W9-AYV-437

ALL THE HELP, RESOURCES, AND PERSONAL SUPPORT YOU AND YOUR STUDENTS NEED!

2-Minute Tutorials and all of the resources you & your students need to get started
www.wileyplus.com/firstday

Student support from an experienced student user
Ask your local representative for details!

Collaborate with your colleagues, find a mentor, attend virtual and live events, and view resources
www.WhereFacultyConnect.com

Pre-loaded, ready-to-use assignments and presentations
www.wiley.com/college/quickstart

Technical Support 24/7
FAQs, online chat, and phone support
www.wileyplus.com/support

Your *WileyPLUS* Account Manager
Training and implementation support
www.wileyplus.com/accountmanager

MAKE IT YOURS!

Introduction to

Organic Chemistry

Introduction to Organic Chemistry

FOURTH EDITION

William H. Brown

Beloit College

Thomas Poon

Claremont McKenna College
Scripps College
Pitzer College

John Wiley & Sons, Inc.

VP & Publisher: Kaye Pace
Associate Publisher & Acquisition Editor: Petra Recter
Project Editor: Jennifer Yee
Marketing Manager: Kristine Ruff
Production Manager: Dorothy Sinclair
Senior Production Editor: Sandra Dumas
Senior Designer: Kevin Murphy
Senior Media Editor: Thomas Kulesa
Media Editor: Marc Wezdecki
Photo Department Manager: Hilary Newman
Photo Editor: Lisa Gee
Production Management Services: Preparé

This book was typeset in 10/12 New Baskerville at Preparé and printed and bound by Courier/Kendallville. The cover was printed by Courier/Kendallville.

The paper in this book was manufactured by a mill whose forest management programs include sustained yield harvesting of its timberlands. Sustained yield harvesting principles ensure that the number of trees cut each year does not exceed the amount of new growth.

This book is printed on acid-free paper. ∞

ISBN 13 978-0470-12923-4
ISBN 13 978-0470-60153-2

Printed in the United States of America.

10 9 8 7 6 5 4 3 2

To Carolyn,
with whom life is a joy

Bill Brown

To Cathy,
the poetry to my prose

Thomas Poon

About the Authors

William H. Brown is Professor Emeritus at Beloit College, where he was twice named Teacher of the Year. He is also the author of two other college textbooks: *Organic Chemistry* 5/e, coauthored with Chris Foote, Brent Iverson, and Eric Anslyn, published in 2009, and *General, Organic, and Biochemistry* 9/e, coauthored with Fred Bettelheim, Mary Campbell, and Shawn Farrell, published in 2010. He received his Ph.D. from Columbia University under the direction of Gilbert Stork and did postdoctoral work at California Institute of Technology and the University of Arizona. Twice he was Director of a Beloit College World Affairs Center seminar at the University of Glasgow, Scotland. In 1999, he retired from Beloit College to devote more time to writing and development of educational materials. Although officially retired, he continues to teach Special Topics in Organic Synthesis on a yearly basis.

Bill and his wife Carolyn enjoy hiking in the canyon country of the Southwest. In addition, they both enjoy quilting and quilts.

Thomas Poon is Associate Professor of Chemistry in the Joint Science Department of Claremont McKenna, Pitzer, and Scripps Colleges, three of the five undergraduate institutions that make up the Claremont Colleges in Claremont, California. He received his B.S. degree from Fairfield University (CT) and his Ph.D. from the University of California, Los Angeles under the direction of Christopher S. Foote. Poon was a Camille and Henry Dreyfus Postdoctoral Fellow under Bradford P. Mundy at Colby College (ME) before joining the faculty at Randolph-Macon College (VA) where he received the Thomas Branch Award for Excellence in Teaching in 1999. He was a visiting scholar at Columbia University (NY) in 2002 (and again in 2004) where he worked on projects in both research and education with his friend and mentor, Nicholas J. Turro. He has taught organic chemistry, forensic chemistry, upper-level courses in advanced laboratory techniques, and a first-year seminar class titled *Science of Identity*. His favorite activity is working alongside undergraduates in the laboratory on research problems involving the investigation of synthetic methodology in zeolites, zeolite photochemistry, natural products isolation, and reactions of singlet oxygen.

When not in the lab, he likes to play guitar and sing funny chemistry songs to his daughter Sophie.

Contents Overview

Contents

Preface

Goals of This Text

Our goals with this text are threefold. The first is to help students understand the logic of organic chemistry and the connectivity of ideas within organic chemistry. To this end, the organization and the order of presentation of topics within the text stress the connections between what students taking this course have learned already and the topics they are currently studying. The second goal is to enable students to see the connections between organic chemistry and the world around them. To this end, we provide a myriad of examples throughout the text to highlight the application of organic chemistry to the health and biological sciences, and, even more broadly, to the world around us. Our third goal is to provide students with all of the latest tools that educational researchers have found to be effective in learning and comprehension. We hope that this logical development of topics, pedagogy, and connectivity to the world around us will help students to understand organic chemistry and to appreciate the contributions organic chemistry has made and will continue to make to our lives.

New to This Edition

How To: Have your students ever wished for an easy-to-follow, step-by-step guide to understanding a problem or concept? We have identified topics in nearly every chapter that often give students a difficult time and created step-by-step *How To* guides for approaching them.

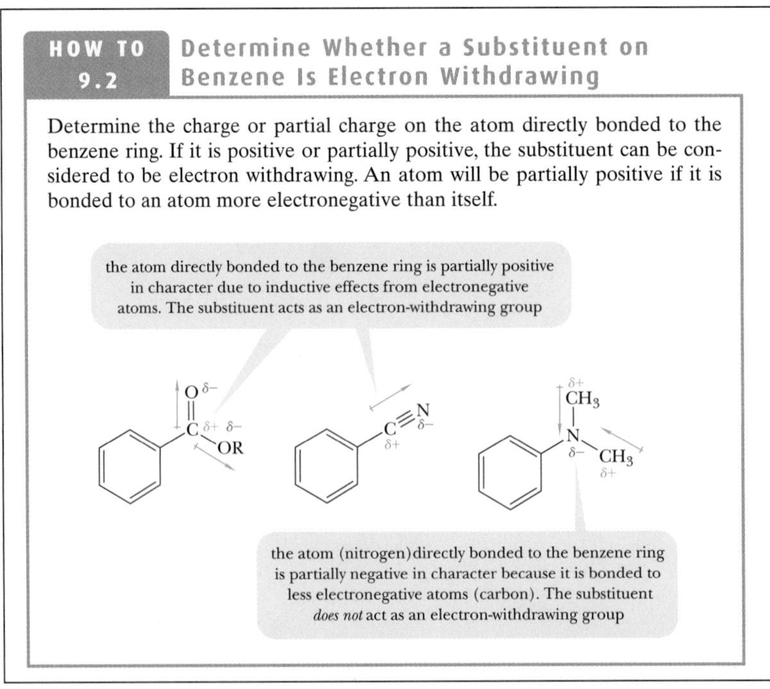

HOW TO 9.2
Determine Whether a Substituent on Benzene Is Electron Withdrawing
Determine the charge or partial charge on the atom directly bonded to the benzene ring. If it is positive or partially positive, the substituent can be considered to be electron withdrawing. An atom will be partially positive if it is bonded to an atom more electronegative than itself.
the atom directly bonded to the benzene ring is partially positive in character due to inductive effects from electronegative atoms. The substituent acts as an electron-withdrawing group
O δ−
C δ+ δ−
OR
C≡N
δ+ δ−
δ+
CH3
N
δ−
CH3
δ+
the atom (nitrogen) directly bonded to the benzene ring is partially negative in character because it is bonded to less electronegative atoms (carbon). The substituent does not act as an electron-withdrawing group

Cumulative Review Questions: In this text, end-of-chapter problems are organized by section, allowing students to easily refer back to the chapter if difficulties arise. This way of organizing practice problems is very useful for learning new material. Wouldn't it be helpful for students to know whether they could do a problem that wasn't categorized for them (i.e., to know whether they could recognize that problem under a different context, such as an exam setting)? To help students in this regard, we have added a section called *Putting It Together* (PIT) at the end of Chapters 3, 6, 10, 15, and 18. Each PIT section is structured much like an exam would be organized, with questions of varying type (multiple choice, short answer, naming, mechanism problems, predict the products, synthesis problems, etc.) and difficulty (often requiring knowledge of concepts from two or more previous chapters). Students' performance on the PIT questions will aid them in assessing their knowledge of the concepts from these groupings of chapters. The answers to the Putting It Together questions appear in the answer appendix of the text.

Putting It Together

The following problems bring together concepts and material from Chapters 13–15. Although the focus may be on these chapters, the problems will also build upon concepts discussed throughout the text thus far.

Choose the best answer for each of the following questions.

1. Which of the following statements is true concerning the following two carboxylic acid derivatives?

A   B

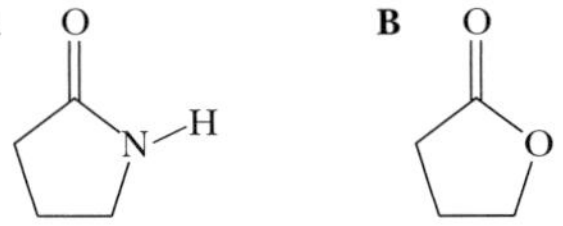

(a) Only molecule **A** can be hydrolyzed.
(b) Only molecule **B** can be hydrolyzed.
(c) Both molecules are hydrolyzable, but **A** will react more quickly than **B**.
(d) Both molecules are hydrolyzable, but **B** will react more quickly than **A**.
(e) **A** and **B** are hydrolyzed at roughly the same rate.

2. How many unique reaction products are formed from the following reaction?

$$CH_3CH_2-\overset{O}{\overset{\|}{C}}-CH_2CH_2-\overset{O}{\overset{\|}{C}}-OCH_3 \xrightarrow[H_2O]{H^+}$$

(a) one (b) two (c) three (d) four (e) five

3. What sequence of reagents will accomplish the following transformation?

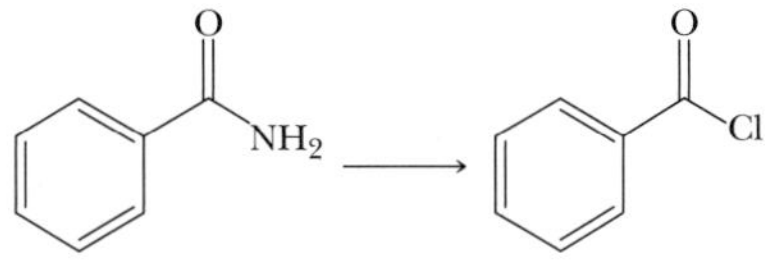

(a) 1. $SOCl_2$
(b) 1. H_2O_2, $SOCl_2$
(c) 1. H_2O_2, HCl
(d) 1. H^+/H_2O_2 2. $SOCl_2$
(e) All of the above

4. Which of the following reactions will not yield butanamide?

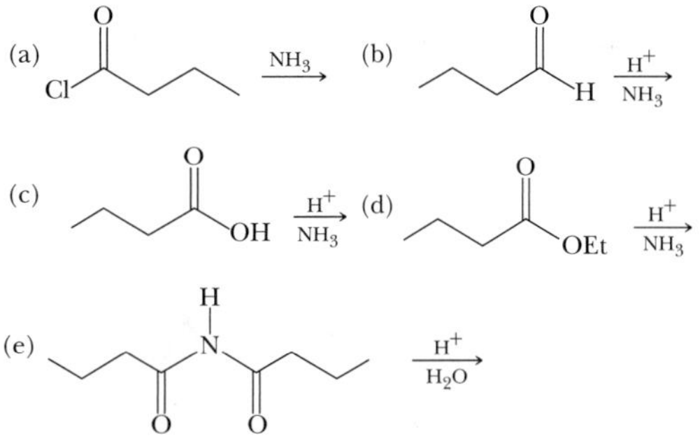

(a) $\xrightarrow{NH_3}$ (b) $\xrightarrow[NH_3]{H^+}$

(c) $\xrightarrow[NH_3]{H^+}$ (d) $\xrightarrow[NH_3]{H^+}$

(e) $\xrightarrow[H_2O]{H^+}$

Problem-Solving Strategies: One of the greatest difficulties students often encounter when attempting to solve problems is knowing where to begin. To help students overcome this challenge, we have added a *Strategy* step for every worked example in the text. The strategy step will help students to determine the starting point for each of the example problems. Once students are familiar with the strategy, they can apply it to all problems of that type.

Example 7.3

Complete these nucleophilic substitution reactions:

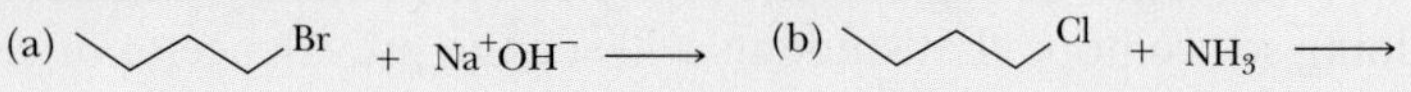

(a) + Na^+OH^- ⟶ (b) + NH_3 ⟶

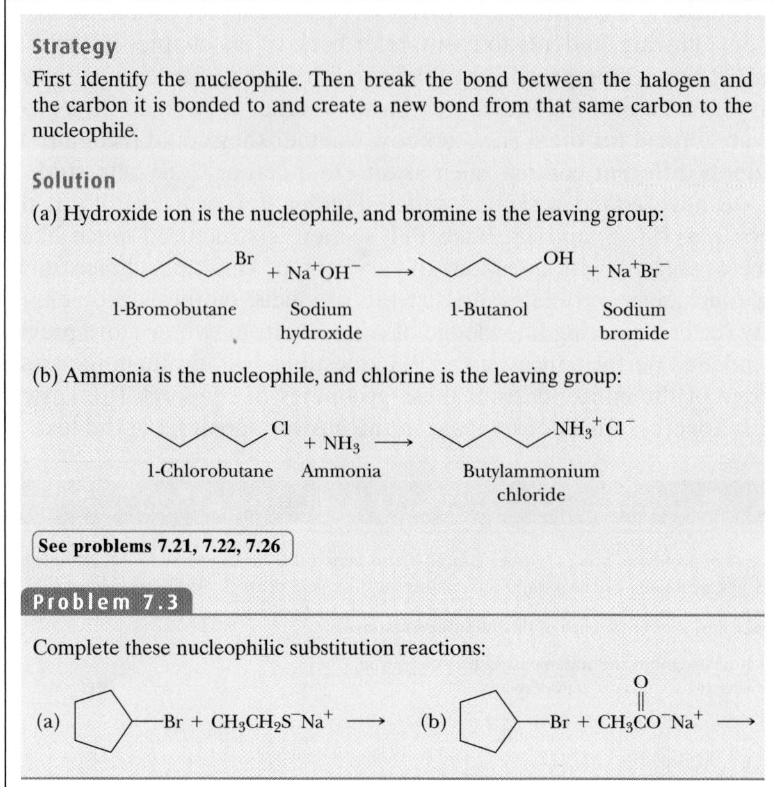

Strategy

First identify the nucleophile. Then break the bond between the halogen and the carbon it is bonded to and create a new bond from that same carbon to the nucleophile.

Solution

(a) Hydroxide ion is the nucleophile, and bromine is the leaving group:

1-Bromobutane + Na^+OH^- (Sodium hydroxide) ⟶ 1-Butanol + Na^+Br^- (Sodium bromide)

(b) Ammonia is the nucleophile, and chlorine is the leaving group:

1-Chlorobutane + NH_3 (Ammonia) ⟶ $NH_3^+\,Cl^-$ Butylammonium chloride

See problems 7.21, 7.22, 7.26

Problem 7.3

Complete these nucleophilic substitution reactions:

(a) cyclopentyl—Br + $CH_3CH_2S^-Na^+$ ⟶ (b) cyclopentyl—Br + $CH_3CO^-Na^+$ (C=O) ⟶

Quick Quizzes: Research on reading comprehension has shown that good readers self-monitor their understanding of what they have just read. We have provided a new tool that will allow students to do this, called the *Quick Quiz*. Quick quizzes are a set of true or false questions at the end of every chapter designed to test students' understanding of the basic concepts presented in the chapter. The questions are not designed to be an indicator of their readiness for an exam. Rather, they are provided for students to assess whether they have the bare minimum of knowledge needed to begin approaching the end-of-chapter problems. The answers to the quizzes are provided at the bottom of the page, so that students can quickly check their progress, and if necessary, return to the appropriate section in the chapter to review the material.

Quick Quiz

Answer true or false to the following questions to assess your general knowledge of the concepts in this chapter. If you have difficulty with any of them, you should review the appropriate section in the chapter (shown in parentheses) before attempting the more challenging end-of-chapter problems.

1. The mechanism of electrophilic aromatic substitution involves three steps: generation of the electrophile, attack of the electrophile on the benzene ring, and proton transfer to regenerate the ring. (9.6)
2. The C—C double bonds in benzene do not undergo the same addition reactions that the C=C double bonds in alkenes undergo. (9.1)
3. Friedel–Crafts acylation is not subject to rearrangements. (9.5)
4. An aromatic compound is planar, possesses a 2*p* orbital on every atom of the ring, and contains either 4, 8, 12, 16, and so on, pi electrons. (9.2)

Putting It in Context—Biochemistry and Health: More examples have been added that show the relevance of organic chemistry to biochemistry and health-related fields. In doing so, we hope that students will see the prevalence of organic chemistry in these fields and realize the benefit of knowing the role organic chemistry plays in other scientific disciplines.

One important example of leaving group stability is found in the methylation of DNA, a process common in all mammals and involved in a variety of biological processes including X-chromosome inactivation in females, inheritance, and carcinogenesis. In DNA methylation, enzymes catalyze the attack of a cytosine unit of DNA on the methyl group of *S*-adenosylmethionine (SAM). All but the methyl group of SAM acts as a leaving group and is able to do so because the positive sulfur atom initially bonded to the methyl group becomes uncharged, making the sulfur atom more stable than before.

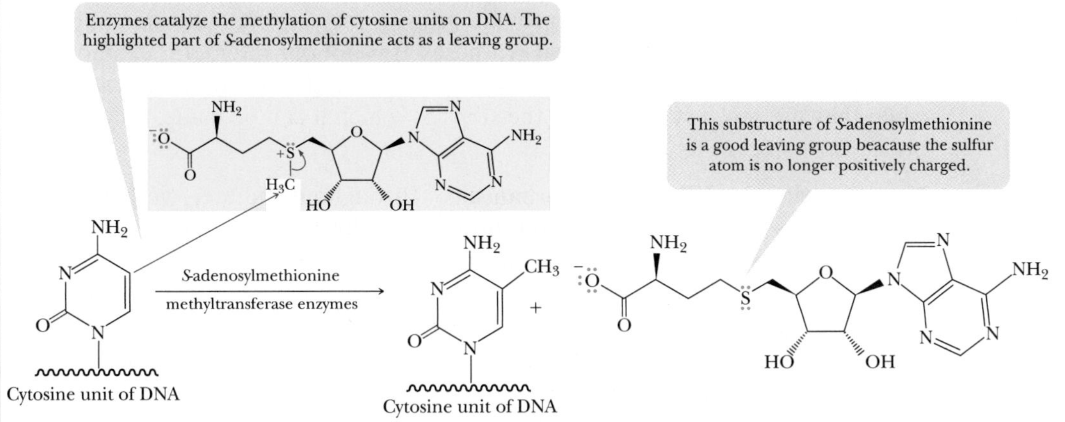

Enzymes catalyze the methylation of cytosine units on DNA. The highlighted part of *S*-adenosylmethionine acts as a leaving group.

This substructure of *S*-adenosylmethionine is a good leaving group beacause the sulfur atom is no longer positively charged.

S-adenosylmethionine methyltransferase enzymes

Cytosine unit of DNA

Cytosine unit of DNA

More Practice Problems: It is widely agreed that one of the best ways to learn the material in organic chemistry is to have students do as many of the practice problems available as possible. We have increased the number of practice problems in the text by over 25%, providing students with even more opportunities to learn the material. For example, we've included a section called *Chemical Transformations* in nearly every chapter, which will help students to familiarize themselves with the reactions covered both in that chapter and in previous chapters. These problems provide a constructivist approach to learning organic chemistry. That is, they illustrate how concepts constantly build on each other throughout the course.

Chemical Transformations

15.46 Test your cumulative knowledge of the reactions learned thus far by completing the following chemical transformations. *Note*: Some will require more than one step. **(See Examples 15.7, 15.8)**

(a) 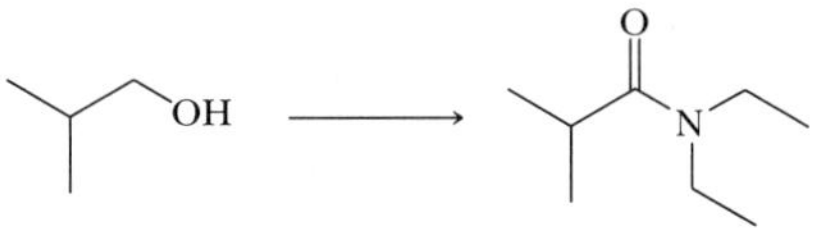 (b) 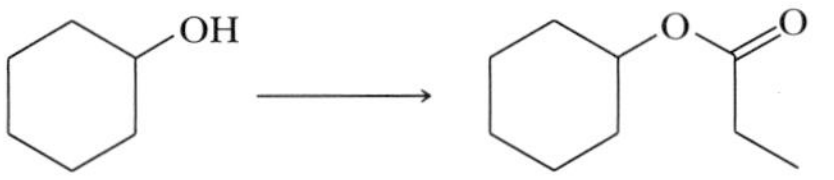 (c)  (d) 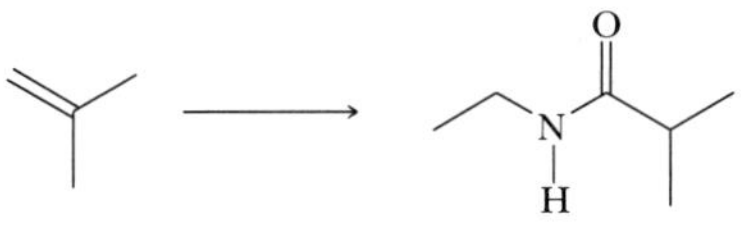 (e) 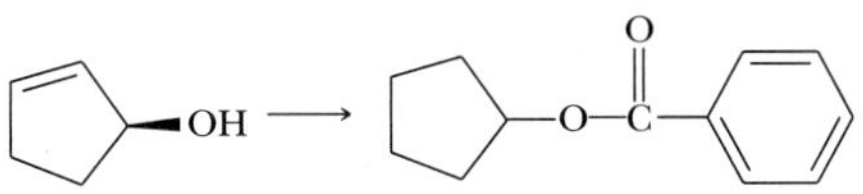 (i) 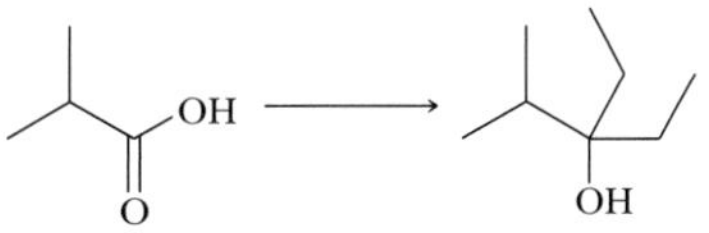 (j) 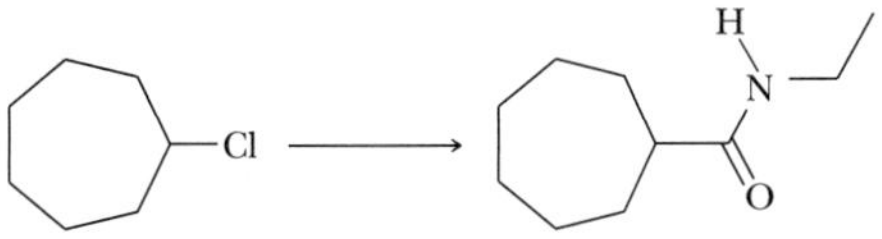 (k) 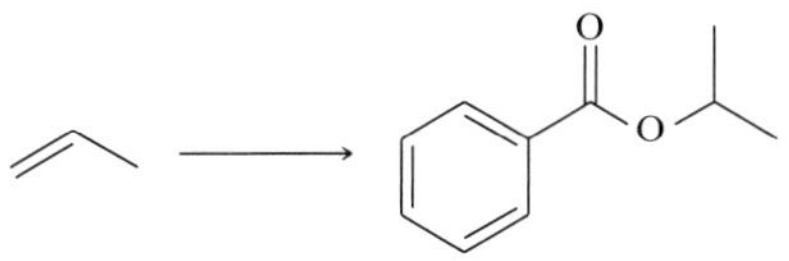 (l)

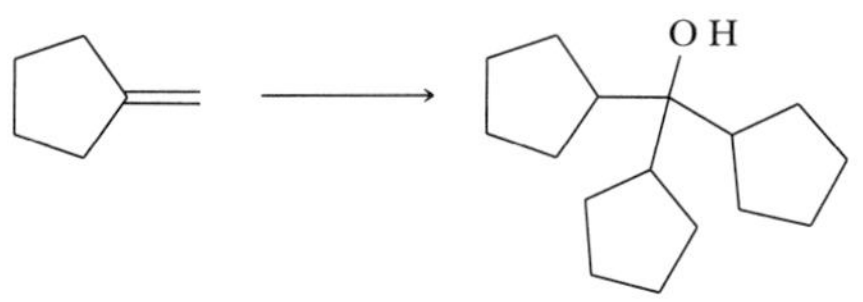

Organic Synthesis: In this text, we treat organic synthesis and all of the challenges it presents as a teaching tool. We recognize that the majority of students taking this course are intending to pursue careers in the health and biological sciences, and that very few intend to become synthetic organic chemists. We also recognize that what organic chemists do best is to synthesize new compounds; that is, they make things. Furthermore, we recognize that one of the keys to mastering organic chemistry is extensive problem solving. To this end, we have developed a large number of synthetic problems in which the target molecule is one with an applied, real-world use. Our purpose in this regard is to provide drills in recognizing and using particular reactions within the context of real syntheses. It is not our intent, for example, that students be able to propose a synthesis for procaine (Novocaine), but rather that when they are given an outline of the steps by which it can be made, they can supply necessary reagents.

New Reactions and Reaction Mechanisms: Throughout the text, we have identified reactions that may have fallen out of favor with chemists and removed them from the text (e.g., the osmylation reaction due to the extreme toxicity of its main reagent). We have also added important reactions that have been significant in the history of organic chemistry and drug development (e.g., hydroboration–oxidation, reactions of alkynes, and others). These reactions will certainly come in handy as students attempt organic synthesis problems. Furthermore, since mechanisms provide a theoretical framework within which students can organize a great deal of organic chemistry, we have paid particular attention to providing mechanisms for nearly all of the reactions in the text. Using this insight into how reagents add to particular compounds, we can make generalizations and then predict how the same reagents might add to similar compounds.

Greater Attention to Visual Learning: Research in knowledge and cognition has shown that visualization and organization can greatly enhance learning. We have more than doubled the number of callouts (short dialog bubbles) to highlight important features of many of the illustrations throughout the text. This places most of the important information in one location. When students try to recall a concept or attempt to solve a problem, we hope that they will try to visualize the relevant illustration from the text. They may be pleasantly surprised to find that the visual cues provided by the callouts help them to remember the content as well as the context of the illustration.

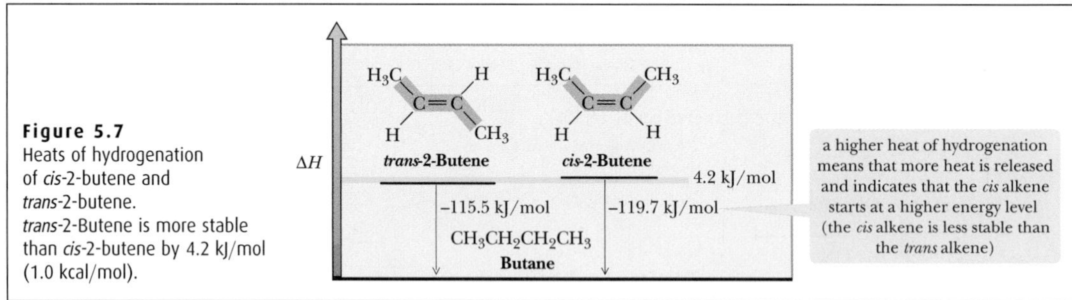

ΔH
trans-2-Butene
cis-2-Butene
4.2 kJ/mol
−115.5 kJ/mol
−119.7 kJ/mol
CH3CH2CH2CH3
Butane
a higher heat of hydrogenation means that more heat is released and indicates that the cis alkene starts at a higher energy level (the cis alkene is less stable than the trans alkene)

Figure 5.7 Heats of hydrogenation of *cis*-2-butene and *trans*-2-butene. *trans*-2-Butene is more stable than *cis*-2-butene by 4.2 kJ/mol (1.0 kcal/mol).

Linking It Together: Many of the end-of-chapter problems are accompanied by a reference to an in-text *Worked Example* that uses the same concept or technique. This allows students to refer back to the worked example if they are having difficulty with a homework problem. Conversely, each worked example contains a reference to end-of-chapter problems that require the same concept or technique to solve. This allows students to pursue more problems of that type and category for immediate practice.

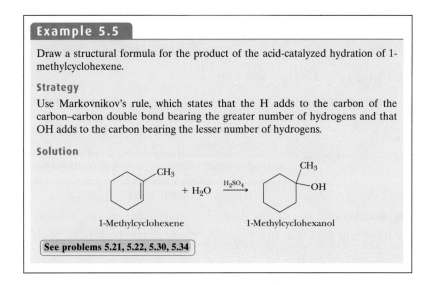

Example 5.5

Draw a structural formula for the product of the acid-catalyzed hydration of 1-methylcyclohexene.

Strategy

Use Markovnikov's rule, which states that the H adds to the carbon of the carbon–carbon double bond bearing the greater number of hydrogens and that OH adds to the carbon bearing the lesser number of hydrogens.

Solution

$$\text{1-Methylcyclohexene} + H_2O \xrightarrow{H_2SO_4} \text{1-Methylcyclohexanol}$$

See problems 5.21, 5.22, 5.30, 5.34

Section 5.3 Electrophilic Additions to Alkenes

5.17 From each pair, select the more stable carbocation: **(See Example 5.3)**

(a) $CH_3CH_2CH_2^+$ or $CH_3\overset{+}{C}HCH_3$

(b) $CH_3\underset{+}{C}H(CH_3)CHCH_3$ or $CH_3\underset{+}{C}(CH_3)CH_2CH_3$

5.18 From each pair, select the more stable carbocation: **(See Example 5.3)**

(a) 1-methylcyclohexyl cation ($-CH_3$) or 2-methylcyclohexyl cation ($-CH_3$)

(b) 1-methylcyclohexyl cation ($-CH_3$) or cyclohexylmethyl cation ($-\overset{+}{C}H_2$)

Key Terms and Concepts: The key terms and concepts from every chapter are now listed at the end of the chapter along with a page number for easy reference.

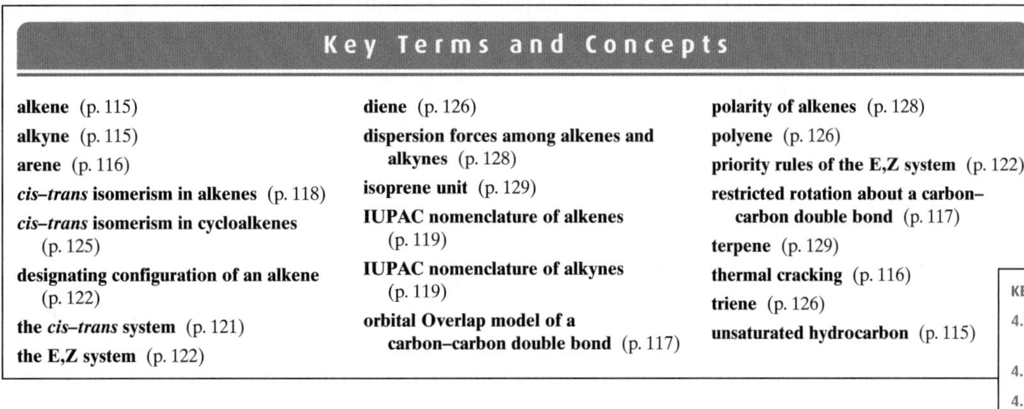

Key Terms and Concepts

alkene (p. 115)
alkyne (p. 115)
arene (p. 116)
***cis–trans* isomerism in alkenes** (p. 118)
***cis–trans* isomerism in cycloalkenes** (p. 125)
designating configuration of an alkene (p. 122)
the *cis–trans* system (p. 121)
the E,Z system (p. 122)
diene (p. 126)
dispersion forces among alkenes and alkynes (p. 128)
isoprene unit (p. 129)
IUPAC nomenclature of alkenes (p. 119)
IUPAC nomenclature of alkynes (p. 119)
orbital Overlap model of a carbon–carbon double bond (p. 117)
polarity of alkenes (p. 128)
polyene (p. 126)
priority rules of the E,Z system (p. 122)
restricted rotation about a carbon–carbon double bond (p. 117)
terpene (p. 129)
thermal cracking (p. 116)
triene (p. 126)
unsaturated hydrocarbon (p. 115)

KEY QUESTIONS

4.1 What Are the Structures and Shapes of Alkenes and Alkynes?
4.2 How Do We Name Alkenes and Alkynes?
4.3 What Are the Physical Properties of Alkenes and Alkynes?
4.4 What Are Terpenes?

HOW TO

4.1 How to Name an Alkene

CHEMICAL CONNECTIONS

4A Ethylene, a Plant Growth Regulator
4B Why Plants Emit Isoprene

Key Questions and Summary of Key Questions: The practice of asking questions prior to, during, or after reading a section of text is present in nearly every reading strategy that has been developed for training effective readers. We have changed all the section headers in the text to appear as questions and have placed them in three places throughout each chapter: (1) in the chapter outline at the beginning of the chapter, (2) within the chapter as the *Section Headers*, and (3) in a section at the

Summary of Key Questions

4.1 What Are the Structures and Shapes of Alkenes and Alkynes?

- An **alkene** is an unsaturated hydrocarbon that contains a carbon–carbon double bond.
- Alkenes have the general formula C_nH_{2n}.
- An **alkyne** is an unsaturated hydrocarbon that contains a carbon–carbon triple bond.
- Alkynes have the general formula C_nH_{2n-2}.
- According to the **orbital Overlap model**, a carbon–carbon double bond consists of one sigma bond formed by the overlap of sp^2 hybrid orbitals and one pi bond formed by the overlap of parallel $2p$ atomic orbitals. It takes approximately 264 kJ/mol (63 kcal/mol) to break the pi bond in ethylene.
- A carbon–carbon triple bond consists of one sigma bond formed by the overlap of *sp* hybrid orbitals and two pi bonds formed by the overlap of pairs of parallel $2p$ orbitals.
- The structural feature that makes ***cis–trans*** **isomerism** possible in alkenes is restricted rotation about the two carbons of the double bond.
- To date, *trans*-cyclooctene is the smallest *trans*-cycloalkene that has been prepared in pure form and is stable at room temperature.

end of the chapter called *Summary of Key Questions*. The latter placement also coincides with bulleted lists of statements that clarify or answer the section header questions. Furthermore, all of the *Chemical Connections Boxes*, a feature of the text that illustrates an application of organic chemistry to an everyday setting, are now followed by one or more questions that relate the application back to a topic or concept from the chapter. In reading the text, students should be encouraged to stop, read these questions, and ponder their meaning. Students may be surprised to see how much more information they can retain and how much easier the material is to understand by doing so.

Molecular Models: Graphics for more than 275 molecular models have been created and incorporated into this text. Further emphasis has been given by adding electrostatic potential maps, which show calculated values for regions of positive and negative charge within molecules and ions. The purpose of these two types of models is to assist students in visualizing organic molecules as three-dimensional objects and to appreciate that the attraction between unlike charges is a major factor in directing molecular reactivity.

The Audience

This book provides an introduction to organic chemistry for students who intend to pursue careers in the sciences and who require a grounding in organic chemistry. For this reason, we make a special effort throughout to show the interrelation between organic chemistry and other areas of science, particularly the biological and health sciences. While studying with this book, we hope that students will see that organic chemistry is a tool for these many disciplines, and that organic compounds, both natural and synthetic, are all around them—in pharmaceuticals, plastics, fibers, agrochemicals, surface coatings, toiletry preparations and cosmetics, food additives, adhesives, and elastomers. Furthermore, we hope that students will recognize that organic chemistry is a dynamic and ever-expanding area of science waiting openly for those who are prepared, both by training and an inquisitive nature, to ask questions and explore.

Organization: An Overview

Chapters 1–10 begin a study of organic compounds by first reviewing the fundamentals of covalent bonding, the shapes of molecules, and acid–base chemistry. The structures and typical reactions of the important classes of organic compounds are then discussed: alkanes, alkenes and alkynes, haloalkanes, alcohols and ethers, benzene and its derivatives, and amines, aldehydes and ketones, and finally carboxylic acids and their derivatives.

Chapters 11 and 12 introduce IR spectroscopy, and ^{1}H-NMR and ^{13}C-NMR spectroscopy. Discussion of spectroscopy requires no more background than what students receive in general chemistry. These two chapters are freestanding and can be taken up in any order appropriate to a particular course.

Chapters 13–17 continue the study of organic compounds, including aldehydes and ketones, carboxylic acids, and finally carboxylic acids and their derivatives. Chapter 16 concludes with an introduction to the aldol, Claisen, and Michael reactions, all three of which are important means for the formation of new carbon–carbon bonds. Chapter 17 provides a brief introduction to organic polymer chemistry.

Chapters 18–21 present an introduction to the organic chemistry of carbohydrates, amino acids and proteins, nucleic acids, and lipids. Chapter 22, The Organic Chemistry of Metabolism, demonstrates how the chemistry developed to this point can be applied to an understanding of three major metabolic pathways—glycolysis, the β-oxidation of fatty acids, and the citric acid cycle.

Chapter by Chapter

Chapter 1 begins with a review of the electronic structure of atoms and molecules, and uses the VSEPR model to predict the shapes of molecules and polyatomic ions. The theory of resonance is introduced midway through Chapter 1 and, with it, the use of curved arrows and electron pushing. A knowledge of resonance theory combined with a facility for pushing electrons gives you two powerful tools for writing reaction mechanisms and understanding chemical reactivity. The discussion of resonance is followed by an introduction to the molecular orbital molecular overlap description of covalent bonding. Chapter 1 concludes with an introduction to the concept of a functional group, and in particular to hydroxyl, carbonyl, and carboxyl groups.

Chapter 2 provides a general introduction to acid–base chemistry and concentrates on two major themes: the quantitative determination of the position of equilibrium in an acid–base reaction, and the qualitative relationship between structure and acidity.

Chapter 3 opens with a description of the structure, nomenclature, and conformations of alkanes and cycloalkanes. The IUPAC system is introduced in Section 3.3 through the naming of alkanes, and, in Section 3.5, the IUPAC system is presented as a general system of nomenclature. The concept of stereoisomerism is introduced in this chapter with a discussion of *cis–trans* isomerism in cycloalkanes.

Chapter 4 presents the structure, nomenclature, and physical properties of alkenes and alkynes.

Chapter 5 opens with an introduction to chemical energetics and the concept of a reaction mechanism. The focus of the chapter is then on the reactions of alkenes, organized in the following order: electrophilic addition, oxidation, and reduction. The twin concepts of regioselectivity and stereoselectivity are introduced in the context of electrophilic additions to alkenes. Reactions of alkynes are also introduced in this chapter.

Chapter 6 introduces the concepts of chirality, enantiomers, and diastereoisomers. This material is coupled with the liberal use of molecular models and encourages students to think of organic molecules as three-dimensional objects and to treat them as such in order to gain a deeper understanding of organic and biochemical reactions.

Chapter 7 introduces haloalkanes and uses them as a vehicle for the discussion of nucleophilic substitution and β-elimination. The concepts of one-step and two-step nucleophilic substitutions along with the S_N1 and S_N2 terminology are introduced first, followed by the concepts of E1 and E2 reactions of haloalkanes. The chapter concludes with a discussion of nucleophilic substitution versus β-elimination.

Chapter 8 concentrates on the structure and characteristic reactions of alcohols, including their conversion to alkyl halides, acid-catalyzed dehydration, and oxidation of primary and secondary alcohols. There then follows a brief introduction to the structure, preparation, and acid-catalyzed ring opening of epoxides. This chapter concludes with a discussion of the acidity of thiols and their oxidation to disulfides.

Chapter 9 opens with the structure and nomenclature of benzene and its derivatives, and then presents several important aromatic heterocyclic compounds. The mechanism of electrophilic aromatic substitution, including the theory of directing effects, is presented in detail. The chapter concludes with a discussion of the structure and acidity of phenols.

Chapter 10 presents the structure and nomenclature of amines and then concentrates on the basicity of amines. New to this edition is a discussion of the reaction of primary aromatic amines with nitrous acid and the ways in which the resulting arenediazonium ions are useful in the synthesis of more complex molecules.

Chapters 11–12 present the fundamentals of IR, ^{1}H-NMR, and ^{13}C-NMR spectroscopy, and how these tools are used in the elucidation of molecular structure.

Chapters 13–16 develop the chemistry of carbonyl-containing compounds. First is the chemistry of aldehydes and ketones in Chapter 13, followed by the chemistry of carboxylic acids in Chapter 14 and the chemistry of carboxylic acid derivatives in Chapter 15. A major theme in these chapters is the addition of nucleophiles to a carbonyl carbon to form tetrahedral carbonyl addition intermediates. Chapter 16 introduces the concept of an enolate anion and its involvement in aldol, Claisen, and Dieckmann reactions to form new carbon–carbon bonds. Included in Chapter 16 is the Michael reaction, an important reaction to create new carbon–carbon bonds.

Chapter 17 provides a brief introduction to the organic chemistry of polymers and emphasizes their prevalence and importance in our world.

Chapters 18–22 present the chemistry of carbohydrates (Chapter 18), amino acids and proteins (Chapter 19), nucleic acids (Chapter 20), and lipids (Chapter 21). Chapter 22 presents a discussion of three key metabolic pathways, namely, glycolysis, β-oxidation of fatty acids, and the citric acid cycle. The purpose of this material is to show that the reactions of these pathways are the biochemical equivalents of organic functional group reactions already studied in previous chapters.

Special Features

1. Chemical Connection Boxes

Each of the 38 boxes illustrates an application of organic chemistry to an everyday setting. Following is a complete list of these boxes by chapter:

1A Buckyball: A New Form of Carbon
3A The Poisonous Puffer Fish
3B Octane Rating: What Those Numbers at the Pump Mean
4A Ethylene, a Plant Growth Regulator
4B Why Plants Emit Isoprene
5A *Cis–Trans* Isomerism in Vision

6A Chiral Drugs
7A The Environmental Impact of Chlorofluorocarbons
7B The Effect of Chlorofluorocarbon Legislation on Asthma Sufferers
8A Nitroglycerin: An Explosive and a Drug
8B Blood Alcohol Screening
8C Ethylene Oxide: A Chemical Sterilant
9A Carcinogenic Polynuclear Aromatics and Cancer
9B Capsaicin, for Those Who Like It Hot
10A Morphine as a Clue in the Design and Discovery of Drugs
10B The Poison Dart Frogs of South America: Lethal Amines
11A Infrared Spectroscopy: A Window on Brain Activity
12A Magnetic Resonance Imaging
13A A Green Synthesis of Adipic Acid
14A From Willow Bark to Aspirin and Beyond
14B Esters as Flavoring Agents
14C Ketone Bodies and Diabetes
15A Ultraviolet Sunscreens and Sunblocks
15B From Moldy Clover to a Blood Thinner
15C The Penicillins and Cephalosporins: β-Lactam Antibiotics
15D The Pyrethrins: Natural Insecticides of Plant Origin
15E Systematic Acquired Resistance in Plants
16A Drugs That Lower Plasma Levels of Cholesterol
16B Antitumor Compounds: The Michael Reaction in Nature
17A Stitches That Dissolve
17B Paper or Plastic?
18A Relative Sweetness of Carbohydrate and Artificial Sweeteners
18B A, B, AB, and O Blood-Group Substances
19A Spider Silk: A Chemical and Engineering Wonder of Nature
20A The Search for Antiviral Drugs
20B DNA Fingerprinting
21A Snake Venom Phospholipases
21B Nonsteroidal Estrogen Antagonists

2. Applications of organic chemistry to the world around us, particularly to the biochemical, health, and biological sciences. The following compounds, listed according to the chapter in which they occur, represent real-world applications of organic chemistry. Discussed within the chapter sections, in Chemical Connections boxes, or as topics of end-of-chapter problems, these compounds illustrate to students the relevance of organic chemistry to their world and to many of their future careers.

Chapter 1 Covalent Bonding and Shapes of Molecules

Tetrafluoroethylene
Freon-11 and Freon-12
Dihydroxyacetone
Propylene glycol
Ephedrine
Lactic acid
Alanine
Peroxyacetyl nitrate (PAN)

Chapter 2 Acids and Bases

Citric acid
Acetic acid
Alanine
Glutamic acid

Chapter 3 Alkanes and Cycloalkanes

Tetrodotoxin
Paraffin wax
Asphalt
Gasohol

Petrolatum
Mineral oil
Vaseline
Natural gas
Petroleum
Gasoline
Naphthas
Butyric acid
Limonene
Germacrene A
Glucose
Cholic acid
Cholestanol
Taxol

Chapter 4 Alkenes and Alkynes

Ethylene
Isoprene
Vitamin A (retinol)
Terpenes
Myrcene
Geraniol
Limonene
Menthol
Farnesol
β-Ocimene
Oleic acid
Elaidic acid
Lycopene
Ethephon (Ethryl)
β-Carotene
Santonin
Periplanone
Gossypol
Zoapatanol
Pyrethrins
Cuparene
Herbertene
Linoleic acid
Linolenic acid
Fumaric acid
Aconitic acid

Chapter 5 Reactions of Alkenes and Alkynes

Limonene
Terpin hydrate
Vitamin A and vision
Rhodopsin
Muscular

Chapter 6 Chirality: The Handedness of Molecules

Ibuprofen
Erythrose
Threose
Tartaric acid
Menthol
Cholesterol
Chymotrypsin
Glyceraldehyde
Naproxen
Captopril
L-DOPA
Quinapril (Accupril)
Ramipril (Vasotec)
Enalapril (Altace)
Dopamine
Carvone
Ephedrine
Amoxicillin
Loratidine (Claritin)
Fexophenadine (Allegra)
Fluoxetine (Prozac)
Sertraline (Zoloft)
Paroxetine (Paxil)
Triamcinolone acetonide (Azmacort)

Chapter 7 Haloalkanes

1,1,1,2-Tetrafluoroethane (HFA-134a)
Triclor
Methylchloroform
Chlorofluorocarbons (CFCs)
Hydrofluorocarbons (HFCs)
Hydrochlorofluorocarbons (HCFCs)
Cytosine unit of DNA
S-Adenosylmethionine (SAM)
Immunostimulants in *Echinacea*

Chapter 8 Alcohols, Ethers, and Thiols

Isoflurane
Nitroglycerin
Procaine (Novocaine)
Diphenhydramine (Benadryl)

Ethylene glycol
Ethanol (drinking alcohol)
Ethylene oxide
Methyl *tert*-butyl ether (MTBE)
Disparlure
Adenine unit of DNA
2-Butene-1-thiol
3-Methyl-1-butanethiol
Bombykol

Chapter 9 Benzene and Its Derivatives

Indole
Purine
Adenine
Serotonin
Benzo[a]pyrene
Phenol
Thymol
Vanillin
Eugenol
Hexylresorcinol
Urushiol
Capsaicin
Vitamin E
Butylated hydroxytoluene (BHT)
Butylated hydroxyanisole (BHA)
Bisphenol A
2,4-Dichlorophenoxyacetic acid (2,4-D)
4-Isopropylacetophenone
Musk ambrette
Hydroquinone
p-Quinone
Bupropion (Wellbutrin, Zyban)

Chapter 10 Amines

Coniine
Nicotine
Cocaine
Morphine
Codeine
Heroin
Dextromethorphan
Levomethorphan
Meperidine (Demerol)
Amphetamine
Cetylpyridinium chloride
Batrachotoxin
Batrachotoxinin A
Guanidine group
Norepinephrine
Alanine
Benzocaine
Chloroquin
Epinephrine (adrenaline)
Albuterol (Proventil)
1,4-Butanediamine (putrescine)
Pyridoxamine (Vitamin B6)
Epibatadine
Procaine
Acetyl choline
Metformin (Glucophage)
Acetaminophen
Propoxycaine
Propranolol (Cardinol)
Phenacetin
Triiodoaromatics
Propofol
Wyerone

Chapter 11 Infrared Spectroscopy

Methyl salicylate
Acetylsalicylic acid (aspirin)
Functional Near Infrared Spectroscopy
Isopentyl acetate
L-Tryptophan

Chapter 12 Nuclear Magnetic Resonance Spectroscopy

Anethole
Magnetic Resonance Imaging (MRI)
Citric acid
Prenol
Phenacetin

Chapter 13 Aldehydes and Ketones

Lactic acid
Pyruvic acid
γ-Aminobutyric acid
Rimantidine
Methenamine
Tamoxifen

Rhodopsin
Lysine
Vanillin
Adipic acid
Amphetamine
Methamphetamine
Bupropion (Wellbutrin)
Chlorpromazine
Amitriptyline
Diphenhydramine (Benadryl)
Venlafazine

Chapter 14 Carboxylic Acids

Oxalic acid
Adipic acid
Salicylic acid
Terephthalic acid
Palmitic acid
Stearic acid
Arachidic acid
γ-Aminobutanoic acid (GABA)
Glyceric acid
Maleic acid
Mevalonic acid
Sodium benzoate
Salicin
Acetylsalicylic acid (Aspirin)
Ibuprofen
Naproxen
Cyclooxygenase
Ketoprofen
Ethyl formate
Isopentyl acetate
Octyl acetate
Methyl butanoate
Ethyl butanoate
Methyl 2-aminobenzoate
Oxalosuccinic acid
α-ketoglutaric acid
3-Oxobutanoic acid
3-Hydroxybutanoic acid
Megatomoic acid
Monopotassium oxalate
Potassium sorbate
Zinc 10-undecenoate
Lactic acid
Alanine
Formic acid
Methyl salicylate
Benzocaine
Methylparaben
Propylparaben
Procaine
Meclizine (Bonine)
Albuterol (Proventil)

Chapter 15 Functional Derivatives of Carboxylic Acids

Octyl *p*-methoxycinnamate
Homosalate
Padimate A
Coumarin
Dicoumarol
Warfarin
Glucose 6-phosphate
Pyridoxal phosphate
Amoxicillin
Barbiturates
Barbital
Meprobamate
Phenobarbital
N,*N*-Diethyl-*m*-toluamide (Deet)
Procaine
Keflex
Acetylsalicylic acid (Aspirin)
Methyl salicylate
Pyrethrin I
Permethrin
Spermaceti
Phenacetin
Acetaminophen
Niacin
Propanil
Lidocaine (Xylocaine)
Etidocaine (Duranest)
Mepivacaine (Carbocaine)
Diethylcarbamazine
Methylparaben

Chapter 16 Enolate Anions

Acetyl-CoA
Mevastatin
Lovastatin (Mevacor)
Isopentenyl pyrophosphate
Calicheamicin
Cortisone
Pindone
Fentanyl

Simvastatin (Zocor)
Oxanamide
Meclizine
Octyl *p*-methoxycinnamate

Chapter 17 Organic Polymer Chemistry

Polyamides
Nylon-66
Nylon-6
Kevlar
Polyesters
Poly(ethylene terephthalate)
Dacron polyester
Mylar
Polycarbonates
Lexan
Polyurethanes
Spandex
Lycra
Epoxy resins
Polyethylene
Polypropylene
Poly(vinyl chloride) (PVC)
Polyacrylonitrile (Orlon)
Polytetrafluoroethylene (Teflon)
Polystyrene
Poly(ethyl acrylate)
Poly(methyl methacrylate) (Plexiglas)
Low-density polyethylene (LDPE)
High-density polyethylene (HDPE)
Nylon 6,10
Nomex
Poly(vinyl acetate)

WileyPLUS for Organic Chemistry – *A Powerful Teaching and Learning Solution*

This online teaching and learning environment integrates the **entire digital textbook** with the most effective instructor and student resources to fit every learning style. With *WileyPLUS* (www.wileyplus.com):

- Students achieve concept mastery in a rich, structured environment that's available 24/7.
- Instructors personalize and manage their course more effectively with assessment, assignments, grade tracking, and more.

WileyPLUS can complement your current textbook or replace the printed text altogether. The problem types and resources in *WileyPLUS* are designed to enable and support problem-solving skill development and conceptual understanding. **Three unique repositories of assessment are offered that provide breadth, depth, and flexibility:**

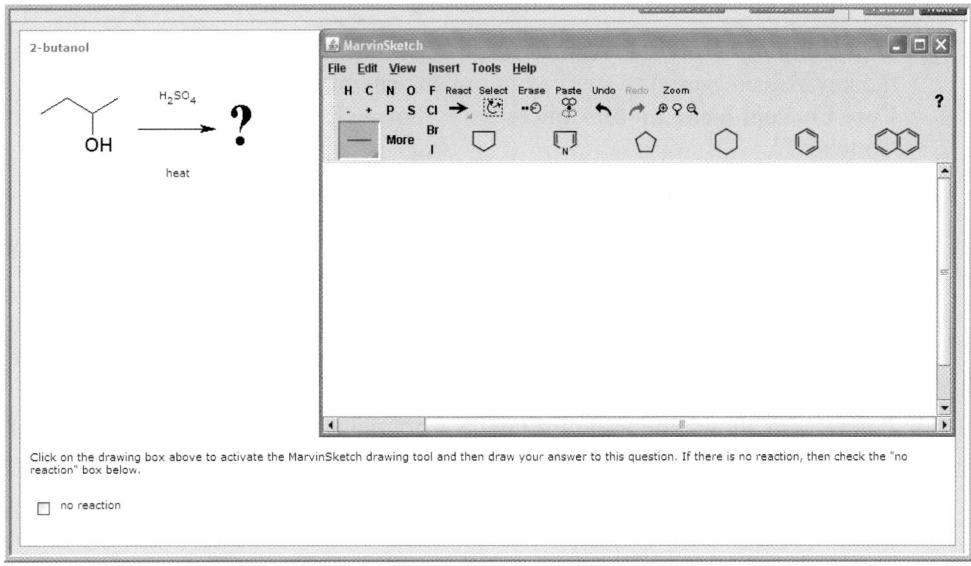

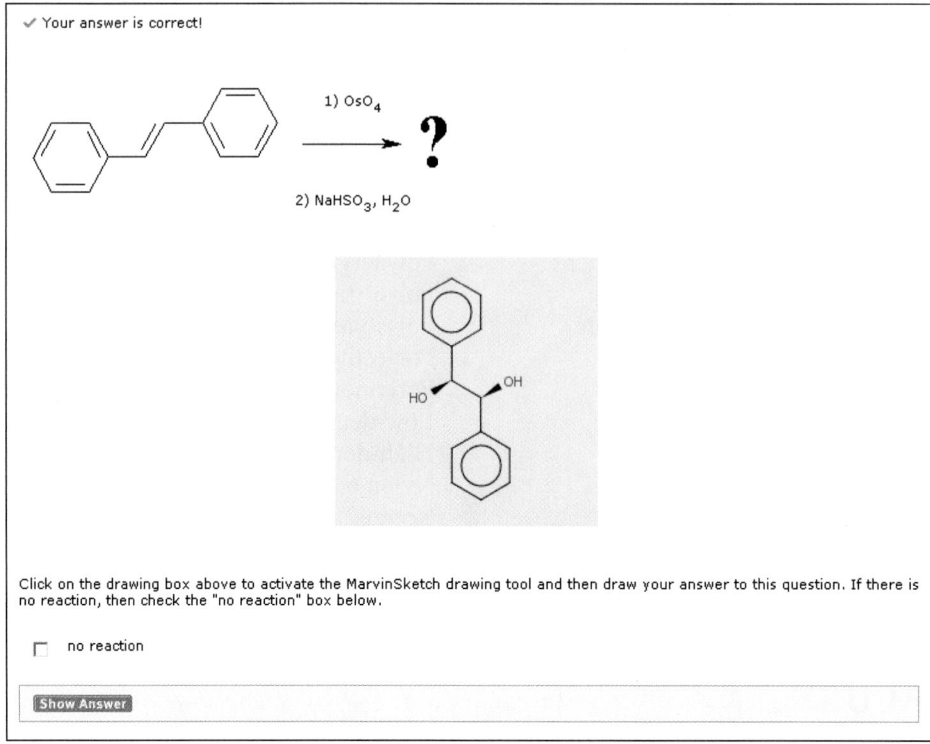

1. **End-of-chapter exercises,** many of which are algorithmic, feature structure drawing and assessment using MarvinSketch, and provide immediate answer feedback. A subset of these end-of-chapter questions is linked to **Guided Online Tutorials,** which are stepped-out problem-solving tutorials that walk the student through the problem, offering individualized feedback at each step.
2. **Prebuilt concept mastery assignments,** organized by topic and concept, feature robust answer feedback.
3. **Test Bank questions** consist of over 2,000 unique questions.

WileyPLUS for Students

Different learning styles, different levels of proficiency, different levels of preparation—each of your students is unique. ***WileyPLUS* offers a myriad of rich multimedia resources for students to facilitate learning.** These include:

- **Skill Building Exercises:** Animated exercises, with instant feedback, reinforce the key skills required to succeed in organic chemistry.
- **Core Concept Animations:** Concepts are thoroughly explained using audio and whiteboard.

WileyPLUS for Instructors

WileyPLUS empowers you with the tools and resources to make your teaching even more effective:

- You can customize your classroom presentation with a wealth of resources and functionality, from PowerPoint slides to a database of rich visuals. You can even add your own materials to your *WileyPLUS* course.
- *WileyPLUS* allows you to help students who might fall behind by tracking their progress and offering assistance easily, even before they come to office hours.
- *WileyPLUS* simplifies and automates such tasks as obsessing student performance, making assignments, scoring student work, keeping grades, and more.

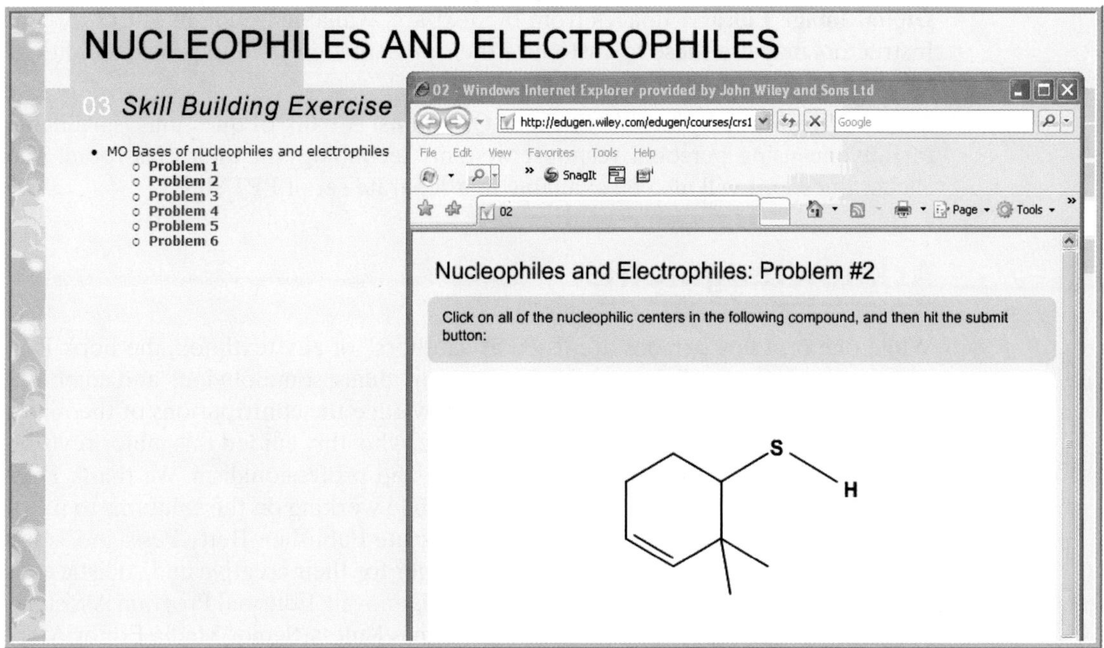

Support Package for Students

Student Solutions Manual: Authored by Felix Lee, of the University of Western Toronto, the Student Study Guide contains detailed solutions to all problems, including the Quick Quiz questions and the Putting It Together questions.

Multimedia Tutors: Several short movies will be available within *WileyPLUS*. Some movies contain complex animations and text annotations that help to explain the more difficult concepts encountered in organic chemistry. Other movies are short online lectures given by one of the authors. Students may find these lectures useful as an alternative presentation of the materials in the text.

Student Flash Cards: Multimedia flash cards will be available within *WileyPLUS*. These virtual flash cards will organize the key reactions and concepts from the course in a useful manner for learning the material. The mechanisms of each reaction will be provided as well as examples of their use.

Support Package for Instructors

All Instructor Resources are available within *WileyPLUS* or they can be accessed by contacting your local Wiley Sales Representative.

PowerPoint Presentations: Authored by William Brown, the PPT lecture slides provide a prebuilt set of approximately 700 slides corresponding to every chapter in the text. The slides include examples and illustrations that help reinforce and test students' grasp of organic chemistry concepts. An additional set of PPT slides, featuring the illustrations, figures, and tables from the text, will also be available. All PPT slide presentations are customizable to fit your course.

Test Bank: Authored by Stefan Bossmann of Kansas State University, the Test Bank for this edition has been revised and updated to include over 2,000 short-answer, multiple-choice, and true-false questions. It is available in both a printed and computerized version.

Digital Image Library: Images from the text are available online in JPEG format. Instructors may use these to customize their presentations and to provide additional visual support for quizzes and exams.

Personal Response Systems/ "Clicker" Questions: A bank of questions is available for anyone using personal response systems technology in their classroom. The clicker questions will also be available in a separate set of PPT slides.

Acknowledgments

While one or a few persons are listed as "authors" of any textbook, the book is in fact the product of collaboration of many individuals, some obvious and some not so obvious. It is with gratitude that we acknowledge the contributions of the many. We begin with our project editor, Jennifer Yee, who ably guided this major revision from beginning to end and did so with grace and professionalism. We thank Felix Lee for his keen eye and attention to detail while working on the solutions to problems in the text. We thank Petra Recter, Associate Publisher; Betty Pessagno, Copy Editor; and Francesca Monaco, Project Manager for their creative and stylistic contributions to the text. We also thank Catherine Donovan, Editorial Program Assistant; Sandra Dumas, Senior Production Editor; Thomas Kulesa, Senior Media Editor; Marc Wezdecki, Media Editor; Evelyn Levich, Assistant Media Editor; Kristine Ruff, Marketing Manager; Lisa Gee, Photo Editor; and Lucy Parkinson, Marketing Assistant. Finally, we thank all our students, both past and present, for their many positive interactions over the years that have guided us in creating this textbook.

List of Reviewers

The authors gratefully acknowledge the following reviewers for their valuable critiques of this book in its many stages as we were developing the Fourth Edition:

Debbie Beard, *Mississippi State University*
Stefan Bossmann, *Kansas State University*
Dana Chatellier, *University of Delaware*
Patricia Chernovitz, *Grantham University*
Wendi David, *Texas State University—San Marcos*
Sushama Dandekar, *University of North Texas*
Maria Gallardo-Williams, *North Carolina State University*
Joseph Gandler, *California State University—Fresno*
Michel Gravel, *University of Saskatchewan*
John Grutzner, *Purdue University*
Ben Gung, *Miami University*
Bettina Heinz, *Palomar College*
James Hershberger, *Miami University*
Richard P. Johnson, *University of New Hampshire*
Spencer Knapp, *Rutgers University*
Felix Lee, *University of Western Ontario*
Douglas Linebarrier, *University of North Carolina at Greensboro*
Brian A. Logue, *South Dakota State University*
Brian Love, *East Carolina University*
Any Pollock, *Michigan State University*
Ginger Powe-McNair, *Louisiana State University*
Michael Rathke, *Michigan State University*
Christian Ray, *University of Illinois at Urbana—Champaign*
Michelle Richards-Babb, *West Virginia University*
Jason Serin, *Glendale Community College*
Eric Trump, *Emporia State University*

We are also grateful to the many people who provided reviews that guided preparation of the earlier editions of our book:

Jennifer Batten, *Grand Rapids Community College*
Dana S. Chatellier, *University of Delaware*
Peter Hamlet, *Pittsburg State University*
John F. Helling, *University of Florida—Gainesville*
Klaus Himmeldirk, *Ohio University—Athens*
Dennis Neil Kevill, *Northern Illinois University*
Dalila G. Kovacs, *Michigan State University—East Lansing*
Tom Munson, *Concordia University*
Robert H. Paine, *Rochester Institute of Technology*
Jeff Piquette, *University of Southern Colorado—Pueblo*
Joe Saunders, *Pennsylvania State University*
K. Barbara Schowen, *University of Kansas—Lawrence*
Robert P. Smart, *Grand Valley State University*
Joshua R. Smith, *Humboldt State University*
Richard T. Taylor, *Miami University—Oxford*
Kjirsten Wayman, *Humboldt State University*

1 Covalent Bonding and Shapes of Molecules

KEY QUESTIONS

HOW TO

CHEMICAL CONNECTIONS

A model of the structure of diamond, one form of pure carbon. Each carbon atom in diamond is bonded to four other carbon atoms at the corners of a tetrahedron. Inset: A model of buckyball, a form of carbon with a molecular formula of C_{60}. *(Charles D. Winters)*

According to the simplest definition, **organic chemistry** is the study of the compounds of carbon. As you study this text, you will realize that organic compounds are everywhere around us—in our foods, flavors, and fragrances; in our medicines, toiletries, and cosmetics; in our plastics, films, fibers, and resins; in our paints and varnishes; in our glues and adhesives; and, of course, in our bodies and in all living things.

Perhaps the most remarkable feature of organic chemistry is that it is the chemistry of carbon and only a few other elements—chiefly hydrogen, oxygen, and nitrogen. Chemists have discovered or made well over 10 million organic compounds.

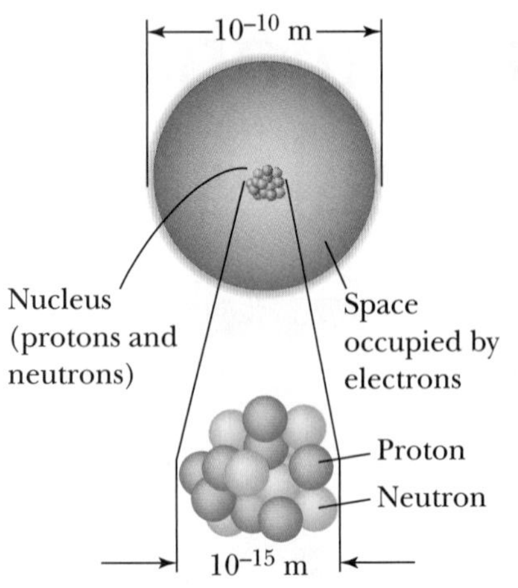

Figure 1.1
A schematic view of an atom. Most of the mass of an atom is concentrated in its small, dense nucleus.

While the majority of them contain carbon and just those three elements, many also contain sulfur, phosphorus, and a halogen (fluorine, chlorine, bromine, or iodine).

Let us begin our study of organic chemistry with a review of how carbon, hydrogen, oxygen, and nitrogen combine by sharing electron pairs to form molecules.

1.1 How Do We Describe the Electronic Structure of Atoms?

You are already familiar with the fundamentals of the electronic structure of atoms from a previous study of chemistry. Briefly, an atom contains a small, dense nucleus made of neutrons and positively charged protons. Most of the mass of an atom is contained in its nucleus, which is surrounded by a much larger extranuclear space that contains the negatively charged electrons. The nucleus of an atom has a diameter of 10^{-14} to 10^{-15} meter (m). The extranuclear space, where the electrons are found, is a much larger area, with a diameter of approximately 10^{-10} m (Figure 1.1).

Shell A region of space around a nucleus where electrons are found.

Electrons do not move freely in the space around a nucleus, but rather are confined to regions of space called **principal energy levels** or, more simply, **shells**. We number these shells 1, 2, 3, and so forth from the inside out. Each shell can contain up to $2n^2$ electrons, where n is the number of the shell. Thus, the first shell can hold 2 electrons, the second 8 electrons, the third 18, the fourth 32, and so on (Table 1.1). Electrons in the first shell are nearest to the positively charged nucleus and are held most strongly by it; these electrons are said to be the lowest in energy. Electrons in higher numbered shells are farther from the nucleus and are held less strongly to it; these electrons are said to be higher in energy.

Orbital A region of space where an electron or pair of electrons spends 90 to 95% of its time.

Shells are divided into subshells designated by the letters $s, p, d,$ and f, and within these subshells, electrons are grouped in orbitals (Table 1.2). An **orbital** is a region of space that can hold 2 electrons. The first shell contains a single orbital called a 1*s*

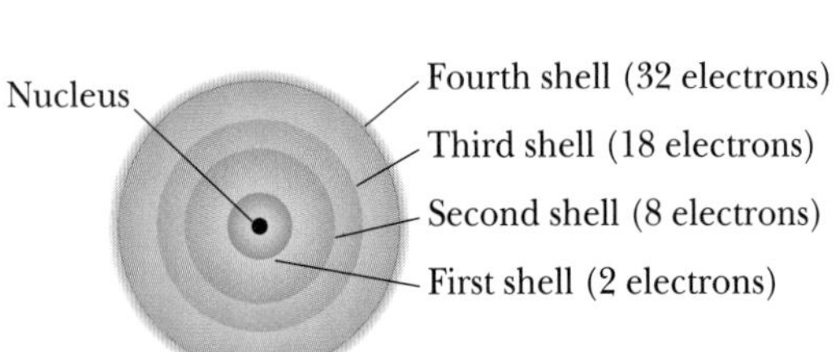

TABLE 1.1 Distribution of Electrons in Shells

Shell	Number of Electrons Shell Can Hold	Relative Energies of Electrons in Each Shell
4	32	Higher
3	18	↑
2	8	
1	2	Lower

TABLE 1.2 Distribution of Orbitals within Shells

Shell	Orbitals Contained in Each Shell	Maximum Number of Electrons Shell Can Hold
4	One 4*s*, three 4*p*, five 4*d*, and seven 4*f* orbitals	2 + 6 + 10 + 14 = 32
3	One 3*s*, three 3*p*, and five 3*d* orbitals	2 + 6 + 10 = 18
2	One 2*s* and three 2*p* orbitals	2 + 6 = 8
1	One 1*s* orbital	2

TABLE 1.3 Ground-State Electron Configurations for Elements 1–18

First Period*	Second Period	Third Period
H 1 $1s^1$	Li 3 [He] $2s^1$	Na 11 [Ne] $3s^1$
He 2 $1s^2$	Be 4 [He] $2s^2$	Mg 12 [Ne] $3s^2$
	B 5 [He] $2s^2 2p_x^{\,1}$	Al 13 [Ne] $3s^2 3p_x^{\,1}$
	C 6 [He] $2s^2 2p_x^{\,1} 2p_y^{\,1}$	Si 14 [Ne] $3s^2 3p_x^{\,1} 3p_y^{\,1}$
	N 7 [He] $2s^2 2p_x^{\,1} 2p_y^{\,1} 2p_z^{\,1}$	P 15 [Ne] $3s^2 3p_x^{\,1} 3p_y^{\,1} 3p_z^{\,1}$
	O 8 [He] $2s^2 2p_x^{\,2} 2p_y^{\,1} 2p_z^{\,1}$	S 16 [Ne] $3s^2 3p_x^{\,2} 3p_y^{\,1} 3p_z^{\,1}$
	F 9 [He] $2s^2 2p_x^{\,2} 2p_y^{\,2} 2p_z^{\,1}$	Cl 17 [Ne] $3s^2 3p_x^{\,2} 3p_y^{\,2} 3p_z^{\,1}$
	Ne 10 [He] $2s^2 2p_x^{\,2} 2p_y^{\,2} 2p_z^{\,2}$	Ar 18 [Ne] $3s^2 3p_x^{\,2} 3p_y^{\,2} 3p_z^{\,2}$

*Elements are listed by symbol, atomic number, and ground-state electron configuration, in that order.

orbital. The second shell contains one $2s$ orbital and three $2p$ orbitals. All p orbitals come in sets of three and can hold up to 6 electrons. The third shell contains one $3s$ orbital, three $3p$ orbitals, and five $3d$ orbitals. All d orbitals come in sets of five and can hold up to 10 electrons. All f orbitals come in sets of seven and can hold up to 14 electrons. In this course, we focus on compounds of carbon with hydrogen, oxygen, and nitrogen, all of which use only electrons in s and p orbitals for covalent bonding. Therefore, we are concerned primarily with s and p orbitals.

A. Electron Configuration of Atoms

The electron configuration of an atom is a description of the orbitals the electrons in the atom occupy. Every atom has an infinite number of possible electron configurations. At this stage, we are concerned only with the **ground-state electron configuration**—the electron configuration of lowest energy. Table 1.3 shows ground-state electron configurations for the first 18 elements of the Periodic Table. We determine the ground-state electron configuration of an atom with the use of the following three rules:

Ground-state electron configuration The electron configuration of lowest energy for an atom, molecule, or ion.

Rule 1. *Orbitals fill in order of increasing energy from lowest to highest.*
Example: In this course, we are concerned primarily with elements of the first, second, and third periods of the Periodic Table. Orbitals in these elements fill in the order $1s$, $2s$, $2p$, $3s$, and $3p$ (Figure 1.2).

Rule 2. *Each orbital can hold up to two electrons with their spins paired.*
Example: With four electrons, the $1s$ and $2s$ orbitals are filled, and we write them as $1s^2 2s^2$. With an additional six electrons, a set of three $2p$ orbitals is filled. We can write them in an expanded form as $2p_x^{\,2} 2p_y^{\,2} 2p_z^{\,2}$. Alternatively, we can group the three filled $2p$ orbitals and write them in a condensed form as $2p^6$. *Spin pairing* means that each electron spins in a direction opposite that of its partner (Figure 1.3). We show this pairing by writing two arrows, one with its head up and the other with its head down.

Rule 3. *When orbitals of equivalent energy are available, but there are not enough electrons to fill them completely, then we add one electron to each equivalent orbital before we add a second electron to any one of them.*
Example: After the $1s$ and $2s$ orbitals are filled with four electrons, we add a fifth electron to the $2p_x$ orbital, a sixth to the $2p_y$ orbital, and a seventh to

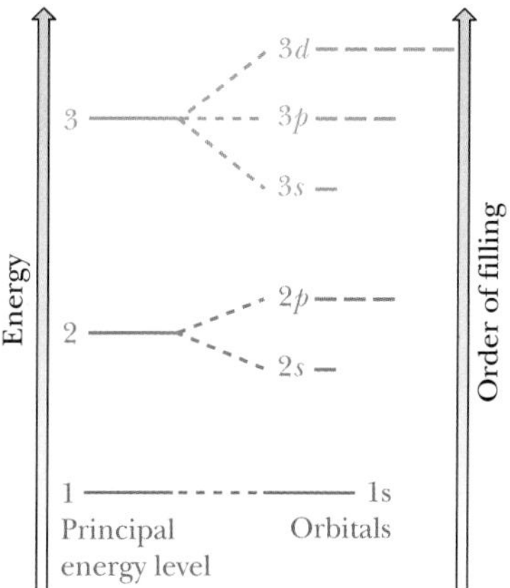

Figure 1.2
Relative energies and order of filling of orbitals through the 3*d* orbitals.

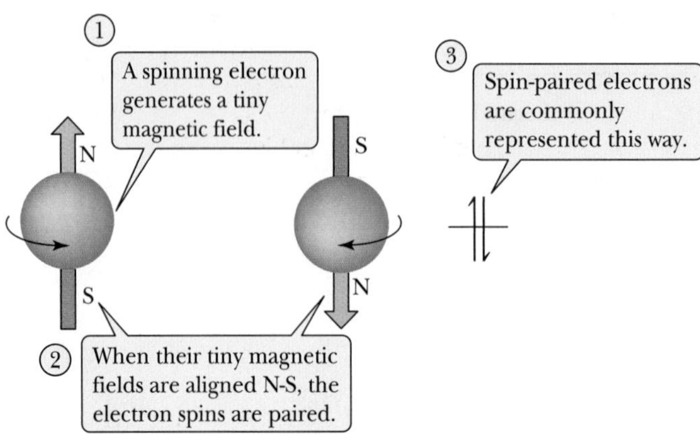

Figure 1.3
The pairing of electron spins.

the $2p_z$ orbital. Only after each $2p$ orbital contains one electron do we add a second electron to the $2p_x$ orbital. For eight electrons, we write the ground-state electron configuration as $1s^2 2s^2 2p_x^{\,2} 2p_y^{\,1} 2p_z^{\,1}$ or, alternatively, $1s^2 2s^2 2p^4$.

Table 1.3 gives the ground-state electron configurations for Elements 1–18. In this table, we use the symbol of the noble gas immediately preceding the particular element to indicate the electron configuration of all filled shells. The electron configuration of helium, for example, is $1s^2$, and that of neon is $2s^2 2p^6$. We show neon's configuration with the shorthand notation [He] $2s^2 2p^6$.

Example 1.1

Write ground-state electron configurations for these elements:

(a) Lithium (b) Oxygen (c) Chlorine

Strategy

Locate each atom in the Periodic Table and determine its atomic number. The order of filling of orbitals is $1s$, $2s$, $2p_x$, $2p_y$, $2p_z$, etc.

Solution

(a) Lithium (atomic number 3): $1s^2 2s^1$. Alternatively, we can write the ground-state electron configuration as [He] $2s^1$.

(b) Oxygen (atomic number 8): $1s^2 2s^2 2p_x^{\,2} 2p_y^{\,1} 2p_z^{\,1}$. Alternatively, we can group the four electrons of the $2p$ orbitals together and write the ground-state electron configuration as $1s^2 2s^2 2p^4$. We can also write it as [He] $2s^2 2p^4$.

(c) Chlorine (atomic number 17): $1s^2 2s^2 2p^6 3s^2 3p^5$. Alternatively, we can write it as [Ne] $3s^2 3p^5$.

See problems 1.17–1.20

Problem 1.1

Write and compare the ground-state electron configurations for the elements in each set.

(a) Carbon and silicon (b) Oxygen and sulfur
(c) Nitrogen and phosphorus

B. Lewis Structures

In discussing the physical and chemical properties of an element, chemists often focus on the outermost shell of its atoms, because electrons in this shell are the ones involved in the formation of chemical bonds and in chemical reactions. We call outer-shell electrons **valence electrons**, and we call the energy level in which they are found the **valence shell**. Carbon, for example, with a ground-state electron configuration of $1s^2 2s^2 2p^2$, has four valence (outer-shell) electrons.

To show the outermost electrons of an atom, we commonly use a representation called a **Lewis structure**, after the American chemist Gilbert N. Lewis (1875–1946), who devised this notation. A Lewis structure shows the symbol of the element, surrounded by a number of dots equal to the number of electrons in the outer shell of an atom of that element. In Lewis structures, the atomic symbol represents the nucleus and all filled inner shells. Table 1.4 shows Lewis structures for the first 18 elements of the Periodic Table. As you study the entries in the table, note that, with the exception of helium, the number of valence electrons of the element corresponds to the group number of the element in the Periodic Table; for example, oxygen, with six valence electrons, is in Group 6A.

Gilbert N. Lewis (1875–1946) introduced the theory of the electron pair that extended our understanding of covalent bonding and of the concept of acids and bases. It is in his honor that we often refer to an "electron dot" structure as a Lewis structure. *(UPI/Bettmann)*

At this point, me must say a word about the numbering of the columns (families or groups) in the Periodic Table. Dmitri Mendeleev gave them numerals and added the letter A for some columns and B for others. This pattern remains in common use in the United States today. In 1985, however, the International Union of Pure and Applied Chemistry (IUPAC) recommended an alternative system in which the columns are numbered 1 to 18 beginning on the left and without added letters. Although we use the original Mendeleev system in this text, the Periodic Table on the inside back cover of the text shows both.

The noble gases helium and neon have filled valence shells. The valence shell of helium is filled with two electrons; that of neon is filled with eight electrons. Neon and argon have in common an electron configuration in which the s and p orbitals of their valence shells are filled with eight electrons. The valence shells of all other elements shown in Table 1.4 contain fewer than eight electrons.

Valence electrons Electrons in the valence (outermost) shell of an atom.

Valence shell The outermost electron shell of an atom.

Lewis structure of an atom The symbol of an element surrounded by a number of dots equal to the number of electrons in the valence shell of the atom.

Compare the Lewis structures given in Table 1.4 with the ground-state electron configurations listed in Table 1.3. Table 1.4 shows the Lewis structure of boron (B), for example, with three valence electrons: the two paired $2s$ electrons and the single $2p_x$ electron listed in Table 1.3. Table 1.4 shows the Lewis structure of carbon (C) with four valence electrons: the two paired $2s$ electrons and the single $2p_x$ and $2p_y$ electrons listed in Table 1.3.

Notice also from Table 1.4 that, for C, N, O, and F in period 2 of the Periodic Table, the valence electrons belong to the second shell. It requires 8 electrons to fill this shell. For Si, P, S, and Cl in period 3 of the Periodic Table, the valence electrons belong to the third shell. With 8 electrons, this shell is only partially filled: the $3s$ and $3p$ orbitals are fully occupied, but the five $3d$ orbitals can accommodate an additional 10 valence electrons. Because of the differences in number and kind of valence shell orbitals available to elements of the second and third periods, significant

TABLE 1.4 Lewis Structures for Elements 1–18 of the Periodic Table

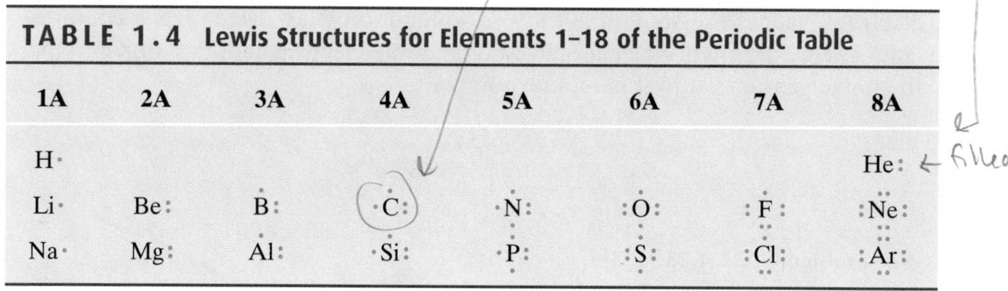

1A	2A	3A	4A	5A	6A	7A	8A
H·							He:
Li·	Be:	·B:	·C:	·N:	:O:	:F:	:Ne:
Na·	Mg:	·Al:	·Si:	·P:	:S:	:Cl:	:Ar:

differences exist in the covalent bonding of oxygen and sulfur and of nitrogen and phosphorus. For example, although oxygen and nitrogen can accommodate no more than 8 electrons in their valence shells, many phosphorus-containing compounds have 10 electrons in the valence shell of phosphorus, and many sulfur-containing compounds have 10 and even 12 electrons in the valence shell of sulfur.

Noble Gas	Noble Gas Notation
He	$1s^2$
Ne	[He] $2s^22p^6$
Ar	[Ne] $3s^23p^6$
Kr	[Ar] $4s^24p^6$
Xe	[Kr] $5s^25p^6$

1.2 What Is the Lewis Model of Bonding?

A. Formation of Ions

In 1916, Lewis devised a beautifully simple model that unified many of the observations about chemical bonding and reactions of the elements. He pointed out that the chemical inertness of the noble gases (Group 8A) indicates a high degree of stability of the electron configurations of these elements: helium with a valence shell of two electrons ($1s^2$), neon with a valence shell of eight electrons ($2s^22p^6$), argon with a valence shell of eight electrons ($3s^23p^6$), and so forth.

The tendency of atoms to react in ways that achieve an outer shell of eight valence electrons is particularly common among elements of Groups 1A–7A (the main-group elements). We give this tendency the special name, the **octet rule**. An atom with almost eight valence electrons tends to gain the needed electrons to have eight electrons in its valence shell and an electron configuration like that of the noble gas nearest it in atomic number. In gaining electrons, the atom becomes a negatively charged ion called an **anion**. An atom with only one or two valence electrons tends to lose the number of electrons required to have the same electron configuration as the noble gas nearest it in atomic number. In losing one or more electrons, the atom becomes a positively charged ion called a **cation**.

Octet rule The tendency among atoms of Group 1A–7A elements to react in ways that achieve an outer shell of eight valence electrons.

Anion An atom or group of atoms bearing a negative charge.

Cation An atom or group of atoms bearing a positive charge.

Example 1.2

Show how the loss of one electron from a sodium atom to form a sodium ion leads to a stable octet:

$$Na \longrightarrow Na^+ + e^-$$

A sodium atom — A sodium ion — An electron

Strategy

To see how this chemical change leads to a stable octet, write the condensed ground-state electron configuration for a sodium atom and for a sodium ion, and then compare the two to that of neon, the noble gas nearest to sodium in atomic number.

Solution

A sodium atom has one electron in its valence shell. The loss of this one valence electron changes the sodium atom to a sodium ion, Na^+, which has a complete octet of electrons in its valence shell and the same electron configuration as neon, the noble gas nearest to it in atomic number.

Na (11 electrons): $1s^22s^22p^6\,3s^1$

Na^+ (10 electrons): $1s^22s^22p^6$

Ne (10 electrons): $1s^22s^22p^6$

See problems 1.22, 1.23

Problem 1.2

Show how the gain of two electrons by a sulfur atom to form a sulfide ion leads to a stable octet:

$$S + 2e^- \longrightarrow S^{2-}$$

B. Formation of Chemical Bonds

According to the Lewis model of bonding, atoms interact with each other in such a way that each atom participating in a chemical bond acquires a valence-shell electron configuration the same as that of the noble gas closest to it in atomic number. Atoms acquire completed valence shells in two ways:

1. An atom may lose or gain enough electrons to acquire a filled valence shell. An atom that gains electrons becomes an anion, and an atom that loses electrons becomes a cation. A chemical bond between an anion and a cation is called an **ionic bond.**
2. An atom may share electrons with one or more other atoms to acquire a filled valence shell. A chemical bond formed by sharing electrons is called a **covalent bond.**

Ionic bond A chemical bond resulting from the electrostatic attraction of an anion and a cation.

Covalent bond A chemical bond resulting from the sharing of one or more pairs of electrons.

Electronegativity A measure of the force of an atom's attraction for electrons it shares in a chemical bond with another atom.

We now ask how we can find out whether two atoms in a compound are joined by an ionic bond or a covalent bond. One way to answer this question is to consider the relative positions of the two atoms in the Periodic Table. Ionic bonds usually form between a metal and a nonmetal. An example of an ionic bond is that formed between the metal sodium and the nonmetal chlorine in the compound sodium chloride, Na^+Cl^-. By contrast, when two nonmetals or a metalloid and a nonmetal combine, the bond between them is usually covalent. Examples of compounds containing covalent bonds between nonmetals include Cl_2, H_2O, CH_4, and NH_3. Examples of compounds containing covalent bonds between a metalloid and a nonmetal include BF_3, $SiCl_4$, and AsH_4.

Another way to identify the type of bond is to compare the electronegativities of the atoms involved, which is the subject of the next subsection.

C. Electronegativity and Chemical Bonds

Electronegativity is a measure of the force of an atom's attraction for electrons that it shares in a chemical bond with another atom. The most widely used scale of electronegativities (Table 1.5) was devised by Linus Pauling in the 1930s. On the Pauling scale, fluorine, the most electronegative element, is assigned an electronegativity of 4.0, and all other elements are assigned values in relation to fluorine.

As you study the electronegativity values in this table, note that they generally increase from left to right within a period of the Periodic Table and generally increase from bottom to top within a group. Values increase from left to right because of the increasing positive charge on the nucleus, which leads to a stronger attraction for electrons in the valence shell. Values increase going up a column because of the decreasing distance of the valence electrons from the nucleus, which leads to stronger attraction between a nucleus and its valence electrons.

Note that the values given in Table 1.5 are only approximate. The electronegativity of a particular element depends not only on its position in the Periodic Table, but also on its oxidation state. The electronegativity of Cu(I) in Cu_2O, for example, is 1.8, whereas the electronegativity of Cu(II) in CuO is 2.0. In spite of these variations, electronegativity is still a useful guide to the distribution of electrons in a chemical bond.

Linus Pauling (1901–1994) was the first person ever to receive two unshared Nobel Prizes. He received the Nobel Prize for Chemistry in 1954 for his contributions to the nature of chemical bonding. He received the Nobel Prize for Peace in 1962 for his efforts on behalf of international control of nuclear weapons and against nuclear testing. *(©UPI/Corbis)*

Electronegativity increases →

Electronegativity increases ↑

Li	Be	B	C	N	O	F
Na	Mg	Al	Si	P	S	Cl
K	Ca					Br
						I

Partial Periodic Table showing commonly encountered elements in organic chemistry. Electronegativity generally increases from left to right within a period and from bottom to top within a group. Hydrogen is less electronegative than the elements in red and more electronegative than those in blue. Hydrogen and phosphorus have the same electronegativity on the Pauling scale.

TABLE 1.5 Electronegativity Values for Some Atoms (Pauling Scale)

1A	2A	3B	4B	5B	6B	7B	8B			1B	2B	3A	4A	5A	6A	7A
								H 2.1								
Li 1.0	Be 1.5											B 2.0	C 2.5	N 3.0	O 3.5	F 4.0
Na 0.9	Mg 1.2											Al 1.5	Si 1.8	P 2.1	S 2.5	Cl 3.0
K 0.8	Ca 1.0	Sc 1.3	Ti 1.5	V 1.6	Cr 1.6	Mn 1.5	Fe 1.8	Co 1.8	Ni 1.8	Cu 1.9	Zn 1.6	Ga 1.6	Ge 1.8	As 2.0	Se 2.4	Br 2.8
Rb 0.8	Sr 1.0	Y 1.2	Zr 1.4	Nb 1.6	Mo 1.8	Tc 1.9	Ru 2.2	Rh 2.2	Pd 2.2	Ag 1.9	Cd 1.7	In 1.7	Sn 1.8	Sb 1.9	Te 2.1	I 2.5
Cs 0.7	Ba 0.9	La 1.1	Hf 1.3	Ta 1.5	W 1.7	Re 1.9	Os 2.2	Ir 2.2	Pt 2.2	Au 2.4	Hg 1.9	Tl 1.8	Pb 1.8	Bi 1.9	Po 2.0	At 2.2

<1.0   1.5 – 1.9   2.5 – 2.9

1.0 – 1.4   2.0 – 2.4   3.0 – 4.0

Example 1.3

Judging from their relative positions in the Periodic Table, which element in each pair has the larger electronegativity?

(a) Lithium or carbon (b) Nitrogen or oxygen
(c) Carbon or oxygen

Strategy

Determine whether the pair resides in the same period (row) or group (column) of the Periodic Table. For those in the same period, electronegativity increases from left to right. For those in the same group, electronegativity increases from bottom to top.

Solution

The elements in these pairs are all in the second period of the Periodic Table. Electronegativity in this period increases from left to right.

(a) C > Li (b) O > N (c) O > C

See problem 1.24

Problem 1.3

Judging from their relative positions in the Periodic Table, which element in each pair has the larger electronegativity?

(a) Lithium or potassium (b) Nitrogen or phosphorus
(c) Carbon or silicon

Ionic Bonds

An ionic bond forms by the transfer of electrons from the valence shell of an atom of lower electronegativity to the valence shell of an atom of higher electronegativity. The more electronegative atom gains one or more valence electrons and becomes an anion; the less electronegative atom loses one or more valence electrons and becomes a cation.

As a guideline, we say that this type of electron transfer to form an ionic compound is most likely to occur if the difference in electronegativity between two atoms is approximately 1.9 or greater. A bond is more likely to be covalent if this difference is less than 1.9. Note that the value 1.9 is somewhat arbitrary: Some chemists prefer a slightly larger value, others a slightly smaller value. The essential point is that the value 1.9 gives us a guidepost against which to decide whether a bond is more likely to be ionic or more likely to be covalent.

An example of an ionic bond is that formed between sodium (electronegativity 0.9) and fluorine (electronegativity 4.0). The difference in electronegativity between these two elements is 3.1. In forming Na^+F^-, the single 3*s* valence electron of sodium is transferred to the partially filled valence shell of fluorine:

$$Na(1s^22s^22p^6\,3s^1) + F(1s^22s^2\,2p^5) \longrightarrow Na^+(1s^22s^22p^6) + F^-(1s^22s^2\,2p^6)$$

As a result of this transfer of one electron, both sodium and fluorine form ions that have the same electron configuration as neon, the noble gas closest to each in atomic number. In the following equation, we use a single-headed curved arrow to show the transfer of one electron from sodium to fluorine:

$$Na\cdot + \cdot\ddot{\underset{..}{F}}\!: \longrightarrow Na^+ \; :\ddot{\underset{..}{F}}\!:^-$$

Covalent Bonds

A covalent bond forms when electron pairs are shared between two atoms whose difference in electronegativity is 1.9 or less. According to the Lewis model, an electron pair in a covalent bond functions in two ways simultaneously: It is shared by two atoms, and, at the same time, it fills the valence shell of each atom.

The simplest example of a covalent bond is that in a hydrogen molecule, H_2. When two hydrogen atoms bond, the single electrons from each atom combine to form an electron pair with the release of energy. A bond formed by sharing a pair of electrons is called a *single bond* and is represented by a single line between the two atoms. The electron pair shared between the two hydrogen atoms in H_2 completes the valence shell of each hydrogen. Thus, in H_2, each hydrogen has two electrons in its valence shell and an electron configuration like that of helium, the noble gas nearest to it in atomic number:

$$H\cdot + \cdot H \longrightarrow H{-}H \qquad \Delta H^0 = -435\text{ kJ/mol } (-104\text{ kcal/mol})$$

The Lewis model accounts for the stability of covalently bonded atoms in the following way: In forming a covalent bond, an electron pair occupies the region between two nuclei and serves to shield one positively charged nucleus from the repulsive force of the other positively charged nucleus. At the same time, an electron pair attracts both nuclei. In other words, an electron pair in the space between two nuclei bonds them together and fixes the internuclear distance to within very narrow limits. The distance between nuclei participating in a chemical bond is called a **bond length**. Every covalent bond has a definite bond length. In H—H, it is 74 pm, where 1 pm $= 10^{-12}$ m.

Although all covalent bonds involve the sharing of electrons, they differ widely in the degree of sharing. We classify covalent bonds into two categories—nonpolar covalent and polar covalent—depending on the difference in electronegativity between the bonded atoms. In a **nonpolar covalent bond**, electrons are shared equally. In a **polar covalent bond**, they are shared unequally. It is important to realize that no sharp line divides these two categories, nor, for that matter, does a sharp line divide polar covalent bonds and ionic bonds. Nonetheless, the rule-of-thumb guidelines in Table 1.6 will help you decide whether a given bond is more likely to be nonpolar covalent, polar covalent, or ionic.

Nonpolar covalent bond A covalent bond between atoms whose difference in electronegativity is less than approximately 0.5.

Polar covalent bond A covalent bond between atoms whose difference in electronegativity is between approximately 0.5 and 1.9.

TABLE 1.6 Classification of Chemical Bonds

Difference in Electronegativity between Bonded Atoms	Type of Bond	Most Likely Formed Between
Less than 0.5	Nonpolar covalent	Two nonmetals or a nonmetal and a metalloid
0.5 to 1.9	Polar covalent	(Two nonmetals or a nonmetal and a metalloid)
Greater than 1.9	Ionic	A metal and a nonmetal

A covalent bond between carbon and hydrogen, for example, is classified as nonpolar covalent because the difference in electronegativity between these two atoms is 2.5 − 2.1 = 0.4 unit. An example of a polar covalent bond is that of H—Cl. The difference in electronegativity between chlorine and hydrogen is 3.0 − 2.1 = 0.9 unit.

Example 1.4

Classify each bond as nonpolar covalent, polar covalent, or ionic:

(a) O—H (b) N—H (c) Na—F (d) C—Mg

Strategy

Use the difference in electronegativity between the two atoms and compare this value with the range of values given in Table 1.6.

Solution

On the basis of differences in electronegativity between the bonded atoms, three of these bonds are polar covalent and one is ionic:

Bond	Difference in Electronegativity	Type of Bond
(a) O—H	3.5 − 2.1 = 1.4	polar covalent
(b) N—H	3.0 − 2.1 = 0.9	polar covalent
(c) Na—F	4.0 − 0.9 = 3.1	ionic
(d) C—Mg	2.5 − 1.2 = 1.3	polar covalent

See problem 1.25

Problem 1.4

Classify each bond as nonpolar covalent, polar covalent, or ionic:

(a) S—H (b) P—H (c) C—F (d) C—Cl

An important consequence of the unequal sharing of electrons in a polar covalent bond is that the more electronegative atom gains a greater fraction of the shared electrons and acquires a partial negative charge, which we indicate by the symbol $\delta-$ (read "delta minus"). The less electronegative atom has a lesser fraction of the shared electrons and acquires a partial positive charge, which we indicate by the symbol $\delta+$ (read "delta plus"). This separation of charge produces a **dipole** (two poles). We can also show the presence of a bond dipole by an arrow, with the head of the arrow near the negative end of the dipole and a cross on the tail of the arrow near the positive end (Figure 1.4).

We can display the polarity of a covalent bond by a type of molecular model called an *electron density model.* In this type of model, a blue color shows the presence of a $\delta+$ charge, and a red color shows the presence of a $\delta-$ charge. Figure 1.4 shows an electron density model of HCl. The ball-and-stick model in the center shows the orientation of the two atoms in space. The transparent surface surrounding the ball-and-stick model shows the relative sizes of the atoms (equivalent to the size shown by a space-filling model). Colors on the surface show the distribution of electron density. We see by the blue color that hydrogen bears a $\delta+$ charge and by the red color that chlorine bears a $\delta-$ charge.

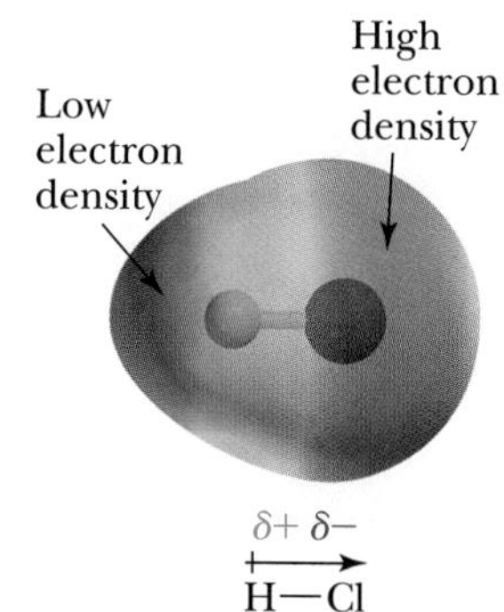

Figure 1.4
An electron density model of HCl. Red indicates a region of high electron density, and blue indicates a region of low electron density.

Example 1.5

Using the symbols $\delta-$ and $\delta+$, indicate the direction of polarity in these polar covalent bonds:

(a) C—O (b) N—H (c) C—Mg

Strategy

To determine the polarity of a covalent bond and the direction of the polarity, compare the electronegativities of the bonded atoms.

Solution

For (a), carbon and oxygen are both in period 2 of the Periodic Table. Because oxygen is farther to the right than carbon, it is more electronegative. For (b), nitrogen is more electronegative than hydrogen. For (c), magnesium is a metal located at the far left of the Periodic Table, and carbon is a nonmetal located at the right. All nonmetals, including hydrogen, have a greater electronegativity than do the metals in columns 1A and 2A. The electronegativity of each element is given below the symbol of the element:

(a) $\overset{\delta+}{\text{C}}—\overset{\delta-}{\text{O}}$ (b) $\overset{\delta-}{\text{N}}—\overset{\delta+}{\text{H}}$ (c) $\overset{\delta-}{\text{C}}—\overset{\delta+}{\text{Mg}}$

2.5 3.5 3.0 2.1 2.5 1.2

See problems 1.26, 1.38, 1.40

Problem 1.5

Using the symbols $\delta-$ and $\delta+$, indicate the direction of polarity in these polar covalent bonds:

(a) C—N (b) N—O (c) C—Cl

In summary, the twin concepts of electronegativity and the polarity of covalent bonds will be very helpful in organic chemistry as a guide to locating centers of chemical reactions. In many of the reactions we will study, reaction is initiated by the attraction between a center of partial positive charge and a center of partial negative charge.

HOW TO 1.1 Draw Lewis Structures of Molecules and Ions

The ability to draw Lewis structures for molecules and ions is a fundamental skill in the study of organic chemistry. The following steps will help you to do this (as you study these guidelines, look at the examples in Table 1.7). As an example, let us draw a Lewis structure of acetic acid, molecular formula $C_2H_4O_2$. It's structural formula, CH_3COOH, gives a hint of the connectivity.

Step 1: *Determine the number of valence electrons in the molecule or ion.*
To do so, add the number of valence electrons contributed by each atom. For ions, add one electron for each negative charge on the ion, and subtract one electron for each positive charge on the ion. For example, the Lewis structure of the water molecule, H_2O, must show eight valence electrons: one from each hydrogen and six from oxygen. The Lewis structure for the hydroxide ion, OH^-, must also show eight valence electrons: one from hydrogen, six from oxygen, plus one for the negative charge on the ion. For acetic acid the molecular formula is $C_2H_4O_2$. The Lewis structure must show 8(2 carbons) + 4(4 hydrogens) + 12(2 oxygens) = 24 valence electrons.

Step 2: *Determine the arrangement of atoms in the molecule or ion.*
This step is the most difficult part of drawing a Lewis structure. Fortunately, the structural formula of a compound can provide valuable information about connectivity. The order in which the atoms are listed in a structural formula is a guide. For example, the CH_3 part of the structural formula of acetic acid tells you that three hydrogen atoms are bonded to the carbon written on the left, and the COOH part tells you that both oxygens are bonded to the same carbon and a hydrogen is bonded to one of the oxygens.

```
     H    O
     |   /
  H—C—C
     |   \
     H    O—H
```

Except for the simplest molecules and ions, the connectivity must be determined experimentally. For some molecules and ions we give as examples, we ask you to propose a connectivity of the atoms. For most, however, we give you the experimentally determined arrangement.

Bonding electrons Valence electrons shared in a covalent bond.

Nonbonding electrons Valence electrons not involved in forming covalent bonds, that is, unshared electrons.

Step 3: *Arrange the remaining electrons in pairs so that each atom in the molecule or ion has a complete outer shell. Show a pair of* ***bonding electrons*** *as a single line between the bonded atoms; show a pair of* ***nonbonding electrons*** *as a pair of Lewis dots.*
To accomplish this, connect the atoms with single bonds. Then arrange the remaining electrons in pairs so that each atom in the molecule or ion has a complete outer shell. Each hydrogen atom must be surrounded by two electrons. Each atom of carbon, oxygen, and nitrogen, as well as each halogen, must be surrounded by eight electrons (per the octet rule). Recall that each neutral carbon atom has four valence electrons and each neutral oxygen atom has six valence electrons. The structure here shows the required 24 valence electrons. The left carbon has four single bonds and a complete valence shell. Each hydrogen also has a complete valence shell. The lower oxygen has

```
     H   ·Ö:
     |   /
  H—C—C
     |  :\
     H    Ö—H
```

two single bonds and two unshared pairs of electrons and, therefore, has a complete valence shell. The original six valence electrons of the upper oxygen are accounted for, but it does not yet have a filled valence shell. Similarly, the original four valence electrons of the right carbon atom are accounted for but it still does not have a complete valence shell.

Notice that in the structure so far, we have accounted for all valence electrons but two atoms do not yet have completed valence shells. Furthermore, one carbon atom and one oxygen atom each have a single unpaired electron.

Step 4: *Use multiple bonds where necessary to eliminate unpaired electrons.* In a **single bond**, two atoms share one pair of electrons. It is sometimes necessary for atoms to share more than one pair of electrons. In a **double bond**, they share two pairs of electrons; we show a double bond by drawing two parallel lines between the bonded atoms. In a **triple bond**, two atoms share three pairs of electrons; we show a triple bond by three parallel lines between the bonded atoms. The following structure combines the unpaired electrons on carbon and oxygen and creates a double bond between the carbon and oxygen. The Lewis structure is now complete.

```
    H    ..
    |    O:
 H—C—C//
    |    \..
    H     O—H
          ..
```

From the study of the compounds in Table 1.7 and other organic compounds, we can make the following generalizations: In neutral (uncharged) organic compounds,

- H has one bond
- C has four bonds
- N has three bonds and one unshared pair of electrons
- O has two bonds and two unshared pair of electrons
- F, Cl, Br, and I have one bond and three unshared pairs of electrons

TABLE 1.7 Lewis Structures for Several Compounds. The number of valence electrons in each molecule is given in parentheses after the molecule's molecular formula.

H—Ö—H (with two unshared pairs on O)	H—N̈—H, with H below N	H—C—H, with H above and below C	H—C̈l:
H_2O (8) Water	NH_3 (8) Ammonia	CH_4 (8) Methane	HCl (8) Hydrogen chloride
H₂C=CH₂ (H, H on each C)	H—C≡C—H	H₂C=Ö	H—Ö—C(=Ö)—Ö—H
C_2H_4 (12) Ethylene	C_2H_2 (10) Acetylene	CH_2O (12) Formaldehyde	H_2CO_3 (24) Carbonic acid

Example 1.6

Draw Lewis structures, showing all valence electrons, for these molecules:

(a) H_2O_2 (b) CH_3OH (c) CH_3Cl

Strategy

Determine the number of valence electrons and the connectivity of the atoms in each molecule. Connect the bonded atoms by single bonds and then arrange the remaining valence electrons so that each atom has a filled valence shell.

Solution

(a) A Lewis structure for hydrogen peroxide, H_2O_2, must show 6 valence electrons from each oxygen and 1 from each hydrogen, for a total of $12 + 2 = 14$ valence electrons. We know that hydrogen forms only one covalent bond, so the connectivity of the atoms must be as follows:

H—O—O—H

The three single bonds account for 6 valence electrons. We place the remaining 8 valence electrons on the oxygen atoms to give each a complete octet:

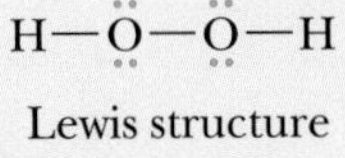

Lewis structure

Ball-and-stick models show only nuclei and covalent bonds; they do not show unshared pairs of electrons

(b) A Lewis structure for methanol, CH_3OH, must show 4 valence electrons from carbon, 1 from each hydrogen, and 6 from oxygen, for a total of $4 + 4 + 6 = 14$ valence electrons. The connectivity of the atoms in methanol is given on the left. The five single bonds in this partial structure account for 10 valence electrons. We place the remaining 4 valence electrons on oxygen as two Lewis dot pairs to give it a complete octet.

```
    H                H
    |                |
H—C—O—H          H—C—Ö—H
    |                |
    H                H
```

The order of attachment of atoms

Lewis structure

(c) A Lewis structure for chloromethane, CH_3Cl, must show 4 valence electrons from carbon, 1 from each hydrogen, and 7 from chlorine, for a total of $4 + 3 + 7 = 14$. Carbon has four bonds, one to each of the hydrogens and one to chlorine. We place the remaining 6 valence electrons on chlorine as three Lewis dot pairs to complete its octet.

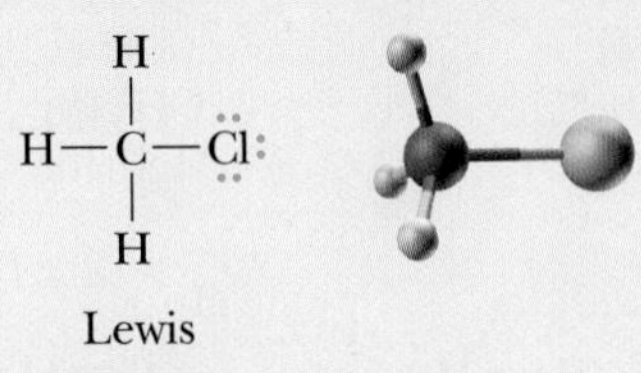

Lewis structure

See problems 1.27, 1.28

Problem 1.6

Draw Lewis structures, showing all valence electrons, for these molecules:

(a) C_2H_6 (b) CS_2 (c) HCN

E. Formal Charge

Throughout this course, we deal not only with molecules, but also with polyatomic cations and polyatomic anions. Examples of polyatomic cations are the hydronium ion, H_3O^+, and the ammonium ion, NH_4^+. An example of a polyatomic anion is the bicarbonate ion, HCO_3^-. It is important that you be able to determine which atom or atoms in a molecule or polyatomic ion bear the positive or negative charge. The charge on an atom in a molecule or polyatomic ion is called its **formal charge**. To derive a formal charge,

Formal charge The charge on an atom in a molecule or polyatomic ion.

Step 1: *Write a correct Lewis structure for the molecule or ion.*

Step 2: *Assign to each atom all its unshared (nonbonding) electrons and one-half its shared (bonding) electrons.*

Step 3: *Compare the number arrived at in Step 2 with the number of valence electrons in the neutral, unbonded atom.* If the number of electrons assigned to a bonded atom is less than that assigned to the unbonded atom, then more positive charges are in the nucleus than counterbalancing negative charges, and the atom has a positive formal charge. Conversely, if the number of electrons assigned to a bonded atom is greater than that assigned to the unbonded atom, then the atom has a negative formal charge.

$$\text{Formal charge} = \begin{matrix}\text{Number of valence}\\ \text{electrons in neutral}\\ \text{unbonded atom}\end{matrix} - \left(\begin{matrix}\text{All unshared}\\ \text{electrons}\end{matrix} + \begin{matrix}\text{One-half of all}\\ \text{shared electrons}\end{matrix}\right)$$

Example 1.7

Draw Lewis structures for these ions, and show which atom in each bears the formal charge:

(a) H_3O^+ (b) CH_3O^-

Strategy

Draw a correct Lewis structure molecule showing all valence electrons on each atom. Then determine the location of the formal charge.

Solution

(a) The Lewis structure for the hydronium ion must show 8 valence electrons: 3 from the three hydrogens, 6 from oxygen, minus 1 for the single positive charge. A neutral, unbonded oxygen atom has 6 valence electrons. To the oxygen atom in H_3O^+, we assign two unshared electrons and one from each shared pair of electrons, giving it a formal charge of $6 - (2 + 3) = +1$.

assigned 5 valence electrons: formal charge of +1

H—Ö⁺—H
|
H

do you have to draw it this way?

(b) The Lewis structure for the methoxide ion, CH_3O^-, must show 14 valence electrons: 4 from carbon, 6 from oxygen, 3 from the hydrogens, plus 1 for the single negative charge. To carbon, we assign 1 electron from each shared pair, giving it a formal charge of $4 - 4 = 0$. To oxygen, we assign 7 valence electrons, giving it a formal charge of $6 - 7 = -1$.

assigned 7 valence electrons: formal charge of −1

H—C(H)(H)—Ö:⁻

See problems 1.30–1.32, 1.34

Problem 1.7

Draw Lewis structures for these ions, and show which atom in each bears the formal charge(s):

(a) $CH_3NH_3^+$ (b) CH_3^+

In writing Lewis structures for molecules and ions, you must remember that elements of the second period, including carbon, nitrogen, and oxygen, can accommodate no more than eight electrons in the four orbitals ($2s, 2p_x, 2p_y$, and $2p_z$) of their valence shells. Following are two Lewis structures for nitric acid, HNO_3, each with the correct number of valence electrons, namely, 24; one structure is acceptable and the other is not:

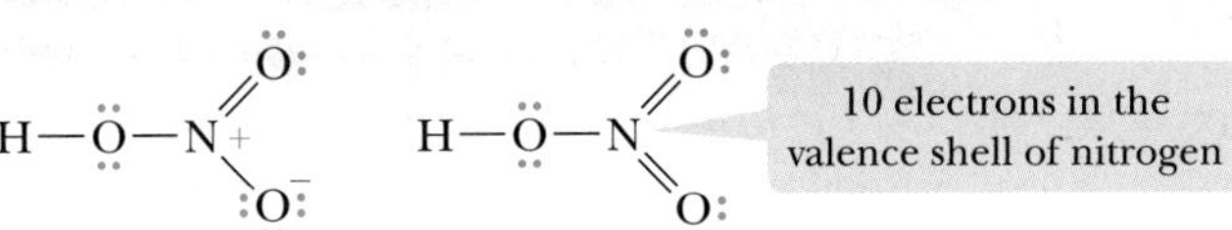

An acceptable Lewis structure / Not an acceptable Lewis structure

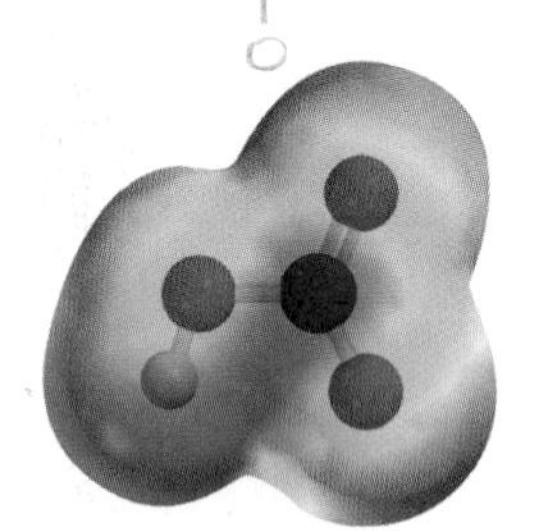

The Lewis structure of HNO_3 shows the negative formal charge localized on one of the oxygen atoms. The electron density model, on the other hand, shows that the negative charge is distributed equally over the two oxygen atoms on the right. The concept of resonance can explain this phenomenon and will be discussed in Section 1.6. Notice also the intense blue color on nitrogen, which is due to its positive formal charge.

The structure on the left is an acceptable Lewis structure. It shows the required 24 valence electrons, and each oxygen and nitrogen has a completed valence shell of 8 electrons. Further, the structure on the left shows a positive formal charge on nitrogen and a negative formal charge on one of the oxygens. An acceptable Lewis structure must show these formal charges. The structure on the right is *not* an acceptable Lewis structure. Although it shows the correct number of valence electrons, it places 10 electrons in the valence shell of nitrogen, yet the four orbitals of the second shell ($2s, 2p_x, 2p_y$, and $2p_z$) can hold no more than 8 valence electrons!

1.3 How Do We Predict Bond Angles and the Shapes of Molecules?

In Section 1.2, we used a shared pair of electrons as the fundamental unit of a covalent bond and drew Lewis structures for several small molecules containing various combinations of single, double, and triple bonds. (See, for example, Table 1.7.) We can predict bond angles in these and other molecules in a very straightforward way by using the concept of **valence-shell electron-pair repulsion (VSEPR)**. According

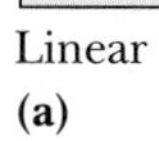

Linear (a) Trigonal planar (b) Tetrahedral (c)

Figure 1.5
Balloon models used to predict bond angles. (a) Two balloons assume a linear shape with a bond angle of 180° about the tie point. (b) Three balloons assume a trigonal planar shape with bond angles of 120° about the tie point. (c) Four balloons assume a tetrahedral shape with bond angles of 109.5° about the tie point. *(Charles D. Winters)*

(a)

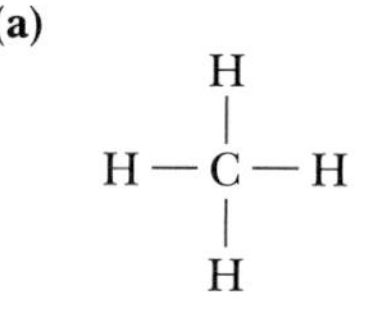

(b)

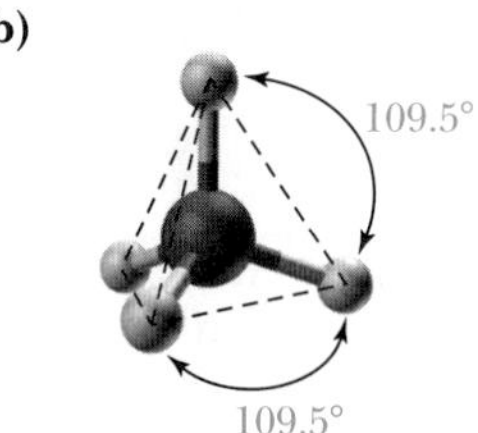

Figure 1.6
The shape of a methane molecule, CH_4. (a) Lewis structure and (b) shape. The hydrogens occupy the four corners of a regular tetrahedron, and all H—C—H bond angles are 109.5°.

to this concept, the valence electrons of an atom may be involved in the formation of single, double, or triple bonds, or they may be unshared. Each combination creates a negatively charged region of space. Because like charges repel each other, the various regions of electron density around an atom spread so that each is as far away from the others as possible.

You can demonstrate the bond angles predicted by VSEPR in a very simple way. Imagine that a balloon represents a region of electron density. If you tie two balloons together by their ends, they assume the shapes shown in Figure 1.5. The point where they are tied together represents the atom about which you want to predict a bond angle, and the balloons represent regions of electron density about that atom.

We use VSEPR in the following way to predict the shape of methane, CH_4. The Lewis structure for CH_4 shows a carbon atom surrounded by four regions of electron density. Each region contains a single pair of electrons forming a bond to a hydrogen atom. According to VSEPR, the four regions radiate from carbon so that they are as far away from each other as possible. The maximum separation occurs when the angle between any two pairs of electrons is 109.5°. Therefore, we predict that the H—C—H bond angles are 109.5° and that the shape of the molecule is **tetrahedral** (Figure 1.6). The H—C—H bond angles in methane have been measured experimentally and found to be 109.5°. Thus, the bond angles and shape of methane predicted by VSEPR are identical to those observed.

We can predict the shape of an ammonia molecule, NH_3, in the same manner. The Lewis structure of NH_3 shows nitrogen surrounded by four regions of electron density. Three regions contain single pairs of electrons forming covalent bonds with hydrogen atoms. The fourth region contains an unshared pair of electrons (Figure 1.7). Using VSEPR, we predict that the four regions of electron density around nitrogen are arranged in a tetrahedral manner and that all H—N—H bond angles are 109.5°. The observed bond angles are 107.3°. We account for this small difference between the predicted and observed angles by proposing that the unshared pair of electrons on nitrogen repels adjacent electron pairs more strongly than bonding pairs repel each other.

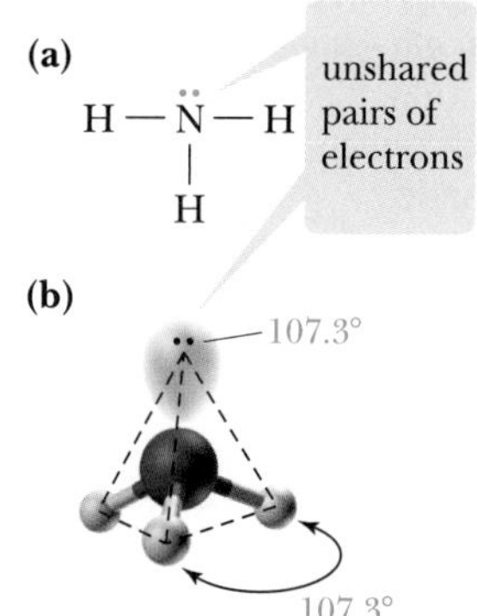

Figure 1.7
The shape of an ammonia molecule, NH_3. (a) Lewis structure and (b) ball-and-stick model. We describe the geometry of an ammonia molecule as **pyramidal**; that is, the molecule has a shape like a triangular-based pyramid with the three hydrogens at the base and nitrogen at the apex.

Figure 1.8 shows a Lewis structure and a ball-and-stick model of a water molecule. In H_2O, oxygen is surrounded by four regions of electron density. Two of these regions contain single pairs of electrons forming covalent bonds to two hydrogens; the remaining two contain unshared electron pairs. Using VSEPR, we predict that the four regions of electron density around oxygen are arranged in a tetrahedral manner and that the H—O—H bond angle is 109.5°. Experimental measurements show that the actual H—O—H bond angle is 104.5°, a value smaller than that predicted. We explain this difference between the predicted and observed bond angle by proposing, as we did for NH_3, that unshared pairs of electrons repel adjacent

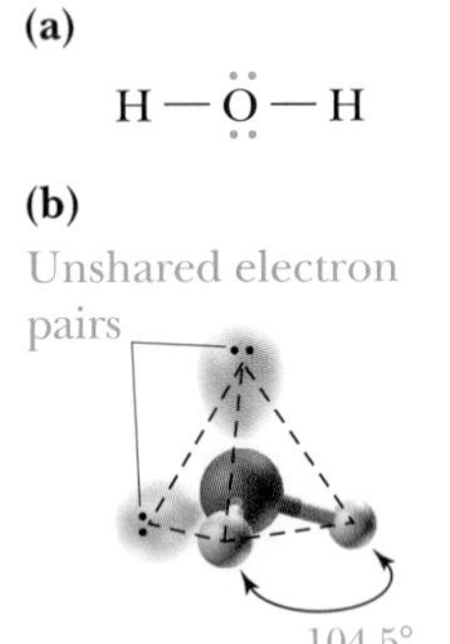

Figure 1.8
The shape of a water molecule, H_2O. (a) A Lewis structure and (b) a ball-and-stick model.

pairs more strongly than do bonding pairs. Note that the distortion from 109.5° is greater in H_2O, which has two unshared pairs of electrons, than it is in NH_3, which has only one unshared pair.

A general prediction emerges from this discussion of the shapes of CH_4, NH_3, and H_2O molecules. If a Lewis structure shows four regions of electron density around an atom, then VSEPR predicts a tetrahedral distribution of electron density and bond angles of approximately 109.5°.

In many of the molecules we encounter, an atom is surrounded by three regions of electron density. Figure 1.9 shows Lewis structures for formaldehyde (CH_2O) and ethylene (C_2H_4).

Using VSEPR, we treat a double bond as a single region of electron density. In formaldehyde, carbon is surrounded by three regions of electron density: Two regions contain single pairs of electrons, which form single bonds to hydrogen atoms; the third region contains two pairs of electrons, which form a double bond to oxygen. In ethylene, each carbon atom is also surrounded by three regions of electron density; two contain single pairs of electrons, and the third contains two pairs of electrons.

Three regions of electron density about an atom are farthest apart when they lie in a plane and make angles of 120° with each other. Thus, we predict that the H—C—H and H—C—O bond angles in formaldehyde and the H—C—H and H—C—C bond angles in ethylene are all 120°.

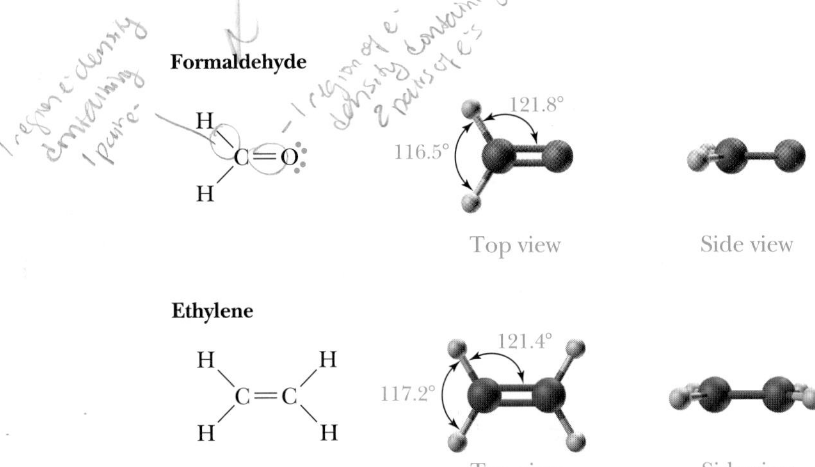

Figure 1.9
Shapes of formaldehyde (CH_2O) and ethylene (C_2H_4). We describe the geometry about each carbon atom as **trigonal planar**.

In still other types of molecules, a central atom is surrounded by only two regions of electron density. Figure 1.10 shows Lewis structures and ball-and-stick models of carbon dioxide (CO_2) and acetylene (C_2H_2).

In carbon dioxide, carbon is surrounded by two regions of electron density, each containing two pairs of electrons and forming a double bond to an oxygen atom. In acetylene, each carbon is also surrounded by two regions of electron density, one containing a single pair of electrons and forming a single bond to a hydrogen atom

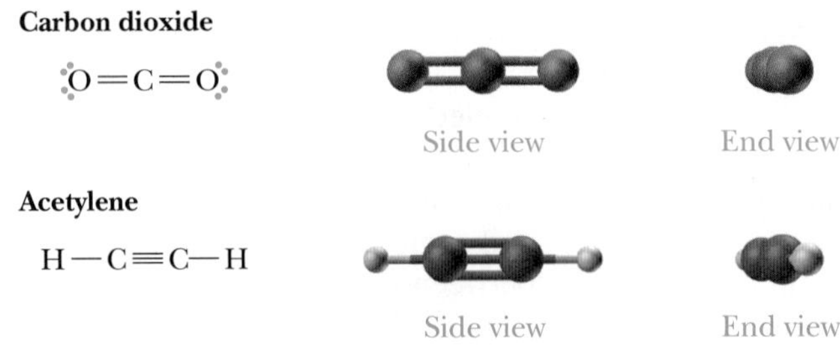

Figure 1.10
Shapes of (a) carbon dioxide (CO_2) and (b) acetylene (C_2H_2).

TABLE 1.8 Predicted Molecular Shapes (VSEPR)

Regions of Electron Density around Central Atom	Predicted Distribution of Electron Density	Predicted Bond Angles	Examples (Shape of the Molecule)
4	Tetrahedral	109.5°	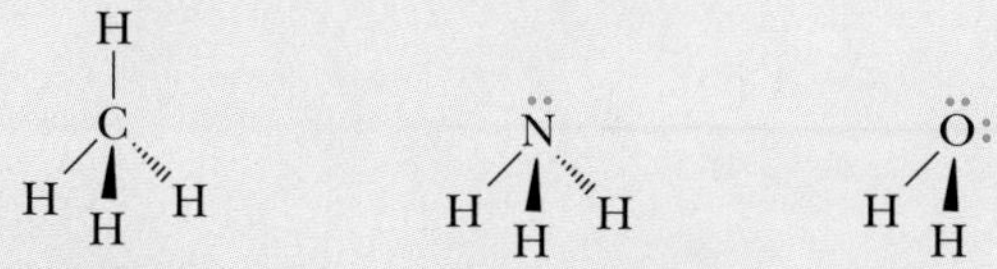 Methane (tetrahedral); Ammonia (pyramidal); Water (bent)
3	Trigonal planar	120°	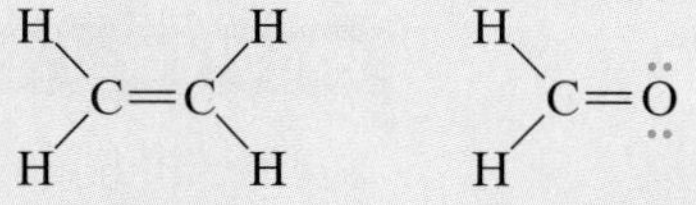 Ethylene (planar); Formaldehyde (planar)
2	Linear	180°	$\ddot{O}=C=\ddot{O}$ Carbon dioxide (linear); H—C≡C—H Acetylene (linear)

Is trigonal planar just called planar?

and the other containing three pairs of electrons and forming a triple bond to a carbon atom. In each case, the two regions of electron density are farthest apart if they form a straight line through the central atom and create an angle of 180°. Both carbon dioxide and acetylene are linear molecules. Table 1.8 summarizes the predictions of VSEPR.

Example 1.8

Predict all bond angles in these molecules:

(a) CH_3Cl (b) $CH_2{=}CHCl$

Strategy

To predict bond angles, first draw a correct Lewis structure for the molecule. Be certain to show all unpaired electrons. Then determine the number of regions of electron density (either 2, 3, or 4) around each atom and use that number to predict bond angles (either 180°, 120°, or 109.5°).

Solution

(a) The Lewis structure for CH_3Cl shows carbon surrounded by four regions of electron density. Therefore, we predict that the distribution of electron pairs about carbon is tetrahedral, that all bond angles are 109.5°, and that the shape of CH_3Cl is tetrahedral:

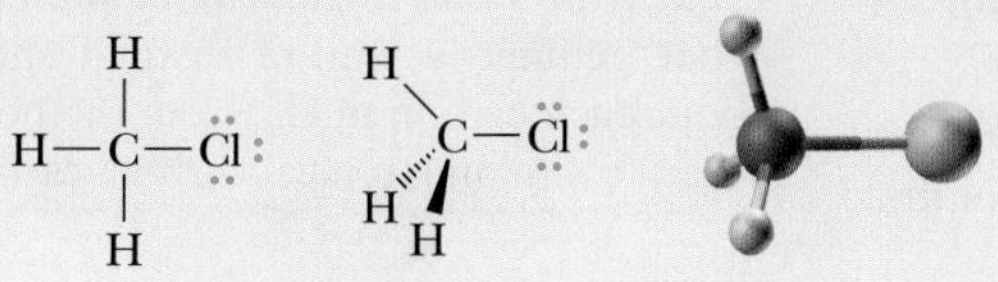

(b) The Lewis structure for $CH_2{=}CHCl$ shows each carbon surrounded by three regions of electron density. Therefore, we predict that all bond angles are 120°.

H :C̈l:
C=C
H H

(Top view)

(Viewed along the C=C bond)

See problems 1.41–1.43

Problem 1.8

Predict all bond angles for these molecules:

(a) CH_3OH (b) CH_2Cl_2 (c) H_2CO_3 (carbonic acid)

CHEMICAL CONNECTIONS 1A

Buckyball: A New Form of Carbon

Many elements in the pure state can exist in different forms. We are all familiar with the fact that pure carbon is found in two forms: graphite and diamond. These forms have been known for centuries, and it was generally believed that they were the only forms of carbon having extended networks of C atoms in well-defined structures.

But that is not so! The scientific world was startled in 1985 when Richard Smalley of Rice University and Harry W. Kroto of the University of Sussex, England, and their coworkers announced that they had detected a new form of carbon with a molecular formula C_{60}. They suggested that the molecule has a structure resembling a soccer ball: 12 five-membered rings and 20 six-membered rings arranged such that each five-membered ring is surrounded by six-membered rings. This structure reminded its discoverers of a geodesic dome, a structure invented by the innovative American engineer and philosopher R. Buckminster Fuller. Therefore, the official name of the new allotrope of carbon has become fullerene. Kroto, Smalley, and Robert F. Curl were awarded the Nobel Prize for Chemistry in 1996 for their work with fullerenes. Many higher fullerenes, such as C_{70} and C_{84}, have also been isolated and studied.

QUESTIONS

Predict the bond angles about the carbon atoms in C_{60}. What geometric feature distinguishes the bond angles about each carbon in C_{60} from the bond angles of a compound containing typical carbon–carbon bonds?

1.4 How Do We Predict If a Molecule Is Polar or Nonpolar?

In Section 1.2C, we used the terms *polar* and *dipole* to describe a covalent bond in which one atom bears a partial positive charge and the other bears a partial negative charge. We also saw that we can use the difference in electronegativity between bonded atoms to determine the polarity of a covalent bond and the direction of its polarity. We can now combine our understanding of bond polarity and molecular geometry (Section 1.3) to predict the polarity of molecules.

A molecule will be polar if (1) it has polar bonds and (2) its centers of partial positive and partial negative charge lie at different places within the molecule. Consider first carbon dioxide, CO_2, a molecule with two polar carbon–oxygen double bonds. Because carbon dioxide is a linear molecule, the centers of negative and positive partial charge coincide; therefore, this molecule is nonpolar.

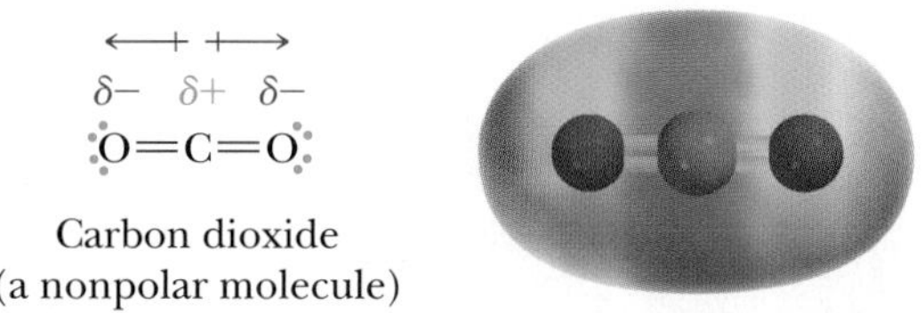
δ− δ+ δ−
:O=C=O:
Carbon dioxide
(a nonpolar molecule)

In a water molecule, each O—H bond is polar, with oxygen, the more electronegative atom, bearing a partial negative charge and each hydrogen bearing a partial positive charge. Because water is a bent molecule, the center of its partial positive charge is between the two hydrogen atoms, and the center of its partial negative charge is on oxygen. Thus, water has polar bonds and, because of its geometry, is a polar molecule.

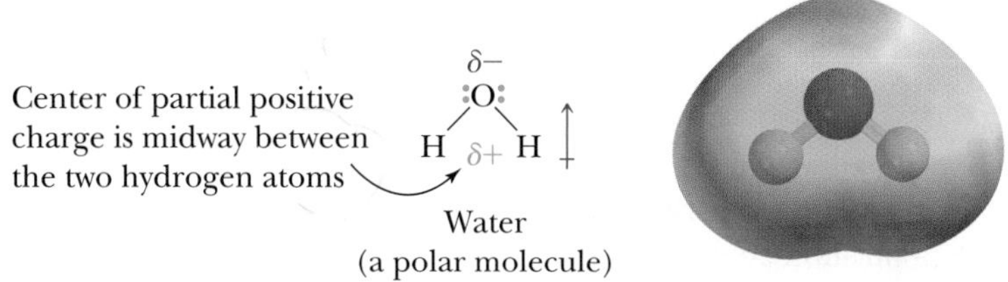
Center of partial positive charge is midway between the two hydrogen atoms
δ−
O
H δ+ H
Water
(a polar molecule)

Ammonia has three polar N—H bonds, and because of its geometry, the centers of partial positive and negative charges are at different places within the molecule. Thus, ammonia has polar bonds and, because of its geometry, is a polar molecule.

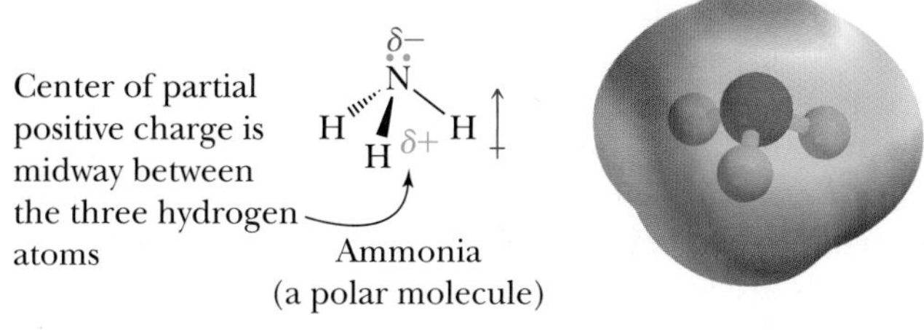
Center of partial positive charge is midway between the three hydrogen atoms
δ−
N
H H δ+ H
Ammonia
(a polar molecule)

Example 1.9

Which of these molecules are polar? For each that is polar, specify the direction of its polarity.

(a) CH_3Cl (b) CH_2O (c) C_2H_2

Strategy

To determine whether a molecule is polar, first determine if it has polar bonds, and if it does, determine whether the centers of positive and negative charge lie at the same or different places within the molecule. If they lie at the same place, the molecule is nonpolar, if they lie at different places, the molecule is polar.

Solution

Both chloromethane (CH_3Cl) and formaldehyde (CH_2O) have polar bonds and, because of their geometry, are polar molecules. Because acetylene (C_2H_2) is linear, and each of its C—H bonds is nonpolar covalent, the molecule is nonpolar.

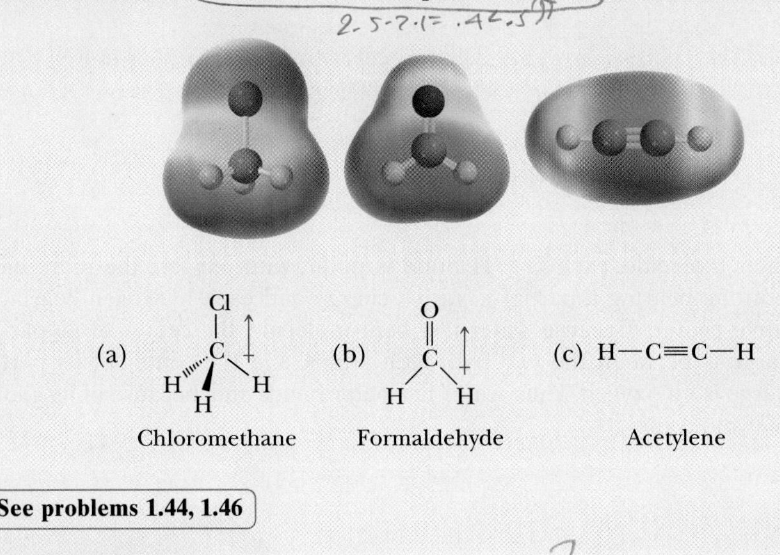

See problems 1.44, 1.46

Problem 1.9

Both carbon dioxide (CO_2) and sulfur dioxide (SO_2) are triatomic molecules. Account for the fact that carbon dioxide is a nonpolar molecule, whereas sulfur dioxide is a polar molecule.

(a) (b) (c)

Figure 1.11
Three Lewis structures for the carbonate ion.

1.5 What Is Resonance?

As chemists developed a better understanding of covalent bonding in organic compounds, it became obvious that, for a great many molecules and ions, no single Lewis structure provides a truly accurate representation. For example, Figure 1.11 shows three Lewis structures for the carbonate ion, CO_3^{2-}, each of which shows carbon bonded to three oxygen atoms by a combination of one double bond and two single bonds. Each Lewis structure implies that one carbon–oxygen bond is different from the other two. This, however, is not the case; it has been shown that all three carbon–oxygen bonds are identical.

To describe the carbonate ion, as well as other molecules and ions for which no single Lewis structure is adequate, we turn to the theory of resonance.

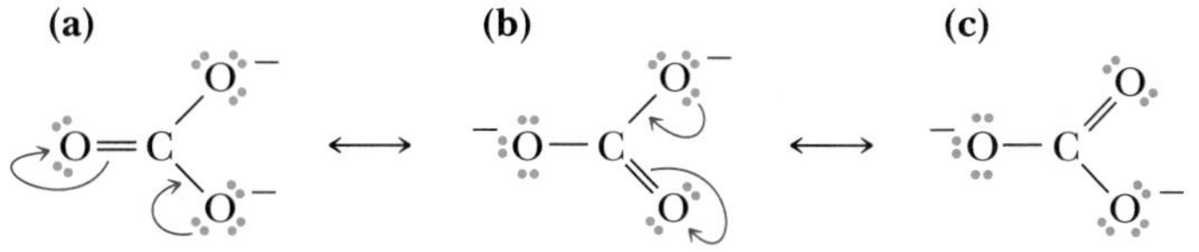

Figure 1.12
The carbonate ion represented as a hybrid of three equivalent contributing structures. Curved arrows show the redistribution of valence electrons between one contributing structure and the next.

A. The Theory of Resonance

The theory of resonance was developed by Linus Pauling in the 1930s. According to this theory, many molecules and ions are best described by writing two or more Lewis structures and considering the real molecule or ion to be a composite of these structures. We call individual Lewis structures **resonance contributing structures**. We show that the real molecule or ion is a **resonance hybrid** of the various contributing structures by interconnecting them with **double-headed arrows**.

Resonance contributing structures Representations of a molecule or ion that differ only in the distribution of valence electrons.

Resonance hybrid A molecule or ion that is best described as a composite of a number of contributing structures.

Double-headed arrow A symbol used to connect contributing structures.

Figure 1.12 shows three resonance contributing structures for the carbonate ion. The three are equivalent, meaning that they have identical patterns of covalent bonding and are of equal energy.

Use of the term *resonance* for this theory of covalent bonding might suggest to you that bonds and electron pairs constantly change back and forth from one position to another over time. This notion is not at all correct. The carbonate ion, for example, has one and only one real structure. The problem is ours: How do we draw that one real structure? The resonance method is a way to describe the real structure and at the same time retain Lewis structures with electron-pair bonds. Thus, although we realize that the carbonate ion is not accurately represented by any one contributing structure shown in Figure 1.12, we continue to represent it as one of these for convenience. We understand, of course, that what is intended is the resonance hybrid.

A final note. Do not confuse resonance contributing structures with equilibration among different species. A molecule described as a resonance hybrid is not equilibrating among individual electron configurations. Rather, the molecule has only one structure, which is best described as a hybrid of its various contributing structures. The colors of the color wheel provide a good analogy. Green is not a primary color; the colors yellow and blue are mixed to make green. You can think of molecules represented by resonance hybrids as being green. Green is not sometimes yellow and sometimes blue. Green is green! In an analogous way, a molecule described as a resonance hybrid is not sometimes one contributing structure and sometimes another. It is a single structure all of the time—the resonance hybrid.

B. Curved Arrows and Electron Pushing

Notice in Figure 1.12 that the only change from resonance contributing structure (a) to (b) and then from (b) to (c) is a redistribution of valence electrons. To show how this redistribution of valence electrons occurs, chemists use a symbol called a **curved arrow**, which shows the repositioning of an electron pair from its origin (the tail of the arrow) to its destination (the head of the arrow). The repositioning may be from an atom to an adjacent bond or from a bond to an adjacent atom.

Curved arrow A symbol used to show the redistribution of valence electrons.

A curved arrow is nothing more than a bookkeeping symbol for keeping track of electron pairs or, as some call it, **electron pushing**. Do not be misled by its simplicity. Electron pushing will help you see the relationship between contributing structures. Furthermore, it will help you follow bond-breaking and bond-forming steps in organic reactions. Understanding this type of electron pushing is a survival skill in organic chemistry.

C. Rules for Writing Acceptable Resonance Contributing Structures

You must follow these four rules in writing acceptable resonance contributing structures:

1. All contributing structures must have the same number of valence electrons.
2. All contributing structures must obey the rules of covalent bonding; thus, no contributing structure may have more than 2 electrons in the valence shell of hydrogen or more than 8 electrons in the valence shell of a second-period element. Third-period elements, such as sulfur and phosphorus, may have up to 12 electrons in their valence shells.
3. The positions of all nuclei must be the same; that is, contributing structures differ only in the distribution of valence electrons.
4. All contributing structures must have the same total number of paired and unpaired electrons.

Example 1.10

Which sets are pairs of acceptable resonance contributing structures?

(a) 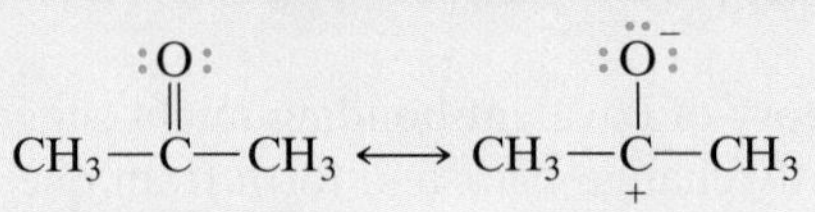 $CH_3-\overset{\overset{\large :O:}{\|}}{C}-CH_3 \longleftrightarrow CH_3-\overset{\overset{\large :\ddot{O}:^-}{|}}{\underset{+}{C}}-CH_3$

(b) 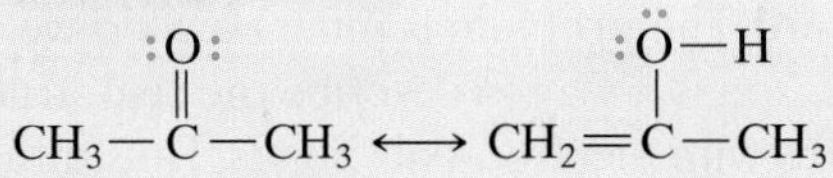 $CH_3-\overset{\overset{\large :O:}{\|}}{C}-CH_3 \longleftrightarrow CH_2=\overset{\overset{\large :\ddot{O}-H}{|}}{C}-CH_3$

Strategy

The concept being examined here is that resonance involves the redistribution of valence electrons; the connectivity of atoms does not change.

Solution

(a) A pair of resonance contributing structures. They differ only in the distribution of valence electrons.

(b) Not a pair of resonance contributing structures. They differ in the arrangement of their atoms. Oxygen is bonded to a hydrogen atom in the Lewis structure on the right but the other structure contains no such bond.

See problem 1.47

Problem 1.10

Which sets are pairs of resonance contributing structures?

(a) 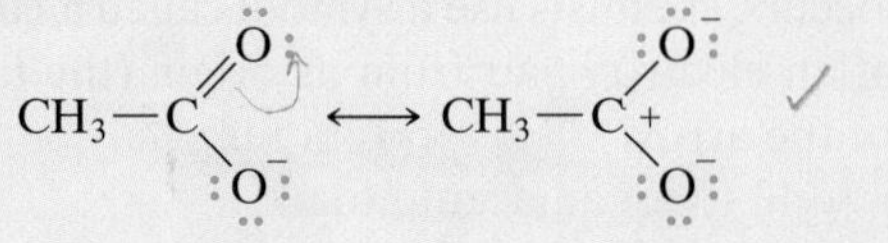 $CH_3-C\begin{matrix}/\!\!/\ddot{O}: \\ \backslash :\ddot{O}:^-\end{matrix} \longleftrightarrow CH_3-C^+\begin{matrix}/ :\ddot{O}:^- \\ \backslash :\ddot{O}:^-\end{matrix}$

(b) 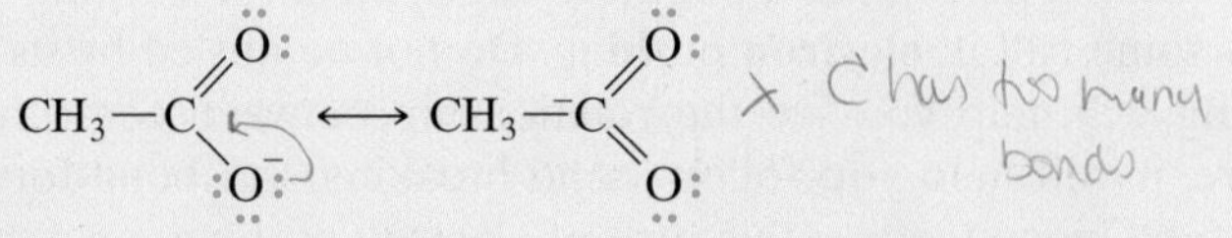 $CH_3-C\begin{matrix}/\!\!/\ddot{O}: \\ \backslash :\ddot{O}:^-\end{matrix} \longleftrightarrow CH_3=C\begin{matrix}/ \ddot{O}: \\ \backslash\!\!\backslash \ddot{O}:\end{matrix}$

Example 1.11

Draw the resonance contributing structure indicated by the curved arrows. Be certain to show all valence electrons and all formal charges.

(a) 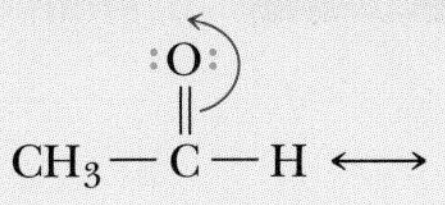 $CH_3-C(=O)-H \longleftrightarrow$

(b) $H-\overset{-}{C}H-C(=O)-H \longleftrightarrow$

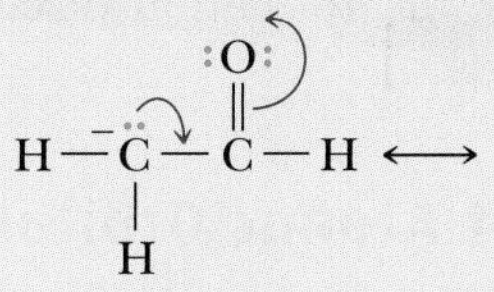

(c) $CH_3-\ddot{O}-\overset{+}{C}H-H \longleftrightarrow$

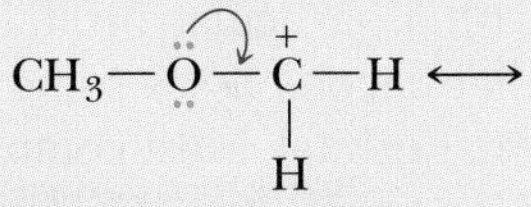

Strategy

Any curved arrow that points to an atom will generate a lone pair of electrons. Any curved arrow that points to a bond will result in an additional bond on top of the original bond. That is, a single bond will become a double bond and a double bond will become a triple bond.

Solution

(a) 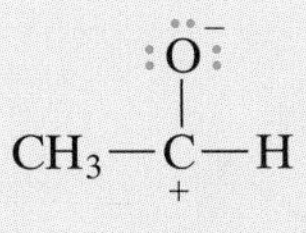 $CH_3-\overset{+}{C}(-O^-)-H$

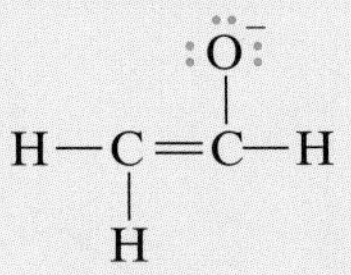

(b) $H-C(H)=C(-O^-)-H$

(c) $CH_3-\overset{+}{O}=C(H)-H$

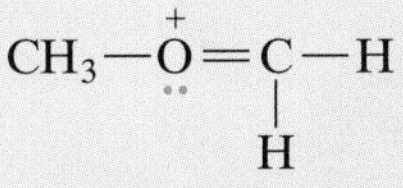

See problems 1.48, 1.50

Problem 1.11

Use curved arrows to show the redistribution of valence electrons in converting resonance contributing structure (a) to (b) and then (b) to (c). Also show, using curved arrows, how (a) can be converted to (c) without going through (b).

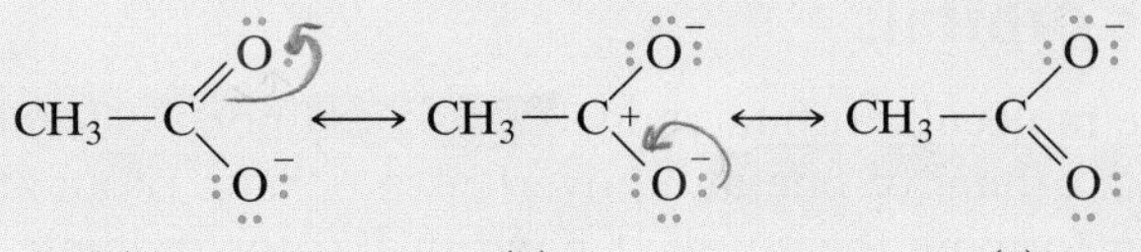

$CH_3-C(=O)-O^- \longleftrightarrow CH_3-C^+(-O^-)-O^- \longleftrightarrow CH_3-C(-O^-)=O$

(a) (b) (c)

1.6 What Is the Molecular Orbital Model of Covalent Bonding?

As much as the Lewis and VSEPR models help us to understand covalent bonding and the geometry of molecules, they leave many questions unanswered. The most important of these questions is the relation between molecular structure and chemical reactivity. For example, carbon–carbon double bonds are different in chemical reactivity from carbon–carbon single bonds. Most carbon–carbon single bonds are quite unreactive but carbon–carbon double bonds, as we will see in Chapter 5, react with a wide variety of reagents. The Lewis model and VSEPR give us no way to account for these differences. Therefore, let us turn to a newer model of covalent bonding, namely, the formation of covalent bonds by the overlap of atomic and molecular orbitals.

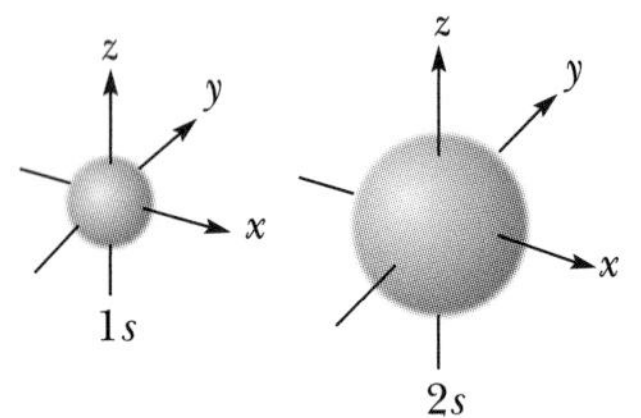

Figure 1.13
Shapes of 1*s* and 2*s* atomic orbitals.

A. Shapes of Atomic Orbitals

One way to visualize the electron density associated with a particular orbital is to draw a boundary surface around the region of space that encompasses some arbitrary percentage of the negative charge associated with that orbital. Most commonly, we draw the boundary surface at 95%. Drawn in this manner, all *s* orbitals have the shape of a sphere with its center at the nucleus (Figure 1.13). Of the various *s* orbitals, the sphere representing the 1*s* orbital is the smallest. A 2*s* orbital is a larger sphere, and a 3*s* orbital is an even larger sphere.

Figure 1.14 shows the three-dimensional shapes of the three 2*p* orbitals, combined in one diagram to illustrate their relative orientations in space. Each 2*p* orbital consists of two lobes arranged in a straight line with the nucleus in the middle. The three 2*p* orbitals are mutually perpendicular and are designated $2p_x$, $2p_y$, and $2p_z$.

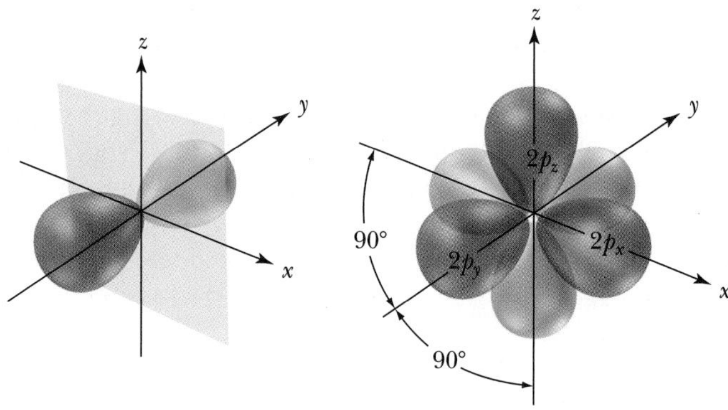

Figure 1.14
Shapes of $2p_x$, $2p_y$, and $2p_z$ atomic orbitals. The three 2*p* orbitals are mutually perpendicular. One lobe of each orbital is shown in red, the other in blue.

B. Formation of a Covalent Bond by the Overlap of Atomic Orbitals

Sigma (σ) bond A covalent bond in which the overlap of atomic orbitals is concentrated along the bond axis.

According to the molecular orbital model, a covalent bond is formed when a portion of an atomic orbital of one atom overlaps a portion of an atomic orbital of another atom. In forming the covalent bond in H_2, for example, two hydrogens approach each other so that their 1*s* atomic orbitals overlap to form a sigma covalent bond (Figure 1.15). A **sigma (σ) bond** is a covalent bond in which orbitals overlap along the axis joining the two nuclei.

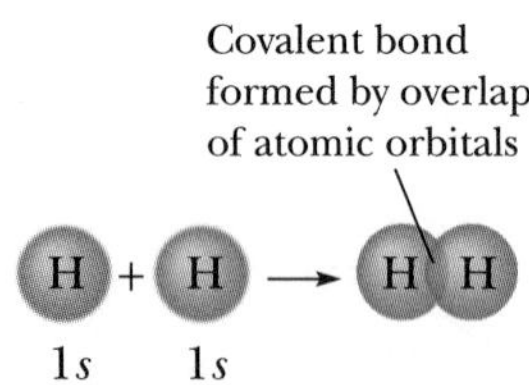

Figure 1.15
Formation of the covalent bond in H_2 by the overlap of the 1*s* atomic orbitals of each hydrogen.

C. Hybridization of Atomic Orbitals

The formation of a covalent bond between two hydrogen atoms is straightforward. The formation of covalent bonds with second-period elements, however, presents the following problem: In forming covalent bonds, atoms of carbon, nitrogen, and oxygen (all second-period elements), use 2*s* and 2*p* atomic orbitals. The three 2*p* atomic orbitals are at angles of 90° to one another (Figure 1.14), and if atoms of second-period elements used these orbitals to form covalent bonds, the bond angles around each would be approximately 90°. Bond angles of 90°, however, are rarely observed in organic molecules. What we find, instead, are bond angles of approximately 109.5° in molecules with only single bonds, 120° in molecules with double bonds, and 180° in molecules with triple bonds:

CH_4 (109.5°) $H_2C{=}CH_2$ (120° (approx)) $H{-}C{\equiv}C{-}H$ (180°)

To account for these observed bond angles, Pauling proposed that atomic orbitals combine to form new orbitals, called **hybrid orbitals**. The number of hybrid orbitals formed is equal to the number of atomic orbitals combined. Elements of the second period form three types of hybrid orbitals, designated sp^3, sp^2, and sp, each of which can contain up to two electrons.

Hybrid orbital An orbital produced from the combination of two or more atomic orbitals.

D. sp^3 Hybrid Orbitals: Bond Angles of Approximately 109.5°

The combination of one 2*s* atomic orbital and three 2*p* atomic orbitals forms four equivalent ***sp*³ hybrid orbitals**. Because they are derived from four atomic orbitals, sp^3 hybrid orbitals always occur in sets of four. Each sp^3 hybrid orbital consists of a larger lobe pointing in one direction and a smaller lobe pointing in the opposite direction. The axes of the four sp^3 hybrid orbitals point toward the corners of a regular tetrahedron, and sp^3 hybridization results in bond angles of approximately 109.5° (Figure 1.16).

sp^3 Hybrid orbital An orbital produced by the combination of one *s* atomic orbital and three *p* atomic orbitals.

Keep in mind that superscripts in the designation of hybrid orbitals tell you how many atomic orbitals have been combined to form the hybrid orbitals. The designation sp^3, for example, tells you that *one s* atomic orbital and *three p* atomic orbitals are combined in forming the hybrid orbital. Do not confuse this use of superscripts with how we use superscripts in writing a ground-state electron configuration—for example, $1s^2 2s^2 2p^5$ for fluorine. In the case of an electron configuration, superscripts tell you the number of electrons in each orbital or set of orbitals.

In Section 1.2, we described the covalent bonding in CH_4, NH_3, and H_2O in terms of the Lewis model, and in Section 1.3 we used VSEPR to predict bond angles of approximately 109.5° in each molecule. Now let us consider the bonding in these molecules in terms of the overlap of atomic orbitals. To bond with four other atoms with bond angles of 109.5°, carbon uses a set of four sp^3 hybrid orbitals. Carbon has four valence electrons, and one electron is placed in each sp^3 hybrid orbital. Each partially filled sp^3 hybrid orbital then overlaps with a partially filled 1*s* atomic

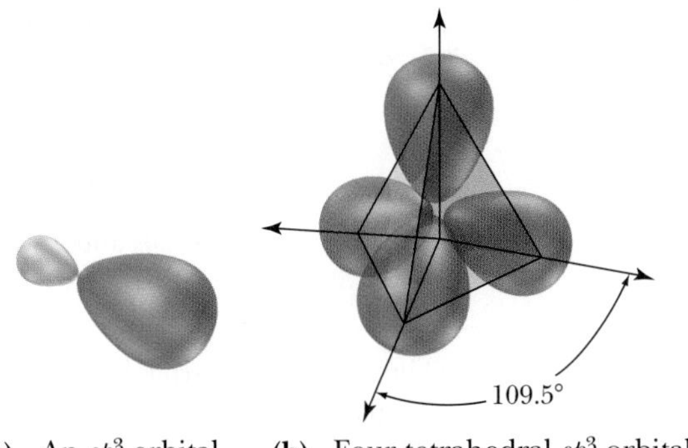

Figure 1.16 sp^3 Hybrid orbitals. (a) Representation of a single sp^3 hybrid orbital showing two lobes of unequal size. (b) Three-dimensional representation of four sp^3 hybrid orbitals, which point toward the corners of a regular tetrahedron. The smaller lobes of each sp^3 hybrid orbital are hidden behind the larger lobes.

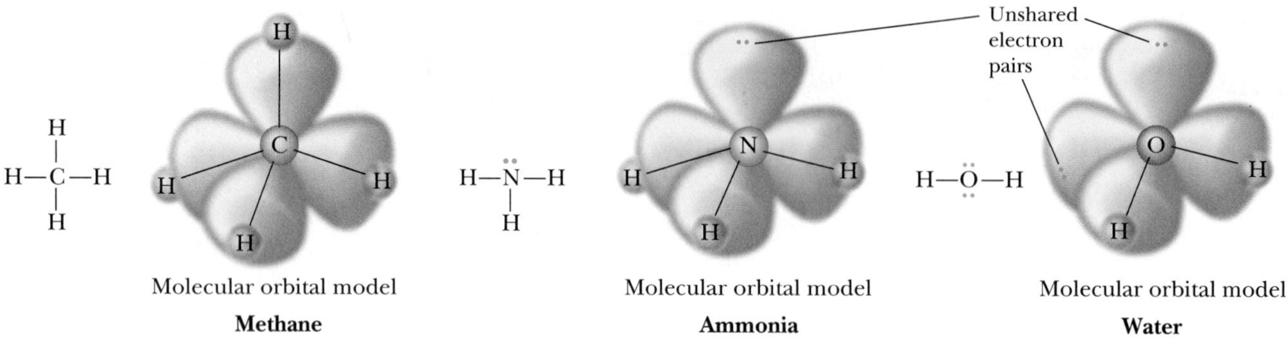

Figure 1.17
Molecular orbital pictures of methane, ammonia, and water.

orbital of hydrogen to form a sigma (σ) bond, and hydrogen atoms occupy the corners of a regular tetrahedron (Figure 1.17).

In bonding with three other atoms, the five valence electrons of nitrogen are distributed so that one sp^3 hybrid orbital is filled with a pair of electrons and the remaining three sp^3 hybrid orbitals have one electron each. Overlapping each partially filled sp^3 hybrid orbital with a 1*s* atomic orbital of a hydrogen atom produces the NH_3 molecule (Figure 1.17).

In bonding with two other atoms, the six valence electrons of oxygen are distributed so that two sp^3 hybrid orbitals are filled and the remaining two have one electron each. Each partially filled sp^3 hybrid orbital overlaps with a 1*s* atomic orbital of hydrogen, and hydrogen atoms occupy two corners of a regular tetrahedron. The remaining two corners of the tetrahedron are occupied by unshared pairs of electrons (Figure 1.17).

E. sp^2 Hybrid Orbitals: Bond Angles of Approximately 120°

sp^2 Hybrid orbital An orbital produced by the combination of one *s* atomic orbital and two *p* atomic orbitals.

The combination of one 2*s* atomic orbital and two 2*p* atomic orbitals forms three equivalent **sp^2 hybrid orbitals**. Because they are derived from three atomic orbitals, sp^2 hybrid orbitals always occur in sets of three. Each sp^2 hybrid orbital consists of two lobes, one larger than the other. The three sp^2 hybrid orbitals lie in a plane and are directed toward the corners of an equilateral triangle; the angle between sp^2 hybrid orbitals is 120°. The third 2*p* atomic orbital (remember, $2p_x$, $2p_y$, and $2p_z$) is not involved in hybridization and consists of two lobes lying perpendicular to the plane of the sp^2 hybrid orbitals. Figure 1.18 shows three equivalent sp^2 hybrid orbitals, along with the remaining unhybridized 2*p* atomic orbital.

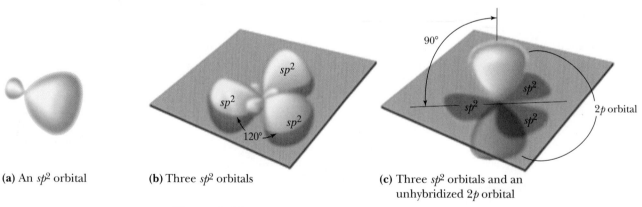

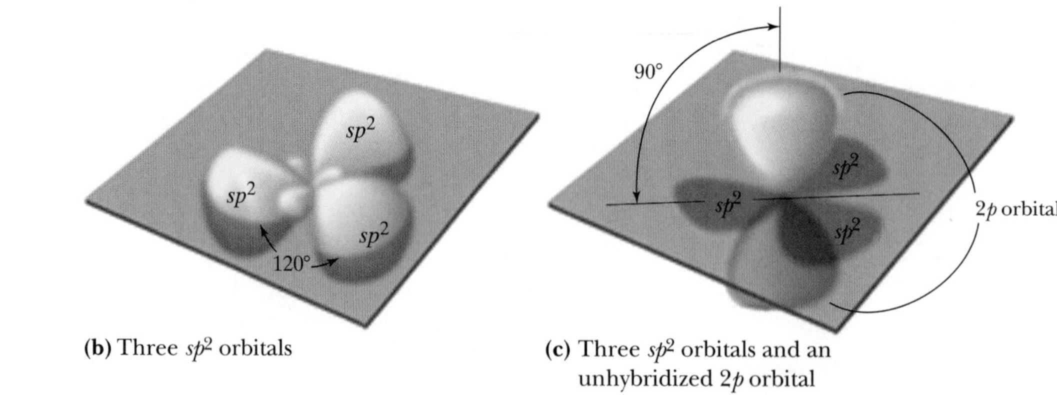

Figure 1.18
sp^2 Hybrid orbitals. (a) A single sp^2 hybrid orbital showing two lobes of unequal size. (b) The three sp^2 hybrid orbitals with their axes in a plane at angles of 120°. (c) The unhybridized 2*p* atomic orbital perpendicular to the plane created by the three sp^2 hybrid orbitals.

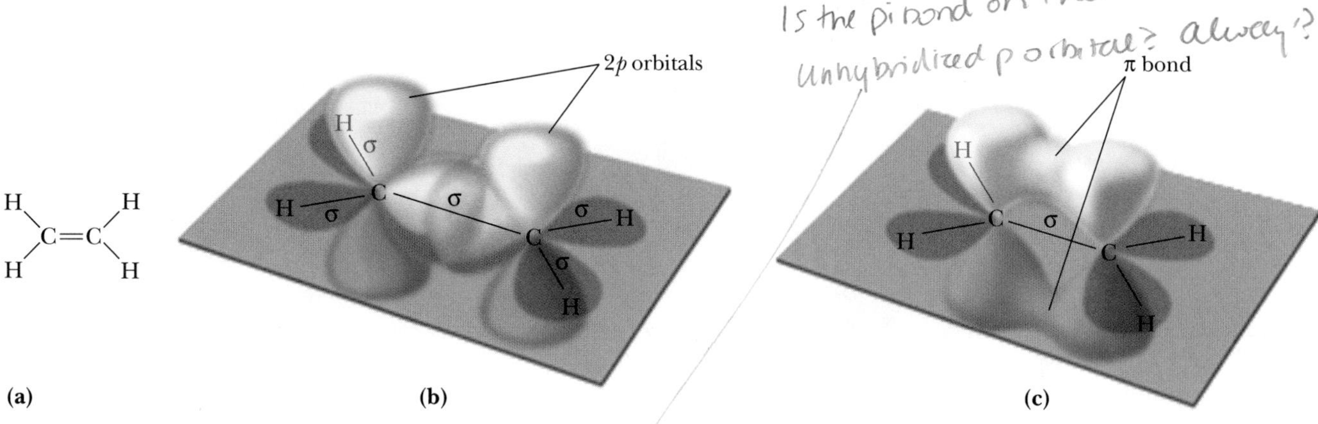

Figure 1.19
Covalent bond formation in ethylene. (a) Lewis structure, (b) overlap of sp^2 hybrid orbitals forms a sigma (σ) bond between the carbon atoms, and (c) overlap of parallel $2p$ orbitals forms a pi (π) bond. Ethylene is a planar molecule; that is, the two carbon atoms of the double bond and the four atoms bonded to them all lie in the same plane.

Second-period elements use sp^2 hybrid orbitals to form double bonds. Figure 1.19(a) shows a Lewis structure for ethylene, C_2H_4. A sigma bond between the carbons in ethylene forms by the overlap of sp^2 hybrid orbitals along a common axis [Figure 1.19(b)]. Each carbon also forms sigma bonds to two hydrogens. The remaining $2p$ orbitals on adjacent carbon atoms lie parallel to each other and overlap to form a pi bond [Figure 1.19(c)]. A **pi (π) bond** is a covalent bond formed by the overlap of parallel p orbitals. Because of the lesser degree of overlap of orbitals forming pi bonds compared with those forming sigma bonds, pi bonds are generally weaker than sigma bonds.

Pi (π) bond A covalent bond formed by the overlap of parallel p orbitals.

The molecular orbital model describes all double bonds in the same way that we have described a carbon–carbon double bond. In formaldehyde, CH_2O, the simplest organic molecule containing a carbon–oxygen double bond, carbon forms sigma bonds to two hydrogens by the overlap of an sp^2 hybrid orbital of carbon and the $1s$ atomic orbital of each hydrogen. Carbon and oxygen are joined by a sigma bond formed by the overlap of sp^2 hybrid orbitals and a pi bond formed by the overlap of unhybridized $2p$ atomic orbitals (Figure 1.20).

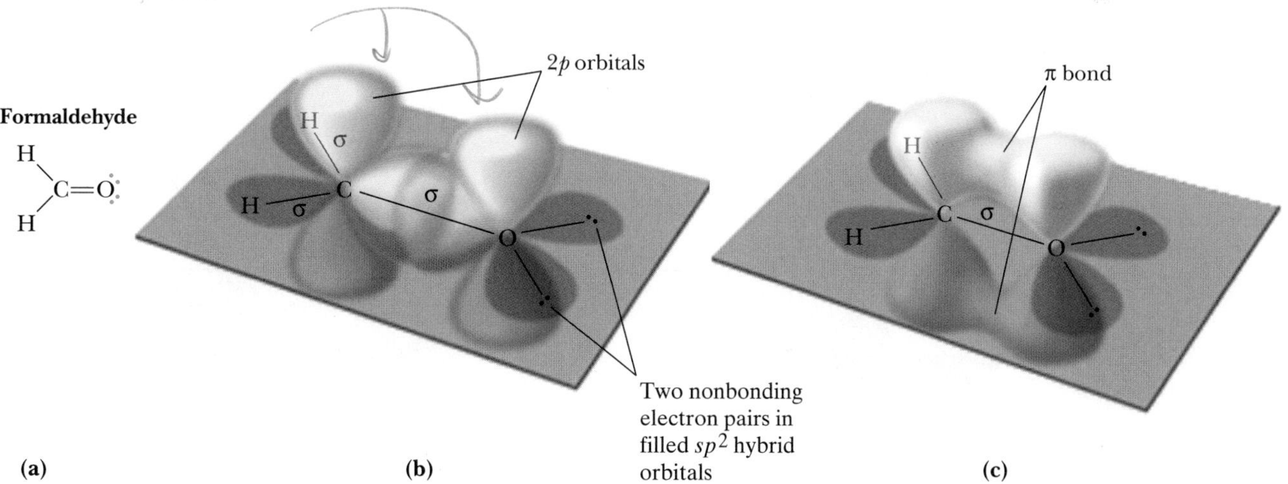

Figure 1.20
A carbon–oxygen double bond. (a) Lewis structure of formaldehyde, CH_2O, (b) the sigma (σ) bond framework and nonoverlapping parallel $2p$ atomic orbitals, and (c) overlap of parallel $2p$ atomic orbitals to form a pi (π) bond.

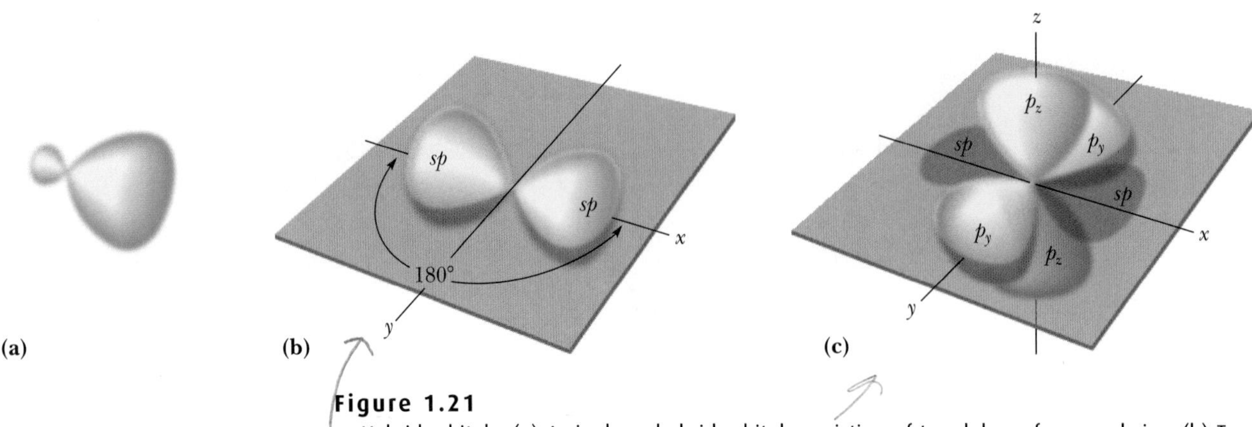

Figure 1.21
sp Hybrid orbitals. (a) A single *sp* hybrid orbital consisting of two lobes of unequal size. (b) Two *sp* hybrid orbitals in a linear arrangement. (c) Unhybridized $2p$ atomic orbitals are perpendicular to the line created by the axes of the two *sp* hybrid orbitals.

F. *sp* Hybrid Orbitals: Bond Angles of Approximately 180°

***sp* Hybrid orbital** A hybrid atomic orbital produced by the combination of one *s* atomic orbital and one *p* atomic orbital.

The combination of one $2s$ atomic orbital and one $2p$ atomic orbital forms two equivalent ***sp* hybrid orbitals**. Because they are derived from two atomic orbitals, *sp* hybrid orbitals always occur in sets of two. The two *sp* hybrid orbitals lie at an angle of 180°. The axes of the unhybridized $2p$ atomic orbitals are perpendicular to each other and to the axis of the two *sp* hybrid orbitals. Figure 1.21 shows the two *sp* hybrid orbitals on the x-axis and the unhybridized $2p$ orbitals on the y-axis and z-axis.

Figure 1.22 shows a Lewis structure and an orbital overlap diagram for acetylene, C_2H_2. A carbon–carbon triple bond consists of one sigma bond and two pi bonds. The sigma bond is formed by the overlap of *sp* hybrid orbitals. One pi bond is formed by the overlap of a pair of parallel $2p$ atomic orbitals. The second pi bond is formed by the overlap of a second pair of parallel $2p$ atomic orbitals.

Table 1.9 summarizes the relationship among the number of groups bonded to carbon, orbital hybridization, and the types of bonds involved.

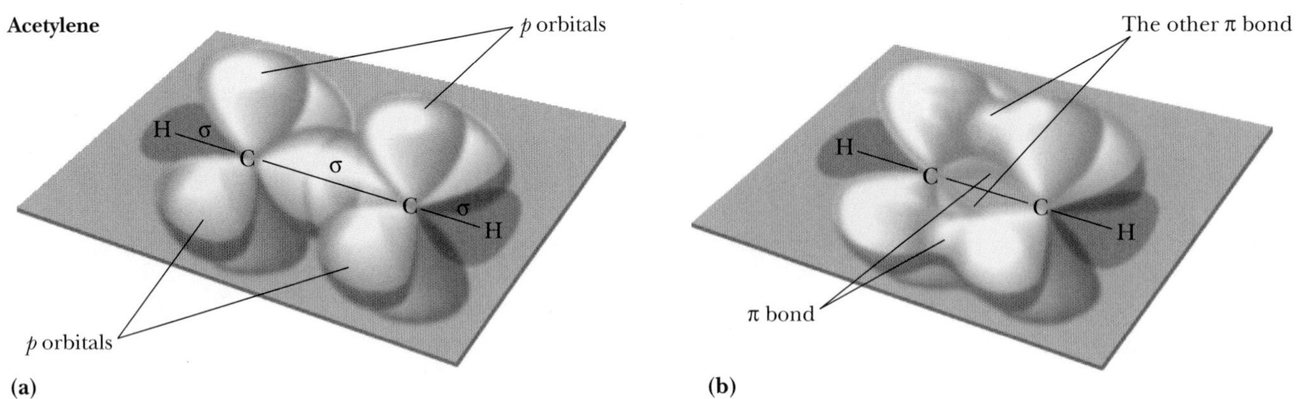

Figure 1.22
Covalent bonding in acetylene. (a) The sigma bond framework shown along with nonoverlapping $2P$ atomic orbitals. (b) Formation of two pi bonds by overlap of two sets of parallel $2P$ atomic orbitals.

TABLE 1.9 Covalent Bonding of Carbon

Groups Bonded to Carbon	Orbital Hybrid-ization	Predicted Bond Angles	Types of Bonds to Carbon	Example	Name
4	sp^3	109.5°	four sigma bonds	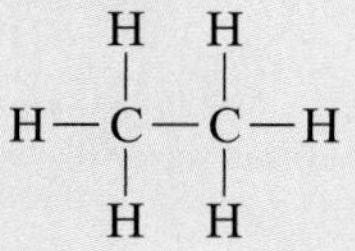 $CH_3{-}CH_3$ (H₃C–CH₃)	ethane
3	sp^2	120°	three sigma bonds and one pi bond	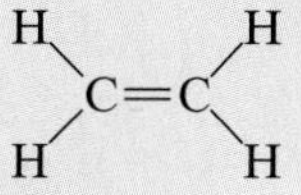 $H_2C{=}CH_2$	ethylene
2	sp	180°	two sigma bonds and two pi bonds	$H{-}C{\equiv}C{-}H$	acetylene

Example 1.12

Describe the bonding in acetic acid, CH_3COOH, in terms of the orbitals involved, and predict all bond angles.

Strategy

First draw a Lewis structure for acetic acid and then determine the number of regions of electron density about each atom.

Solution

The following are three identical Lewis structures. Labels on the first structure point to atoms and show hybridization. Labels on the second structure point to bonds and show the type of bond, either sigma or pi. Labels on the third structure point to atoms and show bond angles about each atom as predicted by valence-shell électron-pair repulsion.

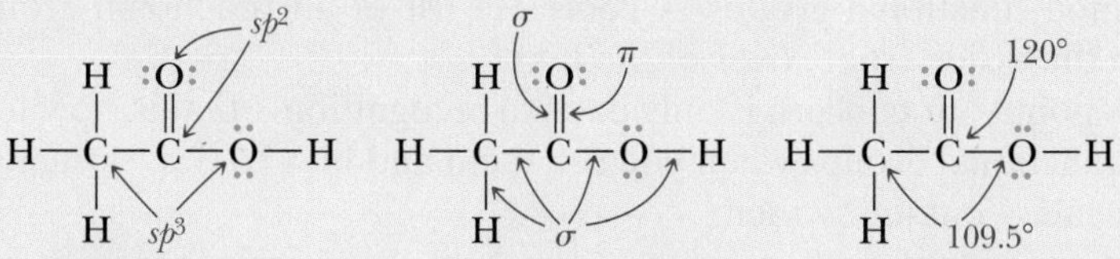

sp² sp³ σ π σ 120° 109.5°

See problem 1.51

Problem 1.12

Describe the bonding in these molecules in terms of the atomic orbitals involved, and predict all bond angles:

(a) $CH_3CH{=}CH_2$ (b) CH_3NH_2

1.7 What Are Functional Groups?

There are over 10 million organic compounds that have been discovered or made by organic chemists! Surely it would seem to be an almost impossible task to learn the physical and chemical properties of this many compounds. Fortunately, the study

TABLE 1.10 Five Common Functional Groups

Functional Group	Name of Group	Present In	Example	Name of Example
—OH	hydroxyl	alcohols	CH_3CH_2OH	Ethanol
$—NH_2$	amino	amines	$CH_3CH_2NH_2$	Ethanamine
O ‖ —C—H	carbonyl	aldehydes	O ‖ CH_3CH	Ethanal
O ‖ —C—	carbonyl	ketones	O ‖ CH_3CCH_3	Acetone
O ‖ —C—OH	carboxyl	carboxylic acids	O ‖ CH_3COH	Acetic acid

Functional group An atom or a group of atoms within a molecule that shows a characteristic set of physical and chemical properties.

of organic compounds is not as formidable a task as you might think. While organic compounds can undergo a wide variety of chemical reactions, only certain portions of their structure are changed in any particular reaction. The part of an organic molecule that undergoes chemical reactions is called a **functional group**, and, as we will see, the same functional group, in whatever organic molecule we find it, undergoes the same types of chemical reactions. Therefore, you do not have to study the chemical reactions of even a fraction of the 10 million known organic compounds. Instead you need only to identify a few characteristic types of functional groups and then study the chemical reactions that each undergoes.

Functional groups are also important because they are the units by which we divide organic compounds into families of compounds. For example, we group those compounds which contain an —OH (hydroxyl) group bonded to a tetrahedral carbon into a family called alcohols, and compounds containing a —COOH (carboxyl) group into a family called carboxylic acids. In Table 1.10, we introduce five of the most common functional groups. A complete list of all functional groups we will study is on the inside front cover of the text.

At this point, our concern is only pattern recognition—that is, how to recognize these five functional groups when you see them and how to draw structural formulas of molecules containing them.

Finally, functional groups serve as the basis for naming organic compounds. Ideally, each of the 10 million or more organic compounds must have a name that is different from every other compound.

To summarize, functional groups

- are sites of chemical reaction; a particular functional group, in whatever compound we find it, undergoes the same types of chemical reactions.
- determine, in large measure, the physical properties of a compound.
- are the units by which we divide organic compounds into families.
- serve as a basis for naming organic compounds.

A. Alcohols

Hydroxyl group An —OH group.

The functional group of an **alcohol** is an **—OH (hydroxyl)** group bonded to a tetrahedral (sp^3 hybridized) carbon atom. In the general formula that follows, we use the symbol R to indicate either a hydrogen or another carbon group. The important point in the general structure is that the —OH group is bonded to a tetrahedral carbon atom:

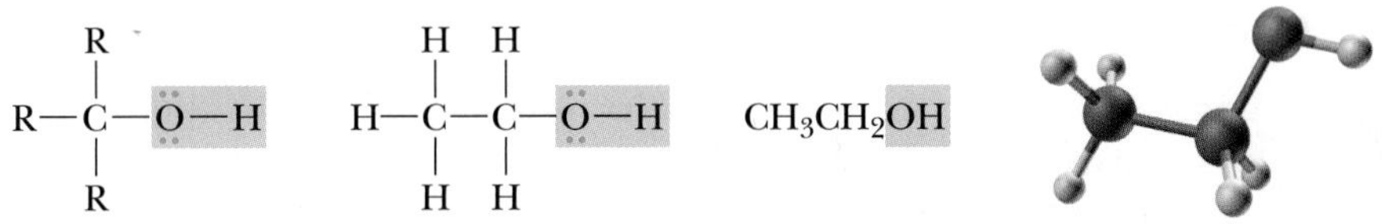

$$\mathrm{R}-\underset{\mathrm{R}}{\overset{\mathrm{R}}{\underset{|}{\overset{|}{\mathrm{C}}}}}-\ddot{\mathrm{O}}-\mathrm{H}$$

Functional group (R=H or carbon group)

$$\mathrm{H}-\underset{\mathrm{H}}{\overset{\mathrm{H}}{\underset{|}{\overset{|}{\mathrm{C}}}}}-\underset{\mathrm{H}}{\overset{\mathrm{H}}{\underset{|}{\overset{|}{\mathrm{C}}}}}-\ddot{\mathrm{O}}-\mathrm{H}$$

Structural formula

CH_3CH_2OH

Condensed structural formula

The rightmost representation of this alcohol is a **condensed structural formula**, CH_3CH_2OH. In a condensed structural formula, CH_3 indicates a carbon bonded to three hydrogens, CH_2 indicates a carbon bonded to two hydrogens, and CH indicates a carbon bonded to one hydrogen. We generally do not show unshared pairs of electrons in a condensed structural formula.

Alcohols are classified as **primary (1°)**, **secondary (2°)**, or **tertiary (3°)**, depending on the number of carbon atoms bonded to the carbon bearing the —OH group:

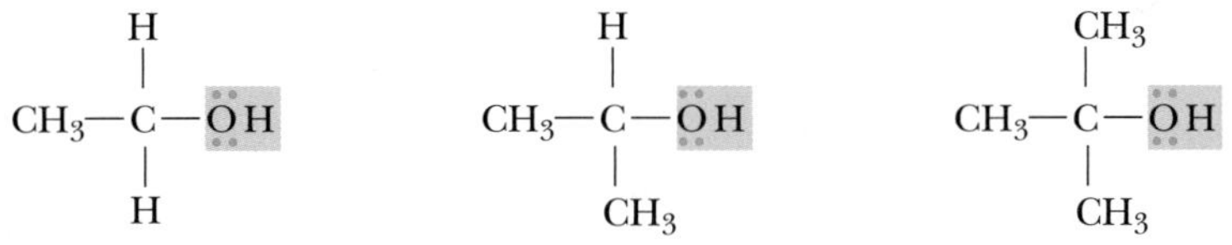

$$CH_3-\underset{H}{\overset{H}{\underset{|}{\overset{|}{C}}}}-\ddot{O}H \qquad CH_3-\underset{CH_3}{\overset{H}{\underset{|}{\overset{|}{C}}}}-\ddot{O}H \qquad CH_3-\underset{CH_3}{\overset{CH_3}{\underset{|}{\overset{|}{C}}}}-\ddot{O}H$$

A 1° alcohol — A 2° alcohol — A 3° alcohol

Example 1.13

Write condensed structural formulas for the two alcohols with the molecular formula C_3H_8O. Classify each as primary, secondary, or tertiary.

Strategy

First, bond the three carbon atoms in a chain with the —OH (hydroxyl) group bonded to either an end carbon or the middle carbon of the chain. Then, to complete each structural formula, add seven hydrogens so that each carbon has four bonds to it:

Solution

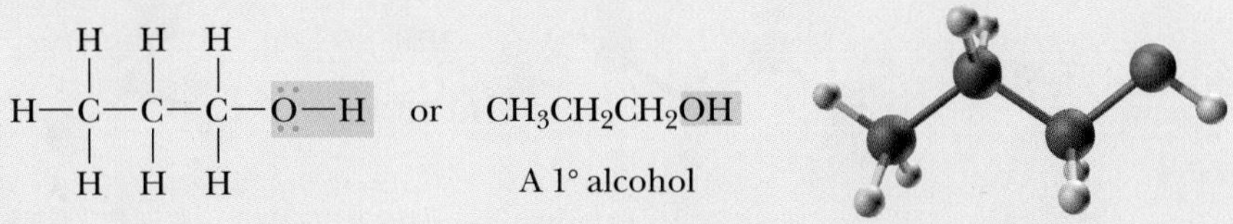

$$\mathrm{H}-\underset{\mathrm{H}}{\overset{\mathrm{H}}{\underset{|}{\overset{|}{\mathrm{C}}}}}-\underset{\mathrm{H}}{\overset{\mathrm{H}}{\underset{|}{\overset{|}{\mathrm{C}}}}}-\underset{\mathrm{H}}{\overset{\mathrm{H}}{\underset{|}{\overset{|}{\mathrm{C}}}}}-\ddot{\mathrm{O}}-\mathrm{H} \quad \text{or} \quad CH_3CH_2CH_2OH$$

A 1° alcohol

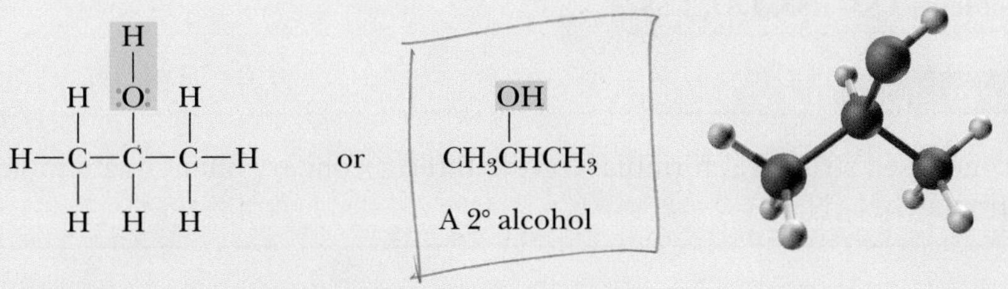

$$\mathrm{H}-\underset{\mathrm{H}}{\overset{\mathrm{H}}{\underset{|}{\overset{|}{\mathrm{C}}}}}-\underset{\mathrm{H}}{\overset{\mathrm{H}-\ddot{\mathrm{O}}}{\underset{|}{\overset{|}{\mathrm{C}}}}}-\underset{\mathrm{H}}{\overset{\mathrm{H}}{\underset{|}{\overset{|}{\mathrm{C}}}}}-\mathrm{H} \quad \text{or} \quad CH_3\overset{OH}{\overset{|}{C}}HCH_3$$

A 2° alcohol

See problems 1.53–1.55, 1.57, 1.58

Problem 1.13

Write condensed structural formulas for the four alcohols with the molecular formula $C_4H_{10}O$. Classify each as primary, secondary, or tertiary.

B. Amines

Amino group An sp^3 hybridized nitrogen atom bonded to one, two, or three carbon groups.

The functional group of an amine is an **amino group**—a nitrogen atom bonded to one, two, or three carbon atoms. In a **primary (1°) amine**, nitrogen is bonded to one carbon atom. In a **secondary (2°) amine**, it is bonded to two carbon atoms, and in a **tertiary (3°) amine**, it is bonded to three carbon atoms. The second and third structural formulas below can be written in a more abbreviated form by collecting the CH_3 groups and writing them as $(CH_3)_2NH$ and $(CH_3)_3N$, respectively.

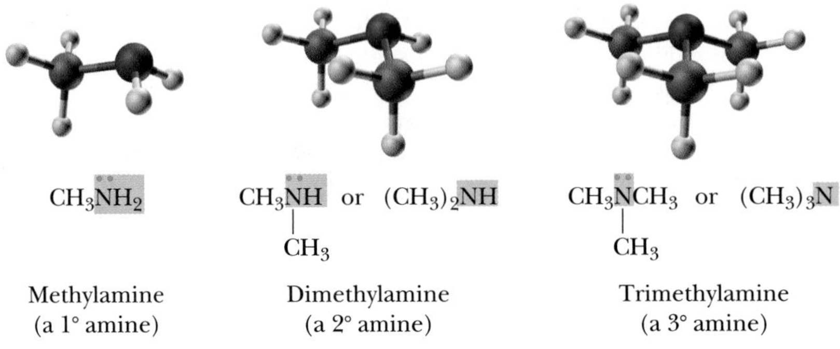

Methylamine (a 1° amine) | Dimethylamine (a 2° amine) | Trimethylamine (a 3° amine)

Example 1.14

Write condensed structural formulas for the two primary (1°) amines with the molecular formula C_3H_9N.

Strategy

For a primary amine, draw a nitrogen atom bonded to two hydrogens and one carbon. The nitrogen may be bonded to the three-carbon chain in two different ways. Then add the seven hydrogens to give each carbon four bonds and give the correct molecular formula:

Solution

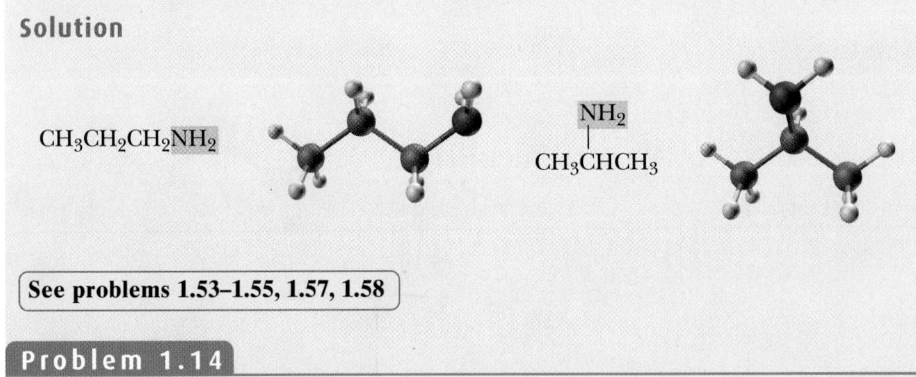

See problems 1.53–1.55, 1.57, 1.58

Problem 1.14

Write condensed structural formulas for the three secondary amines with molecular formula $C_4H_{11}N$.

C. Aldehydes and Ketones

Carbonyl group A C=O group.

Both aldehydes and ketones contain a **C=O (carbonyl)** group. The **aldehyde** functional group contains a carbonyl group bonded to a hydrogen. In formaldehyde, CH_2O, the simplest aldehyde, the carbonyl carbon is bonded to two hydrogen atoms. In a condensed structural formula, the aldehyde group may be written showing the carbon–oxygen double bond as CH═O, or, alternatively, it may be written —CHO. The functional group of a **ketone** is a carbonyl group bonded to two carbon atoms.

Functional group

An aldehyde ($CH_3CH{=}O$)

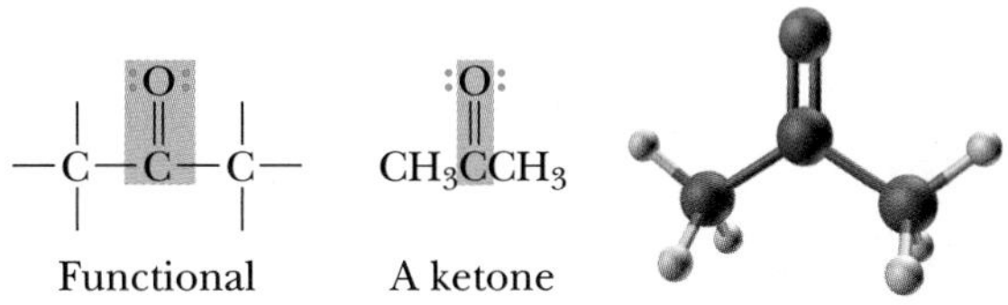

Functional group

A ketone ($CH_3C({=}O)CH_3$)

Example 1.15

Write condensed structural formulas for the two aldehydes with the molecular formula C_4H_8O.

Strategy

First, draw the functional group of an aldehyde and add the remaining carbons, which in this case may be bonded in two different ways. Then, add seven hydrogens to complete the four bonds of each carbon and give the correct molecular formula: Note that the aldehyde group may be written showing the carbon–oxygen double bond as C═O, or, alternatively, it may be written —CHO.

Solution

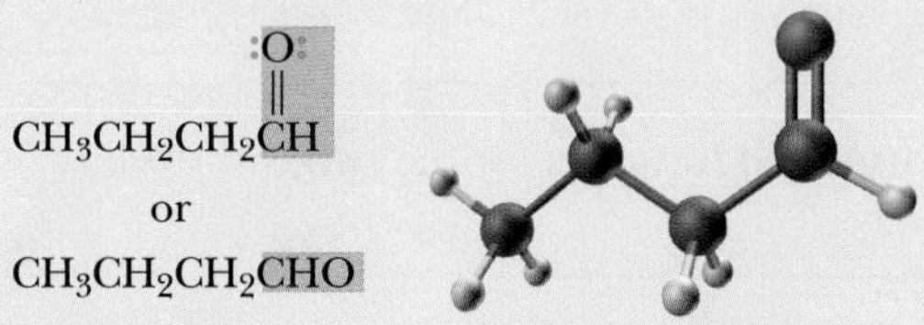

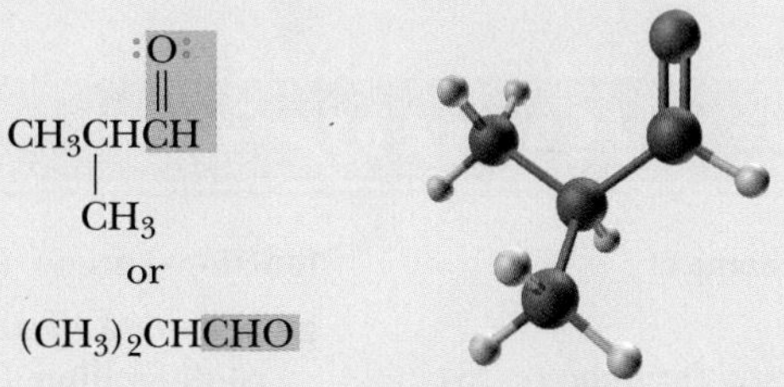

$CH_3CH_2CH_2CH{=}O$

or

$CH_3CH_2CH_2CHO$

$CH_3CH(CH_3)CH{=}O$

or

$(CH_3)_2CHCHO$

See problems 1.53–1.55, 1.57, 1.58

Problem 1.15

Write condensed structural formulas for the three ketones with the molecular formula $C_5H_{10}O$.

D. Carboxylic Acids

Carboxyl group A —COOH group.

The functional group of a carboxylic acid is a **—COOH (carboxyl:** *carb*onyl + hydr*oxyl*) group:

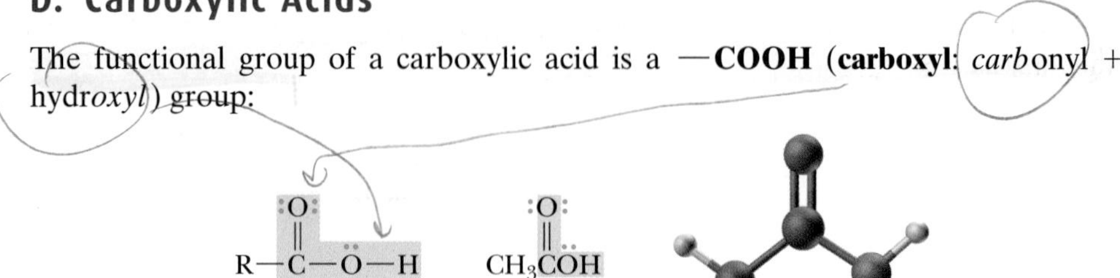

Example 1.16

Write a condensed structural formula for the single carboxylic acid with the molecular formula $C_3H_6O_2$.

Strategy

First draw the carboxyl group and add the two remaining carbons. Finally, add the five remaining hydrogens in such a way that each carbon in the molecule has four bonds to it.

Solution

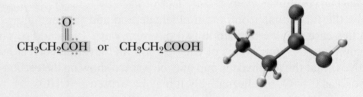

CH_3CH_2COH or CH_3CH_2COOH

See problems 1.53–1.55, 1.57, 1.58

Problem 1.16

Write condensed structural formulas for the two carboxylic acids with the molecular formula $C_4H_8O_2$.

Key Terms and Concepts

alcohol (p. 32)
aldehyde (p. 34)
amino group (p. 34)
anion (p. 6)
atomic orbital (p. 26)
bond length (p. 9)
bonding electrons (p. 12)
carbonyl group (p. 34)
carboxyl group (p. 36)
cation (p. 6)
classification of chemical bonds (p. 9)
condensed structural formula (p. 33)
covalent bond (p. 7)
curved arrow (p. 23)
dipole (p. 11)
double bond (p. 13)
double-headed arrow (p. 23)
electron pushing (p. 23)
electronegativity (p. 7)
formal charge (p. 15)
functional group (p. 32)
ground-state electron configuration (p. 3)
hybrid orbital (p. 27)
hydroxyl group (p. 32)
ionic bond (p. 7)
ketone (p. 35)
Lewis model of bonding (p. 6)
Lewis structure (of an element) (p. 5)
Lewis structure (of molecules and ions) (p. 12)

Molecular orbital model of bonding (p. 25)
Molecular polarity (p. 21)
Nonbonding electrons (p. 12)
Nonpolar covalent bond (p. 9)
Octet rule (p.6)
Orbital (p. 2)
Organic chemistry (p. 1)
Pi (π) bond (p. 29)
Polar covalent bond (p. 9)
Primary (1°) Alcohol (p. 33)
Primary (1°) Amine (p. 34)
Principal energy level (p. 2)
Pyramidal geometry (p. 17)
Resonance (p. 23)
Resonance contributing structure (p. 23)
Resonance hybrid (p. 23)
Secondary (2°) Alcohol (p. 33)
Secondary (2°) Amine (p. 34)
Shapes of atomic *s* and *p* orbitals (p. 26)
Shell (p. 2)
Sigma (σ) bond (p. 26)
Single bond (p. 13)
***sp* hybrid orbitals** (p. 30)
sp^2 hybrid orbitals (p. 28)
sp^3 hybrid orbitals (p. 27)
Tertiary (3°) Alcohol (p. 33)
Tertiary (3°) Amine (p. 34)
Tetrahedral geometry (p. 17)
Trigonal planar geometry (p. 18)
Triple bond (p. 13)
Valence electron (p. 5)
Valence shell (p. 5)
Valence-shell electron-pair repulsion (VSEPR) (p. 16)

Summary of Key Questions

1.1 How Do We Describe the Electronic Structure of Atoms?

- An atom consists of a small, dense nucleus and electrons concentrated about the nucleus in regions of space called **shells**.
- Each shell can contain as many as $2n^2$ electrons, where *n* is the number of the shell. Each shell is subdivided into regions of space called **orbitals**.
- The first shell ($n = 1$) has a single *s* orbital and can hold $2 \times 1^2 = 2$ electrons.
- The second shell ($n = 2$) has one *s* orbital and three *p* orbitals and can hold $2 \times 2^2 = 8$ electrons.
- The **Lewis structure** of an element shows the symbol of the element surrounded by a number of dots equal to the number of electrons in its **valence shell**.

1.2 What Is the Lewis Model of Bonding?

- According to the **Lewis model of bonding**, atoms bond together in such a way that each atom participating in a chemical bond acquires a completed valence-shell electron configuration resembling that of the noble gas nearest it in atomic number.
- Atoms that lose sufficient electrons to acquire a completed valence shell become **cations**; atoms that gain sufficient electrons to acquire a completed valence shell become **anions**.
- An **ionic bond** is a chemical bond formed by the attractive force between an anion and a cation.
- A **covalent bond** is a chemical bond formed by the sharing of electron pairs between atoms.
- The tendency of main-group elements (those of Groups 1A–7A) to achieve an outer shell of eight valence electrons is called the **octet rule**.
- **Electronegativity** is a measure of the force of attraction by an atom for electrons it shares in a chemical bond with another atom. Electronegativity increases from left to right and from bottom to top in the Periodic Table.
- A **Lewis structure** for a molecule or an ion must show (1) the correct arrangement of atoms, (2) the correct number of valence electrons, (3) no more than two electrons in the outer shell of hydrogen, (4) no more than eight electrons in the outer shell of any second-period element, and (5) all formal charges.
- **Formal charge** is the charge on an atom in a molecule or polyatomic ion.

1.3 How Do We Predict Bond Angles and the Shapes of Molecules?

- Valence-shell electron repulsion (VSEPR) predicts bond angles of 109.5° about atoms surrounded by four regions of electron density, bond angles of 120° about atoms surrounded by three regions of electron density, and bond angles of 180° about atoms surrounded by two regions of electron density.

1.4 How Do We Predict If a Molecule Is Polar or Nonpolar?

- As a rough guideline, we say that a **nonpolar covalent bond** is a covalent bond in which the difference in electronegativity between the bonded atoms is less than 0.5 unit.
- A **polar covalent bond** is a covalent bond in which the difference in electronegativity between the bonded atoms is between 0.5 and 1.9 units. In a polar covalent bond, the more electronegative atom bears a partial negative charge ($\delta-$) and the less electronegative atom bears a partial positive charge ($\delta+$).
- A molecule is polar if it has one or more polar bonds and its centers of partial positive and negative charge are at different places within the molecule.

1.5 What Is Resonance?

- According to the **theory of resonance**, a molecule or ion for which no single Lewis structure is adequate is best described by writing two or more **resonance contributing structures** and considering the real molecule or ion to be a **hybrid** of the various contributing structures.
- Resonance contributing structures are interconnected by **double-headed arrows**.
- We show how valence electrons are redistributed from one contributing structure to the next by **curved arrows**. A curved arrow extends from where the electrons are initially shown (on an atom or in a covalent bond) to their new location (an adjacent atom or an adjacent covalent bond).
- The use of curved arrows in this way is commonly referred to as **electron pushing**.

1.6 What Is the Molecular Orbital Model of Covalent Bonding?

- According to the **molecular orbital model**, the formation of a covalent bond results from the overlap of atomic orbitals.
- The greater the overlap, the stronger is the resulting covalent bond.
- The combination of atomic orbitals is called **hybridization**, and the resulting orbitals are called **hybrid orbitals**.
- The combination of one 2*s* atomic orbital and three 2*p* atomic orbitals produces four equivalent ***sp*³ hybrid orbitals**, each pointing toward a corner of a regular tetrahedron at angles of 109.5°.
- The combination of one 2*s* atomic orbital and two 2*p* atomic orbitals produces three equivalent ***sp*² hybrid orbitals**, the axes of which lie in a plane at angles of 120°. Most C=C, C=O, and C=N double bonds are a combination of one **sigma (σ) bond** formed by the overlap of ***sp*² hybrid orbitals** and one **pi (π) bond** formed by the overlap of parallel 2*p* atomic orbitals.
- The combination of one 2*s* atomic orbital and one 2*p* atomic orbital produces two equivalent ***sp* hybrid orbitals**, the axes of which lie in a plane at an angle of 180°.
- All C=C triple bonds are a combination of one sigma bond formed by the overlap of *sp* hybrid orbitals and two pi bonds formed by the overlap of two pairs of parallel 2*p* atomic orbitals.

1.7 What Are Functional Groups?

- **Functional groups** are characteristic structural units by which we divide organic compounds into classes and that serve as a basis for nomenclature. They are also sites of chemical reactivity; a particular functional group, in whatever compound we find it, undergoes the same types of reactions.
- Important functional groups for us at this stage in the course are
  - the **hydroxyl group** of 1°, 2°, and 3° alcohols
  - the **amino group** of 1°, 2°, and 3° amines
  - the **carbonyl group** of aldehydes and ketones
  - the **carboxyl group** of carboxylic acids

Quick Quiz

Answer true or false to the following questions to assess your general knowledge of the concepts in this chapter. If you have difficulty with any of them, you should review the appropriate section in the chapter (shown in parentheses) before attempting the more challenging end-of-chapter problems.

1. These bonds are arranged in order of increasing polarity C—H < N—H < O—H. (1.2)
2. All atoms in a contributing structure must have complete valence shells. (1.5)
3. An electron in a 1*s* orbital is held closer to the nucleus than an electron in a 2*s* orbital. (1.1)
4. A sigma bond and a pi bond have in common that each can result from the overlap of atomic orbitals. (1.6)
5. The molecular formula of the smallest aldehyde is C_3H_6O, and that of the smallest ketone is also C_3H_6O. (1.7)
6. To predict whether a covalent molecule is polar or nonpolar, you must know both the polarity of each covalent bond and the geometry (shape) of the molecule. (1.4)
7. An orbital is a region of space that can hold two electrons. (1.1)
8. In the ground-state electron configuration of an atom, only the lowest-energy orbitals are occupied. (1.1)
9. Electronegativity generally increases with atomic number. (1.2)
10. Paired electron spins means that the two electrons are aligned with their spins North Pole to North Pole and South Pole to South Pole. (1.1)
11. According to the Lewis model of bonding, atoms bond together in such a way that each atom participating in the bond acquires an outer-shell electron configuration matching that of the noble gas nearest it in atomic number. (1.2)
12. A primary amine contains one N—H bond, a secondary amine contains two N—H bonds, and a tertiary amine contains three N—H bonds. (1.7)
13. All bond angles in sets of resonance contributing structures must be the same. (1.5)
14. Electronegativity is a measure of an atom's attraction for electrons it shares in a chemical bond with another atom. (1.2)
15. An orbital can hold a maximum of two electrons with their spins paired. (1.1)
16. Fluorine in the upper right corner of the Periodic Table is the most electronegative element; hydrogen, in the upper left corner, is the least electronegative element. (1.2)
17. A primary alcohol has one —OH group, a secondary alcohol has two —OH groups, and a tertiary alcohol has three —OH groups. (1.7)
18. H_2O and NH_3 are polar molecules, but CH_4 is nonpolar. (1.4)
19. Electronegativity generally increases from top to bottom in a column of the Periodic Table. (1.2)
20. All contributing structures must have the same number of valence electrons. (1.5)
21. A carbon–carbon double bond is formed by the overlap of sp^2 hybrid orbitals, and a triple bond is formed by the overlap of sp^3 hybrid orbitals. (1.6)
22. A covalent bond formed by sharing two electrons is called a double bond. (1.2)
23. The functional groups of an alcohol, an aldehyde, and a ketone have in common the fact that each contains a single oxygen atom. (1.7)
24. Electrons in atoms are confined to regions of space called principal energy levels. (1.1)
25. In a single bond, two atoms share one pair of electrons; in a double bond, they share two pairs of electrons; and in a triple bond, they share three pairs of electrons. (1.2)
26. The Lewis structure for ethene, C_2H_4, must show eight valence electrons. (1.2)
27. The Lewis structure for formaldehyde, CH_2O, must show eight valence electrons. (1.2)
28. The letters VSEPR stand for valence-shell electron pair repulsion. (1.3)
29. In predicting bond angles about a central atom in a covalent bond, VSEPR considers only shared pairs (pairs of electrons involved in forming covalent bonds. (1.3)
30. An *sp* hybrid orbital may contain a maximum of four electrons, an sp^2 hybrid orbital may contain a maximum of six valence electrons, and an sp^3 hybrid orbital may contain a maximum of eight electrons. (1.6)
31. For a central atom surrounded by three regions of electron density, VSEPR predicts bond angles of $360°/3 = 120°$. (1.3)
32. The three 2*p* orbitals are aligned parallel to each other. (1.1.)
33. All molecules with polar bonds are polar. (1.4)
34. Electronegativity generally increases from left to right across a period of the Periodic Table. (1.2)
35. A compound with the molecular formula C_3H_6O may be an aldehyde, a ketone, or a carboxylic acid. (1.7)

36. Dichloromethane, CH_2Cl_2 is polar, but tetrachloromethane, CCl_4, is nonpolar. (1.4)

37. A covalent bond is formed between atoms whose difference in electronegativity is less than 1.9. (1.2)

38. Each principal energy level can hold two electrons. (1.1)

39. Atoms that share electrons to achieve filled valence shells form covalent bonds. (1.2)

40. Contributing structures differ only in the distribution of valence electrons. (1.5)

41. In creating hybrid orbitals (*sp*, sp^2, and sp^3), the number of hybrid orbitals created is equal to the number of atomic orbitals hybridized. (1.6)

42. VSEPR treats the two electron pairs of a double bond and the three electron pairs of a triple bond as one region of electron density. (1.3)

43. If the difference in electronegativity between two atoms is zero (they have identical electronegativities), then the two atoms will not form a covalent bond. (1.2)

44. A carbon–carbon triple bond is a combination of one sigma bond and two pi bonds. (1.6)

45. A carbon–carbon double bond is a combination of two sigma bonds. (1.6)

46. An *s* orbital has the shape of a sphere with the center of the sphere at the nucleus. (1.1)

47. A functional group is a group of atoms in an organic molecule that undergoes a predictable set of chemical reactions. (1.7)

48. In a polar covalent bond, the more electronegative atom has a partial negative charge ($\delta-$) and the less electronegative atom has a partial positive charge ($\delta+$). (1.2)

49. Electronegativity depends on both the nuclear charge and the distance of the valence electrons from the nucleus. (1.2)

50. There are two alcohols with the molecular formula C_3H_8O. (1.7)

51. In methanol, CH_3OH, the O–H bond is more polar than the C—O bond. (1.4)

52. The molecular formula of the smallest carboxylic acid is $C_2H_6O_2$. (1.7)

53. Each 2*p* orbital has the shape of a dumbbell with the nucleus at the midpoint of the dumbbell. (1.1)

54. Atoms that lose electrons to achieve a filled valence shell become cations and form ionic bonds with anions. (1.1)

55. There are three amines with the molecular formula C_3H_9N. (1.7)

Answers: (1) T (2) F (3) T (4) T (5) F (6) T (7) T (8) T (9) F (10) F (11) T (12) F (13) F (14) T (15) T (16) F (17) F (18) T (19) F (20) T (21) F (22) F (23) T (24) F (25) T (26) F (27) F (28) T (29) F (30) F (31) T (32) F (33) F (34) T (35) F (36) T (37) T (38) F (39) F (40) T (41) T (42) T (43) F (44) T (45) F (46) T (47) T (48) T (49) F (50) T (51) T (52) F (53) T (54) T (55) F

Problems

A problem marked with an asterisk indicates an applied "real world" problem. Answers to problems whose numbers are printed in blue are given in Appendix D.

Section 1.1 Electronic Structure of Atoms

1.17 Write the ground-state electron configuration for each element. The atomic number for each is provided in parentheses. **(See Example 1.1)**

(a) Sodium (11) (b) Magnesium (12)

(c) Oxygen (8) (d) Nitrogen (7)

1.18 Write the ground-state electron configuration for each element: **(See Example 1.1)**

(a) Potassium (b) Aluminum

(c) Phosphorus (d) Argon

1.19 Which element has the ground-state electron configuration **(See Example 1.1)**

(a) $1s^22s^22p^63s^23p^4$

(b) $1s^22s^22p^4$

1.20 Which element or ion does not have the ground-state electron configuration $1s^22s^22p^63s^23p^6$? **(See Example 1.1)**

(a) S^{2-} (b) Cl^- (c) Ar (d) Ca^{2+} (e) K

1.21 Define *valence shell* and *valence electron.*

1.22 How many electrons are in the valence shell of each element? **(See Example 1.2)**

(a) Carbon (b) Nitrogen (c) Chlorine

(d) Aluminum (e) Oxygen

1.23 How many electrons are in the valence shell of each ion? **(See Example 1.2)**

(a) H^+ (b) H^-

Section 1.2 Lewis Structures

1.24 Judging from their relative positions in the Periodic Table, which element in each set is more electronegative? **(See Example 1.3)**

(a) Carbon or nitrogen (b) Chlorine or bromine

(c) Oxygen or sulfur

1.25 Which compounds have nonpolar covalent bonds, which have polar covalent bonds, and which have ionic bonds? **(See Example 1.4)**

(a) LiF (b) CH_3F (c) $MgCl_2$ (d) HCl

1.26 Using the symbols $\delta-$ and $\delta+$, indicate the direction of polarity, if any, in each covalent bond: **(See Example 1.5)**

(a) C—Cl (b) S—H (c) C—S (d) P—H

1.27 Write Lewis structures for each of the following compounds, showing all valence electrons (none of the compounds contains a ring of atoms): **(See Example 1.6)**

(a) Hydrogen peroxide, H_2O_2

(b) Hydrazine, N_2H_4

(c) Methanol, CH_3OH

(d) Methanethiol, CH_3SH

(e) Methanamine, CH_3NH_2

(f) Chloromethane, CH_3Cl

(g) Dimethyl ether, CH_3OCH_3

(h) Ethane, C_2H_6

(i) Ethylene, C_2H_4

(j) Acetylene, C_2H_2

(k) Carbon dioxide, CO_2

(l) Formaldehyde, CH_2O

(m) Acetone, CH_3COCH_3

(n) Carbonic acid, H_2CO_3

(o) Acetic acid, CH_3COOH

1.28 Write Lewis structures for these ions: **(See Example 1.6)**

(a) Bicarbonate ion, HCO_3^-

(b) Carbonate ion, CO_3^{2-}

(c) Acetate ion, CH_3COO^-

(d) Chloride ion, Cl^-

1.29 Why are the following molecular formulas impossible?

(a) CH_5 (b) C_2H_7

1.30 Following the rule that each atom of carbon, oxygen, and nitrogen reacts to achieve a complete outer shell of eight valence electrons, add unshared pairs of electrons as necessary to complete the valence shell of each atom in the following ions. Then, assign formal charges as appropriate: **(See Example 1.7)**

(a)
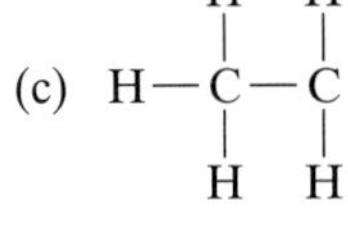

(b)
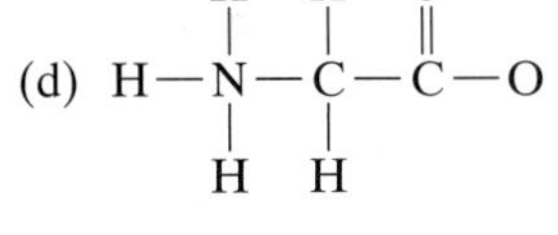

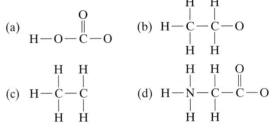

```
        H   H                 H   H   O
        |   |                 |   |   ‖
(c) H — C — C        (d) H — N — C — C — O
        |   |                 |   |
        H   H                 H   H
```

1.31 The following Lewis structures show all valence electrons. Assign formal charges in each structure as appropriate. **(See Example 1.7)**

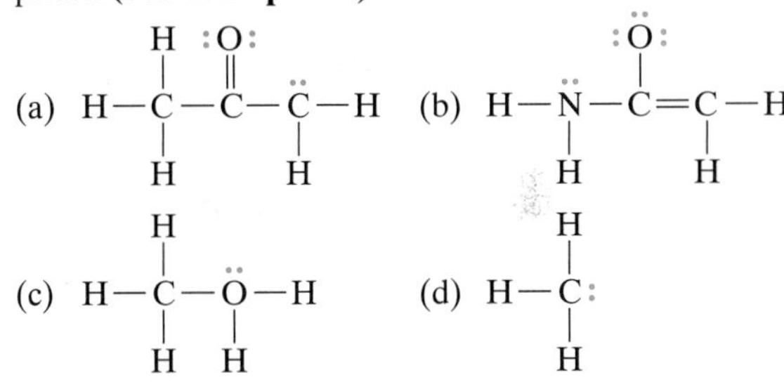

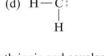

1.32 Each compound contains both ionic and covalent bonds. Draw a Lewis structure for each, and show by charges which bonds are ionic and by dashes which bonds are covalent. **(See Example 1.7)**

(a) NaOH (b) $NaHCO_3$ (c) NH_4Cl

(d) CH_3COONa (e) CH_3ONa

1.33 Silver and oxygen can form a stable compound. Predict the formula of this compound, and state whether the compound consists of ionic or covalent bonds.

1.34 Draw Lewis structures for the following molecule and ions: **(See Example 1.7)**

(a) NH_3 (b) NH_4^+ (c) NH_2^-

Section 1.2 Polarity of Covalent Bonds

1.35 Which statement is true about electronegativity?

(a) Electronegativity increases from left to right in a period of the Periodic Table.

(b) Electronegativity increases from top to bottom in a column of the Periodic Table.

(c) Hydrogen, the element with the lowest atomic number, has the smallest electronegativity.

(d) The higher the atomic number of an element, the greater is its electronegativity.

1.36 Why does fluorine, the element in the upper right corner of the Periodic Table, have the largest electronegativity of any element?

1.37 Arrange the single covalent bonds within each set in order of increasing polarity:

(a) C—H, O—H, N—H

(b) C—H, C—Cl, C—I

(c) C—C, C—O, C—N

(d) C—Li, C—Hg, C—Mg

1.38 Using the values of electronegativity given in Table 1.5, predict which indicated bond in each set is more polar and, using the symbols $\delta+$ and $\delta-$, show the direction of its polarity: **(See Example 1.5)**

(a) CH_3—OH or CH_3O—H

(b) H—NH_2 or CH_3—NH_2

(c) CH_3—SH or CH_3S—H

(d) CH_3—F or H—F

1.39 Identify the most polar bond in each molecule:

(a) $HSCH_2CH_2OH$

(b) $CHCl_2F$

(c) $HOCH_2CH_2NH_2$

1.40 Predict whether the carbon–metal bond in each of these organometallic compounds is nonpolar covalent, polar covalent, or ionic. For each polar covalent bond, show its direction of polarity using the symbols $\delta+$ and $\delta-$. **(See Example 1.5)**

(a) CH_3CH_2—Pb(CH_2CH_3)(CH_2CH_3)—CH_2CH_3
Tetraethyllead

(b) CH_3—Mg—Cl
Methylmagnesium chloride

(c) CH_3—Hg—CH_3
Dimethylmercury

Section 1.3 Bond Angles and Shapes of Molecules

1.41 Using VSEPR, predict bond angles about each highlighted atom: **(See Example 1.8)** # Don't forget to add unbonded e-!

(a) H—C(H)(H)—C(H)(H)—O—H

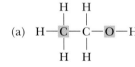

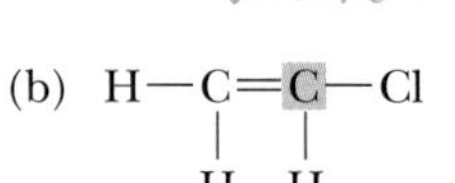

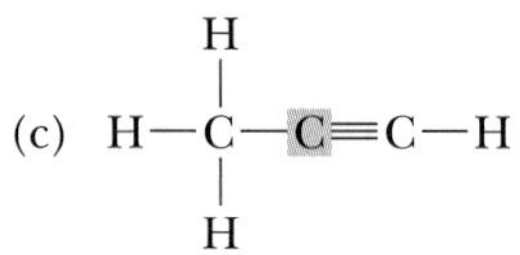

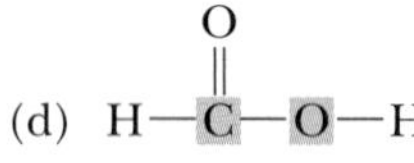

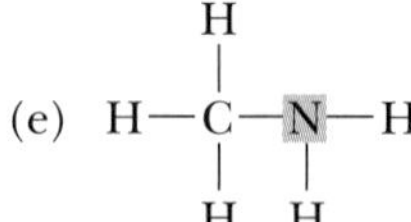

(f) H—O—N=O

1.42 Using VSEPR, predict bond angles about each atom of carbon, nitrogen, and oxygen in these molecules. (*Hint*: First add unshared pairs of electrons as necessary to complete the valence shell of each atom, and then make your predictions of bond angles.) **(See Example 1.8)**

(a) CH_3—CH_2—CH_2—OH

(b) CH_3—CH_2—C(=O)—H

(c) CH_3—CH=CH_2

(d) CH_3—C≡C—CH_3

(e) CH_3—C(=O)—O—CH_3

(f) CH_3—N(CH_3)—CH_3

1.43 Silicon is immediately below carbon in the Periodic Table. Predict the C—Si—C bond angle in tetramethylsilane, $(CH_3)_4Si$. **(See Example 1.8)**

Section 1.4 Polar and Nonpolar Molecules

1.44 Draw a three-dimensional representation for each molecule. Indicate which molecules are polar and the direction of their polarity: **(See Example 1.9)**

(a) CH_3F (b) CH_2Cl_2 (c) $CHCl_3$

(d) CCl_4 (e) $CH_2{=}CCl_2$ (f) $CH_2{=}CHCl$

(g) $CH_3C{\equiv}N$ (h) $(CH_3)_2C{=}O$

*__1.45__ Tetrafluoroethylene, C_2F_4, is the starting material for the synthesis of the polymer poly(tetrafluoroethylene), commonly known as Teflon. Molecules of tetrafluoroethylene are nonpolar. Propose a structural formula for this compound.

*__1.46__ Until several years ago, the two chlorofluorocarbons (CFCs) most widely used as heat-transfer media for refrigeration systems were Freon-11 (trichlorofluoromethane, CCl_3F) and Freon-12 (dichlorodifluoromethane, CCl_2F_2). Draw a three-dimensional representation of each molecule, and indicate the direction of its polarity. **(See Example 1.9)**

Section 1.5 Resonance Contributing Structures

1.47 Which of these statements are true about resonance contributing structures? **(See Example 1.10)**

(a) All contributing structures must have the same number of valence electrons.

(b) All contributing structures must have the same arrangement of atoms.

(c) All atoms in a contributing structure must have complete valence shells.

(d) All bond angles in sets of contributing structures must be the same.

(e) The following pair represents acceptable resonance contributing structures:

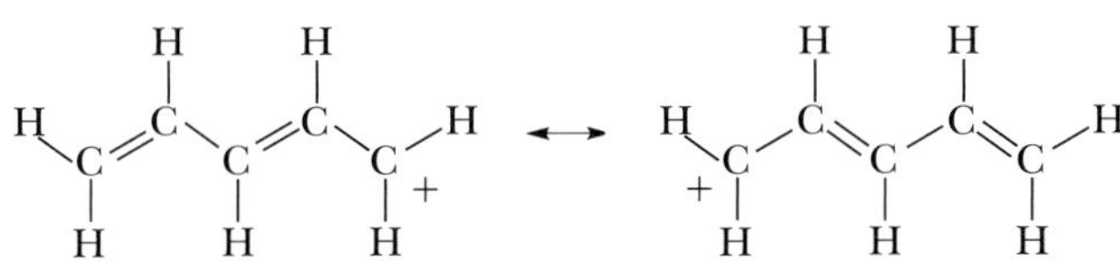

(f) The following pair represents acceptable resonance contributing structures:

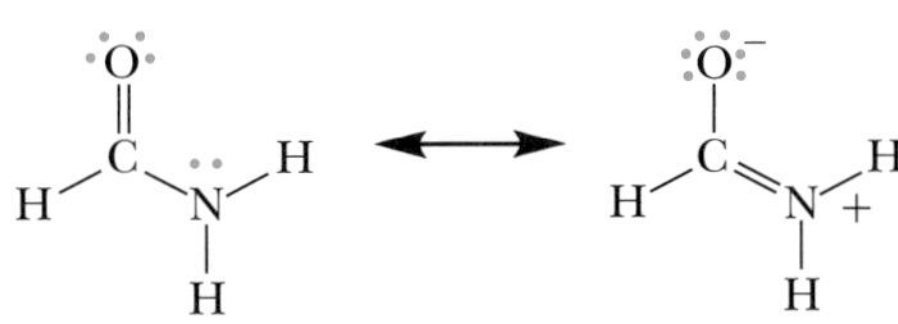

(g) The following pair represents acceptable resonance contributing structures:

$\ddot{O}{=}C{=}\ddot{N}H \longleftrightarrow {}^{-}:\ddot{O}{-}C{\equiv}\overset{+}{N}H$

1.48 Draw the resonance contributing structure indicated by the curved arrow(s), and assign formal charges as appropriate: **(See Example 1.11)**

(a) $H{-}\ddot{O}{-}C(=\ddot{O}:){-}\ddot{O}:^{-} \longleftrightarrow$ (b) $H_2C{=}\ddot{O} \longleftrightarrow$

(c) $CH_3{-}\ddot{O}{-}C(=\ddot{O}:){-}\ddot{O}:^{-} \longleftrightarrow$

1.49 Using VSEPR, predict the bond angles about the carbon atom in each pair of contributing structures in Problem 1.48. In what way do the bond angles change from one contributing structure to the other?

1.50 Draw acceptable resonance contributing structure(s) for each of the compounds shown. **(See Example 1.11)**

(a) $O{=}\overset{+}{O}{-}O^{-}$ (b) $H{-}\overset{+}{N}{\equiv}C{-}\overset{-}{\ddot{N}}{-}H$

(c) 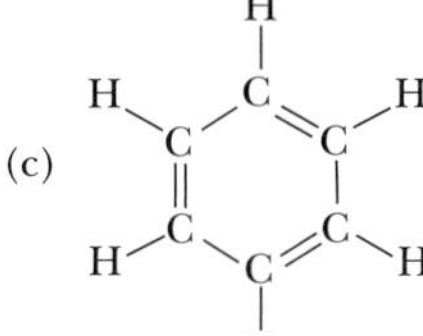

(d) $HO{-}CH(H){-}CH{=}CH{-}\overset{-}{C}H{-}H$

Section 1.6 Hybridization of Atomic Orbitals

1.51 State the hybridization of each highlighted atom:

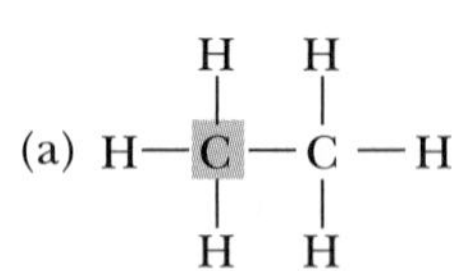

(b) $H_2C{=}CH_2$ (highlighted: C)

(c) H—C≡C—H (highlighted: C)

(d) H—C(H)(H)—N(H)—H (highlighted: N)

(e) H—C(=O)—O—H (highlighted: C, O)

(f) $H_2C{=}O$ (highlighted: C=O)

1.52 Describe each highlighted bond in terms of the overlap of hybrid orbitals: **(See Example 1.12)**

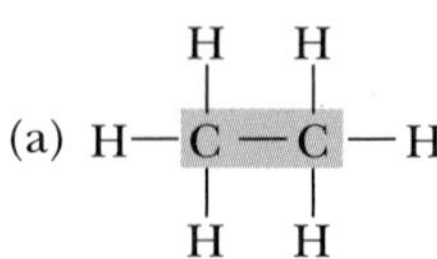

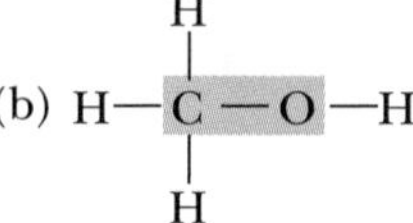

(c) H—C(H)(H)—N(H)—H (highlighted: C—N)

(d) $H_2C{=}O$ (highlighted: C=O)

(e) HC(H)=C(H)H (highlighted: C=C)

(f) H—C(=O)—O—H (highlighted: C—O)

Section 1.7 Functional Groups

1.53 Draw Lewis structures for these functional groups. Be certain to show all valence electrons on each: **(See Examples 1.13–1.16)**

(a) Carbonyl group (b) Carboxyl group

(c) Hydroxyl group (d) Primary amino group

1.54 Draw the structure for a compound with the molecular formula **(See Examples 1.13–1.16)**

(a) C_2H_6O that is an alcohol.

(b) C_3H_6O that is an aldehyde.

(c) C_3H_6O that is a ketone.

(d) $C_3H_6O_2$ that is a carboxylic acid.

(e) $C_4H_{11}N$ that is a tertiary amine.

1.55 Draw condensed structural formulas for all compounds with the molecular formula C_4H_8O that contain **(See Examples 1.13–1.16)**

(a) a carbonyl group. (There are two aldehydes and one ketone.)

(b) a carbon–carbon double bond and a hydroxyl group. (There are eight.)

1.56 Draw structural formulas for **(See Examples 1.13–1.16)**

(a) the eight alcohols with the molecular formula $C_5H_{12}O$.

(b) the eight aldehydes with the molecular formula $C_6H_{12}O$.

(c) the six ketones with the molecular formula $C_6H_{12}O$.

(d) the eight carboxylic acids with the molecular formula $C_6H_{12}O_2$.

(e) the three tertiary amines with the molecular formula $C_5H_{13}N$.

***1.57** Identify the functional groups in each compound (we study each compound in more detail in the indicated section):

(a) $CH_3—CH(OH)—C(=O)—OH$
Lactic acid
(Section 22.4A)

(b) $HO—CH_2—CH_2—OH$
Ethylene glycol
(Section 8.1B)

(c) $CH_3—CH(NH_2)—C(=O)—OH$
Alanine
(Section 19.2B)

(d) $HO—CH_2—CH(OH)—C(=O)—H$
Glyceraldehyde
(Section 18.2A)

(e) $CH_3—C(=O)—CH_2—C(=O)—OH$
Acetoacetic acid
(Section 14.2B)

(f) $H_2NCH_2CH_2CH_2CH_2CH_2CH_2NH_2$
1,6-Hexanediamine
(Section 17.4A)

***1.58** Dihydroxyacetone, $C_3H_6O_3$, the active ingredient in many sunless tanning lotions, contains two 1° hydroxyl groups, each on a different carbon, and one ketone group. Draw a structural formula for dihydroxyacetone. **(See Examples 1.13–1.16)**

***1.59** Propylene glycol, $C_3H_8O_2$, commonly used in airplane deicers, contains a 1° alcohol and a 2° alcohol. Draw a structural formula for propylene glycol. **(See Examples 1.13–1.16)**

***1.60** Ephedrine is a molecule found in the dietary supplement ephedra, which has been linked to adverse health reactions such as heart attacks, strokes, and heart palpitations. The use of ephedra in dietary supplements is now banned by the FDA.

(a) Identify at least two functional groups in ephedrine.

(b) Would you predict ephedrine to be polar or nonpolar?

Ephedrine

***1.61** Ozone (O_3) and carbon dioxide (CO_2) are both known as greenhouse gases. Compare and contrast their shapes, and indicate the hybridization of each atom in the two molecules.

***1.62** In the lower atmosphere that is also contaminated with unburned hydrocarbons, NO_2 participates in a series of reactions. One product of these reactions is peroxyacetyl nitrate (PAN). The connectivity of the atoms in PAN appears below.

(a) Determine the number of valence electrons in this molecule, and then complete its Lewis structure.

(b) Give the approximate values of the bond angles around each atom indicated with an arrow.

Looking Ahead

1.63 Allene, C_3H_4, has the structural formula $H_2C{=}C{=}CH_2$. Determine the hybridization of each carbon in allene and predict the shape of the molecule.

1.64 Dimethylsulfoxide, $(CH_3)_2SO$, is a common solvent used in organic chemistry.

(a) Write a Lewis structure for dimethylsulfoxide.

(b) Predict the hybridization of the sulfur atom in the molecule.

(c) Predict the geometry of dimethylsulfoxide.

(d) Is dimethylsulfoxide a polar or a nonpolar molecule?

1.65 In Chapter 5, we study a group of organic cations called carbocations. Following is the structure of one such carbocation, the *tert*-butyl cation:

tert-Butyl cation

(a) How many electrons are in the valence shell of the carbon bearing the positive charge?

(b) Predict the bond angles about this carbon.

(c) Given the bond angles you predicted in (b), what hybridization do you predict for this carbon?

1.66 We also study the isopropyl cation, $(CH_3)_2CH^+$, in Chapter 5.

(a) Write a Lewis structure for this cation. Use a plus sign to show the location of the positive charge.

(b) How many electrons are in the valence shell of the carbon bearing the positive charge?

(c) Use VSEPR to predict all bond angles about the carbon bearing the positive charge.

(d) Describe the hybridization of each carbon in this cation.

1.67 In Chapter 9, we study benzene, C_6H_6, and its derivatives.

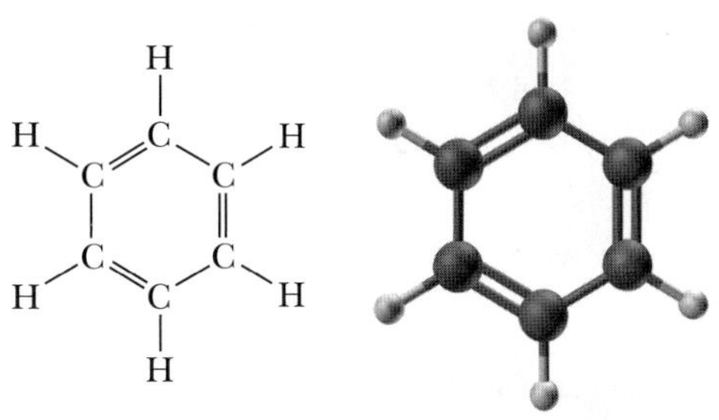

(a) Predict each H—C—C and each C—C—C bond angle on benzene.

(b) State the hybridization of each carbon in benzene.

(c) Predict the shape of a benzene molecule.

1.68 Explain why all the carbon–carbon bonds in benzene are equal in length.

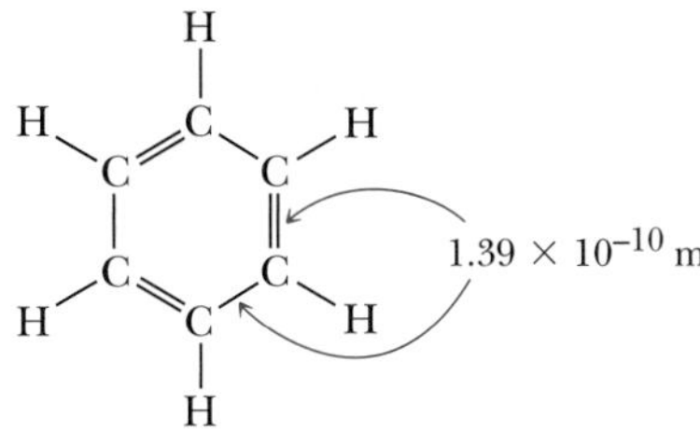

2 Acids and Bases

KEY QUESTIONS

2.1 What Are Arrhenius Acids and Bases?

2.2 What Are Brønsted–Lowry Acids and Bases?

2.3 How Do We Measure the Strength of an Acid or Base?

2.4 How Do We Determine the Position of Equilibrium in an Acid–Base Reaction?

2.5 What Are the Relationships between Acidity and Molecular Structure?

2.6 What Are Lewis Acids and Bases?

HOW TO

2.1 How to Determine the Position of Equilibrium in an Acid–Base Reaction

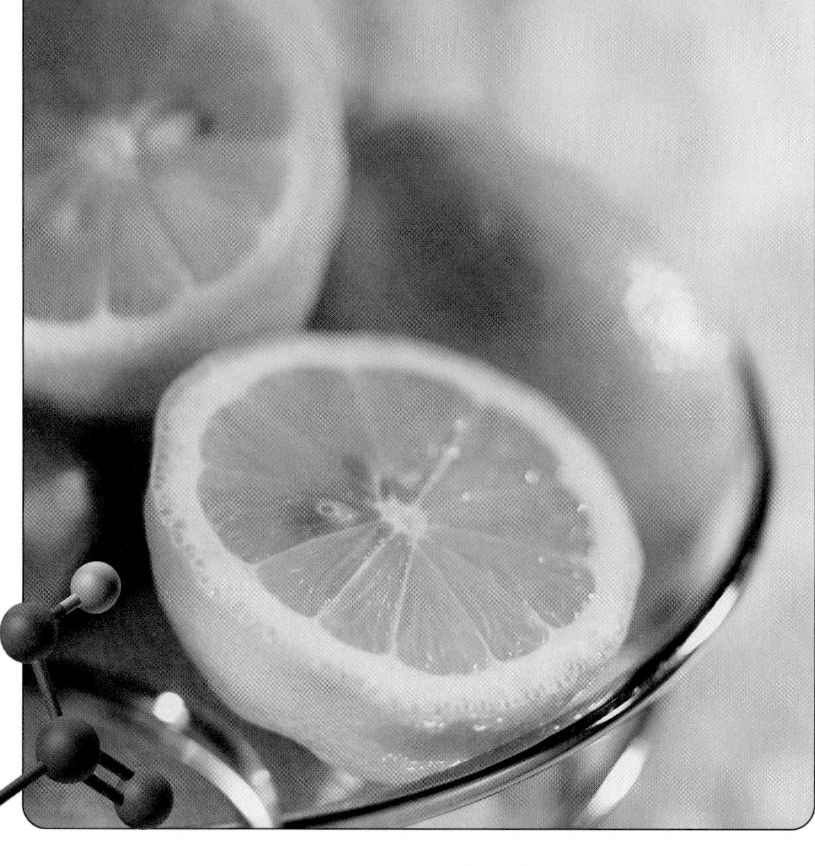

Citrus fruits are sources of citric acid. Lemon juice, for example, contains 5–8% citric acid. Inset: A model of citric acid. *(Corbis Digital Stock)*

A great many organic reactions are acid–base reactions. In this and later chapters, we will study the acid–base properties of the major classes of organic compounds, including alcohols, phenols, carboxylic acids, carbonyl compounds containing α-hydrogens, amines, amino acids, proteins, and, finally, nucleic acids. Furthermore, many organic reactions are catalyzed by proton-donating acids, such as H_3O^+ and $CH_3OH_2{}^+$. Others are catalyzed by Lewis acids, such as $AlCl_3$. It is essential, therefore, that you have a good grasp of the fundamentals of acid–base chemistry.

2.1 What Are Arrhenius Acids and Bases?

Arrhenius acid A substance that dissolves in water to produce H^+ ions.

Arrhenius base A substance that dissolves in water to produce OH^- ions.

The first useful definitions of an acid and a base were put forward by Svante Arrhenius (1859–1927) in 1884; Nobel Prize in Chemistry 1903. According to the original Arrhenius definitions, an acid is a substance that dissolves in water to produce H^+ ions, and a base is a substance that dissolves in water to produce OH^- ions. Today we know that a H^+ ion does not exist in water because it reacts immediately with an H_2O molecule to give a hydronium ion, H_3O^+:

$$H^+(aq) + H_2O(l) \longrightarrow \underset{\text{Hydronium ion}}{H_3O^+(aq)}$$

Apart from this modification, the Arrhenius definitions of acid and base are still valid and useful today, as long as we are talking about aqueous solutions.

When an acid dissolves in water, it reacts with the water to produce H_3O^+. For example, when HCl dissolves in water, it reacts with a water molecule to give hydronium ion and chloride ion:

$$H_2O(l) + HCl(aq) \longrightarrow H_3O^+(aq) + Cl^-(aq)$$

We can show the transfer of a proton from an acid to a base by using a symbol called a **curved arrow**. First we write the Lewis structure of each reactant and product, showing all valence electrons on reacting atoms. Then we use curved arrows to show the change in position of electron pairs during the reaction. The tail of the curved arrow is located at an electron pair. The head of the curved arrow shows the new position of the electron pair. A change in position of an electron pair originating from an atom will form a new bond to that atom, while a change in position of an electron pair originating from a bond will result in breaking that bond.

this pair of electrons is given to chlorine to form chloride ion

$$\text{H—}\ddot{\text{O}}\text{:(—H)} + \text{H—}\ddot{\text{Cl}}\text{:} \longrightarrow \text{H—}\ddot{\text{O}}^+\text{—H(—H)} + \text{:}\ddot{\text{Cl}}\text{:}^-$$

this electron pair is used to form a new O—H bond

In this equation, the leftmost curved arrow shows that an unshared pair of electrons on oxygen changes position to form a new covalent bond with hydrogen. The rightmost curved arrow shows that the pair of electrons of the H—Cl bond is given entirely to chlorine to form a chloride ion. Thus, in the reaction of HCl with H_2O, a proton is transferred from HCl to H_2O, and in the process, an O—H bond forms and an H—Cl bond breaks.

With bases, the situation is slightly different. Many bases are metal hydroxides, such as KOH, NaOH, $Mg(OH)_2$, and $Ca(OH)_2$. These compounds are ionic solids, and when they dissolve in water, their ions merely separate, with each ion solvated by water molecules, as illustrated in the following reaction:

$$NaOH(s) \xrightarrow{H_2O} Na^+(aq) + OH^-(aq)$$

Other bases are not hydroxides. Instead, they produce OH^- ions in water by reacting with water molecules. The most important examples of this kind of base are ammonia, NH_3, and amines (Section 1.7B). When ammonia dissolves in water, it reacts with the water to produce ammonium ions and hydroxide ions:

$$NH_3(aq) + H_2O(l) \rightleftharpoons NH_4^+(aq) + OH^-(aq)$$

As we will see in Section 2.4, ammonia is a weak base, and the position of the equilibrium for its reaction with water lies considerably toward the left. In a 1.0-M solution of NH_3 in water, for example, only about four molecules of NH_3 out of every thousand react with the water to form NH_4^+ and OH^-. Thus, when ammonia dissolves in water, it exists primarily as NH_3 molecules. Nevertheless, some OH^- ions are produced; therefore, NH_3 is a base.

We indicate how the reaction of ammonia with water takes place by using curved arrows to show the transfer of a proton from a water molecule to an ammonia molecule: A key point to remember is that curved arrows always originate from a source of electrons—that is, from either a bonding pair or a lone pair. A curved arrow that originates at a lone pair will form a new covalent bond. A curved arrow that originates at a bonding pair and that points to an atom will always yield a new lone pair.

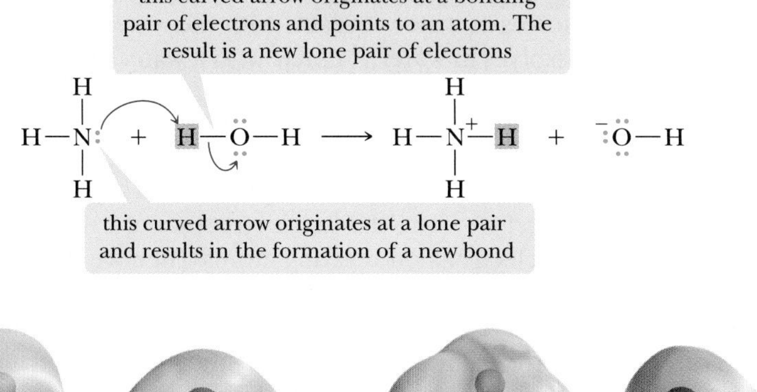

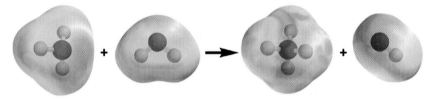

Notice the increase in charge distribution upon protonation of NH_3 and deprotonation of H_2O. The nitrogen of NH_4^+ shows a more intense blue than the nitrogen of NH_3, and the oxygen of OH^- shows a more intense red than the oxygen of H_2O.

Here, the left curved arrow shows that the unshared pair of electrons on nitrogen changes position to form a new covalent bond with a hydrogen of a water molecule. At the same time as the new N—H bond forms, an O—H bond of a water molecule breaks, as indicated by the right arrow, and the pair of electrons forming the H—O bond moves entirely to oxygen, forming OH^-. Thus, ammonia produces an OH^- ion by a proton-transfer reaction from a water molecule and leaves OH^- behind.

The Arrhenius concept of acids and bases is so intimately tied to reactions that take place in water that it has no good way to deal with acid–base reactions in nonaqueous solutions. For this reason, we concentrate in this chapter on the Brønsted–Lowry definitions of acids and bases, which are more useful to us in our discussion of reactions of organic compounds.

2.2 What Are Brønsted–Lowry Acids and Bases?

Brønsted–Lowry acid A proton donor.

Brønsted–Lowry base A proton acceptor.

Conjugate base The species formed when an acid donates a proton.

In 1923, the Danish chemist Johannes Brønsted and the English chemist Thomas Lowry independently proposed the following definitions: An **acid** is a **proton donor**, a **base** is a **proton acceptor**, and an acid–base reaction is a **proton-transfer reaction**. Furthermore, according to the Brønsted–Lowry definitions, any pair of molecules or ions that can be interconverted by the transfer of a proton is called a **conjugate acid–base pair**. When an acid transfers a proton to a base, the acid is converted to its **conjugate base**. When a base accepts a proton, the base is converted to its **conjugate acid**.

Conjugate acid The species formed when a base accepts a proton.

We can illustrate these relationships by examining the reaction of hydrogen chloride with water to form chloride ion and hydronium ion:

conjugate acid–base pair

conjugate acid–base pair

$$\underset{\substack{\text{Hydrogen}\\\text{chloride}\\\text{(Acid)}}}{HCl\ (aq)} + \underset{\substack{\text{Water}\\ \\\text{(Base)}}}{H_2O\ (l)} \longrightarrow \underset{\substack{\text{Chloride}\\\text{ion}\\\text{(Conjugate}\\\text{base of HCl)}}}{Cl^-\ (aq)} + \underset{\substack{\text{Hydronium}\\\text{ion}\\\text{(Conjugate}\\\text{acid of water)}}}{H_3O^+\ (aq)}$$

In this reaction, the acid HCl donates a proton and is converted to its conjugate base, Cl^-. The base H_2O accepts a proton and is converted to its conjugate acid, H_3O^+.

We have illustrated the application of the Brønsted–Lowry definitions with water as a reactant. These definitions, however, do not require water as a reactant. Consider the following reaction between acetic acid and ammonia:

conjugate acid–base pair

conjugate acid–base pair

$$\underset{\substack{\text{Acetic}\\\text{acid}\\\text{(Acid)}}}{CH_3COOH} + \underset{\substack{\text{Ammonia}\\ \\\text{(Base)}}}{NH_3} \rightleftharpoons \underset{\substack{\text{Acetate}\\\text{ion}\\\text{(Conjugate base}\\\text{of acetic acid)}}}{CH_3COO^-} + \underset{\substack{\text{Hydronium}\\\text{ion}\\\text{(Conjugate acid}\\\text{of ammonia)}}}{NH_4^+}$$

We can use curved arrows to show how this reaction takes place:

this electron pair is given to oxygen to form acetate ion

this electron pair is used to form a new N—H bond

$$\underset{\substack{\text{Acetic acid}\\\text{(proton donor)}}}{CH_3-\overset{\overset{\large :O:}{\|}}{C}-\ddot{O}-H} + \underset{\substack{\text{Ammonia}\\\text{(proton acceptor)}}}{:\overset{H}{\overset{|}{N}}\underset{\underset{H}{|}}{}-H} \rightleftharpoons \underset{\text{Acetate ion}}{CH_3-\overset{\overset{\large :O:}{\|}}{C}-\ddot{\underset{..}{O}}:^-} + \underset{\text{Ammonium ion}}{H-\overset{H}{\overset{|}{N^+}}\underset{\underset{H}{|}}{}-H}$$

The rightmost curved arrow shows that the unshared pair of electrons on nitrogen becomes shared between N and H to form a new H—N bond. At the same time that the H—N bond forms, the O—H bond breaks, and the electron pair of the O—H bond moves entirely to oxygen to form the $-O^-$ of the acetate ion. The result of these two electron-pair shifts is the transfer of a proton from an acetic acid molecule to an ammonia molecule. Table 2.1 gives examples of common acids and their conjugate bases. As you study the examples of conjugate acid–base pairs in the table, note the following points:

1. An acid can be positively charged, neutral, or negatively charged. Examples of these charge types are H_3O^+, H_2CO_3, and $H_2PO_4^-$.

TABLE 2.1 Some Acids and Their Conjugate Bases

	Acid	Name	Conjugate Base	Name	
Strong Acids	HI	Hydroiodic acid	I^-	Iodide ion	Weak Bases
↑	HCl	Hydrochloric acid	Cl^-	Chloride ion	↓
	H_2SO_4	Sulfuric acid	HSO_4^-	Hydrogen sulfate ion	
	HNO_3	Nitric acid	NO_3^-	Nitrate ion	
	H_3O^+	Hydronium ion	H_2O	Water	
	HSO_4^-	Hydrogen sulfate ion	SO_4^{2-}	Sulfate ion	
	H_3PO_4	Phosphoric acid	$H_2PO_4^-$	Dihydrogen phosphate ion	
	CH_3COOH	Acetic acid	CH_3COO^-	Acetate ion	
	H_2CO_3	Carbonic acid	HCO_3^-	Bicarbonate ion	
	H_2S	Hydrogen sulfide	HS^-	Hydrogen sulfide ion	
	$H_2PO_4^-$	Dihydrogen phosphate ion	HPO_4^{2-}	Hydrogen phosphate ion	
	NH_4^+	Ammonium ion	NH_3	Ammonia	
	HCN	Hydrocyanic acid	CN^-	Cyanide ion	
	C_6H_5OH	Phenol	$C_6H_5O^-$	Phenoxide ion	
	HCO_3^-	Bicarbonate ion	CO_3^{2-}	Carbonate ion	
	HPO_4^{2-}	Hydrogen phosphate ion	PO_4^{3-}	Phosphate ion	
Weak Acids	H_2O	Water	OH^-	Hydroxide ion	Strong Bases
	C_2H_5OH	Ethanol	$C_2H_5O^-$	Ethoxide ion	

2. A base can be negatively charged or neutral. Examples of these charge types are Cl^- and NH_3.
3. Acids are classified as monoprotic, diprotic, or triprotic, depending on the number of protons each may give up. Examples of **monoprotic acids** include HCl, HNO_3, and CH_3COOH. Examples of **diprotic acids** include H_2SO_4 and H_2CO_3. An example of a **triprotic acid** is H_3PO_4. Carbonic acid, for example, loses one proton to become bicarbonate ion and then a second proton to become carbonate ion:

$$\underset{\text{Carbonic acid}}{H_2CO_3} + H_2O \rightleftharpoons \underset{\text{Bicarbonate ion}}{HCO_3^-} + H_3O^+$$

$$\underset{\text{Bicarbonate ion}}{HCO_3^-} + H_2O \rightleftharpoons \underset{\text{Carbonate ion}}{CO_3^{2-}} + H_3O^+$$

4. Several molecules and ions appear in both the acid and conjugate base columns; that is, each can function as either an acid or a base. The bicarbonate ion, HCO_3^-, for example, can give up a proton to become CO_3^{2-} (in which case it is an acid) or can accept a proton to become H_2CO_3 (in which case it is a base).
5. There is an inverse relationship between the strength of an acid and the strength of its conjugate base. The stronger the acid, the weaker is its conjugate base. HI, for example, is the strongest acid listed in Table 2.1, and I^-, its conjugate base, is the weakest base. As another example, CH_3COOH (acetic acid) is a stronger acid than H_2CO_3 (carbonic acid); conversely, CH_3COO^- (acetate ion) is a weaker base than HCO_3^- (bicarbonate ion).

2, 4

Example 2.1

Write the following acid–base reaction as a proton-transfer reaction. Label which reactant is the acid and which the base, as well as which product is the conjugate base of the original acid and which is the conjugate acid of the original base. Use curved arrows to show the flow of electrons in the reaction.

$$\underset{\text{Acetic acid}}{CH_3\overset{\overset{O}{\|}}{C}OH} + \underset{\text{Bicarbonate ion}}{HCO_3^-} \longrightarrow \underset{\text{Acetate ion}}{CH_3\overset{\overset{O}{\|}}{C}O^-} + \underset{\text{Carbonic acid}}{H_2CO_3}$$

Strategy

First, write a Lewis structure for each reactant by showing all valence electrons on the reacting atoms: Acetic acid is the acid (proton donor), and bicarbonate ion is the base (proton acceptor). The members of a conjugate acid–base pair differ only by a proton, with the acid having the greater number of protons. To find the formula of the conjugate base, we remove one proton from the acid.

Solution

From Table 2.1, we see that acetic acid is a stronger acid and, therefore, is the proton donor in this reaction.

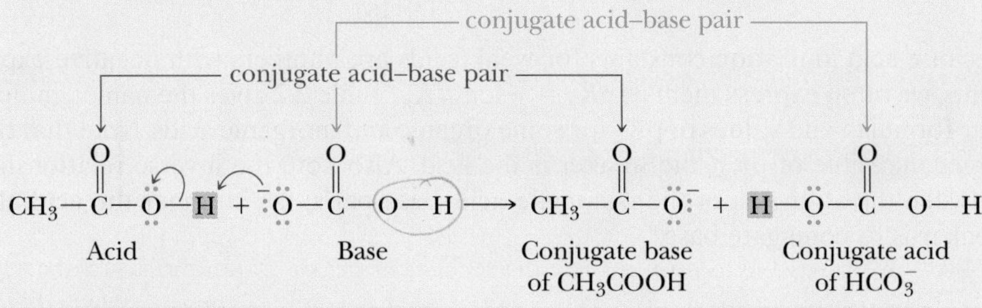

See problems 2.7, 2.8

Problem 2.1

Write each acid–base reaction as a proton-transfer reaction. Label which reactant is the acid and which product is the base, as well as which product is the conjugate base of the original acid and which is the conjugate acid of the original base. Use curved arrows to show the flow of electrons in each reaction.

(a) $CH_3SH + OH^- \longrightarrow CH_3S^- + H_2O$

(b) $CH_3OH + NH_2^- \longrightarrow CH_3O^- + NH_3$

2.3 How Do We Measure the Strength of an Acid or Base?

Strong acid An acid that is completely ionized in aqueous solution.

Strong base A base that is completely ionized in aqueous solution.

A **strong acid** or **strong base** is one that ionizes completely in aqueous solution. When HCl is dissolved in water, a proton is transferred completely from HCl to H_2O to form Cl^- and H_3O^+. There is no tendency for the reverse reaction to occur—for the transfer of a proton from H_3O^+ to Cl^- to form HCl and H_2O. Therefore, when we compare the relative acidities of HCl and H_3O^+, we conclude

memorize

that HCl is the stronger acid and H_3O^+ is the weaker acid. Similarly, H_2O is the stronger base and Cl^- is the weaker base.

Examples of strong acids in aqueous solution are HCl, HBr, HI, HNO_3, $HClO_4$, and H_2SO_4. Examples of strong bases in aqueous solution are LiOH, NaOH, KOH, $Ca(OH)_2$, and $Ba(OH)_2$.

Weak acid An acid that only partially ionizes in aqueous solution.

Weak base A base that only partially ionizes in aqueous solution.

A **weak acid** or **weak base** is one that only partially ionizes in aqueous solution. Most organic acids and bases are weak. Among the most common organic acids we deal with are the carboxylic acids, which contain a carboxyl group, —COOH (Section 1.7D), as shown in the following reaction:

$$\underset{\substack{\text{Acid} \\ \text{(weaker acid)}}}{CH_3\overset{\overset{O}{\|}}{C}OH} + \underset{\substack{\text{Base} \\ \text{(weaker base)}}}{H_2O} \rightleftharpoons \underset{\substack{\text{Conjugate base} \\ \text{of } CH_3COOH \\ \text{(stronger base)}}}{CH_3\overset{\overset{O}{\|}}{C}O^-} + \underset{\substack{\text{Conjugate acid} \\ \text{of } H_2O \\ \text{(stronger acid)}}}{H_3O^+}$$

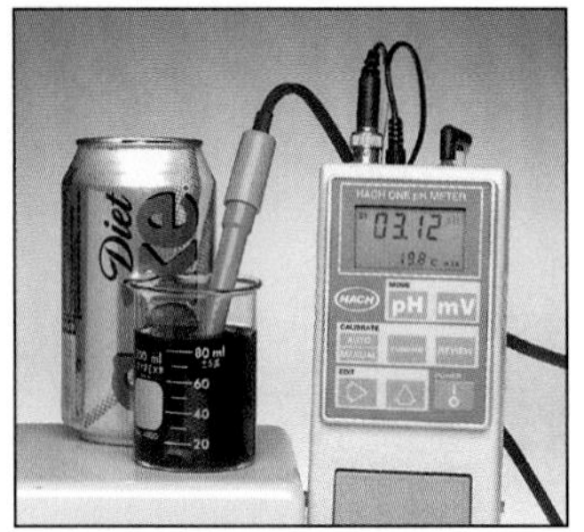

The pH of this soft drink is 3.12. Soft drinks are often quite acidic.
(Charles D. Winters)

The equation for the ionization of a weak acid, HA, in water and the acid ionization constant K_a for this equilibrium are, respectively,

$$HA + H_2O \rightleftharpoons A^- + H_3O^+$$

and

$$K_a = K_{eq}[H_2O] = \frac{[H_3O^+][A^-]}{[HA]}$$

Because acid ionization constants for weak acids are numbers with negative exponents, we often express them as $\mathbf{p}\boldsymbol{K}_\mathbf{a} = -\log_{10} K_a$. Table 2.2 gives the names, molecular formulas, and values of pK_a for some organic and inorganic acids. Note that the larger the value of pK_a, the weaker is the acid. Also note the inverse relationship between the strengths of the conjugate acid–base pairs; the stronger the acid, the weaker is its conjugate base.

TABLE 2.2 pK_a Values for Some Organic and Inorganic Acids

	Acid	Formula	pK_a	Conjugate Base	
Weaker acid	ethane	CH_3CH_3	51	$CH_3CH_2^-$	Stronger base
	ammonia	NH_3	38	NH_2^-	
	ethanol	CH_3CH_2OH	15.9	$CH_3CH_2O^-$	
	water	H_2O	15.7	HO^-	
	methylammonium ion	$CH_3NH_3^+$	10.64	CH_3NH_2	
	bicarbonate ion	HCO_3^-	10.33	CO_3^{2-}	
	phenol	C_6H_5OH	9.95	$C_6H_5O^-$	
	ammonium ion	NH_4^+	9.24	NH_3	
	carbonic acid	H_2CO_3	6.36	HCO_3^-	
	acetic acid	CH_3COOH	4.76	CH_3COO^-	
	benzoic acid	C_6H_5COOH	4.19	$C_6H_5COO^-$	
	phosphoric acid	H_3PO_4	2.1	$H_2PO_4^-$	
	hydronium ion	H_3O^+	−1.74	H_2O	
	sulfuric acid	H_2SO_4	−5.2	HSO_4^-	
	hydrogen chloride	HCl	−7	Cl^-	
Stronger acid	hydrogen bromide	HBr	−8	Br^-	Weaker base
	hydrogen iodide	HI	−9	I^-	

Example 2.2

For each value of pK_a, calculate the corresponding value of K_a. Which compound is the stronger acid?

(a) Ethanol, pK_a = 15.9 (b) Carbonic acid, pK_a = 6.36

Strategy

The stronger acid has the smaller value of pK_a (the larger value of K_a).

Solution

(a) For ethanol, $K_a = 1.3 \times 10^{-16}$ (b) For carbonic acid, $K_a = 4.4 \times 10^{-7}$

Because the value of pK_a for carbonic acid is smaller than that for ethanol, carbonic acid is the stronger acid and ethanol is the weaker acid.

See problem 2.16

Problem 2.2

For each value of K_a, calculate the corresponding value of pK_a. Which compound is the stronger acid?

(a) Acetic acid, $K_a = 1.74 \times 10^{-5}$ (b) Water, $K_a = 2.00 \times 10^{-16}$

Caution: In exercises such as Example 2.2 and Problem 2.2, we ask you to select the stronger acid. You must remember that these and all other acids with ionization constants considerably less than 1.00 are *weak* acids. Thus, although acetic acid is a considerably stronger acid than water, it still is only slightly ionized in water. The ionization of acetic acid in a 0.1 M solution, for example, is only about 1.3%; the major form of this weak acid that is present in a 0.1 M solution is the un-ionized acid!

Forms present in 0.1 M acetic acid

$$\underset{98.7\%}{CH_3\overset{\overset{O}{\|}}{C}OH} + H_2O \rightleftharpoons \underset{1.3\%}{CH_3\overset{\overset{O}{\|}}{C}O^-} + H_3O^+$$

2.4 How Do We Determine the Position of Equilibrium in an Acid–Base Reaction?

We know that HCl reacts with H_2O according to the following equilibrium:

$$HCl + H_2O \longrightarrow Cl^- + H_3O^+$$

We also know that HCl is a strong acid, which means that the position of this equilibrium lies very far to the right.

As we have seen, acetic acid reacts with H_2O according to the following equilibrium:

$$\underset{\text{Acetic acid}}{CH_3COOH} + H_2O \rightleftharpoons \underset{\text{Acetate ion}}{CH_3COO^-} + H_3O^+$$

Acetic acid is a weak acid. Only a few acetic acid molecules react with water to give acetate ions and hydronium ions, and the major species present in equilibrium in aqueous solution is CH_3COOH. The position of this equilibrium, therefore, lies very far to the left.

In the preceding two acid–base reactions, water was the base (proton acceptor). But what if we have a base other than water as the proton acceptor? How can we determine which are the major species present at equilibrium? That is, how can we determine whether the position of equilibrium lies toward the left or toward the right?

As an example, let us examine the acid–base reaction between acetic acid and ammonia to form acetate ion and ammonium ion:

$$\underset{\substack{\text{Acetic acid}\\ \text{(Acid)}}}{CH_3COOH} + \underset{\substack{\text{Ammonia}\\ \text{(Base)}}}{NH_3} \overset{?}{\rightleftharpoons} \underset{\substack{\text{Acetate ion}\\ \text{(Conjugate base}\\ \text{of } CH_3COOH)}}{CH_3COO^-} + \underset{\substack{\text{Ammonium ion}\\ \text{(Conjugate acid}\\ \text{of } NH_3)}}{NH_4^+}$$

Vinegar (which contains acetic acid) and baking soda (sodium bicarbonate) react to produce sodium acetate, carbon dioxide, and water. The carbon dioxide inflates the balloon. *(Charles D. Winters)*

As indicated by the question mark over the equilibrium arrow, we want to determine whether the position of this equilibrium lies toward the left or toward the right. There are two acids present: acetic acid and ammonium ion. There are also two bases present: ammonia and acetate ion. From Table 2.2, we see that CH_3COOH (pK_a 4.76) is the stronger acid, which means that CH_3COO^- is the weaker conjugate base. Conversely, NH_4^+ (pK_a 9.24) is the weaker acid, which means that NH_3 is the stronger conjugate base. We can now label the relative strengths of each acid and base in the equilibrium:

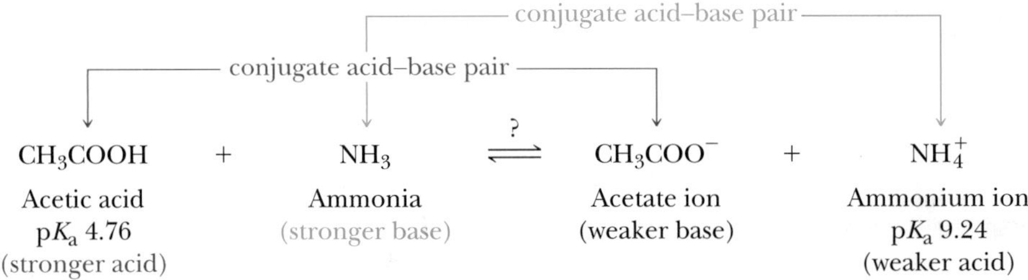

In an acid–base reaction, the position of equilibrium always favors reaction of the stronger acid and stronger base to form the weaker acid and weaker base. Thus, at equilibrium, the major species present are the weaker acid and weaker base. In the reaction between acetic acid and ammonia, therefore, the equilibrium lies to the right, and the major species present are acetate ion and ammonium ion:

$$\underset{\substack{\text{Acetic acid}\\ \text{(stronger acid)}}}{CH_3COOH} + \underset{\substack{\text{Ammonia}\\ \text{(stronger base)}}}{NH_3} \rightleftharpoons \underset{\substack{\text{Acetate ion}\\ \text{(weaker base)}}}{CH_3COO^-} + \underset{\substack{\text{Ammonium ion}\\ \text{(weaker acid)}}}{NH_4^+}$$

HOW TO 2.1 Determine the Position of Equilibrium in an Acid–Base Reaction

1. **Identify** the two acids in the equilibrium; one is on the left side of the equilibrium, the other on the right side.
2. Using the information in Table 2.2, determine which acid is the stronger and which the weaker.
3. Identify the stronger base and weaker base in each equilibrium. Remember that the stronger acid gives the weaker conjugate base and the weaker acid gives the stronger conjugate base.
4. The stronger acid and stronger base react to give the weaker acid and weaker base, and the position of equilibrium lies on the side of the weaker acid and weaker base.

Example 2.3

For each acid–base equilibrium, label the stronger acid, the stronger base, the weaker acid, and the weaker base. Then predict whether the position of equilibrium lies toward the right or toward the left.

(a) $H_2CO_3 + OH^- \rightleftharpoons HCO_3^- + H_2O$

Carbonic acid; Bicarbonate ion

(b) $C_6H_5OH + HCO_3^- \rightleftharpoons C_6H_5O^- + H_2CO_3$

Phenol; Bicarbonate ion; Phenoxide ion; Carbonic acid

Strategy

Identify the two acids in the equilibrium and their relative strengths and the two bases and their relative strengths. The position of the equilibrium lies toward the weaker acid and the weaker base.

Solution

Arrows over each equilibrium show the conjugate acid–base pairs. The position of equilibrium in (a) lies toward the right. In (b) it lies toward the left.

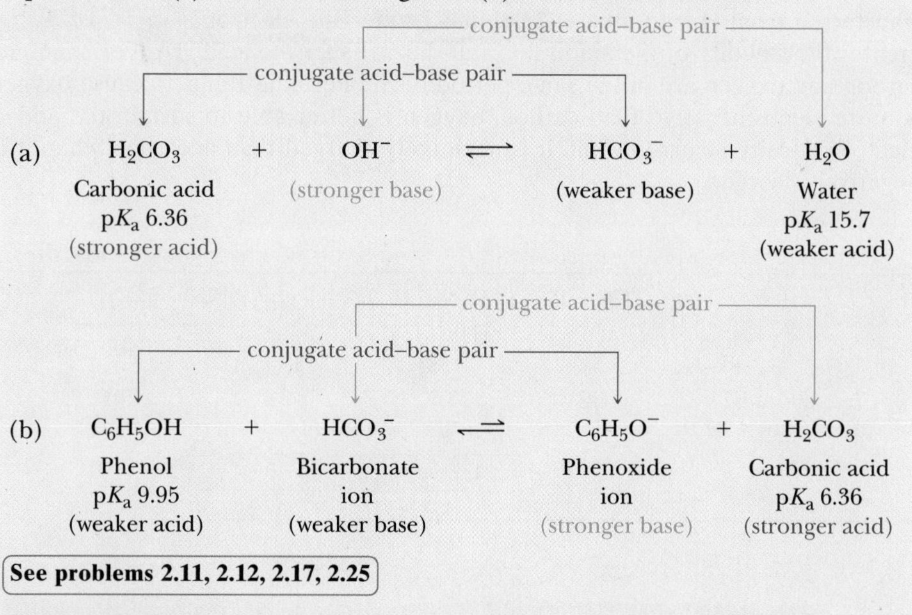

See problems 2.11, 2.12, 2.17, 2.25

Problem 2.3

For each acid–base equilibrium, label the stronger acid, the stronger base, the weaker acid, and the weaker base. Then predict whether the position of equilibrium lies toward the right or the left.

(a) $CH_3NH_2 + CH_3COOH \rightleftharpoons CH_3NH_3^+ + CH_3COO^-$

Methylamine; Acetic acid; Methylammonium ion; Acetate ion

(b) $CH_3CH_2O^- + NH_3 \rightleftharpoons CH_3CH_2OH + NH_2^-$

Ethoxide ion; Ammonia; Ethanol; Amide ion

2.5 What Are the Relationships between Acidity and Molecular Structure?

Now let us examine the relationship between the acidity of organic compounds and their molecular structure. The most important factor in determining the relative acidities of organic acids is the relative stability of the anion, A^-, formed when the acid, HA, transfers a proton to a base. We can understand the relationship involved by considering (A) the electronegativity of the atom bonded to H, (B) resonance, (C) the inductive effect, and (D) the size and delocalization of charge on A^-. We will look at each of these factors briefly in this chapter. We will study them more fully in later chapters when we deal with particular functional groups.

A. Electronegativity: Acidity of HA within a Period of the Periodic Table

Recall that electronegativity is a measure of an atom's attraction for electrons it shares in a covalent bond with another atom. The more electronegative an atom is, the greater its ability to sustain electron density around itself. The relative acidity of the hydrogen acids within a period of the Periodic Table is determined by the stability of A^-, that is, by the stability of the anion that forms when a proton is transferred from HA to a base. Thus, the greater the electronegativity of A, the greater the stability of the anion A^-, and the stronger the acid HA. For example, carbon and oxygen are in the same period of the Periodic Table. Because oxygen is more electronegative than carbon, oxygen is better able to sustain the added electron density incurred when it is negatively charged than is carbon when it is negatively charged.

	$H_3C—H$	$H_2N—H$	$HO—H$	$F—H$
pK_a	51	38	15.7	3.5
Electronegativity of A in A–H	2.5	3.0	3.5	4.0

Increasing acid strength →

Caution: Electronegativity is the major factor when comparing the stability of negatively charged atoms in the same period of the Periodic Table. Other factors, which we will discuss in Section 2.5D and in future chapters, will become significant when comparing atoms in the same group (vertical column) of the Periodic Table.

B. Resonance Effect: Delocalization of the Charge in A^-

Carboxylic acids are weak acids: Values of pK_a for most unsubstituted carboxylic acids fall within the range from 4 to 5. The value of pK_a for acetic acid, for example, is 4.76:

$$\underset{\text{A carboxylic acid}}{CH_3COOH} + H_2O \rightleftharpoons \underset{\text{A carboxylate anion}}{CH_3COO^-} + H_3O^+ \qquad pK_a = 4.76$$

Values of pK_a for most alcohols, compounds that also contain an —OH group, fall within the range from 15 to 18; the value of pK_a for ethanol, for example, is 15.9:

$$\underset{\text{An alcohol}}{CH_3CH_2O{-}H} + H_2O \rightleftharpoons \underset{\text{An alkoxide ion}}{CH_3CH_2O^-} + H_3O^+ \qquad pK_a = 15.9$$

Thus, alcohols are slightly weaker acids than water (pK_a = 15.7) and are much weaker acids than carboxylic acids.

We account for the greater acidity of carboxylic acids compared with alcohols in part by using the resonance model and looking at the relative stabilities of the alkoxide ion and the carboxylate ion. Our guideline is this: The more stable the anion, the farther the position of equilibrium is shifted toward the right and the more acidic is the compound.

There is no resonance stabilization in an alkoxide anion. The ionization of a carboxylic acid, however, gives an anion for which we can write two equivalent contributing structures in which the negative charge of the anion is delocalized; that is, it is spread evenly over the two oxygen atoms:

these contributing structures are equivalent; the carboxylate anion is stabilized by delocalization of the negative charge across the two oxygen atoms

$$CH_3{-}C(=O){-}O{-}H + H_2O \rightleftharpoons CH_3{-}C(=O){-}O^- \longleftrightarrow CH_3{-}C(-O^-){=}O + H_3O^+$$

$$CH_3{-}O{-}H + H_2O \rightleftharpoons CH_3{-}O^- + H_3O^+$$

the alkoxide ion has no resonance contributing structures. Its negative charge is localized on the sole oxygen atom

Because of the delocalization of its charge, a carboxylate anion is significantly more stable than an alkoxide anion. Therefore, the equilibrium for the ionization of a carboxylic acid is shifted to the right relative to that for the ionization of an alcohol, and a carboxylic acid is a stronger acid than an alcohol.

C. The Inductive Effect: Withdrawal of Electron Density from the HA Bond

Inductive effect The polarization of electron density transmitted through covalent bonds caused by a nearby atom of higher electronegativity.

The **inductive effect** is the polarization of electron density transmitted through covalent bonds by a nearby atom of higher electronegativity. We see the operation of the inductive effect when we compare the acidities of acetic acid (pK_a 4.76) and trifluoroacetic acid (pK_a 0.23). Fluorine is more electronegative than carbon and polarizes the electrons of the C—F bond, creating a partial positive charge on the carbon of the $-CF_3$ group. The partial positive charge, in turn, withdraws electron density from the negatively charged $-CO_2^-$ group. The withdrawal of electron density delocalizes the negative charge and makes the conjugate base of trifluoroacetic acid more stable than the conjugate base of acetic acid. The delocalizing effect is apparent when the electron density map of each conjugate base is compared.

$$CH_3COOH + H_2O \rightleftharpoons H_3O^+ + CH_3COO^-$$

Acetic acid
pK_a = 4.76

$$CF_3COOH + H_2O \rightleftharpoons H_3O^+ + CF_3COO^-$$

Trifluoroacetic acid
pK_a = 0.23

Notice that the oxygen atoms on the trifluoroacetate ion are less negative (represented by a lighter shade of red). Thus, the equilibrium for ionization of trifluoroacetic acid is shifted more to the right relative to the ionization of acetic acid, making trifluoroacetic acid more acidic than acetic acid.

D. Size and the Delocalization of Charge in A^-

An important principle in determining the relative stability of unionized acids, HA, is the stability of the conjugate base anion, A^-, resulting from the loss of a proton. The more stable the anion, the greater the acidity of the acid. For example, the relative acidity of the hydrogen halides, HX, is related to the size of the atom bearing the negative charge. The principles of physics teach us that a system bearing a charge (either negative or positive) is more stable if the charge is delocalized. The larger the volume over which the charge of an anion (or cation) is delocalized, the greater the stability of the anion.

Recall from general chemistry that atomic size is a periodic property.

1. **For main group elements, atomic radii increase going down a group in the Periodic Table and increase going across a period.** From the top to the bottom of a group in the Periodic Table, the atomic radii increase because electrons occupy orbitals that are successively larger as the value of *n*, the principal quantum number, increases. Thus, for the halogens (Group 8A elements), iodine has the largest atomic radius and fluorine has the smallest (I > Br > Cl > F).
2. **Anions are always larger than the atoms from which they are derived.** For anions, the nuclear charge is unchanged, but the added electron(s) introduce new repulsions and electron clouds swell. Among the halide ions, I^- has the largest atomic radius and F^- has the smallest ($I^- > Br^- > Cl^- > F^-$).

Thus, when considering the relative acidities of the hydrogen halide acids, we need to consider the relative stabilities of the resulting halide ions formed by ionization of the acid. We know that HI is the strongest acid and that HF is the weakest. We account for this trend by the fact that the negative charge on iodide ion is delocalized over a larger area than is the negative charge on bromide ion. The negative charge on bromide ion in turn is delocalized over a larger area than is the negative charge on chloride ion, and so forth. Thus, HI is the strongest acid in the series because iodide ion is the most stable anion, and HF is the weakest acid because fluoride ion is the least stable ion.

We should note here that a result of resonance (Section 2.5B) is also the delocalization of charge density. Through resonance, the negative charge associated with a carboxylate ion, for example, is more delocalized than is the negative charge associated with an alkoxide ion. Thus a carboxylate ion is more stable than is an alkoxide ion and, therefore, a carboxylic acid is a stronger acid than an alcohol.

Example 2.4

(a) Arrange the following compounds in the order from most acidic to least acidic.

CH_3NH_2 $\quad$ CH_3OH $\quad$ CH_3BH_2

(b) Arrange the following compounds in order from most basic to least basic.

$Cl_3C{-}CH_2O^-$ $\quad$ $H_3C{-}CH_2O^-$ $\quad$ $F_3C{-}CH_2O^-$

Strategy

When determining acidity, assess the stability of the conjugate base—the compound formed after the acid has transferred its proton to a base. Any feature (e.g., electronegativity, resonance, inductive effect, or anion size) that helps to stabilize the conjugate base will make the original acid more acidic. When determining the basicity of a negatively charged species, assess the stability of the base. Any feature (e.g., electronegativity, resonance, or inductive effect) that helps to stabilize the base will make it less basic. That is, a more stable base will be less reactive.

Solution

Because the elements compared in this example (B, C, and O) are all in the same period of the Periodic Table, the stability of the conjugate bases can be compared based on the electronegativity of the element bearing the negative charge. Oxygen, the most electronegative element, is most able to bear the electron density and negative charge. Boron, the least electronegative element, is least able to bear the electron density and negative charge.

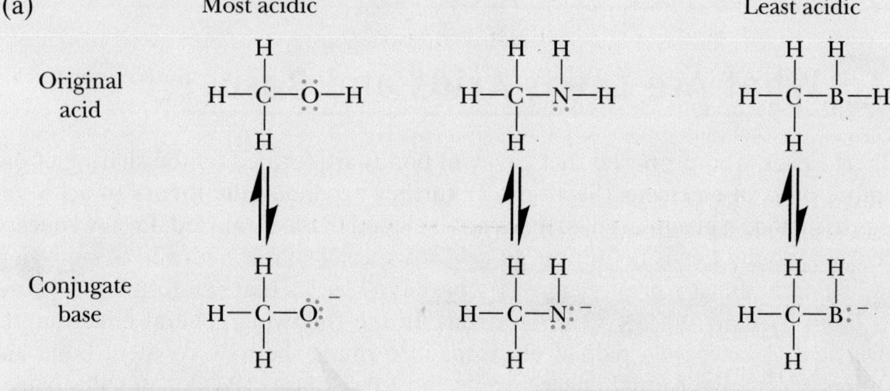

most electronegative atom and therefore most able to bear the electron density and negative charge

least electronegative atom and therefore least able to bear the electron density and negative charge

(b) The stability of bases can be compared based on the extent of an inductive effect within each compound. Fluorine, the most electronegative element, will exert a strong inductive effect on the negatively charged oxygen, thereby delocalizing the negative charge to some extent. This delocalization of negative charge makes the $F_3CCH_2O^-$ ion more stable than either of the other two. Thus the fluorinated ion is the most stable and the least reactive in an acid–base reaction—that is, the least basic:

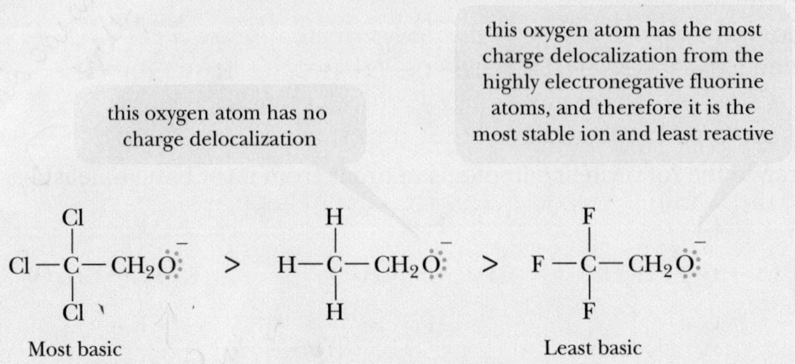

this oxygen atom has no charge delocalization

this oxygen atom has the most charge delocalization from the highly electronegative fluorine atoms, and therefore it is the most stable ion and least reactive

$$CCl_3CH_2O^- > CH_3CH_2O^- > CF_3CH_2O^-$$

Most basic — Least basic

See problems 2.20–2.22

Problem 2.4

(a) Arrange the following compounds in the order from most acidic to least acidic.

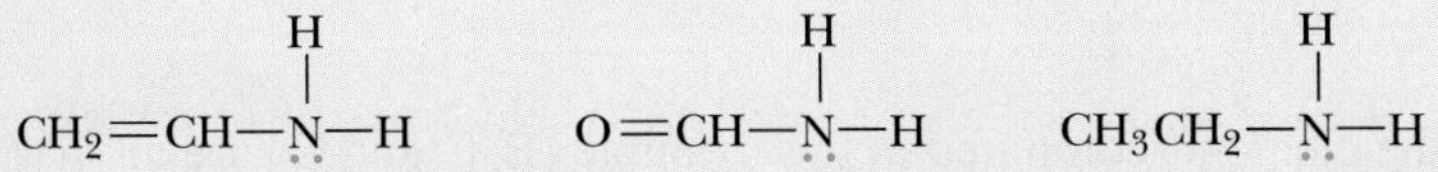

$CH_2{=}CH{-}NH_2$ $\quad$ $O{=}CH{-}NH_2$ $\quad$ $CH_3CH_2{-}NH_2$

(b) Arrange the following compounds in the order from most basic to least basic.

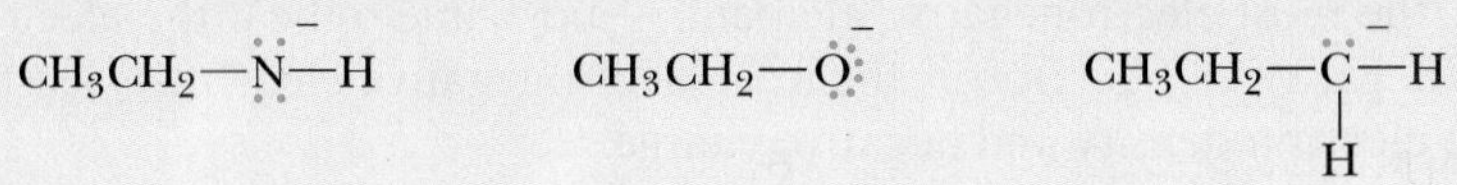

$CH_3CH_2{-}NH^-$ $\quad$ $CH_3CH_2{-}O^-$ $\quad$ $CH_3CH_2{-}CH_2^-$

2.6 What Are Lewis Acids and Bases?

Lewis acid Any molecule or ion that can form a new covalent bond by accepting a pair of electrons.

Lewis base Any molecule or ion that can form a new covalent bond by donating a pair of electrons.

Gilbert Lewis, who proposed that covalent bonds are formed by the sharing of one or more pairs of electrons (Section 1.2), further expanded the theory of acids and bases to include a group of substances not included in the Brønsted–Lowry concept. According to the Lewis definition, an **acid** is a species that can form a new covalent bond by accepting a pair of electrons; a **base** is a species that can form a new covalent bond by donating a pair of electrons. In the following general equation, the Lewis acid, A, accepts a pair of electrons in forming the new covalent bond and acquires a negative formal charge, while the Lewis base, :B, donates the pair of electrons in forming the new covalent bond and acquires a positive formal charge:

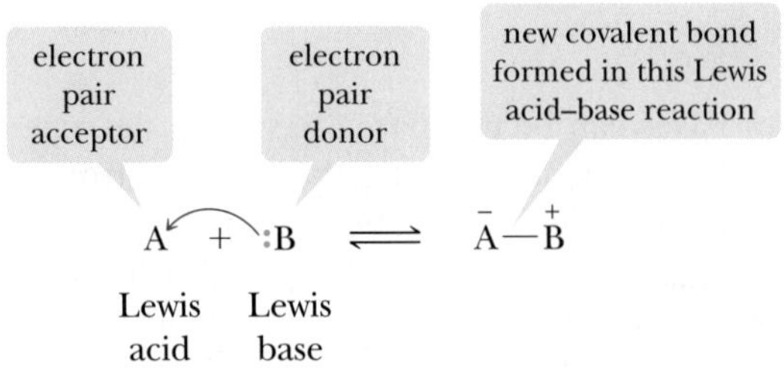

electron pair acceptor

electron pair donor

new covalent bond formed in this Lewis acid–base reaction

$$A + :B \rightleftharpoons \overset{-}{A}{-}\overset{+}{B}$$

Lewis acid — Lewis base

Note that, although we speak of a Lewis base as "donating" a pair of electrons, the term is not fully accurate. "Donating" in this case does not imply that the electron pair under consideration is removed completely from the valence shell of the base. Rather, "donating" means that the electron pair is shared with another atom to form a covalent bond.

As we will see in the chapters that follow, a great many organic reactions can be interpreted as Lewis acid–base reactions. Perhaps the most important (but not the only) Lewis acid is the proton. Isolated protons, of course, do not exist in solution; rather, a proton attaches itself to the strongest available Lewis base. When HCl is dissolved in water, for example, the strongest available Lewis base is an H_2O molecule, and the following proton-transfer reaction takes place:

$$H{-}\ddot{O}{:}(H) + H{-}\ddot{Cl}{:} \longrightarrow H{-}\dot{\dot{O}}^{+}{-}H(H) + {:}\ddot{Cl}{:}^{-}$$

Hydronium ion

When HCl is dissolved in methanol, the strongest available Lewis base is a CH_3OH molecule, and the following proton-transfer reaction takes place. An **oxonium** ion is an ion that contains an oxygen atom with three bonds and bears a positive charge.

$$CH_3{-}\ddot{O}{:}(H) + H{-}\ddot{Cl}{:} \longrightarrow CH_3{-}\dot{\dot{O}}^{+}{-}H(H) + {:}\ddot{Cl}{:}^{-}$$

Methanol (a Lewis base) — An oxonium ion

Table 2.3 gives examples of the most important types of Lewis bases we will encounter in this text arranged in order of their increasing strength in proton-transfer reactions. Note that each of the Lewis bases has at least one atom with an unshared pair of electrons. It is this atom that functions as the Lewis base. Ethers are organic derivatives of water in which both hydrogens of water are replaced by carbon groups. We study the properties of ethers along with those of alcohols in Chapter 8. We study the properties of amines in Chapter 10.

TABLE 2.3 Some Organic Lewis Bases and Their Relative Strengths in Proton-Transfer Reactions

Halide Ions	Water, Alcohols, and Ethers	Ammonia and Amines	Hydroxide Ion and Alkoxide Ions	Amide Ions
$:\ddot{Cl}:^-$	$H{-}\ddot{O}{-}H$	$H{-}\ddot{N}{-}H$ (H)	$H{-}\ddot{O}:^-$	$H{-}\ddot{N}:^-$ (H)
$:\ddot{Br}:^-$	$CH_3{-}\ddot{O}{-}H$	$CH_3{-}\ddot{N}{-}H$ (H)	$CH_3{-}\ddot{O}:^-$	$CH_3{-}\ddot{N}:^-$ (H)
$:\ddot{I}:^-$	$CH_3{-}\ddot{O}{-}CH_3$	$CH_3{-}\ddot{N}{-}H$ (CH_3)		$CH_3{-}\ddot{N}:^-$ (CH_3)
		$CH_3{-}\ddot{N}{-}CH_3$ (CH_3)		
Very weak	Weak	Strong	Stronger	Very strong

Example 2.5

Complete this acid–base reaction. Use curved arrows to show the redistribution of electrons in the reaction. In addition, predict whether the position of this equilibrium lies toward the left or the right.

$$\mathrm{CH_3{-}\underset{\underset{\displaystyle H}{|}}{O^+}{-}H + CH_3{-}\underset{\underset{\displaystyle H}{|}}{N}{-}H \rightleftharpoons}$$

Strategy

First, add unshared pairs of electrons on the reacting atoms to give each a complete octet. Then identify the Lewis base (the electron-pair donor) and the Lewis acid (the electron-pair acceptor). The position of equilibrium lies on the side of the weaker acid and weaker base.

Solution

Proton transfer takes place to form an alcohol and an ammonium ion. We know from Table 2.3 that amines are stronger bases than alcohols. We also know that the weaker the base, the stronger its conjugate acid, and vice versa. From this analysis, we conclude that the position of this equilibrium lies to the right, on the side of the weaker acid and the weaker base.

$$\underset{\text{Stronger acid}}{\mathrm{CH_3{-}\underset{\underset{\displaystyle H}{|}}{\ddot{O}^+}{-}H}} + \underset{\text{Stronger base}}{\mathrm{CH_3{-}\underset{\underset{\displaystyle H}{|}}{\ddot{N}}{-}H}} \rightleftharpoons \underset{\text{Weaker base}}{\mathrm{CH_3{-}\underset{\underset{\displaystyle H}{|}}{\ddot{O}}\!:}} + \underset{\text{Weaker acid}}{\mathrm{CH_3{-}\overset{\overset{\displaystyle H}{|}}{\underset{\underset{\displaystyle H}{|}}{N^+}}{-}H}}$$

See problems 2.7, 2.8, 2.11, 2.26, 2.28–2.30

Problem 2.5

Complete this acid–base reaction. First add unshared pairs of electrons on the reacting atoms to give each atom a complete octet. Use curved arrows to show the redistribution of electrons in the reaction. In addition, predict whether the position of the equilibrium lies toward the left or the right.

$$\mathrm{CH_3{-}O^- + CH_3{-}\overset{\overset{\displaystyle H}{|}}{\underset{\underset{\displaystyle CH_3}{|}}{N^+}}{-}CH_3 \rightleftharpoons}$$

Another type of Lewis acid we will encounter in later chapters is an organic cation in which a carbon is bonded to only three atoms and bears a positive formal charge. Such carbon cations are called carbocations. Consider the reaction that occurs when the following organic cation reacts with a bromide ion:

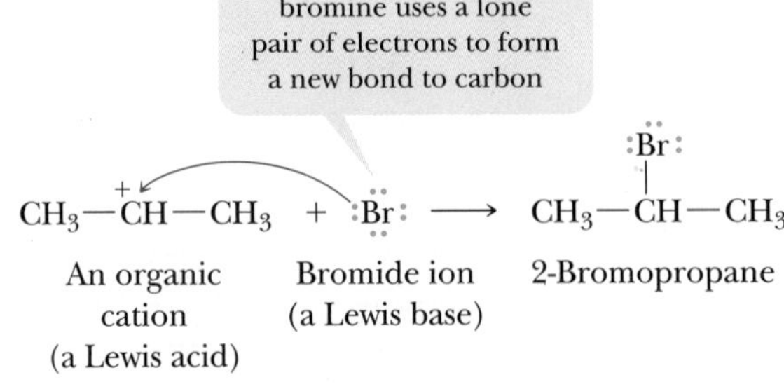

In this reaction, the organic cation is the electron-pair acceptor (the Lewis acid), and bromide ion is the electron-pair donor (the Lewis base).

Example 2.6

Complete the following Lewis acid–base reaction. Show all electron pairs on the reacting atoms and use curved arrows to show the flow of electrons in the reaction:

$$CH_3-\overset{+}{C}H-CH_3 + H_2O \longrightarrow$$

Strategy

Determine which compound will be the electron-pair donor and which will be the electron-pair acceptor. *Hint:* Compounds with empty orbitals in their valence shell usually act as Lewis acids.

Solution

The trivalent carbon atom in the organic cation has an empty orbital in its valence shell and, therefore, is the Lewis acid. Water is the Lewis base.

oxygen uses a lone pair of electrons to form a new bond to carbon

$$\underset{\text{Lewis acid}}{CH_3-\overset{+}{\underset{|\atop H}{C}}-CH_3} + \underset{\text{Lewis base}}{H-\ddot{\underset{..}{O}}-H} \longrightarrow \underset{\text{An oxonium ion}}{CH_3-\overset{H\ \ H\atop \diagdown\ddot{O}^{+}\diagup \atop |}{\underset{|\atop H}{C}}-CH_3}$$

See problems 2.7, 2.8, 2.11, 2.26, 2.28–2.30

Problem 2.6

Write an equation for the reaction between each Lewis acid–base pair, showing electron flow by means of curved arrows. (*Hint:* Aluminum is in Group 3A of the Periodic Table, just under boron. Aluminum in $AlCl_3$ has only six electrons in its valence shell and thus has an incomplete octet.)

(a) $Cl^- + AlCl_3 \longrightarrow$ (b) $CH_3Cl + AlCl_3 \longrightarrow$

Key Terms and Concepts

Arrhenius acid (p. 47)
Arrhenius base (p. 47)
Brønsted–Lowry acid (p. 48)
Brønsted–Lowry base (p. 48)
conjugate acid (p. 49)
conjugate acid–base pair (p. 48)
conjugate base (p. 49)
curved arrow (p. 47)
diprotic acid (p. 50)
Equilibrium in an acid–base reaction favors formation of the weaker acid and weaker base (p. 54)
inductive effect (p. 57)
Lewis acid (p. 60)
Lewis base (p. 60)
monoprotic acid (p. 50)
oxonium ion (p. 61)
Position of equilibrium in an acid–base reaction (p. 54)
proton acceptor (p. 48)
proton donor (p. 48)
proton–transfer reaction (p. 48)
Relationship between molecular structure and acidity (p. 56)
strong acid (p. 51)
strong base (p. 51)
The greater the electronegativity A, the stronger the acid HA (p. 56)
The inductive effect (delocalization of charge) and acidity of an acid HA (p. 57)
The more stable the anion A^-, the greater the acidity of the acid, HA (p. 58)
The resonance effect (delocalization of charge) in the anion A^- and the strength of an acid HA (p. 56)
triprotic acid (p. 50)
weak acid (p. 52)
weak base (p. 52)

Summary of Key Questions

2.1 What Are Arrhenius Acids and Bases?

- An **Arrhenius acid** is a substance that dissolves in aqueous solution to produce H_3O^+ ions.
- An **Arrhenius base** is a substance that dissolves in aqueous solution to produce OH^- ions.

2.2 What Are Brønsted-Lowry Acids and Bases?

- A **Brønsted–Lowry acid** is a proton donor.
- A **Brønsted–Lowry base** is a proton acceptor.
- Neutralization of an acid by a base is a **proton-transfer reaction** in which the acid is transformed into its **conjugate base**, and the base is transformed into its **conjugate acid**.

2.3 How Do We Measure the Strength of an Acid or Base?

- A **strong acid** or **strong base** is one that completely ionizes in water.
- A **weak acid** or **weak base** is one that only partially ionizes in water.
- The strength of a weak acid is expressed by its **ionization constant, K_a**.
- The larger the value of K_a, the stronger the acid, $pK_a = -\log K_a$.

2.4 How Do We Determine the Position of Equilibrium in an Acid–Base Reaction?

- In an acid–base reaction, the position of equilibrium favors the reaction of the stronger acid and the stronger base to form the weaker acid and the weaker base.

2.5 What Are the Relationships between Acidity and Molecular Structure?

The relative acidities of the organic acids, HA, are determined by

- the electronegativity of A.
- the resonance stabilization of the conjugate base, A^-.
- the electron-withdrawing inductive effect, which also stabilizes the conjugate base.
- the size of the atom with the negative charge on the conjugate base.

2.6 What Are Lewis Acids and Bases?

- A **Lewis acid** is a species that forms a new covalent bond by accepting a pair of electrons (an electron-pair acceptor).
- A **Lewis base** is a species that forms a new covalent bond by donating a pair of electrons (an electron-pair donor).

Quick Quiz

Answer true or false to the following questions to assess your general knowledge of the concepts in this chapter. If you have difficulty with any of them, you should review the appropriate section in the chapter (shown in parentheses) before attempting the more challenging end-of-chapter problems.

1. If NH_3 were to behave as an acid, its conjugate base would be NH_2^-. (2.2)
2. Delocalization of electron density is a stabilizing factor. (2.5)
3. Amide ion, NH_2^-, is a Lewis base. (2.6)
4. H_3O^+ is a stronger acid than NH_4^+ and, therefore, NH_3 is a stronger base than H_2O. (2.3)

5. Inductive effects can be used to describe electron delocalization. (2.5)

6. The direction of equilibrium in an acid–base reaction favors the side containing the stronger acid and stronger base. (2.4)

7. The conjugate base of CH_3CH_2OH is $CH_3CH_2O^-$. (2.2)

8. CH_3^+ and NH_4^+ are Lewis acids. (2.6)

9. When an acid, HA, dissolves in water, the solution becomes acidic because of the presence of H^+ ions. (2.1)

10. Between a strong acid and a weak acid, the weak acid will give rise to the stronger conjugate base. (2.4)

11. H_2O can function as an acid (proton donor) and as a base (proton acceptor). (2.2)

12. The strongest base that can exist in aqueous solution is OH^-. (2.4)

13. A strong acid is one that completely ionizes in aqueous solution. (2.3)

14. NH_3 is a Lewis base. (2.6)

15. A Brønsted–Lowry acid is a proton donor. (2.2)

16. When comparing the relative strength of acids, the stronger acid has the smaller value of pK_a. (2.3)

17. When comparing the relative strengths of acids, the stronger acid has the smaller value of K_a. (2.3)

18. The formulas of a conjugate acid–base pair differ only by a proton. (2.2)

19. A Lewis base is an electron donor. (2.6)

20. Acetic acid, CH_3COOH, is a stronger acid than carbonic acid, H_2CO_3, and, therefore, acetate ion, CH_3COO^-, is a stronger base than bicarbonate ion, HCO_3^-. (2.2)

21. The strongest acid that can exist in aqueous solution is H_3O^+. (2.4)

22. A Lewis acid–base reaction results in the formation of a new covalent bond between the Lewis acid and the Lewis base. (2.6)

23. When a base accepts a proton in an acid–base reaction, it is converted into its conjugate base. (2.2)

24. When a metal hydroxide, MOH, dissolves in water, the solution becomes basic because of the presence of hydroxide ions, OH^-. (2.1)

25. A Lewis acid is a proton acceptor. (2.6)

26. If NH_3 were to behave as a base, its conjugate acid would be NH_4^+. (2.2)

27. Resonance effects can be used to describe electron delocalization. (2.5)

28. All Lewis acid–base reactions involve transfer of a proton from the acid to the base.

29. BF_3 is a Lewis acid. (2.6)

30. When HCl dissolves in water, the major ions present are H^+ and Cl^-. (2.3)

31. According to the Arrhenius definitions, acids and bases are limited substances that dissolve in water. (2.1)

32. Acid–base reactions take place only in aqueous solution. (2.6)

33. The conjugate acid of HCO_3^- is H_2CO_3. (2.2)

Answers: (1) T (2) T (3) T (4) T (5) T (6) F (7) T (8) F (9) F (10) T (11) T (12) T (13) T (14) T (15) T (16) T (17) F (18) T (19) T (20) F (21) T (22) T (23) F (24) T (25) F (26) T (27) T (28) F (29) T (30) F (31) T (32) F (33) T

Key Reactions

1. Proton-Transfer Reaction (Section 2.2)

This reaction involves the transfer of a proton from a proton donor (a Brønsted–Lowry acid) to a proton acceptor (a Brønsted–Lowry base):

$$CH_3COOH + :NH_3 \rightleftharpoons CH_3COO^- + H{-}NH_3^+$$

Acetic acid (proton donor) | Ammonia (proton acceptor) | Acetate ion | Ammonium ion

2. Position of Equilibrium in an Acid–Base Reaction (Section 2.4)

Equilibrium favors reaction of the stronger acid with the stronger base to give the weaker acid and the weaker base:

$$CH_3COOH + NH_3 \rightleftharpoons CH_3COO^- + NH_4^+$$

Acetic acid pK_a 4.76 (stronger acid) | (stronger base) | (weaker base) | Ammonium ion pK_a 9.24 (weaker acid)

3. Lewis Acid–Base Reaction (Section 2.6)

A Lewis acid–base reaction involves sharing an electron pair between an electron-pair donor (a Lewis base) and an electron-pair acceptor (a Lewis acid):

$$CH_3{-}\overset{+}{C}H{-}CH_3 + H{-}\ddot{O}{-}H \longrightarrow CH_3{-}CH(\overset{+}{O}H_2){-}CH_3$$

A Lewis acid | A Lewis base | An oxonium ion

Problems

A problem marked with an asterisk indicates an applied "real world" problem. Answers to problems whose numbers are printed in blue are given in Appendix D.

Section 2.1 Arrhenius Acids and Bases

2.7 Complete the net ionic equation for each acid placed in water. Use curved arrows to show the flow of electron pairs in each reaction. Also, for each reaction, determine the direction of equilibrium, using Table 2.2 as a reference for the pK_a values of proton acids. **(See Examples 2.1, 2.8)**

(a) $NH_4^+ + H_2O \rightleftharpoons$ (b) $HCO_3^- + H_2O \rightleftharpoons$

(c) $CH_3-\overset{\overset{O}{\|}}{C}-OH + H_2O \rightleftharpoons$

2.8 Complete the net ionic equation for each base placed in water. Use curved arrows to show the flow of electron pairs in each reaction. Also, for each reaction, determine the direction of equilibrium, using Table 2.2 as a reference for the pK_a values of proton acids formed. **(See Examples 2.1, 2.5)**

(a) $CH_3NH_2 + H_2O \rightleftharpoons$ (b) $HSO_4^- + H_2O \rightleftharpoons$

(c) $Br^- + H_2O \rightleftharpoons$ (d) $CO_3^{2-} + H_2O \rightleftharpoons$

Section 2.2 Brønsted–Lowry Acids and Bases

2.9 How are the formulas of the members of a conjugate acid–base pair related to each other? Within a pair, how can you tell which is the acid?

2.10 Write the structural formula for the conjugate acids of the following structures.

(a) $CH_3-CH_2-\ddot{N}(H)-H$

(b) $H-\ddot{N}(H)-CH_2-CH_2-\ddot{O}H$ (c) 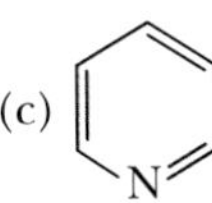

2.11 Complete a net ionic equation for each proton-transfer reaction, using curved arrows to show the flow of electron pairs in each reaction. In addition, write Lewis structures for all starting materials and products. Label the original acid and its conjugate base; label the original base and its conjugate acid. If you are uncertain about which substance in each equation is the proton donor, refer to Table 2.2 for the pK_a values of proton acids. **(See Examples 2.3, 2.5)**

(a) $NH_3 + HCl \longrightarrow$

(b) $CH_3CH_2O^- + HCl \longrightarrow$

(c) $HCO_3^- + OH^- \longrightarrow$

(d) $CH_3COO^- + NH_4^+ \longrightarrow$

(e) $NH_4^+ + OH^- \longrightarrow$

(f) $CH_3COO^- + CH_3NH_3^+ \longrightarrow$

(g) $CH_3CH_2O^- + NH_4^+ \longrightarrow$

(h) $CH_3NH_3^+ + OH^- \longrightarrow$

***2.12** One kind of baking powder contains sodium bicarbonate and calcium dihydrogen phosphate: When water is added, the following reaction occurs. **(See Example 2.3)**

$$HCO_3^-(aq) + H_2PO_4^-(aq) \longrightarrow H_2CO_3(aq) + HPO_4^{2-}(aq)$$

Identify the two acids and the two bases in this reaction. (The H_2CO_3 decomposes to release CO_2, which causes the cake to rise.)

2.13 Each of these molecules and ions can function as a base. Complete the Lewis structure of each base, and write the structural formula of the conjugate acid formed by its reaction with HCl.

(a) CH_3CH_2OH (b) $H\overset{\overset{O}{\|}}{C}H$

(c) $(CH_3)_2NH$ (d) HCO_3^-

2.14 Offer an explanation for the following observations:

(a) H_3O^+ is a stronger acid than NH_4^+.

(b) Nitric acid, HNO_3, is a stronger acid than nitrous acid, HNO_2 (pK_a 3.7).

(c) Ethanol, CH_3CH_2OH, and water have approximately the same acidity.

(d) Trichloroacetic acid, CCl_3COOH (pK_a 0.64), is a stronger acid than acetic acid, CH_3COOH (pK_a 4.74).

(e) Trifluoroacetic acid, CF_3COOH (pK_a 0.23), is a stronger acid than trichloroacetic acid, CCl_3COOH (pK_a 0.64).

2.15 Select the most acidic proton in the following compounds:

(a) $H_3C-\overset{\overset{O}{\|}}{C}-CH_2-\overset{\overset{O}{\|}}{C}-CH_3$

(b) $H_2N-\overset{\overset{NH_2^+}{\|}}{C}-NH_2$

Section 2.3 Quantitative Measure of Acid Strength

2.16 Which has the larger numerical value? **(See Example 2.2)**

(a) The pK_a of a strong acid or the pK_a of a weak acid?

(b) The K_a of a strong acid or the K_a of a weak acid?

*__2.17__ In each pair, select the stronger acid: **(See Example 2.3)**

(a) Pyruvic acid (pK_a 2.49) or lactic acid (pK_a 3.85)

(b) Citric acid (pK_{a1} 3.08) or phosphoric acid (pK_{a1} 2.10)

(c) Nicotinic acid (niacin, K_a 1.4×10^{-5}) or acetylsalicylic acid (aspirin, K_a 3.3×10^{-4})

(d) Phenol (K_a 1.12×10^{-10}) or acetic acid (K_a 1.74×10^{-5})

2.18 Arrange the compounds in each set in order of increasing acid strength. Consult Table 2.2 for pK_a values of each acid.

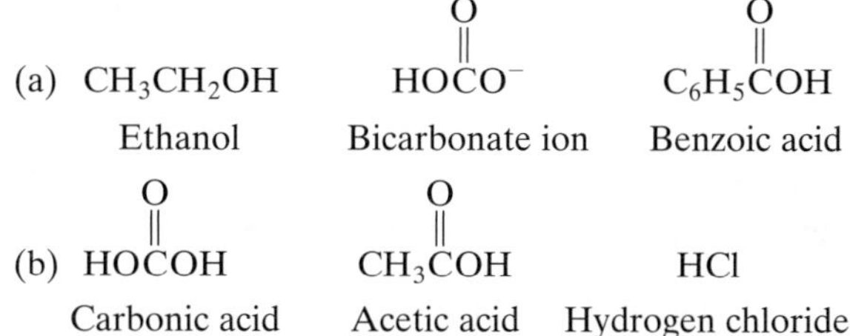

2.19 Arrange the compounds in each set in order of increasing base strength. Consult Table 2.2 for pK_a values of the conjugate acid of each base. (*Hint:* The stronger the acid, the weaker is its conjugate base, and vice versa.)

(a) NH_3 $HOC(=O)O^-$ $CH_3CH_2O^-$

(b) OH^- $HOC(=O)O^-$ $CH_3C(=O)O^-$

(c) H_2O NH_3 $CH_3C(=O)O^-$

(d) NH_2^- $CH_3C(=O)O^-$ OH^-

2.20 Using only the Periodic Table, choose the stronger acid of each pair **(See Example 2.4)**

(a) H_2Se or HBr (b) H_2Se or H_2Te

(c) CH_3OH or CH_3SH

2.21 Explain why H_2S is a stronger acid than H_2O. **(See Example 2.4)**

2.22 Which is the stronger Brønsted–Lowry base, $CH_3CH_2O^-$ or $CH_3CH_2S^-$? What is the basis for your selection? **(See Example 2.4)**

Section 2.4 Position of Equilibrium in Acid–Base Reactions

2.23 Unless under pressure, carbonic acid in aqueous solution breaks down into carbon dioxide and water, and carbon dioxide is evolved as bubbles of gas. Write an equation for the conversion of carbonic acid to carbon dioxide and water.

2.24 For each of the following compounds, will carbon dioxide be evolved when sodium bicarbonate is added to an aqueous solution of the compound?

(a) H_2SO_4 (b) CH_3CH_2OH (c) NH_4Cl

2.25 Acetic acid, CH_3COOH, is a weak organic acid, pK_a 4.76. Write equations for the equilibrium reactions of acetic acid with each base. Which equilibria lie considerably toward the left? Which lie considerably toward the right? **(See Example 2.3)**

(a) $NaHCO_3$ (b) NH_3 (c) H_2O (d) $NaOH$

2.26 The amide ion, NH_2^-, is a very strong base; it is even stronger than OH^-. Write an equation for the reaction that occurs when amide ion is placed in water. Use this equation to show why the amide ion cannot exist in aqueous solution. **(See Example 2.5)**

2.27 For an acid–base reaction, one way to indicate the predominant species at equilibrium is to say that the reaction arrow points to the acid with the higher value of pK_a. For example **(See Example 2.5)**

$$\underset{pK_a\ 9.24}{NH_4^+} + H_2O \longleftarrow NH_3 + \underset{pK_a\ -1.74}{H_3O^+}$$

$$\underset{pK_a\ 9.24}{NH_4^+} + OH^- \longrightarrow NH_3 + \underset{pK_a\ 15.7}{H_2O}$$

Explain why this rule works.

Section 2.6 Lewis Acids and Bases

2.28 Complete the following acid–base reactions, using curved arrow notation to show the flow of electron pairs. In solving these problems, it is essential that you show all valence electrons for the atoms participating directly in each reaction. **(See Example 2.5)**

(a) BF_3 + (tetrahydrofuran ring: O, H_2C, CH_2, $H_2C—CH_2$) $\longrightarrow$

(b) $CH_3—C(CH_3)_2—Cl$ + $Cl—Al(Cl)—Cl$ $\longrightarrow$

2.29 Complete equations for these reactions between Lewis acid–Lewis base pairs. Label which starting material is the Lewis acid and which is the Lewis base, and use a curved arrow to show the flow of the electron pair in each reaction. In solving these problems, it is essential that you show all valence electrons for the atoms participating directly in each reaction. **(see Examples 2.5, 2.6)**

(a) $CH_3{-}\overset{+}{C}H{-}CH_3 + CH_3{-}O{-}H \longrightarrow$

(b) $CH_3{-}\overset{+}{C}H{-}CH_3 + Br^- \longrightarrow$

(c) 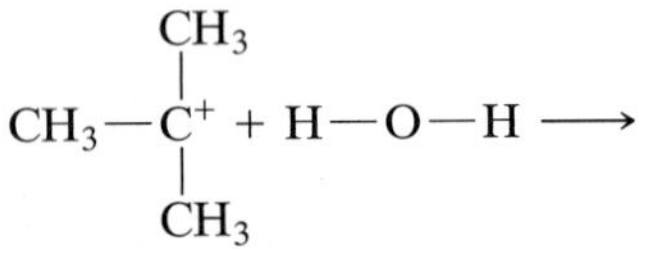 $CH_3{-}\underset{CH_3}{\overset{CH_3}{C^+}} + H{-}O{-}H \longrightarrow$

2.30 Use curved arrow notation to show the flow of electron pairs in each Lewis acid–base reaction. Be certain to show all valence electron pairs on each atom participating in the reaction. **(see Examples 2.5, 2.6)**

(a) 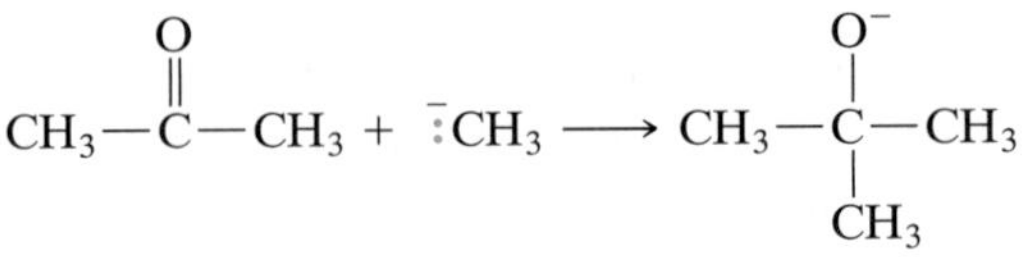 $CH_3{-}\overset{O}{\overset{\|}{C}}{-}CH_3 + \bar{:}CH_3 \longrightarrow CH_3{-}\underset{CH_3}{\overset{O^-}{C}}{-}CH_3$

(b) 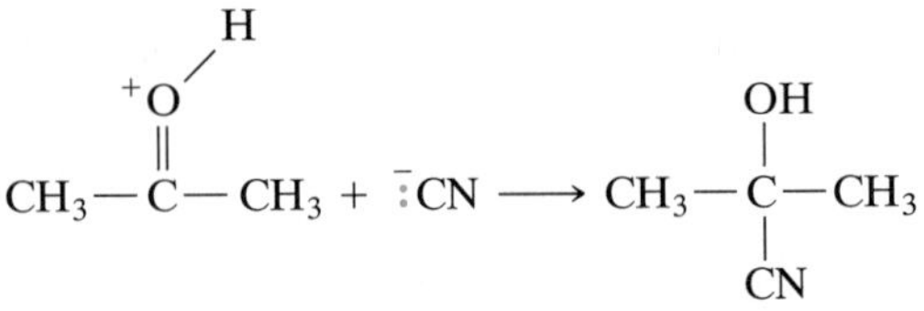 $CH_3{-}\overset{\overset{H}{^{+}O}}{\overset{\|}{C}}{-}CH_3 + \bar{:}CN \longrightarrow CH_3{-}\underset{CN}{\overset{OH}{C}}{-}CH_3$

(c) $CH_3O^- + CH_3{-}Br \longrightarrow CH_3{-}O{-}CH_3 + Br^-$

Looking Ahead

2.31 Alcohols (Chapter 8) are weak organic acids, pK_a15–18. The pK_a of ethanol, CH_3CH_2OH, is 15.9. Write equations for the equilibrium reactions of ethanol with each base. Which equilibria lie considerably toward the right? Which lie considerably toward the left?

(a) $NaHCO_3$ (b) NaOH (c) $NaNH_2$ (d) NH_3

2.32 Phenols (Chapter 9) are weak acids, and most are insoluble in water. Phenol, C_6H_5OH (pK_a 9.95), for example, is only slightly soluble in water, but its sodium salt, $C_6H_5O^-Na^+$, is quite soluble in water. In which of these solutions will phenol dissolve?

(a) Aqueous NaOH (b) Aqueous $NaHCO_3$

(c) Aqueous Na_2CO_3

2.33 Carboxylic acids (Chapter 14) of six or more carbons are insoluble in water, but their sodium salts are very soluble in water. Benzoic acid, C_6H_5COOH (pK_a 4.19), for example, is insoluble in water, but its sodium salt, $C_6H_5COO^-Na^+$, is quite soluble in water. In which of these solutions will benzoic acid dissolve?

(a) Aqueous NaOH (b) Aqueous $NaHCO_3$

(c) Aqueous Na_2CO_3

2.34 As we shall see in Chapter 16, hydrogens on a carbon adjacent to a carbonyl group are far more acidic than those not adjacent to a carbonyl group. The highlighted H in propanone, for example, is more acidic than the highlighted H in ethane:

$CH_3\overset{O}{\overset{\|}{C}}CH_2{-}H$	$CH_3CH_2{-}H$
Propanone	Ethane
$pK_a = 22$	$pK_a = 51$

Account for the greater acidity of propanone in terms of (a) the inductive effect and (b) the resonance effect.

2.35 Explain why the protons in dimethyl ether, $CH_3{-}O{-}CH_3$, are not very acidic.

2.36 Predict whether sodium hydride, NaH, will act as a base or an acid, and provide a rationale for your decision.

***2.37** Alanine is one of the 20 amino acids (it contains both an amino and a carboxyl group) found in proteins (Chapter 19). Is alanine better represented by the structural formula A or B? Explain.

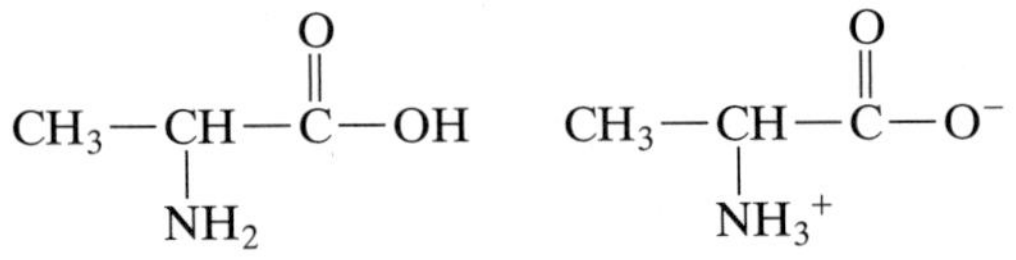

$$\underset{A}{CH_3{-}\underset{NH_2}{CH}{-}\overset{O}{\overset{\|}{C}}{-}OH} \qquad \underset{B}{CH_3{-}\underset{NH_3^+}{CH}{-}\overset{O}{\overset{\|}{C}}{-}O^-}$$

***2.38** Glutamic acid is another of the amino acids found in proteins (Chapter 19):

$$\text{Glutamic acid} \quad HO{-}\overset{O}{\overset{\|}{C}}{-}CH_2{-}CH_2{-}\underset{NH_3^+}{CH}{-}\overset{O}{\overset{\|}{C}}{-}OH$$

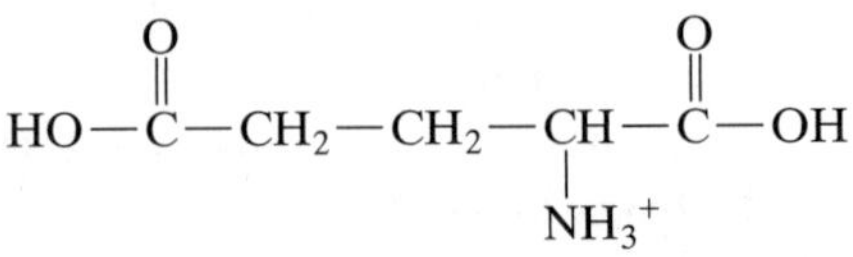

Glutamic acid has two carboxyl groups, one with pK_a 2.10, the other with pK_a 4.07.

(a) Which carboxyl group has which pK_a?

(b) Account for the fact that one carboxyl group is a considerably stronger acid than the other.

3 Alkanes and Cycloalkanes

KEY QUESTIONS

HOW TO

CHEMICAL CONNECTIONS

Bunsen burners burn natural gas, which is primarily methane with small amounts of ethane, propane, butane, and 2-methylbutane. Inset: A model of methane. *(Charles D. Winters)*

In this chapter, we begin our study of organic compounds with the physical and chemical properties of alkanes, the simplest types of organic compounds. Actually, alkanes are members of a larger class of organic compounds called **hydrocarbons.** A hydrocarbon is a compound composed of only carbon and hydrogen. Figure 3.1 shows the four classes of hydrocarbons, along with the characteristic type of bonding between carbon atoms in each.

Alkanes are **saturated hydrocarbons;** that is, they contain only carbon–carbon single bonds. In this context, "saturated" means that each carbon has the maximum number of hydrogens bonded to it. We often refer to alkanes as **aliphatic hydrocarbons,** because the physical properties of the higher members of this class resemble those of the long carbon-chain molecules we find in animal fats and plant oils (Greek: *aleiphar*, fat or oil).

Hydrocarbon A compound that contains only carbon atoms and hydrogen atoms.

Alkane A saturated hydrocarbon whose carbon atoms are arranged in an open chain.

Saturated hydrocarbon A hydrocarbon containing only carbon–carbon single bonds.

Aliphatic hydrocarbon An alternative term to describe an alkane.

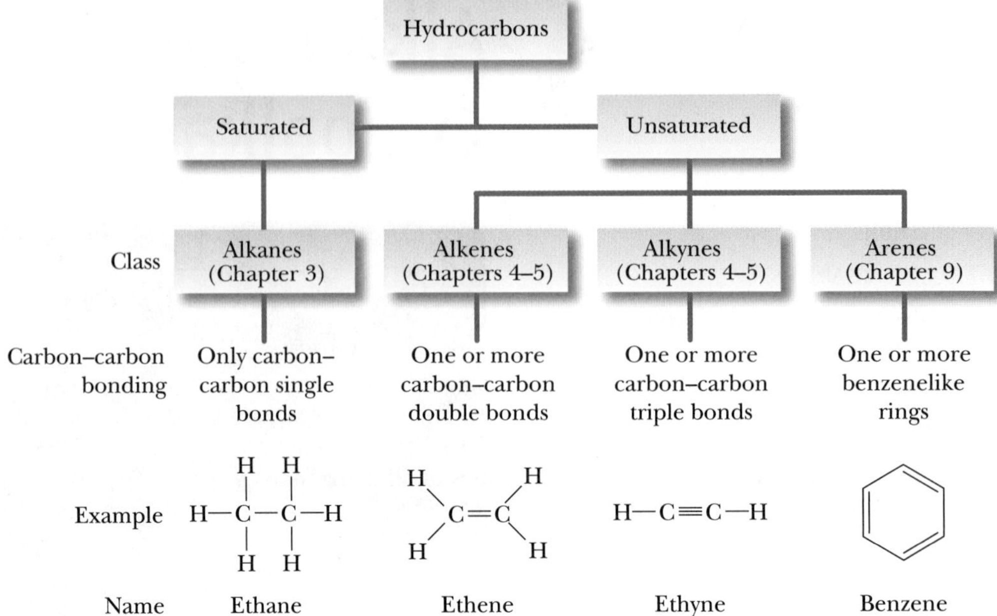

Figure 3.1 The four classes of hydrocarbons.

Butane is the fuel in this lighter. Butane molecules are present in the liquid and gaseous states in the lighter. *(Charles D. Winters)*

A hydrocarbon that contains one or more carbon–carbon double bonds, triple bonds, or benzene rings is classified as an **unsaturated hydrocarbon.** We study alkanes (saturated hydrocarbons) in this chapter. We study alkenes and alkynes (both unsaturated hydrocarbons) in Chapters 4 and 5, and arenes (also unsaturated hydrocarbons) in Chapter 9.

3.1 What Are Alkanes?

Methane (CH_4) and ethane (C_2H_6) are the first two members of the alkane family. Figure 3.2 shows Lewis structures and ball-and-stick models for these molecules. The shape of methane is tetrahedral, and all H—C—H bond angles are 109.5°. Each carbon atom in ethane is also tetrahedral, and all bond angles are approximately 109.5°.

Although the three-dimensional shapes of larger alkanes are more complex than those of methane and ethane, the four bonds about each carbon atom are still arranged in a tetrahedral manner, and all bond angles are still approximately 109.5°.

The next members of the alkane family are propane, butane, and pentane. In the representations that follow, these hydrocarbons are drawn first as condensed structural formulas that show all carbons and hydrogens. They are then drawn in an even more abbreviated form called a **line-angle formula.** In this type of representation, a line represents a carbon-carbon bond, and an angle represents a carbon atom. A line ending represents a $—CH_3$ group. Although hydrogen atoms are not shown in line-angle formulas, they are assumed to be there in sufficient numbers to give each carbon four bonds.

Line-angle formula An abbreviated way to draw structural formulas in which each line ending represents a carbon atom and a line represents a bond.

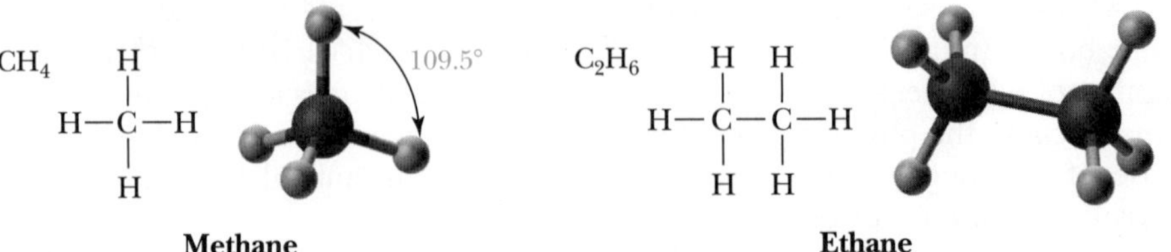

Figure 3.2 Methane and ethane.

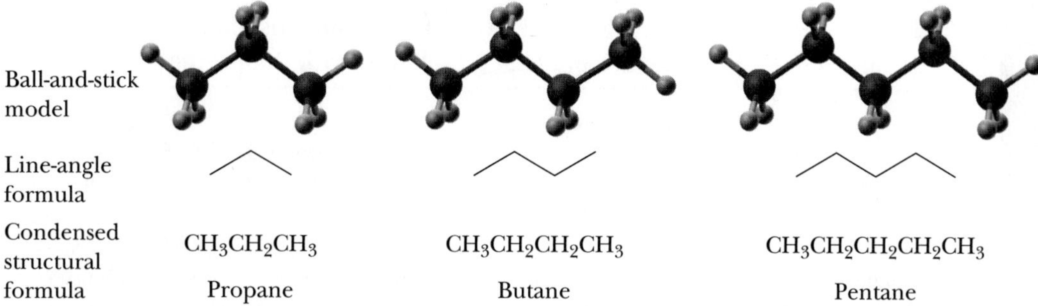
Ball-and-stick model
Line-angle formula
Condensed structural formula
CH3CH2CH3
Propane
CH3CH2CH2CH3
Butane
CH3CH2CH2CH2CH3
Pentane

We can write structural formulas for alkanes in still another abbreviated form. The structural formula of pentane, for example, contains three CH_2 (methylene) groups in the middle of the chain. We can collect these groups together and write the structural formula as $CH_3(CH_2)_3CH_3$. Table 3.1 gives the names and molecular formulas of the first 20 alkanes. Note that the names of all these alkanes end in *-ane*. We will have more to say about naming alkanes in Section 3.3.

HOW TO 3.1 Interpret a Line-Angle Formulas

Use the mnemonic "a carbon at every *bend* and at every *end*. . . ."

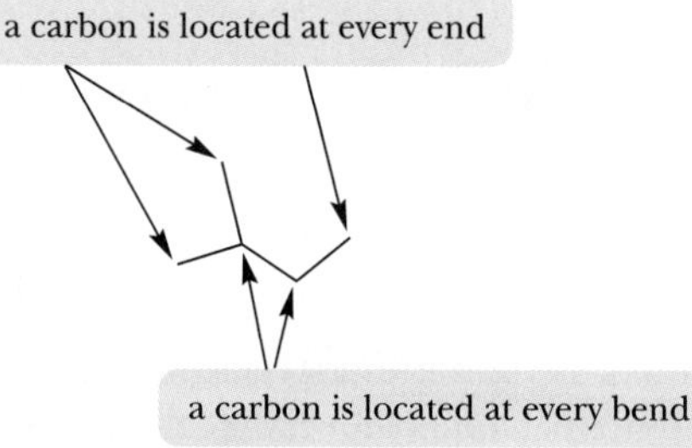
a carbon is located at every end
a carbon is located at every bend

Because carbon requires four bonds to satisfy its valency, count the number of visible bonds to each carbon and then subtract this number from 4 to determine the number of hydrogens also bonded to that carbon (but not shown).

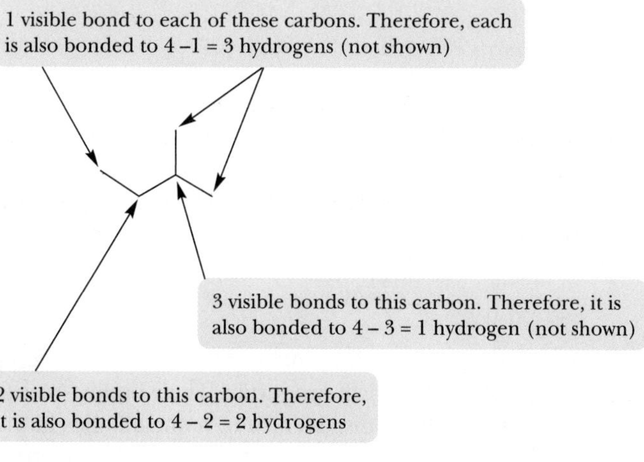
1 visible bond to each of these carbons. Therefore, each is also bonded to 4 –1 = 3 hydrogens (not shown)
3 visible bonds to this carbon. Therefore, it is also bonded to 4 – 3 = 1 hydrogen (not shown)
2 visible bonds to this carbon. Therefore, it is also bonded to 4 – 2 = 2 hydrogens

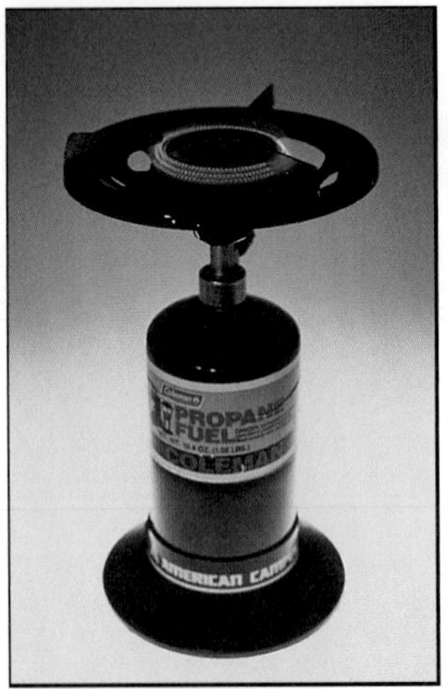

A tank for propane fuel.
(Charles D. Winters)

TABLE 3.1 Names, Molecular Formulas, and Condensed Structural Formulas for the First 20 Alkanes with Unbranched Chains

Name	Molecular Formula	Condensed Structural Formula	Name	Molecular Formula	Condensed Structural Formula
methane	CH_4	CH_4	undecane	$C_{11}H_{24}$	$CH_3(CH_2)_9CH_3$
ethane	C_2H_6	CH_3CH_3	dodecane	$C_{12}H_{26}$	$CH_3(CH_2)_{10}CH_3$
propane	C_3H_8	$CH_3CH_2CH_3$	tridecane	$C_{13}H_{28}$	$CH_3(CH_2)_{11}CH_3$
butane	C_4H_{10}	$CH_3(CH_2)_2CH_3$	tetradecane	$C_{14}H_{30}$	$CH_3(CH_2)_{12}CH_3$
pentane	C_5H_{12}	$CH_3(CH_2)_3CH_3$	pentadecane	$C_{15}H_{32}$	$CH_3(CH_2)_{13}CH_3$
hexane	C_6H_{14}	$CH_3(CH_2)_4CH_3$	hexadecane	$C_{16}H_{34}$	$CH_3(CH_2)_{14}CH_3$
heptane	C_7H_{16}	$CH_3(CH_2)_5CH_3$	heptadecane	$C_{17}H_{36}$	$CH_3(CH_2)_{15}CH_3$
octane	C_8H_{18}	$CH_3(CH_2)_6CH_3$	octadecane	$C_{18}H_{38}$	$CH_3(CH_2)_{16}CH_3$
nonane	C_9H_{20}	$CH_3(CH_2)_7CH_3$	nonadecane	$C_{19}H_{40}$	$CH_3(CH_2)_{17}CH_3$
decane	$C_{10}H_{22}$	$CH_3(CH_2)_8CH_3$	eicosane	$C_{20}H_{42}$	$CH_3(CH_2)_{18}CH_3$

Alkanes have the general molecular formula C_nH_{2n+2}. Thus, given the number of carbon atoms in an alkane, it is easy to determine the number of hydrogens in the molecule and also its molecular formula. For example, decane, with 10 carbon atoms, must have $(2 \times 10) + 2 = 22$ hydrogens and the molecular formula $C_{10}H_{22}$.

3.2 What Is Constitutional Isomerism in Alkanes?

Constitutional isomers Compounds with the same molecular formula, but a different order of attachment of their atoms.

Constitutional isomers are compounds that have the same molecular formula, but different structural formulas. By "different structural formulas," we mean that these compounds differ in the kinds of bonds they have (single, double, or triple) or in their connectivity (the order of attachment among their atoms).

For the molecular formulas CH_4, C_2H_6, and C_3H_8, only one order of attachment of atoms is possible. For the molecular formula C_4H_{10}, two orders of attachment of atoms are possible. In one of these, named butane, the four carbons are bonded in a chain; in the other, named 2-methylpropane, three carbons are bonded in a chain, with the fourth carbon as a branch on the middle carbon of the chain.

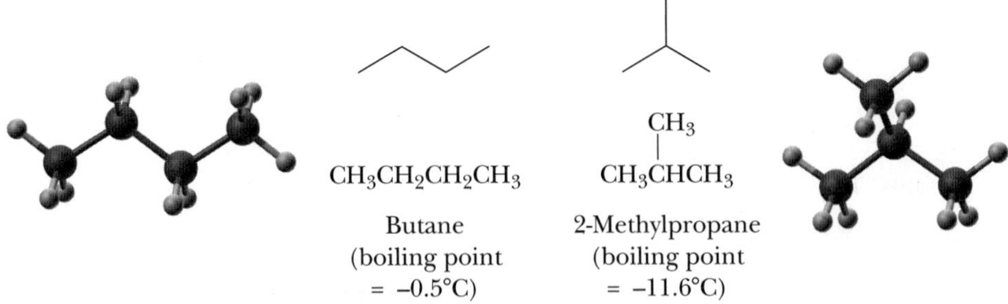

Butane and 2-methylpropane are constitutional isomers; they are different compounds and have different physical and chemical properties. Their boiling points, for example, differ by approximately 11°C. We will discuss how to name alkanes in the next section.

In Section 1.7, we encountered several examples of constitutional isomers, although we did not call them that at the time. We saw that there are two alcohols with the molecular formula C_3H_8O, two aldehydes with the molecular formula C_4H_8O, and two carboxylic acids with the molecular formula $C_4H_8O_2$.

To find out whether two or more structural formulas represent constitutional isomers, write the molecular formula of each and then compare them. All compounds that have the same molecular formula, but different structural formulas, are constitutional isomers.

Example 3.1

Do the structural formulas in each pair represent the same compound or constitutional isomers?

(a) $CH_3CH_2CH_2CH_2CH_2CH_3$ and $CH_3CH_2CH_2$–$CH_2CH_2CH_3$ (the second $CH_2CH_2CH_3$ drawn below, bonded to the last CH_2) (each is C_6H_{14})

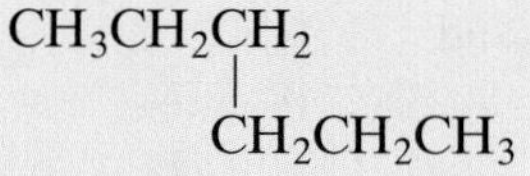

(b) 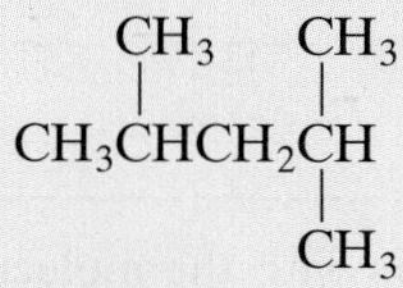 $CH_3CH(CH_3)CH_2CH(CH_3)CH_3$ and $CH_3CH_2CH(CH_3)CH(CH_3)CH_3$ (each is C_7H_{16})

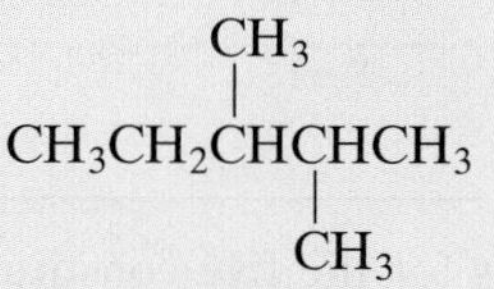

Strategy

To determine whether these structural formulas represent the same compound or constitutional isomers, first find the longest chain of carbon atoms in each. Note that it makes no difference whether the chain is drawn straight or bent. Second, number the longest chain from the end nearest the first branch. Third, compare the lengths of each chain and the sizes and locations of any branches. Structural formulas that have the same connectivity of atoms represent the same compound; those that have a different connectivity of atoms represent constitutional isomers.

Solution

(a) Each structural formula has an unbranched chain of six carbons. The two structures are identical and represent the same compound:

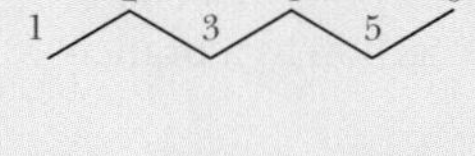

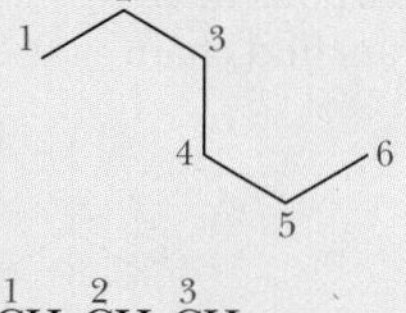

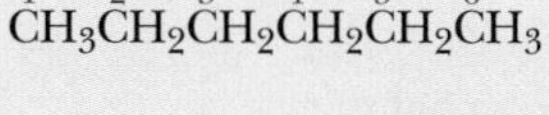

1 2 3 4 5 6 (line-angle chain) and 1 2 3 4 5 6 (bent line-angle chain)

$\overset{1}{C}H_3\overset{2}{C}H_2\overset{3}{C}H_2\overset{4}{C}H_2\overset{5}{C}H_2\overset{6}{C}H_3$ and $\overset{1}{C}H_3\overset{2}{C}H_2\overset{3}{C}H_2$–$\overset{4}{C}H_2\overset{5}{C}H_2\overset{6}{C}H_3$

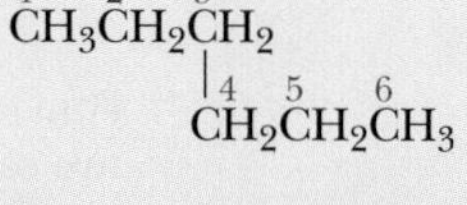

(b) Each structural formula has a chain of five carbons with two CH_3 branches. Although the branches are identical, they are at different locations on the chains. Therefore, these structural formulas represent constitutional isomers:

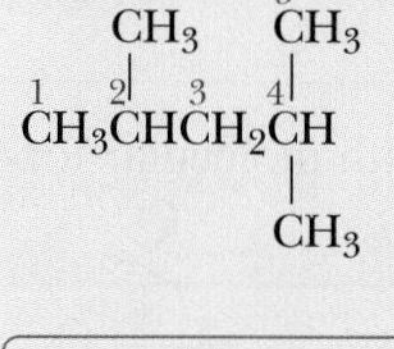

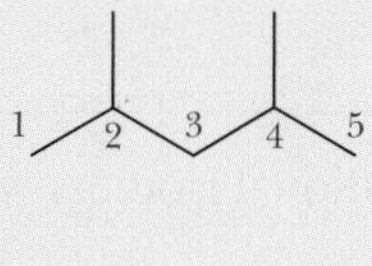

$\overset{1}{C}H_3\overset{2}{C}H(CH_3)\overset{3}{C}H_2\overset{4}{C}H(\overset{5}{C}H_3)CH_3$ (line-angle chain numbered 1 2 3 4 5) and $\overset{5}{C}H_3\overset{4}{C}H_2\overset{3}{C}H(CH_3)\overset{2}{C}H(CH_3)\overset{1}{C}H_3$ (line-angle chain numbered 5 4 3 2 1)

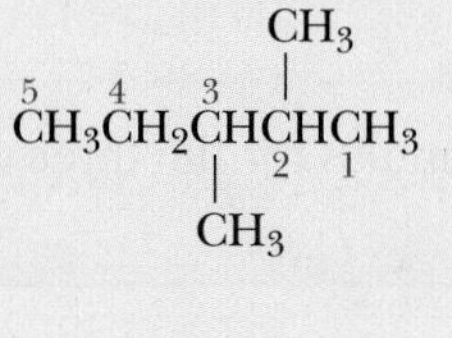

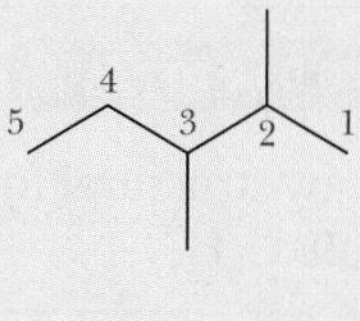

See problems 3.17–3.20, 3.22

Problem 3.1

Do the structural formulas in each pair represent the same compound or constitutional isomers?

(a) 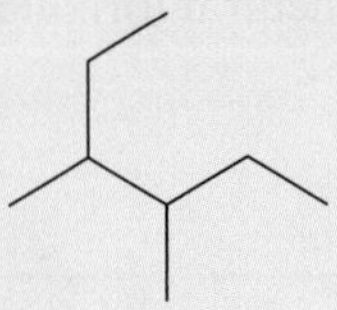 and

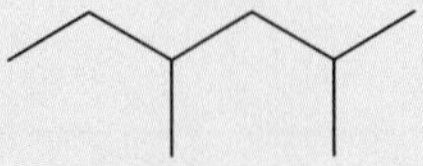

(b) 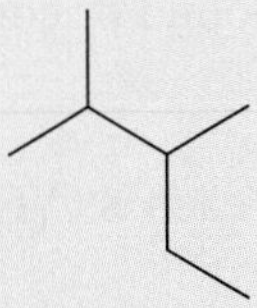 and

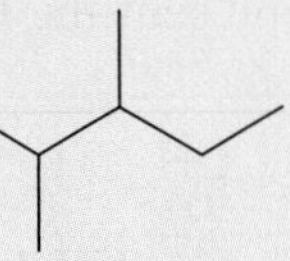

Example 3.2

Draw structural formulas for the five constitutional isomers with the molecular formula C_6H_{14}.

Strategy

In solving problems of this type, you should devise a strategy and then follow it. Here is one such strategy: First, draw a line-angle formula for the constitutional isomer with all six carbons in an unbranched chain. Then, draw line-angle formulas for all constitutional isomers with five carbons in a chain and one carbon as a branch on the chain. Finally, draw line-angle formulas for all constitutional isomers with four carbons in a chain and two carbons as branches.

Solution

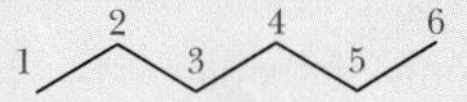

Six carbons in an unbranched chain

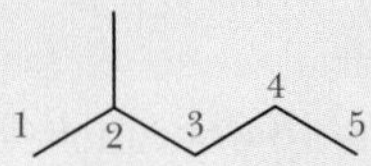

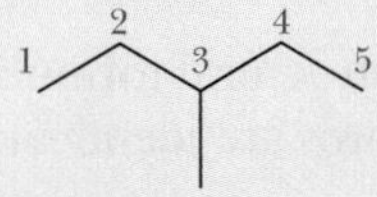

Five carbons in a chain; one carbon as a branch

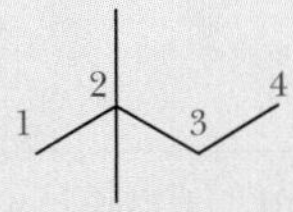

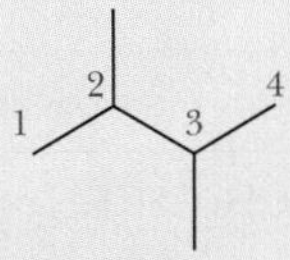

Four carbons in a chain; two carbons as branches

No constitutional isomers with only three carbons in the longest chain are possible for C_6H_{14}.

See problems 3.21, 3.23

Problem 3.2

Draw structural formulas for the three constitutional isomers with molecular formula C_5H_{12}.

The ability of carbon atoms to form strong, stable bonds with other carbon atoms results in a staggering number of constitutional isomers. As the following table shows, there are 3 constitutional isomers with the molecular formula C_5H_{12}, 75 constitutional isomers with the molecular formula $C_{10}H_{22}$, and almost 37 million constitutional isomers with the molecular formula $C_{25}H_{52}$:

Carbon Atoms	Constitutional Isomers
1	0
5	3
10	75
15	4,347
25	36,797,588

Thus, for even a small number of carbon and hydrogen atoms, a very large number of constitutional isomers is possible. In fact, the potential for structural and functional group individuality among organic molecules made from just the basic building blocks of carbon, hydrogen, nitrogen, and oxygen is practically limitless.

3.3 How Do We Name Alkanes?

A. The IUPAC System

Ideally, every organic compound should have a name from which its structural formula can be drawn. For this purpose, chemists have adopted a set of rules established by an organization called the International Union of Pure and Applied Chemistry (IUPAC).

The IUPAC name of an alkane with an unbranched chain of carbon atoms consists of two parts: (1) a prefix that indicates the number of carbon atoms in the chain and (2) the ending **-ane** to show that the compound is a saturated hydrocarbon. Table 3.2 gives the prefixes used to show the presence of 1 to 20 carbon atoms.

The first four prefixes listed in Table 3.2 were chosen by the IUPAC because they were well established in the language of organic chemistry. In fact, they were well established even before there were hints of the structural theory underlying the discipline. For example, the prefix *but-* appears in the name *butyric acid*, a compound of four carbon atoms formed by the air oxidation of butter fat (Latin: *butyrum*, butter). Prefixes to show five or more carbons are derived from Greek or Latin numbers. (See Table 3.1 for the names, molecular formulas, and condensed structural formulas for the first 20 alkanes with unbranched chains.)

TABLE 3.2 Prefixes Used in the IUPAC System to Show the Presence of 1 to 20 Carbons in an Unbranched Chain

Prefix	Number of Carbon Atoms	Prefix	Number of Carbon Atoms
meth-	1	undec-	11
eth-	2	dodec-	12
prop-	3	tridec-	13
but-	4	tetradec-	14
pent-	5	pentadec-	15
hex-	6	hexadec-	16
hept-	7	heptadec-	17
oct-	8	octadec-	18
non-	9	nonadec-	19
dec-	10	eicos-	20

TABLE 3.3 Names of the Most Common Alkyl Groups

Name	Condensed Structural Formula	Name	Condensed Structural Formula
methyl	$-CH_3$	isobutyl	$-CH_2CHCH_3$ CH_3
ethyl	$-CH_2CH_3$		
propyl	$-CH_2CH_2CH_3$	*sec*-butyl	$-CHCH_2CH_3$ CH_3
isopropyl	$-CHCH_3$ CH_3		
butyl	$-CH_2CH_2CH_2CH_3$	*tert*-butyl	CH_3 $-CCH_3$ CH_3

The IUPAC name of an alkane with a branched chain consists of a **parent name** that indicates the longest chain of carbon atoms in the compound and substituent names that indicate the groups bonded to the parent chain.

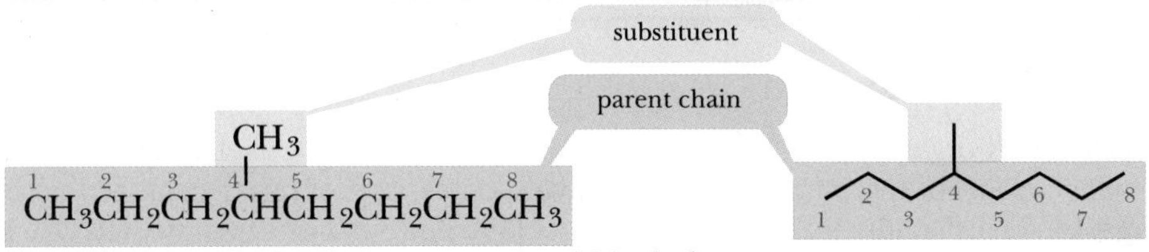

4-Methyloctane

Alkyl group A group derived by removing a hydrogen from an alkane; given the symbol R-.

R- A symbol used to represent an alkyl group.

A substituent group derived from an alkane by the removal of a hydrogen atom is called an **alkyl group** and is commonly represented by the symbol **R-**. We name alkyl groups by dropping the *-ane* from the name of the parent alkane and adding the suffix *-yl*. Table 3.3 gives the names and structural formulas for eight of the most common alkyl groups. The prefix *sec-* is an abbreviation for secondary, meaning a carbon bonded to two other carbons. The prefix *tert-* is an abbreviation for tertiary, meaning a carbon bonded to three other carbons. Note that when these two prefixes are part of a name, they are always italicized.

The rules of the IUPAC system for naming alkanes are as follows:

1. The name for an alkane with an unbranched chain of carbon atoms consists of a prefix showing the number of carbon atoms in the chain and the ending *-ane*.
2. For branched-chain alkanes, take the longest chain of carbon atoms as the parent chain, and its name becomes the root name.
3. Give each substituent on the parent chain a name and a number. The number shows the carbon atom of the parent chain to which the substituent is bonded. Use a hyphen to connect the number to the name:

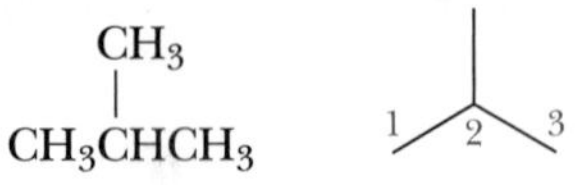

2-Methylpropane

4. If there is one substituent, number the parent chain from the end that gives it the lower number:

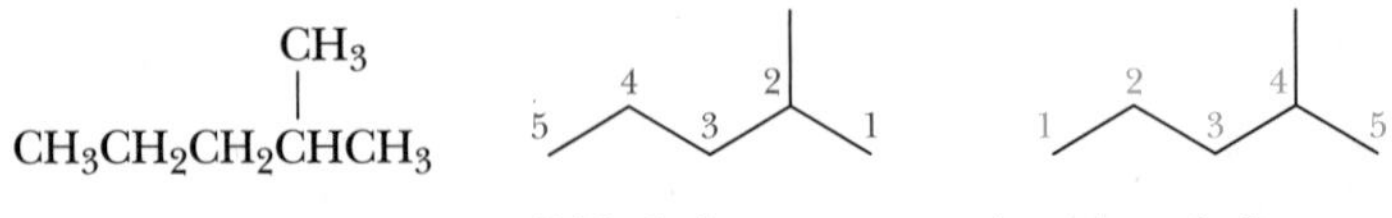

2-Methylpentane (not 4-methylpentane)

5. If there are two or more identical substituents, number the parent chain from the end that gives the lower number to the substituent encountered first. The

number of times the substituent occurs is indicated by the prefix *di-*, *tri-*, *tetra-*, *penta-*, *hexa-*, and so on. A comma is used to separate position numbers:

$$\begin{array}{c}\quad CH_3 \quad\;\; CH_3\\ CH_3CH_2\overset{|}{C}HCH_2\overset{|}{C}HCH_3\end{array}$$

2,4-Dimethylhexane (not 3,5-dimethylhexane)

6. If there are two or more different substituents, list them in alphabetical order, and number the chain from the end that gives the lower number to the substituent encountered first. If there are different substituents in equivalent positions on opposite ends of the parent chain, the substituent of lower alphabetical order is given the lower number:

$$\begin{array}{c}CH_3\\ CH_3CH_2CHCH_2CHCH_2CH_3\\ CH_2CH_3\end{array}$$

3-Ethyl-5-methylheptane (not 3-methyl-5-ethylheptane)

7. The prefixes *di-*, *tri-*, *tetra-*, and so on are not included in alphabetizing. Neither are the hyphenated prefixes *sec-* and *tert-*. Alphabetize the names of the substituents first, and then insert the prefix. In the following example, the alphabetizing parts are ethyl and methyl, not ethyl and dimethyl:

$$\begin{array}{c}CH_3\;\; CH_2CH_3\\ CH_3CCH_2CHCH_2CH_3\\ CH_3\end{array}$$

4-Ethyl-2,2-dimethylhexane
(not 2,2-dimethyl-4-ethylhexane)

Example 3.3

Write IUPAC names for these alkanes:

(a) (b)

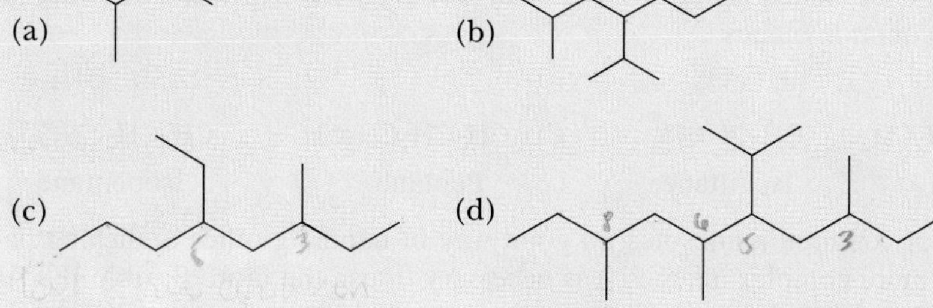

(c) (d)

Strategy

First determine the root name of the alkane. Then name the substituents and place them in alphabetical order. Number the parent chain so as to give the lower number to the substituents encountered first. If substituents have equivalent positions, the lower number is assigned to the substituents with the lower alphabetical order.

Solution

(a) 2-Methylbutane

(b) 4-Isopropyl-2-methylheptane

(c)

5-Ethyl-3-methyloctane

(d)

5-Isopropyl-3,6,8-trimethyldecane
(not 6-isopropyl-3,5,8-trimethyldecane)

See problems 3.24, 3.25, 3.28

Problem 3.3

Write IUPAC names for these alkanes:

(a)

(b)

(c)

(d)

B. Common Names

In the older system of common nomenclature, the total number of carbon atoms in an alkane, regardless of their arrangement, determines the name. The first three alkanes are methane, ethane, and propane. All alkanes of formula C_4H_{10} are called butanes, all those of formula C_5H_{12} are called pentanes, and all those of formula C_6H_{14} are called hexanes. For alkanes beyond propane, *iso* indicates that one end of an otherwise unbranched chain terminates in a $(CH_3)_2CH—$group. Following are examples of common names:

$CH_3CH_2CH_2CH_3$	$CH_3\underset{}{CH}(CH_3)CH_3$	$CH_3CH_2CH_2CH_2CH_3$	$CH_3CH_2CH(CH_3)CH_3$
Butane	Isobutane	Pentane	Isopentane

This system of common names has no good way of handling other branching patterns, so, for more complex alkanes, it is necessary to use the more flexible IUPAC system of nomenclature.

In this text, we concentrate on IUPAC names. However, we also use common names, especially when the common name is used almost exclusively in the everyday discussions of chemists and biochemists. When both IUPAC and common names are given in the text, we always give the IUPAC name first, followed by the common name in parentheses. In this way, you should have no doubt about which name is which.

C. Classification of Carbon and Hydrogen Atoms

We classify a carbon atom as primary (1°), secondary (2°), tertiary (3°), or quaternary (4°), depending on the number of carbon atoms bonded to it. A carbon bonded to one carbon atom is a primary carbon; a carbon bonded to two carbon atoms is a secondary carbon, and so forth. For example, propane contains two primary carbons and one secondary carbon, 2-methylpropane contains three primary carbons and one tertiary carbon, and 2,2,4-trimethylpentane contains five primary carbons, one secondary carbon, one tertiary carbon, and one quaternary carbon:

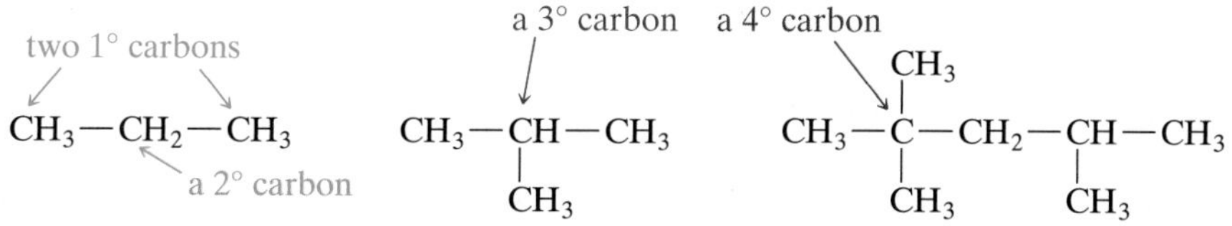

two 1° carbons; a 2° carbon
$CH_3-CH_2-CH_3$
Propane

a 3° carbon
$CH_3-CH(CH_3)-CH_3$
2-Methylpropane

a 4° carbon
$CH_3-C(CH_3)_2-CH_2-CH(CH_3)-CH_3$
2,2,4-Trimethylpentane

Similarly, hydrogens are also classified as primary, secondary, or tertiary, depending on the type of carbon to which each is bonded. Those bonded to a primary carbon are classified as primary hydrogens, those on a secondary carbon are secondary hydrogens, and those on a tertiary carbon are tertiary hydrogens.

Example 3.4

Classify each carbon atom in the following compounds as 1°, 2°, 3°, or 4°.

(a) 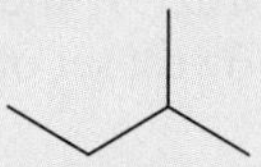 (b)

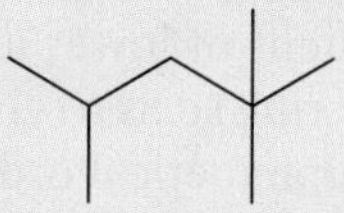

Strategy

To classify carbons, determine whether each is bonded to 1 carbon (1°), 2 carbons (2°), 3 carbons (3°), or 4 carbons (4°).

Solution

(a) 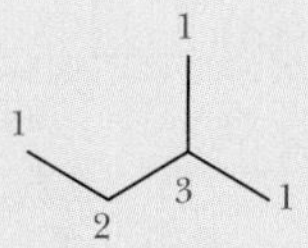 1, 2, 3, 1, 1

(b) 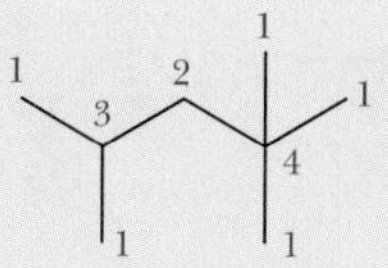 1, 3, 1, 2, 4, 1, 1, 1

See problems 3.26, 3.27

Problem 3.4

Classify each hydrogen atom in the following compounds as 1°, 2°, or 3°.

(a) $CH_3CH(CH_3)CH_2CH_2CH_3$

(b) $CH_3-CH_2-C(CH_3)(CH_3)-CH(CH_3)(CH_3)$

3.4 What Are Cycloalkanes?

A hydrocarbon that contains carbon atoms joined to form a ring is called a *cyclic hydrocarbon*. When all carbons of the ring are saturated, we call the hydrocarbon a **cycloalkane.** Cycloalkanes of ring sizes ranging from 3 to over 30 abound in nature, and, in principle, there is no limit to ring size. Five-membered (cyclopentane) and six-membered (cyclohexane) rings are especially abundant in nature and have received special attention.

Cycloalkane A saturated hydrocarbon that contains carbon atoms joined to form a ring.

Figure 3.3
Examples of cycloalkanes.

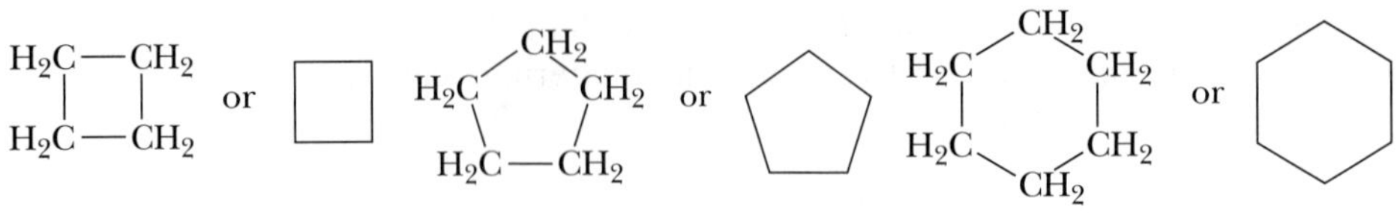

$H_2C—CH_2$ / $H_2C—CH_2$ or [square] — Cyclobutane

CH_2, H_2C, CH_2, $H_2C—CH_2$ or [pentagon] — Cyclopentane

CH_2, H_2C, CH_2, H_2C, CH_2, CH_2 or [hexagon] — Cyclohexane

Figure 3.3 shows the structural formulas of cyclobutane, cyclopentane, and cyclohexane. When writing structural formulas for cycloalkanes, chemists rarely show all carbons and hydrogens. Rather, they use line-angle formulas to represent cycloalkane rings. Each ring is represented by a regular polygon having the same number of sides as there are carbon atoms in the ring. For example, chemists represent cyclobutane by a square, cyclopentane by a pentagon, and cyclohexane by a hexagon.

Cycloalkanes contain two fewer hydrogen atoms than an alkane with the same number of carbon atoms. For instance, compare the molecular formulas of cyclohexane (C_6H_{12}) and hexane (C_6H_{14}). The general formula of a cycloalkane is C_nH_{2n}.

To name a cycloalkane, prefix the name of the corresponding open-chain hydrocarbon with *cyclo*-, and name each substituent on the ring. If there is only one substituent, there is no need to give it a number. If there are two substituents, number the ring by beginning with the substituent of lower alphabetical order. If there are three or more substituents, number the ring so as to give them the lowest set of numbers, and then list the substituents in alphabetical order.

Example 3.5

Write the molecular formula and IUPAC name for each cycloalkane.

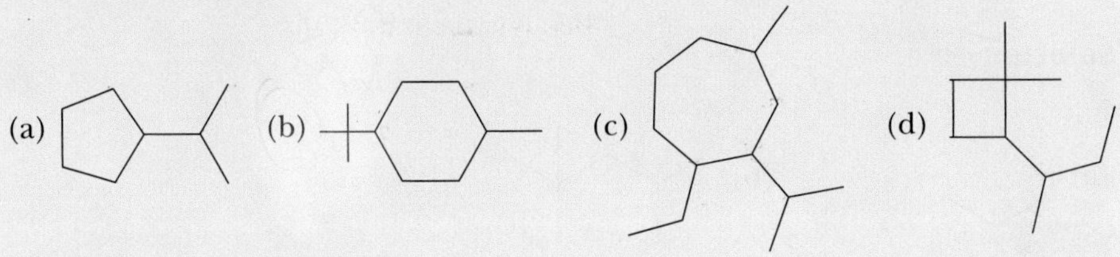

(a) (b) (c) (d)

Strategy

First determine the root name of the cycloalkane. Then name the substituents and place them in alphabetical order. Number the parent chain so as to give the lower number to the substituent encountered first. If substituents have equivalent positions, the lower number is assigned to the substituent with the lower alphabetical order.

Solution

(a) The molecular formula of this cycloalkane is C_8H_{16}. Because there is only one substituent on the ring, there is no need to number the atoms of the ring. The IUPAC name of this compound is isopropylcyclopentane.

(b) Number the atoms of the cyclohexane ring by beginning with *tert*-butyl, the substituent of lower alphabetical order. The compound's name is 1-*tert*-butyl-4-methylcyclohexane, and its molecular formula is $C_{11}H_{22}$.

(c) The molecular formula of this cycloalkane is $C_{13}H_{26}$. The compound's name is 1-ethyl-2-isopropyl-4-methylcycloheptane. The ethyl group is numbered 1 because this allows the isopropyl group to be encountered sooner than if the methyl group were numbered 1.

(d) The molecular formula of this cycloalkane is $C_{10}H_{20}$. The compound's name is 2-*sec*-butyl-1,1-dimethylcyclobutane. This example illustrates that "*sec*" and "di" are not used in alphabetizing for nomenclature.

See problems 3.24, 3.25, 3.28

Problem 3.5

Write the molecular formula and IUPAC name for each cycloalkane:

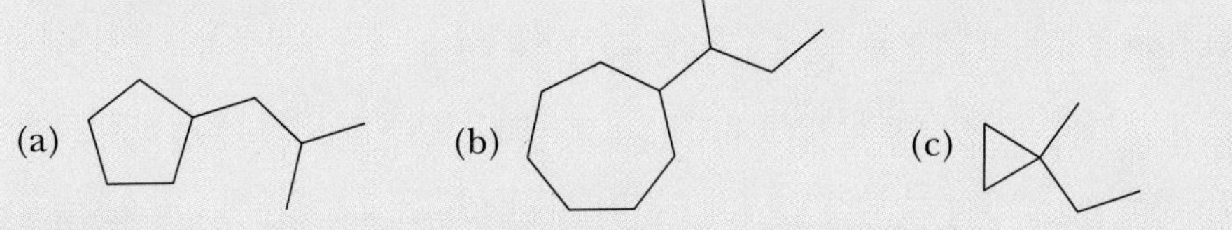

3.5 What Is the IUPAC System of Nomenclature?

The naming of alkanes and cycloalkanes in Sections 3.3 and 3.4 illustrates the application of the IUPAC system of nomenclature to two specific classes of organic compounds. Now let us describe the general approach of the IUPAC system. The name we give to any compound with a chain of carbon atoms consists of three parts: a prefix, an infix (a modifying element inserted into a word), and a suffix. Each part provides specific information about the structural formula of the compound.

1. The prefix shows the number of carbon atoms in the parent chain. Prefixes that show the presence of 1 to 20 carbon atoms in a chain were given in Table 3.2.
2. The infix shows the nature of the carbon–carbon bonds in the parent chain;

Infix	Nature of Carbon–Carbon Bonds in the Parent Chain
-an-	all single bonds
-en-	one or more double bonds
-yn-	one or more triple bonds

3. The suffix shows the class of compound to which the substance belongs;

Suffix	Class of Compound
-e	hydrocarbon
-ol	alcohol
-al	aldehyde
-one	ketone
-oic acid	carboxylic acid

Example 3.6

Following are IUPAC names and structural formulas for four compounds:

(a) $CH_2{=}CHCH_3$ Propene (b) CH_3CH_2OH Ethanol (c) $CH_3CH_2CH_2CH_2\overset{\overset{\displaystyle O}{\|}}{C}OH$ Pentanoic acid (d) $HC{\equiv}CH$ Ethyne

Divide each name into a prefix, an infix, and a suffix, and specify the information about the structural formula that is contained in each part of the name.

Strategy

First look at the first few letters of the name (meth, eth, prop, but, etc.). This is the prefix that tells the number of carbons in the parent chain. Next look at "an," "en," or "yn." These infixes indicate the nature of the carbon–carbon bonds in the parent chain. The letters that follow the infix are part of the suffix, which determines the class of compound to which the molecule belongs.

Solution

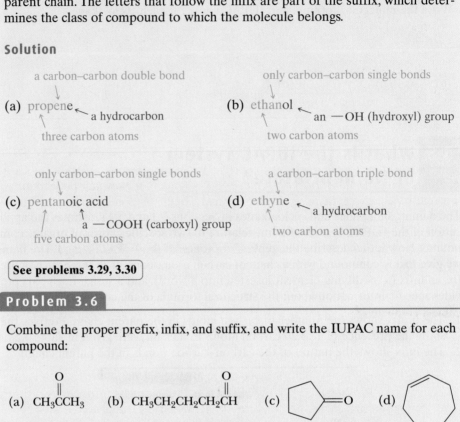

(a) propene — "prop": three carbon atoms; "en": a carbon–carbon double bond; "e": a hydrocarbon

(b) ethanol — "eth": two carbon atoms; "an": only carbon–carbon single bonds; "ol": an —OH (hydroxyl) group

(c) pentanoic acid — "pent": five carbon atoms; "an": only carbon–carbon single bonds; "oic acid": a —COOH (carboxyl) group

(d) ethyne — "eth": two carbon atoms; "yn": a carbon–carbon triple bond; "e": a hydrocarbon

See problems 3.29, 3.30

Problem 3.6

Combine the proper prefix, infix, and suffix, and write the IUPAC name for each compound:

(a) 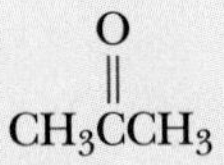 $CH_3C(=O)CH_3$ (b) 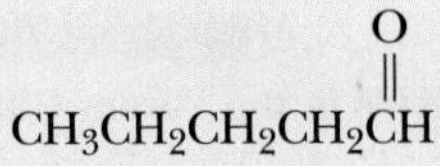 $CH_3CH_2CH_2CH_2CH(=O)$ (c) 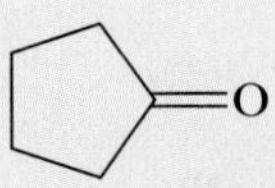 cyclopentane ring with =O (d) 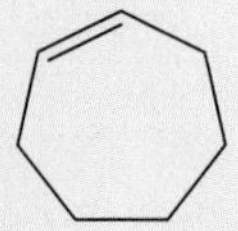 cycloheptene ring

3.6 What Are the Conformations of Alkanes and Cycloalkanes?

Even though structural formulas are useful for showing the order of attachment of atoms, they do not show three-dimensional shapes. As chemists try to understand more and more about the relationships between structure and the chemical and physical properties of molecules, it becomes increasingly important to know more about the three-dimensional shapes of molecules.

In this section, we concentrate on ways to visualize molecules as three-dimensional objects and to visualize not only bond angles within molecules, but also distances between various atoms and groups of atoms not bonded to each other. We also describe strain, which we divide into three types: torsional strain, angle strain, and steric strain. We urge you to build models and to study and manipulate them. Organic molecules are three-dimensional objects, and it is essential that you become comfortable in dealing with them as such.

A. Alkanes

Conformation Any three-dimensional arrangement of atoms in a molecule that results by rotation about a single bond.

Alkanes of two or more carbons can be twisted into a number of different three-dimensional arrangements of their atoms by rotating about one or more carbon–carbon bonds. Any three-dimensional arrangement of atoms that results from rotation about a single bond is called a **conformation**. Figure 3.4(a) shows a ball-and-

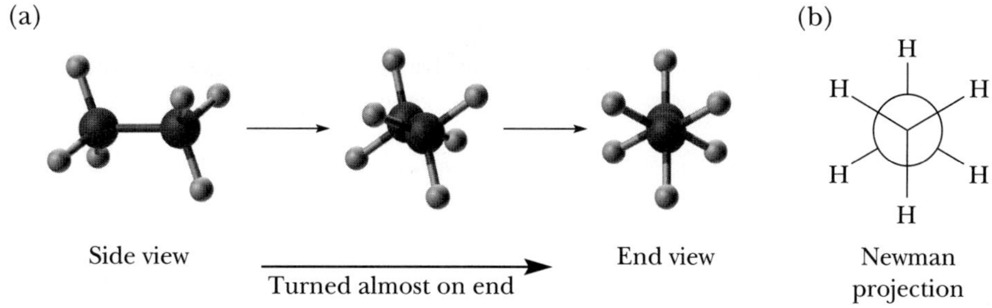

(a) Side view → Turned almost on end → End view

(b) Newman projection

H H H H H H

Figure 3.4
A staggered conformation of ethane. (a) Ball-and-stick model and (b) Newman projection.

stick model of a **staggered conformation** of ethane. In this conformation, the three C—H bonds on one carbon are as far apart as possible from the three C—H bonds on the adjacent carbon. Figure 3.4(b), called a **Newman projection**, is a shorthand way of representing the staggered conformation of ethane. In a Newman projection, we view a molecule along the axis of a C—C bond. The three atoms or groups of atoms nearer your eye appear on lines extending from the center of the circle at angles of 120°. The three atoms or groups of atoms on the carbon farther from your eye appear on lines extending from the circumference of the circle at angles of 120°. Remember that bond angles about each carbon in ethane are approximately 109.5° and not 120°, as this Newman projection might suggest.

Staggered conformation A conformation about a carbon–carbon single bond in which the atoms on one carbon are as far apart as possible from the atoms on the adjacent carbon.

Newman projection A way to view a molecule by looking along a carbon–carbon bond.

Eclipsed conformation A conformation about a carbon–carbon single bond in which the atoms on one carbon are as close as possible to the atoms on the adjacent carbon.

Figure 3.5 shows a ball-and-stick model and a Newman projection of an **eclipsed conformation** of ethane. In this conformation, the three C—H bonds on one carbon are as close as possible to the three C—H bonds on the adjacent carbon. In other words, hydrogen atoms on the back carbon are eclipsed by the hydrogen atoms on the front carbon.

For a long time, chemists believed that rotation about the C—C single bond in ethane was completely free. Studies of ethane and other molecules, however, have shown that a potential energy difference exists between its staggered and eclipsed conformations and that rotation is not completely free. In ethane, the potential energy of the eclipsed conformation is a maximum and that of the staggered conformation is a minimum. The difference in potential energy between these two conformations is approximately 12.6 kJ/mol (3.0 kcal/mol).

The strain induced in the eclipsed conformation of ethane is an example of torsional strain. **Torsional strain** (also called eclipsed interaction strain) is strain that arises when nonbonded atoms separated by three bonds are forced from a staggered conformation to an eclipsed conformation.

Torsional strain (also called eclipsed interaction strain) Strain that arises when atoms separated by three bonds are forced from a staggered conformation to an eclipsed conformation.

Figure 3.5
An eclipsed conformation of ethane. (a, b) Ball-and-stick models and (c) Newman projection.

Example 3.7

Draw Newman projections for one staggered conformation and one eclipsed conformation of propane.

Strategy

Draw the line-angle formula of propane and choose a bond along which to view for the Newman projection. Keep track of the carbons in the line-angle formula and in the Newman projection (numbering them helps). Draw the staggered and eclipsed Newman projections and complete them by adding in the carbons and hydrogens.

Solution

Following are Newman projections and ball-and-stick models of these conformations:

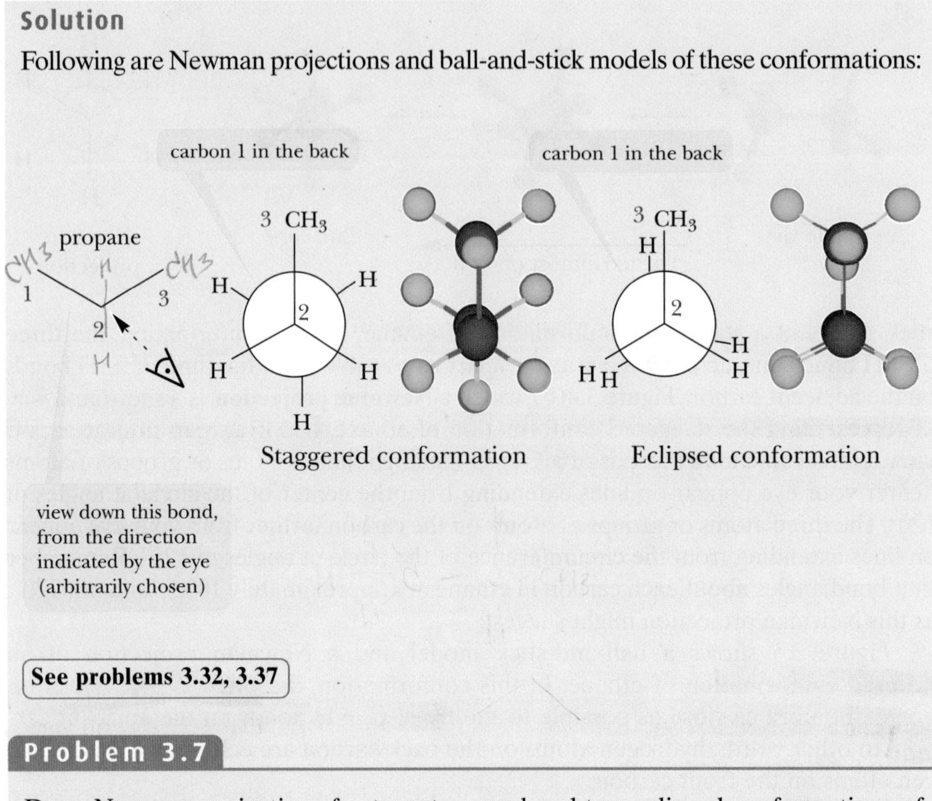

See problems 3.32, 3.37

Problem 3.7

Draw Newman projections for two staggered and two eclipsed conformations of 1,2-dichloroethane.

B. Cycloalkanes

We limit our discussion to the conformations of cyclopentanes and cyclohexanes because these are the most common carbon rings in the molecules of nature.

Cyclopentane

Angle strain The strain that arises when a bond angle is either compressed or expanded compared with its optimal value.

We can draw cyclopentane [Figure 3.6(a)] as a planar conformation with all C—C—C bond angles equal to 108° [Figure 3.6(b)]. This angle differs only slightly from the tetrahedral angle of 109.5°; consequently, there is little angle strain in the planar conformation of cyclopentane. **Angle strain** results when a bond angle in a molecule is either expanded or compressed compared with its optimal values. There are 10 fully eclipsed C—H bonds creating a torsional strain of approximately 42 kJ/mol (10 kcal/mol). To relieve at least a part of this strain, the atoms of the ring twist into the "envelope" conformation [Figure 3.6(c)]. In this conformation, four carbon atoms are in a plane, and the fifth is bent out of the plane, rather like an envelope with its flap bent upward.

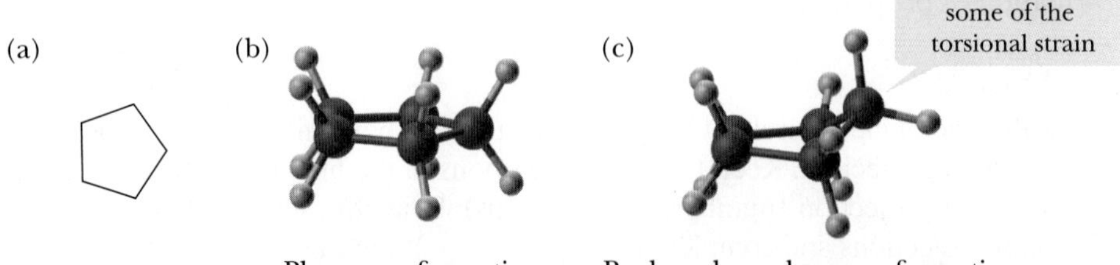

Figure 3.6 Cyclopentane. (a) Structural formula. (b) In the planar conformation, there are 10 pairs of eclipsed C—H interactions. (c) The most stable conformation is a puckered "envelope" conformation.

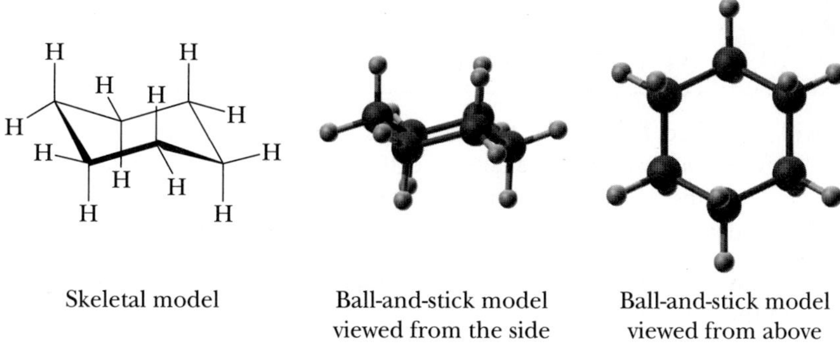

Figure 3.7 Cyclohexane. The most stable conformation is the puckered "chair" conformation.

In the envelope conformation, the number of eclipsed hydrogen interactions is reduced, thereby decreasing torsional strain. The C—C—C bond angles, however, are also reduced, which increases angle strain. The observed C—C—C bond angles in cyclopentane are 105°, indicating that, in its conformation of lowest energy, cyclopentane is slightly puckered. The strain energy in cyclopentane is approximately 23.4 kJ/mol (5.6 kcal/mol).

Cyclohexane

Cyclohexane adopts a number of puckered conformations, the most stable of which is a **chair conformation**. In this conformation (Figure 3.7), all C—C—C bond angles are 109.5° (minimizing angle strain), and hydrogens on adjacent carbons are staggered with respect to one another (minimizing torsional strain). Thus, there is very little strain in a chair conformation of cyclohexane.

In a chair conformation, the C—H bonds are arranged in two different orientations. Six C—H bonds are called **equatorial bonds**, and the other six are called **axial bonds**. One way to visualize the difference between these two types of bonds is to imagine an axis through the center of the chair, perpendicular to the floor [Figure 3.8(a)]. Equatorial bonds are approximately perpendicular to our imaginary axis and alternate first slightly up and then slightly down as you move from one carbon of the ring to the next. Axial bonds are parallel to the imaginary axis. Three axial bonds point up; the other three point down. Notice that axial bonds alternate also, first up and then down as you move from one carbon of the ring to the next. Notice further that if the axial bond on a carbon points upward, then the equatorial bond on that carbon points slightly downward. Conversely, if the axial bond on a particular carbon points downward, then the equatorial bond on that carbon points slightly upward.

There are many other nonplanar conformations of cyclohexane, one of which is the **boat conformation**. You can visualize the interconversion of a chair conformation

Chair conformation The most stable puckered conformation of a cyclohexane ring; all bond angles are approximately 109.5°, and bonds to all adjacent carbons are staggered.

Equatorial bond A bond on a chair conformation of a cyclohexane ring that extends from the ring roughly perpendicular to the imaginary axis of the ring.

Axial bond A bond on a chair conformation of a cyclohexane ring that extends from the ring parallel to the imaginary axis of the ring.

Boat conformation A puckered conformation of a cyclohexane ring in which carbons 1 and 4 of the ring are bent toward each other.

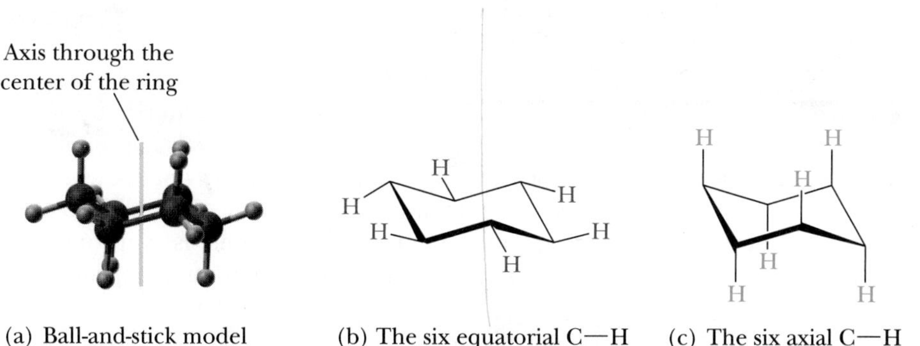

Figure 3.8 Chair conformation of cyclohexane, showing axial and equatorial C—H bonds.

HOW TO 3.2 Draw Alternative Chair Conformations of Cyclohexane

You will be asked frequently to draw three-dimensional representations of chair conformations of cyclohexane and to show spatial relationships among atoms and groups of atoms bonded to the ring. Here are four steps that will help you to draw them. With a little practice you will find them easy to draw.

Step 1: Draw two sets of parallel lines, one line in each set offset from the other in the set as shown.

Step 2: Complete the chair by drawing the head and foot pieces, one up and the other down.

Step 3: Draw the equatorial bonds using ring bonds as a guide. Remember that each equatorial bond is parallel to two ring bonds, and that equatorial bonds on opposite carbons of the ring are parallel to one another. Sets of parallel equatorial and ring bonds are shown here in color.

Equatorial bonds

Step 4: Draw the six axial bonds, as vertical lines. Remember that all axial bonds are parallel to each other. Sets of parallel axial bonds are shown in color.

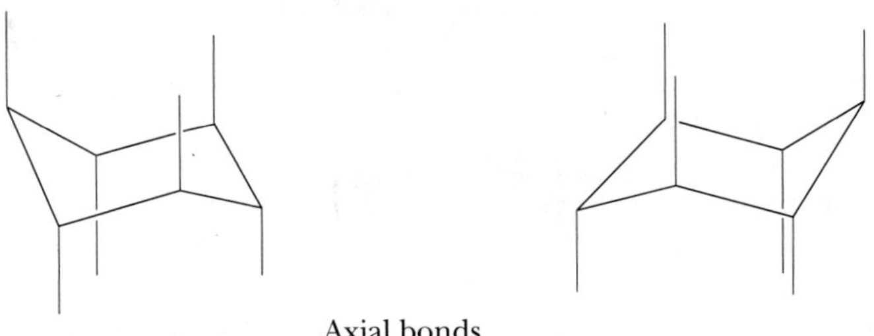

Axial bonds

to a boat conformation by twisting the ring as illustrated in Figure 3.9. A boat conformation is considerably less stable than a chair conformation. In a boat conformation, torsional strain is created by four sets of eclipsed hydrogen interactions, and steric strain is created by the one set of flagpole interactions. **Steric strain** (also called nonbonded interaction strain) results when nonbonded atoms separated by four or more bonds are forced abnormally close to each other—that is, when they are forced closer than their atomic (contact) radii allow. The difference in potential energy between chair and boat conformations is approximately 27 kJ/mol (6.5 kcal/mol), which means that, at room temperature, approximately 99.99% of all cyclohexane molecules are in the chair conformation.

Steric strain The strain that arises when atoms separated by four or more bonds are forced abnormally close to one another.

For cyclohexane, the two equivalent chair conformations can interconvert by one chair twisting first into a boat and then into the other chair. When one chair is

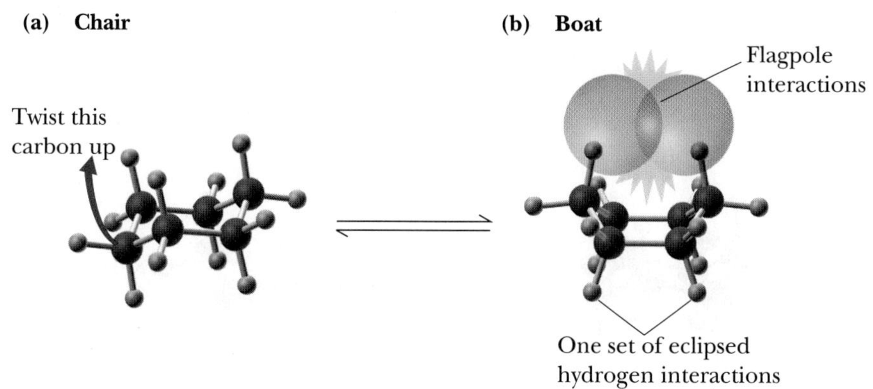

Figure 3.9
Conversion of (a) a chair conformation to (b) a boat conformation. In the boat conformation, there is torsional strain due to the four sets of eclipsed hydrogen interactions and steric strain due to the one set of flagpole interactions. A chair conformation is more stable than a boat conformation.

Example 3.8

Following is a chair conformation of cyclohexane showing a methyl group and one hydrogen:

CH_3

H

(a) Indicate by a label whether each group is equatorial or axial.
(b) Draw the other chair conformation, and again label each group as equatorial or axial.

Strategy

A chair-to-chair interconversion is most often done by changing the orientation of the rightmost and leftmost carbons in the chair conformation of cyclohexane. Remember that after such an interconversion, all prior axial substituents become equatorial and all prior equatorial substituents become axial.

Solution

CH_3 (axial)

H (equatorial)

(a)

CH_3 (equatorial)

H (axial)

(b)

See problem 3.47

Problem 3.8

Following is a chair conformation of cyclohexane with carbon atoms numbered 1 through 6:

5 6 1

4 3 2

(a) Draw hydrogen atoms that are above the plane of the ring on carbons 1 and 2 and below the plane of the ring on carbon 4.
(b) Which of these hydrogens are equatorial? Which are axial?
(c) Draw the other chair conformation. Now which hydrogens are equatorial? Which are axial? Which are above the plane of the ring, and which are below it?

Figure 3.10 Interconversion of chair cyclohexanes. All C—H bonds that are equatorial in one chair are axial in the alternative chair, and vice versa.

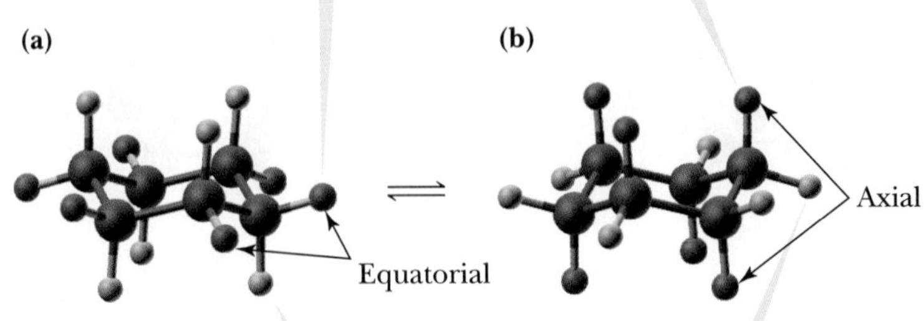

Hydrogen switches from equatorial to axial upon chair-to-chair interconversion, but remains pointing upward
(a)
(b)
Equatorial
Axial
Hydrogen switches from axial to equatorial upon chair-to-chair interconversion, but remains pointing downward

converted to the other, a change occurs in the relative orientations in space of the hydrogen atoms bonded to each carbon: All hydrogen atoms equatorial in one chair become axial in the other, and vice versa (Figure 3.10). The interconversion of one chair conformation of cyclohexane to the other occurs rapidly at room temperature.

If we replace a hydrogen atom of cyclohexane by an alkyl group, the group occupies an equatorial position in one chair and an axial position in the other chair. This means that the two chairs are no longer equivalent and no longer of equal stability.

Diaxial interactions Interactions between groups in parallel axial positions on the same side of a chair conformation of a cyclohexane ring.

A convenient way to describe the relative stabilities of chair conformations with equatorial or axial substituents is in terms of a type of steric strain called **axial–axial (diaxial) interaction**. *Axial–axial interaction* refers to the steric strain existing between an axial substituent and an axial hydrogen (or other group) on the same side of the ring. Consider methylcyclohexane (Figure 3.11). When the $—CH_3$ is equatorial, it is staggered with respect to all other groups on its adjacent carbon atoms. When the $—CH_3$ is axial, it is parallel to the axial C—H bonds on carbons 3 and 5. Thus, for axial methylcyclohexane, there are two unfavorable methyl–hydrogen axial–axial interactions. For methylcyclohexane, the equatorial methyl conformation is favored over the axial methyl conformation by approximately 7.28 kJ/mol (1.74 kcal/mol). At equilibrium at room temperature, approximately 95% of all methylcyclohexane molecules have their methyl group equatorial, and less than 5% have their methyl group axial.

As the size of the substituent increases, the relative amount of the conformation with the group equatorial increases. When the group is as large as *tert*-butyl, the equatorial conformation is approximately 4,000 times more abundant at room temperature than the axial conformation, and, in effect, the ring is "locked" into a chair conformation with the *tert*-butyl group equatorial.

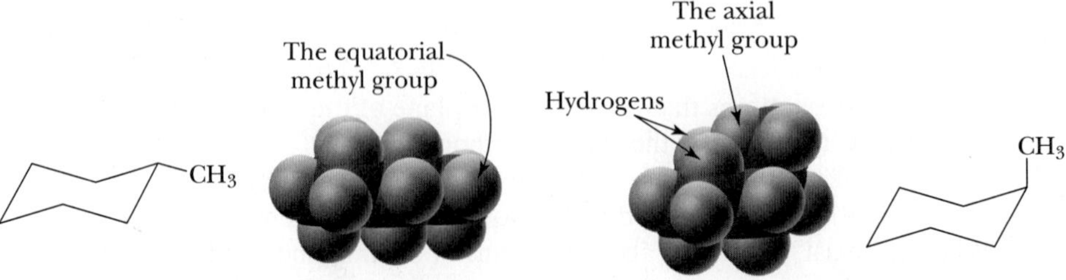

CH_3
The equatorial methyl group
Hydrogens
The axial methyl group
CH_3

(a) Equatorial methylcyclohexane (b) Axial methylcyclohexane

Figure 3.11 Two chair conformations of methylcyclohexane. The two axial–axial interactions (steric strain) make conformation (b) less stable than conformation (a) by approximately 7.28 kJ/mol (1.74 kcal/mol).

Example 3.9

Label all axial–axial interactions in the following chair conformation:

Strategy

Find each axial group. Those on the same side of the chair conformation (either above or below the ring) will participate in axial–axial interactions. Remember that the equatorial substituents do not participate in axial–axial interactions.

Solution

There are four axial–axial interactions: Each axial methyl group has two sets of axial–axial interactions with parallel hydrogen atoms on the same side of the ring. The equatorial methyl group has no axial–axial interactions.

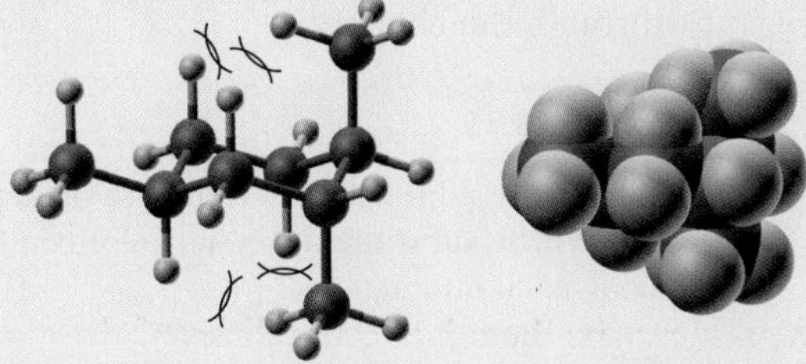

See problem 3.38

Problem 3.9

The conformational equilibria for methyl-, ethyl-, and isopropylcyclohexane are all about 95% in favor of the equatorial conformation, but the conformational equilibrium for *tert*-butylcyclohexane is almost completely on the equatorial side. Explain why the conformational equilibria for the first three compounds are comparable, but that for *tert*-butylcyclohexane lies considerably farther toward the equatorial conformation.

3.7 What Is *Cis-Trans* Isomerism in Cycloalkanes?

Cycloalkanes with substituents on two or more carbons of the ring show a type of isomerism called ***cis–trans* isomerism**. *Cis-trans* isomers have (1) the same molecular formula, (2) the same order of attachment of atoms, and (3) an arrangement of atoms that cannot be interchanged by rotation about sigma bonds under ordinary conditions. By way of comparison, the potential energy difference between conformations is so small that they can be interconverted easily at or near room temperature by rotation about single bonds, while *cis-trans* isomers can only be interconverted at extremely high temperatures or not at all.

***Cis-trans* isomers** Isomers that have the same order of attachment of their atoms, but a different arrangement of their atoms in space, due to the presence of either a ring or a carbon–carbon double bond.

Cis A prefix meaning "on the same side."

Trans A prefix meaning "across from."

We can illustrate *cis–trans* isomerism in cycloalkanes using 1,2-dimethylcyclopentane as an example. In the following structural formula, the cyclopentane ring is drawn as a planar pentagon viewed edge on (in determining the number of *cis–trans*

CHEMICAL CONNECTIONS 3A

The Poisonous Puffer Fish

Nature is by no means limited to carbon in six-membered rings. Tetrodotoxin, one of the most potent toxins known, is composed of a set of interconnected six-membered rings, each in a chair conformation. All but one of these rings have atoms other than carbon in them. Tetrodotoxin is produced in the liver and ovaries of many species of *Tetraodontidae*, especially the puffer fish, so called because it inflates itself to an almost spherical spiny ball when alarmed. The puffer is evidently a species that is highly preoccupied with defense, but the Japanese are not put off. They regard the puffer, called *fugu* in Japanese, as a delicacy. To serve it in a public restaurant, a chef must be registered as sufficiently skilled in removing the toxic organs so as to make the flesh safe to eat.

A puffer fish inflated. *(Dave Fleetham/Stockphotopro, Inc)*

Symptoms of tetrodotoxin poisoning begin with attacks of severe weakness, progressing to complete paralysis and eventual death. Tetrodotoxin exerts its severe poisonous effect by blocking Na^+ ion channels in excitable membranes. The $=NH_2^+$ end of tetrodotoxin lodges in the mouth of a Na^+ ion channel, thus blocking further transport of Na^+ ions through the channel.

QUESTIONS

How many chair conformations are present in tetrodotoxin? Which substituents in tetrodotoxin are involved in axial–axial interactions?

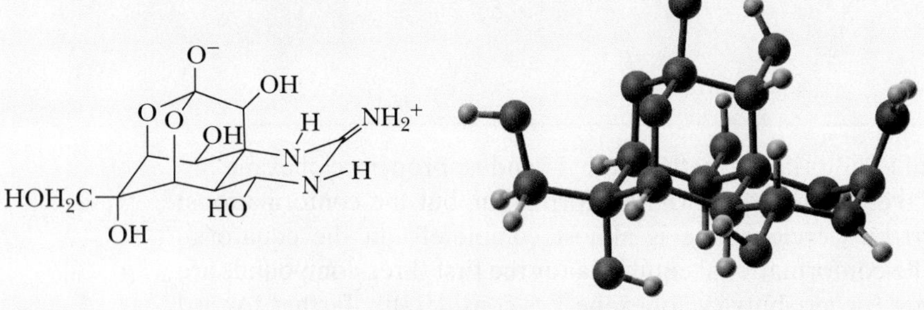

Tetrodotoxin

isomers in a substituted cycloalkane, it is adequate to draw the cycloalkane ring as a planar polygon):

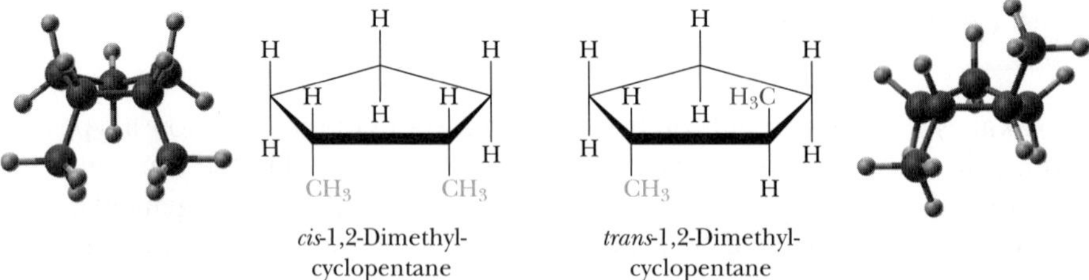

cis-1,2-Dimethylcyclopentane   *trans*-1,2-Dimethylcyclopentane

Carbon–carbon bonds of the ring that project forward are shown as heavy lines. When viewed from this perspective, substituents bonded to the cyclopentane ring project above and below the plane of the ring. In one isomer of 1,2-dimethylcyclopentane, the methyl groups are on the same side of the ring (either both above or both below the plane of the ring); in the other isomer, they are on opposite sides of the ring (one above and one below the plane of the ring).

Alternatively, the cyclopentane ring can be viewed from above, with the ring in the plane of the paper. Substituents on the ring then either project toward you (that is, they project up above the page) and are shown by solid wedges, or they project away from you (they project down below the page) and are shown by broken wedges. In the following structural formulas, only the two methyl groups are shown (hydrogen atoms of the ring are not shown):

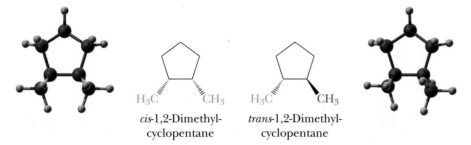

H_3C CH_3

cis-1,2-Dimethyl-cyclopentane

H_3C CH_3

trans-1,2-Dimethyl-cyclopentane

Example 3.10

Which cycloalkanes show *cis–trans* isomerism? For each that does, draw both isomers.

(a) Methylcyclopentane
(b) 1,1-Dimethylcyclobutane
(c) 1,3-Dimethylcyclobutane

Strategy

In order to exhibit *cis–trans* isomerism, a cyclic compound must have at least two substituents on the ring and there must be two possible arrangements (*cis* and *trans*) for any pair of substituents.

Solution

(a) Methylcyclopentane does not show *cis–trans* isomerism: It has only one substituent on the ring.
(b) 1,1-Dimethylcyclobutane does not show *cis–trans* isomerism: Only one arrangement is possible for the two methyl groups on the ring, and they must be *trans*.
(c) 1,3-Dimethylcyclobutane shows *cis–trans* isomerism. Note that, in these structural formulas, we show only the hydrogens on carbons bearing the methyl groups.

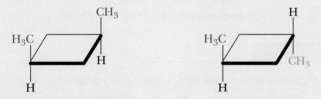

cis-1,3-Dimethylcyclobutane

trans-1,3-Dimethylcyclobutane

See problems 3.41–3.44, 3.46

Problem 3.10

Which cycloalkanes show *cis–trans* isomerism? For each that does, draw both isomers.

(a) 1,3-Dimethylcyclopentane
(b) Ethylcyclopentane
(c) 1-Ethyl-2-methylcyclobutane

Two *cis–trans* isomers exist for 1,4-dimethylcyclohexane. For the purposes of determining the number of *cis–trans* isomers in substituted cycloalkanes, it is adequate to draw the cycloalkane ring as a planar polygon, as is done in the following disubstituted cyclohexanes:

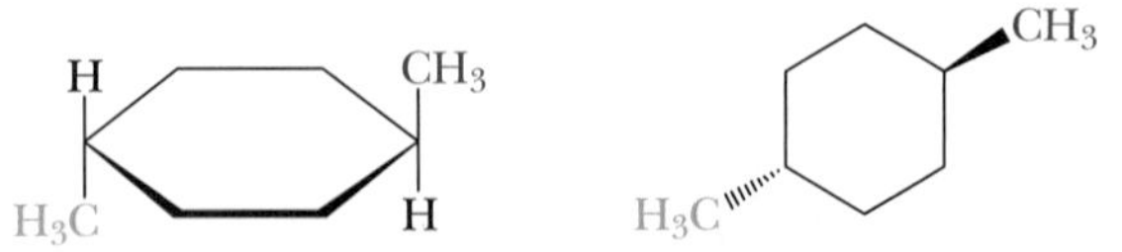

trans-1,4-Dimethylcyclohexane

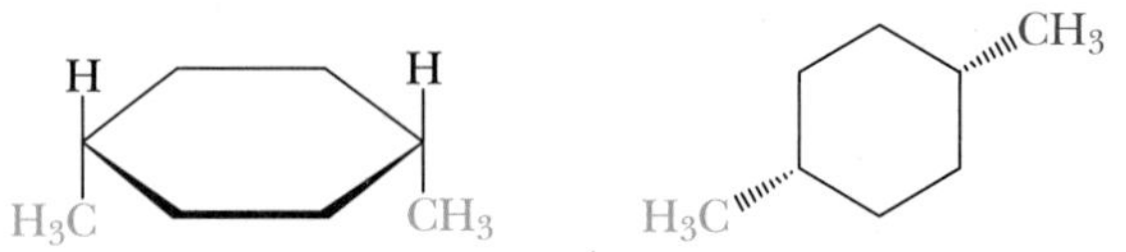

cis-1,4-Dimethylcyclohexane

We can also draw the *cis* and *trans* isomers of 1,4-dimethylcyclohexane as nonplanar chair conformations. In working with alternative chair conformations, it is helpful to remember that all groups axial on one chair are equatorial in the alternative chair, and vice versa. In one chair conformation of *trans*-1,4-dimethylcyclohexane, the two methyl groups are axial; in the alternative chair conformation, they are equatorial. Of these chair conformations, the one with both methyls equatorial is considerably more stable.

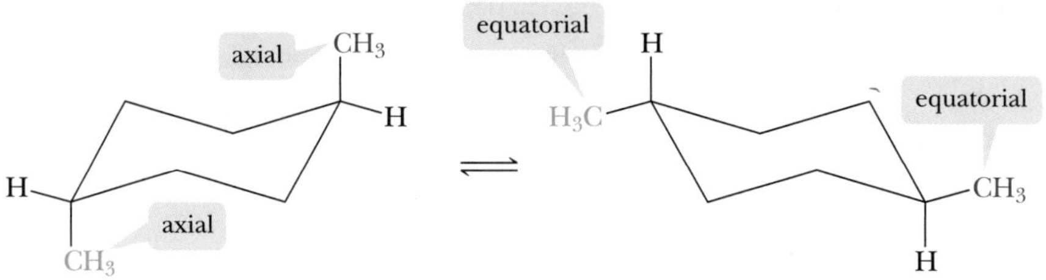

(less stable) (more stable)

trans-1,4-Dimethylcyclohexane

The alternative chair conformations of *cis*-1,4-dimethylcyclohexane are of equal energy. In each chair conformation, one methyl group is equatorial and the other is axial.

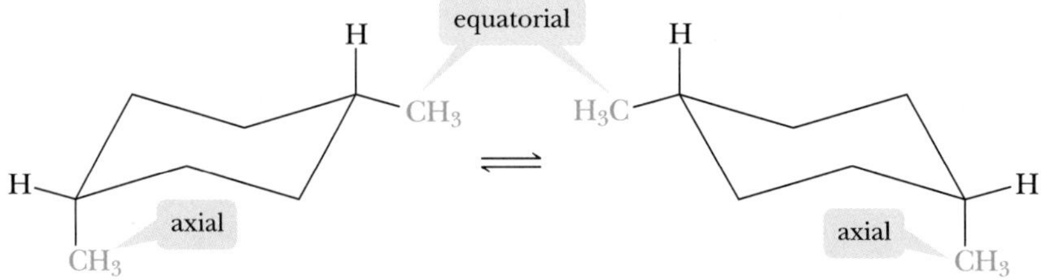

cis-1,4-Dimethylcyclohexane
(these conformations are of equal stability)

Example 3.11

Following is a chair conformation of 1,3-dimethylcyclohexane:

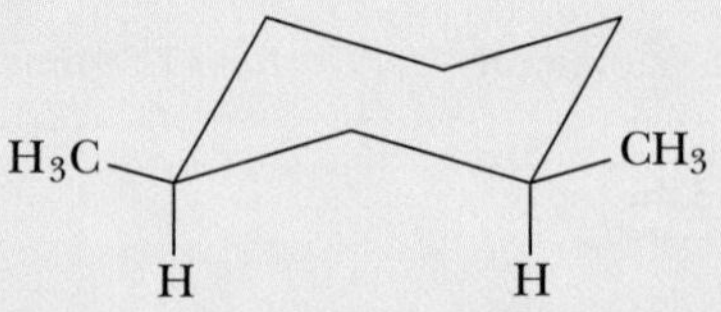

(a) Is this a chair conformation of *cis*-1,3-dimethylcyclohexane or of *trans*-1,3-dimethylcyclohexane?
(b) Draw the alternative chair conformation. Of the two chair conformations, which is the more stable?
(c) Draw a planar hexagon representation of the isomer shown in this example.

Strategy

Determine whether substituents are on the same or different sides of the ring to determine *cis* or *trans*. To perform a chair-to-chair interconversion, change the orientation of the rightmost and leftmost carbons in the chair conformation of cyclohexane. Remember that after such an interconversion, all prior axial substituents become equatorial and all prior equatorial substituents become axial. When converting to planar representations of cyclohexane, show substituents above the ring as coming out of the page (wedges) and those below the ring as going behind the page (dashes).

Solution

(a) The isomer shown is *cis*-1,3-dimethylcyclohexane; the two methyl groups are on the same side of the ring.

(b) H_3C CH_3 H H (more stable) ⇌ CH_3 CH_3 H H (less stable)

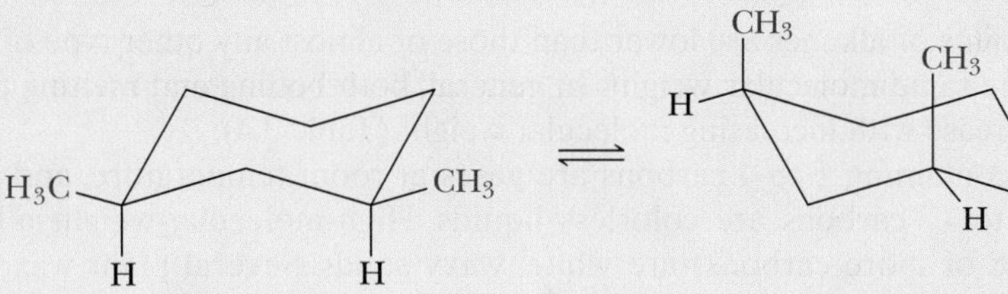

(c) H_3C H_3C H H or H_3C CH_3

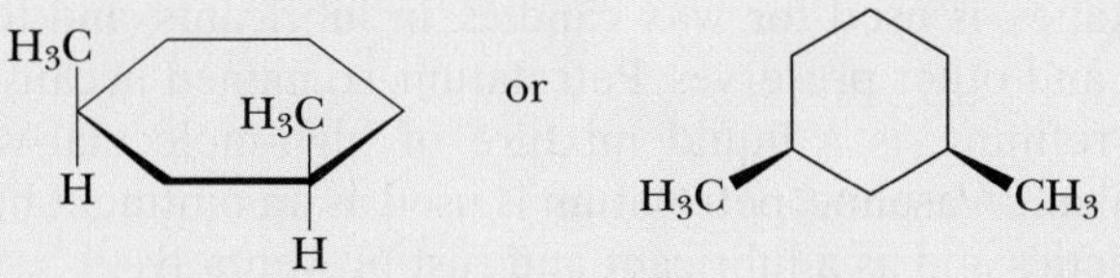

See problems 3.41–3.44, 3.46, 3.47

Problem 3.11

Following is a planar hexagon representation of one isomer of 1,2,4-trimethylcyclohexane. Draw the alternative chair conformations of this compound, and state which is the more stable.

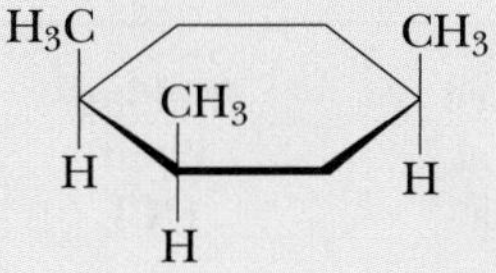

H_3C CH_3 CH_3 H H H

3.8 What Are the Physical Properties of Alkanes and Cycloalkanes?

The most important property of alkanes and cycloalkanes is their almost complete lack of polarity. As we saw in Section 1.2C, the difference in electronegativity between carbon and hydrogen is 2.5 − 2.1 = 0.4 on the Pauling scale, and given this small difference, we classify a C—H bond as nonpolar covalent. Therefore, alkanes are nonpolar compounds, and there are only weak interactions between their molecules.

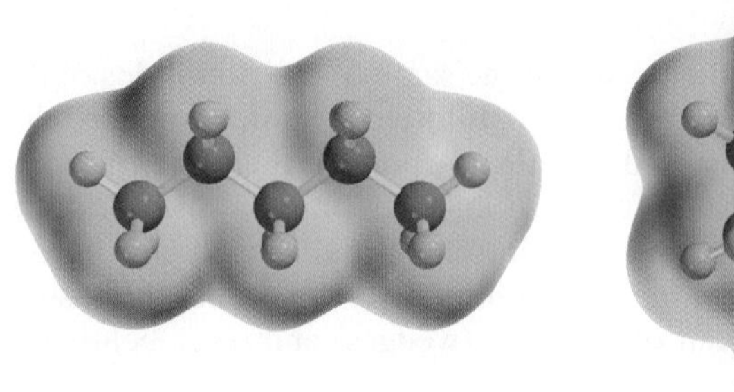

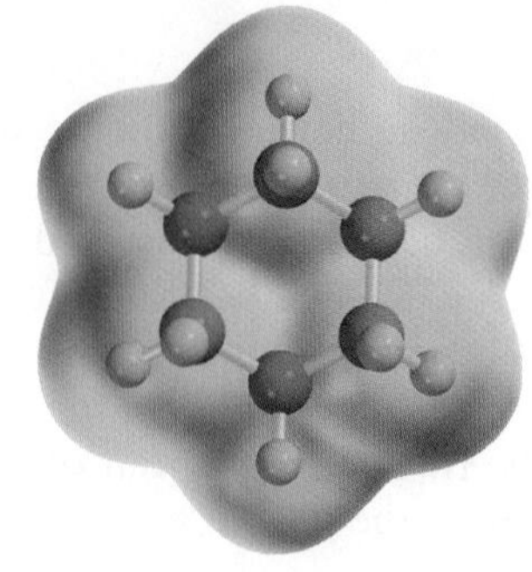

Pentane and cyclohexane. The electron density models show no evidence of any polarity in alkanes and cycloalkanes.

A. Boiling Points

The boiling points of alkanes are lower than those of almost any other type of compound with the same molecular weight. In general, both boiling and melting points of alkanes increase with increasing molecular weight (Table 3.4).

Alkanes containing 1 to 4 carbons are gases at room temperature, and those containing 5 to 17 carbons are colorless liquids. High-molecular-weight alkanes (those with 18 or more carbons) are white, waxy solids. Several plant waxes are high-molecular-weight alkanes. The wax found in apple skins, for example, is an unbranched alkane with the molecular formula $C_{27}H_{56}$. Paraffin wax, a mixture of high-molecular-weight alkanes, is used for wax candles, in lubricants, and to seal home canned jams, jellies, and other preserves. Petrolatum, so named because it is derived from petroleum refining, is a liquid mixture of high-molecular-weight alkanes. Sold as mineral oil and Vaseline, petrolatum is used as an ointment base in pharmaceuticals and cosmetics and as a lubricant and rust preventative.

B. Dispersion Forces and Interactions between Alkane Molecules

Methane is a gas at room temperature and atmospheric pressure. It can be converted to a liquid if cooled to −164°C and to a solid if further cooled to −182°C. The fact that methane (or any other compound, for that matter) can exist as a

TABLE 3.4 Physical Properties of Some Unbranched Alkanes

Name	Condensed Structural Formula	Melting Point (°C)	Boiling Point (°C)	*Density of Liquid (g/mL at 0°C)
methane	CH_4	−182	−164	(a gas)
ethane	CH_3CH_3	−183	−88	(a gas)
propane	$CH_3CH_2CH_3$	−190	−42	(a gas)
butane	$CH_3(CH_2)_2CH_3$	−138	0	(a gas)
pentane	$CH_3(CH_2)_3CH_3$	−130	36	0.626
hexane	$CH_3(CH_2)_4CH_3$	−95	69	0.659
heptane	$CH_3(CH_2)_5CH_3$	−90	98	0.684
octane	$CH_3(CH_2)_6CH_3$	−57	126	0.703
nonane	$CH_3(CH_2)_7CH_3$	−51	151	0.718
decane	$CH_3(CH_2)_8CH_3$	−30	174	0.730

*For comparison, the density of H_2O is 1 g/mL at 4°C.

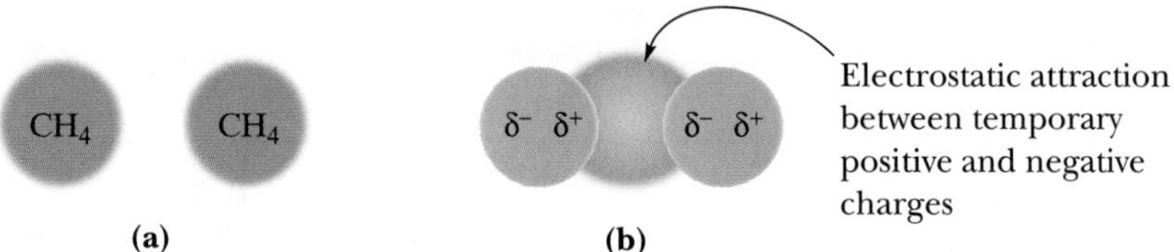

Figure 3.12
Dispersion forces. (a) The average distribution of electron density in a methane molecule is symmetrical, and there is no polarity. (b) Temporary polarization of one molecule induces temporary polarization in an adjacent molecule. Electrostatic attractions between temporary partial positive and partial negative charges are called *dispersion forces*.

liquid or a solid depends on the existence of forces of attraction between particles of the pure compound. Although the forces of attraction between particles are all electrostatic in nature, they vary widely in their relative strengths. The strongest attractive forces are between ions—for example, between Na^+ and Cl^- in NaCl (787 kJ/mol, 188 kcal/mol). Hydrogen bonding is a weaker attractive force (8–42 kJ/mol, 2–10 kcal/mol). We will have more to say about hydrogen bonding in Chapter 8 when we discuss the physical properties of alcohols—compounds containing polar O—H groups.

Dispersion forces (0.08–8 kJ/mol, 0.02–2 kcal/mol) are the weakest intermolecular attractive forces. It is the existence of dispersion forces that accounts for the fact that low-molecular-weight, nonpolar substances such as methane can be liquefied. When we convert methane from a liquid to a gas at −164°C, for example, the process of separating its molecules requires only enough energy to overcome the very weak dispersion forces.

To visualize the origin of dispersion forces, it is necessary to think in terms of instantaneous distributions of electron density rather than average distributions. Over time, the distribution of electron density in a methane molecule is symmetrical [Figure 3.12(a)], and there is no separation of charge. However, at any instant, there is a nonzero probability that the electron density is polarized (shifted) more toward one part of a methane molecule than toward another. This temporary polarization creates temporary partial positive and partial negative charges, which in turn induce temporary partial positive and negative charges in adjacent methane molecules [Figure 3.12(b)]. **Dispersion forces** are weak electrostatic attractive forces that occur between temporary partial positive and partial negative charges in adjacent atoms or molecules.

Dispersion forces Very weak intermolecular forces of attraction resulting from the interaction of temporary induced dipoles.

Because interactions between alkane molecules consist only of these very weak dispersion forces, the boiling points of alkanes are lower than those of almost any other type of compound with the same molecular weight. As the number of atoms and the molecular weight of an alkane increase, the strength of the dispersion forces among alkane molecules increases, and consequently, boiling points increase.

C. Melting Point and Density

The melting points of alkanes increase with increasing molecular weight. The increase, however, is not as regular as that observed for boiling points, because the ability of molecules to pack into ordered patterns of solids changes as the molecular size and shape change.

The average density of the alkanes listed in Table 3.4 is about 0.7 g/mL; that of higher-molecular-weight alkanes is about 0.8 g/mL. All liquid and solid alkanes are less dense than water (1.0 g/mL); therefore, they float on water.

D. Constitutional Isomers Have Different Physical Properties

Alkanes that are constitutional isomers are different compounds and have different physical properties. Table 3.5 lists the boiling points, melting points, and densities of the five constitutional isomers with the molecular formula C_6H_{14}. The boiling point of each of its branched-chain isomers is lower than that of hexane itself, and the more branching there is, the lower is the boiling point. These differences in boiling points

more surface area, an increase in dispersion forces, and a higher boiling point

Hexane

smaller surface area, a decrease in dispersion forces, and a lower boiling point

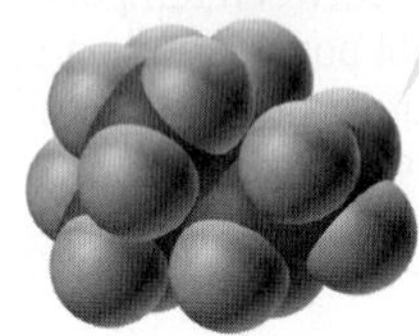

2,2-Dimethylbutane

TABLE 3.5 Physical Properties of the Isomeric Alkanes with the Molecular Formula C_6H_{14}

Name	Melting Point (°C)	Boiling Point (°C)	Density (g/mL)
hexane	−95	69	0.659
3-methylpentane	−6	64	0.664
2-methylpentane	−23	62	0.653
2,3-dimethylbutane	−129	58	0.662
2,2-dimethylbutane	−100	50	0.649

are related to molecular shape in the following way: The only forces of attraction between alkane molecules are dispersion forces. As branching increases, the shape of an alkane molecule becomes more compact, and its surface area decreases. As the surface area decreases, the strength of the dispersion forces decreases, and boiling points decrease. Thus, for any group of alkane constitutional isomers, it is usually observed that the least-branched isomer has the highest boiling point and the most-branched isomer has the lowest boiling point. The trend in melting points is less obvious, but as previously mentioned, it correlates with a molecule's ability to pack into ordered patterns of solids.

Example 3.12

Arrange the alkanes in each set in order of increasing boiling point:

(a) Butane, decane, and hexane
(b) 2-Methylheptane, octane, and 2,2,4-trimethylpentane

Strategy

When determining relative boiling points, remember that as the number of carbon atoms in the chain increases, the dispersion forces among molecules increase and the boiling points increase. Boiling point is also dependent on the degree of branching. For constitutional isomers, the most highly branched isomer has the smallest surface area and the lowest boiling point.

Solution

(a) All of the compounds are unbranched alkanes. As the number of carbon atoms in the chain increases, the dispersion forces among molecules increase, and the boiling points increase. Decane has the highest boiling point, butane the lowest:

Butane (bp −0.5°C) | Hexane (bp 69°C) | Decane (bp 174°C)

(b) These three alkanes are constitutional isomers with the molecular formula C_8H_{18}. Their relative boiling points depend on the degree of branching. 2,2,4-Trimethylpentane, the most highly branched isomer, has the smallest surface area and the lowest boiling point. Octane, the unbranched isomer, has the largest surface area and the highest boiling point.

2,2,4-Trimethylpentane (bp 99°C)   2-Methylheptane (bp 118°C)   Octane (bp 125°C)

See problem 3.49

Problem 3.12

Arrange the alkanes in each set in order of increasing boiling point:

(a) 2-Methylbutane, 2,2-dimethylpropane, and pentane
(b) 3,3-Dimethylheptane, 2,2,4-trimethylhexane, and nonane

3.9 What Are the Characteristic Reactions of Alkanes?

The most important chemical property of alkanes and cycloalkanes is their inertness. They are quite unreactive toward most reagents, a behavior consistent with the fact that they are nonpolar compounds containing only strong sigma bonds. Under certain conditions, however, alkanes and cycloalkanes do react, with oxygen, O_2. By far their most important reaction with oxygen is oxidation (combustion) to form carbon dioxide and water. The oxidation of saturated hydrocarbons is the basis for their use as energy sources for heat [natural gas, liquefied petroleum gas (LPG), and fuel oil] and power (gasoline, diesel fuel, and aviation fuel). Following are balanced equations for the complete combustion of methane, the major component of natural gas, and for propane, the major component of LPG:

$$\underset{\text{Methane}}{CH_4} + 2O_2 \longrightarrow CO_2 + 2H_2O \qquad \Delta H^\circ = -886\ \text{kJ/mol}\ (-212\ \text{kcal/mol})$$

$$\underset{\text{Propane}}{CH_3CH_2CH_3} + 5O_2 \longrightarrow 3CO_2 + 4H_2O \qquad \Delta H^\circ = -2{,}220\ \text{kJ/mol}\ (-530\ \text{kcal/mol})$$

3.10 What Are the Sources of Alkanes?

The three major sources of alkanes throughout the world are the fossil fuels: natural gas, petroleum, and coal. Fossil fuels account for approximately 90% of the total energy consumed in the United States. Nuclear electric power and hydroelectric power make up most of the remaining 10%. In addition, fossil fuels provide the bulk of the raw material for the organic chemicals consumed worldwide.

A. Natural Gas

Natural gas consists of approximately 90–95% methane, 5–10% ethane, and a mixture of other relatively low-boiling alkanes—chiefly propane, butane, and 2-methylpropane. The current widespread use of ethylene as the organic chemical industry's most important building block is largely the result of the ease with which ethane can be separated from natural gas and cracked into ethylene. Cracking is a process whereby a saturated hydrocarbon is converted into an unsaturated hydrocarbon plus H_2. Heating it in a furnace at 800 to 900°C for a fraction of a second cracks ethane. The global production of ethylene in the United States in 2005 was 107 billion kg (235 billion pounds), making it the number-one organic compound produced by the U.S. chemical industry, on a weight basis. The bulk of the ethylene produced is used to create organic polymers, as described in Chapter 17.

$$\underset{\text{Ethane}}{CH_3CH_3} \xrightarrow[\text{(thermal cracking)}]{800\text{–}900°C} \underset{\text{Ethylene}}{CH_2{=}CH_2} + H_2$$

B. Petroleum

A petroleum refinery. *(K. Straiton/Photo Researchers, Inc.)*

Petroleum is a thick, viscous liquid mixture of literally thousands of compounds, most of them hydrocarbons, formed from the decomposition of marine plants and animals. Petroleum and petroleum-derived products fuel automobiles, aircraft, and trains. They provide most of the greases and lubricants required for the machinery of our highly industrialized society. Furthermore, petroleum, along with natural gas, provides close to 90% of the organic raw materials used in the synthesis and manufacture of synthetic fibers, plastics, detergents, drugs, dyes, and a multitude of other products.

It is the task of a petroleum refinery to produce usable products, with a minimum of waste, from the thousands of different hydrocarbons in this liquid mixture. The various physical and chemical processes for this purpose fall into two broad categories: separation processes, which separate the complex mixture into various fractions, and re-forming processes, which alter the molecular structure of the hydrocarbon components themselves.

The fundamental separation process utilized in refining petroleum is fractional distillation (Figure 3.13). Practically all crude oil that enters a refinery goes to distillation units, where it is heated to temperatures as high as 370 to 425°C and separated into fractions. Each fraction contains a mixture of hydrocarbons that boils within a particular range:

1. Gases boiling below 20°C are taken off at the top of the distillation column. This fraction is a mixture of low-molecular-weight hydrocarbons, predominantly propane, butane, and 2-methylpropane, substances that can be liquefied under pressure at room temperature. The liquefied mixture, known as liquefied petroleum gas (LPG), can be stored and shipped in metal tanks and is a convenient source of gaseous fuel for home heating and cooking.
2. Naphthas, bp 20 to 200°C, are a mixture of C_5 to C_{12} alkanes and cycloalkanes. Naphthas also contain small amounts of benzene, toluene, xylene, and other aromatic hydrocarbons (Chapter 9). The light naphtha fraction, bp 20 to 150°C, is the source of straight-run gasoline and averages approximately 25% of crude petroleum. In a sense, naphthas are the most valuable distillation fractions, because they are useful not only as fuel, but also as sources of raw materials for the organic chemical industry.
3. Kerosene, bp 175 to 275°C, is a mixture of C_9 to C_{15} hydrocarbons.
4. Fuel oil, bp 250 to 400°C, is a mixture of C_{15} to C_{18} hydrocarbons. Diesel fuel is obtained from this fraction.
5. Lubricating oil and heavy fuel oil distill from the column at temperatures above 350°C.
6. Asphalt is the black, tarry residue remaining after the removal of the other volatile fractions.

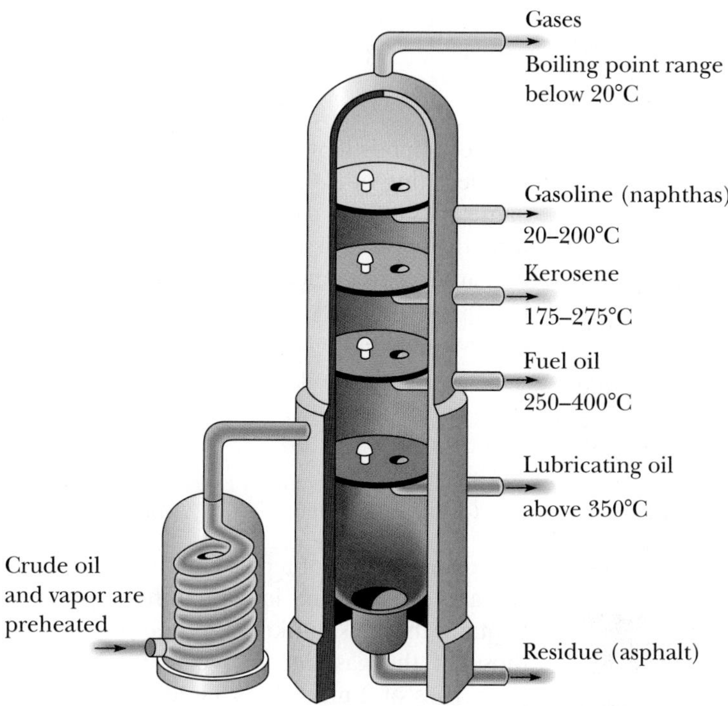

Figure 3.13
Fractional distillation of petroleum. The lighter, more volatile fractions are removed from higher up the column and the heavier, less volatile fractions from lower down.

The two most common re-forming processes are cracking, illustrated by the thermal conversion of ethane to ethylene (Section 3.10A), and catalytic re-forming, illustrated by the conversion of hexane first to cyclohexane and then to benzene:

$$CH_3CH_2CH_2CH_2CH_2CH_3 \xrightarrow[-H_2]{\text{catalyst}} \text{Cyclohexane} \xrightarrow[-3H_2]{\text{catalyst}} \text{Benzene}$$

Hexane　　Cyclohexane　　Benzene

C. Coal

To understand how coal can be used as a raw material for the production of organic compounds, it is necessary to discuss synthesis gas. Synthesis gas is a mixture of carbon monoxide and hydrogen in varying proportions, depending on the means by which it is manufactured. Synthesis gas is prepared by passing steam over coal. It is also prepared by the partial oxidation of methane with oxygen.

$$\underset{\text{Coal}}{C} + H_2O \xrightarrow{\text{heat}} CO + H_2$$

$$\underset{\text{Methane}}{CH_4} + \frac{1}{2}O_2 \xrightarrow{\text{catalyst}} CO + 2H_2$$

Two important organic compounds produced today almost exclusively from carbon monoxide and hydrogen are methanol and acetic acid. In the production of

CHEMICAL CONNECTIONS 3B

Octane Rating: What Those Numbers at the Pump Mean

Gasoline is a complex mixture of C_6 to C_{12} hydrocarbons. The quality of gasoline as a fuel for internal combustion engines is expressed in terms of an *octane rating*. Engine knocking occurs when a portion of the air–fuel mixture explodes prematurely (usually as a result of heat developed during compression) and independently of ignition by the spark plug. Two compounds were selected as reference fuels. One of these, 2,2,4-trimethylpentane (isooctane), has very good antiknock properties (the fuel–air mixture burns smoothly in the combustion chamber) and was assigned an octane rating of 100. (The name *isooctane* is a trivial name; its only relation to the name 2,2,4-trimethylpentane is that both names show eight carbon atoms.) Heptane, the other reference compound, has poor antiknock properties and was assigned an octane rating of 0.

Typical octane ratings of commonly available gasolines. *(Charles D. Winters)*

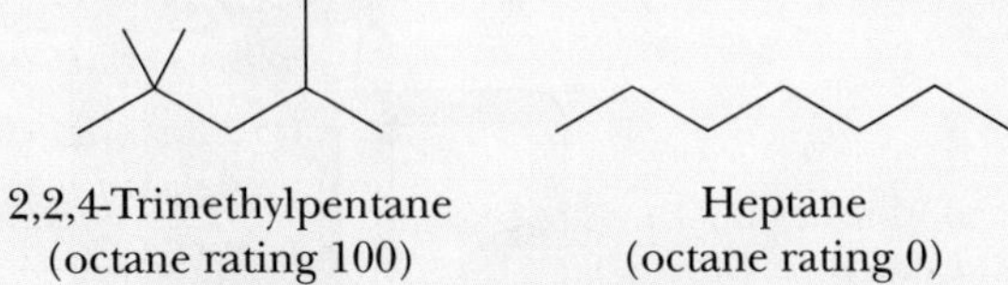

2,2,4-Trimethylpentane (octane rating 100)   Heptane (octane rating 0)

The octane rating of a particular gasoline is that percentage of isooctane in a mixture of isooctane and heptane that has antiknock properties equivalent to those of the gasoline. For example, the antiknock properties of 2-methylhexane are the same as those of a mixture of 42% isooctane and 58% heptane; therefore, the octane rating of 2-methylhexane is 42. Octane itself has an octane rating of −20, which means that it produces even more engine knocking than heptane. Ethanol, the additive to gasohol, has an octane rating of 105. Benzene and toluene have octane ratings of 106 and 120, respectively.

QUESTION

Which would you expect to have a higher boiling point, octane or isooctane (2,2,4-trimethylpentane)?

methanol, the ratio of carbon monoxide to hydrogen is adjusted to 1:2, and the mixture is passed over a catalyst at elevated temperature and pressure:

$$CO + 2H_2 \xrightarrow{\text{catalyst}} \underset{\text{methanol}}{CH_3OH}$$

The treatment of methanol, in turn, with carbon monoxide over a different catalyst gives acetic acid:

$$\underset{\text{Methanol}}{CH_3OH} + CO \xrightarrow{\text{catalyst}} \underset{\text{Acetic acid}}{CH_3\overset{\overset{\displaystyle O}{\|}}{C}OH}$$

Because the processes for making methanol and acetic acid directly from carbon monoxide are commercially proven, it is likely that the decades ahead will see the development of routes to other organic chemicals from coal via methanol.

Key Terms and Concepts

1° Carbon (p. 78)
2° Carbon (p. 78)
3° Carbon (p. 78)
4° Carbon (p. 78)
aliphatic hydrocarbon (p. 69)
alkane (p. 69)
alkyl group (p. 76)
angle strain (p. 84)
asphalt (p. 98)
axial bond (p. 85)
axial–axial (diaxial) interactions in cyclohexanes (p. 88)
boat conformation of cyclohexane (p. 85)
chair conformation of cyclohexane (p. 85)
cis (p. 89)
***cis–trans* isomerism in cycloalkanes** (p. 89)
conformation (p. 82)
constitutional isomers (p. 72)
cycloalkane (p. 79)
dispersion forces (p. 95)
eclipsed conformation (p. 83)
envelope conformation of cyclopentane (p. 84)
equatorial bond (p. 85)
hydrocarbon (p. 69)
International Union of Pure and Applied Chemistry (IUPAC) (p. 75)
IUPAC system of nomenclature (p. 75)
kerosene (p. 98)
line-angle formula (p. 70)
liquefied petroleum gas (LPG) (p. 98)
naphthas (p. 98)
natural gas (p. 97)
Newman projection (p. 83)
octane number (p. 100)
oxidation (combustion) of alkanes (p. 97)
parent chain (p. 76)
petroleum (p. 98)
polarity of alkane molecules (p. 93)
prefix, infix, and suffix (p. 81)
r- (p. 76)
relative stability of axial versus equatorial-substituted cyclohexanes (p. 88)
saturated hydrocarbon (p. 69)
staggered conformation (p. 83)
steric strain (p. 86)
torsional strain (p. 83)
trans (p. 89)
unsaturated hydrocarbon (p. 70)

Summary of Key Questions

3.1 What Are Alkanes?

- A hydrocarbon is a compound that contains only carbon and hydrogen. An alkane is a saturated hydrocarbon and contains only single bonds. Alkanes have the general formula C_nH_{2n+2}.

3.2 What Is Constitutional Isomerism in Alkanes?

- Constitutional isomers have the same molecular formula but a different connectivity (a different order of attachment) of their atoms.

3.3 How Do We Name Alkanes?

- Alkanes are named according to a set of rules developed by the International Union of Pure and Applied Chemistry (IUPAC).
- A carbon atom is classified as primary (1°), secondary (2°), tertiary (3°), or quaternary (4°), depending on the number of carbon atoms bonded to it.
- A hydrogen atom is classified as primary (1°), secondary (2°), or tertiary (3°), depending on the type of carbon to which it is bonded.

3.4 What Are Cycloalkanes?

- A cycloalkane is an alkane that contains carbon atoms bonded to form a ring.
- To name a cycloalkane, prefix the name of the open-chain hydrocarbon with "cyclo."
- Five-membered rings (cyclopentanes) and six-membered rings (cyclohexanes) are especially abundant in the biological world.

3.5 What Is the IUPAC System of Nomenclature?

- The IUPAC system is a general system of nomenclature. The IUPAC name of a compound consists of three parts:

  (1) A prefix that indicates the number of carbon atoms in the parent chain,

  (2) An infix that indicates the nature of the carbon–carbon bonds in the parent chain, and

  (3) A suffix that indicates the class to which the compound belongs.

- Substituents derived from alkanes by the removal of a hydrogen atom are called alkyl groups and are given the symbol R. The name of an alkyl group is formed by dropping the suffix -ane from the name of the parent alkane and adding -yl in its place.

3.6 What Are the Conformations of Alkanes and Cycloalkanes?

- A conformation is any three-dimensional arrangement of the atoms of a molecule that results from rotation about a single bond.
- One convention for showing conformations is the Newman projection. Staggered conformations are lower in energy (more stable) than eclipsed conformations.
- There are three types of molecular strain:

  Torsional strain (also called eclipsed interaction strain) that results when nonbonded atoms separated by three bonds are forced from a staggered conformation to an eclipsed conformation

  Angle strain that results when a bond angle in a molecule is either expanded or compressed compared with its optimal values, and

  Steric strain (also called nonbonded interaction strain) that results when nonbonded atoms separated by four or more bonds are forced abnormally close to each other—that is, when they are forced closer than their atomic (contact) radii would otherwise allow.

- Cyclopentanes, cyclohexanes, and all larger cycloalkanes exist in dynamic equilibrium between a set of puckered conformations. The lowest energy conformation of cyclopentane is an envelope conformation. The lowest energy conformations of cyclohexane are two interconvertible chair conformations. In a chair conformation, six bonds are axial and six are equatorial. Bonds axial in one chair are equatorial in the alternative chair, and vice versa. A boat conformation is higher in energy than chair conformations. The more stable conformation of a substituted cyclohexane is the one that minimizes axial–axial interactions.

3.7 What Is *Cis–Trans* Isomerism in Cycloalkanes?

- *Cis–trans* isomers have the same molecular formula and the same order of attachment of atoms, but arrangements of atoms in space that cannot be interconverted by rotation about single bonds. *Cis* means that substituents are on the same side of the ring; *trans* means that they are on opposite sides of the ring. Most cycloalkanes with substituents on two or more carbons of the ring show *cis–trans* isomerism.

3.8 What Are the Physical Properties of Alkanes and Cycloalkanes?

- Alkanes are nonpolar compounds, and the only forces of attraction between their molecules are dispersion forces, which are weak electrostatic interactions between temporary partial positive and negative charges of atoms or molecules. Low-molecular-weight alkanes, such as methane, ethane, and propane, are gases at room temperature and atmospheric pressure.
- Higher-molecular-weight alkanes, such as those in gasoline and kerosene, are liquids.
- Very high-molecular-weight alkanes, such as those in paraffin wax, are solids.
- Among a set of alkane constitutional isomers, the least branched isomer generally has the highest boiling point; the most branched isomer generally has the lowest boiling point.

3.9 What Are the Characteristic Reactions of Alkanes?

- The most important chemical property of alkanes and cycloalkanes is their inertness. Because they are nonpolar compounds containing only strong sigma bonds, they are quite unreactive toward most reagents.
- By far, their most important reaction is combustion to form carbon dioxide and water. The oxidation of saturated hydrocarbons is the basis for their use as energy sources for heat and power.

3.10 What Are the Sources of Alkanes?

- Natural gas consists of 90–95% methane with lesser amounts of ethane and other lower-molecular-weight hydrocarbons.
- Petroleum is a liquid mixture of literally thousands of different hydrocarbons.
- Synthesis gas, a mixture of carbon monoxide and hydrogen, can be derived from natural gas and coal.

Quick Quiz

Answer true or false to the following questions to assess your general knowledge of the concepts in this chapter. If you have difficulty with any of them, you should review the appropriate section in the chapter (shown in parentheses) before attempting the more challenging end-of-chapter problems.

1. Combustion of alkanes is an endothermic process. (3.9)
2. All alkanes that are liquid at room temperature are more dense than water. (3.8)
3. The two main sources of alkanes the world over are petroleum and natural gas. (3.10)
4. There are four alkyl groups with the molecular formula C_4H_9. (3.3)
5. Sets of constitutional isomers have the same molecular formula and the same physical properties. (3.2)
6. A hydrocarbon is composed of only carbon and hydrogen. (3.1)
7. Cycloalkanes are saturated hydrocarbons. (3.4)
8. The products of complete combustion of an alkane are carbon dioxide and water. (3.9)
9. Alkanes and cycloalkanes show *cis–trans* isomerism. (3.6)
10. Alkenes and alkynes are unsaturated hydrocarbons. (3.1)
11. There are two constitutional isomers with the molecular formula C_4H_{10}. (3.2)
12. Hexane and cyclohexane are constitutional isomers. (3.4)
13. The propyl and isopropyl groups are constitutional isomers. (3.3)
14. There are five constitutional isomers with the molecular formula C_5H_{12}. (3.2)
15. Boiling points among alkanes with unbranched carbon chains increase as the number of carbons in the chain increases. (3.8)
16. In a cyclohexane ring, if an axial bond is above the plane of the ring on a particular carbon atom, axial bonds on the two adjacent carbons are below the plane of the ring. (3.5)
17. Fractional distillation of petroleum separates hydrocarbons based on their melting points (3.10)
18. Among alkane constitutional isomers, the least branched isomer generally has the lowest boiling point. (3.8)
19. The parent name of a cycloalkane is the name of the unbranched alkane with the same number of carbon atoms as are in the cycloalkane ring. (3.4)
20. Octane and 2,2,4-trimethylpentane are constitutional isomers and have the same octane number. (3.10)
21. Liquid alkanes and cycloalkanes are soluble in each other. (3.8)
22. Alkanes and cycloalkanes are insoluble in water. (3.8)
23. The more stable chair conformation of a substituted cyclohexane has the greater number of substituents in equatorial positions. (3.5)
24. The parent name of an alkane is the name of the longest chain of carbon atoms. (3.3)
25. Alkanes are saturated hydrocarbons. (3.1)
26. The general formula of an alkane is C_nH_{2n}, where n is the number of carbon atoms in the alkane. (3.1)
27. The octane number of a particular gasoline is the number of grams of octane per liter. (3.10)

28. *Cis* and *trans* isomers have the same molecular formula, the same connectivity, and the same physical properties. (3.8)

29. A *cis* isomer of a disubstituted cycloalkane can be converted to a *trans* isomer by rotation about an appropriate carbon–carbon single bond. (3.6)

30. All cycloalkanes with two substituents on the ring show *cis–trans* isomerism. (3.6)

31. In all conformations of ethane, propane, butane and higher alkanes, all C—C—C and C-C-H bond angles are approximately 109.5°. (3.5)

32. Conformations have the same molecular formula and the same connectivity, but differ in the three-dimensional arrangement of their atoms in space. (3.5)

33. Constitutional isomers have the same molecular formula and the same connectivity of their atoms. (3.2)

Answers: (1) F (2) F (3) T (4) T (5) F (6) T (7) T (8) T (9) F (10) T (11) T (12) F (13) T (14) F (15) T (16) T (17) F (18) F (19) T (20) F (21) T (22) T (23) T (24) T (25) T (26) F (27) F (28) F (29) F (30) F (31) T (32) T (33) F

Key Reactions

1. Oxidation of Alkanes (Section 3.9)

The oxidation of alkanes to carbon dioxide and water is the basis for their use as energy sources of heat and power:

$$CH_3CH_2CH_3 + 5O_2 \longrightarrow 3CO_2 + 4H_2O + \text{energy}$$

Problems

A problem marked with an asterisk indicates an applied "real world" problem. Answers to problems whose numbers are printed in blue are given in Appendix D.

Section 3.1 Structure of Alkanes

3.13 For each condensed structural formula, write a line-angle formula:

(a) $CH_3CH_2CH(CH_2CH_3)CH(CH(CH_3)_2)CH_2CH(CH_3)CH_3$ — i.e., $CH_3CH_2CHCHCH_2CHCH_3$ with CH_2CH_3 on C3, $CH(CH_3)_2$ on C4, and CH_3 on C6

(b) $CH_3C(CH_3)_2CH_3$ — i.e., CH_3CCH_3 with CH_3 above and below the central C

(c) $(CH_3)_2CHCH(CH_3)_2$

(d) $CH_3CH_2C(CH_2CH_3)_2CH_2CH_3$ — i.e., $CH_3CH_2CCH_2CH_3$ with CH_2CH_3 above and below the central C

(e) $(CH_3)_3CH$

(f) $CH_3(CH_2)_3CH(CH_3)_2$

3.14 Write a condensed structural formula and the molecular formula of each alkane:

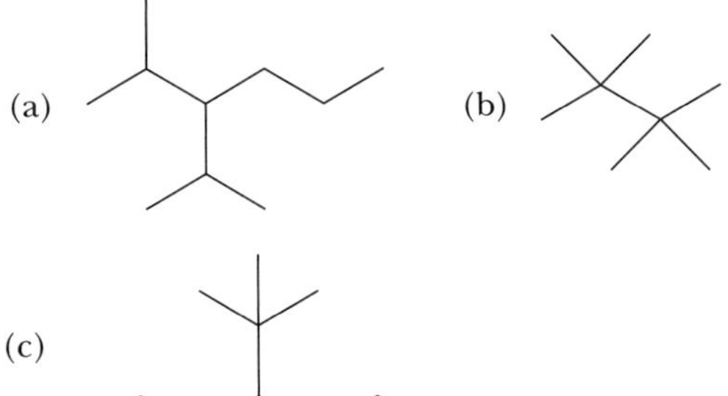

(a) (b) (c)

3.15 For each of the following condensed structural formulas, provide an even more abbreviated formula, using parentheses and subscripts:

(a) $CH_3CH_2CH_2CH_2CH_2CHCH_3$ with CH_3 on the second-to-last C

(b) $HCCH_2CH_2CH_3$ with $CH_2CH_2CH_3$ above and below the first C

(c) $CH_3CCH_2CH_2CH_2CH_2CH_3$ with $CH_2CH_2CH_3$ above and below the second C

Section 3.2 Constitutional Isomerism

3.16 Which statements are true about constitutional isomers?

(a) They have the same molecular formula.

(b) They have the same molecular weight.

(c) They have the same order of attachment of atoms.

(d) They have the same physical properties.

3.17 Each member of the following set of compounds is an alcohol; that is, each contains an —OH (hydroxyl group, Section 1.7A): **(See Example 3.1)**

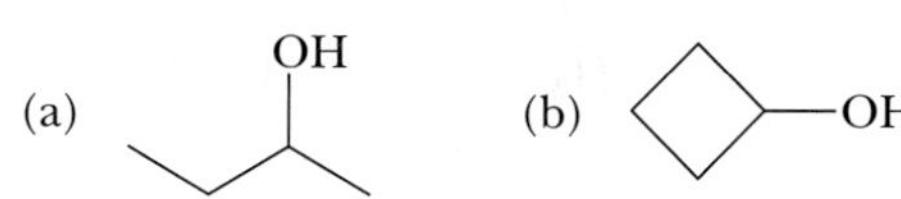

(c) 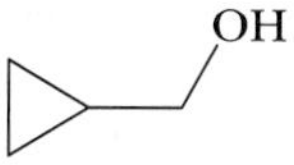 OH (d) 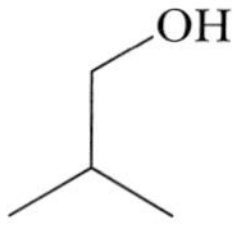 OH

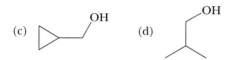

(e) 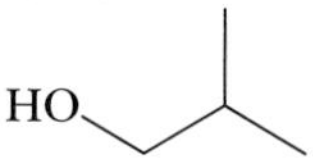 HO (f)  OH

(g) 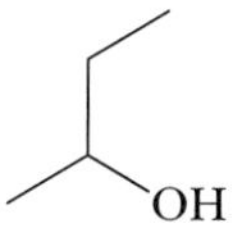 OH (h) 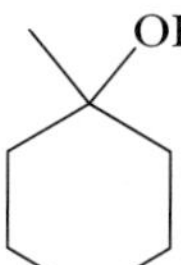 OH

Which structural formulas represent (1) the same compound, (2) different compounds that are constitutional isomers, or (3) different compounds that are not constitutional isomers?

3.18 Each member of the following set of compounds is an amine; that is, each contains a nitrogen atom bonded to one, two, or three carbon groups (Section 1.7B): **(See Example 3.1)**

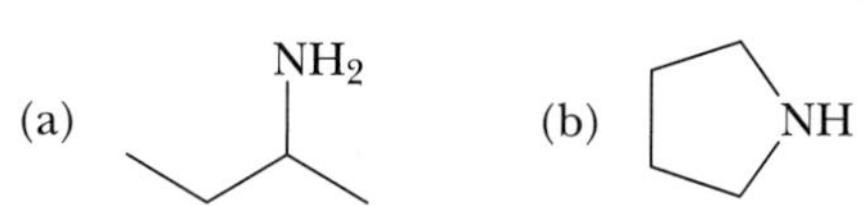

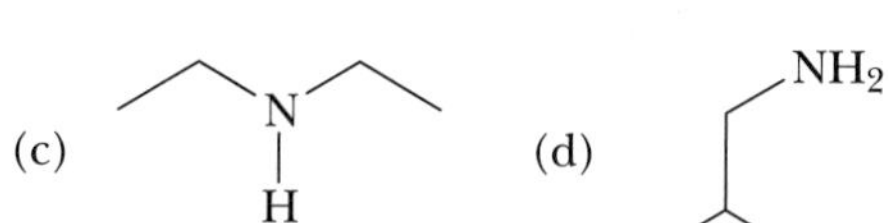

(e) 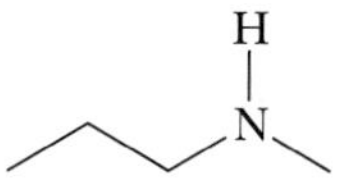 H N (f) 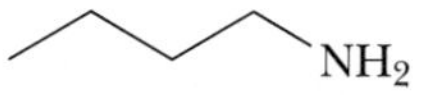 NH_2

(g) 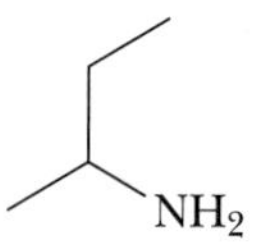 NH_2 (h) N H

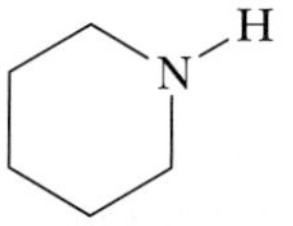

Which structural formulas represent (1) the same compound, (2) different compounds that are constitutional isomers, or (3) different compounds that are not constitutional isomers?

3.19 Each member of the following set of compounds is either an aldehyde or a ketone (Section 1.7C): **(See Example 3.1)**

(a) 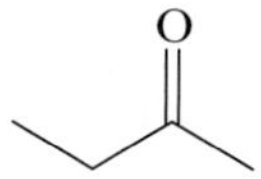 O (b) 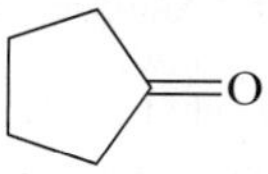 O

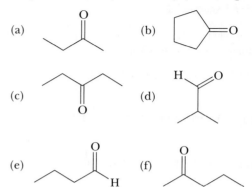

(c) 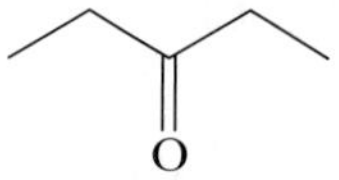 O (d) 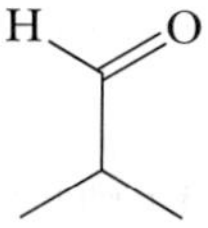 H O

(e) 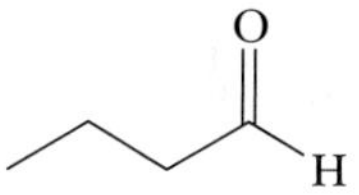 O H (f) 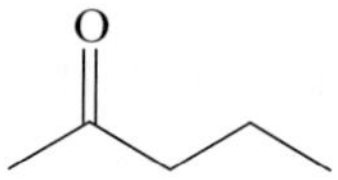 O

(g) 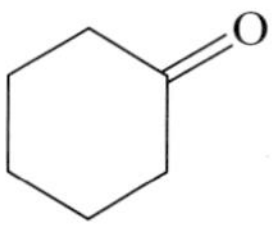 O (h) O H

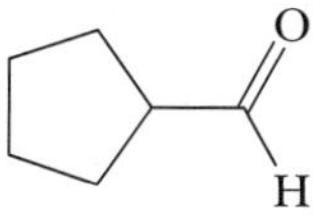

Which structural formulas represent (1) the same compound, (2) different compounds that are constitutional isomers, or (3) different compounds that are not constitutional isomers?

3.20 For each pair of compounds, tell whether the structural formulas shown represent **(See Example 3.1)**

(1) the same compound,

(2) different compounds that are constitutional isomers, or

(3) different compounds that are not constitutional isomers:

(a) 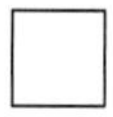 and

(b) 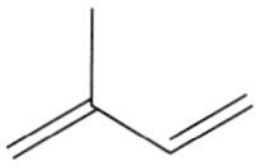 and

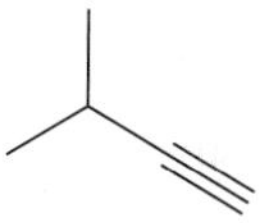

(c) 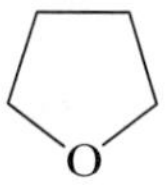 O and O

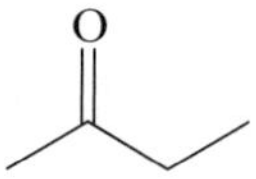

(d)  O and O

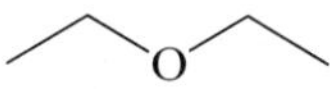

(e) 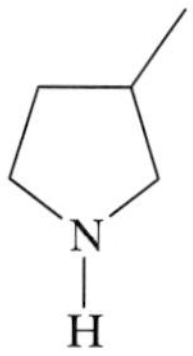 N H and NH_2

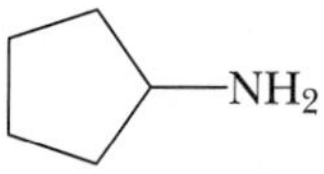

(f)  O and OH

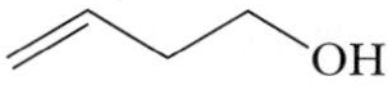

3.21 Name and draw line-angle formulas for the nine constitutional isomers with the molecular formula C_7H_{16}. **(See Example 3.2)**

3.22 Tell whether the compounds in each set are constitutional isomers: **(See Example 3.1)**

(a) CH_3CH_2OH and CH_3OCH_3

(b) $CH_3\overset{\overset{\displaystyle O}{\|}}{C}CH_3$ and $CH_3CH_2\overset{\overset{\displaystyle O}{\|}}{C}H$

(c) $CH_3\overset{\overset{\displaystyle O}{\|}}{C}OCH_3$ and $CH_3CH_2\overset{\overset{\displaystyle O}{\|}}{C}OH$

(d) $CH_3\overset{\overset{\displaystyle OH}{|}}{C}HCH_2CH_3$ and $CH_3\overset{\overset{\displaystyle O}{\|}}{C}CH_2CH_3$

(e) [cyclopentane] and $CH_3CH_2CH_2CH_2CH_3$

(f) [cyclopentane] and $CH_2{=}CHCH_2CH_2CH_3$

3.23 Draw line-angle formulas for **(See Example 3.2)**

(a) The four alcohols with the molecular formula $C_4H_{10}O$.

(b) The two aldehydes with the molecular formula C_4H_8O.

(c) The one ketone with the molecular formula C_4H_8O.

(d) The three ketones with the molecular formula $C_5H_{10}O$.

(e) The four carboxylic acids with the molecular formula $C_5H_{10}O_2$.

Sections 3.2 & 3.3 Nomenclature of Alkanes and Cycloalkanes

3.24 Write IUPAC names for these alkanes and cycloalkanes: **(See Examples 3.3, 3.5)**

(a) $CH_3\underset{\underset{\displaystyle CH_3}{|}}{C}HCH_2CH_2CH_3$ (b) $CH_3\underset{\underset{\displaystyle CH_3}{|}}{C}HCH_2CH_2\underset{\underset{\displaystyle CH_3}{|}}{C}HCH_3$

(c) $CH_3(CH_2)_4\underset{\underset{\displaystyle CH_2CH_3}{|}}{C}HCH_2CH_3$ (d)

(e) (f)

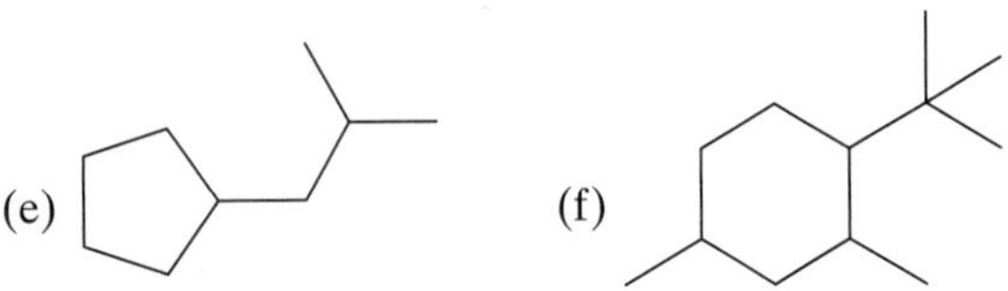

3.25 Write line-angle formulas for these alkanes: **(See Examples 3.3, 3.5)**

(a) 2,2,4-Trimethylhexane

(b) 2,2-Dimethylpropane

(c) 3-Ethyl-2,4,5-trimethyloctane

(d) 5-Butyl-2,2-dimethylnonane

(e) 4-Isopropyloctane

(f) 3,3-Dimethylpentane

(g) *trans*-1,3-Dimethylcyclopentane

(h) *cis*-1,2-Diethylcyclobutane

***3.26** Following is the structure of limonene, the chemical component of oranges that is partly responsible for their citrus scent. Draw the hydrogens present in limonene and classify those bonded to sp^3 hybridized carbons as 1°, 2°, or 3°. **(See Example 3.4)**

Limonene

***3.27** Following is the structure of Germacrene A, a hydrocarbon synthesized in plants and studied for its insecticidal properties. Classify each of the sp^3 hybridized carbons on Germacrene A as 1°, 2°, 3°, or 4°. **(See Example 3.4)**

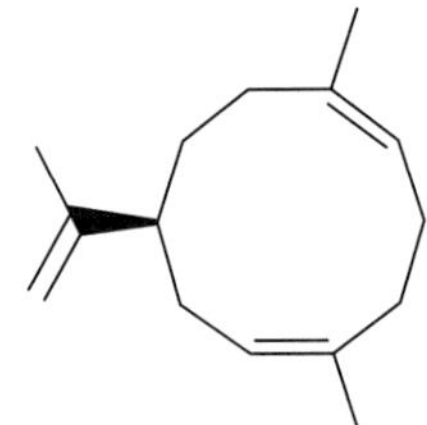

Germacrene A

3.28 Explain why each of the following names is an incorrect IUPAC name and write the correct IUPAC name for the intended compound: **(See Examples 3.3, 3.5)**

(a) 1,3-Dimethylbutane

(b) 4-Methylpentane

(c) 2,2-Diethylbutane

(d) 2-Ethyl-3-methylpentane

(e) 2-Propylpentane

(f) 2,2-Diethylheptane

(g) 2,2-Dimethylcyclopropane

(h) 1-Ethyl-5-methylcyclohexane

3.29 Draw a structural formula for each compound: **(See Example 3.6)**

(a) Ethanol (b) Ethanal
(c) Ethanoic acid (d) Butanone
(e) Butanal (f) Butanoic acid
(g) Propanal (h) Cyclopropanol
(i) Cyclopentanol (j) Cyclopentene
(k) Cyclopentanone

3.30 Write the IUPAC name for each compound: **(See Example 3.6)**

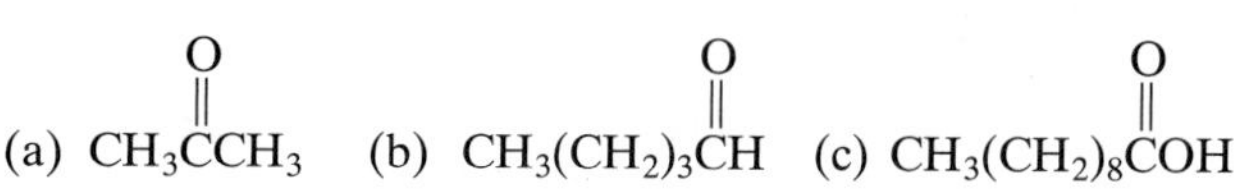

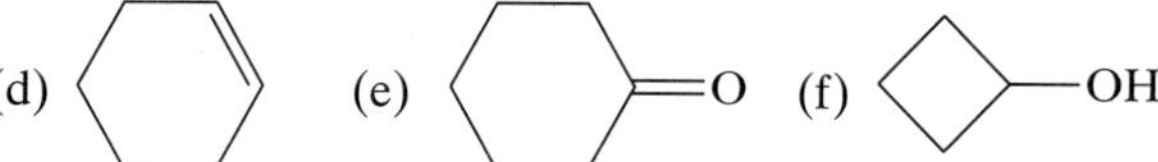

Section 3.6 Conformations of Alkanes and Cycloalkanes

3.31 How many different staggered conformations are there for 2-methylpropane? How many different eclipsed conformations are there? **(See Example 3.6)**

3.32 Looking along the bond between carbons 2 and 3 of butane, there are two different staggered conformations and two different eclipsed conformations. Draw Newman projections of each, and arrange them in order from the most stable conformation to the least stable conformation. **(See Example 3.7)**

3.33 Explain why each of the following Newman projections might not represent the most stable conformation of that molecule:

(a) 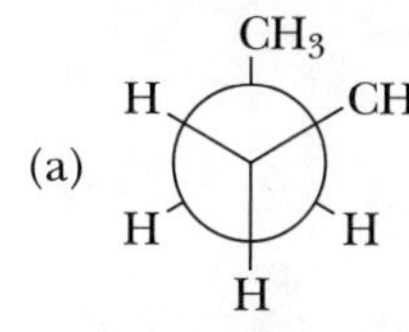

(b) 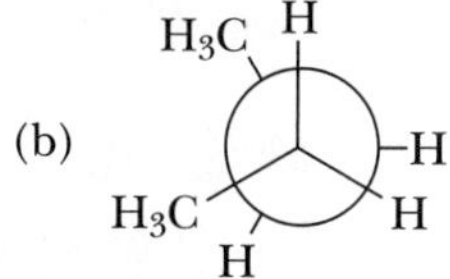

(c) 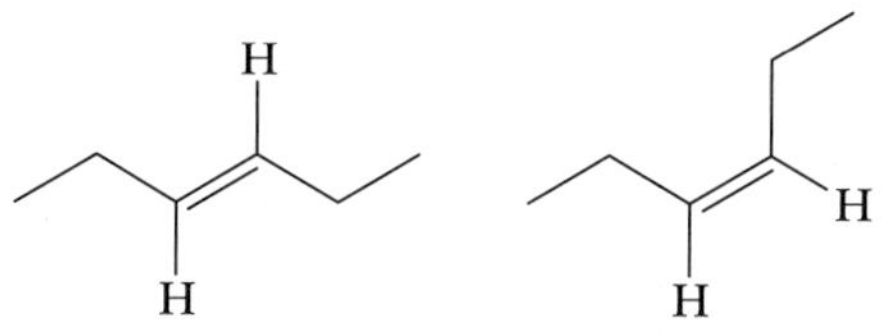

3.34 Explain why the following are not different conformations of 3-hexene:

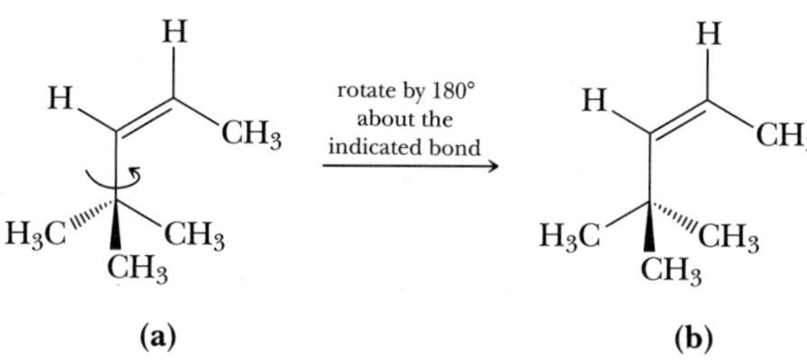

3.35 Which of the following two conformations is the more stable? (*Hint:* Use molecular models to compare structures):

3.36 Determine whether the following pairs of structures in each set represent the same molecule or constitutional isomers, and if they are the same molecule, determine whether they are in the same or different conformations:

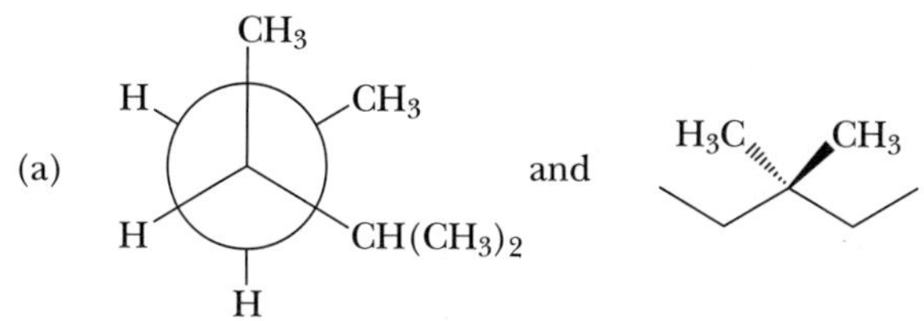

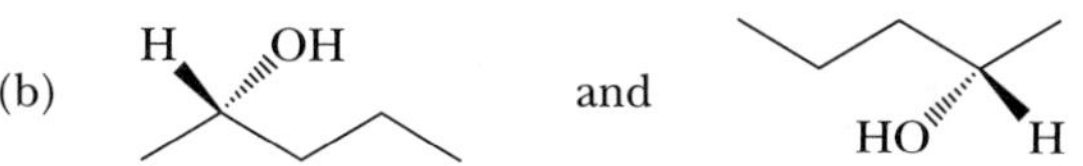

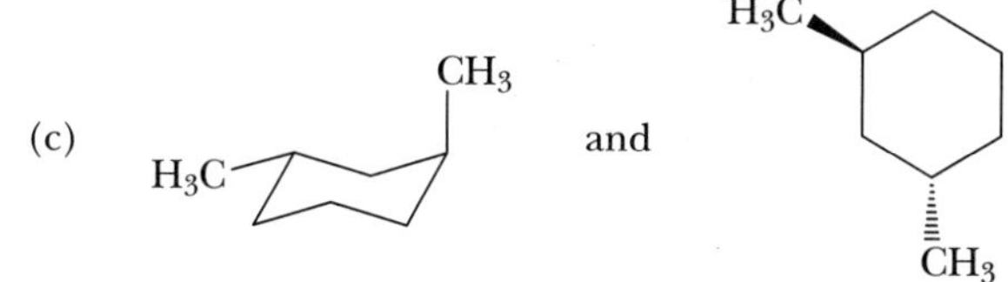

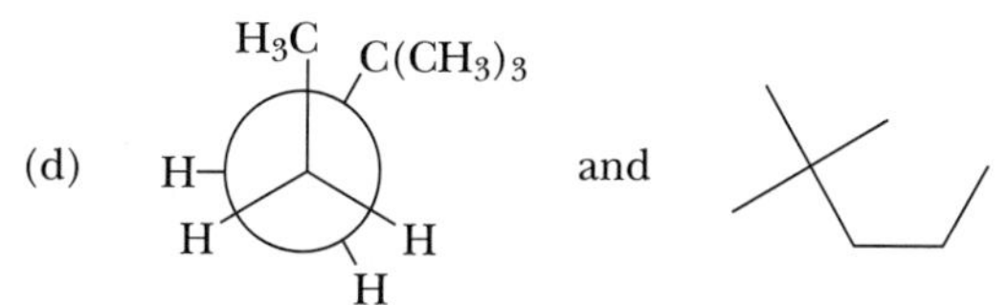

3.37 Draw Newman projections for the most stable conformation of each of the following compounds looking down the indicated bond. **(See Example 3.7)**

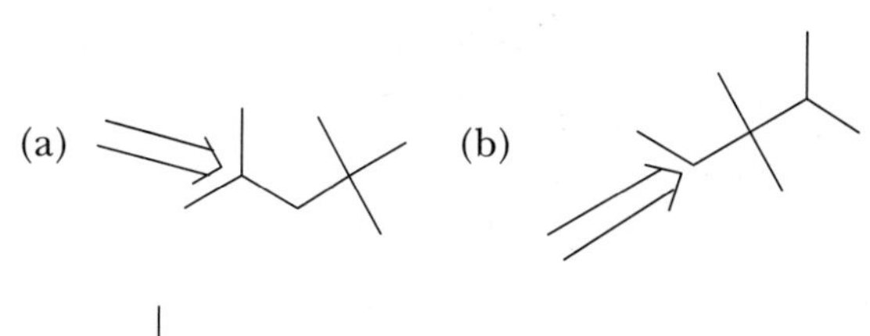

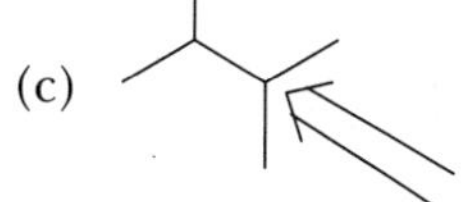

3.38 Draw both chair forms of each of the following compounds and indicate the more stable conformation. **(See Examples 3.8, 3.9)**

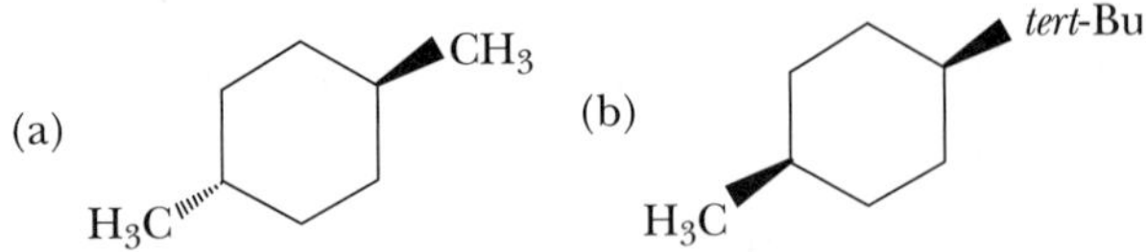

(c) CH_3 H_3C CH_3

(d) CH_3

Section 3.7 *Cis-Trans* Isomerism in Cycloalkanes

3.39 What structural feature of cycloalkanes makes *cis–trans* isomerism in them possible?

3.40 Is *cis–trans* isomerism possible in alkanes?

3.41 Name and draw structural formulas for the *cis* and *trans* isomers of 1,2 dimethylcyclopropane. **(See Examples 3.10, 3.11)**

3.42 Name and draw structural formulas for all cycloalkanes with the molecular formula C_5H_{10}. Be certain to include *cis–trans* isomers, as well as constitutional isomers. **(See Examples 3.10, 3.11)**

3.43 Using a planar pentagon representation for the cyclopentane ring, draw structural formulas for the *cis* and *trans* isomers of **(See Examples 3.10, 3.11)**

(a) 1,2-Dimethylcyclopentane

(b) 1,3-Dimethylcyclopentane

3.44 Draw the alternative chair conformations for the *cis* and *trans* isomers of 1,2-dimethyl cyclohexane, 1,3-dimethylcyclohexane, and 1,4 dimethylcyclohexane. **(See Examples 3.10, 3.11)**

(a) Indicate by a label whether each methyl group is axial or equatorial.

(b) For which isomer(s) are the alternative chair conformations of equal stability?

(c) For which isomer(s) is one chair conformation more stable than the other?

3.45 Use your answers from Problem 3.44 to complete the following table, showing correlations between *cis, trans* isomers and axial, equatorial positions for disubstituted derivatives of cyclohexane:

Position of Substitution	cis	trans
1,4-	a,e or e,a	e,e or a,a
1,3-	___ or ___	___ or ___
1,2-	___ or ___	___ or ___

***3.46** There are four *cis–trans* isomers of 2-isopropyl-5-methylcyclohexanol: **(See Examples 3.10, 3.11)**

2-Isopropyl-5-methylcyclohexanol

(a) Using a planar hexagon representation for the cyclohexane ring, draw structural formulas for these four isomers.

(b) Draw the more stable chair conformation for each of your answers in Part (a).

(c) Of the four *cis–trans* isomers, which is the most stable? If you answered this part correctly, you picked the isomer found in nature and given the name menthol.

Peppermint plant *(Mentha piperita)*, a source of menthol, is a perennial herb with aromatic qualities used in candies, gums, hot and cold beverages, and garnish for punch and fruit. *(John Kaprielian/Photo Researchers, Inc.)*

3.47 Draw alternative chair conformations for each substituted cyclohexane, and state which chair is the more stable: **(See Examples 3.8, 3.11)**

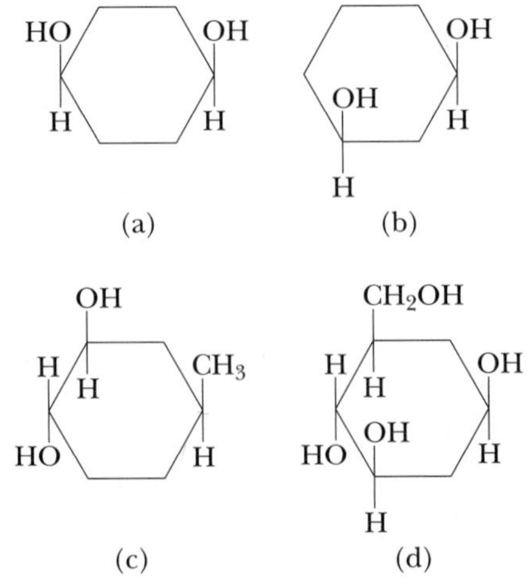

3.48 What kinds of conformations do the six-membered rings exhibit in adamantane?

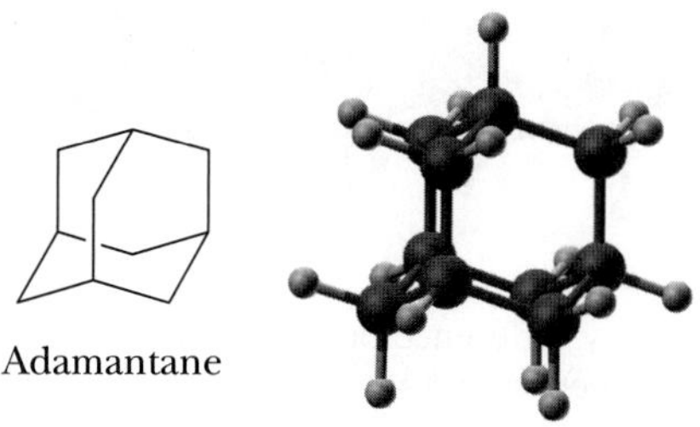
Adamantane

Section 3.8 Physical Properties of Alkanes and Cycloalkanes

3.49 In Problem 3.21, you drew structural formulas for all constitutional isomers with the molecular formula C_7H_{16}. Predict which isomer has the lowest boiling point and which has the highest. **(See Example 3.12)**

3.50 What generalizations can you make about the densities of alkanes relative to that of water?

3.51 What unbranched alkane has about the same boiling point as water? (See Table 3.4.) Calculate the molecular weight of this alkane, and compare it with that of water.

3.52 As you can see from Table 3.4, each CH_2 group added to the carbon chain of an alkane increases the boiling point of the alkane. The increase is greater going from CH_4 to C_2H_6 and from C_2H_6 to C_3H_8 than it is from C_8H_{18} to C_9H_{20} or from C_9H_{20} to $C_{10}H_{22}$. What do you think is the reason for this trend?

3.53 Dodecane, $C_{12}H_{26}$, is an unbranched alkane. Predict the following:

(a) Will it dissolve in water?

(b) Will it dissolve in hexane?

(c) Will it burn when ignited?

(d) Is it a liquid, solid, or gas at room temperature and atmospheric pressure?

(e) Is it more or less dense than water?

***3.54** As stated in Section 3.8A, the wax found in apple skins is an unbranched alkane with the molecular formula $C_{27}H_{56}$. Explain how the presence of this alkane prevents the loss of moisture from within an apple.

Section 3.9 Reactions of Alkanes

3.55 Write balanced equations for the combustion of each hydrocarbon. Assume that each is converted completely to carbon dioxide and water.

(a) Hexane (b) Cyclohexane (c) 2-Methylpentane

***3.56** Following are heats of combustion of methane and propane:

Hydrocarbon	Component of	$\Delta H°$ [kJ/mol (kcal/mol)]
CH_4	natural gas	−886 (−212)
$CH_3CH_2CH_3$	LPG	−2220 (−530)

On a gram-for-gram basis, which of these hydrocarbons is the better source of heat energy?

***3.57** When ethanol is added to gasoline to produce gasohol, the ethanol promotes more complete combustion of the gasoline and is an octane booster (Section 3.10 B). Compare the heats of combustion of 2,2,4-trimethylpentane 5460 kJ/mol (1304 kcal/mol) and ethanol 1369 kJ/mol (327 kcal/mol). Which has the higher heat of combustion in kJ/mol? in kJ/g?

Looking Ahead

3.58 Explain why 1,2-dimethylcyclohexane can exist as *cis–trans* isomers, while 1,2-dimethylcyclododecane cannot.

***3.59** On the left is a representation of the glucose molecule (we discuss the structure and chemistry of glucose in Chapter 18):

Glucose

(a) (b)

(a) Convert this representation to a planar hexagon representation.

(b) Convert this representation to a chair conformation. Which substituent groups in the chair conformation are equatorial? Which are axial?

***3.60** Following is the structural formula of cholic acid (Section 21.4A), a component of human bile whose function is to aid in the absorption and digestion of dietary fats:

Cholic acid

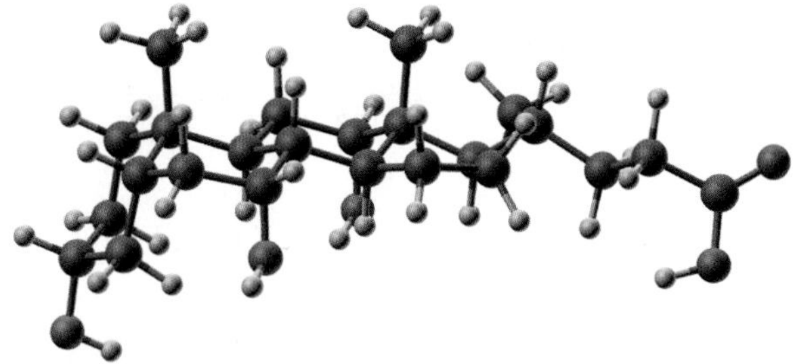

(a) What are the conformations of rings A, B, C, and D?

(b) There are hydroxyl groups on rings A, B, and C. Tell whether each is axial or equatorial.

(c) Is the methyl group at the junction of rings A and B axial or equatorial to ring A? Is it axial or equatorial to ring B?

(d) Is the methyl group at the junction of rings C and D axial or equatorial to ring C?

***3.61** Following is the structural formula and ball-and-stick model of cholestanol:

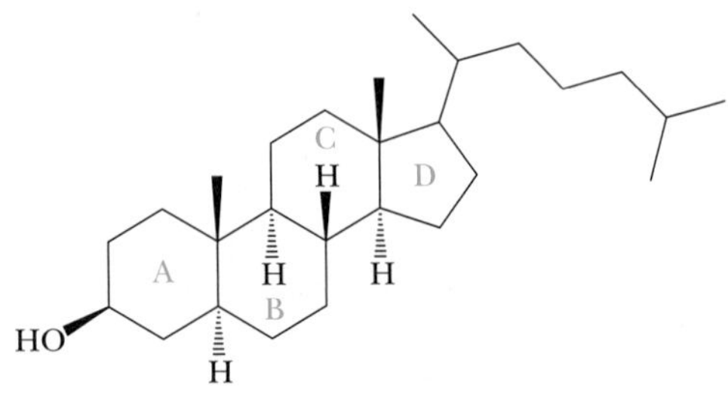

Cholestanol

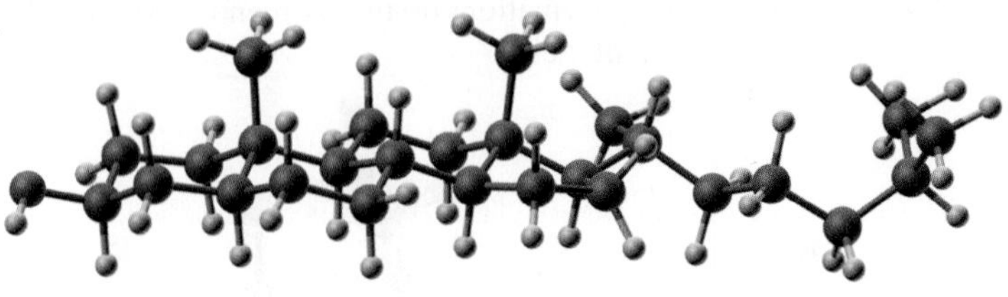

The only difference between this compound and cholesterol (Section 21.4A) is that cholesterol has a carbon–carbon double bond in ring B.

(a) Describe the conformation of rings A, B, C, and D in cholestanol.

(b) Is the hydroxyl group on ring A axial or equatorial?

(c) Consider the methyl group at the junction of rings A and B. Is it axial or equatorial to ring A? Is it axial or equatorial to ring B?

(d) Is the methyl group at the junction of rings C and D axial or equatorial to ring C?

3.62 As we have seen in Section 3.4, the IUPAC system divides the name of a compound into a prefix (showing the number of carbon atoms), an infix (showing the presence of carbon–carbon single, double, or triple bonds), and a suffix (showing the presence of an alcohol, amine, aldehyde, ketone, or carboxylic acid). Assume for the purposes of this problem that, to be alcohol (-ol) or amine (-amine), the hydroxyl or amino group must be bonded to a tetrahedral (sp^3 hybridized) carbon atom.

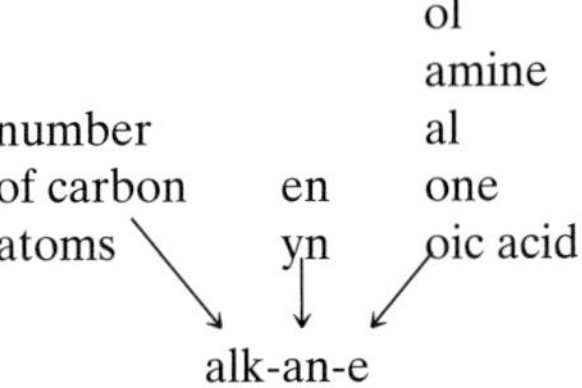

Given this information, write the structural formula of a compound with an unbranched chain of four carbon atoms that is an:

(a) Alkane	(b) Alkene	(c) Alkyne
(d) Alkanol	(e) Alkenol	(f) Alkynol
(g) Alkanamine	(h) Alkenamine	(i) Alkynamine
(j) Alkanal	(k) Alkenal	(l) Alkynal
(m) Alkanone	(n) Alkenone	(o) Alkynone
(p) Alkanoic acid	(q) Alkenoic acid	(r) Alkynoic acid

(*Note:* There is only one structural formula possible for some parts of this problem. For other parts, two or more structural formulas are possible. Where two are more are possible, we will deal with how the IUPAC system distinguishes among them when we come to the chapters on those particular functional groups.)

Putting It Together

The following problems bring together concepts and material from Chapters 1–3.

Choose the best answer for each of the following questions.

1. Which of the following molecules has a net charge of +1?
   (a) CH_2CHCH_3 (b) CH_3CHCH_3
   (c) $CHCCH_3$ (d) $(CH_3)_3CH$
   (e) CH_2CH_2

2. Which of the following statements is true concerning the following compound?

   H–N=C=O (with lone pairs on N and O)

   (a) The central carbon is sp^2 hybridized, and the molecule is planar in geometry.
   (b) The central carbon is sp^2 hybridized, and the molecule is nonplanar in geometry.
   (c) The central carbon is *sp* hybridized, and the molecule is planar in geometry.
   (d) The central carbon is *sp* hybridized, and the molecule is nonplanar in geometry.
   (e) None of these statements is true.

3. Which of the following statements is false concerning *p* orbitals?
   (a) They consist of two equivalent lobes.
   (b) They are absent from the first shell of atomic orbitals.
   (c) They can form π bonds.
   (d) They only participate in bonding on carbon atoms.
   (e) They can hold a maximum of two electrons.

4. Which base (A or B) is stronger and why?

   A: phenoxide ion ($C_6H_5O^-$); B: cyclohexyloxide ion ($C_6H_{11}O^-$)

   (a) **A** is stronger because it has fewer protons for the acid to compete with in acid–base reactions.
   (b) **A** is stronger because inductive effects increase the negative character of its oxygen.
   (c) **B** is stronger because inductive effects increase the negative character of its oxygen.
   (d) **B** is stronger because resonance effects can delocalize its negative charge throughout the molecule.
   (e) **B** is stronger because it has no resonance or inductive effects that can delocalize its negative charge throughout the molecule.

5. Which of the following is the initial product of the reaction between $(CH_3)_3C^+$ and CH_3OH?
   (a) $(H_3C)_3C-\overset{+}{O}(CH_3)-H$
   (b) $(H_3C)_3C-\ddot{O}-H$
   (c) $(H_3C)_3C-\ddot{O}:^- + H_2$
   (d) $(H_3C)_3C-CH_2 + H_2\ddot{O}:$
   (e) $(H_3C)_3C-H + CH_3-\ddot{O}:^-$

6. Select the statement that is false concerning the following acid–base reaction.

   $$H_3C-C(=O)-OH + NaCl \rightleftharpoons H_3C-C(=O)-ONa + HCl$$

   (a) The equilibrium lies on the product side of the reaction.
   (b) The carboxylic acid does not possess a positive charge.
   (c) The chloride ion acts as a Lewis base.
   (d) The chloride ion acts as a Brønsted–Lowry base.
   (e) The carboxylic acid is a weaker acid than HCl.

7. Which of the following statements is false?
   (a) Nonbonded interaction (steric) strain contributes to the energy of butane in the eclipsed conformation.
   (b) All staggered conformations possess zero strain.
   (c) A Newman projection is the picture of a molecule viewed down at least one of its bonds.
   (d) Bonds represented by Newman projections do not freely rotate because they must overcome an energy barrier to rotation.
   (e) Ring strain contributes to the instability of cyclopropane.

8. Which of the following statements is true concerning the isomers *cis*-1,2-dimethylcyclohexane and *cis*-1,3-dimethylcyclohexane?
   (a) They are not constitutional isomers.
   (b) They are conformers.
   (c) The favored conformer of the 1,3-isomer is more stable than that of the 1,2-isomer.
   (d) The favored conformer of the 1,3-isomer and that of the 1,2-isomer are equal in energy.
   (e) The relative stability of the two molecules cannot be predicted.

9. Select the correct order of stability (least stable ---> most stable) for the following conformations.

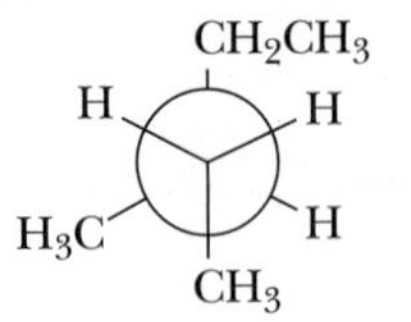

(a) 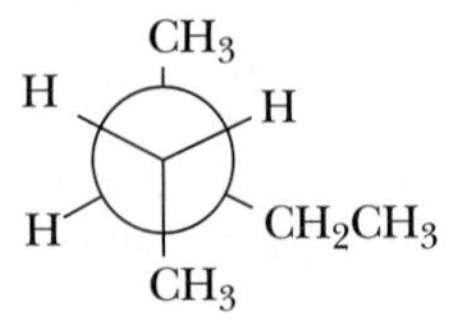 (b) 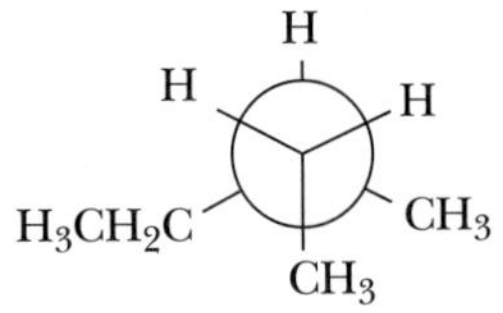 (c)

(a) a, b, c (b) a, c, b (c) b, a, c
(d) c, a, b (e) c, b, a

10. Select the most stable conformation of those shown for 1-*tert*-butyl-3,5-dimethylcyclohexane.

(a) (b) 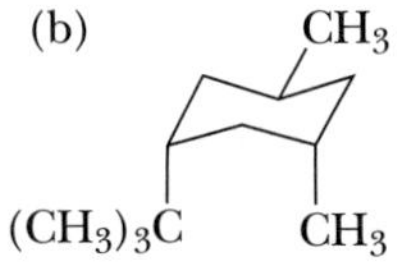 (c) 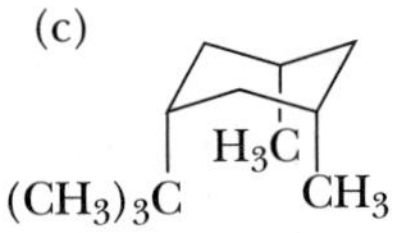 (d) 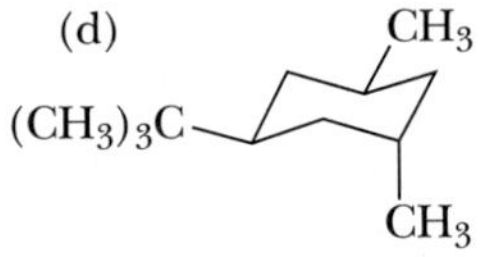 (e)

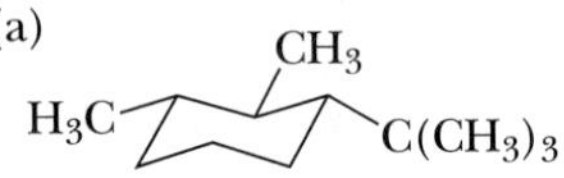

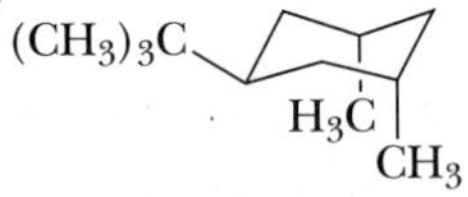

11. Answer the questions that follow regarding the structure of paclitaxel (trade name Taxol®), a compound first isolated from the Pacific Yew tree, which is now used to treat ovarian, breast, and non–small cell lung cancer.

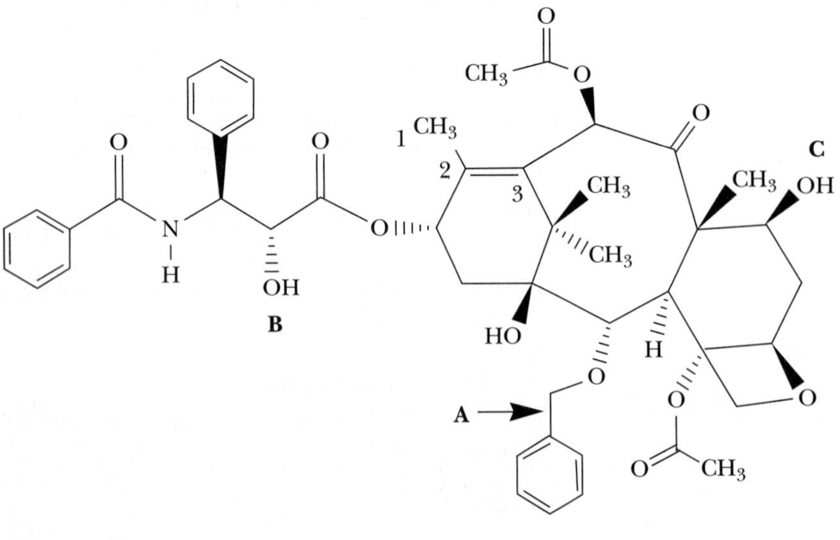

(a) Identify all the hydroxy groups and classify them as 1°, 2°, or 3°.
(b) Identify all the carbonyl groups. Are any of them part of an aldehyde, a ketone, or a carboxylic acid?
(c) What atomic or hybridized orbitals participate in the bond labeled **A**?
(d) Are there any quaternary carbons in paclitaxel?
(e) Explain why hydroxyl group **B** is more acidic than hydroxyl group **C**.
(f) What is the angle of the bond containing atoms 1–2–3?

12. Draw Newman projections of the three most stable conformations of the following compound viewed down the indicated bond and in the indicated direction. Indicate the most favorable conformation. You should be able to briefly describe or illustrate why your choice is the most favorable conformation.

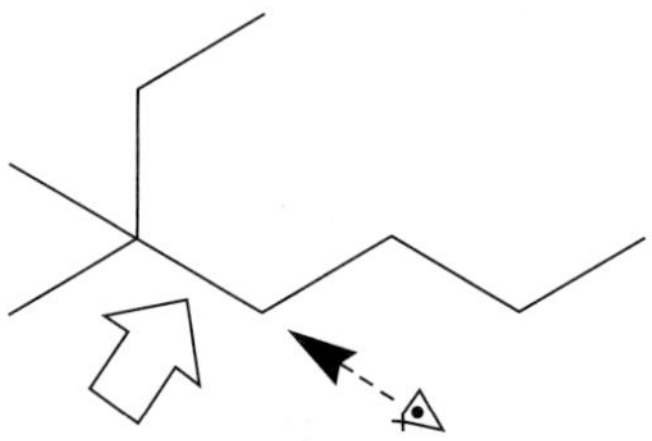

13. Provide IUPAC names for the following compounds.

(a) (b)

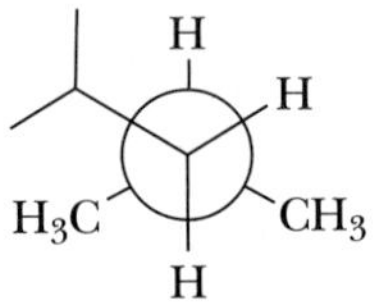

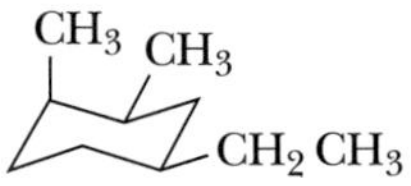

(c) $CH_3CH_2CH(CH_3)CH(CH_2CH_3)CH_2CH(CH_3)_2$

14. For each pair of molecules, select the one that best fits the accompanying description. Provide a concise but thorough rationale for each of your decisions using words and/or pictures.

(a) The higher boiling point?

OH vs. O

A B

(b) The more acidic hydrogen?

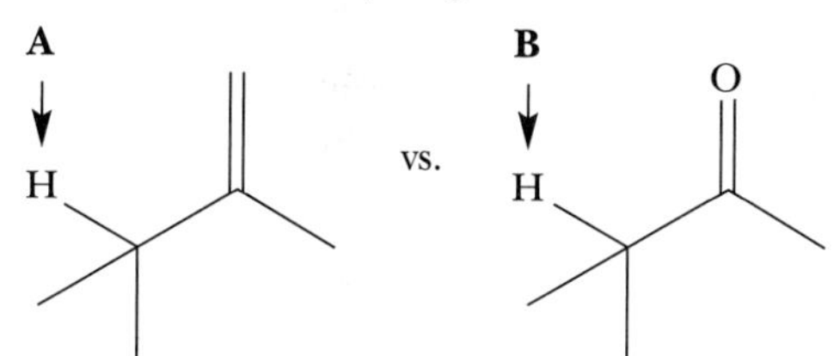

(c) The more basic atom?

A → O

H O H ← B

(d) The more acidic set of protons?

$H_3C—C{\equiv}C—CH_3$ vs. $H_3C—C{\equiv}N{:}$

A B

(e) The stronger base?

O O

vs.

HO O^-Na^+ H_3C O^-Na^+

A B

(f) Possesses the least nonbonding interaction (steric) strain

t-Bu *t*-Bu

vs.

H_3C H_3C

CH_3 CH_3

15. Glutamic acid is one of the common amino acids found in nature. Draw the predominant structure of glutamic acid when placed in a solution of $pH = 3.2$ and indicate its overall charge.

O + NH_3 ← pK_a= 9.67

HO

OH ← pK_a= 4.25

pK_a= 2.19

O

16. Use atomic and hybridized orbitals to illustrate (see example using H_2O) the location of bonding and nonbonding electrons in ethenimine. Do all of the atoms in ethenimine lie in the same plane?

HN=C=CH_2
ethenimine

Example orbital illustration for H_2O

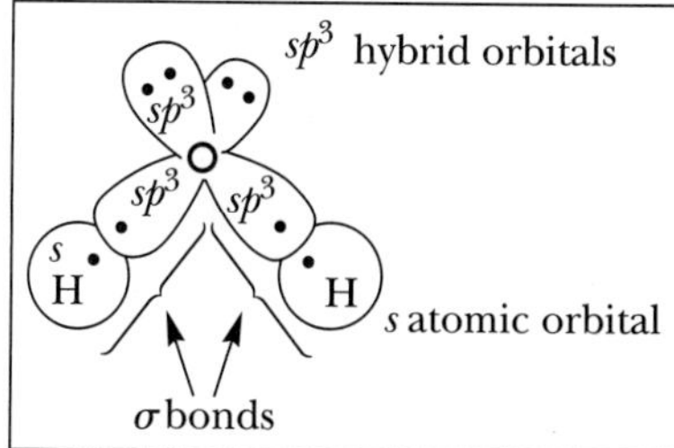

17. The following values have been determined for the amount of energy it takes to place a substituent in the axial position. As shown in the table, going from H to CH_3 causes a drastic increase in free energy (7.28 kJ/mol). However, increasing the size of the R group results in only a minor change in ΔG even when the R group is isopropyl (this only increases ΔG by 1.72 kJ/mol over methyl). *Using perspective (dash-wedge) drawings,* illustrate and explain why the increase in ΔG is only gradual up to isopropyl but increases drastically when the R group is *t*-butyl.

R H ⇌ R H

R	**ΔG kJ/mol (kcal/mol)**
H	0
CH_3	7.28 (1.74)
CH_2CH_3	7.32 (1.75)
$CH(CH_3)_2$	9.00 (2.15)
$C(CH_3)_3$	20.92 (5.00)

18. (a) Draw the two possible products that can form from the Lewis acid–base reaction between methyl formate and BF_3. Indicate the major product and use curved arrow notation to illustrate its formation. Show all charges and nonbonded electrons *in your products.* (b) Use pictures and words to explain why the product you indicated is favored over the other product.

O

H O CH_3 + F B F F ⇌

Methyl formate

19. Use resonance theory to predict whether $[CNO]^-$ or $[NCO]^-$ is the more stable ion. Use pictures *and* words to explain your decision.

20. Provide structures as indicated:

    (a) All compounds of formula C_5H_{10} that exhibit *cis–trans* isomerism.

    (b) Lewis structures and any resonance structures for the ion with formula CH_2NO_2. Show all formal charges and lone pairs of electrons.

    (c) All compounds that upon combustion with 6 mol of O_2 would yield 4 mol of CO_2 and 4 mol of H_2O.

4 Alkenes and Alkynes

KEY QUESTIONS

HOW TO

CHEMICAL CONNECTIONS

Carotene and carotene-like molecules are partnered with chlorophyll in nature to assist in the harvest of sunlight. In autumn, green chlorophyll molecules are destroyed and the yellows and reds of carotene and related molecules become visible. The red color of tomatoes comes from lycopene, a molecule closely related to carotene. See Problems 4.38 and 4.39. Inset: A model of β-carotene. *(Charles D. Winters)*

In this chapter, we begin our study of unsaturated hydrocarbons. An unsaturated hydrocarbon is a hydrocarbon that has fewer hydrogens bonded to carbon than an alkane has. There are three classes of unsaturated hydrocarbons: alkenes, alkynes, and arenes. **Alkenes** contain one or more carbon–carbon double bonds, and **alkynes** contain one or more carbon–carbon triple bonds. Ethene (ethylene) is the simplest alkene, and ethyne (acetylene) is the simplest alkyne:

Alkene An unsaturated hydrocarbon that contains a carbon–carbon double bond.

Alkyne An unsaturated hydrocarbon that contains a carbon–carbon triple bond.

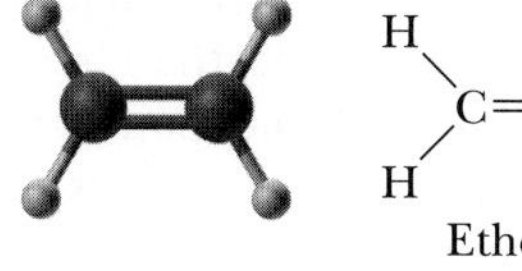

$H_2C{=}CH_2$ (H, H / C=C / H, H)

Ethene (an alkene)

H—C≡C—H

Ethyne (an alkyne)

CHEMICAL CONNECTIONS 4A

Ethylene, a Plant Growth Regulator

As we have noted, ethylene occurs only in trace amounts in nature. Still, scientists have discovered that this small molecule is a natural ripening agent for fruits. Thanks to this knowledge, fruit growers can pick fruit while it is green and less susceptible to bruising. Then, when they are ready to pack the fruit for shipment, the growers can treat it with ethylene gas to induce ripening. Alternatively, the fruit can be treated with ethephon (Ethrel), which slowly releases ethylene and initiates ripening.

Ethephon $Cl{-}CH_2{-}CH_2{-}P(=O)(OH){-}OH$

The next time you see ripe bananas in the market, you might wonder when they were picked and whether their ripening was artificially induced.

QUESTION

Explain the basis for the saying "A rotten apple can spoil the barrel."

Arene A compound containing one or more benzene rings.

Arenes are the third class of unsaturated hydrocarbons. The characteristic structural feature of an arene is the presence of one or more benzene rings. The simplest arene is benzene:

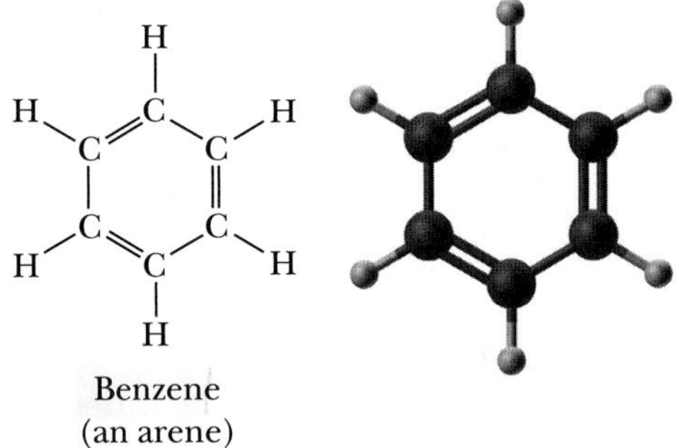

Benzene (an arene)

The chemistry of benzene and its derivatives is quite different from that of alkenes and alkynes. Even though we do not study the chemistry of arenes until Chapter 9, we will show structural formulas of compounds containing benzene rings in earlier chapters. What you need to remember at this point is that a benzene ring is not chemically reactive under any of the conditions we describe in Chapters 4–8.

Compounds containing carbon–carbon double bonds are especially widespread in nature. Furthermore, several low-molecular-weight alkenes, including ethylene and propene, have enormous commercial importance in our modern, industrialized society. The organic chemical industry produces more pounds of ethylene worldwide than any other chemical. Annual production in the United States alone exceeds 20 billion kg (45 billion pounds).

What is unusual about ethylene is that it occurs only in trace amounts in nature. The enormous amounts of it required to meet the needs of the chemical industry are derived the world over by thermal cracking of hydrocarbons. In the United States and other areas of the world with vast reserves of natural gas, the major process for the production of ethylene is thermal cracking of the small quantities of ethane extracted from natural gas. In **thermal cracking**, a saturated hydrocarbon is converted to an unsaturated hydrocarbon plus H_2. Heating ethane in a furnace to 800–900°C for a fraction of a second cracks it to ethylene and hydrogen.

$$\underset{\text{Ethane}}{CH_3CH_3} \xrightarrow[\text{(thermal cracking)}]{800-900°C} \underset{\text{Ethylene}}{CH_2{=}CH_2} + H_2$$

Europe, Japan, and other areas of the world with limited supplies of natural gas depend almost entirely on thermal cracking of petroleum for their ethylene.

The crucial point to recognize is that ethylene and all of the commercial and industrial products made from it are derived from either natural gas or petroleum—both nonrenewable natural resources!

4.1 What Are the Structures and Shapes of Alkenes and Alkynes?

A. Shapes of Alkenes

Using valence-shell electron-pair repulsion VSEPR (Section 1.3), we predict a value of 120° for the bond angles about each carbon in a double bond. The observed H—C—C bond angle in ethylene is 121.7°, a value close to that predicted by VSEPR. In other alkenes, deviations from the predicted angle of 120° may be somewhat larger as a result of strain between groups bonded to one or both carbons of the double bond. The C—C—C bond angle in propene, for example, is 124.7°.

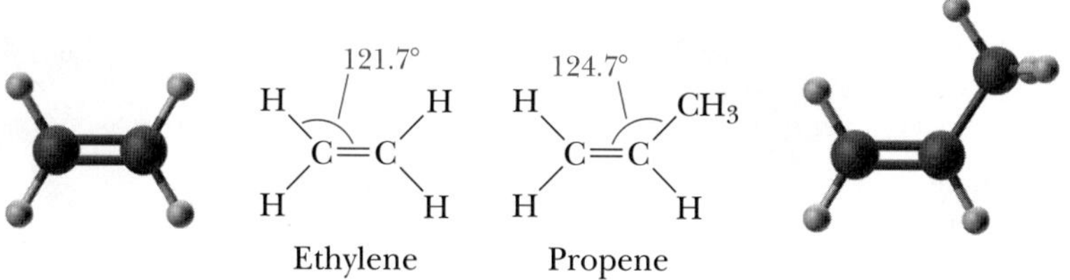

B. Orbital Overlap Model of a Carbon-Carbon Double Bond

In Section 1.6D, we described the formation of a carbon–carbon double bond in terms of the overlap of atomic orbitals. A carbon–carbon double bond consists of one sigma bond and one pi bond. Each carbon of the double bond uses its three sp^2 hybrid orbitals to form sigma bonds to three atoms. The unhybridized $2p$ atomic orbitals, which lie perpendicular to the plane created by the axes of the three sp^2 hybrid orbitals, combine to form the pi bond of the carbon–carbon double bond.

It takes approximately 264 kJ/mol (63 kcal/mol) to break the pi bond in ethylene; that is, to rotate one carbon by 90° with respect to the other so that no overlap occurs between $2p$ orbitals on adjacent carbons (Figure 4.1). This energy is considerably greater than the thermal energy available at room temperature, and, as a consequence, rotation about a carbon–carbon double bond is severely restricted. You might compare rotation about a carbon–carbon double bond, such as the bond in ethylene, with that about a carbon–carbon single bond, such as the bond in ethane (Section 3.6A). Whereas rotation about the carbon–carbon single bond in ethane is relatively free (the energy barrier is approximately 13 kcal/mol), rotation about the carbon–carbon double bond in ethylene is restricted (the energy barrier is approximately 264 kJ/mol).

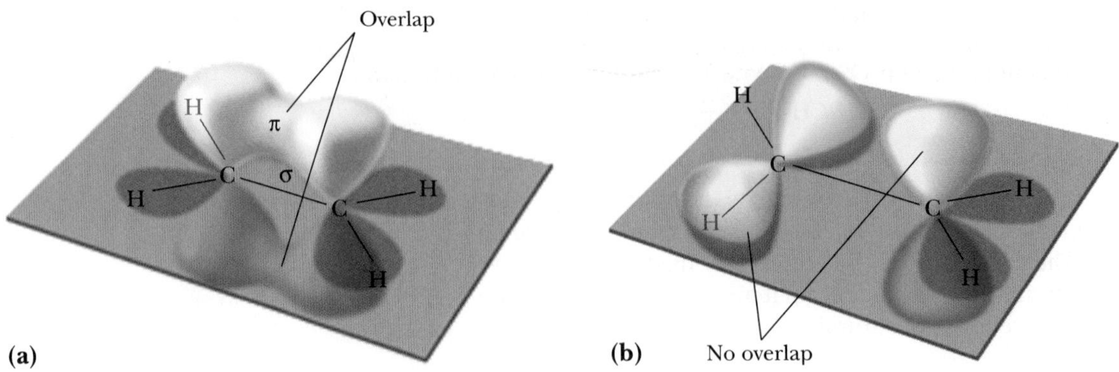

Figure 4.1
Restricted rotation about the carbon–carbon double bond in ethylene. (a) Orbital overlap model showing the pi bond. (b) The pi bond is broken by rotating the plane of one H—C—H group by 90° with respect to the plane of the other H—C—H group.

***Cis–trans* isomerism** Isomers that have the same order of attachment of their atoms, but a different arrangement of their atoms in space due to the presence of either a ring (Chapter 3) or a carbon–carbon double bond (Chapter 4).

C. *Cis–Trans* Isomerism in Alkenes

Because of restricted rotation about a carbon–carbon double bond, an alkene in which each carbon of the double bond has two different groups bonded to it shows *cis–trans* isomerism.

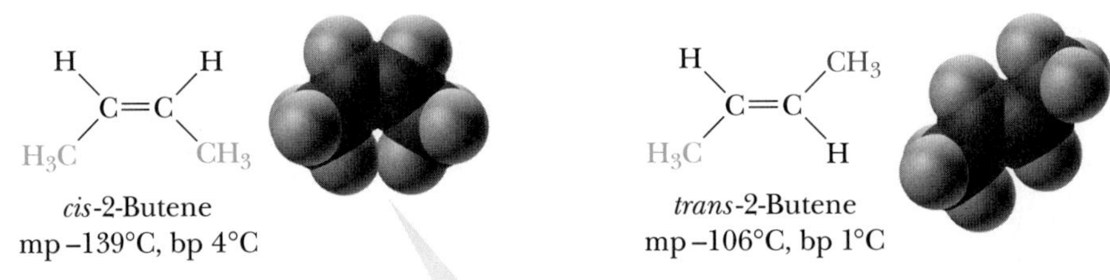

The combustion of acetylene yields energy that produces the very hot temperatures of an oxyacetylene torch. *(Charles D. Winters)*

Consider, for example, 2-butene: In *cis*-2-butene, the two methyl groups are on the same side of the double bond; in *trans*-2-butene, the two methyl groups are on opposite sides of the double bond. These two compounds cannot be converted into one another at room temperature because of the restricted rotation about the double bond; they are different compounds, with different physical and chemical properties.

It takes approximately 264 kJ (63 kcal/mol) to break the π bond of ethylene, that is, to rotate one carbon by 90° with respect to the other where there is zero overlap between the 2*p* orbitals of adjacent carbons (Figure 4.1). This energy is considerably greater than the thermal energy available at room temperature; consequently, rotation about a carbon–carbon double bond does not occur under normal conditions.

Cis alkenes are less stable than their *trans* isomers because of nonbonded interaction strain between alkyl substituents on the same side of the double bond in the *cis* isomer, as can be seen in space-filling models of the *cis* and *trans* isomers of 2-butene. This is the same type of steric strain that results in the preference for equatorial methylcyclohexane over axial methylcyclohexane (Section 3.6B).

D. Structure of Alkynes

The functional group of an alkyne is a **carbon–carbon triple bond**. The simplest alkyne is ethyne, C_2H_2. Ethyne is a linear molecule; all of its bond angles are 180° (Figure 1.10).

According to the molecular orbital model (Section 1.6E), a triple bond is described in terms of the overlap of *sp* hybrid orbitals of adjacent carbons to form a sigma bond, the overlap of parallel $2p_y$ orbitals to form one pi bond, and the overlap of parallel $2p_z$ orbitals to form the second pi bond. In ethyne, each carbon forms a bond to a hydrogen by the overlap of an *sp* hybrid orbital of carbon with a 1*s* atomic orbital of hydrogen.

4.2 How Do We Name Alkenes and Alkynes?

Alkenes are named using the IUPAC system, but, as we shall see, some are still referred to by their common names.

A. IUPAC Names

We form IUPAC names of alkenes by changing the **-an-** infix of the parent alkane to **-en-** (Section 3.5). Hence, $CH_2{=}CH_2$ is named ethene, and $CH_3CH{=}CH_2$ is named propene. In higher alkenes, where isomers exist that differ in the location of the double bond, we use a numbering system. We number the longest carbon chain that contains the double bond in the direction that gives the carbon atoms of the double bond the lower set of numbers. We then use the number of the first carbon of the double bond to show its location. We name branched or substituted alkenes in a manner similar to the way we name alkanes (Section 3.3). We number the carbon atoms, locate the double bond, locate and name substituent groups, and name the main (parent) chain.

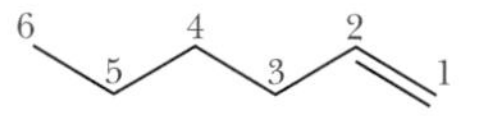

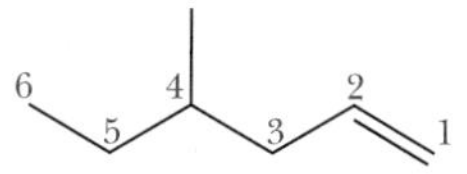

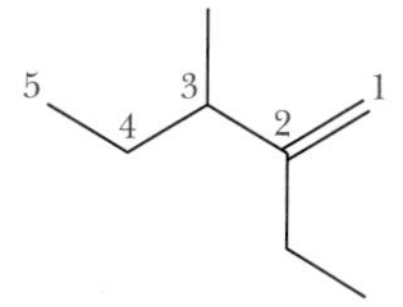

6 5 4 3 2 1
$CH_3CH_2CH_2CH_2CH{=}CH_2$

1-Hexene

CH_3 on C4
6 5 4 3 2 1
$CH_3CH_2CHCH_2CH{=}CH_2$

4-Methyl-1-hexene

CH_3 on C3; CH_2CH_3 on C2
5 4 3 2 1
$CH_3CH_2CHC{=}CH_2$

2-Ethyl-3-methyl-1-pentene

Note that there is a six-carbon chain in 2-ethyl-3-methyl-1-pentene. However, because the longest chain that contains the carbon–carbon double bond has only five carbons, the parent hydrocarbon is pentane, and we name the molecule as a disubstituted 1-pentene.

We form IUPAC names of alkynes by changing the **-an-** infix of the parent alkane to **-yn-** (Section 3.5). Thus, $HC{\equiv}CH$ is named ethyne, and $CH_3C{\equiv}CH$ is named propyne. The IUPAC system retains the name *acetylene*; therefore, there are two acceptable names for $HC{\equiv}CH$: *ethyne* and *acetylene*. Of these two names, *acetylene* is used much more frequently. For larger molecules, we number the longest carbon chain that contains the triple bond from the end that gives the triply bonded carbons the lower set of numbers. We indicate the location of the triple bond by the number of the first carbon of the triple bond.

$$\underset{\displaystyle CH_3}{\overset{4\ \ 3\ \ 2\ \ 1}{CH_3\underset{|}{C}HC{\equiv}CH}}$$

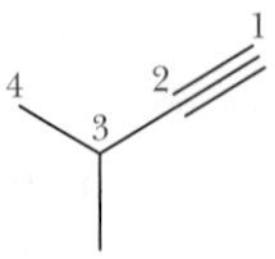

3-Methyl-1-butyne

$$CH_3CH_2C{\equiv}CCH_2C(CH_3)_2CH_3$$

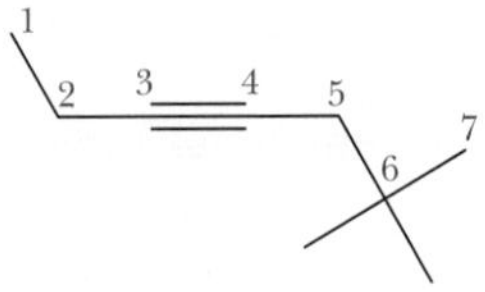

6,6-Dimethyl-3-heptyne

Example 4.1

Write the IUPAC name of each unsaturated hydrocarbon:

(a) $CH_2{=}CH(CH_2)_5CH_3$ (b) 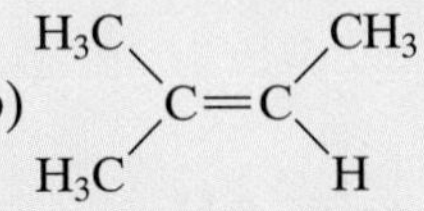 $(H_3C)_2C{=}C(CH_3)H$ (c) $CH_3(CH_2)_2C{\equiv}CCH_3$

Strategy

First look for the longest carbon chain that contains the multiple bond. This chain determines the root name. Number the carbon chain to give the placement of the multiple bond the lowest possible set of numbers. Then identify substituents and give each a name and a number. Locate the position of the multiple bond by the number of its first carbon.

Solution

(a) 1-Octene (b) 2-Methyl-2-butene (c) 2-Hexyne

See problems 4.15–4.20

Problem 4.1

Write the IUPAC name of each unsaturated hydrocarbon:

(a) 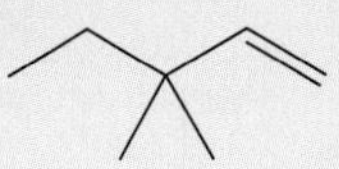 (b) 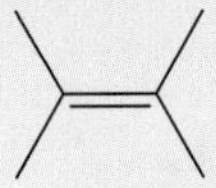 (c)

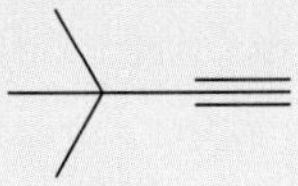

B. Common Names

Despite the precision and universal acceptance of IUPAC nomenclature, some alkenes, particularly those with low molecular weight, are known almost exclusively by their common names, as illustrated by the common names of these alkenes:

	$CH_2{=}CH_2$	$CH_3CH{=}CH_2$	$CH_3C(CH_3){=}CH_2$
IUPAC name:	Ethene	Propene	2-Methylpropene
Common name:	Ethylene	Propylene	Isobutylene

Furthermore, the common names **methylene** (a CH_2 group), **vinyl**, and **allyl** are often used to show the presence of the following alkenyl groups.

Alkenyl Group	Common Name	Example	Common Name
$CH_2{=}CH{-}$	Vinyl	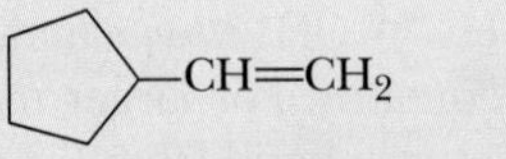 cyclopentyl–$CH{=}CH_2$	Vinylcyclopentane
$CH_2{=}CHCH_2{-}$	Allyl	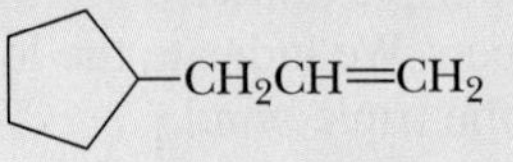 cyclopentyl–$CH_2CH{=}CH_2$	Allylcyclopentane

C. Systems for Designating Configuration in Alkenes

The *Cis–Trans* System

The most common method for specifying the configuration of a disubstituted alkene uses the prefixes *cis* and *trans*. In this system, the orientation of the atoms of the parent chain determines whether the alkene is *cis* or *trans*. Following is a structural formula for the *cis* isomer of 4-methyl-2-pentene:

$H_3C(1)$, H on C^2 $=$ C^3 with H and $CH(CH_3)_2$ (4)

cis-4-Methyl-2-pentene

In this example, carbon atoms of the main chain (carbons 1 and 4) are on the same side of the double bond; therefore, the configuration of this alkene is *cis*.

Example 4.2

Name each alkene, and, using the *cis–trans* system, show the configuration about each double bond:

Strategy

Locate the longest carbon chain that contains the multiple bond and number it from the end that gives the lower set of numbers to the carbon atoms of the multiple bond. Indicate the location of the multiple bond by the number of its first carbon atom. Configuration of a carbon–carbon double bond (*cis* or *trans*) in a common name is determined by the orientation of the carbon atoms of the parent chain relative to each other.

Solution

(a) The chain contains seven carbon atoms and is numbered from the end that gives the lower number to the first carbon of the double bond. The carbon atoms of the parent chain are on opposite sides of the double bond. The compound's name is *trans*-3-heptene.

(b) The longest chain contains seven carbon atoms and is numbered from the right, so that the first carbon of the double bond is carbon 3 of the chain. The carbon atoms of the parent chain are on the same side of the double bond. The compound's name is *cis*-6-methyl-3-heptene.

See problems 4.21, 4.22

Problem 4.2

Name each alkene, and, using the *cis–trans* system, specify its configuration:

E,Z system A system used to specify the configuration of groups about a carbon–carbon double bond.

Z From the German *zusammen*, meaning together; specifies that groups of higher priority on the carbons of a double bond are on the same side.

E From the German *entgegen*, meaning opposite; specifies that groups of higher priority on the carbons of a double bond are on opposite sides.

The E,Z System

The **E,Z system** must be used for tri- and tetrasubstituted alkenes. This system uses a set of rules to assign priorities to the substituents on each carbon of a double bond. If the groups of higher priority are on the same side of the double bond, the configuration of the alkene is **Z** (German: *zusammen*, together). If the groups of higher priority are on opposite sides of the double bond, the configuration is **E** (German: *entgegen*, opposite).

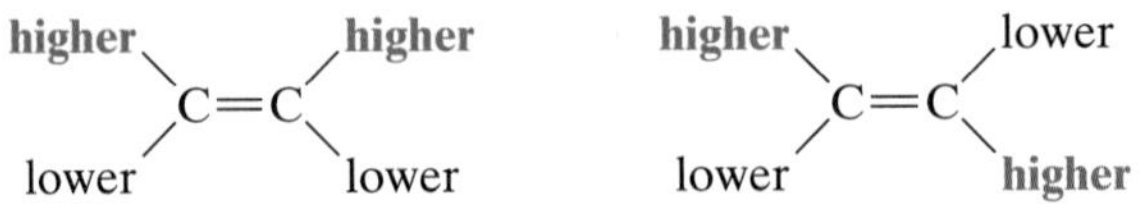

higher, higher / C=C / lower, lower — **Z** (*zusammen*)

higher, lower / C=C / lower, higher — **E** (*entgegen*)

The first step in assigning an E or a Z configuration to a double bond is to label the two groups bonded to each carbon in order of priority.

Priority Rules

1. Priority is based on atomic number: The higher the atomic number, the higher is the priority. Following are several substituents arranged in order of increasing priority (the atomic number of the atom determining priority is shown in parentheses):

$$\underset{(1)}{-H},\ \underset{(6)}{-CH_3},\ \underset{(7)}{-NH_2},\ \underset{(8)}{-OH},\ \underset{(16)}{-SH},\ \underset{(17)}{-Cl},\ \underset{(35)}{-Br},\ \underset{(53)}{-I}$$

Increasing priority →

2. If priority cannot be assigned on the basis of the atoms that are bonded directly to the double bond, look at the next set of atoms, and continue until a priority can be assigned. Priority is assigned at the first point of difference. Following is a series of groups, arranged in order of increasing priority (again, numbers in parentheses give the atomic number of the atom on which the assignment of priority is based):

$$\underset{(1)}{-CH_2-H}\quad \underset{(6)}{-CH_2-CH_3}\quad \underset{(7)}{-CH_2-NH_2}\quad \underset{(8)}{-CH_2-OH}\quad \underset{(17)}{-CH_2-Cl}$$

Increasing priority →

3. In order to compare carbons that are not sp^3 hybridized, the carbons must be manipulated in a way that allows us to maximize the number of groups bonded to them. Thus, we treat atoms participating in a double or triple bond as if they are bonded to an equivalent number of similar atoms by single bonds; that is, atoms of a double bond are replicated. Accordingly,

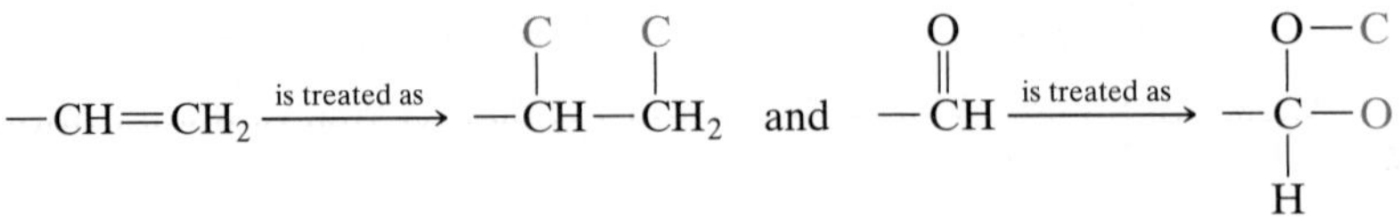

$$-CH{=}CH_2 \xrightarrow{\text{is treated as}} -\overset{\overset{C}{|}}{C}H-\overset{\overset{C}{|}}{C}H_2 \quad \text{and} \quad -\overset{\overset{O}{||}}{C}H \xrightarrow{\text{is treated as}} -\underset{\underset{H}{|}}{\overset{\overset{O-C}{|}}{C}}-O$$

Example 4.3

Assign priorities to the groups in each set:

(a) $-\overset{\overset{O}{||}}{C}OH$ and $-\overset{\overset{O}{||}}{C}H$ (b) $-CH_2NH_2$ and $-\overset{\overset{O}{||}}{C}OH$

Strategy

Priority is based on atomic number; the higher the atomic number the higher the priority. If priority cannot be determined on the basis of the atoms bonded directly to the carbon–carbon double bond, continue to the next set of atoms and continue in this manner until a priority can be assigned.

Solution

(a) The first point of difference is the O of the —OH in the carboxyl group, compared with the —H in the aldehyde group. The carboxyl group is higher in priority:

$$-\overset{\overset{\displaystyle O}{\|}}{C}-O-H \qquad\qquad -\overset{\overset{\displaystyle O}{\|}}{C}-H$$

Carboxyl group (higher priority) — Aldehyde group (lower priority)

(b) Oxygen has a higher priority (higher atomic number) than nitrogen. Therefore, the carboxyl group has a higher priority than the primary amino group:

$$-CH_2NH_2 \qquad\qquad -\overset{\overset{\displaystyle O}{\|}}{C}OH$$

lower priority — higher priority

See problems 4.23, 4.27, 4.28, 4.32

Example 4.4

Name each alkene and specify its configuration by the E,Z system:

(a) $(H)(H_3C)C{=}C(CH_3)(CH(CH_3)_2)$ — H and CH_3 on the upper side; H_3C and $CH(CH_3)_2$ on the lower side

(b) $(Cl)(H_3C)C{=}C(H)(CH_2CH_3)$ — Cl and H on the upper side; H_3C and CH_2CH_3 on the lower side

Strategy

Assign a priority to each atom or group of atoms on the carbon–carbon double bond. If the groups of higher priority are on the same side of the double bond, the alkene has the Z configuration; if they are on opposite sides, the alkene has the E configuration.

Solution

(a) The group of higher priority on carbon 2 is methyl; that of higher priority on carbon 3 is isopropyl. Because the groups of higher priority are on the same side of the carbon–carbon double bond, the alkene has the Z configuration. Its name is (*Z*)-3,4-dimethyl-2-pentene.

(b) Groups of higher priority on carbons 2 and 3 are —Cl and $-CH_2CH_3$. Because these groups are on opposite sides of the double bond, the configuration of this alkene is E, and its name is (*E*)-2-chloro-2-pentene.

See problems 4.23, 4.27, 4.28, 4.32

Problem 4.3

Name each alkene and specify its configuration by the E,Z system:

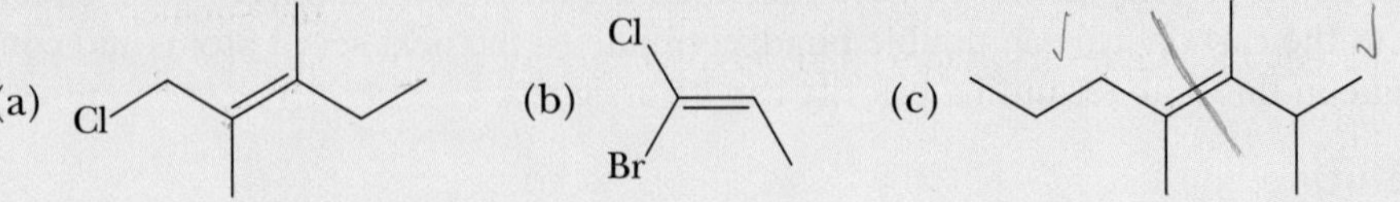

HOW TO 4.1 Name an Alkene

As an example of how to name an alkene, consider the following alkene, drawn here as a line-angle formula.

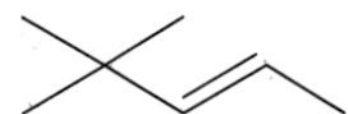

1. *Determine the parent chain, that is, the longest chain of carbon atoms that contains the functional group.*

   In this example, the parent chain is five carbon atoms, making the compound a disubstituted pentene.

2. *Number the parent chain from the end that gives the carbon atoms of the double bond the lower set of numbers.*

   In this example, the parent chain is a disubstituted 2-pentene.

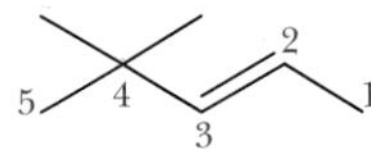

3. *Name and locate the substituents on the parent chain.*

   There are two methyl substituents on carbon 4 of the parent chain, and they are named 4,4-dimethyl-. The name to this point is 4,4-dimethyl-2-pentene.

4. *Determine whether the molecule shows* cis–trans *isomerism. If it does, use either the* cis–trans *or the E,Z system to specify the configuration.*

   In this example, the molecule shows *cis–trans* isomerism, and the double bond has the *trans* configuration. Therefore, the IUPAC name is *trans*-4,4-dimethyl-2-pentene.

   Note that the double bond locator may be placed either before the parent name, as in the name just given, or immediately before the infix specifying the double bond to give the name *trans*-4,4-dimethylpent-2-ene.

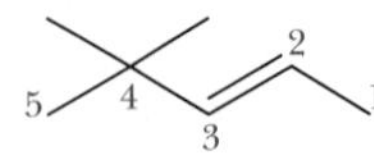

***trans*-4,4-Dimethyl-2-pentene**
or
***trans*-4,4-Dimethylpent-2-ene**
or
***(E)* 4,4-Dimethyl-2-pentene**

D. Naming Cycloalkenes

In naming **cycloalkenes**, we number the carbon atoms of the ring double bond 1 and 2 in the direction that gives the substituent encountered first the smaller number. We name and locate substituents and list them in alphabetical order, as in the following compounds:

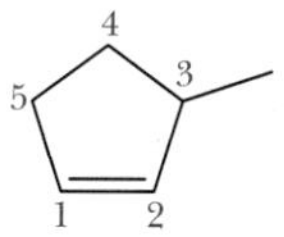

3-Methylcyclopentene
(not 5-methylcyclopentene)

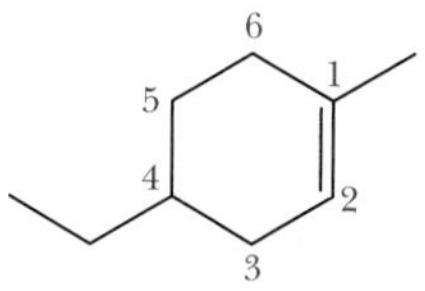

4-Ethyl-1-methylcyclohexene
(not 5-ethyl-2-methylcyclohexene)

Example 4.5

Write the IUPAC name for each cycloalkene:

(a) 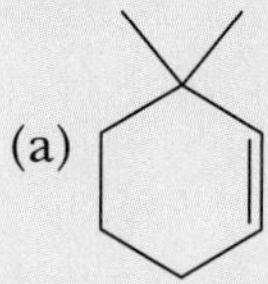

(b) 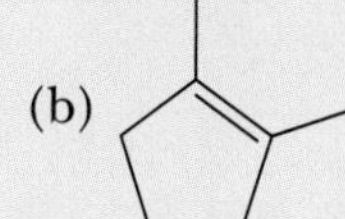

(c) 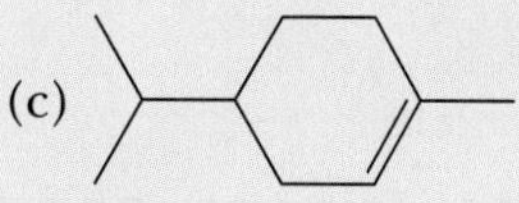

Strategy

The parent name of a cycloalkene is derived from the name of the unbranched alkene with the same number of carbon atoms as are in the ring of the cycloalkene. Number the carbon atoms of the ring 1 and 2 in the direction that gives the substituent encountered first the smaller number. Finally, name and number all substituents and list them in alphabetical order.

Solution

(a) 3,3-Dimethylcyclohexene
(b) 1,2-Dimethylcyclopentene
(c) 4-Isopropyl-1-methylcyclohexene

See problems 4.15–4.20

Problem 4.4

Write the IUPAC name for each cycloalkene:

(a) 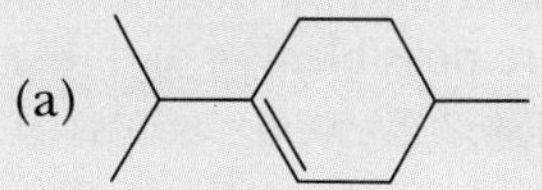

(b) 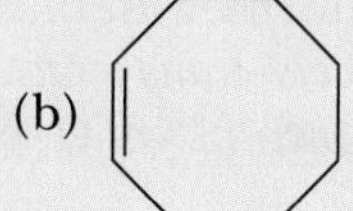

(c) 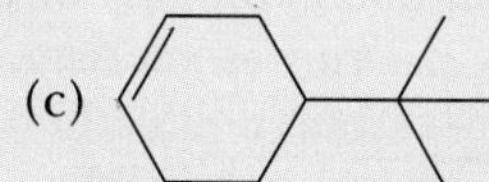

E. *Cis–Trans* Isomerism in Cycloalkenes

Following are structural formulas for four cycloalkenes:

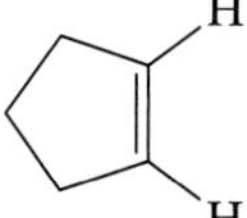

Cyclopentene

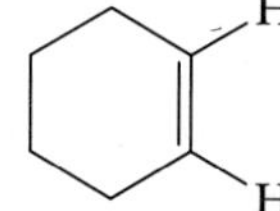

Cyclohexene

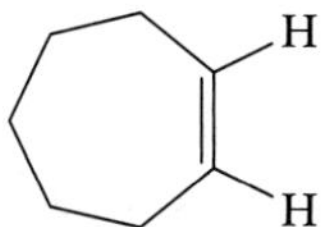

Cycloheptene

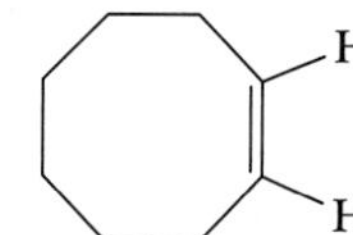

Cyclooctene

In these representations, the configuration about each double bond is *cis*. Because of angle strain, it is not possible to have a *trans* configuration in cycloalkenes of seven or fewer carbons. To date, *trans*-cyclooctene is the smallest *trans*-cycloalkene that has been prepared in pure form and is stable at room temperature. Yet, even in this *trans*-cycloalkene, there is considerable intramolecular strain. *cis*-Cyclooctene is more stable than its *trans* isomer by 38 kJ/mol (9.1 kcal/mol).

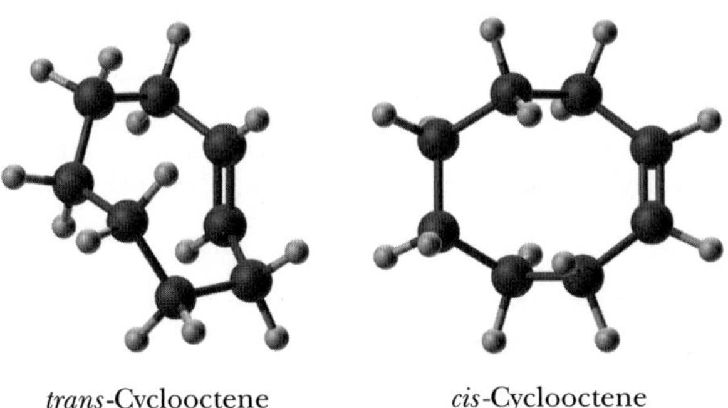

trans-Cyclooctene   *cis*-Cyclooctene

F. Dienes, Trienes, and Polyenes

We name alkenes that contain more than one double bond as alkadienes, alkatrienes, and so forth. We refer to those that contain several double bonds more generally as polyenes (Greek: *poly*, many). Following are three examples of dienes:

$CH_2{=}CHCH_2CH{=}CH_2$

1,4-Pentadiene

$CH_2{=}\overset{\overset{\displaystyle CH_3}{|}}{C}CH{=}CH_2$

2-Methyl-1,3-butadiene (Isoprene)

1,3-Cyclopentadiene

G. *Cis–Trans* Isomerism in Dienes, Trienes, and Polyenes

Thus far, we have considered *cis–trans* isomerism in alkenes containing only one carbon–carbon double bond. For an alkene with one carbon–carbon double bond that can show *cis–trans* isomerism, two *cis–trans* isomers are possible. For an alkene with $\boldsymbol{n}$ carbon–carbon double bonds, each of which can show *cis–trans* isomerism, $\mathbf{2^{n}}$ *cis–trans* isomers are possible.

Example 4.6

How many *cis–trans* isomers are possible for 2,4-heptadiene?

Strategy

Determine which of the carbon–carbon double bonds can show *cis–trans* isomerism. The number of *cis–trans* isomers possible is 2^n. n is the number of double bonds that may exhibit this type of isomerism.

Solution

This molecule has two carbon–carbon double bonds, each of which exhibits *cis–trans* isomerism. As the following table shows, $2^2 = 4$ *cis–trans* isomers are possible (to the right of the table are line-angle formulas for two of these isomers):

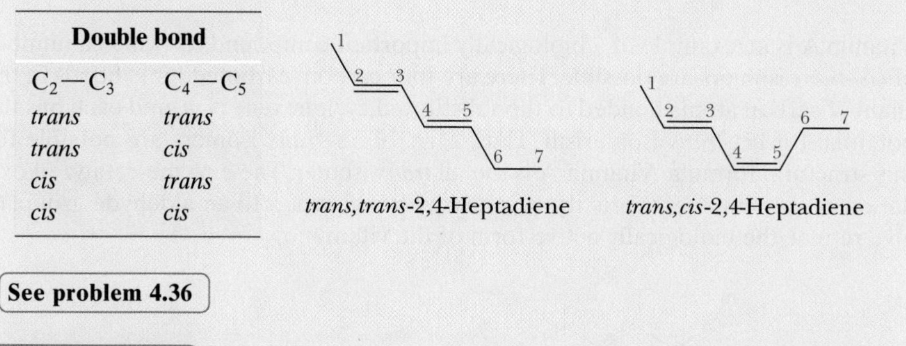

Double bond	
$C_2{-}C_3$	$C_4{-}C_5$
trans	*trans*
trans	*cis*
cis	*trans*
cis	*cis*

trans,trans-2,4-Heptadiene   *trans,cis*-2,4-Heptadiene

See problem 4.36

Problem 4.5

Draw structural formulas for the other two *cis–trans* isomers of 2,4-heptadiene.

Example 4.7

Draw all possible *cis–trans* isomers for the following unsaturated alcohol:

$$\begin{array}{c} \quad CH_3 \qquad\qquad\qquad\quad CH_3 \\ \quad | \qquad\qquad\qquad\qquad | \\ CH_3C{=}CHCH_2CH_2C{=}CHCH_2OH \end{array}$$

Strategy

Cis–trans isomerism is possible only about the double bond between carbons 2 and 3 of the chain. It is not possible for the other double bond because carbon 7 has two identical groups on it. Thus, $2^1 = 2$ *cis-trans* isomers are possible.

Solution

Each isomer may be named by the *cis/trans* system, but as noted earlier, for structures containing a tri- or tetrasubstituted double bond, it is preferable to use the E/Z system.

trans · 7 · 6 · 3 · 2 · OH   cis · 3 · 2 · OH

The *E* isomer
(The *trans* isomer)

The *Z* isomer
(The *cis* isomer)

The *E* isomer of this alcohol, named geraniol, is a major component of the oils of rose, citronella, and lemongrass.

See problem 4.36

Problem 4.6

How many *cis–trans* isomers are possible for the following unsaturated alcohol?

$$\begin{array}{c} \quad CH_3 \qquad\qquad\qquad CH_3 \qquad\qquad\qquad CH_3 \\ CH_3\overset{|}{C}{=}CHCH_2CH_2\overset{|}{C}{=}CHCH_2CH_2\overset{|}{C}{=}CHCH_2OH \end{array}$$

Vitamin A is an example of a biologically important compound for which a number of *cis–trans* isomers are possible. There are four carbon–carbon double bonds in the chain of carbon atoms bonded to the substituted cyclohexene ring, and each has the potential for *cis–trans* isomerism. Thus, $2^4 = 16$ *cis–trans* isomers are possible for this structural formula. Vitamin A is the all *trans* isomer. The enzyme-catalyzed oxidation of vitamin A converts the primary hydroxyl group to an aldehyde group, to give retinal, the biologically active form of the vitamin:

OH

Vitamin A (retinol)

enzyme-catalyzed oxidation →

O

H

Vitamin A aldehyde (retinal)

4.3 What Are the Physical Properties of Alkenes and Alkynes?

Alkenes and alkynes are nonpolar compounds, and the only attractive forces between their molecules are dispersion forces (Section 3.8B). Therefore, their physical properties are similar to those of alkanes (Section 3.8) with the same carbon skeletons. Alkenes and alkynes that are liquid at room temperature have densities less than 1.0 g/mL. Thus, they are less dense than water. Like alkanes, alkenes and alkynes are nonpolar and are soluble in each other. Because of their contrasting polarity with water, they do not dissolve in water. Instead, they form two layers when mixed with water or another polar organic liquid such as ethanol.

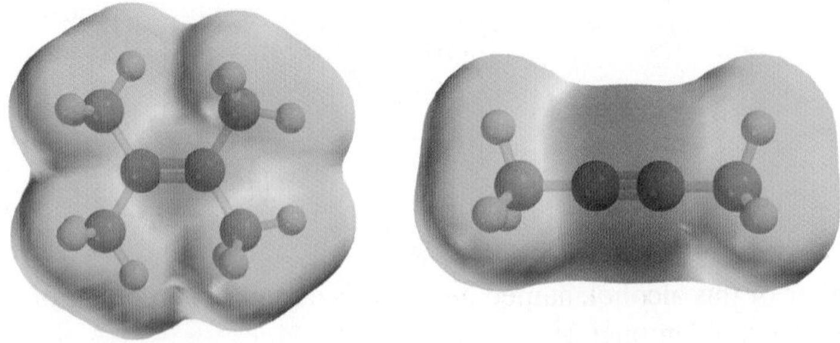

Tetramethylethylene and dimethylacetylene. Both a carbon–carbon double bond and a carbon–carbon triple bond are sites of high electron density and, therefore, sites of chemical reactivity.

Example 4.8

Describe what will happen when 1-nonene is added to the following compounds:

(a) Water (b) 8-Methyl-1-nonyne

Strategy

First determine the polarity of the solvent and the solute. If both are nonpolar liquids, then the solute will be soluble in the solvent. Polar solutes will not dissolve in nonpolar solvents.

Solution

(a) 1-Nonene is an alkene and, therefore, nonpolar. It will not dissolve in a polar solvent such as water. Water and 1-nonene will form two layers; water, which has the higher density, will be the lower layer, and 1-nonene will be the upper layer.

(b) Because alkenes and alkynes are both nonpolar, they will dissolve in one another.

4.4 What Are Terpenes?

Among the compounds found in the essential oils of plants is a group of substances called **terpenes**, all of which have in common the property that their carbon skeletons can be divided into two or more carbon units that are identical with the carbon skeleton of isoprene. Carbon 1 of an **isoprene unit** is called the head, and carbon 4 is called the tail. A terpene is a compound in which the tail of one isoprene unit becomes bonded to the head of another isoprene unit.

Terpene A compound whose carbon skeleton can be divided into two or more units identical with the carbon skeleton of isoprene.

$$CH_2{=}\underset{}{\overset{CH_3}{\overset{|}{C}}}{-}CH{=}CH_2$$

2-Methyl-1,3-butadiene
(Isoprene)

head → C¹—C²(—C)—C³—C⁴ ← tail

Isoprene unit

Terpenes are among the most widely distributed compounds in the biological world, and a study of their structure provides a glimpse of the wondrous diversity that nature can generate from a simple carbon skeleton. Terpenes also illustrate an important principle of the molecular logic of living systems, namely, that, in building large molecules, small subunits are strung together enzymatically by an iterative process and are then chemically modified by precise enzyme-catalyzed reactions. Chemists use the same principles in the laboratory, but our methods do not have the precision and selectivity of the enzyme-catalyzed reactions of cellular systems.

Probably the terpenes most familiar to you, at least by odor, are components of the so-called essential oils obtained by steam distillation or ether extraction of various parts of plants. Essential oils contain the relatively low-molecular-weight substances that are in large part responsible for characteristic plant fragrances. Many essential oils, particularly those from flowers, are used in perfumes.

One example of a terpene obtained from an essential oil is myrcene, $C_{10}H_{16}$, a component of bayberry wax and oils of bay and verbena. Myrcene is a triene with a parent chain of eight carbon atoms and two one-carbon branches (Figure 4.2).

CHEMICAL CONNECTIONS 4B

Why Plants Emit Isoprene

Names like Virginia's *Blue Ridge,* Jamaica's *Blue Mountain Peak,* and Australia's *Blue Mountains* remind us of the bluish haze that hangs over wooded hills in the summertime. In the 1950s, it was discovered that this haze is rich in isoprene, which means that isoprene is far more abundant in the atmosphere than anyone thought. The haze is caused by the scattering of light from an aerosol produced by the photooxidation of isoprene and other hydrocarbons. Scientists now estimate that the global emission of isoprene by plants is 3×10^{11} kg/yr (3.3×10^{8} ton/yr), which represents approximately 2% of all carbon fixed by photosynthesis. A recent study of hydrocarbon emissions in the Atlanta area revealed that plants are by far the largest emitters of hydrocarbons, with plant-derived isoprene accounting for almost 60% of the total.

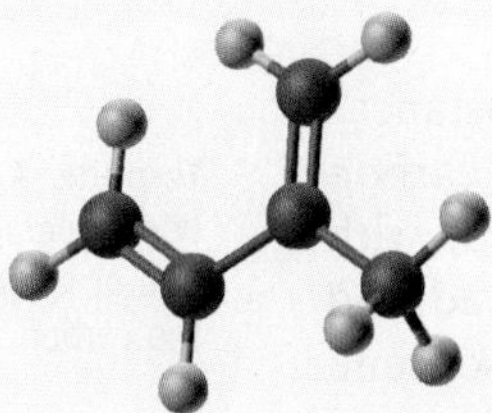

Why do plants emit so much isoprene into the atmosphere rather than use it for the synthesis of terpenes and other natural products? Tom Starkey, a University of Wisconsin plants physiologist, found that the emission of isoprene is extremely sensitive to temperature. Plants grown at 20°C do not emit isoprene, but they begin to emit it when the temperature of their leaves increases to 30°C. In certain plants, isoprene emission can increase as much as tenfold for a 10°C increase in leaf temperature. Starkey studied the relationship between temperature-induced leaf damage and isoprene concentration in leaves of the kudzu plant, a nonnative invasive vine. He discovered that leaf damage, as measured by the destruction of chlorophyll, begins to occur at 37.5°C in the absence of isoprene, but not until 45°C in its presence. Starkey speculates that isoprene dissolves in leaf membranes and in some way increases their tolerance to heat stress. Because isoprene is made rapidly and is also lost rapidly, its concentration correlates with temperature throughout the day.

The haze of the Smoky Mountains is caused by light-scattering from the aerosol produced by the photooxidation of isoprene and other hydrocarbons. *(Digital Vision)* Inset: A model of isoprene

QUESTION

Based on the information in this Chemical Connections what can you deduce about the physical properties of leaf cell membranes?

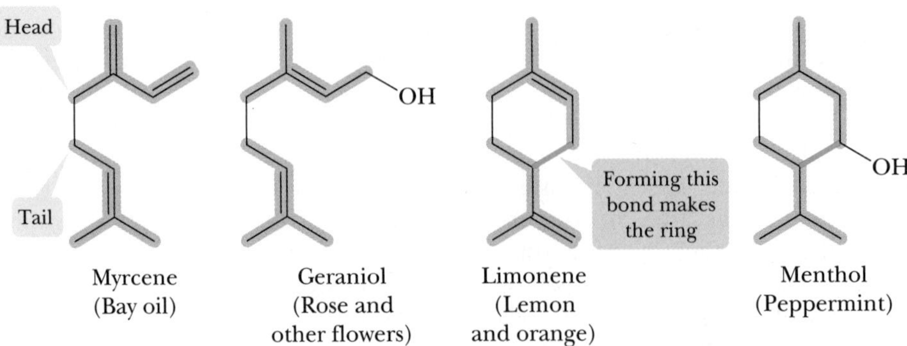

Figure 4.2 Four terpenes, each derived from two isoprene units (highlighted), bonded from the tail of the first unit to the head of the second unit. In limonene and menthol, the formation of an additional carbon–carbon bond creates a six-membered ring.

Farnesol, a terpene with the molecular formula $C_{15}H_{26}O$, includes three isoprene units. The head-to-tail bonds between the isoprene units are shown in black.

Farnesol
(Lily-of-the-valley)

Derivatives of both farnesol and geraniol are intermediates in the biosynthesis of cholesterol (Section 21.4B).

Vitamin A (Section 4.2G), a terpene with the molecular formula $C_{20}H_{30}O$, consists of four isoprene units linked head-to-tail and cross-linked at one point to form a six- membered ring.

Key Terms and Concepts

alkene (p. 115)
alkyne (p. 115)
arene (p. 116)
***cis–trans* isomerism in alkenes** (p. 118)
***cis–trans* isomerism in cycloalkenes** (p. 125)
designating configuration of an alkene (p. 122)
the *cis–trans* system (p. 121)
the E,Z system (p. 122)
diene (p. 126)
dispersion forces among alkenes and alkynes (p. 128)
isoprene unit (p. 129)
IUPAC nomenclature of alkenes (p. 119)
IUPAC nomenclature of alkynes (p. 119)
orbital Overlap model of a carbon–carbon double bond (p. 117)
polarity of alkenes (p. 128)
polyene (p. 126)
priority rules of the E,Z system (p. 122)
restricted rotation about a carbon–carbon double bond (p. 117)
terpene (p. 129)
thermal cracking (p. 116)
triene (p. 126)
unsaturated hydrocarbon (p. 115)

Summary of Key Questions

4.1 What Are the Structures and Shapes of Alkenes and Alkynes?

- An **alkene** is an unsaturated hydrocarbon that contains a carbon–carbon double bond.
- Alkenes have the general formula C_nH_{2n}.
- An **alkyne** is an unsaturated hydrocarbon that contains a carbon–carbon triple bond.
- Alkynes have the general formula C_nH_{2n-2}.
- According to the **orbital Overlap model**, a carbon–carbon double bond consists of one sigma bond formed by the overlap of sp^2 hybrid orbitals and one pi bond formed by the overlap of parallel $2p$ atomic orbitals. It takes approximately 264 kJ/mol (63 kcal/mol) to break the pi bond in ethylene.
- A carbon–carbon triple bond consists of one sigma bond formed by the overlap of sp hybrid orbitals and two pi bonds formed by the overlap of pairs of parallel $2p$ orbitals.
- The structural feature that makes ***cis–trans* isomerism** possible in alkenes is restricted rotation about the two carbons of the double bond.
- To date, *trans*-cyclooctene is the smallest *trans*-cycloalkene that has been prepared in pure form and is stable at room temperature.

4.2 How Do We Name Alkenes and Alkynes?

- According to the IUPAC system, we show the presence of a **carbon–carbon double bond** by changing the infix of the parent hydrocarbon from **-an-** to **-en-**.
- The names *vinyl* and *allyl* are commonly used to show the presence of $-CH{=}CH_2$ and $-CH_2CH{=}CH_2$ groups.
- We show the presence of a **carbon–carbon triple bond** by changing the infix of the parent alkane from **-an-** to **-yn-**.
- The orientation of the carbon atoms of the parent chain about the double bond determines whether an alkene is *cis* or *trans*. If atoms of the parent are on the same side of the double bond, the configuration of the alkene is *cis*; if they are on opposite sides, the configuration is *trans*.
- Using a set of priority rules, we can also specify the configuration of a carbon–carbon double bond by the **E,Z system**.
- If the two groups of higher priority are on the same side of the double bond, the configuration of the alkene is **Z** (German: *zusammen*, together); if they are on opposite sides, the configuration is **E** (German: *entgegen*, opposite).
- To name an alkene containing two or more double bonds, we change the infix to -adien-, -atrien-, and so forth. Compounds containing several double bonds are called polyenes.

4.3 What Are the Physical Properties of Alkenes and Alkynes?

- Alkenes and alkynes are nonpolar compounds, and the only interactions between their molecules are **dispersion forces**.
- The physical properties of alkenes and alkynes are similar to those of alkanes.

4.4 What Are Terpenes?

- The characteristic structural feature of a **terpene** is a carbon skeleton that can be divided into two or more **isoprene units**.
- Terpenes illustrate an important principle of the molecular logic of living systems, namely, that in building large molecules, small subunits are strung together by an iterative process and are then chemically modified by precise enzyme-catalyzed reactions.

Quick Quiz

Answer true or false to the following questions to assess your general knowledge of the concepts in this chapter. If you have difficulty with any of them, you should review the appropriate section in the chapter (shown in parentheses) before attempting the more challenging end-of-chapter problems.

1. Ethylene and acetylene are constitutional isomers. (4.2)
2. Alkanes that are liquid at room temperature are insoluble in water and when added to water will float on water. (4.3)
3. The bulk of the ethylene used by the chemical industry worldwide is obtained from nonrenewable resources. (4.1)
4. Alkenes and alkynes are nonpolar molecules. (4.3)
5. The IUPAC name of $CH_3CH{=}CHCH_3$ is 1,2-dimethylethylene. (4.2)
6. Cyclohexane and 1-hexene are constitutional isomers. (4.1)
7. The IUPAC name of an alkene is derived from the name of the longest chain of carbon atoms that contains the double bond. (4.2)
8. There are two classes of unsaturated hydrocarbons, alkenes and alkynes. (4.1)
9. Both geraniol and menthol (Figure 4.2) show *cis–trans* isomerism. (4.4)
10. Terpenes are identified by their carbon skeletons, namely, that one that can be divided into five-carbon isoprene units. (4.4)
11. 1,2-Dimethylcyclohexene shows *cis–trans* isomerism. (4.2)
12. 2-Methyl 2-butene shows *cis–trans* isomerism. (4.2)
13. Both ethylene and acetylene are planar molecules. (4.1)
14. The physical properties of alkenes are similar to those of alkanes with the same carbon skeletons. (4.3)
15. Isoprene is the common name for 2-methyl-1,3-butadiene. (4.4)

Answers: (1) F (2) T (3) T (4) T (5) F (6) T (7) T (8) F (9) T (10) T (11) F (12) F (13) T (14) T (15) T

Problems

A problem marked with an asterisk indicates an applied "real world" problem. Answers to problems whose numbers are printed in blue are given in Appendix D.

Section 4.1 Structure of Alkenes and Alkynes

4.7 Describe what will happen when *trans*-3-heptene is added to the following compounds:

(a) Cyclohexane (b) Ammonia (*l*)

4.8 Each carbon atom in ethane and in ethylene is surrounded by eight valence electrons and has four bonds to it. Explain how VSEPR (Section 1.3) predicts a bond angle of 109.5° about each carbon in ethane, but an angle of 120° about each carbon in ethylene.

4.9 Explain the difference between saturated and unsaturated.

4.10 Use valence-shell electron-pair repulsion (VSEPR) to predict all bond angles about each of the following highlighted carbon atoms.

(a) 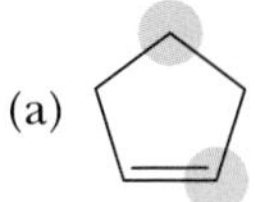

(b) 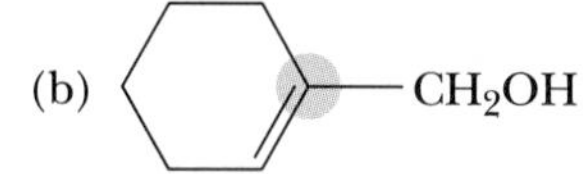

(c) $HC{\equiv}C{-}CH{=}CH_2$

(d) 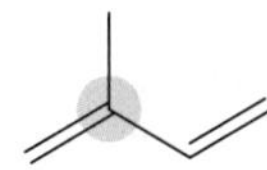

4.11 For each highlighted carbon atom in Problem 4.10, identify which orbitals are used to form each sigma bond and which are used to form each pi bond.

4.12 Predict all bond angles about each highlighted carbon atom:

(a) 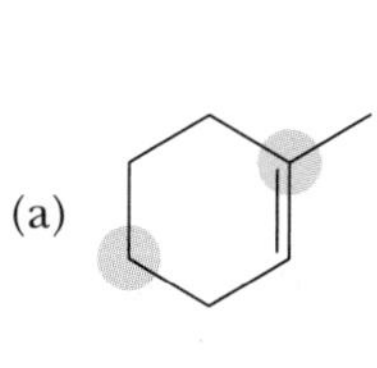

(b) 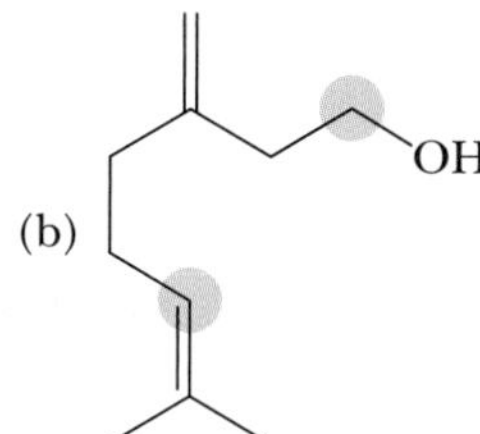

(c) 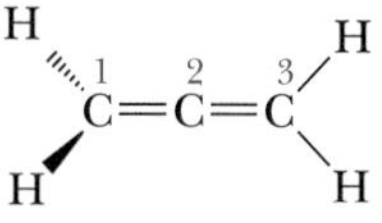

(d) 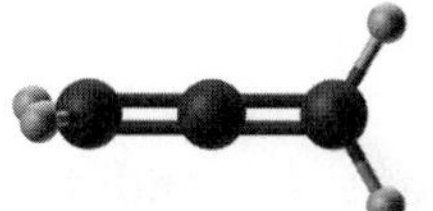

4.13 For each highlighted carbon atom in Problem 4.12, identify which orbitals are used to form each sigma bond and which are used to form each pi bond.

4.14 Following is the structure of 1,2-propadiene (allene). In it, the plane created by H—C—H of carbon 1 is perpendicular to that created by H—C—H of carbon 3.

H, H (on C1) — $\overset{1}{C}{=}\overset{2}{C}{=}\overset{3}{C}$ — H, H (on C3)

1,2-Propadiene (Allene)

Ball-and-stick model

(a) State the orbital hybridization of each carbon in allene.

(b) Account for the molecular geometry of allene in terms of the orbital overlap model. Specifically, explain why all four hydrogen atoms are not in the same plane.

Section 4.2 Nomenclature of Alkenes and Alkynes

4.15 Draw a structural formula for each compound: **(See Examples 4.1, 4.5)**

(a) *trans*-2-Methyl-3-hexene
(b) 2-Methyl-3-hexyne
(c) 2-Methyl-1-butene
(d) 3-Ethyl-3-methyl-1-pentyne
(e) 2,3-Dimethyl-2-butene
(f) *cis*-2-Pentene
(g) (*Z*)-1-Chloropropene
(h) 3-Methylcyclohexene

4.16 Draw a structural formula for each compound: **(See Examples 4.1, 4.5)**

(a) 1-Isopropyl-4-methylcyclohexene
(b) (6*E*)-2,6-Dimethyl-2,6-octadiene
(c) *trans*-1,2-Diisopropylcyclopropane
(d) 2-Methyl-3-hexyne
(e) 2-Chloropropene
(f) Tetrachloroethylene

4.17 Write the IUPAC name for each compound: **(See Examples 4.1, 4.5)**

(a) 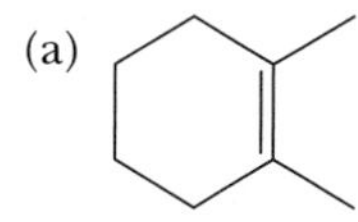

(b) 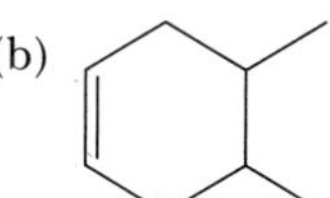

(c) 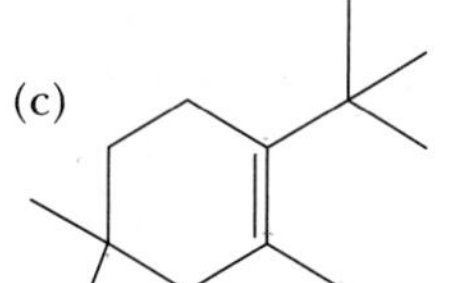

(d) 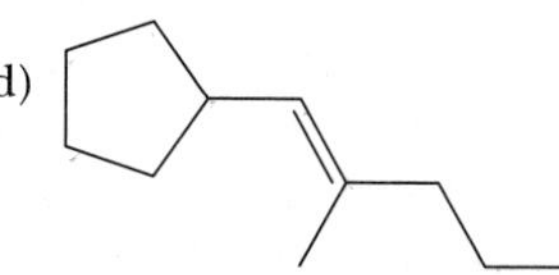

4.18 Write the IUPAC name for each compound: **(See Examples 4.1, 4.5)**

(a) 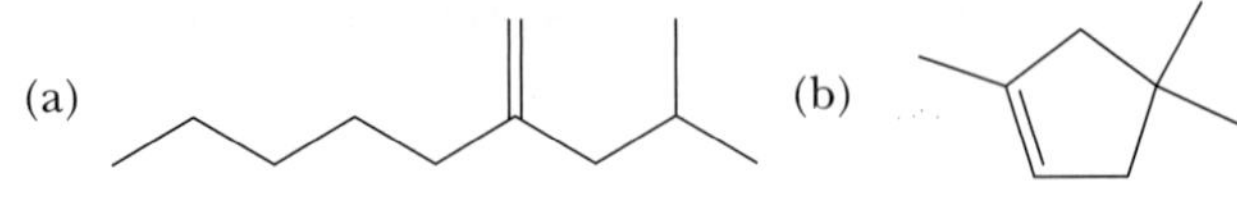 (b)

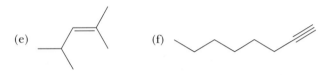

(c) 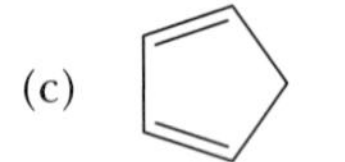 (d)

(e) (f)

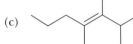

(g) (h)

4.19 Explain why each name is incorrect, and then write a correct name for the intended compound: **(See Examples 4.1, 4.5)**

(a) 1-Methylpropene (b) 3-Pentene
(c) 2-Methylcyclohexene (d) 3,3-Dimethylpentene
(e) 4-Hexyne (f) 2-Isopropyl-2-butene

4.20 Explain why each name is incorrect, and then write a correct name for the intended compound: **(See Examples 4.1, 4.5)**

(a) 2-Ethyl-1-propene (b) 5-Isopropylcyclohexene
(c) 4-Methyl-4-hexene (d) 2-*sec*-Butyl-1-butene
(e) 6,6-Dimethylcyclohexene (f) 2-Ethyl-2-hexene

Sections 4.2 and 4.3 *Cis–Trans* (E/Z) Isomerism in Alkenes and Cycloalkenes

4.21 Which of these alkenes show *cis–trans* isomerism? For each that does, draw structural formulas for both isomers. **(See Example 4.2)**

(a) 1-Hexene (b) 2-Hexene
(c) 3-Hexene (d) 2-Methyl-2-hexene
(e) 3-Methyl-2-hexene (f) 2,3-Dimethyl-2-hexene

4.22 Which of these alkenes show *cis–trans* isomerism? For each that does, draw structural formulas for both isomers. **(See Example 4.2)**

(a) 1-Pentene
(b) 2-Pentene
(c) 3-Ethyl-2-pentene
(d) 2,3-Dimethyl-2-pentene
(e) 2-Methyl-2-pentene
(f) 2,4-Dimethyl-2-pentene

4.23 Which alkenes can exist as pairs of E/Z isomers? For each alkene that does, draw both isomers. **(See Examples 4.3, 4.4)**

(a) $CH_2{=}CHBr$ (b) $CH_3CH{=}CHBr$
(c) $(CH_3)_2C{=}CHCH_3$ (d) $(CH_3)_2CHCH{=}CHCH_3$

4.24 There are three compounds with the molecular formula $C_2H_2Br_2$. Two of these compounds have a dipole greater than zero, and one has no dipole. Draw structural formulas for the three compounds, and explain why two have dipole moments but the third one has none.

4.25 Name and draw structural formulas for all alkenes with the molecular formula C_5H_{10}. As you draw these alkenes, remember that *cis* and *trans* isomers are different compounds and must be counted separately.

4.26 Name and draw structural formulas for all alkenes with the molecular formula C_6H_{12} that have the following carbon skeletons (remember *cis* and *trans* isomers):

(a) C—C(—C)—C—C—C (b) C—C(—C)—C(—C)—C

(c) C—C(—C)(—C)—C—C

4.27 Arrange the groups in each set in order of increasing priority: **(See Examples 4.3, 4.4)**

(a) $-CH_3$, $-Br$, $-CH_2CH_3$
(b) $-OCH_3$, $-CH(CH_3)_2$, $-CH_2CH_2NH_2$
(c) $-CH_2OH$, $-COOH$, $-OH$
(d) $-CH{=}CH_2$, $-CH{=}O$, $-CH(CH_3)_2$

4.28 Name each alkene and specify its configuration using the E,Z system. **(See Examples 4.3, 4.4)**

(a) 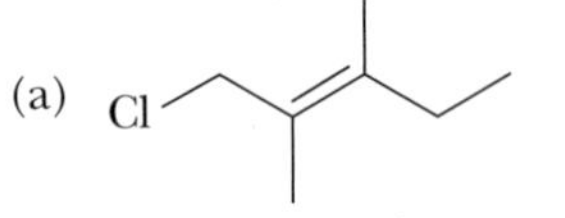

(b) 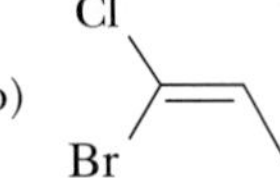

(c)

4.29 Draw the structural formula for at least one bromoalkene with molecular formula C_5H_9Br that **(a)** shows E,Z isomerism and **(b)** does not show E,Z isomerism.

4.30 Is *cis–trans* isomerism possible in alkanes? Explain.

4.31 For each molecule that shows *cis–trans* isomerism, draw the *cis* isomer:

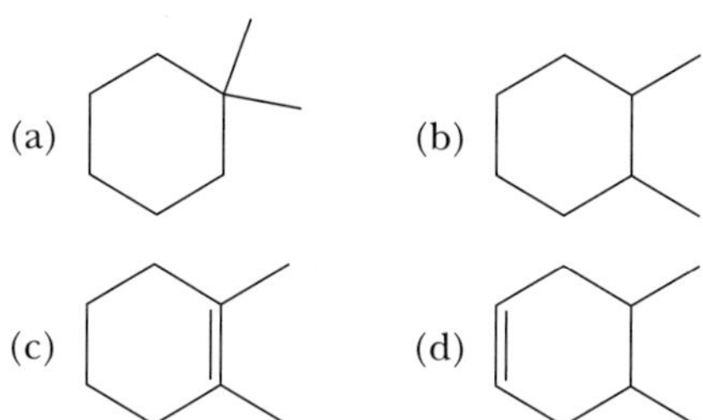

4.32 Explain why each name is incorrect or incomplete, and then write a correct name: **(See Examples 4.3, 4.4)**

(a) (Z)-2-Methyl-1-pentene

(b) (E)-3,4-Diethyl-3-hexene

(c) *trans*-2,3-Dimethyl-2-hexene

(d) (1Z,3Z)-2,3-Dimethyl-1,3-butadiene

4.33 Draw structural formulas for all compounds with the molecular formula C_5H_{10} that are

(a) Alkenes that do not show *cis–trans* isomerism.

(b) Alkenes that do show *cis–trans* isomerism.

(c) Cycloalkanes that do not show *cis–trans* isomerism.

(d) Cycloalkanes that do show *cis–trans* isomerism.

***4.34** β-Ocimene, a triene found in the fragrance of cotton blossoms and several essential oils, has the IUPAC name (3Z)-3,7-dimethyl-1,3,6-octatriene. Draw a structural formula for β-ocimene.

***4.35** Oleic acid and elaidic acid are, respectively, the *cis* and *trans* isomers of 9-octadecenoic acid. One of these fatty acids, a colorless liquid that solidifies at 4°C, is a major component of butterfat. The other, a white solid with a melting point of 44–45°C, is a major component of partially hydrogenated vegetable oils. Which of these two fatty acids is the *cis* isomer and which is the *trans* isomer? (*Hint*: Think about the geometry of packing and the relative strengths of the resulting dispersion forces.)

4.36 Determine whether the structures in each set represent the same molecule, *cis–trans* isomers, or constitutional isomers. If they are the same molecule, determine whether they are in the same or different conformations as a result of rotation about a carbon–carbon single bond. **(See Examples 4.6, 4.7)**

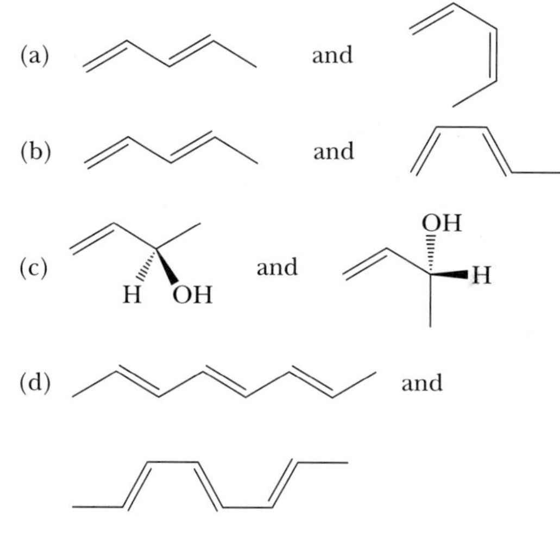

Section 4.4 Terpenes

***4.37** Show that the structural formula of vitamin A (Section 4.2G) can be divided into four isoprene units joined by head-to-tail linkages and cross-linked at one point to form the six-membered ring.

***4.38** Following is the structural formula of lycopene, a deep-red compound that is partially responsible for the red color of ripe fruits, especially tomatoes:

Lycopene

Approximately 20 mg of lycopene can be isolated from 1 kg of fresh, ripe tomatoes.

(a) Show that lycopene is a terpene; that is, show that lycopene's carbon skeleton can be divided into two sets of four isoprene units with the units in each set joined head-to-tail.

(b) How many of the carbon–carbon double bonds in lycopene have the possibility for *cis*–trans isomerism? Lycopene is the all-*trans* isomer.

***4.39** As you might suspect, β-carotene, a precursor of vitamin A, was first isolated from carrots. Dilute solutions of β-carotene are yellow—hence its use as a food coloring. In plants, it is almost always present in combination with chlorophyll to assist in the harvesting of the energy of sunlight. As tree leaves die in the fall, the green of their chlorophyll molecules is replaced by the yellows and reds of carotene and carotene-related molecules.

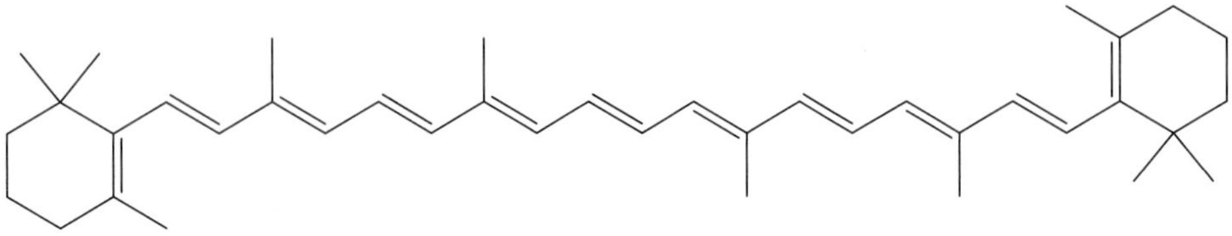

β-Carotene

(a) Compare the carbon skeletons of β-carotene and lycopene. What are the similarities? What are the differences?

(b) Show that β-carotene is a terpene.

***4.40** α-Santonin, isolated from the flower heads of certain species of artemisia, is an anthelmintic—that is, a drug used to rid the body of worms (helminths). It has been estimated that over one-third of the world's population is infested with these parasites. Farnesol is an alcohol with a florid odor:

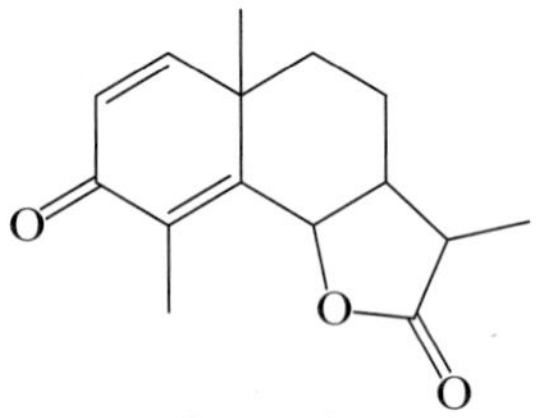

Santonin

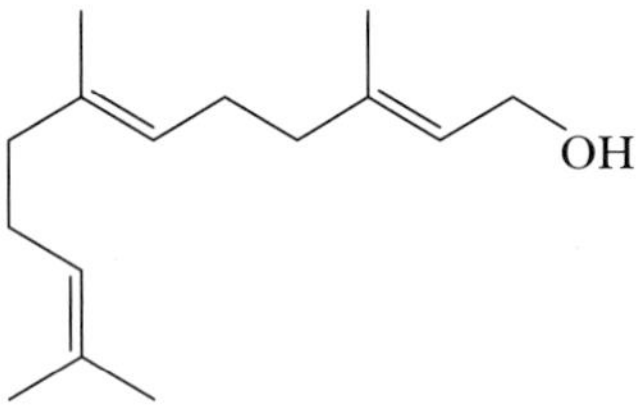

Farnesol

Locate the three isoprene units in santonin, and show how the carbon skeleton of farnesol might be coiled and then cross-linked (a carbon–carbon bond formed between two carbons) to give santonin. Two different coiling patterns of the carbon skeleton of farnesol can lead to santonin. Try to find them both.

***4.41** Periplanone is a pheromone (a chemical sex attractant) isolated from a species of cockroach. Show that the carbon skeleton of periplanone classifies it as a terpene:

Periplanone

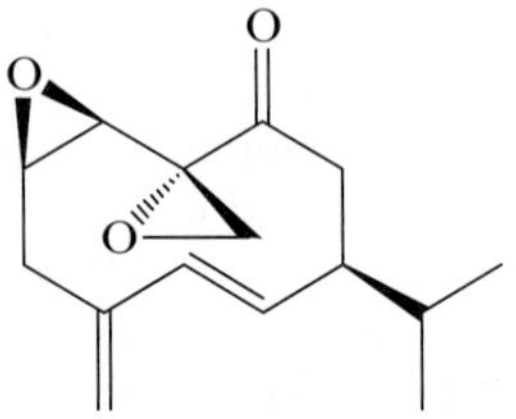

***4.42** Gossypol, a compound found in the seeds of cotton plants, has been used as a male contraceptive in overpopulated countries such as China. Show that gossypol is a terpene:

Gossypol

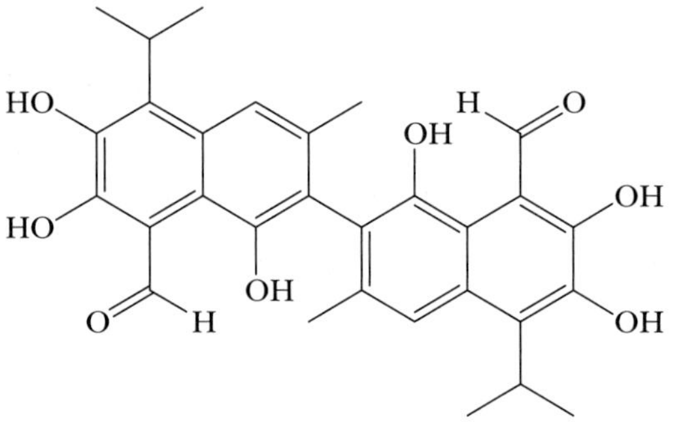

***4.43** In many parts of South America, extracts of the leaves and twigs of *Montanoa tomentosa* are used as a contraceptive, to stimulate menstruation, to facilitate labor, and as an abortifacient. The compound responsible for these effects is zoapatanol:

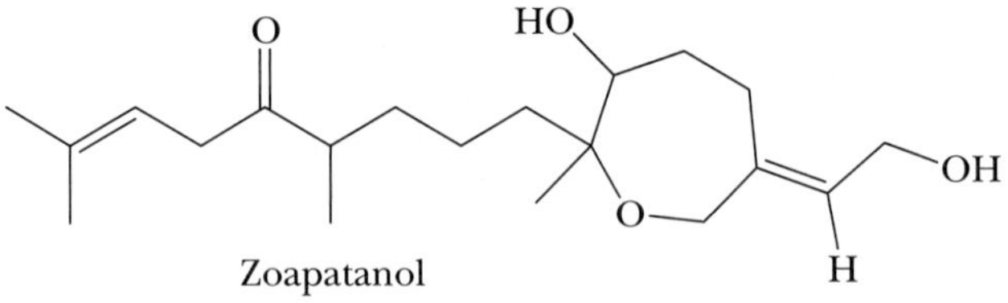

Zoapatanol

(a) Show that the carbon skeleton of zoapatanol can be divided into four isoprene units.

(b) Specify the configuration about the carbon–carbon double bond to the seven-membered ring, according to the E,Z system.

(c) How many *cis–trans* isomers are possible for zoap#atanol? Consider the possibilities for *cis–trans* isomerism in cyclic compounds and about carbon–carbon double bonds.

*4.44 Pyrethrin II and pyrethrosin are natural products isolated from plants of the chrysanthemum family:

Pyrethrin II   Pyrethrosin

Chrysanthemum blossoms.
(Scott Camazine/Photo Researchers, Inc.)

Pyrethrin II is a natural insecticide and is marketed as such.

(a) Label all carbon–carbon double bonds in each about which *cis–trans* isomerism is possible.

(b) Why are *cis–trans* isomers possible about the three-membered ring in pyrethrin II, but not about its five-membered ring?

(c) Show that the ring system of pyrethrosin is composed of three isoprene units.

*4.45 Cuparene and herbertene are naturally occurring compounds isolated from various species of lichen:

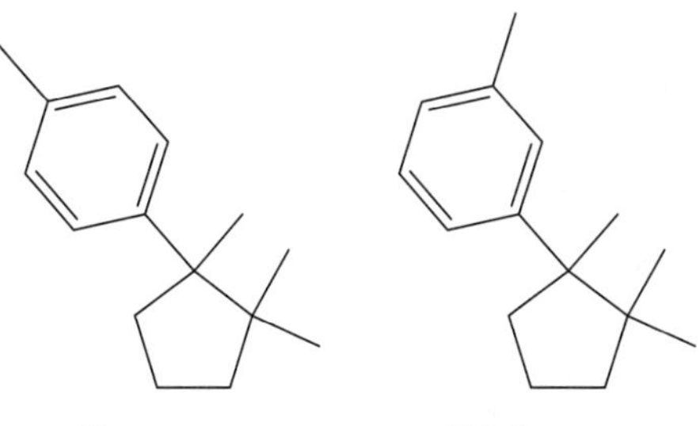

Cuparene   Herbertene

Determine whether one or both of these compounds can be classified as terpenes.

Looking Ahead

4.46 Explain why the central carbon–carbon single bond in 1,3-butadiene is slightly shorter than the central carbon–carbon single bond in 1-butene:

1.47 Å   1.51 Å

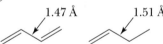

1,3-butadiene   1-butene

4.47 What effect might the ring size in the following cycloalkenes have on the reactivity of the C=C double bond in each?

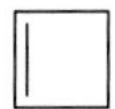 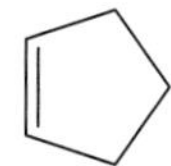 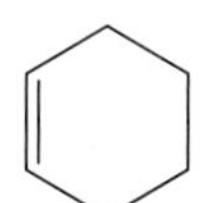

4.48 What effect might each substituent have on the electron density surrounding the alkene C=C bond; that is, how does each substituent affect whether each carbon of the C–C double bond is partially positive or partially negative?

(a) OCH_3 (b) CN (c) $Si(CH_3)_3$

*4.49 In Section 21.1 on the biochemistry of fatty acids, we will study the following three long-chain unsaturated carboxylic acids:

Oleic acid

$CH_3(CH_2)_7CH{=}CH(CH_2)_7COOH$

Linoleic acid

$CH_3(CH_2)_4CH{=}CHCH_2CH{=}CH(CH_2)_7COOH$

Linolenic acid

$CH_3CH_2CH{=}CHCH_2CH{=}CHCH_2CH{=}CH(CH_2)_7COOH$

Each has 18 carbons and is a component of animal fats, vegetable oils, and biological membranes. Because of their presence in animal fats, they are called fatty acids.

(a) How many *cis–trans* isomers are possible for each fatty acid?

(b) These three fatty acids occur in biological membranes almost exclusively in the *cis* configuration. Draw line-angle formulas for each fatty acid, showing the *cis* configuration about each carbon–carbon double bond.

*4.50 Assign an E or a Z configuration and a *cis* or a *trans* configuration to these carboxylic acids, each of which is an intermediate in the citric acid cycle. Under each is given its common name.

(a) Fumaric acid   (b) Aconitic acid

5 Reactions of Alkenes and Alkynes

KEY QUESTIONS

5.1 What Are the Characteristic Reactions of Alkenes?

5.2 What Is a Reaction Mechanism?

5.3 What Are the Mechanisms of Electrophilic Additions to Alkenes?

5.4 What Are Carbocation Rearrangements?

5.5 What Is Hydroboration–Oxidation of an Alkene?

5.6 What Is Ozonolysis of an Alkene?

5.7 How Can an Alkene Be Reduced to an Alkane?

5.8 Why Are 1-Alkynes (Terminal Alkynes) Weak Acids?

5.9 How Can an Acetylide Anion Be Used to Create a New Carbon–Carbon Bond?

5.10 What Is the Product of the Acid-Catalyzed Hydration of an Alkyne?

5.11 How Can Alkynes Be Reduced to Alkenes and Alkanes?

HOW TO

5.1 How to Draw Mechanisms

5.2 How to Predict the Products of an Ozonolysis Reaction

CHEMICAL CONNECTIONS

5A *Cis–Trans* Isomerims in Vision

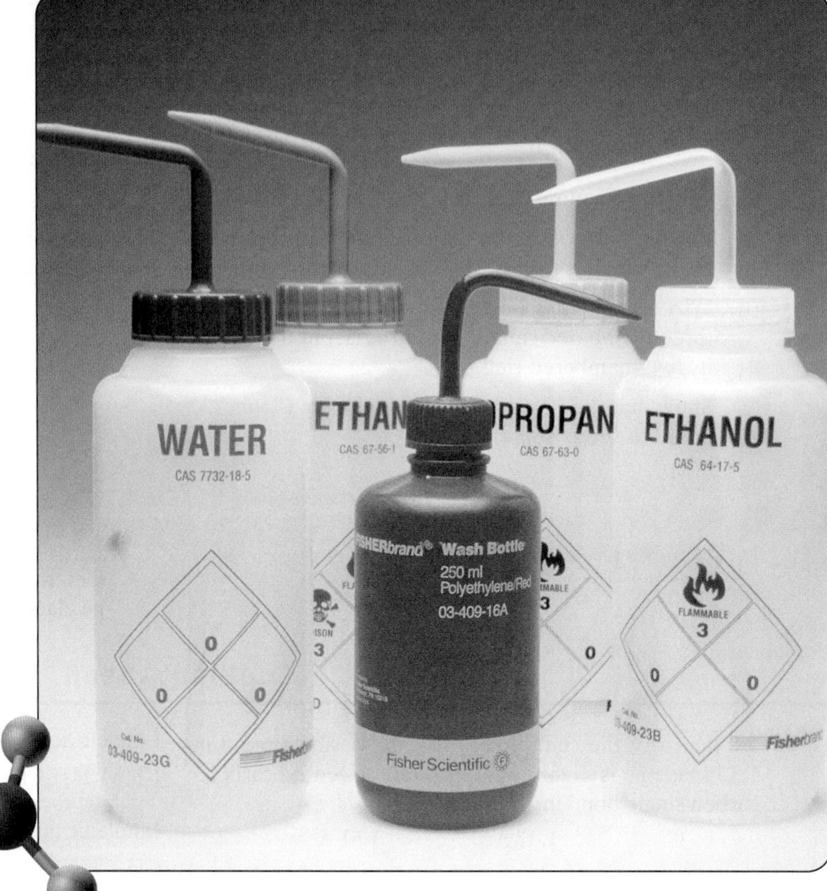

These wash bottles are made of polyethylene. Inset: A model of ethylene. *(Charles D. Winters)*

In this chapter, we begin our systematic study of reaction mechanisms, one of the most important unifying concepts in organic chemistry. We use the reactions of alkenes as the vehicle to introduce this concept.

5.1 What Are the Characteristic Reactions of Alkenes?

The most characteristic reaction of alkenes is **addition to the carbon–carbon double bond** in such a way that the pi bond is broken and, in its place, sigma bonds are formed to two new atoms or groups of atoms. Several examples of reactions at the carbon–carbon double bond are shown in Table 5.1, along with the descriptive name(s) associated with each.

Table 5.1 Characteristic Reactions of Alkenes

Reaction	Descriptive Name(s)
$>C{=}C< + HX \longrightarrow -\underset{H}{\overset{\vert}{C}}-\underset{Cl\ (X)}{\overset{\vert}{C}}-$; X = Cl, Br, I	Hydrochlorination (hydrohalogenation)
$>C{=}C< + H_2O \longrightarrow -\underset{H}{\overset{\vert}{C}}-\underset{OH}{\overset{\vert}{C}}-$	Hydration
$>C{=}C< + X_2 \longrightarrow -\underset{\vert}{\overset{(X)\ Br}{C}}-\underset{Br\ (X)}{\overset{\vert}{C}}-$; $X_2 = Cl_2, Br_2$	Bromination (halogenation)
$>C{=}C< + BH_3 \longrightarrow -\underset{H}{\overset{\vert}{C}}-\underset{BH_2}{\overset{\vert}{C}}-$	Hydroboration
$>C{=}C< + H_2 \longrightarrow -\underset{H}{\overset{\vert}{C}}-\underset{H}{\overset{\vert}{C}}-$	Hydrogenation (reduction)

From the perspective of the chemical industry, the single most important reaction of ethylene and other low-molecular-weight alkenes is the production of **chain-growth polymers** (Greek: *poly*, many, and *meros*, part). In the presence of certain catalysts called *initiators*, many alkenes form polymers by the addition of **monomers** (Greek: *mono*, one, and *meros*, part) to a growing polymer chain, as illustrated by the formation of polyethylene from ethylene:

$$nCH_2 = CH_2 \xrightarrow{\text{initiator}} -\!\!\left(CH_2CH_2\right)\!\!-_n$$

In alkene polymers of industrial and commercial importance, n is a large number, typically several thousand. We discuss this alkene reaction in Chapter 17.

5.2 What Is a Reaction Mechanism?

A **reaction mechanism** describes in detail how a chemical reaction occurs. It describes which bonds break and which new ones form, as well as the order and relative rates of the various bond-breaking and bond-forming steps. If the reaction takes place in solution, the reaction mechanism describes the role of the solvent; if the reaction involves a catalyst, the reaction mechanism describes the role of the catalyst.

Reaction mechanism A step-by-step description of how a chemical reaction occurs.

A. Energy Diagrams and Transition States

To understand the relationship between a chemical reaction and energy, think of a chemical bond as a spring. As a spring is stretched from its resting position, its energy increases. As it returns to its resting position, its energy decreases. Similarly, during a chemical reaction, bond breaking corresponds to an increase in energy, and bond forming corresponds to a decrease in energy. We use an **energy diagram** to show the changes in energy that occur in going from reactants to products. Energy is measured along the vertical axis, and the change in position of the atoms during a reaction is measured on the horizontal axis, called the **reaction coordinate**. The reaction coordinate indicates how far the reaction has progressed, from no reaction to a completed reaction.

Energy diagram A graph showing the changes in energy that occur during a chemical reaction; energy is plotted on the y-axis, and the progress of the reaction is plotted on the x-axis.

Reaction coordinate A measure of the progress of a reaction, plotted on the x-axis in an energy diagram.

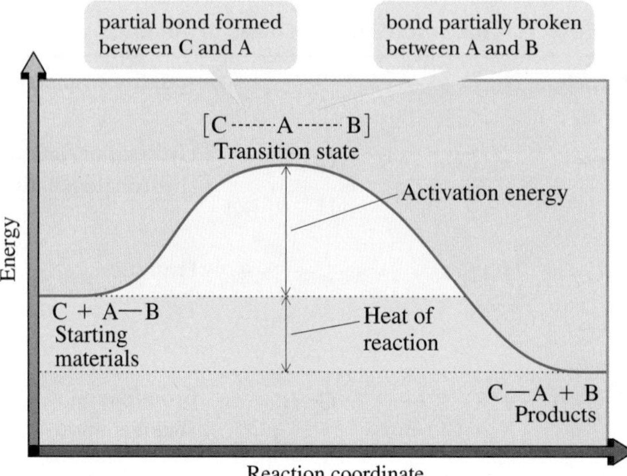

Figure 5.1 An energy diagram for a one-step reaction between C and A—B. The dashed lines in the transition state indicate that the new C—A bond is partially formed and the A—B bond is partially broken. The energy of the reactants is higher than that of the products—the reaction is exothermic.

Figure 5.1 shows an energy diagram for the reaction of C + A—B to form C—A + B. This reaction occurs in one step, meaning that bond breaking in reactants and bond forming in products occur simultaneously.

Heat of reaction The difference in energy between reactants and products.

Exothermic reaction A reaction in which the energy of the products is lower than the energy of the reactants; a reaction in which heat is liberated.

Endothermic reaction A reaction in which the energy of the products is higher than the energy of the reactants; a reaction in which heat is absorbed.

Transition state An unstable species of maximum energy formed during the course of a reaction; a maximum on an energy diagram.

The difference in energy between the reactants and products is called the **heat of reaction, ΔH**. If the energy of the products is lower than that of the reactants, heat is released and the reaction is called **exothermic**. If the energy of the products is higher than that of the reactants, heat is absorbed and the reaction is called **endothermic**. The one-step reaction shown in Figure 5.1 is exothermic.

A **transition state** is the point on the reaction coordinate at which the energy is at a maximum. At the transition state, sufficient energy has become concentrated in the proper bonds so that bonds in the reactants break. As they break, energy is redistributed and new bonds form, giving products. Once the transition state is reached, the reaction proceeds to give products, with the release of energy.

A transition state has a definite geometry, a definite arrangement of bonding and nonbonding electrons, and a definite distribution of electron density and charge. Because a transition state is at an energy maximum on an energy diagram, we cannot isolate it and we cannot determine its structure experimentally. Its lifetime is on the order of a picosecond (the duration of a single bond vibration). As we will see, however, even though we cannot observe a transition state directly by any experimental means, we can often infer a great deal about its probable structure from other experimental observations.

For the reaction shown in Figure 5.1, we use dashed lines to show the partial bonding in the transition state. At the same time, as C begins to form a new covalent bond with A, the covalent bond between A and B begins to break. Upon completion of the reaction, the A—B bond is fully broken and the C—A bond is fully formed.

Activation energy The difference in energy between reactants and the transition state.

The difference in energy between the reactants and the transition state is called the **activation energy**. The activation energy is the minimum energy required for a reaction to occur; it can be considered an energy barrier for the reaction. The activation energy determines the rate of a reaction—that is, how fast the reaction occurs. If the activation energy is large, a very few molecular collisions occur with sufficient energy to reach the transition state, and the reaction is slow. If the activation energy is small, many collisions generate sufficient energy to reach the transition state and the reaction is fast.

Reaction intermediate An unstable species that lies in an energy minimum between two transition states.

In a reaction that occurs in two or more steps, each step has its own transition state and activation energy. Shown in Figure 5.2 is an energy diagram for the conversion of reactants to products in two steps. A **reaction intermediate** corresponds to an energy minimum between two transition states, in this case an intermediate between transition states 1 and 2. Note that because the energies of the reaction intermediates we describe are higher than the energies of either the reactants or the products, these intermediates are highly reactive, and rarely, if ever, can one be isolated.

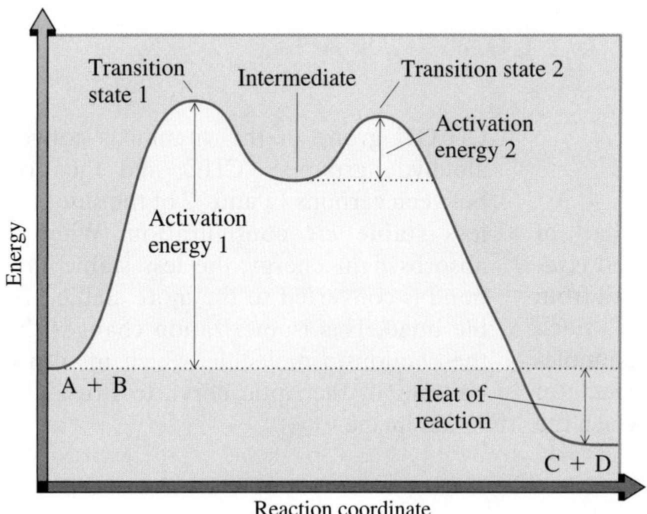

Figure 5.2
Energy diagram for a two-step reaction involving the formation of an intermediate. The energy of the reactants is higher than that of the products, and energy is released in the conversion of A + B to C + D.

The slowest step in a multistep reaction, called the **rate-determining step**, is the step that crosses the highest energy barrier. In the two-step reaction shown in Figure 5.2, Step 1 crosses the higher energy barrier and is, therefore, the rate-determining step.

Rate-determining step The step in a reaction sequence that crosses the highest energy barrier; the slowest step in a multistep reaction.

Example 5.1

Draw an energy diagram for a two-step exothermic reaction in which the second step is rate determining.

Strategy

A two-step reaction involves the formation of an intermediate. In order for the reaction to be exothermic, the products must be lower in energy than the reactants. In order for the second step to be rate determining, it must cross the higher energy barrier.

Solution

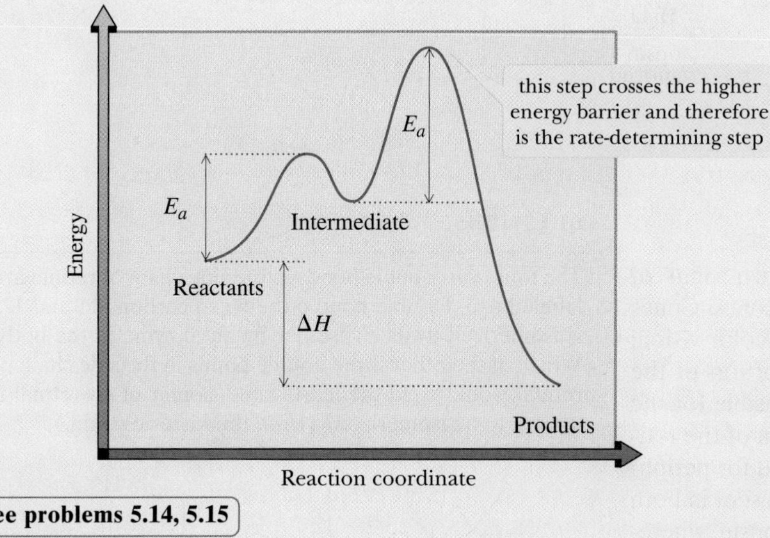

See problems 5.14, 5.15

Problem 5.1

In what way would the energy diagram drawn in Example 5.1 change if the reaction were endothermic?

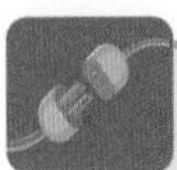

CHEMICAL CONNECTIONS 5A

Cis-Trans *Isomerism in Vision*

The retina—the light-detecting layer in the back of our eyes—contains reddish compounds called *visual pigments*. Their name, *rhodopsin*, is derived from the Greek word meaning "rose colored." Each rhodopsin molecule is a combination of one molecule of a protein called opsin and one molecule of 11-*cis*-retinal, a derivative of vitamin A in which the CH_2OH group of the vitamin is converted to an aldehyde group, —CHO, and the double bond between carbons 11 and 12 of the side chain is in the less stable *cis* configuration. When rhodopsin absorbs light energy, the less stable 11-*cis* double bond is converted to the more stable 11-*trans* double bond. This isomerization changes the shape of the rhodopsin molecule, which in turn causes the neurons of the optic nerve to fire and produce a visual image.

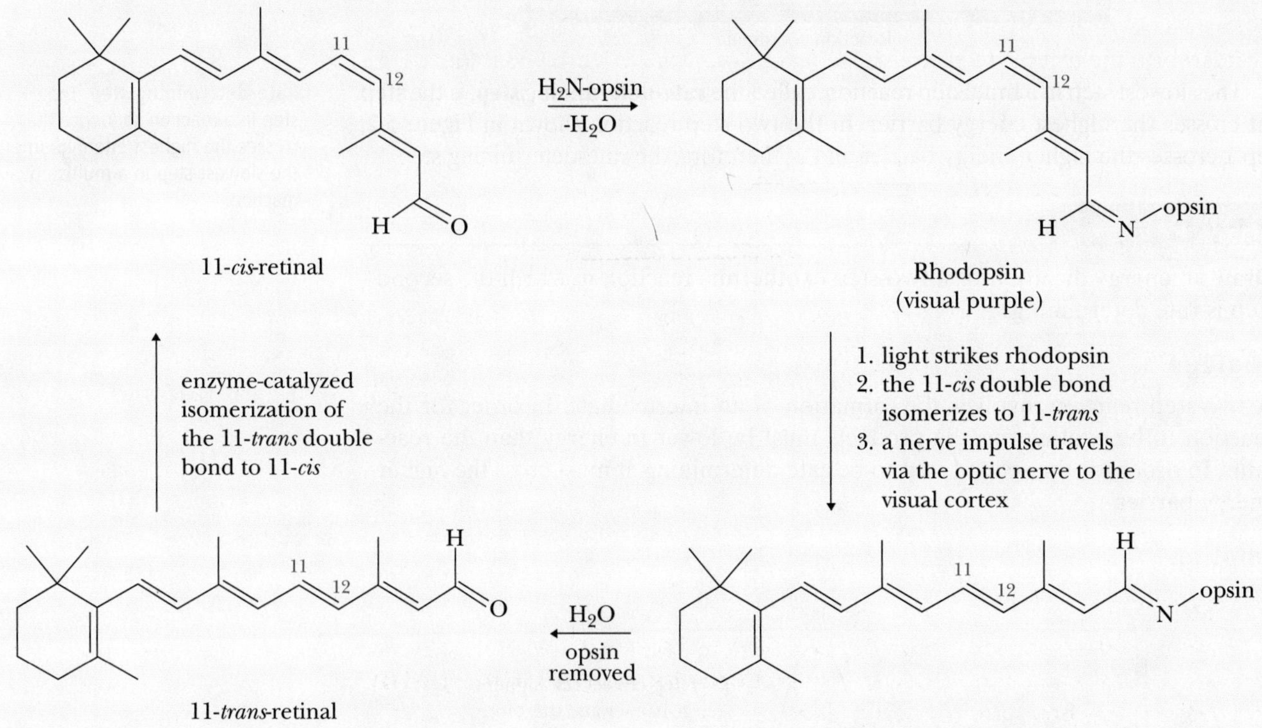

11-*cis*-retinal → (H_2N-opsin, $-H_2O$) → Rhodopsin (visual purple)

1. light strikes rhodopsin
2. the 11-*cis* double bond isomerizes to 11-*trans*
3. a nerve impulse travels via the optic nerve to the visual cortex

H_2O, opsin removed → 11-*trans*-retinal

enzyme-catalyzed isomerization of the 11-*trans* double bond to 11-*cis*

The retina of vertebrates contains two kinds of cells that contain rhodopsin: rods and cones. Cones function in bright light and are used for color vision; they are concentrated in the central portion of the retina, called the *macula*, and are responsible for the greatest visual acuity. The remaining area of the retina consists mostly of rods, which are used for peripheral and night vision. 11-*cis*-Retinal is present in both cones and rods. Rods have one kind of opsin, whereas cones have three kinds—one for blue, one for green, and one for red color vision.

QUESTIONS

The four *trans* double bonds in the side chain of retinal are labeled a–d. Double bond c (between carbons 11 and 12) is isomerized to its *cis* isomer by an enzyme in the body. Which of the other three double bonds in the side chain of retinal would yield the least stable isomer of *cis* retinal if it were to be isomerized? (*Hint:* think steric strain.)

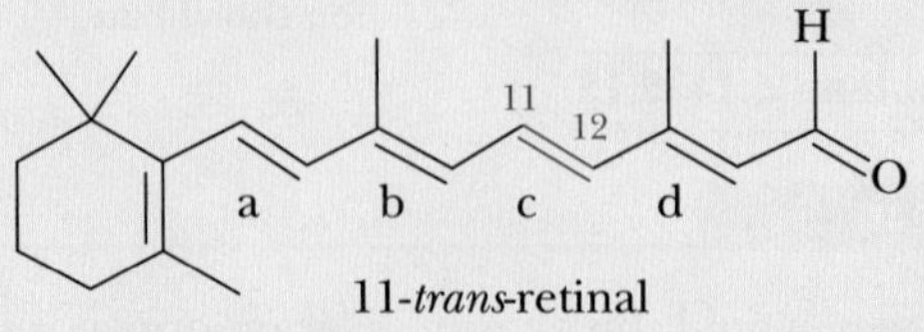

11-*trans*-retinal

B. Developing a Reaction Mechanism

To develop a reaction mechanism, chemists begin by designing experiments that will reveal details of a particular chemical reaction. Next, through a combination of experience and intuition, they propose one or more sets of steps or mechanisms, each of which might account for the overall chemical transformation. Finally, they test each proposed mechanism against the experimental observations to exclude those mechanisms that are not consistent with the facts.

A mechanism becomes generally established by excluding reasonable alternatives and by showing that it is consistent with every test that can be devised. This, of course, does not mean that a generally accepted mechanism is a completely accurate description of the chemical events, but only that it is the best chemists have been able to devise. It is important to keep in mind that, as new experimental evidence is obtained, it may be necessary to modify a generally accepted mechanism or possibly even discard it and start all over again.

Before we go on to consider reactions and reaction mechanisms, we might ask why it is worth the trouble to establish them and your time to learn about them. One reason is very practical. Mechanisms provide a theoretical framework within which to organize a great deal of descriptive chemistry. For example, with insight into how reagents add to particular alkenes, it is possible to make generalizations and then predict how the same reagents might add to other alkenes. A second reason lies in the intellectual satisfaction derived from constructing models that accurately reflect the behavior of chemical systems. Finally, to a creative scientist, a mechanism is a tool to be used in the search for new knowledge and new understanding. A mechanism consistent with all that is known about a reaction can be used to make predictions about chemical interactions as yet unexplored, and experiments can be designed to test these predictions. Thus, reaction mechanisms provide a way not only to organize knowledge, but also to extend it.

5.3 What Are the Mechanisms of Electrophilic Additions to Alkenes?

We begin our introduction to the chemistry of alkenes with an examination of three types of addition reactions: the addition of hydrogen halides (HCl, HBr, and HI), water (H_2O), and halogens (Cl_2, Br_2). We first study some of the experimental observations about each addition reaction and then its mechanism. By examining these particular reactions, we develop a general understanding of how alkenes undergo addition reactions.

As we will show for the addition reactions of alkenes and for the reactions of many other classes of organic compounds, high-electron-density regions of molecules or ions react with low-electron-density regions of other molecules or ions, often resulting in the formation of a new covalent bond. We call an electron-rich species a **nucleophile** (nucleus loving), meaning that it seeks a region of low electron density. We call a low-electron-density species an **electrophile** (electron loving), meaning that it seeks a region of high electron density. Note that nucleophiles are Lewis bases and electrophiles are Lewis acids (Section 2.6).

A. Addition of Hydrogen Halides

The hydrogen halides HCl, HBr, and HI add to alkenes to give haloalkanes (alkyl halides). These additions may be carried out either with the pure reagents or in the presence of a polar solvent such as acetic acid. The addition of HCl to ethylene gives chloroethane (ethyl chloride):

$$\underset{\text{Ethylene}}{CH_2{=}CH_2} + HCl \longrightarrow \underset{\text{Chloroethane}}{\overset{H}{\overset{|}{C}}H_2{-}\overset{Cl}{\overset{|}{C}}H_2}$$

The addition of HCl to propene gives 2-chloropropane (isopropyl chloride); hydrogen adds to carbon 1 of propene and chlorine adds to carbon 2. If the orientation of addition were reversed, 1-chloropropane (propyl chloride) would be formed. The observed result is that 2-chloropropane is formed to the virtual exclusion of 1-chloropropane:

$$\underset{\text{Propene}}{CH_3CH{=}CH_2} + HCl \longrightarrow \underset{\text{2-Chloropropane}}{CH_3\overset{Cl}{\overset{|}{C}}H{-}\overset{H}{\overset{|}{C}}H_2} + \underset{\substack{\text{1-Chloropropane}\\ \text{(not observed)}}}{CH_3\overset{H}{\overset{|}{C}}H{-}\overset{Cl}{\overset{|}{C}}H_2}$$

Regioselective reaction A reaction in which one direction of bond forming or bond breaking occurs in preference to all other directions.

Markovnikov's rule In the addition of HX or H_2O to an alkene, hydrogen adds to the carbon of the double bond having the greater number of hydrogens.

We say that the addition of HCl to propene is highly regioselective and that 2-chloropropane is the major product of the reaction. A **regioselective reaction** is a reaction in which one direction of bond forming or breaking occurs in preference to all other directions.

Vladimir Markovnikov observed this regioselectivity and made the generalization, known as **Markovnikov's rule**, that, in the addition of HX to an alkene, hydrogen adds to the doubly bonded carbon that has the greater number of hydrogens already bonded to it. Although Markovnikov's rule provides a way to predict the product of many alkene addition reactions, it does not explain why one product predominates over other possible products.

Example 5.2

Name and draw a structural formula for the major product of each alkene addition reaction:

(a) $CH_3\overset{CH_3}{\overset{|}{C}}{=}CH_2 + HI \longrightarrow$ (b) 1-methylcyclopentene (CH_3 on the double-bond carbon) $+ HCl \longrightarrow$

Strategy

Use Markovnikov's rule, which predicts that H adds to the least substituted carbon of the double bond and halogen adds to the more substituted carbon.

Solution

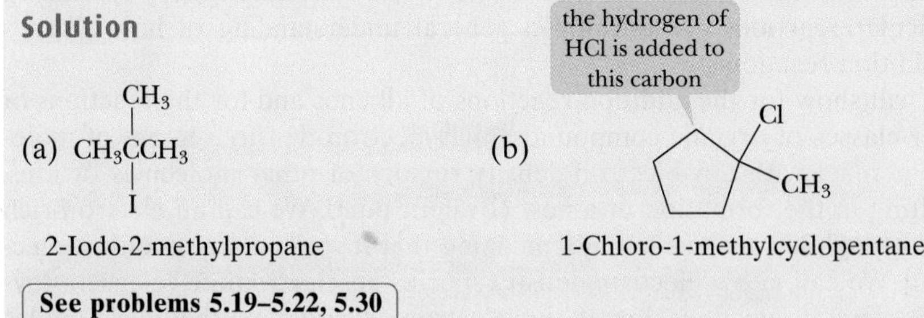

See problems 5.19–5.22, 5.30

Problem 5.2

Name and draw a structural formula for the major product of each alkene addition reaction:

(a) $CH_3CH{=}CH_2 + HI \longrightarrow$ (b) methylenecyclohexane ($=CH_2$) $+ HI \longrightarrow$

Chemists account for the addition of HX to an alkene by a two-step mechanism, which we illustrate by the reaction of 2-butene with hydrogen chloride to give 2-chlorobutane. Let us first look at this two-step mechanism in general and then go back and study each step in detail.

Mechanism: Electrophilic Addition of HCl to 2-Butene

Step 1: The reaction begins with the transfer of a proton from HCl to 2-butene, as shown by the two curved arrows on the left side of Step 1:

$$CH_3CH{=}CHCH_3 + \overset{\delta+}{H}{-}\overset{\delta-}{\ddot{C}l}: \xrightarrow[\text{determining}]{\text{slow, rate}} CH_3\overset{+}{C}H{-}\underset{}{\overset{H}{C}}HCH_3 + :\ddot{C}l:^-$$

sec-Butyl cation (a 2° carbocation intermediate)

The first curved arrow shows the breaking of the pi bond of the alkene and its electron pair now forming a new covalent bond with the hydrogen atom of HCl. In this step, the carbon–carbon double bond of the alkene is the nucleophile (the electron-rich, nucleus-seeking species) and HCl is the electrophile (the electron-poor, electron-seeking species). The second curved arrow shows the breaking of the polar covalent bond in HCl and this electron pair being given entirely to chlorine, forming chloride ion. Step 1 in this mechanism results in the formation of an organic cation and chloride ion.

Step 2: The reaction of the *sec*-butyl cation (an electrophile and a Lewis acid) with chloride ion (a nucleophile and a Lewis base) completes the valence shell of carbon and gives 2-chlorobutane:

$$:\ddot{C}l:^- + CH_3\overset{+}{C}HCH_2CH_3 \xrightarrow{\text{fast}} CH_3\overset{:\ddot{C}l:}{C}HCH_2CH_3$$

Chloride ion (a Lewis base) — *sec*-Butyl cation (a Lewis acid) — 2-Chlorobutane

HOW TO 5.1 Draw Mechanisms

Mechanisms show how bonds are broken and formed. Although individual atoms may change positions in a reaction, the curved arrows used in a mechanism are for the purpose of showing electron movement. Therefore, it is important to remember that curved arrow notation always shows the arrow originating from a bond or from an unshared electron pair (not the other way around).

Correct use of curved arrows...

$$\text{alkene} + H{-}\ddot{B}r: \longrightarrow \text{carbocation} + :\ddot{B}r:^-$$

Incorrect use of curved arrows...

$$\text{alkene} + H{-}\ddot{B}r: \longrightarrow \text{carbocation} + :\ddot{B}r:^-$$

a common mistake is to use curved arrows to indicate the movement of atoms rather than electrons

Figure 5.3
The structure of the *tert*-butyl cation. (a) Lewis structure and (b) an orbital picture.

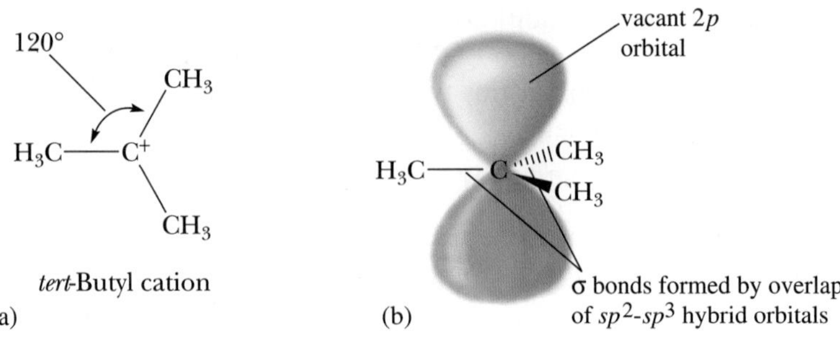

Now let us go back and look at the individual steps in more detail. There is a great deal of important organic chemistry embedded in these two steps, and it is crucial that you understand it now.

Carbocation A species containing a carbon atom with only three bonds to it and bearing a positive charge.

Electrophile Any molecule or ion that can accept a pair of electrons to form a new covalent bond; a Lewis acid.

Step 1 results in the formation of an organic cation. One carbon atom in this cation has only six electrons in its valence shell and carries a charge of +1. A species containing a positively charged carbon atom is called a **carbocation** (*carbo*n + *cation*). Carbocations are classified as primary (1°), secondary (2°), or tertiary (3°), depending on the number of carbon atoms bonded directly to the carbon bearing the positive charge. All carbocations are Lewis acids (Section 2.6) and **electrophiles.**

In a carbocation, the carbon bearing the positive charge is bonded to three other atoms, and, as predicted by valence-shell electron-pair repulsion (VSEPR), the three bonds about that carbon are coplanar and form bond angles of approximately 120°. According to the molecular orbital model of bonding, the electron-deficient carbon of a carbocation uses its sp^2 hybrid orbitals to form sigma bonds to three groups. The unhybridized 2*p* orbital lies perpendicular to the sigma bond framework and contains no electrons. A Lewis structure and an orbital overlap diagram for the *tert*-butyl cation are shown in Figure 5.3.

Figure 5.4 shows an energy diagram for the two-step reaction of 2-butene with HCl. The slower, rate-determining step (the one that crosses the higher energy barrier) is Step 1, which leads to the formation of the 2° carbocation intermediate. This intermediate lies in an energy minimum between the transition states for Steps 1 and 2. As soon as the carbocation intermediate (a Lewis acid) forms, it reacts with chloride ion (a Lewis base) in a Lewis acid–base reaction to give 2-chlorobutane. Note that the energy level for 2-chlorobutane (the product) is lower than the energy level for 2-butene and HCl (the reactants). Thus, in this alkene addition reaction, heat is released; the reaction is, accordingly, exothermic.

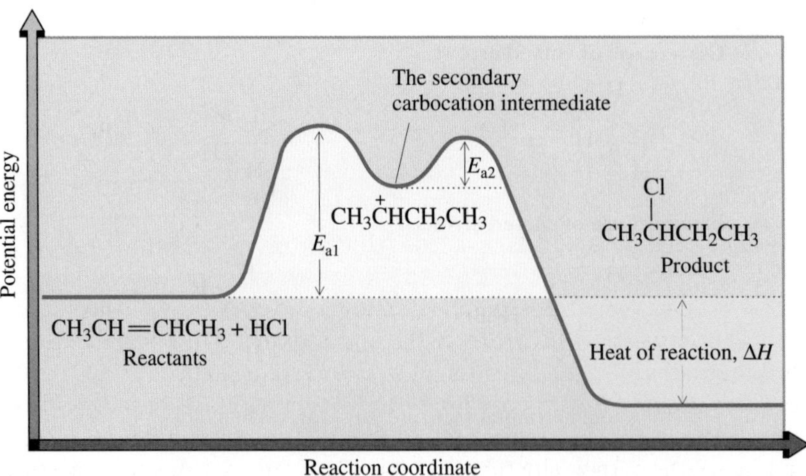

Figure 5.4
Energy diagram for the two-step addition of HCl to 2-butene. The reaction is exothermic.

Relative Stabilities of Carbocations: Regioselectivity and Markovnikov's Rule

The reaction of HX and an alkene can, at least in principle, give two different carbocation intermediates, depending on which of the doubly bonded carbon atoms forms a bond with H^+, as illustrated by the reaction of HCl with propene:

this carbon forms the bond to hydrogen

$$CH_3CH{=}CH_2 + H{-}Cl \longrightarrow CH_3CH_2\overset{+}{C}H_2 \xrightarrow{Cl^-} CH_3CH_2CH_2Cl$$

Propene — Propyl cation (a 1° carbocation) — 1-Chloropropane (not formed)

this carbon forms the bond to hydrogen

$$CH_3CH{=}CH_2 + H{-}Cl \longrightarrow CH_3\overset{+}{C}HCH_3 \xrightarrow{Cl^-} CH_3CH(Cl)CH_3$$

Propene — Isopropyl cation (a 2°carbocation) — 2-Chloropropane (product formed)

The observed product is 2-chloropropane. Because carbocations react very quickly with chloride ions, the absence of 1-chloropropane as a product tells us that the 2° carbocation is formed in preference to the 1° carbocation.

Similarly, in the reaction of HCl with 2-methylpropene, the transfer of a proton to the carbon–carbon double bond might form either the isobutyl cation (a 1° carbocation) or the *tert*-butyl cation (a 3° carbocation):

this carbon forms the bond to hydrogen

$$CH_3C(CH_3){=}CH_2 + H{-}Cl \longrightarrow CH_3CH(CH_3)\overset{+}{C}H_2 \xrightarrow{Cl^-} CH_3CH(CH_3)CH_2Cl$$

2-Methylpropene — Isobutyl cation (a 1° carbocation) — 1-Chloro-2-methylpropane (not formed)

this carbon forms the bond to hydrogen

$$CH_3C(CH_3){=}CH_2 + H{-}Cl \longrightarrow CH_3\overset{+}{C}(CH_3)CH_3 \xrightarrow{Cl^-} CH_3C(CH_3)(Cl)CH_3$$

2-Methylpropene — *tert*-Butyl cation (a 3° carbocation) — 2-Chloro-2-methylpropane (product formed)

In this reaction, the observed product is 2-chloro-2-methylpropane, indicating that the 3° carbocation forms in preference to the 1° carbocation.

From such experiments and a great amount of other experimental evidence, we learn that a 3° carbocation is more stable and requires a lower activation energy for its formation than a 2° carbocation. A 2° carbocation, in turn, is more stable and requires a lower activation energy for its formation than a 1° carbocation. In fact, 1° carbocations are so unstable and so difficult to form that they are never observed in solution; they should never be proposed as a reaction intermediate, when other more stable carbocations are an option. It follows that a more stable carbocation intermediate forms faster than a less stable carbocation intermediate. Following is the order of stability of four types of alkyl carbocations:

$H-\overset{+}{C}(H)-H$	$H_3C-\overset{+}{C}(H)-H$	$H_3C-\overset{+}{C}(CH_3)-H$	$H_3C-\overset{+}{C}(CH_3)-CH_3$
Methyl cation (methyl)	Ethyl cation (1°)	Isopropyl cation (2°)	*tert*-Butyl cation (3°)

Order of increasing carbocation stability →

Although the concept of the relative stabilities of carbocations had not been developed in Markovnikov's time, their relative stabilities is the underlying basis for his rule; that is, the proton of H—X adds to the less substituted carbon of a double bond because this mode of addition produces the more stable carbocation intermediate.

Now that we know the order of stability of carbocations, how do we account for it? The principles of physics teach us that a system bearing a charge (either positive or negative) is more stable if the charge is delocalized. Using this principle, we can explain the order of stability of carbocations if we assume that alkyl groups bonded to a positively charged carbon release electrons toward the cationic carbon and thereby help delocalize the charge on the cation. The electron-releasing ability of alkyl groups bonded to a cationic carbon is accounted for by the **inductive effect** (Section 2.5C).

Inductive effect The polarization of electron density transmitted through covalent bends caused by a nearby atom of higher electronegativity.

The inductive effect operates in the following way: The electron deficiency of the carbon atom bearing a positive charge exerts an electron-withdrawing inductive effect that polarizes electrons from adjacent sigma bonds toward it. Thus, the positive charge of the cation is not localized on the trivalent carbon, but rather is delocalized over nearby atoms as well. The larger the volume over which the positive charge is delocalized, the greater is the stability of the cation. Thus, as the number of alkyl groups bonded to the cationic carbon increases, the stability of the cation increases as well. Figure 5.5 illustrates the electron-withdrawing inductive effect of the positively charged carbon and the resulting delocalization of charge. According to quantum mechanical calculations, the charge on carbon in the methyl cation is approximately +0.645, and the charge on each of the hydrogen atoms is +0.118. Thus, even in the methyl cation, the positive charge is not localized on carbon. Rather, it is delocalized over the volume of space occupied by the entire ion. The polarization of electron density and the delocalization of charge are even more extensive in the *tert*-butyl cation.

the methyl groups donate electron density toward the carbocation carbon, thus delocalizing the positive charge

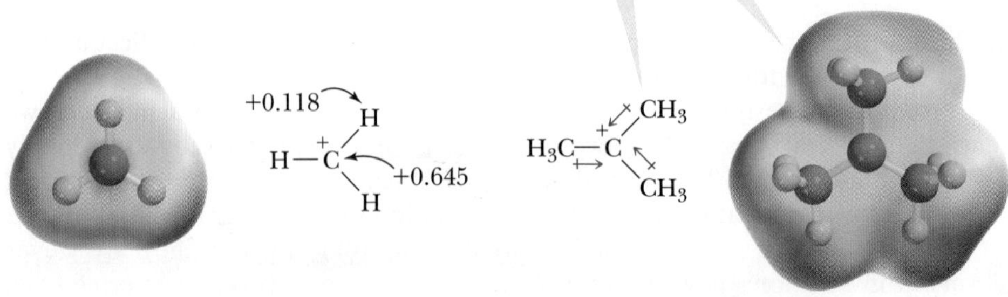

Figure 5.5
Methyl and *tert*-butyl cations. Delocalization of positive charge by the electron-withdrawing inductive effect of the trivalent, positively charged carbon according to molecular orbital calculations.

Example 5.3

Arrange these carbocations in order of increasing stability:

(a) (b) (c)

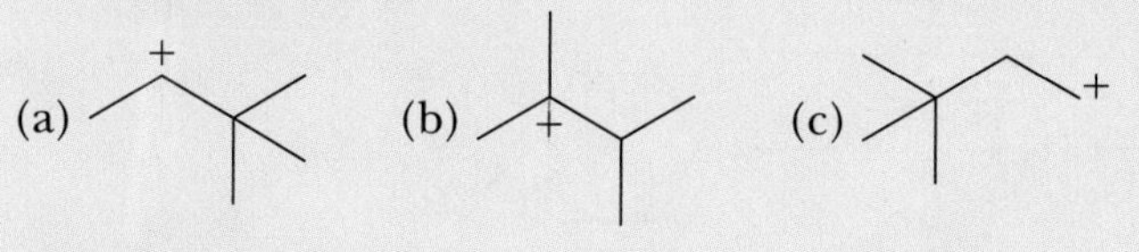

Strategy

Determine the degree of substitution of the positively charged carbon and then consider the order of decreasing stability of alkyl carbocations is 3° > 2° > 1°.

Solution

Carbocation (a) is secondary, (b) is tertiary, and (c) is primary. In order of increasing stability, they are *c* < *a* < *b*.

See problems 5.17, 5.18

Problem 5.3

Arrange these carbocations in order of increasing stability:

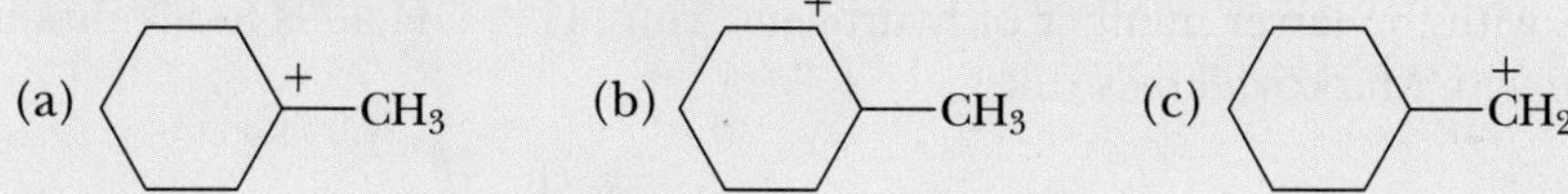

Example 5.4

Propose a mechanism for the addition of HI to methylenecyclohexane to give 1-iodo-1-methylcyclohexane:

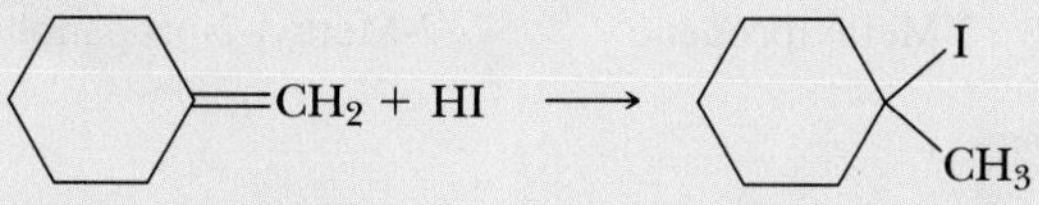

$=CH_2$ + HI ⟶ I, CH_3

Methylenecyclohexane 1-Iodo-1-methylcyclohexane

Which step in your mechanism is rate determining?

Strategy

Propose a two-step mechanism similar to that proposed for the addition of HCl to propene. Formation of the carbocation intermediate is rate determining.

Solution

Step 1: A rate-determining proton transfer from HI to the carbon–carbon double bond gives a 3° carbocation intermediate:

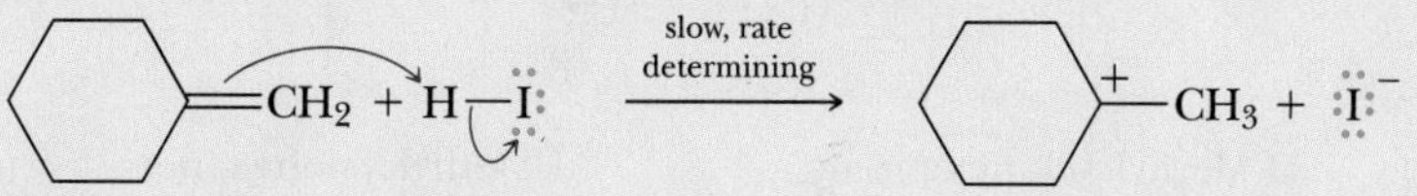

$=CH_2$ + H—I: —slow, rate determining→ + $-CH_3$ + :I:⁻

Methylenecyclohexane A 3° carbocation intermediate

Step 2: Reaction of the 3° carbocation intermediate (a Lewis acid) with iodide ion (a Lewis base) completes the valence shell of carbon and gives the product:

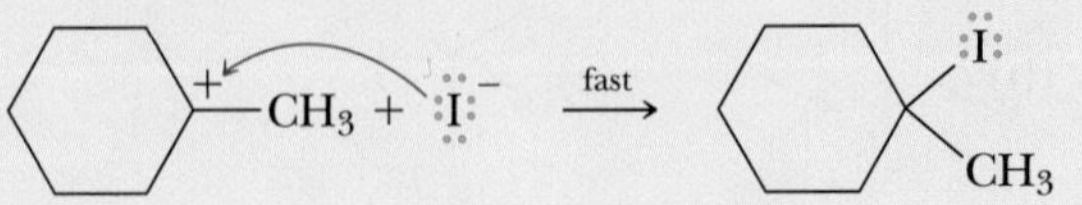

$$\text{(cyclohexyl cation)}^{+}\text{—}CH_3 + :\ddot{I}:^{-} \xrightarrow{\text{fast}} \text{1-iodo-1-methylcyclohexane}$$

1-Iodo-1-methylcyclohexane

See problem 5.31

Problem 5.4

Propose a mechanism for the addition of HI to 1-methylcyclohexene to give 1-iodo-1-methylcyclohexane. Which step in your mechanism is rate determining?

B. Addition of Water: Acid-Catalyzed Hydration

Hydration Addition of water.

In the presence of an acid catalyst—most commonly, concentrated sulfuric acid—water adds to the carbon–carbon double bond of an alkene to give an alcohol. The addition of water is called **hydration**. In the case of simple alkenes, H adds to the carbon of the double bond with the greater number of hydrogens and OH adds to the carbon with the lesser number of hydrogens. Thus, H—OH adds to alkenes in accordance with Markovnikov's rule:

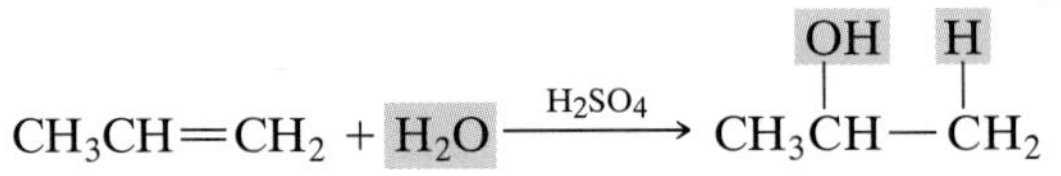

$$\underset{\text{Propene}}{CH_3CH{=}CH_2} + H_2O \xrightarrow{H_2SO_4} \underset{\text{2-Propanol}}{CH_3\overset{OH}{CH}{-}\overset{H}{CH_2}}$$

$$\underset{\text{2-Methylpropene}}{CH_3\overset{CH_3}{C}{=}CH_2} + H_2O \xrightarrow{H_2SO_4} \underset{\text{2-Methyl-2-propanol}}{CH_3\underset{HO}{\overset{CH_3}{C}}{-}\underset{H}{CH_2}}$$

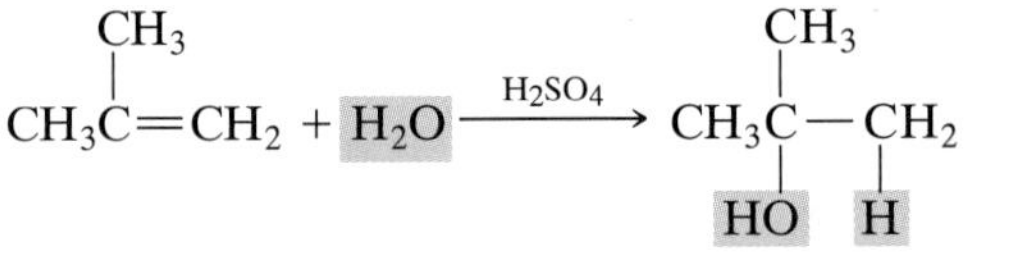

Example 5.5

Draw a structural formula for the product of the acid-catalyzed hydration of 1-methylcyclohexene.

Strategy

Use Markovnikov's rule, which states that the H adds to the carbon of the carbon–carbon double bond bearing the greater number of hydrogens and that OH adds to the carbon bearing the lesser number of hydrogens.

Solution

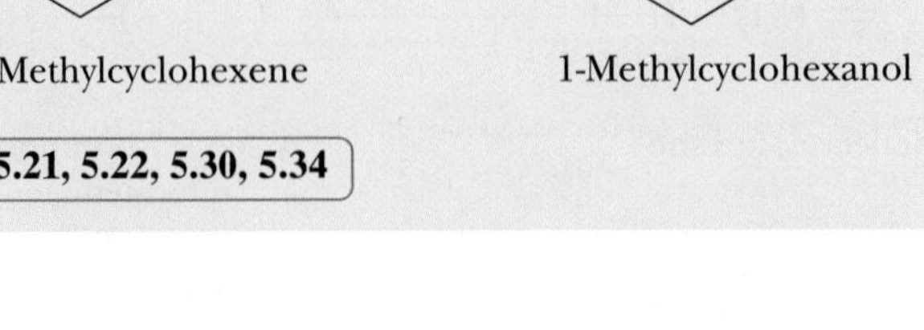

$$\text{1-Methylcyclohexene} + H_2O \xrightarrow{H_2SO_4} \text{1-Methylcyclohexanol}$$

1-Methylcyclohexene — 1-Methylcyclohexanol

See problems 5.21, 5.22, 5.30, 5.34

Problem 5.5

Draw a structural formula for the product of each alkene hydration reaction:

(a) [alkene] + H_2O $\xrightarrow{H_2SO_4}$ (b) [alkene] + H_2O $\xrightarrow{H_2SO_4}$

The mechanism for the acid-catalyzed hydration of alkenes is quite similar to what we have already proposed for the addition of HCl, HBr, and HI to alkenes and is illustrated by the hydration of propene to 2-propanol. This mechanism is consistent with the fact that acid is a catalyst. An H_3O^+ is consumed in Step 1, but another is generated in Step 3.

Mechanism: Acid-Catalyzed Hydration of Propene

Step 1: Proton transfer from the acid catalyst, in this case, the hydronium ion, to propene gives a 2° carbocation intermediate (a Lewis acid):

$$CH_3CH{=}CH_2 + H{-}\overset{+}{O}(H){-}H \xrightleftharpoons[\text{determining}]{\text{slow, rate}} CH_3\overset{+}{C}HCH_3 + :\ddot{O}(H){-}H$$

A 2° carbocation intermediate

Step 2: Reaction of the carbocation intermediate (a Lewis acid) with water (a Lewis base) completes the valence shell of carbon and gives an **oxonium ion:**

Oxonium ion An ion in which oxygen is bonded to three other atoms and bears a positive charge.

$$CH_3\overset{+}{C}HCH_3 + :\ddot{O}(H){-}H \xrightleftharpoons{\text{fast}} CH_3CH(\overset{+}{O}H_2)CH_3$$

An oxonium ion

Step 3: Proton transfer from the oxonium ion to water gives the alcohol and generates a new molecule of the catalyst:

$$CH_3CH(\overset{+}{O}H_2)CH_3 + H{-}\ddot{O}{-}H \xrightleftharpoons{\text{fast}} CH_3CH(\ddot{O}H)CH_3 + H{-}\overset{+}{O}(H){-}H$$

Example 5.6

Propose a mechanism for the acid-catalyzed hydration of methylenecyclohexane to give 1-methylcyclohexanol. Which step in your mechanism is rate determining?

Strategy

Propose a three-step mechanism similar to that for the acid-catalyzed hydration of propene.

Solution

The formation of the 3° carbocation intermediate in Step 1 is rate determining.

Step 1: Proton transfer from the acid catalyst to the alkene gives a 3° carbocation intermediate (a Lewis acid):

$$\text{(cyclohexylidene)}{=}CH_2 + H{-}\ddot{O}^+{-}H \ (\text{with } H) \xrightleftharpoons[]{\text{slow, rate determining}} \text{(cyclohexyl)}^+{-}CH_3 + :\ddot{O}{-}H \ (\text{with } H)$$

A 3° carbocation intermediate

Step 2: Reaction of the carbocation intermediate (a Lewis acid) with water (a Lewis base) completes the valence shell of carbon and gives an oxonium ion:

$$\text{(cyclohexyl)}^+{-}CH_3 + :\ddot{O}{-}H \ (\text{with } H) \xrightleftharpoons{\text{fast}} \text{(cyclohexyl)}(CH_3){-}\overset{+}{O}(H){-}H$$

An oxonium ion

Step 3: Proton transfer from the oxonium ion to water gives the alcohol and regenerates the acid catalyst:

$$\text{(cyclohexyl)}(CH_3){-}\overset{+}{O}(H){-}H + :\ddot{O}{-}H \ (\text{with } H) \xrightleftharpoons{\text{fast}} \text{(cyclohexyl)}(CH_3){-}\ddot{O}{-}H + H{-}\ddot{O}^+{-}H \ (\text{with } H)$$

See problems 5.31–5.33, 5.40

Problem 5.6

Propose a mechanism for the acid-catalyzed hydration of 1-methylcyclohexene to give 1-methylcyclohexanol. Which step in your mechanism is rate determining?

C. Addition of Bromine and Chlorine

Chlorine (Cl_2) and bromine (Br_2) react with alkenes at room temperature by the addition of halogen atoms to the two carbon atoms of the double bond, forming two new carbon–halogen bonds:

$$CH_3CH{=}CHCH_3 + Br_2 \xrightarrow[CH_2Cl_2]{} CH_3\underset{}{\overset{Br}{C}}H{-}\overset{Br}{C}HCH_3$$

2-Butene 2,3-Dibromobutane

Fluorine, F_2, also adds to alkenes, but because its reactions are very fast and difficult to control, addition of fluorine is not a useful laboratory reaction. Iodine, I_2, also adds, but the reaction is not preparatively useful.

The addition of bromine and chlorine to a cycloalkene gives a *trans* dihalocycloalkane. For example, the addition of bromine to cyclohexene gives *trans*-1,2-dibromocyclohexane; the *cis* isomer is not formed. Thus, the addition of a halogen to a cycloalkene is stereoselective. A **stereoselective reaction** is a reaction in which one stereoisomer is formed or destroyed in preference to all others that might be formed or destroyed. We say that addition of bromine to an alkene occurs with **anti stereoselectivity**.

Stereoselective reaction A reaction in which one stereoisomer is formed or destroyed in preference to all others that might be formed or destroyed.

Anti stereoselectivity Addition of atoms or groups of atoms from opposite sides or faces of a carbon–carbon double bond.

$$\text{Cyclohexene} + Br_2 \xrightarrow[CH_2Cl_2]{} \textit{trans}\text{-1,2-Dibromocyclohexane}$$

Cyclohexene *trans*-1,2-Dibromocyclohexane

trans

The reaction of bromine with an alkene is a particularly useful qualitative test for the presence of a carbon–carbon double bond. If we dissolve bromine in dichloromethane, the solution turns red. Both alkenes and dibromoalkanes are colorless. If we now mix a few drops of the bromine solution with an alkene, a dibromoalkane is formed, and the solution becomes colorless.

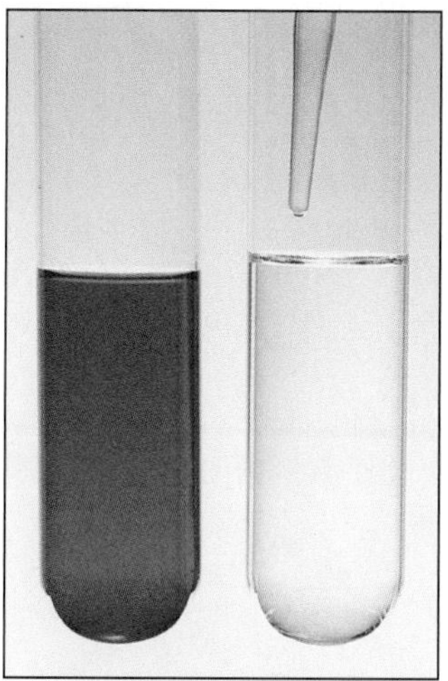

A solution of bromine in dichloromethane is red. Add a few drops of an alkene and the red color disappears. *(Charles D. Winters)*

Example 5.7

Complete these reactions, showing the stereochemistry of each product:

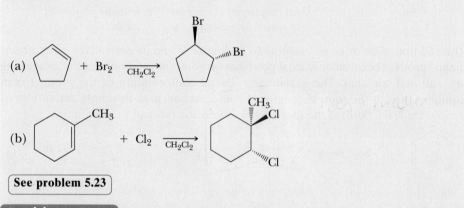

(a) cyclopentene + Br_2 $\xrightarrow{CH_2Cl_2}$ (b) 1-methylcyclohexene (CH_3) + Cl_2 $\xrightarrow{CH_2Cl_2}$

Strategy

The addition of both Br_2 and Cl_2 to cycloalkenes occurs with anti stereoselectivity; the two halogen atoms are *trans* to each other in the product.

Solution

(a) cyclopentene + Br_2 $\xrightarrow{CH_2Cl_2}$ trans-1,2-dibromocyclopentane (Br, Br)

(b) 1-methylcyclohexene (CH_3) + Cl_2 $\xrightarrow{CH_2Cl_2}$ trans-1,2-dichloro-1-methylcyclohexane (CH_3, Cl, Cl)

See problem 5.23

Problem 5.7

Complete these reactions:

(a) $CH_3C(CH_3)_2CH{=}CH_2 + Br_2 \xrightarrow{CH_2Cl_2}$ (b) methylenecyclohexane (CH_2) + Cl_2 $\xrightarrow{CH_2Cl_2}$

Stereoselectivity and Bridged Halonium Ion Intermediates

We explain the addition of bromine and chlorine to cycloalkenes, as well as their anti stereoselectivity (they always add *trans* to each other), by a two-step mechanism that involves a halogen atom bearing a positive charge, called a **halonium ion**. The cyclic structure of which this ion is a part is called a **bridged halonium ion**. The bridged bromonium ion shown in the mechanism that follows might look odd to you, but it is an acceptable Lewis structure. A calculation of formal charge places a positive charge on bromine. Then, in Step 2, a bromide ion reacts with the bridged intermediate from the side opposite that occupied by the bromine atom, giving the dibromoalkane. Thus, bromine atoms add from opposite faces of the carbon–carbon double bond.

Halonium ion An ion in which a halogen atom bears a positive charge.

Mechanism: Addition of Bromine with Anti Selectivity

Step 1: Reaction of the pi electrons of the carbon–carbon double bond with bromine forms a bridged bromonium ion intermediate in which bromine bears a positive formal charge:

A bridged bromonium ion intermediate

Step 2: A bromide ion (a Lewis base) attacks carbon (a Lewis acid) from the side opposite the bridged bromonium ion, opening the three-membered ring:

Anti (coplanar) orientation of added bromine atoms

A Newman projection of the product

The addition of chlorine or bromine to cyclohexene and its derivatives gives a *trans* diaxial product because only axial positions on adjacent atoms of a cyclohexane ring are anti and coplanar. The initial *trans* diaxial conformation of the product is in equilibrium with the *trans* diequatorial conformation, and, in simple derivatives of cyclohexane, the latter is the more stable conformation and predominates.

$+ Br_2 \longrightarrow$

trans Diaxial

trans Diequatorial (more stable)

5.4 What Are Carbocation Rearrangements?

Rearrangement A reaction which the connectivity of atoms in a product is different from that in the starting material.

As we have seen in the preceding discussion, the expected product of electrophilic addition to a carbon–carbon double bond involves rupture of the π bond and formation of two new σ bonds in its place. In the addition of HCl to 3,3-dimethyl-1-butene, however, only 17% of 2-chloro-3,3-dimethylbutane, the expected product, is formed. The major product is 2-chloro-2,3-dimethylbutane, a compound with a different connectivity of its carbon atoms than that in the starting material. We say that the formation of 2-chloro-2,3-dimethylbutane involves a **rearrangement**. Typically, either an alkyl group or a hydrogen atom migrates, with its bonding pair of electrons, from an adjacent atom to an electron-deficient atom. In the rearrangements we examine in this chapter, migration is to an adjacent electron-deficient carbon atom bearing a positive charge. In other words, rearrangement is to the positively charged carbon of a carbocation.

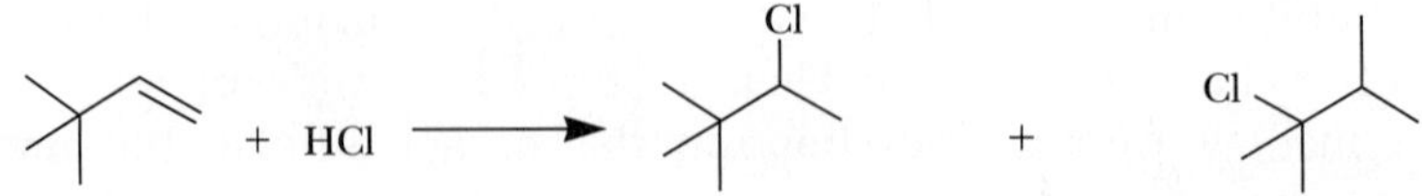

3,3-Dimethyl-1-butene

2-Chloro-3,3-dimethylbutane (the expected product 17%)

2-Chloro-2,3-dimethylbutane (the major product 83%)

Formation of the rearranged product in this reaction can be accounted for by the following mechanism, the key step of which is a type of rearrangement called a **1,2-shift**. In the rearrangement shown in Step 2, the migrating group is a methyl group with its pair of bonding electrons.

Step 1: Proton transfer from the HCl to the alkene gives a 2° carbocation intermediate.

+ H—Cl: ⟶ + :Cl:⁻

3,3-Dimethyl-1-butene

A 2° carbocation intermediate

Step 2: Migration of a methyl group with its bonding electrons from an adjacent carbon gives a more stable 3° carbocation intermediate. The major movement is that of the bonding electron pair with the methyl group following.

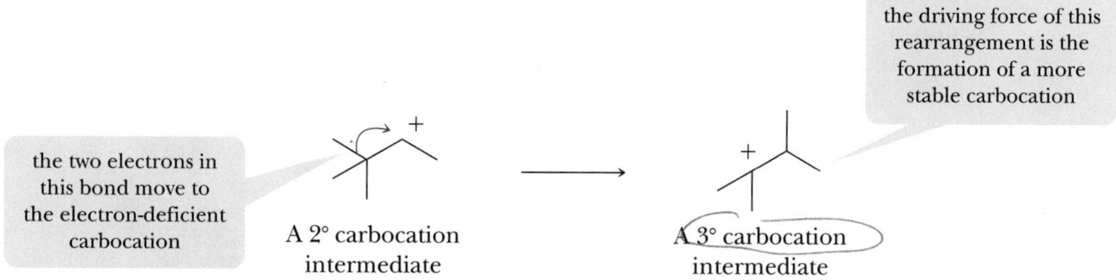
the driving force of this rearrangement is the formation of a more stable carbocation
the two electrons in this bond move to the electron-deficient carbocation
A 2° carbocation intermediate
A 3° carbocation intermediate

Step 3: Reaction of the 3° carbocation intermediate (an electrophile and a Lewis acid) with chloride ion (a nucleophile and a Lewis base) gives the rearranged product.

:Cl:⁻ + ⟶ Cl

A 3° carbocation intermediate

The driving force for this rearrangement is the fact that the less stable 2° carbocation is converted to a more stable 3° carbocation. From the study of this and other carbocation rearrangements, we find that 2° carbocations rearrange to 3° carbocations. 1° Carbocations are never observed for reactions taking place in solution and should not be proposed as reaction intermediates.

Rearrangements also occur in the acid-catalyzed hydration of alkenes, especially when a carbocation formed in the first step can rearrange to a more stable carbocation. For example, the acid-catalyzed hydration of 3-methyl-1-butene gives 2-methyl-2-butanol. In this example, the group that migrates is a hydrogen with its bonding pair of electrons, in effect, a hydride ion H:⁻.

this H migrates to an adjacent carbon

$$CH_3CH(CH_3)CH{=}CH_2 + H_2O \longrightarrow CH_3C(CH_3)(OH)CH_2CH_3$$

3-Methyl-1-butene ⟶ 2-Methyl-2-butanol

Example 5.8

Propose a mechanism for the acid-catalyzed hydration of 3-methyl-1-butene to give 2-methyl-2-butanol.

Strategy

Propose a mechanism similar to that proposed for the acid-catalyzed hydration of an alkene involving proton transfer from the acid catalyst to form a carbocation intermediate, rearrangement of the carbocation intermediate to a more stable intermediate, reaction of the more stable carbocation with water to form an oxonium ion, and finally proton transfer from the oxonium ion to water to give the product and regenerate the acid catalyst. Lest you be tempted to use H^+ to initiate the reaction, remember that ionization of a strong acid in water generates a hydronium ion and an anion. Hydronium ion and not H^+ is the true catalyst in this reaction.

Solution

Step 1: Proton transfer from the hydronium ion (the acid catalyst) to the carbon–carbon double bond gives a 2° carbocation intermediate.

2-Methyl-1-butene Hydronium ion A 2° carbocation intermediate

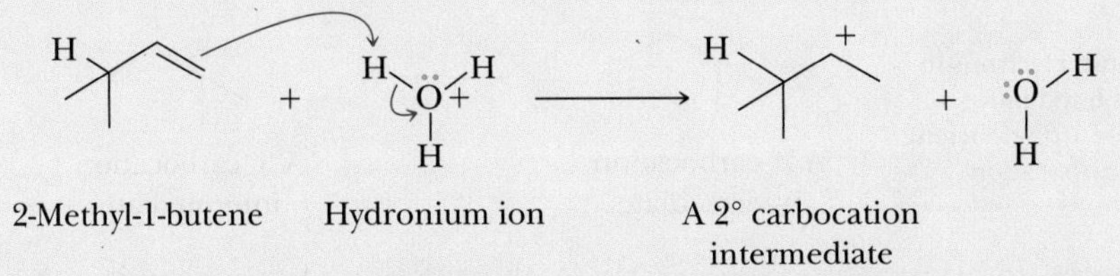

Step 2: A 1,2-shift of a hydrogen from an adjacent carbon with its bonding pair of electrons to the positively charged carbon gives a more stable 3° carbocation intermediate.

A 2° carbocation intermediate A 3° carbocation intermediate

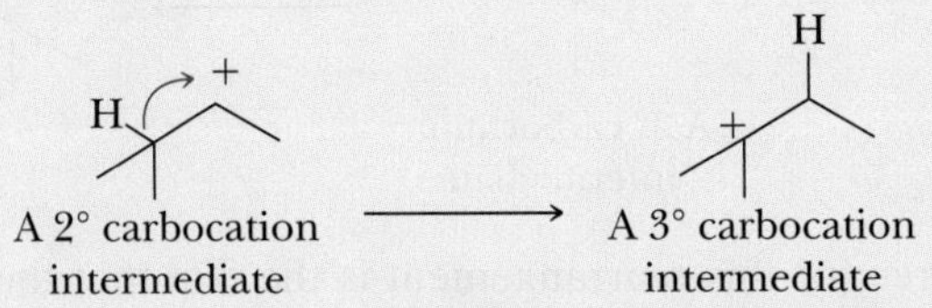

Step 3: Reaction of the 3° carbocation (an electrophile and a Lewis acid) with a water molecule (a nucleophile and a Lewis base) completes the valence shell of carbon and gives an oxonium ion.

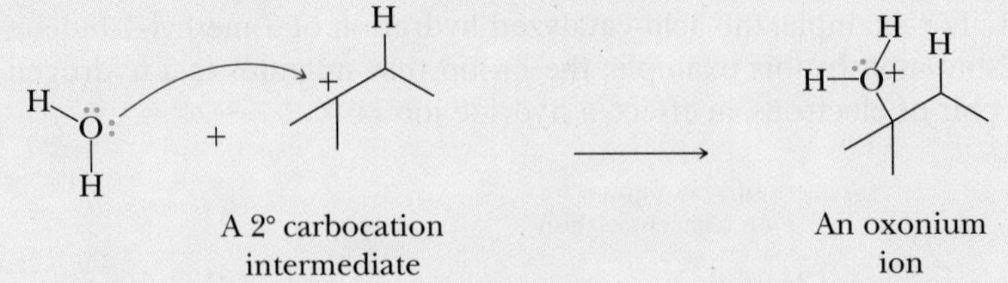

A 2° carbocation intermediate An oxonium ion

Step 4: Proton transfer from the oxonium ion to water gives the alcohol and regenerates the acid catalyst.

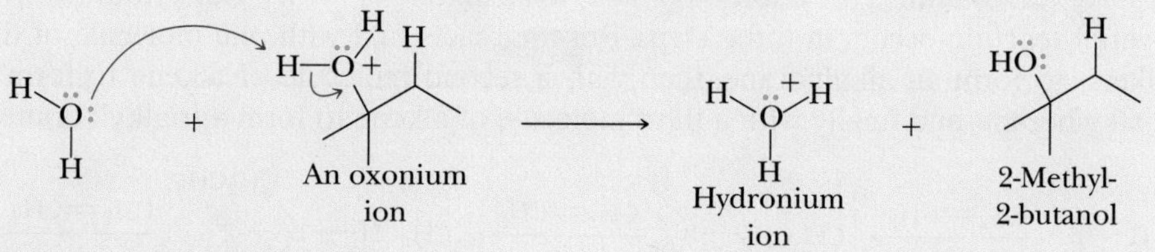

H–Ö:–H + H_2O^+–(oxonium) ⟶ H_3O^+ + 2-Methyl-2-butanol

An oxonium ion

Hydronium ion

2-Methyl-2-butanol

See problems 5.31–5.33, 5.40

Problem 5.8

The acid-catalyzed hydration of 3,3-dimethyl-1-butene gives 2,3-dimethyl-2-butanol as the major product. Propose a mechanism for the formation of this alcohol.

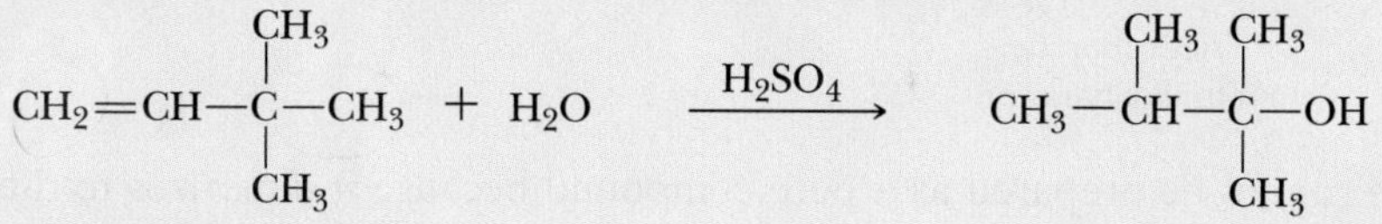

$$CH_2{=}CH{-}C(CH_3)_2{-}CH_3 + H_2O \xrightarrow{H_2SO_4} CH_3{-}CH(CH_3){-}C(CH_3)(OH){-}CH_3$$

3,3-Dimethyl-1-butene → 2,3-Dimethyl-2-butanol

5.5 What Is Hydroboration-Oxidation of an Alkene?

The result of hydroboration and subsequent oxidation of an alkene is hydration of the carbon–carbon double bond, here illustrated by the hydroboration–oxidation of 1-hexene to give 1-hexanol.

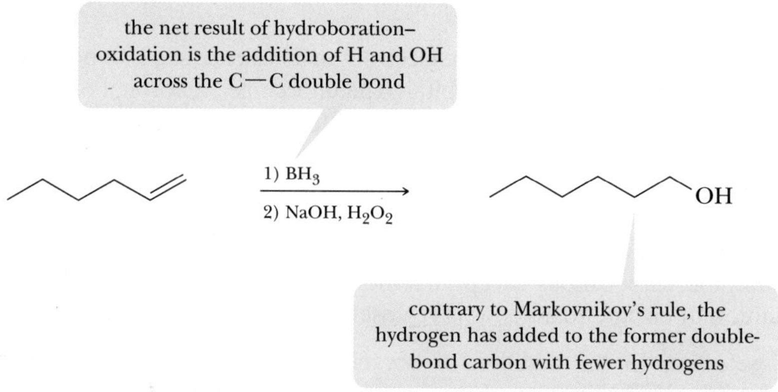

the net result of hydroboration–oxidation is the addition of H and OH across the C—C double bond

$$\text{1-hexene} \xrightarrow[\text{2) NaOH, } H_2O_2]{\text{1) } BH_3} \text{1-hexanol (OH)}$$

contrary to Markovnikov's rule, the hydrogen has added to the former double-bond carbon with fewer hydrogens

Because hydrogen is added to the more substituted carbon of the double bond and —OH to the less substituted carbon, we refer to the regiochemistry of hydroboration and subsequent oxidation as **anti-Markovnikov hydration.**

Note by way of comparison that acid-catalyzed hydration of 1-hexene follows Markovnikov's rule and gives 2-hexanol.

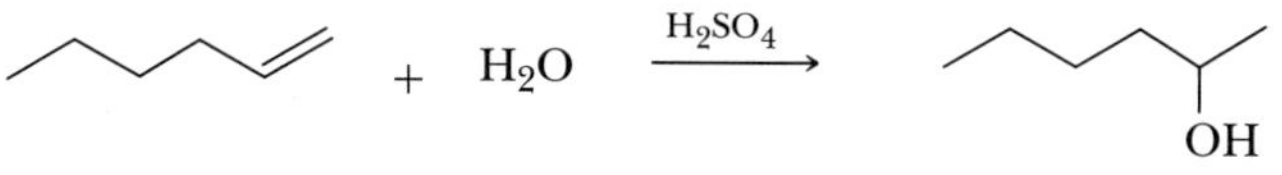

$$\text{1-Hexene} + H_2O \xrightarrow{H_2SO_4} \text{2-Hexanol}$$

The special value of hydration of an alkene by the combination of hydroboration–oxidation is that its regioselectivity is opposite that of acid-catalyzed hydration.

=Borane

Hydroboration is the addition of BH_3 to an alkene to form a trialkylborane. The overall reaction occurs in three steps. Borane reacts first with one molecule of the alkene to form an alkylborane, then with a second molecule of alkene to form a dialkylborane, and finally with a third molecule of alkene to form a trialkylborane.

$$H-BH_2 \xrightarrow{CH_2=CH_2} CH_3CH_2-BH_2 \xrightarrow{CH_2=CH_2} (CH_3CH_2)_2B-H \xrightarrow{CH_2=CH_2} (CH_3CH_2)_3B$$

Borane

Ethylborane (an alkylborane)

Diethylborane (a dialkylborane)

Triethylborane (a trialkylborane)

Borane cannot be prepared as a pure compound because it dimerizes to diborane B_2H_6, a toxic gas that ignites spontaneously in air.

$$2BH_3 \rightleftharpoons B_2H_6$$

Borane Diborane

However, BH_3 forms a stable Lewis acid–base complex with ethers. Borane is most commonly used as a commercially available solution of BH_3 in tetrahydrofuran (THF).

$$2\ \text{THF}(:O:) + B_2H_6 \rightleftharpoons 2\ :\overset{+}{O}-\overset{-}{B}H_3$$

Tetrahydrofuran (THF) $BH_3 \bullet THF$

Boron, atomic number 5, has three electrons in its valence shell. To bond with three other atoms, boron uses sp^2 hybrid orbitals. The unoccupied $2p$ orbital of boron is perpendicular to the plane created by boron and the three other atoms to which it is bonded. BH_3 is a planar molecule with H–B–H bond angles of 120°. An example of a stable, trivalent boron compound is boron trifluoride, BF_3, a planar molecule with F–B–F bond angles of 120° (Section 1.3E). Because of the vacant $2p$ orbital in the valence shell of boron, BH_3, BF_3, and all other tricovalent compounds of boron are electrophiles and closely resemble carbocations, except that they are electrically neutral.

Addition of borane to alkenes is regioselective and stereoselective.

- Regioselective: In the addition of borane to an unsymmetrical alkene, boron becomes bonded predominantly to the less substituted carbon of the double bond.
- Stereoselective: Hydrogen and boron add from the same face of the double bond; that is, the reaction is **syn** (from the same side) **stereoselective**.

Both the regioselectivity and syn stereoselectivity are illustrated by hydroboration of 1-methylcyclopentene.

$$\text{1-Methylcyclopentene} + BH_3 \longrightarrow \text{(two syn-addition enantiomers, } CH_3,\ H,\ BR_2,\ H)$$

1-Methylcyclopentene

(Syn addition of BH_3)
(R = 2-methylcyclopentyl)

Mechanism: Hydroboration of an Alkene

The addition of borane to an alkene is initiated by coordination of the vacant 2*p* orbital of boron (an electrophile) with the electron pair of the pi bond (a nucleophile). Chemists account for the stereoselectivity of hydroboration by proposing the formation of a cyclic, four-center transition state. Boron and hydrogen add simultaneously and from the same face of the double bond, with boron adding to the less substituted carbon atom of the double bond. As shown in the mechanism, there is a slight polarity (about 5%) to the B—H bond because hydrogen (2.1) is slightly more electronegative than boron (2.0).

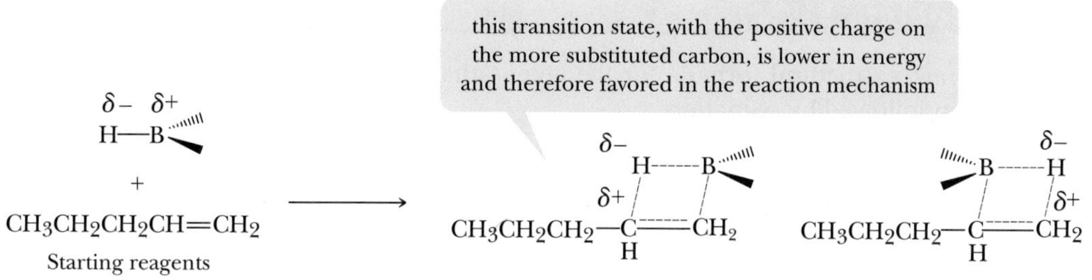

this transition state, with the positive charge on the more substituted carbon, is lower in energy and therefore favored in the reaction mechanism

$\overset{\delta-}{H}$—$\overset{\delta+}{B}$ + $CH_3CH_2CH_2CH{=}CH_2$ ⟶ transition states

Starting reagents

$CH_3CH_2CH_2$—$\overset{\delta+}{C}$H---CH_2 (with $\overset{\delta-}{H}$---B) $CH_3CH_2CH_2$—CH---$\overset{\delta+}{C}H_2$ (with B---$\overset{\delta-}{H}$)

Two possible transition states (some carbocation character exists in both transition states)

We account for the regioselectivity by steric factors. Boron, the larger part of the reagent, adds selectively to the less hindered carbon of the double bond, and hydrogen, the smaller part of the reagent, adds to the more hindered carbon. It is believed that the observed regioselectivity is due largely to steric effects.

Trialkylboranes are rarely isolated. Rather, they are converted directly to other products formed by substitution of another atom (H, O, N, C, or halogen) for boron. One of the most important reactions of trialkylboranes is with hydrogen peroxide in aqueous sodium hydroxide during which an OH group is substituted for boron. In this reaction, hydrogen peroxide is an oxidizing agent and, under these conditions, oxidizes a trialkylborane to an alcohol and sodium borate, Na_3BO_3.

$$(RO)_3B + 3NaOH \longrightarrow 3ROH + Na_3BO_3$$

A trialky lborane   Sodium borate

Hydrogen peroxide oxidation of a trialkylborane is stereoselective in that the configuration of the alkyl group is retained; whatever the position of boron in relation to other groups in the trialkylborane, the OH group by which it is replaced occupies the same position. Thus, the net result of hydroboration–oxidation of an alkene is syn stereoselective addition of H and OH to a carbon–carbon double bond.

Example 5.9

Draw structural formulas for the alcohol formed by hydroboration–oxidation of each alkene.

(a) $CH_3\overset{CH_3}{\overset{|}{C}}{=}CHCH_3$ (b) 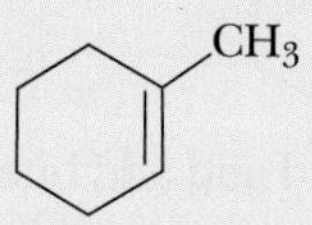 CH_3

Strategy

Hydroboration–oxidation is regioselective (—OH adds to the less substituted carbon of the carbon–carbon double bond, and —H adds to the more substituted carbon). It is also stereoselective (—H and —OH add to the same face of the double bond).

Solution

(a) 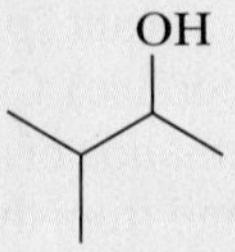 3-Methyl-2-butanol

(b) 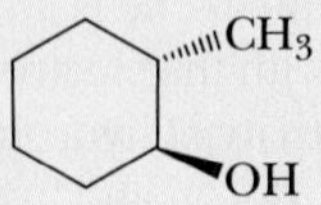 *trans*-2-Methylcylohexanol

See problem 5.38

Problem 5.9

Draw a structural formula for the alkene that gives each alcohol on hydroboration followed by oxidation.

(a)

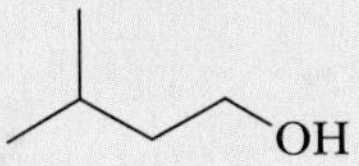

(b)

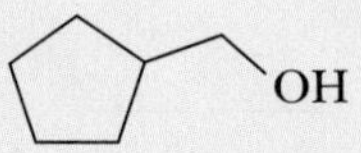

5.6 What Is Ozonolysis of an Alkene?

Treating an alkene with ozone, O_3, followed by a suitable workup, cleaves the carbon–carbon double bond and forms two carbonyl (C=O) groups in its place. The alkene in this reaction has undergone **oxidation**. This reaction is noteworthy because it is one of the very few organic reactions that break carbon–carbon bonds.

The alkene is dissolved in an inert solvent, such as dichloromethane, CH_2Cl_2, and a stream of ozone is bubbled through the solution. The products isolated depend on the reaction conditions. Hydrolysis of the reaction mixture with water yields hydrogen peroxide, an oxidizing agent that can bring about further oxidation. To prevent side reactions caused by reactive peroxide intermediates, a weak reducing agent, most commonly dimethylsulfide, $(CH_3)_2S$, is added during the workup to reduce peroxides to water. Ozonolysis of 2-methyl-2-pentene, for example, gives propanone and propanal.

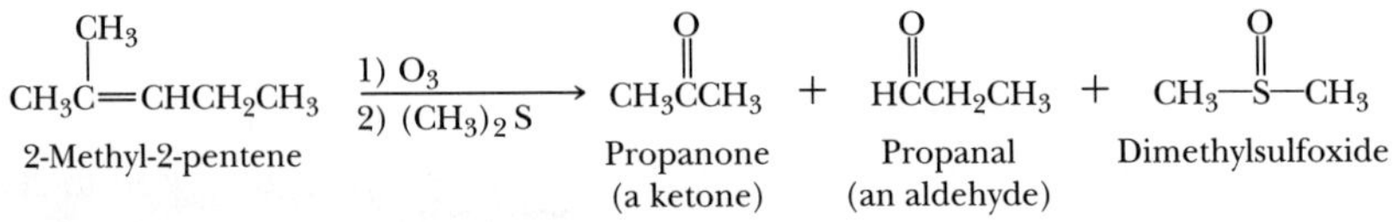

$$CH_3\overset{CH_3}{\overset{|}{C}}{=}CHCH_2CH_3 \xrightarrow[2)\ (CH_3)_2S]{1)\ O_3} CH_3\overset{O}{\overset{\|}{C}}CH_3 + H\overset{O}{\overset{\|}{C}}CH_2CH_3 + CH_3{-}\overset{O}{\overset{\|}{S}}{-}CH_3$$

2-Methyl-2-pentene → Propanone (a ketone) + Propanal (an aldehyde) + Dimethylsulfoxide

HOW TO 5.2 Predict the Products of an Ozonolysis Reaction

1. Identify all C–C double bonds in the reactant.

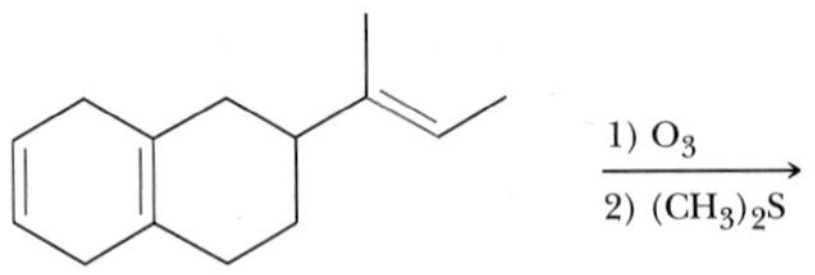

1) O_3 2) $(CH_3)_2S$

2. Cleave each C–C double bond and add one oxygen atom to each carbon.

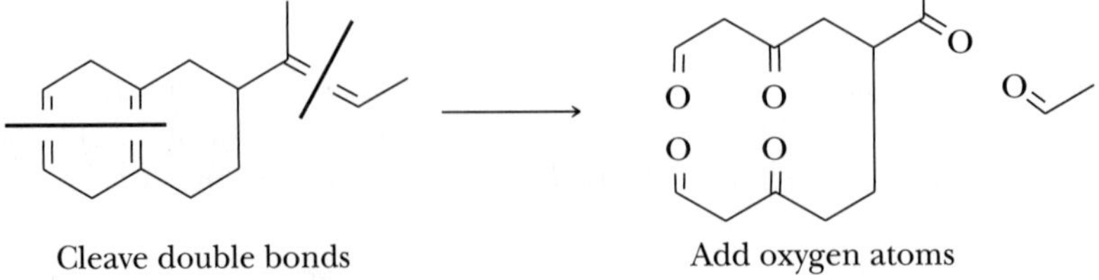

Cleave double bonds → Add oxygen atoms

3. Clean up the chemical structure by drawing in the hydrogens for all aldehyde groups.

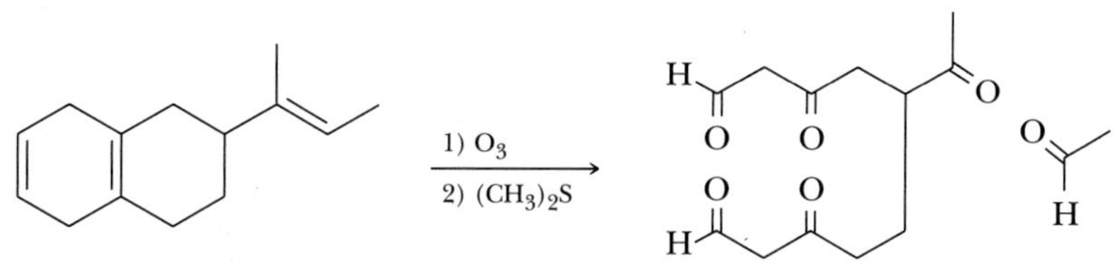

1) O_3
2) $(CH_3)_2S$

:O:
||
C–H
aldehyde
(carbonyl group
bonded to a H)

Example 5.10

Draw structural formulas for the products of the following ozonolysis reactions, and name the new functional groups formed in the oxidation.

(a) 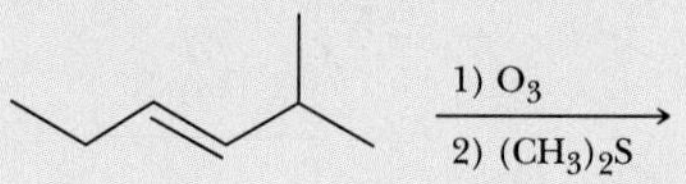 1) O_3 2) $(CH_3)_2S$ (b) 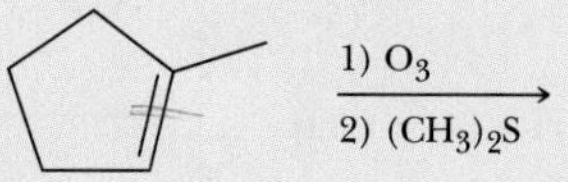 1) O_3 2) $(CH_3)_2S$

Strategy

The carbon–carbon double bond is cleaved, and in its place two carbonyl groups are formed.

Solution

(a) 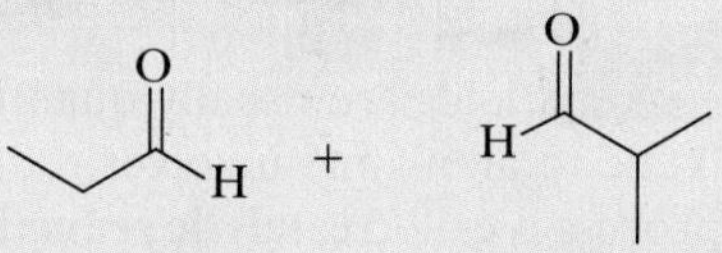 (an aldehyde) + (an aldehyde)

(b) 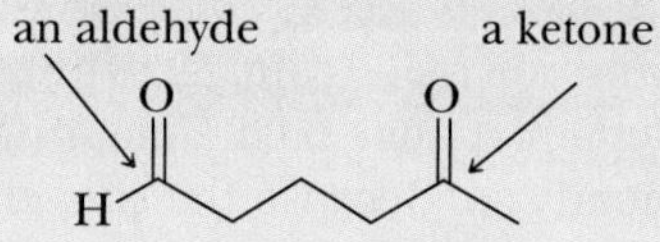 an aldehyde a ketone (a ketoaldehyde)

See problem 5.46

Problem 5.10

What alkene with the molecular formula C_6H_{10}, when treated with ozone and then dimethylsulfide, gives the following products?

(a) 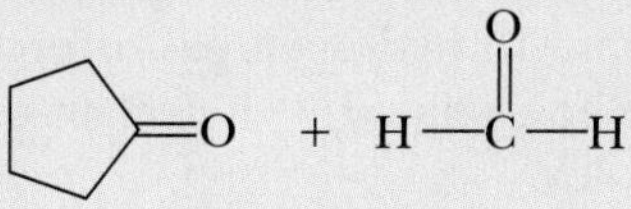 cyclopentanone =O + H—C(=O)—H

(b) $CH_3\overset{O}{\overset{\|}{C}}CH_2CH_2CH_2\overset{O}{\overset{\|}{C}}H$

To understand how ozonolysis occurs, let us first look at the structure of ozone. Following are four contributing structures to the ozone hybrid. Each shows the required 3 × 6 = 18 valence electrons. Like other reagents that add to alkenes, ozone is strongly electrophilic, as can be seen by examining Lewis structures for the four contributing structures for its resonance hybrid.

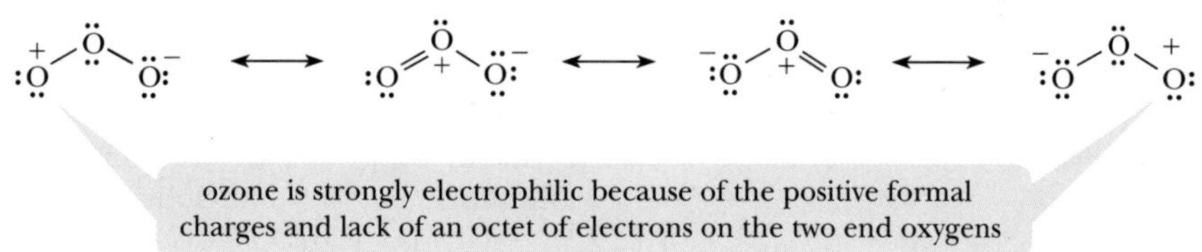

ozone is strongly electrophilic because of the positive formal charges and lack of an octet of electrons on the two end oxygens

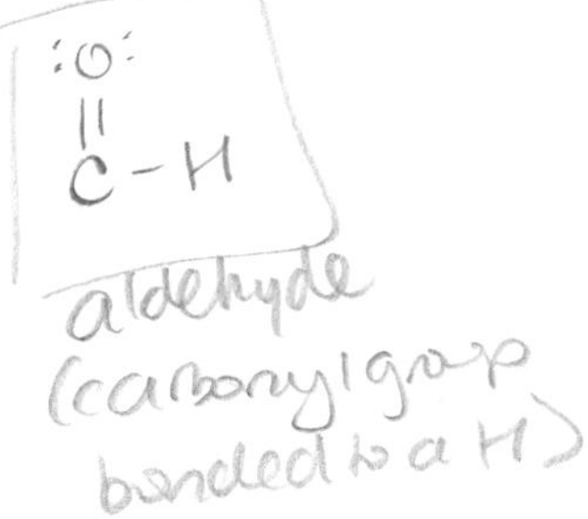

Initial reaction of an alkene with ozone gives an intermediate called a molozonide.

A molozonide

The molozonide then undergoes a rearrangement of atoms and valence electrons to give an ozonide, which, when treated with dimethylsulfide, gives the final products.

$$CH_3CH{=}CHCH_3 \xrightarrow{O_3} [\text{A molozonide}] \longrightarrow \text{An ozonide} \xrightarrow{(CH_3)_2S} 2\,CH_3CH{=}O$$

2-Butene — A molozonide — An ozonide — Acetaldehyde

5.7 How Can an Alkene Be Reduced to an Alkane?

Most alkenes react quantitatively with molecular hydrogen, H_2, in the presence of a transition metal catalyst to give alkanes. Commonly used transition metal catalysts include platinum, palladium, ruthenium, and nickel. Yields are usually quantitative or nearly so. Because the conversion of an alkene to an alkane involves reduction by hydrogen in the presence of a catalyst, the process is called **catalytic reduction** or, alternatively, **catalytic hydrogenation**.

$$\text{Cyclohexene} + H_2 \xrightarrow[25°C,\ 3\ atm]{Pd} \text{Cyclohexane}$$

Cyclohexene — Cyclohexane

The metal catalyst is used as a finely powdered solid, which may be supported on some inert material such as powdered charcoal or alumina. The reaction is carried out by dissolving the alkene in ethanol or another nonreacting organic solvent, adding the solid catalyst, and exposing the mixture to hydrogen gas at pressures from 1 to 100 atm. Alternatively, the metal may be chelated with certain organic molecules and used in the form of a soluble complex.

Catalytic reduction is stereoselective, the most common pattern being the **syn addition** of hydrogens to the carbon–carbon double bond. The catalytic reduction of 1,2-dimethylcyclohexene, for example, yields *cis*-1,2-dimethylcyclohexane along with lesser amounts of *trans*-1,2-dimethylcyclohexane.

$$\text{1,2-Dimethylcyclohexene} + H_2 \xrightarrow{Pt} \textit{cis}\text{-1,2-Dimethylcyclohexane} + \textit{trans}\text{-1,2-Dimethylcyclohexane}$$

1,2-Dimethylcyclohexene — 70% to 85% *cis*-1,2-Dimethylcyclohexane — 30% to 15% *trans*-1,2-Dimethylcyclohexane

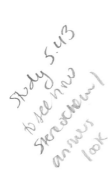

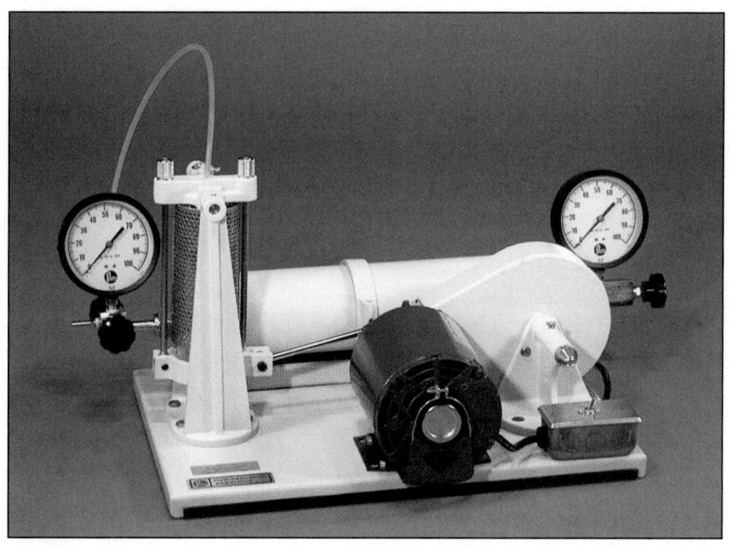

A Paar shaker-type hydrogenation apparatus. *(Paar Instrument Co., Moline, IL)*

The transition metals used in catalytic reduction are able to adsorb large quantities of hydrogen onto their surfaces, probably by forming metal–hydrogen sigma bonds. Similarly, these transition metals adsorb alkenes on their surfaces, with the formation of carbon–metal bonds [Figure 5.6(a)]. Hydrogen atoms are added to the alkene in two steps.

Heats of Hydrogenation and the Relative Stabilities of Alkenes

The **heat of hydrogenation** of an alkene is defined as its heat of reaction, $\Delta H°$, with hydrogen, to form an alkane. Table 5.2 lists the heats of hydrogenation of several alkenes.

Three important points follow from the information given in the table.

1. The reduction of an alkene to an alkane is an exothermic process. This observation is consistent with the fact that, during hydrogenation, there is net conversion of a weaker pi bond to a stronger sigma bond; that is, one sigma bond (H—H) and one pi bond (C═C) are broken, and two new sigma bonds (C—H) are formed.
2. The heat of hydrogenation depends on the degree of substitution of the carbon–carbon double bond: The greater the substitution, the lower is the heat of hydrogenation. Compare, for example, the heats of hydrogenation of ethylene (no substituents), propene (one substituent), 1-butene (one substituent), and the *cis* and *trans* isomers of 2-butene (two substituents each).
3. The heat of hydrogenation of a *trans* alkene is lower than that of the isomeric *cis* alkene. Compare, for example, the heats of hydrogenation of *cis*-2-butene and

(a)

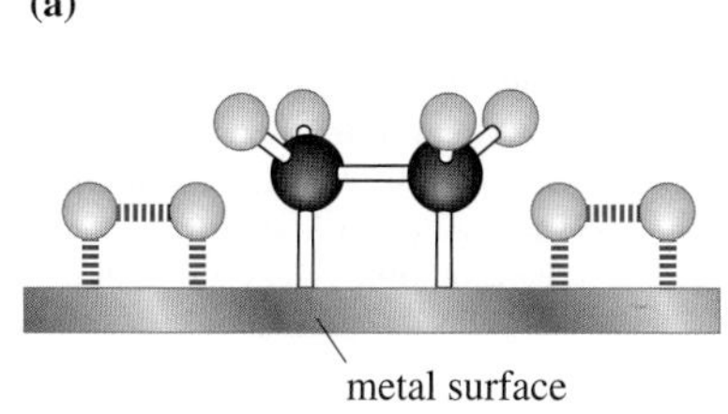

(b)

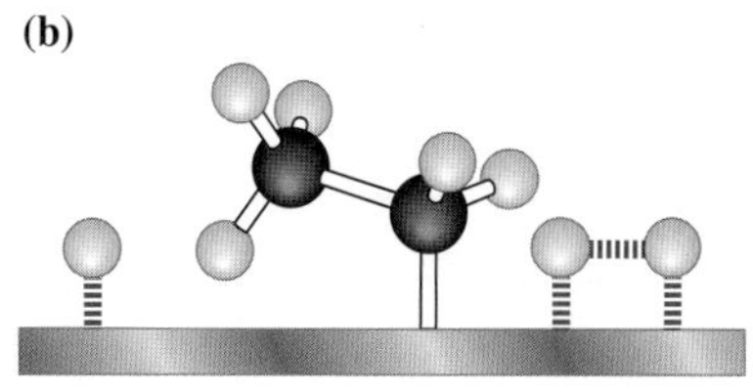

(c)

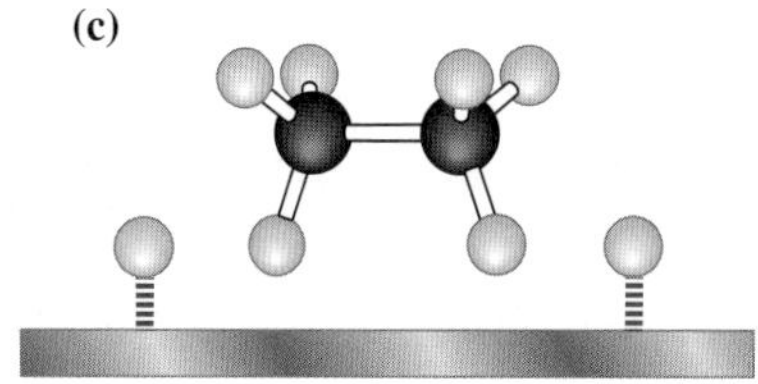

Figure 5.6
Syn addition of hydrogen to an alkene involving a transition metal catalyst. (a) Hydrogen and the alkene are adsorbed on the metal surface, and (b) one hydrogen atom is transferred to the alkene, forming a new C—H bond. The other carbon remains adsorbed on the metal surface. (c) A second C—H bond forms, and the alkane is desorbed.

TABLE 5.2 Heats of Hydrogenation of Several Alkenes

Name	Structural Formula	ΔH [kJ (kcal)/mol]	
Ethylene	$CH_2{=}CH_2$	−137 (−32.8)	
Propene	$CH_3CH{=}CH_2$	−126 (−30.1)	
1-Butene	$CH_3CH_2CH{=}CH_2$	−127 (−30.3)	Ethylene
cis-2-Butene	$(H_3C)(H)C{=}C(CH_3)(H)$	−120 (−28.6)	
trans-2-Butene	$(H_3C)(H)C{=}C(H)(CH_3)$	−115 (−27.6)	*trans*-2-Butene
2-Methyl-2-butene	$(H_3C)_2C{=}C(CH_3)(H)$	−113 (−26.9)	
2,3-Dimethyl-2-butene	$(H_3C)_2C{=}C(CH_3)_2$	−111 (−26.6)	2,3-Dimethyl-2-butene

trans-2-butene. Because the reduction of each alkene gives butane, any difference in their heats of hydrogenation must be due to a difference in relative energy between the two alkenes (Figure 5.7). The alkene with the lower (less negative) value of $\Delta H°$ is the more stable alkene.

These features of a hydrogenation reaction allow us to compare the stabilities and reactivities of any two alkenes that would yield the same product upon hydrogenation. Thus we explain the greater stability of *trans* alkenes relative to *cis* alkenes in terms of nonbonded interaction strain. In *cis*-2-butene, the two $-CH_3$ groups are sufficiently close to each other that there is repulsion between their electron clouds. This repulsion is reflected in the larger heat of hydrogenation (decreased stability) of *cis*-2-butene compared with that of *trans*-2-butene (approximately 4.2 kJ/mol).

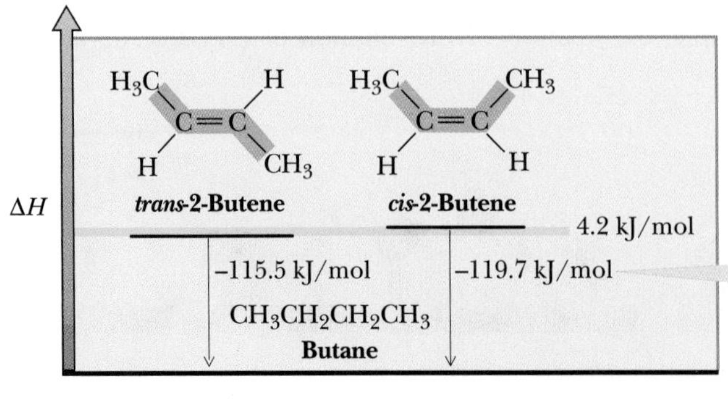

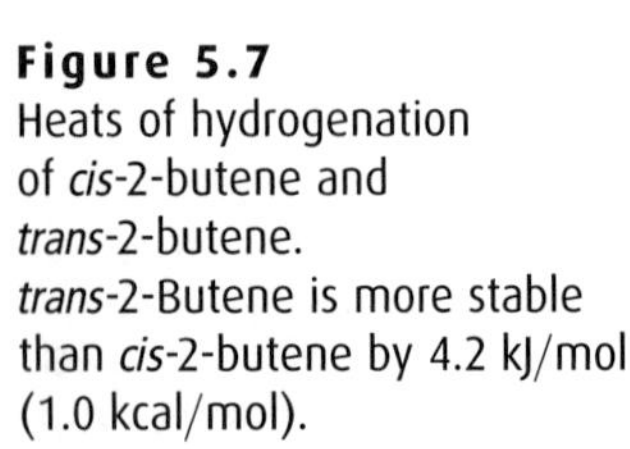

Figure 5.7
Heats of hydrogenation of *cis*-2-butene and *trans*-2-butene. *trans*-2-Butene is more stable than *cis*-2-butene by 4.2 kJ/mol (1.0 kcal/mol).

5.8 Why Are 1-Alkynes (Terminal Alkynes) Weak Acids?

One of the major differences between the chemistry of alkynes and that of alkenes and alkanes is that a hydrogen bonded to a carbon atom of a terminal alkyne is sufficiently acidic (pK_a 25) that it can be removed by a strong base, such as sodium amide, $NaNH_2$, to give an acetylide anion.

$$H-C{\equiv}C-H + {}^{-}{:}\ddot{N}H_2 \rightleftharpoons H-C{\equiv}C{:}^{-} + \ddot{N}H_3 \qquad K_{eq}=10^{13}$$

Acetylene	Amide	Acetylide	Ammonia
pK_a 25	anion	anion	pK_a 38
(stronger acid)	(stronger base)	(weaker base)	(weaker acid)

In this equilibrium, acetylene is the stronger acid and sodium amide is the stronger base, and the position of equilibrium lies considerably toward the right and favors formation of the acetylide anion and ammonia (Section 2.4). Table 5.3 gives pK_a values for an alkane, alkene, and an alkyne hydrogen. Also given for comparison is the value for water.

Because water (pK_a 15.7) is a stronger acid than acetylene (pK_a 25), the hydroxide ion is not a strong enough base to convert a terminal alkyne to an alkyne anion. The position of equilibrium for this acid–base reaction lies toward the left.

$$H-C{\equiv}C-H + {}^{-}{:}\ddot{O}H \rightleftharpoons H-C{\equiv}C{:}^{-} + H-\ddot{O}H$$

pK_a 25			pK_a 15.7
(weaker acid)	(weaker base)	(stronger base)	(stronger acid)

The pK_a values for alkene hydrogens (pK_a approximately 44) and alkane hydrogens (pK_a approximately 51) are so large (they are so weakly acidic) that neither the commonly used alkali metal hydroxides nor sodium amide are strong enough bases to remove a proton from an alkene or an alkane.

Why is the acidity of a hydrogen bonded to a triple-bonded carbon so much more acidic than one bonded to a double-bonded carbon of an alkene or to an alkane? We explain these relative acidities in the following way. The lone pair of electrons on a carbon anion lies in a hybrid orbital: an sp^3 hybrid orbital for an

Table 5.3 Acidity of Alkanes, Alkenes, and Alkynes

Weak Acid		Conjugate Base	pK_a
Water	HO–H	HO^-	15.7
Alkyne	HC≡C–H	$HC{\equiv}C^-$	25
Alkene	CH_2=CH–H	$CH_2{=}CH^-$	44
Alkane	CH_3CH_2–H	$CH_3CH_2^-$	51

(Arrow: Increasing acidity, toward water)

alkane, an sp^2 hybrid orbital for an alkene, and an sp hybrid orbital for an alkyne. An sp hybrid orbital has 50% s character, an sp^2 hybrid orbital has 33% s character, and an sp^3 hybrid orbital has 25% s character. Recall from your course in general chemistry and from Chapter 1 of this text that a $2s$ orbital is lower in energy than a $2p$ orbital. Consequently, electrons in a $2s$ orbital are held more tightly to the nucleus than those in a $2p$ orbital. The more s character in a hybrid orbital of carbon, the more electronegative the carbon atom will be, resulting in a greater stability of the anion and thus a more acidic hydrogen. Of the three types of organic compounds in the series alkyne, alkene, and alkane, the carbon in an alkyne (sp hybridized with 50% s character) is the most electronegative. Therefore, an alkyne anion is the most stable of the series, and an alkyne is the strongest acid of the series. By similar reasoning, the alkane carbon (sp^3 hybridized and 25% s character) is the least electronegative, and an alkane is the weakest acid of the series. An alkene, with 33% s character, is intermediate. Finally, it is only the hydrogen of a 1-alkyne that shows this type of acidity. No other hydogens of an alkyne have comparable acidity, and no other hydrogens are removed by $NaNH_2$.

these hydrogens are much lower in acidity and are not deprotonated by $NaNH_2$

only this hydrogen is acidic enough to be deprotonated by $NaNH_2$

$$CH_3{-}CH_2{-}CH_2{-}C{\equiv}C{-}H$$

5.9 How Can an Acetylide Anion Be Used to Create a New Carbon–Carbon Bond?

As we have seen, an acetylide anion is a strong base. It is also a nucleophile—it has an unshared pair of electrons that it can donate to an electrophilic carbon atom to form a new carbon–carbon bond.

To see how the use of an acetylide anion can lead to the formation of a new carbon–carbon bond, consider chloromethane, CH_3Cl. The C—Cl bond of chloromethane is polar covalent, with carbon bearing a partial positive charge because of the difference in electronegativity between carbon and chlorine.

$$H{-}\overset{H}{\underset{H}{\overset{|}{\underset{|}{C}}}}{}^{\delta+}{-}Cl^{\delta-}$$

In this instance, an acetylide anion donates its unshared pair of electrons to the carbon of chloromethane and in so doing displaces the halogen atom.

$$H{-}C{\equiv}C{:}^{-}\,Na^{+} + H{-}\overset{H}{\underset{H}{\overset{|}{\underset{|}{C}}}}{}^{\delta+}{-}\ddot{\underset{..}{Cl}}{:}^{\delta-} \longrightarrow H{-}C{\equiv}C{-}CH_3 + Na^{+}Cl^{-}$$

The important result is the formation of a new carbon–carbon bond. As is the case with so many organic reactions, this is an instance where reaction is brought about by the interaction of positive and negative charges of interacting molecules.

Because an alkyl group is added to the original alkyne molecule, this type of reaction is called an **alkylation reaction**. We limit our discussion in this chapter to reactions of acetylide anions with methyl and primary haloalkanes. We will discuss the scope and limitation of this type of nucleophilic substitution in more detail in Chapter 7. For reasons we will discuss there, alkylation of nucleophilic acetylide anions is practical only for methyl and primary halides. While this alkylation reaction can be used with limited success with secondary haloalkanes, it fails altogether for tertiary haloalkanes.

Because of the ready availability of acetylene and the ease with which it is converted to a nucleophile, alkylation of acetylide anions is the most convenient laboratory method used for the synthesis of other alkynes. The process can be repeated, and a terminal alkyne in turn can be converted to an internal alkyne. An important feature of this reaction is that a new carbon–carbon skeleton can be made, allowing for the construction of larger carbon skeletons from smaller ones. In the following scheme, the carbon skeleton of 3-heptyne is constructed from acetylene and two lower-molecular-weight haloalkanes.

$$\underset{\text{Acetylene}}{HC{\equiv}CH} \xrightarrow[\text{2) } CH_3CH_2Br]{\text{1) } NaNH_2} \underset{\text{1-Butyne}}{CH_3CH_2C{\equiv}CH} \xrightarrow[\text{4) } CH_3CH_2CH_2Br]{\text{3) } NaNH_2} \underset{\text{3-Heptyne}}{CH_3CH_2C{\equiv}CCH_2CH_2CH_3}$$

Example 5.11

Propose a synthesis for each alkyne starting with acetylene and any necessary organic and inorganic reagents.

(a) cyclohexyl–$CH_2C{\equiv}CH$ (b) $CH_3C{\equiv}CCH_2CH(CH_3)CH_3$ (c) $CH_3C{\equiv}CCH_2CH_2CH_2CH_3$

Strategy

Each alkyne can be synthesized by alkylation of an appropriate alkyne anion. First decide which new carbon–carbon bond or bonds must be formed by alkylation and which alkyne anion nucleophile and haloalkane pair is required to give the desired product. Synthesis of a terminal alkyne from acetylene requires only one nucleophilic substitution, and synthesis of an internal alkyne from acetylene requires two nucleophilic substitutions.

Solution

(a) $HC{\equiv}CH \xrightarrow[\text{2) cyclohexyl–}CH_2Br]{\text{1) } NaNH_2}$ cyclohexyl–$CH_2C{\equiv}CH$

(b) $HC{\equiv}CH \xrightarrow[\text{2) } BrCH_3]{\text{1) } NaNH_2} CH_3C{\equiv}CH \xrightarrow[\text{4) } BrCH_2CH(CH_3)CH_3]{\text{3) } NaNH_2} CH_3C{\equiv}CCH_2CH(CH_3)CH_3$

(c) $HC{\equiv}CH \xrightarrow[\text{2) } BrCH_3]{\text{1) } NaNH_2} CH_3C{\equiv}CH \xrightarrow[\text{4) } BrCH_2CH_2CH_2CH_3]{\text{3) } NaNH_2} CH_3C{\equiv}CCH_2CH_2CH_2CH_3$

See problem 5.56

Problem 5.11

Propose a synthesis for each alkyne starting with acetylene and any necessary organic and inorganic reagents.

(a) 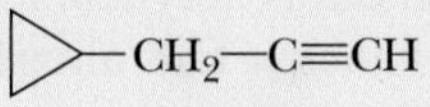 (cyclopropyl)$-CH_2-C\equiv CH$ (b) 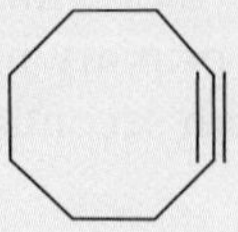 (cyclooctyne)

5.10 What Is the Product of the Acid-Catalyzed Hydration of an Alkyne?

In the presence of concentrated sulfuric acid and Hg(II) salts, alkynes undergo the addition of water in a reaction that follows Markovnikov's rule—that is, it corresponds to the addition of H to the less substituted carbon of the triple bond and —OH to the more substituted carbon, as illustrated by the hydration of propyne.

$$\underset{\text{Propyne}}{CH_3C\equiv CH} + H_2O \xrightarrow[HgSO_4]{H_2SO_4} \underset{\substack{\text{Propen-2-ol} \\ \text{(an enol)}}}{CH_3\overset{\overset{\displaystyle OH}{|}}{C}=CH_2} \rightleftharpoons \underset{\substack{\text{Propanone} \\ \text{(acetone)}}}{CH_3\overset{\overset{\displaystyle O}{\|}}{C}CH_3}$$

The initial product of hydration of an alkyne is an **enol**, a compound containing a hydroxyl group bonded to a carbon of a carbon–carbon double bond. The name "enol" is derived from the fact that the compound is both an alkene (-en-) and an alcohol (-ol).

Enols are in equilibrium with a constitutional isomer formed by migration of a hydrogen atom from oxygen to carbon and rearrangement of the carbon–carbon double bond to form a carbon–oxygen double bond. Propanone (the keto form) is more stable than the enol form. Ketone forms in general are more stable than enol forms, primarily because a carbon–oxygen π bond is stronger than a carbon–carbon π bond.

The keto and enol forms of propanone are called tautomers. **Tautomers** are constitutional isomers that are in equilibrium with each other and differ only in the position of a hydrogen atom and double bond. We will have more to say about tautomerism and its significance in Section 13.8.

Example 5.12

Draw the structural formula of the product of acid-catalyzed hydration of each alkyne.

(a) 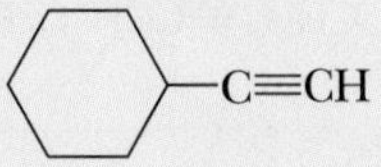 (cyclohexyl)$-C\equiv CH$ (b) $HC\equiv CCH_2\overset{\overset{\displaystyle CH_3}{|}}{C}HCH_3$

Strategy

As per Markovnikov's rule, H adds to the terminal carbon and OH adds to the internal carbon of the carbon–carbon triple bond. The immediate product of reaction is an enol that is in equilibrium with a more stable keto form.

Solution

(a) $C_6H_{11}-C{\equiv}CH + H_2O \xrightarrow[HgSO_4]{H_2SO_4} C_6H_{11}-\overset{OH}{\overset{|}{C}}{=}CH_2 \rightleftharpoons C_6H_{11}-\overset{O}{\overset{\|}{C}}-CH_3$

(b) $HC{\equiv}CCH_2\overset{CH_3}{\overset{|}{C}}HCH_3 + H_2O \xrightarrow[HgSO_4]{H_2SO_4} CH_2{=}\overset{OH}{\overset{|}{C}}CH_2\overset{CH_3}{\overset{|}{C}}HCH_3 \rightleftharpoons CH_3\overset{O}{\overset{\|}{C}}CH_2\overset{CH_3}{\overset{|}{C}}HCH_3$

See problem 5.49

Problem 5.12

Acid-catalyzed hydration of 2-pentyne gives a mixture of two ketones, each with the molecular formula $C_5H_{10}O$. Propose structural formulas for these two ketones and for the enol from which each is derived.

5.11 How Can Alkynes Be Reduced to Alkenes and Alkanes?

Treatment of an alkyne with H_2 in the presence of a transition metal catalyst, most commonly Pd, Pt, or Ni, results in the addition of two moles of H_2 to the alkyne and its conversion to an alkane. Catalytic reduction of an alkyne can be brought about at or slightly above room temperature and with moderate pressures of hydrogen gas.

$$\underset{\text{2-Butyne}}{CH_3C{\equiv}CCH_3} + 2H_2 \xrightarrow[\text{3 atm}]{\text{Pd, Pt, or Ni}} \underset{\text{Butane}}{CH_3CH_2CH_2CH_3}$$

Reduction of an alkyne occurs in two stages: first, addition of one mole of H_2 to form an alkene and then addition of the second mole of H_2 to the alkene to form the alkane. In most cases, it is not possible to stop the reaction at the alkene stage. However, by careful choice of catalyst, it is possible to stop the reaction at the addition of one mole of hydrogen. The catalyst most commonly used for this purpose consists of finely powdered palladium metal deposited on solid calcium carbonate that has been specially modified with lead salts. This combination is known as the **Lindlar catalyst**. Reduction (hydrogenation) of alkynes over a Lindlar catalyst is stereoselective: **syn addition** of two hydrogen atoms to the carbon–carbon triple bond gives a *cis* alkene:

$$\underset{\text{2-Pentyne}}{H_3C-C{\equiv}C-CH_2CH_3} \xrightarrow[\text{Lindlar catalyst}]{H_2} \underset{\textit{cis}\text{-2-Pentene}}{(H_3C)(H)C{=}C(H)(CH_2-CH_3)}$$

Because addition of hydrogen in the presence of the Lindlar catalyst is stereoselective for syn addition, it has been proposed that reduction proceeds by simultaneous or nearly simultaneous transfer of two hydrogen atoms from the surface of the metal catalyst to the alkyne. Earlier we presented a similar mechanism for the catalytic reduction of an alkene to an alkane (Section 5.7).

Note that catalytic reduction using the Lindlar catalyst gives a *cis* alkene. The isomeric *trans* alkene can be generated by reduction using sodium or lithium metal dissolved in liquid ammonia. In this reaction, the alkali metal is oxidized to Li^+ or Na^+, and the alkyne is reduced to an alkene.

The mechanism of this reaction is beyond the scope of this course. What is important about this pair of reduction methods is that we now have a way to convert an alkyne to either a *cis* alkene or a *trans* alkene.

$$H_3C{-}C{\equiv}C{-}CH_2CH_3 \xrightarrow[NH_3(l)]{Li} \text{trans-}H_3C(H)C{=}C(H)CH_2CH_3$$

2-Pentyne → *trans*-2-Penene

Organic chemistry is the foundation for the synthesis of new compounds such as medicines, agrochemicals, and plastics, to name just a few. In order to make these compounds, organic chemists must rely on a vast collection of reactions. The reactions presented in this chapter will already allow you to achieve the synthesis of complex molecules that may require multiple steps to make. As you continue your studies of organic chemistry, new reactions will be presented, the same reactions that have allowed for the creation of the millions of compounds that have contributed to the progress of civilization.

Key Terms and Concepts

Summary of Key Questions

5.1 What Are the Characteristic Reactions of Alkenes?

- A characteristic reaction of alkenes is **addition**, during which a pi bond is broken and sigma bonds are formed to two new atoms or groups of atoms. Alkene addition reactions include addition of halogen acids, H—Cl, acid-catalyzed addition of H_2O to form an alcohol, addition of halogens, X_2, hydroboration followed by oxidation to give an alcohol, and transition metal-catalyzed addition of H_2 to form an alkane.

5.2 What Is a Reaction Mechanism?

- A **reaction mechanism** is a description of (1) how and why a chemical reaction occurs, (2) which bonds break and which new ones form, (3) the order and relative rates in which the various bond-breaking and bond-forming steps take place, and (4) the role of the catalyst if the reaction involves a catalyst.
- **Transition state theory** provides a model for understanding the relationships among reaction rates, molecular structure, and energetics.
- A key postulate of transition state theory is that a **transition state** is formed.
- The difference in energy between reactants and the transition state is called the **activation energy**.
- An **intermediate** is an energy minimum between two transition states.
- The slowest step in a multistep reaction, called the **rate-determining step**, is the one that crosses the highest energy barrier.

5.3 What Are the Mechanisms of Electrophilic Additions to Alkenes?

- An **electrophile** is any molecule or ion that can accept a pair of electrons to form a new covalent bond. All electrophiles are Lewis acids.
- The rate-determining step in **electrophilic addition** to an alkene is reaction of an electrophile with a carbon–carbon double bond to form a **carbocation**, an ion that contains a carbon with only six electrons in its valence shell and has a positive charge.
- Carbocations are planar with bond angles of 120° about the positive carbon.
- The order of stability of carbocations is $3° > 2° > 1° >$ methyl. Primary carbocations, however, are so unstable and have such a high energy of activation for their formation that they are never formed in solution.

5.4 What Are Carbocation Rearrangements?

- The driving force for a carbocation rearrangement is conversion to a more stable 2° or 3° carbocation.
- Rearrangement is by a 1,2-shift in which an atom or group of atoms with its bonding electrons moves from an adjacent carbon to an electron-deficient carbon.

5.5 What Is Hydroboration–Oxidation of an Alkene?

- Hydroboration of an alkene is the addition of BH_2 and H across a C—C double bond.
- Hydroboration occurs with anti-Markovnikov regioselectivity with the H adding to the carbon with the fewer number of hydrogens.
- Oxidation of the hydroboration product results in the replacement of the boron group with an —OH group.
- Hydroboration–oxidation is syn stereoselective.

5.6 What Is Ozonolysis of an Alkene?

- The reaction of ozone, O_3, with a C—C double bond results in cleavage of the double bond and the formation of two carbonyl groups in its place.

5.7 How Can an Alkene Be Reduced to an Alkane?

- The reaction of an alkene with H_2 in the presence of a transition metal catalyst converts all C—C double bonds in the alkene to C—C single bonds via the syn stereoselective addition of a hydrogen to each carbon of the former double bond.
- The heats of reaction, ΔH, of hydrogenation reactions can be used to compare the relative stabilities of alkenes.

5.8 Why Are 1-Alkynes (Terminal Alkynes) Weak Acids?

- Terminal alkynes are weakly acidic (pK_a 25) and can be converted to acetylide anions by strong bases such as sodium amide, $NaNH_2$.

5.9 How Can an Acetylide Anion Be Used to Create a New Carbon–Carbon Bond?

- Acetylide anions are both strong bases and nucleophiles. As nucleophiles, they can be alkylated by treatment with a methyl, primary, or secondary haloalkane. In this way, acetylene serves as a two-carbon building block for the synthesis of larger carbon skeletons.

5.10 What Is the Product of the Acid-Catalyzed Hydration of an Alkyne?

- In the presence of concentrated sulfuric acid and Hg(II) salts, alkynes undergo the addition of water in a reaction that follows Markovnikov's rule.

5.11 How Can Alkynes Be Reduced to Alkenes and Alkanes?

- Treatment of alkyne with H_2 in the presence of a transition metal catalyst, most commonly Pd, Pt, or Ni, results in the addition of two moles of H_2 to the alkyne and its conversion to an alkane.
- Reduction of an alkyne using the Lindlar catalyst results in syn-stereoselective addition of one mole of H_2 to an alkyne. With this reagent, a disubstituted alkyne can be reduced to a *cis*-alkene.
- Reduction of an alkyne using sodium or lithium metal dissolved in liquid ammonia results in anti-stereoselective addition of one mole of H_2 to an alkyne.

Quick Quiz

Answer true or false to the following questions to assess your general knowledge of the concepts in this chapter. If you have difficulty with any of them, you should review the appropriate section in the chapter (shown in parentheses) before attempting the more challenging end-of-chapter problems.

1. Catalytic reduction of an alkene is syn stereoselective. (5.7)
2. Borane, BH_3, is a Lewis acid. (5.5)
3. All electrophiles are positively charged. (5.3)
4. Catalytic hydrogenation of cyclohexene gives hexane. (5.7)
5. A rearrangement will occur in the reaction of 2-methyl-2-pentene with HBr. (5.4)
6. All nucleophiles are negatively charged. (5.3)
7. In hydroboration, BH_3 behaves as an electrophile. (5.5)
8. In catalytic hydrogenation of an alkene, the reducing agent is the transition metal catalyst. (5.7)
9. Alkene addition reactions involve breaking a pi bond and forming two new sigma bonds in its place. (5.3)
10. When ozone is reacted with an alkene followed by treatment with dimethylsulfide, the final product will contain two carbonyl groups for every C—C double bond present in the original alkene. (5.6)
11. The foundation for Markovnikov's rule is the relative stability of carbocation intermediates. (5.3)
12. Acid-catalyzed hydration of an alkene is regioselective. (5.3)
13. Water adds twice to an alkyne; the first molecule of water adds to form an enol, followed by a second molecule of water to form a ketone or an aldehyde. (5.10)
14. The mechanism for addition of HBr to an alkene involves one transition state and two reactive intermediates. (5.3)
15. Hydroboration of an alkene is regioselective and stereoselective. (5.5)
16. According to the mechanism given in the text for acid-catalyzed hydration of an alkene, the —H and —OH groups added to the double bond both arise from the same molecule of H_2O. (5.3)
17. Acid-catalyzed addition of H_2O to an alkene is called *hydration*. (5.3)
18. If a compound fails to react with Br_2, it is unlikely that the compound contains a carbon–carbon double bond. (5.3)
19. The conversion of ethylene, $CH_2{=}CH_2$, to ethanol, CH_3CH_2OH, is an oxidation reaction. (5.3)
20. Addition of Br_2 and Cl_2 to cyclohexene is anti-stereoselective. (5.3)
21. A carbocation is a carbon that has four bonds to it and bears a positive charge. (5.3)
22. The geometry about the positively charged carbon of a carbocation is best described as trigonal planar. (5.3)
23. The carbocation derived by proton transfer to ethylene is $CH_3CH_2^+$. (5.3)

24. Alkyl carbocations are stabilized by the electron-withdrawing inductive effect of the positively charged carbon of the carbocation. (5.3)

25. The oxygen atom of an oxonium ion obeys the octet rule. (5.3)

26. Hydroxide ion (HO^-) is a stronger base than amide ion (H_2N^-). (5.8)

27. Markovnikov's rule refers to the regioselectivity of addition reactions to carbon–carbon double bonds. (5.3)

28. A rearrangement, in which a hydride ion shifts, will occur in the reaction of 3-methyl-1-pentene with HCl. (5.4)

29. Acid-catalyzed hydration of 1-butene gives 1-butanol, and acid-catalyzed hydration of 2-butene gives 2-butanol. (5.3)

30. Alkynes are always more acidic than alkenes. (5.8)

31. Alkenes are good starting materials for reactions in which it is necessary to form a C—C bond. (5.9)

32. Alkynes can be reduced to *cis* alkenes but not to *trans* alkenes. (5.11)

Detailed explanations for many of these answers can be found in the accompanying Solutions Manual.

(1) T (2) T (3) F (4) F (5) F (6) F (7) T (8) F (9) T (10) T
(11) T (12) T (13) F (14) F (15) T (16) F (17) T (18) T (19) F
(20) T (21) F (22) T (23) T (24) T (25) T (26) F (27) T (28) T
(29) F (30) F (31) F (32) F

Key Reactions

1. Addition of H—X to an Alkene (Section 5.3A)

The addition of H—X is regioselective and follows Markovnikov's rule. Reaction occurs in two steps and involves the formation of a carbocation intermediate:

$$\text{1-methylcyclopentene} + HCl \longrightarrow \text{1-chloro-1-methylcyclopentane}$$

2. Acid-Catalyzed Hydration of an Alkene (Section 5.3B)

Hydration of an alkene is regioselective and follows Markovnikov's rule. Reaction occurs in two steps and involves the formation of a carbocation intermediate:

$$CH_3C(CH_3){=}CH_2 + H_2O \xrightarrow{H_2SO_4} CH_3C(CH_3)(OH)CH_3$$

3. Addition of Bromine and Chlorine to an Alkene (Section 5.3C)

Addition of halogen occurs in two steps and involves anti-stereoselective addition by way of a bridged bromonium or chloronium ion intermediate:

$$\text{cyclohexene} + Br_2 \xrightarrow{CH_2Cl_2} \textit{trans}\text{-1,2-dibromocyclohexane}$$

4. Carbocation Rearrangements (Section 5.4)

Rearrangement is from a less stable carbocation intermediate to a more stable one by a 1,2-shift. Rearrangements often occur during the hydrochlorination and acid-catalyzed hydration of alkenes.

3,3-Dimethy-1-butene + HCl ⟶ 2-Chloro-2,3-dimethylbutane

5. Hydroboration–Oxidation of an Alkene (Section 5.5)

Addition of BH_3 to an alkene is syn-stereoselective and regioselective: boron adds to the less substituted carbon of the double bond, and hydrogen adds to the more substituted carbon. Hydroboration–oxidation results in anti-Markovnikov hydration of the alkene.

$$\text{1-methylcyclohexene} \xrightarrow[2)\ NaOH,\ H_2O_2]{1)\ BH_3} \textit{trans}\text{-2-methylcyclohexanol}$$

6. Ozonolysis of an Alkene (Section 5.6)

Treating an alkene with ozone, followed by dimethylsulfide, cleaves the carbon–carbon double bond of the alkene and gives two carbonyl groups in its place. The mechanism involves formation of a molozonide intermediate that rearranges to an ozonide intermediate, which is reduced using dimethylsulfide to give the carbonyl products.

$$\text{1-methylcyclohexene} \xrightarrow[2)\ (CH_3)_2S]{1)\ O_3} \text{6-oxoheptanal}$$

7. Reduction of an Alkene: Formation of Alkanes (Section 5.7)

Catalytic reduction involves predominantly the syn-stereoselective addition of hydrogen:

$$\text{1,2-dimethylcyclohexene} + H_2 \xrightarrow{\text{transition metal catalyst}} \textit{cis}\text{-1,2-dimethylcyclohexane}$$

8. Acidity of Terminal Alkynes (Section 5.8)
Treatment of a terminal alkyne (pK_a 25) with a strong base, most commonly $NaNH_2$, gives an acetylide anion salt.

$$H{-}C{\equiv}C{-}H + Na^+NH_2^- \longrightarrow H{-}C{\equiv}C{:}^-Na^+ + NH_3$$

9. Alkylation on an Acetylide Anion (Section 5.9)
Acetylide anions are nucleophiles and displace halogen from methyl and 1° haloalkanes. Alkylation of acetylide anions is a valuable way to assemble a larger carbon skeleton.

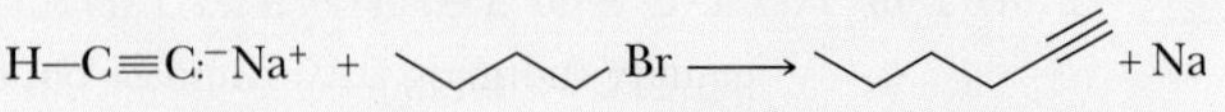

$$H{-}C{\equiv}C{:}^-Na^+ + \text{(bromoalkane)}\ Br \longrightarrow \text{(alkyne)} + Na$$

10. Acid-Catalyzed Hydration of an Alkyne (Section 5.10)
Acid-catalyzed addition of water in the presence of Hg^{2+} salts is regioselective. Keto-enol tautomerism of the resulting enol gives a ketone.

$$CH_3C{\equiv}CH + H_2O \xrightarrow[HgSO_4]{H_2SO_4} CH_3\overset{HO}{C}{=}CH_2 \rightleftharpoons CH_3\overset{O}{\overset{\|}{C}}CH_3$$

Propen-2-ol (an enol) — Propanone (acetone)

11. Reduction of an Alkyne (Section 5.11)
Several different reagents reduce alkynes. Catalytic reduction using a transition metal catalyst gives an alkane. Catalytic reduction using a specially prepared catalyst called the Lindlar catalyst gives a *cis* alkene. Reduction using lithium or sodium metal dissolved in liquid ammonia gives a *trans* alkene.

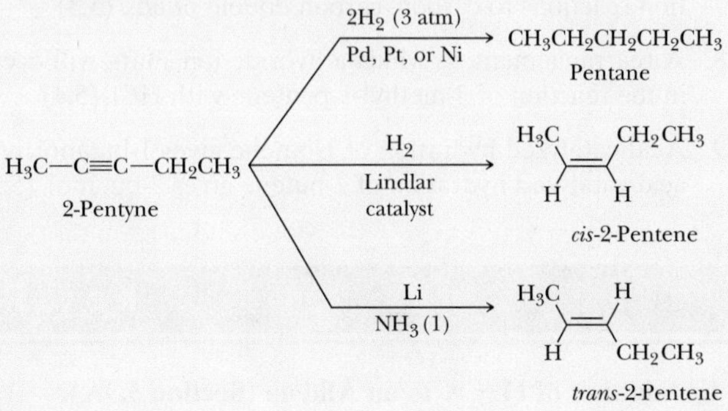

$H_3C{-}C{\equiv}C{-}CH_2CH_3$ (2-Pentyne)

$\xrightarrow[Pd, Pt, or Ni]{2H_2 (3 atm)} CH_3CH_2CH_2CH_2CH_3$ (Pentane)

$\xrightarrow[Lindlar catalyst]{H_2}$ cis-2-Pentene (H_3C and CH_2CH_3 on the same side, H and H)

$\xrightarrow[NH_3 (l)]{Li}$ trans-2-Pentene (H_3C and CH_2CH_3 on opposite sides)

Problems

A problem marked with an asterisk indicates an applied "real world" problem. Answers to problems whose numbers are printed in blue are given in Appendix D.

Section 5.2 Energy Diagrams

5.13 Describe the differences between a transition state and a reaction intermediate.

5.14 Sketch an energy diagram for a one-step reaction that is very slow and only slightly exothermic. How many transition states are present in this reaction? How many intermediates are present? **(See Example 5.1)**

5.15 Sketch an energy diagram for a two-step reaction that is endothermic in the first step, exothermic in the second step, and exothermic overall. How many transition states are present in this two-step reaction? How many intermediates are present? **(See Example 5.1)**

5.16 Determine whether each of the following statements is true or false, and provide a rationale for your decision:
(a) A transition state can never be lower in energy than the reactants from which it was formed.
(b) An endothermic reaction cannot have more than one intermediate.
(c) An exothermic reaction cannot have more than one intermediate.

Section 5.3 Electrophilic Additions to Alkenes

5.17 From each pair, select the more stable carbocation: **(See Example 5.3)**

(a) $CH_3CH_2CH_2^+$ or $CH_3\overset{+}{C}HCH_3$

(b) $CH_3\underset{+}{C}H(CH_3)CHCH_3$ or $CH_3\underset{+}{C}(CH_3)CH_2CH_3$

5.18 From each pair, select the more stable carbocation: **(See Example 5.3)**

(a) 1-methylcyclohexyl cation ($\overset{+}{C}$–CH_3) or 2-methylcyclohexyl cation (+ on ring carbon adjacent to C–CH_3)

(b) 1-methylcyclohexyl cation (ring $\overset{+}{C}$–CH_3) or cyclohexyl–$\overset{+}{C}H_2$

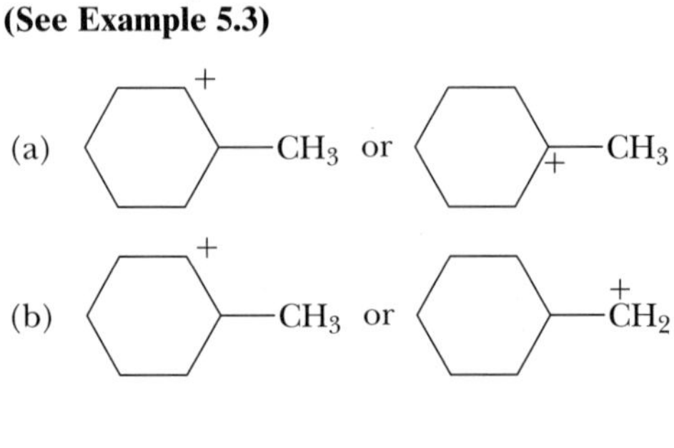

5.19 Draw structural formulas for the isomeric carbocation intermediates formed by the reaction of each alkene with HCl. Label each carbocation as primary, secondary, or tertiary, and state which, if either, of the isomeric carbocations is formed more readily. **(See Example 5.2)**

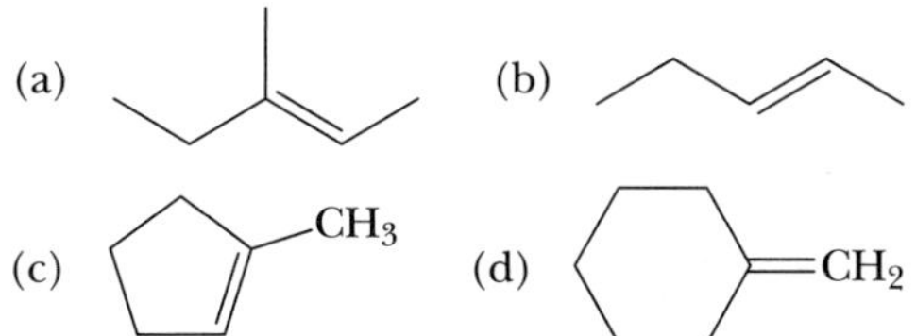

5.20 From each pair of compounds, select the one that reacts more rapidly with HI, draw the structural formula of the major product formed in each case, and explain the basis for your ranking: **(See Example 5.2)**

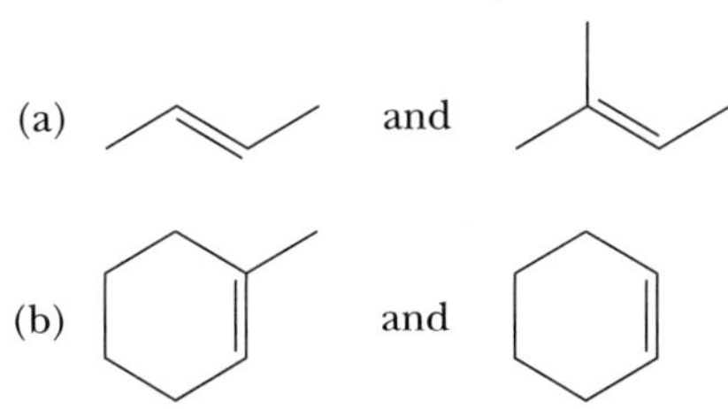

5.21 Complete these equations by predicting the major product formed in each reaction: **(See Examples 5.2, 5.5)**

(a) + HCl ⟶

(b) + H_2O $\xrightarrow{H_2SO_4}$

(c) + HI ⟶

(d) + HCl ⟶

(e) + H_2O $\xrightarrow{H_2SO_4}$

(f) + H_2O $\xrightarrow{H_2SO_4}$

5.22 The reaction of 2-methyl-2-pentene with each reagent is regioselective. Draw a structural formula for the product of each reaction, and account for the observed regioselectivity. **(See Examples 5.2, 5.5, 5.9)**

(a) HI (b) H_2O in the presence of H_2SO_4

(c) BH_3 followed by H_2O_2, NaOH

5.23 The addition of bromine and chlorine to cycloalkenes is stereoselective. Predict the stereochemistry of the product formed in each reaction: **(See Example 5.7)**

(a) 1-Methylcyclohexene + Br_2

(b) 1,2-Dimethylcyclopentene + Cl_2

5.24 Draw a structural formula for an alkene with the indicated molecular formula that gives the compound shown as the major product. Note that more than one alkene may give the same compound as the major product.

(a) C_5H_{10} + H_2O $\xrightarrow{H_2SO_4}$ OH

(b) C_5H_{10} + Br_2 ⟶ Br Br

(c) C_7H_{12} + HCl ⟶ Cl

5.25 Draw the structural formula for an alkene with the molecular formula C_5H_{10} that reacts with Br_2 to give each product:

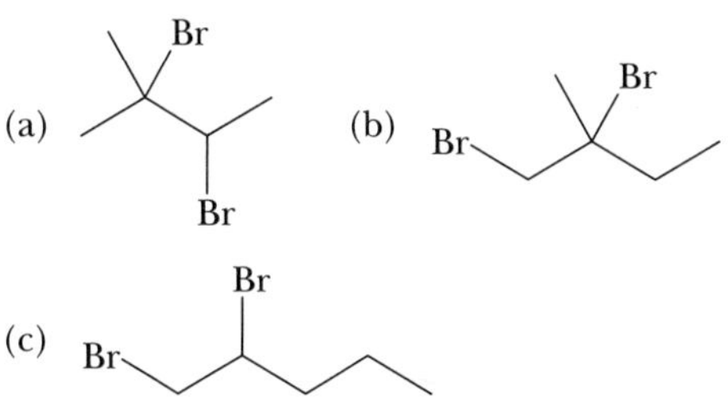

5.26 Draw the structural formula for a cycloalkene of molecular formula C_6H_{10} that reacts with Cl_2 to give each compound:

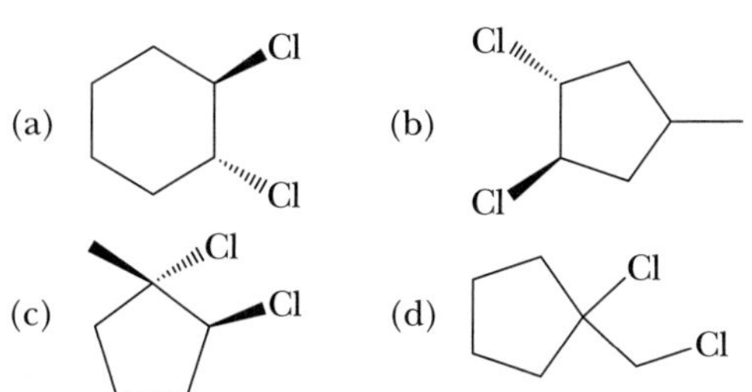

5.27 Draw the structural formula for an alkene with the molecular formula C_5H_{10} that reacts with HCl to give the indicated chloroalkane as the major product:

(a) Cl (b) Cl (c) Cl

5.28 Draw the structural formula of an alkene that undergoes acid-catalyzed hydration to give the indicated alcohol as the major product. More than one alkene may give each compound as the major product.

(a) 3-Hexanol (b) 1-Methylcyclobutanol
(c) 2-Methyl-2-butanol (d) 2-Propanol

5.29 Draw the structural formula of an alkene that undergoes acid-catalyzed hydration to give each alcohol as the major product. More than one alkene may give each compound as the major product.

(a) Cyclohexanol
(b) 1,2-Dimethylcyclopentanol
(c) 1-Methylcyclohexanol
(d) 1-Isopropyl-4-methylcyclohexanol

5.30 Complete these equations by predicting the major product formed in each reaction. Note that certain of these reactions involve rearrangements. **(See Examples 5.2, 5.5)**

(a) 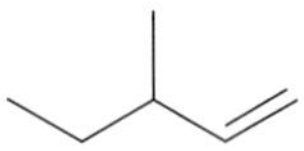 + HBr ⟶

(b) 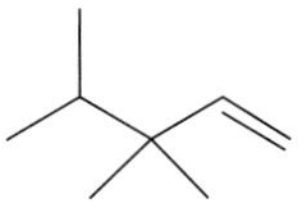 + H_2O $\xrightarrow{H_2SO_4}$

(c) 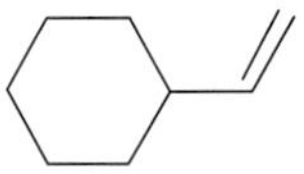 + H_2O $\xrightarrow{H_2SO_4}$

(d)  + HI ⟶

(e) 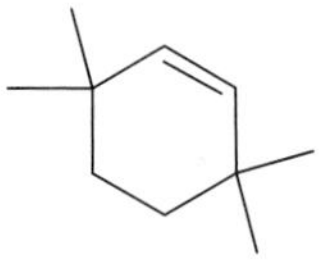 + HCl ⟶

(f) 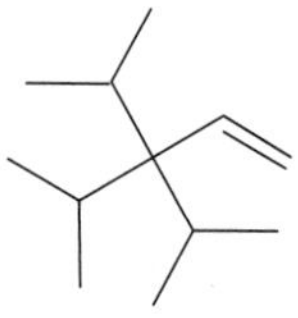 + H_2O $\xrightarrow{H_2SO_4}$

5.31 Propose a mechanism for each reaction in Problem 5.30. **(See Examples 5.4, 5.6, 5.8)**

5.32 Propose a mechanism for the following acid-catalyzed dehydration. **(See Examples 5.6, 5.8)**

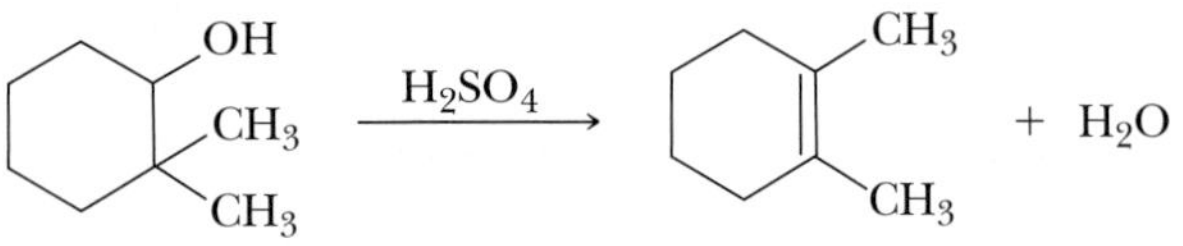

OH, CH_3, CH_3 $\xrightarrow{H_2SO_4}$ CH_3, CH_3 + H_2O

5.33 Propose a mechanism for each of the following transformations. **(See Examples 5.4, 5.6, 5.8)**

(a) + HBr ⟶ Br

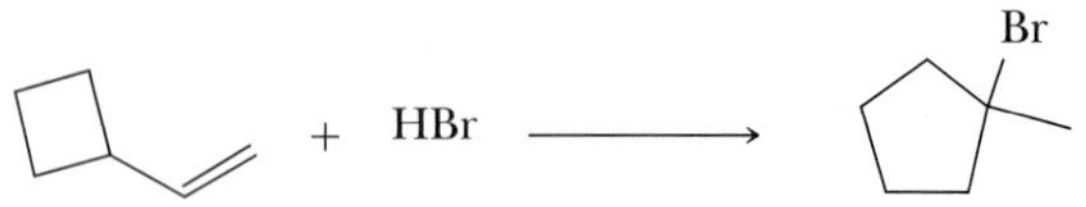

(b) + H_2O $\xrightarrow{H_2SO_4}$ HO

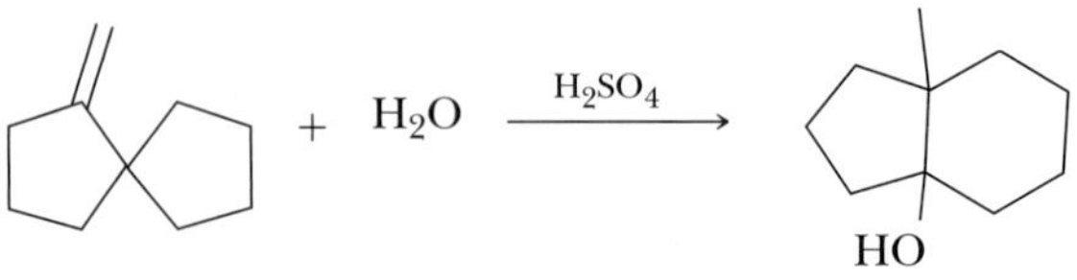

5.34 Terpin is prepared commercially by the acid-catalyzed hydration of limonene: **(See Example 5.5)**

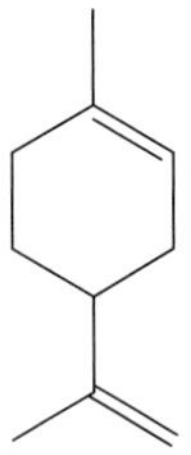

Limonene + $2H_2O$ $\xrightarrow{H_2SO_4}$ $C_{10}H_{20}O_2$ Terpin

(a) Propose a structural formula for terpin and a mechanism for its formation.

(b) How many *cis–trans* isomers are possible for the structural formula you propose?

(c) Terpin hydrate, the isomer in terpin in which the one-carbon and three-carbon substituents are *cis* to each other, is used as an expectorant in cough medicines. Draw the alternative chair conformations for terpin hydrate, and state which of the two is the more stable.

5.35 Propose a mechanism for this reaction and account for its regioselectivity.

$$CH_3-\overset{CH_3}{\overset{|}{C}}=CH_2 + ICl \longrightarrow -CH_3-\underset{Cl}{\underset{|}{\overset{CH_3}{\overset{|}{C}}}}-CH_2I$$

5.36 The treatment of 2-methylpropene with methanol in the presence of a sulfuric acid catalyst gives *tert*-butyl methyl ether:

$$CH_3\overset{CH_3}{\overset{|}{C}}=CH_2 + CH_3OH \xrightarrow{H_2SO_4} CH_3\underset{CH_3}{\underset{|}{\overset{CH_3}{\overset{|}{C}}}}-OCH_3$$

2-Methylpropene Methanol *tert*-Butyl methyl ether

Propose a mechanism for the formation of this ether.

5.37 Treating cyclohexene with HBr in the presence of acetic acid gives a mixture of bromocyclohexane and cyclohexyl acetate.

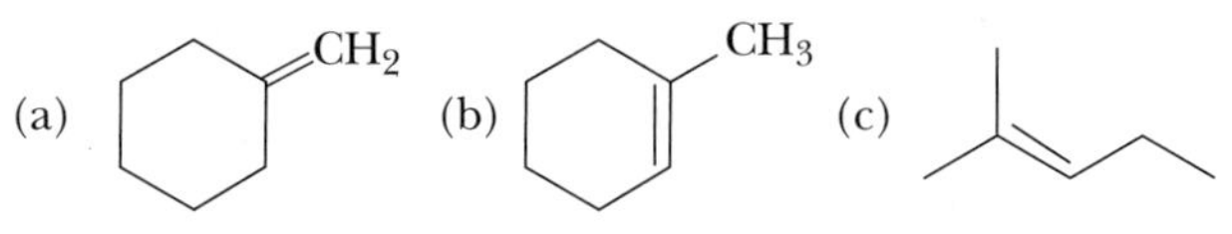

Account for the formation of each product but do not be concerned with the relative percentages of each.

5.38 Draw a structural formula for the alcohol formed by treating each alkene with borane in tetrahydrofuran (THF), followed by hydrogen peroxide in aqueous sodium hydroxide, and specify the stereochemistry where appropriate. **(See Example 5.9)**

(a) (b) (c)

(d) (e)

5.39 Treatment of 1-methylcyclohexene with methanol in the presence of a sulfuric acid catalyst gives a compound of molecular formula $C_8H_{16}O$. Propose a structural formula for this compound and a mechanism for its formation.

$$\text{1-Methylcyclohexene} + CH_3OH \xrightarrow{H_2SO_4} C_8H_{16}O$$

1-Methylcyclohexene Methanol

5.40 *cis*-3-Hexene and *trans*-3-hexene are different compounds and have different physical and chemical properties. Yet, when treated with H_2O/H_2SO_4, each gives the same alcohol. What is the alcohol, and how do you account for the fact that each alkene gives the same one? **(See Examples 5.6, 5.8)**

Sections 5.5–5.7 Oxidation-Reduction

5.41 Which of these transformations involve oxidation, which involve reduction, and which involve neither oxidation nor reduction?

(a) $CH_3CH(OH)CH_3 \longrightarrow CH_3C(=O)CH_3$

(b) $CH_3CH(OH)CH_3 \longrightarrow CH_3CH{=}CH_2$

(c) $CH_3CH{=}CH_2 \longrightarrow CH_3CH_2CH_3$

5.42 Write a balanced equation for the combustion of 2-methylpropene in air to give carbon dioxide and water. The oxidizing agent is O_2, which makes up approximately 20% of air.

5.43 Draw the product formed by treating each alkene with H_2/Ni:

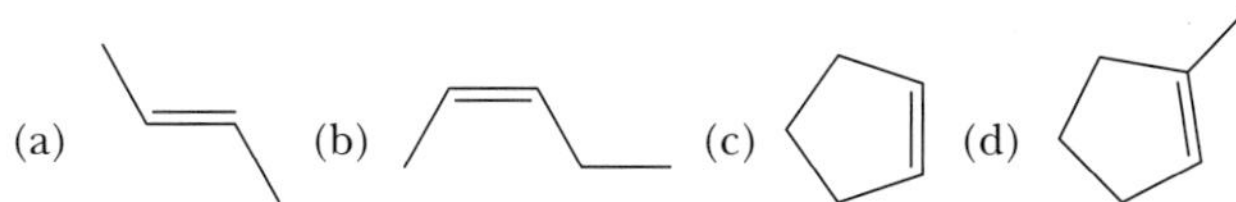

5.44 Hydrocarbon A, C_5H_8, reacts with 2 moles of Br_2 to give 1,2,3,4-tetrabromo-2-methylbutane. What is the structure of hydrocarbon A?

5.45 Two alkenes, A and B, each have the formula C_5H_{10}. Both react with H_2/Pt and with HBr to give identical products. What are the structures of A and B?

5.46 Draw the structural formula of the alkene that reacts with ozone followed by dimethylsulfide to give each product or set of products. **(See Example 5.10)**

(a) $C_7H_{12} \xrightarrow[\text{2) } (CH_3)_2S]{\text{1) } O_3}$

(b) $C_{10}H_{18} \xrightarrow[\text{2) } (CH_3)_2S]{\text{1) } O_3}$

(c) $C_{10}H_{18} \xrightarrow[\text{2) } (CH_3)_2S]{\text{1) } O_3}$

5.47 Consider the following reaction.

$C_8H_{12} \xrightarrow[\text{2) } (CH_3)_2S]{\text{1) } O_3}$ HC(=O)–C₆H₁₀–CH(=O)

Cyclohexane-1,4-dicarbaldehyde

(a) Draw a structural formula for the compound with the molecular formula C_8H_{12}.

(b) Do you predict the product to be the *cis* isomer, the *trans* isomer, or a mixture of *cis* and *trans* isomers? Explain.

(c) Draw a suitable stereorepresentation for the more stable chair conformation of the dicarbaldehyde formed in this oxidation.

5.48 The hydrocarbon muscalure, $C_{23}H_{46}$, is the sex attractant of the common housefly. Ozonolysis followed by treatment with dimethylsulfide gives a mixture of the following two aldehydes.

$$CH_3(CH_2)_{12}\overset{O}{\overset{\|}{C}}H + CH_3(CH_2)_7\overset{O}{\overset{\|}{C}}H$$

Propose a structural formula of muscalure. How many stereoisomers are possible for the structure you have proposed? Muscalure has been synthesized as follows.

$$HC\equiv CH \xrightarrow[\text{2) } CH_3(CH_2)_{11}CH_2Cl]{\text{1) } NaNH_2} A \xrightarrow[\text{2) } CH_3(CH_2)_6CH_2Cl]{\text{1) } NaNH_2}$$

$$B \xrightarrow[\text{Lindlar catalyst}]{H_2} \text{Muscalure } (C_{23}H_{46})$$

Based on this synthesis, assign a configuration to the carbon–carbon double bond in muscalure.

Sections 5.8–5.11 Reactions of Alkynes

5.49 Complete these equations by predicting the major products formed in each reaction. If more than one product is equally likely, draw both products. **(See Example 5.12)**

(a) [alkyne] + H_2O $\xrightarrow[HgSO_4]{H_2SO_4}$

(b) [alkyne] + H_2O $\xrightarrow[HgSO_4]{H_2SO_4}$

(c) [alkyne] + H_2O $\xrightarrow[HgSO_4]{H_2SO_4}$

(d) [alkyne] + H_2O $\xrightarrow[HgSO_4]{H_2SO_4}$

5.50 Complete these equations by predicting the major product formed in each reaction. If more than one product is equally likely, draw both products.

(a) [alkyne] $\xrightarrow[Ni]{2H_2}$

(b) [alkyne] $\xrightarrow[\text{Lindlar catalyst}]{H_2}$

Synthesis

5.51 Show how to convert ethylene into these compounds:

(a) Ethane (b) Ethanol
(c) Bromoethane (d) 1,2-Dibromoethane
(e) Chloroethane

5.52 Show how to convert cyclopentene into these compounds:

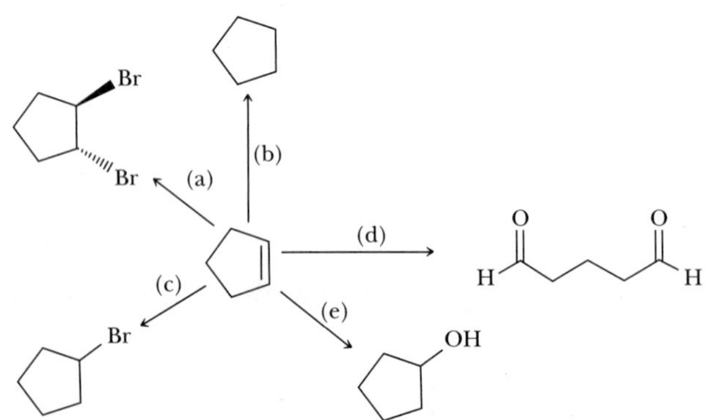

5.53 Show how to convert methylenecyclohexane into each of these compounds.

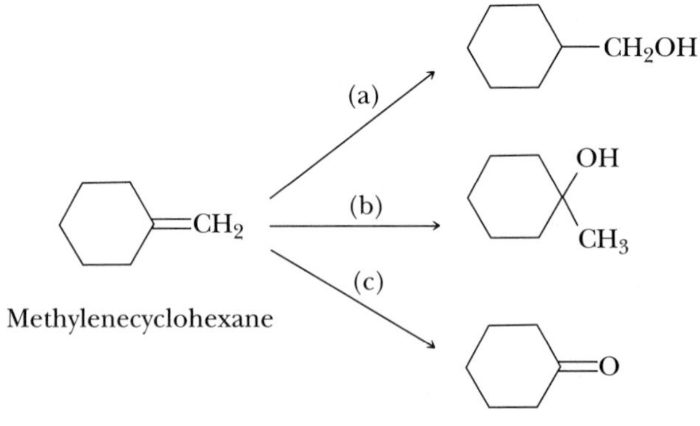

5.54 Show how to convert 1-butene into these compounds:

(a) Butane (b) 2-Butanol
(c) 1-Butanol (d) 2-Bromobutane
(e) 1,2-Dibromobutane

5.55 Show how the following compounds can be synthesized in good yields from an alkene:

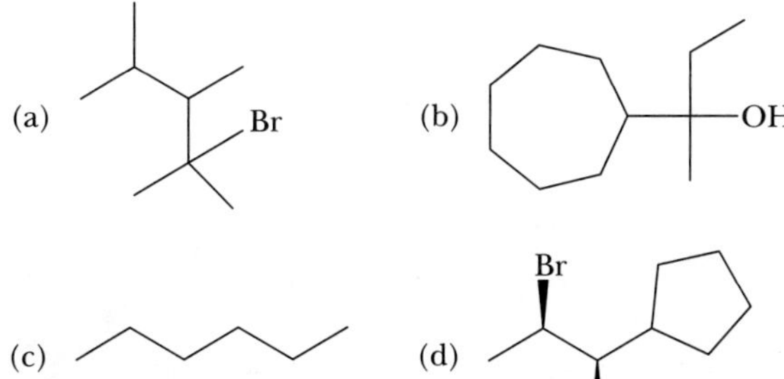

(c) (d) Br Br

5.56 How would you prepare both *cis*-3-hexene and *trans*-3-hexene using only acetylene as the source of carbon atoms, and using any necessary inorganic regents? **(See Example 5.11)**

5.57 Test your cumulative knowledge of the reactions learned thus far by completing the following chemical transformations. Note that some will require more than one step.

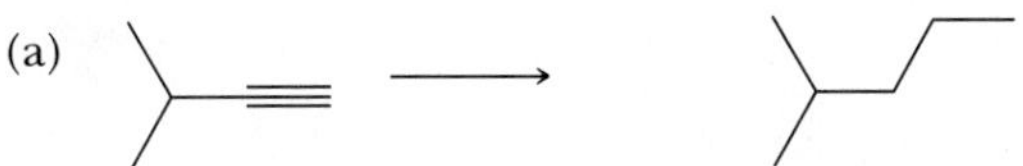

(b) H—≡—H ⟶

(c) Br ⟶ HO

(d) Br ⟶

(e) Cl ⟶ OH

(f) ⟶ Br Br

(g) $H_2C{=}CH_2$ ⟶ O H + O H H

(h) H—≡—H ⟶ O H

Looking Ahead

5.58 Each of the following 2° carbocations is more stable than the tertiary-butyl carbocation shown:

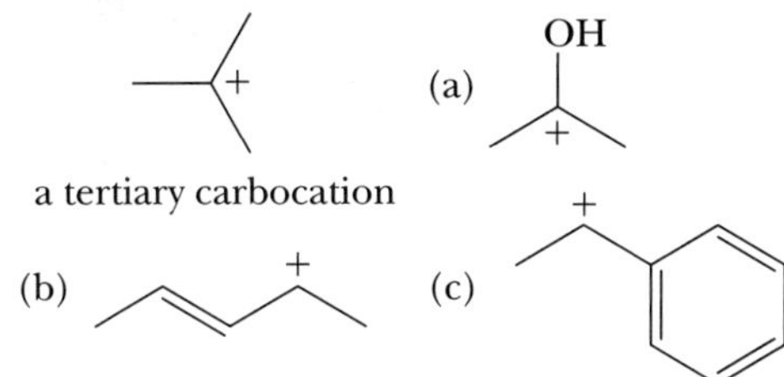

Provide an explanation for each cation's enhanced stability.

5.59 Recall that an alkene possesses a π cloud of electrons above and below the plane of the C=C bond. Any reagent can therefore react with either face of the double bond. Determine whether the reaction of each of the given reagents with the top face of *cis*-2-butene will produce the same product as the reaction of the same reagent with the bottom face. (*Hint:* Build molecular models of the products and compare them.)

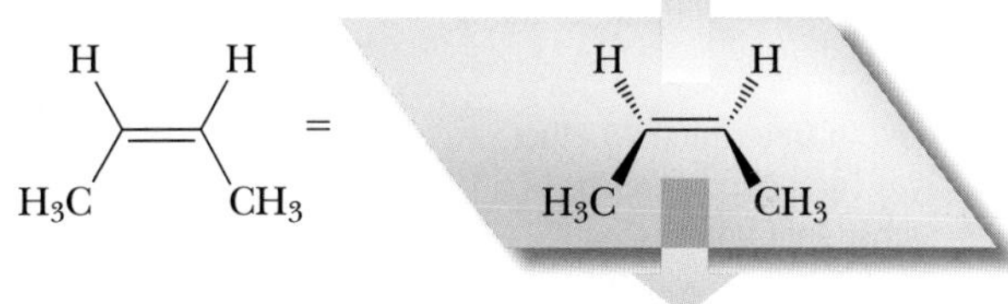

(a) H_2/Pt (b) $\dfrac{1)\ BH_3}{2)\ NaOH,\ HOOH}$ (c) Br_2/CH_2Cl_2

5.60 This reaction yields two products in differing amounts:

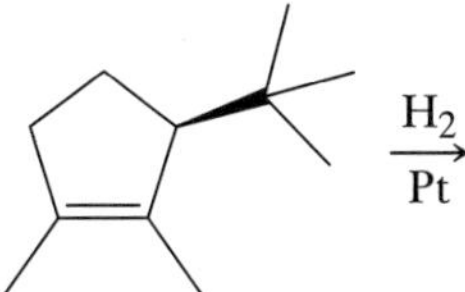

$\xrightarrow[Pt]{H_2}$

Draw the two products and predict which product is favored.

6 Chirality: The Handedness of Molecules

KEY QUESTIONS

6.1 What Are Stereoisomers?

6.2 What Are Enantiomers?

6.3 How Do We Designate the Configuration of a Stereocenter?

6.4 What Is the 2^n Rule?

6.5 How Do We Describe the Chirality of Cyclic Molecules with Two or More Stereocenters?

6.6 How Do We Describe the Chirality of Molecules with Three or More Stereocenters?

6.7 What Are the Properties of Stereoisomers?

6.8 How Is Chirality Detected in the Laboratory?

6.9 What Is the Significance of Chirality in the Biological World?

6.10 How Can Enantiomers Be Resolved?

HOW TO

6.1 How to Draw Enantiomers

6.2 How to Determine the R & S Configuration without Rotating the Molecule

6.3 How to Determine Whether Two Compounds Are the Same, Enantiomers, or Diastereomers without the Need to Spatially Manipulate the Molecule

CHEMICAL CONNECTIONS

6A Chiral Drugs

Tartaric acid is found in grapes and other fruits, both free and as its salts (see Section 6.4B). Inset: A model of tartaric acid. *(Pierre-Louis Martin/Photo Researchers, Inc.)*

Mirror image The reflection of an object in a mirror.

In this chapter, we will explore the relationships between three-dimensional objects and their mirror images. When you look in a mirror, you see a reflection, or **mirror image**, of yourself. Now, suppose your mirror image becomes a three-dimensional object. We could then ask, "What is the relationship between you and your mirror image?" By relationship, we mean "Can your reflection be superposed on the original 'you' in such a way that every detail of the reflection corresponds exactly to the original?" The answer is that you and your mirror image are not superposable. If you have a ring on the little finger of your right hand, for example, your mirror image has the ring on the little finger of its left hand. If you part your hair on your right side, the part will be on the left side in your mirror image. Simply stated, you and your reflection are different objects. You cannot superpose one on the other.

An understanding of relationships of this type is fundamental to an understanding of organic chemistry and biochemistry. In fact, the ability to visualize molecules as three-dimensional objects is a survival skill in organic chemistry and biochemistry. We suggest that you purchase a set of molecular models. Alternatively you may have access to a computer lab with a modeling program. We urge you to use molecular models frequently as an aid to visualizing the spatial concepts in this and later chapters.

The horns of this African gazelle show chirality and are mirror images of each other. *(William H. Brown)*

6.1 What Are Stereoisomers?

Stereoisomers have the same molecular formula and the same connectivity of atoms in their molecules, but different three-dimensional orientations of their atoms in space. The one example of stereoisomers we have seen thus far is that of *cis–trans* isomers in cycloalkanes (Section 3.7) and alkenes (Section 4.1C):

Stereoisomers Isomers that have the same molecular formula and the same connectivity, but different orientations of their atoms in space.

CH_3 / CH_3 and CH_3 / CH_3 H_3C, CH_3, C=C, H, H and H_3C, H, C=C, H, CH_3

cis-1,2-Dimethylcyclohexane *trans*-1,2-Dimethylcyclohexane *cis*-2-Butene *trans*-2-Butene

In this chapter, we study enantiomers and diastereomers (Figure 6.1).

6.2 What Are Enantiomers?

Enantiomers are stereoisomers that are nonsuperposable mirror images of each other. The significance of enantiomerism is that, except for inorganic and a few simple organic compounds, the vast majority of molecules in the biological world show this type of isomerism, including carbohydrates (Chapter 18), lipids (Chapter 21), amino acids and proteins (Chapter 19), and nucleic acids (DNA and RNA, Chapter 20). Further, approximately one-half of the medications used in human medicine also show this type of isomerism.

Enantiomers Stereoisomers that are nonsuperposable mirror images; the term refers to a relationship between pairs of objects.

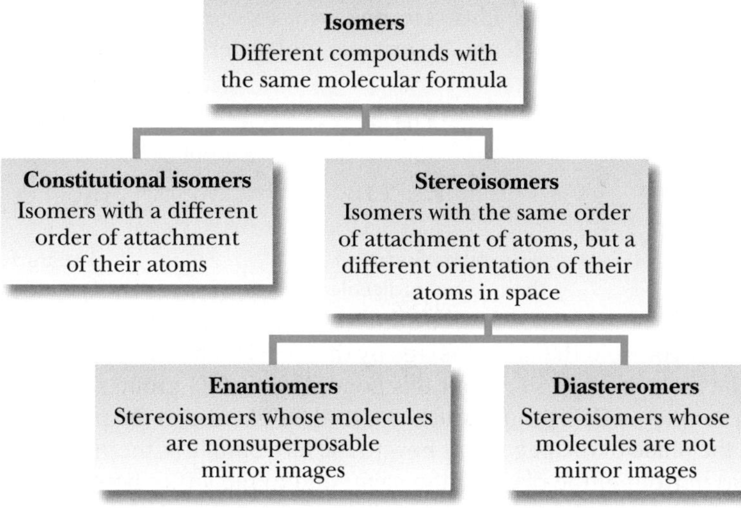

Figure 6.1
Relationships among isomers.

As an example of a molecule that exhibits enantiomerism, let us consider 2-butanol. As we go through the discussion of this molecule, we focus on carbon 2, the carbon bearing the —OH group. What makes this carbon of interest is that it has four different groups bonded to it. The most common cause of enantiomerism among organic molecules is a carbon bonded to four different groups.

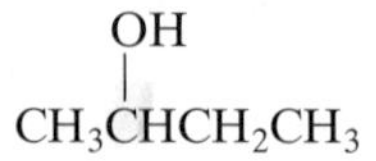

$$CH_3\underset{}{\overset{OH}{|}}{C}HCH_2CH_3$$

2-Butanol

The structural formula we have just drawn does not show the shape of 2-butanol or the orientation of its atoms in space. To do this, we must consider the molecule as a three-dimensional object. On the left are a ball-and-stick model of 2-butanol and a perspective drawing of what we will call the "original" molecule. In this drawing, the —OH and —CH_3 groups on carbon-2 are in the plane of the paper; the —H is behind the plane and the —CH_2CH_3 group is in front of the plane.

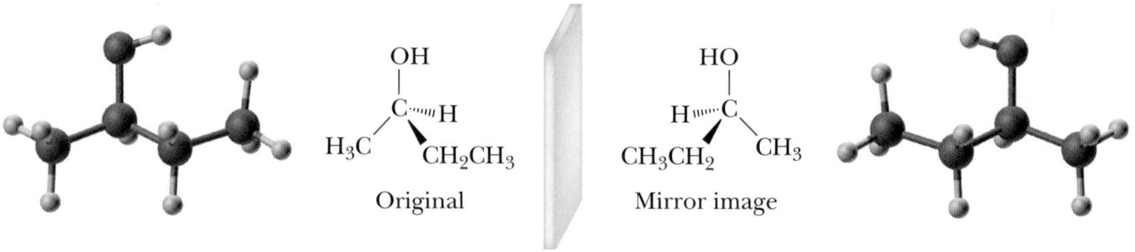

OH, C, H, H_3C, CH_2CH_3

Original

HO, H, C, CH_3CH_2, CH_3

Mirror image

To the right in the preceding diagram is the mirror image of the original molecule. Every molecule and, in fact, every object in the world around us, has a mirror image. The question we now need to ask is "What is the relationship between the original of 2-butanol and its mirror image?" To answer this question, you need to imagine that you can pick up the mirror image and move it in space in any way you wish. If you can move the mirror image in space and find that it fits over the original so that every bond, atom, and detail of the mirror image exactly matches the bonds, atoms, and details of the original, then the two are **superposable**. In this case, the mirror image and the original represent the same molecule; they are only oriented differently in space. If, however, no matter how you turn the mirror image in space, it will not fit exactly on the original with every detail matching, then the two are **nonsuperposable**; they are different molecules.

The key point here is that either an object is superposable on its mirror image or it isn't. Now let us look at 2-butanol and its mirror image and ask, "Are they or are they not superposable?"

The following drawings illustrate one way to see that the mirror image of 2-butanol is not superposable on the original molecule:

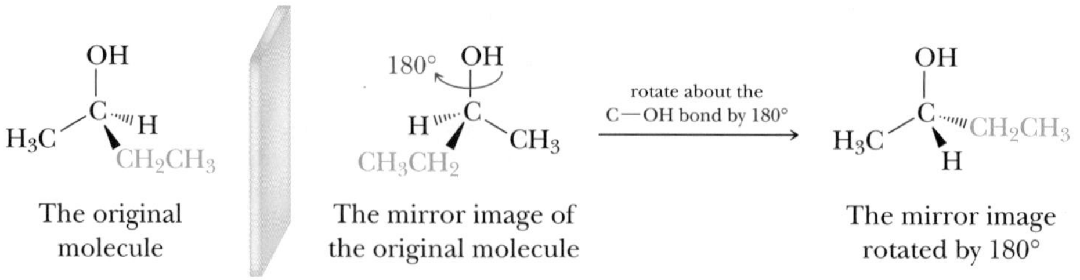

OH, C, H, H_3C, CH_2CH_3

The original molecule

180°, OH, H, C, CH_3, CH_3CH_2

The mirror image of the original molecule

rotate about the C—OH bond by 180°

OH, C, CH_2CH_3, H_3C, H

The mirror image rotated by 180°

Imagine that you hold the mirror image by the C—OH bond and rotate the bottom part of the molecule by 180° about this bond. The —OH group retains its position in space, but the —CH_3 group, which was to the right and in the plane of the paper, is still in the plane of the paper, but now to the left. Similarly, the —CH_2CH_3 group, which was in front of the plane of the paper and to the left, is now behind the plane and to the right.

Now move the rotated mirror image in space, and try to fit it on the original so that all bonds and atoms match:

CH_2CH_3 pointing away

The mirror image rotated by 180° → OH, C, CH_2CH_3, H_3C, H

H pointing toward you

H pointing away

The original molecule → OH, C, H, H_3C, CH_2CH_3

CH_2CH_3 pointing toward you

Left- and right-handed sea shells. If you cup a right-handed shell in your right hand with your thumb pointing from the narrow end to the wide end, the opening will be on your right. *(Charles D. Winters)*

By rotating the mirror image as we did, its $—OH$ and $—CH_3$ groups now fit exactly on top of the $—OH$ and $—CH_3$ groups of the original. But the $—H$ and $—CH_2CH_3$ groups of the two do not match: The $—H$ is away from you in the original, but toward you in the mirror image; the $—CH_2CH_3$ group is toward you in the original, but away from you in the mirror image. We conclude that the original of 2-butanol and its mirror image are nonsuperposable and, therefore, are different compounds.

To summarize, we can rotate the mirror image of 2-butanol in space in any way we want, but as long as no bonds are broken or rearranged, only two of the four groups bonded to carbon-2 of the mirror image can be made to coincide with those on the original. Because 2-butanol and its mirror image are not superposable, they are enantiomers. Like gloves, enantiomers always occur in pairs.

Objects that are not superposable on their mirror images are said to be **chiral** (pronounced ki′-ral, rhymes with spiral; from the Greek: *cheir*, hand); that is, they show handedness. Chirality is encountered in three-dimensional objects of all sorts. Your left hand is chiral, and so is your right hand. A spiral binding on a notebook is chiral. A machine screw with a right-handed twist is chiral. A ship's propeller is chiral. As you examine the objects in the world around you, you will undoubtedly conclude that the vast majority of them are chiral.

Chiral From the Greek *cheir*, meaning hand; objects that are not superposable on their mirror images.

As we said before we examined the original and the mirror image of 2-butanol, the most common cause of enantiomerism in organic molecules is the presence of a carbon with four different groups bonded to it. Let us examine this statement further by considering a molecule such as 2-propanol, which has no such carbon. In this molecule, carbon-2 is bonded to three different groups, but no carbon is bonded to four different groups. The question we ask is, "Is the mirror image of 2-propanol superposable on the original, or isn't it?"

In the following diagram, on the left is a three-dimensional representation of 2-propanol, and on the right is its mirror image:

OH, C, H, H_3C, CH_3 — Original

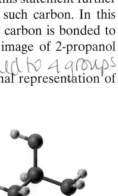

OH, C, H, H_3C, CH_3 — Mirror image

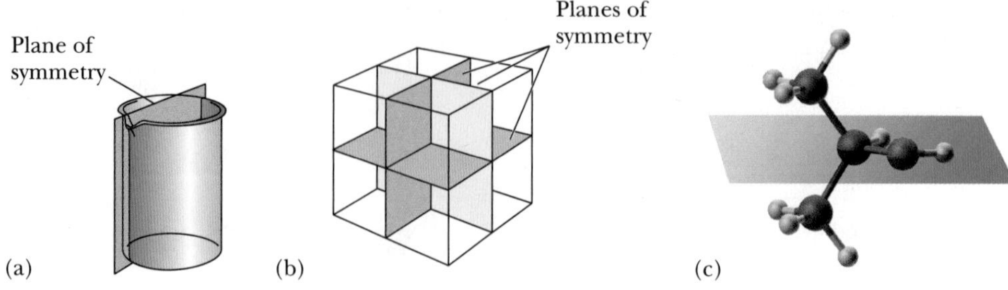

Figure 6.2
Planes of symmetry in (a) a beaker, (b) a cube, and (c) 2-propanol. The beaker and 2-propanol each have one plane of symmetry; the cube has several planes of symmetry, only three of which are shown in the figure.

The question we now ask is "What is the relationship of the mirror image to the original?" This time, let us rotate the mirror image by 120° about the C—OH bond and then compare it with the original. When we do this rotation, we see that all atoms and bonds of the mirror image fit exactly on the original. This means that the structures we first drew for the original and its mirror image are, in fact, the same molecule viewed from different perspectives:

OH
H_3C—C—H
CH_3
The original

180° OH
H—C—CH_3
H_3C
The mirror image

rotate about the C—OH bond by 120° →

OH
H_3C—C—H
CH_3
The mirror image rotated by 180°

Achiral An object that lacks chirality; an object that has no handedness.

Plane of symmetry An imaginary plane passing through an object and dividing it such that one half is the mirror image of the other half.

Stereocenter An atom at which the interchange of two atoms or groups of atoms bonded to it produces a different stereoisomer.

If an object and its mirror image are superposable, then the object and its mirror image are identical, and there is no possibility of enantiomerism. We say that such an object is **achiral** (without chirality).

An achiral object has at least one plane of symmetry. A **plane of symmetry** (also called a *mirror plane*) is an imaginary plane passing through an object and dividing it so that one-half of the object is the reflection of the other half. The beaker shown in Figure 6.2 has a single plane of symmetry, whereas a cube has several planes of symmetry. 2-Propanol also has a single plane of symmetry.

To repeat, the most common cause of chirality in organic molecules is a tetrahedral carbon atom with four different groups bonded to it. We call such a carbon atom a **stereocenter**. 2-Butanol has one stereocenter; 2-propanol has none.

As another example of a molecule with a stereocenter, consider 2-hydroxypropanoic acid, more commonly named lactic acid. Lactic acid is a product of anaerobic glycolysis and is what gives sour cream its sour taste. Figure 6.3 shows three-dimensional representations of lactic acid and its mirror image. In these representations, all bond angles about the central carbon atom are approximately 109.5°, and the four bonds projecting from it are directed toward the corners of a regular tetrahedron. Lactic acid shows enantiomerism; that is, it and its mirror image are not superposable, but rather are different molecules.

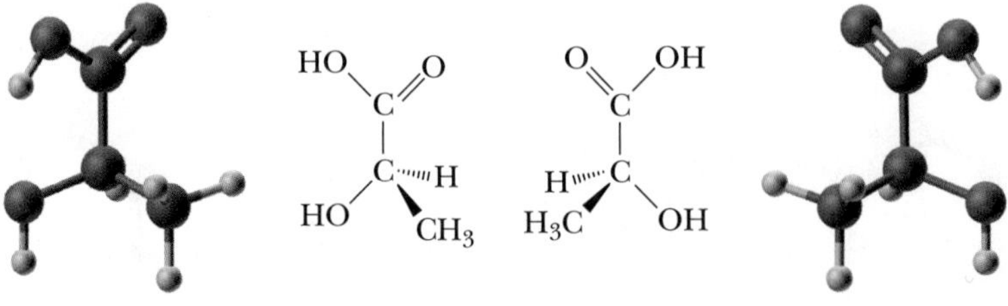

Figure 6.3
Three-dimensional representations of lactic acid and its mirror image.

HOW TO 6.1 Draw Enantiomers

Now that we know what enantiomers are, we can think about how to represent their three-dimensional structures on a two-dimensional page. Let us take one of the enantiomers of 2-butanol as an example. Following are four different representations of this enantiomer:

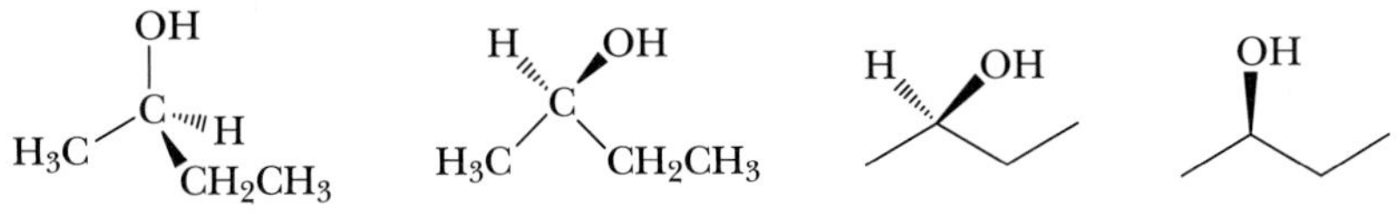

(1) (2) (3) (4)

In our initial discussions of 2-butanol, we used (1) to show the tetrahedral geometry of the stereocenter; in it, two groups are in the plane of the paper, a third is coming out of the plane toward us, and the fourth is behind the plane, away from us. We can turn (1) slightly in space and tip it a bit to place the carbon framework in the plane of the paper. Doing so gives us representation (2), in which we still have two groups in the plane of the paper, one coming toward us and one going away from us. For an even more abbreviated representation of this enantiomer of 2-butanol, we can turn (2) into the line-angle formula (3). Although we don't normally show hydrogens in a line-angle formula, we do so in (3) just to remind ourselves that the fourth group on this stereocenter is really there and that it is H. Finally, we can carry the abbreviation a step further and write 2-butanol as (4). Here, we omit the H on the stereocenter, but we know that it must be there (carbon needs four bonds), and we know that it must be behind the plane of the paper. Clearly, the abbreviated formulas (3) and (4) are the easiest to draw, and we will rely on these representations throughout the remainder of the text. When you have to draw three-dimensional representations of stereocenters, try to keep the carbon framework in the plane of the paper and the other two atoms or groups of atoms on the stereocenter toward and away from you, respectively. Using representation (4) as a model, we get the following two different representations of its enantiomer:

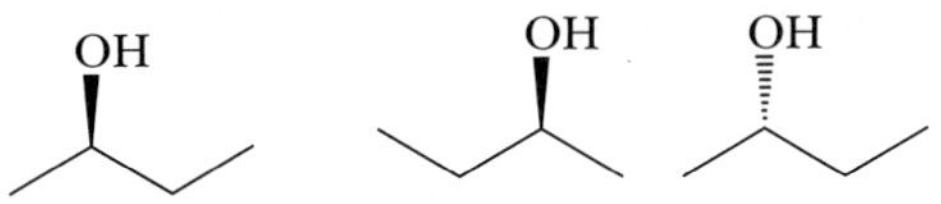

One enantiomer of 2-butanol — Alternative representations for its mirror image

Notice that in the first alternative, the carbon skeleton has been reversed.

Example 6.1

Each of the following molecules has one stereocenter:

(a) $CH_3CH(Cl)CH_2CH_3$ (b)

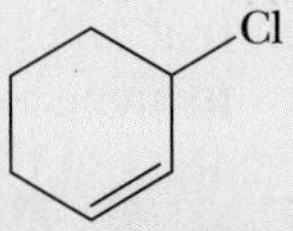

Identify the stereocenter in each and draw stereorepresentations of the enantiomers of each.

Strategy

You will find it helpful to study models of each pair of enantiomers and to view them from different perspectives. As you work with these models, notice that each enantiomer has a carbon atom bonded to four different groups, which makes the molecule chiral. Translate what you see in each model by using perspective drawings.

Solution

The hydrogen at the stereocenter is shown in (a) but not in (b).

(a) 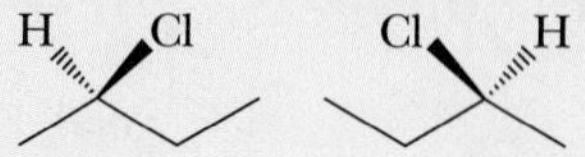 H Cl Cl H (b) 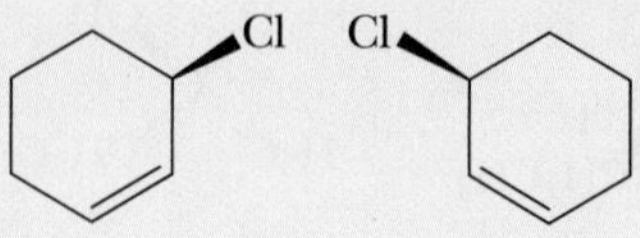 Cl Cl

See problems 6.15, 6.19–6.22

Problem 6.1

Each of the following molecules has one stereocenter:

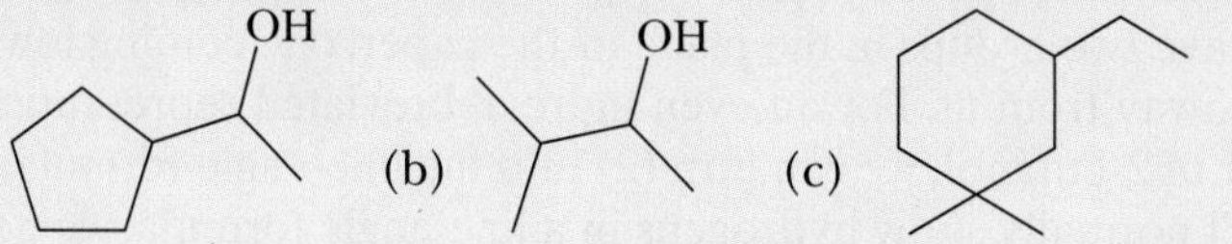

(a) OH (b) OH (c)

Identify the stereocenter in each and draw stereorepresentations of the enantiomers of each.

6.3 How Do We Designate the Configuration of a Stereocenter?

Because enantiomers are different compounds, each must have a different name. The over-the-counter drug ibuprofen, for example, shows enantiomerism and can exist as the pair of enantiomers shown here:

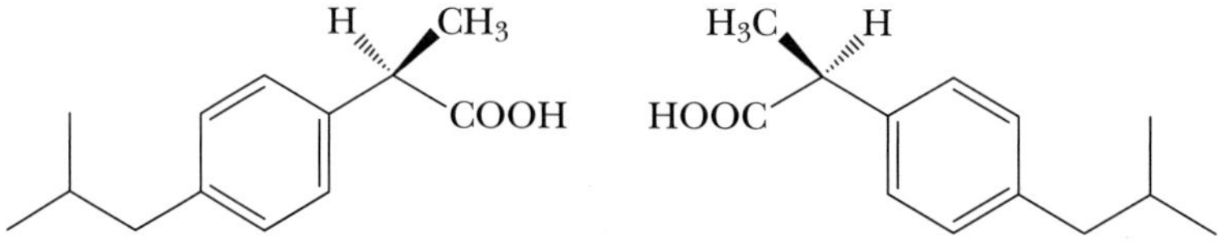

H CH_3 COOH | H_3C H HOOC

The inactive enantiomer of ibuprofen | The active enantiomer

Only one enantiomer of ibuprofen is biologically active. This enantiomer reaches therapeutic concentrations in the human body in approximately 12 minutes. However, in this case, the inactive enantiomer is not wasted. The body converts it to the active enantiomer, but that takes time.

What we need is a way to name each enantiomer of ibuprofen (or any other pair of enantiomers for that matter) so that we can refer to them in conversation or in writing. To do so, chemists have developed the **R,S system**. The first step in assigning an R or S configuration to a stereocenter is to arrange the groups bonded to it in order of priority. For this, we use the same set of **priority rules** we used in Section 4.2C to assign an E,Z configuration to an alkene.

R,S system A set of rules for specifying the configuration about a stereocenter.

To assign an R or S configuration to a stereocenter,

1. Locate the stereocenter, identify its four substituents, and assign a priority from 1 (highest) to 4 (lowest) to each substituent.

2. Orient the molecule in space so that the group of lowest priority (4) is directed away from you, as would be, for instance, the steering column of a car. The three groups of higher priority (1–3) then project toward you, as would the spokes of a steering wheel.
3. Read the three groups projecting toward you in order, from highest priority (1) to lowest priority (3).
4. If reading the groups proceeds in a clockwise direction, the configuration is designated **R** (Latin: *rectus*, straight, correct); if reading proceeds in a counterclockwise direction, the configuration is **S** (Latin: *sinister*, left). You can also visualize this situation as follows: Turning the steering wheel to the right equals R, and turning it to the left equals S.

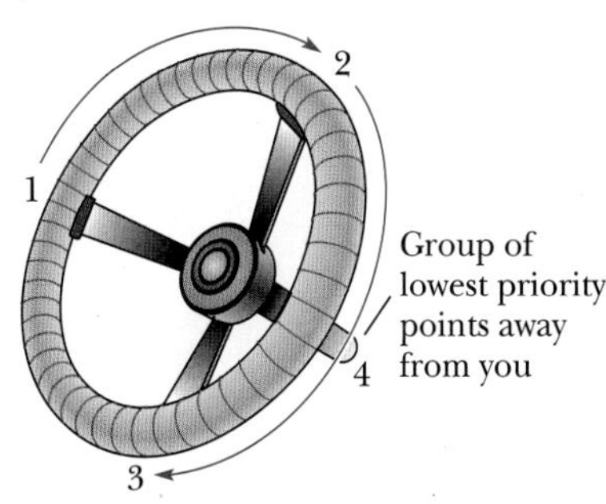

R From the Latin *rectus*, meaning right; used in the R,S system to show that the order of priority of groups on a stereocenter is clockwise.

S From the Latin *sinister*, meaning left; used in the R,S system to show that the order of priority of groups on a stereocenter is counterclockwise.

Example 6.2

Assign an R or S configuration to each stereocenter:

Strategy

First determine the priorities of the groups bonded to the stereocenter. If necessary reorient the molecule so that the group of lowest priority is away from you. Then read the R/S configuration by going from highest to lowest priority.

Solution

View each molecule through the stereocenter and along the bond from the stereocenter toward the group of lowest priority.

(a) The order of priority is $—Cl > —CH_2CH_3 > —CH_3 > —H$. The group of lowest priority, H, points away from you. Reading the groups in the order 1, 2, 3 occurs in the counterclockwise direction, so the configuration is S.

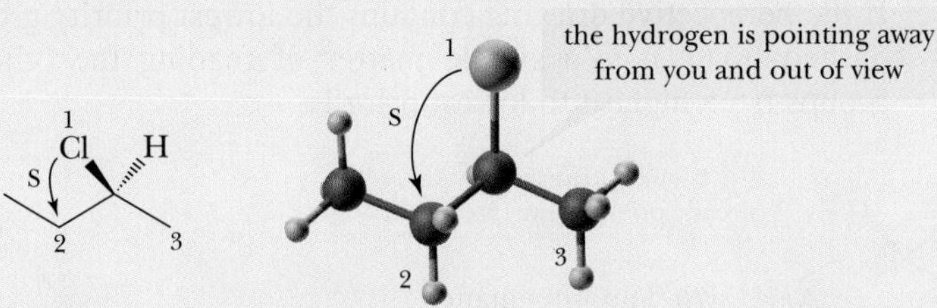

(b) The order of priority is $—OH > —CH{=}CH > —CH_2—CH_2 > —H$. With hydrogen, the group of lowest priority, pointing away from you, reading the groups in the order 1, 2, 3 occurs in the clockwise direction, so the configuration is R.

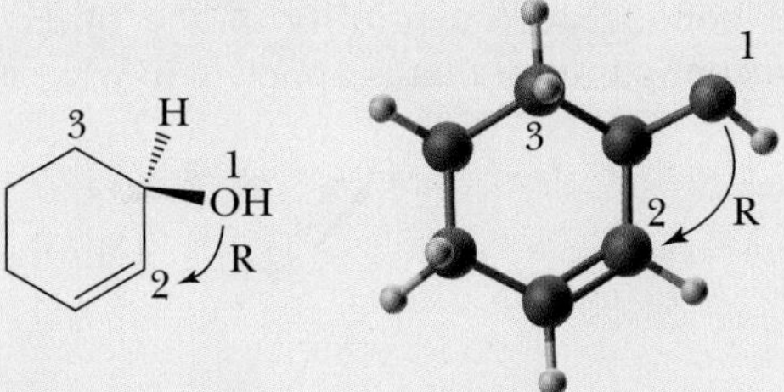

See problems 6.24–6.27, 6.29, 6.39

Problem 6.2

Assign an R or S configuration to each stereocenter:

(a) (b) (c)

Now let us return to our three-dimensional drawing of the enantiomers of ibuprofen and assign each an R or S configuration. In order of decreasing priority, the groups bonded to the stereocenter are $-COOH > -C_6H_4 > -CH_3 > H$. In the enantiomer on the left, reading the groups on the stereocenter in order of priority occurs clockwise. Therefore, this enantiomer is (*R*)-ibuprofen, and its mirror image is (*S*)-ibuprofen:

(*R*)-Ibuprofen
(the inactive enantiomer)

(*S*)-Ibuprofen
(the active enantiomer)

HOW TO 6.2 Determine the R & S Configuration without Rotating the Molecule

If you are having difficulty visualizing the spatial rotation of perspective drawings, the following techniques may be of use.

Scenario 1: *The lowest priority group is already directed away from you.*
If the perspective drawing contains the lowest priority group on a dashed bond, it is a simple matter of reading the other three groups from highest to lowest priority.

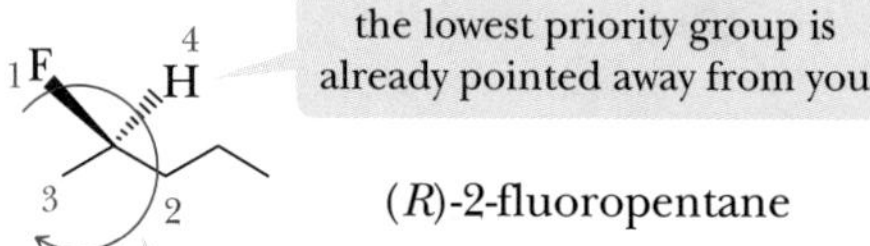
1F
4
H
3
2
the lowest priority group is already pointed away from you

(*R*)-2-fluoropentane

the R/S configuration is read without the need for any spatial manipulation

Scenario 2: *The lowest priority group is directed toward you.*
If the perspective drawing contains the lowest priority group on a wedged bond, read the priority of the other three groups, but assign a configuration that is opposite to what is actually read.

the lowest priority group is pointed toward you

(*S*)-2-fluoropentane

the R/S configuration is read, but then the opposite configuration is chosen. In this example, the priority reading appears to be *R*, but we switch it to *S* because the lowest priority group is pointed toward you.

Scenario 3: *The lowest priority group is in the plane of the page.*
If the perspective drawing contains the lowest priority group in the plane of the page, view down the bond connecting the group to the stereocenter and draw a Newman projection (Section 3.6A).

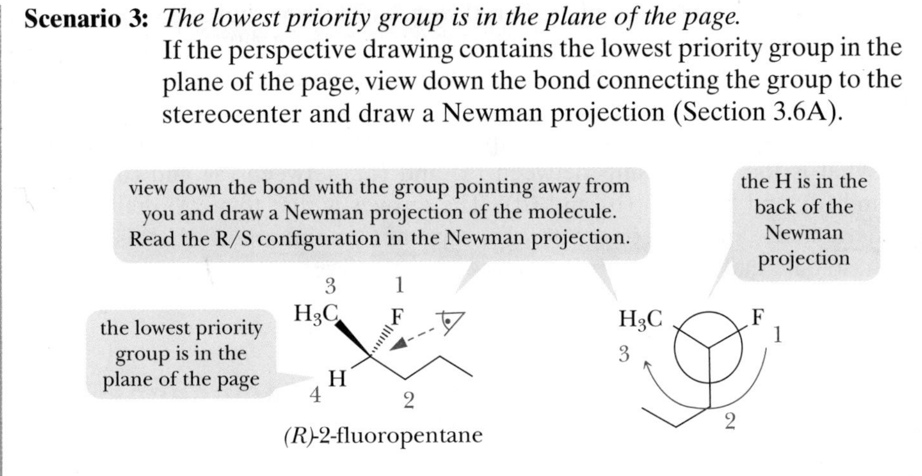
view down the bond with the group pointing away from you and draw a Newman projection of the molecule. Read the R/S configuration in the Newman projection.
the H is in the back of the Newman projection
the lowest priority group is in the plane of the page
(R)-2-fluoropentane

6.4 What Is the 2^n Rule?

Now let us consider molecules with two stereocenters. To generalize, for a molecule with ***n*** stereocenters, the maximum number of stereoisomers possible is $\mathbf{2^n}$. We have already verified that, for a molecule with one stereocenter, $2^1 = 2$ stereoisomers (one pair of enantiomers) are possible. For a molecule with two stereocenters, $2^2 = 4$ stereoisomers are possible; for a molecule with three stereocenters, $2^3 = 8$ stereoisomers are possible, and so forth.

A. Enantiomers and Diastereomers

We begin our study of molecules with two stereocenters by considering 2,3,4-trihydroxybutanal. Its two stereocenters are marked with asterisks:

$$HOCH_2-\overset{*}{C}H(OH)-\overset{*}{C}H(OH)-CH=O$$

2,3,4-Trihydroxybutanal

The maximum number of stereoisomers possible for this molecule is $2^2 = 4$, each of which is drawn in Figure 6.4.

Stereoisomers (a) and (b) are nonsuperposable mirror images and are, therefore, a pair of enantiomers. Stereoisomers (c) and (d) are also nonsuperposable mirror images and are a second pair of enantiomers. We describe the four stereoisomers of 2,3,4-trihydroxybutanal by saying that they consist of two pairs of enantiomers.

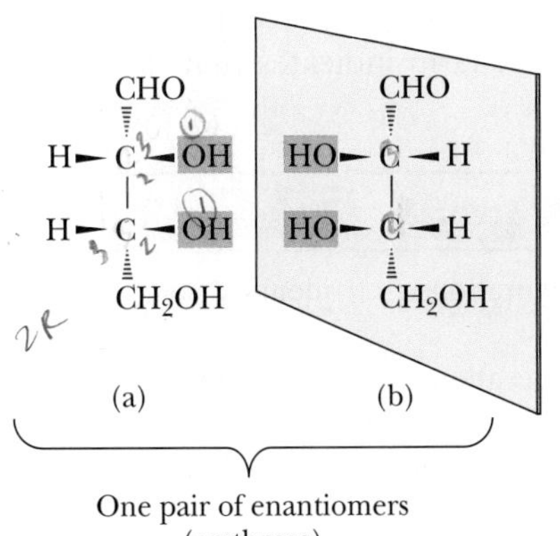

One pair of enantiomers (erythrose)

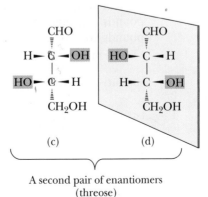

(c) (d)

A second pair of enantiomers (threose)

Figure 6.4
The four stereoisomers of 2,3,4-trihydroxybutanal, a compound with two stereocenters. Configurations (a) and (b) are (2*R*,3*R*) and (2*S*,3*S*), respectively. Configurations (c) and (d) are (2*R*,3*S*) and (2*S*,3*R*), respectively.

Enantiomers (a) and (b) are named **erythrose**, which is synthesized in erythrocytes (red blood cells)—hence the name. Enantiomers (c) and (d) are named **threose**. Erythrose and threose belong to the class of compounds called carbohydrates, which we discuss in Chapter 18.

We have specified the relationship between (a) and (b) and between (c) and (d). What is the relationship between (a) and (c), between (a) and (d), between (b) and (c), and between (b) and (d)? The answer is that they are diastereomers. **Diastereomers** are stereoisomers that are not enantiomers; that is, they are stereoisomers that are not mirror images of each other.

Diastereomers Stereoisomers that are not mirror images of each other; the term refers to relationships among objects.

HOW TO 6.3 Determine Whether Two Compounds Are the Same, Enantiomers, or Diastereomers without the Need to Spatially Manipulate the Molecule

If you are having difficulty visualizing the spatial rotation of perspective drawings, the following technique may be of use.

Step 1: *Verify that the compounds are stereoisomers.*
Make sure that the two compounds in question have the same molecular formula and the same connectivity of atoms.

Br OH Br OH CH_3

C chain is bent

Chemical Formula for both: $C_6H_{13}BrO$
Both have a 6-carbon chain with Br at the 5 position and OH at the 2 position

Step 2: *Assign R/S configurations to each stereocenter in both compounds.*
See How To 6.2 for instructions.

Br (R) (S) OH Br (R) OH (R) CH_3

Step 3: *Compare the configuration at corresponding stereocenters.*
If the configurations match, the compounds are identical. If the configurations are opposite at each corresponding stereocenter, the compounds are enantiomers. Any other scenario indicates that the compounds are diastereomers.

different configuration

Br (R) (S) OH Br (R) OH (R) CH_3

same configuration

Possible Scenario	Relationship
all configurations the same	identical compounds
all configurations opposite	enantiomers
any other scenario	diastereomers

Example 6.3

Following are stereorepresentations of the four stereoisomers of 1,2,3-butanetriol:

(1) CH_2OH / H—C—OH (S) / HO—C—H (S) / CH_3

(2) CH_2OH / H—C—OH / H—C—OH / CH_3

(3) CH_2OH / HO—C—H / HO—C—H / CH_3

(4) CH_2OH / HO—C—H (R) / H—C—OH (R) / CH_3

Configurations are given for the stereocenters in (1) and (4).

(a) Which compounds are enantiomers?
(b) Which compounds are diastereomers?

Strategy

Determine the R/S configuration of the stereocenters in each compound and compare corresponding stereocenters to determine their relationship (see How To 6.3).

Solution

(a) Compounds (1) and (4) are one pair of enantiomers, and compounds (2) and (3) are a second pair of enantiomers. Note that the configurations of the stereocenters in (1) are the opposite of those in (4), its enantiomer.
(b) Compounds (1) and (2), (1) and (3), (2) and (4), and (3) and (4) are diastereomers.

See problem 6.23

Problem 6.3

Following are stereorepresentations of the four stereoisomers of 3-chloro-2-butanol:

(1) CH_3 / H—C—OH / Cl—C—H / CH_3

(2) CH_3 / H—C—OH / H—C—Cl / CH_3

(3) CH_3 / HO—C—H / H—C—Cl / CH_3

(4) CH_3 / HO—C—H / Cl—C—H / CH_3

(a) Which compounds are enantiomers?
(b) Which compounds are diastereomers?

B. Meso Compounds

Certain molecules containing two or more stereocenters have special symmetry properties that reduce the number of stereoisomers to fewer than the maximum number predicted by the 2^n rule. One such molecule is 2,3-dihydroxybutanedioic acid, more commonly named tartaric acid:

HOC(=O)—*CH(OH)—*CH(OH)—C(=O)OH

2,3-Dihydroxybutanedioic acid
(tartaric acid)

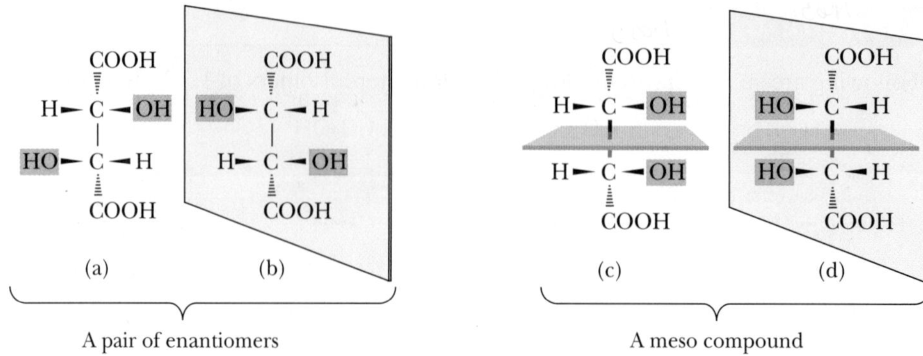

Figure 6.5 Stereoisomers of tartaric acid. One pair of enantiomers and one meso compound. The presence of an internal plane of symmetry indicates that the molecule is achiral.

Tartaric acid is a colorless, crystalline compound occurring largely in the vegetable kingdom, especially in grapes. During the fermentation of grape juice, potassium bitartrate (one —COOH group is present as a potassium salt, $—COO^-K^+$) deposits as a crust on the sides of wine casks. Then, collected and purified, it is sold commercially as cream of tartar.

Carbons 2 and 3 of tartaric acid are stereocenters, and, from the 2^n rule, the maximum number of stereoisomers possible is $2^2 = 4$. Figure 6.5 shows the two pairs of mirror images of this compound. Structures (a) and (b) are nonsuperposable mirror images and, therefore, are a pair of enantiomers. Structures (c) and (d) are also mirror images, but they are superposable. To see this, imagine that you rotate (d) by 180° in the plane of the paper, lift it out of the plane of the paper, and place it on top of (c). If you do this mental manipulation correctly, you will find that (d) is superposable on (c). Therefore, (c) and (d) are *not* different molecules; they are the same molecule, just oriented differently. Because (c) and its mirror image are superposable, (c) is achiral.

Another way to verify that (c) is achiral is to see that it has a plane of symmetry that bisects the molecule in such a way that the top half is the reflection of the bottom half. Thus, even though (c) has two stereocenters, it is achiral. The stereoisomer of tartaric acid represented by (c) or (d) is called a **meso compound**, defined as an achiral compound that contains two or more stereocenters.

Meso compound An achiral compound possessing two or more stereocenters.

We can now return to the original question: How many stereoisomers are there of tartaric acid? The answer is three: one meso compound and one pair of enantiomers. Note that the meso compound is a diastereomer of each of the other stereoisomers.

Example 6.4

Following are stereorepresentations of the three stereoisomers of 2,3-butanediol:

(1)	(2)	(3)
CH_3	CH_3	CH_3
H—C—OH	H—C—OH	HO—C—H
HO—C—H	H—C—OH	H—C—OH
CH_3	CH_3	CH_3

(a) Which are enantiomers?
(b) Which is the meso compound?

Strategy

Enantiomers are nonsuperposable mirror images. A meso compound is an achiral compound with two or more stereocenters, that is, a compound with two or more stereocenters that has a superposable mirror image.

Solution

(a) Compounds (1) and (3) are enantiomers.
(b) Compound (2) has an internal plane of symmetry and, therefore, is a meso compound.

See problems 6.23, 6.36, 6.38

Problem 6.4

Following are four Newman projection formulas for tartaric acid:

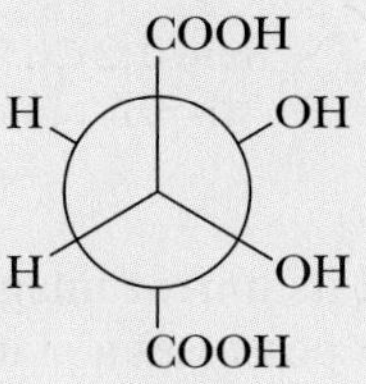

(1)

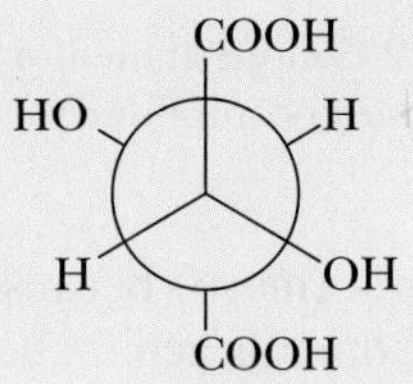

(2)

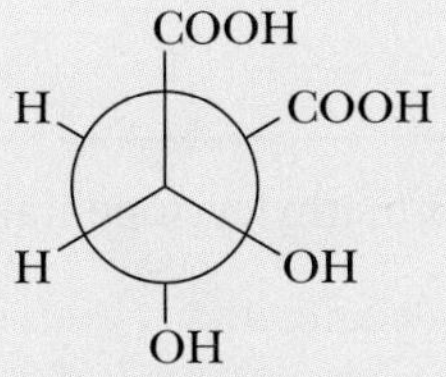

(3)

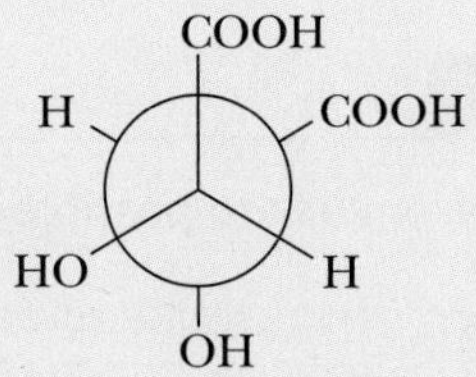

(4)

(a) Which represent the same compound?
(b) Which represent enantiomers?
(c) Which represent(s) meso tartaric acid?

6.5 How Do We Describe the Chirality of Cyclic Molecules with Two or More Stereocenters?

In this section, we concentrate on derivatives of cyclopentane and cyclohexane that contain two stereocenters. We can analyze chirality in these cyclic compounds in the same way we analyzed it in acyclic compounds.

A. Disubstituted Derivatives of Cyclopentane

Let us start with 2-methylcyclopentanol, a compound with two stereocenters. Using the 2^n rule, we predict a maximum of $2^2 = 4$ stereoisomers. Both the *cis* isomer and the *trans* isomer are chiral. The *cis* isomer exists as one pair of enantiomers, and the *trans* isomer exists as a second pair:

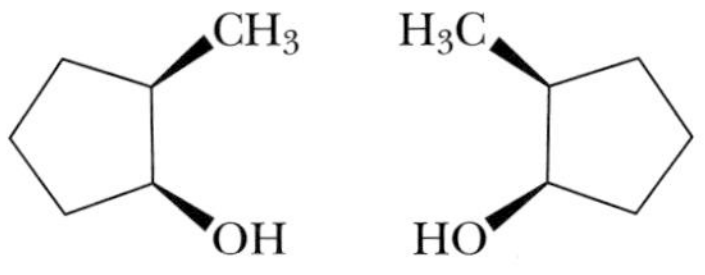

cis-2-Methylcyclopentanol
(a pair of enantiomers)

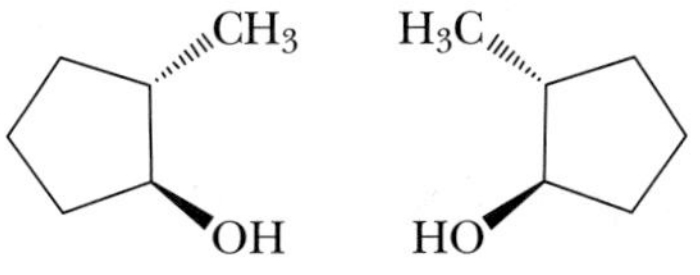

trans-2-Methylcyclopentanol
(a pair of enantiomers)

1,2-Cyclopentanediol also has two stereocenters; therefore, the 2^n rule predicts a maximum of $2^2 = 4$ stereoisomers. As seen in the following stereodrawings, only three stereoisomers exist for this compound:

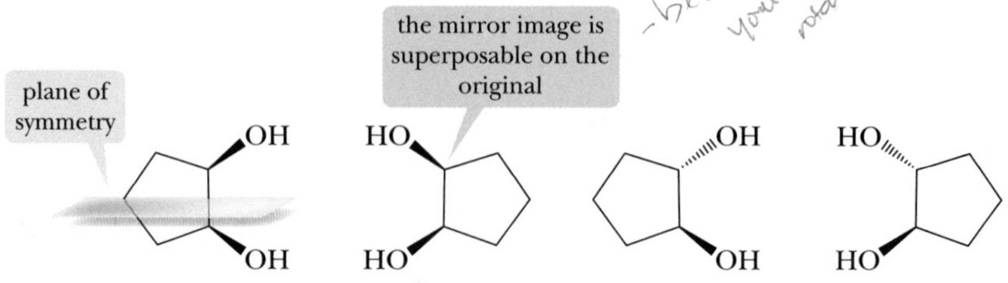

plane of symmetry

the mirror image is superposable on the original

cis-1,2-Cyclopentanediol
(a meso compound)

trans-1,2-Cyclopentanediol
(a pair of enantiomers)

The *cis* isomer is achiral (meso) because it and its mirror image are superposable. Alternatively, the *cis* isomer is achiral because it possesses a plane of symmetry that bisects the molecule into two mirror-image halves. The *trans* isomer is chiral and exists as a pair of enantiomers.

Example 6.5

How many stereoisomers are possible for 3-methylcyclopentanol?

Strategy

First identify all possible stereocenters, draw all possible pairs of stereoisomers, and determine which, if any, of the possible pairs of stereoisomers are meso compounds.

Solution

There are two stereocenters in this compound and, therefore, four stereoisomers of 3-methylcyclopentanol. The *cis* isomer exists as one pair of enantiomers and the *trans* isomer as a second pair:

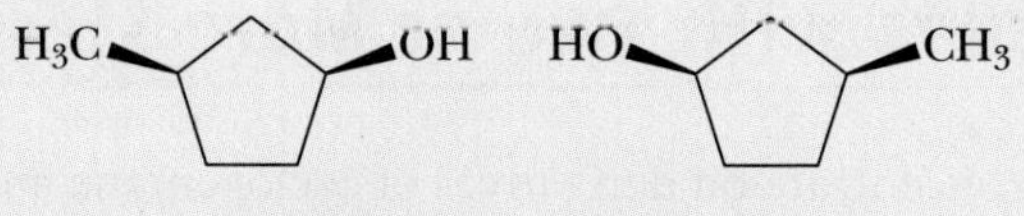

cis-3-Methylcyclopentanol
(a pair of enantiomers)

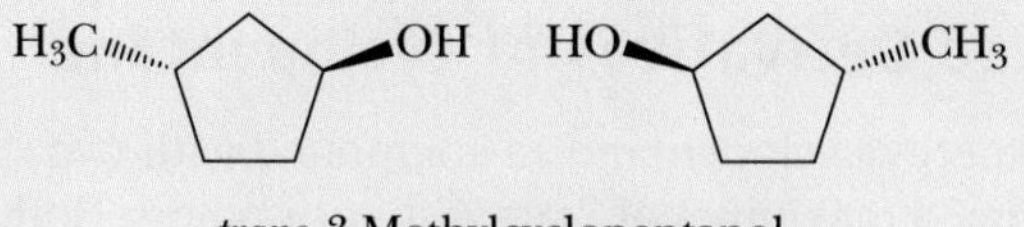

trans-3-Methylcyclopentanol
(a pair of enantiomers)

See problems 6.31, 6.33–6.35, 6.38, 6.39

Problem 6.5

How many stereoisomers are possible for 1,3-cyclopentanediol?

B. Disubstituted Derivatives of Cyclohexane

As an example of a disubstituted cyclohexane, let us consider the methylcyclohexanols. 4-Methylcyclohexanol can exist as two stereoisomers—a pair of *cis–trans* isomers:

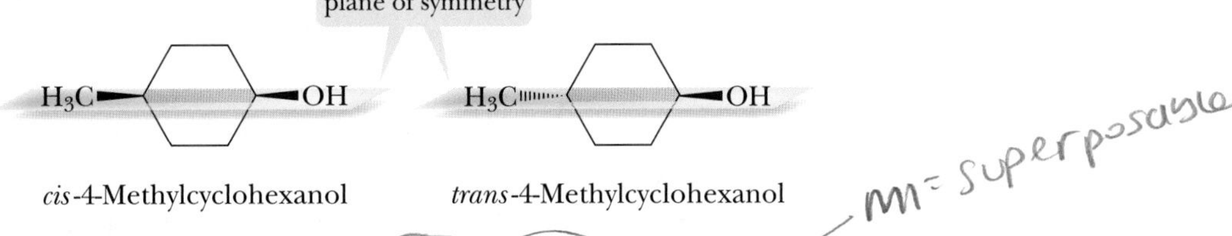

cis-4-Methylcyclohexanol   *trans*-4-Methylcyclohexanol

Both the *cis* and the *trans* isomers are achiral. In each, a plane of symmetry runs through the CH_3 and OH groups and the two attached carbons.

3-Methylcyclohexanol has two stereocenters and exists as $2^2 = 4$ stereoisomers, with the *cis* isomer existing as one pair of enantiomers and the *trans* isomer as a second pair:

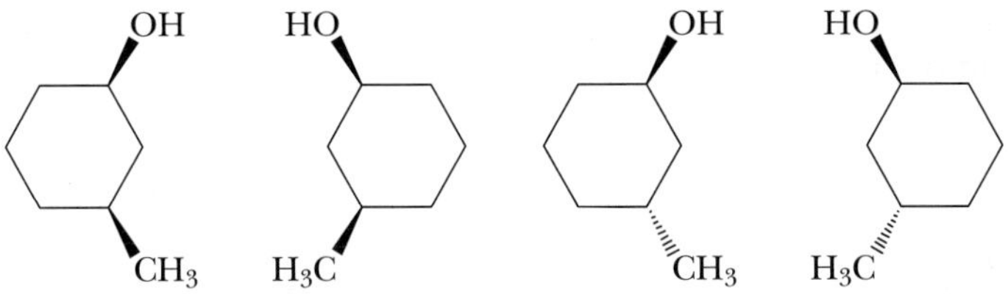

cis-3-Methylcyclohexanol (a pair of enantiomers)

trans-3-Methylcyclohexanol (a pair of enantiomers)

Similarly, 2-methylcyclohexanol has two stereocenters and exists as $2^2 = 4$ stereoisomers, with the *cis* isomer existing as one pair of enantiomers and the *trans* isomer as a second pair:

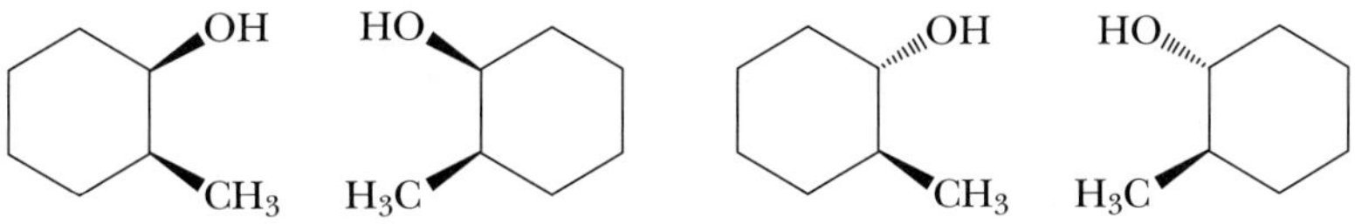

cis-2-Methylcyclohexanol (a pair of enantiomers)

trans-2-Methylcyclohexanol (a pair of enantiomers)

Example 6.6

How many stereoisomers exist for 1,3-cyclohexanediol?

Strategy

Locate all stereocenters and use the 2^n rule to determine the maximum number of stereoisomers possible. Determine which, if any, of the possible stereoisomers are meso compounds.

Solution

1,3-Cyclohexanediol has two stereocenters, and, according to the 2^n rule, a maximum of $2^2 = 4$ stereoisomers is possible. The *trans* isomer of this compound exists as a pair of enantiomers. The *cis* isomer has a plane of symmetry and is a meso compound. Therefore, although the 2^n rule predicts a maximum of four stereoisomers for 1,3-cyclohexanediol, only three exist—one pair of enantiomers and one meso compound:

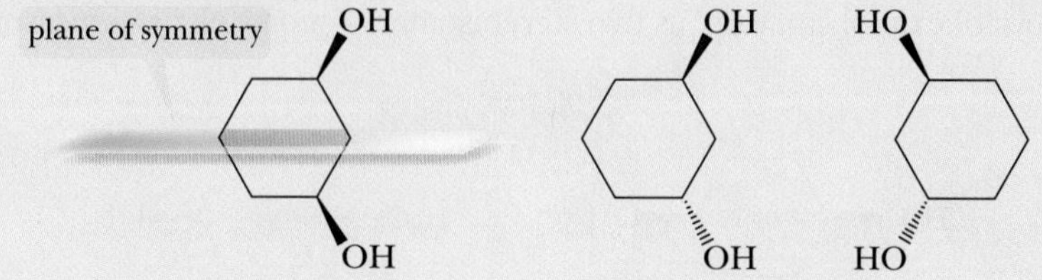

plane of symmetry

OH OH HO

OH OH HO

cis-1,3-Cyclohexanediol
(a meso compound)

trans-1,3-Cyclohexanediol
(a pair of enantiomers)

See problems 6.31, 6.33–6.35, 6.38, 6.39

Problem 6.6

How many stereoisomers exist for 1,4-cyclohexanediol?

6.6 How Do We Describe the Chirality of Molecules with Three or More Stereocenters?

The 2^n rule applies equally well to molecules with three or more stereocenters. Here is a disubstituted cyclohexanol with three stereocenters, each marked with an asterisk:

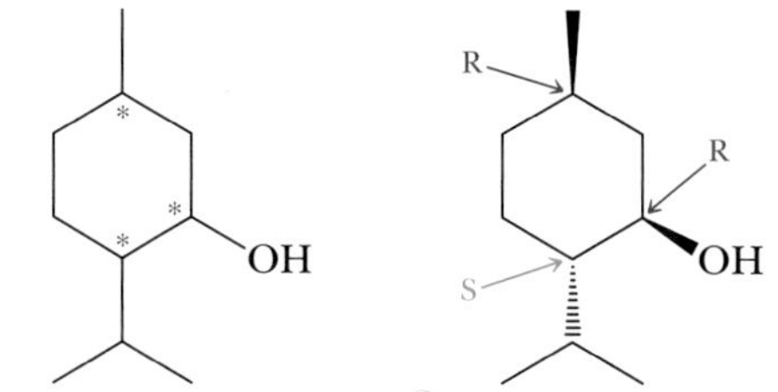

OH R R OH S

2-Isopropyl-5-methyl-cyclohexanol

Menthol

There is a maximum of $2^3 = 8$ stereoisomers possible for this molecule. Menthol, one of the eight, has the configuration shown on the right. The configuration at each stereocenter is indicated. Menthol is present in peppermint and other mint oils.

Cholesterol, a more complicated molecule, has eight stereocenters:

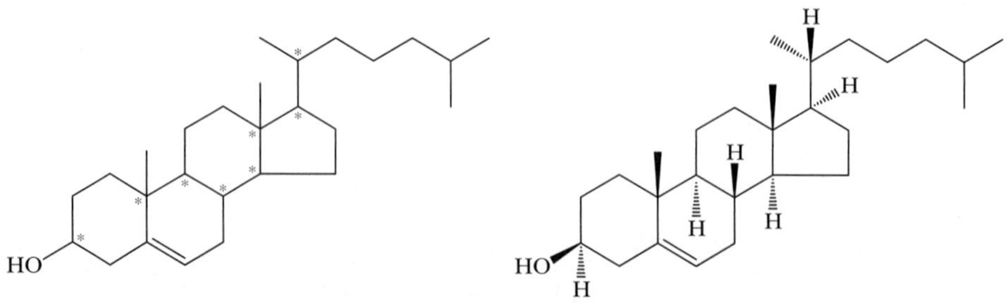

HO HO H H H H H H

Cholesterol has 8 stereocenters;
256 stereoisomers are possible

This is the stereoisomer found in
human metabolism

TABLE 6.1 Some Physical Properties of the Stereoisomers of Tartaric Acid

	(*R,R*)-Tartaric acid	(*S,S*)-Tartaric acid	Meso-tartaric acid
	COOH; H–C–OH; HO–C–H; COOH	COOH; HO–C–H; H–C–OH; COOH	COOH; H–C–OH; H–C–OH; COOH
Specific rotation*	+12.7	−12.7	0
Melting point (°C)	171–174	171–174	146–148
Density at 20°C (g/cm^3)	1.7598	1.7598	1.660
Solubility in water at 20°C (g/100 mL)	139	139	125
pK_1 (25°C)	2.98	2.98	3.23
pK_2 (25°C)	4.34	4.34	4.82

* Specific rotation is discussed in the next section.

To identify the stereocenters, remember to add an appropriate number of hydrogens to complete the tetravalence of each carbon you think might be a stereocenter.

6.7 What Are the Properties of Stereoisomers?

Enantiomers have identical physical and chemical properties in achiral environments. The enantiomers of tartaric acid (Table 6.1), for example, have the same melting point, the same boiling point, the same solubilities in water and other common solvents, and the same values of pK_a (the acid ionization constant), and they all undergo the same acid–base reactions. The enantiomers of tartaric acid do, however, differ in optical activity (the ability to rotate the plane of polarized light), a property that is discussed in the next section.

Diastereomers have different physical and chemical properties, even in achiral environments. Meso-tartaric acid has different physical properties from those of the enantiomers.

6.8 How Is Chirality Detected in the Laboratory?

As we have already established, enantiomers are different compounds, and we must expect, therefore, that they differ in some property or properties. One property that differs between enantiomers is their effect on the plane of polarized light. Each member of a pair of enantiomers rotates the plane of polarized light, and for this reason, enantiomers are said to be **optically active**. To understand how optical activity is detected in the laboratory, we must first understand plane-polarized light and a polarimeter, the instrument used to detect optical activity.

Optically active Showing that a compound rotates the plane of polarized light.

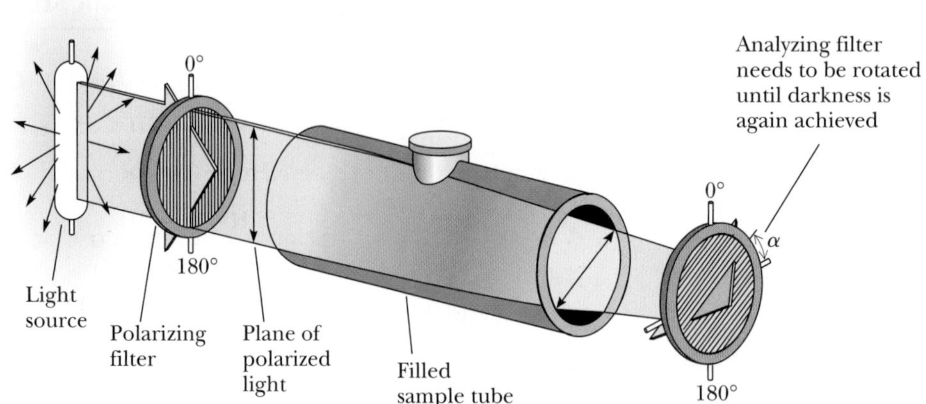

Figure 6.6
Schematic diagram of a polarimeter with its sample tube containing a solution of an optically active compound. The analyzing filter has been turned clockwise by α degrees to restore the dark field.

A. Plane-Polarized Light

Ordinary light consists of waves vibrating in all planes perpendicular to its direction of propagation (Figure 6.6). Certain materials, such as calcite and Polaroid™ sheet (a plastic film containing properly oriented crystals of an organic substance embedded in it), selectively transmit light waves vibrating in parallel planes. Electromagnetic radiation vibrating in only parallel planes is said to be **plane polarized**.

Plane-polarized light Light vibrating only in parallel planes.

B. A Polarimeter

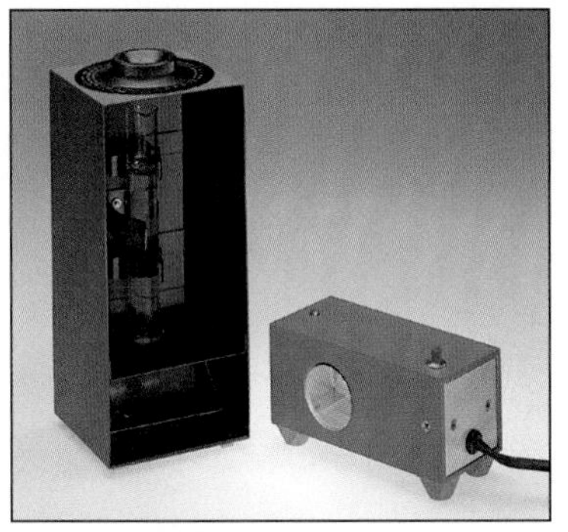

A polarimeter is used to measure the rotation of plane-polarized light as it passes through a sample. *(Richard Magna, 1992, Fundamental Photographs)*

A **polarimeter** consists of a light source, a polarizing filter and an analyzing filter (each made of calcite or Polaroid™ film), and a sample tube (Figure 6.6). If the sample tube is empty, the intensity of light reaching the detector (in this case, your eye) is at its maximum when the polarizing axes of the two filters are parallel. If the analyzing filter is turned either clockwise or counterclockwise, less light is transmitted. When the axis of the analyzing filter is at right angles to the axis of the polarizing filter, the field of view is dark. This position of the analyzing filter is taken to be 0° on the optical scale.

The ability of molecules to **rotate the plane of polarized light** can be observed with the use of a polarimeter in the following way: First, a sample tube filled with solvent is placed in the polarimeter, and the analyzing filter is adjusted so that no light passes through to the observer; that is, the filter is set to 0°. Then we place a solution of an optically active compound in the sample tube. When we do so, we find that a certain amount of light now passes through the analyzing filter. We also find that the plane of polarized light from the polarizing filter has been rotated so that it is no longer at an angle of 90° to the analyzing filter. Consequently, we rotate the analyzing filter to restore darkness in the field of view. The number of degrees, α, through which we must rotate the analyzing filter to restore darkness to the field of view is called the **observed rotation**. If we must turn the analyzing filter to the right (clockwise) to restore the dark field, we say that the compound is **dextrorotatory** (Latin: *dexter*, on the right side); if we must turn it to the left (counterclockwise), we say that the compound is **levorotatory** (Latin: *laevus*, on the left side).

Polarimeter An instrument for measuring the ability of a compound to rotate the plane of polarized light.

Observed rotation The number of degrees through which a compound rotates the plane of polarized light.

Dextrorotatory Rotating the plane of polarized light in a polarimeter to the right.

Levorotatory Rotating the plane of polarized light in a polarimeter to the left.

The magnitude of the observed rotation for a particular compound depends on its concentration, the length of the sample tube, the temperature, the solvent, and the wavelength of the light used. The **specific rotation**, $[\alpha]$, is defined as the observed rotation at a specific cell length and sample concentration expressed in grams per milliliter.

Specific rotation Observed rotation of the plane of polarized light when a sample is placed in a tube 1.0 dm long at a concentration of 1.0 g/mL.

$$\text{Specific rotation} = [\alpha]_{\lambda}^{T} = \frac{\text{Observed rotation (degrees)}}{\text{Length (dm)} \times \text{Concentration}}$$

The standard cell length is 1 decimeter (1 dm = 0.1 m). For a pure liquid sample, the concentration is expressed in grams per milliliter (g/mL; density). The temperature (T, in degrees centigrade) and wavelength (λ, in nanometers) of light are designated, respectively, as superscripts and subscripts. The light source most commonly used in polarimetry is the sodium D line (λ = 589 nm), the same line responsible for the yellow color of sodium-vapor lamps.

In reporting either observed or specific rotation, it is common to indicate a dextrorotatory compound with a plus sign in parentheses, (+), and a levorotatory compound with a minus sign in parentheses, (−). For any pair of enantiomers, one enantiomer is dextrorotatory and the other is levorotatory. For each member, the value of the specific rotation is exactly the same, but the sign is opposite. Following are the specific rotations of the enantiomers of 2-butanol at 25°C, observed with the D line of sodium:

H OH HO H

(S)-(+)-2-Butanol (R)-(−)-2-Butanol

$[\alpha]_D^{25}$ + 13.52° $[\alpha]_D^{25}$ −13.52°

C. Racemic Mixtures

An equimolar mixture of two enantiomers is called a **racemic mixture**, a term derived from the name "racemic acid" (Latin: *racemus*, a cluster of grapes), originally given to an equimolar mixture of the enantiomers of tartaric acid (Table 6.1). Because a racemic mixture contains equal numbers of the dextrorotatory and the levorotatory molecules, its specific rotation is zero. Alternatively, we say that a racemic mixture is **optically inactive**. A racemic mixture is indicated by adding the prefix (±) to the name of the compound.

Racemic mixture A mixture of equal amounts of two enantiomers.

Optically inactive Showing that a compound or mixture of compounds does not rotate the plane of polarized light.

6.9 What Is the Significance of Chirality in the Biological World?

Except for inorganic salts and a relatively few low-molecular-weight organic substances, the molecules in living systems, both plant and animal, are chiral. Although these molecules can exist as a number of stereoisomers, almost invariably only one stereoisomer is found in nature. Of course, instances do occur in which more than one stereoisomer is found, but these rarely exist together in the same biological system.

A. Chirality in Biomolecules

Perhaps the most conspicuous examples of chirality among biological molecules are the enzymes, all of which have many stereocenters. An example is chymotrypsin, an enzyme found in the intestines of animals. This enzyme catalyzes the digestion of proteins (Section 19.5). Chymotrypsin has 251 stereocenters. The maximum number of stereoisomers possible is thus 2^{251}, a staggeringly large number, almost beyond comprehension. Fortunately, nature does not squander its precious energy and resources unnecessarily: Only one of these stereoisomers is produced and used by any given organism.

Because enzymes are chiral substances, most either produce or react with only substances that match their stereochemical requirements.

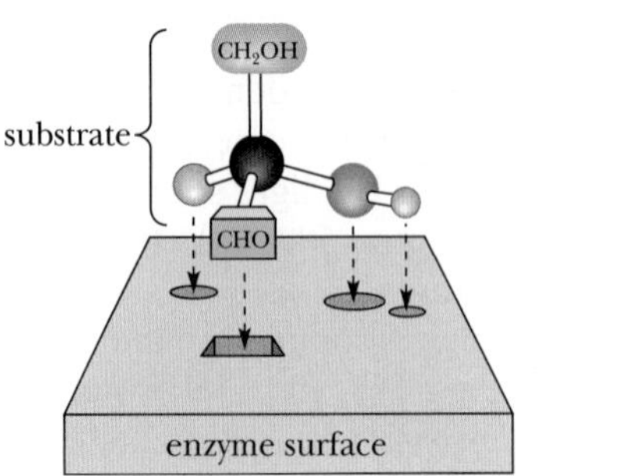

This enantiomer of glyceraldehyde fits the three specific binding sites on the enzyme surface

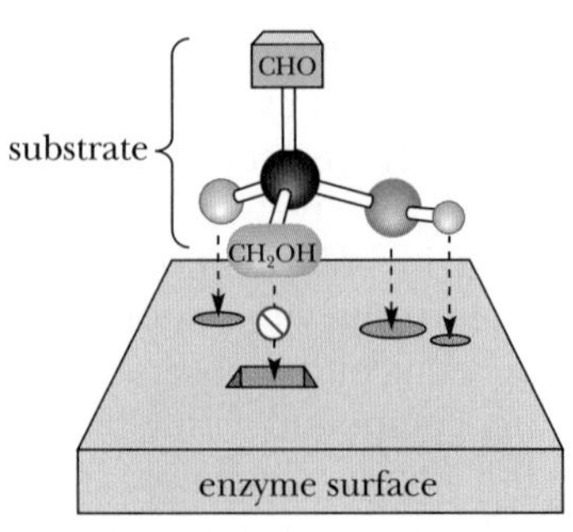

This enantiomer of glyceraldehyde does not fit the same binding sites

Figure 6.7
A schematic diagram of an enzyme surface capable of interacting with (*R*)-(+)-glyceraldehyde at three binding sites, but with (*S*)-(−)-glyceraldehyde at only two of these sites.

B. How an Enzyme Distinguishes between a Molecule and Its Enantiomer

An enzyme catalyzes a biological reaction of a molecule by first positioning it at a **binding site** on the enzyme's surface. An enzyme with binding sites specific for three of the four groups on a stereocenter can distinguish between a molecule and its enantiomer or one of its diastereomers. Assume, for example, that an enzyme involved in catalyzing a reaction of glyceraldehyde has on its surface a binding site specific for —H, a second specific for —OH, and a third specific for —CHO. Assume further that the three sites are arranged on the enzyme surface as shown in Figure 6.7. The enzyme can distinguish (*R*)-(+)-glyceraldehyde (the natural, or biologically active, form) from its enantiomer because the natural enantiomer can be absorbed, with three groups interacting with their appropriate binding sites; for the S enantiomer, at best only two groups can interact with these binding sites.

Because interactions between molecules in living systems take place in a chiral environment, it should come as no surprise that a molecule and its enantiomer or one of its diastereomers elicit different physiological responses. As we have already seen, (*S*)-ibuprofen is active as a pain and fever reliever, whereas its R enantiomer is inactive. The S enantiomer of the closely related analgesic naproxen is also the active pain reliever of this compound, but its R enantiomer is a liver toxin!

(*S*)-Ibuprofen

(*S*)-Naproxen

6.10 How Can Enantiomers Be Resolved?

Resolution Separation of a racemic mixture into its enantiomers.

Resolution is the separation of a racemic mixture into its enantiomers. Because two enantiomers have the same physical properties, separating them, in general, is difficult, but scientists have developed a number of ways to do it. In this section, we illustrate just two of the several laboratory methods for resolution, the use of enzymes as chiral catalysts and the use of solid chiral materials to differentiate between enantiomers made to come in contact with these materials.

CHEMICAL CONNECTIONS 6A

Chiral Drugs

Some of the common drugs used in human medicine (for example, aspirin, Section 14.4B) are achiral. Others are chiral and are sold as single enantiomers. The penicillin and erythromycin classes of antibiotics and the drug Captopril are all chiral drugs. Captopril, which is highly effective for the treatment of high blood pressure and congestive heart failure, was developed in a research program designed to discover effective inhibitors of angiotensin-converting enzyme (ACE). Captopril is manufactured and sold as the (*S*,*S*)-stereoisomer. A large number of chiral drugs, however, are sold as racemic mixtures. The popular analgesic ibuprofen (the active ingredient in Motrin®, Advil®, and many other nonaspirin analgesics) is an example. Only the S enantiomer of the pain reliever ibuprofen is biologically active.

Captopril

(*S*)-Ibuprofen

For racemic drugs, most often only one enantiomer exerts the beneficial effect, whereas the other enantiomer either has no effect or may exert a detrimental effect. Thus, enantiomerically pure drugs should, more often than not, be more effective than their racemic counterparts. A case in point is 3,4-dihydroxyphenylalanine, which is used in the treatment of Parkinson's disease. The active drug is dopamine. Unfortunately, this compound does not cross the blood–brain barrier to the required site of action in the brain. Consequently, what is administered instead is the prodrug, a compound that is not active by itself, but is converted in the body to an active drug. 3,4-Dihydroxyphenylalanine is such a prodrug; it crosses the blood–brain barrier and then undergoes decarboxylation, catalyzed by the enzyme dopamine decarboxylase, to give dopamine. Decarboxylation is the loss of carbon dioxide from a carboxyl group ($R-CO_2H$).

$$R-CO_2H \xrightarrow{\text{decarboxylation}} R-H + CO_2$$

(*S*)-(−)-3,4-Dihydroxyphenylalanine (L-DOPA) $[\alpha]^{13}_{D}$ −13.1° $\xrightarrow{\text{enzyme-catalyzed decarboxylation}}$ Dopamine + CO_2

Dopamine decarboxylase is specific for the S enantiomer, which is commonly known as L-DOPA. It is essential, therefore, to administer the enantiomerically pure prodrug. Were the prodrug to be administered in a racemic form, there could be a dangerous buildup of the R enantiomer, which cannot be metabolized by the enzymes present in the brain.

QUESTION

Draw all possible stereoisomers of Captopril and determine the relationship between each. Following are structural formulas for three other angiotensin-converting enzyme (ACE) inhibitors, all members of the "*pril*" family. Which are chiral? For each that is chiral, determine the number of stereoisomers possible for it. List the similarities in structure among each of these four drugs.

Quinapril (Accupril)

Ramipril (Vasotec)

Enalapril (Altace)

A. Enzymes as Resolving Agents

One class of enzymes that has received particular attention in this regard is the esterases, which catalyze the hydrolysis of esters (Section 15.1C) to give an alcohol and a carboxylic acid. We illustrate this method by describing the resolution of (*R,S*)-naproxen. The ethyl esters of both (*R*)- and (*S*)-naproxen are solids with very low solubilities in water. Chemists then use an esterase in alkaline solution to selectively hydrolyze the (*S*)-ester, which goes into the aqueous solution as the sodium salt of the (*S*)-carboxylic acid. The (*R*)-ester is unaffected by these conditions. Filtering the alkaline solution recovers the crystals of the (*R*)-ester. After the crystals are removed, the alkaline solution is acidified to precipitate pure (*S*)-naproxen. The recovered (*R*)-ester can be racemized (converted to an *R,S*-mixture) and again treated with the esterase. Thus, by recycling the (*R*)-ester, all the racemic ester is converted to (*S*)-naproxen.

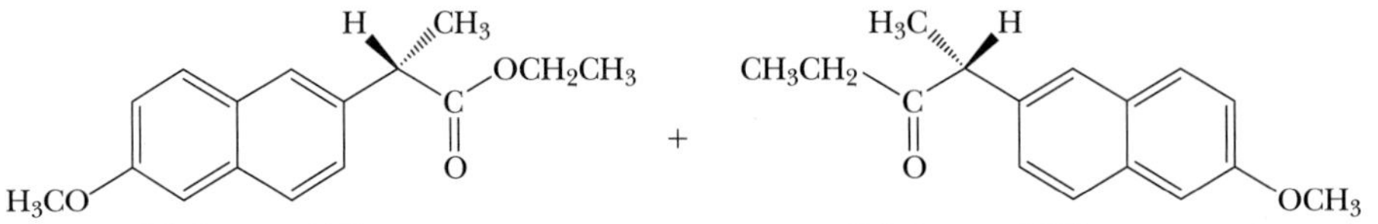

H, CH_3, OCH_2CH_3, C, O, H_3CO

Ethyl ester of (*S*)-naproxen

+

H_3C, H, CH_3CH_2, C, O, OCH_3

Ethyl ester of (*R*)-naproxen (not affected by the esterase)

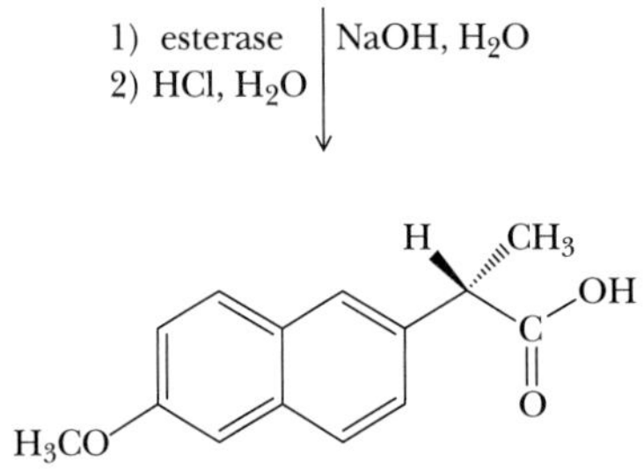

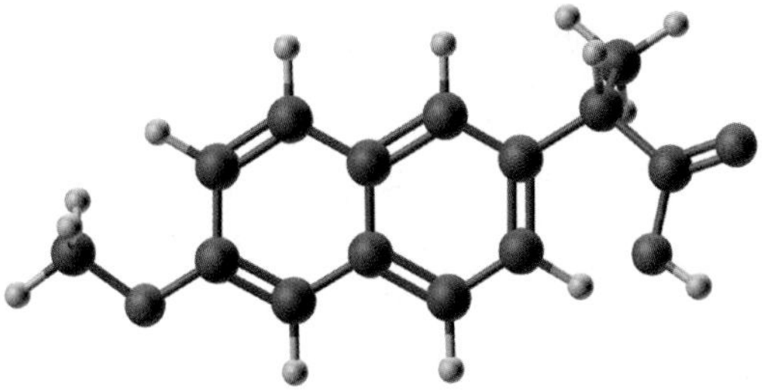

1) esterase / NaOH, H_2O
2) HCl, H_2O

H, CH_3, OH, C, O, H_3CO

(*S*)-Naproxen

The sodium salt of (*S*)-naproxen is the active ingredient in Aleve® and a score of other over-the-counter nonsteroidal anti-inflammatory preparations.

B. Resolution by Means of Chromatography on a Chiral Substrate

Chromatography is a term used to describe the purification of substances in which a sample to be purified interacts with a solid material, and different components of the sample separate based on differences in their interactions with the solid material. The solid material is packed into a column, and a solution of the substance dissolved in a suitable solvent is passed down the column. The more weakly bound components of the mixture pass through the column more quickly than the more tightly bound components.

A common method for resolving enantiomers today is chromatography using a chiral packing material in the column. Each enantiomer in principle interacts differently with the chiral packing material, and the elution time will be different for each enantiomer. A wide variety of chiral column packing materials have been developed for this purpose.

Recently, the U.S. Food and Drug Administration established new guidelines for the testing and marketing of chiral drugs. After reviewing these guidelines, many drug companies have decided to develop only single enantiomers of new chiral drugs. In addition to regulatory pressure, there are patent considerations: If a company has patents on a racemic drug, a new patent can often be taken out on one of its enantiomers.

Key Terms and Concepts

Summary of Key Questions

6.1 What Are Stereoisomers?

- **Stereoisomers** have the same connectivity of their atoms, but a different three-dimensional orientation of their atoms in space.
- A **mirror image** is the reflection of an object in a mirror.

6.2 What Are Enantiomers?

- **Enantiomers** are a pair of stereoisomers that are nonsuperposable mirror images. A molecule that is not superposable on its mirror image is said to be **chiral**.
- Chirality is a property of an object as a whole, not of a particular atom.
- An **achiral** object possesses a **plane of symmetry**—an imaginary plane passing through the object and dividing it such that one half is the reflection of the other half.
- A **stereocenter** is an atom at which the interchange of two atoms or groups of atoms bonded to it produces a different stereoisomer.
- The most common type of stereocenter among organic compounds is a tetrahedral carbon atom with four different groups bonded to it.

6.3 How Do We Designate the Configuration of a Stereocenter?

- The **configuration** at a stereocenter can be specified by the **R,S convention**.
- To apply this convention, (1) each atom or group of atoms bonded to the stereocenter is assigned a priority and numbered from highest priority to lowest priority, (2) the molecule is oriented in space so that the group of lowest priority is directed away from the observer, and (3) the remaining three groups are read in order, from highest priority to lowest priority.
- If the reading of groups is clockwise, the configuration is **R** (Latin: *rectus*, right). If the reading is counterclockwise, the configuration is **S** (Latin: *sinister*, left).

6.4 What Is the 2^n Rule?

- For a molecule with ***n*** stereocenters, the maximum number of stereoisomers possible is $\mathbf{2^n}$.
- **Diastereomers** are stereoisomers that are not mirror images.
- Certain molecules have special symmetry properties that reduce the number of stereoisomers to fewer than that predicted by the 2^n rule.

- A **meso** compound contains two or more stereocenters assembled in such a way that its molecules are achiral.
- Enantiomers have identical physical and chemical properties in achiral environments.
- Diastereomers have different physical and chemical properties.

6.5 How Do We Describe the Chirality of Cyclic Molecules with Two or More Stereocenters?

- When evaluating the symmetry of cyclic structures, such as derivatives of cyclohexane and cyclopentane, it is helpful to evaluate planar representations.

6.6 How Do We Describe the Chirality of Molecules with Three or More Stereocenters?

- For a molecule with n stereocenters, the maximum number of stereoisomers possible is 2^n.

6.7 What Are the Properties of Stereoisomers?

- Enantiomers have identical physical and chemical properties in achiral environments.
- Diastereomers have different physical and chemical properties.

6.8 How Is Chirality Detected in the Laboratory?

- Light that vibrates in only parallel planes is said to be **plane polarized**.
- A **polarimeter** is an instrument used to detect and measure the magnitude of optical activity. **Observed rotation** is the number of degrees the plane of polarized light is rotated.
- **Specific rotation** is the observed rotation measured with a cell 1 dm long and a solution with a concentration of 1.00 g/mL.
- If the analyzing filter must be turned clockwise to restore the zero point, the compound is **dextrorotatory**. If the analyzing filter must be turned counterclockwise to restore the zero point, the compound is **levorotatory**.
- A compound is said to be **optically active** if it rotates the plane of polarized light. Each member of a pair of enantiomers rotates the plane of polarized light an equal number of degrees, but opposite in direction.
- A **racemic mixture** is a mixture of equal amounts of two enantiomers and has a specific rotation of zero.
- A meso compound is optically active.

6.9 What Is the Significance of Chirality in the Biological World?

- An enzyme catalyzes the biological reactions of molecules by first positioning them at a binding site on its surface. An enzyme with a binding site specific for three of the four groups on a stereocenter can distinguish between a molecule and its enantiomer.

6.10 How Can Enantiomers Be Resolved?

- **Resolution** is the experimental process of separating a mixture of enantiomers into two pure enantiomers.
- One means of resolution is to treat the racemic mixture with an enzyme that catalyzes a specific reaction of one enantiomer, but not the other.
- A common method for resolving enantiomers is chromatography using a chiral packing material in the column. Each enantiomer in principle interacts differently with the chiral packing material and the elution time will be different for each enantiomer.

Quick Quiz

Answer true or false to the following questions to assess your general knowledge of the concepts in this chapter. If you have difficulty with any of them, you should review the appropriate section in the chapter (shown in parentheses) before attempting the more challenging end-of-chapter problems.

1. Enantiomers are always chiral. (6.2)
2. An unmarked cube is chiral. (6.1)
3. Stereocenters can be designated using E and Z. (6.3)
4. A chiral molecule will always have a diastereomer. (6.2)
5. Every object in nature has a mirror image. (6.1)
6. A molecule that possesses an internal plane of symmetry can never be chiral. (6.2)
7. Pairs of enantiomers have the same connectivity. (6.1)
8. Enantiomers, like gloves, occur in pairs. (6.2)
9. A cyclic molecule with two stereocenters will always have only three stereoisomers. (6.5)
10. An achiral molecule will always have a diastereomer. (6.2)
11. The *cis* and *trans* isomers of 2-butene are chiral. (6.1)
12. A human foot is chiral. (6.1)
13. A compound with n stereocenters will always have 2^n stereoisomers. (6.4)
14. A molecule with three or more stereocenters cannot be meso. (6.6)
15. A molecule with three or more stereocenters must be chiral. (6.6)
16. Each member of a pair of enantiomers will have the same boiling point. (6.7)
17. If a molecule is not superposable on its mirror image, it is chiral. (6.1)
18. For a molecule with two tetrahedral stereocenters, four stereoisomers are possible. (6.2)
19. Constitutional isomers have the same connectivity. (6.1)
20. Enantiomers can be separated by interacting them with the same chiral environment or chemical agent. (6.10)
21. Enzymes are achiral molecules that can differentiate chiral molecules. (6.9)
22. *Cis* and *trans* stereoisomers of a cyclic compound can be classified as diastereomers. (6.5)
23. 3-Pentanol is the mirror image of 2-pentanol. (6.2)
24. Diastereomers do not have a mirror image. (6.2)
25. The most common cause of chirality in organic molecules is the presence of a tetrahedral carbon atom with four different groups bonded to it. (6.1)
26. Each member of a pair of enantiomers will have the same density. (6.7)
27. The carbonyl carbon of an aldehyde or a ketone cannot be a stereocenter. (6.1)
28. For a molecule with three stereocenters, $3^2 = 9$ stereoisomers are possible. (6.2)
29. Diastereomers can be resolved using traditional methods such as distillation. (6.10)
30. A racemic mixture is optically inactive. (6.8)
31. 2-Pentanol and 3-pentanol are chiral and show enantiomerism. (6.2)
32. A diastereomer of a chiral molecule must also be chiral. (6.2)
33. In order to designate the configuration of a stereocenter, the priority of groups must be read in a clockwise or counterclockwise fashion after the lowest priority group is placed facing toward the viewer. (6.3)
34. A compound with n stereocenters will always be one of the 2^n stereoisomers of that compound. (6.4)
35. Each member of a pair of enantiomers could react differently in a chiral environment. (6.7)
36. A chiral molecule will always have an enantiomer. (6.2)
37. Each member of a pair of diastereomers will have the same melting point. (6.7)
38. If a chiral compound is dextrorotatory, its enantiomer is levorotatory by the same number of degrees. (6.8)
39. All stereoisomers are optically active. (6.8)
40. There are usually equal amounts of each enantiomer of a chiral biological molecule in a living organism. (6.9)

Answers: (1) T (2) F (3) F (4) F (5) T (6) T (7) T (8) T
(9) F (10) F (11) F (12) T (13) F (14) F (15) F (16) T
(17) T (18) T (19) F (20) T (21) F (22) T (23) F (24) F
(25) T (26) T (27) T (28) F (29) T (30) T (31) F (32) F
(33) F (34) T (35) T (36) T (37) F (38) T (39) F (40) F

Problems

A problem marked with an asterisk indicates an applied "real world" problem. Answers to problems whose numbers are printed in blue are given in Appendix D.

Section 6.1 Chirality

6.7 Define the term *stereoisomer*. Name four types of stereoisomers.

6.8 In what way are constitutional isomers different from stereoisomers? In what way are they the same?

6.9 Compare and contrast the meaning of the terms *conformation* and *configuration*.

***6.10** Which of these objects are chiral (assume that there is no label or other identifying mark)?

(a) A pair of scissors (b) A tennis ball

(c) A paper clip (d) A beaker

(e) The swirl created in water as it drains out of a sink or bathtub

***6.11** Think about the helical coil of a telephone cord or the spiral binding on a notebook, and suppose that you view the spiral from one end and find that it has a left-handed twist. If you view the same spiral from the other end, does it have a right-handed twist or a left-handed twist from that end as well?

***6.12** Next time you have the opportunity to view a collection of augers or other seashells that have a helical twist, study the chirality of their twists. Do you find an equal number of left-handed and right-handed augers, or, for example, do they all have the same handedness? What about the handedness of augers compared with that of other spiral shells?

Median cross section through the shell of a chambered nautilus found in the deep waters of the Pacific Ocean. The shell shows handedness; this cross section is a right-handed spiral. *(Photo Disc. Inc./Getty Images)*

***6.13** Next time you have an opportunity to examine any of the seemingly endless varieties of spiral pasta (rotini, fusilli, radiatori, tortiglioni), examine their twist. Do the twists of any one kind all have a right-handed twist, do they all have a left-handed twist, or are they a racemic mixture?

6.14 One reason we can be sure that sp^3-hybridized carbon atoms are tetrahedral is the number of stereoisomers that can exist for different organic compounds.

(a) How many stereoisomers are possible for $CHCl_3$, CH_2Cl_2, and CHBrClF if the four bonds to carbon have a tetrahedral geometry?

(b) How many stereoisomers are possible for each of the compounds if the four bonds to the carbon have a square planar geometry?

Section 6.2 Enantiomers

6.15 Which compounds contain stereocenters? **(See Example 6.1)**

(a) 2-Chloropentane (b) 3-Chloropentane

(c) 3-Chloro-1-pentene (d) 1,2-Dichloropropane

6.16 Using only C, H, and O, write a structural formula for the lowest-molecular-weight chiral molecule of each of the following compounds:

(a) Alkane (b) Alcohol

(c) Aldehyde (d) Ketone

(e) Carboxylic acid

6.17 Which alcohols with the molecular formula $C_5H_{12}O$ are chiral?

6.18 Which carboxylic acids with the molecular formula $C_6H_{12}O_2$ are chiral?

6.19 Draw the enantiomer for each molecule: **(See Example 6.1)**

(a) OH, H_3C, C, COOH, H

(b) CHO, H, C, OH, CH_2OH

(c) COOH, H_2N, C, H, CH_3

(d) O, OH

(e) OH

(f) OH, NH_2

(g) O, COOH

(h) CH_3, H, OH, H, OH, CH_3

(i) 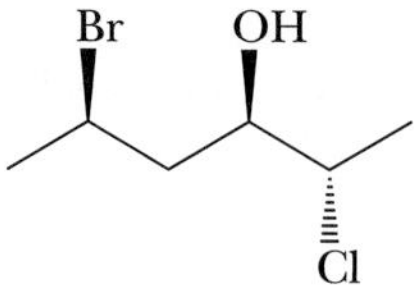 (j) 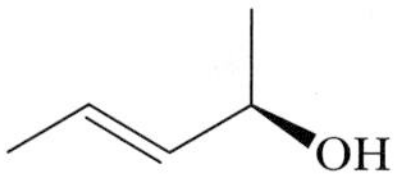 (k) 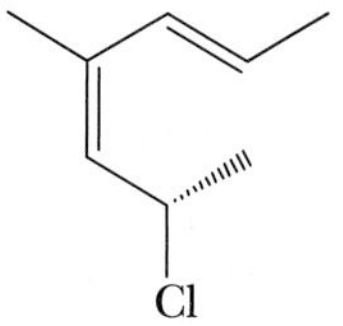 (l) 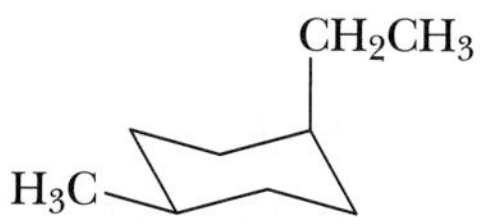 H_3C — CH_2CH_3

6.20 Mark each stereocenter in these molecules with an asterisk (note that not all contain stereocenters): **(See Example 6.1)**

(a) 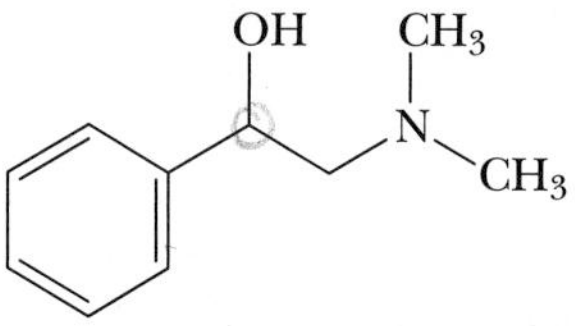 (b) 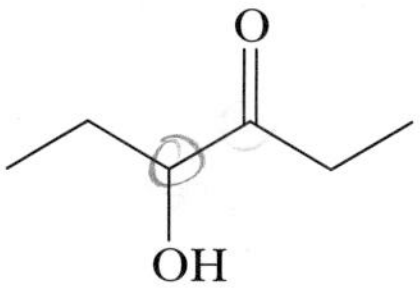 (c) 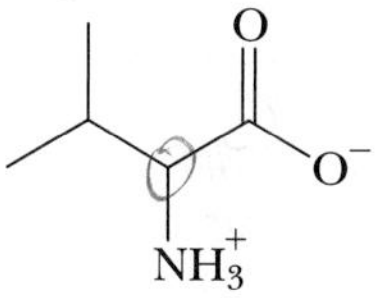 (d)

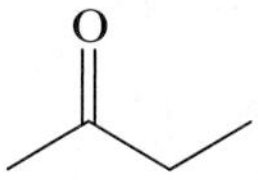

6.21 Mark each stereocenter in these molecules with an asterisk (note that not all contain stereocenters): **(See Example 6.1)**

(a) 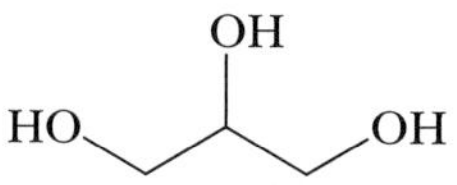 (b) 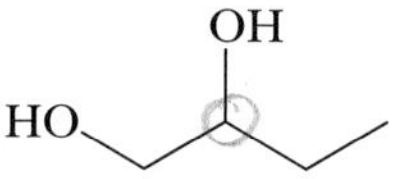 (c) 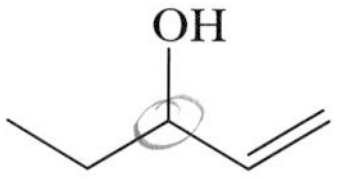 (d)

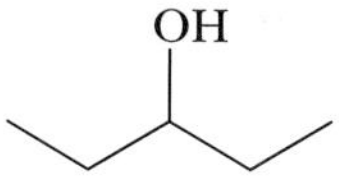

6.22 Mark each stereocenter in these molecules with an asterisk (note that not all contain stereocenters): **(See Example 6.1)**

(a) 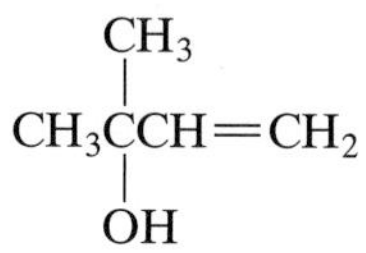 $CH_3C(CH_3)(OH)CH{=}CH_2$

(b) 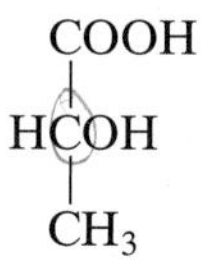 $HC(COOH)(OH)CH_3$

(c) 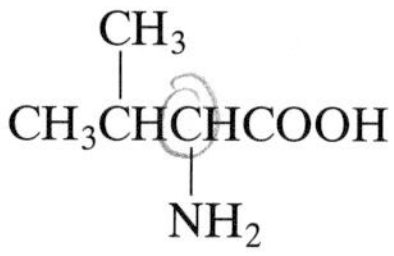 $CH_3CH(CH_3)CH(NH_2)COOH$

(d) 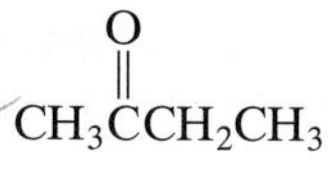 $CH_3CCH_2CH_3$ (C=O)

(e) 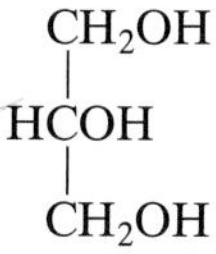 $HOCH(CH_2OH)CH_2OH$

(f) 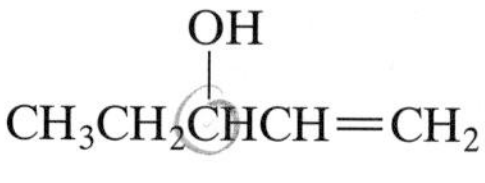 $CH_3CH_2CH(OH)CH{=}CH_2$

(g) 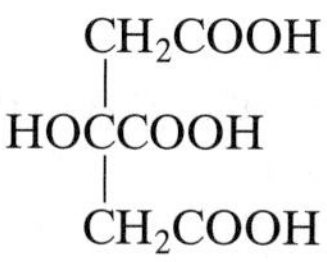 $HOC(COOH)(CH_2COOH)CH_2COOH$

6.23 Following are eight stereorepresentations of lactic acid: **(See Examples 6.3, 6.4)**

(a) 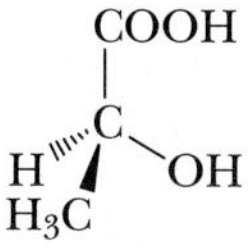 (b) 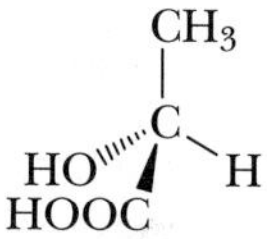 (c) 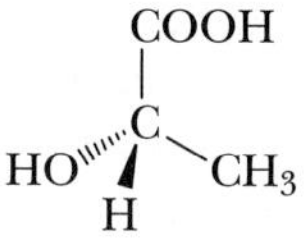 (d) 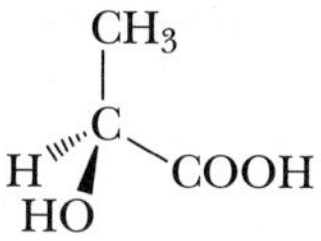 (e) 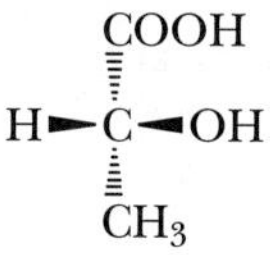 (f) 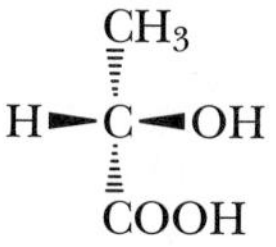 (g) 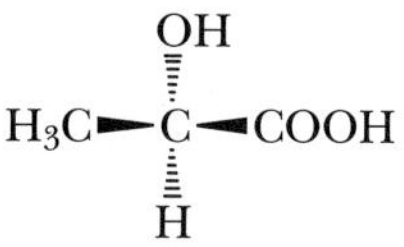 (h)

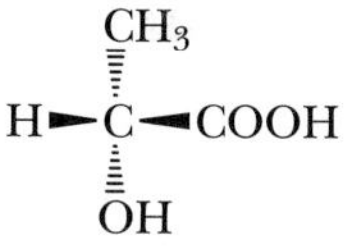

Take (a) as a reference structure. Which stereorepresentations are identical with (a) and which are mirror images of (a)?

Section 6.3 Designation of Configuration: The R,S Convention

6.24 Assign priorities to the groups in each set: **(See Example 6.2)**

(a) $-H \quad -CH_3 \quad -OH \quad -CH_2OH$

(b) $-CH_2CH{=}CH_2 \quad -CH{=}CH_2 \quad -CH_3 \quad -CH_2COOH$

(c) $-CH_3 \quad -H \quad -COO^- \quad -NH_3^+$

(d) $-CH_3 \quad -CH_2SH \quad -NH_3^+ \quad -COO^-$

6.25 Which molecules have R configurations? **(See Example 6.2)**

(a) 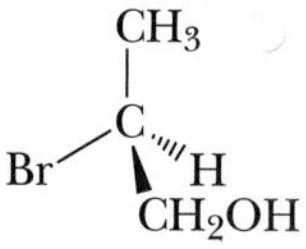 (b) 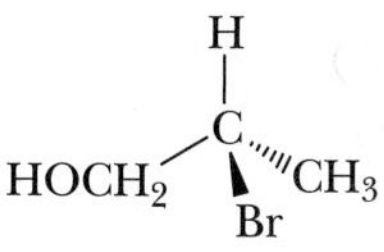 (c) 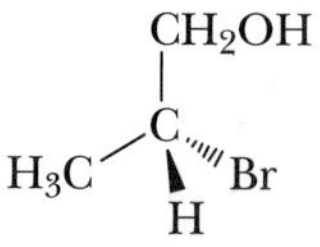 (d)

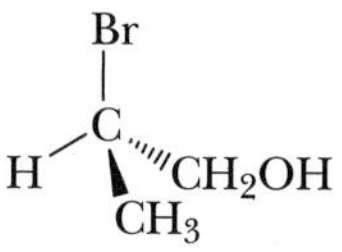

***6.26** Following are structural formulas for the enantiomers of carvone: **(See Example 6.2)**

(−)-Carvone (Spearmint oil)

(+)-Carvone (Caraway and dillseed oil)

Each enantiomer has a distinctive odor characteristic of the source from which it can be isolated. Assign an R or S configuration to the stereocenter in each. How can they have such different properties when they are so similar in structure?

6.27 Following is a staggered conformation of one of the stereoisomers of 2-butanol: **(See Example 6.2)**

(a) Is this (R)-2-butanol or (S)-2-butanol?

(b) Draw a Newman projection for this staggered conformation, viewed along the bond between carbons 2 and 3.

(c) Draw a Newman projection for one more staggered conformations of this molecule. Which of your conformations is the more stable? Assume that —OH and —CH_3 are comparable in size.

Sections 6.5 and 6.6 Molecules with Two or More Stereocenters

6.28 Write the structural formula of an alcohol with molecular formula $C_6H_{14}O$ that contains two stereocenters.

***6.29** For centuries, Chinese herbal medicine has used extracts of *Ephedra sinica* to treat asthma. Investigation of this plant resulted in the isolation of ephedrine, a potent dilator of the air passages of the lungs. The naturally occurring stereoisomer is levorotatory and has the following structure: **(See Example 6.2)**

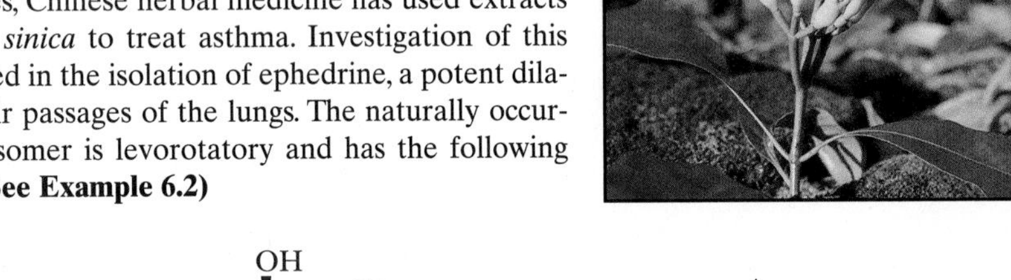

Ephedra sinica, a source of ephedrine, a potent bronchodilator. *(Paolo Koch/Photo Researchers, Inc.)*

Ephedrine

Assign an R or S configuration to each stereocenter.

***6.30** The specific rotation of naturally occurring ephedrine, shown in Problem 6.29, is −41°. What is the specific rotation of its enantiomer?

6.31 Label each stereocenter in these molecules with an asterisk and tell how many stereoisomers exist for each. **(See Examples 6.5, 6.6)**

(a) $CH_3CH(OH)CH(OH)COOH$

(b) $HOOC—CH_2—CH(COOH)—CH(OH)—COOH$

(c)

(d)

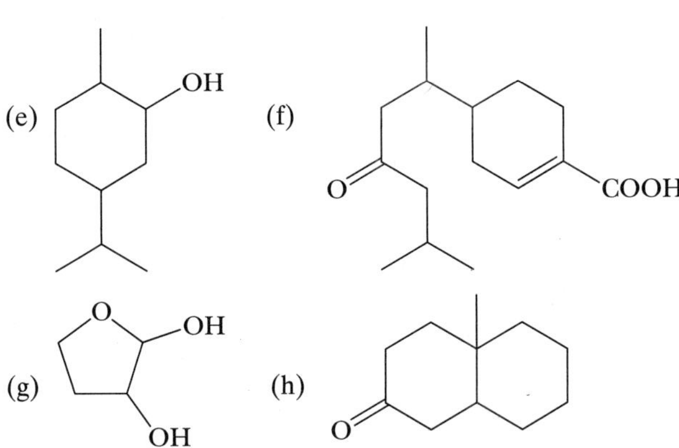

(e)

(f)

(g)

(h)

How many stereoisomers are possible for each molecule?

***6.32** Label the four stereocenters in amoxicillin, which belongs to the family of semisynthetic penicillins:

Amoxicillin

***6.33** Label all stereocenters in loratadine (Claritin®) and fexofenadine (Allegra®), now the top-selling antihistamines in the United States. Tell how many stereoisomers are possible for each. **(See Examples 6.5, 6.6)**

(a)

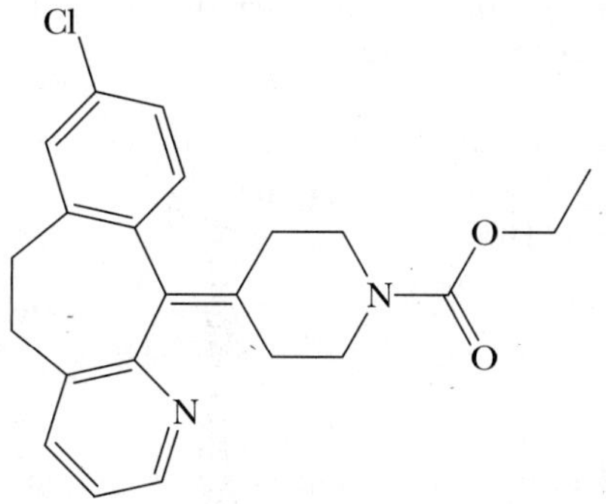

Cl N O O

Loratadine
(Claritin)

(b)

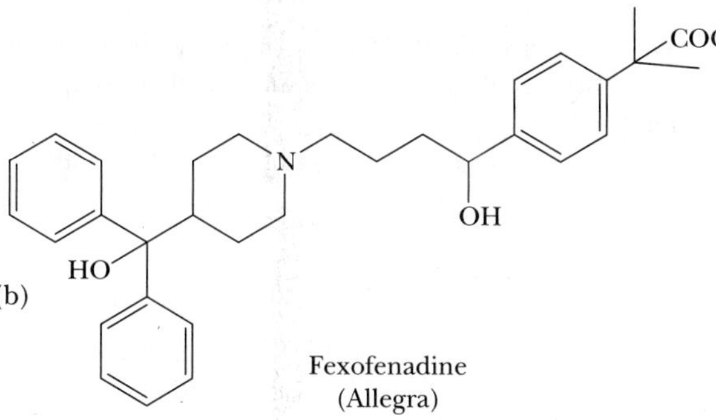

HO N OH COOH

Fexofenadine
(Allegra)

How many stereoisomers are possible for each compound?

***6.34** Following are structural formulas for three of the most widely prescribed drugs used to treat depression. Label all stereocenters in each compound and tell how many stereoisomers are possible for each compound. **(See Examples 6.5, 6.6)**

(a)

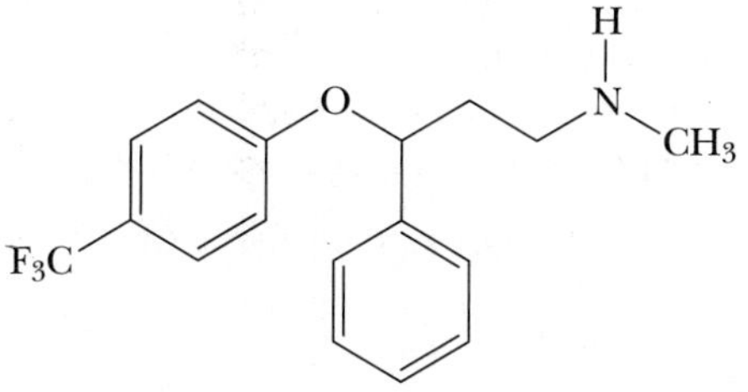

F_3C O H N CH_3

Fluoxetine
(Prozac®)

(b)

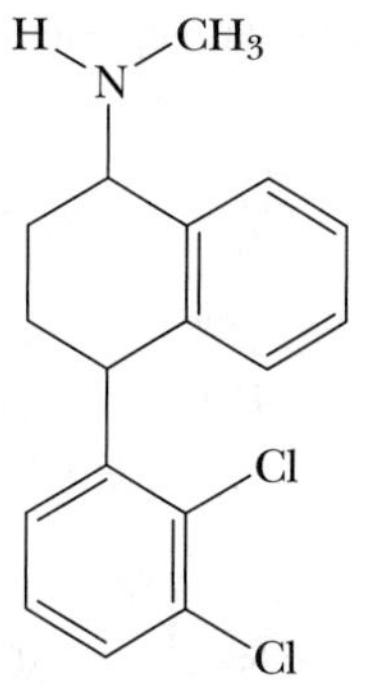

H N CH_3 Cl Cl

Sertraline
(Zoloft®)

(c)

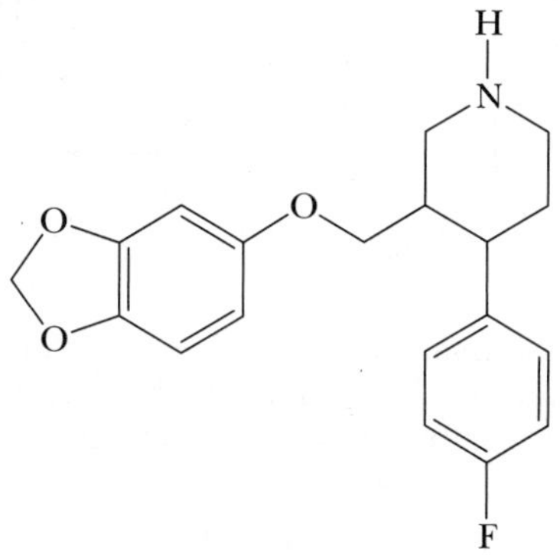

O O O H N F

Paroxetine
(Paxil®)

***6.35** Triamcinolone acetonide, the active ingredient in Azmacort® Inhalation Aerosol, is a steroid used to treat bronchial asthma: **(See Examples 6.5, 6.6)**

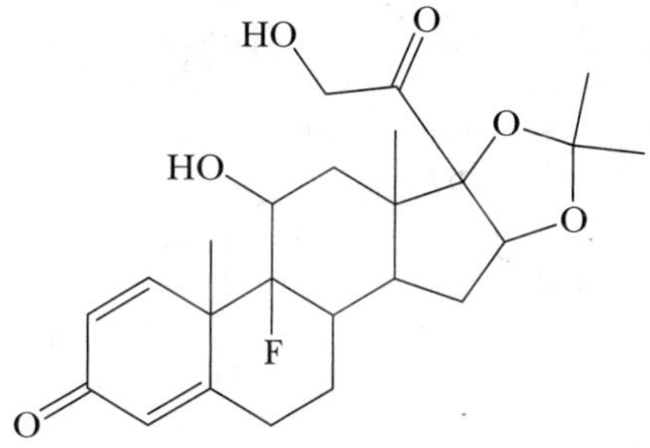

HO O HO O O F O

Triamcinolone acetonide

(a) Label the eight stereocenters in this molecule.

(b) How many stereoisomers are possible for the molecule? (Of this number, only one is the active ingredient in Azmacort.)

6.36 Which of these structural formulas represent meso compounds? **(See Example 6.4)**

(a) 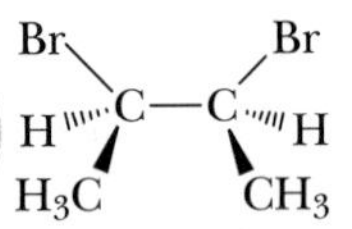 (b) 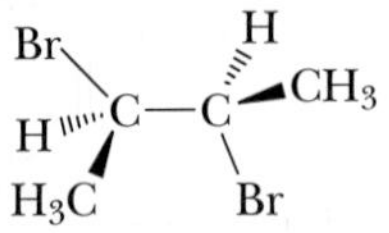 (c) 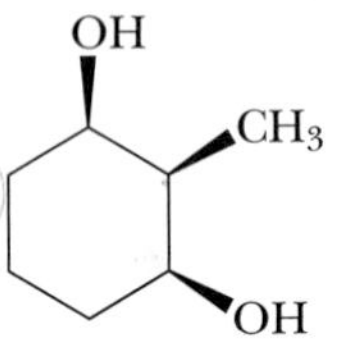 (d) 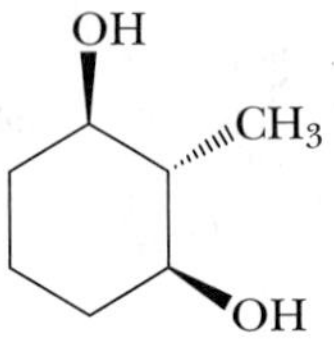 (e) 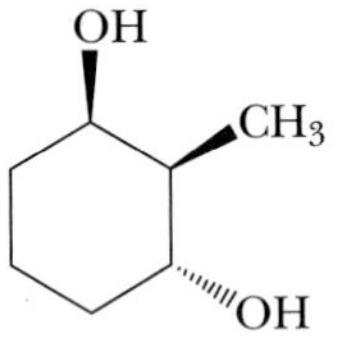 (f)

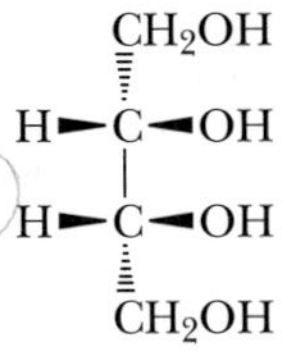

6.37 Draw a Newman projection, viewed along the bond between carbons 2 and 3, for both the most stable and the least stable conformations of meso-tartaric acid:

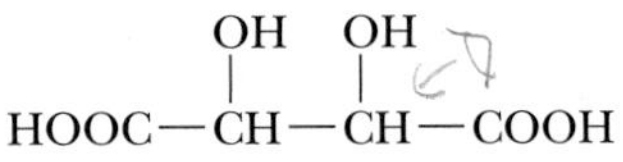

HOOC—CH(OH)—CH(OH)—COOH

6.38 How many stereoisomers are possible for 1,3-dimethylcyclopentane? Which are pairs of enantiomers? Which are meso compounds? **(See Examples 6.4–6.6)**

6.39 In Problem 3.59, you were asked to draw the more stable chair conformation of glucose, a molecule in which all groups on the six-membered ring are equatorial: **(See Examples 6.2, 6.5, 6.6)**

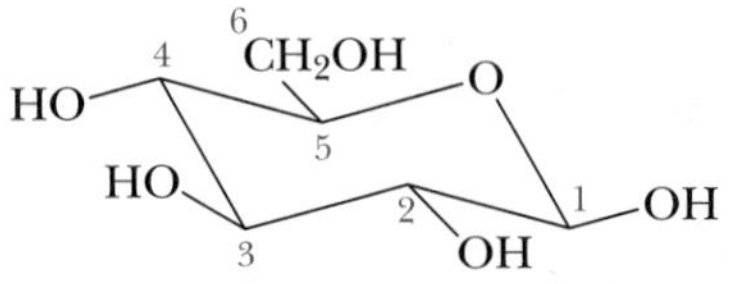

(a) Identify all stereocenters in this molecule.

(b) How many stereoisomers are possible?

(c) How many pairs of enantiomers are possible?

(d) What is the configuration (*R* or *S*) at carbons 1 and 5 in the stereoisomer shown?

6.40 What is a racemic mixture? Is a racemic mixture optically active? That is, will it rotate the plane of polarized light?

Chemical Transformations

6.41 Test your cumulative knowledge of the reactions learned so far by completing the following chemical transformations. Pay particular attention to the stereochemistry in the product. Where more than one stereoisomer is possible, show each stereoisomer. *Note that some transformations will require more than one step.*

(a)

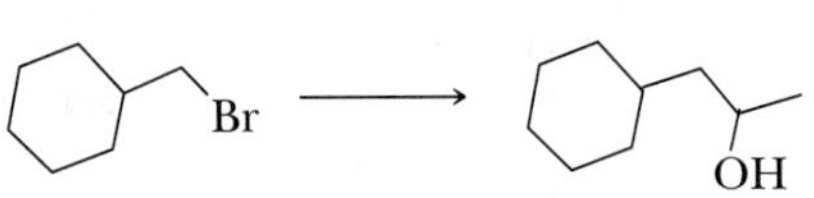

(b) 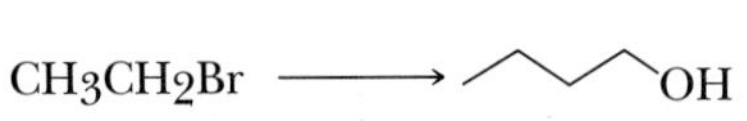 CH_3CH_2Br ⟶

(c)  H—≡—H ⟶

(d)

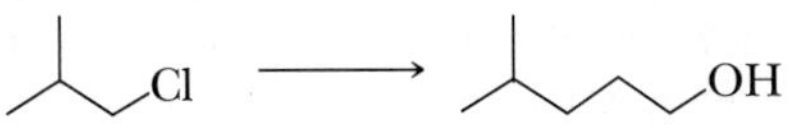

(e)

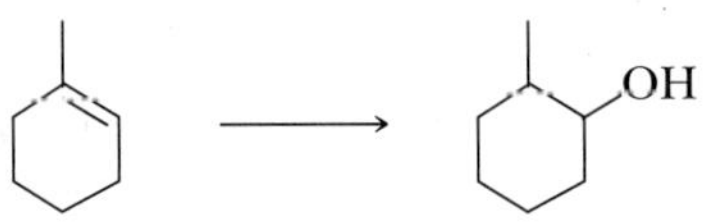

(f)

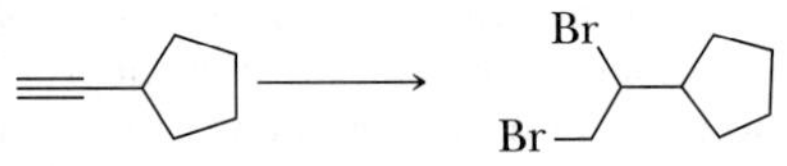

(g)  H_3C ⟶ H_3C

(h) $H_2C{=}CH_2$ ⟶

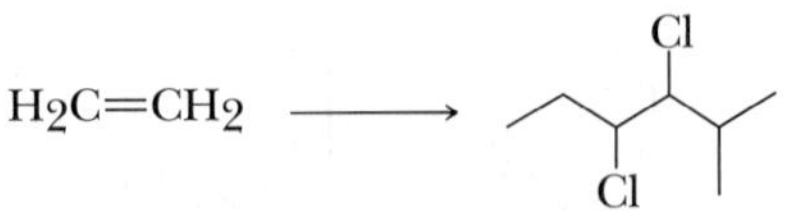

Looking Ahead

6.42 Predict the product(s) of the following reactions (in cases where more than one stereoisomer is possible, show each stereoisomer):

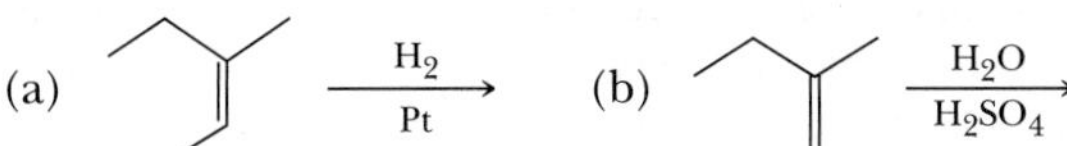

(e) CH_3 $\xrightarrow{HBr}$ (f) $\xrightarrow[Pt]{H_2}$

6.43 What alkene, when treated with H_2/Pd, will ensure a 100% yield of the stereoisomer shown?

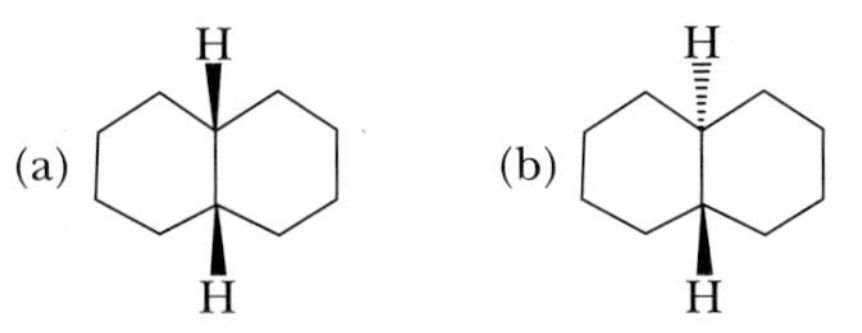

6.44 Which of the following reactions will yield a racemic mixture of products?

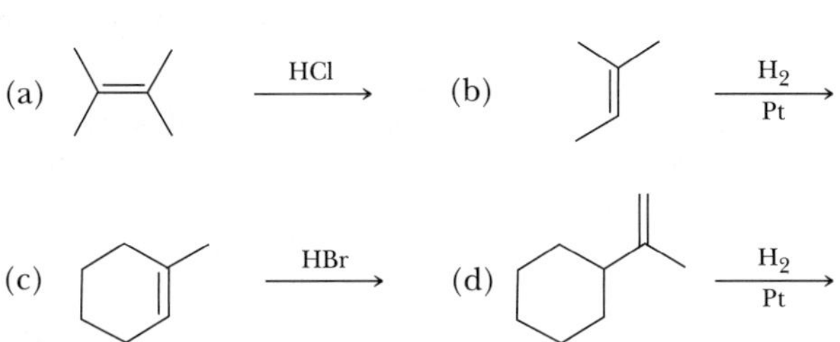

6.45 Draw all the stereoisomers that can be formed in the following reaction:

$\xrightarrow{HCl}$

Comment on the utility of this particular reaction as a synthetic method.

6.46 Explain why the product of the following reaction does not rotate the plane of polarized light:

(a) $\xrightarrow[CH_2Cl_2]{Br_2}$ (b) $\xrightarrow[Pt]{H_2}$

Putting It Together

The following problems bring together concepts and material from Chapters 4–6. Although the focus may be on these chapters, the problems will also build on concepts discussed thus far.

Choose the best answer for each of the following questions.

1. Which of the following will *not* rotate the plane of polarized light?
   (a) A 50:50 ratio of (*R*)-2-butanol and *cis*-2-butene.
   (b) A 70:20 ratio of (*R*)-2-butanol and *S*-2-butanol.
   (c) A 50:25:25 ratio of (*S*)-2-butanol, *cis*-2-butene, and *trans*-2-butene.
   (d) A 20:70 ratio of *trans*-2-butene and *cis*-2-butene.
   (e) None of the above (i.e., all of them will rotate plane-polarized light)

2. Which of the following *cis* isomers of dimethylcyclohexane is *not* meso?
   (a) *cis*-1,4-dimethylcyclohexane
   (b) *cis*-1,3-dimethylcyclohexane
   (c) *cis*-1,2-dimethylcyclohexane
   (d) All of the above (i.e., none of them is meso)
   (e) None of the above (i.e., all of them are meso)

3. How many products are possible in the following Lewis acid–base reaction?

   F—B(CH_3)(CH_2CH_3) + $:\ddot{O}H^-$ ⟶

   (a) One (b) Two (c) Three (d) Four
   (e) None (no reaction will take place)

4. What is the relationship between the following two molecules?

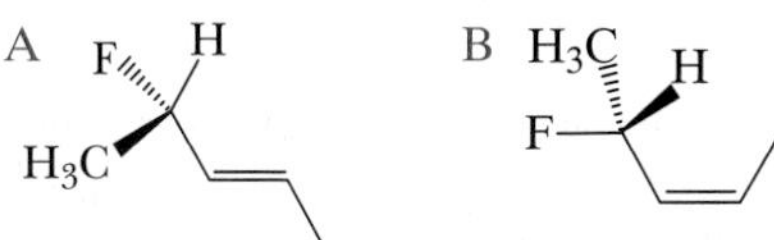

(a) They are identical.
(b) They are enantiomers.
(c) They are diastereomers.
(d) They are constitutional isomers.
(e) They are nonisomers.

5. Which stereoisomer of 2,4-hexadiene is the *least* stable?
(a) *Z,Z*-2,4-hexadiene
(b) *Z,E*-2,4-hexadiene
(c) *E,Z*-2,4-hexadiene
(d) *E,E*-2,4-hexadiene
(e) All are equal in stability.

6. Select the shortest C—C single bond in the following molecule.

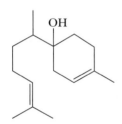

(a) a (b) b (c) c (d) d (e) e

7. Which of the following statements is true of β-bisabolol?

β-Bisabolol

(a) The compound is a terpene and possesses one isoprene unit.
(b) The compound is a terpene and possesses two isoprene units.
(c) The compound is a terpene and possesses three isoprene units.
(d) The compound is a terpene and possesses four isoprene units.
(e) The compound is not a terpene.

8. How many products are formed in the following reaction?

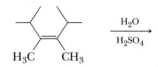

$\xrightarrow[H_2SO_4]{H_2O}$

(a) 1 (b) 2 (c) 3 (d) 4 (e) 5

9. Which of the following is *true* when two isomeric alkenes are treated with H_2/Pt?
(a) The alkene that releases more energy in the reaction is the more stable alkene.
(b) The alkene with the lower melting point will release less energy in the reaction.
(c) The alkene with the lower boiling point will release less energy in the reaction.
(d) Both alkenes will release equal amounts of energy in the reaction.
(e) None of these statements is true.

10. An unknown compound reacts with two equivalents of H_2 catalyzed by Ni. The unknown also yields 5 CO_2 and 4 H_2O upon combustion. Which of the following could be the unknown compound?

(a) (b) 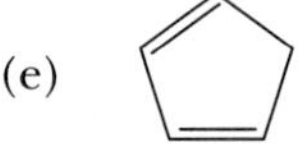

(c) (d)

(e)

11. Provide structures for all possible compounds of formula C_5H_6 that would react quantitatively with $NaNH_2$.

12. Answer the questions that follow regarding the following compound, which has been found in herbal preparations of *Echinacea*, the genus name for a variety of plants marketed for their immunostimulant properties.

(a) How many stereoisomers exist for the compound shown?
(b) Would you expect the compound to be soluble in water?
(c) Is the molecule chiral?
(d) What would be the product formed in the reaction of this compound with an excess amount of H_2/Pt?

13. Provide IUPAC names for the following compounds.

(a) (b)

(c)

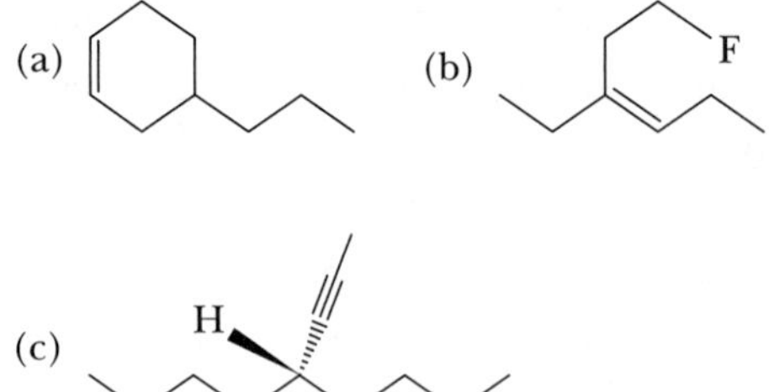

14. Compound **A** is an optically inactive compound with a molecular formula of C_5H_8. Catalytic hydrogenation of **A** gives an optically inactive compound, **B** (M.F. = C_5H_{10}), as the sole product. Furthermore, reaction of **A** with HBr results in a single compound, **C**, with a molecular formula of C_5H_9Br. Provide structures for **A**, **B**, and **C**.

15. An optically active compound, **A**, has a molecular formula of C_6H_{12}. Hydroboration–oxidation of **A** yields an optically active product, **B**, with a molecular formula of $C_6H_{14}O$. Catalytic hydrogenation of **A** yields an optically inactive product, **C**, with a molecular formula of C_6H_{14}. Propose structures for **A**, **B**, and **C**.

16. Based on the following hydrogenation data, which is more stable, the alkene **(A)** with the double bond outside of the ring or the alkene **(B)** with the double bond inside the ring? Use a reaction energy diagram to illustrate your point.

$\Delta H = -23.84$ kcal/mol

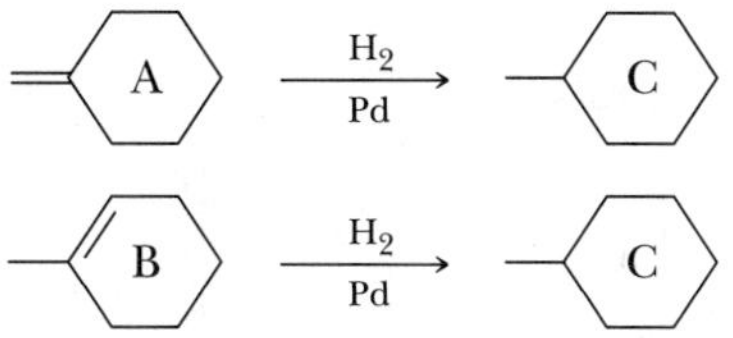

A $\xrightarrow[\text{Pd}]{H_2}$ C

B $\xrightarrow[\text{Pd}]{H_2}$ C

$\Delta H = -20.69$ kcal/mol

17. Explain whether the following pairs of compounds could be separated by resolution of enantiomers. If such separation is not possible, indicate so and explain your answer.

(a)

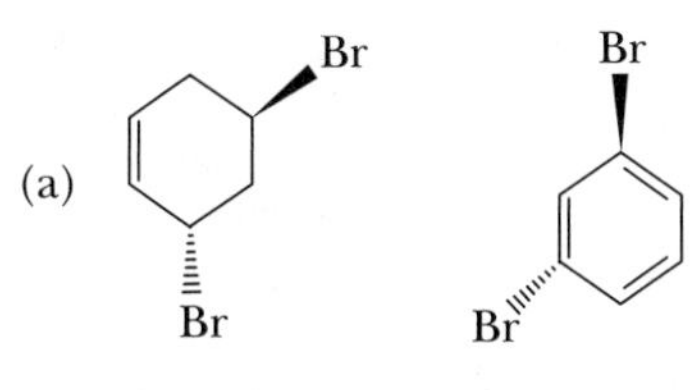

(b)

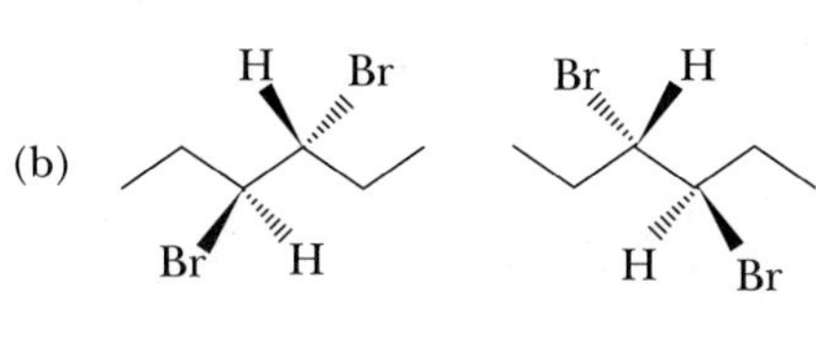

(c)

18. Predict whether solutions containing equal amounts of each pair of the structures shown would rotate the plane of polarized light.

(a)

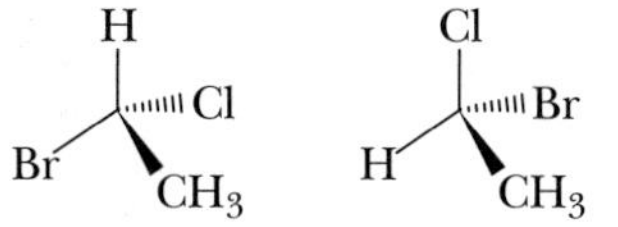

(b)

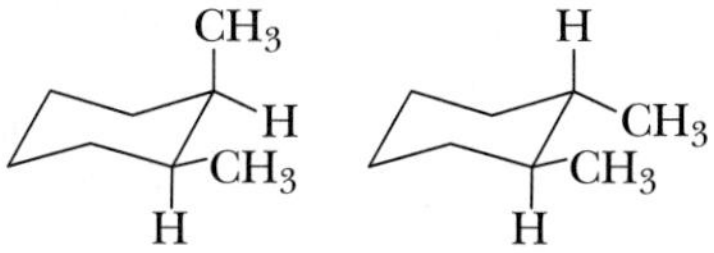

(c)

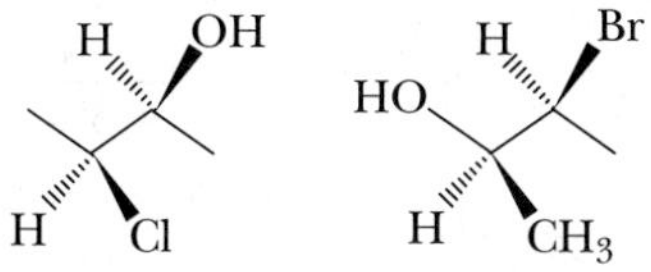

19. Complete the following chemical transformations.

(a)

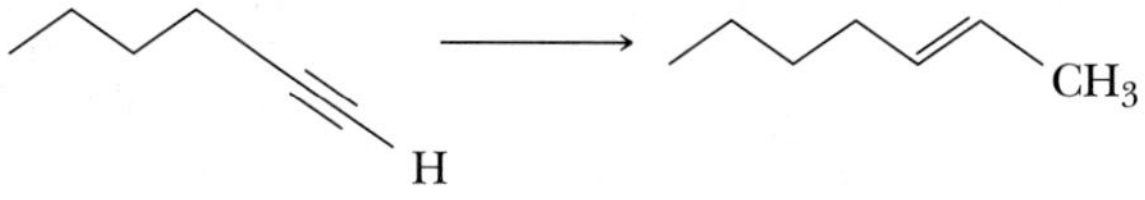

(b) Any alkyne 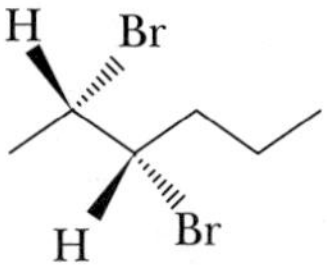 → + enantiomer

(c) Any alkene → +

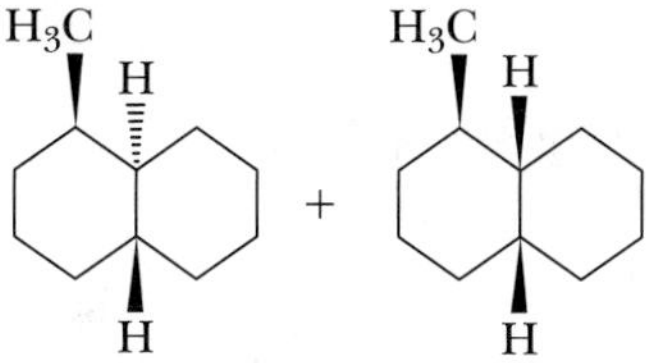

20. Provide a mechanism for the following series of reactions. Show all charges and lone pairs of electrons in your structures as well as the structures of all intermediates.

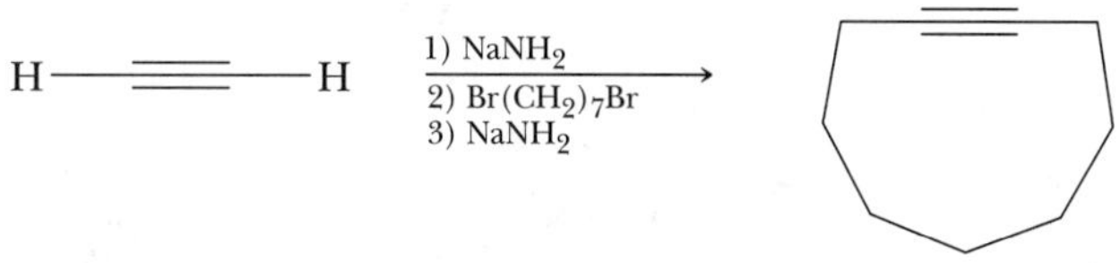

H—C≡C—H $\xrightarrow{\text{1) } NaNH_2 \text{ 2) } Br(CH_2)_7Br \text{ 3) } NaNH_2}$

21. Predict the major product or products of each of the following reactions. Be sure to consider stereochemistry in your answers.

(a) 1) $BH_3 \cdot THF$; 2) HOOH, NaOH, H_2O

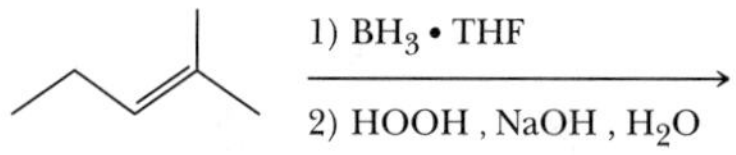

(b) H_2, Pt

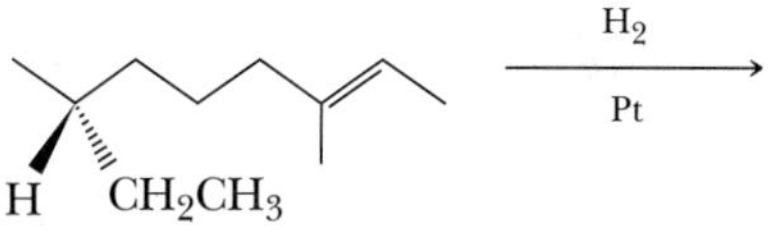

(c) Lindlar's Pd, H_2

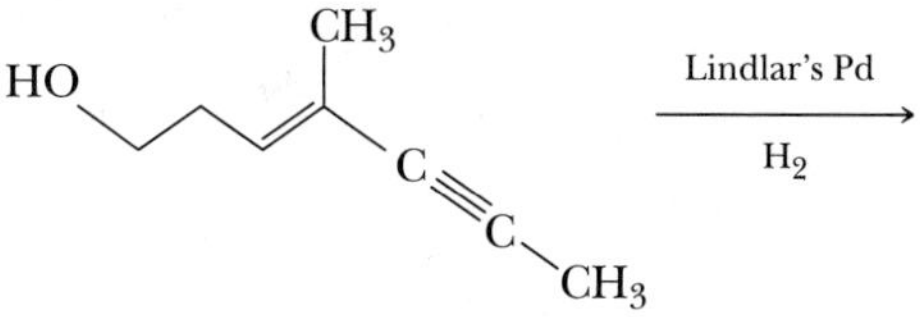

(d) HI

(e) 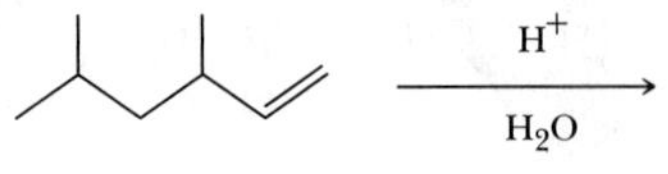 $\xrightarrow[H_2O]{H^+}$

(f) 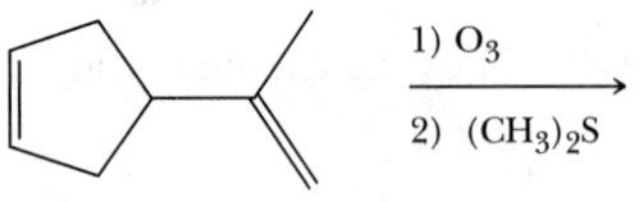 $\xrightarrow[2)\ (CH_3)_2S]{1)\ O_3}$

22. Provide a mechanism for the following reaction. Show all charges and lone pairs of electrons in your structures as well as the structures of all intermediates.

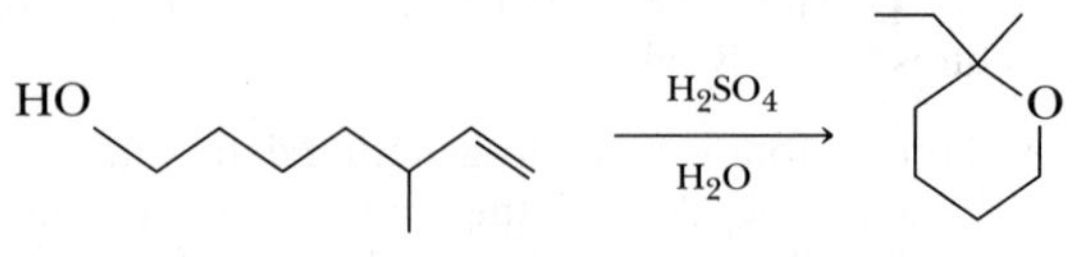

HO $\xrightarrow[H_2O]{H_2SO_4}$ O

7 Haloalkanes

KEY QUESTIONS

7.1 How Are Haloalkanes Named?

7.2 What Are the Characteristic Reactions of Haloalkanes?

7.3 What Are the Products of Nucleophilic Aliphatic Substitution Reactions?

7.4 What Are the S_N2 and S_N1 Mechanisms for Nucleophilic Substitution?

7.5 What Determines Whether S_N1 or S_N2 Predominates?

7.6 How Can S_N1 and S_N2 Be Predicted Based on Experimental Conditions?

7.7 What Are the Products of β-Elimination?

7.8 What Are the E1 and E2 Mechanisms for β-Elimination?

7.9 When Do Nucleophilic Substitution and β-Elimination Compete?

HOW TO

7.1 How to Name Cyclic Haloalkanes

7.2 How to Recognize Substitution and β-Elimination Reactions

7.3 How to Complete a Substitution Reaction

7.4 How to Predict the Type of Substitution Reaction a Haloalkane Will Undergo

7.5 How to Complete an Elimination Reaction

7.6 How to Draw Mechanisms

7.7 How to Predict the Type of β-Elimination Reaction a Haloalkane Will Undergo

CHEMICAL CONNECTIONS

7A The Environmental Impact of Chlorofluorocarbons

7B The Effect of Chlorofluorocarbon Legislation on Asthma Sufferers

Asthma patient using a metered-dose inhaler to deliver the drug albuterol. The drug is propelled by haloalkanes such as 1,1,1,2-tetrafluoroethane. Inset: A molecule of 1,1,1,2-tetrafluoroethane (HFA-134a). *(Carolyn A. McKeone/Photo Researchers, Inc.)*

ompounds containing a halogen atom covalently bonded to an sp^3 hybridized carbon atom are named **haloalkanes** or, in the common system of nomenclature, *alkyl halides*. The general symbol for an **alkyl halide** is R—X, where X may be F, Cl, Br, or I:

Alkyl halide A compound containing a halogen atom covalently bonded to an alkyl group; given the symbol RX.

$$R-\ddot{\underset{\cdot\cdot}{X}}:$$

A haloalkane (An alkyl halide)

In this chapter, we study two characteristic reactions of haloalkanes: nucleophilic substitution and β-elimination. Haloalkanes are useful molecules because they can be converted to alcohols, ethers, thiols, amines, and alkenes and are thus versatile molecules. Indeed, haloalkanes are often used as starting materials for the synthesis of many useful compounds encountered in medicine, food chemistry, and agriculture (to name a few).

7.1 How Are Haloalkanes Named?

A. IUPAC Names

IUPAC names for haloalkanes are derived by naming the parent alkane according to the rules given in Section 3.3A:

- Locate and number the parent chain from the direction that gives the substituent encountered first the lower number.
- Show halogen substituents by the prefixes *fluoro*-, *chloro*-, *bromo*-, and *iodo*-, and list them in alphabetical order along with other substituents.
- Use a number preceding the name of the halogen to locate each halogen on the parent chain.
- In haloalkenes, the location of the double bond determines the numbering of the parent hydrocarbon. In molecules containing functional groups designated by a suffix (for example, *-ol*, *-al*, *-one*, *-oic acid*), the location of the functional group indicated by the suffix determines the numbering:

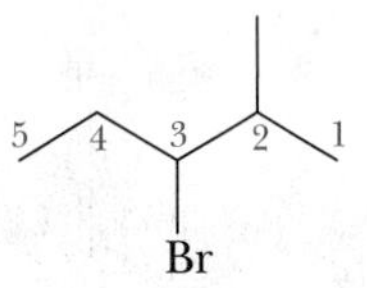

3-Bromo-2-methylpentane 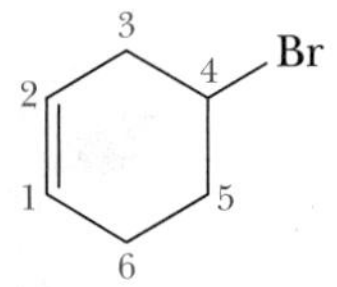   4-Bromocyclohexene 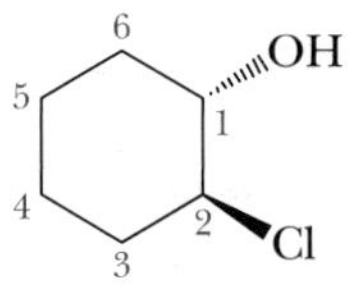   *trans*-2-Chlorocyclohexanol

B. Common Names

Common names of haloalkanes consist of the common name of the alkyl group, followed by the name of the halide as a separate word. Hence, the name **alkyl halide** is a common name for this class of compounds. In the following examples, the IUPAC name of the compound is given first, followed by its common name, in parentheses:

$CH_3CHFCH_2CH_3$	$CH_2{=}CHCl$
2-Fluorobutane	Chloroethene
(*sec*-Butyl fluoride)	(Vinyl chloride)

Several of the polyhalomethanes are common solvents and are generally referred to by their common, or trivial, names. Dichloromethane (methylene chloride) is the most widely used haloalkane solvent. Compounds of the type CHX_3 are called **haloforms**. The common name for $CHCl_3$, for example, is *chloroform*. The common name for CH_3CCl_3 is *methyl chloroform*. Methyl chloroform and trichloroethylene are solvents for commercial dry cleaning.

CH_2Cl_2	$CHCl_3$	CH_3CCl_3	$CCl_2{=}CHCl$
Dichloromethane	Trichloromethane	1,1,1-Trichloroethane	Trichloroethene
(Methylene chloride)	(Chloroform)	(Methyl chloroform)	(Trichlor)

HOW TO 7.1 Name Cyclic Haloalkanes

(a) First, determine the root name of the cycloalkane.

the root name of a 5-carbon ring is "cyclopentane"
Cl
Br

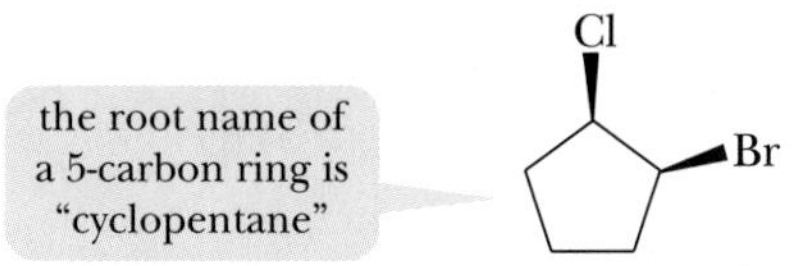

Cyclopentane

(b) Name and number the halogen substituents.

in rings, multiple halogens are listed alphabetically and numbered in alphabetical order
Cl
Br

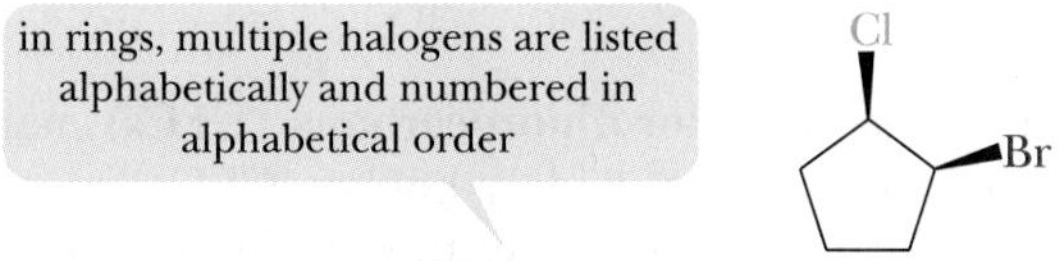

1 -Bromo-2-chlorocyclopentane

(c) Don't forget to indicate stereochemistry.

see Chapter 6 for rules on assigning R/S configurations
Cl
R
S
Br

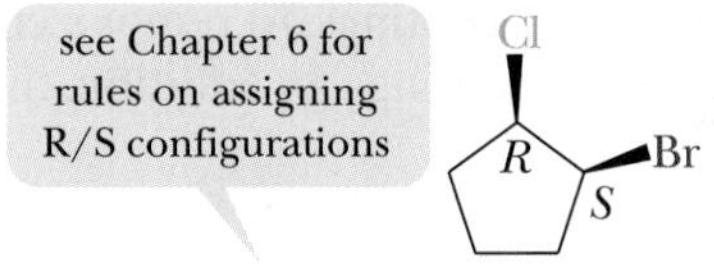

(1*S*,2*R*)-1-Bromo-2-chlorocyclopentane

or

cis-1-Bromo-2-chlorocyclopentene

Example 7.1

Write the IUPAC name for each compound:

(a) Br (b) Br (c) H Br (d) F I

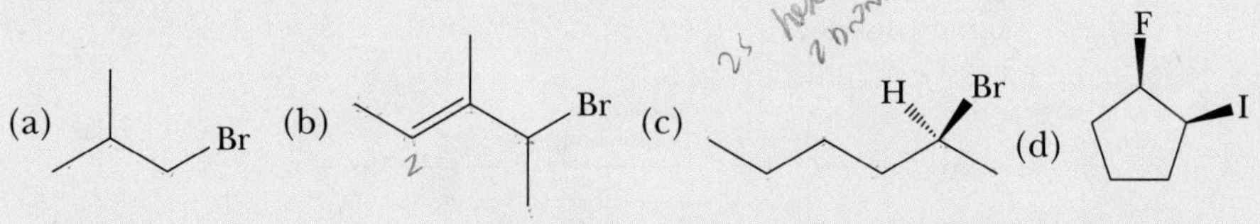

Strategy

First look for the longest chain of carbons. This will allow you to determine the root name. Then identify the atoms or groups of atoms that are not part of that chain of carbons. These are your substituents.

Solution

(a) 1-Bromo-2-methylpropane. Its common name is isobutyl bromide.
(b) (*E*)-4-Bromo-3-methyl-2-pentene.
(c) (*S*)-2-Bromohexane.
(d) (1*R*,2*S*)-1-Fluoro-2-iodocyclopentane or *cis*-1-fluoro-2-iodocyclopentanol.

See problems 7.9–7.12

Problem 7.1

Write the IUPAC name for each compound:

(a) 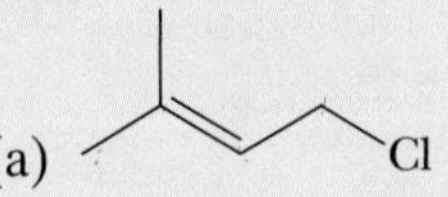 [structure: Cl] (b) 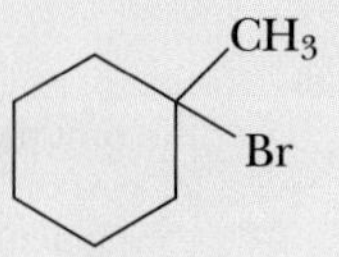 [structure: CH_3, Br]

(c) CH_3CHCH_2Cl with Cl on C-2 (d) 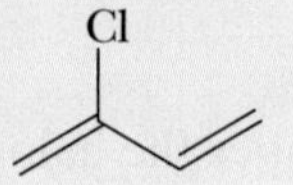 [structure: Cl]

Of all the haloalkanes, the **chlorofluorocarbons (CFCs)** manufactured under the trade name Freon® are the most widely known. CFCs are nontoxic, nonflammable, odorless, and noncorrosive. Originally, they seemed to be ideal replacements for the hazardous compounds such as ammonia and sulfur dioxide formerly used as heat-transfer agents in refrigeration systems. Among the CFCs most widely used for this purpose were trichlorofluoromethane (CCl_3F, Freon-11) and dichlorodifluoromethane (CCl_2F_2, Freon-12). The CFCs also found wide use as industrial cleaning solvents to prepare surfaces for coatings, to remove cutting oils and waxes from millings, and to remove protective coatings. In addition, they were employed as propellants in aerosol sprays.

7.2 What Are the Characteristic Reactions of Haloalkanes?

Nucleophile An atom or a group of atoms that donates a pair of electrons to another atom or group of atoms to form a new covalent bond.

Nucleophilic substitution A reaction in which one nucleophile is substituted for another.

A **nucleophile** (nucleus-loving reagent) is any reagent that donates an unshared pair of electrons to form a new covalent bond. **Nucleophilic substitution** is any reaction in which one nucleophile is substituted for another. In the following general equations, Nu:⁻ is the nucleophile, X is the leaving group, and substitution takes place on an sp^3 hybridized carbon atom:

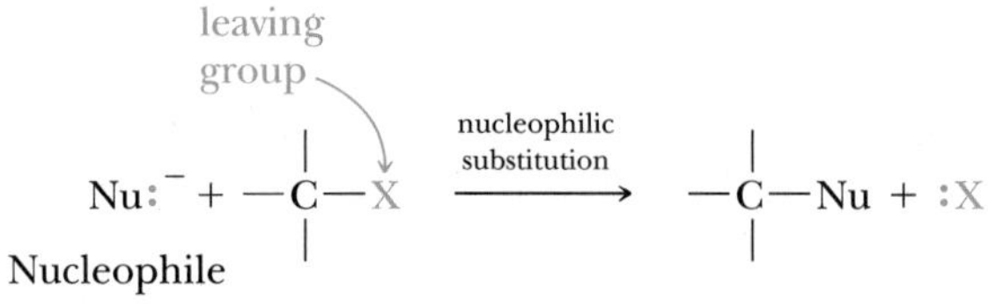

$$\text{Nu:}^- + -\overset{|}{\underset{|}{C}}-X \xrightarrow{\text{nucleophilic substitution}} -\overset{|}{\underset{|}{C}}-\text{Nu} + :X^-$$

Nucleophile; leaving group

Halide ions are among the best and most important leaving groups. Recall from Section 5.9 that nucleophilic substitution occurs in the alkylation of an acetylide ion:

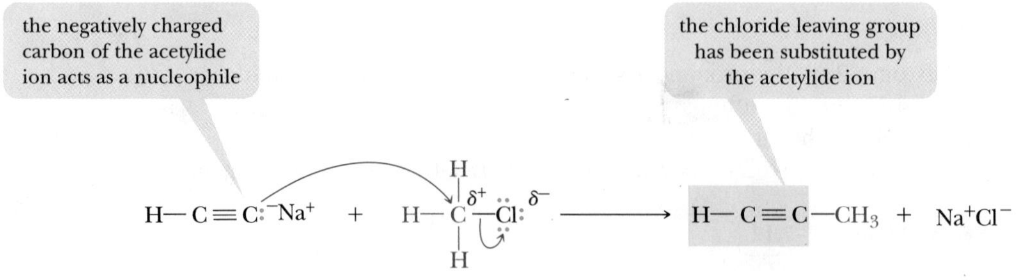

the negatively charged carbon of the acetylide ion acts as a nucleophile

the chloride leaving group has been substituted by the acetylide ion

$$H-C\equiv C:^-Na^+ + H-\overset{H}{\underset{H}{C}}{}^{\delta+}-\ddot{Cl}:^{\delta-} \longrightarrow H-C\equiv C-CH_3 + Na^+Cl^-$$

CHEMICAL CONNECTIONS 7A

The Environmental Impact of Chlorofluorocarbons

Concern about the environmental impact of CFCs arose in the 1970s when researchers found that more than 4.5×10^5 kg/yr of these compounds were being emitted into the atmosphere. In 1974, Sherwood Rowland and Mario Molina announced their theory, which has since been amply confirmed, that CFCs catalyze the destruction of the stratospheric ozone layer. When released into the air, CFCs escape to the lower atmosphere. Because of their inertness, however, they do not decompose there. Slowly, they find their way to the stratosphere, where they absorb ultraviolet radiation from the sun and then decompose. As they do so, they set up a chemical reaction that leads to the destruction of the stratospheric ozone layer, which shields the Earth against short-wavelength ultraviolet radiation from the sun. An increase in short-wavelength ultraviolet radiation reaching the Earth is believed to promote the destruction of certain crops and agricultural species and even to increase the incidence of skin cancer in light-skinned individuals.

The concern about CFCs prompted two conventions, one in Vienna in 1985 and one in Montreal in 1987, held by the United Nations Environmental Program. The 1987 meeting produced the Montreal Protocol, which set limits on the production and use of ozone-depleting CFCs and urged the complete phaseout of their production by the year 1996. Only two members of the UN have failed to ratify the protocol in its original form.

Rowland, Molina, and Paul Crutzen (a Dutch chemist at the Max Planck Institute for Chemistry in Germany) were awarded the 1995 Nobel Prize for chemistry. As the Royal Swedish Academy of Sciences noted in awarding the prize, "By explaining the chemical mechanisms that affect the thickness of the ozone layer, these three researchers have contributed to our salvation from a global environmental problem that could have catastrophic consequences."

The chemical industry responded to the crisis by developing replacement refrigerants that have a much lower ozone-depleting potential. The most prominent replacements are the hydrofluorocarbons (HFCs) and hydrochlorofluorocarbons (HCFCs), such as the following:

CF_3CH_2F (F–C(F)(F)–C(F)(H)–H)
HFC-134a

CH_3CCl_2F (H–C(H)(H)–C(Cl)(Cl)–F)
HCFC-141b

These compounds are much more chemically reactive in the atmosphere than the Freons are and are destroyed before they reach the stratosphere. However, they cannot be used in air conditioners in 1994 and earlier model cars.

QUESTION

Provide IUPAC names for HFC-134a and HCFC-141b.

Because all nucleophiles are also bases, nucleophilic substitution and base-promoted **β-elimination** are competing reactions. The ethoxide ion, for example, is both a nucleophile and a base. With bromocyclohexane, it reacts as a nucleophile (pathway shown in red) to give ethoxycyclohexane (cyclohexyl ethyl ether) and as a base (pathway shown in blue) to give cyclohexene and ethanol:

β-Elimination reaction The removal of atoms or groups of atoms from two adjacent carbon atoms, as for example, the removal of H and X from an alkyl halide or H and OH from an alcohol to form a carbon–carbon double bond.

as a nucleophile, ethoxide ion attacks this carbon

Bromocyclohexane + $CH_3CH_2O^-Na^+$ (a nucleophile and a base)

as a base, ethoxide ion attacks this hydrogen

nucleophilic substitution, ethanol → cyclohexyl–OCH_2CH_3 + Na^+Br^-

β-elimination, ethanol → cyclohexene + CH_3CH_2OH + Na^+Br^-

HOW TO 7.2 Recognize Substitution and β-Elimination Reactions

(a) Substitution reactions always result in the replacement of one atom or group of atoms in a reactant with another atom or group of atoms.

(b) β-Elimination reactions always result in the removal of a hydrogen and an atom or group of atoms on adjacent carbon atoms and in the formation of a C—C double bond.

In this chapter, we study both of these organic reactions. Using them, we can convert haloalkanes to compounds with other functional groups including alcohols, ethers, thiols, sulfides, amines, nitriles, alkenes, and alkynes. Thus, an understanding of nucleophilic substitution and β-elimination opens entirely new areas of organic chemistry.

Example 7.2

Determine whether the following haloalkanes underwent substitution, elimination, or both substitution and elimination:

(a) Br $\xrightarrow[H_2S]{\overset{+}{Na}\overset{-}{S}H}$ SH + NaBr

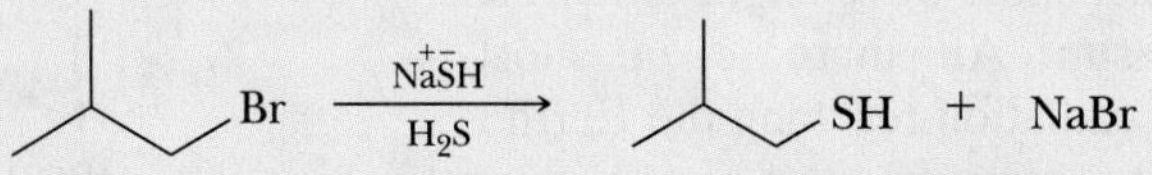

(b) Cl $\xrightarrow{CH_3CH_2OH}$ OCH_2CH_3 + OCH_2CH_3 + HCl

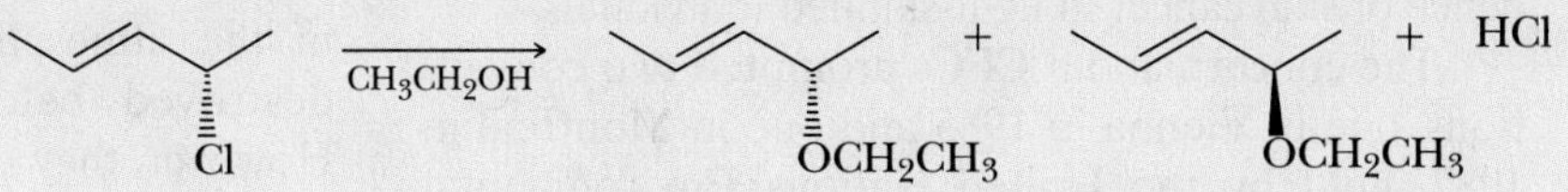

(c) I $\xrightarrow[\text{acetone}]{\overset{+}{Na}\overset{-}{O}H}$ + H_2O + NaI

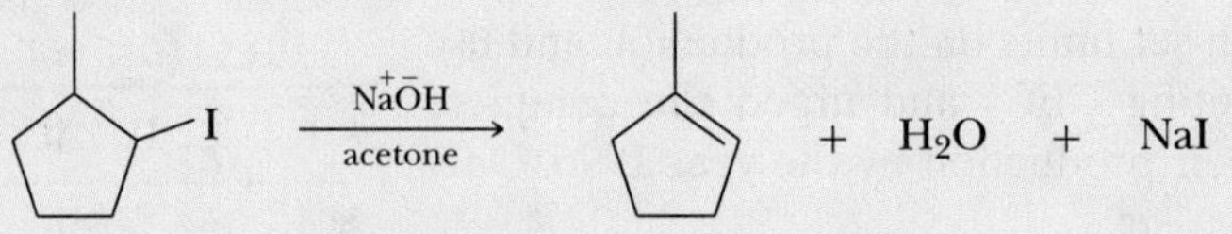

Strategy

Look for the halogen in the reactant. Has it been replaced by a different atom or group of atoms in the product(s)? If so, the reaction was a substitution reaction. If the carbon that was once attached to the halogen is now part of a C—C double bond in the product(s), an elimination reaction occurred.

Solution

(a) Substitution; the bromine was replaced by a thiol group.
(b) Substitution; in both products, an ethoxyl group replaces chlorine.
(c) β-Elimination; a hydrogen atom and an iodo group have been removed, and an alkene forms as a result.

Problem 7.2

Determine whether the following haloalkanes underwent substitution, elimination, or both substitution and elimination:

(a) Br; $\overset{+\ -}{K}\overset{O}{\overset{\|}{OCCH_3}}$ / CH_3COOH → O, O; + KBr

(b) Cl; $\overset{+\ -}{NaOCH_2CH_3}$ / CH_3CH_2OH → OCH_2CH_3; + ; + NaCl

7.3 What Are the Products of Nucleophilic Aliphatic Substitution Reactions?

Nucleophilic substitution is one of the most important reactions of haloalkanes and can lead to a wide variety of new functional groups, several of which are illustrated in Table 7.1. As you study the entries in this table, note the following points:

1. If the nucleophile is negatively charged, as, for example, OH^- and RS^-, then the atom donating the pair of electrons in the substitution reaction becomes neutral in the product.
2. If the nucleophile is uncharged, as, for example, NH_3 and CH_3OH, then the atom donating the pair of electrons in the substitution reaction becomes positively charged in the product. The products then often undergo a second step involving proton transfer to yield a neutral substitution product.

TABLE 7.1 Some Nucleophilic Substitution Reactions

Reaction: $Nu:^- + CH_3X \longrightarrow CH_3Nu + :X^-$

Nucleophile		Product	Class of Compound Formed
$HO:^-$	⟶	CH_3OH	An alcohol
$RO:^-$	⟶	CH_3OR	An ether
$HS:^-$	⟶	CH_3SH	A thiol (a mercaptan)
$RS:^-$	⟶	CH_3SR	A sulfide (a thioether)
$:I:^-$	⟶	$CH_3I:$	An alkyl iodide
$:NH_3$	⟶	$CH_3NH_3^+$	An alkylammonium ion
HOH	⟶	$CH_3\overset{+}{O}{-}H$ (with H on O)	An alcohol (after proton transfer)
CH_3OH	⟶	$CH_3\overset{+}{O}{-}CH_3$ (with H on O)	An ether (after proton transfer)

notice that a nucleophile does not need to be negatively charged

HOW TO 7.3 Complete a Substitution Reaction

(a) Identify the leaving group.

leaving group

Cyclohexyl–Br $\xrightarrow[\text{HOCH}_3]{\overset{+\,-}{\text{NaOCH}_3}}$

(b) Identify the nucleophile and its nucleophilic atom. The nucleophilic atom will be the negatively charged atom or the atom with a lone pair of electrons to donate. If both a negatively charged atom and an uncharged atom with a lone pair of electrons exist, the negatively charged atom will be the more nucleophilic atom. In the following example, $NaOCH_3$ is a better nucleophile than $HOCH_3$.

Cyclohexyl–Br $\xrightarrow[\text{HOCH}_3]{\overset{+\,-}{\text{NaOCH}_3}}$

this oxygen is uncharged

(c) Replace the leaving group in the reactant with the nucleophilic atom. Any groups connected to the nucleophilic atom through covalent bonds will remain bonded to that atom in the product.

sodium was not covalently bound to the oxygen and is not involved in the substitution reaction

the oxygen and any groups covalently bound to it replace the bromine

Cyclohexyl–Br $\xrightarrow[\text{HOCH}_3]{\overset{+\,-}{\text{NaOCH}_3}}$ Cyclohexyl–OCH_3

(d) Spectator ions will usually be shown as part of an ion pair with the negatively charged leaving group.

show the negatively charged leaving group and the spectator cation as an ion pair

Cyclohexyl–Br $\xrightarrow[\text{HOCH}_3]{\overset{+\,-}{\text{NaOCH}_3}}$ Cyclohexyl–OCH_3 + NaBr

Example 7.3

Complete these nucleophilic substitution reactions:

(a) $\sim\sim$Br + Na^+OH^- $\longrightarrow$ (b) $\sim\sim$Cl + NH_3 $\longrightarrow$

Strategy

First identify the nucleophile. Then break the bond between the halogen and the carbon it is bonded to and create a new bond from that same carbon to the nucleophile.

Solution

(a) Hydroxide ion is the nucleophile, and bromine is the leaving group:

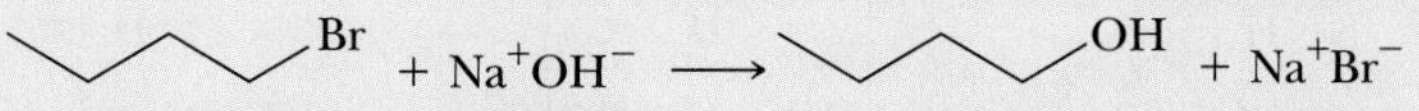

1-Bromobutane + Na^+OH^- (Sodium hydroxide) $\longrightarrow$ 1-Butanol + Na^+Br^- (Sodium bromide)

(b) Ammonia is the nucleophile, and chlorine is the leaving group:

1-Chlorobutane + NH_3 (Ammonia) $\longrightarrow$ Butylammonium chloride ($NH_3^+\ Cl^-$)

? why is Cl⁻ still attached?

See problems 7.21, 7.22, 7.26

Problem 7.3

Complete these nucleophilic substitution reactions:

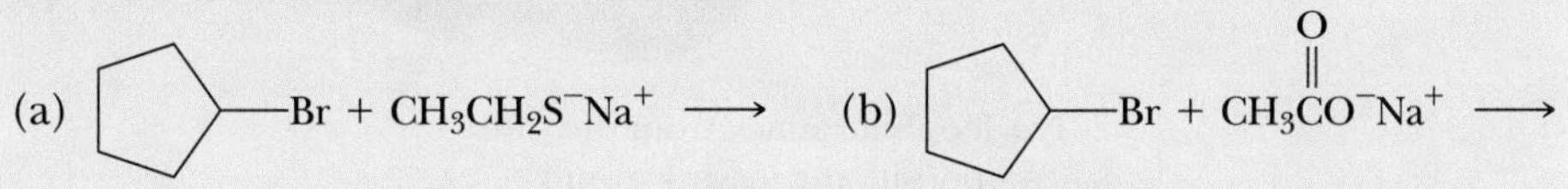

(a) Cyclopentyl—Br + $CH_3CH_2S^-Na^+$ $\longrightarrow$ (b) Cyclopentyl—Br + $CH_3CO^-Na^+$ (with C=O) $\longrightarrow$

7.4 What Are the S_N2 and S_N1 Mechanisms for Nucleophilic Substitution?

On the basis of a wealth of experimental observations developed over a 70-year period, chemists have proposed two limiting mechanisms for nucleophilic substitutions. A fundamental difference between them is the timing of bond breaking between carbon and the leaving group and of bond forming between carbon and the nucleophile.

A. S_N2 Mechanism

At one extreme, the two processes are *concerted*, meaning that bond breaking and bond forming occur simultaneously. Thus, the departure of the leaving group is assisted by the incoming nucleophile. This mechanism is designated **S_N2**, where *S* stands for *S*ubstitution, *N* for *N*ucleophilic, and 2 for a ***bi*molecular reaction**. This type of substitution reaction is classified as bimolecular because both the haloalkane and the nucleophile are involved in the rate-determining step. That is, both species contribute to the rate law of the reaction:

Bimolecular reaction A reaction in which two species are involved in the reaction leading to the transition state of the rate-determining step.

$$\text{Rate} = k[\text{haloalkane}][\text{nucleophile}]$$

Following is an S_N2 mechanism for the reaction of hydroxide ion and bromomethane to form methanol and bromide ion:

Inversion of configuration The reversal of the arrangement of atoms or groups of atoms about a reaction center in an S_N2 reaction.

Mechanism: An S_N2 Reaction

The nucleophile attacks the reactive center from the side opposite the leaving group; that is, an S_N2 reaction involves a backside attack by the nucleophile.

$$HO^- + CH_3Br \longrightarrow \left[\overset{\delta-}{HO}\cdots CH_3 \cdots \overset{\delta-}{Br} \right] \longrightarrow HO{-}CH_3 + Br^-$$

note inversion of the reactive center

Reactants — Transition state with simultaneous bond breaking and bond forming — Products

Figure 7.1 shows an energy diagram for an S_N2 reaction. There is a single transition state and no reactive intermediate.

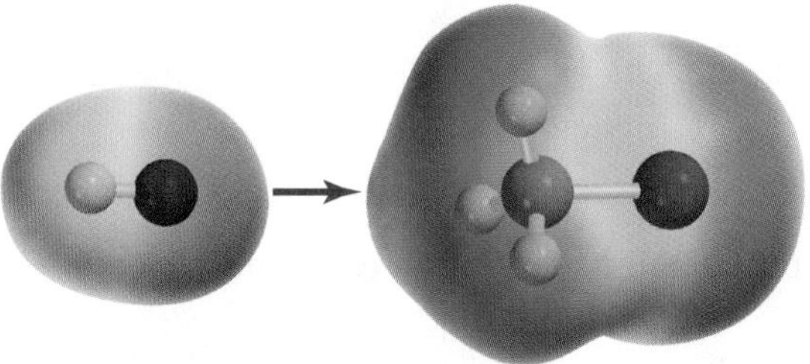

Nucleophilic attack from the side opposite the leaving group

An S_N2 reaction is driven by the attraction between the negative charge of the nucleophile (in this case the negatively charged oxygen of the hydroxide ion) and the center of positive charge of the electrophile (in this case the partial positive charge on the carbon bearing the bromine leaving group).

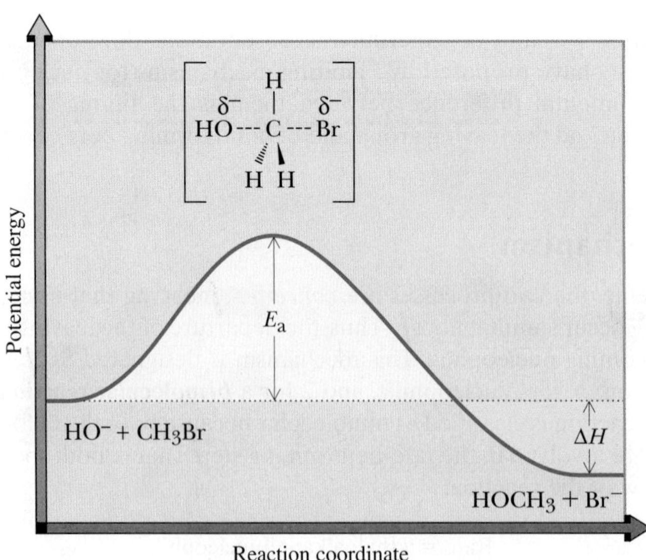

Figure 7.1
An energy diagram for an S_N2 reaction. There is one transition state and no reactive intermediate.

B. S_N1 Mechanism

In the other limiting mechanism, called S_N1, bond breaking between carbon and the leaving group is completed before bond forming with the nucleophile begins. In the designation $\mathbf{S_N1}$, *S* stands for *S*ubstitution, *N* stands for *N*ucleophilic, and *1* stands for a ***uni*molecular reaction**. This type of substitution is classified as unimolecular because only the haloalkane is involved in the rate-determining step; that is, only the haloalkane contributes to the rate law governing the rate-determining step:

Unimolecular reaction A reaction in which only one species is involved in the reaction leading to the transition state of the rate-determining step.

$$\text{Rate} = k[\text{haloalkane}]$$

An S_N1 reaction is illustrated by the **solvolysis** reaction of 2-bromo-2-methylpropane (*tert*-butyl bromide) in methanol to form 2-methoxy-2-methylpropane (*tert*-butyl methyl ether). You may notice that the second step of the mechanism is identical to the second step of the mechanism for the addition of hydrogen halides (H—X) to alkenes (Section 5.3A) and the acid-catalyzed hydration of alkenes (Section 5.3B).

Solvolysis A nucleophilic substitution reaction in which the solvent is the nucleophile.

Mechanism: An S_N1 Reaction

Step 1: The ionization of a C—X bond forms a 3° carbocation intermediate:

$(CH_3)_3C-Br \xrightarrow{\text{slow, rate determining}} (CH_3)_3C^+ + Br^-$

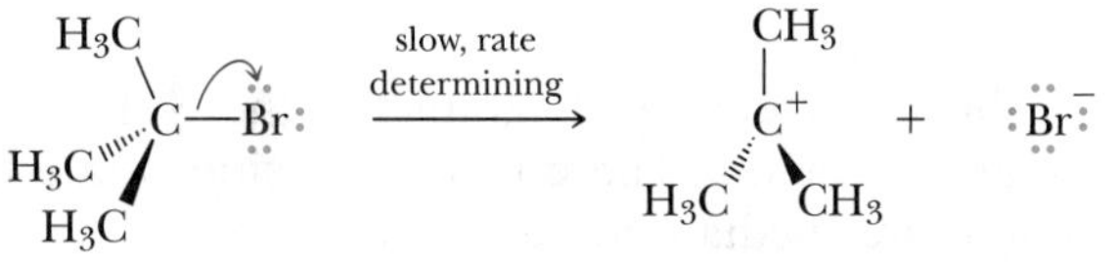

A carbocation intermediate; carbon is trigonal planar

Step 2: Reaction of the carbocation intermediate (an electrophile) with methanol (a nucleophile) gives an oxonium ion. Attack by the nucleophile occurs with equal probability from either face of the planar carbocation intermediate.

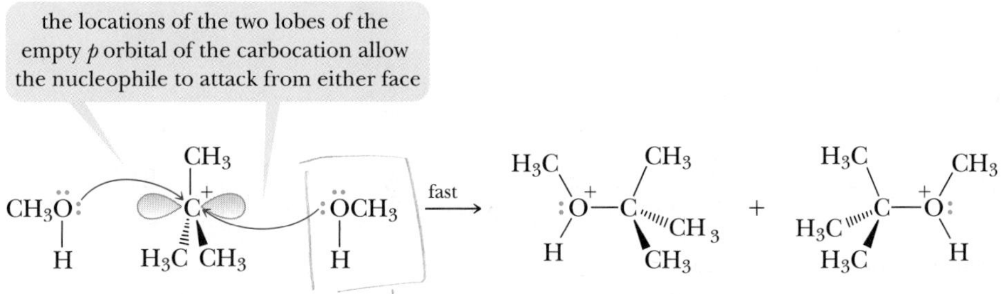

the locations of the two lobes of the empty *p* orbital of the carbocation allow the nucleophile to attack from either face

$CH_3OH + (CH_3)_3C^+ + HOCH_3 \xrightarrow{\text{fast}} (CH_3)_3C-\overset{+}{O}(H)CH_3 + (CH_3)_3C-\overset{+}{O}(H)CH_3$

Step 3: Proton transfer from the oxonium ion to methanol (the solvent) completes the reaction and gives *tert*-butyl methyl ether:

$(CH_3)_3C-\overset{+}{O}(H)CH_3 + H-O-CH_3 \overset{\text{fast}}{\rightleftharpoons} (CH_3)_3C-O-CH_3 + H-\overset{+}{O}(H)-CH_3$

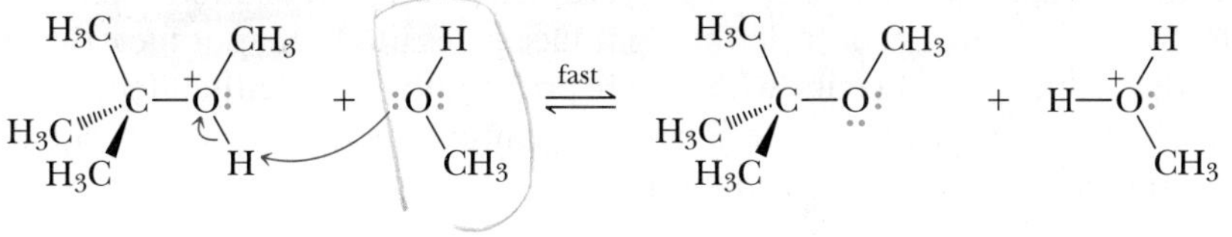

Figure 7.2
An energy diagram for the S_N1 reaction of 2-bromo-2-methylpropane and methanol. There is one transition state leading to formation of the carbocation intermediate in Step 1 and a second transition state for the reaction of the carbocation intermediate with methanol in Step 2. Step 1 crosses the higher energy barrier and is, therefore, rate determining.

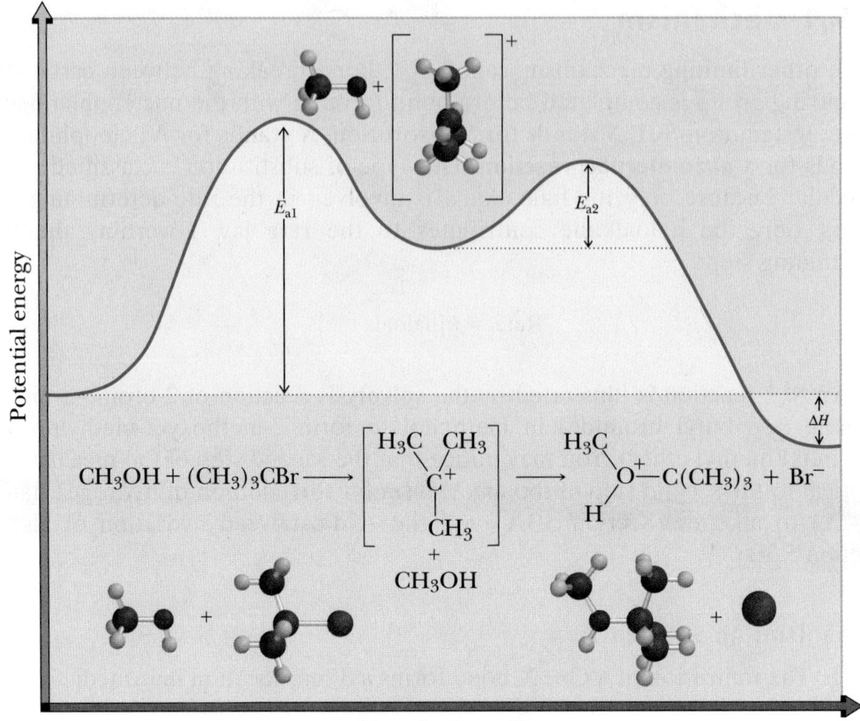

Figure 7.2 shows an energy diagram for the S_N1 reaction of 2-bromo-2-methylpropane and methanol. There is one transition state leading to formation of the carbocation intermediate in Step 1 and a second transition state for reaction of the carbocation intermediate with methanol in Step 2 to give the oxonium ion. The reaction leading to formation of the carbocation intermediate crosses the higher energy barrier and is, therefore, the rate-determining step.

If an S_N1 reaction is carried out on a 2° haloalkane, a 2° carbocation is formed as an intermediate. Recall from Section 5.4 that a 2° carbocation can undergo a rearrangement to form a more stable 3° carbocation. This is illustrated in the solvolysis reaction of 2-bromo-3,3-dimethylbutane in ethanol.

the 2° carbocation intermediate rearranges before the nucleophile has a chance to attack

Br — $\xrightarrow{CH_3CH_2OH}$ — A 2° carbocation — $\longrightarrow$ — A 3° carbocation — $\xrightarrow{CH_3CH_2\ddot{O}H}$ — $CH_3CH_2—\overset{+}{O}$(H) — $\xrightarrow{CH_3CH_2\ddot{O}H}$ — $CH_3CH_2—\ddot{O}$: — Major product

2-Bromo-3,3-dimethylbutane

If an S_N1 reaction is carried out at a tetrahedral stereocenter, the major product is a racemic mixture. We can illustrate this result with the following example: Upon ionization, the R enantiomer forms an achiral carbocation intermediate. Attack by the nucleophile from the left face of the carbocation intermediate gives the S enantiomer; attack from the right face gives the R enantiomer. Because attack by the nucleophile occurs with equal probability from either face of the planar carbocation intermediate, the R and S enantiomers are formed in equal amounts, and the product is a racemic mixture.

Ionization, $-Cl^-$; (1) CH_3OH, (2) $(-H^+)$ Proton transfer to solvent; $H_3CO—C$; $C—OCH_3$; + HCl

R enantiomer Planar carbocation (achiral) S enantiomer R enantiomer

A racemic mixture

7.5 What Determines Whether S_N1 or S_N2 Predominates?

Let us now examine some of the experimental evidence on which these two contrasting mechanisms are based. As we do so, we consider the following questions:

1. What effect does the structure of the nucleophile have on the rate of reaction?
2. What effect does the structure of the haloalkane have on the rate of reaction?
3. What effect does the structure of the leaving group have on the rate of reaction?
4. What is the role of the solvent?

A. Structure of the Nucleophile

Nucleophilicity is a kinetic property, which we measure by relative rates of reaction. We can establish the relative nucleophilicities for a series of nucleophiles by measuring the rate at which each displaces a leaving group from a haloalkane—for example, the rate at which each displaces bromide ion from bromoethane in ethanol at 25°C:

$$CH_3CH_2Br + NH_3 \longrightarrow CH_3CH_2NH_3^+ + Br^-$$

From these studies, we can then make correlations between the structure of the nucleophile and its **relative nucleophilicity**. Table 7.2 lists the types of nucleophiles we deal with most commonly in this text.

Relative nucleophilicity The relative rates at which a nucleophile reacts in a reference nucleophilic substitution reaction.

Because the nucleophile participates in the rate-determining step in an S_N2 reaction, the better the nucleophile, the more likely it is that the reaction will occur by that mechanism. The nucleophile does not participate in the rate-determining step for an S_N1 reaction. Thus, an S_N1 reaction can, in principle, occur at approximately the same rate with any of the common nucleophiles, regardless of their relative nucleophilicities.

B. Structure of the Haloalkane

S_N1 reactions are governed mainly by **electronic factors**, namely, the relative stabilities of carbocation intermediates. S_N2 reactions, by contrast, are governed mainly by **steric factors**, and their transition states are particularly sensitive to crowding about the site of reaction. The distinction is as follows:

Steric hindrance The ability of groups, because of their size, to hinder access to a reaction site within a molecule.

1. *Relative stabilities of carbocations.* As we learned in Section 5.3A, 3° carbocations are the most stable carbocations, requiring the lowest activation energy for their formation, whereas 1° carbocations are the least stable, requiring the highest

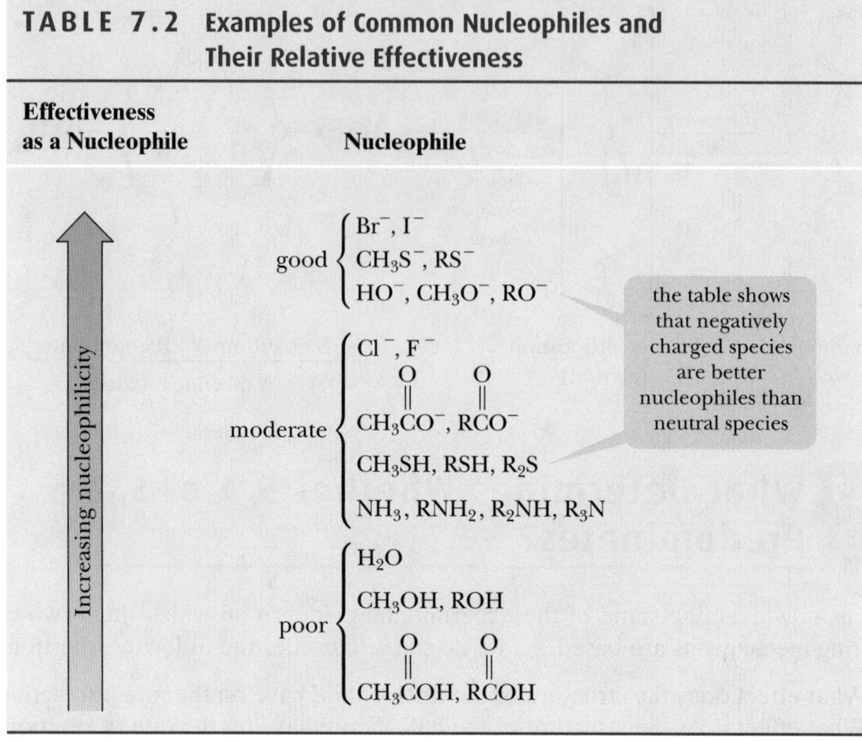

TABLE 7.2 Examples of Common Nucleophiles and Their Relative Effectiveness

Effectiveness as a Nucleophile	Nucleophile
good	Br^-, I^- CH_3S^-, RS^- HO^-, CH_3O^-, RO^-
moderate	Cl^-, F^- $CH_3C(=O)O^-$, $RC(=O)O^-$ CH_3SH, RSH, R_2S NH_3, RNH_2, R_2NH, R_3N
poor	H_2O CH_3OH, ROH $CH_3C(=O)OH$, $RC(=O)OH$

Increasing nucleophilicity
the table shows that negatively charged species are better nucleophiles than neutral species

activation energy for their formation. In fact, 1° carbocations are so unstable that they have never been observed in solution. Therefore, 3° haloalkanes are most likely to react by carbocation formation; 2° haloalkanes are less likely to react in this manner, and methyl and 1° haloalkanes never react in that manner.

2. *Steric hindrance.* To complete a substitution reaction, the nucleophile must approach the substitution center and begin to form a new covalent bond to it. If we compare the ease of approach by the nucleophile to the substitution center of a 1° haloalkane with that of a 3° haloalkane, we see that the approach is considerably easier in the case of the 1° haloalkane. Two hydrogen atoms and one alkyl group screen the backside of the substitution center of a 1° haloalkane. In contrast, three alkyl groups screen the backside of the substitution center of a 3° haloalkane. This center in bromoethane is easily accessible to a nucleophile, while there is extreme crowding around the substitution center in 2-bromo-2-methylpropane:

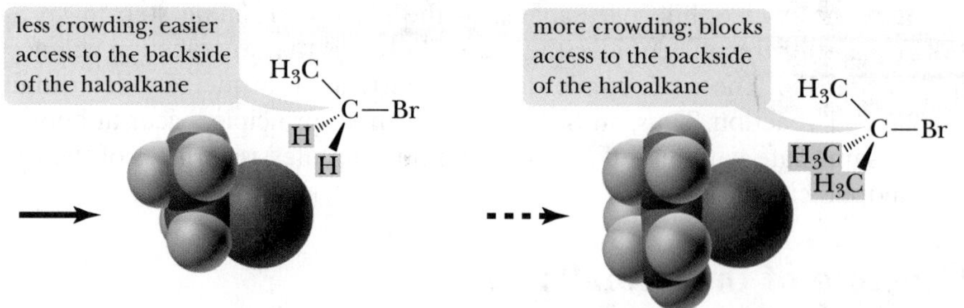

less crowding; easier access to the backside of the haloalkane
H_3C
C—Br
H
H
more crowding; blocks access to the backside of the haloalkane
H_3C
C—Br
H_3C
H_3C

Bromoethane (Ethyl bromide) 2-Bromo-2-methylpropane (*tert*-Butyl bromide)

Given the competition between electronic and steric factors, we find that 3° haloalkanes react by an S_N1 mechanism because 3° carbocation intermediates are particularly stable and because the backside approach of a nucleophile to the substitution center in a 3° haloalkane is hindered by the three groups surrounding it; 3° haloalkanes never react by an S_N2 mechanism. Halomethanes and 1° haloalkanes have little crowding around the substitution center and react by an S_N2 mechanism;

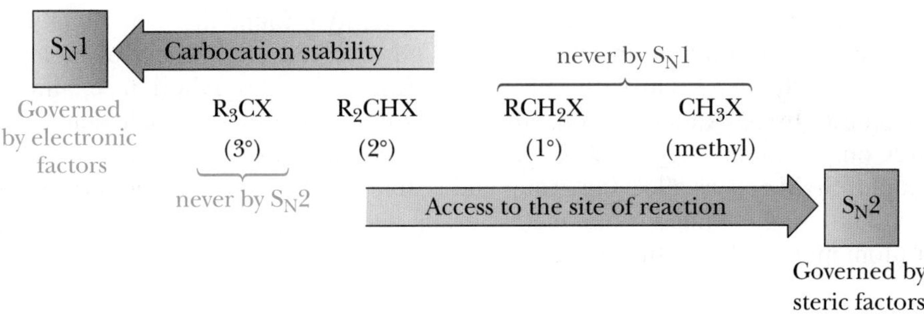

Figure 7.3
Effect of electronic and steric factors in competition between S_N1 and S_N2 reactions of haloalkanes.

they never react by an S_N1 mechanism, because methyl and primary carbocations are so unstable. Secondary haloalkanes may react by either an S_N1 or an S_N2 mechanism, depending on the nucleophile and solvent. The competition between electronic and steric factors and their effects on relative rates of nucleophilic substitution reactions of haloalkanes are summarized in Figure 7.3.

C. The Leaving Group

In the transition state for nucleophilic substitution on a haloalkane, the halogen leaving group develops a partial negative charge in both S_N1 and S_N2 reactions. The halogens Cl^-, Br^-, and I^- make good leaving groups because their size and electronegativity help to stabilize the resulting negative charge. Thus, the ability of a group to function as a leaving group is related to how stable it is as an anion. The most stable anions and the best leaving groups are the conjugate bases of strong acids. We can use the information on the relative strengths of organic and inorganic acids in Table 2.1 to determine which anions are the best leaving groups:

Greater ability to act as leaving group ←

Rarely act as leaving groups in nucleophilic substitution and β-elimination reactions

$$I^- > Br^- > Cl^- > H_2O \gg CH_3\overset{\overset{\large O}{\|}}{C}O^- > HO^- > CH_3O^- > NH_2^-$$

← Greater stability of anion; greater strength of conjugate acid

The best leaving groups in this series are the halogens I^-, Br^-, and Cl^-. Hydroxide ion (OH^-), methoxide ion (CH_3O^-), and amide ion (NH_2^-) are such poor leaving groups that they rarely, if ever, are displaced in nucleophilic aliphatic substitution reactions. H_2O can act as a leaving group if an —OH group of an alcohol is first protonated by an acid.

an —OH group is a poor leaving group and does not undergo substitution

$$(CH_3)_2C(OH)CH_2CH_3 \xrightarrow[H_2O]{Na^+Cl^-} \text{no reaction}$$

$$(CH_3)_2C(\ddot{O}H)CH_2CH_3 \xrightarrow{H-\ddot{C}l:} (CH_3)_2C(\overset{+}{O}H_2)CH_2CH_3 + :\ddot{C}l:^- \longrightarrow H_2\ddot{O} + (CH_3)_2\overset{+}{C}CH_2CH_3 \xrightarrow{:\ddot{C}l:^-} (CH_3)_2C(\ddot{C}l:)CH_2CH_3$$

a strong acid can protonate the —OH group

H_2O is a good leaving group because it is a stable, neutral molecule

One important example of leaving group stability is found in the methylation of DNA, a process common in all mammals and involved in a variety of biological processes including X-chromosome inactivation in females, inheritance, and carcinogenesis. In DNA methylation, enzymes catalyze the attack of a cytosine unit of DNA on the methyl group of *S*-adenosylmethionine (SAM). All but the methyl group of SAM acts as a leaving group and is able to do so because the positive sulfur atom initially bonded to the methyl group becomes uncharged, making the sulfur atom more stable than before.

Enzymes catalyze the methylation of cytosine units on DNA. The highlighted part of *S*-adenosylmethionine acts as a leaving group.

This substructure of *S*-adenosylmethionine is a good leaving group beacause the sulfur atom is no longer positively charged.

S-adenosylmethionine methyltransferase enzymes

Cytosine unit of DNA

Cytosine unit of DNA

We will have more to say about leaving groups other than the halides in subsequent chapters.

D. The Solvent

Solvents provide the medium in which reactants are dissolved and in which nucleophilic substitution reactions take place. Common solvents for these reactions are divided into two groups: **protic** and **aprotic**.

Protic solvent A hydrogen bond donor solvent, as for example water, ethanol, and acetic acid.

Protic solvents contain —OH groups and are hydrogen-bond donors. Common protic solvents for nucleophilic substitution reactions are water, low-molecular-weight alcohols, and low-molecular-weight carboxylic acids (Table 7.3).

TABLE 7.3 Common Protic Solvents

Protic Solvent	Structure	Polarity of Solvent	Notes
Water	H_2O	Increasing (↑)	These solvents favor S_N1 reactions. The greater the polarity of the solvent, the easier it is to form carbocations in it because both the carbocation and the negatively charged leaving group can be solvated.
Formic acid	HCOOH		
Methanol	CH_3OH		
Ethanol	CH_3CH_2OH		
Acetic acid	CH_3COOH		

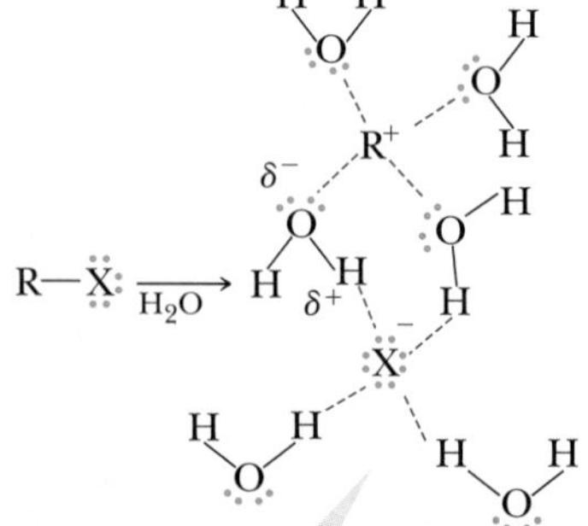

polar protic solvents can solvate both the anion and the cation components of the S_N1 reaction

TABLE 7.4 Common Aprotic Solvents

Aprotic Solvent	Structure	Polarity of Solvent	Notes
Dimethyl sulfoxide (DMSO)	$CH_3\overset{O}{\overset{\|\|}{S}}CH_3$	Increasing (top is most polar)	These solvents favor S_N2 reactions. Although solvents at the top of this list are polar, the formation of carbocations in them is far more difficult than in protic solvents because the anionic leaving group cannot be solvated by these solvents.
Acetone	$CH_3\overset{O}{\overset{\|\|}{C}}CH_3$		
Dichloromethane	CH_2Cl_2		
Diethyl ether	$(CH_3CH_2)_2O$		

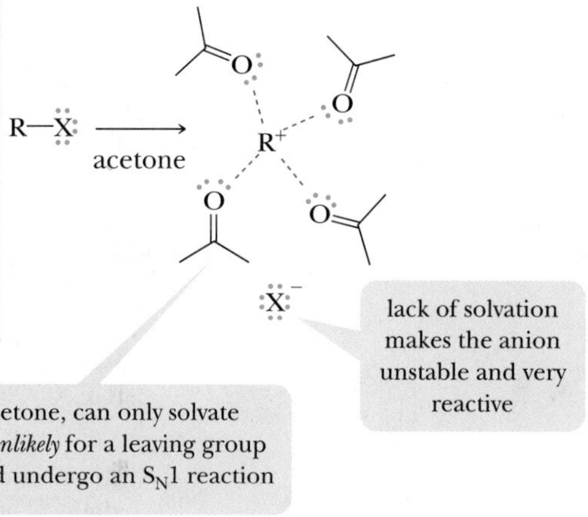

R—X: → (acetone) R^+ solvated by acetone; :X:⁻

polar aprotic solvents, like acetone, can only solvate cations effectively, making it *unlikely* for a leaving group to break its bond to carbon and undergo an S_N1 reaction

lack of solvation makes the anion unstable and very reactive

Each is able to solvate both the anionic and cationic components of ionic compounds by electrostatic interaction between its partially negatively charged oxygen(s) and the cation and between its partially positively charged hydrogen and the anion. These same properties aid in the ionization of C—X bonds to give an X^- anion and a carbocation; thus, protic solvents are good solvents in which to carry out S_N1 reactions.

Aprotic solvents do not contain —OH groups and cannot function as hydrogen-bond donors. They are unable to promote the formation of a carbocation because the leaving group would be unsolvated. Therefore aprotic solvents cannot be used in S_N1 reactions. Table 7.4 lists the aprotic solvents most commonly used for nucleophilic substitution reactions. Dimethyl sulfoxide and acetone are polar aprotic solvents; dichloromethane and diethyl ether are nonpolar aprotic solvents. The aprotic solvents listed in the table are particularly good ones in which to carry out S_N2 reactions. Because polar aprotic solvents are able to solvate only cations and not anions, they allow for "naked" and highly reactive anions as nucleophiles when used with ionic nucleophiles such as Na^+CN^-, Na^+OH^-, and so on.

Aprotic solvent A solvent that cannot serve as a hydrogen bond donor, as for example, acetone, diethyl ether, and dichloromethane.

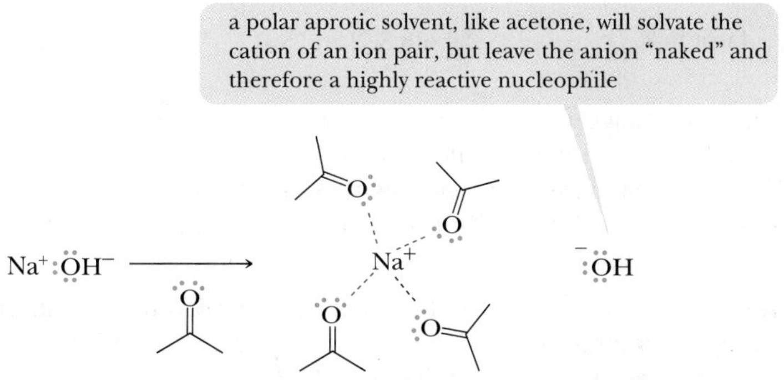

a polar aprotic solvent, like acetone, will solvate the cation of an ion pair, but leave the anion "naked" and therefore a highly reactive nucleophile

Na^+ :ÖH⁻ → (acetone) Na^+ solvated by acetone; :ÖH⁻

Table 7.5 summarizes the factors favoring S_N1 or S_N2 reactions; it also shows the change in configuration when nucleophilic substitution takes place at a stereocenter.

TABLE 7.5 Summary of S_N1 versus S_N2 Reactions of Haloalkanes

Type of Haloalkane	S_N2	S_N1
Methyl CH_3X	**S_N2 is favored.**	**S_N1 does not occur.** The methyl cation is so unstable that it is never observed in solution.
Primary RCH_2X	**S_N2 is favored.**	**S_N1 does not occur.** Primary carbocations are so unstable that they are not observed in solution.
Secondary R_2CHX	**S_N2 is favored** in aprotic solvents with good nucleophiles.	**S_N1 is favored** in protic solvents with poor nucleophiles.
Tertiary R_3CX	**S_N2 does not occur,** because of steric hindrance around the substitution center.	**S_N1 is favored** because of the ease of formation of tertiary carbocations.
Substitution at a stereocenter	**Inversion of configuration.** The nucleophile attacks the stereocenter from the side opposite the leaving group.	**Racemization.** The carbocation intermediate is planar, and attack by the nucleophile occurs with equal probability from either side.

HOW TO 7.4 Predict the Type of Substitution Reaction a Haloalkane Will Undergo

(a) Identify and assess the stability of the potential leaving group. A substitution reaction will not occur unless there is a good leaving group.

(b) Classify the structure of the haloalkane. Methyl and primary haloalkanes do not undergo S_N1 reactions, while tertiary haloalkanes do not undergo S_N2 reactions.

methyl	1°	2°	3°
$H-C(H)(H)-X$	$H-C(H)(R)-X$	$R-C(H)(R)-X$	$R-C(R)(R)-X$

favor S_N2 (methyl, 1°) — both S_N2 and S_N1 (2°) — favor S_N1 (3°)

(c) Identify the nucleophile and assess its relative nucleophilicity. S_N2 reactions are favored with good nucleophiles and rarely occur with poor nucleophiles. S_N1 reactions can occur with both poor and moderate nucleophiles, but receive competition from the S_N2 mechanism in the presence of good nucleophiles.

(d) Identify and classify the solvent. A polar protic solvent is required for an S_N1 reaction. Polar aprotic solvents favor S_N2 reactions, although S_N2 reactions can also occur in protic solvents.

(e) If no single factor eliminates or mandates either substitution mechanism, try to determine whether these factors collectively favor the predominance of one mechanism over the other.

Example 7.4

Answer the following questions:

(a) The rate of a substitution reaction of a haloalkane is unchanged when the nucleophile is switched from hydroxide to ammonia. What type of substitution reaction does this haloalkane likely undergo?
(b) When (*R*)-2-bromobutane is reacted with diethylamine, $(CH_3CH_2)_2NH$, the reaction solution gradually loses optical activity. What type of substitution mechanism is in operation in this reaction?

Strategy

It is important to remember the details that go along with each of the two substitution mechanisms. In an S_N1 reaction, look for (1) the rate of the reaction to be unaffected by the concentration of the nucleophile, (2) the rate of the reaction to be unaffected by the type of nucleophile, or (3) the formation of two products (often enantiomers if stereochemistry exists at the reacting carbon). In an S_N2 reaction, look for (1) the rate of the reaction to be dependent on the concentration of the nucleophile, (2) the rate of the reaction to be dependent on the type of nucleophile, or (3) the formation of only one product.

Solution

(a) S_N1. Hydroxide ion is a better nucleophile than ammonia. S_N1 reactions are unaffected by the effectiveness of the nucleophile. If the reaction were to occur by an S_N2 mechanism, we would expect reaction with the better nucleophile to result in a faster reaction.
(b) S_N1. Diethylamine is a moderate nucleophile and likely favors the S_N1 mechanism. This is confirmed by the stereochemical data, because an S_N2 reaction would yield only the S enantiomer because of the required backside attack. The loss of optical activity most likely indicates that a carbocation intermediate is formed, followed by the attack of the nucleophile to form equal amounts of the enantiomers shown.

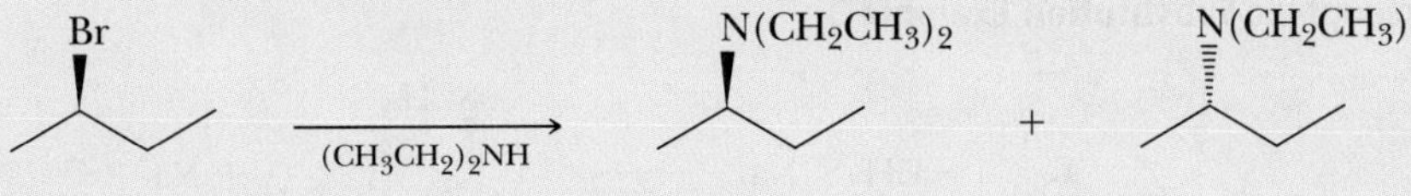

Br — $(CH_3CH_2)_2NH$ → $N(CH_2CH_3)_2$ + $N(CH_2CH_3)_2$

(*R*)-2-bromobutane (*R*) (*S*)

See problems 7.20, 7.23, 7.27

Problem 7.4

Answer the following questions:

(a) Potassium cyanide, KCN, reacts faster than trimethylamine, $(CH_3)_3N$, with 1-chloropentane. What type of substitution mechanism does this haloalkane likely undergo?
(b) Compound A reacts faster with dimethylamine, $(CH_3)_2NH$, than compound B. What does this reveal about the relative ability of each haloalkane to undergo S_N2? S_N1?

A Br B Br

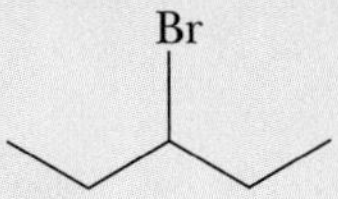

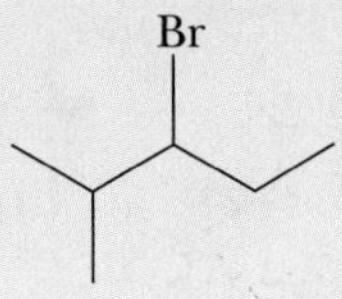

7.6 How Can S_N1 and S_N2 Be Predicted Based on Experimental Conditions?

Predictions about the mechanism for a particular nucleophilic substitution reaction must be based on considerations of the structure of the haloalkane, the nucleophile, and the solvent. Following are analyses of three such reactions:

Nucleophilic Substitution Example 1

Cl + CH_3OH ⟶ OCH_3 + HCl

R enantiomer

Methanol is a polar protic solvent and a good one in which to form carbocations. 2-Chlorobutane ionizes in methanol to form a 2° carbocation intermediate. Methanol is a weak nucleophile. From this analysis, we predict that reaction is by an S_N1 mechanism. The 2° carbocation intermediate (an electrophile) then reacts with methanol (a nucleophile) followed by proton transfer to give the observed product. The product is formed as a 50:50 mixture of R and S configurations; that is, it is formed as a racemic mixture.

Nucleophilic Substitution Example 2

Br + Na^+I^- $\xrightarrow[\text{DMSO}]{}$ I + Na^+Br^-

This is a 1° bromoalkane in the presence of iodide ion, a good nucleophile. Because 1° carbocations are so unstable, they never form in solution, and an S_N1 reaction is not possible. Dimethyl sulfoxide (DMSO), a polar aprotic solvent, is a good solvent in which to carry out S_N2 reactions. From this analysis, we predict that reaction is by an S_N2 mechanism.

Nucleophilic Substitution Example 3

Br + $CH_3S^-Na^+$ $\xrightarrow[\text{acetone}]{}$ SCH_3 + Na^+Br^-

S enantiomer

Bromine ion is a good leaving group on a 2° carbon. The methylsulfide ion is a good nucleophile. Acetone, a polar aprotic solvent, is a good medium in which to carry out S_N2 reactions, but a poor medium in which to carry out S_N1 reactions. We predict that reaction is by an S_N2 mechanism and that the product formed has the R configuration.

Example 7.5

Write the expected product for each nucleophilic substitution reaction, and predict the mechanism by which the product is formed:

(a) I + CH_3OH $\xrightarrow[\text{methanol}]{}$

(b) Br + $CH_3\overset{O}{\overset{\|}{C}}O^- Na^+$ $\xrightarrow[\text{DMSO}]{}$

Strategy

Determine whether the electrophile's reaction center is 1°, 2°, or 3°. Then assess the nucleophilicity of the nucleophile. If it is poor, then the reaction will most likely proceed by an S_N1 mechanism, provided that there exists a polar protic solvent and the reaction center is 2° or 3°. If there is a good nucleophile, then the reaction will most likely proceed by an S_N2 mechanism, provided that the reaction center is 1° or 2°. If the nucleophile is moderate, focus on the solvent polarity and reaction center of the electrophile. Remember that S_N1 mechanisms only occur in polar protic solvents.

Solution

(a) Methanol is a poor nucleophile. It is also a polar protic solvent that is able to solvate carbocations. Ionization of the carbon–iodine bond forms a 2° carbocation intermediate. We predict an S_N1 mechanism:

$$\text{(iodocyclopentane)} + CH_3OH \xrightarrow[\text{methanol}]{S_N1} \text{(methoxycyclopentane, } OCH_3\text{)} + HI$$

(b) Bromide is a good leaving group on a 2° carbon. Acetate ion is a moderate nucleophile. DMSO is a particularly good solvent for S_N2 reactions. We predict substitution by an S_N2 mechanism with inversion of configuration at the stereocenter:

? How do you know?

because attacked from backside?

$$\text{(2-bromohexane, Br)} + CH_3\overset{O}{\overset{\|}{C}}O^- Na^+ \xrightarrow[\text{DMSO}]{S_N2} \text{(product, } O\overset{O}{\overset{\|}{C}}CH_3\text{)} + Na^+Br^-$$

S vs R?

See problems 7.23, 7.25–7.28, 7.34, 7.35

Problem 7.5

Write the expected product for each nucleophilic substitution reaction, and predict the mechanism by which the product is formed:

(a) (tert-butyl bromocyclohexane, Br) $+ Na^+SH^- \xrightarrow{\text{acetone}}$

(b) $CH_3\overset{Cl}{\overset{|}{C}}HCH_2CH_3 + H\overset{O}{\overset{\|}{C}}OH \xrightarrow{\text{formic acid}}$

7.7 What Are the Products of β-Eliminations?

In this section, we study a type of β-elimination called **dehydrohalogenation**. In the presence of a strong base, such as hydroxide ion or ethoxide ion, halogen can be removed from one carbon of a haloalkane and hydrogen from an adjacent carbon to form a carbon–carbon double bond:

Dehydrohalogenation Removal of —H and —X from adjacent carbons; a type of β-elimination.

$$\underset{\text{A haloalkane}}{-\overset{|\beta}{\underset{H}{C}}-\overset{|\alpha}{\underset{X}{C}}-} + \underset{\text{Base}}{CH_3CH_2O^-Na^+} \xrightarrow{CH_3CH_2OH} \underset{\text{An alkene}}{>C=C<} + CH_3CH_2OH + Na^+X^-$$

ethoxide (?)

As the equation shows, we call the carbon bearing the halogen the *α-carbon* and the adjacent carbon the *β-carbon.*

Because most nucleophiles can also act as bases and vice versa, it is important to keep in mind that β-elimination and nucleophilic substitution are competing reactions. In this section, we concentrate on β-elimination. In Section 7.9, we examine the results of competition between the two.

Common strong bases used for β-elimination are OH^-, OR^-, and NH_2^-. Following are three examples of base-promoted β-elimination reactions:

1-Bromooctane + t-BuO^-K^+ (Potassium *tert*-butoxide) ⟶ 1-Octene + t-BuOH + Na^+Br^-

2-Bromo-2-methylbutane $\xrightarrow[CH_3CH_2OH]{CH_3CH_2O^-Na^+}$ 2-Methyl-2-butene (major product) + 2-Methyl-1-butene

1-Bromo-1-methylcyclopentane $\xrightarrow[CH_3OH]{CH_3O^-Na^+}$ 1-Methylcyclopentene (major product) + Methylenecyclopentane

In the first example, the base is shown as a reactant. In the second and third examples, the base is a reactant, but is shown over the reaction arrow. Also in the second and third examples, there are nonequivalent β-carbons, each bearing a hydrogen; therefore, two alkenes are possible from each β-elimination reaction. In each case, the major product of these and most other β-elimination reactions is the more substituted (and therefore the more stable—see Section 5.7B) alkene. We say that each reaction follows **Zaitsev's rule** or, alternatively, that each undergoes Zaitsev elimination, to honor the chemist who first made this generalization.

Zaitsev's rule A rule stating that the major product from a β-elimination reaction is the most stable alkene; that is, the major product is the alkene with the greatest number of substituents on the carbon–carbon double bond.

HOW TO 7.5 Complete an Elimination Reaction

(a) Identify and assess the leaving group. An elimination reaction will not occur unless there is a good leaving group.

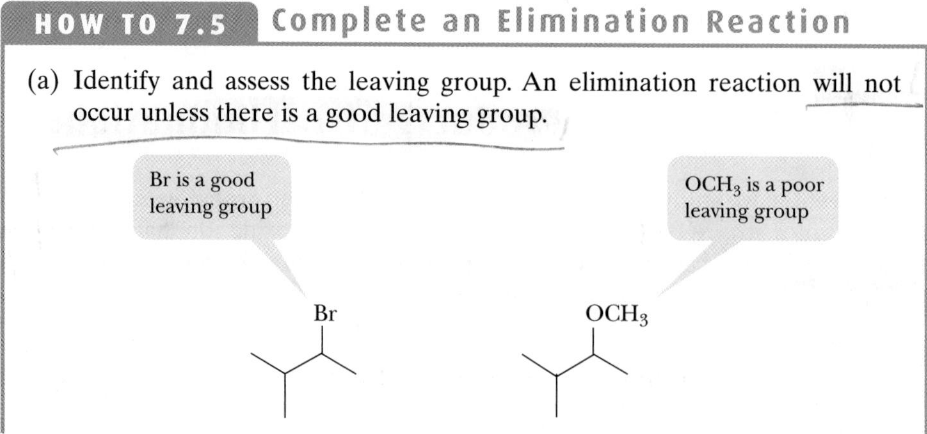
Br is a good leaving group
OCH3 is a poor leaving group
Br
OCH3

(b) Label the carbon bonded to the leaving group as "α" (alpha).

(c) Label any carbon bonded to the α-carbon as "β" (beta). *Note:* Only do so if the β-carbon is bonded to a hydrogen atom.

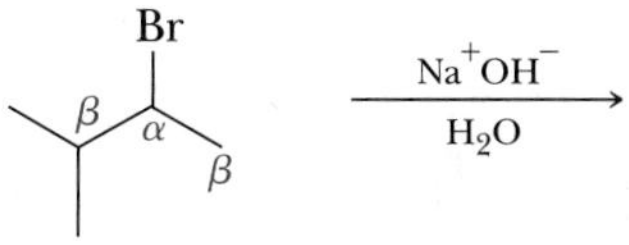

Br β α β $\xrightarrow[H_2O]{Na^+OH^-}$

(d) Remove the leaving group and a β-hydrogen from the molecule and place a new double bond between the α and β carbons. This forms a β-elimination product.

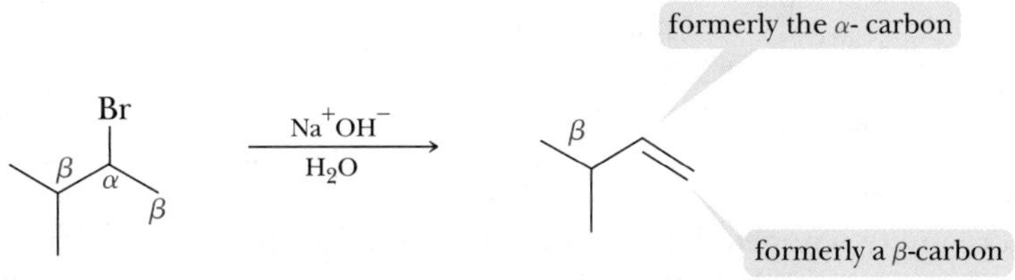

Br β α β $\xrightarrow[H_2O]{Na^+OH^-}$ β

formerly the α- carbon

formerly a β-carbon

(e) Repeat Step (d) above for any other β-carbons to form a different β-elimination product.

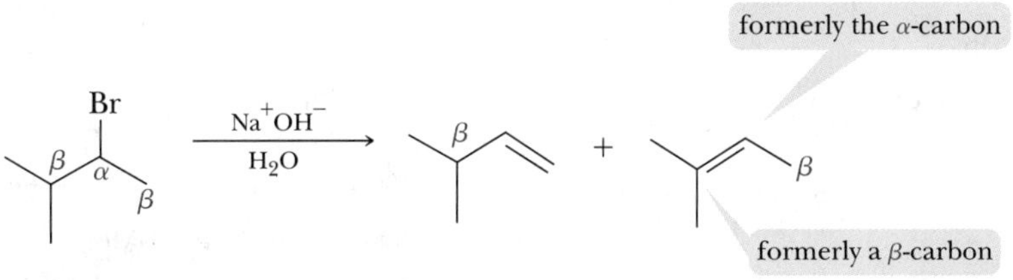

Br β α β $\xrightarrow[H_2O]{Na^+OH^-}$ β + β

formerly the α-carbon

formerly a β-carbon

Example 7.6

Predict the β-elimination product(s) formed when each bromoalkane is treated with sodium ethoxide in ethanol (if two might be formed, predict which is the major product):

(a) 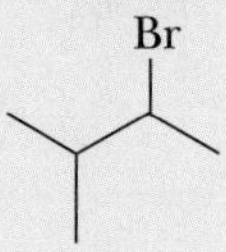 Br (b) 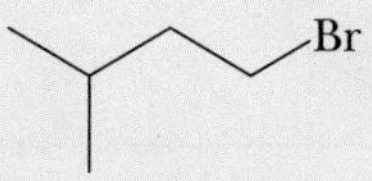 Br (c) 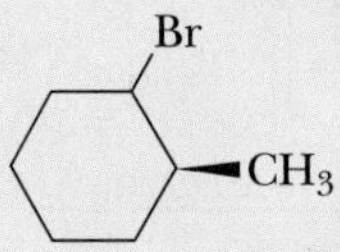 Br CH_3

Strategy

Label the carbon attached to the halogen as α. Then label any carbons next to the α-carbon as β. If the β-carbon is attached to at least one hydrogen, remove that hydrogen and the halogen and draw a C—C double bond between the α-and β-carbons. Start over and repeat this process for any other β-carbons that meet this criteria. Each time you are able to do this will result in an elimination product.

Solution

(a) There are two nonequivalent β-carbons in this bromoalkane, and two alkenes are possible. 2-Methyl-2-butene, the more substituted alkene, is the major product:

Br β β $\xrightarrow[\text{EtOH}]{\text{EtO}^-\text{Na}^+}$ +

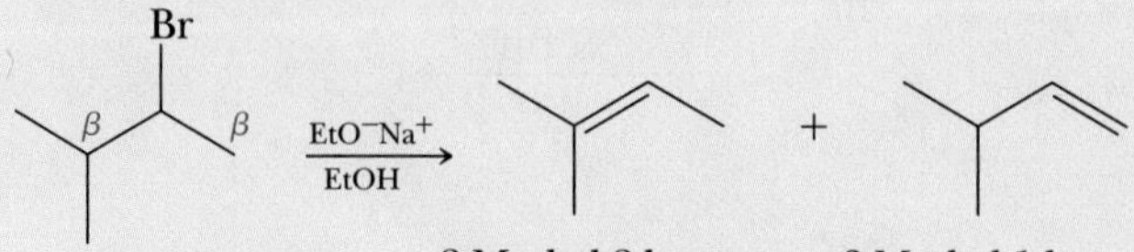

2-Methyl-2-butene (major product) 3-Methyl-1-butene

(b) There is only one β-carbon in this bromoalkane, and only one alkene is possible.

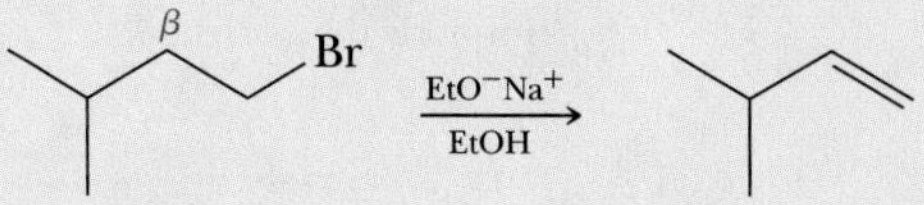

β Br $\xrightarrow[\text{EtOH}]{\text{EtO}^-\text{Na}^+}$

3-Methyl-1-butene

(c) There are two nonequivalent β-carbons in this cyclic bromoalkane, and two alkenes are possible. 1-Methylcyclohexene, the more substituted alkene, is the major product:

the stereocenter remains unchanged in the minor product

the stereocenter is lost in the major product

Br β β CH_3 $\xrightarrow[\text{EtOH}]{\text{EtO}^-\text{Na}^+}$ CH_3 + CH_3

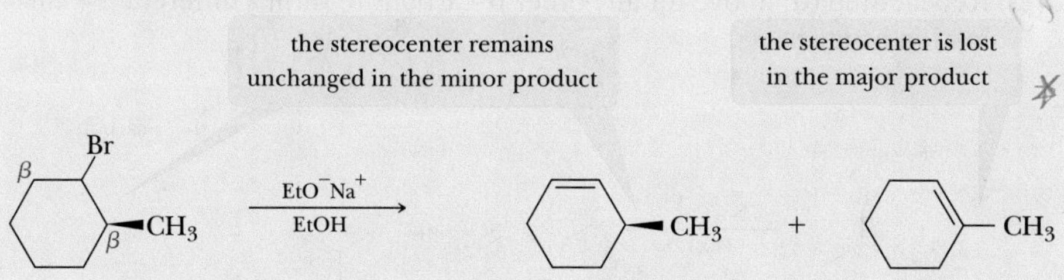

(*S*)-3-methylcyclohexene 1-Methylcyclohexene (major product)

See problems 7.36–7.38

Problem 7.6

Predict the β-elimination products formed when each chloroalkane is treated with sodium ethoxide in ethanol (if two products might be formed, predict which is the major product):

(a) Cl CH_3 (b) CH_2Cl (c) Cl CH_3

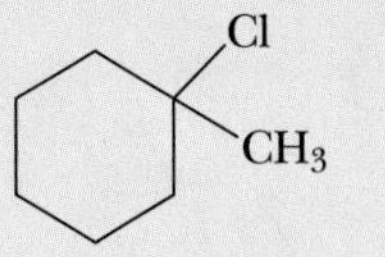

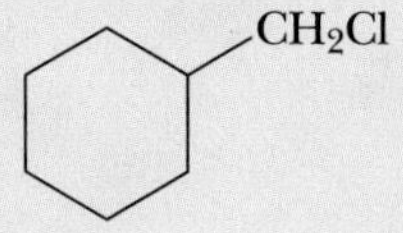

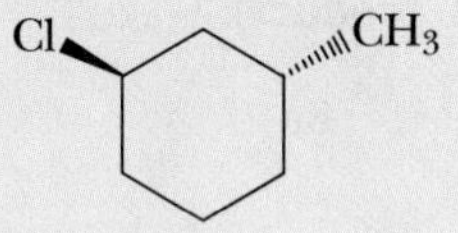

7.8 What Are the E1 and E2 Mechanisms for β-Elimination?

There are two limiting mechanisms of β-elimination reactions. A fundamental difference between them is the timing of the bond-breaking and bond-forming steps. Recall that we made this same statement about the two limiting mechanisms for nucleophilic substitution reactions in Section 7.4.

HOW TO 7.6 Draw Mechanisms

Remember that curved arrow notation always shows the arrow originating from a bond or from a lone pair of electrons.

Correct use of curved arrows...

$$H_3N{:} + H{-}C(H)(H)CH_3{-}C(H)(H){-}Cl{:} \longrightarrow NH_4^+Cl^- + H_3C(H)C{=}C(H)H$$

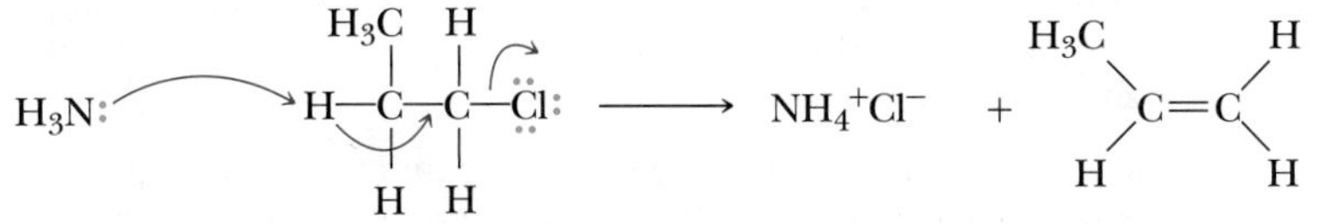

Incorrect use of curved arrows...

$$H_3N{:} \leftarrow H{-}C(H)(H)CH_3{-}C(H)(H){-}Cl{:} \longrightarrow NH_4^+Cl^- + H_3C(H)C{=}C(H)H$$

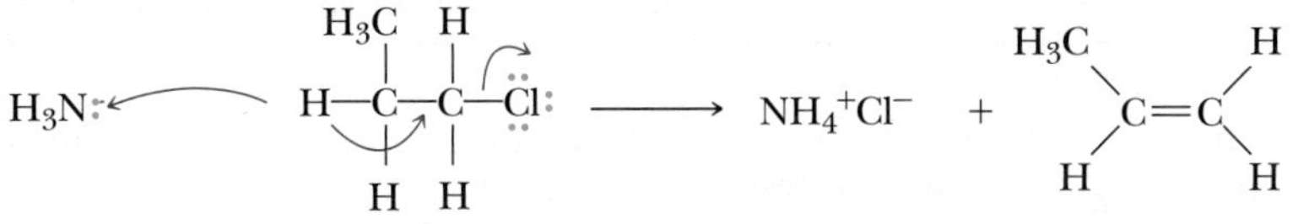

A. E1 Mechanism

At one extreme, breaking of the C—X bond is complete before any reaction occurs with base to lose a hydrogen and before the carbon–carbon double bond is formed. This mechanism is designated **E1**, where *E* stands for *e*limination and *1* stands for a *uni*molecular reaction; only *one* species, in this case the haloalkane, is involved in the rate-determining step. The rate law for an E1 reaction has the same form as that for an S_N1 reaction:

$$\text{Rate} = k[\text{haloalkane}]$$

The mechanism for an E1 reaction is illustrated by the reaction of 2-bromo-2-methylpropane to form 2-methylpropene. In this two-step mechanism, the rate-determining step is the ionization of the carbon–halogen bond to form a carbocation intermediate (just as it is in an S_N1 mechanism).

Mechanism: E1 Reaction of 2-Bromo-2-methylpropane

Step 1: Rate-determining ionization of the C—Br bond gives a carbocation intermediate:

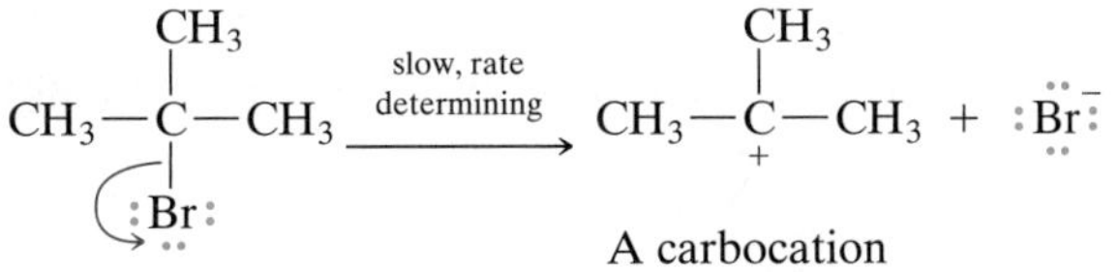

$$CH_3{-}\underset{:Br:}{\overset{CH_3}{C}}{-}CH_3 \xrightarrow[\text{determining}]{\text{slow, rate}} CH_3{-}\overset{CH_3}{\underset{+}{C}}{-}CH_3 + :\ddot{Br}:^-$$

A carbocation intermediate

Step 2: Proton transfer from the carbocation intermediate to methanol (which in this instance is both the solvent and a reactant) gives the alkene:

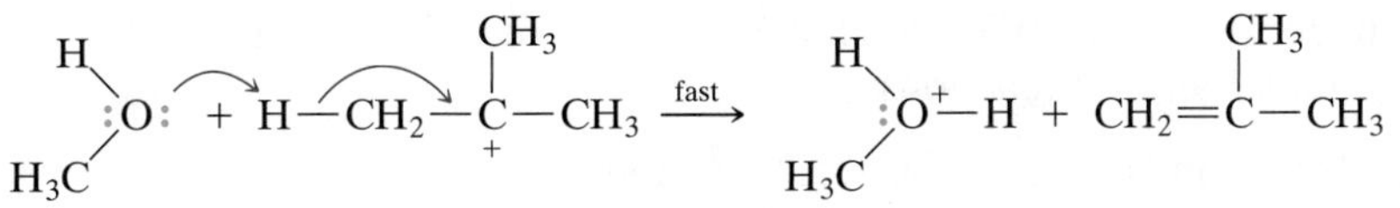

$$(H)(H_3C)\ddot{O}{:} + H{-}CH_2{-}\overset{CH_3}{\underset{+}{C}}{-}CH_3 \xrightarrow{\text{fast}} (H)(H_3C){:}\overset{+}{O}{-}H + CH_2{=}\overset{CH_3}{C}{-}CH_3$$

B. E2 Mechanism

At the other extreme is a concerted process. In an **E2** reaction, *E* stands for *e*limination, and *2* stands for *bi*molecular. Because the base removes a β-hydrogen at the

TABLE 7.6 Summary of E1 versus E2 Reactions of Haloalkanes

Haloalkane	E1	E2
Primary RCH_2X	E1 does not occur. Primary carbocations are so unstable that they are never observed in solution.	E2 is favored.
Secondary R_2CHX	Main reaction with weak bases such as H_2O and ROH.	Main reaction with strong bases such as OH^- and OR^-.
Tertiary R_3CX	Main reaction with weak bases such as H_2O and ROH.	Main reaction with strong bases such as OH^- and OR^-.

same time the C—X bond is broken to form a halide ion, the rate law for the rate-determining step is dependent on both the haloalkane and the base:

$$\text{Rate} = k[\text{haloalkane}][\text{base}]$$

The stronger the base, the more likely it is that the E2 mechanism will be in operation. We illustrate an E2 mechanism by the reaction of 1-bromopropane with sodium ethoxide.

Mechanism: E2 Reaction of 1-Bromopropane

In this mechanism, proton transfer to the base, formation of the carbon–carbon double bond, and the ejection of bromide ion occur simultaneously; that is, all bond-forming and bond-breaking steps occur at the same time.

$$CH_3CH_2\ddot{O}:^- + H{-}\underset{}{\overset{CH_3}{\overset{|}{CH}}}{-}CH_2{-}\ddot{Br}: \longrightarrow CH_3CH_2\ddot{O}{-}H + CH_3CH{=}CH_2 + :\ddot{Br}:^-$$

For both E1 and E2 reactions, the major product is that formed in accordance with Zaitsev's rule (Section 7.7) as illustrated by this E2 reaction:

2-Bromohexane $\xrightarrow[CH_3OH]{CH_3O^-Na^+}$ 2-Hexene (74%) + 1-Hexene (26%)

Table 7.6 summarizes these generalizations about β-elimination reactions of haloalkanes.

HOW TO 7.7 Predict the Type of β-Elimination Reaction a Haloalkane Will Undergo

(a) Classify the structure of the haloalkane. Primary haloalkanes will not undergo E1 reactions. Secondary and tertiary haloalkanes will undergo both E1 and E2 reactions.

(b) Identify and assess the base. E2 reactions are favored with strong bases and rarely occur with weak bases. E2 reactions can occur in any solvent. E1 reactions can occur with both weak and strong bases, but require polar protic solvents to stabilize the carbocation formed in the first step of the reaction.

Example 7.7

Predict whether each β-elimination reaction proceeds predominantly by an E1 or E2 mechanism, and write a structural formula for the major organic product:

(a) $CH_3C(CH_3)(Cl)CH_2CH_3 + Na^+OH^- \xrightarrow[H_2O]{80°C}$ (b) $CH_3C(CH_3)(Cl)CH_2CH_3 \xrightarrow{CH_3COOH}$

Strategy

Identify the solvent and the base. If the base is strong, an E2 mechanism is favored to occur. If the base is weak and the solvent is polar protic, then an E1 mechanism is favored to occur.

Solution

(a) A 3° chloroalkane is heated with NaOH, a strong base. Elimination by an E2 reaction predominates, giving 2-methyl-2-butene as the major product:

$$CH_3C(CH_3)(Cl)CH_2CH_3 + Na^+OH^- \xrightarrow[H_2O]{80°C} CH_3C(CH_3){=}CHCH_3 + NaCl + H_2O$$

(b) A 3° chloroalkane dissolved in acetic acid, a solvent that promotes the formation of carbocations, forms a 3° carbocation that then loses a proton to give 2-methyl-2-butene as the major product. The reaction is by an E1 mechanism:

$$CH_3C(CH_3)(Cl)CH_2CH_3 \xrightarrow{CH_3COOH} CH_3C(CH_3){=}CHCH_3 + HCl$$

See problems 7.36–7.38

Problem 7.7

Predict whether each elimination reaction proceeds predominantly by an E1 or E2 mechanism, and write a structural formula for the major organic product:

(a) 3-bromopentane (Br) $+ CH_3O^- Na^+ \xrightarrow{\text{methanol}}$

(b) substituted chlorocyclohexane (Cl) $+ Na^+OH^- \xrightarrow{\text{acetone}}$

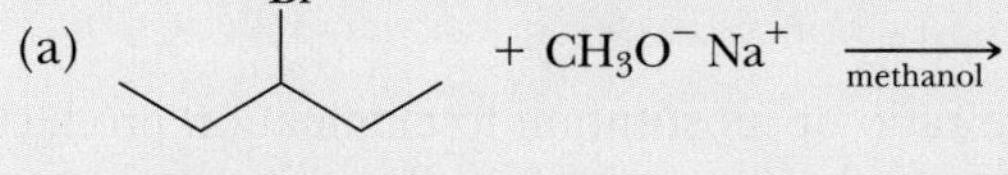

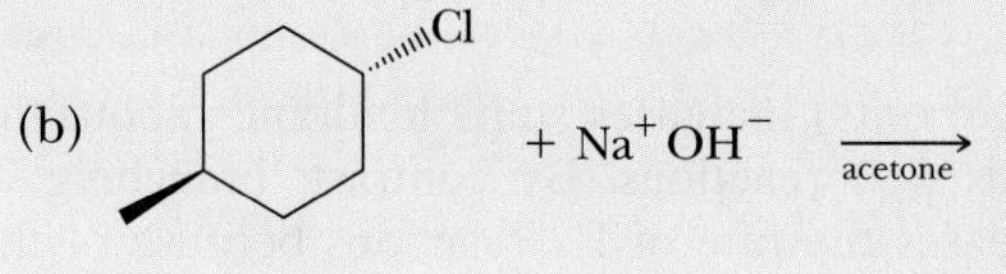

7.9 When Do Nucleophilic Substitution and β-Elimination Compete?

Thus far, we have considered two types of reactions of haloalkanes: nucleophilic substitution and β-elimination. Many of the nucleophiles we have examined—for example, hydroxide ion and alkoxide ions—are also strong bases. Accordingly, nucleophilic substitution and β-elimination often compete with each other, and the ratio of products formed by these reactions depends on the relative rates of the two reactions:

$$\text{H—C—C—X} + :\text{Nu}^- \begin{cases} \xrightarrow{\text{nucleophilic substitution}} \text{H—C—C—Nu} + :\text{X}^- \\ \xrightarrow{\beta\text{-elimination}} \text{C=C} + \text{H—Nu} + :\text{X}^- \end{cases}$$

A. S_N1-versus-E1 Reactions

Reactions of secondary and tertiary haloalkanes in polar protic solvents give mixtures of substitution and elimination products. In both reactions, Step 1 is the formation of a carbocation intermediate. This step is then followed by either (1) the loss of a hydrogen to give an alkene (E1) or (2) reaction with solvent to give a substitution product (S_N1). In polar protic solvents, the products formed depend only on the structure of the particular carbocation. For example, *tert*-butyl chloride and *tert*-butyl iodide in 80% aqueous ethanol both react with solvent, giving the same mixture of substitution and elimination products:

$$(CH_3)_3C\text{—I} \xrightarrow{-I^-} \quad (CH_3)_3C\text{—Cl} \xrightarrow{-Cl^-} \quad CH_3\text{—}\overset{+}{C}(CH_3)\text{—}CH_3$$

$$CH_3\overset{+}{C}(CH_3)_2 \xrightarrow{\text{E1}} CH_2\text{=}C(CH_3)_2 + H^+$$

$$CH_3\overset{+}{C}(CH_3)_2 \xrightarrow[H_2O]{S_N1} CH_3\text{—}C(CH_3)_2\text{—OH} + H^+$$

$$CH_3\overset{+}{C}(CH_3)_2 \xrightarrow[CH_3CH_2OH]{S_N1} CH_3\text{—}C(CH_3)_2\text{—}OCH_2CH_3 + H^+$$

Because iodide ion is a better leaving group than chloride ion, *tert*-butyl iodide reacts over 100 times faster than *tert*-butyl chloride. Yet the ratio of products is the same.

B. S_N2-versus-E2 Reactions

It is considerably easier to predict the ratio of substitution to elimination products for reactions of haloalkanes with reagents that act as both nucleophiles and bases. The guiding principles are as follows:

1. Branching at the α-carbon or β-carbon(s) increases steric hindrance about the α-carbon and significantly retards S_N2 reactions. By contrast, branching at the α-carbon or β-carbon(s) increases the rate of E2 reactions because of the increased stability of the alkene product.

2. The greater the nucleophilicity of the attacking reagent, the greater is the S_N2-to-E2 ratio. Conversely, the greater the basicity of the attacking reagent, the greater is the E2-to-S_N2 ratio.

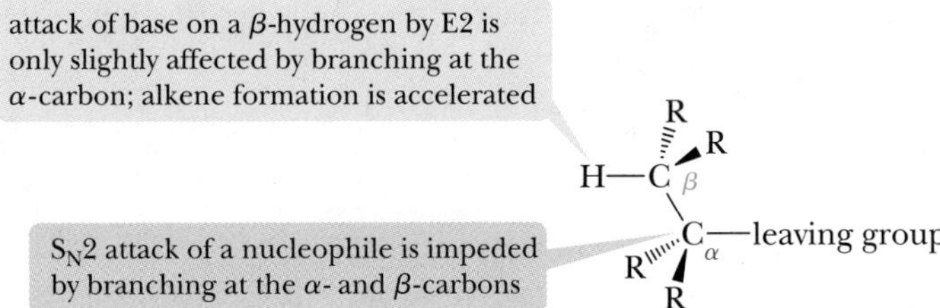
attack of base on a β-hydrogen by E2 is only slightly affected by branching at the α-carbon; alkene formation is accelerated
R
R
H—C
β
C—leaving group
α
R
R
S_N2 attack of a nucleophile is impeded by branching at the α- and β-carbons

Primary halides react with bases/nucleophiles to give predominantly substitution products. With strong bases, such as hydroxide ion and ethoxide ion, a percentage of the product is formed by an E2 reaction, but it is generally small compared with that formed by an S_N2 reaction. With strong, bulky bases, such as *tert*-butoxide ion, the E2 product becomes the major product. Tertiary halides react with all strong bases/good nucleophiles to give only elimination products.

Secondary halides are borderline, and substitution or elimination may be favored, depending on the particular base/nucleophile, solvent, and temperature at which the reaction is carried out. Elimination is favored with strong bases/good nucleophiles—for example, hydroxide ion and ethoxide ion. Substitution is favored with weak bases/poor nucleophiles—for example, acetate ion. Table 7.7 summarizes these generalizations about substitution versus elimination reactions of haloalkanes.

TABLE 7.7 Summary of Substitution versus Elimination Reactions of Haloalkanes

Halide	Reaction	Comments
Methyl	S_N2	The only substitution reactions observed.
CH_3X	~~S_N1~~	S_N1 reactions of methyl halides are never observed. The methyl cation is so unstable that it is never formed in solution.
Primary RCH_2X	S_N2	The main reaction with strong bases such as OH^- and EtO^-. Also, the main reaction with good nucleophiles/weak bases, such as I^- and CH_3COO^-.
	E2	The main reaction with strong, bulky bases, such as potassium *tert*-butoxide.
	~~S_N1/E1~~	Primary cations are never formed in solution; therefore, S_N1 and E1 reactions of primary halides are never observed.
Secondary R_2CHX	S_N2	The main reaction with weak bases/good nucleophiles, such as I^- and CH_3COO^-.
	E2	The main reaction with strong bases/good nucleophiles, such as OH^- and $CH_3CH_2O^-$.
	S_N1/E1	Common in reactions with weak nucleophiles in polar protic solvents, such as water, methanol, and ethanol.
Tertiary R_3CX	~~S_N2~~	S_N2 reactions of tertiary halides are never observed because of the extreme crowding around the 3° carbon.
	E2	Main reaction with strong bases, such as HO^- and RO^-.
	S_N1/E1	Main reactions with poor nucleophiles/weak bases.

Example 7.8

Predict whether each reaction proceeds predominantly by substitution (S_N1 or S_N2) or elimination (E1 or E2) or whether the two compete, and write structural formulas for the major organic product(s):

(a) [structure: 2-chloro-2-methylbutane] + Na^+OH^- $\xrightarrow[H_2O]{80°C}$ (b) [structure: 1-bromo-3-methylbutane] + $(C_2H_5)_3N$ $\xrightarrow[CH_2Cl_2]{30°C}$

Strategy

First, determine whether the reagent acts predominantly as a base or a nucleophile. If it is a weak base but a good nucleophile, substitution is more likely to occur. If the reagent is a strong base but a poor nucleophile, elimination is more likely to occur. When the reagent can act equally as both a base and a nucleophile, use other factors to decide whether substitution or elimination predominates. These include the degree of substitution about the reacting center (1° haloalkanes will not undergo E1 or S_N1 reactions, 3° haloalkanes will not undergo S_N2 reactions) or type of solvent (E1 and S_N1 reactions require polar protic solvents).

Solution

(a) A 3° halide is heated with a strong base/good nucleophile. Elimination by an E2 reaction predominates to give 2-methyl-2-butene as the major product:

[structure: 2-chloro-2-methylbutane] + Na^+OH^- $\xrightarrow[H_2O]{80°C}$ [structure: 2-methyl-2-butene] + NaCl + H_2O

(b) Reaction of a 1° halide with triethylamine, a moderate nucleophile/weak base, gives substitution by an S_N2 reaction:

[structure: 1-bromo-3-methylbutane] + $(C_2H_5)_3N$ $\xrightarrow[CH_2Cl_2]{30°C}$ [structure: 3-methylbutyl]$\overset{+}{N}(C_2H_5)_3Br^-$

See problem 7.43

Problem 7.8

Predict whether each reaction proceeds predominantly by substitution (S_N1 or S_N2) or elimination (E1 or E2) or whether the two compete, and write structural formulas for the major organic product(s):

(a) [structure: 3-bromopentane] + $CH_3O^-\ Na^+$ $\xrightarrow{\text{methanol}}$

(b) [structure: substituted cyclohexane with Cl (hashed wedge) and methyl (wedge)] + $Na^+\ OH^-$ $\xrightarrow{\text{acetone}}$

CHEMICAL CONNECTIONS 7B

The Effect of Chlorofluorocarbon Legislation on Asthma Sufferers

The Montreal Protocol on Substances That Deplete the Ozone Layer was proposed in 1987 and enacted in 1989. As a result of this treaty, and its many revisions, the phaseout of CFCs and many other substances harmful to the ozone layer has been achieved in many industrialized nations. However, the Montreal Protocol provided exceptions for products in which the use of CFCs was essential because no viable alternatives existed. One such product was albuterol metered-dose inhalers, which use CFCs as propellants to deliver the drug and are used by asthma patients worldwide. In the United States, this exemption from the Montreal Protocol expired in December 2008, thanks to the Clean Air Act and the availability of another type of propellant known as hydrofluoroalkanes (HFAs). One drawback of HFA-equipped inhalers is that of cost; HFA inhalers cost three to six times as much as CFC-enabled inhalers because generic versions do not yet exist. This has sparked concerns from patients, physicians, and patients' rights groups over the ability of the nearly 23 million people in the United States who suffer from asthma to obtain treatment. Other differences include taste, smell, temperature of inhalant upon ejection, and effectiveness in colder climates and higher altitudes (HFAs are more effective under these conditions than CFCs). These practical differences are a result of the absence of chlorine in HFAs versus in CFCs, and are an excellent example of how changes in chemical structure can affect the properties of molecules and their ultimate applications in society.

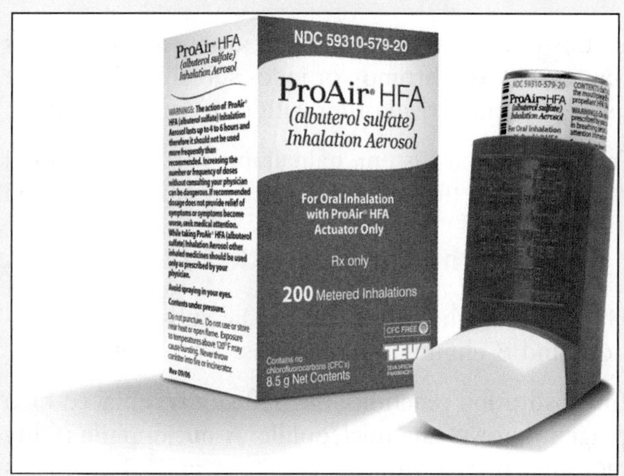

(Courtesy Teva USA)

```
    F  F               F  F  F
    |  |               |  |  |
 F—C—C—H          F—C—C—C—F
    |  |               |  |  |
    F  H               F  H  F
  HFA-134a            HFA-227
```

Hydrofluoroalkanes used in CFC-free medical inhalers

QUESTION

Would you expect HFA-134a or HFA-227 to undergo an S_N1 reaction? An S_N2 reaction? Why or why not?

Key Terms and Concepts

alkyl halide (p. 215)
α-carbon (p. 236)
aprotic solvent (p. 230)
backside attack (p. 224)
β-carbon (p. 236)
β-elimination reaction (p. 219)
bimolecular reaction (p. 223)
chlorofluorocarbons (p. 218)
concerted (p. 223)
dehydrohalogenation (p. 235)
DMSO (p. 231)
E1 (p. 239)
E2 (p. 239)
electronic factors (p. 227)
haloform (p. 216)
haloalkane (p. 215)
inversion of configuration (p. 224)
leaving group (p. 218)
naked ions (p. 231)
nucleophile (p. 218)
nucleophilic substitution (p. 218)
nucleophilicity (p. 227)
protic solvent (p. 230)
racemization (p. 226)
relative nucleophilicity (p. 227)
S_N1 (p. 225)
S_N2 (p. 223)
solvolysis (p. 225)
steric factors (p. 227)
steric hindrance (p. 227)
unimolecular reaction (p. 225)
Zaitsev's rule (p. 236)

Summary of Key Questions

- Haloalkanes are compounds that contain a halogen covalently bonded to an sp^3-hybridized carbon.
- Haloalkanes can be converted to a variety of other functional groups through substitution and elimination reactions.

7.1 How Are Haloalkanes Named?

- In the IUPAC system, halogen atoms are named as fluoro-, chloro-, bromo-, or iodo- substituents and are listed in alphabetical order with other substituents.
- In the common system, haloalkanes are named alkyl halides, where the name is derived by naming the alkyl group followed by the name of the halide as a separate word (e.g., methyl chloride).
- Compounds of the type CHX_3 are called haloforms.

7.2 What Are the Characteristic Reactions of Haloalkanes?

- Haloalkanes undergo nucleophilic substitution reactions and β-elimination reactions.
- In substitution reactions, the halogen is replaced by a reagent known as a nucleophile. A nucleophile is any molecule or ion with an unshared pair of electrons that can be donated to another atom or ion to form a new covalent bond; alternatively, a nucleophile is a Lewis base.
- In elimination reactions, the halogen and an adjacent hydrogen are removed to form an alkene.

7.3 What Are the Products of Nucleophilic Aliphatic Substitution Reactions?

- The product of a nucleophilic substitution reaction varies depending on the nucleophile used in the reaction. For example, when the nucleophile is hydroxide (HO^-), the product will be an alcohol (ROH).
- Nucleophilic substitution reactions can be used to transform haloalkanes into alcohols, ethers, thiols, sulfides, alkyl iodides, and alkyl ammonium ions, to name a few.

7.4 What Are the S_N2 and S_N1 Mechanisms for Nucleophilic Substitution?

- An S_N2 reaction occurs in one step. The departure of the leaving group is assisted by the incoming nucleophile, and both nucleophile and leaving group are involved in the transition state. S_N2 reactions are stereoselective; reaction at a stereocenter proceeds with inversion of configuration.
- An S_N1 reaction occurs in two steps. Step 1 is a slow, rate-determining ionization of the C—X bond to form a carbocation intermediate, followed in Step 2 by its rapid reaction with a nucleophile to complete the substitution. For S_N1 reactions taking place at a stereocenter, the major reaction occurs with racemization.

7.5 What Determines Whether S_N1 or S_N2 Predominates?

- The stability of the leaving group. The ability of a group to function as a leaving group is related to its stability as an anion. The most stable anions and the best leaving groups are the conjugate bases of strong acids.
- The nucleophilicity of a reagent. Nucelophilicity is measured by the rate of its reaction in a reference nucleophilic substitution.
- The structure of the haloalkane. S_N1 reactions are governed by electronic factors, namely, the relative stabilities of carbocation intermediates. S_N2 reactions are governed by steric factors, namely, the degree of crowding around the site of substitution.
- The nature of the solvent. Protic solvents contain —OH groups, interact strongly with polar molecules and ions, and are good solvents in which to form carbocations. Protic solvents favor S_N1 reactions. Aprotic solvents do not contain —OH groups. Common aprotic solvents are dimethyl sulfoxide, acetone, diethyl ether, and dichloromethane. Aprotic solvents do not interact as strongly with polar molecules and ions, and carbocations are less likely to form in them. Aprotic solvents favor S_N2 reactions.

- A nonhalogenated compound with a good leaving group can, like haloalkanes, undergo substitution reactions.
- Halogens make good leaving groups because either their size (as in I^- or Br^-) or electronegativity (Cl^-) help to stabilize the resulting negative charge. F^- is not a good leaving group because HF is a weak acid. HCl, HBr, and HI are strong acids, making their halide anions weak bases.
- The ability of a group to function as a leaving group is related to how stable it is as an anion.
- The most stable anions and the best leaving groups are the conjugate bases of strong acids.

7.6 How Can S_N1 and S_N2 Be Predicted Based on Experimental Conditions?

- Predictions about the mechanism for a particular nucleophilic substitution reaction must be based on considerations of the structure of the haloalkane, the nucleophile, the leaving group, and the solvent.

7.7 What Are the Products of β-Elimination?

- Dehydrohalogenation, a type of β-elimination reaction, is the removal of H and X from adjacent carbon atoms, resulting in the formation of a carbon–carbon double bond.
- A β-elimination that gives the most highly substituted alkene is called Zaitsev elimination.

7.8 What Are the E1 and E2 Mechanisms for β-Elimination?

- An E1 reaction occurs in two steps: breaking the C—X bond to form a carbocation intermediate, followed by the loss of an H^+ to form the alkene.
- An E2 reaction occurs in one step: reaction with the base to remove an H^+, formation of the alkene, and departure of the leaving group, all occurring simultaneously.

7.9 When Do Nucleophilic Substitution and β-Elimination Compete?

- Many of the nucleophiles we have examined—for example, hydroxide ion and alkoxide ions—are also strong bases. As a result, nucleophilic substitution and β-elimination often compete with each other, and the ratio of products formed by these reactions depends on the relative rates of the two reactions.

Quick Quiz

Answer true or false to the following questions to assess your general knowledge of the concepts in this chapter. If you have difficulty with any of them, you should review the appropriate section in the chapter (shown in parentheses) before attempting the more challenging end-of-chapter problems.

1. An S_N1 reaction can result in two products that are stereoisomers. (7.4)
2. In naming halogenated compounds, "haloalkane" is the IUPAC form of the name while "alkyl halide" is the common form of the name. (7.1)
3. A substitution reaction results in the formation of an alkene. (7.3)
4. Sodium ethoxide ($CH_3CH_2O^-$ Na^+) can act as a base and as a nucleophile in its reaction with bromocyclohexane. (7.2)
5. The rate law of the E2 reaction is dependent on just the haloalkane concentration. (7.8)
6. The mechanism of the S_N1 reaction involves the formation of a carbocation intermediate. (7.4)
7. Polar protic solvents are required for E1 or S_N1 reactions to occur. (7.9)
8. OH^- is a better leaving group than Cl^-. (7.5)
9. When naming haloalkanes with more than one type of halogen, numbering priority is given to the halogen with the higher mass. (7.1)

10. S_N2 reactions prefer good nucleophiles, while S_N1 reactions proceed with most any nucleophile. (7.5)
11. The stronger the base, the better is the leaving group. (7.5)
12. S_N2 reactions are more likely to occur with 2° haloalkanes than with 1° haloalkanes. (7.9)
13. A solvolysis reaction is a reaction performed without solvent. (7.4)
14. The degree of substitution at the reaction center affects the rate of an S_N1 reaction but not an S_N2 reaction. (7.5)
15. A reagent must possess a negative charge to react as a nucleophile. (7.3)
16. Elimination reactions favor the formation of the more substituted alkene. (7.7)
17. The best leaving group is one that is unstable as an anion. (7.5)
18. In the S_N2 reaction, the nucleophile attacks the carbon from the side opposite that of the leaving group. (7.4)
19. Only haloalkanes can undergo substitution reactions. (7.5)
20. All of the following are polar aprotic solvents; acetone, DMSO, ethanol. (7.5)

Answers: (1) T (2) T (3) F (4) T (5) F (6) T (7) T (8) F (9) F (10) T (11) F (12) F (13) F (14) F (15) F (16) T (17) F (18) T (19) F (20) F

Key Reactions

1. Nucleophilic Aliphatic Substitution: S_N2 (Section 7.4A)

S_N2 reactions occur in one step, and both the nucleophile and the leaving group are involved in the transition state of the rate-determining step. The nucleophile may be negatively charged or neutral. S_N2 reactions result in an inversion of configuration at the reaction center. They are accelerated in polar aprotic solvents, compared with polar protic solvents. S_N2 reactions are governed by steric factors, namely, the degree of crowding around the site of reaction.

$$I^- + (CH_3CH_2)(H)(H_3C)C{-}Cl \longrightarrow I{-}C(CH_2CH_3)(H)(CH_3) + Cl^-$$

$$(CH_3)_3N + (CH_3CH_2)(H)(H_3C)C{-}Cl \longrightarrow (CH_3)_3\overset{+}{N}{-}C(CH_2CH_3)(H)(CH_3) + Cl^-$$

2. Nucleophilic Aliphatic Substitution: S_N1 (Section 7.4B)

An S_N1 reaction occurs in two steps. Step 1 is a slow, rate-determining ionization of the C—X bond to form a carbocation intermediate, followed in Step 2 by its rapid reaction with a nucleophile to complete the substitution. Reaction at a stereocenter gives a racemic product. S_N1 reactions are governed by electronic factors, namely, the relative stabilities of carbocation intermediates:

$$\text{1-chloro-1-methylcyclohexane } (CH_3, Cl) + CH_3CH_2OH \longrightarrow \text{1-ethoxy-1-methylcyclohexane } (CH_3, OCH_2CH_3) + HCl$$

3. β-Elimination: E1 (Section 7.8A)

E1 reactions involve the elimination of atoms or groups of atoms from adjacent carbons. Reaction occurs in two steps and involves the formation of a carbocation intermediate:

$$\text{(Cl-substituted alkane)} \xrightarrow[CH_3COOH]{E1} \text{(alkene)} + HCl$$

4. β-Elimination: E2 (Section 7.8B)

An E2 reaction occurs in one step: reaction with base to remove a hydrogen, formation of the alkene, and departure of the leaving group, all occurring simultaneously:

$$\text{(Br-substituted alkane)} \xrightarrow[CH_3OH]{CH_3O^-Na^+} \text{(internal alkene) (74\%)} + \text{(terminal alkene) (26\%)}$$

Problems

A problem marked with an asterisk indicates an applied "real world" problem. Answers to problems whose numbers are printed in blue are given in Appendix D.

Section 7.1 Nomenclature

7.9 Write the IUPAC name for each compound: **(See Example 7.1)**

(a) $CH_2{=}CF_2$

(b) [cyclopentene ring bearing Br]

(c) [skeletal structure with Cl]

(d) $Cl(CH_2)_6Cl$

(e) CF_2Cl_2

(f) [skeletal structure with Br]

7.10 Write the IUPAC name for each compound (be certain to include a designation of configuration, where appropriate, in your answer): **(See Example 7.1)**

(a) [structure with H, Br]

(b) H_3C–[cyclohexane]–Br

(c) [cyclohexene with Cl]

(d) [structure with Cl]

(e) [structure with Br, Cl]

(f) [structure with Br, H, CH_3, H, Br, H_3C]

7.11 Draw a structural formula for each compound (given are IUPAC names): **(See Example 7.1)**

(a) 3-Bromopropene

(b) (*R*)-2-Chloropentane

(c) meso-3,4-Dibromohexane

(d) *trans*-1-Bromo-3-isopropylcyclohexane

(e) 1,2-Dichloroethane

(f) Bromocyclobutane

7.12 Draw a structural formula for each compound (given are common names): **(See Example 7.1)**

(a) Isopropyl chloride

(b) *sec*-Butyl bromide

(c) Allyl iodide

(d) Methylene chloride

(e) Chloroform

(f) *tert*-Butyl chloride

(g) Isobutyl chloride

7.13 Which compounds are 2° alkyl halides?

(a) Isobutyl chloride

(b) 2-Iodooctane

(c) *trans*-1-Chloro-4-methylcyclohexane

Synthesis of Alkyl Halides

7.14 What alkene or alkenes and reaction conditions give each alkyl halide in good yield? (*Hint*: Review Chapter 5.) **(See Example 5.2)**

(a) [cyclopentane with Br]

(b) $CH_3C(CH_3)(Br)CH_2CH_2CH_3$

(c) [cyclohexane with CH_3 and Cl]

7.15 Show reagents and conditions that bring about these conversions: **(See Example 5.2)**

(a) [alkene] ⟶ [alkyl chloride with Cl]

(b) $CH_3CH_2CH{=}CH_2 \longrightarrow CH_3CH_2CH(I)CH_3$

(c) $CH_3CH{=}CHCH_3 \longrightarrow CH_3CH(Cl)CH_2CH_3$

(d) [1-methylcyclopentene, CH_3] ⟶ [cyclopentane with CH_3 and Br]

Sections 7.2–7.6 Nucleophilic Aliphatic Substitution

7.16 Write structural formulas for these common organic solvents:

(a) Dichloromethane (b) Acetone

(c) Ethanol (d) Diethyl ether

(e) Dimethyl sulfoxide

7.17 Arrange these protic solvents in order of increasing polarity:

(a) H_2O (b) CH_3CH_2OH

(c) CH_3OH

7.18 Arrange these aprotic solvents in order of increasing polarity:

(a) Acetone (b) Pentane

(c) Diethyl ether

7.19 From each pair, select the better nucleophile:

(a) H_2O or OH^-

(b) CH_3COO^- or OH^-

(c) CH_3SH or CH_3S^-

7.20 Which statements are true for S_N2 reactions of haloalkanes? **(See Example 7.4)**

(a) Both the haloalkane and the nucleophile are involved in the transition state.

(b) The reaction proceeds with inversion of configuration at the substitution center.

(c) The reaction proceeds with retention of optical activity.

(d) The order of reactivity is 3° > 2° > 1° > methyl.

(e) The nucleophile must have an unshared pair of electrons and bear a negative charge.

(f) The greater the nucleophilicity of the nucleophile, the greater is the rate of reaction.

7.21 Complete these S_N2 reactions: **(See Examples 7.3, 7.5)**

(a) $Na^+I^- + CH_3CH_2CH_2Cl \xrightarrow{\text{acetone}}$

(b) NH_3 + (cyclohexyl)–Br $\xrightarrow{\text{ethanol}}$

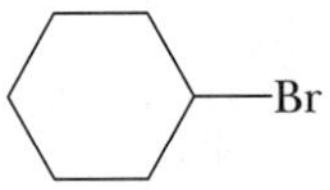

(c) $CH_3CH_2O^-Na^+ + CH_2{=}CHCH_2Cl \xrightarrow{\text{ethanol}}$

7.22 Complete these S_N2 reactions: **(See Examples 7.3, 7.5)**

(a) 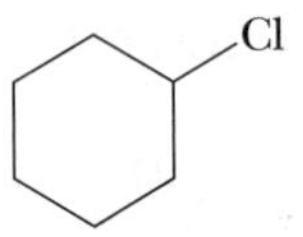 (cyclohexane ring bearing Cl) $+ CH_3\overset{O}{\overset{\|}{C}}O^-Na^+ \xrightarrow{\text{ethanol}}$

(b) $CH_3\overset{I}{\overset{|}{C}}HCH_2CH_3 + CH_3CH_2S^-Na^+ \xrightarrow{\text{acetone}}$

(c) $CH_3\overset{CH_3}{\overset{|}{C}}HCH_2CH_2Br + Na^+I^- \xrightarrow{\text{acetone}}$

(d) $(CH_3)_3N + CH_3I \xrightarrow{\text{acetone}}$

(e) 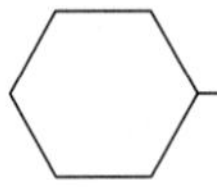 (cyclohexyl)–$CH_2Br + CH_3O^-Na^+ \xrightarrow{\text{methanol}}$

(f) 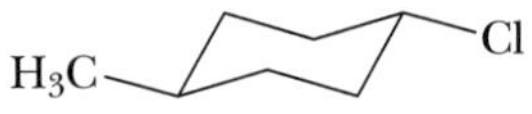 H_3C–(cyclohexane chair)–Cl $+ CH_3S^-Na^+ \xrightarrow{\text{ethanol}}$

(g) 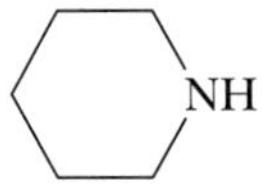 (piperidine ring) NH $+ CH_3(CH_2)_6CH_2Cl \xrightarrow{\text{ethanol}}$

(h) 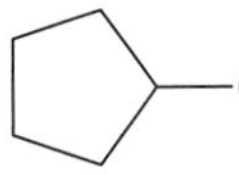 (cyclopentyl)–$CH_2Cl + NH_3 \xrightarrow{\text{ethanol}}$

7.23 You were told that each reaction in Problem 7.22 proceeds by an S_N2 mechanism. Suppose you were not told the mechanism. Describe how you could conclude, from the structure of the haloalkane, the nucleophile, and the solvent, that each reaction is in fact an S_N2 reaction. **(See Examples 7.4, 7.5)**

7.24 In the following reactions, a haloalkane is treated with a compound that has two nucleophilic sites. Select the more nucleophilic site in each part, and show the product of each S_N2 reaction:

(a) $HOCH_2CH_2NH_2 + CH_3I \xrightarrow{\text{ethanol}}$

(b) 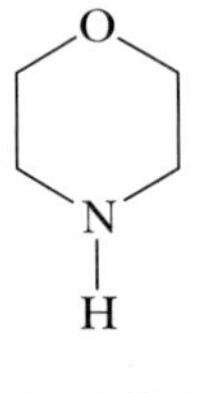 (morpholine ring: O, N–H) $+ CH_3I \xrightarrow{\text{ethanol}}$

(c) $HOCH_2CH_2SH + CH_3I \xrightarrow{\text{ethanol}}$

7.25 Which statements are true for S_N1 reactions of haloalkanes? **(See Example 7.5)**

(a) Both the haloalkane and the nucleophile are involved in the transition state of the rate-determining step.

(b) The reaction at a stereocenter proceeds with retention of configuration.

(c) The reaction at a stereocenter proceeds with loss of optical activity.

(d) The order of reactivity is 3° > 2° > 1° > methyl.

(e) The greater the steric crowding around the reactive center, the lower is the rate of reaction.

(f) The rate of reaction is greater with good nucleophiles compared with poor nucleophiles.

7.26 Draw a structural formula for the product of each S_N1 reaction: **(See Examples 7.3, 7.5)**

(a) $CH_3CHClCH_2CH_3 + CH_3CH_2OH \xrightarrow{\text{ethanol}}$
S enantiomer

(b) 1-chloro-1-methylcyclopentane $+ CH_3OH \xrightarrow{\text{methanol}}$

(c) $(CH_3)_3CCl + CH_3COOH \xrightarrow{\text{acetic acid}}$

(d) 3-bromocyclohexene $-Br + CH_3OH \xrightarrow{\text{methanol}}$

(e) Br $+ CH_3CH_2OH \xrightarrow{\text{ethanol}}$

(f) Cl $+ CH_3COOH \xrightarrow{\text{acetic acid}}$

7.27 You were told that each substitution reaction in Problem 7.26 proceeds by an S_N1 mechanism. Suppose that you were not told the mechanism. Describe how you could conclude, from the structure of the haloalkane, the nucleophile, and the solvent, that each reaction is in fact an S_N1 reaction. **(See Examples 7.4, 7.5)**

7.28 Select the member of each pair that undergoes nucleophilic substitution in aqueous ethanol more rapidly: **(See Example 7.5)**

(a) Cl or Cl

(b) Br or Br

(c) Br or CH_3 Br

7.29 Propose a mechanism for the formation of the products (but not their relative percentages) in this reaction:

$$(CH_3)_3CCl \xrightarrow[25°C]{20\% H_2O,\ 80\% CH_3CH_2OH} \underbrace{(CH_3)_3COCH_2CH_3 + (CH_3)_3COH}_{85\%} + \underset{15\%}{CH_3C(CH_3){=}CH_2} + HCl$$

7.30 The rate of reaction in Problem 7.29 increases by 140 times when carried out in 80% water to 20% ethanol, compared with 40% water to 60% ethanol. Account for this difference.

7.31 Select the member of each pair that shows the greater rate of S_N2 reaction with KI in acetone:

(a) Cl or Cl

(b) Cl or Br

(c) Cl or Cl

(d) Br or Br

7.32 What hybridization best describes the reacting carbon in the S_N2 transition state?

7.33 Haloalkenes such as vinyl bromide, $CH_2{=}CHBr$, undergo neither S_N1 nor S_N2 reactions. What factors account for this lack of reactivity?

7.34 Show how you might synthesize the following compounds from a haloalkane and a nucleophile: **(See Example 7.5)**

(a) CN (b) CN

(c) O O (d) SH

(e) 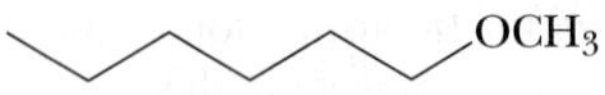 (f)

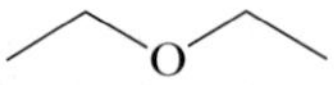

(g)

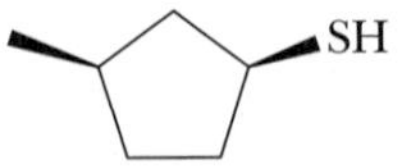

7.35 Show how you might synthesize each compound from a haloalkane and a nucleophile: **(See Example 7.5)**

(a) 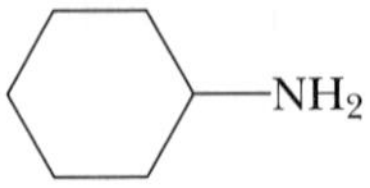 $-NH_2$ (b) 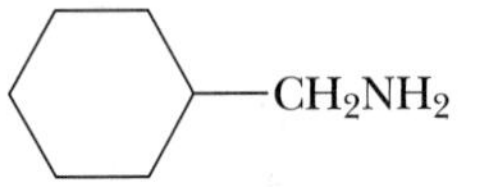 $-CH_2NH_2$

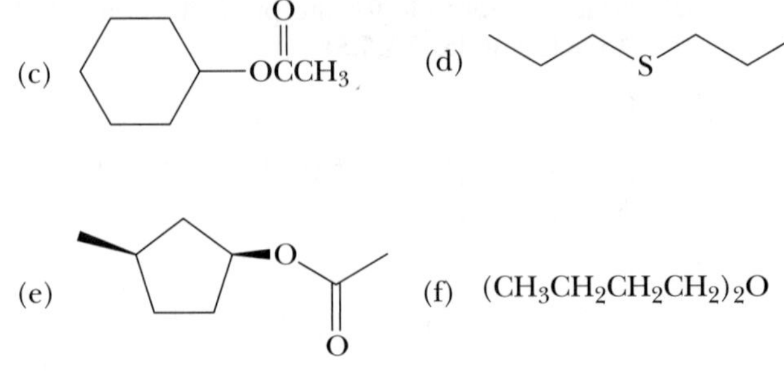

(e) (f) $(CH_3CH_2CH_2CH_2)_2O$

Sections 7.7–7.8 β-Eliminations

7.36 Draw structural formulas for the alkene(s) formed by treating each of the following haloalkanes with sodium ethoxide in ethanol. Assume that elimination is by an E2 mechanism. Where two alkenes are possible, use Zaitsev's rule to predict which alkene is the major product: **(See Examples 7.6, 7.7)**

(a) 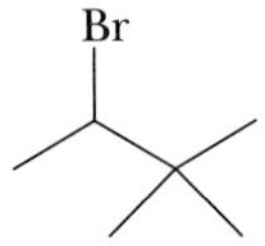 (b) 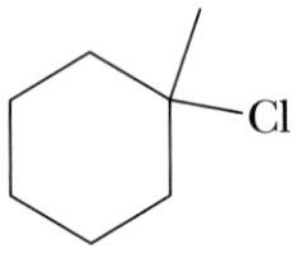 (c) 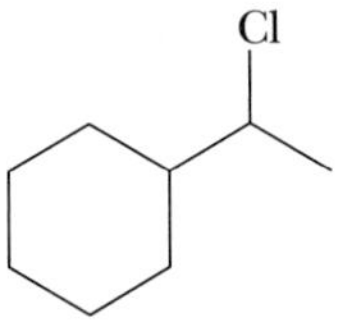 (d)

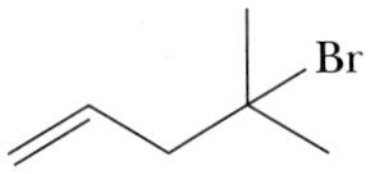

7.37 Which of the following haloalkanes undergo dehydrohalogenation to give alkenes that do not show *cis–trans* isomerism? **(See Examples 7.6, 7.7)**

(a) 2-Chloropentane (b) 2-Chlorobutane
(c) Chlorocyclohexane (d) Isobutyl chloride

7.38 How many isomers, including *cis–trans* isomers, are possible for the major product of dehydrohalogenation of each of the following haloalkanes? **(See Examples 7.6, 7.7)**

(a) 3-Chloro-3-methylhexane
(b) 3-Bromohexane

7.39 What haloalkane might you use as a starting material to produce each of the following alkenes in high yield and uncontaminated by isomeric alkenes?

(a) 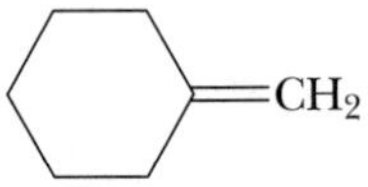 $=CH_2$ (b) $CH_3\overset{CH_3}{\overset{|}{C}}HCH_2CH{=}CH_2$

7.40 For each of the following alkenes, draw structural formulas of all chloroalkanes that undergo dehydrohalogenation when treated with KOH to give that alkene as the major product (for some parts, only one chloroalkane gives the desired alkene as the major product; for other parts, two chloroalkanes may work):

(a) 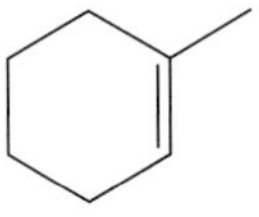 (b) 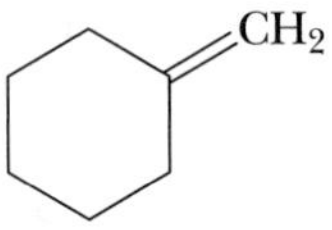 (c) 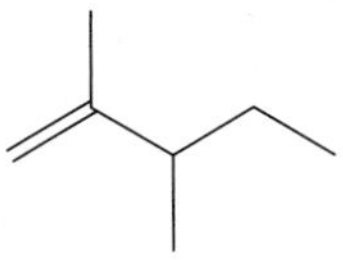 (d) 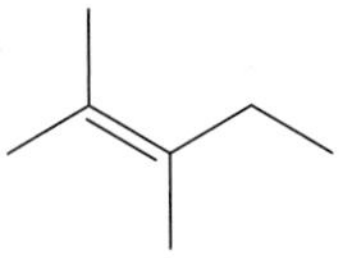 (e)

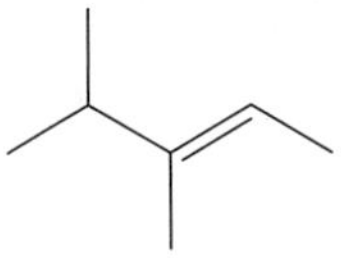

7.41 When *cis*-4-chlorocyclohexanol is treated with sodium hydroxide in ethanol, it gives only the substitution product *trans*-1,4-cyclohexanediol (1). Under the same experimental conditions, *trans*-4-chlorocyclohexanol gives 3-cyclohexenol (2) and product (3):

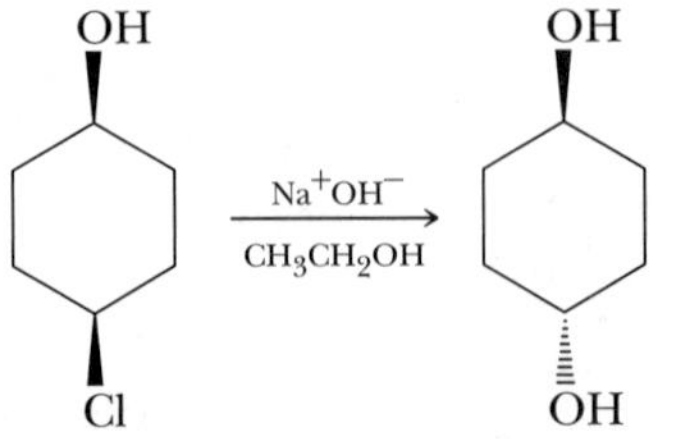

$\xrightarrow[CH_3CH_2OH]{Na^+OH^-}$

cis-4-Chloro-cyclohexanol (1)

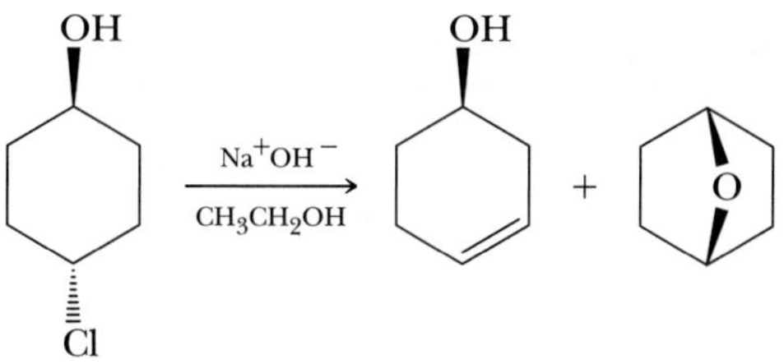

OH OH

Na$^+$OH$^-$ / CH_3CH_2OH → (2) + (3)

Cl

trans-4-Chloro-cyclohexanol (2) (3)

(a) Propose a mechanism for the formation of product (1), and account for its configuration.

(b) Propose a mechanism for the formation of product (2).

(c) Account for the fact that the product (3) is formed from the *trans* isomer, but not from the *cis* isomer.

Section 7.9 Synthesis and Predict the Product

7.42 Show how to convert the given starting material into the desired product (note that some syntheses require only one step, whereas others require two or more steps):

use Ch. 5 rxns

(a) 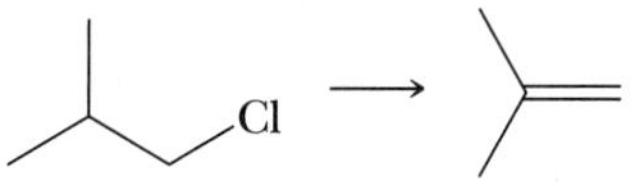 Cl →

(b) 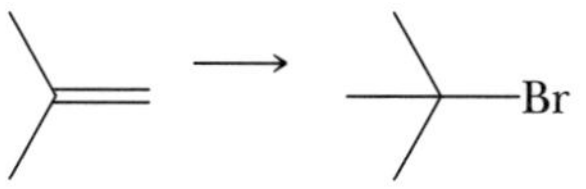 → Br

(c) 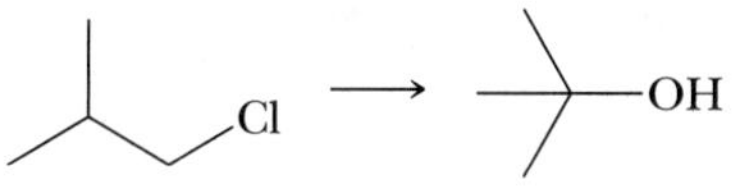 Cl → OH

(d) 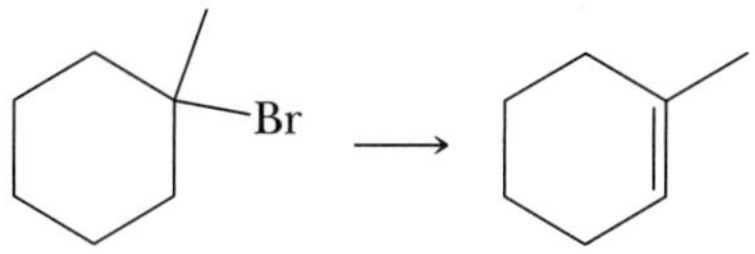 Br →

(e) 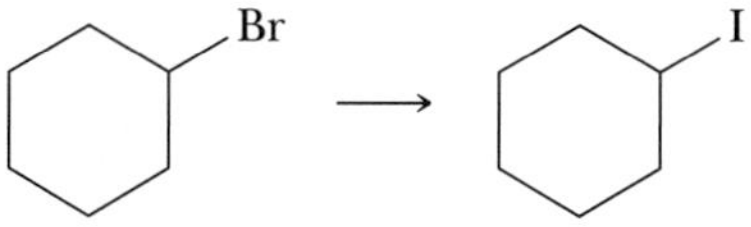 Br → I

(f) 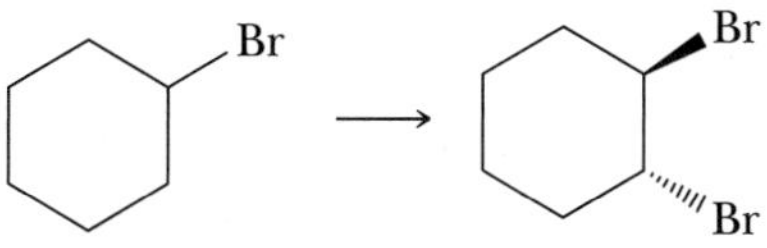 Br → Br, Br

7.43 Complete these reactions by determining the type of reaction and mechanism (S_N1, S_N2, E1, or E2) that they undergo. **(See Example 7.8)**

(a) 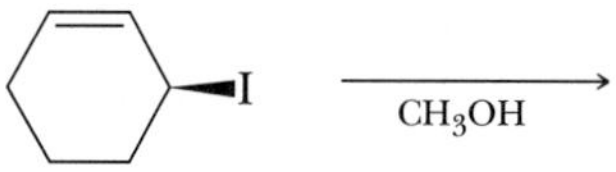 I → CH_3OH

(b) 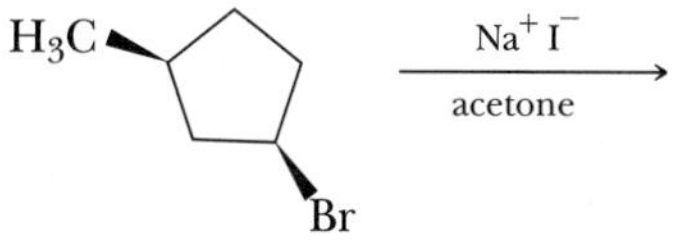 H_3C, Br → Na$^+$ I$^-$ / acetone

(c) 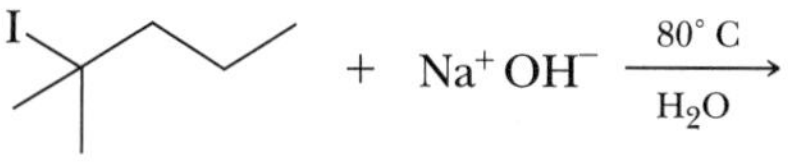 I + Na$^+$ OH$^-$ → 80° C / H_2O

(d) 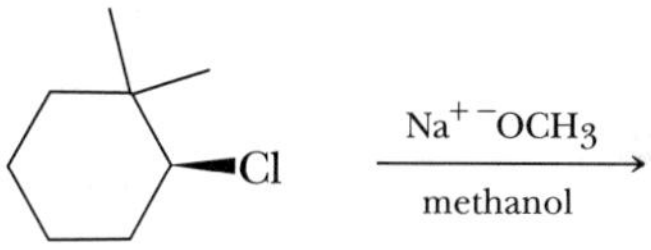 Cl → Na$^+$ $^-OCH_3$ / methanol

(e) 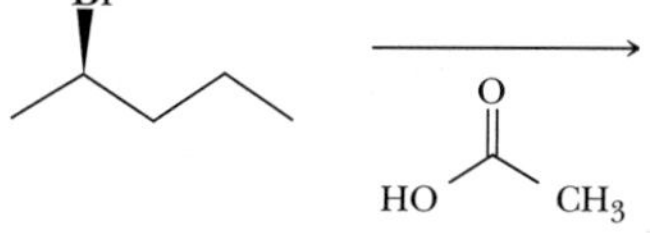 Br → HO(C=O)CH_3 (acetic acid)

(f) 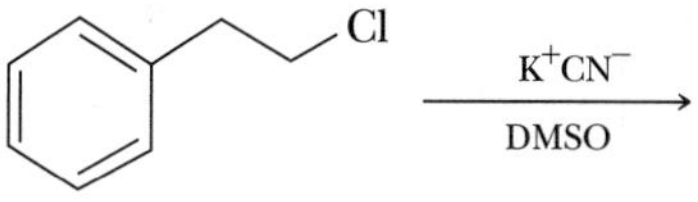 Cl → K$^+$CN$^-$ / DMSO

Chemical Transformations

7.44 Test your cumulative knowledge of the reactions learned thus far by completing the following chemical transformations. *Note: Some will require more than one step.*

(a) 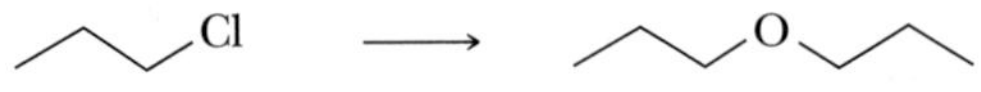 Cl ⟶ O

(b) 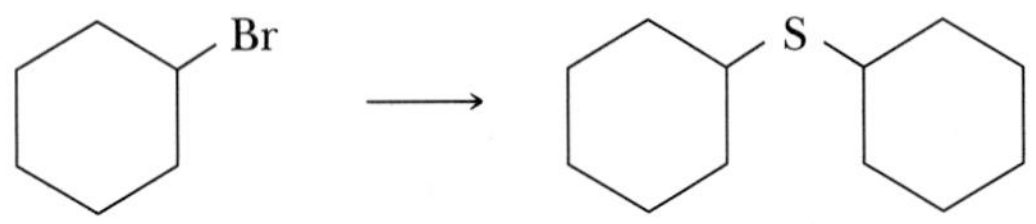 Br ⟶ S

(c)  I ⟶ Br

(d) 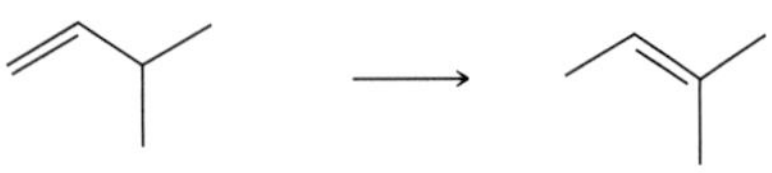 ⟶

(e) 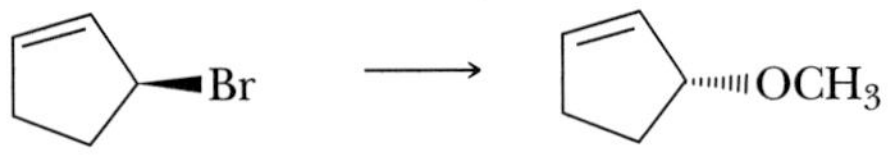 Br ⟶ OCH_3

(f) 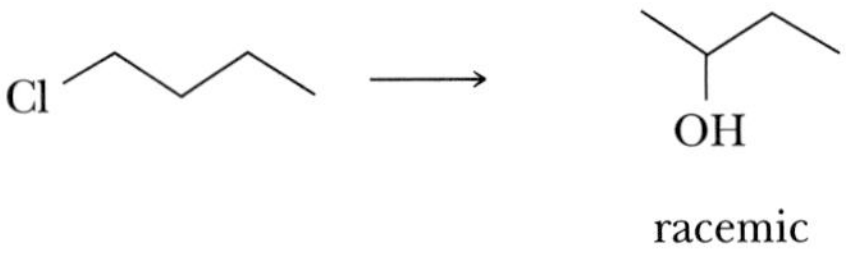 Cl ⟶ OH
racemic

(g) 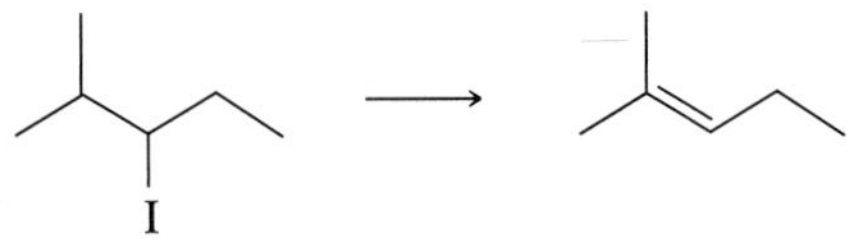 I ⟶

(h)  Cl ⟶ Br

(i) Br ⟶ CH_3 H OH
racemic

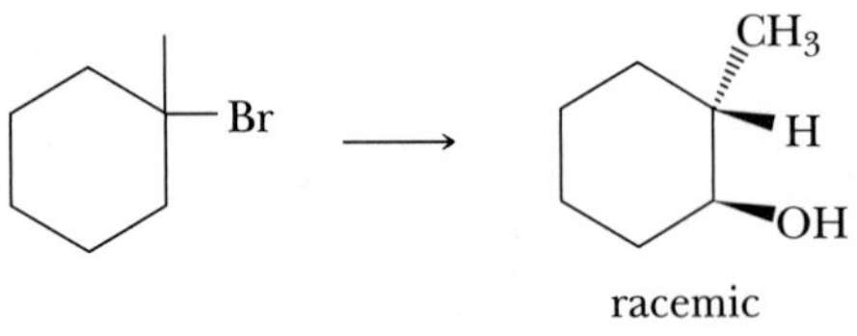

(j) 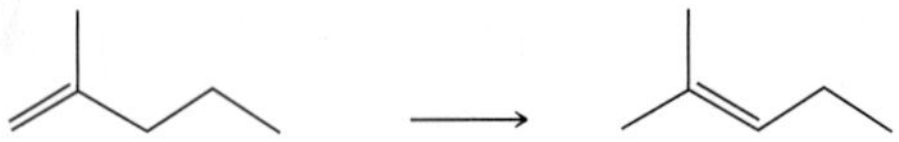 ⟶

(k)  I ⟶ HO

(l) 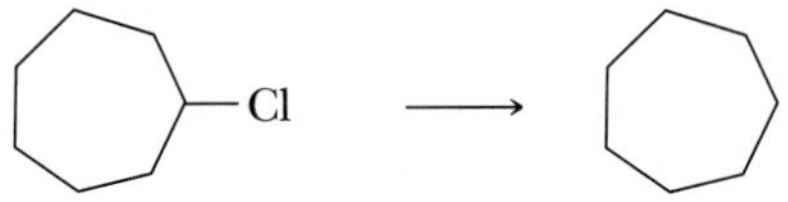 Cl ⟶

(m) 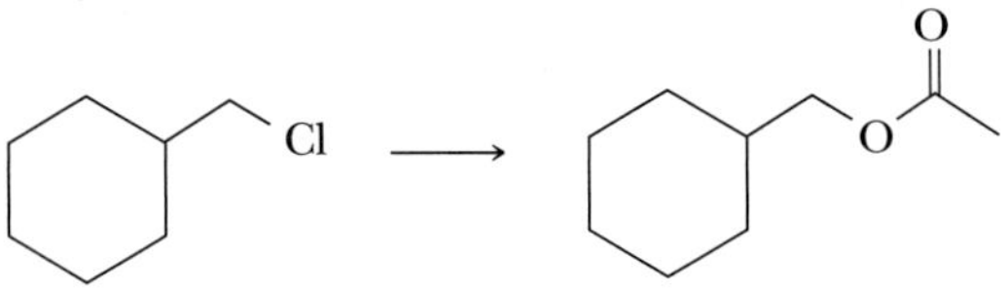 Cl ⟶ O O

(n) 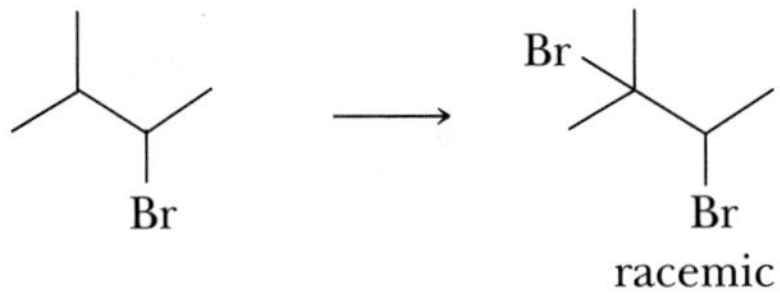 Br ⟶ Br Br
racemic

(o) ⟶ OCH_2CH_3

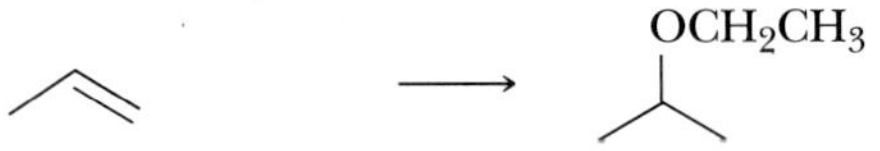

(p) 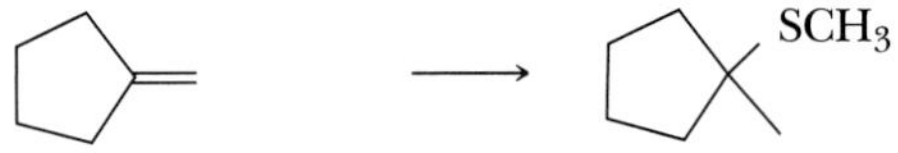 ⟶ SCH_3

(q)  Cl ⟶ Br Br

(r) 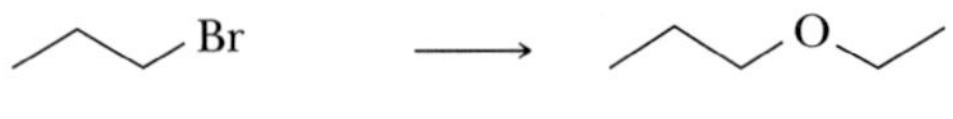 Br ⟶ O

Looking Ahead

7.45 The Williamson ether synthesis involves treating a haloalkane with a metal alkoxide. Following are two reactions intended to give benzyl *tert*-butyl ether. One reaction gives the ether in good yield, the other does not. Which reaction gives the ether? What is the product of the other reaction, and how do you account for its formation?

(a) $(CH_3)_3CO^- K^+ + C_6H_5CH_2Cl \xrightarrow{DMSO} (CH_3)_3COCH_2C_6H_5 + KCl$

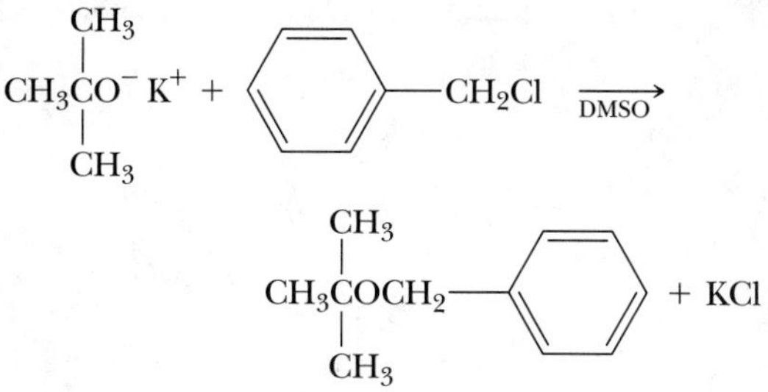

(b) $C_6H_5CH_2O^- K^+ + (CH_3)_3CCl \xrightarrow{DMSO} (CH_3)_3COCH_2C_6H_5 + KCl$

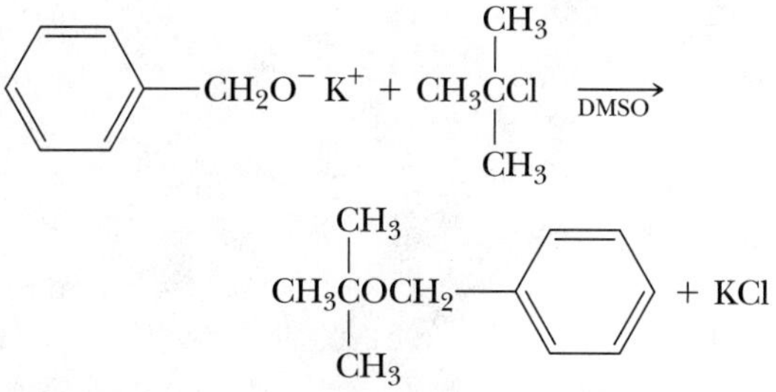

7.46 The following ethers can, in principle, be synthesized by two different combinations of haloalkane or halocycloalkane and metal alkoxide. Show one combination that forms ether bond (1) and another that forms ether bond (2). Which combination gives the higher yield of ether?

(a) (1) (2)

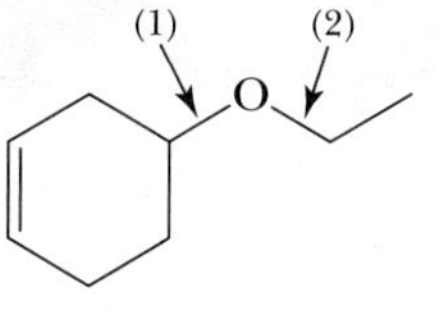

(b) (1) (2)

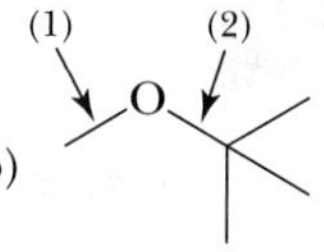

(c) (1) (2)

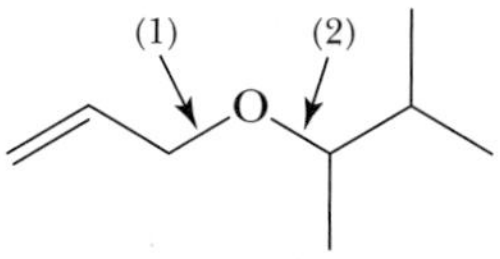

7.47 Propose a mechanism for this reaction:

$Cl—CH_2—CH_2—OH \xrightarrow{Na_2CO_3,\ H_2O} H_2C—CH_2$ (with O bridging)

2-Chloroethanol — Ethylene oxide

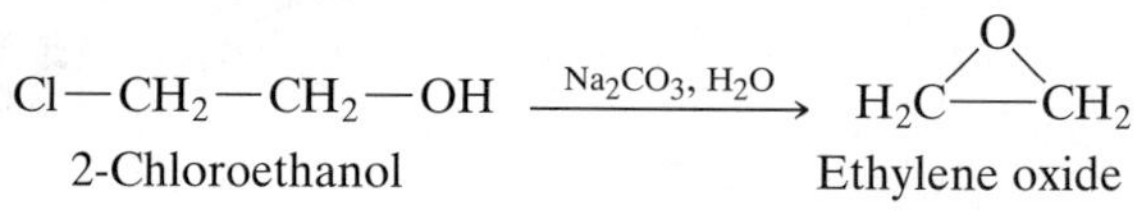

7.48 An OH group is a poor leaving group, and yet substitution occurs readily in the following reaction. Propose a mechanism for this reaction that shows how OH overcomes its limitation of being a poor leaving group.

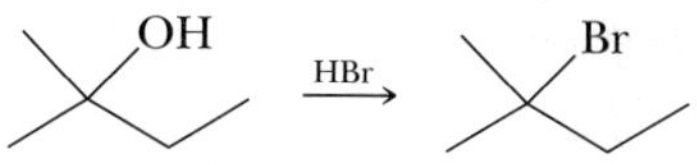

OH → (HBr) → Br

7.49 Explain why (*S*)-2-bromobutane becomes optically inactive when treated with sodium bromide in DMSO:

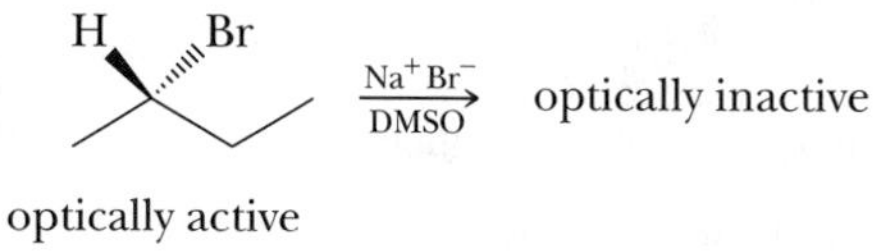

H Br → $\xrightarrow[DMSO]{Na^+Br^-}$ optically inactive

optically active

7.50 Explain why phenoxide is a much poorer nucleophile than cyclohexoxide:

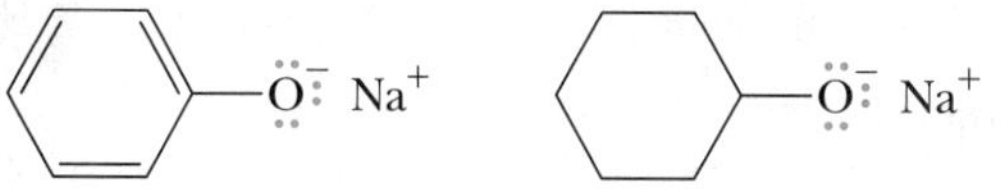

$\ddot{O}^{-}$ Na^+ — $\ddot{O}^{-}$ Na^+

Sodium phenoxide — Sodium cyclohexoxide

7.51 In ethers, each side of the oxygen is essentially an OR group and is thus a poor leaving group. Epoxides are three-membered ring ethers. Explain why an epoxide reacts readily with a nucleophile despite being an ether.

$R—O—R + :Nu^- \longrightarrow$ no reaction

An ether

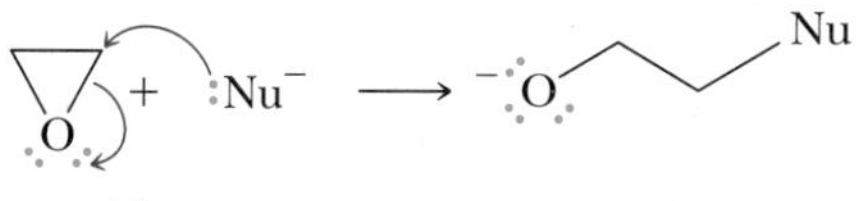

An epoxide $+ :Nu^- \longrightarrow {}^-:\ddot{O}$ ... Nu

An epoxide

8 Alcohols, Ethers, and Thiols

KEY QUESTIONS

8.1 What Are Alcohols?

8.2 What Are the Characteristic Reactions of Alcohols?

8.3 What Are Ethers?

8.4 What Are Epoxides?

8.5 What Are Thiols?

8.6 What Are the Characteristic Reactions of Thiols?

HOW TO

8.1 How to Name Cyclic Alcohols

8.2 How to Predict Relative Boiling Points of Compounds of Similar Molecular Weight

8.3 How to Predict the Position of Equilibrium of an Acid–Base Reaction

8.4 How to Complete a Dehydration Reaction

8.5 How to Predict the Product of an Epoxidation Reaction

CHEMICAL CONNECTIONS

8A Nitroglycerin: An Explosive and a Drug

8B Blood Alcohol Screening

8C Ethylene Oxide: A Chemical Sterilant

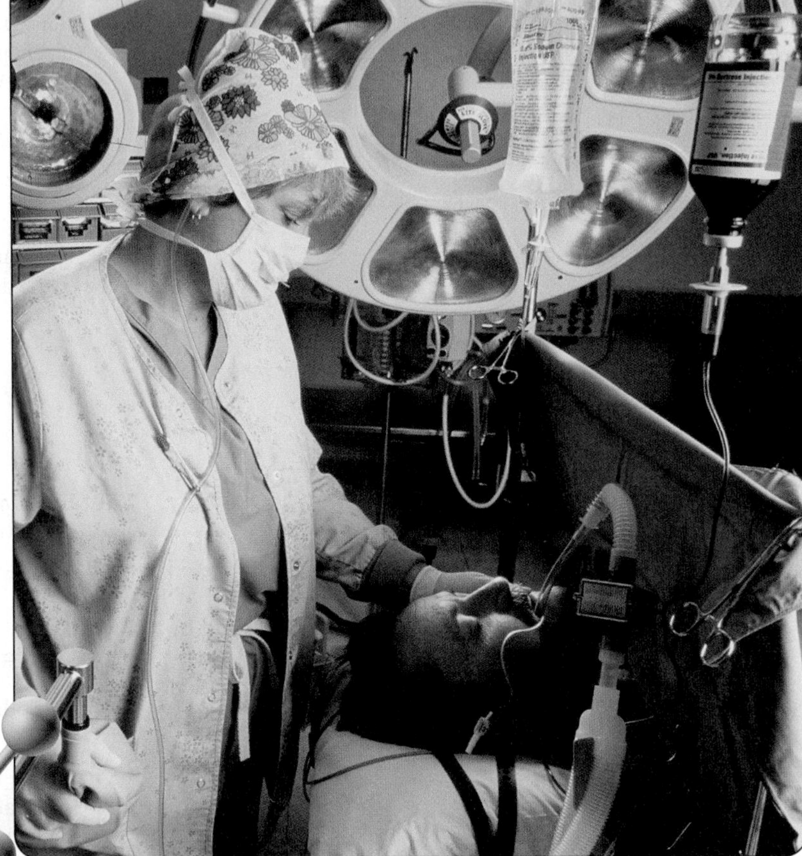

An anesthesiologist administers isoflurane to a patient before surgery. The discovery that inhaling ethers could make a patient insensitive to pain revolutionized the practice of medicine. Inset: A model of isoflurane, $CF_3CHClOCHF_2$, a halogenated ether widely used as an inhalation anesthetic in both human and veterinary medicine.

In this chapter, we study the physical and chemical properties of alcohols and ethers, two classes of oxygen-containing compounds we have seen as products of chemical reactions (Sections 5.3B and 7.4). We also study thiols, a class of sulfur-containing compounds. Thiols are like alcohols in structure, except that they contain an —SH group rather than an —OH group.

CH_3CH_2OH	$CH_3CH_2OCH_2CH_3$	CH_3CH_2SH
Ethanol (an alcohol)	Diethyl ether (an ether)	Ethanethiol (a thiol)

These three compounds are certainly familiar to you. Ethanol is the fuel additive in gasohol, the alcohol in alcoholic beverages, and an important industrial and laboratory solvent. Diethyl ether was the first inhalation anesthetic used in general surgery. It is also an important industrial and laboratory solvent. Ethanethiol, like other low-molecular-weight thiols, has a stench. Smells such as those from skunks, rotten eggs, and sewage are caused by thiols.

Alcohols are particularly important in both laboratory and biochemical transformations of organic compounds. They can be converted into other types of compounds, such as alkenes, haloalkanes, aldehydes, ketones, carboxylic acids, and esters. Not only can alcohols be converted to these compounds, but they also can be prepared from them. Thus, alcohols play a central role in the interconversion of organic functional groups.

8.1 What Are Alcohols?

A. Structure

The functional group of an **alcohol** is an **—OH (hydroxyl) group** bonded to an sp^3 hybridized carbon atom (Section 1.7A). The oxygen atom of an alcohol is also sp^3 hybridized. Two sp^3 hybrid orbitals of oxygen form sigma bonds to atoms of carbon and hydrogen. The other two sp^3 hybrid orbitals of oxygen each contain an unshared pair of electrons. Figure 8.1 shows a Lewis structure and ball-and-stick model of methanol, CH_3OH, the simplest alcohol.

Alcohol A compound containing an —OH (hydroxyl) group bonded to an sp^3 hybridized carbon.

(a)

H
|
:O:
|
H—C—H
|
H

(b)

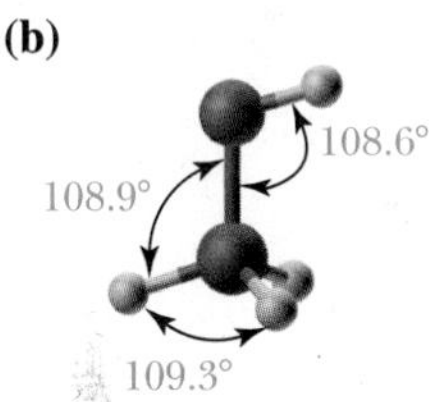

Figure 8.1
Methanol, CH_3OH.
(a) Lewis structure and
(b) ball-and-stick model.
The measured H—O—C bond angle in methanol is 108.6°, very close to the tetrahedral angle of 109.5°.

B. Nomenclature

We derive the IUPAC names for alcohols in the same manner as those for alkanes, with the exception that the ending of the parent alkane is changed from *-e* to *-ol*. The ending *-ol* tells us that the compound is an alcohol.

1. Select, as the parent alkane, the longest chain of carbon atoms that contains the —OH, and number that chain from the end closer to the —OH group. In numbering the parent chain, the location of the —OH group takes precedence over alkyl groups and halogens.
2. Change the suffix of the parent alkane from *-e* to *-ol* (Section 3.5), and use a number to show the location of the —OH group. For cyclic alcohols, numbering begins at the carbon bearing the —OH group.
3. Name and number substituents and list them in alphabetical order.

To derive common names for alcohols, we name the alkyl group bonded to —OH and then add the word *alcohol.* Following are the IUPAC names and, in parentheses, the common names of eight low-molecular-weight alcohols:

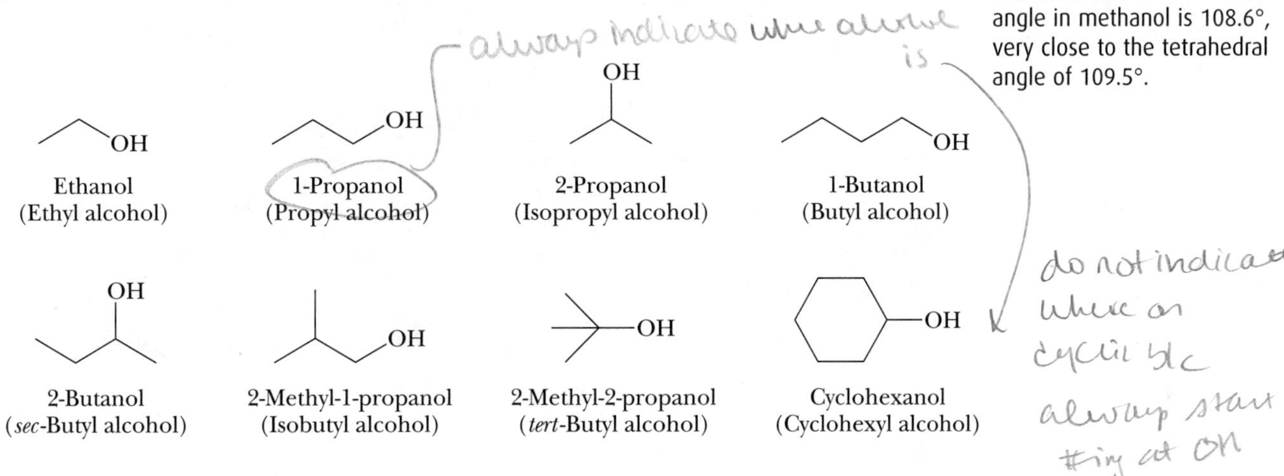

HOW TO 8.1 Name Cyclic Alcohols

(a) First, determine the root name of the cycloalkane, and replace the suffix *-e* with *-ol.*

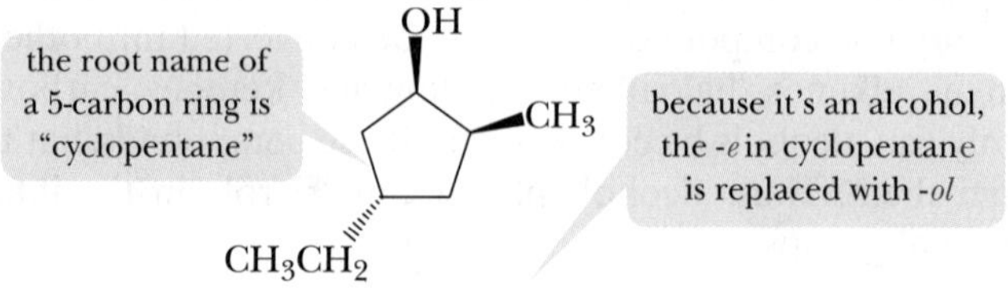

the root name of a 5-carbon ring is "cyclopentane"

OH

CH_3

CH_3CH_2

because it's an alcohol, the *-e* in cyclopentane is replaced with *-ol*

Cyclopentanol

(b) Name and number the substituents. Numbering begins at the carbon bearing the —OH group and proceeds in the direction that gives the lowest *total* for all substituents.

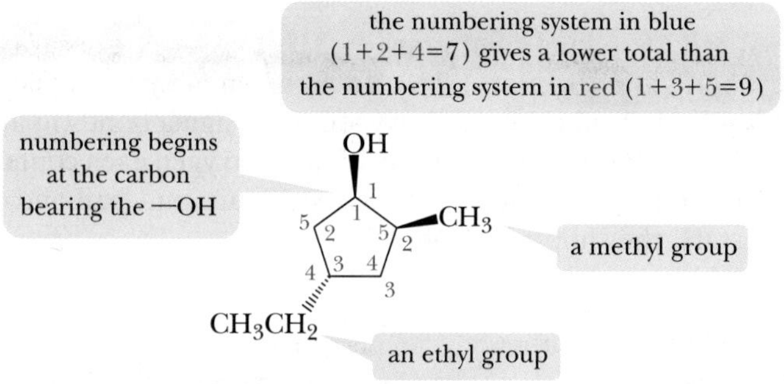

the numbering system in blue (1+2+4=7) gives a lower total than the numbering system in red (1+3+5=9)

numbering begins at the carbon bearing the —OH

OH

CH_3

a methyl group

CH_3CH_2

an ethyl group

(c) Place substituents in alphabetical order preceded by its position on the ring. Position 1 for the —OH group is assumed.

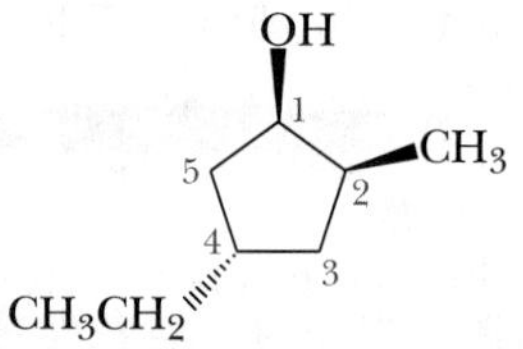

OH

CH_3

CH_3CH_2

4-Ethyl-2-methylcyclopentanol

(d) Don't forget to indicate stereochemistry.

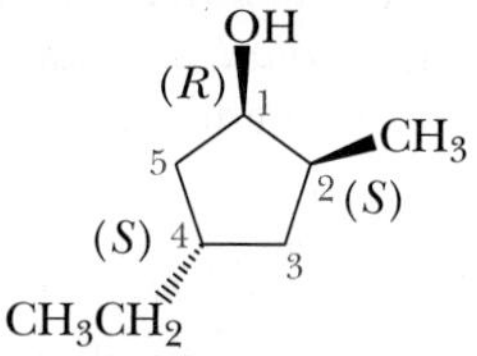

OH

(*R*) 1

CH_3

2 (*S*)

(*S*) 4

CH_3CH_2

(1*R*,2*S*,4*S*)-4-Ethyl-2-methylcyclopentanol

Example 8.1

Write the IUPAC name for each alcohol:

(a) $CH_3(CH_2)_6CH_2OH$

(b) 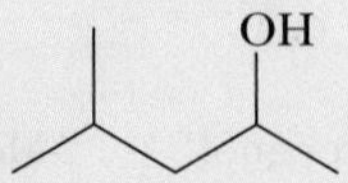 OH

(c) 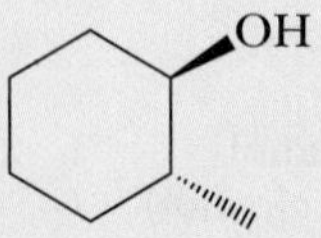 OH

Strategy

First look for the longest chain of carbons that contains the —OH group. This will allow you to determine the root name. Then identify the atoms or groups of atoms that are not part of that chain of carbons. These are your substituents.

Solution

(a) 1-Octanol (b) 4-Methyl-2-pentanol
(c) *trans*-2-Methylcyclohexanol or (1*R*,2*R*)-2-Methylcyclohexanol

See problems 8.14, 8.15, 8.17

Problem 8.1

Write the IUPAC name for each alcohol:

(a) 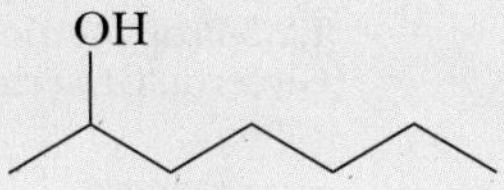 OH (b) 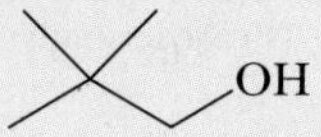 OH (c) 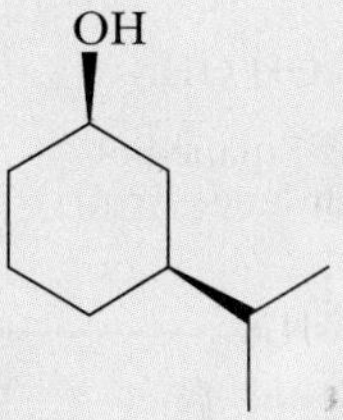 OH

We classify alcohols as **primary** (1°), **secondary** (2°), or **tertiary** (3°), depending on whether the —OH group is on a primary, secondary, or tertiary carbon (Section 1.7A).

Example 8.2

Classify each alcohol as primary, secondary, or tertiary:

(a) 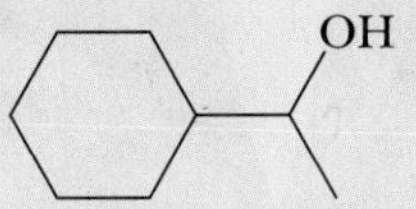 OH (b) CH_3COH with CH_3 above and CH_3 below (c) 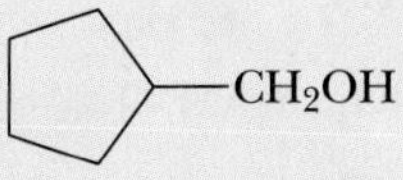 $—CH_2OH$

Strategy

Determine how many carbons are bonded to the carbon bonded to the —OH group (1 carbon = 1°, 2 carbons = 2°, and 3 carbons = 3°).

Solution

(a) Secondary (2°) (b) Tertiary (3°) (c) Primary (1°)

See problem 8.13

Problem 8.2

Classify each alcohol as primary, secondary, or tertiary:

(a) 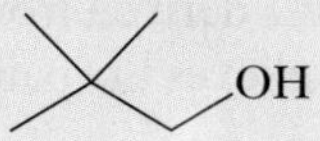 OH (b) 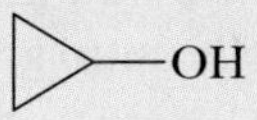 —OH

(c) $CH_2{=}CHCH_2OH$ (d) 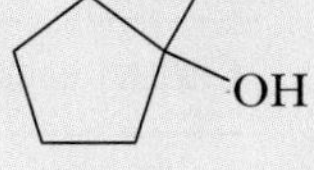 OH

Glycol A compound with two hydroxyl (—OH) groups on different carbons.

Ethylene glycol is a polar molecule and dissolves readily in water, a polar solvent. *(Charles D. Winter)*

In the IUPAC system, a compound containing two hydroxyl groups is named as a **diol**, one containing three hydroxyl groups is named as a **triol**, and so on. In IUPAC names for diols, triols, and so on, the final *-e* of the parent alkane name is retained, as for example, in 1,2-ethanediol.

As with many organic compounds, common names for certain diols and triols have persisted. Compounds containing two hydroxyl groups on different carbons are often referred to as **glycols**. Ethylene glycol and propylene glycol are synthesized from ethylene and propylene, respectively—hence their common names:

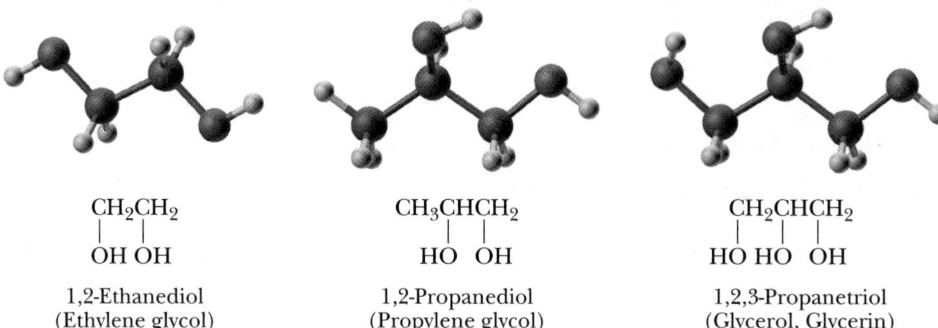

CH_2CH_2 (OH OH)	CH_3CHCH_2 (HO OH)	CH_2CHCH_2 (HO HO OH)
1,2-Ethanediol (Ethylene glycol)	1,2-Propanediol (Propylene glycol)	1,2,3-Propanetriol (Glycerol, Glycerin)

CHEMICAL CONNECTIONS 8A

Nitroglycerin: An Explosive and a Drug

In 1847, Ascanio Sobrero (1812–1888) discovered that 1,2,3-propanetriol, more commonly named glycerin, reacts with nitric acid in the presence of sulfuric acid to give a pale yellow, oily liquid called nitroglycerin:

$$\begin{array}{l} CH_2-OH \\ | \\ CH-OH \\ | \\ CH_2-OH \end{array} + 3\,HNO_3 \xrightarrow{H_2SO_4} \begin{array}{l} CH_2-ONO_2 \\ | \\ CH-ONO_2 \\ | \\ CH_2-ONO_2 \end{array} + 3\,H_2O$$

1,2,3-Propanetriol (Glycerol, Glycerin) → 1,2,3-Propanetriol trinitrate (Nitroglycerin)

Sobrero also discovered the explosive properties of the compound: When he heated a small quantity of it, it exploded! Soon, nitroglycerin became widely used for blasting in the construction of canals, tunnels, roads, and mines and, of course, for warfare.

One problem with the use of nitroglycerin was soon recognized: It was difficult to handle safely, and accidental explosions occurred frequently. The Swedish chemist Alfred Nobel (1833–1896) solved the problem: He discovered that a claylike substance called diatomaceous earth absorbs nitroglycerin so that it will not explode without a fuse. He gave the name *dynamite* to this mixture of nitroglycerine, diatomaceous earth, and sodium carbonate.

Surprising as it may seem, nitroglycerin is used in medicine to treat angina pectoris, the symptoms of which are sharp chest pains caused by a reduced flow of blood in the coronary artery. Nitroglycerin, which is available in liquid (diluted with alcohol to render it nonexplosive), tablet, or paste form, relaxes the smooth muscles of blood vessels, causing dilation of the coronary artery. This dilation, in turn, allows more blood to reach the heart.

The fortune of Alfred Nobel, 1833–1896, built on the manufacture of dynamite, now funds the Nobel Prizes. *(Bettmann/Corbis)*

When Nobel became ill with heart disease, his physicians advised him to take nitroglycerin to relieve his chest pains. He refused, saying he could not understand how the explosive could relieve chest pains. It took science more than 100 years to find the answer. We now know that it is nitric oxide, NO, derived from the nitro groups of nitroglycerin, that relieves the pain.

QUESTION

Classify each hydroxyl group in glycerol as either 1°, 2°, or 3°.

We often refer to compounds containing —OH and C=C groups as unsaturated alcohols. To name an unsaturated alcohol,

1. Number the parent alkane so as to give the —OH group the lowest possible number.
2. Show the double bond by changing the infix of the parent alkane from *-an-* to *-en-* (Section 3.5), and show the alcohol by changing the suffix of the parent alkane from *-e* to *-ol*.
3. Use numbers to show the location of both the carbon–carbon double bond and the hydroxyl group.

Example 8.3

Write the IUPAC name for each alcohol:

(a) $CH_2{=}CHCH_2OH$ (b) HO OH

(c) OH

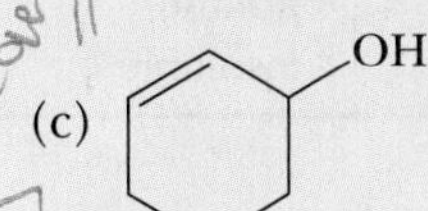

(d) HO

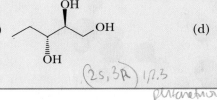

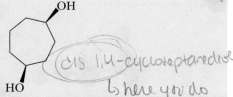

Strategy

First look for the longest chain of carbons. This will allow you to determine the root name. If the alcohol is unsaturated, the name will follow the general form alken-#-ol. If there are two —OH groups, name the compound as an alkanediol.

Solution

(a) 2-Propen-1-ol. Its common name is allyl alcohol.
(b) 2,2-Dimethyl-1,4-butanediol.
(c) 2-Cyclohexenol.
(d) *cis*-3-Hexen-1-ol. This unsaturated alcohol is sometimes called leaf alcohol because of its occurrence in leaves of fragrant plants, including trees and shrubs.

See problems 8.14, 8.15, 8.17

Problem 8.3

Write the IUPAC name for each alcohol:

(a) OH (b) OH

(c) OH OH OH (d) OH HO

C. Physical Properties

The most important physical property of alcohols is the polarity of their —OH groups. Because of the large difference in electronegativity (Table 1.5) between oxygen and carbon (3.5 − 2.5 = 1.0) and between oxygen and hydrogen (3.5 − 2.1 = 1.4), both the C—O and O—H bonds of an alcohol are polar covalent, and alcohols are polar molecules, as illustrated in Figure 8.2 for methanol.

(a)

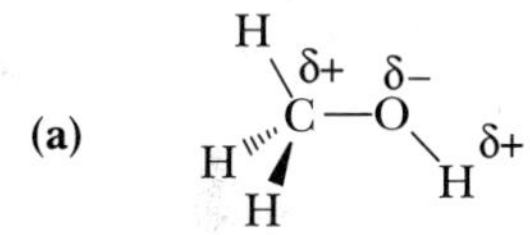

(b)

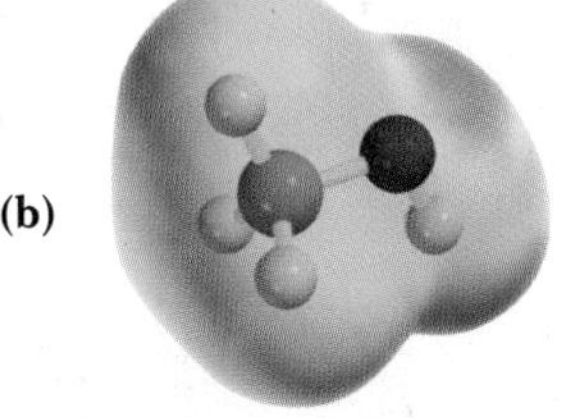

Figure 8.2 Polarity of the C—O—H bond in methanol. (a) There are partial positive charges on carbon and hydrogen and a partial negative charge on oxygen. (b) An electron density map showing the partial negative charge (in red) around oxygen and a partial positive charge (in blue) around hydrogen of the —OH group.

TABLE 8.1 Boiling Points and Solubilities in Water of Five Groups of Alcohols and Alkanes of Similar Molecular Weight

Structural Formula	Name	Molecular Weight	Boiling Point (°C)	Solubility in Water
CH_3OH	methanol	32	65	infinite
CH_3CH_3	ethane	30	−89	insoluble
CH_3CH_2OH	ethanol	46	78	infinite
$CH_3CH_2CH_3$	propane	44	−42	insoluble
$CH_3CH_2CH_2OH$	1-propanol	60	97	infinite
$CH_3CH_2CH_2CH_3$	butane	58	0	insoluble
$CH_3CH_2CH_2CH_2OH$	1-butanol	74	117	8 g/100 g
$CH_3CH_2CH_2CH_2CH_3$	pentane	72	36	insoluble
$CH_3CH_2CH_2CH_2CH_2OH$	1-pentanol	88	138	2.3 g/100 g
$HOCH_2CH_2CH_2CH_2OH$	1,4-butanediol	90	230	infinite
$CH_3CH_2CH_2CH_2CH_2CH_3$	hexane	86	69	insoluble

Table 8.1 lists the boiling points and solubilities in water for five groups of alcohols and alkanes of similar molecular weight. Notice that, of the compounds compared in each group, the alcohol has the higher boiling point and is the more soluble in water.

Hydrogen bonding The attractive force between a partial positive charge on hydrogen and partial negative charge on a nearby oxygen, nitrogen, or fluorine atom.

Alcohols have higher boiling points than alkanes of similar molecular weight, because alcohols are polar molecules and can associate in the liquid state by a type of dipole–dipole intermolecular attraction called **hydrogen bonding** (Figure 8.3). The strength of hydrogen bonding between alcohol molecules is approximately 8.4 to 21 kJ/mol (2 to 5 kcal/mol). For comparison, the strength of the O—H covalent bond in an alcohol molecule is approximately 460 kJ/mol (110 kcal/mol). As we see by comparing these numbers, an O----H hydrogen bond is considerably weaker than an O—H covalent bond. Nonetheless, it is sufficient to have a dramatic effect on the physical properties of alcohols.

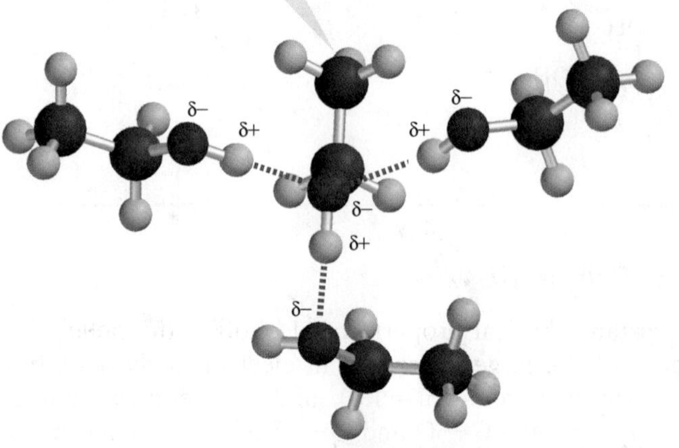
this molecule is shown participating in three hydrogen bonds (shown as ········): two through oxygen and one through hydrogen
δ−
δ+
δ+
δ−
δ−
δ+
δ−

Figure 8.3
The association of ethanol molecules in the liquid state. Each O—H can participate in up to three hydrogen bonds (one through hydrogen and two through oxygen). Only two of these three possible hydrogen bonds per molecule are shown in the figure.

HOW TO 8.2 Predict Relative Boiling Points of Compounds of Similar Molecular Weight

(a) Look for features that make a compound's boiling point higher than another's such as greater polarity, the ability to form hydrogen bonds with that compound's molecules (true for compounds containing N—H or O—H bonds), and greater surface area.

(b) Boiling will typically follow the trend:

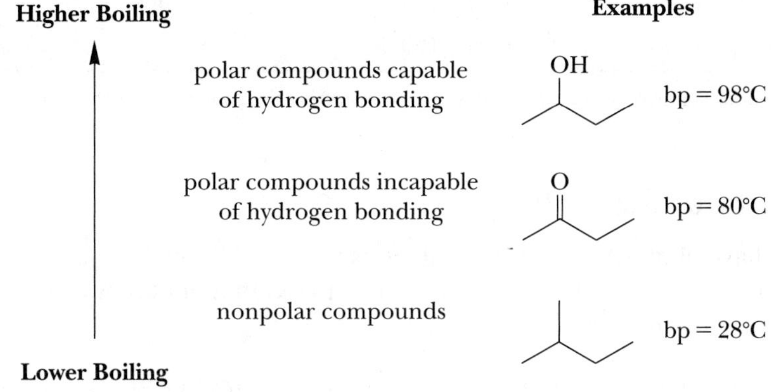

Because of hydrogen bonding between alcohol molecules in the liquid state, extra energy is required to separate each hydrogen-bonded alcohol molecule from its neighbors—hence the relatively high boiling points of alcohols compared with those of alkanes. The presence of additional hydroxyl groups in a molecule further increases the extent of hydrogen bonding, as can be seen by comparing the boiling points of 1-pentanol (138°C) and 1,4-butanediol (230°C), both of which have approximately the same molecular weight.

Because of increased dispersion forces (Section 3.8B) between larger molecules, boiling points of all types of compounds, including alcohols, increase with increasing molecular weight. (Compare, for example, the boiling points of ethanol, 1-propanol, 1-butanol, and 1-pentanol.)

Alcohols are much more soluble in water than are alkanes, alkenes, and alkynes of comparable molecular weight. Their increased solubility is due to hydrogen bonding between alcohol molecules and water. Methanol, ethanol, and 1-propanol are soluble in water in all proportions. As molecular weight increases, the physical properties of alcohols become more like those of hydrocarbons of comparable molecular weight. Alcohols of higher molecular weight are much less soluble in water because of the increase in size of the hydrocarbon portion of their molecules.

8.2 What Are the Characteristic Reactions of Alcohols?

In this section, we study the acidity and basicity of alcohols, their dehydration to alkenes, their conversion to haloalkanes, and their oxidation to aldehydes, ketones, or carboxylic acids.

TABLE 8.2 pK_a Values for Selected Alcohols in Dilute Aqueous Solution*

Compound	Structural Formula	pK_a	
hydrogen chloride	HCl	−7	Stronger acid
acetic acid	CH_3COOH	4.8	↑
methanol	CH_3OH	15.5	
water	H_2O	15.7	
ethanol	CH_3CH_2OH	15.9	
2-propanol	$(CH_3)_2CHOH$	17	
2-methyl-2-propanol	$(CH_3)_3COH$	18	Weaker acid

*Also given for comparison are pK_a values for water, acetic acid, and hydrogen chloride.

A. Acidity of Alcohols

Alcohols have about the same pK_a values as water (15.7), which means that aqueous solutions of alcohols have about the same pH as that of pure water. The pK_a of methanol, for example, is 15.5:

$$CH_3\ddot{\underset{..}{O}}-H + :\ddot{O}(H)-H \rightleftharpoons CH_3\ddot{\underset{..}{O}}:^- + H-\ddot{O}^+(H)-H \qquad K_a = \frac{[CH_3O^-][H_3O^+]}{[CH_3OH]} = 3.2 \times 10^{-16}$$

$$pK_a = 15.5$$

Table 8.2 gives the acid ionization constants for several low-molecular-weight alcohols. Methanol and ethanol are about as acidic as water. Higher-molecular-weight, water-soluble alcohols are slightly weaker acids than water. Even though alcohols have some slight acidity, they are not strong enough acids to react with weak bases such as sodium bicarbonate or sodium carbonate. (At this point, it would be worthwhile to review Section 2.4 and the discussion of the position of equilibrium in acid–base reactions.) Note that, although acetic acid is a "weak acid" compared with acids such as HCl, it is still 10^{10} times stronger as an acid than alcohols are.

Methanol reacts with sodium metal with the evolution of hydrogen gas.
(Charles D. Winters)

B. Basicity of Alcohols

In the presence of strong acids, the oxygen atom of an alcohol is a weak base and reacts with an acid by proton transfer to form an oxonium ion:

$$CH_3CH_2-\ddot{\underset{..}{O}}-H + H-\ddot{O}^+(H)-H \xrightarrow{H_2SO_4} CH_3CH_2-\overset{+}{\ddot{O}}(H)-H + :\ddot{O}(H)-H$$

Ethanol; Hydronium ion (pK_a − 1.7); Ethyloxonium ion (pK_a − 2.4)

Thus, alcohols can function as both weak acids and weak bases.

C. Reaction with Active Metals

Like water, alcohols react with Li, Na, K, Mg, and other active metals to liberate hydrogen and to form metal alkoxides. In the following oxidation–reduction reaction, Na is oxidized to Na^+ and H^+ is reduced to H_2:

$$2\,CH_3OH + 2\,Na \longrightarrow 2\,CH_3O^-Na^+ + H_2$$

Sodium methoxide

To name a metal alkoxide, name the cation first, followed by the name of the anion. The name of an alkoxide ion is derived from a prefix showing the number of carbon atoms and their arrangement (*meth*-, *eth*-, *isoprop*-, tert-*but*-, and so on) followed by the suffix -*oxide*.

Alkoxide ions are somewhat stronger bases than is the hydroxide ion. In addition to sodium methoxide, the following metal salts of alcohols are commonly used in organic reactions requiring a strong base in a nonaqueous solvent; sodium ethoxide in ethanol and potassium *tert*-butoxide in 2-methyl-2-propanol (*tert*-butyl alcohol):

$$CH_3CH_2O^-Na^+ \qquad CH_3C(CH_3)_2O^-K^+$$

Sodium ethoxide   Potassium *tert*-butoxide

As we saw in Chapter 7, alkoxide ions can also be used as nucleophiles in substitution reactions.

HOW TO 8.3 Predict the Position of Equilibrium of an Acid–Base Reaction

(a) Identify the two acids and two bases in the equilibrium.

(b) The position of equilibrium lies on the side with the weaker acid and weaker base.

(c) Following are common types of compounds encountered in organic chemistry and their relative acidities.

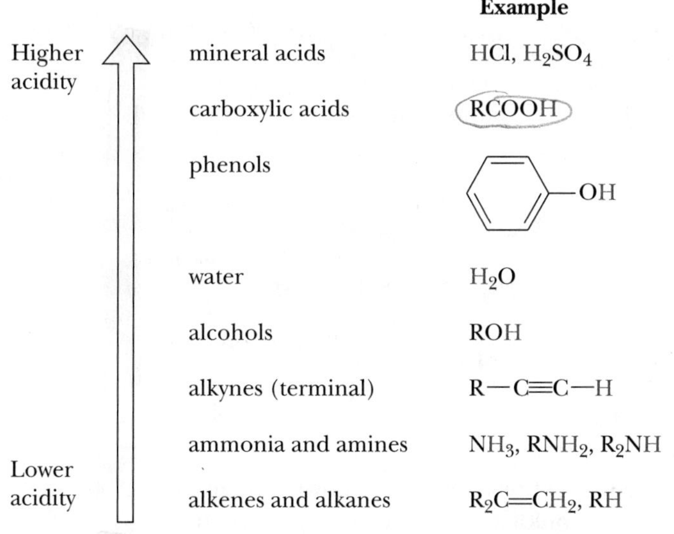

		Example
Higher acidity	mineral acids	HCl, H_2SO_4
	carboxylic acids	RCOOH
	phenols	C_6H_5—OH
	water	H_2O
	alcohols	ROH
	alkynes (terminal)	$R{-}C{\equiv}C{-}H$
	ammonia and amines	NH_3, RNH_2, R_2NH
Lower acidity	alkenes and alkanes	$R_2C{=}CH_2$, RH

Example 8.4

Write balanced equations for the following reactions. If the reaction is an acid–base reaction, predict its position of equilibrium.

(a) C_6H_{11}—OH + Na $\longrightarrow$

(b) $Na^+ NH_2^-$ + (2-methyl-1-propanol) OH $\rightleftharpoons$

(c) $CH_3CH_2O^- Na^+$ + $H_3C{-}\overset{O}{\overset{\|}{C}}{-}OH$ $\rightleftharpoons$

Strategy

First determine what type of reaction is occurring. When elemental sodium is used, an oxidation–reduction reaction takes place, producing a sodium alkoxide and hydrogen gas. In acid–base reactions, the position of the equilibrium resides on the side with the weaker acid and weaker base (i.e., the more stable species).

Solution

(a) 2 (cyclohexyl)—OH + 2 Na $\longrightarrow$ 2 (cyclohexyl)—$O^- Na^+$ + H_2

(b) $Na^+ \ddot{N}H_2^-$ + (2-methyl-1-propanol) $\ddot{O}H$ $\rightleftharpoons$ (2-methyl-1-propoxide) $\ddot{O}{:}^- Na^+$ + $\ddot{N}H_3$

Stronger base	$pK_a = 15.8$ Stronger acid	Weaker base	$pK_a = 38$ Weaker acid

the right side of the equation contains the more stable species, especially when comparing the alkoxide anion with NH_2^- because the oxygen in the alkoxide anion is more electronegative and better able to hold a negative charge than the nitrogen in NH_2^-

(c) $CH_3CH_2\ddot{O}{:}^- Na^+$ + $H_3C{-}\overset{:O:}{\overset{\|}{C}}{-}\ddot{O}H$ $\rightleftharpoons$ $CH_3CH_2\ddot{O}H$ + $H_3C{-}\overset{:O:}{\overset{\|}{C}}{-}\ddot{O}{:}^- Na^+$

Stronger base	$pK_a = 4.76$ Stronger acid	$pK_a = 15.9$ Weaker acid	Weaker base

$\updownarrow$

$H_3C{-}\overset{:\ddot{O}:^-}{\overset{|}{C}}{=}\ddot{O}\ Na^+$

the right side of the equation contains the more stable species, especially when comparing the conjugate base of the carboxylic acid (known as a carboxylate anion) with the ethoxide anion because the negative charge in the carboxylate anion can be delocalized by resonance

See problems 8.28–8.32, 8.34, 8.35

Problem 8.4

Write balanced equations for the following reactions. If the reaction is an acid–base reaction, predict its position of equilibrium.

(a) (cyclopentyl)—OH + $HO{-}\overset{O}{\overset{\|}{C}}{-}OH$ $\rightleftharpoons$

(b) (1-butanol) OH + Na $\longrightarrow$

(c) CH_3CH_2OH + $H_3C{-}\overset{O}{\overset{\|}{C}}{-}O^- Na^+$ $\rightleftharpoons$

D. Conversion to Haloalkanes

The conversion of an alcohol to an alkyl halide involves substituting halogen for —OH at a saturated carbon. The most common reagents for this conversion are the halogen acids and $SOCl_2$.

Reaction with HCl, HBr, and HI

Water-soluble tertiary alcohols react very rapidly with HCl, HBr, and HI. Mixing a tertiary alcohol with concentrated hydrochloric acid for a few minutes at room temperature converts the alcohol to a water-insoluble chloroalkane that separates from the aqueous layer.

$$CH_3C(CH_3)_2OH + HCl \xrightarrow{25^\circ C} CH_3C(CH_3)_2Cl + H_2O$$

2-Methyl-2-propanol → 2-Chloro-2-methylpropane

Low-molecular-weight, water-soluble primary and secondary alcohols do not react under these conditions.

Water-insoluble tertiary alcohols are converted to tertiary halides by bubbling gaseous HX through a solution of the alcohol dissolved in diethyl ether or tetrahydrofuran (THF):

$$\text{1-Methylcyclohexanol} + HCl \xrightarrow[\text{ether}]{0^\circ C} \text{1-Chloro-1-methylcyclohexane} + H_2O$$

1-Methyl-cyclohexanol → 1-Chloro-1-methyl cyclohexane

Water-insoluble primary and secondary alcohols react only slowly under these conditions.

Primary and secondary alcohols are converted to bromoalkanes and iodoalkanes by treatment with concentrated hydrobromic and hydroiodic acids. For example, heating 1-butanol with concentrated HBr gives 1-bromobutane:

$$\text{1-Butanol (OH)} + HBr \longrightarrow \text{1-Bromobutane (Br)} + H_2O$$

1-Butanol → 1-Bromobutane (Butyl bromide)

On the basis of observations of the relative ease of reaction of alcohols with HX ($3^\circ > 2^\circ > 1^\circ$), it has been proposed that the conversion of tertiary and secondary alcohols to haloalkanes by concentrated HX occurs by an S_N1 mechanism (Section 7.4) and involves the formation of a carbocation intermediate. *Note:* recall that secondary carbocations are subject to rearrangement to more stable tertiary carbocations (Section 5.4).

Mechanism: Reaction of a Tertiary Alcohol with HCl: An S_N1 Reaction

Step 1: Rapid and reversible proton transfer from the acid to the OH group gives an oxonium ion. The result of this proton transfer is to convert the leaving group from OH^-, a poor leaving group, to H_2O, a better leaving group:

the OH^- group is converted to H_2O, a better leaving group

$$(CH_3)_3C{-}\ddot{O}{-}H + H{-}\overset{+}{O}H_2 \underset{\text{reversible}}{\overset{\text{rapid and}}{\rightleftharpoons}} (CH_3)_3C{-}\overset{+}{O}H_2 + H_2\ddot{O}$$

2-Methyl-2-propanol (*tert*-Butyl alcohol) — An oxonium ion

Step 2: Loss of water from the oxonium ion gives a 3° carbocation intermediate:

$$(CH_3)_3C{-}\overset{+}{O}H_2 \xrightarrow[S_N1]{\text{slow, rate determining}} (CH_3)_3C^+ + H_2\ddot{O}$$

An oxonium ion — A 3° carbocation intermediate

Step 3: Reaction of the 3° carbocation intermediate (an electrophile) with chloride ion (a nucleophile) gives the product:

$$(CH_3)_3C^+ + :\ddot{Cl}:^- \xrightarrow{\text{fast}} (CH_3)_3C{-}\ddot{Cl}:$$

2-Chloro-2-methylpropane (*tert*-Butyl chloride)

chloride ion is produced in the initial reaction of H_2O with HCl

$$H{-}\ddot{O}{-}H + H{-}\ddot{Cl}: \rightarrow H{-}\overset{+}{O}H{-}H + :\ddot{Cl}:^-$$

1) Protonate the alcohol in both

Primary alcohols react with HX by an S_N2 mechanism. In the rate-determining step, the halide ion displaces H_2O from the carbon bearing the oxonium ion. The displacement of H_2O and the formation of the C—X bond are simultaneous.

Mechanism: Reaction of a Primary Alcohol with HBr: An S_N2 Reaction

Step 1: Rapid and reversible proton transfer to the OH group which converts the leaving group from OH^-, a poor leaving group, to H_2O, a better leaving group:

$$CH_3CH_2CH_2CH_2{-}\ddot{O}H + H{-}\overset{+}{O}H_2 \underset{\text{reversible}}{\overset{\text{rapid and}}{\rightleftharpoons}} CH_3CH_2CH_2CH_2{-}\overset{+}{O}H_2 + H_2\ddot{O}$$

An oxonium ion

Step 2: The nucleophilic displacement of H_2O by Br^- gives the bromoalkane:

$$:\ddot{Br}:^- + CH_3CH_2CH_2CH_2{-}\overset{+}{O}H_2 \xrightarrow[S_N2]{\text{slow, rate determining}} CH_3CH_2CH_2CH_2{-}\ddot{Br}: + H_2\ddot{O}$$

Why do tertiary alcohols react with HX by formation of carbocation intermediates, whereas primary alcohols react by direct displacement of —OH (more accurately, by displacement of $—OH_2^+$)? The answer is a combination of the same two factors involved in nucleophilic substitution reactions of haloalkanes (Section 7.5B):

1. *Electronic factors* Tertiary carbocations are the most stable (require the lowest activation energy for their formation), whereas primary carbocations are the least stable (require the highest activation energy for their formation). Therefore, tertiary alcohols are most likely to react by carbocation formation; secondary alcohols are intermediate, and primary alcohols rarely, if ever, react by carbocation formation.
2. *Steric factors* To form a new carbon–halogen bond, halide ion must approach the substitution center and begin to form a new covalent bond to it. If we compare the ease of approach to the substitution center of a primary oxonium ion with that of a tertiary oxonium ion, we see that approach is considerably easier in the case of a primary oxonium ion. Two hydrogen atoms and one alkyl group screen the back side of the substitution center of a primary oxonium ion, whereas three alkyl groups screen the back side of the substitution center of a tertiary oxonium ion.

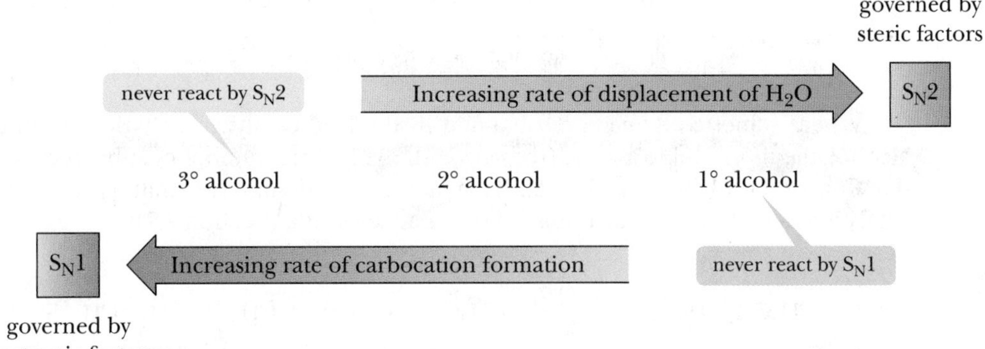

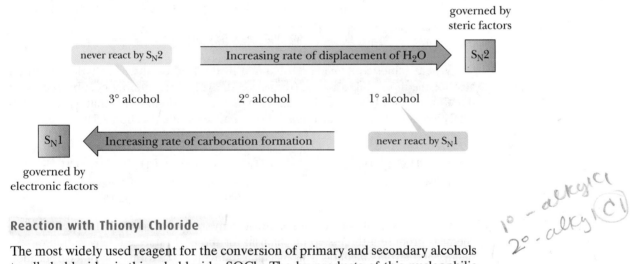

Reaction with Thionyl Chloride

The most widely used reagent for the conversion of primary and secondary alcohols to alkyl chlorides is thionyl chloride, $SOCl_2$. The by-products of this nucleophilic substitution reaction are HCl and SO_2, both given off as gases. Often, an organic base such as pyridine (Section 10.1) is added to react with and neutralize the HCl by-product:

$$\text{1-Heptanol (OH)} + \underset{\text{Thionyl chloride}}{SOCl_2} \xrightarrow{\text{pyridine}} \text{1-Chloroheptane (Cl)} + SO_2 + HCl$$

E. Acid-Catalyzed Dehydration to Alkenes

An alcohol can be converted to an alkene by **dehydration**—that is, by the elimination of a molecule of water from adjacent carbon atoms. In the laboratory, the dehydration of an alcohol is most often brought about by heating it with either 85% phosphoric acid or concentrated sulfuric acid. Primary alcohols are the most difficult to dehydrate and generally require heating in concentrated sulfuric acid at temperatures as high as 180°C. Secondary alcohols undergo acid-catalyzed dehydration at somewhat lower

Dehydration Elimination of a molecule of water from a compound.

temperatures. The acid-catalyzed dehydration of tertiary alcohols often requires temperatures only slightly above room temperature:

$$CH_3CH_2OH \xrightarrow[180°C]{H_2SO_4} CH_2{=}CH_2 + H_2O$$

$$\text{Cyclohexanol} \xrightarrow[140°C]{H_2SO_4} \text{Cyclohexene} + H_2O$$

$$(CH_3)_3COH \xrightarrow[50°C]{H_2SO_4} CH_3C(CH_3){=}CH_2 + H_2O$$

2-Methyl-2-propanol (*tert*-Butyl alcohol) → 2-Methylpropene (Isobutylene)

Thus, the ease of acid-catalyzed dehydration of alcohols occurs in this order:

1° alcohol < 2° alcohol < 3° alcohol

Ease of dehydration of alcohols →

When isomeric alkenes are obtained in the acid-catalyzed dehydration of an alcohol, the more stable alkene (the one with the greater number of substituents on the double bond; see Section 5.3B) generally predominates; that is, the acid-catalyzed dehydration of alcohols follows Zaitsev's rule (Section 7.7):

$$CH_3CH_2CH(OH)CH_3 \xrightarrow[heat]{85\%\ H_3PO_4} CH_3CH{=}CHCH_3 + CH_3CH_2CH{=}CH_2$$

2-Butanol → 2-Butene (80%) + 1-Butene (20%)

2°, 3° = E1

On the basis of the relative ease of dehydration of alcohols (3° > 2° > 1°), chemists propose a three-step mechanism for the acid-catalyzed dehydration of secondary and tertiary alcohols. This mechanism involves the formation of a carbocation intermediate in the rate-determining step and therefore is an E1 mechanism.

again, protonate the OH

Mechanism: Acid-Catalyzed Dehydration of 2-Butanol: An E1 Mechanism

Step 1: Proton transfer from H_3O^+ to the OH group of the alcohol gives an oxonium ion. A result of this step is to convert OH^-, a poor leaving group, into H_2O, a better leaving group:

$$CH_3CH(\ddot{O}H)CH_2CH_3 + H{-}\ddot{O}^+H_2 \underset{\text{reversible}}{\overset{\text{rapid and}}{\rightleftharpoons}} CH_3CH(\ddot{O}^+H_2)CH_2CH_3 + :\ddot{O}H_2$$

An oxonium ion

Step 2: Breaking of the C—O bond gives a 2° carbocation intermediate and H_2O:

H₂O is a good leaving group

$$CH_3CH(\ddot{O}^+H_2)CH_2CH_3 \underset{\text{determining}}{\overset{\text{slow, rate}}{\rightleftharpoons}} CH_3\overset{+}{C}HCH_2CH_3 + H_2\ddot{O}:$$

A 2° carbocation intermediate

Step 3: Proton transfer from the carbon adjacent to the positively charged carbon to H_2O gives the alkene and regenerates the catalyst. The sigma electrons of a C—H bond become the pi electrons of the carbon–carbon double bond:

$$CH_3-\overset{+}{C}H-CH(H)-CH_3 + :\ddot{O}(H)-H \xrightarrow{\text{rapid}} CH_3-CH=CH-CH_3 + H-\overset{+}{\ddot{O}}(H)-H$$

Because the rate-determining step in the acid-catalyzed dehydration of secondary and tertiary alcohols is the formation of a carbocation intermediate, the relative ease of dehydration of these alcohols parallels the ease of formation of carbocations.

Primary alcohols react by the following two-step mechanism, in which Step 2 is the rate-determining step.

Mechanism: Acid-Catalyzed Dehydration of a Primary Alcohol: An E2 Mechanism

Step 1: Proton transfer from H_3O^+ to the OH group of the alcohol gives an oxonium ion:

$$CH_3CH_2-\ddot{O}-H + H-\overset{+}{\ddot{O}}(H)-H \underset{\text{reversible}}{\overset{\text{rapid and}}{\rightleftharpoons}} CH_3CH_2-\overset{+}{O}(H)(H): + :\ddot{O}(H)-H$$

Step 2: Simultaneous proton transfer to solvent and loss of H_2O gives the alkene:

$$H-\ddot{O}(H): + H-C(H)(H)-CH_2-\overset{+}{O}(H)(H): \xrightarrow[\text{E2}]{\text{slow, rate determining}} H-\overset{+}{\ddot{O}}(H)-H + (H)(H)C=C(H)(H) + :\ddot{O}(H)-H$$

In Section 5.3B, we discussed the acid-catalyzed hydration of alkenes to give alcohols. In the current section, we discussed the acid-catalyzed dehydration of alcohols to give alkenes. In fact, hydration–dehydration reactions are reversible. Alkene hydration and alcohol dehydration are competing reactions, and the following equilibrium exists:

$$\underset{\text{An alkene}}{>C=C<} + H_2O \overset{\text{acid catalyst}}{\rightleftharpoons} \underset{\text{An alcohol}}{-\underset{H}{C}-\underset{OH}{C}-}$$

How, then, do we control which product will predominate? Recall that LeChâtelier's principle states that a system in equilibrium will respond to a stress in the equilibrium by counteracting that stress. This response allows us to control these two reactions to give the desired product. Large amounts of water (achieved with the use of dilute aqueous acid) favor alcohol formation, whereas a scarcity of water (achieved with the use of concentrated acid) or experimental conditions by which water is removed (for example, heating the reaction mixture above 100°C) favor alkene formation. Thus, depending on the experimental conditions, it is possible to use the hydration–dehydration equilibrium to prepare either alcohols or alkenes, each in high yields.

HOW TO 8.4 Complete a Dehydration Reaction

(a) A dehydration reaction is very similar to a dehydrohalogenation reaction (Section 7.7) except that the hydroxyl group must be protonated to generate a better leaving group.

OH^- is a poor leaving group

HOH is a better leaving group

H_2SO_4

(b) Label the carbon bonded to the leaving group as "α" (alpha).

(c) Label any carbon bonded to the α-carbon as "β" (beta). *Note*: Only do so if the β-carbon is bonded to a hydrogen atom.

(d) Remove the leaving group (H_2O) and a β-hydrogen from the molecule and place a new double bond between the α and β carbons. This forms a dehydration product.

formerly the α-carbon

formerly a β-carbon

(e) Repeat step (d) above for any other β-carbons to form a different dehydration product.

formerly the α-carbon

formerly a β-carbon

(f) When forming an alkene by E1 elimination, consider the fact that both *cis* and *trans* isomers of the alkene are possible.

both *cis* and *trans* alkenes are formed

Example 8.5

For each of the following alcohols, draw structural formulas for the alkenes that form upon acid-catalyzed dehydration, and predict which alkene is the major product from each alcohol. Be aware that rearrangements may occur because carbocations are formed in the reactions.

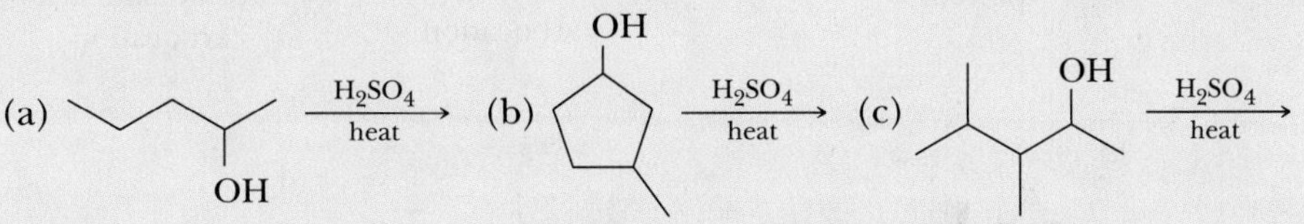

(a) OH $\xrightarrow[\text{heat}]{H_2SO_4}$ (b) OH $\xrightarrow[\text{heat}]{H_2SO_4}$ (c) OH $\xrightarrow[\text{heat}]{H_2SO_4}$

Strategy

Label the carbon bonded to the —OH group as α. This is where the carbocation will form in the mechanism of the reaction. Consider whether a rearrangement (Section 5.4) will occur, and if so, relabel the new carbocation as α. Then label any carbons next to the α-carbon as β. If the β-carbon is bonded to at least one hydrogen, remove that hydrogen and the —OH and draw a C—C double bond between the α- and β-carbons. Start over and repeat this process for any other β-carbons that meet this criteria. Each time you are able to do this will result in an elimination product.

Solution

(a) The elimination of H_2O from carbons 2 and 3 gives 2-pentene, which can form as *cis–trans* isomers; the elimination of H_2O from carbons 1 and 2 gives 1-pentene. *trans*-2-Pentene, with two alkyl groups (an ethyl and a methyl) on the double bond and with *trans* being more stable than *cis* (Section 5.7B), is the major product. 1-Pentene, with only one alkyl group (a propyl group) on the double bond, is a minor product:

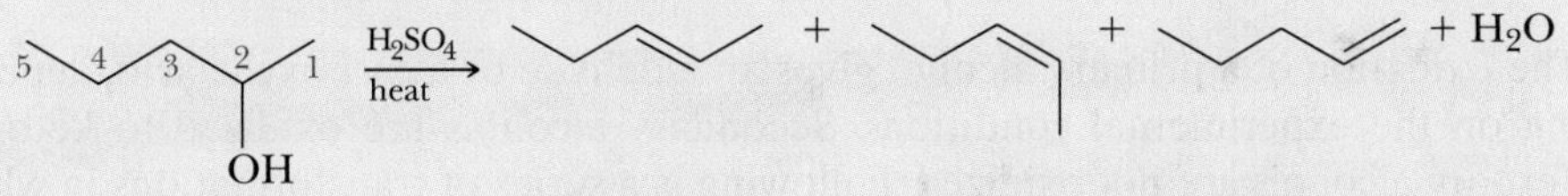

5 4 3 2 1 OH $\xrightarrow[\text{heat}]{H_2SO_4}$ + + $+ H_2O$

2-Pentanol | *trans*-2-Pentene (major product) | *cis*-2-Pentene | 1-Pentene

(b) The elimination of H_2O from carbons 1 and 2 gives 3-methylcyclopentene; the elimination of H_2O from carbons 1 and 5 gives 4-methylcyclopentene. Because both products are disubstituted alkenes (two carbons bonded to each C—C double bond), they will be formed in approximately equal amounts.

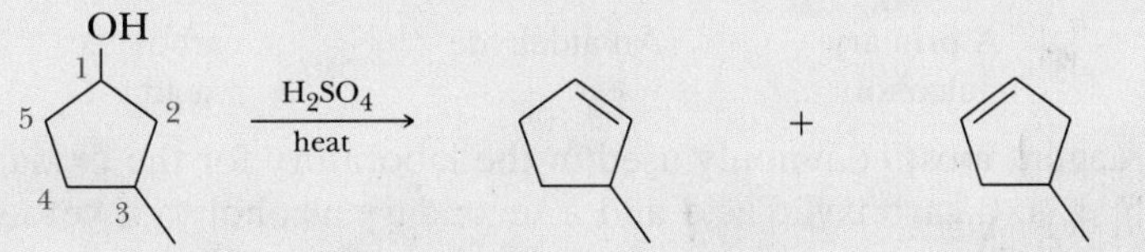

OH 1 2 3 4 5 $\xrightarrow[\text{heat}]{H_2SO_4}$ +

3-Methylcyclopentanol | 3-Methylcyclopentene | 4-Methylcyclopentene

(c) This reaction initially forms a 2° carbocation intermediate, which rearranges via a 1,2-hydride shift (Section 5.4) to form the more stable 3° carbocation. This new carbocation has three β hydrogens and a C—C double bond and can form in three places. 2,3-Dimethyl-2-pentene is the product with the more substituted double bond and is therefore the most stable and major product.

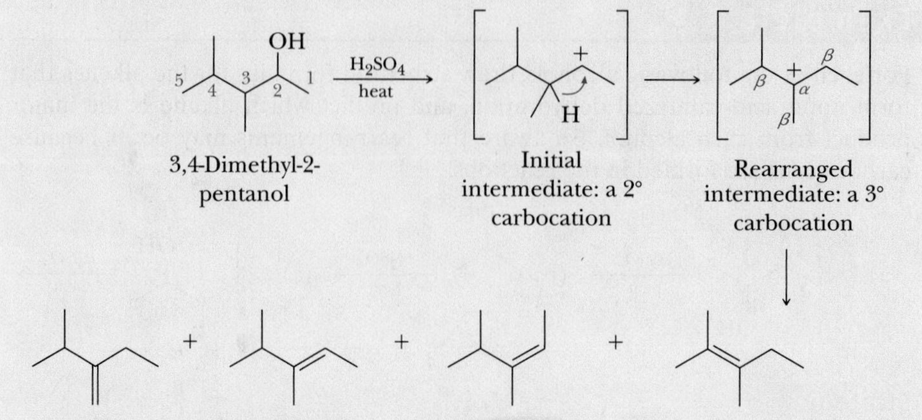

OH

5 4 3 2 1

$\xrightarrow[\text{heat}]{H_2SO_4}$

3,4-Dimethyl-2-pentanol

Initial intermediate: a 2° carbocation

Rearranged intermediate: a 3° carbocation

2-Ethyl-3-methyl-1-butene + (*E*)-3,4-Dimethyl-2-pentene + (*Z*)-3,4-Dimethyl-2-pentene + 2,3-Dimethyl-2-pentene (major product)

See problems 8.41–8.45

Problem 8.5

For each of the following alcohols, draw structural formulas for the alkenes that form upon acid-catalyzed dehydration of that alcohol, and predict which alkene is the major product from each alcohol:

(a) 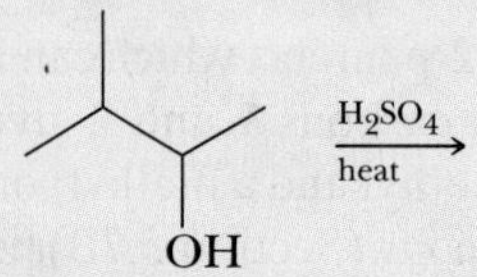 OH $\xrightarrow[\text{heat}]{H_2SO_4}$ (b) 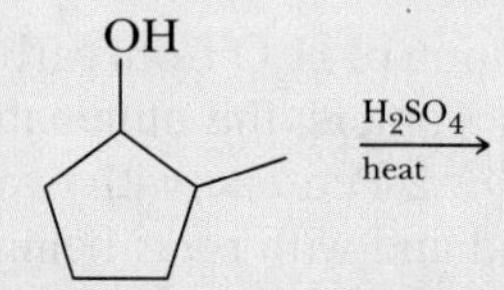 OH $\xrightarrow[\text{heat}]{H_2SO_4}$

F. Oxidation of Primary and Secondary Alcohols

The oxidation of a primary alcohol gives an aldehyde or a carboxylic acid, depending on the experimental conditions. Secondary alcohols are oxidized to ketones. Tertiary alcohols are not oxidized. Following is a series of transformations in which a primary alcohol is oxidized first to an aldehyde and then to a carboxylic acid. The fact that each transformation involves oxidation is indicated by the symbol O in brackets over the reaction arrow:

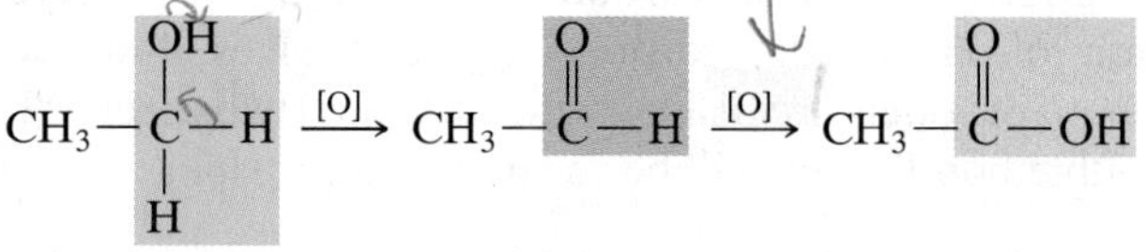

$$CH_3-\underset{H}{\overset{OH}{C}}-H \xrightarrow{[O]} CH_3-\overset{O}{\overset{\|}{C}}-H \xrightarrow{[O]} CH_3-\overset{O}{\overset{\|}{C}}-OH$$

A primary alcohol — An aldehyde — A carboxylic acid

The reagent most commonly used in the laboratory for the oxidation of a primary alcohol to a carboxylic acid and a secondary alcohol to a ketone is chromic acid, H_2CrO_4. Chromic acid is prepared by dissolving either chromium(VI) oxide or potassium dichromate in aqueous sulfuric acid:

$$CrO_3 + H_2O \xrightarrow{H_2SO_4} H_2CrO_4$$

Chromium(VI) oxide — Chromic acid

$$K_2Cr_2O_7 \xrightarrow{H_2SO_4} H_2Cr_2O_7 \xrightarrow{H_2O} 2\,H_2CrO_4$$

Potassisum dichromate — Chromic acid

The oxidation of 1-octanol by chromic acid in aqueous sulfuric acid gives octanoic acid in high yield. These experimental conditions are more than sufficient to oxidize the intermediate aldehyde to a carboxylic acid:

$$CH_3(CH_2)_6CH_2OH \xrightarrow[H_2SO_4,\ H_2O]{CrO_3} \left[CH_3(CH_2)_6\overset{O}{\overset{\|}{C}}H \right] \longrightarrow CH_3(CH_2)_6\overset{O}{\overset{\|}{C}}OH$$

1-Octanol — Octanal (not isolated) — Octanoic acid

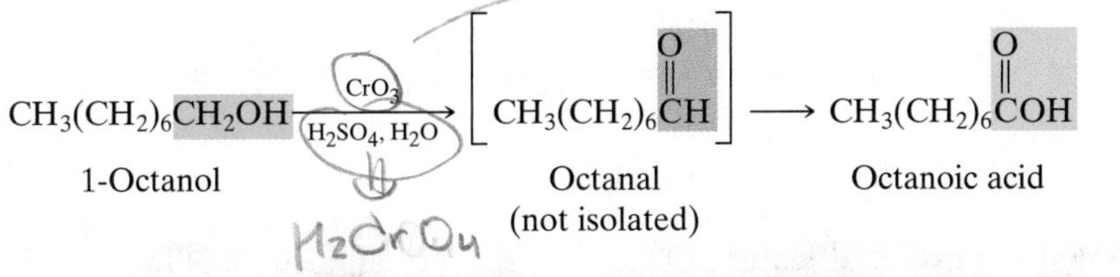

The form of Cr(VI) commonly used for the oxidation of a primary alcohol to an aldehyde is prepared by dissolving CrO_3 in aqueous HCl and adding pyridine to precipitate **pyridinium chlorochromate (PCC)** as a solid. PCC oxidations are carried out in aprotic solvents, most commonly dichloromethane, CH_2Cl_2:

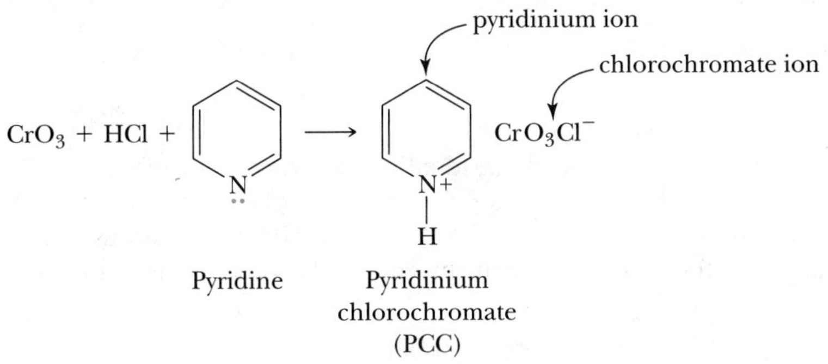

$$CrO_3 + HCl + \text{Pyridine} \longrightarrow \text{Pyridinium } CrO_3Cl^-$$

pyridinium ion

chlorochromate ion

Pyridine — Pyridinium chlorochromate (PCC)

PCC is selective for the oxidation of primary alcohols to aldehydes. It is less reactive than the previously discussed oxidation with chromic acid in aqueous sulfuric acid, and the reaction is run stoichiometrically so that no PCC remains once all the alcohol molecules have been converted to aldehyde. PCC also has little effect on carbon–carbon double bonds or other easily oxidized functional groups. In the following example, geraniol is oxidized to geranial without affecting either carbon–carbon double bond:

$$\text{Geraniol (OH)} \xrightarrow[CH_2Cl_2]{PCC} \text{Geranial (CHO)}$$

Geraniol — Geranial

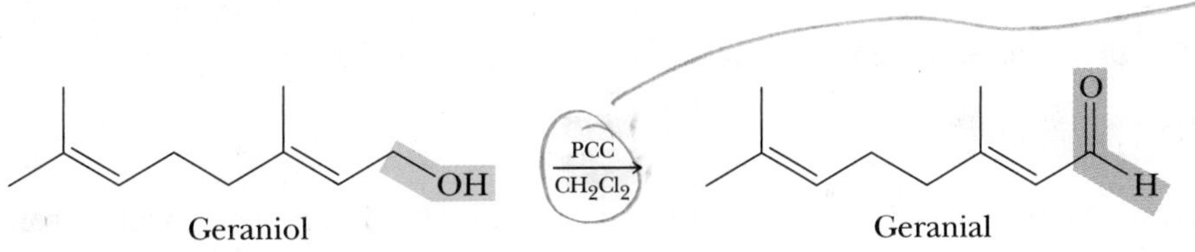

Secondary alcohols are oxidized to ketones by both chromic acid and PCC:

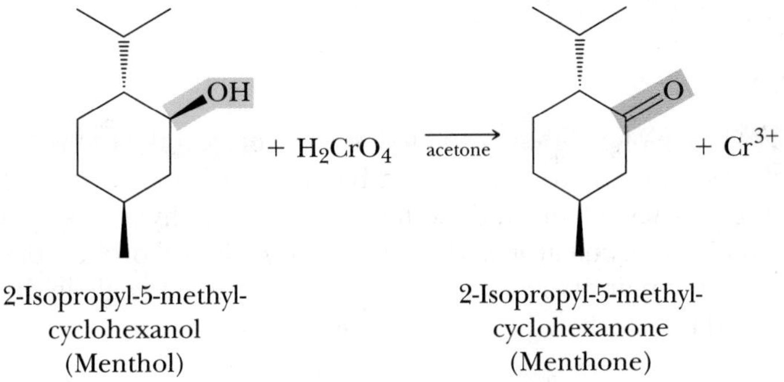

$$\text{Menthol (OH)} + H_2CrO_4 \xrightarrow{\text{acetone}} \text{Menthone (=O)} + Cr^{3+}$$

2-Isopropyl-5-methyl-cyclohexanol (Menthol) — 2-Isopropyl-5-methyl-cyclohexanone (Menthone)

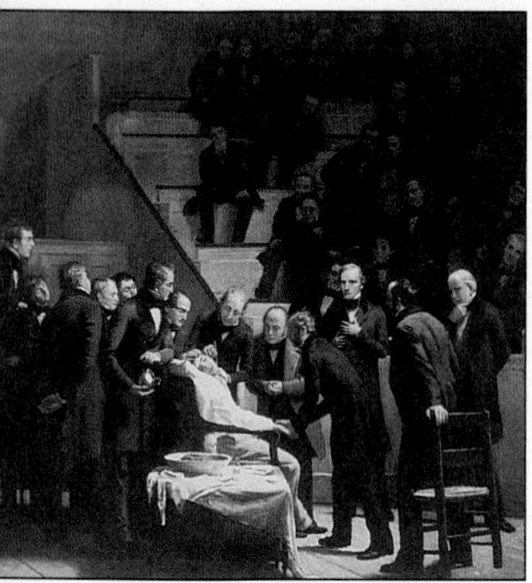

This painting by Robert Hinckley shows the first use of diethyl ether as an anesthetic in 1846. Dr. Robert John Collins was removing a tumor from the patient's neck, and the dentist W. T. G. Morton—who discovered its anesthetic properties—administered the ether. *(Boston Medical Library in the Francis A. Courtney Library of Medicine)*

Tertiary alcohols are resistant to oxidation, because the carbon bearing the —OH is bonded to three carbon atoms and therefore cannot form a carbon–oxygen double bond:

$$\text{1-Methylcyclopentanol (cyclopentane with } CH_3 \text{ and OH)} + H_2CrO_4 \xrightarrow[\text{acetone}]{H^+} \text{(no oxidation)}$$

1-Methylcyclopentanol

Note that the essential feature of the oxidation of an alcohol is the presence of at least one hydrogen on the carbon bearing the OH group. Tertiary alcohols lack such a hydrogen; therefore, they are not oxidized.

Example 8.6

Draw the product of the treatment of each of the following alcohols with PCC:

(a) 1-Hexanol (b) 2-Hexanol (c) Cyclohexanol

Strategy

In oxidation reactions of alcohols, identify the type of alcohol as 1°, 2°, or 3°. Tertiary alcohols remain unreactive. Secondary alcohols are oxidized to ketones. Primary alcohols are oxidized to aldehydes when PCC is used as the oxidizing agent, and to carboxylic acids when chromic acid is used as the oxidizing agent.

Solution

1-Hexanol, a primary alcohol, is oxidized to hexanal. 2-Hexanol, a secondary alcohol, is oxidized to 2-hexanone. Cyclohexanol, a secondary alcohol, is oxidized to cyclohexanone.

(a) Hexanal (b) 2-Hexanone (c) Cyclohexanone

See problems 8.34, 8.35, 8.41–8.46

Problem 8.6

Draw the product of the treatment of each alcohol in Example 8.6 with chromic acid.

(a)

```
    H       H
    |  ..   |
H — C — O — C — H
    |  ..   |
    H       H
```

(b)

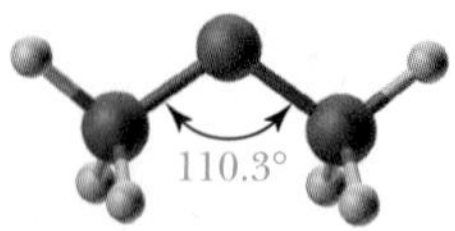

Figure 8.4
Dimethyl ether, CH_3OCH_3. (a) Lewis structure and (b) ball-and-stick model.

8.3 What Are Ethers?

A. Structure

Ether A compound containing an oxygen atom bonded to two carbon atoms.

The functional group of an **ether** is an atom of oxygen bonded to two carbon atoms. Figure 8.4 shows a Lewis structure and a ball-and-stick model of dimethyl ether, CH_3OCH_3, the simplest ether. In dimethyl ether, two sp^3 hybrid orbitals of oxygen form sigma bonds to carbon atoms. The other two sp^3 hybrid orbitals of oxygen each contain an unshared pair of electrons. The C—O—C bond angle in dimethyl ether is 110.3°, close to the predicted tetrahedral angle of 109.5°.

CHEMICAL CONNECTIONS 8B

Blood Alcohol Screening

Potassium dichromate oxidation of ethanol to acetic acid is the basis for the original breath alcohol screening test used by law enforcement agencies to determine a person's blood alcohol content. The test is based on the difference in color between the dichromate ion (reddish orange) in the reagent and the chromium(III) ion (green) in the product. Thus, color change can be used as a measure of the quantity of ethanol present in a breath sample:

$$\underset{\text{Ethanol}}{CH_3CH_2OH} + \underset{\substack{\text{Dichromate ion}\\ \text{(reddish orange)}}}{Cr_2O_7^{2-}} \xrightarrow[H_2O]{H_2SO_4} \underset{\text{Acetic acid}}{CH_3\overset{\overset{\displaystyle O}{\|}}{C}OH} + \underset{\substack{\text{Chromium(III)}\\ \text{ion (green)}}}{Cr^{3+}}$$

In its simplest form, a breath alcohol screening test consists of a sealed glass tube containing a potassium dichromate–sulfuric acid reagent impregnated on silica gel. To administer the test, the ends of the tube are broken off, a mouthpiece is fitted to one end, and the other end is inserted into the neck of a plastic bag. The person being tested then blows into the mouthpiece until the plastic bag is inflated.

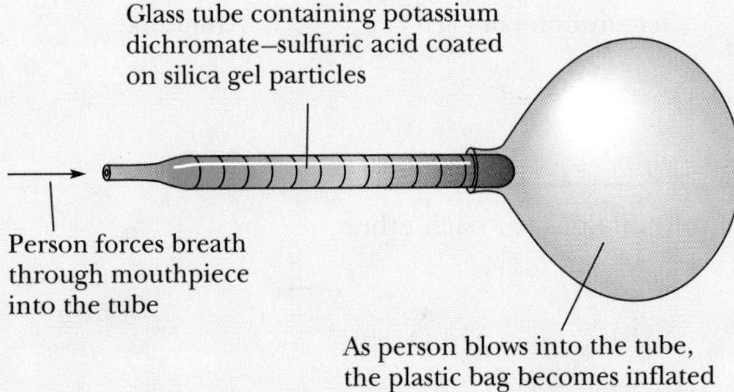
Glass tube containing potassium dichromate–sulfuric acid coated on silica gel particles
Person forces breath through mouthpiece into the tube
As person blows into the tube, the plastic bag becomes inflated

As breath containing ethanol vapor passes through the tube, reddish-orange dichromate ion is reduced to green chromium(III) ion. The concentration of ethanol in the breath is then estimated by measuring how far the green color extends along the length of the tube. When it extends beyond the halfway point, the person is judged as having a sufficiently high blood alcohol content to warrant further, more precise testing.

The Breathalyzer, a more precise testing device, operates on the same principle as the simplified screening test. In a Breathalyzer test, a measured volume of breath is bubbled through a solution of potassium dichromate in aqueous sulfuric acid, and the color change is measured spectrophotometrically.

Both tests measure alcohol in the breath. The legal definition of being under the influence of alcohol is based on *blood* alcohol content, not breath alcohol content. The chemical correlation between these two measurements is that air deep within the lungs is in equilibrium with blood passing through the pulmonary arteries, and an equilibrium is established between blood alcohol and breath alcohol. It has been determined by tests in persons drinking alcohol that 2100 mL of breath contains the same amount of ethanol as 1.00 mL of blood.

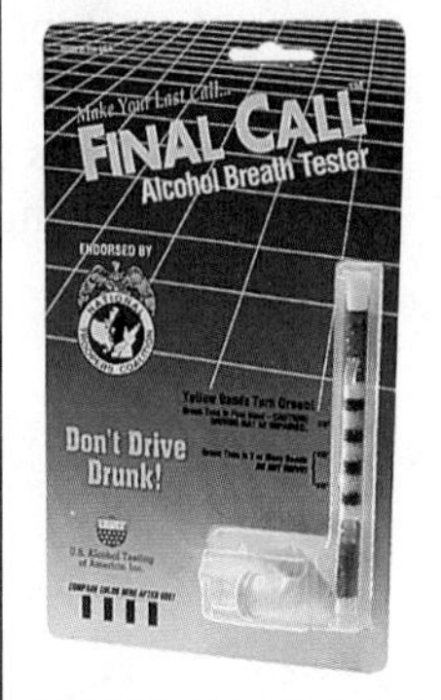
FINAL CALL
Alcohol Breath Tester
Don't Drive Drunk!

A device for testing the breath for the presence of ethanol. When ethanol is oxidized by potassium dichromate, the reddish-orange color of dichromate ion turns to green as it is reduced to chromium(III) ion. *(Charles D. Winters)*

QUESTIONS

Although methanol* and isopropyl alcohol are much more toxic than ethanol and would rarely be found in one's breath, would these two compounds also give a positive alcohol screening test? If so, what would be the products of these reactions?

*Methanol is indeed much more toxic than ethanol, as many found out during Prohibition when they drank wood alcohol instead of ethanol. Methanol causes damage to the nerve sheaths, and one symptom of methanol poisoning is intense pain in response to light.

In ethyl vinyl ether, the ether oxygen is bonded to one sp^3 hybridized carbon and one sp^2 hybridized carbon:

$CH_3CH_2{-}O{-}CH{=}CH_2$

Ethyl vinyl ether

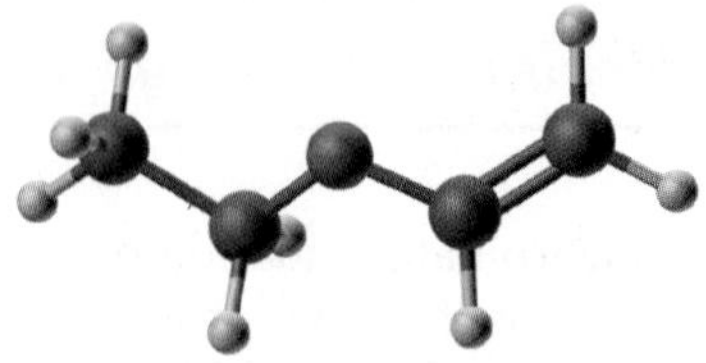

B. Nomenclature

Alkoxy group An —OR group, where R is an alkyl group.

In the IUPAC system, ethers are named by selecting the longest carbon chain as the parent alkane and naming the —OR group bonded to it as an **alkoxy** (*alk*yl + *oxy*gen) group. Common names are derived by listing the alkyl groups bonded to oxygen in alphabetical order and adding the word *ether*.

$CH_3CH_2OCH_2CH_3$ — Ethoxyethane (Diethyl ether)

$CH_3OC(CH_3)_3$ — 2-Methoxy-2-methylpropane (methyl *tert*-butyl ether, MTBE)

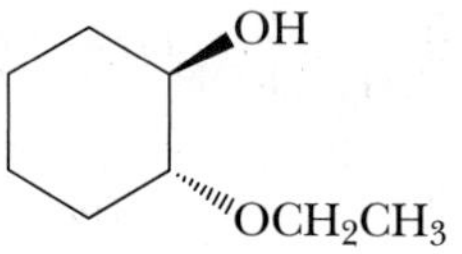

OH / OCH_2CH_3 on cyclohexane — (1*R*,2*R*)-2-Ethoxycyclohexanol (*trans*-2-Ethoxycyclohexanol)

Chemists almost invariably use common names for low-molecular-weight ethers. For example, although ethoxyethane is the IUPAC name for $CH_3CH_2OCH_2CH_3$, it is rarely called that, but rather is called diethyl ether, ethyl ether, or, even more commonly, simply ether. The abbreviation for *tert*-butyl methyl ether, used at one time as an octane-improving additive to gasolines, is *MTBE*, after the common name of methyl *tert*-butyl ether.

Cyclic ether An ether in which the oxygen is one of the atoms of a ring.

Cyclic ethers are heterocyclic compounds in which the ether oxygen is one of the atoms in a ring. These ethers are generally known by their common names:

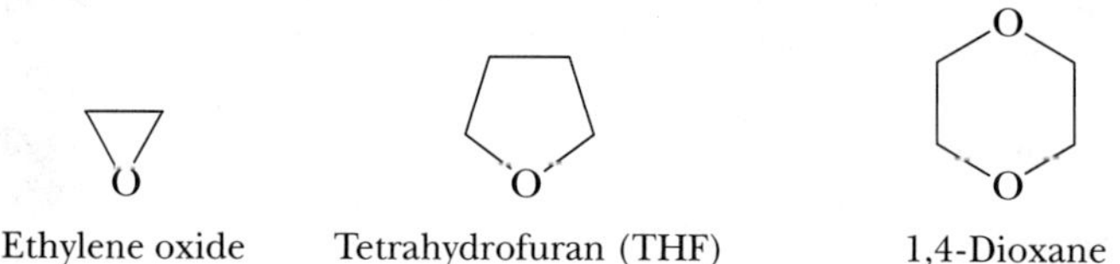

Ethylene oxide   Tetrahydrofuran (THF)   1,4-Dioxane

Example 8.7

Write the IUPAC and common names for each ether:

(a) $CH_3C(CH_3)_2OCH_2CH_3$

(b) cyclohexyl–O–cyclohexyl

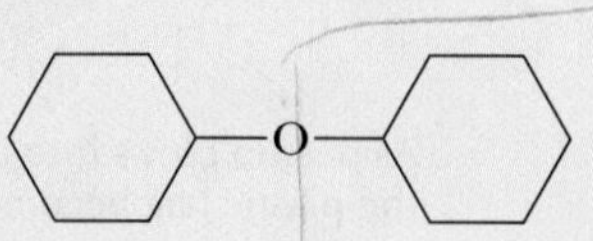

Strategy

As with all nomenclature problems, first determine the root name of the compound. In the IUPAC system, —OR groups are named as alkoxy groups. In the common nomenclature system, the alkyl groups bonded to oxygen are named in alphabetical order, followed by the word "ether."

Solution

(a) 2-Ethoxy-2-methylpropane. Its common name is *tert*-butyl ethyl ether.
(b) Cyclohexoxycyclohexane. Its common name is dicyclohexyl ether.

See problem 8.16

Problem 8.7

Write the IUPAC and common names for each ether:

(a) $CH_3CH(CH_3)CH_2OCH_2CH_3$ (b) cyclopentyl–OCH_3

C. Physical Properties

Ethers are polar compounds in which oxygen bears a partial negative charge and each carbon bonded to it bears a partial positive charge (Figure 8.5). Because of steric hindrance, however, only weak forces of attraction exist between ether molecules in the pure liquid. Consequently, boiling points of ethers are much lower than those of alcohols of comparable molecular weight (Table 8.3). Boiling points of ethers are close to those of hydrocarbons of comparable molecular weight (compare Tables 3.4 and 8.3).

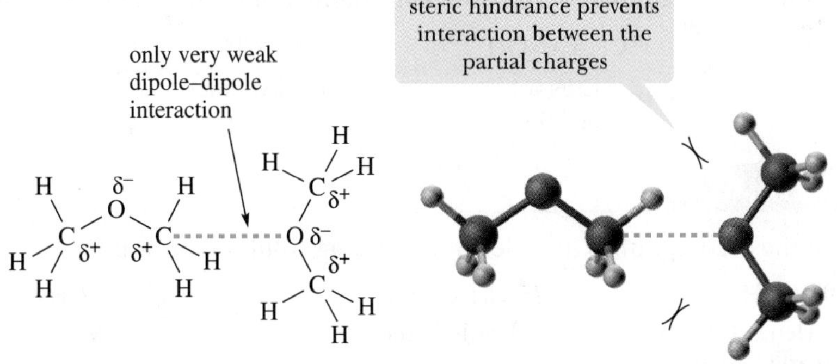
only very weak
dipole–dipole
interaction
steric hindrance prevents
interaction between the
partial charges

Figure 8.5 Ethers are polar molecules, but because of steric hindrance, only weak attractive interactions exist between their molecules in the pure liquid.

TABLE 8.3 Boiling Points and Solubilities in Water of Some Alcohols and Ethers of Comparable Molecular Weight

Structural Formula	Name	Molecular Weight	Boiling Point (°C)	Solubility in Water
CH_3CH_2OH	ethanol	46	78	infinite
CH_3OCH_3	dimethyl ether	46	−24	7.8 g/100 g
$CH_3CH_2CH_2CH_2OH$	1-butanol	74	117	7.4 g/100 g
$CH_3CH_2OCH_2CH_3$	diethyl ether	74	35	8 g/100 g
$CH_3CH_2CH_2CH_2CH_2OH$	1-pentanol	88	138	2.3 g/100 g
$HOCH_2CH_2CH_2CH_2OH$	1,4-butanediol	90	230	infinite
$CH_3CH_2CH_2CH_2OCH_3$	butyl methyl ether	88	71	slight
$CH_3OCH_2CH_2OCH_3$	ethylene glycol dimethyl ether	90	84	infinite

Figure 8.6
Ethers are hydrogen-bond acceptors only. They are not hydrogen-bond donors.

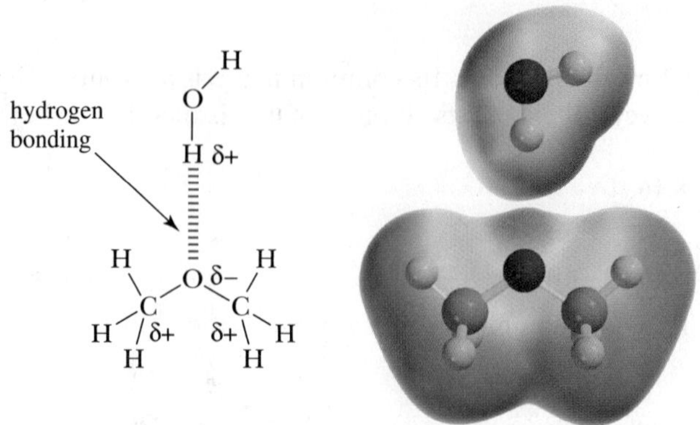

Dimethyl ether in water. The partially negative oxygen of the ether is the hydrogen bond acceptor, and a partially positive hydrogen of a water molecule is the hydrogen bond donor.

Because the oxygen atom of an ether carries a partial negative charge, ethers form hydrogen bonds with water (Figure 8.6) and are more soluble in water than are hydrocarbons of comparable molecular weight and shape (compare data in Tables 3.4 and 8.3).

The effect of hydrogen bonding is illustrated dramatically by comparing the boiling points of ethanol (78°C) and its constitutional isomer dimethyl ether (−24°C). The difference in boiling points between these two compounds is due to the polar O—H group in the alcohol, which is capable of forming intermolecular hydrogen bonds. This hydrogen bonding increases the attractive force between molecules of ethanol; thus, ethanol has a higher boiling point than dimethyl ether:

CH_3CH_2OH	CH_3OCH_3
Ethanol	Dimethyl ether
bp 78°C	bp −24°C

Example 8.8

Arrange these compounds in order of increasing solubility in water:

$CH_3OCH_2CH_2OCH_3$	$CH_3CH_2OCH_2CH_3$	$CH_3CH_2CH_2CH_2CH_2CH_3$
Ethylene glycol dimethyl ether	Diethyl ether	Hexane

Strategy

Look for features that make organic compounds more soluble in water. These are, from most significant to least significant, (1) the ability to form hydrogen bonds with water, (2) polarity, and (3) low molecular weight.

Solution

Water is a polar solvent. Hexane, a nonpolar hydrocarbon, has the lowest solubility in water. Both diethyl ether and ethylene glycol dimethyl ether are polar compounds, due to the presence of their polar C—O—C groups, and each interacts with water as a hydrogen-bond acceptor. Because ethylene glycol dimethyl ether has more sites within its molecules for hydrogen bonding, it is more soluble in water than diethyl ether:

$CH_3CH_2CH_2CH_2CH_2CH_3$	$CH_3CH_2OCH_2CH_3$	$CH_3OCH_2CH_2OCH_3$
Insoluble	8 g/100 g water	Soluble in all proportions

See problems 8.23–8.25

Problem 8.8

Arrange these compounds in order of increasing boiling point:

$CH_3OCH_2CH_2OCH_3$ $\quad$ $HOCH_2CH_2OH$ $\quad$ $CH_3OCH_2CH_2OH$

D. Reactions of Ethers

Ethers, R—O—R, resemble hydrocarbons in their resistance to chemical reaction. They do not react with oxidizing agents, such as potassium dichromate or potassium permanganate. They are not affected by most acids or bases at moderate temperatures. Because of their good solvent properties and general inertness to chemical reaction, ethers are excellent solvents in which to carry out many organic reactions.

8.4 What Are Epoxides?

A. Structure and Nomenclature

An **epoxide** is a cyclic ether in which oxygen is one atom of a three-membered ring:

Epoxide A cyclic ether in which oxygen is one atom of a three-membered ring.

Functional group of an epoxide

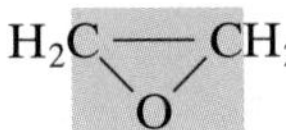

Ethylene oxide

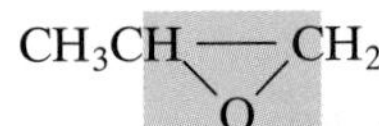

Propylene oxide

Although epoxides are technically classed as ethers, we discuss them separately because of their exceptional chemical reactivity compared with other ethers.

Common names for epoxides are derived by giving the common name of the alkene from which the epoxide might have been derived, followed by the word *oxide*; an example is ethylene oxide.

B. Synthesis from Alkenes

Ethylene oxide, one of the few epoxides manufactured on an industrial scale, is prepared by passing a mixture of ethylene and air (or oxygen) over a silver catalyst:

$$2\,CH_2{=}CH_2 + O_2 \xrightarrow[\text{heat}]{Ag} 2\,H_2C\overset{O}{—}CH_2$$

Ethylene $\qquad$ Ethylene oxide

In the United States, the annual production of ethylene oxide by this method is approximately 10^9 kg.

The most common laboratory method for the synthesis of epoxides from alkenes is oxidation with a peroxycarboxylic acid (a peracid), RCO_3H. One peracid used for this purpose is peroxyacetic acid:

$$CH_3\overset{\overset{O}{\|}}{C}OOH$$

Peroxyacetic acid
(Peracetic acid)

Following is a balanced equation for the epoxidation of cyclohexene by a peroxycarboxylic acid. In the process, the peroxycarboxylic acid is reduced to a carboxylic acid:

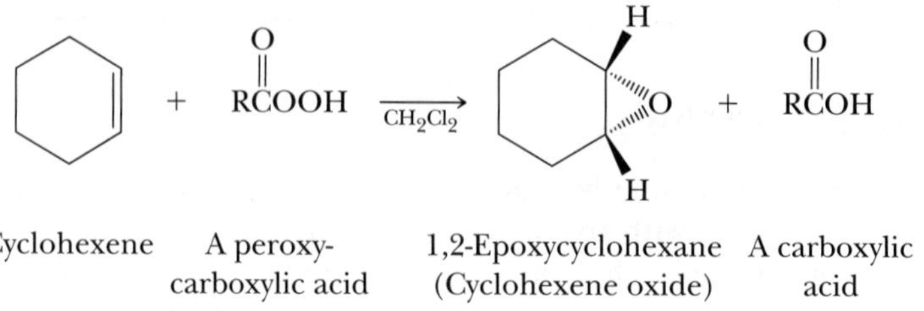

$$\text{Cyclohexene} + \text{RCOOH} \xrightarrow{CH_2Cl_2} \text{1,2-Epoxycyclohexane} + \text{RCOH}$$

Cyclohexene | A peroxycarboxylic acid | 1,2-Epoxycyclohexane (Cyclohexene oxide) | A carboxylic acid

The epoxidation of an alkene is stereoselective. The epoxidation of *cis*-2-butene, for example, yields only *cis*-2-butene oxide:

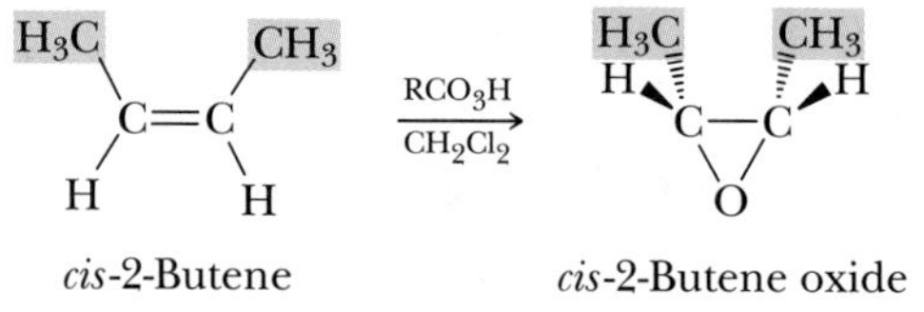

$$\text{cis-2-Butene} \xrightarrow[CH_2Cl_2]{RCO_3H} \text{cis-2-Butene oxide}$$

cis-2-Butene | *cis*-2-Butene oxide

HOW TO 8.5 Predict the Product of an Epoxidation Reaction

The key feature of an epoxidation reaction of an alkene and a peroxycarboxylic acid is the formation of an epoxide with retention of stereochemistry about the reacting C—C double bond. This means that the relative relationship of all groups about the double bond must be the same in the product epoxide as shown in the acyclic and cyclic examples.

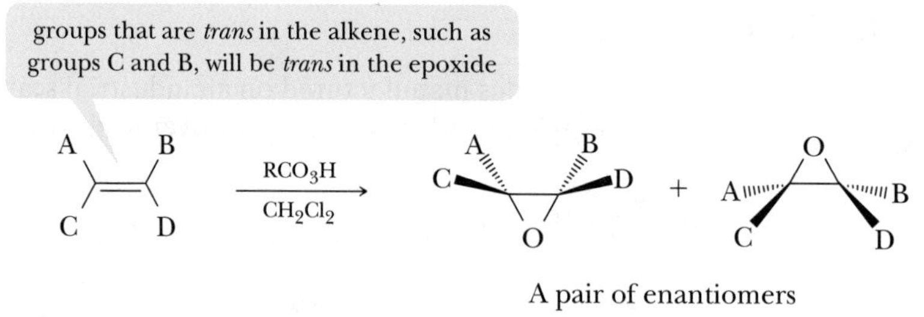

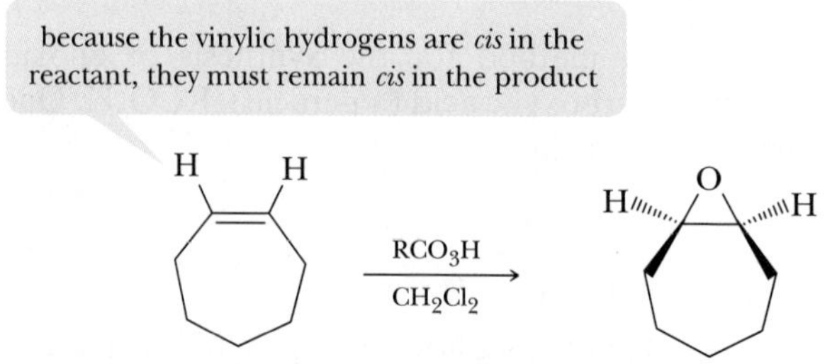

groups that are *trans* in the alkene, such as groups C and B, will be *trans* in the epoxide

$$\xrightarrow[CH_2Cl_2]{RCO_3H}$$

A pair of enantiomers

because the vinylic hydrogens are *cis* in the reactant, they must remain *cis* in the product

$$\xrightarrow[CH_2Cl_2]{RCO_3H}$$

Example 8.9

Draw a structural formula of the epoxide formed by treating *trans*-2-butene with a peroxycarboxylic acid.

Strategy

To predict the product of a peroxycarboxylic acid and an alkene, convert its C—C double bond to a C—C single bond in which both carbons are bonded to the same oxygen in a three-membered ring.

Solution

The oxygen of the epoxide ring is added by forming both carbon–oxygen bonds from the same side of the carbon–carbon double bond:

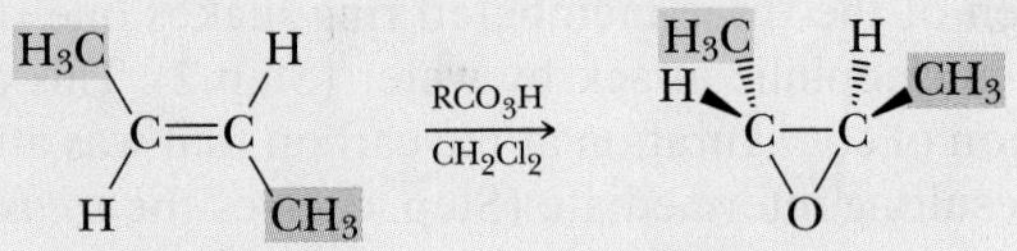

$$\underset{\textit{trans}\text{-2-Butene}}{H_3C(H)C{=}C(H)CH_3} \xrightarrow[CH_2Cl_2]{RCO_3H} \underset{\textit{trans}\text{-2-Butene oxide}}{H_3C(H)C{-}C(H)CH_3 \text{ (epoxide, O bridging both C)}}$$

See problems 8.43–8.45, 8.47

Problem 8.9

Draw the structural formula of the epoxide formed by treating 1,2-dimethylcyclopentene with a peroxycarboxylic acid.

C. Ring-Opening Reactions

Ethers are not normally susceptible to reaction with aqueous acid (Section 8.3D). Epoxides, however, are especially reactive because of the angle strain in the three-membered ring. The normal bond angle about an sp^3 hybridized carbon or oxygen atom is 109.5°. Because of the strain associated with the compression of bond angles in the three-membered epoxide ring from the normal 109.5° to 60°, epoxides undergo ring-opening reactions with a variety of reagents.

In the presence of an acid catalyst—most commonly, perchloric acid—epoxides are hydrolyzed to glycols. As an example, the acid-catalyzed hydrolysis of ethylene oxide gives 1,2-ethanediol:

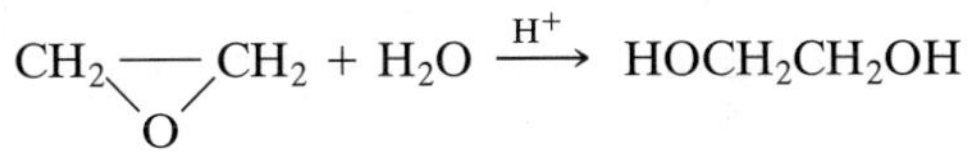

$$\underset{\text{Ethylene oxide}}{CH_2{-}CH_2 \text{ (bridged by O)}} + H_2O \xrightarrow{H^+} \underset{\substack{\text{1,2-Ethanediol}\\ \text{(Ethylene glycol)}}}{HOCH_2CH_2OH}$$

Annual production of ethylene glycol in the United States is approximately 10^{10} kg. Two of its largest uses are in automotive antifreeze and as one of the two starting materials for the production of polyethylene terephthalate (PET), which is fabricated into such consumer products as Dacron® polyester, Mylar®, and packaging films (Section 17.4B).

The acid-catalyzed ring opening of epoxides shows a stereoselectivity typical of S_N2 reactions: The nucleophile attacks anti to the leaving hydroxyl group, and the

—OH groups in the glycol thus formed are anti. As a result, the hydrolysis of an epoxycycloalkane yields a *trans*-1,2-cycloalkanediol:

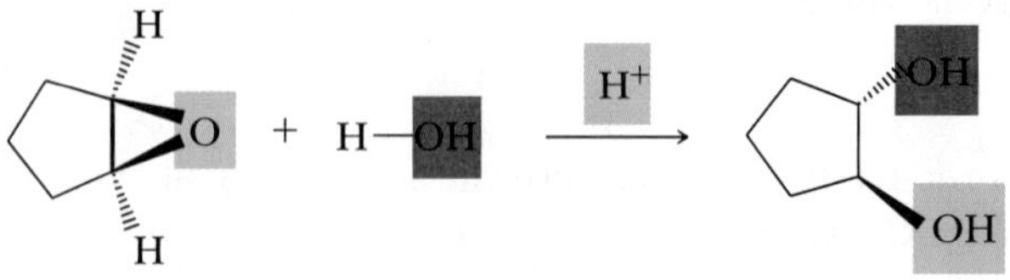

Normally, epoxides will not react with H_2O because water is a poor nucleophile. As shown below, the reaction is made possible because the acid catalyst protonates the epoxide oxygen (Step 1), generating a highly reactive oxonium ion. The positive charge on the oxygen of the three-membered ring makes one of the epoxide carbons susceptible to nucleophilic attack by water (Step 2). This opens the epoxide (Step 3) with inversion of configuration at the carbon that was attacked. Transfer of a proton from the resulting intermediate (Step 4) gives the *trans* glycol and regenerates the acid.

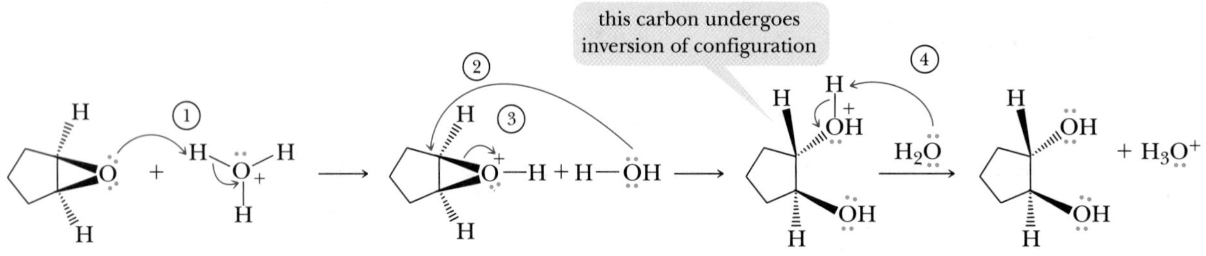

this carbon undergoes inversion of configuration

Oxonium ion

Example 8.10

Draw the structural formula of the product formed by treating cyclohexene oxide with aqueous acid. Be certain to show the stereochemistry of the product.

Strategy

The acid-catalyzed ring opening of an epoxide always results in a *trans*-1,2,-diol, with the two carbons formerly part of the epoxide bonded to each of the two hydroxyl groups.

Solution

The acid-catalyzed hydrolysis of the three-membered epoxide ring gives a *trans* glycol:

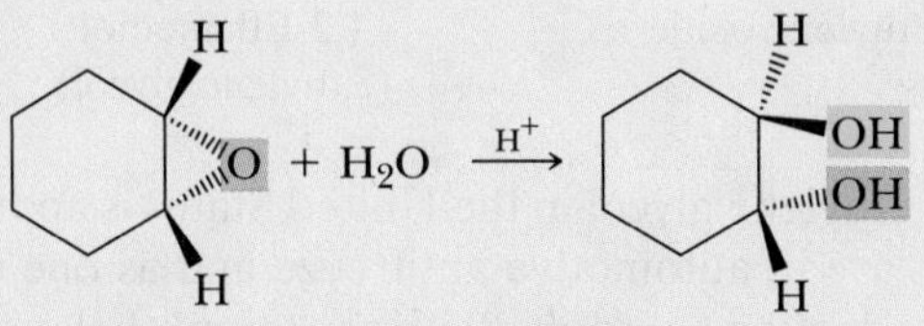

trans-1,2-cyclohexanediol

See problems 8.26, 8.43–8.45

Problem 8.10

Show how to convert 1,2-dimethylcyclohexene to *trans*-1,2-dimethylcyclohexane-1,2-diol.

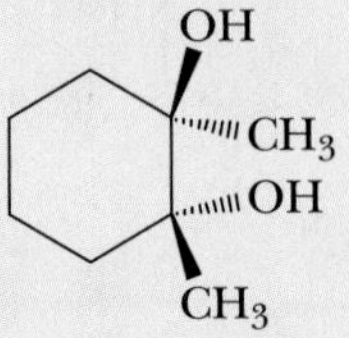

trans-1,2-Dimethylcyclohexane-1,2-diol

Just as ethers are not normally susceptible to reaction with electrophiles, neither are they normally susceptible to reaction with nucleophiles. Because of the strain associated with the three-membered ring, however, epoxides undergo ring-opening reactions with good nucleophiles such as ammonia and amines (Chapter 10), alkoxide ions, and thiols and their anions (Section 8.6). Good nucleophiles attack the ring by an S_N2 mechanism and show a stereoselectivity for attack of the nucleophile at the less hindered carbon of the three-membered ring. The result is an alcohol with the former nucleophile bonded to a carbon β to the newly formed hydroxyl group. An illustration is the reaction of 1-methylcyclohexene oxide with ammonia to give the stereoisomer of 2-amino-1-methylcyclohexanol in which the hydroxyl group and the amino group are *trans*:

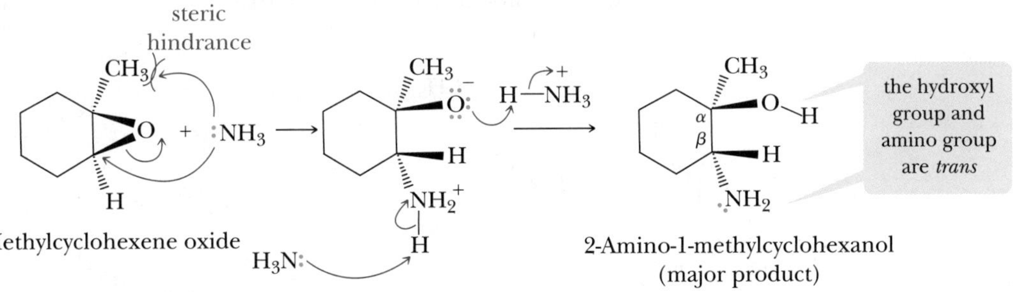

steric hindrance

1-Methylcyclohexene oxide

2-Amino-1-methylcyclohexanol (major product)

the hydroxyl group and amino group are *trans*

The value of epoxides lies in the number of nucleophiles that bring about ring opening and the combinations of functional groups that can be prepared from them. The following chart summarizes the three most important of these nucleophilic ring-opening reactions (the characteristic structural feature of each ring-opening product is shown in color):

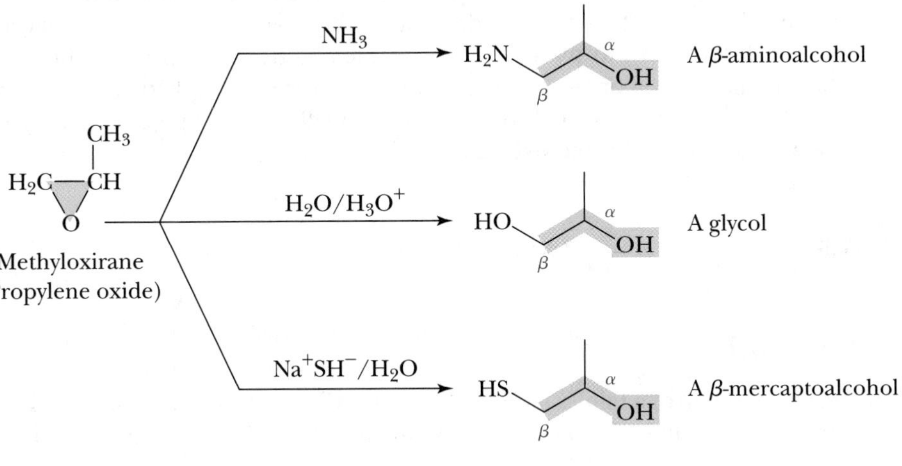

Methyloxirane (Propylene oxide)

NH_3 → A β-aminoalcohol

H_2O/H_3O^+ → A glycol

Na^+SH^-/H_2O → A β-mercaptoalcohol

CHEMICAL CONNECTIONS 8C

Ethylene Oxide: A Chemical Sterilant

Because ethylene oxide is such a highly strained molecule, it reacts with the types of nucleophilic groups present in biological materials. At sufficiently high concentrations, ethylene oxide reacts with enough molecules in cells to cause the death of microorganisms. This toxic property is the basis for using ethylene oxide as a chemical sterilant. In hospitals, surgical instruments and other items that cannot be made disposable are now sterilized by exposure to ethylene oxide.

NH_2 ... ~ DNA ~ Adenine + (ethylene oxide) ⟶

QUESTION

One of the ways that ethylene oxide has been found to kill microorganisms is by reacting with the adenine components of their DNA at the atom indicated in red. Propose a mechanism and an initial product for this reaction. *Hint:* first draw in any lone pairs of electrons in adenine.

Ethylene oxide and substituted ethylene oxides are valuable building blocks for the synthesis of larger organic molecules. Following are structural formulas for two common drugs, each synthesized in part from ethylene oxide:

Procaine (Novocaine)

Diphenhydramine (Benadryl)

Novocaine was the first injectable local anesthetic. Benadryl was the first synthetic antihistamine. The portion of the carbon skeleton of each that is derived from the reaction of ethylene oxide with a nitrogen–nucleophile is shown in color.

In later chapters, after wc have developed the chemistry of more functional groups, we will show how to synthesize Novocaine and Benadryl from readily available starting materials. For the moment, however, it is sufficient to recognize that the unit —O—C—C—Nu can be derived by nucleophilic opening of ethylene oxide or a substituted ethylene oxide.

8.5 What Are Thiols?

A. Structure

Thiol A compound containing an —SH (sulfhydryl) group.

The functional group of a **thiol** is an —SH (sulfhydryl) group. Figure 8.7 shows a Lewis structure and a ball-and-stick model of methanethiol, CH_3SH, the simplest thiol.

H
|
H—C—S̈—H
|
H

(a)

100.3°

(b)

Figure 8.7
Methanethiol, CH_3SH.
(a) Lewis structure and
(b) ball-and-stick model.
The C—S—H bond angle is 100.9°, somewhat smaller than the tetrahedral angle of 109.5°.

Methanethiol. The electronegativities of carbon and sulfur are virtually identical (2.5 each), while sulfur is slightly more electronegative than hydrogen (2.5 versus 2.1). The electron density model shows some slight partial positive charge on hydrogen of the S—H group and some slight partial negative charge on sulfur.

The most outstanding property of low-molecular-weight thiols is their stench. They are responsible for the unpleasant odors such as those from skunks, rotten eggs, and sewage. The scent of skunks is due primarily to two thiols:

$CH_3CH{=}CHCH_2SH$ — 2-Butene-1-thiol

$CH_3CH(CH_3)CH_2CH_2SH$ — 3-Methyl-1-butanethiol

The scent of skunks is a mixture of two thiols, 3-methyl-1-butanethiol and 2-butene-1-thiol. *(Stephen J. Krasemann/ Photo Researchers, Inc.)*

A blend of low-molecular-weight thiols is added to natural gas as an odorant. The most common of these odorants is 2-methyl-2-propanethiol (*tert*-butyl mercaptan), because it is the most resistant to oxidation and has the greatest soil penetration. 2-Propanethiol is also used for this purpose, usually as a blend with *tert*-butyl mercaptan.

Mercaptan A common name for any molecule containing an —SH group.

Natural gas odorants:

$CH_3—C(CH_3)_2—SH$ — 2-Methyl-2-propanethiol (*tert*-Butyl mercaptan)

$CH_3—CH(SH)—CH_3$ — 2-Propanethiol (Isopropyl mercaptan)

B. Nomenclature

The sulfur analog of an alcohol is called a thiol (thi- from the Greek: *theion*, sulfur) or, in the older literature, a **mercaptan**, which literally means "mercury capturing." Thiols react with Hg^{2+} in aqueous solution to give sulfide salts as insoluble precipitates. Thiophenol, C_6H_5SH, for example, gives $(C_6H_5S)_2Hg$.

In the IUPAC system, thiols are named by selecting as the parent alkane the longest chain of carbon atoms that contains the —SH group. To show that the compound is a thiol, we add *-thiol* to the name of the parent alkane and number the parent chain in the direction that gives the —SH group the lower number.

Common names for simple thiols are derived by naming the alkyl group bonded to —SH and adding the word *mercaptan*. In compounds containing other functional groups, the presence of an —SH group is indicated by the prefix **mercapto-**. According to the IUPAC system, —OH takes precedence over —SH in both numbering and naming:

CH_3CH_2SH — Ethanethiol (Ethyl mercaptan)

$CH_3CH(CH_3)CH_2SH$ — 2-Methyl-1-propanethiol (Isobutyl mercaptan)

$HSCH_2CH_2OH$ — 2-Mercaptoethanol

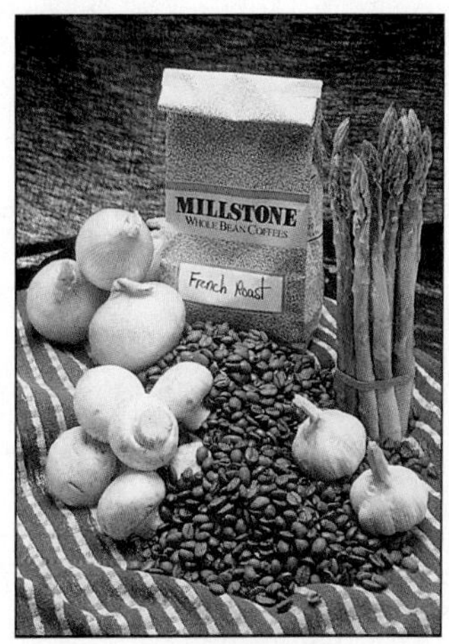

Mushrooms, onions, garlic, and coffee all contain sulfur compounds. One of these present in coffee is

O SH

(Charles D. Winters)

Sulfur analogs of ethers (thioethers) are named by using the word *sulfide* to show the presence of the —S— group. Following are common names of two sulfides:

CH_3SCH_3 — Dimethyl sulfide

$CH_3CH_2SCH(CH_3)CH_3$ — Ethyl isopropyl sulfide

Example 8.11

Write the IUPAC name for each compound:

(a) [structure: SH]
(b) [structure: SH]
(c) [structure: OH, CH_3, SH]

Strategy

Identify the root name of the compound. If the compound only contains an —SH group, name it as an alkanethiol. If the compound contains both an —OH group and an —SH group, name the compound as an alcohol with a mercapto substituent. Remember that priority for numbering is given to the —OH group.

Solution

(a) The parent alkane is pentane. We show the presence of the —SH group by adding *thiol* to the name of the parent alkane. The IUPAC name of this thiol is 1-pentanethiol. Its common name is pentyl mercaptan.
(b) The parent alkane is butane. The IUPAC name of this thiol is 2-butanethiol. Its common name is *sec*-butyl mercaptan.
(c) The parent alkane is pentane. Because —OH receives priority over —SH, the compound is named as an alcohol, with the —OH group receiving priority for numbering as well.

[structure: OH, CH_3, SH; numbered 1 2 3 4 5] not, [structure: OH, CH_3, SH; numbered 5 4 3 2 1]

(2*R*,4*R*)-5-mercapto-4-methylpentan-2-ol

See problems 8.14, 8.15

Problem 8.11

Write the IUPAC name for each thiol:

(a) [structure: SH]
(b) [structure: SH]

C. Physical Properties

Because of the small difference in electronegativity between sulfur and hydrogen (2.5 − 2.1 = 0.4), we classify the S—H bond as nonpolar covalent. Because of this lack of polarity, thiols show little association by hydrogen bonding. Consequently,

TABLE 8.4 Boiling Points of Three Thiols and Three Alcohols with the Same Number of Carbon Atoms

Thiol	Boiling Point (°C)	Alcohol	Boiling Point (°C)
methanethiol	6	methanol	65
ethanethiol	35	ethanol	78
1-butanethiol	98	1-butanol	117

they have lower boiling points and are less soluble in water and other polar solvents than are alcohols of similar molecular weight. Table 8.4 gives the boiling points of three low-molecular-weight thiols. For comparison, the table also gives the boiling points of alcohols with the same number of carbon atoms.

Earlier, we illustrated the importance of hydrogen bonding in alcohols by comparing the boiling points of ethanol (78°C) and its constitutional isomer dimethyl ether (24°C). By comparison, the boiling point of ethanethiol is 35°C, and that of its constitutional isomer dimethyl sulfide is 37°C:

CH_3CH_2SH	CH_3SCH_3
Ethanethiol	Dimethyl sulfide
bp 35°C	bp 37°C

The fact that the boiling points of these constitutional isomers are almost identical indicates that little or no association by hydrogen bonding occurs between thiol molecules.

8.6 What Are the Characteristic Reactions of Thiols?

In this section, we discuss the acidity of thiols and their reaction with strong bases, such as sodium hydroxide, and with molecular oxygen.

A. Acidity

Hydrogen sulfide is a stronger acid than water:

$$H_2O + H_2O \rightleftharpoons HO^- + H_3O^+ \qquad pK_a = 15.7$$

$$H_2S + H_2O \rightleftharpoons HS^- + H_3O^+ \qquad pK_a = 7.0$$

Similarly, thiols are stronger acids than alcohols. Compare, for example, the pK_a's of ethanol and ethanethiol in dilute aqueous solution:

$$CH_3CH_2OH + H_2O \rightleftharpoons CH_3CH_2O^- + H_3O^+ \qquad pK_a = 15.9$$

$$CH_3CH_2SH + H_2O \rightleftharpoons CH_3CH_2S^- + H_3O^+ \qquad pK_a = 8.5$$

Thiols are sufficiently strong acids that, when dissolved in aqueous sodium hydroxide, they are converted completely to alkylsulfide salts:

$$CH_3CH_2SH + Na^+OH^- \longrightarrow CH_3CH_2S^-Na^+ + H_2O$$

CH_3CH_2SH	Na^+OH^-	$CH_3CH_2S^-Na^+$	H_2O
pK_a 8.5			pK_a 15.7
Stronger acid	Stronger base	Weaker base	Weaker acid

To name salts of thiols, give the name of the cation first, followed by the name of the alkyl group to which the suffix *-sulfide* is added. For example, the sodium salt derived from ethanethiol is named sodium ethylsulfide.

B. Oxidation to Disulfides

Many of the chemical properties of thiols stem from the fact that the sulfur atom of a thiol is oxidized easily to several higher oxidation states. The most common reaction of thiols in biological systems is their oxidation to disulfides, the functional group of which is a **disulfide** (—S—S—) bond. Thiols are readily oxidized to disulfides by molecular oxygen. In fact, they are so susceptible to oxidation that they must be protected from contact with air during storage. Disulfides, in turn, are easily reduced to thiols by several reagents. This easy interconversion between thiols and disulfides is very important in protein chemistry, as we will see in Chapter 20:

$$2\,HOCH_2CH_2SH \underset{\text{reduction}}{\overset{\text{oxidation}}{\rightleftharpoons}} HOCH_2CH_2S{-}SCH_2CH_2OH$$

A thiol　　　　A disulfide

We derive common names of simple disulfides by listing the names of the groups bonded to sulfur and adding the word *disulfide*, as, for example, $CH_3S{-}SCH_3$, which is named dimethyldisulfide.

Example 8.12

Predict the products of the following reactions. If the reaction is an acid–base reaction, predict its position of equilibrium.

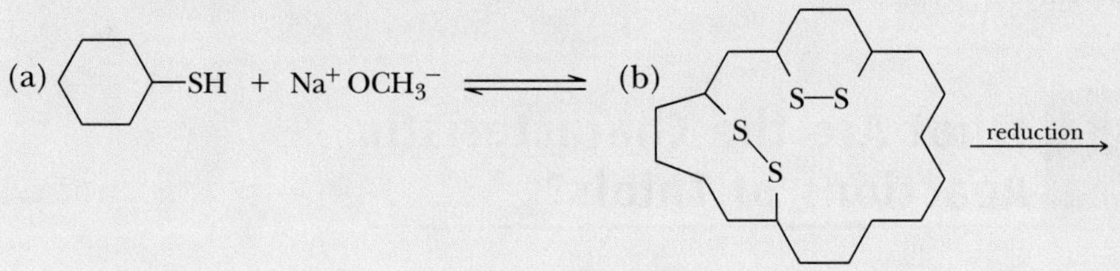

(a) $C_6H_{11}{-}SH + Na^+\,OCH_3^- \rightleftharpoons$

(b) [macrocyclic ring containing two disulfide (S—S) bonds] $\xrightarrow{\text{reduction}}$

Strategy

First determine what type of reaction is occurring. An oxidation reaction of a thiol produces a disulfide bond (—S—S—). A reduction of a disulfide bond produces two mercapto groups. Thiols can also act as weak acids (although at a pK_a of 8.5, they are relatively strong for an organic acid).

Solution

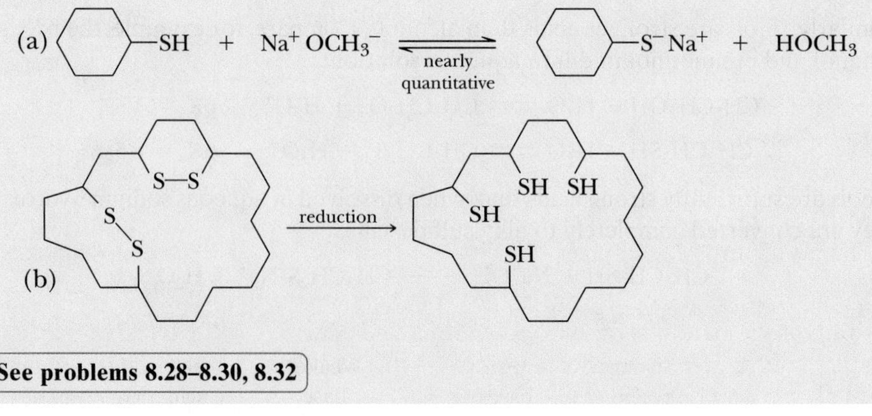

(a) $C_6H_{11}{-}SH + Na^+\,OCH_3^- \underset{\text{quantitative}}{\overset{\text{nearly}}{\rightleftharpoons}} C_6H_{11}{-}S^-Na^+ + HOCH_3$

(b) [macrocycle with S—S and S—S bonds] $\xrightarrow{\text{reduction}}$ [ring-opened product with SH SH, SH, SH]

See problems 8.28–8.30, 8.32

Problem 8.12

Predict the products of the following reactions. If the reaction is an acid–base reaction, predict its position of equilibrium.

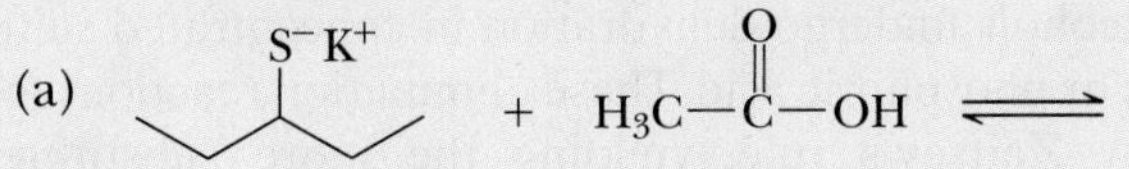

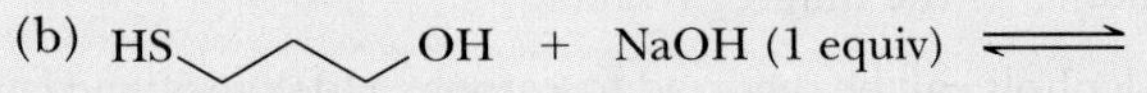

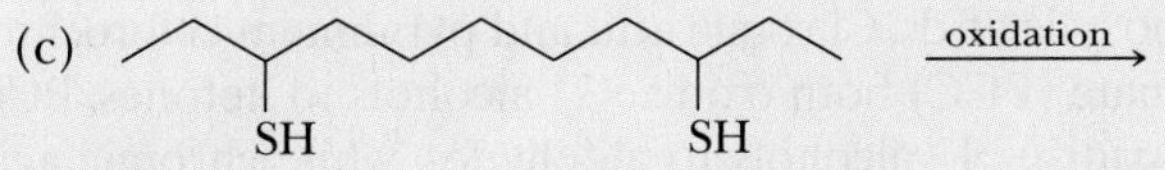

(a) (3-pentanethiolate, S^-K^+) + $H_3C-\overset{O}{\overset{\|}{C}}-OH$ ⇌

(b) $HS\sim\sim OH$ + NaOH (1 equiv) ⇌

(c) (dithiol with two SH groups) $\xrightarrow{\text{oxidation}}$

Key Terms and Concepts

Summary of Key Questions

8.1 What Are Alcohols?

- The functional group of an alcohol is an —OH (hydroxyl) group bonded to an sp^3 hybridized carbon. Alcohols have many uses. For example, they are used as fuel additives, solvents, and deicers. They are used in medicine as a sterilizer, and ethanol is prevalent in the beverage industry. They can be transformed to a variety of other functional groups, making them useful in organic synthesis.
- Alcohols are classified as 1°, 2°, or 3°, depending on whether the —OH group is bonded to a primary, secondary, or tertiary carbon.
- IUPAC names of alcohols are derived by changing the suffix of the parent alkane from *-e* to *-ol*. The chain is numbered to give the carbon bearing —OH the lower number.
- Common names for alcohols are derived by naming the alkyl group bonded to —OH and adding the word *alcohol*.
- Alcohols are polar compounds with oxygen bearing a partial negative charge and both the carbon and hydrogen bonded to it bearing partial positive charges.
- Because of intermolecular association by hydrogen bonding, the boiling points of alcohols are higher than those of hydrocarbons of comparable molecular weight.
- Because of increased dispersion forces, the boiling points of alcohols increase with increasing molecular weight.
- Alcohols interact with water by hydrogen bonding and therefore are more soluble in water than are hydrocarbons of comparable molecular weight.

8.2 What Are the Characteristic Reactions of Alcohols?

- Alcohols undergo acid–base reactions, acting both as weak acids and weak bases. The two smallest alcohols, methanol and ethanol, are comparable to water in acidity, while most 2° and 3° alcohols are less acidic than water.
- Alcohols react with active metals (e.g., Li, Na, K, etc.) to give alkoxides.
- Alcohols react with hydrogen halides (HCl, HBr, and HI) to give haloalkanes via substitution reactions. The mechanism of the reaction is either S_N1 or S_N2 depending on the classification (1°, 2°, or 3°) of the alcohol.
- Alcohols react with thionyl chloride, $SOCl_2$, to give chloroalkanes.
- Alcohols undergo dehydration in concentrated sulfuric or phosphoric acid. These elimination reactions follow Zaitsev's rule, yielding the more substituted alkene as the major product.
- Alcohols can be oxidized to ketones, aldehydes, and carboxylic acids. Chromic acid and pyridinium chlorochromate (PCC) both oxidize 2° alcohols to ketones. PCC oxidizes 1° alcohols to aldehydes, while chromic acid oxidizes 1° alcohols to carboxylic acids.

8.3 What are Ethers?

- The functional group of an ether is an atom of oxygen bonded to two carbon atoms. Ethers are used as solvents and in medicine as anesthetics.
- In the IUPAC name of an ether, the parent alkane is named, and then the —OR group is named as an alkoxy substituent. Common names are derived by naming the two groups bonded to oxygen followed by the word ether. Ethers are weakly polar compounds. Their boiling points are close to those of hydrocarbons of comparable molecular weight. Because ethers are hydrogen-bond acceptors, they are more soluble in water than are hydrocarbons of comparable molecular weight.
- Ethers are relatively resistant to chemical transformation and, for this reason, are often employed as solvents in chemical reactions.

8.4 What Are Epoxides?

- An epoxide is a three-membered cyclic ether in which oxygen is one of the atoms of the three-membered ring.
- Epoxides can be synthesized from the reaction of an alkene with a peroxycarboxylic acid (RCO_3H). The reaction proceeds such that the relative stereochemistry about the C—C double bond is retained in the product epoxide.
- Epoxides undergo ring-opening reactions due to the strain of their three-membered rings. In acid-catalyzed hydrolysis, epoxides made from cyclic alkenes are transformed into *trans*-glycols. Good nucleophiles can also open the epoxide ring via nucleophilic attack at the least substituted carbon of the three-membered ring.

8.5 What Are Thiols?

- A thiol is the sulfur analog of an alcohol; it contains an —SH (sulfhydryl) group in place of an —OH group. Thiols are important compounds in several biological processes.
- Thiols are named in the same manner as alcohols, but the suffix *-e* is retained, and *-thiol* is added. Common names for thiols are derived by naming the alkyl group bonded to —SH and adding the word "mercaptan." In compounds containing functional groups of higher precedence, the presence of —SH is indicated by the prefix *mercapto-*. For thioethers, name the two groups bonded to sulfur, followed by the word "sulfide."
- The S—H bond is nonpolar, and the physical properties of thiols are more like those of hydrocarbons of comparable molecular weight.

8.6 What Are the Characteristic Reactions of Thiols?

- Thiols ($pK_a \approx 8.5$) are stronger acids than alcohols and are quantitatively deprotonated by hydroxide.
- Thiols can be oxidized to give a disulfide (—S—S—) bond. This process is reversible through reduction.

Quick Quiz

Answer true or false to the following questions to assess your general knowledge of the concepts in this chapter. If you have difficulty with any of them, you should review the appropriate section in the chapter (shown in parentheses) before attempting the more challenging end-of-chapter problems.

1. Dehydration of an alcohol proceeds either by an E1 or an E2 mechanism. (8.2)
2. Epoxides are more reactive than acyclic ethers. (8.3, 8.4)
3. Attack of an electrophile on the carbon of an epoxide ring results in opening of the ring. (8.4)
4. A hydrogen bond is a form of dipole–dipole interaction. (8.1)
5. Alcohols have higher boiling points than thiols of the same molecular weight. (8.1, 8.5)
6. Thiols are more acidic than alcohols. (8.2, 8.5)
7. Alcohols can act as hydrogen-bond donors but not as hydrogen-bond acceptors. (8.1)
8. Alcohols can function as both acids and bases. (8.2)
9. Ethers can act as hydrogen-bond donors but not as hydrogen-bond acceptors. (8.3)
10. Reduction of a thiol produces a disulfide. (8.6)
11. Ethers are more reactive than alcohols. (8.2, 8.3)
12. $(CH_3CH_2)_2CHOH$ is classified as a 3° alcohol. (8.1)
13. PCC will oxidize a secondary alcohol to a ketone. (8.2)
14. PCC will oxidize a primary alcohol to a carboxylic acid. (8.2)
15. Alcohols have higher boiling points than ethers of the same molecular weight. (8.1, 8.3)
16. A dehydration reaction yields an epoxide as the product. (8.2)
17. Alcohols can be converted to alkenes. (8.2)
18. Alcohols can be converted to haloalkanes. (8.2)
19. In naming alcohols, "alkyl alcohol" is the IUPAC form of the name, while "alkanol" is the common form of the name. (8.1)
20. —OH is a poor leaving group. (8.2)
21. A glycol is any alcohol with at least two hydroxyl groups bonded to different carbons. (8.1)

Answers: (1) T (2) T (3) F (4) T (5) T (6) T (7) F (8) T (9) F (10) F (11) F (12) F (13) T (14) F (15) T (16) F (17) T (18) T (19) F (20) T (21) T

Key Reactions

1. Acidity of Alcohols (Section 8.2A)

In dilute aqueous solution, methanol and ethanol are comparable in acidity to water. Secondary and tertiary alcohols are weaker acids than water.

$$CH_3OH + H_2O \rightleftharpoons CH_3O^- + H_3O^+ \qquad pK_a = 15.5$$

2. Reaction of Alcohols with Active Metals (Section 8.2C)

Alcohols react with Li, Na, K, and other active metals to form metal alkoxides, which are somewhat stronger bases than NaOH and KOH:

$$2\,CH_3CH_2OH + 2\,Na \longrightarrow 2\,CH_3CH_2O^-Na^+ + H_2$$

3. Reaction of Alcohols with HCl, HBr, and HI (Section 8.2D)

Primary alcohols react with HBr and HI by an S_N2 mechanism:

$$CH_3CH_2CH_2CH_2OH + HBr \longrightarrow CH_3CH_2CH_2CH_2Br + H_2O$$

Tertiary alcohols react with HCl, HBr, and HI by an S_N1 mechanism, with the formation of a carbocation intermediate:

$$CH_3\overset{CH_3}{\underset{CH_3}{C}}OH + HCl \xrightarrow{25^\circ C} CH_3\overset{CH_3}{\underset{CH_3}{C}}Cl + H_2O$$

Secondary alcohols may react with HCl, HBr, and HI by an S_N2 or an S_N1 mechanism, depending on the alcohol and experimental conditions.

4. Reaction of Alcohols with $SOCl_2$ (Section 8.2D)

This is often the method of choice for converting an alcohol to an alkyl chloride:

$$CH_3(CH_2)_5OH + SOCl_2 \longrightarrow CH_3(CH_2)_5Cl + SO_2 + HCl$$

5. Acid-Catalyzed Dehydration of Alcohols (Section 8.2E)

When isomeric alkenes are possible, the major product is generally the more substituted alkene (Zaitsev's rule):

$$CH_3CH_2CH(OH)CH_3 \xrightarrow[\text{heat}]{H_3PO_4}$$

$$CH_3CH{=}CHCH_3 + CH_3CH_2CH{=}CH_2 + H_2O$$

Major product

6. Oxidation of a Primary Alcohol to an Aldehyde (Section 8.2F)

This oxidation is most conveniently carried out by using pyridinium chlorochromate (PCC):

$$\text{cyclopentyl}{-}CH_2OH \xrightarrow[CH_2Cl_2]{PCC} \text{cyclopentyl}{-}\overset{O}{\overset{\|}{C}}H$$

7. Oxidation of a Primary Alcohol to a Carboxylic Acid (Section 8.2F)

A primary alcohol is oxidized to a carboxylic acid by chromic acid:

$$CH_3(CH_2)_4CH_2OH + H_2CrO_4 \xrightarrow[\text{acetone}]{H_2O}$$

$$CH_3(CH_2)_4\overset{O}{\overset{\|}{C}}OH + Cr^{3+}$$

8. Oxidation of a Secondary Alcohol to a Ketone (Section 8.2F)

A secondary alcohol is oxidized to a ketone by chromic acid and by PCC:

$$CH_3(CH_2)_4CH(OH)CH_3 + H_2CrO_4 \longrightarrow CH_3(CH_2)_4\overset{O}{\overset{\|}{C}}CH_3 + Cr^{3+}$$

9. Oxidation of an Alkene to an Epoxide (Section 8.4B)

The most common method for the synthesis of an epoxide from an alkene is oxidation with a peroxycarboxylic acid, such as peroxyacetic acid:

$$\text{cyclohexene} + R\overset{O}{\overset{\|}{C}}OOH \longrightarrow \text{cyclohexene oxide (H, H cis)} + R\overset{O}{\overset{\|}{C}}OH$$

10. Acid-Catalyzed Hydrolysis of Epoxides (Section 8.4C)

Acid-catalyzed hydrolysis of an epoxide derived from a cycloalkene gives a *trans* glycol (hydrolysis of cycloalkene oxide is stereoselective, giving the *trans* glycol):

$$\text{cyclopentene oxide} + H_2O \xrightarrow{H^+} \textit{trans}\text{-1,2-cyclopentanediol}$$

11. Nucleophilic Ring Opening of Epoxides (Section 8.4C)

Good nucleophiles, such as ammonia and amines, open the highly strained epoxide ring by an S_N2 mechanism and show a regioselectivity for attack of the nucleophile at the less hindered carbon of the three-membered ring. The reaction favors the stereoselective formation of the *trans* product:

$$\text{cyclohexene oxide} + NH_3 \longrightarrow \textit{trans}\text{-2-aminocyclohexanol}$$

Cyclohexene oxide *trans*-2-Aminocyclohexanol

12. Acidity of Thiols (Section 8.6A)

Thiols are weak acids, pK_a 8–9, but are considerably stronger acids than alcohols, pK_a 16–18.

$$CH_3CH_2SH + H_2O \rightleftharpoons CH_3CH_2S^- + H_3O^+ \quad pK_a = 8.5$$

13. Oxidation to Disulfides (Section 8.6B)

Oxidation of a thiol by O_2 gives a disulfide:

$$2\,RSH + \tfrac{1}{2}O_2 \longrightarrow RS{-}SR + H_2O$$

Problems

A problem marked with an asterisk indicates an applied "real world" problem. Answers to problems whose numbers are printed in blue are given in Appendix D.

Structure and Nomenclature

8.13 Classify the alcohols as primary, secondary, or tertiary. **(See Example 8.2)**

(a) 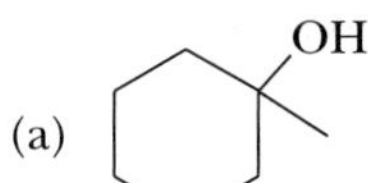

(b) $(CH_3)_3COH$

(c) 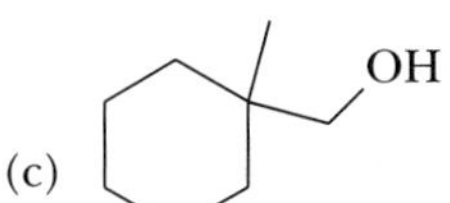

(d) 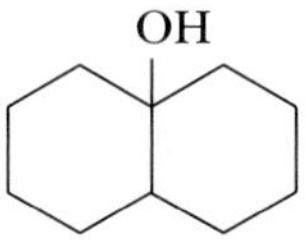 OH

(e) 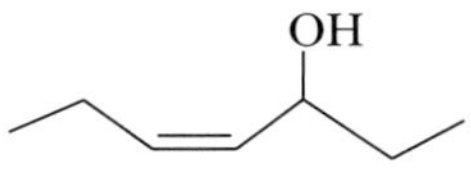 OH

(f) 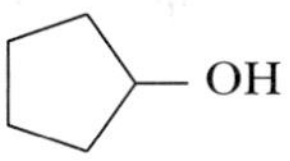 OH

(g) 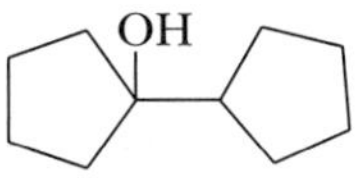 OH

(h) 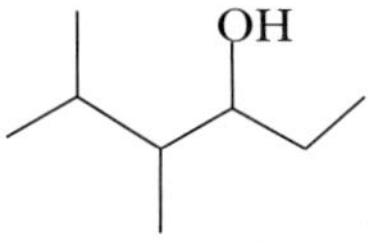 OH

8.14 Provide an IUPAC or common name for these compounds: **(See Examples 8.1, 8.3, 8.11)**

(a) 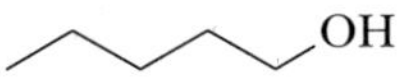 OH

(b) 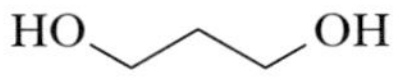 HO OH

(c) 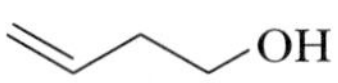 OH

(d) 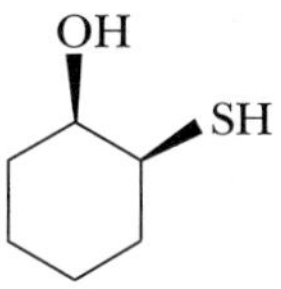 OH SH

(e) 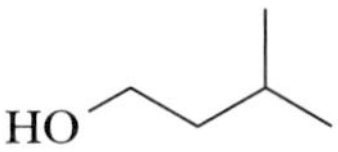 HO

(f) 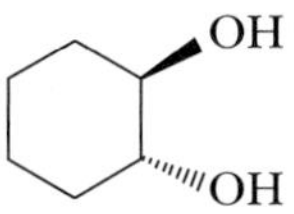 OH OH

(g) 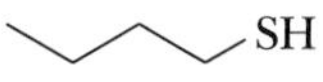 SH

(h) 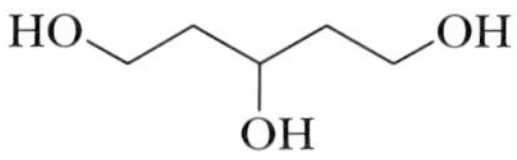 HO OH OH

8.15 Draw a structural formula for each alcohol: **(See Examples 8.1, 8.3, 8.11)**

(a) Isopropyl alcohol

(b) Propylene glycol

(c) (*R*)-5-Methyl-2-hexanol

(d) 2-Methyl-2-propyl-1,3-propanediol

(e) 2,2-Dimethyl-1-propanol

(f) 2-Mercaptoethanol

(g) 1,4-Butanediol

(h) (*Z*)-5-Methyl-2-hexen-1-ol

(i) *cis*-3-Penten-1-ol

(j) *trans*-1,4-Cyclohexanediol

8.16 Write names for these ethers: **(See Example 8.7)**

(a) 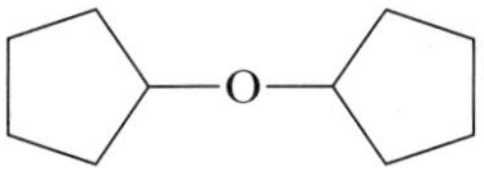 O

(b) 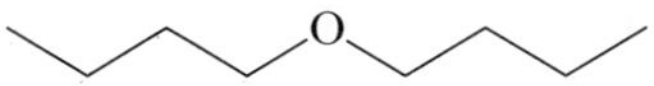 O

(c) 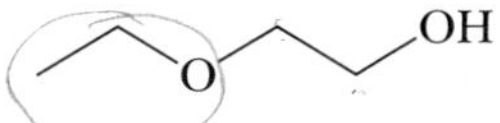 O OH

8.17 Name and draw structural formulas for the eight isomeric alcohols with molecular formula $C_5H_{12}O$. Which are chiral? **(See Example 8.1, 8.3)**

Physical Properties

8.18 Arrange these compounds in order of increasing boiling point (values in °C are −42, 78, 117, and 198):

(a) $CH_3CH_2CH_2CH_2OH$ (b) CH_3CH_2OH

(c) $HOCH_2CH_2OH$ (d) $CH_3CH_2CH_3$

8.19 Arrange these compounds in order of increasing boiling point (values in °C are −42, −24, 78, and 118):

(a) CH_3CH_2OH (b) CH_3OCH_3

(c) $CH_3CH_2CH_3$ (d) CH_3COOH

8.20 Propanoic acid and methyl acetate are constitutional isomers, and both are liquids at room temperature:

$CH_3CH_2\overset{O}{\overset{\|}{C}}OH$ — Propanoic acid

$CH_3\overset{O}{\overset{\|}{C}}OCH_3$ — Methyl acetate

One of these compounds has a boiling point of 141°C; the other has a boiling point of 57°C. Which compound has which boiling point?

8.21 Draw all possible staggered conformations of ethylene glycol ($HOCH_2CH_2OH$). Can you explain why the gauche conformation is more stable than the anti conformation by approximately 4.2 kJ/mol (1 kcal/mol)? **(See Example 3.7)**

8.22 Following are structural formulas for 1-butanol and 1-butanethiol:

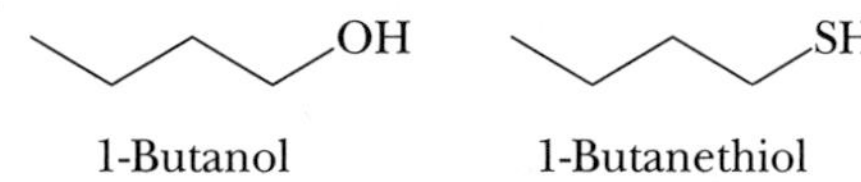

1-Butanol 1-Butanethiol

One of these compounds has a boiling point of 98.5°C; the other has a boiling point of 117°C. Which compound has which boiling point?

8.23 From each pair of compounds, select the one that is more soluble in water: **(See Example 8.8)**

(a) CH_2Cl_2 or CH_3OH

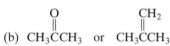

(b) $CH_3\overset{O}{\overset{\|}{C}}CH_3$ or $CH_3\overset{CH_2}{\overset{\|}{C}}CH_3$

(c) CH_3CH_2Cl or NaCl

(d) $CH_3CH_2CH_2SH$ or $CH_3CH_2CH_2OH$

(e) $CH_3CH_2\overset{OH}{\overset{|}{C}}HCH_2CH_3$ or $CH_3CH_2\overset{O}{\overset{\|}{C}}CH_2CH_3$

8.24 Arrange the compounds in each set in order of decreasing solubility in water: **(See Example 8.8)**

(a) Ethanol; butane; diethyl ether

(b) 1-Hexanol; 1,2-hexanediol; hexane

8.25 Each of the following compounds is a common organic solvent: **(See Example 8.8)**

(a) CH_2Cl_2 or CH_3CH_2OH

(b) $CH_3CH_2OCH_2CH_3$ or CH_3CH_2OH

(c) $CH_3\overset{O}{\overset{\|}{C}}CH_3$ or $CH_3CH_2OCH_2CH_3$

(d) $CH_3CH_2OCH_2CH_3$ or $CH_3(CH_2)_3CH_3$

From each pair of compounds, select the solvent with the greater solubility in water.

Synthesis of Alcohols

8.26 Give the structural formula of an alkene or alkenes from which each alcohol or glycol can be prepared: **(See Examples 5.5, 8.10)**

(a) 2-Butanol (b) 1-Methylcyclohexanol

(c) 3-Hexanol (d) 2-Methyl-2-pentanol

(e) Cyclopentanol (f) 1,2-Propanediol

8.27 The addition of bromine to cyclopentene and the acid-catalyzed hydrolysis of cyclopentene oxide are both stereoselective; each gives a *trans* product. Compare the mechanisms of these two reactions, and show how each mechanism accounts for the formation of the *trans* product.

Acidity of Alcohols and Thiols

8.28 From each pair, select the stronger acid, and, for each stronger acid, write a structural formula for its conjugate base: **(See Examples 8.4, 8.12)**

(a) H_2O or H_2CO_3

(b) CH_3OH or CH_3COOH

(c) CH_3COOH or CH_3CH_2SH

8.29 Arrange these compounds in order of increasing acidity (from weakest to strongest): **(See Examples 8.4, 8.12)**

$CH_3CH_2CH_2OH$ $CH_3CH_2\overset{O}{\overset{\|}{C}}OH$ $CH_3CH_2CH_2SH$

8.30 From each pair, select the stronger base, and, for each stronger base, write the structural formula of its conjugate acid: **(See Examples 8.4, 8.12)**

(a) OH^- or CH_3O^-

(b) $CH_3CH_2S^-$ or $CH_3CH_2O^-$

(c) $CH_3CH_2O^-$ or NH_2^-

8.31 Label the stronger acid, stronger base, weaker acid, and weaker base in each of the following equilibria, and then predict the position of each equilibrium (for pK_a values, see Table 2.1): **(See Example 8.4)**

(a) $CH_3CH_2O^- + HCl \rightleftharpoons CH_3CH_2OH + Cl^-$

(b) $CH_3\overset{O}{\overset{\|}{C}}OH + CH_3CH_2O^- \rightleftharpoons CH_3\overset{O}{\overset{\|}{C}}O^- + CH_3CH_2OH$

8.32 Predict the position of equilibrium for each acid–base reaction; that is, does each lie considerably to the left, does each lie considerably to the right, or are the concentrations evenly balanced? **(See Examples 8.4, 8.12)**

(a) $CH_3CH_2OH + Na^+OH^- \rightleftharpoons CH_3CH_2O^-Na^+ + H_2O$

(b) $CH_3CH_2SH + Na^+OH^- \rightleftharpoons CH_3CH_2S^-Na^+ + H_2O$

(c) $CH_3CH_2OH + CH_3CH_2S^-Na^+ \rightleftharpoons CH_3CH_2O^-Na^+ + CH_3CH_2SH$

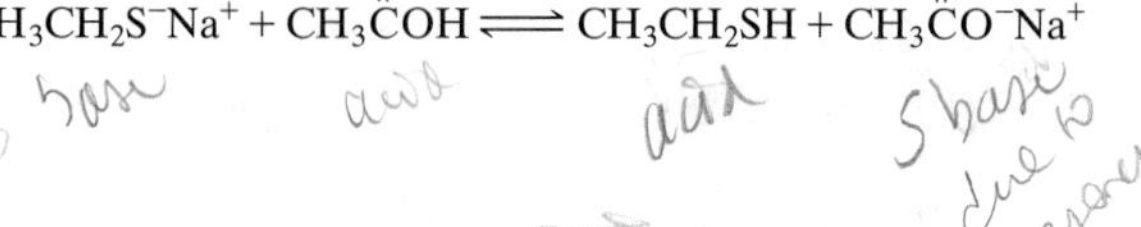

(d) $CH_3CH_2S^-Na^+ + CH_3\overset{O}{\overset{\|}{C}}OH \rightleftharpoons CH_3CH_2SH + CH_3\overset{O}{\overset{\|}{C}}O^-Na^+$

Reactions of Alcohols

8.33 Show how to distinguish between cyclohexanol and cyclohexene by a simple chemical test. (*Hint:* Treat each with Br_2 in CCl_4 and watch what happens.)

8.34 Write equations for the reaction of 1-butanol, a primary alcohol, with these reagents: **(See Examples 8.4, 8.6)**

(a) Na metal
(b) HBr, heat
(c) $K_2Cr_2O_7$, H_2SO_4, heat
(d) $SOCl_2$
(e) Pyridinium chlorochromate (PCC)

8.35 Write equations for the reaction of 2-butanol, a secondary alcohol, with these reagents: **(See Examples 8.4, 8.6)**

(a) Na metal (b) H_2SO_4, heat
(c) HBr, heat (d) $K_2Cr_2O_7$, H_2SO_2, heat
(e) $SOCl_2$ (f) Pyridinium chlorochromate (PCC)

8.36 When (*R*)-2-butanol is left standing in aqueous acid, it slowly loses its optical activity. When the organic material is recovered from the aqueous solution, only 2-butanol is found. Account for the observed loss of optical activity.

8.37 What is the most likely mechanism of the following reaction?

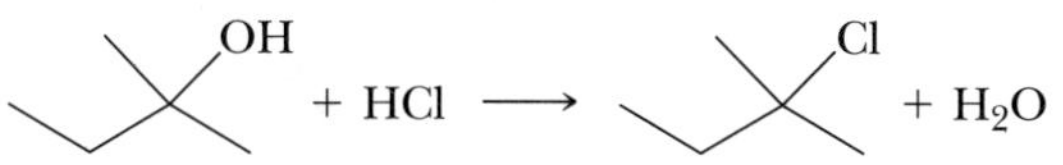

OH + HCl ⟶ Cl + H_2O

Draw a structural formula for the intermediate(s) formed during the reaction.

8.38 Complete the equations for these reactions: **(See Examples 8.6, 8.9)**

(a) OH + H_2CrO_4 ⟶

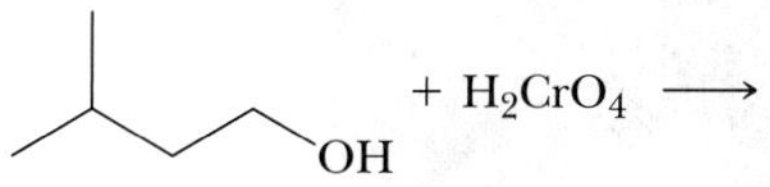

(b) OH + $SOCl_2$ ⟶

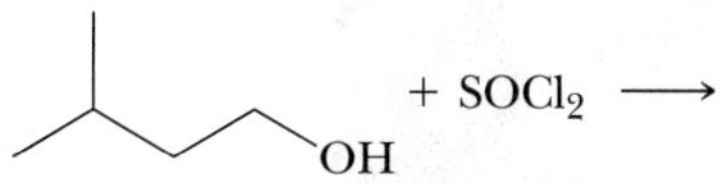

(c) OH HCl ⟶

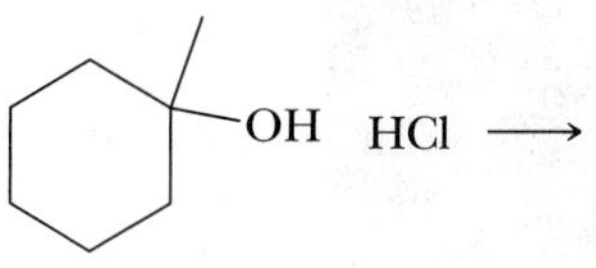

(d) HO OH + HBr (excess) ⟶

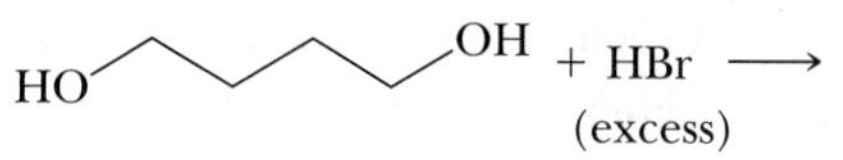

(e) OH + H_2CrO_4 ⟶

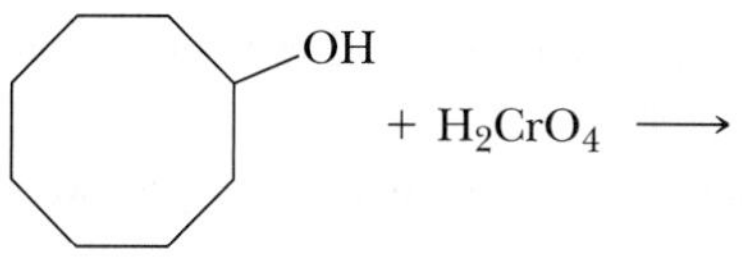

(f) 1) RCOOH (O) 2) H^+/H_2O ⟶

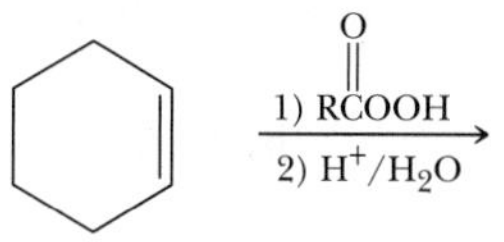

***8.39** In the commercial synthesis of methyl *tert*-butyl ether (MTBE), once used as an antiknock, octane-improving gasoline additive, 2-methylpropene and methanol are passed over an acid catalyst to give the ether: **(See Examples 5.5, 5.6)**

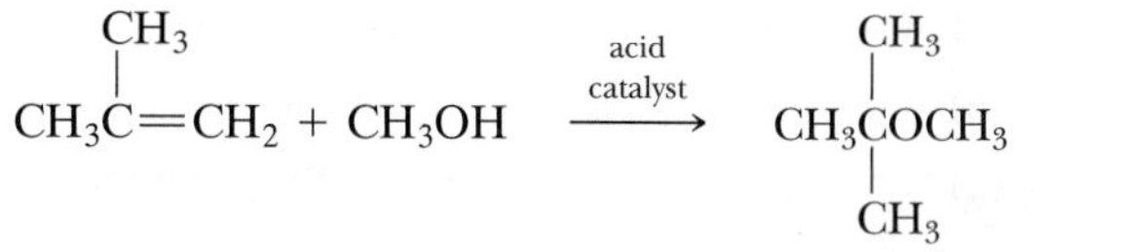

$$CH_3C(CH_3){=}CH_2 + CH_3OH \xrightarrow{\text{acid catalyst}} CH_3C(CH_3)_2OCH_3$$

2-Methylpropene (Isobutylene) Methanol 2-Methoxy-2-methylpropane (Methyl *tert*-butyl ether, MTBE)

Propose a mechanism for this reaction.

8.40 Cyclic bromoalcohols, upon treatment with base, can sometimes undergo intramolecular S_N2 reactions to form the ethers shown in reactions (a) and (b). Provide a mechanism for reactions (a) and (b). Indicate why equation (c) does not yield a similar reaction.

(a) OH Br base ⟶ O

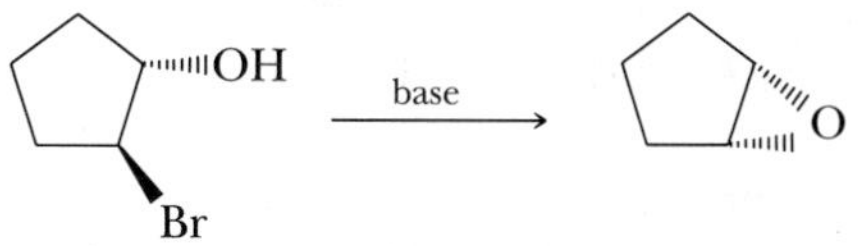

(b) Br OH base ⟶ O

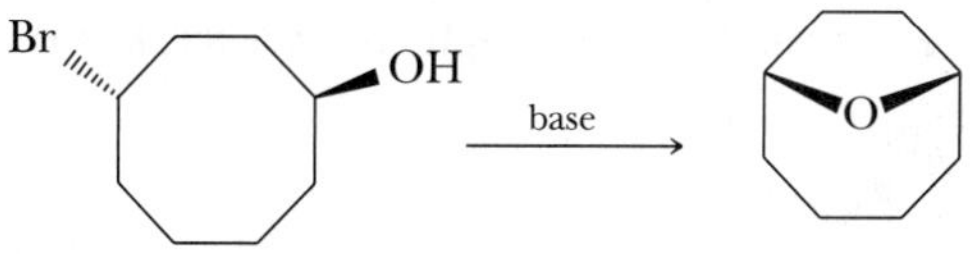

(c) Br OH base ⟶ no reaction

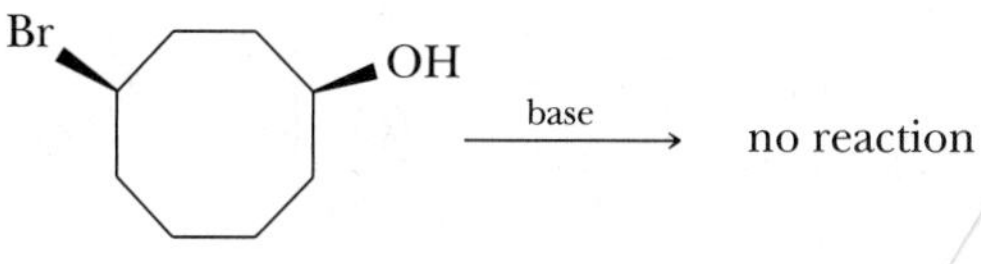

Syntheses

8.41 Show how to convert **(See Examples 8.5, 8.6, 8.10)**

(a) 1-Propanol to 2-propanol in two steps.

(b) Cyclohexene to cyclohexanone in two steps.

(c) Cyclohexanol to *trans*-1,2-cyclohexanediol in three steps.

(d) Propene to propanone (acetone) in two steps.

8.42 Show how to convert cyclohexanol to these compounds: **(See Examples 8.5, 8.6)**

(a) Cyclohexene (b) Cyclohexane

(c) Cyclohexanone

8.43 Show reagents and experimental conditions that can be used to synthesize these compounds from 1-propanol (any derivative of 1-propanol prepared in an earlier part of this problem may be used for a later synthesis): **(See Examples 8.5, 8.6, 8.9, 8.10)**

(a) Propanal (b) Propanoic acid

(c) Propene (d) 2-Propanol

(e) 2-Bromopropane (f) 1-Chloropropane

(g) Propanone (h) 1,2-Propanediol

8.44 Show how to prepare each compound from 2-methyl-1-propanol (isobutyl alcohol): **(See Examples 8.5, 8.6, 8.9, 8.10)**

(a) $CH_3C(CH_3){=}CH_2$ (b) $CH_3C(CH_3)(OH)CH_3$

(c) $CH_3C(CH_3)(OH)-CH_2OH$ (d) $CH_3CH(CH_3)COOH$

For any preparation involving more than one step, show each intermediate compound formed.

8.45 Show how to prepare each compound from 2-methylcyclohexanol: **(See Examples 8.5, 8.6, 8.9, 8.10)**

(a) 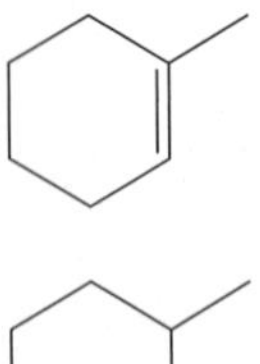 (b) 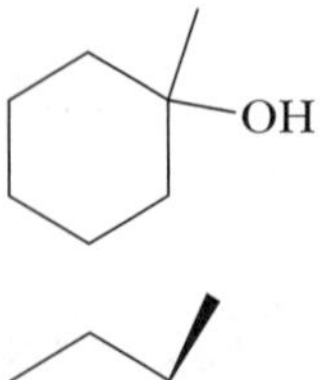 OH

(c) 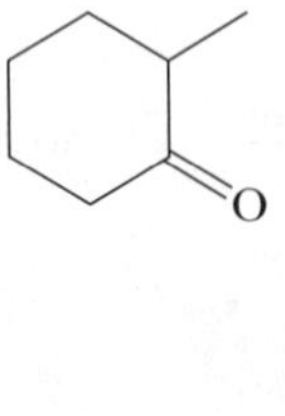 O (d) 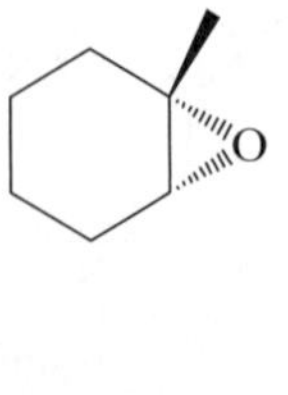 O

(e) 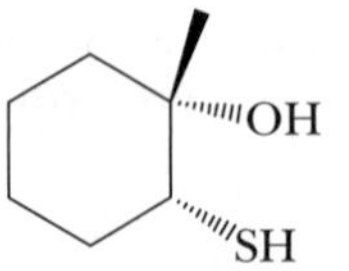 OH SH (f) 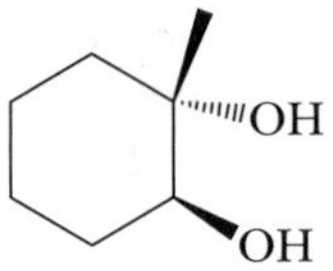 OH OH

For any preparation involving more than one step, show each intermediate compound formed.

8.46 Show how to convert the alcohol on the left to compounds (a), (b), and (c). **(See Example 8.6)**

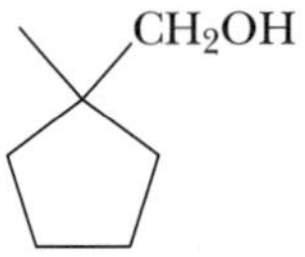

CH_2OH

(a) 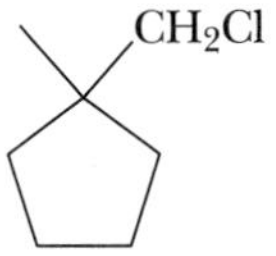 CH_2Cl (b) 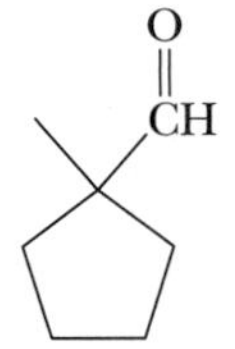 $\overset{O}{\|}CH$ (c) 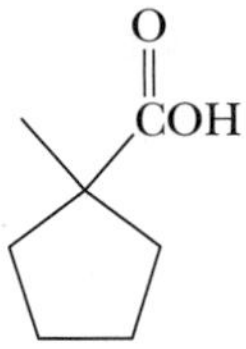 $\overset{O}{\|}COH$

***8.47** Disparlure, a sex attractant of the gypsy moth (*Porthetria dispar*), has been synthesized in the laboratory from the following (*Z*)-alkene: **(See Example 8.9)**

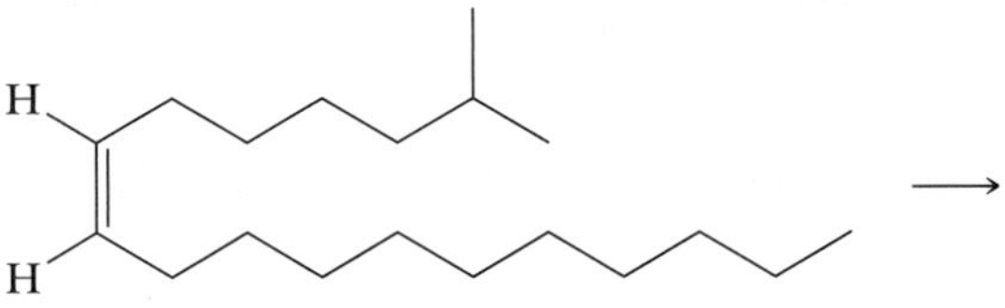

H H →

(*Z*)-2-Methyl-7-octadecene

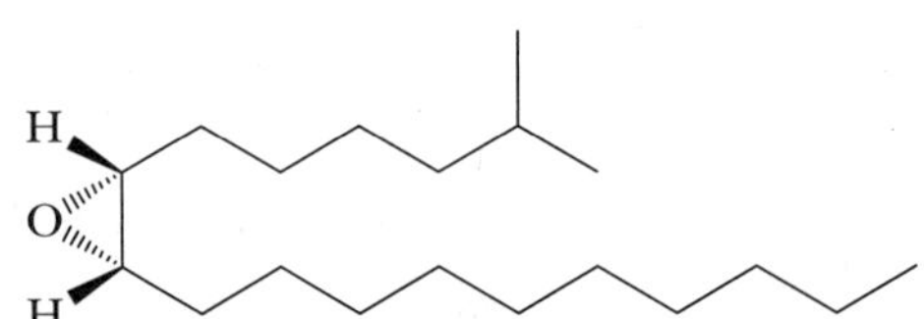

Disparlure

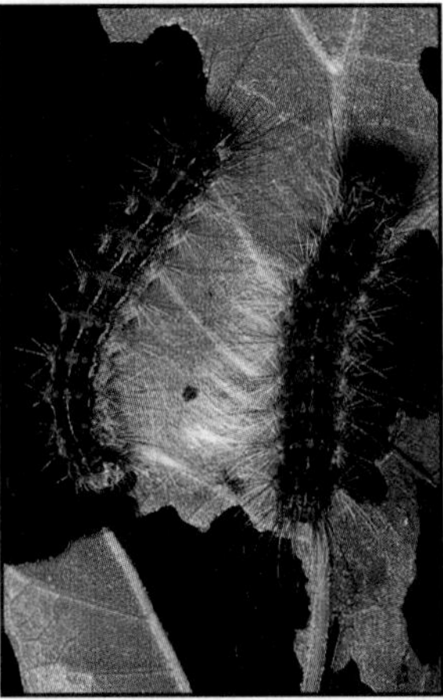

Gypsy moth caterpillars. *(William D. Griffin/Animals, Animals)*

(a) How might the (*Z*)-alkene be converted to disparlure?

(b) How many stereoisomers are possible for disparlure? How many are formed in the sequence you chose?

***8.48** The chemical name for bombykol, the sex pheromone secreted by the female silkworm moth to attract male silkworm moths, is *trans*-10-*cis*-12-hexadecadien-1-ol. (The compound has one hydroxyl group and two carbon–carbon double bonds in a 16-carbon chain.)

(a) Draw a structural formula for bombykol, showing the correct configuration about each carbon–carbon double bond.

(b) How many *cis–trans* isomers are possible for the structural formula you drew in part (a)? All possible *cis–trans* isomers have been synthesized in the laboratory, but only the one named bombykol is produced by the female silkworm moth, and only it attracts male silkworm moths.

Chemical Transformations

8.49 Test your cumulative knowledge of the reactions learned thus far by completing the following chemical transformations. *Note: Some will require more than one step.*

(a) 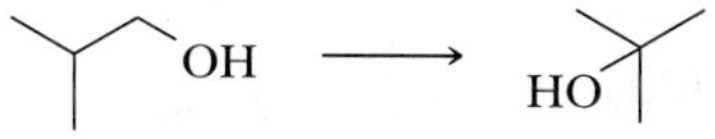 OH → HO

(b) 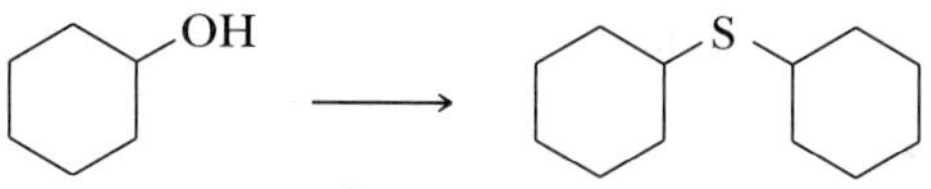 OH → S

(c) 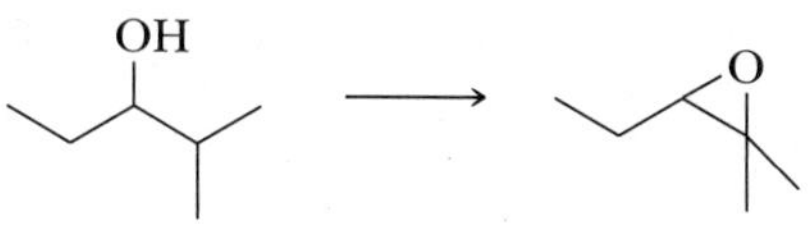 OH → O

(d) → OH, OH

(e) OH → OCH$_3$

(f) → OH

(g) Br → CH$_3$, H, O

(h) HO → O

(i) OH →

(j) → O, OH

(k) → OH, H$_2$N

(l) Cl → OH, OH

(m) → HO, H, N

(n) Br → O

(o) HOCH$_2$CH$_3$ → OCH$_2$CH$_3$

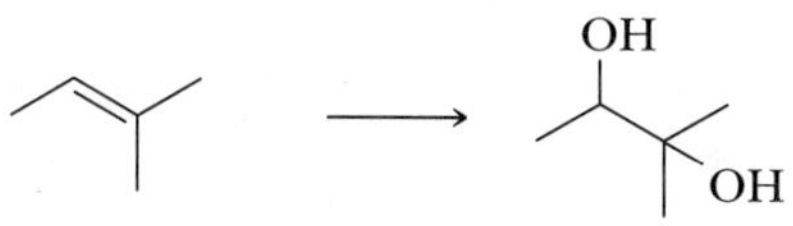

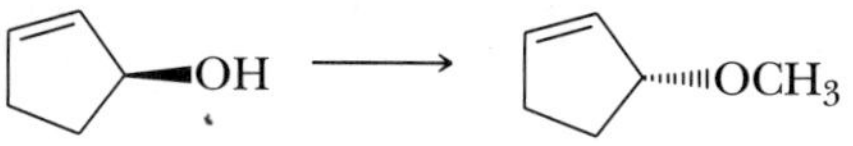

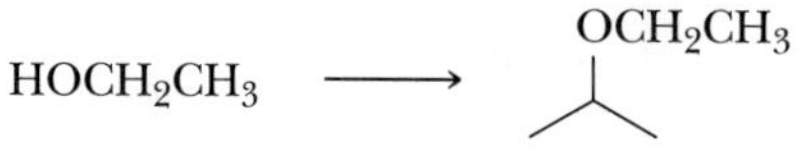

(p) 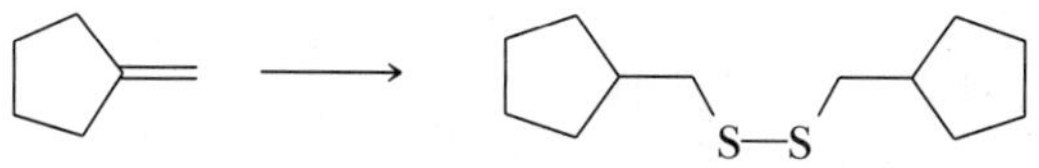 → S–S

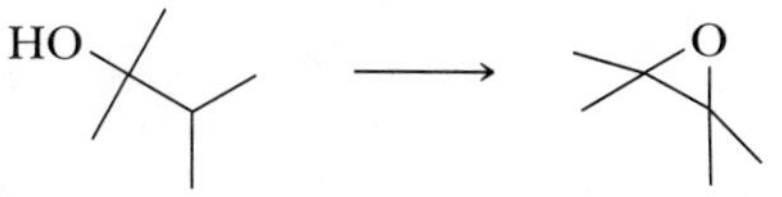

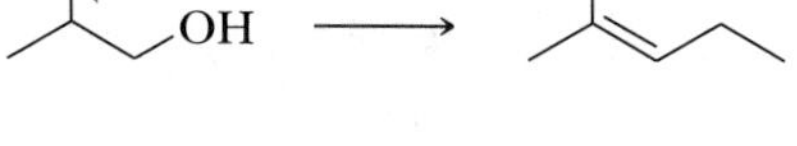

(q) 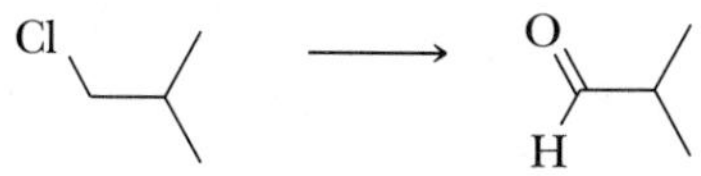 Cl → O, H

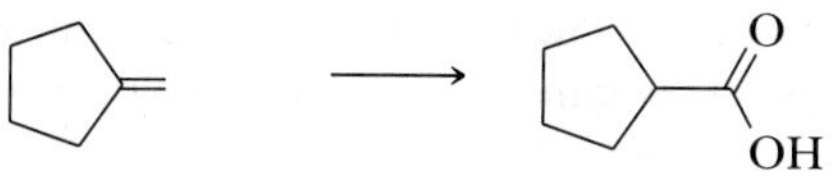

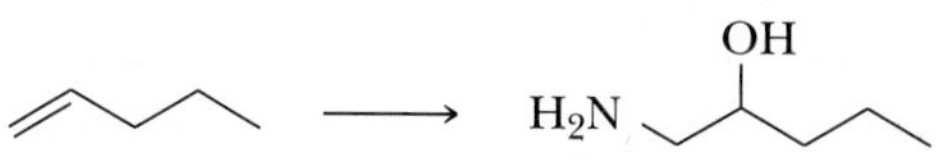

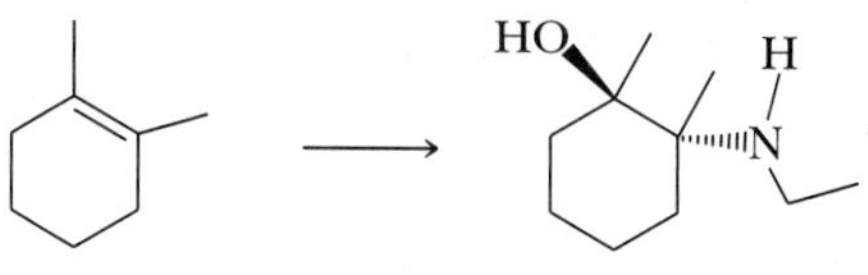

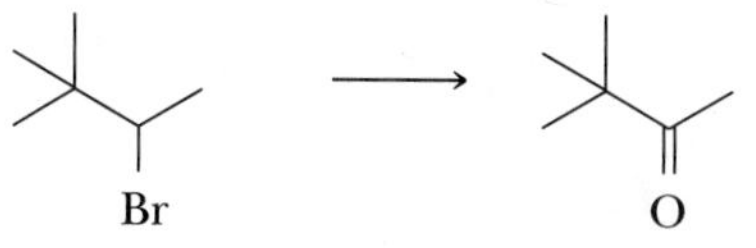

(r) 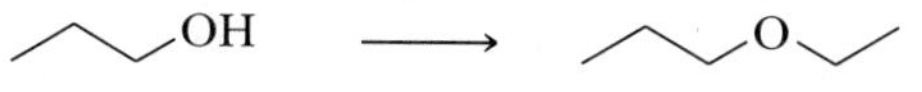 OH → O

Looking Ahead

8.50 Compounds that contain an N—H group associate by hydrogen bonding.

(a) Do you expect this association to be stronger or weaker than that between compounds containing an O—H group?

(b) Based on your answer to part (a), which would you predict to have the higher boiling point, 1-butanol or 1-butanamine?

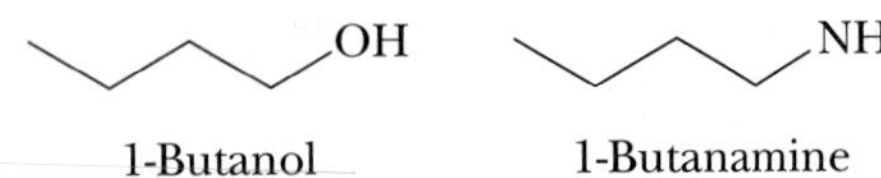

8.51 Draw a resonance structure for methyl vinyl ether in which the oxygen is positively charged. Compared with ethyl methyl ether, how does the resonance structure for methyl vinyl ether influence the reactivity of its oxygen toward an electrophile?

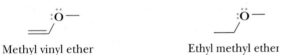

8.52 Rank the members in each set of reagents from most to least nucleophilic:

(a) [acetate anion, $CH_3C(=O)O^-$] [N-methylacetamide anion, $CH_3C(=O)N^-CH_3$] [thioacetate anion, $CH_3C(=O)S^-$]

(b) $R—\ddot{O}^-$ $R—\ddot{N}H^-$ $R—\ddot{C}H_2^-$

8.53 In Chapter 15 we will see that the reactivity of the following carbonyl compounds is directly proportional to the stability of the leaving group. Rank the order of reactivity of these carbonyl compounds from most reactive to least reactive based on the stability of the leaving group.

A: $RC(=O)OCH_3$ B: $RC(=O)NH_2$ C: $RC(=O)Cl$

9 Benzene and Its Derivatives

KEY QUESTIONS

9.1 What Is the Structure of Benzene?

9.2 What Is Aromaticity?

9.3 How Are Benzene Compounds Named, and What Are Their Physical Properties?

9.4 What Is the Benzylic Position, and How Does It Contribute to Benzene Reactivity?

9.5 What Is Electrophilic Aromatic Substitution?

9.6 What Is the Mechanism of Electrophilic Aromatic Substitution?

9.7 How Do Existing Substituents on Benzene Affect Electrophilic Aromatic Substitution?

9.8 What Are Phenols?

HOW TO

9.1 How to Determine Whether a Lone Pair of Electrons Will Be Part of an Aromatic Pi System

9.2 How to Determine Whether a Substituent on Benzene Is Electron Withdrawing

CHEMICAL CONNECTIONS

9A Carcinogenic Polynuclear Aromatics and Cancer

9B Capsaicin, for Those Who Like It Hot

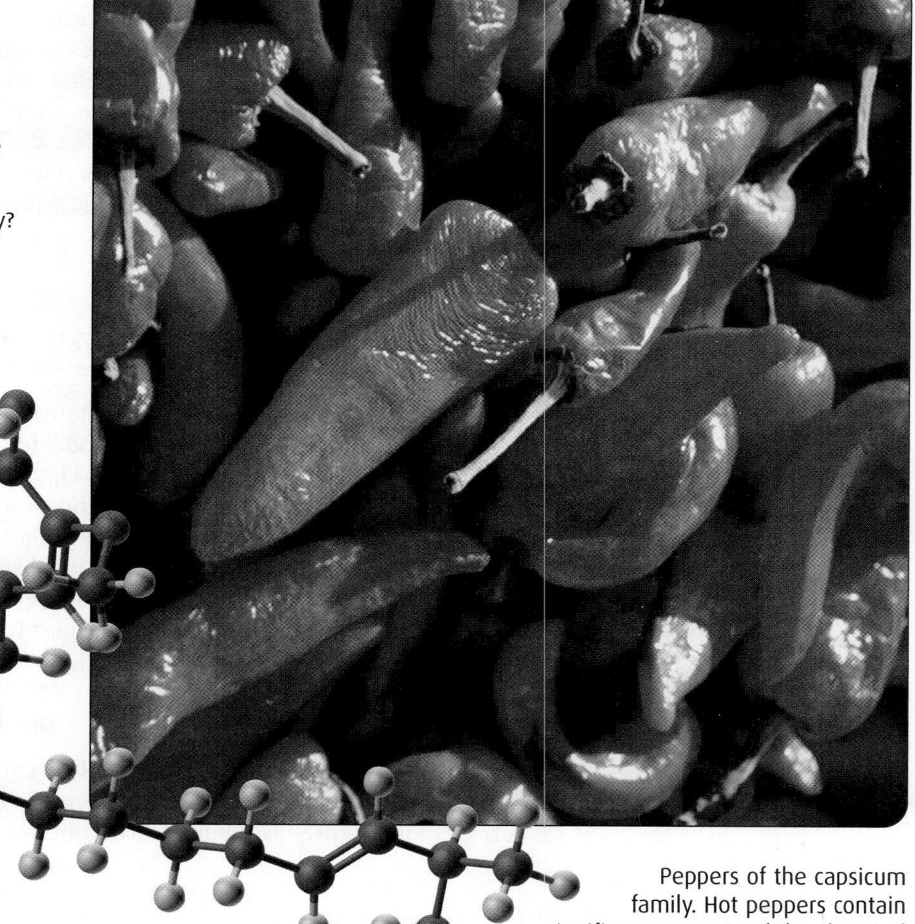

Peppers of the capsicum family. Hot peppers contain significant amounts of the chemical capsaicin, which is used for medicinal purposes as well as for tantalizing taste buds (see Chemical Connections, "Capsaicin, for Those Who Like It Hot"). Inset: A model of capsaicin. *(Douglas Brown)*

Benzene, a colorless liquid, was first isolated by Michael Faraday in 1825 from the oily residue that collected in the illuminating gas lines of London. Benzene's molecular formula, C_6H_6, suggests a high degree of unsaturation. For comparison, an alkane with six carbons has a molecular formula of C_6H_{14}, and a cycloalkane with six carbons has a molecular formula of C_6H_{12}. Considering benzene's high degree of unsaturation, it might be expected to show many of the reactions characteristic of alkenes. Yet, benzene is remarkably *un*reactive! It does not undergo the addition, oxidation, and reduction reactions characteristic of alkenes. For example, benzene does not react with bromine, hydrogen chloride, or other reagents that usually add to carbon–carbon double bonds. Nor is benzene oxidized by chromic acid or ozone under conditions that readily oxidize alkenes. When benzene reacts, it does so by substitution in which a hydrogen atom is replaced by another atom or a group of atoms.

The term *aromatic* was originally used to classify benzene and its derivatives because many of them have distinctive odors. It became clear, however, that a sounder classification for these compounds would be one based on structure and chemical reactivity, not aroma. As it is now used, the term **aromatic** refers instead to the fact that benzene and its derivatives are highly unsaturated compounds that are unexpectedly stable toward reagents that react with alkenes.

Aromatic compound A term used to classify benzene and its derivatives.

Arene An aromatic hydrocarbon.

We use the term **arene** to describe aromatic hydrocarbons, by analogy with alkane and alkene. Benzene is the parent arene. Just as we call a group derived by the removal of an H from an alkane an alkyl group and give it the symbol R—, we call a group derived by the removal of an H from an arene an **aryl group** and give it the symbol **Ar—**.

Aryl group A group derived from an aromatic compound (an arene) by the removal of an H; given the symbol Ar—.

Ar— The symbol used for an aryl group, by analogy with R— for an alkyl group.

9.1 What Is the Structure of Benzene?

Let us imagine ourselves in the mid-nineteenth century and examine the evidence on which chemists attempted to build a model for the structure of benzene. First, because the molecular formula of benzene is C_6H_6, it seemed clear that the molecule must be highly unsaturated. Yet benzene does not show the chemical properties of alkenes, the only unsaturated hydrocarbons known at that time. Benzene does undergo chemical reactions, but its characteristic reaction is substitution rather than addition. When benzene is treated with bromine in the presence of ferric chloride as a catalyst, for example, only one compound with the molecular formula C_6H_5Br forms:

$$\underset{\text{Benzene}}{C_6H_6} + Br_2 \xrightarrow{FeCl_3} \underset{\text{Bromobenzene}}{C_6H_5Br} + HBr$$

Chemists concluded, therefore, that all six carbons and all six hydrogens of benzene must be equivalent. When bromobenzene is treated with bromine in the presence of ferric chloride, three isomeric dibromobenzenes are formed:

$$\underset{\text{Bromobenzene}}{C_6H_5Br} + Br_2 \xrightarrow{FeCl_3} \underset{\text{Dibromobenzene}}{C_6H_4Br_2} + HBr$$

(formed as a mixture of three constitutional isomers)

For chemists in the mid-nineteenth century, the problem was to incorporate these observations, along with the accepted tetravalence of carbon, into a structural formula for benzene. Before we examine their proposals, we should note that the problem of the structure of benzene and other aromatic hydrocarbons has occupied the efforts of chemists for over a century. It was not until 1930s that chemists developed a general understanding of the unique chemical properties of benzene and its derivatives.

A. Kekulé's Model of Benzene

The first structure for benzene, proposed by August Kekulé in 1872, consisted of a six-membered ring with alternating single and double bonds and with one hydrogen bonded to each carbon. Kekulé further proposed that the ring contains three double bonds which shift back and forth so rapidly that the two forms cannot be separated. Each structure has become known as a **Kekulé structure**.

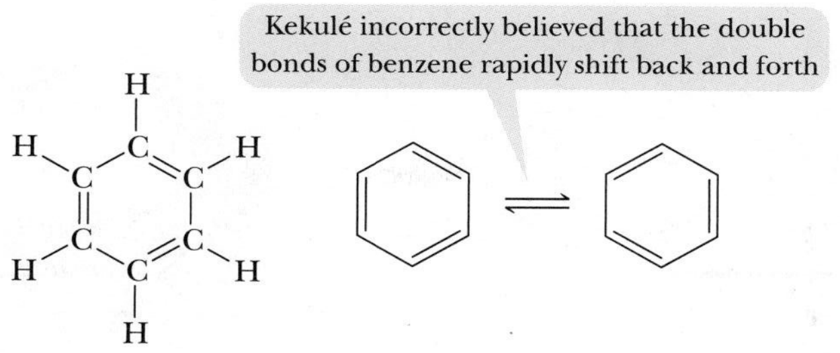

Kekulé incorrectly believed that the double bonds of benzene rapidly shift back and forth

A Kekulé structure, showing all atoms

Kekulé structures as line-angle formulas

Because all of the carbons and hydrogens of Kekulé's structure are equivalent, substituting bromine for any one of the hydrogens gives the same compound. Thus, Kekulé's proposed structure was consistent with the fact that treating benzene with bromine in the presence of ferric chloride gives only one compound with the molecular formula C_6H_5Br.

His proposal also accounted for the fact that the bromination of bromobenzene gives three (and only three) isomeric dibromobenzenes:

$$C_6H_5Br + Br_2 \xrightarrow{FeCl_3} \text{(ortho)} + \text{(meta)} + \text{(para)} + HBr$$

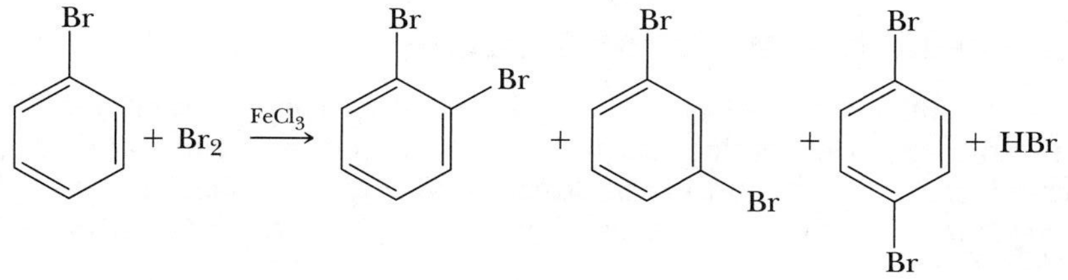

The three isomeric dibromobenzenes

Although Kekulé's proposal was consistent with many experimental observations, it was contested for years. The major objection was that it did not account for the unusual chemical behavior of benzene. If benzene contains three double bonds, why, his critics asked, doesn't it show the reactions typical of alkenes? Why doesn't it add three moles of bromine to form 1,2,3,4,5,6-hexabromocyclohexane? Why, instead, does benzene react by substitution rather than addition?

B. The Molecular Orbital Model of Benzene

The concepts of the **hybridization of atomic orbitals** and the **theory of resonance**, developed by Linus Pauling in the 1930s, provided the first adequate description of the structure of benzene. The carbon skeleton of benzene forms a regular hexagon with C—C—C and H—C—C bond angles of 120°. For this type of bonding, carbon uses sp^2 hybrid orbitals (Section 1.6E). Each carbon forms sigma bonds to two adjacent carbons by the overlap of sp^2–sp^2 hybrid orbitals and one sigma bond to hydrogen by the overlap of sp^2–$1s$ orbitals. As determined experimentally, all carbon–carbon bonds in benzene are the same length, 1.39 Å, a value almost midway between the length of a single bond between sp^3 hybridized carbons (1.54 Å) and that of a double bond between sp^2 hybridized carbons (1.33 Å):

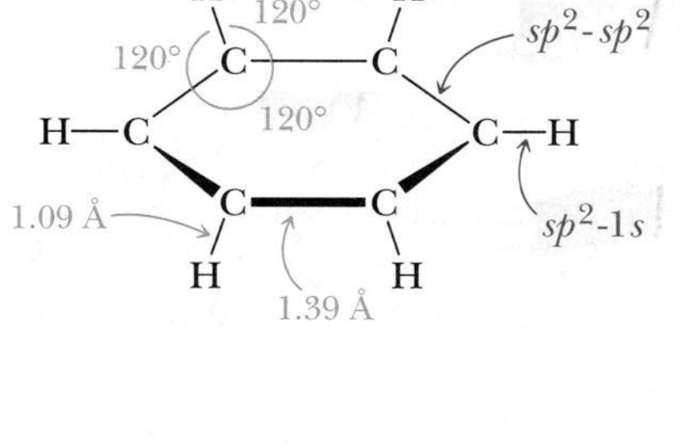

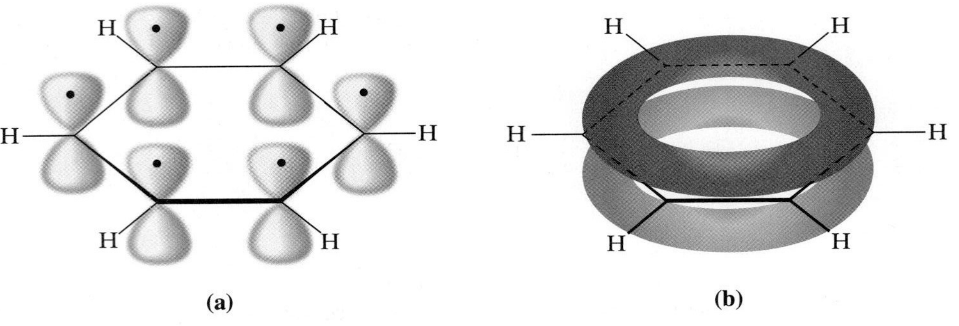

Figure 9.1
Molecular orbital model of bonding in benzene. (a) The carbon, hydrogen framework. The six $2p$ orbitals, each with one electron, are shown uncombined. (b) The overlap of parallel $2p$ orbitals forms a continuous pi cloud, shown by one torus above the plane of the ring and a second below the plane of the ring.

Each carbon also has a single unhybridized $2p$ orbital that contains one electron. These six $2p$ orbitals lie perpendicular to the plane of the ring and overlap to form a continuous pi cloud encompassing all six carbons. The electron density of the pi system of a benzene ring lies in one torus (a doughnut-shaped region) above the plane of the ring and a second torus below the plane (Figure 9.1).

C. The Resonance Model of Benzene

One of the postulates of resonance theory is that, if we can represent a molecule or ion by two or more contributing structures, then that molecule cannot be adequately represented by any single contributing structure. We represent benzene as a hybrid of two equivalent contributing structures, often referred to as *Kekulé structures*:

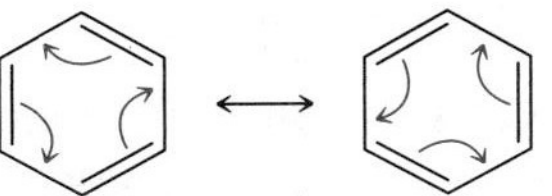

Benzene as a hybrid of two equivalent contributing structures

Each Kekulé structure makes an equal contribution to the hybrid; thus, the C—C bonds are neither single nor double bonds, but something intermediate. We recognize that neither of these contributing structures exists (they are merely alternative ways to pair $2p$ orbitals with no reason to prefer one over the other) and that the actual structure is a superposition of both. Nevertheless, chemists continue to use a single contributing structure to represent this molecule because it is as close as we can come to an accurate structure within the limitations of classical valence bond structures and the tetravalence of carbon.

D. The Resonance Energy of Benzene

Resonance energy The difference in energy between a resonance hybrid and the most stable of its hypothetical contributing structures.

Resonance energy is the difference in energy between a resonance hybrid and its most stable hypothetical contributing structure. One way to estimate the resonance energy of benzene is to compare the heats of hydrogenation of cyclohexene and benzene. In the presence of a transition metal catalyst, hydrogen readily reduces cyclohexene to cyclohexane (Section 5.7):

$$\text{cyclohexene} + H_2 \xrightarrow[\text{1–2 atm}]{\text{Ni}} \text{cyclohexane} \qquad \Delta H^0 = -120 \text{ kJ/mol} \; (-28.6 \text{ kcal/mol})$$

By contrast, benzene is reduced only very slowly to cyclohexane under these conditions. It is reduced more rapidly when heated and under a pressure of several hundred atmospheres of hydrogen:

$$\text{benzene} + 3\,H_2 \xrightarrow[\text{200–300 atm}]{\text{Ni}} \text{cyclohexane} \qquad \Delta H^0 = -209 \text{ kJ/mol } (-49.8 \text{ kcal/mol})$$

The catalytic reduction of an alkene is an exothermic reaction (Section 5.7B). The heat of hydrogenation per double bond varies somewhat with the degree of substitution of the double bond; for cyclohexene $\Delta H^0 = -120$ kJ/mol (−28.6 kcal/mol). If we imagine benzene in which the $2p$ electrons do not overlap outside of their original C—C double bonds, a hypothetical compound with alternating single and double bonds, we might expect its heat of hydrogenation to be $3 \times -120 = -359$ kJ/mol (−85.8 kcal/mol). Instead, the heat of hydrogenation of benzene is only −209 kJ/mol (−49.8 kcal/mol). The difference of 151 kJ/mol (36.0 kcal/mol) between the expected value and the experimentally observed value is the **resonance energy of benzene**. Figure 9.2 shows these experimental results in the form of a graph.

For comparison, the strength of a carbon–carbon single bond is approximately 333–418 kJ/mol (80–100 kcal/mol), and that of hydrogen bonding in water and low-molecular-weight alcohols is approximately 8.4–21 kJ/mol (2–5 kcal/mol). Thus, although the resonance energy of benzene is less than the strength of a carbon–carbon single bond, it is considerably greater than the strength of hydrogen bonding in water and alcohols. In Section 8.1C, we saw that hydrogen bonding has a dramatic effect on the physical properties of alcohols compared with those of alkanes. In this chapter, we see that the resonance energy of benzene and other aromatic hydrocarbons has a dramatic effect on their chemical reactivity.

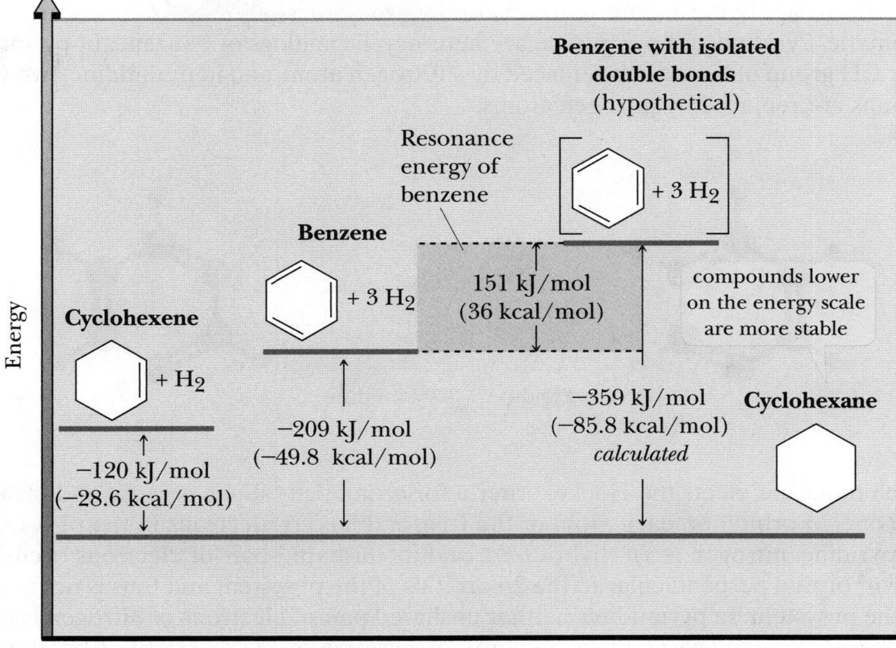

Figure 9.2 The resonance energy of benzene, as determined by a comparison of the heats of hydrogenation of cyclohexene, benzene, and the hypothetical benzene.

Following are resonance energies for benzene and several other aromatic hydrocarbons:

Resonance energy [kJ/mol (kcal/mol)]	Benzene	Naphthalene	Anthracene	Phenanthrene
	151 (36)	255 (61)	347 (83)	381 (91)

9.2 What Is Aromaticity?

Many other types of molecules besides benzene and its derivatives show aromatic character; that is, they contain high degrees of unsaturation, yet fail to undergo characteristic alkene addition and oxidation–reduction reactions. What chemists had long sought to understand were the principles underlying aromatic character. The German chemical physicist Erich Hückel solved this problem in the 1930s.

Hückel's criteria are summarized as follows. To be aromatic, a ring must

1. Have one $2p$ orbital on each of its atoms.
2. Be planar or nearly planar, so that there is continuous overlap or nearly continuous overlap of all $2p$ orbitals of the ring.
3. Have 2, 6, 10, 14, 18, and so forth pi electrons in the cyclic arrangement of $2p$ orbitals.

Benzene meets these criteria. It is cyclic, planar, has one $2p$ orbital on each carbon atom of the ring, and has 6 pi electrons (an aromatic sextet) in the cyclic arrangement of its $2p$ orbitals.

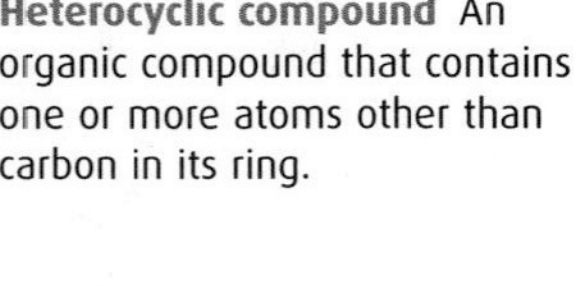

Heterocyclic compound An organic compound that contains one or more atoms other than carbon in its ring.

Let us apply these criteria to several **heterocyclic compounds**, all of which are aromatic. Pyridine and pyrimidine are heterocyclic analogs of benzene. In pyridine, one CH group of benzene is replaced by a nitrogen atom, and in pyrimidine, two CH groups are replaced by nitrogen atoms:

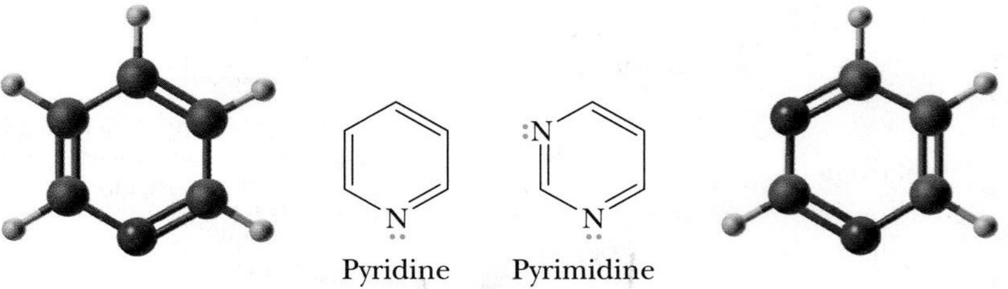

Each molecule meets the Hückel criteria for aromaticity: Each is cyclic and planar, has one $2p$ orbital on each atom of the ring, and has six electrons in the pi system. In pyridine, nitrogen is sp^2 hybridized, and its unshared pair of electrons occupies an sp^2 orbital perpendicular to the $2p$ orbitals of the pi system and thus is not a part of the pi system. In pyrimidine, neither unshared pair of electrons of nitrogen is part of the pi system. The resonance energy of pyridine is 134 kJ/mol (32 kcal/mol), slightly less than that of benzene. The resonance energy of pyrimidine is 109 kJ/mol (26 kcal/mol).

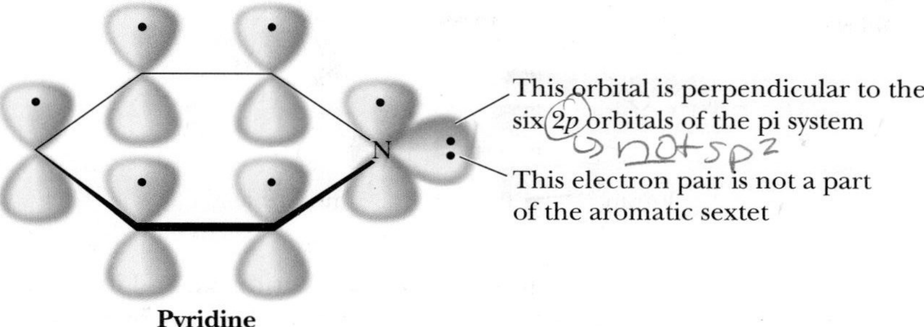
N
This orbital is perpendicular to the six 2*p* orbitals of the pi system
This electron pair is not a part of the aromatic sextet

Pyridine

HOW TO 9.1 Determine Whether a Lone Pair of Electrons Will Be Part of an Aromatic Pi System

(a) First, determine whether the atom containing the lone pair of electrons is part of a double bond. If it is part of a double bond, it is not possible for the lone pair to be part of the aromatic pi system.

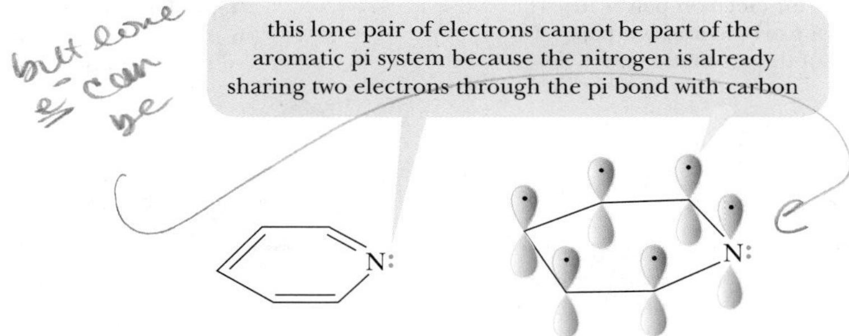
this lone pair of electrons cannot be part of the aromatic pi system because the nitrogen is already sharing two electrons through the pi bond with carbon
N:
N:

(b) If the atom containing the lone pair of electrons is not part of a double bond, it is possible for the lone pair of electrons to be part of the pi system. Determine this by placing the atom in a hybridization state that places the lone pair of electrons in a *p* orbital. If this increases the number of aromatic pi electrons to either 2, 6, 10, 14, and so on, then the lone pair of electrons will become part of the pi aromatic system. If placing the lone pair of electrons in the pi system changes the total number of pi electrons to any other number (e.g., 3–5, 7–9, etc.), the lone pair will not be part of the aromatic pi system.

a nitrogen atom with three single bonds is normally sp^3 hybridized. However, to determine if the lone pair of electrons belongs in the pi system, we must change the hybridization of nitrogen to sp^2 so that the electrons can reside in a *p* orbital

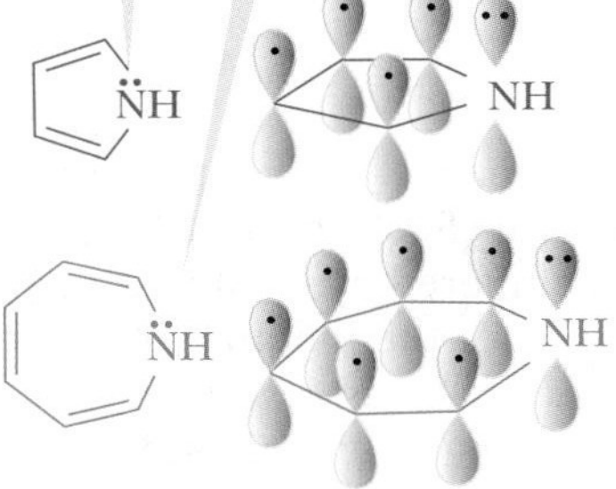
NH
NH
NH
NH

The lone pair gives the pi system six electrons. Therefore, the nitrogen should be sp^2 hybridized.

The lone pair gives the pi system eight electrons. Therefore, the nitrogen should not be sp^2 hybridized.

The five-membered-ring compounds furan, pyrrole, and imidazole are also aromatic:

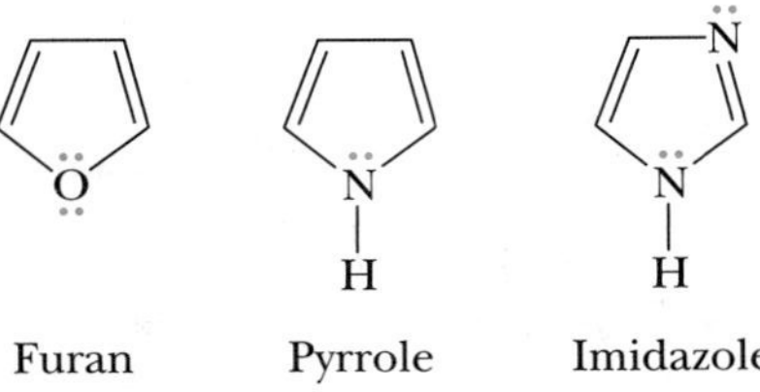

In these planar compounds, each heteroatom is sp^2 hybridized, and its unhybridized $2p$ orbital is part of a continuous cycle of five $2p$ orbitals. In furan, one unshared pair of electrons of the heteroatom lies in the unhybridized $2p$ orbital and is a part of the pi system (Figure 9.3). The other unshared pair of electrons lies in an sp^2 hybrid orbital, perpendicular to the $2p$ orbitals, and is not a part of the pi system. In pyrrole, the unshared pair of electrons on nitrogen is part of the aromatic sextet. In imidazole, the unshared pair of electrons on one nitrogen is part of the aromatic sextet; the unshared pair on the other nitrogen is not.

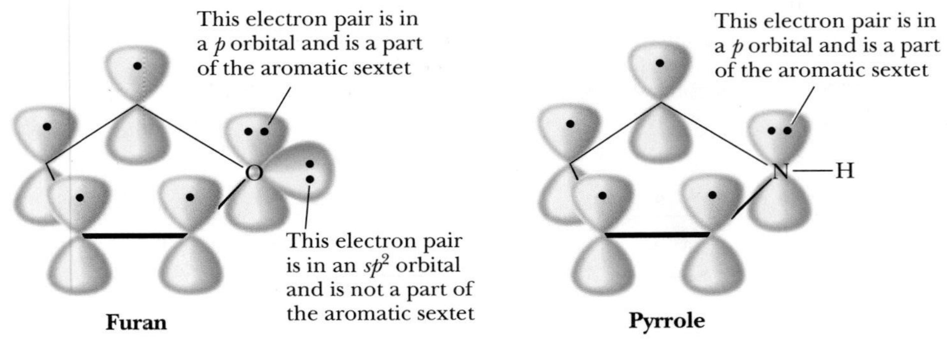

Figure 9.3
Origin of the six pi electrons (the aromatic sextet) in furan and pyrrole. The resonance energy of furan is 67 kJ/mol (16 kcal/mol); that of pyrrole is 88 kJ/mol (21 kcal/mol).

Nature abounds with compounds having a heterocyclic ring fused to one or more other rings. Two such compounds especially important in the biological world are indole and purine:

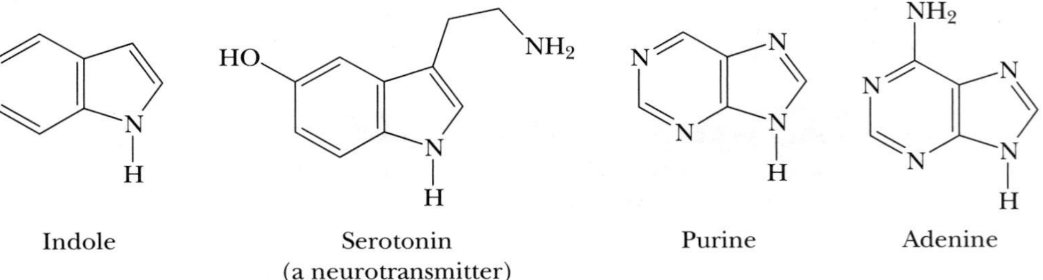

Indole contains a pyrrole ring fused with a benzene ring. Compounds derived from indole include the amino acid L-tryptophan (Section 19.2C) and the neurotransmitter serotonin. Purine contains a six-membered pyrimidine ring fused with a five-membered imidazole ring. Adenine is one of the building blocks of deoxyribonucleic acids (DNA) and ribonucleic acids (RNA), as described in Chapter 20. It is also a component of the biological oxidizing agent nicotinamide adenine dinucleotide, abbreviated NAD^+ (Section 22.1B).

Example 9.1

Which of the following compounds are aromatic?

(a) 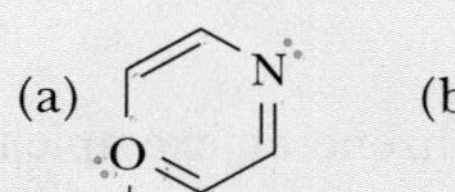 (b) (c)

Strategy

Determine whether each atom of the ring contains a $2p$ orbital and whether the molecule is planar. If these criteria are met, determine the number of pi electrons. Those having 2, 6, 10, 14, and so on electrons are aromatic.

Solution

(a) 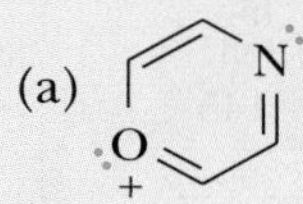

This molecule is planar, and each atom of the ring contains a $2p$ orbital. There is a total of 6 pi electrons. The molecule is aromatic.

(b) 

This molecule is planar, and each atom of the ring contains a $2p$ orbital. There is a total of 4 pi electrons. The molecule is not aromatic.

(c) 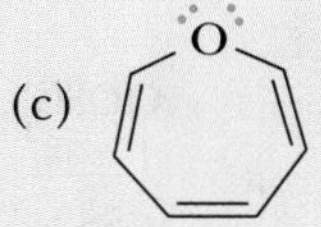

Treat the molecule as planar for the purposes of determining aromaticity. Also, treat each carbon atom in the ring as containing a $2p$ orbital. That is, treat the oxygen atom as sp^2 hybridized, so that one of its lone pairs of electrons will enter the pi electron system (if we do not do this, the molecule cannot be aromatic because an oxygen atom with two lone pairs of electrons and two single bonds is normally sp^3 hybridized). Despite these special considerations, the molecule ends up with a total of eight pi electrons, so the molecule is not aromatic. Because it is not aromatic, the oxygen has no driving force to be sp^2 hybridized and is, in fact, sp^3 hybridized. Also, the molecule has no driving force to be planar, and in fact, the molecule is nonplanar.

See problem 9.11

Problem 9.1

Which of the following compounds are aromatic?

(a) H + (b) H B (c) O +

9.3 How Are Benzene Compounds Named, and What Are Their Physical Properties?

A. Monosubstituted Benzenes

Monosubstituted alkylbenzenes are named as derivatives of benzene; an example is ethylbenzene. The IUPAC system retains certain common names for several of the simpler monosubstituted alkylbenzenes. Examples are **toluene** (rather than methylbenzene) and **styrene** (rather than phenylethylene):

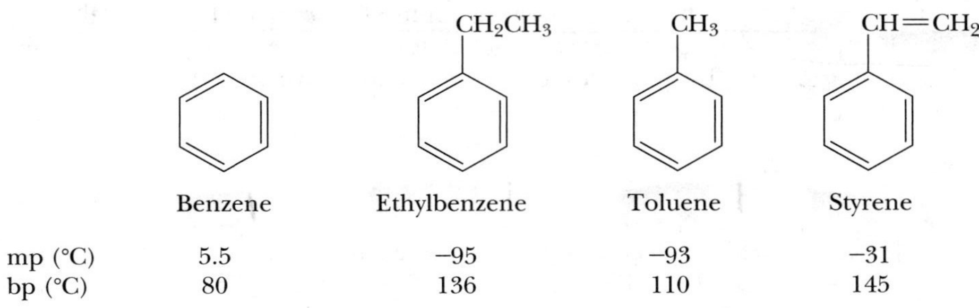

The common names **phenol**, **aniline**, **benzaldehyde**, **benzoic acid**, and **anisole** are also retained by the IUPAC system:

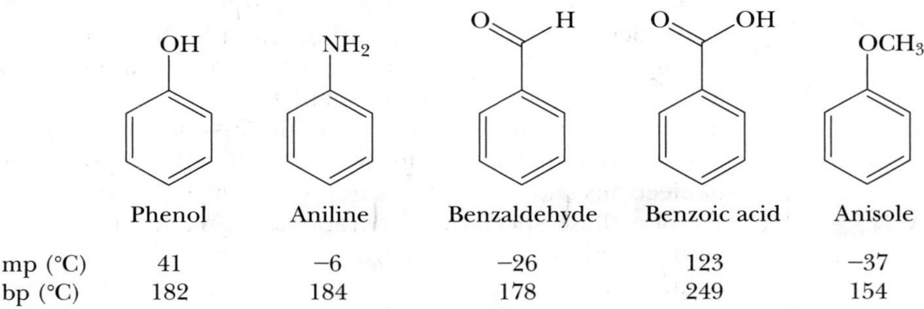

The physical properties of substituted benzenes vary depending on the nature of the substituent. Alkylbenzenes, like other hydrocarbons, are nonpolar and thus have lower boiling points than benzenes with polar substituents such as phenol, aniline, and benzoic acid. The melting points of substituted benzenes depend on whether or not their molecules can be packed close together. Benzene, which has no substituents and is flat, can pack its molecules very closely, giving it a considerably higher melting point than many substituted benzenes.

As noted in the introduction to Chapter 5, the substituent group derived by the loss of an H from benzene is a **phenyl** group (Ph); that derived by the loss of an H from the methyl group of toluene is a **benzyl group** (Bn):

Phenyl group C_6H_5—, the aryl group derived by removing a hydrogen from benzene.

Benzyl group $C_6H_5CH_2$—, the alkyl group derived by removing a hydrogen from the methyl group of toluene.

H | — | CH_3 | CH_2—

Benzene | Phenyl group (Ph) | Toluene | Benzyl group (Bn)

In molecules containing other functional groups, phenyl groups and benzyl groups are often named as substituents:

(Z)-2-Phenyl-2-butene — $CH_3(Ph)C{=}C(H)CH_3$

2-Phenylethanol — $PhCH_2CH_2OH$

Benzyl chloride — $PhCH_2Cl$

B. Disubstituted Benzenes

When two substituents occur on a benzene ring, three constitutional isomers are possible. We locate substituents either by numbering the atoms of the ring or by using the locators **ortho**, **meta**, and **para**. The numbers 1,2- are equivalent to *ortho* (Greek: straight); 1,3- to *meta* (Greek: after); and 1,4- to *para* (Greek: beyond).

When one of the two substituents on the ring imparts a special name to the compound, as, for example, toluene, phenol, and aniline, then we name the compound as a derivative of that parent molecule. In this case, the special substituent occupies ring position number 1. The IUPAC system retains the common name **xylene** for the three isomeric dimethylbenzenes. When neither group imparts a special name, we locate the two substituents and list them in alphabetical order before the ending *-benzene*. The carbon of the benzene ring with the substituent of lower alphabetical ranking is numbered C-1.

Ortho (*o*) Refers to groups occupying positions 1 and 2 on a benzene ring.

Meta (*m*) Refers to groups occupying positions 1 and 3 on a benzene ring.

Para (*p*) Refers to groups occupying positions 1 and 4 on a benzene ring.

CH_3 / Br	NH_2 / Cl	CH_3 / CH_3	CH_2CH_3 / Cl
4-Bromotoluene (*p*-Bromotoluene)	3-Chloroaniline (*m*-Chloroaniline)	1,3-Dimethylbenzene (*m*-Xylene)	1-Chloro-4-ethylbenzene (*p*-Chloroethylbenzene)

C. Polysubstituted Benzenes

When three or more substituents are present on a ring, we specify their locations by numbers. If one of the substituents imparts a special name, then the molecule is named as a derivative of that parent molecule. If none of the substituents imparts a special name, we number them to give the smallest set of numbers and list them in alphabetical order before the ending *-benzene*. In the following examples, the first compound is a derivative of toluene, and the second is a derivative of phenol. Because there is no special name for the third compound, we list its three substituents in alphabetical order, followed by the word *benzene*:

1-CH_3, 2-NO_2, 4-Cl	1-OH, 2-Br, 4-Br, 6-Br	1-CH_2CH_3, 2-Br, 4-NO_2
4-Chloro-2-nitrotoluene	2,4,6-Tribromophenol	2-Bromo-1-ethyl-4-nitrobenzene

Example 9.2

Write names for these compounds:

H_3C I COOH Br Br Cl NO_2 NO_2

(a) (b) (c) (d)

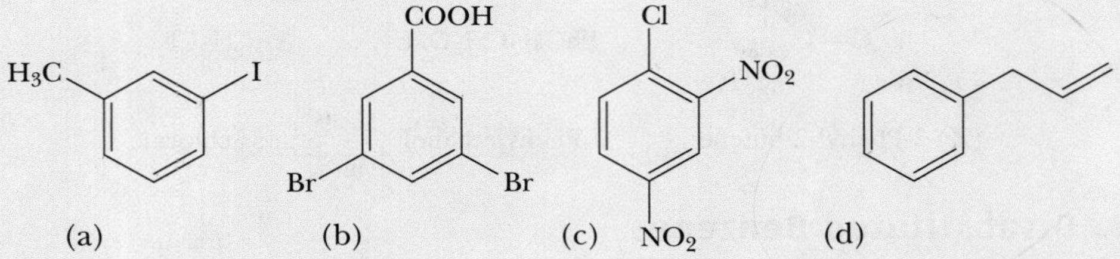

Strategy

First, determine whether one of the substituents imparts a special name to the benzene compound (e.g., toluene, phenol, aniline, etc.). Identify all substituents and list them in alphabetical order. Use numbers to indicate relative position. The locators *ortho*, *meta*, or *para* can be used for disubstituted benzenes.

Solution

(a) 3-Iodotoluene or *m*-iodotoluene (b) 3,5-Dibromobenzoic acid
(c) 1-Chloro-2,4-dinitrobenzene (d) 3-Phenylpropene

See problems 9.13, 9.14

Problem 9.2

Write names for these compounds:

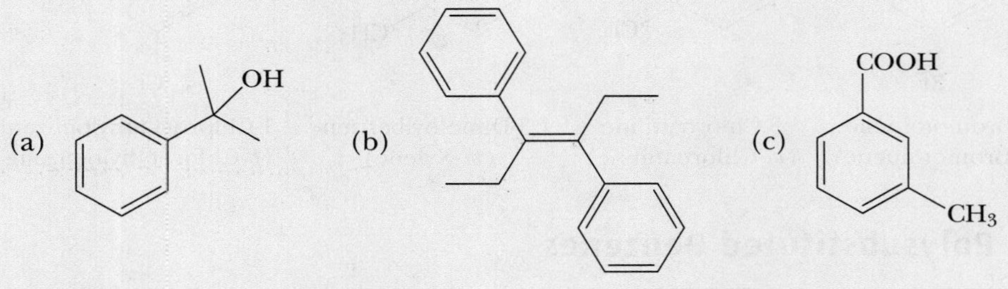

(a) OH (b) (c) COOH CH_3

Polynuclear aromatic hydrocarbon A hydrocarbon containing two or more fused aromatic rings.

Polynuclear aromatic hydrocarbons (PAHs) contain two or more aromatic rings, each pair of which shares two ring carbon atoms. Naphthalene, anthracene, and phenanthrene, the most common PAHs, and substances derived from them are found in coal tar and high-boiling petroleum residues. At one time, naphthalene was used as a moth repellent and insecticide in protecting woolens and furs, but its use has decreased due to the introduction of chlorinated hydrocarbons such as *p*-dichlorobenzene. Also found in coal tar are lesser amounts of benzo[a]pyrene. This compound is found as well in the exhausts of gasoline-powered internal combustion engines (for example, automobile engines) and in cigarette smoke. Benzo[a]pyrene is a very potent carcinogen and mutagen.

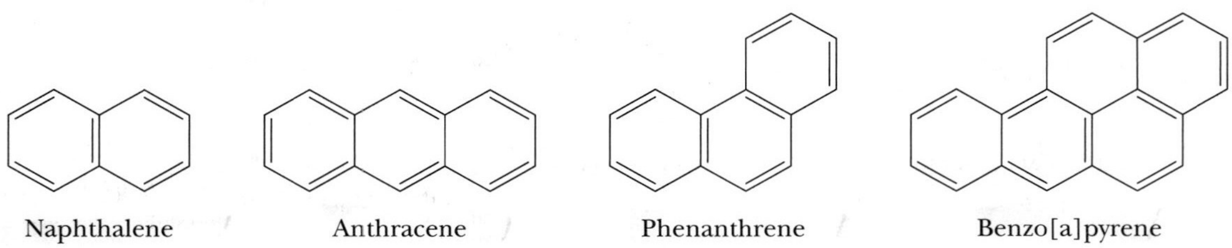

Naphthalene Anthracene Phenanthrene Benzo[a]pyrene

9.4 What Is the Benzylic Position, and How Does It Contribute to Benzene Reactivity?

As we have mentioned, benzene's aromaticity causes it to resist many of the reactions that alkenes typically undergo. However, chemists have been able to react benzene in other ways. This is fortunate, because benzene rings are abundant in many of the compounds that society depends upon, including various medications, plastics, and preservatives for food. We begin our discussion of benzene reactions with processes that take place not on the ring itself, but at the carbon immediately bonded to the benzene ring. This carbon is known as a **benzylic carbon.**

Benzylic carbon An sp^3 hybridized carbon bonded to a benzene ring.

Benzene is unaffected by strong oxidizing agents, such as H_2CrO_4 and $KMnO_4$. When we treat toluene with these oxidizing agents under vigorous conditions, the side-chain methyl group is oxidized to a carboxyl group to give benzoic acid:

CH_3 (Toluene) + H_2CrO_4 ⟶ $COOH$ (Benzoic acid) + Cr^{3+}

The fact that the side-chain methyl group is oxidized, but the aromatic ring is unchanged, illustrates the remarkable chemical stability of the aromatic ring. Halogen

CHEMICAL CONNECTIONS 9A

Carcinogenic Polynuclear Aromatics and Cancer

A **carcinogen** is a compound that causes cancer. The first carcinogens to be identified were a group of polynuclear aromatic hydrocarbons, all of which have at least four aromatic rings. Among them is benzo[a]pyrene, one of the most carcinogenic of the aromatic hydrocarbons. It forms whenever there is incomplete combustion of organic compounds. Benzo[a]pyrene is found, for example, in cigarette smoke, automobile exhaust, and charcoal-broiled meats.

Benzo[a]pyrene causes cancer in the following way: Once it is absorbed or ingested, the body attempts to convert it into a more soluble compound that can be excreted easily. To this end, a series of enzyme-catalyzed reactions transforms benzo[a]pyrene into a **diol epoxide**, a compound that can bind to DNA by reacting with one of its amino groups, thereby altering the structure of DNA and producing a cancer-causing mutation:

Benzo[a]pyrene —(enzyme-catalyzed oxidation)⟶ A diol epoxide (O, HO, OH)

QUESTION

Show how the outer perimeter of benzo[a]pyrene satisfies Hückel's criteria for aromaticity. Is the outer perimeter of the highlighted portion of the diol epoxide product of benzo[a]pyrene also aromatic?

and nitro substituents on an aromatic ring are unaffected by these oxidations. For example, chromic acid oxidizes 2-chloro-4-nitrotoluene to 2-chloro-4-nitrobenzoic acid. Notice that in this oxidation, the nitro and chloro groups remain unaffected:

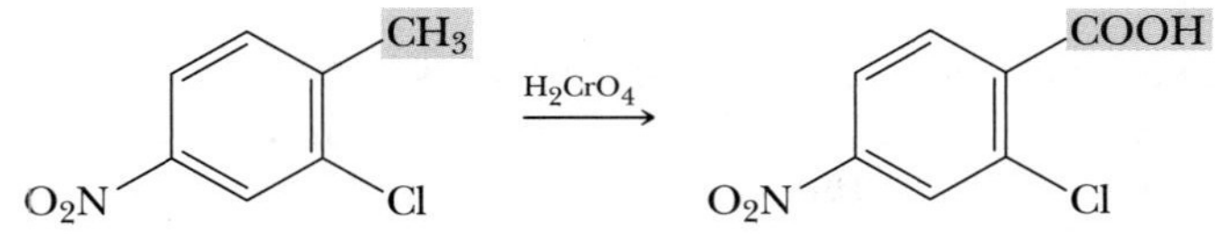

CH_3 $\xrightarrow{H_2CrO_4}$ COOH

O_2N Cl O_2N Cl

2-Chloro-4-nitrotoluene 2-Chloro-4-nitrobenzoic acid

Ethylbenzene and isopropylbenzene are also oxidized to benzoic acid under these conditions. The side chain of *tert*-butylbenzene, which has no benzylic hydrogen, is not affected by these oxidizing conditions.

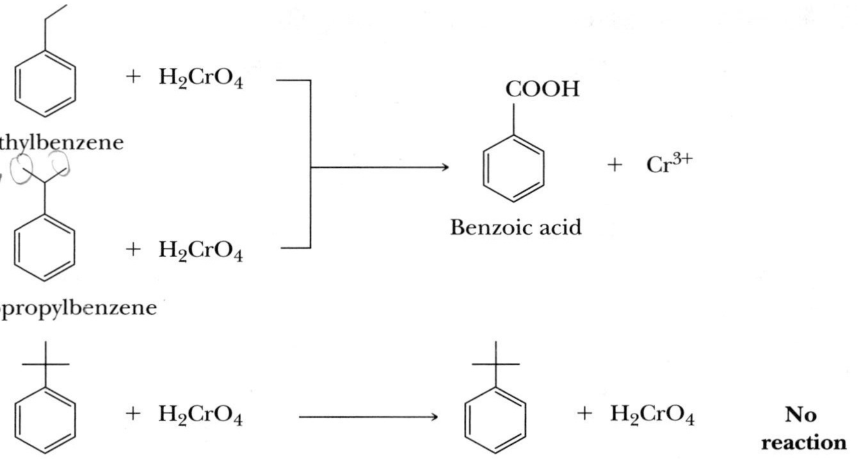

+ H_2CrO_4

Ethylbenzene

COOH

$\longrightarrow$ + Cr^{3+}

Benzoic acid

+ H_2CrO_4

Isopropylbenzene

+ H_2CrO_4 $\longrightarrow$ + H_2CrO_4 **No reaction**

tert-Butylbenzene *tert*-Butylbenzene

From these observations, we conclude that, if a benzylic hydrogen exists, then the benzylic carbon (Section 9.3A) is oxidized to a carboxyl group and all other carbons of the side chain are removed. If no benzylic hydrogen exists, as in the case of *tert*-butylbenzene, then the side chain is not oxidized.

If more than one alkyl side chain exists, each is oxidized to —COOH. Oxidation of *m*-xylene gives 1,3-benzenedicarboxylic acid, more commonly named isophthalic acid:

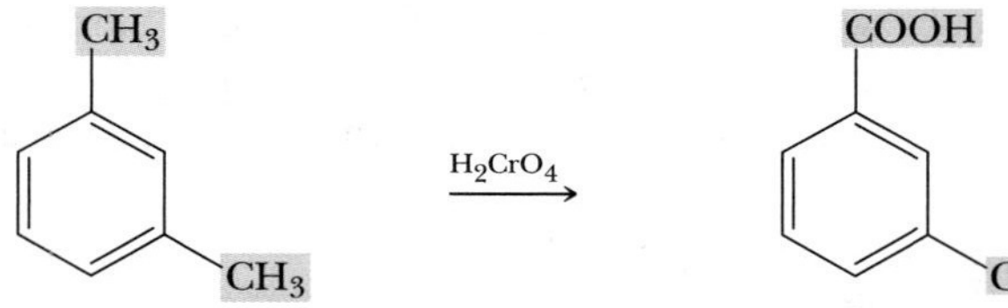

CH_3 COOH

$\xrightarrow{H_2CrO_4}$

CH_3 COOH

m-Xylene 1,3-Benzenedicarboxylic acid (Isophthalic acid)

Example 9.3

Predict the products resulting from vigorous oxidation of each compound by H_2CrO_4:

(a) 1,4-dimethylbenzene (*p*-xylene) (b)

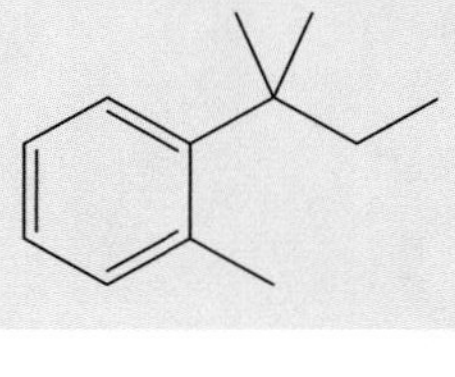

Strategy

Identify all the alkyl groups in the reactant. If a benzylic hydrogen exists on an alkyl group, chromic acid will oxidize it to a —COOH group. The various by-products that are formed from benzylic oxidation reactions are usually not specified.

Solution

chromic acid oxidizes both alkyl groups to —COOH groups, and the product is terephthalic acid, one of two compounds required for the synthesis of Dacron polyester and Mylar (Section 17.4B)

(a) 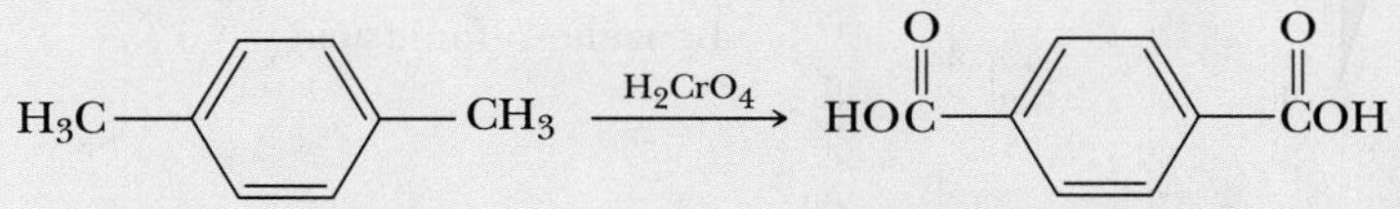 H_3C—C₆H₄—CH_3 $\xrightarrow{H_2CrO_4}$ HOC(=O)—C₆H₄—C(=O)OH

1,4-Dimethylbenzene (*p*-Xylene)

1,4-Benzenedicarboxylic acid (Terephthalic acid)

(b) 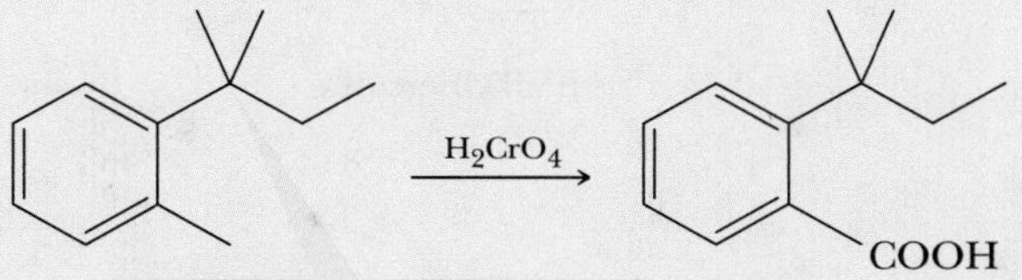 $\xrightarrow{H_2CrO_4}$ COOH

this alkyl group has no benzylic hydrogens and is not oxidized

See problem 9.30

Problem 9.3

Predict the products resulting from vigorous oxidation of each compound by H_2CrO_4:

(a)

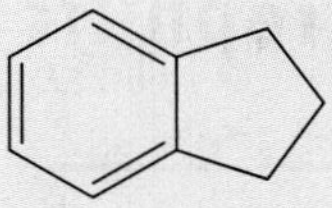

(b) 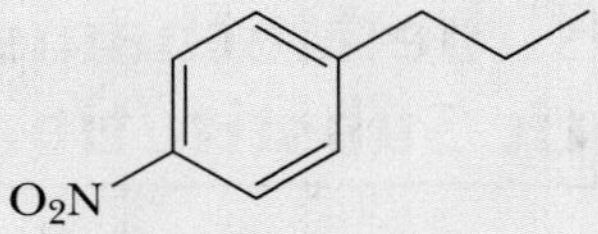 O_2N

9.5 What Is Electrophilic Aromatic Substitution?

Although benzene is resistant to most of the reactions presented thus far for alkenes, it is not completely unreactive. By far the most characteristic reaction of aromatic compounds is substitution at a ring carbon. Some groups that can be introduced directly onto the ring are the halogens, the nitro (—NO_2) group, the sulfonic acid (—SO_3H) group, alkyl (—R) groups, and acyl (RCO—) groups.

Halogenation:

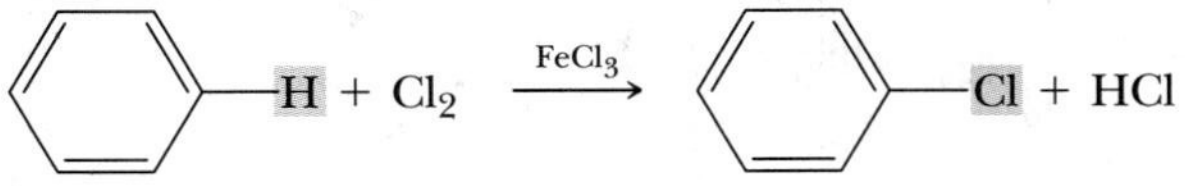

$$C_6H_5\text{—H} + Cl_2 \xrightarrow{FeCl_3} C_6H_5\text{—Cl} + HCl$$

Chlorobenzene

Nitration:

C_6H_5–H + HNO_3 $\xrightarrow{H_2SO_4}$ C_6H_5–NO_2 + H_2O

Nitrobenzene

Sulfonation:

C_6H_5–H + H_2SO_4 ⟶ C_6H_5–SO_3H + H_2O

Benzenesulfonic acid

Alkylation:

C_6H_5–H + RX $\xrightarrow{AlCl_3}$ C_6H_5–R + HX

An alkylbenzene

Acylation:

C_6H_5–H + R–C(=O)–X $\xrightarrow{AlCl_3}$ C_6H_5–C(=O)R + HX

An acyl halide

An acylbenzene

9.6 What Is the Mechanism of Electrophilic Aromatic Substitution?

Electrophilic aromatic substitution A reaction in which an electrophile, E^+, substitutes for a hydrogen on an aromatic ring.

In this section, we study several types of **electrophilic aromatic substitution** reactions—that is, reactions in which a hydrogen of an aromatic ring is replaced by an electrophile, E^+. The mechanisms of these reactions are actually very similar. In fact, they can be broken down into three common steps:

Step 1: Generation of the electrophile:

$$\text{Reagent(s)} \longrightarrow E^+$$

Step 2: Attack of the electrophile on the aromatic ring to give a resonance-stabilized cation intermediate:

C_6H_5–H + E^+ $\xrightarrow{\text{slow}}$ [cation intermediate] ⟷ [cation intermediate] ⟷ [cation intermediate]

Resonance-stabilized cation intermediate

Step 3: Proton transfer to a base to regenerate the aromatic ring:

$$\text{(cation intermediate with H and E)} + :\text{Base} \xrightarrow{\text{fast}} \text{(C}_6\text{H}_5\text{–E)} + \text{Base-H}$$

The reactions we are about to study differ only in the way the electrophile is generated and in the base that removes the proton to re-form the aromatic ring. You should keep this principle in mind as we explore the details of each reaction.

A. Chlorination and Bromination

Chlorine alone does not react with benzene, in contrast to its instantaneous addition to cyclohexene (Section 5.3C). However, in the presence of a Lewis acid catalyst, such as ferric chloride or aluminum chloride, chlorine reacts to give chlorobenzene and HCl. Chemists account for this type of electrophilic aromatic substitution by the following three-step mechanism:

Mechanism: Electrophilic Aromatic Substitution—Chlorination

Step 1: *Formation of the Electrophile:* Reaction between chlorine (a Lewis base) and $FeCl_3$ (a Lewis acid) gives an ion pair containing a chloronium ion (an electrophile):

$$:\ddot{Cl}-\ddot{Cl}: + FeCl_3 \rightleftharpoons :\ddot{Cl}^{+}-\ddot{Cl}-Fe^{-}Cl_3 \rightleftharpoons :\ddot{Cl}^{+}\ :\ddot{Cl}-Fe^{-}Cl_3$$

Chlorine (a Lewis base) | Ferric chloride (a Lewis acid) | A molecular complex with a positive charge on chlorine and a negative charge on iron | An ion pair containing a chloronium ion (the electrophile)

Step 2: *Attack of the Electrophile on the Ring:* Reaction of the Cl_2–$FeCl_3$ ion pair with the pi electron cloud of the aromatic ring forms a resonance-stabilized cation intermediate, represented here as a hybrid of three contributing structures:

$$\text{benzene} + :\ddot{Cl}^{+} \xrightarrow{\text{slow, rate determining}} \text{(three contributing structures)}$$

Resonance-stabilized cation intermediate

The positive charge on the resonance-stabilized intermediate is distributed approximately equally on the carbon atoms 2, 4, and 6 of the ring relative to the point of substitution.

Step 3: *Proton Transfer:* Proton transfer from the cation intermediate to $FeCl_4^-$ forms HCl, regenerates the Lewis acid catalyst, and gives chlorobenzene:

$$\text{Cation intermediate} + :\ddot{Cl}-Fe^{-}Cl_3 \xrightarrow{\text{fast}} \text{Chlorobenzene} + H-\ddot{Cl}: + FeCl_3$$

Treatment of benzene with bromine in the presence of ferric chloride or aluminum chloride gives bromobenzene and HBr. The mechanism for this reaction is the same as that for chlorination of benzene.

The major difference between the addition of halogen to an alkene and substitution by halogen on an aromatic ring is the fate of the cation intermediate formed after the halogen is added to the compound. Recall from Section 5.3C that the addition of chlorine to an alkene is a two-step process, the first and slower step of which is the formation of a bridged chloronium ion intermediate. This intermediate then reacts with chloride ion to complete the addition. With aromatic compounds, the cation intermediate loses H^+ to regenerate the aromatic ring and regain its large resonance stabilization. There is no such resonance stabilization to be regained in the case of an alkene.

B. Nitration and Sulfonation

The sequence of steps for the nitration and sulfonation of benzene is similar to that for chlorination and bromination. For nitration, the electrophile is the **nitronium ion**, NO_2^+, generated by the reaction of nitric acid with sulfuric acid. In the following equations nitric acid is written $HONO_2$ to show more clearly the origin of the nitronium ion.

Mechanism: Formation of the Nitronium Ion

Step 1: Proton transfer from sulfuric acid to the OH group of nitric acid gives the conjugate acid of nitric acid:

$$\underset{\text{Nitric acid}}{H-\ddot{O}-NO_2} + H-\ddot{O}-SO_3H \rightleftharpoons \underset{\text{Conjugate acid of nitric acid}}{H-\overset{+}{\underset{H}{O}}-NO_2} + HSO_4^-$$

(H_2SO_4)

Step 2: Loss of water from this conjugate acid gives the nitronium ion, NO_2^+:

$$H-\overset{+}{\underset{H}{O}}-NO_2 \rightleftharpoons H-\underset{H}{\ddot{O}}: + \underset{\text{The nitronium ion}}{NO_2^+}$$

the electrophile

Mechanism: Formation of the Sulfonium Ion

The sulfonation of benzene is carried out using hot, concentrated sulfuric acid. The electrophile under these conditions is either SO_3 or HSO_3^+, depending on the experimental conditions. The HSO_3^+ electrophile is formed from sulfuric acid in the following way:

the electrophile

$$\underset{\text{Sulfuric acid}}{HO-\overset{O}{\underset{O}{S}}-\ddot{O}H} + H^+ \rightleftharpoons HO-\overset{O}{\underset{O}{S}}-\overset{+}{O}H_2 \rightleftharpoons \underset{\text{The sulfonium ion}}{HO-\overset{O}{\underset{O}{S^+}}} + :\ddot{O}H_2$$

(H_2SO_4)

Example 9.4

Write a stepwise mechanism for the nitration of benzene.

Strategy

Keep in mind that the mechanisms of electrophilic aromatic substitution reactions are all very similar. After the formation of the electrophile, attack of the electrophile on the aromatic ring occurs to give a resonance-stabilized cation intermediate. The last step of the mechanism is proton transfer to a base to regenerate the aromatic ring. The base in nitration is water, which was generated in the formation of the electrophile.

Solution

Step 1: Reaction of the nitronium ion (an electrophile) with the benzene ring (a nucleophile) gives a resonance-stabilized cation intermediate.

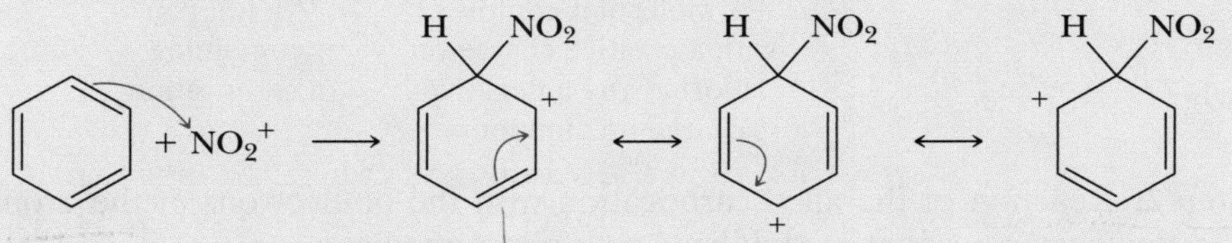

Step 2: Proton transfer from this intermediate to H_2O regenerates the aromatic ring and gives nitrobenzene:

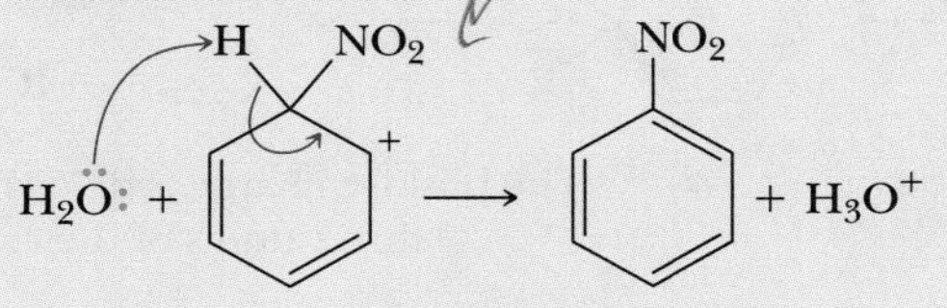

Nitrobenzene

See problems 9.21, 9.22

Problem 9.4

Write a stepwise mechanism for the sulfonation of benzene. Use HSO_3^+ as the electrophile.

C. Friedel–Crafts Alkylation

Alkylation of aromatic hydrocarbons was discovered in 1877 by the French chemist Charles Friedel and a visiting American chemist, James Crafts. They discovered that mixing benzene, a haloalkane, and $AlCl_3$ results in the formation of an alkylbenzene and HX. **Friedel–Crafts alkylation** forms a new carbon–carbon bond between benzene and an alkyl group, as illustrated by reaction of benzene with 2-chloropropane in the presence of aluminum chloride:

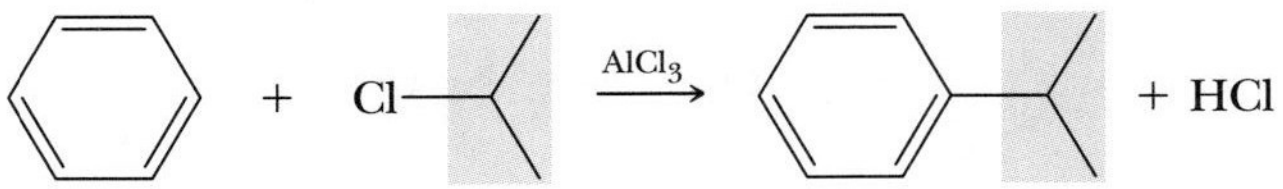

Benzene + 2-Chloropropane (Isopropyl chloride) $\xrightarrow{AlCl_3}$ Isopropylbenzene (Cumene) + HCl

Friedel–Crafts alkylation is among the most important methods for forming new carbon–carbon bonds to aromatic rings.

Mechanism: Friedel–Crafts Alkylation

Step 1: Reaction of a haloalkane (a Lewis base) with aluminum chloride (a Lewis acid) gives a molecular complex in which aluminum has a negative formal charge and the halogen of the haloalkane has a positive formal charge.

Redistribution of electrons in this complex then gives an alkyl carbocation as part of an ion pair:

$$R-\ddot{Cl}: + AlCl_3 \rightleftharpoons R-\overset{+}{Cl}-\overset{-}{Al}Cl_3 \rightleftharpoons R^+ :\ddot{Cl}-\overset{-}{Al}Cl_3$$

the electrophile

A molecular complex with a positive charge on chlorine and a negative charge on aluminum

An ion pair containing a carbocation

Step 2: Reaction of the alkyl carbocation with the pi electrons of the aromatic ring gives a resonance-stabilized cation intermediate:

The positive charge is delocalized onto three atoms of the ring

Step 3: Proton transfer regenerates the aromatic character of the ring and the Lewis acid catalyst:

$$+ :\ddot{Cl}-\overset{-}{Al}Cl_3 \longrightarrow C_6H_5-R + AlCl_3 + H-\ddot{Cl}:$$

There are two major limitations on Friedel–Crafts alkylations. The first is that it is practical only with stable carbocations, such as 3° carbocations, resonance–stabilized carbocations, or 2° carbocations that cannot undergo rearrangement (Section 5.4). Primary carbocations will undergo rearrangement, resulting in multiple products as well as bonding of the benzene ring to unexpected carbons in the former haloalkane.

The second limitation on Friedel–Crafts alkylation is that it fails altogether on benzene rings bearing one or more strongly electron-withdrawing groups. The following table shows some of these groups:

$$C_6H_5Y + RX \xrightarrow{AlCl_3} \text{No reaction}$$

When Y Equals Any of These Groups, the Benzene Ring Does Not Undergo Friedel–Crafts Alkylation

$-CH(=O)$	$-CR(=O)$	$-COH(=O)$	$-COR(=O)$	$-CNH_2(=O)$
$-SO_3H$	$-C\equiv N$	$-NO_2$	$-NR_3^+$	
$-CF_3$	$-CCl_3$			

HOW TO 9.2 Determine Whether a Substituent on Benzene Is Electron Withdrawing

Determine the charge or partial charge on the atom directly bonded to the benzene ring. If it is positive or partially positive, the substituent can be considered to be electron withdrawing. An atom will be partially positive if it is bonded to an atom more electronegative than itself.

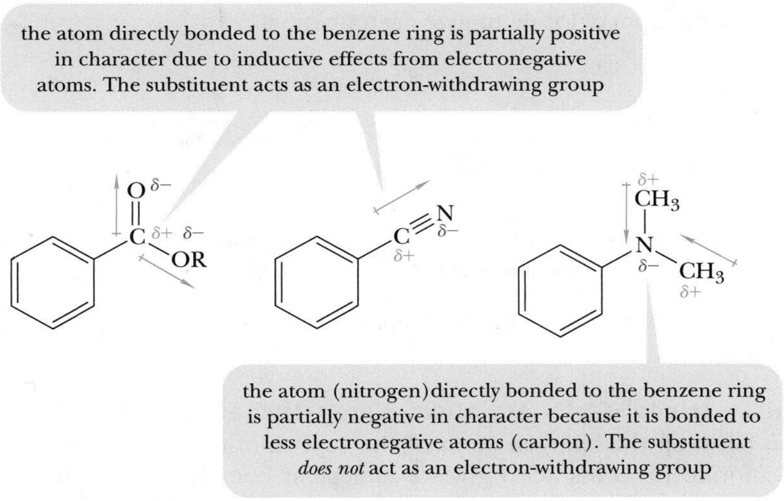

the atom directly bonded to the benzene ring is partially positive in character due to inductive effects from electronegative atoms. The substituent acts as an electron-withdrawing group

the atom (nitrogen) directly bonded to the benzene ring is partially negative in character because it is bonded to less electronegative atoms (carbon). The substituent *does not* act as an electron-withdrawing group

A common characteristic of the groups listed in the preceding table is that each has either a full or partial positive charge on the atom bonded to the benzene ring. For carbonyl-containing compounds, this partial positive charge arises because of the difference in electronegativity between the carbonyl oxygen and carbon. For $-CF_3$ and $-CCl_3$ groups, the partial positive charge on carbon arises because of the difference in electronegativity between carbon and the halogens bonded to it. In both the nitro group and the trialkylamonium group, there is a positive charge on nitrogen:

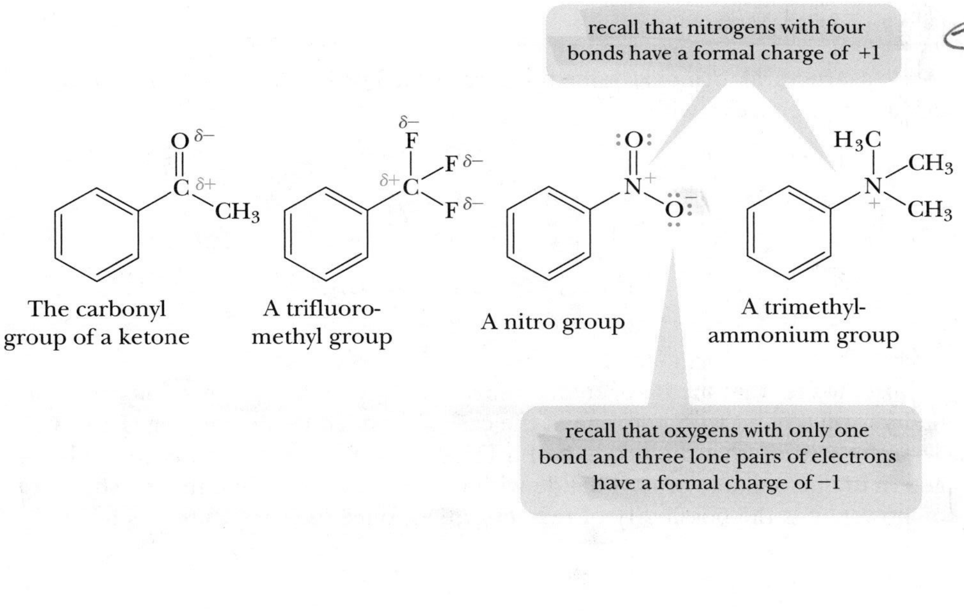

recall that nitrogens with four bonds have a formal charge of +1

don't forget these rules

The carbonyl group of a ketone

A trifluoromethyl group

A nitro group

A trimethylammonium group

recall that oxygens with only one bond and three lone pairs of electrons have a formal charge of −1

D. Friedel–Crafts Acylation

Acyl halide A derivative of a carboxylic acid in which the —OH of the carboxyl group is replaced by a halogen—most commonly, chlorine.

Friedel and Crafts also discovered that treating an aromatic hydrocarbon with an acyl halide (Section 15.1A) in the presence of aluminum chloride gives a ketone. An **acyl halide** is a derivative of a carboxylic acid in which the —OH of the carboxyl group is replaced by a halogen, most commonly chlorine. Acyl halides are also referred to as acid halides. An RCO— group is known as an acyl group; hence, the reaction of an acyl halide with an aromatic hydrocarbon is known as **Friedel–Crafts acylation**, as illustrated by the reaction of benzene and acetyl chloride in the presence of aluminum chloride to give acetophenone:

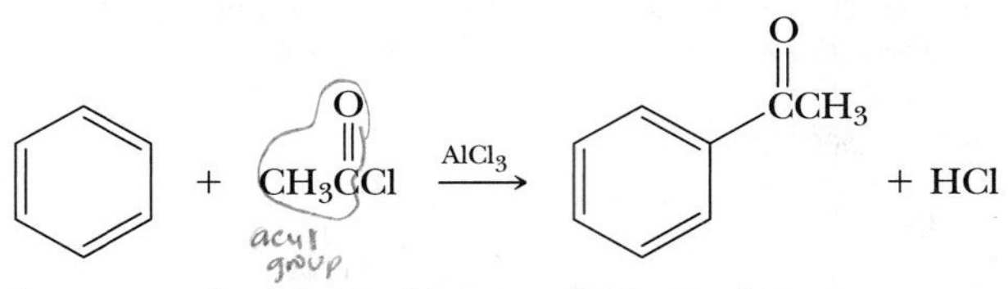

$$C_6H_6 + CH_3\overset{O}{\overset{\|}{C}}Cl \xrightarrow{AlCl_3} C_6H_5\overset{O}{\overset{\|}{C}}CH_3 + HCl$$

Benzene — Acetyl chloride (an acyl halide) — Acetophenone (a ketone)

In Friedel–Crafts acylations, the electrophile is an acylium ion, generated in the following way:

Mechanism: Friedel–Crafts Acylation—Generation of an Acylium Ion

Reaction between the halogen atom of the acyl chloride (a Lewis base) and aluminum chloride (a Lewis acid) gives a molecular complex. The redistribution of valence electrons in turn gives an ion pair containing an **acylium ion**:

$$R-\overset{O}{\overset{\|}{C}}-\ddot{Cl}: + AlCl_3 \rightleftharpoons R-\overset{O}{\overset{\|}{C}}-\overset{+}{\ddot{Cl}}-\overset{-}{Al}Cl_3 \rightleftharpoons R-\overset{O}{\overset{\|}{C^+}}\ :\ddot{Cl}-\overset{-}{Al}Cl_3$$

the electrophile

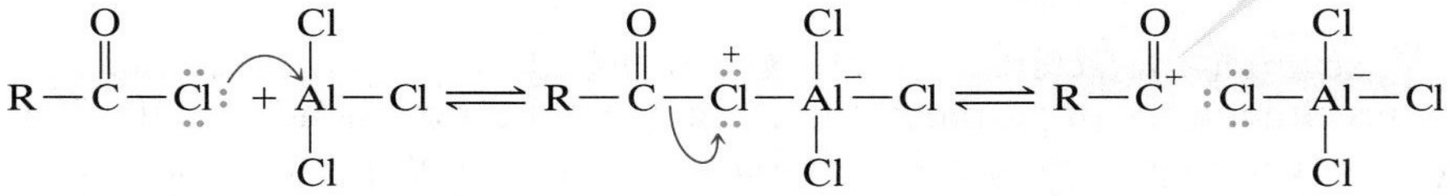

An acyl chloride (a Lewis base) — Aluminum chloride (a Lewis acid) — A molecular complex with a positive charge on chlorine and a negative charge on aluminum — An ion pair containing an acylium ion

Example 9.5

Write a structural formula for the product formed by Friedel–Crafts alkylation or acylation of benzene with

(a) 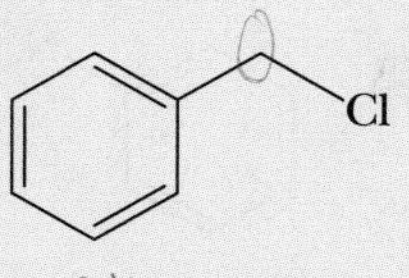 $C_6H_5CH_2Cl$ (b) C_6H_5COCl (c) $(CH_3)_2CHCHClCH_3$

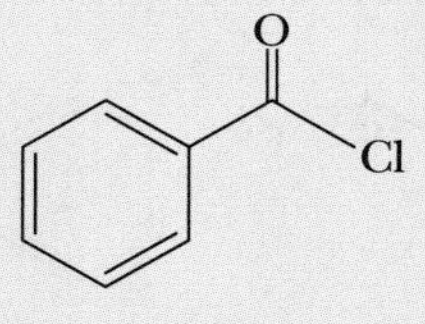

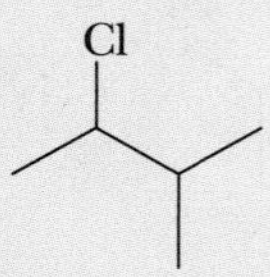

Strategy

Utilize the fact that the halogenated reagent in Friedel–Crafts reactions will normally form a bond with benzene at the carbon bonded to the halogen (Br or Cl). Therefore, to predict the product of a Friedel–Crafts reaction, replace the halogen in the haloalkane or acyl halide with the benzene ring. One thing to be wary of, however, is the possibility of rearrangement once the carbocation is formed.

Solution

(a) Treatment of benzyl chloride with aluminum chloride gives the resonance-stabilized benzyl cation. Reaction of this cation with benzene, followed by loss of H^+, gives diphenylmethane:

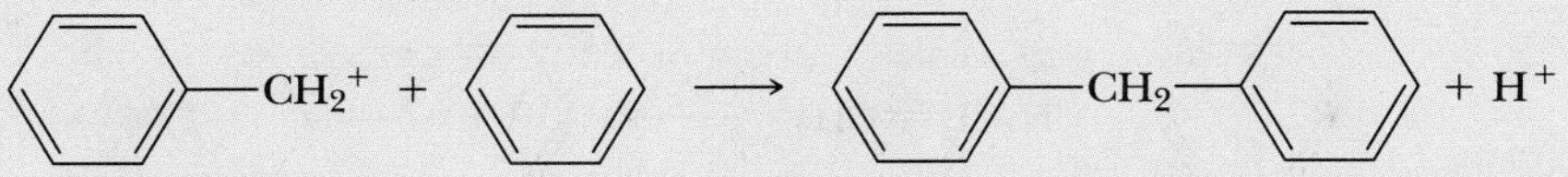

$-CH_2^+$ + $\longrightarrow$ $-CH_2-$ + H^+

Benzyl cation — Diphenylmethane

(b) Treatment of benzoyl chloride with aluminum chloride gives an acyl cation. Reaction of this cation with benzene, followed by loss of H^+, gives benzophenone:

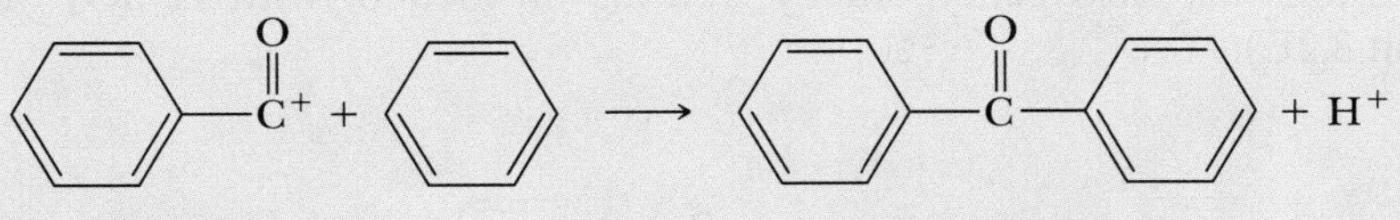

$\overset{O}{\overset{\|}{C^+}}$ + $\longrightarrow$ $\overset{O}{\overset{\|}{C}}$ + H^+

Benzoyl cation — Benzophenone

(c) Treatment of 2-chloro-3-methylbutane with aluminum chloride gives a 2° carbocation. Since there is an adjacent 3° carbon, a 1,2-hydride shift can occur to form the more stable 3° carbon. It is this carbon that reacts with benzene, followed by loss of H^+, to give 2-methyl-2-phenylbutane.

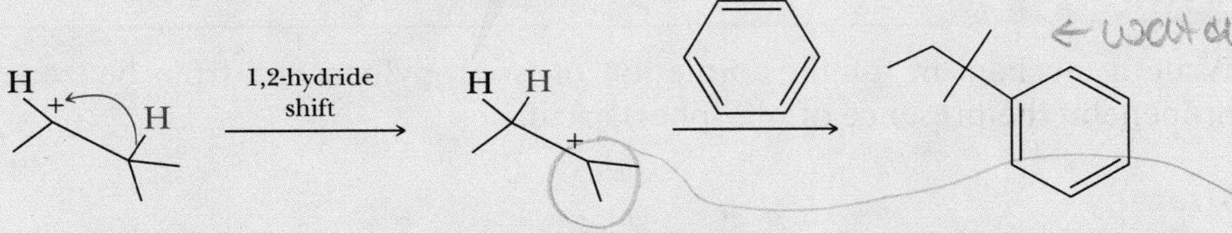

See problems 9.18, 9.19

Problem 9.5

Write a structural formula for the product formed from Friedel–Crafts alkylation or acylation of benzene with

(a) 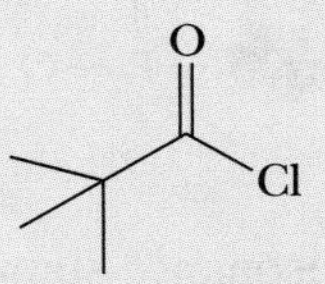 O, Cl (b) 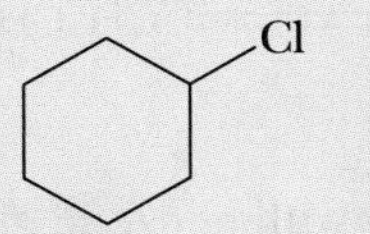 Cl (c) 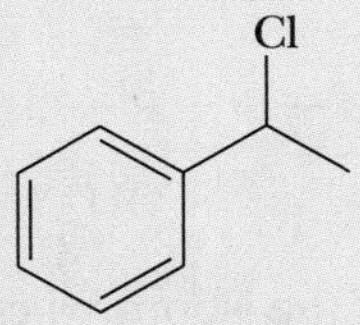 Cl

E. Other Electrophilic Aromatic Alkylations

Once it was discovered that Friedel–Crafts alkylations and acylations involve cationic intermediates, chemists realized that other combinations of reagents and catalysts could give the same products. We study two of these reactions in this section: the generation of carbocations from alkenes and from alcohols.

As we saw in Section 5.3B, treatment of an alkene with a strong acid, most commonly H_2SO_4 or H_3PO_4, generates a carbocation. Isopropylbenzene is synthesized industrially by reacting benzene with propene in the presence of an acid catalyst:

$$\text{Benzene} + CH_3CH{=}CH_2 \xrightarrow{H_3PO_4} \text{Isopropylbenzene}$$

Benzene Propene Isopropylbenzene (Cumene)

Carbocations are also generated by treating an alcohol with H_2SO_4 or H_3PO_4 (Section 8.2E):

$$\text{Benzene} + HO\text{–}C(CH_3)_3 \xrightarrow{H_3PO_4} C_6H_5C(CH_3)_3 + H_2O$$

Benzene 2-Methyl-2-phenylpropane (*tert*-Butylbenzene)

Example 9.6

Write a mechanism for the formation of isopropylbenzene from benzene and propene in the presence of phosphoric acid.

Strategy

Draw the mechanism for the formation of the carbocation. This step constitutes the generation of the electrophile. The remaining steps in the mechanism are the usual: attack of the electrophile on benzene and proton transfer to rearomatize the ring.

Solution

Step 1: Proton transfer from phosphoric acid to propene gives the isopropyl cation:

$$CH_3CH{=}CH_2 + H\text{–}\ddot{O}\text{–}P(=O)(OH)\text{–}O\text{–}H \underset{\text{reversible}}{\overset{\text{fast and}}{\rightleftharpoons}} CH_3\overset{+}{C}HCH_3 + {}^{-}\!:\ddot{O}\text{–}P(=O)(OH)\text{–}O\text{–}H$$

Step 2: Reaction of the isopropyl cation with benzene gives a resonance-stabilized carbocation intermediate:

$$\text{Benzene} + {}^{+}CH(CH_3)_2 \underset{\text{limiting}}{\overset{\text{slow, rate}}{\rightleftharpoons}} \text{cyclohexadienyl cation bearing H and } CH(CH_3)_2$$

Step 3: Proton transfer from this intermediate to dihydrogen phosphate ion gives isopropylbenzene:

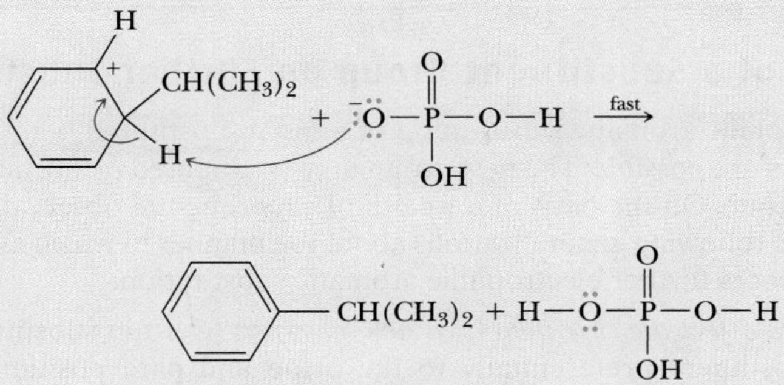

$CH(CH_3)_2$ + $^{-}\ddot{O}-P(=O)(OH)-O-H$ $\xrightarrow{\text{fast}}$ $C_6H_5-CH(CH_3)_2 + H-\ddot{O}-P(=O)(OH)-O-H$

Isopropylbenzene

See problems 9.18, 9.19, 9.33, 9.34

Problem 9.6

Write a mechanism for the formation of *tert*-butylbenzene from benzene and *tert*-butyl alcohol in the presence of phosphoric acid.

F. Comparison of Alkene Addition and Electrophilic Aromatic Substitution (EAS)

Electrophilic aromatic substitution represents the second instance in which we have encountered a C=C double bond attacking an electrophile. The first instance was in our discussion of alkene addition reactions in Section 5.3. Notice the similarities in the first step where a C=C double bond attacks an electrophilic atom (H^+ or E^+). In Step 2, however, alkene addition results in the attack of a nucleophile on the carbocation, while EAS results in abstraction of a hydrogen by base. In one reaction, the C=C double bond is destroyed, while in the other, the C=C double bond is regenerated.

Addition to an Alkene

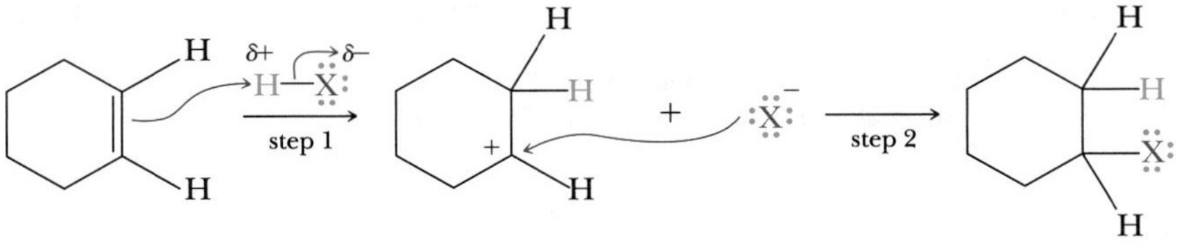

H, H, $\delta+$ H—X: $\delta-$, step 1, H, H, +, + :X:$^-$, step 2, H, H, X:, H

Electrophilic Aromatic Substitution

H, H, E^+, step 1, H, E, +, H, + :Base, step 2, E, H

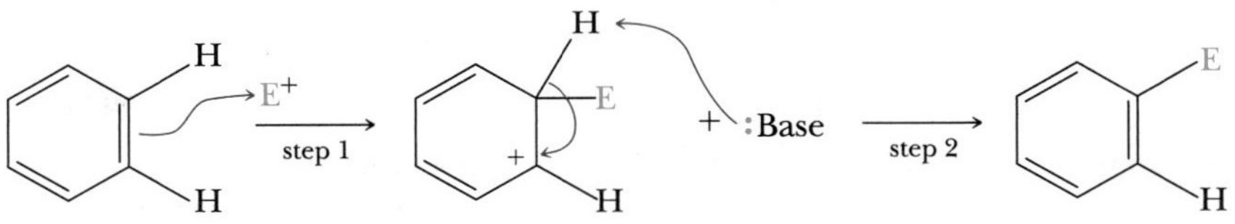

9.7 How Do Existing Substituents on Benzene Affect Electrophilic Aromatic Substitution?

A. Effects of a Substituent Group on Further Substitution

In the electrophilic aromatic substitution of a monosubstituted benzene, three isomeric products are possible: The new group may be oriented ortho, meta, or para to the existing group. On the basis of a wealth of experimental observations, chemists have made the following generalizations about the manner in which an existing substituent influences further electrophilic aromatic substitution:

1. *Substituents affect the orientation of new groups.* Certain substituents direct a second substituent preferentially to the ortho and para positions; other substituents direct it preferentially to a meta position. In other words, we can classify substituents on a benzene ring as **ortho–para directing** or **meta directing**.
2. *Substituents affect the rate of further substitution.* Certain substituents cause the rate of a second substitution to be greater than that of benzene itself, whereas other substituents cause the rate of a second substitution to be lower than that of benzene. In other words, we can classify groups on a benzene ring as **activating** or **deactivating** toward further substitution.

Ortho–para director Any substituent on a benzene ring that directs electrophilic aromatic substitution preferentially to ortho and para positions.

Meta director Any substituent on a benzene ring that directs electrophilic aromatic substitution preferentially to a meta position.

Activating group Any substituent on a benzene ring that causes the rate of electrophilic aromatic substitution to be greater than that for benzene.

Deactivating group Any substituent on a benzene ring that causes the rate of electrophilic aromatic substitution to be lower than that for benzene.

To see the operation of these directing and activating–deactivating effects, compare, for example, the products and rates of bromination of anisole and nitrobenzene. Bromination of anisole proceeds at a rate 1.8×10^9 greater than that of bromination of benzene (the methoxy group is activating), and the product is a mixture of *o*-bromoanisole and *p*-bromoanisole (the methoxy group is ortho–para directing):

the bromination of anisole proceeds many times faster than the bromination of benzene

OCH_3 + Br_2 $\xrightarrow{CH_3COOH}$ OCH_3, Br + OCH_3, Br + HBr

Anisole | *o*-Bromoanisole (4%) | *p*-Bromoanisole (96%)

We see quite another situation in the nitration of nitrobenzene, which proceeds 10,000 times slower than the nitration of benzene itself. (A nitro group is strongly deactivating.) Also, the product consists of approximately 93% of the meta isomer and less than 7% of the ortho and para isomers combined (the nitro group is meta directing):

the nitration of nitrobenzene proceeds many times slower than the nitration of benzene

NO_2 + HNO_3 $\xrightarrow[100°C]{H_2SO_4}$ NO_2, NO_2 + NO_2, NO_2 + NO_2, NO_2 + H_2O

Nitrobenzene | *m*-Dinitrobenzene (93%) | *o*-Dinitrobenzene | *p*-Dinitrobenzene

Less than 7% combined

TABLE 9.1 Effects of Substituents on Further Electrophilic Aromatic Substitution

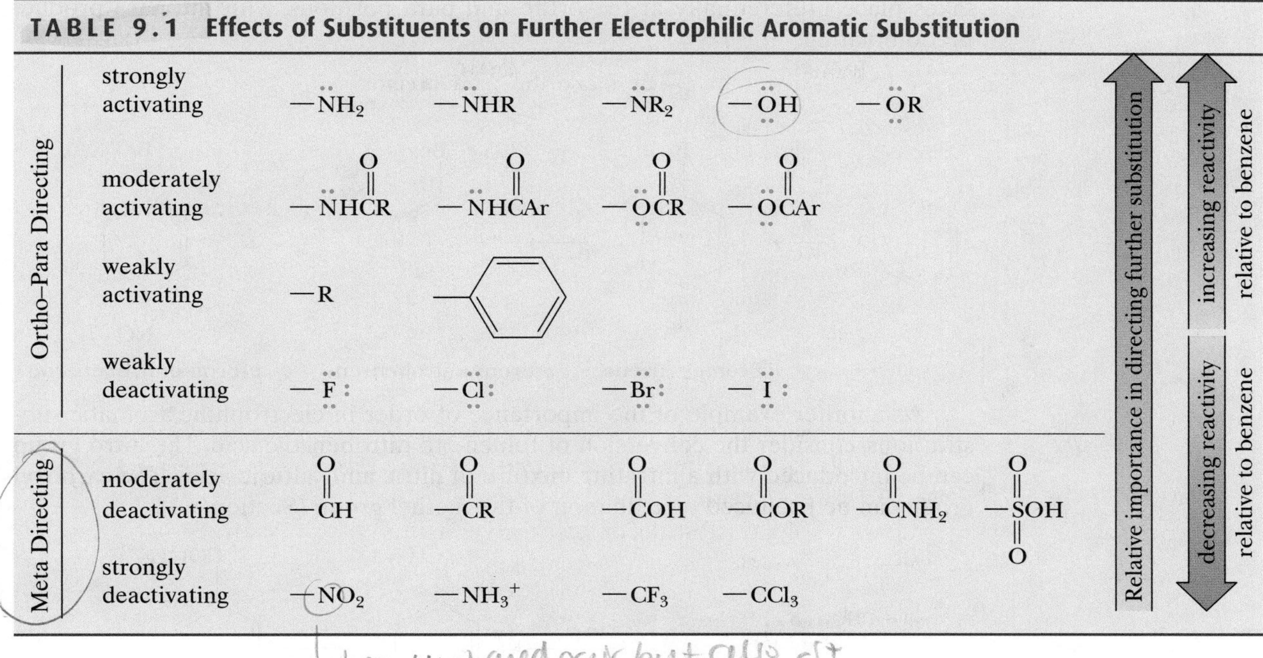

Directing	Effect							Relative importance in directing further substitution
Ortho–Para Directing	strongly activating	$-\ddot{N}H_2$	$-\ddot{N}HR$	$-\ddot{N}R_2$	$-\ddot{O}H$	$-\ddot{O}R$		increasing reactivity relative to benzene
	moderately activating	$-\ddot{N}HCR$ (C=O)	$-\ddot{N}HCAr$ (C=O)	$-\ddot{O}CR$ (C=O)	$-\ddot{O}CAr$ (C=O)			
	weakly activating	$-R$	$-C_6H_5$					
	weakly deactivating	$-\ddot{F}:$	$-\ddot{Cl}:$	$-\ddot{Br}:$	$-\ddot{I}:$			decreasing reactivity relative to benzene
Meta Directing	moderately deactivating	$-CH$ (C=O)	$-CR$ (C=O)	$-COH$ (C=O)	$-COR$ (C=O)	$-CNH_2$ (C=O)	$-SOH$ (SO_2)	
	strongly deactivating	$-NO_2$	$-NH_3^+$	$-CF_3$	$-CCl_3$			

Table 9.1 lists the directing and activating–deactivating effects for the major functional groups with which we are concerned in this text.

If we compare these ortho–para and meta directors for structural similarities and differences, we can make the following generalizations:

1. Alkyl groups, phenyl groups, and substituents in which the atom bonded to the ring has an unshared pair of electrons are ortho–para directing. All other substituents are meta directing.
2. Except for the halogens, all ortho–para directing groups are activating toward further substitution. The halogens are weakly deactivating.
3. All meta directing groups carry either a partial or full positive charge on the atom bonded to the ring.

We can illustrate the usefulness of these generalizations by considering the synthesis of two different disubstituted derivatives of benzene. Suppose we wish to prepare *m*-bromonitrobenzene from benzene. This conversion can be carried out in two steps: nitration and bromination. If the steps are carried out in just that order, the major product is indeed *m*-bromonitrobenzene. The nitro group is a meta director and directs bromination to a meta position:

$-NO_2$ is a meta director

$$\text{Benzene} \xrightarrow[H_2SO_4]{HNO_3} \text{Nitrobenzene} \xrightarrow[FeCl_3]{Br_2} m\text{-Bromonitrobenzene}$$

Nitrobenzene　　*m*-Bromonitrobenzene

If, however, we reverse the order of the steps and first form bromobenzene, we now have an ortho–para directing group on the ring. Nitration of bromobenzene then takes place preferentially at the ortho and para positions, with the para product predominating:

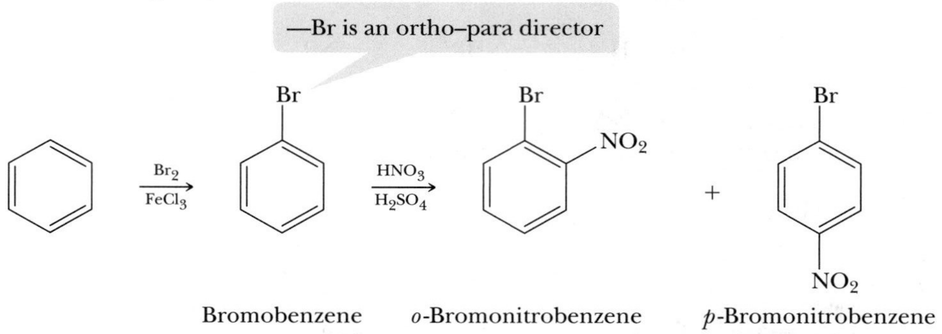

—Br is an ortho–para director

Br$_2$ / FeCl$_3$; HNO$_3$ / H$_2$SO$_4$

Bromobenzene — *o*-Bromonitrobenzene + *p*-Bromonitrobenzene

As another example of the importance of order in electrophilic aromatic substitutions, consider the conversion of toluene to nitrobenzoic acid. The nitro group can be introduced with a nitrating mixture of nitric and sulfuric acids. The carboxyl group can be produced by oxidation of the methyl group (Section 9.4).

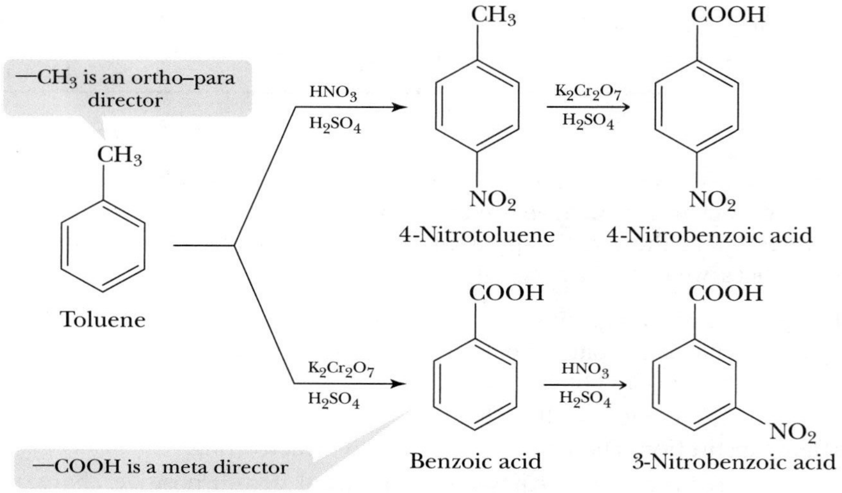

—CH_3 is an ortho–para director

Toluene → (HNO$_3$ / H$_2$SO$_4$) → 4-Nitrotoluene → (K$_2$Cr$_2$O$_7$ / H$_2$SO$_4$) → 4-Nitrobenzoic acid

Toluene → (K$_2$Cr$_2$O$_7$ / H$_2$SO$_4$) → Benzoic acid → (HNO$_3$ / H$_2$SO$_4$) → 3-Nitrobenzoic acid

—COOH is a meta director

Nitration of toluene yields a product with the two substituents para to each other, whereas nitration of benzoic acid yields a product with the substituents meta to each other. Again, we see that the order in which the reactions are performed is critical.

Note that, in this last example, we show nitration of toluene producing only the para isomer. In practice, because methyl is an ortho–para directing group, both ortho and para isomers are formed. In problems in which we ask you to prepare one or the other of these isomers, we assume that both form and that there are physical methods by which you can separate them and obtain the desired isomer.

Example 9.7

Complete the following electrophilic aromatic substitution reactions. Where you predict meta substitution, show only the meta product. Where you predict ortho–para substitution, show both products:

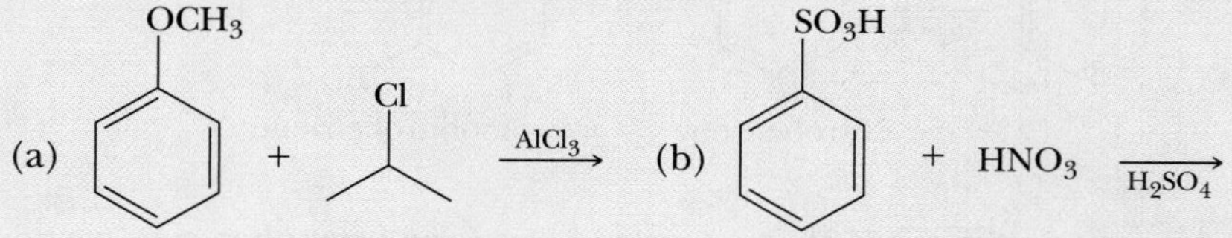

(a) $C_6H_5OCH_3$ + 2-chloropropane (Cl) $\xrightarrow{AlCl_3}$ (b) $C_6H_5SO_3H$ + HNO_3 $\xrightarrow{H_2SO_4}$

Strategy

Determine whether the existing substituent is ortho–para or meta directing prior to completing the reaction.

Solution

The methoxyl group in (a) is ortho–para directing and strongly activating. The sulfonic acid group in (b) is meta directing and moderately deactivating.

$-OCH_3$ is an ortho–para director

(a) OCH_3 benzene + Cl isopropyl chloride $\xrightarrow{AlCl_3}$ 2- Isopropylanisole (ortho-isopropylanisole) + 4-Isopropylanisole (para-isopropylanisole)

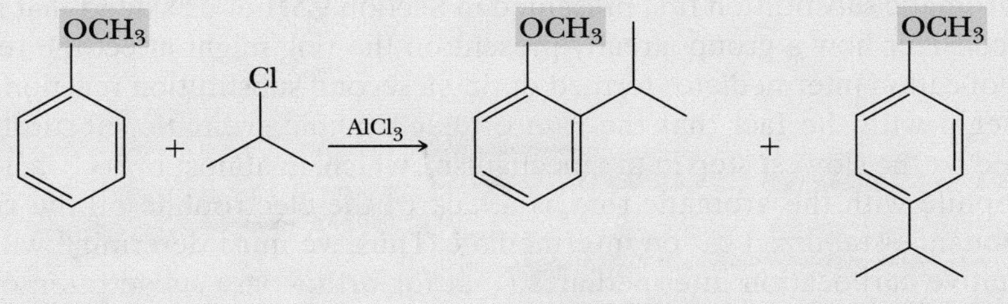

$-SO_3H$ is a meta director

(b) SO_3H benzene + HNO_3 $\xrightarrow{H_2SO_4}$ 3-Nitrobenzenesulfonic acid (meta-nitrobenzenesulfonic acid)

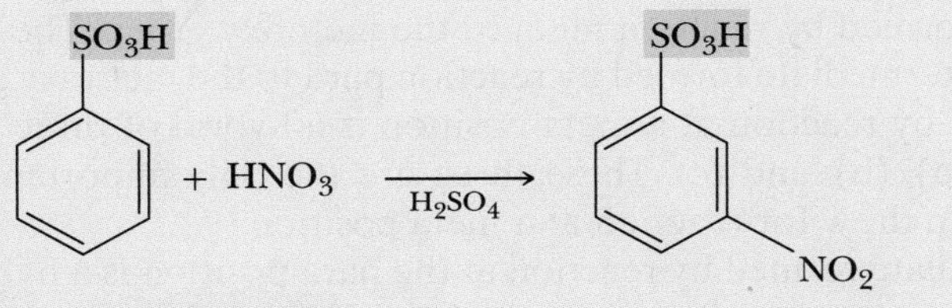

See problems 9.24–9.26, 9.31, 9.32, 9.42–9.44, 9.46, 9.47

Problem 9.7

Complete the following electrophilic aromatic substitution reactions. Where you predict meta substitution, show only the meta product. Where you predict ortho–para substitution, show both products:

(a) methyl benzoate + HNO_3 $\xrightarrow{H_2SO_4}$

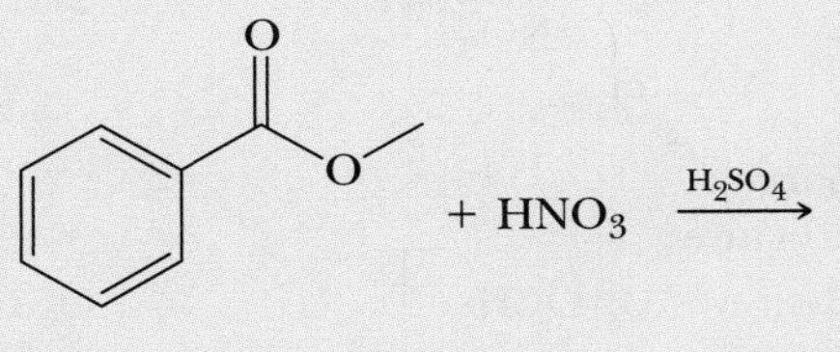

(b) phenyl acetate + HNO_3 $\xrightarrow{H_2SO_4}$

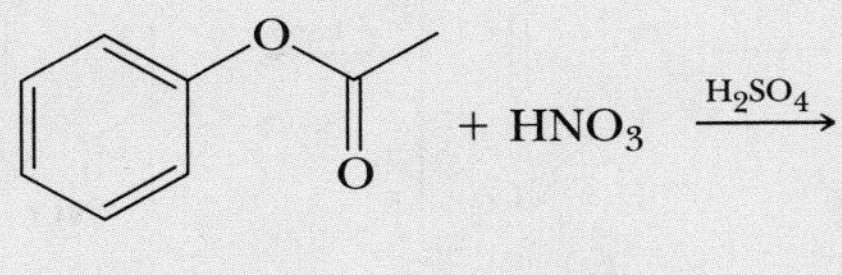

B. Theory of Directing Effects

As we have just seen, a group on an aromatic ring exerts a major effect on the patterns of further substitution. We can make these three generalizations:

1. If there is a lone pair of electrons on the atom bonded to the ring, the group is an ortho–para director.
2. If there is a full or partial positive charge on the atom bonded to the ring, the group is a meta director.
3. Alkyl groups are ortho–para directors.

We account for these patterns by means of the general mechanism for electrophilic aromatic substitution first presented in Section 9.5. Let us extend that mechanism to consider how a group already present on the ring might affect the relative stabilities of cation intermediates formed during a second substitution reaction.

We begin with the fact that the rate of electrophilic aromatic substitution is determined by the slowest step in the mechanism, which, in almost every reaction of an electrophile with the aromatic ring, is attack of the electrophile on the ring to give a resonance-stabilized cation intermediate. Thus, we must determine which of the alternative carbocation intermediates (that for ortho–para substitution or that for meta substitution) is the more stable. That is, we need to show which of the alternative cationic intermediates has the lower activation energy for its formation.

Nitration of Anisole

The rate-determining step in nitration is reaction of the nitronium ion with the aromatic ring to produce a resonance-stabilized cation intermediate. Figure 9.4 shows the cation intermediate formed by reaction meta to the methoxy group. The figure also shows the cationic intermediate formed by reaction para to the methoxy group. The intermediate formed by reaction at a meta position is a hybrid of three major contributing structures: (a), (b), and (c). These three are the only important contributing structures we can draw for reaction at a meta position.

The cationic intermediate formed by reaction at the para position is a hybrid of four major contributing structures: (d), (e), (f), and (g). What is important about structure (f) is that all atoms in it have complete octets, which means that this structure contributes more to the hybrid than structures (d), (e), or (g). Because the cation formed by reaction at an ortho or para position on anisole has a greater resonance stabilization and, hence, a lower activation energy for its formation, nitration of anisole occurs preferentially in the ortho and para positions.

meta attack

OCH_3 + NO_2^+ slow → [(a) ⟷ (b) ⟷ (c)] fast, $-H^+$ → product with OCH_3 and NO_2

para attack

OCH_3 + NO_2^+ slow → [(d) ⟷ (e) ⟷ (f) ⟷ (g)] fast, $-H^+$ → product with OCH_3 and NO_2

The most important contributing structure

Figure 9.4
Nitration of anisole. Reaction of the electrophile meta and para to a methoxy group. Regeneration of the aromatic ring is shown from the right-most contributing structure in each case.

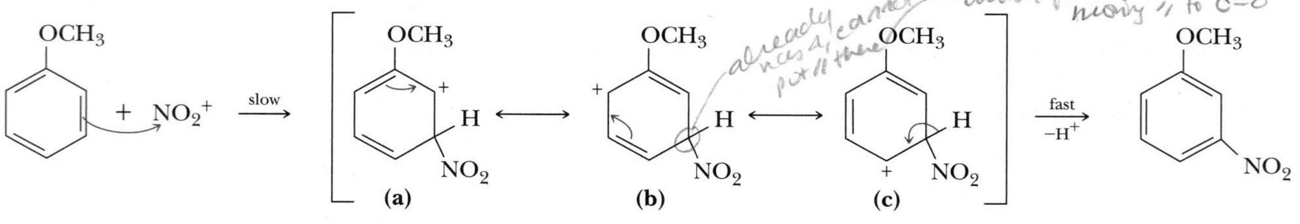

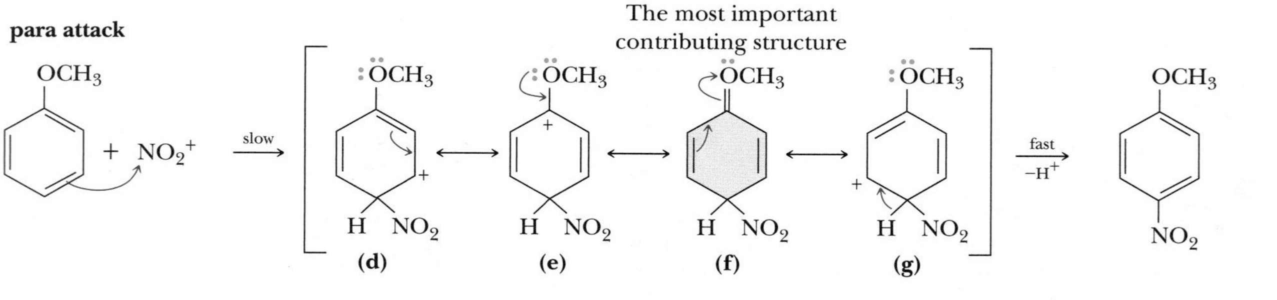

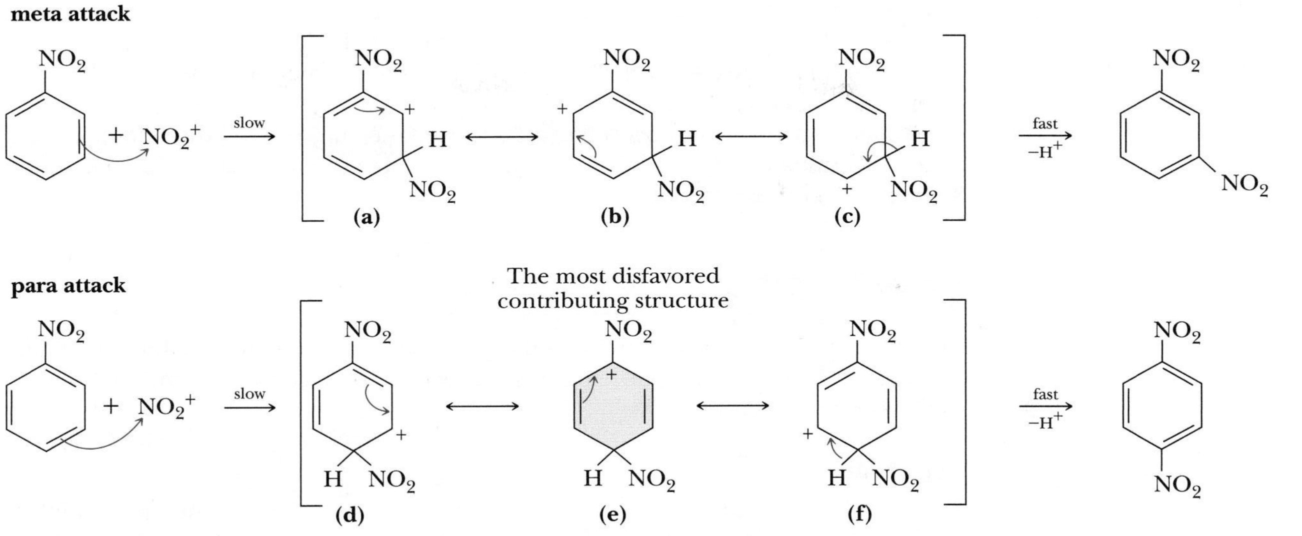

Figure 9.5 Nitration of nitrobenzene. Reaction of the electrophile meta and para to a nitro group. Regeneration of the aromatic ring is shown from the rightmost contributing structure in each case.

Nitration of Nitrobenzene

Figure 9.5 shows the resonance-stabilized cation intermediates formed by reaction of the nitronium ion meta to the nitro group and also para to it.

Each cation in the figure is a hybrid of three contributing structures; no additional ones can be drawn. Now we must compare the relative resonance stabilizations of each hybrid. If we draw a Lewis structure for the nitro group showing the positive formal charge on nitrogen, we see that contributing structure (e) places positive charges on adjacent atoms:

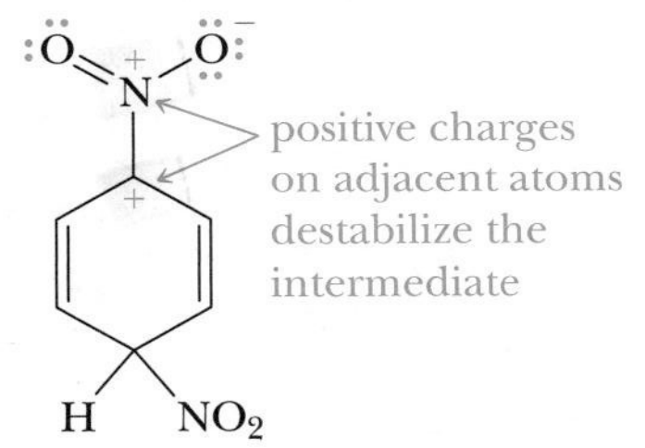

Because of the electrostatic repulsion thus generated, structure (e) makes only a negligible contribution to the hybrid. None of the contributing structures for reaction at a meta position places positive charges on adjacent atoms. As a consequence, resonance stabilization of the cation formed by reaction at a meta position is greater than that for the cation formed by reaction at a para (or ortho) position. Stated alternatively, the activation energy for reaction at a meta position is less than that for reaction at a para position.

A comparison of the entries in Table 9.1 shows that almost all ortho–para directing groups have an unshared pair of electrons on the atom bonded to the aromatic ring. Thus, the directing effect of most of these groups is due primarily to the ability of the atom bonded to the ring to delocalize further the positive charge on the cation intermediate.

The fact that alkyl groups are also ortho–para directing indicates that they, too, help to stabilize the cation intermediate. In Section 5.3A, we saw that alkyl groups stabilize carbocation intermediates and that the order of stability of carbocations is $3° > 2° > 1° >$ methyl. Just as alkyl groups stabilize the cation intermediates formed in reactions of alkenes, they also stabilize the carbocation intermediates formed in electrophilic aromatic substitutions.

To summarize, any substituent on an aromatic ring that further stabilizes the cation intermediate directs ortho–para, and any group that destabilizes the cation intermediate directs meta.

Example 9.8

Draw contributing structures formed during the para nitration of chlorobenzene, and show how chlorine participates in directing the incoming nitronium ion to ortho–para positions.

Strategy

Draw the intermediate that is formed initially from attack of the electrophile. Then draw a contributing structure by moving electrons from the pi bond adjacent to the positive charge. Repeat for all contributing structures until all resonance possibilities have been exhausted. *Note:* Be sure to look for resonance possibilities outside of the benzene ring.

Solution

Contributing structures (a), (b), and (d) place the positive charge on atoms of the ring, while contributing structure (c) places it on chlorine and thus creates additional resonance stabilization for the cation intermediate:

:Cl: :Cl: :Cl: :Cl:

H NO_2 H NO_2 H NO_2 H NO_2

(a) (b) (c) (d)

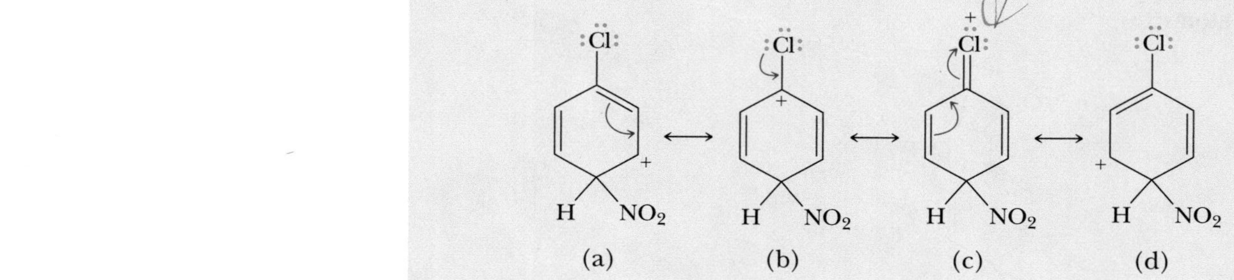

See problem 9.29

Problem 9.8

Because the electronegativity of oxygen is greater than that of carbon, the carbon of a carbonyl group bears a partial positive charge, and its oxygen bears a partial negative charge. Using this information, show that a carbonyl group is meta directing:

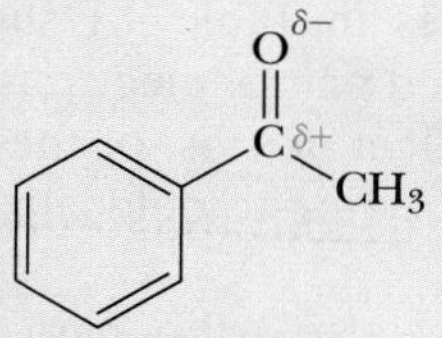

$O^{\delta-}$ $C^{\delta+}$ CH_3

C. Theory of Activating–Deactivating Effects

We account for the activating–deactivating effects of substituent groups by a combination of resonance and inductive effects:

1. Any resonance effect, such as that of $-NH_2$, $-OH$, and $-OR$, which delocalizes the positive charge of the cation intermediate lowers the activation energy for its formation and is activating toward further electrophilic aromatic substitution. That is, these groups increase the rate of electrophilic aromatic substitution, compared with the rate at which benzene itself reacts.
2. Any resonance or inductive effect, such as that of $-NO_2$, $-C{=}O$, $-SO_3H$, $-NR_3^+$, $-CCl_3$, and $-CF_3$, which decreases electron density on the ring, deactivates the ring to further substitution. That is, these groups decrease the rate of further electrophilic aromatic substitution, compared with the rate at which benzene itself reacts.
3. Any inductive effect (such as that of $-CH_3$ or another alkyl group), which releases electron density toward the ring, activates the ring toward further substitution.

In the case of the halogens, the resonance and inductive effects operate in opposite directions. As Table 9.1 shows, the halogens are ortho–para directing, but, unlike other ortho–para directors listed in the table, the halogens are weakly deactivating. These observations can be accounted for in the following way.

similar e⁻ w/drawing effect

1. *The inductive effect of halogens.* The halogens are more electronegative than carbon and have an electron-withdrawing inductive effect. Aryl halides, therefore, react more slowly in electrophilic aromatic substitution than benzene does.
2. *The resonance effect of halogens.* A halogen ortho or para to the site of electrophilic attack stabilizes the cation intermediate by delocalization of the positive charge:

:Cl–C₆H₅ + E⁺ ⟶ [:Cl–(cation, H, E) ⟷ ⁺Cl=(ring, H, E)]

Example 9.9

Predict the product of each electrophilic aromatic substitution.

(a) 3-nitrophenol (OH, NO_2) + Br_2 $\xrightarrow{FeCl_3}$ (b) 4-methylbenzoic acid (COOH, CH_3) + HNO_3 $\xrightarrow{H_2SO_4}$

Strategy

Determine the activating and deactivating effect of each group. The key to predicting the orientation of further substitution on a disubstituted arene is that ortho–para directing groups are always better at activating the ring

toward further substitution than meta directing groups (Table 9.1). This means that, when there is competition between ortho–para directing and meta directing groups, the ortho–para group wins.

Solution

(a) The ortho–para directing and activating —OH group determines the position of bromination. Bromination between the —OH and —NO_2 groups is only a minor product because of steric hindrance to attack of bromine at this position:

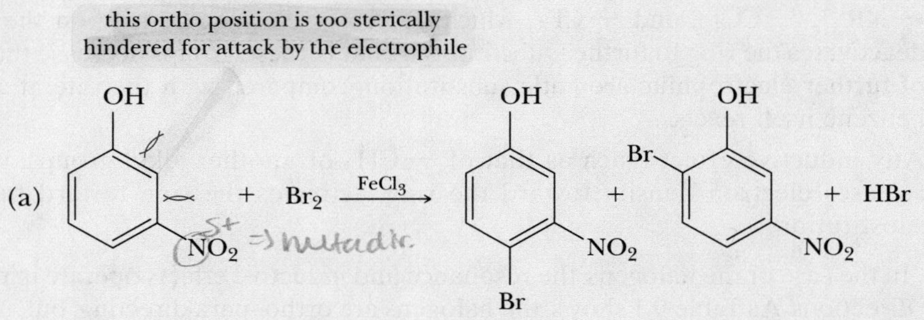

this ortho position is too sterically hindered for attack by the electrophile

(a) OH, NO_2 + Br_2 $\xrightarrow{FeCl_3}$ OH, NO_2, Br + OH, Br, NO_2 + HBr

(b) The ortho–para directing and activating methyl group determines the position of nitration:

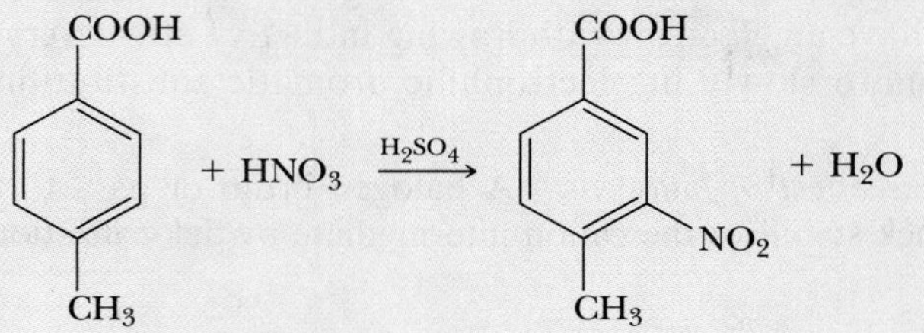

COOH, CH_3 + HNO_3 $\xrightarrow{H_2SO_4}$ COOH, NO_2, CH_3 + H_2O

See problems 9.24–9.26, 9.31, 9.32, 9.42–9.44, 9.46, 9.47

Problem 9.9

Predict the product of treating each compound with HNO_3/H_2SO_4:

9.8 What Are Phenols?

A. Structure and Nomenclature

Phenol A compound that contains an —OH bonded to a benzene ring.

The functional group of a **phenol** is a hydroxyl group bonded to a benzene ring. We name substituted phenols either as derivatives of phenol or by common names:

OH OH OH OH OH

OH CH$_3$ OH OH

Phenol	3-Methylphenol (*m*-Cresol)	1,2-Benzenediol (Catechol)	1,3-Benzenediol (Resorcinol)	1,4-Benzenediol (Hydroquinone)

Thymol is a constituent of garden thyme, Thymus vulgaris. *(Wally Eberhart/Visuals Unlimited)*

Phenols are widely distributed in nature. Phenol itself and the isomeric cresols (*o*-, *m*-, and *p*-cresol) are found in coal tar. Thymol and vanillin are important constituents of thyme and vanilla beans, respectively:

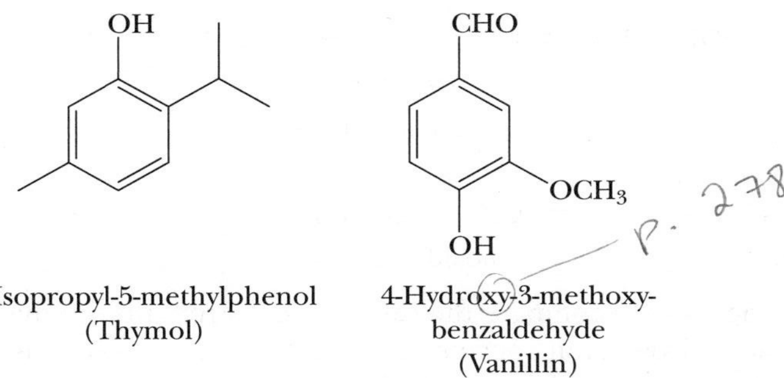

2-Isopropyl-5-methylphenol (Thymol)

4-Hydroxy-3-methoxybenzaldehyde (Vanillin)

Poison ivy. *(Charles D. Winters)*

Phenol, or carbolic acid, as it was once called, is a low-melting solid that is only slightly soluble in water. In sufficiently high concentrations, it is corrosive to all kinds of cells. In dilute solutions, phenol has some antiseptic properties and was introduced into the practice of surgery by Joseph Lister, who demonstrated his technique of aseptic surgery in the surgical theater of the University of Glasgow School of Medicine in 1865. Nowadays, phenol has been replaced by antiseptics that are both more powerful and have fewer undesirable side effects. Among these is hexylresorcinol, which is widely used in nonprescription preparations as a mild antiseptic and disinfectant.

OH OH OCH$_3$ OH OH OH

Hexylresorcinol	Eugenol	Urushiol

Eugenol, which can be isolated from the flower buds (cloves) of *Eugenia aromatica*, is used as a dental antiseptic and analgesic. Urushiol is the main component in the irritating oil of poison ivy.

B. Acidity of Phenols

Phenols and alcohols both contain an —OH group. We group phenols as a separate class of compounds, however, because their chemical properties are quite different from those of alcohols. One of the most important of these differences is that

TABLE 9.2 Relative Acidities of 0.1-M Solutions of Ethanol, Phenol, and HCl

Acid Ionization Equation	$[H^+]$	pH
$CH_3CH_2OH + H_2O \rightleftharpoons CH_3CH_2O^- + H_3O^+$	1×10^{-7}	7.0
$C_6H_5OH + H_2O \rightleftharpoons C_6H_5O^- + H_3O^+$	3.3×10^{-6}	5.4
$HCl + H_2O \rightleftharpoons Cl^- + H_3O^+$	0.1	1.0

phenols are significantly more acidic than are alcohols. Indeed, the acid ionization constant for phenol is 10^6 times larger than that of ethanol!

$$C_6H_5\ddot{O}H + H_2O \rightleftharpoons C_6H_5\ddot{O}:^- + H_3O^+ \quad K_a = 1.1 \times 10^{-10} \quad pK_a = 9.95$$

Phenol Phenoxide ion

$$CH_3CH_2\ddot{O}H + H_2O \rightleftharpoons CH_3CH_2\ddot{O}:^- + H_3O^+ \quad K_a = 1.3 \times 10^{-16} \quad pK_a = 15.9$$

Ethanol Ethoxide ion

Another way to compare the relative acid strengths of ethanol and phenol is to look at the hydrogen ion concentration and pH of a 0.1-M aqueous solution of each (Table 9.2). For comparison, the hydrogen ion concentration and pH of 0.1 M HCl are also included.

In aqueous solution, alcohols are neutral substances, and the hydrogen ion concentration of 0.1 M ethanol is the same as that of pure water. A 0.1-M solution of phenol is slightly acidic and has a pH of 5.4. By contrast, 0.1 M HCl, a strong acid (completely ionized in aqueous solution), has a pH of 1.0.

The greater acidity of phenols compared with alcohols results from the greater stability of the phenoxide ion compared with an alkoxide ion. The negative charge on the phenoxide ion is delocalized by resonance. The two contributing structures on the left for the phenoxide ion place the negative charge on oxygen, while the three on the right place the negative charge on the ortho and para positions of the ring. Thus, in the resonance hybrid, the negative charge of the phenoxide ion is delocalized over four atoms, which stabilizes the phenoxide ion realtive to an alkoxide ion, for which no delocalization is possible:

These two Kekulé structures are equivalent

These three contributing structures delocalize the negative charge onto carbon atoms of the ring

Note that, although the resonance model gives us a way of understanding why phenol is a stronger acid than ethanol, it does not provide us with any quantitative means of predicting just how much stronger an acid it might be. To find out how much stronger one acid is than another, we must determine their pK_a values experimentally and compare them.

Ring substituents, particularly halogen and nitro groups, have marked effects on the acidities of phenols through a combination of inductive and resonance effects. Because the halogens are more electronegative than carbon, they withdraw electron density from the negatively charged oxygen in the conjugate base, stabilizing the phenoxide ion. Nitro groups have greater electron-withdrawing ability than halogens and thus have a greater stabilizing effect on the phenoxide ion, making nitrophenol even more acidic than chlorophenol.

Increasing acid strength

Phenol
pK_a 9.95

4-Chlorophenol
pK_a 9.18

4-Nitrophenol
pK_a 7.15

electron-withdrawing groups withdraw electron density from the negatively charged oxygen of the conjugate base, delocalizing the charge, and thus stabilizing the ion

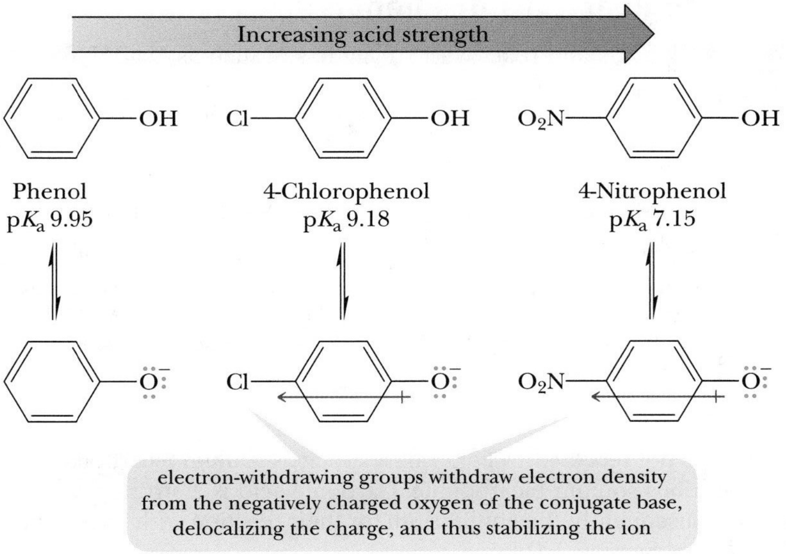

Example 9.10

Arrange these compounds in order of increasing acidity: 2,4-dinitrophenol, phenol, and benzyl alcohol.

Strategy

Draw each conjugate base. Then determine which conjugate base is more stable using the principles of resonance and inductive effects. The more stable the conjugate base, the more acidic the acid from which it was generated.

Solution

Benzyl alcohol, a primary alcohol, has a pK_a of approximately 16–18 (Section 8.2A). The pK_a of phenol is 9.95. Nitro groups are electron withdrawing and increase the acidity of the phenolic —OH group. In order of increasing acidity, these compounds are:

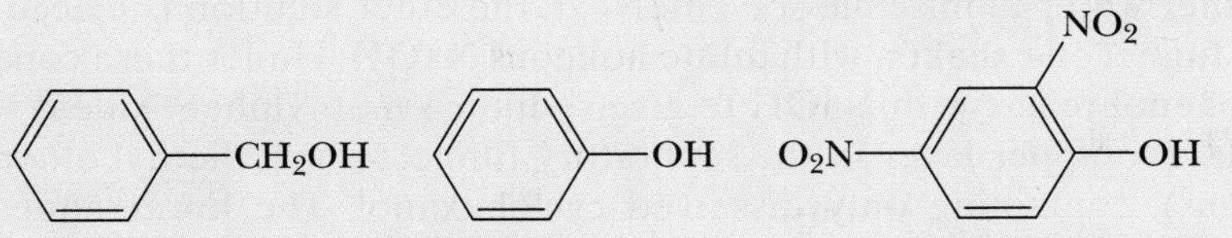

Benzyl alcohol
pK_a 16–18

Phenol
pK_a 9.95

2,4-Dinitrophenol
pK_a 3.96

See problems 9.36–9.38

Problem 9.10

Arrange these compounds in order of increasing acidity: 2,4-dichlorophenol, phenol, cyclohexanol.

C. Acid–Base Reactions of Phenols

Phenols are weak acids and react with strong bases, such as NaOH, to form water-soluble salts:

$$C_6H_5\text{—}OH + NaOH \longrightarrow C_6H_5\text{—}O^-Na^+ + H_2O$$

Phenol	Sodium hydroxide	Sodium phenoxide	Water
pK_a 9.95			pK_a 15.7
(stronger acid)	(stronger base)	(weaker base)	(weaker acid)

Most phenols do not react with weaker bases, such as sodium bicarbonate, and do not dissolve in aqueous sodium bicarbonate. Carbonic acid is a stronger acid than most phenols, and, consequently, the equilibrium for their reaction with bicarbonate ion lies far to the left (see Section 2.4):

$$C_6H_5\text{—}OH + NaHCO_3 \rightleftharpoons C_6H_5\text{—}O^-Na^+ + H_2CO_3$$

Phenol	Sodium bicarbonate	Sodium phenoxide	Carbonic acid
pK_a 9.95			pK_a 6.36
(weaker acid)	(weaker base)	(stronger base)	(stronger acid)

The fact that phenols are weakly acidic, whereas alcohols are neutral, provides a convenient way to separate phenols from water-insoluble alcohols. Suppose that we want to separate 4-methylphenol from cyclohexanol. Each is only slightly soluble in water; therefore, they cannot be separated on the basis of their water solubility. They can be separated, however, on the basis of their difference in acidity. First, the mixture of the two is dissolved in diethyl ether or some other water-immiscible solvent. Next, the ether solution is placed in a separatory funnel and shaken with dilute aqueous NaOH. Under these conditions, 4-methylphenol reacts with NaOH to give sodium 4-methylphenoxide, a water-soluble salt. The upper layer in the separatory funnel is now diethyl ether (density 0.74 g/cm^3), containing only dissolved cyclohexanol. The lower aqueous layer contains dissolved sodium 4-methylphenoxide. The layers are separated, and distillation of the ether (bp 35°C) leaves pure cyclohexanol (bp 161°C). Acidification of the aqueous phase with 0.1 M HCl or another strong acid converts sodium 4-methylphenoxide to 4-methylphenol, which is insoluble in water and can be extracted with ether and recovered in pure form. The following flowchart summarizes these experimental steps:

CHEMICAL CONNECTIONS 9B

Capsaicin, for Those Who Like It Hot

Capsaicin, the pungent principle from the fruit of various peppers (*Capsicum* and Solanaceae), was isolated in 1876, and its structure was determined in 1919:

Capsaicin
(from various types of peppers)

The inflammatory properties of capsaicin are well known; the human tongue can detect as little as one drop of it in 5 L of water. Many of us are familiar with the burning sensation in the mouth and sudden tearing in the eyes caused by a good dose of hot chili peppers. Capsaicin-containing extracts from these flaming foods are also used in sprays to ward off dogs or other animals that might nip at your heels while you are running or cycling.

Ironically, capsaicin is able to cause pain and relieve it as well. Currently, two capsaicin-containing creams, Mioton and Zostrix®, are prescribed to treat the burning pain associated with postherpetic neuralgia, a complication of shingles. They are also prescribed for diabetics, to relieve persistent foot and leg pain.

The mechanism by which capsaicin relieves pain is not fully understood. It has been suggested that, after it is applied, the nerve endings in the area responsible for the transmission of pain remain temporarily numb. Capsaicin remains bound to specific receptor sites on these pain-transmitting neurons, blocking them from further action. Eventually, capsaicin is removed from the receptor sites, but in the meantime, its presence provides needed relief from pain.

QUESTIONS

Would you predict capsaicin to be more soluble in water or more soluble in 1-octanol?

Would your prediction remain the same if capsaicin were first treated with a molar equivalent of NaOH?

D. Phenols as Antioxidants

An important reaction in living systems, foods, and other materials that contain carbon–carbon double bonds is **autoxidation**—that is, oxidation requiring oxygen and no other reactant. If you open a bottle of cooking oil that has stood for a long time, you will notice a hiss of air entering the bottle. This sound occurs because the consumption of oxygen by autoxidation of the oil creates a negative pressure inside the bottle.

Cooking oils contain esters of polyunsaturated fatty acids. You need not worry now about what esters are; we will discuss them in Chapter 15. The important point here is that all vegetable oils contain fatty acids with long hydrocarbon chains, many of which have one or more carbon–carbon double bonds. (See Problem 4.49 for the structures of three of these fatty acids.) Autoxidation takes place at a carbon adjacent to a double bond—that is, at an **allylic carbon**.

Autoxidation is a radical chain process that converts an R—H group into an R—O—O—H group, called a *hydroperoxide*. The process begins when energy in the form of heat or light causes a molecule with a weak bond to form two **radicals**, atoms, or molecules with an unpaired electron. This step is known as **chain initiation**. In the laboratory, small amounts of compounds such as peroxides, ROOR, are used as initiators because they are easily converted to RO· radicals by light or heat. Scientists are still unsure precisely what compounds act as initiators in nature. Once a radical is generated, it reacts with a molecule by removing the hydrogen atom together with one of its electrons (H·) from an allylic carbon. The carbon losing the H· now has only seven electrons in its valence shell, one of which is unpaired.

Step 1: *Chain Initiation—Formation of a Radical from a Nonradical Compound*
The radical generated from the exposure of the initiator to light or heat causes the removal of a hydrogen atom (H·) adjacent to a C=C double bond to give an allylic radical:

section of a fatty acid hydrocarbon chain

$$\text{RO—OR} \xrightarrow[\text{or heat}]{\text{light}} 2\,\text{RO}\cdot \xrightarrow{\text{—CH}_2\text{CH=CH—CH(H)—}} \text{—CH}_2\text{CH=CH—}\dot{\text{C}}\text{H—} + \text{ROH}$$

Radical initiator — Radical — An allylic radical

Step 2a: *Chain Propagation—Reaction of a Radical to Form a New Radical*
The allylic radical reacts with oxygen, itself a diradical, to form a hydroperoxy radical. The new covalent bond of the hydroperoxy radical forms by the combination of one electron from the allylic radical and one electron from the oxygen diradical:

$$\text{—CH}_2\text{CH=CH—}\dot{\text{C}}\text{H—} + \cdot\text{O—O}\cdot \longrightarrow \text{—CH}_2\text{CH=CH—CH(O—O}\cdot\text{)—}$$

Oxygen is a diradical — A hydroperoxy radical

Step 2b: *Chain Propagation—Reaction of a Radical to Form a New Radical*
The hydroperoxy radical removes an allylic hydrogen atom (H·) from a new fatty acid hydrocarbon chain to complete the formation of a hydroperoxide and, at the same time, produce a new allylic radical:

$$\text{—CH}_2\text{CH=CH—CH(O—O}\cdot\text{)—} + \text{—CH}_2\text{CH=CH—CH(H)—} \longrightarrow$$

Section of a new fatty acid hydrocarbon chain

O=O (a) ·O—O· (b)

Although we represent molecular oxygen with the Lewis structure shown in (a), oxygen has long been known to exist and behave as a diradical, as shown in (b).

$$-CH_2CH{=}CH-\underset{|}{CH}(O-O-H)-CH_2- \quad + \quad -CH_2CH{=}CH-\dot{C}H-$$

A hydroperoxide   A new allylic radical

The most important point about the pair of chain propagation steps is that they form a continuous cycle of reactions. The new radical formed in Step 2b next reacts with another molecule of O_2 in Step 2a to give a new hydroperoxy radical, which then reacts with a new hydrocarbon chain to repeat Step 2b, and so forth. This cycle of propagation steps repeats over and over in a chain reaction. Thus, once a radical is generated in Step 1, the cycle of propagation steps may repeat many thousands of times, generating thousands and thousands of hydroperoxide molecules. The number of times the cycle of chain propagation steps repeats is called the **chain length**.

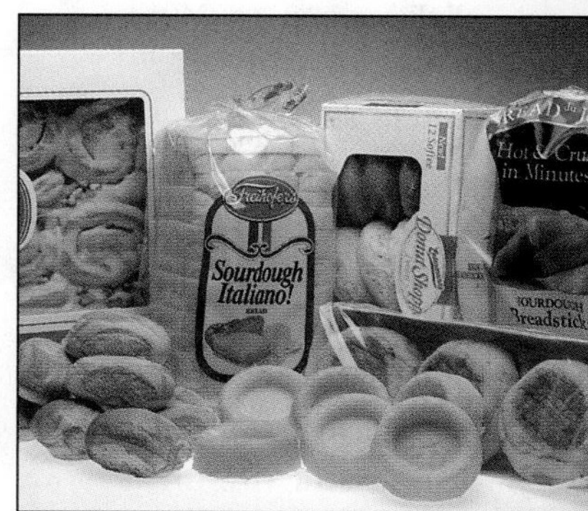

Butylated hydroxytoluene (BHT) is often used as an antioxidant in baked goods to "retard spoilage." *(Charles D. Winters)*

Hydroperoxides themselves are unstable and, under biological conditions, degrade to short-chain aldehydes and carboxylic acids with unpleasant "rancid" smells. These odors may be familiar to you if you have ever smelled old cooking oil or aged foods that contain polyunsaturated fats or oils. A similar formation of hydroperoxides in the low-density lipoproteins deposited on the walls of arteries leads to cardiovascular disease in humans. In addition, many effects of aging are thought to be the result of the formation and subsequent degradation of hydroperoxides.

Fortunately, nature has developed a series of defenses, including the phenol vitamin E, ascorbic acid (vitamin C), and glutathione, against the formation of destructive hydroperoxides. The compounds that defend against hydroperoxides are "nature's scavengers." Vitamin E, for example, inserts itself into either Step 2a or 2b, donates an H· from its phenolic —OH group to the allylic radical, and converts the radical to its original hydrocarbon chain. Because the vitamin E radical is stable, it breaks the cycle of chain propagation steps, thereby preventing the further formation of destructive hydroperoxides. While some hydroperoxides may form, their numbers are very small and they are easily decomposed to harmless materials by one of several enzyme-catalyzed reactions.

Unfortunately, vitamin E is removed in the processing of many foods and food products. To make up for this loss, phenols such as BHT and BHA are added to foods to "retard [their] spoilage" (as they say on the packages) by autoxidation:

[Structures: Vitamin E (HO, O, subscript 3, H); Butylated hydroxytoluene (OH); Butylated hydroxyanisole (OH, OCH_3)]

Vitamin E

*B*utylated *h*ydroxy*t*oluene (BHT)

*B*utylated *h*ydroxy*a*nisole (BHA)

Similar compounds are added to other materials, such as plastics and rubber, to protect them against autoxidation. The protective properties of phenols may explain why the health benefits of foods such as green tea, wine, and blueberries (each of which contains large amounts of phenolic compounds) have been lauded by nutritionists and others in the medical community.

Key Terms and Concepts

acidity of phenols (p. 335)
activating (p. 326)
acylium ion (p. 322)
allylic carbon (p. 340)
aniline (p. 310)
anisole (p. 310)
Ar— (p. 302)
arene (p. 302)
aromatic compound (p. 302)
aryl group (p. 302)
autoxidation (p. 340)
benzaldehyde (p. 310)
benzoic acid (p. 310)
benzyl group (p. 310)
benzylic carbon (p. 313)
carcinogen (p. 313)
deactivating (p. 326)
electrophilic aromatic substitution (p. 316)
Friedel–Crafts acylation (p. 322)
Friedel–Crafts alkylation (p. 319)
heterocyclic compound (p. 306)
Kekulé structure (p. 302)
meta (p. 311)
meta directing (p. 326)
molecular orbital model of benzene (p. 303)
nitronium ion (p. 318)
ortho (p. 311)
ortho–para directing (p. 326)
para (p. 311)
phenol (p. 334)
phenyl (p. 310)
polynuclear aromatic hydrocarbons (p. 312)
radical (p. 340)
resonance energy (p. 304)
resonance energy benzene (p. 304)
styrene (p. 310)
toluene (p. 310)
xylene (p. 311)

Summary of Key Questions

9.1 What Is the Structure of Benzene?

- Benzene is a molecule with a high degree of unsaturation possessing the molecular formula C_6H_6. Each carbon has a single unhybridized $2p$ orbital that contains one electron. These six $2p$ orbitals lie perpendicular to the plane of the ring and overlap to form a continuous pi cloud encompassing all six carbons.
- Benzene and its alkyl derivatives are classified as **aromatic hydrocarbons,** or **arenes**.

9.2 What Is Aromaticity?

- According to the Hückel criteria for aromaticity, a five- or six-membered ring is aromatic if it (1) has one *p* orbital on each atom of the ring, (2) is planar so that overlap of all *p* orbitals of the ring is continuous or nearly so, and (3) has 6, 10, 14, etc. pi electrons in the overlapping system of *p* orbitals.
- A **heterocyclic aromatic compound** contains one or more atoms other than carbon in an aromatic ring.

9.3 How Are Benzene Compounds Named, and What Are Their Physical Properties?

- Aromatic compounds are named by the IUPAC system. The common names toluene, xylene, styrene, phenol, aniline, benzaldehyde, and benzoic acid are retained.
- The $C_6H_5—$ group is named **phenyl**, and the $C_6H_5CH_2—$ group is named **benzyl**.
- To locate two substituents on a benzene ring, either number the atoms of the ring or use the locators **ortho (*o*), meta (*m*),** and **para (*p*)**.
- **Polynuclear aromatic hydrocarbons** contain two or more fused benzene rings.

9.4 What Is the Benzylic Position, and How Does It Contribute to Benzene Reactivity?

- The **benzylic position** is the carbon of an alkyl substituent immediately bonded to the benzene ring.
- The benzylic position of a benzene ring can be oxidized by chromic acid without affecting any of the benzene ring atoms.

9.5 What Is Electrophilic Aromatic Substitution?

- A characteristic reaction of aromatic compounds is **electrophilic aromatic substitution**, which involves the substitution of one of the ring hydrogens of benzene for an electrophilic reagent.
- The five types of electrophilic aromatic substitution discussed here are nitration, halogenation, sulfonation, Friedel–Crafts alkylation, and Friedel–Crafts acylation.

9.6 What Is the Mechanism of Electrophilic Aromatic Substitution?

- The mechanism of electrophilic aromatic substitution can be broken down into three common steps: (1) generation of the electrophile, (2) attack of the electrophile on the aromatic ring to give a resonance-stabilized cation intermediate, and (3) proton transfer to a base to regenerate the aromatic ring.
- The five electrophilic aromatic substitution reactions studied here differ in their mechanism of formation of the electrophile (Step 1) and the specific base used to effect the proton transfer to regenerate the aromatic ring (Step 3).

9.7 How Do Existing Substituents on Benzene Affect Electrophilic Aromatic Substitution?

- Substituents on an aromatic ring influence both the rate and site of further substitution.
- Substituent groups that direct an incoming group preferentially to the ortho and para positions are called **ortho–para directors**. Those that direct an incoming group preferentially to the meta positions are called **meta directors**.
- **Activating groups** cause the rate of further substitution to be faster than that for benzene; **deactivating groups** cause it to be slower than that for benzene.
- A mechanistic rationale for directing effects is based on the degree of resonance stabilization of the possible cation intermediates formed upon reaction of the aromatic ring and the electrophile.
- Groups that stabilize the cation intermediate are ortho–para directors; groups that destabilize it are deactivators and meta directors.

9.8 What Are Phenols?

- The functional group of a **phenol** is an —OH group bonded to a benzene ring.
- Phenol and its derivatives are weak acids, with pK_a approximately 10.0, but are considerably stronger acids than alcohols, with pK_a 16–18.
- Various phenols are used to prevent autoxidation, a radical chain process that converts an R—H group into an R—O—O—H group and causes spoilage in foods.

Quick Quiz

Answer true or false to the following questions to assess your general knowledge of the concepts in this chapter. If you have difficulty with any of them, you should review the appropriate section in the chapter (shown in parentheses) before attempting the more challenging end-of-chapter problems.

1. The mechanism of electrophilic aromatic substitution involves three steps: generation of the electrophile, attack of the electrophile on the benzene ring, and proton transfer to regenerate the ring. (9.6)
2. The C—C double bonds in benzene do not undergo the same addition reactions that the C═C double bonds in alkenes undergo. (9.1)
3. Friedel–Crafts acylation is not subject to rearrangements. (9.5)
4. An aromatic compound is planar, possesses a 2*p* orbital on every atom of the ring, and contains either 4, 8, 12, 16, and so on, pi electrons. (9.2)

5. When naming disubstituted benzenes, the locators para, meta, and ortho refer to substituents that are 1,2, 1,3, and 1,4, respectively. (9.3)

6. The electrophile in the chlorination or bromination of benzene is an ion pair containing a chloronium or bromonium ion. (9.6)

7. An ammonium group ($-NH_3^+$) on a benzene ring will direct an attacking electrophile to a meta position. (9.7)

8. Reaction of chromic acid, H_2CrO_4, with a substituted benzene always oxidizes every alkyl group at the benzylic position to a new carboxyl group. (9.4)

9. Benzene consists of two resonance structures that rapidly interconvert between each other. (9.1)

10. The electrophile in the nitration of benzene is the nitrate ion. (9.6)

11. A benzene ring with an $-OH$ bonded to it is referred to as "phenyl." (9.3)

12. Friedel–Crafts alkylation of a primary haloalkane with benzene will always result in a new bond between benzene and the carbon that was bonded to the halogen. (9.5)

13. Resonance energy is the energy a ring contains due to the stability afforded it by its resonance structures. (9.1)

14. A phenol will react quantitatively with NaOH. (9.8)

15. The use of a haloalkane and $AlCl_3$ is the only way to synthesize an alkylbenzene. (9.6)

16. Phenols are more acidic than alcohols. (9.8)

17. Substituents of polysubstituted benzene rings can be numbered according to their distance from the substituent that imparts a special name to the compound. (9.3)

18. If a benzene ring contains both a weakly activating group and a strongly deactivating group, the strongly deactivating group will direct the attack of an electrophile. (9.7)

19. Oxygen, O_2, can be considered a diradical. (9.8)

20. The contributing structures for the attack of an electrophile to the ortho position of aniline are more stable than those for the attack at the meta position. (9.7)

21. A deactivating group will cause its benzene ring to react slower than benzene itself. (9.7)

22. Friedel–Crafts alkylation is promoted by the presence of electron-withdrawing groups. (9.5)

23. Autooxidation takes place at allylic carbons. (9.8)

24. The contributing structures for the attack of an electrophile to the meta position of nitrobenzene are more stable than those for the attack at the ortho or para position. (9.7)

Answers: (1) T (2) T (3) T (4) F (5) F (6) T (7) T (8) F (9) F (10) F (11) F (12) F (13) F (14) T (15) F (16) T (17) T (18) F (19) T (20) T (21) T (22) F (23) T (24) T

Key Reactions

1. Oxidation at a Benzylic Position (Section 9.4)
A benzylic carbon bonded to at least one hydrogen is oxidized to a carboxyl group:

$$H_3C-C_6H_4-CH(CH_3)_2 \xrightarrow[H_2SO_4]{K_2Cr_2O_7} HOOC-C_6H_4-COOH$$

2. Chlorination and Bromination (Section 9.6A)
The electrophile is a halonium ion, Cl^+ or Br^+, formed by treating Cl_2 or Br_2 with $AlCl_3$ or $FeCl_3$:

$$C_6H_6 + Cl_2 \xrightarrow{AlCl_3} C_6H_5-Cl + HCl$$

3. Nitration (Section 9.6B)
The electrophile is the nitronium ion, NO_2^+, formed by treating nitric acid with sulfuric acid:

$$C_6H_5Br + HNO_3 \xrightarrow{H_2SO_4} o\text{-}BrC_6H_4NO_2 + p\text{-}BrC_6H_4NO_2 + H_2O$$

4. Sulfonation (Section 9.6B)
The electrophile is HSO_3^+:

$$C_6H_6 + H_2SO_4 \longrightarrow C_6H_5{-}SO_3H + H_2O$$

5. Friedel–Crafts Alkylation (Section 9.6C)
The electrophile is an alkyl carbocation formed by treating an alkyl halide with a Lewis acid:

$$C_6H_6 + (CH_3)_2CHCl \xrightarrow{AlCl_3} C_6H_5{-}CH(CH_3)_2 + HCl$$

6. Friedel–Crafts Acylation (Section 9.6D)
The electrophile is an acyl cation formed by treating an acyl halide with a Lewis acid:

$$C_6H_6 + CH_3\overset{O}{\overset{\|}{C}}Cl \xrightarrow{AlCl_3} C_6H_5{-}\overset{O}{\overset{\|}{C}}CH_3 + HCl$$

7. Alkylation Using an Alkene (Section 9.6E)
The electrophile is a carbocation formed by treating an alkene with H_2SO_4 or H_3PO_4:

$$\text{4-}CH_3C_6H_4OH + 2\ CH_3\overset{CH_3}{\overset{|}{C}}{=}CH_2 \xrightarrow{H_3PO_4} \text{2,6-}((CH_3)_3C)_2\text{-4-}CH_3C_6H_2OH$$

8. Alkylation Using an Alcohol (Section 9.6E)
The electrophile is a carbocation formed by treating an alcohol with H_2SO_4 or H_3PO_4:

$$C_6H_6 + (CH_3)_3COH \xrightarrow{H_3PO_4} C_6H_5{-}C(CH_3)_3 + H_2O$$

9. Acidity of Phenols (Section 9.8B)
Phenols are weak acids:

$$C_6H_5{-}OH + H_2O \rightleftharpoons C_6H_5{-}O^- + H_3O^+ \qquad K_a = 1.1 \times 10^{-10} \quad pK_a = 9.95$$

Phenol; Phenoxide ion

Substitution by electron-withdrawing groups, such as the halogens and the nitro group, increases the acidity of phenols.

10. Reaction of Phenols with Strong Bases (Section 9.8C)
Water-insoluble phenols react quantitatively with strong bases to form water-soluble salts:

$$C_6H_5{-}OH + NaOH \longrightarrow C_6H_5{-}O^-Na^+ + H_2O$$

Phenol pK_a 9.95 (stronger acid); Sodium hydroxide (stronger base); Sodium phenoxide (weaker base); Water pK_a 15.7 (weaker acid)

Problems

A problem marked with an asterisk indicates an applied "real-world" problem. Answers to problems whose numbers are printed in blue are given in Appendix D.

Section 9.2 Aromaticity

9.11 Which of the following compounds or chemical entities are aromatic? **(See Example 9.1)**

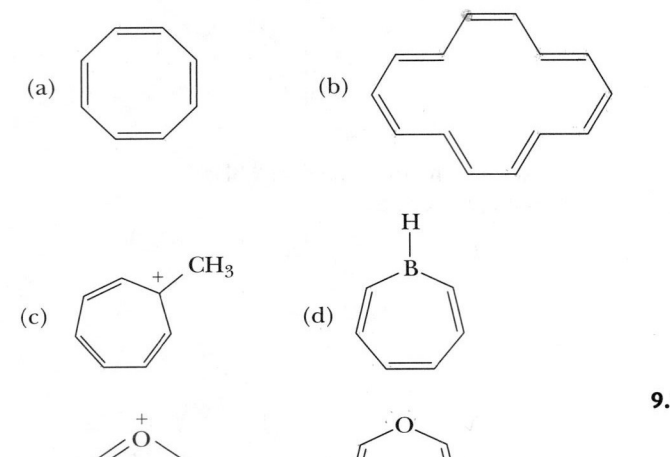

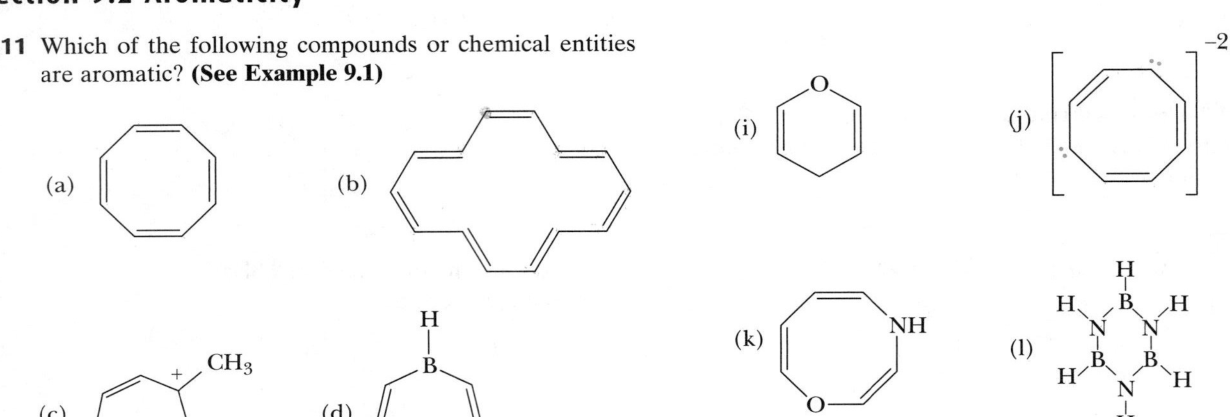

(e) 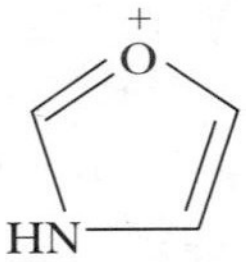 HN, O^+ (f) 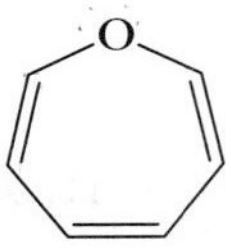 O (g) 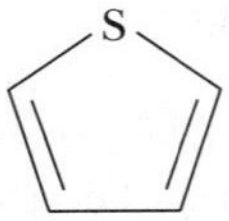 S (h) 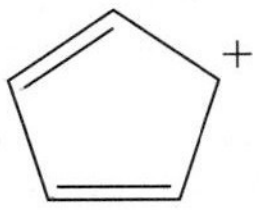 +

9.12 Explain why cyclopentadiene (pK_a 16) is many orders of magnitude more acidic than cyclopentane (pK_a > 50). (*Hint:* Draw the structural formula for the anion formed by removing one of the protons on the $-CH_2-$ group, and then apply the Hückel criteria for aromaticity.)

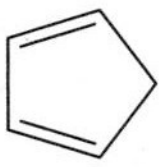

Cyclopentadiene 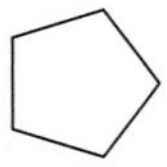 Cyclopentane

Section 9.3 Nomenclature and Structural Formulas

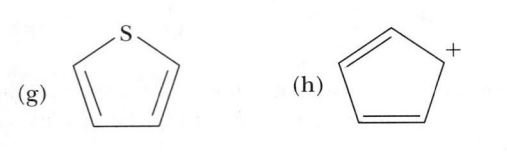

9.13 Name these compounds: **(See Example 9.2)**

(a)  NO_2, Cl

(b) 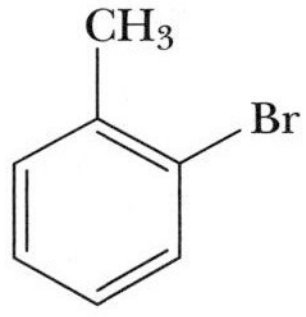 CH_3, Br

(c) 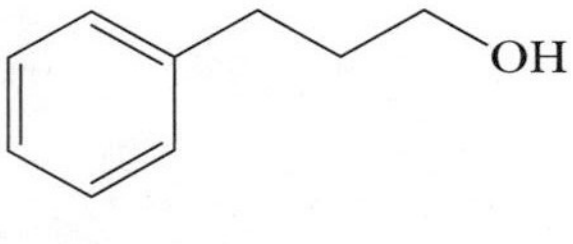 OH

(d) 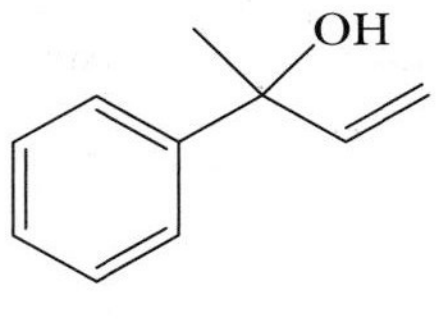 OH

(e) 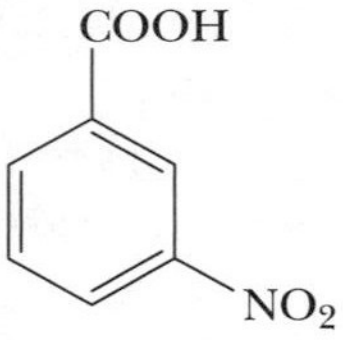 COOH, NO_2

(f) 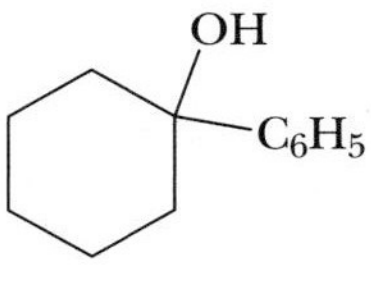 OH, C_6H_5

(g) 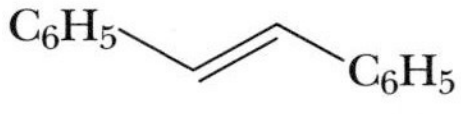 C_6H_5, C_6H_5

(h) 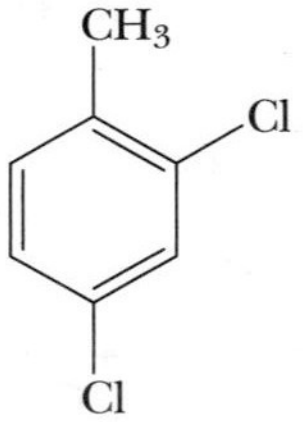 CH_3, Cl, Cl

(i) 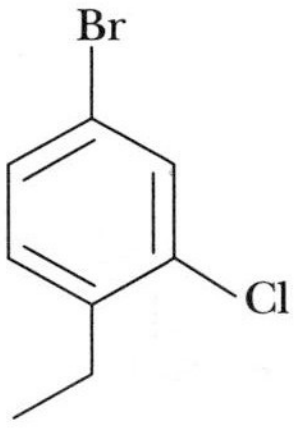 Br, Cl

(j) 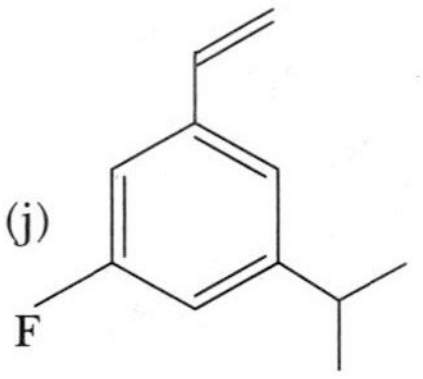 F

9.14 Draw structural formulas for these compounds: **(See Example 9.2)**

(a) 1-Bromo-2-chloro-4-ethylbenzene
(b) 4-Iodo-1,2-dimethylbenzene
(c) 2,4,6-Trinitrotoluene
(d) 4-Phenyl-2-pentanol
(e) *p*-Cresol
(f) 2,4-Dichlorophenol
(g) 1-Phenylcyclopropanol
(h) Styrene (phenylethylene)
(i) *m*-Bromophenol
(j) 2,4-Dibromoaniline
(k) Isobutylbenzene
(l) *m*-Xylene
(m) 4-Bromo-1,2-dichlorobenzene
(n) 5-Fluoro-2-methylphenol
(o) 1-Cyclohexyl-3-ethylbenzene
(p) *m*-Phenylaniline
(q) 3-Methyl-2-vinylbenzoic acid
(r) 2,5-Dimethylanisole

9.15 Show that pyridine can be represented as a hybrid of two equivalent contributing structures.

9.16 Show that naphthalene can be represented as a hybrid of three contributing structures. Show also, by the use of curved arrows, how one contributing structure is converted to the next.

9.17 Draw four contributing structures for anthracene.

Section 9.5 Electrophilic Aromatic Substitution: Monosubstitution

9.18 Draw a structural formula for the compound formed by treating benzene with each of the following combinations of reagents: **(See Examples 9.5, 9.6)**

(a) $CH_3CH_2Cl/AlCl_3$ (b) $CH_2{=}CH_2/H_2SO_4$
(c) CH_3CH_2OH/H_2SO_4

9.19 Show three different combinations of reagents you might use to convert benzene to isopropylbenzene. **(See Examples 9.5, 9.6)**

9.20 How many monochlorination products are possible when naphthalene is treated with $Cl_2/AlCl_3$?

9.21 Write a stepwise mechanism for the following reaction, using curved arrows to show the flow of electrons in each step: **(See Example 9.4)**

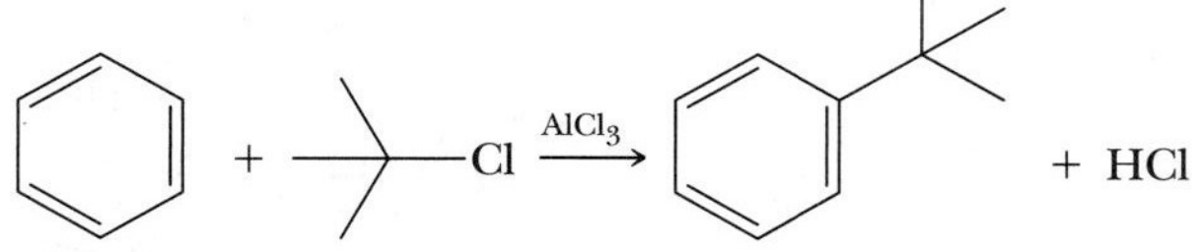

benzene + (CH₃)₃C–Cl $\xrightarrow{AlCl_3}$ *tert*-butylbenzene + HCl

9.22 Write a stepwise mechanism for the preparation of diphenylmethane by treating benzene with dichloromethane in the presence of an aluminum chloride catalyst. **(See Example 9.4)**

9.23 The following alkylation reactions do not yield the compounds shown as the major product. Predict the major product for each reaction and provide a mechanism for their formation.

(a) benzene + 2-chloro-3-methylbutane (Cl) $\xrightarrow{AlCl_3}$ [crossed-out product] + HCl

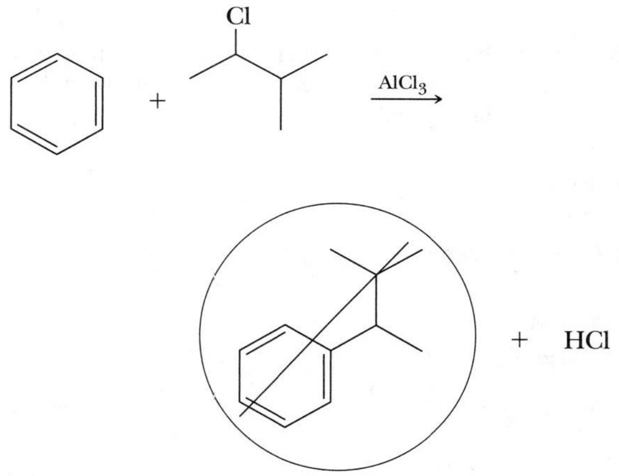

(b) benzene + 3,3-dimethyl-1-butene $\xrightarrow{H_2SO_4}$ [crossed-out product] + HCl

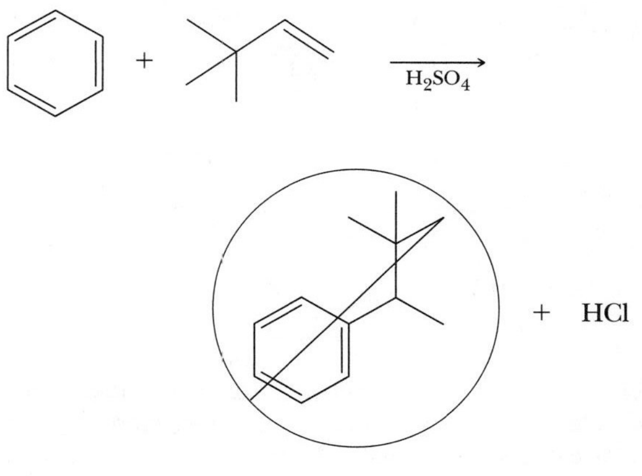

(c) 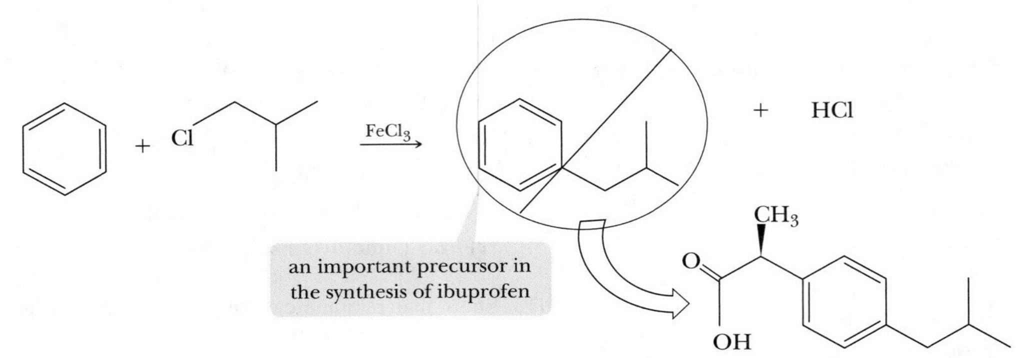 [benzene] + Cl–CH$_2$CH(CH$_3$)$_2$ $\xrightarrow{FeCl_3}$ [isobutylbenzene] + HCl

an important precursor in the synthesis of ibuprofen

CH$_3$ O OH

Ibuprofen

Section 9.7 Electrophilic Aromatic Substitution: Substitution Effects

9.24 When treated with $Cl_2/AlCl_3$, 1,2-dimethylbenzene (*o*-xylene) gives a mixture of two products. Draw structural formulas for these products. **(See Examples 9.7, 9.9)**

9.25 How many monosubstitution products are possible when 1,4-dimethylbenzene (*p*-xylene) is treated with $Cl_2/AlCl_3$? When *m*-xylene is treated with $Cl_2/AlCl_3$? **(See Examples 9.7, 9.9)**

9.26 Draw the structural formula for the major product formed upon treating each compound with $Cl_2/AlCl_3$: **(See Examples 9.7, 9.9)**

(a) Toluene (b) Nitrobenzene

(c) Chlorobenzene (d) *tert*-Butylbenzene

(e) 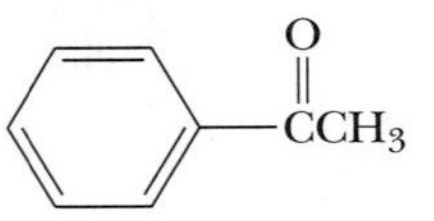 C_6H_5–$\overset{O}{\overset{\|}{C}}CH_3$ (f) 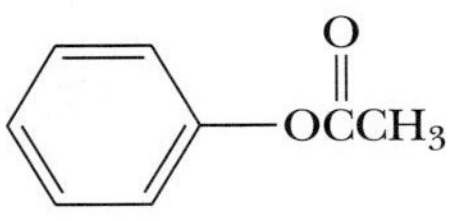 C_6H_5–$O\overset{O}{\overset{\|}{C}}CH_3$

(g) 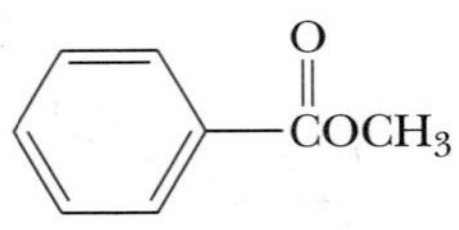 C_6H_5–$\overset{O}{\overset{\|}{C}}OCH_3$ (h) 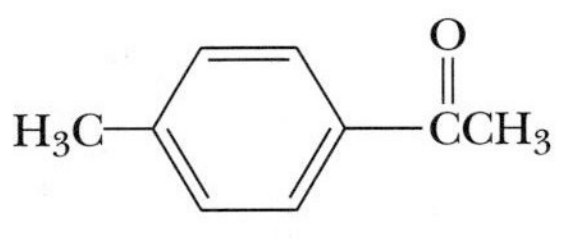 H_3C–C_6H_4–$\overset{O}{\overset{\|}{C}}CH_3$

(i) 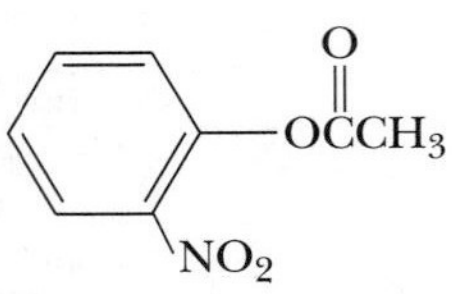 $C_6H_4(NO_2)$–$O\overset{O}{\overset{\|}{C}}CH_3$

9.27 Which compound, chlorobenzene or toluene, undergoes electrophilic aromatic substitution more rapidly when treated with $Cl_2/AlCl_3$? Explain and draw structural formulas for the major product(s) from each reaction.

9.28 Arrange the compounds in each set in order of decreasing reactivity (fastest to slowest) toward electrophilic aromatic substitution:

(a) 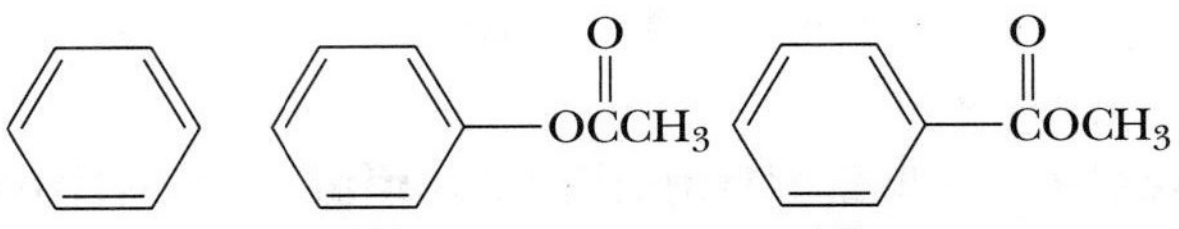 (A) benzene (B) C_6H_5–$O\overset{O}{\overset{\|}{C}}CH_3$ (C) C_6H_5–$\overset{O}{\overset{\|}{C}}OCH_3$

(b) (A) C_6H_5–NO_2 (B) C_6H_5–COOH (C) benzene

(c) (A) C_6H_5–NH_2 (B) C_6H_5–$NH\overset{O}{\overset{\|}{C}}CH_3$ (C) C_6H_5–$\overset{O}{\overset{\|}{C}}NHCH_3$

(d) (A) benzene (B) C_6H_5–CH_3 (C) C_6H_5–OCH_3

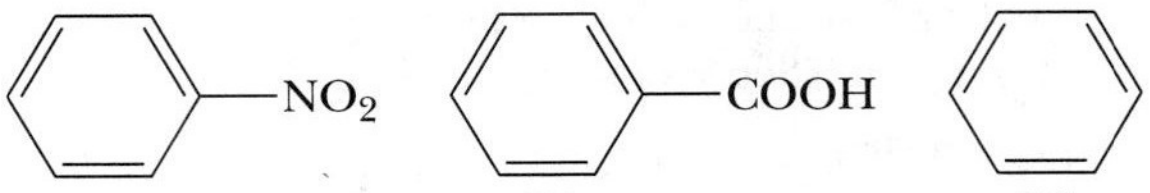

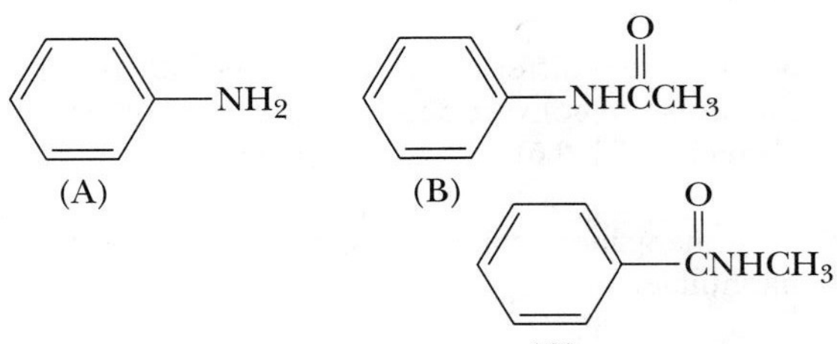

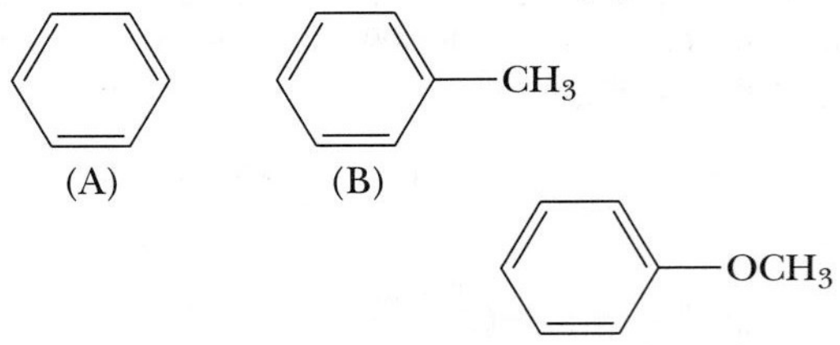

9.29 Account for the observation that the trifluoromethyl group is meta directing, as shown in the following example: **(See Example 9.8)**

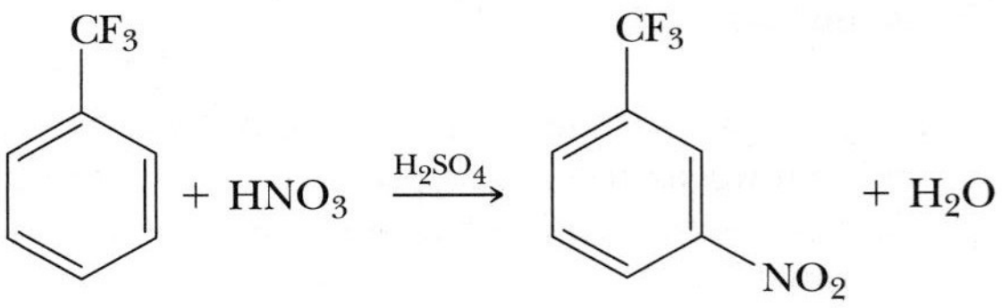

$$C_6H_5CF_3 + HNO_3 \xrightarrow{H_2SO_4} m\text{-}CF_3C_6H_4NO_2 + H_2O$$

9.30 Show how to convert toluene to these carboxylic acids: **(See Example 9.3)**

(a) 4-Chlorobenzoic acid (b) 3-Chlorobenzoic acid

9.31 Show reagents and conditions that can be used to bring about these conversions: **(See Examples 9.7, 9.9)**

(a) CH_3 (toluene) → CH_3, $C(=O)CH_2CH_3$ (para)

(b) OH (phenol) → OH, O_2N, NO_2

(c) OCH_3 → OCH_3, CCH_3 with =O (para)

(d) O-acetyl benzene → O-acetyl, Cl, NO_2

Go back & do - like prob 9.22

9.32 Propose a synthesis of triphenylmethane from benzene as the only source of aromatic rings. Use any other necessary reagents. **(See Examples 9.7, 9.9)**

*__9.33__ Reaction of phenol with acetone in the presence of an acid catalyst gives bisphenol A, a compound used in the production of polycarbonate and epoxy resins (Sections 17.4C and 17.4E): **(See Example 9.6)**

2 (phenyl)–OH + $CH_3\overset{O}{\overset{\|}{C}}CH_3$ $\xrightarrow{H_3PO_4}$

Acetone

HO–(phenylene)–C(CH_3)(CH_3)–(phenylene)–OH + H_2O

Bisphenol A

Propose a mechanism for the formation of bisphenol A. (*Hint:* The first step is a proton transfer from phosphoric acid to the oxygen of the carbonyl group of acetone.)

*__9.34__ 2,6-Di-*tert*-butyl-4-methylphenol, more commonly known as butylated hydroxytoluene, or BHT, is used as an antioxidant in foods to "retard spoilage." BHT is synthesized industrially from 4-methylphenol (*p*-cresol) by reaction with 2-methylpropene in the presence of phosphoric acid: **(See Example 9.6)**

OH (4-Methylphenol) + 2 (2-Methylpropene) $\xrightarrow{H_3PO_4}$

4-Methylphenol 2-Methylpropene

OH

2,6-Di-*tert*-butyl-4-methylphenol
(Butylated hydroxytoluene, BHT)

Propose a mechanism for this reaction.

*__9.35__ The first herbicide widely used for controlling weeds was 2,4-dichlorophenoxyacetic acid (2,4-D). Show how this compound might be synthesized from 2,4-dichlorophenol and chloroacetic acid, $ClCH_2COOH$:

Cl–(ring, Cl)–OH → Cl–(ring, Cl)–$OCH_2\overset{O}{\overset{\|}{C}}OH$

2,4-Dichlorophenol 2,4-Dichlorophenoxyacetic acid (2,4-D)

Section 9.8 Acidity of Phenols

9.36 Use the resonance theory to account for the fact that phenol (pK_a 9.95) is a stronger acid than cyclohexanol (pK_a 18). **(See Example 9.10)**

9.37 Arrange the compounds in each set in order of increasing acidity (from least acidic to most acidic): **(See Example 9.10)**

(a) C_6H_5–OH C_6H_{11}–OH CH_3COOH

(b) C_6H_5–OH $NaHCO_3$ H_2O

(c) O_2N–C_6H_4–OH C_6H_5–OH C_6H_5–CH_2OH

9.38 From each pair, select the stronger base: **(See Example 9.10)**

(a) C_6H_5–O^- or OH^-

(b) C_6H_5–O^- or C_6H_{11}–O^-

(c) C_6H_5–O^- or HCO_3^-

(d) C_6H_5–O^- or CH_3COO^-

9.39 Account for the fact that water-insoluble carboxylic acids (pK_a 4–5) dissolve in 10% sodium bicarbonate with the evolution of a gas, but water-insoluble phenols (pK_a 9.5–10.5) do not show this chemical behavior.

9.40 Describe a procedure for separating a mixture of 1-hexanol and 2-methylphenol (*o*-cresol) and recovering each in pure form. Each is insoluble in water, but soluble in diethyl ether.

Syntheses

9.41 Using styrene, $C_6H_5CH{=}CH_2$, as the only aromatic starting material, show how to synthesize these compounds. In addition to styrene, use any other necessary organic or inorganic chemicals. Any compound synthesized in one part of this problem may be used to make any other compound in the problem:

(a) C_6H_5–COOH (C=O)   (b) C_6H_5–CH(Br)CH_3

(c) C_6H_5–CH(OH)CH_3 (d) C_6H_5–C(=O)CH_3

(e) C_6H_5–CH_2CH_3 (f) C_6H_5–CH(OH)CH_2OH

9.42 Show how to synthesize these compounds, starting with benzene, toluene, or phenol as the only sources of aromatic rings. Assume that, in all syntheses, you can separate mixtures of ortho–para products to give the desired isomer in pure form: **(See Examples 9.7, 9.9)**

(a) *m*-Bromonitrobenzene

(b) 1-Bromo-4-nitrobenzene

(c) 2,4,6-Trinitrotoluene (TNT)

(d) *m*-Bromobenzoic acid

(e) *p*-Bromobenzoic acid

(f) *p*-Dichlorobenzene

(g) *m*-Nitrobenzenesulfonic acid

(h) 1-Chloro-3-nitrobenzene

9.43 Show how to synthesize these aromatic ketones, starting with benzene or toluene as the only sources of aromatic rings. Assume that, in all syntheses, mixtures of ortho–para products can be separated to give the desired isomer in pure form: **(See Examples 9.7, 9.9)**

(a) 4-methylacetophenone (O)   (b) 4-bromoacetophenone (O, Br)

(c) 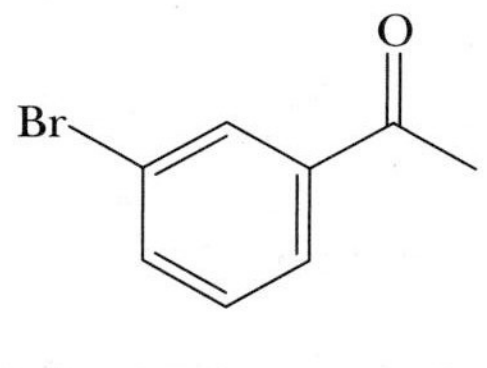 Br O

***9.44** The following ketone, isolated from the roots of several members of the iris family, has an odor like that of violets and is used as a fragrance in perfumes. Describe the synthesis of this ketone from benzene. **(See Examples 9.7, 9.9)**

4-Isopropylacetophenone

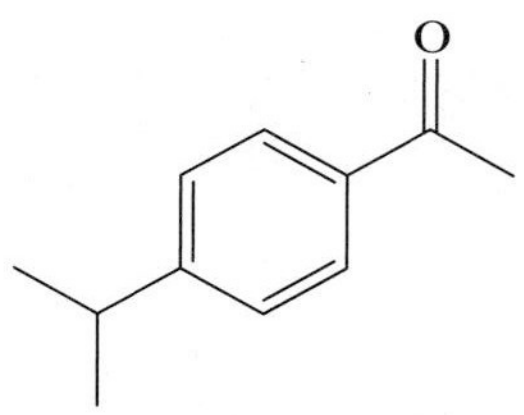

***9.45** The bombardier beetle generates p-quinone, an irritating chemical, by the enzyme-catalyzed oxidation of hydroquinone, using hydrogen peroxide as the oxidizing agent. Heat generated in this oxidation produces superheated steam, which is ejected, along with p-quinone, with explosive force.

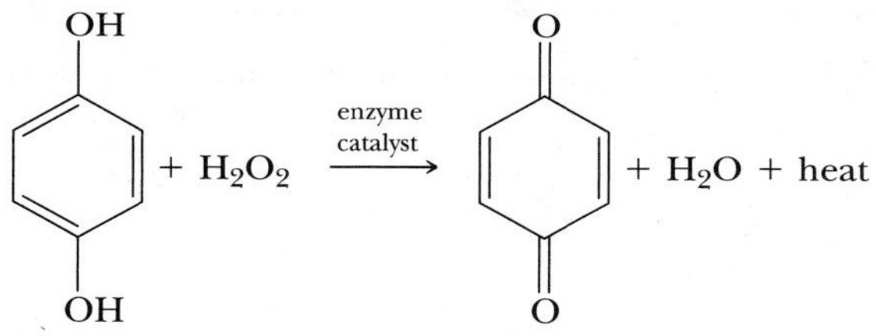

Hydroquinone + H_2O_2 $\xrightarrow{\text{enzyme catalyst}}$ p-Quinone + H_2O + heat

(a) Balance the equation.

(b) Show that this reaction of hydroquinone is an oxidation.

***9.46** Following is a structural formula for musk ambrette, a synthetic musk used in perfumes to enhance and retain fragrance: **(See Examples 9.7, 9.9)**

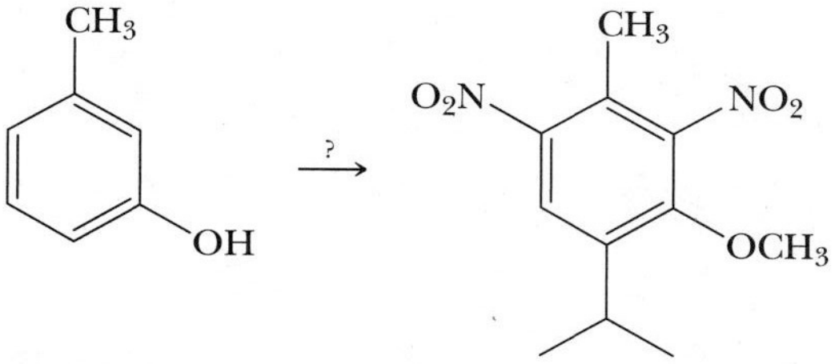

m-Cresol $\xrightarrow{?}$ Musk ambrette

Propose a synthesis for musk ambrette from m-cresol.

***9.47** 1-(3-Chlorophenyl)propanone is a building block in the synthesis of bupropion, the hydrochloride salt of which is the antidepressant Wellbutrin. During clinical trials, researchers discovered that smokers reported a diminished craving for tobacco after one to two weeks on the drug. Further clinical trials confirmed this finding, and the drug is also marketed under the trade name Zyban® as an aid in smoking cessation. Propose a synthesis for this building block from benzene. (We will see in Section 13.9 how to complete the synthesis of bupropion.) **(See Examples 9.7, 9.9)**

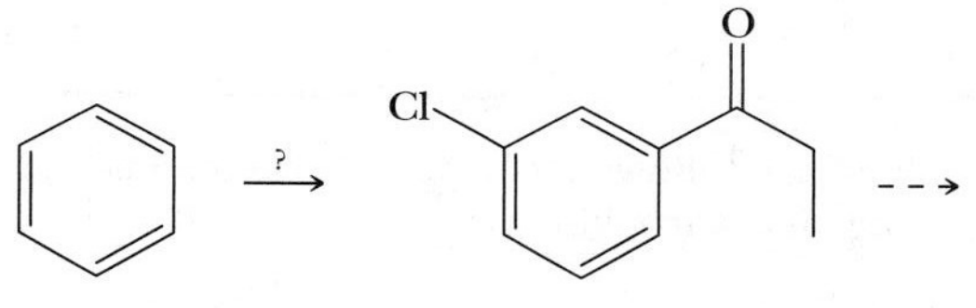

Benzene $\xrightarrow{?}$ 1-(3-Chlorophenyl)-1-propanone - - →

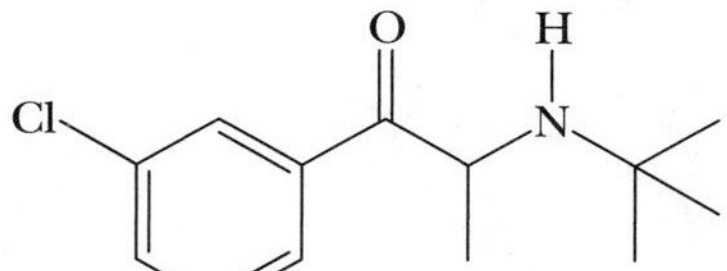

Bupropion (Wellbutrin, Zyban)

Chemical Transformations

9.48 Test your cumulative knowledge of the reactions learned thus far by completing the following chemical transformations. *Note: Some will require more than one step.*

(a) 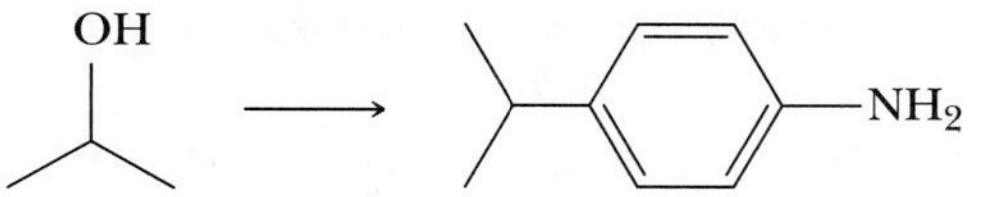 2-propanol (OH) ⟶ 4-isopropylaniline (NH_2)

(b) 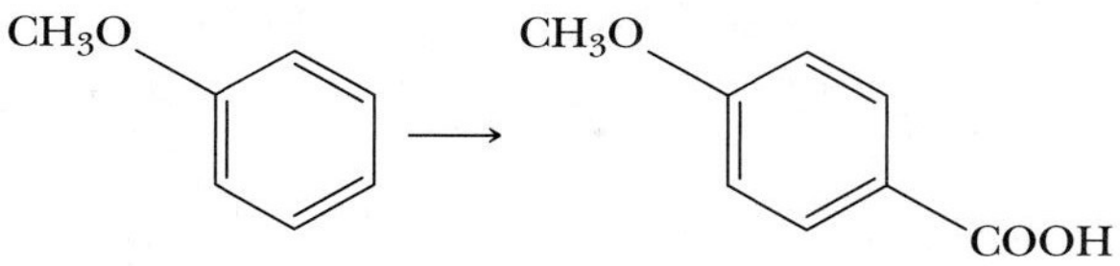 CH_3O-benzene ⟶ CH_3O-C6H4-COOH (para)

(c) 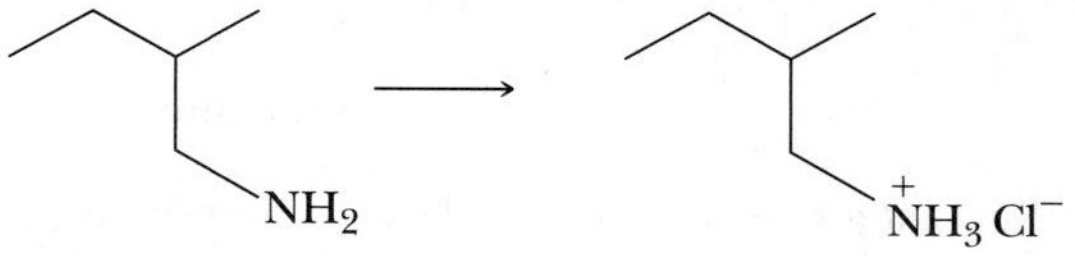 2-ethyl-butylamine (NH_2) ⟶ ammonium salt ($\overset{+}{N}H_3\ Cl^-$)

(d) 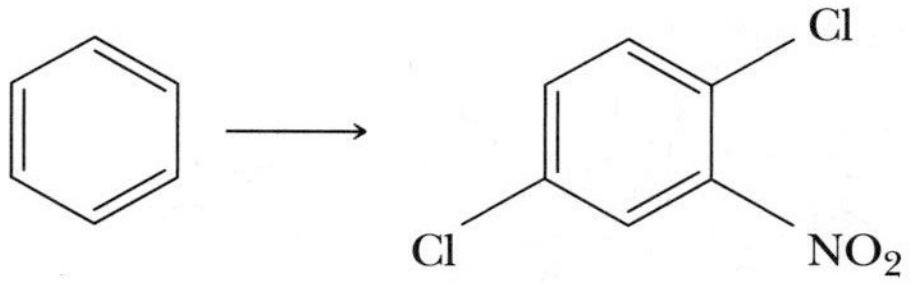 benzene ⟶ 2,4-dichloronitrobenzene (Cl, Cl, NO_2)

(e) 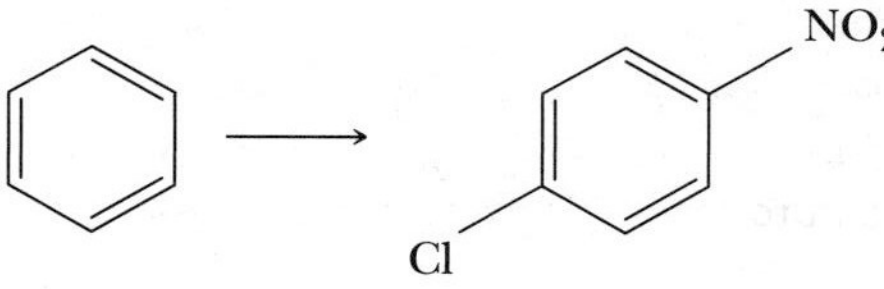 benzene ⟶ 1-chloro-4-nitrobenzene (NO_2, Cl)

(f) 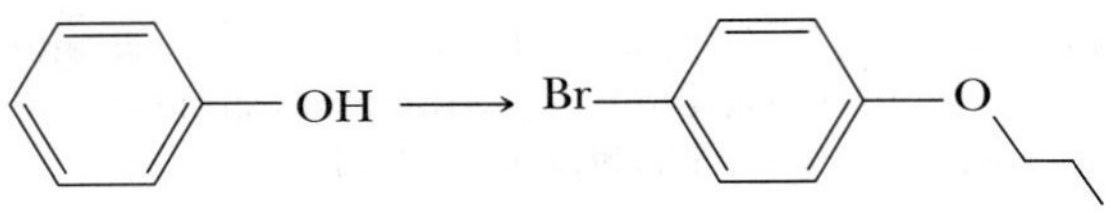 OH ⟶ Br O

(g) 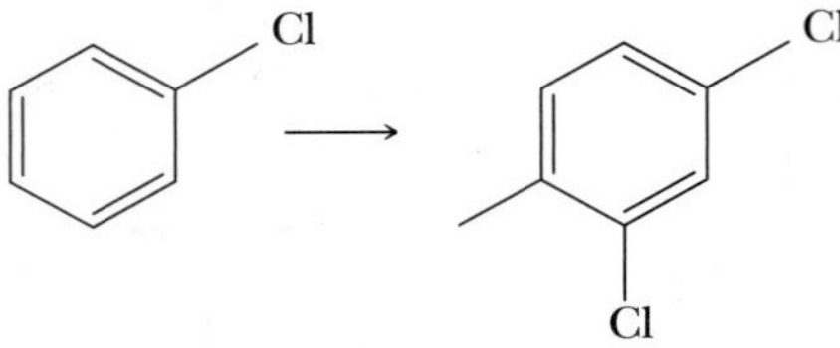 Cl ⟶ Cl Cl

(h) 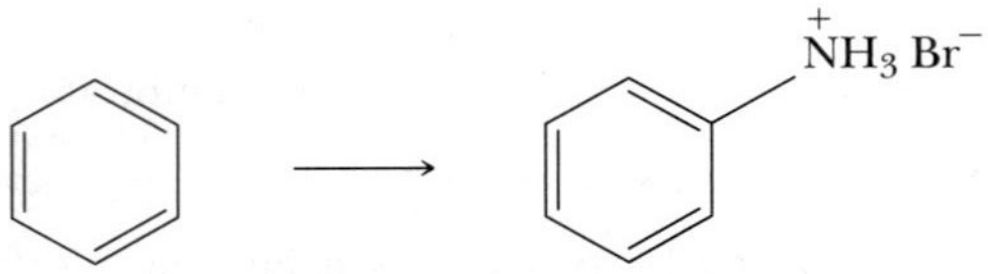 ⟶ $\overset{+}{N}H_3$ Br^-

(i) 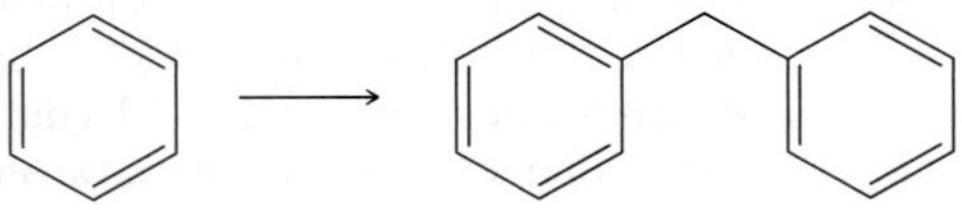 ⟶

(j) 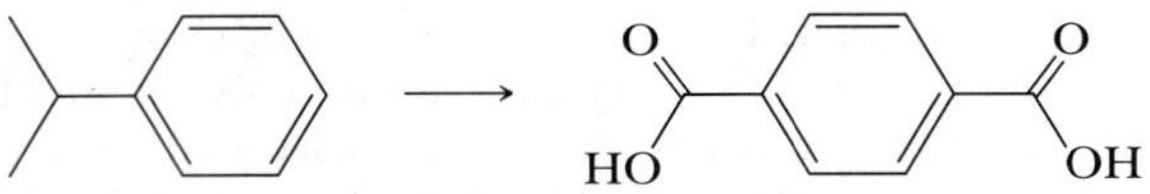 ⟶ O HO O OH

(k) 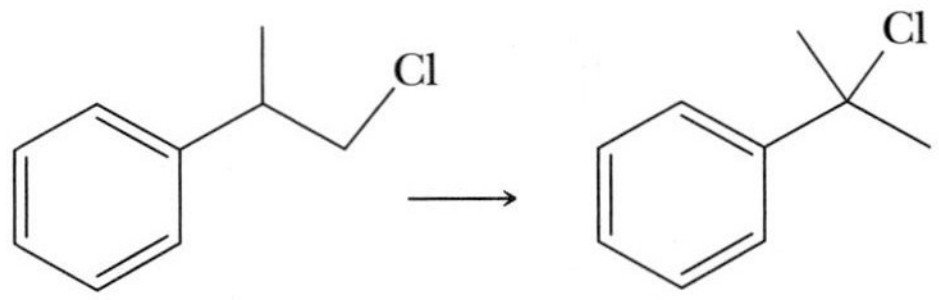 Cl ⟶ Cl

(l) 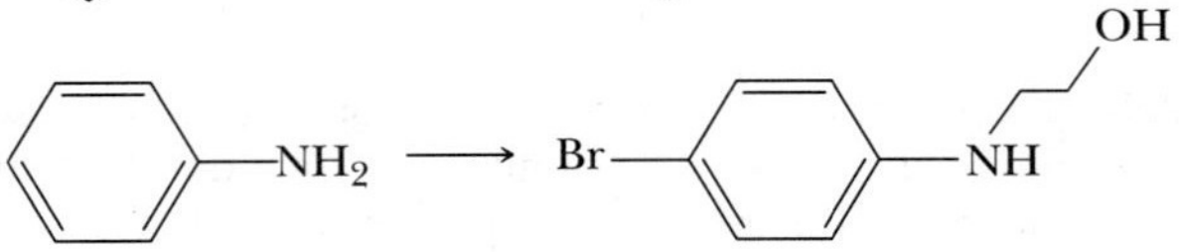 NH_2 ⟶ Br NH OH

(m) 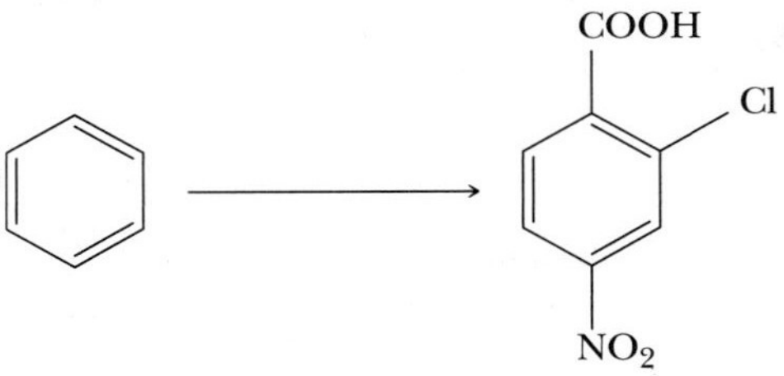 ⟶ COOH Cl NO_2

(n) 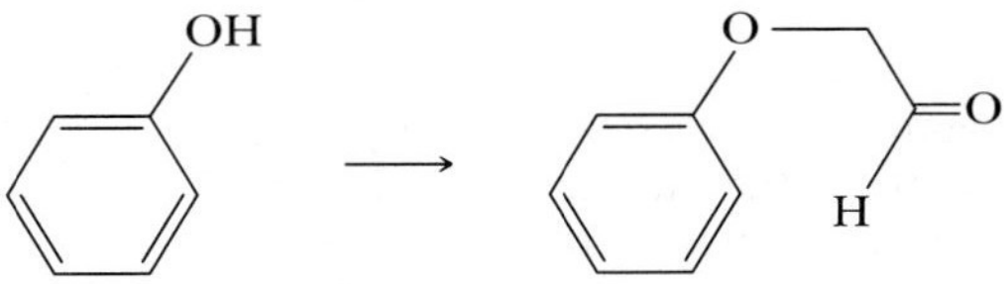 OH ⟶ O O H

Looking Ahead

9.49 Which of the following compounds can be made directly by using an electrophilic aromatic substitution reaction?

(a) 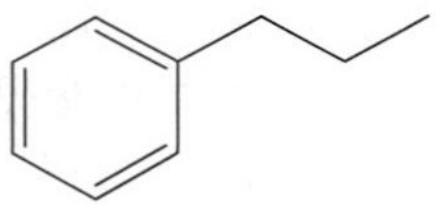 (b)

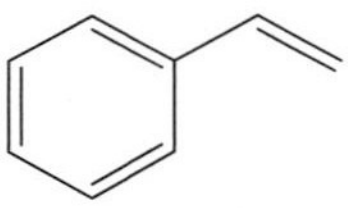

(c) 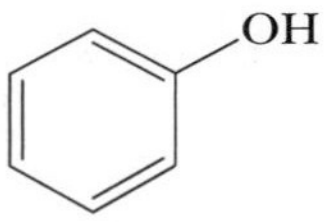 OH (d) 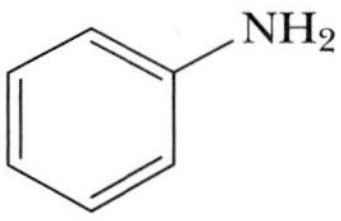 NH_2

9.50 Which compound is a better nucleophile?

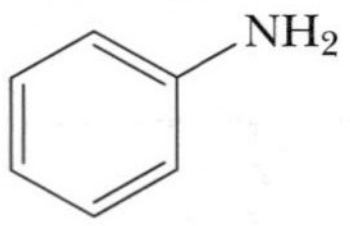

NH_2 or NH_2

Aniline Cyclohexanamine

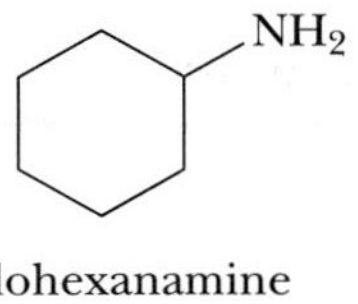

9.51 Suggest a reason that the following arenes do not undergo electrophilic aromatic substitution when $AlCl_3$ is used in the reaction:

(a) 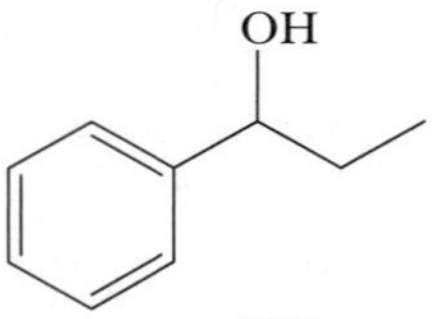 OH (b) 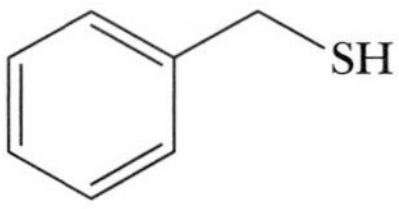 SH

(c) 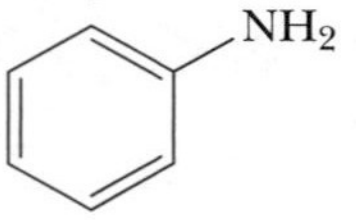 NH_2

9.52 Predict the product of the following acid–base reaction:

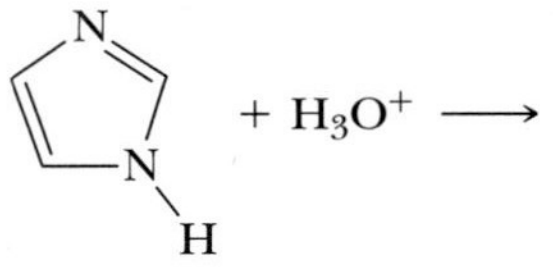

N N H + H_3O^+ ⟶

9.53 Which haloalkane reacts faster in an S_N1 reaction?

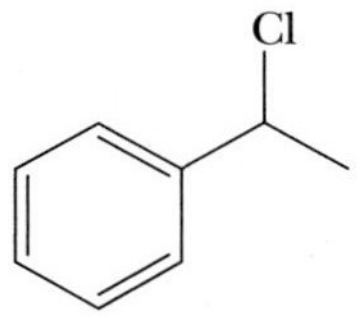

Cl or Cl

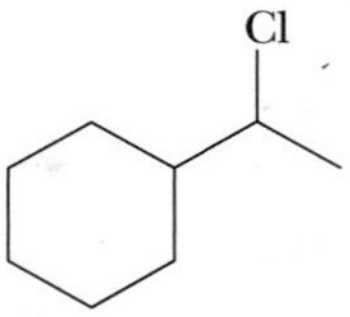

9.54 Which of the following compounds is more basic?

O O

Furan or Tetrahydrofuran

10 Amines

KEY QUESTIONS

HOW TO

CHEMICAL CONNECTIONS

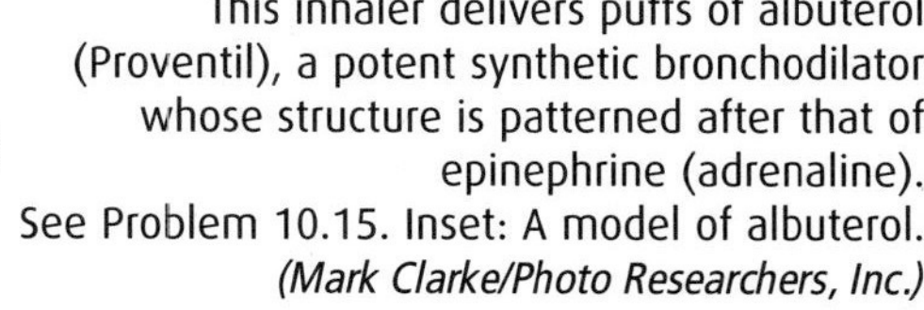

This inhaler delivers puffs of albuterol (Proventil), a potent synthetic bronchodilator whose structure is patterned after that of epinephrine (adrenaline). See Problem 10.15. Inset: A model of albuterol. *(Mark Clarke/Photo Researchers, Inc.)*

Carbon, hydrogen, and oxygen are the three most common elements in organic compounds. Because of the wide distribution of amines in the biological world, nitrogen is the fourth most common component of organic compounds. The most important chemical properties of amines are their basicity and their nucleophilicity.

CHEMICAL CONNECTIONS 10A

Morphine as a Clue in the Design and Discovery of Drugs

The analgesic, soporific, and euphoriant properties of the dried juice obtained from unripe seed pods of the opium poppy *Papaver somniferum* have been known for centuries. By the beginning of the nineteenth century, the active principal, morphine, had been isolated and its structure determined:

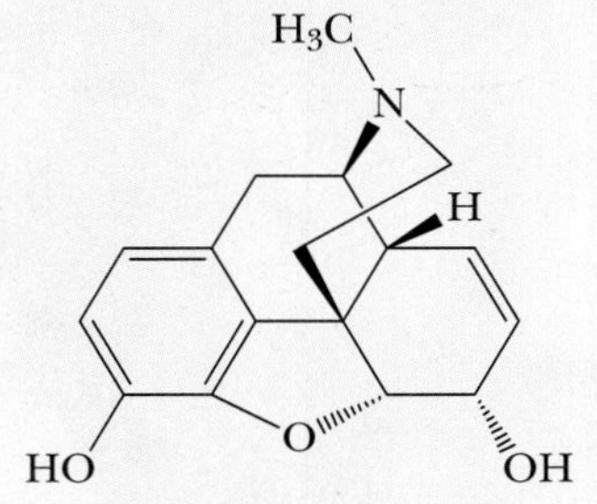

Morphine

Also occurring in the opium poppy is codeine, a monomethyl ether of morphine:

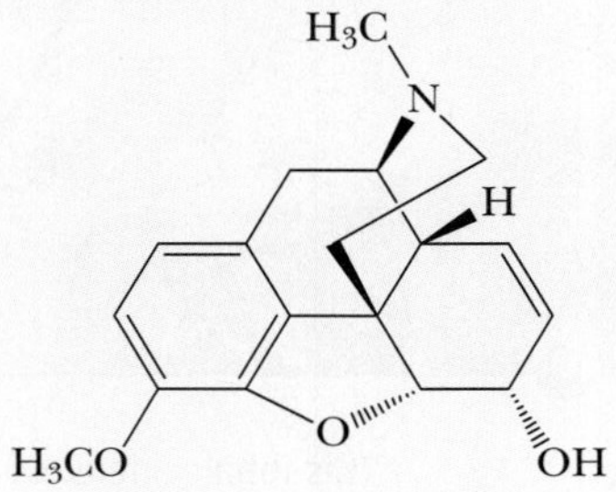

Codeine

Heroin is synthesized by treating morphine with two moles of acetic anhydride:

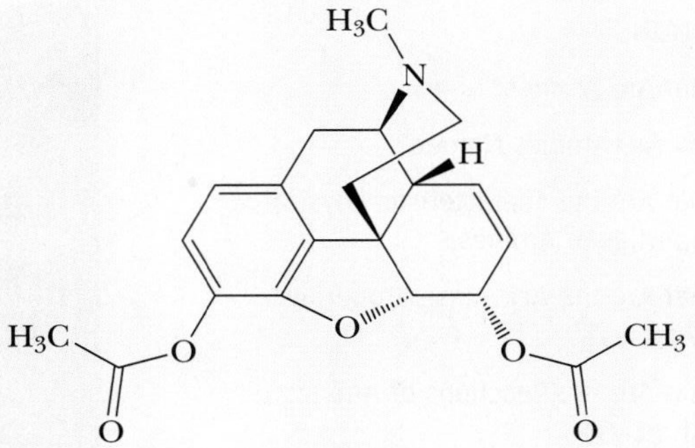

Heroin

Even though morphine is one of modern medicine's most effective painkillers, it has two serious side effects: It is addictive, and it depresses the respiratory control center of the central nervous system. Large doses of morphine (or heroin) can lead to death by respiratory failure. One strategy in the ongoing research to produce painkillers has been to synthesize compounds related in structure to morphine, in the hope that they would be equally effective analgesics, but with diminished side effects. Following are structural formulas for two such compounds that have proven to be clinically useful:

10.1 What Are Amines?

Amines are derivatives of ammonia (NH_3) in which one or more hydrogens are replaced by alkyl or aryl groups. Amines are classified as primary (1°), secondary (2°), or tertiary (3°), depending on the number of hydrogen atoms of ammonia that are replaced by alkyl or aryl groups (Section 1.7B). As we saw with ammonia, the three atoms or groups attached to the nitrogen in amines assume a trigonal pyramidal geometry:

$:NH_3$	$CH_3-\ddot{N}H_2$	$CH_3-\ddot{N}H-CH_3$	$CH_3-\ddot{N}(CH_3)-CH_3$
Ammonia	Methylamine (a 1° amine)	Dimethylamine (a 2° amine)	Trimethylamine (a 3° amine)

H_3CO — (−)-enantiomer = Levomethorphan
(+)-enantiomer = Dextromethorphan

redraw →

Meperidine
(Demerol)

Levomethorphan is a potent analgesic. Interestingly, its dextrorotatory enantiomer, dextromethorphan, has no analgesic activity. It does, however, show approximately the same cough-suppressing activity as morphine and is used extensively in cough remedies.

It has been discovered that there can be even further simplification in the structure of morphine-like analgesics. One such simplification is represented by meperidine, the hydrochloride salt of which is the widely used analgesic Demerol®.

It was hoped that meperidine and related synthetic drugs would be free of many of the morphine-like undesirable side effects. It is now clear, however, that they are not. Meperidine, for example, is definitely addictive. In spite of much determined research, there are as yet no agents as effective as morphine for the relief of severe pain that are absolutely free of the risk of addiction.

How and in what regions of the brain does morphine act? In 1979, scientists discovered that there are specific receptor sites for morphine and other opiates and that these sites are clustered in the brain's limbic system, the area involved in emotion and the perception of pain. Scientists then asked, "Why does the human brain have receptor sites specific for morphine?" Could it be that the brain produces its own opiates? In 1974, scientists discovered that opiate-like compounds are indeed present in the brain; in 1975, they isolated a brain opiate that was named *enkephalin*, meaning "in the brain." Unlike morphine and its derivatives, enkephalin possesses an entirely different structure consisting of a sequence of five peptides (Section 19.4). Scientists have yet to understand the role of these natural brain opiates. Perhaps when we do understand their biochemistry, we will discover clues that will lead to the design and synthesis of more potent, but less addictive, analgesics.

QUESTION

Identify the functional groups in morphine and meperidine. Classify the amino group in these opiates according to type (that is, primary, secondary, tertiary, heterocyclic, aliphatic, or aromatic).

Amines are further divided into aliphatic amines and aromatic amines. In an **aliphatic amine**, all the carbons bonded directly to nitrogen are derived from alkyl groups; in an **aromatic amine**, one or more of the groups bonded directly to nitrogen are aryl groups:

Aliphatic amine An amine in which nitrogen is bonded only to alkyl groups.

Aromatic amine An amine in which nitrogen is bonded to one or more aryl groups.

Aniline (a 1° aromatic amine)	*N*-Methylaniline (a 2° aromatic amine)	Benzyldimethylamine (a 3° aliphatic amine)
$C_6H_5-\ddot{N}H_2$	$C_6H_5-\ddot{N}(CH_3)-H$	$C_6H_5-CH_2-\ddot{N}(CH_3)-CH_3$

not bonded directly to N

An amine in which the nitrogen atom is part of a ring is classified as a **heterocyclic amine**. When the nitrogen is part of an aromatic ring (Section 9.2), the amine

Heterocyclic amine An amine in which nitrogen is one of the atoms of a ring.

Heterocyclic aromatic amine An amine in which nitrogen is one of the atoms of an aromatic ring.

is classified as a **heterocyclic aromatic amine**. Following are structural formulas for two heterocyclic aliphatic amines and two heterocyclic aromatic amines:

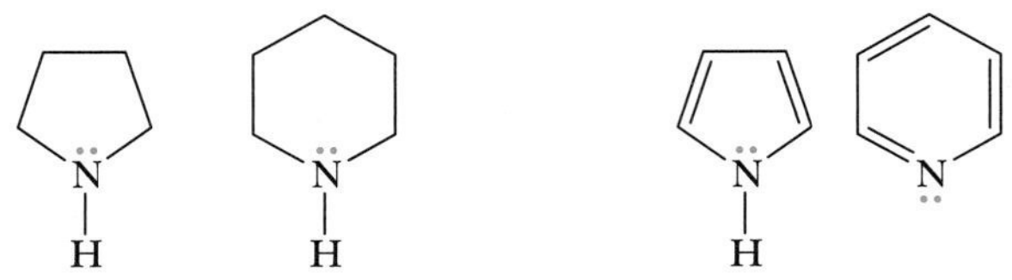

Pyrrolidine Piperidine
(heterocyclic aliphatic amines)

Pyrrole Pyridine
(heterocyclic aromatic amines)

Example 10.1

Alkaloids are basic nitrogen-containing compounds of plant origin, many of which have physiological activity when administered to humans. The ingestion of coniine, present in water hemlock, can cause weakness, labored respiration, paralysis, and, eventually, death. Coniine was the toxic substance in "poison hemlock" that caused the death of Socrates. In small doses, nicotine is an addictive stimulant. In larger doses, it causes depression, nausea, and vomiting. In still larger doses, it is a deadly poison. Solutions of nicotine in water are used as insecticides. Cocaine is a central nervous system stimulant obtained from the leaves of the coca plant. Classify each amino group in these alkaloids according to type (that is, primary, secondary, tertiary, heterocyclic, aliphatic, or aromatic):

(a) 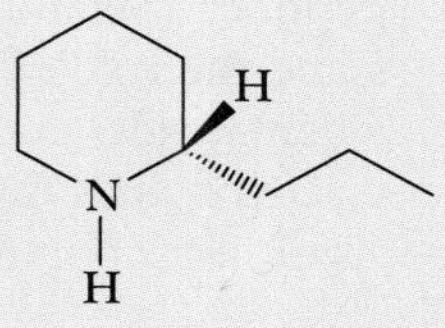 (*S*)-Coniine

(b) 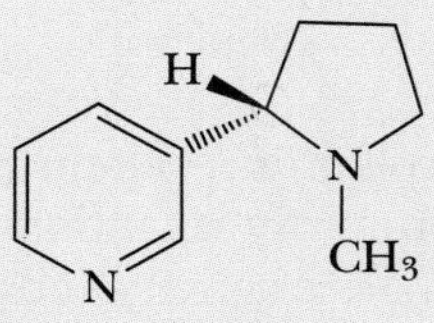 (*S*)-Nicotine

(c) 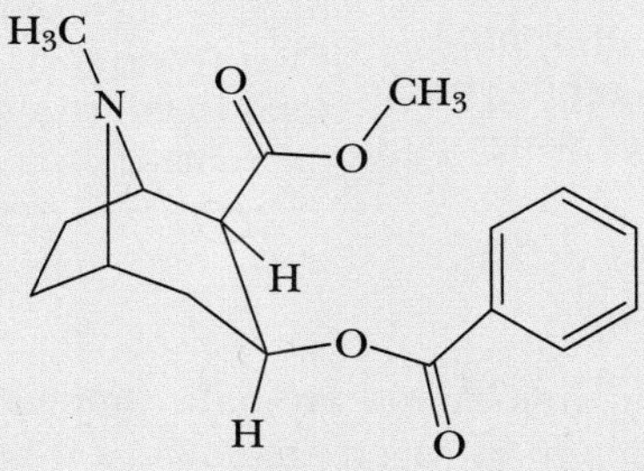 Cocaine

Strategy

Locate each nitrogen in each compound. If a nitrogen is part of a ring, the amine is heterocyclic. If that ring is aromatic, it is classified as a heterocyclic aromatic amine (1°, 2°, or 3° does not apply). If the ring is not aromatic, it is a heterocyclic aliphatic amine that should also be classified as 1°, 2°, or 3°. *Note*: The presence of more than one nitrogen can result in multiple classifications for the molecule, depending on the part of the compound being referred to.

Solution

(a) A secondary heterocyclic aliphatic amine.
(b) One tertiary heterocyclic aliphatic amine and one heterocyclic aromatic amine.
(c) A tertiary heterocyclic aliphatic amine.

See problems 10.13–10.16

Problem 10.1

Identify all carbon stereocenters in coniine, nicotine, and cocaine.

10.2 How Are Amines Named?

A. Systematic Names

Systematic names for aliphatic amines are derived just as they are for alcohols. The suffix *-e* of the parent alkane is dropped and is replaced by *-amine*; that is, they are named alkanamines:

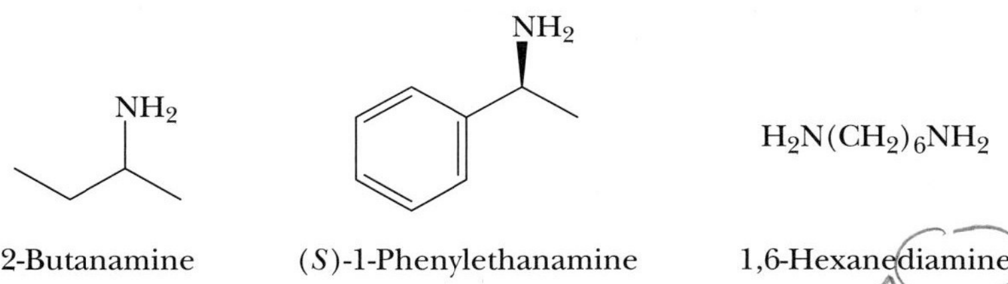

2-Butanamine (*S*)-1-Phenylethanamine 1,6-Hexanediamine

Example 10.2

Write the IUPAC name or provide the structural formula for each amine:

(a) 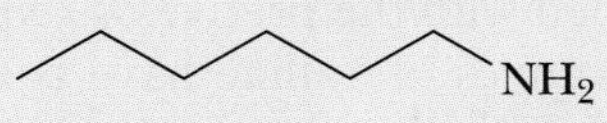 NH$_2$ (b) 2-Methyl-1-propanamine

(c) 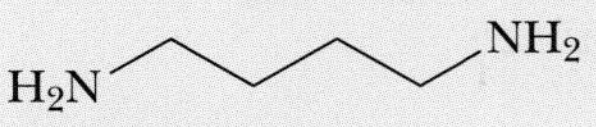 H_2N NH$_2$ (d) *trans*-4-Methylcyclohexanamine

(e)

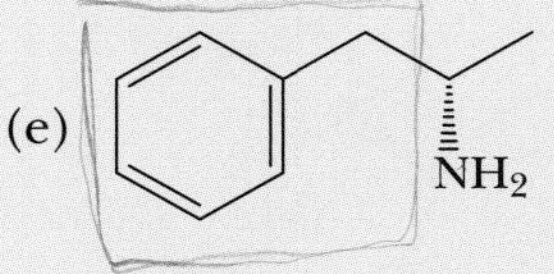

Strategy

When naming, look for the longest chain of carbons that contains the amino group. This will allow you to determine the root name. Then identify and name the substituents, the atoms or groups of atoms that are not part of that chain of carbons.

To translate a name to a structure, identify the carbon chain from the root name and add the substituents to the correct position on the chain.

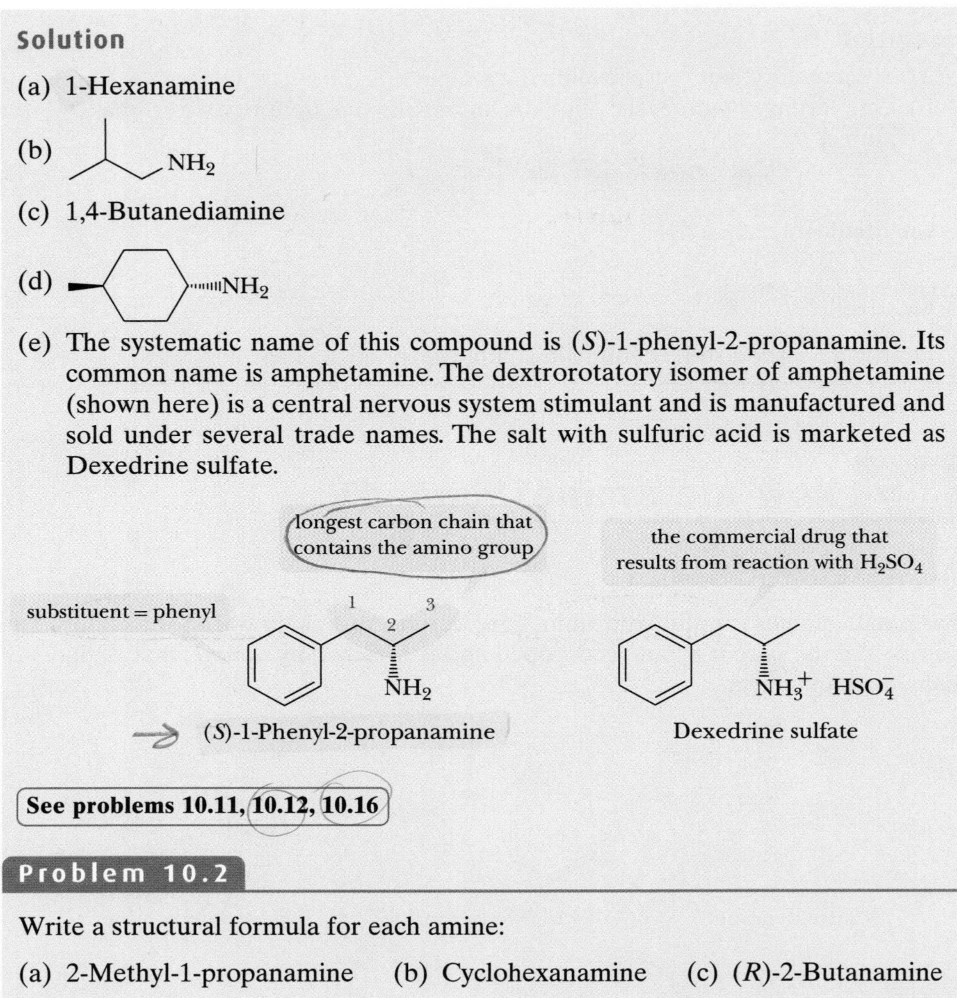

Solution

(a) 1-Hexanamine

(b) NH_2

(c) 1,4-Butanediamine

(d) NH_2

(e) The systematic name of this compound is (*S*)-1-phenyl-2-propanamine. Its common name is amphetamine. The dextrorotatory isomer of amphetamine (shown here) is a central nervous system stimulant and is manufactured and sold under several trade names. The salt with sulfuric acid is marketed as Dexedrine sulfate.

(*S*)-1-Phenyl-2-propanamine

Dexedrine sulfate

See problems 10.11, 10.12, 10.16

Problem 10.2

Write a structural formula for each amine:

(a) 2-Methyl-1-propanamine (b) Cyclohexanamine (c) (*R*)-2-Butanamine

IUPAC nomenclature retains the common name **aniline** for $C_6H_5NH_2$, the simplest aromatic amine. Its simple derivatives are named with the prefixes *o*-, *m*-, and *p*-, or numbers to locate substituents. Several derivatives of aniline have common names that are still widely used. Among these are **toluidine**, for a methyl-substituted aniline, and **anisidine**, for a methoxy-substituted aniline:

NH_2	NH_2, NO_2	NH_2, CH_3	NH_2, OCH_3
Aniline	4-Nitroaniline (*p*-Nitroaniline)	4-Methylaniline (*p*-Toluidine)	3-Methoxyaniline (*m*-Anisidine)

Secondary and tertiary amines are commonly named as *N*-substituted primary amines. For unsymmetrical amines, the largest group is taken as the parent amine; then the smaller group or groups bonded to nitrogen are named, and their location is indicated by the prefix *N* (indicating that they are attached to nitrogen):

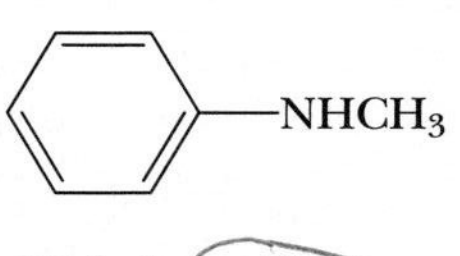

C_6H_5—$NHCH_3$ *N*-Methylaniline

C_5H_9—$N(CH_3)_2$ *N,N*-Dimethylcyclopentanamine

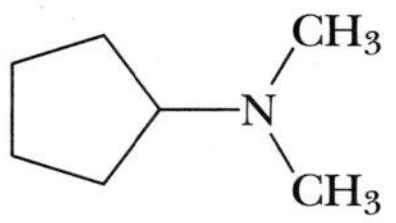

Following are names and structural formulas for four heterocyclic aromatic amines, the common names of which have been retained by the IUPAC:

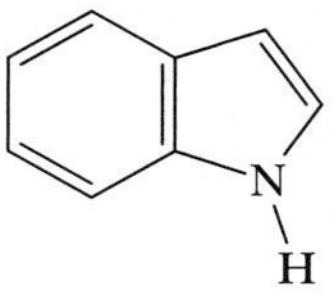

Indole 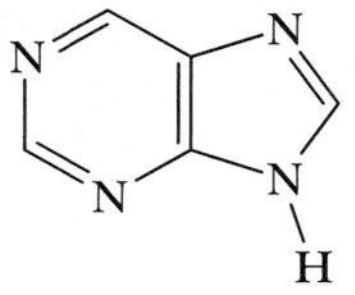 Purine 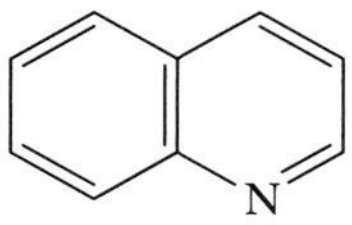 Quinoline 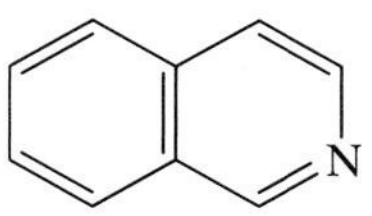 Isoquinoline

Among the various functional groups discussed in this text, the —NH_2 group has one of the lowest priorities. The following compounds each contain a functional group of higher precedence than the amino group, and, accordingly, the amino group is indicated by the prefix *amino*-:

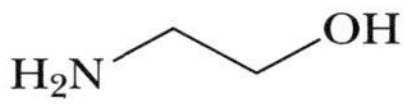

H_2N—CH_2CH_2—OH
2-Aminoethanol (Ethanolamine)

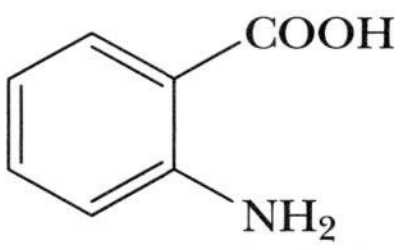

C_6H_4(COOH)(NH_2)
2-Aminobenzoic acid (Anthranilic acid)

B. Common Names

Common names for most aliphatic amines are derived by listing the alkyl groups bonded to nitrogen in alphabetical order in one word ending in the suffix *-amine*; that is, they are named as **alkylamines**:

CH_3NH_2 Methylamine

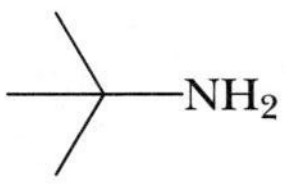

—NH_2 *tert*-Butylamine

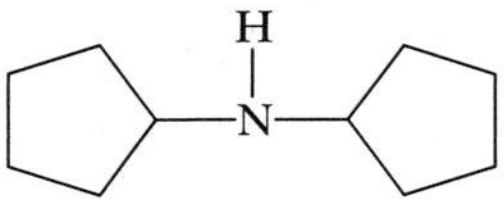

Dicyclopentylamine

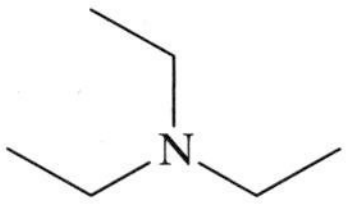

Triethylamine

Example 10.3

Write the IUPAC name or provide the structural formula for each amine:

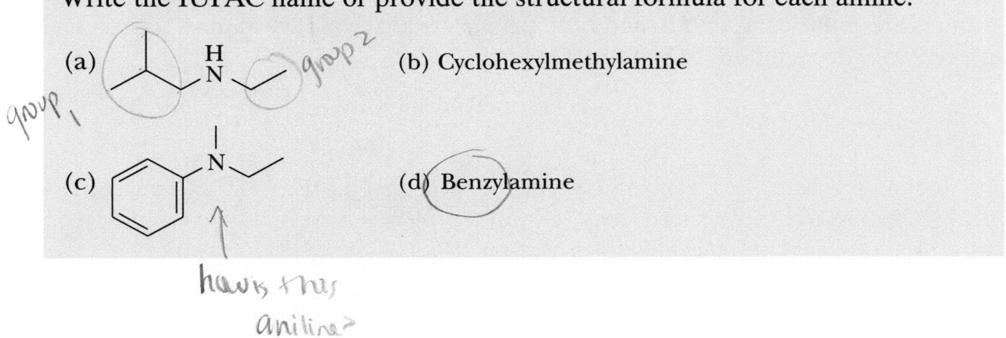

Strategy

When naming, look for the longest chain of carbons that contains the amino group. This will allow you to determine the root name. If the longest chain of carbons is a benzene ring, the amine may be named as an aniline derivative. When identifying the substituents, remember that substitutents bonded to a nitrogen are preceded by "*N*-."

To translate a name to a structure, identify the carbon chain from the root name and add the substituents to the correct position on the molecule.

Solution

(a) *N*-ethyl-2-methyl-1-propanamine

(b)

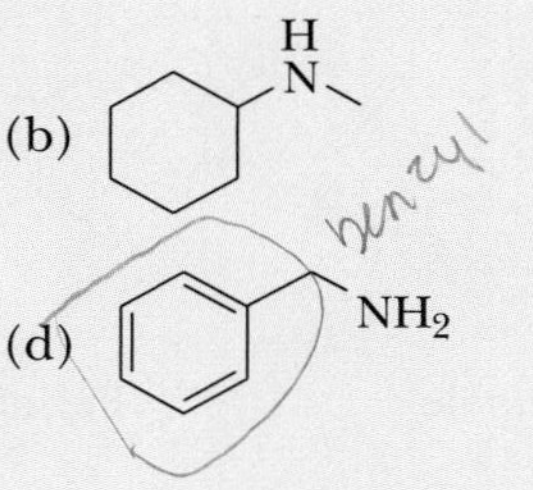

(c) *N*-ethyl-*N*-methylaniline

(d)

See problems 10.11, 10.12, 10.16

Problem 10.3

Write a structural formula for each amine:

(a) Isobutylamine (b) Triphenylamine (c) Diisopropylamine

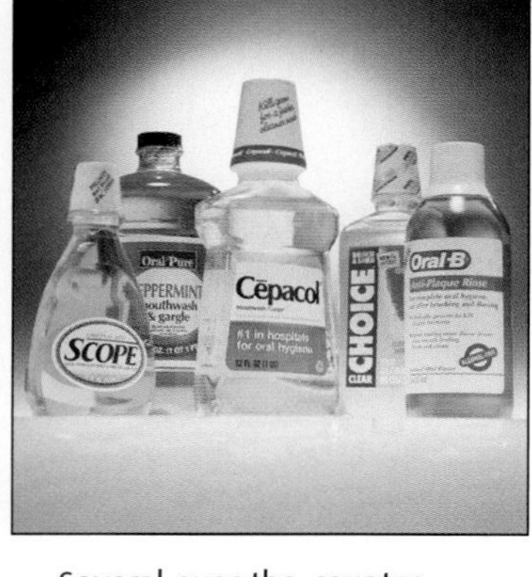

Several over-the-counter mouthwashes contain *N*-alkylatedpyridinium chlorides as an antibacterial agent. *(Charles D. Winters)*

When four atoms or groups of atoms are bonded to a nitrogen atom, we name the compound as a salt of the corresponding amine. We replace the ending *-amine* (or aniline, pyridine, or the like) by *-ammonium* (or *anilinium*, *pyridinium*, or the like) and add the name of the anion (chloride, acetate, and so on). Compounds containing such ions have properties characteristic of salts, such as increased water solubility, high melting points, and high boiling points. Following are three examples (cetylpyridinium chloride is used as a topical antiseptic and disinfectant):

$(CH_3)_4N^+Cl^-$ — Tetramethylammonium chloride

Cl^- $\overset{+}{N}CH_2(CH_2)_{14}CH_3$ — Hexadecylpyridinium chloride (Cetylpyridinium chloride)

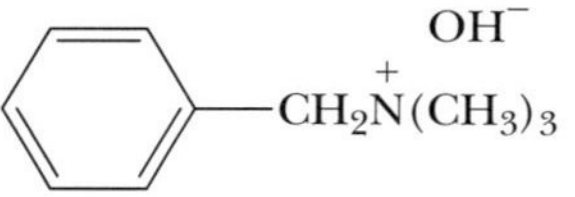

OH^- $-CH_2\overset{+}{N}(CH_3)_3$ — Benzyltrimethylammonium hydroxide

10.3 What Are the Characteristic Physical Properties of Amines?

Amines are polar compounds, and both primary and secondary amines form intermolecular hydrogen bonds (Figure 10.1).

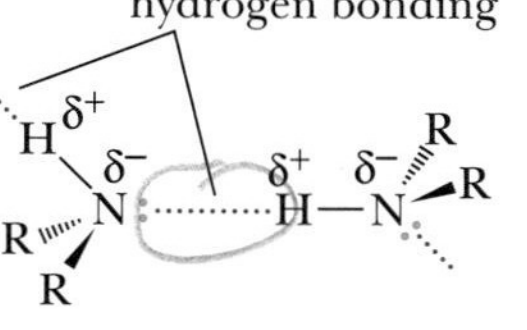

Figure 10.1 Intermolecular association of 1° and 2° amines by hydrogen bonding. Nitrogen is approximately tetrahedral in shape, with the axis of the hydrogen bond along the fourth position of the tetrahedron.

CHEMICAL CONNECTIONS 10B

The Poison Dart Frogs of South America: Lethal Amines

The Noanamá and Embrá peoples of the jungles of western Colombia have used poison blow darts for centuries, perhaps millennia. The poisons are obtained from the skin secretions of several highly colored frogs of the genus *Phyllobates* (*neará* and *kokoi* in the language of the native peoples). A single frog contains enough poison for up to 20 darts. For the most poisonous species (*Phyllobates terribilis*), just rubbing a dart over the frog's back suffices to charge the dart with poison.

Scientists at the National Institutes of Health became interested in studying these poisons when it was discovered that they act on cellular ion channels, which would make them useful tools in basic research on mechanisms of ion transport. A field station was established in western Colombia to collect the relatively common poison dart frogs. From 5,000 frogs, 11 mg of batrachotoxin and batrachotoxinin A were isolated. These names are derived from *batrachos*, the Greek word for frog.

Poison dart frog, *Phyllobates terribilis*. (*Juan M. Renjifo/Animals/Earth Scenes*)

Batrachotoxin and batrachotoxinin A are among the most lethal poisons ever discovered:

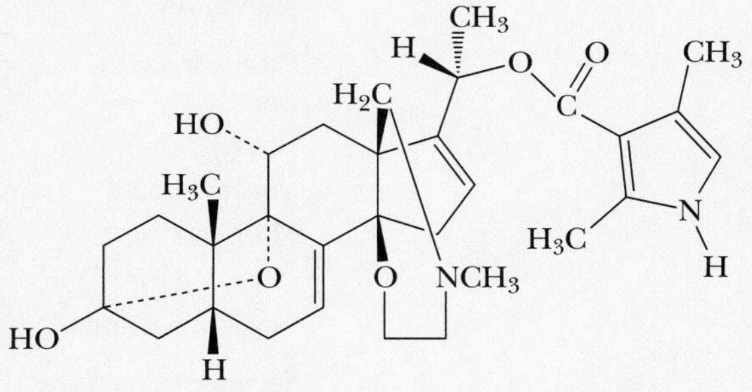

Batrachotoxin

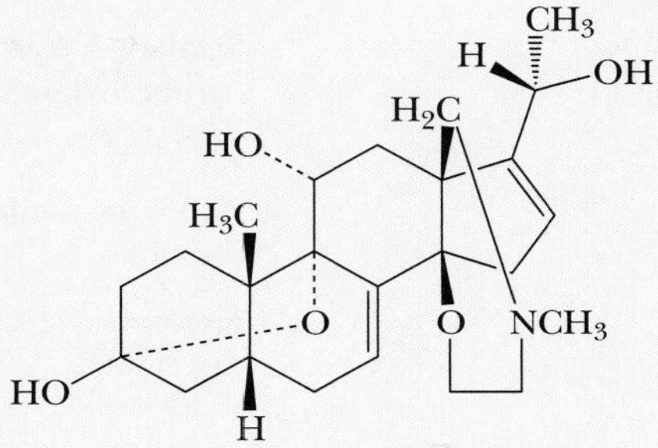

Batrachotoxinin A

It is estimated that as little as 200 μg of batrachotoxin is sufficient to induce irreversible cardiac arrest in a human being. It has been determined that they act by causing voltage-gated Na^+ channels in nerve and muscle cells to be blocked in the open position, which leads to a huge influx of Na^+ ions into the affected cell.

The batrachotoxin story illustrates several common themes in the discovery of new drugs. First, information about the kinds of biologically active compounds and their sources are often obtained from the native peoples of a region. Second, tropical rain forests are a rich source of structurally complex, biologically active substances. Third, an entire ecosystem, not only the plants, is a potential source of fascinating organic molecules.

QUESTIONS

Would you expect batrachotoxin or batrachotoxin A to be more soluble in water? Why?

Predict the product formed from the reaction of batrachotoxin with one equivalent of a weak acid such as acetic acid, CH_3COOH.

TABLE 10.1 Physical Properties of Selected Amines

Name	Structural Formula	Melting Point (°C)	Boiling Point (°C)	Solubility in Water
Ammonia	NH_3	−78	−33	very soluble
Primary Amines				
methylamine	CH_3NH_2	−95	−6	very soluble
ethylamine	$CH_3CH_2NH_2$	−81	17	very soluble
propylamine	$CH_3CH_2CH_2NH_2$	−83	48	very soluble
butylamine	$CH_3(CH_2)_3NH_2$	−49	78	very soluble
benzylamine	$C_6H_5CH_2NH_2$	10	185	very soluble
cyclohexylamine	$C_6H_{11}NH_2$	−17	135	slightly soluble
Secondary Amines				
dimethylamine	$(CH_3)_2NH$	−93	7	very soluble
diethylamine	$(CH_3CH_2)_2NH$	−48	56	very soluble
Tertiary Amines				
trimethylamine	$(CH_3)_3N$	−117	3	very soluble
triethylamine	$(CH_3CH_2)_3N$	−114	89	slightly soluble
Aromatic Amines				
aniline	$C_6H_5NH_2$	−6	184	slightly soluble
Heterocyclic Aromatic Amines				
pyridine	C_5H_5N	−42	116	very soluble

An N—H----N hydrogen bond is weaker than an O—H----O hydrogen bond, because the difference in electronegativity between nitrogen and hydrogen $(3.0 - 2.1 = 0.9)$ is less than that between oxygen and hydrogen $(3.5 - 2.1 = 1.4)$. We can illustrate the effect of intermolecular hydrogen bonding by comparing the boiling points of methylamine and methanol:

	CH_3NH_2	CH_3OH
molecular weight (g/mol)	31.1	32.0
boiling point (°C)	−6.3	65.0

Both compounds have polar molecules and interact in the pure liquid by hydrogen bonding. Methanol has the higher boiling point because hydrogen bonding between its molecules is stronger than that between molecules of methylamine.

All classes of amines form hydrogen bonds with water and are more soluble in water than are hydrocarbons of comparable molecular weight. Most low-molecular-weight amines are completely soluble in water (Table 10.1). Higher-molecular-weight amines are only moderately soluble or insoluble.

Example 10.4

Account for the fact that *n*-butylamine has a higher boiling point than *t*-butylamine.

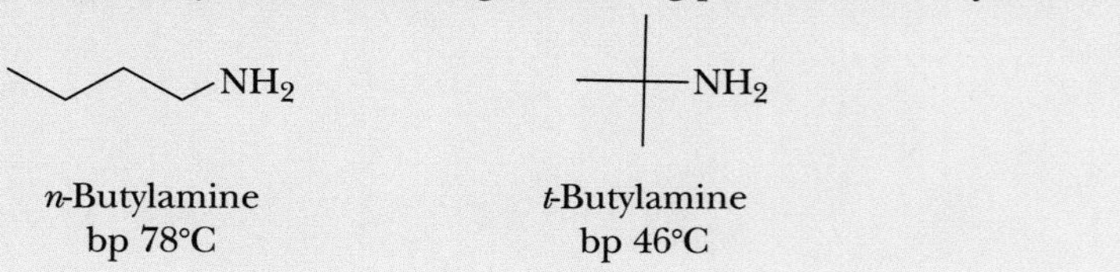

n-Butylamine
bp 78°C

t-Butylamine
bp 46°C

Strategy

Identify structural differences that might affect the intermolecular attractions between the molecules of each compound.

Solution

Both molecules can participate in hydrogen bonding. However, the *t*-butyl group is larger and bulkier, making it more difficult for the molecules of *t*-butylamine to hydrogen bond to each other.

See problems 10.18–10.20

Problem 10.4

Account for the fact that diethylamine has a higher boiling point than diethyl ether.

Diethylamine (H–N between two ethyl groups)	Diethyl ether (O between two ethyl groups)
Diethylamine bp 55°C	Diethyl ether bp 34.6°C

10.4 What Are the Acid–Base Properties of Amines?

Like ammonia, all amines are weak bases, and aqueous solutions of amines are basic. The following acid–base reaction between an amine and water is written using curved arrows to emphasize that, in this proton-transfer reaction, the unshared pair of electrons on nitrogen forms a new covalent bond with hydrogen and displaces hydroxide ion:

$$CH_3{-}\ddot{N}H_2 + H{-}\ddot{O}{-}H \rightleftharpoons CH_3{-}\overset{+}{N}H_3 \;\; {}^{-}\!:\ddot{O}{-}H$$

Methylamine   Methylammonium hydroxide

The equilibrium constant for the reaction of an amine with water, K_{eq}, has the following form, illustrated for the reaction of methylamine with water to give methylammonium hydroxide:

$$K_{eq} = \frac{[CH_3NH_3^+][OH^-]}{[CH_3NH_2][H_2O]}$$

Because the concentration of water in dilute solutions of methylamine in water is essentially a constant ($[H_2O] = 55.5$ mol/L), it is combined with K_{eq} in a new constant called a *base ionization constant*, K_b. The value of K_b for methylamine is 4.37×10^{-4} ($pK_b = 3.36$):

$$K_b = K_{eq}[H_2O] = \frac{[CH_3NH_3^+][OH^-]}{[CH_3NH_2]} = 4.37 \times 10^{-4}$$

It is also common to discuss the basicity of amines by referring to the acid ionization constant of the corresponding conjugate acid, as illustrated for the ionization of the methylammonium ion:

$$CH_3NH_3^+ + H_2O \rightleftharpoons CH_3NH_2 + H_3O^+$$

$$K_a = \frac{[CH_3NH_2][H_3O^+]}{[CH_3NH_3^+]} = 2.29 \times 10^{-11} \qquad pK_a = 10.64$$

Values of pK_a and pK_b for any acid–conjugate base pair are related by the equation

$$pK_a + pK_b = 14.00$$

Values of pK_a and pK_b for selected amines are given in Table 10.2.

TABLE 10.2 Base Strengths (pK_b) of Selected Amines and Acid Strengths (pK_a) of Their Conjugate Acids*

Amine	Structure	pK_b	pK_a
Ammonia	NH_3	4.74	9.26
Primary Amines			
methylamine	CH_3NH_2	3.36	10.64
ethylamine	$CH_3CH_2NH_2$	3.19	10.81
cyclohexylamine	$C_6H_{11}NH_2$	3.34	10.66
Secondary Amines			
dimethylamine	$(CH_3)_2NH$	3.27	10.73
diethylamine	$(CH_3CH_2)_2NH$	3.02	10.98
Tertiary Amines			
trimethylamine	$(CH_3)_3N$	4.19	9.81
triethylamine	$(CH_3CH_2)_3N$	3.25	10.75
Aromatic Amines			
aniline	C_6H_5–NH_2	9.37	4.63
4-methylaniline	H_3C–C_6H_4–NH_2	8.92	5.08
4-chloroaniline	Cl–C_6H_4–NH_2	9.85	4.15
4-nitroaniline	O_2N–C_6H_4–NH_2	13.0	1.0
Heterocyclic Aromatic Amines			
pyridine	(ring with N)	8.75	5.25
imidazole	(ring with N and N–H)	7.05	6.95

* For each amine, $pK_a + pK_b = 14.00$.

Example 10.5

Predict the position of equilibrium for this acid–base reaction:

$$CH_3NH_2 + CH_3COOH \rightleftharpoons CH_3NH_3^+ + CH_3COO^-$$

Strategy

Use the approach we developed in Section 2.4 to predict the position of equilibrium in acid–base reactions. Equilibrium favors reaction of the stronger acid and stronger base to form the weaker acid and the weaker base. It is helpful to remember that even though ammonium ions are positively charged, they are much weaker acids than carboxylic acids.

Solution

In this reaction, equilibrium favors the formation of methylammonium ion and acetate ion, which are the weaker acid and base, respectively:

$$CH_3NH_2 + CH_3COOH \rightleftharpoons CH_3NH_3^+ + CH_3COO^-$$

CH_3NH_2	CH_3COOH	$CH_3NH_3^+$	CH_3COO^-
	$pK_a = 4.76$	$pK_a = 10.64$	
Stronger base	Stronger acid	Weaker acid	Weaker base

See problem 10.25

Problem 10.5

Predict the position of equilibrium for this acid–base reaction:

$$CH_3NH_3^+ + H_2O \rightleftharpoons CH_3NH_2 + H_3O^+$$

Given information such as that in Table 10.2, we can make the following generalizations about the acid–base properties of the various classes of amines:

1. All aliphatic amines have about the same base strength, pK_b 3.0–4.0, and are slightly stronger bases than ammonia.
2. Aromatic amines and heterocyclic aromatic amines are considerably weaker bases than are aliphatic amines. Compare, for example, values of pK_b for aniline and cyclohexylamine:

$$C_6H_{11}NH_2 + H_2O \rightleftharpoons C_6H_{11}NH_3^+OH^- \qquad pK_b = 3.34 \quad K_b = 4.5 \times 10^{-4}$$

Cyclohexylamine → Cyclohexylammonium hydroxide

the lower the value of pK_b, the stronger the base

$$C_6H_5NH_2 + H_2O \rightleftharpoons C_6H_5NH_3^+OH^- \qquad pK_b = 9.37 \quad K_b = 4.3 \times 10^{-10}$$

Aniline → Anilinium hydroxide

The base ionization constant for aniline is smaller (the larger the value of pK_b, the weaker is the base) than that for cyclohexylamine by a factor of 10^6.

Aromatic amines are weaker bases than are aliphatic amines because of the resonance interaction of the unshared pair on nitrogen with the pi system of the aromatic ring. Because no such resonance interaction is possible for an alkylamine, the electron pair on its nitrogen is more available for reaction with an acid:

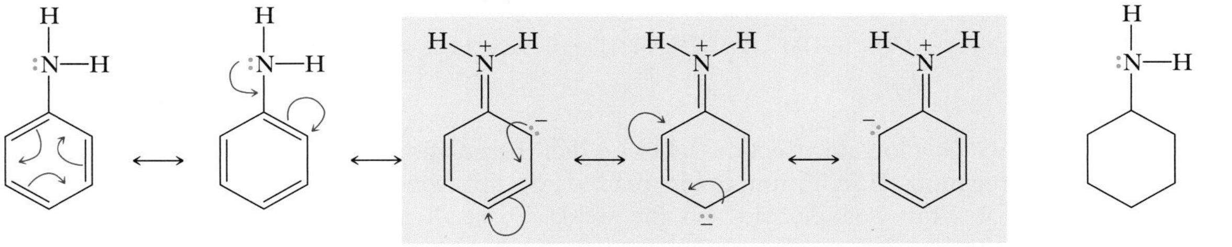

Two Kekulé structures

Interaction of the electron pair on nitrogen with the pi system of the aromatic ring reduces the availability of the electron pair to participate in a reaction with an acid

No resonance is possible with alkylamines

3. Electron-withdrawing groups such as halogen, nitro, and carbonyl decrease the basicity of substituted aromatic amines by decreasing the availability of the electron pair on nitrogen:

$-NO_2$ reduces the availability of the electron pair on nitrogen to participate in a reaction with an acid via both an inductive effect and a resonance effect

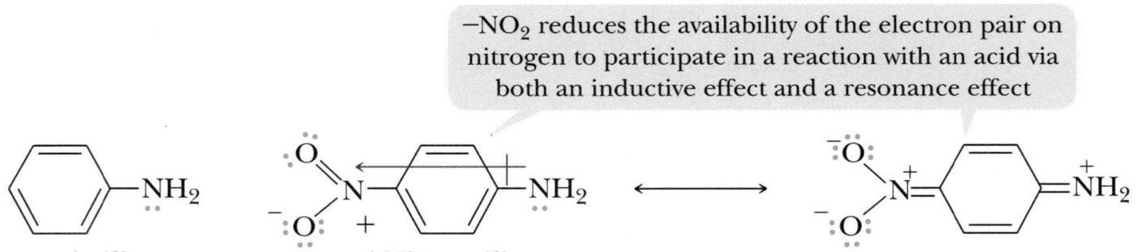

Aniline
pK_b 9.37

4-Nitroaniline
pK_b 13.0

Recall from Section 9.8B that these same substituents increase the acidity of phenols.

HOW TO 10.1 Predict the Relative Basicity of Amines

The basicity of an amine will depend on the ability of its nitrogen atom to donate its lone pair of electrons in an acid–base reaction. When assessing an electron pair's availability, look for the following possibilities:

(a) Resonance contribution

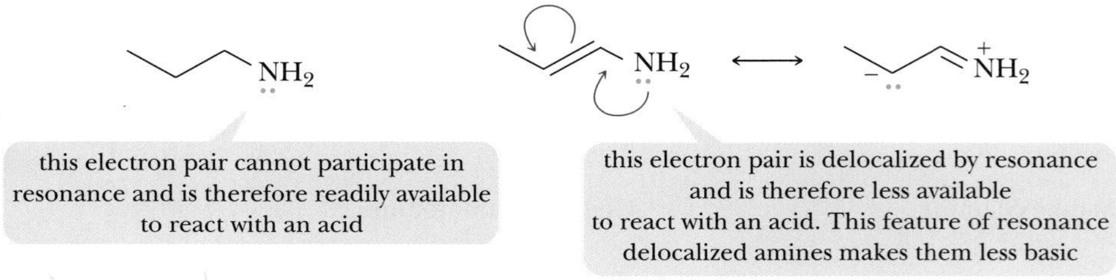

this electron pair cannot participate in resonance and is therefore readily available to react with an acid

this electron pair is delocalized by resonance and is therefore less available to react with an acid. This feature of resonance delocalized amines makes them less basic

(b) Induction

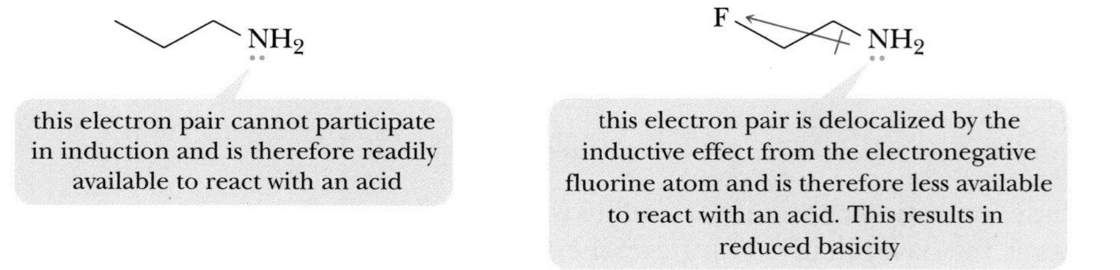

this electron pair cannot participate in induction and is therefore readily available to react with an acid

this electron pair is delocalized by the inductive effect from the electronegative fluorine atom and is therefore less available to react with an acid. This results in reduced basicity

Example 10.6

Select the stronger base in each pair of amines:

(a) 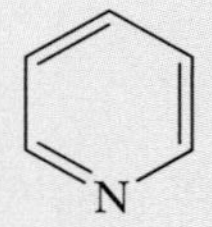 N (pyridine) or O, N–H (morpholine) (b) 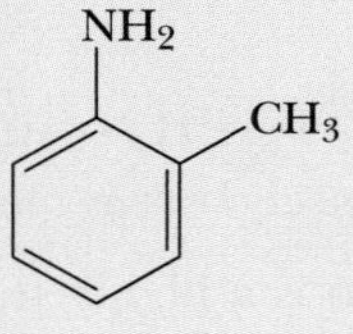 NH_2, CH_3 or CH_2NH_2

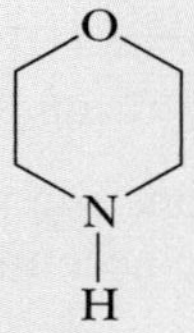

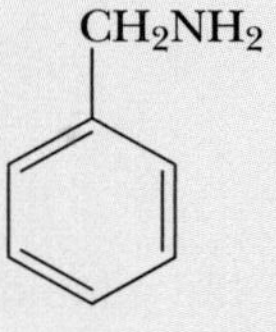

(A) (B) (C) (D)

Strategy

Use Table 10.2 to compare values of pK_b. Alternatively, look for resonance, inductive, or steric effects that might enhance or diminish the availability of a lone pair on the nitrogen of each molecule.

Solution

(a) Morpholine (B) is the stronger base (pK_b 5.79). It has a basicity comparable to that of secondary aliphatic amines. Pyridine (A), a heterocyclic aromatic amine (pK_b 8.75), is considerably less basic than aliphatic amines.
(b) Benzylamine (D), a primary aliphatic amine, is the stronger base (pK_b 3–4). *o*-Toluidine (C), an aromatic amine, is the weaker base (pK_b 9–10). In the absence of Table 10.2, one can see that the electron pair on nitrogen in *o*-toluidine can participate in resonance with the benzene ring, while there are no resonance possibilities in benzylamine. This results in *o*-toluidine's electron pair being less available for reaction with an acid.

See problems 10.21–10.25, 10.29–10.31

Problem 10.6

Select the stronger acid from each pair of ions:

(a) 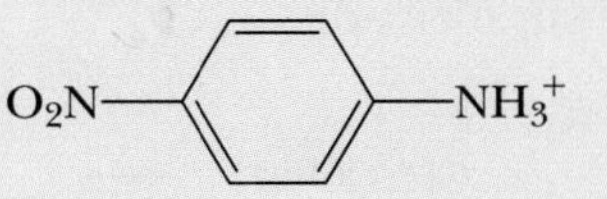 O_2N–C_6H_4–NH_3^+ or H_3C–C_6H_4–NH_3^+

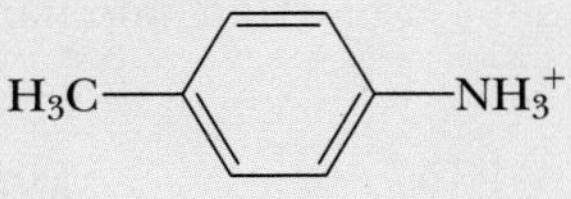

(A) (B)

(b) 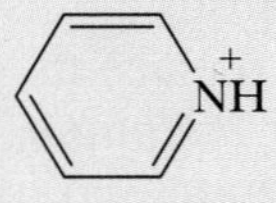 $\overset{+}{N}H$ (pyridinium) or C_6H_{11}–NH_3^+

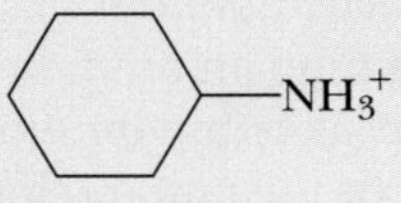

(C) (D)

Guanidine, with pK_b 0.4, is the strongest base among neutral compounds:

$$H_2N-\overset{\overset{\displaystyle NH}{\|}}{C}-NH_2 + H_2O \rightleftharpoons H_2N-\overset{\overset{\displaystyle ^+NH_2}{\|}}{C}-NH_2 + OH^- \qquad pK_b = 0.4$$

Guanidine Guanidinium ion

The remarkable basicity of guanidine is attributed to the fact that the positive charge on the guanidinium ion is delocalized equally over the three nitrogen atoms, as shown by these three equivalent contributing structures:

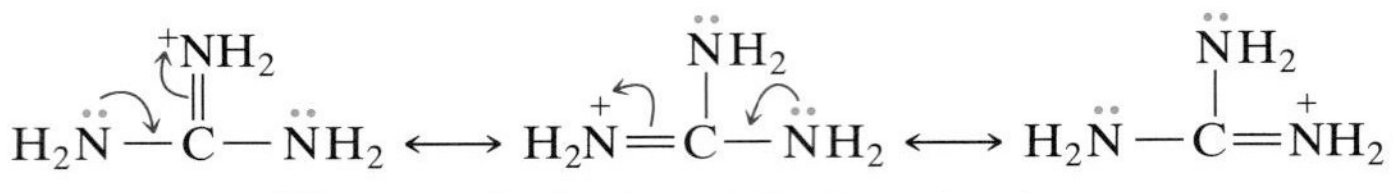

$$H_2\ddot{N}-\overset{\overset{+}{N}H_2}{\overset{\|}{C}}-\ddot{N}H_2 \longleftrightarrow H_2\overset{+}{N}=\overset{\ddot{N}H_2}{\overset{|}{C}}-\ddot{N}H_2 \longleftrightarrow H_2\ddot{N}-\overset{\ddot{N}H_2}{\overset{|}{C}}=\overset{+}{N}H_2$$

Three equivalent contributing structures

Hence, the guanidinium ion is a highly stable cation. The presence of a guanidine group on the side chain of the amino acid arginine accounts for the basicity of its side chain (Section 19.2A).

10.5 What Are the Reactions of Amines with Acids?

Amines, whether soluble or insoluble in water, react quantitatively with strong acids to form water-soluble salts, as illustrated by the reaction of (*R*)-norepinephrine (noradrenaline) with aqueous HCl to form a hydrochloride salt:

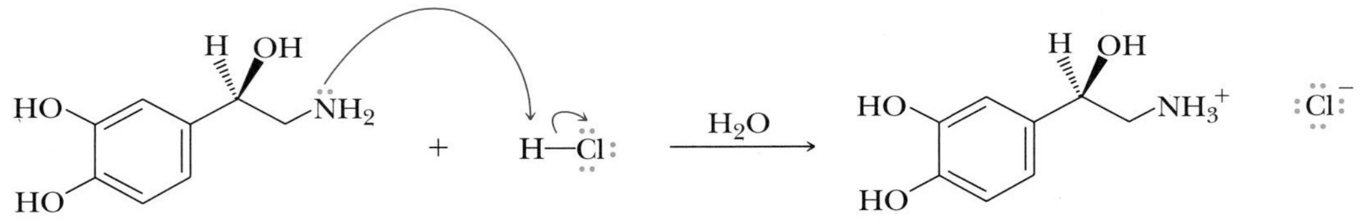

(*R*)-Norepinephrine (only slightly soluble in water) + $H-\ddot{Cl}:$ $\xrightarrow{H_2O}$ (*R*)-Norepinephrine hydrochloride (a water-soluble salt)

Norepinephrine, secreted by the medulla of the adrenal gland, is a neurotransmitter. It has been suggested that it is a neurotransmitter in those areas of the brain that mediate emotional behavior.

Example 10.7

Complete each acid–base reaction, and name the salt formed:

(a) $(CH_3CH_2)_2NH + HCl \longrightarrow$

(b) 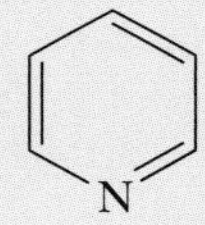 pyridine $+ CH_3COOH \longrightarrow$

Strategy

Identify the acidic proton in the acid. The amine nitrogen will abstract this proton to form the ammonium salt. In naming an ammonium salt, replace the ending *-amine* (or aniline, pyridine, or the like) by *-ammonium* (or *anilinium, pyridinium*, or the like) and add the name of the anion (chloride, acetate, and so on).

Solution

(a) $(CH_3CH_2)_2NH_2^+Cl^-$

Diethylammonium chloride

(b) 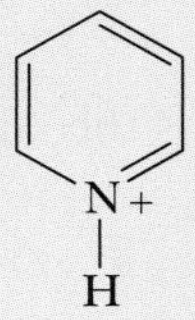 pyridinium (N^+-H) CH_3COO^-

Pyridinium acetate

See problem 10.25

p. 3ue

Problem 10.7

Complete each acid–base reaction and name the salt formed:

(a) $(CH_3CH_2)_3N + HCl \longrightarrow$

(b) (cyclohexane ring)NH + $CH_3COOH \longrightarrow$

The basicity of amines and the solubility of amine salts in water can be used to separate amines from water-insoluble, nonbasic compounds. Shown in Figure 10.2 is a flowchart for the separation of aniline from anisole. Note that aniline is recovered from its salt by treatment with NaOH.

Example 10.8

Following are two structural formulas for alanine (2-aminopropanoic acid), one of the building blocks of proteins (Chapter 19):

$$\underset{\text{(A)}}{CH_3\underset{\displaystyle NH_2}{\underset{|}{C}}H\overset{\displaystyle O}{\overset{\|}{C}}OH} \quad \text{or} \quad \underset{\text{(B)}}{CH_3\underset{\displaystyle NH_3^+}{\underset{|}{C}}H\overset{\displaystyle O}{\overset{\|}{C}}O^-}$$

Is alanine better represented by structural formula (A) or structural formula (B)?

Strategy

Begin by considering the acidity and basicity of the functional groups within alanine. How might they react if they were part of separate molecules?

Solution

Structural formula (A) contains both an amino group (a base) and a carboxyl group (an acid). Proton transfer from the stronger acid ($-COOH$) to the stronger base ($-NH_2$) gives an internal salt; therefore, (B) is the better representation for alanine. Within the field of amino acid chemistry, the internal salt represented by (B) is called a **zwitterion** (Chapter 19).

Problem 10.8

As shown in Example 10.8, alanine is better represented as an internal salt. Suppose that the internal salt is dissolved in water.

(a) In what way would you expect the structure of alanine in aqueous solution to change if concentrated HCl were added to adjust the pH of the solution to 2.0?

(b) In what way would you expect the structure of alanine in aqueous solution to change if concentrated NaOH were added to bring the pH of the solution to 12.0?

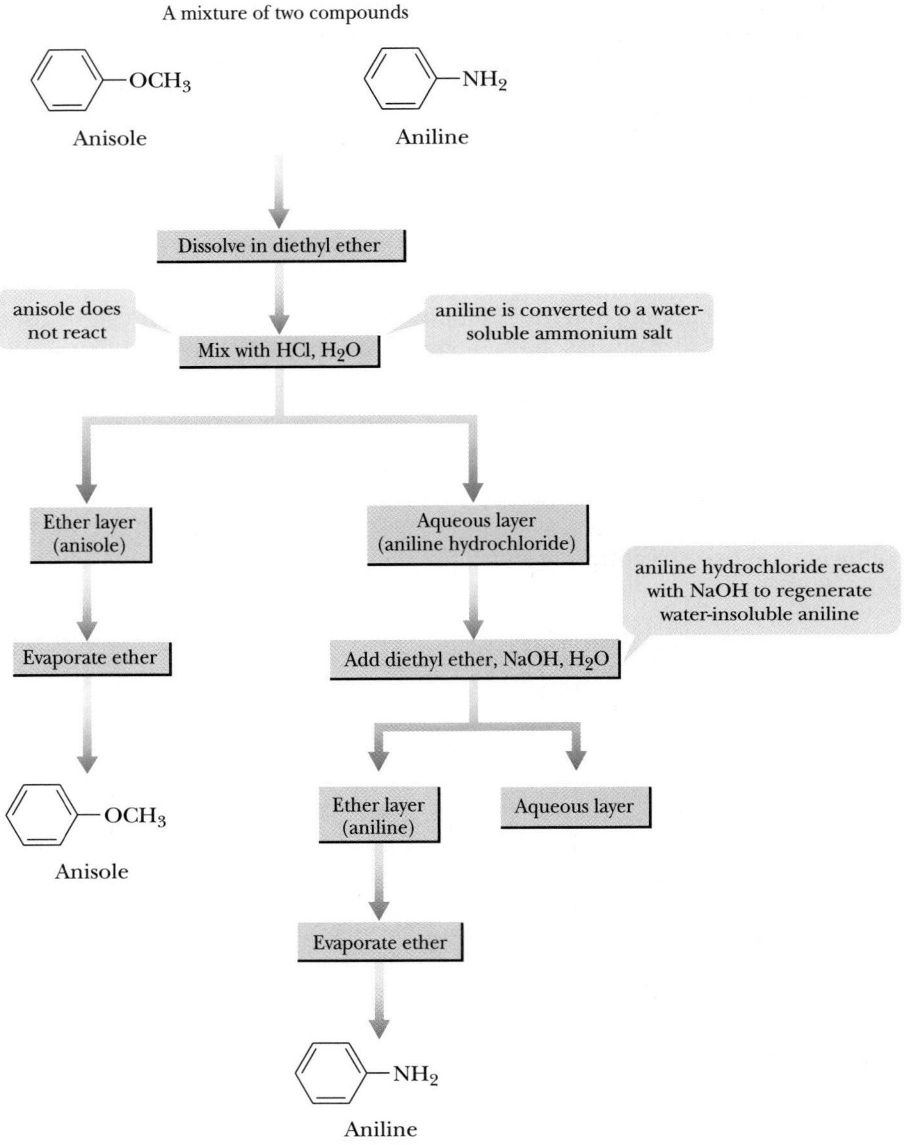

Figure 10.2 Separation and purification of an amine and a neutral compound.

10.6 How Are Arylamines Synthesized?

As we have already seen (Section 9.6B), the nitration of an aromatic ring introduces a NO_2 group. A particular value of nitration is the fact that the resulting nitro group can be reduced to a primary amino group, $-NH_2$, by hydrogenation in the presence of a transition metal catalyst such as nickel, palladium, or platinum:

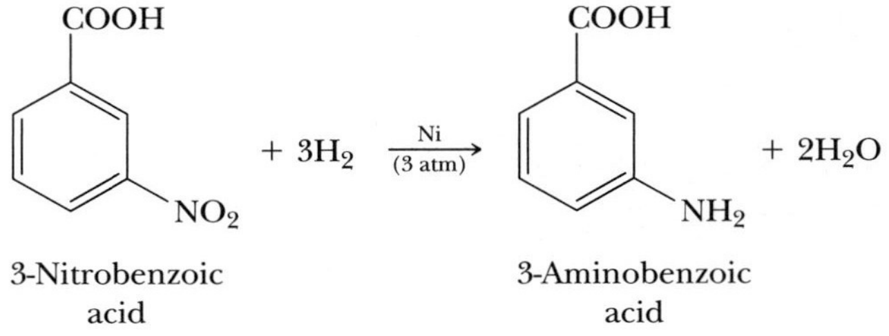

$$\text{3-Nitrobenzoic acid} + 3H_2 \xrightarrow[(3\text{ atm})]{\text{Ni}} \text{3-Aminobenzoic acid} + 2H_2O$$

This method has the potential disadvantage that other susceptible groups, such as a carbon–carbon double bond, and the carbonyl group of an aldehyde or ketone, may also be reduced. Note that neither the —COOH nor the aromatic ring is reduced under these conditions.

Alternatively, a nitro group can be reduced to a primary amino group by a metal in acid:

$$\text{2,4-Dinitrotoluene} \xrightarrow[C_2H_5OH,\ H_2O]{Fe,\ HCl} CH_3C_6H_3(NH_3^+Cl^-)_2 \xrightarrow{NaOH,\ H_2O} \text{2,4-Diaminotoluene}$$

2,4-Dinitrotoluene 2,4-Diaminotoluene

The most commonly used metal-reducing agents are iron, zinc, and tin in dilute HCl. When reduced by this method, the amine is obtained as a salt, which is then treated with a strong base to liberate the free amine.

10.7 How Are Primary Aromatic Amines Used in Synthesis?

Nitrous acid, HNO_2, is an unstable compound that is prepared by adding sulfuric or hydrochloric acid to an aqueous solution of sodium nitrite, $NaNO_2$. Nitrous acid is a weak acid and ionizes according to the following equation:

$$HNO_2 + H_2O \rightleftharpoons H_3O^+ + NO_2^- \qquad K_a = 4.26 \times 10^{-4}$$

Nitrous acid $\qquad pK_a = 3.37$

Nitrous acid reacts with amines in different ways, depending on whether the amine is primary, secondary, or tertiary and whether it is aliphatic or aromatic. We concentrate on the reaction of nitrous acid with primary aromatic amines, because this reaction is useful in organic synthesis.

Treatment of a primary aromatic amine—for example, aniline—with nitrous acid gives a **diazonium salt**:

$$C_6H_5-NH_2 + NaNO_2 + HCl \xrightarrow[0°C]{H_2O} C_6H_5-\overset{+}{N}\equiv N\ Cl^- + H_2O$$

Aniline (a 1° aromatic amine) Sodium nitrite Benzenediazonium chloride

We can also write the equation for this reaction in the following more abbreviated form:

$$C_6H_5-NH_2 \xrightarrow[0°C]{NaNO_2,\ HCl} C_6H_5-N_2^+Cl^-$$

Benzenediazonium chloride

When we warm an aqueous solution of an **arenediazonium salt**, the $—N_2^+$ group is replaced by an —OH group. This reaction is one of the few methods we have for the synthesis of phenols. It enables us to convert an aromatic amine to a phenol by first forming the arenediazonium salt and then heating the solution. In this manner, we can convert 2-bromo-4-methylaniline to 2-bromo-4-methylphenol:

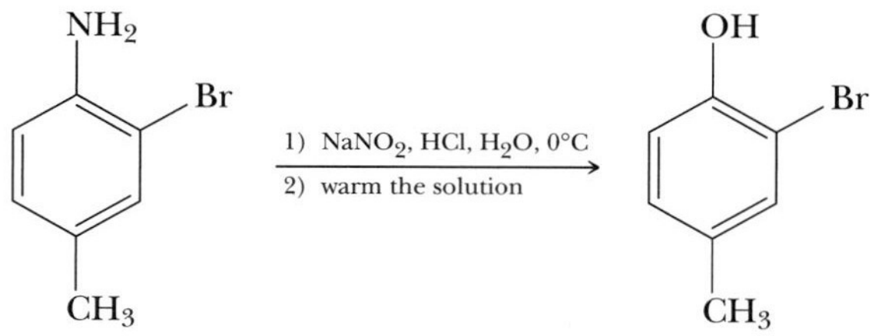

NH_2 Br CH_3 — 1) $NaNO_2$, HCl, H_2O, 0°C 2) warm the solution → OH Br CH_3

2-Bromo-4-methylaniline 2-Bromo-4-methylphenol

Example 10.9

Show the reagents that will bring about each step in this conversion of toluene to 4-hydroxybenzoic acid:

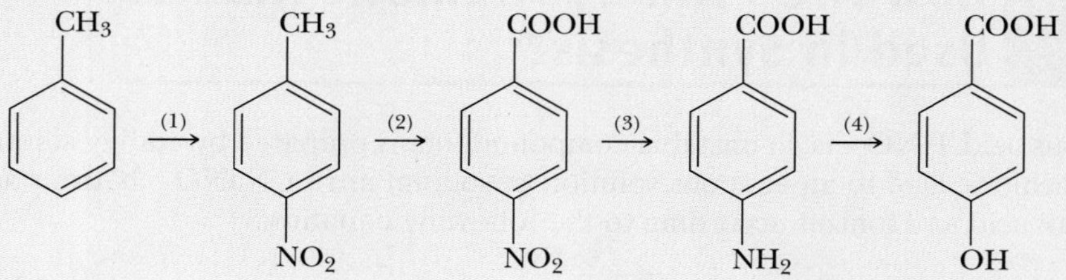

CH_3 (1)→ CH_3/NO_2 (2)→ COOH/NO_2 (3)→ COOH/NH_2 (4)→ COOH/OH

Toluene 4-Hydroxybenzoic acid

Strategy

Use a combination of reactions from this chapter and previous chapters. Remember to consider the regioselectivity of reactions.

Solution

Step 1: Nitration of toluene, using nitric acid/sulfuric acid (Section 9.6B), followed by separation of the ortho and para isomers.

Step 2: Oxidation of the benzylic carbon, using chromic acid (Section 9.4).

Step 3: Reduction of the nitro group, either using H_2 in the presence of a transition metal catalyst or using Fe, Sn, or Zn in the presence of aqueous HCl (Section 10.6).

Step 4: Treatment of the aromatic amine with $NaNO_2$/HCl to form the diazonium ion salt and then warming the solution.

See problems 10.36–10.44

Problem 10.9

Show how you can use the same set of steps in Example 10.9, but in a different order, to convert toluene to 3-hydroxybenzoic acid.

Treatment of an arenediazonium salt with hypophosphorous acid, H_3PO_2, reduces the diazonium group and replaces it with —H, as illustrated by the conversion of aniline to 1,3,5-trichlorobenzene. Recall that the $—NH_2$ group is a powerful activating and ortho–para directing group (Section 9.7A). Treatment of aniline with chlorine requires no catalyst and gives 2,4,6-trichloroaniline (to complete the conversion, we treat the trichloroaniline with nitrous acid followed by hypophosphorous acid):

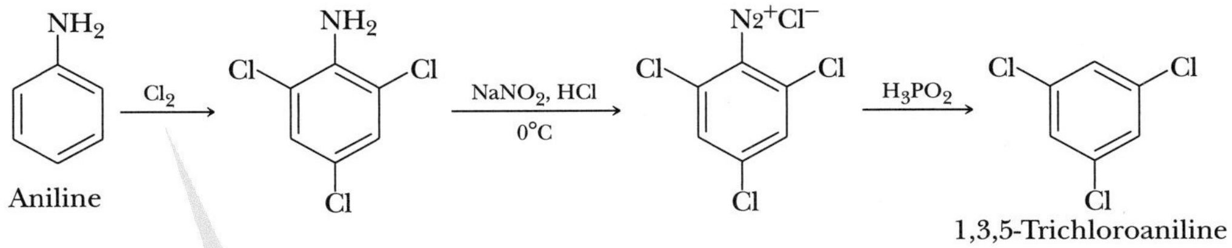

Aniline $\xrightarrow{Cl_2}$ 2,4,6-trichloroaniline $\xrightarrow[0°C]{NaNO_2,\ HCl}$ $N_2^+Cl^-$ (2,4,6-trichlorobenzenediazonium chloride) $\xrightarrow{H_3PO_2}$ 1,3,5-Trichloroaniline

normally, chlorination requires a catalyst such as $AlCl_3$ or $FeCl_3$, but the amino group is such a powerful activating group that the reaction proceeds without a catalyst

10.8 How Do Amines Act as Nucleophiles?

In Chapter 7, we learned that amines are moderate nucleophiles (Table 7.2) due to the presence of a lone pair of electrons on the nitrogen atom. Therefore, they should undergo nucleophilic substitution reactions with haloalkanes and other compounds containing a good leaving group (Section 7.5). In the reaction shown below, the nitrogen atom of an amine displaces chlorine in a haloalkane to yield ammonium chloride.

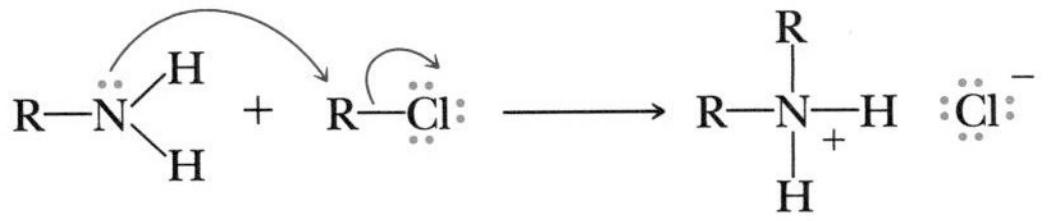

$$R{-}\ddot{N}H_2 + R{-}\ddot{\underset{..}{Cl}}{:} \longrightarrow R_2NH_2^+ \ {:}\ddot{\underset{..}{Cl}}{:}^-$$

At the beginning of this reaction, when only a few product molecules are formed, plenty of amine starting material (a weak base) remains to react with the hydrogen of the ammonium salt to yield a secondary amine and another ammonium chloride ion.

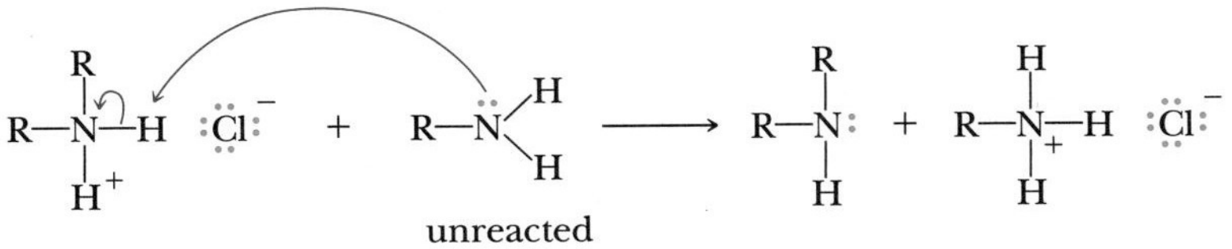

$$R_2NH_2^+ \ {:}\ddot{\underset{..}{Cl}}{:}^- + R{-}\ddot{N}H_2 \longrightarrow R_2\ddot{N}H + RNH_3^+ \ {:}\ddot{\underset{..}{Cl}}{:}^-$$

unreacted starting material

Remaining in the reaction mixture are some initial product, $R_2NH_2^+\ Cl^-$, some of the secondary amine, R_2NH, and lots of unreacted starting material and haloalkane.

The secondary amine is also a nucleophile, and because only a few of the initial R—Cl molecules have reacted at this early stage of the reaction, there are plenty left to react with either amine now in the reaction mixture.

compounds remaining in the reaction mixture

$$R{-}\overset{+}{N}H_3\ :\ddot{Cl}:^- + R_2\overset{+}{N}H_2\ :\ddot{Cl}:^- + R_2\ddot{N}H + R\ddot{N}H_2 + R{-}\ddot{Cl}: \longrightarrow R_3\overset{+}{N}H\ :\ddot{Cl}:^- + \text{all other products previously formed}$$

a small amount; a small amount; a small amount; a large amount; a large amount

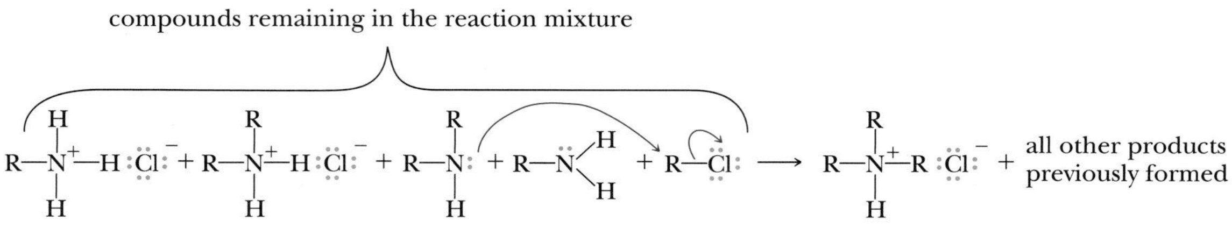

The process can continue to give one other nitrogen-based product, the quaternary ammonium salt. The final composition of the reaction will consist of varying ratios of RNH_2, R_2NH, R_3N, and $R_4N^+Cl^-$. Because the ratio of products is difficult to control or predict, we avoid using an amine (or ammonia) as a nucleophile in nucleophilic aliphatic substitution reactions.

Example 10.10

Determine all possible nitrogen-based products that can be formed in the following reaction:

$$CH_3CH_2CH_2NH_2 \quad + \quad CH_3CH_2Br \longrightarrow$$

Strategy

Keep in mind that the reaction of amines with haloalkanes often results in multiple nitrogen-based products with one or more alkyl groups from the haloalkane forming a bond with the nitrogen atom of the original amine.

Solution

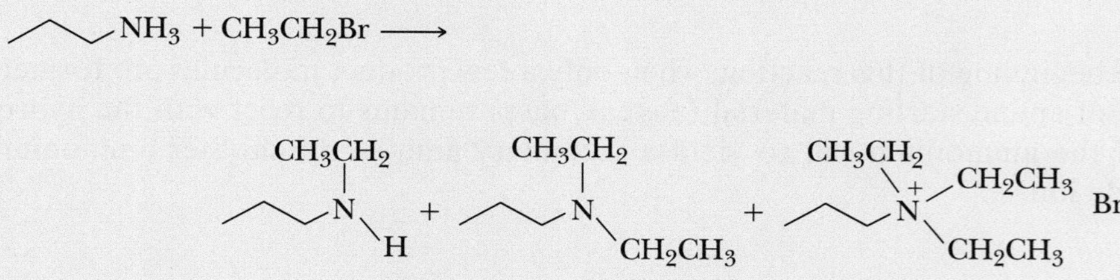

$$CH_3CH_2CH_2NH_3 + CH_3CH_2Br \longrightarrow$$

$$CH_3CH_2CH_2N(H)CH_2CH_3 + CH_3CH_2CH_2N(CH_2CH_3)_2 + CH_3CH_2CH_2\overset{+}{N}(CH_2CH_3)_3\ Br^-$$

Problem 10.10

Determine all possible nitrogen-based products that can be formed in the following reaction:

$$\text{piperidine (NH)} \quad + \quad CH_3Cl \longrightarrow$$

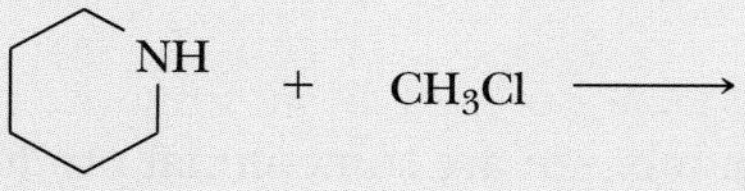

Key Terms and Concepts

aliphatic amine (p. 355)
alkaloid (p. 356)
amines as nucleophiles (p. 373)
ammonium salt (p. 360)
aniline (p. 358)
anisidine (p. 358)
arenediazonium salt (p. 372)
aromatic amine (p. 355)
base ionization constant (K_b) (p. 363)
heterocyclic amine (p. 355)
heterocyclic aromatic amine (p. 356)
pK_b (p. 363)
primary amine (p. 354)
secondary amine (p. 354)
tertiary amine (p. 354)
toluidine (p. 358)
zwitterion (p. 369)

Summary of Key Questions

10.1 What Are Amines?

- Amines are derivatives of ammonia (NH_3) in which one or more hydrogens are replaced by alkyl or aryl groups.
- Amines are classified as **primary**, **secondary**, or **tertiary**, depending on the number of hydrogen atoms of ammonia replaced by alkyl or aryl groups.
- In an **aliphatic amine**, all carbon atoms bonded to nitrogen are derived from alkyl groups.
- In an **aromatic amine**, one or more of the groups bonded to nitrogen are aryl groups.
- A **heterocyclic amine** is an amine in which the nitrogen atom is part of a ring.
- A **heterocyclic aromatic amine** is an amine in which the nitrogen atom is part of an aromatic ring.

10.2 How Are Amines Named?

- In systematic nomenclature, aliphatic amines are named **alkanamines**.
- In the common system of nomenclature, aliphatic amines are named **alkylamines**; the alkyl groups are listed in alphabetical order in one word ending in the suffix *-amine*.
- An ion containing nitrogen bonded to four alkyl or aryl groups is named as a quaternary ammonium ion.

10.3 What Are the Characteristic Physical Properties of Amines?

- Amines are polar compounds, and primary and secondary amines associate by intermolecular hydrogen bonding.
- Because an N—H----N hydrogen bond is weaker than an O—H----O hydrogen bond, amines have lower boiling points than alcohols of comparable molecular weight and structure.
- All classes of amines form hydrogen bonds with water and are more soluble in water than are hydrocarbons of comparable molecular weight.

10.4 What Are the Acid–Base Properties of Amines?

- Amines are weak bases, and aqueous solutions of amines are basic. The base ionization constant for an amine in water is given the symbol K_b.
- It is also common to discuss the acid–base properties of amines by reference to the acid ionization constant, K_a, for the conjugate acid of the amine.
- Acid and base ionization constants for an amine in water are related by the equation $pK_a + pK_b = 14.0$.

10.5 What Are the Reactions of Amines with Acids?

- Amines react quantitatively with strong acids to form water-soluble salts.
- The basicity of amines and the solubility of amine salts in water can be used to separate amines from water-insoluble, nonbasic compounds.

10.6 How Are Arylamines Synthesized?

- Arylamines can be made by reducing the nitro group on a benzene ring.

10.7 How Are Primary Aromatic Amines Used in Synthesis?

- Primary aromatic amines can be converted to diazonium salts, which consist of an $—N_2^+$ group bonded to the benzene ring.
- The $—N_2^+$ group of an arenediazonium salt can be replaced with an —OH group to form a phenol.
- The $—N_2^+$ group of an arenediazonium salt can be replaced with a —H atom.

10.8 How Do Amines Act as Nucleophiles?

- Amines are moderate nucleophiles and can participate in nucleophilic aliphatic substitution reactions.
- Reaction of ammonia or amines with haloalkanes often results in multiple products in varying ratios.

Quick Quiz

Answer true or false to the following questions to assess your general knowledge of the concepts in this chapter. If you have difficulty with any of them, you should review the appropriate section in the chapter (shown in parentheses) before attempting the more challenging end-of-chapter problems.

1. An amine with an $—NH_2$ group bonded to a tertiary carbon is classified as a tertiary amine. (10.1)
2. A hydroxyl group can be directly added to a benzene ring via an electrophilic aromatic substitution reaction. (10.7)
3. An efficient way to make diethylamine is to react ammonia with two equivalents of chloroethane. (10.8)
4. The IUPAC name of $CH_3CH_2CH_2CH_2NHCH_3$ is 2-pentanamine. (10.2)
5. An amino group can be directly added to a benzene ring via an electrophilic aromatic substitution reaction. (10.6)
6. A tertiary amine would be expected to be more water soluble than a secondary amine of the same molecular formula. (10.3)
7. The pK_b of an amine can be determined from the pK_a of its conjugate acid. (10.4)
8. The lower the value of pK_b, the stronger the base. (10.4)
9. The basicity of amines and the solubility of amine salts in water can be used to separate amines from water-insoluble, nonbasic compounds. (10.5)
10. Aromatic amines are more basic than aliphatic amines. (10.4)
11. A heterocyclic aromatic amine must contain one or more aryl groups directly bonded to nitrogen outside of the ring. (10.1)
12. Guanidine is a strong neutral base because its conjugate acid is resonance stabilized. (10.4)
13. Ammonia is a slightly weaker base than most aliphatic amines. (10.4)
14. An amino group forms stronger hydrogen bonds than a hydroxy group. (10.3)
15. A heterocyclic amine must contain a ring and a nitrogen atom as a member of the ring. (10.1)
16. An electron-withdrawing group in an amine decreases its basicity. (10.4)

Answers: (1) F (2) F (3) F (4) F (5) F (6) F (7) T (8) T (9) T (10) F (11) F (12) T (13) T (14) F (15) T (16) T

Key Reactions

1. Basicity of Aliphatic Amines (Section 10.4)
Most aliphatic amines have comparable basicities (pK_b 3.0–4.0) and are slightly stronger bases than ammonia:

$$CH_3NH_2 + H_2O \rightleftharpoons CH_3NH_3^+ + OH^- \quad pK_b = 3.36$$

2. Basicity of Aromatic Amines (Section 10.4)
Aromatic amines (pK_b 9.0–10.0) are considerably weaker bases than are aliphatic amines. Resonance stabilization from interaction of the unshared electron pair on nitrogen with the pi system of the aromatic ring decreases the availability of that electron pair for reaction with an acid. Substitution on the ring by electron-withdrawing groups decreases the basicity of the $—NH_2$ group:

$$C_6H_5—NH_2 + H_2O \rightleftharpoons C_6H_5—NH_3^+ + OH^- \quad pK_b = 9.37$$

3. Reaction of Amines with Strong Acids (Section 10.5)
All amines react quantitatively with strong acids to form water-soluble salts:

$$C_6H_5—N(CH_3)_2 + HCl \longrightarrow C_6H_5—\overset{+}{N}H(CH_3)_2\,Cl^-$$

Insoluble in water — A water-soluble salt

4. Reduction of an Aromatic NO_2 Group (Section 10.6)
An NO_2 group on an aromatic ring can be reduced to an amino group by catalytic hydrogenation or by treatment with a metal and hydrochloric acid, followed by a strong base to liberate the free amine:

$$C_6H_5NO_2 + 3H_2 \xrightarrow[(3\ atm)]{Ni} C_6H_5NH_2$$

$$\text{1,3-}C_6H_4(NO_2)_2 \xrightarrow[C_2H_5OH,H_2O]{Fe,\ HCl} \text{1,3-}C_6H_4(NH_3^+Cl^-)_2 \xrightarrow{NaOH,\ H_2O} \text{1,3-}C_6H_4(NH_2)_2$$

5. Conversion of a Primary Aromatic Amine to a Phenol (Section 10.7)
Treatment of a primary aromatic amine with nitrous acid gives an arenediazonium salt. Heating the aqueous solution of this salt brings about the evolution of N_2 and forms a phenol:

$$\text{2-}CH_3C_6H_4NH_2 \xrightarrow[0°C]{NaNO_2\ HCl} \text{2-}CH_3C_6H_4N_2^+Cl^- \xrightarrow{heat} \text{2-}CH_3C_6H_4OH$$

6. Reduction of an Arenediazonium Salt (Section 10.7)
Treatment of an arenediazonium salt with hypophosphorous acid, H_3PO_2, results in replacement of the N_2^+ group by H:

$$\text{2,4-}Cl_2C_6H_3NH_2 \xrightarrow[0°C]{NaNO_2\ HCl} \text{2,4-}Cl_2C_6H_3N_2^+Cl^- \xrightarrow{H_3PO_2} \text{1,3-}C_6H_4Cl_2$$

Problems

A problem marked with an asterisk indicates an applied "real world" problem. Answers to problems whose numbers are printed in blue are given in Appendix D.

Structure and Nomenclature

10.11 Draw a structural formula for each amine: **(See Examples 10.2, 10.3)**

(a) (R)-2-Butanamine
(b) 1-Octanamine
(c) 2,2-Dimethyl-1-propanamine
(d) 1,5-Pentanediamine
(e) 2-Bromoaniline
(f) Tributylamine

(g) *N,N*-Dimethylaniline
(h) Benzylamine
(i) *tert*-Butylamine
(j) *N*-Ethylcyclohexanamine
(k) Diphenylamine
(l) Isobutylamine

10.12 Draw a structural formula for each amine: **(See Examples 10.2, 10.3)**

(a) 4-Aminobutanoic acid
(b) 2-Aminoethanol (ethanolamine)
(c) 2-Aminobenzoic acid
(d) (*S*)-2-Aminopropanoic acid (alanine)
(e) 4-Aminobutanal
(f) 4-Amino-2-butanone

10.13 Draw examples of 1°, 2°, and 3° amines that contain at least four sp^3 hybridized carbon atoms. Using the same criterion, provide examples of 1°, 2°, and 3° alcohols. How does the classification system differ between the two functional groups? **(See Example 10.1)**

***10.14** Classify each amino group as primary, secondary, or tertiary and as aliphatic or aromatic: **(See Example 10.1)**

(a) H_2N O O

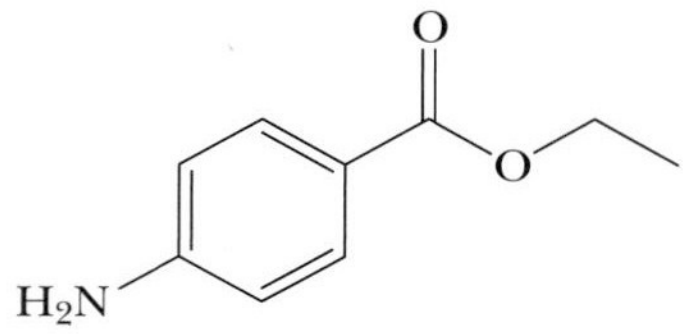

Benzocaine
(a topical anesthetic)

(b) H N N Cl N

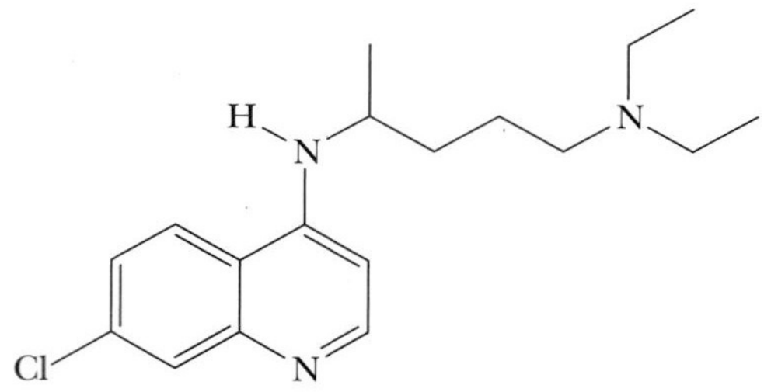

Chloroquine
(a drug for the treatment of malaria)

***10.15** Epinephrine is a hormone secreted by the adrenal medulla. Among epinephrine's actions, it is a bronchodilator. Albuterol, sold under several trade names, including Proventil® and Salbumol®, is one of the most effective and widely prescribed antiasthma drugs. The R enantiomer of albuterol is 68 times more effective in the treatment of asthma than the S enantiomer. **(See Example 10.1)**

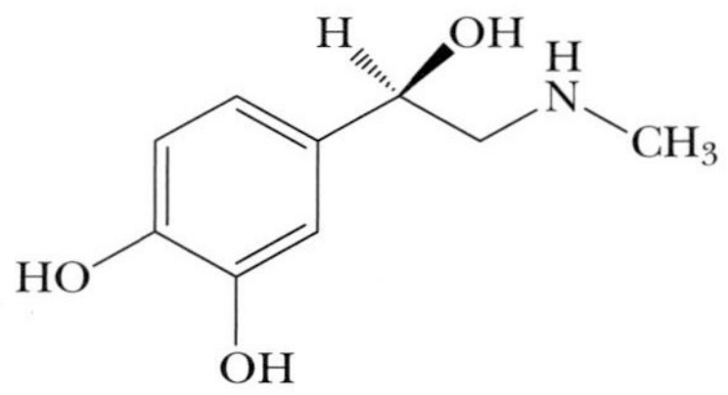

H OH H N CH_3 HO OH

(*R*)-Epinephrine
(Adrenaline)

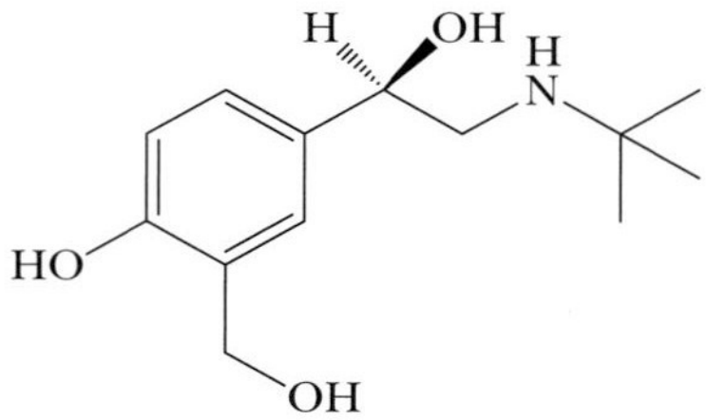

H OH H N HO OH

(*R*)-Albuterol

(a) Classify each amino group as primary, secondary, or tertiary.
(b) List the similarities and differences between the structural formulas of these compounds.

10.16 There are eight constitutional isomers with the molecular formula $C_4H_{11}N$. Name and draw structural formulas for each. Classify each amine as primary, secondary, or tertiary. **(See Examples 10.1–10.3)**

10.17 Draw a structural formula for each compound with the given molecular formula: **(See Example 10.3)**

(a) A 2° arylamine, C_7H_9N
(b) A 3° arylamine, $C_8H_{11}N$
(c) A 1° aliphatic amine, C_7H_9N
(d) A chiral 1° amine, $C_4H_{11}N$
(e) A 3° heterocyclic amine, $C_5H_{11}N$
(f) A trisubstituted 1° arylamine, $C_9H_{13}N$
(g) A chiral quaternary ammonium salt, $C_9H_{22}NCl$

Physical Properties

10.18 Propylamine, ethylmethylamine, and trimethylamine are constitutional isomers with the molecular formula C_3H_9N: **(See Example 10.4)**

$CH_3CH_2CH_2NH_2$	$CH_3CH_2NHCH_3$	$(CH_3)_3N$
bp 48°C	bp 37°C	bp 3°C
Propylamine	Ethylmethylamine	Trimethylamine

Account for the fact that trimethylamine has the lowest boiling point of the three, and propylamine has the highest.

10.19 Account for the fact that 1-butanamine has a lower boiling point than 1-butanol: **(See Example 10.4)**

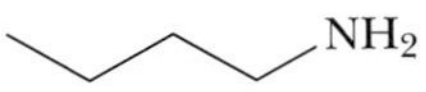

$CH_3CH_2CH_2CH_2NH_2$ — bp 78°C, 1-Butanamine

$CH_3CH_2CH_2CH_2OH$ — bp 117°C, 1-Butanol

***10.20** Account for the fact that putrescine, a foul-smelling compound produced by rotting flesh, ceases to smell upon treatment with two equivalents of HCl: **(See Example 10.4)**

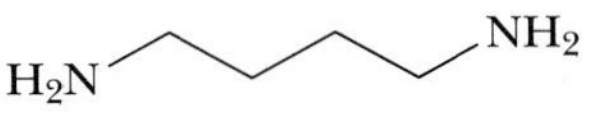

$H_2NCH_2CH_2CH_2CH_2NH_2$

1,4-Butanediamine (Putrescine)

Basicity of Amines

10.21 Account for the fact that amines are more basic than alcohols. **(See Example 10.6)**

10.22 From each pair of compounds, select the stronger base: **(See Example 10.6)**

(a) 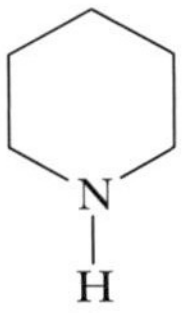 piperidine (N–H) or pyridine

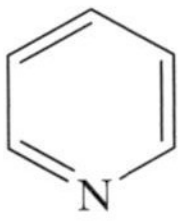

(b) 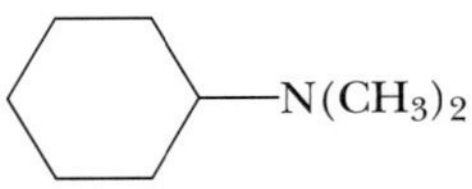 cyclohexyl–$N(CH_3)_2$ or phenyl–$N(CH_3)_2$

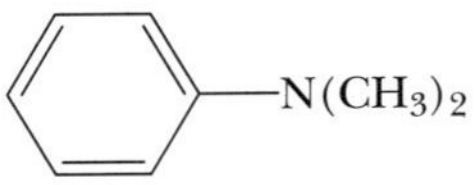

(c) 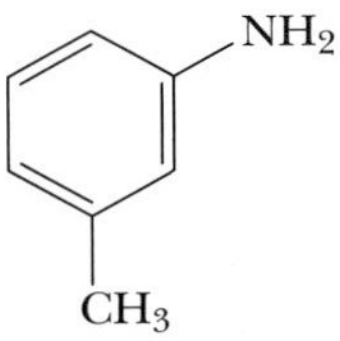 3-methylaniline (NH_2, CH_3) or benzyl amine (CH_2NH_2)

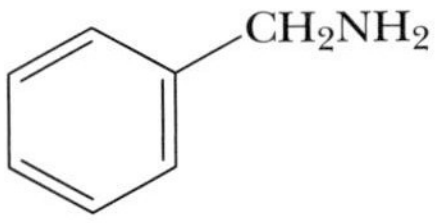

(d) 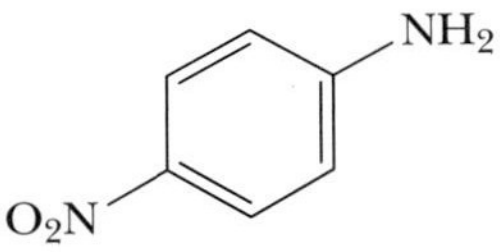 4-nitroaniline (O_2N, NH_2) or 4-methylaniline (H_3C, NH_2)

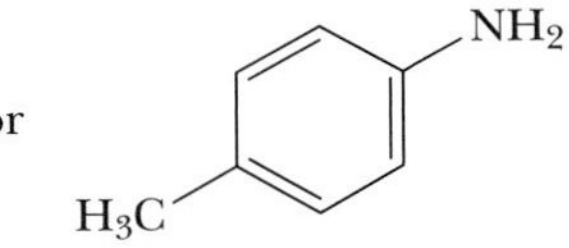

10.23 Account for the fact that substitution of a nitro group makes an aromatic amine a weaker base, but makes a phenol a stronger acid. For example, 4-nitroaniline is a weaker base than aniline, but 4-nitrophenol is a stronger acid than phenol. **(See Example 10.6)**

10.24 Select the stronger base in this pair of compounds: **(See Example 10.6)**

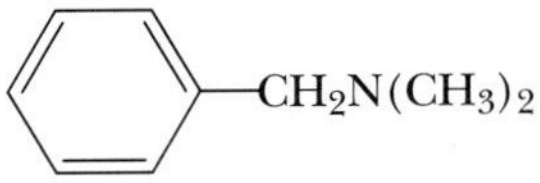

$C_6H_5CH_2N(CH_3)_2$ or $C_6H_5CH_2\overset{+}{N}(CH_3)_3\,OH^-$

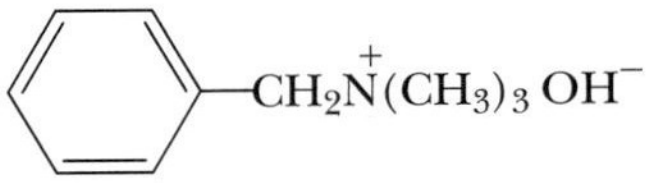

10.25 Complete the following acid–base reactions and predict the position of equilibrium for each. Justify your prediction by citing values of pK_a for the stronger and weaker acid in each equilibrium. For values of acid ionization constants, consult Table 2.2 (pK_a's of some inorganic and organic acids), Table 8.2 (pK_a's of alcohols), Section 9.8B (acidity of phenols), and Table 10.2 (base strengths of amines). Where no ionization constants are given, make the best estimate from aforementioned tables and section. **(See Examples 10.5–10.7)**

(a) CH_3COOH + pyridine $\rightleftharpoons$

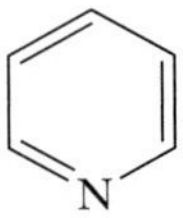

Acetic acid, Pyridine

(b) 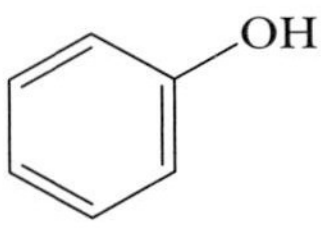 phenol (OH) + $(CH_3CH_2)_3N$ $\rightleftharpoons$

Phenol, Triethylamine

(c) $PhCH_2CH(CH_3)NH_2 + CH_3CH(OH)COOH \rightleftharpoons$

1-Phenyl-2-propanamine (Amphetamine), 2-Hydroxypropanoic acid (Lactic acid)

(d) $PhCH_2CH(CH_3)NHCH_3 + CH_3COOH \rightleftharpoons$

Methamphetamine, Acetic acid

10.26 The pK_a of the morpholinium ion is 8.33:

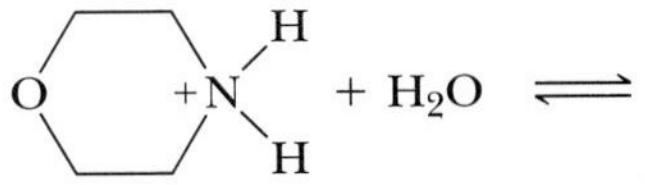

Morpholinium ion (O ring with $^+NH_2$) + H_2O $\rightleftharpoons$ Morpholine (O ring with NH) + H_3O^+ pK_a = 8.33

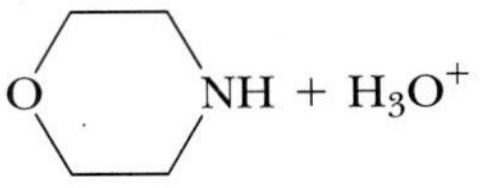

(a) Calculate the ratio of morpholine to morpholinium ion in aqueous solution at pH 7.0.

(b) At what pH are the concentrations of morpholine and morpholinium ion equal?

***10.27** The pK_b of amphetamine (Example 10.2e) is approximately 3.2. Calculate the ratio of amphetamine to its conjugate acid at pH 7.4, the pH of blood plasma.

10.28 Calculate the ratio of amphetamine to its conjugate acid at pH 1.0, such as might be present in stomach acid.

***10.29** Following is a structural formula of pyridoxamine, one form of vitamin B_6: **(See Examples 10.6, 10.7)**

CH_2NH_2, HO, CH_2OH, H_3C, N

Pyridoxamine
(Vitamin B_6)

(a) Which nitrogen atom of pyridoxamine is the stronger base?

(b) Draw the structural formula of the hydrochloride salt formed when pyridoxamine is treated with one mole of HCl.

***10.30** Epibatidine, a colorless oil isolated from the skin of the Ecuadorian poison frog *Epipedobates tricolor*, has several times the analgesic potency of morphine. It is the first chlorine-containing, nonopioid (nonmorphine-like in structure) analgesic ever isolated from a natural source: **(See Example 10.6)**

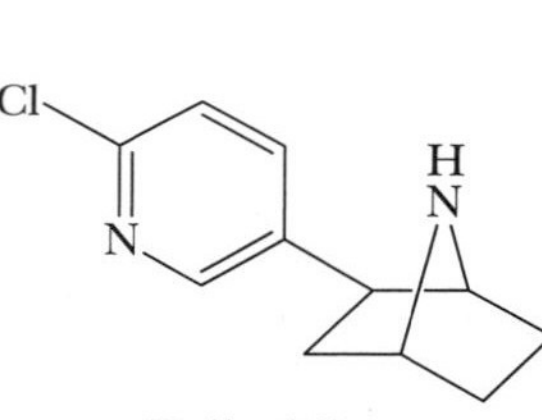

Epibatidine

Poison arrow frog.
(Stephen J. Krasemann/ Photo Researchers, Inc.)

(a) Which of the two nitrogen atoms of epibatidine is the more basic?

(b) Mark all stereocenters in this molecule.

***10.31** Procaine was one of the first local anesthetics for infiltration and regional anesthesia: **(See Examples 10.6, 10.7)**

O, O, N, H_2N

Procaine

The hydrochloride salt of procaine is marketed as Novocaine®.

(a) Which nitrogen atom of procaine is the stronger base?

(b) Draw the formula of the salt formed by treating procaine with one mole of HCl.

(c) Is procaine chiral? Would a solution of Novocaine in water be optically active or optically inactive?

***10.32** Treatment of trimethylamine with 2-chloroethyl acetate gives the neurotransmitter acetylcholine as its chloride salt: **(See Example 10.7)**

$$(CH_3)_3N + CH_3\overset{\overset{\large O}{\|}}{C}OCH_2CH_2Cl \longrightarrow C_7H_{16}ClNO_2$$

Acetylcholine chloride

Propose a structural formula for this quaternary ammonium salt and a mechanism for its formation.

10.33 Aniline is prepared by the catalytic reduction of nitrobenzene:

$$C_6H_5{-}NO_2 \xrightarrow[Ni]{H_2} C_6H_5{-}NH_2$$

Devise a chemical procedure based on the basicity of aniline to separate it from any unreacted nitrobenzene.

10.34 Suppose that you have a mixture of the following three compounds:

H_3C—C_6H_4—NO_2
4-Nitrotoluene
(*p*-Nitrotoluene)

H_3C—C_6H_4—NH_2
4-Methylaniline
(*p*-Toluidine)

H_3C—C_6H_4—OH
4-Methylphenol
(*p*-Cresol)

Devise a chemical procedure based on their relative acidity or basicity to separate and isolate each in pure form.

***10.35** Following is a structural formula for metformin, the hydrochloride salt of which is marketed as the antidiabetic Glucophage®: **(See Example 10.7)**

NH, NH, H_3C, N, N, NH_2, CH_3, H

Metformin

Metformin was introduced into clinical medicine in the United States in 1995 for the treatment of type 2 diabetes. More than 25 million prescriptions for this drug were written in 2000, making it the most commonly prescribed brand-name diabetes medication in the nation.

(a) Draw the structural formula for Glucophage®.

(b) Would you predict Glucophage® to be soluble or insoluble in water? Soluble or insoluble in blood plasma? Would you predict it to be soluble or insoluble in diethyl ether? In dichloromethane? Explain your reasoning.

Synthesis

***10.36** 4-Aminophenol is a building block in the synthesis of the analgesic acetaminophen. Show how this building block can be synthesized in two steps from phenol (in Chapter 15, we will see how to complete the synthesis of acetaminophen): **(See Example 10.9)**

Phenol →(1) 4-Nitrophenol →(2) 4-Aminophenol - - → Acetaminophen

HO, NO_2, NH_2, NH—C(=O)—CH_3

***10.37** 4-Aminobenzoic acid is a building block in the synthesis of the topical anesthetic benzocaine. Show how this building block can be synthesized in three steps from toluene (in Chapter 15, we will see how to complete the synthesis of benzocaine): **(See Example 10.9)**

Toluene (CH_3) →(1) O_2N/CH_3 →(2) O_2N/COOH →(3) 4-Aminobenzoic acid (H_2N/COOH) - - → Ethyl 4-aminobenzoate (Benzocaine) (H_2N/C(=O)—OC_2H_5)

***10.38** The compound 4-aminosalicylic acid is one of the building blocks needed for the synthesis of propoxycaine, one of the family of "caine" anesthetics. Some other members of this family of local anesthetics are procaine (Novocaine®), lidocaine (Xylocaine®), and mepivicaine (Carbocaine®). 4-Aminosalicylic acid is synthesized from salicylic acid in five steps (in Chapter 15, we will see how to complete the synthesis of propoxycaine): **(See Example 10.9)**

Salicylic acid (OH, COOH) →(1) O_2N/OH/COOH →(2) H_2N/OH/COOH →(3) O_2N/H_2N/OH/COOH →(4) O_2N/OH/COOH →(5) 4-Aminosalicylic acid (H_2N/OH/COOH) - - → Propoxycaine

Show reagents that will bring about the synthesis of 4-aminosalicylic acid.

*10.39 A second building block for the synthesis of propoxycaine is 2-diethylaminoethanol: **(See Example 10.9)**

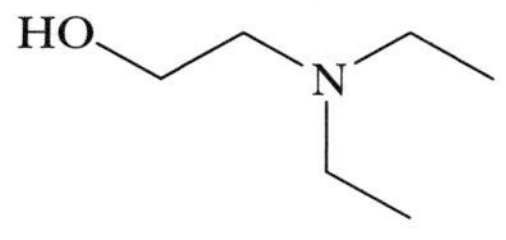

HO N

2-Diethylaminoethanol

Show how this compound can be prepared from ethylene oxide and diethylamine.

*10.40 Following is a two-step synthesis of the antihypertensive drug propranolol, a so-called beta blocker with vasodilating action: **(See Example 10.9)**

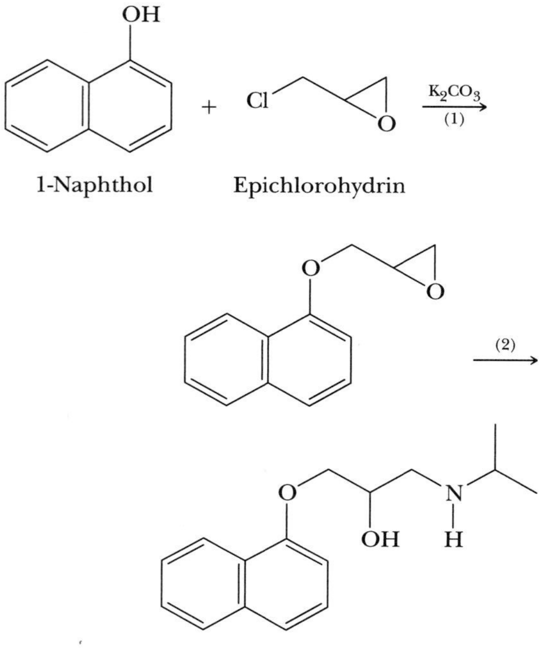

OH + Cl O $\xrightarrow[(1)]{K_2CO_3}$

1-Naphthol Epichlorohydrin

O O $\xrightarrow{(2)}$

O OH N H

Propranolol
(Cardinol)

Propranolol and other beta blockers have received enormous clinical attention because of their effectiveness in treating hypertension (high blood pressure), migraine headaches, glaucoma, ischemic heart disease, and certain cardiac arrhythmias. The hydrochloride salt of propranolol has been marketed under at least 30 brand names, one of which is Cardinol®. (Note the "card-" part of the name, after *cardiac*.)

(a) What is the function of potassium carbonate, K_2CO_3, in Step 1? Propose a mechanism for the formation of the new oxygen–carbon bond in this step.

(b) Name the amine used to bring about Step 2, and propose a mechanism for this step.

(c) Is propranolol chiral? If so, how many stereoisomers are possible for it?

*10.41 The compound 4-ethoxyaniline, a building block of the over-the-counter analgesic phenacetin, is synthesized in three steps from phenol: **(See Example 10.9)**

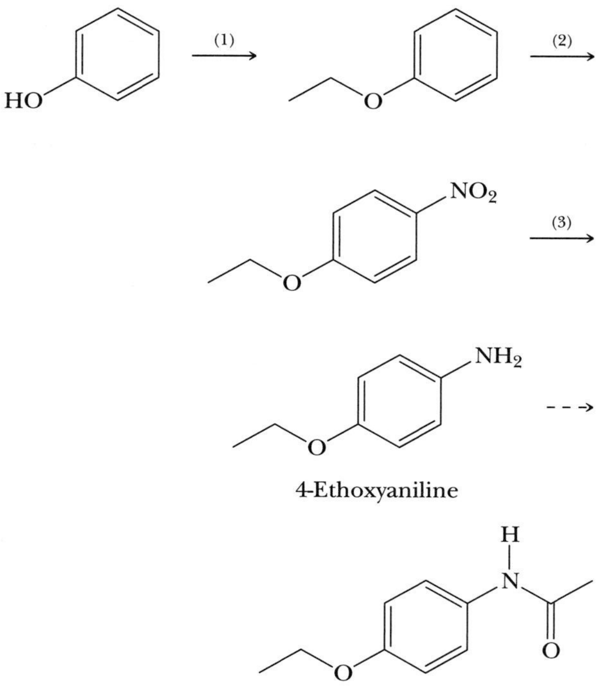

HO $\xrightarrow{(1)}$ O $\xrightarrow{(2)}$

O NO_2 $\xrightarrow{(3)}$

O NH_2 - - →

4-Ethoxyaniline

O N H O

Phenacetin

Show reagents for each step of the synthesis of 4-ethoxyaniline. (In Chapter 15, we will see how to complete this synthesis.)

*10.42 Radiopaque imaging agents are substances administered either orally or intravenously that absorb X rays more strongly than body material does. One of the best known of these agents is barium sulfate, the key ingredient in the "barium cocktail" used for imaging of the gastrointestinal tract. Among other X-ray imaging agents are the so-called triiodoaromatics. You can get some idea of the kinds of imaging for which they are used from the following selection of trade names: Angiografin®, Gastrografin, Cardiografin, Cholografin, Renografin, and Urografin®. The most common of the triiodiaromatics are derivatives of these three triiodobenzenecarboxylic acids: **(See Example 10.9)**

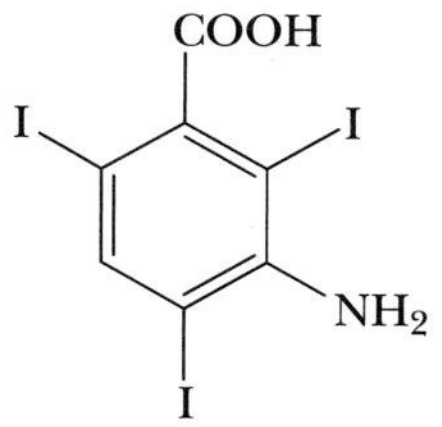

COOH I I NH_2 I

3-Amino-2,4,6-triiodobenzoic acid

COOH I I H_2N NH_2 I

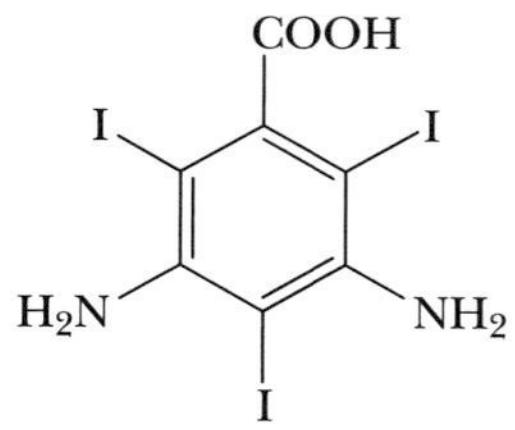

3,5-Diamino-2,4,6-triiodobenzoic acid

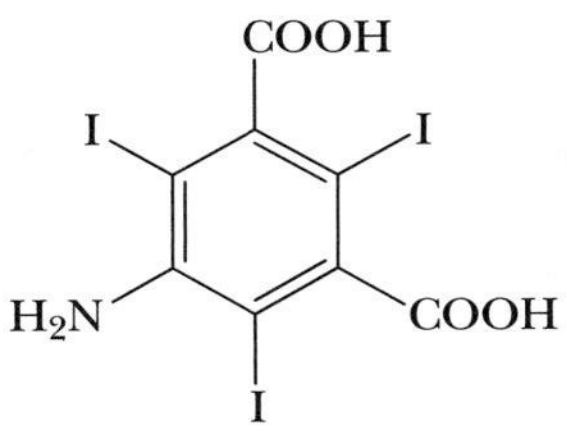

5-Amino-2,4,6-triiodoisophthalic acid

3-Amino-2,4,6-triiodobenzoic acid is synthesized from benzoic acid in three steps:

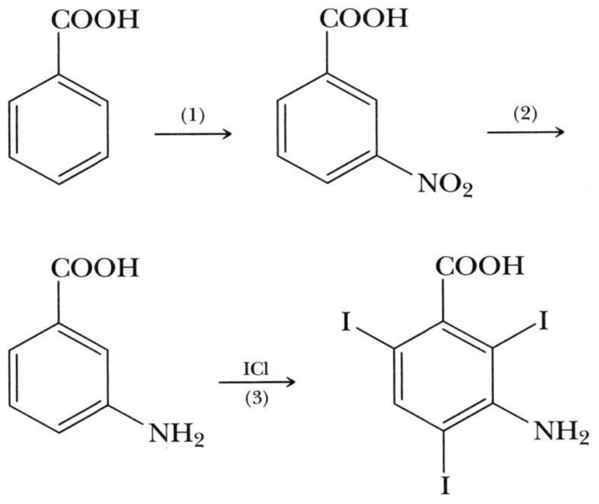

(1) (2) ICl (3)

3-Aminobenzoic acid

3-Amino-2,4,6-triiodobenzoic acid

(a) Show reagents for Steps (1) and (2).

(b) Iodine monochloride, ICl, a black crystalline solid with a melting point of 27.2°C and a boiling point of 97°C, is prepared by mixing equimolar amounts of I_2 and Cl_2. Propose a mechanism for the iodination of 3-aminobenzoic acid by this reagent.

(c) Show how to prepare 3,5-diamino-2,4,6-triiodobenzoic acid from benzoic acid.

(d) Show how to prepare 5-amino-2,4,6-triiodoisophthalic acid from isophthalic acid (1,3-benzenedicarboxylic acid).

***10.43** The intravenous anesthetic propofol is synthesized in four steps from phenol: **(See Example 10.9)**

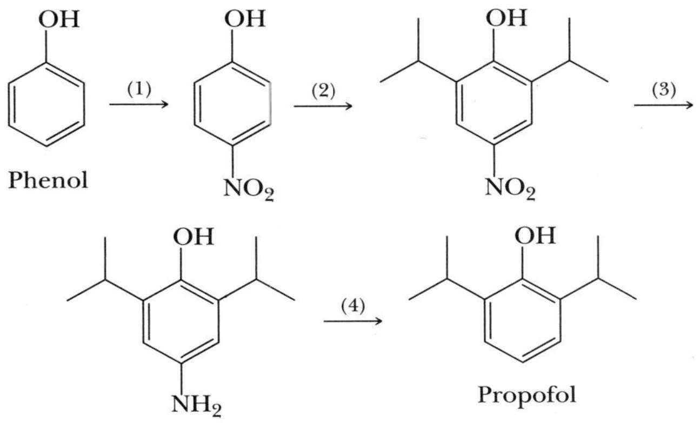

Phenol (1) (2) (3) (4) Propofol

Show reagents to bring about each step.

Chemical Transformations

10.44 Test your cumulative knowledge of the reactions learned thus far by completing the following chemical transformations. *Note: Some will require more than one step.* **(See Example 10.9)**

(a)

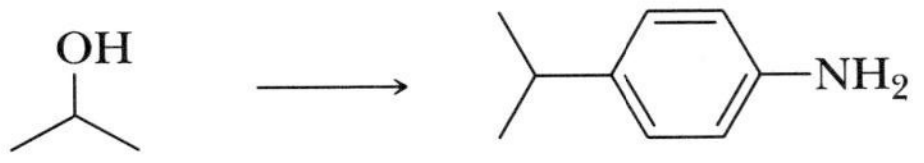

(b)

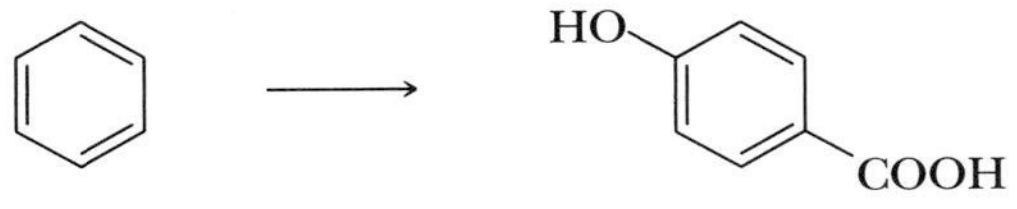

(c)

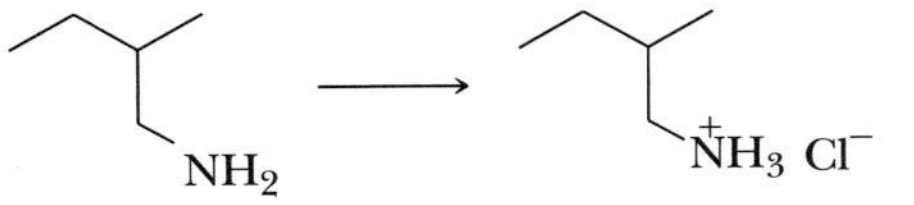

(d)

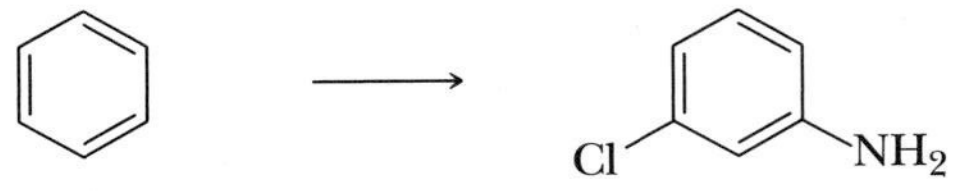

(e)

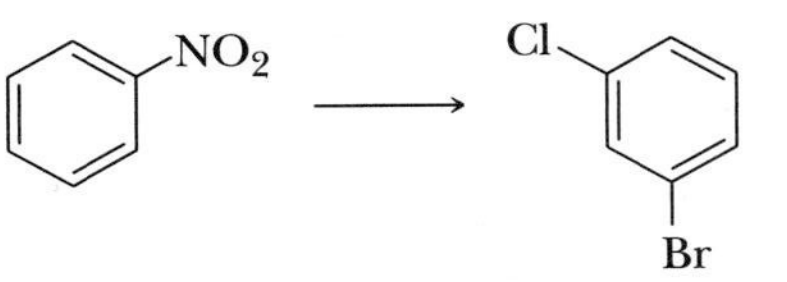

(f)

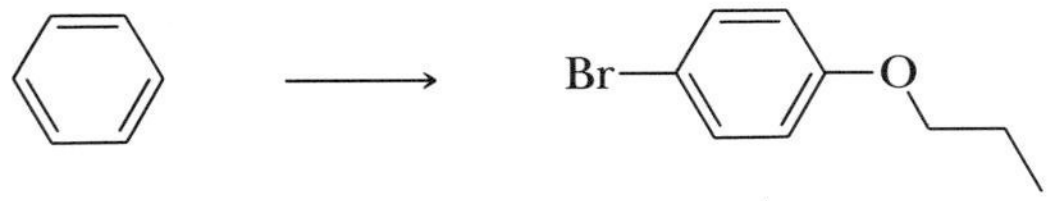

(g)

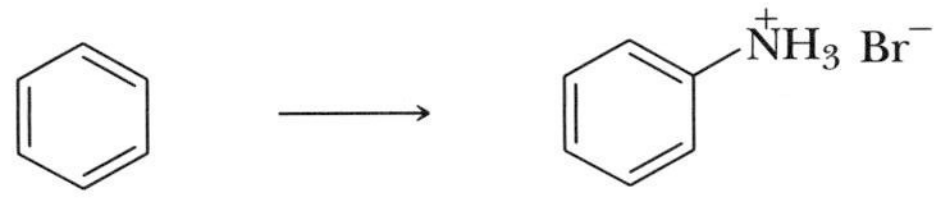

(h)

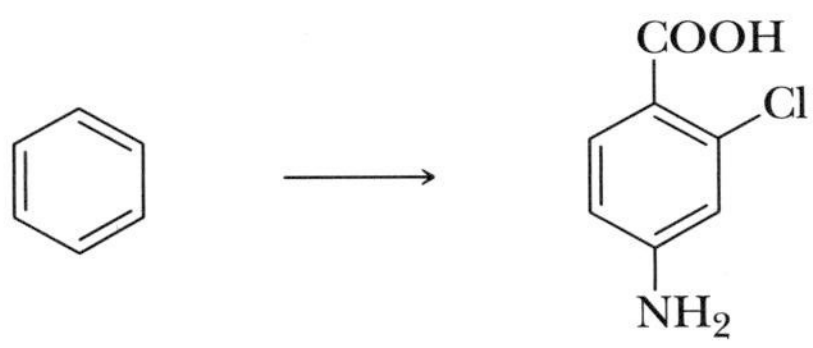

(i)

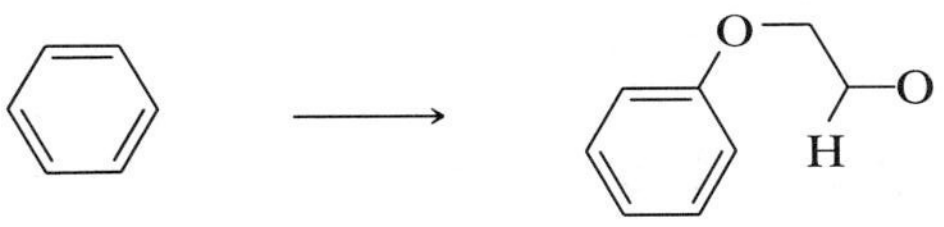

(j)

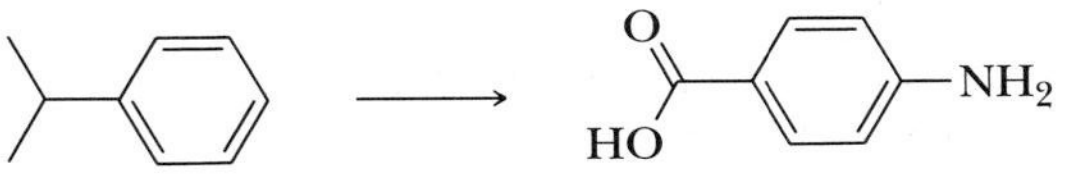

(k) chlorobenzene ⟶ 1,3-dichlorobenzene (Cl, Cl)

(l) benzene ⟶ Br–C_6H_4–NH–CH_2CH_2–OH

Looking Ahead

10.45 State the hybridization of the nitrogen atom in each of the following compounds:

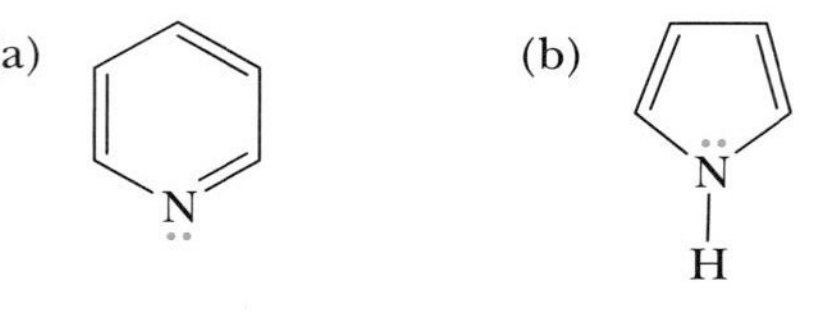

(c) NH_2 (d) H_3C, O, N, CH_3, CH_3

10.46 Amines can act as nucleophiles. For each of the following molecules, circle the most likely atom that would be attacked by the nitrogen of an amine:

(a) (b) O, OCH_3

(c) Cl, Br

10.47 Draw a Lewis structure for a molecule with formula C_3H_7N that does not contain a ring or an alkene (a carbon–carbon double bond).

10.48 Rank the following leaving groups in order from best to worst:

R—Cl R—O—C(=O)—R R—OCH_3 R—$N(CH_3)_2$

Putting It Together

The following problems bring together concepts and material from Chapters 7–10. Although the focus may be on these chapters, the problems will also build on concepts discussed throughout the text thus far.

Choose the best answer for each of the following questions.

1. Arrange the following amines from lowest to highest boiling point.

A CH_3CH_2–N–H with CH_3 **B** $CH_3CH_2CH_2$—NH_2 **C** H_3C–N–CH_3 with CH_3

(a) **A, B, C** (b) **C, B, A** (c) **B, C, A**
(d) **B, A, C** (e) **C, A, B**

2. Which of the following statements is true regarding the following two molecules?

A N—H **B** N—H

(a) Both **A** and **B** are aromatic.
(b) Both **A** and **B** are aliphatic amines.
(c) The nitrogen atoms in **A** and **B** are both sp^3 hybridized.
(d) **B** is more basic than **A**.
(e) Both **A** and **B** are planar molecules.

3. Which series of reagents can be used to achieve the following transformation?

cyclohexene ⟶ cyclohexanone (=O)

(a) 1. HBr 2. H_2SO_4
(b) 1. H_2SO_4, H_2O 2. PCC
(c) 1. HCl 2. $SOCl_2$
(d) 1. H_3PO_4, H_2O 2. H_2CrO_4
(e) More than one of these will achieve the transformation.

4. Arrange the following from strongest to weakest base.

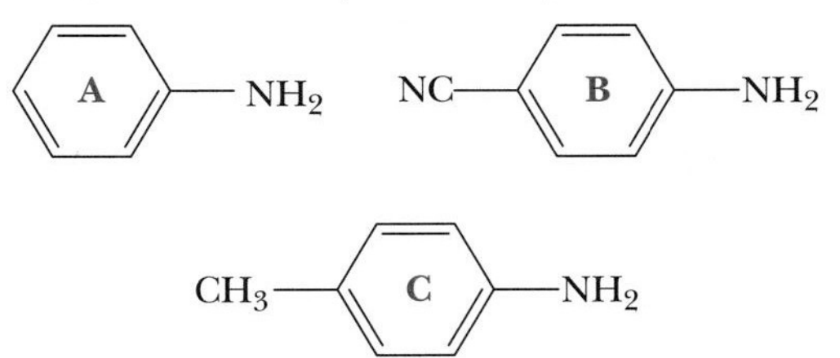

(a) **A, B, C** (b) **B, C, A** (c) **C, A, B** (d) **A, C, B** (e) **B, A, C**

5. How many products are possible from the following elimination reaction?

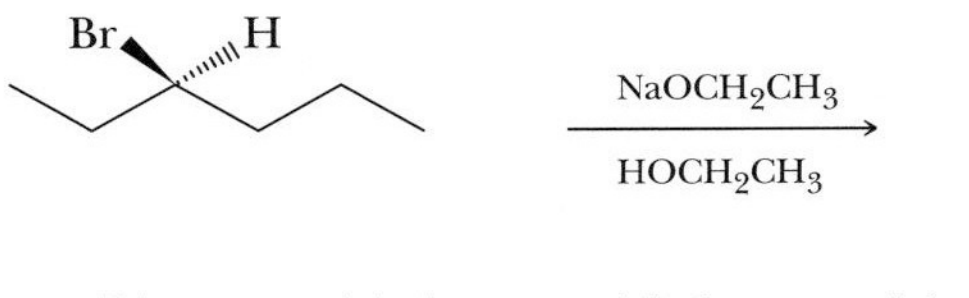

(a) one (b) two (c) three (d) four (e) six

6. Which series of reagents can be used to achieve the following transformation?

(a) 1. HCl	2. RCO_3H
(b) 1. $SOCl_2$	2. RCO_3H
(c) 1. Na	2. RCO_3H
(d) 1. H_3PO_4	2. RCO_3H
(e) 1. H_2CrO_4	2. RCO_3H

7. Consider the following situation: An ether solution containing phenol and a neutral compound is extracted with 30% sodium bicarbonate. Next the ether solution is extracted with 30% NaOH. Finally, the ether solution is extracted with distilled water. Which solution contains the phenol?

(a) The 30% sodium bicarbonate solution.
(b) The 30% NaOH solution.
(c) The ether.
(d) The distilled water.
(e) Not enough information to determine.

8. Which of the following statements is true concerning the following two molecules?

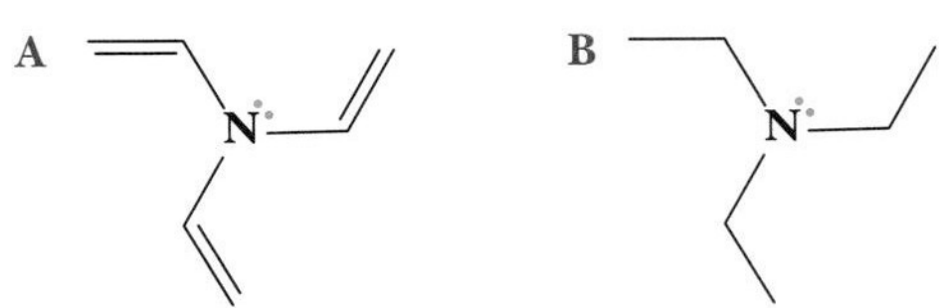

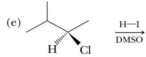

(a) Both are aromatic.
(b) Only one molecule is an amine.
(c) **B** is more polar than **A**.
(d) **A** is more basic than **B**.
(e) All of these statements are true.

9. Which combination of reagents would be most likely to undergo an S_N2 reaction?

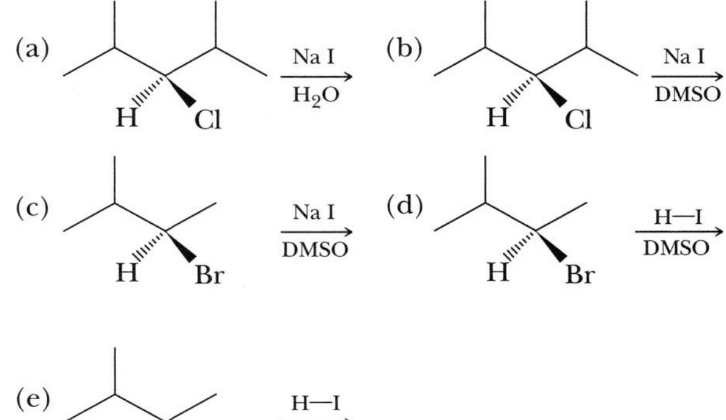

10. Which series of reagents can be used to achieve the following transformation?

(a) 1. $CH_3Br/FeBr_3$ 2. HNO_3/H_2SO_4 3. $H_2SO_4/K_2Cr_2O_4$	(b) 1. HNO_3/H_2SO_4 2. $H_2SO_4/K_2Cr_2O_4$ 3. $CH_3Br/FeBr_3$
(c) 1. $H_2SO_4/K_2Cr_2O_4$ 2. $CH_3Br/FeBr_3$ 3. HNO_3/H_2SO_4	(d) 1. $CH_3Br/FeBr_3$ 2. $H_2SO_4/K_2Cr_2O_4$ 3. HNO_3/H_2SO_4
(e) 1. HNO_3/H_2SO_4 2. $CH_3Br/FeBr_3$ 3. $H_2SO_4/K_2Cr_2O_4$	

11. Determine which aryl amine (**A** or **B**) is more basic and provide a rationale for your determination.

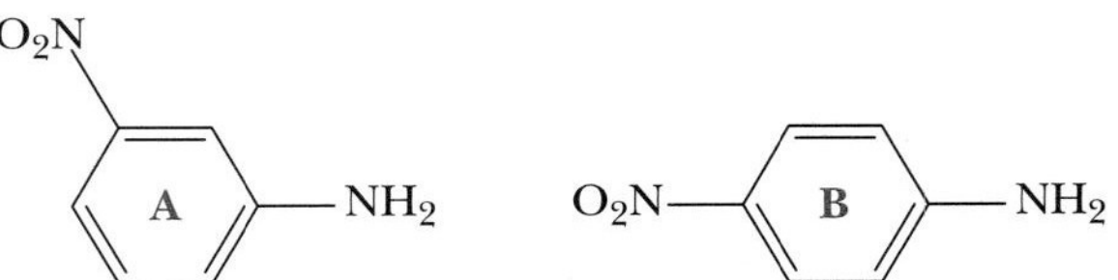

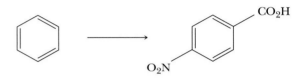

12. Answer the questions that follow regarding the compound Wyerone, which is obtained from fava beans (*Vicia faba*) and has been found to possess antifungal properties.

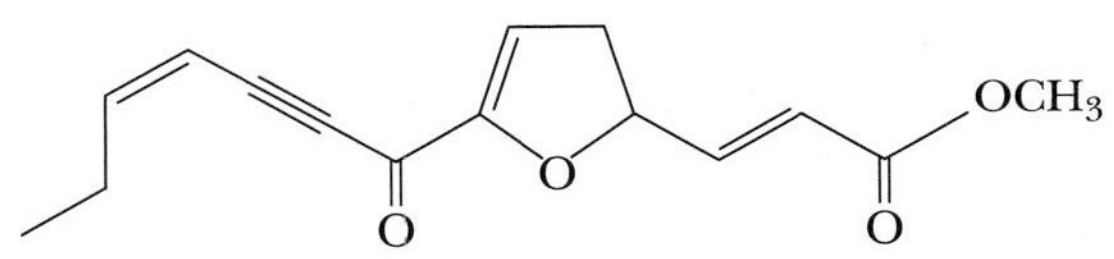

Wyerone

(a) Would you expect the compound to be soluble in water?

(b) How many stereoisomers exist for the compound shown?

(c) Is the molecule chiral?

(d) How many equivalents of Br_2 in CH_2Cl_2 would Wyerone be expected to react with?

13. Provide IUPAC names for the following compounds

(a) N

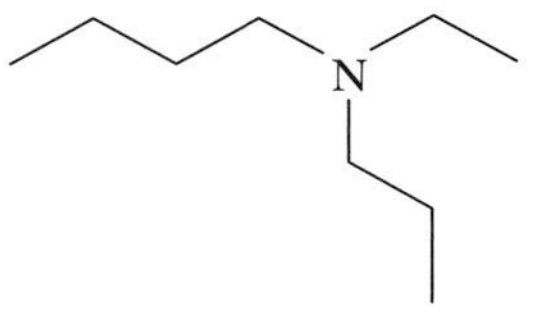

(b) OH

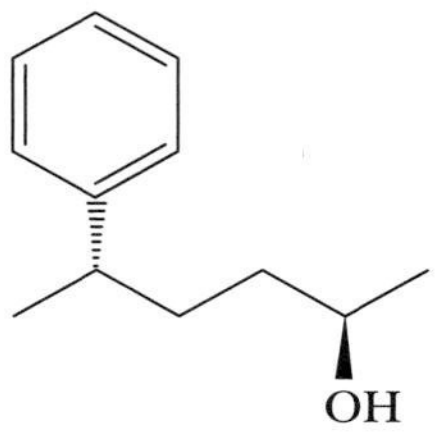

(c) O

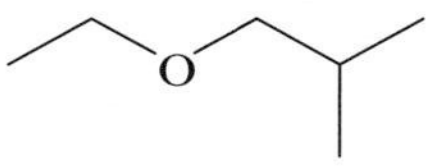

(d) OH

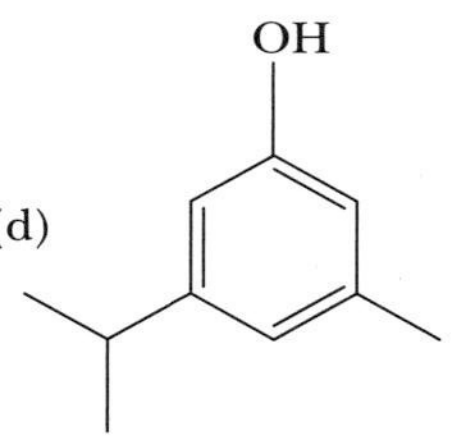

14. Determine whether highlighted proton **A** or **B** is more acidic and provide a rationale for your selection.

(a) H H

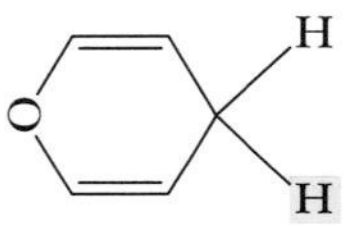

(b) N H H

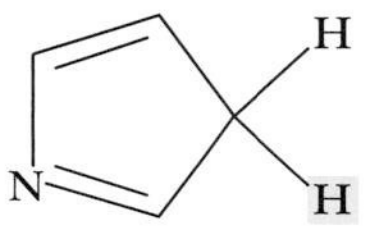

15. Select the answer that best fits each description and provide an explanation for your decision.

(a) The best nucleophile

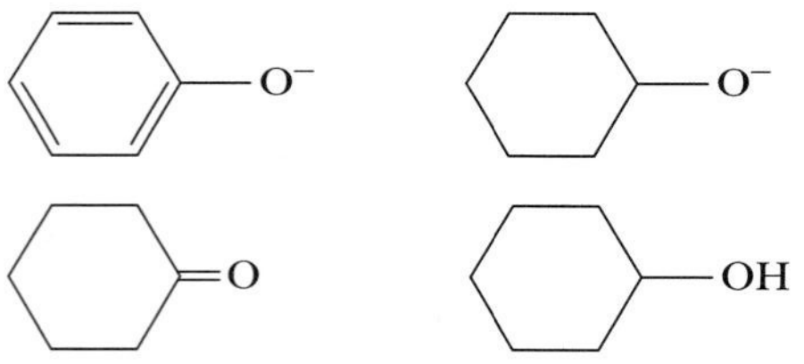

O^- O^- O OH

(b) The best leaving group

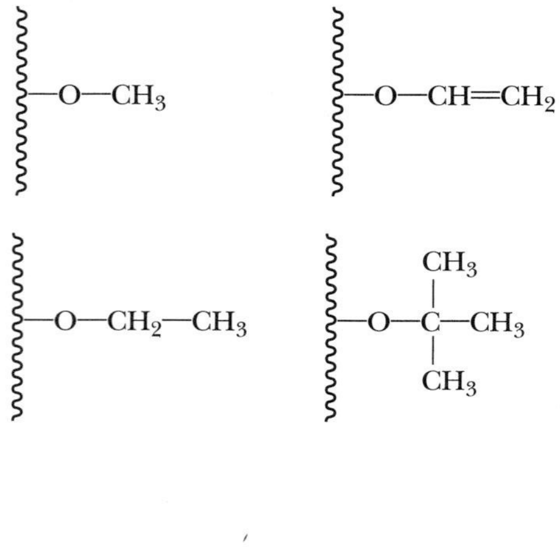

$-O-CH_3$ $-O-CH=CH_2$ $-O-CH_2-CH_3$ $-O-C(CH_3)_2-CH_3$

16. Provide a mechanism for the following reaction. Show all charges and lone pairs of electrons in your structures as well as the structures of all intermediates.

\+ H; H—Br

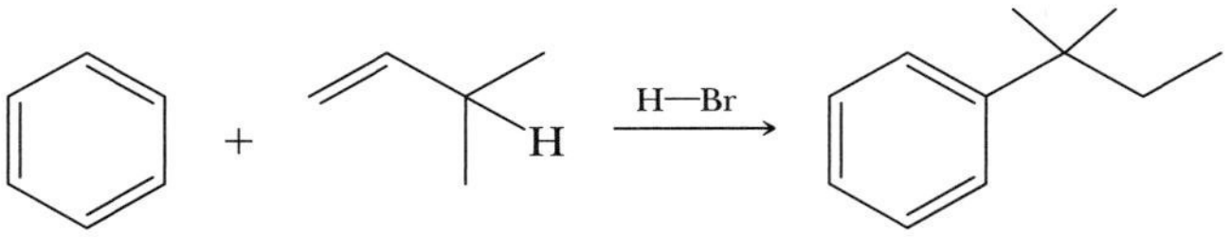

17. When the following nucleophilic substitution reaction was performed, the major product was found to possess the molecular formula $C_{13}H_{30}N$ rather than $C_5H_{13}N$, the formula of the desired product shown below. Provide the structure of the major product and explain why it is formed over the desired product.

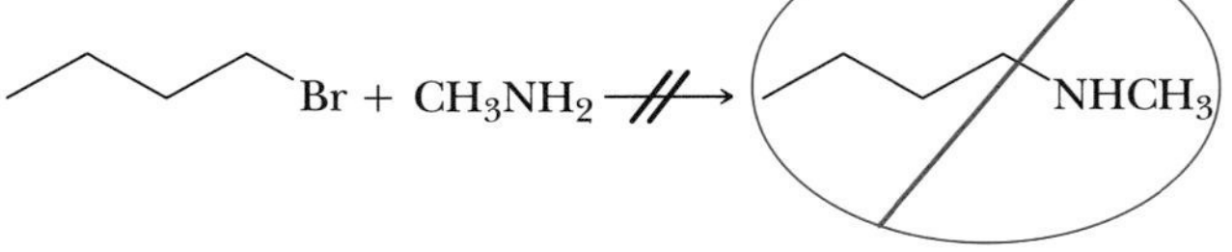

Br + CH_3NH_2 ⇸ $NHCH_3$

18. Complete the following chemical transformations.

(a) Cl → CN

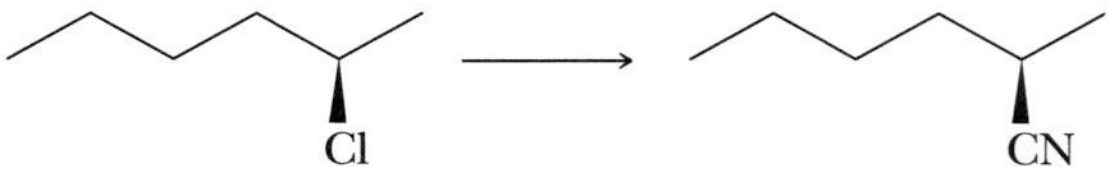

This enantiomer as the *only* product.

(b) OH → OH, O

(c) → OCH_3 (racemic)

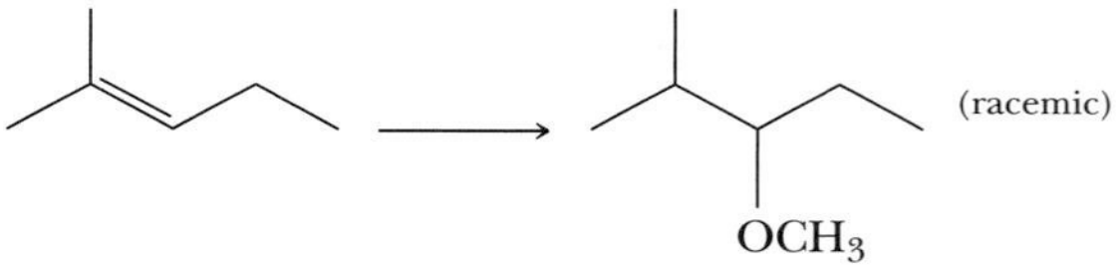

(d) Br → O

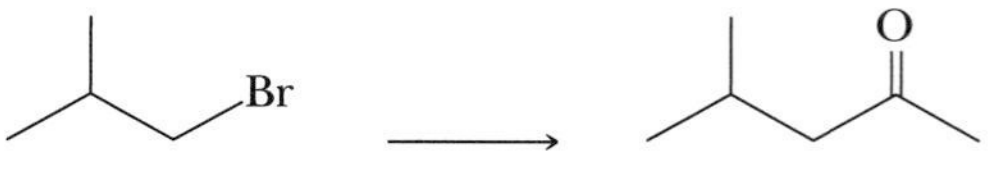

19. Provide a mechanism for the following reaction. Show all charges and lone pairs of electrons in your structures as well as the structures of all intermediates.

H_2SO_4; H_3C CH_3

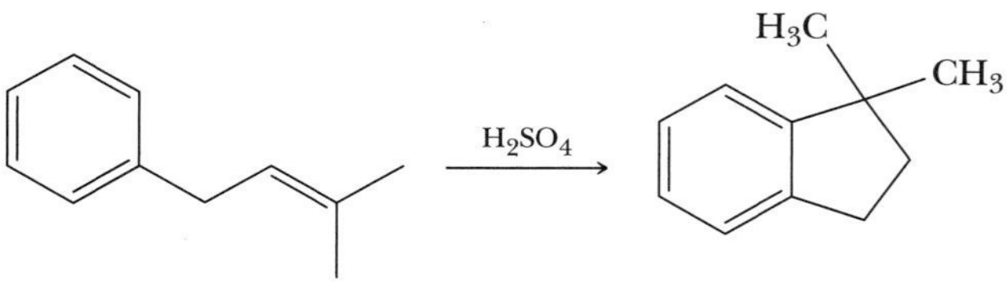

20. Predict the major product or products of each of the following reactions. Be sure to consider stereochemistry in your answers.

(a) 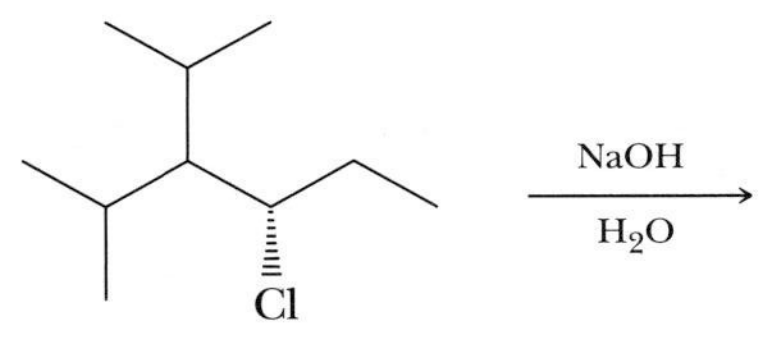 NaOH / H_2O →

Cl

(b) 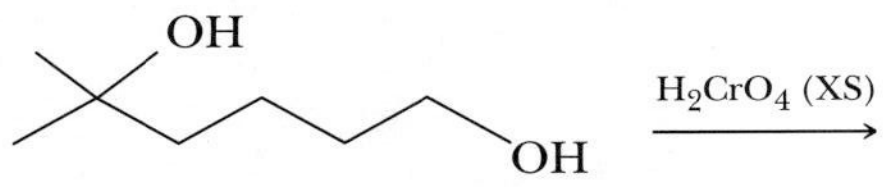 OH OH H_2CrO_4 (XS) →

(c) 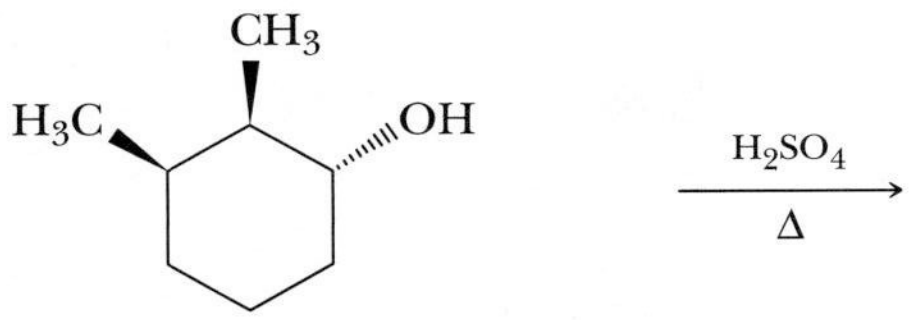 CH_3 H_3C OH H_2SO_4 / Δ →

(d) 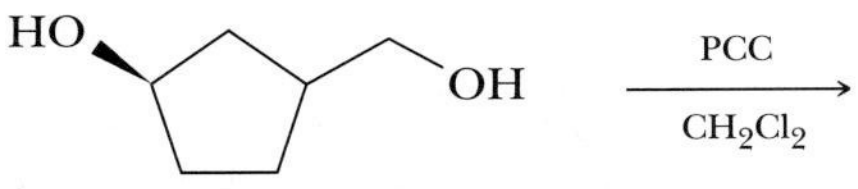 HO OH PCC / CH_2Cl_2 →

(e) 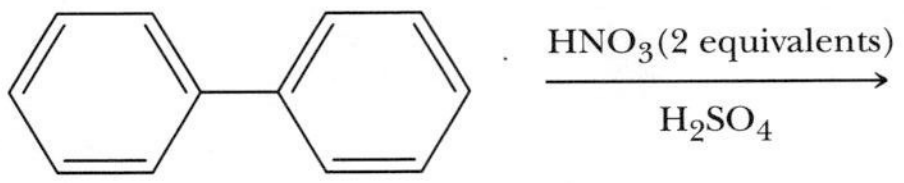 HNO_3(2 equivalents) / H_2SO_4 →

21. Provide a mechanism for the following reaction. Show all charges and lone pairs of electrons in your structures as well as the structures of all intermediates.

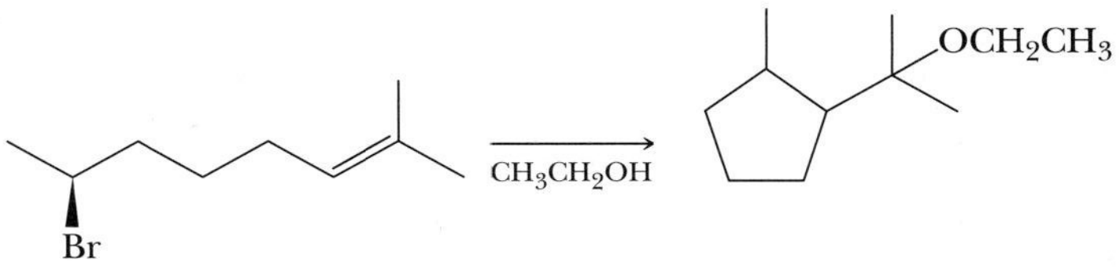

Br → CH_3CH_2OH → OCH_2CH_3

11 Infrared Spectroscopy

KEY QUESTIONS

11.1 What Is Electromagnetic Radiation?

11.2 What Is Molecular Spectroscopy?

11.3 What Is Infrared Spectroscopy?

11.4 How Do We Interpret Infrared Spectra?

HOW TO

11.1 How to Approach Infrared Spectroscopy Structure Determination Problems

CHEMICAL CONNECTIONS

11A Infrared Spectroscopy: A Window on Brain Activity

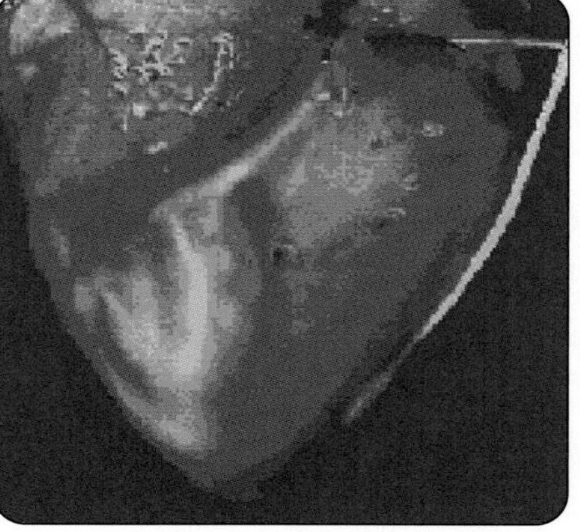
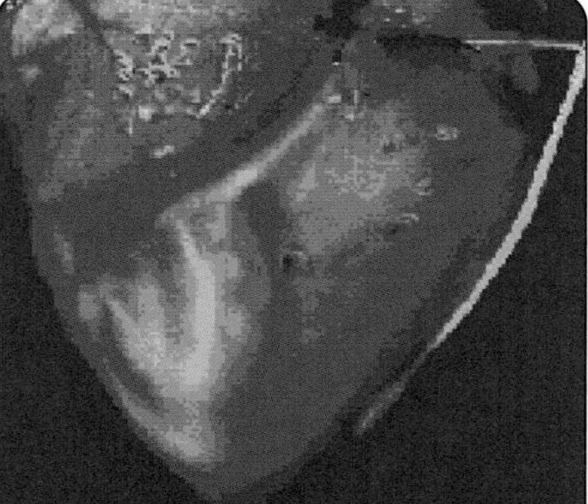

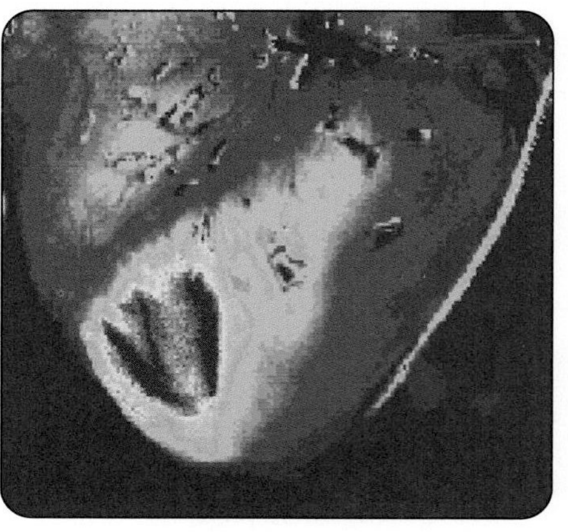

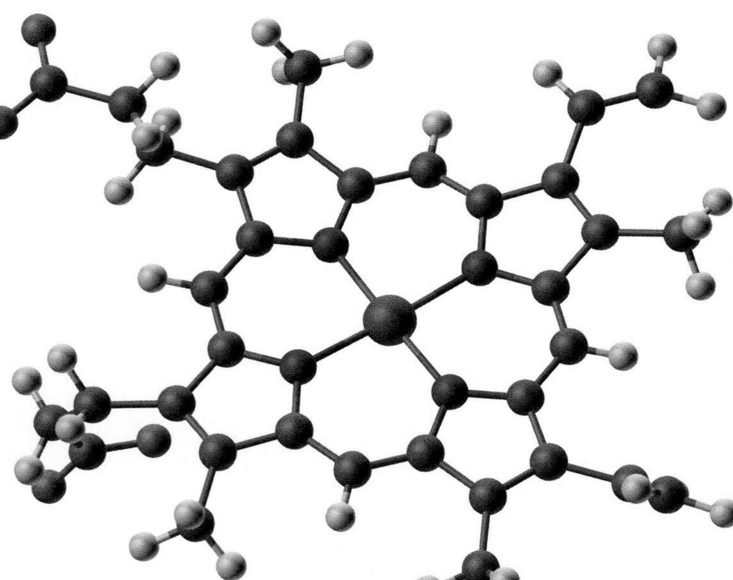

Infrared image of healthy heart (up) and heart with CAD, coronary artery disease (down). Inset: A model of heme, the part of the protein hemoglobin that binds oxygen. Infrared spectroscopy can be used as a noninvasive method for detecting deoxygenated hemoglobin, which tends to occur in high levels in blocked arteries. *(Image courtesy of the National Research Council Institute for Biodiagnostics – 2004.)*

Determining the molecular structure of a compound is a central theme in science. In medicine, for example, the structure of any drug must be known before the drug can be approved for use in patients. In the biotechnology and pharmaceutical industries, knowledge of a compound's structure can provide new leads to promising therapeutics. In organic chemistry, knowledge of the structure of a compound is essential to its use as a reagent or a precursor to other molecules.

Chemists rely almost exclusively on instrumental methods of analysis for structure determination. We begin this chapter with a discussion of infrared (IR) spectroscopy. In the next chapter, we discuss nuclear magnetic resonance (NMR) spectroscopy. These two commonly used techniques involve the interaction of molecules with electromagnetic radiation. Thus, in order to understand the fundamentals of spectroscopy, we must first review some of the fundamentals of electromagnetic radiation.

11.1 What Is Electromagnetic Radiation?

Gamma rays, X rays, ultraviolet light, visible light, infrared radiation, microwaves, and radio waves are all part of the electromagnetic spectrum. Because **electromagnetic radiation** behaves as a wave traveling at the speed of light, it is described in terms of its wavelength and frequency. Table 11.1 summarizes the wavelengths, frequencies, and energies of some regions of the electromagnetic spectrum.

Electromagnetic radiation Light and other forms of radiant energy.

Wavelength is the distance between any two consecutive identical points on the wave. Wavelength is given the symbol λ (Greek lowercase lambda) and is usually expressed in the SI base unit of meters. Other derived units commonly used to express wavelength are given in Table 11.2.

Wavelength (λ) The distance between two consecutive identical points on a wave.

The **frequency** of a wave is the number of full cycles of the wave that pass a given point in a second. Frequency is given the symbol $\boldsymbol{\nu}$ (Greek nu) and is reported in **hertz** (Hz), which has the unit of reciprocal seconds (s^{-1}). Wavelength and frequency are inversely proportional, and we can calculate one from the other from the relationship

Frequency (ν) A number of full cycles of a wave that pass a point in a second.

Hertz (Hz) The unit in which wave frequency is reported; s^{-1} (read *per second*).

$$\nu\lambda = c$$

where ν is frequency in hertz, c is the velocity of light (3.00×10^8 m/s), and λ is the wavelength in meters. For example, consider infrared radiation—or heat

TABLE 11.1 Wavelength, Frequency, and Energy Relationships of Some Regions of the Electromagnetic Spectrum

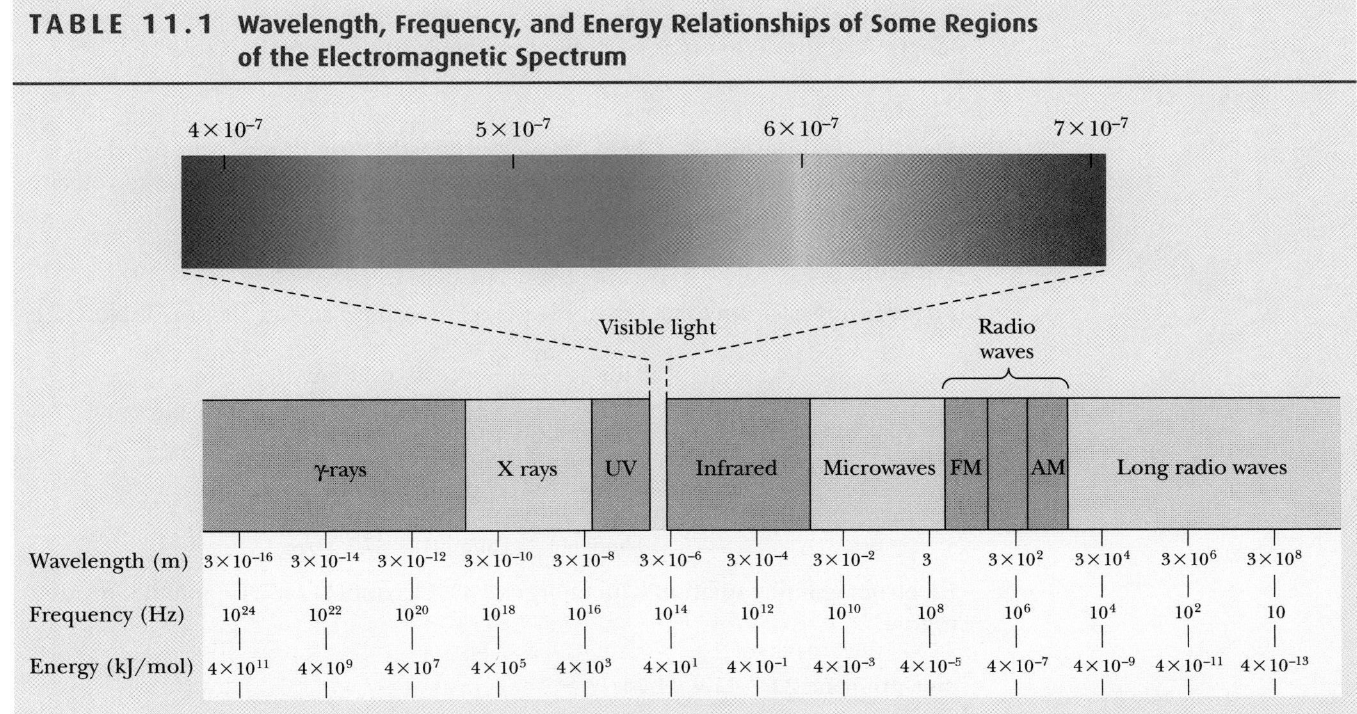

TABLE 11.2 Common Units Used to Express Wavelength (λ)

Unit	Relation to Meter
meter (m)	——
millimeter (mm)	$1\ \text{mm} = 10^{-3}\ \text{m}$
micrometer (μm)	$1\ \mu\text{m} = 10^{-6}\ \text{m}$
nanometer (nm)	$1\ \text{nm} = 10^{-9}\ \text{m}$
Angstrom (Å)	$1\ \text{Å} = 10^{-10}\ \text{m}$

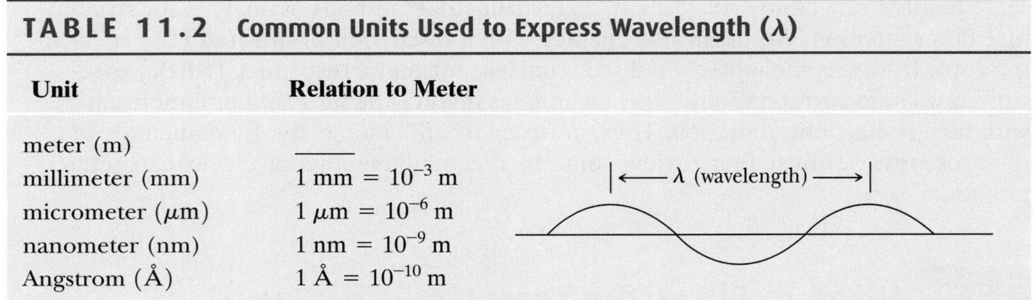

radiation, as it is also called—with wavelength 1.5×10^{-5} m. The frequency of this radiation is

$$\nu = \frac{3.0 \times 10^{8}\ \text{m/s}}{1.5 \times 10^{-5}\ \text{m}} = 2.0 \times 10^{13}\ \text{Hz}$$

An alternative way to describe electromagnetic radiation is in terms of its properties as a stream of particles. We call these particles **photons**. The energy in a mole of photons and the frequency of radiation are related by the equation

$$E = h\nu = h\frac{c}{\lambda}$$

where E is the energy in kJ/mol and h is Planck's constant, $3.99 \times 10^{-13}\ \text{kJ} \cdot \text{s} \cdot \text{mol}^{-1}$ ($9.54 \times 10^{-14}\ \text{kcal} \cdot \text{s} \cdot \text{mol}^{-1}$). This equation tells us that high-energy radiation corresponds to short wavelengths, and vice versa. Thus, ultraviolet light (higher energy) has a shorter wavelength (approximately 10^{-7} m) than infrared radiation (lower energy), which has a wavelength of approximately 10^{-5} m.

Example 11.1

Calculate the energy, in kilocalories per mole of radiation, of a wave with wavelength 2.50 μm. What type of radiant energy is this? (Refer to Table 11.1.)

Strategy

Use the relationship $E = hc/\lambda$. Make certain that the dimensions for distance are consistent: If the dimension of wavelength is meters, then express the velocity of light in meters per second.

Solution

First convert 2.50 μm to meters, using the relationship $1\ \mu\text{m} = 10^{-6}$ m (Table 11.2):

$$2.50\ \mu\text{m} \times \frac{10^{-6}\ \text{m}}{1\ \mu\text{m}} = 2.50 \times 10^{-6}\ \text{m}$$

Now substitute this value into the equation $E = hc/\lambda$:

$$E = \frac{hc}{\lambda} = 3.99 \times 10^{-13}\frac{\text{kJ} \cdot \text{s}}{\text{mol}} \times 3.00 \times 10^{8}\frac{\text{m}}{\text{s}} \times \frac{1}{2.50 \times 10^{-6}\ \text{m}}$$
$$= 47.7\ \text{kJ/mol}\ (11.4\ \text{kcal/mol}) = 47.7\ \text{kJ/mol}$$

Electromagnetic radiation with energy of 47.7 kJ/mol is radiation in the infrared region.

See problems 11.7–11.9, 11.24

Problem 11.1

Calculate the energy of red light (680 nm) in kilocalories per mole. Which form of radiation carries more energy, infrared radiation with wavelength 2.50 μm or red light with wavelength 680 nm?

11.2 What Is Molecular Spectroscopy?

Organic molecules are flexible structures. As we discussed in Chapter 3, atoms and groups of atoms rotate about covalent single bonds. Covalent bonds stretch and bend just as if the atoms are joined by flexible springs. In addition, molecules contain electrons that can move from one electronic energy level to another. We know from experimental observations and from theories of molecular structure that all energy changes within a molecule are quantized; that is, they are subdivided into small, but well-defined, increments. For example, vibrations of bonds within molecules can undergo transitions only between allowed vibrational energy levels.

We can cause an atom or molecule to undergo a transition from energy state E_1 to a higher energy state E_2 by irradiating it with electromagnetic radiation corresponding to the energy difference between states E_1 and E_2, as illustrated schematically in Figure 11.1. When the atom or molecule returns to the ground state E_1, an equivalent amount of energy is emitted.

Molecular spectroscopy is the experimental process of measuring which frequencies of radiation a substance absorbs or emits and then correlating those frequencies with specific types of molecular structures. In **infrared (IR) spectroscopy**, we irradiate a compound with infrared radiation, the absorption of which causes covalent bonds to change from a lower vibrational energy level to a higher one. Because different functional groups have different bond strengths, the energy required to bring about these transitions will vary from one functional group to another. Thus, in infrared spectroscopy, we detect functional groups by the vibrations of their bonds.

Molecular spectroscopy The study of the frequencies of electromagnetic radiation that are absorbed or emitted by substances and the correlation between these frequencies and specific types of molecular structure.

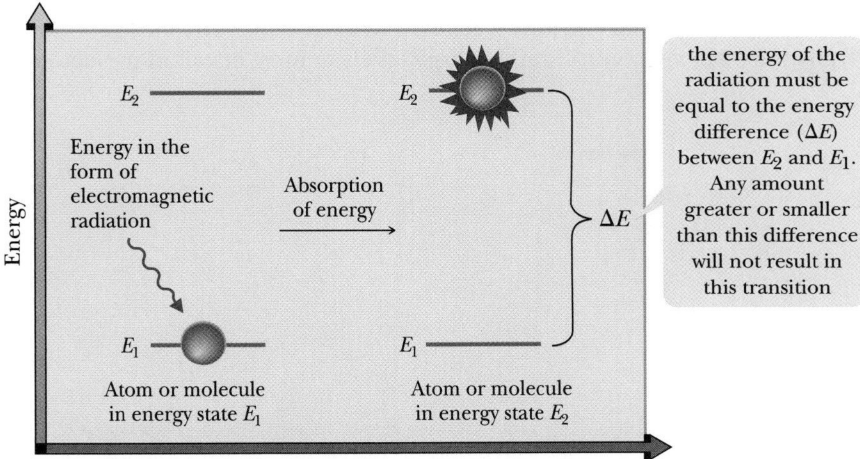

Figure 11.1
Absorption of energy in the form of electromagnetic radiation excites an atom or a molecule in energy state E_1 to a higher energy state E_2.

11.3 What Is Infrared Spectroscopy?

A. The Vibrational Infrared Spectrum

The infrared region of the electromagnetic spectrum (Table 11.1) covers the range from 0.78×10^{-6} m (just above the visible region) to 200×10^{-6} m (just below the microwave region). In organic chemistry, however, we routinely use only the middle portion of this range, called the **vibrational infrared** region. This region extends from 2.5×10^{-6} to 25×10^{-6} m. We commonly refer to radiation in the vibrational infrared region by its **wavenumber** ($\bar{\nu}$), the number of waves per centimeter:

Vibrational infrared The portion of the infrared region that extends from 4000 to 400 cm^{-1}.

Wavenumber ($\bar{\nu}$) A characteristic of electromagnetic radiation equal to the number of waves per centimeter.

$$\bar{\nu} = \frac{1}{\lambda\ (\text{cm})} = \frac{10^{-2}\ (\text{m} \cdot \text{cm}^{-1})}{\lambda\ (\text{m})}$$

Expressed in wavenumbers, the vibrational region of the infrared spectrum extends from 4000 to 400 cm^{-1} (the unit cm^{-1} is read "reciprocal centimeter"):

$$\bar{\nu} = \frac{10^{-2}\ \text{m} \cdot \text{cm}^{-1}}{2.5 \times 10^{-6}\ \text{m}} = 4000\ \text{cm}^{-1} \qquad \bar{\nu} = \frac{10^{-2}\ \text{m} \cdot \text{cm}^{-1}}{25 \times 10^{-6}\ \text{m}} = 400\ \text{cm}^{-1}$$

An advantage of using wavenumbers is that they are directly proportional to energy; the higher the wavenumber, the higher is the energy of radiation.

Figure 11.2 shows an infrared spectrum of aspirin. The horizontal axis at the bottom of the chart is calibrated in wavenumbers (cm^{-1}); that at the top is calibrated in wavelength (micrometers, μm). The wavenumber scale is often divided into two or more linear regions. For all spectra reproduced in this text, it is divided into three linear regions: 4000–2200 cm^{-1}, 2200–1000 cm^{-1}, and 1000–400 cm^{-1}. The vertical axis measures transmittance, with 100% transmittance at the top and 0% transmittance at the bottom. Thus, the baseline for an infrared spectrum (100% transmittance of radiation through the sample = 0% absorption) is at the top of the chart, and the absorption of radiation corresponds to a trough or valley. Strange as it may seem, we commonly refer to infrared absorptions as peaks, even though they are actually troughs.

B. Molecular Vibrations

Atoms joined by covalent bonds are not permanently fixed in one position, but instead undergo continual vibrations relative to each other. The energies associated with transitions between vibrational energy levels in most covalent molecules range

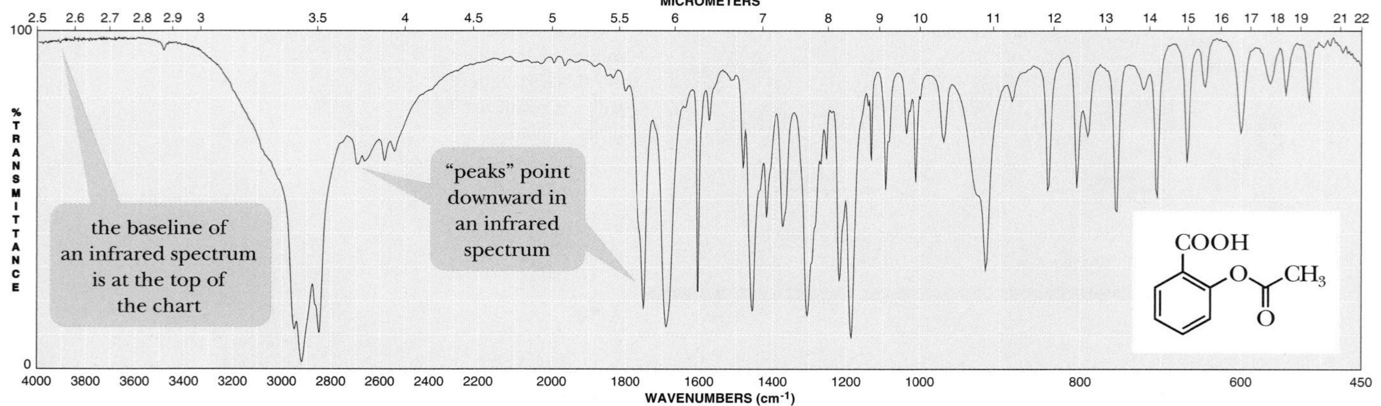

Figure 11.2
Infrared spectrum of aspirin.

from 8.4 to 42 kJ/mol (2 to 10 kcal/mol). Such transitions can be induced by the absorption of radiation in the infrared region of the electromagnetic spectrum.

For a molecule to absorb infrared radiation, the bond undergoing vibration must be polar, and its vibration must cause a periodic change in the bond dipole; the greater the polarity of the bond, the more intense is the absorption. Any vibration that meets this criterion is said to be **infrared active**. Covalent bonds in homonuclear diatomic molecules, such as H_2 and Br_2, and some carbon–carbon double bonds in symmetrical alkenes and alkynes do not absorb infrared radiation because they are not polar bonds. The multiple bonds in the following two molecules, for example, do not have a dipole moment and, therefore, are not infrared active:

Neither of the unsaturated bonds in these molecules is infrared active because the vibrational motions shown do not result in a change in bond dipole (due to the symmetry about these bonds).

$$(H_3C)_2C{=}C(CH_3)_2 \longrightarrow\longleftarrow (H_3C)_2C{=}C(CH_3)_2 \longleftarrow\longrightarrow (H_3C)_2C{=}C(CH_3)_2$$

2,3-Dimethyl-2-butene

$$H_3C{-}C{\equiv}C{-}CH_3 \longrightarrow\longleftarrow H_3C{-}C{\equiv}C{-}CH_3 \longleftarrow\longrightarrow H_3C{-}C{\equiv}C{-}CH_3$$

2-Butyne

For a nonlinear molecule containing n atoms, $3n - 6$ allowed fundamental vibrations exist. For a molecule as simple as ethanol, CH_3CH_2OH, there are 21 fundamental vibrations, and for hexanoic acid, $CH_3(CH_2)_4COOH$, there are 54. Thus, even for relatively simple molecules, a large number of vibrational energy levels exists, and the patterns of energy absorption for these and larger molecules are quite complex.

The simplest vibrational motions in molecules giving rise to the absorption of infrared radiation are **stretching** and **bending** motions. Illustrated in Figure 11.3 are the fundamental stretching and bending vibrations for a methylene group.

To one skilled in the interpretation of infrared spectra, absorption patterns can yield an enormous amount of information about chemical structure. We, however, have neither the time nor the need to develop that level of competence. The value of infrared spectra for us is that we can use them to determine the presence or absence of particular functional groups. A carbonyl group, for example, typically shows strong absorption at approximately 1630–1800 cm^{-1}. The position of absorption for a particular carbonyl group depends on (1) whether it is that of an aldehyde, a ketone, a carboxylic acid, an ester, or an amide, and (2) if the carbonyl carbon is in a ring, the size of the ring.

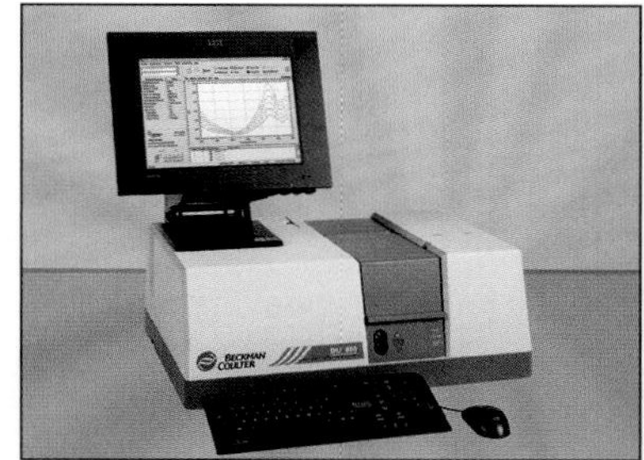

A Beckman Coulter DU 800 infrared spectrophotometer. Spectra are shown in the monitor. *(Courtesy of Beckman Coulter, Inc.)*

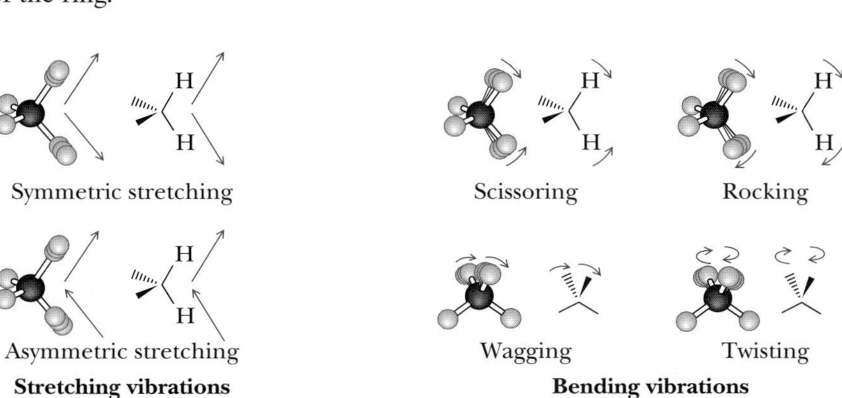

Figure 11.3 Fundamental modes of vibration for a methylene group.

TABLE 11.3 Characteristic IR Absorptions of Selected Functional Groups

Bond	Frequency (cm^{-1})	Intensity
O—H	3200–3500	strong and broad
N—H	3100–3500	medium
C—H	2850–3100	medium to strong
C≡C	2100–2260	weak
C=O	1630–1800	strong
C=C	1600–1680	weak
C—O	1050–1250	strong

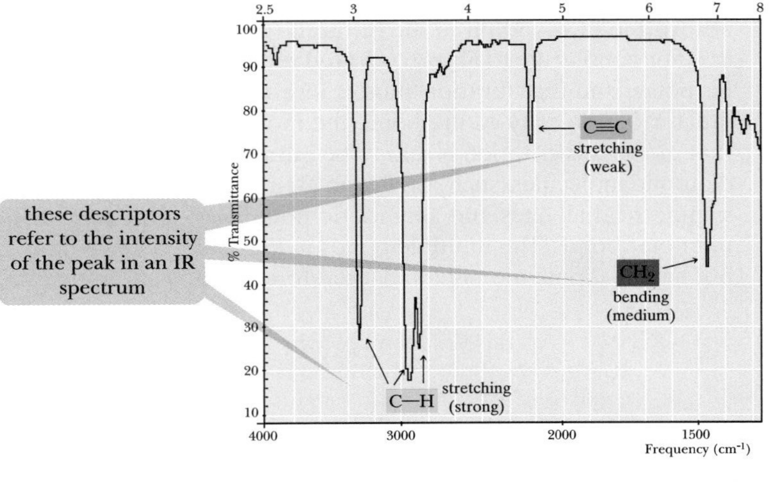

C. Correlation Tables

Data on absorption patterns of selected functional groups are collected in tables called **correlation tables.** Table 11.3 gives the characteristic infrared absorptions for the types of bonds and functional groups we deal with most often. Appendix 5 contains a cumulative correlation table. In these tables, we refer to the intensity of a particular absorption as **strong** (**s**), **medium** (**m**), or **weak** (**w**).

In general, we will pay most attention to the region from 3650 to 1000 cm^{-1} because the characteristic stretching vibrations for most functional groups are found in this region. Vibrations in the region from 1000 to 400 cm^{-1} can arise from phenomena such as combinations of two or more bands or harmonics of fundamental absorption bands. Such vibrations are much more complex and far more difficult to analyze. Because even slight variations in molecular structure lead to differences in absorption patterns in this region, it is often called the **fingerprint region.** If two compounds have even slightly different structures, the differences in their infrared spectra are most clearly discernible in the fingerprint region.

Fingerprint region The portion of the vibrational infrared region that extends from 1000 to 400 cm^{-1} and that is unique to every compound.

Example 11.2

Determine the functional group that is most likely present if a compound shows IR absorption at

(a) 1705 cm^{-1} (b) 2950 cm^{-1}

Strategy

Refer to correlation Table 11.3 or to even more specific correlation tables in this chapter (e.g., Tables 11.4–11.7). Eventually, through the continual practice of problems in this chapter, many of these vibrational stretching frequencies and intensities will become familiar to you.

Solution

(a) A C=O group (b) An aliphatic C—H group

See problems 11.14, 11.15, 11.21–11.23, 11.26

Problem 11.2

A compound shows strong, very broad IR absorption in the region from 3200 to 3500 cm^{-1} and strong absorption at 1715 cm^{-1}. What functional group accounts for both of these absorptions?

Example 11.3

Propanone and 2-propen-1-ol are constitutional isomers. Show how to distinguish between these two isomers by IR spectroscopy.

$$CH_3-\overset{\overset{\large O}{\|}}{C}-CH_3 \qquad\qquad CH_2{=}CH-CH_2-OH$$

Propanone (Acetone) — 2-Propen-1-ol (Allyl alcohol)

Strategy

Because IR spectroscopy distinguishes between characteristic vibrational frequencies of differing *functional groups*, identify the *different* functional groups in the pair of molecules and predict (using correlation tables) the vibrational frequencies that these functional groups would exhibit in an IR spectrum.

Solution

Only propanone shows strong absorption in the C=O stretching region, 1630–1800 cm^{-1}. Alternatively, only 2-propen-1-ol shows strong absorption in the O—H stretching region, 3200–3500 cm^{-1}.

See problems 11.14, 11.15, 11.21–11.23, 11.26

Problem 11.3

Propanoic acid and methyl ethanoate are constitutional isomers. Show how to distinguish between these two compounds by IR spectroscopy.

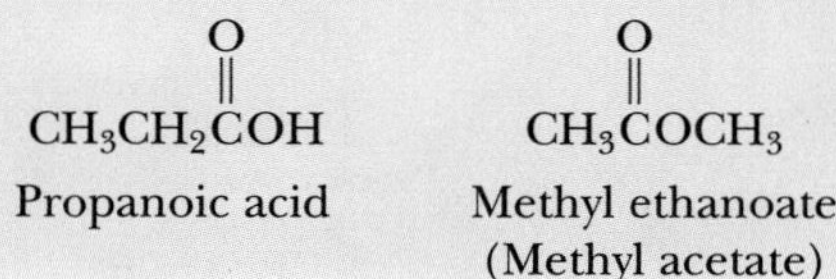

11.4 How Do We Interpret Infrared Spectra?

Interpreting spectroscopic data is a skill that is easy to acquire through practice and exposure to examples. An IR spectrum will reveal not only the functional groups that are present in a sample, but also those that can be excluded from consideration. Often, we can determine the structure of a compound solely from the data in the spectrum of the compound. Other times, we may need additional information, such as the molecular formula of the compound, or knowledge of the reactions used to synthesize the molecule. In this section, we will see specific examples of IR spectra for characteristic functional groups. Familiarizing yourself with them will help you to master the technique of spectral interpretation.

CHEMICAL CONNECTIONS 11A

Infrared Spectroscopy: A Window on Brain Activity

Some of the great advantages of infrared spectroscopy are the relatively low cost, sensitivity, and speed of its instrumentation. The medical and scientific community recognized these benefits, along with the fact that some frequencies of infrared light can harmlessly penetrate human tissue and bone, in creating a technique called functional Near Infrared Spectroscopy (fNIRS). In fNIRS, a patient is fitted with headgear containing many fiber optic cables that allow infrared light in the 700–1000 nm range to be shone through the skull and into the brain. Separate fiber optic cables in the headgear collect the light that reemerges from the brain and direct it to a spectrophotometer, which quantifies the intensity of the light. The instrument measures changes in the concentration of oxygenated and deoxygenated hemoglobin, which absorbs light in the 700–1000 nm range. Because an assortment of tasks that the brain may be asked to carry out result in varying blood flow and oxygenation levels in different parts of the brain, fNIRS can be used to determine how certain thoughts or actions affect brain activity.

QUESTIONS

Could fNIRS be used to detect free oxygen (O_2) levels in the lungs? Why or why not?

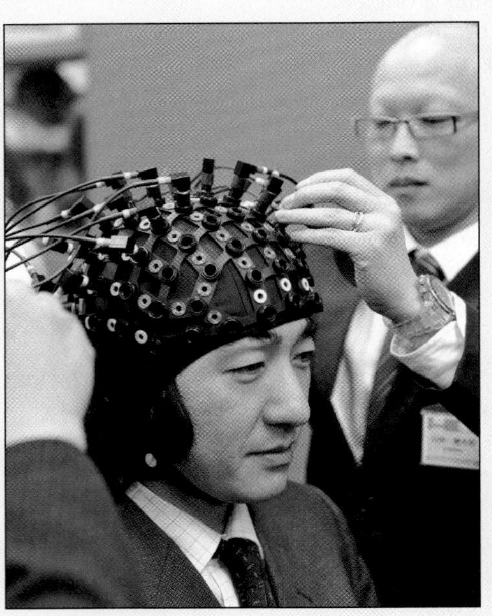

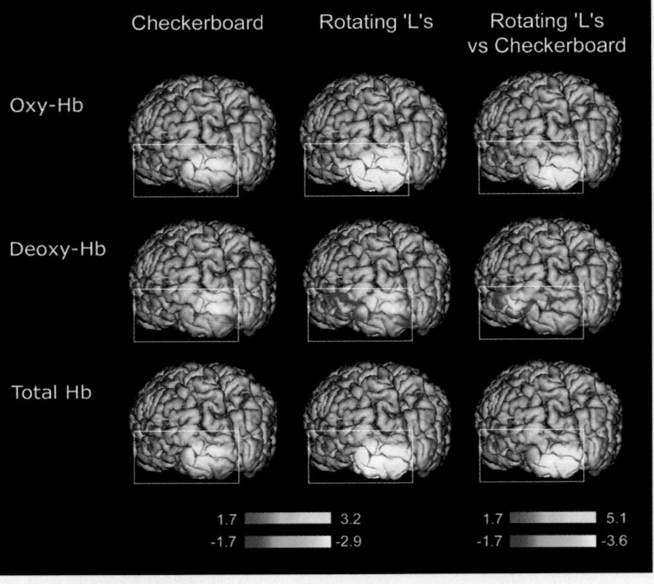

A research subject is asked to perform several mental tasks (left) while fNIRS analysis reveals varying levels of oxy- and deoxyhemoglobin in the blood flowing through the subject's brain (left: *AFP/Getty Images, Inc.*; right: *From Matthias L. Schroeter, NeuroImage 21: 284, Figure 2 (2004). Photo provided courtesy of M. Schroeter.*)

A. Alkanes

Infrared spectra of alkanes are usually simple, with few peaks, the most common of which are given in Table 11.4.

Figure 11.4 shows an infrared spectrum of decane. The strong peak with multiple splittings between 2850 and 3000 cm^{-1} is characteristic of alkane C—H stretching. The C—H peak is strong in this spectrum because there are so many C—H bonds and no other functional groups. The other prominent peaks in the spectrum are a methylene bending absorption at 1465 cm^{-1} and a methyl bending absorption at 1380 cm^{-1}. Because alkane CH, CH_2, and CH_3 groups are present in many organic compounds, these peaks are among the most commonly encountered in infrared spectroscopy.

TABLE 11.4 Characteristic IR Absorptions of Alkanes, Alkenes, and Alkynes

Hydrocarbon	Vibration	Frequency (cm^{-1})	Intensity
Alkane			
C—H	stretching	2850–3000	strong
CH_2	bending	1450–1465	medium
CH_3	bending	1375 and 1450	weak to medium
Alkene			
C—H	stretching	3000–3100	weak to medium
C=C	stretching	1600–1680	weak to medium
Alkyne			
C—H	stretching	3300	weak to medium
C≡C	stretching	2100–2260	weak to medium

B. Alkenes

An easily recognized alkene absorption is the vinylic =C—H stretching band slightly to the left of (at a greater wavenumber than) 3000 cm^{-1}. Also characteristic of alkenes is C=C stretching at 1600–1680 cm^{-1}. This vibration, however, is often weak and difficult to observe. Both vinylic =C—H stretching and C=C stretching can be seen in the infrared spectrum of cyclopentene (Figure 11.5). Also visible are the aliphatic C—H stretching near 2900 cm^{-1} and the methylene bending near 1440 cm^{-1}.

C. Alkynes

Terminal alkynes exhibit C≡C—H stretching at 3300 cm^{-1}. This absorption band is absent in internal alkynes, because the triple bond is not bonded to a proton. All alkynes absorb weakly between 2100 and 2260 cm^{-1}, due to C≡C stretching. This stretching shows clearly in the spectrum of 1-octyne (Figure 11.6).

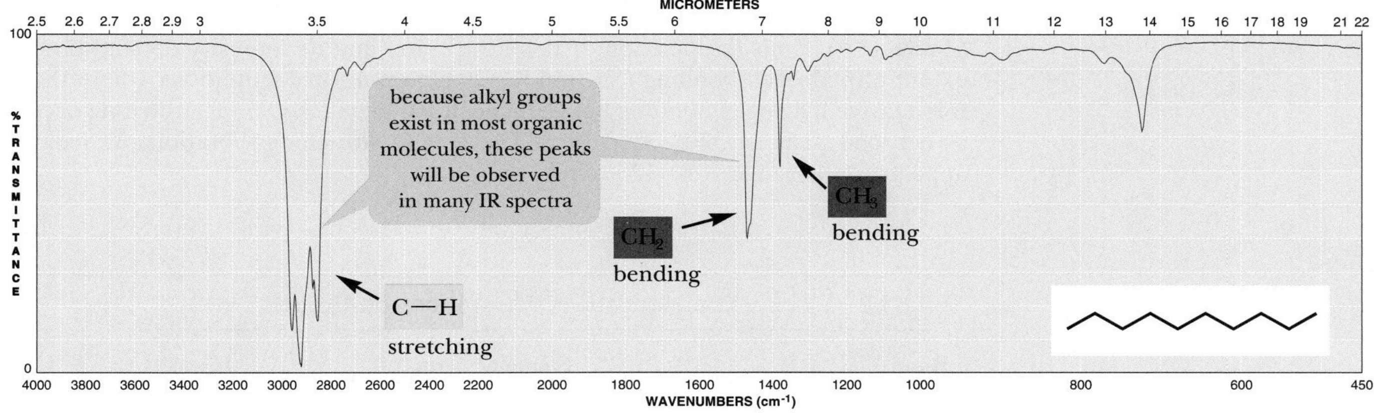

Figure 11.4
Infrared spectrum of decane.

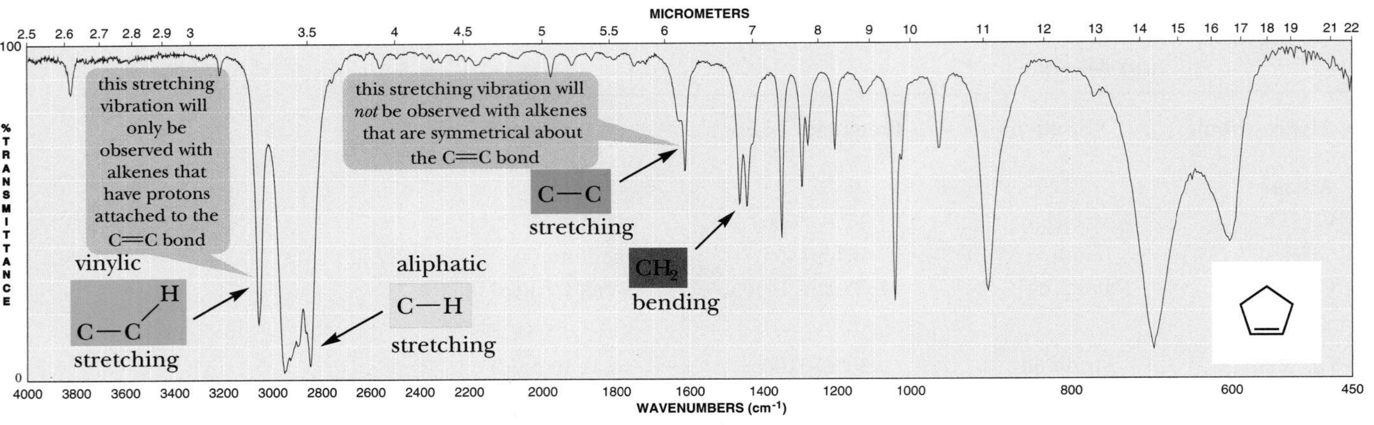

Figure 11.5
Infrared spectrum of cyclopentene.

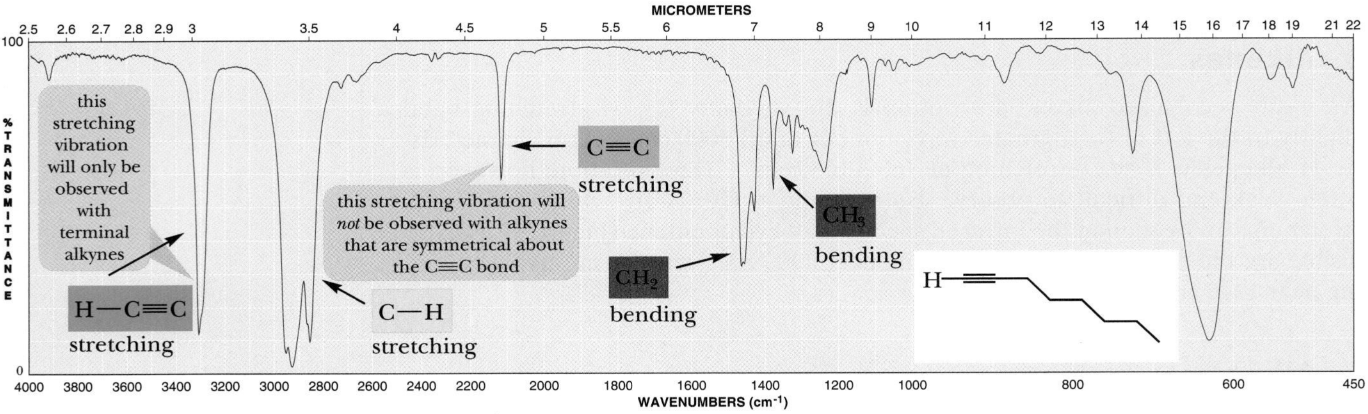

Figure 11.6
Infrared spectrum of 1-octyne.

D. Alcohols

Alcohols are easily recognized by their characteristic O—H stretching absorption (Table 11.5). Both the position of this absorption and its intensity depend on the extent of hydrogen bonding (Section 8.1C). Under normal conditions, where there is extensive hydrogen bonding between alcohol molecules, O—H stretching occurs as a broad peak at 3200–3500 cm^{-1}. The C—O stretching vibration of alcohols appears in the range 1050–1250 cm^{-1}.

TABLE 11.5 Characteristic IR Absorptions of Alcohols

Bond	Frequency (cm^{-1})	Intensity
O—H (hydrogen bonded)	3200–3500	medium, broad
C—O	1050–1250	medium

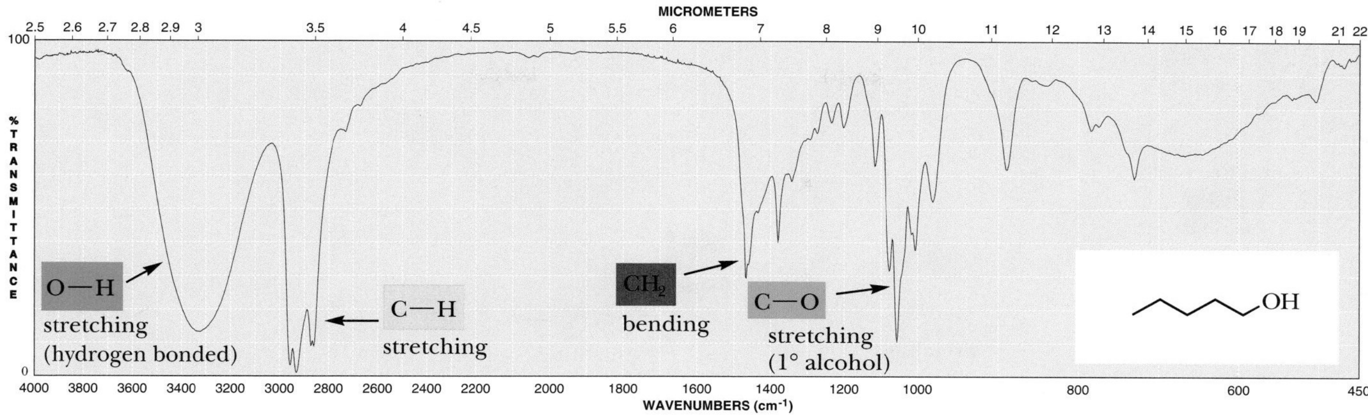

Figure 11.7
Infrared spectrum of 1-pentanol.

Figure 11.7 shows an infrared spectrum of 1-pentanol. The hydrogen-bonded O—H stretching appears as a strong, broad peak centered at 3340 cm^{-1}. The C—O stretching appears near 1050 cm^{-1}, a value characteristic of primary alcohols.

E. Ethers

The C—O stretching frequencies of ethers are similar to those observed in alcohols and esters. Dialkyl ethers typically show a single absorption in this region between 1070 and 1150 cm^{-1}. The presence or absence of O—H stretching at 3200–3500 cm^{-1} for a hydrogen-bonded O—H can be used to distinguish between an ether and an alcohol. The C—O stretching vibration is also present in esters. In this case, we can use the presence or absence of C=O stretching to distinguish between an ether and an ester. Figure 11.8 shows an infrared spectrum of diethyl ether. Notice the absence of O—H stretching.

F. Benzene and Its Derivatives

Aromatic rings show a medium to weak peak in the C—H stretching region at approximately 3030 cm^{-1}, characteristic of sp^2 C—H bonds. They also show several absorptions due to C=C stretching between 1450 and 1600 cm^{-1}. In addition, aromatic rings show strong absorption in the region from 690 to 900 cm^{-1} due to C—H bending. Finally, the presence of weak, broad bands between 1700 and 2000 cm^{-1} is an indicator of a benzene ring (Table 11.6).

TABLE 11.6 Characteristic IR Absorptions of Aromatic Hydrocarbons

Bond	Vibration	Frequency (cm^{-1})	Intensity
C—H	stretching	3030	medium to weak
C—H	bending	690–900	strong
C=C	stretching	1475 and 1600	strong to medium

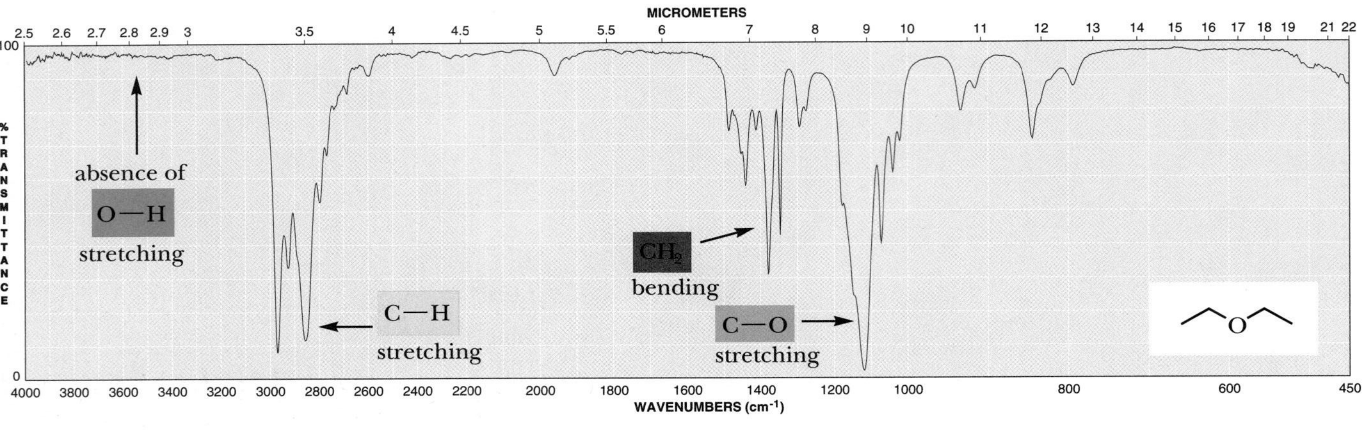

Figure 11.8
Infrared spectrum of diethyl ether.

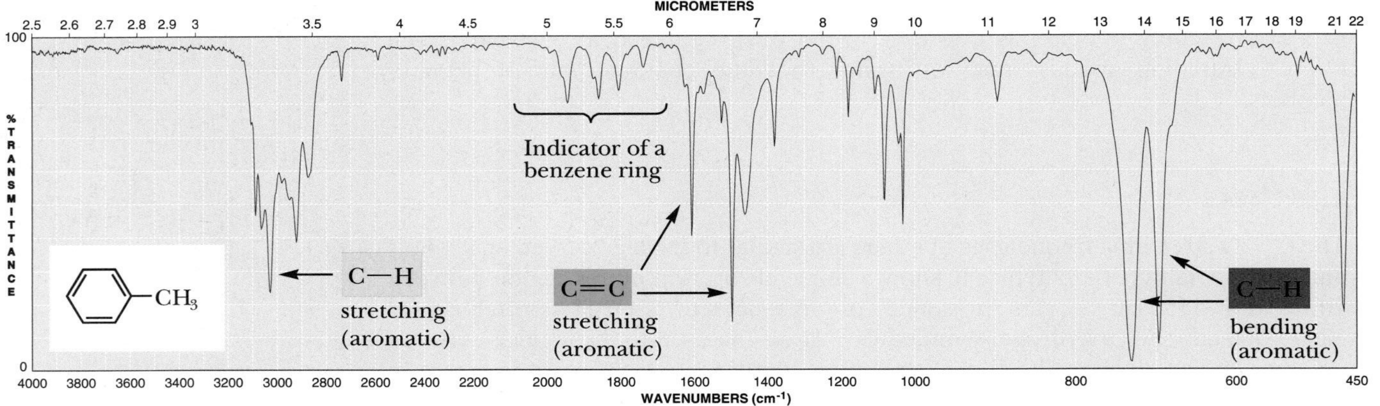

Figure 11.9
Infrared spectrum of toluene.

The C—H and C=C absorption patterns characteristic of aromatic rings can be seen in the infrared spectrum of toluene (Figure 11.9).

G. Amines

The most important and readily observed infrared absorptions of primary and secondary amines are due to N—H stretching vibrations and appear in the region from 3100 to 3500 cm^{-1}. Primary amines have two peaks in this region, one caused by a symmetric stretching vibration and the other by asymmetric stretching. The two N—H stretching absorptions characteristic of a primary amine can be seen in the IR spectrum of butanamine (Figure 11.10). Secondary amines give only one absorption in this region. Tertiary amines have no N—H and therefore are transparent in this region of the infrared spectrum.

H. Aldehydes and Ketones

Aldehydes and ketones (Section 1.7C) show characteristic strong infrared absorption between 1705 and 1780 cm^{-1} associated with the stretching vibration of the

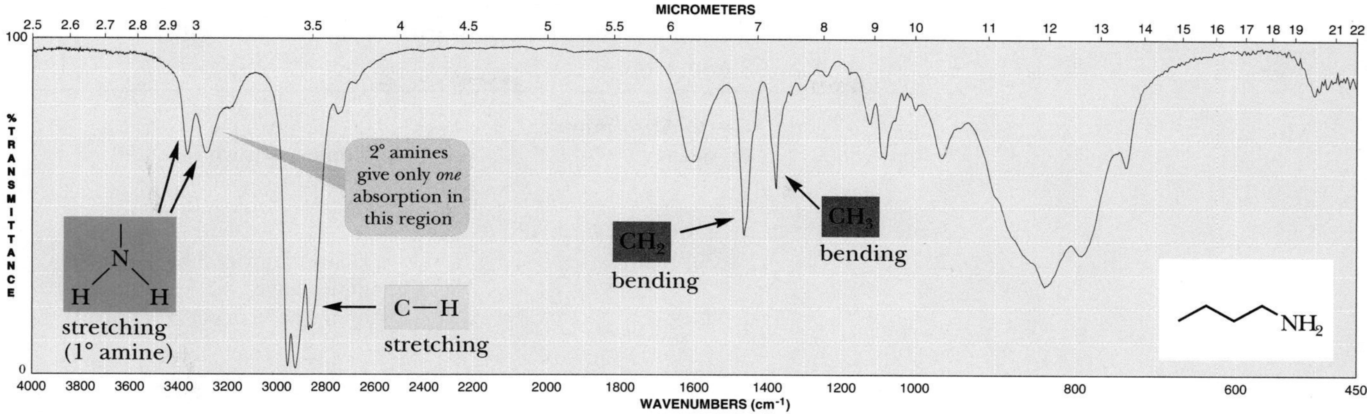

Figure 11.10
Infrared spectrum of butanamine, a primary amine.

carbon–oxygen double bond. The stretching vibration for the carbonyl group of menthone occurs at 1705 cm^{-1} (Figure 11.11).

Because several different functional groups contain a carbonyl group, it is often not possible to tell from absorption in this region alone whether the carbonyl-containing compound is an aldehyde, a ketone, a carboxylic acid, or an ester.

I. Carboxylic Acids and Their Derivatives

The most important infrared absorptions of carboxylic acids and their functional derivatives are due to the C=O stretching vibration; these absorptions are summarized in Table 11.7.

The carboxyl group of a carboxylic acid gives rise to two characteristic absorptions in the infrared spectrum. One of these occurs in the region from 1700 to 1725 cm^{-1} and is associated with the stretching vibration of the carbonyl group. This region is essentially the same as that for the absorption of the carbonyl groups of aldehydes and ketones. The other infrared absorption characteristic of a carboxyl group is a peak between 2400 and 3400 cm^{-1} due to the stretching vibration of the

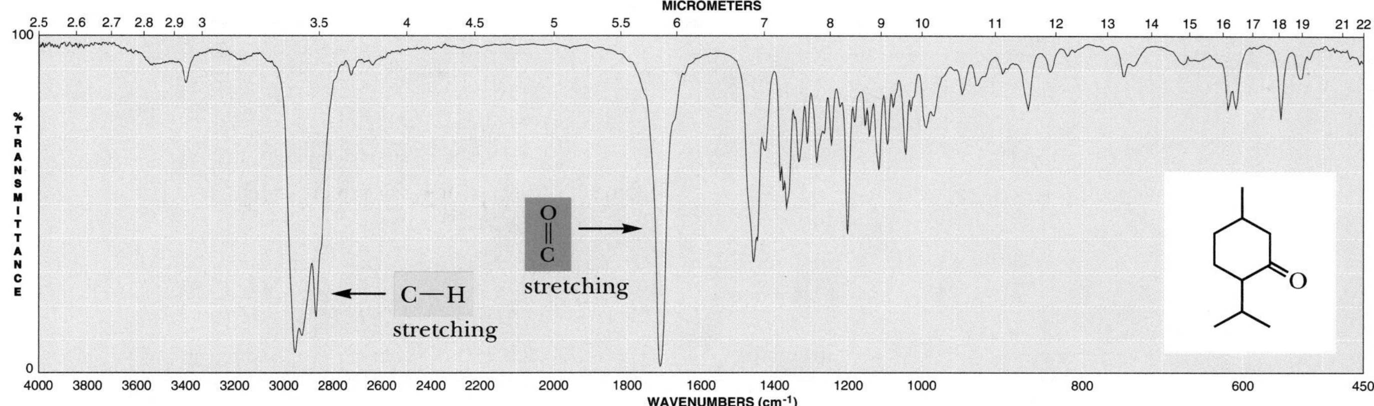

Figure 11.11
Infrared spectrum of menthone.

TABLE 11.7 Characteristic IR Absorptions of Carboxylic Acids, Esters, and Amides

Compound	C=O Absorption Frequency (cm^{-1})	Additional Absorptions (cm^{-1})
$RC(=O)NH_2$	1630–1680	N—H stretching at 3200 and 3400 (1° amides have two N—H peaks) (2° amides have one N—H peak)
RC(=O)OH	1700–1725	O—H stretching at 2400–3400 C—O stretching at 1210–1320
RC(=O)OR	1735–1800	C—O stretching at 1000–1100 and 1200–1250

O—H group. This peak, which often overlaps the C—H stretching absorptions, is generally very broad due to hydrogen bonding between molecules of the carboxylic acid. Both C=O and O—H stretchings can be seen in the infrared spectrum of butanoic acid, shown in Figure 11.12.

Esters display strong C=O stretching absorption in the region between 1735 and 1800 cm^{-1}. In addition, they display strong C—O stretching absorption in the region from 1000 to 1250 cm^{-1} (Figure 11.13).

The carbonyl stretching of amides occurs at 1630–1680 cm^{-1}, a lower series of wavenumbers than for other carbonyl compounds. Primary and secondary amides show N—H stretching in the region from 3200 to 3400 cm^{-1}; primary amides ($RCONH_2$) show two N—H absorptions, whereas secondary amides (RCONHR) show only a single N—H absorption. Tertiary amides, of course, do not show N—H stretching absorptions. See the following three spectra (Figure 11.14).

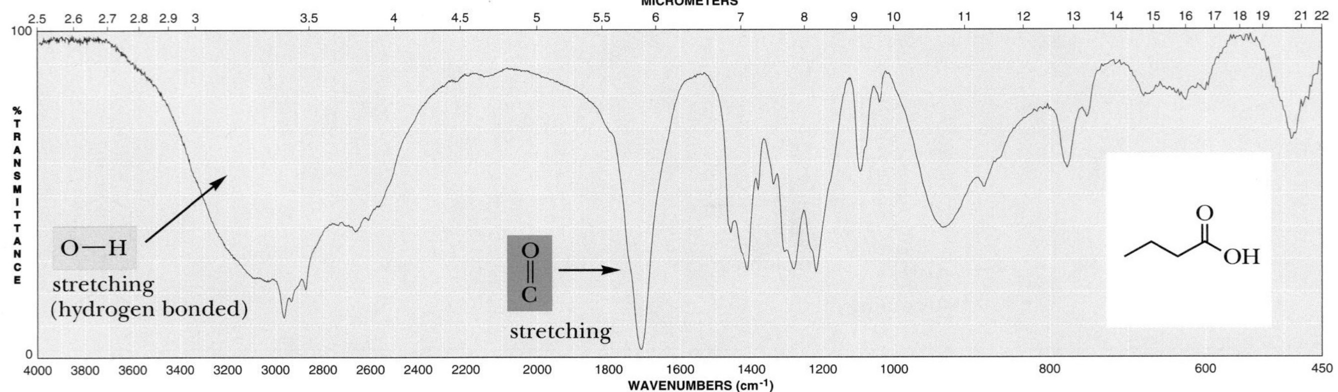

Figure 11.12
Infrared spectrum of butanoic acid.

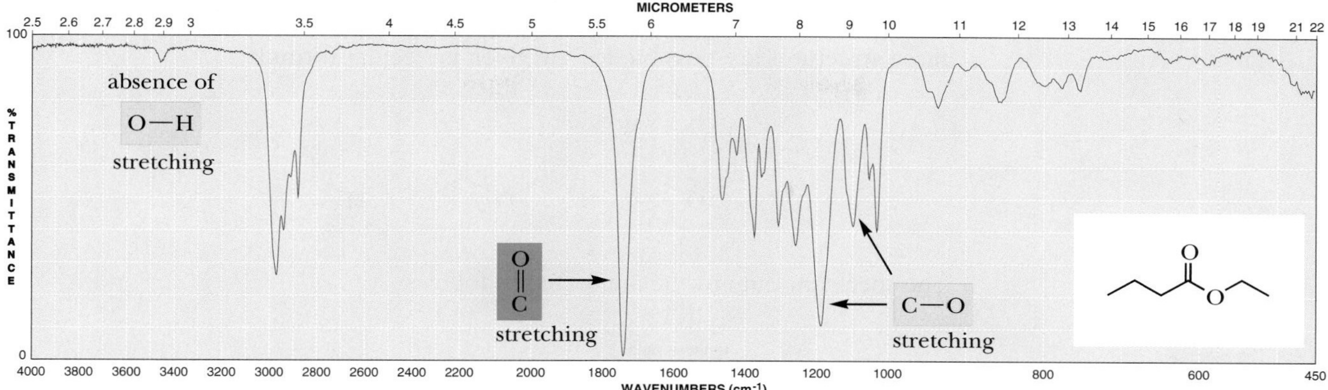

Figure 11.13
Infrared spectrum of ethyl butanoate.

Example 11.4

An unknown compound with the molecular formula $C_3H_6O_2$ yields the following IR spectrum. Draw possible structures for the unknown.

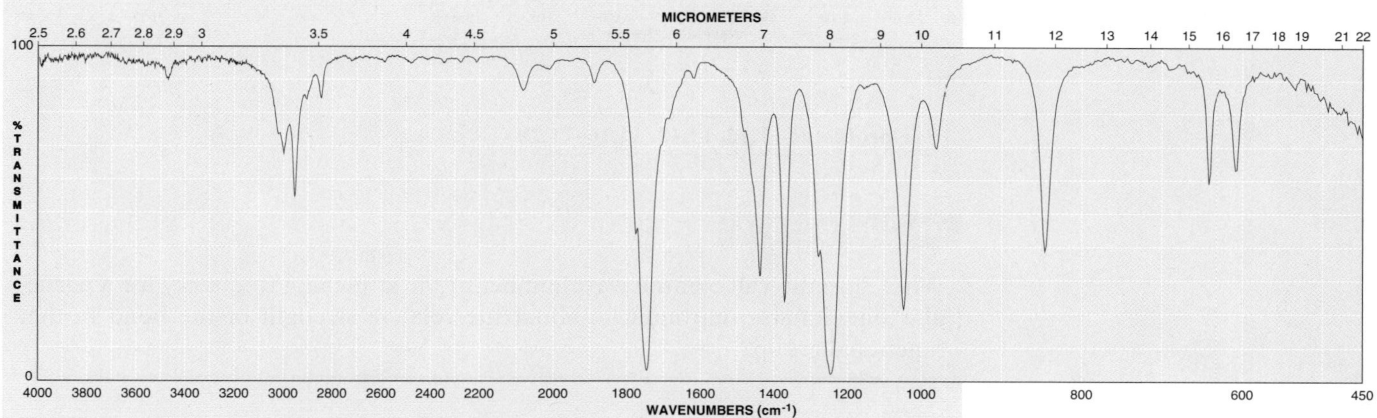

Strategy

Start at 4000 cm^{-1} and move down the wavenumber scale. Make note of characteristic peaks, especially those that are unique to certain functional groups. Observe that the absence of peaks also provides clues for the types of functional groups that cannot be present. Once all the possible functional groups have been identified, propose chemical structures using these functional groups and the elements provided by the molecular formula. In Section 11.4J, we learn the concept of index of hydrogen deficiency, which can also be used in these types of problems to determine the structure of an unknown compound.

Solution

The IR spectrum shows a strong absorption at approximately 1750 cm^{-1}, which is indicative of a C=O group. The spectrum also shows strong C—O absorption peaks at 1250 and 1050 cm^{-1}. Furthermore, there are no peaks above 3000 cm^{-1}, which eliminates the possibility of an O—H group. On the basis of these data, three structures are possible for the given molecular formula:

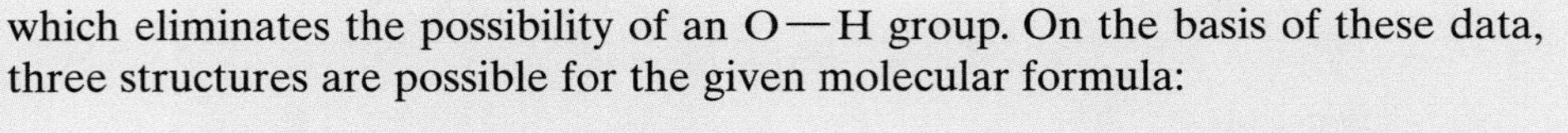

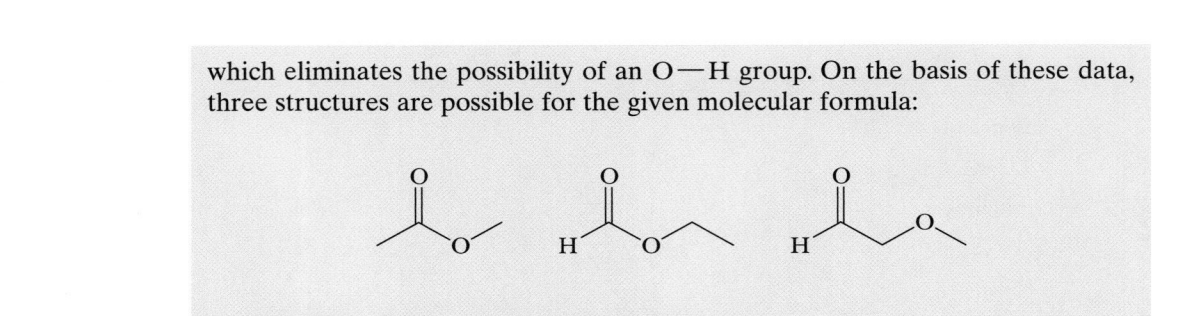

The spectrum can now be annotated as follows:

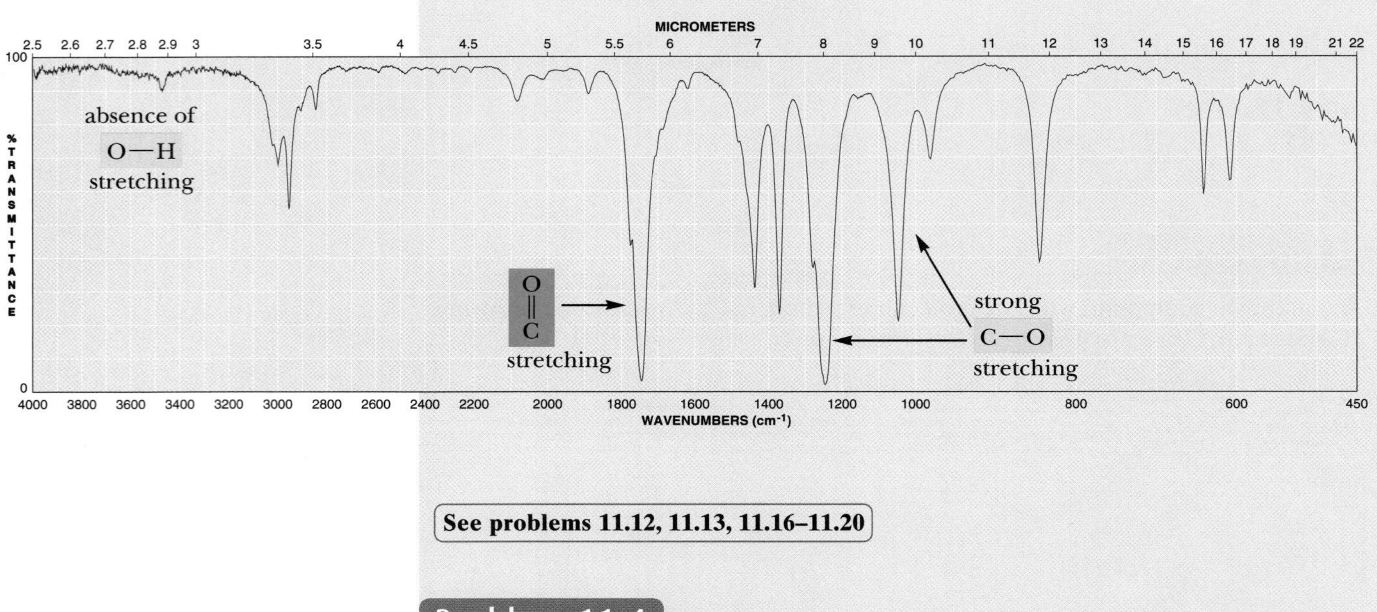

See problems 11.12, 11.13, 11.16–11.20

Problem 11.4

What does the value of the wavenumber of the stretching frequency for a particular functional group indicate about the relative strength of the bond in that functional group?

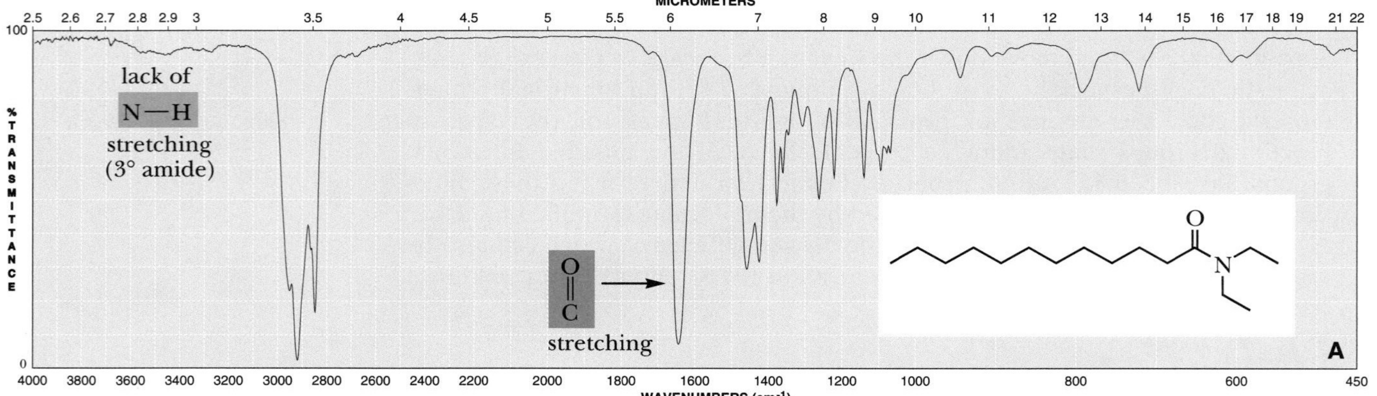

Figure 11.14
Infrared spectra of *N,N*-diethyldodecanamide (**A**, a tertiary amide), *N*-methylbenzamide (**B**, a secondary amide), butanamide (**C**, a primary amide).

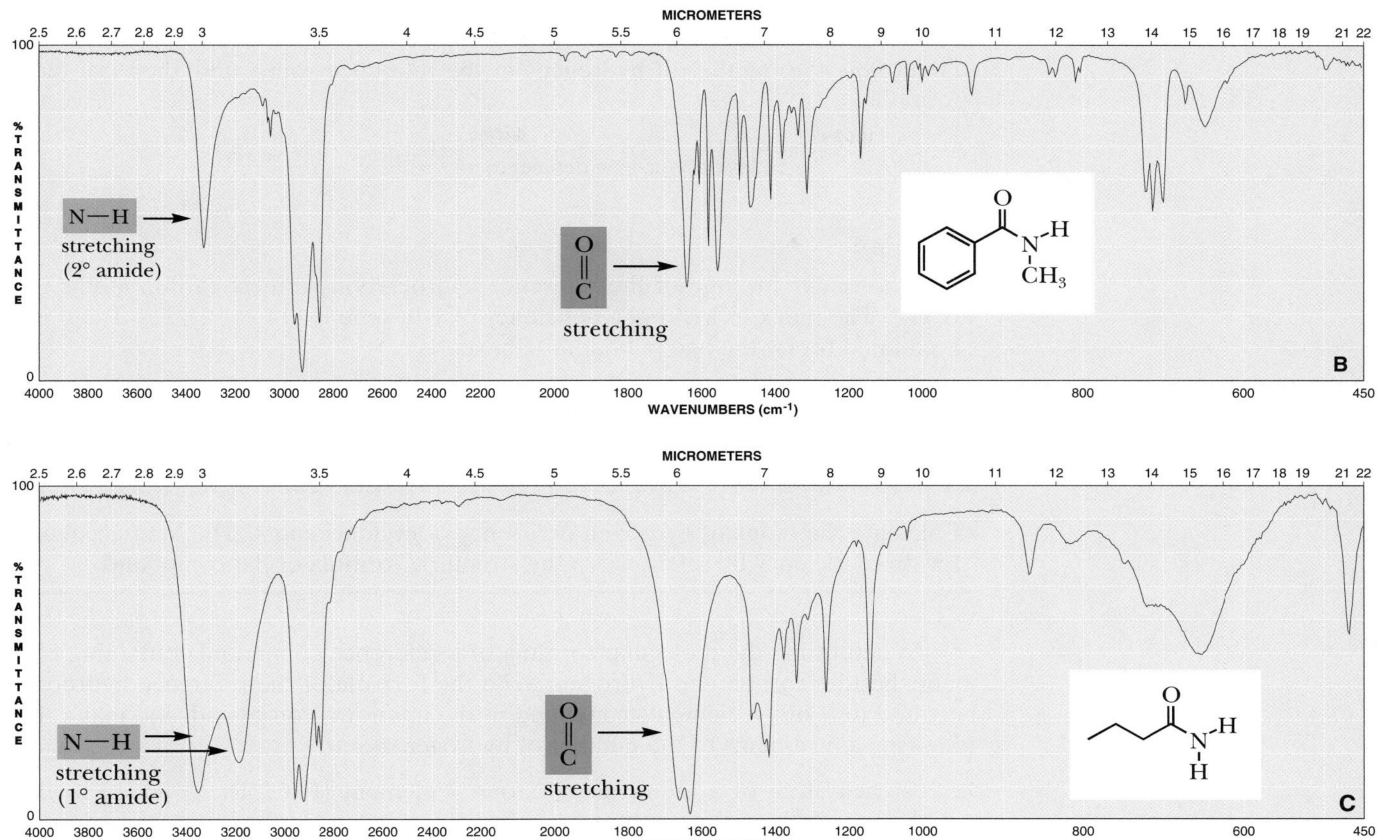

Figure 11.14 ***(Continued)***

J. Index of Hydrogen Deficiency

We can obtain valuable information about the structural formula of an unknown compound by inspecting its molecular formula. In addition to learning the number of atoms of carbon, hydrogen, oxygen, nitrogen, and so forth in a molecule of the compound, we can determine what is called its **index of hydrogen deficiency**, which is the sum of the number of rings and pi bonds in a molecule. We determine this quantity by comparing the number of hydrogens in the molecular formula of a compound of unknown structure with the number of hydrogens in a **reference compound** with the same number of carbon atoms and with no rings or pi bonds. The molecular formula of a reference hydrocarbon is C_nH_{2n+2} (Section 3.1).

Index of hydrogen deficiency The sum of the number of rings and pi bonds in a molecule.

$$\text{Index of hydrogen deficiency} = \frac{(\text{H}_{\text{reference}} - \text{H}_{\text{molecule}})}{2}$$

Example 11.5

Calculate the index of hydrogen deficiency for 1-hexene, with the molecular formula C_6H_{12}, and account for this deficiency by reference to the structural formula of the compound.

Strategy

Determine the number of hydrogens in the reference compound; then use the formula

$$\text{Index of hydrogen deficiency} = \frac{(H_{reference} - H_{molecule})}{2}.$$

Solution

The molecular formula of the reference hydrocarbon with six carbon atoms is C_6H_{14}. The index of hydrogen deficiency of 1-hexene is $(14 - 12)/2 = 1$ and is accounted for by the one pi bond in 1-hexene.

See problems 11.10–11.13, 11.16–11.20

Problem 11.5

Calculate the index of hydrogen deficiency of cyclohexene, C_6H_{10}, and account for this deficiency by reference to the structural formula of the compound.

To determine the molecular formula of a reference compound containing elements besides carbon and hydrogen, write the formula of the reference hydrocarbon, add to it other elements contained in the unknown compound, and make the following adjustments to the number of hydrogen atoms:

1. For each atom of a monovalent Group 7 element (F, Cl, Br, I) added to the reference hydrocarbon, subtract one hydrogen; halogen substitutes for hydrogen and reduces the number of hydrogens by one per halogen. The general formula of an acyclic monochloroalkane, for example, is $C_nH_{2n+1}Cl$.
2. No correction is necessary for the addition of atoms of Group 6 elements (O, S, Se) to the reference hydrocarbon. Inserting a divalent Group 6 element into a reference hydrocarbon does not change the number of hydrogens.
3. For each atom of a trivalent Group 5 element (N and P) added to the formula of the reference hydrocarbon, add one hydrogen. Inserting a trivalent Group 5 element adds one hydrogen to the molecular formula of the reference compound. The general molecular formula for an acyclic alkylamine, for example, is $C_nH_{2n+3}N$.

Example 11.6

Isopentyl acetate, a compound with a bananalike odor, is a component of the alarm pheromone of honeybees. The molecular formula of isopentyl acetate is $C_7H_{14}O_2$. Calculate the index of hydrogen deficiency of this compound.

Strategy

Determine the number of hydrogens in the reference compound then use the formula

$$\text{Index of hydrogen deficiency} = \frac{(H_{reference} - H_{molecule})}{2}.$$

Solution

The molecular formula of the reference hydrocarbon is C_7H_{16}. Adding oxygens to this formula does not require any correction in the number of hydrogens. The molecular formula of the reference compound is $C_7H_{16}O_2$, and the index of hydrogen deficiency is $(16 - 14)/2 = 1$, indicating either one ring or one pi bond. Following is the structural formula of isopentyl acetate, which contains one pi bond, in this case in the carbon–oxygen double bond:

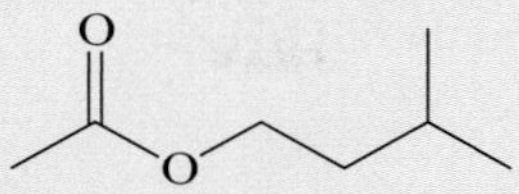

Isopentyl acetate

See problems 11.10–11.13, 11.16–11.20

Problem 11.6

The index of hydrogen deficiency of niacin is 5. Account for this value by reference to the structural formula of niacin.

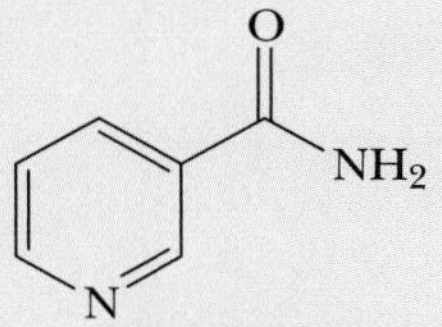

NH_2

Nicotinamide
(Niacin)

HOW TO 11.1 Approach Infrared Spectroscopy Structure Determination Problems

It is useful to develop a systematic approach to problems that ask you to determine a structure given a molecular formula and an infrared spectrum. Following are some guidelines for tackling such problems.

(a) *Determine the index of hydrogen deficiency (IHD).* Knowing the potential number of rings, double bonds, or triple bonds in an unknown compound is of great assistance in solving the structure. For example, if IHD = 1, you know that the unknown can have either a ring or a double bond, but not both. It also cannot have a triple bond because that would require an IHD = 2.

(b) *Move from left to right to identify functional groups in the IR spectrum.* Because the types of transitions in an IR spectrum become less specific as absorptions approach and go lower than 1000 cm^{-1}, it is most useful to start on the left side of an IR spectrum. The shorter the bond, the higher in wavenumber its absorption. This is why O—H, N—H, and C—H bond vibrations occur above 3900 cm^{-1}. Proceeding to the right, we then encounter C—C triple bond stretching vibrations. C═O bonds are shorter than C═C bonds and therefore occur at higher wavenumbers (1800–1650 cm^{-1}) than C═C bonds (1680–1600 cm^{-1}).

(c) *Draw possible structures and verify your structures with the data.* Do not try to think of the answer entirely in your mind. Rather, jot down some structures on paper. Once you have one or more possible structures, verify that these possibilities work with the IHD value and the functional groups indicated by the IR spectrum. Usually, incorrect structures will obviously conflict with one or more of these data items.

Example 11.7

Determine possible structures for a compound that yields the following IR spectrum and has a molecular formula of C_7H_8O:

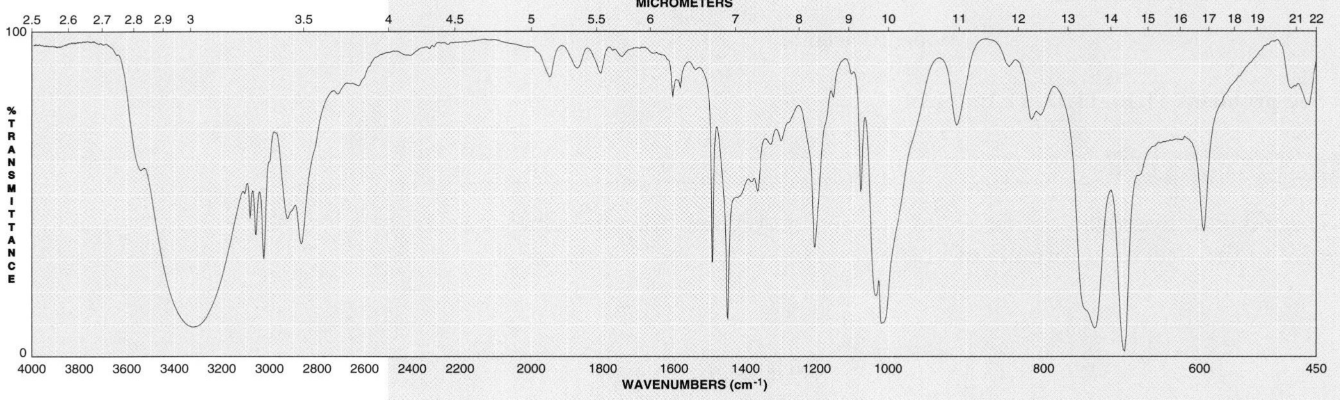

Strategy

Determine the index of hydrogen deficiency and use this value as a guide to the combination of rings, double bonds, or triple bonds possible in the unknown. Analyze the IR spectrum, starting at 4000 cm^{-1} and moving down the wavenumber scale. Make note of characteristic peaks, especially those that are unique to certain functional groups. Note that the absence of peaks also provides clues for the types of functional groups that cannot be present. Once all the possible functional groups have been identified, propose chemical structures using these functional groups and the elements provided by the molecular formula.

Solution

The index of hydrogen deficiency for C_7H_8O is 4, based on the reference formula C_7H_{16}. While accounting for this value may seem daunting at first (consider the possible combinations of four rings or pi bonds), keep in mind that there is one common functional group in organic chemistry that has an index of hydrogen deficiency of 4, namely, a benzene ring. The characteristic aromatic C—H bending bands at 690 and 740 cm^{-1} and the weak, broad bands between 1700 and 2000 cm^{-1} support the existence of a benzene ring. Also, sp^2 C—H stretching is present just above the 3000-cm^{-1} mark. Aromatic C=C stretching absorption bands at 1450 and 1490 cm^{-1} also indicate a benzene ring. Because there is no strong absorption between 1630 and 1800 cm^{-1}, a carbonyl group (C=O) can be excluded. The last piece of evidence is the strong, broad O—H stretching peak at approximately 3310 cm^{-1}. Because we must have an OH group, we cannot propose any structures with an OCH_3 (ether) group. Based on this interpretation of the spectrum, the following four structures are possible:

OH OH OH HO

The given spectrum can now be annotated as follows:

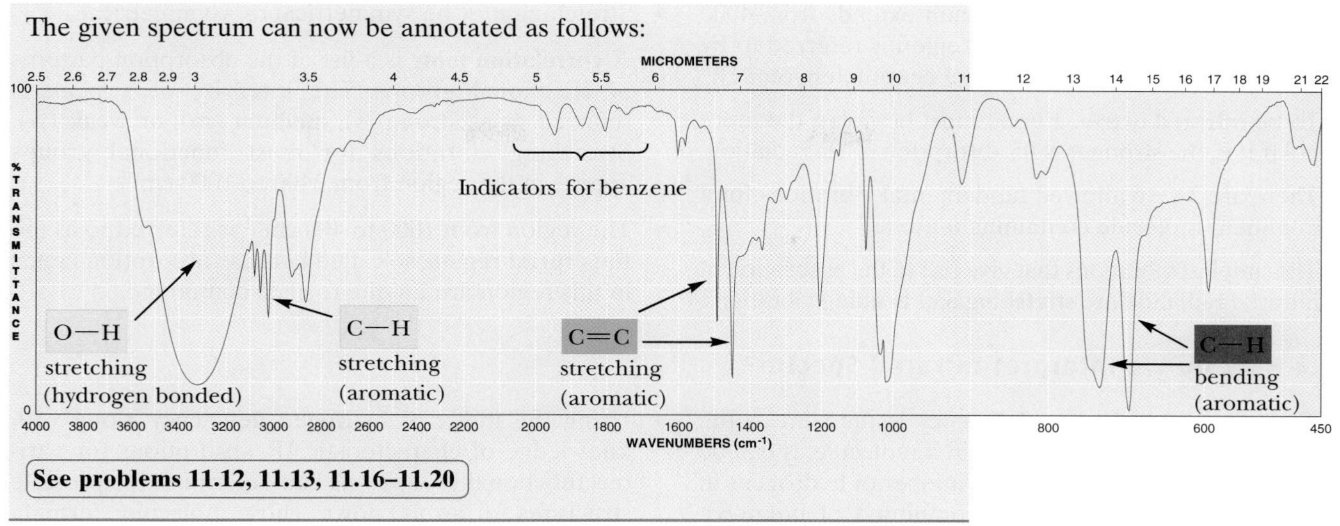

See problems 11.12, 11.13, 11.16–11.20

The preceding example illustrates the power and limitations of IR spectroscopy. The power lies in its ability to provide us with information regarding the functional groups in a molecule. IR spectroscopy does not, however, provide us with information on how those functional groups are connected. Fortunately, another type of spectroscopy—nuclear magnetic resonance (NMR) spectroscopy—does provide us with connectivity information. NMR spectroscopy is the topic of the next chapter.

Key Terms and Concepts

bending vibrational motion (p. 393)
correlation table (p. 394)
electromagnetic radiation (p. 389)
electromagnetic spectrum (p. 389)
fingerprint region (p. 394)
frequency (ν) (p. 389)
hertz (Hz) (p. 389)
index of hydrogen deficiency (p. 405)
infrared active (p. 393)
infrared (IR) spectroscopy (p. 391)
infrared spectrum (p. 392)
intensity (stong, medium, and weak) of absorptions (p. 394)
molecular spectroscopy (p. 391)
photon (p. 390)
reference compound (p. 405)
stretching vibrational motion (p. 393)
vibrational infrared (p. 392)
wavelength (λ) (p. 389)
wavenumber ($\overline{\nu}$) (p. 392)

Summary of Key Questions

11.1 What Is Electromagnetic Radiation?

- **Electromagnetic radiation** is a wave traveling at the speed of light that can be described in terms of its **wavelength** (λ) and its **frequency** (ν).
- Frequency is reported in **hertz (Hz)**.
- An alternative way to describe electromagnetic radiation is in terms of its energy where $E = h\nu$.

11.2 What Is Molecular Spectroscopy?

- **Molecular spectroscopy** is the experimental process of measuring which frequencies of radiation are absorbed or emitted by a substance and correlating these patterns with details of molecular structure.

11.3 What Is Infrared Spectroscopy?

- **Infrared spectroscopy** is molecular spectroscopy applied to frequencies of infrared radiation.
- Interactions of molecules with **infrared radiation** excite covalent bonds to higher vibrational energy levels.

- The **vibrational infrared** spectrum extends from 4000 to 400 cm^{-1}. Radiation in this region is referred to by its wavenumber ($\bar{\nu}$) in reciprocal centimeters (cm^{-1}).
- To be **infrared active**, a bond must be polar; the more polar it is, the stronger is its absorption of IR radiation.
- There are $3n - 6$ allowed fundamental vibrations for a nonlinear molecule containing n atoms.
- The simplest vibrations that give rise to the absorption of infrared radiation are **stretching** and **bending** vibrations.
- Stretching may be symmetrical or asymmetrical.
- A **correlation table** is a list of the absorption patterns of functional groups. The intensity of a peak is referred to as **strong (s)**, **medium (m)**, or **weak (w)**. Stretching vibrations for most functional groups appear in the region from 3400 to 1000 cm^{-1}.
- The region from 1000 to 400 cm^{-1} is referred to as the **fingerprint region**, so called because absorption bands in this region are unique to each compound.

11.4 How Do We Interpret Infrared Spectra?

- The **index of hydrogen deficiency** is the sum of the number of rings and pi bonds in a molecule. It can be determined by comparing the number of hydrogens in the molecular formula of a compound of unknown structure with the number of hydrogens in a reference compound with the same number of carbon atoms and with no rings or pi bonds.
- Using the index of hydrogen deficiency along with knowledge of characteristic IR absorptions for various functional groups, one can determine the possible structures for an unknown whose molecular formula is known.

Quick Quiz

Answer true or false to the following questions to assess your general knowledge of the concepts in this chapter. If you have difficulty with any of them, you should review the appropriate section in the chapter (shown in parentheses) before attempting the more challenging end-of-chapter problems.

1. A weak absorption band in an infrared spectrum can be attributed to, among other things, absorption of infrared light by a low polarity bond. (11.3)
2. Wavelength and frequency are directly proportional. That is, as wavelength increases, frequency increases. (11.1)
3. IR spectroscopy can be used to distinguish between a terminal alkyne and an internal alkyne. (11.4)
4. A transition between two energy states, E_1 and E_2, can be made to occur using light equal to or greater than the energy difference between E_1 and E_2. (11.2)
5. A compound with the molecular formula $C_5H_{10}O$ could contain a C≡C triple bond, two C=O bonds, or two rings. (11.4)
6. A compound with the molecular formula $C_7H_{12}O$ has an index of hydrogen deficiency of 2. (11.4)
7. The number of fundamental vibrations a molecule may undergo can be determined if the geometry and number of atoms are known. (11.3)
8. Electromagnetic radiation can be described as a wave, as a particle, and in terms of energy. (11.1)
9. The collection of absorption peaks in the 1000–400 cm^{-1} region of an IR sprectrum is unique to a particular compound (i.e., no two compounds will yield the same spectrum in this region). (11.3)
10. C—H stretching vibrations occur at higher wavenumbers than C—C stretching vibrations. (11.4)
11. It is not possible to use IR spectroscopy to distinguish between a ketone and a carboxylic acid. (11.4)
12. A wavenumber, $\bar{\nu}$, is directly proportional to frequency. (11.3)
13. IR spectroscopy cannot be used to distinguish between an alcohol and an ether. (11.4)
14. Infrared spectroscopy measures transitions between electronic energy levels. (11.2)
15. The index of hydrogen deficiency can reveal the possible number of rings, double bonds, or triple bonds in a compound based solely on its molecular formula. (11.4)
16. Light of wavelength 400 nm is higher in energy than light of wavelength 600 nm. (11.1)
17. A compound with the molecular formula $C_6H_{14}FN$ has an index of hydrogen deficiency of 1. (11.4)
18. IR spectroscopy can be used to distinguish between 1°, 2°, and 3° amines. (11.4)

Answers: (1) T (2) F (3) T (4) F (5) F (6) T (7) T (8) T (9) T (10) T (11) F (12) T (13) F (14) F (15) T (16) T (17) F (18) T

Problems

A problem marked with an asterisk indicates an applied "real world" problem. Answers to problems whose numbers are printed in blue are given in Appendix D.

Section 11.1 Electromagnetic Radiation

11.7 Which puts out light of higher energy, a green laser pointer or a red laser pointer? **(See Example 11.1)**

11.8 Calculate the energy, in kilocalories per mole of radiation, of a wave with wavelength 2 m. What type of radiant energy is this? **(See Example 11.1)**

11.9 A molecule possesses molecular orbitals that differ in energy by 82 kcal/mol. What wavelength of light would be required to cause a transition between these two energy levels? What region of the electromagnetic spectrum does this energy correspond to? **(See Example 11.1)**

Section 11.4 Index of Hydrogen Deficiency

11.10 Complete the following table: **(See Examples 11.5, 11.6)**

Class of Compound	Molecular Formula	Index of Hydrogen Deficiency	Reason for Hydrogen Deficiency
alkane	C_nH_{2n+2}	0	(reference hydrocarbon)
alkene	C_nH_{2n}	1	one pi bond
alkyne	______	____	________________
alkadiene	______	____	________________
cycloalkane	______	____	________________
cycloalkene	______	____	________________

***11.11** Calculate the index of hydrogen deficiency of each compound: **(See Examples 11.5, 11.6)**

(a) Aspirin, $C_9H_8O_4$

(b) Ascorbic acid (vitamin C), $C_6H_8O_6$

(c) Pyridine, C_5H_5N

(d) Urea, CH_4N_2O

(e) Cholesterol, $C_{27}H_{46}O$

(f) Trichloroacetic acid, $C_2HCl_3O_2$

11.12 Compound A, with the molecular formula C_6H_{10}, reacts with H_2/Ni to give compound B, with the molecular formula C_6H_{12}. The IR spectrum of compound A is provided. From this information about compound A tell **(See Examples 11.4–11.7)**

(a) Its index of hydrogen deficiency.

(b) The number of rings or pi bonds (or both) in compound A.

(c) What structural feature(s) would account for compound A's index of hydrogen deficiency.

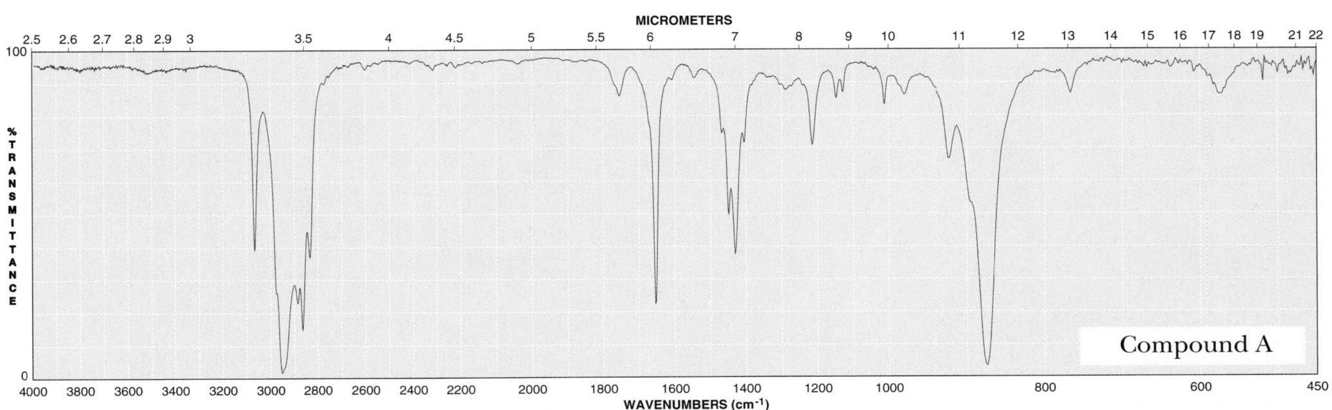

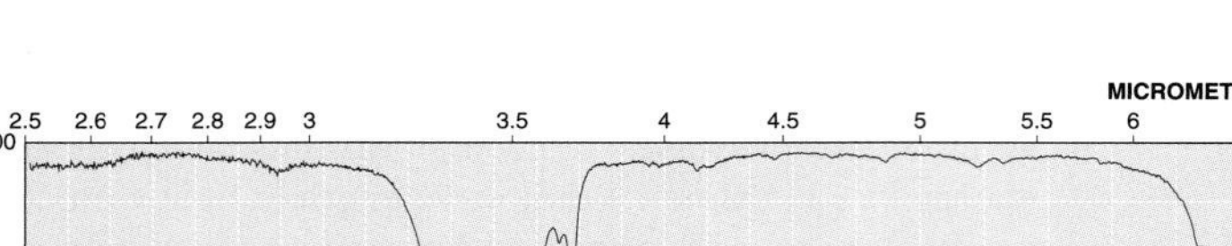

11.13 Compound C, with the molecular formula C_6H_{12}, reacts with H_2/Ni to give compound D, with the molecular formula C_6H_{14}. The IR spectrum of compound C is provided. From this information about compound C, tell **(See Examples 11.4–11.7)**

(a) Its index of hydrogen deficiency.

(b) The number of rings or pi bonds (or both) in compound C.

(c) What structural feature(s) would account for compound C's index of hydrogen deficiency.

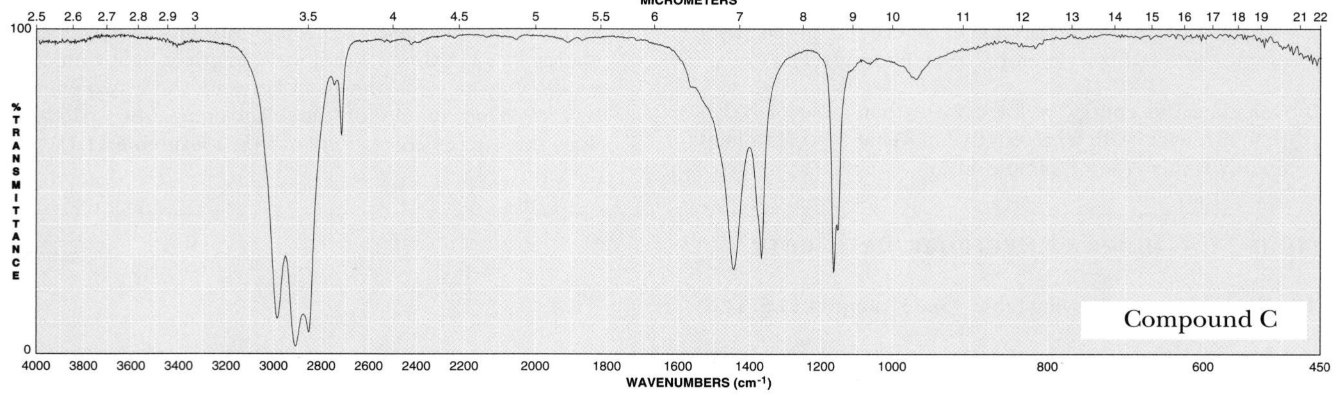

11.14 Following are infrared spectra of compounds E and F: One spectrum is of 1-hexanol, the other of nonane. Assign each compound its correct spectrum. **(See Examples 11.2, 11.3)**

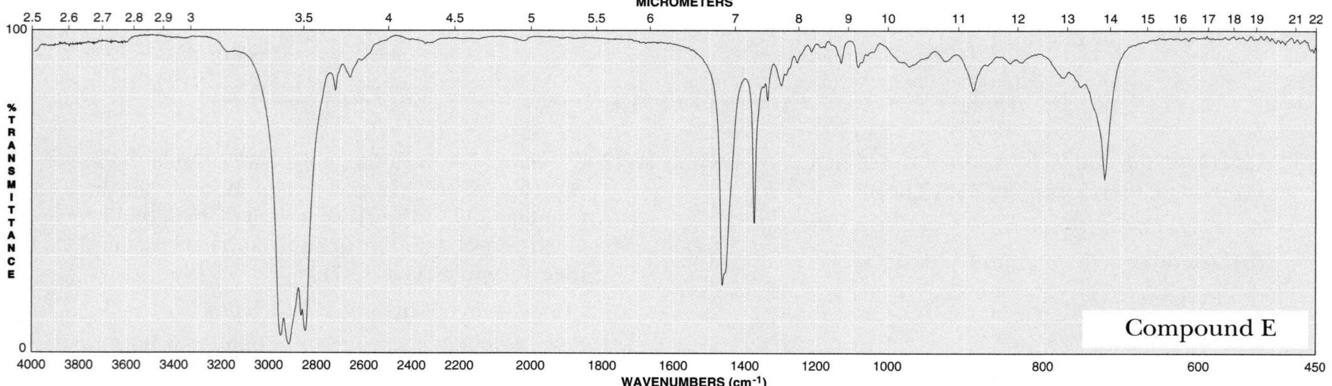

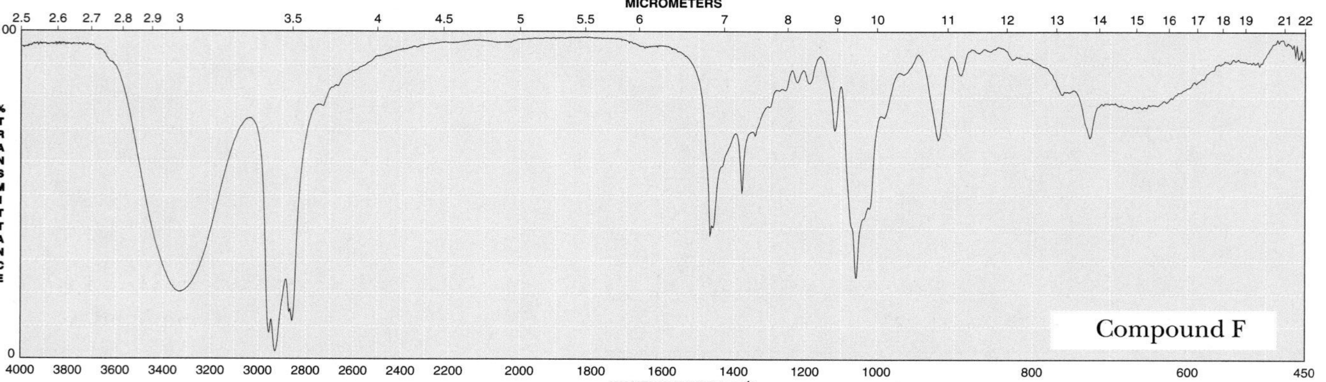

11.15 2-Methyl-1-butanol and *tert*-butyl methyl ether are constitutional isomers with the molecular formula $C_5H_{12}O$. Assign each compound its correct infrared spectrum, G or H: **(See Examples 11.2–11.3)**

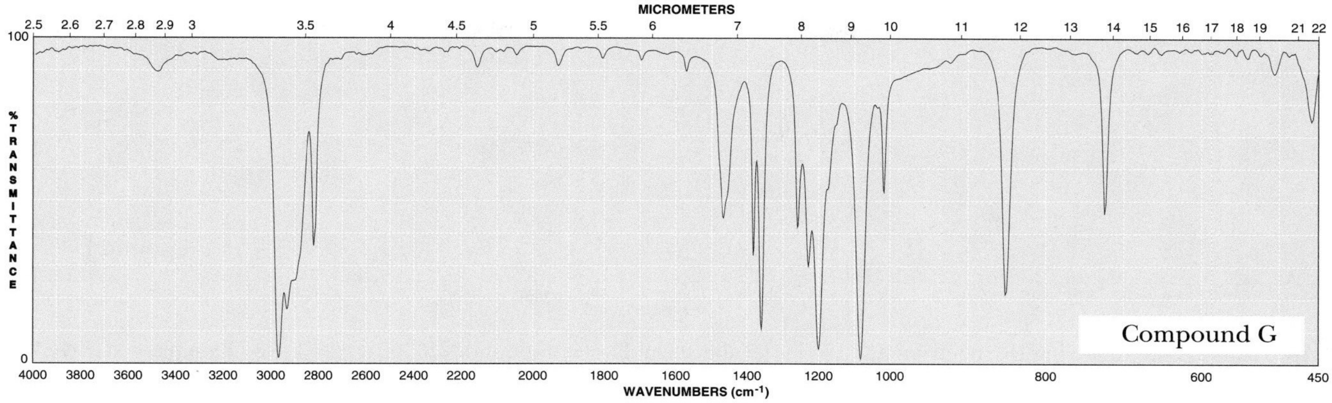

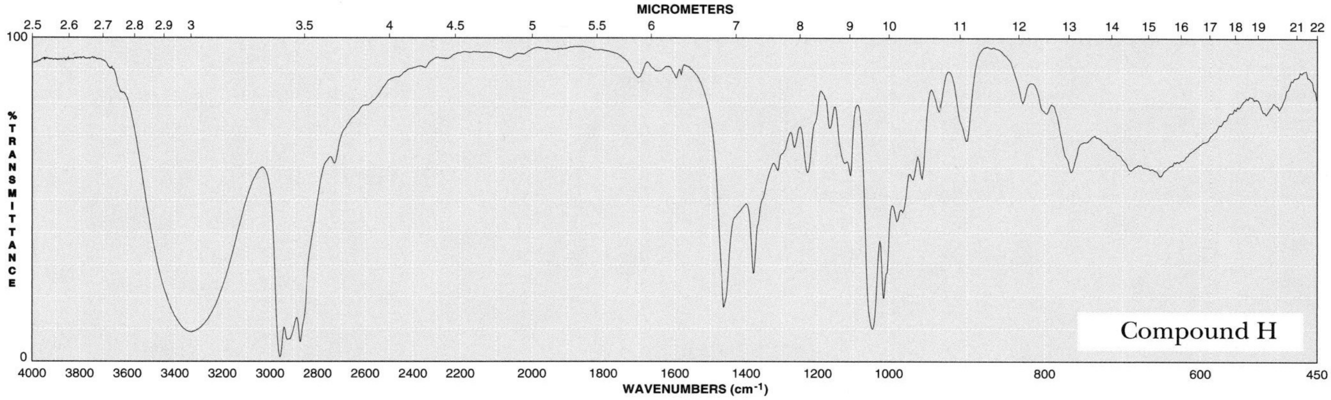

Compound G

Compound H

11.16 Examine the following IR spectrum and the molecular formula of compound I, $C_9H_{12}O$: Tell **(See Examples 11.4–11.7)**

(a) Its index of hydrogen deficiency.

(b) The number of rings or pi bonds (or both) in compound I.

(c) What one structural feature would account for this index of hydrogen deficiency.

(d) What oxygen-containing functional group compound I contains.

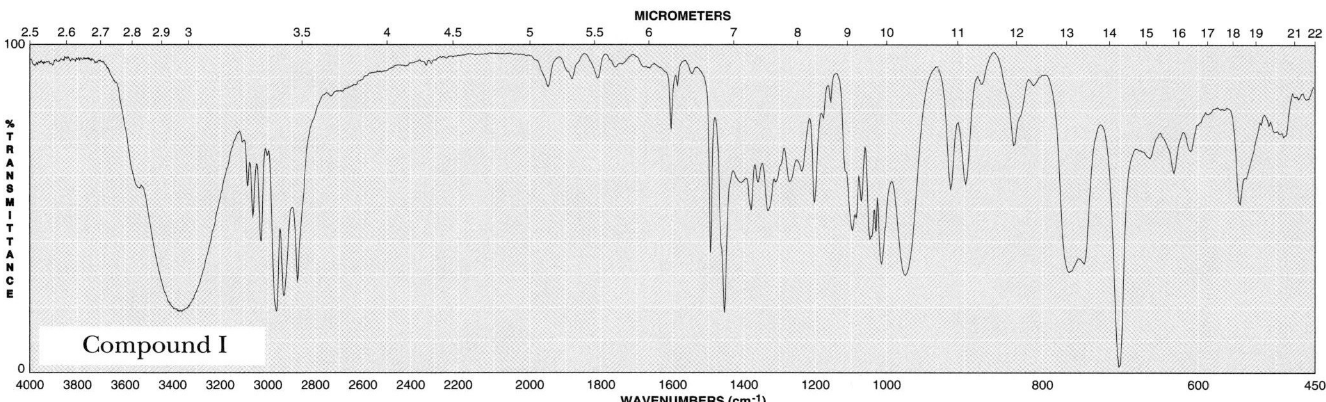

11.17 Examine the following IR spectrum and the molecular formula of compound J, $C_5H_{13}N$: Tell **(See Examples 11.4–11.7)**

(a) Its index of hydrogen deficiency.

(b) The number of rings or pi bonds (or both) in compound J.

(c) The nitrogen-containing functional group(s) compound J might contain.

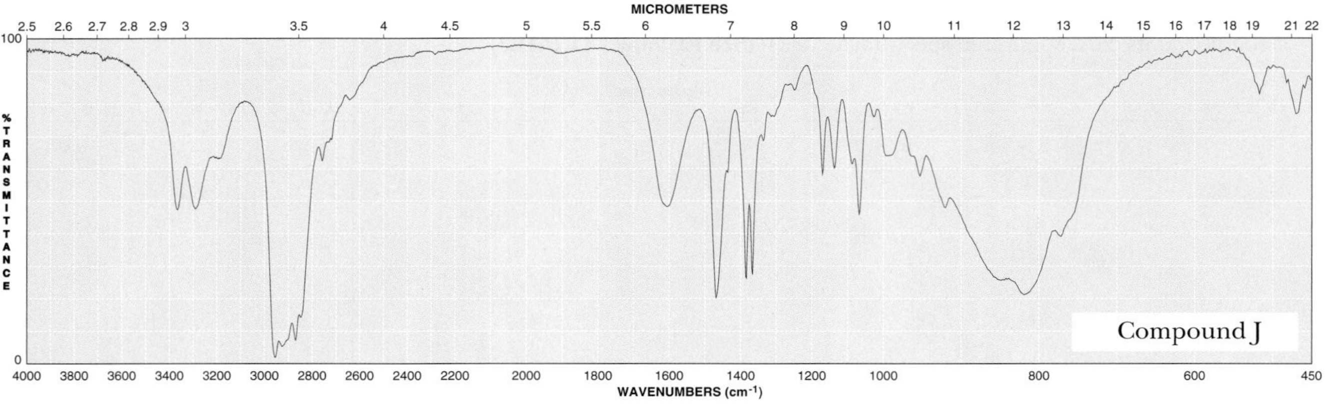

11.18 Examine the following IR spectrum and the molecular formula of compound K, $C_6H_{12}O$: Tell **(See Examples 11.4–11.7)**

(a) Its index of hydrogen deficiency.

(b) The number of rings or pi bonds (or both) in compound K.

(c) What structural features would account for this index of hydrogen deficiency.

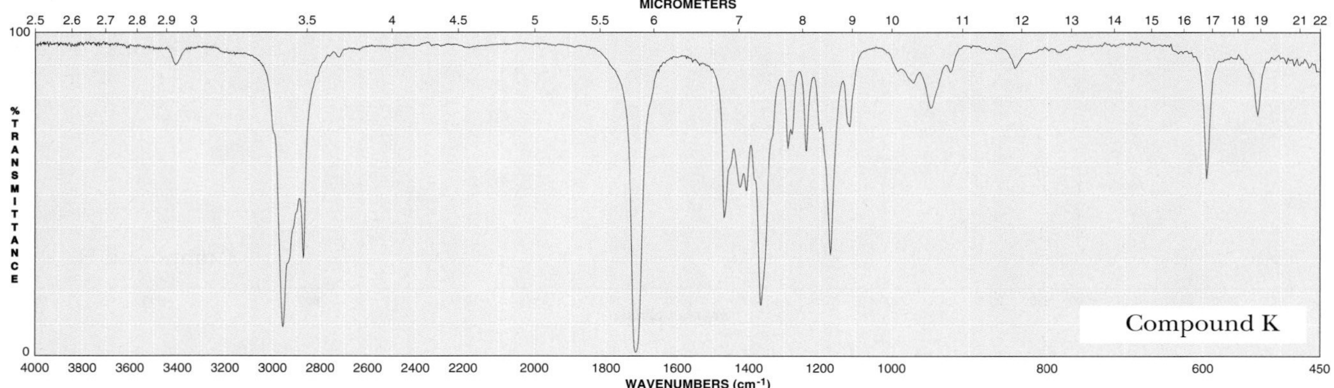

11.19 Examine the following IR spectrum and the molecular formula of compound L, $C_6H_{12}O_2$: Tell **(See Examples 11.4–11.7)**

(a) Its index of hydrogen deficiency.

(b) The number of rings or pi bonds (or both) in compound L.

(c) The oxygen-containing functional group(s) compound L might contain.

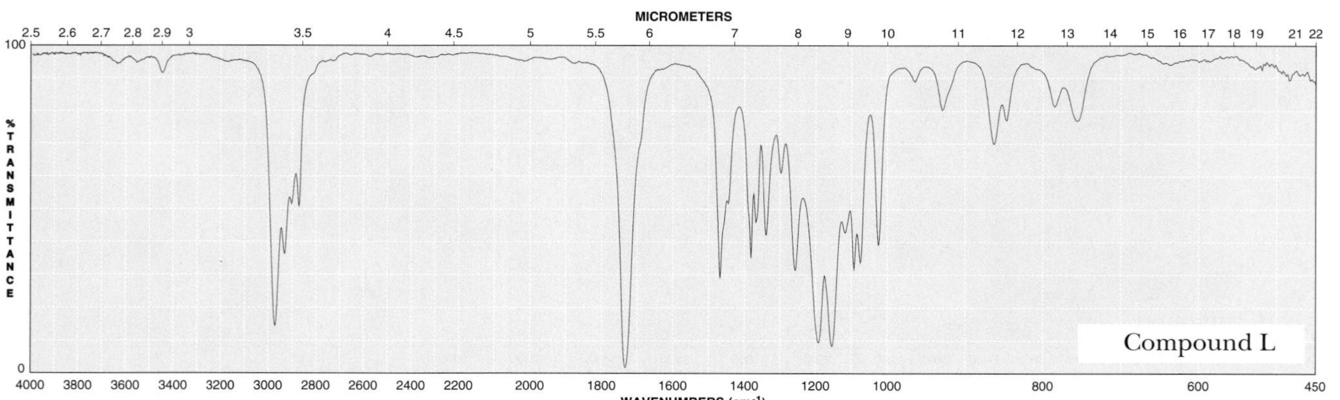

11.20 Examine the following IR spectrum and the molecular formula of compound M, C_3H_7NO: Tell **(See Examples 11.4–11.7)**

(a) Its index of hydrogen deficiency.

(b) The number of rings or pi bonds (or both) in compound M.

(c) The oxygen- and nitrogen-containing functional group(s) in compound M.

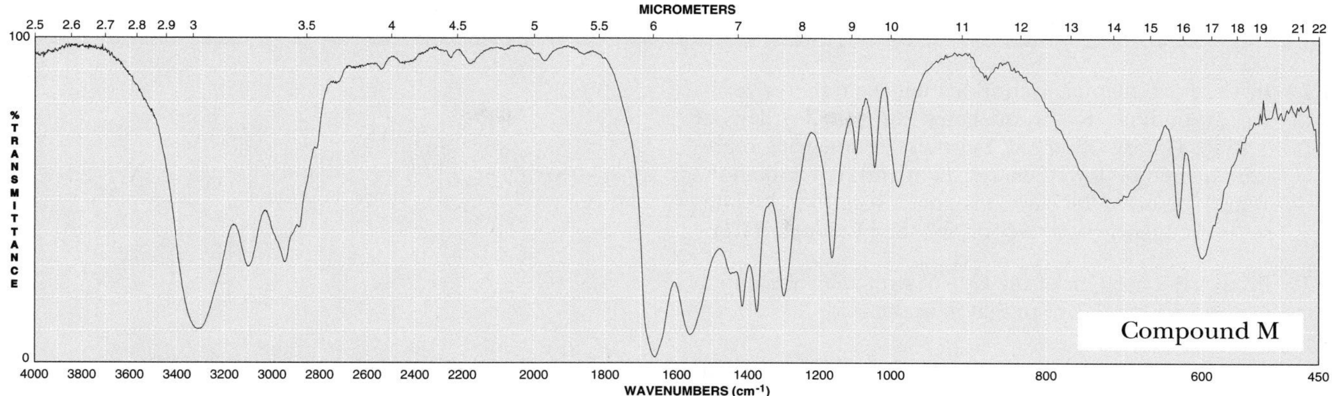

11.21 Show how IR spectroscopy can be used to distinguish between the compounds in each of the following pairs: **(See Examples 11.2, 11.3)**

(a) 1-Butanol and diethyl ether
(b) Butanoic acid and 1-butanol
(c) Butanoic acid and 2-butanone
(d) Butanal and 1-butene
(e) 2-Butanone and 2-butanol
(f) Butane and 2-butene

11.22 For each pair of compounds that follows, list one major feature that appears in the IR spectrum of one compound, but not the other. In your answer, state what type of bond vibration is responsible for the spectral feature you list, and give its approximate position in the IR spectrum. **(See Examples 11.2, 11.3)**

(a) $C_6H_5\overset{O}{\overset{\|}{C}}H$ and $C_6H_5\overset{O}{\overset{\|}{C}}OH$

(b) $C_6H_{11}\overset{O}{\overset{\|}{C}}N(CH_3)_2$ and $C_6H_{11}CH_2N(CH_3)_2$

(c) (six-membered lactone, ring O and C=O) and $HO(CH_2)_4\overset{O}{\overset{\|}{C}}OH$

(d) $C_6H_{11}\overset{O}{\overset{\|}{C}}NH_2$ and $C_6H_{11}\overset{O}{\overset{\|}{C}}N(CH_3)_2$

***11.23** Following are an infrared spectrum and a structural formula for methyl salicylate, the fragrant component of oil of wintergreen. On this spectrum, locate the absorption peak(s) due to **(See Examples 11.2, 11.3)**

(a) O—H stretching of the hydrogen-bonded —OH group (very broad and of medium intensity).
(b) C—H stretching of the aromatic ring (sharp and of weak intensity).
(c) C=O stretching of the ester group (sharp and of strong intensity).
(d) C=C stretching of the aromatic ring (sharp and of medium intensity).

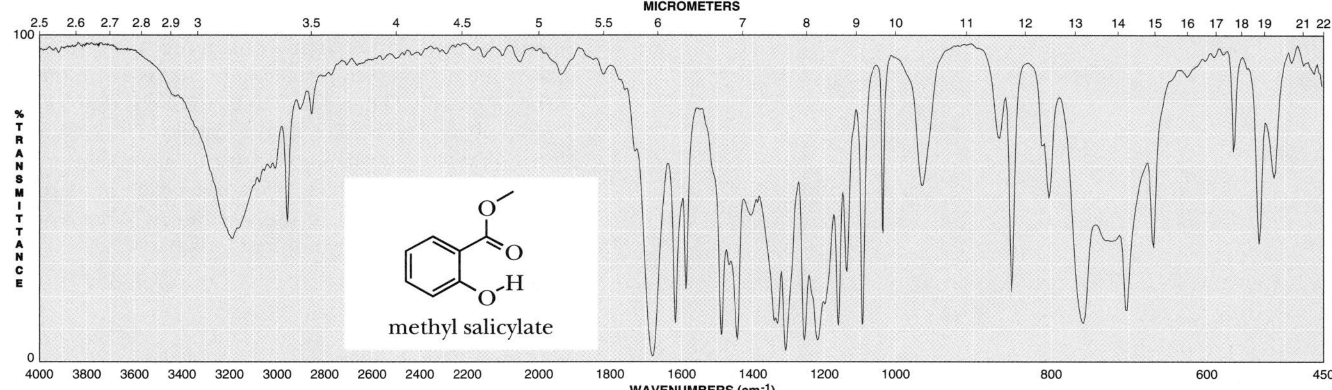

Looking Ahead

11.24 In the next chapter, transitions between energy levels corresponding to frequencies on the order of 3×10^8 Hz are observed. Do these frequencies represent higher or lower energy than infrared radiation? Which region of the electromagnetic spectrum does this set of frequencies correspond to? **(See Example 11.1)**

11.25 Predict the position of the C=O stretching absorption in acetate ion relative to that in acetic acid:

Acetic acid   Acetate ion

11.26 Determine whether IR spectroscopy can be used to distinguish between the following pairs of molecules (assume that you do not have the reference spectrum of either molecule): **(See Examples 11.2, 11.3)**

(a) and

(b) and

(c) and

(d) and

***11.27** Following is the IR spectrum of L-tryptophan, a naturally occurring amino acid that is abundant in foods such as turkey:

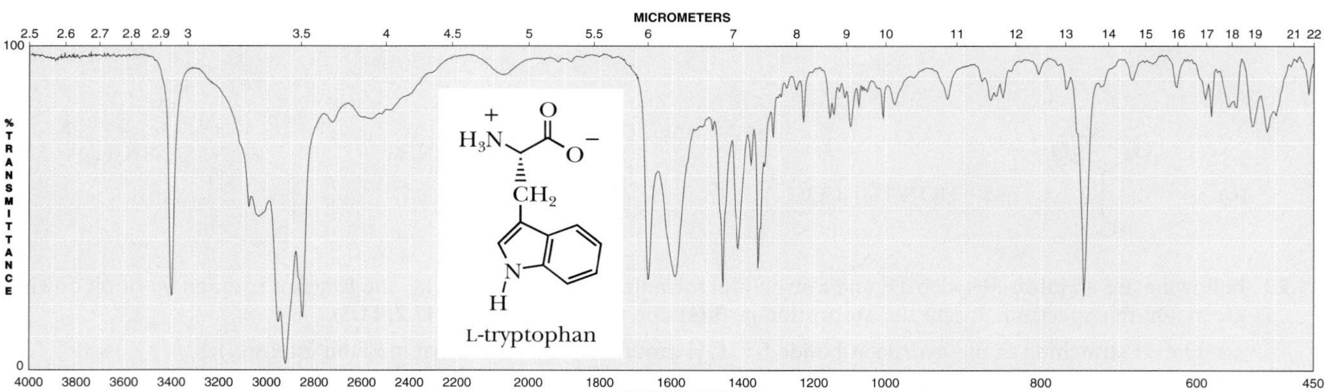

For many years, the L-tryptophan in turkey was believed to make people drowsy after Thanksgiving dinner. Scientists now know that consuming L-tryptophan makes one drowsy only if the compound is taken on an empty stomach. Therefore, it is unlikely that one's Thanksgiving Day turkey is the cause of drowsiness. Notice that L-tryptophan contains one stereocenter. Its enantiomer, D-tryptophan, does not occur in nature, but can be synthesized in the laboratory. What would the IR spectrum of D-tryptophan look like?

12 Nuclear Magnetic Resonance Spectroscopy

KEY QUESTIONS

- **12.1** What Are the Magnetic Properties of Nuclei?
- **12.2** What Is Nuclear Magnetic Resonance?
- **12.3** What Is Shielding?
- **12.4** How Is an NMR Spectrum Obtained?
- **12.5** How Many Resonance Signals Will a Compound Yield in Its NMR Spectrum?
- **12.6** What Is Signal Integration?
- **12.7** What Is Chemical Shift?
- **12.8** What Is Signal Splitting?
- **12.9** What Is ^{13}C-NMR Spectroscopy, and How Does It Differ from ^{1}H-NMR Spectroscopy?
- **12.10** How Do We Interpret NMR Spectra?
- **12.11** How Do We Solve an NMR Problem?

HOW TO

- **12.1** How to Determine Whether an Atomic Nucleus Has a Spin (Behaves as If It Were a Tiny Bar Magnet)
- **12.2** How to Use Symmetry to Determine Equivalency of Hydrogens in a Molecule
- **12.3** How to Use Symmetry to Determine Equivalency of Carbons in a Molecule

CHEMICAL CONNECTIONS

- **12A** Magnetic Resonance Imaging

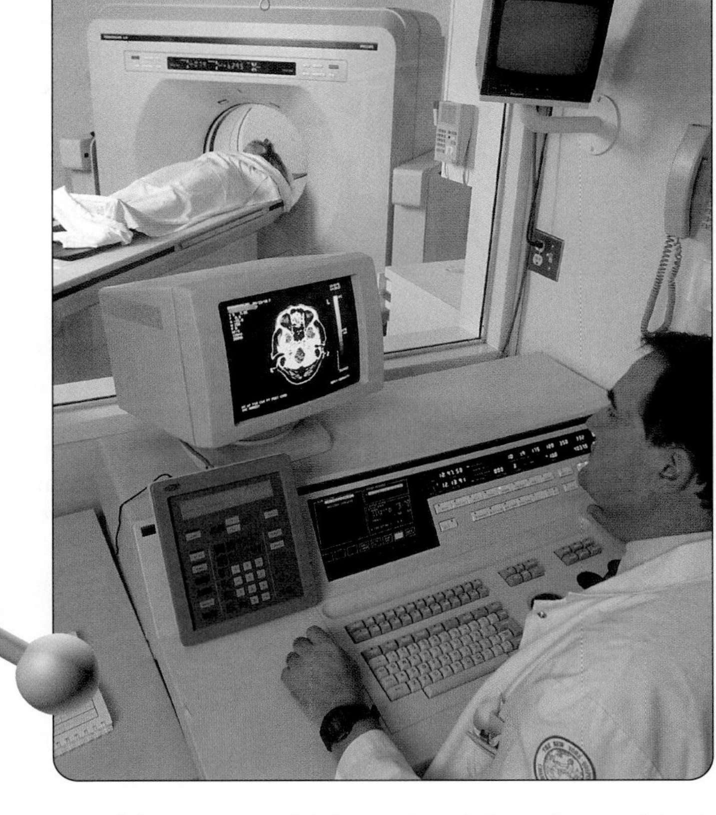

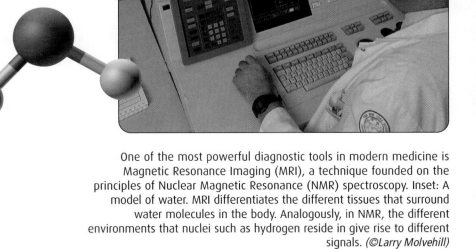

One of the most powerful diagnostic tools in modern medicine is Magnetic Resonance Imaging (MRI), a technique founded on the principles of Nuclear Magnetic Resonance (NMR) spectroscopy. Inset: A model of water. MRI differentiates the different tissues that surround water molecules in the body. Analogously, in NMR, the different environments that nuclei such as hydrogen reside in give rise to different signals. *(©Larry Molvehill)*

In the previous chapter, we discussed infrared spectroscopy and saw how infrared radiation could be used to determine the types of functional groups present in an unknown compound. In this chapter, we concentrate on the absorption of radio-frequency radiation, which causes transitions between nuclear spin energy levels; that is, we concentrate on a technique known as nuclear magnetic resonance (NMR) spectroscopy.

The phenomenon of nuclear magnetic resonance was first detected in 1946 by U.S. scientists Felix Bloch and Edward Purcell, who shared the 1952 Nobel Prize for Physics for their discoveries. The particular value of **nuclear magnetic resonance**

(NMR) spectroscopy is that it gives us information about the number and types of atoms in a molecule, for example, about the number and types of hydrogens using **^{1}H-NMR spectroscopy**, and about the number and types of carbons using **^{13}C-NMR spectroscopy**.

12.1 What Are the Magnetic Properties of Nuclei?

From your study of general chemistry, you may already be familiar with the concept that an electron has a spin and that a spinning charge creates an associated magnetic field. In effect, an electron behaves as if it is a tiny bar magnet. An atomic nucleus that has an odd mass or an odd atomic number also has a **spin** and behaves as if it is a tiny bar magnet. Recall that when designating isotopes, a superscript represents the mass of the element. Thus, the nuclei of ^{1}H and ^{13}C, isotopes of the two elements most common in organic compounds, also have a spin, whereas the nuclei of ^{12}C and ^{16}O do not have a spin and do not behave as tiny bar magnets. Accordingly, in this sense, nuclei of ^{1}H and ^{13}C are quite different from nuclei of ^{12}C and ^{16}O.

HOW TO 12.1 Determine Whether an Atomic Nucleus Has a Spin (Behaves as If It Were a Tiny Bar Magnet)

Determine the mass and atomic number of the atom. If *either* is an odd number, the atom will have a spin and behave as a tiny bar magnet.

atomic mass (may vary for isotopes of the same element)

$^{m}_{n}\text{Atom}$

atomic number (always the same for all isotopes of any element)

Example 12.1

Which of the following nuclei are capable of behaving like tiny bar magnets?

(a) $^{14}_{6}C$ (b) $^{14}_{7}N$

Strategy

Any nucleus that has a spin (those that have either an odd mass or an odd atomic number) will act as a tiny bar magnet.

Solution

(a) $^{14}_{6}C$, a radioactive isotope of carbon, has neither an odd mass number nor an odd atomic number and therefore cannot behave as if it were a tiny bar magnet.

(b) $^{14}_{7}N$, the most common naturally occurring isotope of nitrogen (99.63% of all nitrogen atoms), has an odd atomic number and therefore behaves as if it were a tiny bar magnet.

Problem 12.1

Which of the following nuclei are capable of behaving like tiny bar magnets?

(a) $^{31}_{15}P$ (b) $^{195}_{78}Pt$

Within a collection of ^{1}H and ^{13}C atoms, the spins of their tiny nuclear bar magnets are completely random in orientation. When we place them between the poles of a powerful magnet, however, interactions between their nuclear spins and the **applied magnetic field** are quantized, and only two orientations are allowed (Figure 12.1).

At an applied field strength of 7.05 tesla (T), which is readily available with present-day superconducting electromagnets, the difference in energy between nuclear spin states for ^{1}H is 0.120 J (0.0286 cal)/mol, which corresponds to electromagnetic radiation of approximately 300 MHz (300,000,000 Hz). At the same magnetic field strength, the difference in energy between nuclear spin states for ^{13}C is 0.035 J (0.0072 cal)/mol, which corresponds to electromagnetic radiation of 75 MHz. Thus, we can use electromagnetic radiation in the radio-frequency range (Table 11.1) to detect changes in nuclear spin states for ^{1}H and ^{13}C. In the next several sections, we describe how these measurements are made for nuclear spin states of these two isotopes and then how this information can be correlated with molecular structure.

12.2 What Is Nuclear Magnetic Resonance?

When hydrogen nuclei are placed in an applied magnetic field, a small majority of their nuclear spins align with the applied field in the lower energy state. Irradiation of the nuclei in the lower energy spin state with radio-frequency radiation of the appropriate energy causes them to absorb energy and results in their nuclear spins flipping from the lower energy state to the higher energy state, as illustrated in Figure 12.2. In this context, **resonance** is defined as the absorption of electromagnetic radiation by a spinning nucleus and the resulting flip of its nuclear spin state.

Resonance The absorption of electromagnetic radiation by a spinning nucleus and the resulting "flip" of its spin from a lower energy state to a higher energy state.

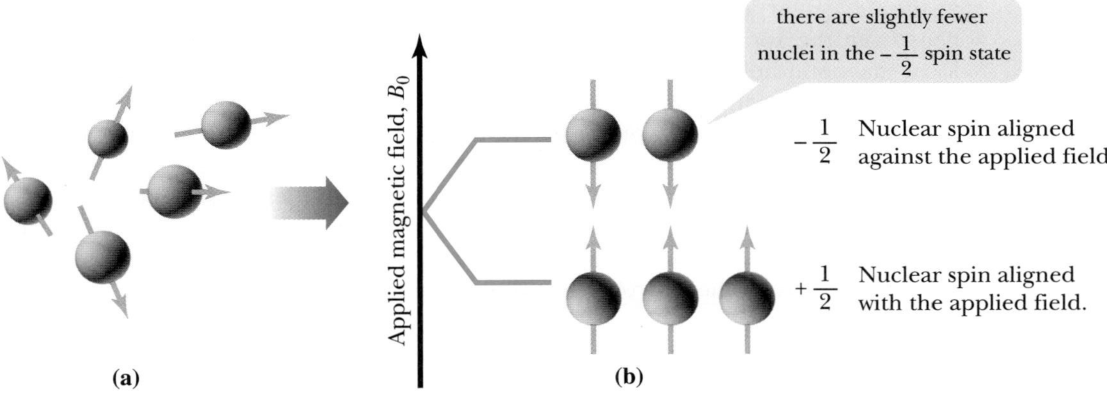
there are slightly fewer nuclei in the $-\frac{1}{2}$ spin state
Applied magnetic field, B_0
$-\frac{1}{2}$ Nuclear spin aligned against the applied field.
$+\frac{1}{2}$ Nuclear spin aligned with the applied field.
(a)
(b)

Figure 12.1
^{1}H and ^{13}C nuclei (a) in the absence of an applied magnetic field and (b) in the presence of an applied field. ^{1}H and ^{13}C nuclei with spin $+\frac{1}{2}$ are aligned with the applied magnetic field and are in the lower spin energy state; those with spin $-\frac{1}{2}$ are aligned against the applied magnetic field and are in the higher spin energy state.

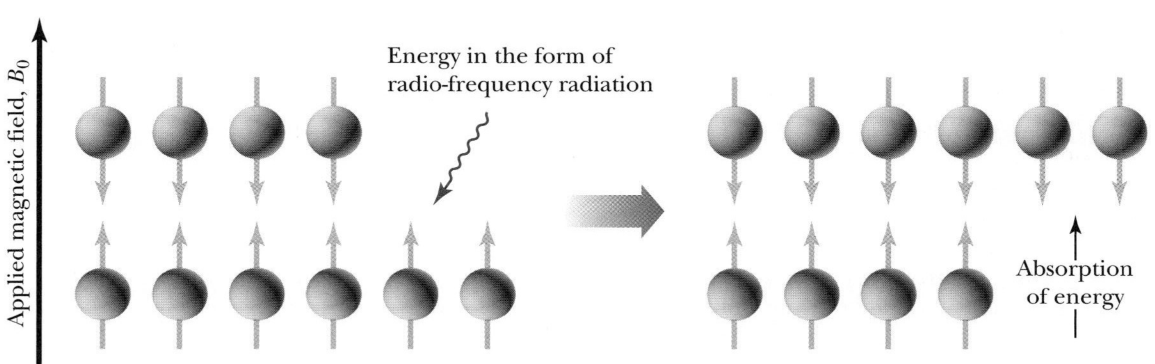

Figure 12.2
An example of resonance for nuclei of spin $\frac{1}{2}$.

Resonance signal A recording of nuclear magnetic resonance in an NMR spectrum.

The instrument we use to detect this absorption and resulting flip of nuclear spin state records it as a **resonance signal**.

12.3 What Is Shielding?

If we were dealing with ^1H nuclei isolated from all other atoms and electrons, any combination of applied field and electromagnetic radiation that produces a resonance signal for one hydrogen nucleus would produce the same resonance signal for all other hydrogen nuclei. In other words, the same amount of energy would cause all hydrogens to resonate, and the hydrogens would be indistinguishable one from another. NMR would then be an ineffective technique for determining the structure of a molecule because all the hydrogens in a compound would resonate at the same frequency, giving rise to one and only one NMR signal.

Fortunately, hydrogens in most organic molecules are surrounded by electrons and by other atoms. The electrons that surround a nucleus also have spin and thereby create **local magnetic fields** that oppose the applied field. Although these local magnetic fields created by electrons are orders of magnitude weaker than the applied magnetic fields used in NMR spectroscopy, they are nonetheless significant at the molecular level. The result of these local magnetic fields is to shield hydrogens from the applied field. The greater the **shielding** of a particular hydrogen by local magnetic fields, the greater is the strength of the applied field necessary to bring that hydrogen into resonance.

Shielding In NMR spectroscopy, electrons around a nucleus create their own local magnetic fields and thereby shield the nucleus from the applied magnetic field.

As we learned in previous chapters, the electron density around a nucleus can be influenced by the atoms that surround the nucleus. For example, the electron density around the hydrogen atoms in fluoromethane is less than that around the hydrogen atoms in chloromethane, due to the greater electronegativity of fluorine relative to chlorine. Thus, we can say that the hydrogen atoms in chloromethane are *more shielded* than the hydrogen atoms in fluoromethane:

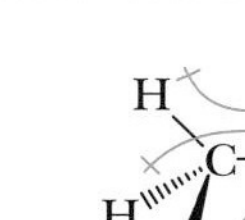

Chlorine is less electronegative than fluorine, resulting in a smaller inductive effect and thereby a greater electron density around each hydrogen. We say that the hydrogens in chloromethane are more **shielded** (by their local environment) than those in fluoromethane.

Fluorine's greater electronegativity produces a larger inductive effect and thereby reduces the electron density around each hydrogen. We say that these hydrogens are **deshielded**.

The differences in resonance frequencies among the various ^1H nuclei within a molecule caused by shielding are generally very small. The difference between the resonance frequencies of hydrogens in chloromethane compared with those in fluoromethane, for example, is only 360 Hz under an applied field of 7.05 tesla. Considering that the radio-frequency radiation used at this applied field is approximately 300 MHz (300×10^6 Hz), the difference in resonance frequencies between these two sets of hydrogens is only slightly greater than 1 **part per million** (1 ppm) compared with the irradiating frequency.

$$\frac{360 \text{ Hz}}{300 \times 10^6 \text{ Hz}} = \frac{1.2}{10^6} = 1.2 \text{ ppm}$$

NMR spectrometers are able to detect these small differences in resonance frequencies. The importance of shielding for elucidating the structure of a molecule will be discussed in Section 12.7.

12.4 How Is an NMR Spectrum Obtained?

The essential elements of an NMR spectrometer are a powerful magnet, a radio-frequency generator, a radio-frequency detector, and a sample tube (Figure 12.3).

The sample is dissolved in a solvent having no hydrogens, most commonly deuterochloroform ($CDCl_3$) or deuterium oxide (D_2O). The sample cell is a small glass tube suspended in the gap between the pole pieces of the magnet and set spinning on its long axis to ensure that all parts of the sample experience a homogeneous applied field. In a typical ^1H-NMR spectrum, the horizontal axis represents the **delta (δ) scale**, with values from 0 on the right to 10 on the left. The vertical axis represents the intensity of the resonance signal.

It is customary to measure the resonance frequencies of individual nuclei relative to the resonance frequency of the same nuclei in a reference compound. The reference compound now universally accepted for ^1H-NMR and ^{13}C-NMR spectroscopy is **tetramethylsilane (TMS)** because it is relatively unreactive and its hydrogen and carbon atoms are highly shielded due to the less electronegative

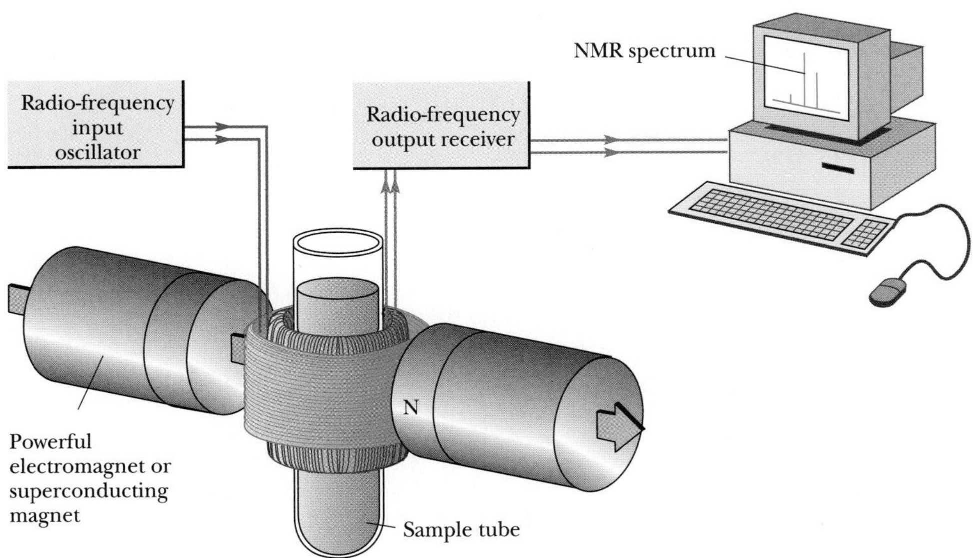

Figure 12.3
Schematic diagram of a nuclear magnetic resonance spectrometer.

silicon atom. The latter fact ensures that most other resonance signals will be less shielded than the signal for TMS.

$$\begin{array}{c} CH_3 \\ | \\ H_3C-Si-CH_3 \\ | \\ CH_3 \end{array}$$

Tetramethylsilane (TMS)

When we determine a ^{1}H-NMR spectrum of a compound, we report how far the resonance signals of its hydrogens are shifted from the resonance signal of the hydrogens in TMS. When we determine a ^{13}C-NMR spectrum, we report how far the resonance signals of its carbons are shifted from the resonance signal of the four carbons in TMS.

Chemical shift, δ The position of a signal on an NMR spectrum relative to the signal of tetramethylsilane (TMS); expressed in delta (δ) units, where 1 δ equals 1 ppm.

To standardize reporting of NMR data, workers have adopted a quantity called the **chemical shift (δ)**. Chemical shift is calculated by dividing the shift in frequency of a signal (relative to that of TMS) by the operating frequency of the spectrometer. Because NMR spectrometers operate at MHz frequencies (i.e., millions of Hz), we express the chemical shift in parts per million. A sample calculation is provided for the signal at 2.05 ppm in the ^{1}H-NMR spectrum for methyl acetate (Figure 12.4), a compound used in the manufacture of artificial leather.

$$\delta = \frac{\text{Shift in frequency of a signal from TMS (Hz)}}{\text{Operating frequency of the spectrometer (Hz)}}$$

the hydrogens on this methyl group cause a signal to occur 615 Hz from the TMS signal in a 300 MHz NMR spectrometer

$$\text{e.g.,}\quad CH_3\overset{\overset{O}{\|}}{C}-OCH_3 \qquad \frac{615\ \text{Hz}}{300 \times 10^6\ \text{Hz}} = \frac{2.05\ \text{Hz}}{\text{million Hz}} = 2.05 \text{ parts per million (ppm)}$$

The small signal at δ 0 in this spectrum represents the hydrogens of the reference compound, TMS. The remainder of the spectrum consists of two signals: one for the hydrogens of the $-OCH_3$ group and one for the hydrogens of the methyl bonded to the carbonyl group. It is not our purpose at the moment to determine why each set of hydrogens give rise to their respective signals, but only to recognize the form in which we record an NMR spectrum and to understand the meaning of the calibration marks.

Figure 12.4
^{1}H-NMR spectrum of methyl acetate.

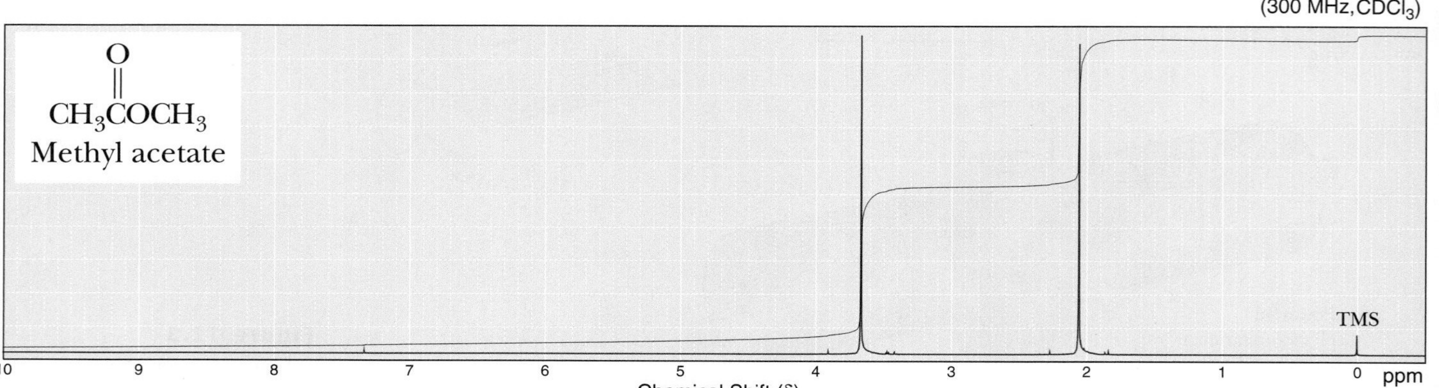

A note on terminology. If a signal is shifted toward the left on the chart paper, we say that it is shifted **downfield**, meaning that nuclei giving rise to that signal are less shielded and come into resonance at a weaker applied field. Conversely, if a signal is shifted toward the right of the spectrum, we say that it is shifted **upfield**, meaning that nuclei giving rise to that signal are more shielded and come into resonance at a stronger applied field.

← downfield upfield →

10 8 6 4 2 0

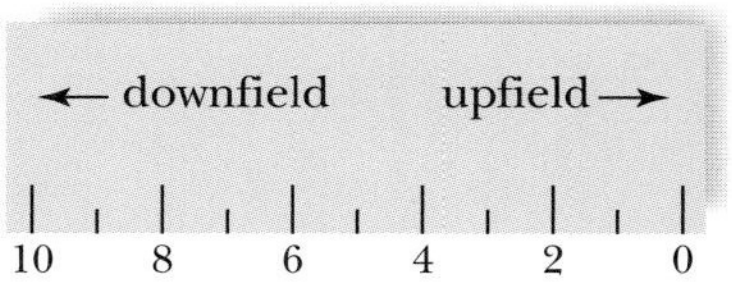

Downfield A term used to refer to the relative position of a signal on an NMR spectrum. Downfield indicates a peak to the left of the spectrum (a weaker applied field).

Upfield A term used to refer to the relative position of a signal on an NMR spectrum. Upfield indicates a peak to the right of the spectrum (a stronger applied field).

Equivalent hydrogens Hydrogens that have the same chemical environment.

12.5 How Many Resonance Signals Will a Compound Yield in Its NMR Spectrum?

Given the structural formula of a compound, how do we know how many signals to expect? The answer is that **equivalent hydrogens** give the same ^{1}H-NMR signal; conversely, nonequivalent hydrogens give different ^{1}H-NMR signals. A direct way to determine which hydrogens in a molecule are equivalent is to replace each in turn by a test atom, such as a halogen atom. Two hydrogens treated in this way are equivalent if their "substituted" versions are the same compound or if they are enantiomers of each other. If replacement gives different compounds, the two hydrogens are nonequivalent.

Using this substitution test, we can show that propane contains two sets of equivalent hydrogens: a set of six equivalent 1° hydrogens and a set of two equivalent 2° hydrogens. Thus we would expect to see two signals, one for the six equivalent $—CH_3$ hydrogens and one for the two equivalent $—CH_2—$ hydrogens:

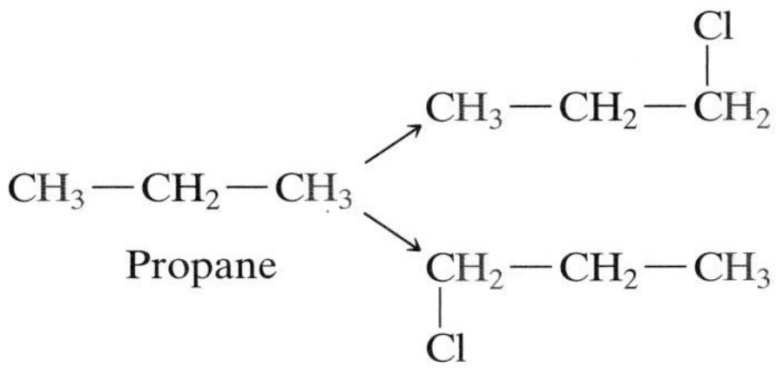

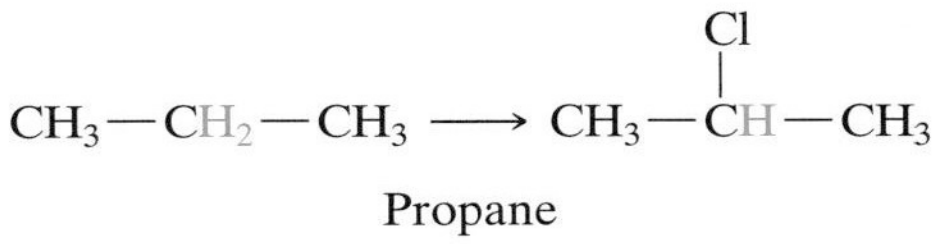

$CH_3—CH_2—CH_3$ (Propane) → $CH_3—CH_2—CH_2Cl$ and $ClCH_2—CH_2—CH_3$

Replacement of any of the red hydrogens by chlorine gives 1-chloropropane; thus, all the red hydrogens are **equivalent**.

$CH_3—CH_2—CH_3$ (Propane) ⟶ $CH_3—CHCl—CH_3$

Replacement of either of the blue hydrogens by chlorine gives 2-chloropropane; thus, both of the blue hydrogens are **equivalent**.

HOW TO 12.2 Use Symmetry to Determine Equivalency of Hydrogens in a Molecule

While it is reliable to determine equivalency of hydrogens by replacing each with a halogen and naming each subsequent compound, the following methods can also be used to spot equivalency:

(a) The hydrogens of a methyl group that can freely rotate will always be equivalent to each other.

methyl group hydrogens will always be equivalent to each other if there is free rotation

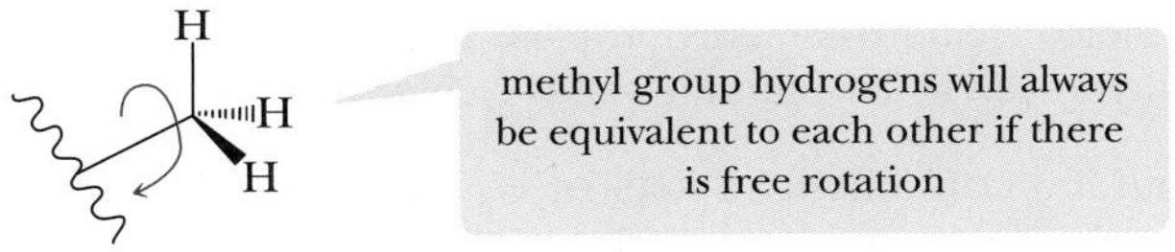

(b) The hydrogens in all three methyl groups of a *tert*-butyl group that can freely rotate will always be equivalent to each other.

CH_3 CH_3 CH_3

all nine hydrogens of a *tert*-butyl group will always be equivalent to each other if there is free rotation

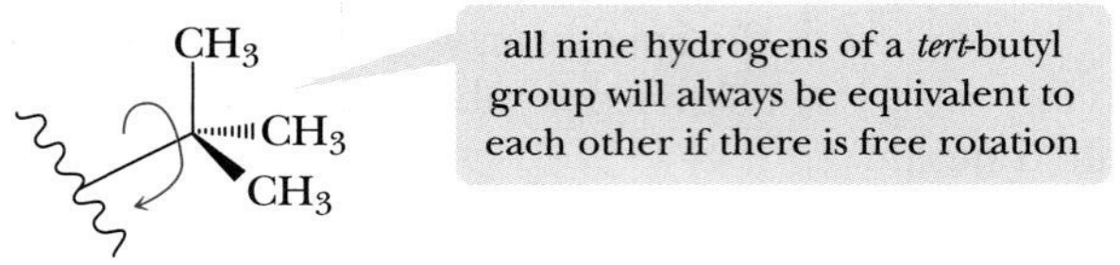

(c) If a plane of symmetry exists in a compound, corresponding hydrogens on each side of the symmetry plane will be equivalent.

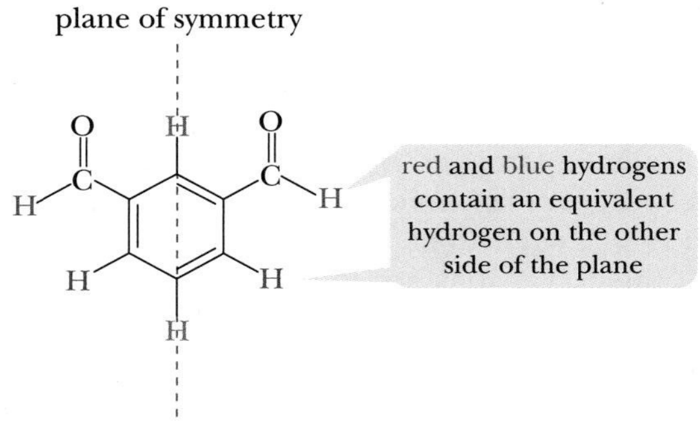

plane of symmetry

red and blue hydrogens contain an equivalent hydrogen on the other side of the plane

Example 12.2

State the number of sets of equivalent hydrogens in each compound and the number of hydrogens in each set:

(a) 2-Methylpropane (b) 2-Methylbutane (c) m-Xylene

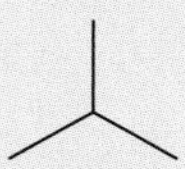

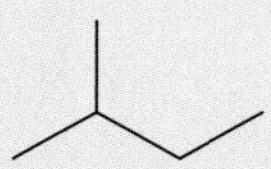

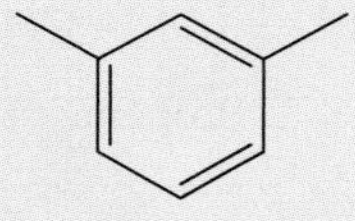

Strategy

A reliable way to determine whether hydrogens are equivalent is to replace each with a halogen and name the resulting compound. Hydrogens are equivalent if the two molecules containing their replacements are the same compound or are enantiomers. See How To 12.2 for other ways to determine equivalency.

Solution

(a) 2-Methylpropane contains two sets of equivalent hydrogens—a set of nine equivalent 1° hydrogens and one 3° hydrogen:

nine equivalent 1° hydrogens → H_3C, H_3C, CH_3; H ← one 3° hydrogen

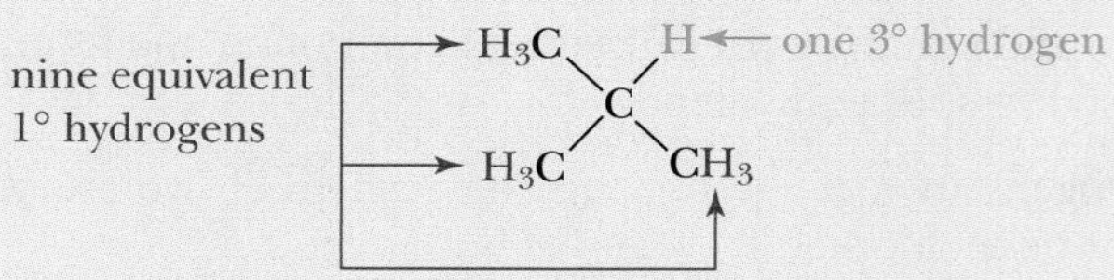

Replacing any one of the red hydrogens with a chlorine yields 1-chloro-2-methylpropane. Replacing the blue hydrogen with a chlorine yields 2-chloro-2-methylpropane.

(b) 2-Methylbutane contains four sets of equivalent hydrogens—two different sets of 1° hydrogens, one set of 2° hydrogens, and one 3° hydrogen:

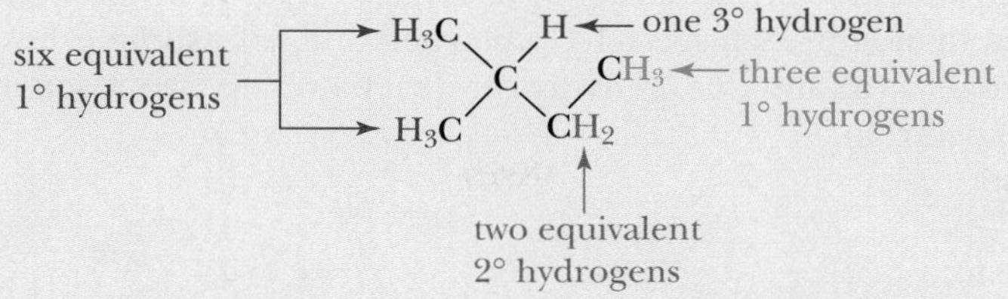

six equivalent 1° hydrogens → H_3C, H_3C; H ← one 3° hydrogen; CH_3 ← three equivalent 1° hydrogens; CH_2 ↑ two equivalent 2° hydrogens

Replacing any one of the red hydrogens with a chlorine yields 1-chloro-2-methylbutane. Replacing the blue hydrogen with a chlorine yields 2-chloro-2-methylbutane. Replacing a purple hydrogen with a chlorine yields 2-chloro-3-methylbutane. Replacing a green hydrogen with chlorine yields 1-chloro-3-methylbutane.

(c) *m*-Xylene contains four sets of equivalent hydrogens—one set of methyl group hydrogens, one set of hydrogens on the benzene ring *ortho* to one methyl group, one hydrogen on the benzene ring *ortho* to both methyl groups, and one hydrogen on the benzene ring *meta* to both methyl groups. In this solution, symmetry is used to illustrate equivalency.

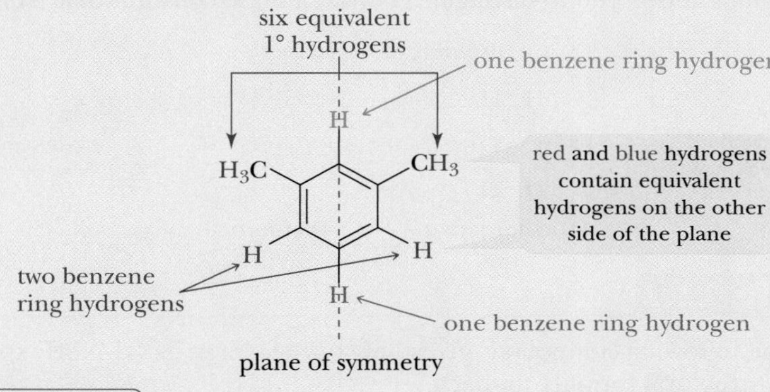

See problem 12.11

Problem 12.2

State the number of sets of equivalent hydrogens in each compound and the number of hydrogens in each set:

(a) 3-Methylpentane

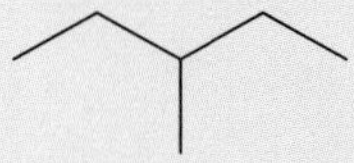

(b) 2,2,4-Trimethylpentane

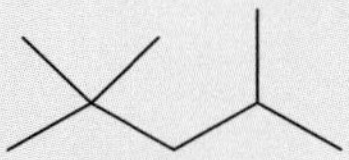

(c) 2,5-Dichloro-4-methyltoluene

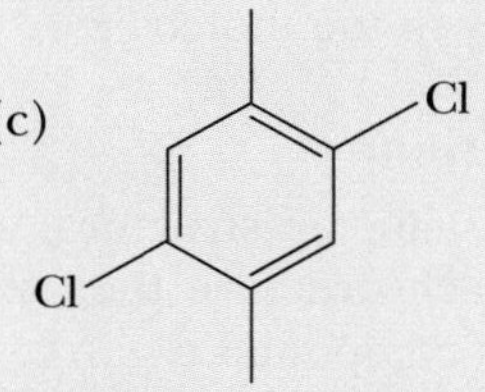

Symmetrical compounds tend to contain a higher amount of equivalent hydrogens, and thus fewer resonance signals in their NMR spectra. Here are four symmetrical organic compounds, each of which has one set of equivalent hydrogens and gives one signal in its ^{1}H-NMR spectrum:

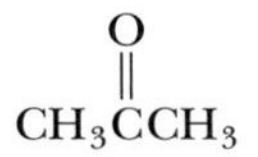

$CH_3C(=O)CH_3$	$ClCH_2CH_2Cl$	(cyclopentane ring)	$(H_3C)_2C{=}C(CH_3)_2$
Propanone (Acetone)	1,2-Dichloroethane	Cyclopentane	2,3-Dimethyl-2-butene

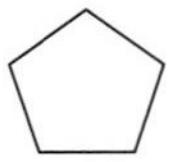

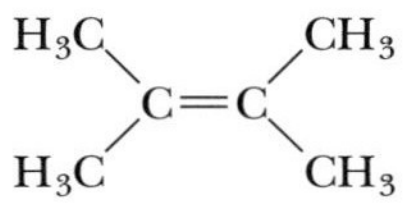

Molecules with two or more sets of equivalent hydrogens give rise to a different resonance signal for each set. 1,1-Dichloroethane, for example, has three equivalent 1° hydrogens (a) and one 2° hydrogen (b); there are two resonance signals in its ^{1}H-NMR spectrum.

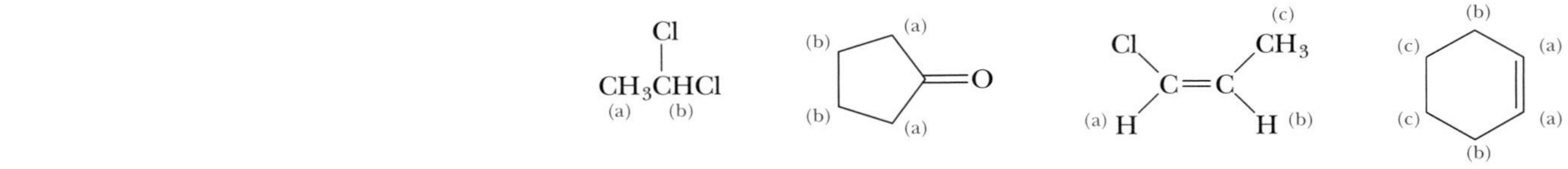

1,1-Dichloroethane (2 signals)	Cyclopentanone (2 signals)	(*Z*)-1-Chloropropene (3 signals)	Cyclohexene (3 signals)

You should see immediately that valuable information about a compound's molecular structure can be obtained simply by counting the number of signals in a ^{1}H-NMR spectrum of that compound. Consider, for example, the two constitutional isomers with the molecular formula $C_2H_4Cl_2$. The compound 1,2-dichloroethane has four equivalent 2° hydrogens and shows one signal in its ^{1}H-NMR spectrum. Its constitutional isomer 1,1-dichloroethane has three equivalent 1° hydrogens and one 2° hydrogen; this isomer shows two signals in its ^{1}H-NMR spectrum. Thus, simply counting signals allows you to distinguish between these constitutional isomers.

Isomers of $C_2H_4Cl_2$

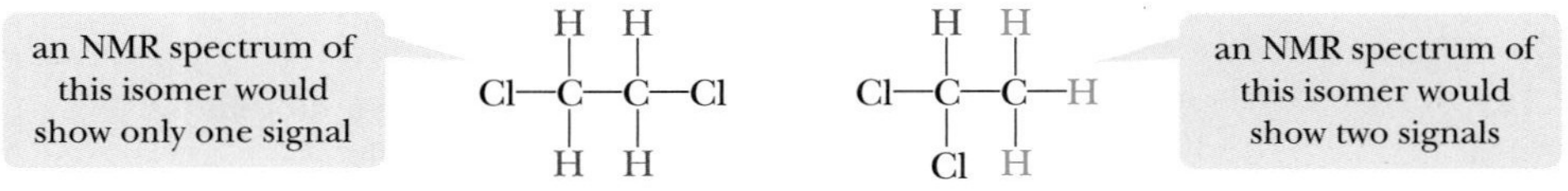

an NMR spectrum of this isomer would show only one signal

1,2-Dichloroethane

1,1-Dichloroethane

an NMR spectrum of this isomer would show two signals

Example 12.3

Each of the following compounds gives only one signal in its ^{1}H-NMR spectrum. Propose a structural formula for each.

(a) C_2H_6O (b) $C_3H_6Cl_2$ (c) C_6H_{12}

Strategy

Use the number of signals in an NMR spectrum to guide your choice of structure. A compound that yields fewer signals than it has hydrogen atoms indicates symmetry in the molecule.

Solution

Following are structural formulas for each of the given compounds. Notice that, for each structure, the replacement of any hydrogen with a chlorine will yield the same compound regardless of the hydrogen being replaced.

(a) CH_3OCH_3 (b) $CH_3CCl_2CH_3$ (c) cyclohexane or $(CH_3)_2C{=}C(CH_3)_2$

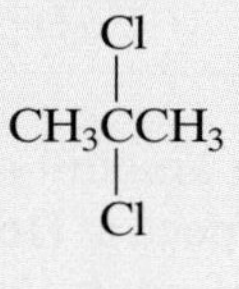

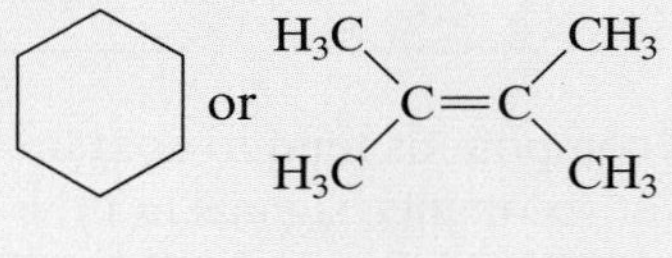

Problem 12.3

Each of the following compounds gives only one signal in its ^{1}H-NMR spectrum. Propose a structural formula for each compound.

(a) C_3H_6O (b) C_5H_{10} (c) C_5H_{12} (d) $C_4H_6Cl_4$

12.6 What Is Signal Integration?

We have just seen that the number of signals in a ^{1}H-NMR spectrum gives us information about the number of sets of equivalent hydrogens. Signal areas in a ^{1}H-NMR spectrum can be measured by a mathematical technique called *integration*. Integration is done in the following way: As a spectrum is being recorded, the instrument's computer numerically adds together the areas under each of the signals. In the spectra shown in this text, this information is displayed in the form of a **line of integration** superposed on the original spectrum. The vertical rise of the line of integration over each signal is proportional to the area under that signal, which, in turn, is proportional to the number of hydrogens giving rise to the signal.

Figure 12.5 shows an integrated ^{1}H-NMR spectrum of the gasoline additive *tert*-butyl acetate ($C_6H_{12}O_2$). The spectrum shows signals at δ 1.44 and 1.95. The integrated height of the upfield (to the right) signal is nearly three times as tall as the height of the downfield (to the left) signal. This relationship corresponds to an **integration ratio** of 3 : 1. We know from the molecular formula that there is a total of 12 hydrogens in the molecule. The ratios obtained from the integration lines are consistent with the presence of one set of 9 equivalent hydrogens and one set of 3 equivalent hydrogens. Alternatively, we could use the horizontal lines on the chart as a unit of measure. (Ten chart divisions are equivalent to the distance between two consecutive solid horizontal lines.) In this fashion, we obtain an integration value of 67 for the upfield signal and 23 for the downfield signal. (Alternatively, we could measure the lines with a ruler.) These values add up to 90 chart divisions for 12 hydrogens. Dividing the total number of chart divisions by the total number of hydrogens gives $90 \div 12 = 7.5$ chart divisions per hydrogen. Thus, the signal at δ 1.44 represents $67 \div 7.5 \approx 9$ equivalent hydrogens, and the signal at δ 1.95 represents $23 \div 7.5 \approx 3$ hydrogens. We will often make use of shorthand notation in referring to an NMR spectrum of a molecule. The notation lists the chemical shift of each signal, beginning with the most deshielded signal and followed by the number of hydrogens that give rise to each signal (based on the integration). The shorthand notation describing the spectrum of *tert*-butyl acetate (Figure 12.5) would be δ 1.95 (3H) and δ 1.44 (9H).

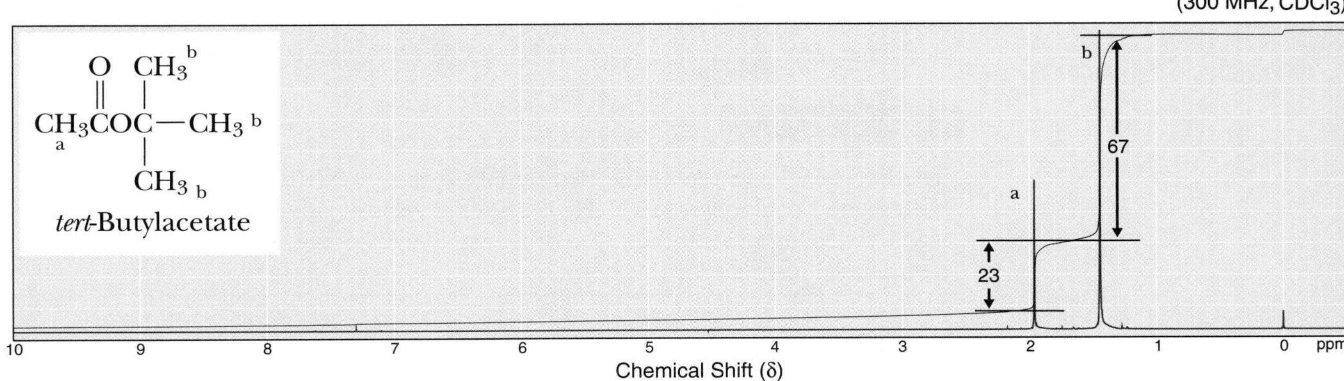

Figure 12.5
^{1}H-NMR spectrum of *tert*-butyl acetate, $C_6H_{12}O_2$, showing a line of integration. The ratio of signal heights for the two peaks is 3 : 1, which, for a molecule possessing 12 hydrogens, corresponds to 9 equivalent hydrogens of one set and 3 equivalent hydrogens of another set.

Example 12.4

Following is a ^{1}H-NMR spectrum for a compound with the molecular formula $C_9H_{10}O_2$. From an analysis of the integration line, calculate the number of hydrogens giving rise to each signal.

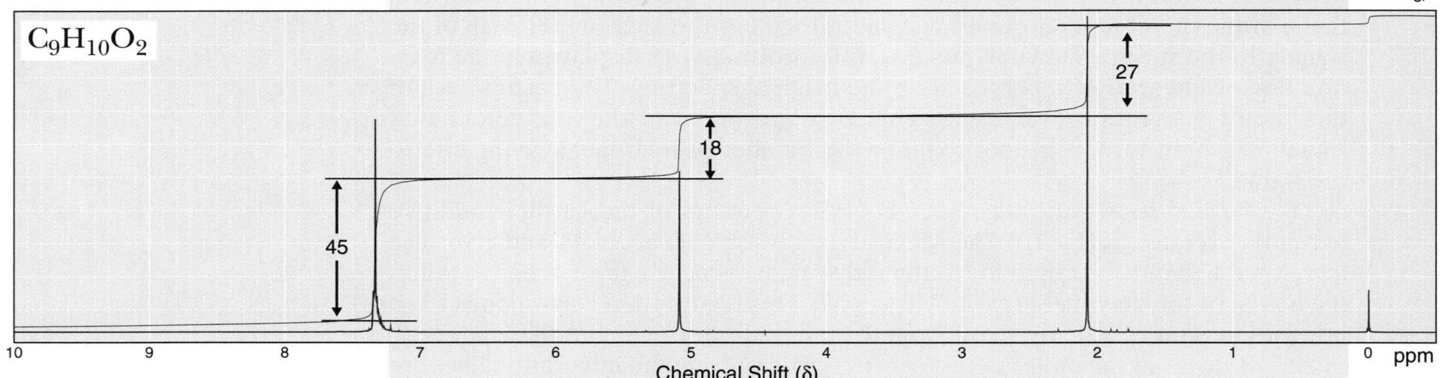

Strategy

Once the lengths of the integration lines are determined, divide all the numbers by the lowest value to obtain a minimized ratio. For example, if the lengths of the integration lines are 4 : 12 : 8, the minimized ratio becomes (4 : 12 : 8)/4 = 1 : 3 : 2. The minimized ratio values can be treated as the true number of hydrogens if the sum of these numbers is equal to the number of hydrogens from the molecular formula. If the molecular formula contains a greater number of hydrogens than the sum of the ratio values, the ratio values need to be multiplied by some factor to bring their sum to that of the total number of hydrogens. For example, if the minimized ratio is 1 : 3 : 2 and the total number of hydrogens is 12, the ratio must be adjusted by a factor of 2 to be 2 : 6 : 4.

Solution

The ratio of the relative signal heights (obtained from the number of horizontal chart divisions) is 5 : 2 : 3 (from downfield to upfield). The molecular formula indicates that there are 10 hydrogens. Thus, the signal at δ 7.34 represents 5 hydrogens, the signal at δ 5.08 represents 2 hydrogens, and the signal at δ 2.06 represents 3 hydrogens. Consequently, the signals and the number of hydrogens each signal represents are δ 7.34 (5H), δ 5.08 (2H), and δ 2.06 (3H).

Problem 12.4

The line of integration of the two signals in the ^{1}H-NMR spectrum of a ketone with the molecular formula $C_7H_{14}O$ shows a vertical rise of 62 and 10 chart divisions. Calculate the number of hydrogens giving rise to each signal, and propose a structural formula for this ketone.

12.7 What Is Chemical Shift?

The position of a signal along the x-axis of an NMR spectrum is known as the **chemical shift** of that signal (Section 12.4). The chemical shift of a signal in a ^{1}H-NMR spectrum can give us valuable information about the type of hydrogens giving rise to that absorption. Hydrogens on methyl groups bonded to sp^3 hybridized carbons, for example, give a signal near δ 0.8–1.0 (compare Figure 12.5). Hydrogens on methyl groups bonded to a carbonyl carbon give signals near δ 2.1–2.3 (compare Figures 12.4 and 12.5), and hydrogens on methyl groups bonded to oxygen give signals near δ 3.7–3.9 (compare Figure 12.4). Table 12.1 lists the average chemical shift for most of the types of hydrogens we deal with in this text. Notice that most of the values shown fall within a rather narrow range from 0 to 13 δ units (ppm). In fact, although the table shows a variety of functional groups and hydrogens bonded to them, we can use the following rules of thumb to remember the chemical shifts of most types of hydrogen:

Chemical Shift (δ)	Type of Hydrogen
0–2	H bonded to an sp^3 carbon.
2–2.8	H bonded to an sp^3 carbon that is at an allylic or benzylic position (i.e., adjacent to a C—C double bond or a benzene ring).
2–4.5	H bonded to an sp^3 carbon that is close to an electronegative element such as N, O, or X. The more electronegative the element, the higher is the chemical shift. Also, the closer the electronegative atom, the higher is the chemical shift.
4.6–5.7	H bonded to an sp^2 carbon in an alkene.
6.5–8.5	H bonded to an sp^2 carbon in an aromatic compound.
9.5–10.1	H bonded to a C=O (an aldehyde hydrogen).
10–13	H of a carboxyl (COOH) group.

TABLE 12.1 Average Values of Chemical Shifts of Representative Types of Hydrogens

Type of Hydrogen (R = alkyl, Ar = aryl)	Chemical Shift (δ)*	Type of Hydrogen (R = alkyl, Ar = aryl)	Chemical Shift (δ)*
$(C\mathbf{H}_3)_4Si$	0 (by definition)	$RCOC\mathbf{H}_3$ (C=O)	3.7–3.9
$RC\mathbf{H}_3$	0.8–1.0	$RCOC\mathbf{H}_2R$ (C=O)	4.1–4.7
$RC\mathbf{H}_2R$	1.2–1.4	$RC\mathbf{H}_2I$	3.1–3.3
$R_3C\mathbf{H}$	1.4–1.7	$RC\mathbf{H}_2Br$	3.4–3.6
$R_2C{=}CRC\mathbf{H}R_2$	1.6–2.6	$RC\mathbf{H}_2Cl$	3.6–3.8
$RC{\equiv}C\mathbf{H}$	2.0–3.0	$RC\mathbf{H}_2F$	4.4–4.5
$ArC\mathbf{H}_3$	2.2–2.5	$ArO\mathbf{H}$	4.5–4.7
$ArC\mathbf{H}_2R$	2.3–2.8	$R_2C{=}C\mathbf{H}_2$	4.6–5.0
$RO\mathbf{H}$	0.5–6.0	$R_2C{=}C\mathbf{H}R$	5.0–5.7
$RC\mathbf{H}_2OH$	3.4–4.0	$Ar\mathbf{H}$	6.5–8.5
$RC\mathbf{H}_2OR$	3.3–4.0	$RC\mathbf{H}$ (C=O)	9.5–10.1
$R_2N\mathbf{H}$	0.5–5.0	$RCO\mathbf{H}$ (C=O)	10–13
$RCC\mathbf{H}_3$ (C=O)	2.1–2.3		
$RCC\mathbf{H}_2R$ (C=O)	2.2–2.6		

*Values are approximate. Other atoms within the molecule may cause the signal to appear outside these ranges.

Example 12.5

Following are two constitutional isomers with the molecular formula $C_6H_{12}O_2$:

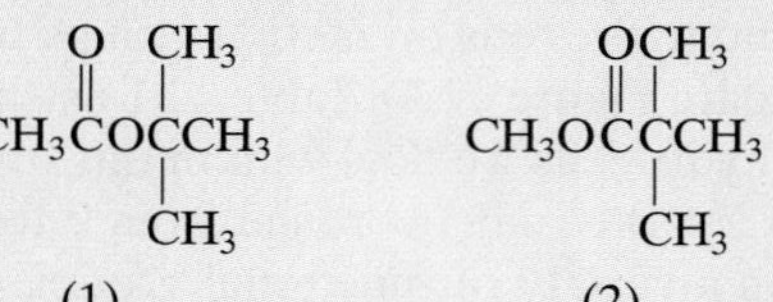

(a) Predict the number of signals in the ^{1}H-NMR spectrum of each isomer.
(b) Predict the ratio of areas of the signals in each spectrum.
(c) Show how to distinguish between these isomers on the basis of chemical shift.

Strategy

This example poses a series of steps that you will repeat often when asked to predict the NMR spectrum of a compound. First, determine the number of signals by determining the number of equivalent hydrogens (Section 12.5 and How To 12.2). Then predict the ratio of areas of the signals by counting the number of hydrogens in each equivalent set (if the set of ratios can be reduced further, divide each number by a value that produces whole number ratios). Finally, predict the chemical shift of each set of equivalent hydrogens. If the rules of thumb for predicting chemical shift do not apply, refer to Table 12.1.

Solution

(a) Each compound contains a set of nine equivalent methyl hydrogens and a set of three equivalent methyl hydrogens.
(b) The ^{1}H-NMR spectrum of each consists of two signals in the ratio 9 : 3, or 3 : 1.
(c) The two constitutional isomers can be distinguished by the chemical shift of the single $-CH_3$ group, or in other words, the signal that integrates to 3 (shown in red for each compound). Using our rules of thumb, we find that the hydrogens of CH_3O are less shielded (appear farther downfield) than the hydrogens of $CH_3C{=}O$. Table 12.1 gives approximate values for each chemical shift. Experimental values are as follows:

δ 1.95 → CH_3 of $CH_3C(=O)OC(CH_3)_3$; δ 1.44 → $C(CH_3)_3$ (1)

δ 3.67 → CH_3O of $CH_3OC(=O)C(CH_3)_3$; δ 1.20 → $C(CH_3)_3$ (2)

Problem 12.5

Following are two constitutional isomers with the molecular formula $C_4H_8O_2$:

(a) Predict the number of signals in the ^{1}H-NMR spectrum of each isomer.
(b) Predict the ratio of areas of the signals in each spectrum.
(c) Show how to distinguish between these isomers on the basis of chemical shift.

12.8 What Is Signal Splitting?

We have now seen three kinds of information that can be derived from an examination of a ^{1}H-NMR spectrum:

1. From the number of signals, we can determine the number of sets of equivalent hydrogens.
2. By integrating over signal areas, we can determine the relative numbers of hydrogens giving rise to each signal.
3. From the chemical shift of each signal, we can derive information about the types of hydrogens in each set.

We can derive a fourth kind of information from the splitting pattern of each signal. Consider, for example, the ^{1}H-NMR spectrum of 1,1,2-trichloroethane (Figure 12.6), a solvent for waxes and natural resins. This molecule contains two 2° hydrogens and one 3° hydrogen, and, according to what we have learned so far, we predict two signals with relative areas 2 : 1, corresponding to the two hydrogens of the $-CH_2-$ group and the one hydrogen of the $-CHCl_2$ group. You see from the spectrum, however, that there are in fact five **peaks**. How can this be, when we predict only two signals? The answer is that a hydrogen's resonance frequency can be affected by the tiny magnetic fields of other hydrogens close by. Those fields cause the signal to be **split** into numerous peaks. Hydrogens split each other if they are separated by no more than three bonds—for example, H—C—C—H or H—C=C—H. (There are three bonds in each case.) If there are more than three bonds, as in H—C—C—C—H, then there is normally no splitting. A signal with just one peak is called a **singlet**. A signal that is split into two peaks is called a **doublet**. Signals that are split into three and four peaks are called **triplets** and **quartets**, respectively.

The grouping of two peaks at δ 3.96 in the ^{1}H-NMR spectrum of 1,1,2-trichloroethane is the signal for the hydrogens of the $-CH_2-$ group, and the grouping of three peaks at δ 5.77 is the signal for the single hydrogen of the $-CHCl_2$ group. We say that the CH_2 signal at δ 3.96 is split into a doublet and that the CH signal at δ 5.77 is split into a triplet. In this phenomenon, called **signal splitting**, the ^{1}H-NMR signal from one set of hydrogens is split by the influence of neighboring nonequivalent hydrogens.

The degree of signal splitting can be predicted on the basis of the **(n + 1) rule**, according to which, if a hydrogen has n hydrogens nonequivalent to it, but

Peak (NMR) The units into which an NMR signal is split—two peaks in a doublet, three peaks in a triplet, and so on.

Singlet A signal that consists of one peak; the hydrogens that give rise to the signal have no neighboring nonequivalent hydrogens.

Doublet A signal that is split into two peaks; the hydrogens that give rise to the signal have one neighboring nonequivalent hydrogen.

Triplet A signal that is split into three peaks; the hydrogens that give rise to the signal have two neighboring nonequivalent hydrogens that are equivalent to each other.

Quartet A signal that is split into four peaks; the hydrogens that give rise to the signal have three neighboring nonequivalent hydrogens that are equivalent to each other.

Signal splitting Splitting of an NMR signal into a set of peaks by the influence of neighboring nuclei.

(n + 1) rule The ^{1}H-NMR signal of a hydrogen or set of equivalent hydrogens with n other hydrogens on neighboring carbons is split into (n + 1) peaks.

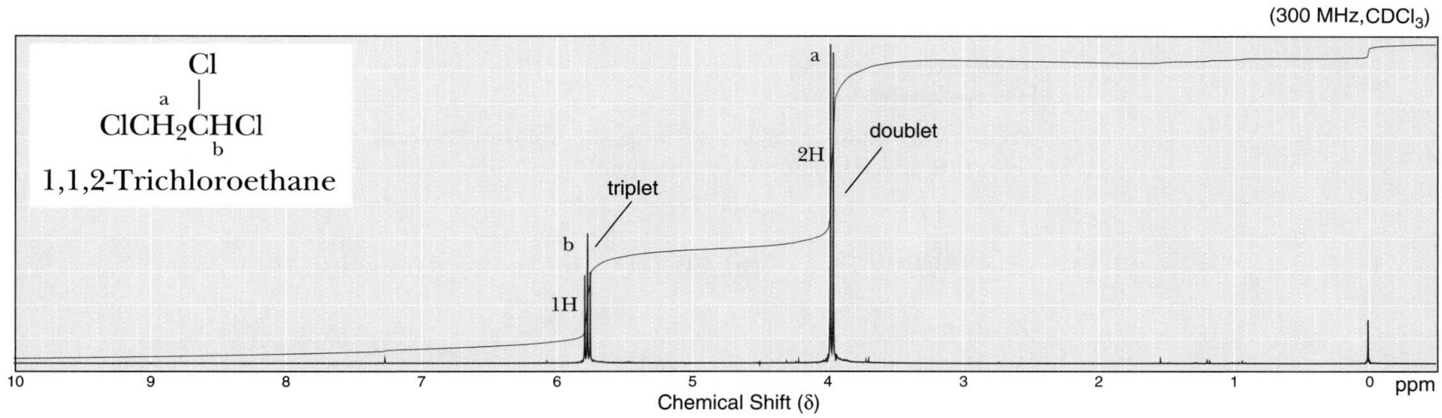

Figure 12.6
^{1}H-NMR spectrum of 1,1,2-trichloroethane.

equivalent among themselves, on the same or adjacent atom(s), then the ¹H-NMR signal of the hydrogen is split into $(n + 1)$ peaks.

Let us apply the $(n + 1)$ rule to the analysis of the spectrum of 1,1,2-trichloroethane. The two hydrogens of the $—CH_2—$ group have one nonequivalent neighboring hydrogen $(n = 1)$; their signal is split into a doublet $(1 + 1 = 2)$. The single hydrogen of the $—CHCl_2$ group has a set of two nonequivalent neighboring hydrogens $(n = 2)$; its signal is split into a triplet $(2 + 1 = 3)$.

For these hydrogens, $n = 1$; their signal is split into $(1 + 1)$ or 2 peaks—a **doublet**

For this hydrogen, $n = 2$; its signal is split into $(2 + 1)$ or 3 peaks—a **triplet**

$$Cl—CH_2—CH(Cl)—Cl$$

It is important to remember that the $(n + 1)$ rule of signal splitting applies only to hydrogens with *equivalent* neighboring hydrogens. When more than one set of neighboring hydrogens exists, the $(n + 1)$ rule no longer applies. An example of where the $(n + 1)$ rule no longer applies is illustrated in the ¹H-NMR spectrum of 1-chloropropane. The two hydrogens on carbon 2 (a CH_2 group) of 1-chloropropane are flanked on one side by a set of 2H on carbon 1, and on the other side by a set of 3H on carbon 3. Because the sets of hydrogen on carbons 1 and 3 are nonequivalent to each other and also nonequivalent to the hydrogens on carbon 2, they cause the signal for the CH_2 group on carbon 2 to be split into a complex pattern, which we will refer to simply as a multiplet.

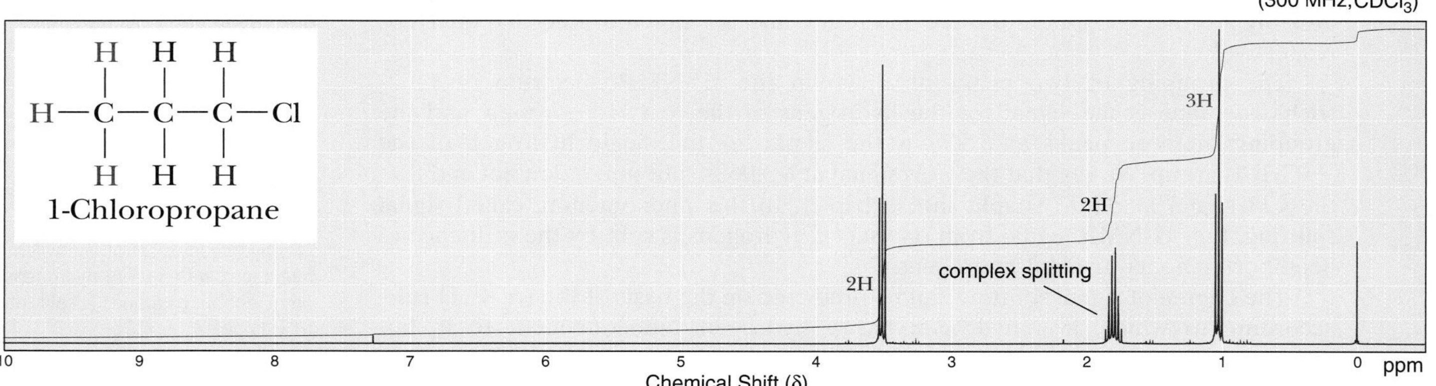

Example 12.6

Predict the number of signals and the splitting pattern of each signal in the ¹H-NMR spectrum of each compound.

(a) $CH_3C(=O)CH_2CH_3$ (b) $CH_3CH_2C(=O)CH_2CH_3$ (c) $CH_3C(=O)CH(CH_3)_2$

Strategy

Determine the number of signals by determining the number of equivalent hydrogens (Section 12.5 and How To 12.2). For each set of equivalent hydrogens, determine the number of equivalent neighbors (n) and apply the $(n + 1)$ rule to determine the splitting pattern.

Solution

The sets of equivalent hydrogens in each molecule are color coded. In molecule (a), the signal for the red methyl group is unsplit (a singlet) because the group is too far (>3 bonds) from any other hydrogens. The blue —CH_2— group has three neighboring hydrogens ($n = 3$) and thus shows a signal split into a quartet ($3 + 1 = 4$). The green methyl group has two neighboring hydrogens ($n = 2$), and its signal is split into a triplet. The integration ratios for these signals would be 3:2:3. Parts (b) and (c) can be analyzed in the same way. Thus, molecule (b) shows a triplet and a quartet in the ratio 3:2. Molecule (c) shows a singlet, a septet ($6 + 1 = 7$), and a doublet in the ratio 3:1:6.

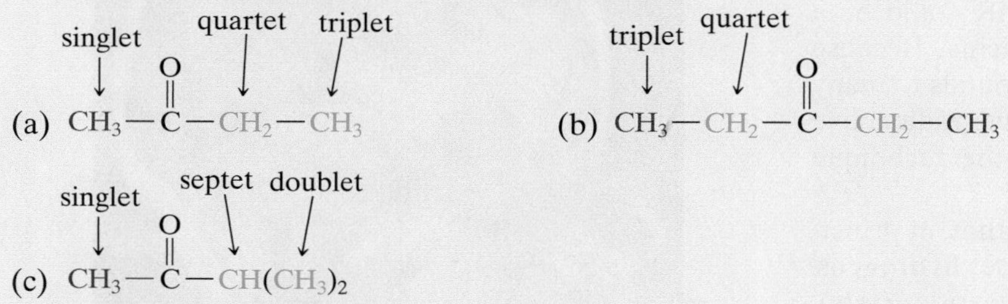

(a) CH_3—C(=O)—CH_2—CH_3 (singlet, quartet, triplet)

(b) CH_3—CH_2—C(=O)—CH_2—CH_3 (triplet, quartet)

(c) CH_3—C(=O)—$CH(CH_3)_2$ (singlet, septet, doublet)

Problem 12.6

Following are pairs of constitutional isomers. Predict the number of signals and the splitting pattern of each signal in the ^{1}H-NMR spectrum of each isomer.

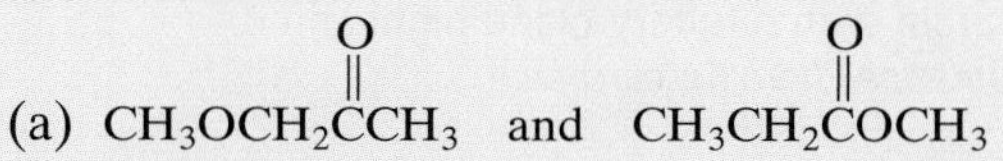

(a) $CH_3OCH_2CCH_3$ (C=O) and $CH_3CH_2COCH_3$ (C=O)

(b) $CH_3CCl_2CH_3$ and $ClCH_2CH_2CH_2Cl$

12.9 What Is ^{13}C-NMR Spectroscopy, and How Does It Differ from ^{1}H-NMR Spectroscopy?

Nuclei of carbon-12, the most abundant (98.89%) natural isotope of carbon, do not have nuclear spin and are not detected by NMR spectroscopy. Nuclei of carbon-13 (natural abundance 1.11%), however, do have nuclear spin and are detected by NMR spectroscopy in the same manner as hydrogens are detected. Thus, NMR can be used to obtain information about 1.11% of all the carbon atoms in a sample. Just as in ^{1}H-NMR spectroscopy, ^{13}C-NMR spectroscopy yields a signal for each set of equivalent carbons in a molecule.

Because both ^{13}C and ^{1}H have spinning nuclei and generate magnetic fields, ^{13}C couples with each ^{1}H bonded to it and gives a signal split according to the $(n + 1)$ rule. In the most common mode for recording a ^{13}C spectrum, this coupling is eliminated by instrumental techniques, so as to simplify the spectrum. In these **hydrogen-decoupled spectra**, all ^{13}C signals appear as singlets. The hydrogen-decoupled ^{13}C-NMR spectrum of citric acid (Figure 12.7), a compound used to increase the solubility of many pharmaceutical drugs in water, consists of four singlets. Again, notice that, as in ^{1}H-NMR, equivalent carbons generate only one signal.

CHEMICAL CONNECTIONS 12A

Magnetic Resonance Imaging

Nuclear magnetic resonance was discovered and explained by physicists in the 1950s, and, by the 1960s, it had become an invaluable analytical tool for chemists. By the early 1970s, it was realized that the imaging of parts of the body via NMR could be a valuable addition to diagnostic medicine. Because the term *nuclear magnetic resonance* sounds to many people as if the technique might involve radioactive material, health care personnel call the technique *magnetic resonance imaging* (MRI).

The body contains several nuclei that, in principle, could be used for MRI. Of these, hydrogens, most of which come from water, triglycerides (fats), and membrane phospholipids, give the most useful signals. Phosphorus MRI is also used in diagnostic medicine.

Recall that, in NMR spectroscopy, energy in the form of radio-frequency radiation is absorbed by nuclei in the sample. The relaxation time is the characteristic time at which excited nuclei give up this energy and relax to their ground state.

In 1971, Raymond Damadian discovered that the relaxation of water in certain cancerous tumors takes much longer than the relaxation of water in normal cells. Thus, it was reasoned that if a relaxation image of the body could be obtained, it might be possible to identify tumors at an early stage. Subsequent work demonstrated that many tumors can be identified in this way.

Another important application of MRI is in the examination of the brain and spinal cord. White and gray matter, the two different layers of the brain, are easily distinguished by MRI, which is useful in the study of such diseases as multiple sclerosis. Magnetic resonance imaging and X-ray imaging are in many cases complementary: The hard, outer layer of bone is essentially invisible to MRI, but shows up extremely well in X-ray images, whereas soft tissue is nearly transparent to X rays, but shows up in MRI.

The key to any medical imaging technique is knowing which part of the body gives rise to which signal. In MRI, the patient is placed in a magnetic field gradient that can be varied from place to place. Nuclei in the weaker magnetic field gradient absorb radiation at a lower frequency. Nuclei elsewhere, in the stronger magnetic field, absorb radiation at a higher frequency. Because a magnetic field gradient along a single axis images a plane, MRI techniques can create views of any part of the body in slicelike sections. In 2003, Paul Lauterbur and Sir Peter Mansfield were awarded the Nobel Prize in Physiology or Medicine for their discoveries that led to the development of these imaging techniques.

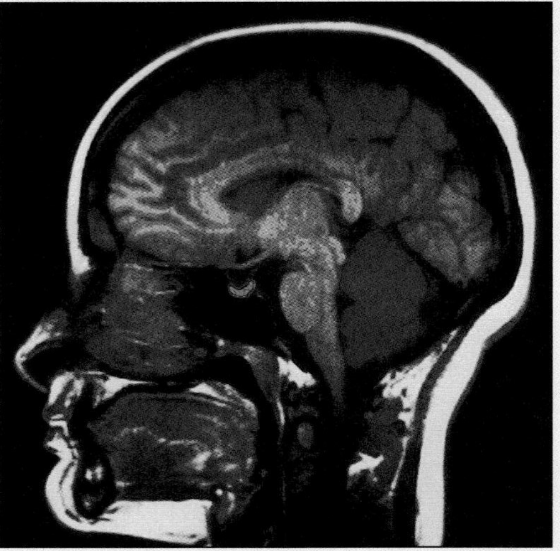

Computer-enhanced MRI scan of a normal human brain with pituitary gland highlighted *(Scott Camazine/Photo Researchers)*

QUESTION

In ^{1}H-NMR spectroscopy, the chemical sample is set spinning on its long axis to ensure that all parts of the sample experience a homogeneous applied field. Homogeneity is also required in MRI. Keeping in mind that the "sample" in MRI is a human being, how do you suppose this is achieved?

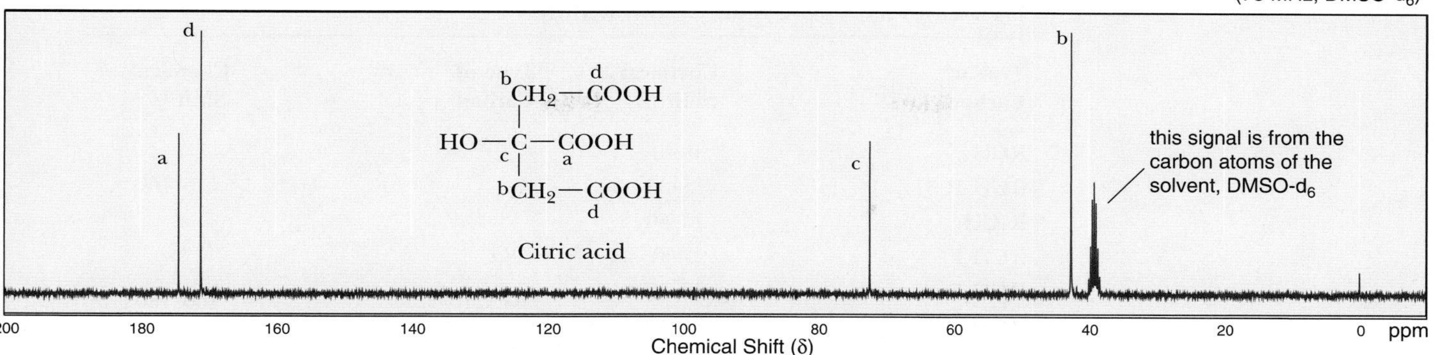

Figure 12.7
Hydrogen-decoupled ^{13}C-NMR spectrum of citric acid.

Table 12.2 shows approximate chemical shifts in ^{13}C-NMR spectroscopy. As with ^{1}H-NMR, we can use the following rules of thumb to remember the chemical shifts of various types of carbons:

Chemical Shift (δ)	Type of Carbon
0–50	sp^3 carbon (3° > 2° > 1°).
50–80	sp^3 carbon bonded to an electronegative element such as N, O, or X. The more electronegative the element, the larger is the chemical shift.
100–160	sp^2 carbon of an alkene or an aromatic compound.
160–180	carbonyl carbon of a carboxylic acid or carboxylic acid derivative (Chapters 14 and 15).
180–210	carbonyl carbon of a ketone or an aldehyde.

Notice how much broader the range of chemical shifts is for ^{13}C-NMR spectroscopy than for ^{1}H-NMR spectroscopy. Whereas most chemical shifts for ^{1}H-NMR spectroscopy fall within a rather narrow range of 0–13 ppm, those for ^{13}C-NMR spectroscopy cover 0–210 ppm. Because of this expanded scale, it is very unusual to find any two nonequivalent carbons in the same molecule with identical chemical shifts. Most commonly, each different type of carbon within a molecule has a distinct signal that is clearly resolved (i.e., separated) from all other signals. Notice further that the chemical shift of carbonyl carbons is quite distinct from the chemical shifts of sp^3 hybridized carbons and other types of sp^2 hybridized carbons. The presence or absence of a carbonyl carbon is quite easy to recognize in a ^{13}C-NMR spectrum.

A great advantage of ^{13}C-NMR spectroscopy is that it is generally possible to count the number of different types of carbon atoms in a molecule. There is one caution here, however: Because of the particular manner in which spin-flipped ^{13}C nuclei return to their lower energy states, integrating signal areas is often unreliable, and it is generally not possible to determine the number of carbons of each type on the basis of the signal areas.

TABLE 12.2 ^{13}C-NMR Chemical Shifts

Type of Carbon	Chemical Shift (δ)	Type of Carbon	Chemical Shift (δ)
RCH_3	0–40	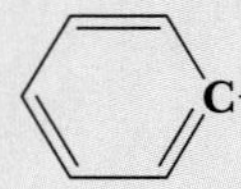 C_6H_5—C—R (aromatic)	110–160
RCH_2R	15–55		
R_3CH	20–60		
RCH_2I	0–40		
RCH_2Br	25–65	RC(=O)OR	160–180
RCH_2Cl	35–80		
R_3COH	40–80		
R_3COR	40–80	$RC(=O)NR_2$	165–180
RC≡CR	65–85		
$R_2C{=}CR_2$	100–150	RC(=O)OH	175–185
		RC(=O)H, RC(=O)R	180–210

HOW TO 12.3 Use Symmetry to Determine Equivalency of Carbons in a Molecule

The following methods can be used to spot equivalency for carbon atoms in a molecule:

(a) The carbons in all three methyl groups of a *tert*-butyl group that can freely rotate will always be equivalent to each other.

all three carbons of a *tert*-butyl group will always be equivalent to each other if there is free rotation

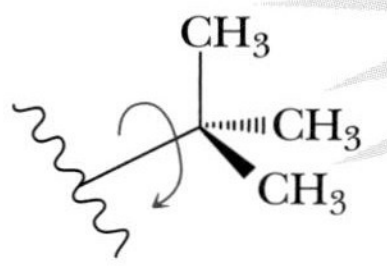

(b) If a plane of symmetry exists in a compound, corresponding carbons on each side of the symmetry plane will be equivalent.

Example 1

Plane of symmetry

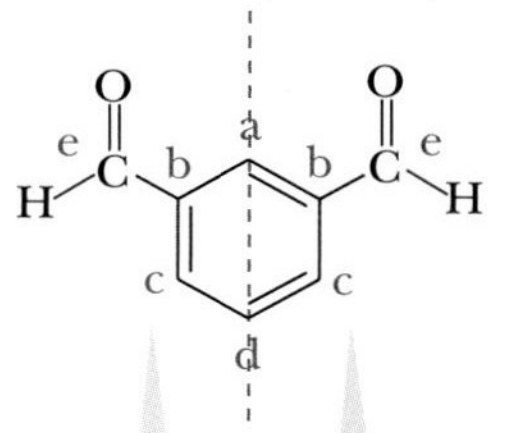

corresponding carbons on opposite sides of the plane of symmetry are equivalent

Example 2

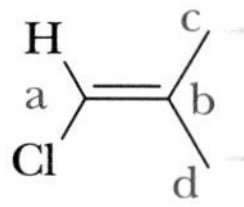

these two methyl carbons are *not* equivalent because there is no plane of symmetry between them

Example 12.7

Predict the number of signals in a proton-decoupled ^{13}C-NMR spectrum of each compound:

(a) CH_3COCH_3 (with C=O) (b) $CH_3CH_2CH_2CCH_3$ (with C=O) (c) $CH_3CH_2CCH_2CH_3$ (with C=O)

Strategy

Because we cannot replace each carbon atom with a halogen (as we did to determine equivalency in ^1H-NMR), inasmuch as a halogen only has a valence of 1, we will need to use symmetry to determine equivalency (see How To 12.3).

Solution

Here is the number of signals in each spectrum, along with the chemical shift of each, color coded to the carbon responsible for that signal. The chemical shifts of the carbonyl carbons are quite distinctive (Table 12.2) and occur at δ 171.37, 208.85, and 211.97 in these examples.

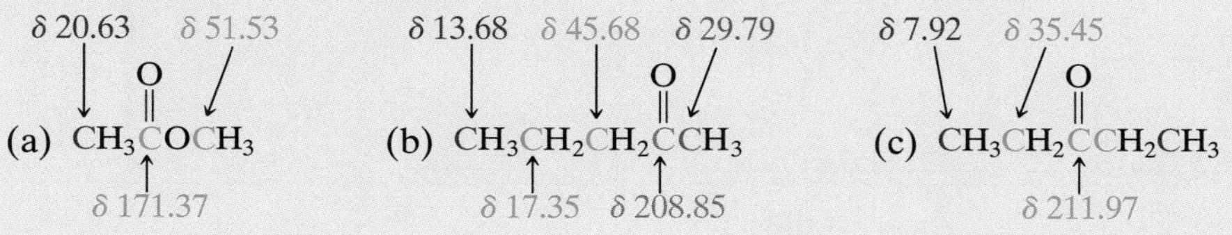

(a) CH_3COCH_3: δ 20.63, δ 51.53, δ 171.37

(b) $CH_3CH_2CH_2CCH_3$: δ 13.68, δ 45.68, δ 29.79, δ 17.35, δ 208.85

(c) $CH_3CH_2CCH_2CH_3$: δ 7.92, δ 35.45, δ 211.97

See problem 12.12

Problem 12.7

Explain how to distinguish between the members of each pair of constitutional isomers, on the basis of the number of signals in the ^{13}C-NMR spectrum of each isomer:

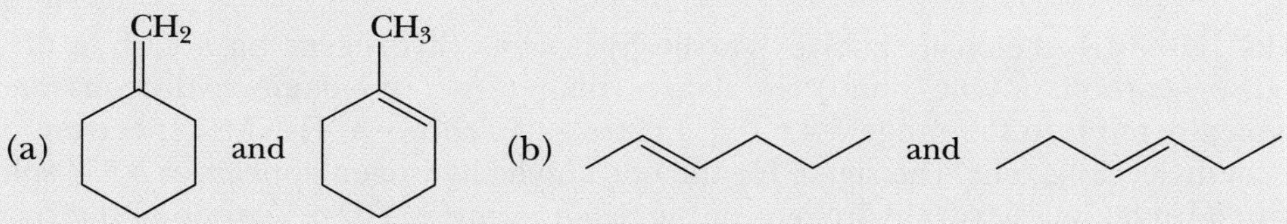

(a) methylenecyclohexane (CH_2) and 1-methylcyclohexene (CH_3) (b) 2-hexene and 3-hexene

12.10 How Do We Interpret NMR Spectra?

A. Alkanes

Because all hydrogens in alkanes are in very similar chemical environments, ^1H-NMR chemical shifts of their hydrogens fall within a narrow range of δ 0.8–1.7. Chemical shifts for alkane carbons in ^{13}C-NMR spectroscopy fall within the considerably wider range of δ 0–60. Notice how it is relatively easy to distinguish all the signals in the ^{13}C-NMR spectrum of 2,2,4-trimethylpentane (Figure 12.8; common name isooctane), a major component in gasoline, compared with the signals in the ^1H-NMR spectrum of isooctane (Figure 12.9). In the ^{13}C-NMR spectrum, we can see all five signals, while in the ^1H-NMR spectrum we expect to see four signals, but in fact see only three. The reason is that nonequivalent hydrogens often have similar chemical shifts, which lead to an overlap of signals.

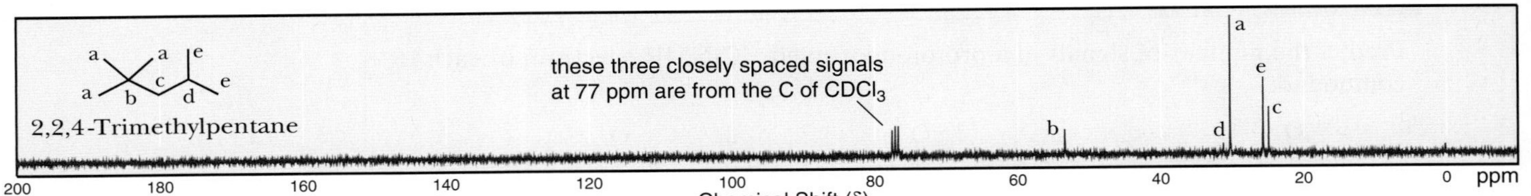

Figure 12.8
Hydrogen-decoupled ^{13}C-NMR spectrum of 2,2,4-trimethylpentane, showing all five signals.

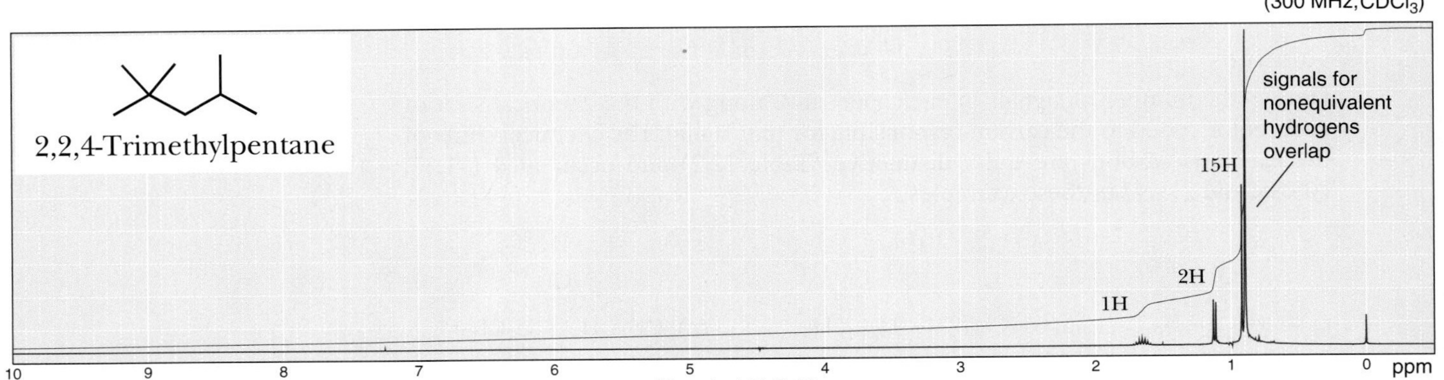

Figure 12.9
^{1}H-NMR spectrum of 2,2,4-trimethylpentane, showing only three signals. Two sets of nonequivalent hydrogens produce signals at δ 0.91, thus giving rise to an integration ratio of 1 : 2 : 15, for the peaks at δ 1.65, 1.12, and 0.91, respectively.

B. Alkenes

The ^{1}H-NMR chemical shifts of vinylic hydrogens (hydrogens on a carbon of a carbon–carbon double bond) are larger than those of alkane hydrogens and typically fall into the range δ 4.6–5.7. Figure 12.10 shows a ^{1}H-NMR spectrum of 1-methylcyclohexene. The signal for the one vinylic hydrogen appears at δ 5.4, split into a triplet by the two hydrogens of the neighboring $—CH_2—$ group of the ring.

Figure 12.10
^{1}H-NMR spectrum of 1-methylcyclohexene.

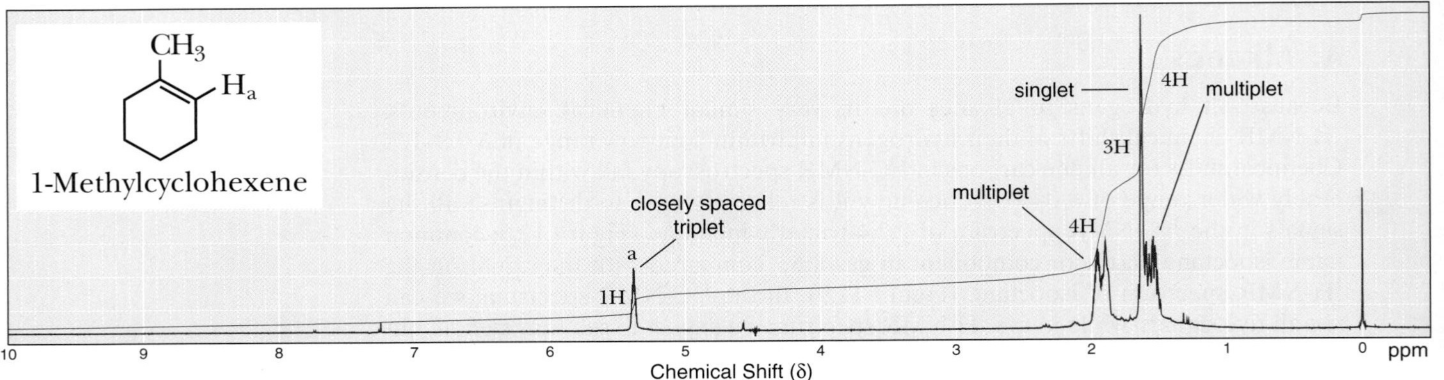

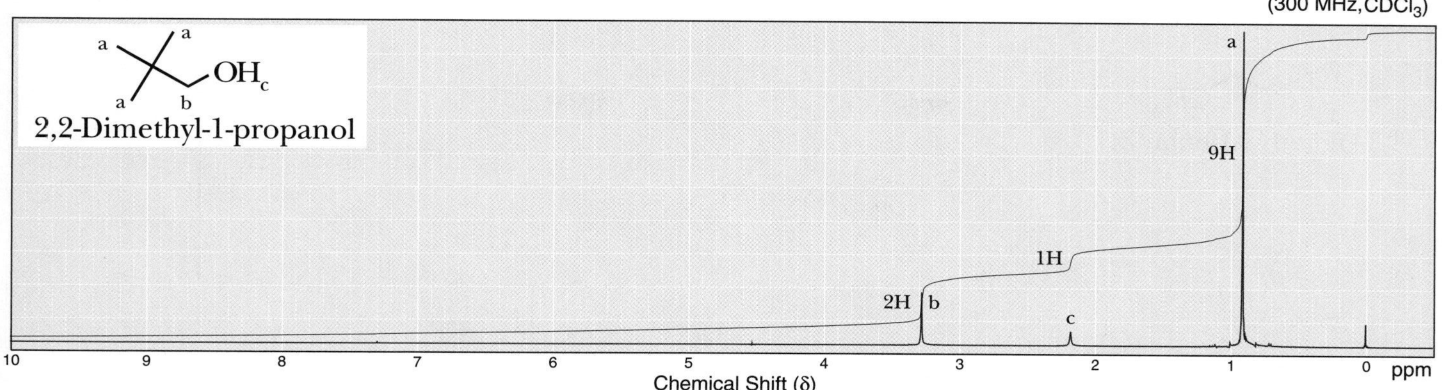

Figure 12.11
^{1}H-NMR spectrum of 2,2-dimethyl-1-propanol.

The sp^2 hybridized carbons of alkenes come into resonance in ^{13}C-NMR spectroscopy in the range δ 100–150 (Table 12.2), which is considerably downfield from resonances of sp^3 hybridized carbons.

C. Alcohols

The chemical shift of a hydroxyl hydrogen in a ^{1}H-NMR spectrum is variable and depends on the purity of the sample, the solvent, and the temperature. Normally, the shift appears in the range δ 3.0–4.5, but, depending on experimental conditions, it may appear as far upfield as δ 0.5. Hydrogens on the carbon bearing the —OH group are deshielded by the electron-withdrawing inductive effect of the oxygen atom, and their absorptions typically appear in the range δ 3.4–4.0. Figure 12.11 shows a ^{1}H-NMR spectrum of 2,2-dimethyl-1-propanol. The spectrum consists of three signals. The hydroxyl hydrogen appears at δ 2.19 as a slightly broad singlet. The signal of the hydrogens on the carbon bearing the hydroxyl group in 2,2-dimethyl-1-propanol appears as a singlet at δ 3.32.

Signal splitting between the hydrogen on O—H and its neighbors on the adjacent $—CH_2—$ group is not normally seen in the ^{1}H-NMR spectrum of 2,2-dimethyl-1-propanol. The reason is that most samples of alcohol contain traces of acid, base, or other impurities that catalyze the transfer of the hydroxyl proton from the oxygen of one alcohol molecule to that of another alcohol molecule. This transfer, which is very fast compared with the time scale required to make an NMR measurement, decouples the hydroxyl proton from all other protons in the molecule. For this same reason, the hydroxyl proton does not usually split the signal of any α-hydrogens. It is, however, possible to slow down or eliminate this proton exchange (e.g., by cooling the sample temperature or diluting the sample), in which case coupling may then be observed.

R–O(δ−)–H(δ+) ··· R–O(δ−)–H(δ+) ··· R–O(δ−)–H(δ+) ⇌(fast) H(δ+)–O(δ−)–R ··· H(δ+)–O(δ−)–R ··· H(δ+)–O(δ−)–R

hydrogen bonding — these protons have been exchanged

Signals from alcohol hydrogens typically appear as broad singlets, as shown in the ^{1}H-NMR spectrum of 3,3-dimethyl-1-butanol (Figure 12.12).

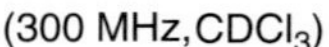

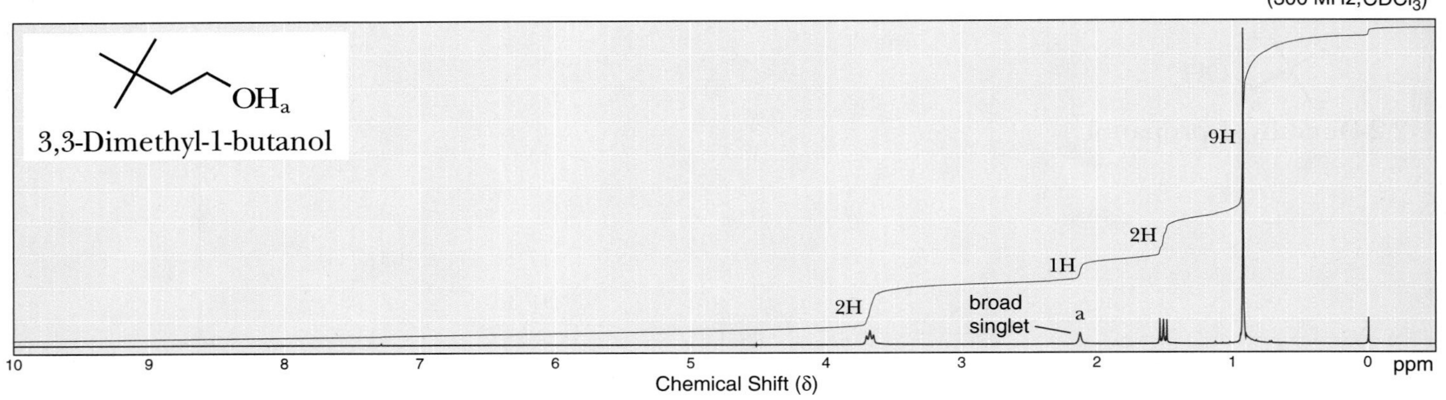

Figure 12.12
^{1}H-NMR spectrum of 3,3-dimethyl-1-butanol, showing the decoupled hydroxyl signal at δ 2.12.

D. Benzene and Its Derivatives

All six hydrogens of benzene are equivalent, and their signal appears in its ^{1}H-NMR spectrum as a sharp singlet at δ 7.27. Hydrogens bonded to a substituted benzene ring appear in the region δ 6.5–8.5. Few other types of hydrogens give signals in this region; thus, aromatic hydrogens are quite easily identifiable by their distinctive chemical shifts.

Recall that vinylic hydrogens are in resonance at δ 4.6–5.7 (Section 12.10B). Hence, aromatic hydrogens absorb radiation even farther downfield than vinylic hydrogens do.

The ^{1}H-NMR spectrum of toluene (Figure 12.13) shows a singlet at δ 2.32 for the three hydrogens of the methyl group and a closely spaced multiplet at δ 7.3 for the five hydrogens of the aromatic ring. Depending on the substituents and substitution pattern on a benzene ring, ^{1}H-NMR signals for aromatic hydrogens can appear anywhere from a broad, overlapping set of peaks to well-resolved sets of peaks.

In ^{13}C-NMR spectroscopy, carbon atoms of aromatic rings appear in the range δ 110–160. Benzene, for example, shows a single signal at δ 128. Because carbon-13 signals for alkene carbons appear in the same range, it is generally not possible to establish the presence of an aromatic ring by ^{13}C-NMR spectroscopy alone. ^{13}C-NMR

Figure 12.13
^{1}H-NMR spectrum of toluene.

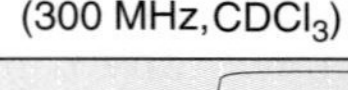

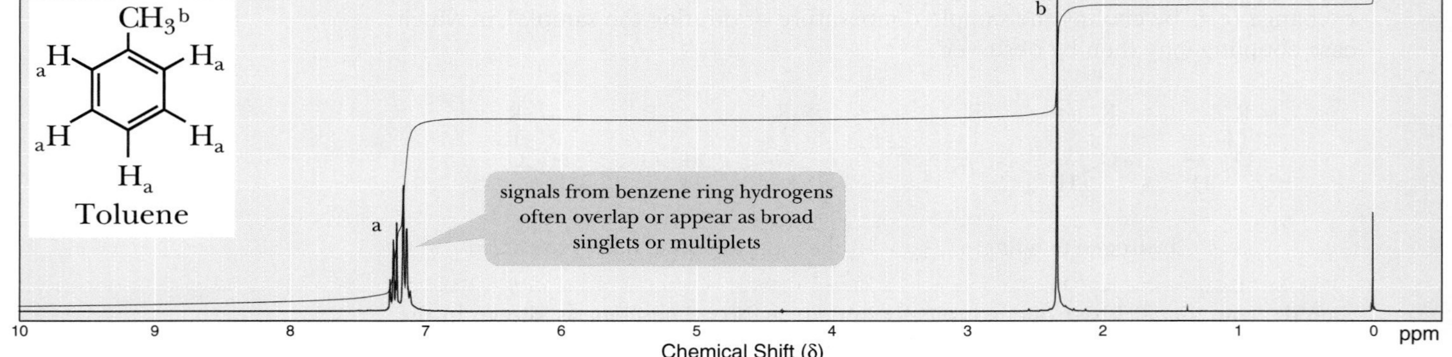

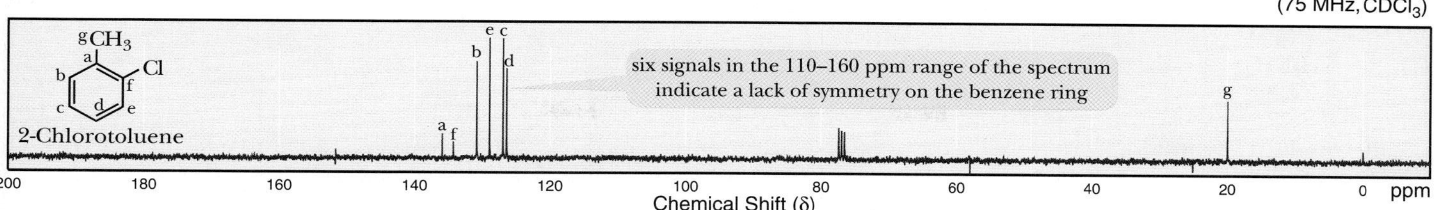

Figure 12.14
^{13}C-NMR spectrum of 2-chlorotoluene.

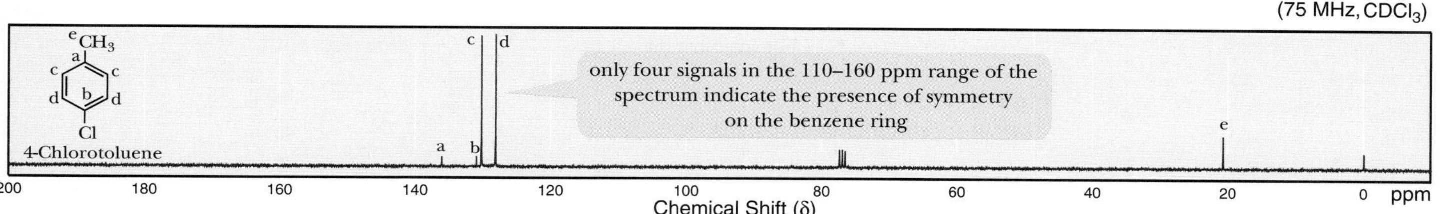

Figure 12.15
^{13}C-NMR spectrum of 4-chlorotoluene.

spectroscopy is particularly useful, however, in establishing substitution patterns of aromatic rings. The ^{13}C-NMR spectrum of 2-chlorotoluene (Figure 12.14) shows six signals in the aromatic region; the compound's more symmetric isomer 4-chlorotoluene (Figure 12.15) shows only four signals in the aromatic region. Thus, all one needs to do is count signals to distinguish between these constitutional isomers.

E. Amines

The chemical shifts of amine hydrogens, like those of hydroxyl hydrogens, are variable and may be found in the region δ 0.5–5.0, depending on the solvent, the concentration, and the temperature. Furthermore, the rate of **intermolecular exchange of hydrogens** is sufficiently rapid, compared with the time scale of an NMR measurement, that signal splitting between amine hydrogens and hydrogens on an adjacent α-carbon is prevented.

fast

hydrogen bonding

these protons have been exchanged

Thus, amine hydrogens generally appear as singlets. The NH_2 hydrogens in benzylamine, $C_6H_5CH_2NH_2$, for example, appear as a singlet at δ 1.40 (Figure 12.16).

F. Aldehydes and Ketones

^{1}H-NMR spectroscopy is an important tool for identifying aldehydes and for distinguishing between aldehydes and other carbonyl-containing compounds. Just as a carbon–carbon double bond causes a downfield shift in the signal of a vinylic hydrogen (Section 12.10B), a carbon–oxygen double bond causes a downfield shift in the signal of an aldehyde hydrogen, typically to δ 9.5–10.1. Signal splitting between this

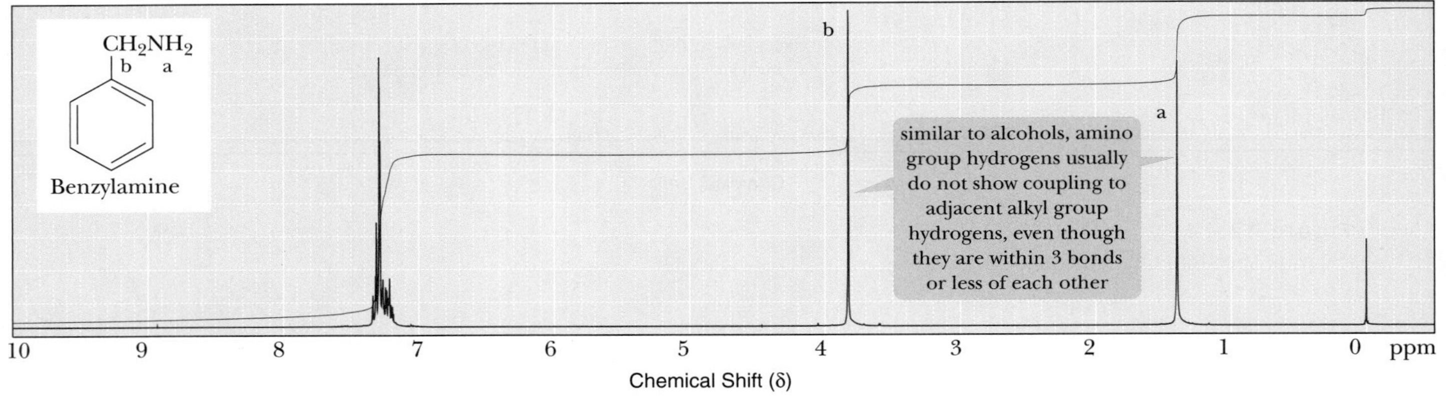

Figure 12.16
^{1}H-NMR spectrum of benzylamine.

hydrogen and those on the adjacent α-carbon is slight; consequently, the aldehyde hydrogen signal often appears as a closely spaced doublet or triplet. In the spectrum of butanal, for example, the triplet signal for the aldehyde hydrogen at δ 9.78 is so closely spaced that it almost looks like a singlet (Figure 12.17).

Just as the signal for an aldehyde hydrogen is only weakly split by the adjacent nonequivalent α-hydrogens, the α-hydrogens are weakly split by the aldehyde hydrogen. Hydrogens on an α-carbon of an aldehyde or ketone typically appear around δ 2.1–2.6. The carbonyl carbons of aldehydes and ketones are readily identifiable in ^{13}C-NMR spectroscopy by the position of their signal between δ 180 and 210.

G. Carboxylic Acids

The hydrogen of the carboxyl group gives a signal in the range δ 10–13. The chemical shift of a carboxyl hydrogen is so large—even larger than the chemical shift of an aldehyde hydrogen (δ 9.5–10.1)—that it serves to distinguish carboxyl hydrogens from most other types of hydrogens. The signal of the carboxyl hydrogen of 2-methylpropanoic acid is at the left of the ^{1}H-NMR spectrum in Figure 12.18 and has

Figure 12.17
^{1}H-NMR spectrum of butanal.

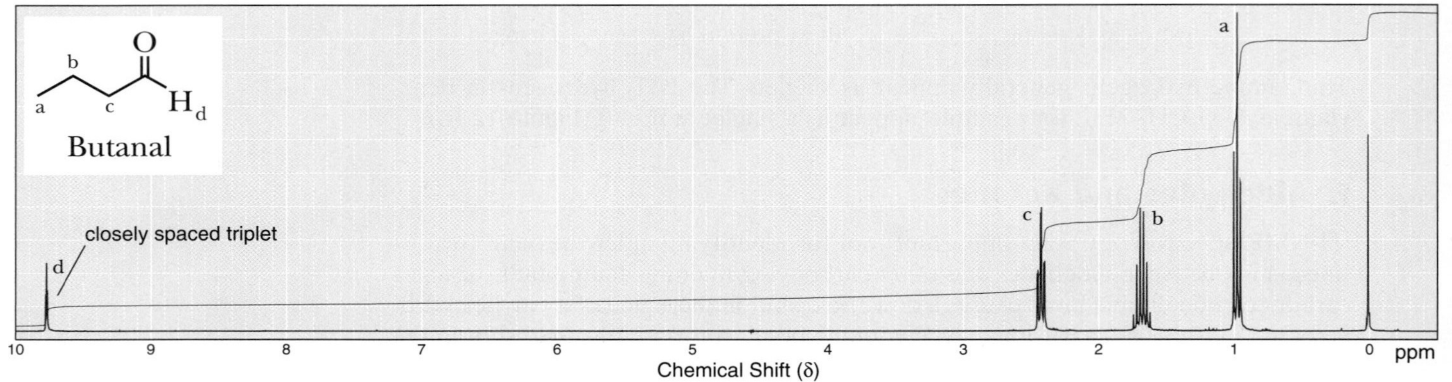

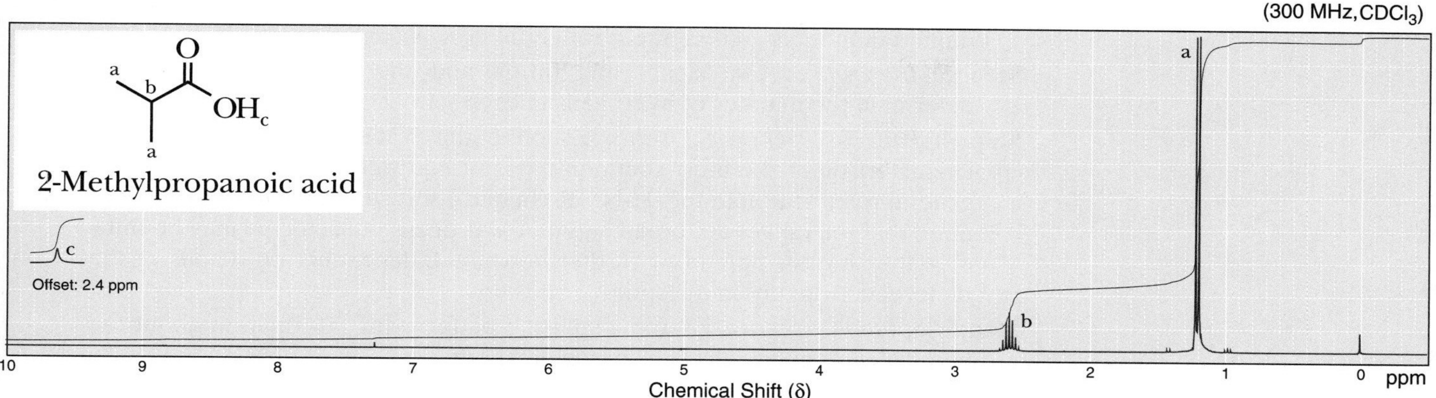

Figure 12.18
^{1}H-NMR spectrum of 2-methylpropanoic acid (isobutyric acid).

been offset by δ 2.4. (Add 2.4 to the position at which the signal appears on the spectrum.) The chemical shift of this hydrogen is δ 12.0.

H. Esters

Hydrogens on the α-carbon of the carbonyl group of an ester are slightly deshielded and give signals at δ 2.1–2.6. Hydrogens on the carbon bonded to the ester oxygen are more strongly deshielded and give signals at δ 3.7–4.7. It is thus possible to distinguish between ethyl acetate and its constitutional isomer, methyl propanoate, by the chemical shifts of either the singlet $—CH_3$ absorption (compare δ 2.04 with 3.68) or the quartet $—CH_2—$ absorption (compare δ 4.11 with 2.33):

δ 2.04(s) δ 4.11(q) δ 2.33(q) δ 3.68(s)

$$CH_3—\overset{\overset{\large O}{\|}}{C}—O—CH_2—CH_3 \qquad CH_3—CH_2—\overset{\overset{\large O}{\|}}{C}—O—CH_3$$

Ethyl acetate Methyl propanoate

12.11 How Do We Solve an NMR Problem?

One of the first steps in determining the molecular structure of a compound is to establish the compound's molecular formula. In the past, this was most commonly done by elemental analysis, combustion to determine the percent composition, and so forth. More commonly today, we determine molecular weight and molecular formula by a technique known as *mass spectrometry*. (An explanation of the technique is beyond the scope of this book.) In the examples that follow, we assume that the molecular formula of any unknown compound has already been determined, and we proceed from there, using spectral analysis to determine a structural formula.

The following steps may prove helpful as a systematic approach to solving ^{1}H-NMR spectral problems:

Step 1: *Molecular formula and index of hydrogen deficiency.* Examine the molecular formula, calculate the index of hydrogen deficiency (Section 11.4J), and deduce what information you can about the presence or absence of rings or pi bonds.

Step 2: *Number of signals.* Count the number of signals, to determine the minimum number of sets of equivalent hydrogens in the compound.

Step 3: *Integration.* Use signal integration and the molecular formula to determine the number of hydrogens in each set.

Step 4: *Pattern of chemical shifts.* Examine the NMR spectrum for signals characteristic of the most common types of equivalent hydrogens. (See the general rules of thumb for ^{1}H-NMR chemical shifts in Section 12.7.) Keep in mind that the ranges are broad and that hydrogens of each type may be shifted either farther upfield or farther downfield, depending on details of the molecular structure in question.

Step 5: *Splitting patterns.* Examine splitting patterns for information about the number of nonequivalent hydrogen neighbors.

Step 6: *Structural formula.* Write a structural formula consistent with the information learned in Steps 1–5.

Example 12.8

Following is a ^{1}H-NMR spectrum for a compound that is a colorless liquid with the molecular formula $C_5H_{10}O$. Propose a structural formula for the compound.

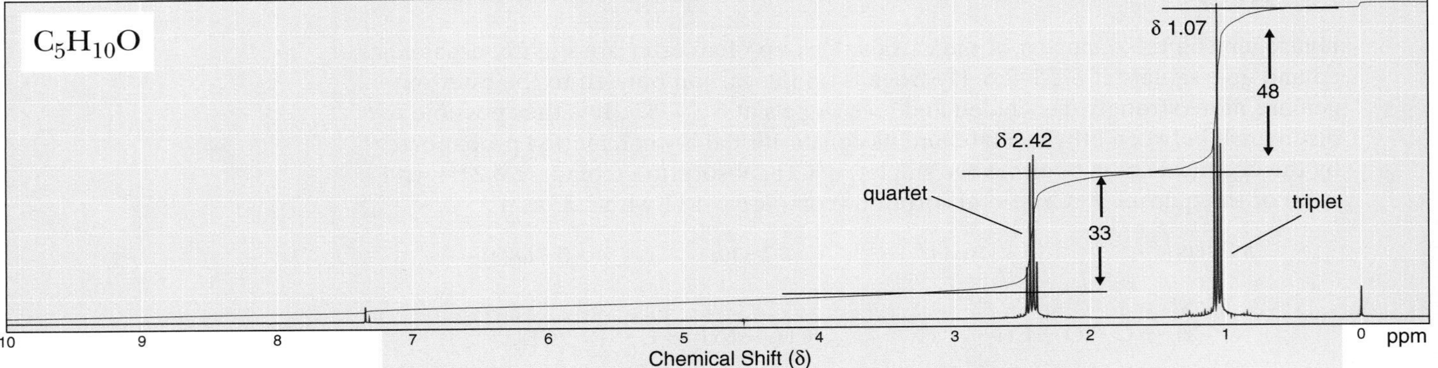

Strategy

^{1}H-NMR spectra can be approached by (1) calculating the index of hydrogen deficiency and deducing what information you can about the presence or absence of rings or pi bonds, (2) counting the number of signals to determine the minimum number of sets of equivalent hydrogens in the compound, (3) using signal integration and the molecular formula to determine the number of hydrogens in each set, (4) examining the NMR spectrum for signals characteristic of the most common types of equivalent hydrogens, (5) examining splitting patterns for information about the number of nonequivalent hydrogen neighbors, and (6) writing a structural formula consistent with the information learned in Steps 1–5.

Solution

Step 1: *Molecular formula and index of hydrogen deficiency.* The reference compound is $C_5H_{12}O$; therefore, the index of hydrogen deficiency is 1. The molecule thus contains either one ring or one pi bond.

Step 2: *Number of signals.* There are two signals (a triplet and a quartet) and therefore two sets of equivalent hydrogens.

Step 3: *Integration.* By signal integration, we calculate that the number of hydrogens giving rise to each signal is in the ratio 3 : 2. Because there are 10 hydrogens, we conclude that the signal assignments are δ 1.07 (6H) and δ 2.42 (4H).

Step 4: *Pattern of chemical shifts.* The signal at δ 1.07 is in the alkyl region and, based on its chemical shift, most probably represents a methyl group. No signal occurs at δ 4.6 to 5.7; thus, there are no vinylic hydrogens. (If a carbon–carbon double bond is in the molecule, no hydrogens are on it; that is, it is tetrasubstituted.)

Step 5: *Splitting pattern.* The methyl signal at δ 1.07 is split into a triplet (t); hence, it must have two neighboring hydrogens, indicating $-CH_2CH_3$. The signal at δ 2.42 is split into a quartet (q); thus, it must have three neighboring hydrogens, which is also consistent with $-CH_2CH_3$. Consequently, an ethyl group accounts for these two signals. No other signals occur in the spectrum; therefore, there are no other types of hydrogens in the molecule.

Step 6: *Structural formula.* Put the information learned in the previous steps together to arrive at the following structural formula. Note that the chemical shift of the methylene group ($-CH_2-$) at δ 2.42 is consistent with an alkyl group adjacent to a carbonyl group.

δ 2.42 (q) δ 1.07 (t)

$$CH_3-CH_2-\overset{\overset{\large O}{\|}}{C}-CH_2-CH_3$$

3-Pentanone

See problems 12.14, 12.8–12.41

Problem 12.8

Following is a ^{1}H-NMR spectrum for prenol, a compound that possesses a fruity odor and that is commonly used in perfumes. Prenol has the molecular formula $C_5H_{10}O$. Propose a structural formula for Prenol.

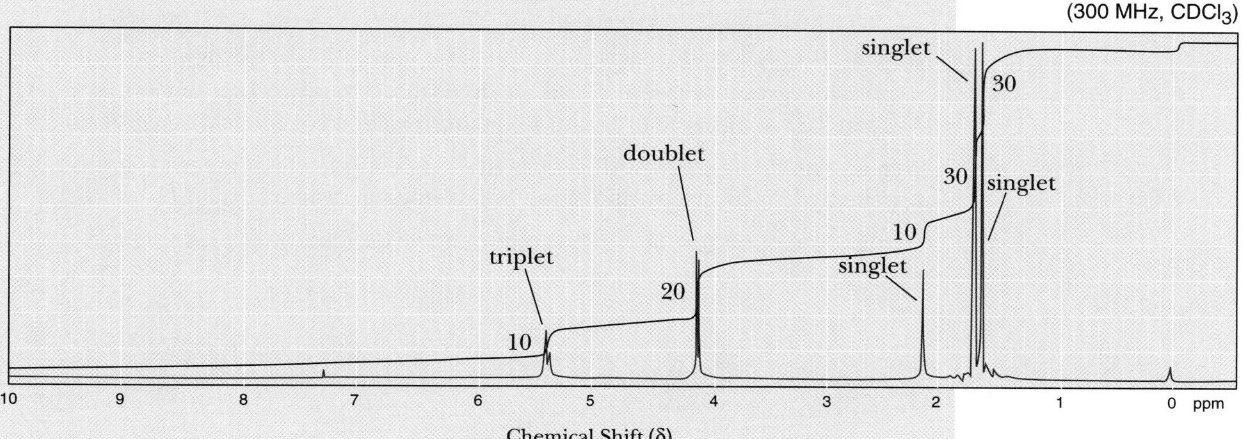

Example 12.9

Following is a ^{1}H-NMR spectrum for a compound that is a colorless liquid with the molecular formula $C_7H_{14}O$. Propose a structural formula for the compound.

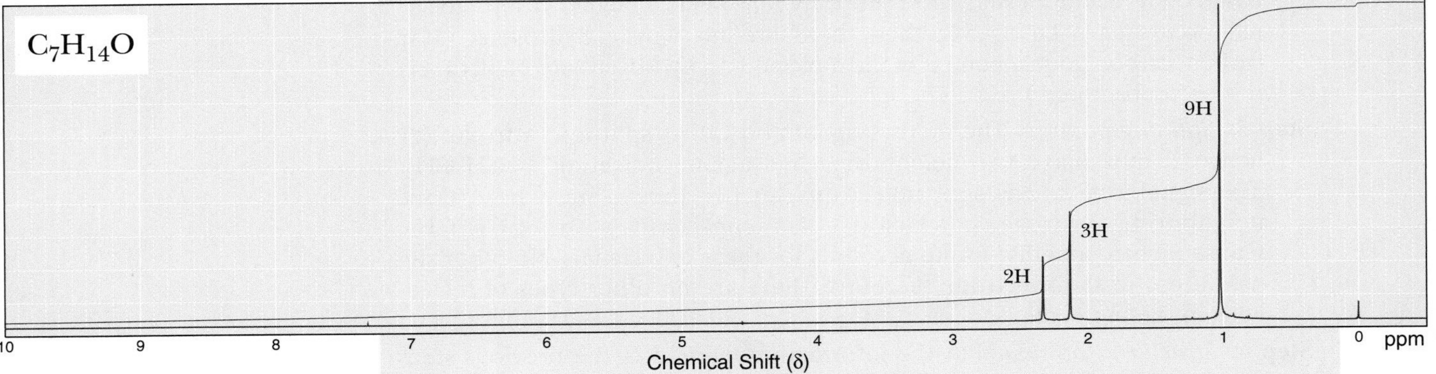

Strategy

^{1}H-NMR spectra can be approached by (1) calculating the index of hydrogen deficiency and deducing what information you can about the presence or absence of rings or pi bonds, (2) counting the number of signals to determine the minimum number of sets of equivalent hydrogens in the compound, (3) using signal integration and the molecular formula to determine the number of hydrogens in each set, (4) examining the NMR spectrum for signals characteristic of the most common types of equivalent hydrogens, (5) examining splitting patterns for information about the number of nonequivalent hydrogen neighbors, and (6) writing a structural formula consistent with the information learned in Steps 1–5.

Solution

Step 1: *Molecular formula and index of hydrogen deficiency.* The index of hydrogen deficiency is 1; thus, the compound contains one ring or one pi bond.

Step 2: *Number of signals.* There are three signals and therefore three sets of equivalent hydrogens.

Step 3: *Integration.* By signal integration, we calculate that the number of hydrogens giving rise to each signal is in the ratio 9 : 3 : 2, reading from left to right.

Step 4: *Pattern of chemical shifts.* The singlet at δ 1.01 is characteristic of a methyl group adjacent to an sp^3 hybridized carbon. The singlets at δ 2.11 and 2.32 are characteristic of alkyl groups adjacent to a carbonyl group.

Step 5: *Splitting pattern.* All signals are singlets (s), which means that none of the hydrogens are within three bonds of each other.

Step 6: *Structural formula.* The compound is 4,4-dimethyl-2-pentanone:

δ 1.01(s) δ 2.32(s) δ 2.11(s)

$$CH_3-\underset{CH_3}{\overset{CH_3}{C}}-CH_2-\overset{O}{\overset{\|}{C}}-CH_3$$

4,4-Dimethyl-2-pentanone

See problems 12.14, 12.8–12.41

Problem 12.9

Following is a ^{1}H-NMR spectrum for a compound that is a colorless liquid with the molecular formula $C_7H_{14}O$. Propose a structural formula for the compound.

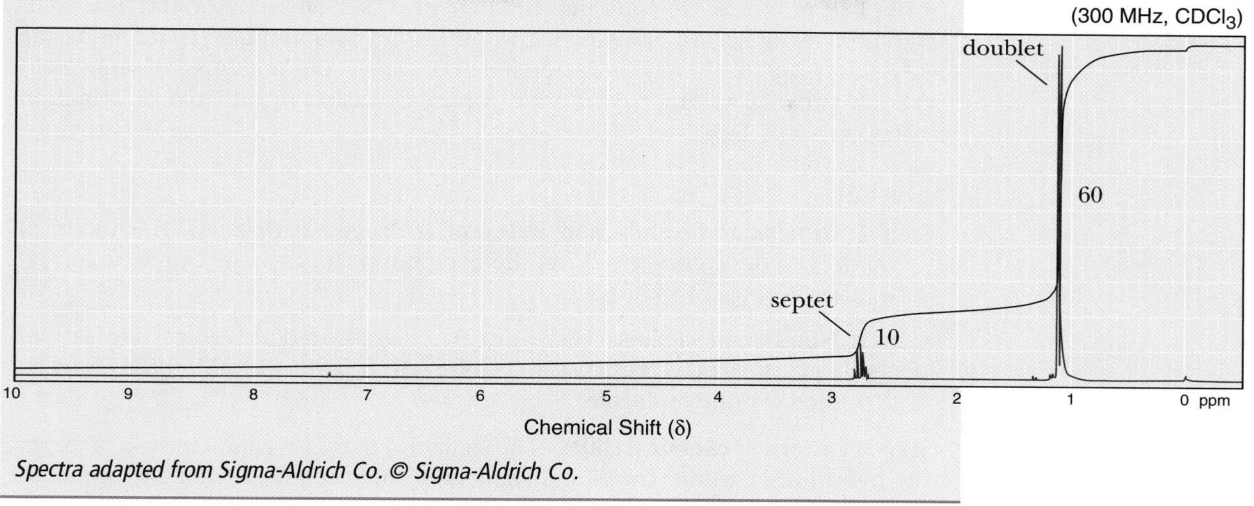

Spectra adapted from Sigma-Aldrich Co. © Sigma-Aldrich Co.

The following steps may prove helpful as a systematic approach to solving ^{13}C-NMR spectral problems:

Step 1: Molecular formula and index of hydrogen deficiency. Examine the molecular formula, calculate the index of hydrogen deficiency (Section 11.4J), and deduce what information you can about the presence or absence of rings or pi bonds.

Step 2: Number of signals. Count the number of signals to determine the minimum number of sets of equivalent carbons in the compound.

Step 3: Pattern of chemical shifts. Examine the NMR spectrum for signals characteristic of the most common types of equivalent carbons (see the general rules of thumb for ^{13}C-NMR chemical shifts in Section 12.9). Keep in mind that these ranges are broad and that carbons of each type may be shifted either farther upfield or farther downfield, depending on details of the molecular structure in question.

Step 4: Structural formula. Write a structural formula consistent with the information learned in Steps 1–3. *Note*: Because ^{13}C-NMR does not provide information about neighboring hydrogens, it may be more difficult to elucidate the structure of a compound based solely on ^{13}C-NMR data.

Example 12.10

Following is a ^{13}C-NMR spectrum for a compound that is a colorless liquid with the molecular formula C_7H_7Cl. Propose a structural formula for the compound.

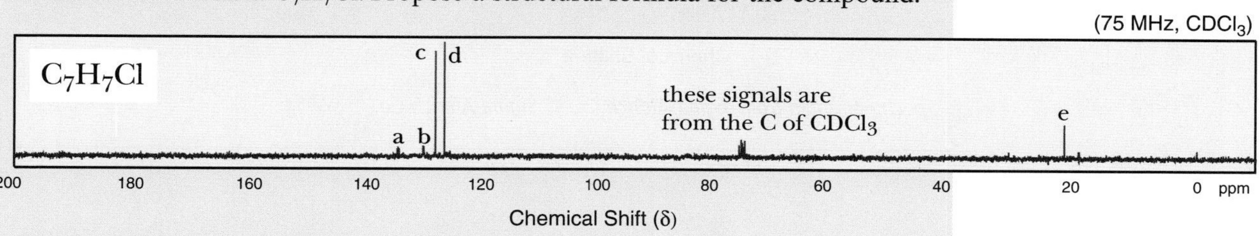

Spectra adapted from Sigma-Aldrich Co. © Sigma-Aldrich Co.

Strategy

^{13}C-NMR spectra can be approached by (1) calculating the index of hydrogen deficiency and deducing what information you can about the presence or absence of rings or pi bonds, (2) counting the number of signals to determine the minimum number of sets of equivalent carbons in the compound, (3) examining the NMR spectrum for signals characteristic of the most common types of equivalent carbons, and (4) writing a structural formula consistent with the information learned in Steps 1–3.

Solution

Step 1: Molecular formula and index of hydrogen deficiency. The index of hydrogen deficiency is 4; thus, the compound can contain a myriad combination of rings or pi bonds.

Step 2: Number of signals. There are five signals and therefore five sets of equivalent carbons. Because there are seven carbons total, there must be symmetry in the molecule.

Step 3: Pattern of chemical shifts. The signal **(e)** at δ 23 is characteristic of an sp^3 hybridized carbon. The four signals **(a–d)** between δ 120 and 140 are characteristic of sp^2 hybridized carbons. Because it would be unlikely for a molecule with only seven carbon atoms to have 4 pi bonds (due to IHD = 4), it is likely that these signals represent the carbons of a benzene ring.

Step 4: Structural formula. Because there must be symmetry in the molecule, the most likely structure of the compound is:

c d
H_3C b a Cl
e
c d

4-Chlorotoluene

See problems 12.13, 12.17, 12.18, 12.22, 12.23, 12.29–12.33

Problem 12.10

Following is a ^{13}C-NMR spectrum for a compound that is a colorless liquid with the molecular formula $C_4H_8Br_2$. Propose a structural formula for the compound.

(75 MHz, $CDCl_3$)

these signals are from the C of $CDCl_3$

200 180 160 140 120 100 80 60 40 20 0 ppm

Chemical Shift (δ)

Spectra adapted from Sigma-Aldrich Co. © Sigma-Aldrich Co.

Key Terms and Concepts

^{1}**H-NMR spectroscopy** (p. 418)
^{13}**C-NMR spectroscopy** (p. 433)
applied magnetic field (p. 419)
chemical shift (δ) (p. 422)
delta (δ) scale (p. 421)
deshielded (p. 420)
doublet (p. 431)
downfield (p. 423)
equivalency (p. 423)
equivalent hydrogens (p. 423)
hydrogen-decoupled spectrum (p. 433)
integration (p. 427)
integration ratio (p. 427)
intermolecular proton exchange (p. 441)
line of integration (p. 427)
local magnetic field (p. 420)
(n + 1) rule (p. 431)
nuclear magnetic resonance (NMR) spectroscopy (p. 417)
part per million (ppm) (p. 421)
peak (NMR) (p. 431)
quartet (p. 431)
resonance (p. 419)
resonance signal (p. 420)
shielding (p. 420)
signal (p. 420)
signal splitting (p. 431)
singlet (p. 431)
spin (p. 418)
tetramethylsilane (TMS) (p. 421)
triplet (p. 431)
upfield (p. 423)

Summary of Key Questions

12.1 What Are the Magnetic Properties of Nuclei?

- An atomic nucleus that has an odd mass or an odd atomic number also has a spin and behaves as if it were a tiny bar magnet.
- When a collection of ^{1}H and ^{13}C atoms is placed between the poles of a powerful magnet, interactions between their nuclear spins and the applied magnetic field are quantized, and only two orientations are allowed.

12.2 What Is Nuclear Magnetic Resonance?

- When placed between the poles of a powerful magnet, the nuclear spins of these elements become aligned either with the applied field or against it.
- Nuclear spins aligned with the applied field are in the lower energy state; those aligned against the applied field are in the higher energy state.
- Resonance is the absorption of electromagnetic radiation by a nucleus and the resulting "flip" of its nuclear spin from a lower energy spin state to a higher energy spin state.

12.3 What Is Shielding?

- The experimental conditions required to cause nuclei to resonate are affected by the local chemical and magnetic environment.
- Electrons around a hydrogen also have spin and create a local magnetic field that shields the hydrogen from the applied field.

12.4 How Is an NMR Spectrum Obtained?

- An NMR spectrometer records resonance as a signal.

12.5 How Many Resonance Signals Will a Compound Yield in Its NMR Spectrum?

- Equivalent hydrogens within a molecule have identical chemical shifts.

12.6 What Is Signal Integration?

- The area of a ^{1}H-NMR signal is proportional to the number of equivalent hydrogens giving rise to that signal. Determination of these areas is termed integration.

12.7 What Is Chemical Shift?

- In a ^{1}H-NMR spectrum, a resonance signal is reported by how far it is shifted from the resonance signal of the 12 equivalent hydrogens in tetramethylsilane (TMS).
- A resonance signal in a ^{13}C-NMR spectrum is reported by how far it is shifted from the resonance signal of the four equivalent carbons in TMS.
- A chemical shift (δ) is the frequency shift from TMS, divided by the operating frequency of the spectrometer.

12.8 What Is Signal Splitting?

- In signal splitting, the ^{1}H-NMR signal from one hydrogen or set of equivalent hydrogens is split by the influence of nonequivalent hydrogens on the same or adjacent carbon atoms.
- According to the $(n + 1)$ rule, if a hydrogen has n hydrogens that are nonequivalent to it, but are equivalent among themselves, on the same or adjacent carbon atom(s), its ^{1}H-NMR signal is split into $(n + 1)$ peaks.
- Complex splitting occurs when a hydrogen is flanked by two or more sets of hydrogens and those sets are nonequivalent.
- Splitting patterns are commonly referred to as singlets, doublets, triplets, quartets, quintets, and multiplets.

12.9 What Is ^{13}C-NMR Spectroscopy, and How Does It Differ from ^{1}H-NMR Spectroscopy?

- A ^{13}C-NMR spectrum normally spans the range δ 0–210 (versus δ 0–13 for ^{1}H-NMR).
- ^{13}C-NMR spectra are commonly recorded in a hydrogen-decoupled instrumental mode. In this mode, all ^{13}C signals appear as singlets.
- Integration is not normally performed in ^{13}C-NMR.

12.10 How Do We Interpret NMR Spectra?

- NMR spectra are interpreted by comparing their signals with those characteristic of various types of compounds (e.g., alkanes, alkenes, alcohols).

12.11 How Do We Solve an NMR Problem?

- ^{1}H-NMR spectra can be approached by (1) calculating the index of hydrogen deficiency and deducing what information you can about the presence or absence of rings or pi bonds, (2) counting the number of signals to determine the minimum number of sets of equivalent hydrogens in the compound, (3) using signal integration and the molecular formula to determine the number of hydrogens in each set, (4) examining the NMR spectrum for signals characteristic of the most common types of equivalent hydrogens, (5) examining splitting patterns for information about the number of nonequivalent hydrogen neighbors, and (6) writing a structural formula consistent with the information learned in Steps 1–5.
- ^{13}C-NMR spectra can be approached by (1) calculating the index of hydrogen deficiency and deducing what information you can about the presence or absence of rings or pi bonds, (2) counting the number of signals to determine the minimum number of sets of equivalent carbons in the compound, (3) examining the NMR spectrum for signals characteristic of the most common types of equivalent carbons, and (4) writing a structural formula consistent with the information learned in Steps 1–3.

Quick Quiz

Answer true or false to the following questions to assess your general knowledge of the concepts in this chapter. If you have difficulty with any of them, you should review the appropriate section in the chapter (shown in parentheses) before attempting the more challenging end-of-chapter problems.

1. Integration reveals the number of neighboring hydrogens in a ^{1}H-NMR spectrum. (12.6)
2. An alkene (vinylic) hydrogen can be distinguished from a benzene ring hydrogen via ^{1}H-NMR spectroscopy. (12.10)
3. The NMR signal of a shielded nucleus appears more upfield than the signal for a deshielded nucleus. (12.3)
4. The chemical shift of a nucleus depends on its resonance frequency. (12.4)
5. A ketone can be distinguished from an aldehyde via ^{13}C-NMR spectroscopy. (12.9)
6. A ^{1}H-NMR spectrum with an integration ratio of 3 : 1 : 2 could represent a compound with the molecular formula C_5H_9O. (12.10)
7. A set of hydrogens are equivalent if replacing each of them with a halogen results in compounds of the same name. (12.5)
8. The area under each peak in a ^{1}H-NMR spectrum can be determined using a technique known as integration. (12.6)
9. All atomic nuclei have a spin, which allows them to be analyzed by NMR spectroscopy. (12.1)
10. The resonance frequency of a nucleus depends on its amount of shielding. (12.3)
11. A carboxylic acid can be distinguished from an aldehyde via ^{1}H-NMR spectroscopy. (12.10)
12. Resonance is the excitation of a magnetic nucleus in one spin state to a higher spin state. (12.2)
13. A compound with an index of hydrogen deficiency of 1 can contain either one ring, one double bond, or one triple bond. (12.10)
14. A set of hydrogens represented by a doublet indicates that there are two neighboring equivalent hydrogens. (12.8)
15. TMS, tetramethylsilane, is a type of solvent used in NMR spectroscopy. (12.4)
16. The methyl carbon of 1-chlorobutane will yield a ^{1}H-NMR signal that appears as a triplet. (12.8)

Answers: (1) F (2) T (3) T (4) T (5) F (6) F (7) T (8) T (9) F (10) T (11) T (12) T (13) F (14) F (15) F (16) T

Problems

A problem marked with an asterisk indicates an applied "real world" problem. Answers to problems whose numbers are printed in blue are given in Appendix D.

Section 12.5 Equivalency of Hydrogens and Carbons

12.11 Determine the number of signals you would expect to see in the ^{1}H-NMR spectrum of each of the following compounds. **(See Example 12.2)**

(a) 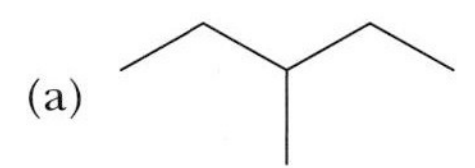

(b) 

(c) 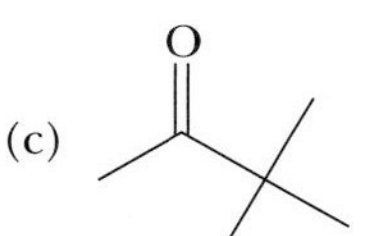

(d) 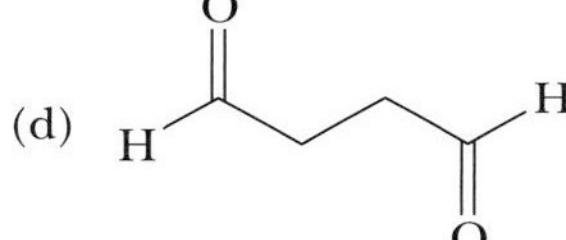

(e) HO

(f) H N

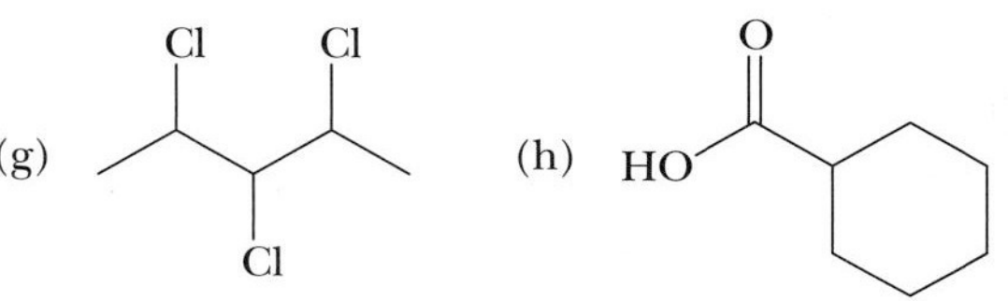

12.12 Determine the number of signals you would expect to see in the ^{13}C-NMR spectrum of each of the compounds in Problem 12.11. **(See Example 12.7)**

Sections 12.10 and 12.11 Interpreting ^{1}H-NMR and ^{13}C-NMR Spectra

12.13 Following are structural formulas for the constitutional isomers of xylene and three sets of ^{13}C-NMR spectra. Assign each constitutional isomer its correct spectrum. **(See Example 12.10)**

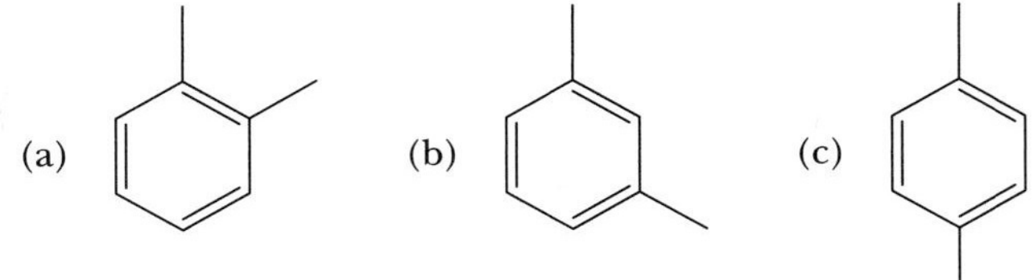

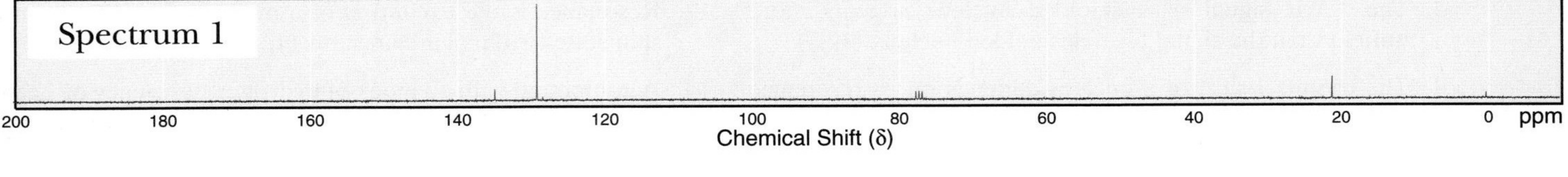

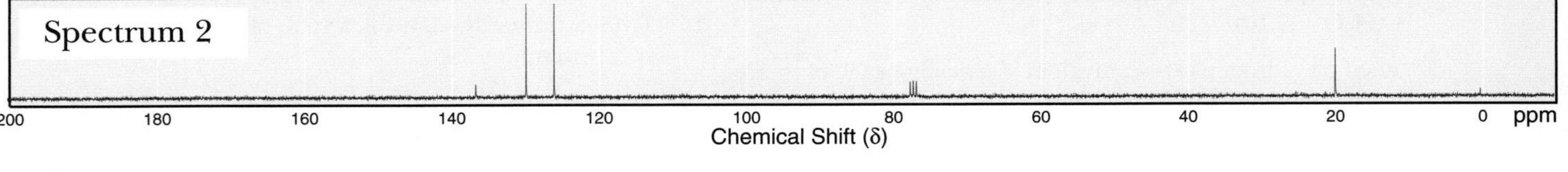

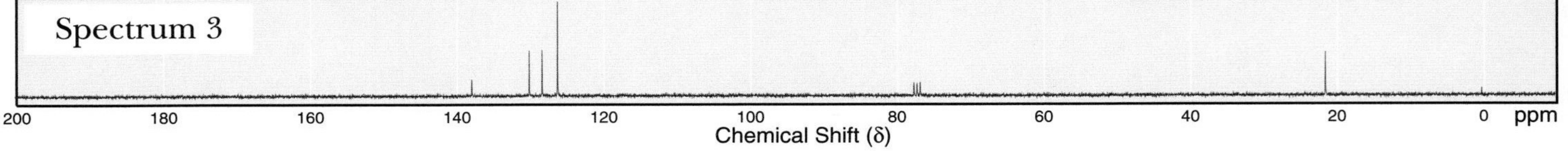

12.14 Following is a ^{1}H-NMR spectrum for compound A, with the molecular formula C_7H_{14}. Compound A decolorizes a solution of bromine in carbon tetrachloride. Propose a structural formula for compound A. **(See Examples 12.8, 12.9)**

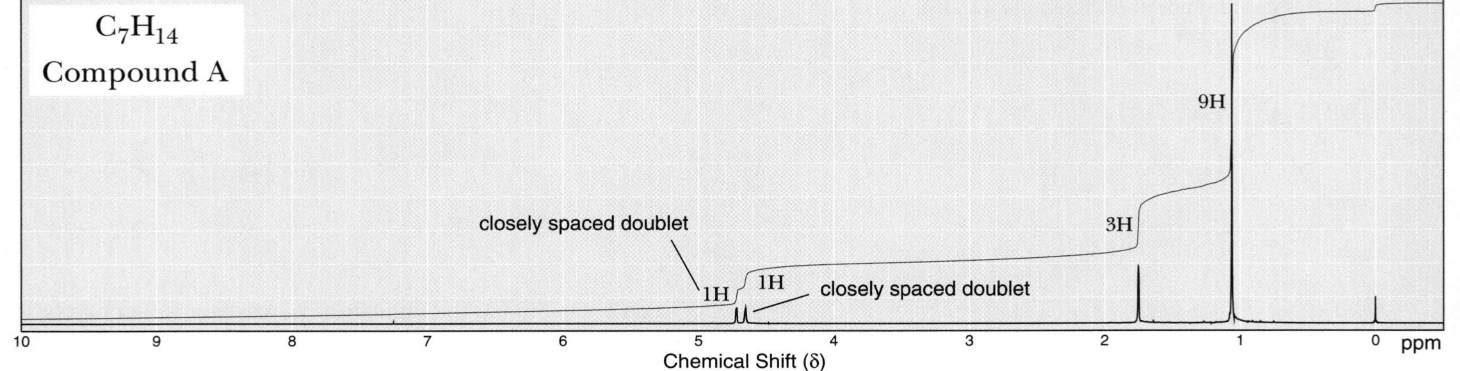

12.15 Following is a ^{1}H-NMR spectrum for compound B, with the molecular formula C_8H_{16}. Compound B decolorizes a solution of Br_2 in CCl_4. Propose a structural formula for compound B. **(See Examples 12.8, 12.9)**

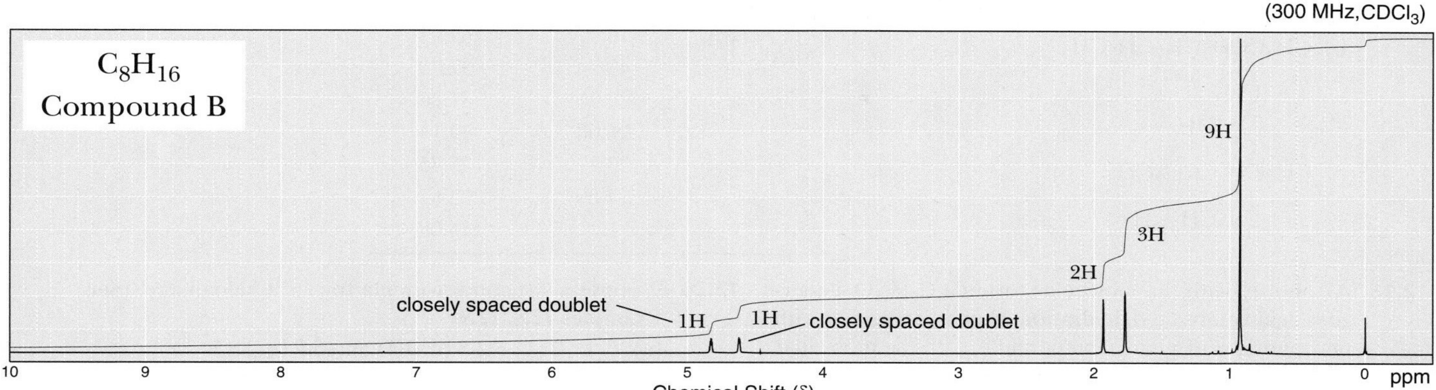

12.16 Following are the ^{1}H-NMR spectra of compounds C and D, each with the molecular formula C_4H_7Cl. Each compound decolorizes a solution of Br_2 in CCl_4. Propose structural formulas for compounds C and D. **(See Examples 12.8, 12.9)**

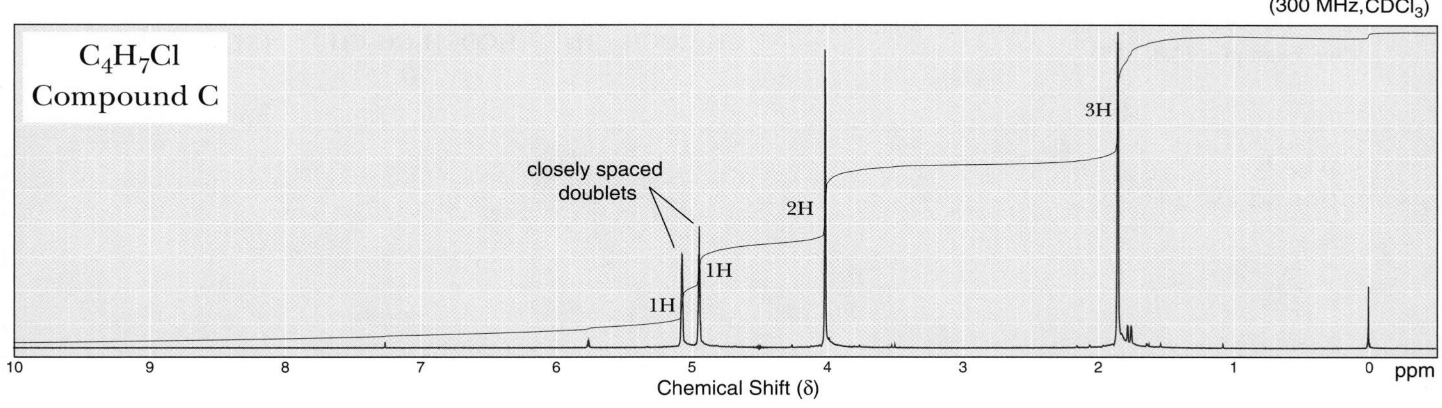

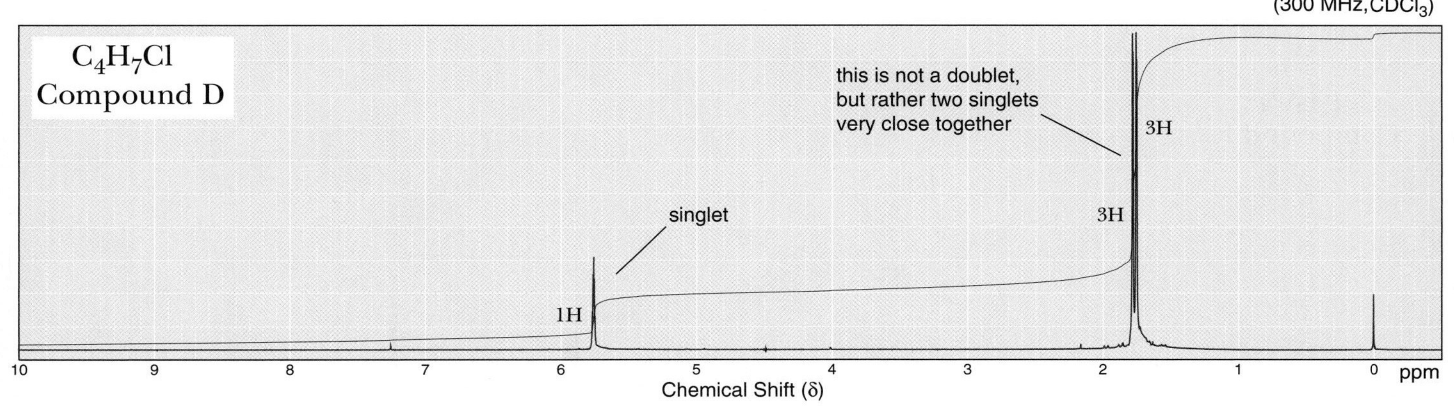

12.17 Following are the structural formulas of three alcohols with the molecular formula $C_7H_{16}O$ and three sets of ^{13}C-NMR spectral data. Assign each constitutional isomer to its correct spectral data. **(See Example 12.10)**

(a) $CH_3CH_2CH_2CH_2CH_2CH_2CH_2OH$

(b) $CH_3C(OH)(CH_3)CH_2CH_2CH_2CH_3$

```
        OH
        |
(b) CH3CCH2CH2CH2CH3
        |
        CH3
```

(c) $CH_3CH_2C(OH)(CH_2CH_3)CH_2CH_3$

```
            OH
            |
(c) CH3CH2CCH2CH3
            |
            CH2CH3
```

Spectrum 1	Spectrum 2	Spectrum 3
74.66	70.97	62.93
30.54	43.74	32.79
7.73	29.21	31.86
	26.60	29.14
	23.27	25.75
	14.09	22.63
		14.08

12.18 Alcohol E, with the molecular formula $C_6H_{14}O$, undergoes acid-catalyzed dehydration when it is warmed with phosphoric acid, giving compound F, with the molecular formula C_6H_{12}, as the major product. A ^1H-NMR spectrum of compound E shows peaks at δ 0.89 (t, 6H), 1.12 (s, 3H), 1.38 (s, 1H), and 1.48 (q, 4H). The ^{13}C-NMR spectrum of compound E shows peaks at δ 72.98, 33.72, 25.85, and 8.16. Propose structural formulas for compounds E and F. **(See Examples 12.8–12.10)**

12.19 Compound G, $C_6H_{14}O$, does not react with sodium metal and does not discharge the color of Br_2 in CCl_4. The ^1H-NMR spectrum of compound G consists of only two signals: a 12H doublet at δ 1.1 and a 2H septet at δ 3.6. Propose a structural formula for compound G. **(See Examples 12.8, 12.9)**

12.20 Propose a structural formula for each haloalkane: **(See Examples 12.8, 12.9)**

(a) $C_2H_4Br_2$ δ 2.5 (d, 3H) and 5.9 (q, 1H)
(b) $C_4H_8Cl_2$ δ 1.67 (d, 6H) and 2.15 (q, 4H)
(c) $C_5H_8Br_4$ δ 3.6 (s, 8H)
(d) C_4H_9Br δ 1.1 (d, 6H), 1.9 (m, 1H), and 3.4 (d, 2H)
(e) $C_5H_{11}Br$ δ 1.1 (s, 9H) and 3.2 (s, 2H)
(f) $C_7H_{15}Cl$ δ 1.1 (s, 9H) and 1.6 (s, 6H)

12.21 Following are structural formulas for esters (1), (2), and (3) and three ^1H-NMR spectra. Assign each compound its correct spectrum (H, I, or J) and assign all signals to their corresponding hydrogens. **(See Examples 12.8, 12.9)**

```
      O                O                   O
      ||               ||                  ||
CH3COCH2CH3     HCOCH2CH2CH3      CH3OCCH2CH3
    (1)              (2)                (3)
```

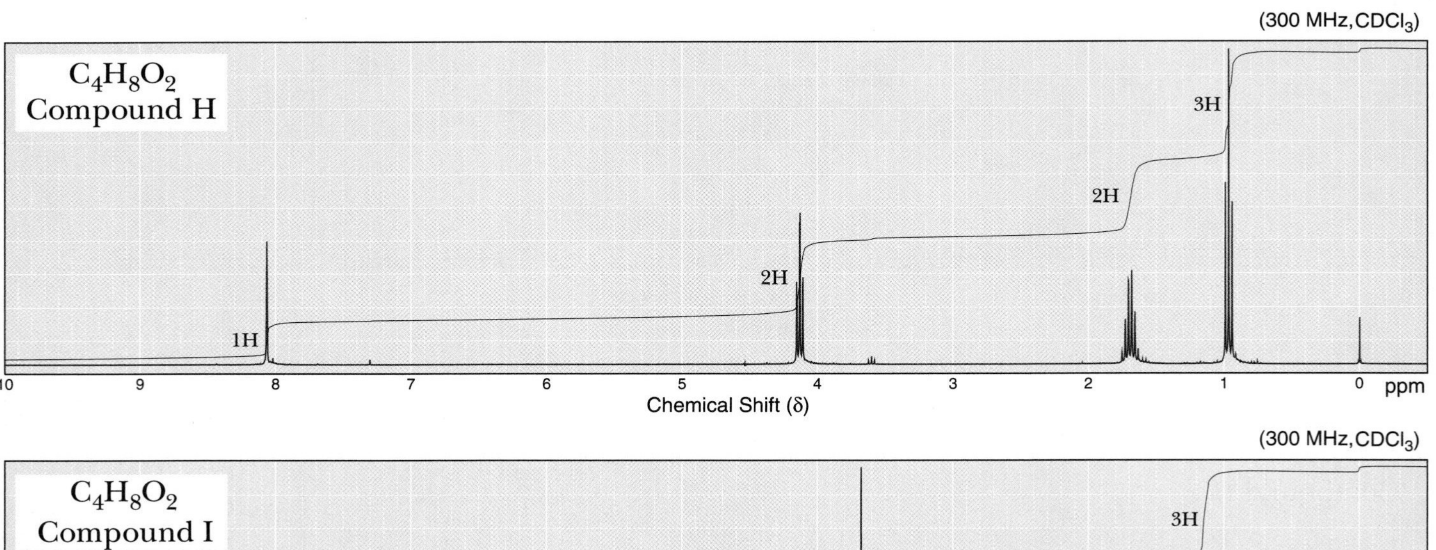

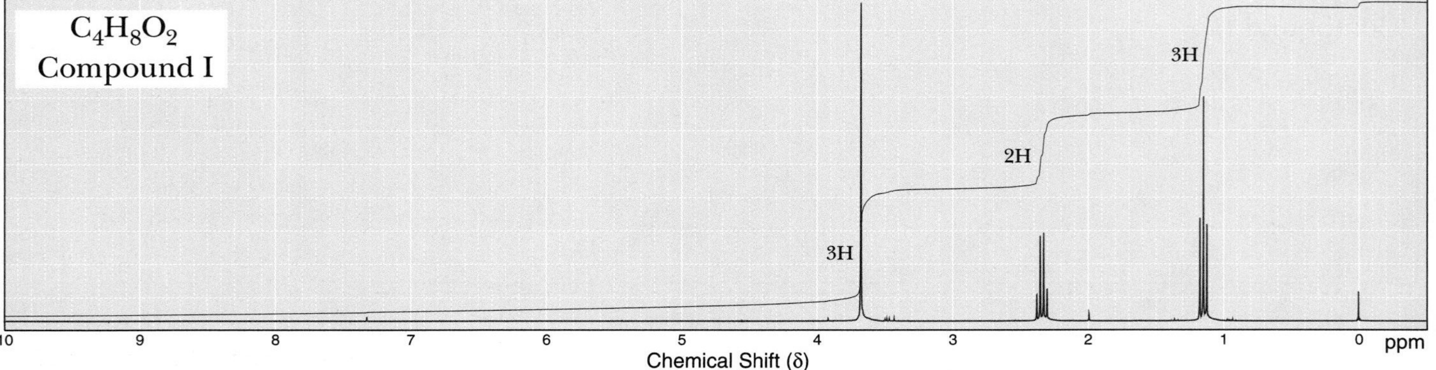

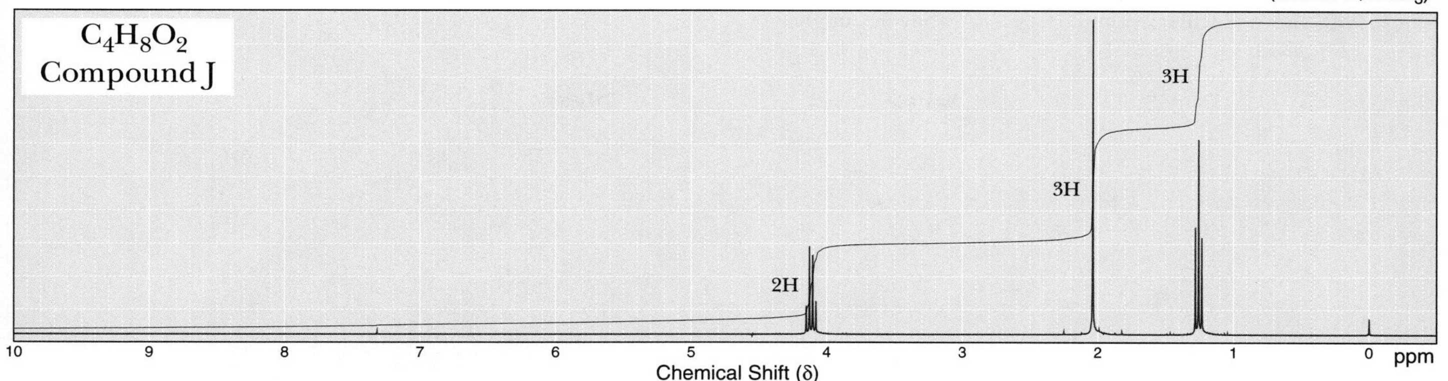

12.22 Compound K, $C_{10}H_{10}O_2$, is insoluble in water, 10% NaOH, and 10% HCl. A ^{1}H-NMR spectrum of compound K shows signals at δ 2.55 (s, 6H) and 7.97 (s, 4H). A ^{13}C-NMR spectrum of compound K shows four signals. From this information, propose a structural formula for K. **(See Examples 12.8–12.10)**

12.23 Compound L, $C_{15}H_{24}O$, is used as an antioxidant in many commercial food products, synthetic rubbers, and petroleum products. Propose a structural formula for compound L based on its ^{1}H-NMR and ^{13}C-NMR spectra. **(Sees Example 12.8–12.10)**

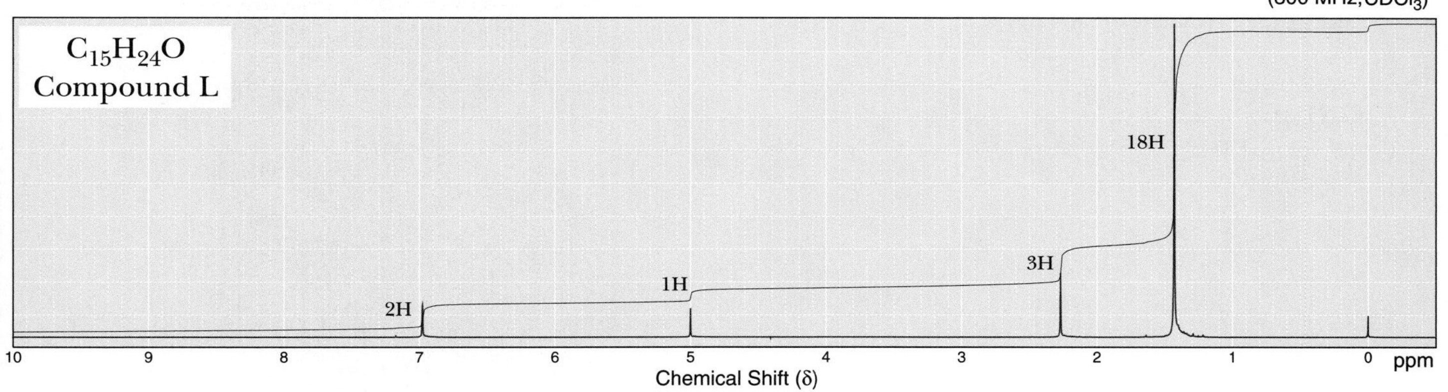

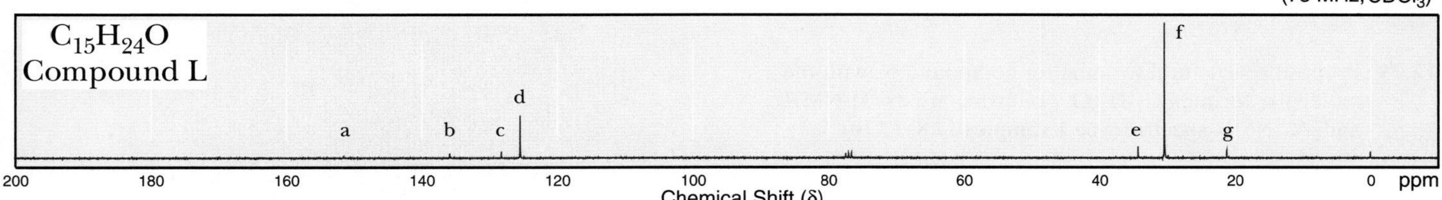

12.24 Propose a structural formula for these compounds, each of which contains an aromatic ring: **(See Examples 12.8, 12.9)**

(a) $C_9H_{10}O$ δ 1.2 (t, 3H), 3.0 (q, 2H), and 7.4-8.0 (m, 5H)

(b) $C_{10}H_{12}O_2$ δ 2.2 (s, 3H), 2.9 (t, 2H), 4.3 (t, 2H), and 7.3 (s, 5H)

(c) $C_{10}H_{14}$ δ 1.2 (d, 6H), 2.3 (s, 3H), 2.9 (septet, 1H), and 7.0 (s, 4H)

(d) C_8H_9Br δ 1.8 (d, 3H), 5.0 (q, 1H), and 7.3 (s, 5H)

12.25 Compound M, with the molecular formula $C_9H_{12}O$, readily undergoes acid-catalyzed dehydration to give compound N, with the molecular formula C_9H_{10}. A ^{1}H-NMR spectrum of compound M shows signals at δ 0.91 (t, 3H), 1.78 (m, 2H), 2.26 (d, 1H), 4.55 (m, 1H), and 7.31 (m, 5H). From this information, propose structural formulas for compounds M and N. **(See Examples 12.8, 12.9)**

12.26 Propose a structural formula for each ketone: **(See Examples 12.8, 12.9)**

(a) C_4H_8O δ 1.0 (t, 3H), 2.1 (s 3H), and 2.4 (q, 2H)

(b) $C_7H_{14}O$ δ 0.9 (t, 6H), 1.6 (sextet, 4H), and 2.4 (t, 4H)

12.27 Propose a structural formula for compound O, a ketone with the molecular formula $C_{10}H_{12}O$: **(See Examples 12.8, 12.9)**

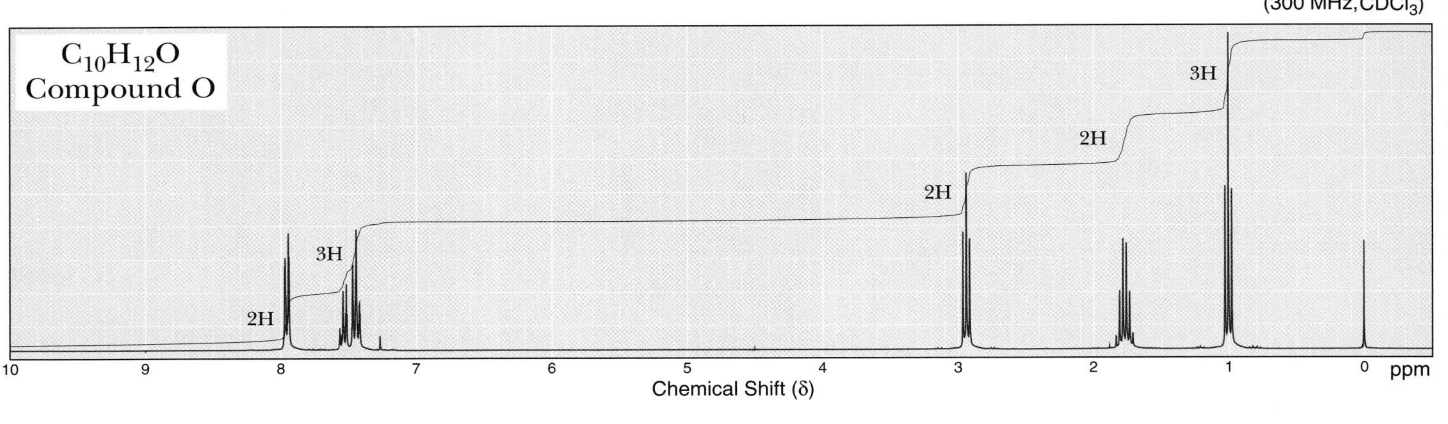

12.28 Following is a ^{1}H-NMR spectrum for compound P, with the molecular formula $C_6H_{12}O_2$. Compound P undergoes acid-catalyzed dehydration to give compound Q, $C_6H_{10}O$. Propose structural formulas for compounds P and Q. **(See Examples 12.8, 12.9)**

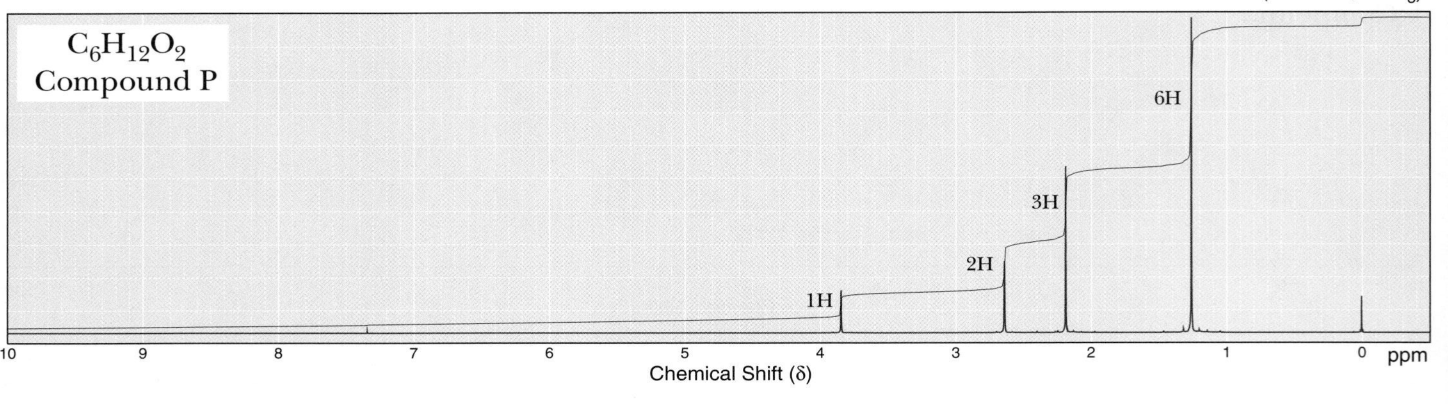

12.29 Propose a structural formula for compound R, with the molecular formula $C_{12}H_{16}O$. Following are its ^{1}H-NMR and ^{13}C-NMR spectra: **(See Examples 12.8–12.10)**

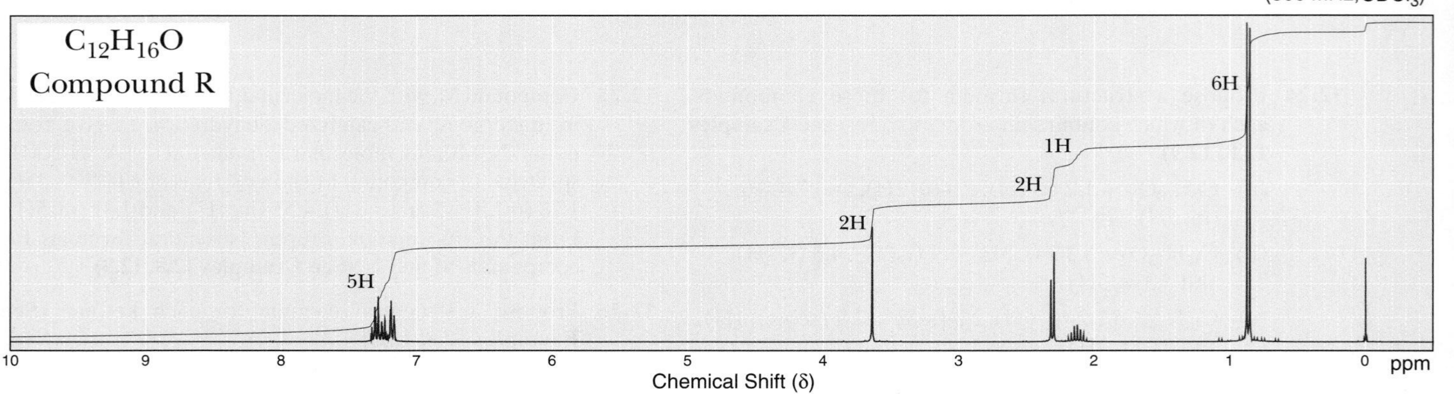

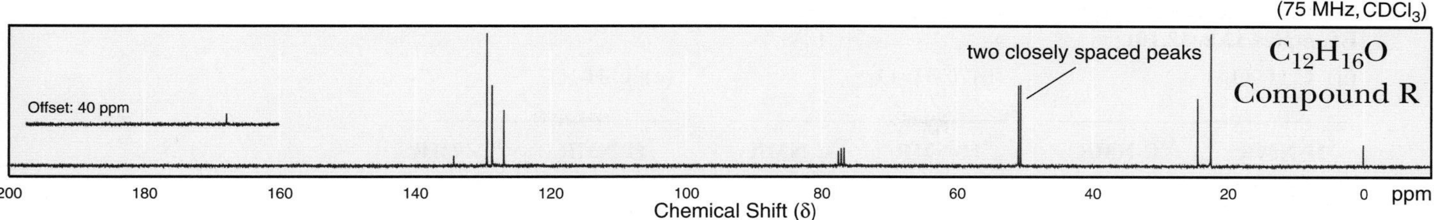

12.30 Propose a structural formula for each carboxylic acid: **(See Examples 12.8–12.10)**

(a) $C_5H_{10}O_2$

^{1}H-NMR	^{13}C-NMR
0.94 (t, 3H)	180.7
1.39 (m, 2H)	33.89
1.62 (m, 2H)	26.76
2.35 (t, 2H)	22.21
12.0 (s, 1H)	13.69

(b) $C_6H_{12}O_2$

^{1}H-NMR	^{13}C-NMR
1.08 (s, 9H)	179.29
2.23 (s, 2H)	46.82
12.1 (s, 1H)	30.62
	29.57

(c) $C_5H_8O_4$

^{1}H-NMR	^{13}C-NMR
0.93 (t, 3H)	170.94
1.80 (m, 2H)	53.28
3.10 (t, 1H)	21.90
12.7 (s, 2H)	11.81

12.31 Following are ^{1}H-NMR and ^{13}C-NMR spectra of compound S, with the molecular formula $C_7H_{14}O_2$. Propose a structural formula for compound S. **(See Examples 12.8–12.10)**

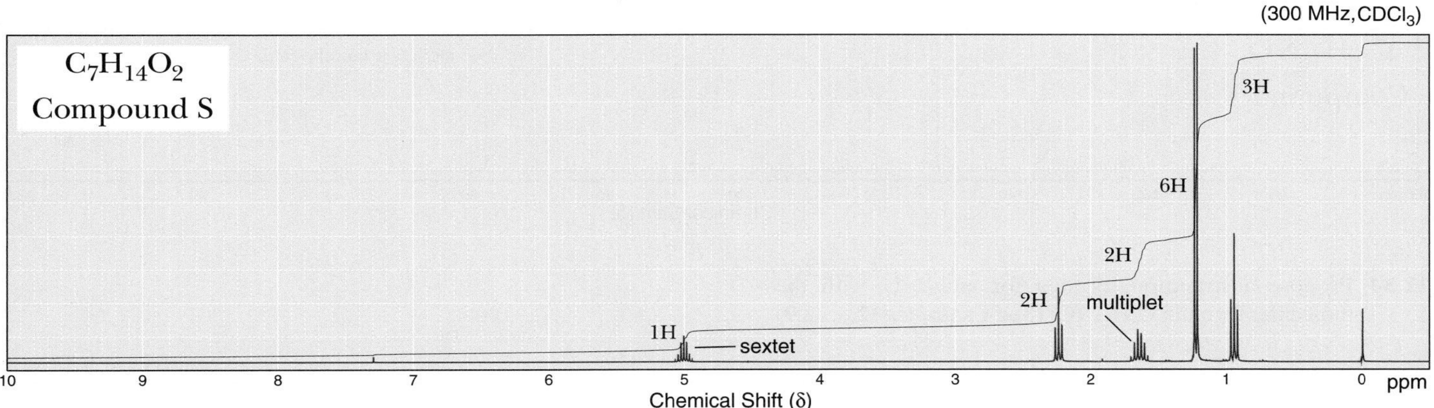

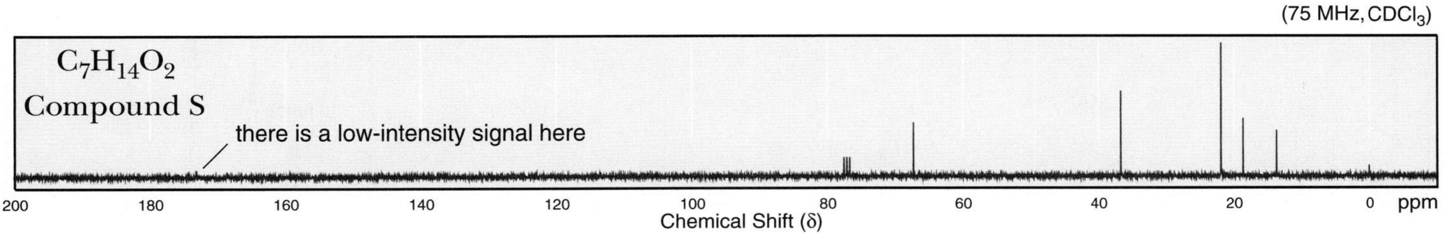

12.32 Propose a structural formula for each ester: **(See Examples 12.8–12.10)**

(a) $C_6H_{12}O_2$

^{1}H-NMR	^{13}C-NMR
1.18 (d, 6H)	177.16
1.26 (t, 3H)	60.17
2.51 (m, 1H)	34.04
4.13 (q, 2H)	19.01
	14.25

(b) $C_7H_{12}O_4$

^{1}H-NMR	^{13}C-NMR
1.28 (t, 6H)	166.52
3.36 (s, 2H)	61.43
4.21 (q, 4H)	41.69
	14.07

(c) $C_7H_{14}O_2$

^{1}H-NMR	^{13}C-NMR
0.92 (d, 6H)	171.15
1.52 (m, 2H)	63.12
1.70 (m, 1H)	37.31
2.09 (s, 3H)	25.05
4.10 (t, 2H)	22.45
	21.06

12.33 Following are ^{1}H-NMR and ^{13}C-NMR spectra of compound T, with the molecular formula $C_{10}H_{15}NO$. Propose a structural formula for this compound. **(See Examples 12.8–12.10)**

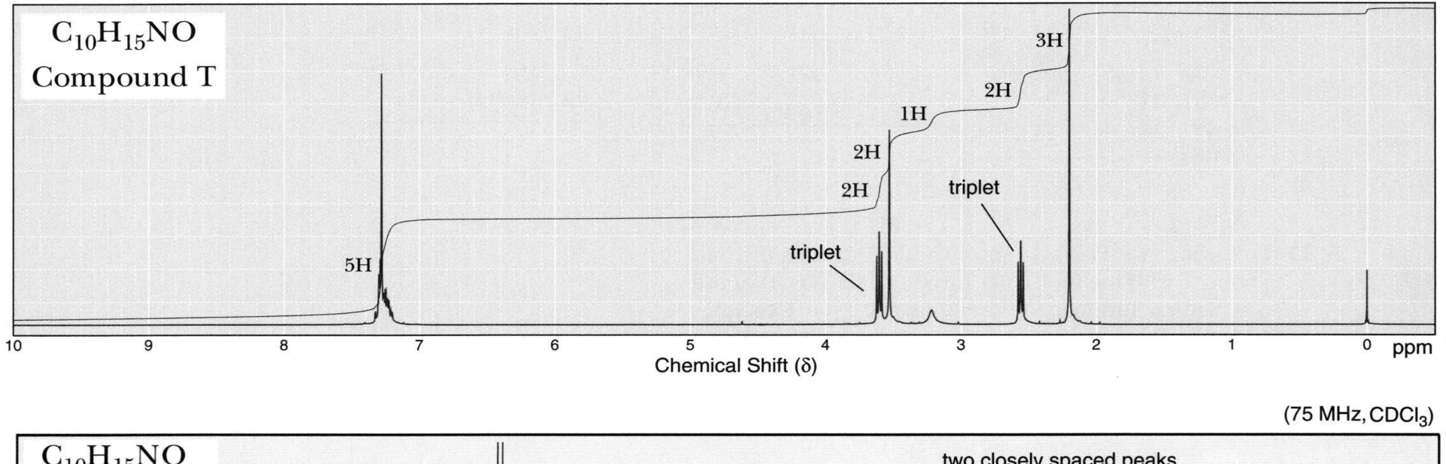

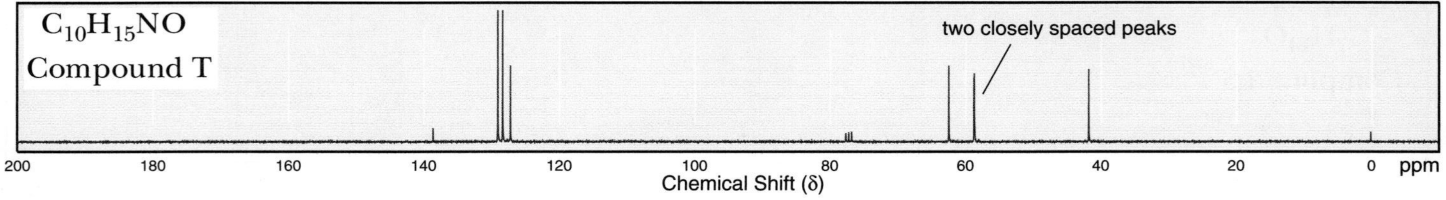

12.34 Propose a structural formula for amide U, with the molecular formula $C_6H_{13}NO$: **(See Examples 12.8, 12.9)**

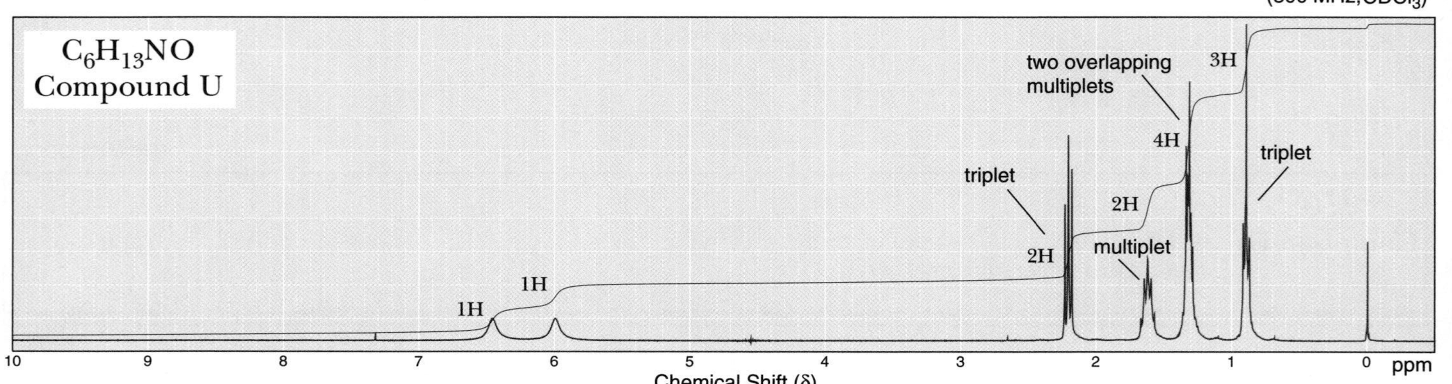

*12.35 Propose a structural formula for the analgesic phenacetin, with molecular formula $C_{10}H_{13}NO_2$, based on its ^{1}H-NMR spectrum: **(See Examples 12.8, 12.9)**

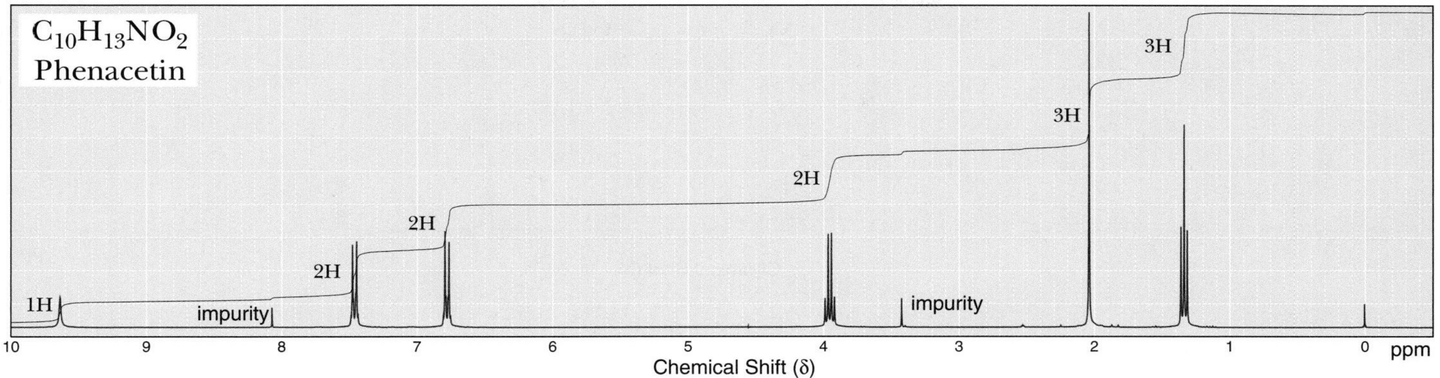

12.36 Propose a structural formula for compound V, an oily liquid with the molecular formula $C_8H_9NO_2$. Compound V is insoluble in water and aqueous NaOH, but dissolves in 10% HCl. When its solution in HCl is neutralized with NaOH, compound V is recovered unchanged. A ^{1}H-NMR spectrum of compound V shows signals at δ 3.84 (s, 3H), 4.18 (s, 2H), 7.60 (d, 2H), and 8.70 (d, 2H). **(See Examples 12.8, 12.9)**

*12.37 Following is a ^{1}H-NMR spectrum and a structural formula for anethole, $C_{10}H_{12}O$, a fragrant natural product obtained from anise. Using the line of integration, determine the number of protons giving rise to each signal. Show that this spectrum is consistent with the structure of anethole. **(See Examples 12.8, 12.9)**

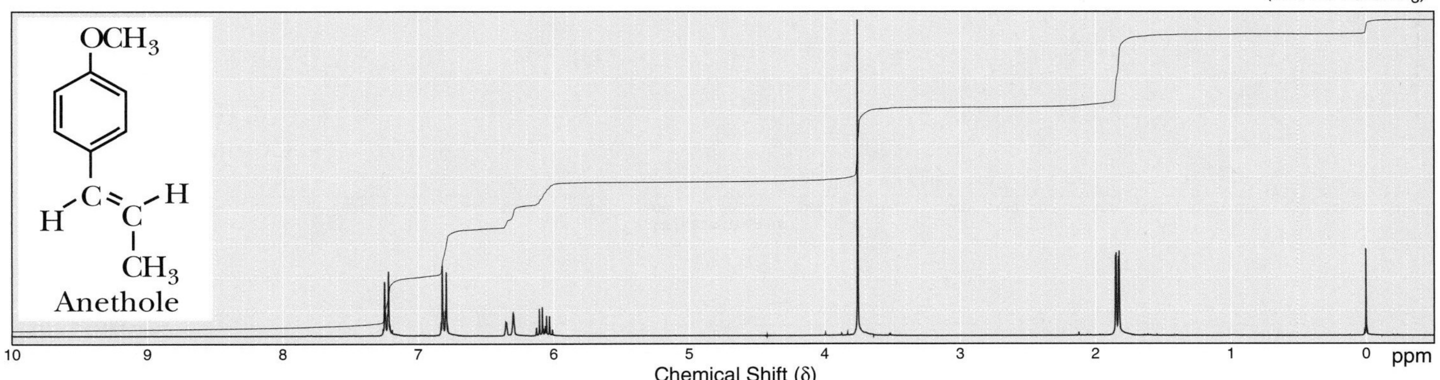

12.38 Propose a structural formula for compound W, with the molecular formula C_4H_6O, based on the following IR and ^{1}H-NMR spectra: **(See Examples 12.8, 12.9)**

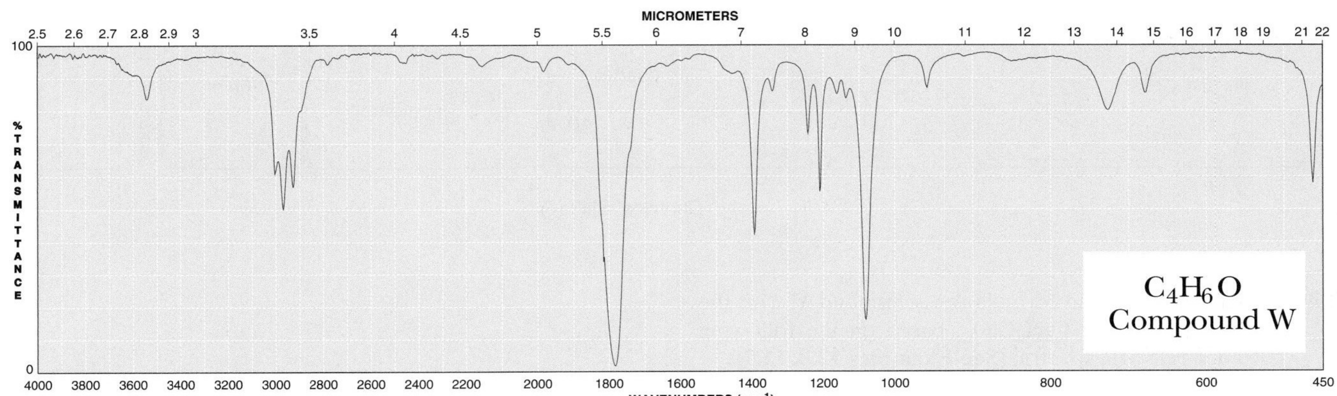

(300 MHz, $CDCl_3$)

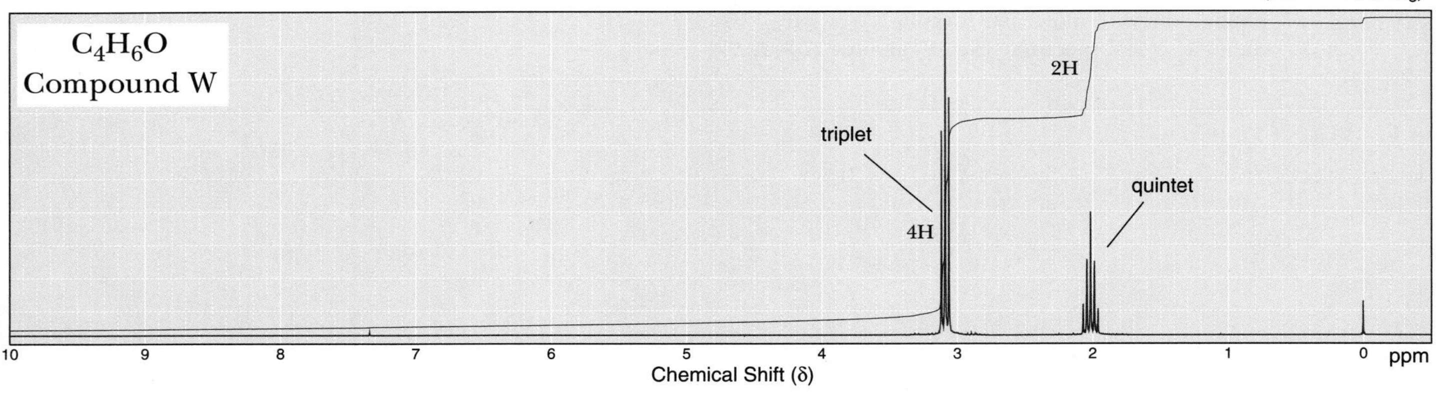

12.39 Propose a structural formula for compound X, with the molecular formula $C_5H_{10}O_2$, based on the following IR and ^1H-NMR spectra: **(See Examples 12.8, 12.9)**

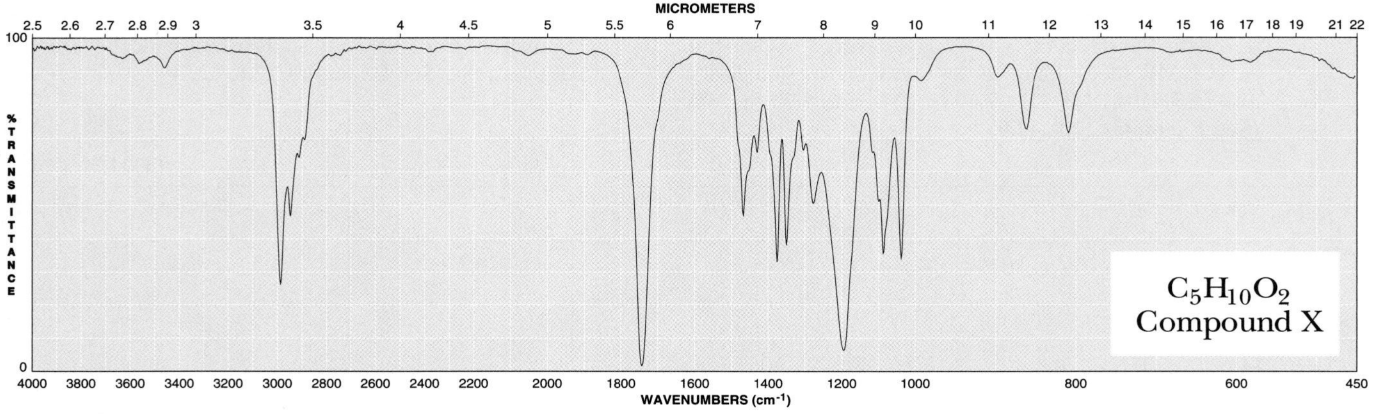

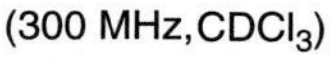

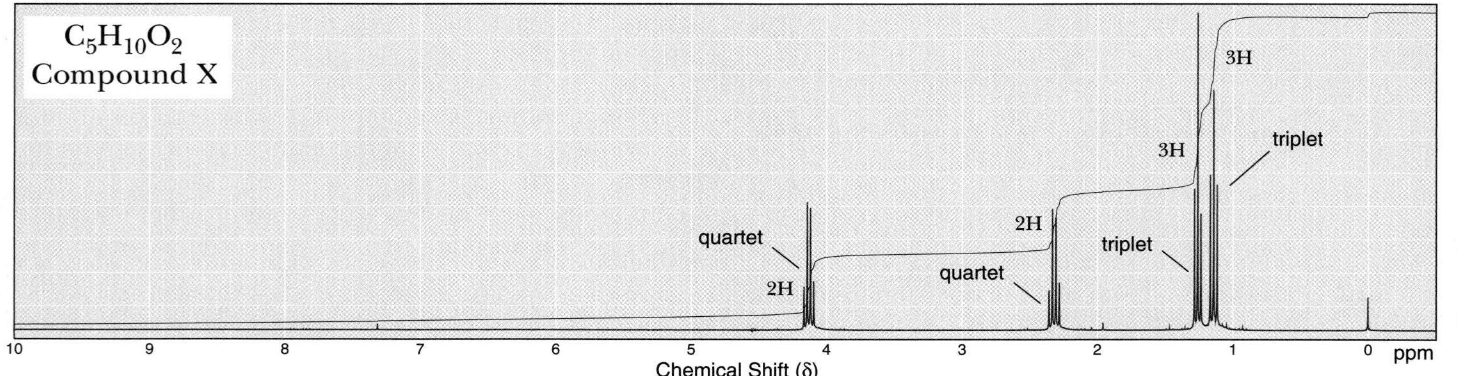

12.40 Propose a structural formula for compound Y, with the molecular formula $C_5H_9ClO_2$, based on the following IR and ^1H-NMR spectra: **(See Examples 12.8, 12.9)**

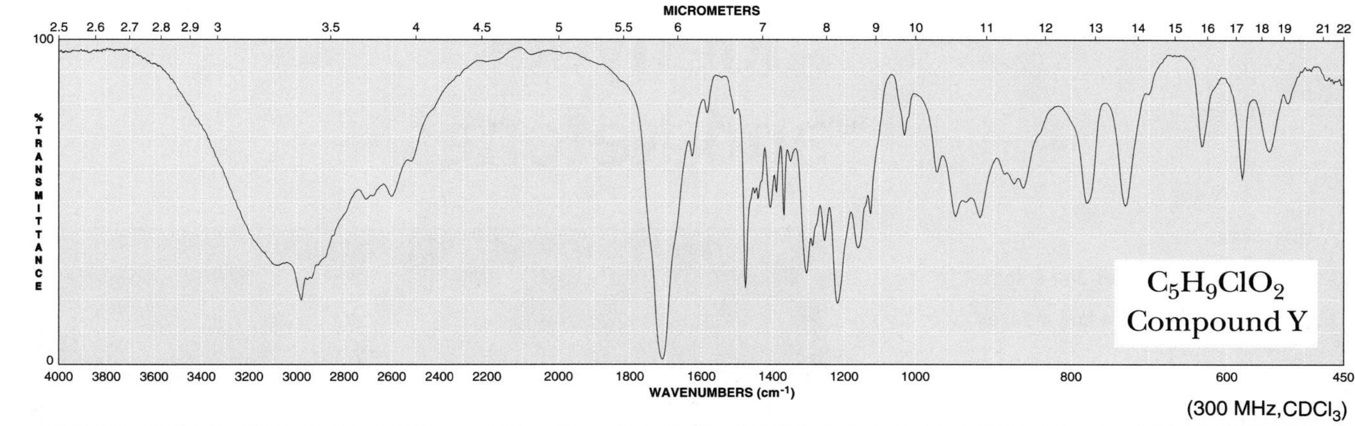

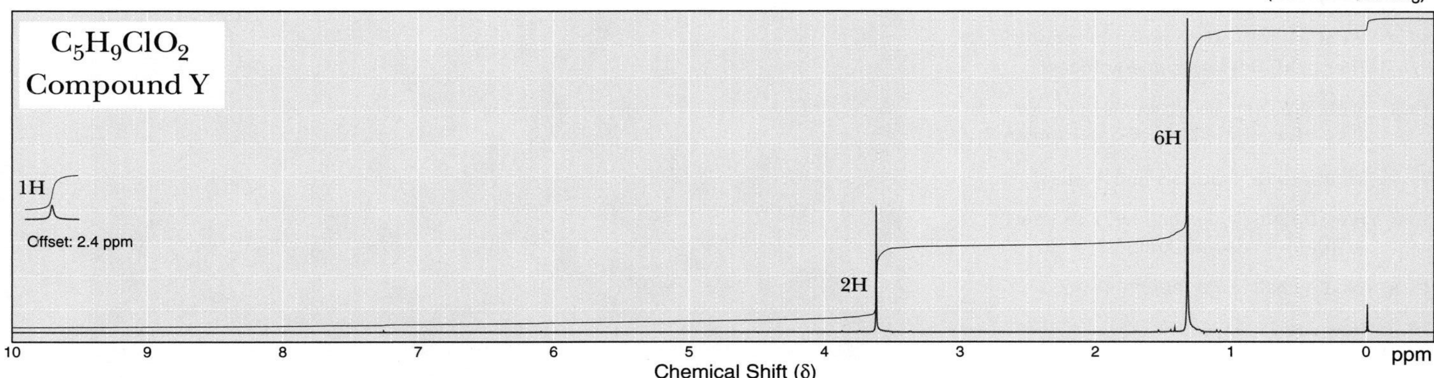

12.41 Propose a structural formula for compound Z, with the molecular formula $C_6H_{14}O$, based on the following IR and ^{1}H-NMR spectra: **(See Examples 12.8, 12.9)**

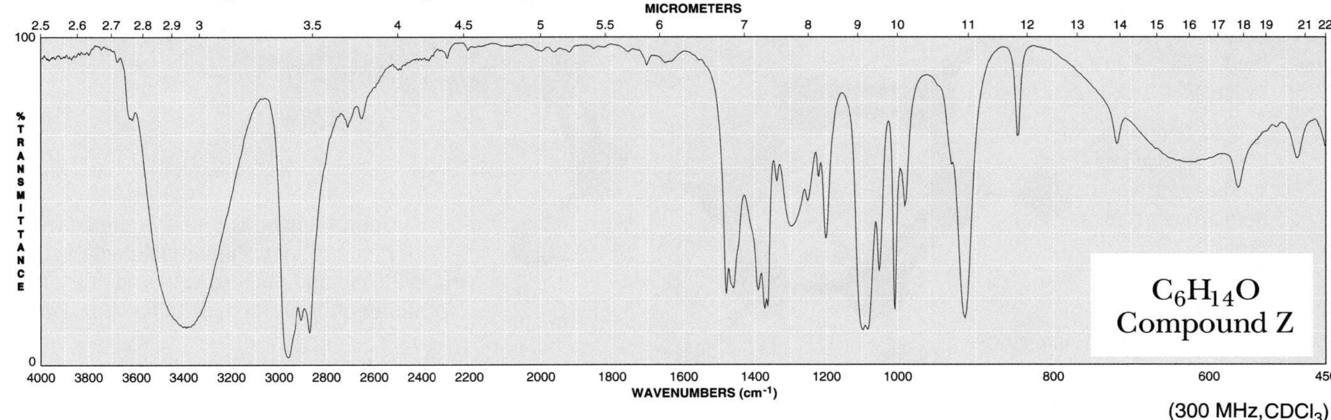

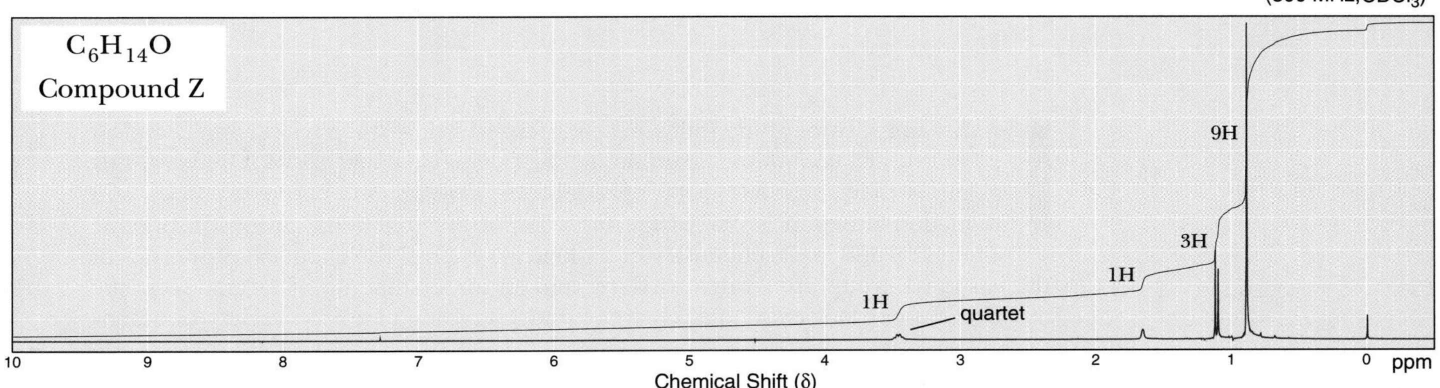

13 Aldehydes and Ketones

KEY QUESTIONS

HOW TO

CHEMICAL CONNECTIONS

Benzaldehyde is found in the kernels of bitter almonds. Cinnamaldehyde is found in Ceylonese and Chinese cinnamon oils. Inset: A model of benzaldehyde. *(Charles D. Winters)*

In this and several of the following chapters, we study the physical and chemical properties of compounds containing the carbonyl group, C=O. Because this group is the functional group of aldehydes, ketones, and carboxylic acids and their derivatives, it is one of the most important functional groups in organic chemistry and in the chemistry of biological systems. The chemical properties of the carbonyl group are straightforward, and an understanding of its characteristic reaction themes leads very quickly to an understanding of a wide variety of organic reactions.

13.1 What Are Aldehydes and Ketones?

The functional group of an **aldehyde** is a carbonyl group bonded to a hydrogen atom (Section 1.7C). In methanal (common name: formaldehyde), the simplest aldehyde, the carbonyl group is bonded to two hydrogen atoms. In other aldehydes, it is bonded to one hydrogen atom and one carbon atom. The functional group of a **ketone** is a carbonyl group bonded to two carbon atoms (Section 1.7C). Following are Lewis structures for the aldehydes methanal and ethanal, and a Lewis structure for propanone, the simplest ketone. Under each in parentheses is its common name:

$\overset{\text{O}}{\overset{\|\|}{\text{HCH}}}$	$\overset{\text{O}}{\overset{\|\|}{CH_3CH}}$	$\overset{\text{O}}{\overset{\|\|}{CH_3CCH_3}}$
Methanal (Formaldehyde)	Ethanal (Acetaldehyde)	Propanone (Acetone)

Aldehyde A compound containing a carbonyl group bonded to hydrogen (a CHO group).

Ketone A compound containing a carbonyl group bonded to two carbons.

A carbon–oxygen double bond consists of one sigma bond formed by the overlap of sp^2 hybrid orbitals of carbon and oxygen and one pi bond formed by the overlap of parallel 2p orbitals. The two nonbonding pairs of electrons on oxygen lie in the two remaining sp^2 hybrid orbitals (Figure 1.20).

Dihydroxyacetone is the active ingredient in several artificial tanning preparations. *(Andy Washnik)*

13.2 How Are Aldehydes and Ketones Named?

A. IUPAC Nomenclature

The IUPAC system of nomenclature for aldehydes and ketones follows the familiar pattern of selecting the longest chain of carbon atoms that contains the functional group as the parent alkane. We show the aldehyde group by changing the suffix *-e* of the parent alkane to *-al*, as in methanal (Section 3.5). Because the carbonyl group of an aldehyde can appear only at the end of a parent chain and numbering must start with that group as carbon-1, its position is unambiguous; there is no need to use a number to locate it.

For **unsaturated aldehydes**, the presence of a carbon–carbon double bond is indicated by the infix *-en-*. As with other molecules with both an infix and a suffix, the location of the suffix determines the numbering pattern.

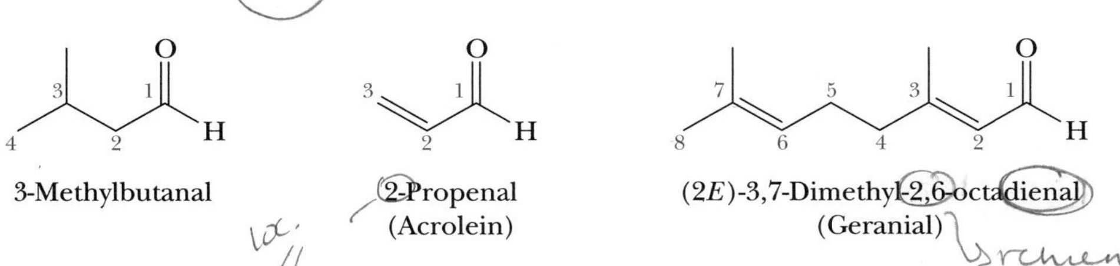

3-Methylbutanal   2-Propenal (Acrolein)   (2*E*)-3,7-Dimethyl-2,6-octadienal (Geranial)

For cyclic molecules in which —CHO is bonded directly to the ring, we name the molecule by adding the suffix *-carbaldehyde* to the name of the ring. We number the atom of the ring bearing the aldehyde group as number 1:

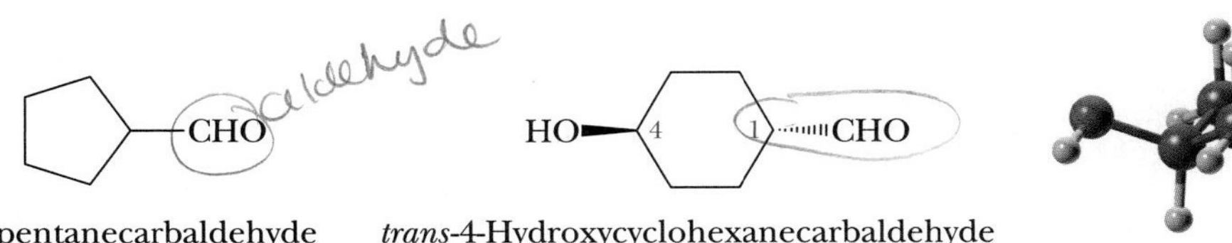

Cyclopentanecarbaldehyde   *trans*-4-Hydroxycyclohexanecarbaldehyde

Among the aldehydes for which the IUPAC system retains common names are benzaldehyde and cinnamaldehyde. Note here the alternative ways of writing the phenyl group. In benzaldehyde, it is written as a line-angle formula and in cinnamaldehyde it is abbreviated C_6H_5—.

Benzaldehyde (—CHO)

trans-3-Phenyl-2-propenal (Cinnamaldehyde): $C_6H_5(H)C{=}C(H)CHO$

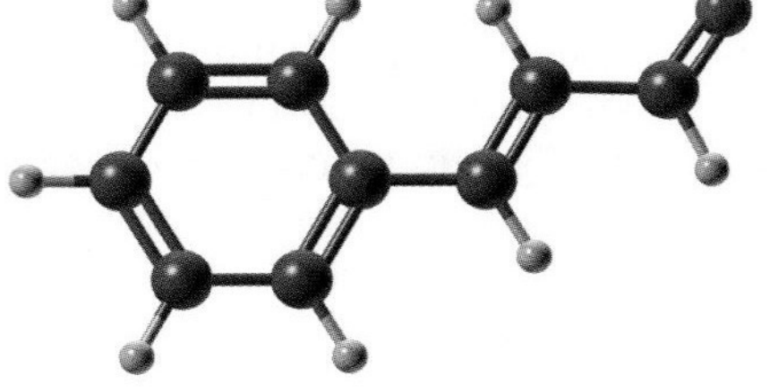

Two other aldehydes whose common names are retained in the IUPAC system are formaldehyde and acetaldehyde.

In the IUPAC system, we name ketones by selecting the longest chain that contains the carbonyl group and making that chain the parent alkane. We indicate the presence of the ketone by changing the suffix from *-e* to *-one* (Section 3.5). We number the parent chain from the direction that gives the carbonyl carbon the smaller number. The IUPAC system retains the common names acetophenone and benzophenone:

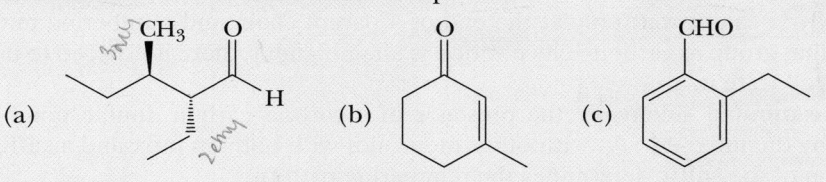

5-Methyl-3-hexanone 2-Methylcyclohexanone Acetophenone Benzophenone

Example 13.1

Write the IUPAC name for each compound:

(a) (b) (c) CHO

Strategy

First determine the root name from the longest chain of carbons that contains the carbonyl group. If the carbonyl is an aldehyde, the suffix will be *-al*. If the carbonyl is a ketone, the suffix will be *-one*. Then identify the atoms or groups of atoms that are not part of that chain of carbons. These are your substituents. If the root name indicates a ring and an aldehyde is bonded to the ring, the suffix *-carbaldehyde* is used. Finally, remember that certain aldehydes and ketones retain their common names in the IUPAC system.

Solution

(a) The longest chain has six carbons, but the longest chain that contains the carbonyl group has five carbons. The IUPAC name of this compound is (2*R*,3*R*)-2-ethyl-3-methylpentanal.
(b) Number the six-membered ring beginning with the carbonyl carbon. The IUPAC name of this compound is 3-methyl-2-cyclohexenone.
(c) This molecule is derived from benzaldehyde. Its IUPAC name is 2-ethylbenzaldehyde.

See problems 13.17, 13.18

Problem 13.1

Write the IUPAC name for each compound:

(a) 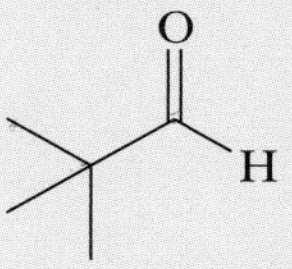 (b) 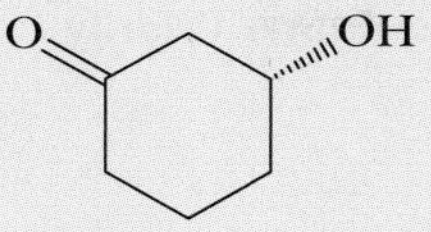 (c) 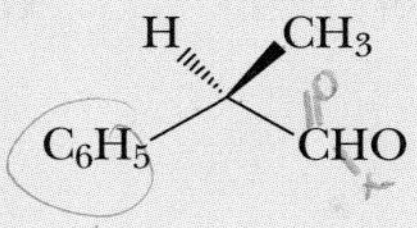 C_6H_5 H CH_3 CHO

Study

= benzyl so one C is not on the ring ⇒ count it as part of the chain, ring = C_5H_5 = phenyl

Example 13.2

Write structural formulas for all ketones with molecular formula $C_6H_{12}O$, and give each its IUPAC name. Which of these ketones are chiral?

Strategy

Start with an unbranched carbon skeleton. Place the carbonyl group, one at a time, at each position (except carbon-1). Next, consider branching possibilities, and repeat the process of placing the carbonyl at different positions. A ketone will be chiral if it has one stereocenter or if it has two or more stereocenters and is not superposable on its mirror image.

Solution

Following are line-angle formulas and IUPAC names for the six ketones with the given molecular formula:

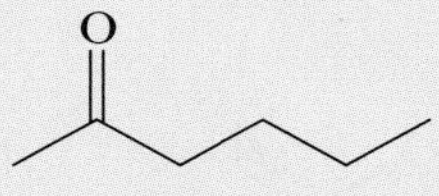

2-Hexanone 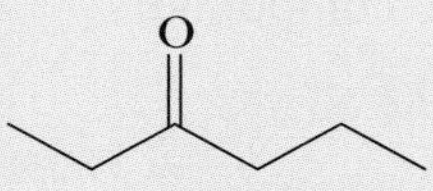 3-Hexanone 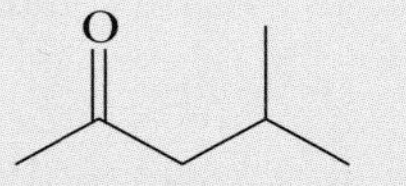 4-Methyl-2-pentanone

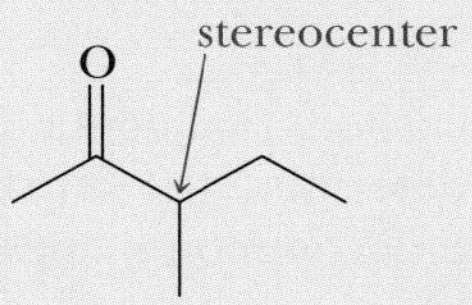

stereocenter

3-Methyl-2-pentanone 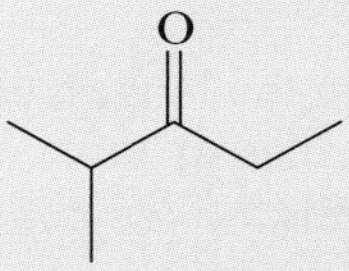 2-Methyl-3-pentanone 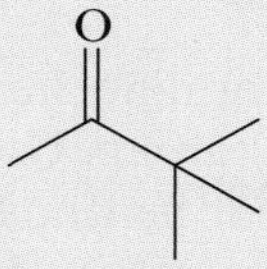 3,3-Dimethyl-2-butanone

Only 3-methyl-2-pentanone has a stereocenter and is chiral.

See problems 13.15, 13.16

Problem 13.2

Write structural formulas for all aldehydes with molecular formula $C_6H_{12}O$, and give each its IUPAC name. Which of these aldehydes are chiral?

B. IUPAC Names for More Complex Aldehydes and Ketones

In naming compounds that contain more than one functional group, the IUPAC has established an **order of precedence of functional groups**. Table 13.1 gives the order of precedence for the functional groups we have studied so far.

Order of precedence of functional groups A system for ranking functional groups in order of priority for the purposes of IUPAC nomenclature.

Table 13.1 Increasing Order of Precedence of Six Functional Groups

Functional Group	Suffix	Prefix	Example of When the Functional Group Has Lower Priority
Carboxyl	-oic acid	—	
Aldehyde	-al	oxo-	3-Oxopropanoic acid 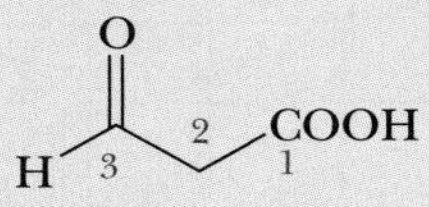
Ketone	-one	oxo-	3-Oxobutanal 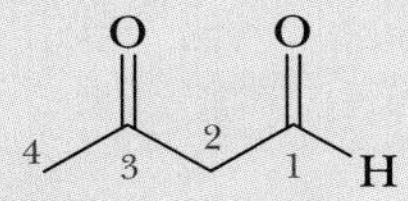
Alcohol	-ol	hydroxy-	4-Hydroxy-2-butanone 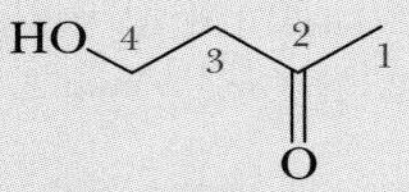
Amino	-amine	amino-	2-Amino-1-propanol 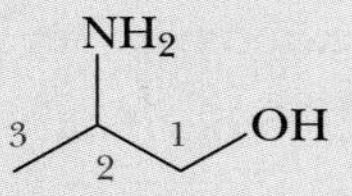
Sulfhydryl	-thiol	mercapto-	2-Mercaptoethanol 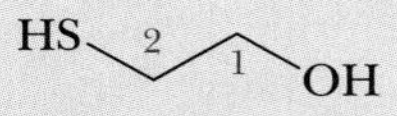

memorize

Example 13.3

Write the IUPAC name for each compound:

(a) 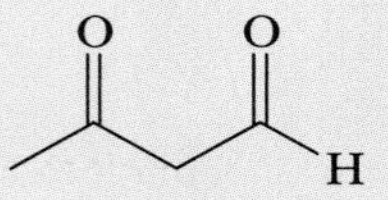 (b) 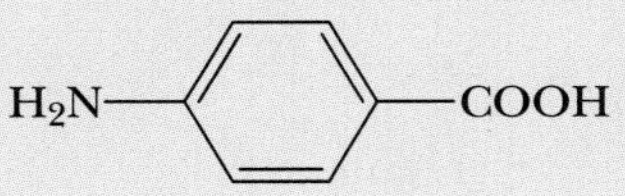 H_2N—C₆H₄—COOH (c)

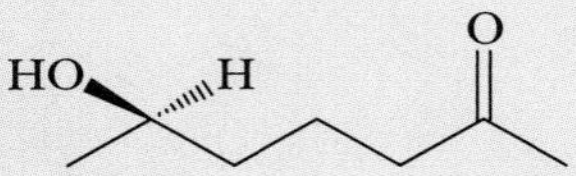

Strategy

First determine the root name from the longest chain of carbons that contains the carbonyl group. Use the priority rules in Table 13.1 to determine the suffix and prefix. For benzene ring compounds, remember to use any common names that have been retained in the IUPAC system.

Solution

(a) An aldehyde has higher precedence than a ketone, so we indicate the presence of the carbonyl group of the ketone by the prefix *oxo-*. The IUPAC name of this compound is 3-oxobutanal.

(b) The carboxyl group has higher precedence, so we indicate the presence of the amino group by the prefix *amino-*. The IUPAC name is 4-aminobenzoic acid. Alternatively, the compound may be named *p*-aminobenzoic acid, abbreviated PABA. PABA, a growth factor of microorganisms, is required for the synthesis of folic acid.

(c) The C=O group has higher precedence than the —OH group, so we indicate the —OH group by the prefix *hydroxy-*. The IUPAC name of this compound is (*R*)-6-hydroxy-2-heptanone.

See problems 13.17, 13.18

Problem 13.3

Write IUPAC names for these compounds, each of which is important in intermediary metabolism:

(a) $CH_3CH(OH)COOH$ — Lactic acid

(b) $CH_3C(=O)COOH$ — Pyruvic acid

(c) H_2N–$CH_2CH_2CH_2$–$C(=O)OH$ — γ-Aminobutyric acid

The name shown is the one by which the compound is more commonly known in the biological sciences.

C. Common Names

The common name for an aldehyde is derived from the common name of the corresponding carboxylic acid by dropping the word *acid* and changing the suffix *-ic* or *-oic* to *-aldehyde*. Because we have not yet studied common names for carboxylic acids, we are not in a position to discuss common names for aldehydes. We can, however, illustrate how they are derived by reference to two common names of carboxylic acids with which you are familiar. The name formaldehyde is derived from formic acid, the name acetaldehyde from acetic acid:

$HC(=O)H$	$HC(=O)OH$	$CH_3C(=O)H$	$CH_3C(=O)OH$
Formaldehyde	Formic acid	Acetaldehyde	Acetic acid

Common names for ketones are derived by naming each alkyl or aryl group bonded to the carbonyl group as a separate word, followed by the word *ketone*. Groups are generally listed in order of increasing atomic weight. (Methyl ethyl ketone, abbreviated MEK, is a common solvent for varnishes and lacquers):

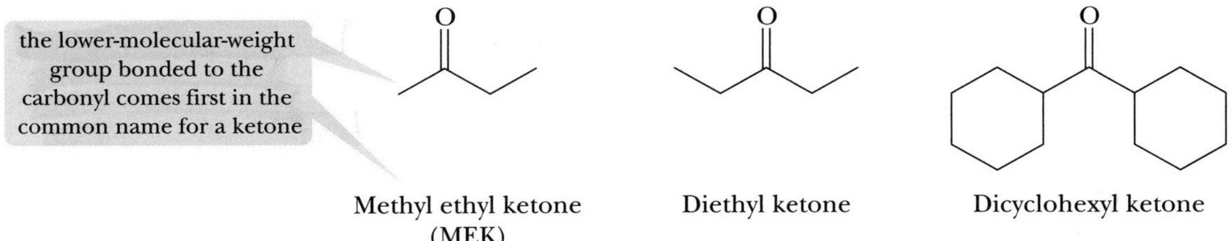
the lower-molecular-weight group bonded to the carbonyl comes first in the common name for a ketone
O
O
O

Methyl ethyl ketone (MEK) Diethyl ketone Dicyclohexyl ketone

13.3 What Are the Physical Properties of Aldehydes and Ketones?

Oxygen is more electronegative than carbon (3.5 compared with 2.5; Table 1.5); therefore, a carbon–oxygen double bond is polar, with oxygen bearing a partial negative charge and carbon bearing a partial positive charge:

the more important contributing structure

$\delta+$ $\delta-$ C=O
Polarity of a carbonyl group

C=O: $\longleftrightarrow$ +C—O:$^-$
A carbonyl group as a resonance hybrid

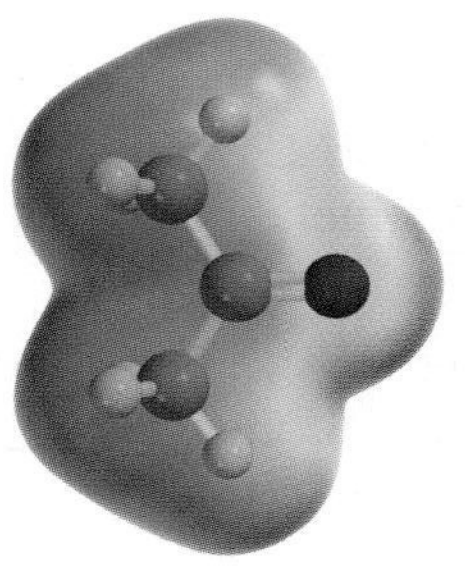

The electron density model shows that the partial positive charge on an acetone molecule is distributed both on the carbonyl carbon and on the two attached methyl groups as well.

In addition, the resonance structure on the right emphasizes that, in reactions of a carbonyl group, carbon acts as an electrophile and a Lewis acid. The carbonyl oxygen, by contrast, acts as a nucleophile and a Lewis base.

Because of the polarity of the carbonyl group, aldehydes and ketones are polar compounds and interact in the liquid state by dipole–dipole interactions. As a result, aldehydes and ketones have higher boiling points than those of nonpolar compounds of comparable molecular weight. Table 13.2 lists the boiling points of six compounds of comparable molecular weight.

Pentane and diethyl ether have the lowest boiling points of these six compounds. Both butanal and 2-butanone are polar compounds, and because of the intermolecular attraction between carbonyl groups, their boiling points are higher than those of pentane and diethyl ether. Alcohols (Section 8.1C) and carboxylic acids (Section 14.3) are polar compounds, and their molecules associate by hydrogen bonding; their boiling points are higher than those of butanal and 2-butanone, compounds whose molecules cannot associate in that manner.

Because the carbonyl groups of aldehydes and ketones interact with water molecules by hydrogen bonding, low-molecular-weight aldehydes and ketones are more soluble in water than are nonpolar compounds of comparable molecular weight. Table 13.3 lists the boiling points and solubilities in water of several low-molecular-weight aldehydes and ketones.

TABLE 13.2 Boiling Points of Six Compounds of Comparable Molecular Weight

Name	Structural Formula	Molecular Weight	Boiling Point (°C)
Diethyl ether	$CH_3CH_2OCH_2CH_3$	74	34
Pentane	$CH_3CH_2CH_2CH_2CH_3$	72	36
Butanal	$CH_3CH_2CH_2CHO$	72	76
2-Butanone	$CH_3CH_2COCH_3$	72	80
1-Butanol	$CH_3CH_2CH_2CH_2OH$	74	117
Propanoic acid	CH_3CH_2COOH	72	141

TABLE 13.3 Physical Properties of Selected Aldehydes and Ketones

IUPAC Name	Common Name	Structural Formula	Boiling Point (°C)	Solubility (g/100 g water)
Methanal	Formaldehyde	HCHO	−21	infinite
Ethanal	Acetaldehyde	CH_3CHO	20	infinite
Propanal	Propionaldehyde	CH_3CH_2CHO	49	16
Butanal	Butyraldehyde	$CH_3CH_2CH_2CHO$	76	7
Hexanal	Caproaldehyde	$CH_3(CH_2)_4CHO$	129	slight
Propanone	Acetone	CH_3COCH_3	56	infinite
2-Butanone	Methyl ethyl ketone	$CH_3COCH_2CH_3$	80	26
3-Pentanone	Diethyl ketone	$CH_3CH_2COCH_2CH_3$	101	5

although the "solubilities" of methanal and ethanal are reported as "infinite," it should be noted that 99% of initial methanal and 57% of initial ethanal are converted to compounds known as hydrates upon addition of water

$H_2C{=}O \overset{H_2O}{\rightleftharpoons} H_2C(OH)_2$

Hydrate of methanal

13.4 What Is the Most Common Reaction Theme of Aldehydes and Ketones?

The partially positive charge on the carbonyl carbon (Section 13.3) is the cause of the most common reaction theme of the carbonyl group, the addition of a nucleophile to form a **tetrahedral carbonyl addition intermediate**. In the following general reaction, the nucleophilic reagent is written as Nu:$^-$ to emphasize the presence of its unshared pair of electrons:

$$Nu{:}^- + R_2C^{\delta+}{=}O^{\delta-} \longrightarrow Nu{-}CR_2{-}O^-$$

Tetrahedral carbonyl addition intermediate

Because the carbon of the carbonyl is trigonal planar in geometry, attack of the nucleophile can occur from either side of the pi bond. This results in two possible stereoisomers if the aldehyde or ketone is not symmetrical:

attack from this face yields this stereoisomer

$$Nu{:}^- + R_1R_2C{=}O \longrightarrow R_1R_2C(Nu)O^- + R_1R_2C(Nu)O^-$$

Tetrahedral carbonyl addition intermediates

attack from this face yields this stereoisomer

13.5 What Are Grignard Reagents, and How Do They React with Aldehydes and Ketones?

From the perspective of the organic chemist, the addition of carbon nucleophiles is the most important type of nucleophilic addition to a carbonyl group because these reactions form new carbon–carbon bonds. In this section, we describe the preparation and reactions of Grignard reagents and their reaction with aldehydes and ketones.

A. Formation and Structure of Organomagnesium Compounds

Organometallic compound A compound containing a carbon–metal bond.

Grignard reagent An organomagnesium compound of the type RMgX or ArMgX.

Alkyl, aryl, and vinylic halides react with Group I, Group II, and certain other metals to form **organometallic compounds**. Within the range of organometallic compounds, organomagnesium compounds are among the most readily available, easily prepared, and easily handled. They are commonly named **Grignard reagents**, after Victor Grignard, who was awarded a 1912 Nobel Prize in Chemistry for their discovery and their application to organic synthesis.

Grignard reagents are typically prepared by the slow addition of a halide to a stirred suspension of magnesium metal in an ether solvent, most commonly diethyl ether or tetrahydrofuran (THF). Organoiodides and bromides generally react rapidly under these conditions, whereas chlorides react more slowly. Butylmagnesium bromide, for example, is prepared by adding 1-bromobutane to an ether suspension of magnesium metal. Aryl Grignards, such as phenylmagnesium bromide, are prepared in a similar manner:

1-Bromobutane + Mg $\xrightarrow{\text{ether}}$ Butylmagnesium bromide

Bromobenzene + Mg $\xrightarrow{\text{ether}}$ Phenylmagnesium bromide

in a Grignard reagent, the magnesium atom has inserted itself between the bromine and carbon atom

Given that the difference in electronegativity between carbon and magnesium is 1.3 units (2.5 − 1.2), the carbon–magnesium bond is best described as polar covalent, with carbon bearing a partial negative charge and magnesium bearing a partial positive charge. In the structural formula on the right, the carbon–magnesium bond is shown as ionic to emphasize its nucleophilic character. Note that although we can write a Grignard reagent as a **carbanion**, a more accurate representation shows it as a polar covalent compound:

Carbanion An anion in which carbon has an unshared pair of electrons and bears a negative charge.

$$CH_3(CH_2)_2\overset{H}{\underset{H}{C}}{}^{\delta-}\!-\!{}^{\delta+}MgBr \qquad CH_3(CH_2)_2\overset{H}{\underset{H}{C}}{:}^{-}\ \overset{+}{M}gBr$$

carbon is a nucleophile

The feature that makes Grignard reagents so valuable in organic synthesis is that the carbon bearing the halogen is now transformed into a nucleophile.

B. Reaction with Protic Acids

Grignard reagents are very strong bases and react readily with a wide variety of acids (proton donors) to form alkanes. Ethylmagnesium bromide, for example, reacts instantly with water to give ethane and magnesium salts. This reaction is an example of a stronger acid and a stronger base reacting to give a weaker acid and a weaker base (Section 2.4):

$$\overset{\delta-}{CH_3CH_2}-\overset{\delta+}{MgBr} + H-OH \longrightarrow CH_3CH_2-H + Mg^{2+} + OH^- + Br^-$$

Stronger base	pK_a 15.7 Stronger acid	pK_a 51 Weaker acid	Weaker base

Any compound containing an O—H, N—H, and S—H bond or a relatively acidic hydrogen will react with a Grignard reagent by proton transfer. Following are examples of compounds containing those functional groups:

HOH	ROH	ArOH	RCOOH	RNH_2	RSH	R—C≡C—H
Water	Alcohols	Phenols	Carboxylic acids	Amines	Thiols	Terminal alkynes

Because Grignard reagents react so rapidly with these proton acids, Grignard reagents cannot be made from any halogen-containing compounds that also contain them.

Example 13.4

Write an equation for the acid–base reaction between ethylmagnesium iodide and an alcohol. Use curved arrows to show the flow of electrons in this reaction. In addition, show that the reaction is an example of a stronger acid and stronger base reacting to form a weaker acid and weaker base.

Strategy

Show the reaction of the Grignard reagent with a generic alcohol (ROH) to form an alkane and a magnesium alkoxide. In drawing the mechanism, remember that the Grignard reagent reacts as a base by donating the electrons in its C—Mg bond to form a new bond to the electrophile (in this case, H^+).

Solution

The alcohol is the stronger acid, and ethyl carbanion is the stronger base:

$$CH_3CH_2-MgI + H-\ddot{O}R \longrightarrow CH_3CH_2-H + R\ddot{O}{:}^- MgI^+$$

Ethylmagnesium iodide (stronger base)	An alcohol pK_a 16–18 (stronger acid)	Ethane pK_a 51 (weaker acid)	A magnesium alkoxide (weaker base)

See problems 13.19, 13.21, 13.22

Problem 13.4

Explain how these Grignard reagents react with molecules of their own kind to "self-destruct":

(a) HO—C_6H_4—MgBr (b) HO(C=O)$CH_2CH_2CH_2CH_2$MgBr

C. Addition of Grignard Reagents to Aldehydes and Ketones

The special value of Grignard reagents is that they provide excellent ways to form new carbon–carbon bonds. In their reactions, Grignard reagents behave as carbanions. A carbanion is a good nucleophile and adds to the carbonyl group of an aldehyde or a ketone to form a tetrahedral carbonyl addition intermediate. The driving force for these reactions is the attraction of the partial negative charge on the carbon of the organometallic compound to the partial positive charge of the carbonyl carbon. In the examples that follow, the magnesium–oxygen bond, which forms after the tetrahedral carbonyl addition intermediate is formed, is written $—O^{-}[MgBr]^{+}$ to emphasize its ionic character. The alkoxide ions formed in Grignard reactions are strong bases (Section 8.2C) and form alcohols when treated with an aqueous acid such as HCl or aqueous NH_4Cl during workup.

Addition to Formaldehyde Gives a 1° Alcohol

Treatment of a Grignard reagent with formaldehyde, followed by hydrolysis in aqueous acid, gives a primary alcohol:

recall that hydronium is the active agent in HCl/H_2O

$$CH_3CH_2—MgBr + H—\underset{}{\overset{O}{\overset{\|}{C}}}—H \xrightarrow{\text{ether}} CH_3CH_2—CH_2—O^{-}[MgBr]^{+} \xrightarrow{H_3O^+} CH_3CH_2—CH_2—OH + Mg^{2+}$$

Formaldehyde — A magnesium alkoxide — 1-Propanol (a 1° alcohol)

Addition to an Aldehyde (Except Formaldehyde) Gives a 2° Alcohol

Treatment of a Grignard reagent with any aldehyde other than formaldehyde, followed by hydrolysis in aqueous acid, gives a secondary alcohol:

$$C_6H_{11}—MgBr + CH_3—\overset{O}{\overset{\|}{C}}—H \xrightarrow{\text{ether}} C_6H_{11}—CHCH_3(O^{-}[MgBr]^{+}) \xrightarrow{H_3O^+} C_6H_{11}—CHCH_3(OH) + Mg^{2+}$$

Acetaldehyde — A magnesium alkoxide — 1-Cyclohexylethanol (a 2° alcohol)

Addition to a Ketone Gives a 3° Alcohol

Treatment of a Grignard reagent with a ketone, followed by hydrolysis in aqueous acid, gives a tertiary alcohol:

$$C_6H_5—MgBr + CH_3—\overset{O}{\overset{\|}{C}}—CH_3 \xrightarrow{\text{ether}} C_6H_5—C(CH_3)_2(O^{-}[MgBr]^{+}) \xrightarrow{H_3O^+} C_6H_5—C(CH_3)_2(OH) + Mg^{2+}$$

Acetone — A magnesium alkoxide — 2-Phenyl-2-propanol (a 3° alcohol)

HOW TO 13.1 Predict the Product of a Grignard Reaction

(a) Using the fact that a Grignard reaction involves the formation of a carbon–carbon bond, identify the nucleophilic carbon (i.e., the carbon bonded to the magnesium atom).

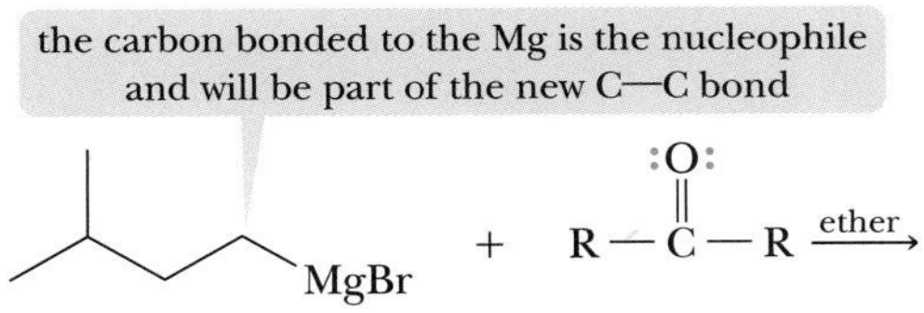

the carbon bonded to the Mg is the nucleophile and will be part of the new C—C bond

MgBr + R—C(=O)—R —ether→

(b) Check to see that there are no O—H, N—H, or S—H bonds in the reagents or solvent. These will undergo proton transfer with the Grignard reagent and prevent the reaction with the carbonyl from occurring.

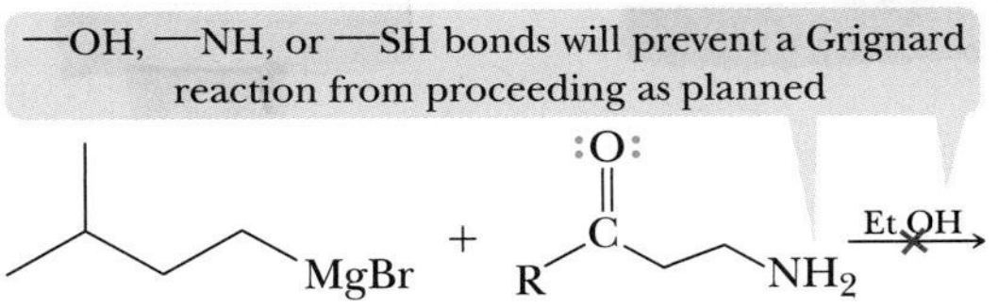

—OH, —NH, or —SH bonds will prevent a Grignard reaction from proceeding as planned

MgBr + R—C(=O)—CH₂—NH₂ —EtOH→

(c) Create a new bond between the carbon identified in Step 1 and the carbonyl carbon. The nucleophilic carbon from the Grignard reagent will no longer be bonded to MgBr. Instead, the MgBr should be shown to be ionically coordinated with the negatively charged oxygen that was part of the carbonyl. If there is a workup step, the magnesium salt is converted to an alcohol.

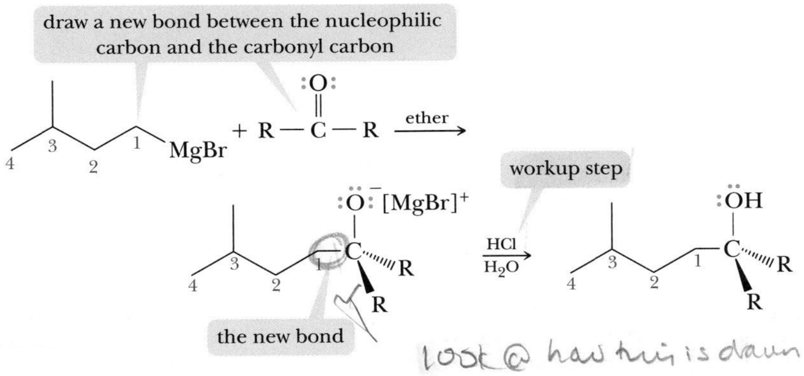

draw a new bond between the nucleophilic carbon and the carbonyl carbon

MgBr + R—C(=O)—R —ether→

workup step

$:\ddot{O}:^-[MgBr]^+$ —HCl, H_2O→ :ÖH

the new bond

Example 13.5

2-Phenyl-2-butanol can be synthesized by three different combinations of a Grignard reagent and a ketone. Show each combination.

Strategy

The Grignard reagent used to synthesize any alcohol can be determined by identifying a C—C bond connecting the alcohol carbon to the continuing carbon chain. Remove this bond, convert the C—OH to C=O and convert the other piece to a Grignard reagent.

Solution

Curved arrows in each solution show the formation of the new carbon–carbon bond and the alkoxide ion, and labels on the final product show which set of reagents forms each bond:

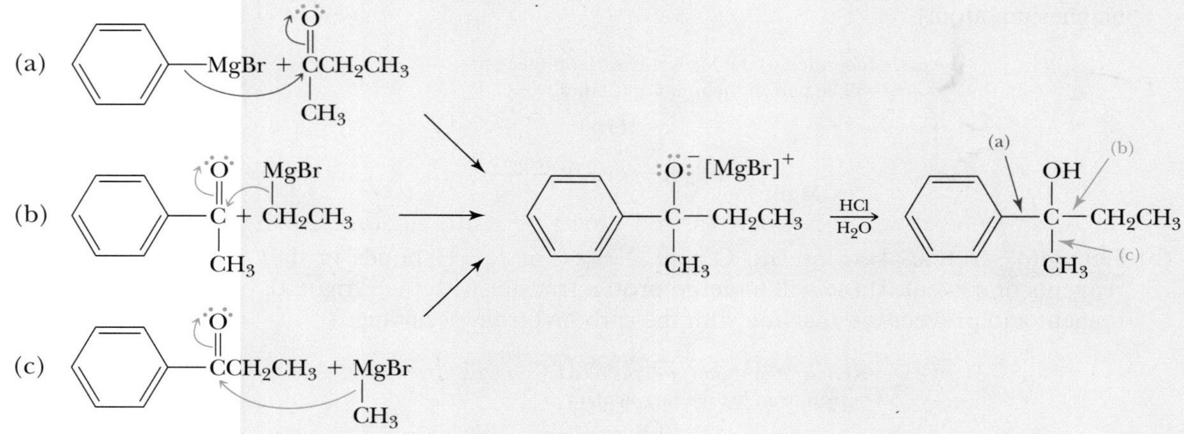

(a) $C_6H_5MgBr + CH_3C(=O)CH_2CH_3$

(b) $C_6H_5C(=O)CH_3 + CH_2CH_3MgBr$

(c) $C_6H_5C(=O)CH_2CH_3 + CH_3MgBr$

$\longrightarrow C_6H_5C(O^-[MgBr]^+)(CH_3)CH_2CH_3 \xrightarrow[H_2O]{HCl} C_6H_5C(OH)(CH_3)CH_2CH_3$

(a), (b), (c)

See problems 13.19, 13.21, 13.22

Problem 13.5

Show how these three compounds can be synthesized from the same Grignard reagent:

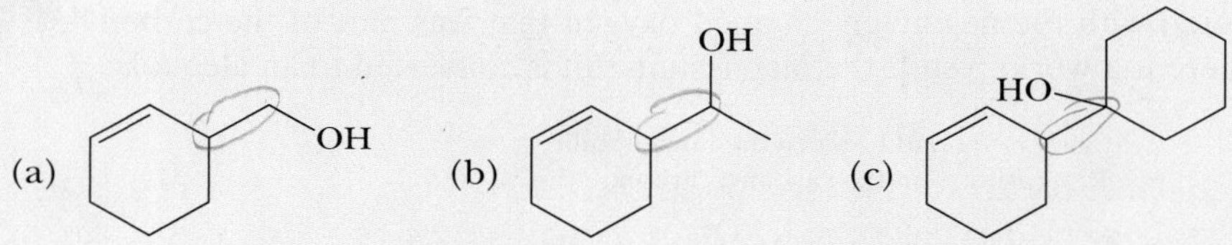

(a) OH (b) OH (c) HO

13.6 What Are Hemiacetals and Acetals?

A. Formation of Acetals

Hemiacetal A molecule containing an —OH and an —OR or —OAr group bonded to the same carbon.

The addition of a molecule of alcohol to the carbonyl group of an aldehyde or a ketone forms a **hemiacetal** (a half-acetal). This reaction is catalyzed by both acid and base: Oxygen adds to the carbonyl carbon and hydrogen adds to the carbonyl oxygen:

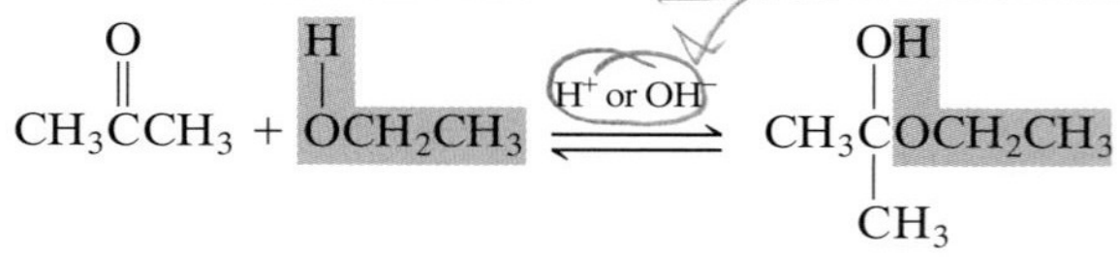

$$CH_3CCH_3(=O) + H-OCH_2CH_3 \xrightleftharpoons{H^+ \text{ or } OH^-} CH_3C(OH)(CH_3)OCH_2CH_3$$

A hemiacetal

The functional group of a hemiacetal is a carbon bonded to an —OH group and an —OR or —OAr group:

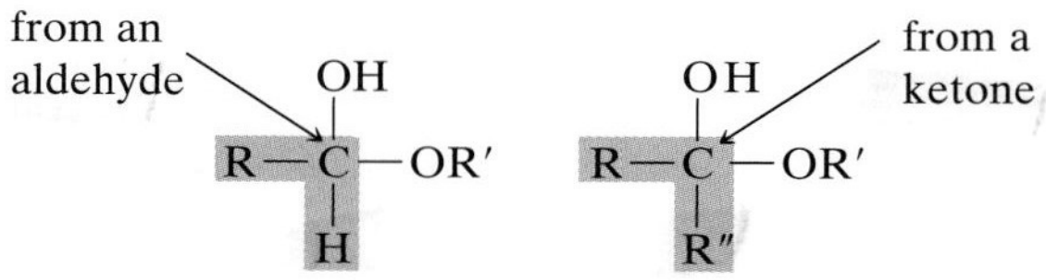

from an aldehyde: $R-C(OH)(H)-OR'$

from a ketone: $R-C(OH)(R'')-OR'$

Hemiacetals

The mechanism for the base-catalyzed conversion of an aldehyde or a ketone to a hemiacetal can be divided into three steps. Note that the base OH^- is a true catalyst in this reaction; it is used in Step 1, but a replacement OH^- is generated in Step 3.

Mechanism: Base-Catalyzed Formation of a Hemiacetal

Step 1: Proton transfer from the alcohol to the base gives an alkoxide ion:

$$CH_3\overset{O}{\overset{\|}{C}}CH_3 + H{-}\ddot{O}CH_2CH_3 + :\ddot{O}H^- \rightleftharpoons CH_3\overset{O}{\overset{\|}{C}}CH_3 + {}^-:\ddot{O}CH_2CH_3 + H_2O$$

alcohol

An alkoxide ion

Step 2: Addition of the alkoxide ion to the carbonyl gives a tetrahedral carbonyl addition intermediate:

$$CH_3\overset{\ddot{O}:}{\overset{\|}{C}}CH_3 + {}^-:\ddot{O}CH_2CH_3 + H_2O \rightleftharpoons CH_3\underset{CH_3}{\underset{|}{\overset{:\ddot{O}:^-}{\overset{|}{C}}}}\ddot{O}CH_2CH_3 + H_2O$$

Tetrahedral carbonyl addition intermediate

Step 3: Proton transfer from water to the tetrahedral carbonyl addition intermediate gives the hemiacetal and regenerates the hydroxide catalyst:

$$CH_3\underset{CH_3}{\underset{|}{\overset{:\ddot{O}:^-}{\overset{|}{C}}}}\ddot{O}CH_2CH_3 + H{-}\ddot{O}{-}H \rightleftharpoons CH_3\underset{CH_3}{\underset{|}{\overset{OH}{\overset{|}{C}}}}OCH_2CH_3 + {}^-:\ddot{O}{-}H$$

The mechanism for the acid-catalyzed conversion of an aldehyde or ketone to a hemiacetal can be divided into three steps. Note that the acid H—A is a true catalyst in this reaction: it is used in Step 1, but a replacement H—A is generated in Step 3.

Mechanism: Acid-Catalyzed Formation of a Hemiacetal

Step 1: Proton transfer from H—A to the carbonyl gives a resonance-stabilized cation. The more significant resonance structure places the positive charge on the carbon:

$$CH_3\overset{\ddot{O}:}{\overset{\|}{C}}CH_3 \underset{}{\overset{H-A}{\rightleftharpoons}} \left[CH_3\overset{\overset{+}{\ddot{O}}-H}{\overset{\|}{C}}-CH_3 \longleftrightarrow CH_3\underset{+}{\overset{:\ddot{O}-H}{\overset{|}{C}}}-CH_3 \right] + A:^-$$

A resonance-stabilized cation

Step 2: Addition of the alcohol to the resonance-stabilized cation gives an oxonium ion. *Note:* The attack of the alcohol can be to either contributing structure:

$$\left[CH_3\overset{\overset{+}{\ddot{O}}-H}{\overset{\|}{C}}-CH_3 \longleftrightarrow CH_3\underset{+}{\overset{:\ddot{O}-H}{\overset{|}{C}}}-CH_3 \right] + H{-}\ddot{O}CH_2CH_3 \rightleftharpoons CH_3\underset{H-\overset{+}{\ddot{O}}CH_2CH_3}{\underset{|}{\overset{OH}{\overset{|}{C}}}}CH_3$$

An oxonium ion

Step 3: Proton transfer from the oxonium ion to A^- gives the hemiacetal and regenerates the acid catalyst:

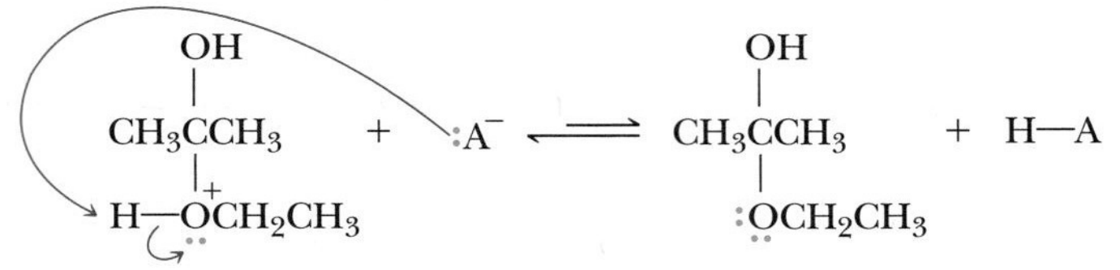

$$CH_3C(OH)(CH_3)\text{—}\overset{+}{O}H\text{—}CH_2CH_3 + :A^- \rightleftharpoons CH_3C(OH)(CH_3)\text{—}\ddot{O}CH_2CH_3 + H\text{—}A$$

Hemiacetals are generally unstable and are only minor components of an equilibrium mixture, except in one very important type of molecule. When a hydroxyl group is part of the same molecule that contains the carbonyl group, and a five- or six-membered ring can form, the compound exists almost entirely in a cyclic hemiacetal form:

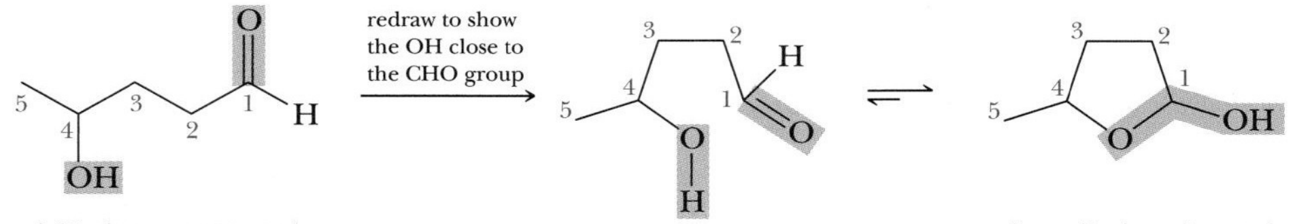

redraw to show the OH close to the CHO group

4-Hydroxypentanal

A cyclic hemiacetal (major form present at equilibrium)

We shall have much more to say about cyclic hemiacetals when we consider the chemistry of carbohydrates in Chapter 18.

Acetal A molecule containing two —OR or —OAr groups bonded to the same carbon.

Hemiacetals can react further with alcohols to form **acetals** plus a molecule of water. This reaction is acid catalyzed:

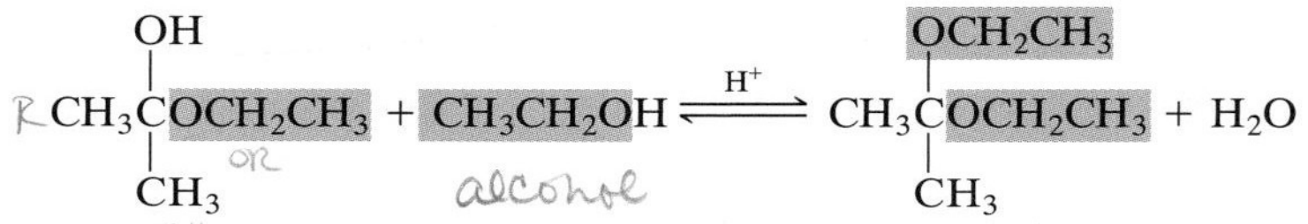

$$CH_3C(OH)(CH_3)OCH_2CH_3 + CH_3CH_2OH \overset{H^+}{\rightleftharpoons} CH_3C(OCH_2CH_3)(CH_3)OCH_2CH_3 + H_2O$$

A hemiacetal — A diethyl acetal

The functional group of an acetal is a carbon bonded to two —OR or —OAr groups:

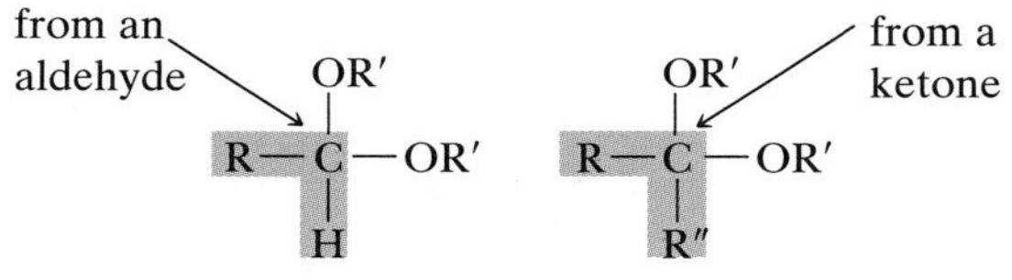

from an aldehyde: $R\text{—}C(OR')(H)\text{—}OR'$ — from a ketone: $R\text{—}C(OR')(R'')\text{—}OR'$

Acetals

The mechanism for the acid-catalyzed conversion of a hemiacetal to an acetal can be divided into four steps. Note that acid H—A is a true catalyst in this reaction; it is used in Step 1, but a replacement H—A is generated in Step 4.

Mechanism: Acid-Catalyzed Formation of an Acetal

Step 1: Proton transfer from the acid, H—A, to the hemiacetal OH group gives an oxonium ion:

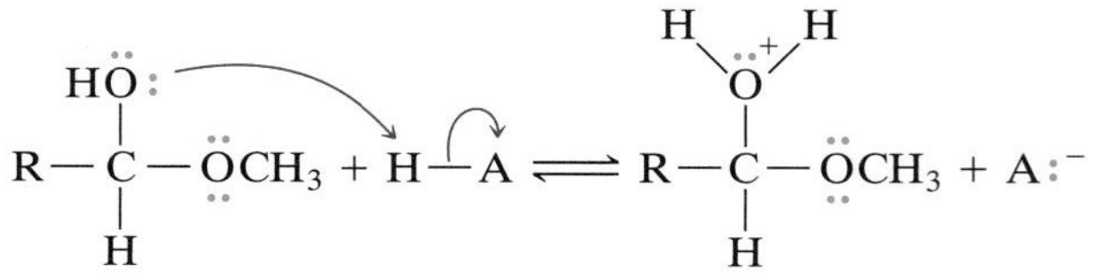

$$R\text{—}C(\ddot{O}H)(H)\text{—}\ddot{O}CH_3 + H\text{—}A \rightleftharpoons R\text{—}C(\overset{+}{O}H_2)(H)\text{—}\ddot{O}CH_3 + A:^-$$

An oxonium ion

Step 2: Loss of water from the oxonium ion gives a resonance-stabilized cation:

$$R-\underset{H}{\overset{H-\overset{+}{O}-H}{C}}-\ddot{O}CH_3 \rightleftharpoons R-\underset{H}{C}=\overset{+}{O}CH_3 \longleftrightarrow R-\underset{H}{\overset{+}{C}}-\ddot{O}CH_3 + H_2\ddot{O}:$$

A resonance-stabilized cation

Step 3: Reaction of the resonance-stabilized cation (an electrophile) with methanol (a nucleophile) gives the conjugate acid of the acetal:

$$CH_3-\underset{}{\overset{H}{\ddot{O}}}: + R-\underset{H}{C}=\overset{+}{O}CH_3 \rightleftharpoons R-\underset{H}{\overset{H-\overset{+}{O}-CH_3}{C}}-\ddot{O}CH_3$$

A protonated acetal

Step 4: Proton transfer from the protonated acetal to A^- gives the acetal and generates a new molecule of H—A, the acid catalyst:

$$A:^- + R-\underset{H}{\overset{H-\overset{+}{O}-CH_3}{C}}-\ddot{O}CH_3 \rightleftharpoons HA + R-\underset{H}{\overset{:\ddot{O}-CH_3}{C}}-\ddot{O}CH_3$$

A protonated acetal — An acetal

Formation of acetals is often carried out using the alcohol as a solvent and dissolving either dry HCl (hydrogen chloride) or arenesulfonic acid (Section 9.6B), $ArSO_3H$, in the alcohol. Because the alcohol is both a reactant and the solvent, it is present in large molar excess, which drives the reaction to the right and favors acetal formation. Alternatively, the reaction may be driven to the right by the removal of water as it is formed:

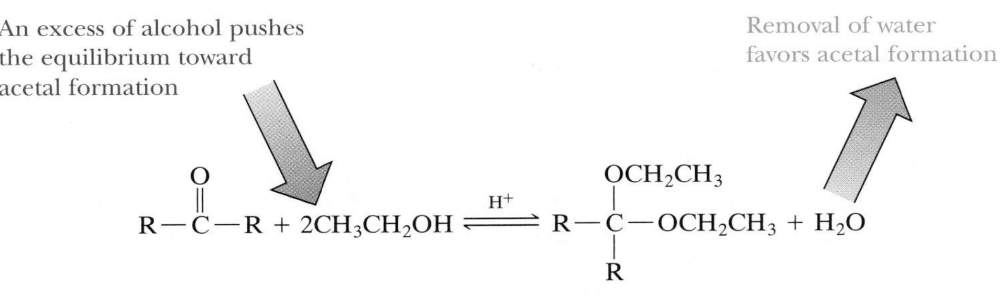
An excess of alcohol pushes the equilibrium toward acetal formation
Removal of water favors acetal formation
R—C(=O)—R + 2CH3CH2OH ⇌(H+) R—C(OCH2CH3)(R)—OCH2CH3 + H2O

A diethyl acetal

HOW TO 13.2 Determine the Reactants Used to Synthesize a Hemiacetal or Acetal

(a) Identify the carbon atom that is bonded to two oxygen atoms. This carbon atom is the carbonyl carbon that was converted to the carbon of the acetal or hemiacetal group.

the carbon that is bonded to two oxygen atoms is the former carbonyl carbon

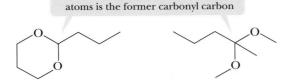

(b) Remove both C—O bonds and add back a hydrogen to each oxygen to obtain the alcohol reagent(s) used. Then convert the carbon identified in Step 1 to a carbonyl group.

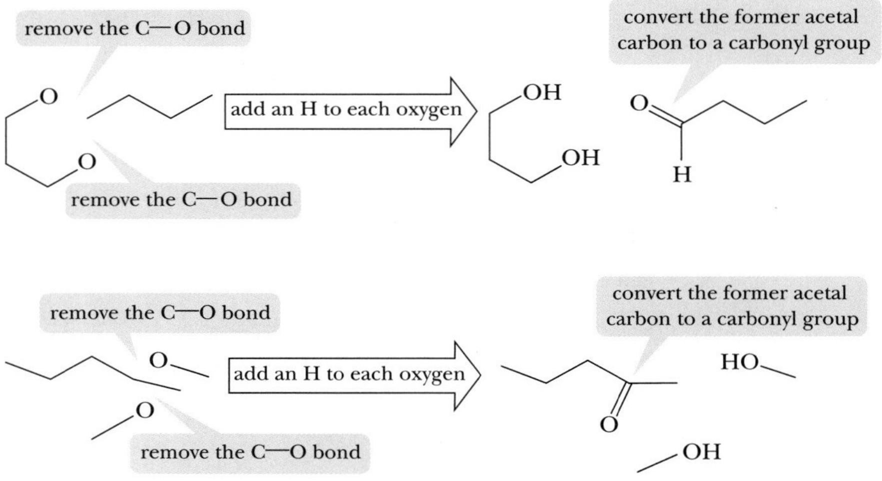

remove the C—O bond

add an H to each oxygen

convert the former acetal carbon to a carbonyl group

remove the C—O bond

remove the C—O bond

add an H to each oxygen

convert the former acetal carbon to a carbonyl group

remove the C—O bond

Example 13.6

Show the reaction of the carbonyl group of each ketone with one molecule of alcohol to form a hemiacetal and then with a second molecule of alcohol to form an acetal (note that, in part (b), ethylene glycol is a diol, and one molecule of it provides both —OH groups):

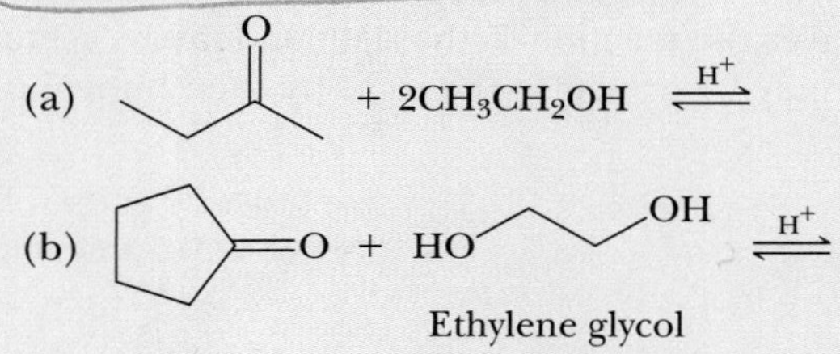

(a) + $2CH_3CH_2OH$ $\overset{H^+}{\rightleftharpoons}$

(b) + HO—OH $\overset{H^+}{\rightleftharpoons}$

Ethylene glycol

Strategy

In forming the hemiacetal, one molecule of the alcohol is added to the carbonyl carbon, resulting in an OR group and an OH group bonded to the carbon that was previously part of the carbonyl. In forming an acetal, two molecules of the alcohol are added to the carbonyl carbon, resulting in two OR groups bonded to the carbon that was previously part of the carbonyl.

Solution

Here are structural formulas of the hemiacetal and then the acetal:

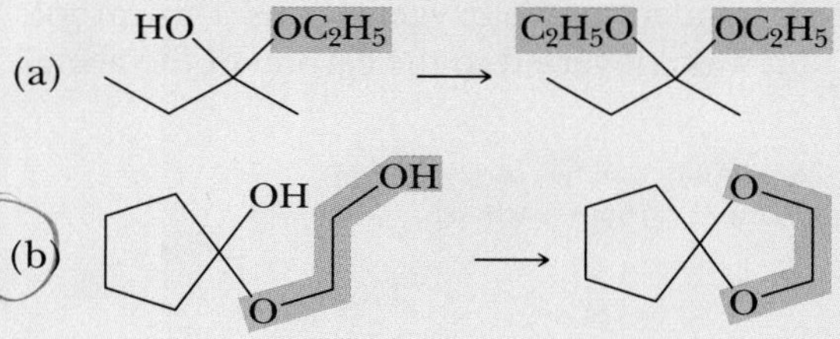

(a) HO OC_2H_5 ⟶ C_2H_5O OC_2H_5

(b) OH OH O ⟶ O O

See problems 13.23–13.26

Problem 13.6

The hydrolysis of an acetal forms an aldehyde or a ketone and two molecules of alcohol. Following are structural formulas for three acetals:

(a) (b) (c)

Draw the structural formulas for the products of the hydrolysis of each in aqueous acid.

Like ethers, acetals are unreactive to bases, to reducing agents such as H_2/M, to Grignard reagents, and to oxidizing agents (except, of course, those which involve aqueous acid). Because of their lack of reactivity toward these reagents, acetals are often used to protect the carbonyl groups of aldehydes and ketones while reactions are carried out on functional groups in other parts of the molecule.

B. Acetals as Carbonyl-Protecting Groups

The use of acetals as carbonyl-protecting groups is illustrated by the synthesis of 5-hydroxy-5-phenylpentanal from benzaldehyde and 4-bromobutanal:

Benzaldehyde + 4-Bromobutanal ⟶?? 5-Hydroxy-5-phenylpentanal

One obvious way to form a new carbon–carbon bond between these two molecules is to treat benzaldehyde with the Grignard reagent formed from 4-bromobutanal. This Grignard reagent, however, would react immediately with the carbonyl group of another molecule of 4-bromobutanal, causing it to self-destruct during preparation (Section 13.4B). A way to avoid this problem is to protect the carbonyl group of 4-bromobutanal by converting it to an acetal. Cyclic acetals are often used because they are particularly easy to prepare.

the carbonyl is protected by converting it to an acetal

Br(CH₂)₃CHO + Ethylene glycol $\overset{H^+}{\rightleftharpoons}$ A cyclic acetal + H_2O

Treatment of the protected bromoaldehyde with magnesium in diethyl ether, followed by the addition of benzaldehyde, gives a magnesium alkoxide:

the protected carbonyl will not react with any of the reagents used in this synthesis

A cyclic acetal $\xrightarrow[\text{Ether}]{\text{Mg}}$ A Grignard reagent $\xrightarrow{\text{Benzaldehyde}}$ A magnesium alkoxide

Treatment of the magnesium alkoxide with aqueous acid accomplishes two things. First, protonation of the alkoxide anion gives the desired hydroxyl group, and then, **hydrolysis** of the cyclic acetal regenerates the aldehyde group:

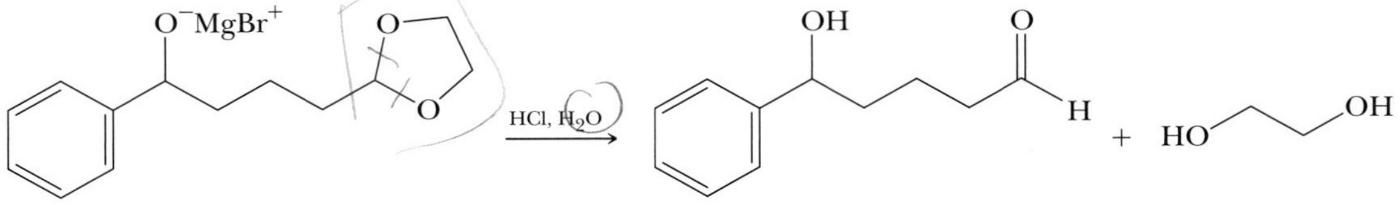

$\xrightarrow{HCl,\ H_2O}$ + HO–CH₂CH₂–OH

Example 13.7

Propose a method for the following transformation. *Note*: Catalytic hydrogenation affects carbonyls as well as C—C double bonds.

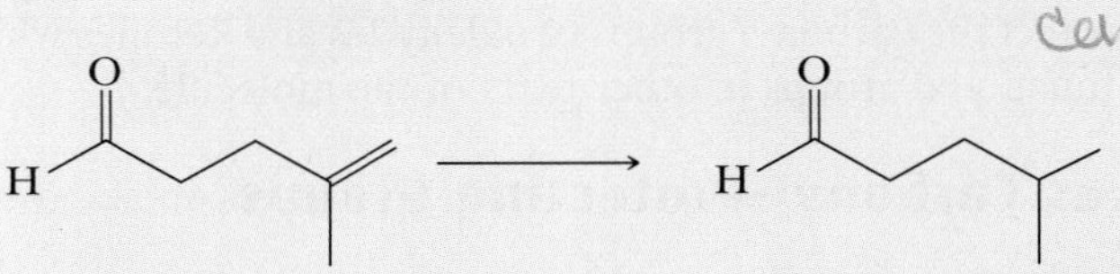

Strategy

Decide which reaction(s) are needed to achieve the interconversion of functional groups. Before applying any reaction to the targeted functional group, determine whether any other functional groups in the compound will react with the reagents proposed. If these other reactions are undesirable, determine whether the functional groups can be protected.

Solution

It is important to protect the carbonyl group. Otherwise, it will be reduced to an alcohol by H_2/Pt (Section 13.10A):

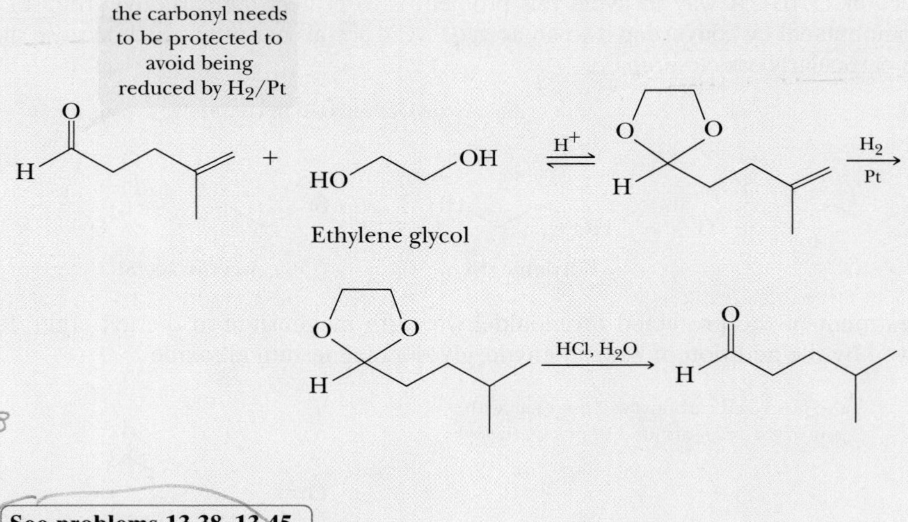

the carbonyl needs to be protected to avoid being reduced by H_2/Pt

Ethylene glycol

$\xrightleftharpoons{H^+}$ $\xrightarrow[Pt]{H_2}$ $\xrightarrow{HCl,\ H_2O}$

See problems 13.38–13.45

Problem 13.7

Propose a method for the following transformation:

$$\text{2-(hydroxymethyl)cyclohexanone} \longrightarrow \text{2-(1-hydroxy-1-phenylmethyl)cyclohexanone}$$

13.7 How Do Aldehydes and Ketones React with Ammonia and Amines?

A. Formation of Imines

Ammonia, primary aliphatic amines (RNH_2), and primary aromatic amines ($ArNH_2$) react with the carbonyl group of aldehydes and ketones in the presence of an acid catalyst to give a product that contains a carbon–nitrogen double bond. A molecule containing a carbon–nitrogen double bond is called an **imine** or, alternatively, a **Schiff base**:

Imine A compound containing a carbon–nitrogen double bond; also called a Schiff base.

Schiff base An alternative name for an imine.

$$\underset{\text{Ethanal}}{CH_3\overset{O}{\overset{\|}{C}}H} + \underset{\text{Aniline}}{H_2N{-}C_6H_5} \underset{}{\overset{H^+}{\rightleftharpoons}} \underset{\substack{\text{An imine}\\\text{(A Schiff base)}}}{CH_3CH{=}N{-}C_6H_5} + H_2O$$

$$\underset{\text{Cyclohexanone}}{C_6H_{10}{=}O} + \underset{\text{Ammonia}}{NH_3} \overset{H^+}{\rightleftharpoons} \underset{\substack{\text{An imine}\\\text{(A Schiff base)}}}{C_6H_{10}{=}NH} + H_2O$$

As with hemiacetal- and acetal-forming reactions, imine formation is reversible; acid-catalyzed hydrolysis of an imine gives a 1° amine and an aldehyde or a ketone. When one equivalent of acid is used, the 1° amine, a weak base, is converted to an ammonium salt.

$$\underset{\substack{\text{An imine}\\\text{(a Schiff base)}}}{C_6H_{10}{=}NCH_3} \xrightarrow[H_2O]{HCl} \underset{\text{Cyclohexanone}}{C_6H_{10}{=}O} + \underset{\text{Ammonium salt}}{\overset{+}{N}H_3CH_3Cl^-}$$

Mechanism: Formation of an Imine from an Aldehyde or a Ketone

Step 1: Addition of the nitrogen atom of ammonia or a primary amine, both good nucleophiles, to the carbonyl carbon, followed by a proton transfer, gives a tetrahedral carbonyl addition intermediate:

$$\mathrm{>C{=}O} + H_2N{-}R \rightleftharpoons -\overset{O^-}{\underset{|}{C}}-\overset{H}{N^+}(H)-R \rightleftharpoons -\overset{OH}{\underset{|}{C}}-N(H)-R$$

A tetrahedral carbonyl addition intermediate

Step 2: Protonation of the OH group, followed by loss of water and proton transfer to solvent gives the imine. Notice that the loss of water and the proton transfer have the characteristics of an E2 reaction. Three things happen simultaneously in this dehydration: a base (in this case a water molecule) removes a proton from N, the carbon–nitrogen double bond forms, and the leaving group (in this case, a water molecule) departs:

$$H_3O^+ + -\overset{OH}{\underset{|}{C}}-N(H)-R \rightleftharpoons -\overset{OH_2^+}{\underset{|}{C}}-N(H)-R \rightleftharpoons \mathrm{>C{=}N}-R + H_2O + H_3O^+$$

An imine

(The flow of electrons here is similar to that in an E2 reaction.)

To give but one example of the importance of imines in biological systems, the active form of vitamin A aldehyde (retinal) is bound to the protein opsin in the human retina in the form of an imine called *rhodopsin* or *visual purple* (see Chemical Connections 5A). The amino acid lysine (see Table 18.1) provides the primary amino group for this reaction:

11-*cis*-Retinal + H_2N—Opsin ⟶ Rhodopsin (Visual purple)

Example 13.8

Predict the products formed in each reaction:

(a) 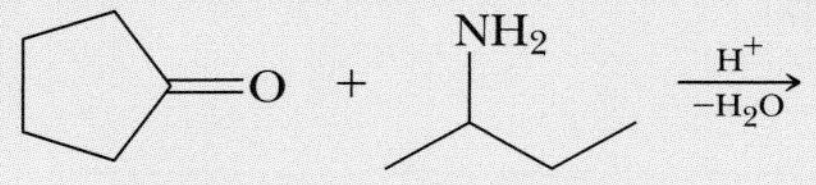 cyclopentanone (=O) + NH_2-substituted sec-butyl amine $\xrightarrow[-H_2O]{H^+}$

(b) 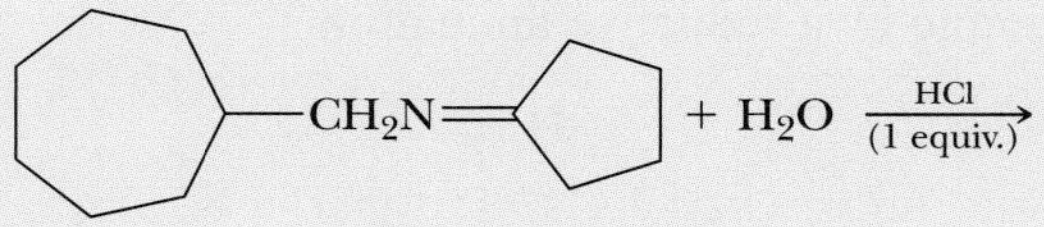 cycloheptyl–CH_2N=cyclopentylidene + H_2O $\xrightarrow[\text{(1 equiv.)}]{HCl}$

Strategy

In an imine-forming reaction, the C=O group is converted to a C=N group and the nitrogen of the former 1° amine loses both of its hydrogens. In the reverse process, the C=N group is converted back to a C=O group and two hydrogens are added back to the nitrogen to form a 1° amine.

Solution

Reaction (a) is an imine-forming reaction, while reaction (b) is the acid-catalyzed hydrolysis of an imine to an ammonium salt and a ketone:

(a) 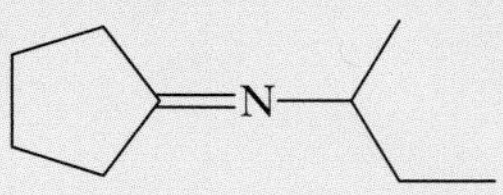 cyclopentylidene=N–(sec-butyl)

(b) 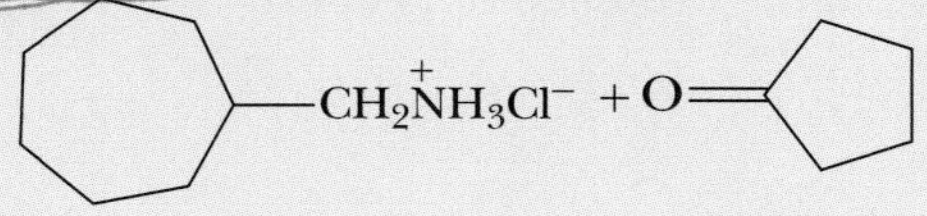 cycloheptyl–$CH_2\overset{+}{N}H_3Cl^-$ + O=cyclopentanone

See problems 13.29–13.32

Problem 13.8

Predict the products formed in each reaction. *Note*: Acid-catalyzed hydrolysis of an imine gives an amine and an aldehyde or a ketone. When one equivalent of acid is used, the amine is converted to its ammonium salt.

(a) 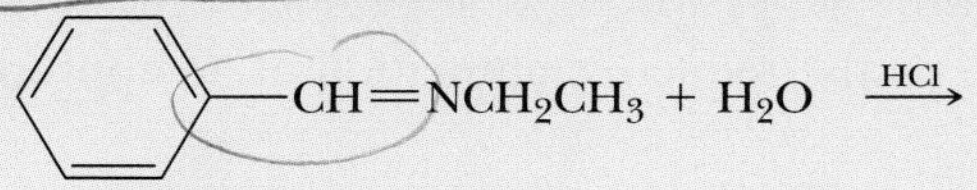 C_6H_5–$CH{=}NCH_2CH_3$ + H_2O $\xrightarrow{HCl}$

(b) 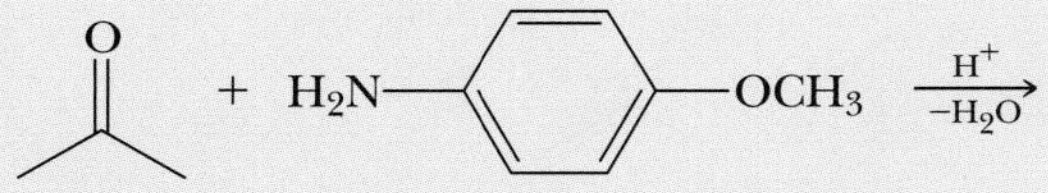 acetone (O=C(CH$_3$)$_2$) + H_2N–C_6H_4–OCH_3 $\xrightarrow[-H_2O]{H^+}$

B. Reductive Amination of Aldehydes and Ketones

One of the chief values of imines is that the carbon–nitrogen double bond can be reduced to a carbon–nitrogen single bond by hydrogen in the presence of a nickel or other transition metal catalyst. By this two-step reaction, called **reductive amination**, a primary amine is converted to a secondary amine by way of an imine, as illustrated by the conversion of cyclohexylamine to dicyclohexylamine:

Reductive amination The formation of an imine from an aldehyde or a ketone, followed by the reduction of the imine to an amine.

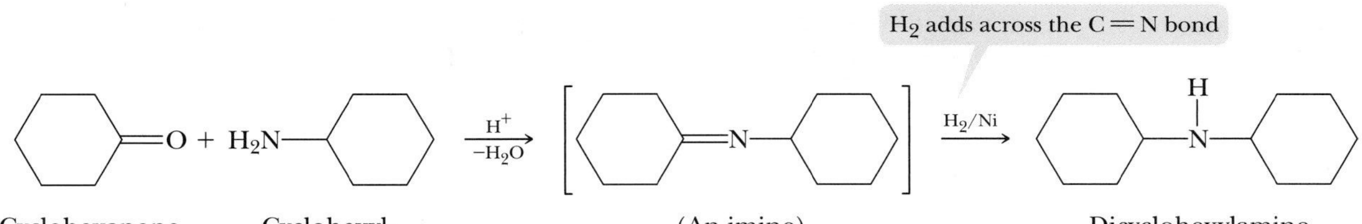

H$_2$ adds across the C=N bond

Cyclohexanone (=O) + H_2N–Cyclohexylamine (a 1° amine) $\xrightarrow[-H_2O]{H^+}$ [(An imine) C$_6$H$_{10}$=N–C$_6$H$_{11}$] $\xrightarrow{H_2/Ni}$ Dicyclohexylamine (a 2° amine) C$_6$H$_{11}$–NH–C$_6$H$_{11}$

Cyclohexanone | Cyclohexylamine (a 1° amine) | (An imine) | Dicyclohexylamine (a 2° amine)

Conversion of an aldehyde or a ketone to an amine is generally carried out in one laboratory operation by mixing together the carbonyl-containing compound, the amine or ammonia, hydrogen, and the transition metal catalyst. The imine intermediate is not isolated.

Example 13.9

Show how to synthesize each amine by a reductive amination:

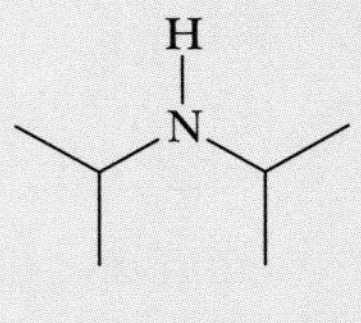

Strategy

Identify the C—N bond formed in the reductive amination. The carbon of the C—N bond is part of the carbonyl starting material, and the nitrogen is part of the 1° amine.

Solution

Treat the appropriate compound, in each case a ketone, with ammonia or an amine in the presence of H_2/Ni:

(a) 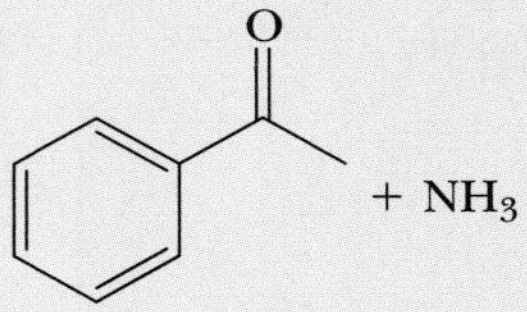 O, + NH_3 (b) 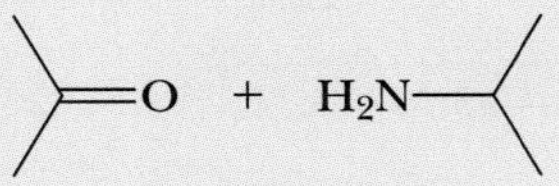 =O + H_2N—

See problems 13.29–13.32

Problem 13.9

Show how to prepare each amine by the reductive amination of an appropriate aldehyde or ketone:

(a) 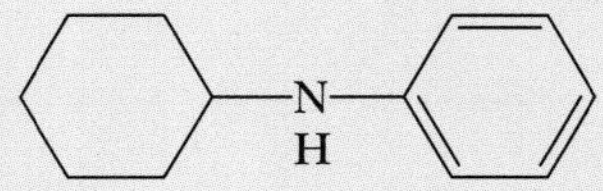 N, H (b) 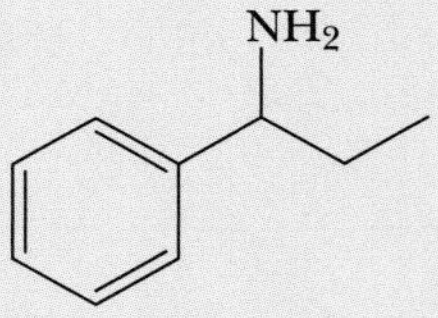 NH_2

13.8 What Is Keto-Enol Tautomerism?

A. Keto and Enol Forms

α-Carbon A carbon atom adjacent to a carbonyl group.

α-Hydrogen A hydrogen on an α-carbon.

A carbon atom adjacent to a carbonyl group is called an ***α*-carbon**, and any hydrogen atoms bonded to it are called ***α*-hydrogens:**

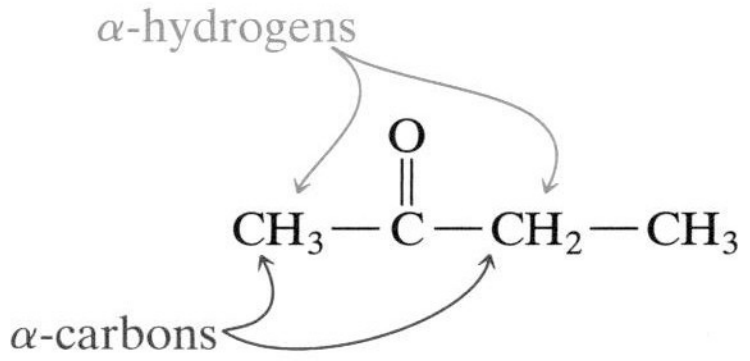

α-hydrogens

$$CH_3-\overset{O}{\overset{\|}{C}}-CH_2-CH_3$$

α-carbons

TABLE 13.4 The Position of Keto-Enol Equilibrium for Four Aldehydes and Ketones*

Keto Form		Enol Form	% Enol at Equilibrium
$CH_3CH{=}O$	⇌	$CH_2{=}CH{-}OH$	6×10^{-5}
$CH_3C({=}O)CH_3$	⇌	$CH_3C(OH){=}CH_2$	6×10^{-7}
Cyclopentanone	⇌	Cyclopentenol	1×10^{-6}
Cyclohexanone	⇌	Cyclohexenol	4×10^{-5}

*Data from J. March, *Advanced Organic Chemistry*, 4th ed. (New York, Wiley Interscience, 1992) p. 70.

An aldehyde or ketone that has at least one α-hydrogen is in equilibrium with a constitutional isomer called an **enol**. The name *enol* is derived from the IUPAC designation of it as both an alkene (*-en-*) and an alcohol (*-ol*):

Enol A molecule containing an —OH group bonded to a carbon of a carbon–carbon double bond.

$$CH_3{-}\overset{\overset{\displaystyle O}{\|}}{C}{-}CH_3 \rightleftharpoons CH_3{-}\overset{\overset{\displaystyle OH}{|}}{C}{=}CH_2$$

Acetone (keto form) — Acetone (enol form)

→ different compounds w/same molecular formula and different order of attachment of their atoms

Keto and enol forms are examples of **tautomers**—constitutional isomers that are in equilibrium with each other and that differ in the location of a hydrogen atom and a double bond relative to a heteroatom, most commonly O, S, or N. This type of isomerism is called **tautomerism**.

Tautomers Constitutional isomers that differ in the location of hydrogen and a double bond relative to O, N, or S.

For most simple aldehydes and ketones, the position of the equilibrium in keto–enol tautomerism lies far on the side of the keto form (Table 13.4), because a carbon–oxygen double bond is stronger than a carbon–carbon double bond. We saw how this trend was utilized in the hydration reactions of alkynes to form ketones (Section 5.10):

$$R{-}C{\equiv}C{-}H \xrightarrow[HgSO_4,\ H_2O]{H_2SO_4} \left[\begin{matrix} HO & & H \\ & C{=}C & \\ R & & H \end{matrix}\right] \underset{\text{taut.}}{\rightleftharpoons} \begin{matrix} O & & H \\ & C{-}C{-}H & \\ R & & H \end{matrix}$$

Enol — Keto

The equilibration of keto and enol forms is catalyzed by acid, as shown in the following two-step mechanism (note that a molecule of H—A is consumed in Step 1, but another is generated in Step 2):

Mechanism: Acid-Catalyzed Equilibration of Keto and Enol Tautomers

Step 1: Proton transfer from the acid catalyst, H—A, to the carbonyl oxygen forms the conjugate acid of the aldehyde or ketone:

$$CH_3{-}\overset{\overset{\displaystyle \ddot{O}:}{\|}}{C}{-}CH_3 + H{-}A \underset{}{\overset{\text{fast}}{\rightleftharpoons}} CH_3{-}\overset{\overset{\displaystyle \overset{+}{\ddot{O}}{-}H}{\|}}{C}{-}CH_3 + :A^-$$

Keto form — The conjugate acid of the ketone

Step 2: Proton transfer from the α-carbon to the base, A^-, gives the enol and generates a new molecule of the acid catalyst, H—A:

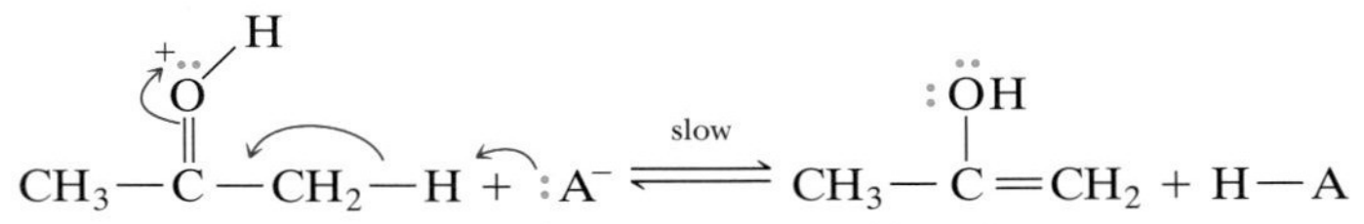

$$CH_3-\overset{\overset{+}{O}-H}{\overset{\|}{C}}-CH_2-H + :A^- \underset{}{\overset{\text{slow}}{\rightleftharpoons}} CH_3-\overset{:\ddot{O}H}{\overset{|}{C}}=CH_2 + H-A$$

Enol form

Example 13.10

Write two enol forms for each compound, and state which enol of each predominates at equilibrium:

(a) 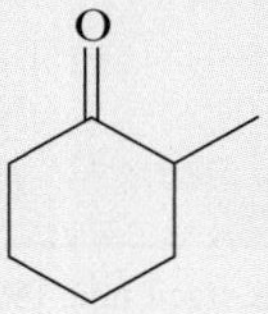 (b)

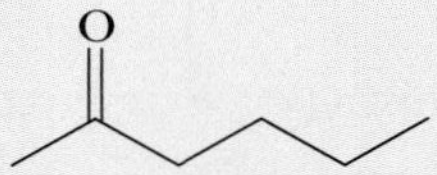

Strategy

An enol can form on either side of the carbonyl as long as there is an α-hydrogen to be abstracted in Step 2 of the mechanism. To decide which enol predominates, recall that the more substituted an alkene is, the more stable it is (see Section 5.7B).

Solution

In each case, the major enol form has the more substituted (the more stable) carbon–carbon double bond:

(a) 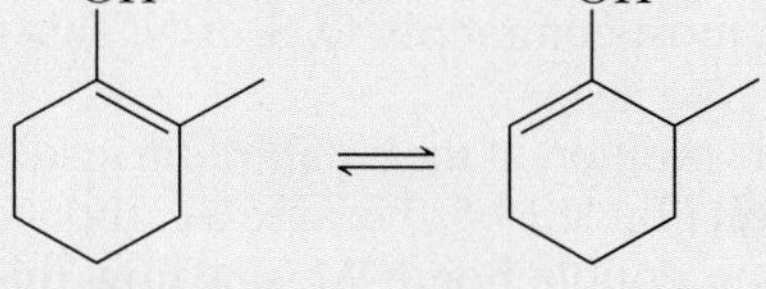 OH ⇌ OH

Major enol

(b) 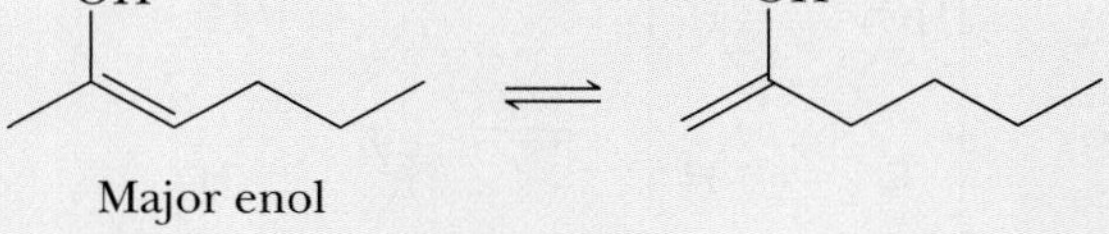 OH ⇌ OH

Major enol

See problem 13.34

Problem 13.10

Draw the structural formula for the keto form of each enol:

(a) 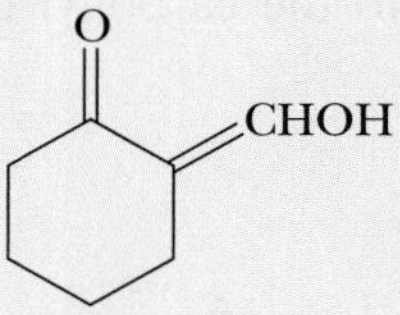 O, CHOH (b) 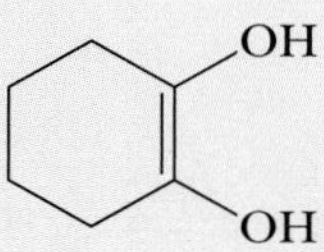 OH, OH (c) 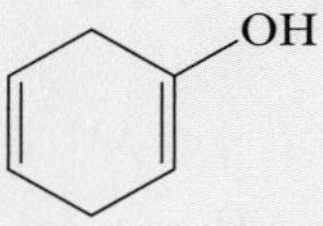 OH

Study this one

B. Racemization at an α-Carbon

Racemization The conversion of a pure enantiomer into a racemic mixture.

When enantiomerically pure (either *R* or *S*) 3-phenyl-2-butanone is dissolved in ethanol, no change occurs in the optical activity of the solution over time. If, however, a trace of acid (for example, HCl) is added, the optical activity of the solution begins to decrease and gradually drops to zero. When 3-phenyl-2-butanone is isolated from this solution, it is found to be a racemic mixture (Section 6.8C). This observation can be explained by the acid-catalyzed formation of an achiral enol intermediate. Tautomerism of the achiral enol to the chiral keto form generates the *R* and *S* enantiomers with equal probability:

(*R*)-3-Phenyl-2-butanone ⇌ An achiral enol ⇌ (*S*)-3-Phenyl-2-butanone

Racemization by this mechanism occurs only at α-carbon stereocenters with at least one α-hydrogen. This process is usually an undesired side effect of acid impurities in a sample, because it is often, in medicine for example, important to have an enantiomerically pure form of a compound rather than a racemic mixture.

C. α-Halogenation

Aldehydes and ketones with at least one α-hydrogen react with bromine and chlorine at the α-carbon to give an α-haloaldehyde or α-haloketone. Acetophenone, for example, reacts with bromine in acetic acid to give an α-bromoketone:

Acetophenone $+ Br_2 \xrightarrow{CH_3COOH}$ α-Bromoacetophenone $+ HBr$

α-Halogenation is catalyzed by both acid and base. For acid-catalyzed halogenation, the HBr or HCl generated by the reaction catalyzes further reaction.

Mechanism: Acid-Catalyzed α-Halogenation of a Ketone

Step 1: Acid-catalyzed keto–enol tautomerism gives the enol:

Keto form ⇌ Enol form

Step 2: Nucleophilic attack of the enol on the halogen molecule gives the α-haloketone after proton transfer to generate HBr:

$+ :Br—Br: \longrightarrow$ $+ :Br:^- \longrightarrow$ $+ HBr$

look how to draw

The value of α-halogenation is that it converts an α-carbon into a center that now has a good leaving group bonded to it and that is therefore susceptible to attack by a variety of good nucleophiles. In the following illustration, diethylamine (a nucleophile) reacts with the α-bromoketone to give an α-diethylaminoketone:

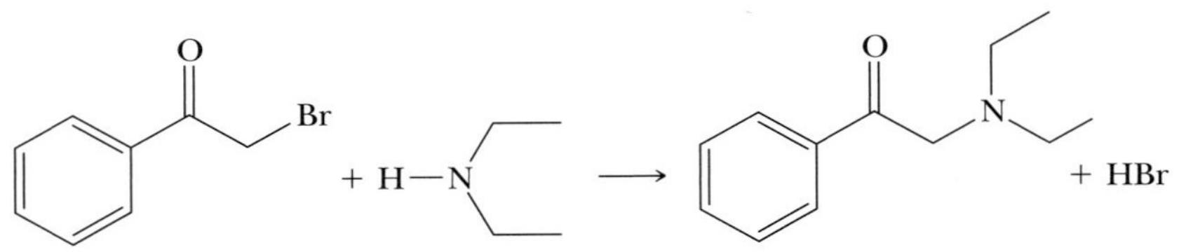

An α-bromoketone + H—N(Et)$_2$ ⟶ An α-diethylaminoketone + HBr

In practice, this type of nucleophilic substitution is generally carried out in the presence of a weak base such as potassium carbonate to neutralize the HX as it is formed.

13.9 How Are Aldehydes and Ketones Oxidized?

A. Oxidation of Aldehydes to Carboxylic Acids

Aldehydes are oxidized to carboxylic acids by a variety of common oxidizing agents, including chromic acid and molecular oxygen. In fact, aldehydes are one of the most easily oxidized of all functional groups. Oxidation by chromic acid (Section 8.2F) is illustrated by the conversion of hexanal to hexanoic acid:

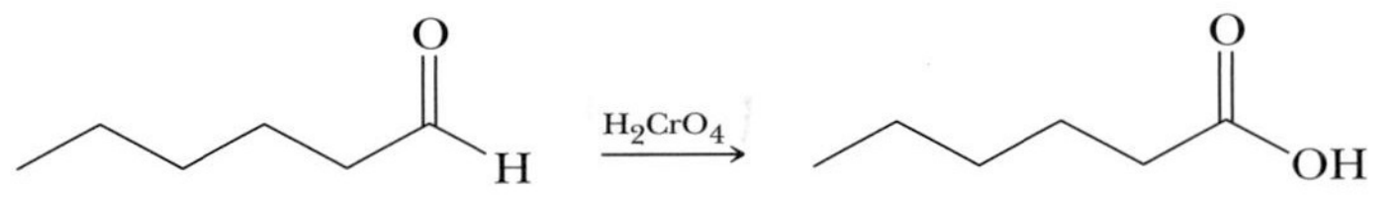

Hexanal $\xrightarrow{H_2CrO_4}$ Hexanoic acid

Aldehydes are also oxidized to carboxylic acids by silver ion. One laboratory procedure is to shake a solution of the aldehyde dissolved in aqueous ethanol or tetrahydrofuran (THF) with a slurry of Ag_2O:

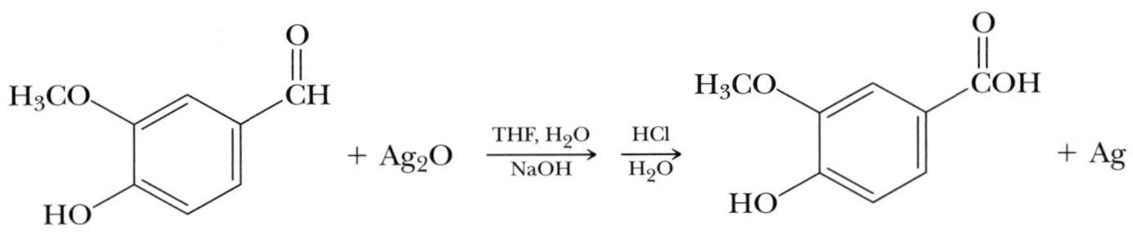

Vanillin (from vanilla) + Ag_2O $\xrightarrow[NaOH]{THF, H_2O}$ $\xrightarrow[H_2O]{HCl}$ Vanillic acid + Ag

Tollens' reagent, another form of silver ion, is prepared by dissolving $AgNO_3$ in water, adding sodium hydroxide to precipitate silver ion as Ag_2O, and then adding aqueous ammonia to redissolve silver ion as the silver–ammonia complex ion:

$$Ag^+NO_3^- + 2NH_3 \xrightleftharpoons{NH_3,\ H_2O} Ag(NH_3)_2{}^+NO_3^-$$

When Tollens' reagent is added to an aldehyde, the aldehyde is oxidized to a carboxylic anion, and Ag^+ is reduced to metallic silver. If this reaction is carried out properly, silver precipitates as a smooth, mirrorlike deposit—hence the name **silver-mirror test**:

$$\underset{\text{}}{\overset{\displaystyle O}{\overset{\|}{RCH}}} + 2Ag(NH_3)_2^+ \xrightarrow{NH_3,\, H_2O} \overset{\displaystyle O}{\overset{\|}{RCO^-}} + \underset{\substack{\text{Precipitates as} \\ \text{silver mirror}}}{2Ag} + 4NH_3$$

A silver mirror has been deposited in the inside of this flask by the reaction of an aldehyde with Tollens' reagent. *(Charles D. Winters)*

Nowadays, Ag^+ is rarely used for the oxidation of aldehydes, because of the cost of silver and because other, more convenient methods exist for this oxidation. The reaction, however, is still used for silvering mirrors. In the process, formaldehyde or glucose is used as the aldehyde to reduce Ag^+.

Aldehydes are also oxidized to carboxylic acids by molecular oxygen and by hydrogen peroxide.

$$2\,C_6H_5\overset{\displaystyle O}{\overset{\|}{C}}H + O_2 \longrightarrow 2\,C_6H_5\overset{\displaystyle O}{\overset{\|}{C}}OH$$

Benzaldehyde → Benzoic acid

Molecular oxygen is the least expensive and most readily available of all oxidizing agents, and, on an industrial scale, air oxidation of organic molecules, including aldehydes, is common. Air oxidation of aldehydes can also be a problem: Aldehydes that are liquid at room temperature are so sensitive to oxidation by molecular oxygen that they must be protected from contact with air during storage. Often, this is done by sealing the aldehyde in a container under an atmosphere of nitrogen.

Example 13.11

Draw a structural formula for the product formed by treating each compound with Tollens' reagent, followed by acidification with aqueous HCl:

(a) Pentanal (b) Cyclopentanecarbaldehyde

Strategy

Aldehydes are oxidized to carboxylic acids by Tollens' reagent.

Solution

The aldehyde group in each compound is oxidized to a carboxyl group:

(a) $CH_3CH_2CH_2CH_2\overset{\displaystyle O}{\overset{\|}{C}}OH$ Pentanoic acid

(b) C_5H_9—COOH Cyclopentanecarboxylic acid

See problems 13.36, 13.37

Problem 13.11

Complete these oxidations:

(a) 3-Oxobutanal + O_2 $\longrightarrow$

(b) 3-Phenylpropanal + Tollens' reagent $\longrightarrow$

CHEMICAL CONNECTIONS 13A

A Green Synthesis of Adipic Acid

The current industrial production of adipic acid relies on the oxidation of a mixture of cyclohexanol and cyclohexanone by nitric acid:

$$4\ \text{Cyclohexanol} + 6HNO_3 \longrightarrow 4\ \text{Hexanedioic acid (Adipic acid)} + 3N_2O\ \text{(Nitrous oxide)} + 3H_2O$$

A by-product of this oxidation is nitrous oxide, a gas considered to play a role in global warming and the depletion of the ozone layer in the atmosphere, as well as contributing to acid rain and acid smog. Given the fact that worldwide production of adipic acid is approximately 2.2 billion metric tons per year, the production of nitrous oxide is enormous. In spite of technological advances that allow for the recovery and recycling of nitrous oxide, it is estimated that approximately 400,000 metric tons escapes recovery and is released into the atmosphere each year.

Recently, Ryoji Noyori and coworkers at Nagoya University in Japan developed a "green" route to adipic acid, one that involves the oxidation of cyclohexene by 30% hydrogen peroxide catalyzed by sodium tungstate, Na_2WO_4:

$$\text{Cyclohexene} + 4H_2O_2 \xrightarrow[{[CH_3(C_8H_{17})_3N]HSO_4}]{Na_2WO_4} \text{Hexanedioic acid (Adipic acid)} + 4H_2O$$

In this process, cyclohexene is mixed with aqueous 30% hydrogen peroxide, and sodium tungstate and methyltrioctylammonium hydrogen sulfate are added to the resulting two-phase system. (Cyclohexene is insoluble in water.) Under these conditions, cyclohexene is oxidized to adipic acid in approximately 90% yield.

While this route to adipic acid is environmentally friendly, it is not yet competitive with the nitric acid oxidation route because of the high cost of 30% hydrogen peroxide. What will make it competitive is either a considerable reduction in the cost of hydrogen peroxide or the institution of more stringent limitations on the emission of nitrous oxide into the atmosphere (or a combination of these).

QUESTION

Using chemistry presented in this and previous chapters, propose a synthesis for adipic acid from cyclohexene.

B. Oxidation of Ketones to Carboxylic Acids

Ketones are much more resistant to oxidation than are aldehydes. For example, ketones are not normally oxidized by chromic acid or potassium permanganate. In fact, these reagents are used routinely to oxidize secondary alcohols to ketones in good yield (Section 8.2F).

Ketones undergo oxidative cleavage, via their enol form, by potassium dichromate and potassium permanganate at higher temperatures and by higher concentrations of nitric acid, HNO_3. The carbon–carbon double bond of the enol is cleaved to form two carboxyl or ketone groups, depending on the substitution pattern of the original ketone. An important industrial application of this reaction is the oxidation of cyclohexanone to hexanedioic acid (adipic acid), one of the two monomers required for the synthesis of the polymer nylon 66 (Section 17.4A):

Cyclohexanone (keto form) ⇌ Cyclohexanone (enol form) $\xrightarrow{HNO_3}$ Hexanedioic acid (adipic acid)

13.10 How Are Aldehydes and Ketones Reduced?

Aldehydes are reduced to primary alcohols and ketones to secondary alcohols:

$$\underset{\text{An aldehyde}}{RCH{=}O} \xrightarrow{\text{reduction}} \underset{\text{A primary alcohol}}{RCH_2OH} \qquad \underset{\text{A ketone}}{RC({=}O)R'} \xrightarrow{\text{reduction}} \underset{\text{A secondary alcohol}}{RCH(OH)R'}$$

A. Catalytic Reduction

The carbonyl group of an aldehyde or a ketone is reduced to a hydroxyl group by hydrogen in the presence of a transition metal catalyst, most commonly finely divided palladium, platinum, nickel, or rhodium. Reductions are generally carried out at temperatures from 25 to 100°C and at pressures of hydrogen from 1 to 5 atm. Under such conditions, cyclohexanone is reduced to cyclohexanol:

Cyclohexanone $+ H_2 \xrightarrow[25°C,\ 2\ atm]{Pt}$ Cyclohexanol

The catalytic reduction of aldehydes and ketones is simple to carry out, yields are generally very high, and isolation of the final product is very easy. A disadvantage is that some other functional groups (for example, carbon–carbon double bonds) are also reduced under these conditions.

trans-2-Butenal (Crotonaldehyde) $\xrightarrow[Ni]{2H_2}$ 1-Butanol

B. Metal Hydride Reductions

By far the most common laboratory reagents used to reduce the carbonyl group of an aldehyde or a ketone to a hydroxyl group are sodium borohydride and lithium aluminum hydride. Each of these compounds behaves as a source of **hydride ion**, a

Hydride ion A hydrogen atom with two electrons in its valence shell; $H:^-$.

very strong nucleophile. The structural formulas drawn here for these reducing agents show formal negative charges on boron and aluminum:

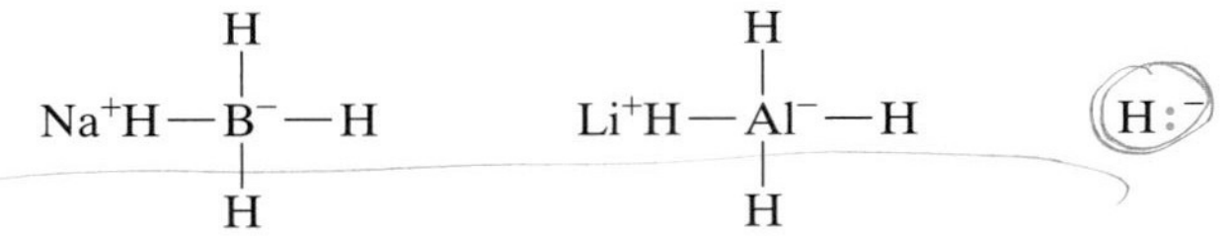

$$Na^+H-\underset{H}{\overset{H}{|}}B^--H \qquad Li^+H-\underset{H}{\overset{H}{|}}Al^--H \qquad H:^-$$

Sodium borohydride — Lithium aluminum hydride — Hydride ion

In fact, hydrogen is more electronegative than either boron or aluminum (H = 2.1, Al = 1.5, and B = 2.0), and the formal negative charge in the two reagents resides more on hydrogen than on the metal.

Lithium aluminum hydride is a very powerful reducing agent; it rapidly reduces not only the carbonyl groups of aldehydes and ketones, but also those of carboxylic acids (Section 14.5) and their functional derivatives (Section 15.8). Sodium borohydride is a much more selective reagent, reducing only aldehydes and ketones rapidly.

Reductions using sodium borohydride are most commonly carried out in aqueous methanol, in pure methanol, or in ethanol. The initial product of reduction is a tetraalkyl borate, which is converted to an alcohol and sodium borate salts upon treatment with water. One mole of sodium borohydride reduces 4 moles of aldehyde or ketone:

$$4R\overset{O}{\overset{\|}{C}}H + NaBH_4 \xrightarrow{CH_3OH} \underset{\text{A tetraalkyl borate}}{(RCH_2O)_4B^-Na^+} \xrightarrow{H_2O} 4RCH_2OH + \text{borate salts}$$

The key step in the metal hydride reduction of an aldehyde or a ketone is the transfer of a hydride ion from the reducing agent to the carbonyl carbon to form a tetrahedral carbonyl addition intermediate. In the reduction of an aldehyde or a ketone to an alcohol, only the hydrogen atom attached to carbon comes from the hydride-reducing agent; the hydrogen atom bonded to oxygen comes from the water added to hydrolyze the metal alkoxide salt.

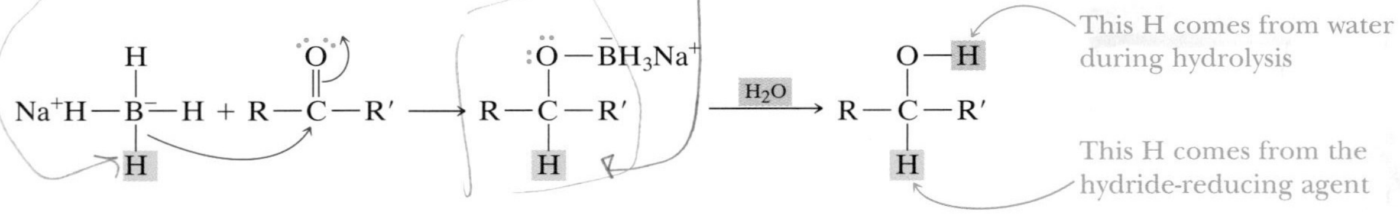

$$Na^+H-\underset{H}{\overset{H}{|}}B^--H + R-\overset{O}{\overset{\|}{C}}-R' \longrightarrow R-\underset{H}{\overset{:\ddot{O}-\bar{B}H_3Na^+}{|}}C-R' \xrightarrow{H_2O} R-\underset{H}{\overset{O-H}{|}}C-R'$$

This H comes from water during hydrolysis

This H comes from the hydride-reducing agent

The next two equations illustrate the selective reduction of a carbonyl group in the presence of a carbon–carbon double bond and, alternatively, the selective reduction of a carbon–carbon double bond in the presence of a carbonyl group.

Selective reduction of a carbonyl group:

$$RCH=CH\overset{O}{\overset{\|}{C}}R' \xrightarrow[2.\ H_2O]{1.\ NaBH_4} RCH=CH\overset{OH}{\overset{|}{C}}HR'$$

A carbon–carbon double bond can be reduced selectively in the presence of a carbonyl group by first protecting the carbonyl group using an acetal.

Selective reduction of a carbon–carbon double bond using a protecting group:

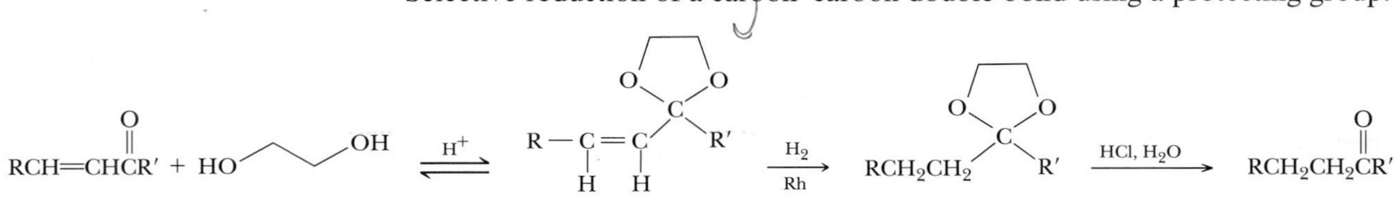

$$RCH=CH\overset{O}{\overset{\|}{C}}R' + HOCH_2CH_2OH \underset{}{\overset{H^+}{\rightleftharpoons}} R-\underset{H}{C}=\underset{H}{C}-\underset{(\text{1,3-dioxolane})}{C}-R' \xrightarrow[Rh]{H_2} RCH_2CH_2-\underset{(\text{1,3-dioxolane})}{C}-R' \xrightarrow{HCl,\ H_2O} RCH_2CH_2\overset{O}{\overset{\|}{C}}R'$$

Example 13.12

Complete these reductions:

(a) 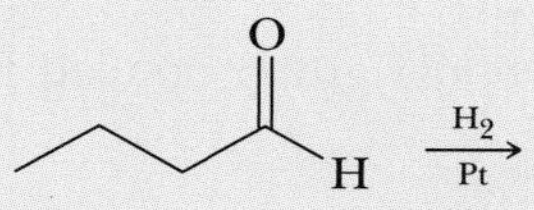 butanal $\xrightarrow[Pt]{H_2}$ (b) 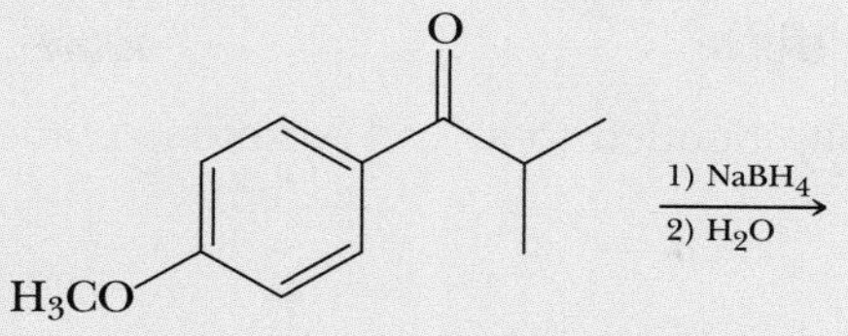 1-(4-methoxyphenyl)-2-methylpropan-1-one (H_3CO–C₆H₄–CO–CH(CH₃)₂) $\xrightarrow[2)\ H_2O]{1)\ NaBH_4}$

Strategy

Consider all the functional groups that can react with each reducing reagent. Alkenes, ketones, aldehydes, and imines are just some examples of functional groups that can be reduced.

Solution

The carbonyl group of the aldehyde in (a) is reduced to a primary alcohol, and that of the ketone in (b) is reduced to a secondary alcohol:

(a)  butan-1-ol (OH) (b) 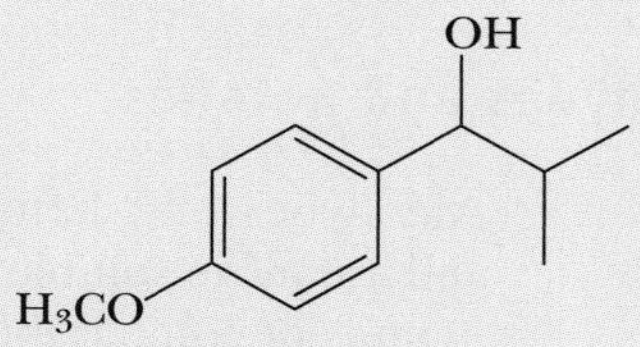 H_3CO–C₆H₄–CH(OH)–CH(CH₃)₂

See problems 13.36, 13.37 - see p. 491 for expl-for 13.37 e

Problem 13.12

What aldehyde or ketone gives each alcohol upon reduction by $NaBH_4$?

(a) 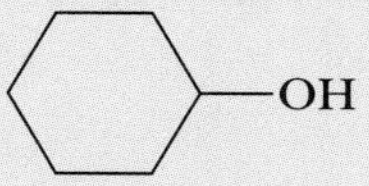 cyclohexanol (–OH) (b) 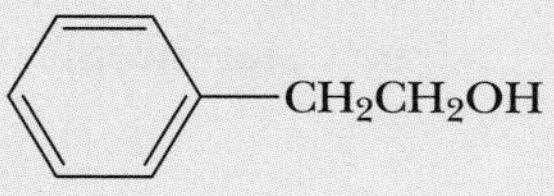 C₆H₅–CH_2CH_2OH

(c) 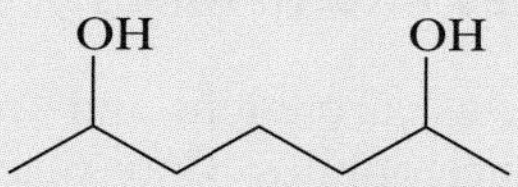 hexane-2,5-diol (OH, OH) study

Key Terms and Concepts

acetal (p. 476)
acetaldehyde (p. 463)
acetone (p. 463)
acetophenone (p. 464)
acid-catalyzed hydrolysis (p. 480)
aldehyde (p. 463)
α-carbon (p. 484)
α-halogenation (p. 487)
α-hydrogen (p. 484)
benzaldehyde (p. 464)
benzophenone (p. 464)
carbanion (p. 470)
carbonyl-protecting group (p. 479)
enol (p. 485)
formaldehyde (p. 463)
Grignard reagent (p. 470)
hemiacetal (p. 474)
hydride ion (p. 491)
imine (p. 481)
IUPAC order of preference of functional groups (p. 465)
keto form (p. 485)
ketone (p. 463)
magnesium alkoxide (p. 472)
organometallic compound (p. 470)
racemization (p. 487)
reductive amination (p. 483)
Schiff base (p. 481)
silver-mirror test (p. 489)
tautomerism (p. 485)
tautomers (p. 485)
tetrahedral carbonyl addition intermediate (p. 469)
Tollens' reagent (p. 488)
unsaturated aldehyde (p. 463)

Summary of Key Questions

13.1 What Are Aldehydes and Ketones?

- An **aldehyde** contains a carbonyl group bonded to a hydrogen atom and a carbon atom.
- A **ketone** contains a carbonyl group bonded to two carbons.

13.2 How Are Aldehydes and Ketones Named?

- An aldehyde is named by changing *-e* of the parent alkane to *-al*.
- A CHO group bonded to a ring is indicated by the suffix *-carbaldehyde*.
- A ketone is named by changing *-e* of the parent alkane to *-one* and using a number to locate the carbonyl group.
- In naming compounds that contain more than one functional group, the IUPAC system has established an **order of precedence of functional groups**. If the carbonyl group of an aldehyde or a ketone is lower in precedence than other functional groups in the molecule, it is indicated by the infix **-oxo-**.

13.3 What Are the Physical Properties of Aldehydes and Ketones?

- Aldehydes and ketones are polar compounds and interact in the pure state by dipole–dipole interactions.
- Aldehydes and ketones have higher boiling points and are more soluble in water than are nonpolar compounds of comparable molecular weight.

13.4 What Is the Most Common Reaction Theme of Aldehdyes and Ketones?

- The common reaction theme of the carbonyl group of aldehydes and ketones is the addition of a nucleophile to form a **tetrahedral carbonyl addition intermediate**.
- Because the carbon of the carbonyl is trigonal planar in geometry, attack of the nucleophile can occur from either side of the pi bond, potentially generating a new stereocenter.

13.5 What Are Grignard Reagents, and How Do They React with Aldehydes and Ketones?

- **Grignard reagents** are organomagnesium compounds of the generic formula RMgX.
- The carbon–metal bond in Grignard reagents has a high degree of partial ionic character.
- Grignard reagents behave as carbanions and are both strong bases and good nucleophiles. They react with aldehydes and ketones by adding to the carbonyl carbon.

13.6 What Are Hemiacetals and Acetals?

- The addition of a molecule of alcohol to the carbonyl group of an aldehyde or a ketone forms a **hemiacetal**.
- Hemiacetals can react further with alcohols to form **acetals** plus a molecule of water.
- Because of their lack of reactivity toward nucleophilic and basic reagents, acetals are often used to protect the carbonyl groups of aldehydes and ketones while reactions are carried out on functional groups in other parts of the molecule.

13.7 How Do Aldehydes and Ketones React with Ammonia and Amines?

- Ammonia, 1° aliphatic amines (RNH_2), and 1° aromatic amines ($ArNH_2$) react with the carbonyl group of aldehydes and ketones in the presence of an acid catalyst to give **imines**, compounds that contain a carbon–nitrogen double bond.

13.8 What Is Keto–Enol Tautomerism?

- A carbon atom adjacent to a carbonyl group is called an **α-carbon**, and any hydrogen atoms bonded to it are called **α-hydrogens**.
- An aldehyde or a ketone, which is said to be in its **keto** form, that has at least one α-hydrogen is in equilibrium with a constitutional isomer called an **enol**. This type of isomerism is called **tautomerism**.
- Tautomerism, catalyzed by trace amounts of acid or base, is the cause of racemization of chiral aldehydes and ketones when a stereocenter exists at an α-carbon.
- The enol form allows aldehydes and ketones to be halogenated at the α-position.

13.9 How Are Aldehydes and Ketones Oxidized?

- Aldehydes are oxidized to carboxylic acids by a variety of common oxidizing agents, including chromic acid, the Tollens' reagent, and molecular oxygen.
- Ketones are much more resistant to oxidation than are aldehydes. However, they undergo oxidative cleavage, via their enol form, by potassium dichromate and potassium permanganate at higher temperatures and by higher concentrations of HNO_3.

13.10 How Are Aldehydes and Ketones Reduced?

- Aldehydes are reduced to primary alcohols and ketones to secondary alcohols by catalytic hydrogenation or through the use of the metal hydrides $NaBH_4$ or $LiAlH_4$.

Quick Quiz

Answer true or false to the following questions to assess your general knowledge of the concepts in this chapter. If you have difficulty with any of them, you should review the appropriate section in the chapter (shown in parentheses) before attempting the more challenging end-of-chapter problems.

1. Catalytic hydrogenation can be performed on both alkenes and aldehydes. (13.10)
2. Nucleophiles react with aldehydes and ketones to form tetrahedral carbonyl addition intermediates. (13.4)
3. The carboxyl group (COOH) has a higher priority in naming than all other functional groups. (13.2)
4. A stereocenter at the α-carbon of an aldehyde or a ketone will undergo racemization over time in the presence of an acid or a base. (13.8)
5. Acetone is the lowest-molecular-weight ketone. (13.3)
6. Aldehydes can be oxidized to ketones and carboxylic acids. (13.9)
7. Ketones are less water soluble than alcohols of comparable molecular weight. (13.3)
8. A Grignard reagent cannot be formed in the presence of an NH, OH, or SH group. (13.5)
9. Ketones have higher boiling points than alkanes of comparable molecular weight. (13.3)
10. An aldehyde has a higher priority in naming than a ketone. (13.2)
11. A Grignard reagent is a good electrophile. (13.5)
12. Any reaction that oxidizes an aldehyde to a carboxylic acid will also oxidize a ketone to a carboxylic acid. (13.9)
13. Aldehyes are more water soluble than ethers of comparable molecular weight. (13.3)
14. Aldehydes react with Grignard reagents (followed by acid workup) to form 1° alcohols. (13.5)
15. An imine can be reduced to an amine through catalytic hydrogenation. (13.7)
16. Sodium borohydride, $NaBH_4$, is more reactive and less selective than lithium aluminum hydride, $LiAlH_4$. (13.10)
17. An acetal can only result from the base-catalyzed addition of an alcohol to a hemiacetal. (13.6)
18. A Grignard reagent is a strong base. (13.5)
19. Acetal formation is reversible. (13.6)
20. An imine is the result of the reaction of a 2° amine with an aldehyde or a ketone. (13.7)
21. Ketones react with Grignard reagents (followed by acid workup) to form 2° alcohols. (13.5)

22. Aldehydes and ketones can undergo tautomerism. (13.8)
23. Acetaldehyde is the lowest-molecular-weight aldehyde. (13.3)
24. A ketone that possesses an α-hydrogen can undergo α-halogenation. (13.8)
25. A carbonyl group is polarized such that the oxygen atom is partially positive and the carbon atom is partially negative. (13.3)
26. Acetals are stable to bases, nucleophiles, and reducing agents. (13.6)
27. A "carbaldehyde" is an aldehyde in which the carbonyl group is adjacent to a C—C double bond. (13.1)
28. A hemiacetal can result from the acid-catalyzed or base-catalyzed addition of an alcohol to an aldehyde or a ketone. (13.6)

Answers: (1) T (2) T (3) T (4) T (5) T (6) F (7) T (8) T (9) T (10) T (11) F (12) F (13) T (14) F (15) T (16) F (17) F (18) T (19) T (20) F (21) F (22) T (23) F (24) T (25) F (26) T (27) F (28) T

Key Reactions

1. Reaction with Grignard Reagents (Section 13.5C)

Treatment of formaldehyde with a Grignard reagent, followed by hydrolysis in aqueous acid, gives a primary alcohol. Similar treatment of any other aldehyde gives a secondary alcohol:

$$CH_3CH{=}O \xrightarrow[\text{2) HCl, } H_2O]{\text{1) } C_6H_5MgBr} C_6H_5CH(OH)CH_3$$

Treatment of a ketone with a Grignard reagent gives a tertiary alcohol:

$$CH_3C({=}O)CH_3 \xrightarrow[\text{2) HCl, } H_2O]{\text{1) } C_6H_5MgBr} C_6H_5C(OH)(CH_3)_2$$

2. Addition of Alcohols to Form Hemiacetals (Section 13.6)

Hemiacetals are only minor components of an equilibrium mixture of aldehyde or ketone and alcohol, except where the —OH and C=O groups are parts of the same molecule and a five- or six-membered ring can form:

$$CH_3CH(OH)CH_2CH_2CH{=}O \rightleftharpoons \text{(cyclic hemiacetal: } H_3C\text{-substituted five-membered ring with O, bearing OH)}$$

4-Hydroxypentanal ⇌ A cyclic hemiacetal

3. Addition of Alcohols to Form Acetals (Section 13.6)

The formation of acetals is catalyzed by acid:

$$\text{cyclopentanone} + HOCH_2CH_2OH \overset{H^+}{\rightleftharpoons} \text{(cyclic acetal, } O{-}CH_2{-}CH_2{-}O\text{ spiro ring)} + H_2O$$

4. Addition of Ammonia and Amines (Section 13.7)

The addition of ammonia or a primary amine to the carbonyl group of an aldehyde or a ketone forms a tetrahedral carbonyl addition intermediate. Loss of water from this intermediate gives an imine (a Schiff base):

$$\text{cyclopentanone} + H_2NCH_3 \overset{H^+}{\rightleftharpoons} \text{cyclopentylidene}{=}NCH_3 + H_2O$$

5. Reductive Amination to Amines (Section 13.7B)

The carbon–nitrogen double bond of an imine can be reduced by hydrogen in the presence of a transition metal catalyst to a carbon–nitrogen single bond:

$$\text{cyclohexanone} + H_2N\text{-cyclohexyl} \xrightarrow{-H_2O} [\text{cyclohexylidene}{=}N\text{-cyclohexyl}] \xrightarrow{H_2/Ni} \text{cyclohexyl-}NH\text{-cyclohexyl}$$

6. Keto–Enol Tautomerism (Section 13.8A)

The keto form generally predominates at equilibrium:

$$CH_3C({=}O)CH_3 \rightleftharpoons CH_3C(OH){=}CH_2$$

Keto form (Approx 99.9%) — Enol form

7. Oxidation of an Aldehyde to a Carboxylic Acid (Section 13.9)

The aldehyde group is among the most easily oxidized functional groups. Oxidizing agents include H_2CrO_4, Tollens' reagent, and O_2:

$$\text{2-hydroxybenzaldehyde (HOC}_6\text{H}_4\text{CHO)} + Ag_2O \xrightarrow[\text{2) } H_2O,\ HCl]{\text{1) THF, } H_2O,\ NaOH} \text{2-hydroxybenzoic acid (HOC}_6\text{H}_4\text{COOH)} + Ag$$

8. Catalytic Reduction (Section 13.10A)
Catalytic reduction of the carbonyl group of an aldehyde or a ketone to a hydroxyl group is simple to carry out, and yields of alcohols are high:

$$\text{cyclohexanone} + H_2 \xrightarrow[25°C,\ 2\ atm]{Pt} \text{cyclohexanol}$$

9. Metal Hydride Reduction (Section 13.10B)
Both $LiAlH_4$ and $NaBH_4$ reduce the carbonyl group of an aldehyde or a ketone to an hydroxyl group. They are selective in that neither reduces isolated carbon–carbon double bonds:

$$\text{cyclohex-3-en-1-one} \xrightarrow[\text{2) } H_2O]{\text{1) } NaBH_4} \text{cyclohex-3-en-1-ol}$$

Problems

A problem marked with an asterisk indicates an applied "real world" problem. Answers to problems whose numbers are printed in blue are given in Appendix D.

Preparation of Aldehydes and Ketones (see Chapters 8 and 9)

13.13 Complete these reactions:

(a) cyclooctanol $\xrightarrow[H_2SO_4]{K_2Cr_2O_7}$

(b) cyclopentyl–CH_2OH $\xrightarrow[CH_2Cl_2]{PCC}$

(c) cyclopentyl–CH_2OH $\xrightarrow[H_2SO_4]{K_2Cr_2O_7}$

(d) benzene + $(CH_3)_2CHCH_2COCl$ $\xrightarrow{AlCl_3}$

13.14 Show how you would bring about these conversions:

(a) 1-Pentanol to pentanal
(b) 1-Pentanol to pentanoic acid
(c) 2-Pentanol to 2-pentanone
(d) 1-Pentene to 2-pentanone
(e) Benzene to acetophenone
(f) Styrene to acetophenone
(g) Cyclohexanol to cyclohexanone
(h) Cyclohexene to cyclohexanone

Sections 13.1 and 13.2 Structure and Nomenclature

13.15 Draw a structural formula for the one ketone with molecular formula C_4H_8O and for the two aldehydes with molecular formula C_4H_8O. **(See Example 13.2)**

13.16 Draw structural formulas for the four aldehydes with molecular formula $C_5H_{10}O$. Which of these aldehydes are chiral? **(See Example 13.2)**

13.17 Name these compounds: **(See Examples 13.1, 13.3)**

(a) 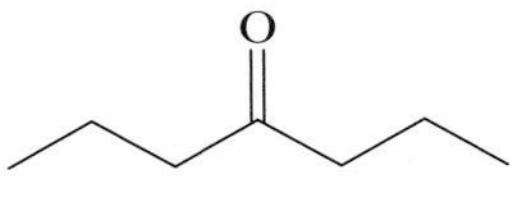 (b) 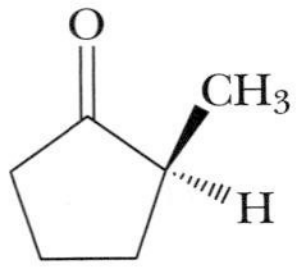 CH_3 H

(c) 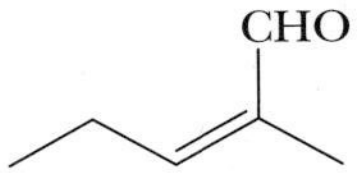 CHO (d) 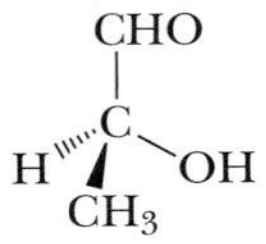 CHO C H OH CH_3

(e) 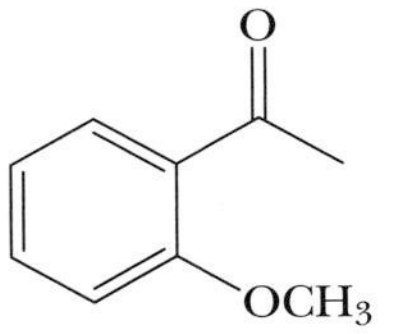 OCH_3 (f) 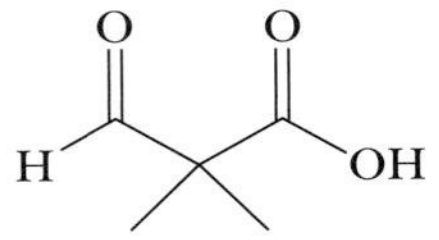 H OH

(g)

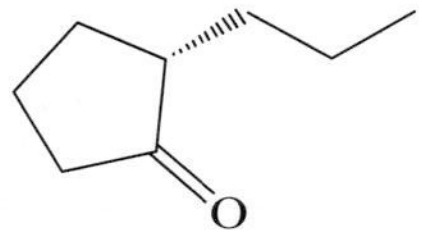

13.18 Draw structural formulas for these compounds: **(See Examples 13.1, 13.3)**

(a) 1-Chloro-2-propanone
(b) 3-Hydroxybutanal
(c) 4-Hydroxy-4-methyl-2-pentanone
(d) 3-Methyl-3-phenylbutanal
(e) *(S)*-3-bromocyclohexanone
(f) 3-Methyl-3-buten-2-one
(g) 5-Oxohexanal
(h) 2,2-Dimethylcyclohexanecarbaldehyde
(i) 3-Oxobutanoic acid

Section 13.5 Addition of Carbon Nucleophiles

13.19 Write an equation for the acid–base reaction between phenylmagnesium iodide and a carboxylic acid. Use curved arrows to show the flow of electrons in this reaction. In addition, show that the reaction is an example of a stronger acid and stronger base reacting to form a weaker acid and weaker base. **(See Examples 13.4, 13.5)**

13.20 Diethyl ether is prepared on an industrial scale by the acid-catalyzed dehydration of ethanol:

$$2CH_3CH_2OH \xrightarrow[180°C]{H_2SO_4} CH_3CH_2OCH_2CH_3 + H_2O$$

Explain why diethyl ether used in the preparation of Grignard reagents must be carefully purified to remove all traces of ethanol and water.

13.21 Draw structural formulas for the product formed by treating each compound with propylmagnesium bromide, followed by hydrolysis in aqueous acid: **(See Examples 13.4, 13.5)**

(a) CH_2O (b)

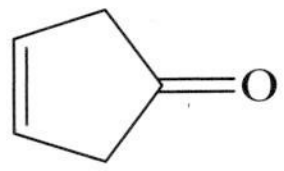

(c) 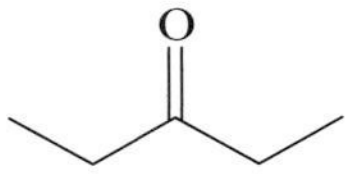 (d) 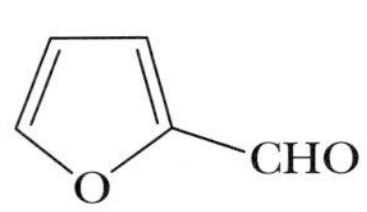 CHO

(e) 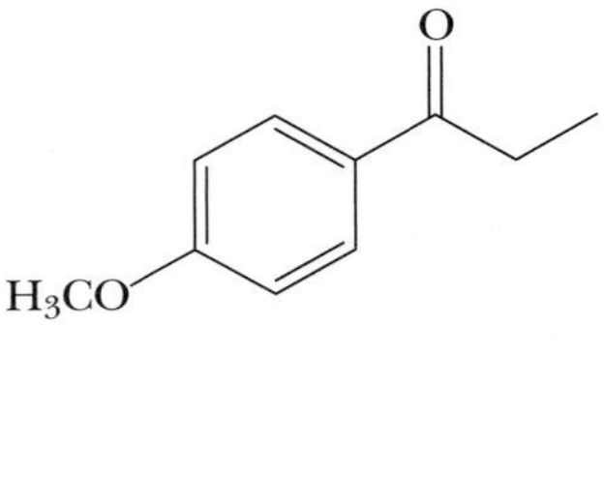 H_3CO

13.22 Suggest a synthesis for each alcohol, starting from an aldehyde or a ketone and an appropriate Grignard reagent (the number of combinations of Grignard reagent and aldehyde or ketone that might be used is shown in parentheses below each target molecule): **(See Examples 13.4, 13.5)**

(a) 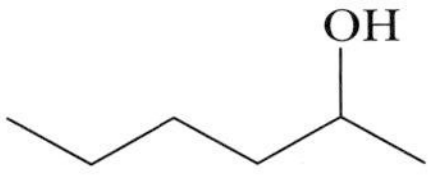 OH

(Two combinations)

(b) 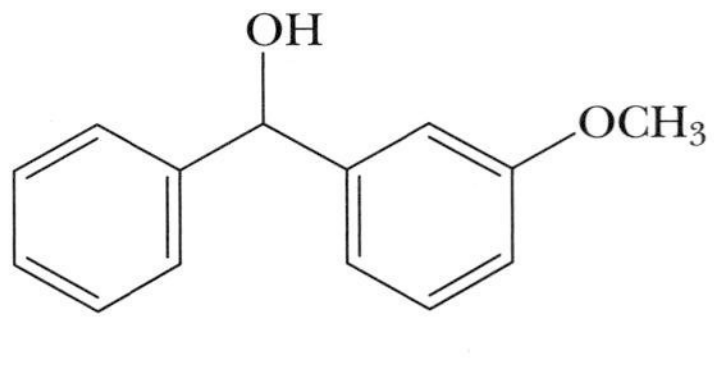 OH OCH_3

(Two combinations)

(c) 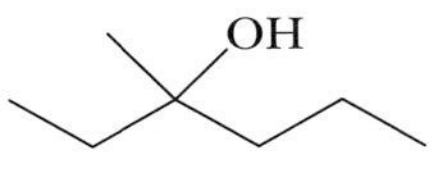 OH

(Three combinations)

Section 13.6 Addition of Oxygen Nucleophiles

13.23 5-Hydroxyhexanal forms a six-membered cyclic hemiacetal that predominates at equilibrium in aqueous solution: **(See Example 13.6)**

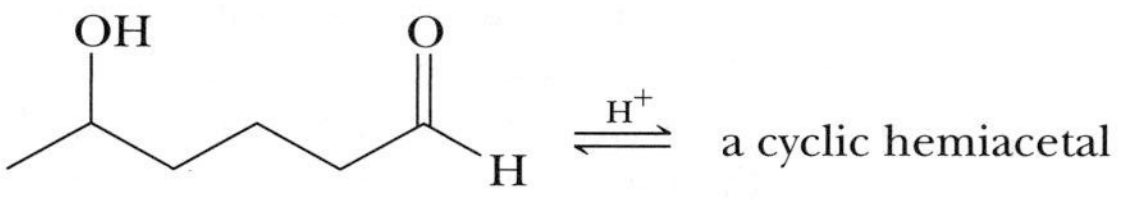

OH O H $\overset{H^+}{\rightleftharpoons}$ a cyclic hemiacetal

5-Hydroxyhexanal

(a) Draw a structural formula for this cyclic hemiacetal.

(b) How many stereoisomers are possible for 5-hydroxyhexanal?

(c) How many stereoisomers are possible for the cyclic hemiacetal?

(d) Draw alternative chair conformations for each stereoisomer.

(e) For each stereoisomer, which alternative chair conformation is the more stable?

13.24 Draw structural formulas for the hemiacetal and then the acetal formed from each pair of reactants in the presence of an acid catalyst: **(See Example 13.6)**

(a) 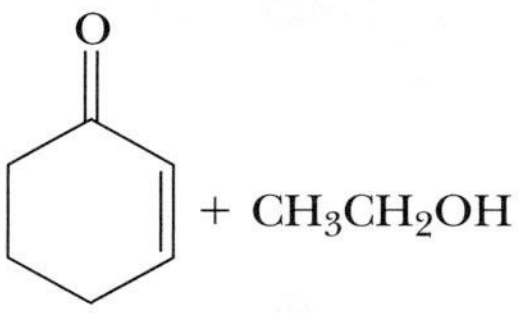 O + CH_3CH_2OH

(b) 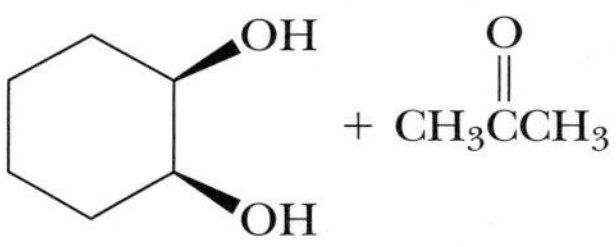 OH OH + $CH_3\overset{O}{\overset{\|}{C}}CH_3$

(c) CHO + CH_3OH

13.25 Draw structural formulas for the products of hydrolysis of each acetal in aqueous acid: **(See Example 13.6)**

(a) 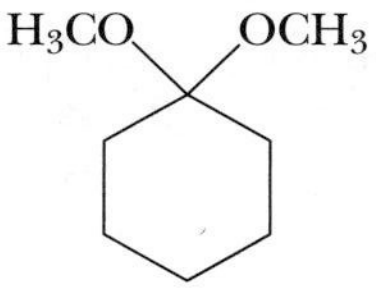 H_3CO OCH_3

(b) 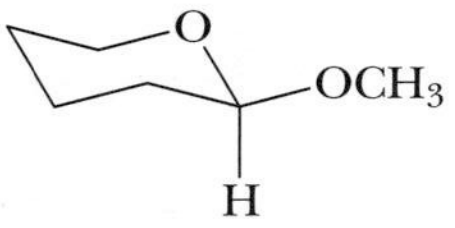 O OCH_3 H

(c) 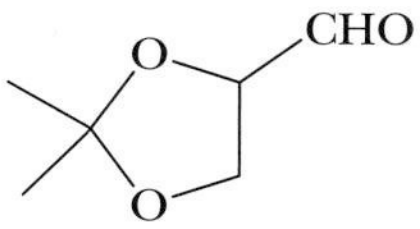 O O CHO

***13.26** The following compound is a component of the fragrance of jasmine: From what carbonyl-containing compound and alcohol is the compound derived? **(See Example 13.6)**

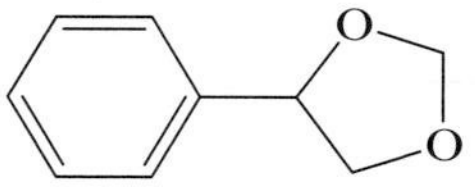

O O

13.27 Propose a mechanism for the formation of the cyclic acetal by treating acetone with ethylene glycol in the presence of an acid catalyst. Make sure that your mechanism is consistent with the fact that the oxygen atom of the water molecule is derived from the carbonyl oxygen of acetone.

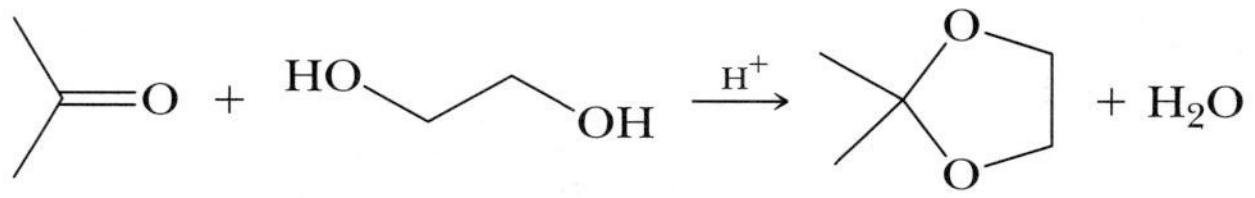

=O + HO OH $\xrightarrow{H^+}$ O O + H_2O

Acetone Ethylene glycol

13.28 Propose a mechanism for the formation of a cyclic acetal from 4-hydroxypentanal and one equivalent of methanol: If the carbonyl oxygen of 4-hydroxypentanal is enriched with oxygen-18, does your mechanism predict that the oxygen label appears in the cyclic acetal or in the water? Explain.

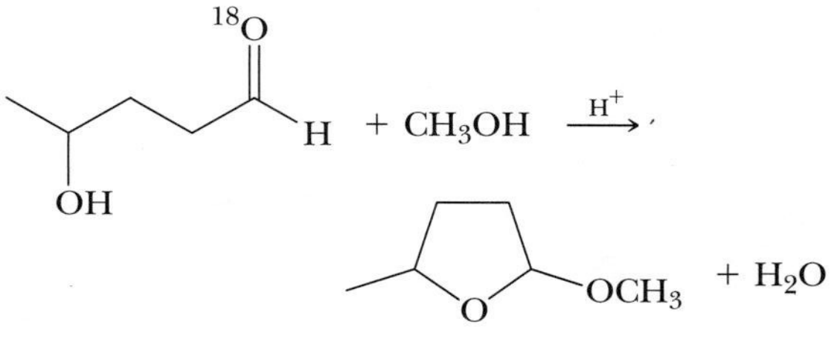

^{18}O H OH + CH_3OH $\xrightarrow{H^+}$

O OCH_3 + H_2O

Section 13.7 Addition of Nitrogen Nucleophiles

13.29 Show how this secondary amine can be prepared by two successive reductive aminations: **(See Examples 13.8, 13.9)**

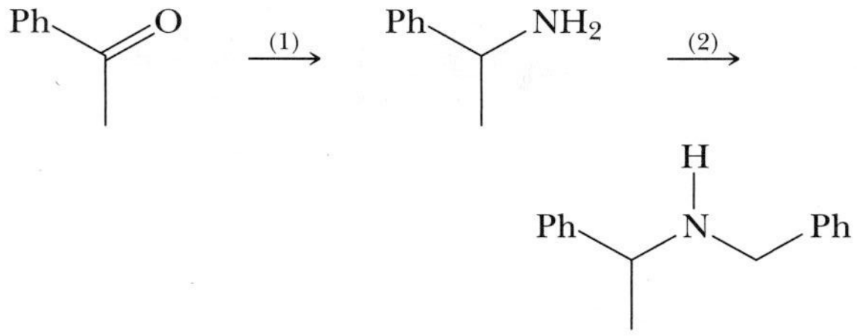

Ph O $\xrightarrow{(1)}$ Ph NH_2 $\xrightarrow{(2)}$

Ph N(H) Ph

13.30 Show how to convert cyclohexanone to each of the following amines: **(See Examples 13.8, 13.9)**

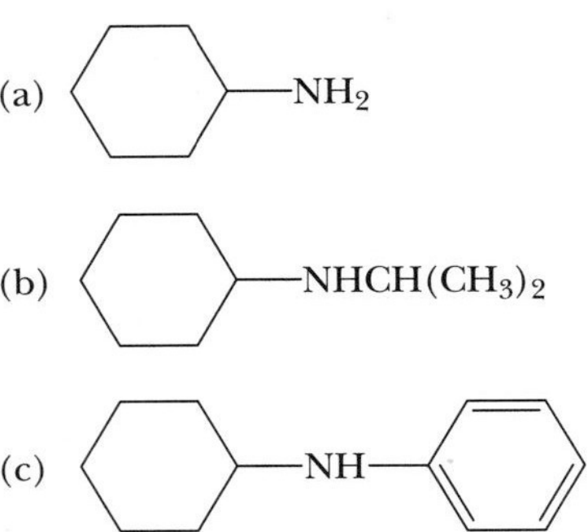

***13.31** Following are structural formulas for amphetamine and methamphetamine: **(See Examples 13.8, 13.9)**

(a) NH_2 Amphetamine (b) $H-N-CH_3$ Methamphetamine

The major central nervous system effects of amphetamine and amphetaminelike drugs are locomotor stimulation, euphoria and excitement, stereotyped behavior, and anorexia. Show how each drug can be synthesized by the reductive amination of an appropriate aldehyde or ketone.

***13.32** Rimantadine is effective in preventing infections caused by the influenza A virus and in treating established illness. The drug is thought to exert its antiviral effect by blocking a late stage in the assembly of the virus. Following is the final step in the synthesis of rimantadine: **(See Examples 13.8, 13.9)**

O → NH_2

Rimantadine
(an antiviral agent)

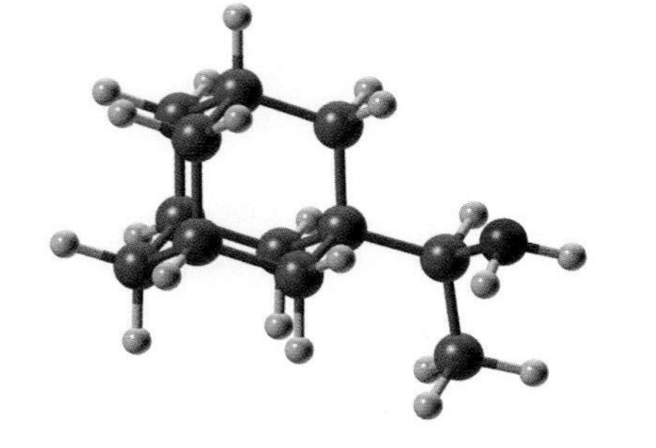

(a) Describe experimental conditions to bring about this conversion.

(b) Is rimantadine chiral?

***13.33** Methenamine, a product of the reaction of formaldehyde and ammonia, is a *prodrug*—a compound that is inactive by itself, but is converted to an active drug in the body by a biochemical transformation. The strategy behind the use of methenamine as a prodrug is that nearly all bacteria are sensitive to formaldehyde at concentrations of 20 mg/mL or higher. Formaldehyde cannot be used directly in medicine, however, because an effective concentration in plasma cannot be achieved with safe doses. Methenamine is stable at pH 7.4 (the pH of blood plasma), but undergoes acid-catalyzed hydrolysis to formaldehyde and ammonium ion under the acidic conditions of the kidneys and the urinary tract:

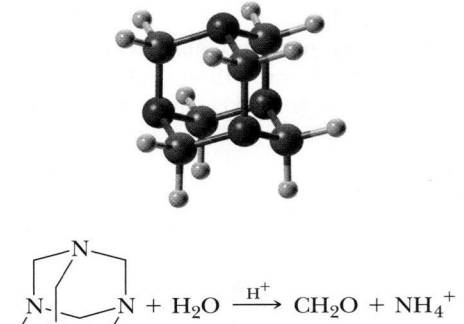

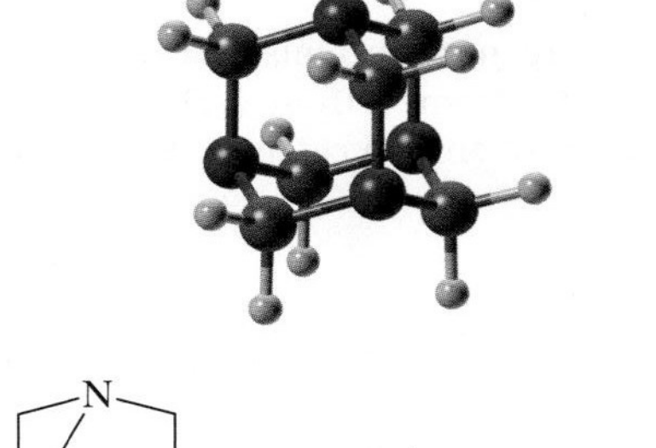

$$\text{Methenamine} + H_2O \xrightarrow{H^+} CH_2O + NH_4^+$$

Methenamine

Thus, methenamine can be used as a site-specific drug to treat urinary infections.

(a) Balance the equation for the hydrolysis of methenamine to formaldehyde and ammonium ion.

(b) Does the pH of an aqueous solution of methenamine increase, remain the same, or decrease as a result of the hydrolysis of the compound? Explain.

(c) Explain the meaning of the following statement: The functional group in methenamine is the nitrogen analog of an acetal.

(d) Account for the observation that methenamine is stable in blood plasma, but undergoes hydrolysis in the urinary tract.

Section 13.8 Keto–Enol Tautomerism

13.34 The following molecule belongs to a class of compounds called enediols: Each carbon of the double bond carries an —OH group:

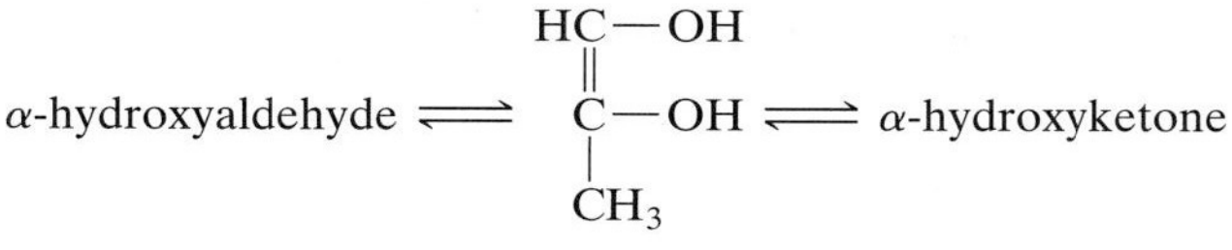

$$\alpha\text{-hydroxyaldehyde} \rightleftharpoons \underset{\substack{| \\ CH_3}}{\overset{\substack{HC-OH \\ \|}}{C}}-OH \rightleftharpoons \alpha\text{-hydroxyketone}$$

An enediol

Draw structural formulas for the α-hydroxyketone and the α-hydroxyaldehyde with which this enediol is in equilibrium. **(See Example 13.10)**

13.35 In dilute aqueous acid, (*R*)-glyceraldehyde is converted into an equilibrium mixture of (*R,S*)-glyceraldehyde and dihydroxyacetone:

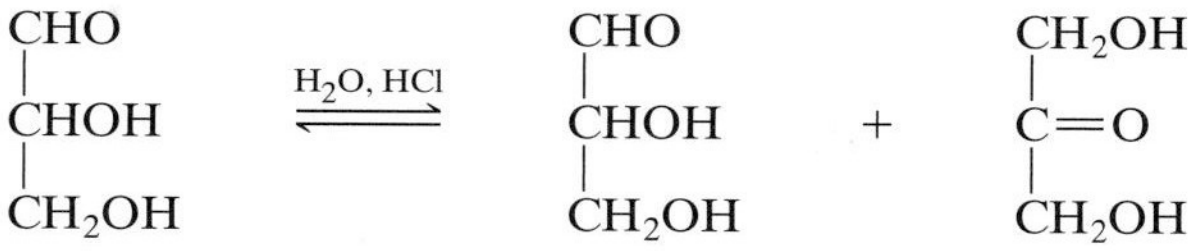

$$\begin{array}{c} CHO \\ | \\ CHOH \\ | \\ CH_2OH \end{array} \xrightleftharpoons{H_2O,\ HCl} \begin{array}{c} CHO \\ | \\ CHOH \\ | \\ CH_2OH \end{array} + \begin{array}{c} CH_2OH \\ | \\ C{=}O \\ | \\ CH_2OH \end{array}$$

(*R*)-Glyceraldehyde (*R,S*)-Glyceraldehyde Dihydroxyacetone

Propose a mechanism for this isomerization.

Section 13.9 Oxidation/Reduction of Aldehydes and Ketones

13.36 Draw a structural formula for the product formed by treating butanal with each of the following sets of reagents: **(See Examples 13.11, 13.12)**

(a) $LiAlH_4$ followed by H_2O

(b) $NaBH_4$ in CH_3OH/H_2O

(c) H_2/Pt

(d) $Ag(NH_3)_2^+$ in NH_3/H_2O and then HCl/H_2O

(e) H_2CrO_4

(f) $C_6H_5NH_2$ in the presence of H_2/Ni

13.37 Draw a structural formula for the product of the reaction of *p*-bromoacetophenone with each set of reagents in Problem 13.36. **(See Examples 13.11, 13.12)**

Synthesis

13.38 Show the reagents and conditions that will bring about the conversion of cyclohexanol to cyclohexanecarbaldehyde: **(See Example 13.7)**

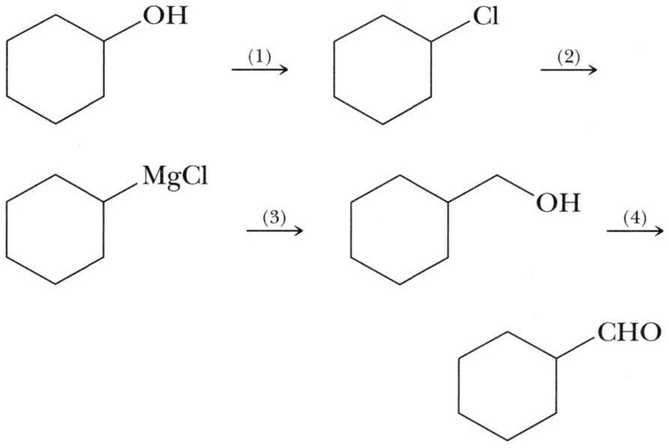

Cyclohexanol (OH) $\xrightarrow{(1)}$ chlorocyclohexane (Cl) $\xrightarrow{(2)}$ cyclohexylmagnesium chloride (MgCl) $\xrightarrow{(3)}$ cyclohexylmethanol (OH) $\xrightarrow{(4)}$ cyclohexanecarbaldehyde (CHO)

13.39 Starting with cyclohexanone, show how to prepare these compounds (in addition to the given starting material, use any other organic or inorganic reagents, as necessary): **(See Example 13.7)**

(a) Cyclohexanol

(b) Cyclohexene

(c) 6-Oxohexanal

(d) 1-Methylcyclohexanol

(e) 1-Methylcyclohexene

(f) 1-Phenylcyclohexanol

(g) 1-Phenylcyclohexene

(h) Cyclohexene oxide

(i) *trans*-1,2-Cyclohexanediol

13.40 Show how to bring about these conversions (in addition to the given starting material, use any other organic or inorganic reagents, as necessary): **(See Example 13.7)**

(a) 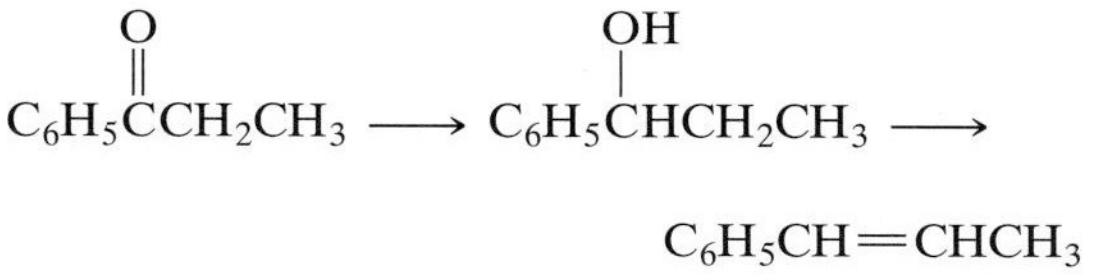 $C_6H_5\overset{O}{\overset{\|}{C}}CH_2CH_3 \longrightarrow C_6H_5\overset{OH}{\overset{|}{C}}HCH_2CH_3 \longrightarrow C_6H_5CH{=}CHCH_3$

(b) cyclopentanone (=O) ⟶ cyclopentanol (—OH) ⟶ chlorocyclopentane (—Cl) ⟶ cyclopentylmethanol (—CH_2OH)

(c) cyclopentanone (=O) ⟶ 1-cyclopentylcyclopentanol (OH)

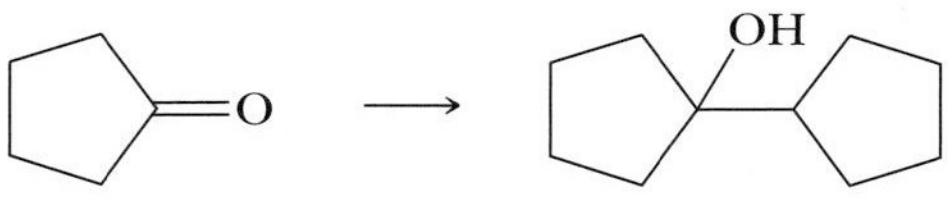

(d) 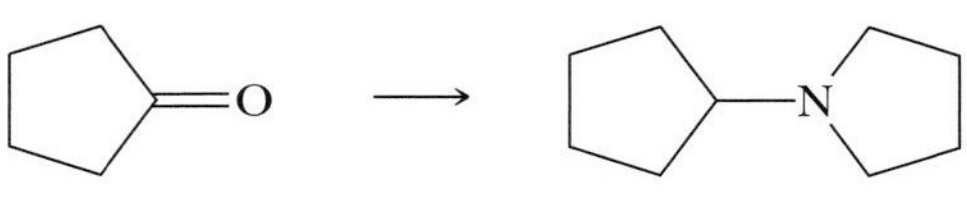 cyclopentanone (=O) ⟶ N-cyclopentylpyrrolidine (N)

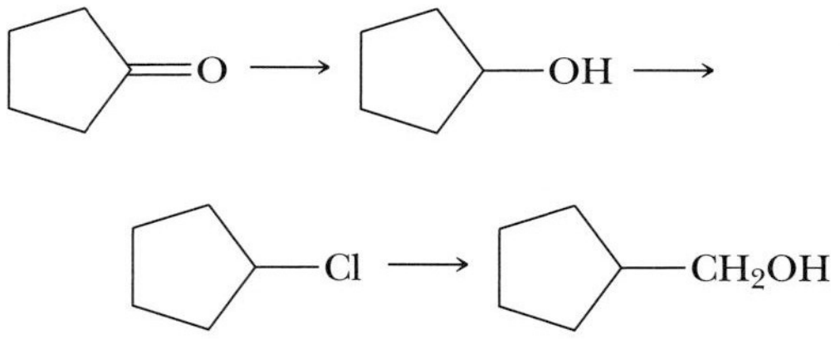

***13.41** Many tumors of the breast are estrogen dependent. Drugs that interfere with estrogen binding have antitumor activity and may even help prevent the occurrence of tumors. A widely used antiestrogen drug is tamoxifen: **(See Example 13.7)**

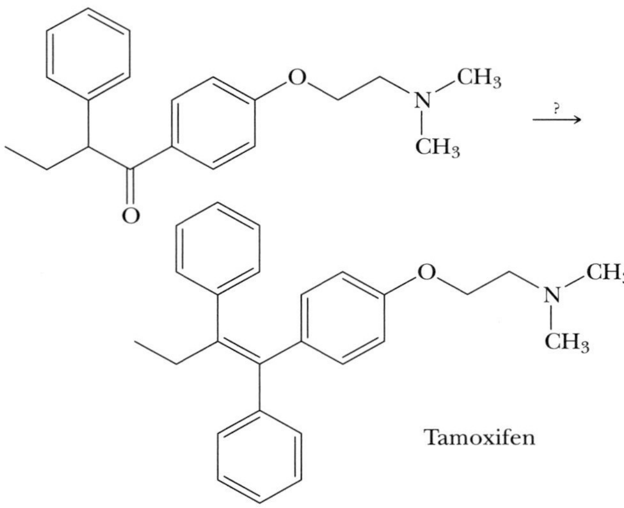

Tamoxifen

(a) How many stereoisomers are possible for tamoxifen?

(b) Specify the configuration of the stereoisomer shown here.

(c) Show how tamoxifen can be synthesized from the given ketone using a Grignard reaction, followed by dehydration.

***13.42** Following is a possible synthesis of the antidepressant bupropion (Wellbutrin®): **(See Example 13.7)**

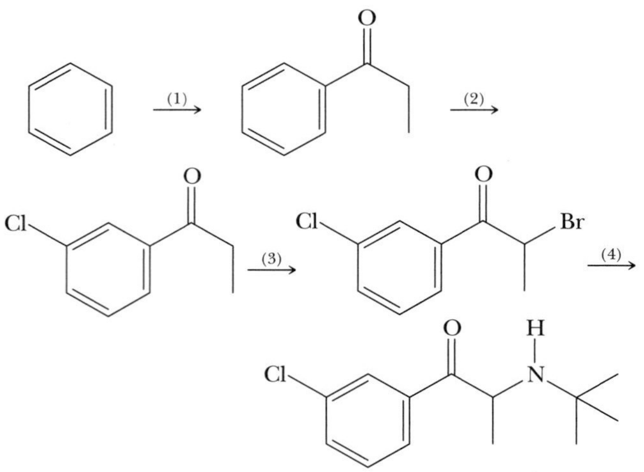

Bupropion (Wellbutrin®)

Show the reagents that will bring about each step in this synthesis.

***13.43** The synthesis of chlorpromazine in the 1950s and the discovery soon thereafter of the drug's antipsychotic activity opened the modern era of biochemical investigations into the pharmacology of the central nervous system. One of the compounds prepared in the search for more effective antipsychotics was amitriptyline. **(See Example 13.7)**

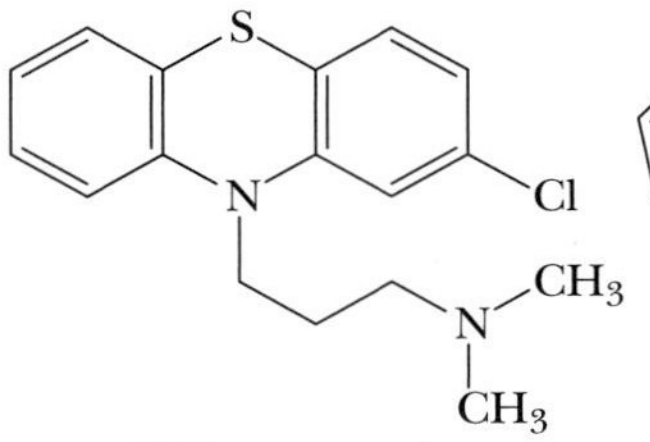

Chlorpromazine 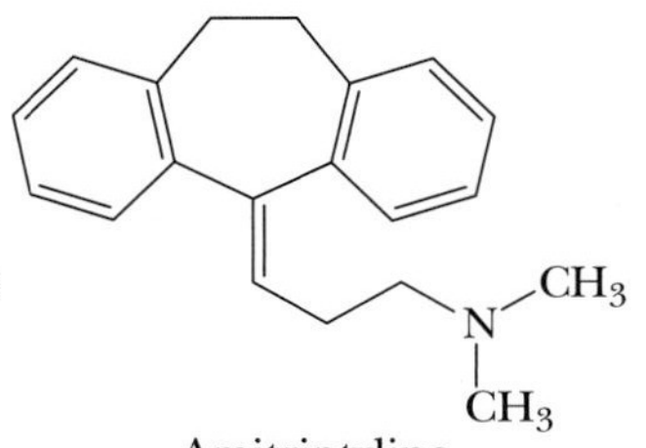   Amitriptyline

Surprisingly, amitriptyline shows antidepressant activity rather than antipsychotic activity. It is now known that amitriptyline inhibits the reuptake of norepinephrine and serotonin from the synaptic cleft. Because the reuptake of these neurotransmitters is inhibited, their effects are potentiated. That is, the two neurotransmitters remain available to interact with serotonin and norepinephrine receptor sites longer and continue to cause excitation of serotonin and norepinephrine-mediated neural pathways. The following is a synthesis for amitriptyline:

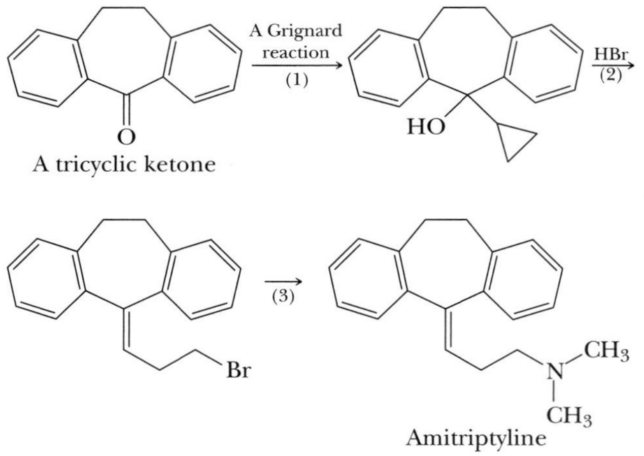

A tricyclic ketone —A Grignard reaction (1)→ —HBr (2)→ —(3)→ Amitriptyline

(a) Propose a reagent for Step 1.

(b) Propose a mechanism for Step 2. (*Note:* It is not acceptable to propose a primary carbocation as an intermediate.)

(c) Propose a reagent for Step 3.

***13.44** Following is a synthesis for diphenhydramine: **(See Example 13.7)**

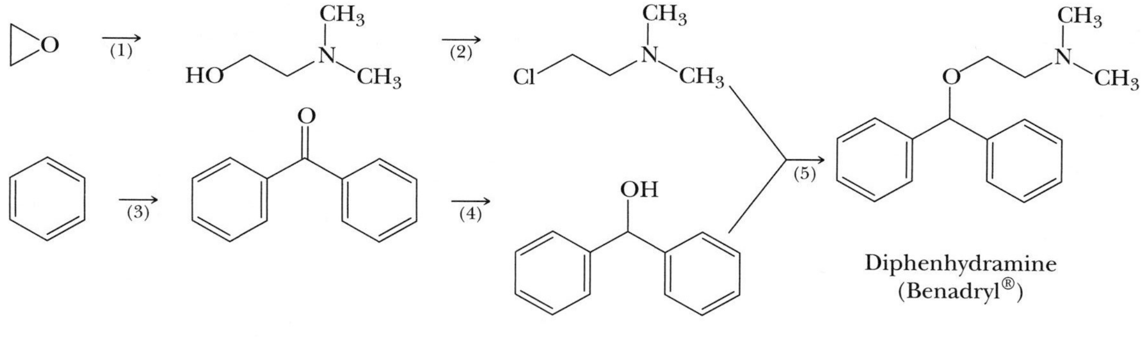

Diphenhydramine
(Benadryl®)

The hydrochloride salt of this compound, best known by its trade name, Benadryl®, is an antihistamine.

(a) Propose reagents for Steps 1 and 2.

(b) Propose reagents for Steps 3 and 4.

(c) Show that Step 5 is an example of nucleophilic aliphatic substitution. What type of mechanism—S_N1 or S_N2—is more likely for this reaction? Explain.

***13.45** Following is a synthesis for the antidepressant venlafaxine: **(See Example 13.7)**

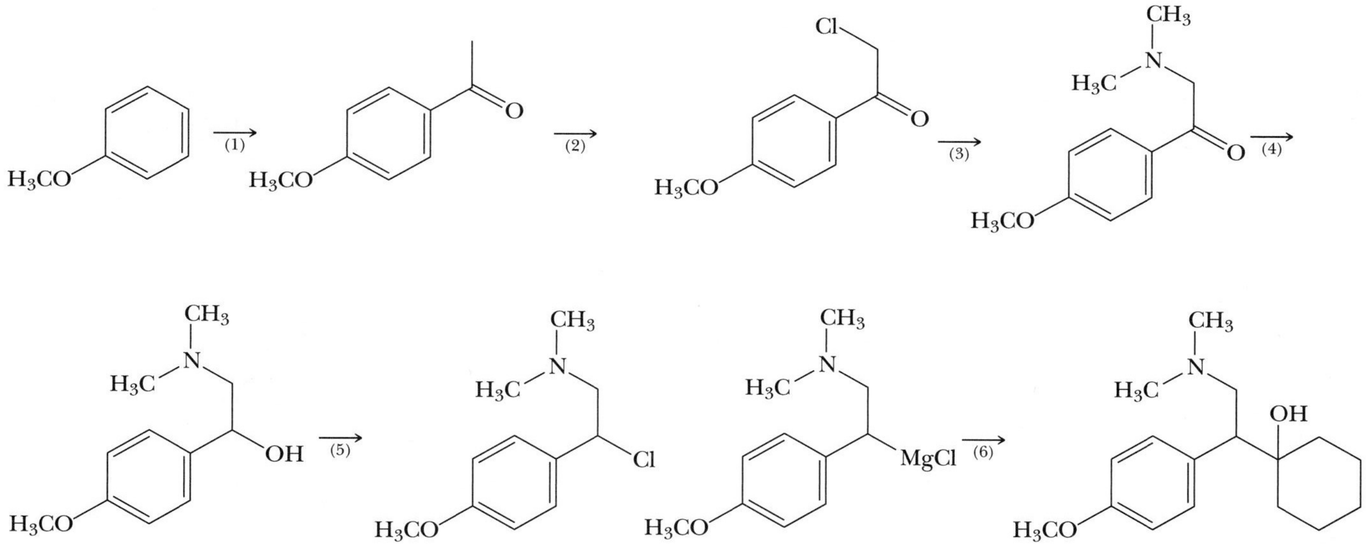

Venlafaxine

(a) Propose a reagent for Step 1, and name the type of reaction that takes place.

(b) Propose reagents for Steps 2 and 3.

(c) Propose reagents for Steps 4 and 5.

(d) Propose a reagent for Step 6, and name the type of reaction that takes place.

Chemical Transformations

13.46 Test your cumulative knowledge of the reactions learned thus far by completing the following chemical transformations. *Note*: Some will require more than one step.

(a) OH ⟶ OH

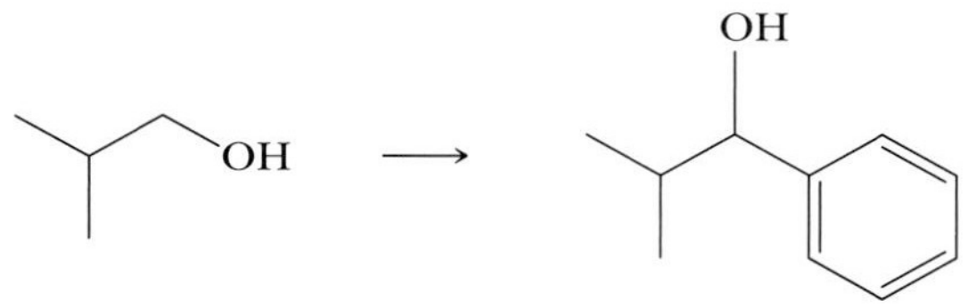

(b) OH ⟶ O O

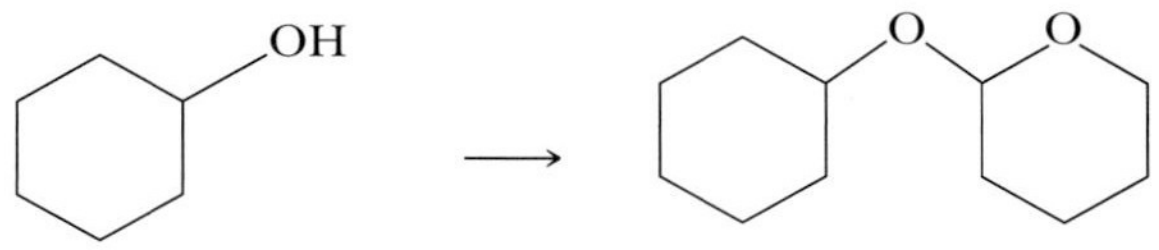

(c) OH ⟶ OH

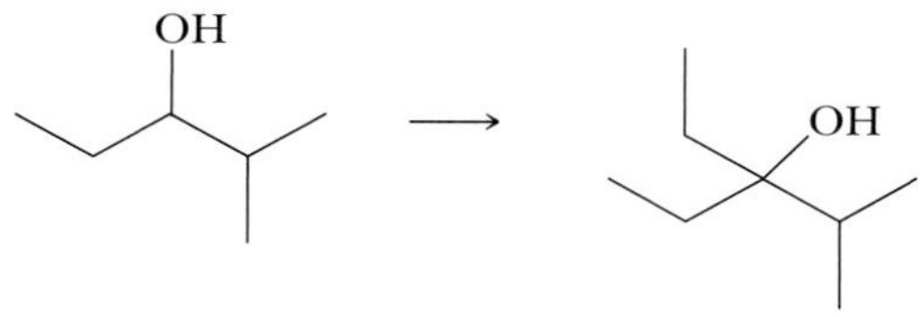

(d) ⟶ O Br

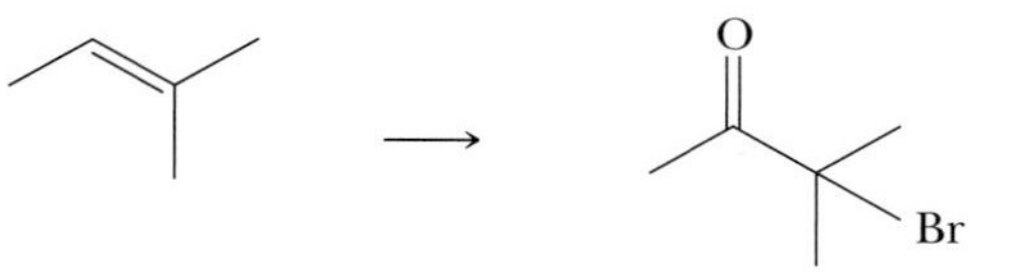

(e) OH ⟶ O

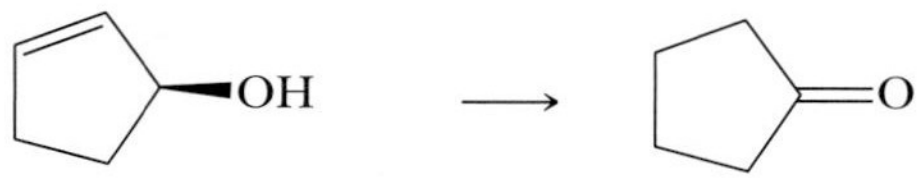

(f) ⟶ HN

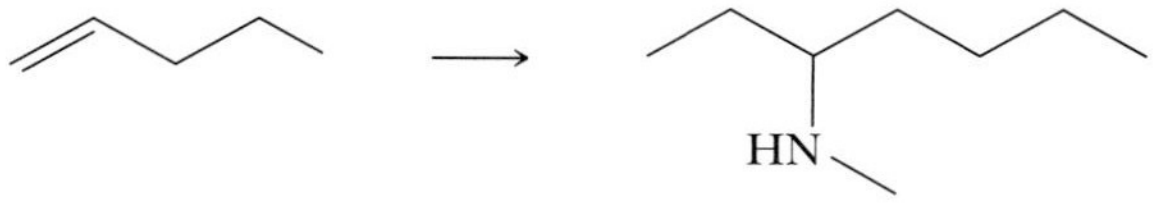

(g) ⟶ NH Cl

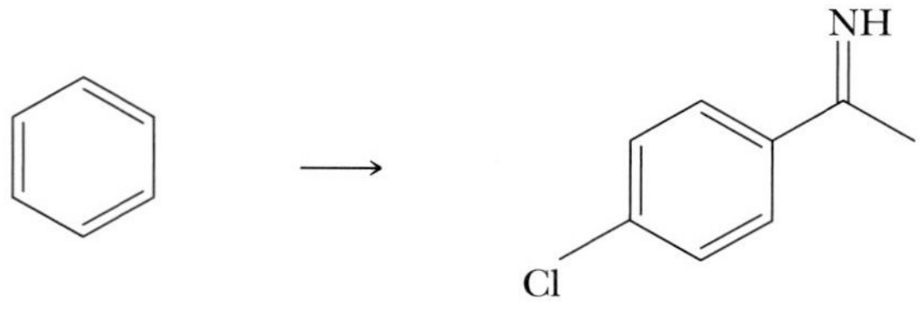

(h) Br H O ⟶ O H O

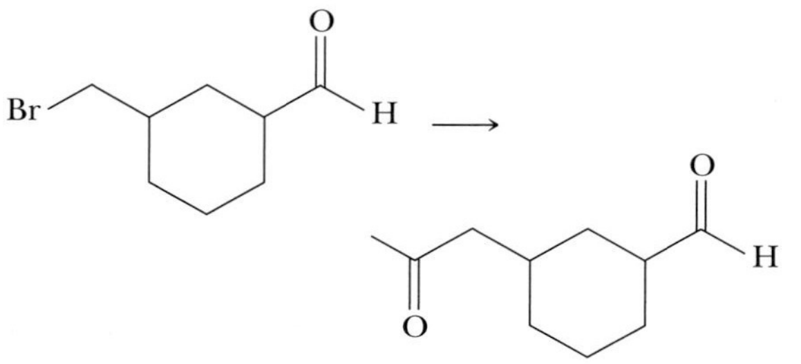

(i) OH ⟶

(j) ⟶ O OH

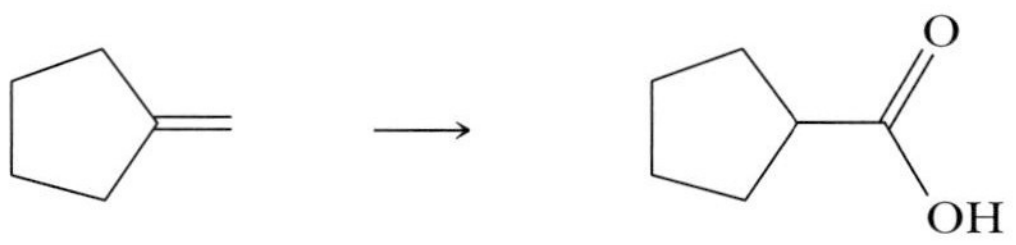

(k) ⟶ OH

(l) Cl ⟶ N H

(m) O ⟶ O O

(n) Br ⟶ NH_2

(o) $H_3C—C≡C—H$ ⟶ O

(p) ⟶

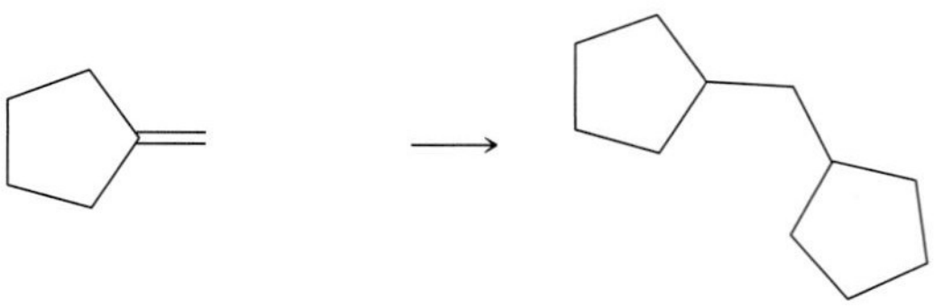

(q) Cl ⟶ O HO

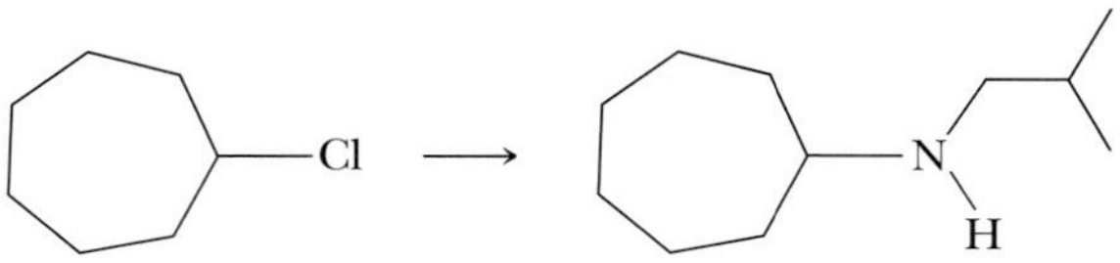

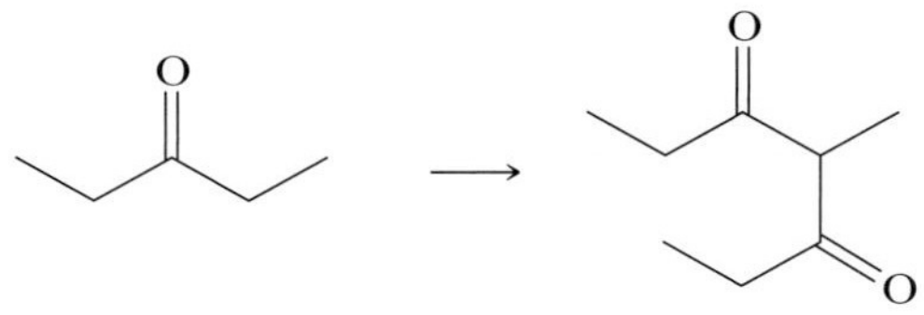

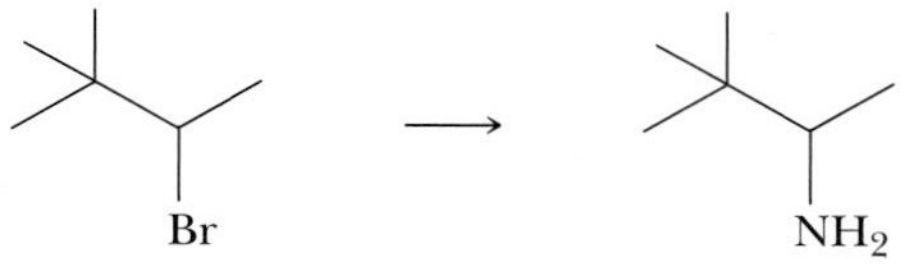

(r) OH ⟶ H N

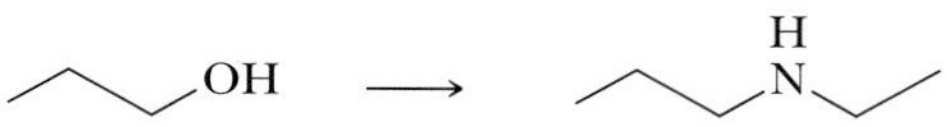

Spectroscopy

13.47 Compound A, $C_5H_{10}O$, is used as a flavoring agent for many foods that possess a chocolate or peach flavor. Its common name is isovaleraldehyde, and it gives ^{13}C-NMR peaks at δ 202.7, 52.7, 23.6, and 22.6. Provide a structural formula for isovaleraldehyde and give its IUPAC name.

13.48 Following are ^{1}H-NMR and IR spectra of compound B, $C_6H_{12}O_2$:

(300 MHz,$CDCl_3$)

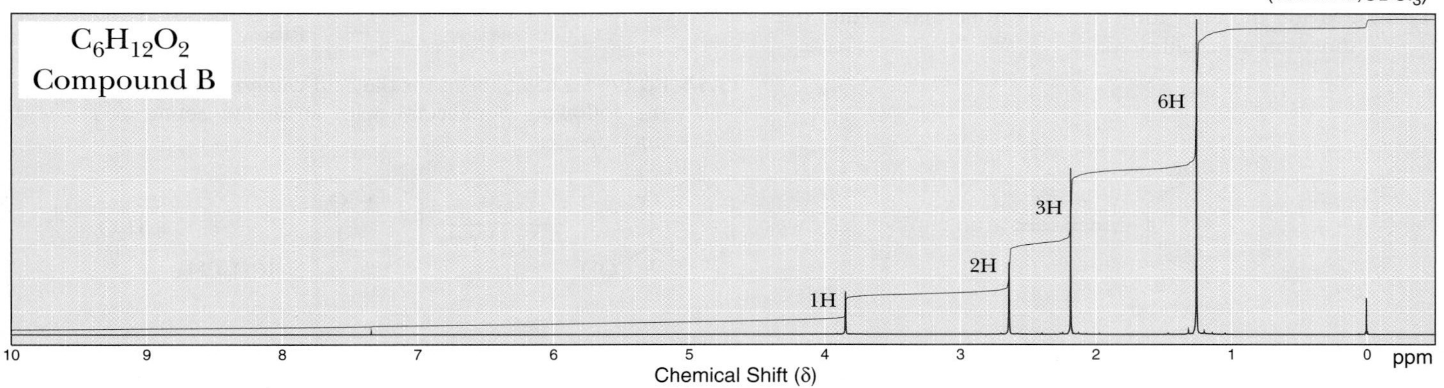

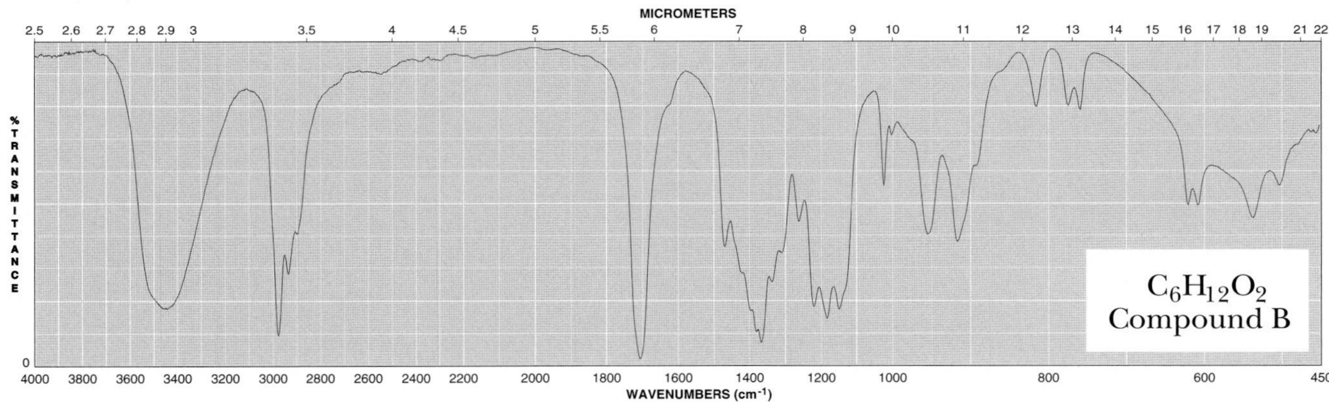

Propose a structural formula for compound B.

13.49 Compound C, $C_9H_{18}O$, is used in the automotive industry to retard the flow of solvent and thus improve the application of paints and coatings. It yields ^{13}C-NMR peaks at δ 210.5, 52.4, 24.5, and 22.6. Provide a structure and an IUPAC name for compound C.

Looking Ahead

13.50 Reaction of a Grignard reagent with carbon dioxide, followed by treatment with aqueous HCl, gives a carboxylic acid. Propose a structural formula for the bracketed intermediate formed by the reaction of phenylmagnesium bromide with CO_2, and propose a mechanism for the formation of this intermediate:

MgBr + CO_2 ⟶ [intermediate (not isolated)] $\xrightarrow{HCl,\ H_2O}$ C(=O)OH

13.51 Rank the following carbonyls in order of increasing reactivity to nucleophilic attack, and explain your reasoning.

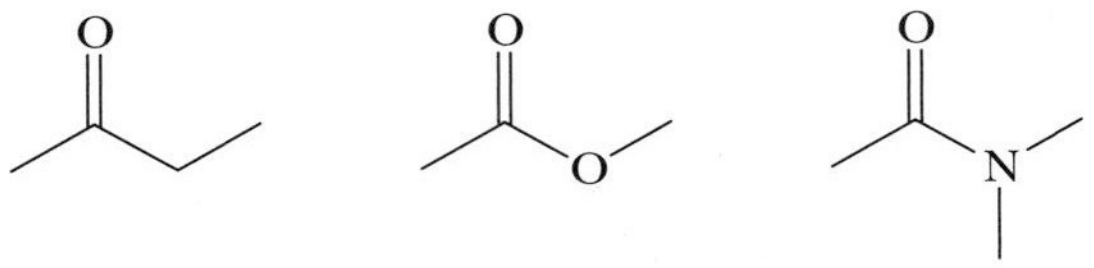

13.52 Provide the enol form of this ketone and predict the direction of equilibrium:

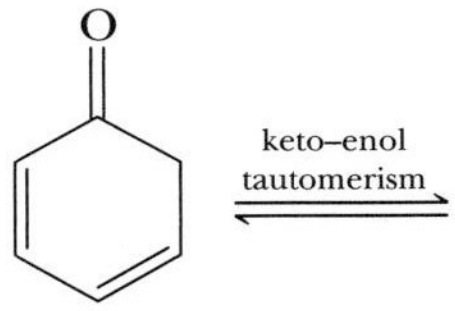

keto–enol tautomerism

an enol

13.53 Draw the cyclic hemiacetal formed by reaction of the highlighted —OH group with the aldehyde group:

(a)

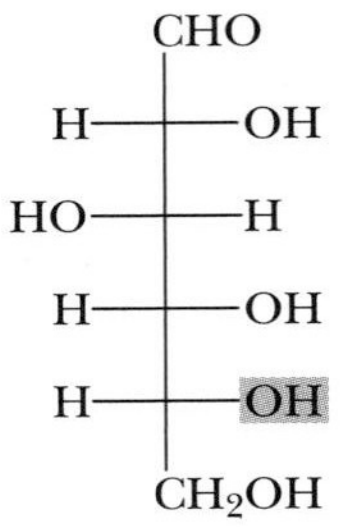

	CHO	
H		OH
HO		H
H		OH
H		**OH**
	CH_2OH	

Glucose

(b)

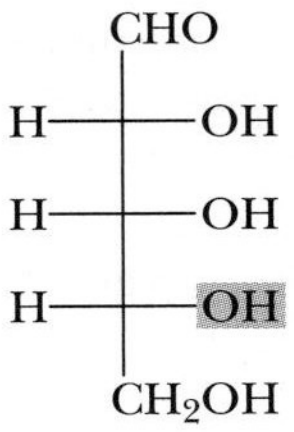

	CHO	
H		OH
H		OH
H		**OH**
	CH_2OH	

Ribose

13.54 Propose a mechanism for the acid-catalyzed reaction of the following hemiacetal, with an amine acting as a nucleophile:

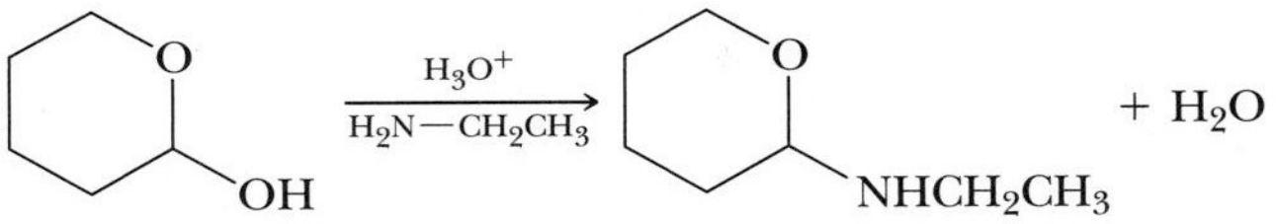

$\xrightarrow[H_2N-CH_2CH_3]{H_3O^+}$ $NHCH_2CH_3$ + H_2O

14 Carboxylic Acids

KEY QUESTIONS

14.1 What Are Carboxylic Acids?

14.2 How Are Carboxylic Acids Named?

14.3 What Are the Physical Properties of Carboxylic Acids?

14.4 What Are the Acid–Base Properties of Carboxylic Acids?

14.5 How Are Carboxyl Groups Reduced?

14.6 What Is Fischer Esterification?

14.7 What Are Acid Chlorides?

14.8 What Is Decarboxylation?

HOW TO

14.1 How to Predict the Product of a Fischer Esterification

14.2 How to Predict the Product of a β-Decarboxylation Reaction

CHEMICAL CONNECTIONS

14A From Willow Bark to Aspirin and Beyond

14B Esters as Flavoring Agents

14C Ketone Bodies and Diabetes

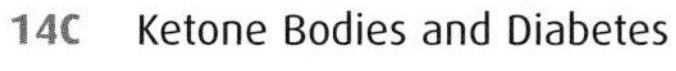

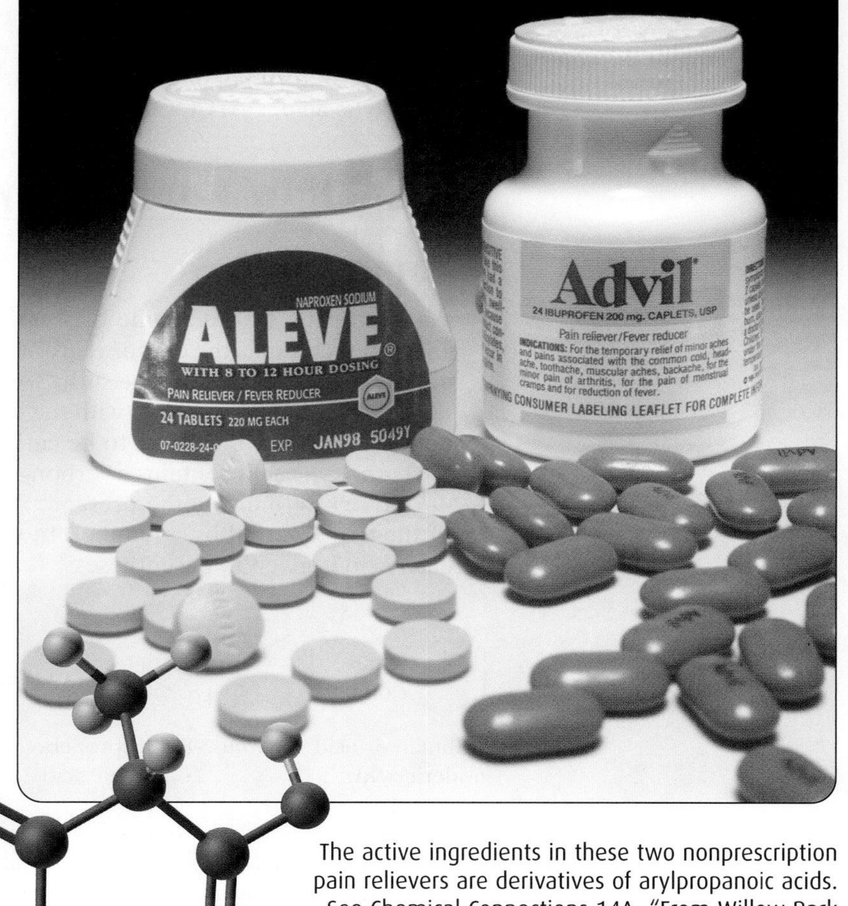

The active ingredients in these two nonprescription pain relievers are derivatives of arylpropanoic acids. See Chemical Connections 14A, "From Willow Bark to Aspirin and Beyond." Inset: A model of (S)-ibuprofen. *(Charles D. Winters)*

Carboxylic acids are another class of organic compounds containing the carbonyl group. Their occurrence in nature is widespread, and they are important components of foodstuffs such as vinegar, butter, and vegetable oils. The most important chemical property of carboxylic acids is their acidity. Furthermore, carboxylic acids form numerous important derivatives, including esters, amides, anhydrides, and acid halides. In this chapter, we study carboxylic acids themselves; in Chapters 15 and 16, we study their derivatives.

14.1 What Are Carboxylic Acids?

Carboxyl group A —COOH group.

The functional group of a carboxylic acid is a **carboxyl group**, so named because it is made up of a **carb**onyl group and a hydr**oxyl** group (Section 1.7D). Following is a Lewis structure of the carboxyl group, as well as two alternative representations of it:

$$-C(=\ddot{O}:)-\ddot{O}-H \qquad -COOH \qquad -CO_2H$$

The general formula of an aliphatic carboxylic acid is RCOOH; that of an aromatic carboxylic acid is ArCOOH.

14.2 How Are Carboxylic Acids Named?

A. IUPAC System

We derive the IUPAC name of a carboxylic acid from that of the longest carbon chain that contains the carboxyl group by dropping the final *-e* from the name of the parent alkane and adding the suffix *-oic*, followed by the word *acid* (Section 3.5). We number the chain beginning with the carbon of the carboxyl group. Because the carboxyl carbon is understood to be carbon 1, there is no need to give it a number. If the carboxylic acid contains a carbon–carbon double bond, we change the infix from *-an-* to *-en-* to indicate the presence of the double bond, and we show the location of the double bond by a number. In the following examples, the common name of each acid is given in parentheses:

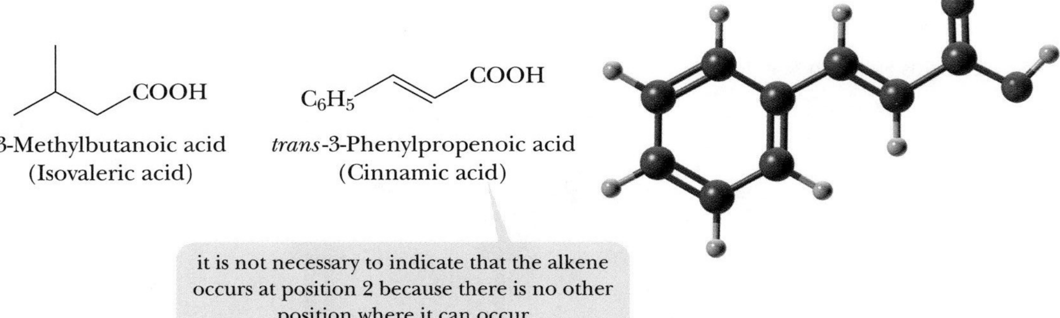
COOH
C_6H_5
COOH
3-Methylbutanoic acid
(Isovaleric acid)
trans-3-Phenylpropenoic acid
(Cinnamic acid)
it is not necessary to indicate that the alkene occurs at position 2 because there is no other position where it can occur

In the IUPAC system, a carboxyl group takes precedence over most other functional groups (Table 13.1), including hydroxyl and amino groups, as well as the carbonyl groups of aldehydes and ketones. As illustrated in the following examples, an —OH group of an alcohol is indicated by the prefix *hydroxy*-, an $-NH_2$ group of an amine by *amino*-, and an =O group of an aldehyde or ketone by *oxo*-:

OH … COOH — 5-Hydroxyhexanoic acid

H_2N … COOH — 4-Aminobutanoic acid

O … COOH — 5-Oxohexanoic acid

Dicarboxylic acids are named by adding the suffix *-dioic*, followed by the word *acid*, to the name of the carbon chain that contains both carboxyl groups. Because the two carboxyl groups can be only at the ends of the parent chain, there is no need to number them. Following are IUPAC names and common names for several important aliphatic dicarboxylic acids:

Ethanedioic acid
(Oxalic acid)

Propanedioic acid
(Malonic acid)

Butanedioic acid
(Succinic acid)

Pentanedioic acid
(Glutaric acid)

Hexanedioic acid
(Adipic acid)

The name *oxalic acid* is derived from one of its sources in the biological world, namely, plants of the genus *Oxalis*, one of which is rhubarb. Oxalic acid also occurs in human and animal urine, and calcium oxalate (the calcium salt of oxalic acid) is a major component of kidney stones. Adipic acid is one of the two monomers required for the synthesis of the polymer nylon 66. The U.S. chemical industry produces approximately 1.8 billion pounds of adipic acid annually, solely for the synthesis of nylon 66 (Section 17.4A).

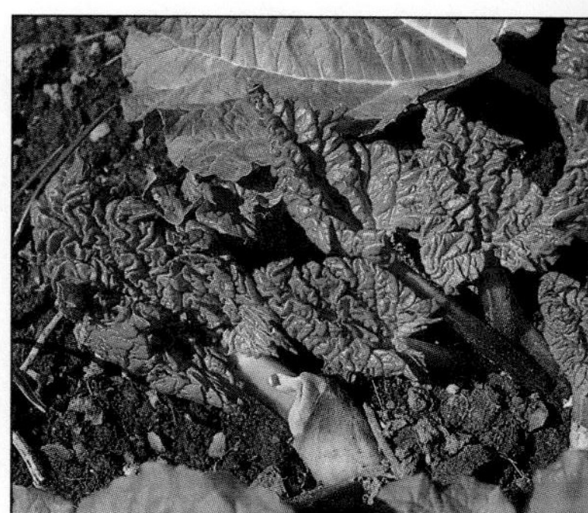

Leaves of the rhubarb plant contain oxalic acid as its potassium and sodium salts. *(Hans Reinhard/OKAPIA/Photo Researchers, Inc.)*

A carboxylic acid containing a carboxyl group bonded to a cycloalkane ring is named by giving the name of the ring and adding the suffix *-carboxylic acid*. The atoms of the ring are numbered beginning with the carbon bearing the —COOH group:

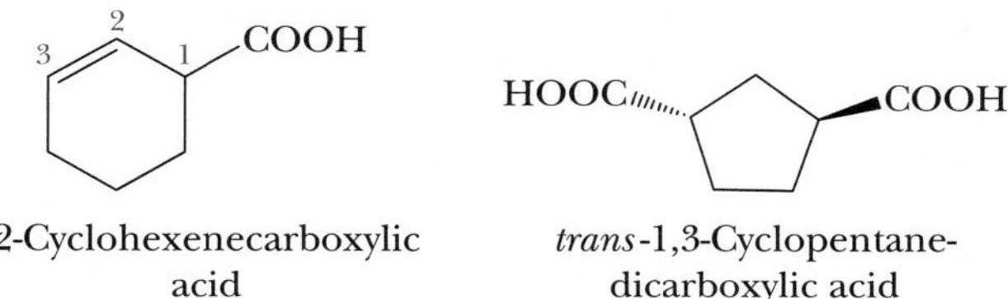

2-Cyclohexenecarboxylic acid

trans-1,3-Cyclopentanedicarboxylic acid

The simplest aromatic carboxylic acid is benzoic acid. Derivatives are named by using numbers and prefixes to show the presence and location of substituents relative to the carboxyl group. Certain aromatic carboxylic acids have common names by which they are more usually known. For example, 2-hydroxybenzoic acid is more often called salicylic acid, a name derived from the fact that this aromatic carboxylic acid was first obtained from the bark of the willow, a tree of the genus *Salix*. Aromatic dicarboxylic acids are named by adding the words *dicarboxylic acid* to *benzene*. Examples are 1,2-benzenedicarboxylic acid and 1,4-benzenedicarboxylic acid. Each is more usually known by its common name: phthalic acid and terephthalic acid, respectively. Terephthalic acid is one of the two organic components required for the synthesis of the textile fiber known as Dacron® polyester (Section 17.4B).

Benzoic acid

2-Hydroxybenzoic acid
(Salicylic acid)

1,2-Benzenedicarboxylic acid
(Phthalic acid)

1,4-Benzenedicarboxylic acid
(Terephthalic acid)

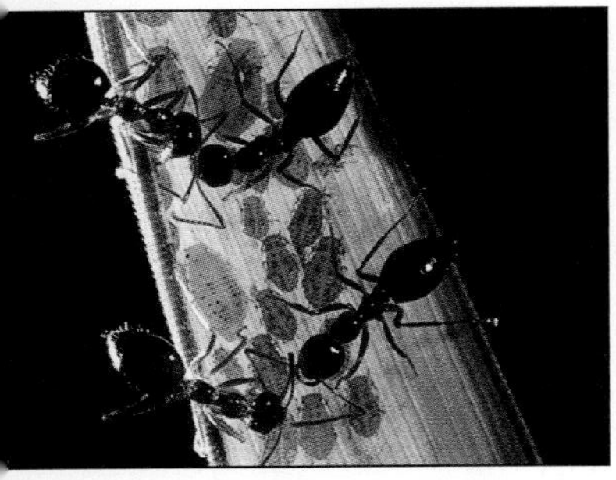

Formic acid was first obtained in 1670 from the destructive distillation of ants, whose genus is *Formica*. It is one of the components of the venom of stinging ants. *(Ted Nelson/Dembinsky Photo Associates)*

TABLE 14.1 Several Aliphatic Carboxylic Acids and Their Common Names

Structure	IUPAC Name	Common Name	Derivation
HCOOH	methanoic acid	formic acid	Latin: *formica*, ant
CH_3COOH	ethanoic acid	acetic acid	Latin: *acetum*, vinegar
CH_3CH_2COOH	propanoic acid	propionic acid	Greek: *propion*, first fat
$CH_3(CH_2)_2COOH$	butanoic acid	butyric acid	Latin: *butyrum*, butter
$CH_3(CH_2)_3COOH$	pentanoic acid	valeric acid	Latin: *valere*, to be strong
$CH_3(CH_2)_4COOH$	hexanoic acid	caproic acid	Latin: *caper*, goat
$CH_3(CH_2)_6COOH$	octanoic acid	caprylic acid	Latin: *caper*, goat
$CH_3(CH_2)_8COOH$	decanoic acid	capric acid	Latin: *caper*, goat
$CH_3(CH_2)_{10}COOH$	dodecanoic acid	lauric acid	Latin: *laurus*, laurel
$CH_3(CH_2)_{12}COOH$	tetradecanoic acid	myristic acid	Greek: *myristikos*, fragrant
$CH_3(CH_2)_{14}COOH$	hexadecanoic acid	palmitic acid	Latin: *palma*, palm tree
$CH_3(CH_2)_{16}COOH$	octadecanoic acid	stearic acid	Greek: *stear*, solid fat
$CH_3(CH_2)_{18}COOH$	icosanoic acid	arachidic acid	Greek: *arachis*, peanut

B. Common Names

Aliphatic carboxylic acids, many of which were known long before the development of structural theory and IUPAC nomenclature, are named according to their source or for some characteristic property. Table 14.1 lists several of the unbranched aliphatic carboxylic acids found in the biological world, along with the common name of each. Those with 16, 18, and 20 carbon atoms are particularly abundant in fats and oils (Section 21.1) and the phospholipid components of biological membranes (Section 21.3).

When common names are used, the Greek letters α, β, γ, δ, and so forth are often added as a prefix to locate substituents. The α-position in a carboxylic acid is the position next to the carboxyl group; an α-substituent in a common name is equivalent to a 2-substituent in an IUPAC name. *GABA*, short for *g*amma-*a*mino*b*utyric *a*cid, is an inhibitory neurotransmitter in the central nervous system of humans:

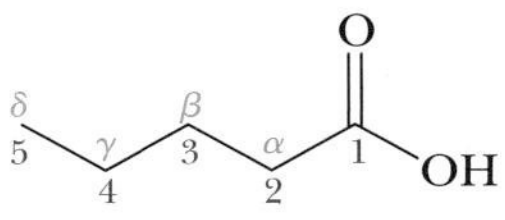

δ 5, γ 4, β 3, α 2, 1 — O, OH

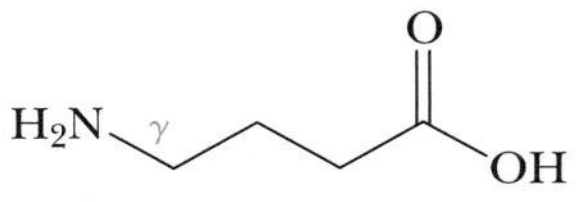

H_2N γ — O, OH

4-Aminobutanoic acid
(γ-Aminobutyric acid, GABA)

In common nomenclature, the prefix *keto-* indicates the presence of a ketone carbonyl in a substituted carboxylic acid (as illustrated by the common name β-ketobutyric acid):

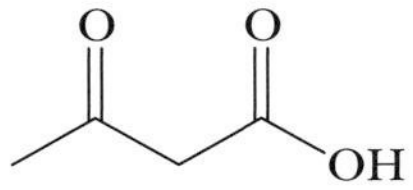

O O OH

3-Oxobutanoic acid
(β-Ketobutyric acid;
Acetoacetic acid)

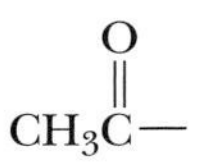

$CH_3\overset{O}{\overset{\|}{C}}—$

Acetyl group
(Aceto group)

An alternative common name for 3-oxobutanoic acid is acetoacetic acid. In deriving this common name, this ketoacid is regarded as a substituted acetic acid, and the $CH_3C(=O)—$ substituent is named an **aceto group**.

Aceto group A $CH_3CO—$ group.

Example 14.1

Write the IUPAC name for each carboxylic acid:

(a) $CH_3(CH_2)_7$(H)C=C(H)$(CH_2)_7COOH$

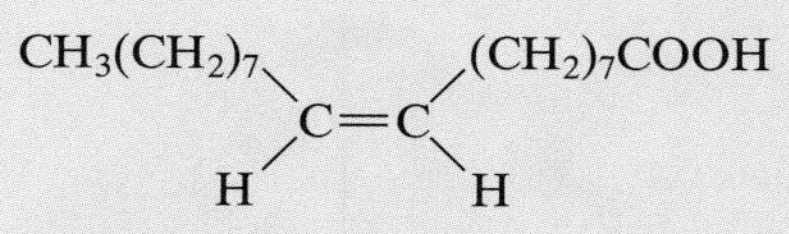

(b) cyclohexane ring bearing COOH and OH

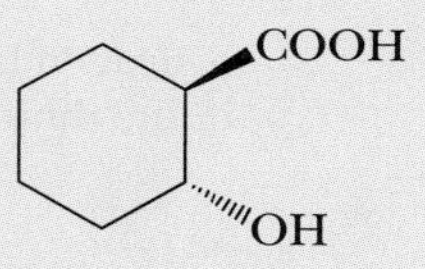

(c) C with OH, H, CH_3, COOH

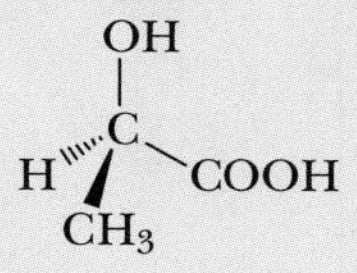

(d) $ClCH_2COOH$

Strategy

Identify the longest chain of carbon atoms that contains the carboxyl group to determine the root name. The suffix *-e* is then changed to *-anoic acid*. For cyclic carboxylic acids, *carboxylic acid* is appended to the name of the cycloalkane (without dropping the suffix *-e*). As usual, remember to note stereochemistry (*E/Z*, *cis/trans*, or *R/S*) where appropriate.

Solution

(a) *cis*-9-Octadecenoic acid (oleic acid)
(b) *trans*-2-Hydroxycyclohexanecarboxylic acid
(c) (*R*)-2-Hydroxypropanoic acid [(*R*)-lactic acid]
(d) Chloroethanoic acid (chloroacetic acid)

See problems 14.9–14.12, 14.15

Problem 14.1

Each of the following compounds has a well-recognized common name. A derivative of glyceric acid is an intermediate in glycolysis (Section 22.3). Maleic acid is an intermediate in the tricarboxylic acid (TCA) cycle. Mevalonic acid is an intermediate in the biosynthesis of steroids (Section 21.4B).

(a) HO, COOH, HO, H — Glyceric acid

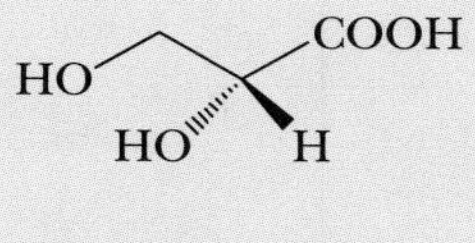

(b) COOH, COOH — Maleic acid

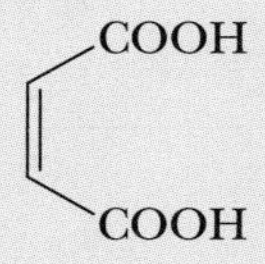

(c) HO, CH3, COOH, HO — Mevalonic acid

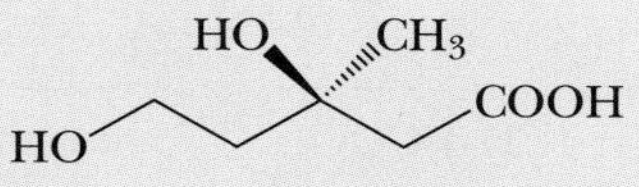

Write the IUPAC name for each compound. Be certain to show the configuration of each.

14.3 What Are the Physical Properties of Carboxylic Acids?

In the liquid and solid states, carboxylic acids are associated by intermolecular hydrogen bonding into dimers, as shown for acetic acid:

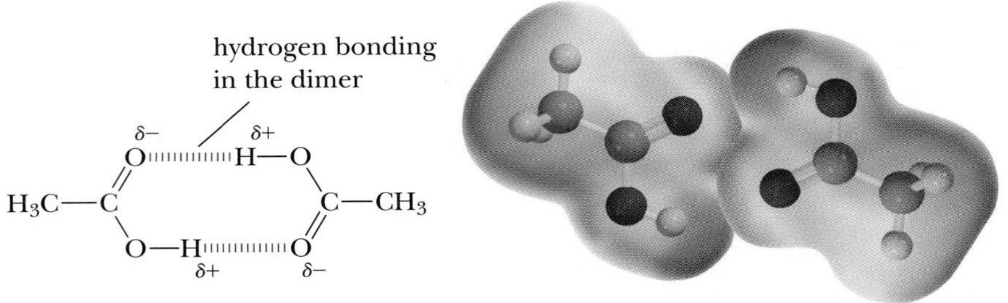

Carboxylic acids have significantly higher boiling points than other types of organic compounds of comparable molecular weight, such as alcohols, aldehydes, and ketones. For example, butanoic acid (Table 14.2) has a higher boiling point than either 1-pentanol or pentanal. The higher boiling points of carboxylic acids result from their polarity and from the fact that they form very strong intermolecular hydrogen bonds.

Carboxylic acids also interact with water molecules by hydrogen bonding through both their carbonyl and hydroxyl groups. Because of these hydrogen-bonding interactions, carboxylic acids are more soluble in water than are alcohols, ethers, aldehydes, and ketones of comparable molecular weight. The solubility of a carboxylic acid in water decreases as its molecular weight increases. We account for this trend in the following way: A carboxylic acid consists of two regions of different polarity—a polar hydrophilic carboxyl group and, except for formic acid, a nonpolar hydrophobic hydrocarbon chain. The **hydrophilic** carboxyl group increases water solubility; the **hydrophobic** hydrocarbon chain decreases water solubility.

Hydrophilic From the Greek, meaning "water loving."

Hydrophobic From the Greek, meaning "water hating."

TABLE 14.2 Boiling Points and Solubilities in Water of Selected Carboxylic Acids, Alcohols, and Aldehydes of Comparable Molecular Weight

Structure	Name	Molecular Weight	Boiling Point (°C)	Solubility (g/100 mL H_2O)
CH_3COOH	acetic acid	60.5	118	infinite
$CH_3CH_2CH_2OH$	1-propanol	60.1	97	infinite
CH_3CH_2CHO	propanal	58.1	48	16
$CH_3(CH_2)_2COOH$	butanoic acid	88.1	163	infinite
$CH_3(CH_2)_3CH_2OH$	1-pentanol	88.1	137	2.3
$CH_3(CH_2)_3CHO$	pentanal	86.1	103	slight
$CH_3(CH_2)_4COOH$	hexanoic acid	116.2	205	1.0
$CH_3(CH_2)_5CH_2OH$	1-heptanol	116.2	176	0.2
$CH_3(CH_2)_5CHO$	heptanal	114.1	153	0.1

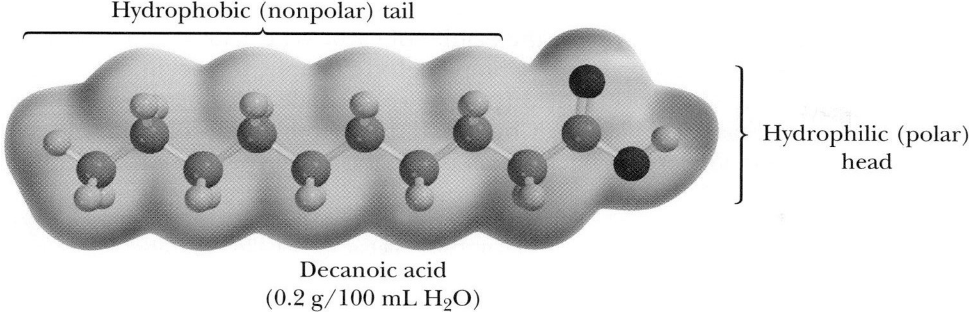

Decanoic acid
(0.2 g/100 mL H_2O)

The first four aliphatic carboxylic acids (formic, acetic, propanoic, and butanoic acids) are infinitely soluble in water because the hydrophilic character of the carboxyl group more than counterbalances the hydrophobic character of the hydrocarbon chain. As the size of the hydrocarbon chain increases relative to the size of the carboxyl group, water solubility decreases. The solubility of hexanoic acid in water is 1.0 g/100 g water; that of decanoic acid is only 0.2 g/100 g water.

One other physical property of carboxylic acids must be mentioned: The liquid carboxylic acids, from propanoic acid to decanoic acid, have extremely foul odors, about as bad as those of thiols, though different. Butanoic acid is found in stale perspiration and is a major component of "locker room odor." Pentanoic acid smells even worse, and goats, which secrete C_6, C_8, and C_{10} acids, are not famous for their pleasant odors.

14.4 What Are the Acid–Base Properties of Carboxylic Acids?

A. Acid Ionization Constants

Carboxylic acids are weak acids. Values of K_a for most unsubstituted aliphatic and aromatic carboxylic acids fall within the range from 10^{-4} to 10^{-5}. The value of K_a for acetic acid, for example, is 1.74×10^{-5}, and the pK_a of acetic acid is 4.76:

$$CH_3COOH + H_2O \rightleftharpoons CH_3COO^- + H_3O^+$$

$$K_a = \frac{[CH_3COO^-][H_3O^+]}{[CH_3COOH]} = 1.74 \times 10^{-5}$$

$$pK_a = 4.76$$

As we discussed in Section 2.5B, carboxylic acids are stronger acids (pK_a 4–5) than alcohols (pK_a 16–18) because resonance stabilizes the **carboxylate** anion by delocalizing its negative charge. No comparable resonance stabilization exists in alkoxide ions.

$$RCOOH \rightleftharpoons H^+ + \left[RC(=O)O^- \longleftrightarrow RC(-O^-)=O \right]$$

pK_a 4–5

$$R{-}\ddot{O}H \rightleftharpoons H^+ + R{-}\ddot{O}^-$$

pK_a 16–18

resonance stabilization delocalizes the negative charge

no resonance stabilization

Substitution at the α-carbon of an atom or a group of atoms of higher electronegativity than carbon increases the acidity of carboxylic acids, often by several orders of magnitude (Section 2.5C). Compare, for example, the acidities of acetic acid (pK_a 4.76) and chloroacetic acid (pK_a 2.86). A single chlorine substituent on the α-carbon increases acid strength by nearly 100! Both dichloroacetic acid and trichloroacetic acid are stronger acids than phosphoric acid (pK_a 2.1):

the inductive effect of an electronegative atom delocalizes the negative charge and stabilizes the carboxylate ion

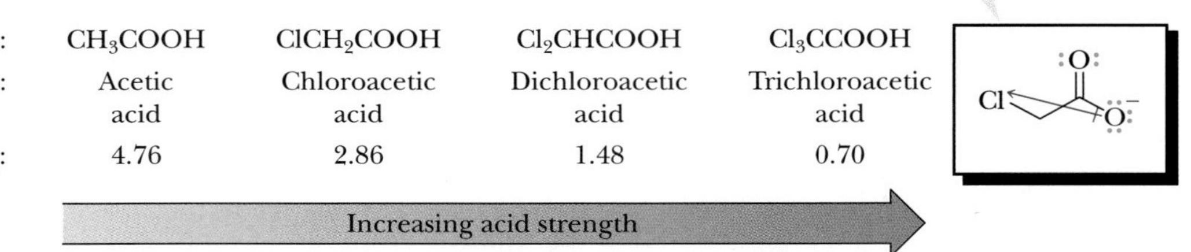

Formula:	CH_3COOH	$ClCH_2COOH$	$Cl_2CHCOOH$	Cl_3CCOOH
Name:	Acetic acid	Chloroacetic acid	Dichloroacetic acid	Trichloroacetic acid
pK_a:	4.76	2.86	1.48	0.70

Increasing acid strength →

The acid-strengthening effect of halogen substitution falls off rather rapidly with increasing distance from the carboxyl group. Although the acid ionization constant for 2-chlorobutanoic acid (pK_a 2.83) is 100 times that for butanoic acid, the acid ionization constant for 4-chlorobutanoic acid (pK_a 4.52) is only about twice that for butanoic acid:

2-Chlorobutanoic acid (pK_a 2.83)	3-Chlorobutanoic acid (pK_a 3.98)	4-Chlorobutanoic acid (pK_a 4.52)	Butanoic acid (pK_a 4.82)

Decreasing acid strength →

Example 14.2

Which acid in each set is the stronger?

(a) Propanoic acid or 2-Hydroxypropanoic acid (Lactic acid)

(b) 2-Hydroxypropanoic acid (Lactic acid) or 2-Oxopropanoic acid (Pyruvic acid)

Strategy

Draw the conjugate base of each acid and look for possible stabilization of the ion via resonance or inductive effects. The conjugate base that is more greatly stabilized will indicate the more acidic carboxylic acid.

Solution

(a) 2-Hydroxypropanoic acid (pK_a 3.85) is a stronger acid than propanoic acid (pK_a 4.87), because of the electron-withdrawing inductive effect of the hydroxyl oxygen.

(b) 2-Oxopropanoic acid (pK_a 2.06) is a stronger acid than 2-hydroxypropanoic acid (pK_a 3.08), because of the greater electron-withdrawing inductive effect of the carbonyl oxygen compared with that of the hydroxyl oxygen.

See problems 14.20–14.22, 14.48

Problem 14.2

Match each compound with its appropriate pK_a value:

$CH_3C(CH_3)_2COOH$ — 2,2-Dimethylpropanoic acid

CF_3COOH — Trifluoroacetic acid

$CH_3CH(OH)COOH$ — 2-Hydroxypropanoic acid (Lactic acid)

pK_a values = 5.03, 3.85, and 0.22.

B. Reaction with Bases

All carboxylic acids, whether soluble or insoluble in water, react with NaOH, KOH, and other strong bases to form water-soluble salts:

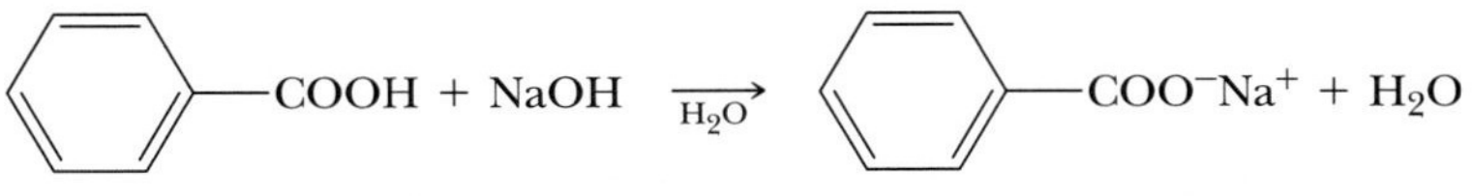

$$C_6H_5COOH + NaOH \xrightarrow{H_2O} C_6H_5COO^-Na^+ + H_2O$$

Benzoic acid (slightly soluble in water) — Sodium benzoate (60 g/100 mL water)

Sodium benzoate, a fungal growth inhibitor, is often added to baked goods "to retard spoilage." Calcium propanoate is used for the same purpose.

Carboxylic acids also form water-soluble salts with ammonia and amines:

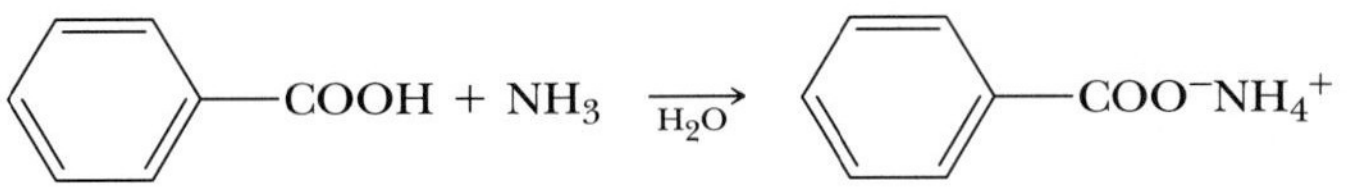

$$C_6H_5COOH + NH_3 \xrightarrow{H_2O} C_6H_5COO^-NH_4^+$$

Benzoic acid (slightly soluble in water) — Ammonium benzoate (20 g/100 mL water)

As described in Section 2.4, carboxylic acids react with sodium bicarbonate and sodium carbonate to form water-soluble sodium salts and carbonic acid (a relatively weak acid). Carbonic acid, in turn, decomposes to give water and carbon dioxide, which evolves as a gas:

$$CH_3COOH + Na^+HCO_3^- \xrightarrow{H_2O} CH_3COO^-Na^+ + H_2CO_3$$

$$H_2CO_3 \longrightarrow CO_2 + H_2O$$

$$CH_3COOH + Na^+HCO_3^- \longrightarrow CH_3COO^-Na^+ + CO_2 + H_2O$$

Salts of carboxylic acids are named in the same manner as are salts of inorganic acids: Name the cation first and then the anion. Derive the name of the anion from the name of the carboxylic acid by dropping the suffix *-ic acid* and adding the suffix *-ate*. For example, the name of $CH_3CH_2COO^-Na^+$ is sodium propanoate, and that of $CH_3(CH_2)_{14}COO^-Na^+$ is sodium hexadecanoate (sodium palmitate).

Example 14.3

Complete each acid–base reaction and name the salt formed:

(a) $CH_3CH_2CH_2COOH + NaOH \longrightarrow$ (b) $CH_3CH(OH)COOH + NaHCO_3 \longrightarrow$

Strategy

Identify the base and the most acidic hydrogen of the acid. Remember that sodium bicarbonate ($NaHCO_3$) typically reacts to yield carbonic acid, which subsequently decomposes to give CO_2 and H_2O.

Solution

Each carboxylic acid is converted to its sodium salt. In (b), carbonic acid is formed (not shown) and decomposes to carbon dioxide and water:

(a) $CH_3CH_2CH_2COOH + NaOH \longrightarrow CH_3CH_2CH_2COO^-Na^+ + H_2O$

Butanoic acid → Sodium butanoate

(b) $CH_3CH(OH)COOH + NaHCO_3 \longrightarrow CH_3CH(OH)COO^-Na^+ + H_2O + CO_2$

2-Hydroxypropanoic acid (Lactic acid) → Sodium 2-hydroxypropanoate (Sodium lactate)

See problems 14.23, 14.34

Problem 14.3

Write an equation for the reaction of each acid in Example 14.3 with ammonia, and name the salt formed.

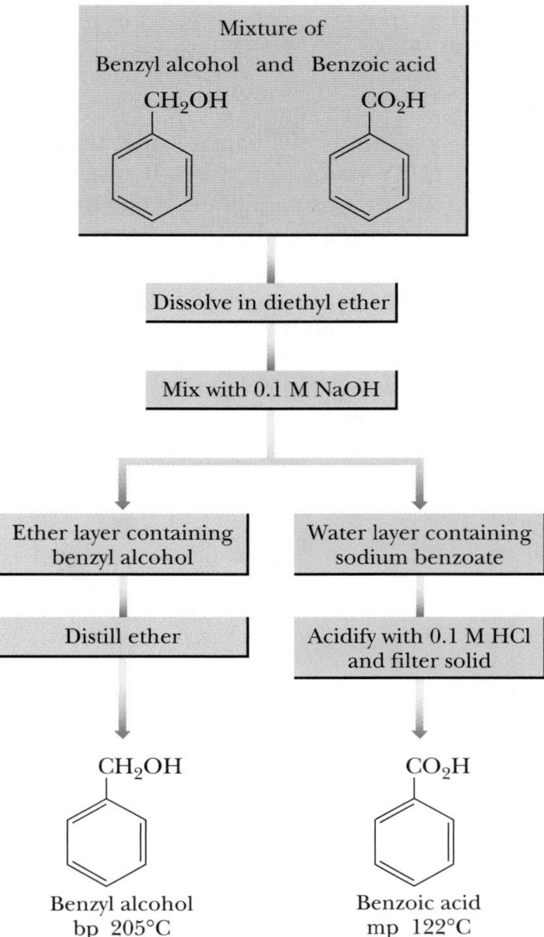

Figure 14.1 Flowchart for separation of benzoic acid from benzyl alcohol.

A consequence of the water solubility of carboxylic acid salts is that we can convert water-insoluble carboxylic acids to water-soluble alkali metal or ammonium salts and then extract them into aqueous solution. In turn, we can transform the salt into the free carboxylic acid by adding HCl, H_2SO_4, or some other strong acid. These reactions allow us to separate water-insoluble carboxylic acids from water-insoluble neutral compounds.

Figure 14.1 shows a flowchart for the separation of benzoic acid, a water-insoluble carboxylic acid, from benzyl alcohol, a water-insoluble nonacidic compound. First, we dissolve the mixture of benzoic acid and benzyl alcohol in diethyl ether. Next, we shake the ether solution with aqueous NaOH to convert benzoic acid to its water-soluble sodium salt. Then we separate the ether from the aqueous phase. Distillation of the ether solution yields first diethyl ether (bp 35°C) and then benzyl alcohol (bp 205°C). When we acidify the aqueous solution with HCl, benzoic acid precipitates as a water-insoluble solid (mp 122°C) and is recovered by filtration. The ability to separate compounds based on their acid–base properties is very important in laboratory and industrial chemistry.

14.5 How Are Carboxyl Groups Reduced?

The carboxyl group is one of the organic functional groups that is most resistant to reduction. It is not affected by catalytic reduction (H_2/M) under conditions that easily reduce aldehydes and ketones to alcohols and that reduce alkenes to alkanes.

CHEMICAL CONNECTIONS 14A

From Willow Bark to Aspirin and Beyond

The first drug developed for widespread use was aspirin, today's most common pain reliever. Americans alone consume approximately 80 billion tablets of aspirin a year! The story of the development of this modern pain reliever goes back more than 2,000 years: In 400 B.C.E., the Greek physician Hippocrates recommended chewing bark of the willow tree to alleviate the pain of childbirth and to treat eye infections.

The active component of willow bark was found to be salicin, a compound composed of salicyl alcohol joined to a unit of β-D-glucose (Section 18.2). Hydrolysis of salicin in aqueous acid gives salicyl alcohol, which can then be oxidized to salicylic acid, an even more effective reliever of pain, fever, and inflammation than salicin and one without its extremely bitter taste:

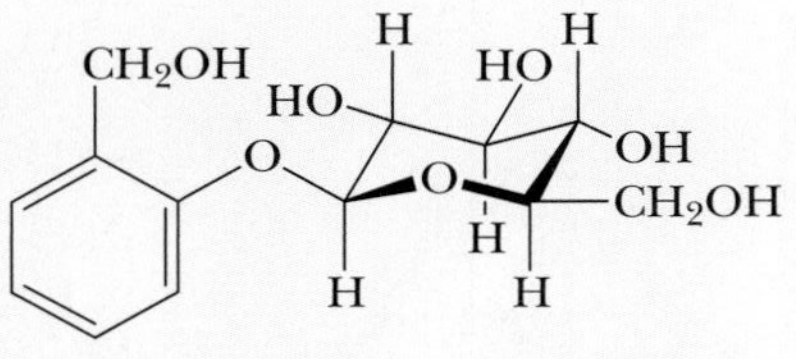

Salicin

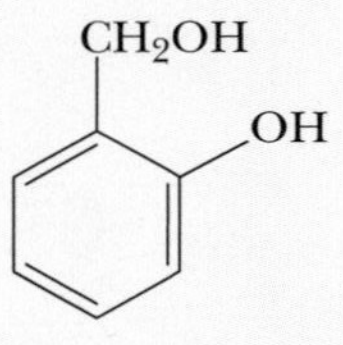

Salicyl alcohol

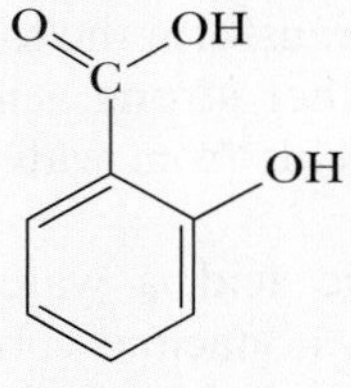

Salicylic acid

Unfortunately, patients quickly recognized salicylic acid's major side effect: It causes severe irritation of the mucous membrane lining the stomach.

In the search for less irritating, but still effective, derivatives of salicylic acid, chemists at the Bayer division of I. G. Farben in Germany prepared acetylsalicylic acid in 1883 and gave it the name *aspirin*, a word derived from the German *spirsäure* (salicylic acid), with the initial *a* for the acetyl group:

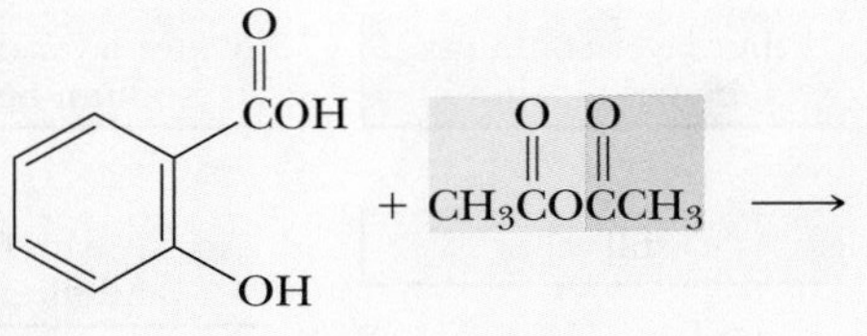

Salicylic acid + CH_3COCCH_3 (Acetic anhydride) 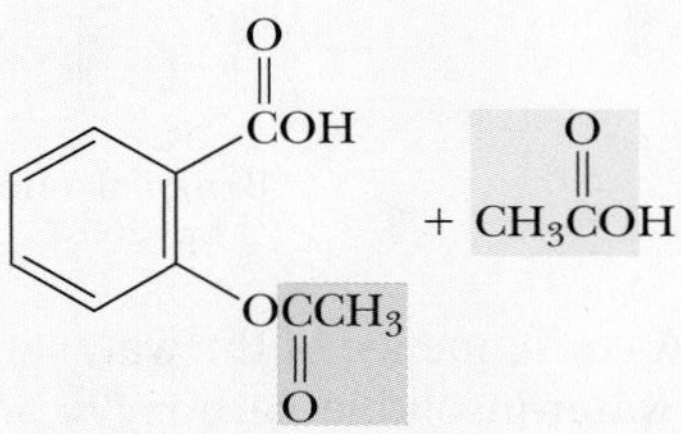 ⟶ Acetyl salicylate (Aspirin) + CH_3COH

Aspirin proved to be less irritating to the stomach than salicylic acid and also more effective in relieving the pain and inflammation of rheumatoid

The most common reagent for the reduction of a carboxylic acid to a primary alcohol is the very powerful reducing agent lithium aluminum hydride (Section 13.10B).

A. Reduction of a Carboxyl Group

Lithium aluminum hydride, $LiAlH_4$, reduces a carboxyl group to a primary alcohol in excellent yield. Reduction is most commonly carried out in diethyl ether or tetrahydrofuran (THF). The initial product is an aluminum alkoxide, which is then treated with water to give the primary alcohol and lithium and aluminum hydroxides:

arthritis. Bayer began large-scale production of aspirin in 1899.

In the 1960s, in a search for even more effective and less irritating analgesics and anti-inflammatory drugs, the Boots Pure Drug Company in England studied compounds related in structure to salicylic acid. They discovered an even more potent compound, which they named ibuprofen, and soon thereafter, Syntex Corporation in the United States developed naproxen and Rhone–Poulenc in France developed ketoprofen:

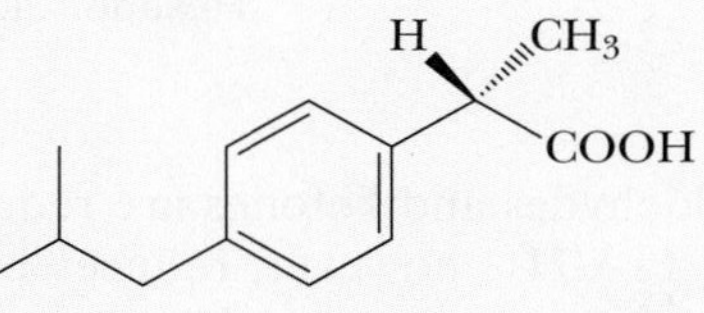

H CH$_3$ COOH

(*S*)-Ibuprofen

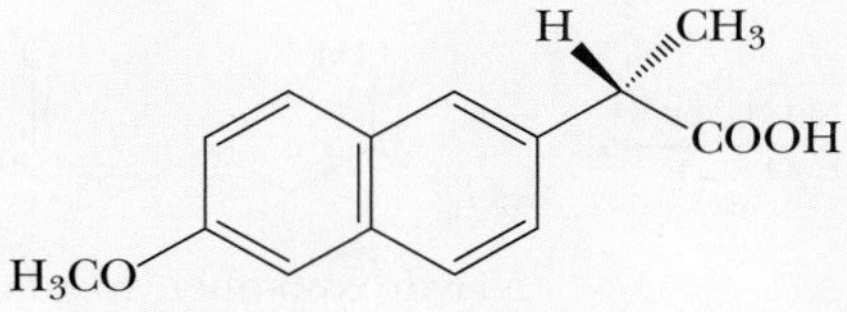

H CH$_3$ COOH H$_3$CO

(*S*)-Naproxen

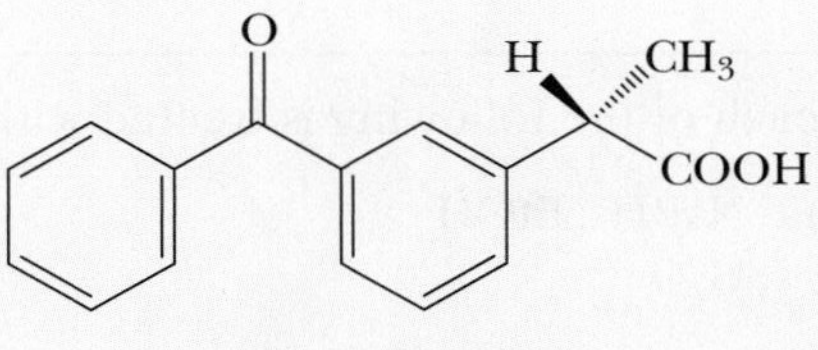

O H CH$_3$ COOH

(*S*)-Ketoprofen

Notice that each compound has one stereocenter and can exist as a pair of enantiomers. For each drug, the physiologically active form is the *S* enantiomer. Even though the *R* enantiomer of ibuprofen has none of the analgesic or anti-inflammatory activity, it is converted in the body to the active *S* enantiomer.

In the 1960s, scientists discovered that aspirin acts by inhibiting cyclooxygenase (COX), a key enzyme in the conversion of arachidonic acid to prostaglandins (Section 21.5). With this discovery, it became clear why only one enantiomer of ibuprofen, naproxen, and ketoprofen is active: Only the *S* enantiomer of each has the correct handedness to bind to COX and inhibit its activity.

The discovery that these drugs owe their effectiveness to the inhibition of COX opened an entirely new avenue for drug research. If we know more about the structure and function of this key enzyme, might it be possible to design and discover even more effective nonsteroidal anti-inflammatory drugs for the treatment of rheumatoid arthritis and other inflammatory diseases?

And so continues the story that began with the discovery of the beneficial effects of chewing willow bark.

QUESTION

Draw the product of the reaction of salicylic acid with (a) one equivalent of NaOH, (b) two equivalents of NaOH, and (c) two equivalents of $NaHCO_3$.

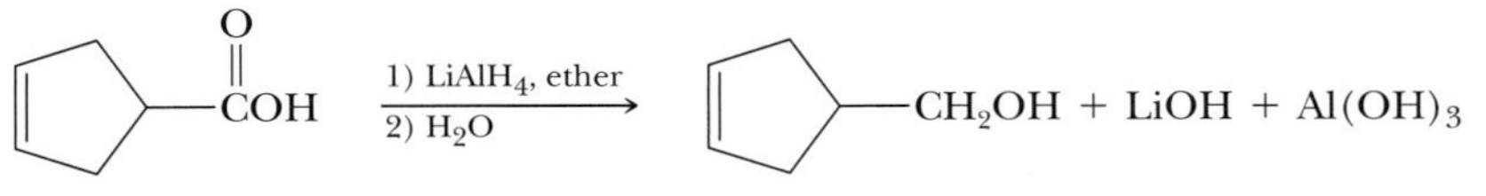

O ‖ COH —1) $LiAlH_4$, ether; 2) H_2O→ $CH_2OH + LiOH + Al(OH)_3$

3-Cyclopentene-carboxylic acid

4-Hydroxymethyl-cyclopentene

These hydroxides are insoluble in diethyl ether or THF and are removed by filtration. Evaporation of the solvent yields the primary alcohol.

Alkenes are generally not affected by metal hydride-reducing reagents. These reagents function as hydride ion donors; that is, they function as nucleophiles, and alkenes are not normally attacked by nucleophiles.

B. Selective Reduction of Other Functional Groups

Catalytic hydrogenation (at least under the same conditions used to reduce ketones and aldehydes) does not reduce carboxyl groups, but does reduce alkenes to alkanes. Therefore, we can use H_2/M to reduce this functional group selectively in the presence of a carboxyl group:

5-Hexenoic acid + H_2 $\xrightarrow[25°C,\ 2\ atm]{Pt}$ Hexanoic acid

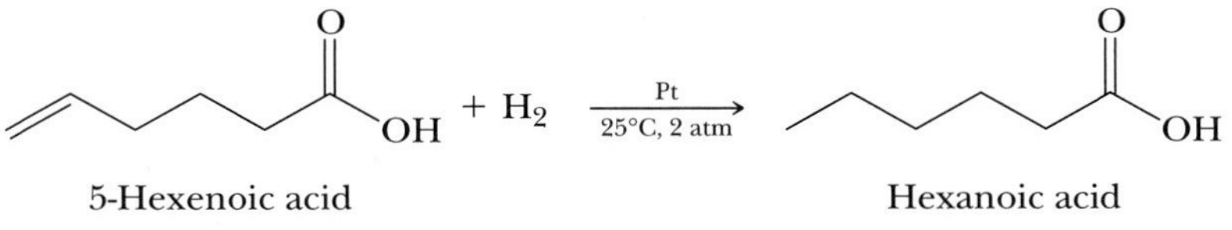

We saw in Section 13.10B that aldehydes and ketones are reduced to alcohols by both $LiAlH_4$ and $NaBH_4$. Only $LiAlH_4$, however, reduces carboxyl groups. Thus, it is possible to reduce an aldehyde or a ketone carbonyl group selectively in the presence of a carboxyl group by using the less reactive $NaBH_4$ as the reducing agent:

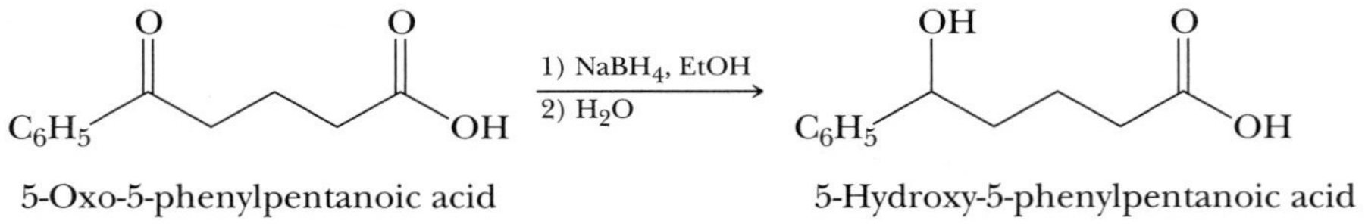

5-Oxo-5-phenylpentanoic acid $\xrightarrow[2)\ H_2O]{1)\ NaBH_4,\ EtOH}$ 5-Hydroxy-5-phenylpentanoic acid

Example 14.4

Provide the product formed when each of the following is treated with:

(i) H_2/Pd (ii) 1. $LiAlH_4$, ether 2. H_2O (iii) 1. $NaBH_4$, EtOH 2. H_2O

In each reaction, assume an excess of reagent is available for reaction.

(a) 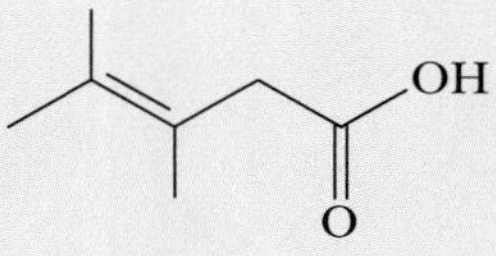 (b)

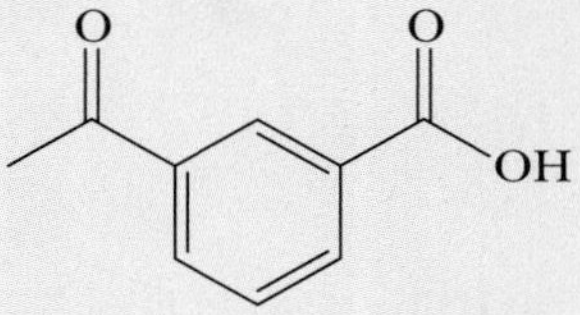

Strategy

Remember that carboxylic acids are only reduced by $LiAlH_4$, alkenes are only reduced by H_2/M, aldehydes and ketones are reduced by all metal hydride reducing agents, and benzene rings are resistant to each of these reducing-reagents. Remember to consider stereochemistry in the outcome of each reaction.

Solution

Here are structural formulas for the major product produced in each reaction:

(a) 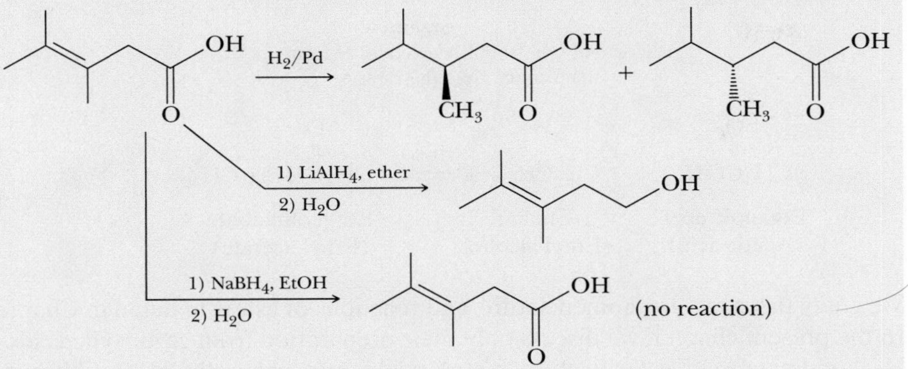 OH, O; H_2/Pd → OH, CH_3, O + OH, CH_3, O

1) $LiAlH_4$, ether 2) H_2O → OH

1) $NaBH_4$, EtOH 2) H_2O → OH, O (no reaction)

(b) 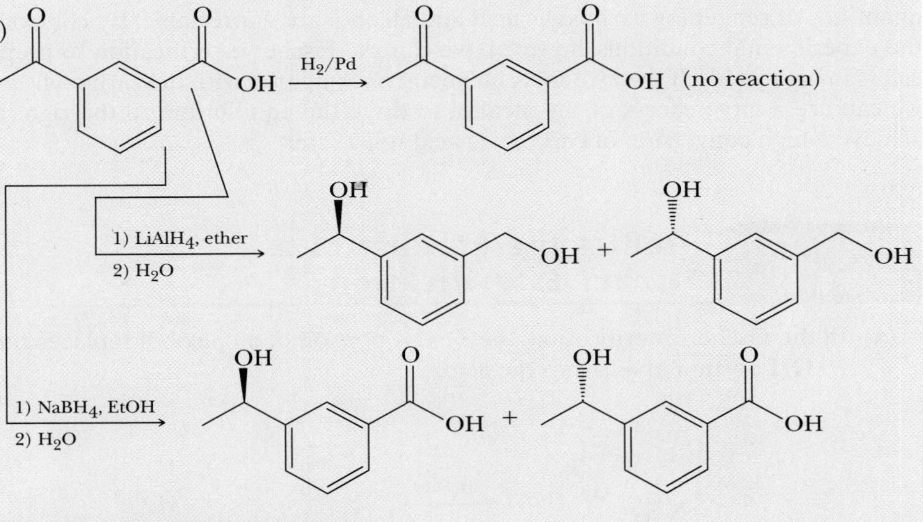 O, O, OH; H_2/Pd → O, O, OH (no reaction)

1) $LiAlH_4$, ether 2) H_2O → OH, OH + OH, OH

1) $NaBH_4$, EtOH 2) H_2O → OH, O, OH + OH, O, OH

See problems 14.30–14.32, 14.47

Problem 14.4

Provide the product formed when each of the following is treated with:

(i) H_2/Pd (ii) 1. $LiAlH_4$, ether 2. H_2O (iii) 1. $NaBH_4$, EtOH 2. H_2O

In each reaction, presume that an excess of reagent is available for each reaction.

(a) 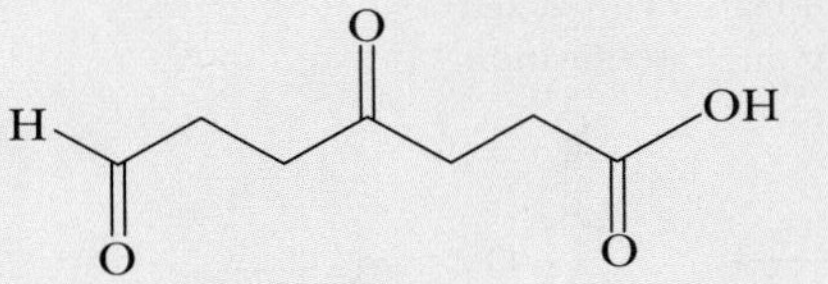 H, O, O, OH, O (b) 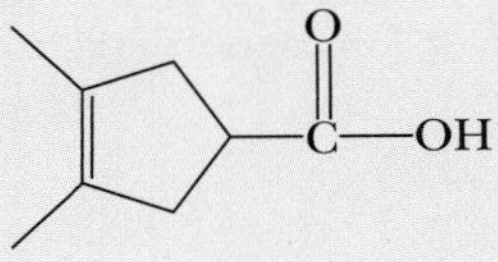 O, C—OH

14.6 What Is Fischer Esterification?

Treatment of a carboxylic acid with an alcohol in the presence of an acid catalyst—most commonly, concentrated sulfuric acid—gives an ester. This method of forming an ester is given the special name **Fischer esterification** after the German chemist

Fischer esterification The process of forming an ester by refluxing a carboxylic acid and an alcohol in the presence of an acid catalyst, commonly sulfuric acid.

These products all contain ethyl acetate as a solvent. *(Charles D. Winter)*

Emil Fischer (1852–1919). As an example of Fischer esterification, treating acetic acid with ethanol in the presence of concentrated sulfuric acid gives ethyl acetate and water:

removal of OH from the acid and H from the alcohol gives the ester

$$\underset{\substack{\text{Ethanoic acid}\\\text{(Acetic acid)}}}{CH_3\overset{O}{\overset{\|}{C}}OH} + \underset{\substack{\text{Ethanol}\\\text{(Ethyl alcohol)}}}{CH_3CH_2OH} \underset{}{\overset{H_2SO_4}{\rightleftharpoons}} \underset{\substack{\text{Ethyl ethanoate}\\\text{(Ethyl acetate)}}}{CH_3\overset{O}{\overset{\|}{C}}OCH_2CH_3} + H_2O$$

We study the structure, nomenclature, and reactions of esters in detail in Chapter 15. In the present chapter, we discuss only their preparation from carboxylic acids.

Acid-catalyzed esterification is reversible, and generally, at equilibrium, the quantities of remaining carboxylic acid and alcohol are appreciable. By controlling the experimental conditions, however, we can use Fischer esterification to prepare esters in high yields. If the alcohol is inexpensive compared with the carboxylic acid, we can use a large excess of the alcohol to drive the equilibrium to the right and achieve a high conversion of carboxylic acid to its ester.

HOW TO 14.1 Predict the Product of a Fischer Esterification

(a) In the Fischer esterification, the —OR portion of an alcohol replaces the —OH portion of a carboxylic acid.

$$R\text{-}C(=O)OH + HOR \xrightarrow{H_2SO_4} R\text{-}C(=O)OR + H_2O$$

(b) This simple fact of the mechanism allows us to predict the product of any type of Fischer esterification. For example, in the following Fischer esterification, the alcohol is already a part of the molecule. In cases of intramolecular Fischer esterifications, it is often helpful to number the atoms in the molecule.

in this case, the R group is ultimately connected to the carboxyl group undergoing esterification

$$\text{HO(6)–C(5)(CH}_3)_2\text{–CH}_2(4)\text{–CH}_2(3)\text{–CH}_2(2)\text{–C(1)(=O)OH} \xrightarrow{H_2SO_4} \text{six-membered lactone (O 6 bonded to C 1)} + H_2O$$

numbering the atoms allows us to see that atom-6 will form a new bond to atom-1, resulting in the formation of a six-membered ring

Example 14.5

Complete these Fischer esterification reactions:

(a) 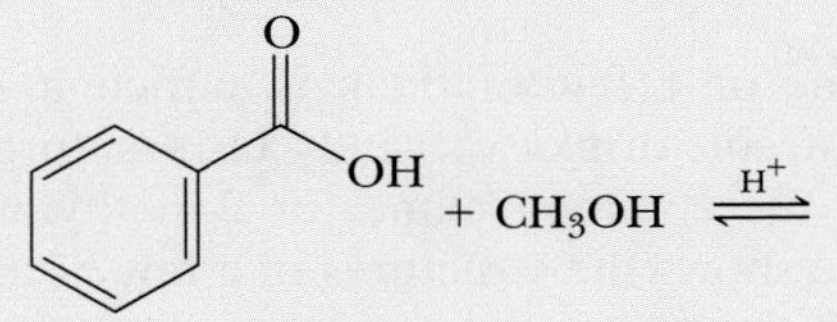 $C_6H_5COOH + CH_3OH \overset{H^+}{\rightleftharpoons}$

(b) 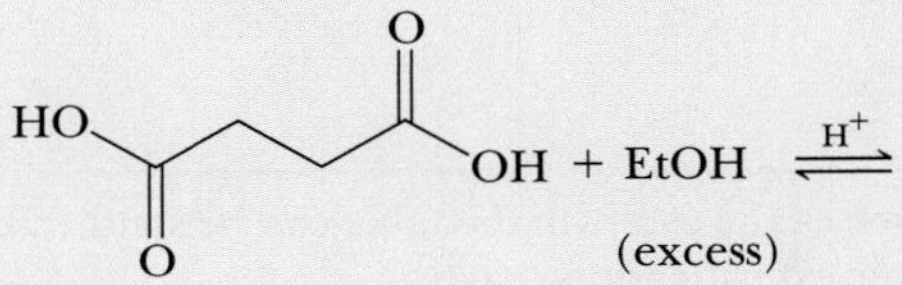 $HOOC{-}CH_2CH_2{-}COOH + EtOH \overset{H^+}{\rightleftharpoons}$
(excess)

Strategy

In a Fischer esterification, each carboxyl group is converted to an ester in which the —OR group originates from the alcohol reagent.

Solution

Here is a structural formula for the ester produced in each reaction:

(a) 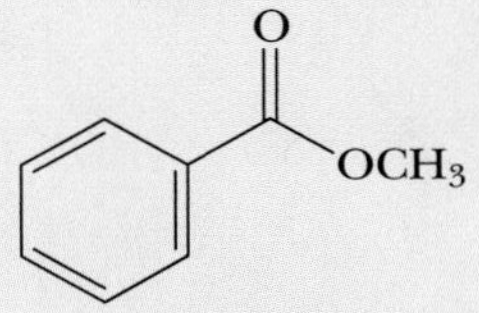 $C_6H_5COOCH_3$

Methyl benzoate

(b) 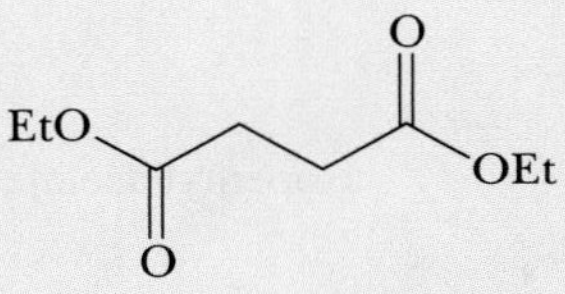 $EtOOC{-}CH_2CH_2{-}COOEt$

Diethyl butanedioate
(Diethyl succinate)

See problems 14.30, 14.33, 14.39, 14.40, 14.47

Problem 14.5

Complete these Fischer esterification reactions:

(a) 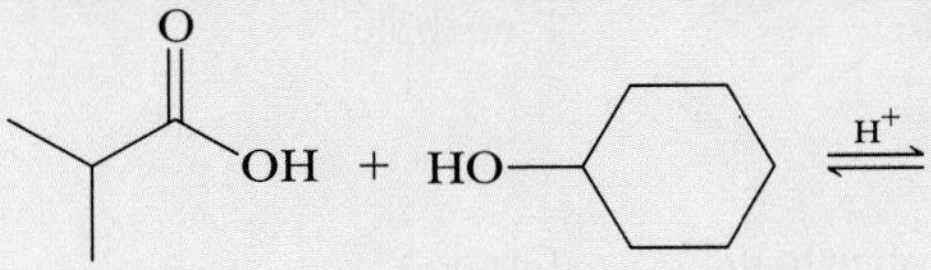 $(CH_3)_2CHCOOH$ + HO—cyclohexyl $\overset{H^+}{\rightleftharpoons}$

(b) 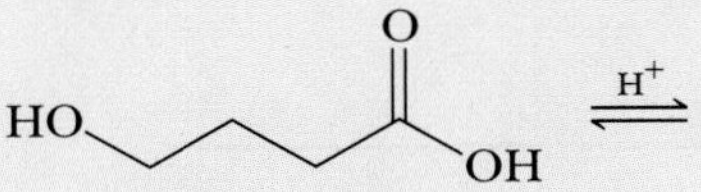 $HOCH_2CH_2CH_2COOH \overset{H^+}{\rightleftharpoons}$ (a cyclic ester)

Following is a mechanism for Fischer esterification, and we urge you to study it carefully. It is important that you understand this mechanism thoroughly, because it is a model for many of the reactions of the functional derivatives of carboxylic acids presented in Chapter 15. Note that, although we show the acid catalyst as H_2SO_4 when we write Fisher esterification reactions, the actual proton-transfer acid that initiates the reaction is the oxonium formed by the transfer of a proton

CHEMICAL CONNECTIONS 14B

Esters as Flavoring Agents

Flavoring agents are the largest class of food additives. At present, over a thousand synthetic and natural flavors are available. The majority of these are concentrates or extracts from the material whose flavor is desired and are often complex mixtures of from tens to hundreds of compounds. A number of ester flavoring agents are synthesized industrially. Many have flavors very close to the target flavor, and adding only one or a few of them is sufficient to make ice cream, soft drinks, or candy taste natural. (Isopentane is the common name for 2-methylbutane.) The table shows the structures of a few of the esters used as flavoring agents:

QUESTION

Show how each of the esters in the table can be synthesized using a Fischer esterification reaction.

Structure	Name	Flavor
$HCOOCH_2CH_3$	Ethyl formate	Rum
$CH_3COOCH_2CH_2CH(CH_3)_2$	Isopentyl acetate	Banana
$CH_3COO(CH_2)_7CH_3$	Octyl acetate	Orange
$CH_3CH_2CH_2COOCH_3$	Methyl butanoate	Apple
$CH_3CH_2CH_2COOCH_2CH_3$	Ethyl butanoate	Pineapple
$2\text{-}H_2NC_6H_4COOCH_3$ (NH2)	Methyl 2-aminobenzoate (Methyl anthranilate)	Grape

from H_2SO_4 (the stronger acid) to the alcohol (the stronger base) used in the esterification reaction:

$$CH_3{-}\ddot{O}{-}H + H{-}\ddot{O}{-}\underset{\underset{O}{\|}}{\overset{\overset{O}{\|}}{S}}{-}O{-}H \rightleftharpoons CH_3{-}\underset{\underset{H}{|}}{\ddot{O}^{+}}{-}H + {}^{-}{:}\ddot{O}{-}\underset{\underset{O}{\|}}{\overset{\overset{O}{\|}}{S}}{-}O{-}H$$

Mechanism: Fischer Esterification

① Proton transfer from the acid catalyst to the carbonyl oxygen increases the electrophilicity of the carbonyl carbon . . .

② which is then attacked by the nucleophilic oxygen atom of the alcohol . . .

③ to form an oxonium ion.

④ Proton transfer from the oxonium ion to a second molecule of alcohol . . .

⑤ gives a tetrahedral carbonyl addition intermediate (TCAI).

⑥ Proton transfer to one of the —OH groups of the TCAI . . .

⑦ gives a new oxonium ion.

⑧ Loss of water from this oxonium ion . . .

⑨ gives the ester and water, and regenerates the acid catalyst.

14.7 What Are Acid Chlorides?

The functional group of an acid halide is a carbonyl group bonded to a halogen atom. Among the acid halides, acid chlorides are the most frequently used in the laboratory and in industrial organic chemistry:

$-\overset{O}{\overset{\|}{C}}-X$ — Functional group of an acid halide

$CH_3\overset{O}{\overset{\|}{C}}Cl$ — Acetyl chloride

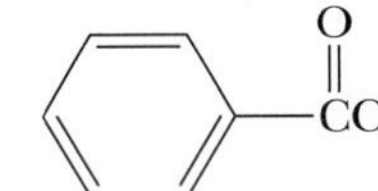

Benzoyl chloride

We study the nomenclature, structure, and characteristic reactions of acid halides in Chapter 15. In this chapter, our concern is only with their synthesis from carboxylic acids.

The most common way to prepare an acid chloride is to treat a carboxylic acid with thionyl chloride, the same reagent that converts an alcohol to a chloroalkane (Section 8.2D):

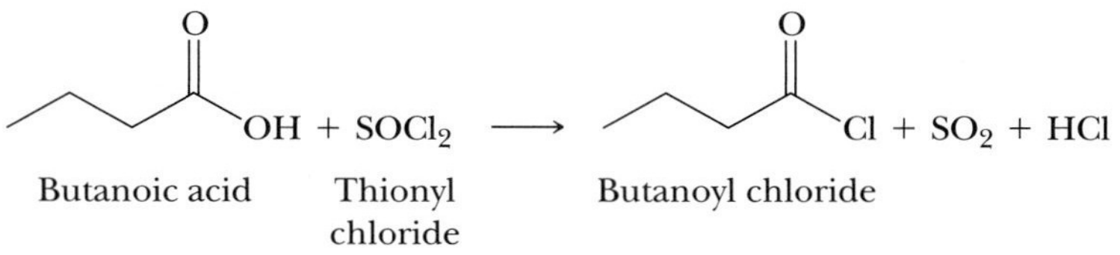

$$\text{Butanoic acid} + SOCl_2 \longrightarrow \text{Butanoyl chloride} + SO_2 + HCl$$

Butanoic acid — Thionyl chloride — Butanoyl chloride

The mechanism of this reaction consists of four steps. In the first step, the carboxyl group adds to the sulfur atom of thionyl chloride to generate a tetrahedral sulfur intermediate. In Step 2, the sulfonyl group is regenerated, and chloride ion is released. The chloride ion attacks the carbonyl carbon in Step 3, forming a tetrahedral carbonyl addition intermediate. The sulfonyl group highlighted in Step 3 is an excellent leaving group. This allows a lone pair of electrons to collapse back toward the C—O bond to regenerate the carbonyl carbon while expelling the leaving group sulfochloridous acid. This sulfur-based acid is unstable and breaks down to yield sulfur dioxide and HCl. The mechanism shown in Step 4 is a common mode of reactivity for functional derivatives of carboxylic acids (Chapter 15).

Step 1:

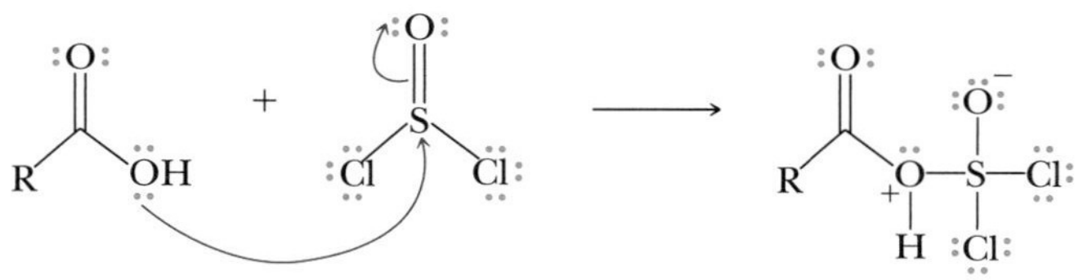

Step 2:

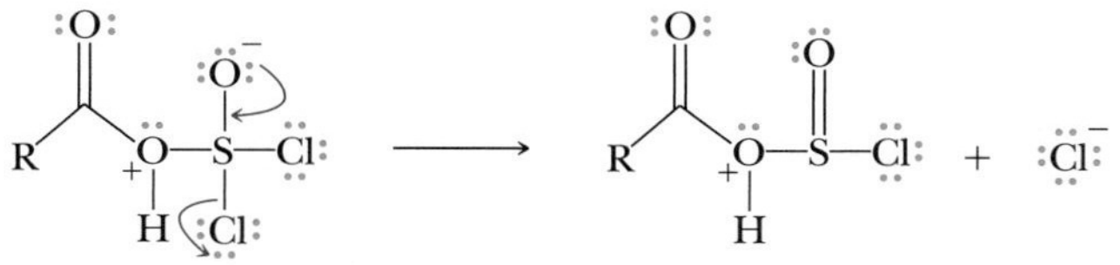

Step 3:

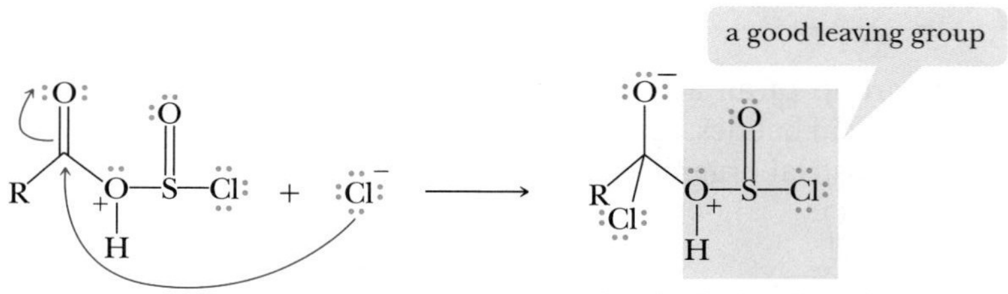

a good leaving group

Tetrahedral carbonyl addition intermediate

Step 4:

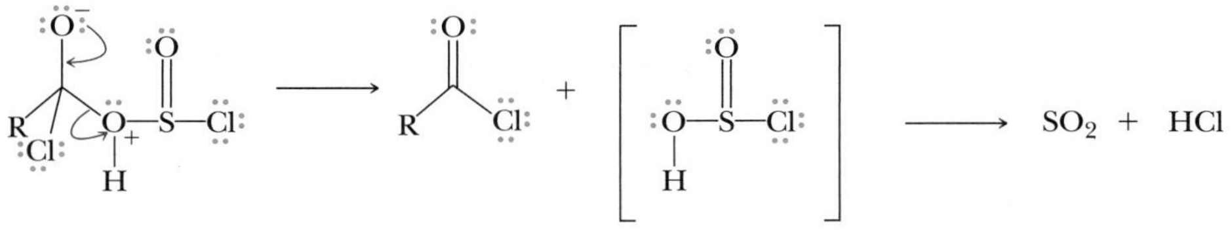

Sulfochloridous acid

$\longrightarrow SO_2 + HCl$

Example 14.6

Complete each equation:

(a) 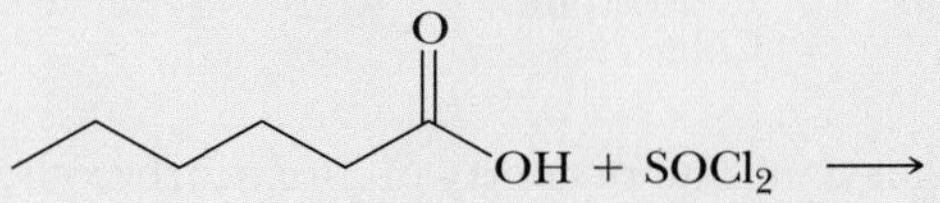 $OH + SOCl_2 \longrightarrow$

(b) 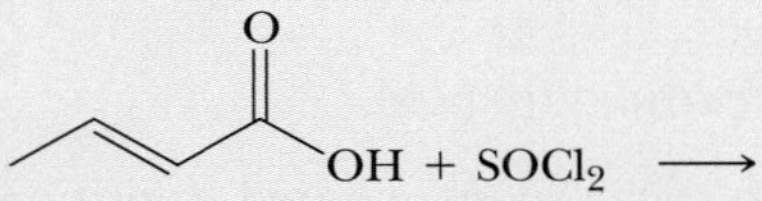 $OH + SOCl_2 \longrightarrow$

Strategy

Thionyl chloride effectively causes —OH groups (for example, those of alcohols and carboxylic acids) to be replaced by Cl. Don't forget to show the by-products of the reaction (SO_2 and HCl).

Solution

Following are the products for each reaction:

(a) 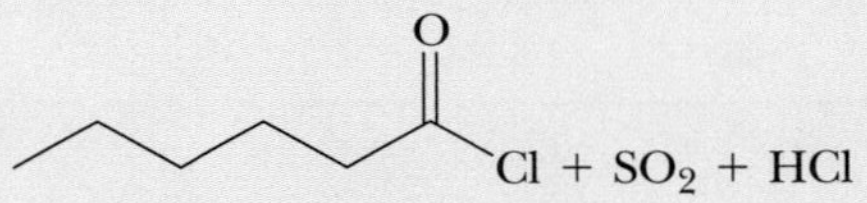 $Cl + SO_2 + HCl$ (b) 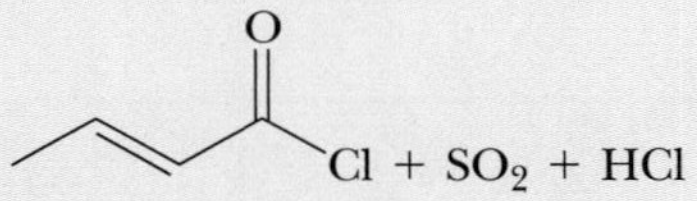 $Cl + SO_2 + HCl$

See problems 14.30, 14.47

Problem 14.6

Complete each equation:

(a) 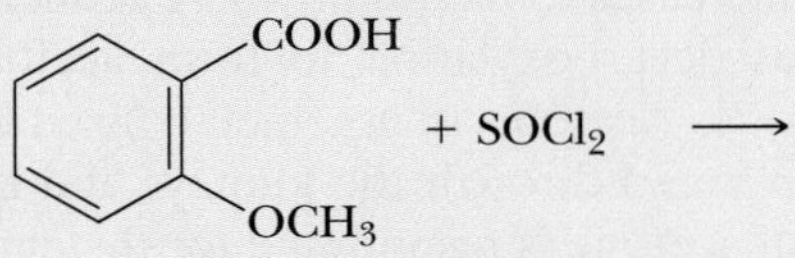 COOH, OCH_3 $+ SOCl_2 \longrightarrow$ (b) 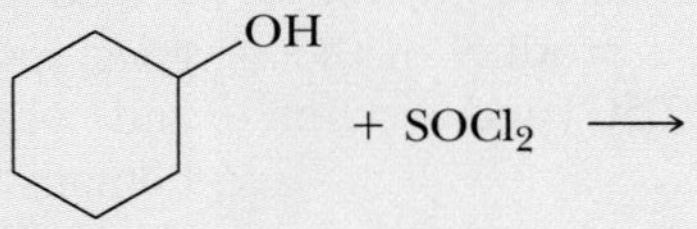 OH $+ SOCl_2 \longrightarrow$

14.8 What Is Decarboxylation?

A. β-Ketoacids

Decarboxylation is the loss of CO_2 from a carboxyl group. Almost any carboxylic acid, heated to a very high temperature, undergoes decarboxylation:

Decarboxylation Loss of CO_2 from a carboxyl group.

$$RCOOH \xrightarrow[\text{(high temperature)}]{\text{decarboxylation}} RH + CO_2$$

Most carboxylic acids, however, are quite resistant to moderate heat and melt or even boil without decarboxylation. Exceptions are carboxylic acids that have a carbonyl group β to the carboxyl group. This type of carboxylic acid undergoes decarboxylation quite readily on mild heating. For example, when 3-oxobutanoic acid (acetoacetic acid) is heated moderately, it undergoes decarboxylation to give acetone and carbon dioxide:

3-Oxobutanoic acid (Acetoacetic acid) $\xrightarrow{\text{warm}}$ Acetone + CO_2

Decarboxylation on moderate heating is a unique property of 3-oxocarboxylic acids (β-ketoacids) and is not observed with other classes of ketoacids.

Mechanism: Decarboxylation of a β-Ketocarboxylic Acid

Step 1: Redistribution of six electrons in a cyclic six-membered transition state gives carbon dioxide and an enol:

(A cyclic six-membered transition state) $\xrightarrow{(1)}$ enol of a ketone + $O=C=O$ $\underset{(2)}{\rightleftharpoons}$ acetone + CO_2

CHEMICAL CONNECTIONS 14C

Ketone Bodies and Diabetes

3-Oxobutanoic acid (acetoacetic acid) and its reduction product, 3-hydroxybutanoic acid, are synthesized in the liver from acetyl-CoA, a product of the metabolism of fatty acids (Section 22.5C) and certain amino acids:

3-Oxobutanoic acid (Acetoacetic acid)

3-Hydroxybutanoic acid (β-Hydroxybutyric acid)

3-Hydroxybutanoic acid and 3-oxobutanoic acid are known collectively as ketone bodies.

The concentration of ketone bodies in the blood of healthy, well-fed humans is approximately 0.01 mM/L. However, in persons suffering from starvation or diabetes mellitus, the concentration of ketone bodies may increase to as much as 500 times normal. Under these conditions, the concentration of acetoacetic acid increases to the point where it undergoes spontaneous decarboxylation to form acetone and carbon dioxide. Acetone is not metabolized by humans and is excreted through the kidneys and the lungs. The odor of acetone is responsible for the characteristic "sweet smell" on the breath of severely diabetic patients.

QUESTION

Show the mechanism for the decarboxylation of acetoacetic acid. Explain why 3-hydroxybutanoic acid cannot undergo decarboxylation.

Step 2: Keto–enol tautomerism (Section 13.8A) of the enol gives the more stable keto form of the product:

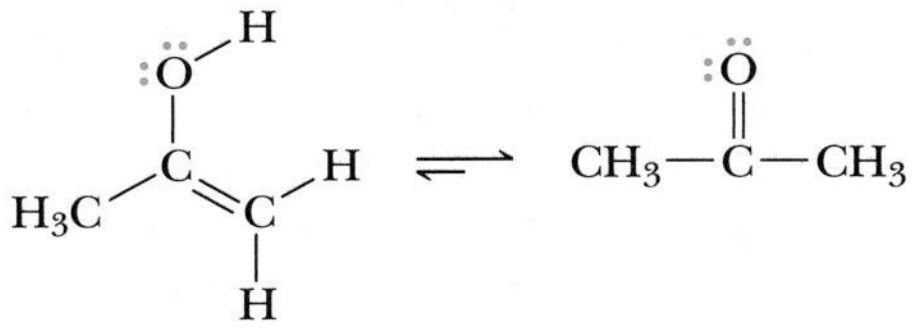

$$H_3C{-}C(OH){=}CH_2 \rightleftharpoons CH_3{-}\overset{O}{\overset{\|}{C}}{-}CH_3$$

HOW TO 14.2 Predict the Product of a β-Decarboxylation Reaction

(a) The most important criterion of a decarboxylation reaction is that the carbonyl group be at the β-position relative to a carboxyl group. Therefore, identify each carboxyl group in a molecule and determine whether a carbonyl is β to it.

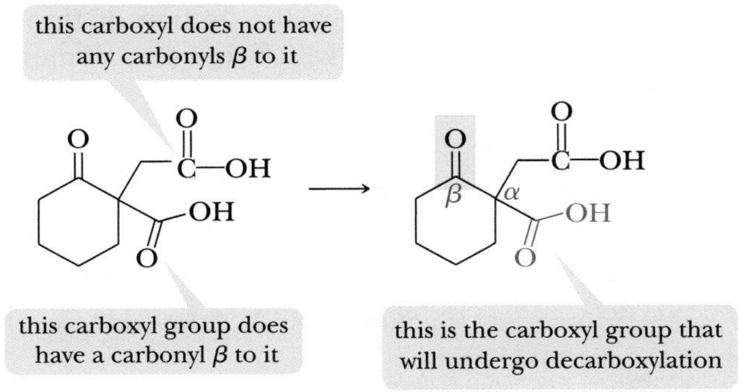

this carboxyl does not have any carbonyls β to it

this carboxyl group does have a carbonyl β to it

this is the carboxyl group that will undergo decarboxylation

(b) Once all the carboxyl groups with β-carbonyls are identified, it is a simple matter of replacing the carboxyl groups with a hydrogen.

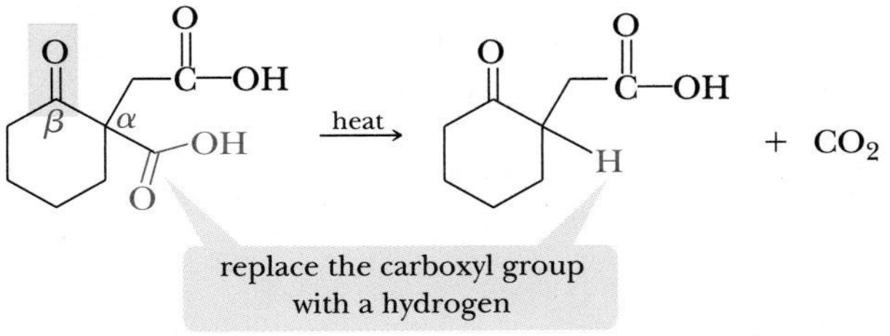

$\xrightarrow{\text{heat}}$ + CO_2

replace the carboxyl group with a hydrogen

An important example of decarboxylation of a β-ketoacid in the biological world occurs during the oxidation of foodstuffs in the tricarboxylic acid (TCA) cycle. Oxalosuccinic acid, one of the intermediates in this cycle, undergoes spontaneous decarboxylation to produce α-ketoglutaric acid. Only one of the three carboxyl groups of oxalosuccinic acid has a carbonyl group in the position β to it, and it is this carboxyl group that is lost as CO_2:

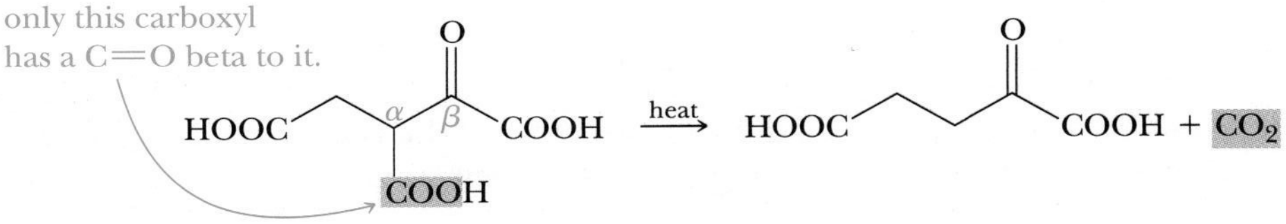

only this carboxyl has a C=O beta to it.

HOOC–CH₂–CH(COOH)–C(=O)–COOH $\xrightarrow{\text{heat}}$ HOOC–CH₂–CH₂–C(=O)–COOH + CO_2

Oxalosuccinic acid α-Ketoglutaric acid

B. Malonic Acid and Substituted Malonic Acids

The presence of a ketone or an aldehyde carbonyl group on the carbon β to the carboxyl group is sufficient to facilitate decarboxylation. In the more general reaction, decarboxylation is facilitated by the presence of any carbonyl group on the β carbon, including that of a carboxyl group or an ester. Malonic acid and substituted malonic acids, for example, undergo decarboxylation on heating, as illustrated by the decarboxylation of malonic acid when it is heated slightly above its melting point of 135–137°C:

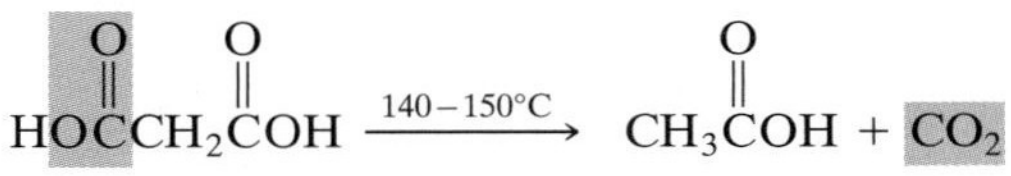

$$HOC(=O)CH_2C(=O)OH \xrightarrow{140-150^\circ C} CH_3C(=O)OH + CO_2$$

Propanedioic acid
(Malonic acid)

The mechanism for decarboxylation of malonic acids is similar to what we have just studied for the decarboxylation of β-ketoacids. The formation of a cyclic, six-membered transition state involving a redistribution of three electron pairs gives the enol form of a carboxylic acid, which, in turn, isomerizes to the carboxylic acid.

Mechanism: Decarboxylation of a β-Dicarboxylic Acid

Step 1: Rearrangement of six electrons in a cyclic six-membered transition state gives carbon dioxide and the enol form of a carboxyl group.

Step 2: Keto–enol tautomerism (Section 13.8A) of the enol gives the more stable keto form of the carboxyl group.

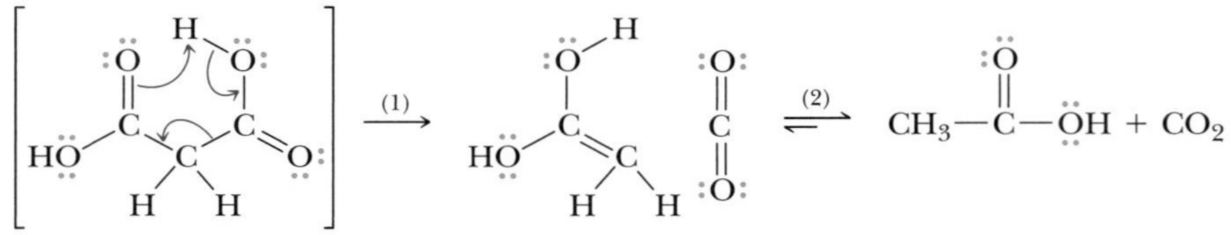

$$\text{[cyclic transition state]} \xrightarrow{(1)} HO-C(OH)=CH_2 + O=C=O \underset{}{\overset{(2)}{\rightleftharpoons}} CH_3-C(=O)-OH + CO_2$$

A cyclic six-membered transition state

Enol of a carboxyl group

Example 14.7

Each of these carboxylic acids undergoes thermal decarboxylation:

(a) 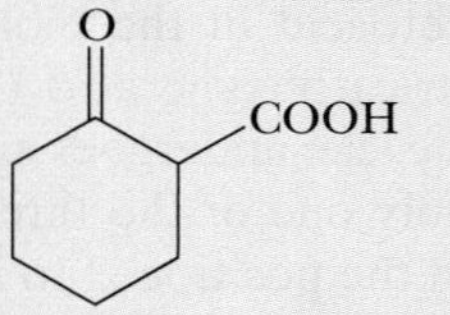 2-oxocyclohexane-1-carboxylic acid (COOH)

(b) 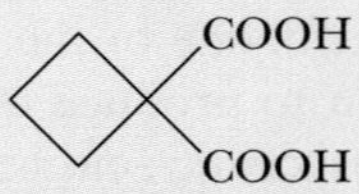 cyclobutane-1,1-dicarboxylic acid (COOH, COOH)

Draw a structural formula for the enol intermediate and final product formed in each reaction.

Strategy

It is often helpful to draw the full Lewis structure of the β-carboxyl group and to position it to allow a cyclic six-membered transition state:

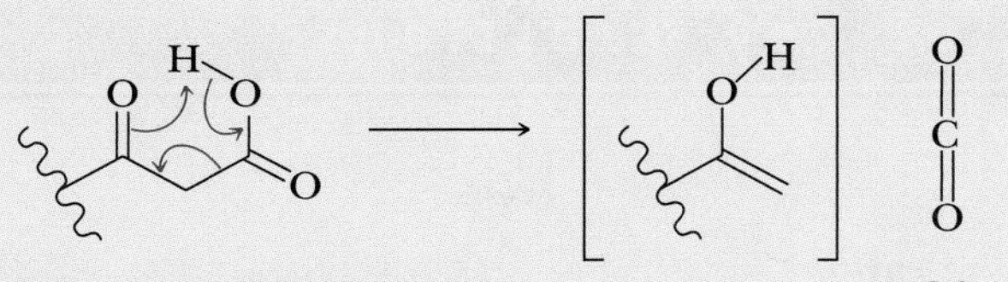

A cyclic six-membered transition state → [Enol] CO_2

By carefully keeping track of the movement of electrons, the bonds made and the bonds broken, one can arrive at the enol intermediate. Predicting the final product is a simple matter of replacing the —COOH group that is β to a carbonyl in the molecule with a hydrogen atom.

Solution

(a) [Enol intermediate] → cyclohexanone + CO_2

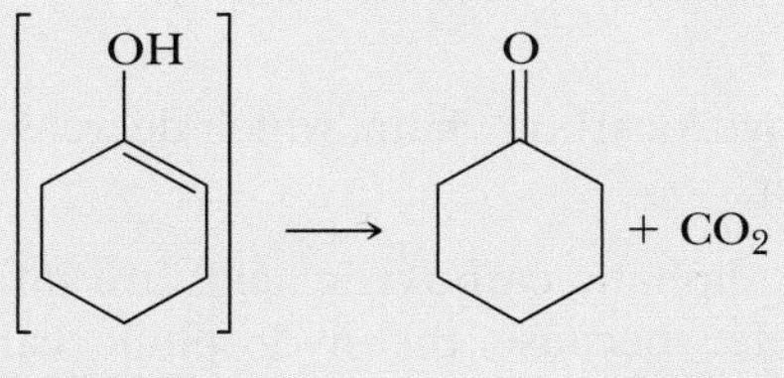

Enol intermediate

(b) [Enol intermediate] → cyclobutane–COOH + CO_2

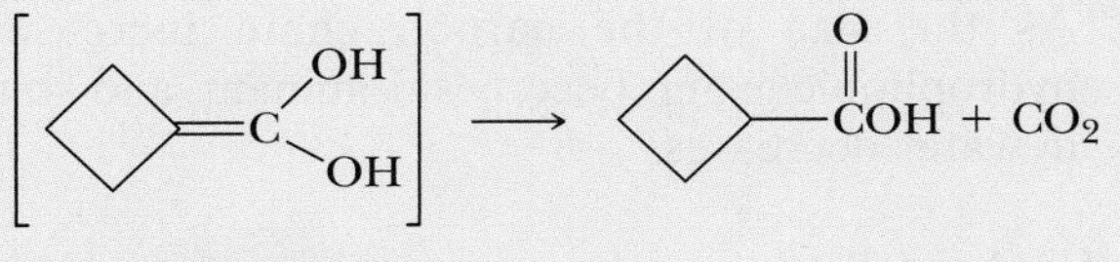

Enol intermediate

See problems 14.41, 14.47

Problem 14.7

Draw the structural formula for the indicated β-ketoacid:

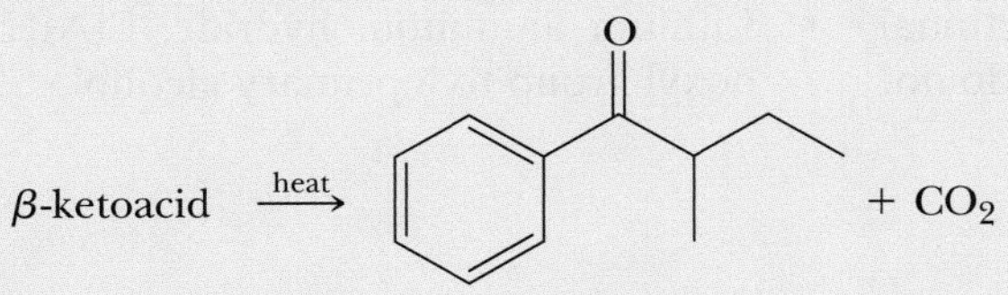

β-ketoacid $\xrightarrow{\text{heat}}$ (product) + CO_2

Key Terms and Concepts

aceto group (p. 511)
acetoacetic acid (p. 510)
acid chloride (p. 525)
acidity of carboxylic acids (p. 513)
aliphatic carboxylic acid (p. 508)
aromatic carboxylic acid (p. 508)
β-ketoacid (p. 528)
carboxylate group (p. 513)
carboxyl group (p. 508)
cyclic six-membered transition state (p. 528)
decarboxylation (p. 527)
ester (p. 522)
Fischer esterification (p. 521)
hydrophilicity (p. 512)
hydrophobicity (p. 512)
malonic acid (p. 530)
reduction of carboxyl groups (p. 517)
thionyl chloride (p. 526)

Summary of Key Questions

14.1 What Are Carboxylic Acids?

- The functional group of a **carboxylic acid** is the **carboxyl group**, **—COOH**.

14.2 How Are Carboxylic Acids Named?

- IUPAC names of carboxylic acids are derived from the parent alkane by dropping the suffix *-e* and adding *-oic acid*.
- Dicarboxylic acids are named as *-dioic acids*.

14.3 What Are the Physical Properties of Carboxylic Acids?

- Carboxylic acids are polar compounds that associate by hydrogen bonding into dimers in the liquid and solid states.
- Carboxylic acids have higher boiling points and are more soluble in water than alcohols, aldehydes, ketones, and ethers of comparable molecular weight.
- A carboxylic acid consists of two regions of different polarity: a polar, **hydrophilic** carboxyl group, which increases solubility in water, and a nonpolar, **hydrophobic** hydrocarbon chain, which decreases solubility in water.
- The first four aliphatic carboxylic acids are infinitely soluble in water because the hydrophilic carboxyl group more than counterbalances the hydrophobic hydrocarbon chain.
- As the size of the carbon chain increases, the hydrophobic group becomes dominant, and solubility in water decreases.

14.4 What Are the Acid–Base Properties of Carboxylic Acids?

- Values of $\mathbf{p}\boldsymbol{K}_\mathbf{a}$ for aliphatic carboxylic acids are in the 4.0 to 5.0 range.
- Electron-withdrawing substituents near the carboxyl group increase acidity in both aliphatic and aromatic carboxylic acids.

14.5 How Are Carboxyl Groups Reduced?

- The carboxyl group is one of the organic functional groups that is most resistant to reduction. They do not react with H_2/M or $NaBH_4$.
- Lithium aluminum hydride, $LiAlH_4$, reduces a carboxyl group to a primary alcohol.

14.6 What Is Fischer Esterification?

- **Fischer esterification** is a method of forming an ester by treatment of a carboxylic acid with an alcohol in the presence of an acid catalyst.

14.7 What Are Acid Chlorides?

- The functional group of an acid chloride is a carbonyl group bonded to a chlorine atom.
- The most common way to prepare an acid chloride is to treat a carboxylic acid with thionyl chloride.

14.8 What Is Decarboxylation?

- **Decarboxylation** is the loss of CO_2 from a carboxyl group.
- Carboxylic acids that have a carbonyl group β to the carboxyl group readily undergo decarboxylation on mild heating.

Quick Quiz

Answer true or false to the following questions to assess your general knowledge of the concepts in this chapter. If you have difficulty with any of them, you should review the appropriate section in the chapter (shown in parentheses) before attempting the more challenging end-of-chapter problems.

1. In naming carboxylic acids, it is always necessary to indicate the position at which the carboxyl group occurs. (14.2)
2. 2-Propylpropanedioic acid can undergo decarboxylation at relatively moderate temperatures. (14.8)
3. Fischer esterification is reversible. (14.6)
4. The hydrophilic group of a carboxylic acid decreases water solubility. (14.3)
5. Both alcohols and carboxylic acids react with $SOCl_2$. (14.7)
6. Fischer esterification involves the reaction of a carboxylic acid with another carboxylic acid. (14.6)
7. An electronegative atom on a carboxylic acid can potentially increase the acid's acidity. (14.4)
8. A carboxyl group is reduced to a 1° alcohol by H_2/Pt. (14.5)
9. A carboxyl group is reduced to a 1° alcohol by $NaBH_4$. (14.5)
10. A carboxyl group that has been deprotonated is called a carboxylate group. (14.4)
11. A carboxyl group is reduced to a 1° alcohol by $LiAlH_4$. (14.5)
12. The conjugate base of a carboxylic acid is resonance-stabilized. (14.4)
13. Carboxylic acids possess both a region of polarity and a region of nonpolarity. (14.3)
14. Carboxylic acids are less acidic than phenols. (14.4)
15. 4-Oxopentanoic acid can undergo decarboxylation at relatively moderate temperatures. (14.8)
16. The γ position of a carboxylic acid refers to carbon-4 of the chain. (14.2)

Answers: (1) F (2) T (3) T (4) F (5) T (6) F (7) T (8) F (9) F (10) T (11) T (12) T (13) T (14) F (15) F (16) T

Key Reactions

1. Acidity of Carboxylic Acids (Section 14.4A)

Values of pK_a for most unsubstituted aliphatic and aromatic carboxylic acids are within the range from 4 to 5:

$$CH_3\overset{O}{\overset{\|}{C}}OH + H_2O \rightleftharpoons CH_3\overset{O}{\overset{\|}{C}}O^- + H_3O^+ \quad pK_a = 4.76$$

Substitution by electron-withdrawing groups decreases pK_a (increases acidity).

2. Reaction of Carboxylic Acids with Bases (Section 14.4B)

Carboxylic acids form water-soluble salts with alkali metal hydroxides, carbonates, and bicarbonates, as well as with ammonia and amines:

$$C_6H_5{-}COOH + NaOH \xrightarrow{H_2O} C_6H_5{-}COO^-Na^+ + H_2O$$

3. Reduction by Lithium Aluminum Hydride (Section 14.5)

Lithium aluminum hydride reduces a carboxyl group to a primary alcohol:

$$\text{(cyclopentenyl)}{-}\overset{O}{\overset{\|}{C}}OH \xrightarrow[2)\ H_2O]{1)\ LiAlH_4} \text{(cyclopentenyl)}{-}CH_2OH$$

4. Fischer Esterification (Section 14.6)

Fischer esterification is reversible:

$$CH_3\overset{O}{\overset{\|}{C}}OH + HOCH_2CH_2CH_3 \underset{}{\overset{H_2SO_4}{\rightleftharpoons}} CH_3\overset{O}{\overset{\|}{C}}OCH_2CH_2CH_3 + H_2O$$

One way to force the equilibrium to the right is to use an excess of the alcohol.

5. Conversion to Acid Halides (Section 14.7)
Acid chlorides, the most common and widely used of the acid halides, are prepared by treating carboxylic acids with thionyl chloride:

$CH_3CH_2CH_2COOH + SOCl_2 \longrightarrow CH_3CH_2CH_2COCl + SO_2 + HCl$

6. Decarboxylation of β-Ketoacids (Section 14.8A)
The mechanism of decarboxylation involves the redistribution of bonding electrons in a cyclic, six-membered transition state:

2-Oxocyclohexanecarboxylic acid $\xrightarrow{\text{warm}}$ cyclohexanone $+ CO_2$

7. Decarboxylation of β-Dicarboxylic Acids (Section 14.8B)
The mechanism of decarboxylation of a β-dicarboxylic acid is similar to that of decarboxylation of a β-ketoacid:

$$HO\overset{O}{\overset{\|}{C}}CH_2\overset{O}{\overset{\|}{C}}OH \xrightarrow{\text{heat}} CH_3\overset{O}{\overset{\|}{C}}OH + CO_2$$

Problems

A problem marked with an asterisk indicates an applied "real world" problem. Answers to problems whose numbers are printed in blue are given in Appendix D.

Section 14.2 Structure and Nomenclature

14.8 Name and draw structural formulas for the four carboxylic acids with molecular formula $C_5H_{10}O_2$. Which of these carboxylic acids is chiral?

14.9 Write the IUPAC name for each compound: **(See Example 14.1)**

(a) cyclohexene ring–COOH

(b) $CH_3CH(OH)CH_2CH_2COOH$

(c) $(CH_3)_2C{=}CHCH_2CH_2C(CH_3){=}CHCOOH$

(d) 1-methylcyclopentane–COOH

(e) $CH_3CH_2CH_2CH_2CH_2COO^-NH_4^+$

(f) $HOOCCH_2CH(OH)COOH$

14.10 Draw a structural formula for each carboxylic acid: **(See Example 14.1)**

(a) 4-Nitrophenylacetic acid
(b) 4-Aminopentanoic acid
(c) 3-Chloro-4-phenylbutanoic acid
(d) *cis*-3-Hexenedioic acid
(e) 2,3-Dihydroxypropanoic acid
(f) 3-Oxohexanoic acid
(g) 2-Oxocyclohexanecarboxylic acid
(h) 2,2-Dimethylpropanoic acid

***14.11** Megatomoic acid, the sex attractant of the female black carpet beetle, has the structure **(See Example 14.1)**

$$CH_3(CH_2)_7CH{=}CHCH{=}CHCH_2COOH$$
Megatomoic acid

(a) What is the IUPAC name of megatomoic acid?
(b) State the number of stereoisomers possible for this compound.

***14.12** The IUPAC name of ibuprofen is 2-(4-isobutylphenyl) propanoic acid. Draw a structural formula of ibuprofen. **(See Example 14.1)**

14.13 Draw structural formulas for these salts:

(a) Sodium benzoate
(b) Lithium acetate
(c) Ammonium acetate
(d) Disodium adipate
(e) Sodium salicylate
(f) Calcium butanoate

*14.14 The monopotassium salt of oxalic acid is present in certain leafy vegetables, including rhubarb. Both oxalic acid and its salts are poisonous in high concentrations. Draw a structural formula of monopotassium oxalate.

*14.15 Potassium sorbate is added as a preservative to certain foods to prevent bacteria and molds from causing spoilage and to extend the foods' shelf life. The IUPAC name of potassium sorbate is potassium (2*E*,4*E*)-2,4-hexadienoate. Draw a structural formula of potassium sorbate. **(See Example 14.1)**

*14.16 Zinc 10-undecenoate, the zinc salt of 10-undecenoic acid, is used to treat certain fungal infections, particularly *tinea pedis* (athlete's foot). Draw a structural formula of this zinc salt.

Section 14.3 Physical Properties

14.17 Arrange the compounds in each set in order of increasing boiling point:

(a) $CH_3(CH_2)_5COOH$ $CH_3(CH_2)_6CHO$ $CH_3(CH_2)_6CH_2OH$

(b) CH_3CH_2COOH $CH_3CH_2CH_2CH_2OH$ $CH_3CH_2OCH_2CH_3$

Section 14.4 Preparation of Carboxylic Acids

14.18 Draw a structural formula for the product formed by treating each compound with warm chromic acid, H_2CrO_4:

(a) $CH_3(CH_2)_4CH_2OH$

(b)

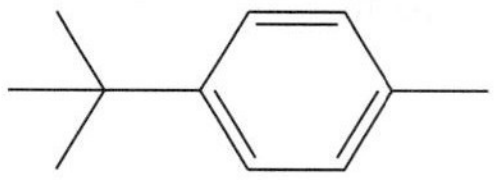

(c) 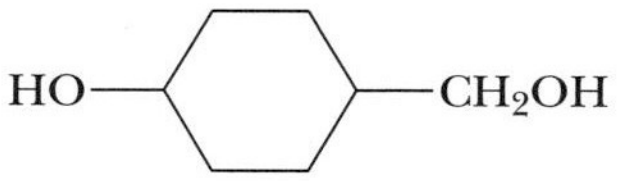 HO— —CH_2OH

14.19 Draw a structural formula for a compound with the given molecular formula that, on oxidation by chromic acid, gives the carboxylic acid or dicarboxylic acid shown:

(a) $C_6H_{14}O$ $\xrightarrow{\text{oxidation}}$ COOH

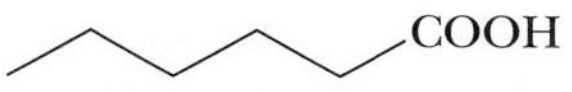

(b) $C_6H_{12}O$ $\xrightarrow{\text{oxidation}}$ COOH

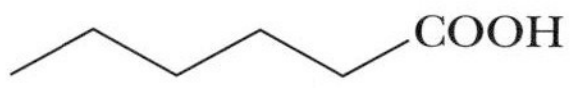

(c) $C_6H_{14}O_2$ $\xrightarrow{\text{oxidation}}$ HOOC COOH

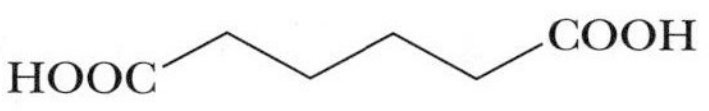

Acidity of Carboxylic Acids

14.20 Which is the stronger acid in each pair? **(See Example 14.2)**

(a) Phenol (pK_a 9.95) or benzoic acid (pK_a 4.17)

(b) Lactic acid (K_a 1.4×10^{-4}) or ascorbic acid (K_a 6.8×10^{-5})

14.21 Arrange these compounds in order of increasing acidity: benzoic acid, benzyl alcohol, and phenol. **(See Example 14.2)**

14.22 Assign the acid in each set its appropriate pK_a: **(See Example 14.2)**

(a) 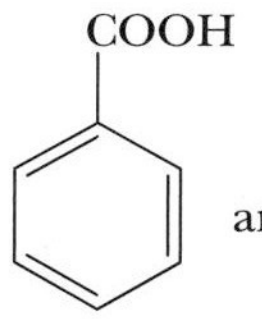 COOH and COOH, NO_2 (pK_a 4.19 and 3.14)

(b) 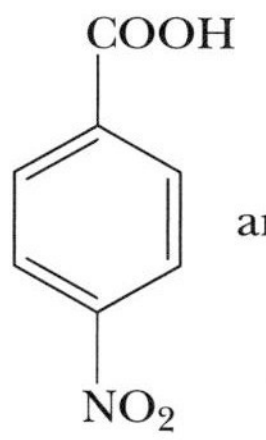 COOH, NO_2 and COOH, NH_2 (pK_a 4.92 and 3.14)

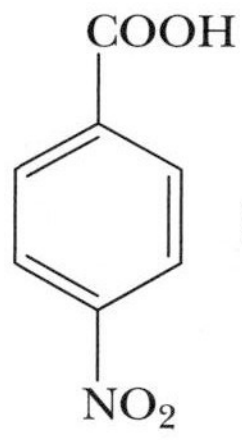

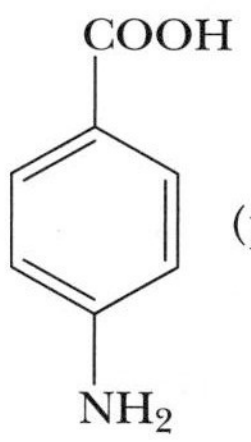

(c) $CH_3\overset{O}{\overset{\|}{C}}CH_2COOH$ and $CH_3\overset{O}{\overset{\|}{C}}COOH$ (pK_a 3.58 and 2.49)

(d) $CH_3\overset{OH}{\overset{|}{C}}HCOOH$ and CH_3CH_2COOH (pK_a 3.85 and 3.08)

14.23 Complete these acid–base reactions: **(See Example 14.3)**

(a) 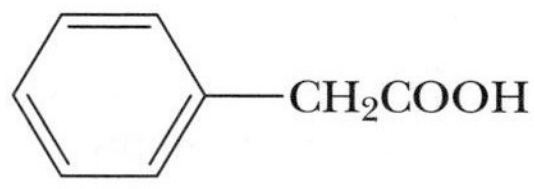 $-CH_2COOH + NaOH \longrightarrow$

(b) $CH_3CH{=}CHCH_2COOH + NaHCO_3 \longrightarrow$

(c) 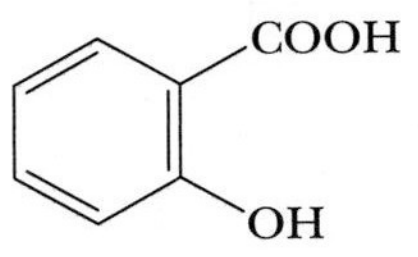 (2-hydroxybenzoic acid: benzene ring with COOH and OH) + $NaHCO_3 \longrightarrow$

(d) $CH_3CH(OH)COOH + H_2NCH_2CH_2OH \longrightarrow$

(e) $CH_3CH{=}CHCH_2COO^-Na^+ + HCl \longrightarrow$

*14.24 The normal pH range for blood plasma is 7.35–7.45. Under these conditions, would you expect the carboxyl group of lactic acid (pK_a 3.85) to exist primarily as a carboxyl group or as a carboxylate anion? Explain.

*14.25 The pK_a of ascorbic acid (Section 18.7) is 4.76. Would you expect ascorbic acid dissolved in blood plasma (pH 7.35–7.45) to exist primarily as ascorbic acid or as ascorbate anion? Explain.

*14.26 Excess ascorbic acid is (pK_a 4.76) excreted in the urine, the pH of which is normally in the range from 4.8 to 8.4. What form of ascorbic acid, ascorbic acid itself or ascorbate anion, would you expect to be present in urine with pH 8.4?

*14.27 The pH of human gastric juice is normally in the range from 1.0 to 3.0. What form of lactic acid (pK_a 3.85), lactic acid itself or its anion, would you expect to be present in the stomach?

*14.28 Following are two structural formulas for the amino acid alanine (Section 19.2):

$$CH_3-CH(NH_2)-C(=O)-OH \quad \text{(A)} \qquad CH_3-CH(NH_3^+)-C(=O)-O^- \quad \text{(B)}$$

Is alanine better represented by structural formula A or B? Explain.

*14.29 In Chapter 19, we discuss a class of compounds called amino acids, so named because they contain both an amino group and a carboxyl group. Following is a structural formula for the amino acid alanine in the form of an internal salt:

$$CH_3CH(NH_3^+)CO^- \text{ (C=O) } \quad \text{Alanine}$$

What would you expect to be the major form of alanine present in aqueous solution at (a) pH 2.0, (b) pH 5–6, and (c) pH 11.0? Explain.

Sections 14.5–14.8 Reactions of Carboxylic Acids

14.30 Give the expected organic products formed when phenylacetic acid, $PhCH_2COOH$, is treated with each of the following reagents: **(See Examples 14.4–14.6)**

(a) $SOCl_2$

(b) $NaHCO_3, H_2O$

(c) $NaOH, H_2O$

(d) NH_3, H_2O

(e) $LiAlH_4$, followed by H_2O

(f) $NaBH_4$, followed by H_2O

(g) $CH_3OH + H_2SO_4$ (catalyst)

(h) H_2/Ni at 25°C and 3 atm pressure

14.31 Show how to convert *trans*-3-phenyl-2-propenoic acid (cinnamic acid) to these compounds: **(See Example 14.4)**

(a) 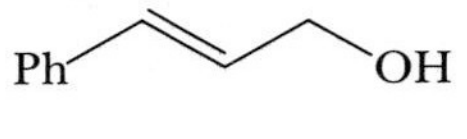 Ph–CH=CH–CH₂OH (skeletal: Ph, OH)

(b) 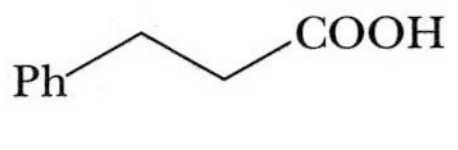 Ph–CH₂CH₂–COOH (skeletal: Ph, COOH)

(c) 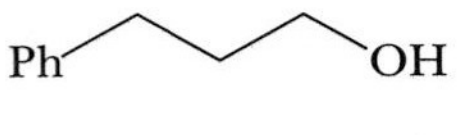 Ph–CH₂CH₂CH₂–OH (skeletal: Ph, OH)

14.32 Show how to convert 3-oxobutanoic acid (acetoacetic acid) to these compounds: **(See Example 14.4)**

(a) $CH_3CH(OH)CH_2COOH$

(b) $CH_3CH(OH)CH_2CH_2OH$

(c) $CH_3CH{=}CHCOOH$

14.33 Complete these examples of Fischer esterification (assume an excess of the alcohol): **(See Example 14.5)**

(a) 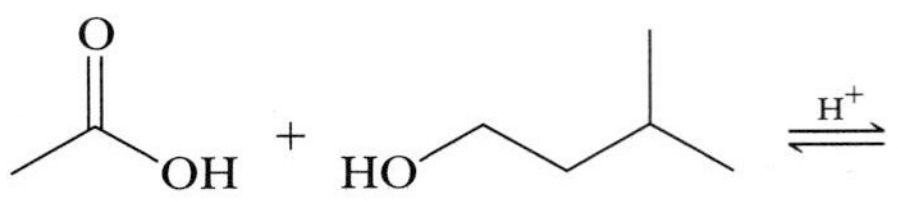 $CH_3C(=O)OH$ + HO–CH₂CH₂CH(CH₃)₂ $\overset{H^+}{\rightleftharpoons}$

(b) 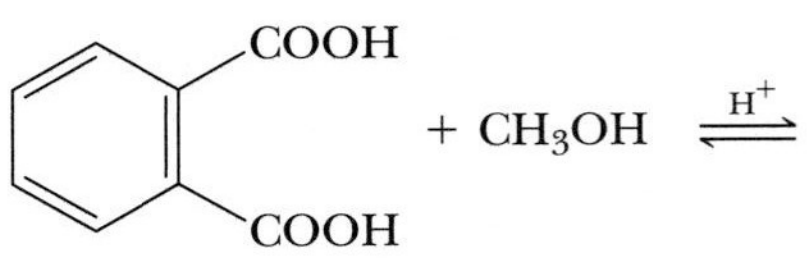 (benzene-1,2-dicarboxylic acid: COOH, COOH) $+ CH_3OH \overset{H^+}{\rightleftharpoons}$

(c) 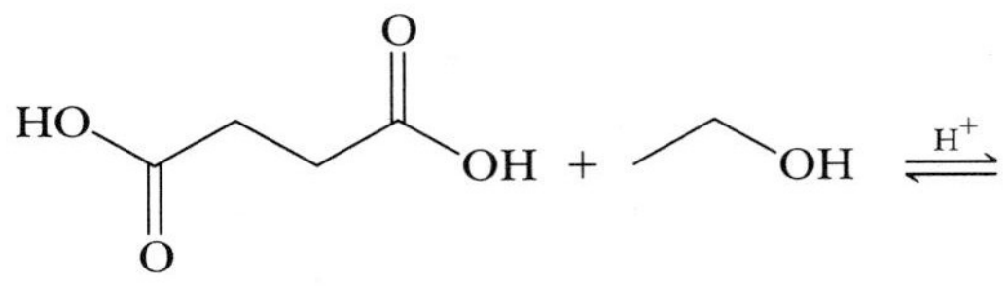 HO–C(=O)–CH₂CH₂–C(=O)–OH + $CH_3CH_2OH \overset{H^+}{\rightleftharpoons}$

*14.34 Formic acid is one of the components responsible for the sting of biting ants and is injected under the skin by bees and wasps. A way to relieve the pain is to rub the area of the sting with a paste of baking soda ($NaHCO_3$) and water, which neutralizes the acid. Write an equation for this reaction. **(See Example 14.5)**

*14.35 Methyl 2-hydroxybenzoate (methyl salicylate) has the odor of oil of wintergreen. This ester is prepared by the Fischer esterification of 2-hydroxybenzoic acid (salicylic acid) with methanol. Draw a structural formula of methyl 2-hydroxybenzoate.

*14.36 Benzocaine, a topical anesthetic, is prepared by treating 4-aminobenzoic acid with ethanol in the presence of an acid catalyst, followed by neutralization. Draw a structural formula of benzocaine.

*14.37 Examine the structural formulas of pyrethrin and permethrin. (See Chemical Connections 15D.)

(a) Locate the ester groups in each compound.

(b) Is pyrethrin chiral? How many stereoisomers are possible for it?

(c) Is permethrin chiral? How many stereoisomers are possible for it?

*14.38 A commercial Clothing & Gear Insect Repellant gives the following information about permethrin, its active ingredient:

Cis/trans ratio: Minimum 35% (+/−) *cis* and maximum 65% (+/−) *trans*

(a) To what does the *cis/trans* ratio refer?

(b) To what does the designation "(+/−)" refer?

14.39 From what carboxylic acid and alcohol is each of the following esters derived? **(See Example 14.5)**

(a) 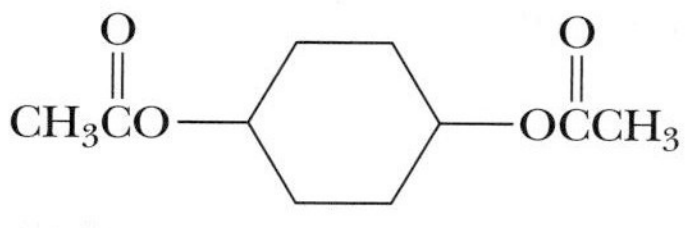 $CH_3\overset{O}{\overset{\|}{C}}O$—[cyclohexane-1,4-diyl]—$O\overset{O}{\overset{\|}{C}}CH_3$

(b) $CH_3O\overset{O}{\overset{\|}{C}}CH_2CH_2\overset{O}{\overset{\|}{C}}OCH_3$

(c) 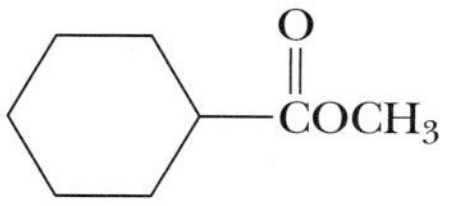 [cyclohexyl]—$\overset{O}{\overset{\|}{C}}OCH_3$

(d) $CH_3CH_2CH{=}CH\overset{O}{\overset{\|}{C}}OCH(CH_3)_2$

14.40 When treated with an acid catalyst, 4-hydroxybutanoic acid forms a cyclic ester (a lactone). Draw the structural formula of this lactone. **(See Example 14.5)**

14.41 Draw a structural formula for the product formed on thermal decarboxylation of each of the following compounds: **(See Example 14.7)**

(a) $C_6H_5\overset{O}{\overset{\|}{C}}CH_2COOH$

(b) 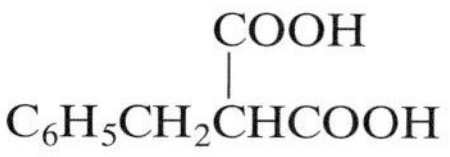 $C_6H_5CH_2\underset{}{\overset{COOH}{\overset{|}{C}}}HCOOH$

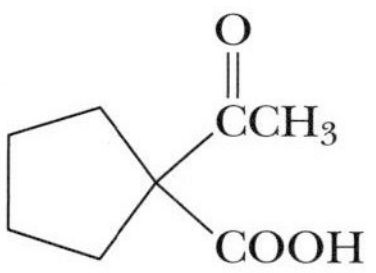

(c) [cyclopentane bearing $\overset{O}{\overset{\|}{C}}CH_3$ and COOH on the same carbon]

Synthesis

*14.42 Methyl 2-aminobenzoate, a flavoring agent with the taste of grapes (see Chemical Connections 14B), can be prepared from toluene by the following series of steps:

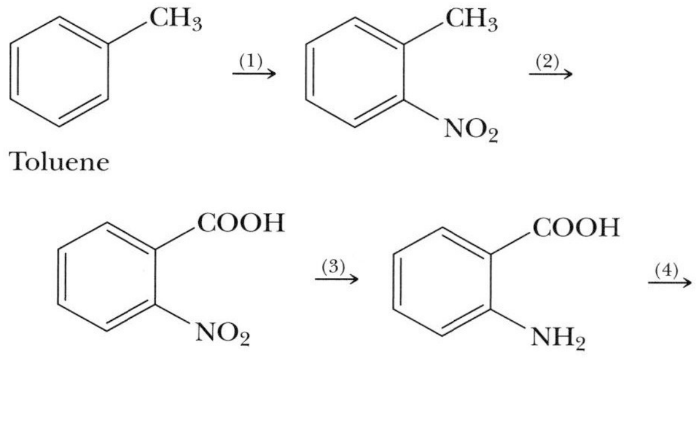

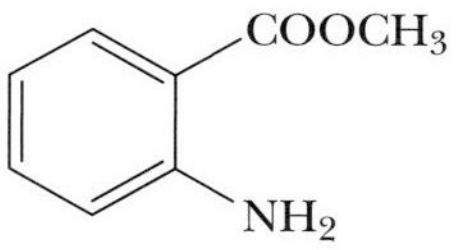

Toluene ($C_6H_5CH_3$) $\xrightarrow{(1)}$ 2-nitrotoluene (CH_3, NO_2) $\xrightarrow{(2)}$ 2-nitrobenzoic acid ($COOH$, NO_2) $\xrightarrow{(3)}$ 2-aminobenzoic acid ($COOH$, NH_2) $\xrightarrow{(4)}$ Methyl 2-aminobenzoate ($COOCH_3$, NH_2)

Methyl 2-amino-benzoate

Show how you might bring about each step in this synthesis.

*14.43 Methylparaben and propylparaben are used as preservatives in foods, beverages, and cosmetics:

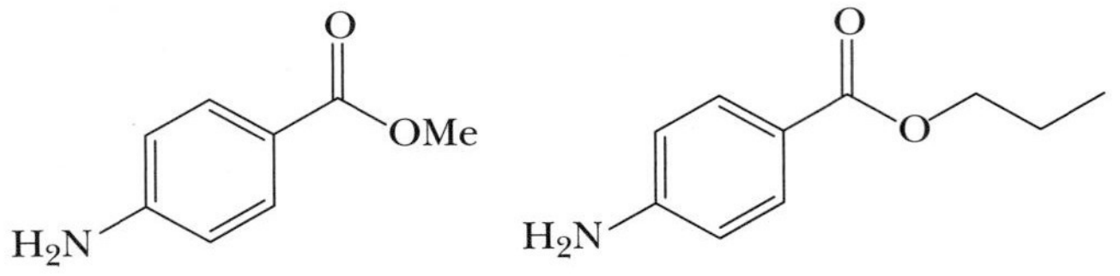

Methyl 4-aminobenzoate (Methylparaben)

Propyl 4-aminobenzoate (Propylparaben)

Show how the synthetic scheme in Problem 14.42 can be modified to give each of these compounds.

*14.44 Procaine (its hydrochloride is marketed as Novocaine®) was one of the first local anesthetics developed for infiltration and regional anesthesia. It is synthesized by the following Fischer esterification:

p-Aminobenzoic acid + 2-Diethylaminoethanol $\xrightarrow{\text{Fischer esterification}}$ Procaine

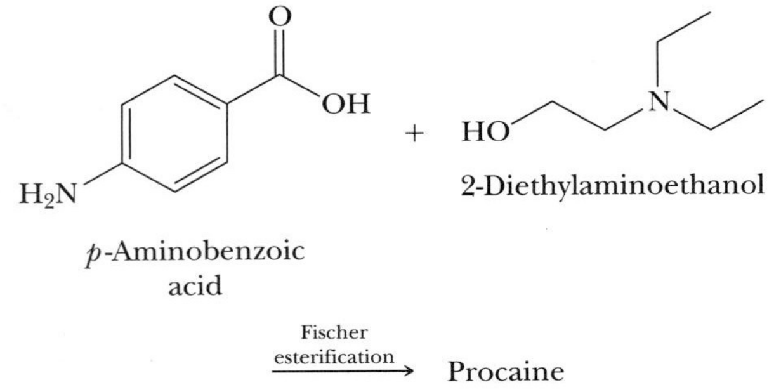

Draw a structural formula for procaine.

*14.45 Meclizine is an antiemetic: It helps prevent, or at least lessen, the vomiting associated with motion sickness, including seasickness. Among the names of the over-the-counter preparations of meclizine are Bonine®, Sea-Legs, Antivert®, and Navicalm®. Meclizine can be synthesized by the following series of steps:

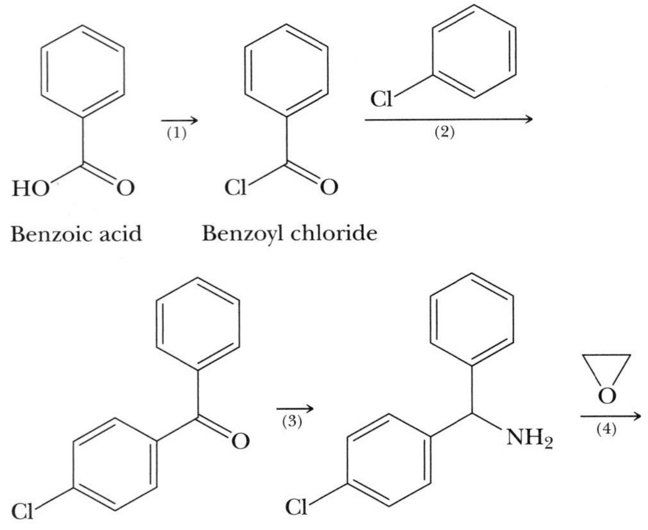

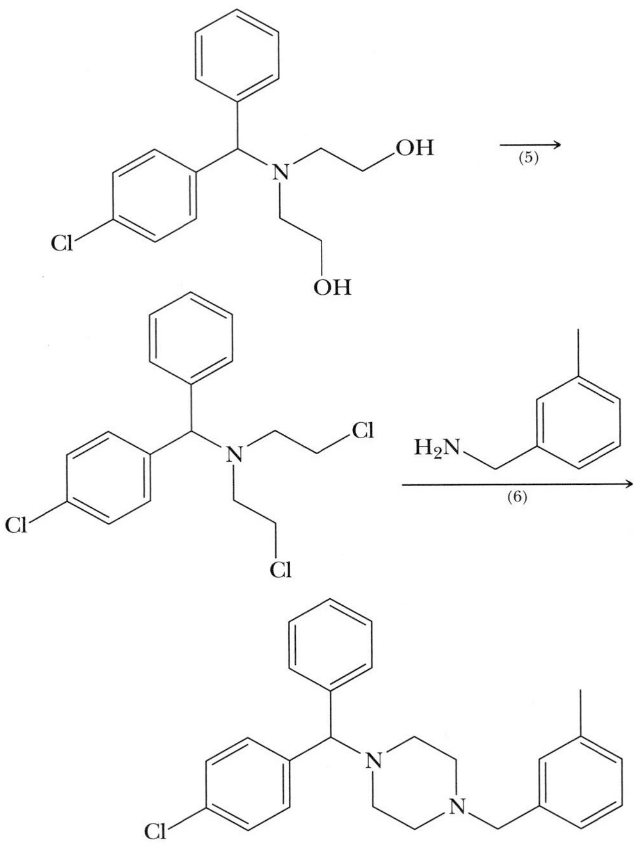

Benzoic acid $\xrightarrow{(1)}$ Benzoyl chloride $\xrightarrow{(2)}$ ... $\xrightarrow{(3)}$ ... $\xrightarrow{(4)}$ ... $\xrightarrow{(5)}$ ... $\xrightarrow{(6)}$ Meclizine

Meclizine

(a) Propose a reagent for Step 1.

(b) The catalyst for Step 2 is $AlCl_3$. Name the type of reaction that occurs in Step 2.

(c) Propose reagents for Step 3.

(d) Propose a mechanism for Step 4, and show that it is an example of nucleophilic aliphatic substitution.

(e) Propose a reagent for Step 5.

(f) Show that Step 6 is also an example of nucleophilic aliphatic substitution.

*14.46 Chemists have developed several syntheses for the antiasthmatic drug albuterol (Proventil). One of these syntheses starts with salicylic acid, the same acid that is the starting material for the synthesis of aspirin:

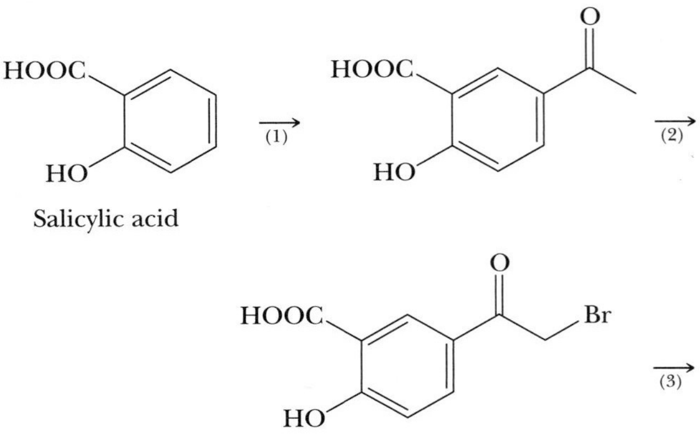

Salicylic acid $\xrightarrow{(1)}$ ... $\xrightarrow{(2)}$ ... $\xrightarrow{(3)}$

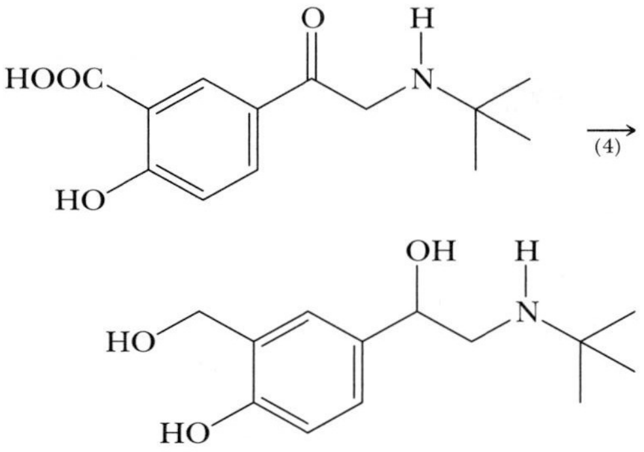

Albuterol

(a) Propose a reagent and a catalyst for Step 1. What name is given to this type of reaction?

(b) Propose a reagent for Step 2.

(c) Name the amine used to bring about Step 3.

(d) Step 4 is a reduction of two functional groups. Name the functional groups reduced and tell what reagent will accomplish the reduction.

Chemical Transformations

14.47 Test your cumulative knowledge of the reactions learned thus far by completing the following chemical transformations. *Note*: Some will require more than one step. **(See Examples 14.4–14.7)**

(a)

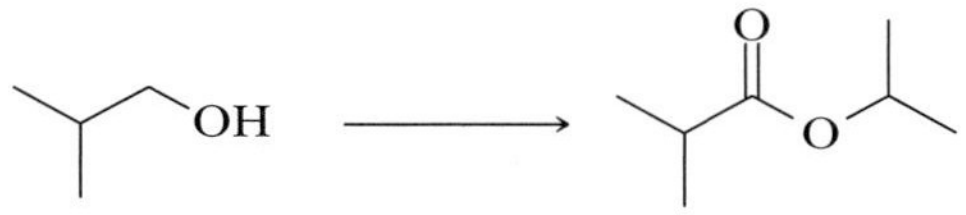

(b)

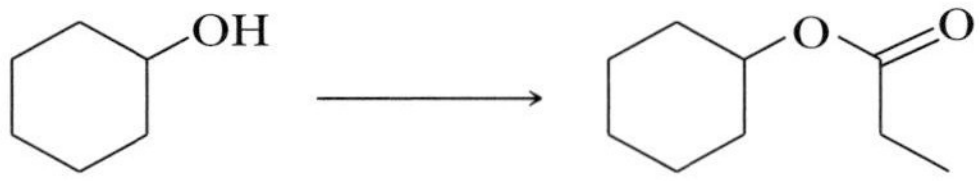

(c)

(d)

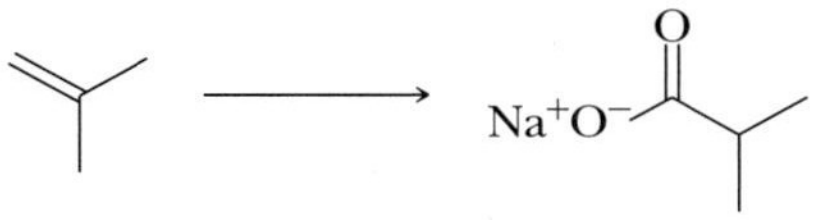

(e)

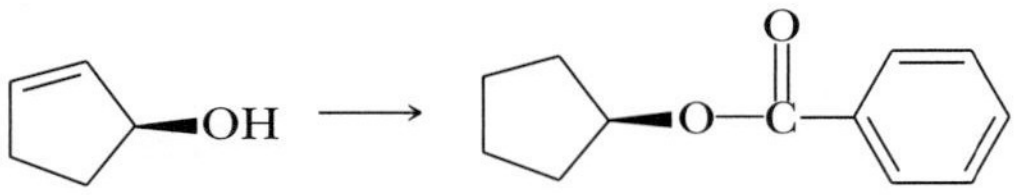

(f)

(g)

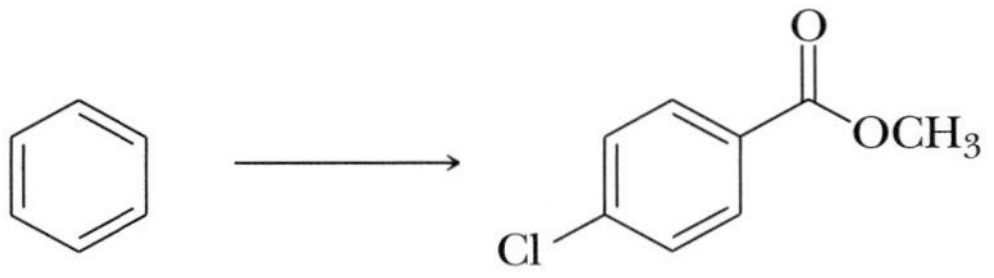

(h)

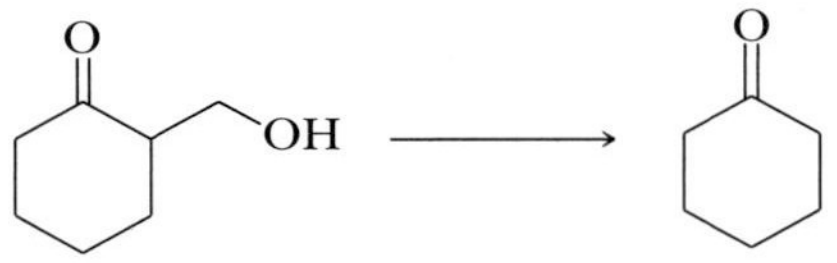

(i)

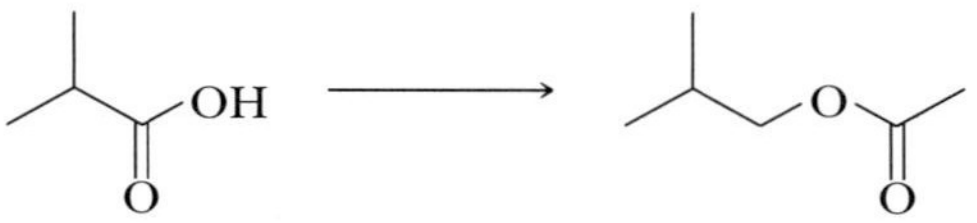

(j)

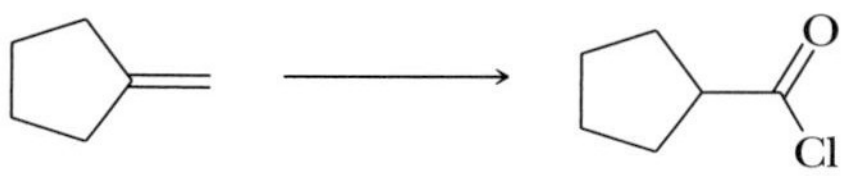

(k)

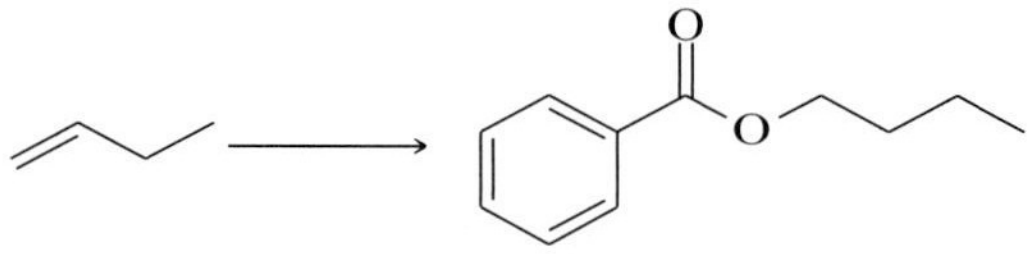

(l)

(m)

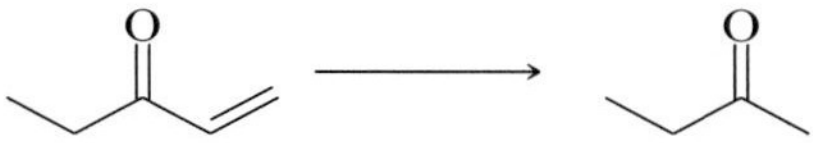

(n)

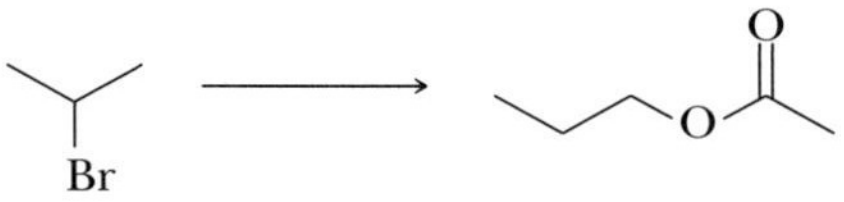

(o) $H_3C—C\equiv C—H$ ⟶

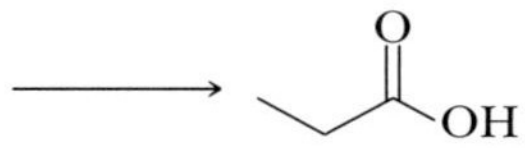

(p)

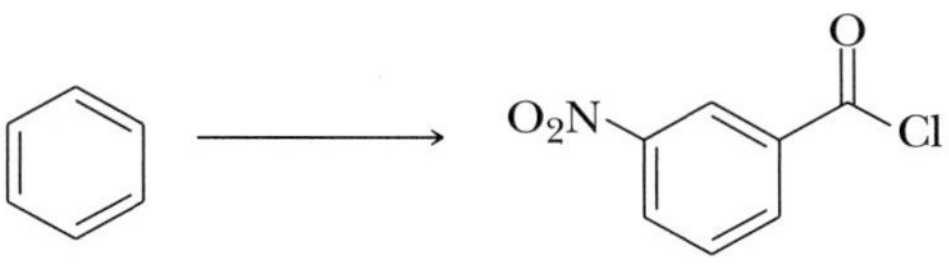

Looking Ahead

14.48 Explain why α-amino acids, the building blocks of proteins (Chapter 20), are nearly a thousand times more acidic than aliphatic carboxylic acids: **(See Example 14.2)**

An α-amino acid
$pK_a \approx 2$

An aliphatic acid
$pK_a \approx 5$

14.49 Which is more difficult to reduce with $LiAlH_4$, a carboxylic acid or a carboxylate ion?

14.50 Show how an ester can react with H^+/H_2O to give a carboxylic acid and an alcohol (*Hint:* This is the reverse of Fischer esterification):

$$RCH_2COOR' \xrightleftharpoons{H^+/H_2O} RCH_2COOH + HOR'$$

14.51 In Chapter 13, we saw how Grignard reagents readily attack the carbonyl carbon of ketones and aldehydes. Should the same process occur with Grignards and carboxylic acids? With esters?

14.52 In Section 14.6, it was suggested that the mechanism for the Fischer esterification of carboxylic acids would be a model for many of the reactions of the functional derivatives of carboxylic acids. One such reaction, the reaction of an acid halide with water, is the following:

$$RCOCl \xrightarrow{H_2O} RCOOH + HCl$$

Suggest a mechanism for this reaction.

15 Functional Derivatives of Carboxylic Acids

KEY QUESTIONS

15.1 What Are Some Derivatives of Carboxylic Acids, and How Are They Named?

15.2 What Are the Characteristic Reactions of Carboxylic Acid Derivatives?

15.3 What Is Hydrolysis?

15.4 How Do Carboxylic Acid Derivatives React with Alcohols?

15.5 How Do Carboxylic Acid Derivatives React with Ammonia and Amines?

15.6 How Can Functional Derivatives of Carboxylic Acids Be Interconverted?

15.7 How Do Esters React with Grignard Reagents?

15.8 How Are Derivatives of Carboxylic Acids Reduced?

HOW TO

15.1 How to Name Functional Derivatives of Carboxylic Acids

15.2 How to Approach Multistep Synthesis Problems

CHEMICAL CONNECTIONS

15A Ultraviolet Sunscreens and Sunblocks

15B From Moldy Clover to a Blood Thinner

15C The Penicillins and Cephalosporins: β-Lactam Antibiotics

15D The Pyrethrins: Natural Insecticides of Plant Origin

15E Systematic Acquired Resistance in Plants

Macrophotograph of the fungus *Penicillium notatum* growing on a petri dish culture of Whickerman's agar. This fungus was used as an early source of the first penicillin antibiotic. Inset: A model of amoxicillin. *(Andrew McClenaghan/Photo Researchers, Inc.)*

In this chapter, we study four classes of organic compounds, all derived from the carboxyl group: acid halides, acid anhydrides, esters, and amides. Under the general formula of each functional group is a drawing to help you see how the group is formally related to a carboxyl group. The loss of —OH from a carboxyl group and H— from H—Cl, for example, gives an acid chloride, and similarly, the loss of —OH from a carboxyl group and H— from ammonia gives an amide:

$RCCl$ (C=O) An acid chloride	$RCOCR'$ (two C=O) An acid anhydride	$RCOR'$ (C=O) An ester	$RCNH_2$ (C=O) An amide
↑ $-H_2O$	↑ $-H_2O$	↑ $-H_2O$	↑ $-H_2O$
$RC(=O)—OH$ H—Cl	$RC(=O)—OH$ H—$OC(=O)R'$	$RC(=O)—OH$ H—OR'	$RC(=O)—OH$ H—NH_2

15.1 What Are Some Derivatives of Carboxylic Acids, and How Are They Named?

A. Acid Halides

Acid halide A derivative of a carboxylic acid in which the —OH of the carboxyl group is replaced by a halogen—most commonly, chlorine.

The functional group of an **acid halide** (acyl halide) is an **acyl group (RCO—)** bonded to a halogen atom (Section 14.7). The most common acid halides are acid chlorides:

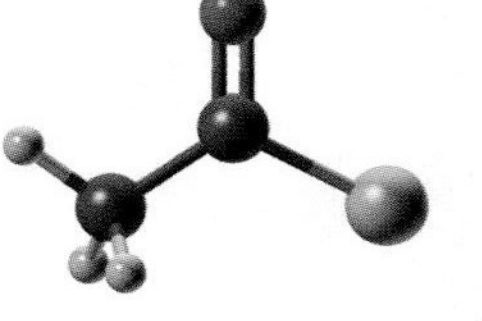

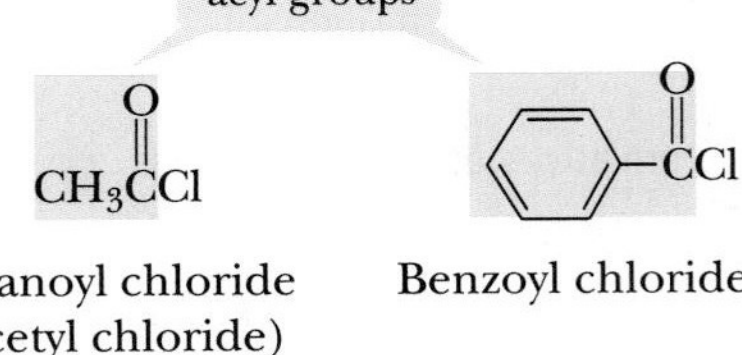

Acid halides are named by changing the suffix *-ic acid* in the name of the parent carboxylic acid to *-yl halide*.

B. Acid Anhydrides

Carboxylic Anhydrides

Carboxylic anhydride A compound in which two acyl groups are bonded to an oxygen.

The functional group of a **carboxylic anhydride** (commonly referred to simply as an anhydride) is two acyl groups bonded to an oxygen atom. The anhydride may be symmetrical (having two identical acyl groups), or it may be mixed (having two different acyl groups). Symmetrical anhydrides are named by changing the suffix *acid* in the name of the parent carboxylic acid to *anhydride*:

$CH_3C(=O)—O—C(=O)CH_3$

Acetic anhydride

$C_6H_5C(=O)—O—C(=O)C_6H_5$

Benzoic anhydride

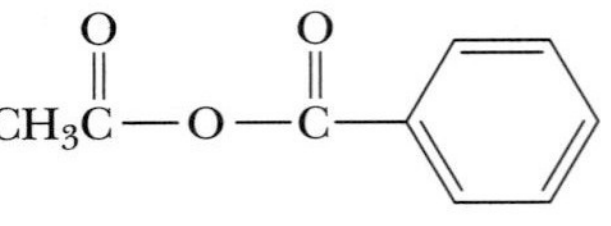

Acetic benzoic anhydride
(a mixed anhydride)

Mixed anhydrides are named by identifying the two parent carboxylic acids from both acyl groups and placing those names in succession, in alphabetical order, without the "acid" part of the name followed by the word *anhydride*.

Phosphoric Anhydrides

Because of the special importance of anhydrides of phosphoric acid in biochemical systems (Chapter 22), we include them here to show the similarity between them and the anhydrides of carboxylic acids. The functional group of a **phosphoric anhydride**

CHEMICAL CONNECTIONS 15A

Ultraviolet Sunscreens and Sunblocks

Ultraviolet (UV) radiation (Section 11.1, Table 11.1) penetrating the earth's ozone layer is arbitrarily divided into two regions: UVB (290–320 nm) and UVA (320–400 nm). UVB, a more energetic form of radiation than UVA, interacts directly with molecules of the skin and eyes, causing skin cancer, aging of the skin, eye damage leading to cataracts, and delayed sunburn that appears 12 to 24 hours after exposure. UVA radiation, by contrast, causes tanning. It also damages skin, albeit much less efficiently than UVB. The role of UVA in promoting skin cancer is less well understood.

Commercial sunscreen products are rated according to their sun protection factor (SPF), which is defined as the minimum effective dose of UV radiation that produces a delayed sunburn on protected skin compared with unprotected skin. Two types of active ingredients are found in commercial sunblocks and sunscreens. The most common sunblock agent is zinc oxide, ZnO, a white crystalline substance that reflects and scatters UV radiation. Sunscreens, the second type of active ingredient, absorb UV radiation and then reradiate it as heat. Sunscreens are most effective in screening out UVB radiation, but they do not screen out UVA radiation. Thus, they allow tanning, but prevent the UVB-associated damage. Given here are structural formulas for three common esters used as UVB-screening agents, along with the name by which each is most commonly listed in the "Active Ingredients" label on commercial products:

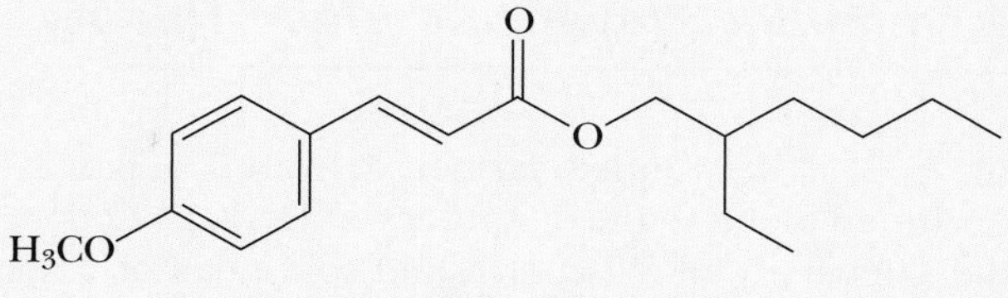

H_3CO O

Octyl *p*-methoxycinnamate

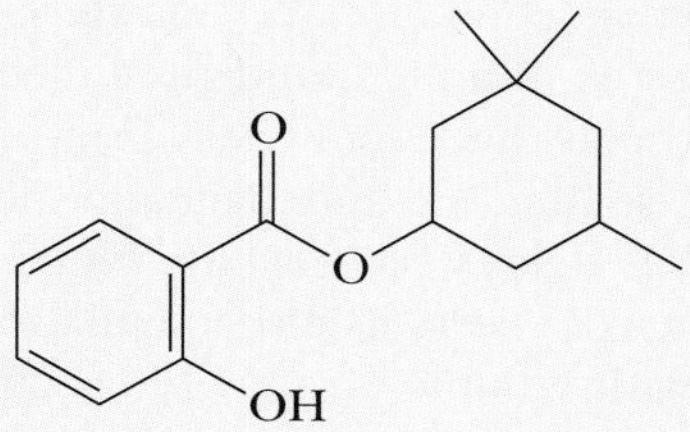

O O OH

Homosalate

O O H_3C N CH_3

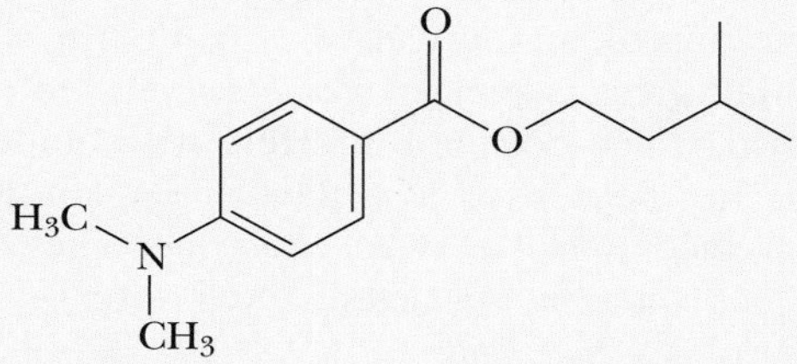

Padimate A

QUESTION

Show how each sunscreen can be synthesized from a carboxylic acid and alcohol using the Fischer esterification reaction (Section 14.6).

is two phosphoryl groups bonded to an oxygen atom. Shown here are structural formulas for two anhydrides of phosphoric acid, H_3PO_4, and the ions derived by ionization of the acidic hydrogens of each:

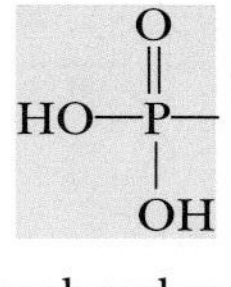

O ‖ HO—P— | OH

phosphoryl group

HO—P(=O)(OH)—O—P(=O)(OH)—OH

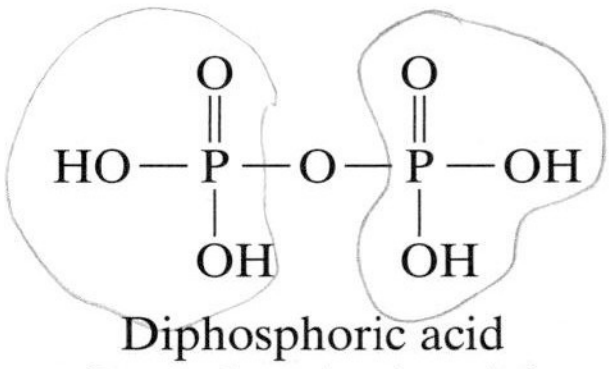

Diphosphoric acid (Pyrophosphoric acid)

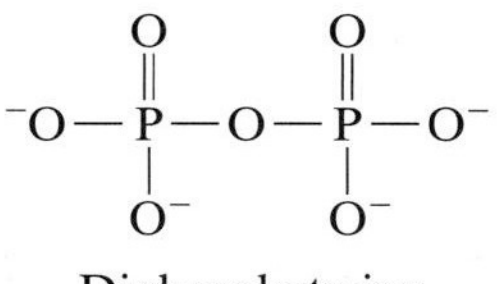

$^-$O—P(=O)(O$^-$)—O—P(=O)(O$^-$)—O$^-$

Diphosphate ion (Pyrophosphate ion)

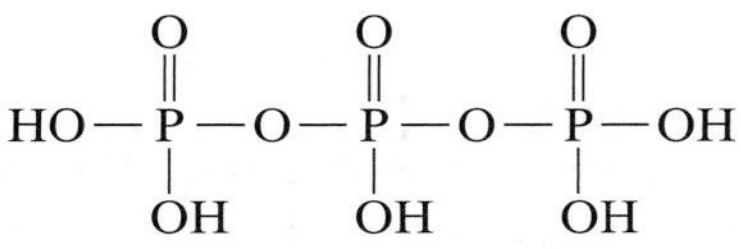

HO—P(=O)(OH)—O—P(=O)(OH)—O—P(=O)(OH)—OH

Triphosphoric acid

$^-$O—P(=O)(O$^-$)—O—P(=O)(O$^-$)—O—P(=O)(O$^-$)—O$^-$

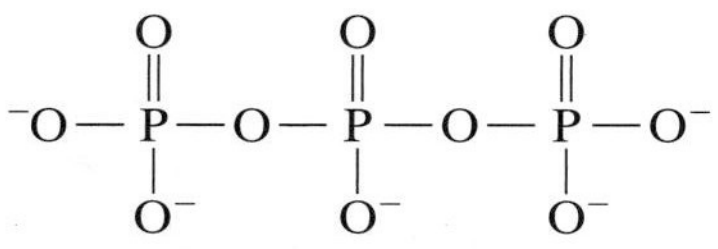

Triphosphate ion

CHEMICAL CONNECTIONS 15B

From Moldy Clover to a Blood Thinner

In 1933, a disgruntled farmer delivered a pail of unclotted blood to the laboratory of Dr. Karl Link at the University of Wisconsin and told tales of cows bleeding to death from minor cuts. Over the next couple of years, Link and his collaborators discovered that when cows are fed moldy clover, their blood clotting is inhibited, and they bleed to death from minor cuts and scratches. From the moldy clover, Link isolated the anticoagulant dicoumarol, a substance that delays or prevents blood from clotting. Dicoumarol exerts its anticoagulation effect by interfering with vitamin K activity (Section 21.6D). Within a few years after its discovery, dicoumarol became widely used to treat victims of heart attack and others at risk for developing blood clots.

Dicoumarol is a derivative of coumarin, a cyclic ester that gives sweet clover its pleasant smell. Coumarin, which does not interfere with blood clotting and has been used as a flavoring agent, is converted to dicoumarol as sweet clover becomes moldy. Notice that coumarin is a lactone (cyclic ester), whereas dicoumarol is a dilactone:

as sweet clover becomes moldy →

Coumarin
(from sweet clover)

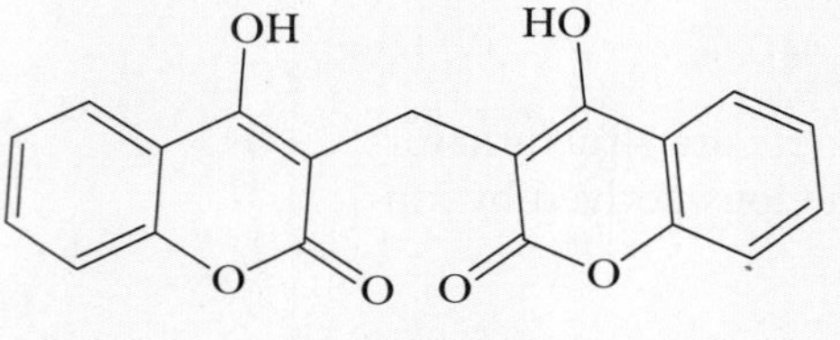

Dicoumarol
(an anticoagulant)

In a search for even more potent anticoagulants, Link developed warfarin (named after the Wisconsin Alumni Research Foundation), now used primarily as a rat poison: When rats consume warfarin, their blood fails to clot, and they bleed to death. Sold under the brand name Coumadin®, warfarin is also used as a blood thinner in humans. The *S* enantiomer is more active than the *R* enantiomer. The commercial product is a racemic mixture.

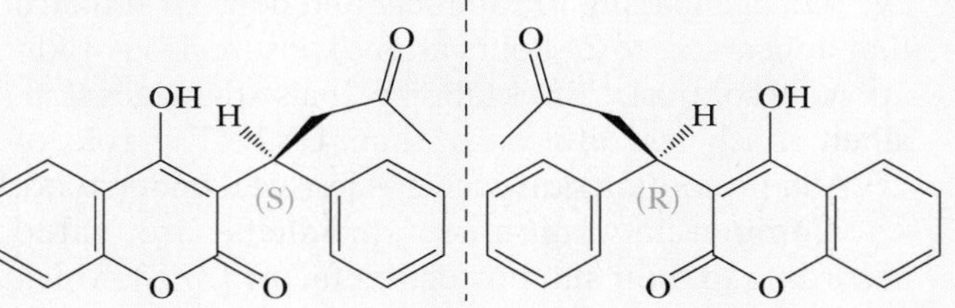

Warfarin
(a synthetic anticoagulant)

The powerful anticoagulant dicoumarol was first isolated from moldy clover. *(Daniel MAR/iStockphoto)*

QUESTION

Identify warfarin as an α, β, γ, etc., lactone. Identify each part of warfarin that can undergo keto–enol tautomerisation and show the tautomer at that position.

C. Esters and Lactones

Esters of Carboxylic Acids

The functional group of a **carboxylic ester** (commonly referred to simply as an ester) is an acyl group bonded to —OR or —OAr. Both IUPAC and common names of esters are derived from the names of the parent carboxylic acids. The alkyl or aryl group bonded to oxygen is named first, followed by the name of the acid, in which the suffix *-ic acid* is replaced by the suffix *-ate*:

the name of the group attached to the —O— comes first

$CH_3COCH_2CH_3$ (C=O)

Ethyl ethanoate
(Ethyl acetate)

Diethyl butanedioate
(Diethyl succinate)

Lactone A cyclic ester.

A cyclic ester is called a **lactone**. The IUPAC name of a lactone is formed by dropping the suffix *-oic acid* from the name of the parent carboxylic acid and adding the suffix *-olactone*. The common name is similarly derived. The location of the oxygen atom in the ring is indicated by a number if the IUPAC name of the acid is used and by a Greek letter α, β, γ, δ, ε, and so forth if the common name of the acid is used.

4-Butanolactone
(A γ-lactone)

Esters of Phosphoric Acid

Phosphoric acid has three —OH groups and forms mono-, di-, and triphosphoric esters, which are named by giving the name(s) of the alkyl or aryl group(s) bonded to oxygen, followed by the word *phosphate*—for example, dimethyl phosphate. In more complex phosphoric esters, it is common to name the organic molecule and then show the presence of the phosphoric ester by using either the word *phosphate* or the prefix *phospho*-. On the right are two phosphoric esters, each of special importance in the biological world. The first reaction in the metabolism of glucose is the formation of a phosphoric ester of D-glucose (Section 22.3), to give D-glucose 6-phosphate. Pyridoxal phosphate is one of the metabolically active forms of vitamin B_6. Each of these esters is shown as it is ionized at p*H* 7.4, the p*H* of blood plasma; the two hydrogens of each phosphate group are ionized, giving the phosphate group a charge of −2:

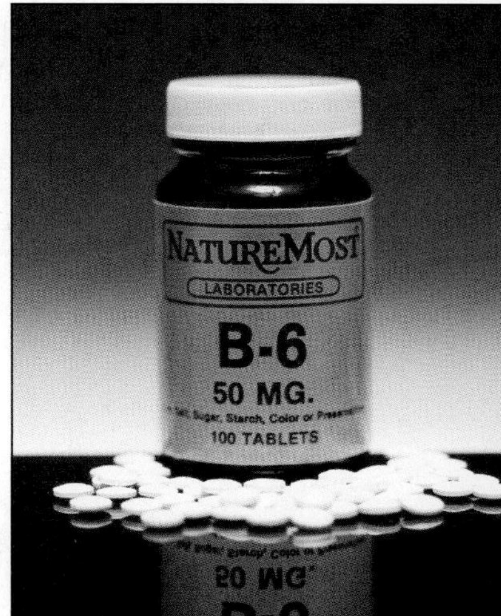

Vitamin B_6, pyridoxal. *(Charles D. Winters)*

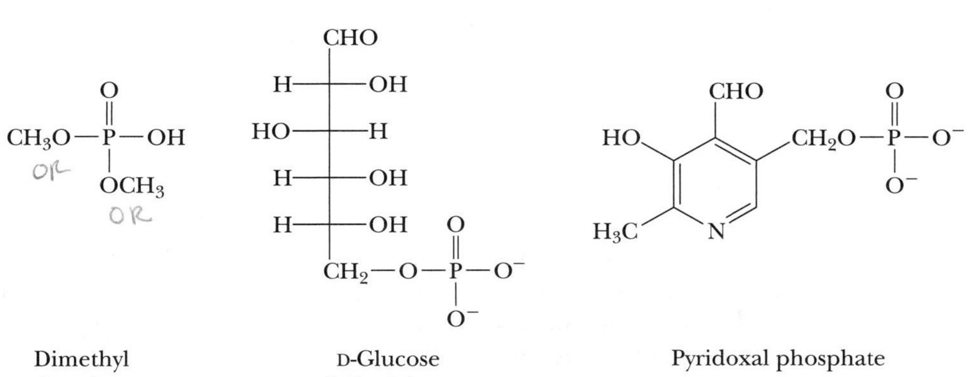

CHEMICAL CONNECTIONS 15C

The Penicillins and Cephalosporins: β-Lactam Antibiotics

The **penicillins** were discovered in 1928 by the Scottish bacteriologist Sir Alexander Fleming. As a result of the brilliant experimental work of Sir Howard Florey, an Australian pathologist, and Ernst Chain, a German chemist who fled Nazi Germany, penicillin G was introduced into the practice of medicine in 1943. For their pioneering work in developing one of the most effective antibiotics of all time, Fleming, Florey, and Chain were awarded the Nobel Prize in Medicine or Physiology in 1945.

The mold from which Fleming discovered penicillin was *Penicillium notatum*, a strain that gives a relatively low yield of penicillin. Commercial production of the antibiotic uses *P. chrysogenum*, a strain cultured from a mold found growing on a grapefruit in a market in Peoria, Illinois. The penicillins owe their antibacterial activity to a common mechanism that inhibits the biosynthesis of a vital part of bacterial cell walls.

The structural feature common to all penicillins is a ***β*-lactam** ring fused to a five-membered ring containing one S atom and one N atom:

The penicillins differ in the group bonded to the carbonyl carbon

HO, O, NH, H H, S, NH_2, N, O, β-lactam, COOH

Amoxicillin
(a β-lactam antibiotic)

Soon after the penicillins were introduced into medical practice, penicillin-resistant strains of bacteria began to appear and have since proliferated. One approach to combating resistant strains is to synthesize newer, more effective penicillins. Among those that have been developed are ampicillin, methicillin, and amoxicillin. Another approach is to search for newer, more effective β-lactam antibiotics. The most effective of these discovered so far are the **cephalosporins**, the first of which was isolated from the fungus *Cephalosporium acremonium*. This class of β-lactam antibiotics has an even broader spectrum of antibacterial activity than the penicillins and is effective against many penicillin-resistant bacterial strains.

The cephalosporins differ in the group bonded to the carbonyl carbon...

...and the group bonded to this carbon of the six-membered ring

O, H H, NH, S, NH_2, N, O, CH_3, COOH

Keflex
(a β-lactam antibiotic)

QUESTION

What would you except to be the major form of amoxicillin present in aqueous solution at (a) pH 2.0, (b) at pH 5–6, and (c) at pH 11.0? Explain.

D. Amides and Lactams

The functional group of an **amide** is an acyl group bonded to a trivalent nitrogen atom. Amides are named by dropping the suffix *-oic acid* from the IUPAC name of the parent acid, or *-ic acid* from its common name, and adding *-amide*. If the nitrogen atom of an amide is bonded to an alkyl or aryl group, the group is named and its location on nitrogen is indicated by *N*-. Two alkyl or aryl groups on nitrogen are indicated by *N,N*-di- if the groups are identical or by *N*-alkyl-*N*-alkyl if they are different:

CH_3CNH_2 (C=O)
Acetamide (a 1° amide)

$CH_3C{-}N(H)CH_3$ (C=O)
N-Methylacetamide (a 2° amide)

$H{-}C{-}N(CH_3)_2$ (C=O)
N,N-Dimethyl-formamide (DMF) (a 3° amide)

Amide bonds are the key structural feature that joins amino acids together to form polypeptides and proteins (Chapter 19).

Cyclic amides are given the special name **lactam**. Their common names are derived in a manner similar to those of lactones, with the difference that the suffix *-olactone* is replaced by *-olactam*:

Lactam A cyclic amide.

3-Butanolactam (A *β*-lactam)

6-Hexanolactam (An *ε*-lactam)

indicates C to which N atom is attached

6-Hexanolactam is a key intermediate in the synthesis of nylon-6 (Section 17.4A).

HOW TO 15.1 Name Functional Derivatives of Carboxylic Acids

The key to naming one of the four main functional derivatives of carboxylic acids is to realize how its name differs from that of the corresponding carboxylic acid. The following table highlights the difference for each derivative in italics.

Functional Derivative	Carboxylic Acid Name	Derivative Name	Example	
acid halide	alkanoic *acid*	alkanoyl *halide*	propanoic acid	propanoyl chloride
acid anhydride	alkanoic *acid*	alkanoic *anhydride*	propanoic acid	propanoic anhydride
ester	alkanoic *acid*	*alkyl* alkanoate	butanoic acid	CH_3O (or OAr) OR methyl butanoate
amide	alkanoic *acid*	alkan*amide*	butanoic acid	H_2N butanamide

Example 15.1

Write the IUPAC name for each compound:

(a) 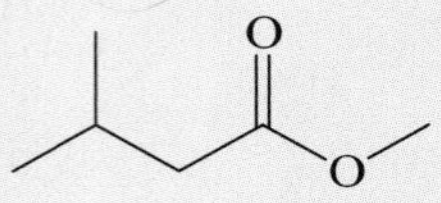 (b)

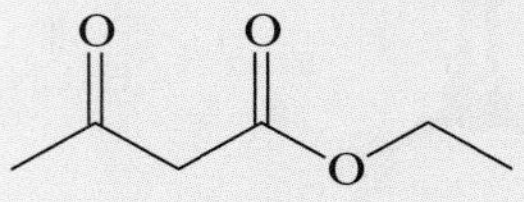

(c) 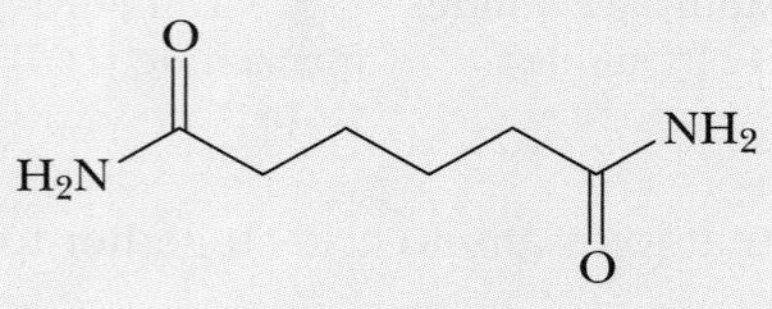 H_2N … NH_2 (d) 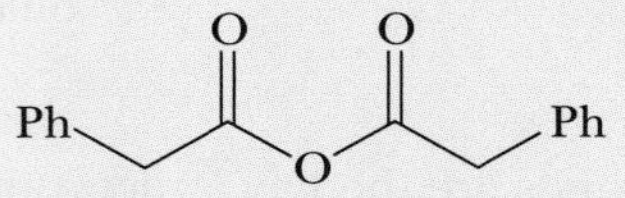 Ph … Ph

Strategy

Identify the longest chain containing the functional derivative to establish the root name. Treat the molecule as if each functional derivative group were a carboxyl group and name it as a carboxylic acid. Then change the suffix of the name to reflect the derivative. See How To 15.1 for examples.

Solution

Given first are IUPAC names and then, in parentheses, common names:

(a) Methyl 3-methylbutanoate (methyl isovalerate, from isovaleric acid)
(b) Ethyl 3-oxobutanoate (ethyl β-ketobutyrate, from β-ketobutyric acid)
(c) Hexanediamide (adipamide, from adipic acid)
(d) Phenylethanoic anhydride (phenylacetic anhydride, from phenylacetic acid)

See problems 15.9–15.11

Problem 15.1

Draw a structural formula for each compound:

(a) *N*-Cyclohexylacetamide
(b) *sec*-Butyl acetate
(c) Cyclobutyl butanoate
(d) *N*-(2-Octyl)benzamide
(e) Diethyl adipate
(f) Propanoic anhydride

15.2 What Are the Characteristic Reactions of Carboxylic Acid Derivatives?

The most common reaction theme of acid halides, anhydrides, esters, and amides is the addition of a nucleophile to the carbonyl carbon to form a tetrahedral carbonyl addition intermediate. To this extent, the reactions of these functional groups are similar to nucleophilic addition to the carbonyl groups in aldehydes and ketones (Section 13.4). The **tetrahedral carbonyl addition intermediate** (TCAI) formed from an aldehyde or a ketone then adds H^+. The result of this reaction is nucleophilic addition to a carbonyl group of an aldehyde or a ketone:

the oxygen of the TCAI abstracts a proton from the acid of the workup step

Nucleophilic acyl addition: An aldehyde or a ketone ($R-CO-R$) $+ :Nu^-$ $\longrightarrow$ [Tetrahedral carbonyl addition intermediate] $\xrightarrow{H^+}$ Addition product

For functional derivatives of carboxylic acids, the fate of the tetrahedral carbonyl addition intermediate is quite different from that of aldehydes and ketones. This intermediate collapses to expel the leaving group and regenerate the carbonyl group. The result of this addition–elimination sequence is **nucleophilic acyl substitution**:

Nucleophilic acyl substitution A reaction in which a nucleophile bonded to a carbonyl carbon is replaced by another nucleophile.

the oxygen of the TCAI releases a pair of electrons to expel Y^- and reform the carbonyl

Nucleophilic acyl substitution: $R-CO-Y$ $+ :Nu^-$ $\longrightarrow$ [Tetrahedral carbonyl addition intermediate] $\longrightarrow$ Substitution product ($R-CO-Nu$) $+ :Y^-$

The major difference between these two types of carbonyl addition reactions is that aldehydes and ketones do not have a group, Y, that can leave as a stable anion. They undergo only nucleophilic acyl addition. The four carboxylic acid derivatives we study in this chapter do have a group, Y, that can leave as a stable anion; accordingly, they undergo nucleophilic acyl substitution.

In this general reaction, we show the nucleophile and the leaving group as anions. That need not be the case, however: Neutral molecules, such as water, alcohols, ammonia, and amines, may also serve as nucleophiles in the acid-catalyzed version of the reaction. We show the leaving groups here as anions to illustrate an important point about leaving groups, namely, that the weaker the base, the better is the leaving group (Section 7.5C):

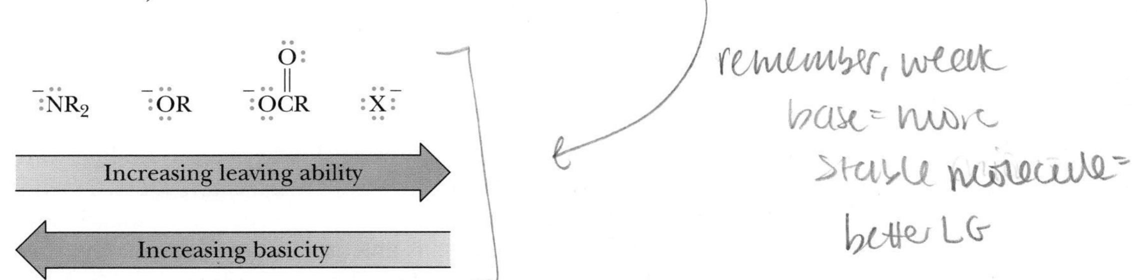

The weakest base in this series, and thus the best leaving group, is halide ion; acid halides are the most reactive toward nucleophilic acyl substitution. The strongest base, and hence the poorest leaving group, is amide ion; amides are the least reactive toward nucleophilic acyl substitution. Acid halides and acid anhydrides are so reactive that they are not found in nature. Esters and amides, however, are universally present.

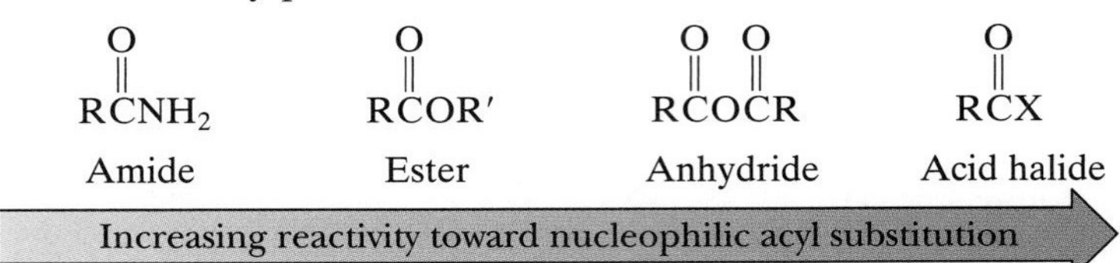

15.3 What Is Hydrolysis?

Hydrolysis (Greek: *hudor*, water; *lyein*, separate) is a chemical process whereby a bond (or bonds) in a molecule is broken by its reaction with water. In hydrolysis, the water molecule is also typically split into H^+ and OH^-.

A. Acid Chlorides

Low-molecular-weight acid chlorides react very rapidly with water to form carboxylic acids and HCl:

this bond is hydrolyzed by the addition of water

$$CH_3\overset{O}{\overset{\|}{C}}{-}Cl + H_2O \longrightarrow CH_3\overset{O}{\overset{\|}{C}}OH + HCl$$

Higher-molecular-weight acid chlorides are less soluble and consequently react less rapidly with water.

B. Acid Anhydrides

Acid anhydrides are generally less reactive than acid chlorides. The lower-molecular-weight anhydrides, however, react readily with water to form two molecules of carboxylic acid:

$$CH_3\overset{O}{\overset{\|}{C}}{-}O{-}\overset{O}{\overset{\|}{C}}CH_3 + H_2O \longrightarrow CH_3\overset{O}{\overset{\|}{C}}OH + HO\overset{O}{\overset{\|}{C}}CH_3$$

one of these C—O bonds is hydrolyzed by the addition of water

C. Esters

Esters are hydrolyzed only very slowly, even in boiling water. Hydrolysis becomes considerably more rapid, however, when esters are refluxed in aqueous acid or base. When we discussed acid-catalyzed (Fischer) esterification in Section 14.6, we pointed out that esterification is an equilibrium reaction. Hydrolysis of esters in aqueous acid is also an equilibrium reaction and proceeds by the same mechanism as esterification, except in reverse. The role of the acid catalyst is to protonate the carbonyl oxygen (Step 1), thereby increasing the electrophilic character of the carbonyl carbon toward attack by water (Step 2) to form a tetrahedral carbonyl addition intermediate. An internal proton transfer to the alkoxy group (Step 3) makes that group a good leaving group and allows the collapse of this intermediate (Step 4) to give a carboxylic acid and an alcohol. In this reaction, acid is a catalyst; it is consumed in the first step, but another is generated at the end of the reaction:

$$RC(=O)OCH_3 + H_2\ddot{O} \overset{H^+}{\rightleftharpoons} RC(=\overset{+}{O}H)\ddot{O}CH_3 \rightleftharpoons \left[R\overset{OH}{\underset{H_3C\ddot{O}:}{C}}\overset{+}{O}H_2 \right]$$

this bond is hydrolyzed by the addition of water

(1) (2) (3)

Tetrahedral carbonyl addition intermediate

⇅ internal proton transfer

$$RC(=O)\ddot{O}H + CH_3OH + H_3O^+ \underset{H_2\ddot{O}}{\rightleftharpoons} \left[R\overset{OH}{\underset{H_3C\overset{+}{O}H}{C}}\ddot{O}H \right]$$

(4)

the acid is regenerated

Tetrahedral carbonyl addition intermediate

Hydrolysis of esters may also be carried out with hot aqueous base, such as aqueous NaOH. Hydrolysis of esters in aqueous base is often called **saponification**, a reference to the use of this reaction in the manufacture of soaps (Section 21.2A). Each mole of ester hydrolyzed requires 1 mole of base, as shown in the following balanced equation:

Saponification Hydrolysis of an ester in aqueous NaOH or KOH to an alcohol and the sodium or potassium salt of a carboxylic acid.

$$R\overset{O}{\overset{\|}{C}}OCH_3 + NaOH \xrightarrow{H_2O} R\overset{O}{\overset{\|}{C}}O^-Na^+ + CH_3OH$$

Mechanism: Hydrolysis of an Ester in Aqueous Base

Step 1: Addition of hydroxide ion to the carbonyl carbon of the ester gives a tetrahedral carbonyl addition intermediate:

$$R-\overset{:\ddot{O}}{\overset{\|}{C}}-\ddot{O}CH_3 + {}^-\!:\ddot{O}H \rightleftharpoons R-\overset{:\ddot{O}:^-}{\underset{:\ddot{O}H}{C}}-\ddot{O}CH_3$$

Step 2: Collapse of this intermediate gives a carboxylic acid and an alkoxide ion:

$$R-\overset{:\ddot{O}:^-}{\underset{:\ddot{O}H}{C}}-\ddot{O}CH_3 \rightleftharpoons R-\overset{\ddot{O}:}{\overset{\|}{C}}-\ddot{O}H + {}^-\!:\ddot{O}CH_3$$

Step 3: Proton transfer from the carboxyl group (an acid) to the alkoxide ion (a base) gives the carboxylate anion. This step is irreversible because the alcohol is not a strong enough nucleophile to attack a carboxylate anion:

$$R-\overset{\overset{\displaystyle O}{\|}}{C}-O-H + {}^{-}OCH_3 \longrightarrow R-\overset{\overset{\displaystyle O}{\|}}{C}-O^{-} + H-OCH_3$$

There are two major differences between the hydrolysis of esters in aqueous acid and that in aqueous base:

1. For hydrolysis in aqueous acid, acid is required in only catalytic amounts. For hydrolysis in aqueous base, base is required in equimolar amounts, because it is a reactant, not just a catalyst.
2. Hydrolysis of an ester in aqueous acid is reversible. Hydrolysis in aqueous base is irreversible, because a carboxylic acid anion is not attacked by ROH.

Example 15.2

Complete and balance equations for the hydrolysis of each ester in aqueous sodium hydroxide, showing all products as they are ionized in aqueous NaOH:

(a) $C_6H_5COOCH(CH_3)_2 + NaOH \xrightarrow{H_2O}$

(b) $CH_3COOCH_2CH_2OOCCH_3 + NaOH \xrightarrow{H_2O}$

Strategy

The hydrolysis of an ester results in a carboxyl group and an alcohol for every ester group in the molecule. In aqueous base, one mole of NaOH is consumed for every ester group in the molecule.

Solution

The products of hydrolysis of (a) are benzoic acid and 2-propanol. In aqueous NaOH, benzoic acid is converted to its sodium salt. Therefore, 1 mole of NaOH is required for the hydrolysis of 1 mole of this ester. Compound (b) is a diester of ethylene glycol. Two moles of NaOH are required for its hydrolysis:

(a) $C_6H_5COO^-Na^+ + HO-CH(CH_3)_2$

Sodium benzoate; 2-Propanol (Isopropyl alcohol)

(b) $2CH_3\overset{\overset{\displaystyle O}{\|}}{C}O^-Na^+ + HOCH_2CH_2OH$

Sodium acetate; 1,2-Ethanediol (Ethylene glycol)

See problems 15.19, 15.20, 15.31

Problem 15.2

Complete and balance equations for the hydrolysis of each ester in aqueous solution, showing each product as it is ionized under the given experimental conditions:

(a) $C_6H_4(COOCH_3)_2$ (1,2-) + NaOH (excess) $\xrightarrow{H_2O}$

(b) $CH_3C(=O)CH_2CH_2C(=O)OCH_2CH_3 + H_2O \xrightarrow{HCl}$

D. Amides

Amides require considerably more vigorous conditions for hydrolysis in both acid and base than do esters. Amides undergo hydrolysis in hot aqueous acid to give a carboxylic acid and ammonia. Hydrolysis is driven to completion by the acid–base reaction between ammonia or the amine and acid to form an ammonium salt. One mole of acid is required per mole of amide:

$$CH_3CH_2CH(Ph)C(=O)NH_2 + H_2O + HCl \xrightarrow{heat} CH_3CH_2CH(Ph)C(=O)OH + NH_4^+Cl^-$$

2-Phenylbutanamide — 2-Phenylbutanoic acid

In aqueous base, the products of amide hydrolysis are a carboxylic acid and ammonia or an amine. Base-catalyzed hydrolysis is driven to completion by the acid–base reaction between the carboxylic acid and base to form a salt. One mole of base is required per mole of amide:

$$CH_3\overset{O}{\overset{\|}{C}}NH{-}C_6H_5 + NaOH \xrightarrow[heat]{H_2O} CH_3\overset{O}{\overset{\|}{C}}O^-Na^+ + H_2N{-}C_6H_5$$

N-Phenylethanamide (*N*-Phenylacetamide, Acetanilide) — Sodium acetate — Aniline

The reactions of these functional groups with water are summarized in Table 15.1. Remember that, although all four functional groups react with water, there are large differences in the rates and experimental conditions under which they undergo hydrolysis.

TABLE 15.1 Summary of Reaction of Acid Chlorides, Anhydrides, Esters, and Amides with Water

$$R{-}\overset{O}{\overset{\|}{C}}{-}Cl + H_2O \longrightarrow R{-}\overset{O}{\overset{\|}{C}}{-}OH + HCl$$

$$R{-}\overset{O}{\overset{\|}{C}}{-}O{-}\overset{O}{\overset{\|}{C}}{-}R + H_2O \longrightarrow R{-}\overset{O}{\overset{\|}{C}}{-}OH + HO{-}\overset{O}{\overset{\|}{C}}{-}R$$

$$R{-}\overset{O}{\overset{\|}{C}}{-}OR' + H_2O \xrightarrow{NaOH} R{-}\overset{O}{\overset{\|}{C}}{-}O^-Na^+ + R'OH$$

$$R{-}\overset{O}{\overset{\|}{C}}{-}OR' + H_2O \xrightarrow{H_2SO_4} R{-}\overset{O}{\overset{\|}{C}}{-}OH + R'OH$$

$$R{-}\overset{O}{\overset{\|}{C}}{-}NH_2 + H_2O \xrightarrow{NaOH} R{-}\overset{O}{\overset{\|}{C}}{-}O^-Na^+ + NH_3$$

$$R{-}\overset{O}{\overset{\|}{C}}{-}NH_2 + H_2O \xrightarrow{HCl} R{-}\overset{O}{\overset{\|}{C}}{-}OH + NH_4^+Cl^-$$

Example 15.3

Write equations for the hydrolysis of these amides in concentrated aqueous HCl, showing all products as they exist in aqueous HCl and showing the number of moles of HCl required for the hydrolysis of each amide:

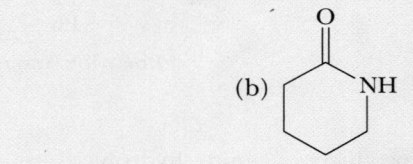

Strategy

The hydrolysis of an amide results in a carboxyl group and an ammonium chloride ion for every amide group in the molecule. Either 1 mole of NaOH (basic conditions) or 1 mole of HCl (acidic conditions) is consumed for every amide group in the molecule.

Solution

(a) Hydrolysis of *N,N*-dimethylacetamide gives acetic acid and dimethylamine. Dimethylamine, a base, is protonated by HCl to form dimethylammonium ion and is shown in the balanced equation as dimethylammonium chloride. Complete hydrolysis of this amide requires 1 mole of HCl for each mole of the amide:

$$CH_3\overset{O}{\overset{\|}{C}}N(CH_3)_2 + H_2O + HCl \xrightarrow[heat]{} CH_3\overset{O}{\overset{\|}{C}}OH + (CH_3)_2NH_2^+Cl^-$$

(b) Hydrolysis of this δ-lactam gives the protonated form of 5-aminopentanoic acid. One mole of acid is required per mole of lactam:

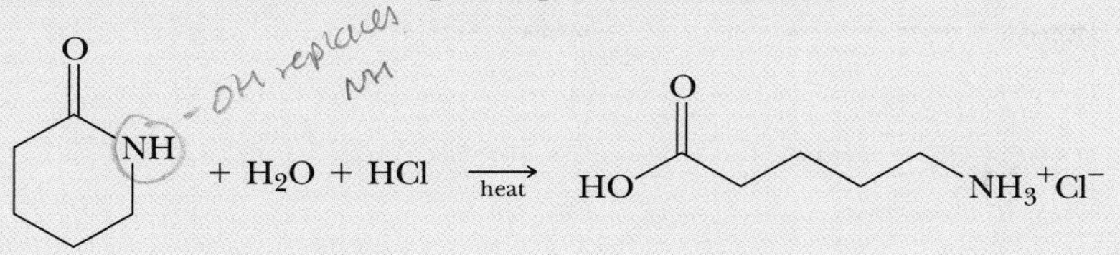

$$\text{(δ-lactam)} + H_2O + HCl \xrightarrow{\text{heat}} HOC(=O)CH_2CH_2CH_2CH_2NH_3^+Cl^-$$

-OH replaces NH

See problems 15.29, 15.32

Problem 15.3

Complete equations for the hydrolysis of the amides in Example 15.3 in concentrated aqueous NaOH. Show all products as they exist in aqueous NaOH, and show the number of moles of NaOH required for the hydrolysis of each amide.

15.4 How Do Carboxylic Acid Derivatives React with Alcohols?

A. Acid Chlorides

Acid chlorides react with alcohols to give an ester and HCl:

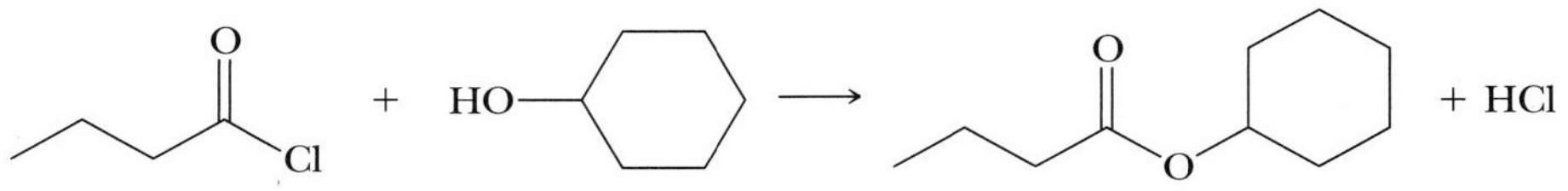

$$CH_3CH_2CH_2COCl + HO\text{–}C_6H_{11} \longrightarrow CH_3CH_2CH_2COO\text{–}C_6H_{11} + HCl$$

Butanoyl chloride | Cyclohexanol | Cyclohexyl butanoate

Because acid chlorides are so reactive toward even weak nucleophiles such as alcohols, no catalyst is necessary for these reactions. Phenol and substituted phenols also react with acid chlorides to give esters.

B. Acid Anhydrides

Acid anhydrides react with alcohols to give 1 mole of ester and 1 mole of a carboxylic acid.

$$\underset{\text{Acetic anhydride}}{CH_3\overset{O}{\overset{\|}{C}}O\overset{O}{\overset{\|}{C}}CH_3} + \underset{\text{Ethanol}}{HOCH_2CH_3} \longrightarrow \underset{\text{Ethyl acetate}}{CH_3\overset{O}{\overset{\|}{C}}OCH_2CH_3} + \underset{\text{Acetic acid}}{CH_3\overset{O}{\overset{\|}{C}}OH}$$

Thus, the reaction of an alcohol with an anhydride is a useful method for synthesizing esters. Aspirin is synthesized on an industrial scale by reacting acetic anhydride with salicylic acid:

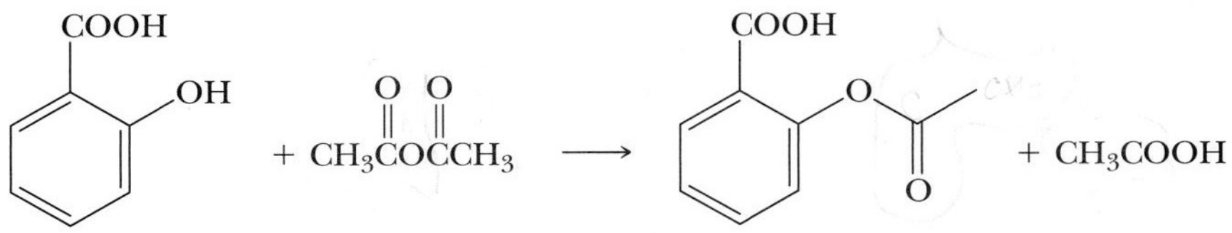

$$\text{2-}HOC_6H_4COOH + CH_3\overset{O}{\overset{\|}{C}}O\overset{O}{\overset{\|}{C}}CH_3 \longrightarrow \text{2-}CH_3COOC_6H_4COOH + CH_3COOH$$

2-Hydroxybenzoic acid (Salicylic acid) | Acetic anhydride | Acetylsalicylic acid (Aspirin) | Acetic acid

TABLE 15.2 Summary of Reaction of Acid Chlorides, Anhydrides, Esters, and Amides with Alcohols

$$R{-}\overset{O}{\overset{\|}{C}}{-}Cl + HOR'' \longrightarrow R{-}\overset{O}{\overset{\|}{C}}{-}OR'' + HCl$$

$$R{-}\overset{O}{\overset{\|}{C}}{-}O{-}\overset{O}{\overset{\|}{C}}{-}R + R''OH \longrightarrow R{-}\overset{O}{\overset{\|}{C}}{-}OR'' + HO{-}\overset{O}{\overset{\|}{C}}{-}R$$

$$R{-}\overset{O}{\overset{\|}{C}}{-}OR' + R''OH \underset{}{\overset{H_2SO_4}{\rightleftharpoons}} R{-}\overset{O}{\overset{\|}{C}}{-}OR'' + R'OH$$

$$R{-}\overset{O}{\overset{\|}{C}}{-}NH_2 + R''OH \longrightarrow \text{No Reaction}$$

C. Esters

When treated with an alcohol in the presence of an acid catalyst, esters undergo an exchange reaction called **transesterification**. In this reaction, the original —OR group of the ester is exchanged for a new —OR group. In the following example, the transesterification can be driven to completion by heating the reaction at a temperature above the boiling point of methanol (65°C) so that methanol distills from the reaction mixture:

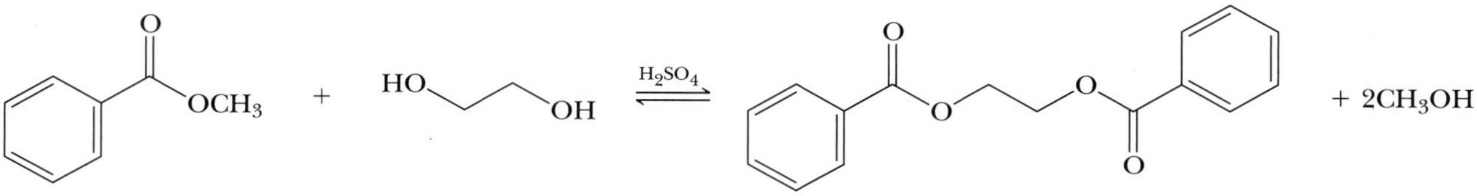

Methyl benzoate + 1,2-Ethanediol (Ethylene glycol) $\overset{H_2SO_4}{\rightleftharpoons}$ (A diester of ethylene glycol) + $2CH_3OH$

D. Amides

Amides do not react with alcohols under any experimental conditions. Alcohols are not strong enough nucleophiles to attack the carbonyl group of an amide.

The reactions of the foregoing functional groups with alcohols are summarized in Table 15.2. As with reactions of these same functional groups with water (Section 15.3), there are large differences in the rates and experimental conditions under which they undergo reactions with alcohols. At one extreme are acid chlorides and anhydrides, which react rapidly; at the other extreme are amides, which do not react at all.

Example 15.4

Complete these equations:

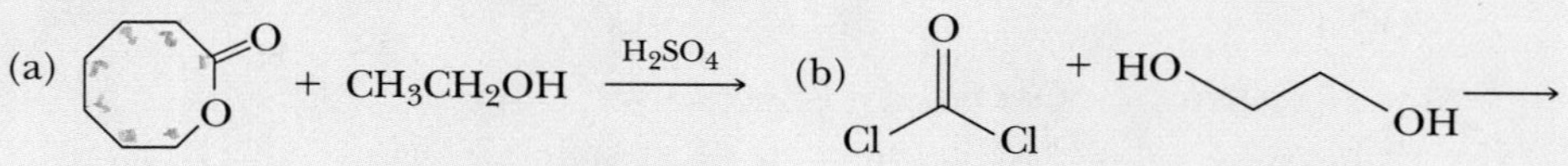

(a) [cyclic ester, C=O, O] + CH_3CH_2OH $\xrightarrow{H_2SO_4}$ (b) $Cl{-}\overset{O}{\overset{\|}{C}}{-}Cl$ + HO~~~OH $\longrightarrow$

Strategy

Acid halides, anhydrides, and esters undergo nucleophilic acyl substitution with alcohols (HOR′), the net result being the replacement of each —X, —OC(O)R, or —OR group with the —OR′ group of the alcohol.

Solution

(a) [cyclic ester, ring atoms numbered 1–7] + CH_3CH_2OH $\xrightarrow{H_2SO_4}$ HO–(C7…C1)–C(=O)OCH_2CH_3

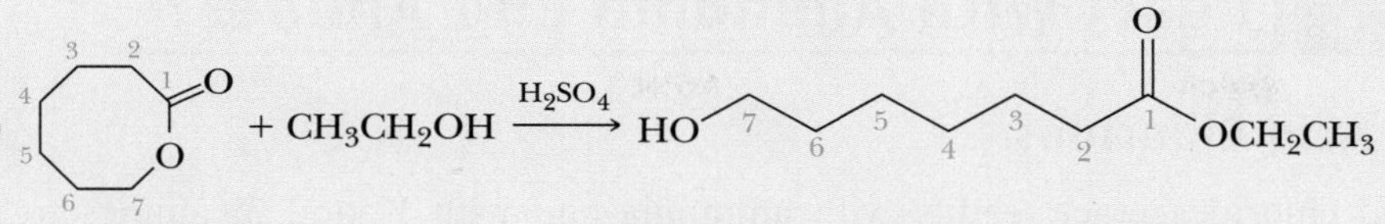

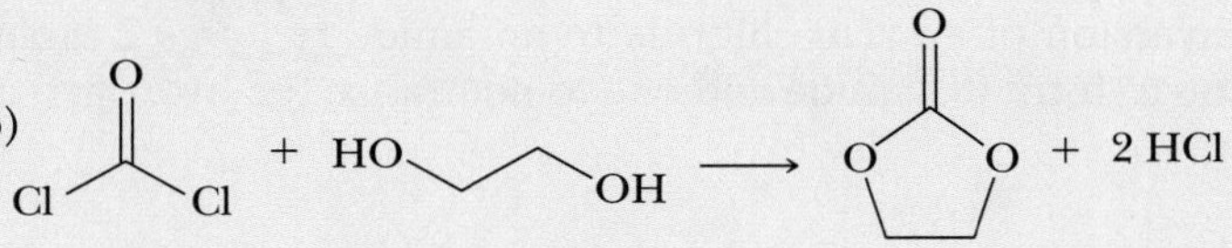

(b) Cl–C(=O)–Cl + HO–CH_2CH_2–OH ⟶ [cyclic carbonate] + 2 HCl

See problems 15.16, 15.17, 15.19–15.22, 15.28

Problem 15.4

Complete these equations (the stoichiometry of each is given in the equation):

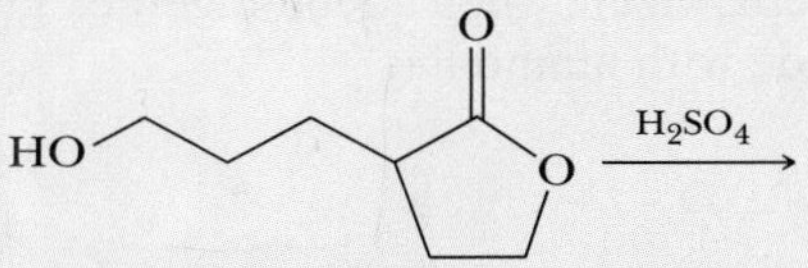

(a) HO–C_6H_4–OH + 2 C_6H_5C(=O)Cl ⟶

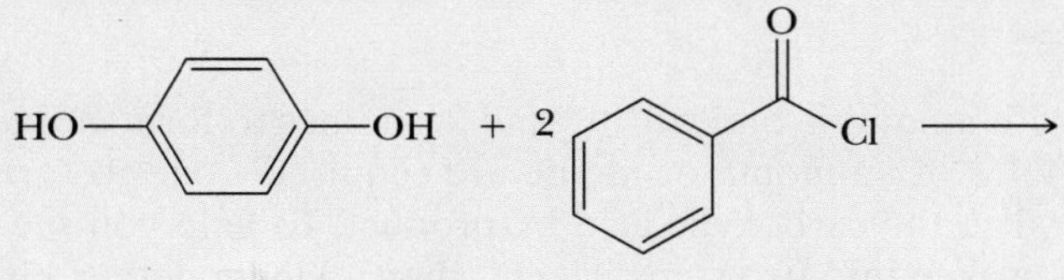

(b) [HO–$CH_2CH_2CH_2$– substituted lactone] $\xrightarrow{H_2SO_4}$

CHEMICAL CONNECTIONS 15D

The Pyrethrins: Natural Insecticides of Plant Origin

Pyrethrum is a natural insecticide obtained from the powdered flower heads of several species of *Chrysanthemum*, particularly *C. cinerariaefolium*. The active substances in pyrethrum, principally pyrethrins I and II, are contact poisons for insects and cold-blooded vertebrates. Because their concentrations in the pyrethrum powder used in chrysanthemum-based insecticides are nontoxic to plants and higher animals, pyrethrum powder is used in household and livestock sprays, as well as in dusts for edible plants. Natural pyrethrins are esters of chrysanthemic acid.

While pyrethrum powders are effective insecticides, the active substances in them are destroyed rapidly in the environment. In an effort to develop synthetic compounds as effective as these natural insecticides but with greater biostability, chemists have prepared a series of esters related in structure to chrysanthemic acid. Permethrin is one of the most commonly used synthetic pyrethrinlike compounds in household and agricultural products.

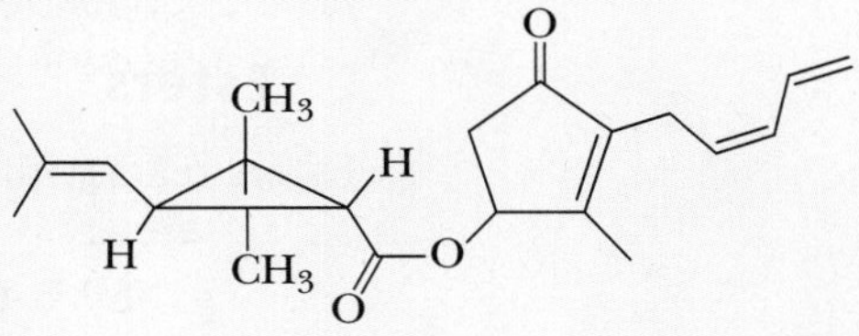

Pyrethrin I

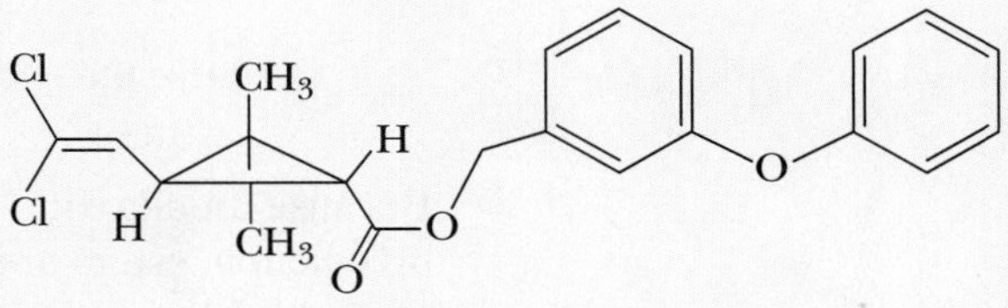

Permethrin

QUESTION

Show the compounds that would result if pyrethrin I and permethrin were to undergo hydrolysis.

15.5 How Do Carboxylic Acid Derivatives React with Ammonia and Amines?

A. Acid Chlorides

Acid chlorides react readily with ammonia and with 1° and 2° amines to form amides. Complete conversion of an acid chloride to an amide requires 2 moles of ammonia or amine: one to form the amide and one to neutralize the hydrogen chloride formed:

$$CH_3(CH_2)_4COCl + 2NH_3 \longrightarrow CH_3(CH_2)_4CONH_2 + NH_4^+Cl^-$$

Hexanoyl chloride | Ammonia | Hexanamide | Ammonium chloride

B. Acid Anhydrides

Acid anhydrides react with ammonia and with 1° and 2° amines to form amides. As with acid chlorides, 2 moles of ammonia or amine are required—one to form the amide and one to neutralize the carboxylic acid by-product. To help you see what happens, this reaction is broken into two steps, which, when added together, give the net reaction for the reaction of an anhydride with ammonia:

$$CH_3\overset{O}{\overset{\|}{C}}O\overset{O}{\overset{\|}{C}}CH_3 + NH_3 \longrightarrow CH_3\overset{O}{\overset{\|}{C}}NH_2 + CH_3\overset{O}{\overset{\|}{C}}OH$$

$$CH_3\overset{O}{\overset{\|}{C}}OH + NH_3 \longrightarrow CH_3\overset{O}{\overset{\|}{C}}O^-NH_4^+$$

$$CH_3\overset{O}{\overset{\|}{C}}O\overset{O}{\overset{\|}{C}}CH_3 + 2NH_3 \longrightarrow CH_3\overset{O}{\overset{\|}{C}}NH_2 + CH_3\overset{O}{\overset{\|}{C}}O^-NH_4^+$$

C. Esters

Esters react with ammonia and with 1° and 2° amines to form amides:

$$PhCH_2COOCH_2CH_3 + NH_3 \longrightarrow PhCH_2CONH_2 + HOCH_2CH_3$$

Ethyl phenylacetate | Phenylacetamide | Ethanol

Because an alkoxide anion is a poor leaving group compared with a halide or carboxylic ion, esters are less reactive toward ammonia, 1° amines, and 2° amines than are acid chlorides or acid anhydrides.

D. Amides

Amides do not react with ammonia or amines.

The reactions of the preceding four functional groups with ammonia and amines are summarized in Table 15.3.

TABLE 15.3 Summary of Reaction of Acid Chlorides, Anhydrides, Esters, and Amides with Ammonia and Amines

$$R-\overset{O}{\overset{\|}{C}}-Cl + 2NH_3 \longrightarrow R-\overset{O}{\overset{\|}{C}}-NH_2 + NH_4^+\,Cl^-$$

$$R-\overset{O}{\overset{\|}{C}}-O-\overset{O}{\overset{\|}{C}}-R + 2NH_3 \longrightarrow R-\overset{O}{\overset{\|}{C}}-NH_2 + R-\overset{O}{\overset{\|}{C}}-O^-NH_4^+$$

$$R-\overset{O}{\overset{\|}{C}}-OR' + NH_3 \rightleftharpoons R-\overset{O}{\overset{\|}{C}}-NH_2 + R'OH$$

$$R-\overset{O}{\overset{\|}{C}}-NH_2 \text{ No reaction with ammonia or amines}$$

Example 15.5

Complete these equations (the stoichiometry of each is given in the equation):

(a) Ethyl butanoate $+ NH_3 \longrightarrow$

(b) Diethyl carbonate $+ 2NH_3 \longrightarrow$

Strategy

Acid halides, anhydrides, and esters undergo nucleophilic acyl substitution with ammonia or amines, the net result being the replacement of each $-X$, $-OC(O)R$, or $-OR$ group with the $-NH_2$ group of ammonia or the $-NHR$ or $-NR_2$ group of the amine.

Solution

(a) Butanamide $+ CH_3CH_2OH$ (b) $H_2N-\overset{O}{\overset{\|}{C}}-NH_2 + 2CH_3CH_2OH$ Urea

See problems 15.18–15.22, 15.24–15.26, 15.31, 15.35

Problem 15.5

Complete these equations (the stoichiometry of each is given in the equation):

(a) $CH_3\overset{O}{\overset{\|}{C}}O-C_6H_4-O\overset{O}{\overset{\|}{C}}CH_3 + 2NH_3 \longrightarrow$ (b) (six-membered cyclic lactone) $+ NH_3 \longrightarrow$

OR'

*Study – when cyclic in all of these probs, keep as one C chain in answer (i.e. just open the ring

15.6 How Can Functional Derivatives of Carboxylic Acids Be Interconverted?

In the last few sections, we have seen that acid chlorides are the most reactive carboxyl derivatives toward nucleophilic acyl substitution and that amides are the least reactive:

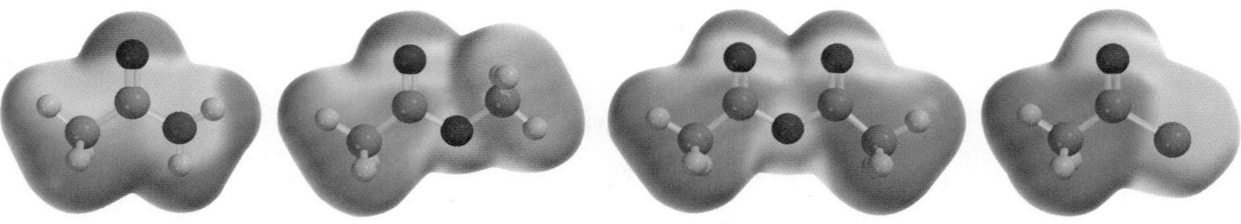

Amide < Ester < Acid anhydride < Acid halide

Increasing reactivity toward nucleophilic acyl substitution →

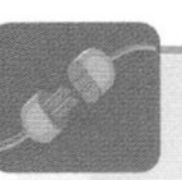

CHEMICAL CONNECTIONS 15E

Systematic Acquired Resistance in Plants

The use of germicides to protect plants from harmful pathogens is common in farming. Recently, plant physiologists discovered that some plant species are able to generate their own defenses against pathogens. The tobacco mosaic virus (TMV), for example, is a particularly devastating pathogen for plants such as tobacco, cucumber, and tomato. Scientists have found that certain strains of these plants produce large amounts of salicylic acid upon being infected with TMV. Accompanying the infection is the appearance of lesions on the leaves of the plants, which help to contain the infection to those localized areas. Furthermore, scientists have discovered that neighboring plants tend to acquire some resistance to TMV. It appears that the infected plant somehow signals neighboring plants of the impending danger by converting salicylic acid to its ester, methyl salicylate:

O, OH, OH → → → O, OCH_3, OH

Salicylic acid — Methyl salicylate

With a lower boiling point and higher vapor pressure than salicylic acid has, the methyl salicylate diffuses into the air from the infected plant, and the surrounding plants use it as a signal to enhance their defenses against TMV.

The tobacco plant, *Nicotiana tobacum*. *(Inga Spence/Index Stock/Photolibrary)*

QUESTION

An early proposal in this research was that the tobacco plant could utilize two molecules of salicylic acid (molar mass 138.12 g/mol) in a nucleophilic acyl substitution reaction to yield a compound with a molar mass of 240.21 g/mol that would be less polar than salicylic acid. Propose a structure for this reaction product.

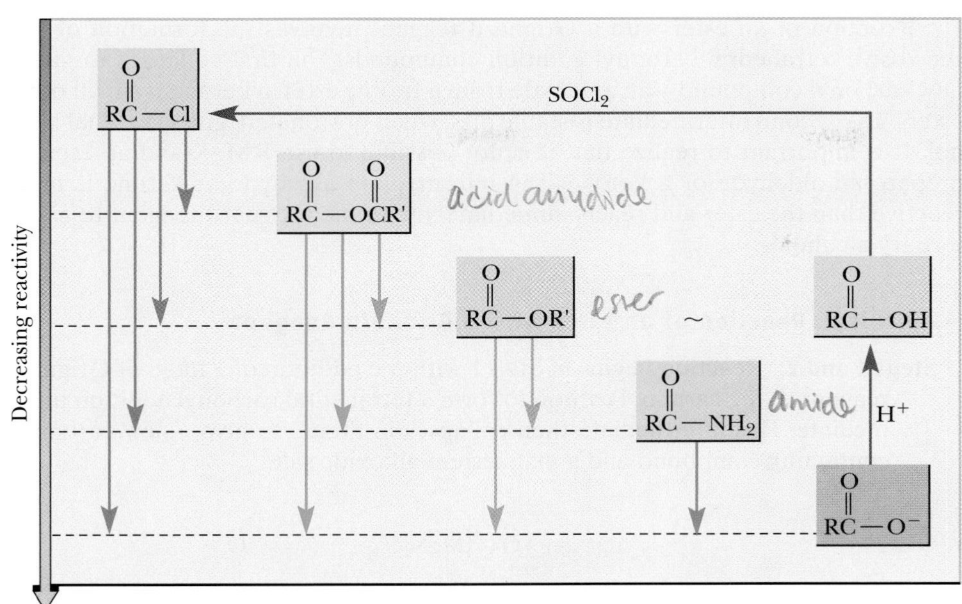

Figure 15.1 Relative reactivities of carboxylic acid derivatives toward nucleophilic acyl substitution. A more reactive derivative may be converted to a less reactive derivative by treatment with an appropriate reagent. Treatment of a carboxylic acid with thionyl chloride converts the carboxylic acid to the more reactive acid chloride. Carboxylic acids are about as reactive as esters under acidic conditions, but are converted to the unreactive carboxylate anions under basic conditions.

Another useful way to think about the relative reactivities of these four functional derivatives of carboxylic acids is summarized in Figure 15.1. Any functional group in this figure can be prepared from any functional group above it by treatment with an appropriate oxygen or nitrogen nucleophile. An acid chloride, for example, can be converted to an acid anhydride, an ester, an amide, or a carboxylic acid. An acid anhydride, ester, or amide, however, does not react with chloride ion to give an acid chloride.

Notice that all carboxylic acid derivatives can be converted to carboxylic acids, which in turn can be converted to acid chlorides. Thus, any acid derivative can be used to synthesize another, either directly or via a carboxylic acid.

15.7 How Do Esters React with Grignard Reagents?

Treating a formic ester with 2 moles of a Grignard reagent, followed by hydrolysis of the magnesium alkoxide salt in aqueous acid, gives a 2° alcohol, whereas treating an ester other than a formate with a Grignard reagent gives a 3° alcohol in which two of the groups bonded to the carbon bearing the —OH group are the same:

$$\underset{\text{An ester of formic acid}}{\mathrm{H\overset{\overset{\large O}{\|}}{C}OCH_3}} + 2\mathrm{RMgX} \longrightarrow \begin{matrix}\text{magnesium}\\ \text{alkoxide}\\ \text{salt}\end{matrix} \xrightarrow{\mathrm{H_2O,\ HCl}} \underset{\text{A 2° alcohol}}{\mathrm{H\overset{\overset{\large OH}{|}}{\underset{\underset{\large R}{|}}{C}}{-}R}} + \mathrm{CH_3OH}$$

$$\underset{\text{An ester other than formic acid}}{\mathrm{CH_3\overset{\overset{\large O}{\|}}{C}OCH_3}} + 2\mathrm{RMgX} \longrightarrow \begin{matrix}\text{magnesium}\\ \text{alkoxide}\\ \text{salt}\end{matrix} \xrightarrow{\mathrm{H_2O,\ HCl}} \underset{\text{A 3° alcohol}}{\mathrm{CH_3\overset{\overset{\large OH}{|}}{\underset{\underset{\large R}{|}}{C}}{-}R}} + \mathrm{CH_3OH}$$

Reaction of an ester with a Grignard reagent involves the formation of two successive tetrahedral carbonyl addition compounds. The first collapses to give a new carbonyl compound—an aldehyde from a formic ester, a ketone from all other esters. The second intermediate is stable and, when protonated, gives the final alcohol. It is important to realize that it is not possible to use RMgX and an ester to prepare an aldehyde or a ketone: The intermediate aldehyde or ketone is more reactive than the ester and reacts immediately with the Grignard reagent to give a tertiary alcohol.

Mechanism: Reaction of an Ester with a Grignard Reagent

Steps 1 and 2: Reaction begins in Step 1 with the addition of 1 mole of Grignard reagent to the carbonyl carbon to form a tetrahedral carbonyl addition intermediate. This intermediate then collapses in Step 2 to give a new carbonyl-containing compound and a magnesium alkoxide salt:

$$CH_3-\overset{O}{\overset{\|}{C}}-OCH_3 + R-MgX \longrightarrow CH_3-\underset{R}{\overset{O^-\ [MgX]^+}{C}}-OCH_3 \longrightarrow CH_3-\underset{R}{\overset{O}{\overset{\|}{C}}} + CH_3O^-\ [MgX]^+$$

A magnesium salt (a tetrahedral carbonyl addition intermediate)

A ketone

Steps 3 and 4: The new carbonyl-containing compound reacts in Step 3 with a second mole of Grignard reagent to form a second tetrahedral carbonyl addition compound, which, after hydrolysis in aqueous acid (Step 4), gives a 3° alcohol (or a 2° alcohol if the starting ester was a formate):

$$CH_3-\underset{R}{\overset{O}{\overset{\|}{C}}} + R-MgX \longrightarrow CH_3-\underset{R}{\overset{O^-\ [MgX]^+}{C}}-R \xrightarrow{H-\overset{+}{O}(H)-H} CH_3-\underset{R}{\overset{OH}{C}}-R$$

A ketone

Magnesium salt

A 3° alcohol

Example 15.6

Complete each Grignard reaction:

(a) $HCOCH_3$ (with C=O) $\xrightarrow[2)\ H_2O,\ HCl]{1)\ 2\ \text{(propyl)MgBr}}$

(b) methyl butanoate ($CH_3CH_2CH_2C(=O)OCH_3$) $\xrightarrow[2)\ H_2O,\ HCl]{1)\ 2PhMgBr}$

Strategy

Reaction of a Grignard reagent with an ester results in an alcohol containing two identical R groups (the R groups from the Grignard reagent) bonded to the former carbonyl carbon.

Solution

Sequence (a) gives a 2° alcohol, and sequence (b) gives a 3° alcohol:

(a) 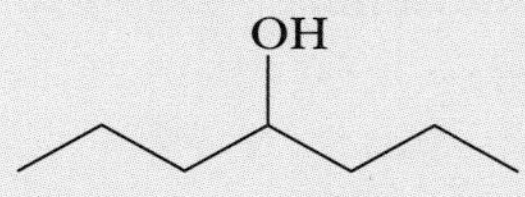 OH (b) 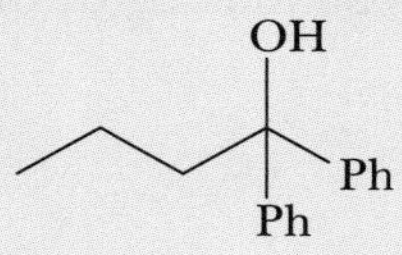 OH Ph Ph

See problems 15.30, 15.33, 15.34

Problem 15.6

Show how to prepare each alcohol by treating an ester with a Grignard reagent:

(a) 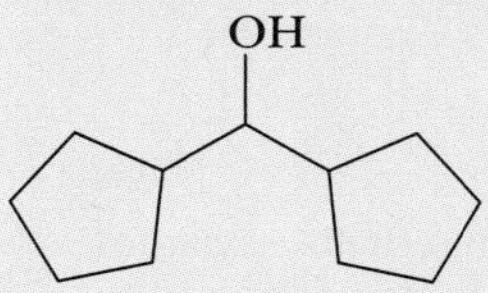 OH (b) 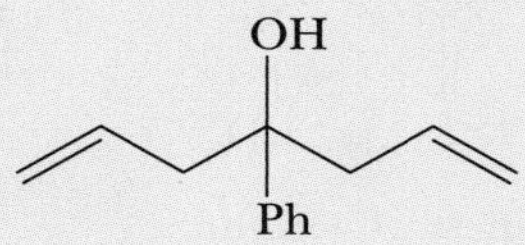 OH Ph

15.8 How Are Derivatives of Carboxylic Acids Reduced?

Most reductions of carbonyl compounds, including aldehydes and ketones, are now accomplished by transferring hydride ions from boron or aluminum hydrides. We have already seen the use of sodium borohydride to reduce the carbonyl groups of aldehydes and ketones to hydroxyl groups (Section 13.10B). We have also seen the use of lithium aluminum hydride to reduce not only the carbonyl groups of aldehydes and ketones, but also carboxyl groups (Section 14.5A), to hydroxyl groups.

A. Esters

An ester is reduced by lithium aluminum hydride to two alcohols. The alcohol derived from the acyl group is primary:

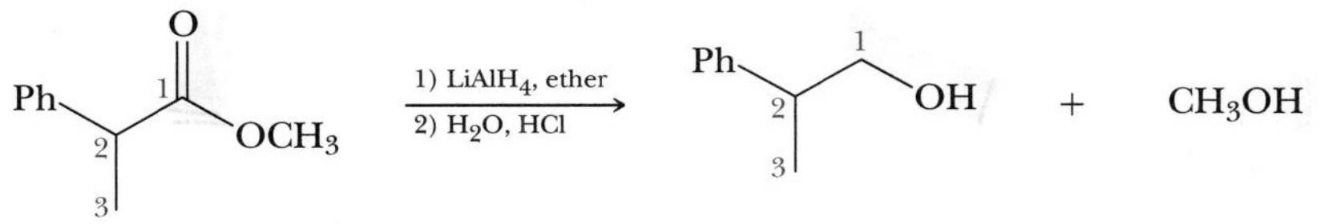

Ph O 1 OCH_3 2 3 — 1) $LiAlH_4$, ether 2) H_2O, HCl → Ph 1 2 3 OH + CH_3OH

Methyl 2-phenyl-propanoate — 2-Phenyl-1-propanol (a 1∞ alcohol) — Methanol

Sodium borohydride is not normally used to reduce esters because the reaction is very slow. Because of this lower reactivity of sodium borohydride toward esters, it is possible to reduce the carbonyl group of an aldehyde or a ketone to a hydroxyl group with this reagent without reducing an ester or a carboxyl group in the same molecule:

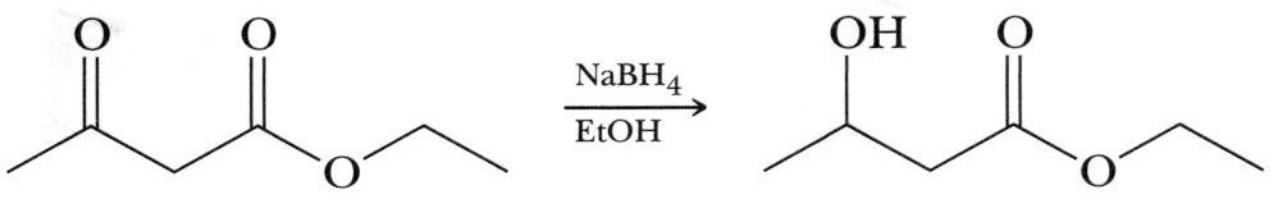

O O O — $NaBH_4$ EtOH → OH O O

B. Amides

Reduction of amides by lithium aluminum hydride can be used to prepare 1°, 2°, or 3° amines, depending on the degree of substitution of the amide:

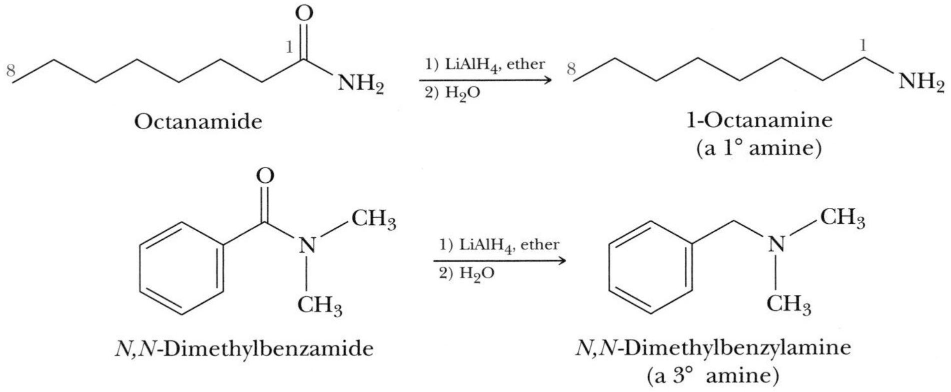

Octanamide $\xrightarrow[\text{2) } H_2O]{\text{1) } LiAlH_4\text{, ether}}$ 1-Octanamine (a 1° amine)

N,N-Dimethylbenzamide $\xrightarrow[\text{2) } H_2O]{\text{1) } LiAlH_4\text{, ether}}$ N,N-Dimethylbenzylamine (a 3° amine)

HOW TO 15.2 Approach Multistep Synthesis Problems

(a) When a given chemical transformation cannot be achieved with a known chemical reaction, it is necessary to use multiple steps to complete the synthesis. One of the most effective ways to accomplish this is through retrosynthetic analysis. The technique, formalized by Harvard Professor and Nobel Laureate E. J. Corey, involves working backwards from a target molecule until the synthesis is achieved. The technique is illustrated using the following transformation:

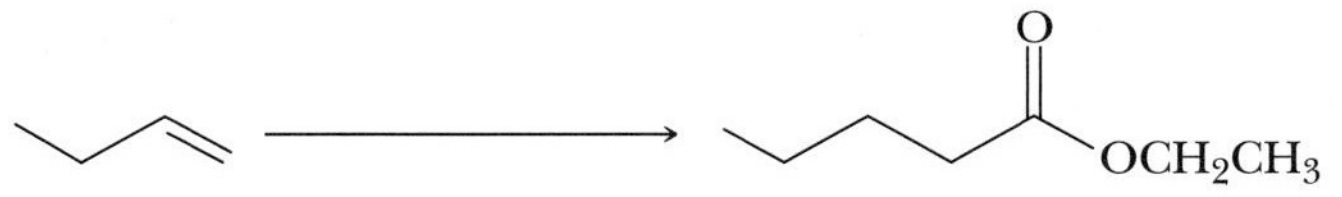

OCH_2CH_3

(b) Because there is no reaction that converts an alkene into an ester while also forming a new C—C bond, we work backwards from the ester. The goal is to identify a reaction (or reactions) that can synthesize esters. One such reaction is the Fischer esterification:

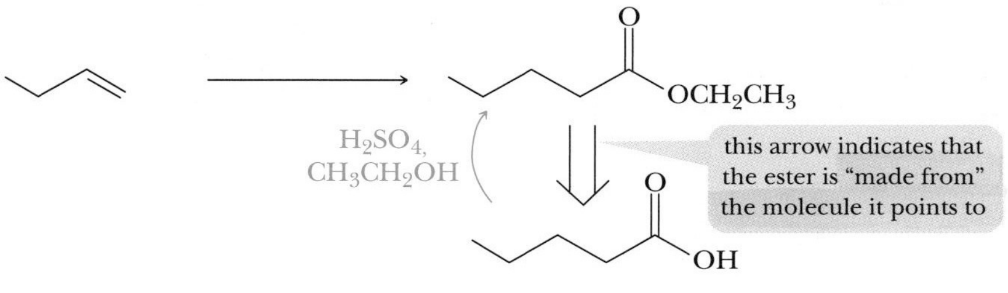

OCH_2CH_3

H_2SO_4, CH_3CH_2OH

this arrow indicates that the ester is "made from" the molecule it points to

OH

(c) Now that we've proposed that the ester can be made from pentanoic acid via Fischer esterification, the next step is to identify a reaction that can produce pentanoic acid. Here we propose oxidation of a 1° alcohol:

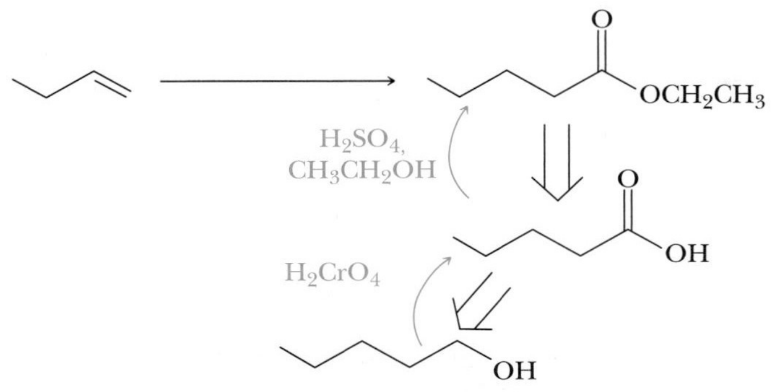

H_2SO_4, CH_3CH_2OH

OCH_2CH_3

H_2CrO_4

OH

OH

(d) The 1° alcohol, in turn, can be made from a Grignard reagent and formaldehyde:

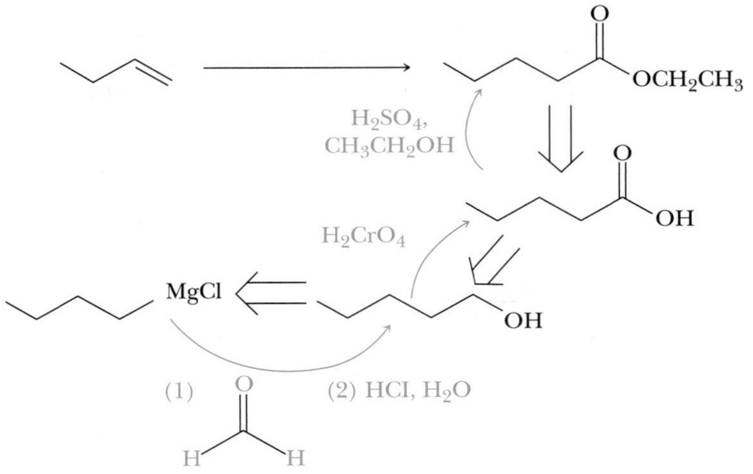

H_2SO_4, CH_3CH_2OH

OCH_2CH_3

H_2CrO_4

OH

MgCl

OH

(1) H–C(=O)–H (2) HCl, H_2O

(e) In this manner, the alkene can be arrived at retrosynthetically:

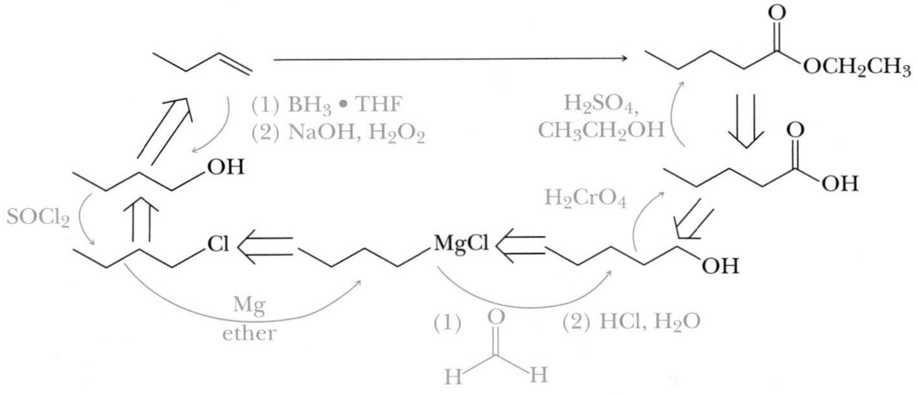

OCH_2CH_3

(1) $BH_3 \bullet THF$ (2) NaOH, H_2O_2

H_2SO_4, CH_3CH_2OH

OH

OH

H_2CrO_4

$SOCl_2$

Cl

MgCl

OH

Mg ether

(1) H–C(=O)–H (2) HCl, H_2O

Example 15.7

Show how to bring about each conversion:

(a) C_6H_5COH (C=O) $\longrightarrow$ $C_6H_5CH_2—N$ (pyrrolidine ring)

(b) cyclohexyl–COH (C=O) $\longrightarrow$ cyclohexyl–CH_2NHCH_3

Strategy

The key in each part is to convert the carboxylic acid to an amide (Section 15.5D) and then reduce the amide with $LiAlH_4$ (Section 15.8B).

Solution

Each amide can be prepared by treating the carboxylic acid with $SOCl_2$ to form the acid chloride (Section 14.7) and then treating the acid chloride with an amine (Section 15.5A). Alternatively, the carboxylic acid can be converted to an ester by Fischer esterification (Section 14.6) and the ester treated with an amine to give the amide. Solution (a) uses the acid chloride route, solution (b) the ester route:

(a) C_6H_5COH (C=O) $\xrightarrow{SOCl_2}$ C_6H_5CCl (C=O) $\xrightarrow{HN\text{(pyrrolidine)}}$

$C_6H_5C—N$ (C=O; pyrrolidine ring) $\xrightarrow[2)\ H_2O]{1)\ LiAlH_4,\ ether}$ $C_6H_5CH_2—N$ (pyrrolidine ring)

(b) cyclohexyl–COH (C=O) $\xrightarrow{CH_3CH_2OH,\ H^+}$ cyclohexyl–$COCH_2CH_3$ (C=O) $\xrightarrow{CH_3NH_2}$

cyclohexyl–$CNHCH_3$ (C=O) $\xrightarrow[2)\ H_2O]{1)\ LiAlH_4,\ ether}$ cyclohexyl–CH_2NHCH_3

See problems 15.27, 15.29, 15.31, 15.32, 15.39, 15.46

Problem 15.7

Show how to convert hexanoic acid to each amine in good yield:

(a) hexyl–$N(CH_3)_2$ (b) hexyl–NH–isopropyl

Example 15.8

Show how to convert phenylacetic acid to these compounds:

(a) 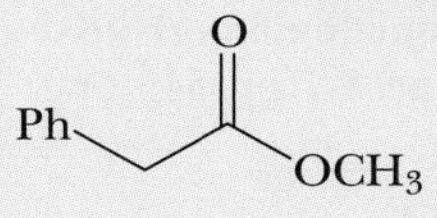 $PhCH_2C(=O)OCH_3$ (b) 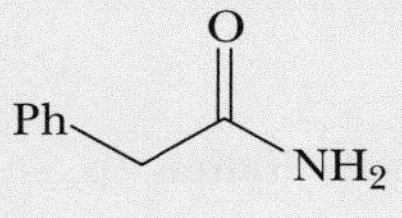 $PhCH_2C(=O)NH_2$

(c)  $PhCH_2CH_2NH_2$ (d) 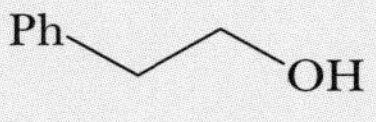 $PhCH_2CH_2OH$

Strategy

Decide whether the functional group interconversion can be done in one step. If not, try to determine what functional group can be converted to the targeted group. For example, a carboxyl group cannot be converted directly to an amine. However, an amide can be converted to an amine. Therefore, one only needs to convert the carboxyl group into an amide to eventually be able to produce the amine.

Solution

Prepare methyl ester (a) by Fischer esterification (Section 14.6) of phenylacetic acid with methanol. Then treat this ester with ammonia to prepare amide (b). Alternatively, treat phenylacetic acid with thionyl chloride (Section 14.7) to give an acid chloride, and then treat the acid chloride with two equivalents of ammonia to give amide (b). Reduction of amide (b) by $LiAlH_4$ gives the 1° amine (c). Similar reduction of either phenylacetic acid or ester (a) gives 1° alcohol (d):

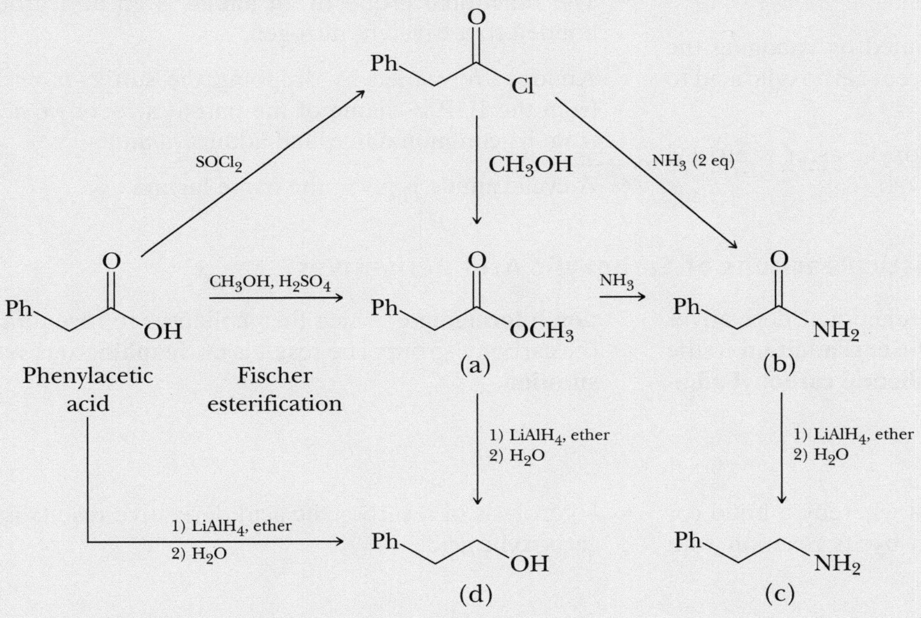

Phenylacetic acid → ($SOCl_2$) → $PhCH_2COCl$

$PhCH_2COCl$ → (CH_3OH) → (a) $PhCH_2CO_2CH_3$

$PhCH_2COCl$ → (NH_3 (2 eq)) → (b) $PhCH_2CONH_2$

Phenylacetic acid → (CH_3OH, H_2SO_4; Fischer esterification) → (a)

(a) → (NH_3) → (b)

(a) → (1) $LiAlH_4$, ether; 2) H_2O) → (d) $PhCH_2CH_2OH$

(b) → (1) $LiAlH_4$, ether; 2) H_2O) → (c) $PhCH_2CH_2NH_2$

Phenylacetic acid → (1) $LiAlH_4$, ether; 2) H_2O) → (d)

See problem 15.46

Problem 15.8

Show how to convert (*R*)-2-phenylpropanoic acid to these compounds:

(a) 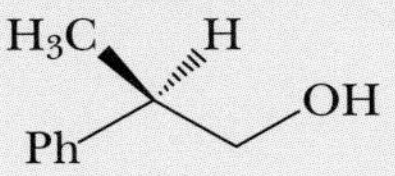 (*R*)-2-Phenyl-1-propanol

(b) 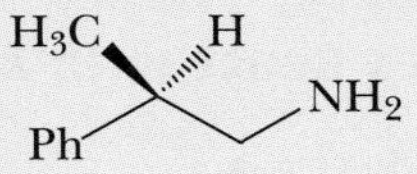 (*R*)-2-Phenyl-1-propanamine

Key Terms and Concepts

acid halide (p. 542)
acyl group (p. 542)
amide (p. 546)
carboxylic anhydride (p. 542)
carboxylic ester (p. 544)
hydrolysis (p. 550)
lactam (p. 547)
lactone (p. 545)
nucleophilic acyl substitution (p. 549)
phosphoric anhydride (p. 542)
phosphoric ester (p. 545)
reactivity of functional derivatives of carboxylic acids (pp. 549, 560)
saponification (p. 551)
tetrahedral carbonyl addition intermediate (p. 548)
transesterification (p. 556)

Summary of Key Questions

15.1 What Are Some Derivatives of Carboxylic Acids, and How Are They Named?

- The functional group of an **acid halide** is an acyl group bonded to a halogen.
- Acid halides are named by changing the suffix *-ic acid* in the name of the parent carboxylic acid to *-yl halide*.
- The functional group of a **carboxylic anhydride** is two acyl groups bonded to an oxygen.
- Symmetrical anhydrides are named by changing the suffix *acid* in the name of the parent carboxylic acid to *anhydride*.
- The functional group of a carboxylic ester is an acyl group bonded to —OR or —OAr.
- An ester is named by giving the name of the alkyl or aryl group bonded to oxygen first, followed by the name of the acid, in which the suffix *-ic acid* is replaced by the suffix *-ate*.
- A cyclic ester is given the name **lactone**.
- The functional group of an **amide** is an acyl group bonded to a trivalent nitrogen.
- Amides are named by dropping the suffix *-oic acid* from the IUPAC name of the parent acid, or *-ic acid* from its common name, and adding *-amide*.
- A cyclic amide is given the name **lactam**.

15.2 What Are the Characteristic Reactions of Carboxylic Acid Derivatives?

- A common reaction theme of functional derivatives of carboxylic acids is **nucleophilic acyl addition** to the carbonyl carbon to form a **tetrahedral carbonyl addition intermediate**, which then collapses to regenerate the carbonyl group. The result is **nucleophilic acyl substitution**.

15.3 What Is Hydrolysis?

- Hydrolysis is a chemical process whereby a bond (or bonds) in a molecule is broken by its reaction with water.
- Hydrolysis of a carboxylic acid derivative results in a carboxylic acid.

15.4 How Do Carboxylic Acid Derivatives React with Alcohols?

- Carboxylic acid derivatives (except for amides) react with alcohols to give esters.
- The reaction conditions required (i.e., neutral, acidic, or basic) depend on the type of derivative.

15.5 How Do Carboxylic Acid Derivatives React with Ammonia and Amines?

- Carboxylic acid derivatives (except for amides) react with ammonia and amines to give amides.

15.6 How Can Functional Derivatives of Carboxylic Acids Be Interconverted?

- Listed in order of increasing reactivity toward nucleophilic acyl substitution, these functional derivatives are:

$\overset{\text{O}}{\underset{\|}{}}$ $RCNH_2$	$RCOR'$	$RCOCR'$	$RCCl$
Amide	Ester	Anhydride	Acid chloride

Reactivity toward nucleophilic acyl substitution →

Less reactive — More reactive

- Any more reactive functional derivative can be directly converted to any less reactive functional derivative by reaction with an appropriate oxygen or nitrogen nucleophile.

15.7 How Do Esters React with Grignard Reagents?

- Reaction of an ester with a Grignard reagent involves the formation of two successive tetrahedral carbonyl addition compounds. The result of the overall reaction is an alcohol containing the two identical alkyl groups from the Grignard reagent.

15.8 How Are Derivatives of Carboxylic Acids Reduced?

- Derivatives of carboxylic acids are resistant to reduction by $NaBH_4$. Therefore, ketones and aldehydes can be selectively reduced in the presence of a carboxylic acid derivative.
- Derivatives of carboxylic acids are resistant to catalytic hydrogenation by H_2/M. Therefore, C—C double and triple bonds can be selectively reduced in the presence of a carboxylic acid derivative.
- $LiAlH_4$ reduces the carboxyl group of acid halides, acid anhydrides, and esters to a 1° alcohol group.
- $LiAlH_4$ reduces amides to amines.

Quick Quiz

Answer true or false to the following questions to assess your general knowledge of the concepts in this chapter. If you have difficulty with any of them, you should review the appropriate section in the chapter (shown in parentheses) before attempting the more challenging end-of-chapter problems.

1. The stronger the base, the better the leaving group. (15.2)
2. Anhydrides can contain C—O double bonds or P—O double bonds. (15.1)
3. Acid anhydrides react with ammonia and amines without the need for acid or base. (15.5)
4. Derivatives of carboxylic acids are reduced by H_2/M. (15.8)
5. Aldehydes and ketones undergo nucleophilic acyl substitution reactions, while derivatives of carboxylic acids undergo nucleophilic addition reactions. (15.2)
6. Esters react with ammonia and amines without the need for acid or base. (15.5)
7. An acyl group is a carbonyl bonded to an alkyl (R) group. (15.1)
8. Hydrolysis is the loss of water from a molecule. (15.3)
9. Esters react with water without the need for acid or base. (15.4)
10. Acid anhydrides react with water without the need for acid or base. (15.3)
11. An acid halide can be converted to an amide in one step. (15.6)
12. An ester can be converted to an acid halide in one step. (15.6)
13. In the hydrolysis of an ester with base, hydroxide ion is a catalyst. (15.3)
14. Derivatives of carboxylic acids are reduced by $NaBH_4$. (15.8)
15. Acid anhydrides react with alcohols without the need for acid or base. (15.4)

16. Acid halides react with water without the need for acid or base. (15.3)
17. An ester of formic acid reacts with Grignard reagents to form 3° alcohols. (15.7)
18. Acid halides react with ammonia and amines without the need for acid or base. (15.5)
19. A cyclic amide is called a lactone. (15.1)
20. The reactivity of a carboxylic acid derivative is dependent on the stability of its leaving group. (15.2)
21. Amides react with ammonia and amines without the need for acid or base. (15.5)
22. An amide can be converted to an ester in one step. (15.6)
23. Amides react with water without the need for acid or base. (15.3)
24. Esters react with alcohols without the need for acid or base. (15.3)
25. Amides react with alcohols under acidic or basic conditions. (15.4)
26. Esters other than formic acid esters react with Grignards to form ketones. (15.7)
27. Acid halides react with alcohols without the need for acid or base. (15.4)
28. An —OR group attached to a P—O double bond is known as an ester. (15.1)

Answers: (1) F (2) T (3) T (4) F (5) F (6) T (7) T (8) F (9) F (10) T (11) T (12) F (13) F (14) F (15) T (16) T (17) F (18) T (19) F (20) T (21) F (22) F (23) F (24) F (25) F (26) F (27) T (28) T

Key Reactions

1. Hydrolysis of an Acid Chloride (Section 15.3A)
Low-molecular-weight acid chlorides react vigorously with water; higher-molecular-weight acid chlorides react less rapidly:

$$CH_3\overset{O}{\overset{\|}{C}}Cl + H_2O \longrightarrow CH_3\overset{O}{\overset{\|}{C}}OH + HCl$$

2. Hydrolysis of an Acid Anhydride (Section 15.3B)
Low-molecular-weight acid anhydrides react readily with water; higher-molecular-weight acid anhydrides react less rapidly:

$$CH_3\overset{O}{\overset{\|}{C}}O\overset{O}{\overset{\|}{C}}CH_3 + H_2O \longrightarrow CH_3\overset{O}{\overset{\|}{C}}OH + HO\overset{O}{\overset{\|}{C}}CH_3$$

3. Hydrolysis of an Ester (Section 15.3C)
Esters are hydrolyzed only in the presence of base or acid; base is required in an equimolar amount, acid is a catalyst:

$$CH_3\overset{O}{\overset{\|}{C}}O\text{-}C_6H_{11} + NaOH \xrightarrow{H_2O} CH_3\overset{O}{\overset{\|}{C}}O^-Na^+ + HO\text{-}C_6H_{11}$$

$$CH_3\overset{O}{\overset{\|}{C}}O\text{-}C_6H_{11} + H_2O \xrightarrow{HCl} CH_3\overset{O}{\overset{\|}{C}}OH + HO\text{-}C_6H_{11}$$

4. Hydrolysis of an Amide (Section 15.3D)
Either acid or base is required in an amount equivalent to that of the amide:

$$CH_3CH_2CH_2\overset{O}{\overset{\|}{C}}NH_2 + H_2O + HCl \xrightarrow[Heat]{H_2O} CH_3CH_2CH_2\overset{O}{\overset{\|}{C}}OH + NH_4^+Cl^-$$

$$CH_3\overset{O}{\overset{\|}{C}}NH\text{-}C_6H_5 + NaOH \xrightarrow[heat]{H_2O} CH_3\overset{O}{\overset{\|}{C}}O^-Na^+ + H_2N\text{-}C_6H_5$$

5. Reaction of an Acid Chloride with an Alcohol (Section 15.4A)
Treatment of an acid chloride with an alcohol gives an ester and HCl:

$$CH_3CH_2CH_2\overset{O}{\overset{\|}{C}}Cl + HOCH_3 \longrightarrow CH_3CH_2CH_2\overset{O}{\overset{\|}{C}}OCH_3 + HCl$$

6. Reaction of an Acid Anhydride with an Alcohol (Section 15.4B)
Treatment of an acid anhydride with an alcohol gives an ester and a carboxylic acid:

$$CH_3\overset{O}{\overset{\|}{C}}O\overset{O}{\overset{\|}{C}}CH_3 + HOCH_2CH_3 \longrightarrow CH_3\overset{O}{\overset{\|}{C}}OCH_2CH_3 + CH_3\overset{O}{\overset{\|}{C}}OH$$

7. Reaction of an Ester with an Alcohol (Section 15.4C)
Treatment of an ester with an alcohol in the presence of an acid catalyst results in transesterification—that is, the replacement of one —OR group by a different —OR group:

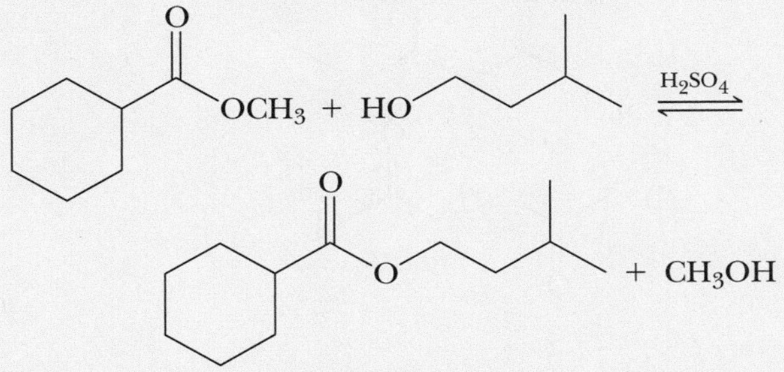

Methyl cyclohexanecarboxylate + HO(CH$_2$)$_2$CH(CH$_3$)$_2$ $\underset{}{\overset{H_2SO_4}{\rightleftharpoons}}$ 3-methylbutyl cyclohexanecarboxylate + CH_3OH

8. Reaction of an Acid Chloride with Ammonia or an Amine (Section 15.5A)
Reaction requires 2 moles of ammonia or amine—1 mole to form the amide and 1 mole to neutralize the HCl by-product:

$$CH_3\overset{O}{\overset{\|}{C}}Cl + 2NH_3 \longrightarrow CH_3\overset{O}{\overset{\|}{C}}NH_2 + NH_4^+Cl^-$$

9. Reaction of an Acid Anhydride with Ammonia or an Amine (Section 15.5B)
Reaction requires 2 moles of ammonia or amine—1 mole to form the amide and 1 mole to neutralize the carboxylic acid by-product:

$$CH_3\overset{O}{\overset{\|}{C}}O\overset{O}{\overset{\|}{C}}CH_3 + 2NH_3 \longrightarrow CH_3\overset{O}{\overset{\|}{C}}NH_2 + CH_3\overset{O}{\overset{\|}{C}}O^-NH_4^+$$

10. Reaction of an Ester with Ammonia or an Amine (Section 15.5C)
Treatment of an ester with ammonia, a 1° amine, or a 2° amine gives an amide:

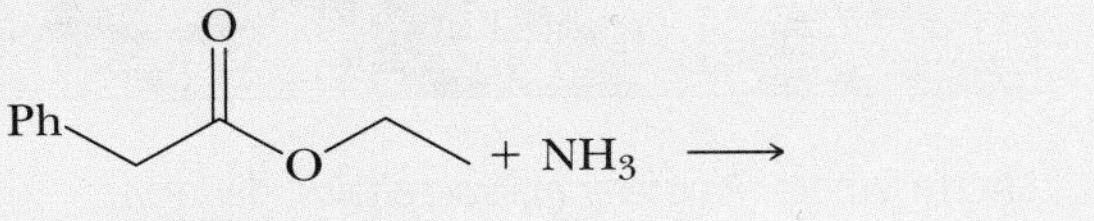

Ethyl phenylacetate + NH_3 ⟶ Phenylacetamide + Ethanol

11. Reaction of an Ester with a Grignard Reagent (Section 15.7)
Treating a formic ester with a Grignard reagent, followed by hydrolysis, gives a 2° alcohol, whereas treating any other ester with a Grignard reagent gives a 3° alcohol:

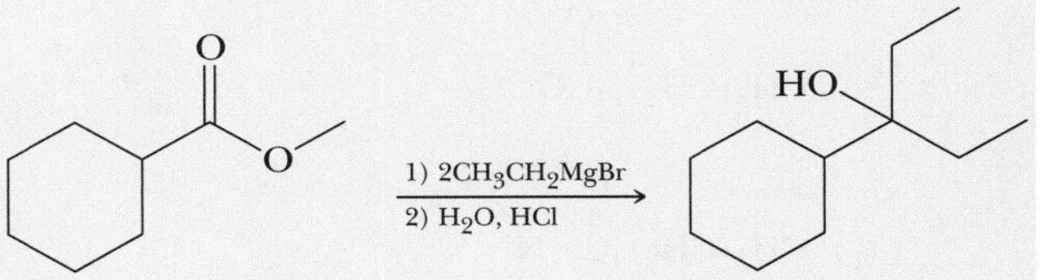

Methyl cyclohexanecarboxylate $\xrightarrow[2)\ H_2O,\ HCl]{1)\ 2CH_3CH_2MgBr}$ 3-cyclohexyl-3-pentanol (HO)

12. Reduction of an Ester (Section 15.8A)
Reduction by lithium aluminum hydride gives two alcohols:

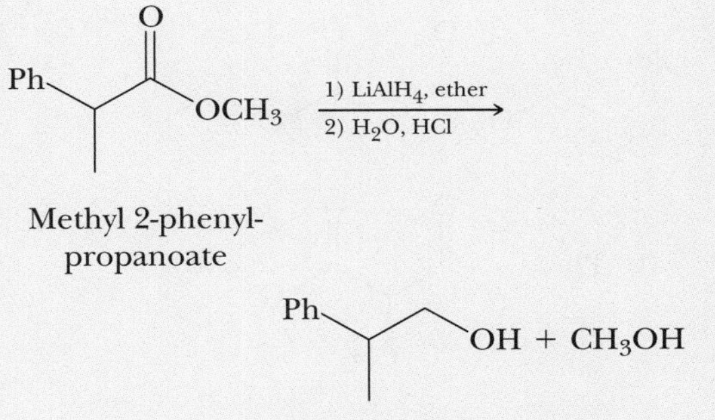

Methyl 2-phenylpropanoate $\xrightarrow[2)\ H_2O,\ HCl]{1)\ LiAlH_4,\ ether}$ 2-Phenyl-1-propanol + CH_3OH (Methanol)

13. Reduction of an Amide (Section 15.8B)
Reduction by lithium aluminum hydride gives an amine:

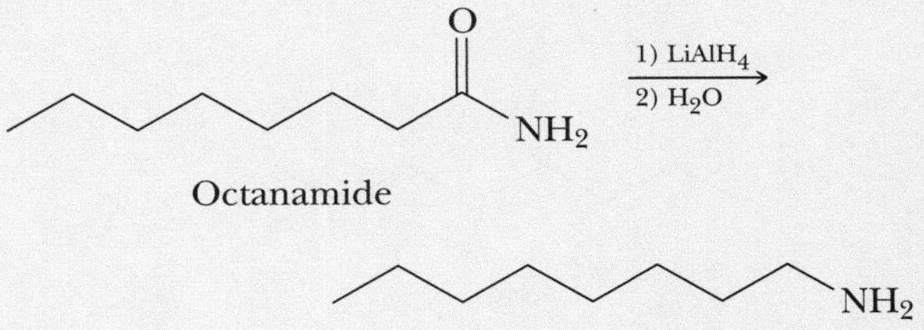

Octanamide $\xrightarrow[2)\ H_2O]{1)\ LiAlH_4}$ 1-Octanamine

Problems

A problem marked with an asterisk indicates an applied "real world" problem. Answers to problems whose numbers are printed in blue are given in Appendix D.

Section 15.1 Structure and Nomenclature

15.9 Draw a structural formula for each compound: **(See Example 15.1)**

(a) Dimethyl carbonate
(b) *p*-Nitrobenzamide
(c) Octanoyl chloride
(d) Diethyl oxalate
(e) Ethyl *cis*-2-pentenoate
(f) Butanoic anhydride
(g) Dodecanamide
(h) Ethyl 3-hydroxybutanoate
(i) Ethyl benzoate
(j) Benzoyl chloride

15.10 Write the IUPAC name for each compound: **(See Example 15.1)**

(a) $C_6H_5-\overset{O}{\overset{\|}{C}}-O-\overset{O}{\overset{\|}{C}}-C_6H_5$

(b) $CH_3(CH_2)_{14}\overset{O}{\overset{\|}{C}}OCH_3$

(c) $CH_3(CH_2)_4\overset{O}{\overset{\|}{C}}-N(CH_3)H$

(d) $H_2N-C_6H_4-\overset{O}{\overset{\|}{C}}-NH_2$

(e)

(f) Ph, CH_3, CH_3

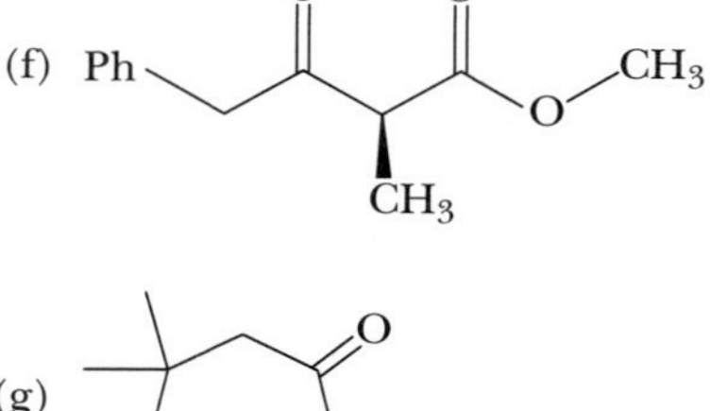

(g)

(h)

(i)

(j)

(k) OH, Cl

(l) Cl

*__15.11__ When oil from the head of the sperm whale is cooled, spermaceti, a translucent wax with a white, pearly luster, crystallizes from the mixture. Spermaceti, which makes up 11% of whale oil, is composed mainly of hexadecyl hexadecanoate (cetyl palmitate). At one time, spermaceti was widely used in the making of cosmetics, fragrant soaps, and candles. Draw a structural formula of cetyl palmitate. **(See Example 15.1)**

Sperm whale, *Physterer macrocephalus,* diving, Kaikoura, NZ. *(Kim Westerskov/Stone/Getty Images)*

Physical Properties

15.12 Acetic acid and methyl formate are constitutional isomers. Both are liquids at room temperature, one with a boiling point of 32°C, the other with a boiling point of 118°C. Which of the two has the higher boiling point?

15.13 Butanoic acid (M. W. 88.11 g/mol) has a boiling point of 162°C, whereas its propyl ester (M.W. 130.18 g/mol) has a boiling point of 142°C. Account for the fact that the boiling point of butanoic acid is higher than that of its propyl ester, even though butanoic acid has a lower molecular weight.

15.14 The constitutional isomers pentanoic acid and methyl butanoate are both slightly soluble in water. One of these compounds has a solubility of 1.5 g/100 ml (25 °C), while the other has a solubility of 4.97 g/100 ml (25 °C). Assign the solubilities to each compound and account for the differences.

Sections 15.2–15.8 Reactions

15.15 Arrange these compounds in order of increasing reactivity toward nucleophilic acyl substitution:

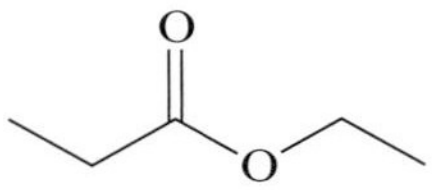

(1) 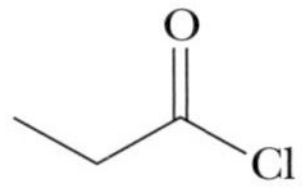 (2) 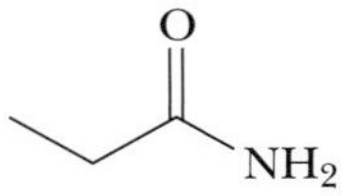 (3) 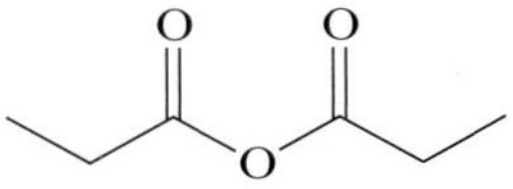 (4)

15.16 A carboxylic acid can be converted to an ester by Fischer esterification. Show how to synthesize each ester from a carboxylic acid and an alcohol by Fischer esterification: **(See Example 15.4)**

(a)

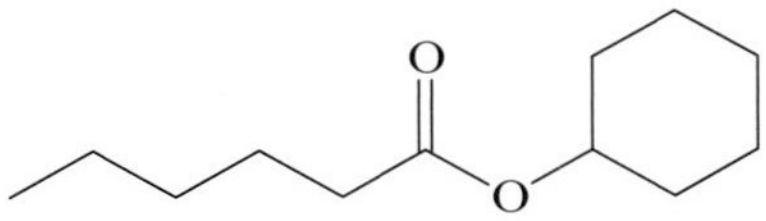

(b)

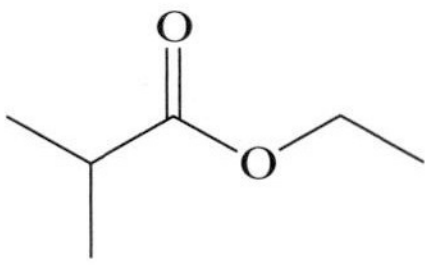

15.17 A carboxylic acid can also be converted to an ester in two reactions by first converting the carboxylic acid to its acid chloride and then treating the acid chloride with an alcohol. Show how to prepare each ester in Problem 15.16 from a carboxylic acid and an alcohol by this two-step scheme. **(See Example 15.4)**

15.18 Show how to prepare these amides by reaction of an acid chloride with ammonia or an amine: **(See Example 15.5)**

(a)

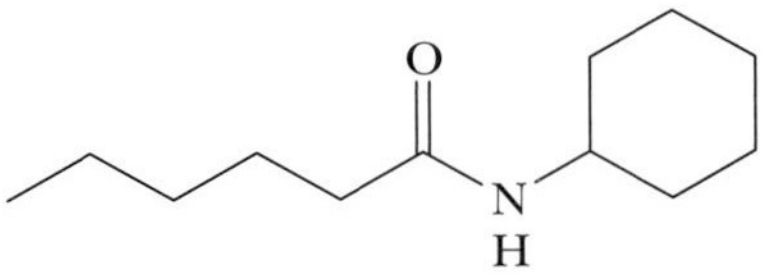

(b)

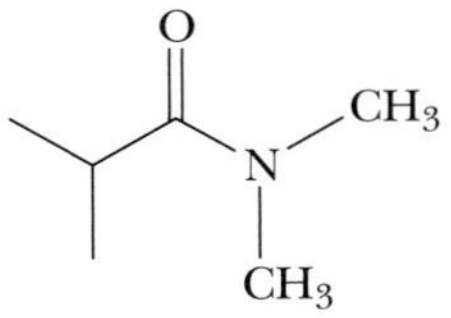

(c)

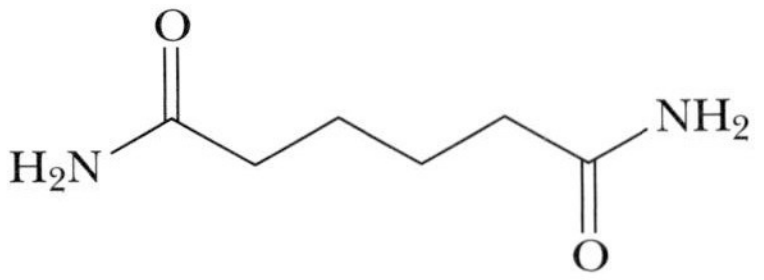

15.19 Balance and write a mechanism for each of the following reactions. **(See Examples 15.2, 15.4, 15.5)**

(a) 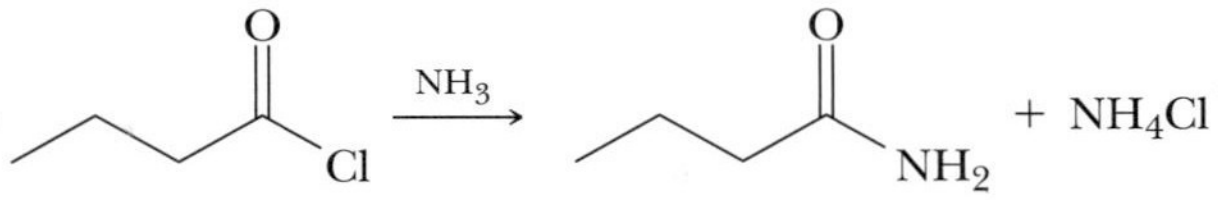 $\xrightarrow{NH_3}$ $+ NH_4Cl$

(b) 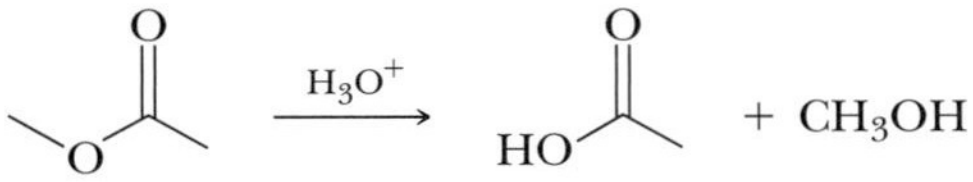 $\xrightarrow{H_3O^+}$ $+ CH_3OH$

(c) 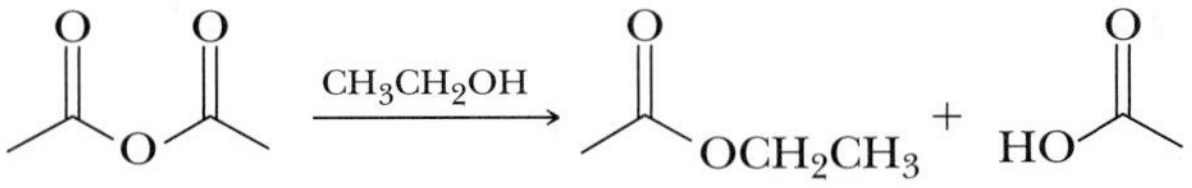 $\xrightarrow{CH_3CH_2OH}$ OCH_2CH_3 + HO

15.20 What product is formed when benzoyl chloride is treated with these reagents? **(See Examples 15.2, 15.4, 15.5)**

(a) C_6H_6, $AlCl_3$
(b) $CH_3CH_2CH_2CH_2OH$
(c) $CH_3CH_2CH_2CH_2SH$
(d) $CH_3CH_2CH_2CH_2NH_2$ (2 equivalents)
(e) H_2O
(f) 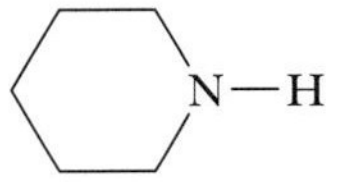 N—H (2 equivalents)

15.21 Write the product(s) of the treatment of propanoic anhydride with each reagent: **(See Examples 15.4, 15.5)**

(a) Ethanol (1 equivalent)
(b) Ammonia (2 equivalents)

15.22 Write the product of the treatment of benzoic anhydride with each reagent: **(See Examples 15.4, 15.5)**

(a) Ethanol (1 equivalent)
(b) Ammonia (2 equivalents)

***15.23** The analgesic phenacetin is synthesized by treating 4-ethoxyaniline with acetic anhydride. Write an equation for the formation of phenacetin.

***15.24** The analgesic acetaminophen is synthesized by treating 4-aminophenol with one equivalent of acetic anhydride. Write an equation for the formation of acetaminophen. (*Hint*: Remember from Section 7.5A that an $—NH_2$ group is a better nucleophile than an —OH group.) **(See Example 15.5)**

***15.25** Nicotinic acid, more commonly named niacin, is one of the B vitamins. Show how nicotinic acid can be converted to ethyl nicotinate and then to nicotinamide: **(See Example 15.5)**

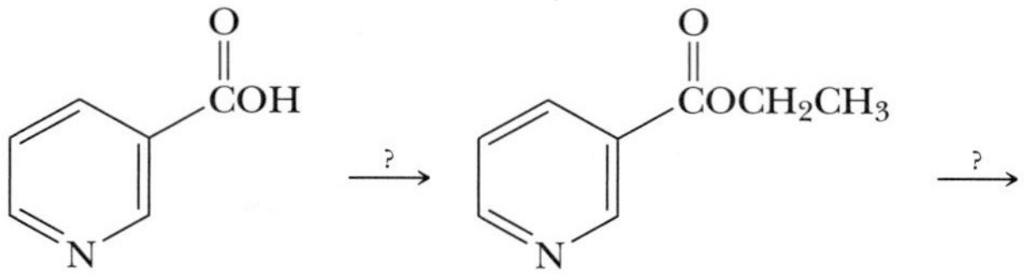

Nicotinic acid (Niacin) $\xrightarrow{?}$ Ethyl nicotinate $\xrightarrow{?}$

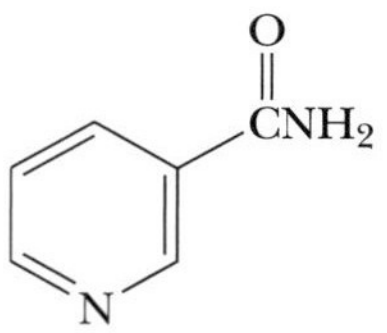

Nicotinamide

15.26 Complete these reactions: **(See Example 15.5)**

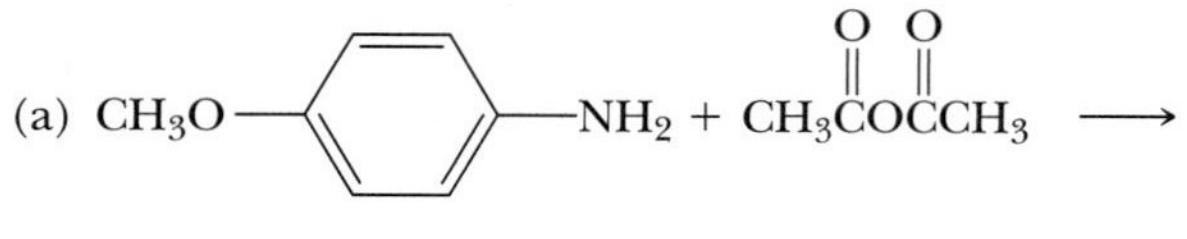

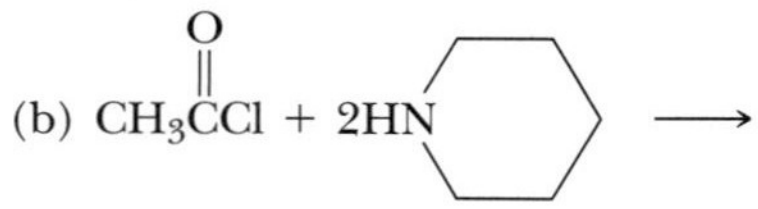

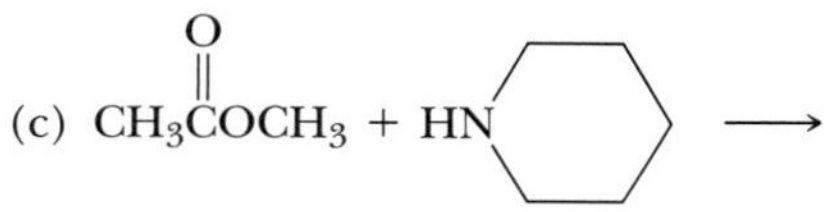

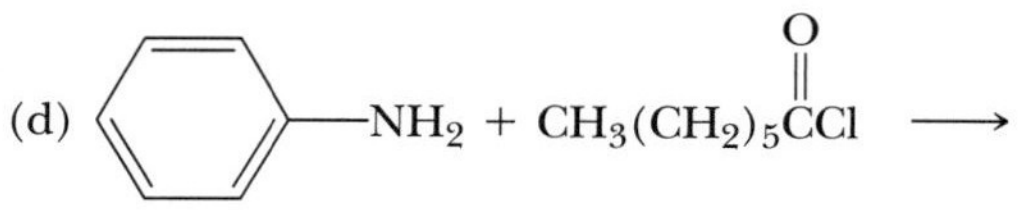

(a) $CH_3O-C_6H_4-NH_2 + CH_3COCCH_3$ (with two C=O) $\longrightarrow$

(b) CH_3CCl (C=O) + 2HN (piperidine) $\longrightarrow$

(c) CH_3COCH_3 (C=O) + HN (piperidine) $\longrightarrow$

(d) $C_6H_5-NH_2 + CH_3(CH_2)_5CCl$ (C=O) $\longrightarrow$

15.27 What product is formed when ethyl benzoate is treated with these reagents? **(See Example 15.7)**

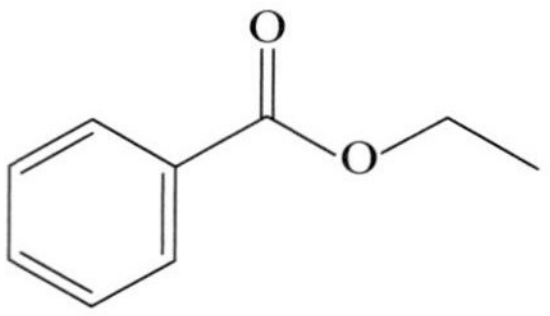

Ethyl benzoate

(a) H_2O, NaOH, heat

(b) $LiAlH_4$, then H_2O

(c) H_2O, H_2SO_4, heat

(d) $CH_3CH_2CH_2CH_2NH_2$

(e) C_6H_5MgBr (2 moles) and then H_2O/HCl

***15.28** Show how to convert 2-hydroxybenzoic acid (salicylic acid) to these compounds: **(See Example 15.4)**

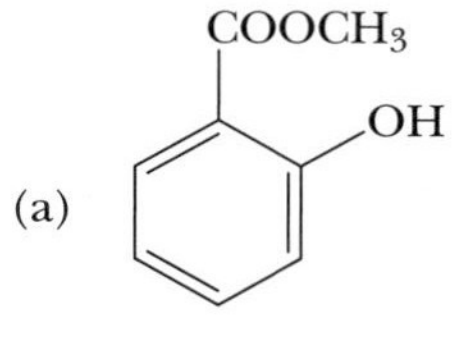

(a) Methyl salicylate (Oil of wintergreen)

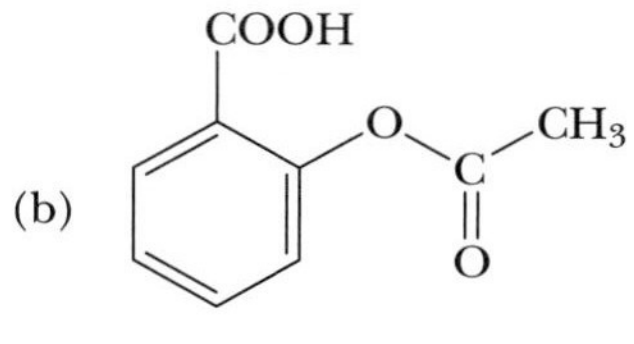

(b) Acetyl salicylic acid (Aspirin)

15.29 What product is formed when benzamide is treated with these reagents? **(See Examples 15.3, 15.7)**

(a) H_2O, HCl, heat

(b) NaOH, H_2O, heat

(c) $LiAlH_4$/ether, then H_2O

15.30 Treating γ-butyrolactone with two equivalents of methylmagnesium bromide, followed by hydrolysis in aqueous acid, gives a compound with the molecular formula $C_6H_{14}O_2$: **(See Example 15.6)**

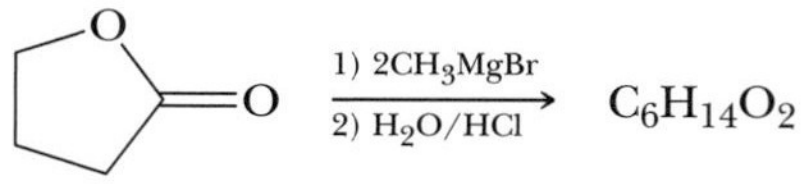

$$\xrightarrow[2)\ H_2O/HCl]{1)\ 2CH_3MgBr} C_6H_{14}O_2$$

Propose a structural formula for this compound.

15.31 Show the product of treating γ-butyrolactone with each reagent: **(See Examples 15.2, 15.5, 15.7)**

(a) NH_3

(b) $LiAlH_4$/ether, then H_2O

(c) NaOH, H_2O, heat

15.32 Show the product of treating *N*-methyl-γ-butyrolactam with each reagent: **(See Examples 15.3, 15.7)**

(a) H_2O, HCl, heat

(b) NaOH, H_2O, heat

(c) $LiAlH_4$/ether, then H_2O

15.33 Complete these reactions: **(See Example 15.6)**

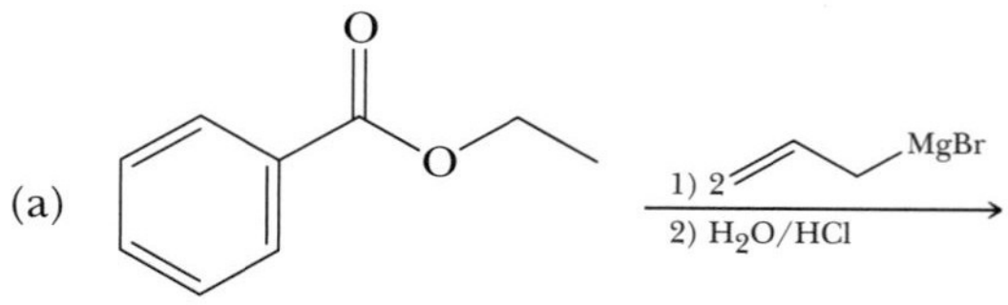

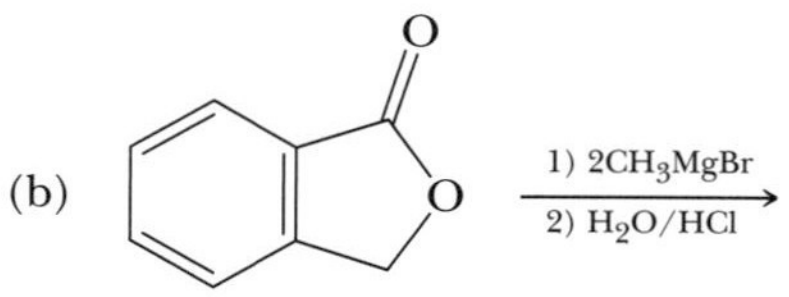

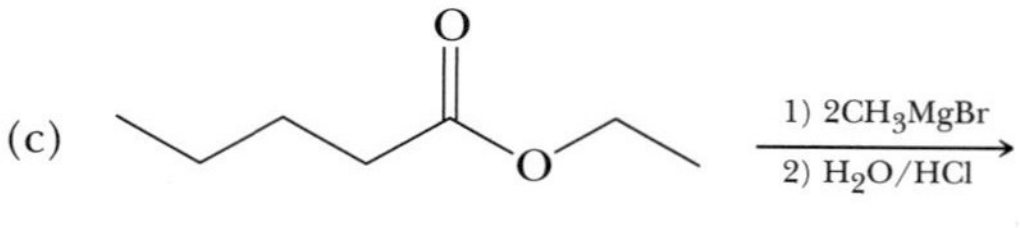

(a) $\xrightarrow[2)\ H_2O/HCl]{1)\ 2\ CH_2{=}CHCH_2MgBr}$

(b) $\xrightarrow[2)\ H_2O/HCl]{1)\ 2CH_3MgBr}$

(c) $\xrightarrow[2)\ H_2O/HCl]{1)\ 2CH_3MgBr}$

15.34 What combination of ester and Grignard reagent can be used to prepare each alcohol? **(See Example 15.6)**

(a) 2-Methyl-2-butanol

(b) 3-Phenyl-3-pentanol

(c) 1,1-Diphenylethanol

15.35 Reaction of a 1° or 2° amine with diethyl carbonate under controlled conditions gives a carbamic ester: **(See Example 15.5)**

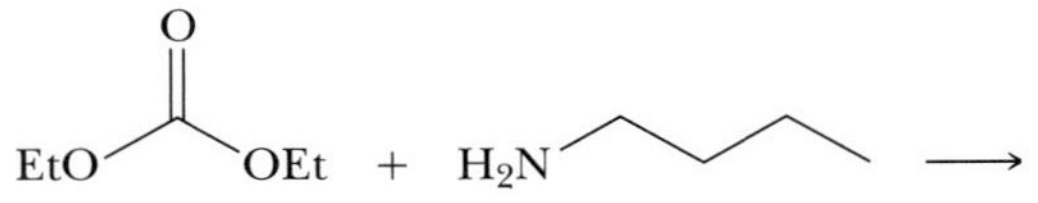

EtO–C(=O)–OEt + H_2N–CH₂CH₂CH₂CH₃ ⟶

Diethyl carbonate   1-Butanamine (Butylamine)

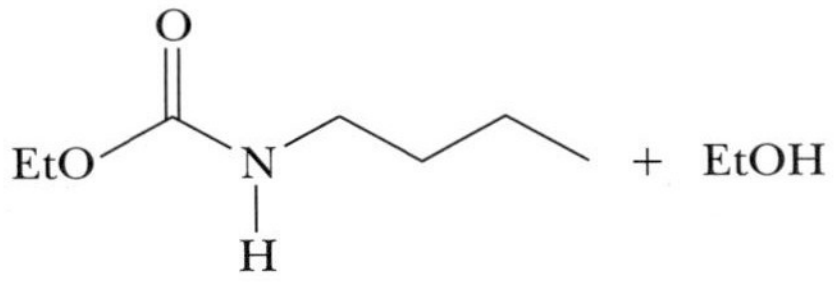

EtO–C(=O)–NH–CH₂CH₂CH₂CH₃ + EtOH

A carbamic ester

Propose a mechanism for this reaction.

15.36 Barbiturates are prepared by treating diethyl malonate or a derivative of diethyl malonate with urea in the presence of sodium ethoxide as a catalyst. Following is an equation for the preparation of barbital from diethyl 2,2-diethylmalonate and urea (barbital, a long-duration hypnotic and sedative, is prescribed under a dozen or more trade names):

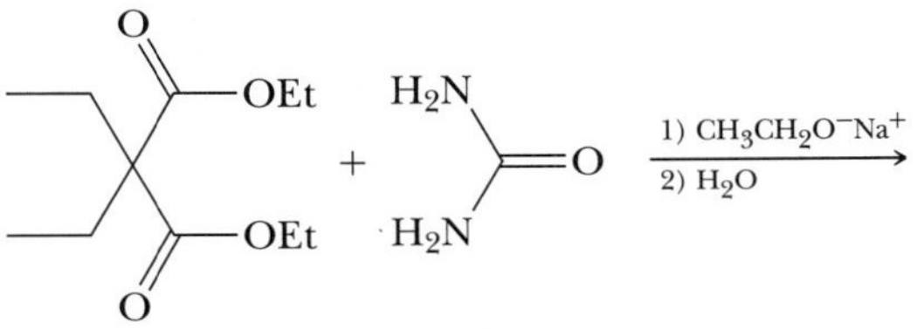

Diethyl 2,2-diethylmalonate + Urea $\xrightarrow[\text{2) } H_2O]{\text{1) } CH_3CH_2O^-Na^+}$ 5,5-Diethylbarbituric acid (Barbital) + 2EtOH

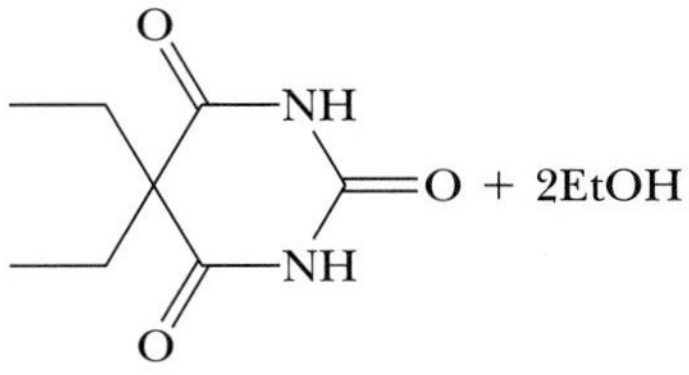

(a) Propose a mechanism for this reaction.

(b) The pK_a of barbital is 7.4. Which is the most acidic hydrogen in this molecule, and how do you account for its acidity?

***15.37** Name and draw structural formulas for the products of the complete hydrolysis of meprobamate and phenobarbital in hot aqueous acid. Meprobamate is a tranquilizer prescribed under 58 different trade names. Phenobarbital is a long-acting sedative, hypnotic, and anticonvulsant. [*Hint*: Remember that, when heated, β-dicarboxylic acids and β-ketoacids undergo decarboxylation (Section 14.8B).]

(a) 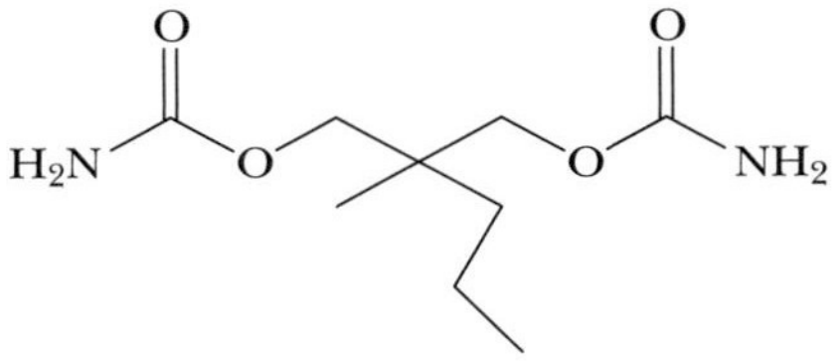 Meprobamate

(b) 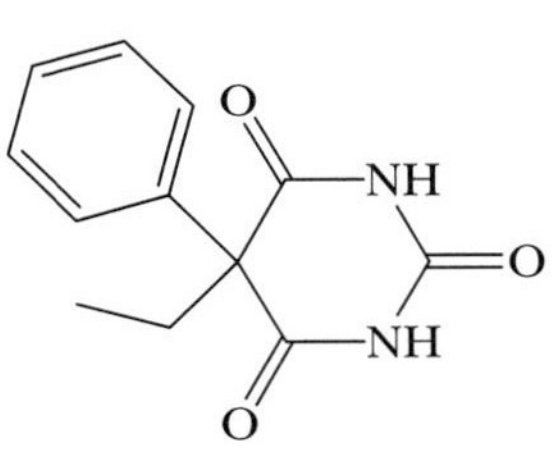 Phenobarbital

Synthesis

***15.38** The active ingredient in several common insect repellents *N,N*-Diethyl-*m*-toluamide (Deet) is synthesized from 3-methylbenzoic acid (*m*-toluic acid) and diethylamine:

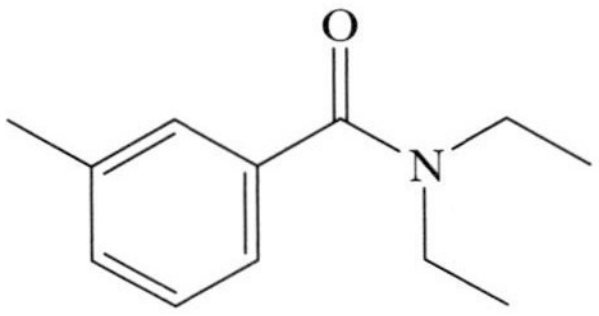

N,N-Diethyl-*m*-toluamide (Deet)

Show how this synthesis can be accomplished.

15.39 Show how to convert ethyl 2-pentenoate into these compounds: **(See Example 15.7)**

(a)

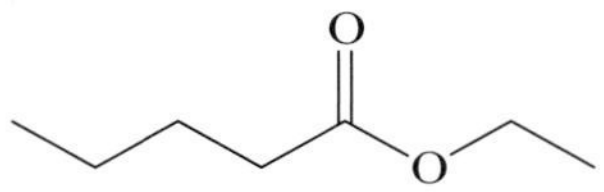

(b) OH

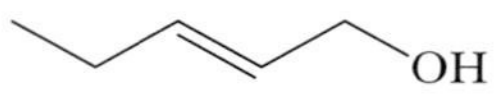

(c) Br O Br

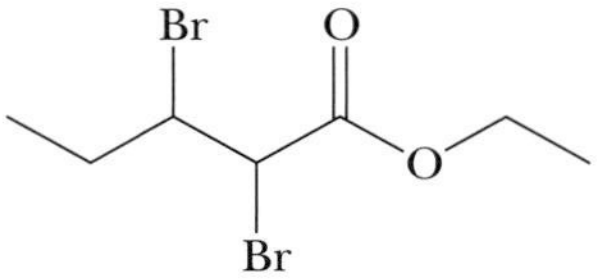

***15.40** Procaine (whose hydrochloride is marketed as Novocaine®) was one of the first local anesthetics for infiltration and regional anesthesia. Show how to synthesize procaine, using the three reagents shown as the sources of carbon atoms:

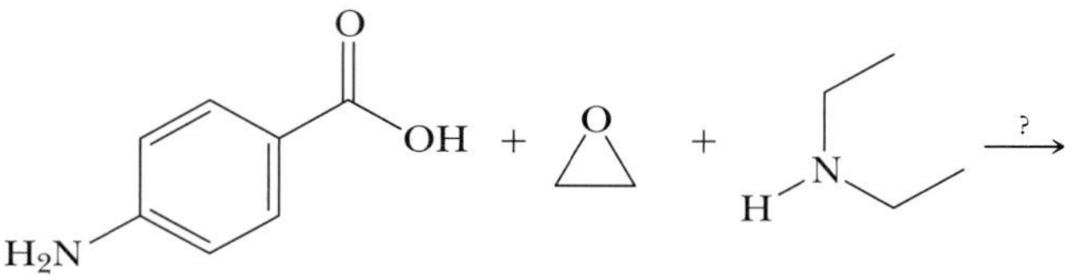

4-Aminobenzoic acid + Ethylene oxide + Diethylamine $\xrightarrow{?}$

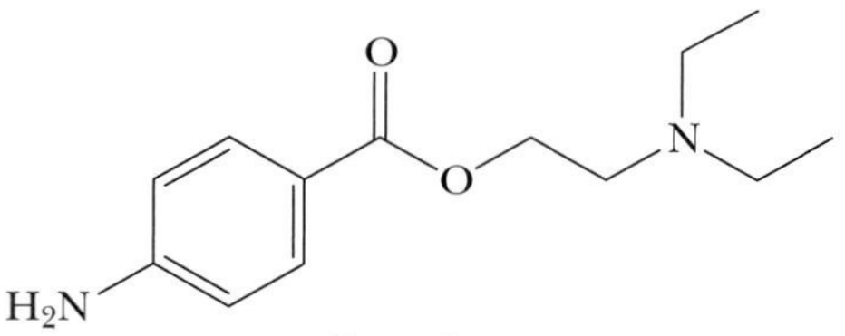

Procaine

***15.41** There are two nitrogen atoms in procaine. Which of the two is the stronger base? Draw the structural formula for the salt that is formed when procaine is treated with 1 mole of aqueous HCl.

***15.42** Starting materials for the synthesis of the herbicide propanil, a weed killer used in rice paddies, are benzene and propanoic acid. Show reagents to bring about this synthesis:

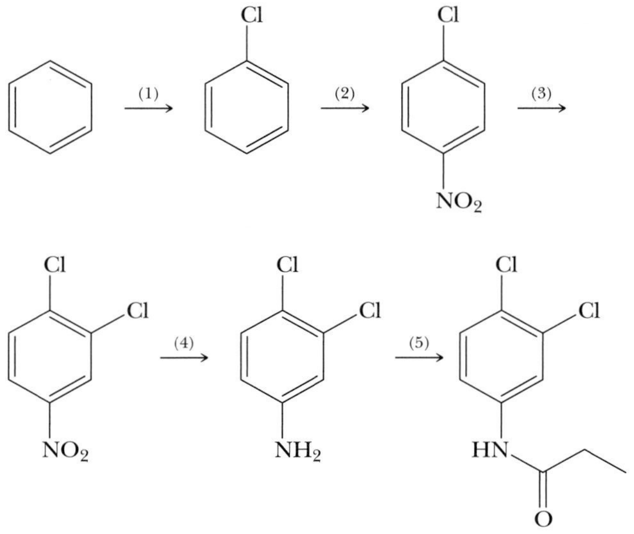

(1) (2) (3) (4) (5)

Propanil

***15.43** Following are structural formulas for three local anesthetics: Lidocaine was introduced in 1948 and is now the most widely used local anesthetic for infiltration and regional anesthesia. Its hydrochloride is marketed under the name Xylocaine®. Etidocaine (its hydrochloride is marketed as Duranest®) is comparable to lidocaine in onset, but its analgesic action lasts two to three times longer. Mepivacaine (its hydrochloride is marketed as Carbocaine®) is faster and somewhat longer in duration than lidocaine.

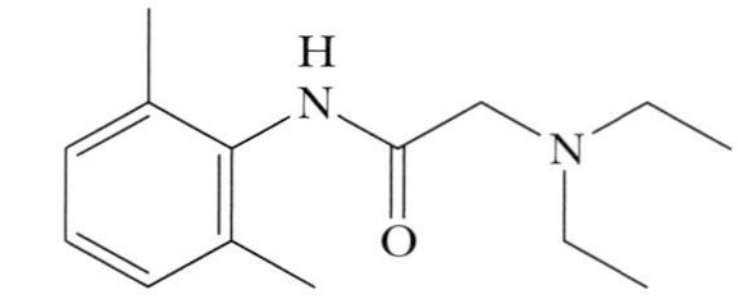

Lidocaine
(Xylocaine®)

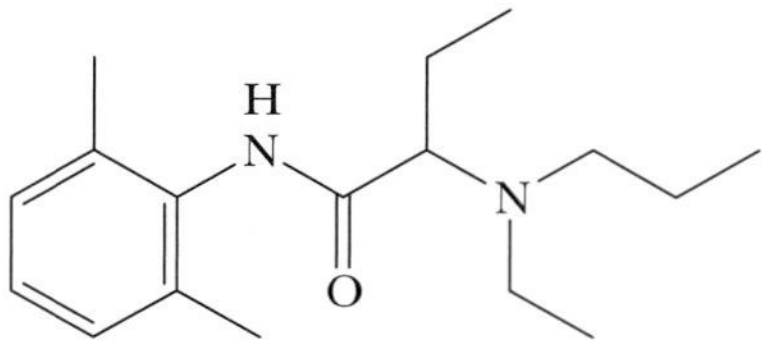

Etidocaine
(Duranest®)

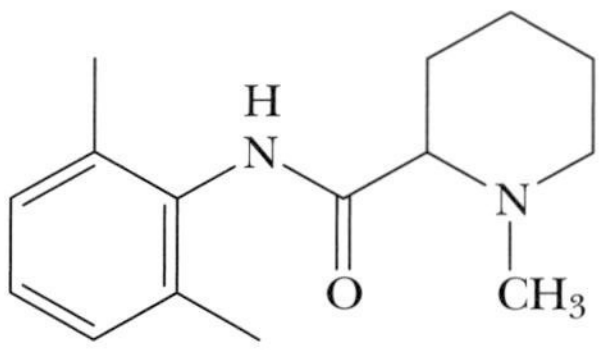

Mepivacaine
(Carbocaine®)

(a) Propose a synthesis of lidocaine from 2,6-dimethylaniline, chloroacetyl chloride ($ClCH_2COCl$), and diethylamine.

(b) Propose a synthesis of etidocaine from 2,6-dimethylaniline, 2-chlorobutanoyl chloride, and ethylpropylamine.

(c) What amine and acid chloride can be reacted to give mepivacaine?

***15.44** Following is the outline of a five-step synthesis for the anthelmintic (against worms) diethylcarbamazine:

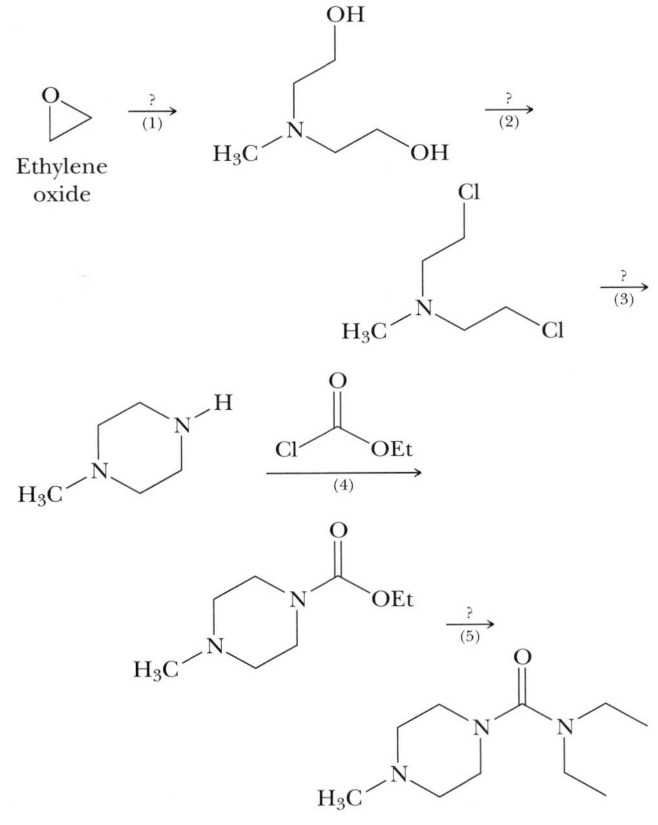

Ethylene oxide $\xrightarrow[(1)]{?}$ $H_3C-N(CH_2CH_2OH)_2$ $\xrightarrow[(2)]{?}$ $H_3C-N(CH_2CH_2Cl)_2$ $\xrightarrow[(3)]{?}$ 1-methylpiperazine (H_3C–N ring N–H) $\xrightarrow[(4)]{Cl-CO-OEt}$ ethyl 4-methylpiperazine-1-carboxylate $\xrightarrow[(5)]{?}$ Diethylcarbamazine

Diethylcarbamazine

Diethylcarbamazine is used chiefly against nematodes, small cylindrical or slender threadlike worms such as the common roundworm, which are parasitic in animals and plants.

(a) Propose a reagent for Step 1. Which mechanism is more likely for this step, S_N1 or S_N2? Explain.

(b) Propose a reagent for Step 2.

(c) Propose a reagent for Step 3.

(d) Ethyl chloroformate, the reagent for Step 4, is both an acid chloride and an ester. Account for the fact that Cl, rather than OCH_2CH_3, is displaced from this reagent.

***15.45** Following is an outline of a five-step synthesis for methylparaben, a compound widely used as a preservative in foods:

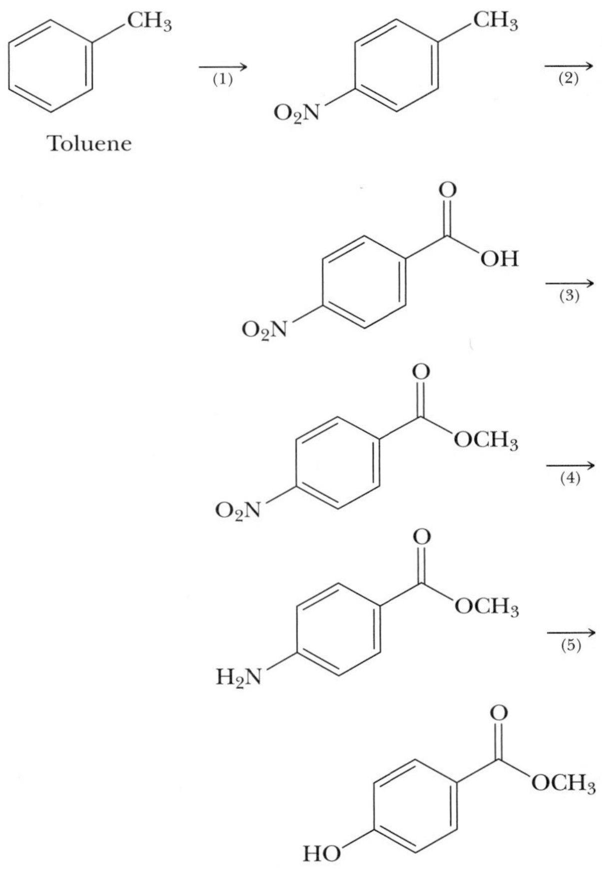

Toluene $\xrightarrow{(1)}$ 4-nitrotoluene (O_2N, CH_3) $\xrightarrow{(2)}$ 4-nitrobenzoic acid (O_2N, COOH) $\xrightarrow{(3)}$ methyl 4-nitrobenzoate (O_2N, $COOCH_3$) $\xrightarrow{(4)}$ methyl 4-aminobenzoate (H_2N, $COOCH_3$) $\xrightarrow{(5)}$ Methyl 4-hydroxybenzoate (HO, $COOCH_3$)

Methyl 4-hydroxybenzoate (Methylparaben)

Propose reagents for each step.

Chemical Transformations

15.46 Test your cumulative knowledge of the reactions learned thus far by completing the following chemical transformations. *Note*: Some will require more than one step. **(See Examples 15.7, 15.8)**

(a) 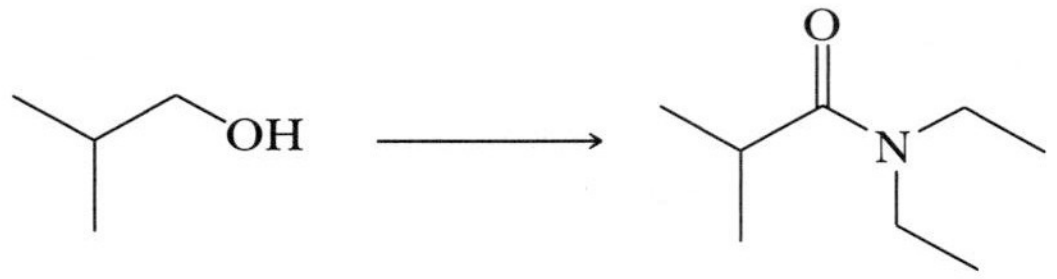 OH → O N

(b) 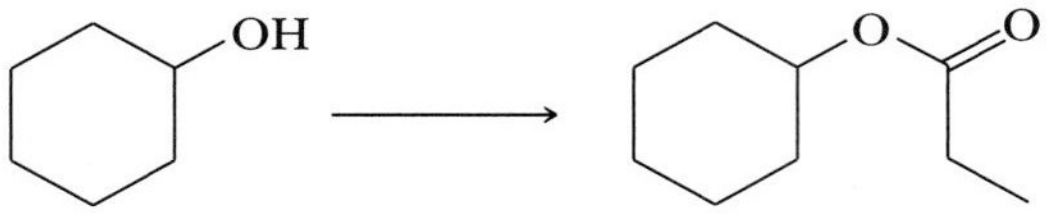 OH → O O

(c)  OH → O Cl

(d) 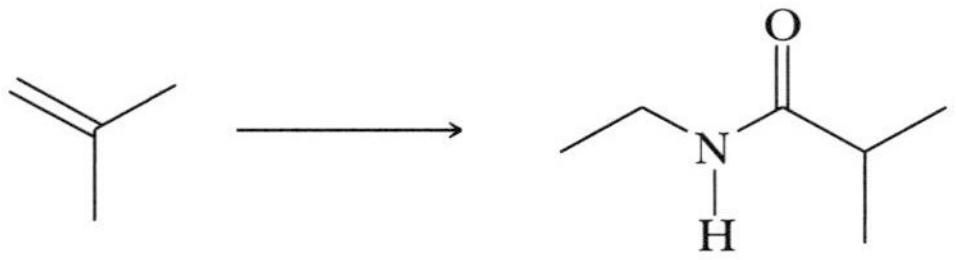 → O N H

(e) OH → O O—C

(f) → O O

(g) → O NH_2 Cl

(h) O OH → O O O

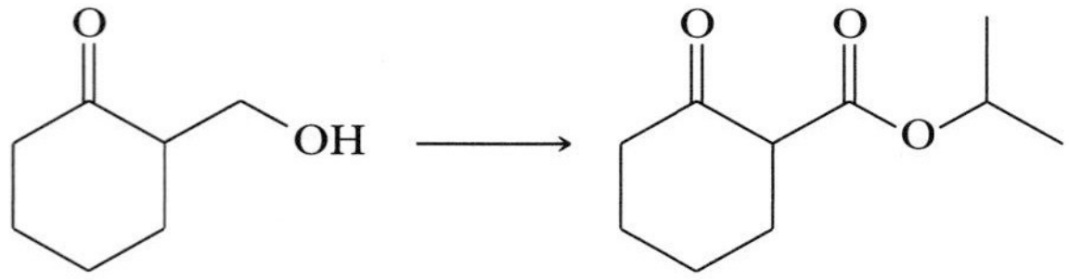

(i) 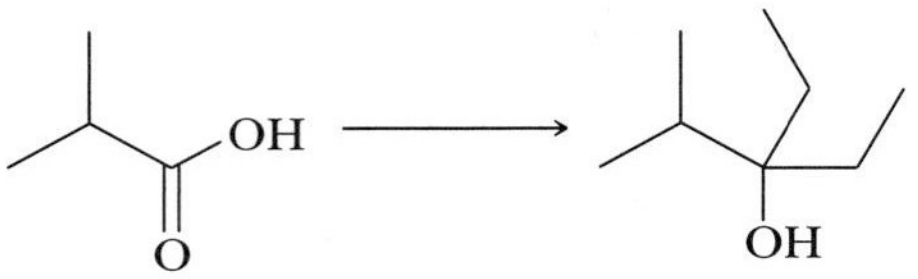 OH O → OH

(j) 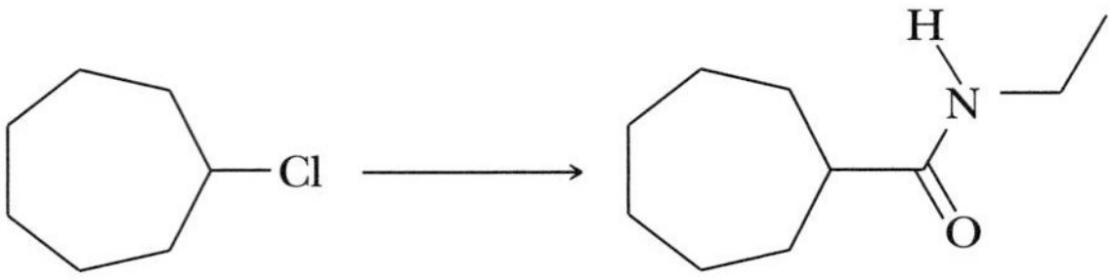 Cl → H N O

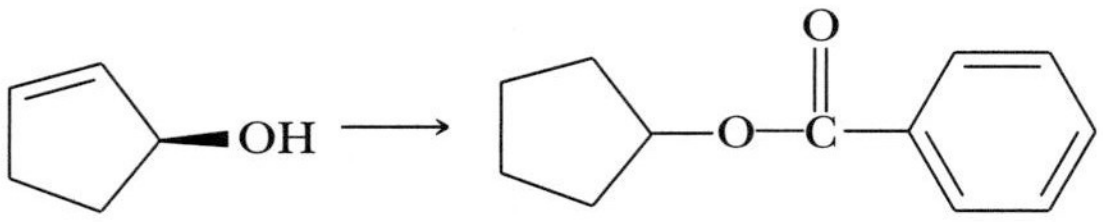

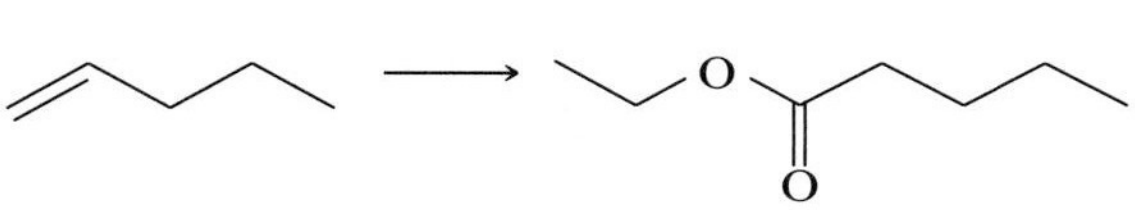

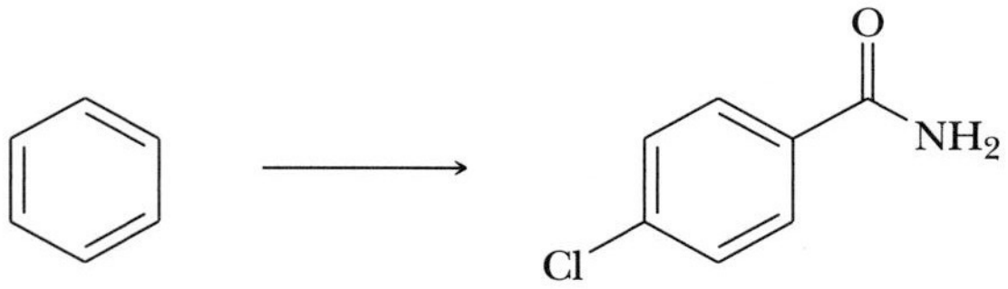

(k) → O O

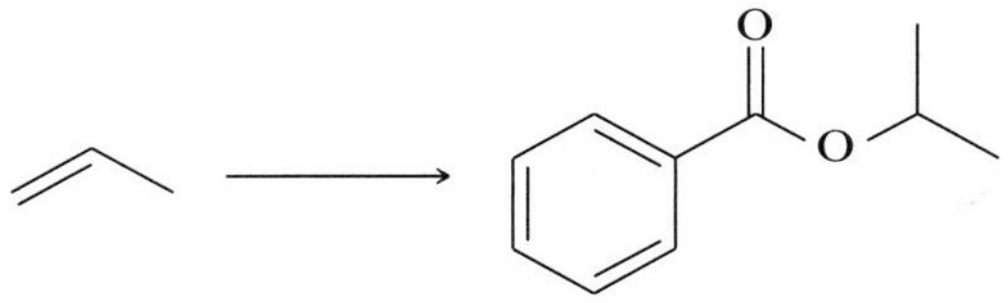

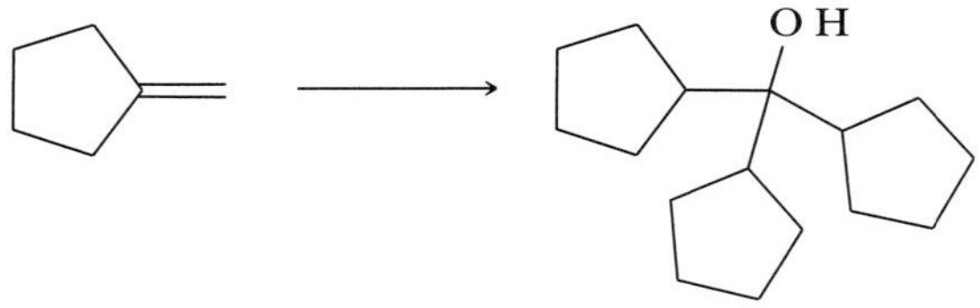

(l) → O H

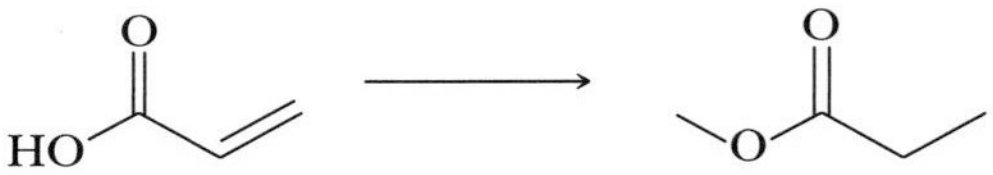

(m) O HO → O O

(n) Br → OH

(o) $H_3C—C≡C—H$ → O NH_2

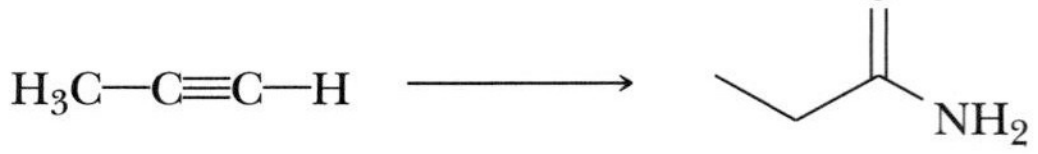

(p) 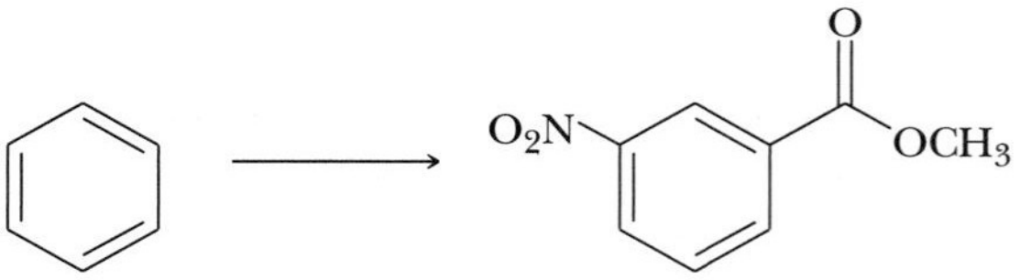 → O_2N O OCH_3

Looking Ahead

15.47 Identify the most acidic proton in each of the following esters:

(a) 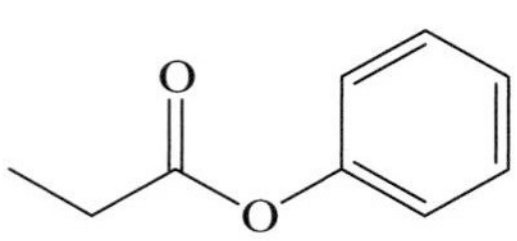 O O (b) 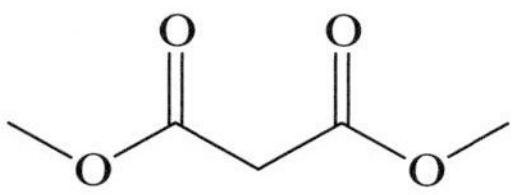 O O O O

15.48 Does a nucleophilic acyl substitution occur between the ester and the nucleophile shown?

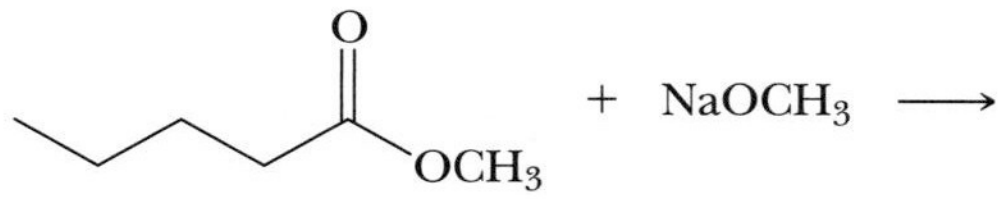

O OCH_3 + $NaOCH_3$ →

Propose an experiment that would verify your answer.

15.49 Explain why a nucleophile, Nu, attacks not only the carbonyl carbon, but also the β-carbon, as indicated in the following α,β-unsaturated ester:

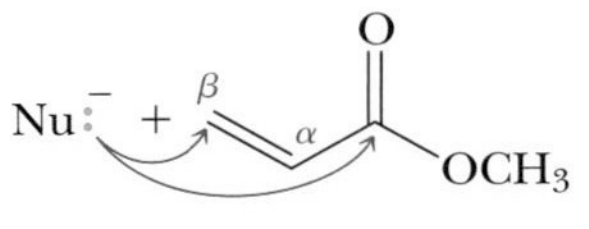

15.50 Explain why a Grignard reagent will not undergo nucleophilic acyl substitution with the following amide:

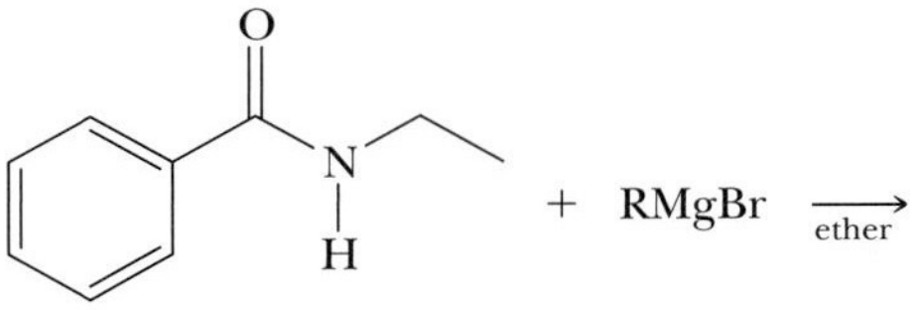

15.51 At low temperatures, the following amide exhibits *cis–trans* isomerism, while at higher temperatures it does not:

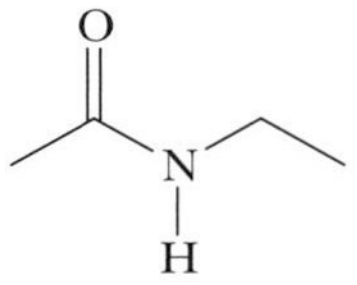

Explain how this is possible.

Putting It Together

The following problems bring together concepts and material from Chapters 13–15. Although the focus may be on these chapters, the problems will also build upon concepts discussed throughout the text thus far.

Choose the best answer for each of the following questions.

1. Which of the following statements is true concerning the following two carboxylic acid derivatives?

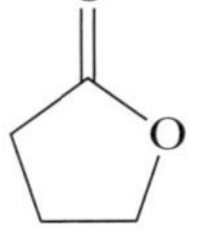

(a) Only molecule **A** can be hydrolyzed.

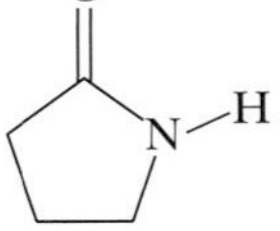

(b) Only molecule **B** can be hydrolyzed.

(c) Both molecules are hydrolyzable, but **A** will react more quickly than **B**.

(d) Both molecules are hydrolyzable, but **B** will react more quickly than **A**.

(e) **A** and **B** are hydrolyzed at roughly the same rate.

2. How many unique reaction products are formed from the following reaction?

$$CH_3CH_2-\overset{O}{\overset{\|}{C}}-CH_2CH_2-\overset{O}{\overset{\|}{C}}-OCH_3 \xrightarrow[H_2O]{H^+}$$

(a) one (b) two (c) three (d) four (e) five

3. What sequence of reagents will accomplish the following transformation?

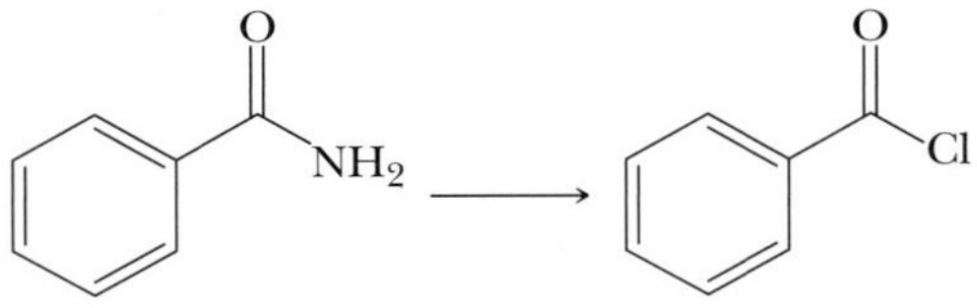

(a) 1. $SOCl_2$

(b) 1. H_2O_2 2. $SOCl_2$

(c) 1. H_2O_2 2. HCl

(d) 1. H^+/H_2O_2 2. $SOCl_2$

(e) All of the above

4. Which of the following reactions will not yield butanamide?

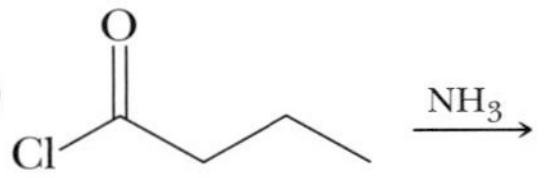

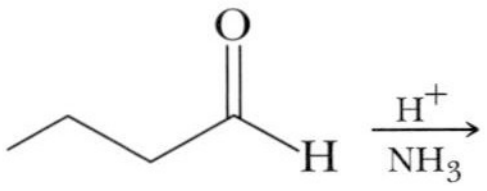

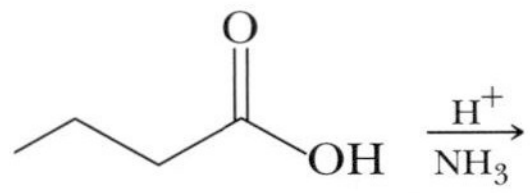

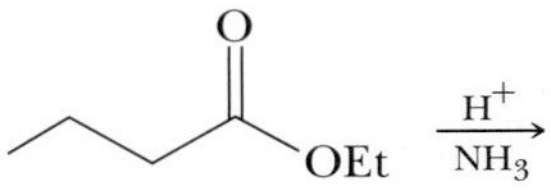

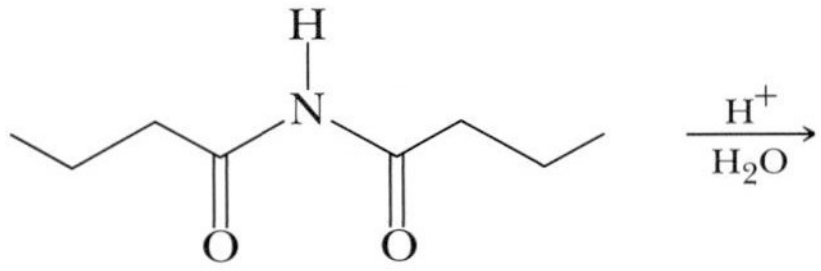

5. Which of the following is the tetrahedral carbonyl addition intermediate (TCAI) for the Fischer esterification of ethanol and benzoic acid?

(a) C_6H_5–C(OH)(OH)(OCH_2CH_3)
(b) C_6H_5–C(OH)(OH)(CH_2CH_3)
(c) C_6H_5–C(OCH_2CH_3)(OCH_2CH_3)(OCH_2CH_3)
(d) C_6H_5–C(OH)(OCH_2CH_3)(OCH_2CH_3)
(e) HO–C(OH)(OCH_2CH_3)(OCH_2CH_3)

6. Which of the following is the enol intermediate in the decarboxylation of ethylpropanedioic acid?

(a) OH, H, H
(b) H, H, OH
(c) OH, HO, H
(d) OH, HO, OH
(e) OH, H, OH

7. Which of the following statements is true concerning the two carboxylic acids shown?

A (3-hydroxybenzoic acid) B (4-hydroxybenzoic acid)

(a) **A** is more acidic than **B** because of an additional resonance effect.
(b) **B** is more acidic than **A** because of an additional resonance effect.
(c) Only the conjugate base of **A** experiences an inductive effect.
(d) Only the conjugate base of **B** experiences an inductive effect.
(e) None of the above.

8. What would be the expected outcome if one equivalent of a Grignard reagent were reacted with the molecule below?

H–C(=O)(A)–$CH_2CH_2CH_2$–C(=O)(B)–CH_3

(a) 100% addition at carbonyl **A**.
(b) 100% addition at carbonyl **B**.
(c) Equal addition at both carbonyls.
(d) Greater distribution of addition at **A**.
(e) Greater distribution of addition at **B**.

9. Which of the following carbonyl carbons would be considered the most electrophilic?

(a) (b) Cl (c) Cl, Cl (d) Cl, Cl, Cl

(e) All are equally electrophilic

10. The following reaction will occur as shown:

HO–(ketone) $\xrightarrow[\text{2) HCl/H}_2\text{O}]{\text{1) CH}_3\text{MgBr/Et}_2\text{O}}$ HO–(tertiary alcohol with HO, CH_3)

(a) True
(b) False

11. Provide a structure for the starting compound needed to produce the product shown. Then show the mechanism of its formation. Show all charges and lone pairs of electrons in your structures.

$\underset{H_2O}{\overset{H^+}{\rightleftharpoons}}$

12. Rank the following from most to least reactive with EtOH. Provide a rationale for your ranking.

(a) Cl, Cl (b) Cl, Cl (c) Cl, F (d) Cl, F (e) Cl

13. Provide IUPAC names for the following compounds.

(a) 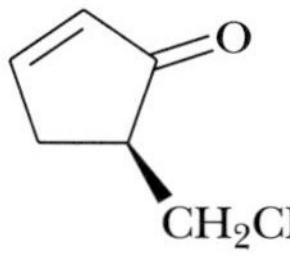
O
CH2CH3

(b) 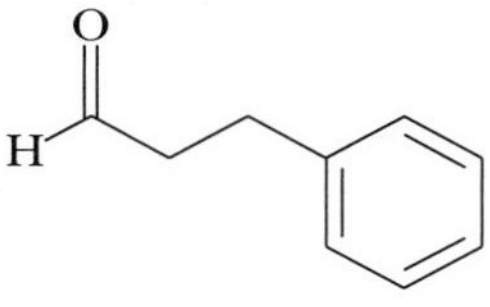
O
H

(c) 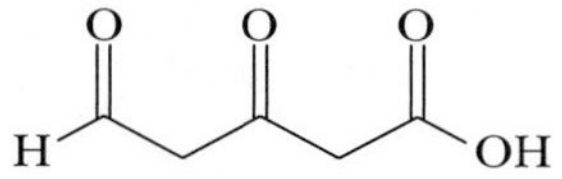
O
O
O
H
OH

14. Provide a mechanism for the following reaction. Show all charges and lone pairs of electrons in your structures, as well as the structures of all intermediates.

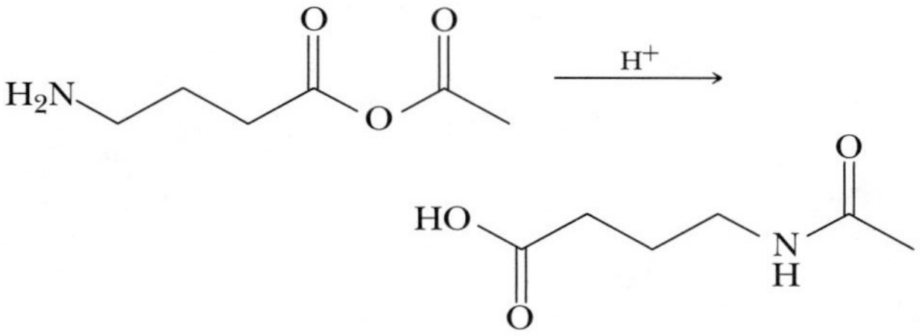
H2N
O
O
O
H+
HO
N
H
O
O

15. Predict the major product of each of the following reactions.

(a) 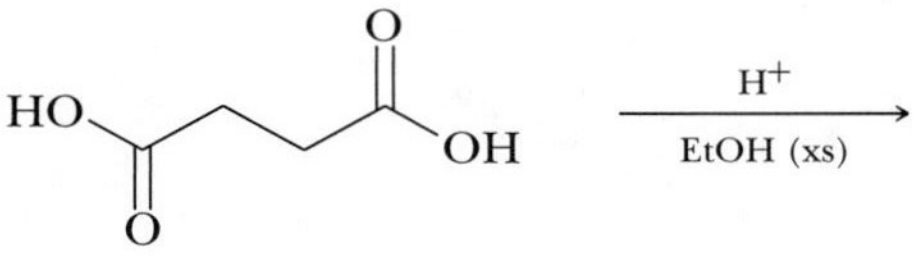
HO
O
O
OH
H+
EtOH (xs)

(b) 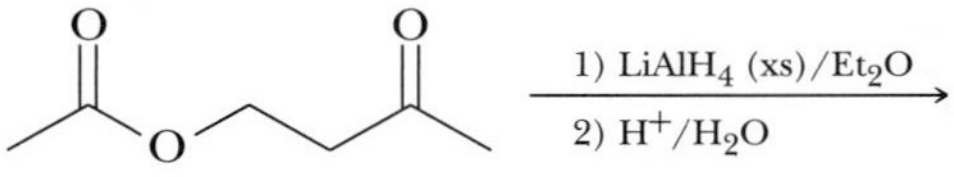
O
O
O
1) LiAlH4 (xs)/Et2O
2) H+/H2O

(c) 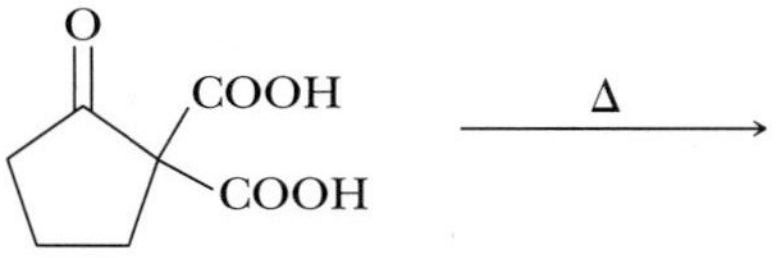
O
COOH
COOH
Δ

(d) 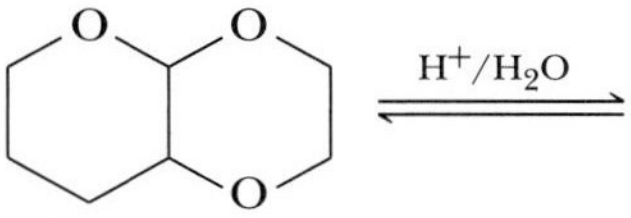
O
O
O
H+/H2O

(e) 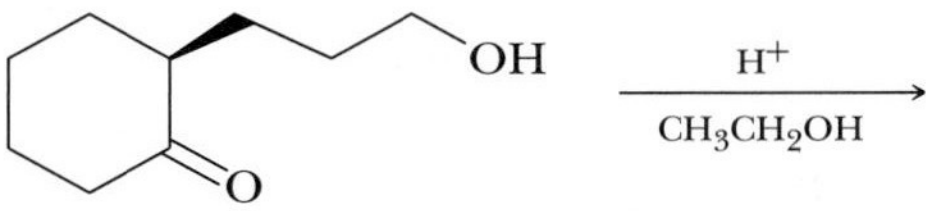
OH
O
H+
CH3CH2OH

16. Complete the following chemical transformations.

(a) 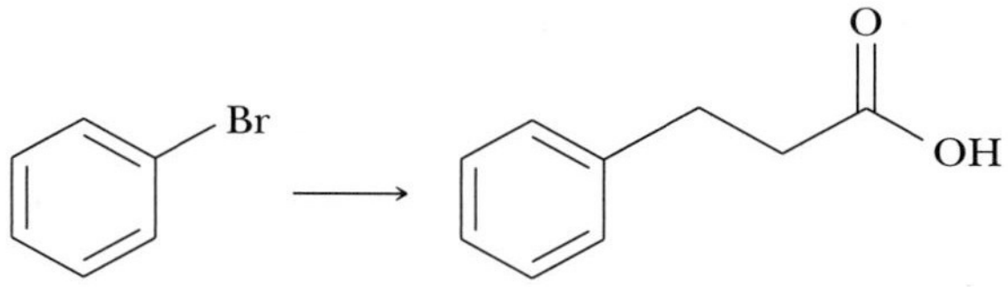
Br
O
OH

(b) 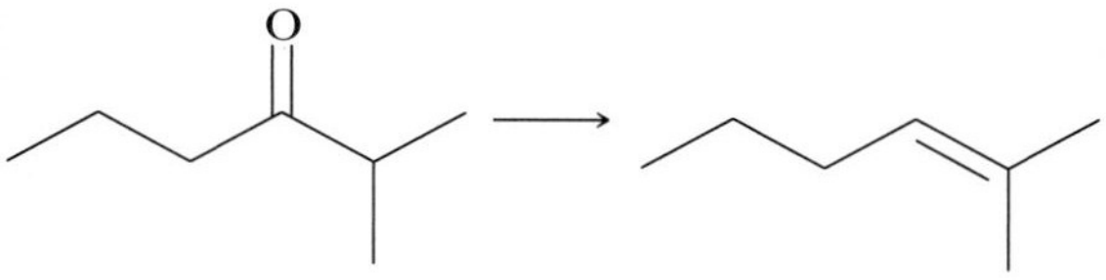
O

(c) 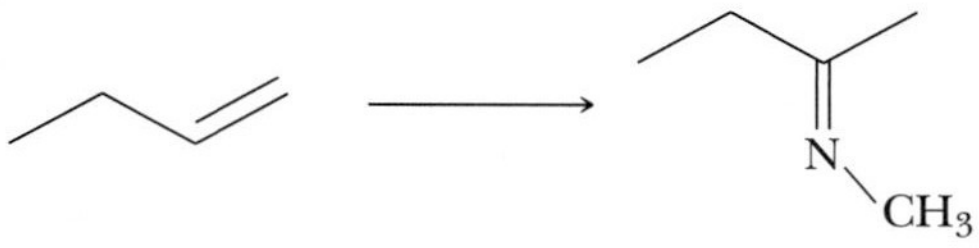
N
CH3

(d) 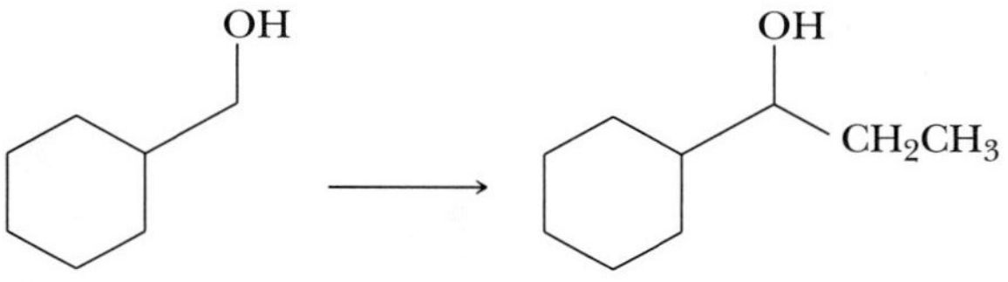
OH
OH
CH2CH3

(e) 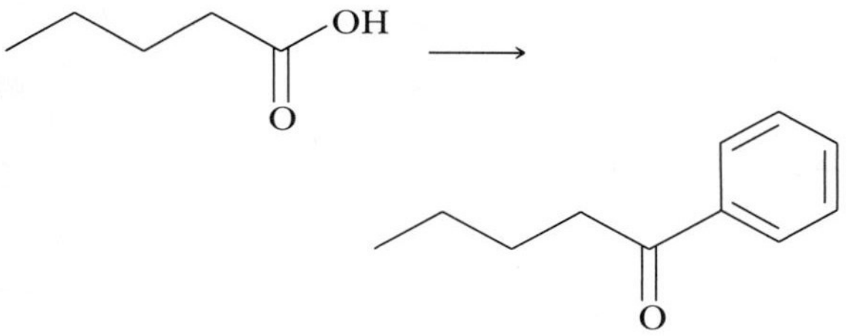
OH
O
O

(f) 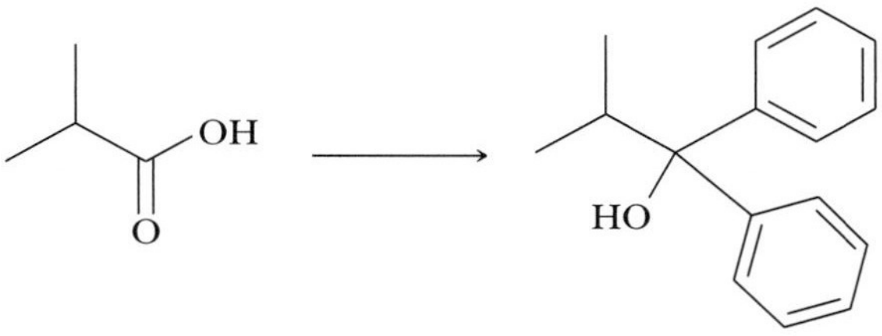
OH
O
HO

(g) 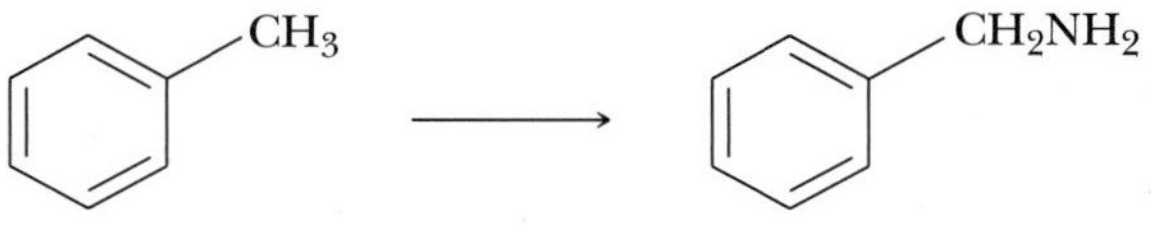
CH3
CH2NH2

(h) 
Br
O
Cl

16 Enolate Anions

KEY QUESTIONS

- 16.1 What Are Enolate Anions, and How Are They Formed?
- 16.2 What Is the Aldol Reaction?
- 16.3 What Are the Claisen and Dieckmann Condensations?
- 16.4 How Are Aldol Reactions and Claisen Condensations Involved in Biological Processes?
- 16.5 What Is the Michael Reaction?

HOW TO

- 16.1 How to Determine the Starting Compounds Used in an Aldol Reaction
- 16.2 How to Determine the Starting Compounds Used in a Claisen Condensation
- 16.3 How to Determine the Starting Compounds Used in a Michael Reaction
- 16.4 How to Recognize the Products of Aldol, Claisen Condensation, and Michael Reactions

CHEMICAL CONNECTIONS

- 16A Drugs That Lower Plasma Levels of Cholesterol
- 16B Antitumor Compounds: The Michael Reaction in Nature

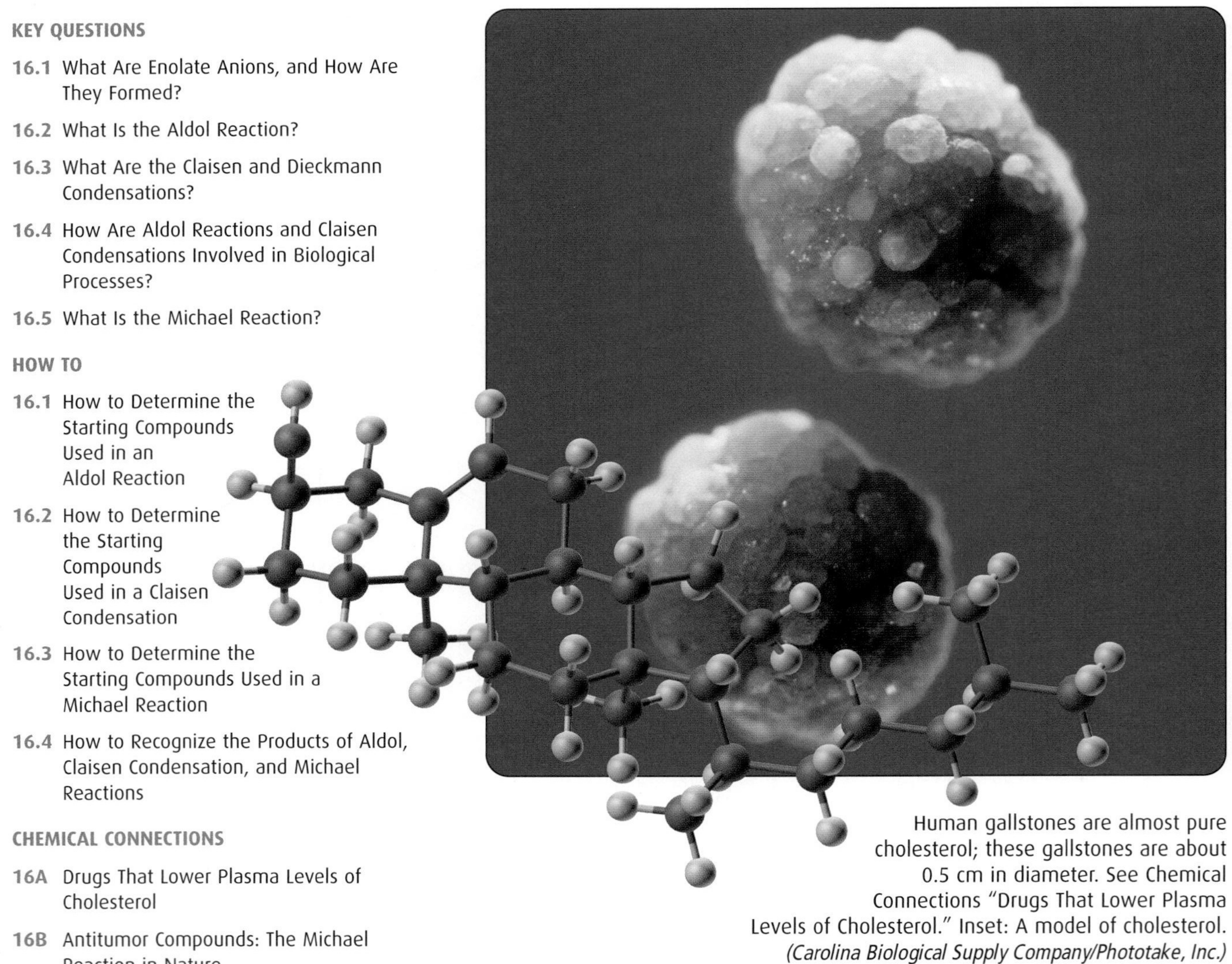

Human gallstones are almost pure cholesterol; these gallstones are about 0.5 cm in diameter. See Chemical Connections "Drugs That Lower Plasma Levels of Cholesterol." Inset: A model of cholesterol. *(Carolina Biological Supply Company/Phototake, Inc.)*

In this chapter, we continue our discussion of the chemistry of carbonyl compounds. In Chapters 13–15, we concentrated on the carbonyl group itself and on nucleophilic additions to it to form tetrahedral carbonyl addition compounds. In the current chapter, we expand on the chemistry of carbonyl-containing compounds and consider the acidity of α-hydrogens and the enolate anions formed by their removal. The reactions presented here represent some of the most important reactions in organic chemistry because they involve the formation of carbon–carbon bonds, allowing the construction of larger molecules from smaller, readily available starting materials.

16.1 What Are Enolates Anions, and How Are They Formed?

A. Acidity of α-Hydrogens

A carbon atom adjacent to a carbonyl group is called an ***α*-carbon**, and a hydrogen atom bonded to it is called an ***α*-hydrogen**:

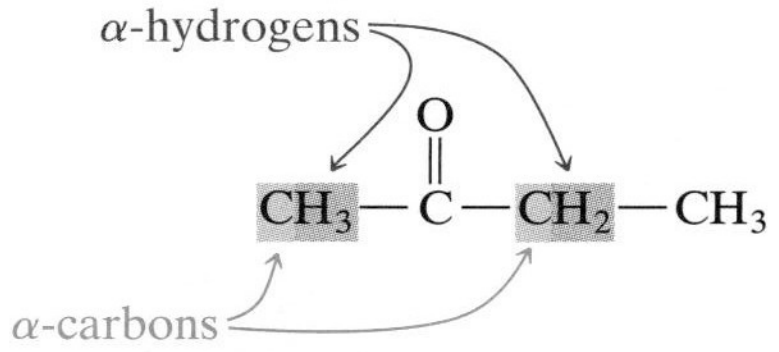

α-hydrogens

$$CH_3-\overset{O}{\overset{\|}{C}}-CH_2-CH_3$$

α-carbons

Because carbon and hydrogen have comparable electronegativities, a C—H bond normally has little polarity, and a hydrogen atom bonded to carbon shows very low acidity (Section 2.3). The situation is different, however, for hydrogens that are alpha to a carbonyl group. As Table 16.1 shows, α-hydrogens of aldehydes, ketones,

TABLE 16.1 The Acidity of Hydrogens Alpha to a Carbonyl Group Relative to Other Organic Hydrogens

Class of Compound	Example	pK_a
a β-diketone	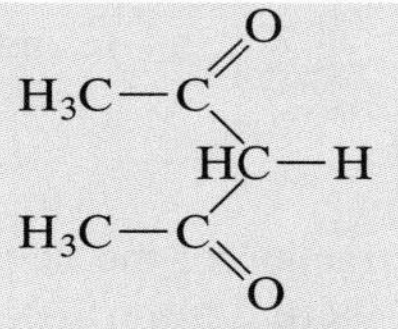 $H_3C-C(=O)-CH(H)-C(=O)-CH_3$	9.5
a phenol	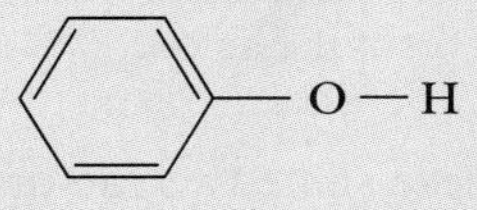 C_6H_5-O-H	10
a β-ketoester (an acetoacetic ester)	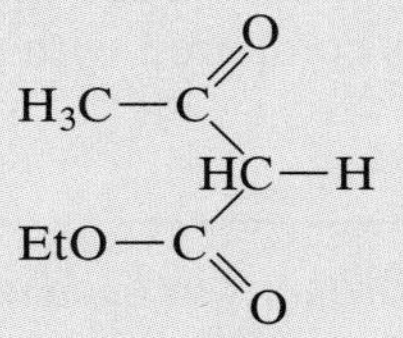 $H_3C-C(=O)-CH(H)-C(=O)-OEt$	10.7
a β-diester (a malonic ester)	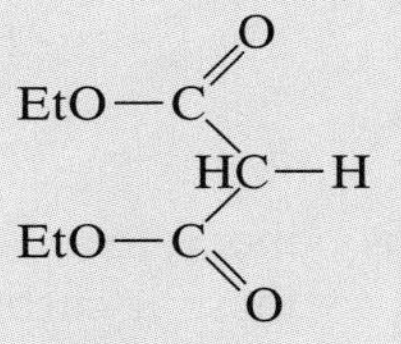 $EtO-C(=O)-CH(H)-C(=O)-OEt$	13
water	$HO-H$	15.7
an alcohol	CH_3CH_2O-H	16
an aldehyde or a ketone	$CH_3\overset{O}{\overset{\|}{C}}CH_2-H$	20
an ester	$EtO\overset{O}{\overset{\|}{C}}CH_2-H$	22
a terminal alkyne	$R-C\equiv C-H$	25
an alkene	$CH_2=CH-H$	44
an alkane	CH_3CH_2-H	51

Increasing acidity (arrow pointing up, from alkane to β-diketone)

and esters are considerably more acidic than alkane and alkene hydrogens, but less acidic than the hydroxyl hydrogen of alcohols. The table also shows that hydrogens that are alpha to two carbonyl groups—for example, in a β-ketoester and a β-diester,—are even more acidic than alcohols.

B. Enolate Anions

Enolate anion An anion formed by the removal of an α-hydrogen from a carbonyl-containing compound.

Carbonyl groups increase the acidity of their alpha hydrogens in two ways. First, the electron-withdrawing inductive effect of the carbonyl group weakens the bond to the alpha hydrogen and promotes its ionization. Second, the negative charge on the resulting **enolate anion** is delocalized by resonance, thus stabilizing it relative to the anion from an alkane or an alkene:

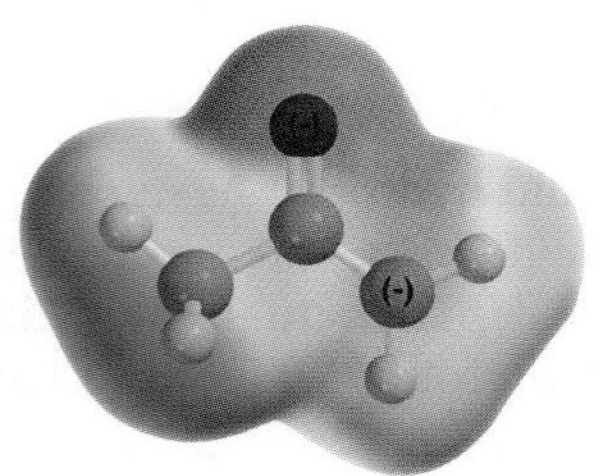

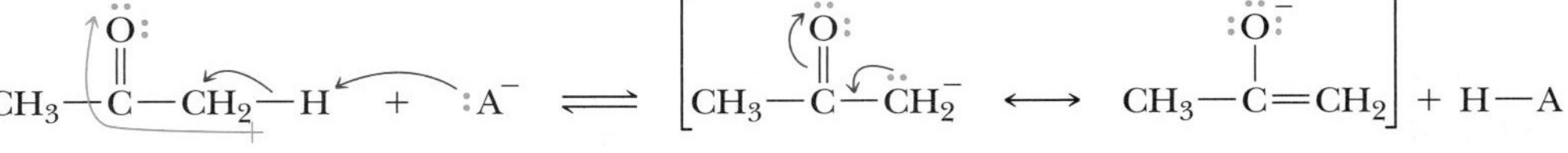

the electron-withdrawing inductive effect of the carbonyl weakens the C—H bond

resonance stabilizes the enolate anion

Although the major fraction of the negative charge of an enolate anion is on the carbonyl oxygen, there is nevertheless a significant partial negative charge on the alpha carbon of the anion.

Recall that we used these same two factors in Section 2.5 to account for the greater acidity of carboxylic acids compared with alcohols.

Example 16.1

Identify the acidic α-hydrogens in each compound:

(a) Butanal (b) 2-Butanone

Strategy

Identify all carbon atoms bonded to a carbonyl group. These are the α-carbons. Any hydrogens bonded to these α-carbon atoms are α-hydrogens and will be more acidic than typical alkane or alkene hydrogens.

Solution

Butanal (a) has one set of acidic α-hydrogens, and 2-butanone (b) has two sets:

(a) $CH_3CH_2CH_2CH(=O)$ (b) $CH_3CH_2C(=O)CH_3$

See problem 16.13

Problem 16.1

Identify the acidic hydrogens in each compound:

(a) 2-methylcyclohexanone (b) Acetophenone

(c) 2,2-diethylcyclohexanone

Enolate anions can be formed either quantitatively or under equilibrium conditions. Use of a base that is much stronger than the ensuing enolate anion results in the quantitative removal of an α-hydrogen.

$$H_3C-\overset{O}{\overset{\|}{C}}-CH_3 + NaNH_2 \longrightarrow H_3C-\overset{O}{\overset{\|}{C}}-\ddot{C}H_2^- + Na^+ + NH_3$$

pK_a 20 (stronger acid)	Sodium amide (stronger base)	Enolate (weaker base)	pK_a 38 (weaker acid)

the left side of each equation contains a much stronger acid and base. For this reason both reactions proceed quantitatively

2,4-pentanedione + NaOH ⟶ enolate + Na^+ + HOH

pK_a 9 (stronger acid)	Sodium hydroxide (stronger base)	Enolate (weaker base)	pK_a 15.7 (weaker acid)

Use of a base that is weaker than the ensuing enolate anion results in an equilibrium in which the enolate exists in very small concentrations.

the stronger acid and base reside on the right side of the equation, causing the equilibrium to favor the reactants

$$H_3C-\overset{O}{\overset{\|}{C}}-CH_3 + NaOH \rightleftharpoons H_3C-\overset{O}{\overset{\|}{C}}-\ddot{C}H_2^- + Na^+ + HOH$$

pK_a 20	Sodium hydroxide	Enolate	pK_a 15.7

C. The Use of Enolate Anions to Form New C—C Bonds

Enolate anions are important building blocks in organic synthesis, and we will study their use as nucleophiles to form new carbon–carbon bonds. In overview, they participate in three types of nucleophilic reactions.

Enolate anions function as nucleophiles in carbonyl addition reactions:

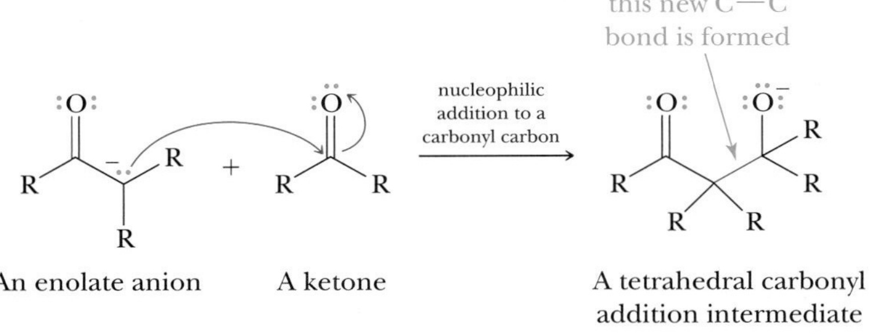

This type of enolate anion reaction is particularly useful among reactions of aldehydes and ketones in the aldol reaction (Section 16.2).

Enolate anions function as nucleophiles in nucleophilic acyl substitution reactions:

this new C—C bond is formed

RO–C(=O)–CH(⁻)–R′ + R–C(=O)–OR ⟶ [tetrahedral intermediate] ⟶ RO–C(=O)–CH(R′)–C(=O)–R + $^{-}$:OR

An enolate anion | An ester | A tetrahedral carbonyl addition intermediate | Product of nucleophilic acyl substitution

This type of enolate anion reaction occurs among esters in the Claisen (Section 16.3A) and Dieckmann condensations (Section 16.3B).

Enolate anions undergo nucleophilic addition to a carbon–carbon double bond if the double bond is conjugated with the carbonyl group of an aldehyde, a ketone, or an ester:

this new C—C bond is formed

$$\text{EtOOC–}\overset{-}{\text{C}}\text{H–COOEt} + \text{CH}_2\text{=CH–C(=O)–CH}_3 \xrightarrow[\text{EtOH}]{\text{EtO}^-\text{Na}^+} (\text{EtOOC})_2\text{CH–CH}_2\text{–CH=C(O}^-\text{)–CH}_3$$

Enolate anion of diethyl propanedioate (Diethyl malonate) | 3-Buten-2-one (Methyl vinyl ketone)

This type of enolate anion reaction is called the Michael reaction (Section 16.5).

16.2 What Is the Aldol Reaction?

A. Formation of Enolate Anions of Aldehydes and Ketones

Treatment of an aldehyde or a ketone containing an acidic α-hydrogen with a strong base, such as sodium hydroxide or sodium ethoxide, gives an enolate anion as a hybrid of two major contributing structures:

$$\underset{\substack{\text{p}K_a\ 20\\ \text{(weaker acid)}}}{\text{CH}_3\overset{\text{O}}{\overset{\|}{\text{C}}}\text{CH}_3} + \text{NaOH} \rightleftharpoons \underset{\text{An enolate anion}}{\left[\text{H–}\overset{-}{\text{C}}\text{H–}\overset{\text{O}}{\overset{\|}{\text{C}}}\text{–CH}_3 \longleftrightarrow \text{H–CH=}\overset{\text{O}^-}{\overset{|}{\text{C}}}\text{CH}_3\right]}\text{Na}^+ + \underset{\substack{\text{p}K_a\ 15.7\\ \text{(stronger acid)}}}{\text{H}_2\text{O}}$$

Given the relative acidities of the two acids in this equilibrium, the position of equilibrium lies considerably to the left. However, the existence of just a small amount of enolate anion is enough to allow the aldol reaction to proceed.

B. The Aldol Reaction

Addition of the enolate anion derived from an aldehyde or a ketone to the carbonyl group of another aldehyde or ketone is illustrated by these examples:

$$\underset{\substack{\text{Ethanal}\\\text{(Acetaldehyde)}}}{CH_3-\overset{O}{\overset{\|}{C}}-H} + \underset{\substack{\text{Ethanal}\\\text{(Acetaldehyde)}}}{\overset{H}{\overset{|}{C}}H_2-\overset{O}{\overset{\|}{C}}-H} \underset{}{\overset{NaOH}{\rightleftharpoons}} \underset{\substack{\text{3-Hydroxybutanal}\\\text{(a }\beta\text{-hydroxyaldehyde)}}}{CH_3-\overset{OH}{\overset{|}{\underset{\beta}{C}}}H-\underset{\alpha}{C}H_2-\overset{O}{\overset{\|}{C}}-H}$$

$$\underset{\substack{\text{Propanone}\\\text{(Acetone)}}}{CH_3-\overset{O}{\overset{\|}{C}}-CH_3} + \underset{\substack{\text{Propanone}\\\text{(Acetone)}}}{\overset{H}{\overset{|}{C}}H_2-\overset{O}{\overset{\|}{C}}-CH_3} \overset{NaOH}{\rightleftharpoons} \underset{\substack{\text{4-Hydroxy-4-methyl-2-pentanone}\\\text{(a }\beta\text{-hydroxyketone)}}}{CH_3-\overset{OH}{\overset{|}{\underset{\underset{CH_3}{|}}{C}}}-\underset{\alpha}{C}H_2-\overset{O}{\overset{\|}{C}}-CH_3}$$

The common name of the product derived from the reaction of acetaldehyde in base is **aldol**, so named because it is both an **ald**ehyde and an alcoh**ol**. *Aldol* is also the generic name given to any product formed in this type of reaction. The functional group of the product of an **aldol reaction** is a β-hydroxyaldehyde or a β-hydroxyketone.

Aldol reaction A carbonyl condensation reaction between two aldehydes or ketones to give a β-hydroxyaldehyde or a β-hydroxyketone.

The key step in a base-catalyzed aldol reaction is nucleophilic addition of the enolate anion from one carbonyl-containing molecule to the carbonyl group of another carbonyl-containing molecule to form a tetrahedral carbonyl addition intermediate. This mechanism is illustrated by the aldol reaction between two molecules of acetaldehyde. Notice that OH^- is a true catalyst: An OH^- is used in Step 1, but another OH^- is generated in Step 3. Notice also the parallel between Step 2 of the aldol reaction and the reaction of Grignard reagents with aldehydes and ketones (Section 13.5) and the first step of their reaction with esters (Section 15.7). Each type of reaction involves the addition of a carbon nucleophile to the carbonyl carbon of another molecule.

Mechanism: Base-Catalyzed Aldol Reaction

Step 1: The removal of an α-hydrogen by base gives a resonance-stabilized enolate anion:

$$H-\ddot{\underset{..}{O}}:^- + H-CH_2-\overset{\ddot{O}:}{\overset{\|}{C}}-H \rightleftharpoons H-\ddot{\underset{..}{O}}-H + \left[\underset{\text{An enolate anion}}{{}^-:CH_2-\overset{\ddot{O}:}{\overset{\|}{C}}-H \longleftrightarrow CH_2=\overset{:\ddot{O}:^-}{\overset{|}{C}}-H} \right]$$

Step 2: Because the equilibrium favors the left in Step 1, there is plenty of unreacted aldehyde (or ketone) remaining in the reaction mixture. Nucleophilic addition of the enolate anion to the carbonyl carbon of an unreacted molecule of aldehyde (or ketone) gives a tetrahedral carbonyl addition intermediate:

$$CH_3-\overset{\ddot{O}:}{\overset{\|}{C}}-H + {}^-:CH_2-\overset{:O:}{\overset{\|}{C}}-H \rightleftharpoons \underset{\substack{\text{A tetrahedral carbonyl}\\\text{addition intermediate}}}{CH_3-\overset{:\ddot{O}:^-}{\overset{|}{C}}H-CH_2-\overset{:O:}{\overset{\|}{C}}-H}$$

Step 3: Reaction of the tetrahedral carbonyl addition intermediate with a proton donor gives the aldol product and generates another hydroxide ion:

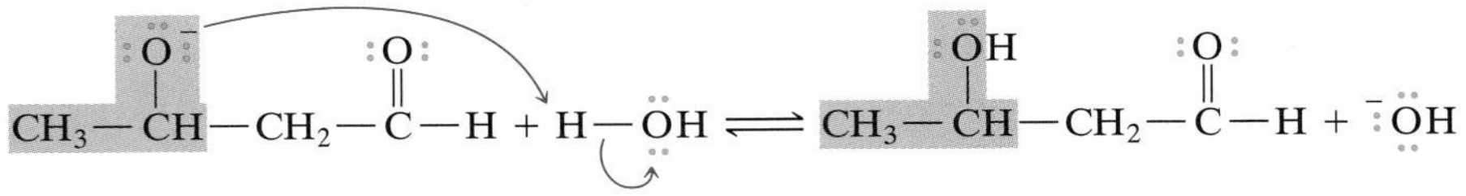

$$CH_3-\underset{\underset{O^-}{|}}{CH}-CH_2-\overset{\overset{O}{\|}}{C}-H + H-OH \rightleftharpoons CH_3-\underset{\underset{OH}{|}}{CH}-CH_2-\overset{\overset{O}{\|}}{C}-H + {}^-OH$$

Example 16.2

Draw the product of the base-catalyzed aldol reaction of each compound:

(a) Butanal (b) Cyclohexanone

Strategy

Draw two molecules of each ketone or aldehyde. Convert one of the molecules to an enolate anion and show it adding to the carbonyl carbon of the other. Be sure to show the regeneration of base accompanied by the formation of the β-hydroxycarbonyl. It often helps to number the atoms in the enolate anion and the ketone or aldehyde.

Solution

The aldol product is formed by the nucleophilic addition of the α-carbon of one compound to the carbonyl carbon of another:

draw two molecules of the aldehyde and show one of them as an enolate anion

(a) [butanal + enolate anion of butanal ⇌ 2-ethyl-3-hydroxyhexanal] + OH^-

new C—C bond

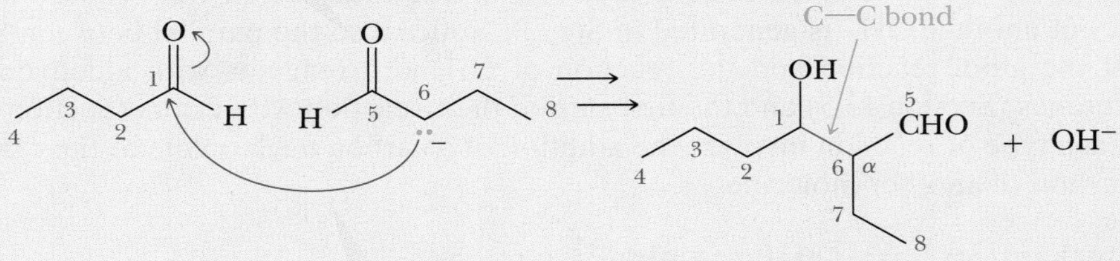

the enolate anion of one molecule attacks the carbonyl carbon of the unreacted aldehyde; a new bond is formed between carbons 1 and 6

draw two molecules of the ketone and show one of them as an enolate anion

(b) [cyclohexanone + enolate anion of cyclohexanone ⇌ aldol product] + OH^-

new C—C bond

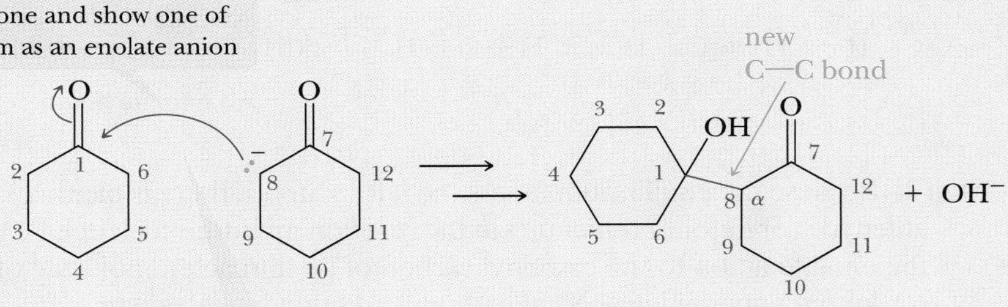

the enolate anion of one molecule attacks the carbonyl carbon of the unreacted ketone; a new bond is formed between carbons 1 and 8

See problems 16.17, 16.20, 16.21

Problem 16.2

Draw the product of the base-catalyzed aldol reaction of each compound:

(a) Acetophenone (b) Cyclopentanone

β-Hydroxyaldehydes and β-hydroxyketones are very easily dehydrated, and often the conditions necessary to bring about an aldol reaction are sufficient to cause dehydration (Section 8.2E). Dehydration can also be brought about by warming the aldol product in dilute acid. The major product from the dehydration of an aldol product is one in which the carbon–carbon double bond is conjugated with the carbonyl group; that is, the product is an **α,β-unsaturated aldehyde or ketone** (so named because the site of unsaturation, the double bond, is between the α and β carbons).

$$\underset{\text{Product of an aldol reaction}}{CH_3CH(OH)CH_2CH{=}O} \xrightarrow{\text{warm in either acid or base}} \underset{\text{An }\alpha,\beta\text{-unsaturated aldehyde}}{CH_3\overset{\beta}{C}H{=}\overset{\alpha}{C}HCH{=}O} + H_2O$$

Base-catalyzed aldol reactions are readily reversible, and generally little aldol product is present at equilibrium. Equilibrium constants for dehydration, however, are usually large, so that, if reaction conditions are sufficiently vigorous to bring about dehydration, good yields of product can be obtained.

Mechanism: Base-Catalyzed Dehydration of an Aldol Product

We can write the following two-step mechanism for the base-catalyzed dehydration of an aldol product:

Step 1: An acid–base reaction removes an α-hydrogen to give an enolate anion.

Step 2: The enolate anion ejects hydroxide ion, regenerating the base and giving the α,β-unsaturated carbonyl compound.

$$CH_3-CH(OH)-CH(H)-CH{=}O + HO^- \xrightarrow{(1)} \underset{\text{An enolate anion}}{CH_3-CH(OH)-CH{=}CH-O^-} \xrightarrow[(2)]{-OH^-} \underset{\text{An }\alpha,\beta\text{-unsaturated aldehyde}}{CH_3-\overset{\beta}{C}H{=}\overset{\alpha}{C}H-CH{=}O}$$

Mechanism: Acid-Catalyzed Dehydration of an Aldol Product

The acid-catalyzed dehydration of an aldol product also takes place in two steps:

Step 1: An acid–base reaction protonates the β-hydroxyl group.

Step 2: Water acts as a base to abstract an α-hydrogen and eject H_2O as a leaving group, giving the α,β-unsaturated carbonyl compound and regenerating the acid.

$$CH_3-CH(OH)-CH(H)-CH{=}O + H_3O^+ \xrightarrow{(1)} CH_3-CH(\overset{+}{O}H_2)-CH(H)-CH{=}O + HOH \xrightarrow[(2)]{-H_2O} \underset{\text{An }\alpha,\beta\text{-unsaturated aldehyde}}{CH_3-\overset{\beta}{C}H{=}\overset{\alpha}{C}H-CH{=}O}$$

Example 16.3

Draw the product of the base-catalyzed dehydration of each aldol product from Example 16.2.

Strategy

The product of dehydration of an aldol product is always an α,β-unsaturated carbonyl compound. The C—C double bond always forms between the α-carbon and the carbon that was once bonded to the —OH group.

Solution

Loss of H_2O from aldol product (a) gives an α,β-unsaturated aldehyde, while loss of H_2O from aldol product (b) gives an α,β-unsaturated ketone:

(a) OH, β, CHO, α → β, CHO, α

the C=C bond will form here

(b) OH, O, β, α → O, β, α

the C=C bond will form here

See problems 16.17, 16.18, 16.20–16.24

Problem 16.3

Draw the product of the base-catalyzed dehydration of each aldol product from Problem 16.2.

C. Crossed Aldol Reactions

Crossed aldol reaction An aldol reaction between two different aldehydes, two different ketones, or an aldehyde and a ketone.

The reactants in the key step of an aldol reaction are an enolate anion and an enolate anion acceptor. In self-reactions, both roles are played by one kind of molecule. **Crossed aldol reactions** are also possible, such as the crossed aldol reaction between acetone and formaldehyde. Because it has no α-hydrogen, formaldehyde cannot form an enolate anion. It is, however, a particularly good enolate anion acceptor because its carbonyl group is unhindered. Acetone forms an enolate anion, but its carbonyl group, which is bonded to two alkyl groups, is less reactive than that of formaldehyde. Consequently, the crossed aldol reaction between acetone and formaldehyde gives 4-hydroxy-2-butanone:

the enolate can react with either (a) unreacted acetone or (b) formaldehdye

$$CH_3\overset{O}{\overset{\|}{C}}CH_3 + H\overset{O}{\overset{\|}{C}}H \underset{}{\overset{NaOH}{\rightleftharpoons}} H_2\ddot{C}^{-}-\overset{O}{\overset{\|}{C}}-CH_3 + H\overset{O}{\overset{\|}{C}}H \rightleftharpoons CH_3\overset{O}{\overset{\|}{C}}CH_2CH_2OH$$

4-Hydroxy-2-butanone

cannot form an enolate

the enolate reacts with formaldehyde because it is less sterically hindered and therefore more reactive than acetone

As this example illustrates, for a crossed aldol reaction to be successful, one of the two reactants should have no α-hydrogen, so that an enolate anion does not form. It also helps if the compound with no α-hydrogen has the more reactive carbonyl—for example, an aldehyde. Following are examples of aldehydes that have no α-hydrogens and that can be used in crossed aldol reactions:

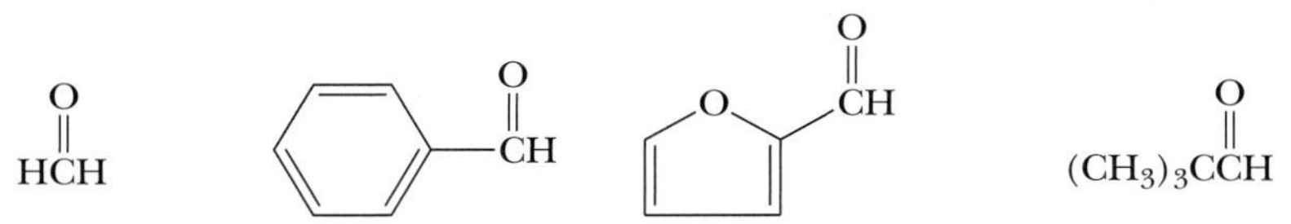

HCH (C=O) — Formaldehyde; C_6H_5CH (C=O) — Benzaldehyde; Furfural; $(CH_3)_3CCH$ (C=O) — 2,2-Dimethylpropanal

Example 16.4

Draw the product of the crossed aldol reaction between furfural and cyclohexanone and the product formed by its base-catalyzed dehydration.

Strategy

Determine which carbonyl compound possesses an abstractable α-hydrogen and draw its enolate anion. Decide which carbonyl compound would be more reactive toward the enolate anion (aldehydes are more reactive than ketones) and show the enolate anion adding to its carbonyl carbon to form a β-hydroxycarbonyl. It often helps to number the atoms in the enolate anion and the ketone or aldehyde. The product of dehydration of an aldol product is always an α,β-unsaturated carbonyl compound. The C—C double bond always forms between the α-carbon and the carbon that was once bonded to the —OH group.

Solution

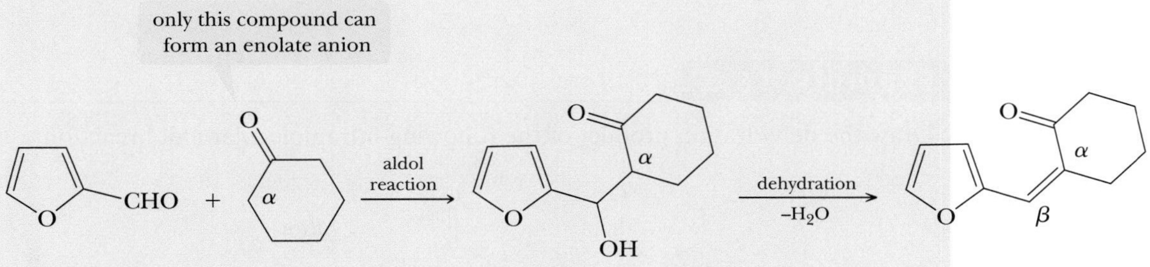

only this compound can form an enolate anion

Furfural (CHO) + Cyclohexanone (α) —aldol reaction→ Aldol product (OH, α) —dehydration, $-H_2O$→ (α, β)

See problems 16.18, 16.19

Problem 16.4

Draw the product of the crossed aldol reaction between benzaldehyde and 3-pentanone and the product formed by its base-catalyzed dehydration.

D. Intramolecular Aldol Reactions

When both the enolate anion and the carbonyl group to which it adds are in the same molecule, aldol reaction results in formation of a ring. This type of **intramolecular aldol reaction** is particularly useful for the formation of five- and six-membered rings. Because they are the most stable rings, and because the equilibrium conditions under which these reactions are performed are driven by stability, five- and six-membered rings form much more readily than four- or seven-and larger-membered rings. Intramolecular aldol reaction of 2,7-octanedione via enolate anion α_3, for example, gives a five-membered ring, whereas intramolecular aldol reaction of this same compound via enolate anion α_1 would give a seven-membered ring. In the case of 2,7-octanedione, the five-membered ring forms in preference to the seven-membered ring:

the five-membered ring is more stable than the seven-membered ring

2,7-Octanedione + KOH

$-H_2O$

(formed)

(not formed)

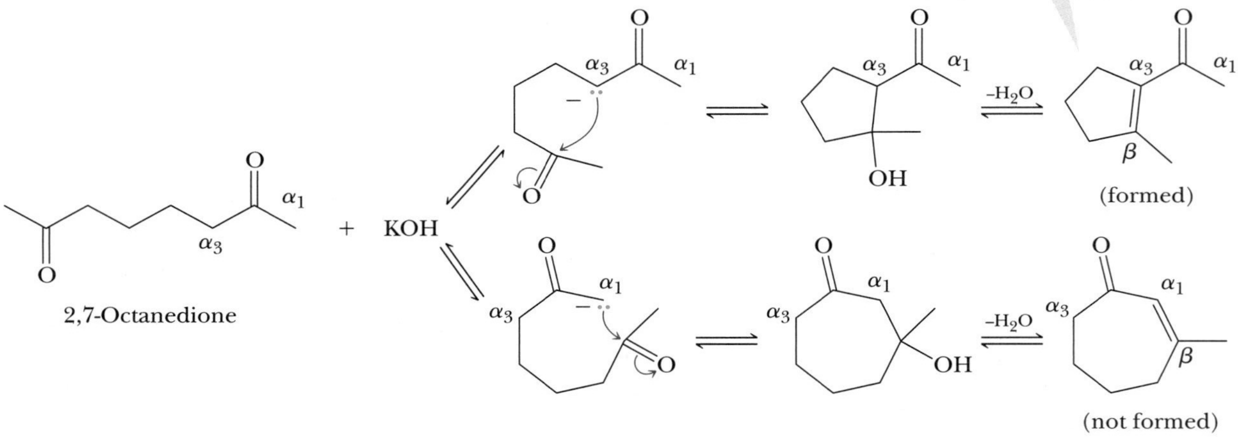

Example 16.5

Draw the dehydration product of the following intramolecular aldol reaction.

+ KOH ⟶

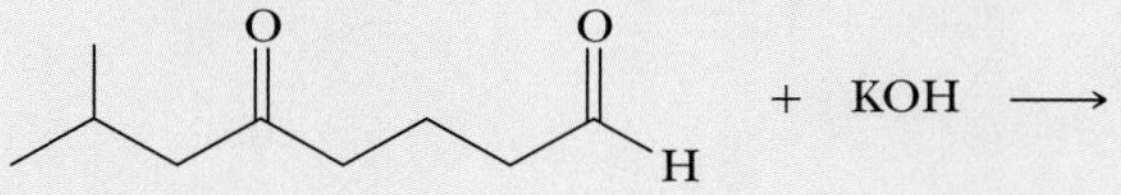

Strategy

Identify all the alpha hydrogens in the molecule, and for each one, form an enolate anion. Then decide which enolate anion would form the more stable ring upon reaction with the other carbonyl in the molecule. It often helps to number the atoms in the ketone or aldehyde. The product of dehydration of any aldol product is always an α,β-unsaturated carbonyl compound. The C—C double bond always forms between the α-carbon and the carbon that was once bonded to the —OH group.

Solution

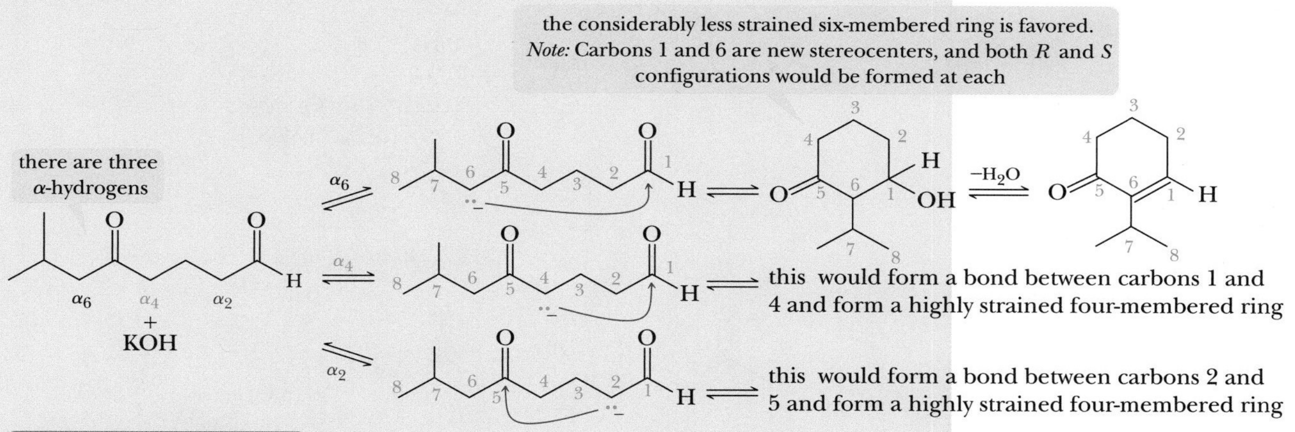

the considerably less strained six-membered ring is favored. *Note:* Carbons 1 and 6 are new stereocenters, and both *R* and *S* configurations would be formed at each

there are three α-hydrogens

KOH

$-H_2O$

this would form a bond between carbons 1 and 4 and form a highly strained four-membered ring

this would form a bond between carbons 2 and 5 and form a highly strained four-membered ring

See problems 16.22–16.24

Problem 16.5

Draw the dehydration product of the following intermolecular aldol reaction.

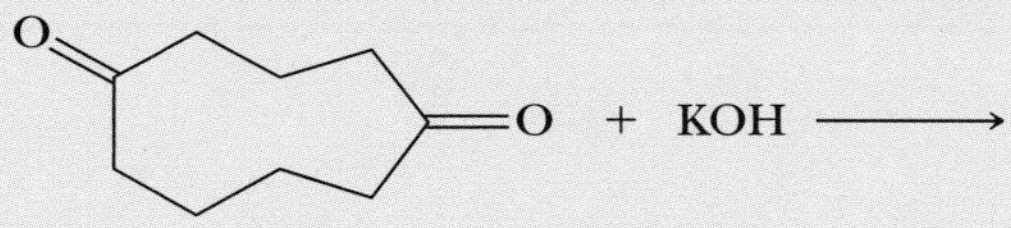

+ KOH ⟶

HOW TO 16.1 Determine the Starting Compounds Used in an Aldol Reaction

It will sometimes be necessary to work retrosynthetically (How To 15.2) from the product of an aldol reaction. Following are the steps for determining the starting compounds used in an aldol reaction.

(a) Locate the carbonyl and identify the α- and β-carbons. An aldol product will contain a carbonyl of a ketone or an aldehyde and either an OH group or a C—C double bond. Once the carbonyl is found, label the carbons using the Greek alphabet in the direction of the OH group or C—C double bond:

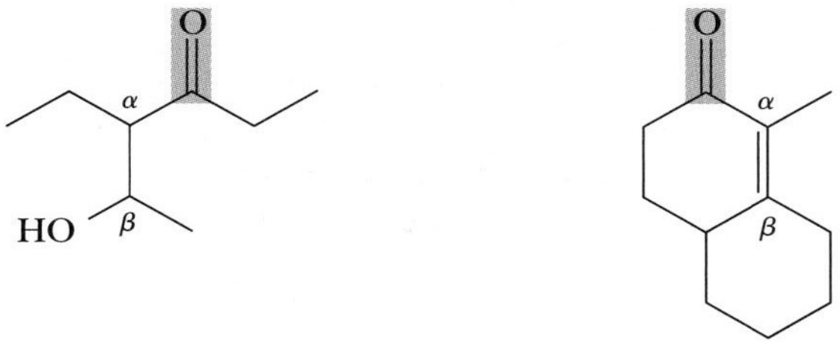

(b) Erase the bond between the α- and β-carbons and convert the carbon labeled β to a carbonyl group (the oxygen of an —OH group attached to the β-carbon becomes the oxygen atom of the carbonyl). Note that a hydrogen is added to the carbon labeled α, although it is usually unnecessary to show this in a line-angle formula:

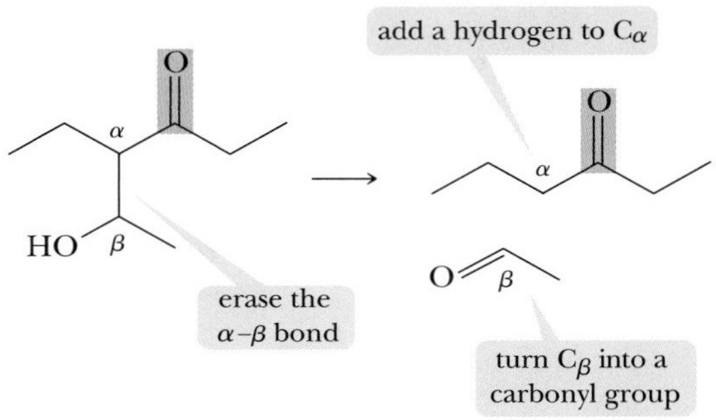

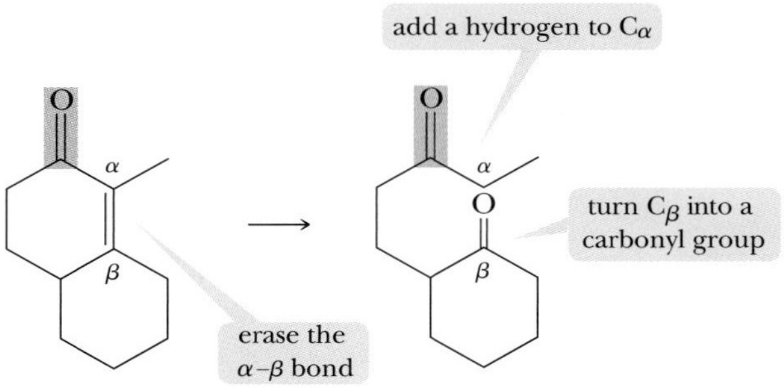

add a hydrogen to C_α

erase the α–β bond

turn C_β into a carbonyl group

add a hydrogen to C_α

turn C_β into a carbonyl group

erase the α–β bond

16.3 What Are the Claisen and Dieckmann Condensations?

A. Claisen Condensation

Claisen condensation A carbonyl condensation reaction between two esters to give a β-ketoester.

In this section, we examine the formation of an enolate anion from one ester, followed by the nucleophilic acyl substitution of the enolate anion at the carbonyl carbon of another ester. One of the first of these reactions discovered was the **Claisen condensation**, named after its discoverer, German chemist Ludwig Claisen (1851–1930). We illustrate a Claisen condensation by the reaction between two molecules of ethyl acetate in the presence of sodium ethoxide, followed by acidification, to give ethyl acetoacetate (note that, in this and many of the equations that follow, we abbreviate the ethyl group as Et):

$$2\,CH_3\overset{O}{\overset{\|}{C}}OEt \xrightarrow[2)\,H_2O,\,HCl]{1)\,EtO^-\,Na^+} CH_3\overset{O}{\overset{\|}{C}}CH_2\overset{O}{\overset{\|}{C}}OEt + EtOH$$

Ethyl ethanoate (Ethyl acetate) — Ethyl 3-oxobutanoate (Ethyl acetoacetate) — Ethanol

The functional group of the product of a Claisen condensation is a ***β*-ketoester**:

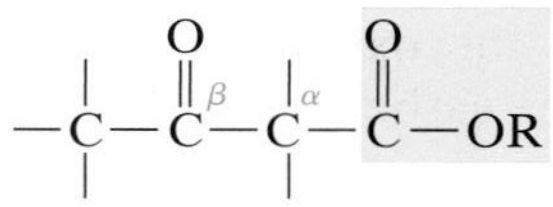

$$-\overset{|}{\underset{|}{C}}-\overset{O}{\overset{\|}{C}}-\overset{|}{\underset{|}{C}}-\overset{O}{\overset{\|}{C}}-OR$$

A β-ketoester

The Claisen condensation of two molecules of ethyl propanoate gives the following β-ketoester:

$$\text{Ethyl propanoate} + \text{Ethyl propanoate} \xrightarrow[\text{2) } H_2O,\ HCl]{\text{1) } EtO^-Na^+} \text{Ethyl 2-methyl-3-oxopentanoate} + EtOH$$

Ethyl propanoate | Ethyl propanoate | Ethyl 2-methyl-3-oxopentanoate

Claisen condensations, like the aldol reaction, require a base. Aqueous bases, such as NaOH, however, cannot be used in Claisen condensations because they would bring about hydrolysis of the ester (saponification, Section 15.3C) instead. Rather, the bases most commonly used in Claisen condensations are nonaqueous bases, such as sodium ethoxide in ethanol and sodium methoxide in methanol. Furthermore, to prevent transesterification (Section 15.4C), the alkyl group (—R) of the base should match the R group in the alkoxyl portion (—OR) of the ester.

Mechanism: Claisen Condensation

As you study this mechanism, note how closely its first two steps resemble the first steps of the aldol reaction (Section 16.1). In each reaction, base removes a proton from an α-carbon in Step 1 to form a resonance-stabilized enolate anion. In Step 2, the enolate anion attacks the carbonyl carbon of another ester molecule to form a tetrahedral carbonyl addition intermediate.

Step 1: Base removes an α-hydrogen from the ester to give a resonance-stabilized enolate anion:

$$EtO^- + H{-}CH_2{-}\overset{O}{\overset{\|}{C}}OEt \rightleftharpoons EtOH + \left[{}^-CH_2{-}\overset{O}{\overset{\|}{C}}OEt \longleftrightarrow CH_2{=}\overset{O^-}{\overset{|}{C}}OEt \right]$$

(weaker base) | pK_a 22 (weaker acid) | pK_a 15.9 (stronger acid) | Resonance-stabilized enolate anion (stronger base)

Because the α-hydrogen of the ester is the weaker acid and ethoxide is the weaker base, the position of this equilibrium lies very much toward the left.

Step 2: Attack of the enolate anion on the carbonyl carbon of another ester molecule gives a tetrahedral carbonyl addition intermediate:

$$CH_3{-}\overset{O}{\overset{\|}{C}}{-}OEt + {}^-CH_2{-}\overset{O}{\overset{\|}{C}}OEt \rightleftharpoons \left[CH_3{-}\underset{OEt}{\underset{|}{\overset{O^-}{\overset{|}{C}}}}{-}CH_2{-}\overset{O}{\overset{\|}{C}}{-}OEt \right]$$

A tetrahedral cabonyl addition intermediate

Step 3: Unlike the tetrahedral carbonyl addition intermediate in the aldol reaction, this intermediate has a leaving group (the ethoxide ion). Collapse of the tetrahedral carbonyl addition intermediate by ejection of the ethoxide ion gives a β-ketoester:

$$CH_3{-}\underset{OEt}{\underset{|}{\overset{O^-}{\overset{|}{C}}}}{-}CH_2{-}\overset{O}{\overset{\|}{C}}{-}OEt \rightleftharpoons CH_3{-}\overset{O}{\overset{\|}{C}}{-}CH_2{-}\overset{O}{\overset{\|}{C}}{-}OEt + EtO^-$$

Step 4: Formation of the enolate anion of the β-ketoester drives the Claisen condensation to the right. The β-ketoester (a stronger acid) reacts with ethoxide ion (a stronger base) to give ethanol (a weaker acid) and the anion of the β-ketoester (a weaker base):

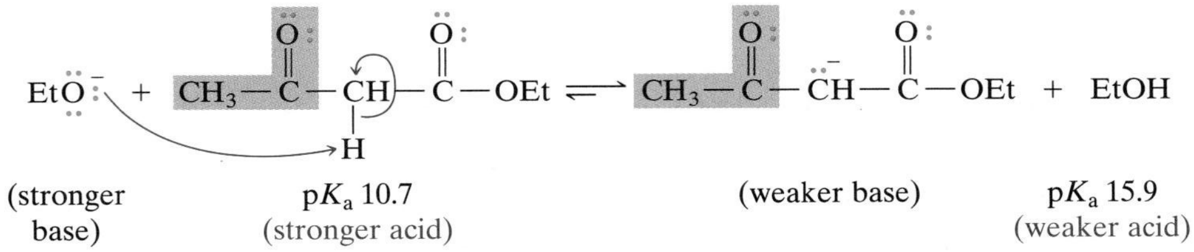

$$EtO^- + CH_3-\overset{O}{\overset{\|}{C}}-CH(H)-\overset{O}{\overset{\|}{C}}-OEt \rightleftharpoons CH_3-\overset{O}{\overset{\|}{C}}-\overset{-}{C}H-\overset{O}{\overset{\|}{C}}-OEt + EtOH$$

(stronger base) — pK_a 10.7 (stronger acid) — (weaker base) — pK_a 15.9 (weaker acid)

The position of equilibrium for this step lies very far toward the right.

Step 5: Acidification of the enolate anion gives the β-ketoester:

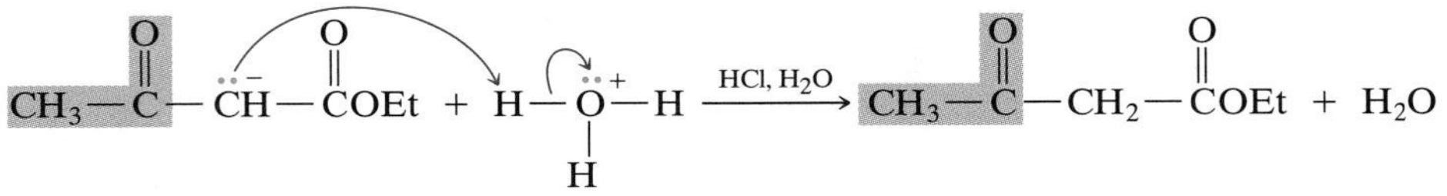

$$CH_3-\overset{O}{\overset{\|}{C}}-\overset{-}{C}H-\overset{O}{\overset{\|}{C}}OEt + H-\overset{+}{O}(H)-H \xrightarrow{HCl,\ H_2O} CH_3-\overset{O}{\overset{\|}{C}}-CH_2-\overset{O}{\overset{\|}{C}}OEt + H_2O$$

Example 16.6

Show the product of the Claisen condensation of ethyl butanoate in the presence of sodium ethoxide followed by acidification with aqueous HCl.

Strategy

Draw two molecules of ethyl butanoate. Convert one of the molecules to an enolate anion and show it adding to the carbonyl carbon of the other. Because Claisen condensations occur with nucleophilic acyl substitution, the —OR group of the carbonyl being attacked is eliminated from the final product. It often helps to number the atoms in the enolate anion and the ester being attacked.

Solution

The new bond formed in a Claisen condensation is between the carbonyl group of one ester and the α-carbon of another:

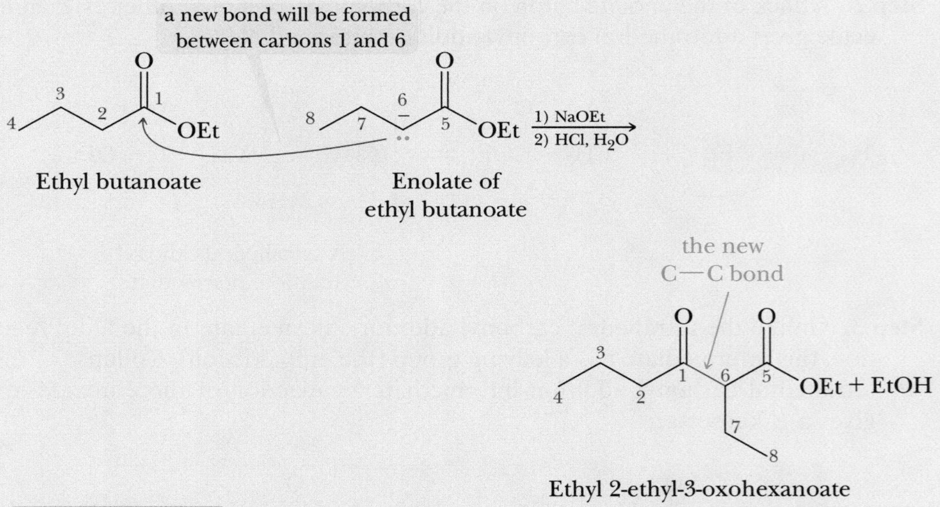

a new bond will be formed between carbons 1 and 6

Ethyl butanoate + Enolate of ethyl butanoate $\xrightarrow[2)\ HCl,\ H_2O]{1)\ NaOEt}$ Ethyl 2-ethyl-3-oxohexanoate + EtOH

(the new C—C bond)

Ethyl 2-ethyl-3-oxohexanoate

See problem 16.27

Problem 16.6

Show the product of the Claisen condensation of ethyl 3-methylbutanoate in the presence of sodium ethoxide.

B. Dieckmann Condensation

An intramolecular Claisen condensation of a dicarboxylic ester to give a five- or six-membered ring is known as a **Dieckmann condensation**. In the presence of one equivalent of sodium ethoxide, diethyl hexanedioate (diethyl adipate), for example, undergoes an intramolecular condensation to form a five-membered ring:

Dieckmann condensation An intramolecular Claisen condensation of an ester of a dicarboxylic acid to give a five- or six-membered ring.

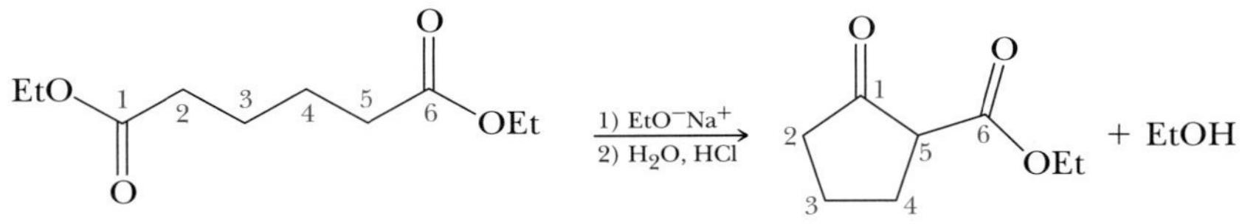

Diethyl hexanedioate (Diethyl adipate) $\xrightarrow[\text{2) } H_2O, HCl]{\text{1) } EtO^-Na^+}$ Ethyl 2-oxocyclopentanecarboxylate + EtOH

The mechanism of a Dieckmann condensation is identical to the mechanism we described for the Claisen condensation. An anion formed at the α-carbon of one ester in Step 1 adds to the carbonyl of the other ester group in Step 2 to form a tetrahedral carbonyl addition intermediate. This intermediate ejects ethoxide ion in Step 3 to regenerate the carbonyl group. Cyclization is followed by formation of the conjugate base of the β-ketoester in Step 4, just as in the Claisen condensation. The β-ketoester is isolated after acidification with aqueous acid.

Example 16.7

Complete the equation for the following Dieckmann condensation:

$\xrightarrow[\text{2) HCl, } H_2O]{\text{1) NaOEt}}$

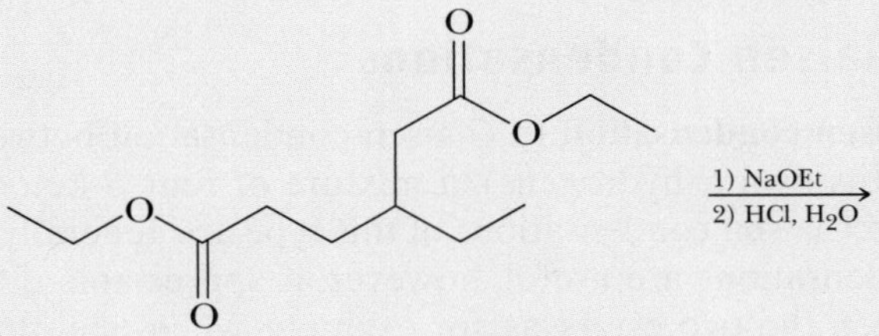

Strategy

Identify the α-carbon for each ester group. Convert one of the α-carbons to an enolate anion and show it adding to the other carbonyl carbon. Because Dieckmann condensations occur with nucleophilic acyl substitution, the —OR group of the carbonyl being attacked is eliminated from the final product. It often helps to number the atoms in the enolate anion and the ester being attacked.

Solution

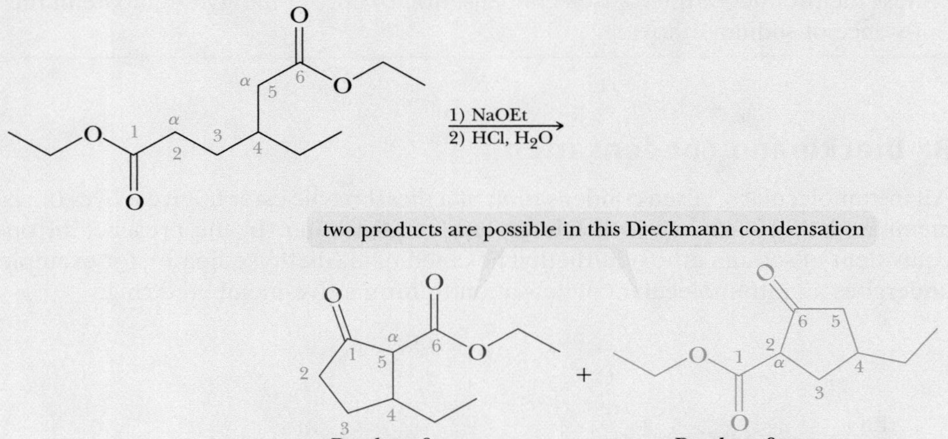

1) NaOEt
2) HCl, H_2O

two products are possible in this Dieckmann condensation

Product from enolate at C_5 + Product from enolate at C_2

See problem 16.33

Problem 16.7

Complete the equation for the following Dieckmann condensation:

OEt, OEt

1) NaOEt
2) HCl, H_2O

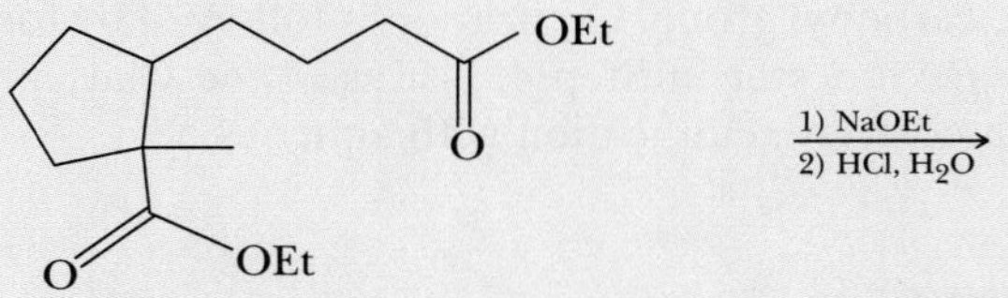

C. Crossed Claisen Condensations

Crossed Claisen condensation A Claisen condensation between two different esters.

In a **crossed Claisen condensation** (a Claisen condensation between two different esters, each with its own α-hydrogens), a mixture of four β-ketoesters is possible; therefore, crossed Claisen condensations of this type are generally not synthetically useful. Such condensations are useful, however, if appreciable differences in reactivity exist between the two esters, as, for example, when one of the esters has no α-hydrogens and can function only as an enolate anion acceptor. These esters have no α-hydrogens:

HCOEt (O)	EtOCOEt (O)	EtOC—COEt (O, O)	C_6H_5—COEt (O)
Ethyl formate	Diethyl carbonate	Diethyl ethanedioate (Diethyl oxalate)	Ethyl benzoate

Crossed Claisen condensations of this type are usually carried out by using the ester with no α-hydrogens in excess. In the following illustration, methyl benzoate is used in excess:

$$\text{PhC(=O)OCH}_3 + \text{CH}_3\text{CH}_2\text{C(=O)OCH}_3 \xrightarrow[\text{2) H}_2\text{O, HCl}]{\text{1) CH}_3\text{O}^-\text{Na}^+} \text{PhC(=O)CH(CH}_3\text{)C(=O)OCH}_3 + \text{CH}_3\text{OH}$$

Methyl benzoate; Methyl propanoate; Methyl 2-methyl-3-oxo-3-phenylpropanoate

Example 16.8

Complete the equation for this crossed Claisen condensation:

$$\text{HCOEt} + \text{CH}_3\text{CH}_2\text{COEt} \xrightarrow[\text{2) H}_2\text{O, HCl}]{\text{1) EtO}^-\text{Na}^+}$$

(each ester shown with C=O)

Strategy

Identify the α-carbon(s) for each ester. Convert one of the α-carbons to an enolate anion and show it adding to the other carbonyl carbon. Repeat this for all α-carbons and ester molecules. Remember that the enolate anion can attack the other ester or an unreacted molecule of itself if equal amounts of ester are used. Because Claisen condensations occur with nucleophilic acyl substitution, the —OR group of the carbonyl being attacked is eliminated from the final product. It often helps to number the atoms in the enolate anion and the ester being attacked.

Solution

$$\text{HCOEt} + \text{CH}_3\text{CH}_2\text{COEt} \xrightarrow[\text{2) H}_2\text{O, HCl}]{\text{1) EtO}^-\text{ Na}^+} \text{HCCH(CH}_3\text{)COEt} + \text{CH}_3\text{CH}_2\text{CCH(CH}_3\text{)COEt} + \text{EtOH}$$

only this ester can become an enolate...

...once it does, it can attack the formyl ester to form this product...

...or it can attack the same starting ester to form this product

See problems 16.29–16.31

Problem 16.8

Complete the equation for this crossed Claisen condensation:

$$\text{C}_6\text{H}_5\text{COEt} + \text{C}_6\text{H}_5\text{CH}_2\text{COEt} \xrightarrow[\text{2) H}_2\text{O, HCl}]{\text{1) EtO}^-\text{Na}^+}$$

HOW TO 16.2 Determine the Starting Compounds Used in a Claisen Condensation

It will sometimes be necessary to work retrosynthetically (How To 15.2) from the product of a Claisen condensation. Following are the steps for determining the starting compounds used in a Claisen condensation.

(a) Locate the carbonyl of an ester and identify the α- and β-carbons. A Claisen condensation product will contain a carbonyl of an ester and another carbonyl of either an aldehyde, a ketone, or another ester. Once the carbonyl of the ester is found, label the carbons using the Greek alphabet in the direction of the other carbonyl group:

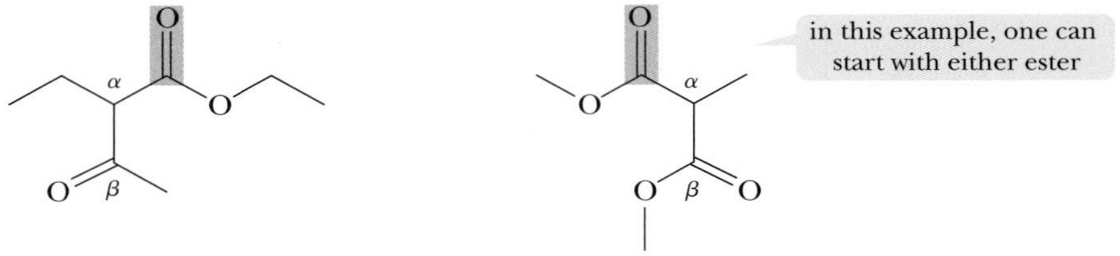

in this example, one can start with either ester

(b) Erase the bond between the α- and β-carbons:

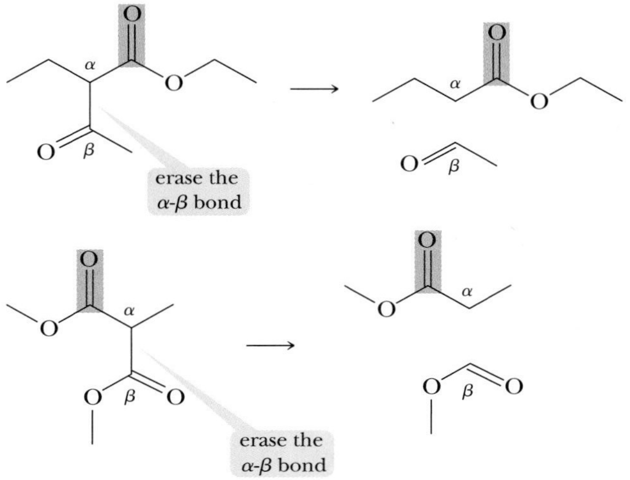

erase the α-β bond

erase the α-β bond

(c) Add a hydrogen to the carbon labeled α and add an —OR group to the carbon labeled β. The type of alkyl group in —OR should match the alkyl group of the other ester:

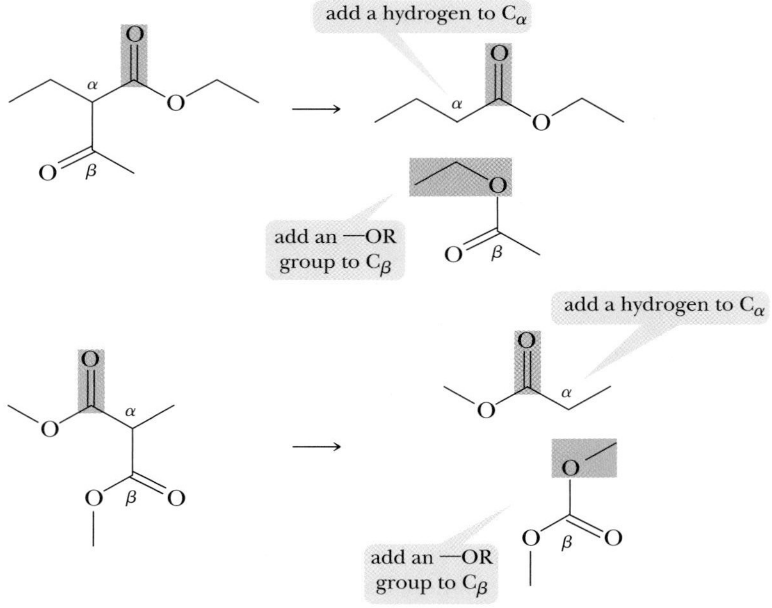

add a hydrogen to C_α

add an —OR group to C_β

add a hydrogen to C_α

add an —OR group to C_β

D. Hydrolysis and Decarboxylation of β-Ketoesters

Recall from Section 15.3C that the hydrolysis of an ester in aqueous sodium hydroxide (saponification), followed by acidification of the reaction mixture with HCl or other mineral acid, converts an ester to a carboxylic acid. Recall also from Section 14.8 that β-ketoacids and β-dicarboxylic acids readily undergo decarboxylation (lose CO_2) when heated. The following equations illustrate the results of a Claisen condensation, followed by saponification, acidification, and decarboxylation:

Claisen condensation:

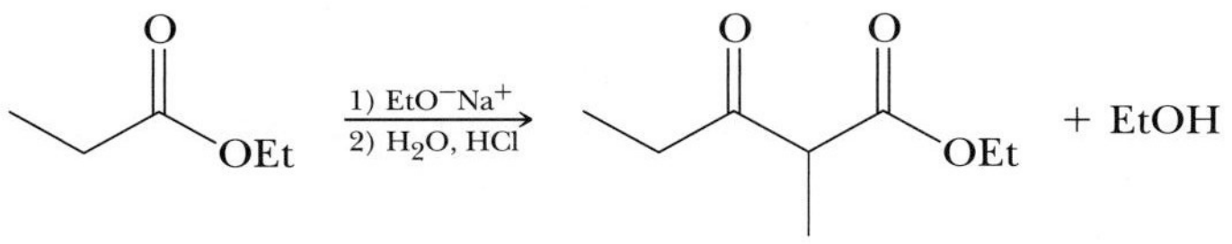

1) EtO^-Na^+ 2) H_2O, HCl; + EtOH

Saponification followed by acidification:

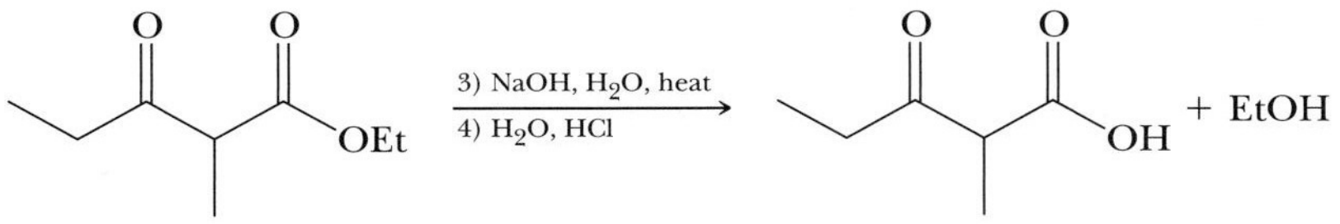

3) NaOH, H_2O, heat 4) H_2O, HCl; + EtOH

Decarboxylation:

5) heat; + CO_2

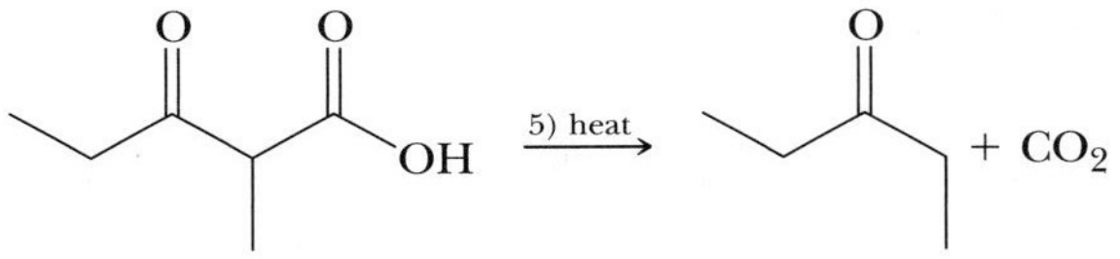

The result of these five steps is a reaction between two molecules of ester, one furnishing a carboxyl group and the other furnishing an enolate anion, to give a ketone and carbon dioxide:

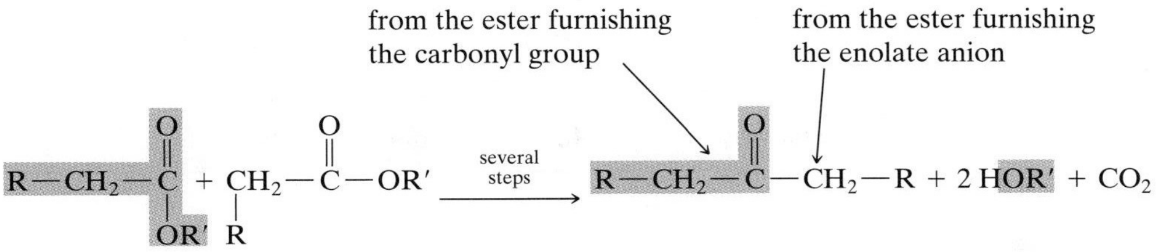

from the ester furnishing the carbonyl group

from the ester furnishing the enolate anion

$$R-CH_2-\overset{O}{\overset{\|}{C}}-OR' + \underset{R}{\underset{|}{CH_2}}-\overset{O}{\overset{\|}{C}}-OR' \xrightarrow{\text{several steps}} R-CH_2-\overset{O}{\overset{\|}{C}}-CH_2-R + 2\,HOR' + CO_2$$

In the general reaction, both ester molecules are the same, and the product is a symmetrical ketone.

Example 16.9

Each set of compounds undergoes (1, 2) Claisen condensation, (3) saponification followed by (4) acidification, and (5) thermal decarboxylation:

(a) $Ph\overset{O}{\overset{\|}{C}}OEt + CH_3\overset{O}{\overset{\|}{C}}OEt$

(b) 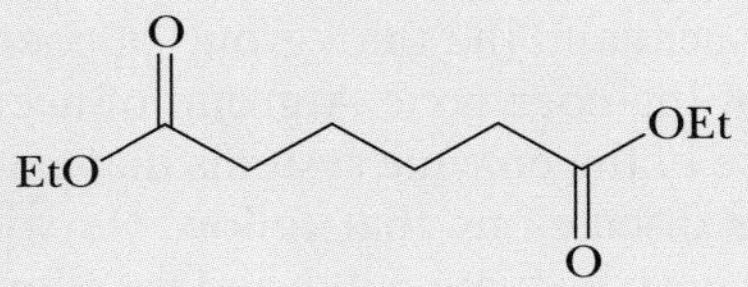 EtO(C=O)–CH₂CH₂CH₂CH₂–(C=O)OEt

Draw a structural formula of the product after completion of this reaction sequence.

Strategy

Proceed with a variation of the Claisen condensation (Ex. 16.6–16.8) and arrive at the final decarboxylated product by removing the carbonyl ester group and replacing it with a hydrogen.

Solution

Steps 1 and 2 bring about a crossed Claisen condensation in (a) and a Dieckmann condensation in (b) to form a β-ketoester. Steps 3 and 4 bring about hydrolysis of the β-ketoester to give a β-ketoacid, and Step 5 brings about decarboxylation to give a ketone:

$$\text{(a)} \xrightarrow{1,2} \text{Ph}\overset{\text{O}}{\overset{\|}{\text{C}}}\text{CH}_2\overset{\text{O}}{\overset{\|}{\text{C}}}\text{OEt} \xrightarrow{3,4} \text{Ph}\overset{\text{O}}{\overset{\|}{\text{C}}}\text{CH}_2\overset{\text{O}}{\overset{\|}{\text{C}}}\text{OH} \xrightarrow{5} \text{Ph}\overset{\text{O}}{\overset{\|}{\text{C}}}\text{CH}_3$$

(b) $\xrightarrow{1,2}$ 2-(COOEt)cyclopentanone $\xrightarrow{3,4}$ 2-(COOH)cyclopentanone $\xrightarrow{5}$ cyclopentanone

See problems 16.28, 16.32, 16.34

Problem 16.9

Show how to convert benzoic acid to 3-methyl-1-phenyl-1-butanone by using a Claisen condensation at some stage in the synthesis:

Benzoic acid (COOH) $\xrightarrow{?}$ 3-Methyl-1-phenyl-1-butanone

16.4 How Are Aldol Reactions and Claisen Condensations Involved in Biological Processes?

Carbonyl condensations are among the most widely used reactions in the biological world for the assembly of new carbon–carbon bonds in such important biomolecules as fatty acids, cholesterol, steroid hormones, and terpenes. One source of carbon atoms for the synthesis of these biomolecules is **acetyl-CoA**, a thioester of acetic acid and the thiol group of coenzyme A (CoA-SH, Section 22.1D). The function of the coenzyme A group of acetyl-CoA is to anchor the acetyl group on the surface of the enzyme systems that catalyze the reactions we examine in this section. In the discussions that follow, we will not be concerned with the mechanism by which each enzyme-catalyzed reaction occurs. Rather, our concern is with recognizing the type of reaction that takes place in each step.

In the Claisen condensation catalyzed by the enzyme thiolase, acetyl-CoA is converted to its enolate anion, which then attacks the carbonyl group of a second molecule of acetyl-CoA to form a tetrahedral carbonyl addition intermediate. Collapse of this intermediate by the loss of CoA-SH gives acetoacetyl-CoA. The mechanism for this condensation reaction is exactly the same as that of the Claisen condensation (Section 16.3A):

Acetyl-CoA + Acetyl-CoA $\xrightarrow[\text{condensation}]{\text{thiolase, Claisen}}$ Acetoacetyl-CoA + CoASH (Coenzyme A)

An enzyme-catalyzed aldol reaction with a third molecule of acetyl-CoA on the ketone carbonyl of acetoacetyl-CoA gives (*S*)-3-hydroxy-3-methylglutaryl-CoA:

the second carbonyl condensation takes place at this carbonyl

Acetoacetyl-CoA + Acetyl-CoA $\xrightarrow{\text{3-hydroxy-3-methyl-glutaryl-CoA synthetase}}$ (*S*)-3-Hydroxy-3-methylglutaryl-CoA + CoASH

Note three features of this reaction. First, the creation of the new stereocenter is stereoselective: Only the S enantiomer is formed. Although the acetyl group of each reactant is achiral, their condensation takes place in a chiral environment created by the enzyme 3-hydroxy-3-methylglutaryl-CoA synthetase. Second, hydrolysis of the thioester group of acetyl-CoA is coupled with the aldol reaction. Third, the carboxyl group is shown as it is ionized at pH 7.4, the approximate pH of blood plasma and many cellular fluids.

Enzyme-catalyzed reduction of the thioester group of 3-hydroxy-3-methylglutaryl-CoA to a primary alcohol gives mevalonic acid, shown here as its anion:

(*S*)-3-Hydroxy-3-methylglutaryl-CoA $\xrightarrow[\text{2NADH} \rightarrow \text{2NAD}^+]{\text{3-hydroxy-3-methyl-glutaryl-CoA reductase}}$ (*R*)-Mevalonate

The reducing agent for this transformation is nicotinamide adenine dinucleotide, abbreviated NADH (Section 22.1B). This reducing agent is the biochemical equivalent of $LiAlH_4$. Each reducing agent functions by delivering a hydride ion ($H:^-$) to the carbonyl carbon of an aldehyde, a ketone, or an ester. Note that, in the reduction, a change occurs in the designation of configuration from S to R, not because of any change in configuration at the stereocenter, but rather because of a change in priority among the four groups bonded to the stereocenter.

Enzyme-catalyzed transfer of a phosphate group from adenosine triphosphate (ATP, Section 20.1) to the 3-hydroxyl group of mevalonate gives a phosphoric ester at carbon 3. Enzyme-catalyzed transfer of a pyrophosphate group (Section 15.1B) from a second molecule of ATP gives a pyrophosphoric ester at

CHEMICAL CONNECTIONS 16A

Drugs That Lower Plasma Levels of Cholesterol

Coronary artery disease is the leading cause of death in the United States and other Western countries, where about one half of all deaths can be attributed to atherosclerosis. Atherosclerosis results from the buildup of fatty deposits called plaque on the inner walls of arteries. A major component of plaque is cholesterol derived from low-density-lipoproteins (LDL), which circulate in blood plasma. Because more than one half of total body cholesterol in humans is synthesized in the liver from acetyl-CoA, intensive efforts have been directed toward finding ways to inhibit this synthesis. The rate-determining step in cholesterol biosynthesis is reduction of 3-hydroxy-3-methylglutaryl-CoA (HMG-CoA) to mevalonic acid. This reduction is catalyzed by the enzyme HMG-CoA reductase and requires two moles of NADPH per mole of HMG-CoA.

Beginning in the early 1970s, researchers at the Sankyo Company in Tokyo screened more than 8000 strains of microorganisms and in 1976 announced the isolation of mevastatin, a potent inhibitor of HMG-CoA reductase, from culture broths of the fungus *Penicillium citrinum*. The same compound was isolated by researchers at Beecham Pharmaceuticals in England from cultures of *Penicillium brevicompactum*. Soon thereafter, a second, more active compound called lovastatin was isolated at the Sankyo Company from the fungus *Monascus ruber*, and at Merck Sharpe & Dohme from *Aspergillus terreus*. Both mold metabolites are extremely effective in lowering plasma concentrations of LDL. The active form of each is the 5-hydroxycarboxylate anion formed by hydrolysis of the δ-lactone.

hydrolysis of the δ-lactone →

$R_1 = R_2 = H$, mevastatin
$R_1 = H, R_2 = CH_3$, lovastatin (Mevacor)
$R_1 = R_2 = CH_3$, simvastatin (Zocor)

The active form of each drug

These drugs and several synthetic modifications now available inhibit HMG-CoA reductase by forming an enzyme-inhibitor complex that prevents further catalytic action of the enzyme. It is reasoned that the 3,5-dihydroxycarboxylate anion part of the active form of each drug binds tightly to the enzyme because it mimics the hemithioacetal intermediate formed by the first reduction of HMG-CoA.

3-Hydroxy-3-methyl glutaryl-CoA (HMG-CoA) —NADPH→ A hemithioacetal intermediate formed by the first NADPH reduction → —NADPH→ Mevalonate

Systematic studies have shown the importance of each part of the drug for effectiveness. It has been found, for example, that the carboxylate anion ($—COO^-$) is essential, as are both the 3-OH and 5-OH groups. It has also been shown that almost any modification of the two fused six-membered rings and their pattern of substitution reduced potency.

QUESTION

Which of the biomolecules in the above reaction scheme could be the product of an aldol reaction?

carbon 5. Enzyme-catalyzed β-elimination from this molecule results in the loss of CO_2 and PO_4^{3-}, both good leaving groups:

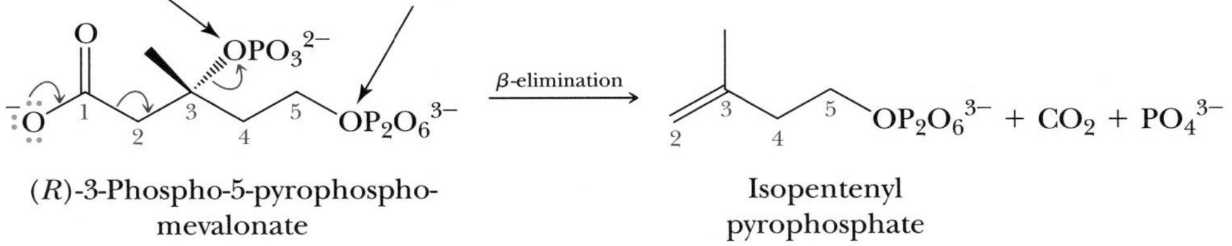

a phosphoric ester
a pyrophosphoric ester
β-elimination
OPO_3^{2-}
$OP_2O_6^{3-}$
$OP_2O_6^{3-} + CO_2 + PO_4^{3-}$

(*R*)-3-Phospho-5-pyrophosphomevalonate

Isopentenyl pyrophosphate

Isopentenyl pyrophosphate has the carbon skeleton of isoprene, the unit into which the carbon skeletons of terpenes can be divided (Section 4.4). This molecule is, in fact, a key intermediate in the biosynthesis of isoprene, terpenes, cholesterol, steroid hormones, and bile acids:

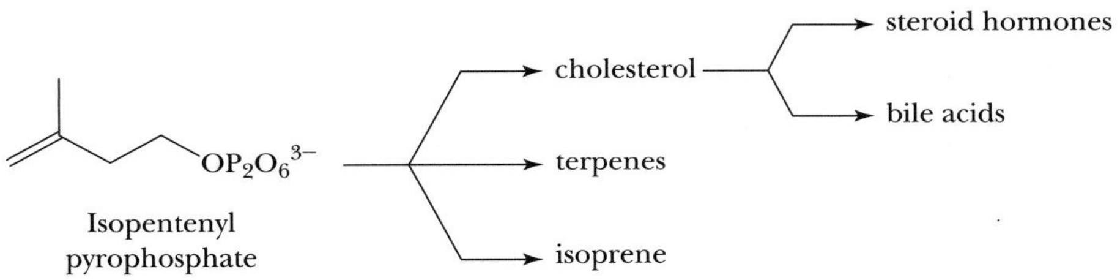

$OP_2O_6^{3-}$

Isopentenyl pyrophosphate

cholesterol → steroid hormones, bile acids
terpenes
isoprene

16.5 What Is the Michael Reaction?

Thus far, we have used carbon nucleophiles in two ways to form new carbon–carbon bonds:

1. Addition of organomagnesium (Grignard) reagents to the carbonyl groups of aldehydes, ketones, and esters.
2. Addition of enolate anions derived from aldehydes or ketones (aldol reactions) and esters (Claisen and Dieckmann condensations) to the carbonyl groups of other aldehydes, ketones, or esters.

Addition of an enolate anion to a carbon–carbon double bond conjugated with a carbonyl group presents an entirely new synthetic strategy. In this section, we study a type of **conjugate addition** involving nucleophilic addition to an electrophilic double bond.

A. Michael Addition of Enolate Anions

Nucleophilic addition of enolate anions to α,β-unsaturated carbonyl compounds was first reported in 1887 by the American chemist Arthur Michael. Following are two examples of **Michael reactions**. In the first example, the nucleophile is the enolate anion of diethyl malonate. In the second example, the nucleophile is the enolate anion of ethyl acetoacetate:

Michael reaction The conjugate addition of an enolate anion or other nucleophile to an α,β-unsaturated carbonyl compound.

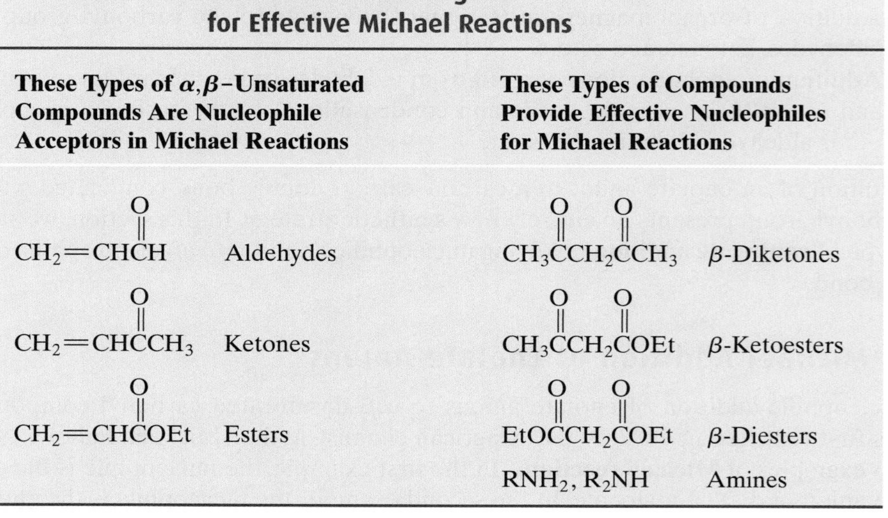

Recall that nucleophiles don't ordinarily add to carbon–carbon double bonds. Rather, they add to electrophiles (Section 5.2). What activates a carbon–carbon double bond for nucleophilic attack in a Michael reaction is the presence of the adjacent carbonyl group. One important resonance structure of α,β-unsaturated carbonyl compounds puts a positive charge on the β-carbon of the double bond, making it electrophilic in its reactivity:

an *activated* C—C double bond due to its partial positive character as shown in one of its resonance contributors

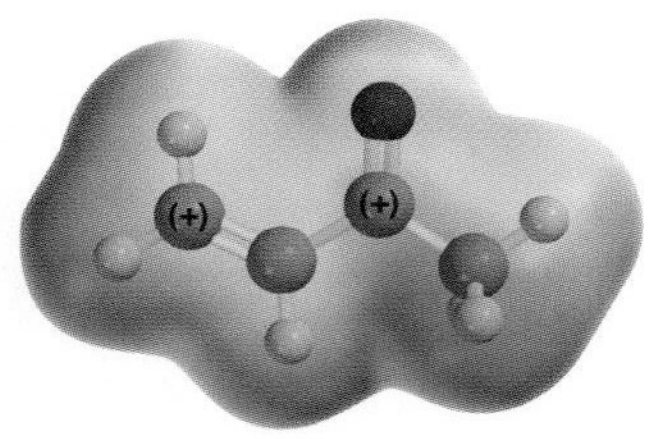

Although the major fraction of the partial positive charge of an α,β-unsaturated aldehyde or ketone is on the carbonyl carbon, there is nevertheless a significant partial positive charge on the beta carbon.

Thus, nucleophiles can add to this type of double bond, which we call "activated" for that reason.

Table 16.2 lists the most common combinations of α,β-unsaturated carbonyl compounds and nucleophiles used in Michael reactions. The most commonly used bases are metal alkoxides, pyridine, and piperidine.

TABLE 16.2 Combinations of Reagents for Effective Michael Reactions

These Types of α,β-Unsaturated Compounds Are Nucleophile Acceptors in Michael Reactions		These Types of Compounds Provide Effective Nucleophiles for Michael Reactions	
$CH_2{=}CHCH{=}O$	Aldehydes	$CH_3C({=}O)CH_2C({=}O)CH_3$	β-Diketones
$CH_2{=}CHC({=}O)CH_3$	Ketones	$CH_3C({=}O)CH_2C({=}O)OEt$	β-Ketoesters
$CH_2{=}CHC({=}O)OEt$	Esters	$EtOC({=}O)CH_2C({=}O)OEt$	β-Diesters
		RNH_2, R_2NH	Amines

We can write the following general mechanism for a Michael reaction:

Mechanism: Michael Reaction—Conjugate Addition of Enolate Anions

Step 1: Treatment of H—Nu with base gives the nucleophile, Nu:⁻.

$$\text{Nu—H} + :\text{B}^- \rightleftharpoons \text{Nu}:^- + \text{H—B}$$

Base

Step 2: Nucleophilic addition of Nu:⁻ to the β-carbon of the conjugated system gives a resonance-stabilized enolate anion:

$$\text{Nu}:^- + \text{—C=C—C(=O)—} \longrightarrow \text{Nu—C—C=C(—O}^-\text{)—} \longleftrightarrow \text{Nu—C—}\bar{\text{C}}\text{—C(=O)—}$$

A resonance-stabilized enolate anion

Step 3: Proton transfer from H—B gives the enol and regenerates the base:

$$\text{Nu—C—C=C(—O}^-\text{)—} + \text{H—B} \longrightarrow \text{Nu—}\overset{4}{\text{C}}\text{—}\overset{3}{\text{C}}\text{=}\overset{2}{\text{C}}\text{(—}\overset{1}{\text{O}}\text{H)—} + :\text{B}^-$$

An enol
(a product of 1,4-addition)

Note that the enol formed in this step corresponds to 1,4-addition to the conjugated system of the α,β-unsaturated carbonyl compound. It is because this intermediate is formed that the Michael reaction is classified as a 1,4-, or conjugate, addition. Note also that, the base, B:⁻, is regenerated, in accordance with the experimental observation that a Michael reaction requires only a catalytic amount of base rather than a molar equivalent.

Step 4: Tautomerism (Section 13.8A) of the less stable enol form gives the more stable keto form:

$$\text{Nu—C—C=C(—OH)—} \rightleftharpoons \text{Nu—C—C(H)—C(=O)—}$$

Enol form (less stable) — Keto form (more stable)

Example 16.10

Draw a structural formula for the product formed by treating each set of reactants with sodium ethoxide in ethanol under conditions of the Michael reaction:

(a) CH_3COCH_2COOEt + $CH_2=CHCOOEt$

(b) $EtOOCCH_2COOEt$ + 2-cyclopentenone

Strategy

The net result of a Michael reaction is the addition of a nucleophile to the β-carbon of an α,β-unsaturated carbonyl compound and the addition of a hydrogen to the α-carbon.

Solution

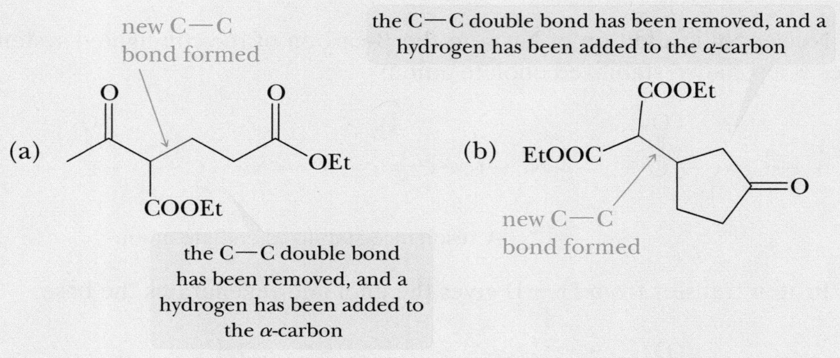

new C—C bond formed

the C—C double bond has been removed, and a hydrogen has been added to the α-carbon

(a) COOEt, OEt

(b) EtOOC, COOEt

new C—C bond formed

the C—C double bond has been removed, and a hydrogen has been added to the α-carbon

See problems 16.35, 16.36

Problem 16.10

Show the product formed from each Michael product in the solution to Example 16.10 after (1) hydrolysis in aqueous NaOH, (2) acidification, and (3) thermal decarboxylation of each β-ketoacid or β-dicarboxylic acid. These reactions illustrate the usefulness of the Michael reaction for the synthesis of 1,5-dicarbonyl compounds.

HOW TO 16.3 Determine the Starting Compounds Used in a Michael Reaction

It will sometimes be necessary to work retrosynthetically (How To 15.2) from the product of a Michael reaction. Following are the steps for determining the starting compounds used in a Michael reaction.

(a) Locate the carbonyl of the ketone, aldehyde, or ester and identify the α- and β-carbons. If there are multiple carbonyls, look for the β-carbon bonded to one of the nucleophiles presented in Table 16.2:

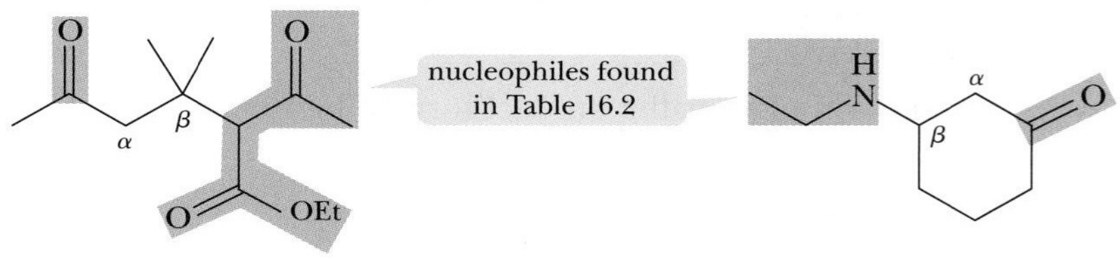

nucleophiles found in Table 16.2

(b) Erase the bond between the β-carbon and the nucleophile:

(c) Draw a C—C double bond between the α- and β-carbons. The nucleophile should be drawn as negatively charged (in the case of carbon) or as a neutral amine (in the case of nitrogen).

draw a C—C double bond and give the carbon nucleophile a negative charge

draw a C—C double bond and make the nitrogen nucleophile neutral in charge

Example 16.11

Show how the series of reactions in Example 16.10 and Problem 16.10 (Michael reaction, hydrolysis, acidification, and thermal decarboxylation) can be used to prepare 2,6-heptanedione.

Strategy

The key is to recognize that a COOH group beta to a ketone can be lost by decarboxylation. Once you find where that COOH group might have been located, you should see which carbons of the target molecule can be derived from the carbon skeleton of ethyl acetoacetate and which carbons can be derived from an α,β-unsaturated carbonyl compound.

Solution

As shown here, the target molecule can be constructed from the carbon skeletons of ethyl acetoacetate and methyl vinyl ketone:

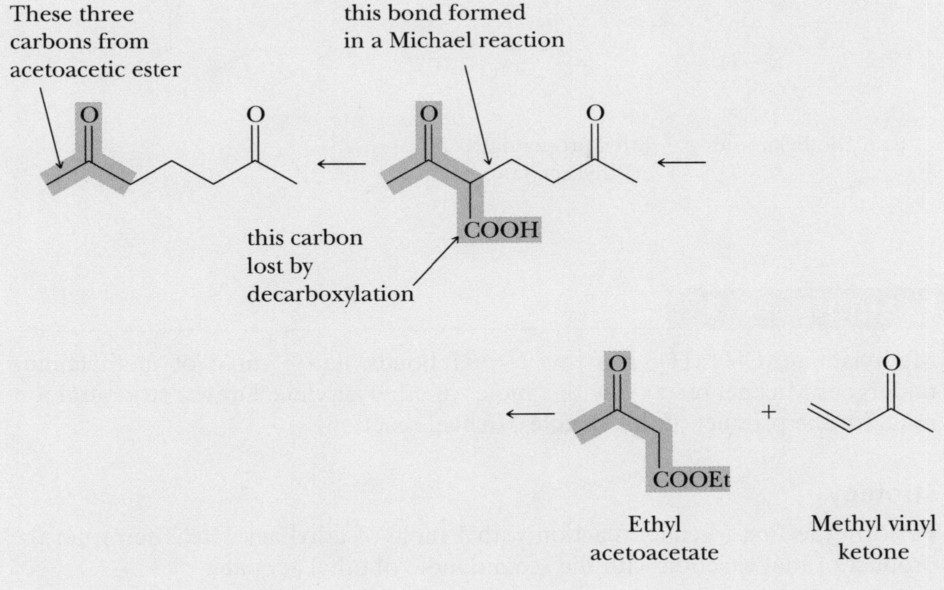

Following are the steps in their conversion to 2,6-heptanedione:

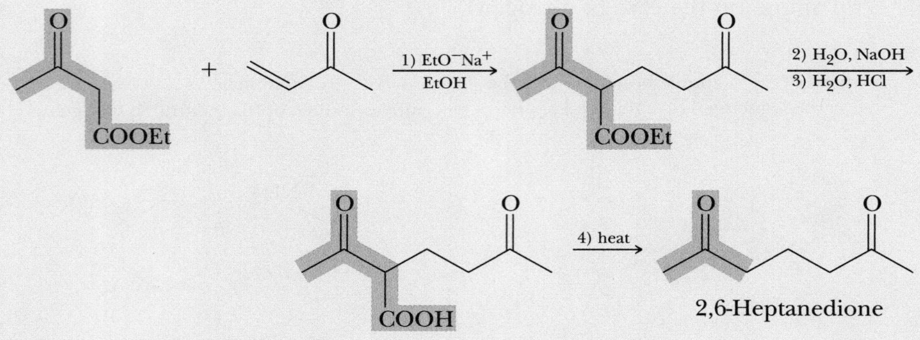

COOEt + → 1) EtO^-Na^+ / EtOH → COOEt → 2) H_2O, NaOH / 3) H_2O, HCl →

COOH → 4) heat → 2,6-Heptanedione

See problem 16.37

Problem 16.11

Show how the sequence consisting of Michael reaction, hydrolysis, acidification, and thermal decarboxylation can be used to prepare pentanedioic acid (glutaric acid).

B. Michael Addition of Amines

As Table 16.2 shows, aliphatic amines also function as nucleophiles in Michael reactions. Diethylamine, for example, adds to methyl acrylate, as shown in the following equation:

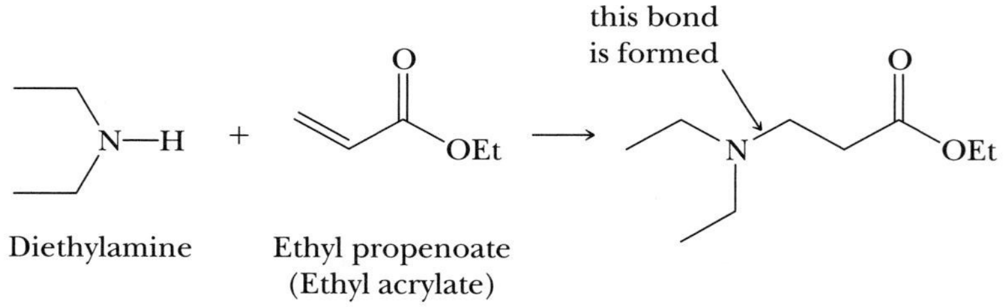

N—H + OEt → N OEt

this bond is formed

Diethylamine

Ethyl propenoate (Ethyl acrylate)

Example 16.12

Methylamine, CH_3NH_2, has two N—H bonds, and 1 mole of methylamine undergoes Michael reaction with 2 moles of ethyl acrylate. Draw a structural formula for the product of this double Michael reaction.

Strategy

Perform the first Michael reaction with 1 mole of ethyl acrylate. Then treat the product of that reaction with the second mole of ethyl acrylate.

Solution

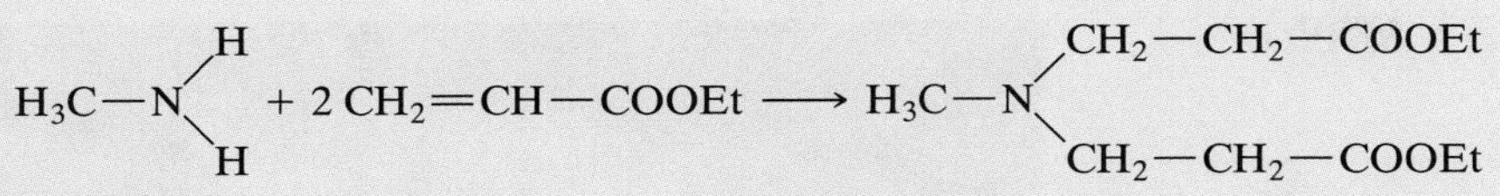

$$H_3C-NH_2 + 2\,CH_2{=}CH-COOEt \longrightarrow H_3C-N(CH_2-CH_2-COOEt)_2$$

See problem 16.36

Problem 16.12

The product of the double Michael reaction in Example 16.12 is a diester that, when treated with sodium ethoxide in ethanol, undergoes a Dieckmann condensation. Draw the structural formula for the product of this Dieckmann condensation followed by acidification with aqueous HCl.

HOW TO 16.4 Recognize the Products of Aldol, Claisen Condensation, and Michael Reactions

Aldol, Claisen condensation, and Michael reactions are some of the most important reactions in organic chemistry because they allow chemists to synthesize larger molecules from smaller, readily available compounds. The following table will help you to recognize when to use each reaction in a synthesis problem.

Reaction to Use	When This Substructure Is Needed in the Final Product	Example
aldol reaction	β-hydroxy carbonyl or α,β-unsaturated carbonyl	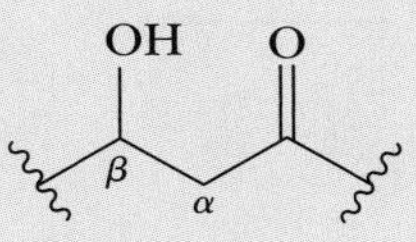 OH O β α or O β α 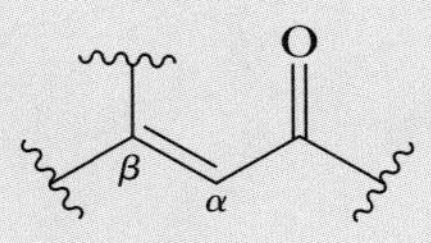
Claisen condensation	β-ketoester	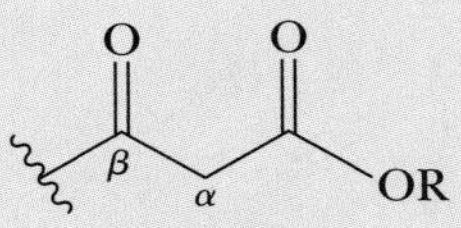 O O β α OR
Michael reaction	β-substituted carbonyl	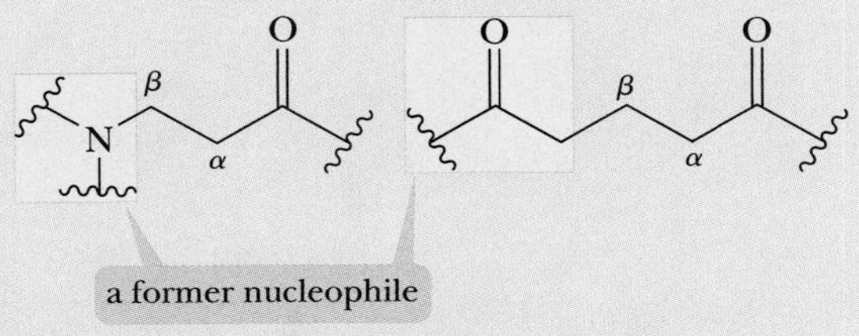 N β α O ; O β α O — a former nucleophile

CHEMICAL CONNECTIONS 16B

Antitumor Compounds: The Michael Reaction in Nature

(Image Courtesy of K.C. Nicolaou, The Scripps Research Institute)

In 1987, a scientist happened upon a red rock (see photo) during a hike while vacationing in Texas. Thinking that there might be interesting chemicals within the organisms growing on the rock, the scientist brought the rock back to his laboratory at Wyeth (formerly Lederle Labs). It was subsequently found that a compound known as calicheamicin could be extracted from the bacteria *Micromonospora echinospora*, which was growing on the rock. The chemical turned out to be bioactive. After the compound's toxicity to cells was further researched, it was found to be among the most potent of all anticancer compounds. The mode of action of calicheamicin was elucidated, and the Michael reaction was found to play a crucial part in the mechanism for reactivity. In Step 1 of the mechanism, a trisulfide group on the molecule (a) is biochemically reduced to the anion of a thiol group. This, in turn, acts as a nucleophile and attacks the α,β-unsaturated ketone of the molecule (b). The product of this intramolecular Michael reaction (c) is thought to place great strain on the enediyne portion of calicheamicin, causing a rearrangement to occur that creates a benzene ring with two unpaired electrons (radicals) that are para to each other on the ring (d). This highly reactive structure acts to cleave both strands of DNA (Section 20.2B), which is the process that makes calicheamicin so damaging to tumor cells. By modifying the sugar unit of the calicheamicin so that it could be attached to cancer-specific antibodies, this chemical, found by a curious chemist on vacation, has become a promising drug for cancer therapy.

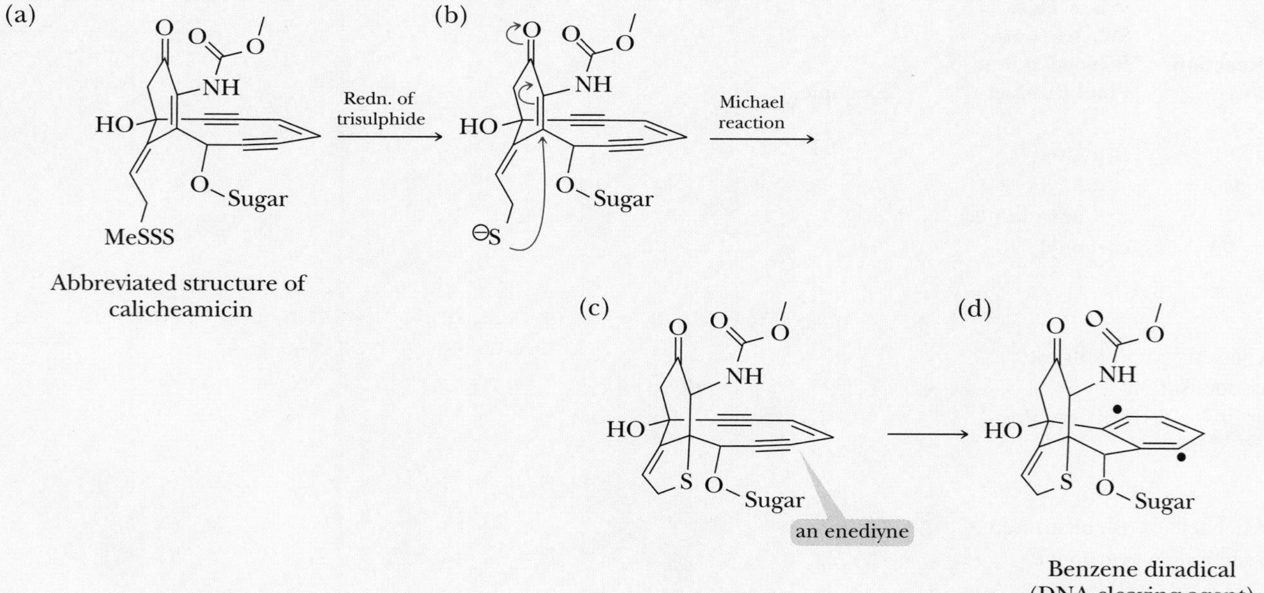

(a) Abbreviated structure of calicheamicin — Redn. of trisulphide → (b) — Michael reaction → (c) an enediyne → (d) Benzene diradical (DNA cleaving agent)

QUESTION

Provide a complete mechanism for the Michael reaction of b to produce c. Using fishhook arrows, provide a mechanism for the rearrangement of c to produce d.

Key Terms and Concepts

Summary of Key Questions

16.1 What Are Enolate Anions, and How Are They Formed?

- An **enolate anion** is an anion formed by the removal of an α-hydrogen from a carbonyl-containing compound.
- Aldehydes, ketones, and esters can be converted to their enolate anions by treatment with a metal alkoxide or other strong base.

16.2 What Is the Aldol Reaction?

- An **aldol reaction** is the addition of an enolate anion from one aldehyde or ketone to the carbonyl carbon of another aldehyde or ketone to form a **β-hydroxyaldehyde** or **β-hydroxyketone**.
- Dehydration of the product of an aldol reaction gives an **α,β-unsaturated aldehyde** or **ketone**.
- **Crossed aldol reactions** are useful only when appreciable differences in reactivity occur between the two carbonyl-containing compounds, such as when one of them has no α-hydrogens and can function only as an enolate anion acceptor.
- When both carbonyl groups are in the same molecule, aldol reaction results in the formation of a ring. These intramolecular aldol reactions are particularly useful for the formation of five- and six-membered rings.

16.3 What Are the Claisen and Dieckmann Condensations?

- A key step in the **Claisen condensation** is the addition of an enolate anion of one ester to a carbonyl group of another ester to form a tetrahedral carbonyl addition intermediate, followed by the collapse of the intermediate to give a β-ketoester.
- The **Dieckmann condensation** is an intramolecular Claisen condensation.

16.4 How Are Aldol Reactions and Claisen Condensations Involved in Biological Processes?

- **Acetyl-CoA** is the source of the carbon atoms for the synthesis of terpenes, cholesterol, steroid hormones, and fatty acids. Various enzymes catalyze biological versions of aldol reactions and Claisen condensations during the syntheses of these compounds.
- Key intermediates in the synthesis of these biomolecules are mevalonic acid and isopentenyl pyrophosphate.

16.5 What Is the Michael Reaction?

- The **Michael reaction** is the addition of a nucleophile to a carbon–carbon double bond activated by an adjacent carbonyl group.
- The Michael reaction results in the formation of a new bond between the nucleophile and the β-carbon of an α,β-unsaturated carbonyl compound. The C—C double bond is converted to a C—C single bond in the reaction.

Quick Quiz

Answer true or false to the following questions to assess your general knowledge of the concepts in this chapter. If you have difficulty with any of them, you should review the appropriate section in the chapter (shown in parentheses) before attempting the more challenging end-of-chapter problems.

1. All ketones and aldehydes with a carbon atom alpha to the carbonyl group can be converted to an enolate anion by treatment with a catalytic amount of base. (16.1)
2. A Dieckmann condensation favors seven- or eight-membered rings over four-, five-, or six- membered rings. (16.3)
3. An intramolecular aldol reaction favors five- or six-membered rings over four-, seven-, or eight-membered rings. (16.2)
4. A hydrogen that is alpha to two carbonyls is less acidic than a hydrogen that is alpha to only one carbonyl. (16.1)
5. The product of a Claisen condensation is a β-hydroxyester. (16.3)
6. The mechanism of a Michael reaction involves enol–keto tautomerization. (16.5)
7. An enolate anion can act as a nucleophile. (16.1)
8. An aldol reaction involves the reaction of an enolate anion with a ketone or an aldehyde. (16.2)
9. The product of an aldol reaction is a β-hydroxyester. (16.2)
10. Aldol reactions and Claisen condensations can be catalyzed by enzymes. (16.4)
11. A crossed aldol reaction is most effective when one of the carbonyl compounds is more reactive toward nucleophilic addition and cannot form an enolate anion. (16.2)
12. Hydrogen atoms alpha to a carbonyl are many times more acidic than vinyl or alkyl hydrogens. (16.1)
13. The Claisen condensation is a reaction between an enolate anion and an ester. (16.3)
14. The α-hydrogen of an ester is more acidic than the α-hydrogen of a ketone. (16.3)
15. An enolate anion is stabilized by resonance. (16.1)
16. All carbonyl compounds with an alpha hydrogen can be converted to an enolate anion by treatment with a catalytic amount of base. (16.1)
17. A crossed Claisen condensation is most effective when one of the carbonyl compounds can only function as an enolate anion acceptor. (16.3)
18. An enolate anion can participate in a Michael reaction. (16.5)
19. The product of an aldol reaction can be dehydrated to yield an α,β-unsaturated carbonyl compound. (16.2)
20. The Michael reaction is the reaction of a nucleophile with the β-carbon of an α,β-unsaturated carbonyl compound. (16.5)
21. An enolate anion can act as a base. (16.1)
22. The product of a Claisen condensation can be hydrolyzed and decarboxylated to form a ketone. (16.3)
23. An amine can participate in a Michael reaction. (16.5)

Answers: (1) F (2) F (3) T (4) F (5) F (6) T (7) T (8) T (9) F (10) T (11) T (12) T (13) T (14) F (15) T (16) F (17) T (18) T (19) T (20) T (21) T (22) T (23) T

Key Reactions

1. The Aldol Reaction (Section 16.2B)

The aldol reaction involves the nucleophilic addition of an enolate anion from one aldehyde or ketone to the carbonyl carbon of another aldehyde or ketone to give a β-hydroxyaldehyde or β-hydroxyketone:

2 [butanal] $\xrightleftharpoons{\text{NaOH}}$ [2-ethyl-3-hydroxyhexanal]

2. Dehydration of the Product of an Aldol Reaction (Section 16.2)

Dehydration of the β-hydroxyaldehyde or ketone from an aldol reaction occurs readily and gives an α,β-unsaturated aldehyde or β-hydroxyketone:

[3-hydroxybutanal] $\xrightarrow{H^+}$ [2-butenal] + H_2O

3. The Claisen Condensation (Section 16.3A)

The product of a Claisen condensation is a β-ketoester:

[ethyl propanoate] $\xrightarrow[\text{2) } H_2O,\ HCl]{\text{1) } EtO^-Na^+}$ [ethyl 2-methyl-3-oxopentanoate] + EtOH

Condensation occurs by nucleophilic acyl substitution in which the attacking nucleophile is the enolate anion of an ester.

4. The Dieckmann Condensation (Section 16.3B)

An intramolecular Claisen condensation is called a Dieckmann condensation:

[diethyl hexanedioate] $\xrightarrow[\text{2) } H_2O,\ HCl]{\text{1) } EtO^-Na^+}$ [ethyl 2-oxocyclopentanecarboxylate] + EtOH

5. Crossed Claisen Condensations (Section 16.3C)

Crossed Claisen condensations are useful only when an appreciable difference exists in the reactivity between the two esters. Such is the case when an ester that has no α-hydrogens can function only as an enolate anion acceptor:

[ethyl benzoate] + [ethyl propanoate] $\xrightarrow[\text{2) } H_2O,\ HCl]{\text{1) } EtO^-Na^+}$ [ethyl 2-methyl-3-oxo-3-phenylpropanoate] + EtOH

6. Hydrolysis and Decarboxylation of β-Ketoesters (Section 16.3D)

Hydrolysis of the ester, followed by decarboxylation of the resulting β-ketoacid, gives a ketone and carbon dioxide:

[COOEt-substituted ketone] $\xrightarrow[\text{3) Heat}]{\text{1) NaOH, } H_2O \text{; 2) } H_2O,\ HCl}$ [ketone] + CO_2

7. The Michael Reaction (Section 16.5)

Attack of a nucleophile at the β-carbon of an α,β-unsaturated carbonyl compound results in conjugate addition:

[ethyl acetoacetate (COOEt)] + [methyl vinyl ketone] $\xrightarrow[\text{EtOH}]{EtO^-Na^+}$ [Michael adduct (COOEt)]

Problems

A problem marked with an asterisk indicates an applied "real world" problem. Answers to problems whose numbers are printed in blue are given in Appendix D.

Sections 16.1 and 16.2 The Aldol Reaction

16.13 Identify the most acidic hydrogen(s) in each compound: **(See Example 16.1)**

(a) 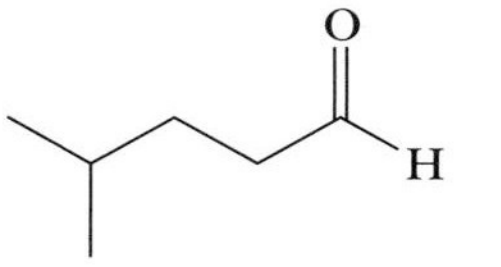 (b) 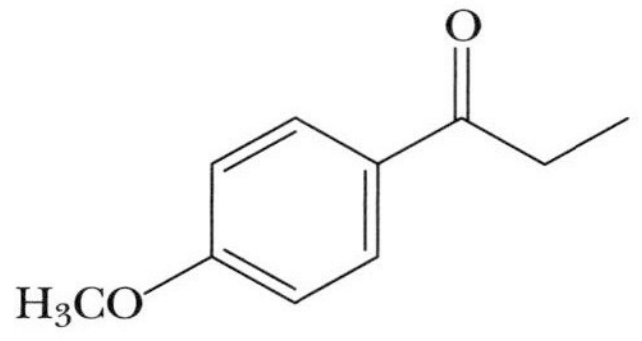 H_3CO (c) 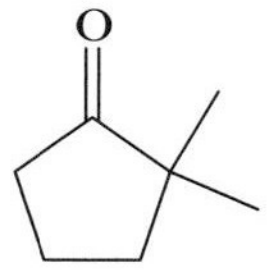 (d) 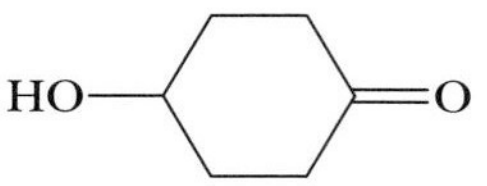 HO— —O (e) 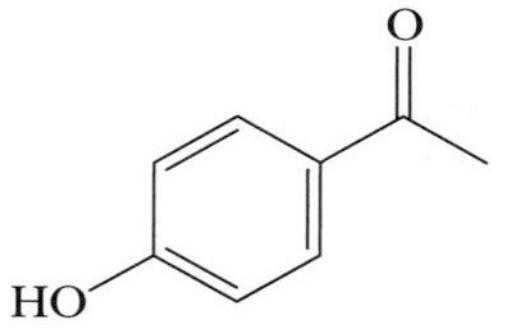 HO (f) 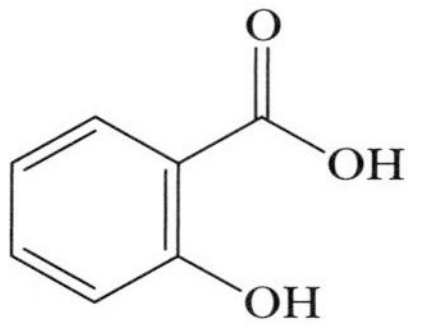 OH OH

16.14 Estimate the pK_a of each compound and arrange them in order of increasing acidity:

(a) $CH_3\overset{O}{\overset{\|}{C}}CH_3$ (b) $CH_3\overset{OH}{\overset{|}{C}}HCH_3$

(c) $CH_3CH_2\overset{O}{\overset{\|}{C}}OH$

16.15 Write a second contributing structure of each anion, and use curved arrows to show the redistribution of electrons that gives your second structure:

(a) $CH_3CH_2\overset{:\ddot{O}:^-}{\overset{|}{C}}{=}CHCH_3$ (b) 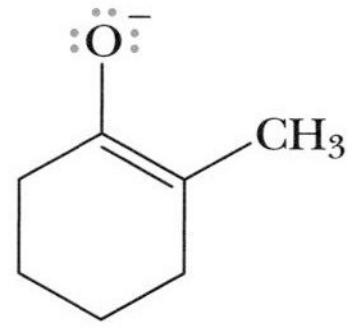 $:\ddot{O}:^-$ CH_3

(c) 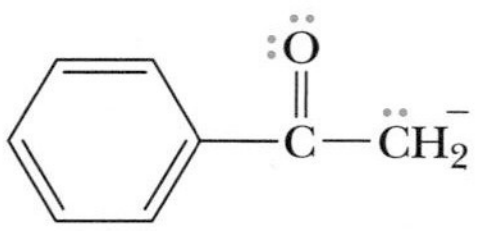 $:\ddot{O}$ $\overset{\|}{C}-\ddot{C}H_2^-$

16.16 Treatment of 2-methylcyclohexanone with base gives two different enolate anions. Draw the contributing structure for each that places the negative charge on carbon.

16.17 Draw a structural formula for the product of the aldol reaction of each compound and for the α,β-unsaturated aldehyde or ketone formed by dehydration of each aldol product: **(See Examples 16.2, 16.3)**

(a) 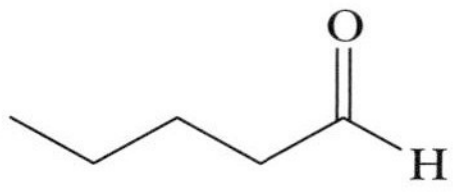 (b) 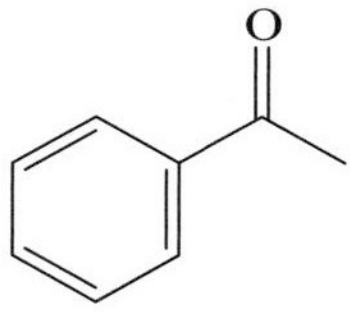 (c) 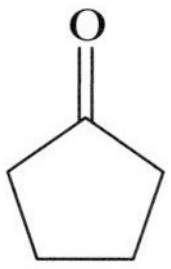 (d) 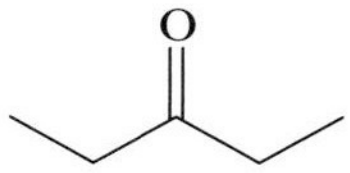 (e) 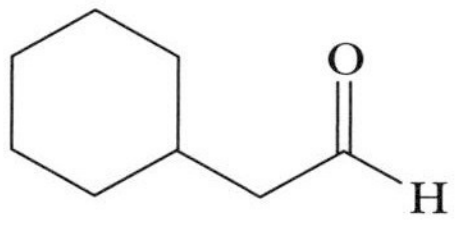 (f)

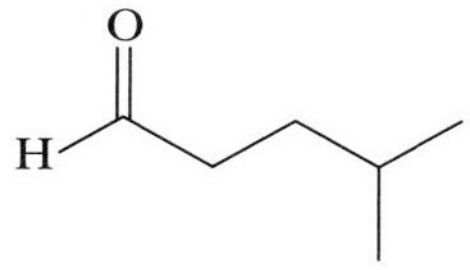

16.18 Draw a structural formula for the product of each crossed aldol reaction and for the compound formed by dehydration of each aldol product: **(See Examples 16.3, 16.4)**

(a) $(CH_3)_3\overset{O}{\overset{\|}{C}}H + CH_3\overset{O}{\overset{\|}{C}}CH_3$

(b) 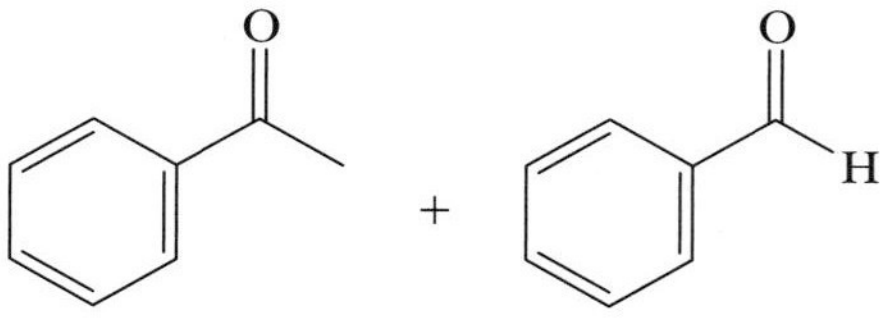 +

(c) 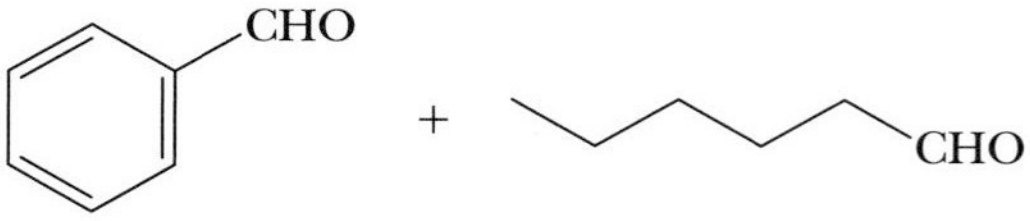 CHO + CHO

(d) 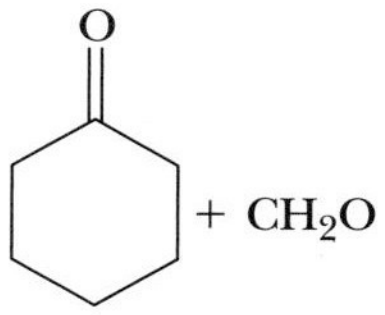 $+ CH_2O$

16.19 When a 1:1 mixture of acetone and 2-butanone is treated with base, six aldol products are possible. Draw a structural formula for each. **(See Example 16.4)**

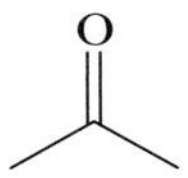

Acetone 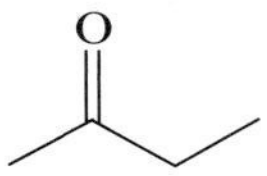   2-Butanone

16.20 Show how to prepare each α,β-unsaturated ketone by an aldol reaction followed by dehydration of the aldol product: **(See Examples 16.2, 16.3)**

(a) 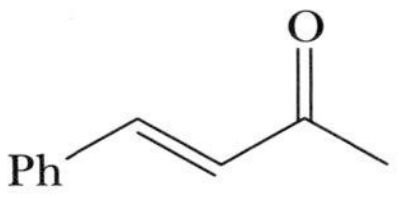 Ph   (b)

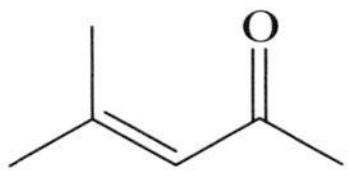

16.21 Show how to prepare each α,β-unsaturated aldehyde by an aldol reaction followed by dehydration of the aldol product: **(See Examples 16.2, 16.3)**

(a) 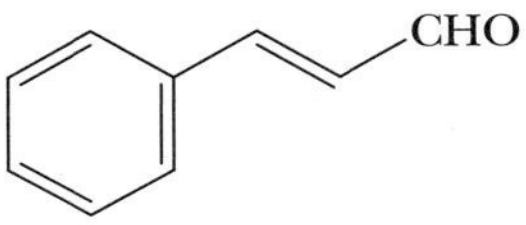 CHO

(b) 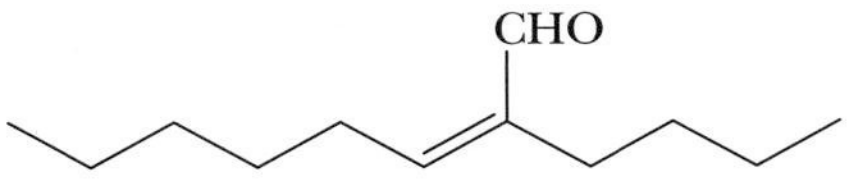 CHO

16.22 When treated with base, the following compound undergoes an intramolecular aldol reaction, followed by dehydration, to give a product containing a ring (yield 78%): **(See Examples 16.3, 16.5)**

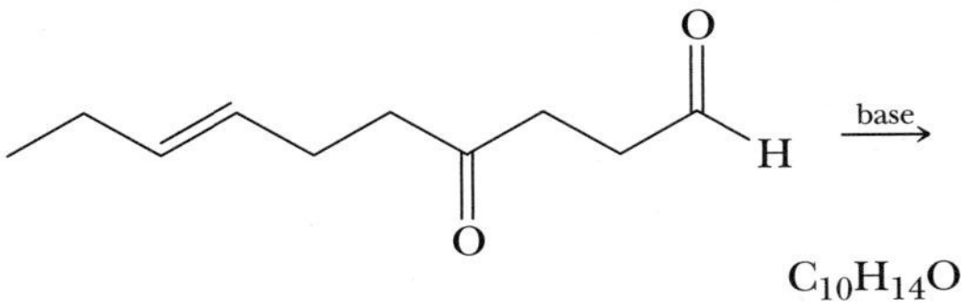

$\xrightarrow{\text{base}}$ $C_{10}H_{14}O + H_2O$

Propose a structural formula for this product.

16.23 Propose a structural formula for the compound with molecular formula $C_6H_{10}O_2$ that undergoes an aldol reaction followed by dehydration to give this α,β-unsaturated aldehyde: **(See Examples 16.3, 16.5)**

$C_6H_{10}O_2 \xrightarrow{\text{base}}$ CHO $+ H_2O$

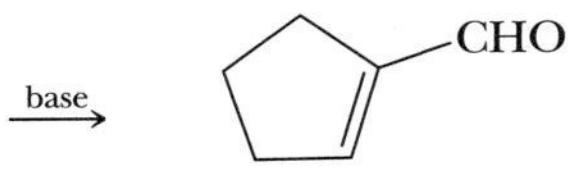

1-Cyclopentenecarbaldehyde

16.24 Show how to bring about this conversion: **(See Examples 16.3, 16.5)**

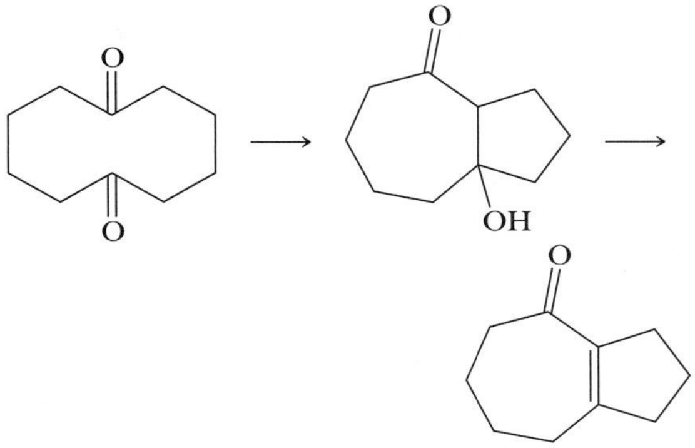

O → O, OH →

***16.25** Oxanamide, a mild sedative, is synthesized from butanal in these five steps:

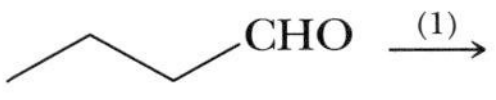

CHO $\xrightarrow{(1)}$ H $\xrightarrow{(2)}$

Butanal 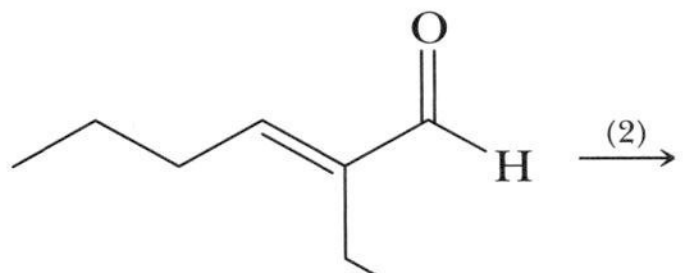   2-Ethyl-2-hexenal

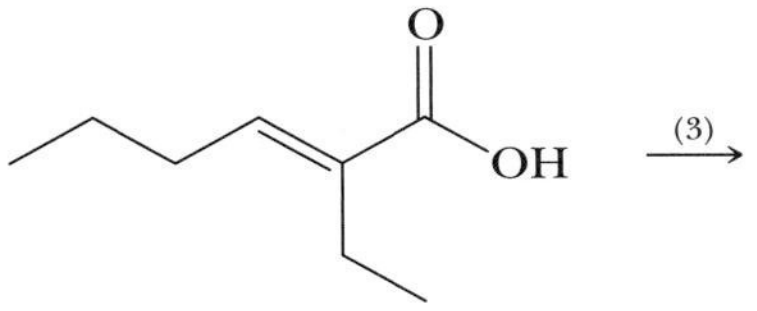

OH $\xrightarrow{(3)}$

2-Ethyl-2-hexenoic acid

Cl $\xrightarrow{(4)}$

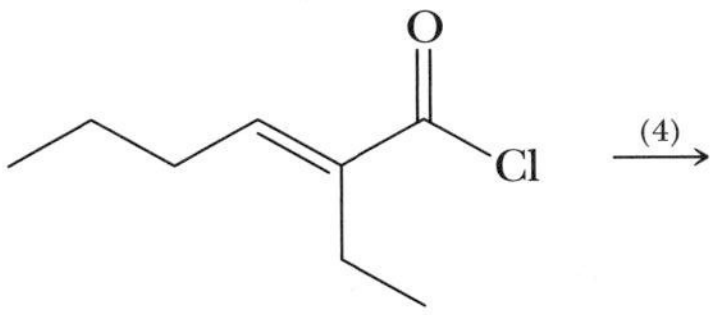

2-Ethyl-2-hexenoyl chloride

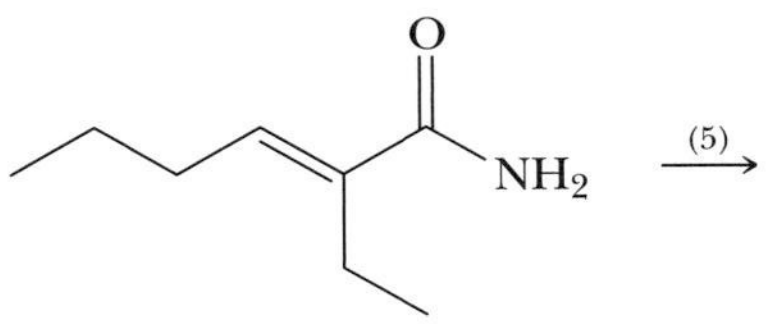

NH_2 $\xrightarrow{(5)}$

2-Ethyl-2-hexenamide

NH_2

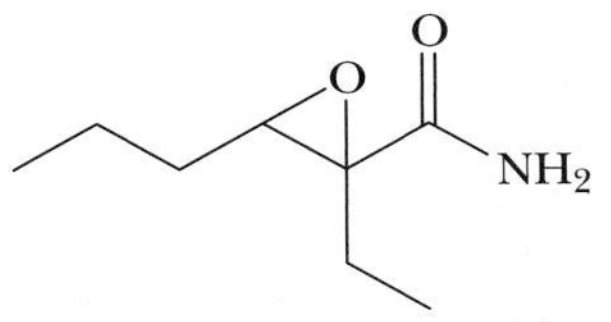

Oxanamide

(a) Show reagents and experimental conditions that might be used to bring about each step in this synthesis.

(b) How many stereocenters are in oxanamide? How many stereoisomers are possible for oxanamide?

16.26 Propose structural formulas for compounds A and B:

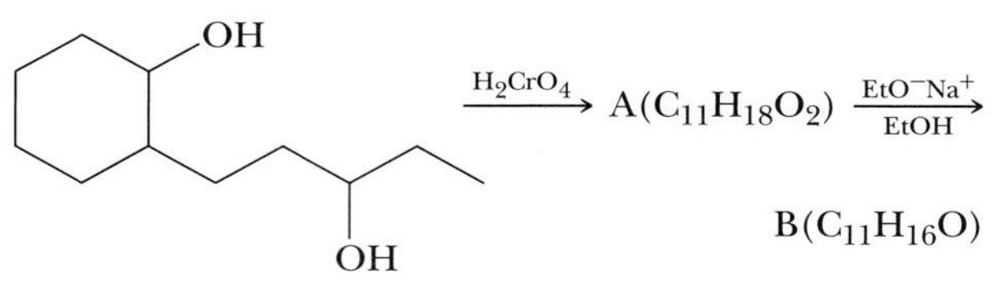

$\xrightarrow{H_2CrO_4}$ A($C_{11}H_{18}O_2$) $\xrightarrow[EtOH]{EtO^-Na^+}$ B($C_{11}H_{16}O$)

Section 16.3 The Claisen and Dieckmann Condensations

16.27 Show the product of the Claisen condensation of each ester: **(See Example 16.6)**

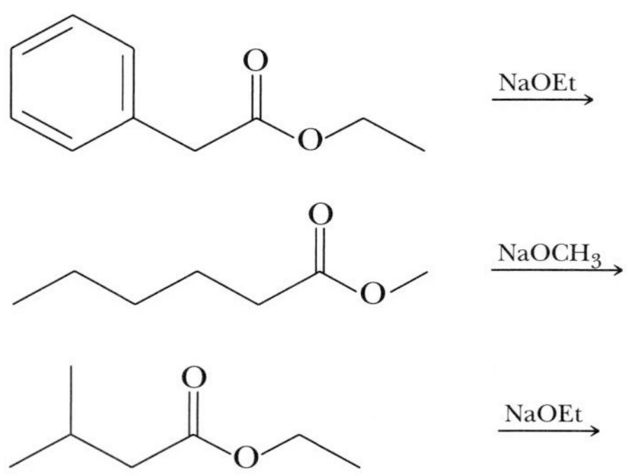

(a) $\xrightarrow{NaOEt}$

(b) $\xrightarrow{NaOCH_3}$

(c) $\xrightarrow{NaOEt}$

16.28 Draw a structural formula for the product of saponification, acidification, and decarboxylation of each β-ketoester formed in Problem 16.27. **(See Example 16.9)**

16.29 When a 1 : 1 mixture of ethyl propanoate and ethyl butanoate is treated with sodium ethoxide, four Claisen condensation products are possible. Draw a structural formula for each product. **(See Example 16.8)**

Ethyl propanoate Ethyl butanoate

16.30 Draw a structural formula for the β-ketoester formed in the crossed Claisen condensation of ethyl propanoate with each ester: **(See Example 16.8)**

(a) EtOC(=O)—C(=O)OEt (b) PhC(=O)OEt (c) HC(=O)OEt

16.31 Complete the equation for this crossed Claisen condensation: **(See Example 16.8)**

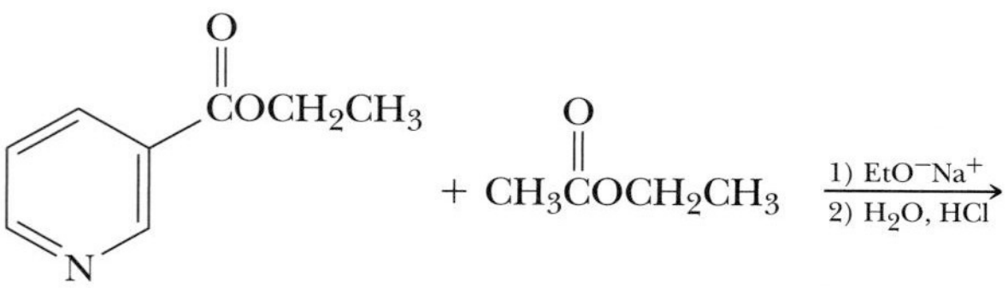

$COCH_2CH_3$ (O) + $CH_3COCH_2CH_3$ (O) $\xrightarrow[2)\ H_2O,\ HCl]{1)\ EtO^-Na^+}$

16.32 The Claisen condensation can be used as one step in the synthesis of ketones, as illustrated by this reaction sequence: **(See Example 16.9)**

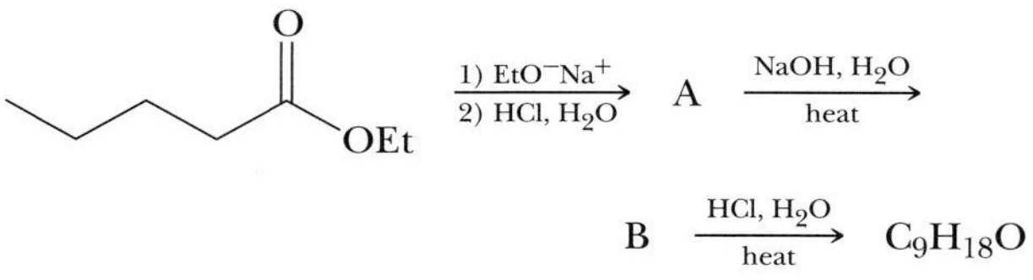

OEt $\xrightarrow[2)\ HCl,\ H_2O]{1)\ EtO^-Na^+}$ A $\xrightarrow[heat]{NaOH,\ H_2O}$ B $\xrightarrow[heat]{HCl,\ H_2O}$ $C_9H_{18}O$

Propose structural formulas for compounds A, B, and the ketone formed in the sequence.

16.33 Draw a structural formula for the product of treating each diester with sodium ethoxide followed by acidification with HCl (*Hint:* These are Dieckmann condensations): **(See Example 16.7)**

(a) COOEt COOEt

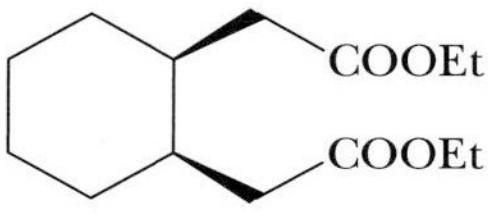

(b) EtO OEt

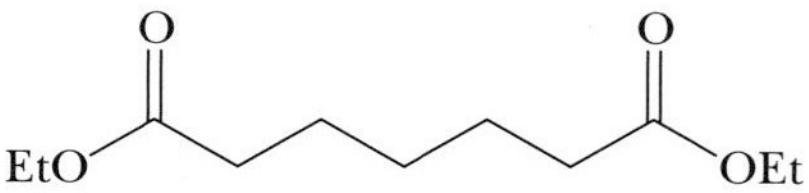

16.34 Claisen condensation between diethyl phthalate and ethyl acetate, followed by saponification, acidification, and decarboxylation, forms a diketone, $C_9H_6O_2$. Propose structural formulas for compounds A, B, and the diketone: **(See Example 16.9)**

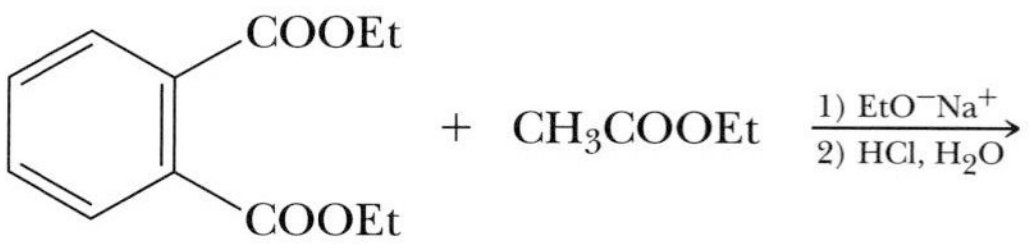

COOEt COOEt + CH_3COOEt $\xrightarrow[2)\ HCl,\ H_2O]{1)\ EtO^-Na^+}$

Diethyl phthalate Ethyl acetate

A $\xrightarrow[heat]{NaOH,\ H_2O}$ B $\xrightarrow[heat]{HCl,\ H_2O}$ $C_9H_6O_2$

***16.35** The rodenticide and insecticide pindone is synthesized by the following sequence of reactions: **(See Example 16.10)**

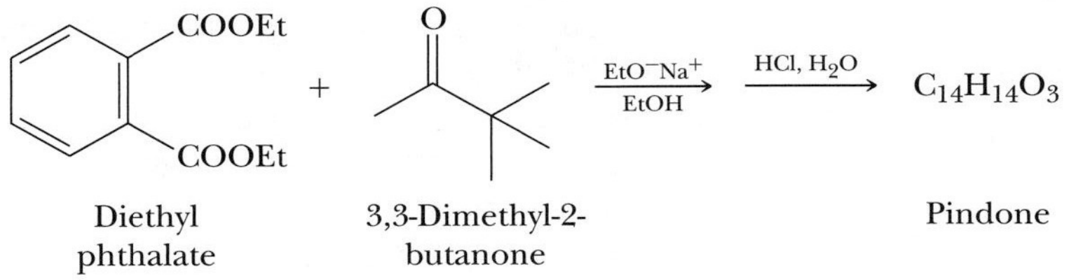

COOEt
COOEt
Diethyl phthalate + O 3,3-Dimethyl-2-butanone $\xrightarrow[\text{EtOH}]{\text{EtO}^-\text{Na}^+}$ $\xrightarrow{\text{HCl, H}_2\text{O}}$ $C_{14}H_{14}O_3$ Pindone

Propose a structural formula for pindone.

Section 16.5 The Michael Reaction

16.36 Show the product of the Michael reaction of each α,β-unsaturated carbonyl compound: **(See Examples 16.10, 16.12)**

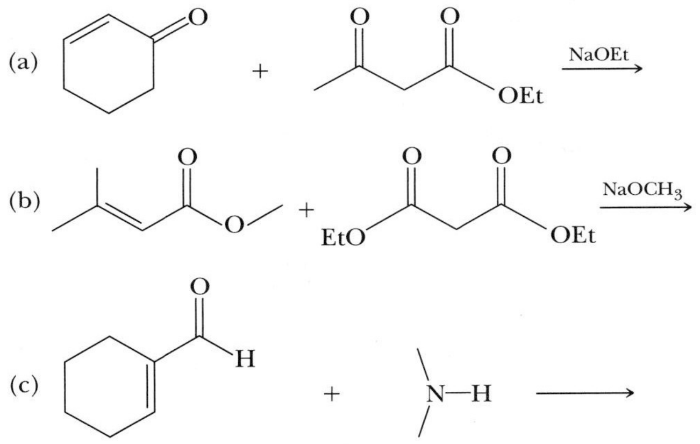

(a) O + O O OEt $\xrightarrow{\text{NaOEt}}$

(b) O O + O O EtO OEt $\xrightarrow{\text{NaOCH}_3}$

(c) O H + N—H $\longrightarrow$

***16.37** Show the outcomes of subjecting the Michael reaction products in Problems 16.36a and 16.36b to hydrolysis, followed by acidification, followed by thermal decarboxylation. **(See Example 16.11)**

***16.38** The classic synthesis of the steroid cortisone, a drug used to treat some types of allergies, involves a Michael reaction in which 1-penten-3-one and the cyclic compound shown are treated with NaOH in the solvent dioxane. Provide a structure for the product of this reaction.

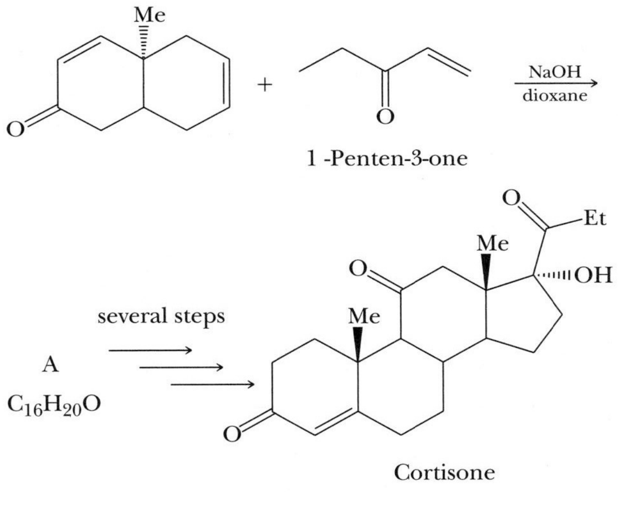

Me O + O 1 -Penten-3-one $\xrightarrow[\text{dioxane}]{\text{NaOH}}$

A $C_{16}H_{20}O$ $\xrightarrow{\text{several steps}}$ Cortisone (O, Et, Me, OH, O, Me, O)

Synthesis

***16.39** Fentanyl is a nonopoid (nonmorphinelike) analgesic used for the relief of severe pain. It is approximately 50 times more potent in humans than morphine itself. One synthesis for fentanyl begins with 2-phenylethanamine:

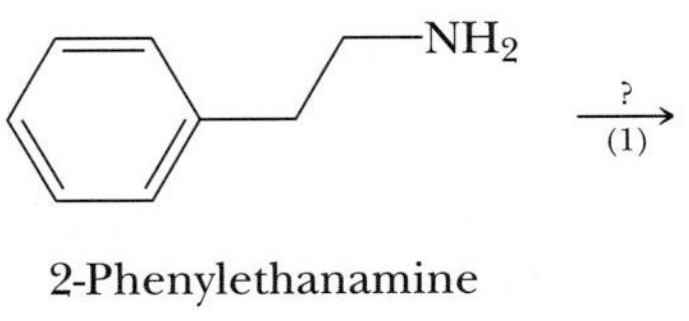

NH_2 2-Phenylethanamine $\xrightarrow[(1)]{?}$

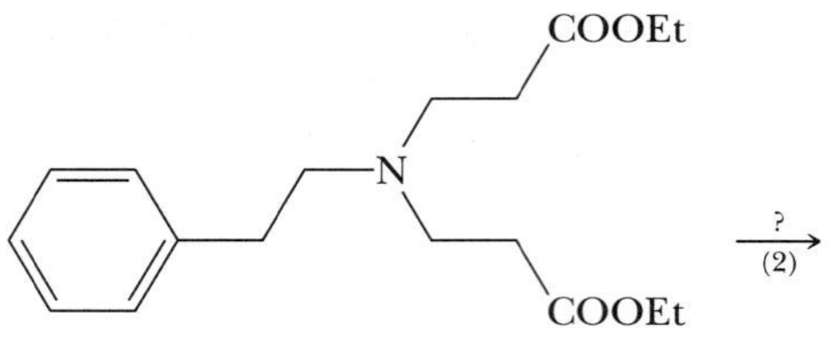

COOEt

N

COOEt

$\xrightarrow[(2)]{?}$

(A)

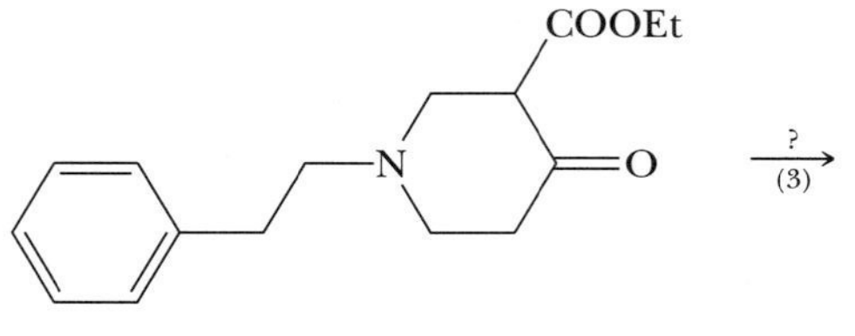

COOEt

N

O

$\xrightarrow[(3)]{?}$

(B)

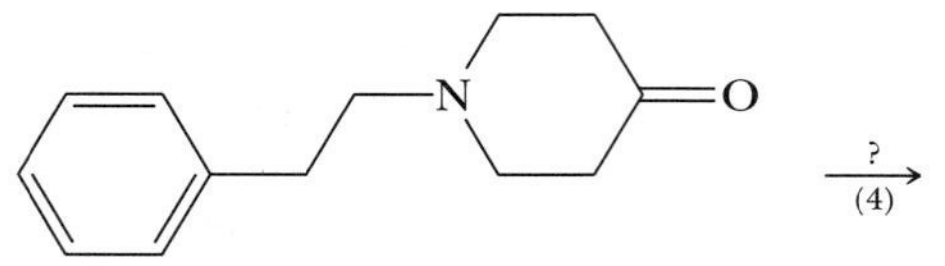

N

O

$\xrightarrow[(4)]{?}$

(C)

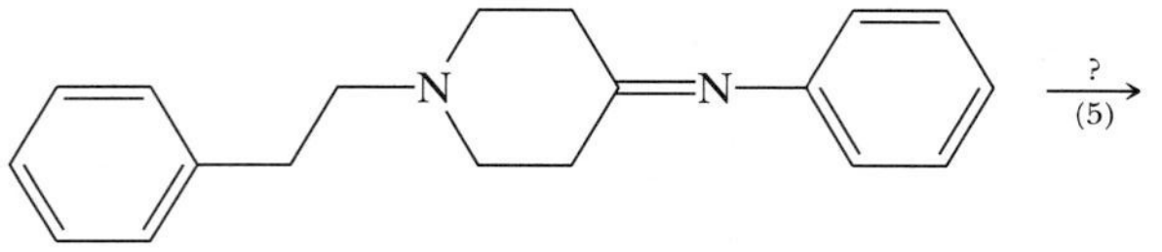

N

N

$\xrightarrow[(5)]{?}$

(D)

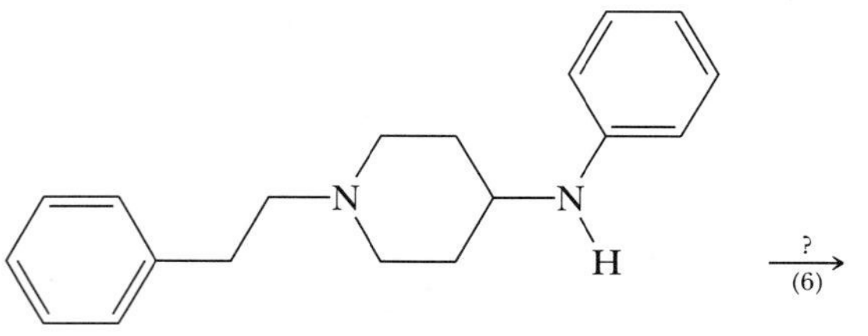

N

N

H

$\xrightarrow[(6)]{?}$

(E)

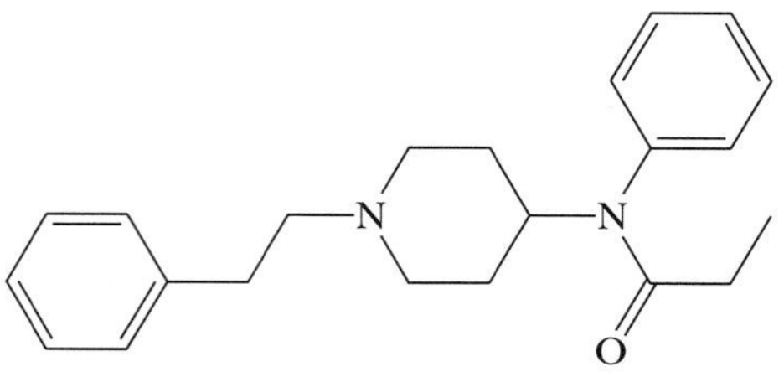

N

N

O

Fentanyl

(a) Propose a reagent for Step 1. Name the type of reaction that occurs in this step.

(b) Propose a reagent to bring about Step 2. Name the type of reaction that occurs in this step.

(c) Propose a series of reagents that will bring about Step 3.

(d) Propose a reagent for Step 4. Identify the imine (Schiff base) part of Compound D.

(e) Propose a reagent to bring about Step 5.

(f) Propose two different reagents, either of which will bring about Step 6.

(g) Is fentanyl chiral? Explain.

***16.40** Meclizine is an antiemetic. (It helps prevent or at least lessen the vomiting associated with motion sickness, including seasickness.) Among the names of the over-the-counter preparations of meclizine are Bonine®, Sea-Legs, Antivert®, and Navicalm®. Meclizine can be produced by the following series of reactions:

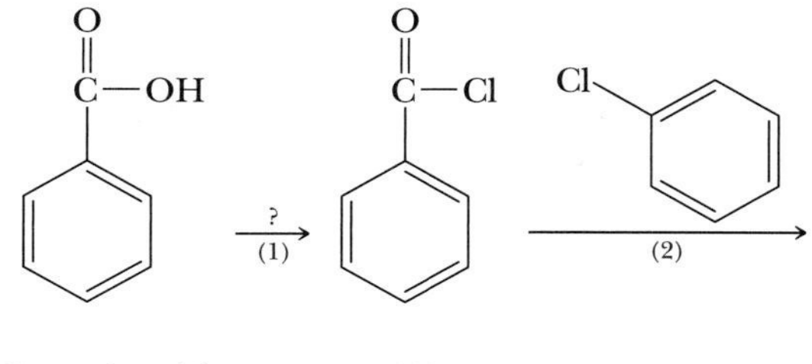

O
‖
C—OH

$\xrightarrow[(1)]{?}$

O
‖
C—Cl

Cl

$\xrightarrow{(2)}$

Benzoic acid

(A)

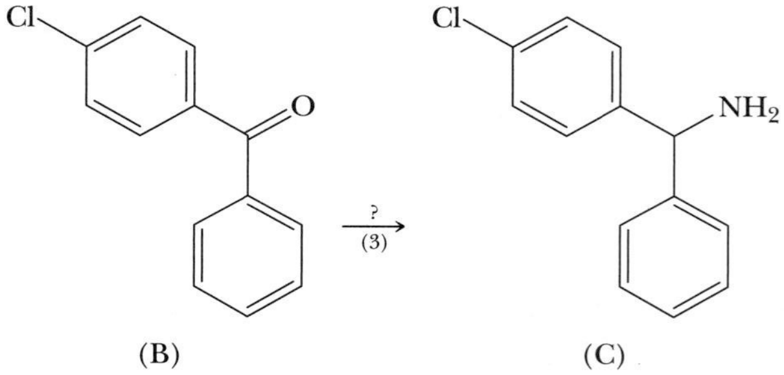

Cl

O

$\xrightarrow[(3)]{?}$

Cl

NH_2

(B)

(C)

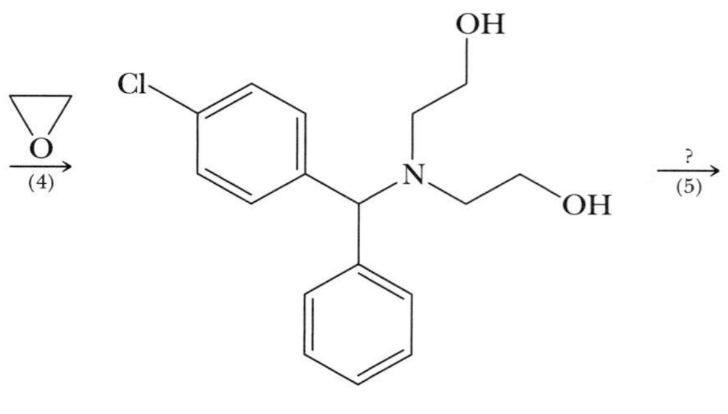

O

$\xrightarrow{(4)}$

Cl

OH

N

OH

$\xrightarrow[(5)]{?}$

(D)

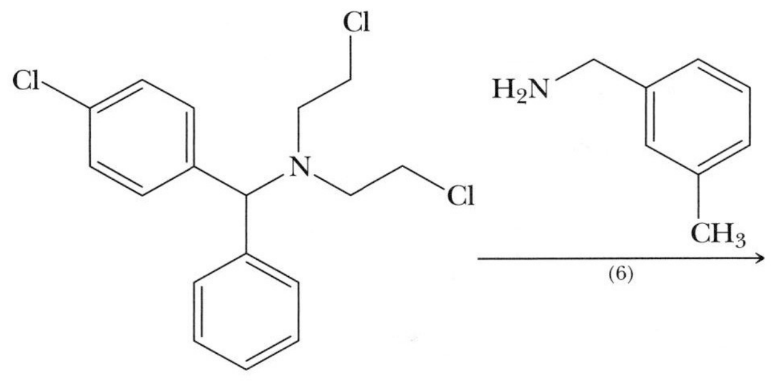

Cl (E) H_2N CH_3 (6)

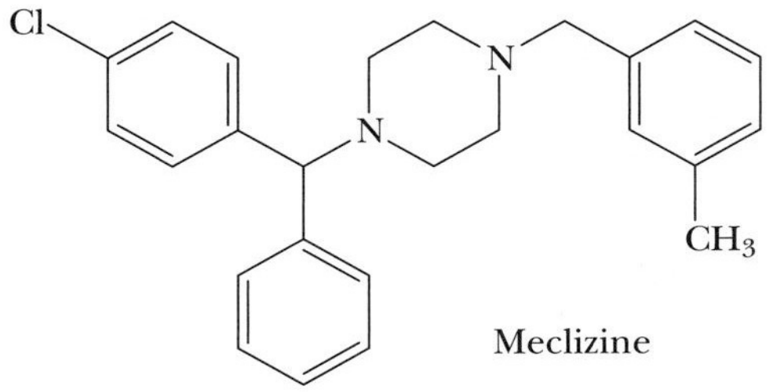

Cl Meclizine CH_3

(a) Name the functional group in (A). What reagent is most commonly used to convert a carboxyl group to this functional group?

(b) The catalyst for Step 2 is aluminum chloride, $AlCl_3$. Name the type of reaction that occurs in this step. The product shown here has the orientation of the new group para to the chlorine atom of chlorobenzene. Suppose you were not told the orientation of the new group. Would you have predicted it to be ortho, meta, or para to the chlorine atom? Explain.

(c) What set of reagents can be used in Step 3 to convert the C=O group to an $-NH_2$ group?

(d) The reagent used in Step 4 is the cyclic ether ethylene oxide. Most ethers are quite unreactive to nucleophiles such as the 1° amine in this step. Ethylene oxide, however, is an exception to this generalization. What is it about ethylene oxide that makes it so reactive toward ring-opening reactions with nucleophiles?

(e) What reagent can be used in Step 5 to convert each 1° alcohol to a 1° halide?

(f) Step 6 is a double nucleophilic displacement. Which mechanism is more likely for this reaction, S_N1 or S_N2? Explain.

***16.41** 2-Ethyl-1-hexanol is used for the synthesis of the sunscreen octyl *p*-methoxycinnamate. (See "Chemical Connections 15A.") This primary alcohol can be synthesized from butanal by the following series of steps:

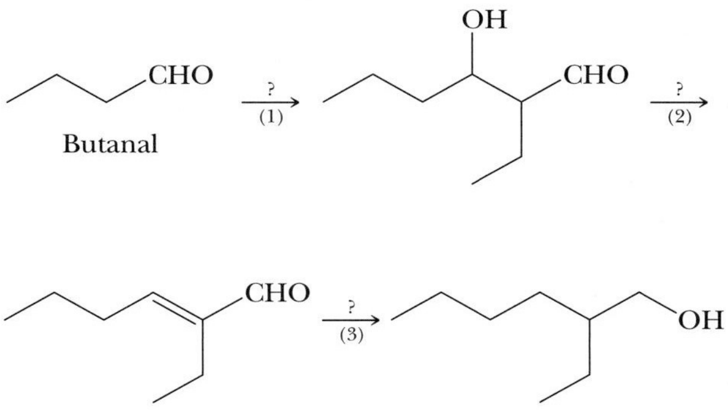

Butanal CHO →? (1) OH CHO →? (2)

CHO →? (3) OH

2-Ethyl-1-hexanol

(a) Propose a reagent to bring about Step 1. What name is given to this type of reaction?

(b) Propose a reagent for Step 2.

(c) Propose a reagent for Step 3.

(d) Following is a structural formula for the commercial sunscreening ingredient:

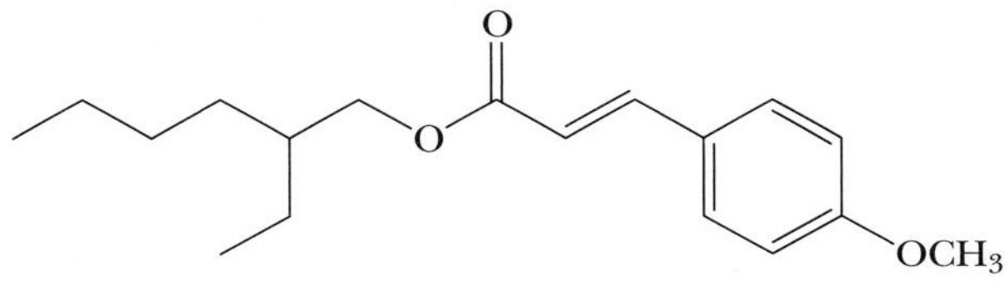

O O OCH_3

Octyl *p*-methoxycinnamate

What carboxylic acid would you use to form this ester? How would you bring about the esterification reaction?

Chemical Transformations

16.42 Test your cumulative knowledge of the reactions learned thus far by completing the following chemical transformations. *Note:* Most will require more than one step.

(a) OH → O H

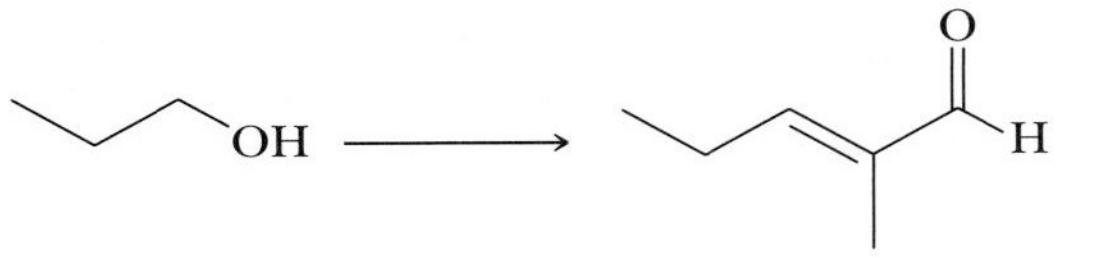

(b) OH → O O OEt

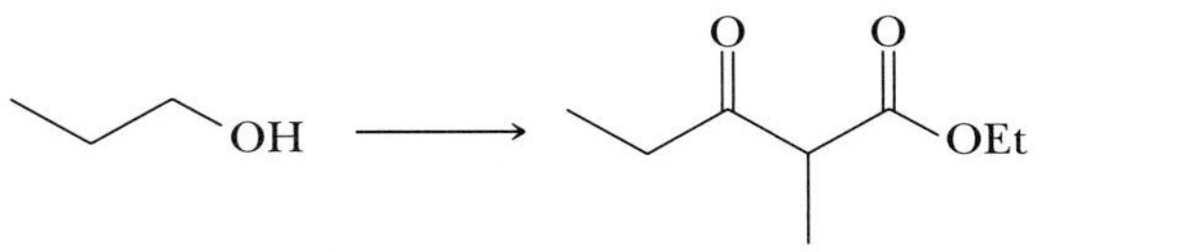

(c) OH → O O H

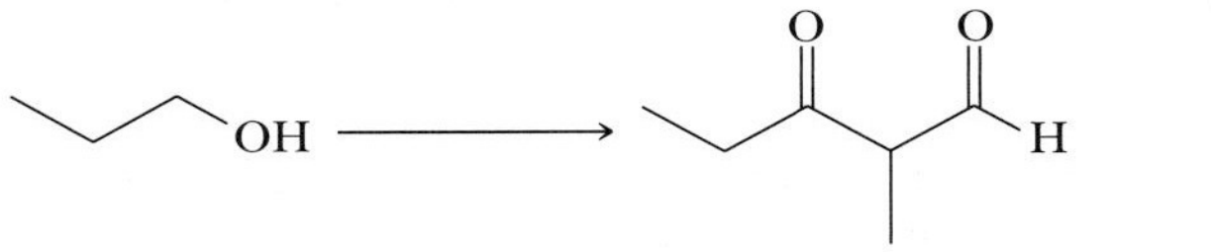

(d) 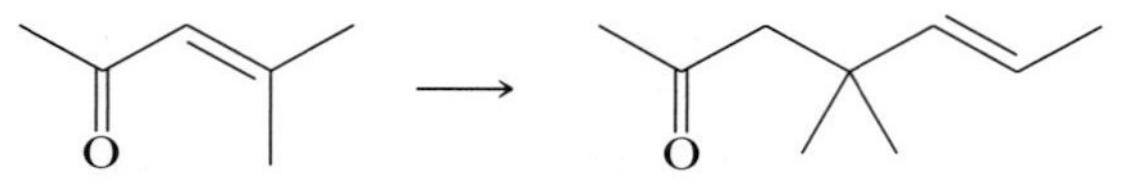 O, O →

(e) 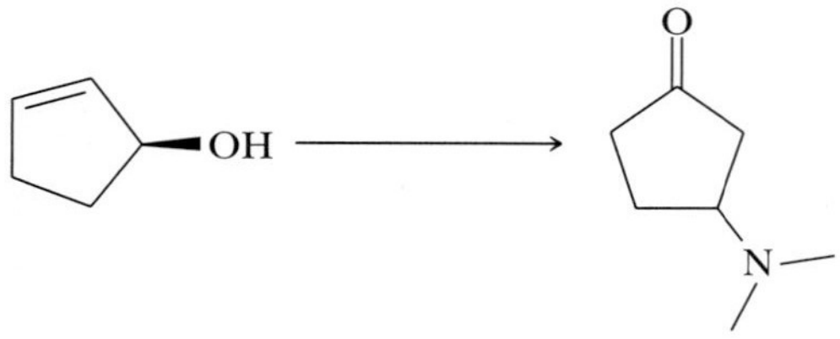 OH → O, N

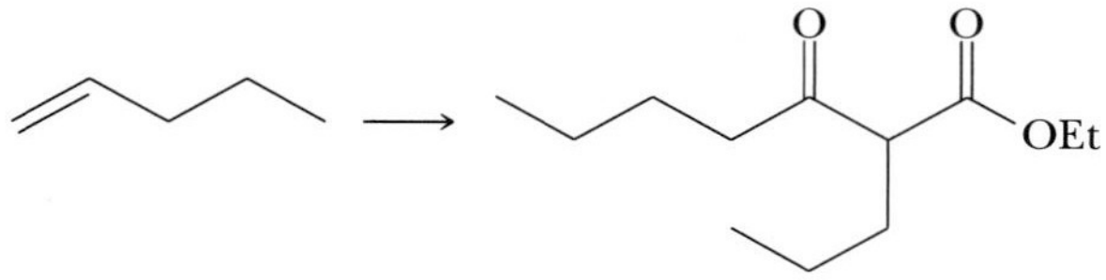

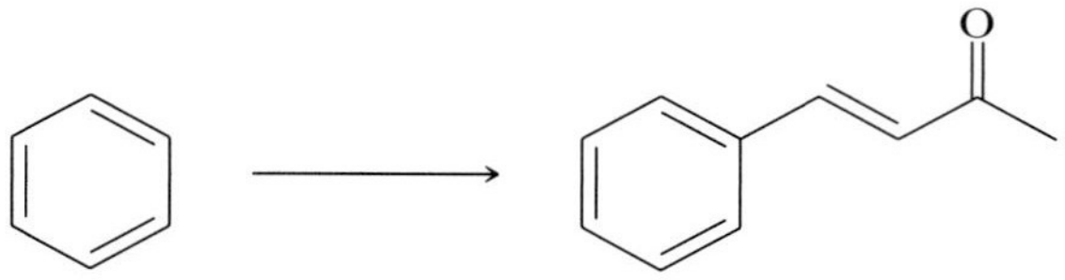

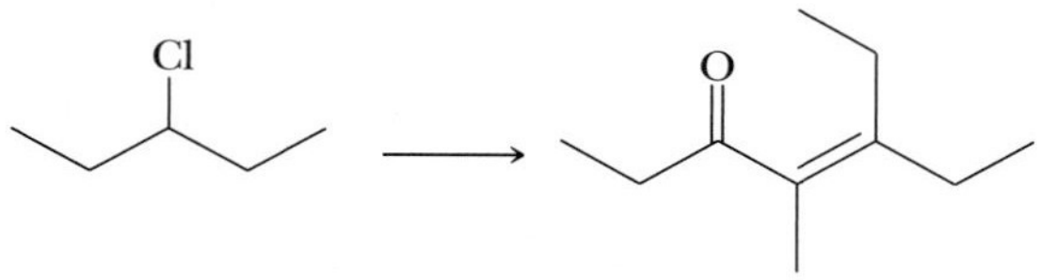

(f) → O, O, OEt

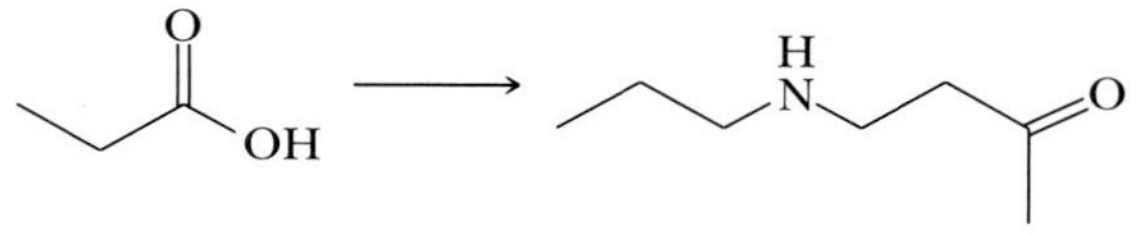

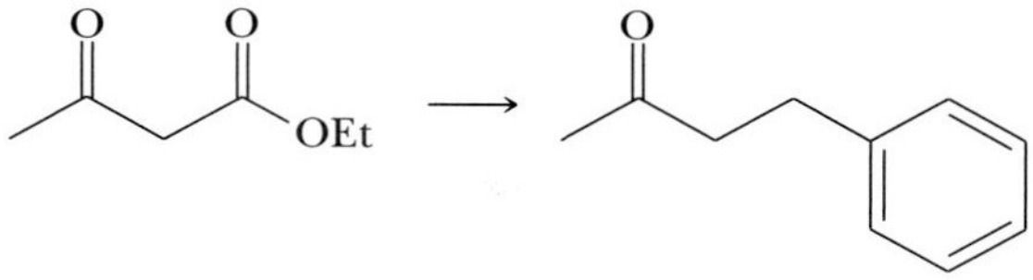

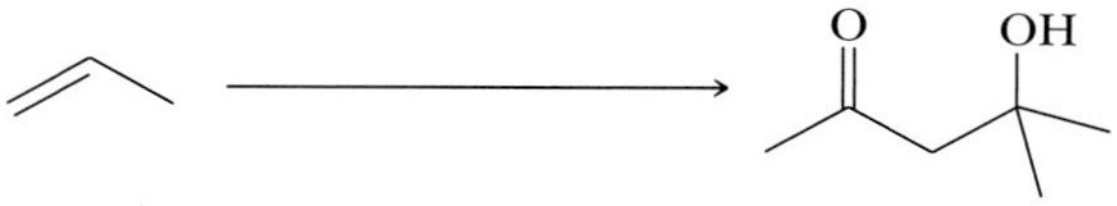

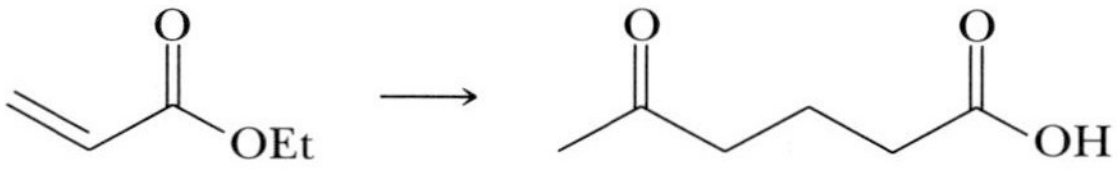

(g) → O

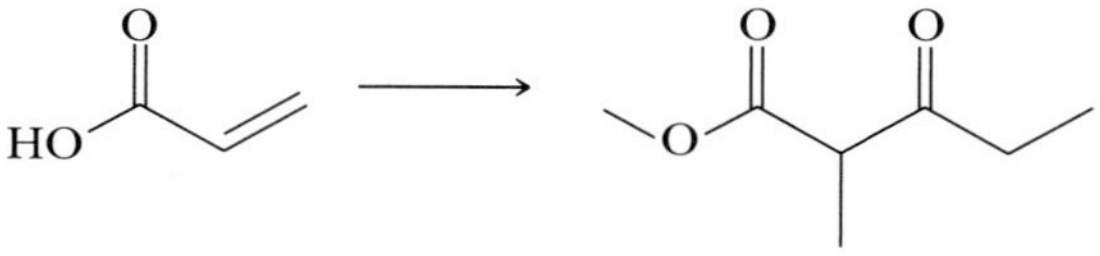

(h) Cl → O

(i) O, OH → H, N, O

(j) O, O, OEt → O

(k) → O, OH

(l) O, OEt → O, O, OH

(m) O, HO → O, O, O

(n) → O, O, OEt

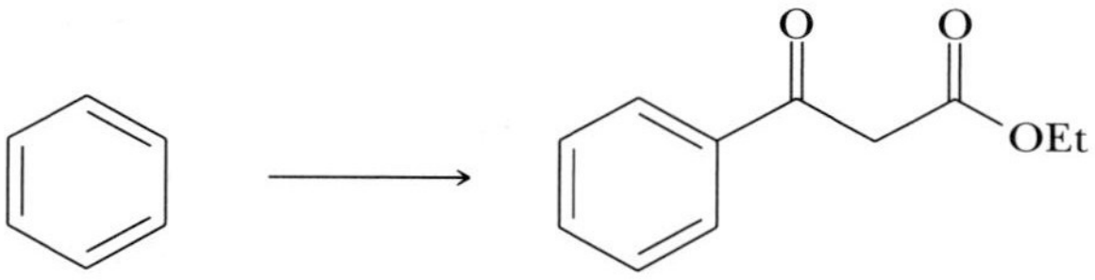

Looking Ahead

*16.43 The following reaction is one of the 10 steps in glycolysis (Section 22.3), a series of enzyme-catalyzed reactions by which glucose is oxidized to two molecules of pyruvate:

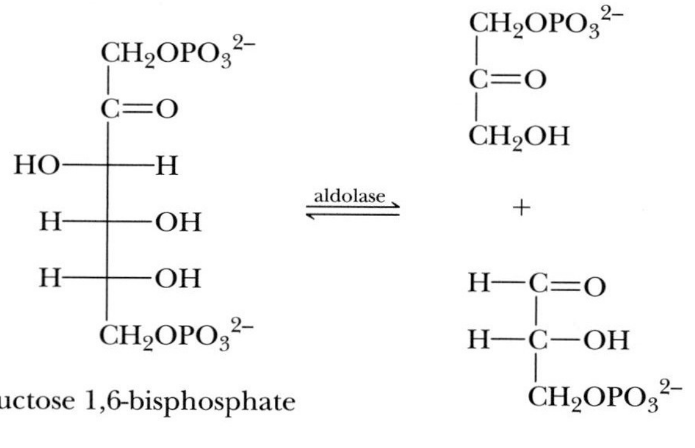

$CH_2OPO_3^{2-}$, C=O, HO—H, H—OH, H—OH, $CH_2OPO_3^{2-}$

Fructose 1,6-bisphosphate

$\xrightleftharpoons{\text{aldolase}}$

$CH_2OPO_3^{2-}$, C=O, CH_2OH Dihydroxyacetone phosphate

+

H—C=O, H—C—OH, $CH_2OPO_3^{2-}$ Glyceraldehyde 3-phosphate

Show that this step is the reverse of an aldol reaction.

***16.44** The following reaction is the fourth in the set of four enzyme-catalyzed steps by which the hydrocarbon chain of a fatty acid (Section 22.5) is oxidized, two carbons at a time, to acetyl-coenzyme A:

$$\underset{\beta\text{-Ketoacyl-CoA}}{R-\overset{O}{\overset{\|}{C}}-CH_2-\overset{O}{\overset{\|}{C}}SCoA} + \underset{\text{Coenzyme A}}{CoA-SH} \longrightarrow$$

$$\underset{\text{An acyl-CoA}}{R-\overset{O}{\overset{\|}{C}}-SCoA} + \underset{\text{Acetyl-CoA}}{CH_3\overset{O}{\overset{\|}{C}}-SCoA}$$

Show that this reaction is the reverse of a Claisen condensation.

***16.45** Steroids are a major type of lipid (Section 21.4) with a characteristic tetracyclic ring system. Show how the **A** ring of the steroid testosterone can be constructed from the indicated precursors, using a Michael reaction followed by an aldol reaction (with dehydration):

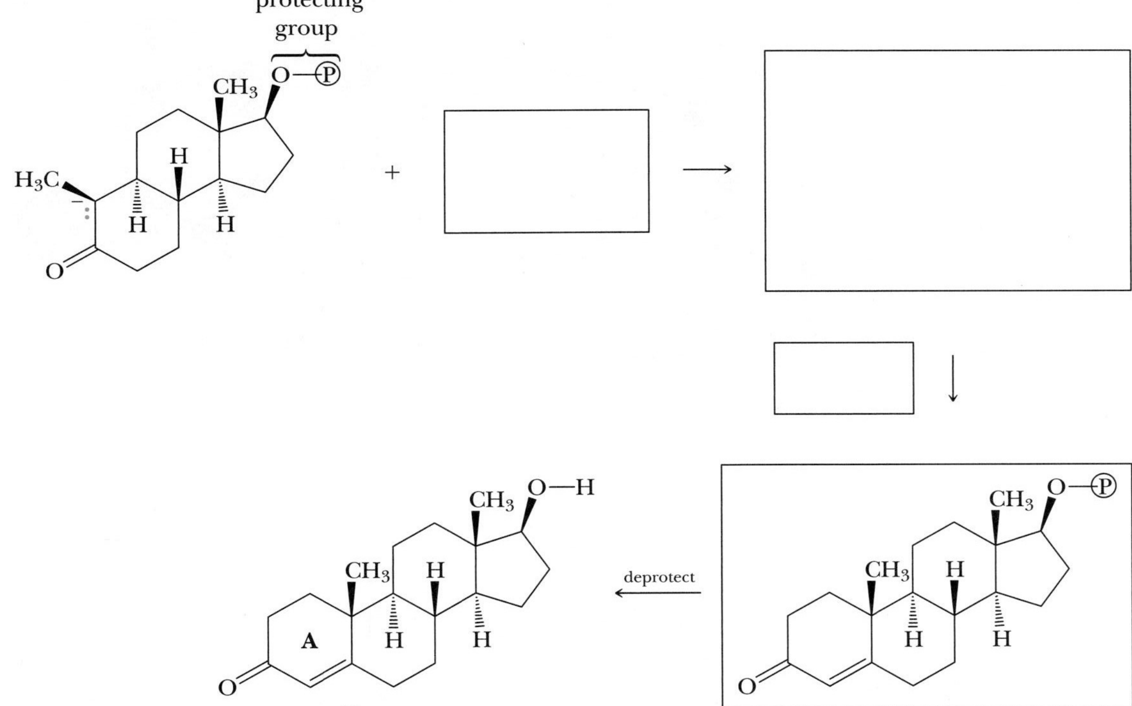

***16.46** The third step of the citric acid cycle (Section 22.6) involves the protonation of one of the carboxylate groups of oxalosuccinate, a β-ketoacid, followed by decarboxylation to form α-ketoglutarate:

$$\underset{\text{Oxalosuccinate}}{\begin{array}{l} H_2C-COO^- \\ \ \ |\\ \ \ CH-COO^- \\ \ \ | \\ O{=}C-COO^- \end{array}} + H^+ \longrightarrow \underset{\alpha\text{-Ketoglutarate}}{\boxed{}} + CO_2$$

Write the structural formula of α-ketoglutarate.

17 Organic Polymer Chemistry

KEY QUESTIONS

17.1 What Is the Architecture of Polymers?

17.2 How Do We Name and Show the Structure of a Polymer?

17.3 What Is Polymer Morphology? Crystalline versus Amorphous Materials

17.4 What Is Step-Growth Polymerization?

17.5 What are Chain-Growth Polymers?

17.6 What Types of Plastics Are Currently Recycled in Large Quantities?

CHEMICAL CONNECTIONS

17A Stitches That Dissolve

17B Paper or Plastic?

Sea of umbrellas on a rainy day in Shanghai, China. Inset: A model of adipic acid, one of the two monomers from which nylon 66 is made. *(Gavin Hellier/Stone/Getty Images)*

The technological advancement of any society is inextricably tied to the materials available to it. Indeed, historians have used the emergence of new materials as a way of establishing a time line to mark the development of human civilization. As part of the search to discover new materials, scientists have made increasing use of organic chemistry for the preparation of synthetic materials known as polymers. The versatility afforded by these polymers allows for the creation and fabrication of materials with ranges of properties unattainable using such materials as wood, metals, and ceramics. Deceptively simple changes in the chemical structure of a given polymer, for example, can change its mechanical properties from those of a sandwich bag to those of a bulletproof vest. Furthermore, structural changes can introduce properties never before imagined in organic polymers. For instance, using well-defined organic reactions, chemists can turn one type of polymer into an insulator (e.g., the rubber sheath that surrounds electrical cords). Treated differently, the same type of polymer can be made into an electrical conductor with a conductivity nearly equal to that of metallic copper!

The years since the 1930s have seen extensive research and development in organic polymer chemistry, and an almost explosive growth in plastics, coatings, and rubber technology has created a worldwide multibillion-dollar industry. A few basic characteristics account for this phenomenal growth. First, the raw materials for synthetic polymers are derived mainly from petroleum. With the development of petroleum-refining processes, raw materials for the synthesis of polymers became generally cheap and plentiful. Second, within broad limits, scientists have learned how to tailor polymers to the requirements of the end use. Third, many consumer products can be fabricated more cheaply from synthetic polymers than from such competing materials as wood, ceramics, and metals. For example, polymer technology created the water-based (latex) paints that have revolutionized the coatings industry, and plastic films and foams have done the same for the packaging industry. The list could go on and on as we think of the manufactured items that are everywhere around us in our daily lives.

17.1 What Is the Architecture of Polymers?

Polymers (Greek: *poly* + *meros*, many parts) are long-chain molecules synthesized by linking **monomers** (Greek: *mono* + *meros*, single part) through chemical reactions. The molecular weights of polymers are generally high compared with those of common organic compounds and typically range from 10,000 g/mol to more than 1,000,000 g/mol. The architectures of these macromolecules can also be quite diverse: There are polymer architectures with linear and branched chains, as well as those with comb, ladder, and star structures (Figure 17.1). Additional structural variations can be achieved by introducing covalent cross-links between individual polymer chains.

Polymer From the Greek *poly*, many and *meros*, parts; any long-chain molecule synthesized by linking together many single parts called monomers.

Monomer From the Greek *mono*, single and *meros*, part; the simplest nonredundant unit from which a polymer is synthesized.

Linear and branched polymers are often soluble in solvents such as chloroform, benzene, toluene, dimethyl sulfoxide (DMSO), and tetrahydrofuran (THF). In addition, many linear and branched polymers can be melted to form highly viscous liquids. In polymer chemistry, the term **plastic** refers to any polymer that can be molded when hot and that retains its shape when cooled. **Thermoplastics** are polymers which, when melted, become sufficiently fluid that they can be molded into shapes that are retained when they are cooled. **Thermosetting plastics**, or thermosets, can be molded when they are first prepared, but once cooled, they harden irreversibly and cannot be remelted. Because of their very different physical characteristics, thermoplastics and thermosets must be processed differently and are used in very different applications.

Plastic A polymer that can be molded when hot and retains its shape when cooled.

Thermoplastic A polymer that can be melted and molded into a shape that is retained when it is cooled.

Thermosetting plastic A polymer that can be molded when it is first prepared, but, once cooled, hardens irreversibly and cannot be remelted.

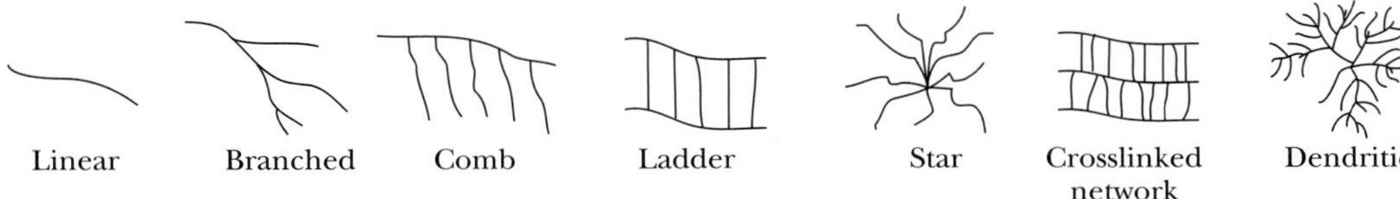

Figure 17.1
Various polymer architectures.

The single most important property of polymers at the molecular level is the size and shape of their chains. A good example of the importance of size is a comparison of paraffin wax, a natural polymer, and polyethylene, a synthetic polymer. These two distinct materials have identical repeat units, namely, $-CH_2-$, but differ greatly in the size of their chains. Paraffin wax has between 25 and 50 carbon atoms per chain, whereas polyethylene has between 1,000 and 3,000 carbon atoms per chain. Paraffin wax, such as that in birthday candles, is soft and brittle, but polyethylene, from which plastic beverage bottles are fabricated, is strong, flexible, and tough. These vastly different properties arise directly from the difference in size and molecular architecture of the individual polymer chains.

17.2 How Do We Name and Show the Structure of a Polymer?

Average degree of polymerization, *n* A subscript placed outside the parentheses of the simplest nonredundant unit of a polymer to indicate that the unit repeats *n* times in the polymer.

We typically show the structure of a polymer by placing parentheses around the **repeating unit**, which is the smallest molecular fragment that contains all the structural features of the chain. A subscript n placed outside the parentheses indicates that the unit repeats n times. Thus, we can reproduce the structure of an entire polymer chain by repeating the enclosed structure in both directions. An example is polypropylene, which is derived from the polymerization of propylene:

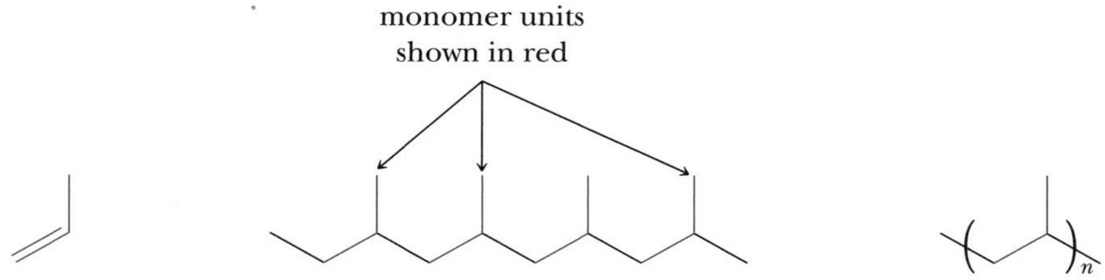

monomer units
shown in red

$CH_2{=}CH(CH_3)$

The monomer (propylene)

$-CH_2CH(CH_3)-CH_2CH(CH_3)-CH_2CH(CH_3)-CH_2CH(CH_3)-$

Part of an extended chain of polypropylene

$\left(CH_2CH(CH_3)\right)_n$

The repeating unit of polypropylene

The most common method of naming a polymer is to add the prefix **poly-** to the name of the monomer from which the polymer is synthesized. Examples are polyethylene and polystyrene. In the case of a more complex monomer or when the name of the monomer is more than one word (e.g., the monomer vinyl chloride), parentheses are used to enclose the name of the monomer:

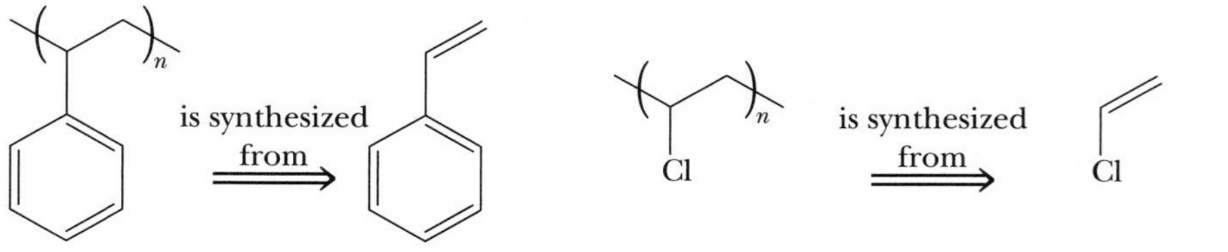

Polystyrene is synthesized from ⟹ Styrene

Poly(vinyl chloride) (PVC) is synthesized from ⟹ Vinyl chloride

Cl Cl

Example 17.1

Given the following structure, determine the polymer's repeating unit; redraw the structure, using the simplified parenthetical notation; and name the polymer:

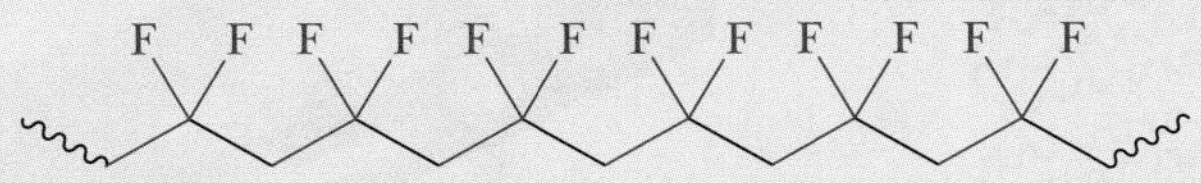

(repeating unit in red)

Strategy

Identify the repeating structural unit of the chain and place parentheses around it. Add the subscript n to indicate that this unit repeats n times.

Solution

The repeating unit is $—CH_2CF_2—$ and the polymer is written $\left(CH_2CF_2\right)_n$. The repeat unit is derived from 1,1-difluoroethylene and the polymer is named poly(1,1-difluoroethylene). This polymer is used in microphone diaphragms.

See problems 17.9, 17.10

Problem 17.1

Given the following structure, determine the polymer's repeat unit; redraw the structure, using the simplified parenthetical notation; and name the polymer:

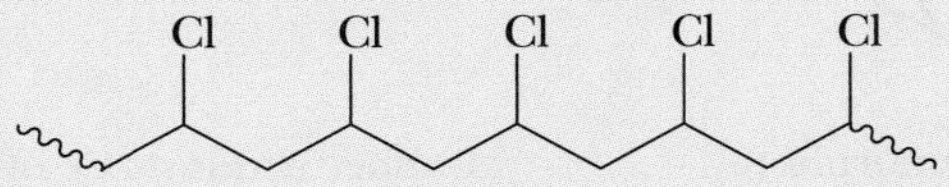

17.3 What Is Polymer Morphology? Crystalline versus Amorphous Materials

Polymers, like small organic molecules, tend to crystallize upon precipitation or as they are cooled from a melt. Acting to inhibit this tendency is the very large size of their molecules, which tends to inhibit diffusion, and their sometimes complicated or irregular structures, which prevent efficient packing of their chains. The result is that polymers in the solid state tend to be composed of both ordered **crystalline domains** (crystallites) and disordered **amorphous domains** (Figure 17.2). The relative amounts of crystalline and amorphous domains differ from polymer to polymer and frequently depend upon the manner in which the material is processed (Figure 17.3).

We often find high degrees of crystallinity in polymers with regular, compact structures and strong intermolecular forces, such as hydrogen bonding. The temperature at which crystallites melt corresponds to the **melt transition temperature (T_m)** of the polymer. As the degree of crystallinity of a polymer increases, its T_m increases, and the polymer becomes more opaque because of the scattering of light by its crystalline domains. With an increase in crystallinity comes a corresponding increase in strength and stiffness. For example, poly(6-aminohexanoic acid), known more commonly as nylon 6 because it is synthesized from a six-carbon monomer, has a

Crystalline domains Ordered crystalline regions in the solid state of a polymer; also called crystallites.

Amorphous domains Disordered, noncrystalline regions in the solid state of a polymer.

Melt transition temperature, T_m The temperature at which crystalline regions of a polymer melt.

Figure 17.2
Examples of various polymer morphologies.

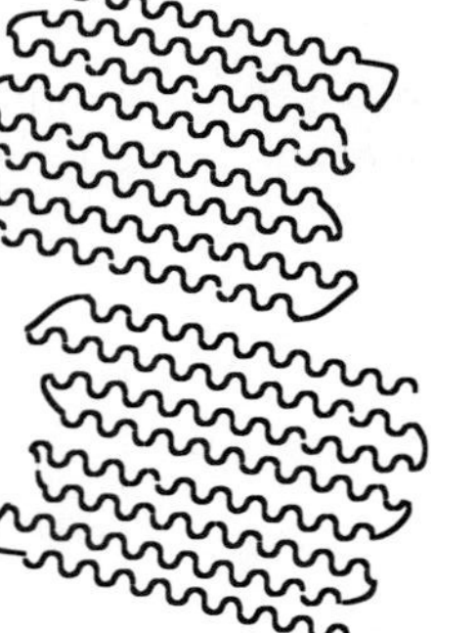

Arrangement of crystalline polymer chains

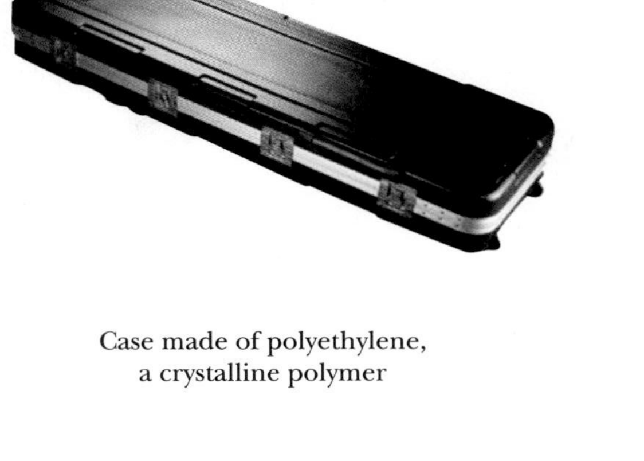

Case made of polyethylene, a crystalline polymer

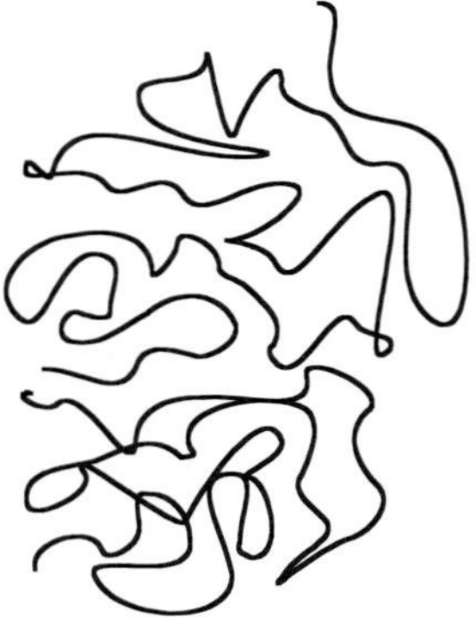
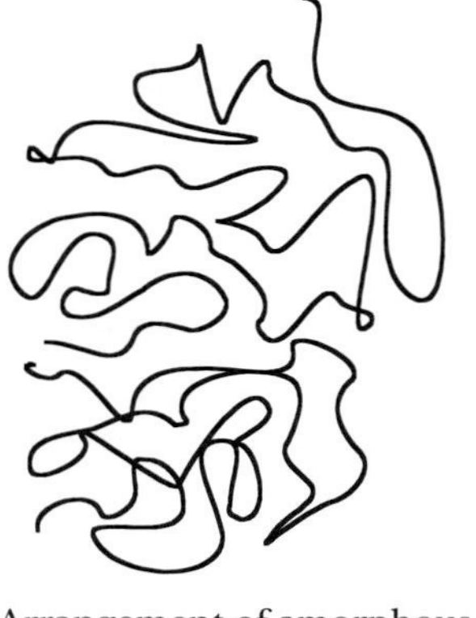
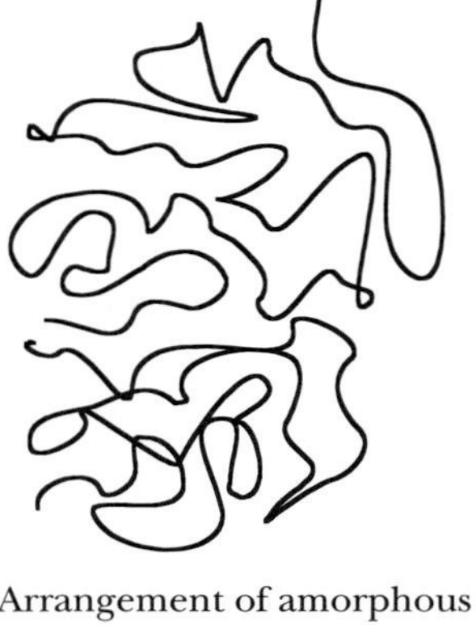

Arrangement of amorphous polymer chains

Rubber bands made of latex rubber, an amorphous polymer

T_m = 223°C. At and well above room temperature, this polymer is a hard, durable material that does not undergo any appreciable change in properties, even on a very hot summer afternoon. Its uses range from textile fibers to the heels of shoes.

Figure 17.3
(a) Example of a polymer chain with both crystalline and amorphous domains, (b) a cup containing a polymer with a greater percentage of crystalline domains, (c) a cup containing a polymer with a greater percentage of amorphous domains.

Amorphous domains have little or no long-range order. Highly amorphous polymers are sometimes referred to as **glassy** polymers. Because they lack crystalline domains that scatter light, amorphous polymers are transparent. In addition, they are typically weak polymers, in terms of both their high flexibility and their low mechanical strength. On being heated, amorphous polymers are transformed from a hard, glassy state to a soft, flexible, rubbery state. The temperature at which this transition occurs is called the **glass transition temperature (T_g)**. Amorphous polystyrene, for example, has a $T_g = 100°C$. At room temperature, it is a rigid solid used for drinking cups, foamed packaging materials, disposable medical wares, tape reels, and so forth. If it is placed in boiling water, it becomes soft and rubbery.

Glass transition temperature, T_g The temperature at which a polymer undergoes the transition from a hard glass to a rubbery state.

This relationship between a polymer's mechanical properties and degree of crystallinity can be illustrated by poly(ethylene terephthalate) (PET):

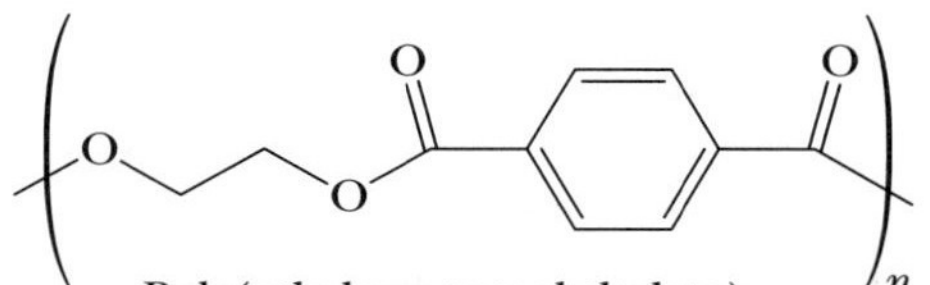

Poly(ethylene terephthalate) (PET)

PET can be made with a percentage of crystalline domains ranging from 0% to about 55%. By cooling the melt quickly, completely amorphous PET forms. By prolonging the cooling time, more molecular diffusion occurs and crystallites form as the chains become more ordered. The differences in mechanical properties between these forms of PET are great. PET with a low degree of crystallinity is used for plastic beverage bottles, whereas fibers drawn from highly crystalline PET are used for textile fibers and tire cords.

Rubber materials must have low T_g values in order to behave as **elastomers (elastic polymers)**. If the temperature drops below its T_g value, then the material is converted to a rigid glassy solid and all elastomeric properties are lost. A poor understanding of this behavior of elastomers contributed to the *Challenger* spacecraft disaster in 1985. The elastomeric O-rings used to seal the solid booster rockets had a T_g value around 0°C. When the temperature dropped to an unanticipated low on the morning of the launch of the craft, the O-ring seals dropped below their T_g value and obediently changed from elastomers to rigid glasses, losing any sealing capabilities. The rest is tragic history. The physicist Richard Feynman sorted this out publicly in a famous televised hearing in which he put a *Challenger*-type O-ring in ice water and showed that its elasticity was lost!

Elastomer A material that, when stretched or otherwise distorted, returns to its original shape when the distorting force is released.

17.4 What Is Step-Growth Polymerization?

Polymerizations in which chain growth occurs in a stepwise manner are called **step-growth** or **condensation polymerizations**. Step-growth polymers are formed by reaction between difunctional molecules, with each new bond created in a separate step. During polymerization, monomers react to form dimers, dimers react with monomers to form trimers, dimers react with dimers to form tetramers, and so on.

Step-growth polymerization A polymerization in which chain growth occurs in a stepwise manner between difunctional monomers, for example, between adipic acid and hexamethylenediamine to form nylon 66. Also referred to as condensation polymerization.

There are two common types of step-growth processes: (1) reaction between A—M—A and B—M—B type monomers to give -(A—M—A—B—M—B)_n polymers and (2) the self-condensation of A—M—B monomers to give -(A—M—B)_n polymers. In this notation, "M" indicates the monomer and "A" and "B" the reactive functional groups on the monomer. In each type of step-growth polymerization, an A functional group reacts exclusively with a B functional group, and a B functional group reacts exclusively with an A functional group. New covalent bonds in step-growth polymerizations are generally formed by polar reactions between A and B functional groups—for example, nucleophilic acyl substitution. In this section, we discuss five types of step-growth polymers: polyamides, polyesters, polycarbonates, polyurethanes, and epoxy resins.

A. Polyamides

Polyamide A polymer in which each monomer unit is joined to the next by an amide bond, as for example nylon 66.

Autoclave An instrument used to sterilize items by subjecting them to steam and high pressure.

In the early 1930s, chemists at E. I. DuPont de Nemours & Company began fundamental research into the reactions between dicarboxylic acids and diamines to form **polyamides**. In 1934, they synthesized the first purely synthetic fiber, nylon 66, so named because it is synthesized from two different monomers, each containing six carbon atoms.

In the synthesis of nylon 66, hexanedioic acid and 1,6-hexanediamine are dissolved in aqueous ethanol, in which they react to form a one-to-one salt called nylon salt. This salt is then heated in an **autoclave** to 250°C and an internal pressure of 15 atm. Under these extreme conditions, $—COO^-$ groups from the diacid and $—NH_3^+$ groups from diamine react by the loss of H_2O to form a polyamide. Nylon 66 formed under these conditions melts at 250 to 260°C and has a molecular weight ranging from 10,000 to 20,000 g/mol:

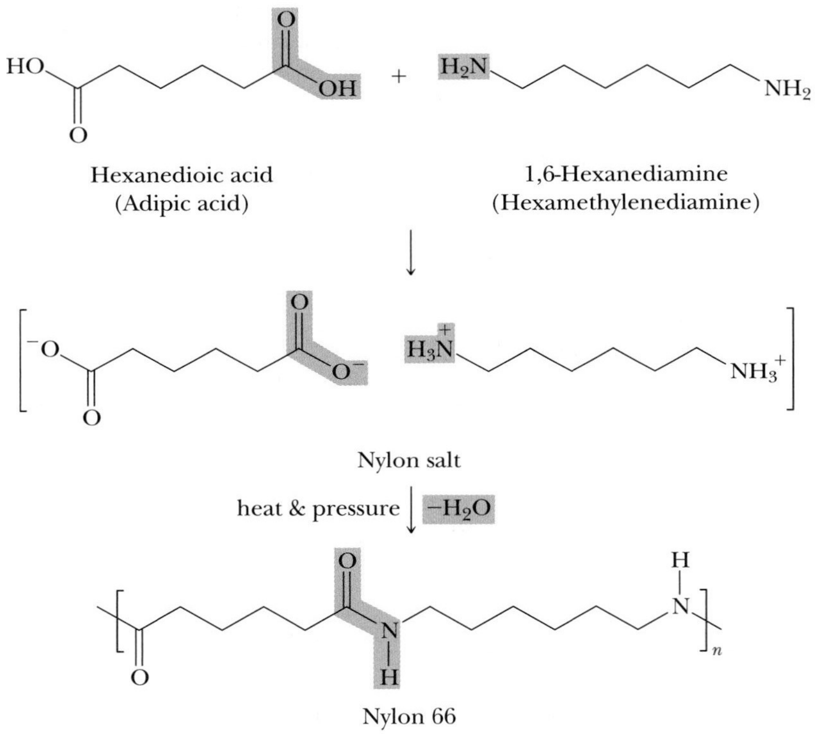

HO, O, OH, +, H_2N, NH_2

Hexanedioic acid (Adipic acid)

1,6-Hexanediamine (Hexamethylenediamine)

^-O, O, O^-, $H_3\overset{+}{N}$, NH_3^+

Nylon salt

heat & pressure, $-H_2O$

O, N, H, N, H, n

Nylon 66

In the first stage of fiber production, crude nylon 66 is melted, spun into fibers, and cooled. Next, the melt-spun fibers are **cold drawn** (drawn at room temperature) to about four times their original length to increase their degree of crystallinity. As the fibers are drawn, individual polymer molecules become oriented in the direction of the fiber axis, and hydrogen bonds form between carbonyl oxygens of one chain and amide hydrogens of another chain (Figure 17.4). The effects of the orientation of polyamide molecules on the physical properties of the fiber are dramatic—both tensile strength and stiffness are increased markedly. Cold drawing is an important step in the production of most synthetic fibers.

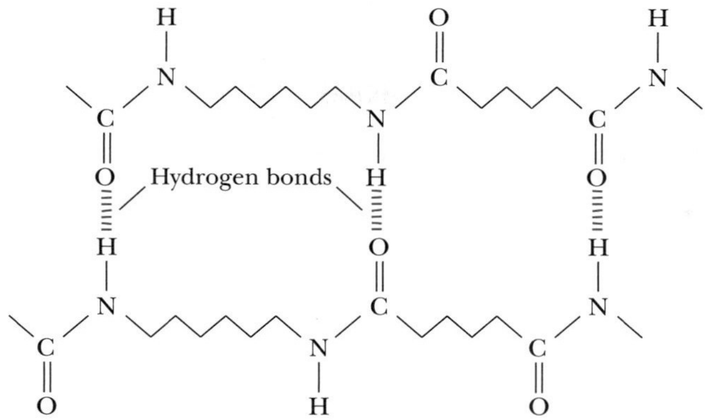

Figure 17.4 The structure of cold-drawn nylon 66. Hydrogen bonds between adjacent polymer chains provide additional tensile strength and stiffness to the fibers.

The current raw-material base for the production of adipic acid is benzene, which is derived almost entirely from catalytic cracking and re-forming of petroleum (Section 3.10B). Catalytic reduction of benzene to cyclohexane (Section 9.1D), followed by catalyzed air oxidation, gives a mixture of cyclohexanol and cyclohexanone. Oxidation of this mixture by nitric acid gives adipic acid:

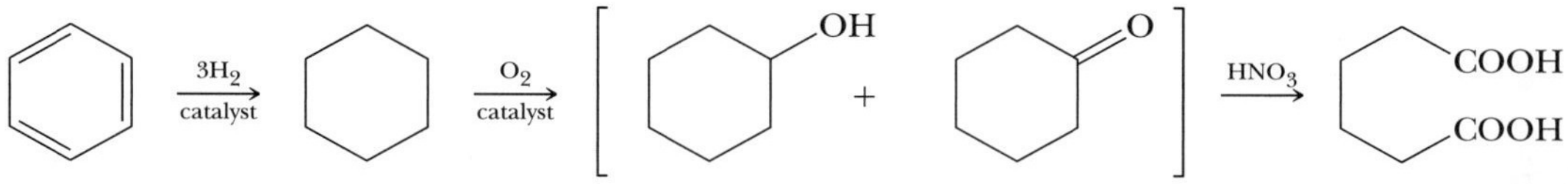

Benzene $\xrightarrow[\text{catalyst}]{3H_2}$ Cyclohexane $\xrightarrow[\text{catalyst}]{O_2}$ [Cyclohexanol + Cyclohexanone] $\xrightarrow{HNO_3}$ Hexanedioic acid (Adipic acid)

Adipic acid, in turn, is a starting material for the synthesis of hexamethylenediamine. Treating adipic acid with ammonia yields an ammonium salt that, when heated, gives adipamide. The catalytic reduction of adipamide then gives hexamethylenediamine:

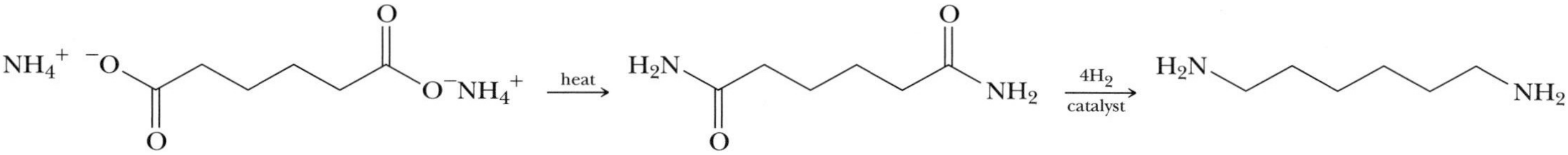

Ammonium hexanedioate (Ammonium adipate) $\xrightarrow{\text{heat}}$ Hexanediamide (Adipamide) $\xrightarrow[\text{catalyst}]{4H_2}$ 1,6-Hexanediamine (Hexamethylenediamine)

Note that carbon sources for the production of nylon 66 are derived entirely from petroleum, which, unfortunately, is not a renewable resource.

The nylons are a family of polymers, the members of which have subtly different properties that suit them to one use or another. The two most widely used members of the family are nylon 66 and nylon 6. Nylon 6 is so named because it is synthesized from caprolactam, a six-carbon monomer. In this synthesis, caprolactam is partially hydrolyzed to 6-aminohexanoic acid and then heated to 250°C to bring about polymerization:

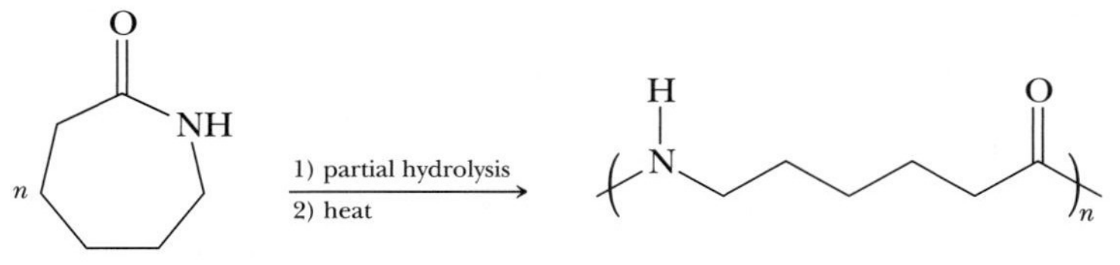

n Caprolactam $\xrightarrow[\text{2) heat}]{\text{1) partial hydrolysis}}$ Nylon 6

Nylon 6 is fabricated into fibers, bristles, rope, high-impact moldings, and tire cords.

Aramid A poly*aromatic amide*; a polymer in which the monomer units are an aromatic diamine and an aromatic dicarboxylic acid.

Based on extensive research into the relationships between molecular structure and bulk physical properties, scientists at DuPont reasoned that a polyamide containing aromatic rings would be stiffer and stronger than either nylon 66 or nylon 6. In early 1960, DuPont introduced Kevlar, a polyaromatic amide **(aramid)** fiber synthesized from terephthalic acid and p-phenylenediamine:

$$n\,HO\overset{O}{\overset{\|}{C}}-C_6H_4-\overset{O}{\overset{\|}{C}}OH + n\,H_2N-C_6H_4-NH_2 \longrightarrow \left(\overset{O}{\overset{\|}{C}}-C_6H_4-\overset{O}{\overset{\|}{C}}NH-C_6H_4-NH\right)_n + 2n\,H_2O$$

1,4-Benzenedicarboxylic acid (Terephthalic acid) — 1,4-Benzenediamine (p-Phenylenediamine) — Kevlar

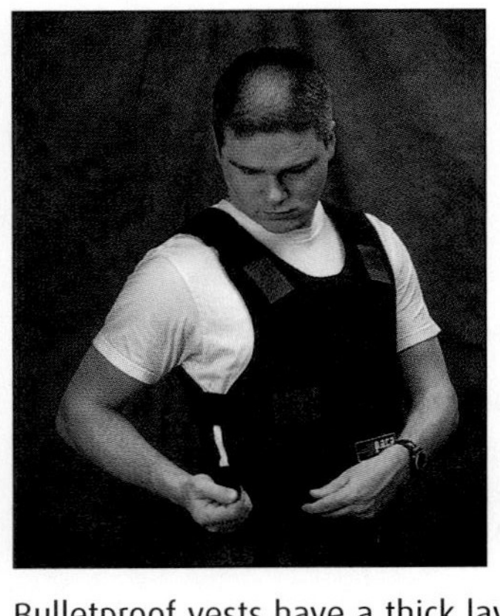

Bulletproof vests have a thick layer of Kevlar. *(Charles D. Winters)*

One of the remarkable features of Kevlar is its light weight compared with that of other materials of similar strength. For example, a 7.6-cm (3-in.) cable woven of Kevlar has a strength equal to that of a similarly woven 7.6-cm (3-in.) steel cable. However, whereas the steel cable weighs about 30 kg/m (20 lb/ft), the Kevlar cable weighs only 6 kg/m (4 lb/ft). Kevlar now finds use in such articles as anchor cables for offshore drilling rigs and reinforcement fibers for automobile tires. Kevlar is also woven into a fabric that is so tough that it can be used for bulletproof vests, jackets, and raincoats.

B. Polyesters

Polyester A polymer in which each monomer unit is joined to the next by an ester bond as, for example, poly(ethylene terephthalate).

The first **polyester**, developed in the 1940s, involved the polymerization of benzene 1,4-dicarboxylic acid (terephthalic acid) with 1,2-ethanediol (ethylene glycol) to give poly(ethylene terephthalate), abbreviated PET. Virtually all PET is now made from the dimethyl ester of terephthalic acid by the following transesterification reaction (Section 15.4C):

Remove CH_3OH

$$CH_3O\overset{O}{\overset{\|}{C}}-C_6H_4-\overset{O}{\overset{\|}{C}}OCH_3 + HOCH_2CH_2OH \xrightarrow[-CH_3OH]{heat} \left(\overset{O}{\overset{\|}{C}}-C_6H_4-\overset{O}{\overset{\|}{C}}OCH_2CH_2O\right)_n$$

Dimethyl terephthalate — 1,2-Ethanediol (Ethylene glycol) — Poly(ethylene terephthalate) (Dacron®, Mylar®)

Because Mylar film has very tiny pores, it is used for balloons that can be inflated with helium; the helium atoms diffuse only slowly through the pores of the film. *(Charles D. Winters)*

The crude polyester can be melted, extruded, and then cold drawn to form the textile fiber Dacron® polyester, the outstanding features of which are its stiffness (about four times that of nylon 66), very high strength, and remarkable resistance to creasing and wrinkling. Because the early Dacron® polyester fibers were harsh to the touch, due to their stiffness, they were usually blended with cotton or wool to make acceptable textile fibers. Newly developed fabrication techniques now produce less harsh Dacron® polyester textile fibers. PET is also fabricated into Mylar® films and recyclable plastic beverage containers.

Ethylene glycol for the synthesis of PET is obtained by the air oxidation of ethylene to ethylene oxide (Section 8.4B), followed by hydrolysis to the glycol (Section 8.4C). Ethylene is, in turn, derived entirely from cracking either petroleum or ethane derived from natural gas (Section 3.10). Terephthalic acid is obtained by the oxidation of p-xylene (Section 9.4), an aromatic hydrocarbon obtained along with benzene and toluene from the catalytic cracking and re-forming of naphtha and other petroleum fractions (Section 3.10B):

$$CH_2{=}CH_2 \xrightarrow[\text{catalyst}]{O_2} H_2C\overset{O}{\diagup\diagdown}CH_2 \xrightarrow{H^+,\ H_2O} HOCH_2CH_2OH$$

Ethylene — Ethylene oxide — 1,2-Ethanediol (Ethylene glycol)

$$H_3C{-}C_6H_4{-}CH_3 \xrightarrow[\text{catalyst}]{O_2} HO\overset{O}{\overset{\|}{C}}{-}C_6H_4{-}\overset{O}{\overset{\|}{C}}OH$$

p-Xylene — Terephthalic acid

C. Polycarbonates

Polycarbonates, the most familiar of which is Lexan®, are a class of commercially important engineering polyesters. Lexan® forms by the reaction between the disodium salt of bisphenol A (Problem 9.33) and phosgene:

Polycarbonate A polyester in which the carboxyl groups are derived from carbonic acid.

remove Na^+Cl^-

$$^+Na\,^-O{-}C_6H_4{-}C(CH_3)_2{-}C_6H_4{-}O^-\,Na^+ + Cl{-}\overset{O}{\overset{\|}{C}}{-}Cl \longrightarrow \left({-}C_6H_4{-}C(CH_3)_2{-}C_6H_4{-}O{-}\overset{O}{\overset{\|}{C}}{-}O{-}\right)_n + NaCl$$

Disodium salt of bisphenol A — Phosgene — Lexan® (a polycarbonate)

Note that phosgene is the diacid chloride (Section 15.1A) of carbonic acid; hydrolysis of phosgene gives H_2CO_3 and 2HCl.

Lexan® is a tough, transparent polymer with high impact and tensile strengths that retains its properties over a wide temperature range. It is used in sporting equipment (for helmets and face masks), to make light, impact-resistant housings for household appliances, and in the manufacture of safety glass and unbreakable windows.

A polycarbonate hockey mask. *(Charles D. Winters)*

D. Polyurethanes

A urethane, or carbamate, is an ester of carbamic acid, H_2NCOOH. Carbamates are most commonly prepared by treating an isocyanate with an alcohol. In this reaction, the H and OR′ of the alcohol add to the C=N bond in a reaction comparable to the addition of an alcohol to a C=O bond:

$$RN{=}C{=}O + R'OH \longrightarrow RNH\overset{O}{\overset{\|}{C}}OR'$$

An isocyanate — A carbamate

Polyurethane A polymer containing the —NHCOO— group as a repeating unit.

Polyurethanes consist of flexible polyester or polyether units (blocks) alternating with rigid urethane units (blocks) derived from a diisocyanate, commonly a mixture of 2,4- and 2,6-toluene diisocyanate:

$$\text{O=C=N–}C_6H_3(CH_3)\text{–N=C=O} + n\text{HO-polymer-OH} \longrightarrow \left[\text{–}\overset{O}{\overset{\|}{C}}\text{NH–}C_6H_3(CH_3)\text{–NH}\overset{O}{\overset{\|}{C}}\text{O-polymer-O–}\right]_n$$

2,6-Toluene diisocyanate — Low-molecular-weight polyester or polyether with OH groups at each end of the chain — A polyurethane

The more flexible blocks are derived from low-molecular-weight (MW 1,000 to 4,000) polyesters or polyethers with —OH groups at each end of their chains. Polyurethane fibers are fairly soft and elastic and have found use as spandex and Lycra®, the "stretch" fabrics used in bathing suits, leotards, and undergarments.

Polyurethane foams for upholstery and insulating materials are made by adding small amounts of water during polymerization. Water reacts with isocyanate groups to form a carbamic acid that undergoes spontaneous decarboxylation to produce gaseous carbon dioxide, which then acts as the foaming agent:

$$\text{RN=C=O} + H_2O \longrightarrow \left[\text{RNH–}\overset{O}{\overset{\|}{C}}\text{–OH}\right] \longrightarrow RNH_2 + CO_2$$

An isocyanate — A carbamic acid (unstable)

Epoxy resin A material prepared by a polymerization in which one monomer contains at least two epoxy groups.

E. Epoxy Resins

Epoxy resins are materials prepared by a polymerization in which one monomer contains at least two epoxy groups. Within this range, a large number of polymeric materials are possible, and epoxy resins are produced in forms ranging from low-viscosity liquids to high-melting solids. The most widely used epoxide monomer is the diepoxide prepared by treating 1 mole of bisphenol A (Problem 9.33) with 2 moles of epichlorohydrin:

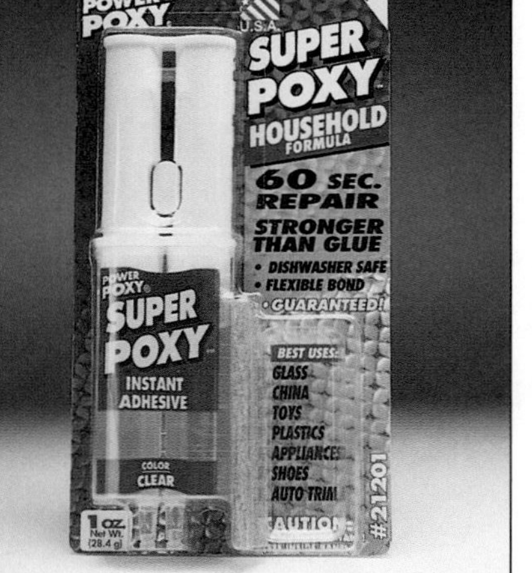

An epoxy resin kit.
(Charles D. Winters)

Epichlorohydrin + Disodium salt of Bisphenol A ($^{+}Na^{-}O–C_6H_4–C(CH_3)_2–C_6H_4–O^{-}Na^{+}$) + Epichlorohydrin

$$\downarrow$$

A diepoxide (glycidyl–O–C_6H_4–$C(CH_3)_2$–C_6H_4–O–glycidyl) + 2NaCl

To prepare the following epoxy resin, the diepoxide monomer is treated with 1,2-ethanediamine (ethylene diamine):

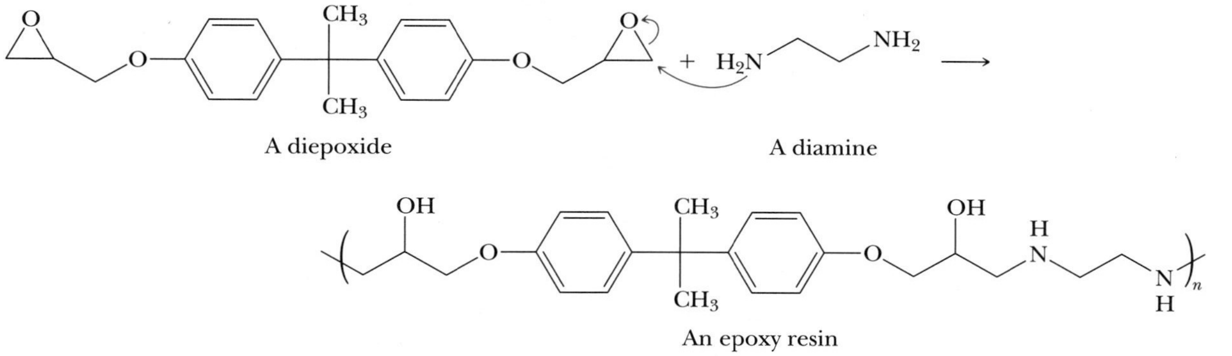

A diepoxide A diamine

An epoxy resin

Ethylene diamine is usually called the catalyst in the two-component formulations that you buy in hardware or craft stores; it is also the component with the acrid smell. The preceding reaction corresponds to nucleophilic opening of the highly strained three-membered epoxide ring (Section 8.4C).

Epoxy resins are widely used as adhesives and insulating surface coatings. They have good electrical insulating properties, which lead to their use in encapsulating electrical components ranging from integrated circuit boards to switch coils and insulators for power transmission systems. Epoxy resins are also used as composites with other materials, such as glass fiber, paper, metal foils, and other synthetic fibers, to create structural components for jet aircraft, rocket motor casings, and so on.

Example 17.2

By what type of mechanism does the reaction between the disodium salt of bisphenol A and epichlorohydrin take place?

Strategy

Examine the mechanism on the previous page and you will see by the curved arrows that an oxygen anion of bisphenol A displaces a chlorine atom from the primary carbon of epichlorohydrin. The phenoxide ion of bisphenol A is a good nucleophile, and chlorine on the primary carbon of epichlorohydrin is the leaving group.

Solution

The mechanism is an S_N2 mechanism.

Problem 17.2

Write the repeating unit of the epoxy resin formed from the following reaction:

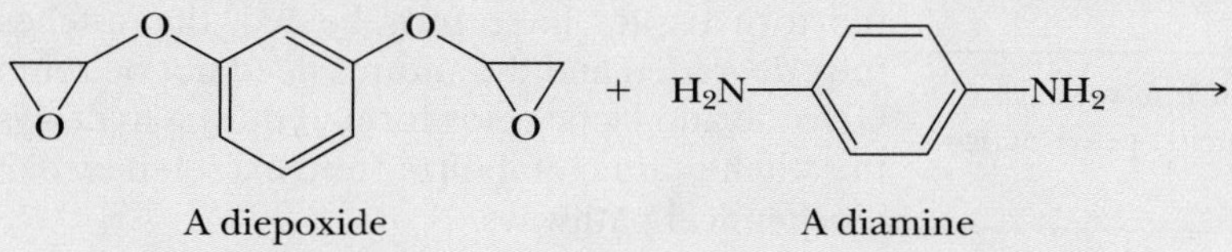

A diepoxide A diamine

17.5 What Are Chain-Growth Polymers?

Chain-growth polymerization A polymerization that involves sequential addition reactions, either to unsaturated monomers or to monomers possessing other reactive functional groups.

From the perspective of the chemical industry, the single most important reaction of alkenes is **chain-growth polymerization**, a type of polymerization in which monomer units are joined together without the loss of atoms. An example is the formation of polyethylene from ethylene:

$$nCH_2{=}CH_2 \xrightarrow{\text{catalyst}} \left(CH_2CH_2 \right)_n$$

Ethylene Polyethylene

The mechanisms of chain-growth polymerization differ greatly from the mechanism of step-growth polymerizations. In the latter, all monomers plus the polymer end groups possess equally reactive functional groups, allowing for all possible combinations of reactions to occur, including monomer with monomer, dimer with dimer, monomer with tetramer, and so forth. In contrast, chain-growth polymerizations involve end groups possessing reactive intermediates that react only with a monomer. The reactive intermediates used in chain-growth polymerizations include radicals, carbanions, carbocations, and organometallic complexes.

CHEMICAL CONNECTIONS 17A

Stitches That Dissolve

As the technological capabilities of medicine have grown, the demand for synthetic materials that can be used inside the body has increased as well. Polymers have many of the characteristics of an ideal biomaterial: They are lightweight and strong, are inert or biodegradable (depending on their chemical structure), and have physical properties (softness, rigidity, elasticity) that are easily tailored to match those of natural tissues. Carbon–carbon backbone polymers are resistant to degradation and are used widely in permanent organ and tissue replacements.

Even though most medical uses of polymeric materials require biostability, applications have been developed that use the biodegradable nature of some macromolecules. An example is the use of glycolic acid/lactic acid copolymers as absorbable sutures:

Glycolic acid + Lactic acid $\xrightarrow[-nH_2O]{\text{copolymerization}}$ A copolymer of poly(glycolic acid)–poly(lactic acid)

Traditional suture materials such as catgut must be removed by a health-care specialist after they have served their purpose. Stitches of these hydroxyester polymers, however, are hydrolyzed slowly over a period of approximately two weeks, and by the time the torn tissues have fully healed, the stitches are fully degraded and the sutures need not be removed. Glycolic and lactic acids formed during hydrolysis of the stitches are metabolized and excreted by existing biochemical pathways.

QUESTION

Propose a mechanism for the hydrolysis of one repeating unit of the copolymer of poly(glycolic acid)-poly(lactic acid).

The number of monomers that undergo chain-growth polymerization is large and includes such compounds as alkenes, alkynes, allenes, isocyanates, and cyclic compounds such as lactones, lactams, ethers, and epoxides. We concentrate on the chain-growth polymerizations of ethylene and substituted ethylenes and show how these compounds can be polymerized by radical and organometallic-mediated mechanisms.

Table 17.1 lists several important polymers derived from ethylene and substituted ethylenes, along with their common names and most important uses.

A. Radical Chain-Growth Polymerization

The first commercial polymerizations of ethylene were initiated by radicals formed by thermal decomposition of organic peroxides, such as benzoyl peroxide. A **radical** is any molecule that contains one or more unpaired electrons. Radicals can be

Radical Any molecule that contains one or more unpaired electrons.

CHEMICAL CONNECTIONS 17B

Paper or Plastic?

Any audiophile will tell you that the quality of any sound system is highly dependent upon its speakers. Speakers create sound by moving a diaphragm in and out to displace air. Most diaphragms are in the shape of a cone, traditionally made of paper. Paper cones are inexpensive, lightweight, rigid, and nonresonant. One disadvantage is their susceptibility to damage by water and humidity. Over time and with exposure, paper cones become weakened, losing their fidelity of sound. Many of the speakers that are available today are made of polypropylene, which is also inexpensive, lightweight, rigid, and nonresonant. Furthermore, not only are polypropylene cones immune to water and humidity, but also their performance is less influenced by heat or cold. Moreover, their added strength makes them less prone to splitting than paper. They last longer and can be displaced more frequently and for longer distances, creating deeper bass notes and higher high notes.

(Photo Courtesy of Crutchfield.com)

QUESTION

Paper speaker cones consist of mostly cellulose, a polymer of the monomer unit known as D-glucose (Chapter 18). Propose why one type of polymer is susceptible to humidity while the other type, polypropylene, is resistant to humidity.

Cellulose

Polypropylene

TABLE 17.1 Polymers Derived from Ethylene and Substituted Ethylenes

Monomer Formula	Common Name	Polymer Name(s) and Common Uses
$CH_2{=}CH_2$	ethylene	polyethylene, Polythene; break-resistant containers and packaging materials
$CH_2{=}CHCH_3$	propylene	polypropylene, Herculon; textile and carpet fibers
$CH_2{=}CHCl$	vinyl chloride	poly(vinyl chloride), PVC; construction tubing
$CH_2{=}CCl_2$	1,1-dichloroethylene	poly(1,1-dichloroethylene); Saran Wrap® is a copolymer with vinyl chloride
$CH_2{=}CHCN$	acrylonitrile	polyacrylonitrile, Orlon®; acrylics and acrylates
$CF_2{=}CF_2$	tetrafluoroethylene	polytetrafluoroethylene, PTFE; Teflon®, nonstick coatings
$CH_2{=}CHC_6H_5$	styrene	polystyrene, Styrofoam™; insulating materials
$CH_2{=}CHCOOCH_2CH_3$	ethyl acrylate	poly(ethyl acrylate); latex paints
$CH_2{=}C(CH_3)COOCH_3$	methyl methacrylate	poly(methyl methacrylate), Lucite®, Plexiglas®; glass substitutes

Fishhook arrow A single-barbed, curved arrow used to show the change in position of a single electron.

formed by the cleavage of a bond in such a way that each atom or fragment participating in the bond retains one electron. In the following equation, **fishhook arrows** are used to show the change in position of single electrons:

$$C_6H_5C(=O)-\ddot{O}-\ddot{O}-C(=O)C_6H_5 \xrightarrow{\text{heat}} C_6H_5C(=O)-\ddot{O}\cdot + \cdot\ddot{O}-C(=O)C_6H_5$$

Benzoyl peroxide — Benzoyloxy radicals

Radical polymerization of ethylene and substituted ethylenes involves three steps: (1) chain initiation, (2) chain propagation, and (3) chain termination. We show these steps here and then discuss each separately in turn.

Chain initiation In radical polymerization, the formation of radicals from molecules containing only paired electrons.

Mechanism: Radical Polymerization of Ethylene

Step 1: Chain initiation—formation of radicals from nonradical compounds:

$$In-In \xrightarrow[\text{or light}]{\text{heat}} 2In\cdot$$

In this equation, In-In represents an initiator that, when heated or irradiated with radiation of a suitable wavelength, cleaves to give two radicals (In·).

Step 2: Chain propagation—reaction of a radical and a molecule to form a new radical:

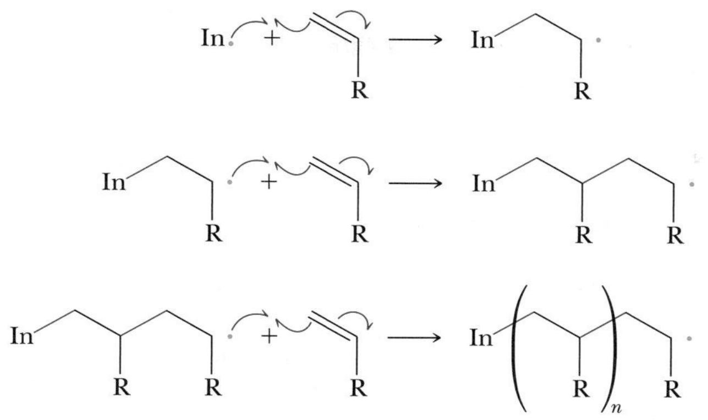

Chain propagation In radical polymerization, a reaction of a radical and a molecule to give a new radical.

Step 3: Chain termination—destruction of radicals:

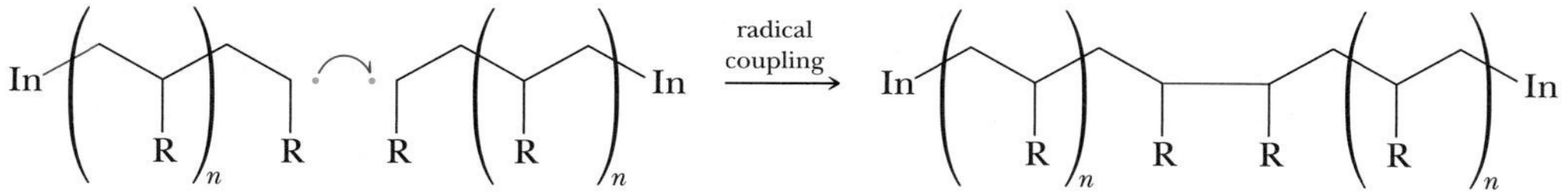

radical coupling

Chain termination In radical polymerization, a reaction in which two radicals combine to form a covalent bond.

The characteristic feature of a chain-initiation step is the formation of radicals from a molecule with only paired electrons. In the case of peroxide-initiated polymerizations of alkenes, chain initiation is by (1) heat cleavage of the O—O bond of a peroxide to give two alkoxy radicals and (2) reaction of an alkoxy radical with a molecule of alkene to give an alkyl radical. In the general mechanism shown, the initiating catalyst is given the symbol In-In and its radical is given the symbol In•.

The structure and geometry of carbon radicals are similar to those of alkyl carbocations. They are planar or nearly so, with bond angles of approximately 120° about the carbon with the unpaired electron. The relative stabilities of alkyl radicals are similar to those of alkyl carbocations because they both possess electron deficient carbons.

methyl < 1° < 2° < 3°

Increasing stability of alkyl radicals

The characteristic feature of a chain-propagation step is the reaction of a radical and a molecule to give a new radical. Propagation steps repeat over and over (propagate), with the radical formed in one step reacting with a monomer to produce a new radical, and so on. The number of times a cycle of chain-propagation steps repeats is called the **chain length** and is given the symbol ***n***. In the polymerization of ethylene, chain-lengthening reactions occur at a very high rate, often as fast as thousands of additions per second, depending on the experimental conditions.

Radical polymerizations of substituted ethylenes almost always give the more stable (more substituted) radical. Because additions are biased in this fashion, the polymerizations of substituted ethylene monomers tend to yield polymers with monomer units joined by the head (carbon 1) of one unit to the tail (carbon 2) of the next unit:

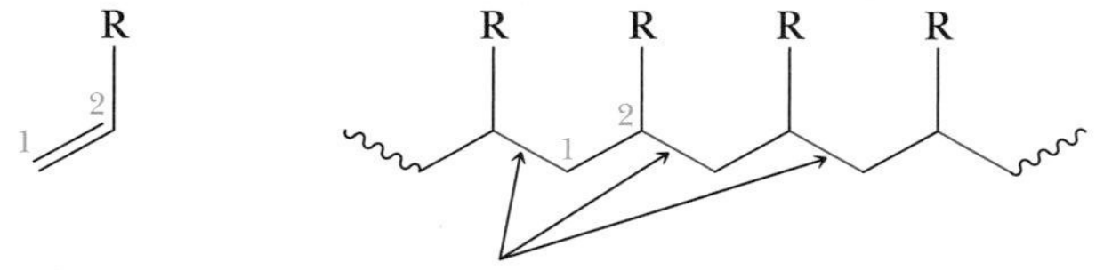

Substituted ethylene monomer

Head-to-tail linkages

In principle, chain-propagation steps can continue until all starting materials are consumed. In practice, they continue only until two radicals react with each other to terminate the process. The characteristic feature of a chain-termination step is the destruction of radicals. In the mechanism shown for radical polymerization of the substituted ethylene, chain termination occurs by the coupling of two radicals to form a new carbon–carbon single bond.

The first commercial process for ethylene polymerization used peroxide catalysts at temperatures of 500°C and pressures of 1,000 atm and produced a soft, tough polymer known as **low-density polyethylene (LDPE)** with a density of between 0.91 and 0.94 g/cm^3 and a melt transition temperature (T_m) of about 115°C. Because LDPE's melting point is only slightly above 100°C, it cannot be used for products that will be exposed to boiling water. At the molecular level, chains of LDPE are highly branched.

Chain-transfer reaction In radical polymerization, the transfer of reactivity of an end group from one chain to another during a polymerization.

The branching on chains of low-density polyethylene results from a "back-biting" reaction in which the radical end group abstracts a hydrogen from the fourth carbon back (the fifth carbon in the chain). Abstraction of this hydrogen is particularly facile because the transition state associated with the process can adopt a conformation like that of a chair cyclohexane. In addition, the less stable 1° radical is converted to a more stable 2° radical. This side reaction is called a **chain-transfer reaction**, because the activity of the end group is "transferred" from one chain to another. Continued polymerization of monomer from this new radical center leads to a branch four carbons long:

$$\xrightarrow{} \quad \xrightarrow{n\,CH_2{=}CH_2}$$

A six-membered transition state leading to 1,5-hydrogen abstraction

Approximately 65% of all LDPE is used for the manufacture of films by a blow-molding technique illustrated in Figure 17.5. LDPE film is inexpensive, which makes it ideal for packaging such consumer items as baked goods, vegetables and other produce and for trash bags.

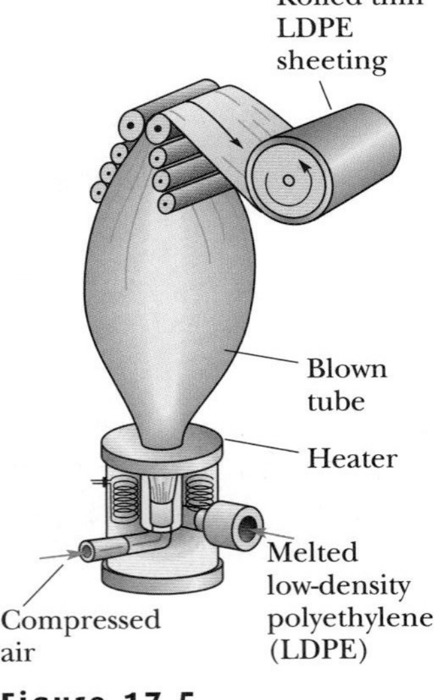

Figure 17.5
Fabrication of an LDPE film. A tube of melted LDPE along with a jet of compressed air is forced through an opening and blown into a giant, thin-walled bubble. The film is then cooled and taken up onto a roller. This double-walled film can be slit down the side to give LDPE film, or it can be sealed at points along its length to make LDPE bags.

B. Ziegler–Natta Chain-Growth Polymerization

In the 1950s, Karl Ziegler of Germany and Giulio Natta of Italy developed an alternative method for the polymerization of alkenes, work for which they shared the Nobel prize in chemistry in 1963. The early Ziegler–Natta catalysts were highly active, heterogeneous materials composed of an $MgCl_2$ support, a Group 4B transition metal halide such as $TiCl_4$, and an alkylaluminum compound—for example, diethylaluminum chloride, $Al(CH_2CH_3)_2Cl$. These catalysts bring about the polymerization of ethylene and propylene at 1–4 atm and at temperatures as low as 60°C.

The catalyst in a Ziegler–Natta polymerization is an alkyltitanium compound formed by reaction between $Al(CH_2CH_3)_2Cl$ and the titanium halide on the surface of a $MgCl_2/TiCl_4$ particle. Once formed, this alkyltitanium species repeatedly inserts ethylene units into the titanium–carbon bond to yield polyethylene.

Mechanism: Ziegler–Natta Catalysis of Ethylene Polymerization

Step 1: Formation of a titanium–ethyl bond:

$$Ti{-}Cl + Al(CH_2CH_3)_2Cl \longrightarrow Ti{-}CH_2CH_3 + Al(CH_2CH_3)Cl_2$$

Step 2: Insertion of ethylene into the titanium–carbon bond:

$$\text{Ti}-CH_2CH_3 + CH_2{=}CH_2 \longrightarrow \text{Ti}-CH_2CH_2CH_2CH_3$$

Over 60 billion pounds of polyethylene are produced worldwide every year with Ziegler–Natta catalysts. Polyethylene from Ziegler–Natta systems, termed **high-density polyethylene (HDPE)**, has a higher density (0.96 g/cm^3) and melt transition temperature (133°C) than low-density polyethylene, is 3 to 10 times stronger, and is opaque rather than transparent. The added strength and opacity are due to a much lower degree of chain branching and a resulting higher degree of crystallinity of HDPE compared with LDPE. Approximately 45% of all HDPE used in the United States is blow molded (Figure 17.6).

Polyethylene films are produced by extruding the molten plastic through a ring-like gap and inflating the film into a balloon. *(Brownie Harris/Corbis)*

Even greater improvements in properties of HDPE can be realized through special processing techniques. In the melt state, HDPE chains have random coiled conformations similar to those of cooked spaghetti. Engineers have developed extrusion techniques that force the individual polymer chains of HDPE to uncoil into linear conformations. These linear chains then align with one another to form highly crystalline materials. HDPE processed in this fashion is stiffer than steel and has approximately four times its tensile strength! Because the density of polyethylene ($\approx$1.0 g/cm^3) is considerably less than that of steel (8.0 g/cm^3), these comparisons of strength and stiffness are even more favorable if they are made on a weight basis.

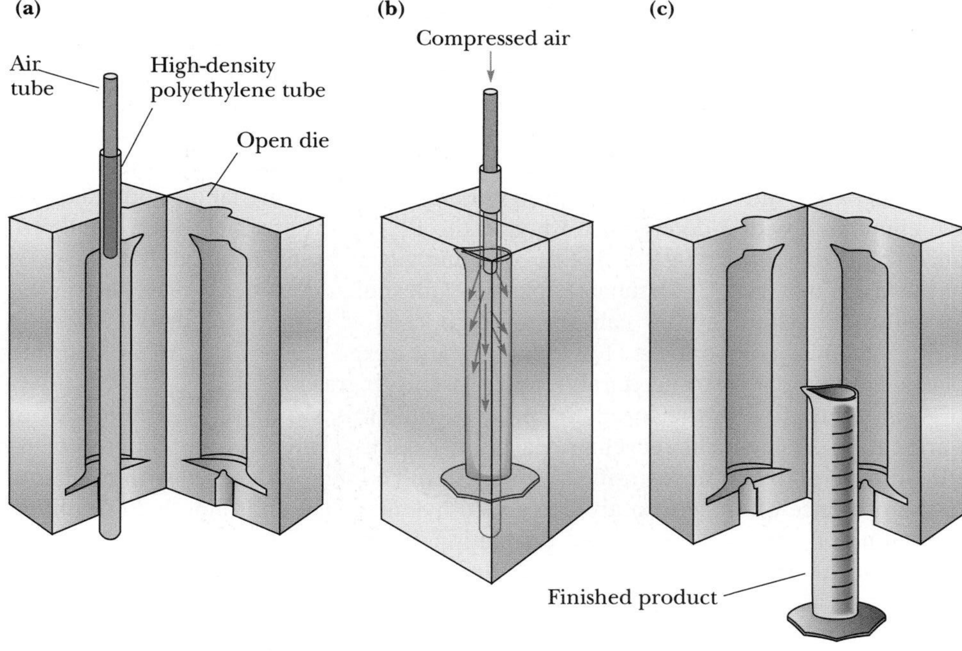

Figure 17.6
Blow molding of an HDPE container. (a) A short length of HDPE tubing is placed in an open die, and the die is closed, sealing the bottom of the tube. (b) Compressed air is forced into the hot polyethylene–die assembly, and the tubing is literally blown up to take the shape of the mold. (c) After cooling, the die is opened, and there is the container!

Some common products packaged in high-density-polyethylene containers. *(Charles D. Winters)*

17.6 What Plastics Are Currently Recycled in Large Quantities?

Polymers in the form of plastics are materials upon which our society is incredibly dependent. Durable and lightweight, plastics are probably the most versatile synthetic materials in existence; in fact, their current production in the United States exceeds that of steel. Plastics have come under criticism, however, for their role in the current trash crisis. They make up 21% of the volume and 8% of the weight of solid waste, most of which is derived from disposable packaging and wrapping. Of the 1.5×10^8 kg of thermoplastic materials produced in the United States per year, less than 2% is recycled.

If the durability and chemical inertness of most plastics make them ideally suited for reuse, why aren't more plastics being recycled? The answer to this question has more to do with economics and consumer habits than with technological obstacles. Because curbside pickup and centralized drop-off stations for recyclables are just now becoming common, the amount of used material available for reprocessing has traditionally been small. This limitation, combined with the need for an additional sorting and separation step, rendered the use of recycled plastics in manufacturing expensive compared with virgin materials. The increase in environmental awareness over the last decade, however, has resulted in a greater demand for recycled products. As manufacturers adapt to satisfy this new market, the recycling of plastics will eventually catch up with that of other materials, such as glass and aluminum.

Six types of plastics are commonly used for packaging applications. In 1988, manufacturers adopted recycling code numbers developed by the Society of the Plastics Industry (Table 17.2). Because the plastics recycling industry still is not fully developed, only PET and HDPE are currently being recycled in large quantities. LDPE, which accounts for about 40% of plastic trash, has been slow in finding acceptance with recyclers. Facilities for the reprocessing of poly(vinyl chloride) (PVC), polypropylene (PP), and polystyrene (PS) exist, but are rare.

The process for the recycling of most plastics is simple, with separation of the desired plastics from other contaminants the most labor-intensive step. For example, PET soft-drink bottles usually have a paper label and adhesive that must be removed before the PET can be reused. The recycling process begins with hand or machine sorting, after which the bottles are shredded into small chips. An air cyclone then removes paper and other lightweight materials. Any remaining labels and adhesives are eliminated with a detergent wash, and the PET chips are then dried. PET produced by this method is 99.9% free of contaminants and sells for about half the price of the virgin material. Unfortunately, plastics with similar densities cannot be separated with this technology, nor can plastics composed of several polymers be broken down into pure components. However, recycled mixed plastics can be molded into plastic lumber that is strong, durable, and resistant to graffiti.

An alternative to the foregoing process, which uses only physical methods of purification, is chemical recycling. Eastman Kodak salvages large amounts of its PET film scrap by a transesterification reaction. The scrap is treated with methanol in the presence of an acid catalyst to give ethylene glycol and dimethyl terephthalate, monomers that are purified by distillation or recrystallization and used as feedstocks for the production of more PET film:

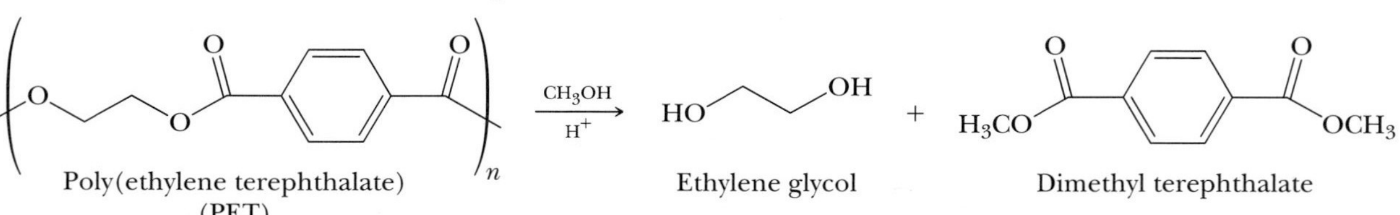

TABLE 17.2 Recycling Codes for Plastics

Recycling Code	Polymer	Common Uses	Uses of Recycled Polymer
1 PET	poly(ethylene terephthalate)	soft-drink bottles, household chemical bottles, films, textile fibers	soft-drink bottles, household chemical bottles, films, textile fibers
2 HDPE	high-density polyethylene	milk and water jugs, grocery bags, bottles	bottles, molded containers
3 V	poly(vinyl chloride), PVC	shampoo bottles, pipes, shower curtains, vinyl siding, wire insulation, floor tiles, credit cards	plastic floor mats
4 LDPE	low-density polyethylene	shrink wrap, trash and grocery bags, sandwich bags, squeeze bottles	trash bags and grocery bags
5 PP	polypropylene	plastic lids, clothing fibers, bottle caps, toys, diaper linings	mixed-plastic components
6 PS	polystyrene	Styrofoam™ cups, egg cartons, disposable utensils, packaging materials, appliances	molded items such as cafeteria trays, rulers, Frisbees™, trash cans, videocasettes
7	all other plastics and mixed plastics	various	plastic lumber, playground equipment, road reflectors

Key Terms and Concepts

amorphous domains (p. 627)
aramid (p. 632)
average degree of polymerization, n (p. 626)
chain-growth polymerization (p. 636)
chain initiation (p. 638)
chain length (p. 639)
chain propagation (p. 639)
chain termination (p. 639)
chain-transfer reaction (p. 640)
cold drawing (p. 630)
crystalline domains (p. 627)
elastomer (p. 629)
epoxy resin (p. 634)
fishhook arrow (p. 638)
glass transition temperature (T_g) (p. 629)
high-density polyethylene (HDPE) (p. 641)
low-density polyethylene (LDPE) (p. 640)
melt transition temperature (T_m) (p. 627)
monomer (p. 625)
plastic (p. 625)
polyamide (p. 630)
polycarbonate (p. 633)
polyester (p. 632)
polymer (p. 625)
polymer morphology (p. 627)
polyurethane (p. 634)
radical (p. 637)
recycling code for plastics (p. 643)
showing the structure of a polymer (p. 626)
step-growth polymerization (p. 629)
thermoplastics (p. 625)
thermosetting plastics (p. 625)

Summary of Key Questions

17.1 What Is the Architecture of Polymers?

- **Polymerization** is the process of joining together many small **monomers** into large, high-molecular-weight **polymers**.
- **Polymers** are long-chain molecules synthesized by linking monomers through chemical reactions. The molecular weight of polymers is generally high compared with those of common organic compounds, and typically range from 10,000 g/mol to more than 1,000,000 g/mol.
- **Thermoplastics** are polymers that can be molded when hot and that retain their shape when cooled.
- **Thermosetting plastics** can be molded when they are first prepared but once cooled, they harden irreversibly and cannot be remelted.

17.2 How Do We Name and Show the Structure of a Polymer?

- The repeat unit of a polymer is the smallest unit that contains all of the structural features of the polymer.
- To show the structure of a polymer, enclose the repeat unit in parentheses and place a subscript n outside the parentheses to show that this structural unit repeats n times in a polymer chain.
- An entire polymer chain can be reproduced by repeating the enclosed structural unit in both directions.
- The most common method of naming a polymer is to add the prefix **poly**- to the name of the monomer from which the polymer is synthesized. If the name of the monomer is two or more words, enclose its name in parentheses.

17.3 What Is Polymer Morphology?

- The properties of polymeric materials depend on the structure of the repeat unit, as well as on the chain architecture and morphology of the material.
- Polymers, like small organic molecules, tend to crystallize upon precipitation or as they are cooled from a melt.
- Acting to inhibit crystallization are the facts that polymer molecules are very large, which inhibits their diffusion, and that their structures are sometimes complicated and irregular.
- Polymers in the solid state tend to be composed of both ordered **crystalline domains** (crystallites) and disordered **amorphous domains**.
- The temperature at which crystallites melt corresponds to the **melt transition temperature (T_m)**.
- As the degree of crystallinity of a polymer increases, its T_m increases, and the polymer becomes more opaque because of the scattering of light by its crystalline domains. With an increase in crystallinity comes an increase in strength and stiffness.
- Amorphous polymers have little or no long-range order. Because they lack crystalline domains that scatter light, amorphous polymers are transparent. In addition they tend to be highly flexible and have low mechanical strength.

17.4 What Is Step-Growth Polymerization?

- **Step-growth polymerizations** involve the stepwise reaction of difunctional monomers.
- The two most common types of step-growth polymerizations involve (1) reaction between A-M-A and B-M-B monomers to give –(A-M-A-B-M-B)$_n$- polymers, where A and B are reactive functional groups, and (2) self-condensations of A-M-B monomers to give –(A-M-B)$_n$- polymers.
- Important commercial polymers synthesized through step-growth processes include polyamides, polyesters, polycarbonates, polyurethanes, and epoxy resins.

17.5 What Is Chain-Growth Polymerization?

- **Chain-growth polymerization** proceeds by the sequential addition of monomer units to an active chain end-group.
- **Radical chain-growth polymerization** consists of three stages: chain initiation, chain propagation, and chain termination.
- Alkyl radicals are planar or almost so with bond angles of approximately 120° about the carbon with the unpaired electron.
- In **chain initiation**, radicals are formed from nonradical molecules.
- In **chain propagation**, a radical and a monomer react to give a new radical.
- **Chain length** is the number of times a cycle of chain propagation steps repeats.
- In **chain termination**, radicals are destroyed.
- **Ziegler–Natta chain-growth polymerization** involves the formation of an alkyl-transition metal compound and then the repeated insertion of alkene monomers into the transition metal-to-carbon bond to yield a saturated polymer chain.

17.6 What Types of Plastics Are Commonly Recycled in Large Quantities?

- The six types of plastics commonly used for packaging applications have been assigned recycling codes with values 1 through 6.
- Currently, only (1) poly(ethylene terephthalate), PET, and (2) high-density polyethylene (HDPE) are recycled in large quantities.

Quick Quiz

Answer true or false to the following questions to assess your general knowledge of the concepts in this chapter. If you have difficulty with any of them, you should review the appropriate section in the chapter (shown in parentheses) before attempting the more challenging end-of-chapter problems.

1. Radicals can undergo chain-growth polymerization. (17.5)
2. A thermosetting plastic cannot be remelted. (17.1)
3. Chain-transfer reactions can lead to branching in polymers. (17.5)
4. Polymers that have low glass transition temperatures can behave as elastomers. (17.3)
5. A highly crystalline polymer will have a glass transition temperature. (17.3)
6. Polymers can be named from the monomeric units from which they are derived. (17.2)
7. Only compounds that have two or more functional groups can undergo step-growth polymerization. (17.4)
8. The propagation step of a radical polymerization mechanism involves the reaction of a radical with another radical. (17.5)
9. A radical is a molecule with an unpaired electron and a positive charge. (17.5)
10. The term *plastics* can be used to refer to all polymers. (17.1)
11. The mechanism of a radical polymerization reaction involves three distinct steps. (17.5)
12. Hydrogen bonding will usually weaken the fibers of a polymer. (17.4)
13. A secondary radical is more stable than a tertiary radical. (17.5)
14. A thermoplastic can be molded multiple times through heating and cooling. (17.1)
15. Ziegler–Natta polymerization uses a titanium catalyst. (17.5)

Answers: (1) T (2) T (3) T (4) T (5) F (6) T (7) T (8) F (9) F (10) F (11) T (12) F (13) F (14) T (15) T

Key Reactions

1. Step-growth polymerization of a dicarboxylic acid and a diamine gives a polyamide (Section 17.4A)

In this equation, M and M′ indicate the remainder of each monomer unit:

$$HO\overset{O}{\overset{\|}{C}}-M-\overset{O}{\overset{\|}{C}}OH + H_2N-M'-NH_2 \xrightarrow{\text{heat}} \left(\overset{O}{\overset{\|}{C}}-M-\overset{O}{\overset{\|}{C}}-\underset{H}{N}-M'-\underset{H}{N} \right)_n + 2nH_2O$$

2. Step-growth polymerization of a dicarboxylic acid and a diol gives a polyester (Section 17.4B)

$$HO\overset{O}{\overset{\|}{C}}-M-\overset{O}{\overset{\|}{C}}OH + HO-M'-OH \xrightarrow[\text{catalyst}]{\text{acid}} \left(\overset{O}{\overset{\|}{C}}-M-\overset{O}{\overset{\|}{C}}-O-M'-O \right)_n + 2nH_2O$$

3. Step-growth polymerization of phosgene and a diol gives a polycarbonate (Section 17.4C)

$$Cl-\overset{O}{\overset{\|}{C}}-Cl + HO-M-OH \longrightarrow \left(O-\overset{O}{\overset{\|}{C}}-O-M \right)_n + 2nHCl$$

4. Step-growth polymerization of a diisocyanate and a diol gives a polyurethane (Section 17.4D)

$$O{=}C{=}N-M-N{=}C{=}O + HO-M'-OH \longrightarrow \left(\overset{O}{\overset{\|}{C}}-\underset{H}{N}-M-\underset{H}{N}-\overset{O}{\overset{\|}{C}}-O-M'-O \right)_n$$

5. Step-growth polymerization of a diepoxide and a diamine gives an epoxy resin (Section 17.4E)

$$\text{(epoxide)}-M-\text{(epoxide)} + H_2N-M'-NH_2 \longrightarrow \left(\underset{H}{N}-CH_2-\underset{OH}{CH}-M-\underset{OH}{CH}-CH_2-\underset{H}{N}-M' \right)_n$$

6. Radical chain-growth polymerization of ethylene and substituted ethylenes (Section 17.5A)

$$nCH_2{=}CHCOOCH_3 \xrightarrow[\text{heat}]{\text{peroxide}} \left(CH_2\underset{}{\overset{COOCH_3}{\overset{|}{C}H}} \right)_n$$

7. Ziegler–Natta chain-growth polymerization of ethylene and substituted ethylenes (Section 17.5B)

$$nCH_2{=}CHCH_3 \xrightarrow[MgCl_2]{TiCl_4/Al(C_2H_5)_2Cl} \left(CH_2\overset{CH_3}{\overset{|}{C}H} \right)_n$$

Problems

A problem marked with an asterisk indicates an applied "real world" problem. Answers to problems whose numbers are printed in blue are given in Appendix D.

Section 17.4 Step-Growth Polymers

*__17.3__ Identify the monomers required for the synthesis of each step-growth polymer:

(a) $\left(\overset{O}{\overset{\|}{C}}-C_6H_4-\overset{O}{\overset{\|}{C}}-OCH_2-C_6H_{10}-CH_2O \right)_n$

Kodel™
(a polyester)

(b) 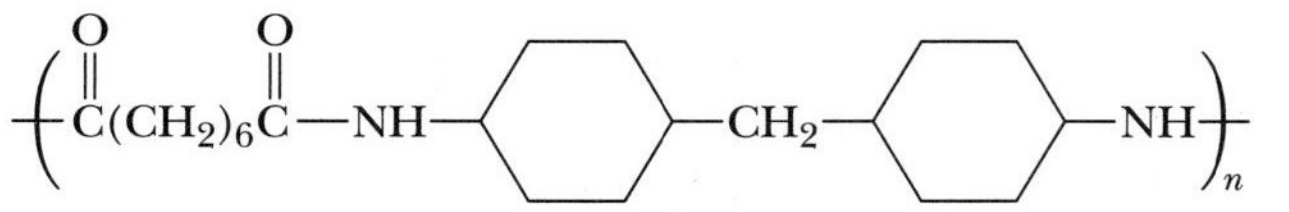 $\left(\overset{O}{\overset{\|}{C}}(CH_2)_6\overset{O}{\overset{\|}{C}}—NH—\text{C}_6\text{H}_{10}—CH_2—\text{C}_6\text{H}_{10}—NH \right)_n$

Quiana™
(a polyamide)

(c) 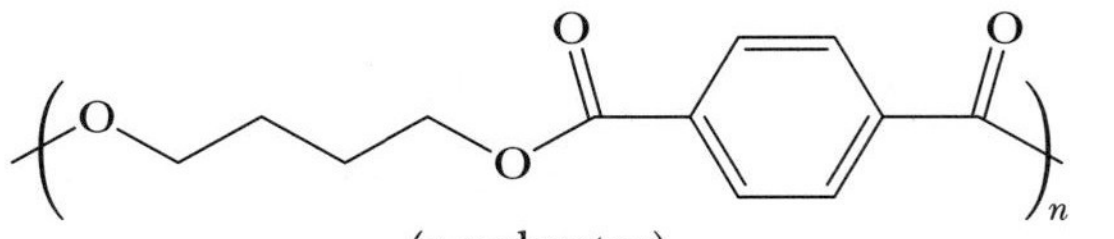 (a polyester)

(d)

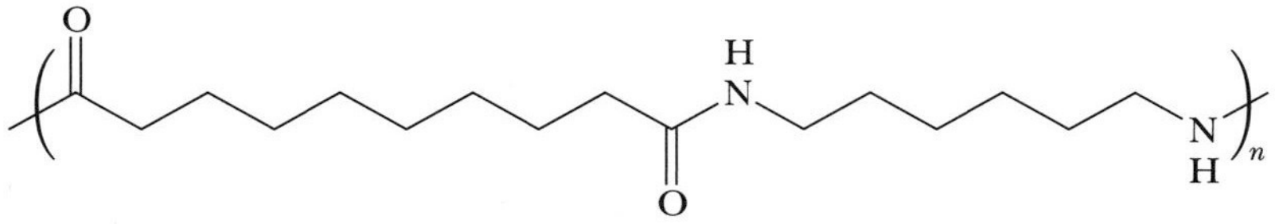

Nylon 6,10
(a polyamide)

17.4 Poly(ethylene terephthalate) (PET) can be prepared by the following reaction:

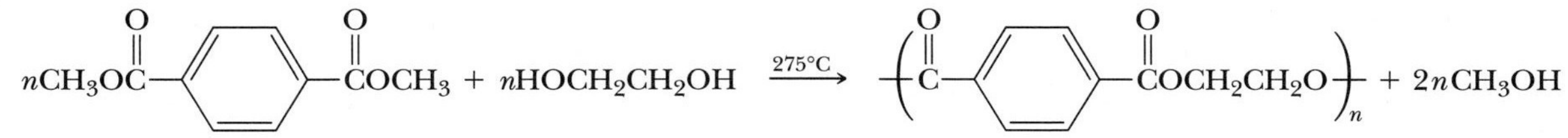

$$nCH_3O\overset{O}{\overset{\|}{C}}—C_6H_4—\overset{O}{\overset{\|}{C}}OCH_3 + nHOCH_2CH_2OH \xrightarrow{275°C} \left(\overset{O}{\overset{\|}{C}}—C_6H_4—\overset{O}{\overset{\|}{C}}OCH_2CH_2O \right)_n + 2nCH_3OH$$

Dimethyl terephthalate | Ethylene glycol | Poly(ethylene terephthalate) | Methanol

Propose a mechanism for the step-growth reaction in this polymerization.

***17.5** Currently, about 30% of PET soft-drink bottles are being recycled. In one recycling process, scrap PET is heated with methanol in the presence of an acid catalyst. The methanol reacts with the polymer, liberating ethylene glycol and dimethyl terephthalate. These monomers are then used as feedstock for the production of new PET products. Write an equation for the reaction of PET with methanol to give ethylene glycol and dimethyl terephthalate.

***17.6** Nomex® is an aromatic polyamide (aramid) prepared from the polymerization of 1,3-benzenediamine and the acid chloride of 1,3-benzenedicarboxylic acid:

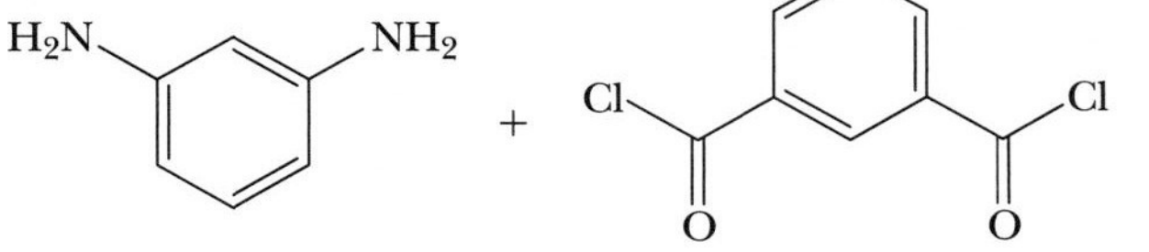

H_2N, NH_2 + Cl, Cl, O, O

1,3-Benzenediamine | 1,3-Benzenedicarbonyl chloride

$\xrightarrow{\text{polymerization}}$ Nomex

The physical properties of the polymer make it suitable for high-strength, high-temperature applications such as parachute cords and jet aircraft tires. Draw a structural formula for the repeating unit of Nomex.

17.7 Nylon 6,10 [Problem 17.3(d)] can be prepared by reacting a diamine and a diacid chloride. Draw the structural formula of each reactant.

Section 17.5 Chain-Growth Polymerization

17.8 Following is the structural formula of a section of polypropylene derived from three units of propylene monomer:

$$-CH_2CH(CH_3)-CH_2CH(CH_3)-CH_2CH(CH_3)-$$

Polypropylene

Draw a structural formula for a comparable section of
(a) Poly(vinyl chloride)
(b) Polytetrafluoroethylene (PTFE)
(c) Poly(methyl methacrylate)

17.9 Following are structural formulas for sections of two polymers: **(See Example 17.1)**

(a) $-CH_2CCl_2CH_2CCl_2CH_2CCl_2-$ (b) $-CH_2CF_2CH_2CF_2CH_2CF_2-$

From what alkene monomer is each polymer derived?

17.10 Draw the structure of the alkene monomer used to make each chain-growth polymer: **(See Example 17.1)**

(a) (b) (c) CH_2Cl (d) $\left(CF(CF_3)-CF_2\right)_n$

17.11 LDPE has a higher degree of chain branching than HDPE. Explain the relationship between chain branching and density.

17.12 Compare the densities of LDPE and HDPE with the densities of the liquid alkanes listed in Table 3.4. How might you account for the differences between them?

17.13 The polymerization of vinyl acetate gives poly(vinyl acetate). Hydrolysis of this polymer in aqueous sodium hydroxide gives poly(vinyl alcohol). Draw the repeat units of both poly(vinyl acetate) and poly(vinyl alcohol):

Vinyl acetate $CH_3-C(=O)-O-CH=CH_2$

17.14 As seen in the previous problem, poly(vinyl alcohol) is made by the polymerization of vinyl acetate, followed by hydrolysis in aqueous sodium hydroxide. Why is poly(vinyl alcohol) not made instead by the polymerization of vinyl alcohol, $CH_2=CHOH$?

17.15 As you know, the shape of a polymer chain affects its properties. Consider the following three polymers:

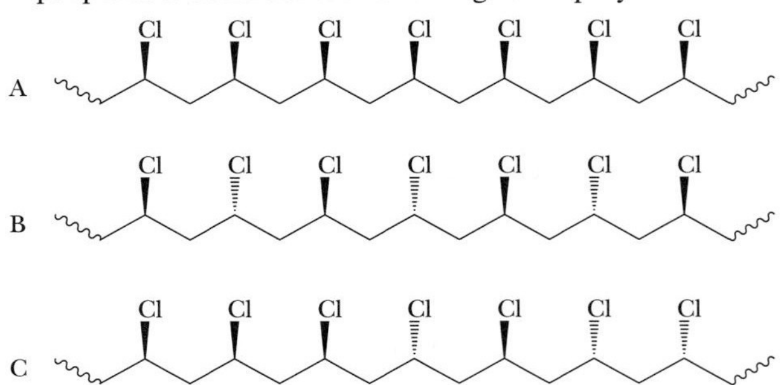

Which do you expect to be the most rigid? Which do you expect to be the most transparent? (Assume the same molecular weights.)

Looking Ahead

17.16 Cellulose, the principle component of cotton, is a polymer of D-glucose in which the monomer unit repeats at the indicated atoms:

D-Glucose

Draw a three-unit section of cellulose.

17.17 Is a repeating unit a requirement for a compound to be called a polymer?

17.18 Proteins are polymers of naturally occurring monomers called amino acids:

a protein

Amino acids differ in the types of R groups available in nature. Explain how the following properties of a protein might be affected upon changing the R groups from $-CH_2CH(CH_3)_2$ to $-CH_2OH$:

(a) solubility in water (b) T_m
(c) crystallinity (d) elasticity

18 Carbohydrates

KEY QUESTIONS

HOW TO

Breads, grains, and pasta are sources of carbohydrates. Inset: A model of glucose. *(Charles D. Winters)*

CHEMICAL CONNECTIONS

18.1 What Are Carbohydrates?

Carbohydrates are the most abundant organic compounds in the plant world. They act as storehouses of chemical energy (glucose, starch, glycogen); are components of supportive structures in plants (cellulose), crustacean shells (chitin), and connective tissues in animals (acidic polysaccharides); and are essential components of nucleic acids (D-ribose and 2-deoxy-D-ribose). Carbohydrates account for approximately three-fourths of the dry weight of plants. Animals (including humans) get their carbohydrates by eating plants, but they do not store much of what they consume. In fact, less than 1 percent of the body weight of animals is made up of carbohydrates.

The word *carbohydrate* means "hydrate of carbon" and derives from the formula $C_n(H_2O)_m$. Two examples of carbohydrates with molecular formulas that can be written alternatively as hydrates of carbon are

- glucose (blood sugar), $C_6H_{12}O_6$, which can be written as $C_6(H_2O)_6$, and
- sucrose (table sugar), $C_{12}H_{22}O_{11}$, which can be written as $C_{12}(H_2O)_{11}$.

Not all carbohydrates, however, have this general formula. Some contain too few oxygen atoms to fit the formula, whereas some contain too many. Some also contain nitrogen. But the term "carbohydrate" has become firmly rooted in chemical nomenclature and, although not completely accurate, it persists as the name for this class of compounds.

Carbohydrate A polyhydroxyaldehyde or polyhydroxyketone or a substance that gives these compounds on hydrolysis.

At the molecular level, most **carbohydrates** are polyhydroxyaldehydes, polyhydroxyketones, or compounds that yield them after hydrolysis. Therefore, the chemistry of carbohydrates is essentially the chemistry of hydroxyl and carbonyl groups and of acetal bonds (Section 13.6A) formed between these two functional groups.

18.2 What Are Monosaccharides?

A. Structure and Nomenclature

Monosaccharide A carbohydrate that cannot be hydrolyzed to a simpler compound.

Monosaccharides have the general formula $C_nH_{2n}O_n$, with one of the carbons being the carbonyl group of either an aldehyde or a ketone. The most common monosaccharides have from three to nine carbon atoms. The suffix *-ose* indicates that a molecule is a carbohydrate, and the prefixes *tri-*, *tetr-*, *pent-*, and so forth, indicate the number of carbon atoms in the chain. Monosaccharides containing an aldehyde group are classified as **aldoses**; those containing a ketone group are classified as **ketoses**.

Aldose A monosaccharide containing an aldehyde group.

Ketose A monosaccharide containing a ketone group.

There are only two **trioses**—glyceraldehyde, which is an aldotriose, and dihydroxyacetone, which is a ketotriose:

CHO
|
CHOH
|
CH_2OH
Glyceraldehyde
(an aldotriose)

CH_2OH
|
C=O
|
CH_2OH
Dihydroxyacetone
(a ketotriose)

Often the designations *aldo-* and *keto-* are omitted, and these molecules are referred to simply as trioses, **tetroses**, and the like. Although these designations do not tell the nature of the carbonyl group, at least they indicate that the monosaccharide contains three and four carbon atoms, respectively.

B. Stereoisomerism

Glyceraldehyde contains one stereocenter and exists as a pair of enantiomers. The stereoisomer shown on the left has the *R* configuration and is named (*R*)-glyceraldehyde, while its enantiomer, shown on the right, is named (*S*)-glyceraldehyde:

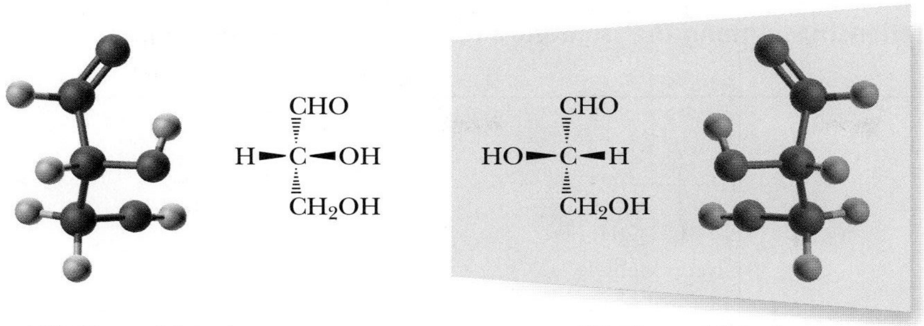

$$\begin{array}{c} CHO \\ H-C-OH \\ CH_2OH \end{array} \qquad \begin{array}{c} CHO \\ HO-C-H \\ CH_2OH \end{array}$$

(*R*)-Glyceraldehyde (*S*)-Glyceraldehyde

C. Fischer Projection Formulas

Chemists commonly use two-dimensional representations called **Fischer projections** to show the configuration of carbohydrates. To draw a Fischer projection, draw a three-dimensional representation with the most oxidized carbon toward the top and the molecule oriented so that the vertical bonds from the stereocenter are directed away from you and the horizontal bonds from it are directed toward you. Then write the molecule as a two-dimensional figure with the stereocenter indicated by the point at which the bonds cross. You now have a Fischer projection.

Fischer projection A two-dimensional representation showing the configuration of a stereocenter; horizontal lines represent bonds projecting forward from the stereocenter, vertical lines represent bonds projecting to the rear.

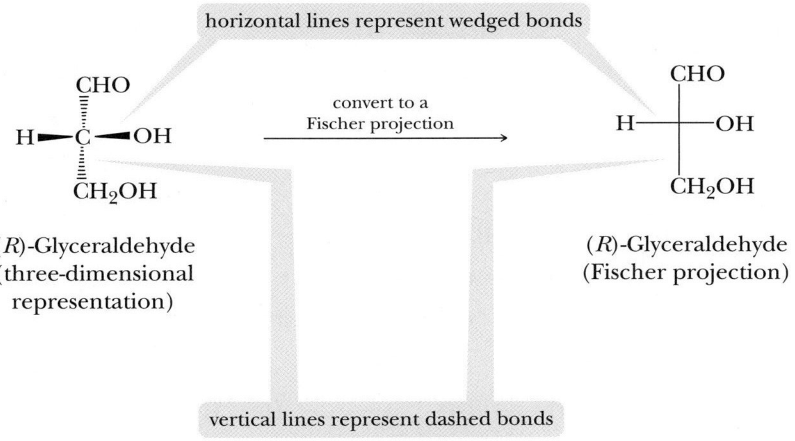

horizontal lines represent wedged bonds

$$\begin{array}{c} CHO \\ H-C-OH \\ CH_2OH \end{array} \xrightarrow{\text{convert to a Fischer projection}} \begin{array}{c} CHO \\ H-\!\!\!+\!\!\!-OH \\ CH_2OH \end{array}$$

(*R*)-Glyceraldehyde (three-dimensional representation)

(*R*)-Glyceraldehyde (Fischer projection)

vertical lines represent dashed bonds

The two horizontal segments of this Fischer projection represent bonds directed toward you, and the two vertical segments represent bonds directed away from you. The only atom in the plane of the paper is the stereocenter.

D. D- and L-Monosaccharides

Even though the *R,S* system is widely accepted today as a standard for designating the configuration of stereocenters, we still commonly designate the configuration of carbohydrates by the D,L system proposed by Emil Fischer in 1891. He assigned the dextrorotatory and levorotatory enantiomers of glyceraldehyde the following configurations and named them D-glyceraldehyde and L-glyceraldehyde, respectively:

$$\begin{array}{c} CHO \\ H-\!\!\!+\!\!\!-OH \\ CH_2OH \end{array} \qquad \begin{array}{c} CHO \\ HO-\!\!\!+\!\!\!-H \\ CH_2OH \end{array}$$

D-Glyceraldehyde $[\alpha]_D^{25} = +13.5°$

L-Glyceraldehyde $[\alpha]_D^{25} = -13.5°$

TABLE 18.1 Configurational Relationships Among the Isomeric D-Aldotetroses, D-Aldopentoses, and D-Aldohexoses

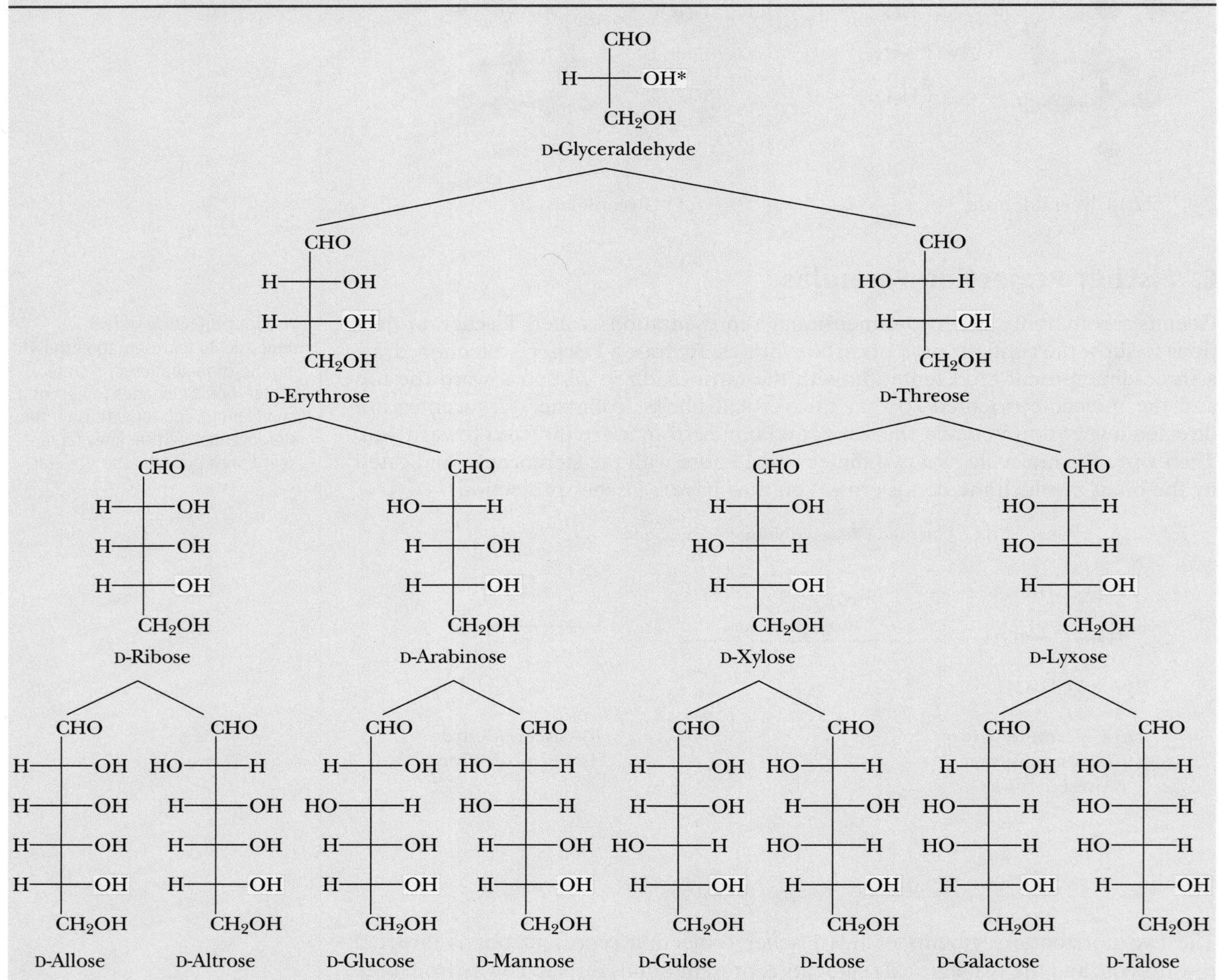

* The configuration of the reference —OH on the penultimate carbon is shown in color.

Penultimate carbon The stereocenter of a monosaccharide farthest from the carbonyl group—for example, carbon 5 of glucose.

D-Monosaccharide A monosaccharide that, when written as a Fischer projection, has the —OH on its penultimate carbon to the right.

L-Monosaccharide A monosaccharide that, when written as a Fischer projection, has the —OH on its penultimate carbon to the left.

D-glyceraldehyde and L-glyceraldehyde serve as reference points for the assignment of relative configurations to all other aldoses and ketoses. The reference point is the stereocenter farthest from the carbonyl group. Because this stereocenter is the next-to-the-last carbon on the chain, it is called the **penultimate carbon**. A **D-monosaccharide** is a monosaccharide that has the same configuration at its penultimate carbon as D-glyceraldehyde (its —OH is on the right in a Fischer projection); an **L-monosaccharide** has the same configuration at its penultimate carbon as L-glyceraldehyde (its —OH is on the left in a Fischer projection). Almost all monosaccharides in the biological world belong to the D series, and the majority of them are either **hexoses** or **pentoses**.

Table 18.1 shows the names and Fischer projection formulas for all D-aldotrioses, tetroses, pentoses, and hexoses. Each name consists of three parts. The letter D specifies the configuration at the stereocenter farthest from the carbonyl group. Prefixes, such as *rib*-, *arabin*-, and *gluc*-, specify the configurations of all other stereocenters relative to one another. The suffix *-ose* shows that the compound is a carbohydrate.

The three most abundant hexoses in the biological world are D-glucose, D-galactose, and D-fructose. The first two are D-aldohexoses; the third, fructose, is a D-2-ketohexose. Glucose, by far the most abundant of the three, is also known as dextrose because it is dextrorotatory. Other names for this monosaccharide include *grape sugar* and *blood sugar*. Human blood normally contains 65–110 mg of glucose/100 mL of blood. D-Fructose is one of the two monosaccharide building blocks of sucrose (table sugar, Section 18.7A):

CH_2OH

C=O

HO—H

H—OH

H—OH

CH_2OH

D-Fructose

this is D-fructose because the OH at the penultimate carbon points to the right as in D-glyceraldehyde

Example 18.1

(a) Draw Fischer projections for the four aldotetroses.
(b) Which of the four aldotetroses are D-monosaccharides, which are L-monosaccharides, and which are enantiomers?
(c) Refer to Table 18.1, and name each aldotetrose you have drawn.

Strategy

Start with the Fischer projection of glyceraldehyde (a triose) and add one more carbon to the chain to give a tetrose. Carbon-1 of an aldotetrose is an aldehyde (as in glyceraldehyde) and carbon 4 is a CH_2OH. The designations D- and L- refer to the configuration of the penultimate carbon, which in the case of a tetrose is carbon 3. In the Fischer projection of a D-aldotetrose, the —OH group on carbon 3 is on the right and in an L-aldotetrose, it is on the left.

Solution

Following are Fischer projections for the four aldotetroses:

D-Erythrose	L-Erythrose	D-Threose	L-Threose
CHO	CHO	CHO	CHO
H—OH	HO—H	HO—H	H—OH
H—³—OH	HO—³—H	H—³—OH	HO—³—H
CH_2OH	CH_2OH	CH_2OH	CH_2OH

(one pair of enantiomers) (a second pair of enantiomers)

See problems 18.13, 18.14

Problem 18.1

(a) Draw Fischer projections for all 2-ketopentoses.
(b) Which of the 2-ketopentoses are D-ketopentoses, which are L-ketopentoses, and which are enantiomers?

E. Amino Sugars

Amino sugars contain an $-NH_2$ group in place of an $-OH$ group. Only three amino sugars are common in nature: D-glucosamine, D-mannosamine, and D-galactosamine. *N*-Acetyl-D-glucosamine, a derivative of D-glucosamine, is a component of many polysaccharides, including connective tissue such as cartilage. It is also a component of chitin, the hard shell-like exoskeleton of lobsters, crabs, shrimp, and other shellfish. Several other amino sugars are components of naturally occurring antibiotics.

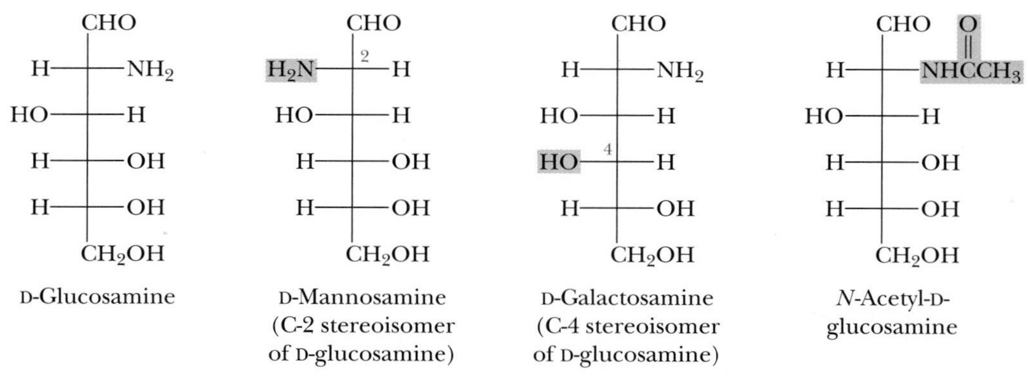

D-Glucosamine — D-Mannosamine (C-2 stereoisomer of D-glucosamine) — D-Galactosamine (C-4 stereoisomer of D-glucosamine) — *N*-Acetyl-D-glucosamine

F. Physical Properties

Monosaccharides are colorless, crystalline solids. Because hydrogen bonding is possible between their polar $-OH$ groups and water, all monosaccharides are very soluble in water. They are only slightly soluble in ethanol and are insoluble in nonpolar solvents such as diethyl ether, dichloromethane, and benzene.

18.3 What Are the Cyclic Structures of Monosaccharides?

In Section 13.6, we saw that aldehydes and ketones react with alcohols to form **hemiacetals**. We also saw that cyclic hemiacetals form very readily when hydroxyl and carbonyl groups are parts of the same molecule and their interaction can form a five- or six-membered ring. For example, 4-hydroxypentanal forms a five-membered cyclic hemiacetal. Note that 4-hydroxypentanal contains one stereocenter and that a second stereocenter is generated at carbon 1 as a result of hemiacetal formation:

redraw to show $-OH$ and $-CHO$ close to each other

new stereocenter

4-Hydroxypentanal — A cyclic hemiacetal

Because monosaccharides have hydroxyl and carbonyl groups in the same molecule, they exist almost exclusively as five- and six-membered cyclic hemiacetals.

A. Haworth Projections

A common way of representing the cyclic structure of monosaccharides is the **Haworth projection**, named after the English chemist Sir Walter N. Haworth, Nobel laureate of 1937. In a Haworth projection, a five- or six-membered cyclic hemiacetal is represented as a planar pentagon or hexagon, respectively, lying roughly perpendicular to the plane of the paper. Groups bonded to the carbons of the ring then lie either above or below the plane of the ring. The new stereocenter created in forming the cyclic hemiacetal is called the **anomeric carbon**. Stereoisomers that differ in configuration only at the anomeric carbon are called **anomers**. The anomeric carbon of an aldose is carbon 1; in D-fructose, the most common ketose, it is carbon 2.

Haworth projection A way of viewing the furanose and pyranose forms of monosaccharides. The ring is drawn flat and viewed through its edge, with the anomeric carbon on the right and the oxygen atom of the ring in the rear to the right.

Anomeric carbon The hemiacetal carbon of the cyclic form of a monosaccharide.

Anomers Monosaccharides that differ in configuration only at their anomeric carbons.

Typically, Haworth projections are written with the anomeric carbon at the right and the hemiacetal oxygen at the back right (Figure 18.1).

As you study the open chain and cyclic hemiacetal forms of D-glucose, note that, in converting from a Fischer projection to a Haworth structure,

- Groups on the right in the Fischer projection point down in the Haworth projection.
- Groups on the left in the Fischer projection point up in the Haworth projection.
- For a D-monosaccharide, the terminal $-CH_2OH$ points up in the Haworth projection.
- The configuration of the anomeric $-OH$ group is relative to the terminal $-CH_2OH$ group: If the anomeric $-OH$ group is on the same side as the terminal $-CH_2OH$, its configuration is β; if the anomeric $-OH$ group is on the opposite side, it is α.

A six-membered hemiacetal ring is shown by the infix **-pyran-**, and a five-membered hemiacetal ring is shown by the infix **-furan-**. The terms **furanose** and **pyranose** are used because monosaccharide five- and six-membered rings correspond to the heterocyclic compounds pyran and furan:

Furanose A five-membered cyclic hemiacetal form of a monosaccharide.

Pyranose A six-membered cyclic hemiacetal form of a monosaccharide.

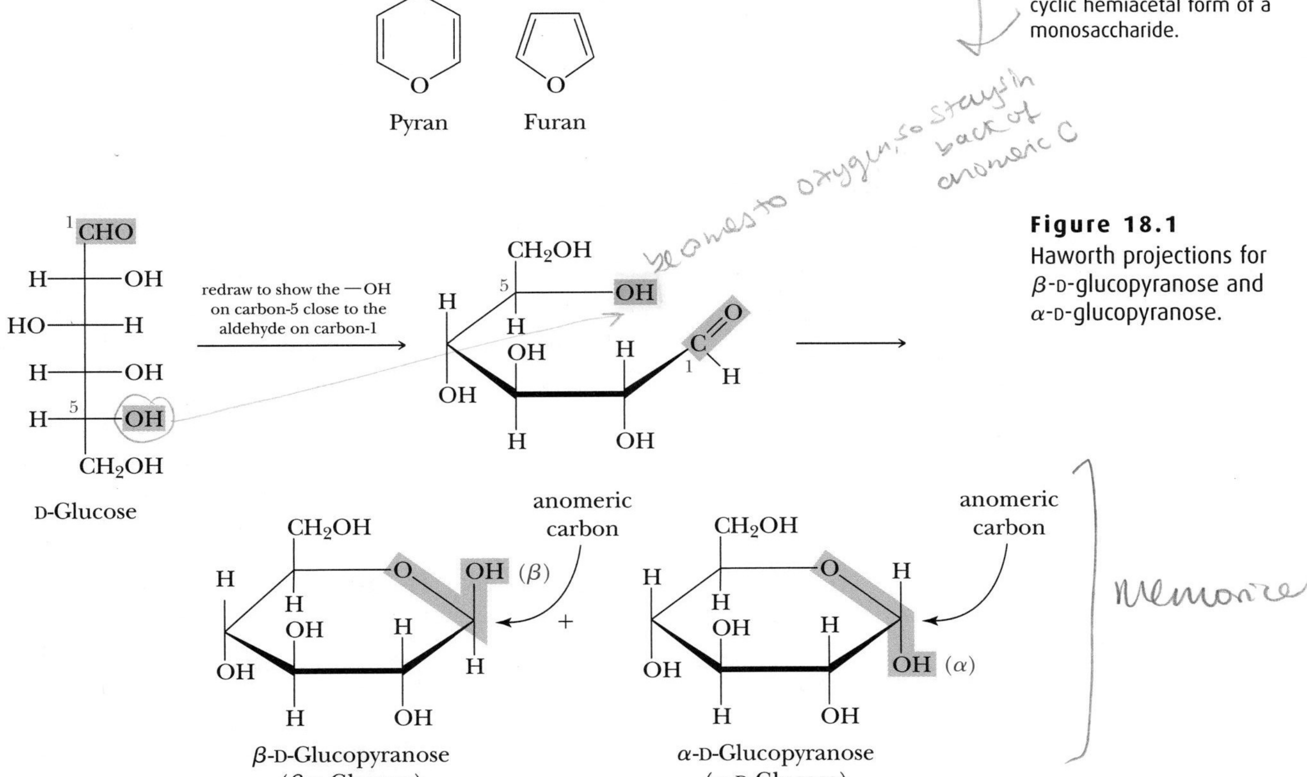

Figure 18.1
Haworth projections for β-D-glucopyranose and α-D-glucopyranose.

Because the α and β forms of glucose are six-membered cyclic hemiacetals, they are named α-D-glucopyranose and β-D-glucopyranose, respectively. The designations *-furan-* and *-pyran-* are not always used, however, in names of monosaccharides. Thus, the glucopyranoses are often named simply α-D-glucose and β-D-glucose.

You would do well to remember the configuration of groups on the Haworth projection of both α-D-glucopyranose and β-D-glucopyranose as reference structures. Knowing how the Fischer projection of any other monosaccharide differs from that of D-glucose, you can then construct the Haworth projection of that other monosaccharide by reference to the Haworth projection of D-glucose.

Example 18.2

Draw Haworth projections for the α and β anomers of D-galactopyranose.

Strategy

One way to arrive at the structures for the α and β anomers of D-galactopyranose is to use the α and β forms of D-glucopyranose as a reference and to remember (or discover by looking at Table 18.1) that D-galactose differs from D-glucose only in the configuration at carbon 4. Thus, you can begin with the Haworth projections shown in Figure 18.1 and then invert the configuration at carbon 4:

Solution

configuration differs from that of D-glucose at C-4

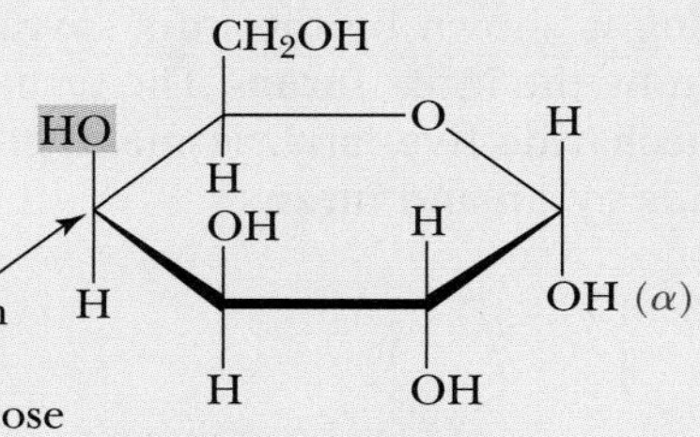

α-D-Galactopyranose
(α-D-Galactose)

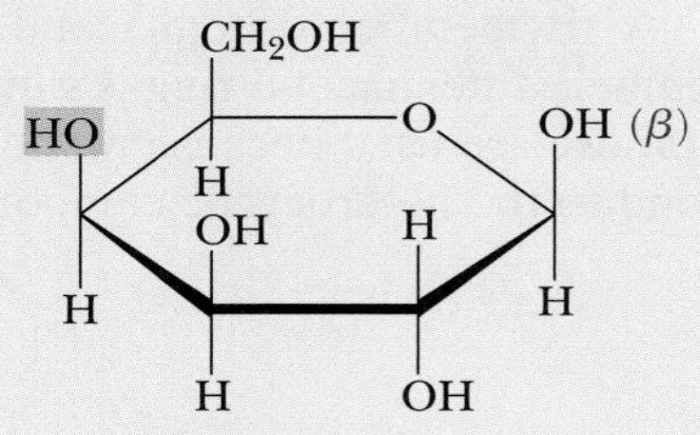

β-D-Galactopyranose
(β-D-Galactose)

See problems 18.23, 18.26

Problem 18.2

Mannose exists in aqueous solution as a mixture of α-D-mannopyranose and β-D-mannopyranose. Draw Haworth projections for these molecules.

Aldopentoses also form cyclic hemiacetals. The most prevalent forms of D-ribose and other pentoses in the biological world are furanoses. Following are Haworth projections for α-D-ribofuranose (α-D-ribose) and β-2-deoxy-D-ribofuranose (β-2-deoxy-D-ribose):

α-D-Ribofuranose
(α-D-Ribose)

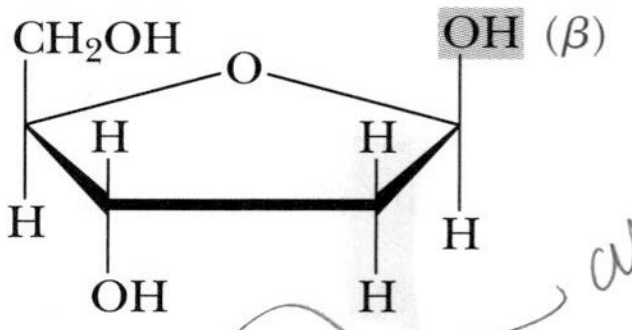

β-2-Deoxy-D-ribofuranose
(β-2-Deoxy-D-ribose)

The prefix 2-deoxy indicates the absence of oxygen at carbon 2. Units of D-ribose and 2-deoxy-D-ribose in nucleic acids and most other biological molecules are found almost exclusively in the β-configuration.

Fructose also forms five-membered cyclic hemiacetals. β-D-Fructofuranose, for example, is found in the disaccharide sucrose (Section 18.7A).

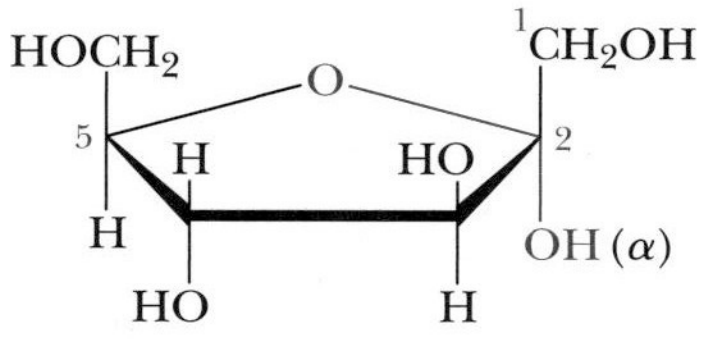

α-D-Fructofuranose
(α-D-Fructose)

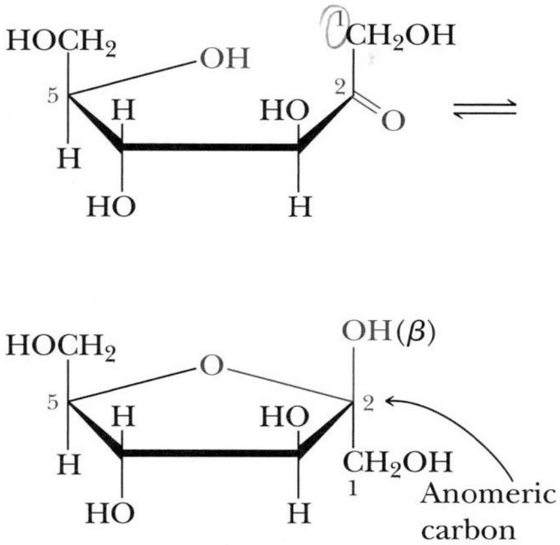

Anomeric carbon

β-D-Fructofuranose
(β-D-Fructose)

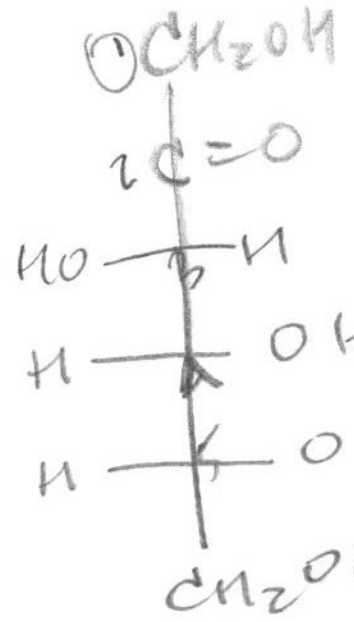

B. Conformation Representations

A five-membered ring is so close to being planar that Haworth projections are adequate to represent furanoses. For pyranoses, however, the six-membered ring is more accurately represented as a **chair conformation** in which strain is a minimum (Section 3.6B). Figure 18.2 shows structural formulas for α-D-glucopyranose and β-D-glucopyranose, both drawn as chair conformations. The figure also shows the open-chain, or free, aldehyde form with which the cyclic hemiacetal forms are in equilibrium in aqueous solution. Notice that each group, including the anomeric —OH, on the chair conformation of β-D-glucopyranose is equatorial. Notice also that the —OH group on the anomeric carbon in α-D-glucopyranose is axial. Because of the equatorial orientation of the —OH on its anomeric carbon, β-D-glucopyranose is more stable than the α-anomer and therefore predominates in aqueous solution.

At this point, you should compare the relative orientations of groups on the D-glucopyranose ring in the Haworth projection and chair conformation:

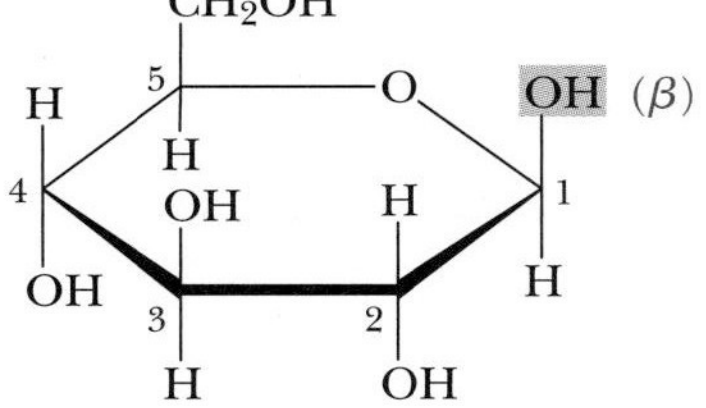

β-D-Glucopyranose
(Haworth projection)

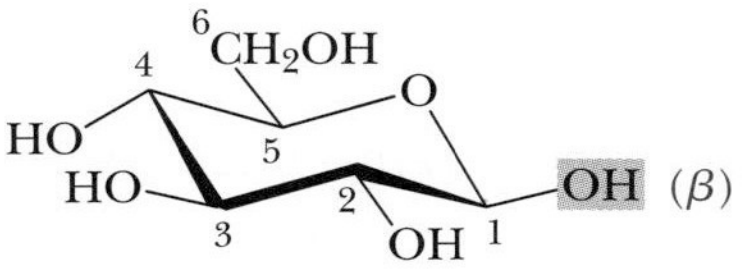

β-D-Glucopyranose
(chair conformation)

Notice that the orientations of groups on carbons 1 through 5 in the Haworth projection of β-D-glucopyranose are up, down, up, down, and up, respectively. The same is the case in the chair conformation.

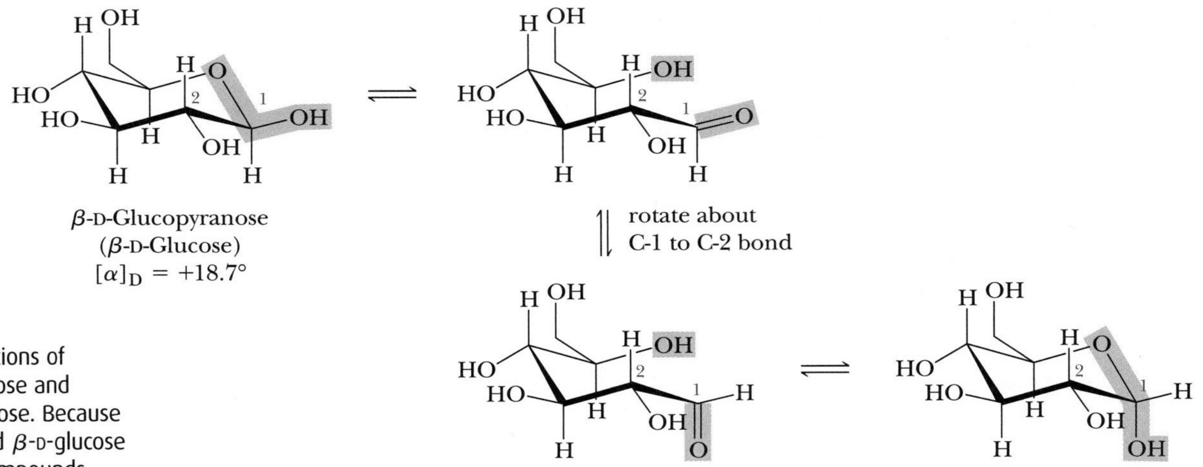

β-D-Glucopyranose (β-D-Glucose) $[\alpha]_D = +18.7°$

rotate about C-1 to C-2 bond

α-D-Glucopyranose (α-D-Glucose) $[\alpha]_D = +112°$

Figure 18.2 Chair conformations of α-D-glucopyranose and β-D-glucopyranose. Because α-D-glucose and β-D-glucose are different compounds (they are anomers), they have different specific rotations.

HOW TO 18.1 Determine the Stereochemistry of OH Groups in Cyclic D-Monosaccharides

There is another way to determine the stereochemistry of OH groups in the D-series of monosaccharides if you forget the structure of glucopyranose:

(a) Draw the Haworth or chair form of the ring and number the corresponding carbon atoms (don't forget to situate the ring oxygen in the correct position):

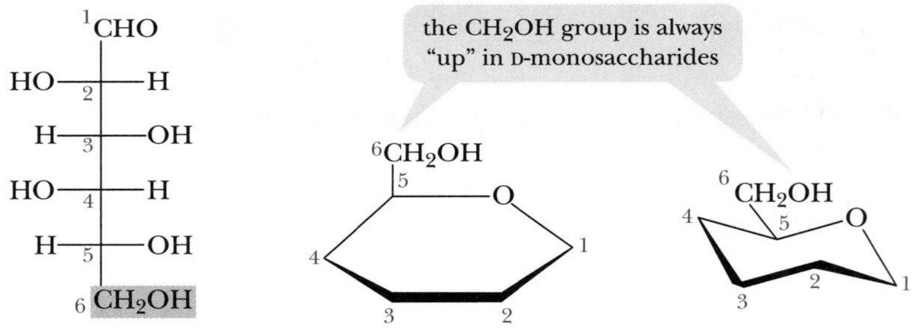

the CH_2OH group is always "up" in D-monosaccharides

D-Idose

(b) Rotate the drawing of the acyclic monosaccharide clockwise by 90°:

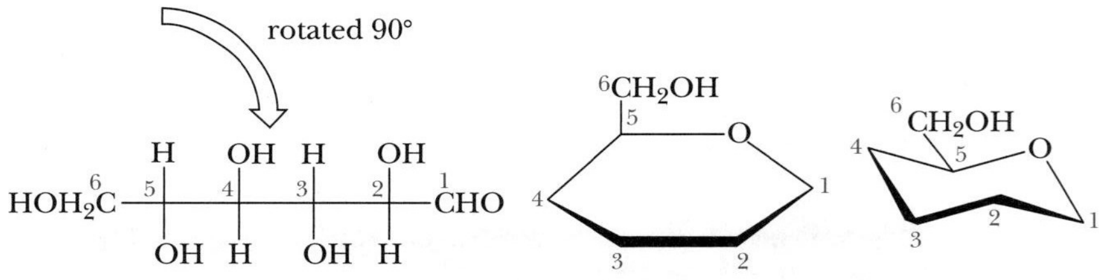

rotated 90°

D-Idose

(c) The direction (up or down) of the OH group on the rotated structure is the direction the OH group should be in on the cyclic monosaccharide:

up OH H OH H HOH$_2$C CHO down OH H OH H down D-Idose

up HO 6 CH$_2$OH O up HO OH down OH

up HO 6 CH$_2$OH HO O up OH down OH

the direction of the anomeric OH depends on whether the monosaccharide is α or β

Example 18.3

Draw chair conformations for α-D-galactopyranose and β-D-galactopyranose. Label the anomeric carbon in each cyclic hemiacetal.

Strategy

The configuration of D-galactose differs from that of D-glucose only at carbon 4. Therefore, draw the α and β, forms of D-glucopyranose and then interchange the positions of the —OH and —H groups on carbon 4.

Solution

Shown are the requested α and β forms of D-galactose. Also shown are the specific rotations of each anomer.

anomeric carbon

β-D-Galactopyranose
(β-D-Galactose)
$[\alpha]_D = +52.8°$

D-Galactose

why is this axial?

α-D-Galactopyranose
(α-D-Galactose)
$[\alpha]_D = +150.7°$

Problem 18.3

Draw chair conformations for α-D-mannopyranose and β-D-mannopyranose. Label the anomeric carbon atom in each.

C. Mutarotation

Mutarotation The change in optical activity that occurs when an α or β form of a carbohydrate is converted to an equilibrium mixture of the two forms.

Mutarotation is the change in specific rotation that accompanies the interconversion of α- and β-anomers in aqueous solution. As an example, a solution prepared by dissolving crystalline α-D-glucopyranose in water shows an initial rotation of $+112°$ (Figure 18.2), which gradually decreases to an equilibrium value of $+52.7°$ as α-D-glucopyranose reaches an equilibrium with β-D-glucopyranose. A solution of β-D-glucopyranose also undergoes mutarotation, during which the specific rotation changes from an initial value of $+18.7°$ to the same equilibrium value of $+52.7°$. The equilibrium mixture consists of 64% β-D-glucopyranose and 36% α-D-glucopyranose and contains only traces (0.003%) of the open-chain form. Mutarotation is common to all carbohydrates that exist in hemiacetal forms.

18.4 What Are the Characteristic Reactions of Monosaccharides?

In this section, we discuss reactions of monosaccharides with alcohols, reducing agents, and oxidizing agents. In addition, we examine how these reactions are useful in our everyday lives.

A. Formation of Glycosides (Acetals)

As we saw in Section 13.6A, treating an aldehyde or a ketone with one molecule of alcohol yields a hemiacetal, and treating the hemiacetal with a molecule of alcohol yields an acetal. Treating a monosaccharide, all forms of which exist as cyclic hemiacetals, with an alcohol gives an acetal, as illustrated by the reaction of β-D-glucopyranose (β-D-glucose) with methanol:

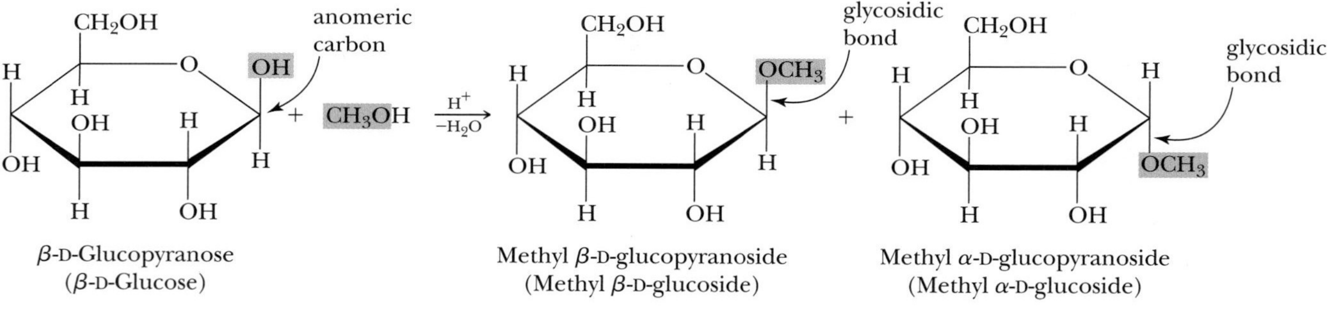

β-D-Glucopyranose (β-D-Glucose)

Methyl β-D-glucopyranoside (Methyl β-D-glucoside)

Methyl α-D-glucopyranoside (Methyl α-D-glucoside)

Glycoside A carbohydrate in which the —OH on its anomeric carbon is replaced by —OR.

Glycosidic bond The bond from the anomeric carbon of a glycoside to an —OR group.

A cyclic acetal derived from a monosaccharide is called a **glycoside**, and the bond from the anomeric carbon to the —OR group is called a **glycosidic bond**. Mutarotation is no longer possible in a glycoside, because, unlike a hemiacetal, an acetal is no longer in equilibrium with the open-chain carbonyl-containing compound in neutral or alkaline solution. Like other acetals (Section 13.6), glycosides are stable in water and aqueous base, but undergo hydrolysis in aqueous acid to an alcohol and a monosaccharide.

We name glycosides by listing the alkyl or aryl group bonded to oxygen, followed by the name of the carbohydrate involved in which the ending **-e** is replaced by **-ide**. For example, glycosides derived from β-D-glucopyranose are named β-D-glucopyranosides; those derived from β-D-ribofuranose are named β-D-ribofuranosides.

Example 18.4

Draw a structural formula for methyl β-D-ribofuranoside (methyl β-D-riboside). Label the anomeric carbon and the glycosidic bond.

Strategy

Furanosides are five-membered cyclic acetals. The anomeric carbon is carbon 1, and the glycosidic bond is formed to carbon 1. For a β-glycosidic bond, the —OR group is above the plane of the ring and on the same side as the terminal —CH_2OH group (carbon 5 of a furanoside).

Solution

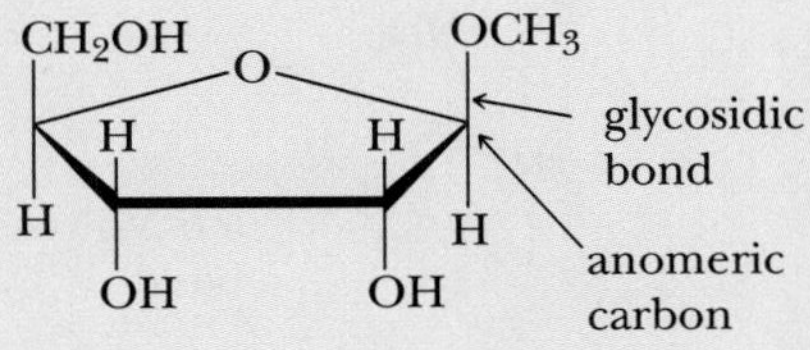

CH_2OH O OCH_3 H H H H OH OH glycosidic bond anomeric carbon

See problems 18.30–18.32

Problem 18.4

Draw a structural formula for the chair conformation of methyl α-D-mannopyranoside (methyl α-D-mannoside). Label the anomeric carbon and the glycosidic bond.

Just as the anomeric carbon of a cyclic hemiacetal undergoes reaction with the —OH group of an alcohol to form a glycoside, it also undergoes reaction with the —NH group of an amine to form an *N*-glycoside. Especially important in the biological world are the *N*-glycosides formed between D-ribose and 2-deoxy-D-ribose (each as a furanose), and the heterocyclic aromatic amines uracil, cytosine, thymine, adenine, and guanine (Figure 18.3). *N*-Glycosides of these compounds are structural units of nucleic acids (Chapter 20).

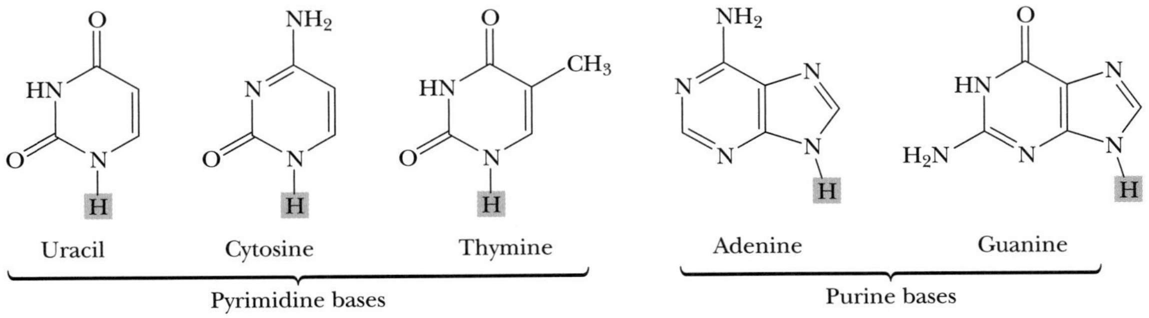

Uracil Cytosine Thymine Adenine Guanine

Pyrimidine bases Purine bases

Figure 18.3
Structural formulas of the five most important purine and pyrimidine bases found in DNA and RNA. The hydrogen atom shown in color is lost in the formation of an *N*-glycoside.

Example 18.5

Draw a structural formula for the β-N-glycoside formed between D-ribofuranose and cytosine. Label the anomeric carbon and the N-glycosidic bond.

Strategy

Start with the Haworth projection of β-D-ribofuranose. Locate the anomeric carbon (carbon 1). Remove the —OH group from carbon 1 and in its place, bond the appropriate nitrogen atom of cytosine (see Figure 18.3) to give the β-N-glycosidic bond.

Solution

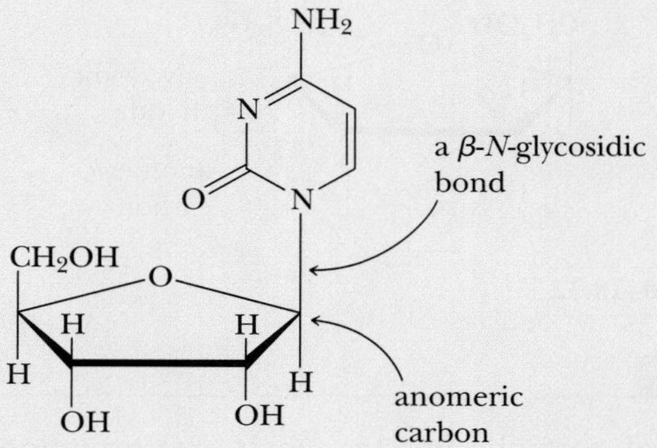

a β-N-glycosidic bond

anomeric carbon

See problem 18.33

Problem 18.5

Draw a structural formula for the β-N-glycoside formed between β-D-ribofuranose and adenine.

B. Reduction to Alditols

Alditol The product formed when the C=O group of a monosaccharide is reduced to a CHOH group.

The carbonyl group of a monosaccharide can be reduced to a hydroxyl group by a variety of reducing agents, including $NaBH_4$ (Section 13.10B). The reduction products are known as **alditols**. Reduction of D-glucose gives D-glucitol, more commonly known as D-sorbitol. Here, D-glucose is shown in the open-chain form, only a small amount of which is present in solution, but, as it is reduced, the equilibrium between the cyclic hemiacetal forms and the open-chain form shifts to replace the D-glucose:

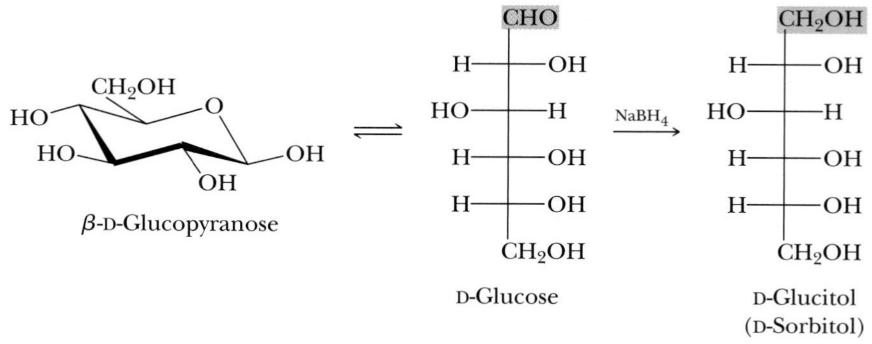

β-D-Glucopyranose ⇌ D-Glucose $\xrightarrow{NaBH_4}$ D-Glucitol (D-Sorbitol)

We name alditols by replacing the *-ose* in the name of the monosaccharide with *-itol*. D-Sorbitol is found in the plant world in many berries and in cherries, plums, pears, apples, seaweed, and algae. It is about 60 percent as sweet as sucrose (table sugar) and is used in the manufacture of candies and as a sugar substitute for diabetics. Among other alditols common in the biological world are erythritol, D-mannitol, and xylitol, the last of which is used as a sweetening agent in "sugarless" gum, candy, and sweet cereals:

Many "sugar-free" products contain sugar alcohols, such as D-sorbitol and xylitol. *(Andy Washnik)*

Erythritol:

$$\begin{array}{c} CH_2OH \\ H-\!\!\!-\!\!\!-OH \\ H-\!\!\!-\!\!\!-OH \\ CH_2OH \end{array}$$

D-Mannitol:

$$\begin{array}{c} CH_2OH \\ HO-\!\!\!-\!\!\!-H \\ HO-\!\!\!-\!\!\!-H \\ H-\!\!\!-\!\!\!-OH \\ H-\!\!\!-\!\!\!-OH \\ CH_2OH \end{array}$$

Xylitol:

$$\begin{array}{c} CH_2OH \\ H-\!\!\!-\!\!\!-OH \\ HO-\!\!\!-\!\!\!-H \\ H-\!\!\!-\!\!\!-OH \\ CH_2OH \end{array}$$

Erythritol   D-Mannitol   Xylitol

Example 18.6

$NaBH_4$ reduces D-glucose to D-glucitol. Do you expect the alditol formed under these conditions to be optically active or optically inactive? Explain.

Strategy

$NaBH_4$ reduces the aldehyde group (—CHO) of D-glucose to a primary alcohol (—CH_2OH). Reduction does not affect any other group in D-glucose. The problem then is to determine if D-glucitol is chiral and, if it is chiral, does it have a plane of symmetry, in which case it would be superposable on its mirror image and, therefore, optically inactive.

Solution

D-Glucitol does not have a plane of symmetry and is chiral. Therefore, we predict that D-glucitol is optically active. Its specific rotation is $-1.7°$.

See problems 18.34–18.39

Problem 18.6

$NaBH_4$ reduces D-erythrose to erythritol. Do you expect the alditol formed under these conditions to be optically active or optically inactive? Explain.

C. Oxidation to Aldonic Acids (Reducing Sugars)

We saw in Section 13.9A that several agents, including O_2, oxidize aldehydes (RCHO) to carboxylic acids (RCOOH). Similarly, under basic conditions, the aldehyde group of an aldose can be oxidized to a carboxylate group. Under these conditions, the cyclic form of the aldose is in equilibrium with the open-chain form, which is then oxidized by the mild oxidizing agent. D-Glucose, for example, is oxidized to D-gluconate (the anion of D-gluconic acid):

β-D-Glucopyranose (β-D-Glucose) ⇌ D-Glucose $\xrightarrow[\text{basic solution}]{\text{oxidizing agent}}$ D-Gluconate (an aldonic acid)

Any carbohydrate that reacts with an oxidizing agent to form an aldonic acid is classified as a **reducing sugar**. (It reduces the oxidizing agent.)

Reducing sugar A carbohydrate that reacts with an oxidizing agent to form an aldonic acid.

HOW TO 18.2 Determine If a Carbohydrate Is a Reducing Sugar

Following are three commonly encountered features in carbohydrates that are reducing sugars.

(1) Any carbohydrate that is an aldehyde is a reducing sugar.

(2) Any ketose that is in equilibrium with its aldehyde is a reducing sugar. This is because under the basic conditions of the Tollens' test, keto–enol tautomerism occurs and converts the ketose to an aldose (it is the aldose that is the actual reducing sugar):

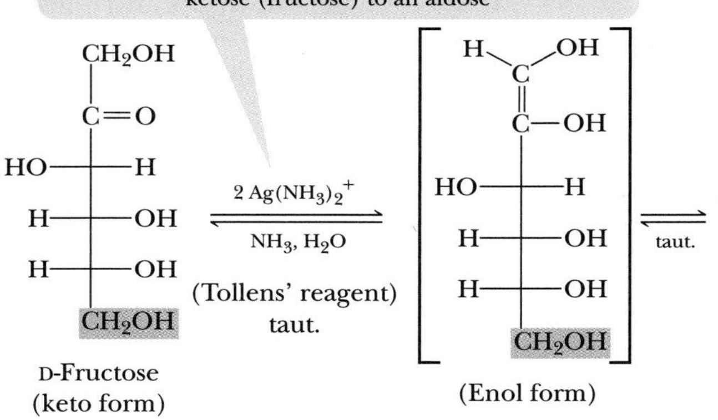

once formed, the Tollens' reagent can oxidize the aldehyde to a carboxyl group

(Keto form)

(3) Any ring carbohydrate that exists as a hemiacetal will be in equilibrium with its acyclic form. The acyclic form will either be an aldose or a ketose, which under the conditions of the Tollens' test will be converted to an aldose (see #2 above). It is the aldose that is the actual reducing sugar:

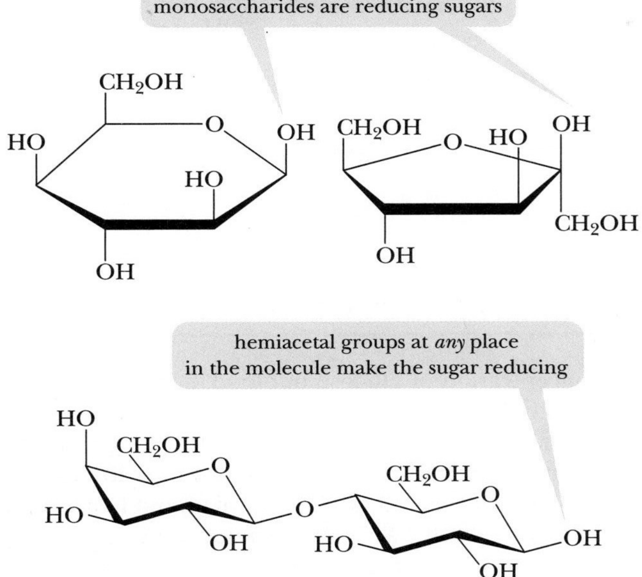
hemiacetal forms of cyclic monosaccharides are reducing sugars
hemiacetal groups at *any* place in the molecule make the sugar reducing

D. Oxidation to Uronic Acids

Enzyme-catalyzed oxidation of the primary alcohol at carbon 6 of a hexose yields a **uronic acid**. Enzyme-catalyzed oxidation of D-glucose, for example, yields D-glucuronic acid, shown here in both its open-chain and cyclic hemiacetal forms:

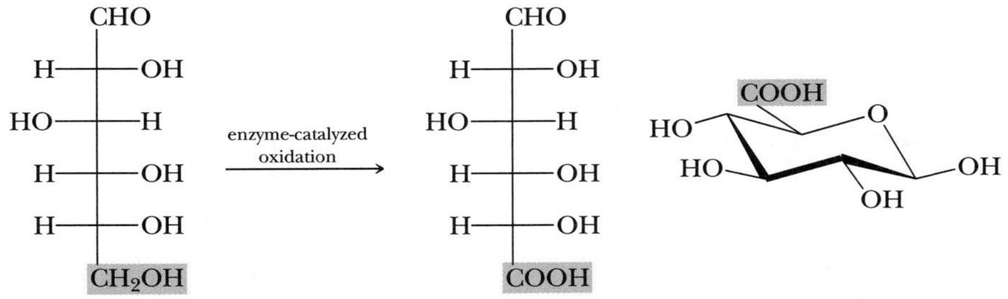

enzyme-catalyzed oxidation

D-Glucose

Fischer projection

Chair conformation

D-Glucuronic acid
(a uronic acid)

D-Glucuronic acid is widely distributed in both the plant and animal worlds. In humans, it is an important component of the acidic polysaccharides of connective tissues. The body also uses it to detoxify foreign phenols and alcohols. In the liver, these compounds are converted to glycosides of glucuronic acid (glucuronides), to be excreted in the urine. The intravenous anesthetic propofol (Problem 10.43), for example, is converted to the following water-soluble glucuronide and then excreted in the urine:

Propofol — A urine-soluble glucuronide

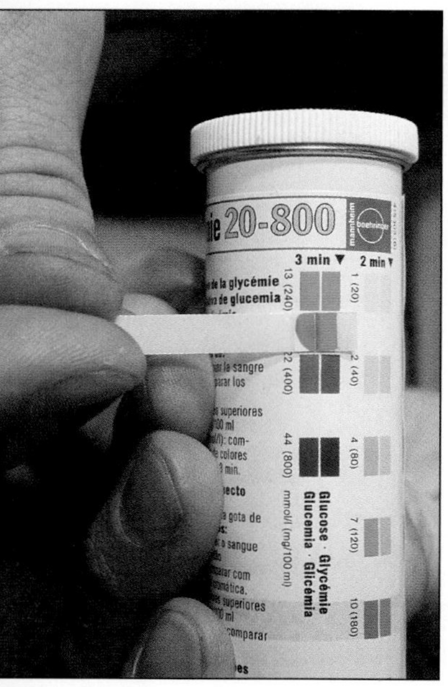

Chemstrip kit for blood glucose. *(Martin Dohrn/SPL/ Photo Researchers, Inc.)*

18.5 How Is the Concentration of Blood Glucose Determined?

Determining the concentration of glucose in blood, urine, or some other biological fluid is among the analytical procedures most often performed in a clinical chemistry laboratory. The need for a rapid and reliable test for glucose stems from the high incidence of diabetes mellitus. Approximately 18 million known diabetics live in the United States, and another 1.3 million are diagnosed each year.

In diabetes mellitus, the body has insufficient levels of the polypeptide hormone insulin (Section 19.4B). If the blood concentration of insulin is too low, muscle and liver cells do not absorb glucose, which, in turn, leads to increased levels of blood glucose (hyperglycemia), impaired metabolism of fats and proteins, ketosis, and, possibly, diabetic coma. Thus, a rapid procedure for the determination of blood glucose levels is critical for early diagnosis and effective management of this disease. In addition to being rapid, a test must be specific for D-glucose; that is, it must give a positive test for glucose, but not react with any other substance that is normally present in biological fluids.

Today, the determination of blood glucose levels is carried out by an enzyme-based procedure that uses the enzyme **glucose oxidase**. This enzyme catalyzes the oxidation of β-D-glucose to D-gluconic acid:

β-D-Glucopyranose (β-D-Glucose) $+ O_2 + H_2O \xrightarrow{\text{glucose oxidase}}$ D-Gluconic acid $+ H_2O_2$ (Hydrogen peroxide)

Glucose oxidase is specific for β-D-glucose. Therefore, complete oxidation of any sample containing both β-D-glucose and α-D-glucose requires the conversion of the α form to the β form. Fortunately, this conversion is rapid and complete in the short time required for the test.

Molecular oxygen, O_2, is the oxidizing agent in this reaction and is reduced to hydrogen peroxide, H_2O_2, the concentration of which can be determined spectrophotometrically. In one procedure, hydrogen peroxide formed in the glucose oxidase–catalyzed reaction oxidizes colorless *o*-toluidine to a colored product in a reaction catalyzed by the enzyme peroxidase:

$$\text{2-Methylaniline } (o\text{-Toluidine}) + H_2O_2 \xrightarrow{\text{peroxidase}} \text{Colored product}$$

2-Methylaniline
(*o*-Toluidine)

The concentration of the colored oxidation product is determined spectrophotometrically and is proportional to the concentration of glucose in the test solution. Alternatively, several commercially available test kits use the glucose oxidase reaction to render a qualitative determination of glucose in urine.

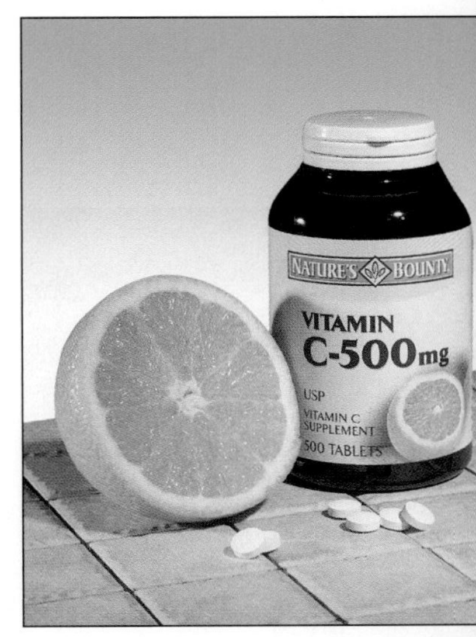

Vitamin C in an orange is identical to its synthetic tablet form. *(Andy Washnik)*

18.6 What Is L-Ascorbic Acid (Vitamin C)?

The structural formula of L-ascorbic acid (vitamin C) resembles that of a monosaccharide. In fact, this vitamin is synthesized both biochemically by plants and some animals and commercially from D-glucose. Humans do not have the enzyme systems required for the synthesis of L-ascorbic acid; therefore, for us, it is a vitamin. Approximately 66 million kilograms of vitamin C are synthesized every year in the United States. L-Ascorbic acid is very easily oxidized to L-dehydroascorbic acid, a diketone:

$$\text{L-Ascorbic acid (Vitamin C)} \underset{\text{reduction}}{\overset{\text{oxidation}}{\rightleftharpoons}} \text{L-Dehydroascorbic acid}$$

L-Ascorbic acid
(Vitamin C)

L-Dehydroascorbic acid

Both L-ascorbic acid and L-dehydroascorbic acid are physiologically active and are found together in most body fluids.

18.7 What Are Disaccharides and Oligosaccharides?

Disaccharide A carbohydrate containing two monosaccharide units joined by a glycosidic bond.

Oligosaccharide A carbohydrate containing from 6 to 10 monosaccharide units, each joined to the next by a glycosidic bond.

Polysaccharide A carbohydrate containing a large number of monosaccharide units, each joined to the next by one or more glycosidic bonds.

Most carbohydrates in nature contain more than one monosaccharide unit. Those that contain two units are called **disaccharides**, those that contain three units are called **trisaccharides**, and so forth. The more general term, **oligosaccharide**, is often used for carbohydrates that contain from 6 to 10 monosaccharide units. Carbohydrates containing larger numbers of monosaccharide units are called **polysaccharides**.

In a disaccharide, two monosaccharide units are joined by a glycosidic bond between the anomeric carbon of one unit and an —OH of the other. Sucrose, lactose, and maltose are three important disaccharides.

A. Sucrose

Sucrose (table sugar) is the most abundant disaccharide in the biological world. It is obtained principally from the juice of sugarcane and sugar beets. In sucrose, carbon 1 of α-D-glucopyranose bonds to carbon 2 of D-fructofuranose by an α-1,2-glycosidic bond:

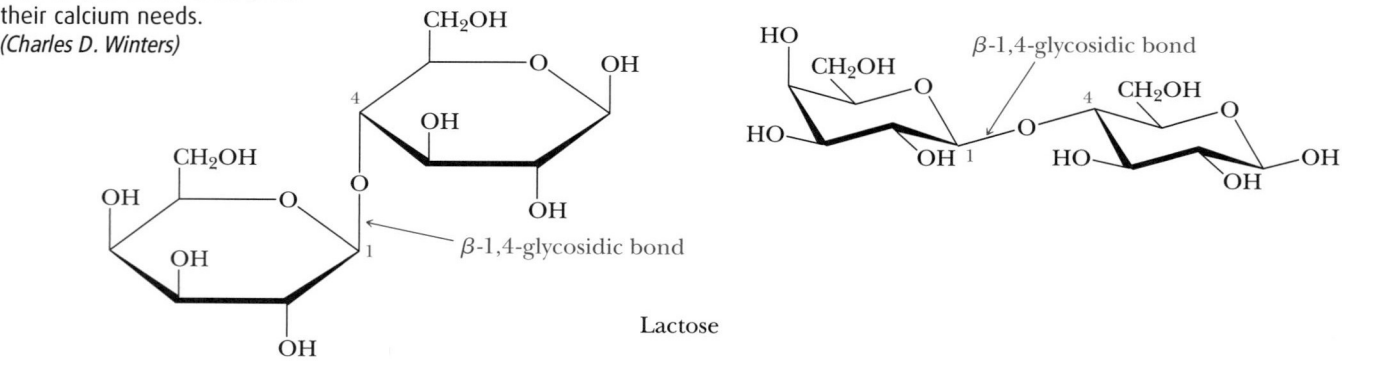

α-1,2-glycosidic bond

a unit of α-D-glucopyranose

a unit of β-D-fructofuranose

Sucrose

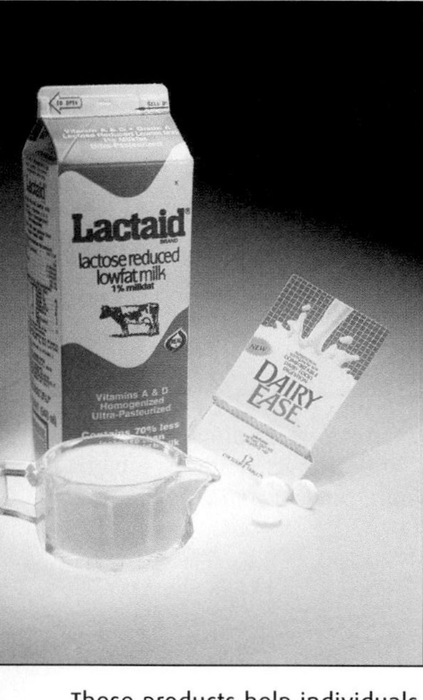

These products help individuals with lactose intolerance meet their calcium needs. *(Charles D. Winters)*

Because the anomeric carbons of both the glucopyranose and fructofuranose units are involved in formation of the glycosidic bond, neither monosaccharide unit is in equilibrium with its open-chain form. Thus, sucrose is a nonreducing sugar.

B. Lactose

Lactose, the principal sugar present in milk, accounts for 5 to 8 percent of human milk and 4 to 6 percent of cow's milk. This disaccharide consists of D-galactopyranose, bonded by a β-1,4-glycosidic bond to carbon 4 of D-glucopyranose:

β-1,4-glycosidic bond

Lactose

Lactose is a reducing sugar, because the cyclic hemiacetal of the D-glucopyranose unit is in equilibrium with its open-chain form and can be oxidized to a carboxyl group.

C. Maltose

Maltose derives its name from its presence in malt, the juice from sprouted barley and other cereal grains. Maltose consists of two units of D-glucopyranose, joined by a glycosidic bond between carbon 1 (the anomeric carbon) of one unit and carbon 4 of the other unit. Because the oxygen atom on the anomeric carbon of the first

CHEMICAL CONNECTIONS 18A

Relative Sweetness of Carbohydrate and Artificial Sweeteners

Among the disaccharide sweetening agents, D-fructose tastes the sweetest—even sweeter than sucrose. The sweet taste of honey is due largely to D-fructose and D-glucose. Lactose has almost no sweetness and is sometimes added to foods as filler. Some people cannot tolerate lactose well, however, and should avoid these foods. The following table lists the sweetness of various carbohydrates and artificial sweeteners relative to that of sucrose:

Carbohydrate	Sweetness Relative to Sucrose	Artificial Sweetener	Sweetness Relative to Sucrose
fructose	1.74	saccharin	450
sucrose (table sugar)	1.00	acesulfame-K	200
honey	0.97	aspartame	180
glucose	0.74	sucralose	600
maltose	0.33		
galactose	0.32		
lactose (milk sugar)	0.16		

QUESTION

Following is the structure of the artificial sweetener sucralose. Indicate all the ways in which it differs from sucrose.

CH_2OH Cl OH O O OH $ClCH_2$ O HO CH_2OH OH

Sucralose

glucopyranose unit is alpha, the bond joining the two units is called an α-1,4-glycosidic bond. Following are a Haworth projection and a chair conformation for β-maltose, so named because the —OH group on the anomeric carbon of the glucose unit on the right are beta:

α-1,4-glycosidic bond

HOH_2C O 1 O 4 CH_2OH OH HO OH OH O OH (β) HO OH — HO HO CH_2OH O 1 HO 4 CH_2OH O O HO OH (β) HO

Maltose

CHEMICAL CONNECTIONS 18B

A, B, AB, and O Blood-Group Substances

Membranes of animal plasma cells have large numbers of relatively small carbohydrates bound to them. In fact, the outsides of most plasma cell membranes are literally sugarcoated. These membrane-bound carbohydrates are part of the mechanism by which the different types of cells recognize each other; in effect, the carbohydrates act as biochemical markers (antigenic determinants). Typically, the membrane-bound carbohydrates contain from 4 to 17 units consisting of just a few different monosaccharides, primarily D-galactose, D-mannose, L-fucose, *N*-acetyl-D-glucosamine, and *N*-acetyl-D-galactosamine. L-Fucose is a 6-deoxyaldohexose:

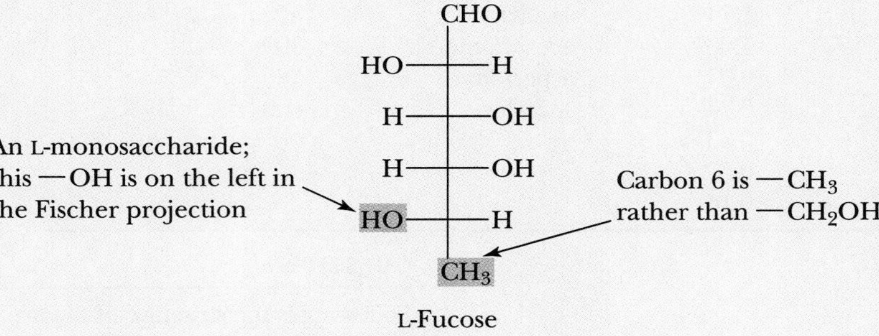

Among the first discovered and best understood of these membrane-bound carbohydrates are those of the ABO blood-group system, discovered in 1900 by Karl Landsteiner (1868–1943). Whether an individual has type A, B, AB, or O blood is genetically determined and depends on the type of trisaccharide or tetrasaccharide bound to the surface of the person's red blood cells. The monosaccharides of each blood group and the type of glycosidic bond joining them are shown in the following figure (the configurations of the glycosidic bonds are in parentheses):

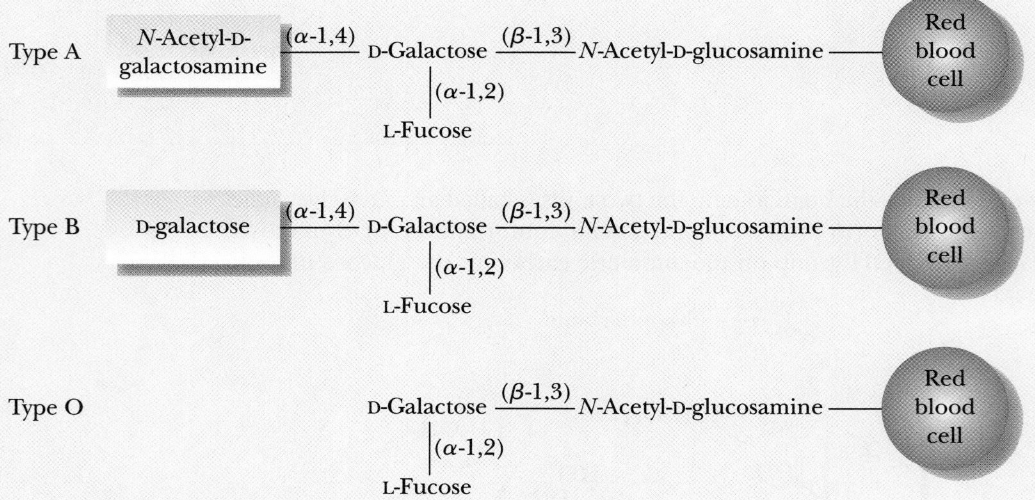

QUESTION

Draw the two pyranose forms of L-fucose.

Example 18.7

Draw a chair conformation for the β anomer of a disaccharide in which two units of D-glucopyranose are joined by an α-1,6-glycosidic bond.

Strategy

First draw a chair conformation of α-D-glucopyranose. Then bond the anomeric carbon of this monosaccharide to carbon 6 of a second D-glucopyranose unit by an α-glycosidic bond. The resulting molecule is either α or β, depending on the orientation of the —OH group on the reducing end of the disaccharide.

Solution

The disaccharide shown here is β:

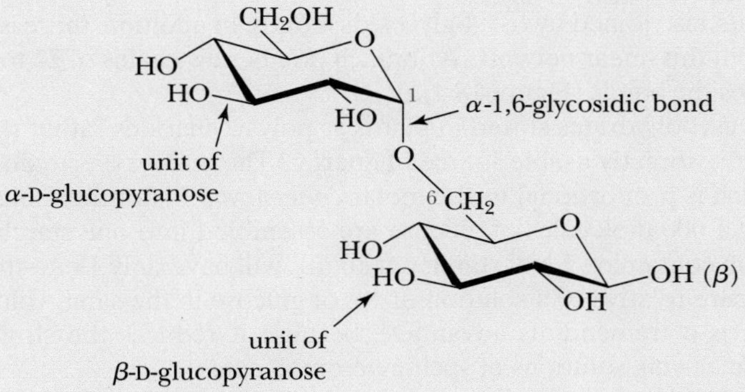

CH_2OH
HO
HO
HO
O
1
α-1,6-glycosidic bond
unit of
α-D-glucopyranose
O
$^6CH_2$
O
HO
HO
OH
OH (β)
unit of
β-D-glucopyranose

See problem 18.48

Problem 18.7

Draw Haworth and chair formulas for the α form of a disaccharide in which two units of D-glucopyranose are joined by a β-1,3-glycosidic bond.

18.8 What Are Polysaccharides?

Polysaccharides consist of a large number of monosaccharide units joined together by glycosidic bonds. Three important polysaccharides, all made up of glucose units, are starch, glycogen, and cellulose.

A. Starch: Amylose and Amylopectin

Starch is found in all plant seeds and tubers and is the form in which glucose is stored for later use. Starch can be separated into two principal polysaccharides: amylose and amylopectin. Although the starch from each plant is unique, most starches contain 20 to 25 percent amylose and 75 to 80 percent amylopectin.

Complete hydrolysis of both amylose and amylopectin yields only D-glucose. Amylose is composed of continuous, unbranched chains of as many as 4,000 D-glucose

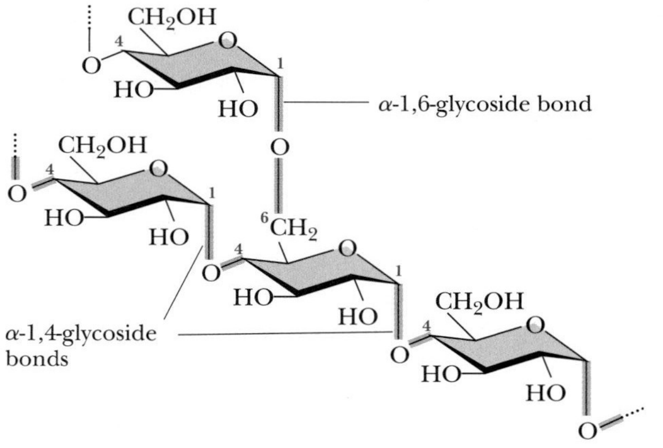

α-1,6-glycoside bond

α-1,4-glycoside bonds

Figure 18.4
Amylopectin is a highly branched polymer of D-glucose. Chains consist of 24 to 30 units of D-glucose, joined by α-1,4-glycosidic bonds, and branches created by α-1,6-glycosidic bonds.

units, joined by α-1,4-glycosidic bonds. Amylopectin contains chains up to 10,000 D-glucose units, also joined by α-1,4-glycosidic bonds. In addition, there is considerable branching from this linear network. At branch points, new chains of 24 to 30 units start by α-1,6-glycosidic bonds (Figure 18.4).

Why are carbohydrates stored in plants as polysaccharides rather than monosaccharides, a more directly usable source of energy? The answer has to do with **osmotic pressure**, which is proportional to the molar *concentration*, not the molecular weight, of a solute. If 1,000 molecules of glucose are assembled into one starch macromolecule, a solution containing 1 g of starch per 10 mL will have only 1 one-thousandth the osmotic pressure relative to a solution of 1 g of glucose in the same volume. This feat of packaging is a tremendous advantage, because it reduces the strain on various membranes enclosing solutions of such macromolecules.

B. Glycogen

Glycogen is the reserve carbohydrate for animals. Like amylopectin, glycogen is a branched polymer of D-glucose containing approximately 10^6 glucose units, joined by α-1,4- and α-1,6-glycosidic bonds. The total amount of glycogen in the body of a well-nourished adult human being is about 350 g, divided almost equally between liver and muscle.

C. Cellulose

Cellulose, the most widely distributed plant skeletal polysaccharide, constitutes almost half of the cell-wall material of wood. Cotton is almost pure cellulose.

Cellulose, a linear polymer of D-glucose units joined by β-1,4-glycosidic bonds (Figure 18.5), has an average molar mass of 400,000 g/mol, corresponding to approximately 2,800 glucose units per molecule.

Cellulose molecules act much like stiff rods, a feature that enables them to align themselves side by side into well-organized, water-insoluble fibers in which the OH groups form numerous intermolecular hydrogen bonds. This arrangement of parallel

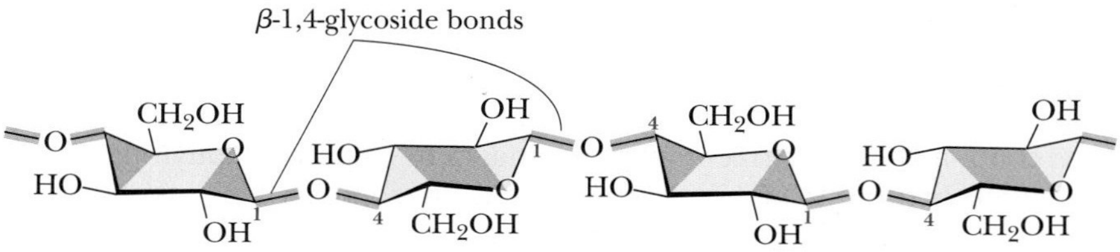

β-1,4-glycoside bonds

Figure 18.5
Cellulose is a linear polymer of D-glucose, joined by β-1,4-glycosidic bonds.

chains in bundles gives cellulose fibers their high mechanical strength and explains why cellulose is insoluble in water. When a piece of cellulose-containing material is placed in water, there are not enough —OH groups on the surface of the fiber to pull individual cellulose molecules away from the strongly hydrogen-bonded fiber.

Humans and other animals cannot use cellulose as food, because our digestive systems do not contain β-glucosidases, enzymes that catalyze the hydrolysis of β-glucosidic bonds. Instead, we have only α-glucosidases; hence, the polysaccharides we use as sources of glucose are starch and glycogen. By contrast, many bacteria and microorganisms do contain β-glucosidases and can digest cellulose. Termites are fortunate (much to our regret) to have such bacteria in their intestines and can use wood as their principal food. Ruminants (cud-chewing animals) and horses can also digest grasses and hay, because β-glucosidase-containing microorganisms are present within their alimentary systems.

D. Textile Fibers from Cellulose

Both **rayon** and acetate rayon are made from chemically modified cellulose and were the first commercially important synthetic textile fibers. In the production of rayon, cellulose fibers are treated with carbon disulfide, CS_2, in aqueous sodium hydroxide. In this reaction, some of the —OH groups on a cellulose fiber are converted to the sodium salt of a xanthate ester, which causes the fibers to dissolve in alkali as a viscous colloidal dispersion:

An —OH group in a cellulose fiber

$$\text{Cellulose—OH} \xrightarrow{\text{NaOH}} \text{Cellulose—O}^-\text{Na}^+ \xrightarrow{\text{S=C=S}} \text{Cellulose—O}\overset{\overset{\text{S}}{\|}}{\text{C}}\text{—S}^-\text{Na}^+$$

Cellulose (insoluble in water)

Sodium salt of a xanthate ester (a viscous colloidal suspension)

The solution of cellulose xanthate is separated from the alkali-insoluble parts of wood and then forced through a spinneret (a metal disc with many tiny holes) into dilute sulfuric acid to hydrolyze the xanthate ester groups and precipitate regenerated cellulose. Regenerated cellulose extruded as a filament is called viscose rayon thread.

In the industrial synthesis of **acetate rayon**, cellulose is treated with acetic anhydride (Section 15.4B):

$$\text{glucose unit (OH, } CH_2\text{, HO, OH)} + 3CH_3\overset{\overset{O}{\|}}{C}O\overset{\overset{O}{\|}}{C}CH_3 \longrightarrow \text{acetylated unit (} \overset{\overset{O}{\|}}{OCCH_3}\text{, } CH_2\text{, } CH_3\underset{\underset{O}{\|}}{C}O\text{, } \underset{\underset{O}{\|}}{OCCH_3}\text{)} + 3CH_3\overset{\overset{O}{\|}}{C}OH$$

A glucose unit in a cellulose fiber

Acetic anhydride

A fully acetylated glucose unit

Acetylated cellulose is then dissolved in a suitable solvent, precipitated, and drawn into fibers known as acetate rayon. The fibers are smooth, soft, resistant to static cling, and dry quickly. One of acetate rayon's most valuable properties is its ability to behave as a thermoplastic (Section 17.1). This allows clothing made from acetate rayon to be heated and permanently pleated upon cooling.

Key Terms and Concepts

A, AB, and O blood group substances (p. 670)
acetate rayon (p. 673)
alditol (p. 662)
aldonic acid (p. 663)
aldose (p. 650)
amino sugar (p. 654)
anomeric carbon (p. 655)
anomers (p. 655)
ascorbic acid (p. 667)
blood sugar (p. 653)
carbohydrates (p. 650)
disaccharide (p. 667)
Fischer projection (p. 651)
furanose (p. 655)
glycogen (p. 672)
glycoside (p. 660)
glycosidic bond (p. 660)
Haworth projection (p. 655)
hexose (p. 652)
D-monosaccharide (p. 652)
ketose (p. 650)
L-monosaccharide (p. 652)
monosaccharide (p. 650)
mutarotation (p. 660)
oligosaccharide (p. 667)
pentose (p. 652)
penultimate carbon (p. 652)
polysaccharide (p. 667)
pyranose (p. 655)
rayon (p. 673)
reducing sugar (p. 664)
starch (p. 671)
tetrose (p. 650)
triose (p. 650)
trisaccharide (p. 667)
uronic acid (p. 665)
vitamin C (p. 667)

Summary of Key Questions

18.1 What Are Carbohydrates?

Carbohydrates are

- the most abundant organic compounds in the plant world.
- storage forms of chemical energy (glucose, starch, glycogen).
- components of supportive structures in plants (cellulose), crustacean shells (chitin), and connective tissues in animals (acidic polysaccharides).
- essential components of nucleic acids (D-ribose and 2-deoxy-D-ribose).

18.2 What Are Monosaccharides?

- **Monosaccharides** are polyhydroxyaldehydes or polyhydroxyketones or compounds that yield them after hydrolysis.
- The most common monosaccharides have the general formula $C_nH_{2n}O_n$ where n varies from three to nine.
- Their names contain the suffix *-ose*.
- The prefixes *tri-*, *tetr-*, *pent-*, and so on show the number of carbon atoms in the chain.
- The prefix *aldo-* shows an aldehyde, the prefix *keto-* a ketone.
- In a **Fischer projection** of a monosaccharide, the carbon chain is written vertically, with the most highly oxidized carbon toward the top. Horizontal lines show groups projecting above the plane of the page and vertical lines show groups projecting behind the plane of the page.
- A monosaccharide that has the same configuration at its penultimate carbon as D-glyceraldehyde is called a **D-monosaccharide**; one that has the same configuration at its penultimate carbon as L-glyceraldehyde is called an **L-monosaccharide**.
- An amino sugar contains an $-NH_2$ group in place of an $-OH$ group.

18.3 What Are the Cyclic Structures of Monosaccharides?

- Monosaccharides exist primarily as cyclic hemiacetals.
- The new stereocenter resulting from cyclic hemiacetal formation is referred to as an **anomeric carbon**. The stereoisomers thus formed are called **anomers**.
- A six-membered cyclic hemiacetal form of a monosaccharide is called a **pyranose**; a five-membered cyclic hemiacetal form is called a **furanose**.
- Furanoses and pyranoses can be drawn as **Haworth projections**.

- Pyranoses can be shown as Haworth projections or as strain-free **chair conformations**.
- The symbol ***β***- indicates that the —OH on the anomeric carbon is on the same side of the ring as the terminal —CH_2OH.
- The symbol ***α***- indicates that —OH on the anomeric carbon is on the opposite side of the ring from the terminal —CH_2OH.
- **Mutarotation** is the change in specific rotation that accompanies the formation of an equilibrium mixture of α- and β-anomers in aqueous solution.

18.4 What Are the Characteristic Reactions of Monosaccharides?

- A **glycoside** is a cyclic acetal derived from a monosaccharide.
- The name of the glycoside is composed of the name of the alkyl or aryl group bonded to the acetal oxygen atom followed by the name of the parent monosaccharide in which the terminal *-e* has been replaced by *-ide*.
- An **alditol** is a polyhydroxy compound formed by reduction of the carbonyl group of a monosaccharide to a hydroxyl group.
- An **aldonic acid** is a carboxylic acid formed by oxidation of the aldehyde group of an aldose.
- Any carbohydrate that reduces an oxidizing agent is called a **reducing sugar**.
- Enzyme-catalyzed oxidation of the terminal —CH_2OH group of a monosaccharide to a COOH group gives a **uronic acid**.

18.6 What Is Ascorbic Acid (Vitamin C)?

- The structure of L-ascorbic acid (vitamin C) resembles that of a monosaccharide.
- **L-Ascorbic acid** is synthesized in nature from D-glucose by a series of enzyme-catalyzed steps.
- Humans do not have the enzymes required to carry out this synthesis and therefore must obtain vitamin C either in the food they eat or as a vitamin supplement.

18.7 What Are Disaccharides and Oligosaccharides?

- A **disaccharide** contains two monosaccharide units joined by a **glycosidic bond**.
- Terms applied to carbohydrates containing larger numbers of monosaccharides are **trisaccharide**, **tetrasaccharide**, etc.
- An **oligosaccharide** is a carbohydrate that contains from six to ten monosaccharide units.
- **Sucrose** is a disaccharide consisting of D-glucose joined to D-fructose by an α-1,2-glycosidic bond.
- **Lactose** is a disaccharide consisting of D-galactose joined to D-glucose by a β-1,4-glycosidic bond.
- **Maltose** is a disaccharide of two molecules of D-glucose joined by an α-1,4-glycosidic bond.

18.8 What Are Polysaccharides?

- **Polysaccharides** consist of a large number of monosaccharide units bonded together by glycosidic bonds.
- **Starch** can be separated into two fractions given the names amylose and amylopectin. **Amylose** is a linear polymer of up to 4,000 units of D-glucopyranose joined by α-1,4-glycosidic bonds. **Amylopectin** is a highly branched polymer of D-glucose joined by α-1,4-glycosidic bonds and, at branch points, by α-1,6-glycosidic bonds.
- **Glycogen**, the reserve carbohydrate of animals, is a highly branched polymer of D-glucopyranose joined by α-1,4-glycosidic bonds and, at branch points, by α-1,6-glycosidic bonds.
- **Cellulose**, the skeletal polysaccharide of plants, is a linear polymer of D-glucopyranose joined by β-1,4-glycosidic bonds.
- **Rayon** is made from chemically modified and regenerated cellulose.
- **Acetate rayon** is made by the acetylation of cellulose.

Quick Quiz

Answer true or false to the following questions to assess your general knowledge of the concepts in this chapter. If you have difficulty with any of them, you should review the appropriate section in the chapter (shown in parentheses) before attempting the more challenging end-of-chapter problems.

1. An acetal of the pyranose or furanose form of a sugar is referred to as a glycoside. (18.4)
2. A monosaccharide can contain the carbonyl of a ketone or the carbonyl of an aldehyde. (18.2)
3. Starch, glycogen, and cellulose are all examples of oligosaccharides. (18.8)
4. An L-sugar and a D-sugar of the same name are enantiomers. (18.2)
5. Alditols are oxidized carbohydrates. (18.4)
6. D-Glucose and D-ribose are diastereomers. (18.2)
7. A pyranoside contains a five-membered ring. (18.3)
8. All monosaccharides dissolve in ether. (18.2)
9. Monosaccharides exist mostly as cyclic hemiacetals. (18.3)
10. A polysaccharide is a glycoside of two monosaccharides. (18.7)
11. α and β in a monosaccharide are used to refer to the positions 1 and 2 carbons away from the carbonyl group. (18.3)
12. Carbohydrates must have the formula $C_n(H_2O)_m$. (18.1)
13. Mutarotation is the establishment of an equilibrium concentration of α and β anomers of a carbohydrate. (18.3)
14. D-Glucose and D-galactose are diastereomers. (18.2)
15. Only acyclic carbohydrates that contain aldehyde groups can act as reducing sugars. (18.4)
16. A glycoside of a monosaccharide cannot act as a reducing sugar. (18.4)
17. The penultimate carbon of an acyclic monosaccharide becomes the anomeric carbon in the cyclic hemiacetal form of the molecule. (18.3)
18. A Fischer projection may be rotated 90°. (18.2)

Answers: (1) T (2) T (3) F (4) T (5) F (6) F (7) F (8) F (9) T (10) F (11) F (12) F (13) T (14) T (15) F (16) F (17) F (18) F

Key Reactions

1. Formation of Cyclic Hemiacetals (Section 18.3)

A monosaccharide existing as a five-membered ring is a furanose; one existing as a six-membered ring is a pyranose. A pyranose is most commonly drawn as a Haworth projection or a chair conformation:

CHO; H—OH; HO—H; H—OH; H—OH; CH_2OH

D-Glucose

anomeric carbon

β-D-Glucopyranose (β-D-Glucose)

2. Mutarotation (Section 18.3C)

Anomeric forms of a monosaccharide are in equilibrium in aqueous solution. Mutarotation is the change in specific rotation that accompanies this equilibration:

β-D-Glucopyranose
$[\alpha]_D^{25}$ +18.7°

Open-chain form

α-D-Glucopyranose
$[\alpha]_D^{25}$ +112°

3. Formation of Glycosides (Section 18.5A)
Treatment of a monosaccharide with an alcohol in the presence of an acid catalyst forms a cyclic acetal called a glycoside:

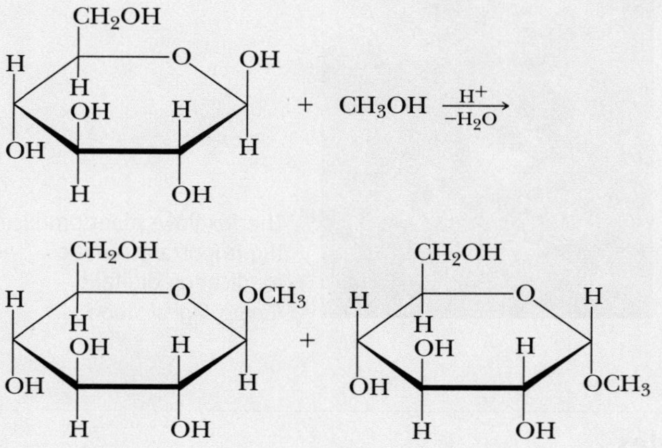

$$\text{monosaccharide} + CH_3OH \xrightarrow[-H_2O]{H^+} \text{glycoside } (\beta) + \text{glycoside } (\alpha)$$

The bond to the new —OR group is called a glycosidic bond.

4. Reduction to Alditols (Section 18.5B)
Reduction of the carbonyl group of an aldose or a ketose to a hydroxyl group yields a polyhydroxy compound called an alditol:

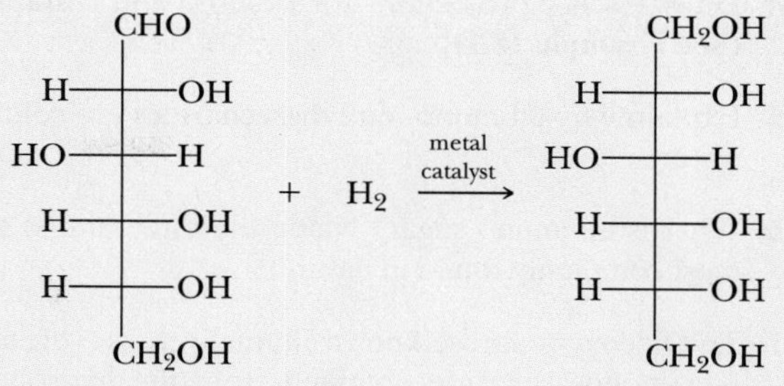

$$\text{D-Glucose} + H_2 \xrightarrow{\text{metal catalyst}} \text{D-Glucitol (D-Sorbitol)}$$

D-Glucose

D-Glucitol
(D-Sorbitol)

5. Oxidation to an Aldonic Acid (Section 18.5C)
Oxidation of the aldehyde group of an aldose to a carboxyl group by a mild oxidizing agent gives a polyhydroxycarboxylic acid called an aldonic acid:

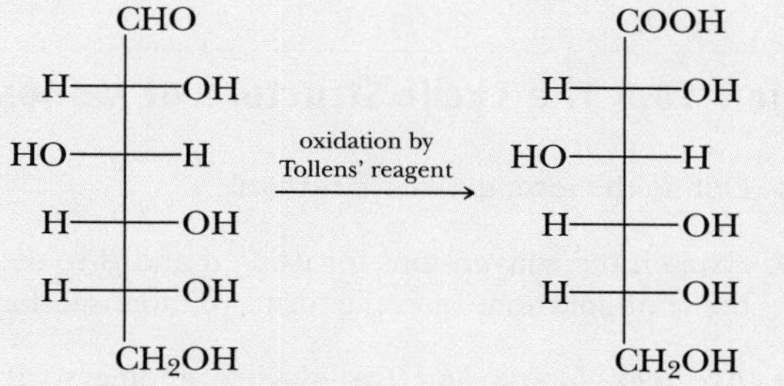

$$\text{D-Glucose} \xrightarrow{\text{oxidation by Tollens' reagent}} \text{D-Gluconic acid}$$

D-Glucose

D-Gluconic acid

Problems

A problem marked with an asterisk indicates an applied "real world" problem. Answers to problems whose numbers are printed in blue are given in Appendix D.

Section 18.2 Monosaccharides

18.8 What is the difference in structure between an aldose and a ketose? Between an aldopentose and a ketopentose?

18.9 Which hexose is also known as dextrose?

18.10 What does it mean to say that D- and L-glyceraldehyde are enantiomers?

18.11 Explain the meaning of the designations D and L as used to specify the configuration of carbohydrates.

18.12 How many stereocenters are present in D-glucose? In D-ribose? How many stereoisomers are possible for each monosaccharide?

18.13 Which compounds are D-monosaccharides and which are L-monosaccharides? **(See Example 18.1)**

(a) CHO; H—OH; HO—H; H—OH; H—OH; CH_2OH

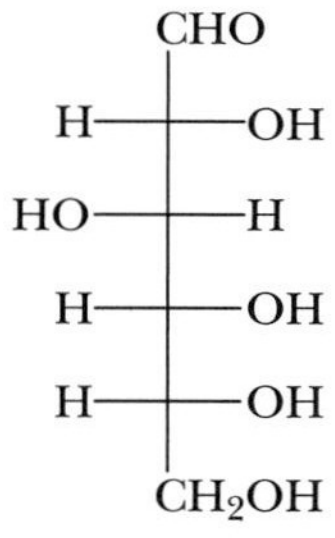

(b) CHO; HO—H; H—OH; HO—H; CH_2OH

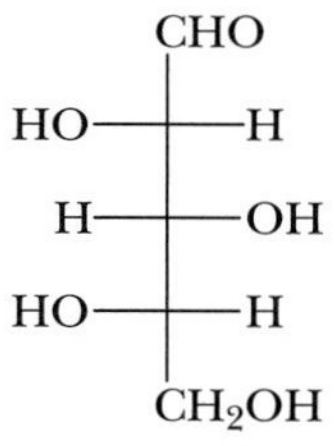

(c) CH_2OH; C=O; H—OH; H—OH; CH_2OH

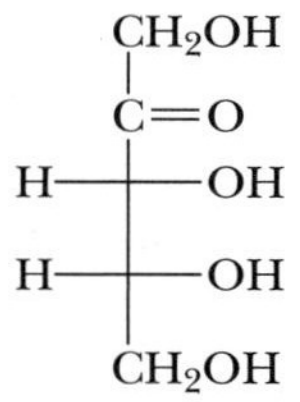

18.14 Draw Fischer projections for L-ribose and L-arabinose. **(See Example 18.1)**

18.15 Explain why all mono- and disaccharides are soluble in water.

18.16 What is an amino sugar? Name the three amino sugars most commonly found in nature.

The foxglove plant produces the important cardiac medication digitalis. *(Corbis Digital Stock)*

***18.17** 2,6-Dideoxy-D-altrose, known alternatively as D-digitoxose, is a monosaccharide obtained from the hydrolysis of digitoxin, a natural product extracted from purple foxglove *(Digitalis purpurea)*. Digitoxin has found wide use in cardiology because it reduces the pulse rate, regularizes heart rhythm, and strengthens the heartbeat. Draw the structural formula of 2,6-dideoxy-D-altrose.

Section 18.3 The Cyclic Structure of Monosaccharides

18.18 Define the term *anomeric carbon.*

18.19 Explain the conventions for using α and β to designate the configurations of cyclic forms of monosaccharides.

18.20 Are α-D-glucose and β-D-glucose anomers? Explain. Are they enantiomers? Explain.

18.21 Are α-D-gulose and α-L-gulose anomers? Explain.

18.22 In what way are chair conformations a more accurate representation of molecular shape of hexopyranoses than are Haworth projections?

18.23 Draw α-D-glucopyranose (α-D-glucose) as a Haworth projection. Now, using only the following information, draw Haworth projections for these monosaccharides: **(See Example 18.2)**

(a) α-D-Mannopyranose (α-D-mannose). The configuration of D-mannose differs from that of D-glucose only at carbon 2.

(b) α-D-Gulopyranose (α-D-gulose). The configuration of D-gulose differs from that of D-glucose at carbons 3 and 4.

18.24 Convert each Haworth projection to an open-chain form and then to a Fischer projection:

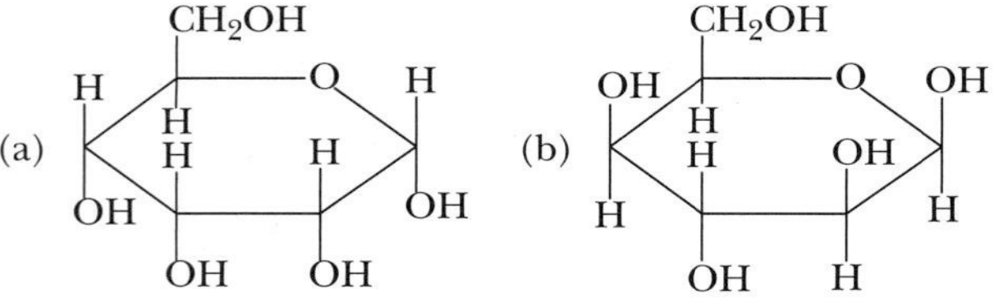

Name the monosaccharides you have drawn.

18.25 Convert each chair conformation to an open-chain form and then to a Fischer projection:

Name the monosaccharides you have drawn.

18.26 The configuration of D-arabinose differs from the configuration of D-ribose only at carbon 2. Using this information, draw a Haworth projection and the chair form for α-D-arabinofuranose (α-D-arabinose). **(See Examples 18.2, 18.3)**

18.27 Explain the phenomenon of mutarotation with reference to carbohydrates. By what means is mutarotation detected?

18.28 The specific rotation of α-D-glucose is $+112.2°$. What is the specific rotation of α-L-glucose?

18.29 When α-D-glucose is dissolved in water, the specific rotation of the solution changes from $+112.2°$ to $+52.7°$. Does the specific rotation of α-L-glucose also change when it is dissolved in water? If so, to what value does it change?

Section 18.4 Reactions of Monosaccharides

18.30 Draw the structural formula for ethyl α-D-galactopyranoside (ethyl α-D-galactoside). Label the anomeric carbon and the glycosidic bond. **(See Example 18.4)**

18.31 Draw the structural formula for methyl β-D-mannopyranoside (methyl β-D-mannoside). Label the anomeric carbon and the glycosidic bond. **(See Example 18.4)**

18.32 Show the two possible products of each reaction. Label the α and β anomers in each reaction. **(See Example 18.4)**

(a) D-Gulose (CHO; H—OH; H—OH; HO—H; H—OH; CH_2OH) + CH_3CH_2OH $\xrightarrow{\text{dilute } H_2SO_4}$

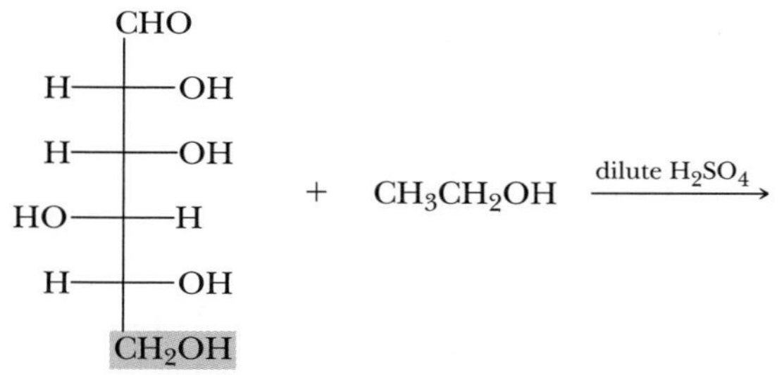

(b) D-Altrose (CHO; HO—H; H—OH; H—OH; H—OH; CH_2OH) + isopropyl alcohol (OH) $\xrightarrow{\text{dilute } H_2SO_4}$

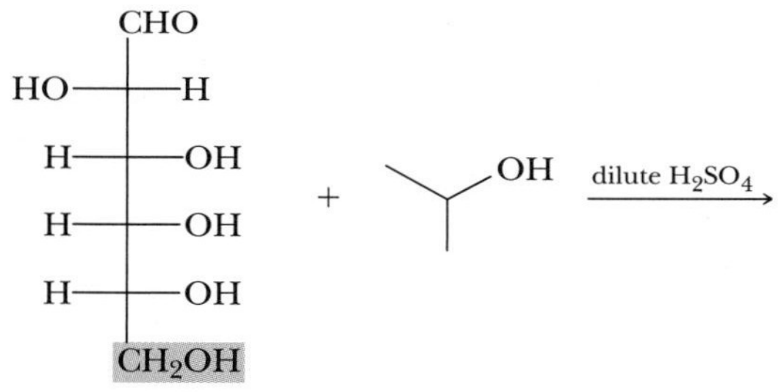

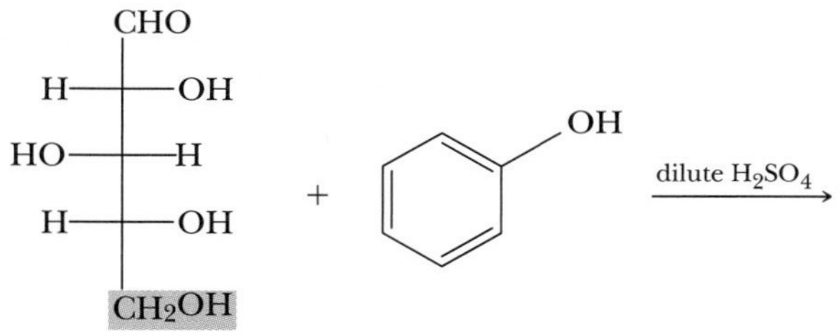

(c) D-Xylose (CHO; H—OH; HO—H; H—OH; CH_2OH) + phenol (OH) $\xrightarrow{\text{dilute } H_2SO_4}$

18.33 Draw a structural formula for the β-*N*-glycoside formed between (a) D-ribofuranose and thymine and (b) D-ribofuranose and guanine. Label the anomeric carbon and the *N*-glycosidic bond. **(See Example 18.5)**

18.34 Draw Fischer projections for the product(s) formed by the reaction of D-galactose with the following compounds, and state whether each product is optically active or optically inactive: **(See Example 18.6)**

(a) $NaBH_4$ in H_2O (b) $AgNO_3$ in NH_3, H_2O

18.35 Repeat Problem 18.34, but using D-ribose in place of D-galactose. **(See Example 18.6)**

18.36 The reduction of D-fructose by $NaBH_4$ gives two alditols, one of which is D-sorbitol. Name and draw a structural formula for the other alditol. **(See Example 18.6)**

18.37 There are four D-aldopentoses (Table 18.1). If each is reduced with $NaBH_4$, which yield optically active alditols? Which yield optically inactive alditols? **(See Example 18.6)**

18.38 Account for the observation that the reduction of D-glucose with $NaBH_4$ gives an optically active alditol, whereas the reduction of D-galactose with $NaBH_4$ gives an optically inactive alditol. **(See Example 18.6)**

18.39 Which two D-aldohexoses give optically inactive (meso) alditols on reduction with $NaBH_4$? **(See Example 18.6)**

***18.40** L-Fucose, one of several monosaccharides commonly found in the surface polysaccharides of animal cells (Chemical Connections 18B), is synthesized biochemically from D-mannose in the following eight steps:

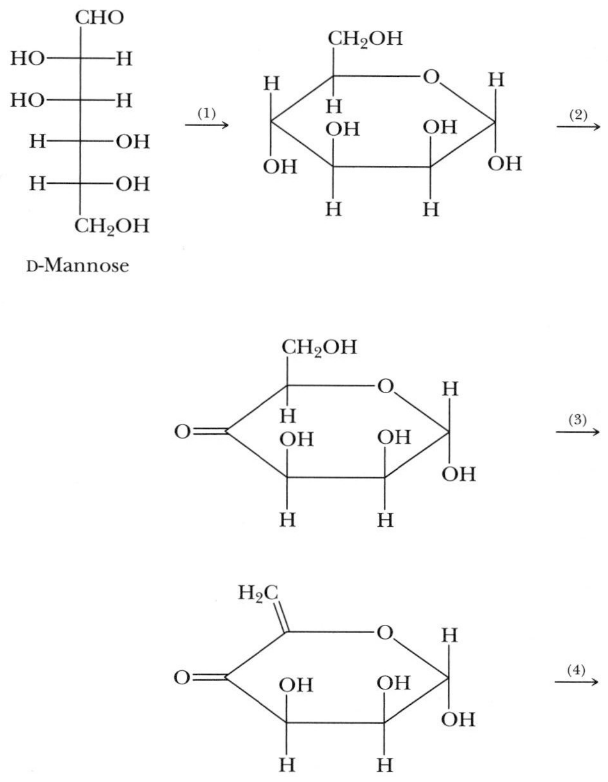

D-Mannose (CHO; HO—H; HO—H; H—OH; H—OH; CH_2OH) $\xrightarrow{(1)}$ pyranose (CH_2OH, O, H, OH, H, OH, OH, OH, H, H) $\xrightarrow{(2)}$ (CH_2OH, O, H, O=, H, OH, OH, OH, H, H) $\xrightarrow{(3)}$ (H_2C, O, H, O=, OH, OH, OH, H, H) $\xrightarrow{(4)}$

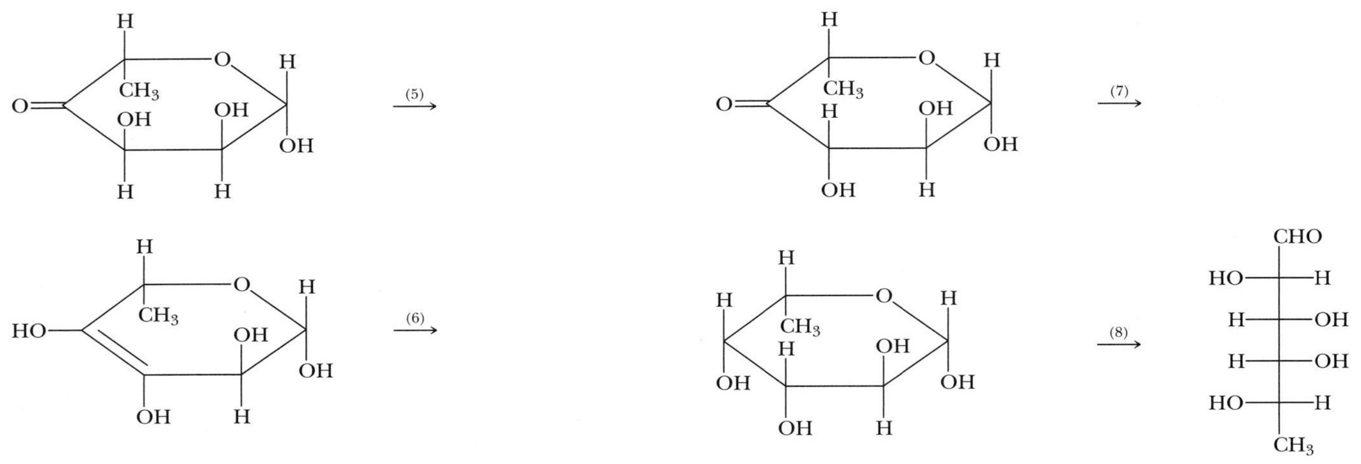

$\xrightarrow{(5)}$ $\xrightarrow{(7)}$ $\xrightarrow{(6)}$ $\xrightarrow{(8)}$

L-Fucose

(a) Describe the type of reaction (oxidation, reduction, hydration, dehydration, or the like) involved in each step.

(b) Explain why this monosaccharide, which is derived from D-mannose, now belongs to the L series.

Section 18.6 Ascorbic Acid

***18.41** Is ascorbic acid a biological oxidizing agent or a biological reducing agent? Explain.

18.42 Ascorbic acid is a diprotic acid with the following acid ionization constants: pK_{a1} 4.10 and pK_{a2} 11.79. The two acidic hydrogens are those connected with the enediol part of the molecule. Which hydrogen has which ionization constant? (*Hint:* Draw separately the anion derived by the loss of one of these hydrogens and that formed by the loss of the other hydrogen. Which anion has the greater degree of resonance stabilization?)

Section 18.7 Disaccharides and Oligosaccharides

18.43 Define the term *glycosidic bond.*

18.44 What is the difference in meaning between the terms *glycosidic bond* and *glucosidic bond*?

18.45 Do glycosides undergo mutarotation?

***18.46** In making candy or syrups from sugar, sucrose is boiled in water with a little acid, such as lemon juice. Why does the product mixture taste sweeter than the starting sucrose solution?

18.47 Which disaccharides are reduced by $NaBH_4$?

(a) Sucrose (b) Lactose (c) Maltose

18.48 Draw Haworth and chair formulas for the β form of a disaccharide in which two units of D-glucopyranose are joined by a β-1,4-glycosidic bond. **(See Example 18.7)**

***18.49** Trehalose is found in young mushrooms and is the chief carbohydrate in the blood of certain insects. Trehalose is a disaccharide consisting of two D-monosaccharide units, each joined to the other by an α-1,1-glycosidic bond:

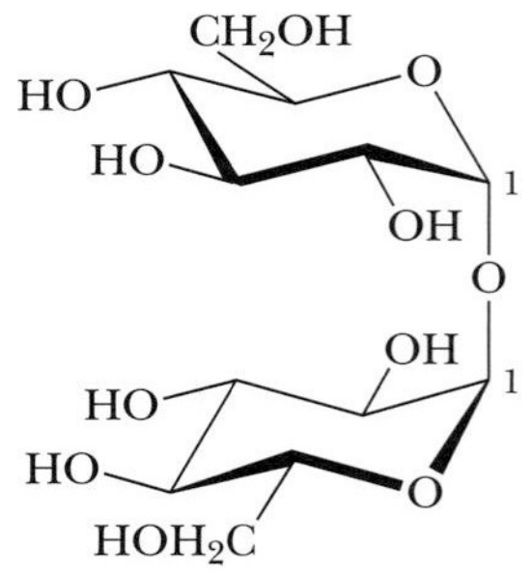

Trehalose

(a) Is trehalose a reducing sugar?

(b) Does trehalose undergo mutarotation?

(c) Name the two monosaccharide units of which trehalose is composed.

***18.50** Hot-water extracts of ground willow bark are an effective pain reliever. Unfortunately, the liquid is so bitter that most persons refuse it. The pain reliever in these infusions is salicin:

Salicin

Name the monosaccharide unit in salicin.

Section 18.8 Polysaccharides

18.51 What is the difference in structure between oligosaccharides and polysaccharides?

18.52 Name three polysaccharides that are composed of units of D-glucose. In which of the three polysaccharides are the glucose units joined by α-glycosidic bonds? In which are they joined by β-glycosidic bonds?

18.53 Starch can be separated into two principal polysaccharides: amylose and amylopectin. What is the major difference in structure between the two?

***18.54** A Fischer projection of N-acetyl-D-glucosamine is given in Section 18.2E.

(a) Draw Haworth and chair structures for the α- and β-pyranose forms of this monosaccharide.

(b) Draw Haworth and chair structures for the disaccharide formed by joining two units of the pyranose form of N-acetyl-D-glucosamine by a β-1,4-glucosidic bond. If your drawing is correct, you have the structural formula for the repeating dimer of chitin, the structural polysaccharide component of the shell of lobsters and other crustaceans.

***18.55** Propose structural formulas for the repeating disaccharide unit in these polysaccharides:

(a) Alginic acid, isolated from seaweed, is used as a thickening agent in ice cream and other foods. Alginic acid is a polymer of D-mannuronic acid in the pyranose form, joined by β-1,4-glycosidic bonds.

(b) Pectic acid is the main component of pectin, which is responsible for the formation of jellies from fruits and berries. Pectic acid is a polymer of D-galacturonic acid in the pyranose form joined by α-1,4-glycosidic bonds.

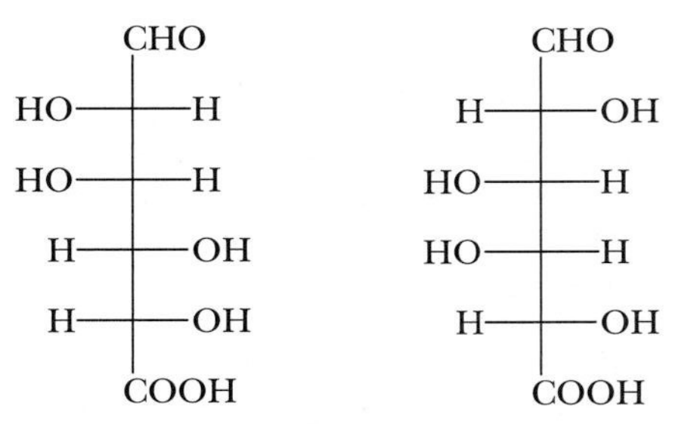

D-Mannuronic acid D-Galacturonic acid

***18.56** The first formula is a Haworth projection and the second is a chair conformation for the repeating disaccharide unit in chondroitin 6-sulfate:

This biopolymer acts as a flexible connecting matrix between the tough protein filaments in cartilage and is available as a dietary supplement, often combined with D-glucosamine sulfate. Some believe that the combination can strengthen and improve joint flexibility.

(a) From what two monosaccharide units is the repeating disaccharide unit of chondroitin 6-sulfate derived?

(b) Describe the glycosidic bond between the two units.

***18.57** Certain complex lipids are constantly being synthesized and decomposed in the body. In several genetic diseases classified as lipid storage diseases, some of the enzymes needed to decompose the complex lipid are defective or missing. As a consequence, the complex lipids accumulate and cause enlarged liver and spleen, mental retardation, blindness, and in certain cases early death. At present no treatment is available for these diseases. The best way to prevent them is genetic counseling. Some of them can be diagnosed during fetal development.

The following is the structure of the lipid that accumulates in Fabray's disease. The genetic defect in this case is that the enzyme α-galactosidase is either missing or defective. This enzyme catalyzes the hydrolysis of the glycosidic bonds formed by α-D-galactopyranose.

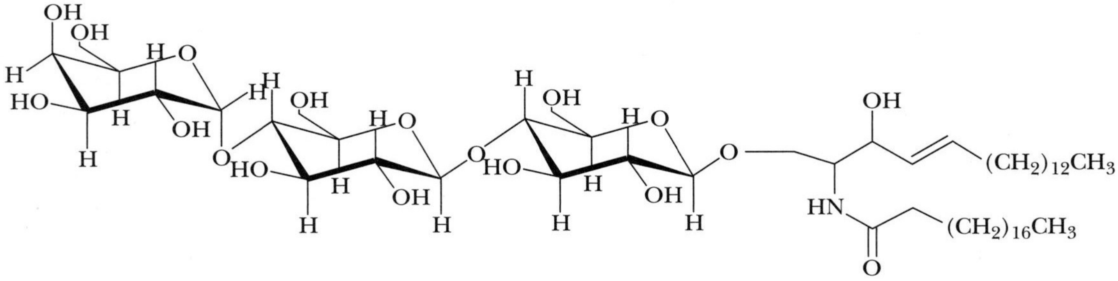

(a) Name the three hexoses present in this lipid

(b) Describe the glycosidic bond between each.

(c) Would you expect this molecule to be soluble or insoluble in water? Explain.

Looking Ahead

***18.58** One step in glycolysis, the pathway that converts glucose to pyruvate (Section 22.3), involves an enzyme-catalyzed conversion of dihydroxyacetone phosphate to D-glyceraldehyde 3-phosphate:

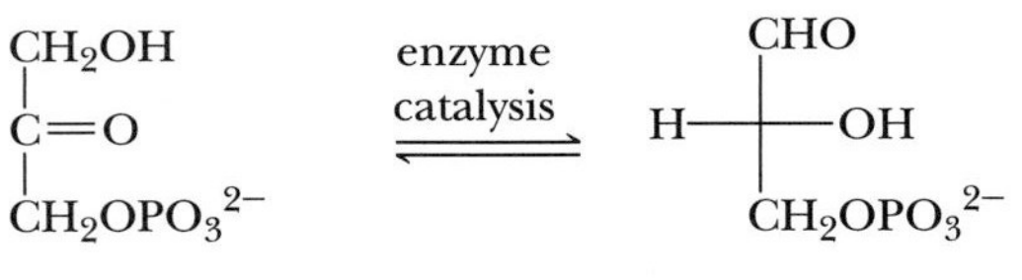

Dihydroxyacetone phosphate ⇌ (enzyme catalysis) D-Glyceraldehyde 3-phosphate

Show that this transformation can be regarded as two enzyme-catalyzed keto–enol tautomerizations (Section 13.8).

***18.59** One pathway for the metabolism of glucose 6-phosphate is its enzyme-catalyzed conversion to fructose 6-phosphate:

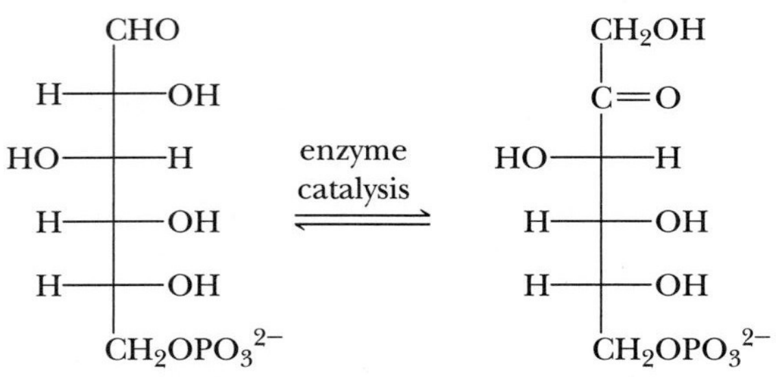

D-Glucose 6-phosphate ⇌ (enzyme catalysis) D-Fructose 6-phosphate

Show that this transformation can be regarded as two enzyme-catalyzed keto–enol tautomerizations.

18.60 Epimers are carbohydrates that differ in configuration at only one stereocenter.

(a) Which of the aldohexoses are epimers of each other?

(b) Are all anomer pairs also epimers of each other? Explain. Are all epimers also anomers? Explain.

***18.61** Oligosaccharides are very valuable therapeutically and are especially difficult to synthesize, even though the starting materials are readily available. Shown is the structure of globotriose, the receptor for a series of toxins synthesized by some strains of *E. coli:*

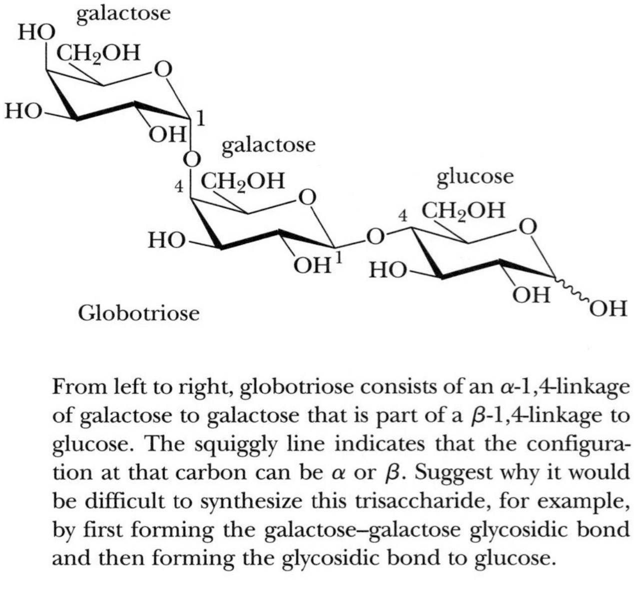

Globotriose

From left to right, globotriose consists of an α-1,4-linkage of galactose to galactose that is part of a β-1,4-linkage to glucose. The squiggly line indicates that the configuration at that carbon can be α or β. Suggest why it would be difficult to synthesize this trisaccharide, for example, by first forming the galactose–galactose glycosidic bond and then forming the glycosidic bond to glucose.

Putting It Together

The following problems bring together concepts and material from Chapters 16–18. Although the focus may be on these chapters, the problems will also build upon concepts discussed throughout the text thus far.

Choose the best answer for each of the following questions.

1. Which carbon on β-maltose will be oxidized by Tollens' reagent?

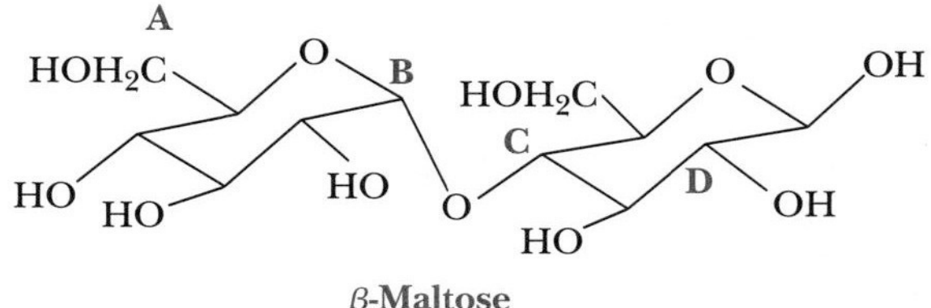

β-**Maltose**

(a) **A** (b) **B** (c) **C** (d) **D** (e) **none of the above**

2. What sequence of reagents will accomplish the following transformation?

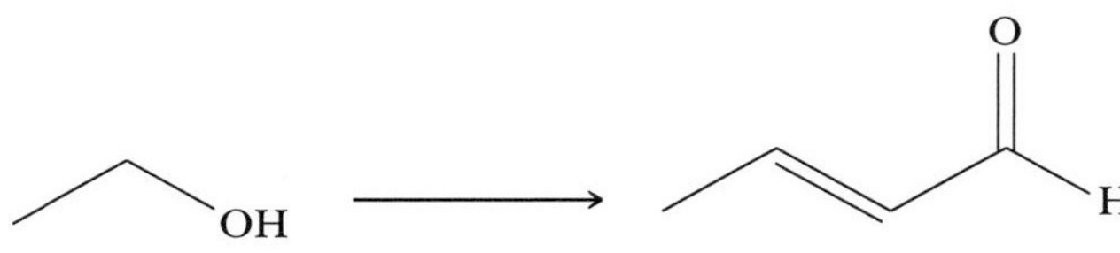

(a) 1. PCC / CH_2Cl_2
2. NaOH / H_2O
3. H^+ / H_2O

(b) 1. H_2CrO_4 / H_2SO_4
2. NaOH / H_2O
3. H^+ / H_2O

(c) 1. NaOH / H_2O
2. H_2CrO_4 / H_2SO_4
3. NaOH / H_2O
4. H^+ / H_2O

(d) 1. $LiAlH_4$ / Et_2O
2. H^+ / H_2O
3. NaOH / H_2O

(e) 1. $NaBH_4$ / EtOH
2. H^+ / H_2O
3. NaOH / H_2O

3. Assuming that the following two polymers are manufactured in similar ways, which of the statements below is *true*?

A B

(a) Polymer **A** will have a lower T_m than polymer **B**.

(b) Polymer **A** will be more amorphous than polymer **B**.

(c) Polymer **A** will be weaker than polymer **B**.

(d) Polymer **A** will have a lower T_g than polymer **B**.

(e) None of the above.

4. Which of the following carbohydrates does *not* undergo mutarotation?

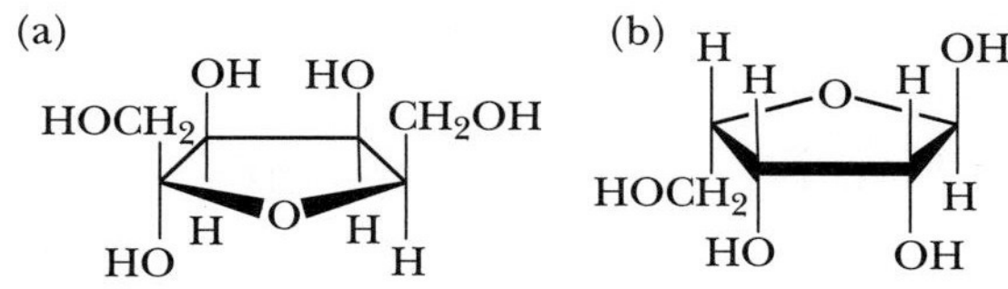

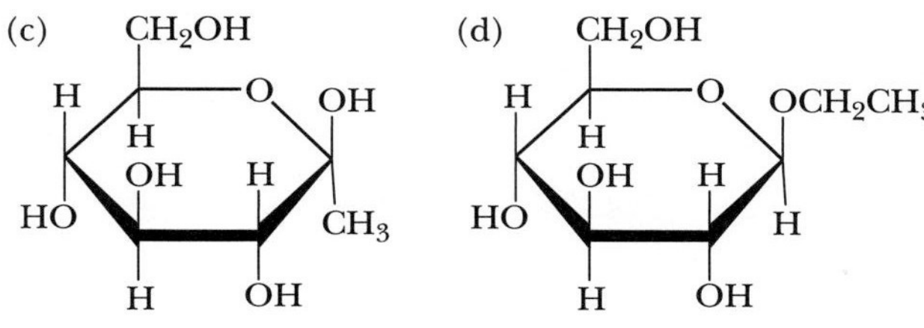

(e) All of these

5. What sequence of reagents will accomplish the following transformation?

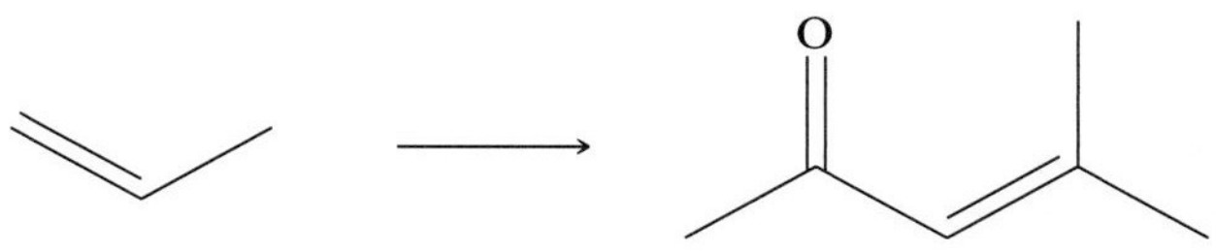

(a) $\xrightarrow{HBr}$ $\xrightarrow[H_2SO_4]{H_2CrO_4}$ $\xrightarrow[H_2O]{NaOH}$

(b) $\xrightarrow[H_2O]{NaOH}$ $\xrightarrow[H_2O]{H^+}$

(c) $\xrightarrow[H_2SO_4]{H_2CrO_4}$ $\xrightarrow[H_2O]{NaOH}$ $\xrightarrow[H_2O]{H^+}$

(d) $\xrightarrow[H_2O]{H_2SO_4}$ $\xrightarrow[H_2SO_4]{H_2CrO_4}$ $\xrightarrow[H_2O]{NaOH}$ $\xrightarrow[H_2O]{H^+}$

(e) $\xrightarrow[H_2O]{H_2SO_4}$ $\xrightarrow[Et_2O]{LiAlH_4}$ $\xrightarrow[H_2O]{H^+}$ $\xrightarrow[H_2O]{NaOH}$ $\xrightarrow[H_2O]{H^+}$

6. Select the most likely product of the following reaction.

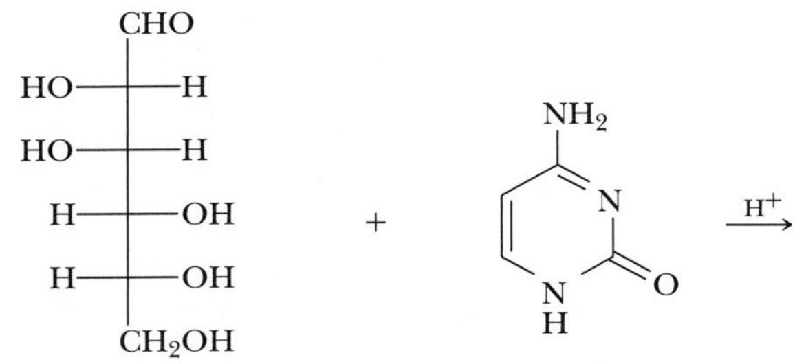

(a)

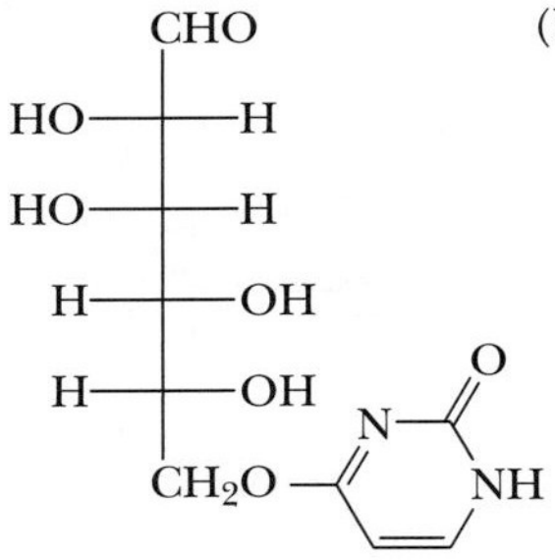

(b)

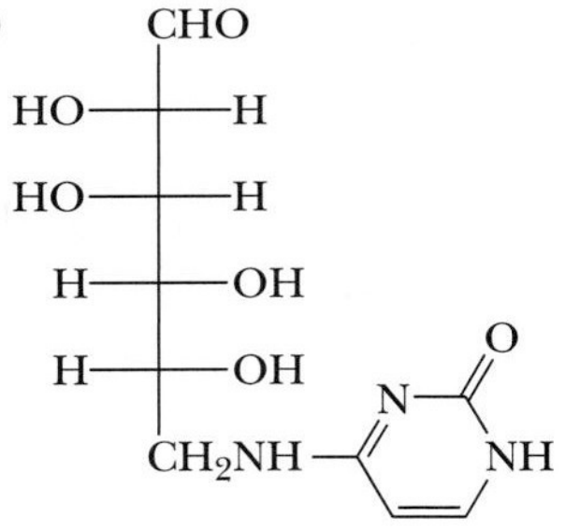

(c)

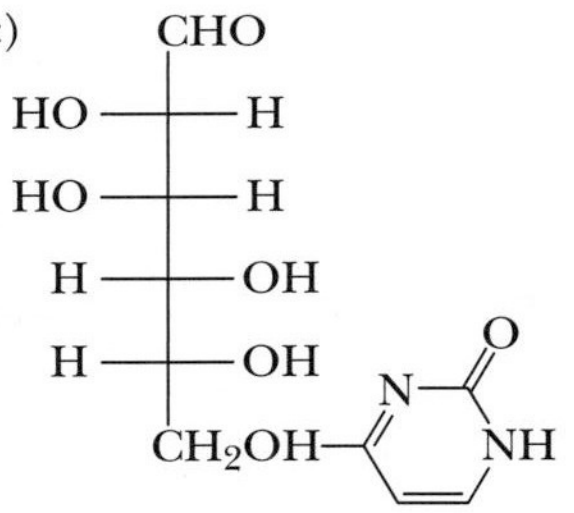

(d)

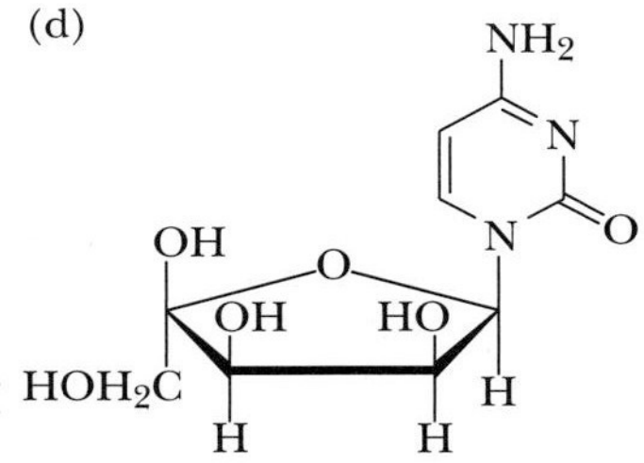

(e) None of the above

7. How many glycosidic bonds exist in the following polysaccharide?

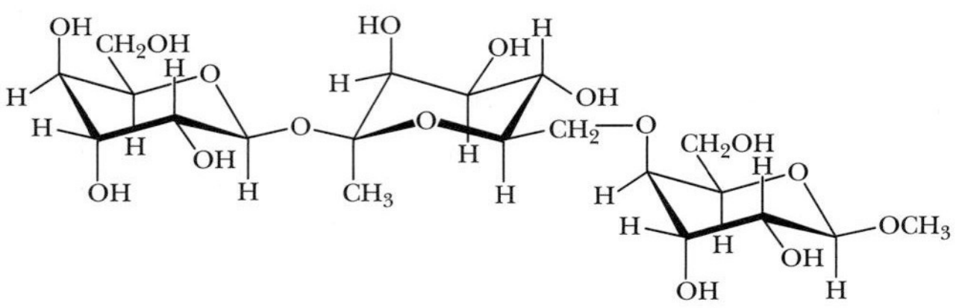

(a) one (b) two (c) three (d) four (e) five

8. Which of the following best classifies the following biological process?

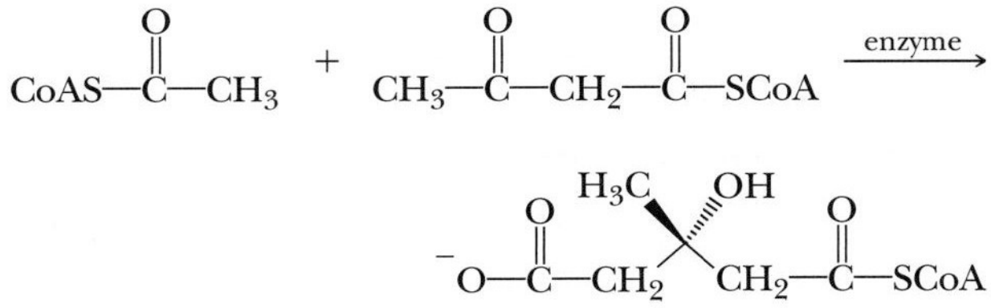

(a) Claisen condensation
(b) Aldol reaction
(c) Nucleophilic acyl substitution
(d) β-elimination
(e) Both **A** and **C**

9. Identify the monomer(s) required for the synthesis of the following polymer:

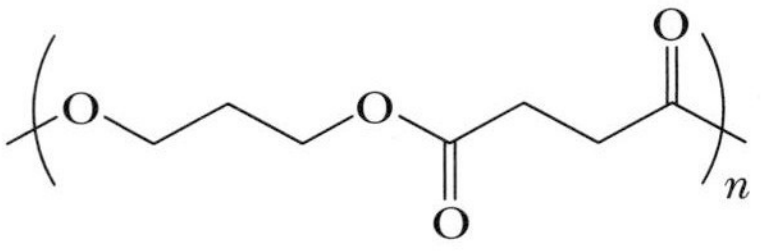

(a)

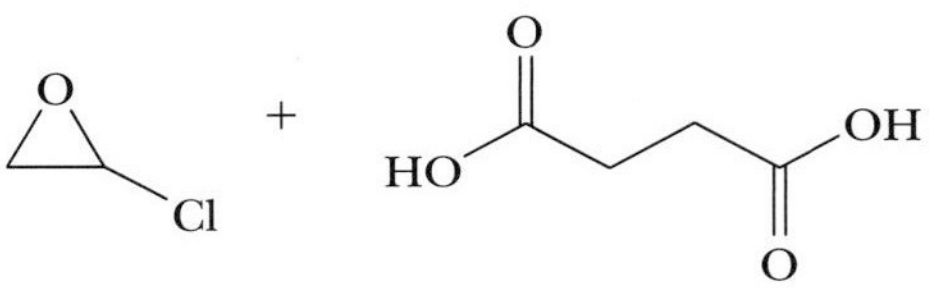

(b)

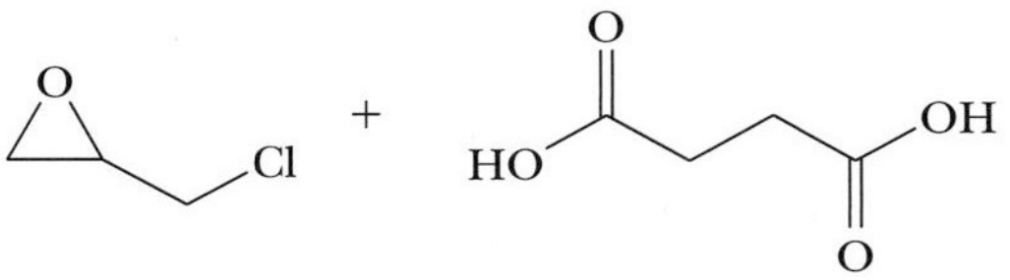

(c)

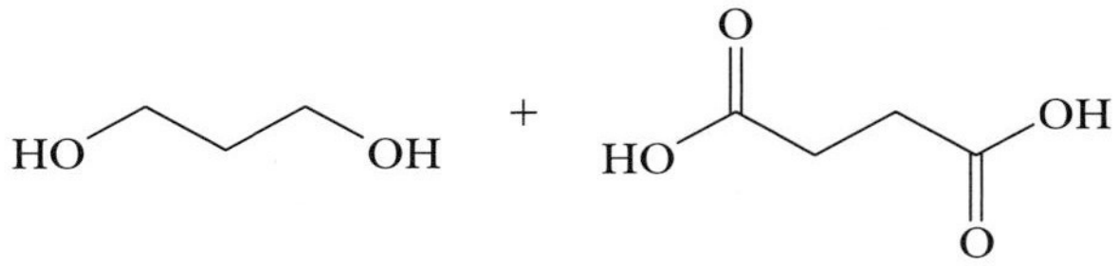

(d)

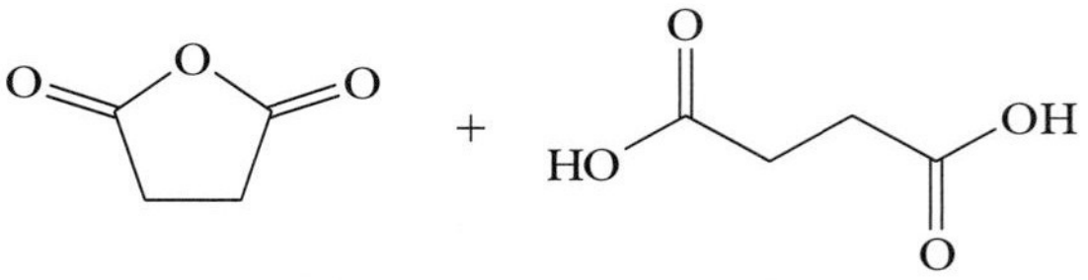

(e) None of these

10. An unknown carbohydrate is placed in a solution of $NaBH_4$/EtOH. After isolating the product, it was discovered that no alditol products were formed. Which of the following carbohydrates could be the unknown?

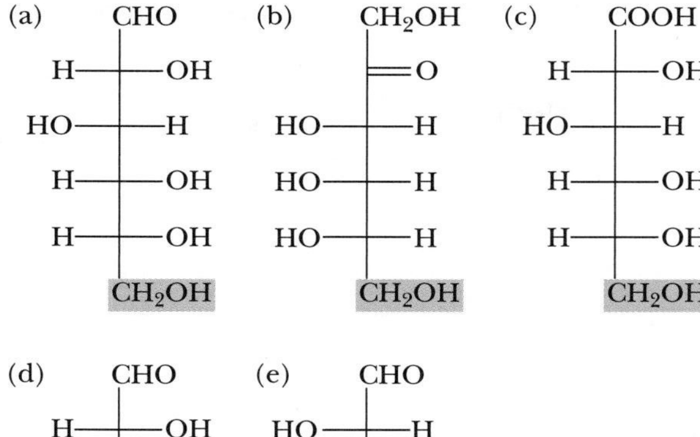

(d) CHO, H–OH, HO–H, CH_2OH (e) CHO, HO–H, HO–H, HO–H, HO–H, CH_2OH

11. Each of the products shown can be made by the reaction indicated under the arrow. Provide a structure for the starting compound(s) needed to produce the product shown. Then show the mechanism of its formation. Show all charges and lone pairs of electrons in your structures.

(a) by an aldol reaction

(b) by a Michael addition

12. NAD^+ is a coenzyme found in all living cells. It acts to carry electrons during biological reactions.

NAD^+

(a) Assign the formal charge in NAD^+ to the appropriate atom.

(b) Is NAD^+ a reducing sugar?

(c) Identify the glycosidic bonds in NAD^+.

(d) From what naturally occurring sugar(s) is NAD^+ derived?

(e) Are the two monosaccharide units α-anomers or β-anomers?

(f) Does NAD^+ undergo mutarotation?

13. Redraw each polymer using the notation in which parentheses are placed around the repeating unit. Identify each as chain-growth or step-growth polymers and identify the monomers required for its synthesis.

(a)

(b)

(c)

14. Provide a mechanism for the following series of reactions. Show all charges and lone pairs of electrons in your structures as well as the structures of all intermediates.

1) $CH_3CH_2NH_2$
2) $CH_3CH_2CH_2Cl$

15. Select the polymer from each pair that would have the higher melt transition temperature and provide an explanation for selection.

(a) $\left(CH_2-CH(Cl)\right)_{1000}$ *Versus* $\left(CH_2-CH(Cl)\right)_{350}$

(b) *Versus*

16. Predict the major product of each of the following reactions.

(a) 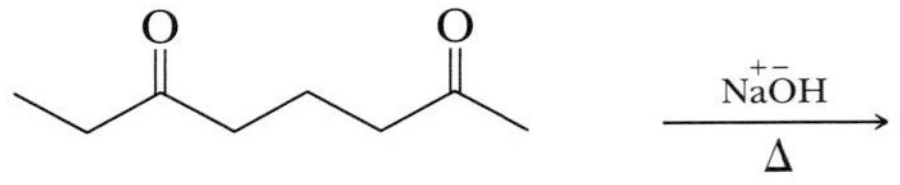 [structure] $\xrightarrow[\Delta]{\overset{+-}{NaOH}}$

(b) 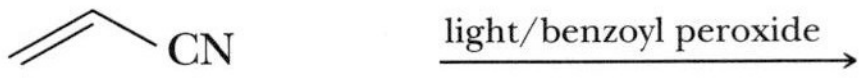 $CH_2{=}CH{-}CN$ $\xrightarrow{\text{light/benzoyl peroxide}}$

(c) 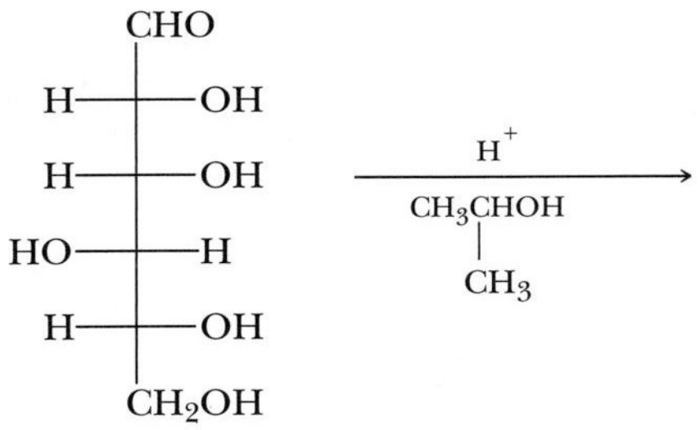 CHO; H—OH; H—OH; HO—H; H—OH; CH_2OH $\xrightarrow[CH_3CHOH\,(CH_3)]{H^+}$

(d) 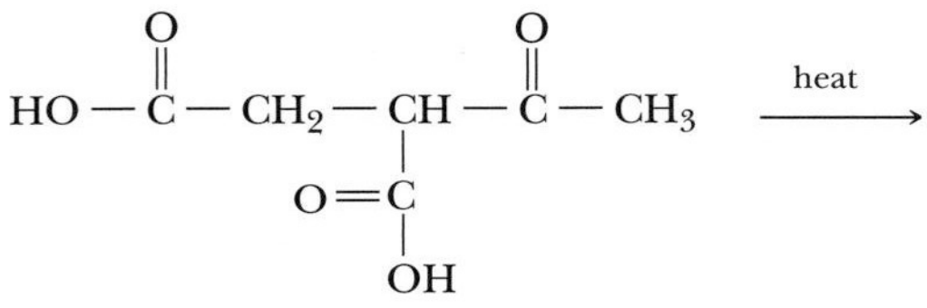 $HO-\overset{O}{\overset{\|}{C}}-CH_2-\underset{\underset{OH}{|}}{\underset{O=C}{|}}{CH}-\overset{O}{\overset{\|}{C}}-CH_3$ $\xrightarrow{\text{heat}}$

(e) 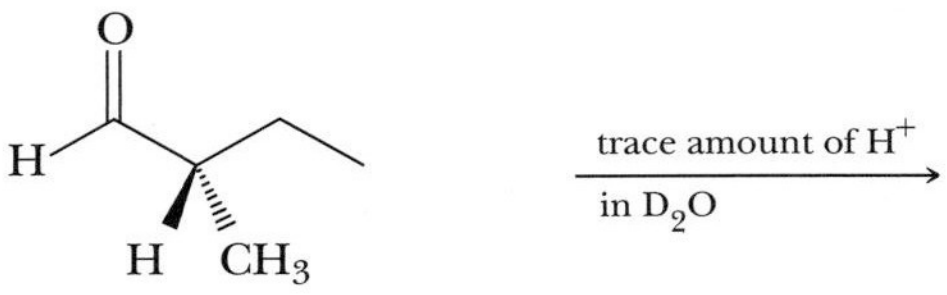 [structure: H, H, CH_3] $\xrightarrow[\text{in } D_2O]{\text{trace amount of } H^+}$

(f) 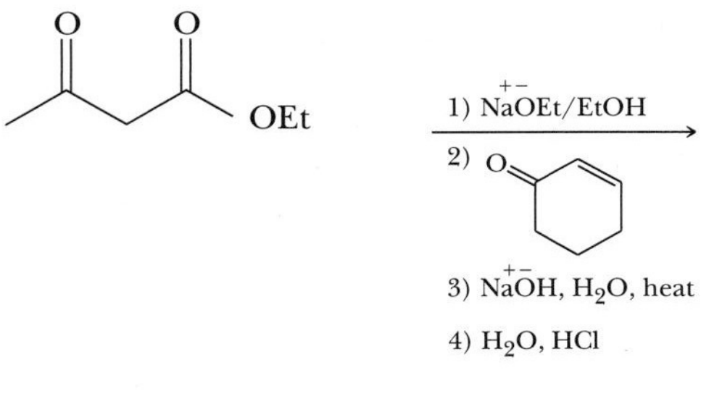 [structure: OEt]
1) $\overset{+-}{NaOEt}$/EtOH
2) [cyclohexenone]
3) $\overset{+-}{NaOH}$, H_2O, heat
4) H_2O, HCl

17. Complete the following chemical transformations.

(a) any hydrocarbon ⟶ $\left(CH_2-\underset{CH_2CH_2CH_3}{\overset{|}{CH}}\right)_n$

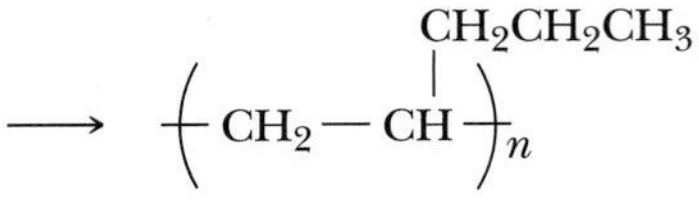

(b) 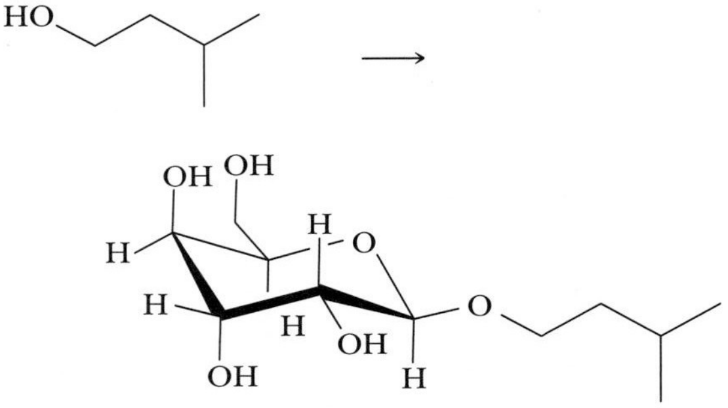 [structure: HO] ⟶ [structure: OH, OH, H, H, H, OH, H, OH, H, O, O]

(c) 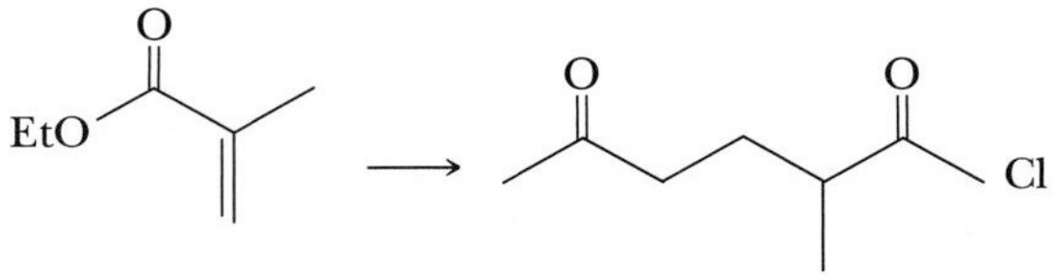 [structure: EtO] ⟶ [structure: Cl]

(d) 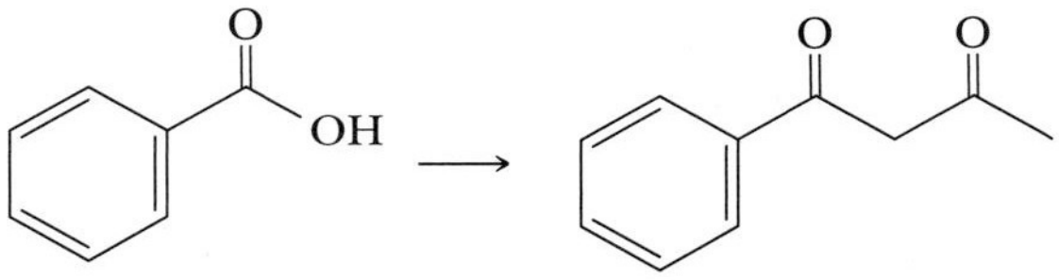 [structure: OH] ⟶ [structure]

(e) any carbonyl compound of 5 carbons or less ⟶ [structure: H, N, H]

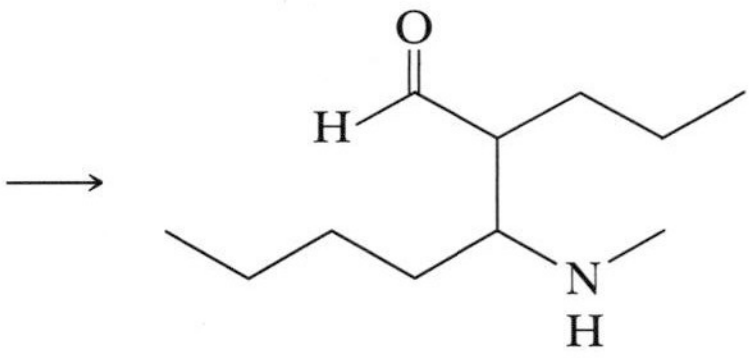

(f) 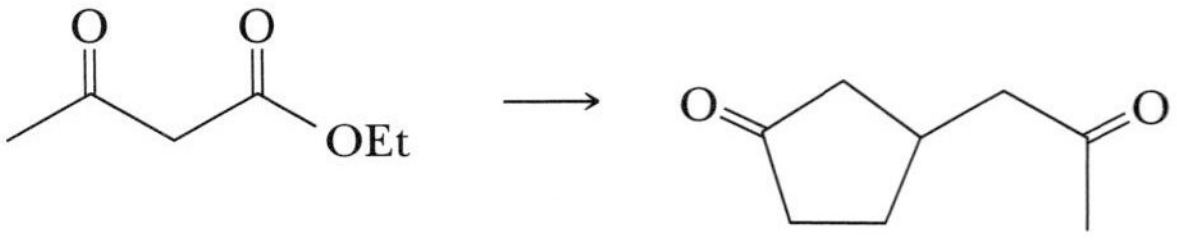 [structure: OEt] ⟶ [structure]

19 Amino Acids and Proteins

KEY QUESTIONS

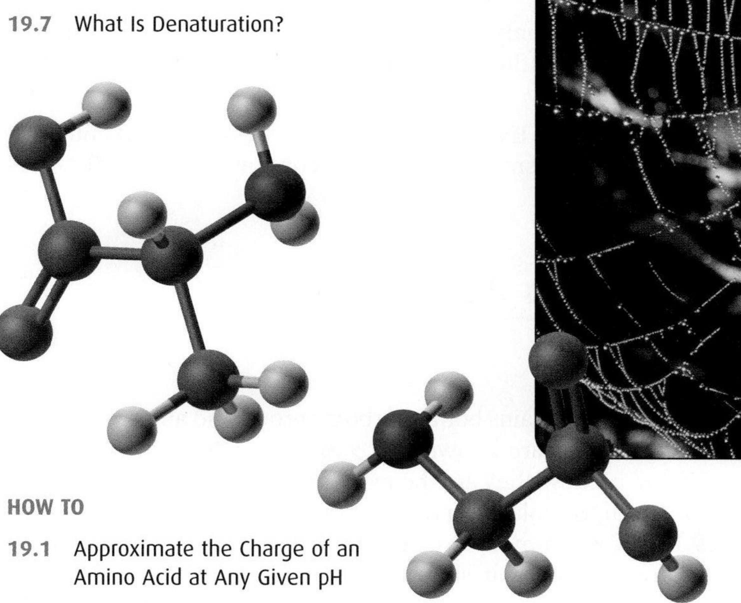

HOW TO

CHEMICAL CONNECTIONS

Spider silk is a fibrous protein that exhibits unmatched strength and toughness. Inset: Models of D-alanine and glycine, the major components of the fibrous protein of silk. *(PhotoDisc Inc./Getty Images)*

We begin this chapter with a study of amino acids, compounds whose chemistry is built on amines (Chapter 10) and carboxylic acids (Chapter 14). We concentrate in particular on the acid–base properties of amino acids because these properties are so important in determining many of the properties of proteins, polymers of amino acids that have many functions in living organisms. With this understanding of the chemistry of amino acids, we then examine the structure of proteins themselves.

19.1 What Are the Many Functions of Proteins?

Proteins are among the most important of all biological compounds. Among the functions performed by these vital molecules are the following:

- *Structure*—Structural proteins such as collagen and keratin are the chief constituents of skin, bones, hair, and nails.
- *Catalysis*—Virtually all reactions that take place in living systems are catalyzed by a special group of proteins called enzymes. Without enzymes, these many reactions would take place so slowly as to be useless.
- *Movement*—Muscle expansion and contraction are involved in every movement we make. Muscle fibers are made of proteins called myosin and actin.
- *Transport*—A large number of proteins perform transport duties. The protein hemoglobin is responsible for the transport of oxygen from the lungs to tissues. Other proteins transport molecules across cell membranes.
- *Hormones*—Many hormones are proteins, including insulin and human growth hormone.
- *Protection*—The production of a group of proteins called antibodies is one of the body's major defenses against disease. The function of the protein fibrinogen is to promote blood clotting.
- *Regulations*—Some proteins not only control the expression of genes, thereby regulating the kind of protein synthesized in a particular cell, but also dictate when such synthesis takes place.

Proteins have other functions as well. Even this brief list, however, should convince you of their vital role in living organisms. A typical cell contains about 9000 different proteins.

19.2 What Are Amino Acids?

A. Structure

Amino acid A compound that contains both an amino group and a carboxyl group.

α-Amino acid An amino acid in which the amino group is on the carbon adjacent to the carboxyl group.

An **amino acid** is a compound that contains both a carboxyl group and an amino group. Although many types of amino acids are known, the ***α*-amino acids** are the most significant in the biological world because they are the monomers from which proteins are constructed. A general structural formula of an α-amino acid is shown in Figure 19.1.

Although Figure 19.1(a) is a common way of writing structural formulas for an amino acid, it is not accurate because it shows an acid ($-COOH$) and a base ($-NH_2$) within the same molecule. These acidic and basic groups react with each other to form an internal salt (a dipolar ion) [Figure 19.1(b)]. This internal salt is given the special name **zwitterion**. Note that a zwitterion has no net charge; it contains one positive charge and one negative charge.

Zwitterion An internal salt of an amino acid.

Because they exist as zwitterions, amino acids have many of the properties associated with salts. They are crystalline solids with high melting points and are fairly soluble in water, but insoluble in nonpolar organic solvents such as ether and hydrocarbon solvents.

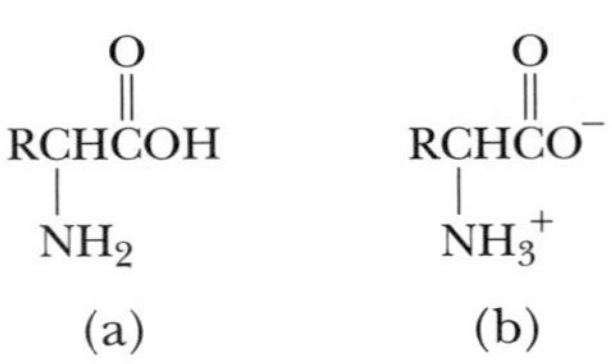

Figure 19.1
An α-amino acid.
(a) Un-ionized form and
(b) internal salt
(zwitterion) form.

B. Chirality

With the exception of glycine, H_2NCH_2COOH, all protein-derived amino acids have at least one stereocenter and therefore are chiral. Figure 19.2 shows Fischer projection formulas for the enantiomers of alanine. The vast majority of carbohydrates in the biological world are of the D-series (Section 18.2), whereas the vast majority of α-amino acids in the biological world are of the L-series.

recall that when the functional group at the penultimate carbon is on the right in a Fischer projection, the compound is part of the D-series

	COO^-	
H—	—	—NH_3^+
	CH_3	

D-Alanine

	COO^-	
$H_3\overset{+}{N}$—	—	—H
	CH_3	

L-Alanine

when the functional group is on the left, it is part of the L-series

Figure 19.2 The enantiomers of alanine. The vast majority of α-amino acids in the biological world have the L-configuration at the α-carbon.

C. Protein-Derived Amino Acids

Table 19.1 gives common names, structural formulas, and standard three-letter and one-letter abbreviations for the 20 common L-amino acids found in proteins. The amino acids shown are divided into four categories: those with nonpolar side chains; those with polar, but un-ionized, side chains; those with acidic side chains; and those with basic side chains. As you study the information in this table, note the following points:

1. All 20 of these protein-derived amino acids are α-amino acids, meaning that the amino group is located on the carbon alpha to the carboxyl group.
2. For 19 of the 20 amino acids, the α-amino group is primary. Proline is different: Its α-amino group is secondary.
3. With the exception of glycine, the α-carbon of each amino acid is a stereocenter. Although not shown in the table, all 19 chiral amino acids have the same relative configuration at the α-carbon. In the D,L convention, all are L-amino acids.
4. Isoleucine and threonine contain a second stereocenter. Four stereoisomers are possible for each amino acid, but only one of the four is found in proteins.
5. The sulfhydryl group of cysteine, the imidazole group of histidine, and the phenolic hydroxyl of tyrosine are partially ionized at pH 7.0, but the ionized form is not the major form present at that pH.

Example 19.1

Of the 20 protein-derived amino acids shown in Table 19.1, how many contain (a) aromatic rings, (b) side-chain hydroxyl groups, (c) phenolic —OH groups, and (d) sulfur?

Strategy

Study the structural formulas of the amino acids given in Table 19.1.

Solution

(a) Phenylalanine, tryptophan, tyrosine, and histidine contain aromatic rings.
(b) Serine and threonine contain side-chain hydroxyl groups.
(c) Tyrosine contains a phenolic —OH group.
(d) Methionine and cysteine contain sulfur.

See problems 19.14, 19.15

Problem 19.1

Of the 20 protein-derived amino acids shown in Table 19.1, (a) which contain no stereocenter and (b) which contain two stereocenters?

TABLE 19.1 The 20 Common Amino Acids Found in Proteins

Nonpolar Side Chains

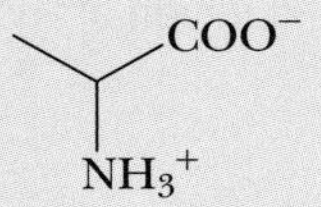

Structure	Name	Structure	Name
COO^-, NH_3^+	Alanine (Ala, A)	COO^-, NH_3^+	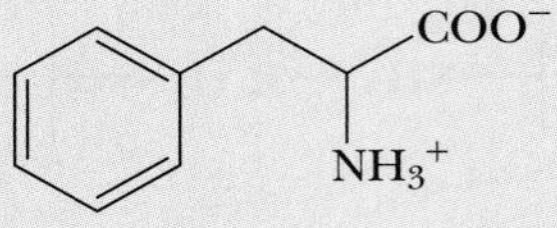 Phenylalanine (Phe, F)
COO^-, NH_3^+	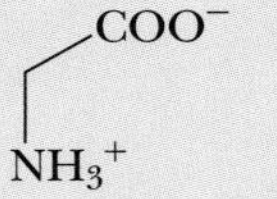 Glycine (Gly, G)	$\overset{+}{N}$, H, H, COO^-	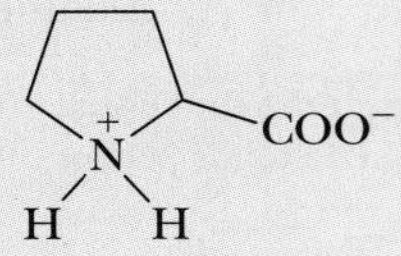 Proline (Pro, P)
COO^-, NH_3^+	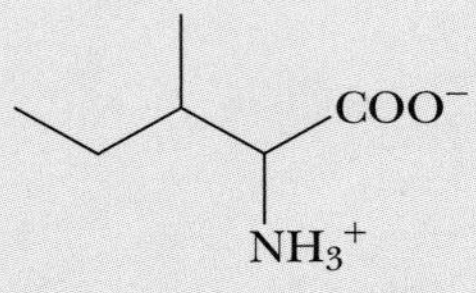 Isoleucine (Ile, I)	COO^-, NH_3^+, N, H	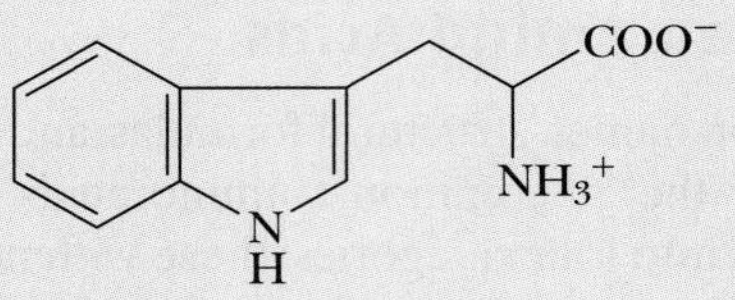 Tryptophan (Trp, W)
COO^-, NH_3^+	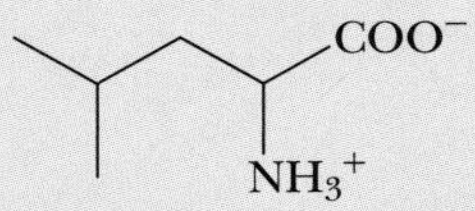 Leucine (Leu, L)		
S, COO^-, NH_3^+	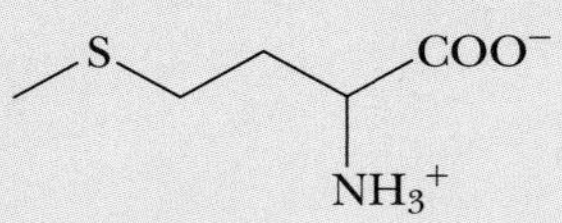 Methionine (Met, M)	COO^-, NH_3^+	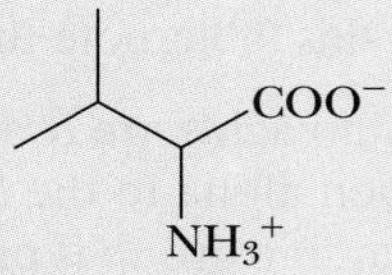 Valine (Val, V)

Polar Side Chains

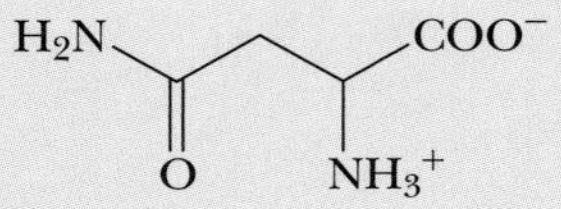

Structure	Name	Structure	Name
H_2N, O, COO^-, NH_3^+	Asparagine (Asn, N)	HO, COO^-, NH_3^+	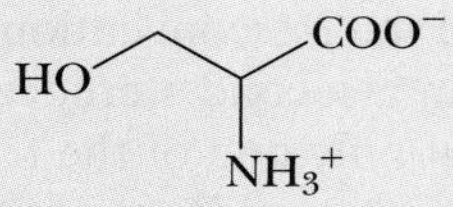 Serine (Ser, S)
O, H_2N, COO^-, NH_3^+	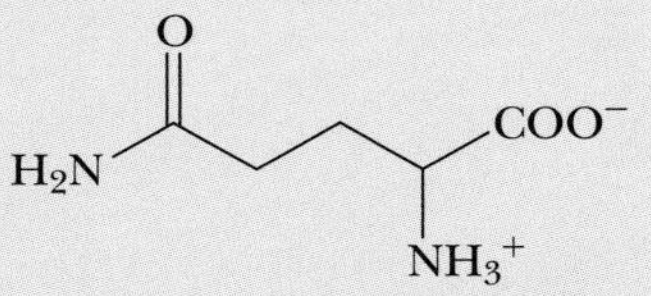 Glutamine (Gln, Q)	OH, COO^-, NH_3^+	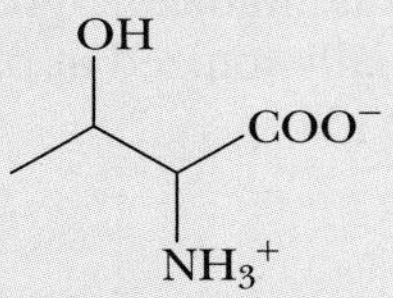 Threonine (Thr, T)

Acidic Side Chains		**Basic Side Chains**	
^-O, O, COO^-, NH_3^+	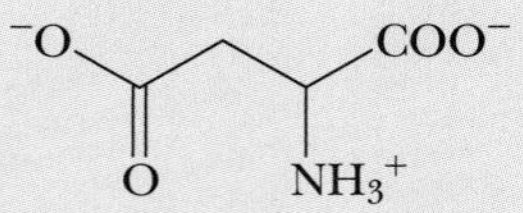 Aspartic acid (Asp, D)	NH_2^+, H_2N, N, H, COO^-, NH_3^+	Arginine (Arg, R)
O, ^-O, COO^-, NH_3^+	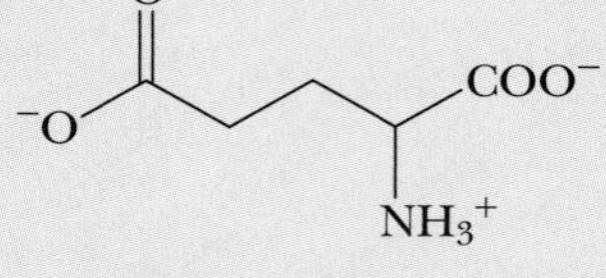 Glutamic acid (Glu, E)	COO^-, NH_3^+, N, N, H	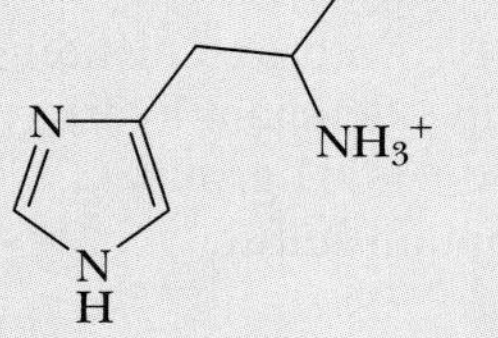 Histidine (His, H)
HS, COO^-, NH_3^+	Cysteine (Cys, C)		
HO, COO^-, NH_3^+	Tyrosine (Tyr, Y)	$H_3\overset{+}{N}$, COO^-, NH_3^+	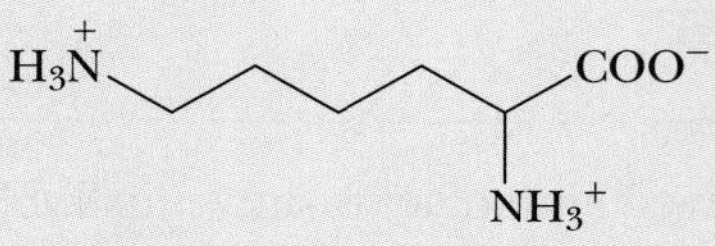 Lysine (Lys, K)

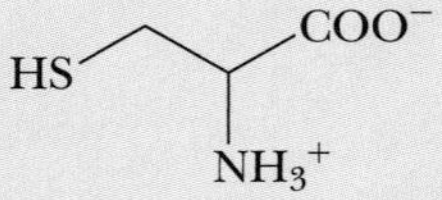

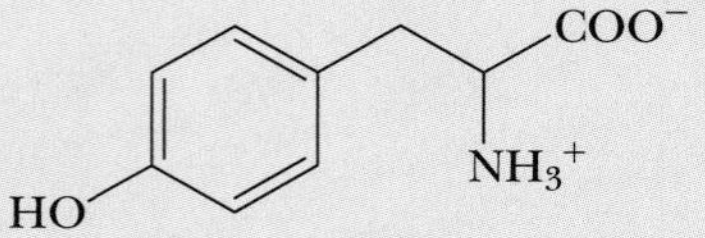

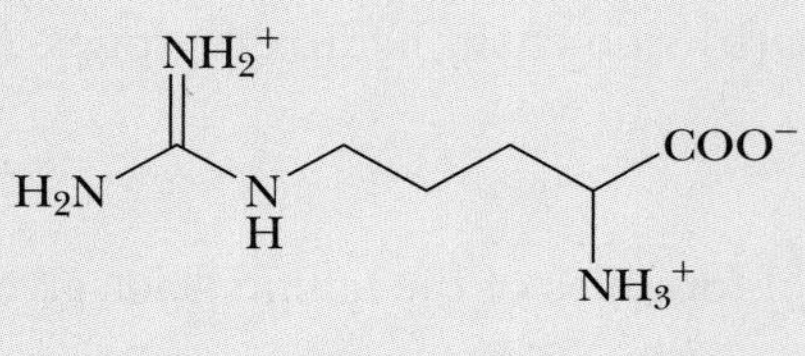

Note: Each ionizable group is shown in the form present in highest concentration in aqueous solution at pH 7.0.

D. Some Other Common L-Amino Acids

Although the vast majority of plant and animal proteins are constructed from just these 20 α-amino acids, many other amino acids are also found in nature. Ornithine and citrulline, for example, are found predominantly in the liver and are an integral part of the urea cycle, the metabolic pathway that converts ammonia to urea:

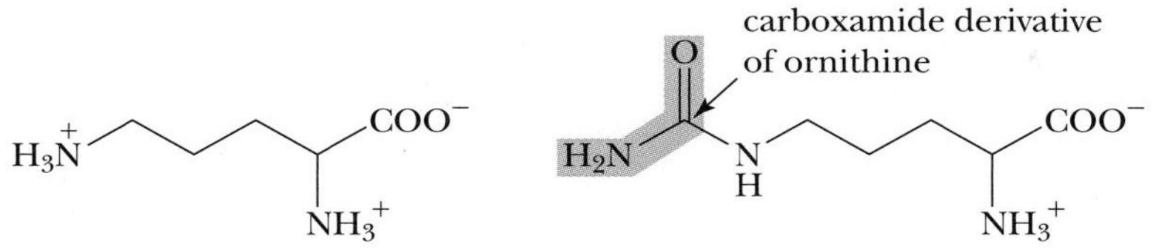

carboxamide derivative of ornithine

Ornithine — Citrulline

Thyroxine and triiodothyronine, two of several hormones derived from the amino acid tyrosine, are found in thyroid tissue:

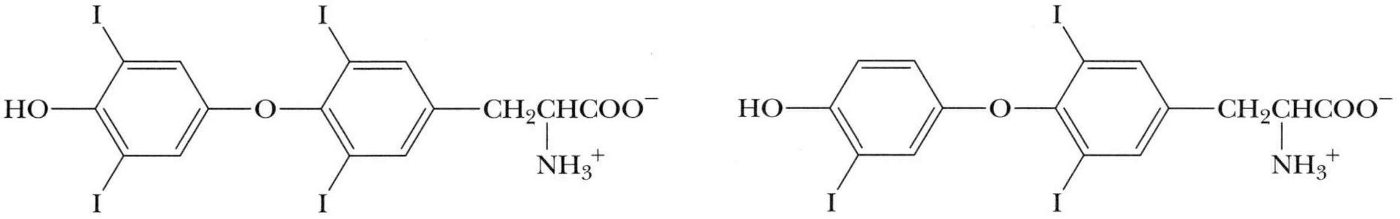

Thyroxine, T_4 — Triiodothyronine, T_3

The principal function of these two hormones is to stimulate metabolism in other cells and tissues.

4-Aminobutanoic acid (γ-aminobutyric acid, or GABA) is found in high concentration (0.8 mM) in the brain, but in no significant amounts in any other mammalian tissue. GABA is synthesized in neural tissue by decarboxylation of the α-carboxyl group of glutamic acid and is a neurotransmitter in the central nervous system of invertebrates and possibly in humans as well:

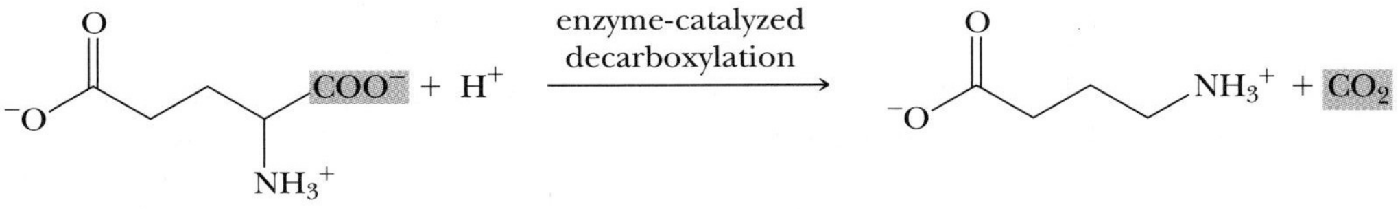

Glutamic acid $+ H^+$ —(enzyme-catalyzed decarboxylation)→ 4-Aminobutanoic acid (γ-Aminobutyric acid, GABA) $+ CO_2$

Only L-amino acids are found in proteins, and only rarely are D-amino acids a part of the metabolism of higher organisms. Several D-amino acids, however, along with their L-enantiomers, are found in lower forms of life. D-Alanine and D-glutamic acid, for example, are structural components of the cell walls of certain bacteria. Several D-amino acids are also found in peptide antibiotics.

19.3 What Are the Acid–Base Properties of Amino Acids?

A. Acidic and Basic Groups of Amino Acids

Among the most important chemical properties of amino acids are their acid–base properties. All are weak polyprotic acids because of the presence of both —COOH and $—NH_3^+$ groups. Given in Table 19.2 are pK_a values for each ionizable group of the 20 protein-derived amino acids.

TABLE 19.2 pK_a Values for Ionizable Groups of Amino Acids

Amino Acid	pK_a of α-COOH	pK_a of α-NH_3^+	pK_a of of Side Chain	Isoelectric Point (pI)
alanine	2.35	9.87	—	6.11
arginine	2.01	9.04	12.48	10.76
asparagine	2.02	8.80	—	5.41
aspartic acid	2.10	9.82	3.86	2.98
cysteine	2.05	10.25	8.00	5.02
glutamic acid	2.10	9.47	4.07	3.08
glutamine	2.17	9.13	—	5.65
glycine	2.35	9.78	—	6.06
histidine	1.77	9.18	6.10	7.64
isoleucine	2.32	9.76	—	6.04
leucine	2.33	9.74	—	6.04
lysine	2.18	8.95	10.53	9.74
methionine	2.28	9.21	—	5.74
phenylalanine	2.58	9.24	—	5.91
proline	2.00	10.60	—	6.30
serine	2.21	9.15	—	5.68
threonine	2.09	9.10	—	5.60
tryptophan	2.38	9.39	—	5.88
tyrosine	2.20	9.11	10.07	5.63
valine	2.29	9.72	—	6.00

Note: Dash indicates no ionizable side chain.

Acidity of α-Carboxyl Groups

The average value of the pK_a for an α-carboxyl group of a protonated amino acid is 2.19. Thus, the α-carboxyl group is a considerably stronger acid than the carboxyl group of acetic acid (pK_a 4.76) and other low-molecular-weight aliphatic carboxylic acids. This greater acidity is accounted for by the electron-withdrawing inductive effect of the adjacent $—NH_3^+$ group (recall that we used similar reasoning in Section 14.4A to account for the relative acidities of acetic acid and its mono-, di-, and trichloroderivatives):

the ammonium group has an electron-withdrawing inductive effect that delocalizes the negative charge and stabilizes the ion

$$RCH(NH_3^+)COOH + H_2O \rightleftharpoons RCH(NH_3^+)COO^- + H_3O^+ \quad pK_a = 2.19$$

$$RCH(R)COOH + H_2O \rightleftharpoons RCH(R)COO^- + H_3O^+ \quad pK_a = 4.5$$

without an ammonium group, there is no electron-withdrawing inductive effect

Acidity of Side-Chain Carboxyl Groups

Due to the electron-withdrawing inductive effect of the α-NH_3^+ group, the side-chain carboxyl groups of protonated aspartic acid and glutamic acid are also stronger acids than acetic acid (pK_a 4.76) Notice that this acid-strengthening inductive effect decreases with increasing distance of the —COOH from the α-NH_3^+. Compare the acidities of the α-COOH of alanine (pK_a 2.35), the γ-COOH of aspartic acid (pK_a 3.86), and the δ-COOH of glutamic acid (pK_a 4.07).

Acidity of α-Ammonium Groups

The average value of pK_a for an α-ammonium group, α-NH_3^+, is 9.47 compared with an average value of 10.76 for primary aliphatic ammonium ions (Section 10.4). Just as the —NH_3^+ group exerts an inductive effect on the carboxylate group, the electronegative oxygen atoms of the carboxylate group exert an electron-withdrawing inductive effect on the —NH_3^+ group. This increases the electron deficiency of the ammonium group, making it more likely to donate a proton to become an uncharged —NH_2 group. Thus, the α-ammonium group of an amino acid is a slightly stronger acid than a primary aliphatic ammonium ion. Conversely, an α-amino group is a slightly weaker base than a primary aliphatic amine.

the electron-withdrawing effect of the carboxylate oxygens increases the positive character of the ammonium group, making the ion less stable and more likely to give up a proton

$$\underset{\substack{|\\NH_3^+}}{RCHCOO^-} + H_2O \rightleftharpoons \underset{\substack{|\\NH_2}}{RCHCOO^-} + H_3O^+ \quad pK_a = 9.47$$

$$R{-}\underset{\substack{|\\NH_3^+}}{CH}{-}R + H_2O \rightleftharpoons R{-}\underset{\substack{|\\NH_2}}{CH}{-}R + H_3O^+ \quad pK_a = 10.60$$

Basicity of the Guanidine Group of Arginine

The side-chain guanidine group of arginine is a considerably stronger base than an aliphatic amine is. As we saw in Section 10.4, guanidine (pK_b 0.4) is the strongest base of any neutral compound. The remarkable basicity of the guanidine group of arginine is attributed to the large resonance stabilization of the protonated form.

$$RNH{-}C(NH_2)({=}NH_2^+) \longleftrightarrow RNH{-}C({=}NH_2^+)(NH_2) \longleftrightarrow R\overset{+}{N}H{=}C(NH_2)(NH_2) + H_2O \rightleftharpoons RN{=}C(NH_2)(NH_2) + H_3O^+ \quad pK_a = 12.48$$

The protonated form of the guanidinium ion side chain of arginine is a hybrid of three contributing structures

the resonance stabilization of the guanidinium ion makes the guanidine nitrogen more basic than an aliphatic amine

no resonance stabilization

$$H{-}\overset{+}{N}{-}R + H_2O \rightleftharpoons :N{-}R + H_3O^+ \quad pK_a = 10.5$$

Ammonium ion

Aliphatic amine

Basicity of the Imidazole Group of Histidine

Because the imidazole group on the side chain of histidine contains six π electrons in a planar, fully conjugated ring, imidazole is classified as a heterocyclic aromatic amine (Section 9.2). The unshared pair of electrons on one nitrogen is a part of the aromatic sextet, whereas that on the other nitrogen is not. It is the pair of electrons that is not part of the aromatic sextet that is responsible for the basic properties of the imidazole ring. Protonation of this nitrogen produces a resonance-stabilized cation:

This lone pair is not part of the aromatic sextet; it is the proton acceptor

$$\text{(imidazolium cation)} \longleftrightarrow \text{(imidazolium cation)} \underset{H_3O^+}{\overset{H_2O}{\rightleftharpoons}} \text{(histidine)} + H_3O^+ \quad pK_a = 6.10$$

Resonance-stabilized imidazolium cation

B. Titration of Amino Acids

Values of pK_a for the ionizable groups of amino acids are most commonly obtained by acid–base titration and by measuring the pH of the solution as a function of added base (or added acid, depending on how the titration is done). To illustrate this experimental procedure, consider a solution containing 1.00 mole of glycine to which has been added enough strong acid so that both the amino and carboxyl groups are fully protonated. Next, the solution is titrated with 1.00 M NaOH; the volume of base added and the pH of the resulting solution are recorded and then plotted as shown in Figure 19.3.

The most acidic group, and the one to react first with added sodium hydroxide, is the carboxyl group. When exactly 0.50 mole of NaOH has been added, the carboxyl group is half neutralized. At this point, the concentration of the zwitterion equals that of the positively charged ion, and the pH of 2.35 equals the pK_a of the carboxyl group (pK_{a1}):

$$\text{At pH} = pK_{a1} \qquad [H_3\overset{+}{N}CH_2COOH] = [H_3\overset{+}{N}CH_2COO^-]$$

Positive ion   Zwitterion

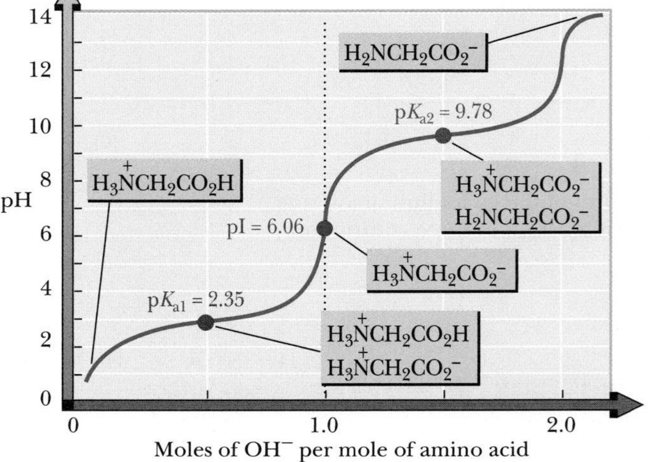

Figure 19.3
Titration of glycine with sodium hydroxide.

The end point of the first part of the titration is reached when 1.00 mole of NaOH has been added. At this point, the predominant species present is the zwitterion, and the observed pH of the solution is 6.06.

The next section of the curve represents titration of the $—NH_3^+$ group. When another 0.50 mole of NaOH has been added (bringing the total to 1.50 moles), half of the $—NH_3^+$ groups are neutralized and converted to $—NH_2$. At this point, the concentrations of the zwitterion and negatively charged ion are equal, and the observed pH is 9.78, the pK_a of the amino group of glycine (pK_{a2}):

$$\text{At pH} = \text{p}K_{a2} \qquad [H_3\overset{+}{N}CH_2COO^-] = [H_2NCH_2COO^-]$$

Zwitterion Negative ion

The second end point of the titration is reached when a total of 2.00 moles of NaOH have been added and glycine is converted entirely to an anion.

C. Isoelectric Point

Titration curves such as that for glycine permit us to determine pK_a values for the ionizable groups of an amino acid. They also permit us to determine another important property: the **isoelectric point, pI**—the pH at which most of the molecules of the amino acid in solution have a net charge of zero. (They are zwitterions.) By examining the titration curve, you can see that the isoelectric point for glycine falls half way between the pK_a values for the carboxyl and amino groups:

Isoelectric point (pI) The pH at which an amino acid, a polypeptide, or a protein has no net charge.

$$\text{pI} = \frac{1}{2}(\text{p}K_a\ \alpha\text{-COOH} + \text{p}K_a\ \alpha\text{-NH}_3^+)$$

$$= \frac{1}{2}(2.35 + 9.78) = 6.06$$

At pH 6.06, the predominant form of glycine molecules is the dipolar ion; furthermore, at this pH, the concentration of positively charged glycine molecules equals the concentration of negatively charged glycine molecules.

When estimating the pI for arginine, aspartic acid, glutamic acid, histidine, and lysine (amino acids that contain either two carboxyl or two ammonium groups), we use the pK_a's of the two groups that are closest in value. For example, the pI of lysine would be determined as follows:

$$\text{pI} = \frac{1}{2}(\text{p}K_a\ \alpha\text{-}\overset{+}{N}H_3 + \text{p}K_a\ \text{side chain-NH}_3^+)$$

$$= \frac{1}{2}(8.95 + 10.53) = 9.74$$

$H_3\overset{+}{N}—CH(C(=O)—OH)—CH_2—CH_2—CH_2—CH_2—\overset{+}{N}H_3$

pK_a 8.95 (α-NH_3^+); pK_a 2.18 (COOH); pK_a 10.53 (side chain NH_3^+)

these two pK_a values are closest

Given a value for the isoelectric point of an amino acid, it is possible to estimate the charge on that amino acid at any pH. For example, the charge on tyrosine at pH 5.63, the isoelectric point of tyrosine, is zero. A small fraction of tyrosine molecules is positively charged at pH 5.00 (0.63 unit less than its pI), and virtually all are positively charged at pH 3.63 (2.00 units less than its pI). As another example, the net charge on lysine is zero at pH 9.74. At pH values smaller than 9.74, an increasing fraction of lysine molecules is positively charged.

At pH values greater than pI, an increasing fraction of its molecules have a net negative charge. To summarize for any amino acid:

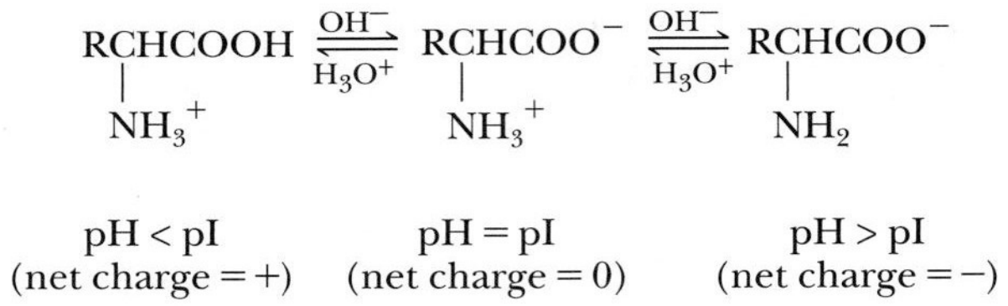

$$\underset{\substack{\text{pH} < \text{pI} \\ (\text{net charge} = +)}}{\underset{NH_3^+}{\overset{|}{RCHCOOH}}} \underset{H_3O^+}{\overset{OH^-}{\rightleftharpoons}} \underset{\substack{\text{pH} = \text{pI} \\ (\text{net charge} = 0)}}{\underset{NH_3^+}{\overset{|}{RCHCOO^-}}} \underset{H_3O^+}{\overset{OH^-}{\rightleftharpoons}} \underset{\substack{\text{pH} > \text{pI} \\ (\text{net charge} = -)}}{\underset{NH_2}{\overset{|}{RCHCOO^-}}}$$

HOW TO 19.1 Approximate the Charge of an Amino Acid at Any Given pH

The pH of a solution is a measure of its acidity. The pK_a value of an ionizable functional group is a measure of the acidity of a proton on that functional group. In your study of general chemistry, you learned that the Henderson-Hasselbalch equation $\left(\text{pH} = \text{p}K_a + \log\frac{[A^-]}{[HA]}\right)$ allows us to relate the pH of a solution to the pK_a of an acid. The equation can be summarized as follows:

(a) If the pH of the solution is lower than the pK_a of an acidic group, the solution behaves as an acid to the group and the group remains protonated.

(b) If the pH of the solution is higher than the pK_a of an acidic group, the acidic group behaves as an acid to the solution and the group is deprotonated.

We use lysine as an example. At pH 5, the charge on lysine will be +1:

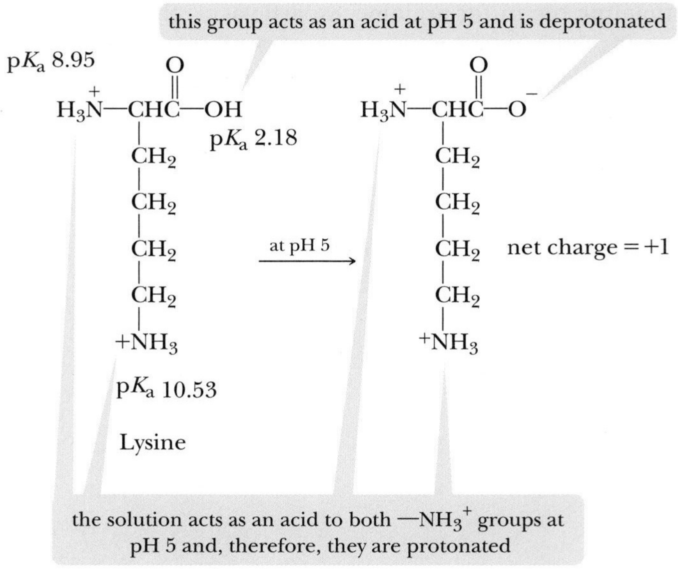

this group acts as an acid at pH 5 and is deprotonated

$$\underset{\text{Lysine}}{\underset{\text{p}K_a\ 10.53}{\overset{\text{p}K_a\ 8.95}{H_3\overset{+}{N}{-}CH(CH_2CH_2CH_2CH_2\overset{+}{N}H_3){-}C(=O){-}OH}}}\ (\text{p}K_a\ 2.18) \xrightarrow{\text{at pH 5}} H_3\overset{+}{N}{-}CH(CH_2CH_2CH_2CH_2\overset{+}{N}H_3){-}C(=O){-}O^- \quad \text{net charge} = +1$$

the solution acts as an acid to both —NH_3^+ groups at pH 5 and, therefore, they are protonated

*Note: This approximation works best when there is a difference of 1 unit between the pH value of the solution and the pK_a value of each ionizable group.

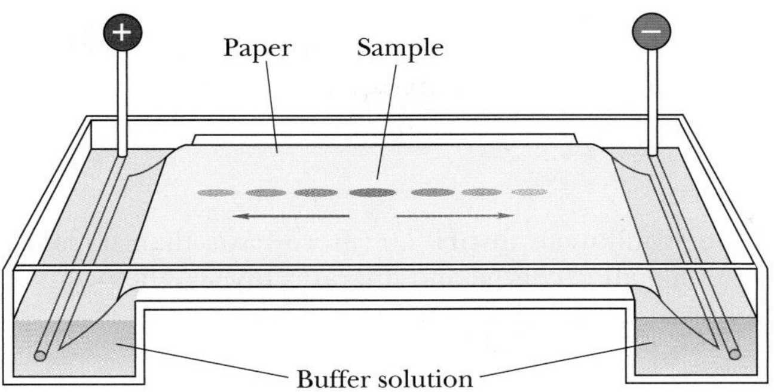

Figure 19.4
Electrophoresis of a mixture of amino acids. Those with a negative charge move toward the positive electrode; those with a positive charge move toward the negative electrode; those with no charge remain at the origin.

D. Electrophoresis

Electrophoresis The process of separating compounds on the basis of their electric charge.

Electrophoresis, a process of separating compounds on the basis of their electric charges, is used to separate and identify mixtures of amino acids and proteins. Electrophoretic separations can be carried out with paper, starch, agar, certain plastics, and cellulose acetate used as solid supports. In paper electrophoresis, a paper strip saturated with an aqueous buffer of predetermined pH serves as a bridge between two electrode vessels (Figure 19.4). Next, a sample of amino acids is applied as a colorless spot on the paper strip. (The amino acid mixture is colorless.) When an electrical potential is then applied to the electrode vessels, amino acids migrate toward the electrode carrying the charge opposite to their own. Molecules having a high charge density move more rapidly than do those with a lower charge density. Any molecule already at its isoelectric point remains at the origin. After the separation is complete, the paper strip is sprayed with a dye that transforms each amino acid into a colored compound, making the separated components visible.

A dye commonly used to detect amino acids is ninhydrin (1,2,3-indanetrione monohydrate). Ninhydrin reacts with α-amino acids to produce an aldehyde, carbon dioxide, and a purple-colored anion:

$$\underset{\text{An } \alpha\text{-amino acid}}{RCH(NH_3^+)CO^-} + 2\ \underset{\text{Ninhydrin}}{\text{Ninhydrin}} \longrightarrow \underset{\text{Purple-colored anion}}{\text{Purple-colored anion}} + RCHO + CO_2 + H_3O^+$$

This reaction is commonly used in both qualitative and quantitative analysis of amino acids. Nineteen of the 20 protein-derived α-amino acids have primary amino groups and give the same purple-colored ninhydrin-derived anion. Proline, a secondary amine, gives a different, orange-colored compound.

Example 19.2

The isoelectric point of tyrosine is 5.63. Toward which electrode does tyrosine migrate during paper electrophoresis at pH 7.0?

Strategy

At its isoelectric point, a molecule of an amino acid has no net charge. In a solution in which pH > pI, its molecules have a net negative charge, and in a solution in which pH < pI, its molecules have a net positive charge. Therefore, in this problem, the important point is to compare the pH of the solution and the pI of the amino acid.

Solution

During paper electrophoresis at pH 7.0 (more basic than its isoelectric point), tyrosine has a net negative charge and migrates toward the positive electrode.

See problems 19.22, 19.23, 19.25, 19.26, 19.32, 19.33

Problem 19.2

The isoelectric point of histidine is 7.64. Toward which electrode does histidine migrate during paper electrophoresis at pH 7.0?

Example 19.3

The electrophoresis of a mixture of lysine, histidine, and cysteine is carried out at pH 7.64. Describe the behavior of each amino acid under these conditions.

Strategy

Compare the pI of each amino acid to the pH of the solution in which it is dissolved and determine the net charge on each at that pH. If the pI of the amino acid is identical to the pH of the solution in which it is dissolved, the amino acid will not migrate from the origin. If the pH of the solution is greater than its pI, an amino acid will migrate toward the positive electrode. If the pH of the solution is less than its pI, an amino acid will migrate toward the positive electrode.

Solution

The isoelectric point of histidine is 7.64. At this pH, histidine has a net charge of zero and does not move from the origin. The pI of cysteine is 5.02; at pH 7.64 (more basic than its isoelectric point), cysteine has a net negative charge and moves toward the positive electrode. The pI of lysine is 9.74; at pH 7.64 (more acidic than its isoelectric point), lysine has a net positive charge and moves toward the negative electrode.

See problem 19.34

Problem 19.3

Describe the behavior of a mixture of glutamic acid, arginine, and valine during paper electrophoresis at pH 6.0.

19.4 What Are Polypeptides and Proteins?

Peptide bond The special name given to the amide bond formed between the α-amino group of one amino acid and the α-carboxyl group of another amino acid.

In 1902, Emil Fischer proposed that proteins were long chains of amino acids joined together by amide bonds formed between the α-carboxyl group of one amino acid and the α-amino group of another. For these amide bonds, Fischer proposed the special name **peptide bond**. Figure 19.5 shows the peptide bond formed between serine and alanine in the dipeptide serylalanine.

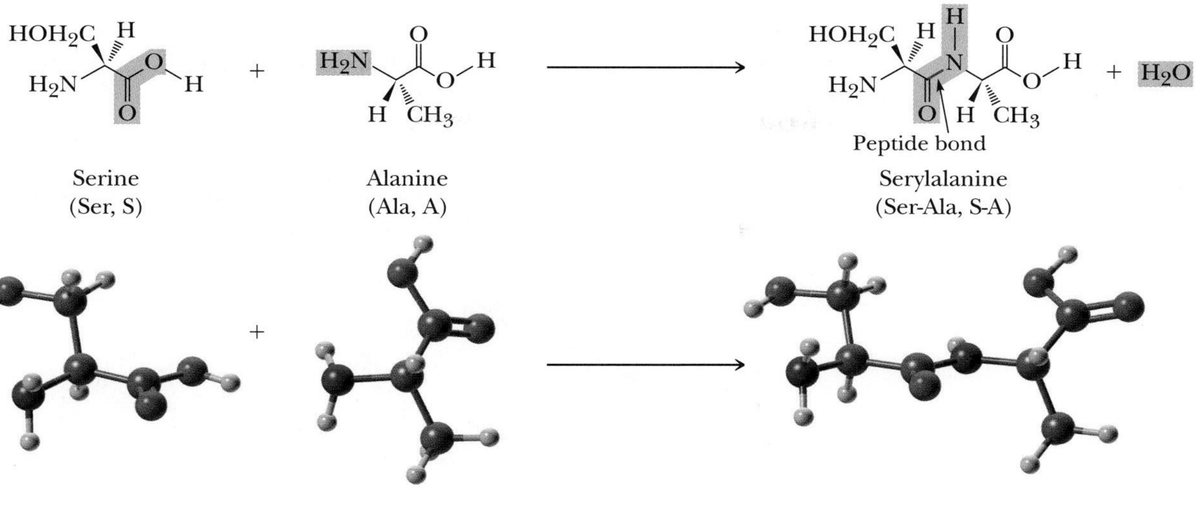

HOH$_2$C H; H$_2$N; O; H + H$_2$N; O; H; H CH$_3$ ⟶ HOH$_2$C H; H; N; O; H$_2$N; O; H; O; H CH$_3$ + H$_2$O

Peptide bond

Serine (Ser, S) — Alanine (Ala, A) — Serylalanine (Ser-Ala, S-A)

Figure 19.5
The peptide bond in serylalanine.

Peptide is the name given to a short polymer of amino acids. We classify peptides by the number of amino acid units in their chains. A molecule containing 2 amino acids joined by an amide bond is called a **dipeptide**. Those containing 3 to 10 amino acids are called **tripeptides**, **tetrapeptides**, **pentapeptides**, and so on. Molecules containing more than 10, but fewer than 20, amino acids are called **oligopeptides**. Those containing several dozen or more amino acids are called **polypeptides**. **Proteins** are biological macromolecules with molecular weight 5,000 or greater and consisting of one or more polypeptide chains. The distinctions in this terminology are not at all precise.

By convention, polypeptides are written from left to right, beginning with the amino acid having the free —NH_3^+ group and proceeding toward the amino acid with the free —COO^- group. The amino acid with the free —NH_3^+ group is called the ***N*-terminal amino acid**, and that with the free —COO^- group is called the ***C*-terminal amino acid**:

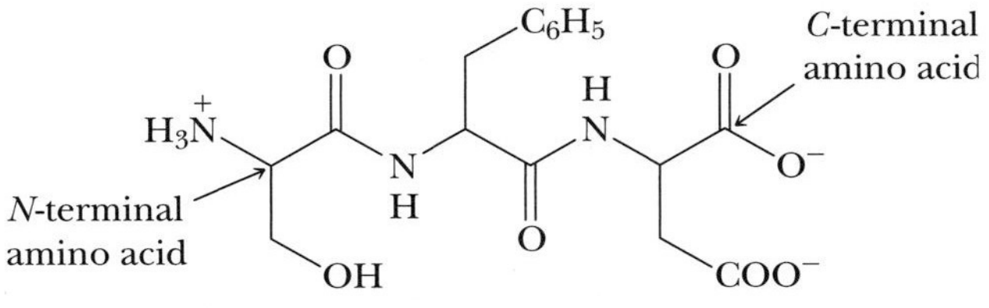

Dipeptide A molecule containing two amino acid units joined by a peptide bond.

Tripeptide A molecule containing three amino acid units, each joined to the next by a peptide bond.

Polypeptide A macromolecule containing 20 or more amino acid units, each joined to the next by a peptide bond.

***N*-Terminal amino acid** The amino acid at the end of a polypeptide chain having the free —NH_3^+ group.

***C*-Terminal amino acid** The amino acid at the end of a polypeptide chain having the free —COO^- group.

$H_3\overset{+}{N}$; O; N H; C_6H_5; O; H N; O; O^-; OH; COO^-

N-terminal amino acid — *C*-terminal amino acid

Ser-Phe-Asp

Example 19.4

Draw a structural formula for Cys-Arg-Met-Asn. Label the *N*-terminal amino acid and the *C*-terminal amino acid. What is the net charge on this tetrapeptide at pH 6.0?

Strategy

Begin by drawing the zwitterion form of each amino acid in order from cysteine to asparagine, each oriented with its α-ammonium group on the left and its α-carboxylate group on the right. Then form peptide bonds by removing a water molecule from between $—COO^-$ and $^+H_3N—$ groups that are next to each other. To determine the net charge on this tetrapeptide, consult Table 19.2 for the pK_a value of the ionizable group on the side chain of each amino acid.

Solution

The backbone of Cys-Arg-Met-Asn, a tetrapeptide, is a repeating sequence of nitrogen-α-carbon-carbonyl. The net charge on this tetrapeptide at pH 6.0 is +1. The following is a structural formula for Cys-Arg-Met-Asn:

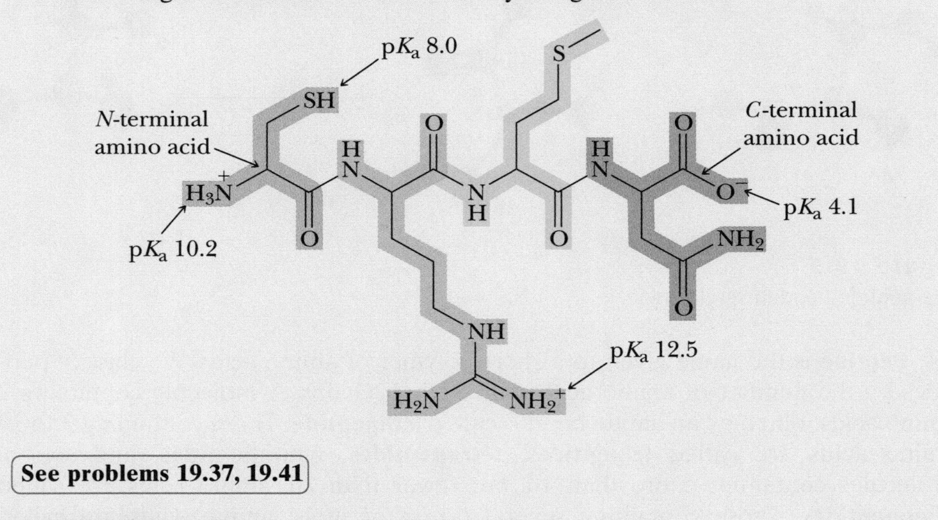

See problems 19.37, 19.41

Problem 19.4

Draw a structural formula for Lys-Phe-Ala. Label the *N*-terminal amino acid and the *C*-terminal amino acid. What is the net charge on this tripeptide at pH 6.0?

19.5 What Is the Primary Structure of a Polypeptide or Protein?

Primary (1°) structure of proteins The sequence of amino acids in the polypeptide chain; read from the *N*-terminal amino acid to the *C*-terminal amino acid.

The **primary (1°) structure** of a polypeptide or protein is the sequence of amino acids in its polypeptide chain. In this sense, the primary structure is a complete description of all covalent bonding in a polypeptide or protein.

In 1953, Frederick Sanger of Cambridge University, England, reported the primary structure of the two polypeptide chains of the hormone insulin. Not only was this a remarkable achievement in analytical chemistry, but also, it clearly established that the molecules of a given protein all have the same amino acid composition and the same amino acid sequence. Today, the amino acid sequences of over 20,000 different proteins are known, and the number is growing rapidly.

A. Amino Acid Analysis

The first step in determining the primary structure of a polypeptide is hydrolysis and quantitative analysis of its amino acid composition. Recall from Section 15.3D that amide bonds are highly resistant to hydrolysis. Typically, a sample of a protein is hydrolyzed in 6 M HCl in a sealed glass vial at 110°C for 24 to 72 hours.

(This hydrolysis can be done in a microwave oven in a shorter time.) After the polypeptide is hydrolyzed, the resulting mixture of amino acids is analyzed by ion-exchange chromatography. In this process, the mixture of amino acids is passed through a specially packed column. Each of the 20 amino acids requires a different time to pass through the column. Amino acids are detected by reaction with ninhydrin as they emerge from the column (Section 19.3D), followed by absorption spectroscopy. Current procedures for the hydrolysis of polypeptides and the analysis of amino acid mixtures have been refined to the point where it is possible to determine the amino acid composition from as little as 50 nanomoles (50×10^{-9} mole) of a polypeptide. Figure 19.6 shows the analysis of a polypeptide hydrolysate by ion-exchange chromatography. Note that, during hydrolysis, the side-chain amide groups of asparagine and glutamine are hydrolyzed, and these amino acids are detected as aspartic acid and glutamic acid. For each glutamine or asparagine hydrolyzed, an equivalent amount of ammonium chloride is formed.

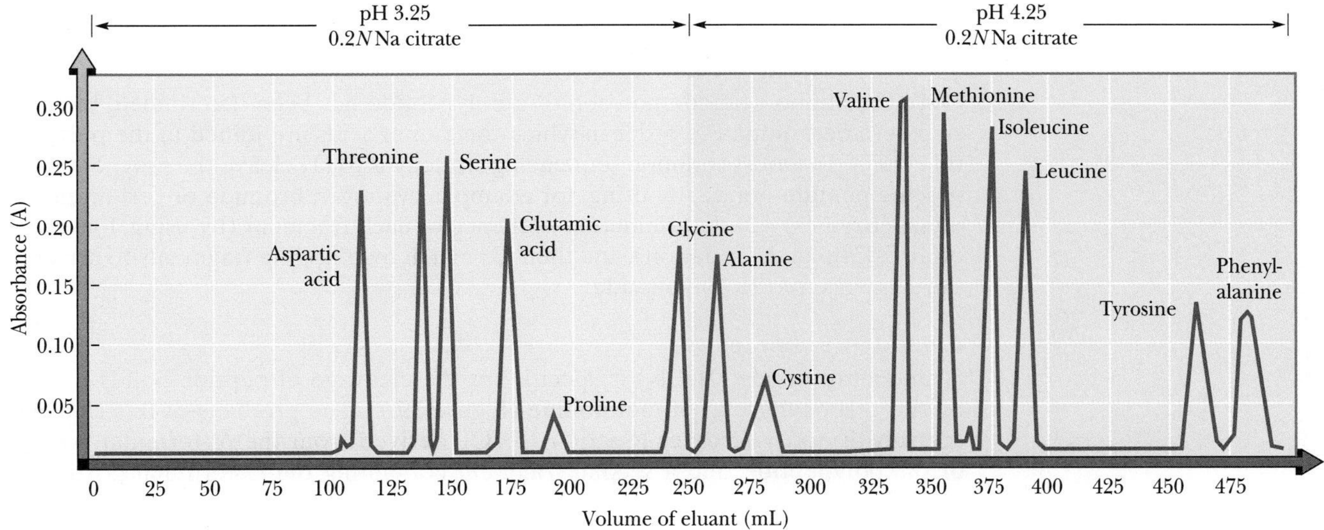

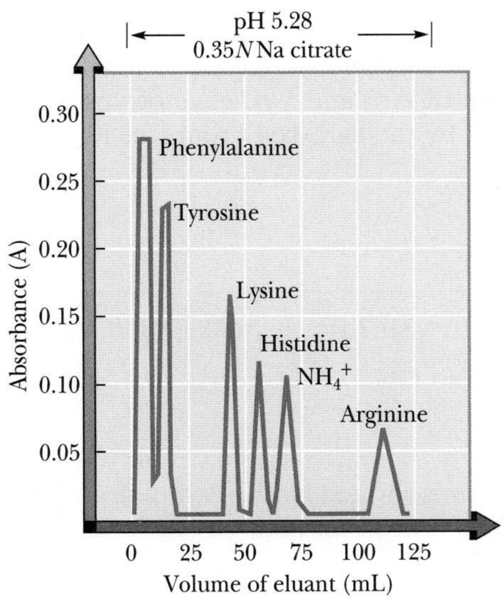

Figure 19.6
Analysis of a mixture of amino acids by ion-exchange chromatography using Amberlite IR-120, a sulfonated polystyrene resin. The resin contains phenyl-SO_3^- Na^+ groups. The amino acid mixture is applied to the column at low pH (3.25), under which conditions the acidic amino acids (Asp, Glu) are weakly bound to the resin and the basic amino acids (Lys, His, Arg) are tightly bound. Sodium citrate buffers of two different concentrations and three different values of pH are used to elute the amino acids from the column. Cysteine is determined as cystine, Cys-S-S-Cys, the disulfide of cysteine.

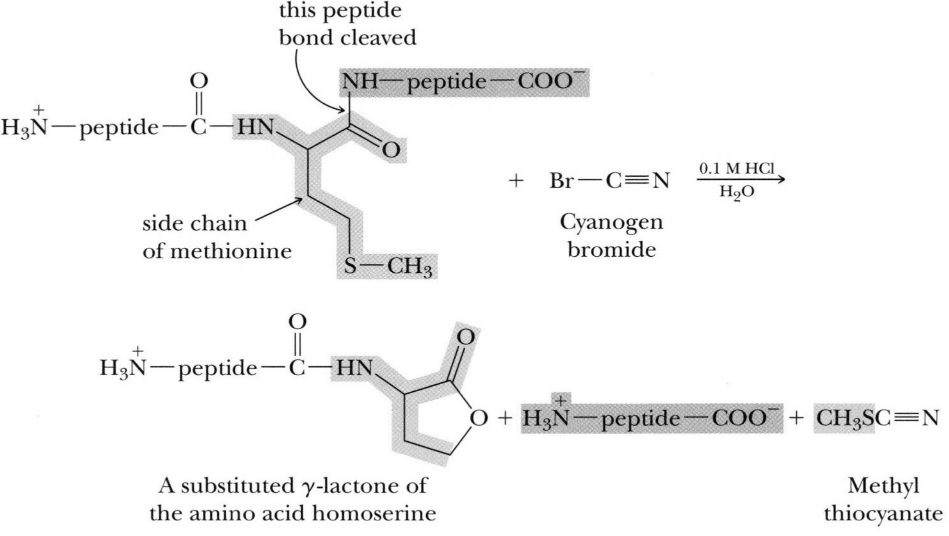

Figure 19.7 Cleavage by cyanogen bromide, BrCN, of a peptide bond formed by the carboxyl group of methionine.

B. Sequence Analysis

Once the amino acid composition of a polypeptide has been determined, the next step is to determine the order in which the amino acids are joined in the polypeptide chain. The most common sequencing strategy is to (1) cleave the polypeptide at specific peptide bonds (by using, for example, cyanogen bromide or certain proteolytic enzymes), (2) determine the sequence of each fragment (by using, for example, the Edman degradation), and then (3) match overlapping fragments to arrive at the sequence of the polypeptide.

Cyanogen Bromide

Cyanogen bromide (BrCN) is specific for the cleavage of peptide bonds formed by the carboxyl group of methionine (Figure 19.7). The products of this cleavage are substituted γ-lactones (Section 15.1C), derived from the *N*-terminal portion of the polypeptide, and a second fragment containing the *C*-terminal portion of the polypeptide.

Enzyme-Catalyzed Hydrolysis of Peptide Bonds

A group of proteolytic enzymes, including trypsin and chymotrypsin, can be used to catalyze the hydrolysis of specific peptide bonds. Trypsin catalyzes the hydrolysis of peptide bonds formed by the carboxyl groups of Arg and Lys; chymotrypsin catalyzes the hydrolysis of peptide bonds formed by the carboxyl groups of Phe, Tyr, and Trp.

Example 19.5

Which of these tripeptides are hydrolyzed by trypsin? By chymotrypsin?

(a) Arg-Glu-Ser (b) Phe-Gly-Lys

Strategy

Trypsin catalyzes the hydrolysis of peptide bonds formed by the carboxyl groups of Lys and Arg. Chymotrypsin catalyzes the hydrolysis of peptide bonds formed by the carboxyl groups of Phe, Tyr, and Trp.

Solution

(a) The peptide bond between Arg and Glu is hydrolyzed in the presence of trypsin.

$$\text{Arg-Glu-Ser} + H_2O \xrightarrow{\text{trypsin}} \text{Arg} + \text{Glu-Ser}$$

Because none of these three aromatic amino acids is present in tripeptide (a), it is not affected by chymotrypsin.

(b) Tripeptide (b) is not affected by trypsin. Although Lys is present, its carboxyl group is at the *C*-terminal end and not involved in peptide bond formation. Tripeptide (b) is hydrolyzed in the presence of chymotrypsin.

$$\text{Phe-Gly-Lys} + H_2O \xrightarrow{\text{chymotrypsin}} \text{Phe} + \text{Gly-Lys}$$

See problem 19.39

Problem 19.5

Which of these tripeptides are hydrolyzed by trypsin? By chymotrypsin?

(a) Tyr-Gln-Val (b) Thr-Phe-Ser (c) Thr-Ser-Phe

Edman Degradation

Of the various chemical methods developed for determining the amino acid sequence of a polypeptide, the one most widely used today is the **Edman degradation**, introduced in 1950 by Pehr Edman of the University of Lund, Sweden. In this procedure, a polypeptide is treated with phenyl isothiocyanate, $C_6H_5N{=}C{=}S$, and then with acid. The effect of Edman degradation is to remove the *N*-terminal amino acid selectively as a substituted phenylthiohydantoin (Figure 19.8), which is then separated and identified.

Edman degradation A method for selectively cleaving and identifying the *N*-terminal amino acid of a polypeptide chain.

The special value of the Edman degradation is that it cleaves the *N*-terminal amino acid from a polypeptide without affecting any other bonds in the chain. Furthermore, Edman degradation can be repeated on the shortened polypeptide, causing the next amino acid in the sequence to be cleaved and identified. In practice, it is possible to sequence as many as the first 20 to 30 amino acids in a polypeptide by this method, using as little as a few milligrams of material.

Most polypeptides in nature are longer than 20 to 30 amino acids, the practical limit on the number of amino acids that can be sequenced by repetitive Edman degradation. The special value of cleavage with cyanogen bromide, trypsin, and chymotrypsin is that, at specific peptide bonds, a long polypeptide chain can be cleaved into smaller polypeptide fragments, and each fragment can then be sequenced separately.

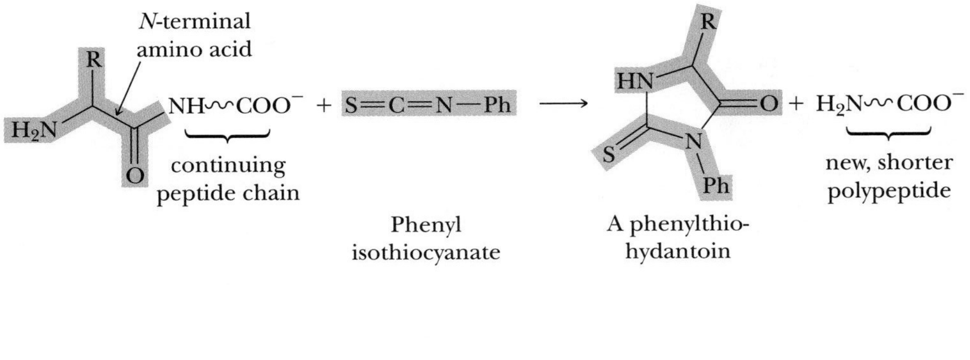

Figure 19.8 Edman degradation. Treatment of a polypeptide with phenyl isothiocyanate followed by acid selectively cleaves the *N*-terminal amino acid as a substituted phenylthiohydantoin.

Example 19.6

Deduce the amino acid sequence of a pentapeptide from the following experimental results (note that, under the column "Amino Acids Determined from Procedure," the amino acids are listed in alphabetical order; in no way does this listing give any information about primary structure):

Experimental Procedure	Amino Acids Determined from Procedure
Amino Acid Analysis of Pentapeptide	Arg, Glu, His, Phe, Ser
Edman Degradation	Glu
Hydrolysis Catalyzed by Chymotrypsin	
Fragment A	Glu, His, Phe
Fragment B	Arg, Ser
Hydrolysis Catalyzed by Trypsin	
Fragment C	Arg, Glu, His, Phe
Fragment D	Ser

Strategy

Review the specificity of each method of degradation:

- Edman degradation: Selectively cleaves the *N*-terminal amino acid.
- Chymotrypsin: Cleaves the peptide bonds formed by the carboxyl groups of Phe, Tyr, and Trp.
- Trypsin: Cleaves the peptide bonds formed by the carboxyl groups of Arg and Lys.

Solution

Edman degradation cleaves Glu from the pentapeptide; therefore, glutamic acid must be the *N*-terminal amino acid, and we have

Glu- (Arg, His, Phe, Ser)

Fragment A from chymotrypsin-catalyzed hydrolysis contains Phe. Because of the specificity of chymotrypsin, Phe must be the *C*-terminal amino acid of fragment A. Fragment A also contains Glu, which we already know is the *N*-terminal amino acid. From these observations, we conclude that the first three amino acids in the chain must be Glu-His-Phe, and we now write the following partial sequence:

Glu-His-Phe-(Arg, Ser)

The fact that trypsin cleaves the pentapeptide means that Arg must be within the pentapeptide chain; it cannot be the *C*-terminal amino acid. Therefore, the complete sequence must be

Glu-His-Phe-Arg-Ser

See problems 19.38, 19.40

Problem 19.6

Deduce the amino acid sequence of an undecapeptide (11 amino acids) from the experimental results shown in the following table:

Experimental Procedure	Amino Acids Determined from Procedure
Amino Acid Analysis of Undecapeptide	Ala, Arg, Glu, Lys_2, Met, Phe, Ser, Thr, Trp, Val
Edman Degradation	Ala
Trypsin-Catalyzed Hydrolysis	
Fragment E	Ala, Glu, Arg
Fragment F	Thr, Phe, Lys
Fragment G	Lys
Fragment H	Met, Ser, Trp, Val
Chymotrypsin-Catalyzed Hydrolysis	
Fragment I	Ala, Arg, Glu, Phe, Thr
Fragment J	Lys_2, Met, Ser, Trp, Val
Treatment with Cyanogen Bromide	
Fragment K	Ala, Arg, Glu, Lys_2, Met, Phe, Thr, Val
Fragment L	Trp, Ser

19.6 What Are the Three-Dimensional Shapes of Polypeptides and Proteins?

A. Geometry of a Peptide Bond

In the late 1930s, Linus Pauling began a series of studies aimed at determining the geometry of a peptide bond. One of his first discoveries was that a peptide bond is planar. As shown in Figure 19.9, the four atoms of a peptide bond and the two α-carbons joined to it all lie in the same plane.

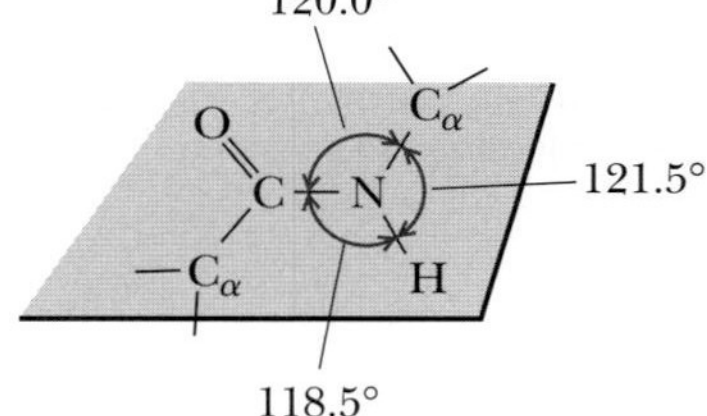

Figure 19.9
Planarity of a peptide bond. Bond angles about the carbonyl carbon and the amide nitrogen are approximately 120°.

Had you been asked in Chapter 1 to describe the geometry of a peptide bond, you probably would have predicted bond angles of 120° about the carbonyl carbon and 109.5° about the amide nitrogen.

Using VSEPR, predict bond angles of 120° about C and 109.5° about N.

An amide bond

This prediction agrees with the observed bond angles of approximately 120° about the carbonyl carbon. It does not agree, however, with the observed bond angles of 120° about the amide nitrogen. To account for the observed geometry, Pauling proposed that a peptide bond is more accurately represented as a resonance hybrid of these two contributing structures:

(1) (2)

Contributing structure (1) shows a carbon–oxygen double bond, and structure (2) shows a carbon–nitrogen double bond. The hybrid, of course, is neither of these.

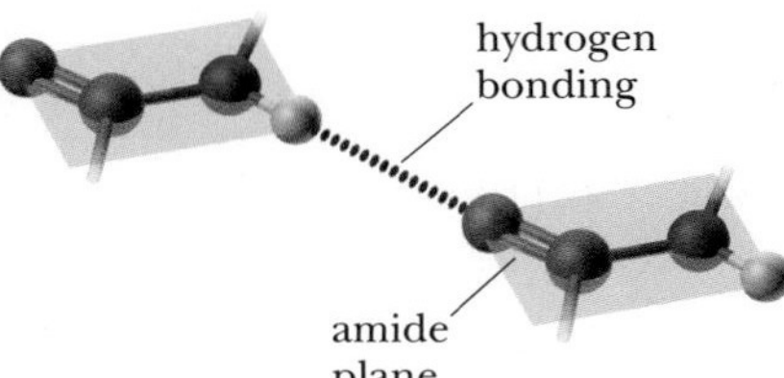

Figure 19.10
Hydrogen bonding between amide groups.

In the real structure, the carbon–nitrogen bond has considerable double-bond character. Accordingly, in the hybrid, the six-atom group of the peptide bond and the two attached α-carbons are planar.

Two configurations are possible for the atoms of a planar peptide bond. In one, the two α-carbons are *cis* to each other; in the other, they are *trans* to each other. The *trans* configuration is more favorable because the α-carbons with the bulky groups bonded to them are farther from each other than they are in the *cis* configuration. Virtually all peptide bonds in naturally occurring proteins studied to date have the *trans* configuration.

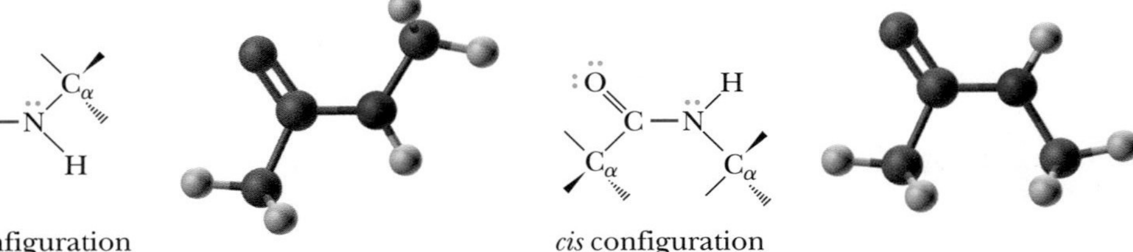

B. Secondary Structure

Secondary (2°) structure of proteins The ordered arrangements (conformations) of amino acids in localized regions of a polypeptide or protein.

Secondary (2°) structure is the ordered arrangement (conformation) of amino acids in localized regions of a polypeptide or protein molecule. The first studies of polypeptide conformations were carried out by Linus Pauling and Robert Corey, beginning in 1939. They assumed that, in conformations of greatest stability, all atoms in a peptide bond lie in the same plane and there is **hydrogen bonding** between the N—H of one peptide bond and the C=O of another, as shown in Figure 19.10.

On the basis of model building, Pauling proposed that two types of secondary structure should be particularly stable: the α-helix and the antiparallel β-pleated sheet.

The α-Helix

α-Helix A type of secondary structure in which a section of polypeptide chain coils into a spiral, most commonly a right-handed spiral.

In the **α-helix** pattern, shown in Figure 19.11, a polypeptide chain is coiled in a spiral. As you study this section of the α-helix, note the following:

1. The helix is coiled in a clockwise, or right-handed, manner. *Right-handed* means that if you turn the helix clockwise, it twists away from you. In this sense, a right-handed helix is analogous to the right-handed thread of a common wood or machine screw.

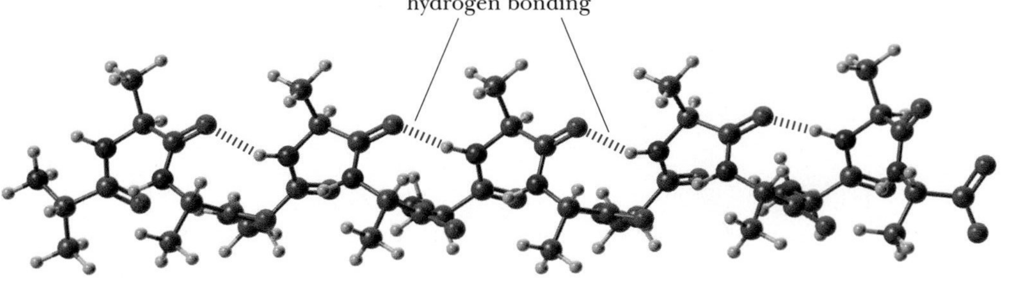

Figure 19.11
An α-helix. The polypeptide chain is repeating units of L-alanine.

2. There are 3.6 amino acids per turn of the helix.
3. Each peptide bond is *trans* and planar.
4. The N—H group of each peptide bond points roughly downward, parallel to the axis of the helix, and the C=O of each peptide bond points roughly upward, also parallel to the axis of the helix.
5. The carbonyl group of each peptide bond is hydrogen bonded to the N—H group of the peptide bond four amino acid units away from it. Hydrogen bonds are shown as dotted lines.
6. All R— groups point outward from the helix.

Almost immediately after Pauling proposed the α-helix conformation, other researchers proved the presence of α-helix conformations in keratin, the protein of hair and wool. It soon became obvious that the α-helix is one of the fundamental folding patterns of polypeptide chains.

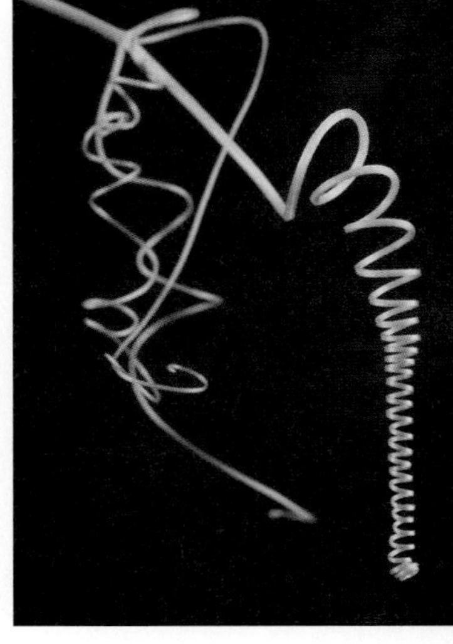

The star cucumber, *Sicyos angulatus,* uses left-handed helical tendrils to attach itself to climbing vines. Its helical pattern is analogous, but in reverse, to the right-handed α-helix of polypeptides. *(Grant Heilman Photography/Alamy)*

The β-Pleated Sheet

An antiparallel **β-pleated sheet** consists of an extended polypeptide chain with neighboring sections of the chain running in opposite (antiparallel) directions. In a parallel β-pleated sheet, the neighboring sections run in the same direction. Unlike the α-helix arrangement, N—H and C=O groups lie in the plane of the sheet and are roughly perpendicular to the long axis of the sheet. The C=O group of each peptide bond is hydrogen bonded to the N—H group of a peptide bond of a neighboring section of the chain (Figure 19.12).

As you study this section of β-pleated sheet in Figure 19.12 note the following:

1. The three sections of the polypeptide chain lie adjacent to each other and run in opposite (antiparallel) directions.
2. Each peptide bond is planar, and the α-carbons are *trans* to each other.
3. The C=O and N—H groups of peptide bonds from adjacent sections point at each other and are in the same plane, so that hydrogen bonding is possible between adjacent sections.
4. The R— groups on any one chain alternate, first above, then below, the plane of the sheet, and so on.

β-Pleated sheet A type of secondary structure in which two sections of polypeptide chain are aligned parallel or antiparallel to one another.

The β-pleated sheet conformation is stabilized by hydrogen bonding between N—H groups of one section of the chain and C=O groups of an adjacent section. By comparison, the α-helix is stabilized by hydrogen bonding between N—H and C=O groups within the same polypeptide chain.

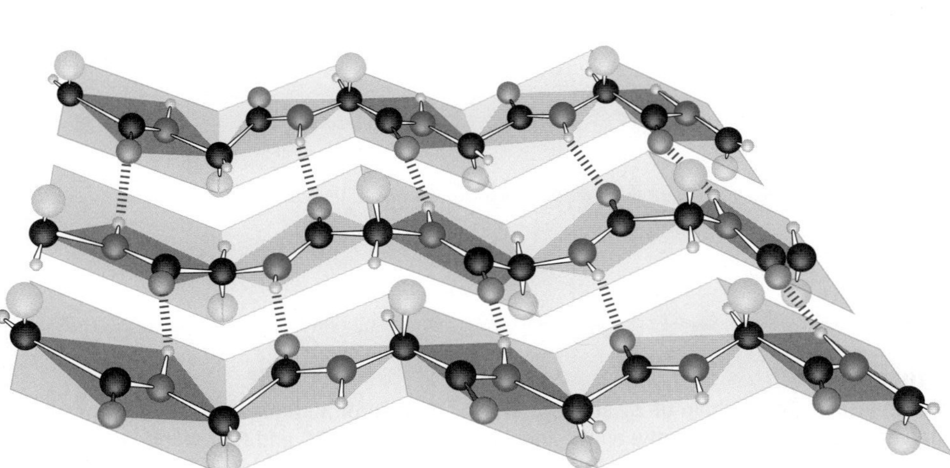

Figure 19.12
β-Pleated sheet conformation with three polypeptide chains running in opposite (antiparallel) directions. Hydrogen bonding between chains is indicated by dashed lines.

C. Tertiary Structure

Tertiary (3°) structure of proteins The three-dimensional arrangement in space of all atoms in a single polypeptide chain.

Tertiary (3°) structure is the overall folding pattern and arrangement in space of all atoms in a single polypeptide chain. No sharp dividing line exists between secondary and tertiary structures. Secondary structure refers to the spatial arrangement of amino acids *close to one another* on a polypeptide chain, whereas tertiary structure refers to the three-dimensional arrangement of *all* atoms in a polypeptide chain. Among the most important factors in maintaining 3° structure are disulfide bonds, hydrophobic interactions, hydrogen bonding, and salt bridges.

Disulfide bond A covalent bond between two sulfur atoms; an —S—S— bond.

Disulfide bonds play an important role in maintaining tertiary structure. Disulfide bonds are formed between side chains of two cysteine units by oxidation of their thiol groups (—SH) to form a disulfide bond (Section 8.6B). Treatment of a disulfide bond with a reducing agent regenerates the thiol groups:

side chains of cysteine
SH
SH
oxidation
reduction
S—S
a disulfide bond

Figure 19.13 shows the amino acid sequence of human insulin. This protein consists of two polypeptide chains: an A chain of 21 amino acids and a B chain of 30 amino acids. The A chain is bonded to the B chain by two interchain disulfide bonds. An intrachain disulfide bond also connects the cysteine units at positions 6 and 11 of the A chain.

As an example of 2° and 3° structure, let us look at the three-dimensional structure of myoglobin, a protein found in skeletal muscle and particularly abundant in diving mammals, such as seals, whales, and porpoises. Myoglobin and its structural relative, hemoglobin, are the oxygen transport and storage molecules of vertebrates. Hemoglobin binds molecular oxygen in the lungs and transports it to myoglobin in muscles. Myoglobin stores molecular oxygen until it is required for metabolic oxidation.

Myoglobin consists of a single polypeptide chain of 153 amino acids. Myoglobin also contains a single heme unit. Heme consists of one Fe^{2+} ion, coordinated in a square planar array with the four nitrogen atoms of a molecule of porphyrin (Figure 19.14).

Figure 19.13
Human insulin. The A chain of 21 amino acids and B chain of 30 amino acids are connected by interchain disulfide bonds between A7 and B7 and between A20 and B19. In addition, a single intrachain disulfide bond occurs between A6 and A11.

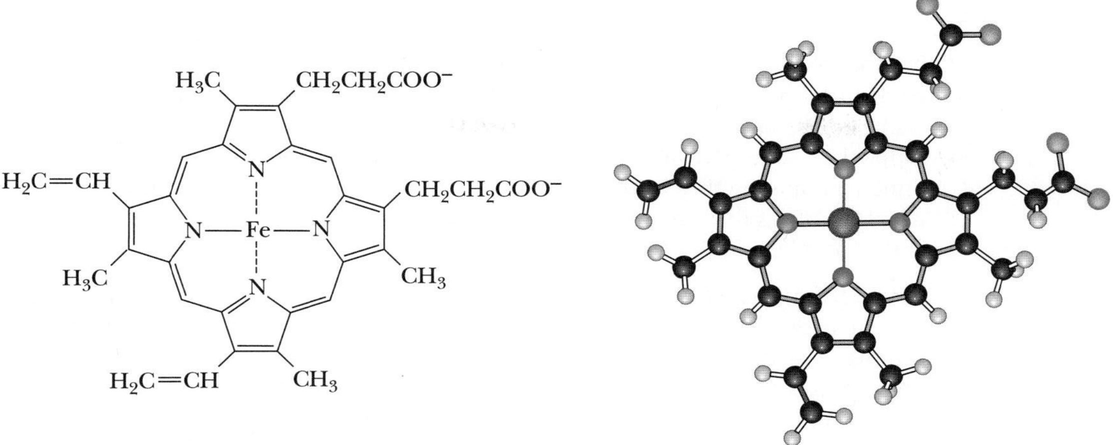

Figure 19.14
The structure of heme, found in myoglobin and hemoglobin.

Determination of the three-dimensional structure of myoglobin represented a milestone in the study of molecular architecture. For their contribution to this research, John C. Kendrew and Max F. Perutz, both of Britain, shared the 1962 Nobel Prize for Chemistry. The secondary and tertiary structures of myoglobin are shown in Figure 19.15. The single polypeptide chain is folded into a complex, almost boxlike shape.

Important structural features of the three-dimensional shape of myoglobin are as follows:

1. The backbone consists of eight relatively straight sections of α-helix, each separated by a bend in the polypeptide chain. The longest section of α-helix has 24 amino acids, the shortest has seven. Some 75% of the amino acids are found in these eight regions of α-helix.
2. Hydrophobic side chains of phenylalanine, alanine, valine, leucine, isoleucine, and methionine are clustered in the interior of the molecule, where they are shielded from contact with water. **Hydrophobic interactions** are a major factor in directing the folding of the polypeptide chain of myoglobin into this compact, three-dimensional shape.

The humpback whale relies on myoglobin as a storage form of oxygen. *(Stuart Westmorland/Stone/Getty Images)*

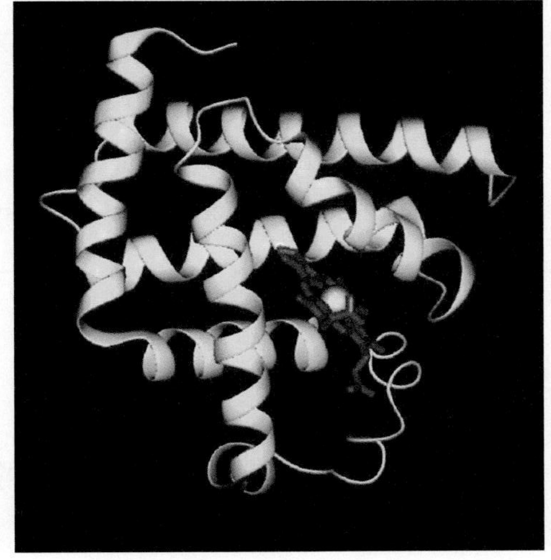

Figure 19.15
Ribbon model of myoglobin. The polypeptide chain is shown in yellow, the heme ligand in red, and the Fe atom as a white sphere.

3. The outer surface of myoglobin is coated with hydrophilic side chains, such as those of lysine, arginine, serine, glutamic acid, histidine, and glutamine, which interact with the aqueous environment by **hydrogen bonding**. The only polar side chains that point to the interior of the myoglobin molecule are those of two histidine units, which point inward toward the heme group.
4. Oppositely charged amino acid side chains close to each other in the three-dimensional structure interact by electrostatic attractions called **salt bridges**. An example of a salt bridge is the attraction of the side chains of lysine ($—NH_3^+$) and glutamic acid ($—COO^-$).

The tertiary structures of hundreds of proteins have also been determined. It is clear that proteins contain α-helix and β-pleated sheet structures, but that wide variations exist in the relative amounts of each. Lysozyme, with 129 amino acids in a single polypeptide chain, has only 25% of its amino acids in α-helix regions. Cytochrome, with 104 amino acids in a single polypeptide chain, has no α-helix structure but does contain several regions of β-pleated sheet. Yet, whatever the proportions of α-helix, β-pleated sheet, or other periodic structure, virtually all nonpolar side chains of water-soluble proteins are directed toward the interior of the molecule, whereas polar side chains are on the surface of the molecule, in contact with the aqueous environment.

Example 19.7

With which of the following amino acid side chains can the side chain of threonine form hydrogen bonds?

(a) Valine (b) Asparagine (c) Phenylalanine
(d) Histidine (e) Tyrosine (f) Alanine

Strategy

Analyze the types of side chains of these amino acids and then look for potential interactions between them by hydrogen bonding.

Solution

The side chain of threonine contains a hydroxyl group that can participate in hydrogen bonding in two ways: (1) Its oxygen has a partial negative charge and can function as a hydrogen bond acceptor; (2) its hydrogen has a partial positive charge and can function as a hydrogen bond donor. Therefore, the side chain of threonine can form hydrogen bonds with the side chains of tyrosine, asparagine, and histidine.

Problem 19.7

At pH 7.4, with what amino acid side chains can the side chain of lysine form salt bridges?

D. Quaternary Structure

Quaternary (4°) structure of proteins The arrangement of polypeptide monomers into a noncovalently bonded aggregation.

Most proteins with molecular weight greater than 50,000 consist of two or more noncovalently linked polypeptide chains. The arrangement of protein monomers into an aggregation is known as **quaternary (4°) structure**. A good example is hemoglobin (Figure 19.16), a protein that consists of four separate polypeptide

CHEMICAL CONNECTIONS 19A

Spider Silk: A Chemical and Engineering Wonder of Nature

Many of society's technological innovations have been inspired by nature. Velcro, for example, is modeled after plant burrs. The water-repellent swimsuits that revolutionized the sport of swimming were modeled after the skin of sharks. And hundreds of medicines are based on natural products. However, one product of nature that for centuries has been difficult to harness or imitate is spider silk. A strand of spider silk is almost five times stronger than a strand of steel of the same diameter. In addition to its strength, a strand of spider silk can be stretched up to 30–40% of its length without breaking. As shown in the following graphic of its secondary structure, a strand of spider silk consists of oriented amorphous regions (**A**), crystalline regions (**B**), and completely amorphous regions (**C**). The oriented amorphous regions are held together by hydrogen bonds and give spider silk its elasticity. The crystalline regions are mostly responsible for the strength of spider silk. They consist of β-pleated sheets and are highly hydrophobic, which also makes spider silk insoluble in water and resistant to rain and dew.

For centuries, spider silk has been highly sought after for its properties. Unfortunately, spiders are solitary animals and, unlike silkworms, cannot be domesticated. This leaves only one option for harnessing the utility of spider silk: its reproduction using artificial means. Research thus far has revealed the detailed structure of spider silk. However, it is a testament to the elegance of nature that despite the fact that spider silk consists mostly of alanine and glycine, researchers have yet to find a way to assemble a strand of spider silk in the laboratory. Incidentally, after a spider web has lost its stickiness, most spiders "recycle" the protein by eating their webs, leaving nary a trace of this chemical and engineering wonder of nature.

QUESTIONS

Draw two pentapeptide strands of polyalanine and show how they can hydrogen bond to each other.

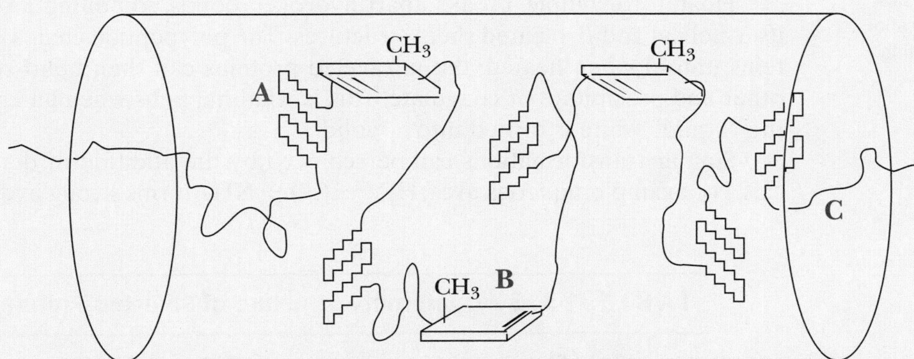

chains: two α-chains of 141 amino acids each and two β-chains of 146 amino acids each.

The major factor stabilizing the aggregation of protein subunits is the **hydrophobic effect**. When separate polypeptide chains fold into compact three-dimensional shapes to expose polar side chains to the aqueous environment and shield nonpolar side chains from water, hydrophobic "patches" may still appear on the surface, in contact with water. These patches can be shielded from water if two

Hydrophobic effect The tendency of nonpolar groups to cluster in such a way as to be shielded from contact with an aqueous environment.

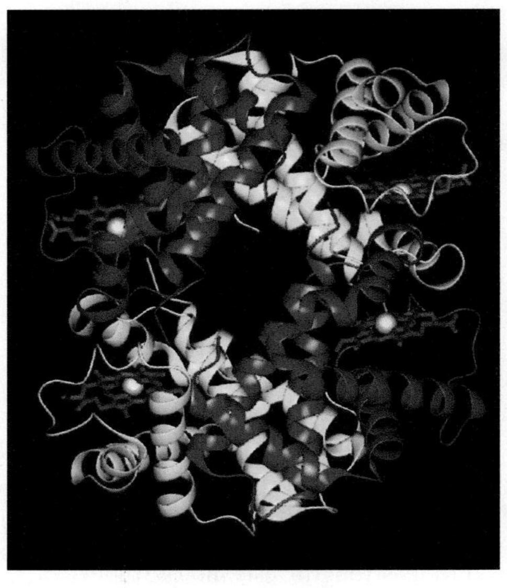

Figure 19.16 Ribbon model of hemoglobin. The α-chains are shown in purple, the β-chains in yellow, the heme ligands in red, and the Fe atoms as white spheres.

or more monomers assemble so that their hydrophobic patches are in contact. The numbers of subunits of several proteins of known quaternary structure are shown in Table 19.3.

19.7 What Is Denaturation?

The functions and properties of a protein arise from a combination of its secondary, tertiary, and quaternary structures that give the protein its particular shape and conformation. Any physical or chemical agent that interferes with these stabilizing structures changes the conformation of the protein and often removes the protein's functionality. We call this process **denaturation**.

Denaturation The loss of secondary, tertiary, and quaternary structure of a protein by a chemical or physical agent and the resulting loss of function.

Heat, for example, breaks apart hydrogen bonds, so boiling a protein destroys its α-helical and β-pleated sheet structures. The polypeptide chains of globular proteins unfold when heated; the unraveled proteins can then bond strongly to each other and precipitate or coagulate. This is what happens when an egg is boiled and the "liquid" white is turned into a "solid."

Similar transformations can be achieved by the addition of denaturing chemicals. For example, aqueous urea, $H_2N—CO—NH_2$, forms strong hydrogen bonds, so

TABLE 19.3 Quaternary Structure of Selected Proteins

Protein	Number of Subunits
alcohol dehydrogenase	2
aldolase	4
hemoglobin	4
lactate dehydrogenase	4
insulin	6
glutamine synthetase	12
tobacco mosaic virus protein disc	17

TABLE 19.4 Protein Denaturing Agents and Their Mode or Site of Action

Denaturing Agent	Affected Groups or Region
heat	hydrogen bonding
6 M urea	hydrogen bonding
detergents	hydrophobic interactions
acids/bases	salt bridges and hydrogen bonding
salts	salt bridges
reducing agents	disulfide bonds
heavy metals	thiol groups
alcohols	hydration layers

it can disrupt other hydrogen bonds and cause globular proteins to unfold. Ethanol denatures proteins by coagulation. Detergents change protein conformation by disturbing the hydrophobic interactions, whereas acids, bases, and salts interfere with ionic salt bridges in the tertiary and quaternary structures. Other chemical agents such as reducing agents can break the disulfide bonds (—S—S) that hold tertiary structures together. In addition, heavy metal ions (Pb^{2+}, Hg^{2+}, Cd^{2+}, etc.) attack thiol (—S—H) groups to form salt bridges. Table 19.4 outlines some common denaturing factors and the regions of the protein that are affected.

Key Terms and Concepts

α-amino acid (p. 688)
amino acid (p. 688)
chirality of α-amino acids (p. 688)
***C*-terminal amino acid** (p. 699)
denaturation (p. 712)
dipeptide (p. 699)
disulfide bond (p. 708)
Edman degradation (p. 703)
electrophoresis (p. 697)
geometry of a peptide bond (p. 705)
α-helix (p. 706)
hydrogen bonding (p. 706)
hydrophobic effect (p. 711)
hydrophobic interaction (p. 709)
isoelectric point (p. 695)
***N*-terminal amino acid** (p. 699)
peptide bond (p. 698)
β-pleated sheet (p. 707)
polypeptide (p. 699)
primary (1°) structure (p. 700)
protein (p. 699)
quaternary (4°) structure (p. 710)
salt bridge (p. 710)
secondary (2°) structure (p. 706)
tertiary (3°) structure (p. 708)
tetrapeptide (p. 699)
tripeptide (p. 699)
zwitterion (p. 688)

Summary of Key Questions

19.1 What Are the Many Functions of Proteins?

Proteins have many roles in growth and metabolism, among which are:

- Structural (collagen)
- Catalytic (trypsin and other digestive enzymes)
- Transport (hemoglobin)
- Movement (myosin and actin)
- Protection (immuloglobulins)
- Hormonal (insulin)

19.2 What Are Amino Acids?

- **α-Amino acids** are compounds that contain an amino group alpha to a carboxyl group.
- Each amino acid has an acid (a —COOH group) and a base (an $—NH_2$ group) that undergo an acid–base reaction to form an internal salt given the special **zwitterion**. A zwitterion has no net charge because it contains one positive charge and one negative charge.
- With the exception of glycine, all protein-derived amino acids are chiral.
- Whereas most monosaccharides in the biological world have the D-configuration, the vast majority of naturally occurring α-amino acids have the L-configurations at the α-carbon. D-amino acids are rare.
- Isoleucine and threonine contain a second stereocenter, and four stereoisomers are possible for each.
- The 20 protein-derived amino acids are commonly divided into four categories: nine with nonpolar side chains, four with polar but un-ionized side chains, four with acidic side chains, and three with basic side chains.

19.3 What Are the Acid–Base Properties of Amino Acids?

- Amino acids are weak polyprotic acids because of their —COOH and $—NH_3^+$ groups.
- The average value of pK_a for an α-carboxyl group of a protonated amino acids is 2.19. Thus the α-carboxyl group is a considerably stronger acid than the carboxyl group of acetic acid (pK_a 4.76), a fact due to the electron-withdrawing inductive effect of the nearby $—NH_3^+$ group of the α-amino acid.
- The average value of pK_a for an α-ammonium group is 9.47, compared to an average value of 10.76 for a primary aliphatic ammonium ion. Thus, the α-ammonium group of an amino acid is a slightly stronger acid than a primary aliphatic amine.
- The side-chain guanidine group of arginine is a considerably stronger base than an aliphatic amine. This remarkable basicity is attributed to the large resonance stabilization of the protonated form relative to the neutral form.
- The **isoelectric point, pI**, of an amino acid, polypeptide, or protein is the pH at which the majority of its molecules have no net charge.
- **Electrophoresis** is the process of separating compounds on the basis of their electric charge. Compounds with a higher charge density move more rapidly than those with a lower charge density.
- Any amino acid or protein in a solution with a pH that equals the pI of the compound remains at the origin. One with a net negative charge moves toward the positive electrode, and one with a net positive charge moves toward the negative electrode.

19.4 What Are Polypeptides and Proteins?

- A **peptide bond** is the special name given to the amide bond formed between α-amino acids.
- A **polypeptide** is a biological macromolecule containing 20 or more amino acids joined by peptide bonds.
- By convention, the sequence of amino acids in a polypeptide is written from the ***N*-terminal amino acid** toward the ***C*-terminal amino acid**.
- A **peptide bond** is planar; that is, the four atoms of the amide bond and the two α-carbons bonded to it lie in the same plane.
- Bond angles about the amide nitrogen and the carbonyl carbon of a peptide bond are approximately 120°.

19.5 What Is the Primary Structure of a Polypeptide or Protein?

- The **primary (1°) structure** of a polypeptide or protein refers to the sequence of amino acids in its polypeptide chain.
- The first step in determination of primary structure is hydrolysis and quantitative analysis of amino acid composition by **ion-exchange chromatography**.
- **Cyanogen bromide** is specific for the cleavage of peptide bonds formed by the carboxyl group of methionine.
- **Trypsin** catalyzes the hydrolysis of peptide bonds formed by the carboxyl groups of arginine and lysine.

- **Chymotrypsin** catalyzes the hydrolysis of peptide bonds formed by the carboxyl groups of phenylalanine, tyrosine, and tryptophan.
- The **Edman degradation** selectively cleaves the *N*-terminal amino acid without affecting any other peptide bonds in a polypeptide or protein.

19.6 What Are the Three-Dimensional Shapes of Polypeptides and Proteins?

- The four atoms of a peptide bond and the two α-carbons bonded to it all lie in the same plane; that is, a peptide bond is planar.
- **Secondary (2°) structure** refers to the ordered arrangement (conformations) of amino acids in localized regions of a polypeptide or protein. The two most prevalent types of secondary structure are the α-helix and the β-pleated sheet, both of which are stabilized by hydrogen bonding.
- In an **α-helix**, the carbonyl group of each peptide bond is hydrogen-bonded to the N—H group of the peptide bond four amino acids away from it.
- In an **antiparallel β-pleated sheet**, neighboring sections of a polypeptide chain run in opposite (antiparallel) directions, and the C=O group of each peptide bond is hydrogen-bonded to the N—H group of a peptide bond in a section of the neighboring antiparallel chain.
- In a **parallel β-sheet**, the neighboring sections of the polypeptide chain run in the same (parallel) directions, and the C=O of each peptide bond is hydrogen-bonded to the N—H group of a peptide bond in a neighboring section of the chain.
- **Tertiary (3°) structure** refers to the overall folding pattern and arrangement in space of all atoms in a single polypeptide chain.
- **Quaternary (4°) structure** is the arrangement of individual polypeptide chains into a noncovalently bonded aggregate. A major factor stabilizing quaternary structure is hydrophobic interaction created when separate polypeptide chains fold into compact three-dimensional shapes that expose their polar side chains to the aqueous environment and shield their nonpolar side chains from the aqueous environment. Any remaining exposed hydrophobic patches can be shielded from water if two or more polypeptide chains assemble so that their hydrophobic patches are in contact.

19.7 What Is Denaturation?

- **Denaturation** is the loss of a protein's properties due to the application of chemical or physical conditions that disrupt its secondary, tertiary, or quaternary structures.

Quick Quiz

Answer true or false to the following questions to assess your general knowledge of the concepts in this chapter. If you have difficulty with any of them, you should review the appropriate section in the chapter (shown in parentheses) before attempting the more challenging end-of-chapter problems.

1. The isoelectric point of an amino acid is the pH at which the majority of molecules in solution have a net charge of -1. (19.3)
2. Proteins can protect an organism against disease. (19.1)
3. Titration of an amino acid can be used to determine both the pK_a of its ionizable groups and its isoelectric point. (19.3)
4. Hydrogen bonding, salt bridges, hydrophobic interactions, and disulfide bonds can each be categorized as stabilizing factors in a protein. (19.6)
5. A polypeptide chain is read from its *C*-terminal end to its *N*-terminal end. (19.5)
6. The majority of naturally occurring amino acids are from the D-series. (19.2)
7. The amino group of an α-amino acid is more basic than the amino group of an aliphatic amine. (19.3)
8. Lysine contains a basic side chain. (19.2)
9. Denaturation is the process of creating a synthetic protein. (19.7)
10. A peptide bond exhibits free rotation at room temperature. (19.6)
11. In Edman degradation, a polypeptide is shortened one amino acid at a time using the reagent phenyl isothiocyanate. (19.5)

12. Phenylalanine contains a polar side chain. (19.2)
13. The side chain of arginine shows enhanced basicity because of resonance stabilization of the ion that results after protonation. (19.3)
14. α-Helices and β-pleated sheets are examples of the tertiary structure of a protein. (19.6)
15. The majority of amino acids in proteins are β-amino acids. (19.2)
16. Proteins can act as catalysts in chemical reactions. (19.1)
17. In electrophoresis, species with a net negative charge will move toward the negative electrode. (19.3)
18. All naturally occurring amino acids are chiral. (19.2)
19. Cyanogen bromide, trypsin, and chymotrypsin each act to cleave peptide bonds at specific amino acids. (19.5)
20. The carboxyl group of an α-amino acid is more acidic than the carboxyl group of an aliphatic carboxylic acid. (19.3)
21. The amino acid sequence of a protein or polypeptide is known as its secondary structure. (19.5)
22. The 20 common, naturally occurring amino acids can be represented by both three- and one-letter abbreviations. (19.2)
23. The side chain of histidine shows enhanced basicity because of the electron-withdrawing inductive effects that stabilize the ion that results after protonation. (19.3)
24. The quaternary structure of a protein describes how smaller, individual protein strands interact to form the overall structure of a protein. (19.6)
25. An amino acid with a net charge of +1 is classified as a zwitterion. (19.2)
26. The side chain in serine is polar and can undergo hydrogen bonding. (19.2)

Answers: (1) F (2) T (3) T (4) T (5) F (6) F (7) F (8) T (9) F (10) F (11) T (12) F (13) T (14) F (15) F (16) T (17) F (18) F (19) T (20) T (21) F (22) T (23) F (24) T (25) F (26) T

Key Reactions

1. Acidity of an α-Carboxyl Group (Section 19.3A)

An α-COOH (pK_a approximately 2.19) of a protonated amino acid is a considerably stronger acid than acetic acid (pK_a 4.76) or other low-molecular-weight aliphatic carboxylic acid, due to the electron-withdrawing inductive effect of the α-NH_3^+ group:

$$RCH(NH_3^+)COOH + H_2O \rightleftharpoons RCH(NH_3^+)COO^- + H_3O^+ \quad pK_a = 2.19$$

2. Acidity of an α-Ammonium Group (Section 19.3A)

An α-NH_3^+ group (pK_a approximately 9.47) is a slightly stronger acid than a primary aliphatic ammonium ion (pK_a approximately 10.76):

$$RCH(NH_3^+)COO^- + H_2O \rightleftharpoons RCH(NH_2)COO^- + H_3O^+ \quad pK_a = 9.47$$

3. Reaction of an α-Amino Acid with Ninhydrin (Section 19.3D)

Treating an α-amino acid with ninhydrin gives a purple-colored solution:

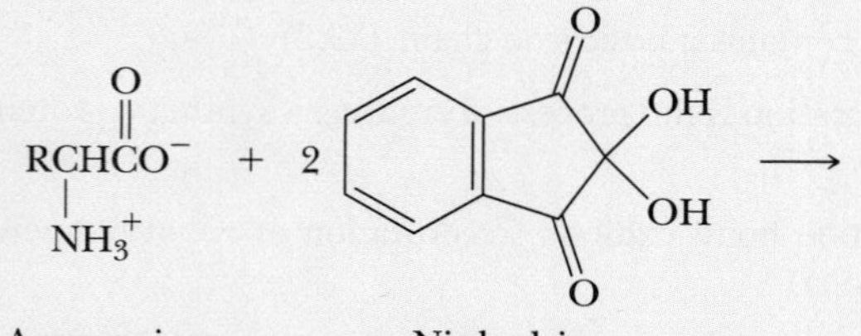

$$RCH(NH_3^+)CO_2^- + 2\ \text{Ninhydrin} \longrightarrow \text{Purple-colored anion} + RCHO + CO_2 + H_3O^+$$

An α-amino acid

Ninhydrin

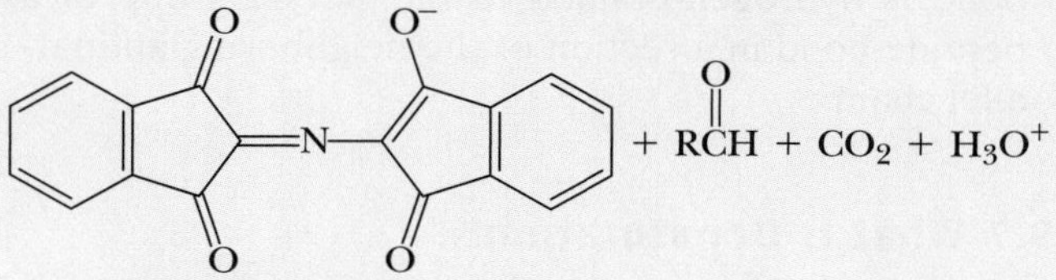

Purple-colored anion

Treating proline with ninhydrin gives an orange-colored solution.

4. Cleavage of a Peptide Bond by Cyanogen Bromide (Section 19.5B)

Cleavage is regioselective for a peptide bond formed by the carboxyl group of methionine:

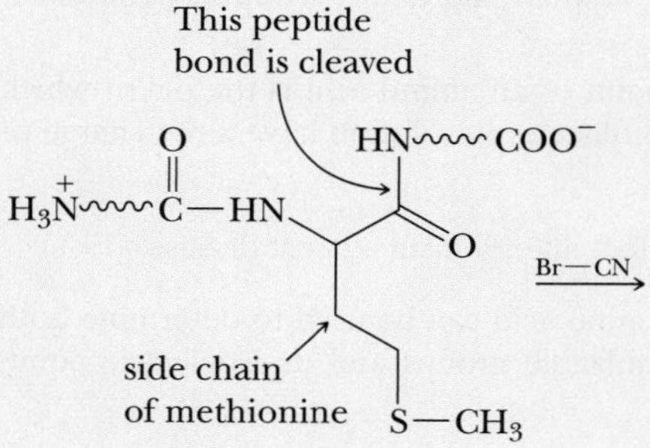

This peptide bond is cleaved

$H_3\overset{+}{N}$~~~C(=O)—HN—CH(side chain)—C(=O)—HN~~~COO^- $\xrightarrow{Br-CN}$

side chain of methionine $S-CH_3$

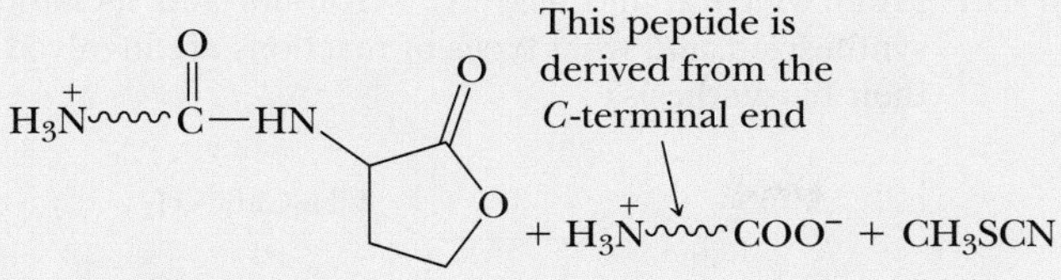

$H_3\overset{+}{N}$~~~$\overset{O}{\overset{\|}{C}}$—HN—(γ-lactone ring) + $H_3\overset{+}{N}$~~~COO^- + CH_3SCN

This peptide is derived from the *C*-terminal end

A substituted γ-lactone of the amino acid homoserine

5. Edman Degradation (Section 19.5B)
Treatment with phenyl isothiocyanate followed by acid removes the *N*-terminal amino acid as a substituted phenylthiohydantoin, which is then separated and identified:

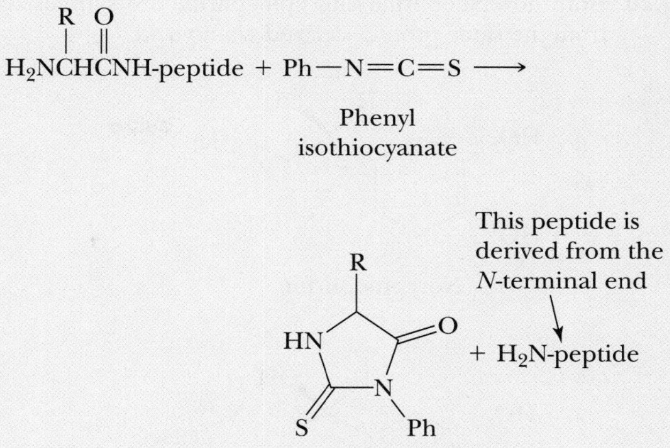

$H_2NCH(R)C(=O)NH$-peptide + Ph—N=C=S ⟶

Phenyl isothiocyanate

A phenylthiohydantoin + H_2N-peptide

This peptide is derived from the *N*-terminal end

Problems

A problem marked with an asterisk indicates an applied "real world" problem. Answers to problems whose numbers are printed in blue are given in Appendix D.

Section 19.2 Amino Acids

19.8 What amino acid does each abbreviation stand for?
(a) Phe (b) Ser (c) Asp (d) Gln
(e) His (f) Gly (g) Tyr

19.9 The configuration of the stereocenter in α-amino acids is most commonly specified using the D,L convention. The configuration can also be identified using the R,S convention (Section 6.3). Does the stereocenter in L-serine have the R or the S configuration?

19.10 Assign an R or S configuration to the stereocenter in each amino acid:
(a) L-Phenylalanine
(b) L-Glutamic acid
(c) L-Methionine

19.11 The amino acid threonine has two stereocenters. The stereoisomer found in proteins has the configuration 2S,3R about the two stereocenters. Draw a Fischer projection of this stereoisomer and also a three-dimensional representation.

19.12 Define the term *zwitterion*.

19.13 Draw zwitterion forms of these amino acids:
(a) Valine
(b) Phenylalanine
(c) Glutamine

19.14 Why are Glu and Asp often referred to as acidic amino acids? **(See Example 19.1)**

19.15 Why is Arg often referred to as a basic amino acid? Which two other amino acids are also referred to as basic amino acids? **(See Example 19.1)**

19.16 What is the meaning of the alpha as it is used in α-amino acid?

***19.17** Several β-amino acids exist. A unit of β-alanine, for example, is contained within the structure of coenzyme A (Section 22.1D). Write the structural formula of β-alanine.

***19.18** Although only L-amino acids occur in proteins, D-amino acids are often a part of the metabolism of lower organisms. The antibiotic actinomycin D, for example, contains a unit of D-valine, and the antibiotic bacitracin A contains units of D-asparagine and D-glutamic acid. Draw Fischer projections and three-dimensional representations for these three D-amino acids.

***19.19** Histamine is synthesized from one of the 20 protein-derived amino acids. Suggest which amino acid is the biochemical precursor of histamine, and name the type of organic reaction(s) (e.g., oxidation, reduction, decarboxylation, nucleophilic substitution) involved in its conversion to histamine.

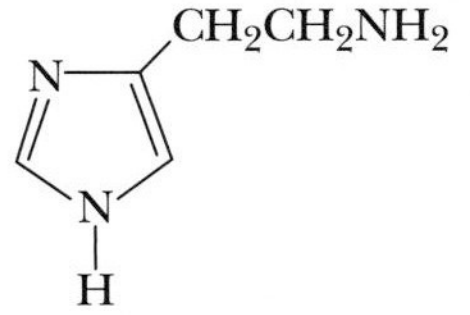

$CH_2CH_2NH_2$ (imidazole ring, N—H)

Histamine

***19.20** Both norepinephrine and epinephrine are synthesized from the same protein-derived amino acid:

(a) Norepinephrine

(b) Epinephrine (Adrenaline)

From which amino acid are the two compounds synthesized, and what types of reactions are involved in their biosynthesis?

***19.21** From which amino acid are serotonin and melatonin synthesized, and what types of reactions are involved in their biosynthesis?

(a) Serotonin

(b) Melatonin

Section 19.3 Acid–Base Behavior of Amino Acids

19.22 Draw a structural formula for the form of each amino acid most prevalent at pH 1.0 **(See Example 19.2)**:

(a) Threonine (b) Arginine
(c) Methionine (d) Tyrosine

19.23 Draw a structural formula for the form of each amino acid most prevalent at pH 10.0 **(See Example 19.2)**:

(a) Leucine (b) Valine
(c) Proline (d) Aspartic acid

19.24 Write the zwitterion form of alanine and show its reaction with:

(a) 1.0 mol NaOH (b) 1.0 mol HCl

19.25 Write the form of lysine most prevalent at pH 1.0, and then show its reaction with each of the following (consult Table 19.2 for pK_a values of the ionizable groups in lysine) **(See Example 19.2)**:

(a) 1.0 mol NaOH (b) 2.0 mol NaOH
(c) 3.0 mol NaOH

19.26 Write the form of aspartic acid most prevalent at pH 1.0, and then show its reaction with the following (consult Table 19.2 for pK_a values of the ionizable groups in aspartic acid) **(See Example 19.2)**:

(a) 1.0 mol NaOH
(b) 2.0 mol NaOH
(c) 3.0 mol NaOH

19.27 Given pK_a values for ionizable groups from Table 19.2, sketch curves for the titration of (a) glutamic acid with NaOH and (b) histidine with NaOH.

19.28 Draw a structural formula for the product formed when alanine is treated with each of the following reagents:

(a) Aqueous NaOH
(b) Aqueous HCl
(c) CH_3CH_2OH, H_2SO_4
(d) $(CH_3CO)_2O$, $CH_3COO^- Na^+$

19.29 Account for the fact that the isoelectric point of glutamine (pI 5.65) is higher than the isoelectric point of glutamic acid (pI 3.08).

19.30 Enzyme-catalyzed decarboxylation of glutamic acid gives 4-aminobutanoic acid (Section 19.2D). Estimate the pI of 4-aminobutanoic acid.

19.31 Guanidine and the guanidino group present in arginine are two of the strongest organic bases known. Account for their basicity.

***19.32** At pH 7.4, the pH of blood plasma, do the majority of protein-derived amino acids bear a net negative charge or a net positive charge? **(See Example 19.2)**

19.33 Do the following compounds migrate to the cathode or the anode on electrophoresis at the specified pH? **(See Example 19.2)**

(a) Histidine at pH 6.8 (b) Lysine at pH 6.8

(c) Glutamic acid at pH 4.0 (d) Glutamine at pH 4.0

(e) Glu-Ile-Val at pH 6.0 (f) Lys-Gln-Tyr at pH 6.0

19.34 At what pH would you carry out an electrophoresis to separate the amino acids in each of the following mixtures? **(See Example 19.3)**

(a) Ala, His, Lys (b) Glu, Gln, Asp (c) Lys, Leu, Tyr

***19.35** Examine the amino acid sequence of human insulin (Figure 19.13), and list each Asp, Glu, His, Lys, and Arg in this molecule. Do you expect human insulin to have an isoelectric point nearer that of the acidic amino acids (pI 2.0–3.0), the neutral amino acids (pI 5.5–6.5), or the basic amino acids (pI 9.5–11.0)?

Section 19.5 Primary Structure of Polypeptides and Proteins

19.36 If a protein contains four different SH groups, how many different disulfide bonds are possible if only a single disulfide bond is formed? How many different disulfides are possible if two disulfide bonds are formed?

19.37 How many different tetrapeptides can be made if **(See Example 19.4)**

(a) The tetrapeptide contains one unit each of Asp, Glu, Pro, and Phe?

(b) All 20 amino acids can be used, but each only once?

19.38 A decapeptide has the following amino acid composition: **(See Example 19.6)**

Ala_2, Arg, Cys, Glu, Gly, Leu, Lys, Phe, Val

Partial hydrolysis yields the following tripeptides:

Cys-Glu-Leu + Gly-Arg-Cys + Leu-Ala-Ala
+ Lys-Val-Phe + Val-Phe-Gly

One round of Edman degradation yields a lysine phenylthiohydantoin. From this information, deduce the primary structure of the given decapeptide.

***19.39** Following is the primary structure of glucagon, a polypeptide hormone with 29 amino acids: **(See Example 19.5)**

1 5 10
His-Ser-Glu-Gly-Thr-Phe-Thr-Ser-Asp-Tyr-Ser-Lys-Tyr-

15 20 25
Leu-Asp-Ser-Arg-Arg-Ala-Gln-Asp-Phe-Val-Gln-Trp-

29
Leu-Met-Asn-Thr

Glucagon

Glucagon is produced in the α-cells of the pancreas and helps maintain the blood glucose concentration within a normal range. Which peptide bonds are hydrolyzed when glucagon is treated with each reagent?

(a) Phenyl isothiocyanate (b) Chymotrypsin

(c) Trypsin (d) Br-CN

19.40 A tetradecapeptide (14 amino acid residues) gives the following peptide fragments on partial hydrolysis **(See Example 19.6)**:

Pentapeptide Fragments	Tetrapeptide Fragments
Phe-Val-Asn-Gln-His	Gln-His-Leu-Cys
His-Leu-Cys-Gly-Ser	His-Leu-Val-Glu
Gly-Ser-His-Leu-Val	Leu-Val-Glu-Ala

From the information shown, deduce the primary structure of the given polypeptide. Fragments are grouped according to size.

19.41 Draw a structural formula of each of the following tripeptides: marking each peptide bond, the *N*-terminal amino acid, and the *C*-terminal amino acid: **(See Example 19.4)**

(a) Phe-Val-Asn (b) Leu-Val-Gln

19.42 Estimate the pI of each tripeptide in Problem 19.41.

***19.43** Glutathione (G-SH), one of the most common tripeptides in animals, plants, and bacteria, is a scavenger of oxidizing agents:

$$\text{Glutathione} \quad H_3\overset{+}{N}\underset{\displaystyle COO^-}{CH}CH_2CH_2\overset{\displaystyle O}{\overset{\|}{C}}NH\underset{\displaystyle CH_2SH}{CH}\overset{\displaystyle O}{\overset{\|}{C}}NHCH_2COO^-$$

In reacting with oxidizing agents, glutathione is converted to G-S-S-G.

(a) Name the amino acids in this tripeptide.

(b) What is unusual about the peptide bond formed by the *N*-terminal amino acid?

(c) Is glutathione a biological oxidizing agent or a biological reducing agent?

(d) Write a balanced equation for the reaction of glutathione with molecular oxygen, O_2, to form G-S-S-G and H_2O. Is molecular oxygen oxidized or reduced in this reaction?

***19.44** Following is a structural formula for the artificial sweetener aspartame:

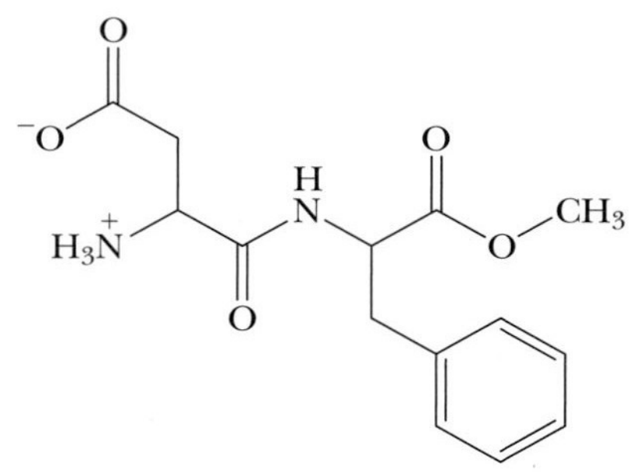

(a) Name the two amino acids in this molecule.

(b) Estimate the isoelectric point of aspartame.

(c) Draw structural formulas for the products of the hydrolysis of aspartame in 1 M HCl.

$^{-}$O O $H_3\overset{+}{N}$ H N O O CH_3 O

Aspartame

Section 19.6 Three-Dimensional Shapes of Polypeptides and Proteins

19.45 Examine the α-helix conformation. Are amino acid side chains arranged all inside the helix, all outside the helix, or randomly?

19.46 Distinguish between intermolecular and intramolecular hydrogen bonding between the backbone groups on polypeptide chains. In what type of secondary structure do you find intermolecular hydrogen bonds? In what type do you find intramolecular hydrogen bonding?

***19.47** Many plasma proteins found in an aqueous environment are globular in shape. Which of the following amino acid side chains would you expect to find on the surface of a globular protein, in contact with the aqueous environment, and which would you expect to find inside, shielded from the aqueous environment?

(a) Leu (b) Arg (c) Ser (d) Lys (e) Phe

Explain.

Section 19.7 Protein Denaturation

***19.48** Enzymes are examples of proteins. Why do enzymes lose their catalytic activity at higher than physiological temperatures?

***19.49** When an egg is boiled, the yolk changes color and consistency but when cooled to its original temperature, it does not regain its original nature. Explain.

19.50 Heating can disrupt the 3° structure of a protein. Explain the chemical processes that occur upon heating a protein.

Looking Ahead

19.51 Some amino acids cannot be incorporated into proteins because they are self-destructive. Homoserine, for example, can use its side-chain OH group in an intramolecular nucleophilic acyl substitution to cleave the peptide bond and form a cyclic structure on one end of the chain:

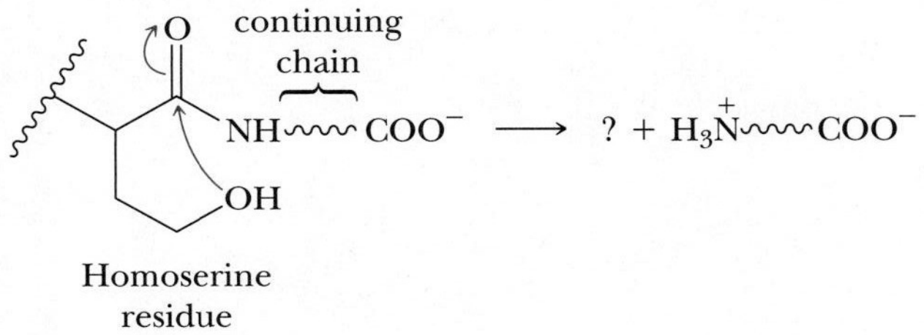

continuing chain

NH~~~COO$^-$ ⟶ ? + $H_3\overset{+}{N}$~~~COO$^-$

OH

Homoserine residue

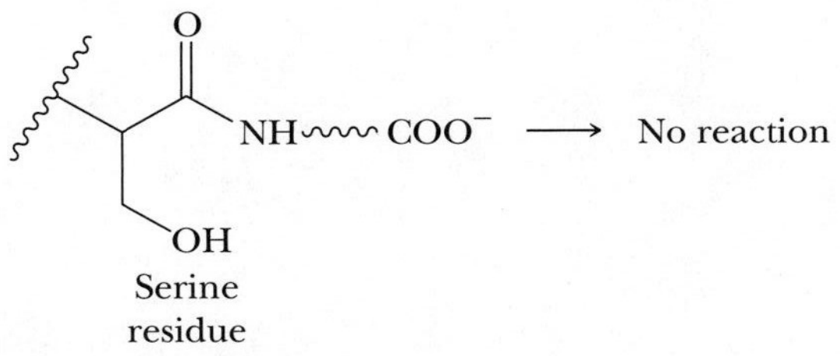

NH~~~COO$^-$ ⟶ No reaction

OH

Serine residue

Draw the cyclic structure formed and explain why serine does not suffer the same fate.

19.52 Would you expect a decapeptide of only isoleucine residues to form an α-helix? Explain.

19.53 Which type of protein would you expect to bring about the following change?

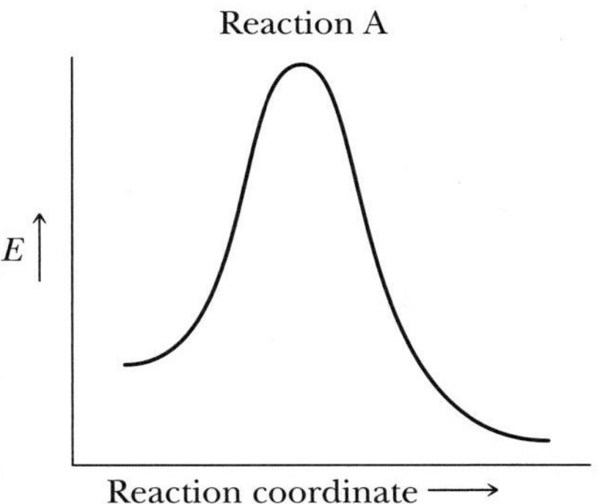

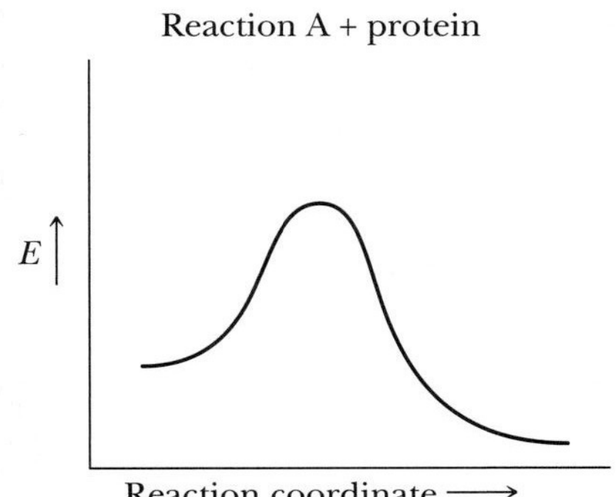

20 Nucleic Acids

KEY QUESTIONS

CHEMICAL CONNECTIONS

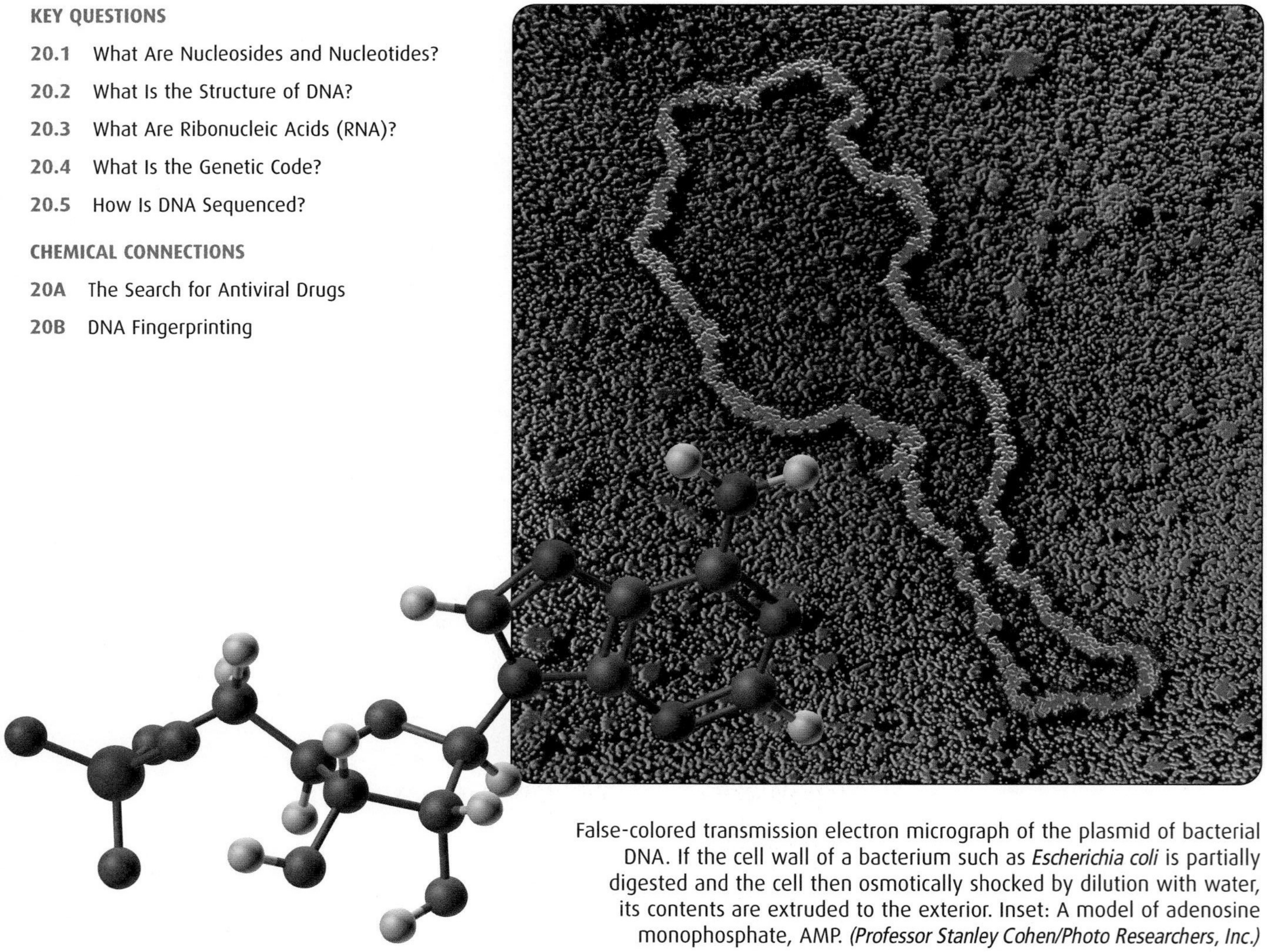

False-colored transmission electron micrograph of the plasmid of bacterial DNA. If the cell wall of a bacterium such as *Escherichia coli* is partially digested and the cell then osmotically shocked by dilution with water, its contents are extruded to the exterior. Inset: A model of adenosine monophosphate, AMP. *(Professor Stanley Cohen/Photo Researchers, Inc.)*

The organization, maintenance, and regulation of cellular function require a tremendous amount of information, all of which must be processed each time a cell is replicated. With very few exceptions, genetic information is stored and transmitted from one generation to the next in the form of **deoxyribonucleic acids (DNA)**. Genes, the hereditary units of chromosomes, are long stretches of double-stranded DNA. If the DNA in a human chromosome in a single cell were uncoiled, it would be approximately 1.8 meters long!

Genetic information is expressed in two stages: transcription from DNA to ribonucleic acids (RNA) and then translation through the synthesis of proteins:

$$\text{DNA} \xrightarrow{\text{transcription}} \text{RNA} \xrightarrow{\text{translation}} \text{proteins}$$

Thus, DNA is an enormous molecule that stores our genetic information, whereas RNA serves in the transcription and translation of this information, which is then expressed through the synthesis of proteins.

In this chapter, we will first examine the DNA molecule in detail to gain an understanding of its structure and function. We start by examining the structure of nucleosides and nucleotides and the manner in which these monomers are covalently bonded to form **nucleic acids**. Then we explore how genetic information is encoded on molecules of DNA, the function of the three types of ribonucleic acids, and, finally, how the primary structure of a DNA molecule is determined.

Nucleic acid A biopolymer containing three types of monomer units: heterocyclic aromatic amine bases derived from purine and pyrimidine, the monosaccharides D-ribose or 2-deoxy-D-ribose, and phosphate.

20.1 What Are Nucleosides and Nucleotides?

Controlled hydrolysis of nucleic acids yields three types of simpler building blocks: heterocyclic aromatic amine bases, the monosaccharide D-ribose or 2-deoxy-D-ribose (Section 18.3), and phosphate ions. Figure 20.1 shows the five heterocyclic aromatic amine bases most common to nucleic acids. Uracil, cytosine, and thymine are referred to as pyrimidine bases after the name of the parent base; adenine and guanine are referred to as purine bases.

Nucleoside A building block of nucleic acids, consisting of D-ribose or 2-deoxy-D-ribose bonded to a heterocyclic aromatic amine base by a β-*N*-glycosidic bond.

A **nucleoside** is a compound containing a pentose, either D-ribose or 2-deoxy-D-ribose bonded to a heterocyclic aromatic amine base by a β-*N*-glycosidic bond (Section 18.4A). Table 20.1 gives the names of the nucleosides derived from the four most common heterocyclic amine bases.

TABLE 20.1 Nomenclature of Nucleosides

Base	Name of Nucleoside
uracil	uridine
thymine	thymidine
cytosine	cytidine
guanine	guanosine
adenine	adenosine

Pyrimidine — Uracil (U) — Cytosine (C) — Thymine (T)

Purine — Adenine (A) — Guanine (G)

Figure 20.1
Names and one-letter abbreviations for the heterocyclic aromatic amine bases most common to DNA and RNA. Bases are numbered according to the patterns of pyrimidine and purine, the parent compounds.

The monosaccharide component of DNA is 2-deoxy-D-ribose (the "2-deoxy" refers to the absence of a hydroxyl group at the 2′ position), whereas that of RNA is D-ribose. The glycosidic bond is between C-1′ (the anomeric carbon) of ribose or 2-deoxyribose and N-1 of a pyrimidine base or N-9 of a purine base. Figure 20.2 shows the structural formula for uridine, a nucleoside derived from ribose and uracil.

Nucleotide A nucleoside in which a molecule of phosphoric acid is esterified with an —OH of the monosaccharide, most commonly either the 3′-OH or the 5′-OH.

A **nucleotide** is a nucleoside in which a molecule of phosphoric acid is esterified with a free hydroxyl of the monosaccharide, most commonly either the 3′-hydroxyl or the 5′-hydroxyl. A nucleotide is named by giving the name of the parent nucleoside, followed by the word "monophosphate." The position of the phosphoric ester is

CHEMICAL CONNECTIONS 20A

The Search for Antiviral Drugs

The search for antiviral drugs has been more difficult than the search for antibacterial drugs primarily because viral replication depends on the metabolic processes of the invaded cell. Thus, antiviral drugs are also likely to cause harm to the cells that harbor the virus. The challenge in developing antiviral drugs is to understand the biochemistry of viruses and to develop drugs that target processes specific to them. Compared with the large number of antibacterial drugs available, there is only a handful of antiviral drugs, and they have nowhere near the effectiveness that antibiotics have in bacterial infections.

Acyclovir was one of the first of a new family of drugs for the treatment of infectious diseases caused by DNA viruses called herpes virus. Herpes infections in humans are of two kinds: herpes simplex type 1, which gives rise to mouth and eye sores, and herpes simplex type 2, which gives rise to serious genital infections. Acyclovir is highly effective against herpes virus–caused genital infections. The structural formula of acyclovir is drawn in the accompanying figure to show its structural relationship to 2-deoxyguanosine. The drug is activated in vivo by the conversion of the primary —OH (which corresponds to the 5′-OH of a riboside or a deoxyriboside) to a triphosphate. Because of its close resemblance to deoxyguanosine triphosphate, an essential precursor of DNA synthesis, acyclovir triphosphate is taken up by viral DNA polymerase to form an enzyme–substrate complex on which no 3′-OH exists for replication to continue. Thus, the enzyme–substrate complex is no longer active (it is a dead-end complex), viral replication is disrupted, and the virus is destroyed.

Perhaps the best known of the HIV-fighting viral antimetabolites is zidovudine (AZT), an analog of deoxythymidine in which the 3′-OH has been replaced by an azido group, N_3. AZT is effective against HIV-1, a retrovirus that is the causative agent of AIDS. AZT is converted in vivo by cellular enzymes to the 5′-triphosphate, is recognized as deoxythymidine 5′-triphosphate by viral RNA-

specified by the number of the carbon to which it is bonded. Figure 20.3 shows a structural formula and a ball-and-stick model of adenosine 5′-monophosphate. Monophosphoric esters are diprotic acids with pK_a values of approximately 1 and 6. Therefore, at a pH of 7, the two hydrogens of a phosphoric monoester are fully ionized, giving a nucleotide a charge of −2.

Nucleoside monophosphates can be further phosphorylated to form nucleoside diphosphates and nucleoside triphosphates. Shown in Figure 20.4 is a structural formula for adenosine 5′-triphosphate (ATP).

Nucleoside diphosphates and triphosphates are also polyprotic acids and are extensively ionized at pH 7.0. pK_a values of the first three ionization steps for adenosine triphosphate are less than 5.0. The value of pK_{a4} is approximately 7.0. Therefore, at pH 7.0, approximately 50% of adenosine triphosphate is present as ATP^{4-} and 50% is present as ATP^{3-}.

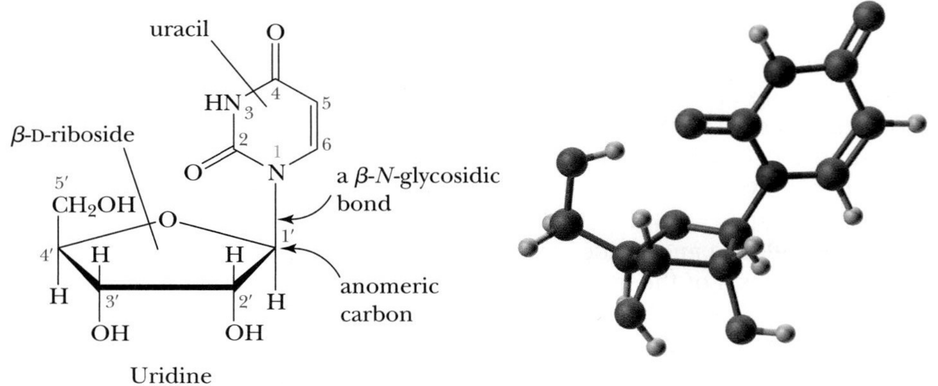

Figure 20.2
Uridine, a nucleoside. Atom numbers on the monosaccharide rings are primed to distinguish them from atom numbers on the heterocyclic aromatic amine bases.

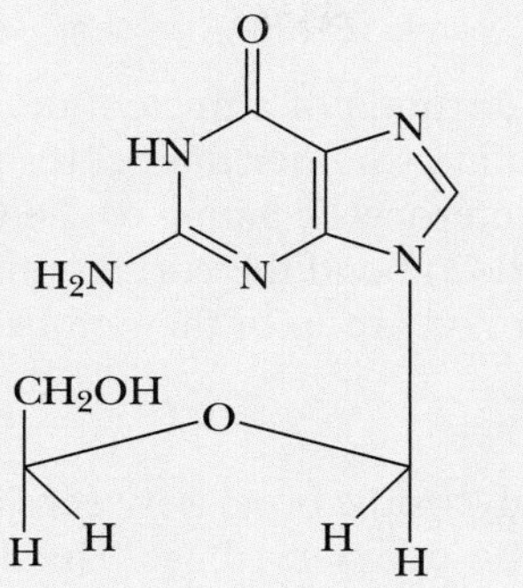

Acyclovir
(drawn to show its structural relationship to 2-deoxyguanosine)

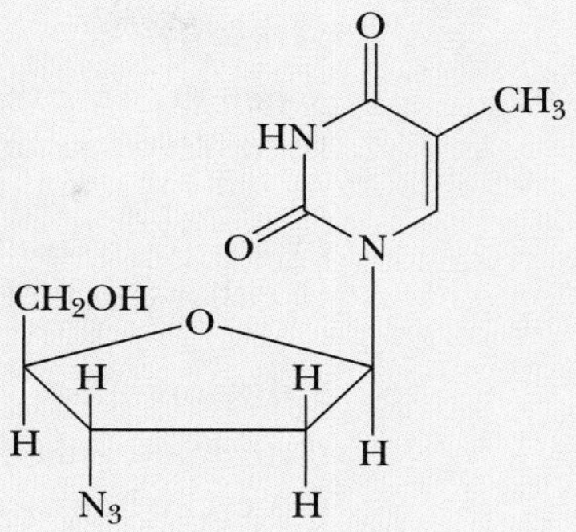

Zidovudine
(Azidothymidine; AZT)

dependent DNA polymerase (reverse transcriptase), and is added to a growing DNA chain. There, it stops chain elongation because no 3′-OH exists on which to add the next deoxynucleotide. AZT owes its effectiveness to the fact that it binds more strongly to viral reverse transcriptase than it does to human DNA polymerase.

QUESTION

Draw the triphosphate of acyclovir as it would exist in solution at pH 7.4.

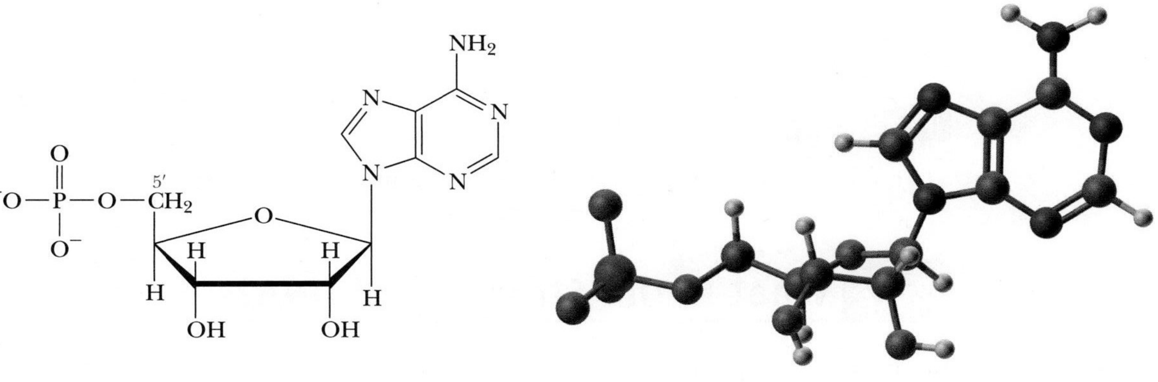

Figure 20.3
Adenosine 5′-monophosphate, a nucleotide. The phosphate group is fully ionized at pH 7.0, giving this nucleotide a charge of −2.

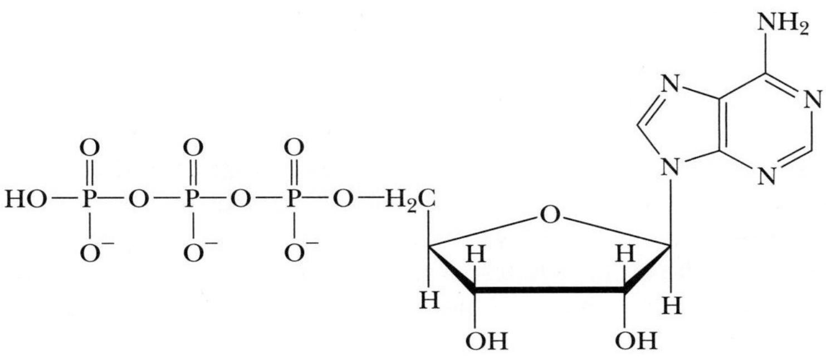

Figure 20.4
Adenosine 5′-triphosphate, ATP.

Example 20.1

Draw a structural formula for 2′-deoxycytidine 5′-diphosphate.

Strategy

When drawing the structural formula of a nucleotide, there are three important things to consider: (1) Determine whether the pentose is D-ribose or 2-deoxy-D-ribose; (2) bond the correct heterocyclic amine to the C1 position of the pentose by a β-N-glycosidic bond; and (3) bond the correct number of phosphate groups to either the 3′ or the 5′ hydroxyl group of the pentose.

Solution

Cytosine is joined by a β-N-glycosidic bond between N-1 of cytosine and C-1′ of the cyclic hemiacetal form of 2-deoxy-D-ribose. The 5′-hydroxyl of the pentose is bonded to a phosphate group by an ester bond, and this phosphate is in turn bonded to a second phosphate group by an anhydride bond:

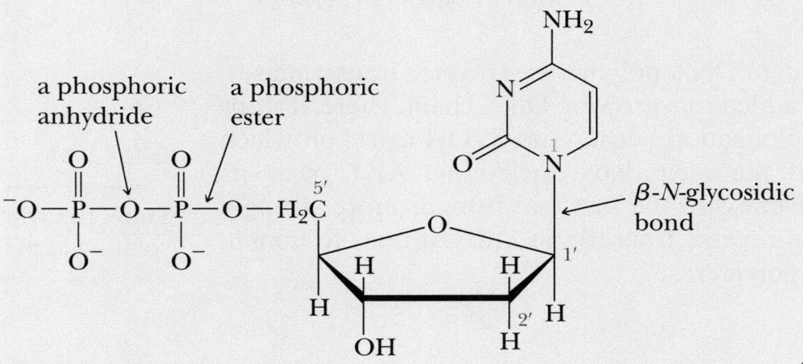

a phosphoric anhydride
a phosphoric ester
β-N-glycosidic bond

See problem 20.11

Problem 20.1

Draw a structural formula for 2′-deoxythymidine 3′-monophosphate.

20.2 What Is the Structure of DNA?

In Chapter 19, we saw that the four levels of structural complexity in polypeptides and proteins are primary, secondary, tertiary, and quaternary. There are three levels of structural complexity in nucleic acids, and although these levels are somewhat comparable to those in polypeptides and proteins, they also differ in significant ways.

Primary structure of nucleic acids The sequence of bases along the pentose–phosphodiester backbone of a DNA or RNA molecule, read from the 5′ end to the 3′ end.

5′ End The end of a polynucleotide at which the 5′-OH of the terminal pentose unit is free.

3′ End The end of a polynucleotide at which the 3′-OH of the terminal pentose unit is free.

A. Primary Structure: The Covalent Backbone

Deoxyribonucleic acids consist of a backbone of alternating units of deoxyribose and phosphate in which the 3′-hydroxyl of one deoxyribose unit is joined by a phosphodiester bond to the 5′-hydroxyl of another deoxyribose unit (Figure 20.5). This pentose–phosphodiester backbone is constant throughout an entire DNA molecule. A heterocyclic aromatic amine base—adenine, guanine, thymine, or cytosine—is bonded to each deoxyribose unit by a β-N-glycosidic bond. The **primary structure** of a DNA molecule is the order of heterocyclic bases along the pentose–phosphodiester backbone. The sequence of bases is read from the **5′ end** to the **3′ end**.

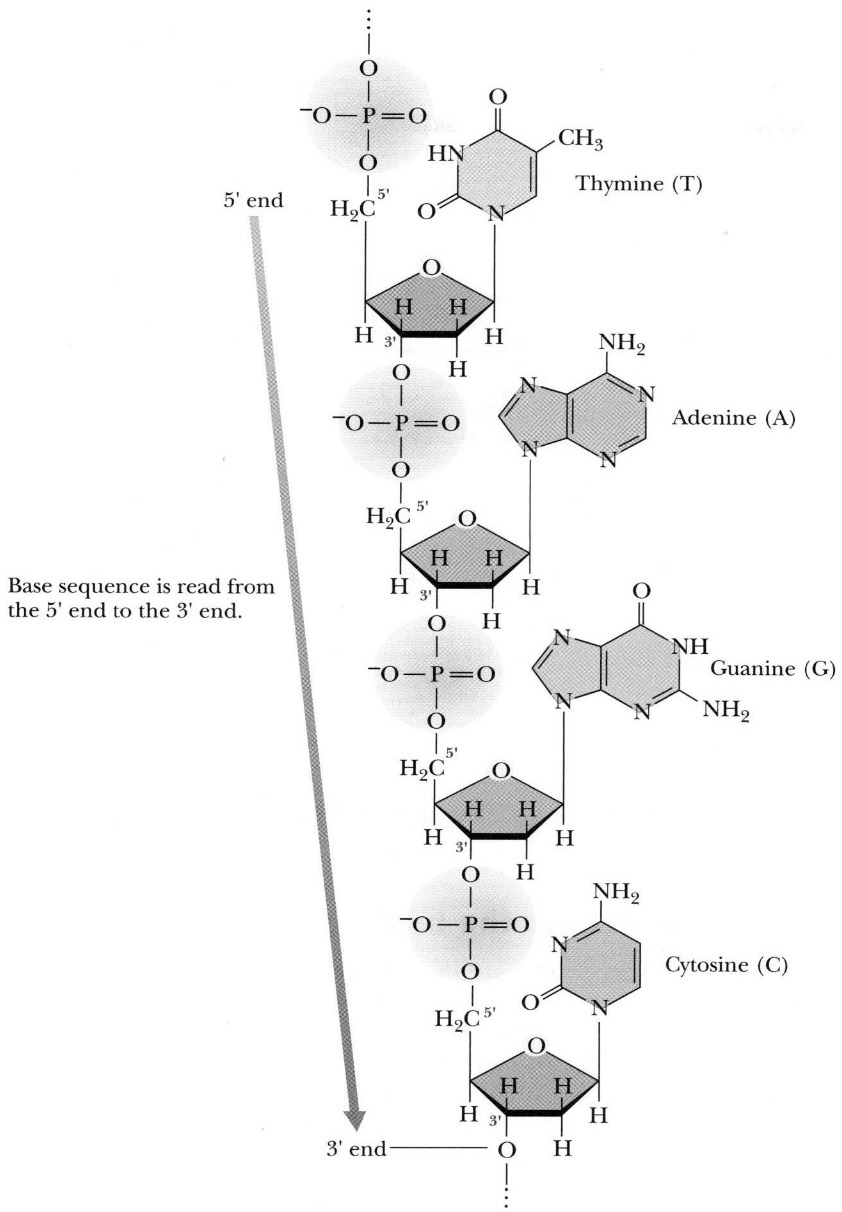

Figure 20.5
A tetranucleotide section of a single-stranded DNA.

Example 20.2

Draw a structural formula for the DNA dinucleotide TG that is phosphorylated at the 5′ end only.

Strategy

Because the sequence of bases in a nucleic acid is always read from the 5′ end to the 3′ end, a thymidine is bonded by a phosphodiester bond to the 5′ end of guanosine. The 3′ end of the guanosine has a free —OH group.

Solution

See problem 20.13

Problem 20.2

Draw a structural formula for the section of DNA that contains the base sequence CTG and is phosphorylated at the 3′ end only.

Secondary structure of nucleic acids The ordered arrangement of strands of nucleic acid.

Rosalind Franklin (1920–1958). In 1951, she joined the Biophysical Laboratory at King's College, London, where she began her studies on the application of X-ray diffraction methods to the study of DNA. She is credited with discoveries that established the density of DNA, its helical conformation, and other significant aspects. Her work was thus important to the model of DNA developed by Watson and Crick. She died in 1958 at the age of 37 and, because a Nobel Prize is never awarded posthumously, she did not share in the 1962 Nobel Prize in Physiology or Medicine with Watson, Crick, and Wilkins. *(Photo Researchers, Inc.)*

B. Secondary Structure: The Double Helix

By the early 1950s, it was clear that DNA molecules consist of chains of alternating units of deoxyribose and phosphate joined by 3′,5′-phosphodiester bonds, with a base attached to each deoxyribose unit by a β-*N*-glycosidic bond. In 1953, the American biologist James D. Watson and the British physicist Francis H. C. Crick proposed a double-helix model for the **secondary structure of DNA**. Watson, Crick, and Maurice Wilkins shared the 1962 Nobel Prize in Physiology or Medicine for "their discoveries concerning the molecular structure of nucleic acids, and its significance for information transfer in living material." Although Rosalind Franklin also took part in this research, her name was omitted from the Nobel list because of her death in 1958 at age 37. The Nobel foundation does not make awards posthumously.

The Watson–Crick model was based on molecular modeling and two lines of experimental observations: chemical analyses of DNA base compositions and mathematical analyses of X-ray diffraction patterns of crystals of DNA.

Base Composition

At one time, it was thought that, in all species, the four principal bases occurred in the same ratios and perhaps repeated in a regular pattern along the pentose–phosphodiester backbone of DNA. However, more precise determinations of their composition by Erwin Chargaff revealed that bases do not occur in the same ratios (Table 20.2).

Researchers drew the following conclusions from the data in the table and related data: To within experimental error,

1. The mole-percent base composition of DNA in any organism is the same in all cells of the organism and is characteristic of the organism.

TABLE 20.2 Comparison of DNA from Several Organisms, Base Composition, in Mole Percent

	Purines		Pyrimidines				
Organism	**A**	**G**	**C**	**T**	**A/T**	**G/C**	**Purines/ Pyrimidines**
human	30.4	19.9	19.9	30.1	1.01	1.00	1.01
sheep	29.3	21.4	21.0	28.3	1.04	1.02	1.03
yeast	31.7	18.3	17.4	32.6	0.97	1.05	1.00
E. coli	26.0	24.9	25.2	23.9	1.09	0.99	1.04

2. The mole percentages of adenine (a purine base) and thymine (a pyrimidine base) are equal. The mole percentages of guanine (a purine base) and cytosine (a pyrimidine base) are also equal.
3. The mole percentages of purine bases (A + G) and pyrimidine bases (C + T) are equal.

Analyses of X-Ray Diffraction Patterns

Additional information about the structure of DNA emerged when X-ray diffraction photographs taken by Rosalind Franklin and Maurice Wilkins were analyzed. The diffraction patterns revealed that, even though the base composition of DNA isolated from different organisms varies, DNA molecules themselves are remarkably uniform in thickness. They are long and fairly straight, with an outside diameter of approximately 20 Å, and not more than a dozen atoms thick. Furthermore, the crystallographic pattern repeats every 34 Å. Herein lay one of the chief problems to be solved: How could the molecular dimensions of DNA be so regular, even though the relative percentages of the various bases differ so widely? With this accumulated information, the stage was set for the development of a hypothesis about DNA structure.

The Watson–Crick Double Helix

The heart of the Watson–Crick model is the postulate that a molecule of DNA is a complementary **double helix** consisting of two antiparallel polynucleotide strands coiled in a right-handed manner about the same axis. As illustrated in the ribbon models in Figure 20.6, chirality is associated with a double helix: Like enantiomers, left-handed and right-handed double helices are related by reflection.

Double helix A type of secondary structure of DNA molecules in which two antiparallel polynucleotide strands are coiled in a right-handed manner about the same axis.

To account for the observed base ratios and uniform thickness of DNA, Watson and Crick postulated that purine and pyrimidine bases project inward toward the axis of the helix and always pair in a specific manner. According to scale models, the dimensions of an adenine–thymine base pair are almost identical to the dimensions of a guanine–cytosine base pair, and the length of each pair is consistent with the core thickness of a DNA strand (Figure 20.7). Thus, if the purine base in one strand is adenine, then its complement in the antiparallel strand must be thymine. Similarly, if the purine in one strand is guanine, its complement in the antiparallel strand must be cytosine.

Figure 20.6
A DNA double helix has a chirality associated with it. Right-handed and left-handed double helices of otherwise identical DNA chains are nonsuperposable mirror images.

A significant feature of Watson and Crick's model is that no other **base pairing** is consistent with the observed thickness of a DNA molecule. A pair of pyrimidine bases is too small to account for the observed thickness, whereas a pair of purine bases is too large. Thus, according to the Watson–Crick model, the repeating units in a double-stranded DNA molecule are not single bases of differing dimensions, but rather base pairs of almost identical dimensions.

To account for the periodicity observed from X-ray data, Watson and Crick postulated that base pairs are stacked one on top of the other, with a distance of 3.4 Å between base pairs and with 10 base pairs in one complete turn of the helix. Thus, there is one complete turn of the helix every 34 Å. Figure 20.8 shows a ribbon model

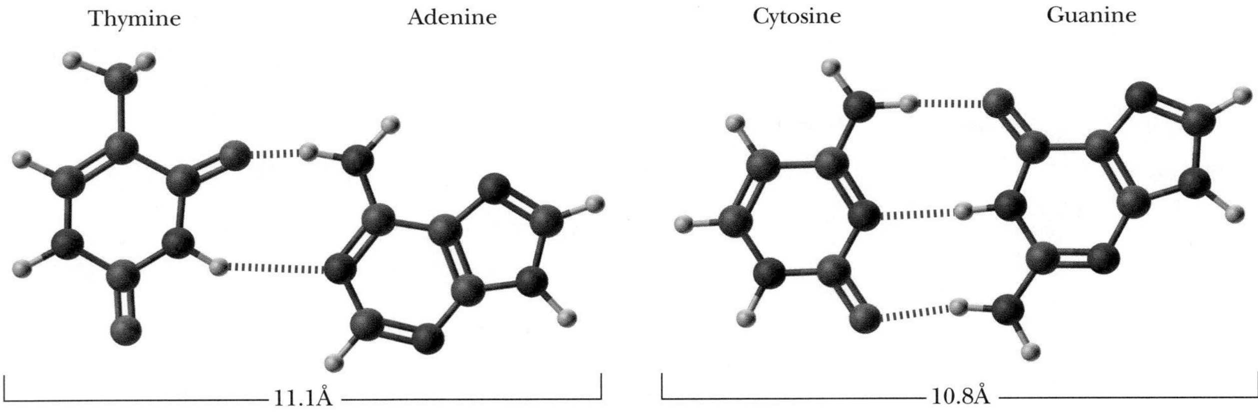

Figure 20.7
Base pairing between adenine and thymine (A-T) and between guanine and cytosine (G-C). An A-T base pair is held by two hydrogen bonds, whereas a G-C base pair is held by three hydrogen bonds.

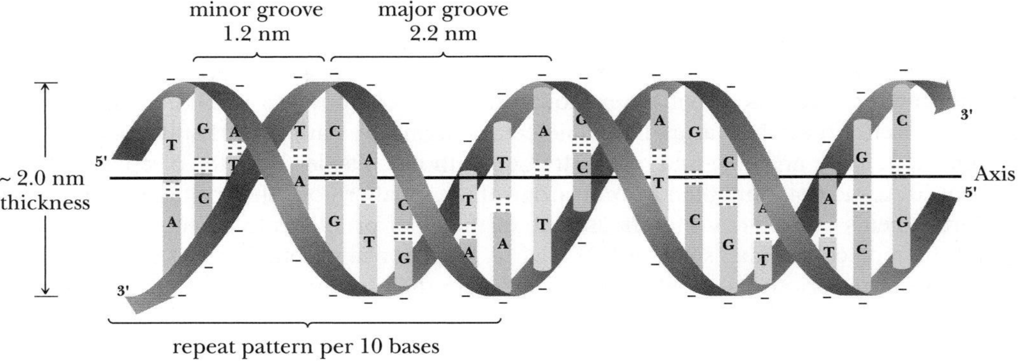

Figure 20.8
Ribbon model of double-stranded B-DNA. Each ribbon shows the pentose–phosphodiester backbone of a single-stranded DNA molecule. The strands are antiparallel, one running to the left from the 5′ end to the 3′ end, the other running to the right from the 5′ end to the 3′ end. Hydrogen bonds are shown by three dotted lines between each G-C base pair and two dotted lines between each A-T base pair.

of double-stranded **B-DNA**, the predominant form of DNA in dilute aqueous solution and thought to be the most common form in nature.

In the double helix, the bases in each base pair are not directly opposite from one another across the diameter of the helix, but rather are slightly displaced. This displacement and the relative orientation of the glycosidic bonds linking each base to the sugar–phosphate backbone lead to two differently sized grooves: a major groove and a minor groove (Figure 20.8). Each groove runs along the length of the cylindrical column of the double helix. The major groove is approximately 22 Å wide, the minor groove approximately 12 Å wide.

Figure 20.9 shows more detail of an idealized B-DNA double helix. The major and minor grooves are clearly recognizable in this model.

Other forms of secondary structure are known that differ in the distance between stacked base pairs and in the number of base pairs per turn of the helix. One of the most common of these, **A-DNA**, also a right-handed helix, is thicker than B-DNA, and has a repeat distance of only 29 Å. There are 10 base pairs per turn of the helix, with a spacing of 2.9 Å between base pairs.

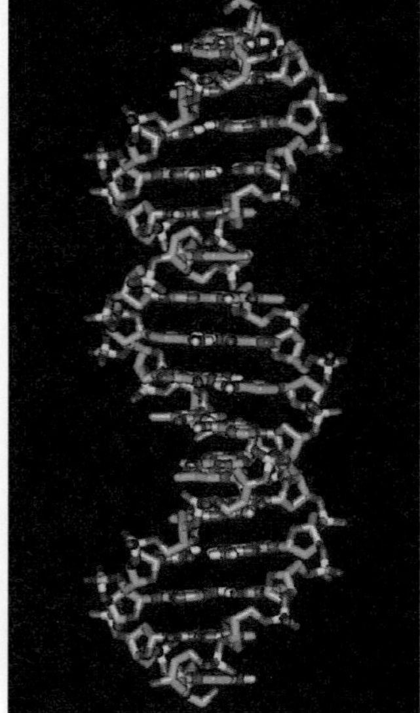

Figure 20.9
B-DNA. An idealized model of B-DNA.

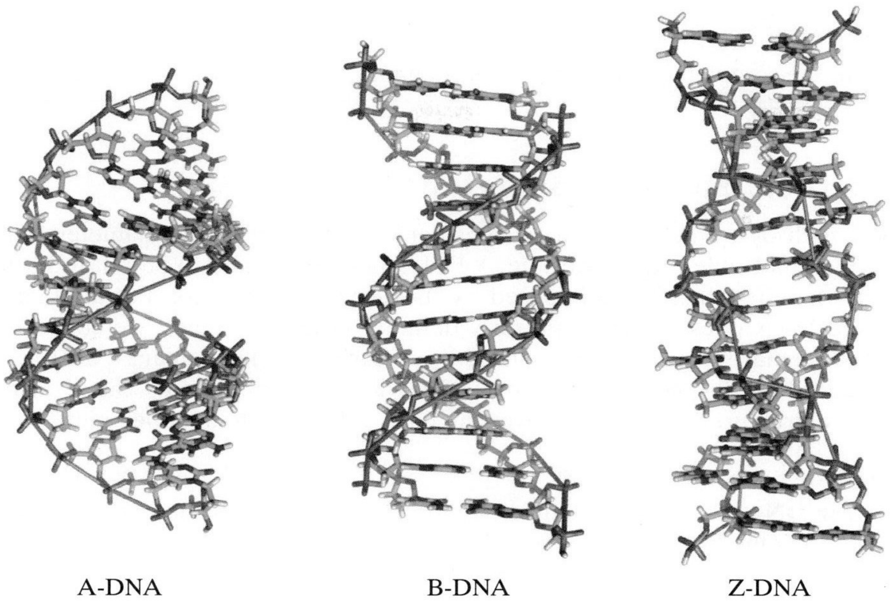

Figure 20.10
Section of A, B, and Z-DNA. *(Richard Wheeler, originally produced for Wikipedia.org)*

A-DNA is normally seen in dehydrated samples of DNA, such as in crystals prepared for X-ray crystallography. Another form of DNA secondary structure is **Z-DNA**, a left-handed helix, which is believed to occur during DNA transcription. There are 12 base pairs per turn of helix, and unlike the other two forms of DNA, the major and minor grooves are nearly identical in width. Figure 20.10 shows the three forms side by side for comparison.

Example 20.3

One strand of a DNA molecule contains the base sequence 5′-ACTTGCCA-3′. Write its complementary base sequence.

Strategy

Remember that a base sequence is always written from the 5′ end of a strand to the 3′ end, and that A pairs with T and that G pairs with C. In double-stranded DNA, the two strands run in opposite (antiparallel) directions, so that the 5′ end of one strand is associated with the 3′ end of the other strand.

Solution

Written from the 5′ end, the complementary strand is 5′-TGGCAAGT-3′.

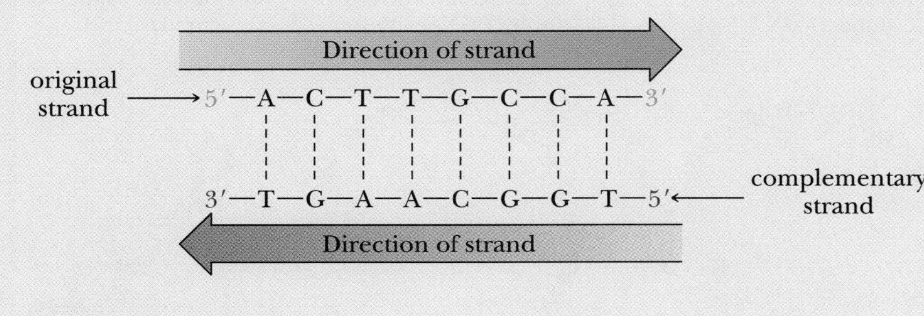

See problem 20.16

Problem 20.3

Write the complementary DNA base sequence for 5′-CCGTACGA-3′.

C. Tertiary Structure: Supercoiled DNA

The length of a DNA molecule is considerably greater than its diameter, and the extended molecule is quite flexible. A DNA molecule is said to be relaxed if it has no twists other than those imposed by its secondary structure. Put another way, relaxed DNA does not have a clearly defined tertiary structure. We consider two types of **tertiary structure**, one induced by perturbations in circular DNA, the other introduced by the coordination of DNA with nuclear proteins called histones. Tertiary structure in DNA, whatever the type, is referred to as **supercoiling**.

Tertiary structure of nucleic acids The three-dimensional arrangement of all atoms of a nucleic acid, commonly referred to as supercoiling.

Supercoiling of Circular DNA

Circular DNA is a type of double-stranded DNA in which the two ends of each strand are joined by phosphodiester bonds [Figure 20.11(a)]. This type of DNA, the most prominent form in bacteria and viruses, is also referred to as *circular duplex* (because it is double-stranded) DNA. One strand of circular DNA may be opened, partially unwound, and then rejoined. The unwound section introduces a strain into the molecule because the nonhelical gap is less stable than hydrogen-bonded, base-paired helical sections. The strain can be localized in the nonhelical gap. Alternatively, it may be spread uniformly over the entire circular DNA by the introduction of **superhelical twists**, one twist for each turn of a helix unwound. The circular DNA shown in Figure 20.11(b) has been unwound by four complete turns of the helix. The strain introduced by this unwinding is spread uniformly over the entire molecule by the introduction of four superhelical twists [Figure 20.11(c)]. Interconversion of relaxed and supercoiled DNA is catalyzed by groups of enzymes called topoisomerases and gyrases.

Circular DNA A type of double-stranded DNA in which the 5′ and 3′ ends of each strand are joined by phosphodiester groups.

Supercoiling of Linear DNA

Supercoiling of linear DNA in plants and animals takes another form and is driven by the interaction between negatively charged DNA molecules and a group of positively charged proteins called **histones** (Figure 20.12). Histones are particularly rich in lysine and arginine and, at the pH of most body fluids, have an abundance of positively charged sites along their length. The complex between negatively charged DNA and positively charged histones is called **chromatin**. Histones associate to form core particles about which double-stranded DNA then wraps. Further coiling of DNA produces the chromatin found in cell nuclei.

Histone A protein, particularly rich in the basic amino acids lysine and arginine, that is found associated with DNA molecules.

Chromatin A complex formed between negatively charged DNA molecules and positively charged histones.

(a) Relaxed: circular (duplex) DNA

(b) Slightly strained: circular (duplex) DNA with four twists of the helix unwound

(c) Strained: supercoiled circular DNA

Figure 20.11 Relaxed and supercoiled DNA. (a) Circular DNA is relaxed. (b) One strand is broken, unwound by four turns, and the ends are then rejoined. The strain of unwinding is localized in the nonhelical gap. (c) Supercoiling by four twists distributes the strain of unwinding uniformly over the entire molecule of circular DNA.

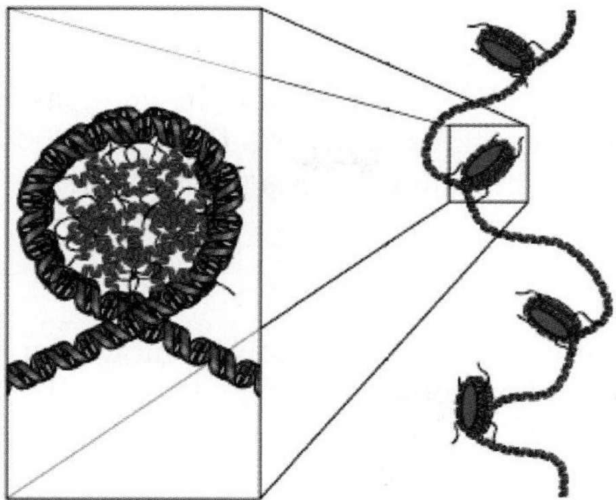

Figure 20.12
A section of a DNA strand showing complexation with histones. *(Richard Wheeler, originally produced for Wikipedia.org)*

Supercoiled DNA from a mitochondrion. *(Don W. Fawcett/ Photo Researchers, Inc.)*

20.3 What Are Ribonucleic Acids (RNA)?

Ribonucleic acids (RNA) are similar to deoxyribonucleic acids (DNA) in that they, too, consist of long, unbranched chains of nucleotides joined by phosphodiester groups between the 3′-hydroxyl of one pentose and the 5′-hydroxyl of the next. There are, however, three major differences in structure between RNA and DNA:

1. The pentose unit in RNA is β-D-ribose rather than β-2-deoxy-D-ribose.
2. The pyrimidine bases in RNA are uracil and cytosine rather than thymine and cytosine.
3. RNA is single stranded rather than double stranded.

Following are structural formulas for the furanose form of D-ribose and for uracil:

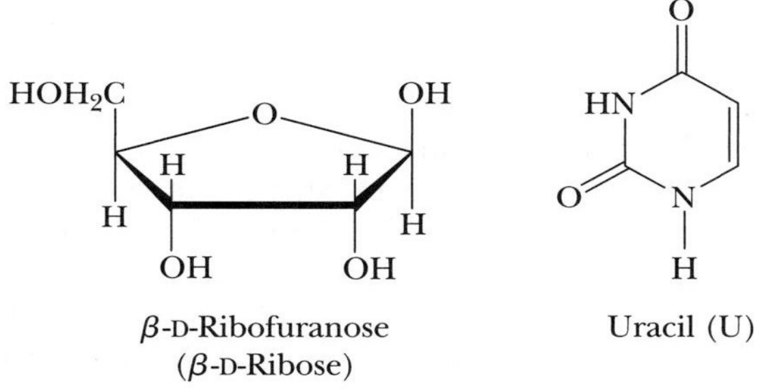

β-D-Ribofuranose (β-D-Ribose)

Uracil (U)

Cells contain up to eight times as much RNA as DNA, and, in contrast to DNA, RNA occurs in different forms and in multiple copies of each form. RNA molecules are classified, according to their structure and function, into three major types: ribosomal RNA, transfer RNA, and messenger RNA. Table 20.3 summarizes the molecular weight, number of nucleotides, and percentage of cellular abundance of the three types of RNA in cells of *E. coli*, one of the best-studied bacteria and a workhorse for cellular study.

TABLE 20.3 Types of RNA Found in Cells of *E. coli*

Type	Molecular Weight Range (amu)	Number of Nucleotides	Percentage of Cell RNA
mRNA	25,000–1,000,000	75–3,000	2
tRNA	23,000–30,000	73–94	16
rRNA	35,000–1,100,000	120–2,904	82

A. Ribosomal RNA

Ribosomal RNA (rRNA) A ribonucleic acid found in ribosomes, the sites of protein synthesis.

The bulk of **ribosomal RNA (rRNA)** is found in the cytoplasm in subcellular particles called ribosomes, which contain about 60% RNA and 40% protein. Ribosomes are the sites in cells at which protein synthesis takes place.

B. Transfer RNA

Transfer RNA (tRNA) A ribonucleic acid that carries a specific amino acid to the site of protein synthesis on ribosomes.

Transfer RNA (tRNA) molecules have the lowest molecular weight of all nucleic acids. They consist of 73 to 94 nucleotides in a single chain. The function of tRNA is to carry amino acids to the sites of protein synthesis on the ribosomes. Each amino acid has at least one tRNA dedicated specifically to this purpose. Several amino acids have more than one. In the transfer process, the amino acid is joined to its specific tRNA by an ester bond between the α-carboxyl group of the amino acid and the 3′ hydroxyl group of the ribose unit at the 3′ end of the tRNA:

amino acid, bound as an ester to its specific tRNA

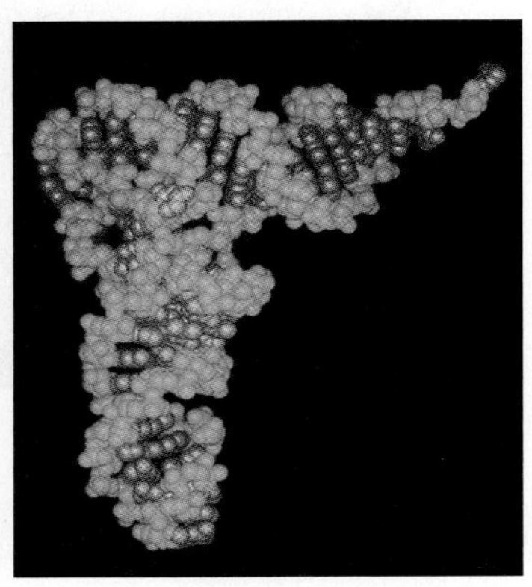

A space-filling model of yeast phenylalanine tRNA. *(After S. H. Kim, in P. Schimmel, D. Söll, and J. N. Abelson, eds., Transfer RNA: Structure, Properties, and Recognition, New York: Cold Spring Harbor Laboratories, 1979)*

C. Messenger RNA

Messenger RNA (mRNA) A ribonucleic acid that carries coded genetic information from DNA to ribosomes for the synthesis of proteins.

Messenger RNA (mRNA) is present in cells in relatively small amounts and is very short-lived. Messenger RNA molecules are single stranded, and their synthesis is directed by information encoded on DNA molecules. Double-stranded DNA is unwound, and a complementary strand of mRNA is synthesized along one strand of the DNA template, beginning from the 3′ end. The synthesis of mRNA from a DNA template is called *transcription* because genetic information contained in a sequence of bases of DNA is transcribed into a complementary sequence of bases on mRNA. The name "messenger" is derived from the function of this type of RNA, which is to carry coded genetic information from DNA to the ribosomes for the synthesis of proteins.

Example 20.4

Following is a base sequence from a portion of DNA:

3′-A-G-C-C-A-T-G-T-G-A-C-C-5′

Write the sequence of bases of the mRNA synthesized with this section of DNA as a template.

Strategy

RNA synthesis begins at the 3′ end of the DNA template and proceeds toward the 5′ end. The complementary mRNA strand is formed using the bases C, G, A, and U. Uracil (U) is the complement of adenine (A) on the DNA template.

Solution

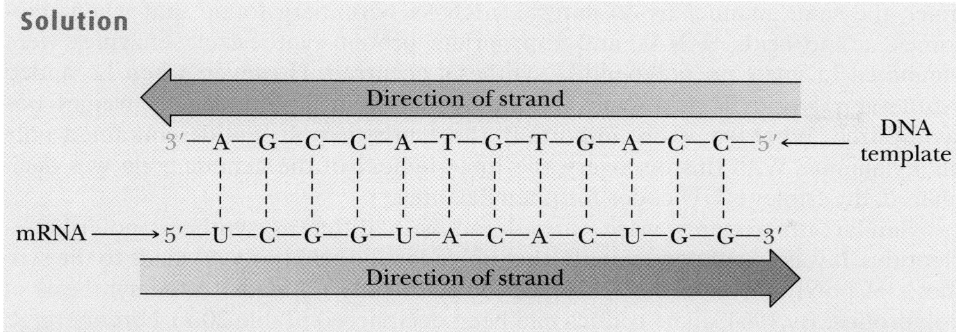

Reading from the 5′ end, we see that the sequence of mRNA is 5′-UCGGUA-CACUGG-3′.

See problems 20.28, 20.29, 20.35, 20.36

Problem 20.4

Here is a portion of the nucleotide sequence in phenylalanine tRNA:

3′-ACCACCUGCUCAGGCCUU-5′

Write the nucleotide sequence of the DNA complement of this sequence.

20.4 What Is the Genetic Code?

A. Triplet Nature of the Code

It was clear by the early 1950s that the sequence of bases in DNA molecules constitutes the store of genetic information and directs the synthesis of messenger RNA, which, in turn, directs the synthesis of proteins. However, the idea that the sequence of bases in DNA directs the synthesis of proteins presents the following problem: How can a molecule containing only four variable units (adenine, cytosine, guanine, and thymine) direct the synthesis of molecules containing up to 20 variable units (the protein-derived amino acids)? How can an alphabet of only four letters code for the order of letters in the 20-letter alphabet that occurs in proteins?

An obvious answer is that there is not one base, but rather a combination of bases coding for each amino acid. If the code consists of nucleotide pairs, there are $4^2 = 16$ combinations; this is a more extensive code, but it is still not extensive enough to code for 20 amino acids. If the code consists of nucleotides in groups of three, there are $4^3 = 64$ combinations; more than enough to code for the primary structure of a protein. This appears to be a very simple solution to the problem for a system that must have taken eons of evolutionary trial and error to develop. Yet proof now exists, from comparisons of gene (nucleic acid) and protein (amino acid) sequences, that nature does indeed use a simple three-letter or triplet code to store genetic information. A triplet of nucleotides is called a **codon**.

Codon A triplet of nucleotides on mRNA that directs the incorporation of a specific amino acid into a polypeptide sequence.

B. Deciphering the Genetic Code

The next question is, Which of the 64 triplets codes for which amino acid? In 1961, Marshall Nirenberg provided a simple experimental approach to the problem, based on the observation that synthetic polynucleotides direct polypeptide synthesis in much the same manner as do natural mRNAs. Nirenberg found that when ribosomes, amino acids, tRNAs, and appropriate protein-synthesizing enzymes were incubated in vitro, no polypeptide synthesis occurred. However, when he added synthetic polyuridylic acid (poly U), a polypeptide of high molecular weight was synthesized. What was more important, the synthetic polypeptide contained only phenylalanine. With this discovery, the first element of the genetic code was deciphered: the triplet UUU codes for phenylalanine.

Similar experiments were carried out with different synthetic polyribonucleotides. It was found, for example, that polyadenylic acid (poly A) leads to the synthesis of polylysine, and that polycytidylic acid (poly C) leads to the synthesis of polyproline. By 1964, all 64 codons had been deciphered (Table 20.4). Nirenberg, R. W. Holley, and H. G. Khorana shared the 1968 Nobel Prize in Physiology or Medicine for their seminal work.

C. Properties of the Genetic Code

Several features of the genetic code are evident from a study of Table 20.4:

1. Only 61 triplets code for amino acids. The remaining three (UAA, UAG, and UGA) are signals for chain termination; they signal to the protein-synthesizing machinery of the cell that the primary sequence of the protein is complete. The three chain termination triplets are indicated in Table 20.4 by "Stop."
2. The code is degenerate, which means that several amino acids are coded for by more than one triplet. Only methionine and tryptophan are coded for by just one triplet. Leucine, serine, and arginine are coded for by six triplets, and the remaining amino acids are coded for by two, three, or four triplets.

TABLE 20.4 The Genetic Code: mRNA Codons and the Amino Acid Each Codon Directs

First Position (5′-end)	Second Position								Third Position (3′-end)
	U		C		A		G		
U	UUU	Phe	UCU	Ser	UAU	Tyr	UGU	Cys	U
	UUC	Phe	UCC	Ser	UAC	Tyr	UGC	Cys	C
	UUA	Leu	UCA	Ser	UAA	Stop	UGA	Stop	A
	UUG	Leu	UCG	Ser	UAG	Stop	UGG	Trp	G
C	CUU	Leu	CCU	Pro	CAU	His	CGU	Arg	U
	CUC	Leu	CCC	Pro	CAC	His	CGC	Arg	C
	CUA	Leu	CCA	Pro	CAA	Gln	CGA	Arg	A
	CUG	Leu	CCG	Pro	CAG	Gln	CGG	Arg	G
A	AUU	Ile	ACU	Thr	AAU	Asn	AGU	Ser	U
	AUC	Ile	ACC	Thr	AAC	Asn	AGC	Ser	C
	AUA	Ile	ACA	Thr	AAA	Lys	AGA	Arg	A
	AUG*	Met	ACG	Thr	AAG	Lys	AGG	Arg	G
G	GUU	Val	GCU	Ala	GAU	Asp	GGU	Gly	U
	GUC	Val	GCC	Ala	GAC	Asp	GGC	Gly	C
	GUA	Val	GCA	Ala	GAA	Glu	GGA	Gly	A
	GUG	Val	GCG	Ala	GAG	Glu	GGG	Gly	G

* AUG also serves as the principal initiation codon.

3. For the 15 amino acids coded for by two, three, or four triplets, it is only the third letter of the code that varies. For example, glycine is coded for by the triplets GGA, GGG, GGC, and GGU.
4. There is no ambiguity in the code, meaning that each triplet codes for one amino acid.

Finally, we must ask one last question about the genetic code: Is the code universal? That is, is it the same for all organisms? Every bit of experimental evidence available today from the study of viruses, bacteria, and higher animals, including humans, indicates that the code is universal. Furthermore, the fact that it is the same for all these organisms means that it has been the same over millions of years of evolution.

Example 20.5

During transcription, a portion of mRNA is synthesized with the following base sequence:

5′-AUG-GUA-CCA-CAU-UUG-UGA-3′

(a) Write the nucleotide sequence of the DNA from which this portion of mRNA was synthesized.
(b) Write the primary structure of the polypeptide coded for by the given section of mRNA.

Strategy

During transcription, mRNA is synthesized from a DNA strand beginning from the 3′ end of the template. The DNA strand must be the complement of the newly synthesized mRNA strand.

Solution

(a)

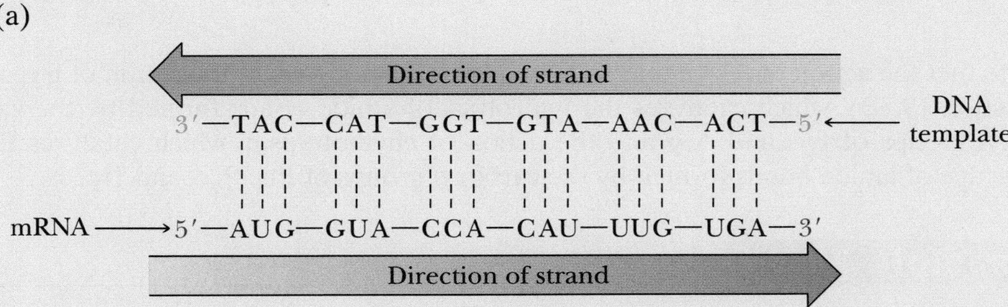

Note that the codon UGA codes for termination of the growing polypeptide chain; therefore, the sequence given in this problem codes for a pentapeptide only.

(b) The sequence of amino acids is shown in the following mRNA strand:

5′-AUG-GUA-CCA-CAU-UUG-UGA-3′
met---val---pro---his---leu---stop

See problem 20.44

Problem 20.5

The following section of DNA codes for oxytocin, a polypeptide hormone:

3′-ACG-ATA-TAA-GTT-TTA-ACG-GGA-GAA-CCA-ACT-5′

(a) Write the base sequence of the mRNA synthesized from this section of DNA.
(b) Given the sequence of bases in part (a), write the primary structure of oxytocin.

20.5 How Is DNA Sequenced?

As recently as 1975, the task of determining the primary structure of a nucleic acid was thought to be far more difficult than determining the primary structure of a protein. Nucleic acids, it was reasoned, contain only 4 different units, whereas proteins contain 20 different units. With only 4 different units, there are fewer specific sites for selective cleavage, distinctive sequences are more difficult to recognize, and there is greater chance of ambiguity in the assignment of sequences. Two breakthroughs reversed this situation. First was the development of a type of electrophoresis called **polyacrylamide gel electrophoresis**, a technique so sensitive that it is possible to separate nucleic acid fragments that differ from one another in only a single nucleotide. The second breakthrough was the discovery of a class of enzymes called **restriction endonucleases**, isolated chiefly from bacteria.

Restriction endonuclease An enzyme that catalyzes the hydrolysis of a particular phosphodiester bond within a DNA strand.

A. Restriction Endonucleases

A restriction endonuclease recognizes a set pattern of four to eight nucleotides and cleaves a DNA strand by hydrolyzing the linking phosphodiester bonds at any site that contains that particular sequence. Molecular biologists have now isolated close to 1,000 restriction endonucleases and characterized their specificities; each cleaves DNA at a different site and produces a different set of restriction fragments. *E. coli*, for example, has a restriction endonuclease EcoRI (pronounced eeko-are-one) that recognizes the hexanucleotide sequence GAATTC and cleaves it between G and A:

cleavage here

$$5'\text{---G-A-A-T-T-C-----}3' \xrightarrow{\text{EcoRI}} 5'\text{---G} + 5'\text{-A-A-T-T-C-----}3'$$

Note that the action of restriction endonucleases is analogous to the action of trypsin (Section 19.5B), which catalyzes the hydrolysis of amide bonds formed by the carboxyl groups of Lys and Arg and the action of chymotrypsin, which catalyzes the cleavage of amide bonds formed by the carboxyl groups of Phe, Tyr, and Trp.

Example 20.6

The following is a section of the gene coding for bovine rhodopsin, along with a table listing several restriction endonucleases, their recognition sequences, and their hydrolysis sites:

5′GTCTACAACCCGGTCATCTACTATCATGATCAACAAGCAGTTCCGGAACT-3′

Enzyme	Recognition Sequence	Enzyme	Recognition Sequence
A*lu*I	AG↓CT	H*pa*II	C↓CGG
B*al*I	TGG↓CCA	M*bo*I	↓GATC
F*nu*DII	CG↓CG	N*ot*I	GC↓GGCCGC
H*ea*III	GG↓CC	S*ac*I	GAGCT↓C

Which endonucleases will catalyze the cleavage of the given section of DNA?

Strategy

Look for each recognition sequence in the DNA strand in the 5′– 3′ direction.

Solution

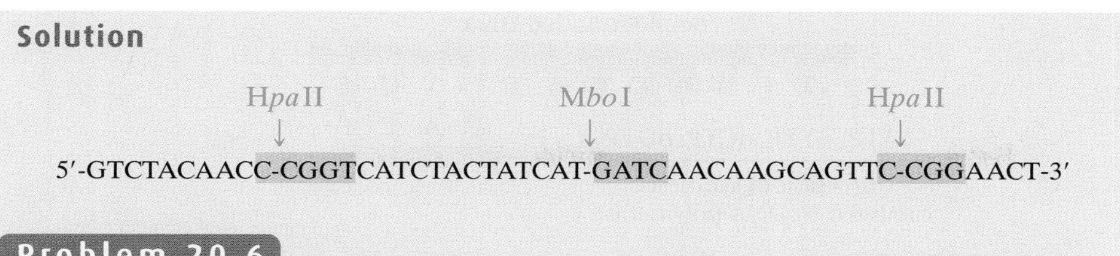

Problem 20.6

The following is another section of the bovine rhodopsin gene:

5′-ACGTCGGGTCGTCGTCCTCTCGCGGTGGTGAGTCTTCCGGCTCTTCT-3′

Which of the endonucleases given in Example 20.6 will catalyze the cleavage of this section?

B. Methods for Sequencing Nucleic Acids

The sequencing of DNA begins with the site-specific cleavage of double-stranded DNA by one or more restriction endonucleases into smaller fragments called **restriction fragments**. Each restriction fragment is then sequenced separately, overlapping base sequences are identified, and the entire sequence of bases is subsequently deduced.

Two methods for sequencing restriction fragments have been devised. The first, developed by Allan Maxam and Walter Gilbert and known as the **Maxam–Gilbert method**, depends on base-specific chemical cleavage. The second method, developed by Frederick Sanger and known as the **chain termination** or **dideoxy method**, depends on the interruption of DNA-polymerase-catalyzed synthesis. Sanger and Gilbert shared the 1980 Nobel Prize in Chemistry for their "development of chemical and biochemical analysis of DNA structure." Sanger's dideoxy method is currently more widely used, so we concentrate on it.

Sanger dideoxy method A method, developed by Frederick Sanger, for sequencing DNA molecules.

C. DNA Replication in Vitro

To appreciate the rationale for the dideoxy method, we must first understand certain aspects of the biochemistry of DNA replication. First, DNA replication takes place when cells divide. During replication, the sequence of nucleotides in one strand is copied as a complementary strand to form the second strand of a double-stranded DNA molecule. Synthesis of the complementary strand is catalyzed by the enzyme DNA polymerase. As shown in the following equation, the DNA chain grows by adding each new unit to the free 3′-OH group of the chain:

The DNA chain grows from the 5′ end to the 3′ end

DNA polymerase

Pyrophosphate

A deoxynucleotide triphosphate (dNTP)

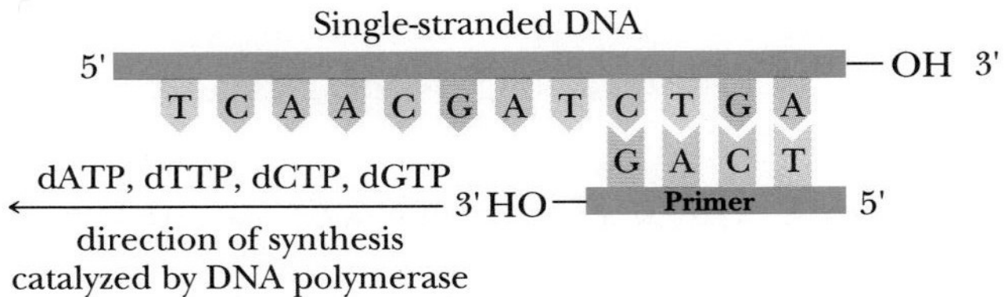

Figure 20.13
DNA polymerase catalyzes the synthesis of the complementary strand of DNA in vitro, using single-stranded DNA as a template, provided that both the four deoxynucleotide triphosphate (dNTP) monomers and a primer are present. The primer provides a short stretch of double-stranded DNA by base pairing with its complement on the single-stranded DNA.

DNA polymerase will also carry out this synthesis in vitro with single-stranded DNA as a template, provided that both the four deoxynucleotide triphosphate (dNTP) monomers and a primer are present. A **primer** is an oligonucleotide that is capable of forming a short section of double-stranded DNA (dsDNA) by base pairing with its complement on a single-stranded DNA (ssDNA). Because a new DNA strand grows from its 5′ to 3′ end, the primer must have a free 3′-OH group to which the first nucleotide of the growing chain is added (Figure 20.13).

D. The Chain Termination, or Dideoxy, Method

The key to the chain termination method is the addition of a 2′,3′-dideoxynucleoside triphosphate (ddNTP) to the synthesizing medium:

A 2′,3′-dideoxynucleoside triphosphate (ddNTP)

Because a ddNTP has no —OH group at the 3′ position, it cannot serve as an acceptor for the next nucleotide to be added to the growing polynucleotide chain. Thus, chain synthesis is terminated at any point where a ddNTP becomes incorporated and hence the designation *chain termination method*.

In the chain termination method, a single-stranded DNA of unknown sequence is mixed with primer and divided into four separate reaction mixtures. To each reaction mixture are added all four deoxynucleoside triphosphates (dNTPs), one of which is labeled in the 5′ phosphoryl group with phosphorus-32 (^{32}P) so that the newly synthesized fragments can be visualized by autoradiography.

$$^{32}_{15}P \longrightarrow ^{32}_{16}S + \text{Beta particle} + \text{Gamma rays}$$

Also added to each reaction mixture are DNA polymerase and one of the four ddNTPs. The ratio of dNTPs to ddNTP in each reaction mixture is adjusted so that a ddNTP is incorporated only infrequently. In each reaction mixture, DNA synthesis takes place, but in a given population of molecules, synthesis is interrupted at every possible site (Figure 20.14).

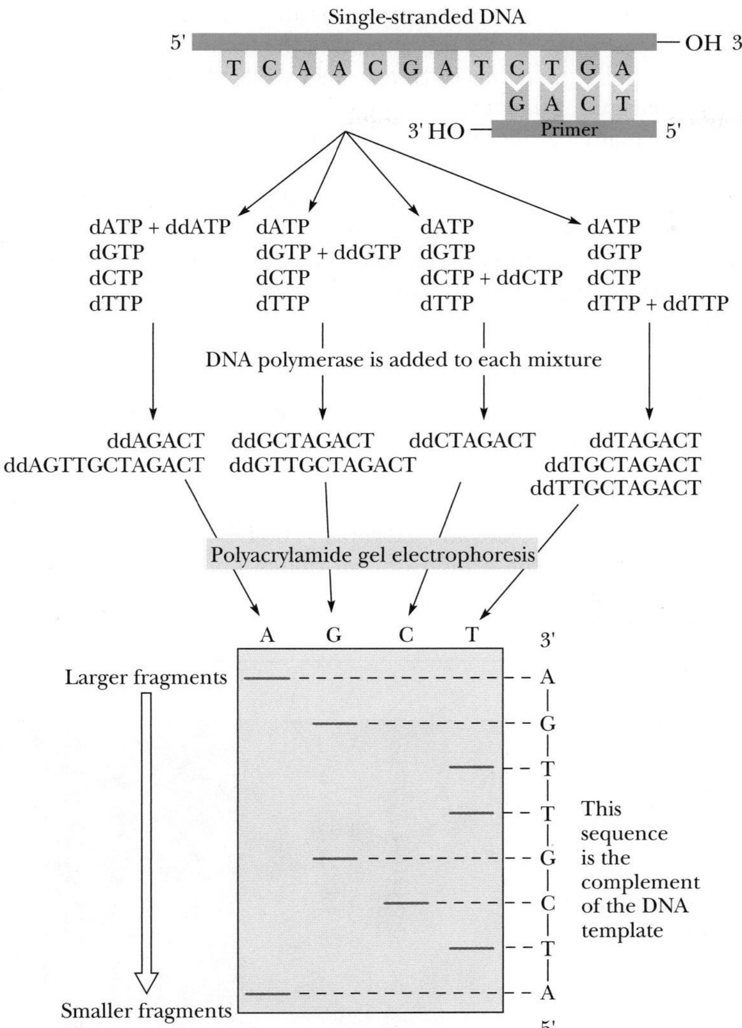

Figure 20.14
The chain termination, or dideoxy, method of DNA sequencing. The primer–DNA template is divided into four separate reaction mixtures, to each of which are added the four dNTPs, DNA polymerase, and one of the four ddNTPs. Synthesis is interrupted at every possible site. The mixtures of oligonucleotides are separated by polyacrylamide gel electrophoresis. The base sequence of the DNA complement is read from the bottom to the top (from the 5′ end to the 3′ end) of the developed gel.

When gel electrophoresis of each reaction mixture is completed, a piece of X-ray film is placed over the gel, and gamma rays released by the radioactive decay of ^{32}P darken the film and create a pattern on it that is an image of the resolved oligonucleotides. The base sequence of the complement of the original single-stranded template is then read directly from bottom to top of the developed film.

A variation on this method is to use a single reaction mixture, with each of the four ddNTPs labeled with a different fluorescent indicator. Each label is then detected by its characteristic spectrum. Automated DNA-sequencing machines using this variation are capable of sequencing up to 10,000 base pairs per day.

E. Sequencing the Human Genome

As nearly everyone knows, the sequencing of the human genome was announced in the spring of 2000 by two competing groups: the so-called Human Genome Project, a loosely linked consortium of publicly funded groups, and a private company called Celera. Actually, this milestone represents, not a complete sequence, but a "rough draft," constituting about 85% of the entire genome. The methodology for sequencing the human genome was based on a refinement of the techniques described earlier that use massively parallel separations of fragments by electrophoresis in capillary tubes. The Celera approach utilized some 300 of the fastest sequencing machines in parallel, each operating on many parallel DNA fragments. Supercomputers were employed to help assemble and compare millions of overlapping sequences.

This achievement represents the beginning of a new era of molecular medicine. From now on, specific genetic deficiencies leading to inherited diseases will be understood on a molecular basis, and new therapies targeted at shutting down undesired genes or turning on desired ones will be developed.

CHEMICAL CONNECTIONS 20B

DNA Fingerprinting

Each human being has a genetic makeup consisting of approximately 3 billion pairs of nucleotides, and, except for identical twins, the base sequence of DNA in one individual is different from that of every other individual. As a result, each person has a unique DNA "fingerprint." To determine a DNA fingerprint, a sample of DNA from a trace of blood, skin, or other tissue is treated with a set of restriction endonucleases, and the 5′ end of each restriction fragment is labeled with phosphorus-32. The resulting ^{32}P-labeled restriction fragments are then separated by polyacrylamide gel electrophoresis and visualized by placing a photographic plate over the developed gel.

In the DNA fingerprint patterns shown in the accompanying figure, lanes 1, 5, and 9 represent internal standards, or control lanes. The lanes contain the DNA fingerprint pattern of a standard virus treated with a standard set of restriction endonucleases. Lanes 2, 3, and 4 were used in a paternity suit. The DNA fingerprint of the mother in lane 4 contains five bands, which match with five of the six bands in the DNA fingerprint of the child in lane 3. The DNA fingerprint of the alleged father in lane 2 contains six bands, three of which match with bands in the DNA fingerprint of the child. Because the child inherits only half of its genes from the father, only half of the child's and father's DNA fingerprints are expected to match. In this instance, the paternity suit was won on the basis of the DNA fingerprint matching.

Lanes 6, 7, and 8 contain DNA fingerprint patterns used as evidence in a rape case. Lanes 7 and 8 are DNA fingerprints of semen obtained from the rape victim. Lane 6 is the DNA fingerprint pattern of the alleged rapist. The DNA fingerprint patterns of the semen do not match and helped to exclude the suspect from the case.

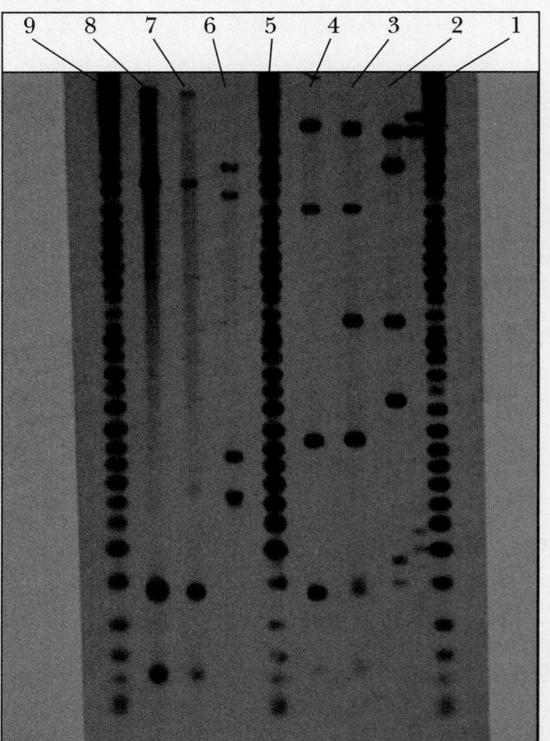

DNA fingerprint. *(Courtesy of Dr. Lawrence Koblinsky)*

QUESTION

Explain the basis of separation for the bands shown on the DNA fingerprinting gel.

Key Terms and Concepts

Summary of Key Questions

20.1 What Are Nucleosides and Nucleotides?

- **Nucleic acids** are composed of three types of monomer units: heterocyclic aromatic amine bases derived from purine and pyrimidine, the monosaccharides D-ribose or 2-deoxy-D-ribose, and phosphate ions.
- A **nucleoside** is a compound containing D-ribose or 2-deoxy-D-ribose bonded to a heterocyclic aromatic amine base by a *β*-*N*-glycosidic bond.
- A **nucleotide** is a nucleoside in which a molecule of phosphoric acid is esterified with an —OH of the monosaccharide, most commonly either the 3′—OH or the 5′—OH.
- Nucleoside mono-, di-, and triphosphates are strong polyprotic acids and are extensively ionized at pH 7.0. At that pH, adenosine triphosphate, for example, is a 50:50 mixture of ATP^{3-} and ATP^{4-}.

20.2 What Is the Structure of DNA?

- The **primary structure of deoxyribonucleic acids (DNA)** consists of units of 2-deoxyribose bonded by 3′,5′-phosphodiester bonds.
- A heterocyclic aromatic amine base is bonded to each deoxyribose unit by a *β*-*N*-glycosidic bond.
- The sequence of bases is read from the 5′ end of the polynucleotide strand to the 3′ end.
- The heart of the **Watson–Crick model** of the structure of DNA is the postulate that a molecule of DNA consists of two antiparallel polynucleotide strands coiled in a right-handed manner about the same axis to form a **double helix**.
- Purine and pyrimidine bases point inward toward the axis of the helix and are always paired G-C and A-T.
- In **B-DNA**, base pairs are stacked one on top of another with a spacing of 2.9 Å and 10 base pairs per 34-Å helical repeat.
- In **A-DNA**, bases are stacked with a spacing of 2.9 Å between base pairs and 11 base pairs per 29-Å helical repeat.
- In Z-DNA, the helix is left-handed and the minor groove is the same width as the major groove. Bases are stacked with a spacing of 3.7 Å between base pairs and 12 base pairs per 45.6-Å helical repeat.
- The tertiary structure of DNA is commonly referred to as **supercoiling**.
- **Circular DNA** is a type of double-stranded DNA in which the ends of each strand are joined by phosphodiester bonds.

- The opening of one strand followed by the partial unwinding and rejoining of the ends introduces strain in the nonhelical gap. The strain can be spread over the entire molecule of circular DNA by the introduction of **superhelical twists**.
- **Histones** are particularly rich in lysine and arginine and, therefore, have an abundance of positive charges.
- The association of DNA with histones produces a pigment called **chromatin**.

20.3 What Are Ribonucleic Acids (RNA)?

- There are two important differences between the primary structure of **ribonucleic acids (RNA)** and DNA:
  (1) The monosaccharide unit in RNA is D-ribose.
  (2) Both RNA and DNA contain the purine bases adenine (A) and guanine (G) and the pyrimidine base cytosine (C). As the fourth base, however, RNA contains uracil (U), whereas DNA contains thymine (T).
- **Messenger RNA (mRNA)** molecules are produced in the process called transcription, and they carry genetic information from the DNA in the nucleus directly to the cytoplasm where protein is synthesized. mRNA consists of a chain of nucleotides whose sequence is exactly complementary to that of one of the DNA strands.
- **Transfer RNA (tRNA)** molecules are relatively small RNA molecules that transport amino acids to the site of protein synthesis in ribosomes.
- **Ribosomal RNA (rRNA)** is a type of RNA that is complexed with proteins and makes up the ribosomes used in the translation of mRNA into protein.

20.4 What Is the Genetic Code?

- A **gene** is a segment of a DNA molecule that carries the sequences of bases that direct the synthesis of one particular protein or RNA molecule.
- The **genetic code** consists of nucleosides in groups of three; that is, it is a triplet code. Only 61 triplets code for amino acids; the remaining three code for the termination of polypeptide synthesis.

20.5 How Is DNA Sequenced?

- **Restriction endonucleases** recognize a set pattern of four to eight nucleotides and cleave a DNA strand by hydrolyzing the linking phosphodiester bonds at any site that contains that particular sequence.
- In the **chain termination or dideoxy method** of DNA sequencing developed by Frederick Sanger, a primer-DNA template is divided into four separate reaction mixtures. To each is added the four dNTPs, one of which is labeled with phosphorus-32. Also added are DNA polymerase and one of the four ddNTPs. Synthesis is interrupted at every possible site. The mixtures of newly synthesized oligonucleotides are separated by polyacrylamide gel electrophoresis and visualized by autoradiography. The base sequence of the DNA complement of the original DNA template is read from the bottom to the top (from the 5′ end to the 3′ end) of the developed photographic plate.

Quick Quiz

Answer true or false to the following questions to assess your general knowledge of the concepts in this chapter. If you have difficulty with any of them, you should review the appropriate section in the chapter (shown in parentheses) before attempting the more challenging end-of-chapter problems.

1. Endonucleases are used to make DNA strands radioactive for sequencing. (20.5)
2. A strand of RNA is the product of transcription. (20.4)
3. There are the same number of hydrogen bonds between a G-C base pair as there are between an A-T base pair. (20.2)

4. There are multiple types of RNA, each serving a different function in a cell. (20.3)
5. The dideoxy method of DNA sequencing uses a nucleoside triphosphate that is deoxygenated at both the 5′ and 3′ positions of the pentose ring. (20.5)
6. The primary structure of DNA is the sequence of its bases and is always read from the 5′ end to the 3′ end. (20.2)
7. A primer is a short section of single-stranded DNA used in the synthesis of DNA. (20.5)
8. There is an equal number of each type of nucleoside (A, T, G, and C) in the DNA of humans. (20.2)
9. The four heterocyclic amine bases of DNA are uracil, cytosine, guanine, and adenine. (20.1)
10. A-DNA is the most common form of DNA in living systems. (20.2)
11. Histones complex with DNA mainly through hydrogen bonding interactions. (20.2)
12. The ionizable oxygen atoms of monophosphate, diphosphate, and triphosphate groups are all protonated at biological pH. (20.2)
13. Messenger RNA (mRNA) molecules carry the amino acids necessary for protein synthesis. (20.3)
14. A codon is the amino acid that is produced by a three-letter sequence of DNA. (20.4)
15. There are an equal number of purines and pyrimidines in the DNA of humans. (20.2)
16. A nucleoside is a compound containing a pentose bonded to a heterocyclic aromatic amine base by a β-*N*-glycosidic bond, where a molecule of phosphoric acid has been esterified with a free hydroxyl of the pentose ring. (20.1)
17. The double helix is an example of the tertiary structure of DNA. (20.2)
18. The carbohydrate ring in DNA is deoxygenated at the 3′ position. (20.2)

Answers: (1) F (2) T (3) F (4) T (5) F (6) T (7) T (8) F (9) F (10) F (11) F (12) F (13) F (14) F (15) T (16) F (17) F (18) F

Problems

A problem marked with an asterisk indicates an applied "real world" problem. Answers to problems whose numbers are printed in blue are given in Appendix D.

Section 20.1 Nucleosides and Nucleotides

20.7 Write the names and structural formulas for the five nitrogen bases found in nucleosides.

20.8 Write the names and structural formulas for the two monosaccharides found in nucleosides.

20.9 Explain the difference in structure between a nucleoside and a nucleotide.

20.10 Following are structural formulas for cytosine and thymine:

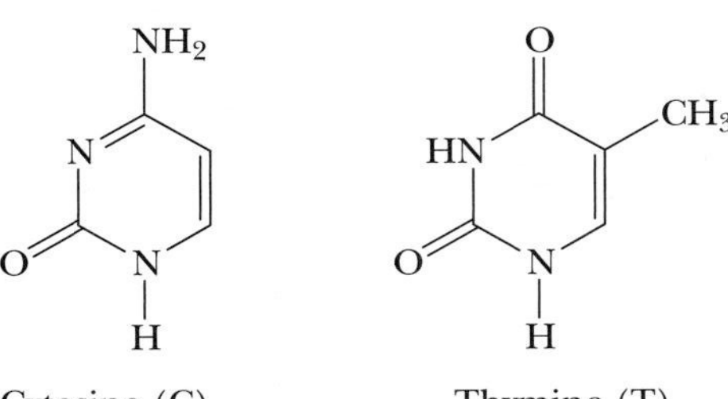

Cytosine (C) Thymine (T)

Draw two additional tautomeric forms for cytosine and three for thymine.

20.11 Draw a structural formula for a nucleoside composed of **(See Example 20.1)**

(a) β-D-Ribose and adenine

(b) β-2-Deoxy-D-ribose and cytosine

20.12 Nucleosides are stable in water and in dilute base. In dilute acid, however, the glycosidic bond of a nucleoside undergoes hydrolysis to give a pentose and a heterocyclic aromatic amine base. Propose a mechanism for this acid-catalyzed hydrolysis.

***20.13** Draw a structural formula for each nucleotide and estimate its net charge at pH 7.4, the pH of blood plasma: **(See Example 20.2)**

(a) 2′-Deoxyadenosine 5′-triphosphate (dATP)

(b) Guanosine 3′-monophosphate (GMP)

(c) 2′-Deoxyguanosine 5′-diphosphate (dGDP)

Section 20.2 The Structure of DNA

20.14 Why are deoxyribonucleic acids called acids? What are the acidic groups in their structure?

***20.15** Human DNA is approximately 30.4% A. Estimate the percentages of G, C, and T and compare them with the values presented in Table 20.2.

20.16 Draw a structural formula for the DNA tetranucleotide 5′-A-G-C-T-3′. Estimate the net charge on this tetranucleotide at pH 7.0. What is the complementary tetranucleotide of this sequence? **(See Example 20.3)**

20.17 List the postulates of the Watson–Crick model of DNA secondary structure.

20.18 The Watson–Crick model is based on certain experimental observations of base composition and molecular dimensions. Describe these observations and show how the model accounts for each.

20.19 List the compositions, abbreviations, and structures of the various nucleotides in DNA.

20.20 What is meant by the term *complementary bases*?

20.21 Discuss the role of hydrophobic interactions in stabilizing double-stranded DNA.

20.22 In terms of hydrogen bonding, which is more stable, an A-T base pair or a G-C base pair?

Section 20.3 Ribonucleic Acids

20.23 Describe the differences between mRNA, tRNA, and rRNA.

20.24 List the compositions, abbreviations, and structural formulas of the most common nucleotides in RNA.

20.25 Compare the degree of hydrogen bonding in the base pair A-T found in DNA with that in the base pair A-U found in RNA.

20.26 Compare DNA and RNA in these ways:

(a) Monosaccharide units

(b) Principal purine and pyrimidine bases

(c) Primary structure

(d) Location in the cell

(e) Function in the cell

20.27 What type of RNA has the shortest lifetime in cells?

20.28 Write the mRNA complement of 5′-ACCGTTAAT-3′. Be certain to label which is the 5′ end and which is the 3′ end of the mRNA strand. **(See Example 20.4)**

20.29 Write the mRNA complement of 5′-TCAACGAT-3′. **(See Example 20.4)**

Section 20.4 The Genetic Code

20.30 What is the genetic code?

20.31 Why are at least three nucleotides needed for one unit of the genetic code?

20.32 Briefly outline the biosynthesis of proteins, starting from DNA.

20.33 How does the protein-synthesizing machinery of the cell know when a genetic sequence is complete?

20.34 At elevated temperatures, nucleic acids become denatured; that is, they unwind into single-stranded DNA. Account for the observation that the higher the G-C content of a nucleic acid, the higher the temperature required for its thermal denaturation.

20.35 Write the DNA complement of 5′-ACCGTTAAT-3′. Be certain to label which is the 5′ end and which is the 3′ end of the complement strand. **(See Example 20.4)**

20.36 Write the DNA complement of 5′-TCAACGAT-3′. **(See Example 20.4)**

20.37 What does it mean to say that the genetic code is degenerate?

20.38 Aspartic acid and glutamic acid have carboxyl groups on their side chains and are called acidic amino acids. Compare the codons for these two amino acids.

20.39 Compare the structural formulas of the aromatic amino acids phenylalanine and tyrosine. Compare the codons for these two amino acids as well.

20.40 Glycine, alanine, and valine are classified as nonpolar amino acids. Compare their codons. What similarities do you find? What differences?

20.41 Codons in the set CUU, CUC, CUA, and CUG all code for the amino acid leucine. In this set, the first and second bases are identical, and the identity of the third base is irrelevant. For what other sets of codons is the third base also irrelevant, and for what amino acid(s) does each set code?

20.42 Compare the amino acids coded for by the codons with a pyrimidine, either U or C, as the second base. Do the majority of the amino acids specified by these codons have hydrophobic or hydrophilic side chains?

20.43 Compare the amino acids coded for by the codons with a purine, either A or G, as the second base. Do the majority of the amino acids specified by these codons have hydrophilic or hydrophobic side chains?

20.44 What polypeptide is coded for by this mRNA sequence? **(See Example 20.5)**

5′-GCU-GAA-GUC-GAG-GUG-UGG-3′

***20.45** The alpha chain of human hemoglobin has 141 amino acids in a single polypeptide chain. Calculate the minimum number of bases on DNA necessary to code for the alpha chain. Include in your calculation the bases necessary for specifying the termination of polypeptide synthesis.

***20.46** In HbS, the human hemoglobin found in individuals with sickle-cell anemia, glutamic acid at position 6 in the beta chain is replaced by valine.

(a) List the two codons for glutamic acid and the four codons for valine.

(b) Show that one of the glutamic acid codons can be converted to a valine codon by a single substitution mutation—that is, by changing one letter in one codon.

Additional Problems

***20.47** Two drugs used in the treatment of acute leukemia are 6-mercaptopurine and 6-thioguanine:

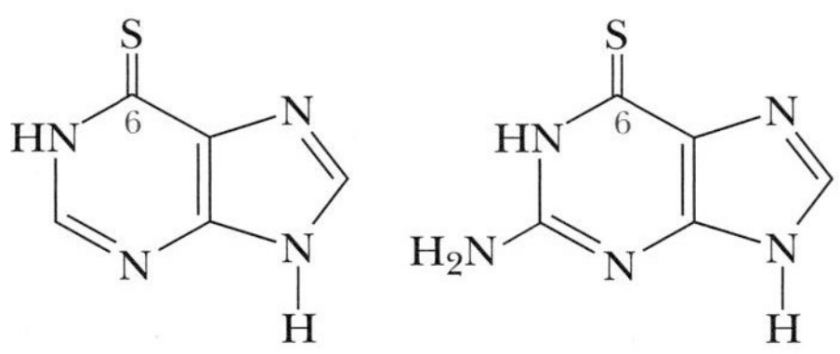

6-Mercaptopurine 6-Thioguanine

Note that, in each drug, the oxygen at carbon 6 of the parent molecule is replaced by divalent sulfur. Draw structural formulas for the enethiol (the sulfur equivalent of an enol) forms of 6-mercaptopurine and 6-thioguanine.

***20.48** Cyclic-AMP, first isolated in 1959, is involved in many diverse biological processes as a regulator of metabolic and physiological activity. In this compound, a single phosphate group is esterified with both the 3′ and 5′ hydroxyls of adenosine. Draw a structural formula of cyclic-AMP.

20.49 Compare the α-helix of proteins and the double helix of DNA in terms of:

(a) The units that repeat in the backbone of the polymer chain.

(b) The projection in space of substituents along the backbone (the R groups in the case of amino acids, purine and pyrimidine bases in the case of double-stranded DNA) relative to the axis of the helix.

***20.50** The loss of three consecutive units of T from the gene that codes for CFTR, a transmembrane conductance regulator protein, results in the disease known as cystic fibrosis. Which amino acid is missing from CFTR to cause this disease?

***20.51** The following compounds have been researched as potential antiviral agents:

(a) Cordycepin (3′-deoxyadenosine)

(b)

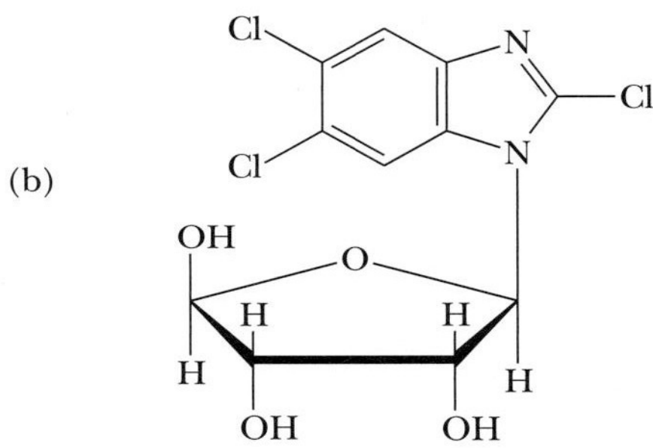

2,5,6-Trichloro-1-(β-D-ribofuranosyl) benzimidazole

(c)

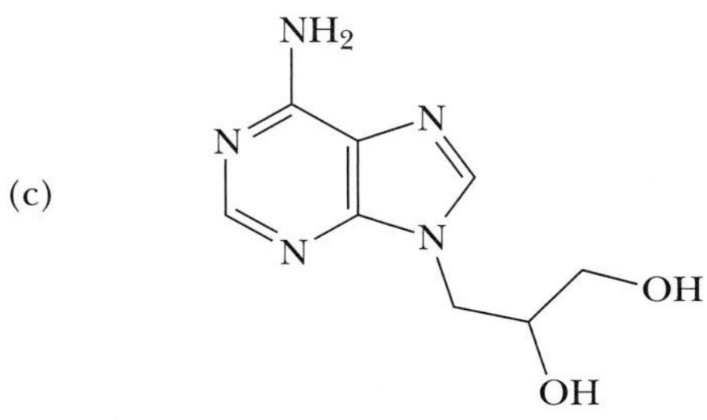

9-(2,3-Dihydroxypropyl) adenine

Suggest how each of these compounds might block the synthesis of RNA or DNA.

20.52 Name the type of covalent bond(s) joining monomers in these biopolymers:

(a) Polysaccharides

(b) Polypeptides

(c) Nucleic acids

***20.53** The ends of chromosomes, called telomeres, can form unique and nonstandard structures. One example is the presence of base pairs between units of guanosine. Show how two guanine bases can pair with each other by hydrogen bonding:

Guanine

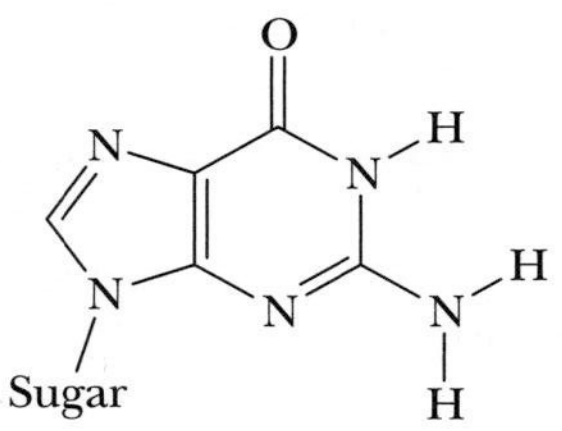

20.54 One synthesis of zidovudine (AZT) involves the following reaction (DMF is the solvent *N,N*-dimethylformamide):

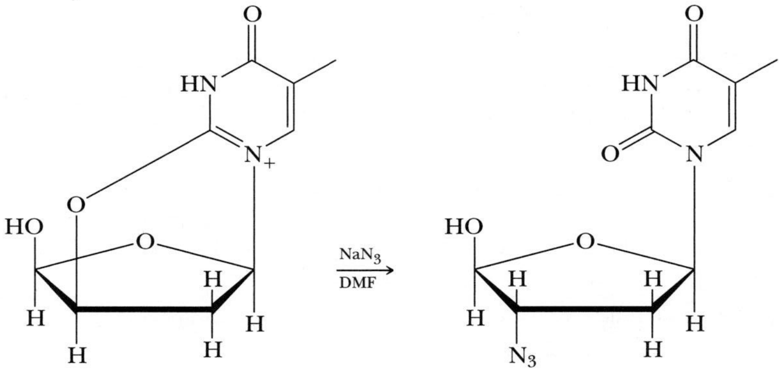

$\xrightarrow[\text{DMF}]{\text{NaN}_3}$

What type of reaction is this?

21 Lipids

KEY QUESTIONS

CHEMICAL CONNECTIONS

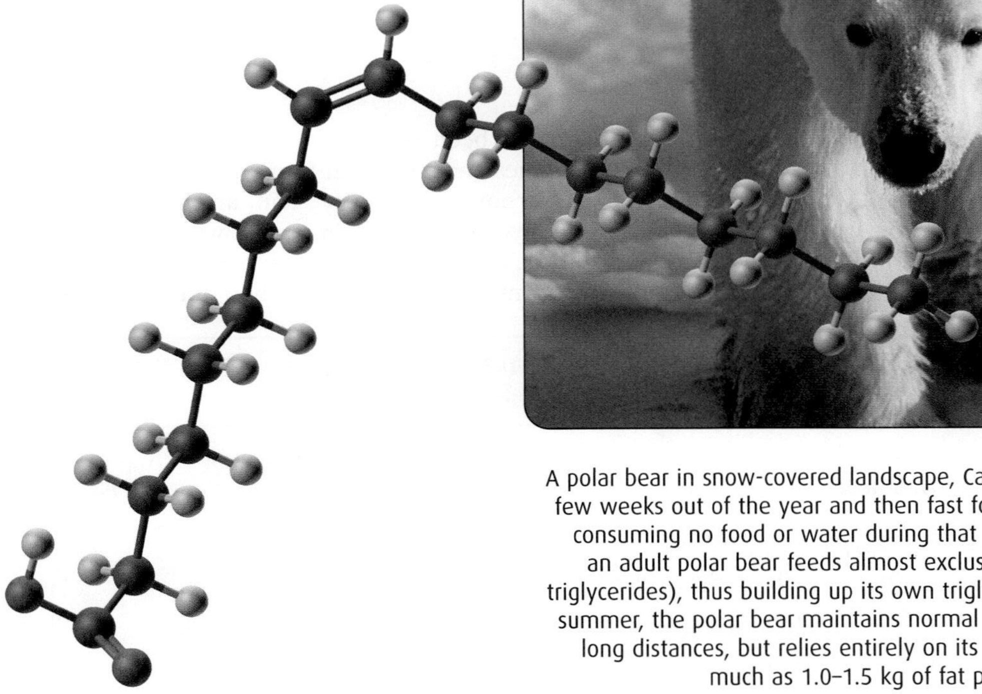

A polar bear in snow-covered landscape, Canada. Polar bears eat only during a few weeks out of the year and then fast for periods of eight months or more, consuming no food or water during that period. Eating mainly in the winter, an adult polar bear feeds almost exclusively on seal blubber (composed of triglycerides), thus building up its own triglyceride reserves. Through the Arctic summer, the polar bear maintains normal physiological activity, roaming over long distances, but relies entirely on its body fat for sustenance, burning as much as 1.0–1.5 kg of fat per day. Inset: A model of oleic acid. *(Daniel J. Cox/Stone/Getty Images)*

Lipids are a heterogeneous group of naturally occurring organic compounds, classified together on the basis of their common solubility properties. Lipids are insoluble in water, but soluble in relatively nonpolar aprotic organic solvents, including diethyl ether, dichloromethane, and acetone.

Lipid A class of biomolecules isolated from plant or animal sources by extraction with nonpolar organic solvents, such as diethyl ether and acetone.

Lipids play three major roles in human biology. First, they are storage depots of chemical energy in the form of triglycerides (fats). While plants store energy in the form of carbohydrates (e.g., starch), humans store it in the form of fat globules in adipose tissue. Second, lipids, in the form of phospholipids, are the water-insoluble components from which biological membranes are constructed. Third, lipids, in the form of steroid hormones, prostaglandins, thromboxanes, and leukotrienes, are chemical messengers. In this chapter, we describe the structures and biological functions of each group of lipids.

21.1 What Are Triglycerides?

Triglyceride (triacylglycerol) An ester of glycerol with three fatty acids.

Animal fats and vegetable oils, the most abundant naturally occurring lipids, are triesters of glycerol and long-chain carboxylic acids. Fats and oils are referred to as **triglycerides** or **triacylglycerols**. Hydrolysis of a triglyceride in aqueous base, followed by acidification, gives glycerol and three fatty acids:

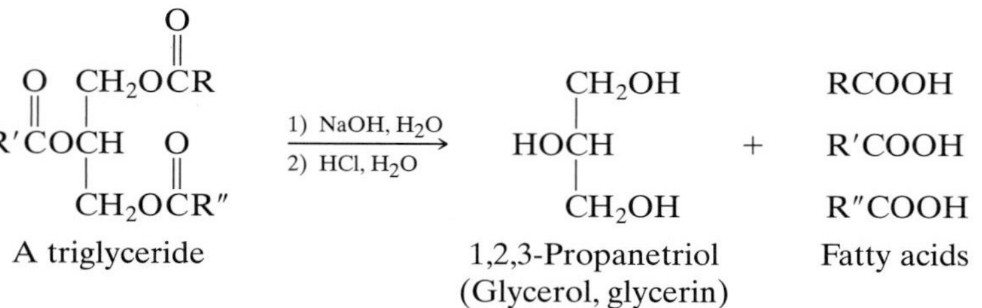

A. Fatty Acids

Fatty acid A long, unbranched-chain carboxylic acid, most commonly of 12 to 20 carbons, derived from the hydrolysis of animal fats, vegetable oils, or the phospholipids of biological membranes.

More than 500 different **fatty acids** have been isolated from various cells and tissues. Given in Table 21.1 are common names and structural formulas for the most abundant of them. The number of carbons in a fatty acid and the number of carbon–carbon double bonds in its hydrocarbon chain are shown by two numbers separated by a colon. In this notation, linoleic acid, for example, is designated as an 18:2 fatty acid; its 18-carbon chain contains two carbon–carbon double bonds.

Some vegetable oils.
(Charles D. Winters)

TABLE 21.1 The Most Abundant Fatty Acids in Animal Fats, Vegetable Oils, and Biological Membranes

Carbon Atoms/ Double Bonds*	Structure	Common Name	Melting Point(°C)
Saturated Fatty Acids			
12:0	$CH_3(CH_2)_{10}COOH$	lauric acid	44
14:0	$CH_3(CH_2)_{12}COOH$	myristic acid	58
16:0	$CH_3(CH_2)_{14}COOH$	palmitic acid	63
18:0	$CH_3(CH_2)_{16}COOH$	stearic acid	70
20:0	$CH_3(CH_2)_{18}COOH$	arachidic acid	77
Unsaturated Fatty Acids			
16:1	$CH_3(CH_2)_5CH{=}CH(CH_2)_7COOH$	palmitoleic acid	1
18:1	$CH_3(CH_2)_7CH{=}CH(CH_2)_7COOH$	oleic acid	16
18:2	$CH_3(CH_2)_4(CH{=}CHCH_2)_2(CH_2)_6COOH$	linoleic acid	−5
18:3	$CH_3CH_2(CH{=}CHCH_2)_3(CH_2)_6COOH$	linolenic acid	−11
20:4	$CH_3(CH_2)_4(CH{=}CHCH_2)_4(CH_2)_2COOH$	arachidonic acid	−49

*The first number is the number of carbons in the fatty acid; the second is the number of carbon–carbon double bonds in its hydrocarbon chain.

Following are several characteristics of the most abundant fatty acids in higher plants and animals:

1. Nearly all fatty acids have an even number of carbon atoms, most between 12 and 20, in an unbranched chain.
2. The three most abundant fatty acids in nature are palmitic acid (16 : 0), stearic acid (18 : 0), and oleic acid (18 : 1).
3. In most unsaturated fatty acids, the *cis* isomer predominates; the *trans* isomer is rare.
4. Unsaturated fatty acids have lower melting points than their saturated counterparts. The greater the degree of unsaturation, the lower is the melting point. Compare, for example, the melting points of these four 18-carbon fatty acids:

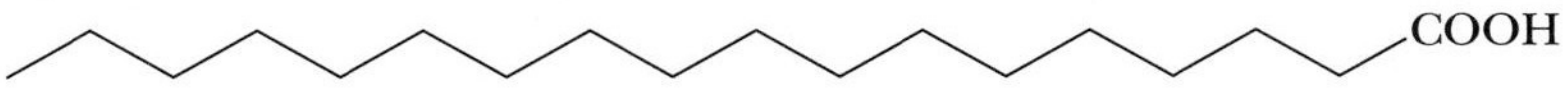

COOH Stearic acid (18 : 0) (mp 70°C)

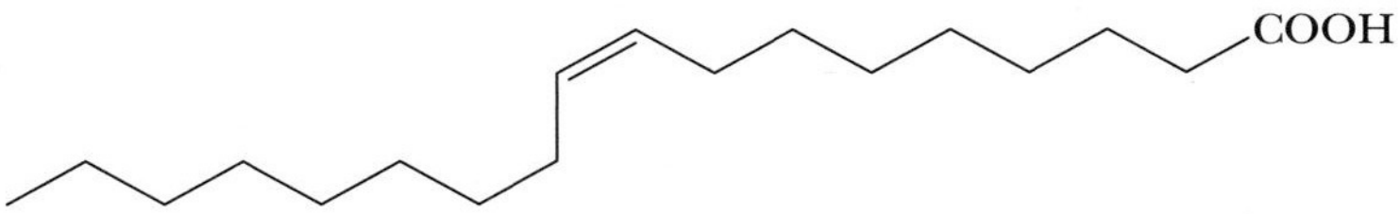

COOH Oleic acid (18 : 1) (mp 16°C)

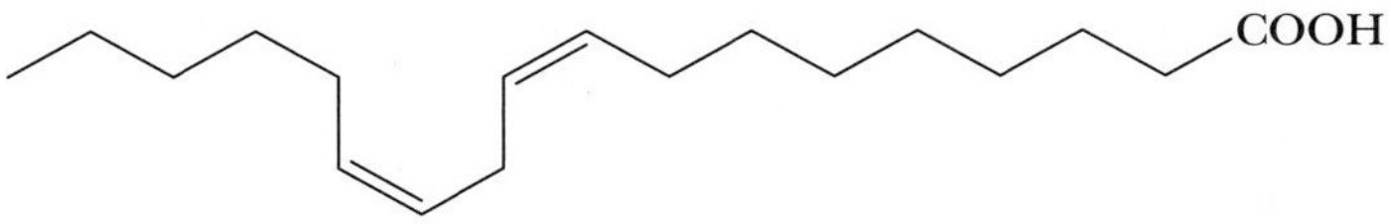

COOH Linoleic acid (18 : 2) (mp –5°C)

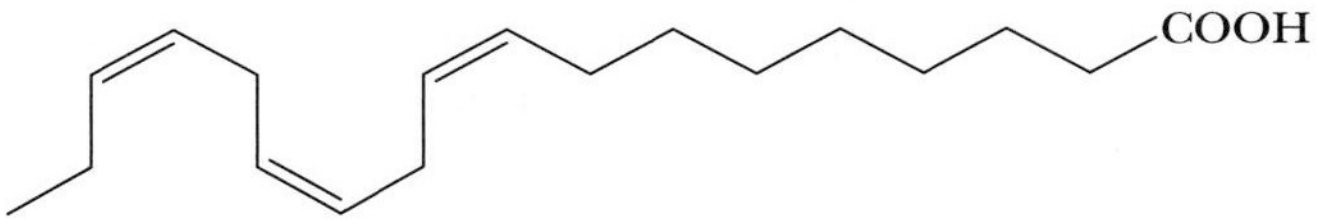

COOH Linolenic acid (18 : 3) (mp –11°C)

Example 21.1

Draw the structural formula of a triglyceride derived from one molecule each of palmitic acid, oleic acid, and stearic acid, the three most abundant fatty acids in the biological world.

Strategy

A triglyceride is a triester of glycerol (a triol) so that each of the three —OH groups of glycerol will form an ester with a carboxylic acid. There are six triglycerides possible using three different fatty acids.

Solution

In this structure, palmitic acid is esterified at carbon 1 of glycerol, oleic acid at carbon 2, and stearic acid at carbon 3:

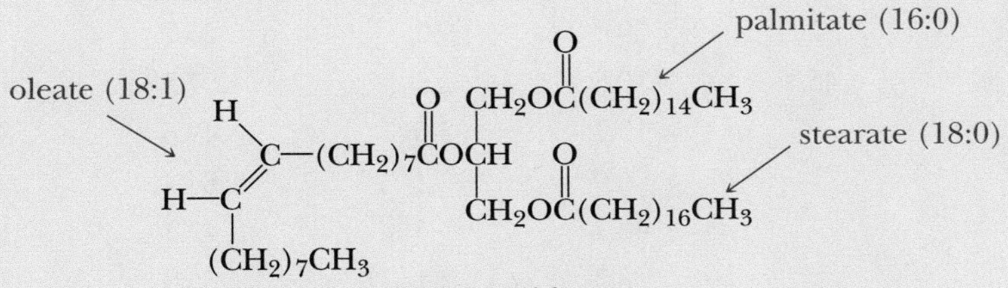

palmitate (16:0)

oleate (18:1)

H–C(=C–H)–$(CH_2)_7COCH$ (O) ; $CH_2OC(CH_2)_{14}CH_3$ (O) ; $CH_2OC(CH_2)_{16}CH_3$ (O) ; $(CH_2)_7CH_3$

stearate (18:0)

A triglyceride

See problem 21.8

Oil A triglyceride that is liquid at room temperature.

Fat A triglyceride that is semisolid or solid at room temperature.

Problem 21.1

(a) How many constitutional isomers are possible for a triglyceride containing one molecule each of palmitic acid, oleic acid, and stearic acid?
(b) Which of the constitutional isomers that you found in Part (a) are chiral?

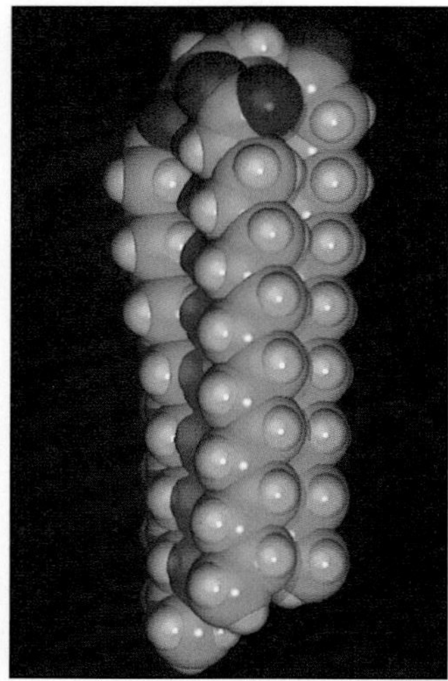

Figure 21.1 Tripalmitin, a saturated triglyceride. *(Brent Iverson, University of Texas)*

B. Physical Properties

The physical properties of a triglyceride depend on its fatty-acid components. In general, the melting point of a triglyceride increases as the number of carbons in its hydrocarbon chains increases and as the number of carbon–carbon double bonds decreases. Triglycerides rich in oleic acid, linoleic acid, and other unsaturated fatty acids are generally liquids at room temperature and are called **oils** (e.g., corn oil and olive oil). Triglycerides rich in palmitic, stearic, and other saturated fatty acids are generally semisolids or solids at room temperature and are called **fats** (e.g., human fat and butter fat). Fats of land animals typically contain approximately 40% to 50% saturated fatty acids by weight (Table 21.2). Most plant oils, on the other hand, contain 20% or less saturated fatty acids and 80% or more unsaturated fatty acids. The notable exception to this generalization about plant oils are the **tropical oils** (e.g., coconut and palm oils), which are considerably richer in low-molecular-weight saturated fatty acids.

The lower melting points of triglycerides that are rich in unsaturated fatty acids are related to differences in the three-dimensional shape between the hydrocarbon chains of their unsaturated and saturated fatty-acid components. Figure 21.1 shows a space-filling model of tripalmitin, a saturated triglyceride. In the model, the hydrocarbon chains lie parallel to each other, giving the molecule an ordered, compact shape. Because of this compact three-dimensional shape and the resulting strength of the dispersion forces (Section 3.8B) between hydrocarbon chains of adjacent molecules, triglycerides that are rich in saturated fatty acids have melting points above room temperature.

The three-dimensional shape of an unsaturated fatty acid is quite different from that of a saturated fatty acid. Recall from Section 21.1A that unsaturated fatty acids of higher organisms are predominantly of the *cis* configuration; *trans* configurations are rare. Figure 21.2 shows a space-filling model of a **polyunsaturated triglyceride**

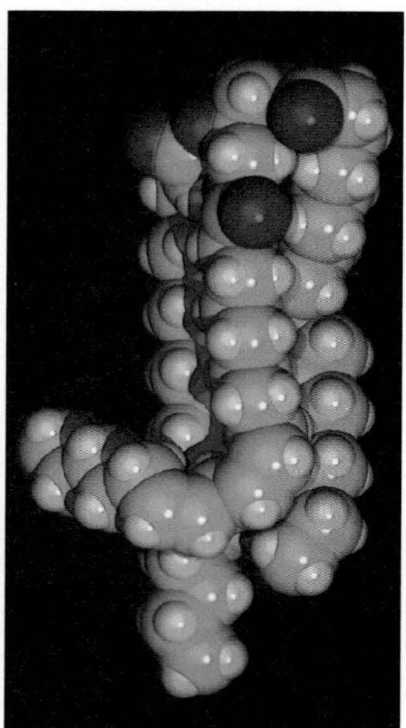

Figure 21.2 A polyunsaturated triglyceride. *(Brent Iverson, University of Texas)*

Polyunsaturated triglyceride A triglyceride having several carbon–carbon double bonds in the hydrocarbon chains of its three fatty acids.

TABLE 21.2 Grams of Fatty Acid per 100 g of Triglyceride of Several Fats and Oils*

	Saturated Fatty Acids			Unsaturated Fatty Acids	
Fat or Oil	**Lauric (12:0)**	**Palmitic (16:0)**	**Stearic (18:0)**	**Oleic (18:1)**	**Linoleic (18:2)**
human fat	—	24.0	8.4	46.9	10.2
beef fat	—	27.4	14.1	49.6	2.5
butter fat	2.5	29.0	9.2	26.7	3.6
coconut oil	45.4	10.5	2.3	7.5	trace
corn oil	—	10.2	3.0	49.6	34.3
olive oil	—	6.9	2.3	84.4	4.6
palm oil	—	40.1	5.5	42.7	10.3
peanut oil	—	8.3	3.1	56.0	26.0
soybean oil	0.2	9.8	2.4	28.9	50.7

*Only the most abundant fatty acids are given; other fatty acids are present in lesser amounts.

derived from one molecule each of stearic acid, oleic acid, and linoleic acid. Each double bond in this polyunsaturated triglyceride has the *cis* configuration.

Polyunsaturated triglycerides have less ordered structures and do not pack together so closely or so compactly as saturated triglycerides. As a consequence, intramolecular and intermolecular dispersion forces are weaker, with the result that polyunsaturated triglycerides have lower melting points than their saturated counterparts.

C. Reduction of Fatty-Acid Chains

For a variety of reasons, in part convenience and in part dietary preference, the conversion of oils to fats has become a major industry. The process is called **hardening** of oils and involves the catalytic reduction (Section 5.8) of some or all of an oil's carbon–carbon double bonds. In practice, the degree of hardening is carefully controlled to produce fats of a desired consistency. The resulting fats are sold for kitchen use (as Crisco®, Spry®, and others). **Margarine** and other butter substitutes are produced by partial hydrogenation of polyunsaturated oils derived from corn, cottonseed, peanut, and soybean oils. To the hardened oil are added β-carotene (to give it a yellow color and make it look like butter), salt, and about 15% milk by volume to form the final emulsion. Vitamins A and D may be added as well. Finally, because the product to this stage is tasteless, acetoin and diacetyl, two compounds that mimic the characteristic flavor of butter, are often added:

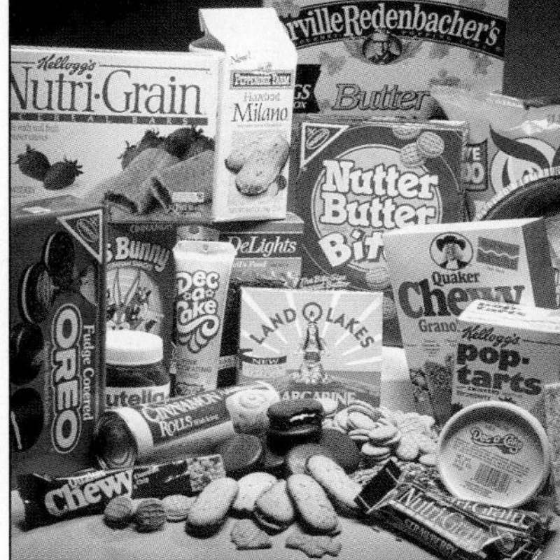

Some common products containing hydrogenated vegetable oils. *(Charles D. Winters)*

$$\underset{\text{3-Hydroxy-2-butanone (Acetoin)}}{CH_3-\overset{\overset{\displaystyle HO}{|}}{CH}-\overset{\overset{\displaystyle O}{\|}}{C}-CH_3} \qquad \underset{\text{2,3-Butanedione (Diacetyl)}}{CH_3-\overset{\overset{\displaystyle O}{\|}}{C}-\overset{\overset{\displaystyle O}{\|}}{C}-CH_3}$$

3-Hydroxy-2-butanone (Acetoin) 2,3-Butanedione (Diacetyl)

21.2 What Are Soaps and Detergents?

A. Structure and Preparation of Soaps

Soap A sodium or potassium salt of a fatty acid.

Natural **soaps** are prepared most commonly from a blend of tallow and coconut oils. In the preparation of tallow, the solid fats of cattle are melted with steam, and the tallow layer that forms on the top is removed. The preparation of soaps begins by boiling these triglycerides with sodium hydroxide. The reaction that takes place is called *saponification* (Latin: *saponem*, soap):

$$\underset{\text{A triglyceride}}{\begin{array}{l} \quad\; CH_2OC(=O)R \\ RC(=O)OCH \\ \quad\; CH_2OC(=O)R \end{array}} + 3NaOH \xrightarrow{\text{saponification}} \underset{\text{1,2,3-Propanetriol (Glycerol; glycerin)}}{\begin{array}{l} CH_2OH \\ CHOH \\ CH_2OH \end{array}} + \underset{\text{Sodium soaps}}{3R\overset{\overset{\displaystyle O}{\|}}{C}O^-Na^+}$$

At the molecular level, saponification corresponds to base-promoted hydrolysis of the ester groups in triglycerides (Section 15.3C). The resulting soaps contain mainly the sodium salts of palmitic, stearic, and oleic acids from tallow and the sodium salts of lauric and myristic acids from coconut oil.

After hydrolysis is complete, sodium chloride is added to precipitate the soap as thick curds. The water layer is then drawn off, and glycerol is recovered by vacuum distillation. The crude soap contains sodium chloride, sodium hydroxide, and other impurities that are removed by boiling the curd in water and reprecipitating with

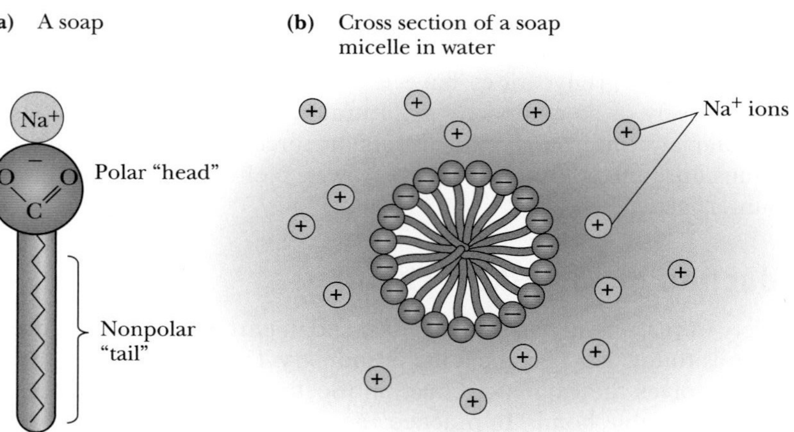

Figure 21.3
Soap micelles. Nonpolar (hydrophobic) hydrocarbon chains are clustered in the interior of the micelle, and polar (hydrophilic) carboxylate groups are on the surface of the micelle. Soap micelles repel each other because of their negative surface charges.

more sodium chloride. After several purifications, the soap can be used as an inexpensive industrial soap without further processing. Other treatments transform the crude soap into pH-controlled cosmetic soaps, medicated soaps, and the like.

B. How Soap Cleans

Soap owes its remarkable cleansing properties to its ability to act as an emulsifying agent. Because the long hydrocarbon chains of natural soaps are insoluble in water, they tend to cluster in such a way as to minimize their contact with surrounding water molecules. The polar carboxylate groups, by contrast, tend to remain in contact with the surrounding water molecules. Thus, in water, soap molecules spontaneously cluster into **micelles** (Figure 21.3). A micelle is a spherical arrangement of organic molecules in water solution clustered so that their **hydrophobic** parts are buried in the sphere and their **hydrophilic** parts are on the surface of the sphere and in contact with water.

Micelle A spherical arrangement of organic molecules in water solution clustered so that their hydrophobic parts are buried inside the sphere and their hydrophilic parts are on the surface of the sphere and in contact with water.

Most of the things we commonly think of as dirt (such as grease, oil, and fat stains) are nonpolar and insoluble in water. When soap and this type of dirt are mixed together, as in a washing machine, the nonpolar hydrocarbon inner parts of the soap micelles "dissolve" the nonpolar dirt molecules. In effect, new soap micelles are formed, this time with nonpolar dirt molecules in the center (Figure 21.4). In this way, nonpolar organic grease, oil, and fat are "dissolved" and washed away in the polar wash water.

Soap micelle with "dissolved" grease
Grease
Soap

Figure 21.4
A soap micelle with a "dissolved" oil or grease droplet.

Soaps, however, have their disadvantages, foremost among which is the fact that they form water-insoluble salts when used in water, which contains Ca(II), Mg(II), or Fe(III) ions (hard water):

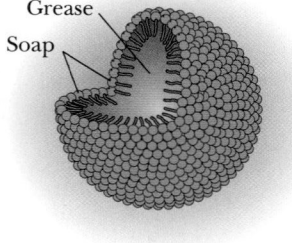

$$\underset{\substack{\text{A sodium soap}\\\text{(soluble in water as micelles)}}}{2CH_3(CH_2)_{14}COO^-Na^+} + Ca^{2+} \longrightarrow \underset{\substack{\text{Calcium salt of a fatty acid}\\\text{(insoluble in water)}}}{[CH_3(CH_2)_{14}COO^-]_2Ca^{2+}} + 2Na^+$$

These calcium, magnesium, and iron salts of fatty acids create problems, including rings around the bathtub, films that spoil the luster of hair, and grayness and roughness that build up on textiles after repeated washings.

C. Synthetic Detergents

After the cleansing action of soaps was understood, chemists were in a position to design a **synthetic detergent**. Molecules of a good detergent, they reasoned, must have a long hydrocarbon chain—preferably, 12 to 20 carbon atoms long—and a polar group at one end of the molecule that does not form insoluble salts with the Ca(II), Mg(II), or Fe(III) ions that are present in hard water. These essential characteristics of natural soaps, they recognized, could be produced in a molecule containing a sulfonate ($—SO_3^-$) group instead of a carboxylate ($—COO^-$) group. Calcium, magnesium, and iron salts of alkylsulfonic acids, $R–SO_3H$, and arylsulfonic acids, $ArSO_3H$, are much more soluble in water than comparable salts of fatty acids.

The most widely used synthetic detergents today are the linear alkylbenzenesulfonates (LAS). One of the most common of these is sodium 4-dodecylbenzenesulfonate. To prepare this type of detergent, a linear alkylbenzene is treated with sulfuric acid to form an alkylbenzenesulfonic acid (Section 9.6B), followed by neutralization of the sulfonic acid with NaOH:

$$CH_3(CH_2)_{10}CH_2—C_6H_5 \xrightarrow[2)\ NaOH]{1)\ H_2SO_4} CH_3(CH_2)_{10}CH_2—C_6H_4—SO_3^-Na^+$$

Dodecylbenzene

Sodium 4-dodecylbenzenesulfonate (an anionic detergent)

The product is mixed with builders and spray dried to give a smooth, flowing powder. The most common builder is sodium silicate. Alkylbenzenesulfonate detergents were introduced in the late 1950s, and today they command close to 90% of the market once held by natural soaps.

Effects of optical bleaches: *(above)* ordinary light; *(below)* black light. *(Charles D. Winters)*

Among the most common additives to detergent preparations are foam stabilizers (to encourage longer-lasting bubbles), bleaches, and optical brighteners. A common foam stabilizer added to liquid soaps, but not laundry detergents (for obvious reasons: think of a top-loading washing machine with foam spewing out of the lid!), is the amide prepared from dodecanoic acid (lauric acid) and 2-aminoethanol (ethanolamine). The most common bleach used in washing powders is sodium perborate tetrahydrate, which decomposes at temperatures above 50°C to give hydrogen peroxide, the actual bleaching agent:

$$CH_3(CH_2)_{10}\overset{O}{\overset{\|}{C}}NHCH_2CH_2OH \qquad O{=}B—O—O^-Na^+ \cdot 4H_2O$$

N-(2-Hydroxyethyl)dodecanamide (a foam stabilizer)

Sodium perborate tetrahydrate (a bleach)

Also added to laundry detergents are optical brighteners (optical bleaches). These substances are absorbed into fabrics and, after absorbing ambient light, fluoresce with a blue color, offsetting the yellow color caused by fabric as it ages. Optical brighteners produce a "whiter-than-white" appearance. You most certainly have observed their effects if you have seen the glow of white T-shirts or blouses when they are exposed to black light (UV radiation).

Phospholipid A lipid containing glycerol esterified with two molecules of fatty acid and one molecule of phosphoric acid.

21.3 What Are Phospholipids?

A. Structure

Phospholipids, or phosphoacylglycerols, as they are more properly named, are the second most abundant group of naturally occurring lipids. They are found almost exclusively in plant and animal membranes, which typically consist of about 40% to 50% phospholipids and 50% to 60% proteins. The most abundant phospholipids are derived from a **phosphatidic acid** (Figure 21.5).

The fatty acids that are most common in phosphatidic acids are palmitic and stearic acids (both fully saturated) and oleic acid (with one double bond in the hydrocarbon chain). Further esterification of a phosphatidic acid with a low-molecular-weight alcohol gives a phospholipid. Several of the most common alcohols found in phospholipids are given in Table 21.3.

All of these products contain lecithin. *(Charles D. Winters)*

nonpolar hydrocarbon tails

A phosphatidate

CH_2—O—$\overset{O}{\parallel}$P(—O^-)—O^-

polar head groups

A phospholipid

CH_2—O—$\overset{O}{\parallel}$P(—O^-)—$OCH_2CH_2\overset{+}{N}(CH_3)_3$

Figure 21.5
A phosphatidic acid and a phospholipid. In a phosphatidic acid, glycerol is esterified with two molecules of fatty acid and one molecule of phosphoric acid. Further esterification of the phosphoric acid group with a low-molecular-weight alcohol gives a phospholipid. Each structural formula shows all functional groups as they are ionized at pH 7.4, the approximate pH of blood plasma and of many biological fluids. Under these conditions, each phosphate group bears a negative charge and each amino group bears a positive charge.

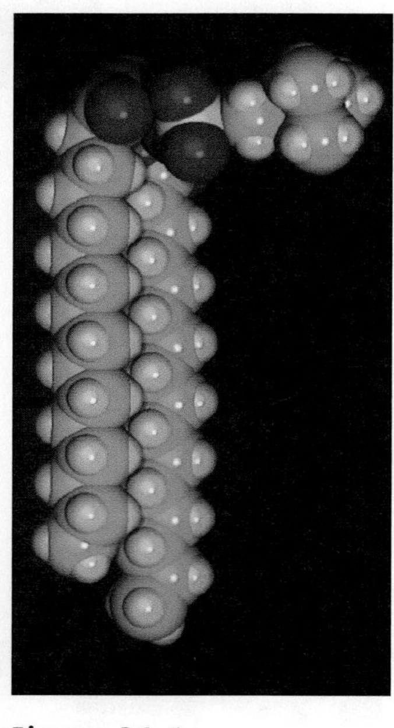

Figure 21.6
Space-filling model of a lecithin. *(Brent Iverson, University of Texas)*

TABLE 21.3 Low-Molecular-Weight Alcohols Most Common to Phospholipids

Alcohols Found in Phospholipids

Structural Formula	Name	Name of Phospholipid
$HOCH_2CH_2NH_2$	ethanolamine	phosphatidylethanolamine (cephalin)
$HOCH_2CH_2\overset{+}{N}(CH_3)_3$	choline	phosphatidylcholine (lecithin)
$HOCH_2CH(NH_3^+)COO^-$	serine	phosphatidylserine
HO, HO, OH, HO, OH, OH (ring)	inositol	phosphatidylinositol

B. Lipid Bilayers

Figure 21.6 shows a space-filling model of a lecithin (a phosphatidycholine). It and other phospholipids are elongated, almost rodlike molecules, with the nonpolar (hydrophobic) hydrocarbon chains lying roughly parallel to one another and the polar (hydrophilic) phosphoric ester group pointing in the opposite direction.

Lipid bilayer A back-to-back arrangement of phospholipid monolayers.

Placed in aqueous solution, phospholipids spontaneously form a **lipid bilayer** (Figure 21.7) in which polar head groups lie on the surface, giving the bilayer an ionic coating. Nonpolar hydrocarbon chains of fatty acids lie buried within the bilayer.

CHEMICAL CONNECTIONS 21A

Snake Venom Phospholipases

The venoms of certain snakes contain enzymes called phospholipases that catalyze the hydrolysis of carboxylic ester bonds of phospholipids. The venom of the eastern diamondback rattlesnake (*Crotalus adamanteus*) and that of the Indian cobra (*Naja naja*) both contain phospholipase PLA_2, which catalyzes the hydrolysis of esters at carbon 2 of phospholipids. The breakdown product of this hydrolysis, a lysolecithin, acts as a detergent and dissolves the membranes of red blood cells, causing them to rupture:

$$\underset{\text{A phospholipid}}{\begin{array}{c} CH_2CH_2NH_3^+ \\ | \\ O \\ | \\ O{=}P{-}O^- \\ | \\ O \\ | \\ CH_2{-}CH{-}CH_2 \\ |\quad\;\; | \\ O\quad\;\; O \\ |\quad\;\; | \\ O{=}C\quad O{=}C \\ |\quad\;\; | \\ R_1\quad\;\; R_2 \end{array}} + H_2O \xrightarrow{PLA_2} \underset{\text{A lysolecithin}}{\begin{array}{c} CH_2CH_2NH_3^+ \\ | \\ O \\ | \\ O{=}P{-}O^- \\ | \\ O \\ | \\ CH_2{-}CH{-}CH_2 \\ |\quad\;\; | \\ O\quad\;\; OH \\ | \\ O{=}C \\ | \\ R_1 \end{array}} + R_2{-}\overset{\overset{\displaystyle O}{\|}}{C}{-}O^- + H^+$$

PLA₂ catalyzes hydrolysis of this ester bond

Indian cobras kill several thousand people each year.

Milking an Indian cobra for its venom. *(Dan McCoy/Rainbow)*

QUESTION

How is a lysolecithin able to act as a detergent?

This self-assembly of phospholipids into a bilayer is a spontaneous process, driven by two types of noncovalent forces:

1. **hydrophobic effects**, which result when nonpolar hydrocarbon chains cluster together and exclude water molecules, and
2. electrostatic interactions, which result when polar head groups interact with water and other polar molecules in the aqueous environment.

Recall from Section 21.2B that the formation of soap micelles is driven by these same noncovalent forces. The polar (hydrophilic) carboxylate groups of soap molecules lie on the surface of the micelle and associate with water molecules, and the nonpolar (hydrophobic) hydrocarbon chains cluster within the micelle and thus are removed from contact with water.

The arrangement of hydrocarbon chains in the interior of a phospholipid bilayer varies from rigid to fluid, depending on the degree of unsaturation of the chains

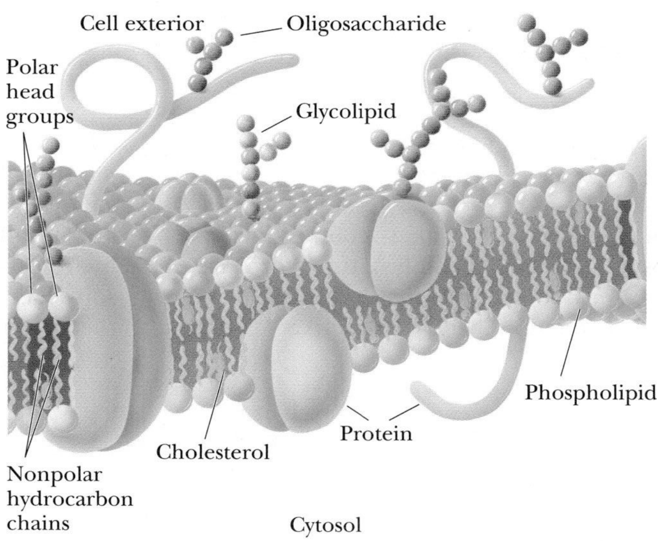

Figure 21.7
Fluid-mosaic model of a biological membrane, showing the lipid bilayer with membrane proteins oriented on the inner and outer surfaces of the membrane, and penetrating the entire thickness of the membrane.

themselves. Saturated hydrocarbon chains tend to lie parallel and closely packed, making the bilayer rigid. Unsaturated hydrocarbon chains, by contrast, have one or more *cis* double bonds, which cause "kinks" in the chains, so they pack neither as closely nor as orderly as saturated chains. The disordered packing of unsaturated hydrocarbon chains leads to fluidity of the bilayer.

Biological membranes are made of lipid bilayers. The most satisfactory model for the arrangement of phospholipids, proteins, and cholesterol in plant and animal membranes is the **fluid-mosaic model** proposed in 1972 by S. J. Singer and G. Nicolson. The term *mosaic* signifies that the various components in the membrane coexist side by side, as discrete units, rather than combining to form new molecules or ions. "Fluid" signifies that the same sort of fluidity exists in membranes that we have already seen in lipid bilayers. Furthermore, the protein components of membranes "float" in the bilayer and can move laterally along the plane of the membrane.

Fluid-mosaic model A model of a biological membrane consisting of a phospholipid bilayer, with proteins, carbohydrates, and other lipids embedded in, and on the surface of, the bilayer.

Steroid A plant or animal lipid having the characteristic tetracyclic ring structure of the steroid nucleus, namely, three six-membered rings and one five-membered ring.

21.4 What Are Steroids?

Steroids are a group of plant and animal lipids that have the tetracyclic ring system shown in Figure 21.8.

The features common to the tetracyclic ring system of most naturally occurring steroids are illustrated in Figure 21.9.

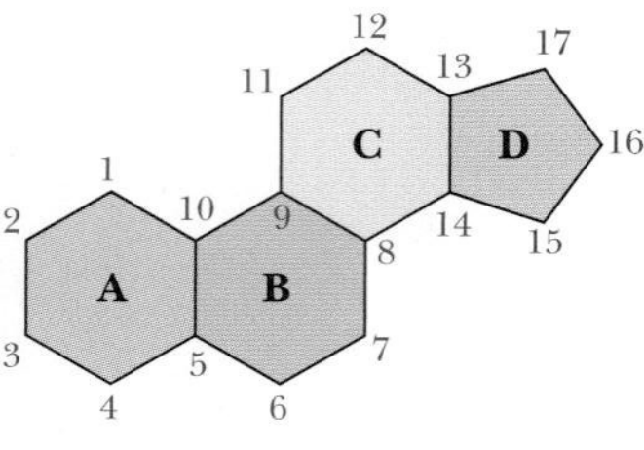

Figure 21.8
The tetracyclic ring system characteristic of steroids.

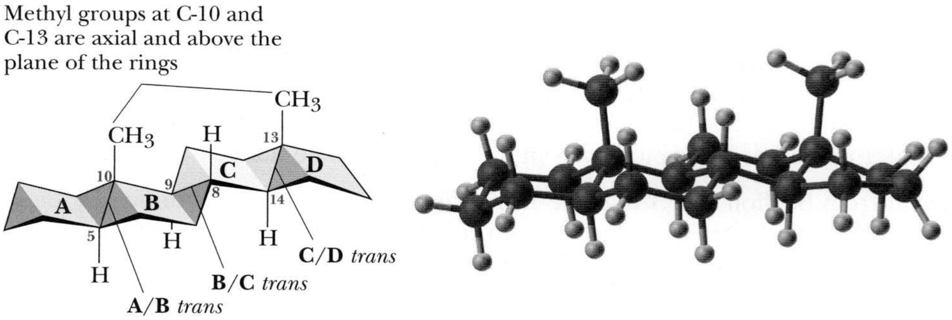

Methyl groups at C-10 and C-13 are axial and above the plane of the rings

Figure 21.9
Features common to the tetracyclic ring system of many steroids.

1. The fusion of the rings is *trans*, and each atom or group of atoms at a ring junction is axial. (Compare, for example, the orientations of —H at carbon 5 and $-CH_3$ at carbon 10.)
2. The pattern of atoms or groups along the points of ring fusion (carbons 5 to 10, 10 to 9, 9 to 8, 8 to 14, and 14 to 13) is nearly always *trans*–anti–*trans*–anti–*trans*.
3. Because of the *trans*–anti–*trans*–anti–*trans* arrangement of atoms or groups along the points of ring fusion, the tetracyclic steroid ring system is nearly flat and quite rigid.
4. Many steroids have axial methyl groups at carbon 10 and carbon 13 of the tetracyclic ring system.

A. Structure of the Major Classes of Steroids

Cholesterol

Cholesterol, a white, water-insoluble, waxy solid found in blood plasma and in all animal tissues, is an integral part of human metabolism in two ways:

1. It is an essential component of biological membranes. The body of a healthy adult contains approximately 140 g of cholesterol, about 120 g of which is present in membranes. Membranes of the central and peripheral nervous systems, for example, contain about 10% cholesterol by weight.
2. It is the compound from which sex hormones, adrenocorticoid hormones, bile acids, and vitamin D are synthesized. Thus, cholesterol is, in a sense, the parent steroid.

Cholesterol has eight stereocenters, and a molecule with this structural feature can exist as 2^8, or 256, stereoisomers (128 pairs of enantiomers). Only the stereoisomer shown here on the right is present in human metabolism:

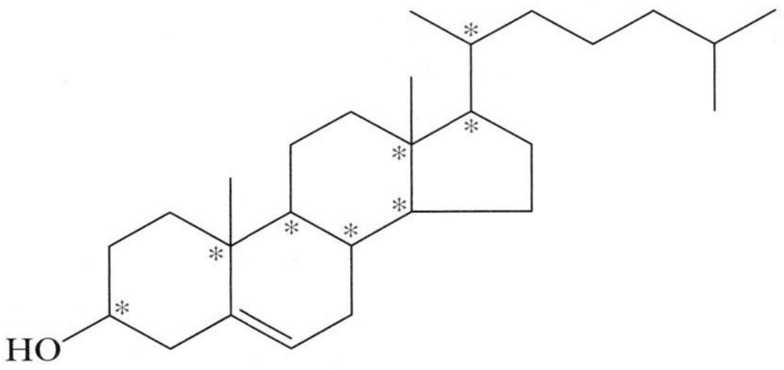

HO

Cholesterol has eight stereocenters; 256 stereoisomers are possible

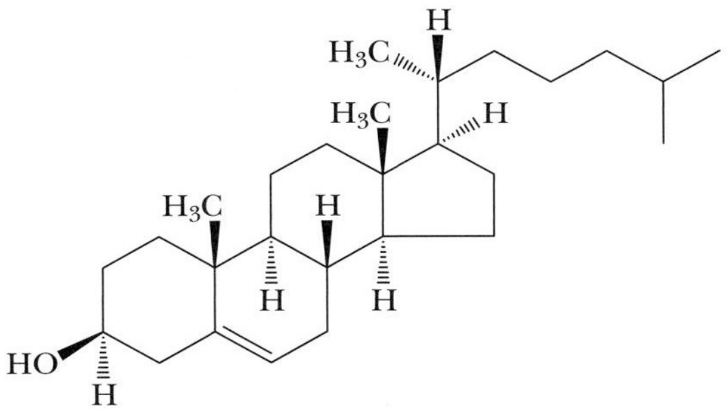

H_3C H H_3C H H_3C H H H HO H

This is the stereoisomer found in human metabolism

Cholesterol is insoluble in blood plasma, but can be transported as a plasma-soluble complex formed by cholesterol with proteins called lipoproteins. **Low-density lipoproteins (LDL)** transport cholesterol from the site of its synthesis in the liver to the various tissues and cells of the body where it is to be used. It is primarily cholesterol associated with LDLs that builds up in atherosclerotic deposits in blood vessels. **High-density lipoproteins (HDL)** transport excess and unused cholesterol from cells back to the liver to be degraded to bile acids and eventual excretion in the feces. It is thought that HDLs retard or reduce atherosclerotic deposits.

Steroid Hormones

Shown in Table 21.4 are representatives of each major class of steroid hormones, along with their principal functions. Female steroid sex hormones are called **estrogens**, and male steroid sex hormones are called **androgens**.

Low-density lipoprotein (LDL) Plasma particles, of density 1.02–1.06 g/mL, consisting of approximately 25% proteins, 50% cholesterol, 21% phospholipids, and 4% triglycerides.

High-density lipoprotein (HDL) Plasma particles, of density 1.06–1.21 g/mL, consisting of approximately 33% proteins, 30% cholesterol, 29% phospholipids, and 8% triglycerides.

Estrogen A steroid hormone, such as estradiol, that mediates the development and sexual characteristics of females.

Androgen A steroid hormone, such as testosterone, that mediates the development and sexual characteristics of males.

TABLE 21.4 Selected Steroid Hormones

Structure	Source and Major Effects
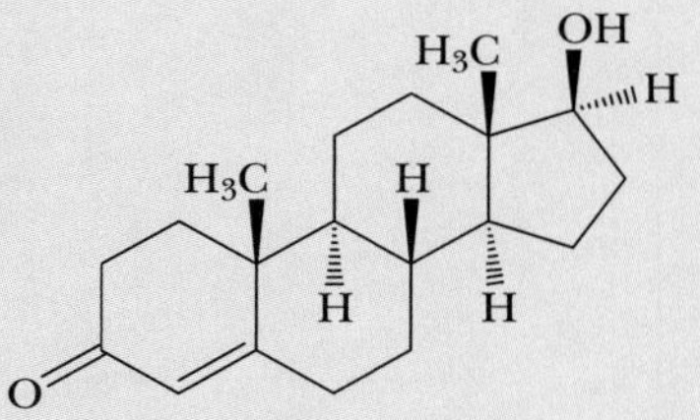 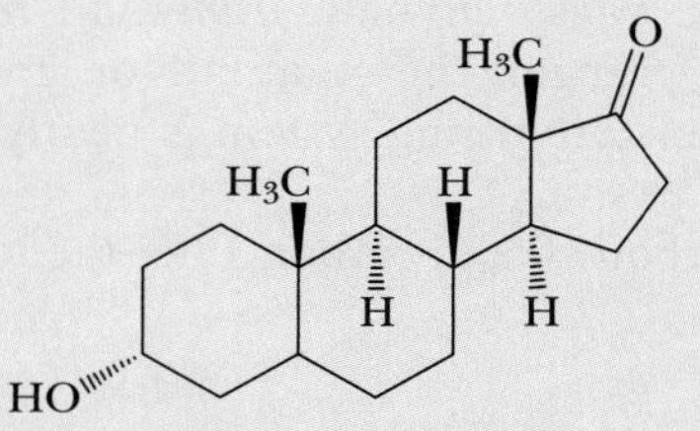 Testosterone; Androsterone	androgens (male sex hormones): synthesized in the testes; responsible for development of male secondary sex characteristics
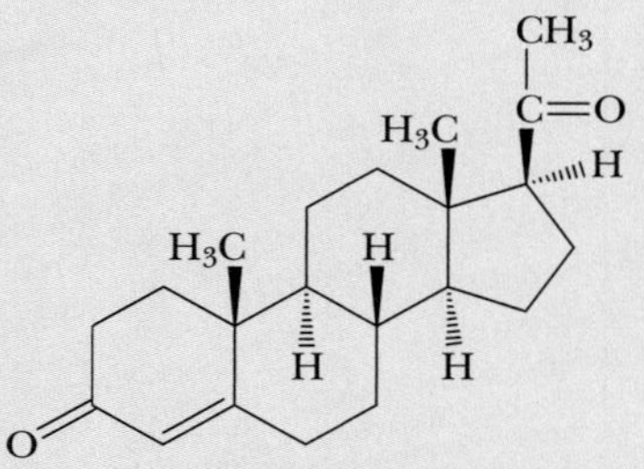 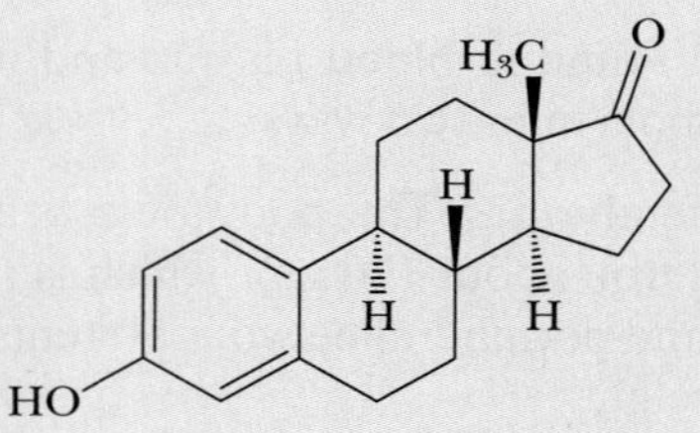 Progesterone; Estrone	estrogens (female sex hormones): synthesized in the ovaries; responsible for development of female secondary sex characteristics and control of the menstrual cycle
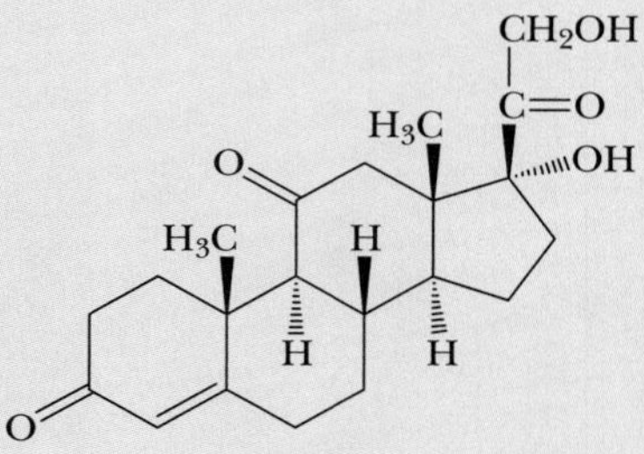 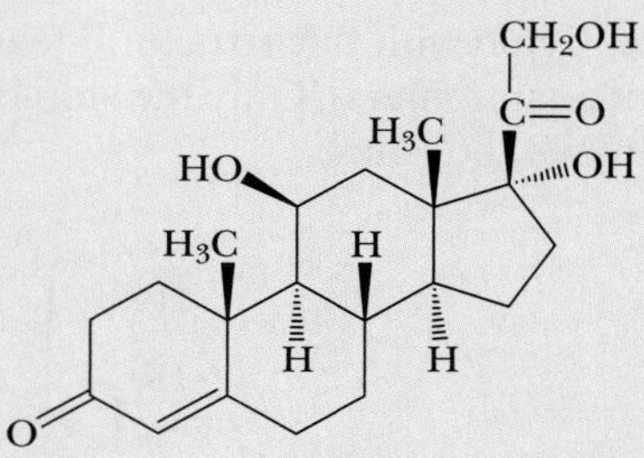 Cortisone; Cortisol	glucocorticoid hormones: synthesized in the adrenal cortex; regulate metabolism of carbohydrates, decrease inflammation, and involved in the reaction to stress
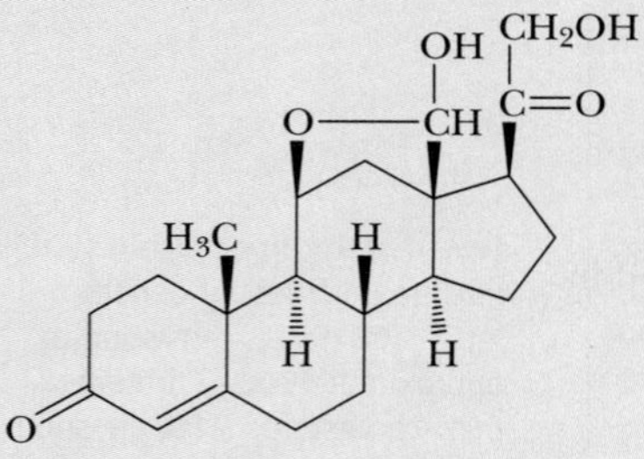 Aldosterone	a mineralocorticoid hormone: synthesized in the adrenal cortex; regulates blood pressure and volume by stimulating the kidneys to absorb Na^+, Cl^-, and HCO_3^-

After scientists understood the role of progesterone in inhibiting ovulation, they realized its potential as a contraceptive. Progesterone itself is relatively ineffective when taken orally. As a result of a massive research program in both industrial and academic laboratories, many synthetic progesterone-mimicking steroids became available in the 1960s. Taken regularly, these drugs prevent ovulation yet allow women to maintain a normal menstrual cycle. Some of the most effective preparations contain a progesterone analog, such as norethindrone, combined with a smaller amount of an estrogenlike material to help prevent irregular menstrual flow during the prolonged use of contraceptive pills.

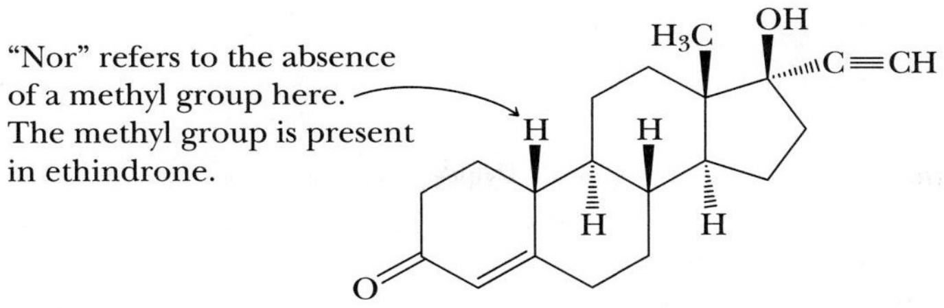

Norethindrone
(a synthetic progesterone analog)

The chief function of testosterone and other androgens is to promote the normal growth of the male reproductive organs (the primary sex characteristics) and the development of the male's characteristic deep voice, pattern of body and facial hair, and musculature (secondary sex characteristics). Although testosterone produces these effects, it is not active when taken orally because it is metabolized in the liver to an inactive steroid. A number of oral **anabolic steroids** have been developed for use in rehabilitation medicine, particularly when muscle atrophy occurs during recovery from an injury. Examples include the following compounds:

Anabolic steroid A steroid hormone, such as testosterone, that promotes tissue and muscle growth and development.

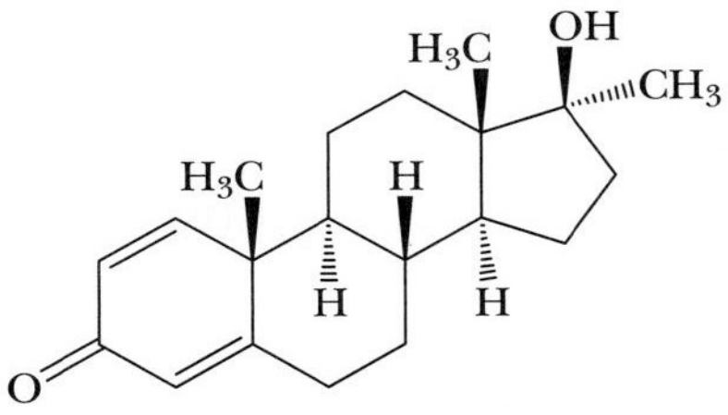

Methandrostenolone

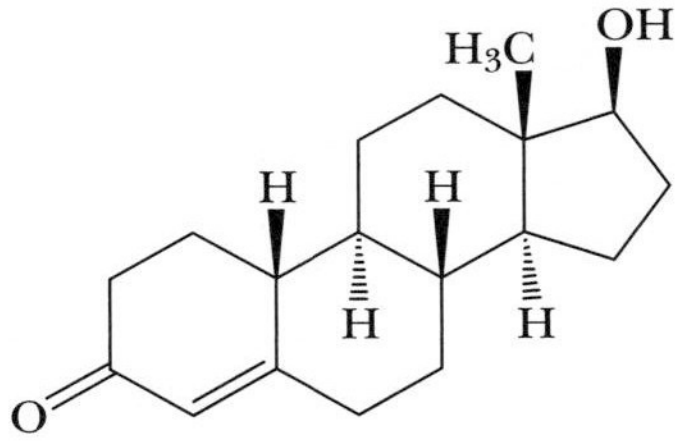

Nandrolone

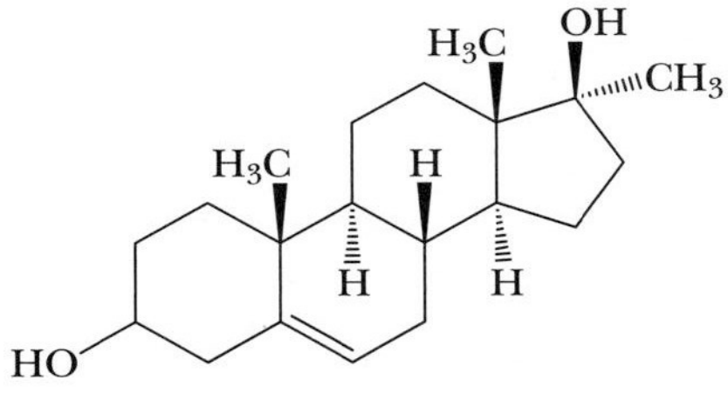

Methandriol

Among certain athletes, the misuse of anabolic steroids to build muscle mass and strength, particularly for sports that require explosive action, is common. The risks associated with abusing anabolic steroids for this purpose are enormous: heightened aggressiveness, sterility, impotence, and premature death from complications of diabetes, coronary artery disease, and liver cancer.

Bile Acids

Figure 21.10 shows a structural formula for cholic acid, a constituent of human bile. The molecule is shown as an anion, because it would be ionized in bile and intestinal fluids. **Bile acids**, or, more properly, bile salts, are synthesized in the liver, stored in the gallbladder, and secreted into the intestine, where their function is to emulsify dietary fats and thereby aid in their absorption and digestion. Furthermore, bile salts are the end products of the metabolism of cholesterol and, thus, are a principal pathway for the elimination of that substance from the body. A characteristic structural feature of bile salts is a *cis* fusion of rings A/B.

Bile acid A cholesterol-derived detergent molecule, such as cholic acid, that is secreted by the gallbladder into the intestine to assist in the absorption of dietary lipids.

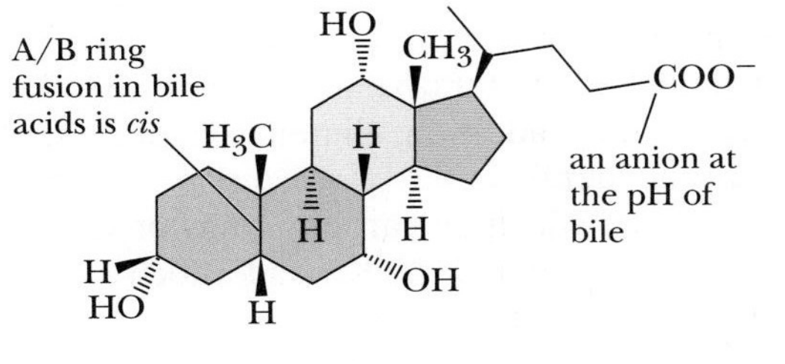

Figure 21.10
Cholic acid, an important constituent of human bile.

CHEMICAL CONNECTIONS 21B

Nonsteroidal Estrogen Antagonists

Estrogens are female sex hormones, the most important of which are estrone, estradiol, and estriol (of the three, β-estradiol is the most potent). (*Note:* As per convention in steroid nomenclature, the designation *beta* means "toward the reader," on the topside as the rings are viewed in the accompanying diagram; *alpha* means "away from the reader," on the bottom side.)

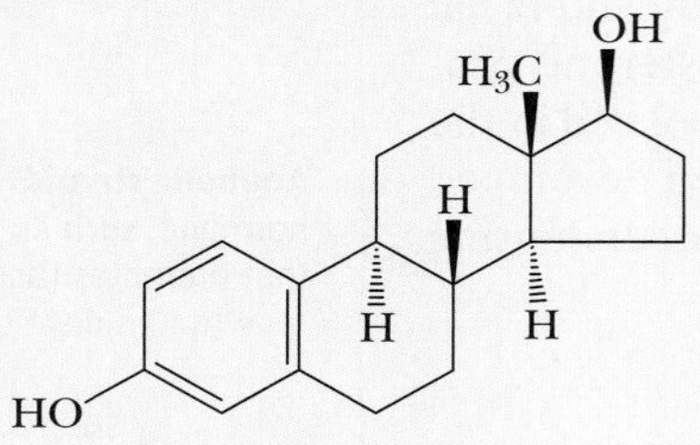

β-Estradiol

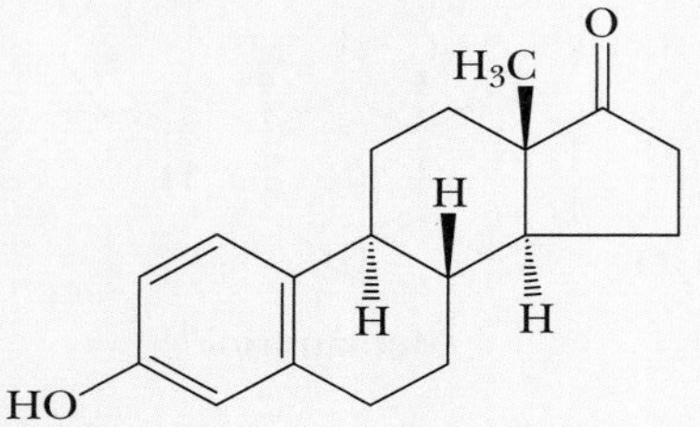

Estrone

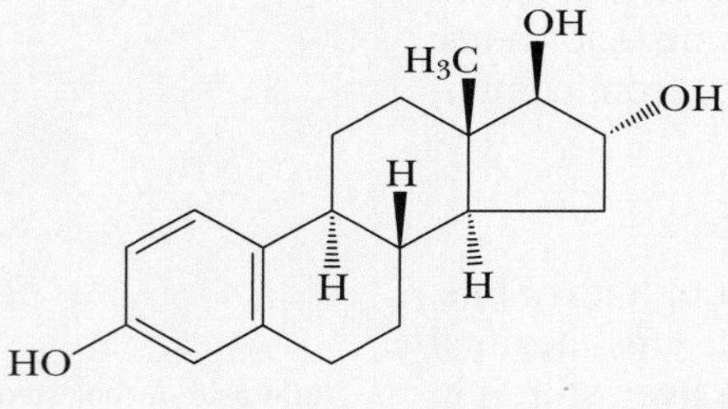

Estriol

As soon as these compounds were isolated in the early 1930s and their pharmacology was studied, it became clear that they are extremely potent. In recent years, there have been intense efforts to design and synthesize molecules that will bind to the estrogen receptor. One target of this research has been nonsteroidal estrogen antagonists—compounds that interact with the estrogen receptor as antagonists (i.e., compounds that will block the effect of endogenous or exogenous estrogens). A feature common to many that have been developed is the presence of a 1,2-diphenylethylene, with one of the benzene rings bearing a dialkylaminoethoxyl substituent. The first nonsteroidal estrogen antagonist of this type to achieve clinical importance was tamoxifen, now an important drug in the prevention and treatment of breast cancer:

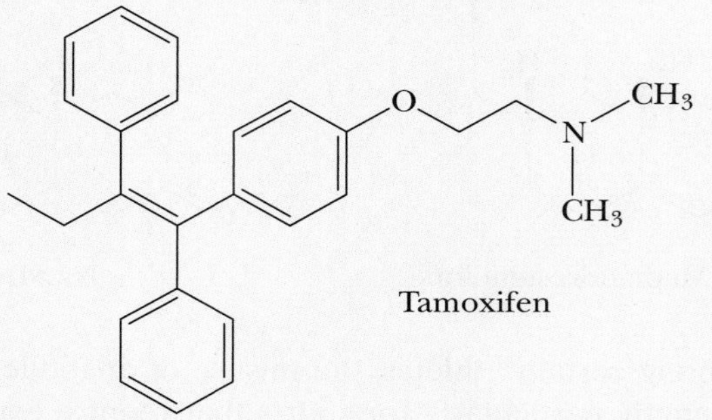

Tamoxifen

QUESTIONS

Determine the E/Z configuration of the carbon–carbon double bond in Tamoxifen.

What is the role of progesterone and similar compounds in contraceptive pills?

B. Biosynthesis of Cholesterol

The biosynthesis of cholesterol illustrates a point we first made in our introduction to the structure of terpenes (Section 4.4): In building large molecules, one of the common patterns in the biological world is to begin with one or more smaller subunits, which are then joined by an iterative process and chemically modified by oxidation, reduction, cross-linking, addition, elimination, or related processes to give a biomolecule with a unique identity.

The building block from which all carbon atoms of steroids are derived is the two-carbon acetyl group of acetyl-CoA. Konrad Bloch of the United States and Feodor Lynen of Germany showed that 15 of the 27 carbon atoms of cholesterol are derived from the methyl group of acetyl-CoA; the remaining 12 carbon atoms

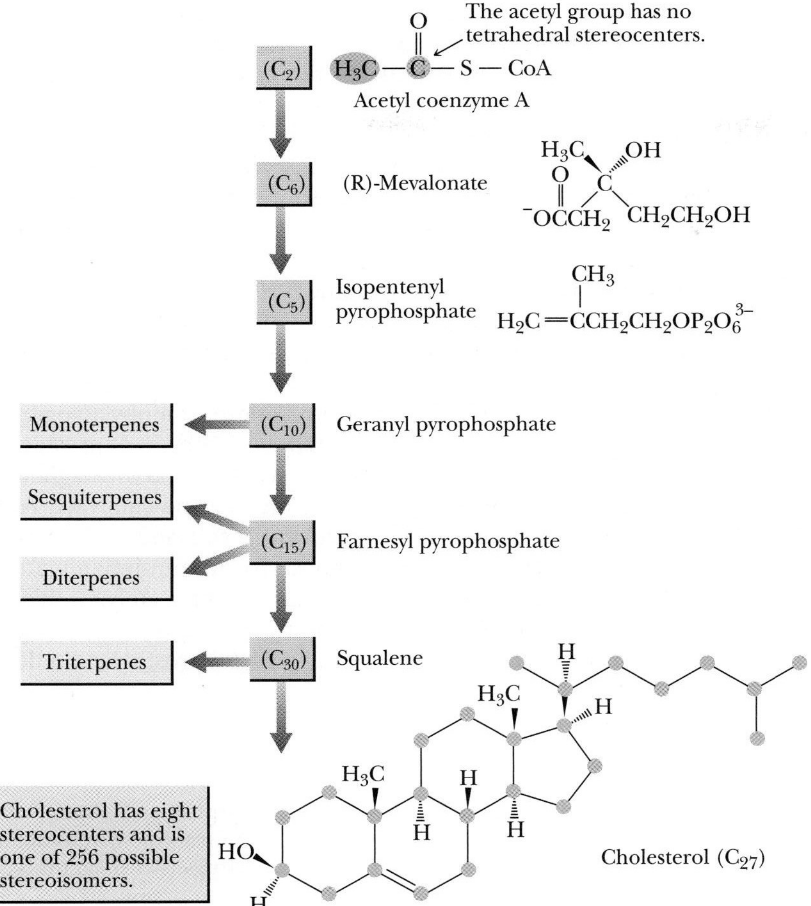

Figure 21.11 Several key intermediates in the synthesis of cholesterol from acetyl groups of acetyl-CoA. Eighteen moles of acetyl-CoA are required for the synthesis of 1 mole of cholesterol.

are derived from the carbonyl group of acetyl-CoA (Figure 21.11). For their discoveries, the two shared the 1964 Nobel Prize in Physiology or Medicine.

A remarkable feature of this synthetic pathway is that the biosynthesis of cholesterol from acetyl-CoA is completely stereoselective; cholesterol is synthesized as only one of 256 possible stereoisomers. We cannot duplicate this exquisite degree of stereoselectivity in the laboratory. Cholesterol is, in turn, the key intermediate in the synthesis of all other steroids:

cholesterol →
- bile acids (e.g., cholic acid)
- sex hormones (e.g., testosterone and estrone)
- mineralocorticoid hormones (e.g., aldosterone)
- glucocorticoid hormones (e.g., cortisone)

21.5 What Are Prostaglandins?

The **prostaglandins** are a family of compounds having in common the 20-carbon skeleton of prostanoic acid:

Prostaglandin A member of the family of compounds having the 20-carbon skeleton of prostanoic acid.

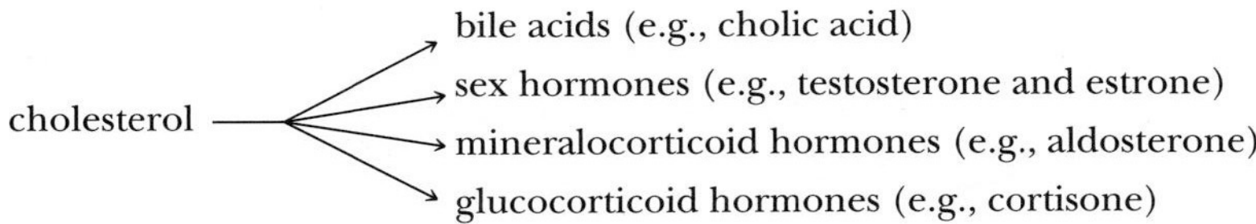

Prostanoic acid

The story of the discovery and structure determination of these remarkable compounds began in 1930 when gynecologists Raphael Kurzrock and Charles Lieb reported that human seminal fluid stimulates the contraction of isolated uterine muscle. A few years later, Ulf von Euler in Sweden confirmed this report and noted that, when injected into the bloodstream, human seminal fluid also stimulates the contraction of intestinal smooth muscle and lowers the blood pressure. Von Euler proposed the name *prostaglandin* for the mysterious substance(s) responsible for these diverse effects, because it was believed at the time that they were synthesized in the prostate gland. Although we now know that prostaglandin production is by no means limited to the prostate gland, the name has stuck.

Prostaglandins are not stored as such in target tissues. Rather, they are synthesized in response to specific physiological triggers. Starting materials for the biosynthesis of prostaglandins are polyunsaturated fatty acids with 20 carbon atoms, stored until they are needed as membrane phospholipid esters. In response to a physiological trigger, the ester is hydrolyzed, the fatty acid is released, and the synthesis of prostaglandins is initiated. Figure 21.12 outlines the steps in the synthesis of several prostaglandins from arachidonic acid. A key step in this biosynthesis is the enzyme-catalyzed reaction of arachidonic acid with two molecules of O_2 to form prostaglandin G_2 (PGG_2). The anti-inflammatory effect of aspirin and other nonsteroidal anti-inflammatory drugs (NSAIDs) results from their ability to inhibit the enzyme that catalyzes this step.

Research on the involvement of prostaglandins in reproductive physiology and the inflammatory process has produced several clinically useful prostaglandin derivatives.

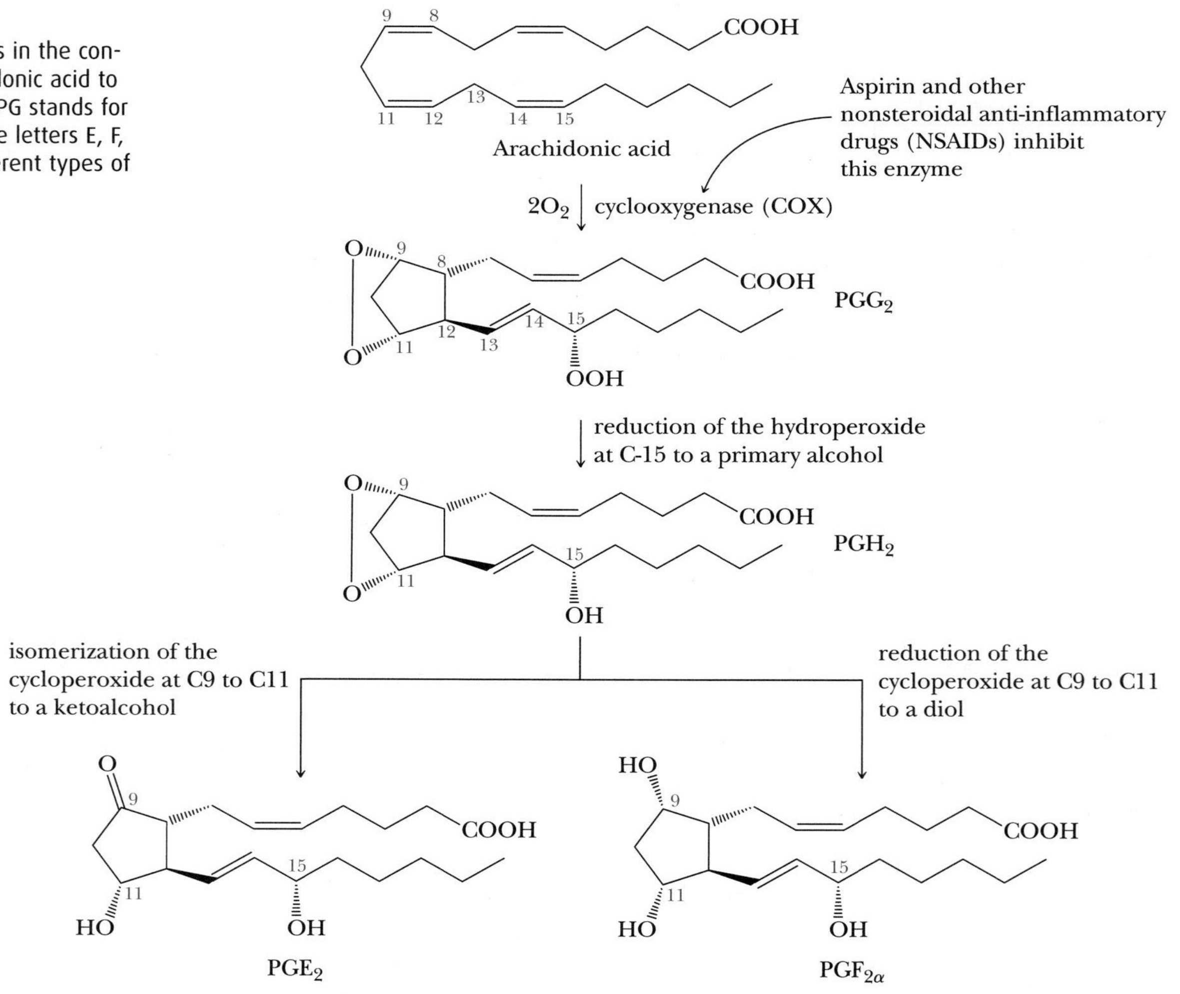

Figure 21.12
Key intermediates in the conversion of arachidonic acid to PGE_2 and $PGF_{2\alpha}$. PG stands for prostaglandin. The letters E, F, G, and H are different types of prostaglandins.

Arachidonic acid

Aspirin and other nonsteroidal anti-inflammatory drugs (NSAIDs) inhibit this enzyme

$2O_2$ ↓ cyclooxygenase (COX)

PGG_2

↓ reduction of the hydroperoxide at C-15 to a primary alcohol

PGH_2

isomerization of the cycloperoxide at C9 to C11 to a ketoalcohol

reduction of the cycloperoxide at C9 to C11 to a diol

PGE_2

$PGF_{2\alpha}$

The observation that $PGF_{2\alpha}$ stimulates contractions of uterine smooth muscle led to a synthetic derivative that is used for therapeutic abortions. A problem with the use of the natural prostaglandins for this purpose is that they are rapidly degraded within the body. In the search for less rapidly degraded prostaglandins, a number of analogs have been prepared, one of the most effective of which is carboprost. This synthetic prostaglandin is 10 to 20 times more potent than the natural $PGF_{2\alpha}$ and is only slowly degraded in the body:

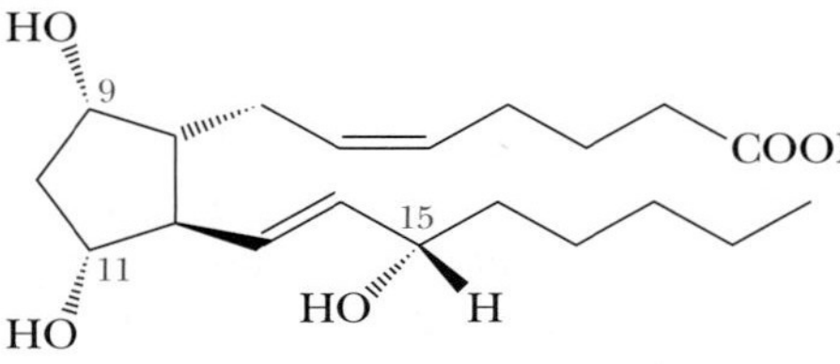

$PGF_{2\alpha}$

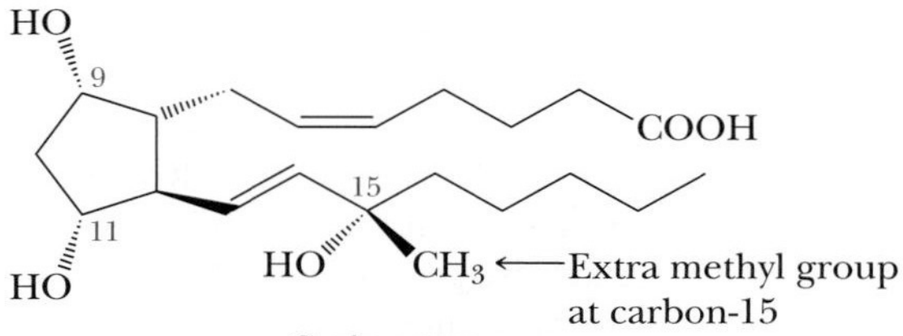

Extra methyl group at carbon-15

Carboprost
(15*S*)-15-Methyl-$PGF_{2\alpha}$

A comparison of these two prostaglandins illustrates how a simple change in the structure of a drug can make a significant change in its effectiveness.

The PGEs, along with several other PGs, suppress gastric ulceration and appear to heal gastric ulcers. The PGE_1 analog, misoprostol, is currently used primarily to prevent ulceration associated with aspirin-like NSAIDs:

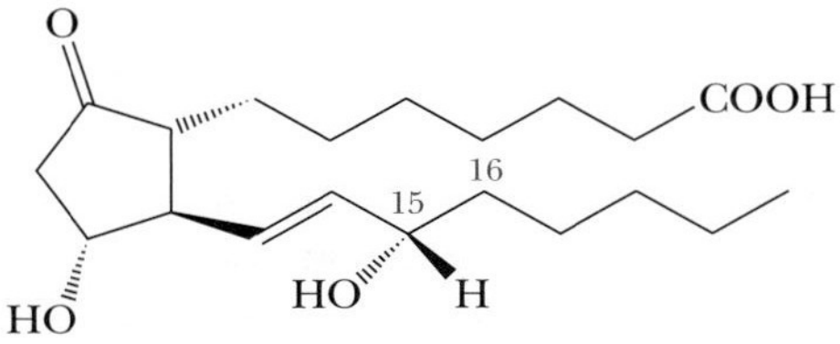

PGE_1

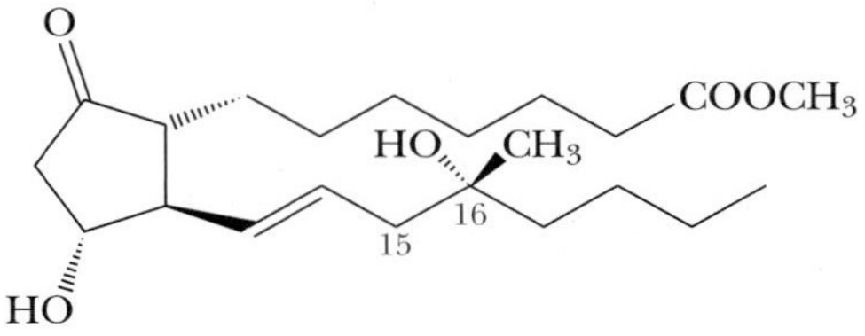

Misoprostol

Prostaglandins are members of an even larger family of compounds called **eicosanoids**, all of which are synthesized from arachidonic acid (20:4). The family includes not only the prostaglandins, but also the leukotrienes, thromboxanes, and prostacyclins:

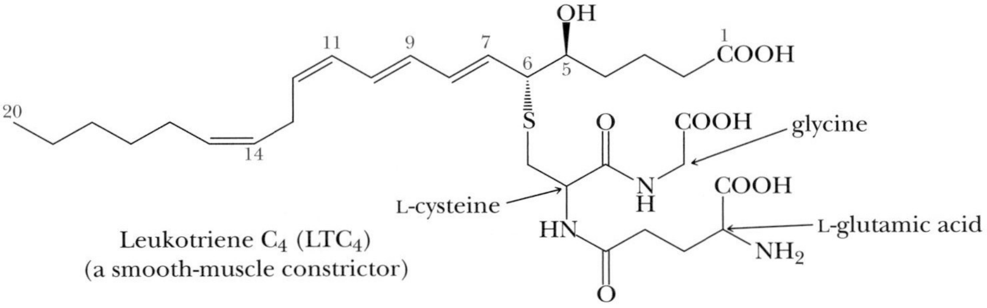

glycine

L-cysteine

L-glutamic acid

Leukotriene C_4 (LTC_4)
(a smooth-muscle constrictor)

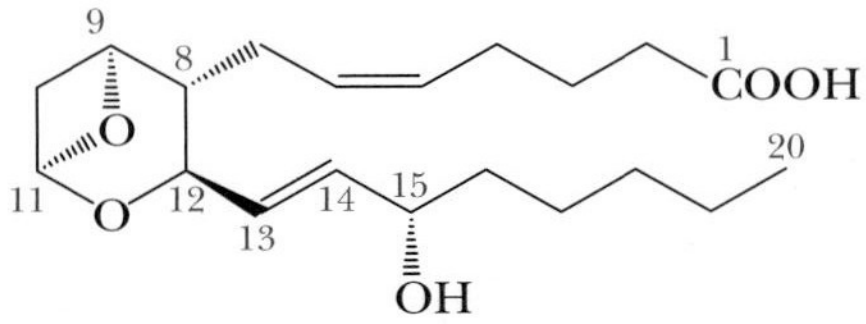

Thromboxane A_2
(a potent vasoconstrictor)

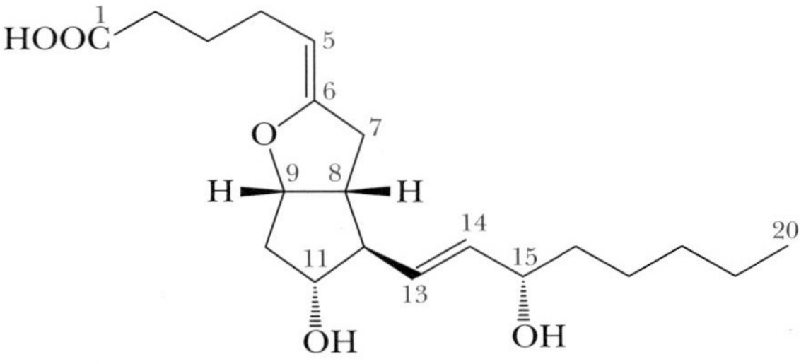

Prostacyclin
(a platelet aggregation inhibitor)

The eicosanoids are extremely widespread, and members of this family of compounds have been isolated from almost every tissue and body fluid.

Leukotrienes are derived from arachidonic acid and are found primarily in leukocytes (white blood cells). Leukotriene C_4 (LTC_4), a typical member of the family, has three conjugated double bonds (hence the suffix *-triene*) and contains the amino acids L-cysteine, glycine, and L-glutamic acid (Section 19.2). An important physiological action of LTC_4 is the constriction of smooth muscles, especially those of the lungs. The synthesis and release of LTC_4 is prompted by allergic reactions. Drugs that inhibit the synthesis of LTC_4 show promise for the treatment of the allergic reactions associated with asthma.

Thromboxane A_2, a potent vasoconstrictor, is also synthesized in the body from arachidonic acid. Its release triggers the irreversible phase of platelet aggregation and the constriction of injured blood vessels. It is thought that aspirin and aspirin-like drugs act as mild anticoagulants because they inhibit cyclooxygenase, the enzyme that initiates the synthesis of thromboxane A_2 from arachidonic acid.

21.6 What Are Fat-Soluble Vitamins?

Vitamins are divided into two broad classes on the basis of their solubility: those that are fat soluble (and hence classed as lipids) and those that are water soluble. The fat-soluble vitamins are A, D, E, and K.

A. Vitamin A

Vitamin A, or retinol, occurs only in the animal world, where the best sources are cod-liver oil and other fish-liver oils, animal liver, and dairy products. Vitamin A in the form of a precursor, or provitamin, is found in the plant world in a group of tetraterpene (C_{40}) pigments called carotenes. The most common of these is β-carotene, abundant in carrots, but also found in some other vegetables, particularly yellow and green ones. β-Carotene has no vitamin A activity; however, after ingestion, it is cleaved at the central carbon–carbon double bond to give retinol (vitamin A):

cleavage of this C=C gives vitamin A

β-Carotene

enzyme-catalyzed cleavage in the liver

2

CH_2OH

Retinol
(Vitamin A)

Probably the best understood role of vitamin A is its participation in the visual cycle in rod cells. In a series of enzyme-catalyzed reactions retinol undergoes (1) a two-electron oxidation to all-*trans*-retinal and (2) isomerization about the carbon 11

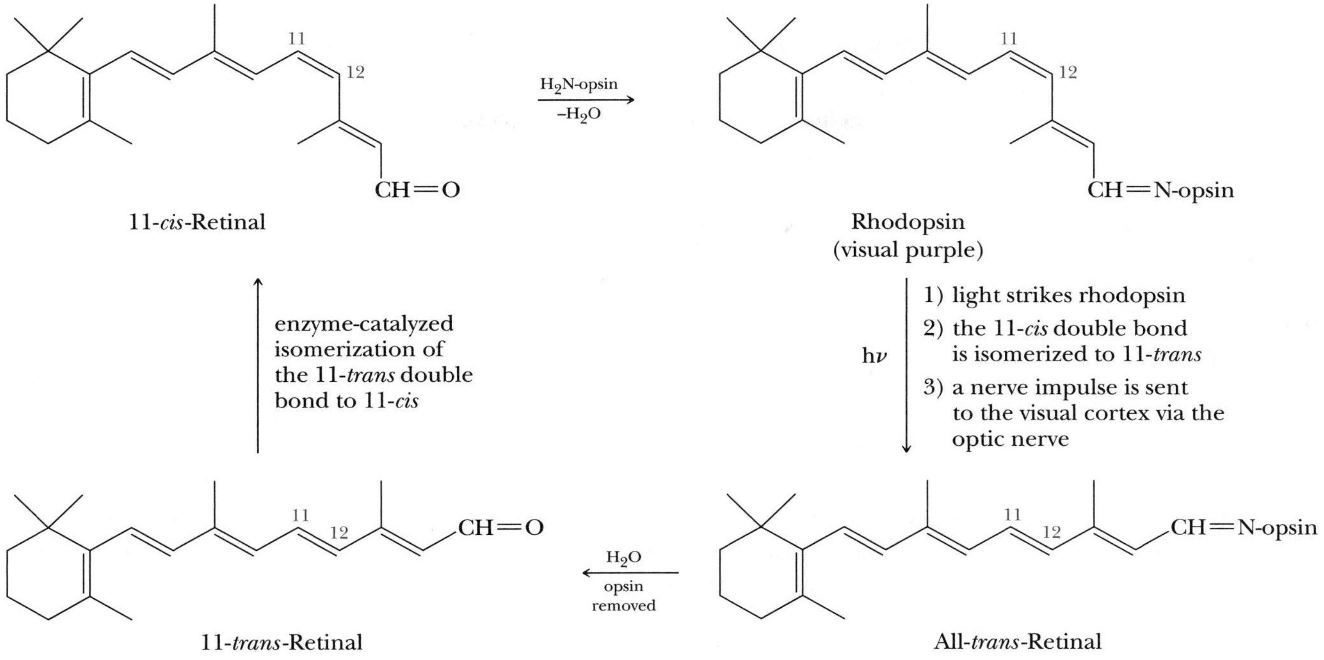

Figure 21.13
The primary chemical reaction of vision in rod cells is the absorption of light by rhodopsin, followed by the isomerization of a carbon–carbon double bond from a *cis* configuration to a *trans* configuration.

to carbon 12 double bond to give 11-*cis*-retinal, (3) thereby forming an imine (Section 13.7A) with the $—NH_2$ from a lysine unit of the protein opsin. The product of these reactions is rhodopsin, a highly conjugated pigment that shows intense absorption in the blue-green region of the visual spectrum.

The primary event in vision is the absorption of light by rhodopsin in rod cells of the retina of the eye to produce an electronically excited molecule. Within several picoseconds (1 picosec = 10^{-12} sec), the excess electronic energy is converted to vibrational and rotational energy, and the 11-*cis* double bond is isomerized to the more stable 11-*trans* double bond. The isomerization triggers a conformational change in the protein opsin that causes neurons in the optic nerve to fire, producing a visual image. Coupled with this light-induced change is the hydrolysis of rhodopsin to give 11-*trans*-retinal and free opsin. At this point, the visual pigment is bleached and in a refractory period. Rhodopsin is regenerated by a series of enzyme-catalyzed reactions that converts 11-*trans*-retinal to 11-*cis*-retinal and then to rhodopsin. The visual cycle is shown in abbreviated form in Figure 21.13.

B. Vitamin D

Vitamin D is the name for a group of structurally related compounds that play a major role in regulating the metabolism of calcium and phosphorus. A deficiency of vitamin D in childhood is associated with rickets, a mineral-metabolism disease that leads to bowlegs, knock-knees, and enlarged joints. Vitamin D_3, the most abundant form of the vitamin in the circulatory system, is produced in the skin of mammals by the action of ultraviolet radiation on 7-dehydrocholesterol (cholesterol with a double bond between carbons 7 and 8). In the liver, vitamin D_3 undergoes an enzyme-catalyzed, two-electron oxidation at carbon 25 of the side chain to form 25-hydroxyvitamin D_3; the oxidizing agent is molecular oxygen, O_2. 25-Hydroxyvitamin

D_3 undergoes further oxidation in the kidneys, also by O_2, to form 1,25-dihydroxyvitamin D_3, the hormonally active form of the vitamin:

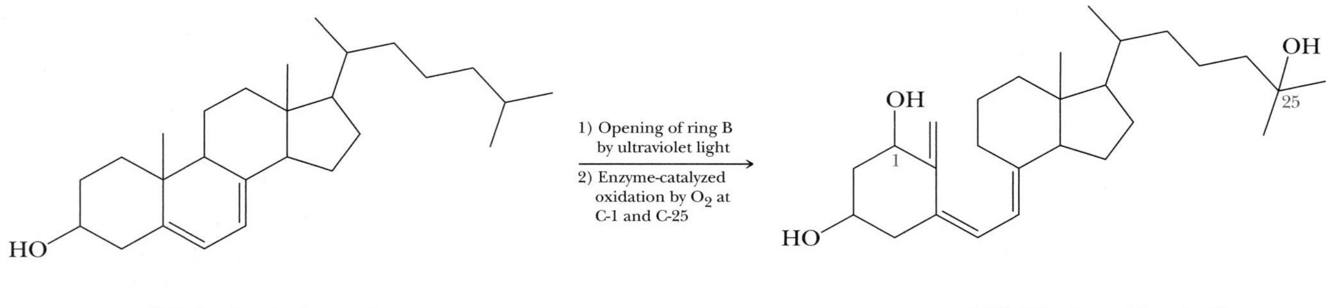

1) Opening of ring B by ultraviolet light
2) Enzyme-catalyzed oxidation by O_2 at C-1 and C-25

7-Dehydrocholesterol

1,25-Dihydroxyvitamin D_3

C. Vitamin E

Vitamin E was first recognized in 1922 as a dietary factor essential for normal reproduction in rats—hence its name *tocopherol*, from the Greek *tocos*, birth, and *pherein*, to bring about. Vitamin E is a group of compounds of similar structure, the most active of which is α-tocopherol:

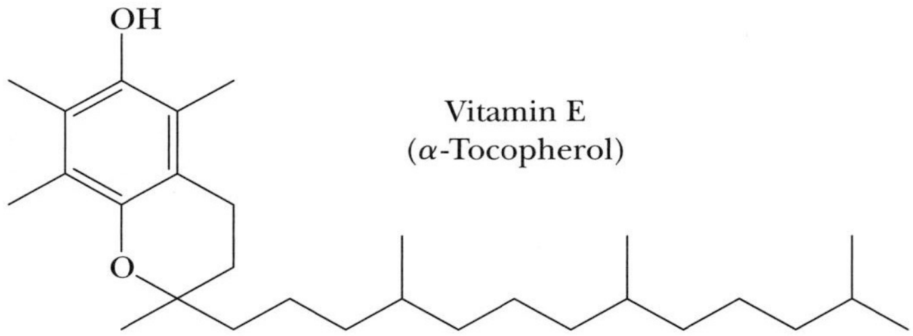

Vitamin E
(α-Tocopherol)

Vitamin E occurs in fish oil; in other oils, such as cottonseed and peanut oil; and in leafy, green vegetables. The richest source is wheat germ oil.

In the body, vitamin E functions as an antioxidant, trapping peroxy radicals of the type HOO• and ROO• formed as a result of enzyme-catalyzed oxidation by molecular oxygen of the unsaturated hydrocarbon chains in membrane phospholipids. There is speculation that peroxy radicals play a role in the aging process and that vitamin E and other antioxidants may retard that process. Vitamin E is also necessary for the proper development and function of the membranes of red blood cells.

D. Vitamin K

The name of this vitamin comes from the German word *koagulation*, signifying its important role in the blood-clotting process. A deficiency of vitamin K results in slowed blood clotting. Natural vitamins of the K family have, for the most part, been replaced in vitamin supplements by synthetic preparations. Menadione, one such synthetic material that exhibits vitamin K activity, has a hydrogen in place of the alkyl chain:

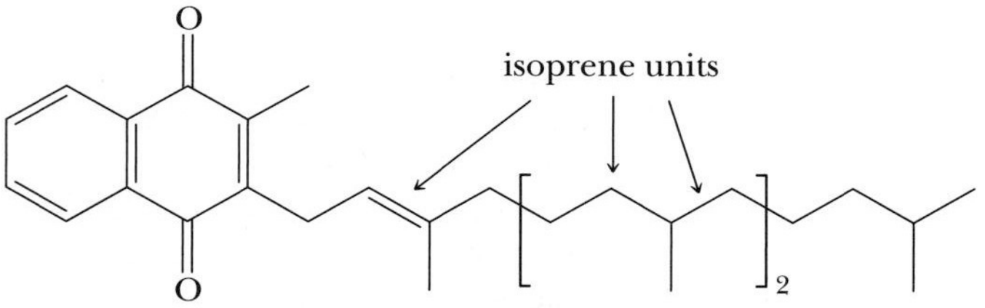

isoprene units

Vitamin K_1

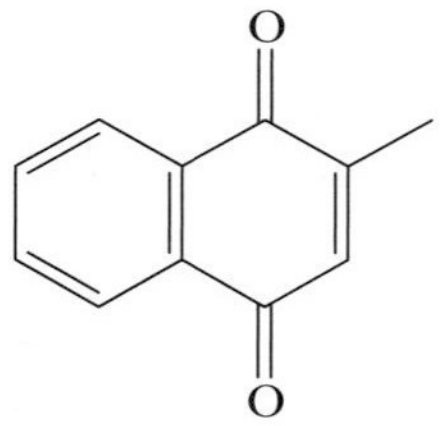

Menadione
(a synthetic vitamin K analog)

Key Terms and Concepts

anabolic steroid (p. 761)
androgen (p. 759)
bile acid (p. 761)
cholesterol (p. 759)
estrogen (p. 759)
fat (p. 752)
fat-soluble vitamin (p. 766)
fatty acid (p. 750)
fluid mosaic model (p. 758)
foam stabilizer (p. 755)
hardening of oils (p. 753)
high-density lipoprotein (HDL) (p. 759)
how soaps clean (p. 754)
hydrophilic effect (p. 754)
hydrophobic effect (p. 757)
leukotriene (p. 766)
lipid (p. 749)
lipid bilayer (p. 756)
low-density lipoprotein (LDL) (p. 759)
margarine (p. 753)
micelle (p. 754)
natural soap (p. 753)
oil (p. 752)
optical bleach (p. 755)
phosphatidic acid (p. 755)
phospholipid (p. 755)
polyunsaturated triglyceride (p. 752)
prostaglandin (p. 763)
reduction of fatty acids (p. 753)
saponification (p. 753)
soap (p. 753)
steroid (p. 758)
steroid ring system (p. 758)
synthetic detergent (p. 754)
triacylglycerol (p. 750)
triglyceride (p. 750)
tropical oil (p. 752)
vitamin A (p. 766)
vitamin A, role in vision (p. 767)
vitamin D (p. 767)
vitamin E (p. 768)
vitamin E as an antioxidant (p. 768)
vitamin K (p. 768)

Summary of Key Questions

- **Lipids** are a heterogeneous class of compounds grouped together on the basis of their solubility properties; they are insoluble in water and soluble in diethyl ether, acetone, and dichloromethane. Carbohydrates, amino acids, and proteins are largely insoluble in these organic solvents.

21.1 What Are Triglycerides?

- **Triglycerides (triacylglycerols)**, the most abundant lipids, are triesters of glycerol and fatty acids.
- **Fatty acids** are long-chain carboxylic acids derived from the hydrolysis of fats, oils, and the phospholipids of biological membranes.
- The melting point of a triglyceride increases as (1) the length of its hydrocarbon chains increases and (2) its degree of saturation increases.
- Triglycerides rich in saturated fatty acids are generally solids at room temperature; those rich in unsaturated fatty acids are generally oils at room temperature.

21.2 What Are Soaps and Detergents?

- **Soaps** are sodium or potassium salts of fatty acids.
- In water, soaps form **micelles**, which "dissolve" nonpolar organic grease and oil.
- Natural soaps precipitate as water-insoluble salts with Mg^{2+}, Ca^{2+}, and Fe^{3+} ions in hard water.
- The most common and most widely used **synthetic detergents** are linear alkylbenzenesulfonates.

21.3 What Are Phospholipids?

- **Phospholipids**, the second most abundant group of naturally occurring lipids, are derived from phosphatidic acids—compounds containing glycerol esterified with two molecules of fatty acid and a molecule of phosphoric acid.
- Further esterification of the phosphoric acid part with a low-molecular-weight alcohol—most commonly, ethanolamine, choline, serine, or inositol—gives a phospholipid.
- When placed in aqueous solution, phospholipids spontaneously form **lipid bilayers**. According to the **fluid-mosaic model**, membrane phospholipids form lipid bilayers with membrane proteins associated with the bilayer as both peripheral and integral proteins.

21.4 What Are Steroids?

- **Steroids** are a group of plant and animal lipids that have a characteristic tetracyclic structure of three six-membered rings and one five-membered ring.
- **Cholesterol**, an integral part of animal membranes, is the compound from which human sex hormones, adrenocorticoid hormones, bile acids, and vitamin D are synthesized.
- **Low-density lipoproteins (LDLs)** transport cholesterol from the site of its synthesis in the liver to tissues and cells where it is to be used.
- **High-density lipoproteins (HDLs)** transport cholesterol from cells back to the liver for its degradation to bile acids and eventual excretion in the feces.
- **Oral contraceptive** pills contain a synthetic progestin (e.g., norethindrone) that prevents ovulation, yet allows women to maintain an otherwise normal menstrual cycle.
- A variety of synthetic **anabolic steroids** are available for use in rehabilitation medicine to treat muscle tissue that has weakened or deteriorated due to injury.
- **Bile acids** differ from most other steroids in that they have a *cis* configuration at the junction of rings A and B.
- The carbon skeleton of cholesterol and those of all biomolecules derived from it originate with the acetyl group (a C_2 unit) of **acetyl-CoA**.

21.5 What Are Prostaglandins?

- **Prostaglandins** are a group of compounds having the 20-carbon skeleton of prostanoic acid.
- Prostaglandins are synthesized from phospholipid-bound arachidonic acid (20:4) and other 20-carbon fatty acids in response to physiological triggers.

21.6 What Are Fat-Soluble Vitamins?

- **Vitamin A** occurs only in the animal world. The carotenes of the plant world are tetraterpenes (C_{40}) and are cleaved into vitamin A after ingestion. The best-understood role of vitamin A is its participation in the **visual cycle**.
- **Vitamin D** is synthesized in the skin of mammals by the action of ultraviolet radiation on 7-dehydrocholesterol. This vitamin plays a major role in the regulation of calcium and phosphorus metabolism.
- **Vitamin E** is a group of compounds of similar structure, the most active of which is α-tocopherol. In the body, vitamin E functions as an antioxidant.
- **Vitamin K** is required for the clotting of blood.

Quick Quiz

Answer true or false to the following questions to assess your general knowledge of the concepts in this chapter. If you have difficulty with any of them, you should review the appropriate section in the chapter (shown in parentheses) before attempting the more challenging end-of-chapter problems.

1. Prostaglandins are formed in response to certain biological triggers. (21.5)
2. A fatty acid consists of a carboxyl group on the end of a long hydrocarbon chain. (21.1)
3. Vitamin D is synthesized from a steroid. (21.6)
4. All steroids have the same tetracyclic ring system in common. (21.4)
5. Fatty acids with high melting points are called oils, while fatty acids with low melting points are called fats. (21.1)
6. Steroids are achiral molecules. (21.4)
7. Synthetic detergents are based on fatty acids where the carboxylate group has been replaced by a sulfonate group. (21.2)
8. According to the fluid-mosaic model, proteins in lipid bilayers are stationary and cannot move around. (21.3)

9. Phospholipids organize into micelles in aqueous solution. (21.2)

10. The greater the degree of unsaturation in a fatty acid, the higher is its melting point. (21.1)

11. The function of both low- and high-density lipoproteins is to transport cholesterol throughout the body. (21.4)

12. A fatty acid that has been deprotonated and coordinated with a sodium ion can act as a soap. (21.2)

13. Isoprene is the building block from which the carbon framework of all steroids is made. (21.4)

14. Treatment of a triglyceride with aqueous base followed by acidification yields glycerol and up to three fatty acids. (21.1)

15. Vitamin A can be synthesized from β-carotene. (21.6)

16. In fatty acids with C—C double bonds, the *trans* form of each double bond predominates. (21.1)

Answers: (1) T (2) T (3) T (4) T (5) F (6) F (7) T (8) F (9) F (10) F (11) T (12) T (13) F (14) T (15) T (16) F

Problems

A problem marked with an asterisk indicates an applied "real world" problem. Answers to problems whose numbers are printed in blue are given in Appendix D.

Section 21.1 Fatty Acids and Triglycerides

21.2 Define the term *hydrophobic*.

21.3 Identify the hydrophobic and hydrophilic region(s) of a triglyceride.

21.4 Explain why the melting points of unsaturated fatty acids are lower than those of saturated fatty acids.

21.5 Oleic acid has a melting point of 16°C. If you were to convert the *cis* double bond to a *trans* double bond, what would happen to the melting point?

21.6 Which would you expect to have the higher melting point, glyceryl trioleate or glyceryl trilinoleate?

21.7 Which animal fat has the highest percentage of unsaturated fatty acids? Which plant oil has the highest percentage of unsaturated fatty acids?

21.8 Draw a structural formula for methyl linoleate. Be certain to show the correct configuration of groups about each carbon–carbon double bond. **(See Example 21.1)**

***21.9** Explain why coconut oil is a liquid triglyceride, even though most of its fatty-acid components are saturated.

***21.10** It is common now to see "contains no tropical oils" on cooking-oil labels, meaning that the oil contains no palm or coconut oil. What is the difference between the composition of tropical oils and that of vegetable oils, such as corn oil, soybean oil, and peanut oil?

***21.11** What is meant by the term *hardening* as applied to vegetable oils?

21.12 How many moles of H_2 are used in the catalytic hydrogenation of 1 mole of a triglyceride derived from glycerol, stearic acid, linoleic acid, and arachidonic acid?

21.13 Characterize the structural features necessary to make a good synthetic detergent.

***21.14** Following are structural formulas for a cationic detergent and a neutral detergent:

$$CH_3(CH_2)_6CH_2\overset{+}{N}(CH_3)_2(CH_2C_6H_5)\quad Cl^-$$

Benzyldimethyloctylammonium chloride
(a cationic detergent)

$$(HOCH_2)_3CCH_2OC(=O)(CH_2)_{14}CH_3$$

Pentaerythrityl palmitate
(a neutral detergent)

Account for the detergent properties of each.

***21.15** Identify some of the detergents used in shampoos and dishwashing liquids. Are they primarily anionic, neutral, or cationic detergents?

21.16 Show how to convert palmitic acid (hexadecanoic acid) into the following:

(a) Ethyl palmitate

(b) Palmitoyl chloride

(c) 1-Hexadecanol (cetyl alcohol)

(d) 1-Hexadecanamine

(e) *N,N*-Dimethylhexadecanamide

*21.17 Palmitic acid (hexadecanoic acid, 16:0) is the source of the hexadecyl (cetyl) group in the following compounds:

Cetylpyridinium chloride: pyridinium N^+–$CH_2(CH_2)_{14}CH_3$ Cl^-

Benzylcetyldimethylammonium chloride: C_6H_5–$CH_2\overset{+}{N}(CH_3)_2CH_2(CH_2)_{14}CH_3$ Cl^-

Each compound is a mild surface-acting germicide and fungicide and is used as a topical antiseptic and disinfectant.

(a) Cetylpyridinium chloride is prepared by treating pyridine with 1-chlorohexadecane (cetyl chloride). Show how to convert palmitic acid to cetyl chloride.

(b) Benzylcetyldimethylammonium chloride is prepared by treating benzyl chloride with *N*,*N*-dimethyl-1-hexadecanamine. Show how this tertiary amine can be prepared from palmitic acid.

Section 21.3 Phospholipids

21.18 Draw the structural formula of a lecithin containing one molecule each of palmitic acid and linoleic acid.

21.19 Identify the hydrophobic and hydrophilic region(s) of a phospholipid.

*21.20 The hydrophobic effect is one of the most important noncovalent forces directing the self-assembly of biomolecules in aqueous solution. The hydrophobic effect arises from tendencies of biomolecules (1) to arrange polar groups so that they interact with the aqueous environment by hydrogen bonding and (2) to arrange nonpolar groups so that they are shielded from the aqueous environment. Show how the hydrophobic effect is involved in directing

(a) The formation of micelles by soaps and detergents.

(b) The formation of lipid bilayers by phospholipids.

(c) The formation of the DNA double helix.

21.21 How does the presence of unsaturated fatty acids contribute to the fluidity of biological membranes?

*21.22 Lecithins can act as emulsifying agents. The lecithin of egg yolk, for example, is used to make mayonnaise. Identify the hydrophobic part(s) and the hydrophilic part(s) of a lecithin. Which parts interact with the oils used in making mayonnaise? Which parts interact with the water?

Sections 21.4 Steroids

21.23 Draw the structural formula for the product formed by treating cholesterol (a) with H_2/Pd; (b) with Br_2.

*21.24 List several ways in which cholesterol is essential for human life. Why do many people find it necessary to restrict their dietary intake of cholesterol?

21.25 Both low-density lipoproteins (LDL) and high-density lipoproteins (HDL) consist of a core of triacylglycerols and cholesterol esters surrounded by a single phospholipid layer. Draw the structural formula of cholesteryl linoleate, one of the cholesterol esters found in this core.

*21.26 Examine the structural formulas of testosterone (a male sex hormone) and progesterone (a female sex hormone). What are the similarities in structure between the two? What are the differences?

*21.27 Examine the structural formula of cholic acid, and account for the ability of this bile salt and others to emulsify fats and oils and thus aid in their digestion.

21.28 Following is a structural formula for cortisol (hydrocortisone):

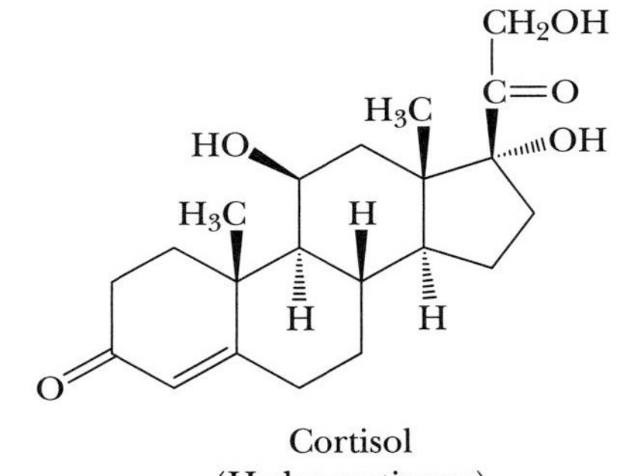

Cortisol
(Hydrocortisone)

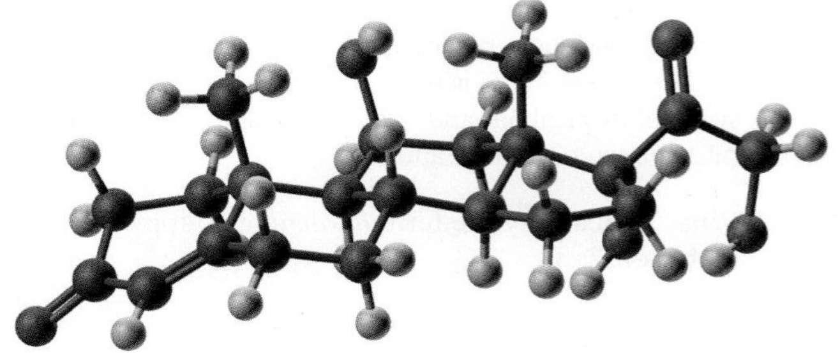

Draw a stereorepresentation of this molecule, showing the conformations of the five- and six-membered rings.

*21.29 How does the oral anabolic steroid methandrostenolone differ structurally from testosterone?

*21.30 Because some types of tumors need estrogen to survive, compounds that compete with the estrogen receptor on tumor cells are useful anticancer drugs. The compound tamoxifen is one such drug. To what part of the estrone molecule is the shape of tamoxifen similar?

Tamoxifen

Estrone

21.31 Estradiol in the body is synthesized from progesterone. What chemical modifications occur when estradiol is synthesized?

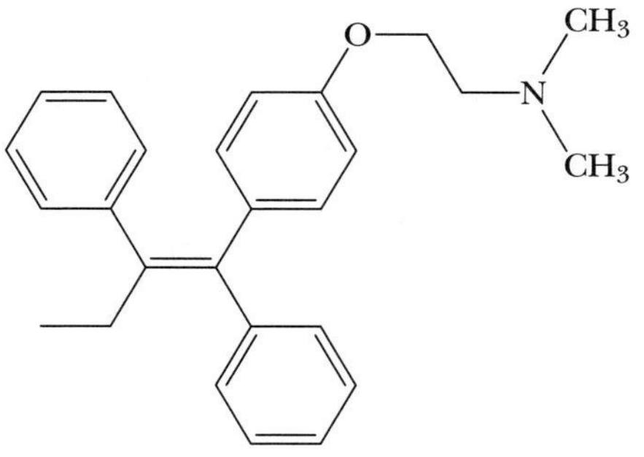

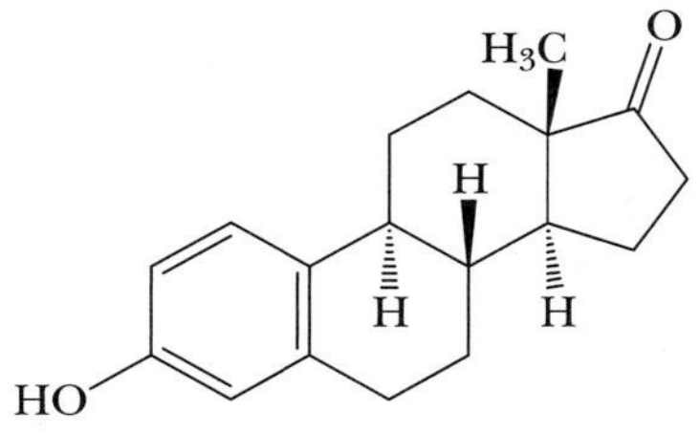

Section 21.5 Prostaglandins

21.32 Examine the structure of $PGF_{2\alpha}$, and

(a) identify all stereocenters.

(b) identify all double bonds about which *cis,trans* isomerism is possible.

(c) state the number of stereoisomers possible for a molecule of this structure.

*21.33 Following is the structure of unoprostone, a compound patterned after the natural prostaglandins (Section 21.5):

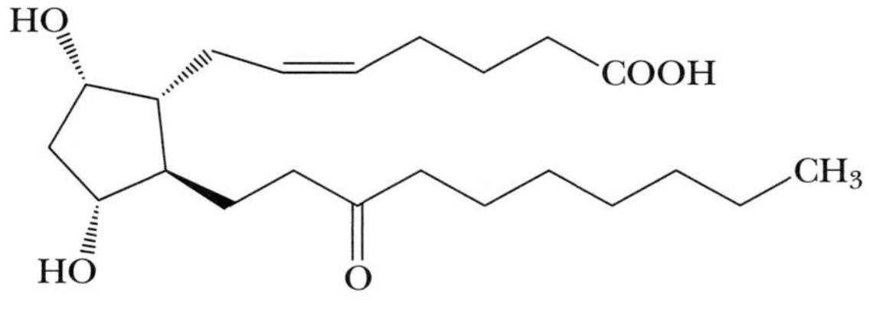

Unoprostone
(antiglaucoma)

Rescula, the isopropyl ester of unoprostone, is an antiglaucoma drug used to treat ocular hypertension. Compare the structural formula of this synthetic prostaglandin with that of $PGF_{2\alpha}$.

*21.34 How does aspirin, an anti-inflammatory drug, prevent strokes caused by blood clots in the brain?

Section 21.6 Fat-Soluble Vitamins

21.35 Examine the structural formula of vitamin A, and state the number of *cis,trans* isomers that are possible for this molecule.

*21.36 The form of vitamin A present in many food supplements is vitamin A palmitate. Draw the structural formula of this molecule.

*21.37 Examine the structural formulas of vitamin A, 1,25-dihydroxy-D_3, vitamin E, and vitamin K_1. Do you expect these vitamins to be more soluble in water or in dichloromethane? Do you expect them to be soluble in blood plasma?

Looking Ahead

***21.38** Here is the structure of a glycolipid, a class of lipid that contains a sugar residue:

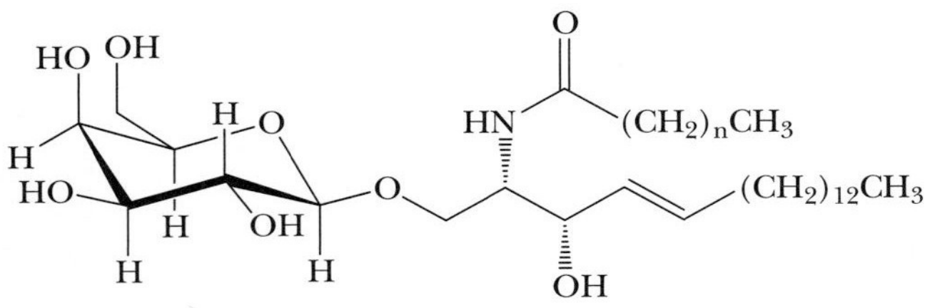

A Glycolipid

Glycolipids are found in cell membranes. (a) Which part of the molecule would you expect to reside on the extracellular side of the membrane? (b) What monosaccharide unit is present in this glycolipid?

21.39 How would you expect temperature to affect fluidity in a cell membrane?

21.40 Which type of lipid movement, A or B, is more favorable in cell membranes? Explain.

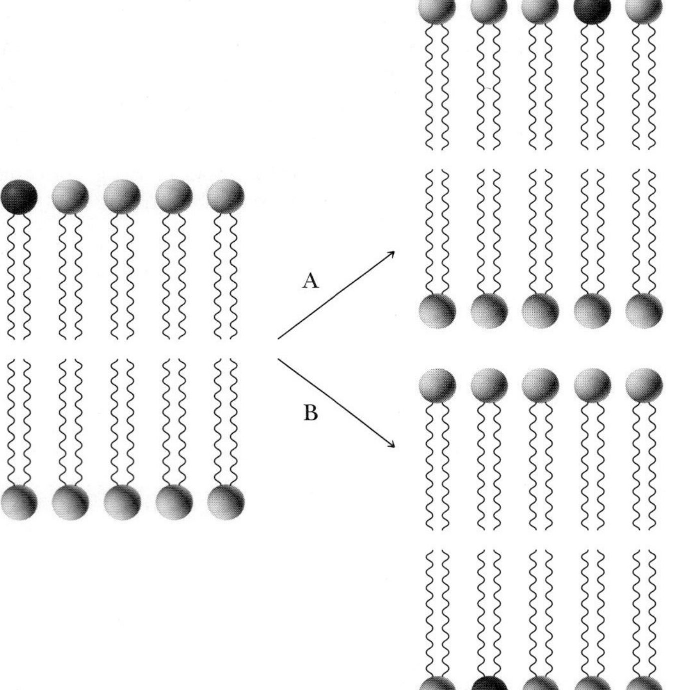

***21.41** Aspirin works by transferring an acetyl group to the side chain of the 530th amino acid in the protein prostaglandin H_2 synthase-1. Draw the product of this reaction:

Residue-530

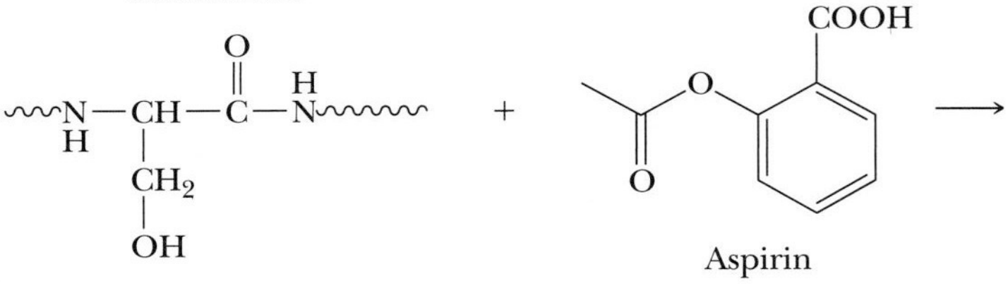

Aspirin

22 The Organic Chemistry of Metabolism

KEY QUESTIONS

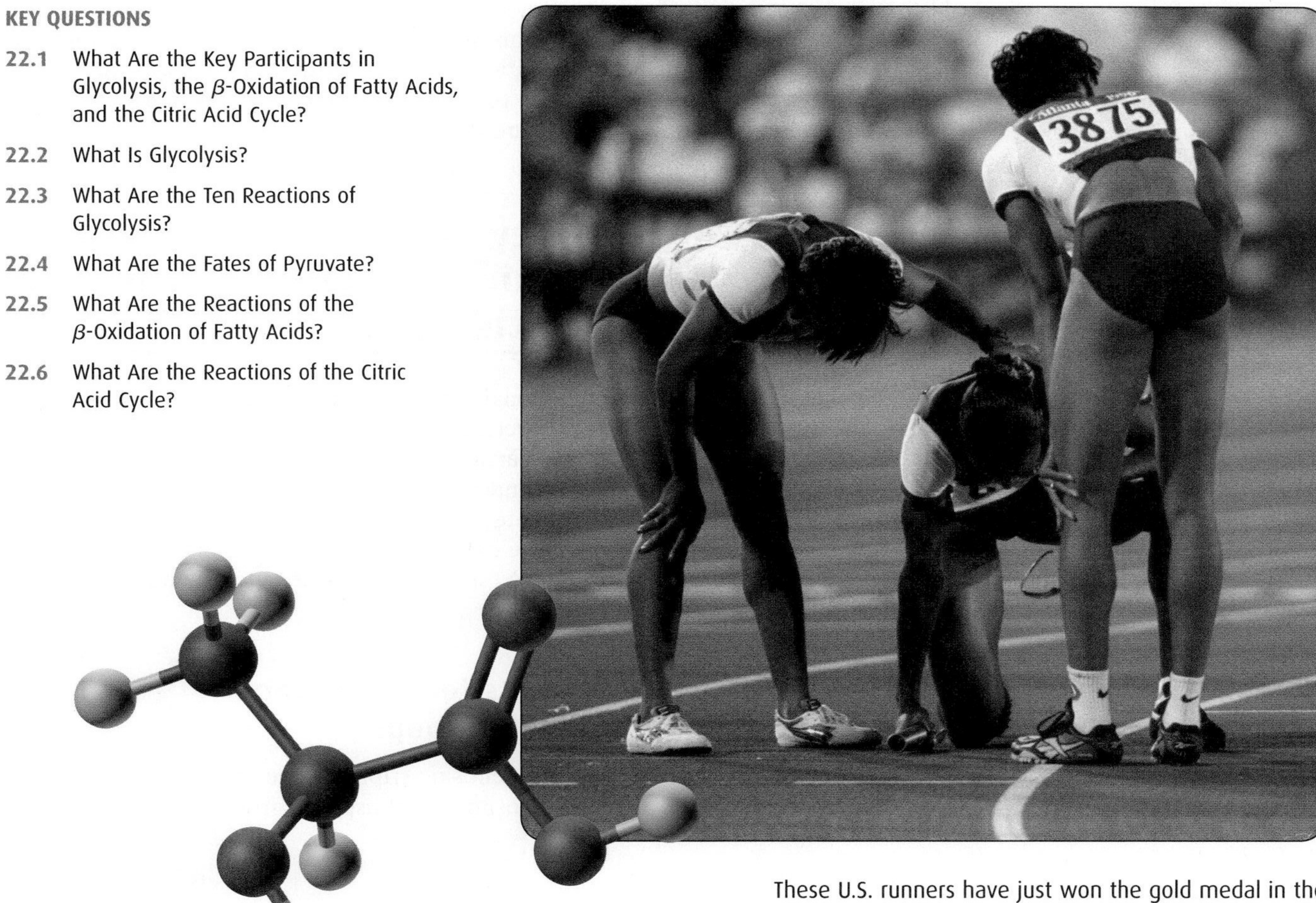

These U.S. runners have just won the gold medal in the 4×400 m relay race at the 1996 Olympic games. The buildup of lactic acid can cause severe muscle pain. Inset: A model of lactic acid. *(J. O. Atlanta 96/Gamma)*

We have now studied the structures and typical reactions of the major types of organic functional groups. Further, we have examined the structures of carbohydrates, amino acids and proteins, nucleic acids, and lipids. Now let us apply this background to the study of the organic chemistry of metabolism. In this chapter, we study three key metabolic pathways: glycolysis, the citric acid cycle, and the β-oxidation of fatty acids. The first is a pathway by which glucose is converted to pyruvate and then to acetyl coenzyme A. The second is a pathway by which the hydrocarbon chains of fatty acids are degraded, two carbons at a time, to acetyl coenzyme A. The third is the pathway by which the carbon skeletons of carbohydrates, fatty acids, and proteins are oxidized to carbon dioxide.

Those of you who go on to courses in biochemistry will undoubtedly study these metabolic pathways in considerable detail, including their role in energy production and conservation, their regulation, and the diseases associated with errors in particular metabolic steps. Our concern in this chapter is more limited. Our goal is to show that the reactions of these pathways are biochemical equivalents of organic functional group reactions we have already studied in detail. For instance, we find examples of keto–enol tautomerism; oxidation of an aldehyde to a carboxylic acid; oxidation of a secondary alcohol to a ketone; an aldol reaction and a reverse aldol reaction; a reverse Claisen condensation; and the formation and hydrolysis of esters, imines, thioesters, and anhydrides. In addition, we will use the mechanisms we have studied earlier to give us insights into the mechanisms of these reactions, all of which are enzyme catalyzed.

22.1 What Are the Key Participants in Glycolysis, β-Oxidation of Fatty Acids, and the Citric Acid Cycle?

To understand the reactions of functional groups that play a role in the β-oxidation of fatty acids, the tricarboxylic acid cycle, and glycolysis, we first need to introduce several of the principal compounds participating in these and a great many other metabolic pathways. Three of these compounds (ATP, ADP, and AMP) are central to the storage and transfer of phosphate groups. Four (NAD^+/NADH and FAD/$FADH_2$) are **coenzymes** involved in the oxidation–reduction of metabolic intermediates. The final compound (coenzyme A) is an agent for the storage and transfer of acetyl groups.

Coenzyme A low-molecular-weight, nonprotein molecule or ion that binds reversibly to an enzyme, functions as a second substrate for the enzyme, and is regenerated by further reaction.

A. ATP, ADP, and AMP: Agents for the Storage and Transfer of Phosphate Groups

Following is a structural formula of adenosine triphosphate (Section 20.1), a compound involved in the storage and transport of phosphate groups:

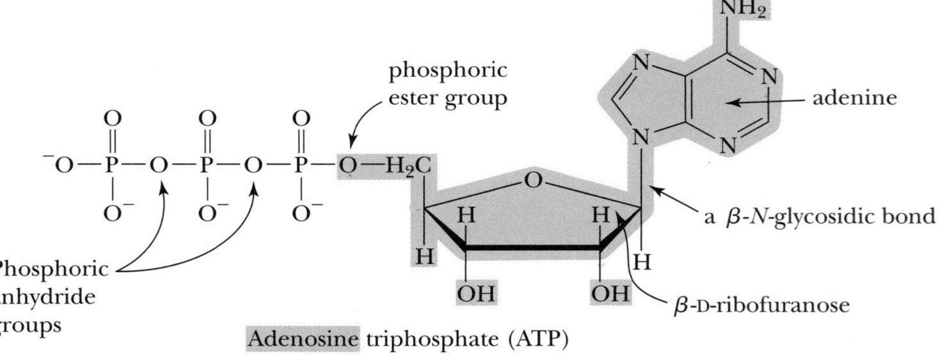

A building block for ATP as well as for five other key participants is adenosine, which consists of a unit of adenine bonded to a unit of D-ribofuranose by a β-N-glycosidic bond. Three phosphate groups are bonded to the terminal $-CH_2OH$ of ribose: one by a phosphoric ester bond, the remaining two by phosphoric anhydride bonds. Hydrolysis of the terminal phosphate group of ATP gives ADP. In the following abbreviated structural formulas, adenosine and its single phosphoric ester group are represented by the symbol AMP (adenosine monophosphate):

$$^{-}O-\overset{\overset{O}{\|}}{\underset{\underset{O^-}{|}}{P}}-O-\overset{\overset{O}{\|}}{\underset{\underset{O^-}{|}}{P}}-O-AMP \quad + \quad H_2O \quad \longrightarrow \quad ^{-}O-\overset{\overset{O}{\|}}{\underset{\underset{O^-}{|}}{P}}-O-AMP \quad + \quad H_2PO_4^{-}$$

Adenosine triphosphate (ATP) — Water (a phosphate acceptor) — Adenosine diphosphate (ADP)

The reaction shown is the hydrolysis of a phosphoric anhydride; the phosphate acceptor is water. In the first two reactions of glycolysis, the phosphate acceptors are —OH groups of glucose and fructose, respectively, and are used to form phosphoric esters of these monosaccharides. In two reactions of glycolysis, ADP is the phosphate acceptor and is converted to ATP.

B. NAD^+/NADH: Agents for Electron Transfer in Biological Oxidation–Reductions

Nicotinamide adenine dinucleotide (NAD^+) is one of the central agents for the transfer of electrons in metabolic oxidations and reductions. NAD^+ is constructed of a unit of ADP, joined by a phosphoric ester bond to the terminal $-CH_2OH$ of *β*-D-ribofuranose, which is in turn joined to the pyridine ring of nicotinamide by a *β*-*N*-glycosidic bond:

Nicotinamide adenine dinucleotide (NAD^+) A biological oxidizing agent. When acting as an oxidizing agent, NAD^+ is reduced to NADH.

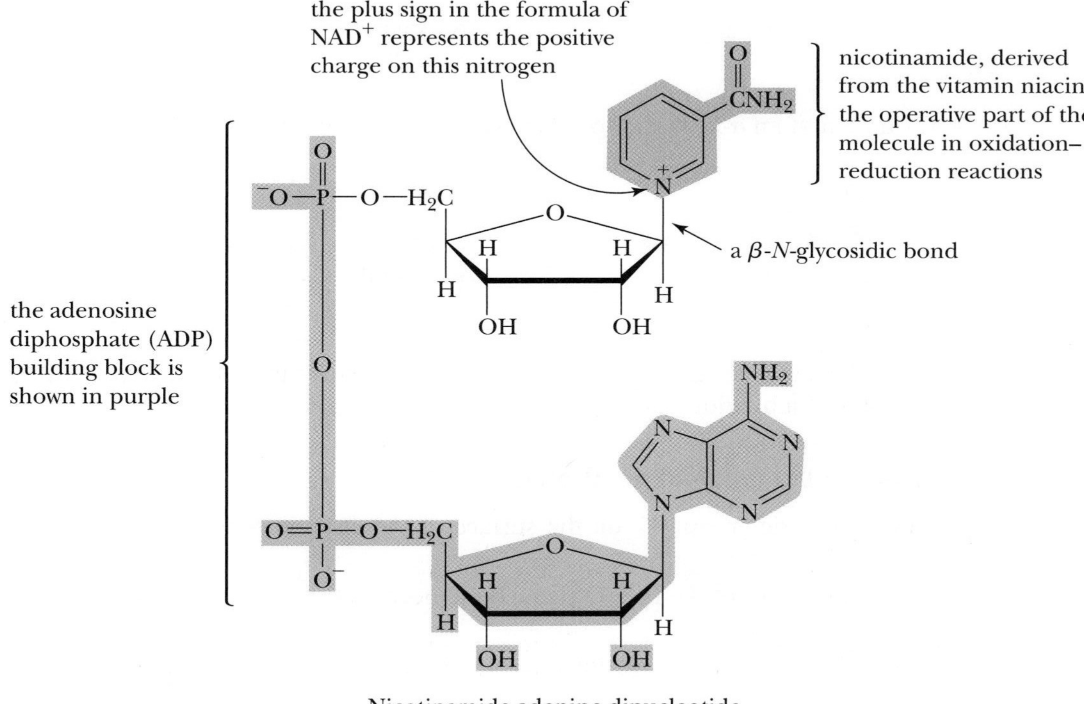

Nicotinamide adenine dinucleotide (NAD^+)

When NAD^+ acts as an oxidizing agent, it is reduced to NADH, which, in turn, is a reducing agent and is oxidized to NAD^+. In these abbreviated structural formulas, the adenine dinucleotide part of each molecule is represented by the symbol Ad:

NAD^+ (oxidized form) $\underset{\text{oxidation}}{\overset{\text{reduction}}{\rightleftharpoons}}$ NADH (reduced form)

NAD⁺ is involved in a variety of enzyme-catalyzed oxidation–reduction reactions. The three types of oxidations we deal with in this chapter are the following:

- Oxidation of a secondary alcohol to a ketone:

$$-\underset{\text{OH}}{\text{CH}}- + NAD^+ \longrightarrow -\overset{\text{O}}{\overset{\|}{\text{C}}}- + NADH + H^+$$

A 2° alcohol → A ketone

- Oxidation of an aldehyde to a carboxylic acid:

$$-\overset{\text{O}}{\overset{\|}{\text{C}}}-H + NAD^+ + H_2O \longrightarrow -\overset{\text{O}}{\overset{\|}{\text{C}}}-OH + NADH + H^+$$

An aldehyde → A carboxylic acid

- Oxidation of an α-ketoacid to a carboxylic acid and carbon dioxide:

$$-\overset{\text{O}}{\overset{\|}{\text{C}}}-COOH + NAD^+ + H_2O \longrightarrow -\overset{\text{O}}{\overset{\|}{\text{C}}}-OH + CO_2 + NADH + H^+$$

An α-ketoacid → A carboxylic acid

As the following mechanism shows, the oxidation of each functional group involves the transfer of a hydride ion to NAD^+.

Mechanism: Oxidation of an Alcohol by NAD^+

Step 1: A basic group, B^-, on the surface of the enzyme removes H^+ from the —OH group.

Step 2: Electrons of the H—O sigma bond become the pi electrons of the C=O bond.

Step 3: A hydride ion is transferred from carbon to NAD^+ to create a new C—H bond.

Step 4: Electrons within the ring flow to the positively charged nitrogen.

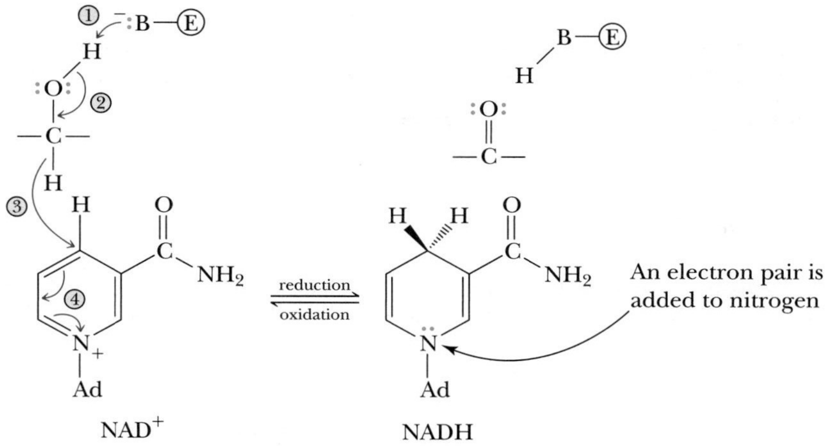

The hydride ion, H:⁻, which is transferred from the secondary alcohol to NAD^+, contains two electrons; thus, NAD^+ and NADH function exclusively in two-electron oxidations and two-electron reductions.

C. FAD/$FADH_2$: Agents for Electron Transfer in Biological Oxidation–Reductions

Flavin adenine dinucleotide (FAD) is also a central component in the transfer of electrons in metabolic oxidations and reductions. In FAD, flavin is bonded to the five-carbon monosaccharide ribitol, which is, in turn, bonded to the terminal phosphate group of ADP:

Flavin adenine dinucleotide (FAD) A biological oxidizing agent. When acting as an oxidizing agent, FAD is reduced to $FADH_2$.

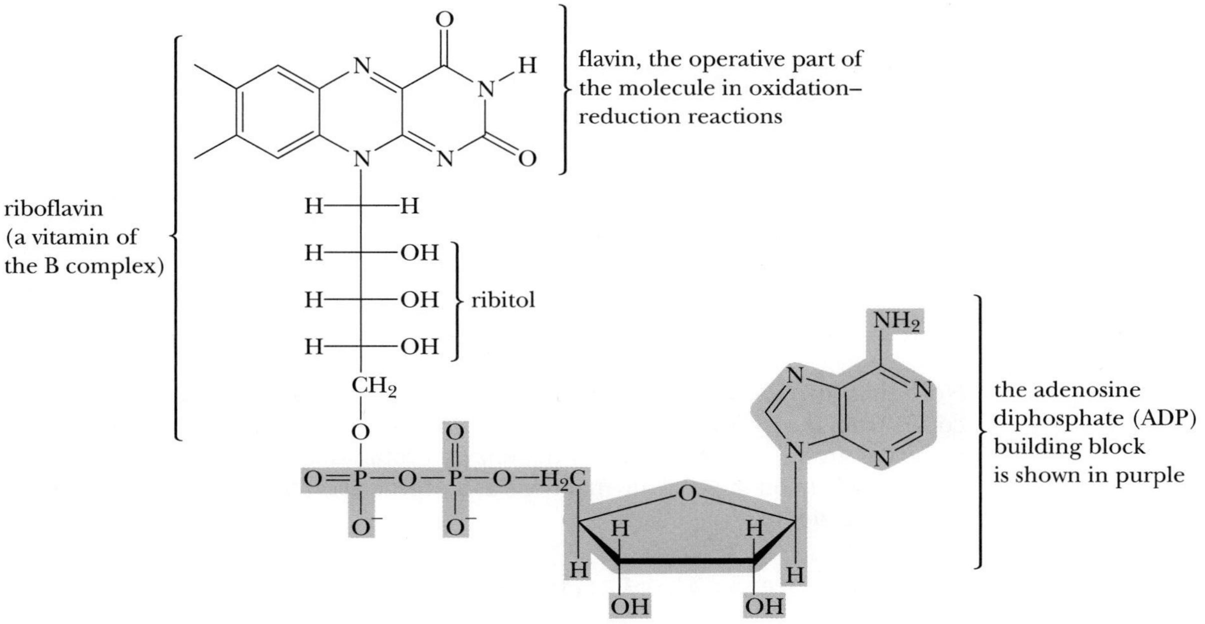

Flavin adenine dinucleotide (FAD)

FAD participates in several types of enzyme-catalyzed oxidation–reduction reactions. Our concern in this chapter is its role in the oxidation of a carbon–carbon single bond in the hydrocarbon chain of a fatty acid to a carbon–carbon double bond, in the process of which FAD is reduced to $FADH_2$:

$$-CH_2-CH_2- + FAD \longrightarrow -CH=CH- + FADH_2$$

a portion of the hydrocarbon chain of a fatty acid

The mechanism by which FAD oxidizes $-CH_2-CH_2-$ to $-CH=CH-$ involves the transfer of a hydride ion from the hydrocarbon chain of the fatty acid to FAD.

Mechanism: Oxidation of a Fatty Acid $-CH_2-CH_2-$ to $-CH=CH-$ by FAD

The individual curved arrows in this mechanism are numbered 1–6 to help you follow the flow of electrons in the transformation.

Step 1: A basic group, $B:^-$, on the surface of the enzyme removes a hydrogen atom from the carbon adjacent to the carboxyl group.

Step 2: Electrons from this C–H sigma bond become the pi electrons of the new C–C double bond.

Step 3: A hydride ion transfers from the carbon beta to the carboxyl group, to a nitrogen atom of flavin.

Step 4: The pi electrons within flavin become redistributed.

Step 5: Electrons of the C=N bond remove a hydrogen atom from the enzyme.

Step 6: A new basic group forms on the surface of the enzyme.

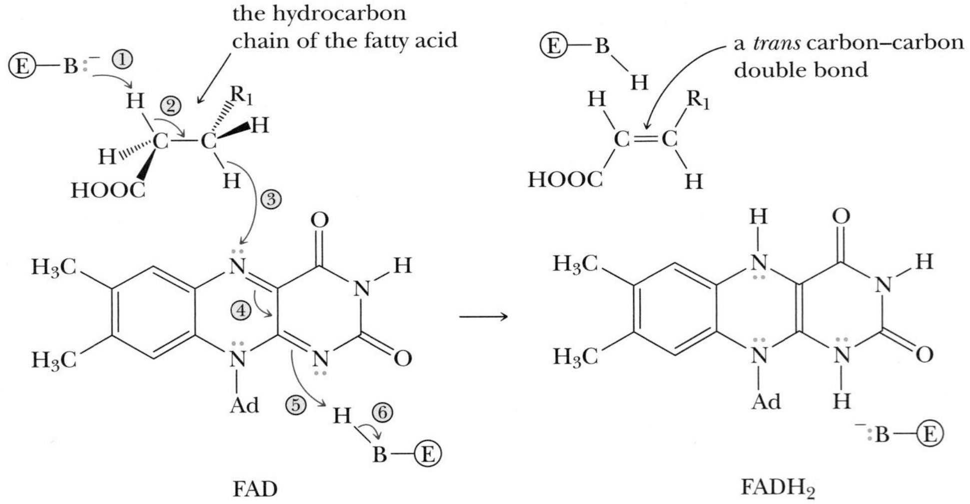

Note that, of the two hydrogen atoms added to FAD to produce $FADH_2$, one comes from the hydrocarbon chain undergoing oxidation and the other comes from an acidic group on the surface of the enzyme catalyzing this oxidation. Note also that one group on the enzyme functions as a proton acceptor and another functions as a proton donor.

D. Coenzyme A: The Carrier of Acetyl Groups

Coenzyme A is derived from four subunits. On the left is a two-carbon unit derived from 2-mercaptoethanamine. This unit is in turn joined by an amide bond to the carboxyl group of 3-aminopropanoic acid (β-alanine). The amino group of β-alanine is joined by an amide bond to the carboxyl group of pantothenic acid, a vitamin of the B complex. Finally, the —OH group of pantothenic acid is joined by a phosphoric ester bond to the terminal phosphate group of ADP:

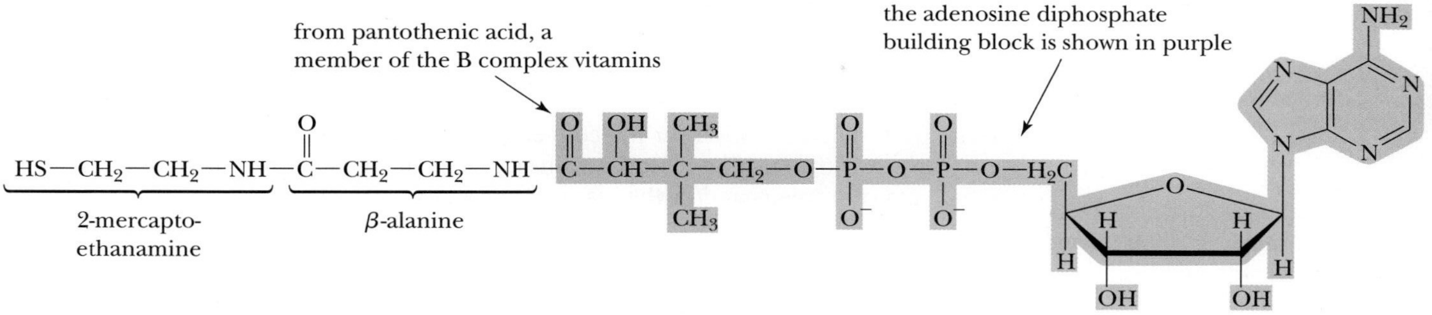

from pantothenic acid, a member of the B complex vitamins

the adenosine diphosphate building block is shown in purple

HS—CH$_2$—CH$_2$—NH—C(=O)—CH$_2$—CH$_2$—NH—C(=O)—CH(OH)—C(CH$_3$)$_2$—CH$_2$—O—P(=O)(O$^-$)—O—P(=O)(O$^-$)—O—H$_2$C—(ribose with OH, OH; adenine with NH$_2$)

2-mercapto-ethanamine

β-alanine

Coenzyme A (Co-A)

A key feature of the structure of coenzyme A is the terminal sulfhydryl (—SH) group. During the degradation of foodstuffs for the production of energy, the carbon skeletons of glucose, fructose, and galactose, along with those of fatty acids, glycerol, and several amino acids, are converted to acetate in the form of a thioester named acetyl coenzyme A, or, more commonly, acetyl-CoA:

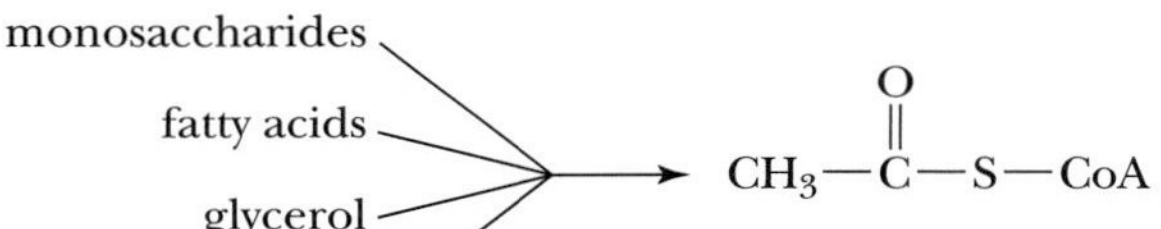

monosaccharides, fatty acids, glycerol, amino acids $\longrightarrow$ CH_3—C(=O)—S—CoA

Acetyl coenzyme A
(Acetyl-CoA)

In Section 22.6, we will see how the two-carbon acetyl group is oxidized to carbon dioxide and water by the reactions of the citric acid cycle.

22.2 What Is Glycolysis?

Nearly every living cell carries out glycolysis. Living things first appeared in an environment lacking O_2, and glycolysis was an early and important pathway for extracting energy from nutrient molecules because it takes place in the absence of oxygen. Glycolysis played a central role in anaerobic metabolic processes for the first billion or so years of biological evolution on earth. Modern organisms still employ it to provide precursor molecules for aerobic pathways, such as the citric acid cycle, and as a short-term energy source when the supply of oxygen is limited.

Glycolysis is a series of 10 enzyme-catalyzed reactions that brings about the oxidation of glucose to two molecules of pyruvate. The oxidizing agent is NAD^+. Furthermore, two molecules of ATP are produced for each molecule of glucose oxidized to pyruvate. Following is the net reaction of glycolysis:

Glycolysis From the Greek *glyko*, sweet, and *lysis*, splitting; a series of 10 enzyme-catalyzed reactions by which glucose is oxidized to two molecules of pyruvate.

$$\underset{\text{Glucose}}{C_6H_{12}O_6} + 2NAD^+ + 2HPO_4^{2-} + 2ADP \xrightarrow[\text{10 enzyme-catalyzed steps}]{\text{glycolysis}} \underset{\text{Pyruvate}}{2CH_3\overset{O}{\overset{\|}{C}}COO^-} + 2H^+ + 2NADH + 2ATP$$

Example 22.1

Show that the conversion of glucose to two molecules of pyruvate is an oxidation. (*Hint:* That it is an oxidation is easiest to see if you take the product to be pyruvic acid; recognize, of course, that, under the pH conditions at which this reaction takes place in cells, pyruvic acid is ionized to pyruvate.)

Strategy

Write a balanced equation for the conversion of glucose to pyruvate, and examine it to see if the starting material gains oxygens or loses hydrogens (oxidation is the gain of oxygens and/or the loss of hydrogens).

Solution

Glucose is $C_6H_{12}O_6$. Two molecules of pyruvic acid are $2(C_3H_4O_3) = C_6H_8O_6$. The number of O atoms remains the same in this conversion but four H are lost. Therefore, the conversion of glucose to pyruvate is an oxidation.

Problem 22.1

Under anaerobic (without oxygen) conditions, glucose is converted to lactate by a metabolic pathway called anaerobic glycolysis or, alternatively, lactate fermentation:

$$\underset{\text{Glucose}}{C_6H_{12}O_6} \xrightarrow[\text{glycolysis}]{\text{anaerobic}} \underset{\text{Lactate}}{2CH_3\overset{\overset{\displaystyle OH}{|}}{C}HCOO^-} + 2H^+$$

Is anaerobic glycolysis a net oxidation, a net reduction, or neither?

22.3 What Are the Ten Reactions of Glycolysis?

Although writing the net reaction of glycolysis is simple, it took several decades of patient, intensive work by scores of scientists to discover the separate reactions by which glucose is converted to pyruvate. Glycolysis is frequently called the *Embden–Meyerhof pathway*, in honor of the two German biochemists, Gustav Embden and Otto Meyerhof, who contributed so greatly to our present knowledge of it. Figure 22.1 shows the 10 reactions of glycolysis.

Reaction 1: Phosphorylation of α-D-Glucose

The transfer of a phosphate group from ATP to glucose gives α-D-glucose 6-phosphate. This conversion is an example of the reaction of an anhydride with an alcohol to form an ester (Section 15.4B); in this case, a phosphoric anhydride reacts with the primary alcohol group of glucose to form a phosphoric ester. In Section 15.6, we saw that a more reactive carboxyl functional group can be transformed into a less reactive carboxyl functional group. The same principle applies to functional derivatives of phosphoric acid. In Reaction 1 of glycolysis, a phosphoric

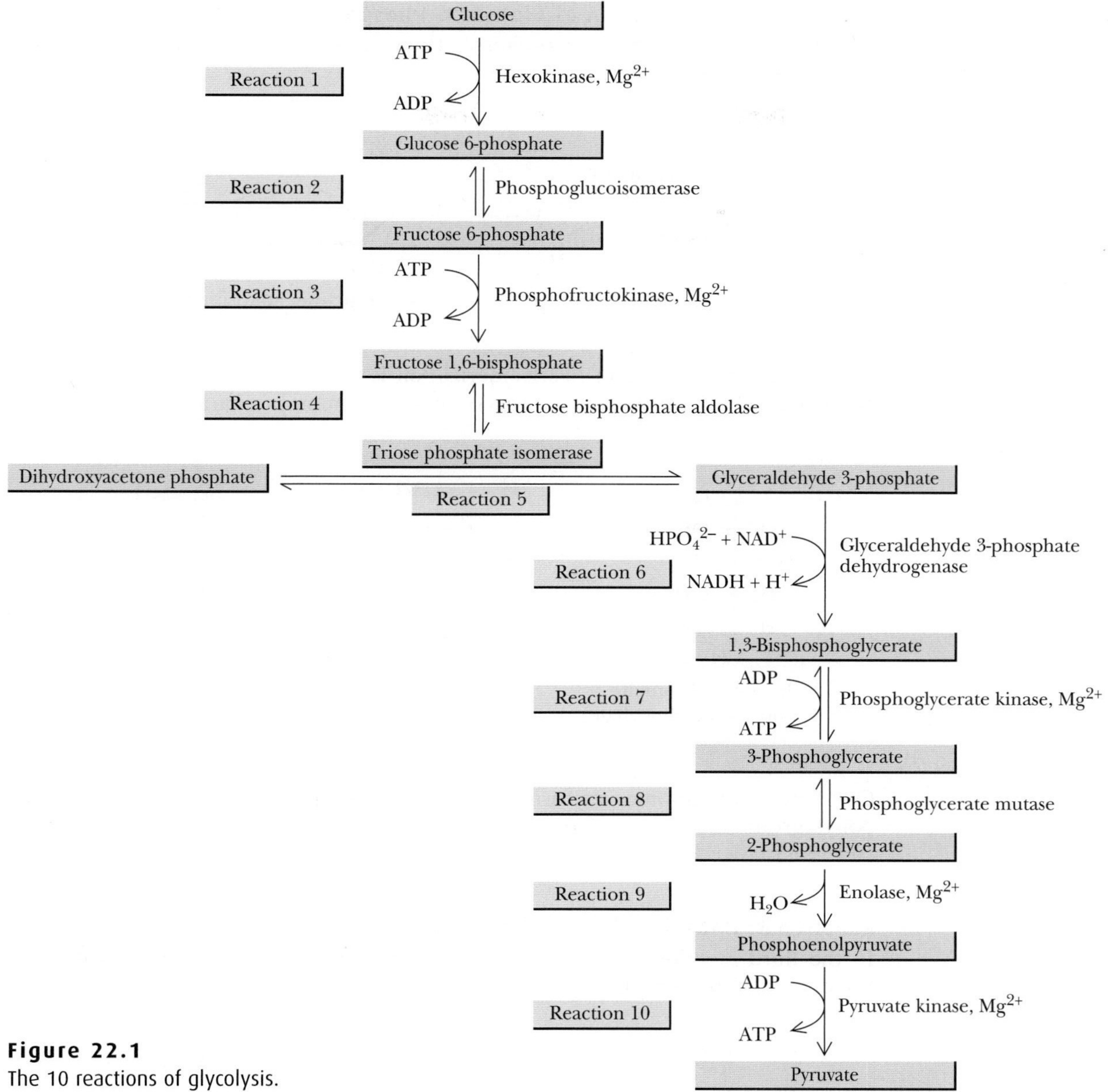

Figure 22.1
The 10 reactions of glycolysis.

anhydride, a more reactive functional group, is converted to a phosphoric ester, a less reactive functional group:

$$\alpha\text{-D-Glucose} + \text{ATP} \xrightarrow[\text{Mg}^{2+}]{\text{hexokinase}} \alpha\text{-D-Glucose 6-phosphate} + \text{ADP}$$

Hexokinase, the enzyme catalyzing this reaction, requires divalent magnesium ion, Mg^{2+}, whose function is to coordinate with two negatively charged oxygens of the terminal phosphate group of ATP and to facilitate attack by the —OH group of glucose on the phosphorus atom of the P=O group.

Reaction 2: Isomerization of Glucose 6-Phosphate to Fructose 6-Phosphate

In this reaction, α-D-glucose 6-phosphate, an aldohexose, is converted to α-D-fructose 6-phosphate, a 2-ketohexose:

$$\alpha\text{-D-Glucose 6-phosphate} \underset{\text{isomerase}}{\overset{\text{phosphogluco-}}{\rightleftharpoons}} \alpha\text{-D-Fructose 6-phosphate}$$

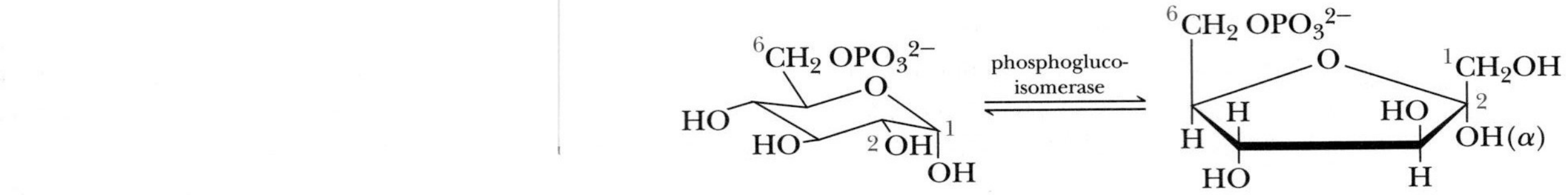

α-D-Glucose 6-phosphate α-D-Fructose 6-phosphate

The chemistry involved in this isomerization is easiest to see by considering the open-chain (Fisher projection) forms of these two monosaccharides:

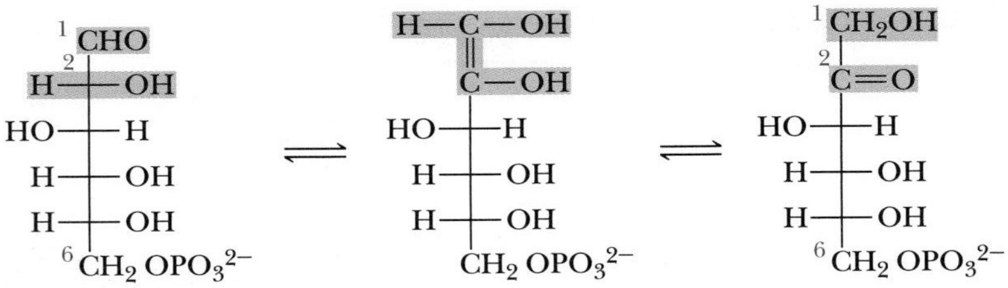

$$\text{Glucose 6-phosphate} \rightleftharpoons \text{An enediol} \rightleftharpoons \text{Fructose 6-phosphate}$$

Glucose 6-phosphate (an aldohexose phosphate) An enediol Fructose 6-phosphate (a 2-ketohexose phosphate)

One keto–enol tautomerism forms an enediol; a second then forms the ketone carbonyl group in fructose 6-phosphate. (See Section 13.8A and Problems 13.34 and 13.35.)

Reaction 3: Phosphorylation of Fructose 6-Phosphate

In the third reaction, a second mole of ATP converts D-fructose 6-phosphate to D-fructose 1,6-bisphosphate:

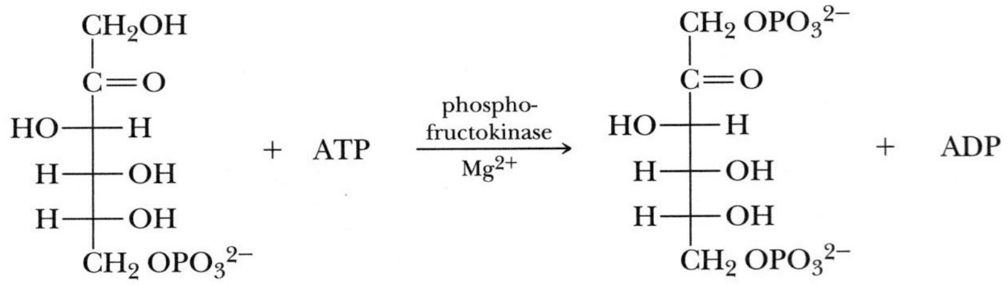

$$\text{D-Fructose 6-phosphate} + \text{ATP} \xrightarrow[\text{Mg}^{2+}]{\text{phosphofructokinase}} \text{D-Fructose 1,6-bisphosphate} + \text{ADP}$$

D-Fructose 6-phosphate D-Fructose 1,6-bisphosphate

Reaction 4: Cleavage of Fructose 1,6-Bisphosphate to Two Triose Phosphates

In the fourth reaction, D-fructose 1,6-bisphosphate is cleaved to dihydroxyacetone phosphate and D-glyceraldehyde 3-phosphate by a reaction that is the reverse of an aldol reaction. Recall from Section 16.2 that an aldol reaction takes place between the α-carbon of one carbonyl-containing compound and the carbonyl carbon of another and that the functional group of the product of an aldol reaction is a β-hydroxyaldehyde or ketone:

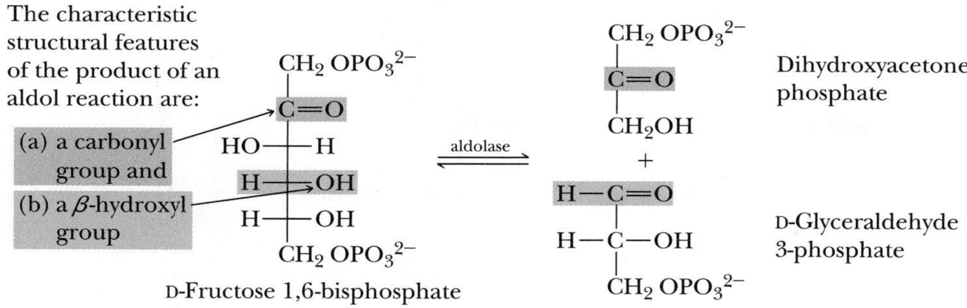

Reaction 5: Isomerization of Dihydroxyacetone Phosphate to D-Glyceraldehyde 3-Phosphate

This interconversion of triose phosphates occurs by the same type of keto–enol tautomerism (Section 13.8A) and enediol intermediate we have already seen in the isomerization of D-glucose 6-phosphate to D-fructose 6-phosphate. Note that D-glyceraldehyde 3-phosphate is chiral and that two enantiomers are possible for it. The enzyme catalyzing this reaction has a high stereospecificity, and only the D enantiomer is formed.

$$CH_2OH-C(=O)-CH_2OPO_3^{2-} \rightleftharpoons CHOH=C(OH)-CH_2OPO_3^{2-} \rightleftharpoons OHC-CH(OH)-CH_2OPO_3^{2-}$$

Dihydroxyacetone phosphate — An enediol intermediate — D-Glyceraldehyde 3-phosphate

Reaction 6: Oxidation of the Aldehyde Group of D-Glyceraldehyde 3-Phosphate

To simplify structural formulas in Reaction 6, D-glyceraldehyde 3-phosphate is abbreviated G—CHO. Two changes occur in this reaction. First, the aldehyde group is oxidized to a carboxyl group, which is, in turn, converted to a mixed anhydride, and second, the oxidizing agent is NAD^+, which is reduced to NADH:

$$G-\overset{O}{\overset{\|}{C}}-H + H_2O + NAD^+ \longrightarrow G-\overset{O}{\overset{\|}{C}}-OH + H^+ + NADH$$

Glyceraldehyde 3-phosphate — Glyceric acid 3-phosphate

The reaction is considerably more complicated than might appear from the balanced equation. As the following mechanism indicates, it involves (1) the formation of a thiohemiacetal, (2) hydride ion transfer to form a thioester, and (3) the conversion of a thioester to a mixed anhydride.

Mechanism: Oxidation of D-Glyceraldehyde 3-Phosphate to 1,3-Bisphosphoglycerate

Step 1: Reaction between D-glyceraldehyde 3-phosphate and a sulfhydryl group of the enzyme gives a thiohemiacetal (Section 13.6):

$$G-\overset{:\ddot{O}}{\overset{\|}{C}}-H + H\ddot{S}-Enz \rightleftharpoons G-\overset{:\ddot{O}H}{\underset{H}{C}}-\ddot{S}-Enz$$

D-Glyceraldehyde 3-phosphate — A thiohemiacetal

Step 2: Oxidation occurs by the transfer of a hydride ion from the thiohemiacetal to NAD^+:

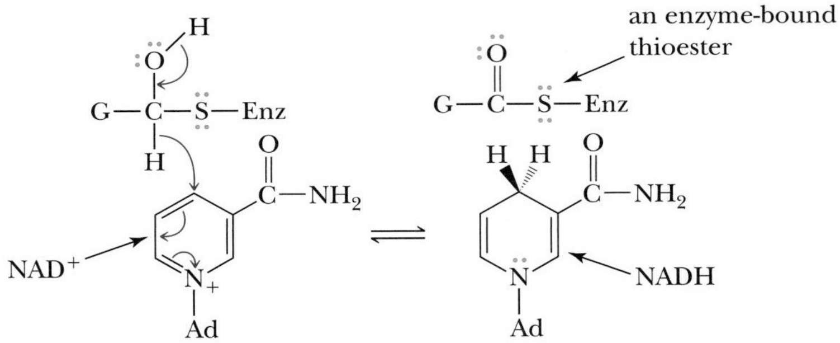

NAD^+

an enzyme-bound thioester

NADH

Step 3: Reaction of the thioester with phosphate ion gives a tetrahedral carbonyl addition intermediate, which then collapses to regenerate the enzyme and give a mixed anhydride of phosphoric acid and glyceric acid:

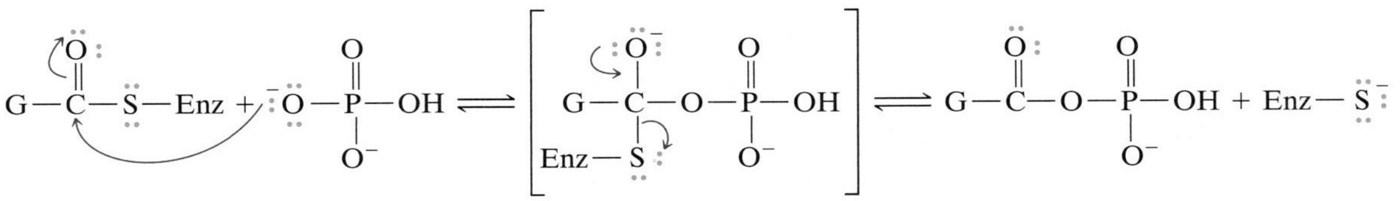

A tetrahedral carbonyl addition intermediate

1,3-Bisphosphoglycerate (a mixed anhydride)

Reaction 7: Transfer of a Phosphate Group from 1,3-Bisphosphoglycerate to ADP

The transfer of a phosphate group in this reaction involves the exchange of one anhydride group for another, namely, the mixed anhydride of 1,3-bisphosphoglycerate for the new phosphoric anhydride in ATP:

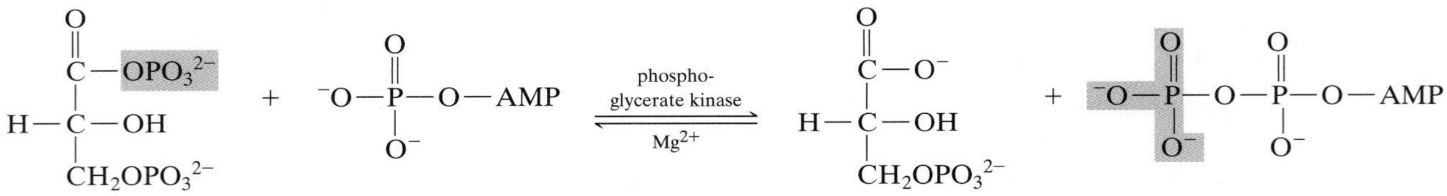

1,3-Bisphosphoglycerate + ADP $\xrightleftharpoons[Mg^{2+}]{\text{phosphoglycerate kinase}}$ 3-Phosphoglycerate + ATP

Reaction 8: Isomerization of 3-Phosphoglycerate to 2-Phosphoglycerate

A phosphate group is transferred from the primary —OH group on carbon 3 to the secondary —OH group on carbon 2:

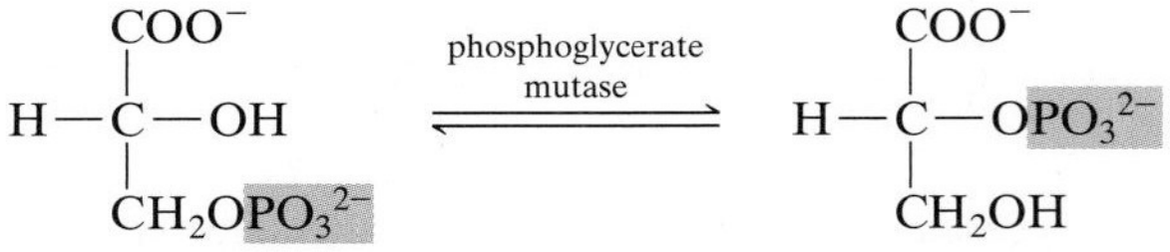

3-Phosphoglycerate $\xrightleftharpoons{\text{phosphoglycerate mutase}}$ 2-Phosphoglycerate

Reaction 9: Dehydration of 2-Phosphoglycerate

Dehydration of the primary alcohol (Section 8.2E) of 2-phosphoglycerate gives phosphoenolpyruvate, which is the ester of phosphoric acid and the enol form of pyruvic acid:

$$\underset{\text{2-Phosphoglycerate}}{\text{H}-\overset{\text{COO}^-}{\underset{\text{CH}_2\text{OH}}{\text{C}}}-\text{OPO}_3^{2-}} \xrightarrow[\text{Mg}^{2+}]{\text{enolase}} \underset{\text{Phosphoenolpyruvate}}{\overset{\text{COO}^-}{\underset{\text{CH}_2}{\overset{|}{\text{C}}\,\Vert}}-\text{OPO}_3^{2-}} + \text{H}_2\text{O}$$

Reaction 10: Transfer of a Phosphate Group from Phosphoenolpyruvate to ADP

Reaction 10 is divided into two steps: the transfer of a phosphate group to ADP to produce ATP and the conversion of the enol form of pyruvate to its keto form by keto–enol tautomerism (Section 13.8A). It is common in writing biochemical reactions to show some reactants and products by a curved arrow set either over or under the main reaction arrow. We use this convention here to show that in this reaction ADP is converted to ATP.

$$\underset{\text{Phosphoenol-pyruvate}}{\text{CH}_2{=}\text{C}(\text{COO}^-)-\text{OPO}_3^{2-}} \xrightarrow[\text{ADP} \frown \text{ATP}]{\text{pyruvate kinase, Mg}^{2+}} \underset{\text{Enol of pyruvate}}{\text{CH}_2{=}\text{C}(\text{COO}^-)-\text{OH}} \rightleftharpoons \underset{\text{Pyruvate}}{\text{CH}_3-\text{C}(\text{COO}^-){=}\text{O}}$$

Summing these 10 reactions gives a balanced equation for the net reaction of glycolysis:

$$\underset{\text{Glucose}}{C_6H_{12}O_6} + 2NAD^+ + 2HPO_4^{2-} + 2ADP \xrightarrow[\text{10 enzyme-catalyzed steps}]{\text{glycolysis}} \underset{\text{Pyruvate}}{2CH_3\overset{O}{\overset{\Vert}{C}}COO^-} + 2H^+ + 2NADH + 2ATP$$

22.4 What Are the Fates of Pyruvate?

Pyruvate does not accumulate in cells, but rather undergoes one of three enzyme-catalyzed reactions, depending on the state of oxygenation and the type of cell in which it is produced. A key to understanding the biochemical logic responsible for two of these possible fates of pyruvate is to recognize that this compound is produced by the oxidation of glucose through the reactions of glycolysis. NAD^+ is the oxidizing agent and is reduced to NADH. For glycolysis to continue, there must be a continuing supply of NAD^+; therefore, under anaerobic conditions (in which there is no oxygen present for the reoxidation of NADH), two of the metabolic pathways we describe use pyruvate in ways that regenerate NAD^+.

A. Reduction to Lactate—Lactate Fermentation

In vertebrates, the most important pathway for the regeneration of NAD^+ under anaerobic conditions is the reduction of pyruvate to lactate, catalyzed by the enzyme lactate dehydrogenase:

$$\underset{\text{Pyruvate}}{CH_3\overset{\overset{O}{\|}}{C}COO^-} + NADH + H_3O^+ \xrightleftharpoons[\text{dehydrogenase}]{\text{lactate}} \underset{\text{Lactate}}{CH_3\overset{\overset{OH}{|}}{C}HCOO^-} + NAD^+ + H_2O$$

Lactate fermentation A metabolic pathway that converts glucose to two molecules of lactate.

Even though **lactate fermentation** allows glycolysis to continue in the absence of oxygen, it also brings about an increase in the concentration of lactate, and, perhaps more importantly, it increases the concentration of hydronium ion, H_3O^+, in muscle tissue and in the bloodstream. This buildup of lactate and H_3O^+ is associated with muscle fatigue. When blood lactate reaches a concentration of about 0.4 mg/100 mL, muscle tissue becomes almost completely exhausted.

Example 22.2

Show that glycolysis, followed by the reduction of pyruvate to lactate (lactate fermentation), leads to an increase in the hydrogen ion concentration in the bloodstream.

Strategy

Write a balanced equation for the conversion of glucose to lactic acid, and then consider the degree of ionization of lactic acid in a solution of pH 7.4, the normal pH of blood plasma.

Solution

Lactate fermentation produces lactic acid, which is completely ionized at pH 7.4, the normal pH of blood plasma. Therefore, the hydronium ion concentration increases:

$$\underset{\text{Glucose}}{C_6H_{12}O_6} + 2H_2O \xrightarrow[\text{fermentation}]{\text{lactate}} \underset{\text{Lactate}}{2CH_3\overset{\overset{OH}{|}}{C}HCOO^-} + 2H_3O^+$$

Problem 22.2

Does lactate fermentation result in an increase or a decrease in blood pH?

B. Reduction to Ethanol—Alcoholic Fermentation

Yeast and several other organisms have an alternative pathway to regenerate NAD^+ under anaerobic conditions. In the first step of this pathway, pyruvate undergoes enzyme-catalyzed decarboxylation to give acetaldehyde:

$$\underset{\text{Pyruvate}}{CH_3\overset{\overset{O}{\|}}{C}COO^-} + H_3O^+ \xrightarrow[\text{decarboxylase}]{\text{pyruvate}} \underset{\text{Acetaldehyde}}{CH_3\overset{\overset{O}{\|}}{C}H} + CO_2 + H_2O$$

The carbon dioxide produced in this reaction is responsible for the foam on beer and the carbonation of naturally fermented wines and champagnes. In the second step, acetaldehyde is reduced by NADH to ethanol:

$$\underset{\text{Acetaldehyde}}{CH_3\overset{\overset{O}{\|}}{C}H} + NADH + H_3O^+ \xrightarrow[\text{dehydrogenase}]{\text{alcoholic}} \underset{\text{Ethanol}}{CH_3CH_2OH} + NAD^+ + H_2O$$

Adding the reactions for the decarboxylation of pyruvate and the reduction of acetaldehyde to the net reaction of glycolysis gives the overall reaction of **alcoholic fermentation**:

Alcoholic fermentation A metabolic pathway that converts glucose to two molecules of ethanol and two molecules of CO_2.

$$\underset{\text{Glucose}}{C_6H_{12}O_6} + 2HPO_4^{2-} + 2ADP + 2H^+ \xrightarrow[\text{fermentation}]{\text{alcoholic}} \underset{\text{Ethanol}}{2CH_3CH_2OH} + 2CO_2 + 2ATP$$

C. Oxidation and Decarboxylation to Acetyl-CoA

Under aerobic conditions, pyruvate undergoes oxidative decarboxylation in which the carboxylate group is converted to carbon dioxide and the remaining two carbons are converted to the acetyl group of acetyl-CoA:

$$\underset{\text{Pyruvate}}{CH_3\overset{O}{\overset{\|}{C}}COO^-} + NAD^+ + CoASH \xrightarrow[\text{decarboxylation}]{\text{oxidative}} \underset{\text{Acetyl-CoA}}{CH_3\overset{O}{\overset{\|}{C}}SCoA} + CO_2 + NADH$$

The oxidative decarboxylation of pyruvate is considerably more complex than is suggested by the preceding equation. In addition to utilizing NAD^+ and coenzyme A, this transformation also requires FAD, thiamine pyrophosphate (which is derived from thiamine, vitamin B1), and lipoic acid:

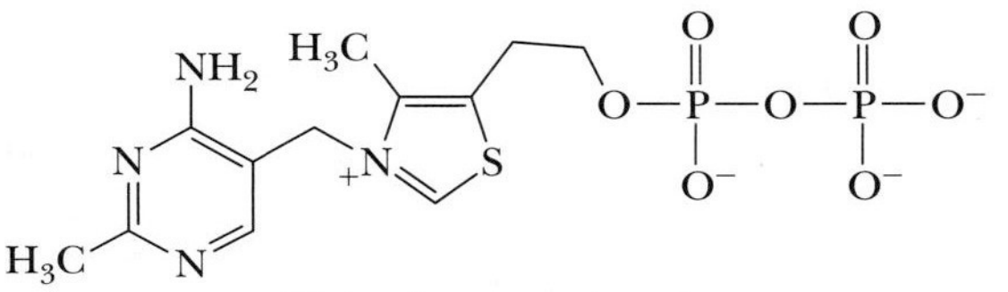

Thiamine pyrophosphate

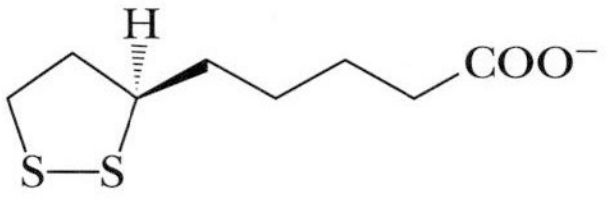

Lipoic acid (as the carboxylate anion)

Acetyl coenzyme A then becomes a fuel for the citric acid cycle, which results in oxidation of the two-carbon chain of the acetyl group to CO_2, with the production of NADH and $FADH_2$. These reduced coenzymes are, in turn, oxidized to NAD^+ and FAD during respiration, with O_2 as the oxidizing agent.

22.5 What Are the Reactions of the β-Oxidation of Fatty Acids?

The first phase in the catabolism of fatty acids involves their release from triglycerides, either those stored in adipose tissue or those ingested from the diet. The hydrolysis of triglycerides is catalyzed by a group of enzymes called lipases.

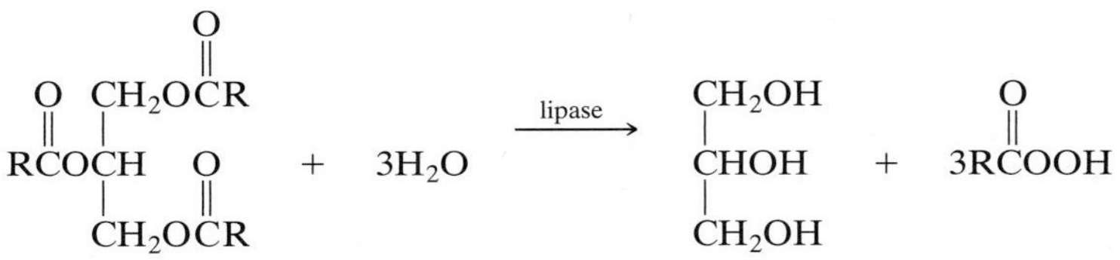

$$\underset{\text{A triglyceride}}{\begin{array}{l} \quad\;\; CH_2O\overset{O}{\overset{\|}{C}}R \\ R\overset{O}{\overset{\|}{C}}OCH \\ \quad\;\; CH_2O\overset{O}{\overset{\|}{C}}R \end{array}} + 3H_2O \xrightarrow{\text{lipase}} \underset{\substack{\text{1,2,3-Propanetriol} \\ \text{(Glycerol; glycerin)}}}{\begin{array}{l} CH_2OH \\ CHOH \\ CH_2OH \end{array}} + \underset{\text{Fatty acids}}{3R\overset{O}{\overset{\|}{C}}OOH}$$

The free fatty acids then pass into the bloodstream and on to cells for oxidation. There are two major stages in the **β-oxidation of fatty acids**: (1) activation of a free fatty acid in the cytoplasm and its transport across the inner mitochondrial membrane, followed by (2) β-oxidation, a repeated sequence of four reactions.

β-Oxidation of fatty acids A series of four enzyme-catalyzed reactions that cleaves carbon atoms, two at a time, from the carboxyl end of a fatty acid.

A. Activation of Fatty Acids: Formation of a Thioester with Coenzyme A

The process of β-oxidation begins in the cytoplasm with the formation of a **thioester** between the carboxyl group of a fatty acid and the sulfhydryl group of coenzyme A. The formation of this acyl-CoA derivative is coupled with the hydrolysis of ATP to AMP and pyrophosphate ion.

Thioester An ester in which the oxygen atom of the —OR group is replaced by an atom of sulfur.

$$R-C(=O)-O^- + HS-CoA \xrightarrow{ATP \quad AMP+P_2O_7^{4-}} R-C(=O)-S-CoA + OH^-$$

Fatty acid (as anion) Coenzyme A An acyl-CoA derivative

The mechanism of this reaction involves attack by the fatty acid carboxylate anion on P=O of a phosphoric anhydride group of ATP to form an intermediate analogous to the tetrahedral carbonyl addition intermediate formed in C=O chemistry. In the intermediate formed in the fatty acid–ATP reaction, the phosphorus attacked by the carboxylate anion becomes bonded to five groups. The collapse of this intermediate gives an acyl-AMP, which is a highly reactive mixed anhydride of the carboxyl group of the fatty acid and the phosphate group of AMP:

$$R-C(=O)-O^- + Ad-O-P(=O)(O^-)-O-P(=O)(O^-)-O-P(=O)(O^-)-O^- \rightleftharpoons \left[R-C(=O)-O-P(O^-)(O^-)(O-Ad)-O-P(=O)(O^-)-O-P(=O)(O^-)-O^- \right]$$

Fatty acid (as anion) ATP intermediate with one phosphorus bonded to five groups

$$\downarrow$$

$$R-C(=O)-O-P(=O)(O^-)-O-Ad + {}^-O-P(=O)(O^-)-O-P(=O)(O^-)-O^-$$

An acyl-AMP (a mixed anhydride) Pyrophosphate ion

This mixed anhydride then undergoes a carbonyl addition reaction with the sulfhydryl group of coenzyme A to form a tetrahedral carbonyl addition intermediate, which collapses to give AMP and an acyl-CoA (a fatty acid thioester of coenzyme A):

$$CoA-SH + R-C(=O)-O-P(=O)(O^-)-O-Ad \rightleftharpoons \left[R-C(O^-)(S-CoA)-O-P(=O)(O^-)-O-Ad \right]$$

Coenzyme A An acyl-AMP tetrahedral carbonyl addition intermediate

$$\downarrow$$

$$R-C(=O)-S-CoA + {}^-O-P(=O)(O^-)-O-Ad$$

An acyl-CoA AMP

At this point, the activated fatty acid is transported into the mitochondrion, where its carbon chain is degraded by the reactions of β-oxidation.

B. The Four Reactions of β-Oxidation

Reaction 1: Oxidation of the Hydrocarbon Chain

The first reaction of β-oxidation is oxidation of the carbon chain between the alpha- and beta-carbons of the fatty acid chain. The oxidizing agent is FAD, which is reduced to $FADH_2$. This reaction is stereoselective: Only the *trans* alkene isomer is formed:

$$\underset{\text{An acyl-CoA}}{R-\overset{\beta}{C}H_2-\overset{\alpha}{C}H_2-\overset{O}{\overset{\|}{C}}-SCoA} + FAD \xrightarrow[\text{dehydrogenase}]{\text{Fatty acyl-CoA}} \underset{\text{A } trans\text{-enoyl-CoA}}{(H)(R)C=C(H)-\overset{O}{\overset{\|}{C}}-SCoA} + FADH_2$$

Reaction 2: Hydration of the Carbon–Carbon Double Bond

Enzyme-catalyzed hydration of the carbon–carbon double bond gives a β-hydroxyacyl-CoA:

$$\underset{\text{A } trans\text{-enoyl-CoA}}{(H)(R)C=C(H)-\overset{O}{\overset{\|}{C}}-SCoA} + H_2O \xrightarrow[\text{hydrase}]{\text{Enoyl-CoA}} \underset{(R)\text{-}\beta\text{-Hydroxyacyl-CoA}}{H\text{,}R\text{,}OH\text{-}C-CH_2-\overset{O}{\overset{\|}{C}}-SCoA}$$

Note that the hydration is regioselective: The —OH is added to carbon 3 of the chain. It is also stereoselective: Only the R enantiomer is formed.

Reaction 3: Oxidation of the β-Hydroxyl Group

In the second oxidation step of β-oxidation, the secondary alcohol is oxidized to a ketone. The oxidizing agent is NAD^+, which is reduced to NADH:

$$\underset{(R)\text{-}\beta\text{-Hydroxyacyl-CoA}}{H\text{,}R\text{,}OH\text{-}C-CH_2-\overset{O}{\overset{\|}{C}}-SCoA} + NAD^+ \xrightarrow[\text{dehydrogenase}]{(R)\text{-}\beta\text{-Hydroxyacyl-CoA}} \underset{\beta\text{-Ketoacyl-CoA}}{R-\overset{O}{\overset{\|}{C}}-CH_2-\overset{O}{\overset{\|}{C}}-SCoA} + NADH$$

Reaction 4: Cleavage of the Carbon Chain

The final step of β-oxidation is cleavage of the carbon chain to give a molecule of acetyl coenzyme A and a new acyl-CoA, the hydrocarbon chain of which is shortened by two carbon atoms:

$$\underset{\beta\text{-Ketoacyl-CoA}}{R-\overset{O}{\overset{\|}{C}}-CH_2-\overset{O}{\overset{\|}{C}}-SCoA} + \underset{\text{Coenzyme A}}{HS-CoA} \xrightarrow{\text{thiolase}} \underset{\text{An acyl-CoA}}{R-\overset{O}{\overset{\|}{C}}-SCoA} + \underset{\text{Acetyl-CoA}}{CH_3\overset{O}{\overset{\|}{C}}-SCoA}$$

Mechanism: A Reverse Claisen Condensation in β-Oxidation of Fatty Acids

Step 1: A sulfhydryl group of the enzyme thiolase attacks the carbonyl carbon of the ketone to form a tetrahedral carbonyl addition intermediate.

Step 2: The addition intermediate collapses to give the enolate anion of acetyl-CoA and an enzyme-bound thioester, which is now shortened by two carbons.

Step 3: The enolate anion reacts with a proton donor to give acetyl-CoA.

Step 4: The enzyme–thioester intermediate undergoes reaction with a molecule of coenzyme A to regenerate a sulfhydryl group on the surface of the enzyme and liberate the fatty acyl-CoA, now shortened by two carbon atoms.

$$R-\overset{O}{\overset{\|}{C}}-CH_2-\overset{O}{\overset{\|}{C}}-SCoA + Enz-S^- \xrightarrow{(1)} \left[R-\underset{Enz-S}{\overset{O^-}{C}}-CH_2-\overset{O}{\overset{\|}{C}}-SCoA \right] \xrightarrow{(2)} R-\overset{O}{\overset{\|}{C}}-S-Enz + CH_2=\overset{O^-}{C}-SCoA$$

Tetrahedral carbonyl addition intermediate; An enzyme-thioester; Enolate anion of acetyl-CoA

$$R-\overset{O}{\overset{\|}{C}}-S-Enz \xrightarrow[(4)]{CoA-SH} Enz-SH + R-\overset{O}{\overset{\|}{C}}-S-CoA$$

A fatty acyl-CoA

$$CH_2=\overset{O^-}{C}-SCoA \xrightarrow[(3)]{\text{proton donor}} CH_3\overset{O}{\overset{\|}{C}}-SCoA$$

Acetyl-CoA

If Steps 1–3 of this mechanism are read in reverse, it is seen as an example of a **Claisen condensation** (Section 16.3A)—the attack by the enolate anion of acetyl-CoA on the carbonyl group of a thioester to form a tetrahedral carbonyl addition intermediate, followed by its collapse to give a β-ketothioester.

The four steps in β-oxidation are summarized in Figure 22.2.

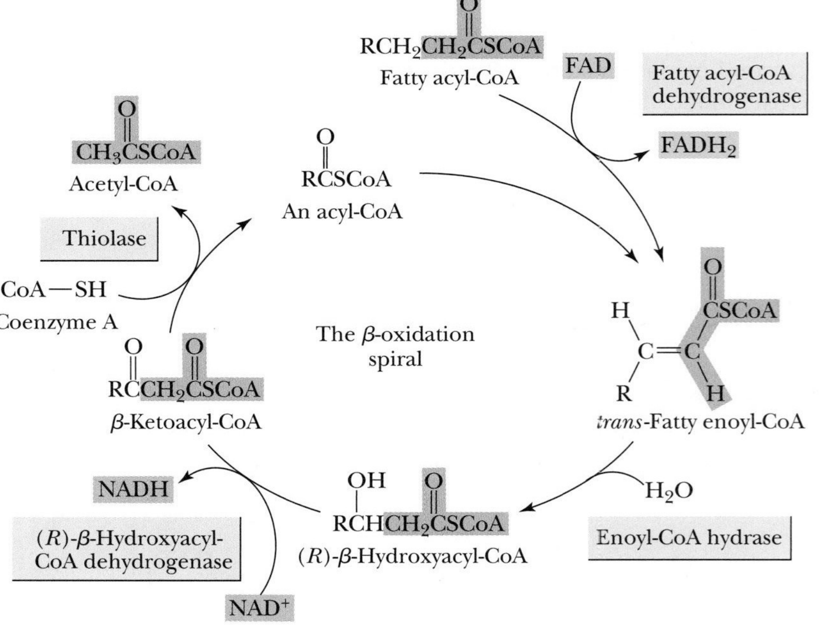

Figure 22.2 The four reactions of β-oxidation. The steps of β-oxidation are called a spiral because, after each series of four reactions, the carbon chain is shortened by two carbon atoms.

C. Repetition of the β-Oxidation Spiral Yields Additional Acetate Units

The series of four reactions of β-oxidation then repeats on the shortened fatty acyl-CoA chain and continues until the entire fatty acid chain is degraded to acetyl-CoA. Seven cycles of β-oxidation of palmitic acid, for example, give eight molecules of acetyl-CoA and involve seven oxidations by FAD and seven oxidations by NAD^+:

$$CH_3(CH_2)_{14}\overset{\overset{\large O}{\|}}{C}OH + 8CoA{-}SH + 7NAD^+ + 7FAD \xrightarrow[]{ATP \quad AMP + P_2O_7^{4-}} 8CH_3\overset{\overset{\large O}{\|}}{C}SCoA + 7NADH + 7FADH_2$$

Hexadecanoic acid (Palmitic acid) — Acetyl coenzyme A

22.6 What Are the Reactions of the Citric Acid Cycle?

Under aerobic conditions, the central metabolic pathway for the oxidation of the carbon skeletons not only of carbohydrates, but also of fatty acids and amino acids, to carbon dioxide is the citric acid cycle, also known as the *tricarboxylic acid* (*TCA*) cycle and Krebs cycle. The last-mentioned name is in honor of Sir Adolph Krebs, the biochemist who first proposed the cyclic nature of this pathway in 1937.

A. Overview of the Cycle

Through the reactions of the citric acid cycle, the carbon atoms of the acetyl group of acetyl-CoA are oxidized to carbon dioxide. There are four separate oxidations in the cycle, three involving NAD^+ and one involving FAD. Figure 22.3 gives an overview of the cycle, showing the four steps.

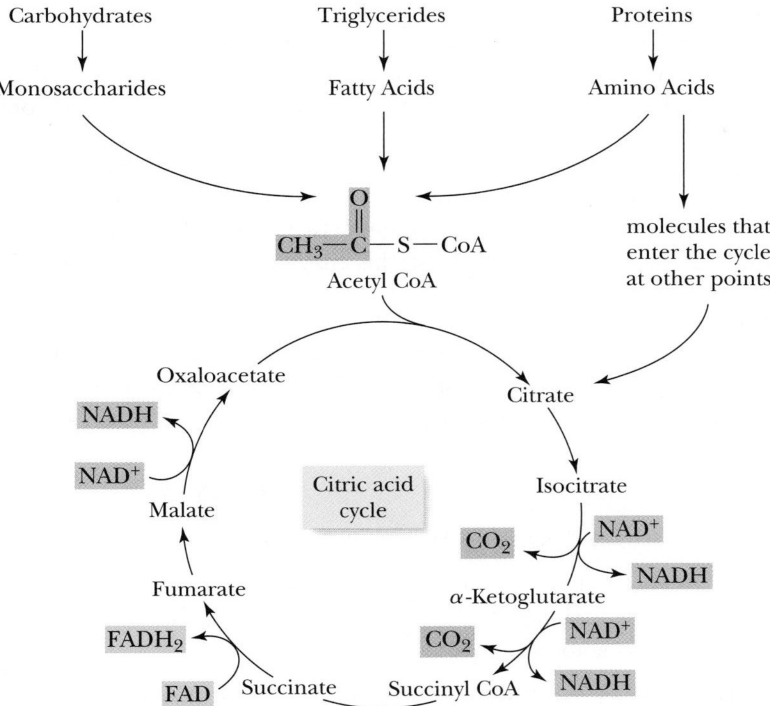

Figure 22.3 The citric acid cycle. Fuel for the cycle is derived from the catabolism (breakdown) of monosaccharides, fatty acids, and amino acids.

B. Reactions of the Citric Acid Cycle

1. Formation of Citrate

The two-carbon acetyl group of acetyl coenzyme A enters the cycle by an enzyme-catalyzed aldol reaction (Section 16.2) between the alpha carbon of acetyl-CoA and the ketone group of oxaloacetate. The product of this reaction is citrate, the tricarboxylic acid from which the cycle derives its name. In the reaction, the carbonyl condensation is coupled with the hydrolysis of the thioester to give free coenzyme A:

$$CH_3C(=O)-S-CoA + O=C(COO^-)-CH_2-COO^- + H_2O \xrightarrow{\text{citrate synthase}} HO-C(COO^-)(CH_2-COO^-)-CH_2-COO^- + CoA-SH$$

Oxaloacetate → Citrate

2. Isomerization of Citrate to Isocitrate

In the second step of the cycle, citrate is converted to the constitutional isomer isocitrate. This isomerization occurs in two steps, both catalyzed by aconitase. First, in a reaction analogous to the acid-catalyzed dehydration of an alcohol (Section 8.2E), citrate undergoes enzyme-catalyzed dehydration of aconitate. Then, in a reaction analogous to acid-catalyzed hydration of an alkene (Section 5.3B), aconitate undergoes enzyme-catalyzed hydration to give isocitrate.

$$\underset{\text{Citrate}}{HO-C(COO^-)(CH_2-COO^-)_2} \xrightarrow[-H_2O]{\text{aconitase}} \underset{\text{Aconitate}}{(CH_2-COO^-)C(COO^-)=CH-COO^-} \xrightarrow[+H_2O]{\text{aconitase}} \underset{\text{Isocitrate}}{(CH_2-COO^-)HC(COO^-)-CH(OH)-COO^-}$$

There are several important features to note about this transformation:

- The dehydration of citrate is completely regioselective: Dehydration is in the direction of the $-CH_2-$ group from the original molecule of oxaloacetate.
- The dehydration of citrate is completely stereoselective: It gives only the *cis* isomer of aconitate.
- The hydration of aconitate is completely regioselective: It gives only isocitrate.
- The hydration of aconitate is completely stereoselective: Isocitrate has two stereocenters, and four possible stereoisomers (two pair of enantiomers). Only one of the four stereoisomers is produced in this enzyme-catalyzed hydration.

3. Oxidation and Decarboxylation of Isocitrate

In Step 3, the secondary alcohol of isocitrate is oxidized to a ketone by NAD^+ in a reaction catalyzed by the enzyme isocitrate dehydrogenase. The product, oxalosuccinate, is a β-ketoacid and undergoes decarboxylation (Section 14.8) to produce α-ketoglutarate:

$$\underset{\substack{\text{Oxalosuccinate}\\(\text{a }\beta\text{-ketoacid})}}{(CH_2-COO^-)HC(\overset{\beta}{COO^-})-C(=O)-COO^-} + H^+ \longrightarrow \underset{\alpha\text{-Ketoglutarate}}{(CH_2-COO^-)CH_2-C(=O)-COO^-} + CO_2$$

Note that only one of the three carboxyl groups in oxalosuccinate is beta to the ketone carbonyl; it is this carboxyl group that undergoes the decarboxylation.

4. Oxidation and Decarboxylation of α-Ketoglutarate

The second molecule of carbon dioxide is generated in the cycle by the same type of oxidative decarboxylation as that for the conversion of pyruvate (also an α-ketoacid) to acetyl-CoA and carbon dioxide (Section 22.4C). In the oxidative decarboxylation of α-ketoglutarate, the carboxyl group is converted to carbon dioxide and the adjacent ketone is oxidized to a carboxyl group in the form of a thioester with coenzyme A:

$$\underset{\alpha\text{-Ketoglutarate}}{\begin{array}{l}CH_2{-}COO^-\\ |\\ CH_2\\ |\\ O{=}C{-}COO^-\end{array}} + NAD^+ + CoA{-}SH \longrightarrow \underset{\text{Succinyl CoA}}{\begin{array}{l}CH_2{-}COO^-\\ |\\ CH_2\\ |\\ O{=}C{-}S{-}CoA\end{array}} + CO_2 + NADH$$

Note: Of the two molecules of carbon dioxide given off in this turn of the citric acid cycle, both carbons are from the carbon skeleton of oxaloacetate; neither is from the acetyl group of acetyl coenzyme A.

5. Conversion of Succinyl CoA to Succinate

Next, in coupled reactions catalyzed by succinyl CoA synthetase, succinyl CoA, HPO_4^{2-}, and guanosine diphosphate (GDP) react to form succinate, guanosine triphosphate (GTP), and coenzyme A:

$$\underset{\text{Succinyl CoA}}{\begin{array}{l}CH_2{-}COO^-\\ |\\ CH_2\\ |\\ O{=}C{-}S{-}CoA\end{array}} + GDP + HPO_4^{2-} \longrightarrow \underset{\text{Succinate}}{\begin{array}{l}COO^-\\ |\\ CH_2\\ |\\ CH_2\\ |\\ COO^-\end{array}} + GTP + CoA{-}SH$$

Observe that to this point in the cycle, the two carbons of the original acetyl group of acetyl-CoA have remained differentiated from the carbon atoms of oxaloacetate. With the production of succinate, however, the two $-CH_2$ groups, as well as the two $-COO^-$ groups, are now indistinguishable.

6. Oxidation of Succinate

In the third oxidation of the cycle, succinate is oxidized to fumarate. The oxidizing agent is FAD, which is reduced to $FADH_2$:

$$\underset{\text{Succinate}}{\begin{array}{l}COO^-\\ |\\ CH_2\\ |\\ CH_2\\ |\\ COO^-\end{array}} + FAD \xrightarrow{\text{succinate dehydrogenase}} \underset{\text{Fumarate}}{(H)(COO^-)C{=}C(H)(^-OOC)} + FADH_2$$

This oxidation is completely stereoselective: Only the *trans* isomer is formed.

7. Hydration of Fumarate

In the second hydration step of the cycle, fumarate is converted to malate:

$$\underset{\text{Fumarate}}{(H)(COO^-)C{=}C(H)(^-OOC)} + H_2O \xrightarrow{\text{fumarase}} \underset{\text{Malate}}{\begin{array}{l}HO{-}CH{-}COO^-\\ \quad\ \ |\\ \quad\ \ CH_2{-}COO^-\end{array}}$$

Fumarase, the enzyme catalyzing this hydration, recognizes only fumarate (and not its *cis* isomer) and gives malate as a single enantiomer.

8. Oxidation of Malate

In the fourth oxidation of the cycle, the secondary alcohol of malate is oxidized to a ketone by NAD^+:

$$\underset{\text{Malate}}{\mathrm{HO{-}CH(CH_2{-}COO^-){-}COO^-}} + NAD^+ \xrightarrow{\text{malate dehydrogenase}} \underset{\text{Oxaloacetate}}{\mathrm{O{=}C(CH_2{-}COO^-){-}COO^-}} + NADH + H^+$$

With the production of oxaloacetate, the reactions of the citric acid cycle are complete. Continued operation of the cycle requires two things: (1) a supply of carbon atoms in the form of acetyl groups from acetyl-CoA and (2) a supply of oxidizing agents in the form of NAD^+ and FAD. For a continuing supply of these two oxidizing agents, the operation of the cycle depends on the reactions of respiration and electron transport, a series of reactions in which the reduced coenzymes NADH and $FADH_2$ are reoxidized by molecular oxygen, O_2.

Another important feature of the cycle is best seen by examining the balanced equation for the cycle:

$$CH_3\overset{\displaystyle O}{\overset{\|}{C}}SCoA + 3NAD^+ + FAD + HPO_4^{2-} + ADP \xrightarrow{\text{citric acid cycle}}$$

$$2CO_2 + 3NADH + FADH_2 + ATP + CoA{-}SH$$

The cycle is truly catalytic: Its intermediates do not enter into the balanced equation for this pathway; they are neither destroyed nor synthesized in the net reaction. The only function of the cycle is to accept acetyl groups from acetyl-CoA, oxidize them to carbon dioxide, and at the same time produce a supply of reduced coenzymes as fuel for electron transport and oxidative phosphorylation. In fact, if any of the intermediates of the cycle are removed, the cycle ceases, because there is no way to regenerate oxaloacetate. Fortunately, the cycle is connected to other metabolic pathways through several of its intermediates. In practice, certain intermediates can be used to synthesize other biomolecules, provided that another intermediate is supplied, which in turn can be converted to oxaloacetate, thus making up for the intermediate withdrawn.

Key Terms and Concepts

acetyl CoA (p. 781)
alcoholic fermentation (p. 789)
ATP, ADP, and AMP (p. 776)
β-oxidation of fatty acids (p. 790)
citric acid cycle (p. 793)
coenzyme (p. 776)
coenzyme A (CoA) (p. 780)
flavin adenine dinucleotide ($FAD/FADH_2$) (p. 779)
glycolysis (p. 781)
lactate fermentation (p. 788)
nicotinamide adenine dinucleotide ($NAD^+/NADH$) (p. 777)
pyruvate (p. 787)
thioester (p. 790)

Summary of Key Questions

Section 22.1 What Are the Key Participants in Glycolysis, β-Oxidation of Fatty Acids, and the Citric Acid Cycle?

ATP, ADP, and **AMP** are agents for the storage and transport of phosphate groups. **Nicotinamide adenine dinucleotide (NAD^+)** and **flavin adenine dinucleotide (FAD)** are agents for the storage and transport of electrons in metabolic oxidations and reductions. NAD^+ is a two-electron oxidizing agent and is reduced to NADH, which, in turn, is a two-electron reducing agent and is oxidized to NAD^+. In the reactions of FAD that are involved in the β-oxidation of fatty acids, FAD is a two-electron oxidizing agent and is reduced to $FADH_2$. **Coenzyme** A is a carrier of acetyl groups.

Section 22.2 What Is Glycolysis?

Glycolysis is a series of 10 enzyme-catalyzed reactions that oxidizes glucose to two molecules of pyruvate.

Section 22.3 What Are the Ten Reactions of Glycolysis?

The 10 reactions of glycolysis can be grouped in the following way:

- Transfer of a phosphate group from ATP to an —OH group of a monosaccharide to form a phosphoric ester (Reactions 1 and 3).
- Interconversion of constitutional isomers by keto–enol tautomerism (Reactions 2 and 5).
- Reverse aldol reaction (Reaction 4).
- Oxidation of an aldehyde group to the mixed anhydride of a carboxylic acid and phosphoric acid (Reaction 6).
- Transfer of a phosphate group from a monosaccharide intermediate to ADP to form ATP (Reactions 7 and 10).
- Transfer of a phosphate group from a 1° alcohol to a 2° alcohol (Reaction 8).
- Dehydration of a 1° alcohol to form a carbon–carbon double bond (Reaction 9).

Section 22.4 What Are the Fates of Pyruvate?

Pyruvate, the product of anaerobic glycolysis, does not accumulate in cells but rather undergoes one of three possible enzyme-catalyzed reactions, depending on the state of oxygenation and the type of cell in which the pyruvate is produced.

- In **lactate fermentation**, pyruvate is reduced to lactate by NADH.
- In **alcoholic fermentation**, pyruvate is converted to acetaldehyde, which is reduced to ethanol by NADH.
- Under aerobic conditions, pyruvate is oxidized to acetyl coenzyme A by NAD^+.

Section 22.5 What Are the Reactions of β-Oxidation of Fatty Acids?

There are two major stages in the metabolism of fatty acids:

- (1) Activation of free fatty acids in the cytoplasm through the formation of thioesters with **coenzyme A** and transport of the activated fatty acids across the inner mitochondrial membrane followed by
- (2) β-Oxidation. **β-Oxidation of fatty acids** is a series of four enzyme-catalyzed reactions by which a fatty acid is degraded to acetyl-CoA.

Section 22.6 What Are the Reactions of the Citric Acid Cycle?

The **citric acid cycle** accepts the two-carbon acetyl group from acetyl-CoA and in a series of steps oxidizes it to two molecules of carbon dioxide. Oxidizing agents are NAD^+ and FAD.

Quick Quiz

Answer true or false to the following questions to assess your general knowledge of the concepts in this chapter. If you have difficulty with any of them, you should review the appropriate section in the chapter (shown in parentheses) before attempting the more challenging end-of-chapter problems.

1. Fatty acids are metabolized through the process of β-oxidation. (22.5)
2. NAD^+, NADH, FAD, and $FADH_2$ are coenzymes that undergo oxidation and reduction during metabolism. (22.1)
3. An end product of glycolysis is pyruvate. (22.2–22.4)
4. The starting point of the citric acid cycle involves acetyl CoA, which can be formed from carbohydrates, triglycerides, or proteins. (22.6)
5. The conversion of a $-CH_2-CH_2-$ unit to $-CH{=}CH-$ by FAD is a reduction reaction. (22.1, 22.5)
6. Lactate fermentation allows pyruvate to be metabolized under aerobic conditions. (22.4)
7. All carbohydrates are directly metabolized through the process of glycolysis. (22.1–22.4)
8. Keto–enol tautomerism is an important reaction in glycolysis. (22.3)
9. Coenzyme A is used in metabolism to store and transfer hydroxyl groups. (22.1)
10. Coenzyme A contains a terminal —OH group, which forms a new bond to acetyl groups in the degradation of monosaccharides, fatty acids, glycerol, and amino acids. (22.1, 22.4–22.6)
11. Alcoholic fermentation is one possible means of pyruvate metabolism. (22.4)
12. ATP, ADP, and AMP are involved in the storage and transfer of adenine groups. (22.1)
13. NAD^+ acts to oxidize both hydroxyl and carbonyl groups. (22.2)

Answers: (1) T (2) T (3) T (4) T (5) F (6) F (7) F (8) T (9) F (10) F (11) T (12) F (13) T

Key Reactions

1. Glycolysis (Sections 22.2 and 22.3)

Glycolysis is a series of 10 enzyme-catalyzed reactions that converts glucose to pyruvate:

$$\underset{\text{Glucose}}{C_6H_{12}O_6} + 2NAD^+ + 2HPO_4^{2-} + 2ADP \xrightarrow{\text{glycolysis}} \underset{\text{Pyruvate}}{2CH_3\overset{O}{\overset{\|}{C}}COO^-} + 2NADH + 2ATP + 2H_3O^+$$

2. Reduction of Pyruvate to Lactate: Lactate Fermentation (Section 22.4A)

$$\underset{\text{Pyruvate}}{CH_3\overset{O}{\overset{\|}{C}}COO^-} + NADH + H^+ \xrightleftharpoons[]{\text{lactate dehydrogenase}} \underset{\text{Lactate}}{CH_3\overset{OH}{\overset{|}{C}}HCOO^-} + NAD^+$$

3. Reduction of Pyruvate to Ethanol: Alcohol Fermentation (Section 22.4B)

The carbon dioxide formed in this reaction is responsible for the foam on beer and the carbonation of naturally fermented wines and champagnes:

$$\underset{\text{Pyruvate}}{CH_3\overset{O}{\overset{\|}{C}}COO^-} + 2H^+ + NADH \xrightarrow{\text{alcoholic fermentation}} \underset{\text{Ethanol}}{CH_3CH_2OH} + CO_2 + NAD^+$$

4. Oxidative Decarboxylation of Pyruvate to Acetyl-CoA (Section 22.4C)

$$\underset{\text{Pyruvate}}{CH_3\overset{O}{\overset{\|}{C}}COO^-} + NAD^+ + CoA{-}SH \xrightarrow{\text{oxidative decarboxylation}} \underset{\text{Acetyl-CoA}}{CH_3\overset{O}{\overset{\|}{C}}SCoA} + CO_2 + NADH$$

5. β-Oxidation of Fatty Acids (Section 22.5)
In this series of four enzyme-catalyzed reactions, the carbon chain of a fatty acid is shortened by two carbon atoms at a time:

$$\underset{\text{Hexadecanoic acid (Palmitic acid)}}{CH_3(CH_2)_{14}\overset{O}{\overset{\|}{C}}OH} + 8CoA{-}SH + 7NAD^+ + 7FAD \xrightarrow{ATP \quad AMP + P_2O_7^{4-}} \underset{\text{Acetyl coenzyme A}}{8CH_3\overset{O}{\overset{\|}{C}}SCoA} + 7NADH + 7FADH_2$$

6. Citric Acid Cycle (Section 22.6)
Through the reactions of the citric acid cycle, the carbon atoms of the acetyl group of acetyl-CoA are oxidized to carbon dioxide:

$$CH_3\overset{O}{\overset{\|}{C}}SCoA + 3NAD^+ + FAD + HPO_4^{2-} + ADP \xrightarrow{\text{citric acid cycle}} 2CO_2 + 3NADH + FADH_2 + ATP + CoA{-}SH$$

There are four separate oxidation steps in the cycle, three involving NAD^+ and one involving FAD.

Problems

A problem marked with an asterisk indicates an applied "real world" problem. Answers to problems whose numbers are printed in blue are given in Appendix D.

Sections 22.2 and 22.3 Glycolysis

***22.3** In many enzyme-catalyzed reactions, a group on the enzyme functions as a proton donor. List several amino acid side chains that could function as proton donors.

***22.4** In many other enzyme-catalyzed reactions, a group on the enzyme surface functions as a proton acceptor. List several amino acid side chains that might function as proton acceptors.

22.5 Name one coenzyme required for glycolysis. From what vitamin is the coenzyme derived?

22.6 Number the carbons of glucose 1 through 6. Which carbons of glucose become the carboxyl groups of the two pyruvates?

22.7 How many moles of lactate are produced from 3 moles of glucose?

***22.8** Although glucose is the principal source of carbohydrates for glycolysis, fructose and galactose are also metabolized for energy.

(a) What is the main dietary source of fructose? Of galactose?

(b) Propose a series of reactions by which fructose might enter glycolysis.

(c) Propose a series of reactions by which galactose might enter glycolysis.

***22.9** How many moles of ethanol are produced per mole of sucrose through the reactions of glycolysis and alcoholic fermentation? How many moles of CO_2 are produced?

***22.10** Glycerol that is derived from the hydrolysis of triglycerides and phospholipids is also metabolized for energy. Propose a series of reactions by which the carbon skeleton of glycerol might enter glycolysis and be oxidized to pyruvate.

22.11 Write a mechanism to show the role of NADH in the reduction of acetaldehyde to ethanol.

***22.12** Ethanol is oxidized in the liver to acetate ion by NAD^+.

(a) Write a balanced equation for this oxidation.

(b) Do you expect the pH of blood plasma to increase, decrease, or remain the same as a result of the metabolism of a significant amount of ethanol?

22.13 When pyruvate is reduced to lactate by NADH, two hydrogens are added to pyruvate: one to the carbonyl carbon, the other to the carbonyl oxygen. Which of these hydrogens is derived from NADH?

22.14 Why is glycolysis called an anaerobic pathway?

22.15 Which carbons of glucose end up in CO_2 as a result of alcoholic fermentation?

22.16 Which steps in glycolysis require ATP? Which steps produce ATP?

Section 22.5 β-Oxidation

22.17 Write structural formulas for palmitic, oleic, and stearic acids, the three most abundant fatty acids.

22.18 A fatty acid must be activated before it can be metabolized in cells. Write a balanced equation for the activation of palmitic acid.

22.19 Name three coenzymes that are necessary for the β-oxidation of fatty acids. From what vitamin is each derived?

***22.20** We have examined β-oxidation of saturated fatty acids, such as palmitic acid and stearic acid. Oleic acid, an unsaturated fatty acid, is also a common component of dietary fats and oils. This unsaturated fatty acid is degraded by β-oxidation, but, at one stage in its degradation, requires an additional enzyme named enoyl-CoA isomerase. Why is this enzyme necessary, and what isomerization does it catalyze? (*Hint:* Consider both the configuration of the carbon–carbon double bond in oleic acid and its position in the carbon chain.)

Section 22.6 Citric Acid Cycle

22.21 What is the main function of the citric acid cycle?

22.22 Which steps in the citric acid cycle involve

(a) The formation of new carbon–carbon bonds

(b) The breaking of carbon–carbon bonds

(c) Oxidation by NAD^+

(d) Oxidation by FAD

(e) Decarboxylation

(f) The creation of new stereocenters

22.23 What does it mean to say that the citric acid cycle is catalytic—that is, that it does not produce any new compounds?

Additional Problems

22.24 Review the oxidation reactions of glycolysis, β-oxidation, and the citric acid cycle, and compare the types of functional groups oxidized by NAD^+ with those oxidized by FAD.

***22.25** The *respiratory quotient* (RQ), used in studies of energy metabolism and exercise physiology, is defined as the ratio of the volume of carbon dioxide produced to the volume of oxygen used:

$$\text{RQ} = \frac{\text{Volume } CO_2}{\text{Volume } O_2}$$

(a) Show that RQ for glucose is 1.00. (*Hint:* Look at the balanced equation for the complete oxidation of glucose to carbon dioxide and water.)

(b) Calculate RQ for triolein, a triglyceride with the molecular formula $C_{57}H_{104}O_6$.

(c) For an individual on a normal diet, RQ is approximately 0.85. Would this value increase or decrease if ethanol were to supply an appreciable portion of the person's caloric needs?

***22.26** Acetoacetate, β-hydroxybutyrate, and acetone are commonly referred to within the health sciences as "ketone bodies," in spite of the fact that one of them is not a ketone at all. All are products of human metabolism and are always present in blood plasma. Most tissues (with the notable exception of the brain) have the enzyme systems necessary to use ketone bodies as energy sources. Ketone bodies are synthesized by the following enzyme-catalyzed reactions:

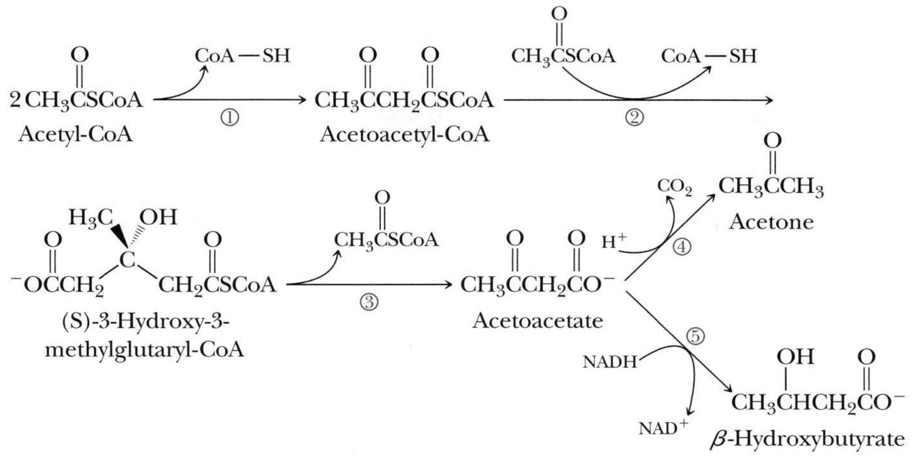

$2\,CH_3CSCoA$ (C=O) → ① ($CoA{-}SH$ released) → CH_3CCH_2CSCoA (two C=O)

Acetyl-CoA → Acetoacetyl-CoA

② CH_3CSCoA (C=O) in, $CoA{-}SH$ out →

$^{-}OCCH_2C(H_3C)(OH)CH_2CSCoA$

(S)-3-Hydroxy-3-methylglutaryl-CoA

③ → CH_3CSCoA released → $CH_3CCH_2CO^-$ (two C=O)

Acetoacetate

④ H^+ in, CO_2 out → CH_3CCH_3 (C=O)

Acetone

⑤ NADH in, NAD^+ out → $CH_3CHCH_2CO^-$ (OH, C=O)

β-Hydroxybutyrate

Describe the type of reaction involved in each step.

22.27 A connecting point between anaerobic glycolysis and β-oxidation is the formation of acetyl-CoA. Which carbon atoms of glucose appear as methyl groups of acetyl-CoA? Which carbon atoms of palmitic acid appear as methyl groups of acetyl-CoA?

22.28 Which of the steps in the following biochemical pathways use molecular oxygen as the oxidizing agent?

(a) Glycolysis

(b) β-Oxidation

(c) The citric acid cycle

Looking Back

22.29 Compare biological (enzyme-catalyzed) reactions with laboratory reactions in terms of

(a) Efficiency of yields

(b) Regiochemical outcome of products

(c) Stereochemical outcome of products

***22.30** Comment on the importance of stereochemistry in the synthesis of new drugs.

***22.31** Of the functional groups that we have studied, which are affected by the acidity of biological environments (biological pH)?

22.32 Can you think of any aspect of your day-to-day life that does not involve or is not affected by organic chemistry? Explain.

Appendix 1

Acid Ionization Constants for the Major Classes of Organic Acids

Class and Example	Typical pK_a	Class and Example	Typical pK_a
Sulfonic acid $C_6H_5SO_2O{-}H$	0–1	Alkylammonium ion $(CH_3CH_2)_3\overset{+}{N}{-}H$	10–12
Carboxylic acid $CH_3\overset{O}{\overset{\Vert}{C}}O{-}H$	3–5	β-ketoester $CH_3{-}\overset{O}{\overset{\Vert}{C}}{-}\overset{H}{\overset{\vert}{C}}H{-}\overset{O}{\overset{\Vert}{C}}OCH_2CH_3$	11
Arylammonium ion $C_6H_5\overset{+}{N}H_2{-}H$	4–5	Water $HO{-}H$	15.7
Thiol $CH_3CH_2S{-}H$	8–12	Alcohol $CH_3CH_2O{-}H$	15–19
Phenol $C_6H_5O{-}H$	9–10	α-Hydrogen of an aldehyde or ketone $CH_3\overset{O}{\overset{\Vert}{C}}CH_2{-}H$	18–20
β-Diketone $CH_3{-}\overset{O}{\overset{\Vert}{C}}{-}\overset{H}{\overset{\vert}{C}}H{-}\overset{O}{\overset{\Vert}{C}}CH_3$	10	α-Hydrogen of an ester $CH_3CH_2O\overset{O}{\overset{\Vert}{C}}CH_2{-}H$	23–25
		Terminal alkyne $CH_3{-}C{\equiv}C{-}H$	25

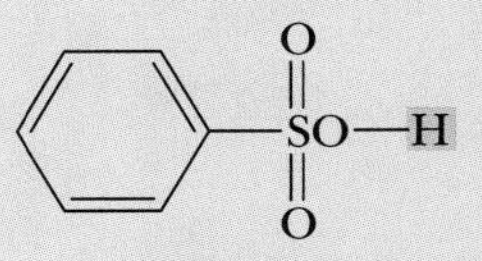

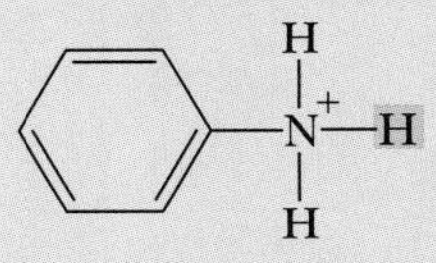

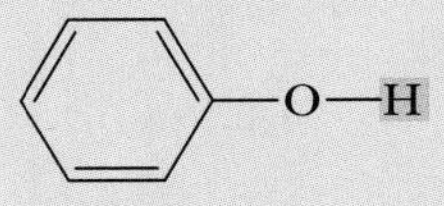

Appendix 2

Characteristic ^{1}H-NMR Chemical Shifts

Type of Hydrogen (R = alkyl, Ar = aryl)	Chemical Shift (δ)*	Type of Hydrogen (R = alkyl, Ar = aryl)	Chemical Shift (δ)*
$(CH_3)_4Si$	0 (by definition)	$RC(=O)CH_2R$	2.2–2.6
RCH_3	0.8–1.0	$RC(=O)OCH_3$	3.7–3.9
RCH_2R	1.2–1.4	$RC(=O)OCH_2R$	4.1–4.7
R_3CH	1.4–1.7	RCH_2I	3.1–3.3
$>C{=}C<CH$	1.6–2.6	RCH_2Br	3.4–3.6
$RC{\equiv}CH$	2.0–3.0	RCH_2Cl	3.6–3.8
$ArCH_3$	2.2–2.5	RCH_2F	4.4–4.5
$ArCH_2R$	2.3–2.8	ArOH	4.5–4.7
ROH	0.5–6.0	$R_2C{=}CH_2$	4.6–5.0
RCH_2OH	3.4–4.0	$R_2C{=}CHR$	5.0–5.7
RCH_2OR	3.3–4.0	ArH	6.5–8.5
R_2NH	0.5–5.0	$RC(=O)H$	9.5–10.1
$RC(=O)CH_3$	2.1–2.3	$RC(=O)OH$	10–13

* Values are relative to tetramethylsilane. Other atoms within the molecule may cause the signal to appear outside these ranges.

Appendix 3

Characteristic ^{13}C-NMR Chemical Shifts

Type of Carbon	Chemical Shift (δ)	Type of Carbon	Chemical Shift (δ)
RCH_3	0–40	C_6H_5–C–R (aromatic C)	110–160
RCH_2R	15–55		
R_3CH	20–60		
RCH_2I	0–40	RC(=O)OR	160–180
RCH_2Br	25–65		
RCH_2Cl	35–80	$RC(=O)NR_2$	165–180
R_3COH	40–80		
R_3COR	40–80	RC(=O)OH	175–185
RC≡CR	65–85		
$R_2C{=}CR_2$	100–150	RC(=O)H, RC(=O)R	180–210

Appendix 4

Characteristic Infrared Absorption Frequencies

Bonding		Frequency (cm^{-1})	Intensity*
C—H	alkane	2850–3000	w–m
	—CH_3	1375 and 1450	w–m
	—CH_2—	1450	m
	alkene	3000–3100	w–m
		650–1000	s
	alkyne	3300	w–m
		1600–1680	w–m
	aromatic	3000–3100	s
		690–900	s
	aldehyde	2700–2800	w
		2800–2900	w
C=C	alkene	1600–1680	w–m
	aromatic	1450 and 1600	w–m
C—O	alcohol, ether,	1050–1100 (sp^3 C—O)	s
	ester, carboxylic		s
	acid, anhydride	1200–1250 (sp^2 C—O)	s
C=O	amide	1630–1680	s
	carboxylic acid	1700–1725	s
	ketone	1705–1780	s
	aldehyde	1705–1740	s
	ester	1735–1800	s
	anhydride	1760 and 1800	s
O—H	alcohol, phenol		
	free	3600–3650	m
	H-bonded	3200–3500	m
	carboxylic acid	2400–3400	m
N—H	amine and amide	3100–3500	m–s

*w = weak, m = medium, s = strong

Glossary

3′ End The end of a polynucleotide at which the 3′-OH of the terminal pentose unit is free.

5′ End The end of a polynucleotide at which the 5′-OH of the terminal pentose unit is free.

α-Amino acid An amino acid in which the amino group is on the carbon adjacent to the carboxyl group.

α-Carbon A carbon atom adjacent to a carbonyl group.

Acetal A molecule containing two —OR or —OAr groups bonded to the same carbon.

Aceto group A CH_3CO— group.

Achiral An object that lacks chirality; an object that has no handedness.

Acid halide A derivative of a carboxylic acid in which the —OH of the carboxyl group is replaced by a halogen—most commonly, chlorine.

Activating group Any substituent on a benzene ring that causes the rate of electrophilic aromatic substitution to be greater than that for benzene.

Activation energy The difference in energy between reactants and the transition state.

Acyl halide A derivative of a carboxylic acid in which the —OH of the carboxyl group is replaced by a halogen—most commonly, chlorine.

α-Helix A type of secondary structure in which a section of polypeptide chain coils into a spiral, most commonly a right-handed spiral.

α-Hydrogen A hydrogen on an α-carbon.

Alcohol A compound containing an —OH (hydroxyl) group bonded to an sp^3 hybridized carbon.

Alcoholic fermentation A metabolic pathway that converts glucose to two molecules of ethanol and two molecules of CO_2.

Aldehyde A compound containing a carbonyl group bonded to hydrogen (a CHO group).

Alditol The product formed when the C=O group of a monosaccharide is reduced to a CHOH group.

Aldol reaction A carbonyl condensation reaction between two aldehydes or ketones to give a β-hydroxyaldehyde or a β-hydroxyketone.

Aldose A monosaccharide containing an aldehyde group.

Aliphatic amine An amine in which nitrogen is bonded only to alkyl groups.

Aliphatic hydrocarbon An alternative term to describe an alkane.

Alkane A saturated hydrocarbon whose carbon atoms are arranged in an open chain.

Alkene An unsaturated hydrocarbon that contains a carbon–carbon double bond.

Alkoxy group An —OR group, where R is an alkyl group.

Aryl group A group derived from an aromatic compound (an arene) by the removal of an H; given the symbol Ar—.

Alkyl halide A compound containing a halogen atom covalently bonded to an alkyl group; given the symbol RX.

Alkyne An unsaturated hydrocarbon that contains a carbon–carbon triple bond.

Amino acid A compound that contains both an amino group and a carboxyl group.

Amino group An sp^3 hybridized nitrogen atom bonded to one, two, or three carbon groups.

Amorphous domains Disordered, noncrystalline regions in the solid state of a polymer.

Anabolic steroid A steroid hormone, such as testosterone, that promotes tissue and muscle growth and development.

Androgen A steroid hormone, such as testosterone, that mediates the development and sexual characteristics of males.

Angle strain The strain that arises when a bond angle is either compressed or expanded compared with its optimal value.

Anion An atom or group of atoms bearing a negative charge.

Anomeric carbon The hemiacetal carbon of the cyclic form of a monosaccharide.

Anomers Monosaccharides that differ in configuration only at their anomeric carbons.

Anti stereoselectivity Addition of atoms or groups of atoms from opposite sides or faces of a carbon–carbon double bond.

Aprotic solvent A solvent that cannot serve as a hydrogen bond donor, as for example, acetone, diethyl ether, and dichloromethane.

Ar— The symbol used for an aryl group, by analogy with R— for an alkyl group.

Aramid A polyaromatic amide; a polymer in which the monomer units are an aromatic diamine and an aromatic dicarboxylic acid.

Arene A compound containing one or more benzene rings.

Arene An aromatic hydrocarbon.

Aromatic amine An amine in which nitrogen is bonded to one or more aryl groups.

Aromatic compound A term used to classify benzene and its derivatives.

Arrhenius acid A substance that dissolves in water to produce H^1 ions.

Arrhenius base A substance that dissolves in water to produce OH^- ions.

Aryl group A group derived from an aromatic compound (an arene) by the removal of an H; given the symbol Ar—.

Autoclave An instrument used to sterilize items by subjecting them to steam and high pressure.

Average degree of polymerization, *n* A subscript placed outside the parentheses of the simplest nonredundant unit of a polymer to indicate that the unit repeats *n* times in the polymer.

Axial bond A bond on a chair conformation of a cyclohexane ring that extends from the ring parallel to the imaginary axis of the ring.

β-Elimination reaction The removal of atoms or groups of atoms from two adjacent carbon atoms, as for example, the removal of H and X from an alkyl halide or H and OH from an alcohol to form a carbon–carbon double bond.

Benzyl group $C_6H_5CH_2$—, the alkyl group derived by removing a hydrogen from the methyl group of toluene.

Benzylic carbon An sp^3 hybridized carbon bonded to a benzene ring.

Bile acid A cholesterol-derived detergent molecule, such as cholic acid, that is secreted by the gallbladder into the intestine to assist in the absorption of dietary lipids.

Bimolecular reaction A reaction in which two species are involved in the reaction leading to the transition state of the rate-determining step.

Boat conformation A puckered conformation of a cyclohexane ring in which carbons 1 and 4 of the ring are bent toward each other.

Bonding electrons Valence electrons shared in a covalent bond.

β-Oxidation of fatty acids A series of four enzyme-catalyzed reactions that cleaves carbon atoms, two at a time, from the carboxyl end of a fatty acid.

β-Pleated sheet A type of secondary structure in which two sections of polypeptide chain are aligned parallel or antiparallel to one another.

Brønsted–Lowry acid A proton donor.

Brønsted–Lowry base A proton acceptor.

Carbanion An anion in which carbon has an unshared pair of electrons and bears a negative charge.

Carbocation A species containing a carbon atom with only three bonds to it and bearing a positive charge.

Carbohydrate A polyhydroxyaldehyde or polyhydroxyketone or a substance that gives these compounds on hydrolysis.

Carbonyl group A C=O group.

Carboxyl group A —COOH group.

Carboxylic anhydride A compound in which two acyl groups are bonded to an oxygen.

Cation An atom or group of atoms bearing a positive charge.

Chain initiation In radical polymerization, the formation of radicals from molecules containing only paired electrons.

Chain propagation In radical polymerization, a reaction of a radical and a molecule to give a new radical.

Chain termination In radical polymerization, a reaction in which two radicals combine to form a covalent bond.

Chain-growth polymerization A polymerization that involves sequential addition reactions, either to unsaturated monomers or to monomers possessing other reactive functional groups.

Chain-transfer reaction In radical polymerization, the transfer of reactivity of an end group from one chain to another during a polymerization.

Chair conformation The most stable puckered conformation of a cyclohexane ring; all bond angles are approximately 109.5°, and bonds to all adjacent carbons are staggered.

Chemical shift, δ The position of a signal on an NMR spectrum relative to the signal of tetramethylsilane (TMS); expressed in delta (δ) units, where 1 δ equals 1 ppm.

Chiral From the Greek *cheir*, meaning hand; objects that are not superposable on their mirror images.

Chromatin A complex formed between negatively charged DNA molecules and positively charged histones.

Circular DNA A type of double-stranded DNA in which the 5′ and 3′ ends of each strand are joined by phosphodiester groups.

Cis A prefix meaning "on the same side."

***Cis–trans* isomerism** Isomers that have the same order of attachment of their atoms, but a different arrangement of their atoms in space due to the presence of either a ring (Chapter 3) or a carbon–carbon double bond (Chapter 4).

***Cis–trans* isomers** Isomers that have the same order of attachment of their atoms, but a different arrangement of their atoms in space, due to the presence of either a ring or a carbon–carbon double bond.

Claisen condensation A carbonyl condensation reaction between two esters to give a β-ketoester.

Codon A triplet of nucleotides on mRNA that directs the incorporation of a specific amino acid into a polypeptide sequence.

Coenzyme A low-molecular-weight, nonprotein molecule or ion that binds reversibly to an enzyme, functions as a second substrate for the enzyme, and is regenerated by further reaction.

Conformation Any three-dimensional arrangement of atoms in a molecule that results by rotation about a single bond.

Conjugate acid The species formed when a base accepts a proton.

Conjugate base The species formed when an acid donates a proton.

Constitutional isomers Compounds with the same molecular formula, but a different order of attachment of their atoms.

Covalent bond A chemical bond resulting from the sharing of one or more pairs of electrons.

Crossed aldol reaction An aldol reaction between two different aldehydes, two different ketones, or an aldehyde and a ketone.

Crossed Claisen condensation A Claisen condensation between two different esters.

Crystalline domains Ordered crystalline regions in the solid state of a polymer; also called crystallites.

***C*-Terminal amino acid** The amino acid at the end of a polypeptide chain having the free $-COO^-$ group.

Curved arrow A symbol used to show the redistribution of valence electrons.

Cyclic ether An ether in which the oxygen is one of the atoms of a ring.

Cycloalkane A saturated hydrocarbon that contains carbon atoms joined to form a ring.

Deactivating group Any substituent on a benzene ring that causes the rate of electrophilic aromatic substitution to be lower than that for benzene.

Decarboxylation Loss of CO_2 from a carboxyl group.

Dehydration Elimination of a molecule of water from a compound.

Dehydrohalogenation Removal of $-H$ and $-X$ from adjacent carbons; a type of β-elimination.

Denaturation The loss of secondary, tertiary, and quaternary structure of a protein by a chemical or physical agent and the resulting loss of function.

Dextrorotatory Rotating the plane of polarized light in a polarimeter to the right.

Diastereomers Stereoisomers that are not mirror images of each other; the term refers to relationships among objects.

Diaxial interactions Interactions between groups in parallel axial positions on the same side of a chair conformation of a cyclohexane ring.

Dieckmann condensation An intramolecular Claisen condensation of an ester of a dicarboxylic acid to give a five- or six-membered ring.

Dipeptide A molecule containing two amino acid units joined by a peptide bond.

Disaccharide A carbohydrate containing two monosaccharide units joined by a glycosidic bond.

Dispersion forces Very weak intermolecular forces of attraction resulting from the interaction of temporary induced dipoles.

Disulfide bond A covalent bond between two sulfur atoms; an $-S-S-$ bond.

D-Monosaccharide A monosaccharide that, when written as a Fischer projection, has the $-OH$ on its penultimate carbon to the right.

Double helix A type of secondary structure of DNA molecules in which two antiparallel polynucleotide strands are coiled in a right-handed manner about the same axis.

Double-headed arrow A symbol used to connect contributing structures.

Doublet A signal that is split into two peaks; the hydrogens that give rise to the signal have one neighboring nonequivalent hydrogen.

Downfield A term used to refer to the relative position of a signal on an NMR spectrum. Downfield indicates a peak to the left of the spectrum (a weaker applied field).

E From the German *entgegen,* meaning opposite; specifies that groups of higher priority on the carbons of a double bond are on opposite sides.

E,Z system A system used to specify the configuration of groups about a carbon–carbon double bond.

Eclipsed conformation A conformation about a carbon–carbon single bond in which the atoms on one carbon are as close as possible to the atoms on the adjacent carbon.

Edman degradation A method for selectively cleaving and identifying the *N*-terminal amino acid of a polypeptide chain.

Elastomer A material that, when stretched or otherwise distorted, returns to its original shape when the distorting force is released.

Electromagnetic radiation Light and other forms of radiant energy.

Electronegativity A measure of the force of an atom's attraction for electrons it shares in a chemical bond with another atom.

Electrophile Any molecule or ion that can accept a pair of electrons to form a new covalent bond; a Lewis acid.

Electrophilic aromatic substitution A reaction in which an electrophile, E^+, substitutes for a hydrogen on an aromatic ring.

Electrophoresis The process of separating compounds on the basis of their electric charge.

Enantiomers Stereoisomers that are nonsuperposable mirror images; the term refers to a relationship between pairs of objects.

Endothermic reaction A reaction in which the energy of the products is higher than the energy of the reactants; a reaction in which heat is absorbed.

Energy diagram A graph showing the changes in energy that occur during a chemical reaction; energy is plotted on the *y*-axis, and the progress of the reaction is plotted on the *x*-axis.

Enol A molecule containing an $-OH$ group bonded to a carbon of a carbon–carbon double bond.

Enolate anion An anion formed by the removal of an α-hydrogen from a carbonyl-containing compound.

Epoxide A cyclic ether in which oxygen is one atom of a three-membered ring.

Epoxy resin A material prepared by a polymerization in which one monomer contains at least two epoxy groups.

Equatorial bond A bond on a chair conformation of a cyclohexane ring that extends from the ring roughly perpendicular to the imaginary axis of the ring.

Equivalent hydrogens Hydrogens that have the same chemical environment.

Estrogen A steroid hormone, such as estradiol, that mediates the development and sexual characteristics of females.

Ether A compound containing an oxygen atom bonded to two carbon atoms.

Exothermic reaction A reaction in which the energy of the products is lower than the energy of the reactants; a reaction in which heat is liberated.

Fat A triglyceride that is semisolid or solid at room temperature.

Fatty acid A long, unbranched-chain carboxylic acid, most commonly of 12 to 20 carbons, derived from the hydrolysis of animal fats, vegetable oils, or the phospholipids of biological membranes.

Fingerprint region The portion of the vibrational infrared region that extends from 1000 to 400 cm^{-1} and that is unique to every compound.

Fischer esterification The process of forming an ester by refluxing a carboxylic acid and an alcohol in the presence of an acid catalyst, commonly sulfuric acid.

Fischer projection A two-dimensional representation showing the configuration of a stereocenter; horizontal lines represent bonds projecting forward from the stereocenter, vertical lines represent bonds projecting to the rear.

Fishhook arrow A single-barbed, curved arrow used to show the change in position of a single electron.

Flavin adenine dinucleotide (FAD) A biological oxidizing agent. When acting as an oxidizing agent, FAD is reduced to $FADH_2$.

Fluid-mosaic model A model of a biological membrane consisting of a phospholipid bilayer, with proteins, carbohydrates, and other lipids embedded in, and on the surface of, the bilayer.

Formal charge The charge on an atom in a molecule or polyatomic ion.

Frequency (ν) A number of full cycles of a wave that pass a point in a second.

Functional group An atom or a group of atoms within a molecule that shows a characteristic set of physical and chemical properties.

Furanose A five-membered cyclic hemiacetal form of a monosaccharide.

Glass transition temperature, T_g The temperature at which a polymer undergoes the transition from a hard glass to a rubbery state.

Glycol A compound with two hydroxyl (—OH) groups on different carbons.

Glycolysis From the Greek *glyko*, sweet, and *lysis*, splitting; a series of 10 enzyme-catalyzed reactions by which glucose is oxidized to two molecules of pyruvate.

Glycoside A carbohydrate in which the —OH on its anomeric carbon is replaced by —OR.

Glycosidic bond The bond from the anomeric carbon of a glycoside to an —OR group.

Grignard reagent An organomagnesium compound of the type RMgX or ArMgX.

Ground-state electron configuration The electron configuration of lowest energy for an atom, molecule, or ion.

Halonium ion An ion in which a halogen atom bears a positive charge.

Haworth projection A way of viewing the furanose and pyranose forms of monosaccharides. The ring is drawn flat and viewed through its edge, with the anomeric carbon on the right and the oxygen atom of the ring in the rear to the right.

Heat of reaction The difference in energy between reactants and products.

Hemiacetal A molecule containing an —OH and an —OR or —OAr group bonded to the same carbon.

Hertz (Hz) The unit in which wave frequency is reported; s^{-1} (read per second).

Heterocyclic amine An amine in which nitrogen is one of the atoms of a ring.

Heterocyclic aromatic amine An amine in which nitrogen is one of the atoms of an aromatic ring.

Heterocyclic compound An organic compound that contains one or more atoms other than carbon in its ring.

High-density lipoprotein (HDL) Plasma particles, of density 1.06–1.21 g/mL, consisting of approximately 33% proteins, 30% cholesterol, 29% phospholipids, and 8% triglycerides.

Histone A protein, particularly rich in the basic amino acids lysine and arginine, that is found associated with DNA molecules.

Hybrid orbital An orbital produced from the combination of two or more atomic orbitals.

Hydration Addition of water.

Hydride ion A hydrogen atom with two electrons in its valence shell; $H\!:^-$.

Hydrocarbon A compound that contains only carbon atoms and hydrogen atoms.

Hydrogen bonding The attractive force between a partial positive charge on hydrogen and partial negative charge on a nearby oxygen, nitrogen, or fluorine atom.

Hydrophilic From the Greek, meaning "water loving."

Hydrophobic effect The tendency of nonpolar groups to cluster in such a way as to be shielded from contact with an aqueous environment.

Hydrophobic From the Greek, meaning "water hating."

Hydroxyl group An —OH group.

Imine A compound containing a carbon–nitrogen double bond; also called a Schiff base.

Index of hydrogen deficiency The sum of the number of rings and pi bonds in a molecule.

Inductive effect The polarization of electron density transmitted through covalent bonds caused by a nearby atom of higher electronegativity.

Inversion of configuration The reversal of the arrangement of atoms or groups of atoms about a reaction center in an S_N2 reaction.

Ionic bond A chemical bond resulting from the electrostatic attraction of an anion and a cation.

Isoelectric point (pI) The pH at which an amino acid, a polypeptide, or a protein has no net charge.

Ketone A compound containing a carbonyl group bonded to two carbons.

Ketose A monosaccharide containing a ketone group.

Lactam A cyclic amide.

Lactate fermentation A metabolic pathway that converts glucose to two molecules of pyruvate.

Lactone A cyclic ester.

Levorotatory Rotating the plane of polarized light in a polarimeter to the left.

Lewis acid Any molecule or ion that can form a new covalent bond by accepting a pair of electrons.

Lewis base Any molecule or ion that can form a new covalent bond by donating a pair of electrons.

Lewis structure of an atom The symbol of an element surrounded by a number of dots equal to the number of electrons in the valence shell of the atom.

Line-angle formula An abbreviated way to draw structural formulas in which each line ending represents a carbon atom and a line represents a bond.

Lipid A class of biomolecules isolated from plant or animal sources by extraction with nonpolar organic solvents, such as diethyl ether and acetone.

Lipid bilayer A back-to-back arrangement of phospholipid monolayers.

L-Monosaccharide A monosaccharide that, when written as a Fischer projection, has the —OH on its penultimate carbon to the left.

Low-density lipoprotein (LDL) Plasma particles, of density 1.02–1.06 g/mL, consisting of approximately 25% proteins, 50% cholesterol, 21% phospholipids, and 4% triglycerides.

Markovnikov's rule In the addition of HX or H_2O to an alkene, hydrogen adds to the carbon of the double bond having the greater number of hydrogens.

Melt transition temperature, T_m The temperature at which crystalline regions of a polymer melt.

Mercaptan A common name for any molecule containing an —SH group.

Meso compound An achiral compound possessing two or more stereocenters.

Messenger RNA (mRNA) A ribonucleic acid that carries coded genetic information from DNA to ribosomes for the synthesis of proteins.

Meta (*m*) Refers to groups occupying positions 1 and 3 on a benzene ring.

Meta director Any substituent on a benzene ring that directs electrophilic aromatic substitution preferentially to a meta position.

Micelle A spherical arrangement of organic molecules in water solution clustered so that their hydrophobic parts are buried inside the sphere and their hydrophilic parts are on the surface of the sphere and in contact with water.

Michael reaction The conjugate addition of an enolate anion or other nucleophile to an α,β-unsaturated carbonyl compound.

Mirror image The reflection of an object in a mirror.

Molecular spectroscopy The study of the frequencies of electromagnetic radiation that are absorbed or emitted by substances and the correlation between these frequencies and specific types of molecular structure.

Monomer From the Greek *mono*, single and *meros*, part; the simplest nonredundant unit from which a polymer is synthesized.

Monosaccharide A carbohydrate that cannot be hydrolyzed to a simpler compound.

Mutarotation The change in optical activity that occurs when an α or β form of a carbohydrate is converted to an equilibrium mixture of the two forms.

(n + 1) rule The ^{1}H-NMR signal of a hydrogen or set of equivalent hydrogens with n other hydrogens on neighboring carbons is split into $(n + 1)$ peaks.

Newman projection A way to view a molecule by looking along a carbon–carbon bond.

Nicotinamide adenine dinucleotide (NAD$^+$) A biological oxidizing agent. When acting as an oxidizing agent, NAD$^+$ is reduced to NADH.

Nonbonding electrons Valence electrons not involved in forming covalent bonds, that is, unshared electrons.

Nonpolar covalent bond A covalent bond between atoms whose difference in electronegativity is less than approximately 0.5.

***N*-Terminal amino acid** The amino acid at the end of a polypeptide chain having the free $-NH_3^+$ group.

Nucleic acid A biopolymer containing three types of monomer units: heterocyclic aromatic amine bases derived from purine and pyrimidine, the monosaccharides D-ribose or 2-deoxy-D-ribose, and phosphate.

Nucleophile An atom or a group of atoms that donates a pair of electrons to another atom or group of atoms to form a new covalent bond.

Nucleophilic acyl substitution A reaction in which a nucleophile bonded to a carbonyl carbon is replaced by another nucleophile.

Nucleophilic substitution A reaction in which one nucleophile is substituted for another.

Nucleoside A building block of nucleic acids, consisting of D-ribose or 2-deoxy-D-ribose bonded to a heterocyclic aromatic amine base by a β-*N*-glycosidic bond.

Nucleotide A nucleoside in which a molecule of phosphoric acid is esterified with an $-OH$ of the monosaccharide, most commonly either the 3′-OH or the 5′-OH.

Observed rotation The number of degrees through which a compound rotates the plane of polarized light.

Octet rule The tendency among atoms of Group 1A–7A elements to react in ways that achieve an outer shell of eight valence electrons.

Oil A triglyceride that is liquid at room temperature.

Oligosaccharide A carbohydrate containing from 6 to 10 monosaccharide units, each joined to the next by a glycosidic bond.

Optically active Showing that a compound rotates the plane of polarized light.

Optically inactive Showing that a compound or mixture of compounds does not rotate the plane of polarized light.

Orbital A region of space where an electron or pair of electrons spends 90 to 95% of its time.

Order of precedence of functional groups A system for ranking functional groups in order of priority for the purposes of IUPAC nomenclature.

Organometallic compound A compound containing a carbon–metal bond.

Ortho (*o*) Refers to groups occupying positions 1 and 2 on a benzene ring.

Ortho–para director Any substituent on a benzene ring that directs electrophilic aromatic substitution preferentially to ortho and para positions.

Oxonium ion An ion in which oxygen is bonded to three other atoms and bears a positive charge.

Para (*p*) Refers to groups occupying positions 1 and 4 on a benzene ring.

Peak (NMR) The units into which an NMR signal is split—two peaks in a doublet, three peaks in a triplet, and so on.

Penultimate carbon The stereocenter of a monosaccharide farthest from the carbonyl group—for example, carbon 5 of glucose.

Peptide bond The special name given to the amide bond formed between the α-amino group of one amino acid and the α-carboxyl group of another amino acid.

Phenol A compound that contains an $-OH$ bonded to a benzene ring.

Phenyl group C_6H_5-, the aryl group derived by removing a hydrogen from benzene.

Phospholipid A lipid containing glycerol esterified with two molecules of fatty acid and one molecule of phosphoric acid.

Pi (π) bond A covalent bond formed by the overlap of parallel p orbitals.

Plane of symmetry An imaginary plane passing through an object and dividing it such that one half is the mirror image of the other half.

Plane-polarized light Light vibrating only in parallel planes.

Plastic A polymer that can be molded when hot and retains its shape when cooled.

Polar covalent bond A covalent bond between atoms whose difference in electronegativity is between approximately 0.5 and 1.9.

Polarimeter An instrument for measuring the ability of a compound to rotate the plane of polarized light.

Polyamide A polymer in which each monomer unit is joined to the next by an amide bond, as for example nylon 66.

Polycarbonate A polyester in which the carboxyl groups are derived from carbonic acid.

Polyester A polymer in which each monomer unit is joined to the next by an ester bond as, for example, poly(ethylene terephthalate).

Polymer From the Greek *poly*, many, and *meros*, parts; any long-chain molecule synthesized by linking together many single parts called monomers.

Polynuclear aromatic hydrocarbon A hydrocarbon containing two or more fused aromatic rings.

Polypeptide A macromolecule containing 20 or more amino acid units, each joined to the next by a peptide bond.

Polysaccharide A carbohydrate containing a large number of monosaccharide units, each joined to the next by one or more glycosidic bonds.

Polyunsaturated triglyceride A triglyceride having several carbon–carbon double bonds in the hydrocarbon chains of its three fatty acids.

Polyurethane A polymer containing the $-NHCOO-$ group as a repeating unit.

Primary (1°) structure of proteins The sequence of amino acids in the polypeptide chain; read from the *N*-terminal amino acid to the *C*-terminal amino acid.

Primary structure of nucleic acids The sequence of bases along the pentose-phosphodiester backbone of a DNA or RNA molecule, read from the 5′ end to the 3′ end.

Prostaglandin A member of the family of compounds having the 20-carbon skeleton of prostanoic acid.

Protic solvent A hydrogen bond donor solvent, as for example water, ethanol, and acetic acid.

Pyranose A six-membered cyclic hemiacetal form of a monosaccharide.

Quartet A signal that is split into four peaks; the hydrogens that give rise to the signal have three neighboring nonequivalent hydrogens that are equivalent to each other.

Quaternary (4°) structure of proteins The arrangement of polypeptide monomers into a noncovalently bonded aggregation.

R From the Latin *rectus*, meaning right; used in the R,S system to show that the order of priority of groups on a stereocenter is clockwise.

R— A symbol used to represent an alkyl group.

R,S system A set of rules for specifying the configuration about a stereocenter.

Racemic mixture A mixture of equal amounts of two enantiomers.

Racemization The conversion of a pure enantiomer into a racemic mixture.

Radical Any molecule that contains one or more unpaired electrons.

Rate-determining step The step in a reaction sequence that crosses the highest energy barrier; the slowest step in a multistep reaction.

Reaction coordinate A measure of the progress of a reaction, plotted on the x-axis in an energy diagram.

Reaction intermediate An unstable species that lies in an energy minimum between two transition states.

Reaction mechanism A step-by-step description of how a chemical reaction occurs.

Rearrangement A reaction which the connectivity of atoms in a product is different from that in the starting material.

Reducing sugar A carbohydrate that reacts with an oxidizing agent to form an aldonic acid.

Reductive amination The formation of an imine from an aldehyde or a ketone, followed by the reduction of the imine to an amine.

Regioselective reaction A reaction in which one direction of bond forming or bond breaking occurs in preference to all other directions.

Relative nucleophilicity The relative rates at which a nucleophile reacts in a reference nucleophilic substitution reaction.

Resolution Separation of a racemic mixture into its enantiomers.

Resonance The absorption of electromagnetic radiation by a spinning nucleus and the resulting "flip" of its spin from a lower energy state to a higher energy state.

Resonance contributing structures Representations of a molecule or ion that differ only in the distribution of valence electrons.

Resonance energy The difference in energy between a resonance hybrid and the most stable of its hypothetical contributing structures.

Resonance hybrid A molecule or ion that is best described as a composite of a number of contributing structures.

Resonance signal A recording of nuclear magnetic resonance in an NMR spectrum.

Restriction endonuclease An enzyme that catalyzes the hydrolysis of a particular phosphodiester bond within a DNA strand.

Ribosomal RNA (rRNA) A ribonucleic acid found in ribosomes, the sites of protein synthesis.

S From the Latin *sinister*, meaning left; used in the R,S system to show that the order of priority of groups on a stereocenter is counterclockwise.

Sanger dideoxy method A method, developed by Frederick Sanger, for sequencing DNA molecules.

Saponification Hydrolysis of an ester in aqueous NaOH or KOH to an alcohol and the sodium or potassium salt of a carboxylic acid.

Saturated hydrocarbon A hydrocarbon containing only carbon–carbon single bonds.

Schiff base An alternative name for an imine.

Secondary (2°) structure of proteins The ordered arrangements (conformations) of amino acids in localized regions of a polypeptide or protein.

Secondary structure of nucleic acids The ordered arrangement of strands of nucleic acid.

Shell A region of space around a nucleus where electrons are found.

Shielding In NMR spectroscopy, electrons around a nucleus create their own local magnetic fields and thereby shield the nucleus from the applied magnetic field.

Sigma (σ) bond A covalent bond in which the overlap of atomic orbitals is concentrated along the bond axis.

Signal splitting Splitting of an NMR signal into a set of peaks by the influence of neighboring nuclei.

Singlet A signal that consists of one peak; the hydrogens that give rise to the signal have no neighboring nonequivalent hydrogens.

Soap A sodium or potassium salt of a fatty acid.

Solvolysis A nucleophilic substitution reaction in which the solvent is the nucleophile.

***sp* Hybrid orbital** A hybrid atomic orbital produced by the combination of one *s* atomic orbital and one *p* atomic orbital.

sp^2 Hybrid orbital An orbital produced by the combination of one *s* atomic orbital and two *p* atomic orbitals.

sp^3 Hybrid orbital An orbital produced by the combination of one *s* atomic orbital and three *p* atomic orbitals.

Specific rotation Observed rotation of the plane of polarized light when a sample is placed in a tube 1.0 dm long at a concentration of 1.0 g/mL.

Staggered conformation A conformation about a carbon–carbon single bond in which the atoms on one carbon are as far apart as possible from the atoms on the adjacent carbon.

Step-growth polymerization A polymerization in which chain growth occurs in a stepwise manner between difunctional monomers, for example, between adipic acid and hexamethylenediamine to form nylon 66. Also referred to as condensation polymerization.

Stereocenter An atom at which the interchange of two atoms or groups of atoms bonded to it produces a different stereoisomer.

Stereoisomers Isomers that have the same molecular formula and the same connectivity, but different orientations of their atoms in space.

Stereoselective reaction A reaction in which one stereoisomer is formed or destroyed in preference to all others that might be formed or destroyed.

Steric hindrance The ability of groups, because of their size, to hinder access to a reaction site within a molecule.

Steric strain The strain that arises when atoms separated by four or more bonds are forced abnormally close to one another.

Steroid A plant or animal lipid having the characteristic tetracyclic ring structure of the steroid nucleus, namely, three six-membered rings and one five-membered ring.

Strong acid An acid that is completely ionized in aqueous solution.

Strong base A base that is completely ionized in aqueous solution.

Tautomers Constitutional isomers that differ in the location of hydrogen and a double bond relative to O, N, or S.

Terpene A compound whose carbon skeleton can be divided into two or more units identical with the carbon skeleton of isoprene.

Tertiary (3°) structure of proteins The three-dimensional arrangement in space of all atoms in a single polypeptide chain.

Tertiary structure of nucleic acids The three-dimensional arrangement of all atoms of a nucleic acid, commonly referred to as supercoiling.

Thermoplastic A polymer that can be melted and molded into a shape that is retained when it is cooled.

Thermosetting plastic A polymer that can be molded when it is first prepared, but, once cooled, hardens irreversibly and cannot be remelted.

Thioester An ester in which the oxygen atom of the —OR group is replaced by an atom of sulfur.

Thiol A compound containing an —SH (sulfhydryl) group.

Torsional strain (also called eclipsed interaction strain) Strain that arises when atoms separated by three bonds are forced from a staggered conformation to an eclipsed conformation.

Trans A prefix meaning "across from."

Transfer RNA (tRNA) A ribonucleic acid that carries a specific amino acid to the site of protein synthesis on ribosomes.

Transition state An unstable species of maximum energy formed during the course of a reaction; a maximum on an energy diagram.

Triglyceride (triacylglycerol) An ester of glycerol with three fatty acids.

Tripeptide A molecule containing three amino acid units, each joined to the next by a peptide bond.

Triplet A signal that is split into three peaks; the hydrogens that give rise to the signal have two neighboring nonequivalent hydrogens that are equivalent to each other.

Unimolecular reaction A reaction in which only one species is involved in the reaction leading to the transition state of the rate-determining step.

Upfield A term used to refer to the relative position of a signal on an NMR spectrum. Upfield indicates a peak to the right of the spectrum (a stronger applied field).

Valence electrons Electrons in the valence (outermost) shell of an atom.

Valence shell The outermost electron shell of an atom.

Vibrational infrared The portion of the infrared region that extends from 4000 to 400 cm^{-1}.

Wavelength (λ) The distance between two consecutive identical points on a wave.

Wavenumber ($\bar{\nu}$) A characteristic of electromagnetic radiation equal to the number of waves per centimeter.

Weak acid An acid that only partially ionizes in aqueous solution.

Weak base A base that only partially ionizes in aqueous solution.

Z From the German *zusammen*, meaning together; specifies that groups of higher priority on the carbons of a double bond are on the same side.

Zaitsev's rule A rule stating that the major product from a β-elimination reaction is the most stable alkene; that is, the major product is the alkene with the greatest number of substituents on the carbon–carbon double bond.

Zwitterion An internal salt of an amino acid.

Answers Section

Chapter 1
Covalent Bonding and Shapes of Molecules

1.1 Each pair possesses the same number of valence electrons (**valence electrons shown in boldface**):
(a) carbon: $1s^2\mathbf{2s^22p^2}$ silicon: $1s^22s^22p^6\mathbf{3s^23p^2}$
(b) oxygen: $1s^2\mathbf{2s^22p^4}$ sulfur: $1s^22s^22p^6\mathbf{3s^23p^4}$
(c) nitrogen: $1s^2\mathbf{2s^22p^3}$ phosphorus: $1s^22s^22p^6\mathbf{3s^23p^3}$

1.2 The electron configuration becomes $1s^22s^22p^6\mathbf{3s^23p^6}$.

1.3 (a) Li (b) N (c) C

1.4 (a) nonpolar covalent (b) nonpolar covalent
(c) polar covalent (d) polar covalent

1.5 (a) $\overset{\delta+}{C}—\overset{\delta-}{N}$ (b) $\overset{\delta+}{N}—\overset{\delta-}{O}$ (c) $\overset{\delta+}{C}—\overset{\delta-}{Cl}$

1.6 (a) H—C(H)(H)—C(H)(H)—H (b) :S̤=C=S̤: (c) H—C≡N:

1.7 (a) H—C(H)(H)—N^+(H)(H)—H (b) H—C^+(H)(H)

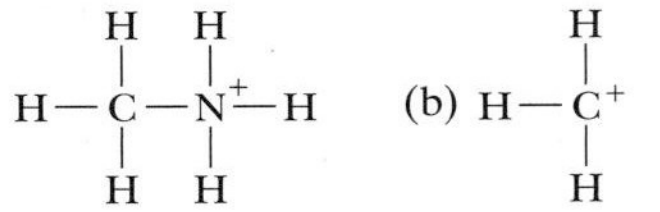

1.8 (a) 109.5° (b) 109.5°
(c) carbon 120° oxygen 109.5°

1.9 SO_2 has a bent shape

1.10 Pair (a) is a set of resonance structures.

1.11 CH_3—C(=O:)(—O^-:) ⟷ CH_3—C^+(—O^-)(—O^-) ⟷ CH_3—C(—O^-)(=O)
(a) (b) (c)

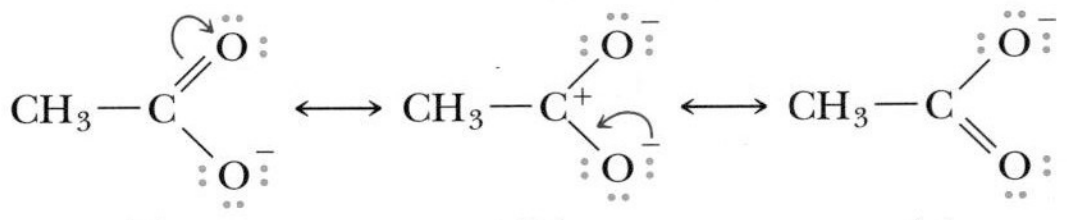

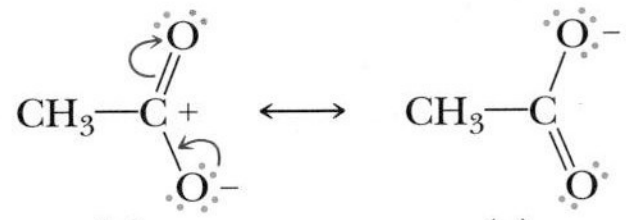

CH_3—C^+(=O)(—O^-) ⟷ CH_3—C(—O^-)(=O)
(a) (c)

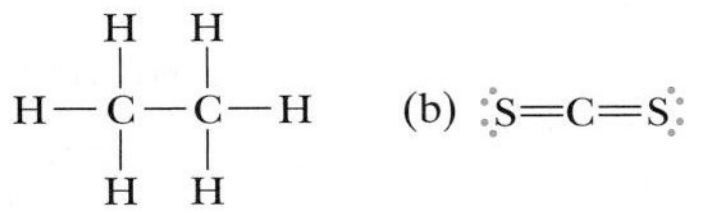

1.12 (a) H—C(H)(H)—C(H)=C(H)—H; bond labels: $\sigma_{s\text{-}sp^3}$, $\sigma_{sp^3\text{-}sp^2}$, $\left[\begin{matrix}\sigma_{sp^2\text{-}sp^2}\\ \pi_{p\text{-}p}\end{matrix}\right.$, $\sigma_{sp^2\text{-}s}$; angles: 109.5°, 120°

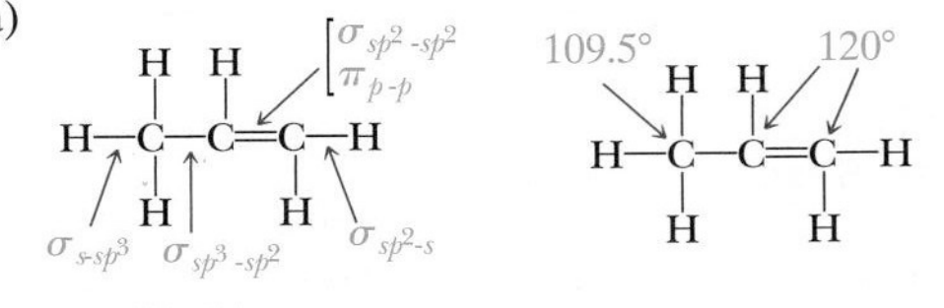

(b) H—C(H)(H)—N̤(H)—H; bond labels: $\sigma_{s\text{-}sp^3}$, $\sigma_{sp^3\text{-}sp^3}$, $\sigma_{sp^3\text{-}s}$; angles: 109.5°, 109.5°

1.13 $CH_3CH_2CH_2CH_2OH$ (1°) $CH_3CH_2CH(OH)CH_3$ (2°)
$CH_3CH(CH_3)CH_2OH$ (1°) $CH_3C(CH_3)(OH)CH_3$ (3°)

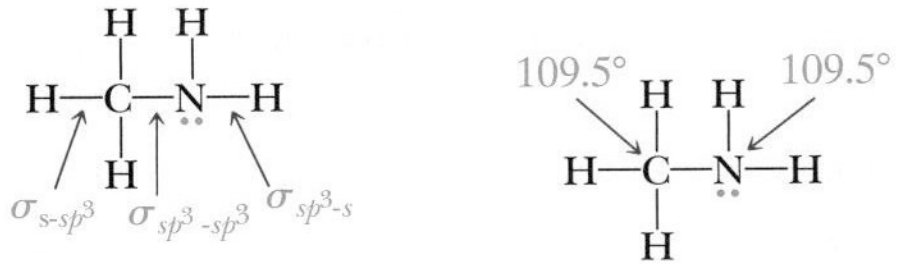

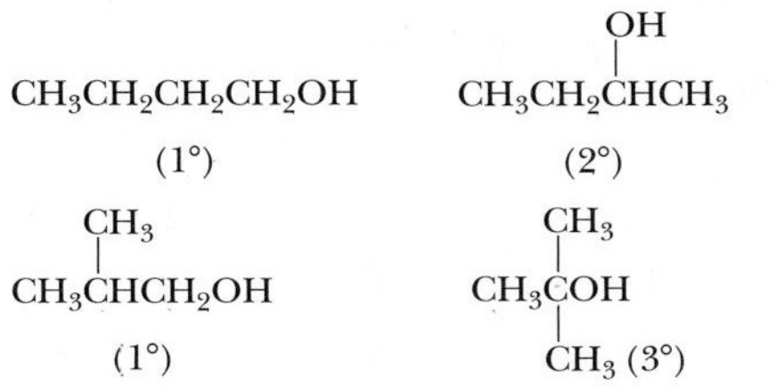

1.14 $CH_3CH_2NHCH_2CH_3$ $CH_3CH_2CH_2NHCH_3$
$CH_3CH(CH_3)NHCH_3$

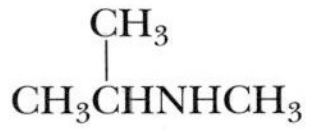

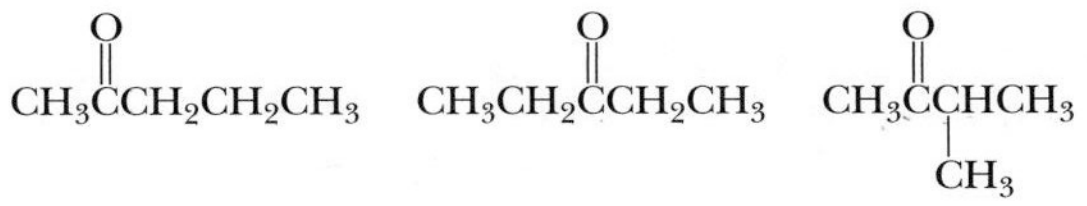

1.15 $CH_3C(=O)CH_2CH_2CH_3$ $CH_3CH_2C(=O)CH_2CH_3$ $CH_3C(=O)CH(CH_3)CH_3$

1.16 $CH_3CH_2CH_2COOH$ $CH_3CH(CH_3)COOH$

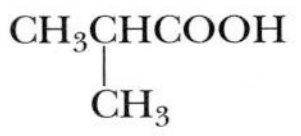

1.17 (a) $1s^22s^22p^63s^1$ (b) $1s^22s^22p^63s^2$
(c) $1s^22s^22p^4$ (d) $1s^22s^22p^3$

1.19 (a) S (b) O

1.21 The *valence shell* of an atom is the outermost shell that can be occupied by electrons in the ground state. The valence shell generally has the highest principal quantum number (n). A *valence electron* is an electron that is situated in the valence shell.

1.23 (a) none (b) two

1.25 (a) ionic (b) C—F polar covalent
C—H nonpolar covalent
(c) polar covalent (d) polar covalent

1.27 (a) H—Ö—Ö—H (b) H—N̈(H)—N̈(H)—H
(c) H—C(H)(H)—Ö—H (d) H—C(H)(H)—S̈—H
(e) H—C(H)(H)—N̈(H)—H (f) H—C(H)(H)—C̈l:
(g) H—C(H)(H)—Ö—C(H)(H)—H (h) H—C(H)(H)—C(H)(H)—H
(i) $H_2C=CH_2$ (j) H—C≡C—H

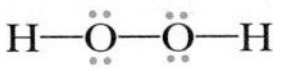

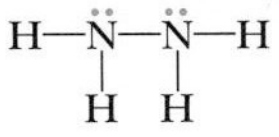

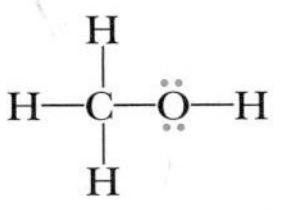

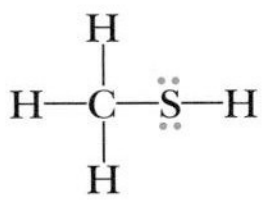

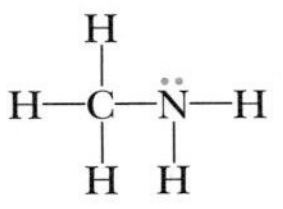

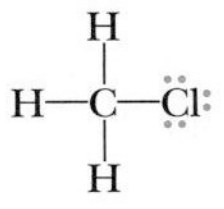

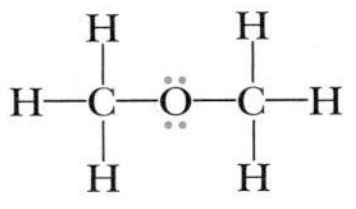

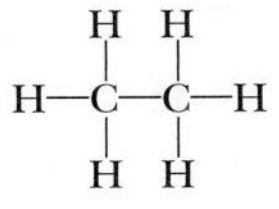

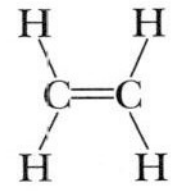

(k) O=C=O (l) $H_2C=O$ (m) $CH_3C(=O)CH_3$

(n) HO–C(=O)–OH (o) $CH_3C(=O)OH$

1.29 (a) If carbon were bonded to five hydrogen atoms, the octet rule would be violated.

(b) Because each hydrogen atom can only bond with one other atom, a single hydrogen atom cannot be bonded to both carbons.

1.31 (a) $H_3C–C(=O)–CH_2^-$ (b) $H_2N–C(O^-)=CH_2$

(c) $CH_3OH_2^+$ (d) CH_3^-

1.33 Ag_2O, and consists of polar covalent bonds

1.35 Only (a) is true

1.37 (a) C—H < N—H < O—H

(b) C—I < C—H < C—Cl

(c) C—C < C—N < C—O

(d) C—Hg < C—Mg < C—Li

1.39 (a) O—H (b) C—F (c) O—H

1.41 (a) $CH_3–CH_2–OH$: 109.5° (C), 109.5° (O) (b) $H_2C=CHCl$: 120°

(c) $CH_3–C≡C–H$: 180° (d) HC(=O)–OH: 120°, 109.5°

(e) $CH_3–NH_2$: 109.5° (f) H–O–N=O: 120°

1.43 109.5°

1.45 $F_2C=CF_2$

1.47 a, b, d, e, f, and g are true

1.49 bond angles do not change

1.51 (a) $CH_3–CH_3$: sp^3 (b) $H_2C=CH_2$: sp^2 (c) H–C≡C–H: sp

(d) $CH_3–NH_2$: N sp^3 (e) HC(=O)–OH: C sp^2, O sp^3 (f) $H_2C=O$: sp^2

1.53 (a) —C(=O)— (b) —C(=O)–O–H

(c) —O–H (d) —NH_2

1.55 (a) $CH_3CH_2CH_2CHO$ $CH_3CH(CH_3)CHO$ $CH_3C(=O)CH_2CH_3$

(b) $CH_2=CHCH_2CH_2OH$ $CH_3CH=CHCH_2OH$

$CH_3CH_2CH=CHOH$ $CH_2=CHCH(OH)CH_3$

$CH_2=C(OH)CH_2CH_3$ $CH_3C(OH)=CHCH_3$

$CH_2=C(CH_3)CH_2OH$ $CH_3C(CH_3)=CHOH$

1.57 (a) 2° hydroxyl group; carboxyl group: $CH_3–CH(OH)–COOH$

(b) 1° hydroxyl groups: HO–CH_2–CH_2–OH

(c) 1° amino group; carboxyl group: $CH_3–CH(NH_2)–COOH$

(d) 1° hydroxyl group; 2° hydroxyl group; carbonyl group (aldehyde): HO–CH_2–CH(OH)–CHO

(e) carbonyl group (ketone); carboxyl group: $CH_3–C(=O)–CH_2–COOH$

(f) 1° amino groups: $H_2N–CH_2CH_2CH_2CH_2CH_2CH_2–NH_2$

1.59 HO—CH_2—CH(OH)—CH_3

1.61 The carbon of CO_2 is *sp* and the molecule is linear. The central O of O_3 is sp^2 and O_3 is bent.

1.63 The central carbon is *sp* and the terminal carbons are sp^2. One CH_2 is perpendicular to the other.

1.65 (a) 6 (b) 120° (c) sp^2

1.67 (a) 120° (b) sp^2 (c) planar

Chapter 2
Acids and Bases

2.1 (a) $CH_3S–H$ + ^-OH ⟶ CH_3S^- + H–OH

acid, base, conjugate base, conjugate acid

(b) $CH_3O–H$ + $^-NH_2$ ⟶ CH_3O^- + H–NH_2

acid, base, conjugate base, conjugate acid

2.2 (a) $pK_a = 4.76$ (b) $pK_a = 15.7$
acetic acid is a stronger acid than water

2.3 (a) $CH_3NH_2 + CH_3COOH \rightleftharpoons CH_3NH_3^+ + CH_3COO^-$

methylamine	acetic acid	methylammonium ion	acetate ion
stronger base	stronger acid	weaker acid	weaker base

(b) $CH_3CH_2O^- + NH_3 \rightleftharpoons CH_3CH_2OH + NH_2^-$

ethoxide ion	ammonia	ethanol	amide ion
weaker base	weaker acid	stronger acid	stronger base

2.4 (a) $2 > 1 > 3$ (b) $3 > 1 > 2$

2.5 $CH_3\ddot{O}^- + H{-}N^+(CH_3)_3 \rightleftharpoons CH_3\ddot{O}{-}H + :N(CH_3)_3$

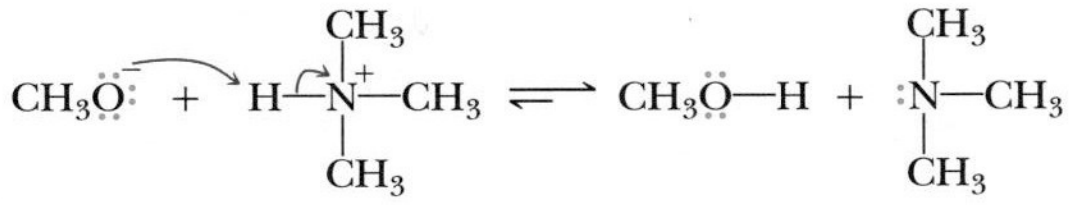

2.6 (a) 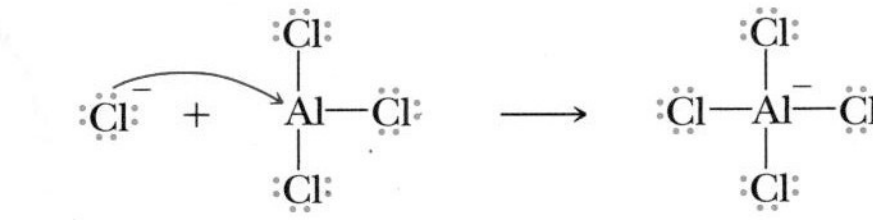 $:\ddot{Cl}:^- + AlCl_3 \longrightarrow Cl{-}Al^-Cl_3$

Lewis base Lewis acid

(b) 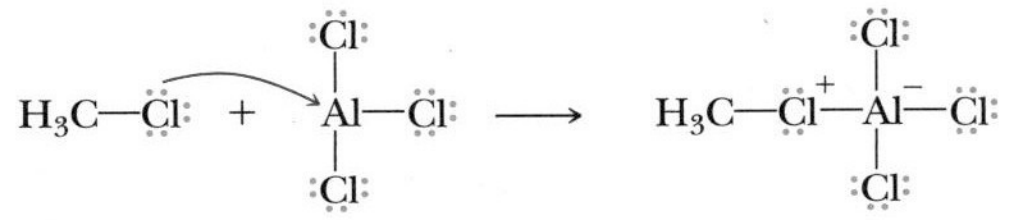 $H_3C{-}\ddot{Cl}: + AlCl_3 \longrightarrow H_3C{-}\overset{+}{Cl}{-}\overset{-}{Al}Cl_3$

Lewis base Lewis acid

2.7

(a) 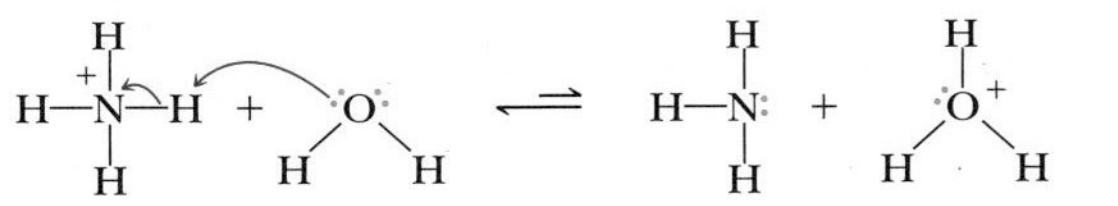 $H{-}N^+H_3 + H_2\ddot{O} \rightleftharpoons H{-}\ddot{N}H_2 + H_3O^+$

(b) 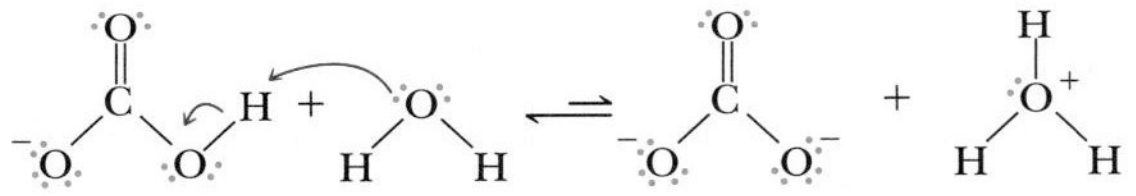 $HOCO_2^- + H_2\ddot{O} \rightleftharpoons CO_3^{2-} + H_3O^+$

(c) 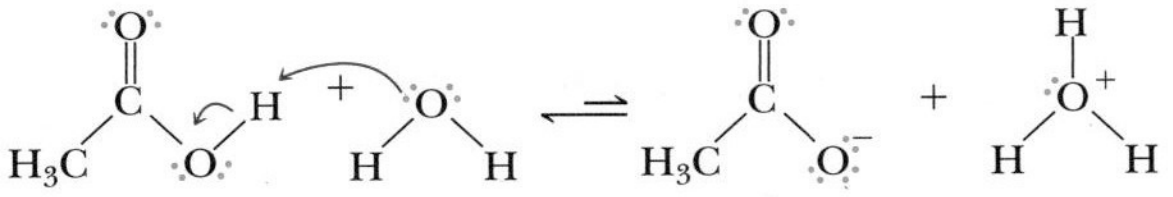 $H_3C{-}C(=O){-}O{-}H + H_2\ddot{O} \rightleftharpoons H_3C{-}C(=O){-}O^- + H_3O^+$

2.9 The formulas of a conjugate acid-base pair differ by one hydrogen atom, as well as one charge. The acid has one hydrogen atom more, but one negative charge less, than the base.

2.11

(a) 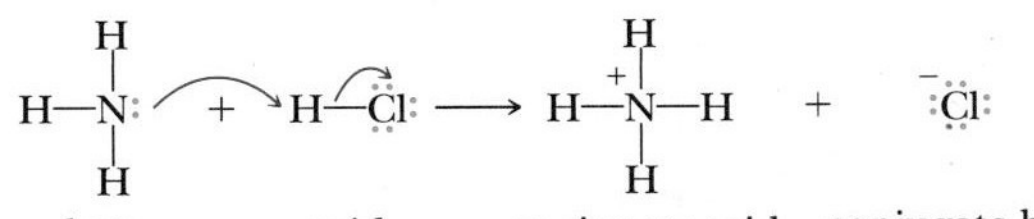 $H{-}\ddot{N}H_2 + H{-}\ddot{Cl}: \longrightarrow H{-}N^+H_3 + :\ddot{Cl}:^-$

base acid conjugate acid conjugate base

(b) 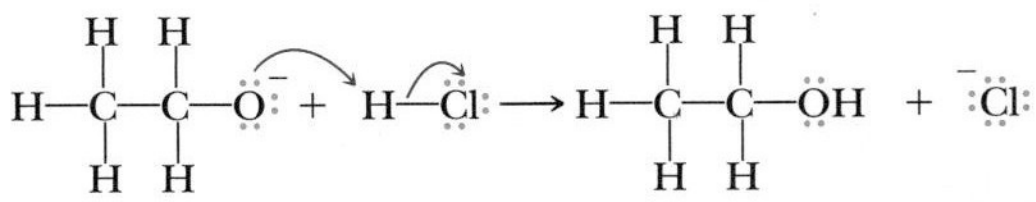 $H{-}CH_2{-}CH_2{-}\ddot{O}:^- + H{-}\ddot{Cl}: \longrightarrow H{-}CH_2{-}CH_2{-}\ddot{O}H + :\ddot{Cl}:^-$

base acid conjugate acid conjugate base

(c) 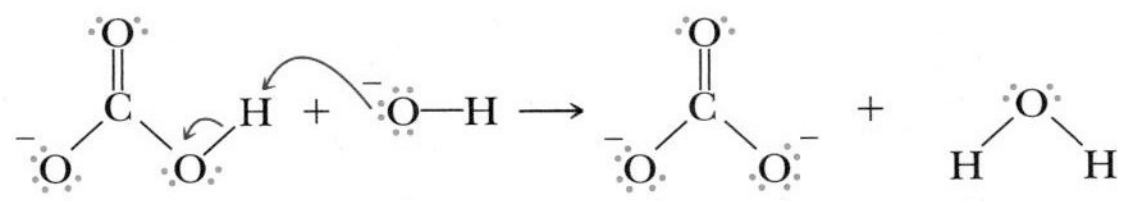 $^-O{-}C(=O){-}O{-}H + {}^-\ddot{O}{-}H \longrightarrow {}^-O{-}C(=O){-}O^- + H_2\ddot{O}$

acid base conjugate base conjugate acid

(d) 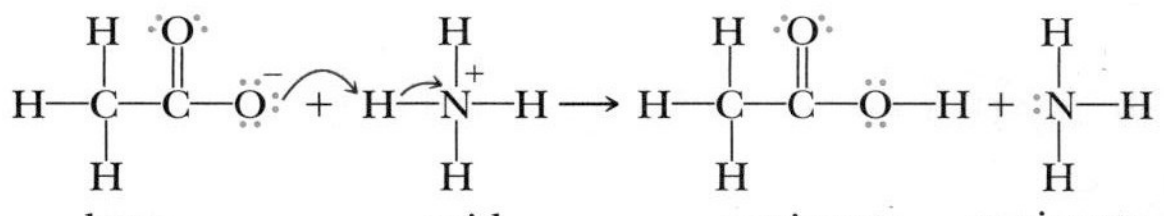 $H{-}CH_2{-}C(=O){-}\ddot{O}:^- + H{-}N^+H_3 \longrightarrow H{-}CH_2{-}C(=O){-}\ddot{O}{-}H + :NH_3$

base acid conjugate acid conjugate base

(e) 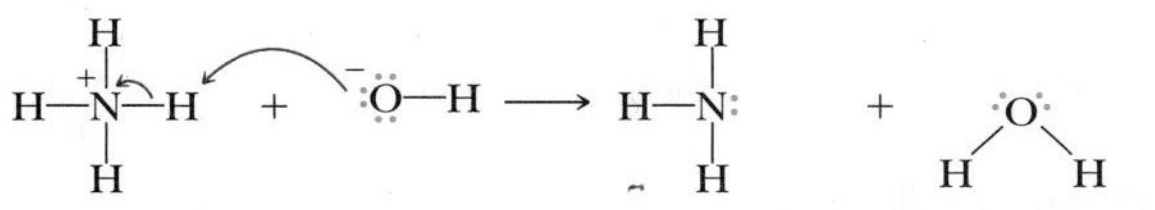 $H{-}N^+H_3 + {}^-\ddot{O}{-}H \longrightarrow H{-}\ddot{N}H_2 + H_2\ddot{O}$

acid base conjugate base conjugate acid

(f) 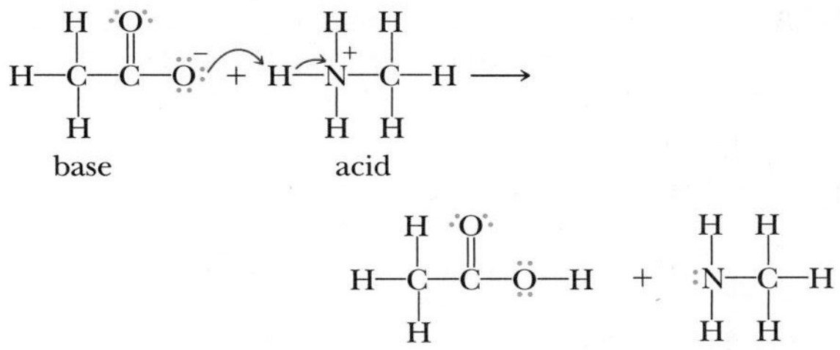 $H{-}CH_2{-}C(=O){-}\ddot{O}:^- + H{-}N^+H_2{-}CH_3 \longrightarrow$

base acid

$H{-}CH_2{-}C(=O){-}\ddot{O}{-}H + :\ddot{N}H{-}CH_3$

conjugate acid conjugate base

(g) 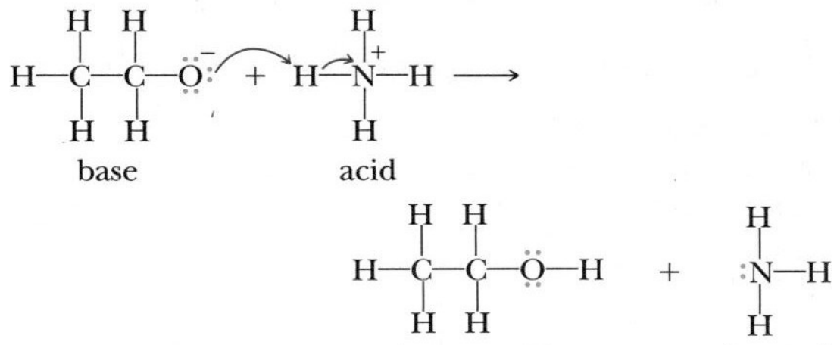 $H{-}CH_2{-}CH_2{-}\ddot{O}:^- + H{-}N^+H_3 \longrightarrow$

base acid

$H{-}CH_2{-}CH_2{-}\ddot{O}{-}H + :NH_3$

conjugate acid conjugate base

(h) 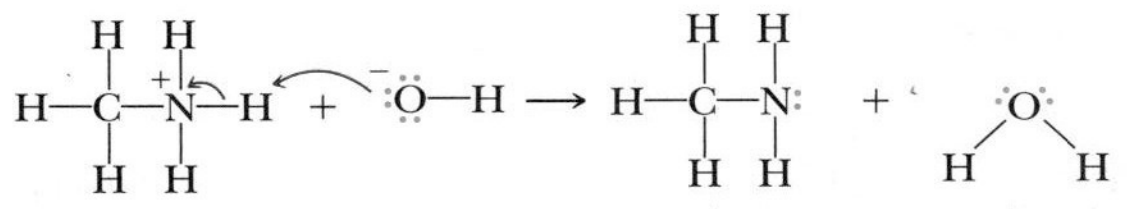 $H{-}CH_2{-}N^+H_3 + {}^-\ddot{O}{-}H \longrightarrow H{-}CH_2{-}\ddot{N}H_2 + H_2\ddot{O}$

acid base conjugate base conjugate acid

2.13

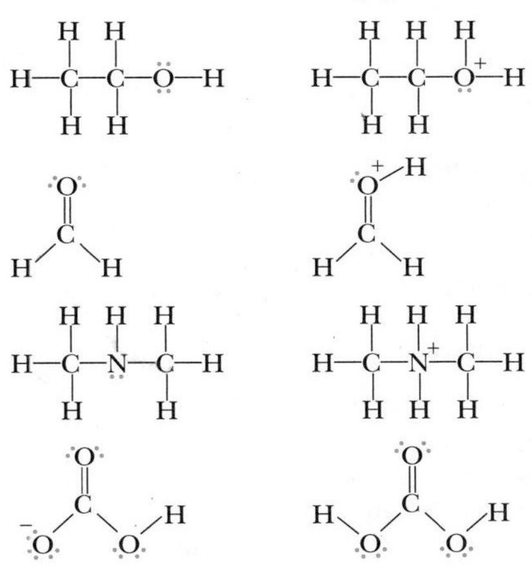

	Bases	Conjugate acids
(a)	$H{-}CH_2{-}CH_2{-}\ddot{O}{-}H$	$H{-}CH_2{-}CH_2{-}\overset{+}{O}H{-}H$
(b)	$H_2C{=}\ddot{O}$	$H_2C{=}\overset{+}{O}{-}H$
(c)	$H{-}CH_2{-}\ddot{N}H{-}CH_2{-}H$	$H{-}CH_2{-}\overset{+}{N}H_2{-}CH_2{-}H$
(d)	$^-O{-}C(=O){-}O{-}H$	$H{-}O{-}C(=O){-}O{-}H$

2.15 (a) 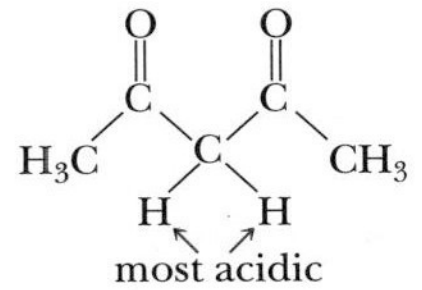 $H_3C{-}C(=O){-}CH_2{-}C(=O){-}CH_3$ — the H atoms on the central C: most acidic

(b) 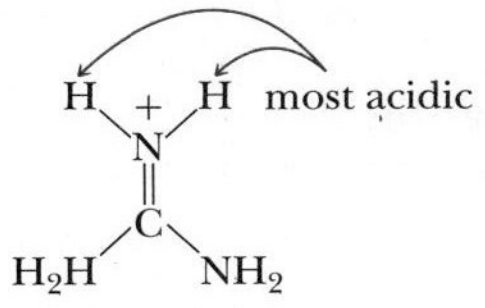 $H_2\overset{+}{N}{=}C(NH_2)_2$ — the H atoms on N^+: most acidic

2.17 (a) pyruvic acid (b) phosphoric acid
(c) acetylsalicylic acid (d) acetic acid

2.19 (a) $HOC(=O)O^-$ < NH_3 < $CH_3CH_2O^-$

(b) $CH_3C(=O)O^-$ < $HOC(=O)O^-$ < HO^-

(c) H_2O < $CH_3C(=O)O^-$ < NH_3

(d) $CH_3C(=O)O^-$ < OH^- < NH_2^-

2.21 As a result of the difference in atomic size between S and O, H_2S ionizes to form a conjugate base (HS^-) that is more stable relative to the conjugate base of H_2O (HO^-).

2.23 $HOC(=O)OH \longrightarrow CO_2 + H_2O$

2.25 (a) $CH_3COOH + HCO_3^- \rightleftharpoons CH_3COO^- + H_2CO_3$

(b) $CH_3COOH + NH_3 \rightleftharpoons CH_3COO^- + NH_4^+$

(c) $CH_3COOH + H_2O \rightleftharpoons CH_3COO^- + H_3O^+$

(d) $CH_3COOH + OH^- \rightleftharpoons CH_3COO^- + H_2O$

2.27 In acid-base equilibria, the position of the equilibrium favors the reaction of the stronger acid and stronger base to give the weaker acid and weaker base.

2.29

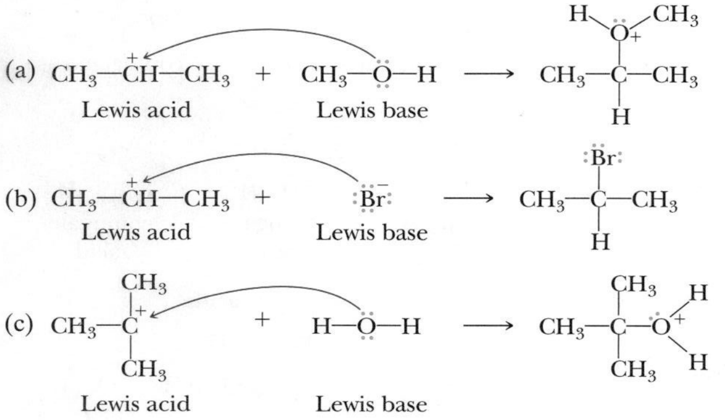

(a) $CH_3-\overset{+}{C}H-CH_3$ + $CH_3-\ddot{O}-H$ $\longrightarrow$ $CH_3-CH(O^+(H)CH_3)-CH_3$
Lewis acid, Lewis base

(b) $CH_3-\overset{+}{C}H-CH_3$ + $:\ddot{Br}:^-$ $\longrightarrow$ $CH_3-CH(Br)-CH_3$
Lewis acid, Lewis base

(c) $CH_3-\overset{+}{C}(CH_3)_2$ + $H-\ddot{O}-H$ $\longrightarrow$ $CH_3-C(CH_3)_2-O^+H_2$
Lewis acid, Lewis base

2.31 (a) $CH_3CH_2OH + HCO_3^- \rightleftharpoons CH_3CH_2O^- + H_2CO_3$

(b) $CH_3CH_2OH + OH^- \rightleftharpoons CH_3CH_2O^- + H_2O$

(c) $CH_3CH_2OH + NH_2^- \rightleftharpoons CH_3CH_2O^- + NH_3$

(d) $CH_3CH_2OH + NH_3 \rightleftharpoons CH_3CH_2O^- + NH_4^+$

2.33 All three of the bases will dissolve benzoic acid.

2.35 The conjugate base of dimethyl ether is a highly unstable C^- ion that is stabilized by only the inductive effect of the electronegative oxygen atom.

2.37 (B) is favored because it's charged functional groups represent the product of a favorable acid–base reaction.

Chapter 3
Alkanes and Cycloalkanes

3.1 (a) constitutional isomers (b) same compound

3.2

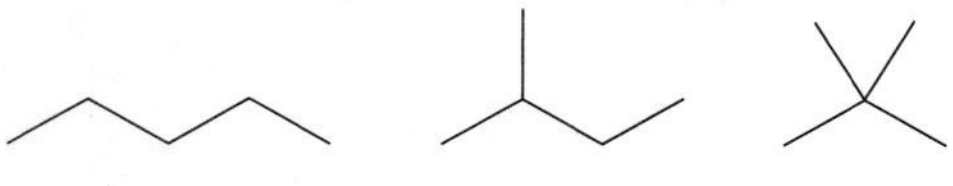

3.3 (a) 5-isopropyl-2-methyloctane
(b) 4-isopropyl-4-propyloctane
(c) 4-ethyl-2,3-dimethylheptane
(d) 4,6-diisopropyl-2-methylnonane

3.4 (a) $CH_3CH(CH_3)CH_2CH_2CH_3$ (1° 3° 2° 2° 1°; CH_3 branch 1°) (b) $CH_3-CH_2-C(CH_3)_2-CH(CH_3)_2$ (1° 2°; CH_3 groups 1°; CH 3°)

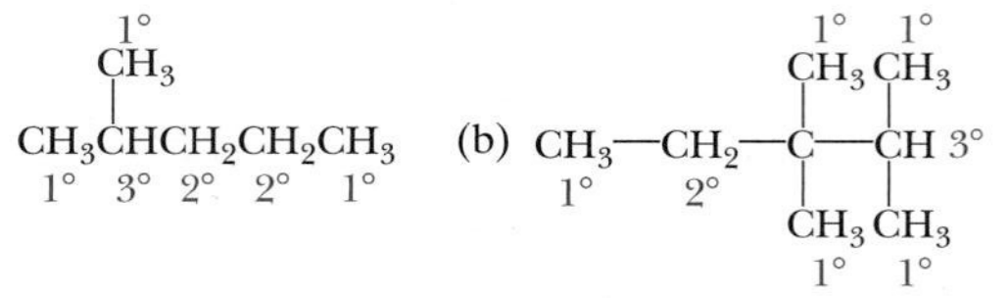

3.5 (a) C_9H_{18} isobutylcyclopentane
(b) $C_{11}H_{22}$ *sec*-butylcycloheptane
(c) C_6H_{12} 1-ethyl-1-methylcyclopropane

3.6 (a) propanone (b) pentanal
(c) cyclopentanone (d) cycloheptene

3.7

Staggered / Eclipsed

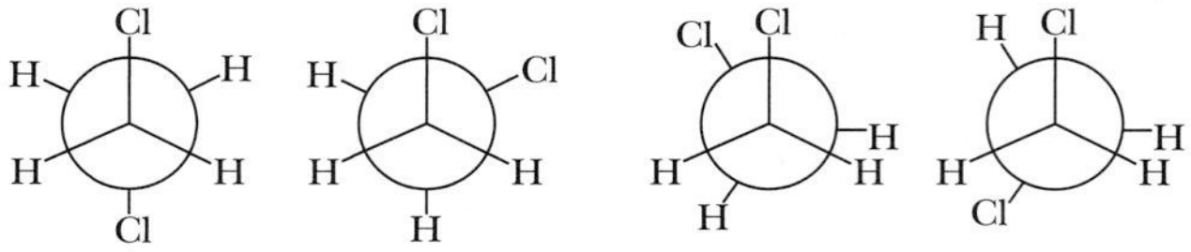

3.8 (a)

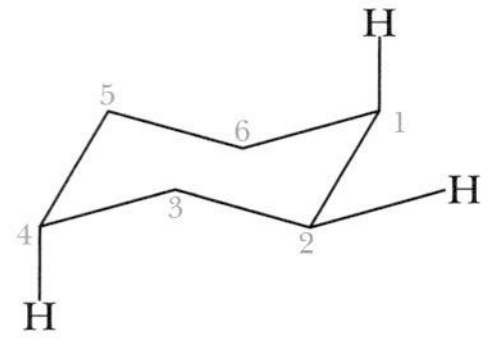

(b) The hydrogen on carbon 2 is equatorial, and the hydrogens on carbons 1 and 4 are axial.

(c)

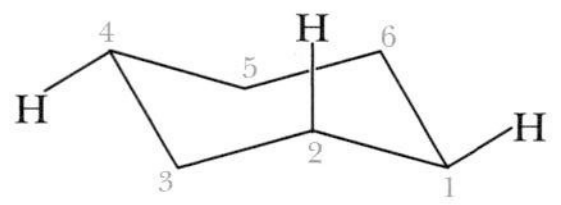

Hydrogens 1 and 4 are equatorial. Hydrogen 2 is axial. Hydrogens 1 and 2 are above, and 4 below, the plane of the ring.

3.9 An axial *tert*-butyl group will always have a $-CH_3$ group directed at the other axial substituents.

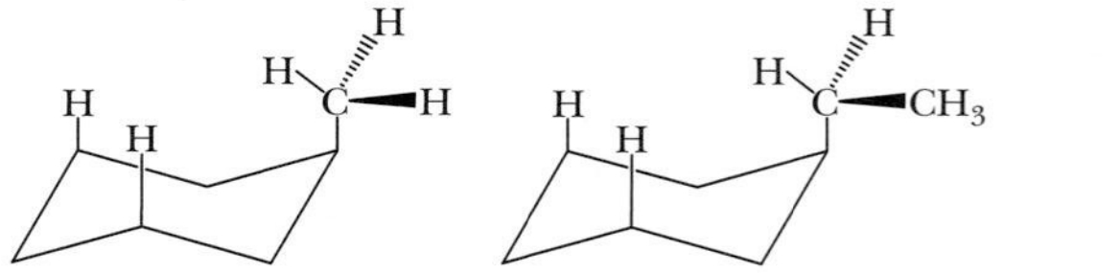

methyl substituent / ethyl substituent

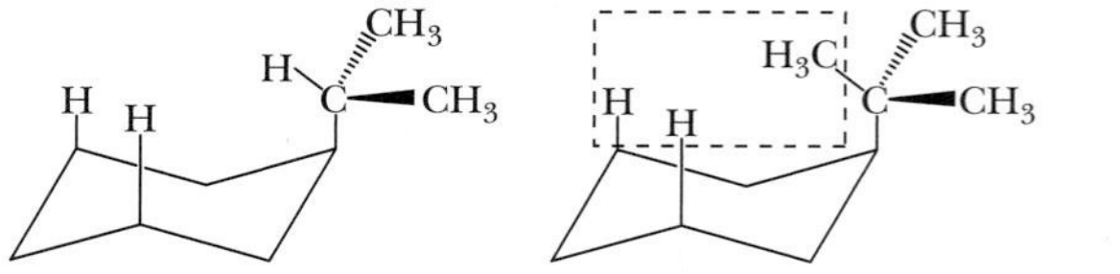

isopropyl substituent / *tert*-butyl substituent

3.10 (a)

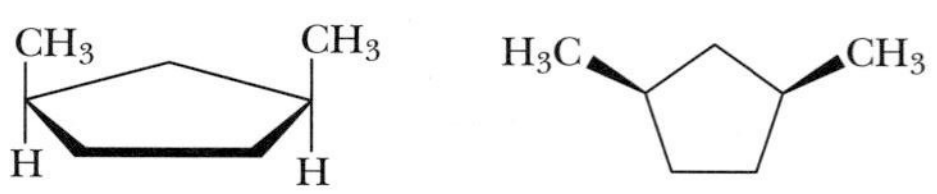

cis-1,3-dimethylcyclopentane

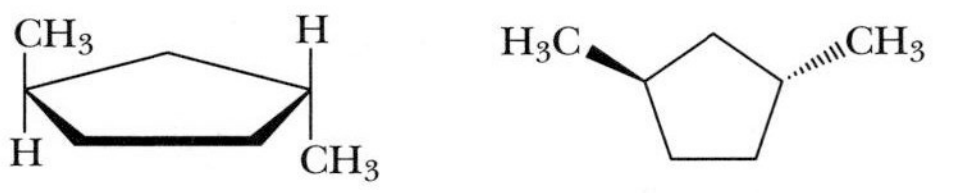

CH_3 H H CH_3 H_3C CH_3

trans-1,3-dimethylcyclopentane

(c) CH_2CH_3 CH_3 H H H_3CH_2C CH_3

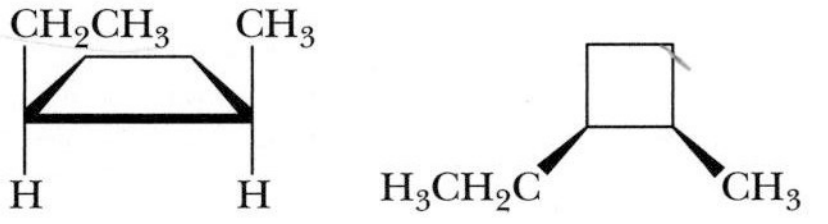

cis-1-ethyl-2-methylcyclobutane

H CH_3 CH_2CH_3 H H_3CH_2C CH_3

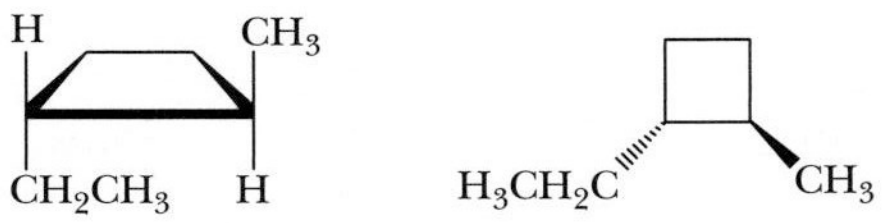

trans-1-ethyl-2-methylcyclobutane

3.11

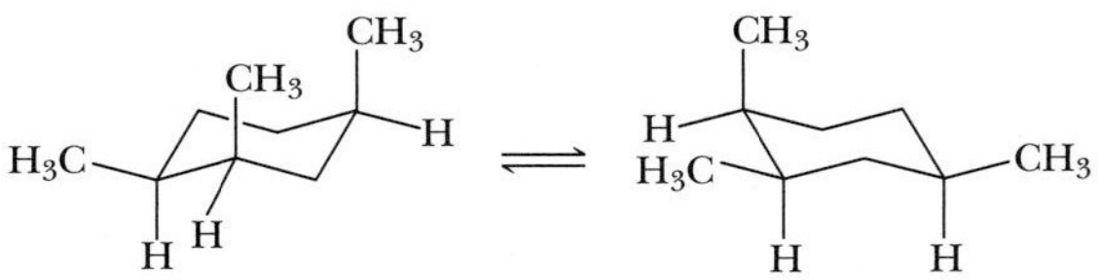

less stable ⇌ more stable

3.12 (a) 2,2-dimethylpropane < 2-methylbutane < pentane
(b) 2,2,4-trimethylhexane < 3,3-dimethylheptane < nonane

3.13 (a) 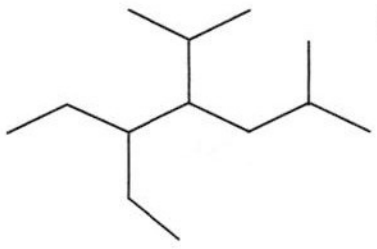 (b)  (c) 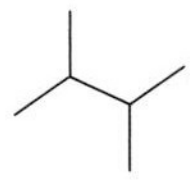 (d) 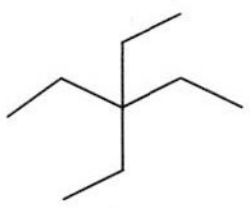 (e)  (f)

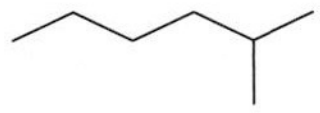

3.15 (a) $CH_3(CH_2)_4CH(CH_3)_2$ (b) $HC((CH_2)_2CH_3)_3$

(c) $CH_3C(CH_2)_4CH_3$ with two $(CH_2)_2CH_3$ substituents

3.17 (1) Compounds (a) and (g) represent the same compound.
Compounds (d) and (e) represent the same compound.
(2) Compounds (a = g), (d = e), and (f) represent constitutional isomers of $C_4H_{10}O$.
Compounds (b) and (c) represent constitutional isomers of C_4H_8O.
(3) The isomers of $C_4H_{10}O$ [(a = g) and (d = e)] are different compounds from the isomers of C_4H_8O [(b) and (c)], and all are different from compound (h).

3.19 (1) None of the compounds are the same.
(2) Compounds (a), (d), and (e) represent constitutional isomers of C_4H_8O.
Compounds (c) and (f) represent constitutional isomers of $C_5H_{10}O$.
Compounds (g) and (h) represent constitutional isomers of $C_6H_{10}O$.
(3) Compound (b), with formula C_5H_8O, is a different compound than all of the isomers listed for each of the respective molecular formulas indicated in (2).

3.21

heptane

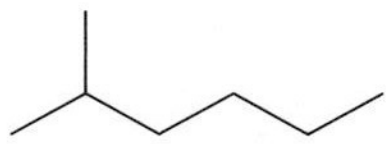

2-methylhexane

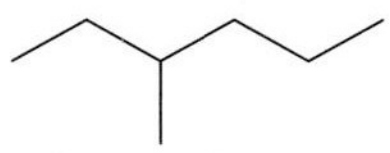

3-methylhexane

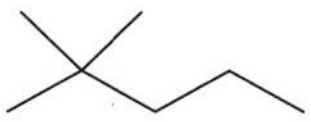

2,2-dimethylpentane

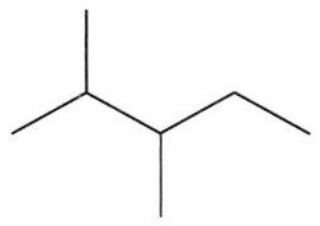

2,3-dimethylpentane

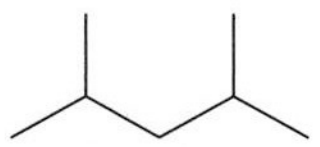

2,4-dimethylpentane

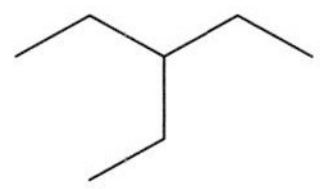

3-ethylpentane

3,3-dimethylpentane

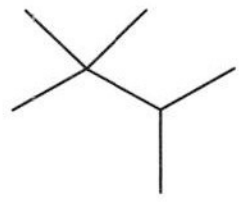

2,2,3-trimethylbutane

3.23

(a)  OH OH OH HO

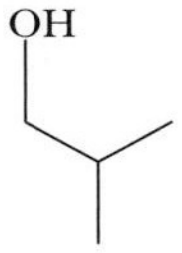

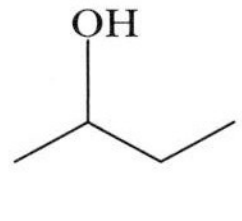

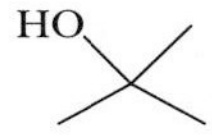

(b) 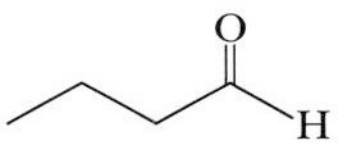 O H O H (c) 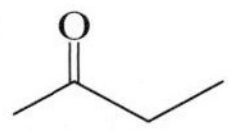 O

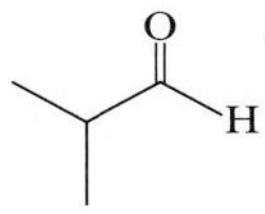

(d) 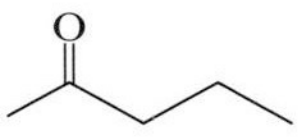 O O O

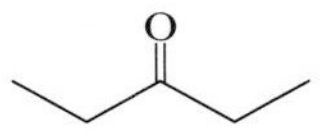

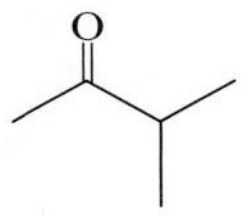

(e) 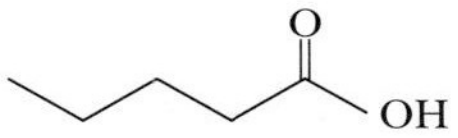 O OH O OH

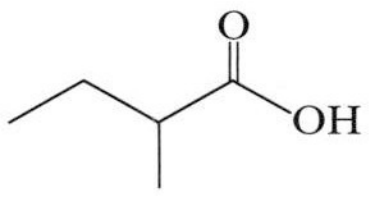

O OH O OH

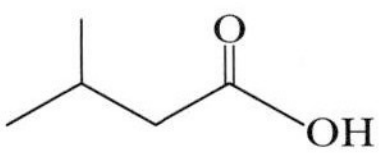

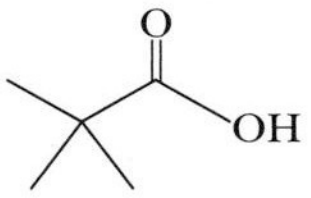

3.25 (a) 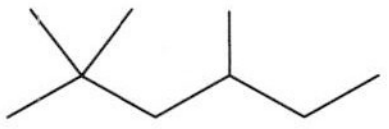 (b)  (c) 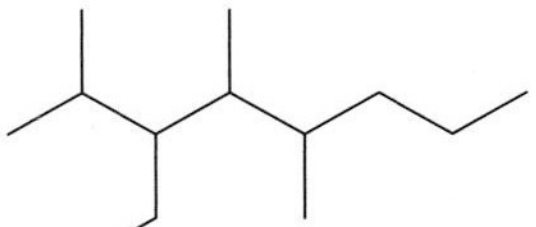 (d) 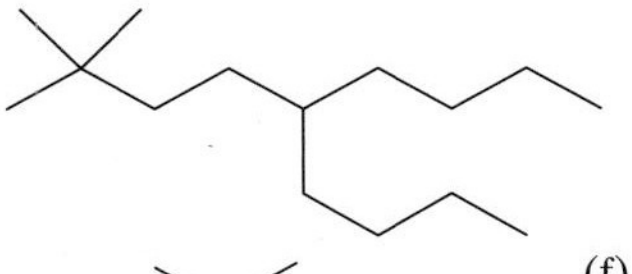 (e) 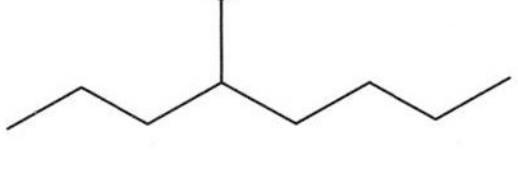 (f)

(g)

(h) CH_2CH_3 / CH_2CH_3

3.27 1°, 2°, 2°, 3°, 1°, 2°, 2°, 2°, 1°, 2°

3.29 (a) CH_3CH_2OH (b) $CH_3\overset{O}{\overset{\|}{C}}H$

(c) $CH_3\overset{O}{\overset{\|}{C}}OH$ (d) $CH_3\overset{O}{\overset{\|}{C}}CH_2CH_3$

(e) $CH_3CH_2CH_2\overset{O}{\overset{\|}{C}}H$ (f) $CH_3CH_2CH_2\overset{O}{\overset{\|}{C}}OH$

(g) $CH_3CH_2\overset{O}{\overset{\|}{C}}H$ (h) —OH

(i) —OH (j)

(k) =O

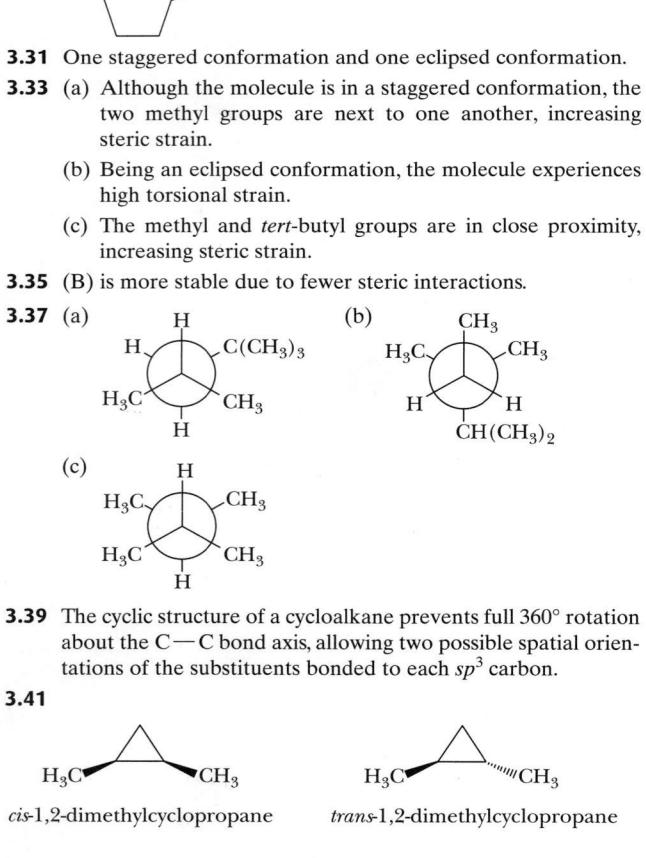

3.31 One staggered conformation and one eclipsed conformation.

3.33 (a) Although the molecule is in a staggered conformation, the two methyl groups are next to one another, increasing steric strain.

(b) Being an eclipsed conformation, the molecule experiences high torsional strain.

(c) The methyl and *tert*-butyl groups are in close proximity, increasing steric strain.

3.35 (B) is more stable due to fewer steric interactions.

3.37 (a) H, H, $C(CH_3)_3$, H_3C, CH_3, H (b) CH_3, H_3C, CH_3, H, H, $CH(CH_3)_2$

(c) H, H_3C, CH_3, H_3C, CH_3, H

3.39 The cyclic structure of a cycloalkane prevents full 360° rotation about the C—C bond axis, allowing two possible spatial orientations of the substituents bonded to each sp^3 carbon.

3.41

H_3C CH_3 — *cis*-1,2-dimethylcyclopropane

H_3C CH_3 — *trans*-1,2-dimethylcyclopropane

3.43

(a)

cis-1,2-dimethylcyclopentane

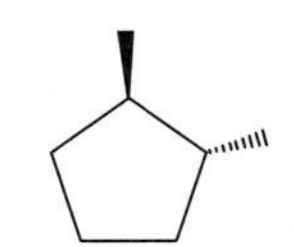

trans-1,2-dimethylcyclopentane

(b)

cis-1,3-dimethylcyclopentane *trans*-1,3-dimethylcyclopentane

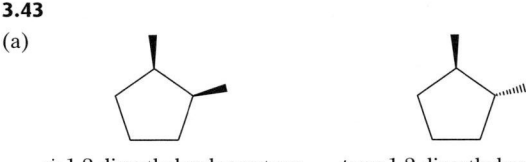

3.45

Position of substitution	*cis*	*trans*
1,2-	a,e or e,a	e,e or a,a
1,3-	e,e or a,a	a,e or e,a
1,4-	a,e or e,a	e,e or a,a

3.47

(a) OH, HO; OH, OH

chairs are of equal stability

(b) OH, OH; HO, OH

more stable

(c) OH, CH_3, OH; HO, HO, CH_3

more stable

(d) CH_2OH OH, OH, OH; CH_2OH, HO, HO, CH_3

more stable

3.49 highest = heptane

lowest = 2,2-dimethylpentane

3.51 Heptane has a boiling point of 98°C. Its molecular formula is C_7H_{16}, which corresponds to a molecular weight of 100 g/mol.

3.53 (a) No (b) Yes (c) Yes (d) liquid (e) less dense

3.55 (a) $2\ CH_3(CH_2)_4CH_3 + 19\ O_2 \longrightarrow 12\ CO_2 + 14\ H_2O$

(b) (cyclohexane) $+\ 9\ O_2 \longrightarrow 6\ CO_2 + 6\ H_2O$

(c) $2\ CH_3\underset{}{C}HCH_2CH_2CH_3$ (with CH_3 on C-2) $+ 19\ O_2 \longrightarrow 12\ CO_2 + 14\ H_2O$

3.57 2,2,4-Trimethylpentane produces more energy both per gram and per mole.

3.59

Glucose

(a)

(b)

All the hydrogens are axial, and all the other substituents are equatorial.

3.61 (a) Rings A, B, and C are in the chair conformation. Ring D, a cyclopentane ring, is in an envelope conformation.
(b) equatorial
(c) axial with respect to both rings
(d) axial

Putting It Together

1. (b) **2.** (c) **3.** (d)

4. Statement (e) is correct because the negative charge in molecule **A** is delocalized by resonance.

5. (a) **6.** (a) **7.** (b) **8.** (c) **9.** (e) **10.** (d)

11. (a)

(b) One of the carbonyl groups is a part of a ketone. The rest are either part of an amide (Chapter 15) or an ester.

(c) The carbon-carbon σ bond involves the overlap of sp^2 and sp^3 hybrid orbitals.

(d) There are two quaternary carbons, one located adjacent to carbon **3** and the other located adjacent to the carbon to which the —OH group **C** is bonded.

(e) The conjugate base of **B** is stabilized by the inductively withdrawing δ^+ charge of the nearby carbonyl (C=O) group.

(f) 120°.

12.

most stable

13. (a) 2,4-dimethylpentane
(b) 4-ethyl-1,2-dimethylcyclohexane
(c) 4-ethyl-2,5-dimethylheptane

14. (a) A (b) B (c) A (d) B (e) A (f) A

15. No net charge

16. Each pair of brackets represents a π bond.

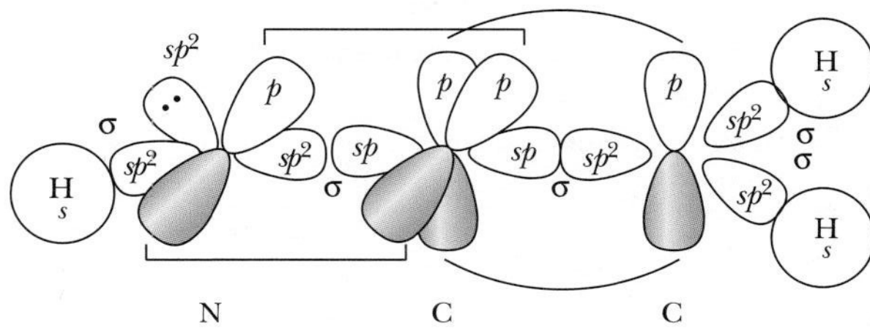

the N—H bond is perpendicular to the two C—H bonds.

17. A *tert*-butyl group has a methyl group that can interact hydrogen atoms via 1,3-diaxial interactions. See Problem 3.9 for more details.

18. The major product is favored because of charge delocalization from a resonance contributing structure.

major product

minor product

19. $:N{\equiv}C{-}\ddot{O}:^- \longleftrightarrow :\ddot{N}^-{=}C{=}\ddot{O}:$ $\quad :C{\equiv}\ddot{N}{-}\ddot{O}:^- \longleftrightarrow :\ddot{C}^-{=}N{=}\ddot{O}:$

$[NCO]^-$ is more stable because the negative charge is delocalized over oxygen and nitrogen atoms. Whereas, the negative charge in $[CNO]^-$ is delocalized over oxygen and carbon atoms.

20.

(a) $CH_3CH_2CH{=}CHCH_3$ 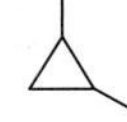

(b)

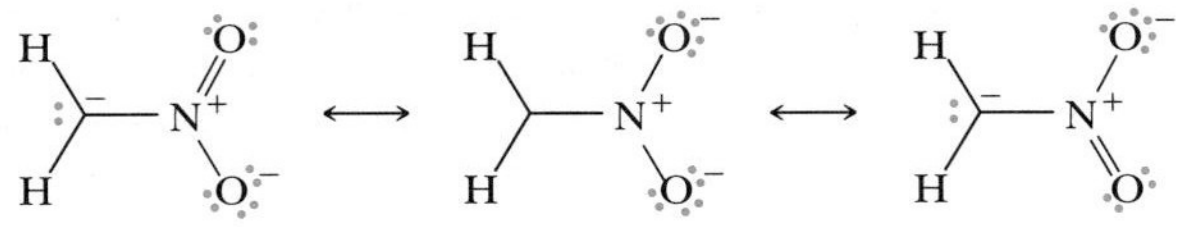

(c)

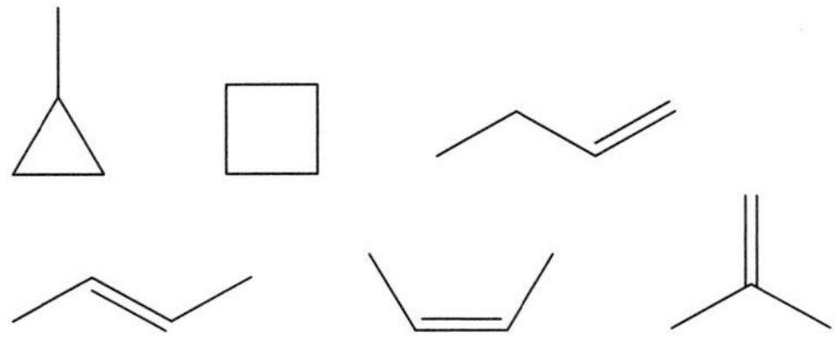

Chapter 4
Alkenes and Alkynes

4.1 (a) 3,3-dimethyl-1-pentene (b) 2,3-dimethyl-2-butene
(c) 3,3-dimethyl-1-butyne

4.2 (a) *cis*-4-methyl-2-pentene (b) *trans*-2,2-dimethyl-3-hexene

4.3 (a) (*E*)-1-chloro-2,3-dimethyl-2-pentene
(b) (*Z*)-1-bromo-1-chloro-1-propene
(c) (*E*)-2,3,4-trimethyl-3-heptene

4.4 (a) 1-isopropyl-4-methylcyclohexene
(b) cyclooctene
(c) 4-*tert*-butylcyclohexene

4.5

4.6 4

4.7 (a) it will dissolve (b) it will not dissolve

4.9 A compound that is *saturated* has the maximum number of hydrogens possible for the number of carbons present in the structure. A compound that is *unsaturated* contains fewer hydrogens than the maximum number that is possible, which is the case when one or more rings or π bonds is present.

4.11 (a) H H sp^3 sp^2 H (b) sp^2 CH_2OH

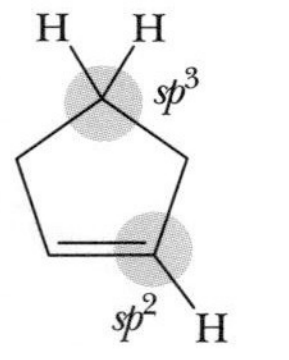

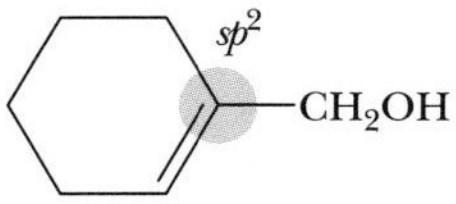

(c) sp sp^2 H H—C≡C—CH=C H (d) sp^2

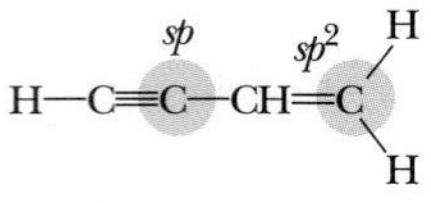

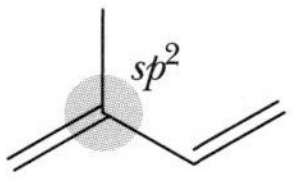

4.13 (a) 3 σ bonds from sp^2 1 π bond from *p*

4 σ bonds from sp^3

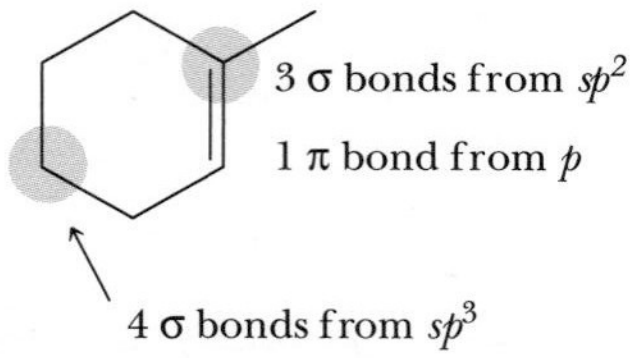

(b) 4 σ bonds from sp^3 OH

3 σ bonds from sp^2 1 π bond from *p*

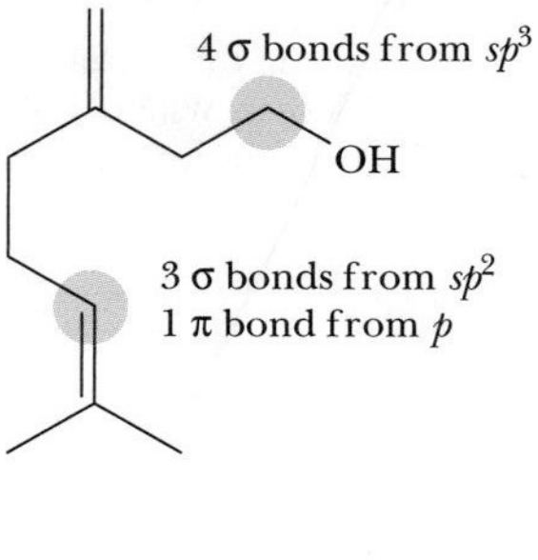

(c) 2 σ bonds from *sp* 2 π bonds from *p*

4 σ bonds from sp^3

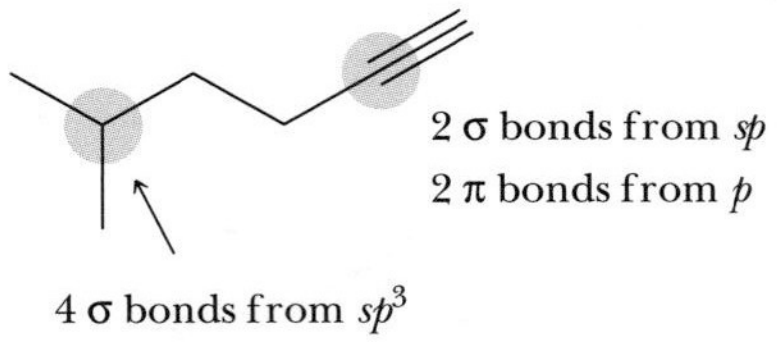

(d) Br 4 σ bonds from sp^3 Br

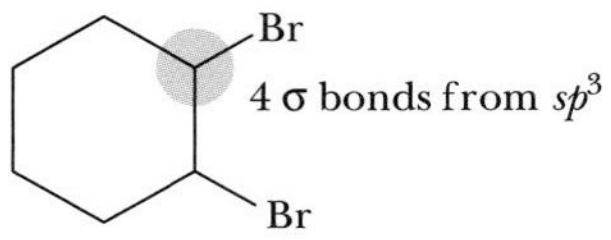

4.15 (a) 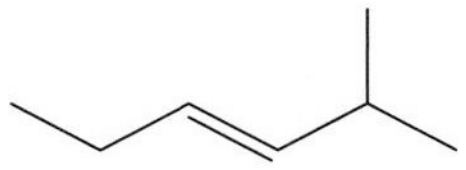 (b)

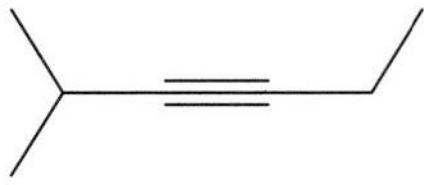

(c) 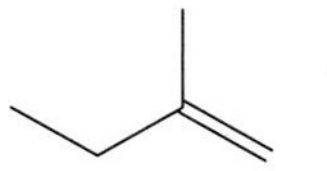 (d)

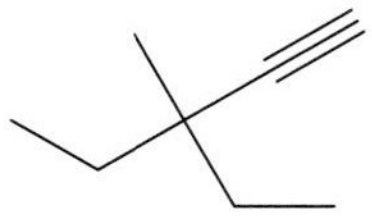

(e) 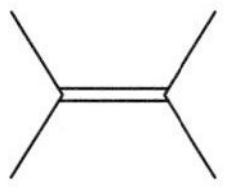 (f)

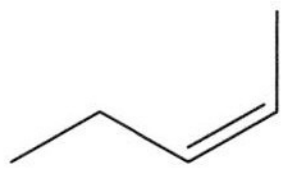

(g)  Cl (h)

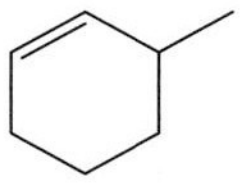

4.17 (a) 1,2-dimethylcyclohexene
(b) 4,5-dimethylcyclohexene
(c) 1-*tert*-butyl-2,4,4-trimethylcyclohexene
(d) (*E*)-1-cyclopentyl-2-methylpentene

4.19 (a) The parent chain is four carbons long, and it is also necessary to indicate the configuration (*E* or *Z*). Correct name: (*E*)-2-butene or (*Z*)-2-butene
(b) The numbering of the chain is incorrect, and it is also necessary to indicate the configuration (*E* or *Z*). Correct name: (*E*)-2-pentene or (*Z*)-2-pentene
(c) The numbering of the ring is incorrect. Correct name: 1-methylcyclohexene
(d) It is necessary to indicate the position of the carbon-carbon double bond. Correct name: 3,3-dimethyl-1-pentene
(e) The numbering of the chain is incorrect. Correct name: 3-hexyne
(f) The parent chain is five carbons long, and it is also necessary to indicate the configuration (*E* or *Z*). Correct name: (*E*)-3,4-dimethyl-2-pentene or (*Z*)-3,4-dimethyl-2-pentene

4.21 (b)

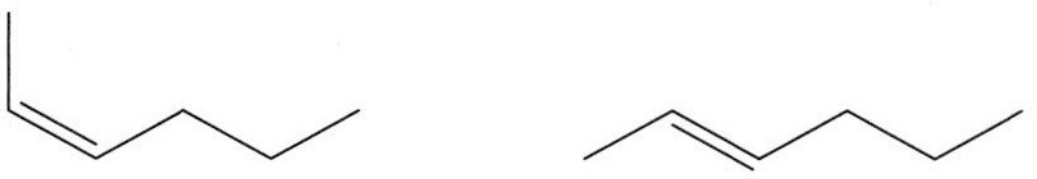

(c)

(e)

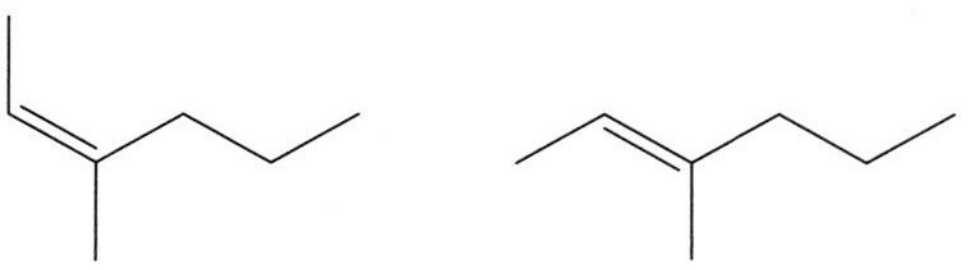

4.23 (b)  Br (d)

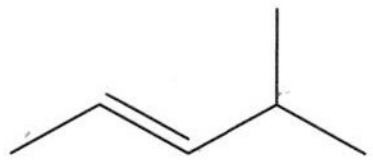

4.25

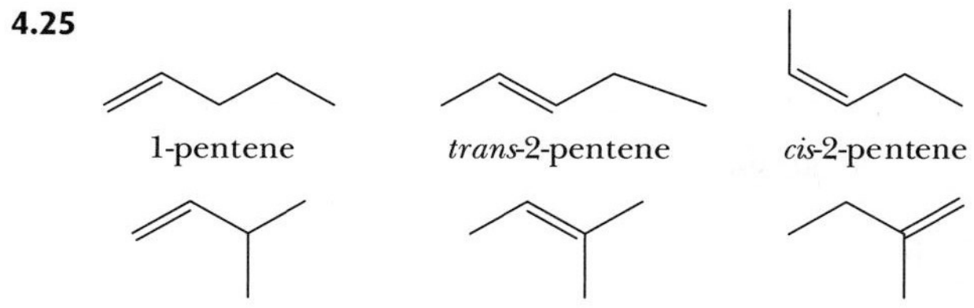

4.27 (a) $-CH_3 < -CH_2CH_3 < -Br$
(b) $-CH_2CH_2NH_2 < -CH(CH_3)_2 < -OCH_3$
(c) $-CH_2OH < -COOH < -OH$
(d) $-CH(CH_3)_2 < -CH{=}CH_2 < -CH{=}O$

4.29 (a)

(b)

4.31 (b) 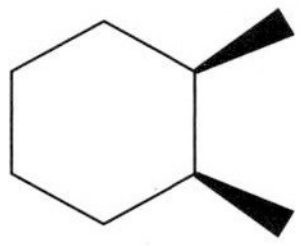 (d)

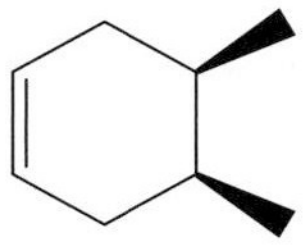

4.33 (a)

(b)

(c)

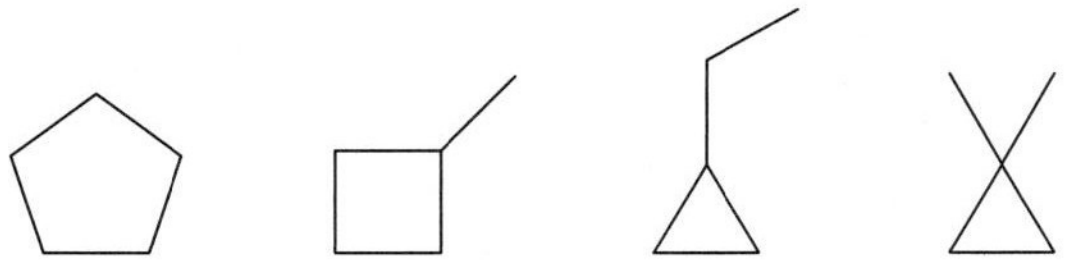

(d)

4.35 Oleic acid is *cis* and elaidic acid is *trans*

4.37

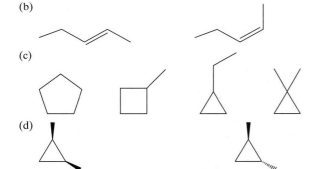

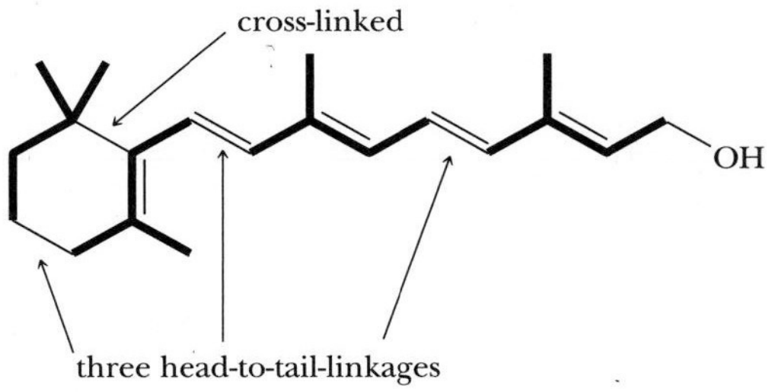

cross-linked

three head-to-tail-linkages

4.39 (a)

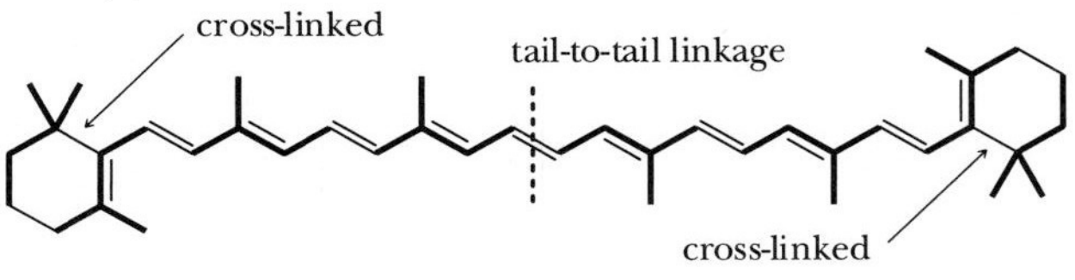

cross-linked

tail-to-tail linkage

cross-linked

The main structural difference between β-carotene and lycopene is that β-carotene has six-membered rings at the two ends of the structure. To form each ring, a two isoprene units are cross-linked. Both β-carotene and lycopene contain two sets of four isoprene units that are joined in the middle via a tail-to-tail linkage. In both molecules, all of the double bonds that can exhibit *cis-trans* isomerism are in the *trans* configuration (with respect to the main chain).

(b) β-Carotene has eight isoprene units.

4.41 The molecule contains three isoprene units, which are highlighted in bold and joined together by two head-to-tail linkages.

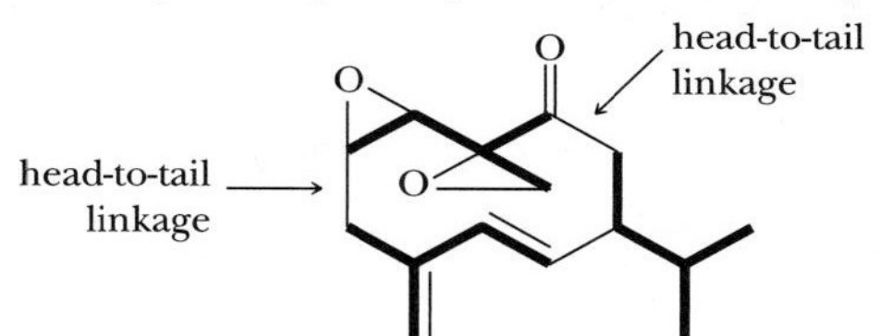

4.43 (a) three head-to-tail linkages

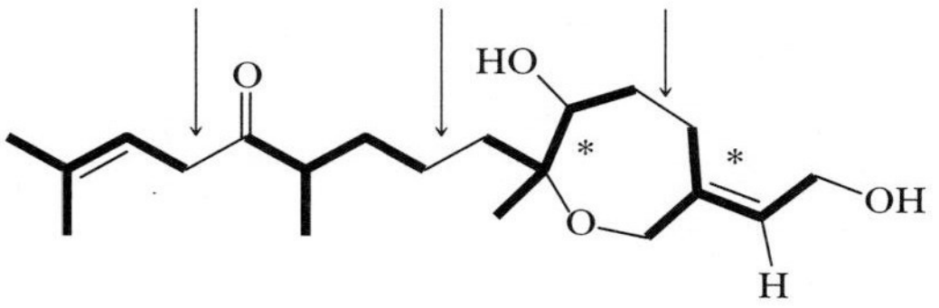

(b) *E* (c) 4

4.45 Both are terpenes

4.47 Smaller rings are more reactive

4.49 (a) Oleic, linoleic, and linolenic acid respectively have 2, 4, and 8 *cis-trans* isomers.

(b)

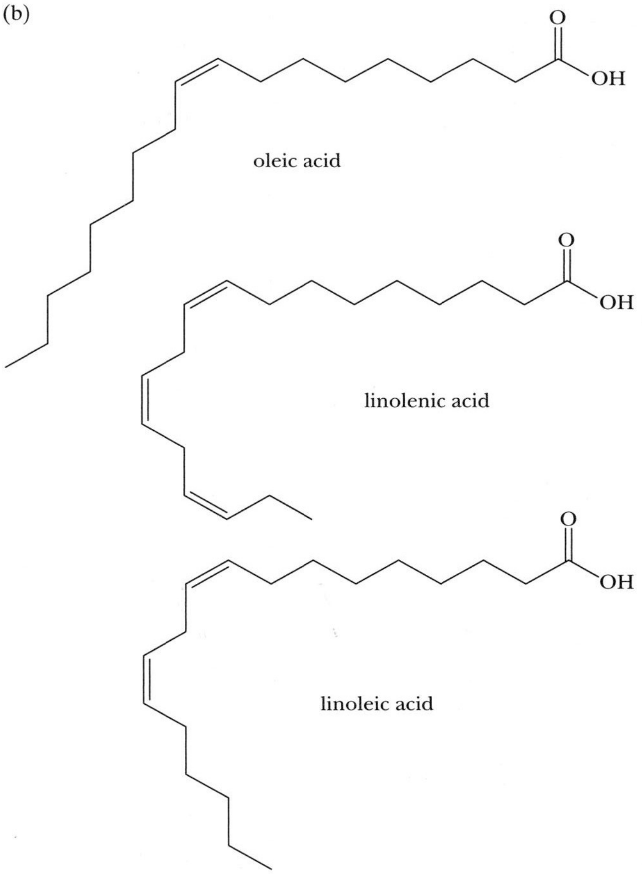

oleic acid

linolenic acid

linoleic acid

Chapter 5 Reactions of Alkenes

5.1 The products would be higher in energy than the reactants

5.2 I

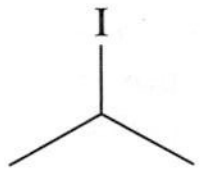

2-iodopropane

CH_3

I

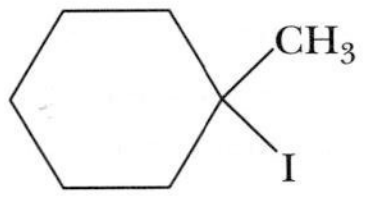

1-iodo-1-methylcyclohexane

5.3 (a) 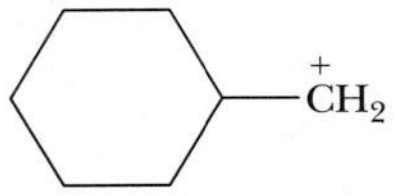 $\overset{+}{C}H_2$ (b) 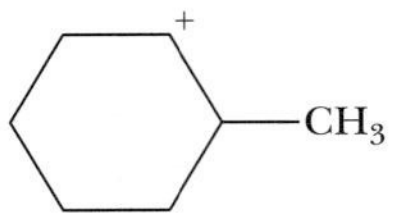 CH_3

least stable

(c) 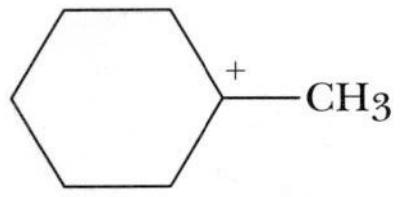 CH_3

most stable

5.4 Step 1:

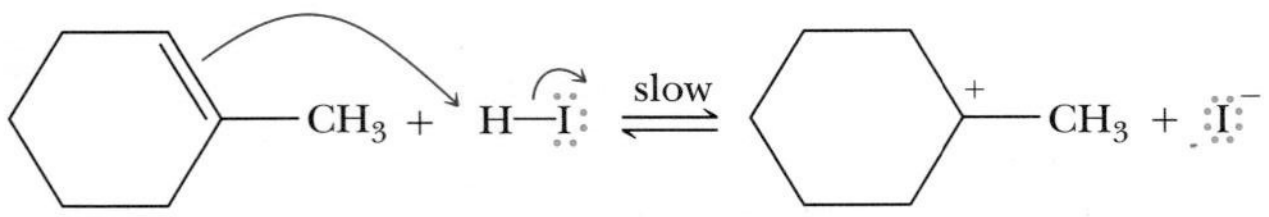

CH_3 + H—I: slow CH_3 + :I:$^-$

Step 2:

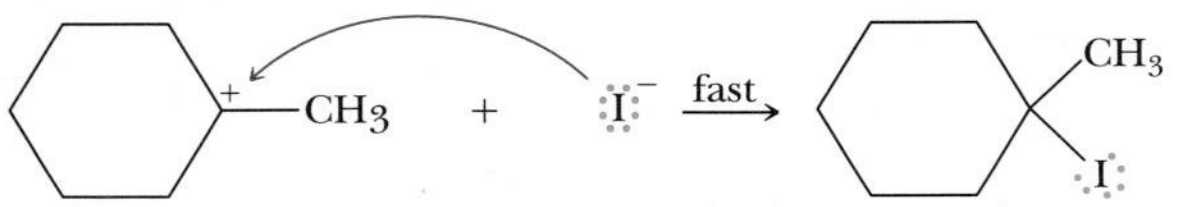

CH_3 + :I:$^-$ fast CH_3 I

5.5 (a) 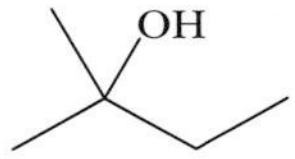 OH (b) 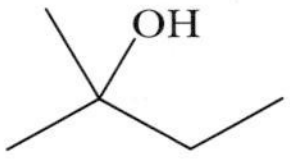 OH

5.6 Step 1:

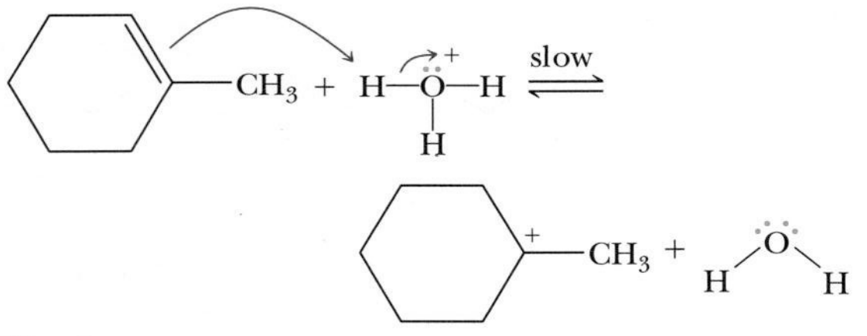

CH_3 + H—O—H (H) slow

CH_3 + H—O—H

Step 2:

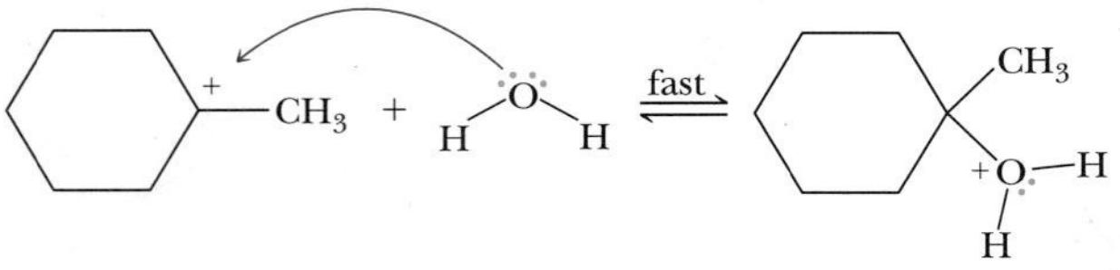

CH_3 + H—O—H fast CH_3 O—H (H)

Step 3:

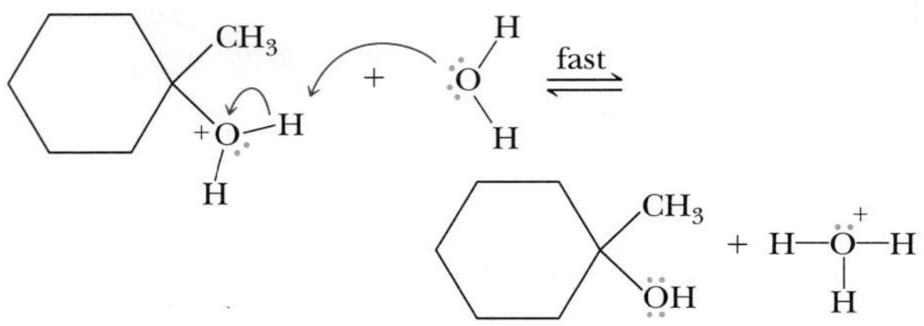

CH_3 O—H (H) + H—O—H fast

CH_3 OH + H—O—H (H)

5.7 (a) 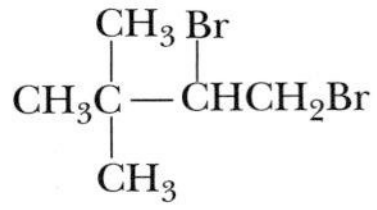 $CH_3C(CH_3)_2$—$CHBrCH_2Br$ (b) 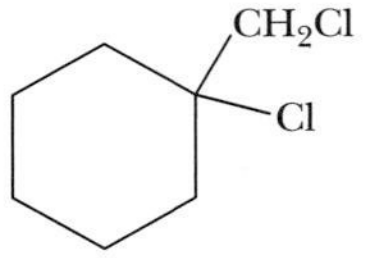 CH_2Cl Cl

5.8 Step 1:

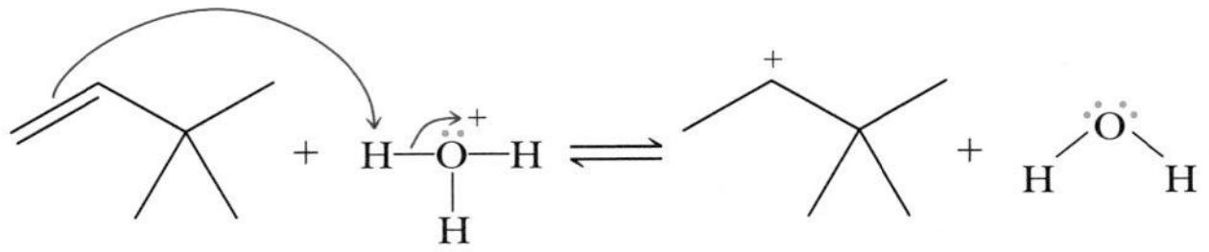

+ H—O—H (H) ⇌ + H—O—H

Step 2:

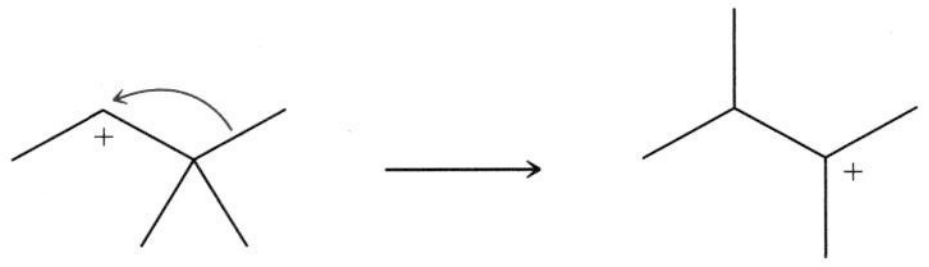

Step 3:

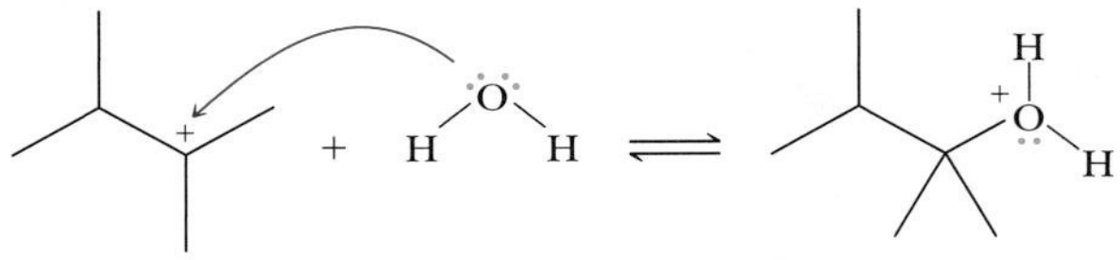

+ H—O—H ⇌ H O H

Step 4:

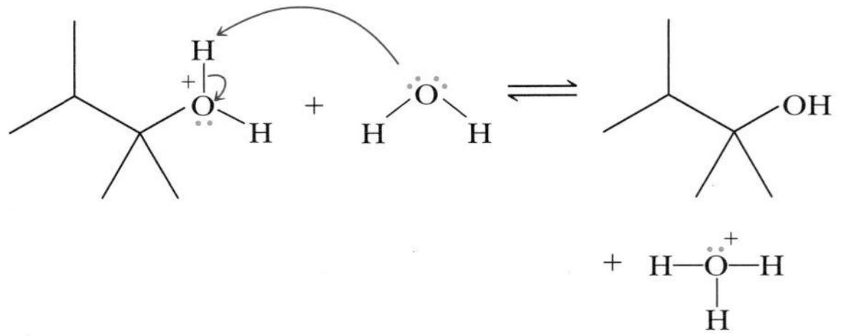

H O H + H—O—H ⇌ OH

+ H—O—H (H)

5.9 (a) 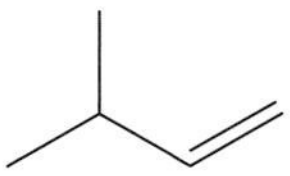 (b)

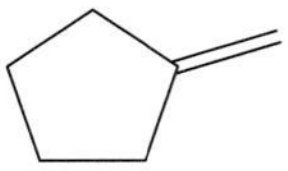

5.10 (a) 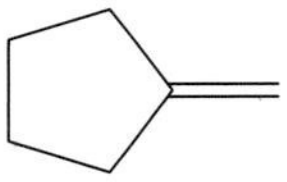 (b)

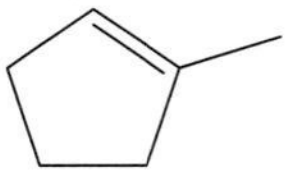

5.11 (a) 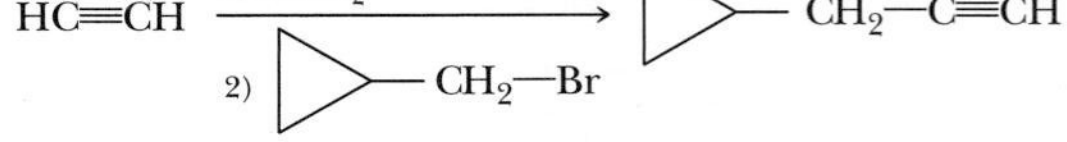 HC≡CH $\xrightarrow[\text{2) cyclopropyl—}CH_2\text{—Br}]{\text{1) }NaNH_2}$ cyclopropyl—CH_2—C≡CH

(b) 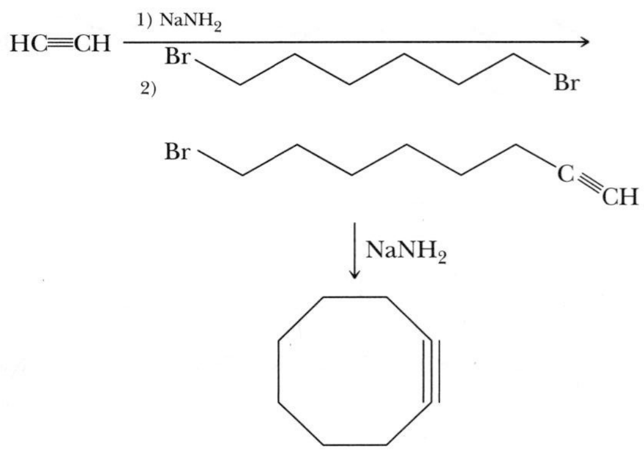 HC≡CH $\xrightarrow[\text{2) Br}\sim\sim\text{Br}]{\text{1) }NaNH_2}$ Br～～C≡CH

$\downarrow NaNH_2$

5.12 + H_2O $\xrightarrow[HgSO_4]{H_2SO_4}$

OH + OH

O O

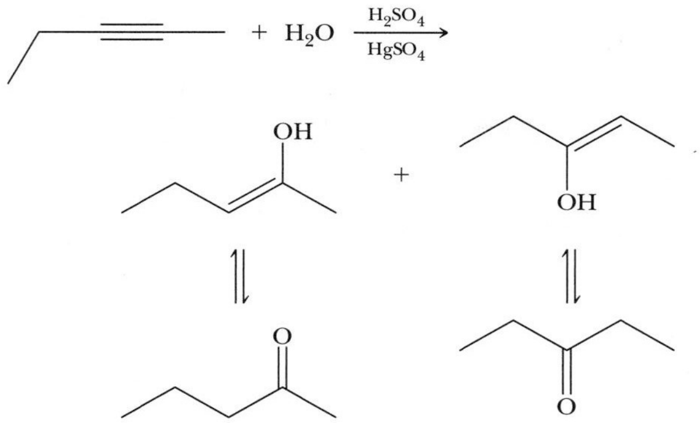

5.13 A transition state is a point on the reaction coordinate where the energy is at a maximum, and it cannot be isolated and its structure can only be postulated. A reaction intermediate corresponds to an energy minimum between two transition states, but the energy of the intermediate is usually higher than the energy of either the products or the reactants.

5.15

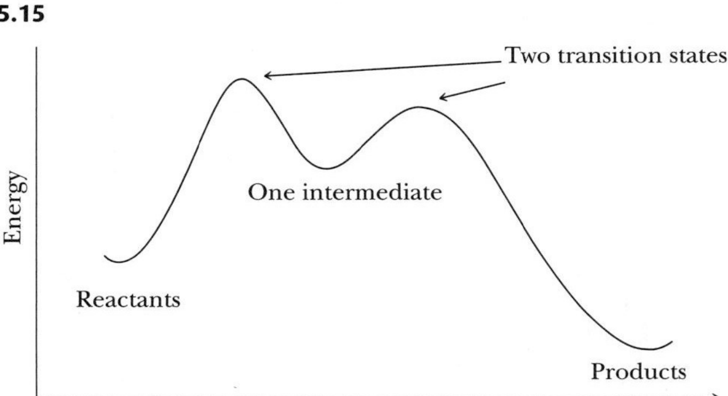

5.17 (a) $CH_3\overset{+}{C}HCH_3$ (b) $CH_3\underset{+}{\overset{CH_3}{\overset{|}{C}}}CH_2CH_3$

5.19 (a) +

3° carbocation (formed more readily) 2° carbocation

(b) +

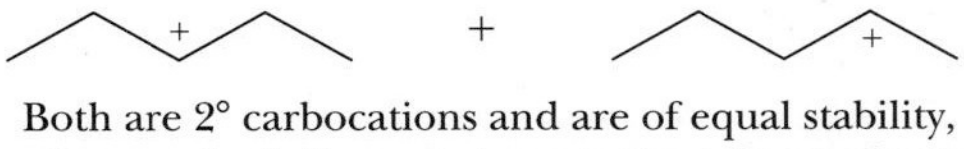

Both are 2° carbocations and are of equal stability, so they are both formed at approximately equal rates

(c) CH_3 + CH_3

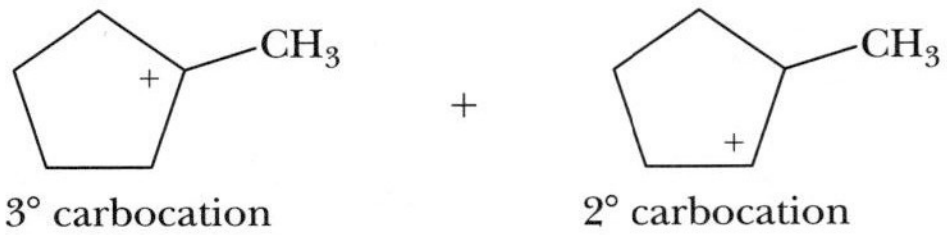

3° carbocation (formed more readily) 2° carbocation

(d) CH_3 + $\overset{+}{C}H_2$

3° carbocation (formed more readily) 1° carbocation

5.21 (a) Cl (b) OH

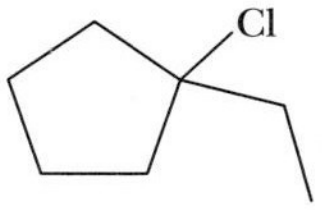

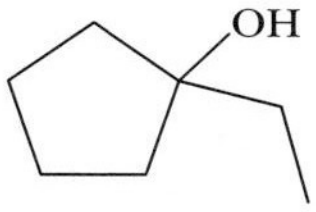

(c) I (d) Cl

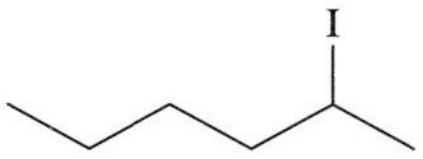

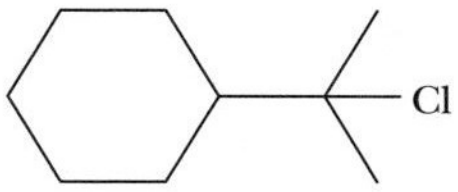

(e) OH (f) OH

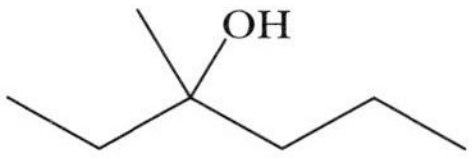

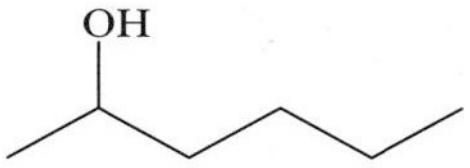

5.23 (a) Br H Br (b) Cl Cl

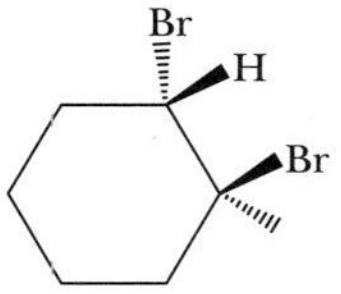

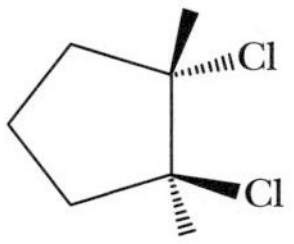

5.25 (a) 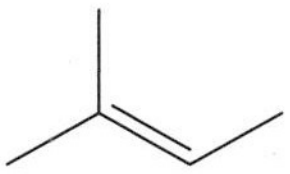 (b)

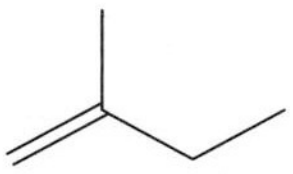

(c)

5.27 (a) 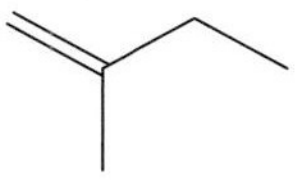 or

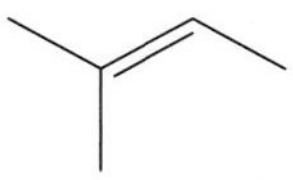

(b) 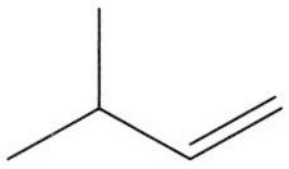 (c)

5.29 (a)

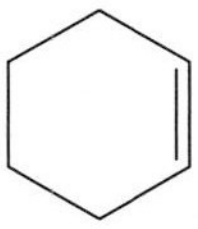

(b) 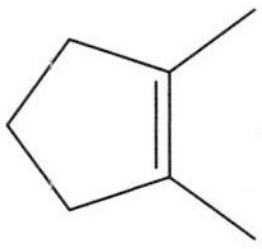 or 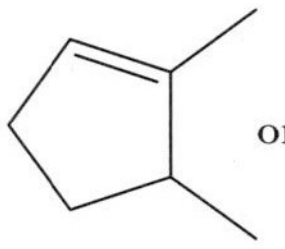 or

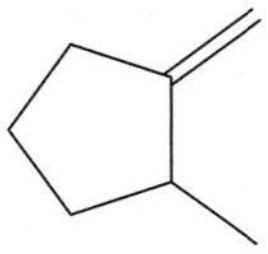

(c) 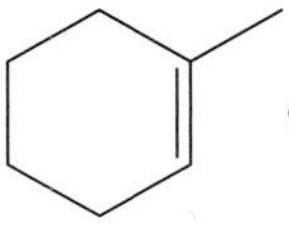 or

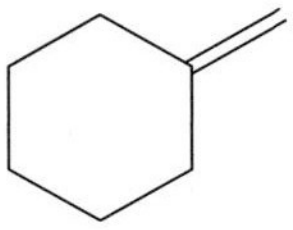

(d)

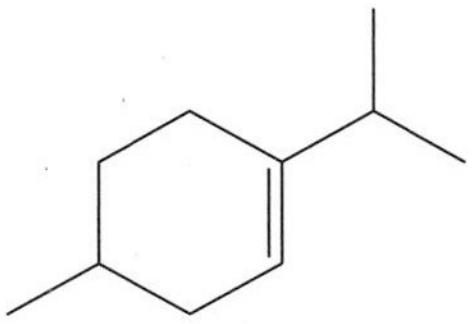

5.31 (a) H H—Br ⟶ H + Br^-

⟶ + $\xrightarrow{Br^-}$ Br

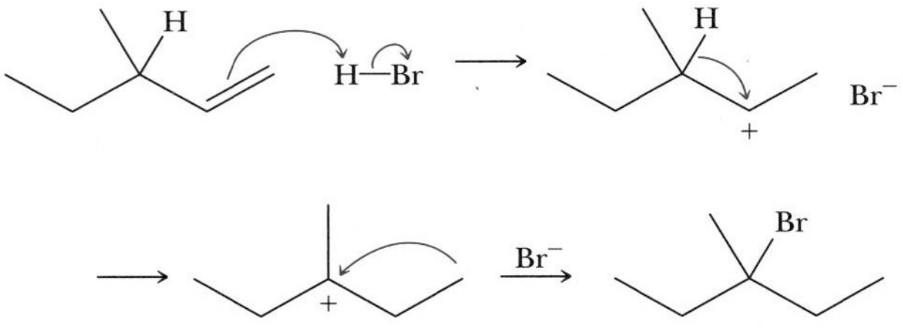

(b)

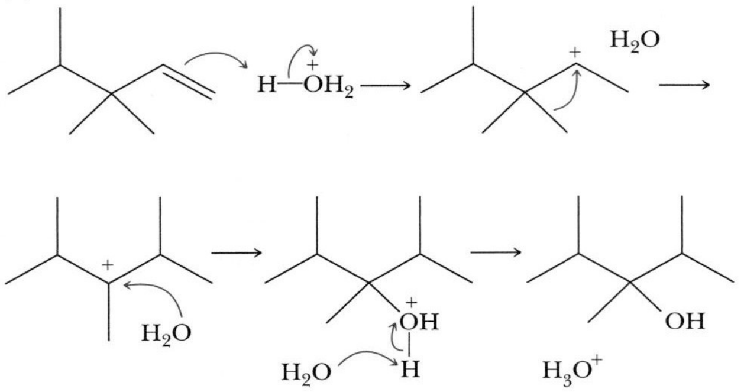

$H-\overset{+}{O}H_2 \longrightarrow$ H_2O $\longrightarrow$

H_2O $\longrightarrow$ $\overset{+}{O}H$ H_2O H $\longrightarrow$ OH H_3O^+

(c)

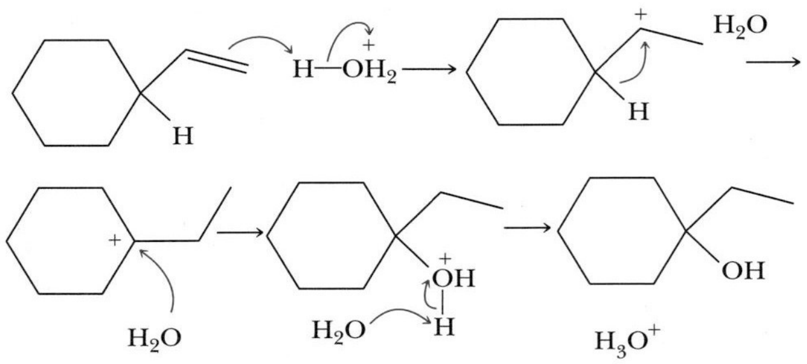

$H-\overset{+}{O}H_2 \longrightarrow$ H H_2O H $\longrightarrow$

H_2O $\longrightarrow$ $\overset{+}{O}H$ H_2O H $\longrightarrow$ OH H_3O^+

(d)

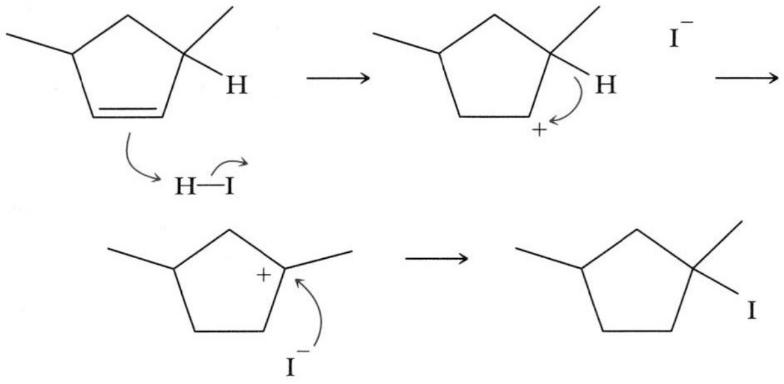

H $H-I$ $\longrightarrow$ H I^- $\longrightarrow$

I^- $\longrightarrow$ I

(e)

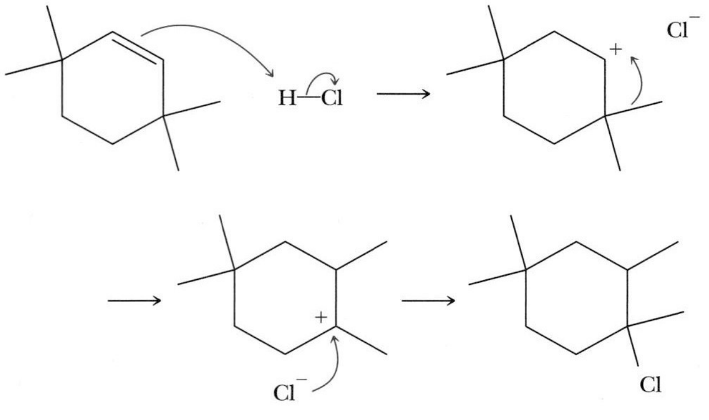

$H-Cl$ $\longrightarrow$ Cl^-

$\longrightarrow$ Cl^- $\longrightarrow$ Cl

(f)

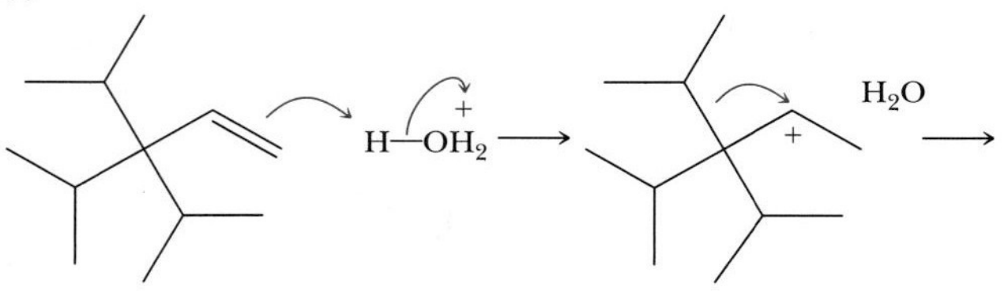

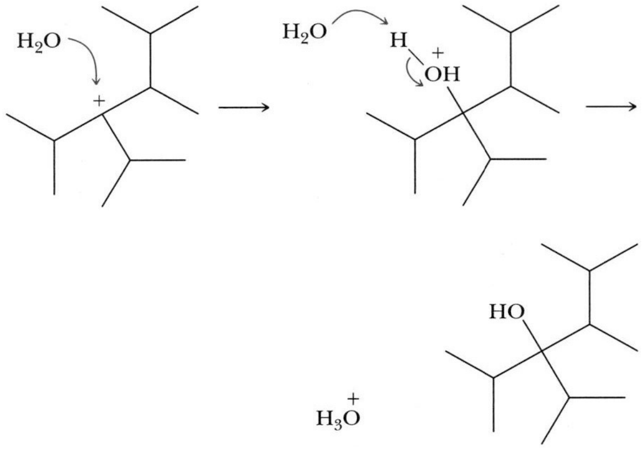

$H-\overset{+}{O}H_2 \longrightarrow$ H_2O $\longrightarrow$

H_2O $\longrightarrow$ H_2O H $\overset{+}{O}H$ $\longrightarrow$

HO $H_3\overset{+}{O}$

5.33

(a)

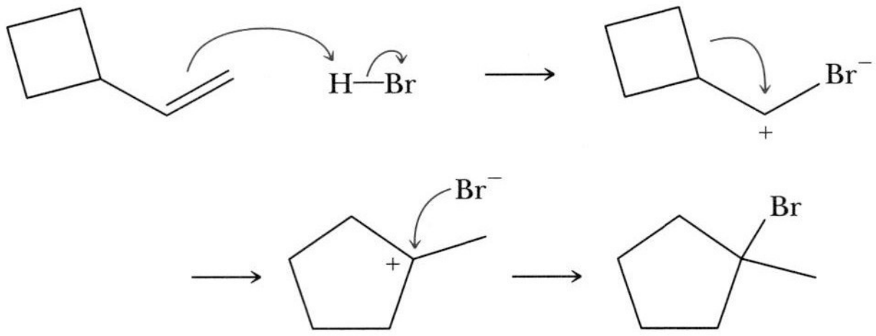

$H-Br$ $\longrightarrow$ Br^-

Br^- $\longrightarrow$ $\longrightarrow$ Br

(b)

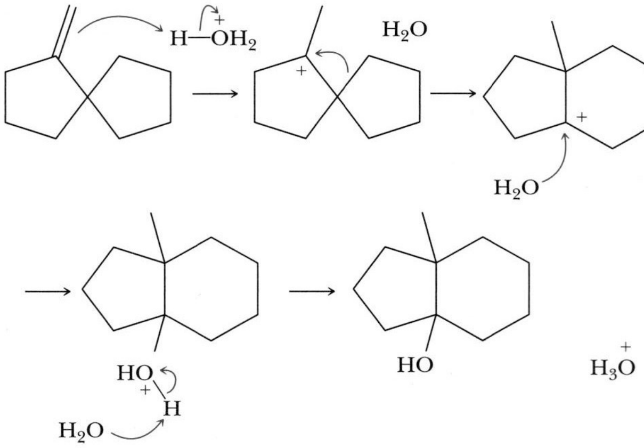

$H-\overset{+}{O}H_2$ $\longrightarrow$ H_2O $\longrightarrow$ H_2O

$\longrightarrow$ $H\overset{+}{O}$ H H_2O $\longrightarrow$ HO $H_3\overset{+}{O}$

5.35

$$CH_3-\underset{}{\overset{CH_3}{\overset{|}{C}}}=CH_2 \;\; I-Cl \longrightarrow CH_3-\overset{CH_3}{\overset{|}{\underset{+}{C}}}-CH_2I \;\; Cl^- \longrightarrow CH_3-\overset{CH_3}{\overset{|}{\underset{Cl}{\underset{|}{C}}}}-CH_2I$$

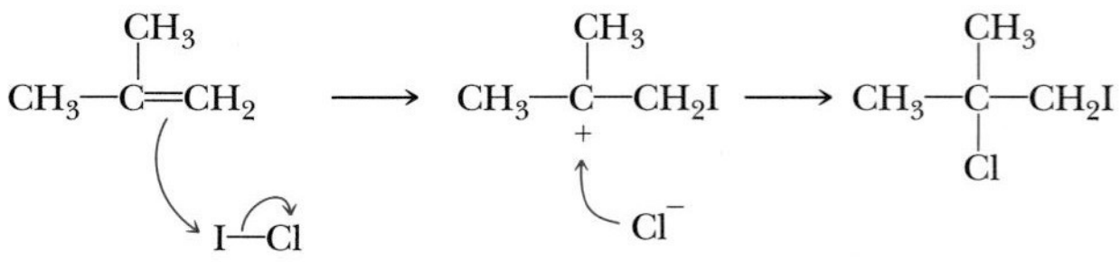

5.37 In the presence of two acids, HBr and acetic acid, the stronger acid (HBr) is more likely to protonate cyclohexene.

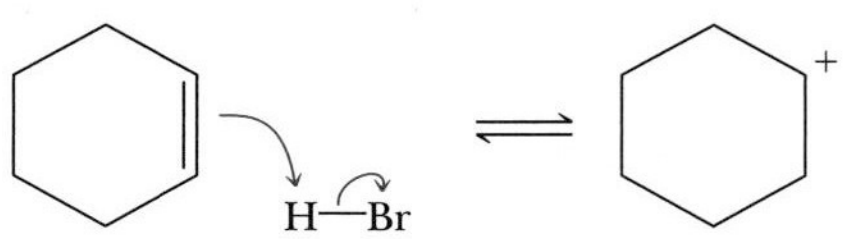

$H-Br$ $\rightleftharpoons$ +

Bromocyclohexane is formed from the nucleophlic attack of the carbocation by Br^-.

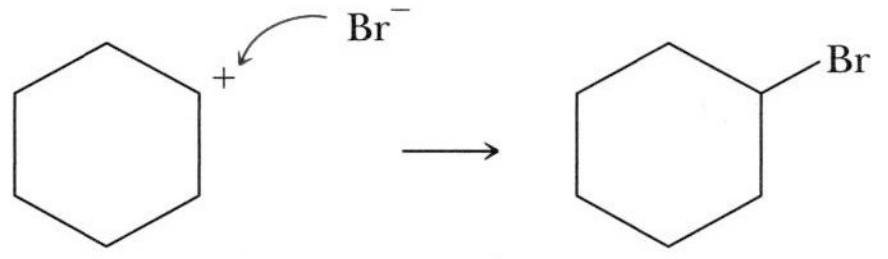

Whereas, nucleophilic attack by acetic acid followed by deprotonation results in the formation of cyclohexyl acetate.

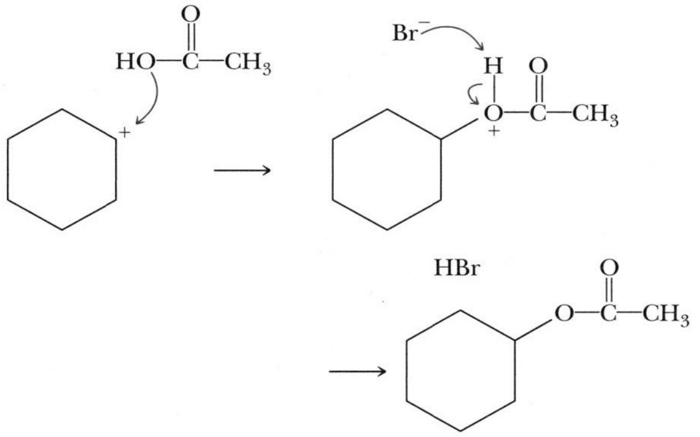

HBr

5.39

$C_8H_{16}O$

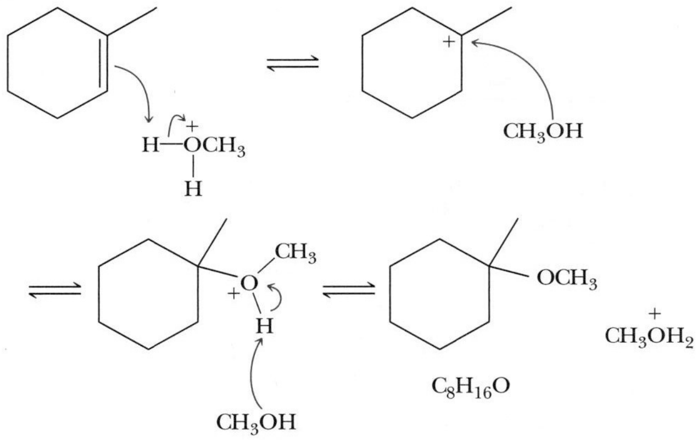

5.41 (a) oxidation (b) neither (c) reduction

5.43 (a)  (b)

(c) 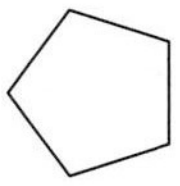 (d)

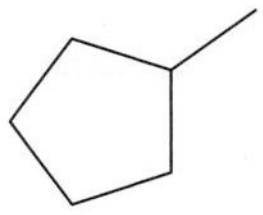

5.45

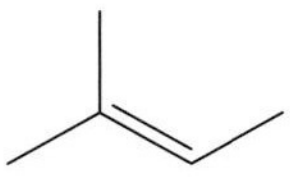

and

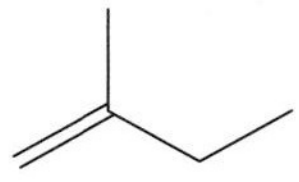

5.47 (a) 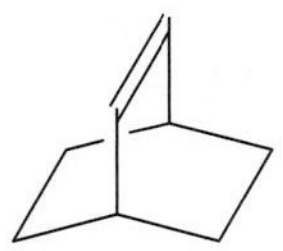 (b) *cis* only

(c)

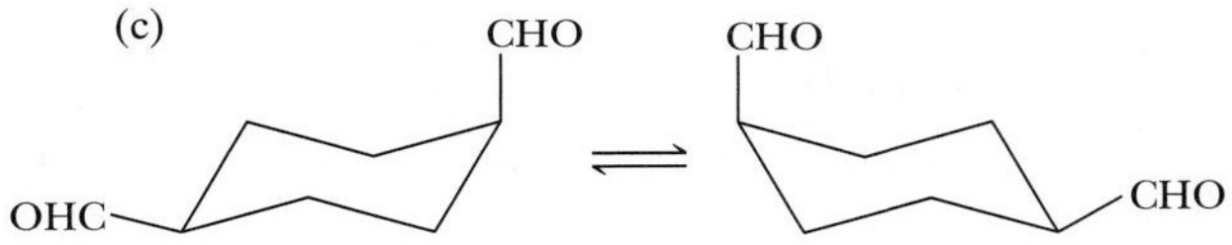

Both chairs are of equal stability

5.49 (a)

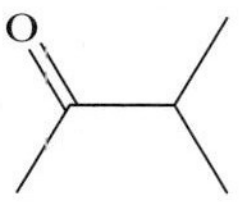

(b)

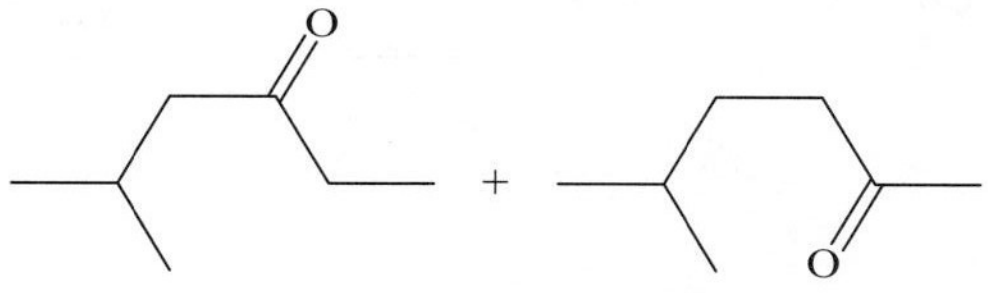

(c)

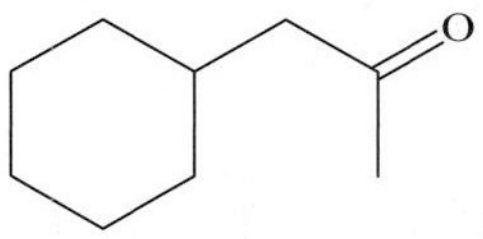

(d)

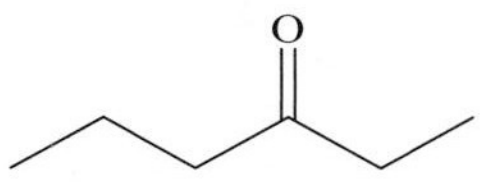

5.51 (a) $H_2C{=}CH_2 \xrightarrow[Ni]{H_2} CH_3CH_3$

(b) $H_2C{=}CH_2 \xrightarrow[H_2SO_4]{H_2O} CH_3CH_2OH$

(c) $H_2C{=}CH_2 \xrightarrow{HBr} CH_3CH_2Br$

(d) $H_2C{=}CH_2 \xrightarrow{Br_2} BrCH_2CH_2Br$

(e) $H_2C{=}CH_2 \xrightarrow{HCl} CH_3CH_2Cl$

5.53

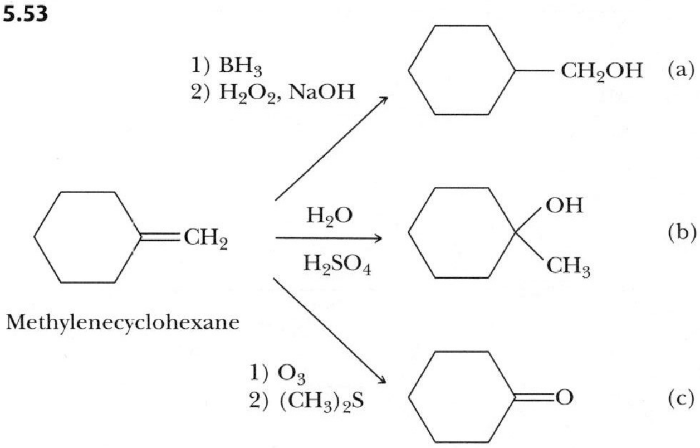

Methylenecyclohexane

1) BH_3
2) H_2O_2, NaOH → (a)

H_2O, H_2SO_4 → (b)

1) O_3
2) $(CH_3)_2S$ → (c)

5.55 (a) 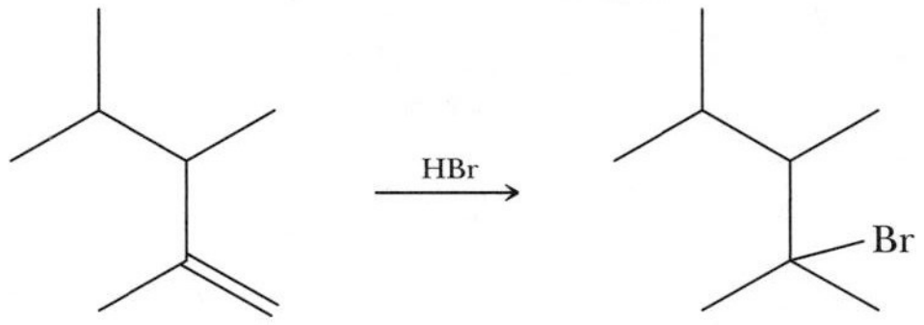 HBr

(b) 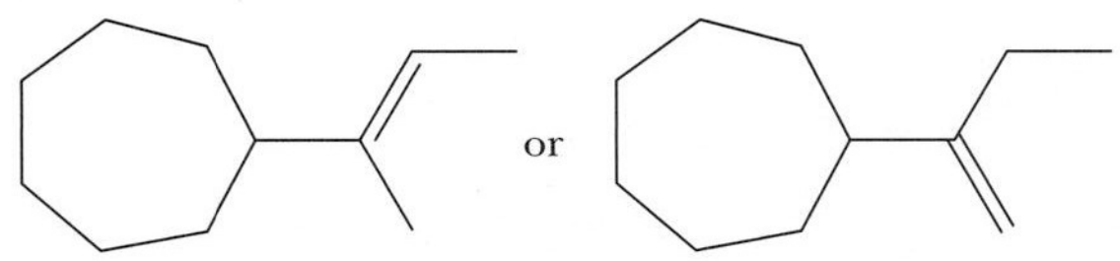 or

$\xrightarrow[H_2SO_4]{H_2O}$

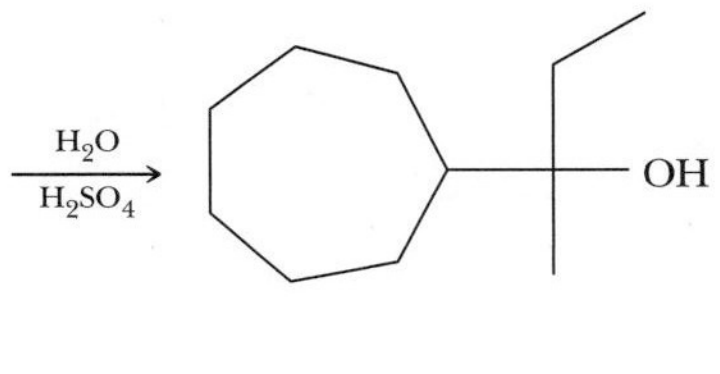

(c)

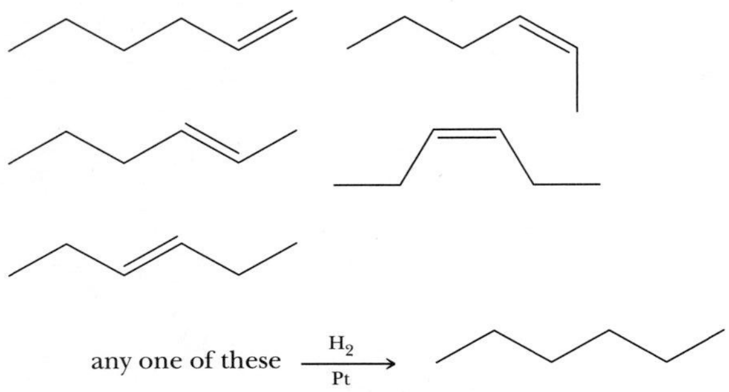

any one of these $\xrightarrow[Pt]{H_2}$

(d) $\xrightarrow{Br_2}$ Br Br

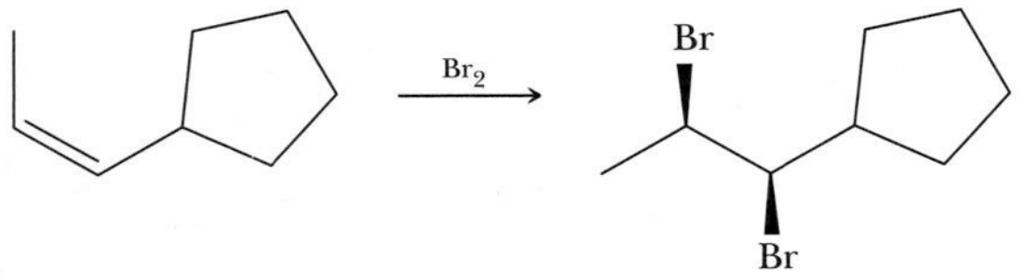

5.57 (a) $\xrightarrow[2)\ CH_3Br]{1)\ NaNH_2}$

$\xrightarrow[Pt]{H_2}$

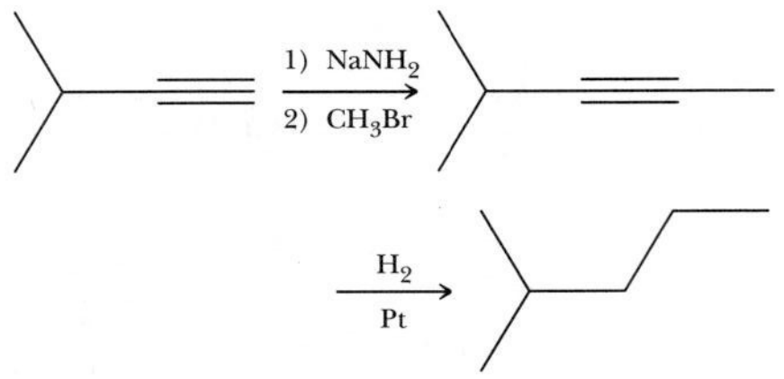

(b)

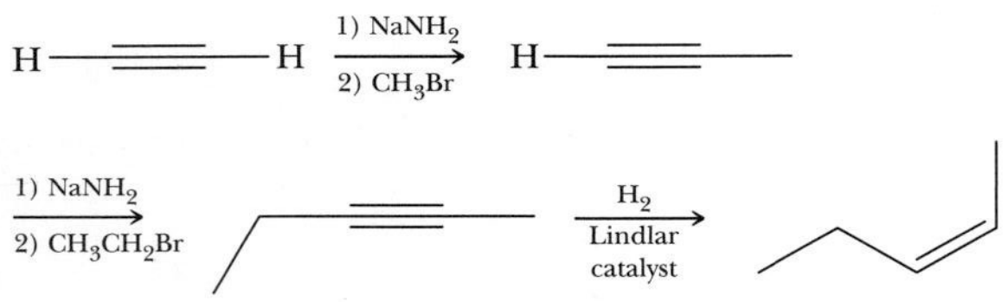

H—≡—H $\xrightarrow[2)\ CH_3Br]{1)\ NaNH_2}$ H—≡—

$\xrightarrow[2)\ CH_3CH_2Br]{1)\ NaNH_2}$ $\xrightarrow[\text{Lindlar catalyst}]{H_2}$

(c)

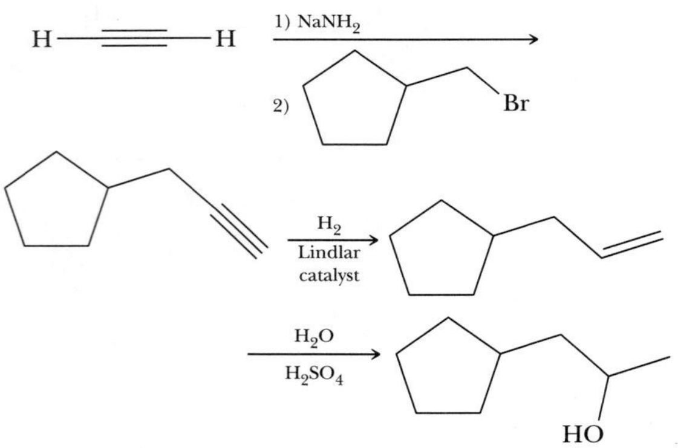

H—≡—H $\xrightarrow{1)\ NaNH_2}$ 2) Br

$\xrightarrow[\text{Lindlar catalyst}]{H_2}$

$\xrightarrow[H_2SO_4]{H_2O}$ HO

(d)

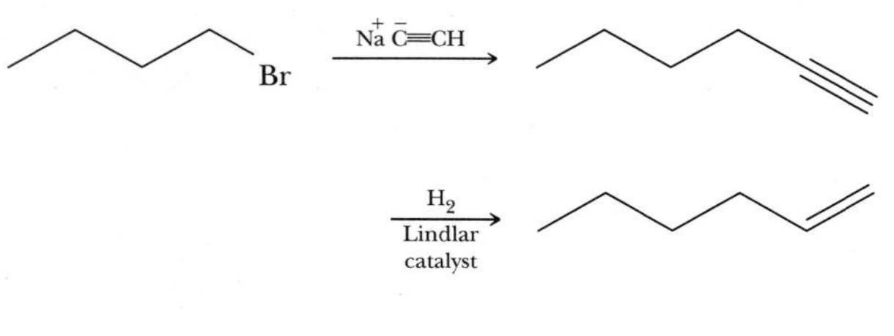

Br $\xrightarrow{\overset{+}{Na}\ \overset{-}{C}{\equiv}CH}$

$\xrightarrow[\text{Lindlar catalyst}]{H_2}$

(e)

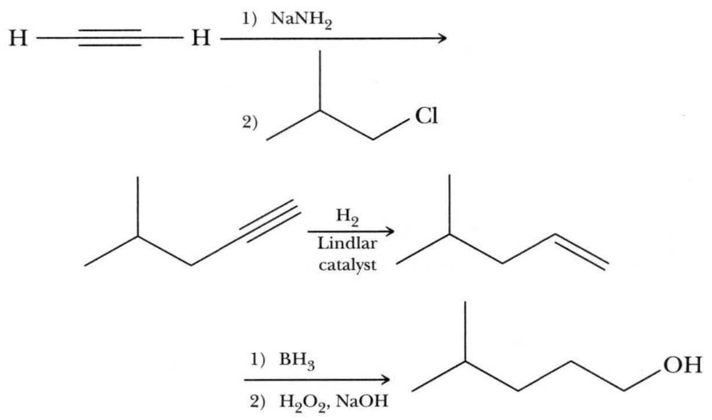

H—≡—H $\xrightarrow{1)\ NaNH_2}$ 2) Cl

$\xrightarrow[\text{Lindlar catalyst}]{H_2}$

$\xrightarrow[2)\ H_2O_2,\ NaOH]{1)\ BH_3}$ OH

(f)

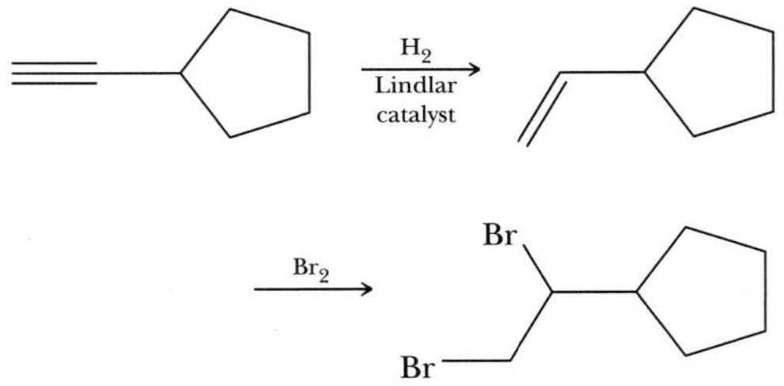

$\xrightarrow[\text{Lindlar catalyst}]{H_2}$

$\xrightarrow{Br_2}$ Br Br

(g)

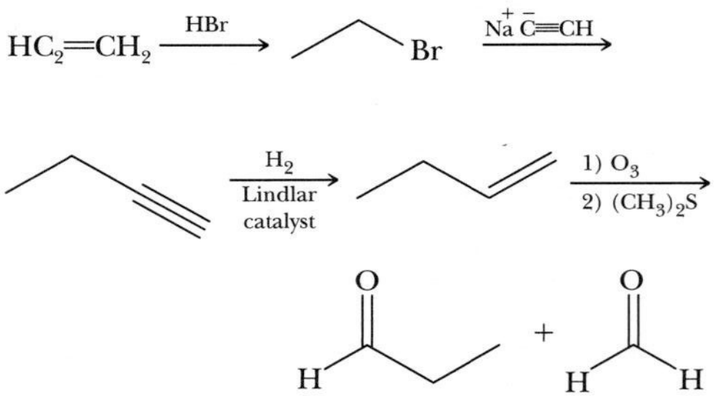

$HC_2{=}CH_2$ $\xrightarrow{HBr}$ Br $\xrightarrow{\overset{+}{Na}\ \overset{-}{C}{\equiv}CH}$

$\xrightarrow[\text{Lindlar catalyst}]{H_2}$ $\xrightarrow[2)\ (CH_3)_2S]{1)\ O_3}$

O H + O H H

(h)

H—≡—H $\xrightarrow[2)\ CH_3CH_2Br]{1)\ NaNH_2}$ —≡—H

$\xrightarrow[2)\ CH_3CH_2Br]{1)\ NaNH_2}$ $\xrightarrow[2)\ (CH_3)_2S]{1)\ O_3}$ O H

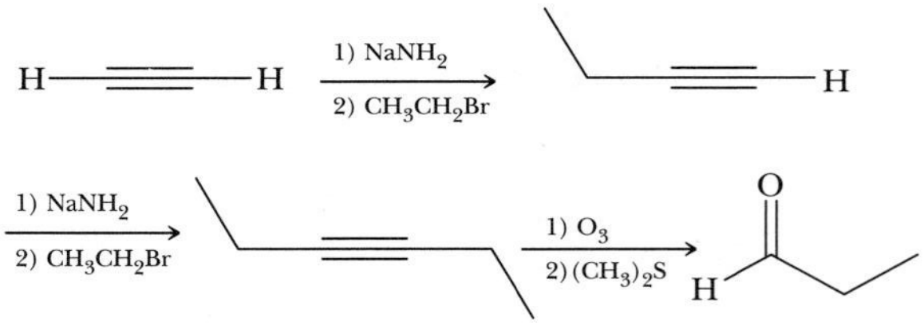

5.59 (a) same (b) different (c) different

Chapter 6 Chirality

6.1 (a) HO H * H OH *

(b) HO H * H OH *

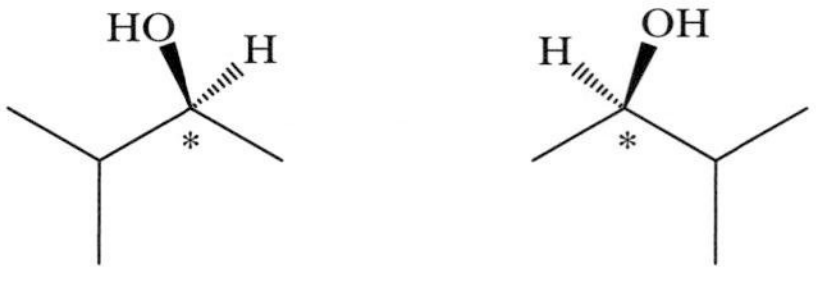

(c)

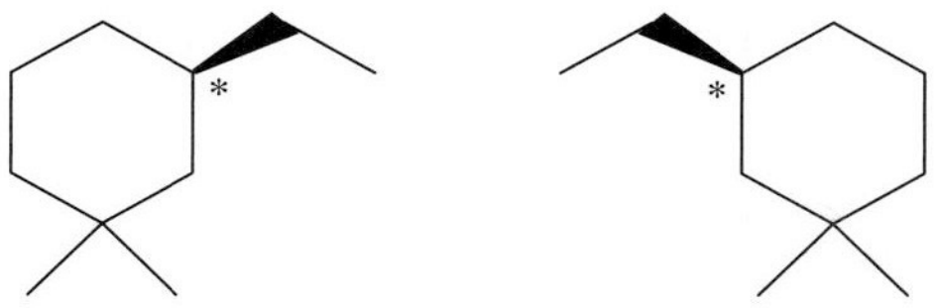

6.2 (a) S (b) S (b) R

6.3 (a) (1) and (3), and (2) and (4).
(b) (1) and (2), (1) and (4), (2) and (3), and (3) and (4).

6.4 (a) (2) and (3) (b) (1) and (4) (c) (2) and (3)

6.5 3

6.6 2

6.7 Stereoisomers are compounds that have the same molecular formula and the same connectivity but have different, noninterconverting orientations of their atoms or groups in three-dimensional space.

Four types of stereoisomers include:

- Enantiomers
- Diastereomers
- *cis-trans* isomers
- meso isomers

6.9 *Conformation* refers to the different, interconverting arrangements of atoms, in three-dimensional space, that are the result of rotation about single bonds. *Configuration* also refers to the different spatial arrangement of atoms, but these spatial arrangements are noninterconverting and cannot be made the same by rotation about single bonds.

6.11 A spiral with a left-handed twist will have a left-handed twist when viewed from either end. For the same reason, viewing a chiral molecule from one direction and in another direction does not mean that you are viewing its mirror image—it is still the same molecule.

6.13 This of course depends on the manufacturer of the pasta! However, in any given box of spiral pasta, usually every piece of pasta is of the same twist because the entire box was manufactured using the same machine. So, the next time you are having spiral pasta, you will most likely be consuming something that is enantiomerically pure!

6.15 All except for 3-chloropropane

6.17

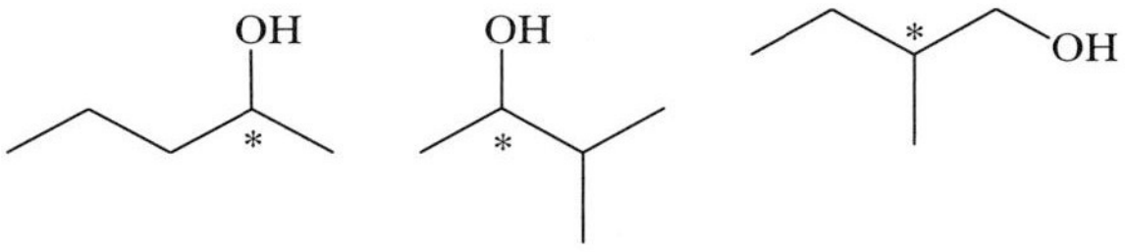

OH OH OH

6.19 (a) OH HOOC CH3 H (b) CHO HO H CH2OH

(c) COOH H NH2 CH3 (d) O OH

(e) OH (f) OH NH2

(g) O COOH (h) CH3 HO H HO H CH3

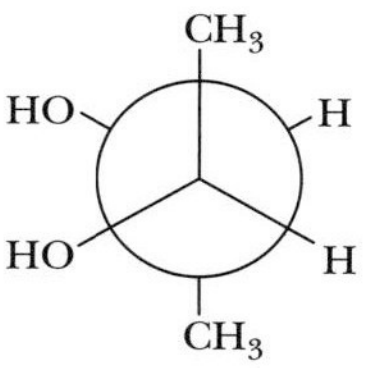

(i) Br OH Cl (j) HO

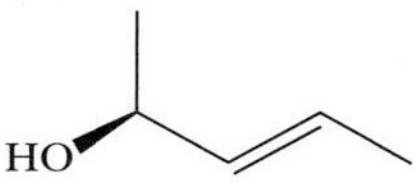

(k) Cl (l) CH2CH3 CH3

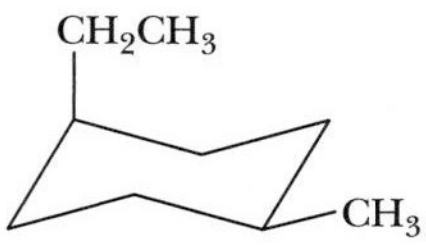

6.21 (b) OH HO (c) OH

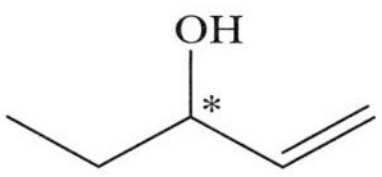

6.23 Structures (b), (c), (d), and (f) are identical to (a). Structures (e), (g), and (h) are mirror images of (a).

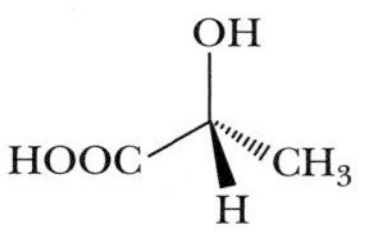

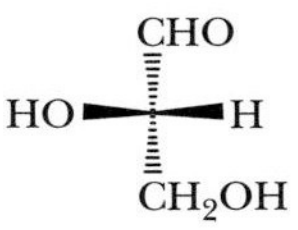

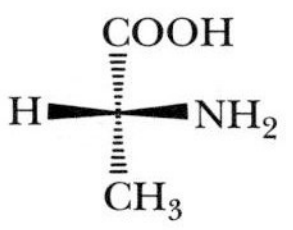

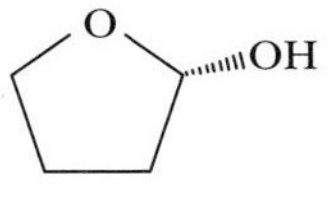

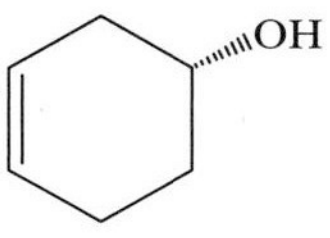

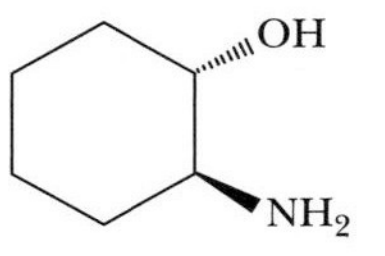

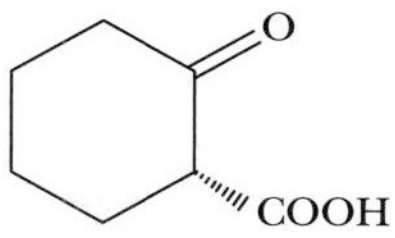

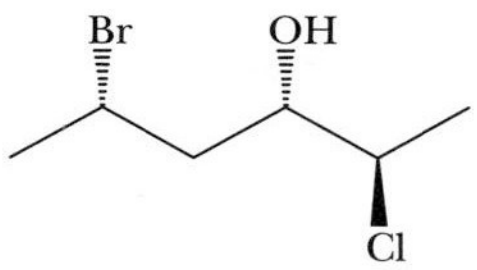

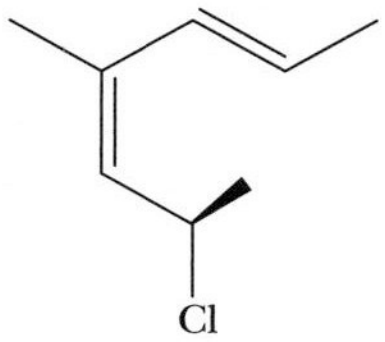

6.25 (c) and (d)

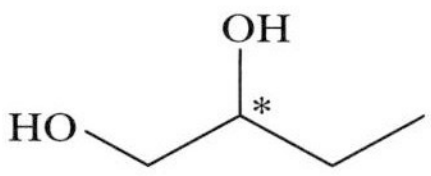

6.27 (a) (*S*)-2-butanol (b) CH3 HO H H H CH3

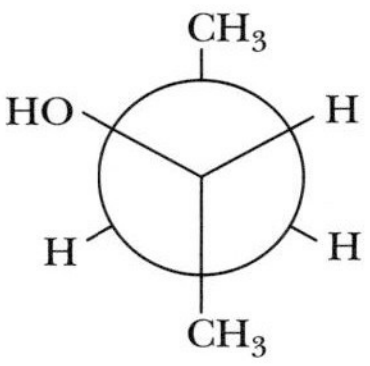

(c)

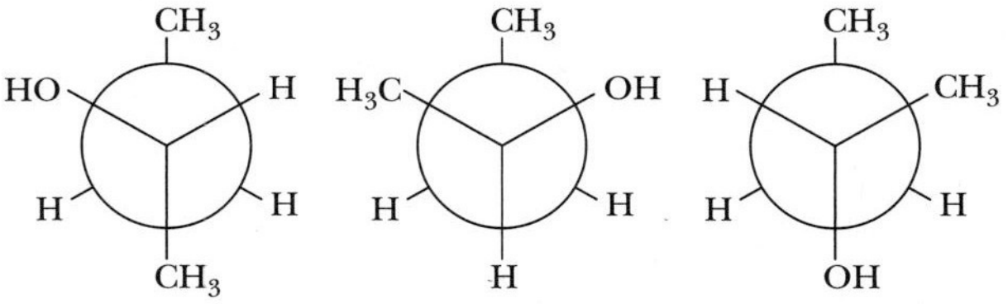

CH3 HO H H H CH3 — more stable

CH3 H3C OH H H H

CH3 H CH3 H H OH — more stable

6.29

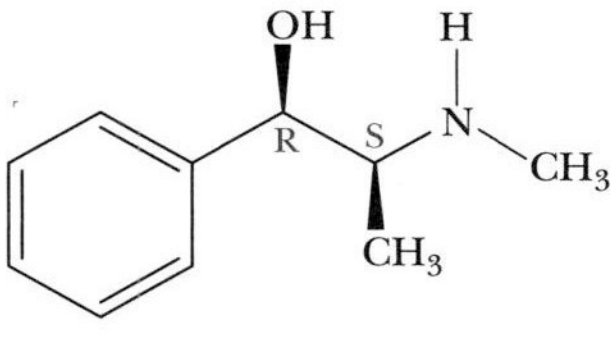

OH H N R S CH3 CH3

6.31

(a) $CH_3\overset{*}{C}H\overset{*}{C}HCOOH$ (HO, OH)

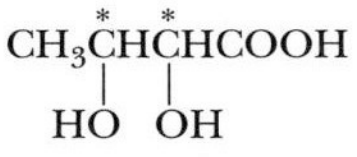

four stereoisomers

(b) $CH_2—COOH$ / $*CH—COOH$ / $HO—\overset{*}{C}H—COOH$

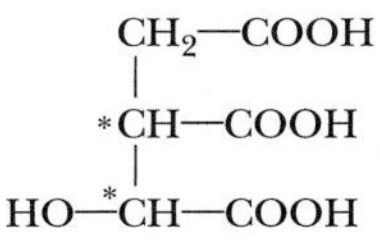

four stereoisomers

(c) OH

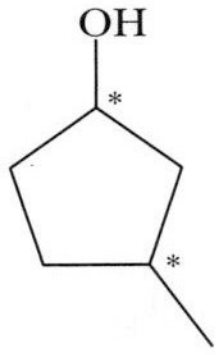

four stereoisomers

(d)

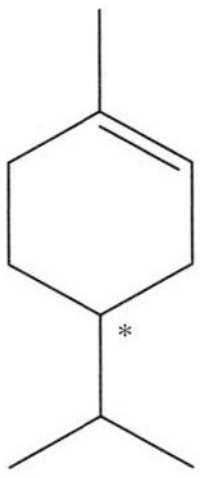

two stereoisomers

(e) OH

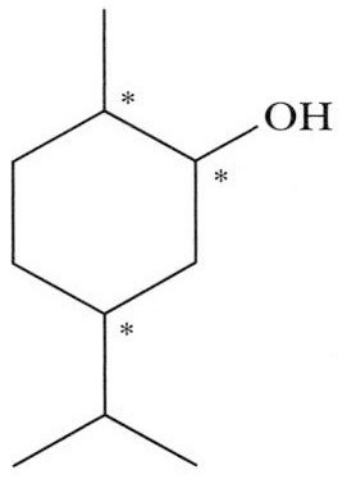

eight stereoisomers

(f) O COOH

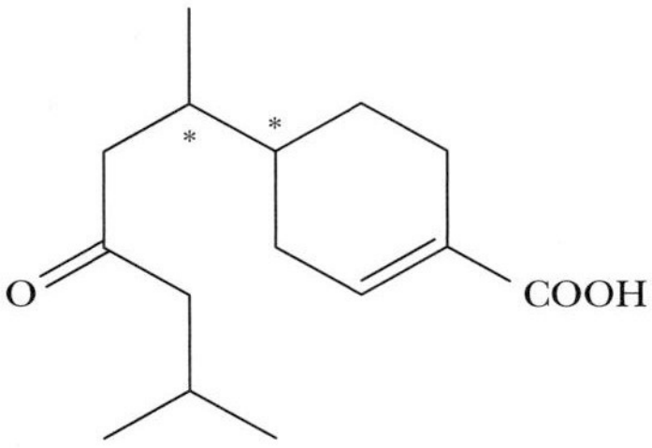

four stereoisomers

(g) O OH OH

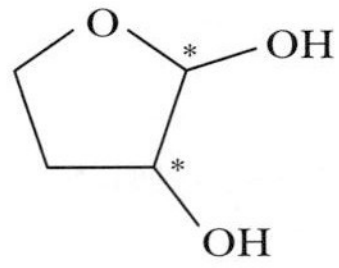

four stereoisomers

(h) O

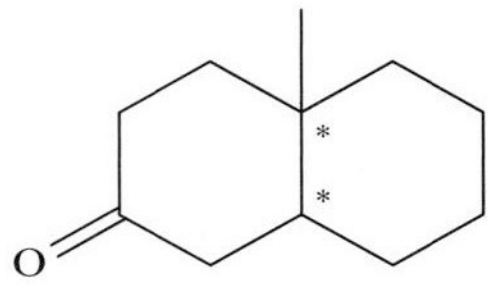

four stereoisomers

6.33 (a) Loratadine has no stereocenters and is achiral. Not all large molecules have stereocenters!

(b) COOH N OH HO

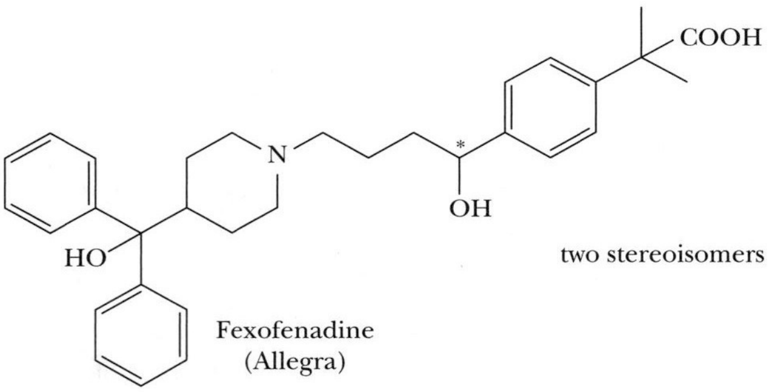

two stereoisomers

Fexofenadine (Allegra)

6.35 (a) HO O HO O O F O

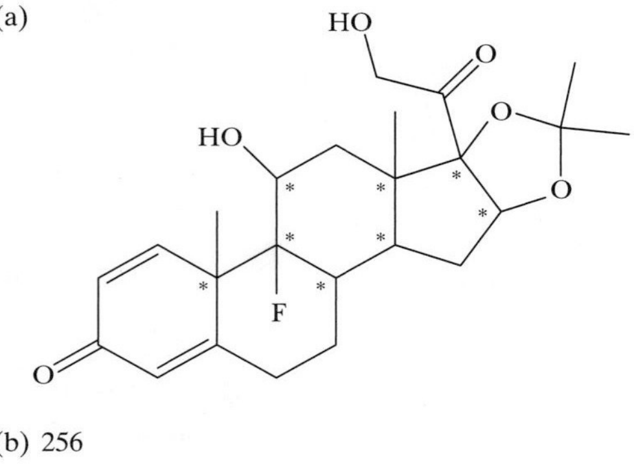

(b) 256

6.37

COOH HO H H OH COOH

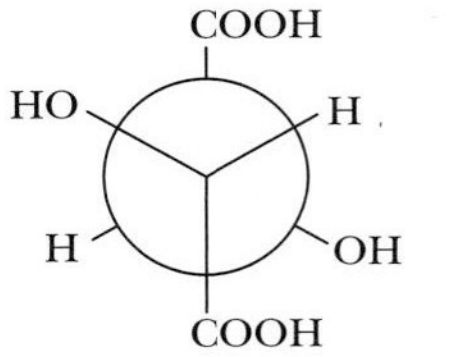

most stable conformation

COOH COOH H H OH OH

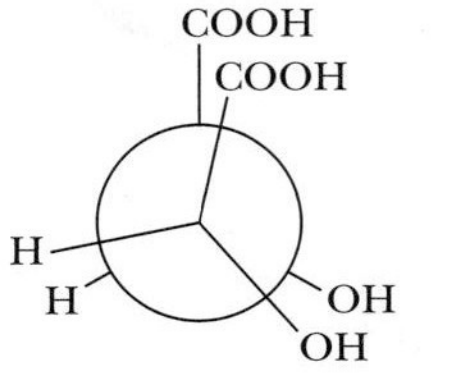

least stable conformation

6.39 (a) HOH_2C O HO HO OH OH

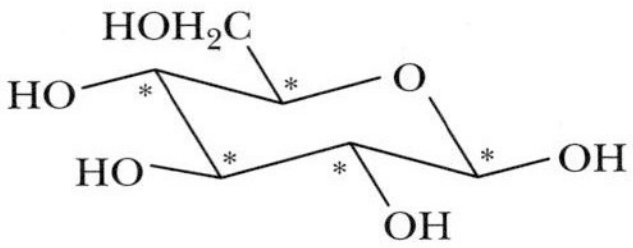

(b) 32

(c) 16

(d) Both 1 and 5 are R

6.41

(a)

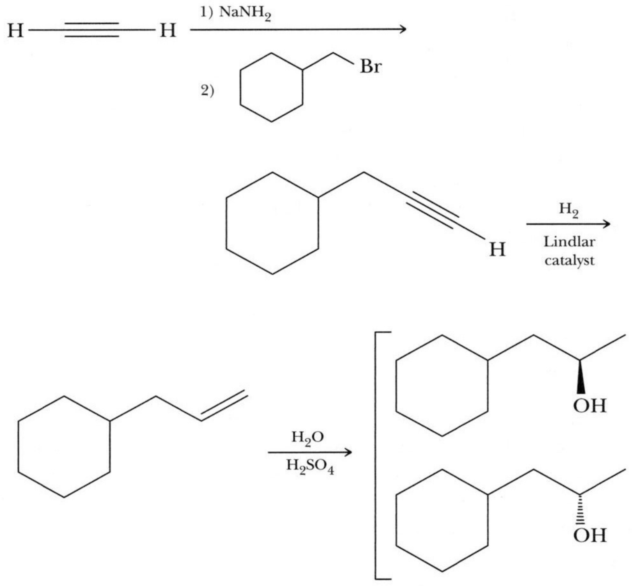

H—≡—H $\xrightarrow{\text{1) } NaNH_2}$ 2) Br

H $\xrightarrow[\text{Lindlar catalyst}]{H_2}$

$\xrightarrow[H_2SO_4]{H_2O}$ OH OH

(b)

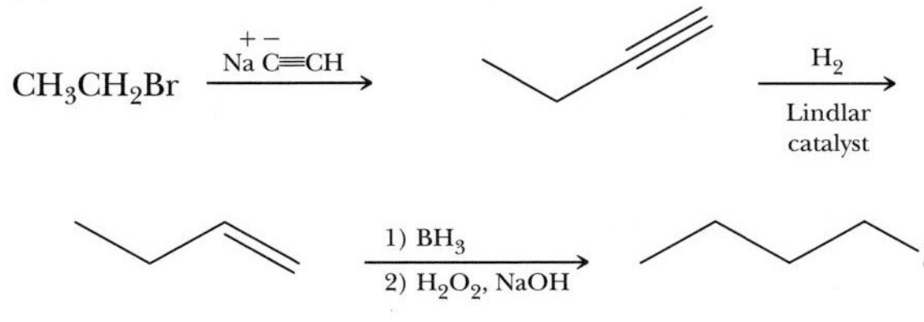

CH_3CH_2Br $\xrightarrow{\overset{+\ -}{Na\ C{\equiv}CH}}$ $\xrightarrow[\text{Lindlar catalyst}]{H_2}$

$\xrightarrow[\text{2) } H_2O_2,\ NaOH]{\text{1) } BH_3}$ OH

(c)

H—≡—H $\xrightarrow[\text{2) } CH_3CH_2Br]{\text{1) } NaNH_2}$ $\xrightarrow[\text{Lindlar catalyst}]{H_2}$

$\xrightarrow{HCl}$ Cl + Cl

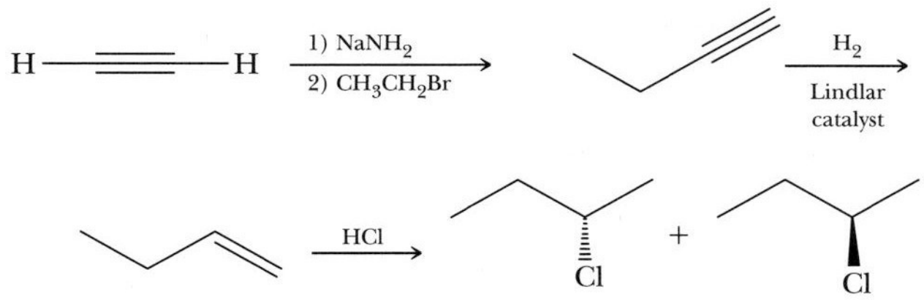

(d)

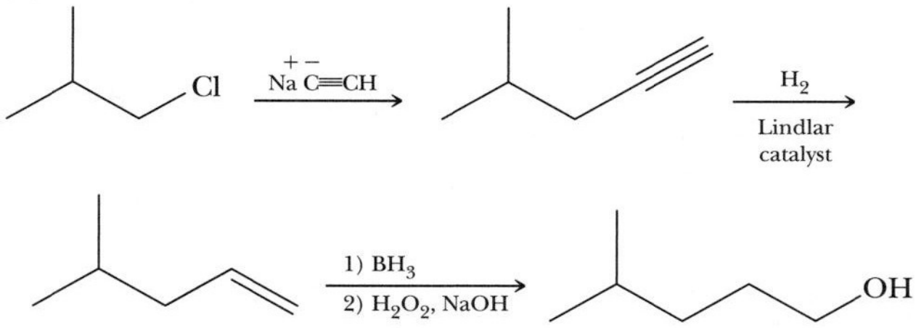

1) BH_3 2) H_2O_2, NaOH

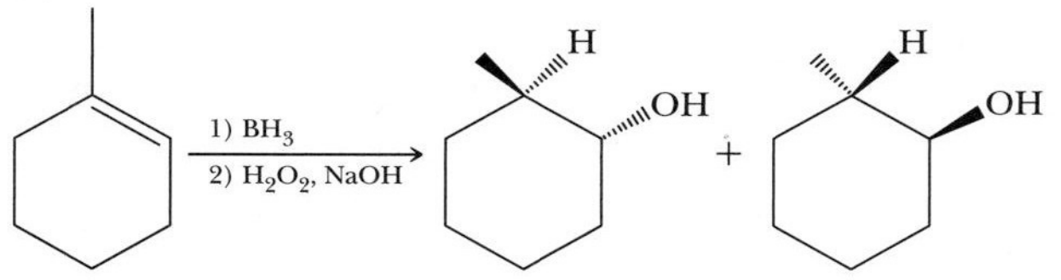

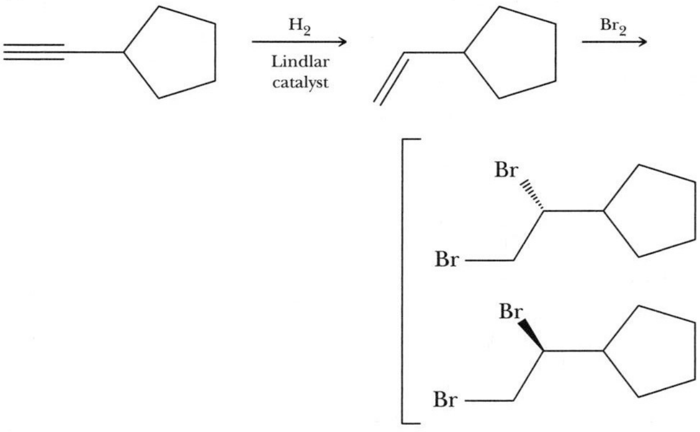

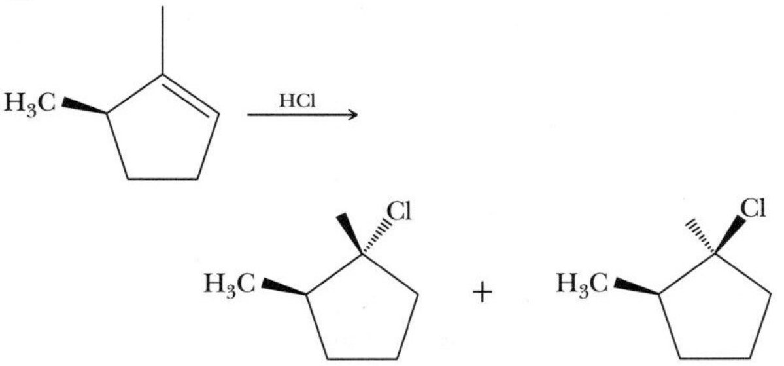

(e)

1) BH_3 2) H_2O_2, NaOH

(f)

H_2 Lindlar catalyst; Br_2

(g)

HCl

(h)

$H_2C{=}CH_2$ —HBr→ ; $\overset{+}{Na}\ \overset{-}{C}{\equiv}CH$; 1) $NaNH_2$ 2) $(CH_3)_2CHBr$; Li, NH_3(l); Cl_2

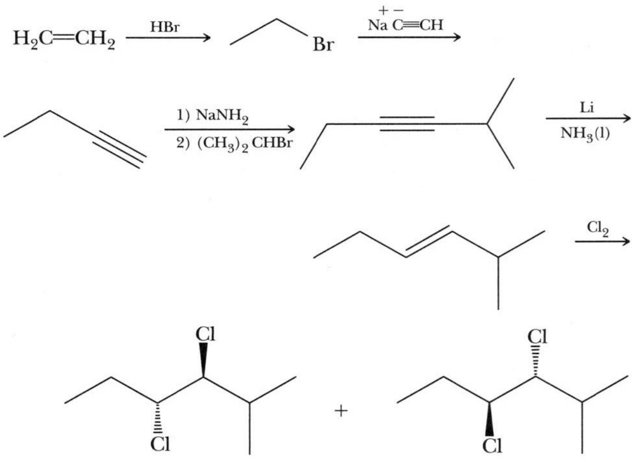

6.43 (a)

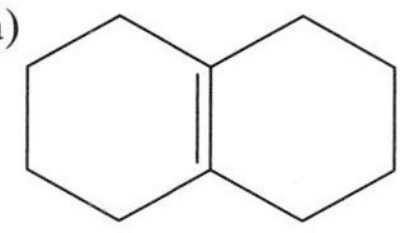

(b) 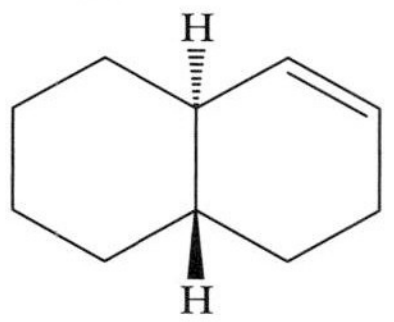 or

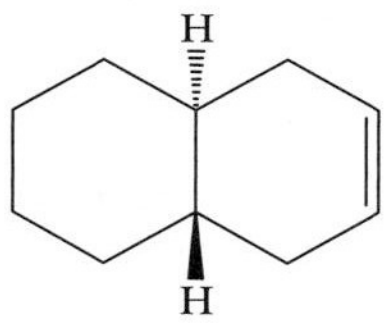

6.45

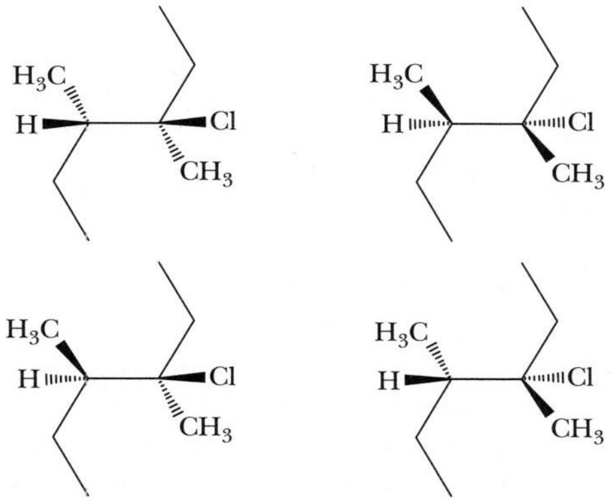

The reaction produces four different stereoisomers. If the goal of the synthesis is to selectively synthesize one of the four stereoisomers, this method is not very useful.

Putting It Together

1. (d) **2.** (a) **3.** (b) **4.** (c) **5.** (a)
6. (c) **7.** (c) **8.** (d) **9.** (e) **10.** (d)

11.

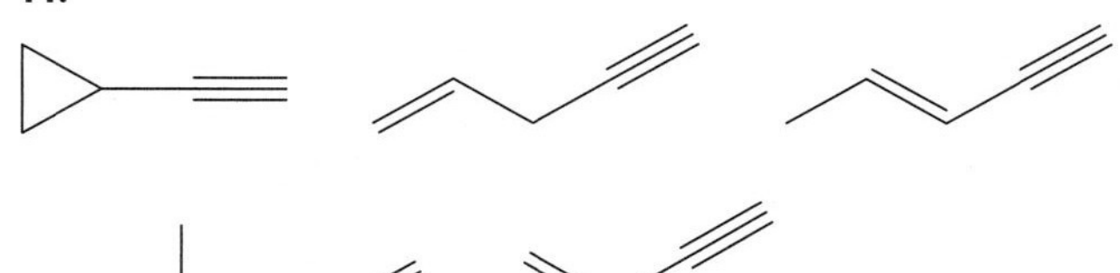

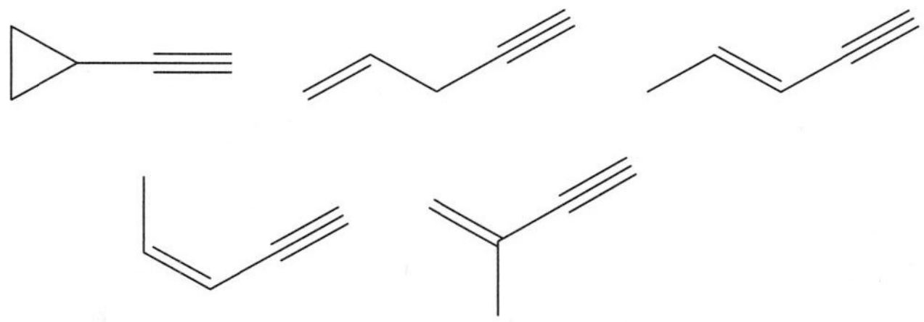

12. (a) 8 (b) no (c) yes
(d) amids are not reduced under these conditions

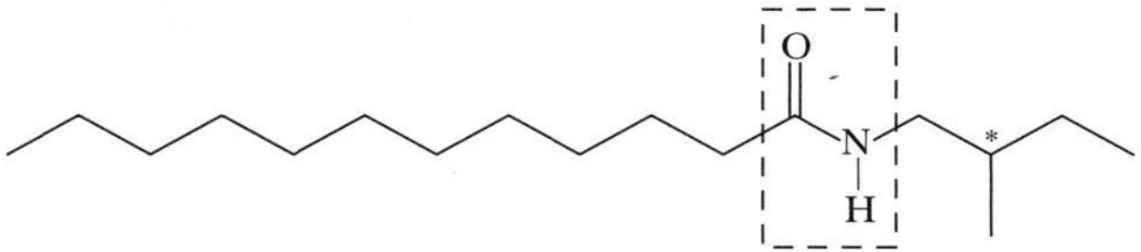

13. (a) 4-propylcyclohexene
(b) (*Z*)-3-ethyl-1-fluoro-3-hexene
(c) (*S*)-4-butyl-7-methyl-2-octyne

14.

A B C

15.

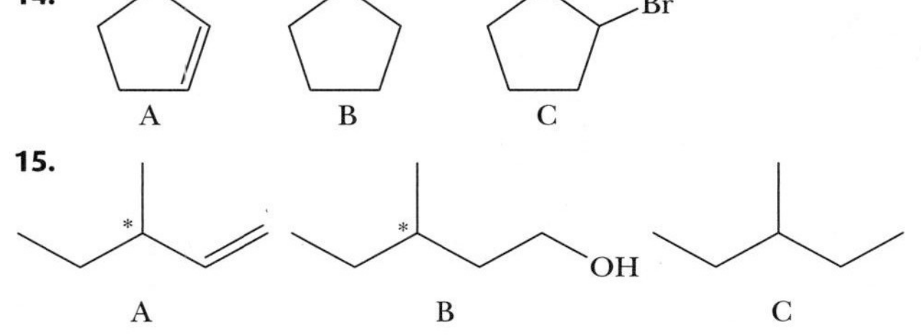

16.

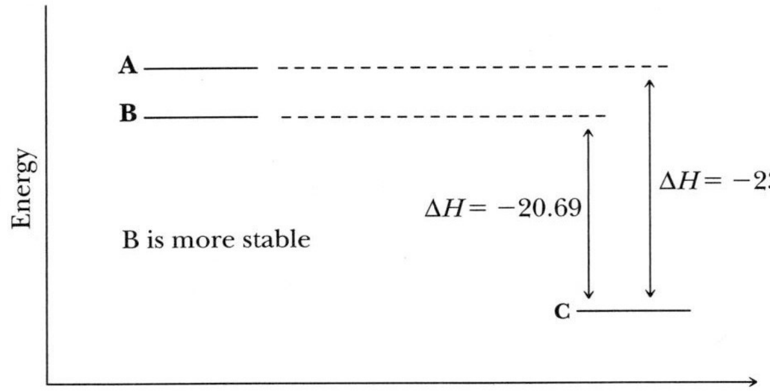

17. (a) The two compounds are enantiomers and have the same physical properties. It is necessary to use a chiral agent, such as chiral substrate chromatography, to separate them.

(b) They are identical.

(c) The two compounds are *cis-trans* isomers and have different properties. For instance, they can be separated by distillation due to their different boiling points.

18. (a) The two structures are identical (both are *S*), so the solution would be enantiomerically pure and will rotate PPL.

(b) The compound on the left is meso and does not rotate PPL. The compound on the right is chiral and does rotate PPL. Thus, the solution does rotate PPL.

(c) The compounds are diastereomers, which do not have equal and opposite optical activites. Overall, the solution would rotate PPL.

19. (a)

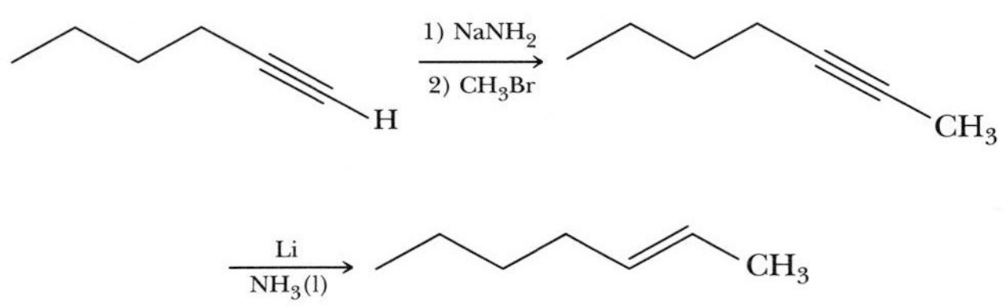

(b)

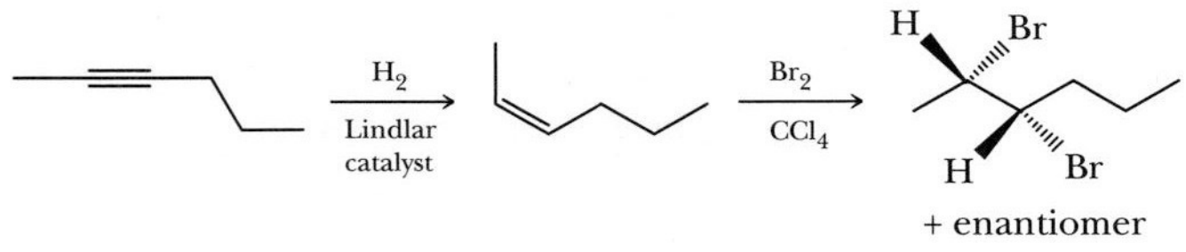

(c)

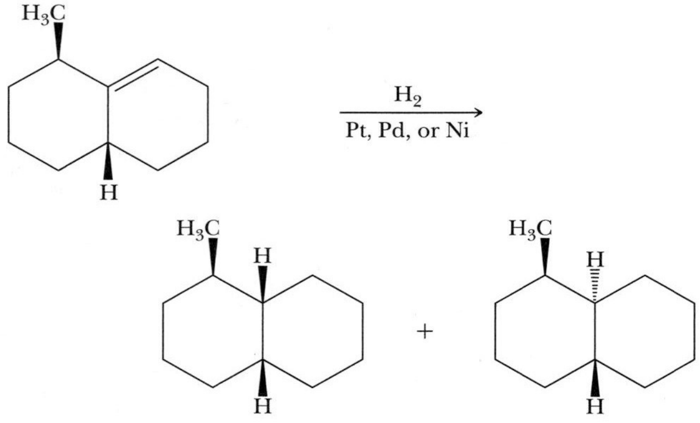

20.

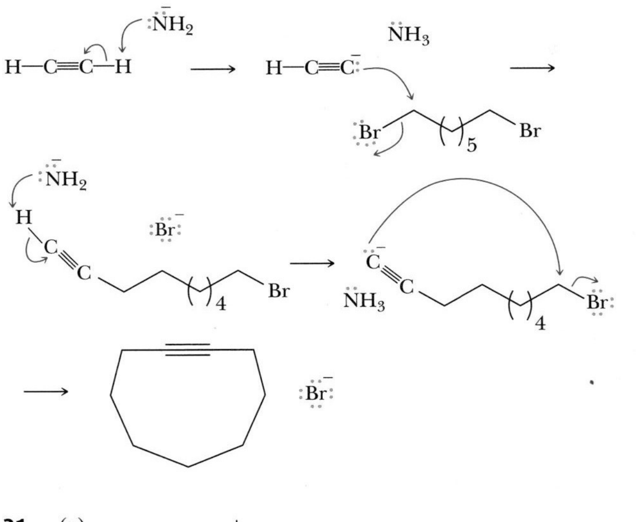

21. (a)

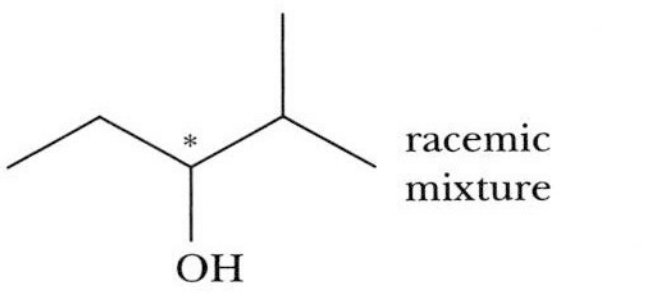

(b)

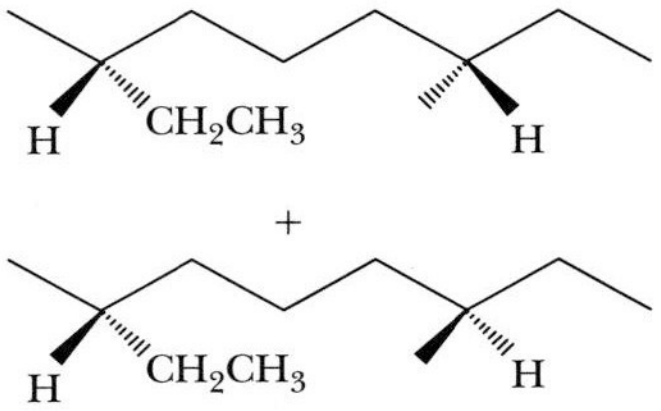

(c)

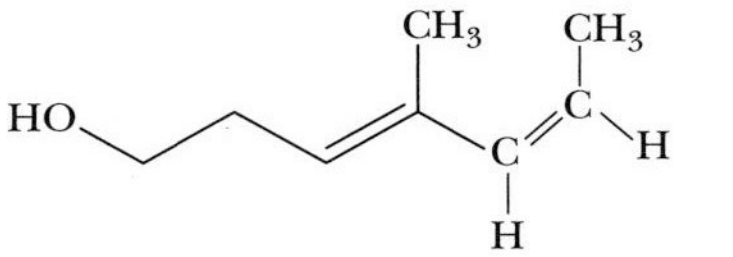

(d)

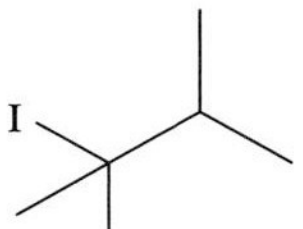

Note: a 1,2-methyl shift has occurred

(e)

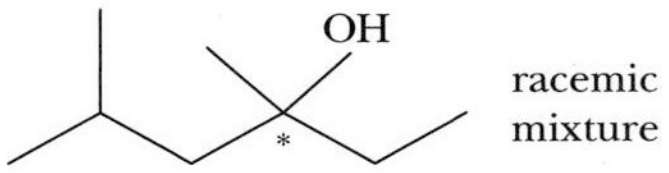

(f)

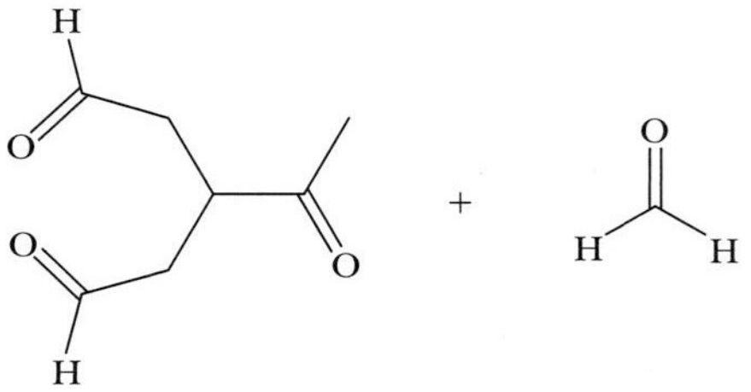

22.

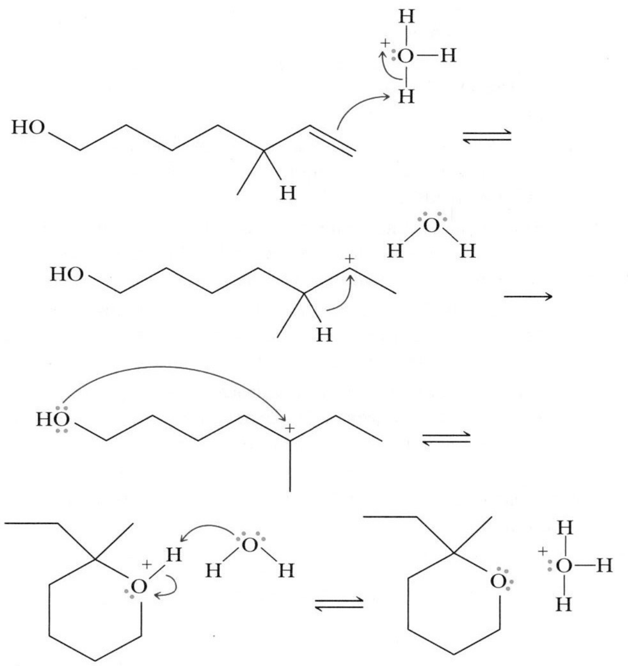

HO ... ⇌ ... → ... ⇌ ...

(b)

$CH_3CH(Cl)CH_2CH_3 + HCOOH \xrightarrow[\text{formic acid}]{S_N1} CH_3CH(OCHO)CH_2CH_3 + HCl$

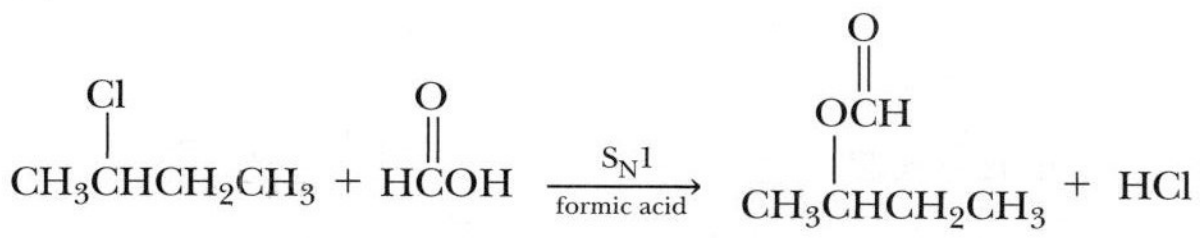

7.6 (a) minor + major

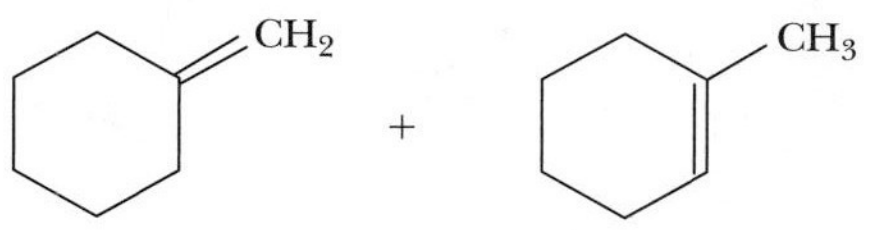

(b)

(c) + equal amounts

7.7 (a) + $CH_3O^-Na^+$ $\xrightarrow[\text{methanol}]{E2}$

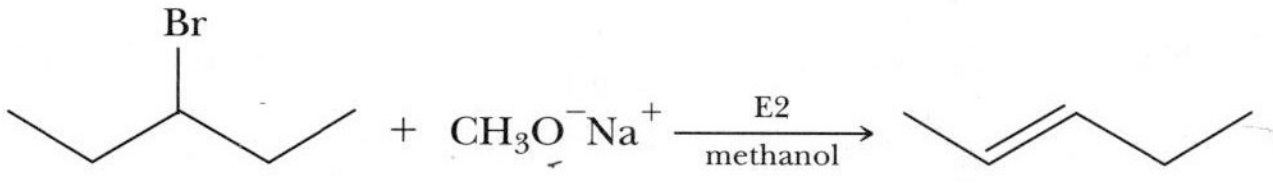

(b) + $Na^+\,OH^-$ $\xrightarrow[\text{acetone}]{E2}$

7.8 (a) + $CH_3O^-Na^+$ $\xrightarrow[\text{methanol}]{E2/S_N2}$ + (OCH_3)

(b) + $Na^+\,OH^-$ $\xrightarrow[\text{acetone}]{S_N2}$ (OH)

Chapter 7 Haloalkanes

7.1 (a) 1-chloro-3-methyl-2-butene
(b) 1-bromo-1-methylcyclohexane
(c) 1,2-dichloropropane
(d) 2-chloro-1,3-butadiene

7.2 (a) substitution
(b) substitution and elimination

7.3 (a) cyclopentyl–Br + $CH_3CH_2S^-Na^+$ ⟶ cyclopentyl–SCH_2CH_3 + Na^+Br^-

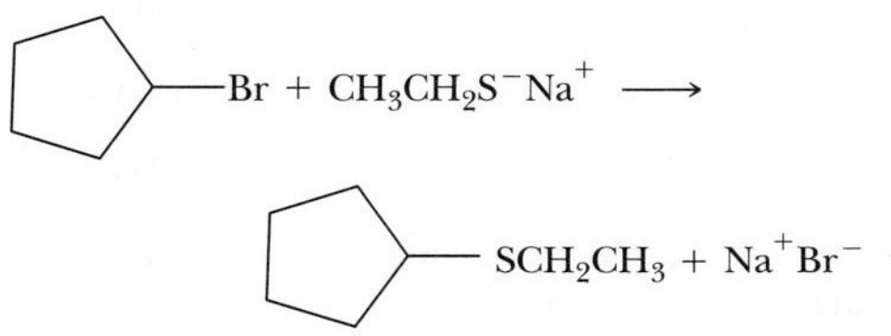

(b) cyclopentyl–Br + $CH_3CO^-Na^+$ (C=O) ⟶ cyclopentyl–$OCCH_3$ (C=O) + $Na^+\,Br^-$

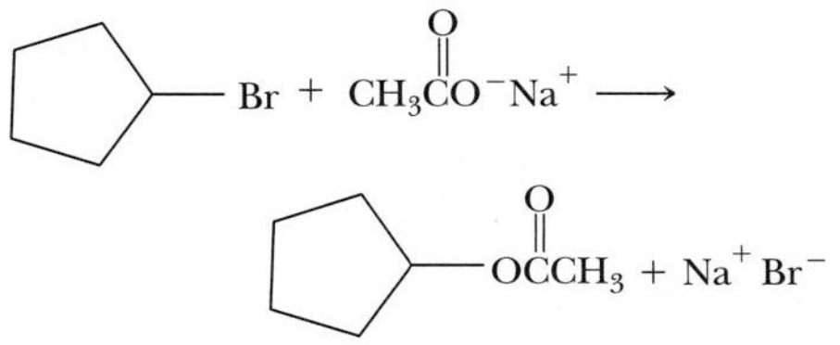

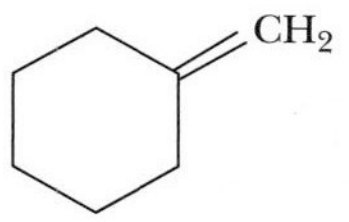

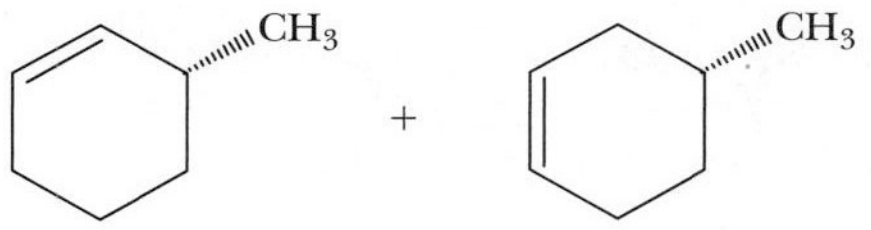

7.4 (a) S_N2
(b) Compound A more reactive by S_N2. Both A and B equally reactive by S_N1.

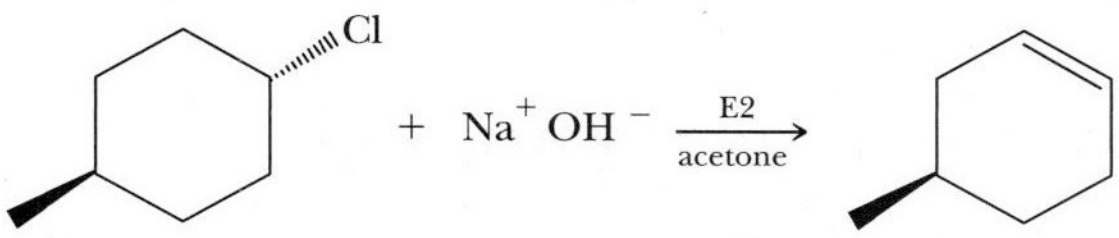

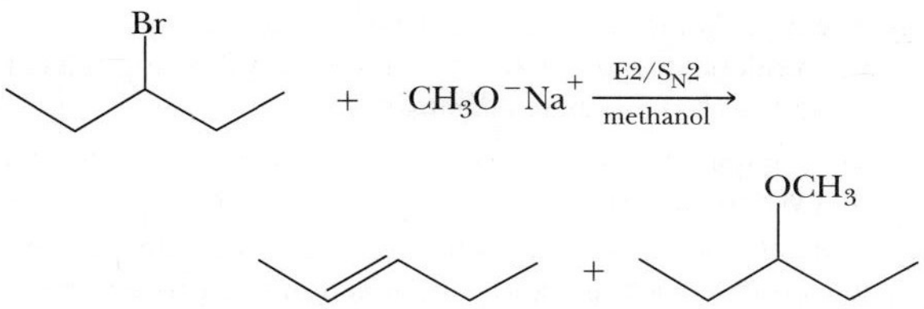

7.5 (a) (Br) + $Na^+\,SH^-$ $\xrightarrow[\text{acetone}]{S_N2}$ Br + Na^+Br^-

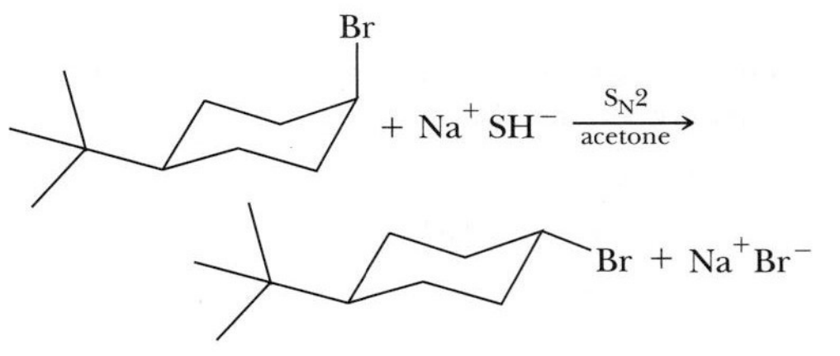

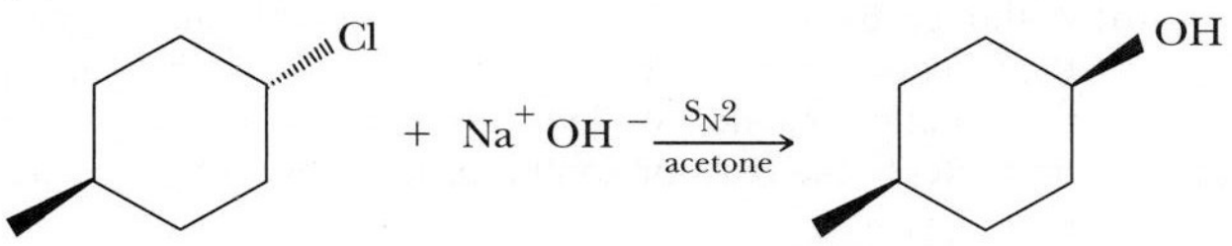

7.9 (a) 1,1-difluoroethene (b) 3-bromocyclopentene
(c) 2-chloro-5-methylhexane (d) 1,6-dichlorohexane
(e) dichlorodifluoromethane (f) 3-bromo-3-ethylpentane

7.11 (a) Br (b) Cl

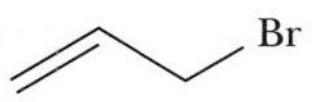

(c) Br Br (d) Br

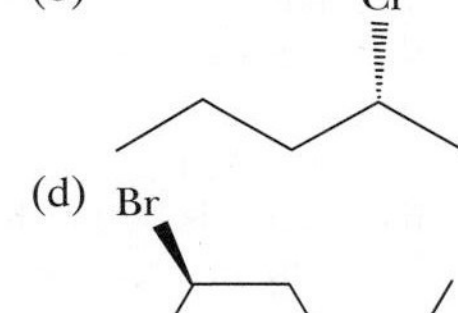

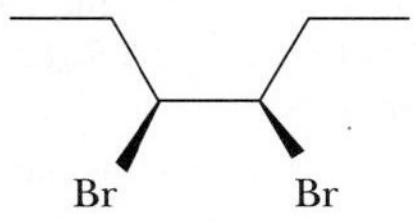

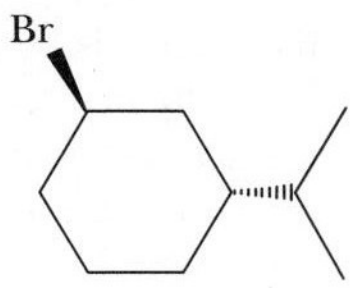

(e) Cl–CH₂CH₂CH₂–Cl   (f) Br–cyclobutane

7.13 (b) and (c) are 2°

7.15 (a) 2-methyl-2-pentene $\xrightarrow{HCl}$ 2-chloro-2-methylpentane

(b) $CH_3CH_2CH{=}CH_2 \xrightarrow{HI} CH_3CH_2CH(I)CH_3$

(c) $CH_3CH{=}CHCH_3 \xrightarrow{HCl} CH_3CH(Cl)CH_2CH_3$

(d) 1-methylcyclopentene $\xrightarrow{HBr}$ 1-bromo-1-methylcyclopentane

7.17 $CH_3CH_2OH < CH_3OH < H_2O$

7.19 (a) OH^- (b) OH^- (c) CH_3S^-

7.21

(a) $Na^+I^- + CH_3CH_2CH_2Cl \xrightarrow{acetone} Na^+Cl^- + CH_3CH_2CH_2I$

(b) NH_3 + bromocyclohexane $\xrightarrow{ethanol}$ cyclohexyl–$NH_3^+Br^-$

(c) $CH_3CH_2O^-Na^+ + CH_2{=}CHCH_2Cl \xrightarrow{ethanol} Na^+Cl^- + CH_2{=}CHCH_2OCH_2CH_3$

7.23 (a) The haloalkane is 2°, acetate is a moderate nucleophile that is a weak base, and ethanol is a moderately ionizing solvent; these favor an S_N2 mechanism.

(b) The haloalkane is 2°, ethyl thiolate is a good nucleophile that is a weak base, and acetone is a weakly ionizing solvent.

(c) The haloalkane is 1°, and when combined with an excellent nucleophile such as iodide, the reaction will proceed by an S_N2 mechanism.

(d) Although the nucleophile is only a moderate nucleophile, the reaction can only proceed by S_N2 because the haloalkane is a methyl halide. Methyl carbocations are extremely unstable, so the reaction cannot proceed by an S_N1 mechanism.

(e) The haloalkane is 1°, and when combined with a good nucleophile such as methoxide, the reaction will proceed by an S_N2 mechanism. Although methoxide is also a good base, 1° haloalkanes will favour substitution over elimination.

(f) The haloalkane is 2°, methyl thiolate is a good nucleophile that is a weak base, and ethanol is a moderately ionizing solvent.

(g) Similar to (d), the amine is a moderate nucleophile. The haloalkane is 1°, which will not react by an S_N1 mechanism. Thus, the favored mechanism is S_N2.

(h) As with (d) and (g), the amine nucleophile (ammonia) is moderate. When combined with a 1° haloalkane, which cannot proceed by S_N1, the favored mechanism is S_N2.

7.25 Only (c) and (d) are true.

7.27 (a) The 2° haloalkane forms a relatively stable carbocation, ethanol is a weak nucleophile, and the solvent (also ethanol) is moderately ionizing.

(b) The 3° haloalkane forms a very stable carbocation, methanol is a weak nucleophile, and the solvent (also methanol) is moderately ionizing.

(c) The 3° haloalkane forms a very stable carbocation, acetic acid is a weak nucleophile, and the solvent (also acetic acid) is strongly ionizing.

(d) The 2° haloalkane forms a carbocation that is stabilized by resonance, methanol is a weak nucleophile, and the solvent (also methanol) is moderately ionizing.

(e) The 2° haloalkane forms a carbocation that can rearrange (hydride shift) to form a 3° carbocation, and like (a), the nucleophile and solvent favor an S_N1 mechanism.

(f) The 2° haloalkane forms a carbocation that can rearrange (alkyl shift) to form a 3° carbocation, and like (c), the nucleophile and solvent favor an S_N1 mechanism.

7.29 Both the S_N1 and E1 mechanisms involve the highly stable *tert*-butyl carbocation, which is generated in the first step of the reaction:

$$(CH_3)_3C{-}Cl \xrightarrow{slow} (CH_3)_3C^+ + Cl^-$$

Attack of the carbocation by ethanol gives the ether product:

$$(CH_3)_3C^+ + HOCH_2CH_3 \longrightarrow (CH_3)_3C{-}O^+(H)CH_2CH_3 \longrightarrow CH_3C(CH_3)_2OCH_2CH_3 + H_3O^+$$

In a similar fashion, nucleophilic attack of the carbocation by water results in the alcohol:

$$(CH_3)_3C^+ + H_2O \longrightarrow (CH_3)_3C{-}O^+H_2 \longrightarrow CH_3C(CH_3)_2OH + H_3O^+$$

Deprotonation of the carbocation by the solvent (water or ethanol) gives the alkene:

$$H_2O + H{-}CH_2{-}C^+(CH_3)_2 \longrightarrow CH_3C(CH_3){=}CH_2 + H_3O^+$$

7.31 Larger alkyl groups bonded to the reacting carbon present greater sterlic hindrance to the attacking nucleophile

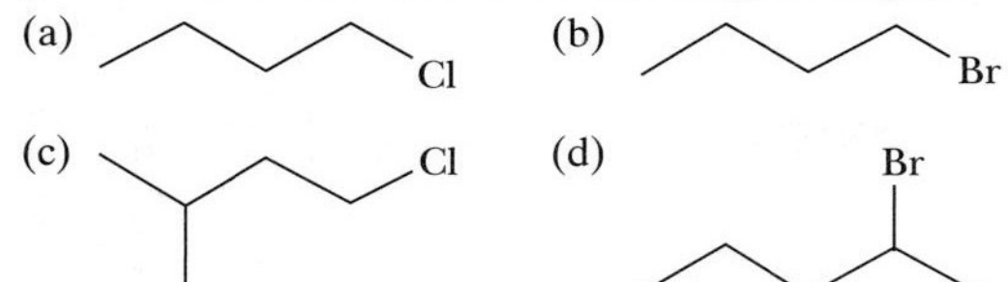

7.33 The alkenyl carbocations produced by ionization of the carbon-halogen bond are too unstable. They fail to undergo S_N2 reactions because the planar geometry of the alkene does not allow backside attack, and the electron-rich nature of the alkene does not attract nucleophiles.

7.35 (a)

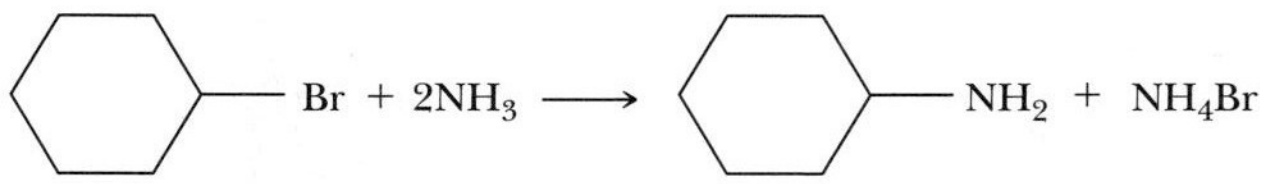

Br + $2NH_3$ ⟶ NH_2 + NH_4Br

(b)

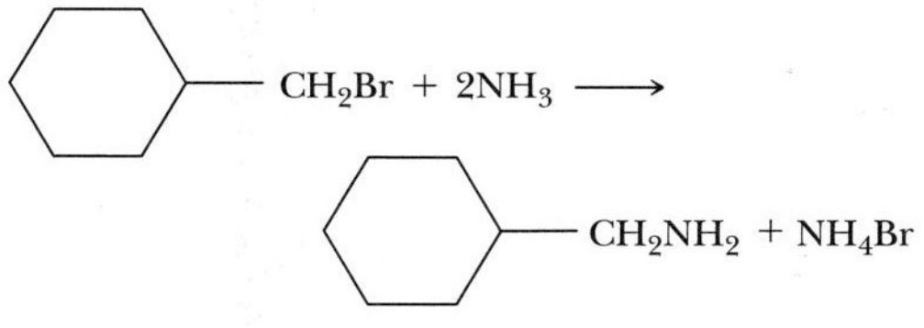

CH_2Br + $2NH_3$ ⟶

CH_2NH_2 + NH_4Br

(c)

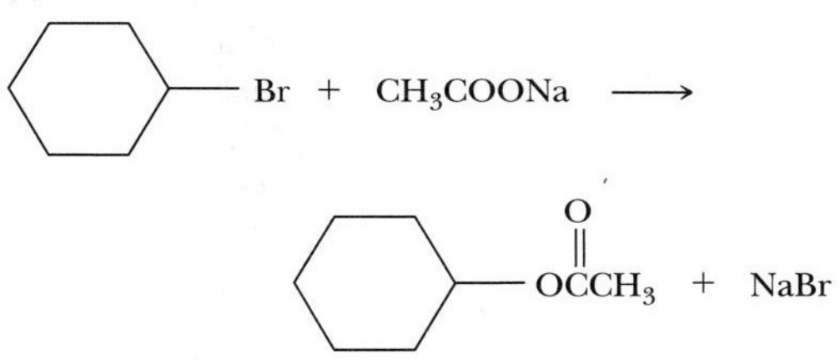

Br + CH_3COONa ⟶

O ‖ $OCCH_3$ + NaBr

(d)

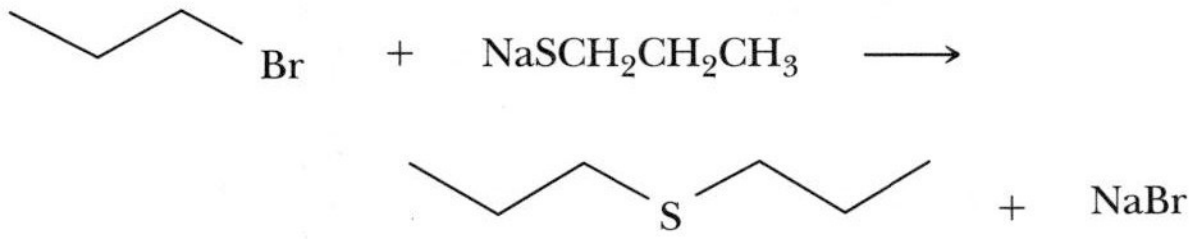

Br + $NaSCH_2CH_2CH_3$ ⟶

S + NaBr

(e)

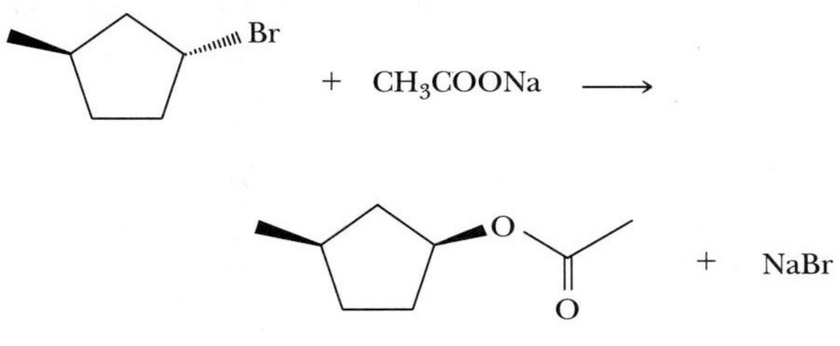

Br + CH_3COONa ⟶

O O + NaBr

(f)

$CH_3(CH_2)_2CH_2Br$ + $NaOCH_2(CH_2)_2CH_3$ ⟶

$(CH_3CH_2CH_2CH_2)_2O$ + NaBr

7.37 Chlorocyclohexane and isobutyl chloride

7.39 (a) CH_2Br (b) CH_3 | $CH_3CHCH_2CH_2CH_2Br$

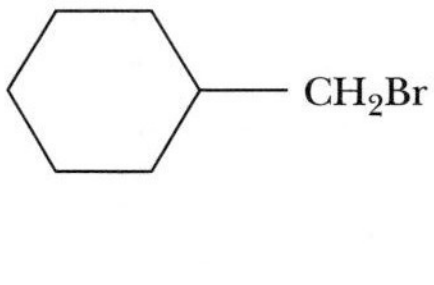

7.41 (a) The inversion of configuration indicates that an S_N2 reaction mechanism occurred.

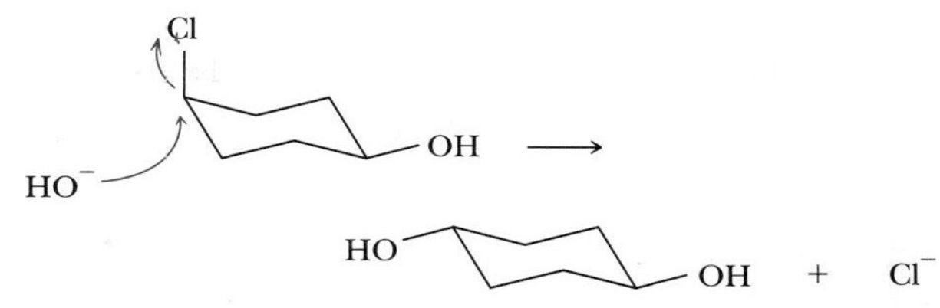

(b)

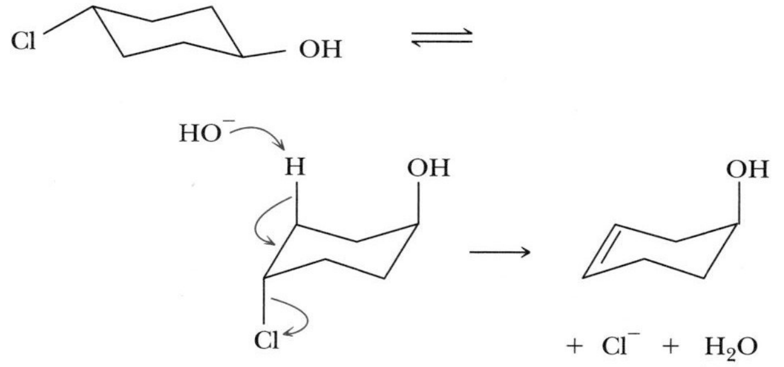

Cl OH ⇌

HO⁻ H OH ⟶ OH

Cl + Cl⁻ + H_2O

(c) With *trans*-4-chlorocyclohexanol, the nucleophilic atom is created by deprotonating the hydroxyl group. The resulting alkoxide is properly oriented to initiate a backside attack. Whereas, the alkoxide nucleophile generated from *cis*-4-chlorocyclohexanol is situated on the same side as the leaving group and cannot initiate a backside attack.

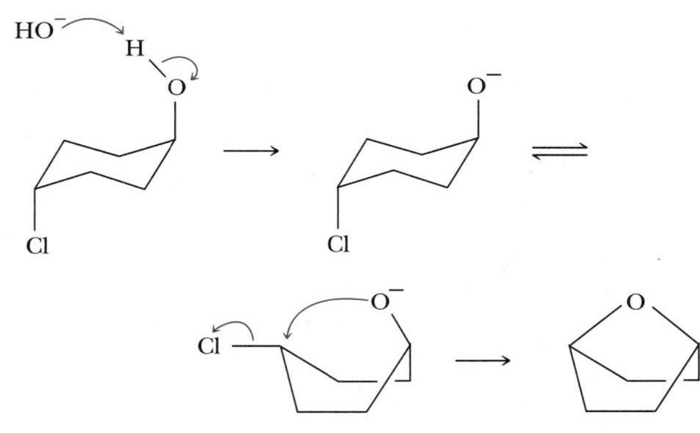

7.43 (a)

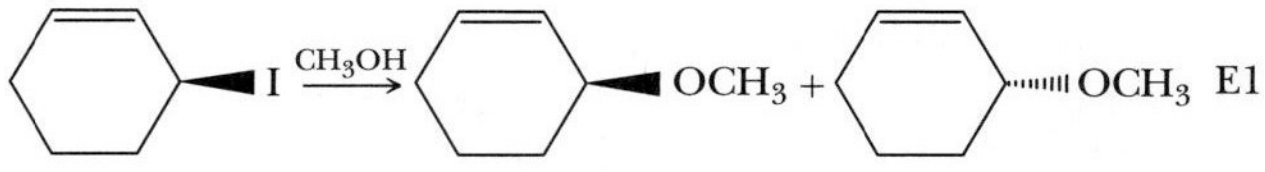

I $\xrightarrow{CH_3OH}$ OCH_3 + OCH_3 E1

(b)

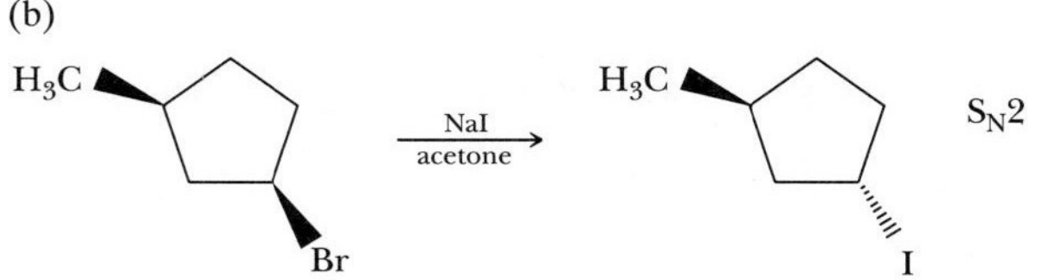

(c)

I + NaOH $\xrightarrow[H_2O]{80°C}$ E2

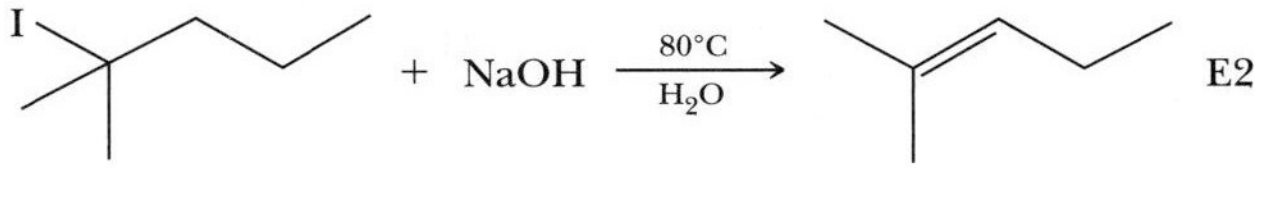

(d)

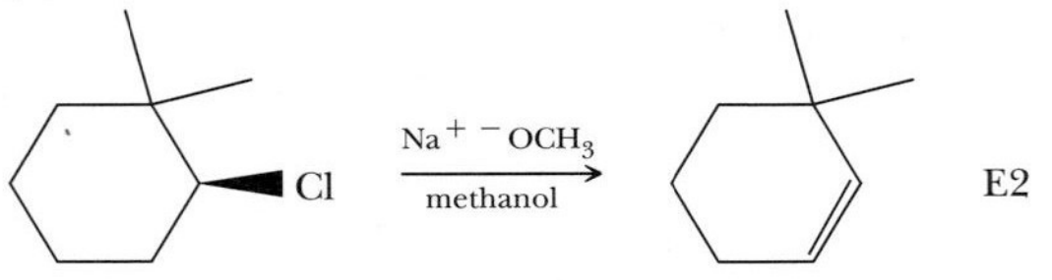

Cl $\xrightarrow[\text{methanol}]{Na^+ \ ^-OCH_3}$ E2

(e)

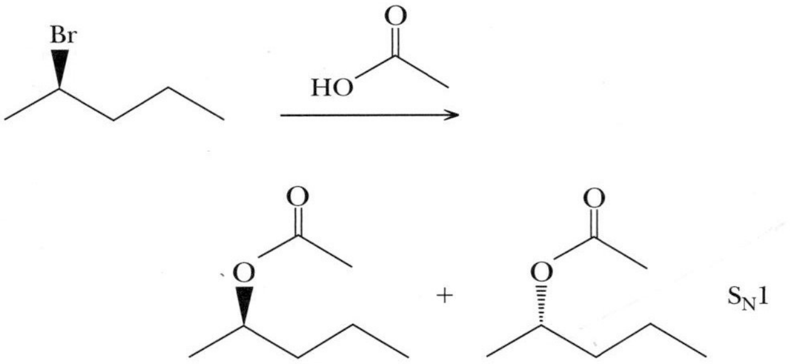

Br + HO(C=O)CH$_3$ ⟶

O(C=O)CH$_3$ + O(C=O)CH$_3$ S_N1

(f)

Cl $\xrightarrow[\text{DMSO}]{\text{KCN}}$ CN S_N2

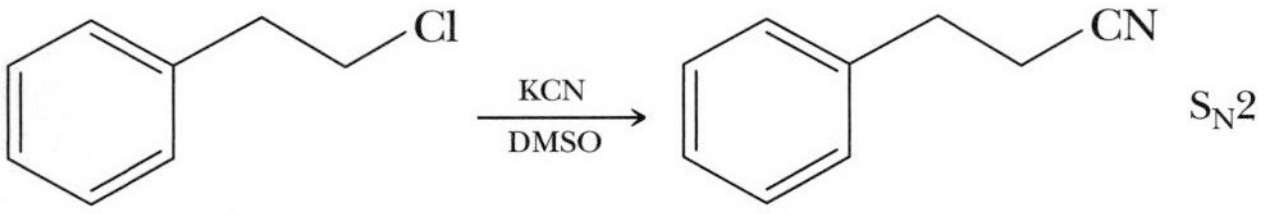

7.45 Reaction (a) gives the ether, but (b) gives 2-methylpropene by an E2 reaction.

7.47

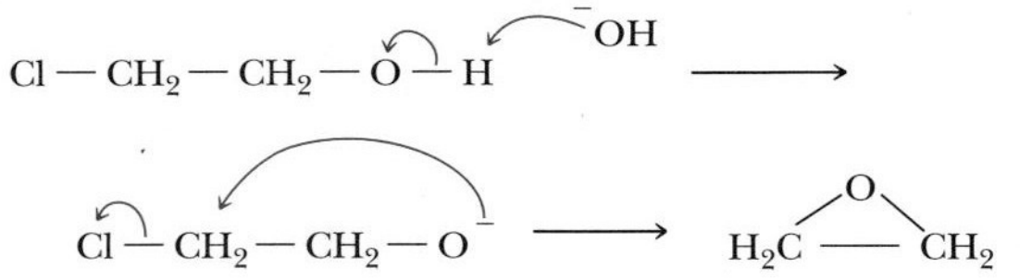

$Cl-CH_2-CH_2-O-H$ + ^-OH ⟶

$Cl-CH_2-CH_2-O^-$ ⟶ H_2C-CH_2 (O bridging)

7.49 Over time, an equilibrium mixture consisting of equal amounts of both enantiomers is formed.

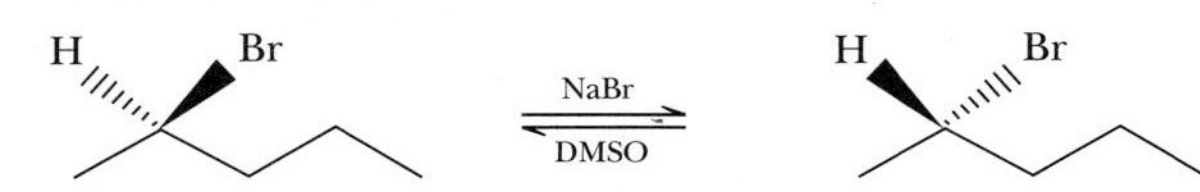

7.51 Although alkoxides are poor leaving group, the opening of the highly strained three-membered ring is the driving force behind the opening of epoxides by nucleophiles.

Chapter 8
Alcohols, Ethers, and Thiols

8.1 (a) 2-heptanol (b) 2,2-dimethyl-1-propanol
(c) (1*R*,3*S*)-3-isopropylcyclohexanol or *cis*-3-isopropylcyclohexanol

8.2 (a) primary (b) secondary
(c) primary (d) tertiary

8.3 (a) *tans*-3-penten-1-ol (b) 2-cyclopentenol
(c) (2*S*,3*R*)-1,2,3-pentanetriol
(d) *cis*-1,4-cycloheptanediol or (1*R*,4*S*)-1,4-cycloheptanediol

8.4 (a)

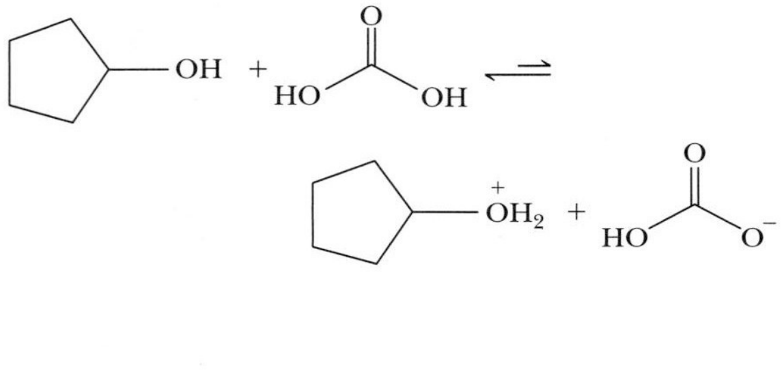

OH + HO(C=O)OH ⇌

$\overset{+}{O}H_2$ + HO(C=O)O$^-$

(b)

2 OH + 2 Na ⟶

2 O^-Na^+ + H_2

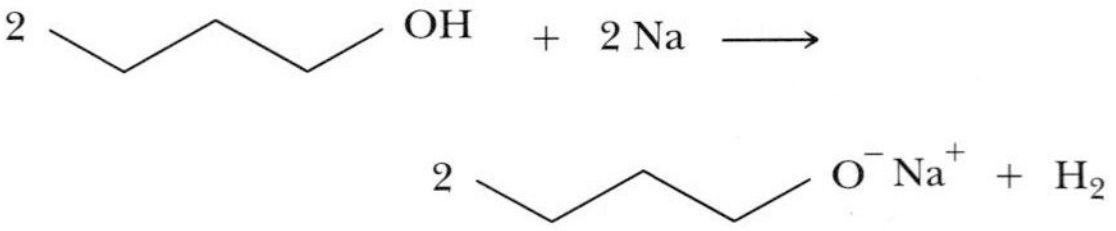

(c)

CH_3CH_2OH + HO(C=O)O^-Na^+ ⇌

$CH_3CH_2O^-Na^+$ + HO(C=O)OH

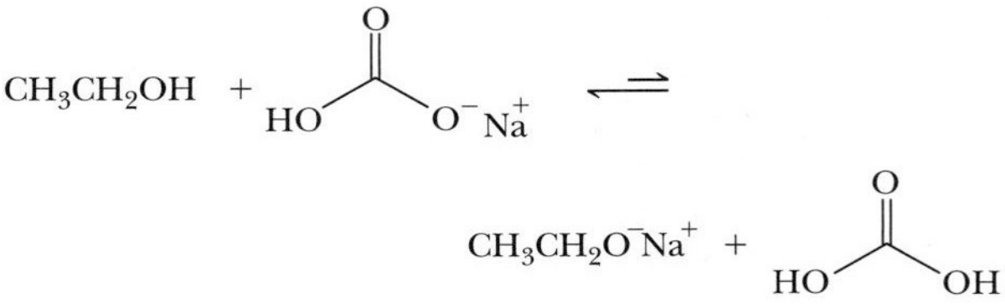

8.5 (a)

major product + minor product

(b)

major product + minor product

8.6 (a)

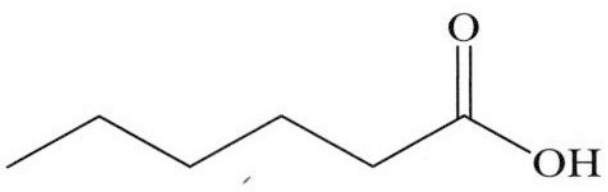

OH

(b)

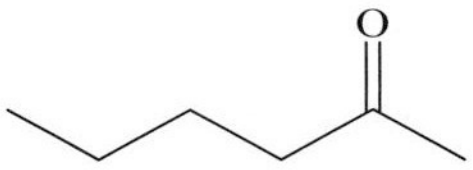

O

(c)

O

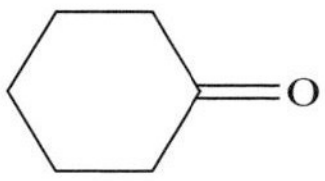

8.7 (a) 1-ethoxy-2-methylpropane or ethyl isobutyl ether
(b) methoxycyclopentane or cyclopentyl methyl ether

8.8 1,2-dimethoxyethane < 2-methoxyethanol < 1,2-ethanediol (ethylene glycol)

8.9

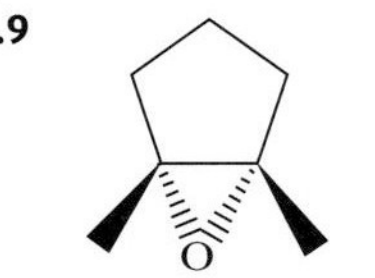

8.10

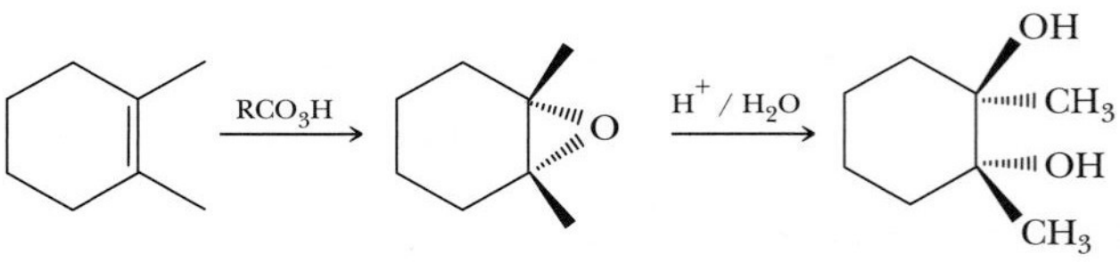

$\xrightarrow{RCO_3H}$ O $\xrightarrow{H^+ / H_2O}$ OH, CH_3, OH, CH_3

8.11 (a) 3-methyl-1-butanethiol
(b) 3-methyl-2-butanethiol

8.12

(a)

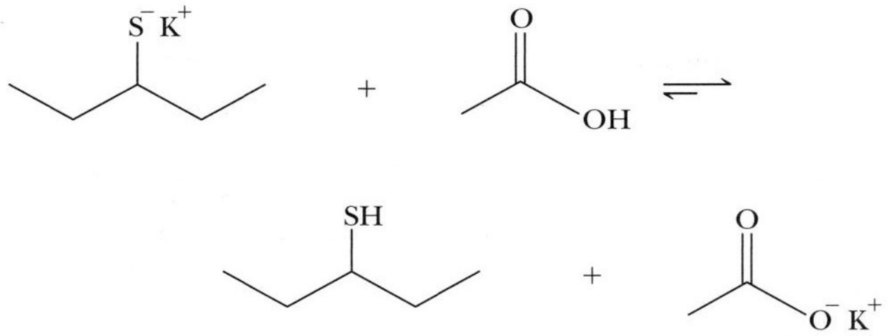

$S^- K^+$ + O, OH ⇌ SH + O, $O^- K^+$

(b)

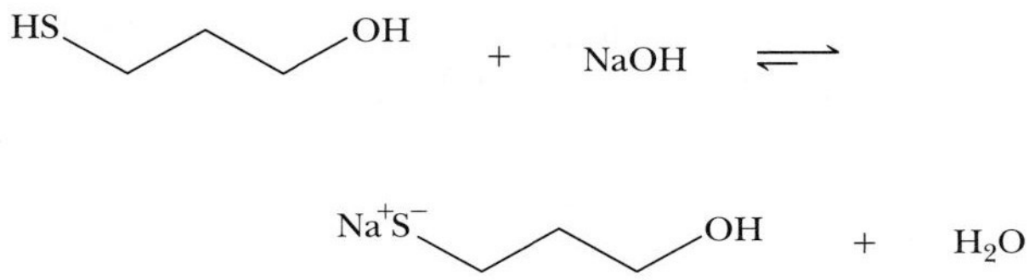

HS OH + NaOH ⇌

Na^+S^- OH + H_2O

(c)

SH SH

oxidation ⟶ S—S

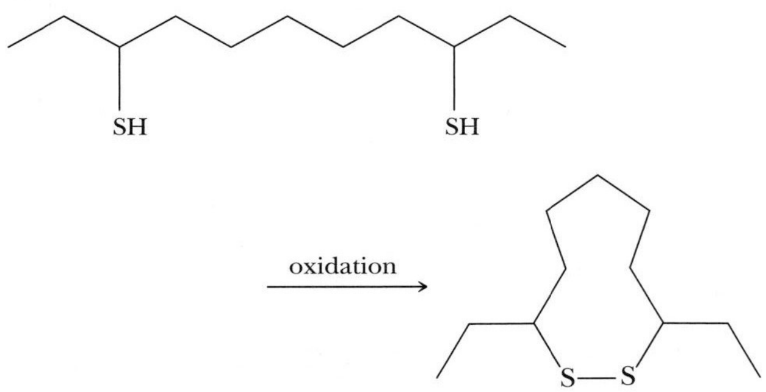

8.13 (a) 3° (b) 3° (c) 1° (d) 3°
(e) 2° (f) 2° (g) 3° (h) 2°

8.15

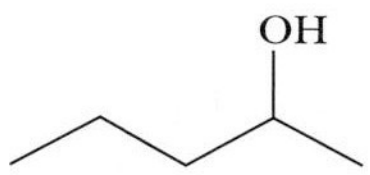

(a) OH

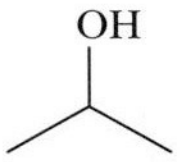

(b) OH OH

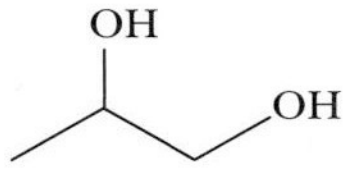

(c) OH

(d) OH HO

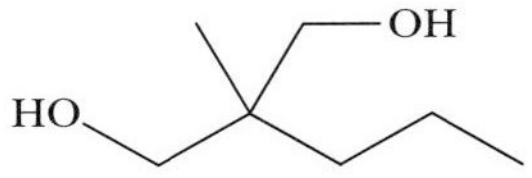

(e) OH

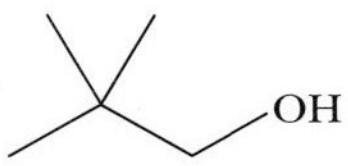

(f) HS OH

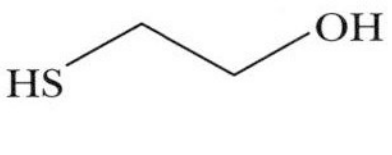

(g) HO OH

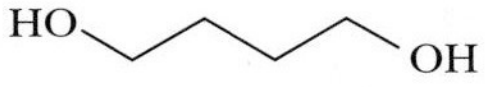

(h) OH

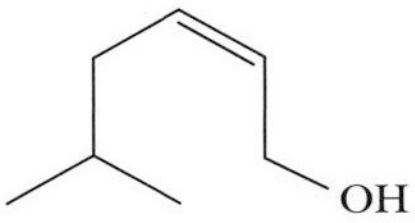

(i) OH

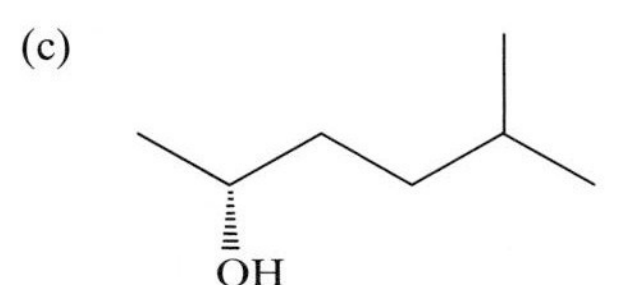

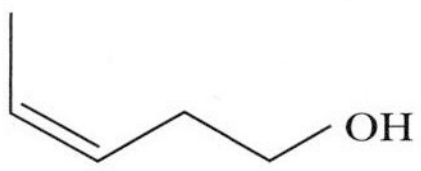

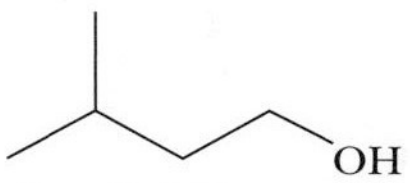

(j) HO OH

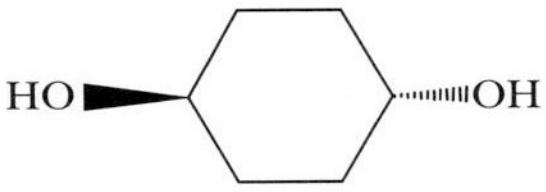

8.17

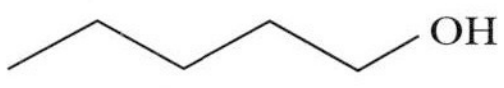

OH 1-pentanol

OH 2-pentanol (chiral)

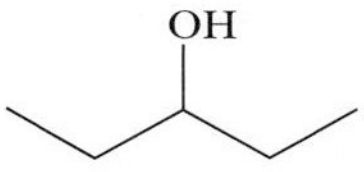

OH 3-pentanol

OH 3-methyl-1-butanol

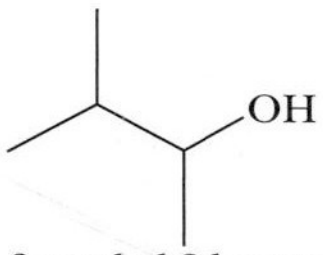

OH 3-methyl-2-butanol (chiral)

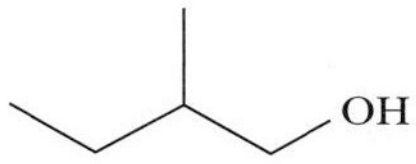

OH 2-methyl-1-butanol (chiral)

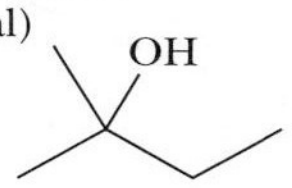

OH 2-methyl-2-butanol

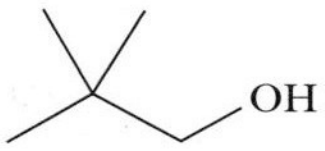

OH 2,2-dimethyl-1-propanol

8.19 $CH_3CH_2CH_3 < CH_3OCH_3 < CH_3CH_2OH < CH_3COOH$

8.21

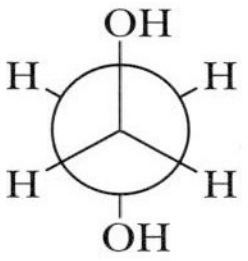

OH, H, H, H, H, OH — anti

H, O, H, H, O, H, H, H — gauche

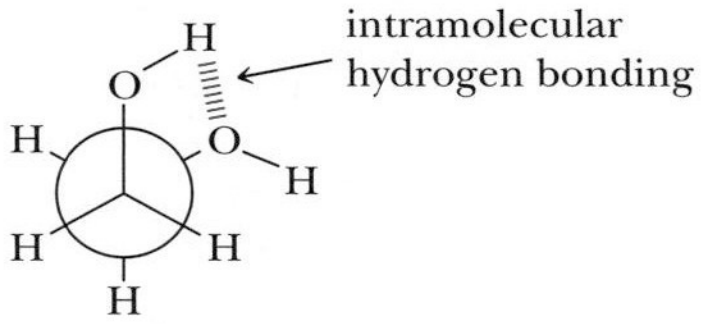

intramolecular hydrogen bonding

8.23 (a) CH_3OH

(b) $CH_3\overset{O}{\overset{\|}{C}}CH_3$

(c) NaCl

(d) $CH_3CH_2CH_2OH$

(e) $CH_3CH_2\overset{OH}{\overset{|}{C}}HCH_2CH_3$

8.25 (a) CH_3CH_2OH

(b) CH_3CH_2OH

(c) $CH_3\overset{O}{\overset{\|}{C}}CH_3$

(d) $CH_3CH_2OCH_2CH_3$

8.27 Acid-catalyzed hydrolysis of epoxide:

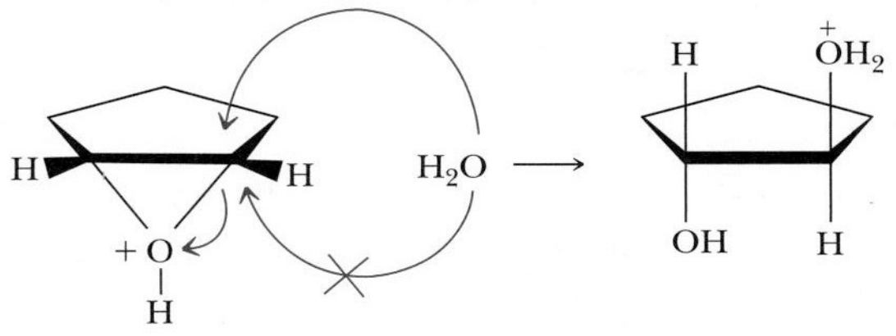

Bromination:

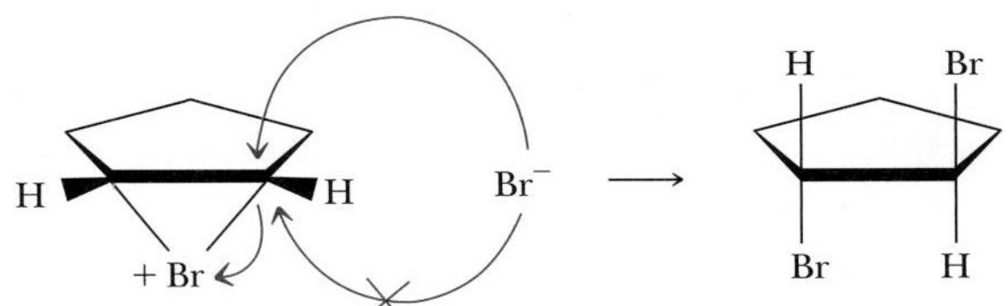

8.29

$CH_3CH_2CH_2OH < CH_3CH_2CH_2SH < CH_3CH_2COH$ (with C=O)

8.31

(a) $CH_3CH_2O^- + HCl \rightleftharpoons CH_3CH_2OH + Cl^-$

stronger base, stronger acid, weaker acid, weaker base

(b) CH_3COH (C=O) $+ CH_3CH_2O^- \rightleftharpoons CH_3CO^-$ (C=O) $+ CH_3CH_2OH$

stronger acid, stronger base, weaker base, weaker acid

8.33 Cyclohexene will decolorize an orange-red solution of bromine dissolved in CCl_4.

8.35

(a) 2 OH … $\xrightarrow{2\,Na}$ 2 O^-Na^+ … $+ H_2$

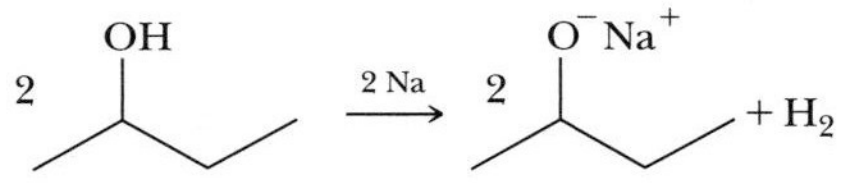

(b) OH … $\xrightarrow[heat]{H_2SO_4}$ … + … + …

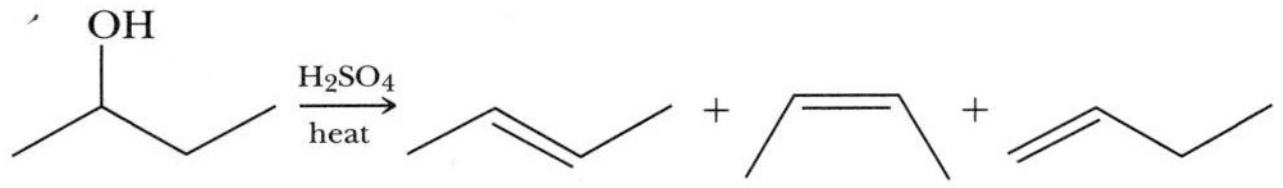

major product

(c) OH … $\xrightarrow[heat]{HBr}$ Br …

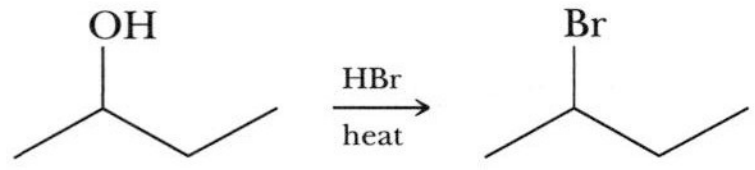

(d) OH … $\xrightarrow[H_2SO_4;\ heat]{K_2Cr_2O_7}$ O …

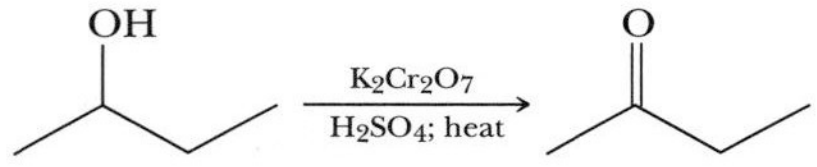

(e) OH … $\xrightarrow{SOCl_2}$ Cl …

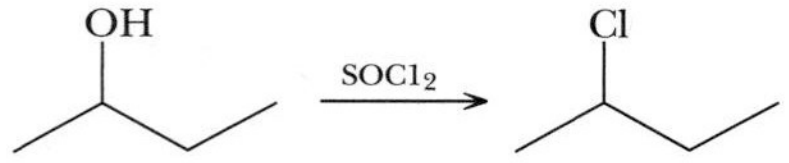

(f) OH … $\xrightarrow{PCC}$ O …

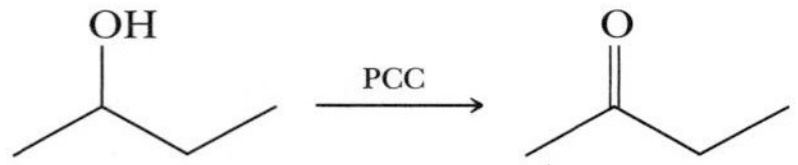

8.37 The reaction most likely proceeds by an S_N1 mechanism. Intermediates:

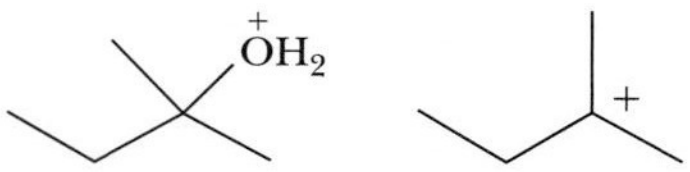

$\overset{+}{O}H_2$ … +

8.39

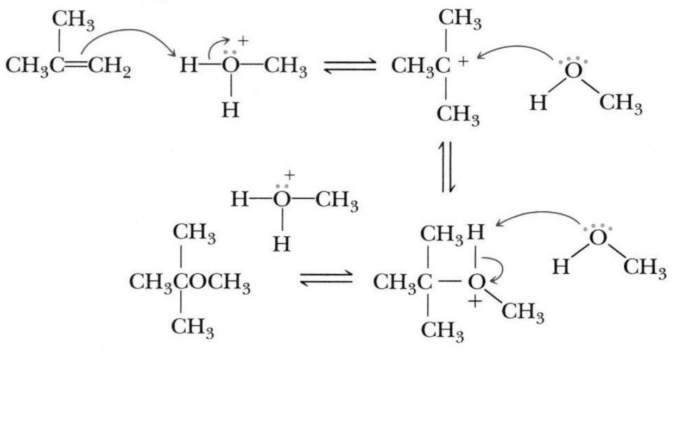

$CH_3C{=}CH_2$ (CH_3) + $H{-}\overset{+}{O}{-}CH_3$ (H) $\rightleftharpoons$ $CH_3\overset{+}{C}$ (CH_3, CH_3) + $H{-}O{-}CH_3$

$\rightleftharpoons$

CH_3COCH_3 (CH_3, CH_3) $\rightleftharpoons$ $CH_3C{-}\overset{+}{O}(H){-}CH_3$ (CH_3, CH_3) + $H{-}\overset{+}{O}{-}CH_3$ (H)

8.41 (a)

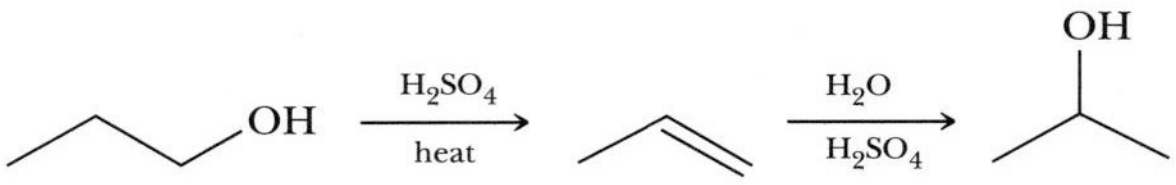

OH $\xrightarrow[heat]{H_2SO_4}$ … $\xrightarrow[H_2SO_4]{H_2O}$ OH

(b)

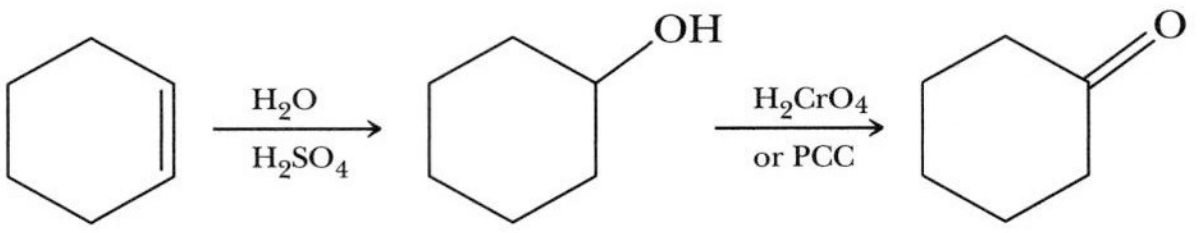

… $\xrightarrow[H_2SO_4]{H_2O}$ OH $\xrightarrow[or\ PCC]{H_2CrO_4}$ O

(c)

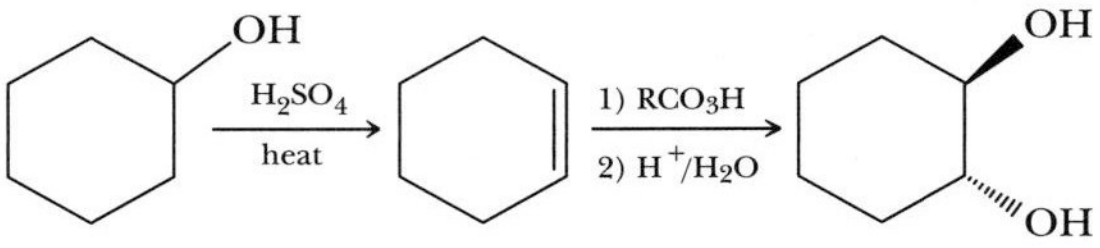

OH $\xrightarrow[heat]{H_2SO_4}$ … $\xrightarrow[2)\ H^+/H_2O]{1)\ RCO_3H}$ OH, OH racemic mixture

(d)

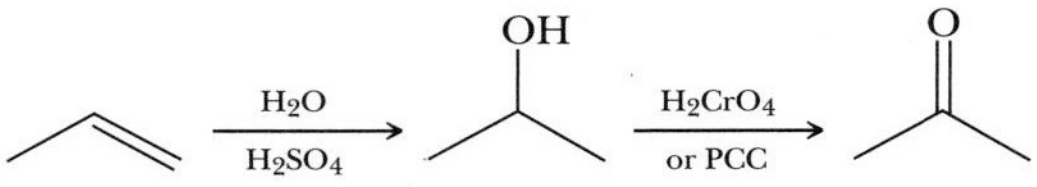

… $\xrightarrow[H_2SO_4]{H_2O}$ OH $\xrightarrow[or\ PCC]{H_2CrO_4}$ O

8.43

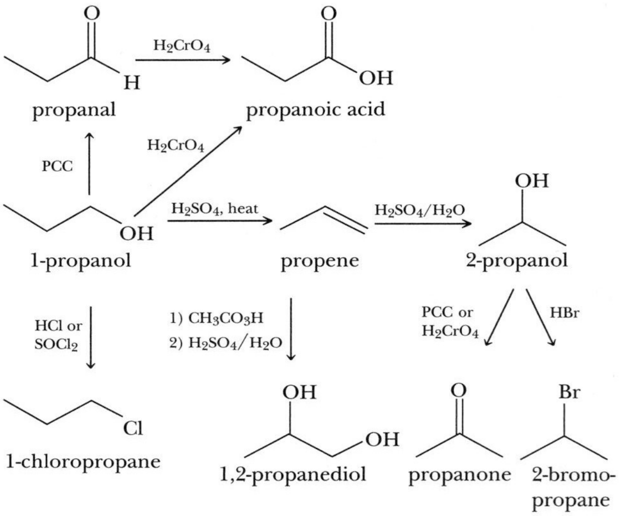

propanal $\xrightarrow{H_2CrO_4}$ propanoic acid

1-propanol $\xrightarrow{PCC}$ propanal; 1-propanol $\xrightarrow{H_2CrO_4}$ propanoic acid

1-propanol $\xrightarrow{H_2SO_4,\ heat}$ propene $\xrightarrow{H_2SO_4/H_2O}$ 2-propanol

1-propanol $\xrightarrow{HCl\ or\ SOCl_2}$ 1-chloropropane

propene $\xrightarrow{1)\ CH_3CO_3H;\ 2)\ H_2SO_4/H_2O}$ 1,2-propanediol

2-propanol $\xrightarrow{PCC\ or\ H_2CrO_4}$ propanone

2-propanol $\xrightarrow{HBr}$ 2-bromo-propane

8.45

OH $\xrightarrow{H_2SO_4,\ heat}$ (a) $\xrightarrow{H_2O/H_2SO_4}$ (b) OH

$\xrightarrow{PCC\ or\ H_2CrO_4}$ (c) O

(a) $\xrightarrow{CH_3CO_3H}$ (d) O

(d) $\xrightarrow{NaSH/H_2O}$ (e) OH, SH

(d) $\xrightarrow{H_2O/H_2SO_4}$ (f) OH, OH

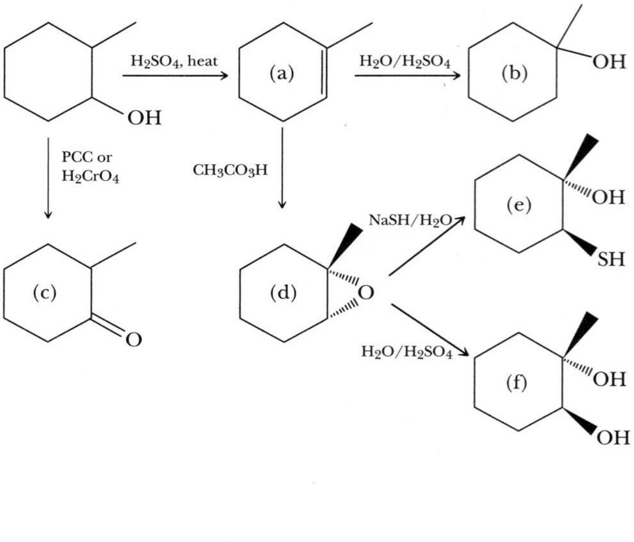

8.47 (a) A peroxycarboxylic acid, such as CH_3CO_3H, will convert the alkene into an epoxide.

(b) There are two stereocenters, giving rise to a maximum of $2^2 = 4$ stereoisomers. However, only two are formed due to the stereoselectivity of the epoxidation reaction.

8.49

(a) OH; H_2SO_4, heat; H_2O, H_2SO_4; HO

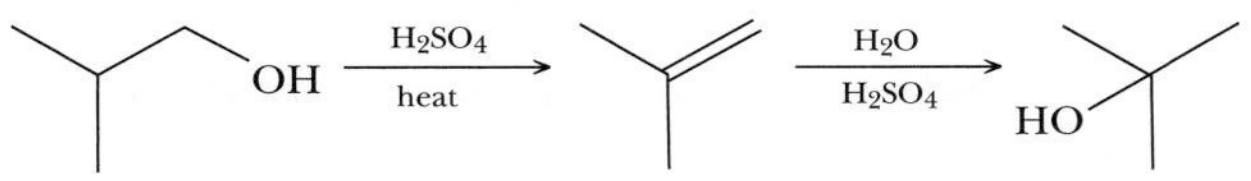

(b) OH; HBr; Br; sodium cyclohexylsulfide; S

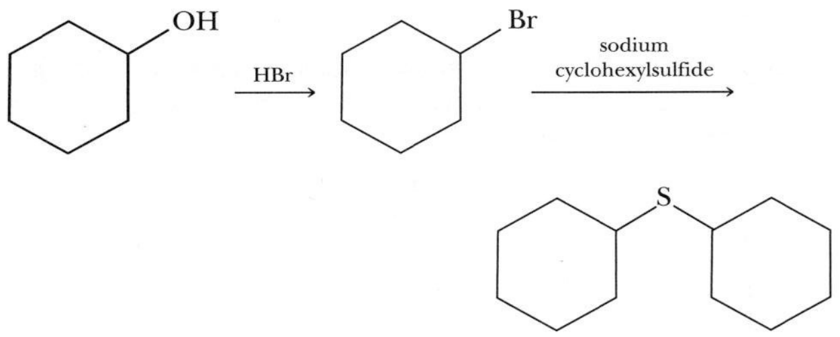

(c) OH; H_2SO_4, heat; RCO_3H; O

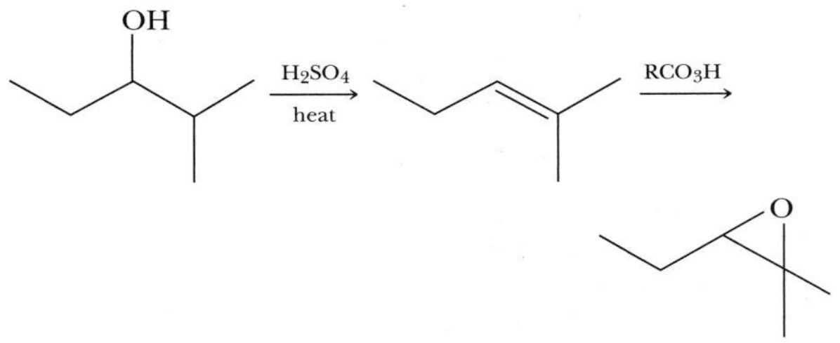

(d) RCO_3H; O; H_2O, H_2SO_4; OH; OH

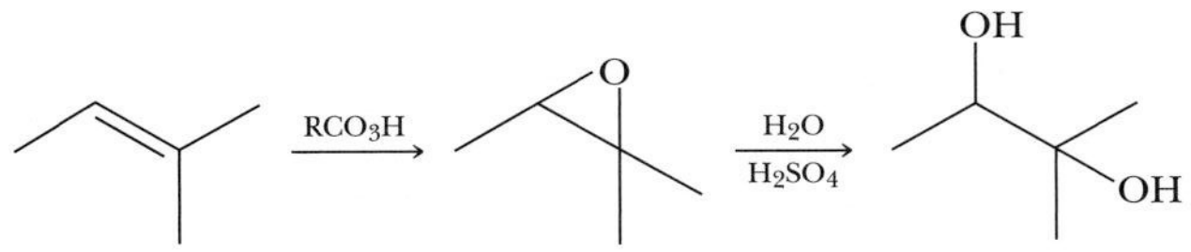

(e) OH; HCl; Cl; NaBr; Br; $NaOCH_3$; OCH_3

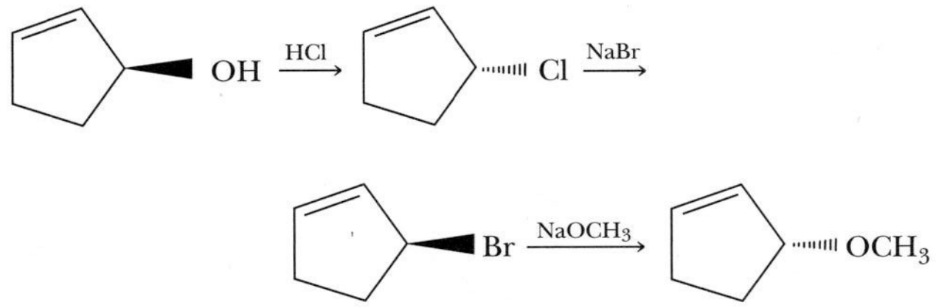

(f) H_2O, H_2SO_4; OH; H_2SO_4, heat; H_2O/H_2SO_4; OH

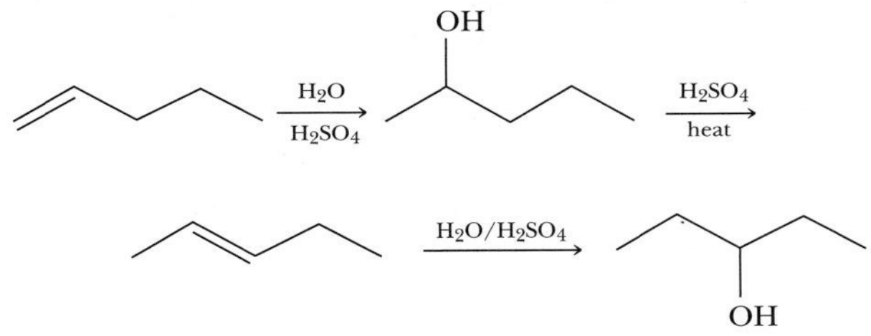

(g) Br; NaOH, heat; 1) BH_3 2) H_2O_2, NaOH; OH; racemic; H_2CrO_4; O; racemic

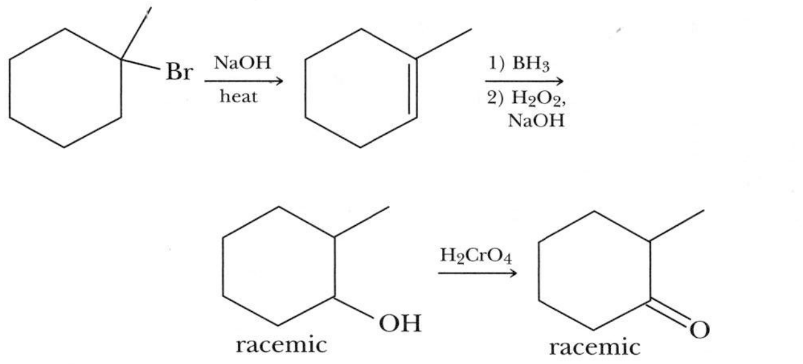

(h) HO; H_2SO_4, heat; RCO_3H; O

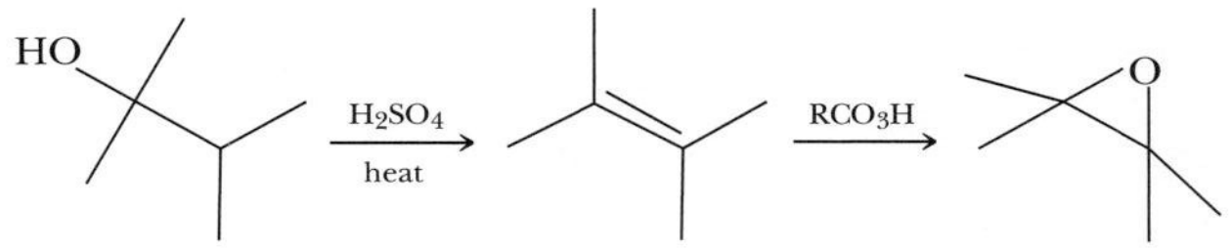

(i) OH; H_2SO_4, heat; RCO_3H; O; 1) $Na^+\,{}^-C{\equiv}CH$ 2) H_3O^+; HO; H_2, Pt; HO; H_2SO_4, heat

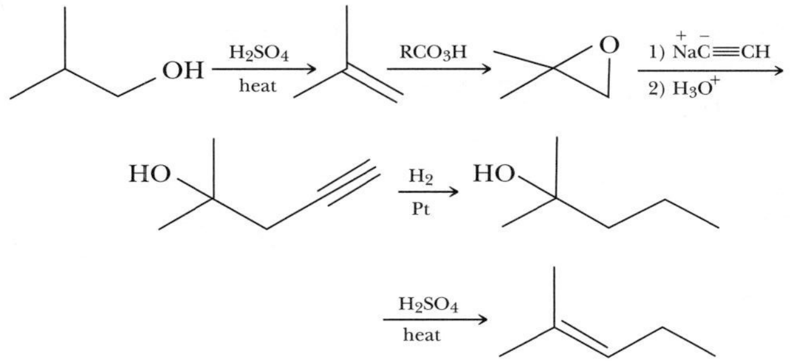

(j) 1) BH_3 2) H_2O_2, NaOH; OH; H_2CrO_4; O; OH

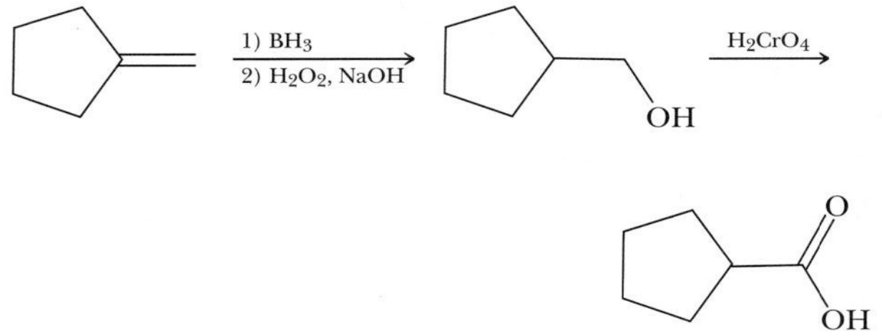

(k) RCO_3H; O; NH_3; OH; H_2N

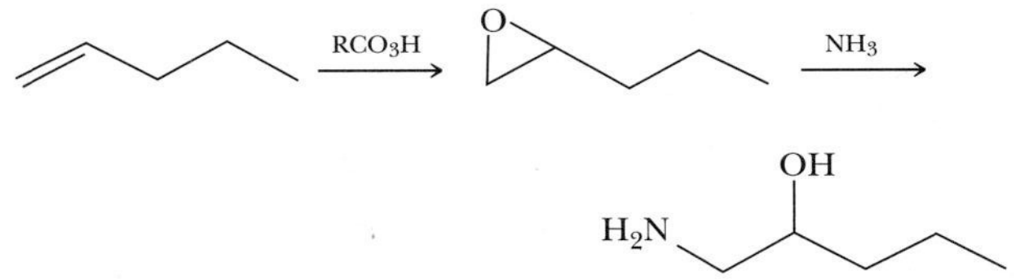

(l) Cl; NaOH, heat; 1) RCO_3H 2) H^+/H_2O; OH; OH; racemic

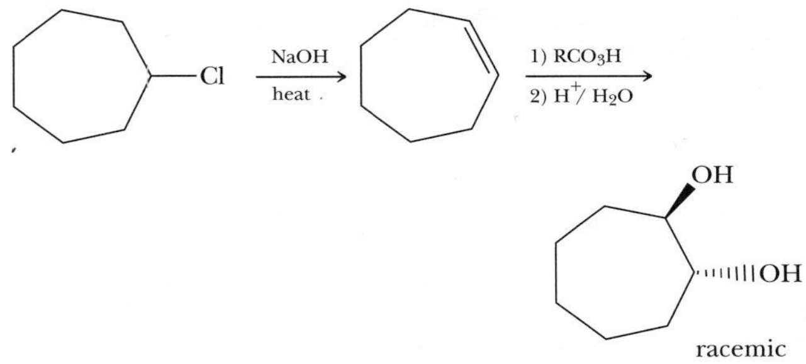

(m)

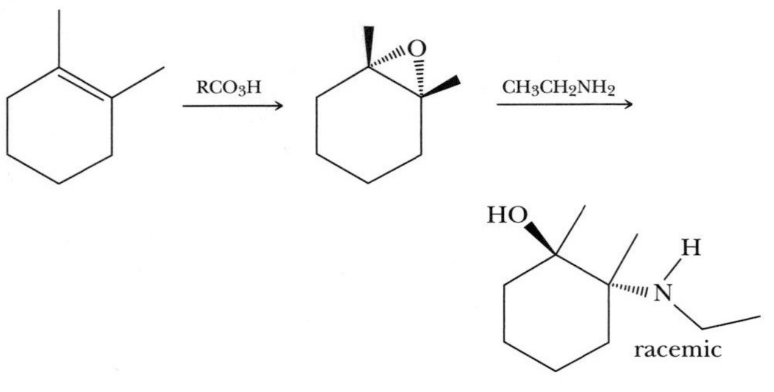

$\xrightarrow{RCO_3H}$ $\xrightarrow{CH_3CH_2NH_2}$

HO H N racemic

(n)

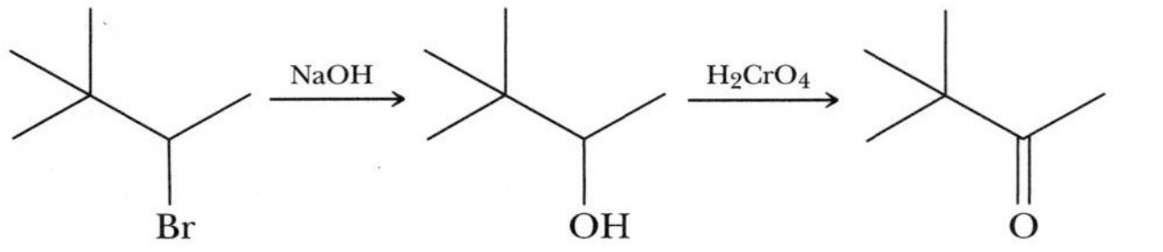

$\xrightarrow{NaOH}$ $\xrightarrow{H_2CrO_4}$

Br OH O

(o)

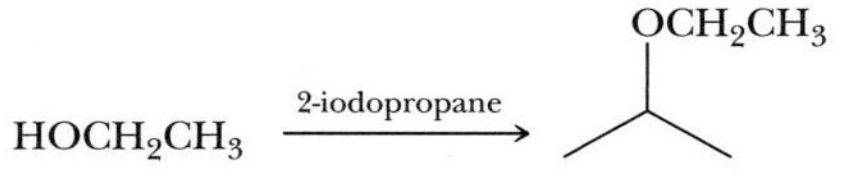

$HOCH_2CH_3$ $\xrightarrow{\text{2-iodopropane}}$ OCH_2CH_3

(p)

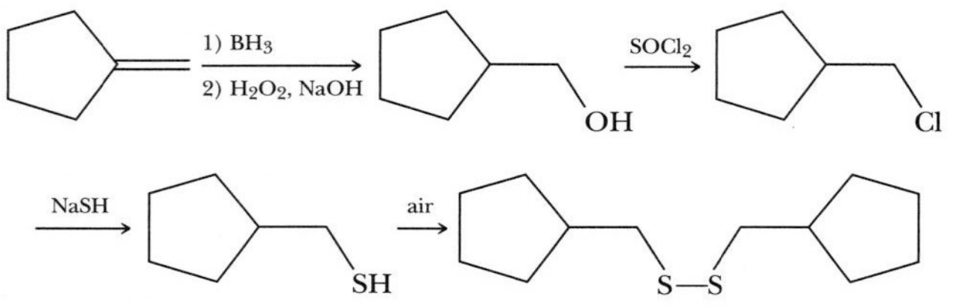

$\xrightarrow{1)\ BH_3;\ 2)\ H_2O_2,\ NaOH}$ OH $\xrightarrow{SOCl_2}$ Cl

$\xrightarrow{NaSH}$ SH $\xrightarrow{air}$ S—S

(q)

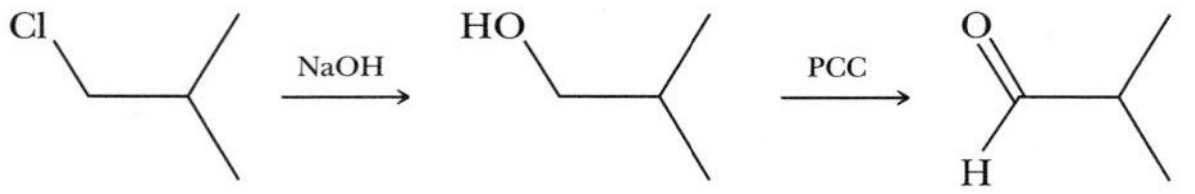

Cl $\xrightarrow{NaOH}$ HO $\xrightarrow{PCC}$ O H

(r)

OH $\xrightarrow{Na}$ ONa $\xrightarrow{bromoethane}$ O

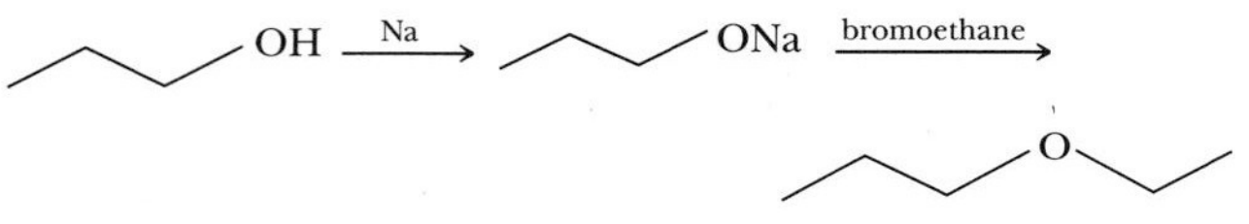

8.51

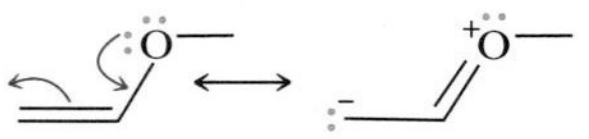

As a result of resonance, the oxygen of methyl vinyl ether carries a partial positive charge, which makes it less reactive towards an electrophile.

8.53

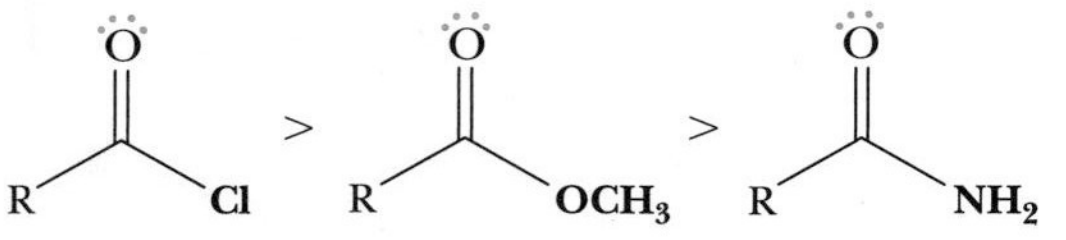

R–C(=O)–Cl > R–C(=O)–OCH_3 > R–C(=O)–NH_2

Chapter 9
Benzene and Its Derivatives

9.1 Compounds (a) and (c) are aromatic

9.2 (a) 2-phenyl-2-propanol
(b) (*E*)-3,4-diphenyl-3-hexene
(c) 3-methylbenzoic acid or *m*-methylbenzoic acid

9.3 (a) COOH COOH (b) COOH O_2N

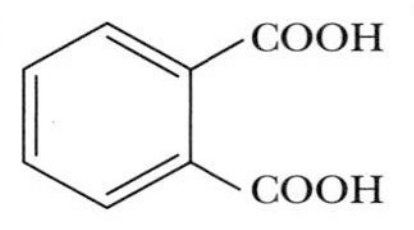

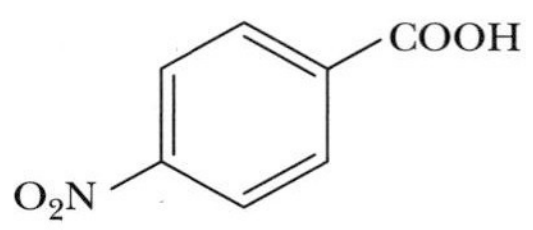

9.4 Step 1: The HSO_3^+ electrophile is generated from H_2SO_4.

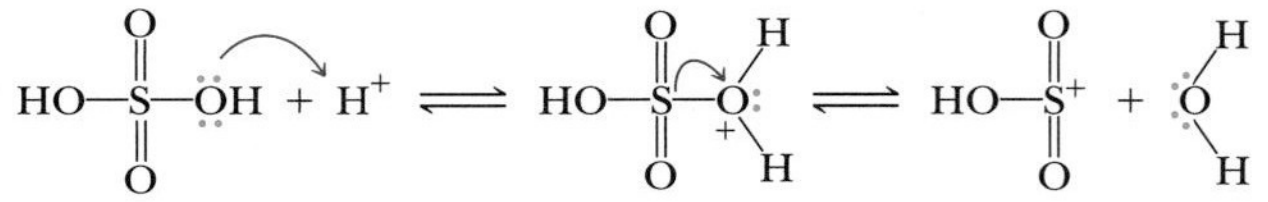

$HO-SO_2-OH + H^+ \rightleftharpoons HO-SO_2-OH_2^+ \rightleftharpoons HO-SO_2^+ + H_2O$

Step 2: The HSO_3^+ electrophile adds to benzene, forming a resonance-stabilized carbocation. This step is slow because aromaticity is destroyed.

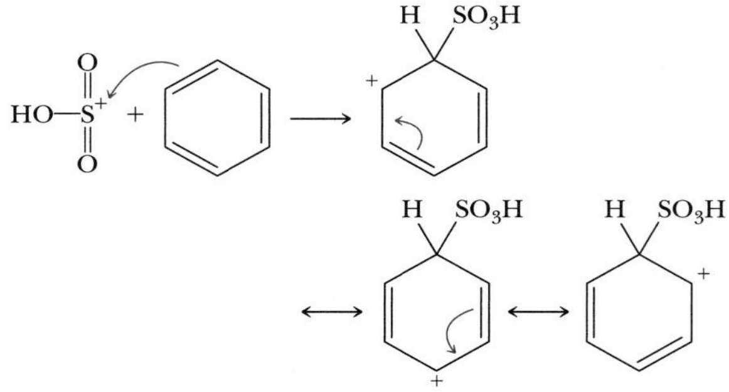

H SO_3H H SO_3H H SO_3H

Step 3: Deprotonation of the carbocation regenerates the benzene ring. This step is fast because aromaticity is regenerated. Note that the proton is removed from the same carbon to which the electrophile (HSO_3^+) was added.

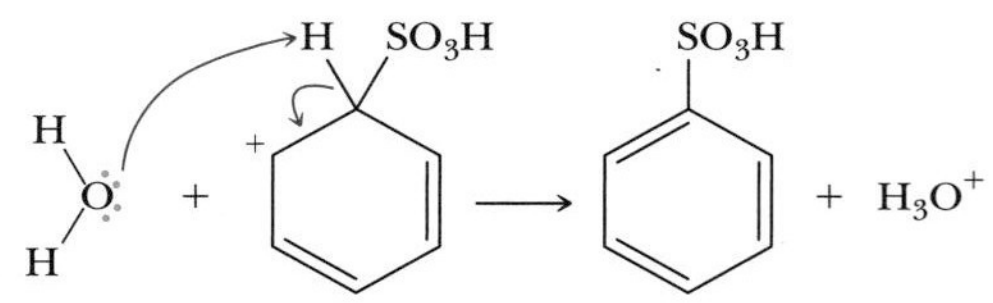

H SO_3H SO_3H + H_3O^+

9.5 (a) O (b)

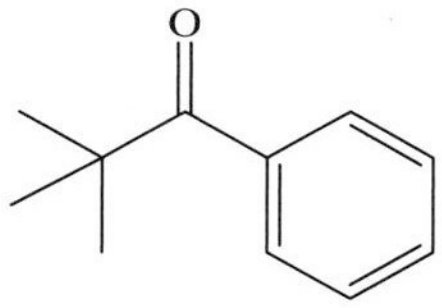

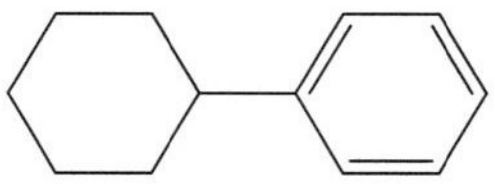

(c)

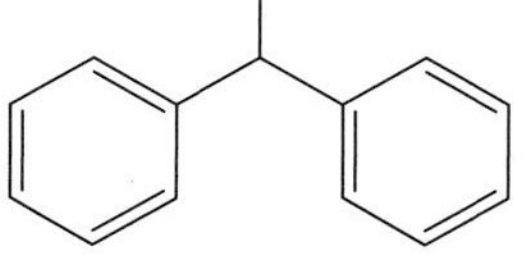

9.6 Step 1: Protonation of the alcohol generates a good leaving group, which departs in the second step. H_2O is a good leaving group because it is neutrally charged and stable.

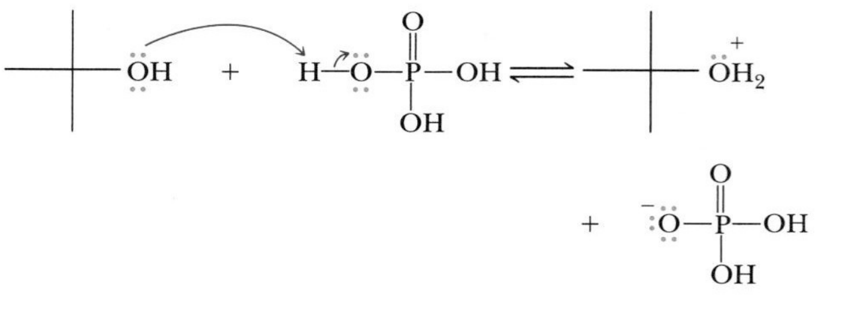

OH + $H-O-P(=O)(OH)-OH \rightleftharpoons$ OH_2^+ + $^-O-P(=O)(OH)-OH$

Step 2: The leaving group leaves as water, generating the *tert*-butyl carbocation.

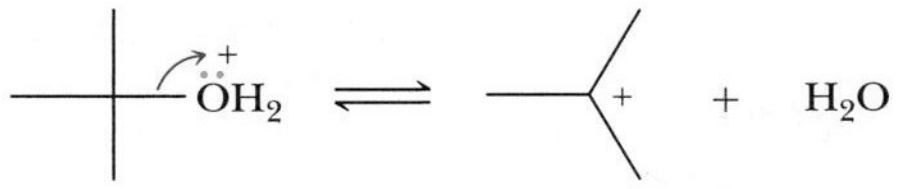

Step 3: The carbocation, an electrophile, reacts with benzene.

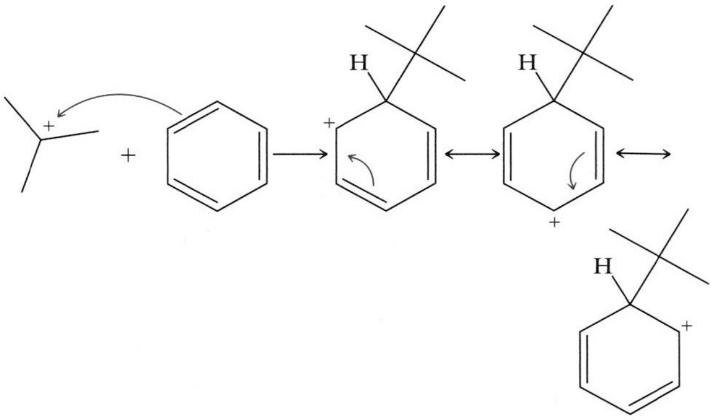

Step 4: Deprotonation regenerates the aromatic ring.

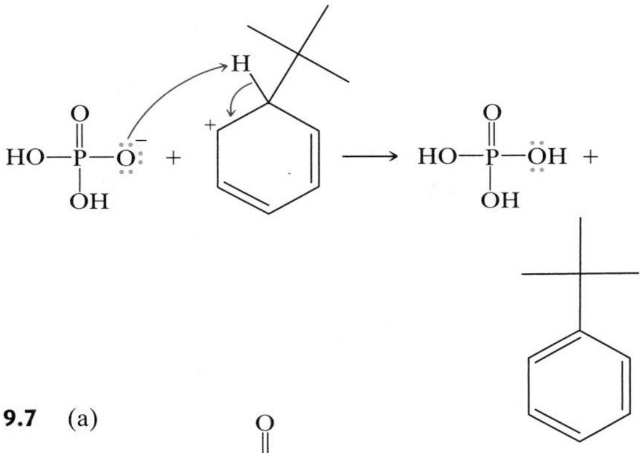

9.7 (a)

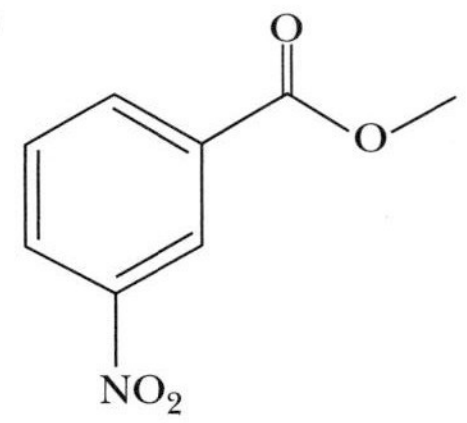

(b)

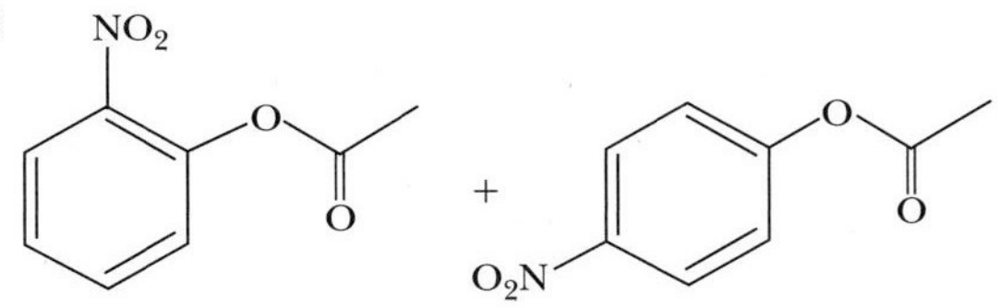

9.8 If the electrophile is added ortho or para, it is possible to draw a resonance structure that places the positive charge directly adjacent to the partially positive carbon atom of the carbonyl group. This interaction destabilizes the carbocation, which in turn disfavors ortho-para attack of the electrophile.

9.9 (a) (b)

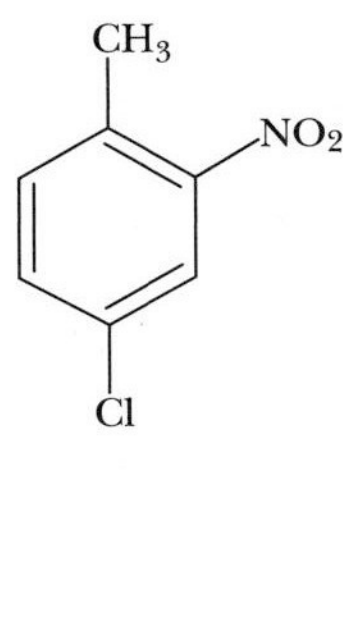

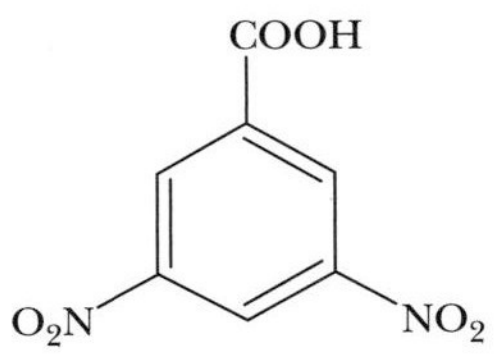

9.10

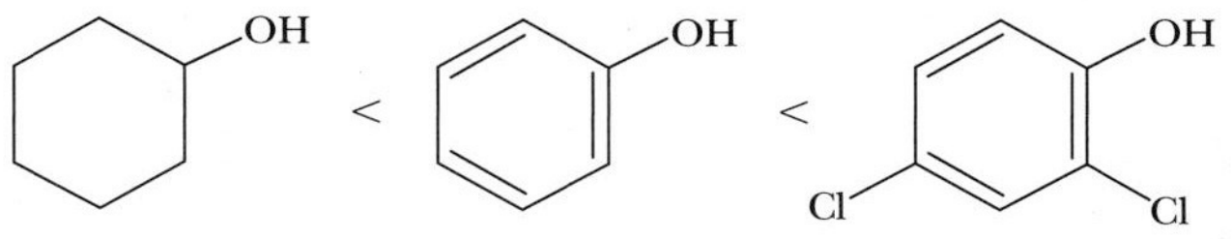

9.11 (c), (d), (e), (g), (j), (k), and (l) are aromatic

9.13 (a) 1-chloro-4-nitrobenzene (*p*-chloronitrobenzene)

(b) 1-bromo-2-methylbenzene or 2-bromotoluene or *o*-bromotoluene

(c) 3-phenyl-1-propanol

(d) 2-phenyl-3-buten-2-ol

(e) 3-nitrobenzoic acid or *m*-nitrobenzoic acid

(f) 1-phenylcyclohexanol

(g) (*E*)-1,2-diphenylethene or *trans*-1,2-diphenylethene

(h) 2,4-dichloro-1-methylbenzene or 2,4-dichlorotoluene

(i) 4-bromo-2-chloro-1-ethylbenzene

(j) 1-fluoro-3-isopropyl-5-vinylbenzene or 3-fluoro-5-isoprpylstyrene

9.15

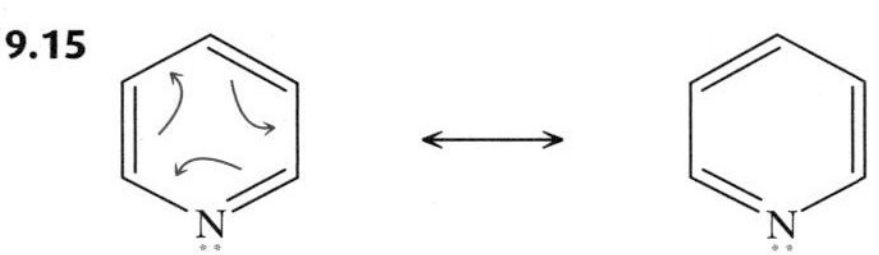

9.17

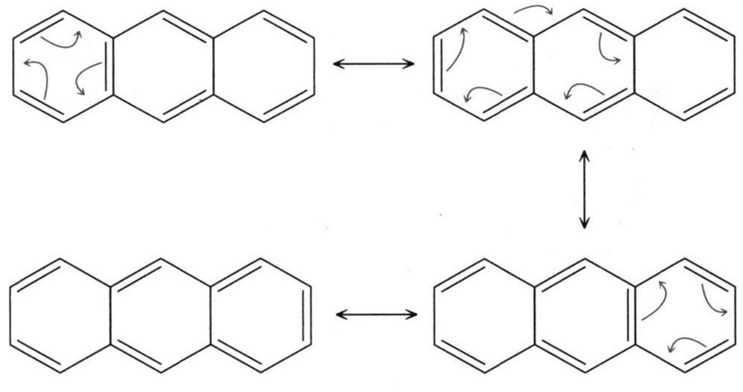

9.19

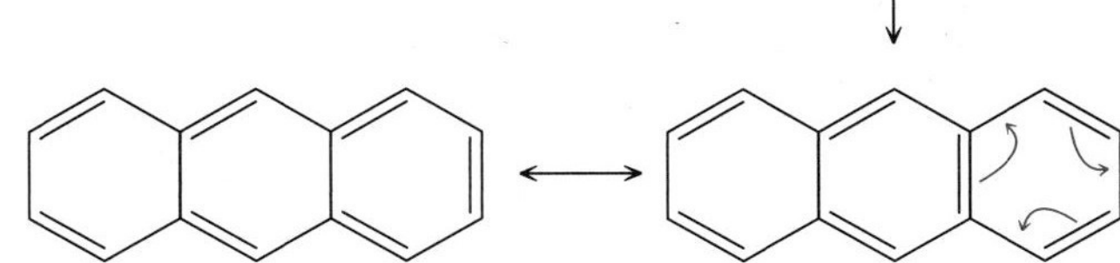

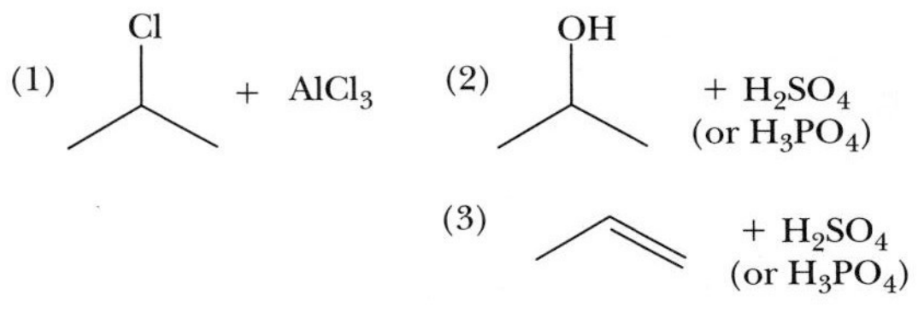

9.21 Step 1: Reaction of *tert*-butyl chloride (Lewis base) with $AlCl_3$ (Lewis acid).

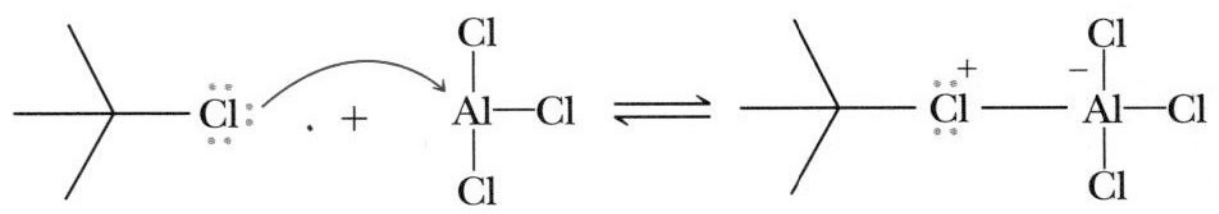

Step 2: Formation of the *tert*-butyl carbocation.

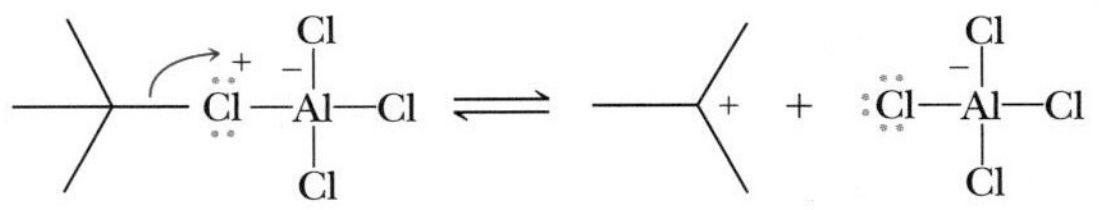

Step 3: Electrophilic attack on the benzene ring by the *tert*-butyl carbocation.

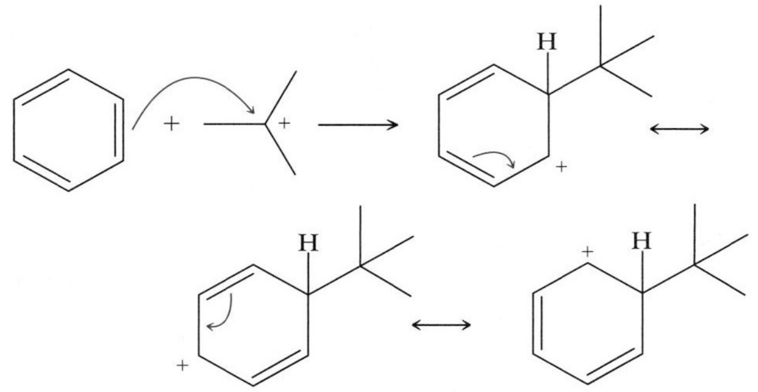

Step 4: Deprotonation to regenerate the aromatic ring, as well as the catalyst.

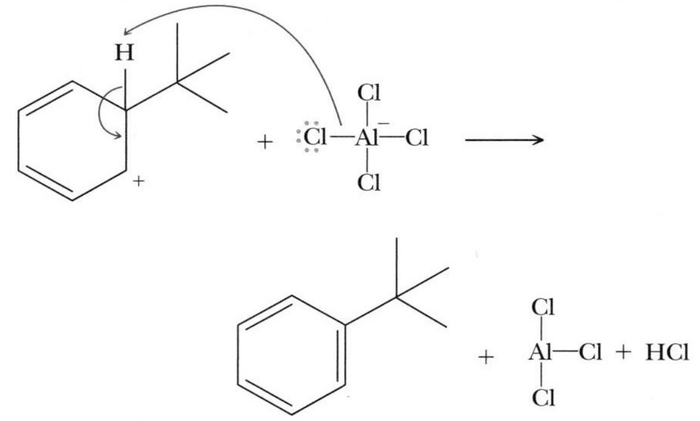

9.23 The carbocations made in these reactions rearrange (1,2-hydride and 1,2-alkyl shifts) to more stable carbocations. These carbocations react with benzene as usual (see Problem 9.21).

(a)

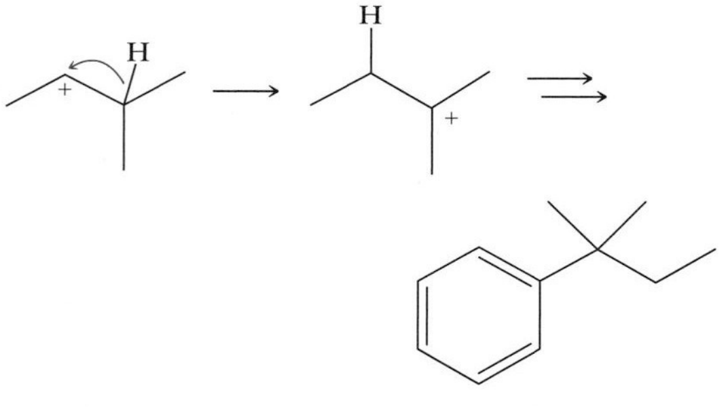

(b)

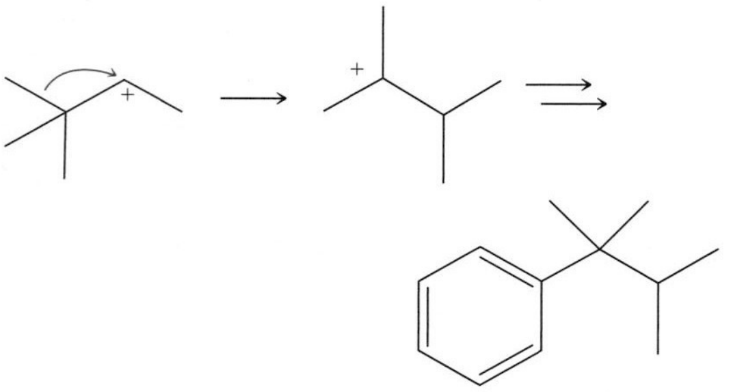

(c)

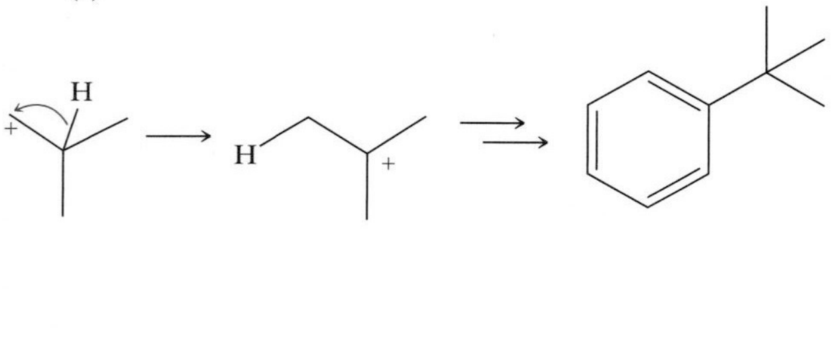

9.25 One product is formed from *p*-xylene, but *m*-xylene forms two products.

9.27 Toluene will react faster than chlorobenzene. Both the methyl and chlorine groups are ortho-para directors, but the electronegative chlorine deactivates the benzene ring towards electrophilic attack.

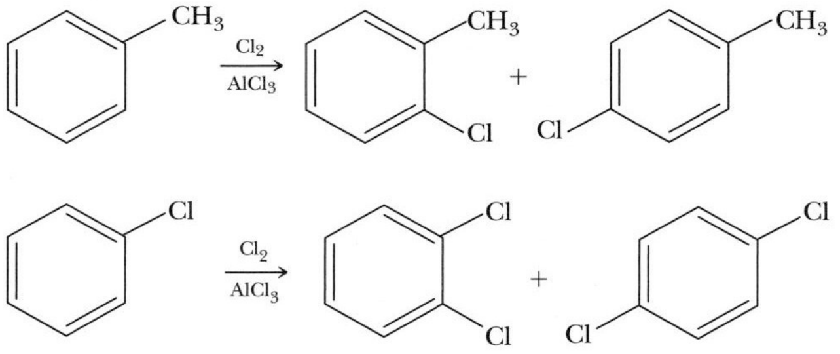

9.29 The trifluoromethyl group is a strong electron-withdrawing group and therefore a meta director. The three fluorine atoms make the carbon atom of the trifluoromethyl group, which is bonded directly to the ring, highly $\delta+$.

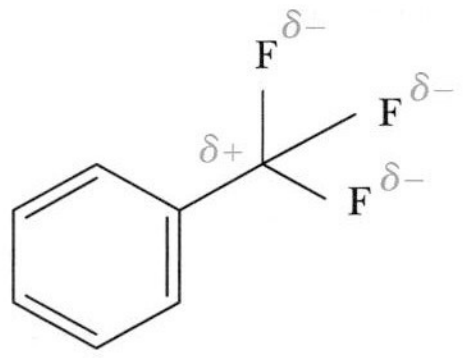

9.31 (a) $Cl/AlCl_3$

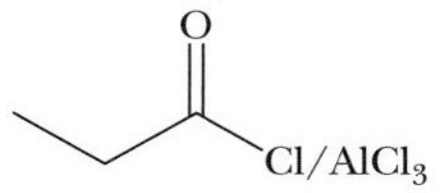

(b) HNO_3 (2 equiv.) / H_2SO_4

(c) $Cl/AlCl_3$

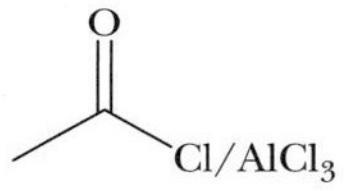

(d) 1) HNO_3/H_2SO_4
2) $Cl_2/AlCl_3$

9.33 Step 1: Protonation of acetone, forming a resonance stabilized cation.

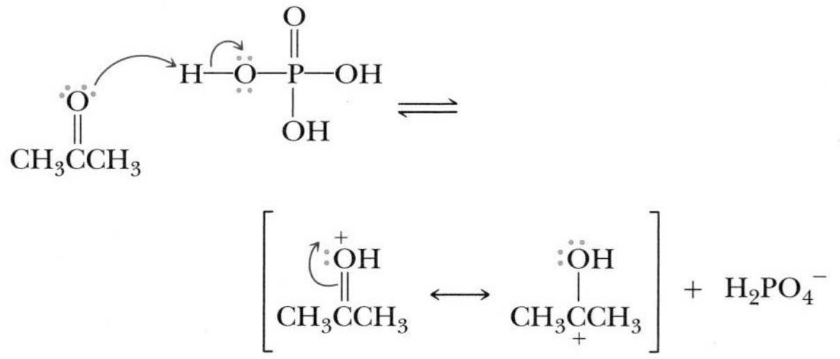

CH_3CCH_3 + $H_2PO_4^-$

Step 2: Electrophilic attack of the benzene ring by the carbocation.

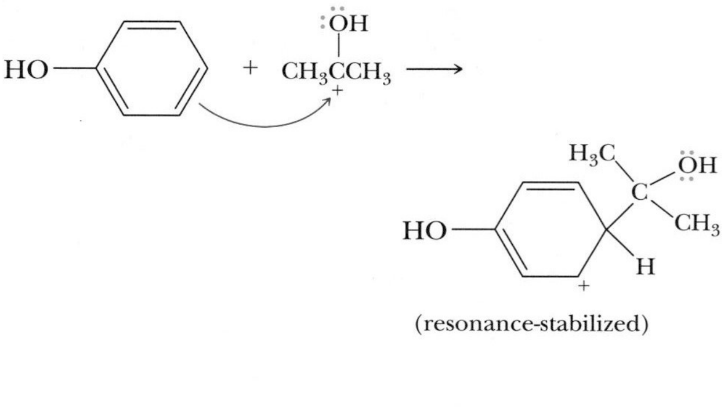

(resonance-stabilized)

Step 3: Deprotonation results in an alcohol.

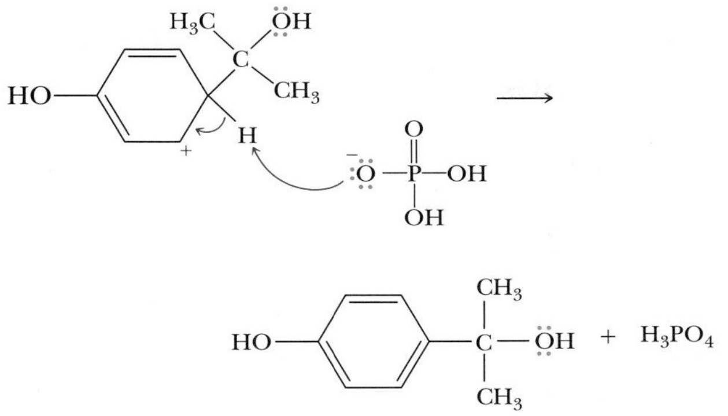

Step 4: In this step, the alcohol is converted into a good leaving group via protonation.

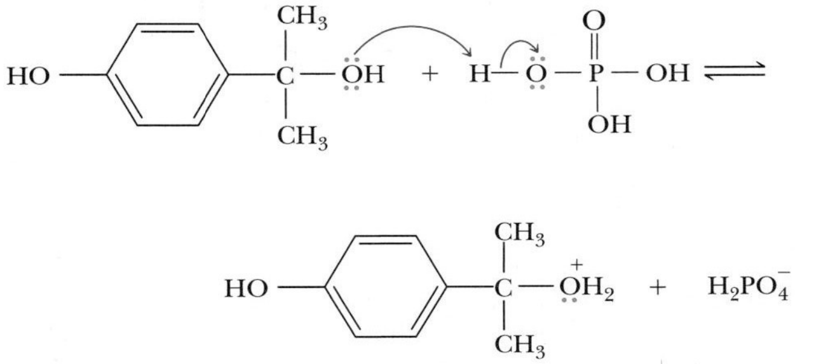

Step 5: Departure of the leaving group results in a resonance-stabilized carbocation.

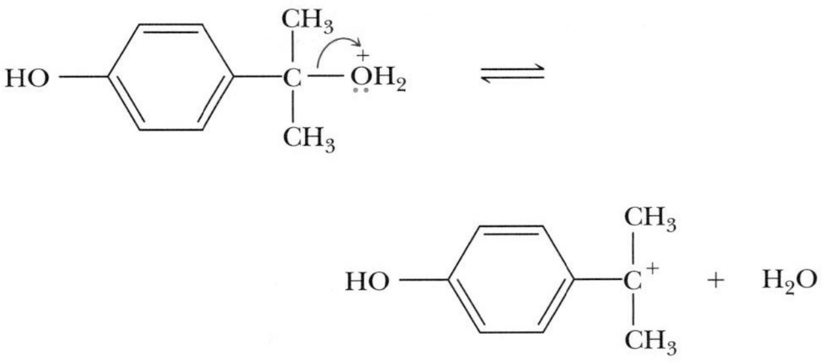

Step 6: This carbocation undergoes an electrophilic aromatic substitution with a second molecule of phenol.

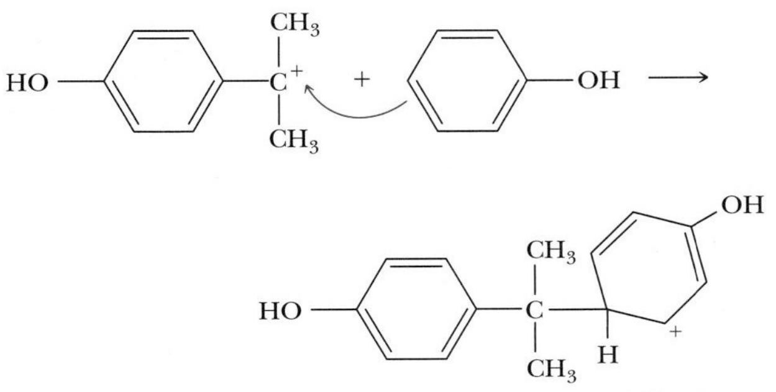

(resonance-stabilized)

Step 7: Deprotonation regenerates the aromatic ring and yields bisphenol A.

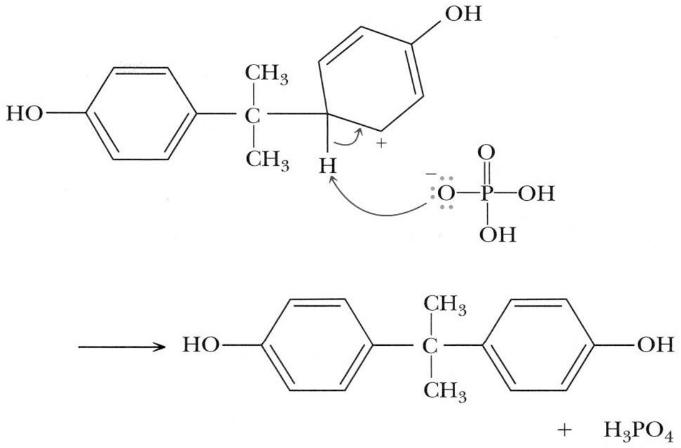

$+ \ H_3PO_4$

9.35 The oxygen atom of 2,4-dichlorophenol undergoes nucleophilic attack on the halogenated carbon of chloroacetic acid.

9.37 (a)

cyclohexanol < phenol < CH_3COOH

(b)

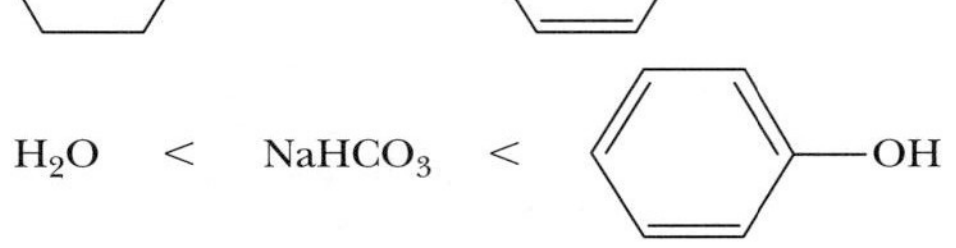

H_2O < $NaHCO_3$ < phenol

(c)

benzyl alcohol ($C_6H_5CH_2OH$) < phenol < O_2N–C_6H_4–OH

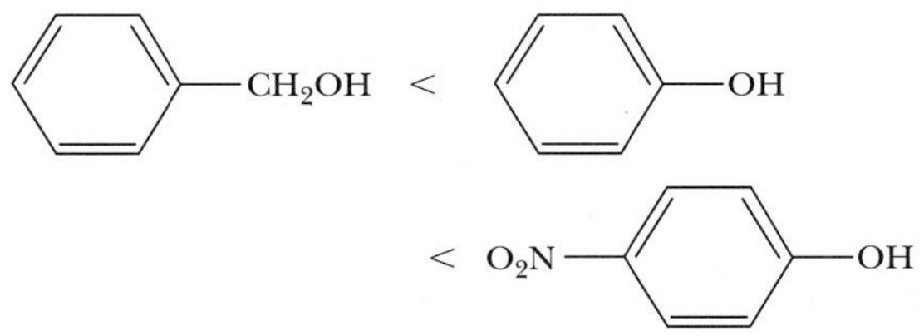

9.39 The gas formed is carbon dioxide, which arises from the decomposition of carbonic acid.
Carbonic acid is formed when HCO_3^- acts as a base. A carboxylic acid is a stronger acid than carbonic acid, so the bicarbonate ion is a strong enough base to deprotonate a carboxylic acid. However, since a phenol is a weaker acid than carbonic acid, bicarbonate is not a strong enough base to deprotonate phenol.

9.41

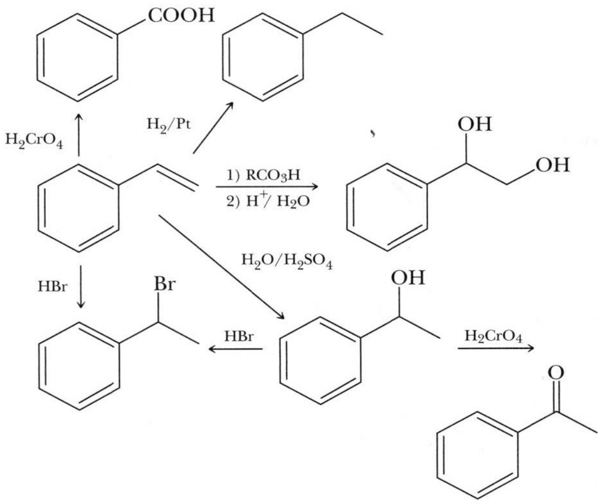

H_2CrO_4; H_2/Pt; 1) RCO_3H 2) H^+/H_2O; H_2O/H_2SO_4; HBr; HBr; H_2CrO_4

9.43 (a) 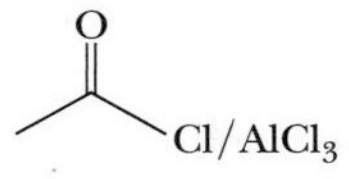 Cl/$AlCl_3$ (from toluene)

(b) 1) $Br_2/FeCl_3$ 2) 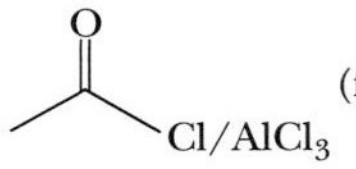 Cl/$AlCl_3$ (from benzene)

(c) 1) 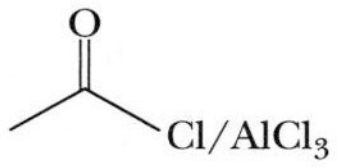 Cl/$AlCl_3$ 2) $Br_2/FeBr_3$ (from benzene)

9.45 (a) OH, OH + H_2O_2 ⟶ O, O + $2H_2O$ + heat

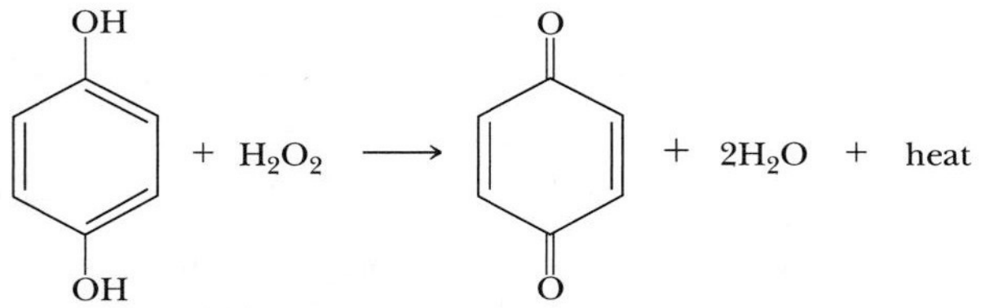

(b) Hydroquinone has lost hydrogen atoms (two), so it has been oxidized.

9.47

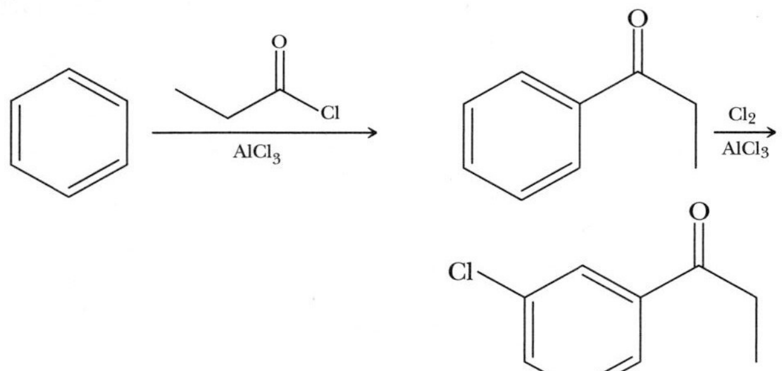

Cl, $AlCl_3$; Cl_2 / $AlCl_3$; O; Cl

9.49 None of these compounds can be made *directly*.

9.51 First, $AlCl_3$ decomposes in the presence of protic acids (ionizable hydrogens) to form HCl. Second, $AlCl_3$ (a Lewis acid) reacts with OH, SH, and NH_2 groups, which are Lewis bases.

9.53 Cl

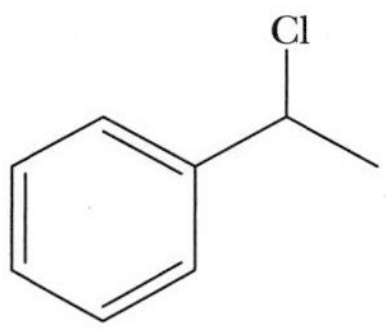

Chapter 10
Amines

10.1 (a) H, N, H, * (b) H, *, N, CH_3, N

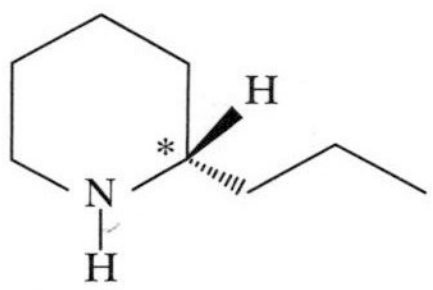

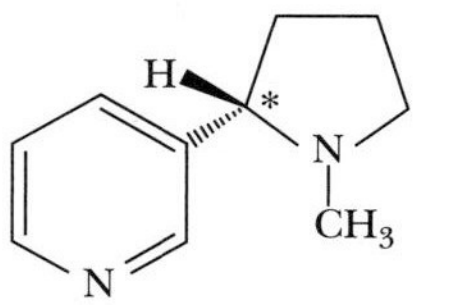

(c) H_3C, N, O, CH_3, O, *, *, *, *, H, O, H, O

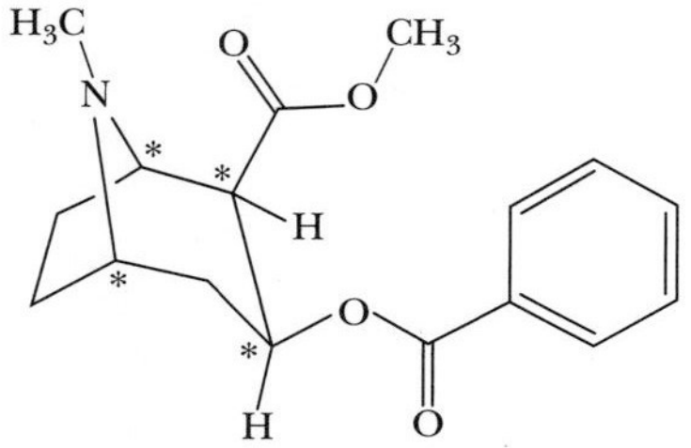

10.2 (a) 2-Methyl-1-propanamine (b) Cyclohexanamine

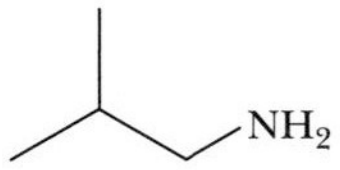

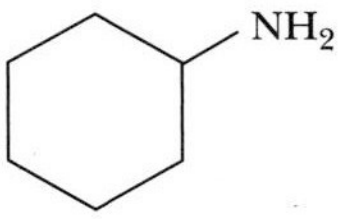

NH_2 NH_2

(c) (*R*)-2-Butanamine

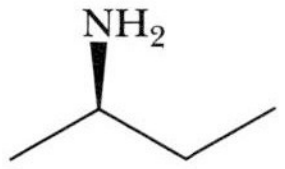

NH_2

10.3 (a) NH_2 (b) N

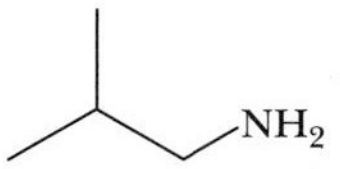

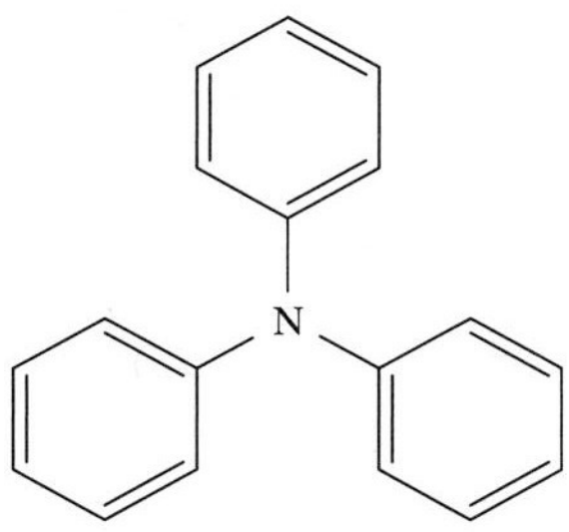

(c) H N

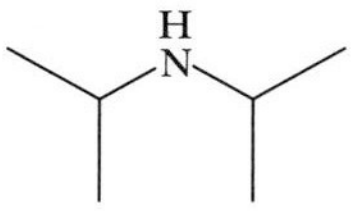

10.4 Diethylamine has an NH group that acts as a hydrogen-bond acceptor and donor. No hydrogen bonding is possible with the ether.

10.5 The equilibrium lies to the left.

10.6 (a) A (b) C

10.7 (a) $(CH_3CH_2)_3NH^+Cl^-$ triethylammonium chloride

(b) 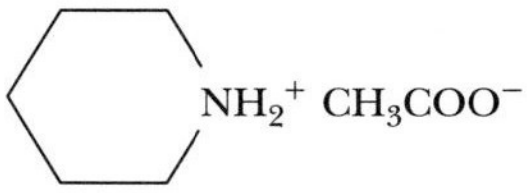 NH_2^+ CH_3COO^- piperidinium acetate

10.8 (a) $CH_3CH(NH_3^+)COH$ with C=O (b) $CH_3CH(NH_2)CO^-$ with C=O

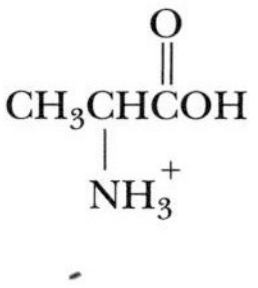

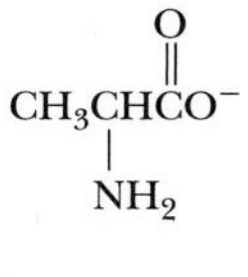

10.9

CH_3 —H_2CrO_4→ COOH —HNO_3/H_2SO_4→ COOH, NO_2 —H_2/Pt→

COOH, NH_2 —HNO_2/HCl, heat→ COOH, OH

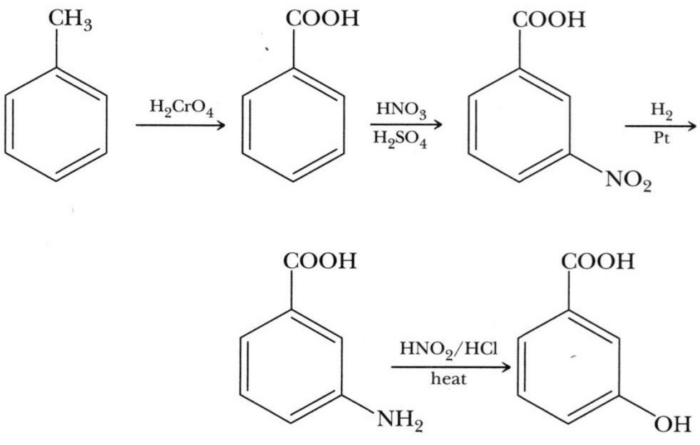

10.10 CH_3, $\overset{+}{N}$—H, Cl^- + CH_3, $\overset{+}{N}$—CH_3, Cl^-

10.11

(a) NH_2

(b) $CH_3(CH_2)_6CH_2NH_2$

(c) NH_2

(d) H_2N, NH_2

(e) NH_2, Br

(f) $(CH_3CH_2CH_2CH_2)_3N$

(g)

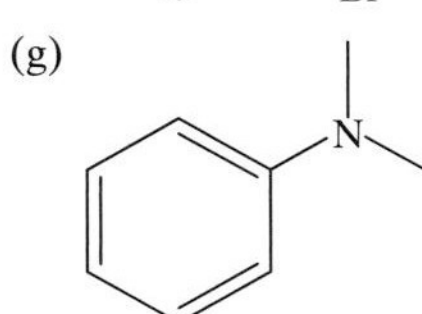

(h)

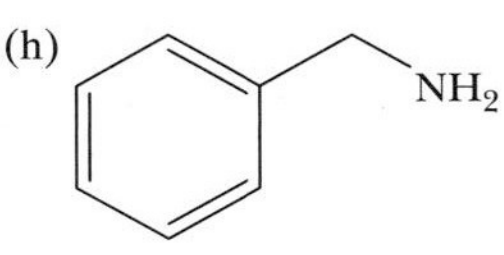

(i)

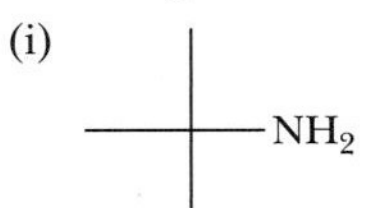

(j) H, N

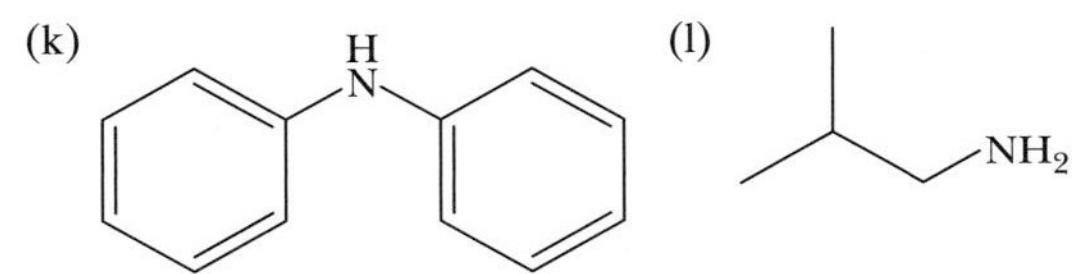

(k)

(l)

10.13 Amines are classified by the number of carbon substituents bonded to the nitrogen atom. Whereas, alcohols are classified by the number of carbon substituents bonded to the carbon bearing the OH group.

NH_2 — 1° amine; N, H — 2° amine; N — 3° amine

OH — 1° alcohol; OH — 2° alcohol; OH — 3° alcohol

10.15 (a) Both amino groups are secondary.

(b) Both compounds contain the same basic framework and have the same configuration at the stereocenter. Differences are the *tert*-butyl and CH_2OH groups in (*R*)-albuterol versus the methyl and OH groups in (*R*)-epinephrine, respectively.

10.17 (a) (b)

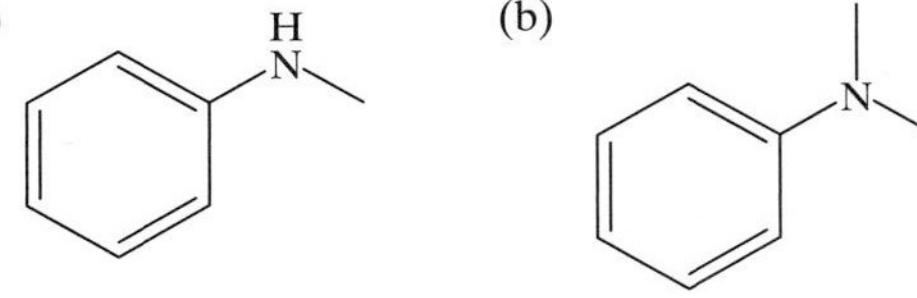

(c) NH_2 (d) NH_2

(e) N (f) NH_2 + isomers

(g) N, +, Cl^- or N, +, Cl^-

10.19 The O—H hydrogen bond is stronger than the N—H hydrogen bond.

10.21 Nitrogen is less electronegative than oxygen, so the lone pair on the nitrogen atom of an amine is less stable (and less tightly held) compared to that on the oxygen atom of an alcohol.

10.23 The nitro group is an electron-withdrawing group. In addition, it is able to stabilize the lone pair of electrons on the nitrogen atom in 4-nitroaniline by resonance.

NH_2 ⟷ $\overset{+}{N}H_2$

Likewise, the nitro group can stabilize the conjugate base of 4-nitrophenol, 4-nitrophenoxide, by both induction and resonance.

O^- ⟷ O

10.25

(a)

CH_3COOH + pyridine ⇌ pyridinium ($\overset{+}{N}$H) + CH_3COO^-

stronger acid $pK_a = 4.76$ | stronger base | weaker acid $pK_a = 5.25$ | weaker base

(b)

phenol (OH) + $(CH_3CH_2)_3N$ ⇌

stronger acid $pK_a = 9.95$ | stronger base

phenoxide (O^-) + $(CH_3CH_2)_3\overset{+}{N}H$

weaker base | weaker acid $pK_a = 10.75$

(c)

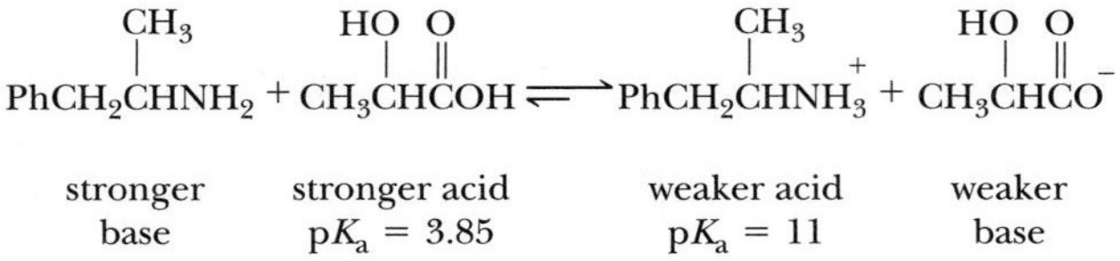

$$\mathrm{PhCH_2CH(CH_3)NH_2 + CH_3CH(OH)COOH \rightleftharpoons PhCH_2CH(CH_3)NH_3^+ + CH_3CH(OH)CO_2^-}$$

stronger base | stronger acid $pK_a = 3.85$ | weaker acid $pK_a = 11$ | weaker base

(d)

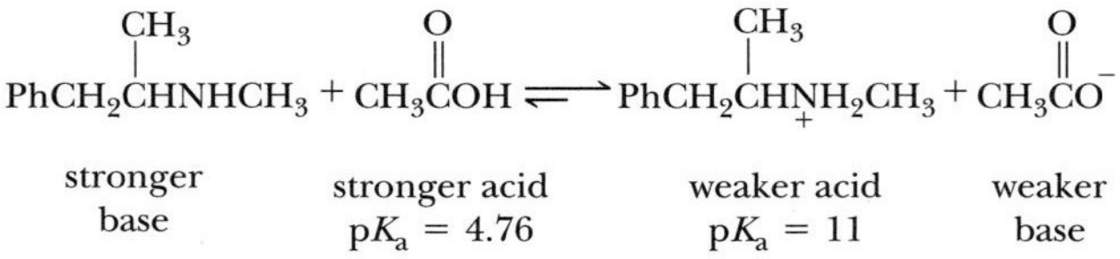

$$\mathrm{PhCH_2CH(CH_3)NHCH_3 + CH_3COOH \rightleftharpoons PhCH_2CH(CH_3)NH_2^+CH_3 + CH_3CO_2^-}$$

stronger base | stronger acid $pK_a = 4.76$ | weaker acid $pK_a = 11$ | weaker base

10.27 1 : 2500

10.29 (a) The N of the CH_2NH_2 group.

(b) $CH_2NH_3^+\ Cl^-$ HO CH_2OH H_3C N

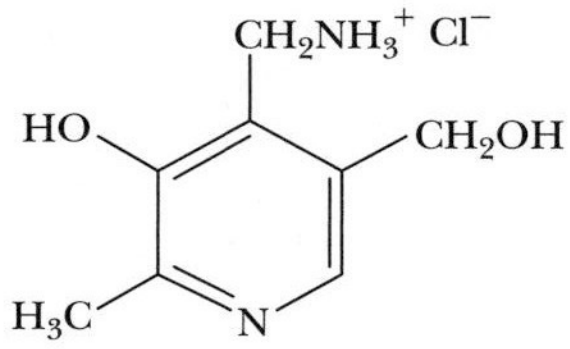

10.31 (a) The tertiary amine.

(b) O Cl^- ^+N O H H_2N

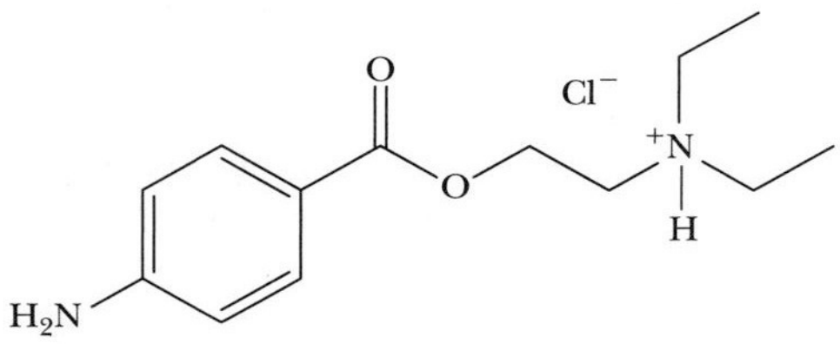

(c) Neither procaine or its salt is chiral. Optically inactive.

10.33 First, dissolve the reaction mixture in an organic solvent such as diethyl ether. The ethereal solution can be extracted with an aqueous acid, such as HCl, which reacts with aniline, a base, to form a water-soluble salt. Thus, the ether layer contains nitrobenzene while the aqueous layer contains the anilinium salt. After separating the two layers, the acidic aqueous layer can be basified with NaOH to deprotonate the anilinium ion. Extraction of the basified aqueous layer with diethyl ether, followed by evaporation of the ether, allows aniline to be isolated.

10.35

(a) $Cl^-\ \overset{+}{N}H_2$ NH N N H NH_2 or NH $\overset{+}{N}H_2\ Cl^-$ N N H NH_2

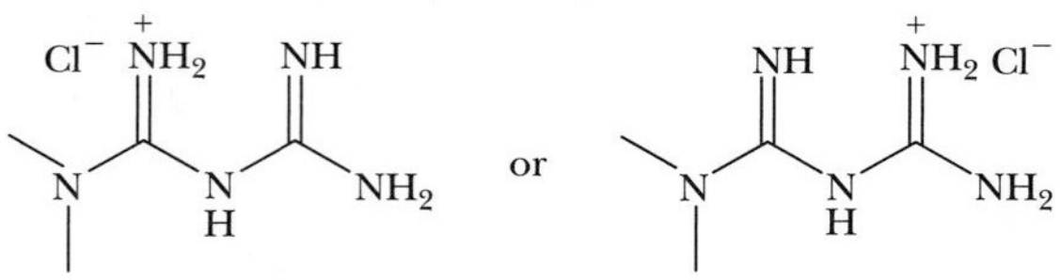

(b) Soluble in both water and blood plasma. Insoluble in diethyl ether and dichloromethane.

10.37 1. HNO_3/H_2SO_4 2. H_2CrO_4 3. H_2/Ni

10.39

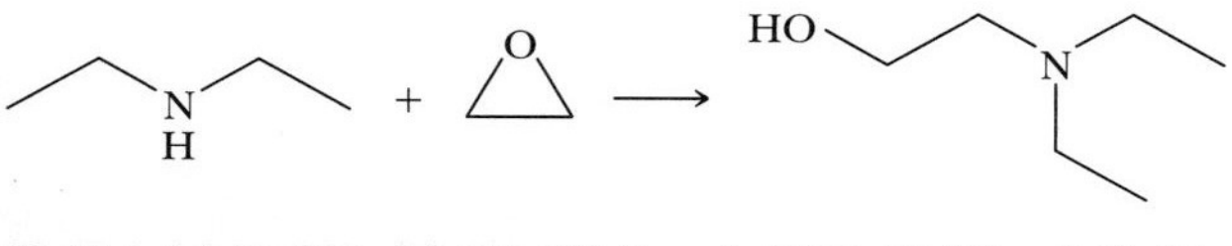

N H + O ⟶ HO N

10.41 1. (a) NaOH (b) CH_3CH_2Br 2. HNO_3/H_2SO_4 3. H_2/Ni

10.43

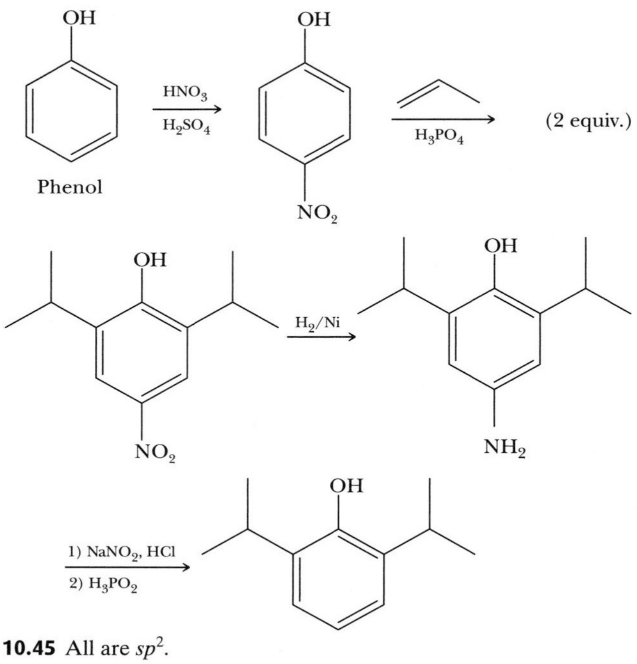

OH Phenol $\xrightarrow[H_2SO_4]{HNO_3}$ OH NO_2 $\xrightarrow{H_3PO_4}$ (2 equiv.)

OH NO_2 $\xrightarrow{H_2/Ni}$ OH NH_2

$\xrightarrow[2)\ H_3PO_2]{1)\ NaNO_2,\ HCl}$ OH

10.45 All are sp^2.

10.47

H N H H H H H H H H H H N H H H H

H H H H N H H H H H H H H N H H H

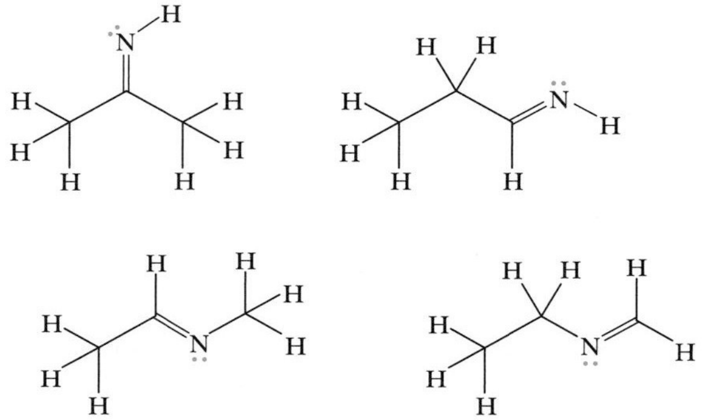

Putting It Together

1.	(e)	**2.**	(d)	**3.**	(e)	**4.**	(c)
5.	(d)	**6.**	(d)	**7.**	(b)	**8.**	(c)
9.	(c)	**10.**	(a)	**11.**	A		

12. (a) No (b) 8 (c) Yes (d) 4

13. (a) *N*-ethyl-*N*-propyl-1-butanamine
(b) (2*R*,5*S*)-5-phenyl-2-hexanol
(c) 1-ethoxy-2-methylpropane (ethyl isobutyl ether)
(d) 3-isopropyl-5-methylphenol

14. The conjugate base of **B** is aromatic, but that of **A** is not. Proton **B** is more acidic than proton **A**.

15. (a) O^-

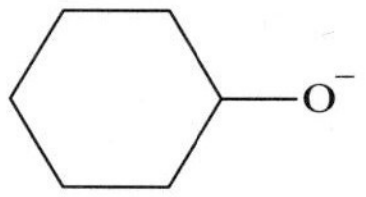

The strength of a nucleophile is inversely related to the stability of the lone pair.

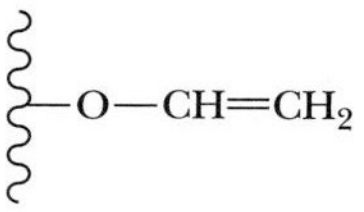

$-O-CH{=}CH_2$

$CH_2{=}CH{-}O^-$ is the most stable leaving group because it is stabilized by resonance.

16.

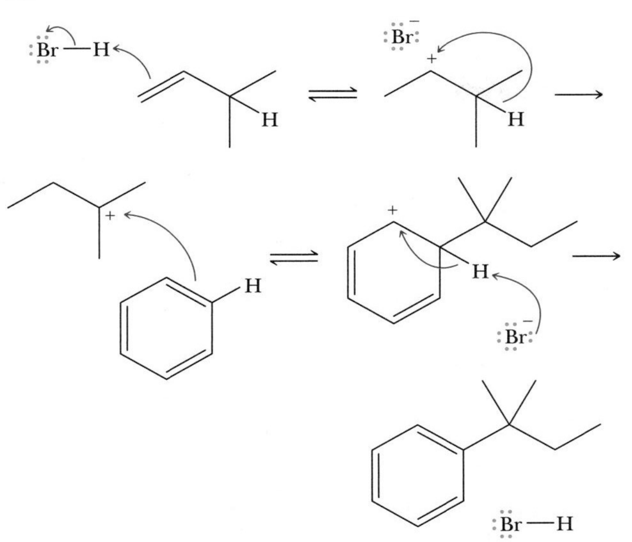

:Br:⁻ + H ⇌ H ⟶ H ⇌ H ⟶ :Br:⁻ Br—H

17. Alkyl groups are electron-donating, so amines with more alkyl groups are more basic and hence better nucleophiles.

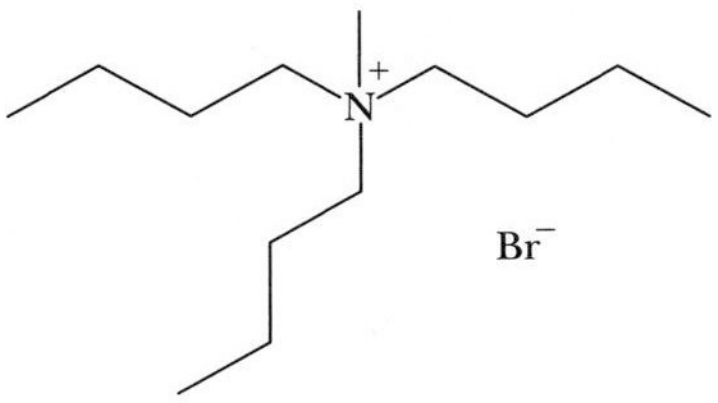

N^+ Br^-

18. (a)

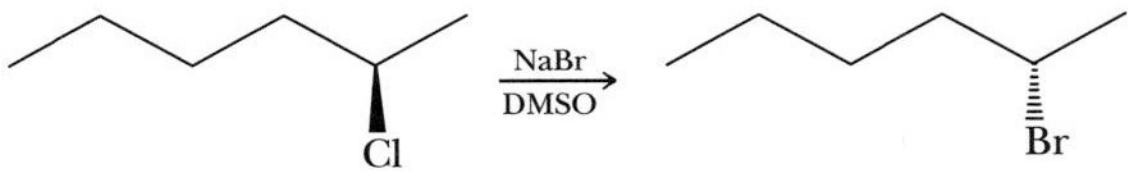

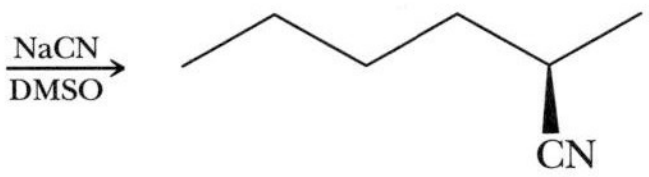

Cl $\xrightarrow[\text{DMSO}]{\text{NaBr}}$ Br

$\xrightarrow[\text{DMSO}]{\text{NaCN}}$ CN This enantiomer as the *only* product

(b)

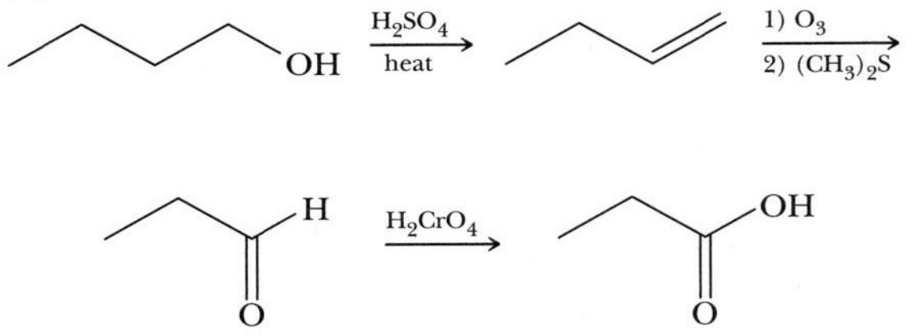

OH $\xrightarrow[\text{heat}]{H_2SO_4}$ $\xrightarrow[2)\ (CH_3)_2S]{1)\ O_3}$

H O $\xrightarrow{H_2CrO_4}$ OH O

(c)

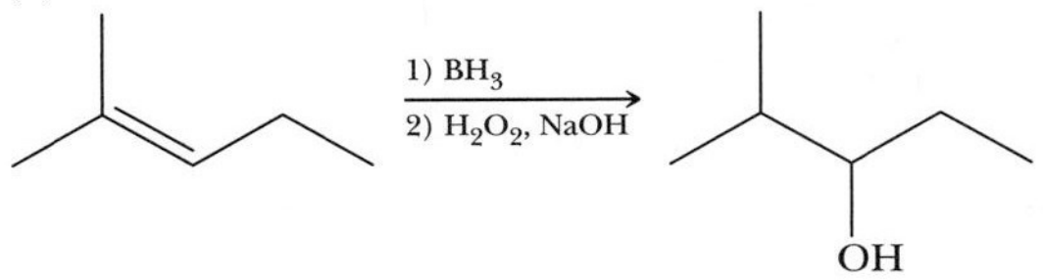

$\xrightarrow[2)\ H_2O_2,\ NaOH]{1)\ BH_3}$ OH racemic mixture

$\xrightarrow[2)\ CH_3I]{1)\ Na}$ OCH_3 racemic mixture

(d)

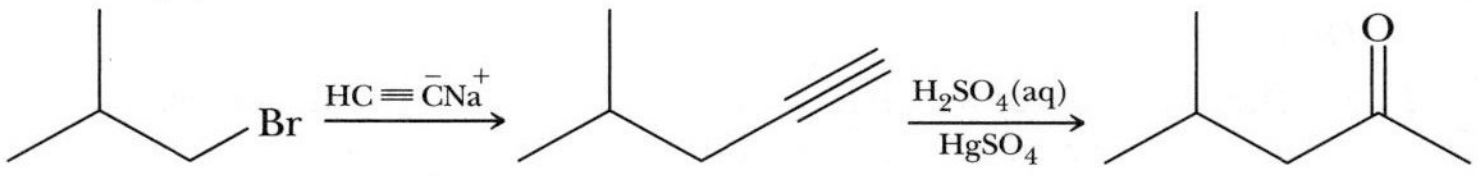

Br $\xrightarrow{HC\equiv \bar{C}Na^+}$ $\xrightarrow[HgSO_4]{H_2SO_4(aq)}$ O

19.

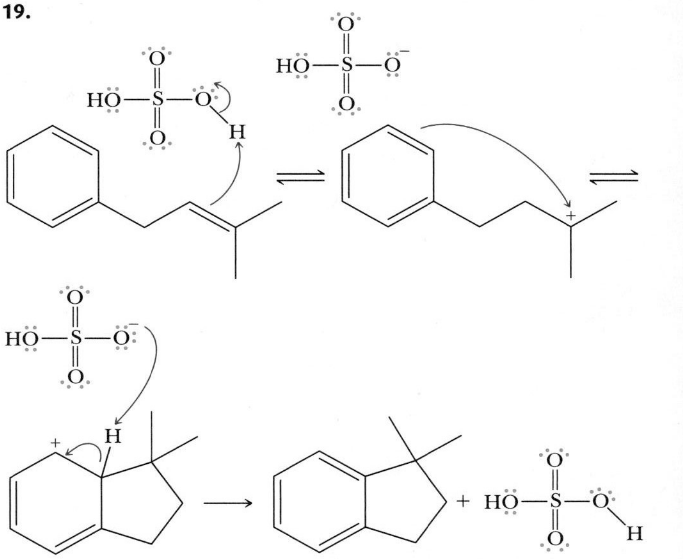

HO—S(=O)(=O)—O—H HO—S(=O)(=O)—O⁻ ⇌ + ⇌

HO—S(=O)(=O)—O⁻ + H ⟶ + HO—S(=O)(=O)—O—H

20.
(a)

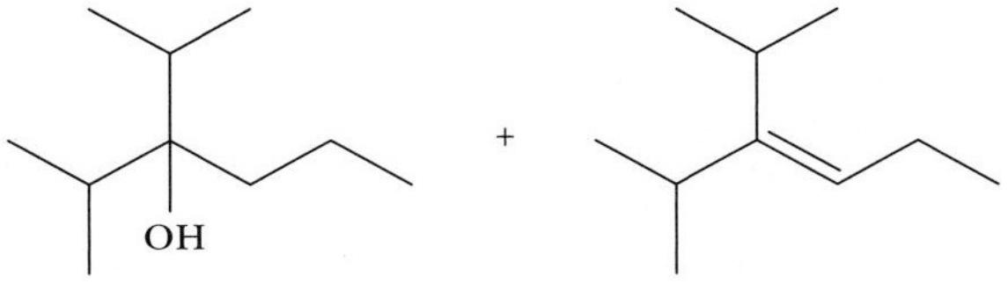

OH +

(via S_N1) (via E2)

(b) OH O OH

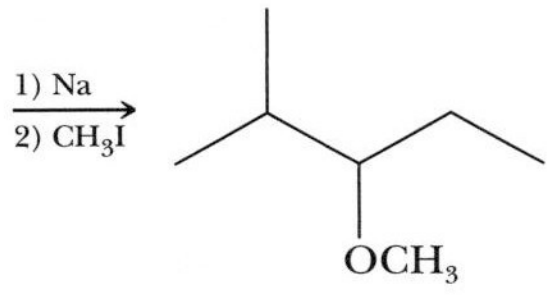

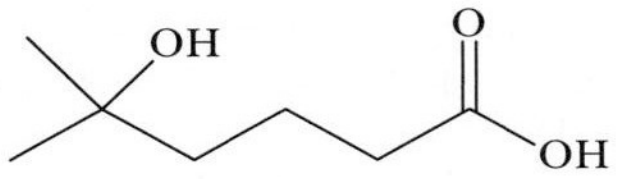

(c) CH_3 H_3C (d) O O H

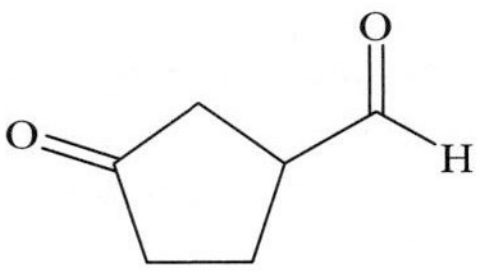

(e) O_2N— —NO_2

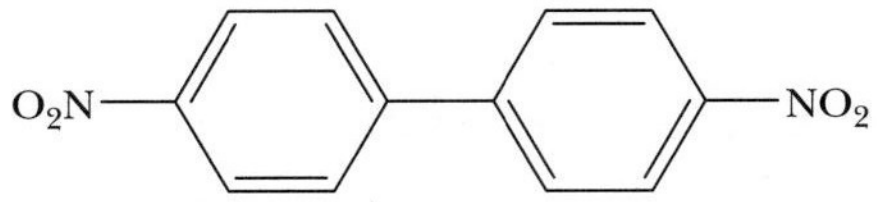

21.

8 7 Br 6 5 4 3 2 1 ⟶

⟶ 8 7+ 6 5 4 3 2 ⟶ 8 7 6 5 4 3 2 + 1

↓ $CH_3CH_2\ddot{O}H$

O—CH_2CH_3 $\xleftarrow{CH_3CH_2\ddot{O}H}$ H +O CH_2CH_3

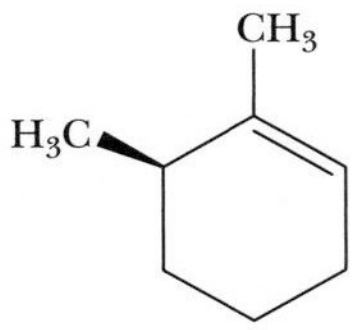

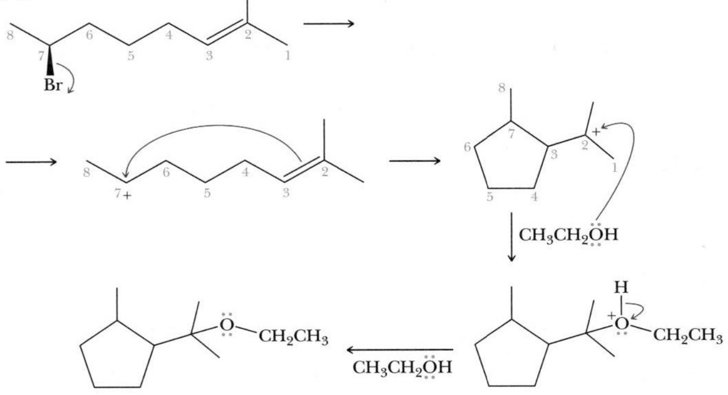

Chapter 11
Infrared Spectroscopy

11.1 The energy of red light (680 nm) is 42.1 kcal/mol and is higher than the energy from infrared radiation of 2500 nm (11.4 kcal/mol).
11.2 A carboxyl group in a carboxylic acid.
11.3 Propanoic acid will have a strong, broad OH absorption between 2400–3400 cm^{-1}.
11.4 The wavenumber is directly related to the energy of the infrared radiation required to stretch the bond. Bonds that require more energy to stretch are stronger bonds.
11.5 IHD = 2. Cyclohexene contains one ring and one C—C double bond.
11.6 Niacin contains one ring and four π bonds.
11.7 Green laser pointer.
11.9 349 nm, UV region.
11.11 (a) 6 (b) 3 (c) 4 (d) 1 (e) 5 (f) 1
11.13 (a) 1
(b) one ring or one π bond
(c) A carbon–carbon double bond that is not bonded to any hydrogen atoms
11.15 G = *tert*-butyl methyl ether
H = 2-methyl-1-butanol
11.17 (a) zero (b) none (c) 1° amine
11.19 (a) one
(b) one ring or one π bond
(c) ester, aldehyde, or ketone
11.21 (a) 1-Butanol will have a strong, broad O—H absorption between 3200 and 3400 cm^{-1}.
(b) Butanoic acid will have a strong C=O absorption near 1700 cm^{-1}.
(c) Butanoic acid will have a broad O—H absorption between 2400 and 3400 cm^{-1}.
(d) Butanal will have a strong C=O absorption near 1700 and 1725 cm^{-1}. 1-Butene will have a vinylic =C—H absorption near 3100 cm^{-1}.
(e) Butanone will have a strong C=O absorption near 1725 cm^{-1}. 2-Butanol will have a strong, broad O—H absorption between 3200 and 3400 cm^{-1}.
(f) 2-Butene will have a vinylic =C—H absorption near 3100 cm^{-1}.
11.23 (a) Centered at 3200 cm^{-1}. (b) 3050 cm^{-1}
(c) 1680 cm^{-1} (d) Between 1450 and 1620 cm^{-1}.
11.25 The C=O stretching absorption for acetate will be at a lower wavenumber.
11.27 The same

Chapter 12
Nuclear Magnetic Resonance Spectroscopy

12.1 Both
12.2 (a) Four sets of equivalent hydrogens. The number of hydrogens in each set are 6, 4, 3, and 1.
(b) Four sets of equivalent hydrogens. The number of hydrogens in each set are 9, 6, 2, and 1.
(c) Two sets of equivalent hydrogens. The number of hydrogens in each set are 6 and 2.
12.3 (a) $H_3C-C(=O)-CH_3$ (b) cyclopentane
(c) $H_3C-C(CH_3)_2-CH_3$ (d) $H_3C-CCl_2-CCl_2-CH_3$
12.4 12 and 2 hydrogens. 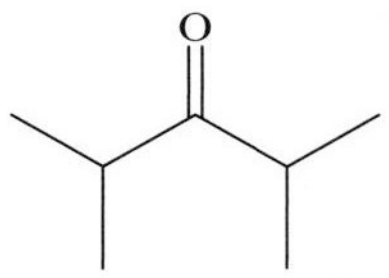
12.5 (a) 3 (b) 3 : 3 : 2
(c) CH_2 hydrogens in (1) are more downfield
12.6 (a) $CH_3OCH_2CCH_3$ (with C=O) three singlets and $CH_3CH_2COCH_3$ (with C=O): quartet, triplet, singlet
(b) $CH_3CCl_2CH_3$ one singlet and $ClCH_2CH_2CH_2Cl$: quintet, triplet
12.7 (a) The compound on the left has a plane of symmetry and will have five ^{13}C signals, while the compound on the right will have seven signals.
(b) The compound on the left will have six ^{13}C signals, while the compound on the right, will have three signals.
12.8 OH
12.9 O
12.10 Br Br
12.11 (a) four (b) five (c) two (d) two
(e) five (f) four (g) three (h) five
12.13 (a) 2 (b) 3 (c) 1
12.15 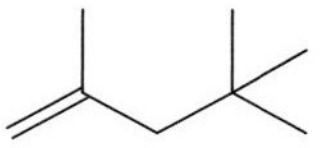
12.17 (a) 3 (b) 2 (c) 1
12.19 O
12.21 (1) J (2) H (3) I
12.23 OH
12.25 OH

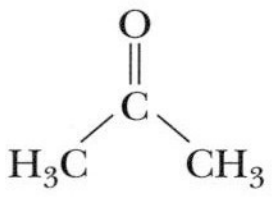

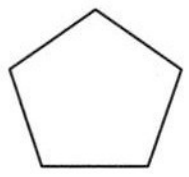

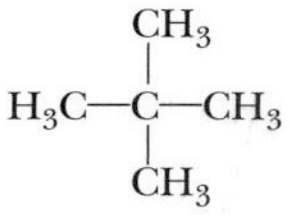

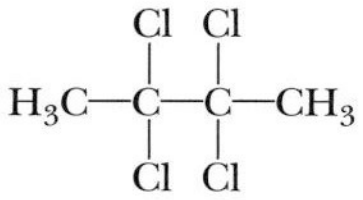

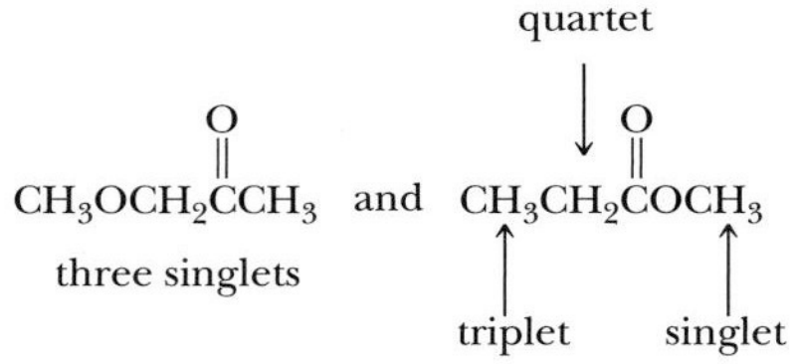

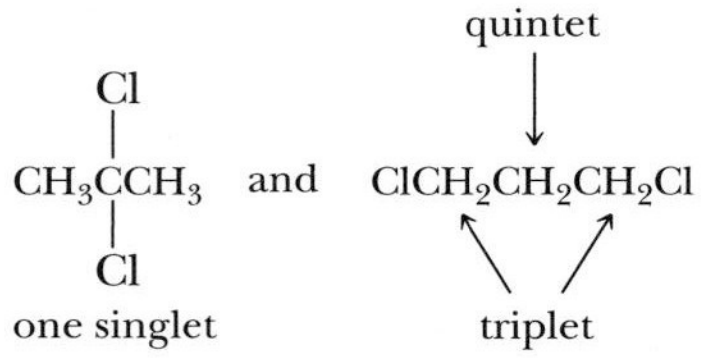

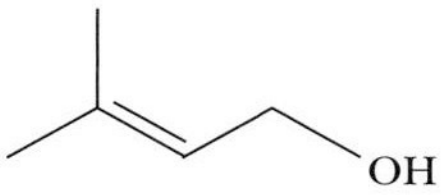

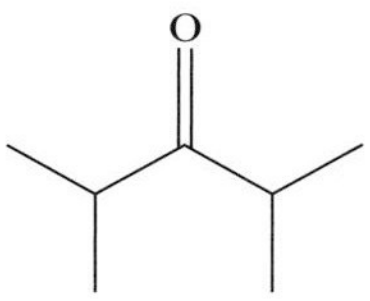

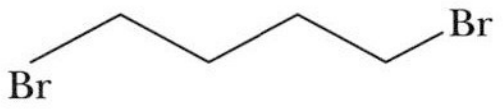

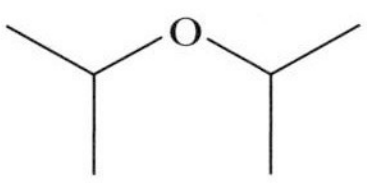

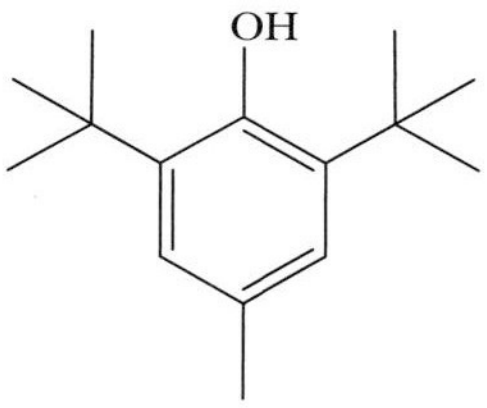

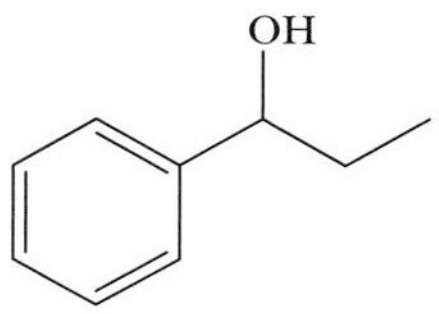

compound M 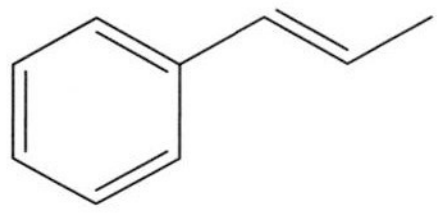 compound N

12.27

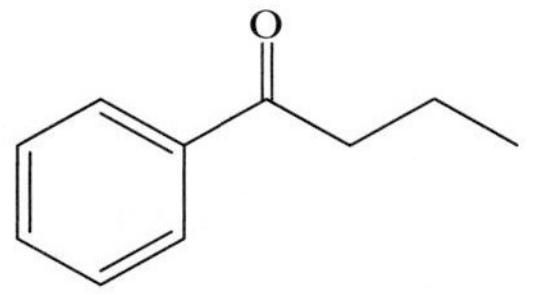

12.29

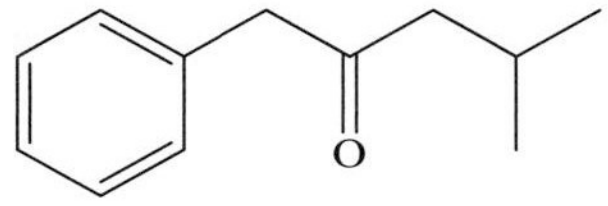

12.31

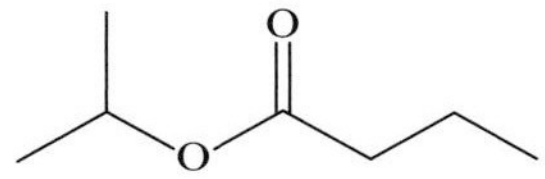

12.33

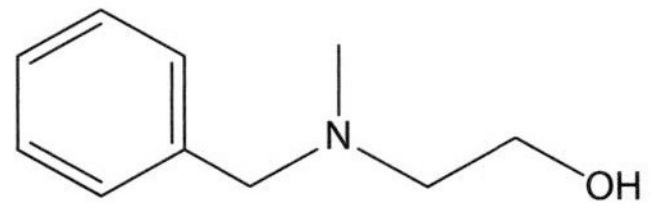

OH

12.35

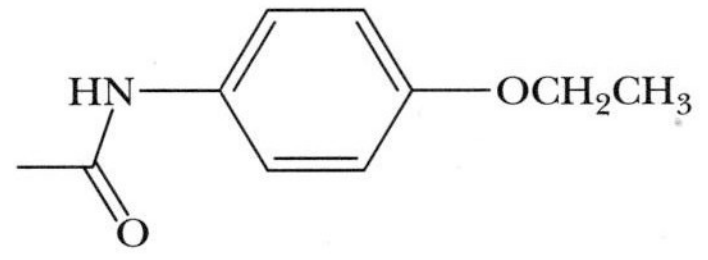

HN— —OCH_2CH_3

12.37

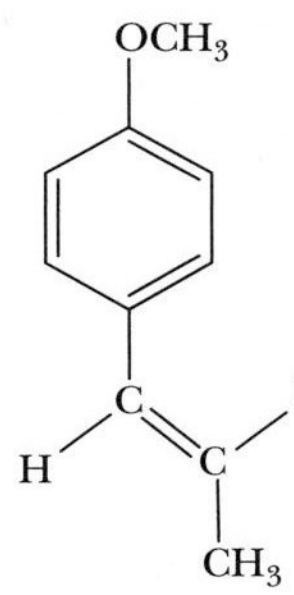

OCH_3

H C H C CH_3

12.39

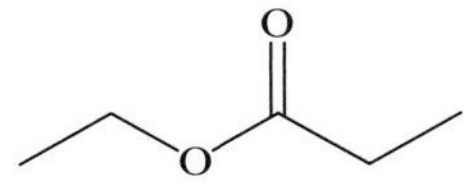

12.41

OH

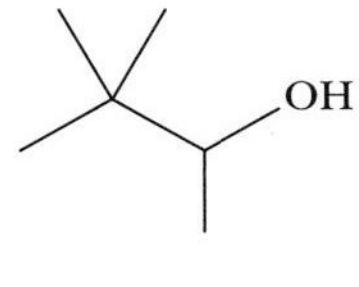

Chapter 13 Aldehydes and Ketones

13.1 (a) 2,2-dimethylpropanal
(b) (*R*)-3-hydroxycyclohexanone
(c) (*R*)-2-phenylpropanal

13.2

CHO hexanal

CHO 4-methylpentanal

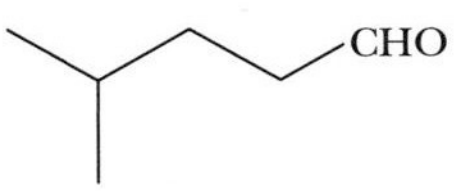

* CHO 3-methylpentanal

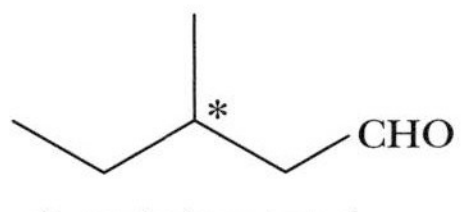

* CHO 2-methylpentanal

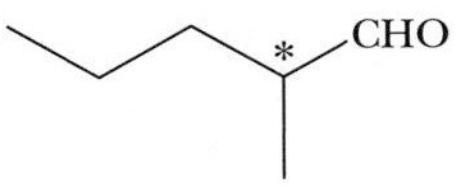

CHO 3,3-dimethylbutanal

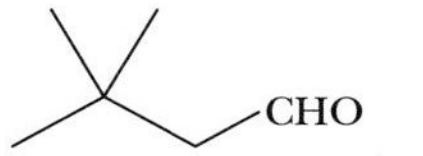

CHO 2,2-dimethylbutanal

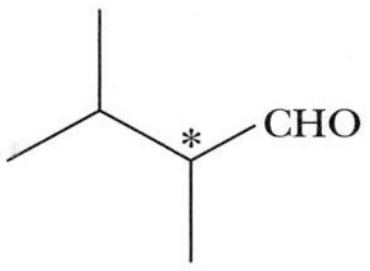

* CHO 2,3-dimethylbutanal

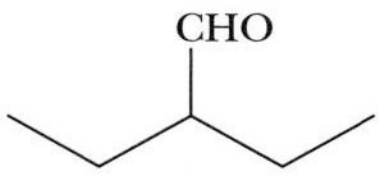

CHO 2-ethylbutanal

13.3 (a) 2-hydroxypropanoic acid
(b) 2-oxopropanoic acid
(c) 4-aminobutanoic acid

13.4 Grignard reagents, which behave as carbanion equivalents, are highly basic and are destroyed by acids. Both of these reagents contain acidic functional groups; (a) contains a phenol and (b) contains a carboxylic acid.

13.5

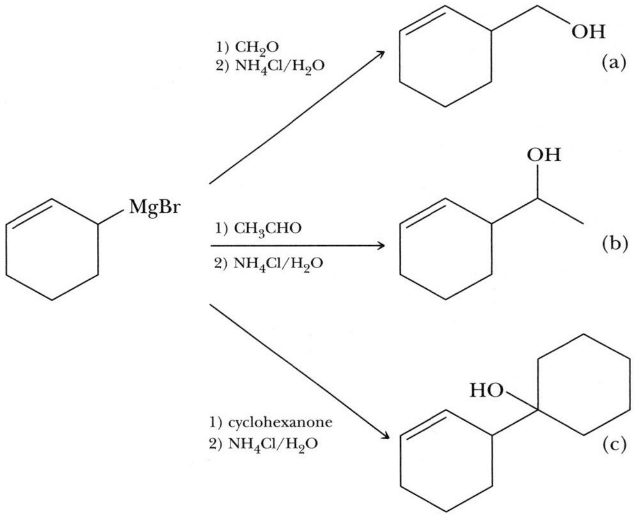

MgBr

1) CH_2O
2) NH_4Cl/H_2O
OH (a)

1) CH_3CHO
2) NH_4Cl/H_2O
OH (b)

1) cyclohexanone
2) NH_4Cl/H_2O
HO (c)

13.6 (a)

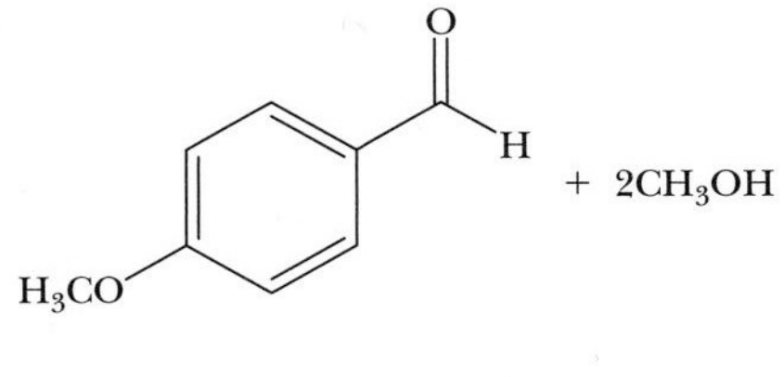

O H H_3CO + $2CH_3OH$

(b) O + HO OH

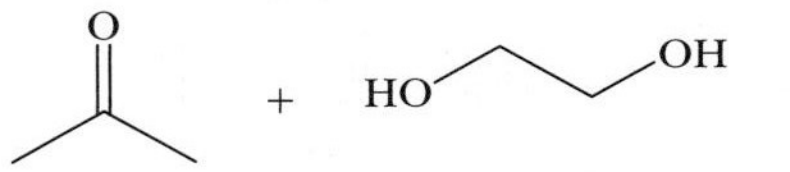

(c) OH O H + CH_3OH

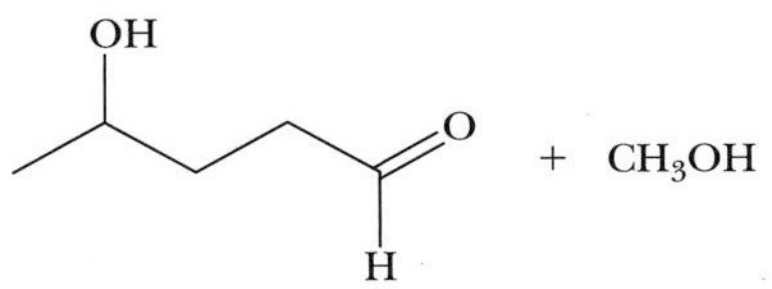

13.7

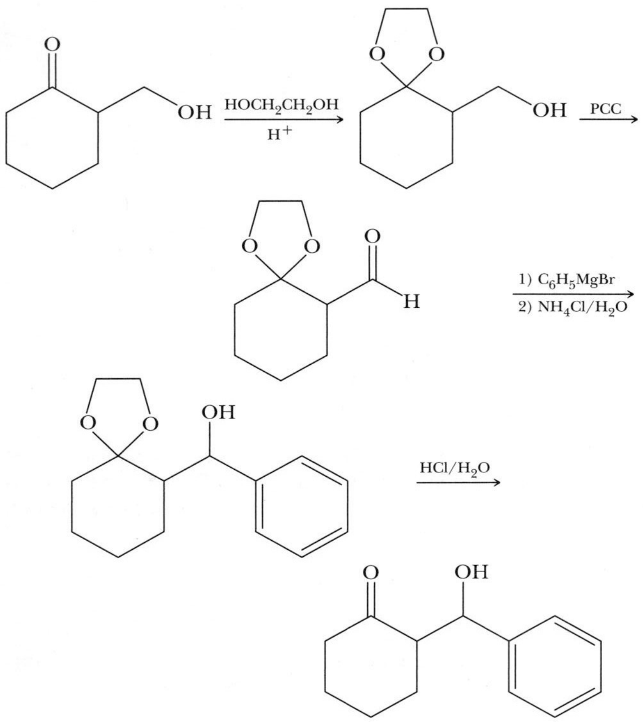

$\xrightarrow[H^+]{HOCH_2CH_2OH}$ OH $\xrightarrow{PCC}$

H $\xrightarrow[2)\ NH_4Cl/H_2O]{1)\ C_6H_5MgBr}$

OH $\xrightarrow{HCl/H_2O}$

O OH

13.8 (a) CHO + $H_3\overset{+}{N}CH_2CH_3$

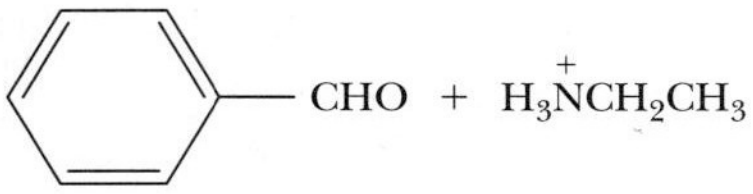

(b) N— —OCH_3 + H_2O

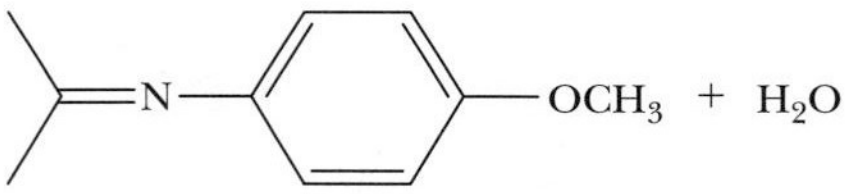

13.9 (a) O + H_2N— $\xrightarrow[H_2O]{H^+}$

N— $\xrightarrow{H_2/Ni}$

H N

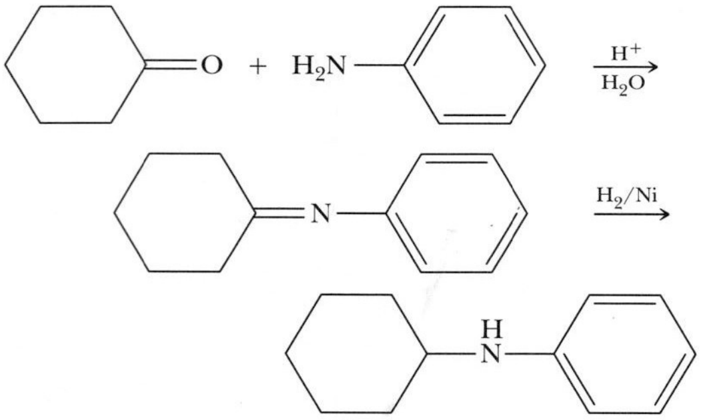

(b) O + NH_3 $\xrightarrow[H_2O]{H^+}$ NH $\xrightarrow{H_2/Ni}$

NH_2

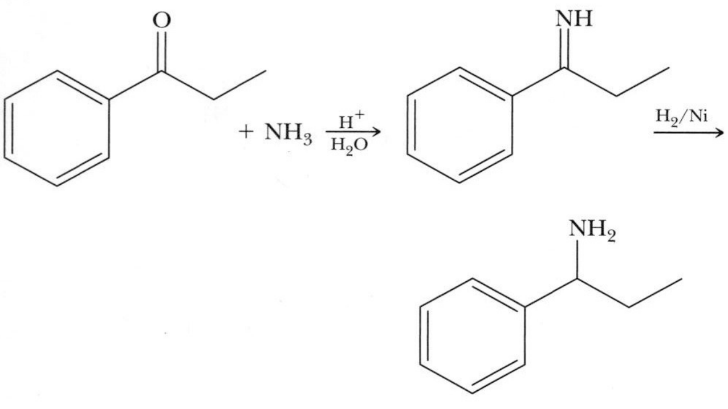

13.10 (a) O O H (b) O OH

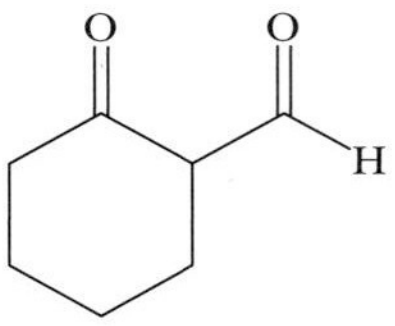

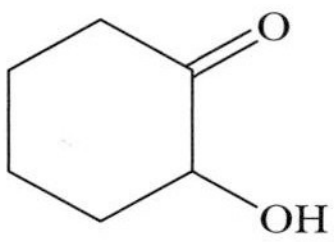

(c) O

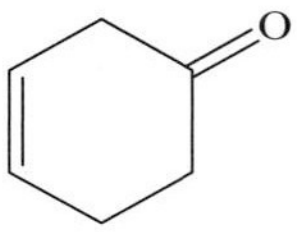

13.11 (a) O O OH

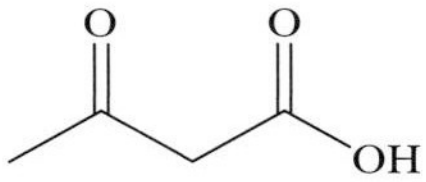

(b) O OH

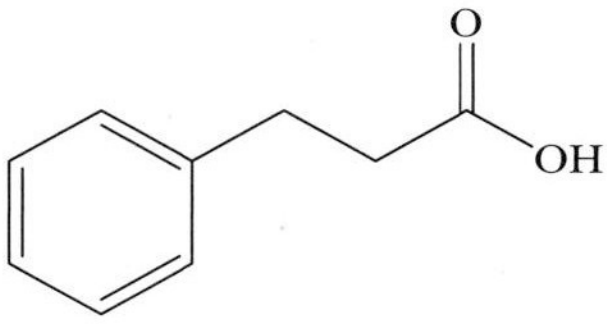

13.12 (a) O (b) O $CH_2\overset{\|}{C}H$

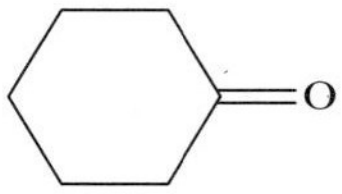

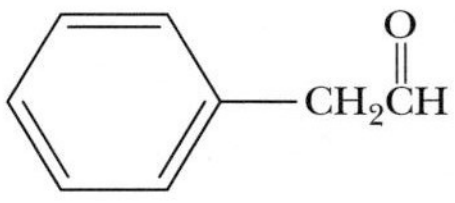

(c) O O

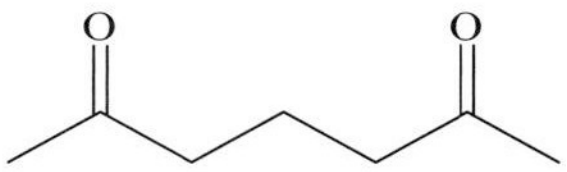

13.13

(a) O (b) CHO

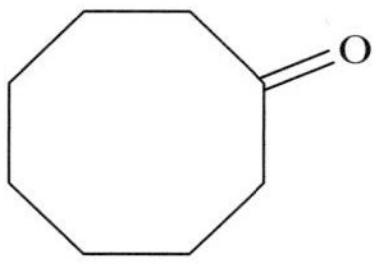

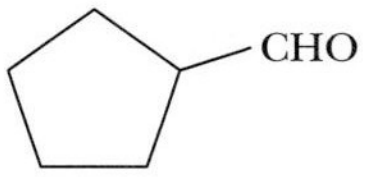

(c) COOH (d) O

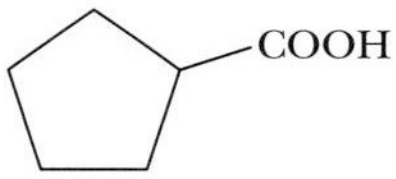

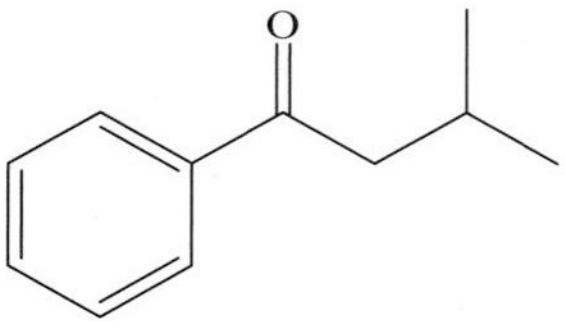

13.15

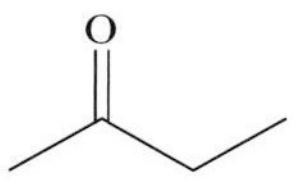

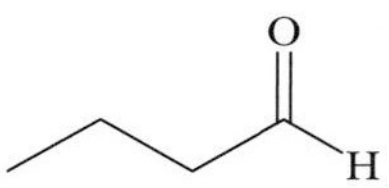

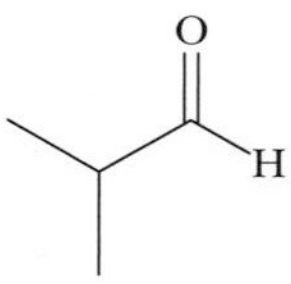

O O H O H

13.17 (a) 4-heptanone
(b) (*S*)-2-methylcyclopentanone
(c) (*Z*)-2-methyl-2-pentenal
(d) (*S*)-2-hydroxypropanal
(e) 2-methoxyacetophenone
(f) 2,2-dimethyl-3-oxopropanoic acid
(g) (*S*)-2-propylcyclopentanone

13.19

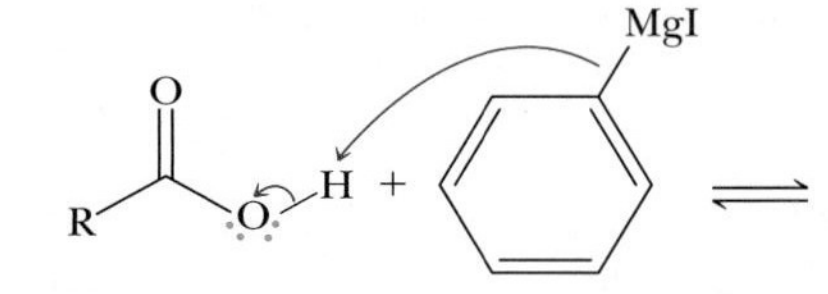

MgI

R O H +

$pK_a = 5$

stronger acid stronger base

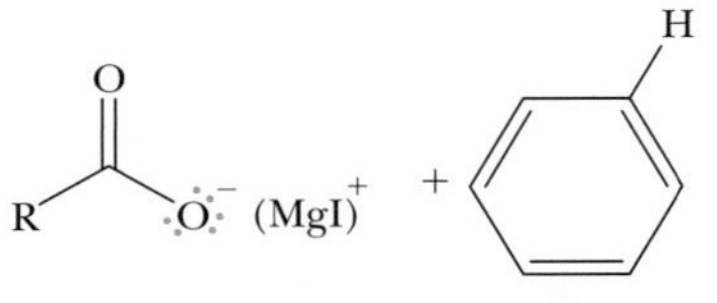

H

R O⁻ $(MgI)^+$ +

$pK_a = 43$

weaker base weaker acid

13.21

(a) OH (b) OH

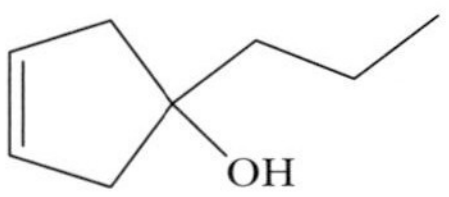

(c) HO racemic (d) racemic OH

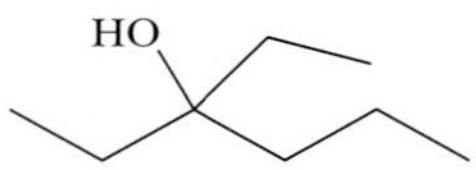

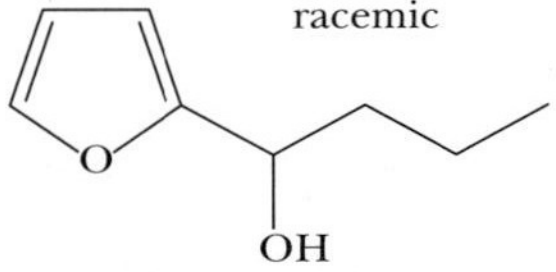

(e) OH racemic H_3CO

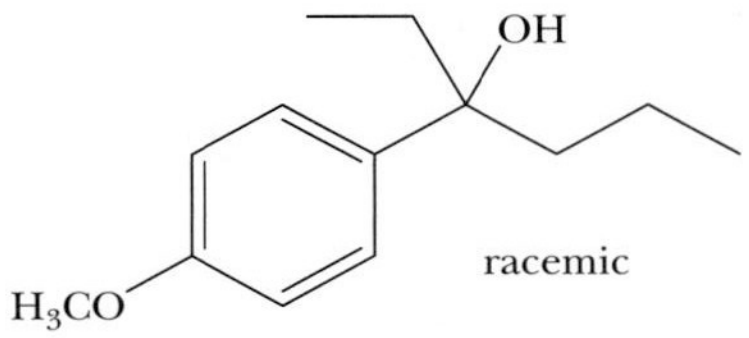

13.23 (a) O OH (b) 2 (c) 4

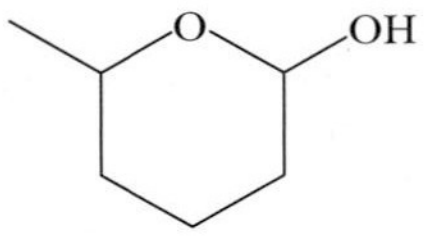

(d)

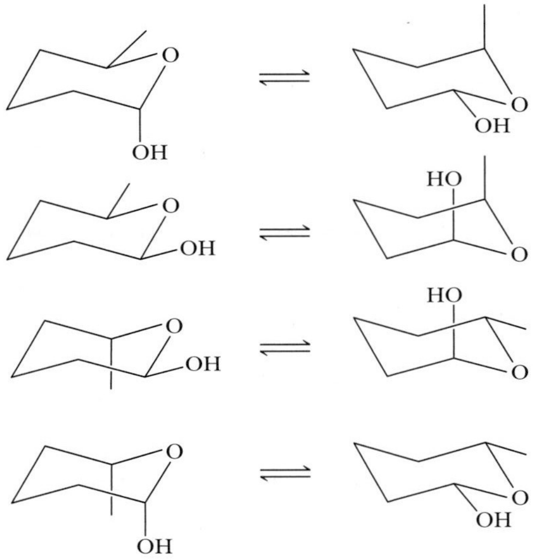

(e)

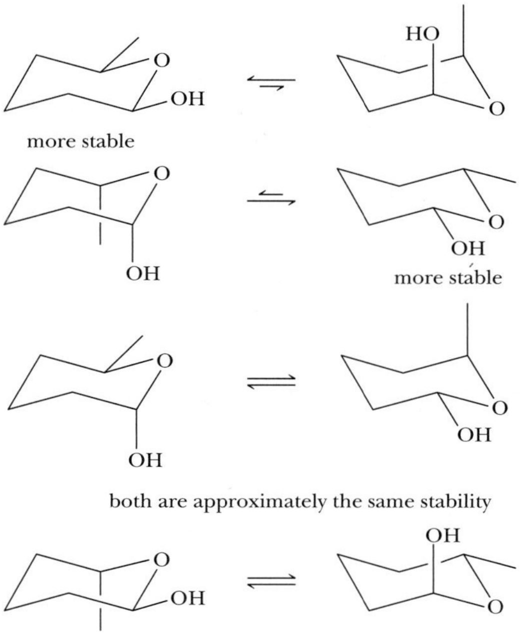

more stable

more stable

both are approximately the same stability

both are approximately the same stability

13.25 (a) O + $2CH_3OH$

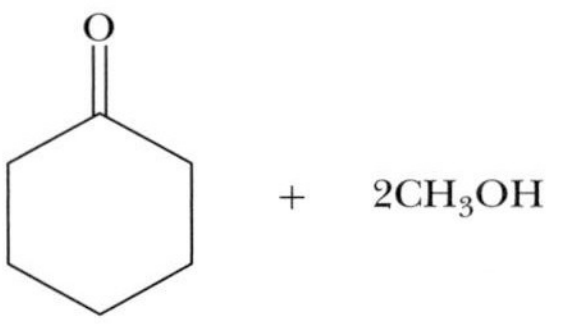

(b) HO H O + CH_3OH

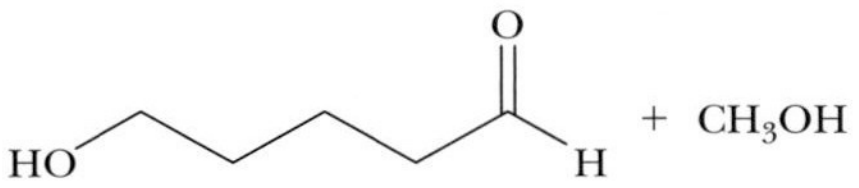

(c) HO H O + O OH

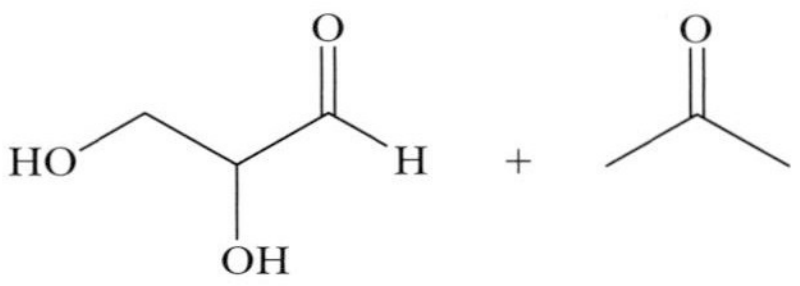

13.27

Step 1:

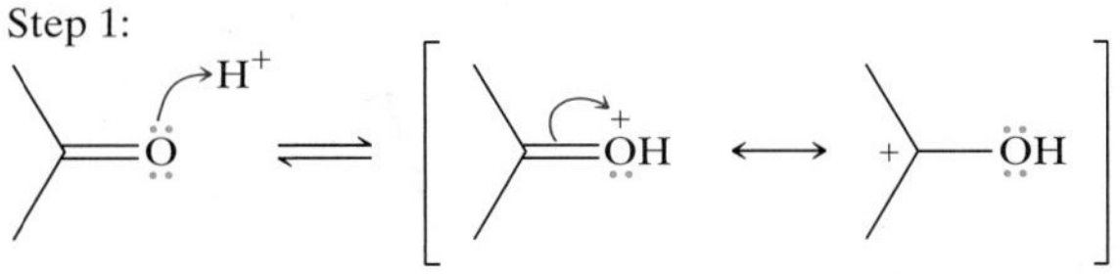

H^+ OH OH

Step 2:

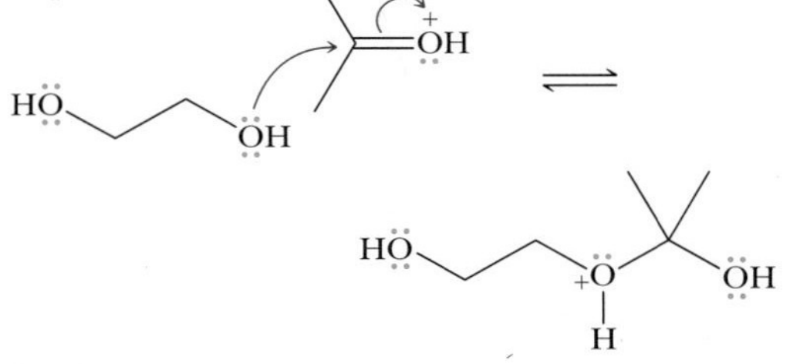

HO OH OH

HO O OH H

Step 3:

HO O OH H ⇌ HO O OH + H^+

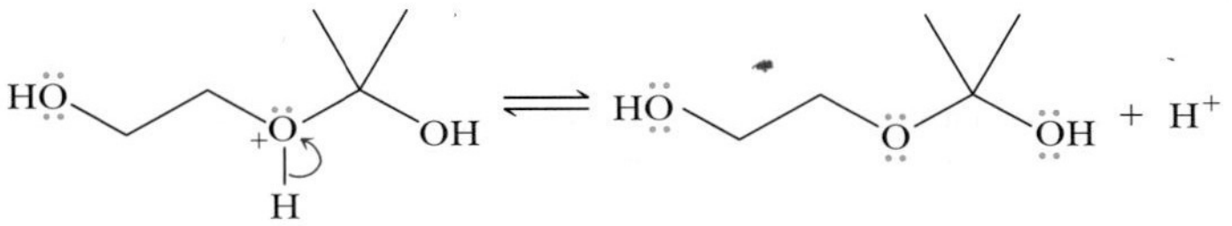

Step 4:

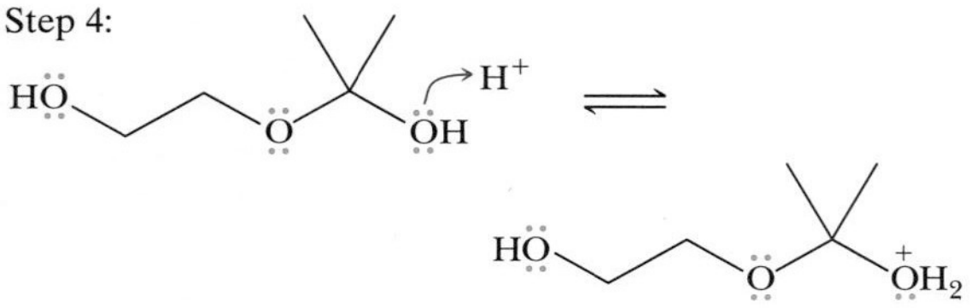

Step 5:

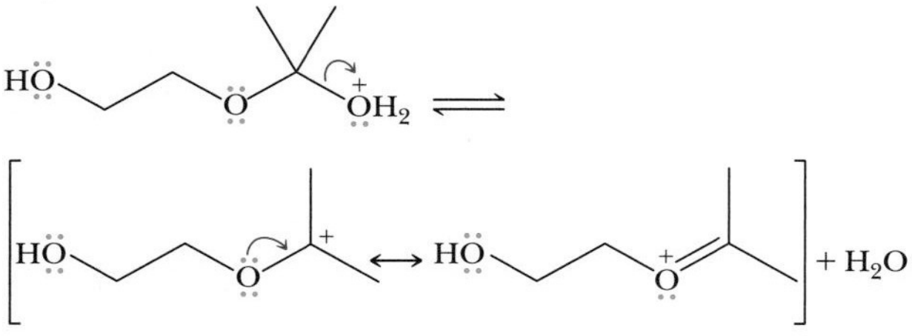

Step 6:

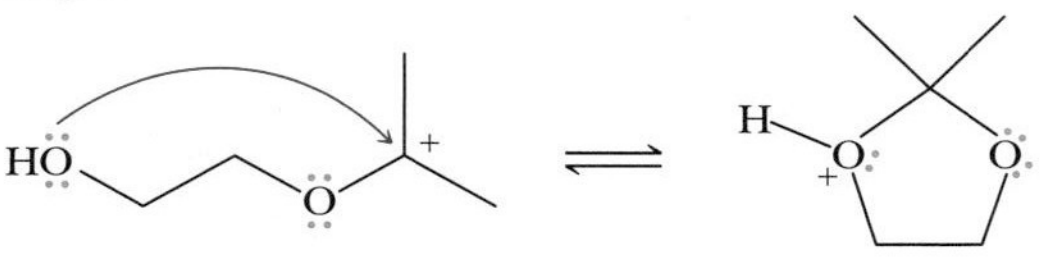

Step 7:

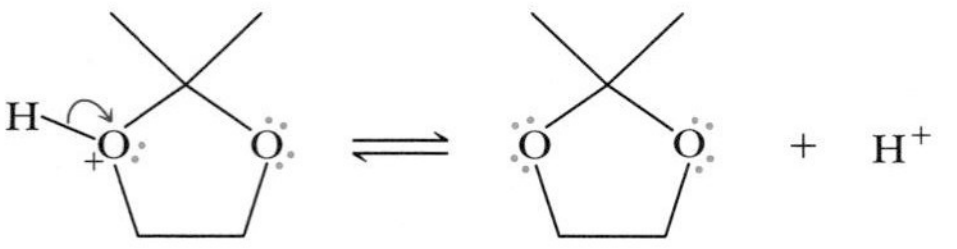

$$\text{(cyclic oxonium)} \rightleftharpoons \text{(acetal)} + H^+$$

13.29 Ph–CO–CH_3 $\xrightarrow[\text{2) } H_2/Ni]{\text{1) } NH_3}$ Ph–CH(CH_3)–NH_2 $\xrightarrow[\text{2) } H_2/Ni]{\text{1) PhCHO}}$ Ph–CH(CH_3)–NH–CH_2–Ph

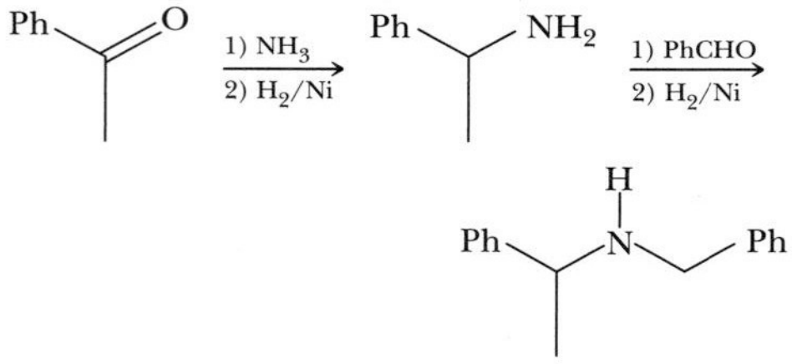

13.31 (a) $+ NH_3 \xrightarrow[Ni]{H_2}$

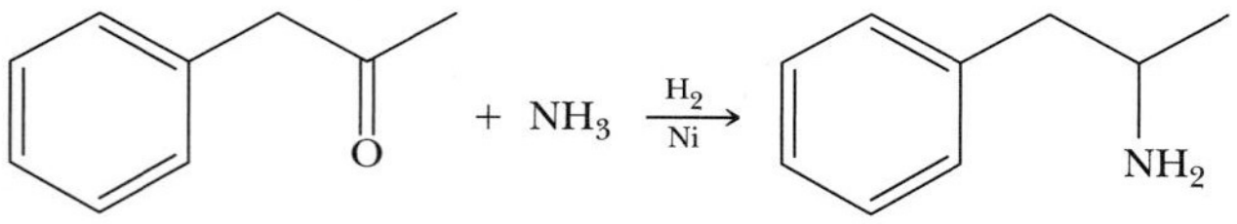

(b) $+ CH_3NH_2 \xrightarrow[Ni]{H_2}$

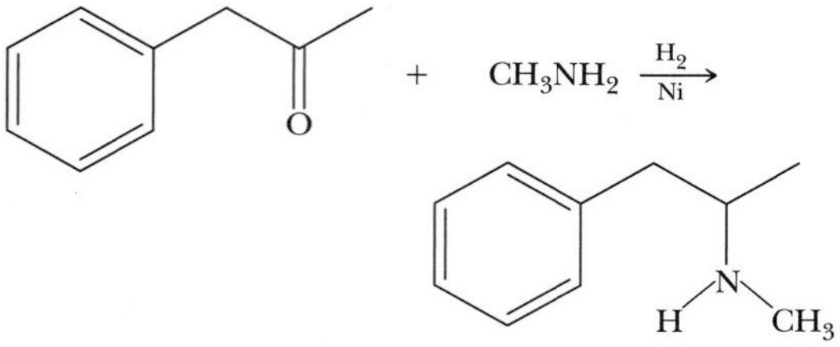

13.33 (a)

$$\text{(methenamine)} + 10H_2O \xrightarrow{H^+} 6CH_2O + 4NH_4^+ + 4OH^-$$

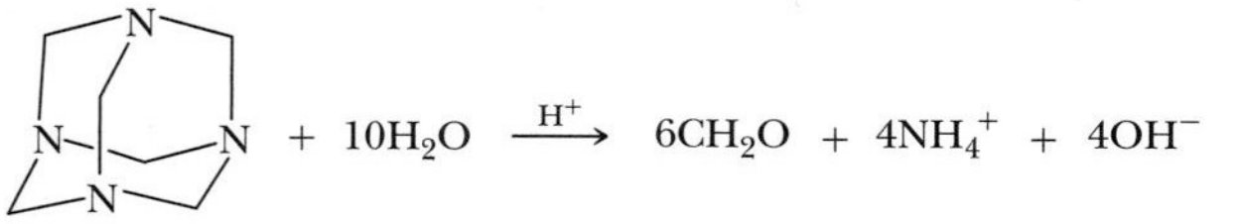

(b) When methenamine is hydrolyzed, ammonium hydroxide is formed. Ammonium hydroxide is a base, so the pH will increase.

(c) Acetals are 1,1-diethers, where a single carbon atom is bonded to two –OR groups. In the case of methenamine, each carbon atom is bonded to two amine functional groups.

(d) The pH of blood plasma is slightly basic (pH of 7.4), but the pH of the urinary tract is acidic. Acetals (and methenamine, which is a nitrogen analog of an acetal) require acidic conditions for hydrolysis to occur, but they are stable under basic conditions.

13.35

Step 1:

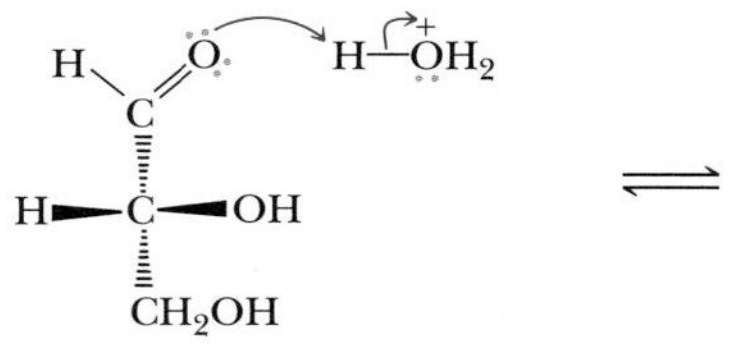

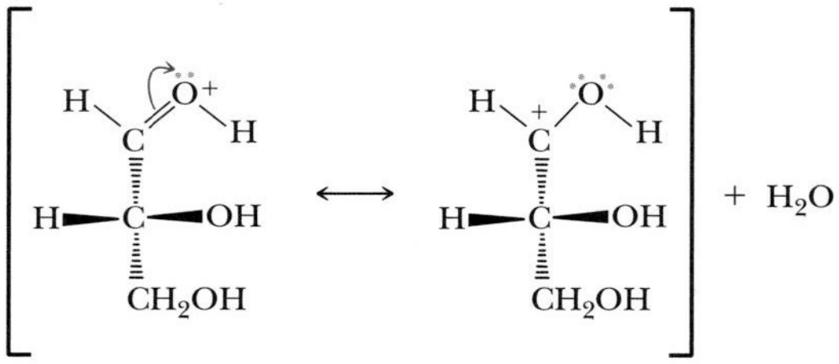

Step 2: Deprotonation of the resonance-stabilized intermediate at the α-carbon (carbon 2) leads to the enediol intermediate. The Z isomer is shown, but the E isomer can be formed as well. Note that the chiral center is destroyed and that the enediol is achiral.

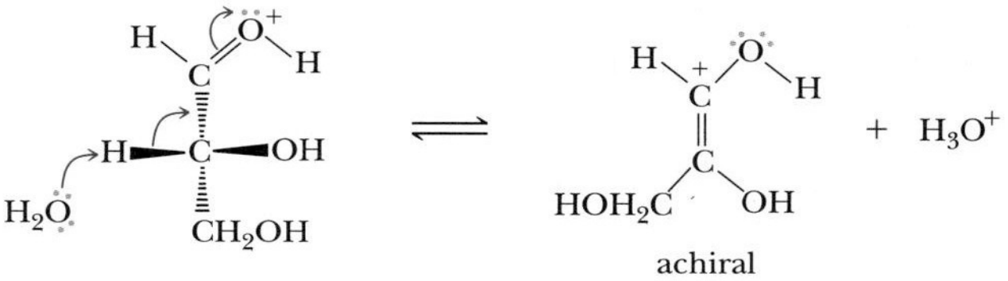

achiral

Step 3: Reprotonation of the same carbon (carbon 2) can occur from either face of the alkene with equal probability. As a result, a racemic mixture is formed.

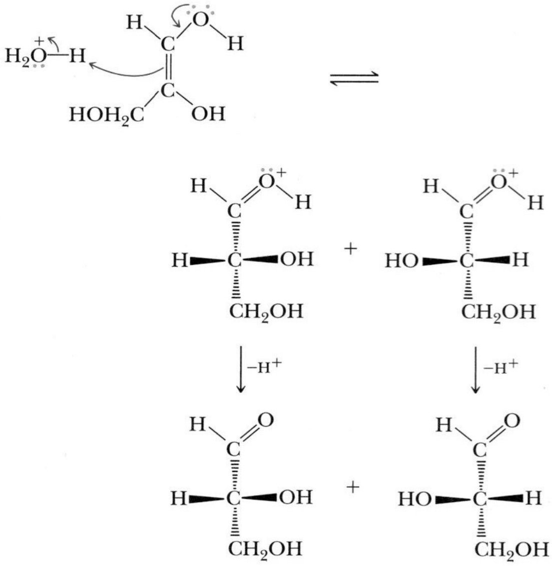

Step 4: If the enediol formed in Step 2 is protonated at carbon 1 instead of carbon 2, the intermediate that is formed can subsequently lead to the formation of dihydroxyacetone.

$H_2\overset{+}{O}$—H   H—C(—O—H)=C(—OH)CH_2OH ⇌

[CH_2OH—C(=$\overset{+}{O}$—H)—CH_2OH ⟷ CH_2OH—$\overset{+}{C}$(—O—H)—CH_2OH] + H_2O

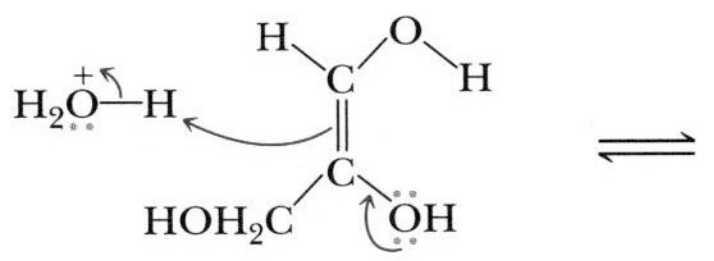

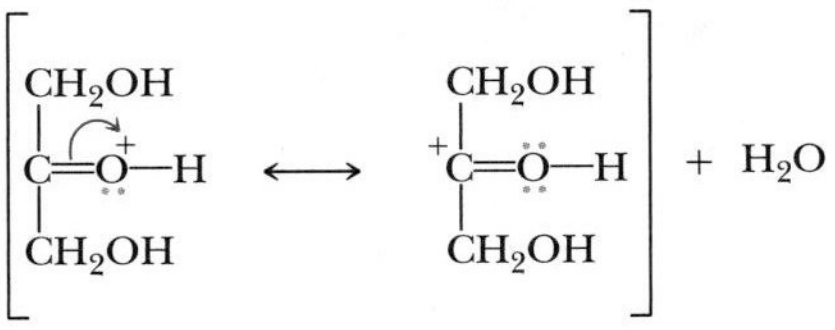

Step 5: Deprotonation results in the formation of dihydroxyacetone.

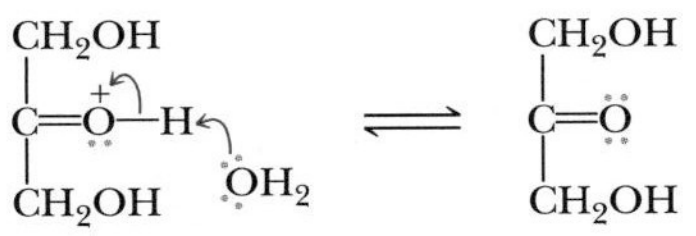

CH_2OH—C(=$\overset{+}{O}$—H)—CH_2OH OH_2 ⇌ CH_2OH—C(=O)—CH_2OH

13.37

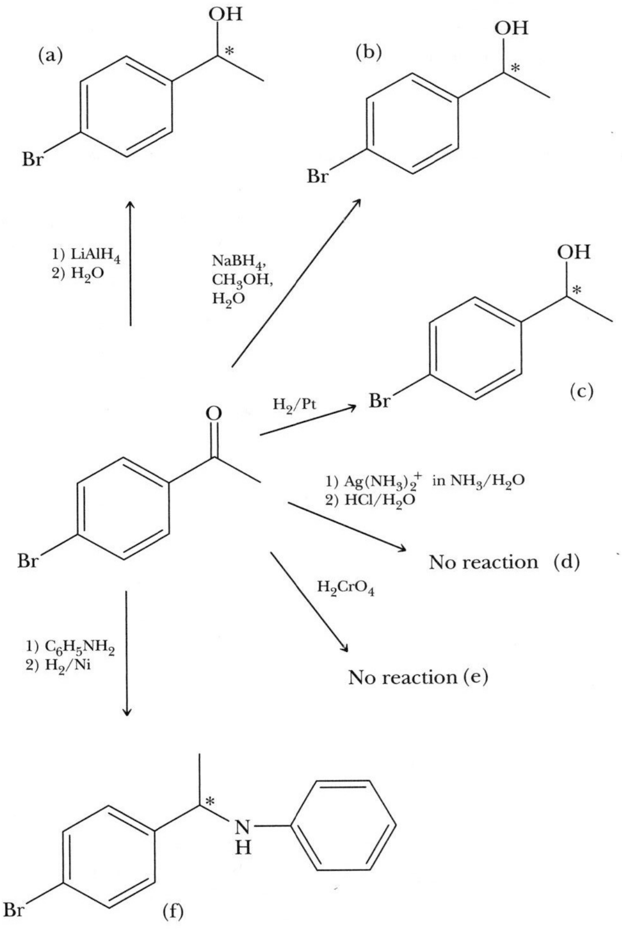

(a) OH, *, Br

(b) OH, *, Br

1) $LiAlH_4$ 2) H_2O

$NaBH_4$, CH_3OH, H_2O

OH, *, Br (c)

H_2/Pt

O, Br

1) $Ag(NH_3)_2^+$ in NH_3/H_2O 2) HCl/H_2O

No reaction (d)

H_2CrO_4

No reaction (e)

1) $C_6H_5NH_2$ 2) H_2/Ni

*, N, H, Br (f)

13.39

(a) OH → H_2SO_4, heat → (b) → 1) O_3 2) $(CH_3)_2S$ → (c) H, O, O, H, 4

RCO_3H

(h) O → H_3O^+ → (i) OH, OH racemic

$NaBH_4$ in MeOH or 1) $LiAlH_4$ 2) H_2O or H_2/Ni

O → 1) CH_3MgBr 2) H_3O^+ → (d) OH → H_2SO_4, heat → (e)

1) PhMgBr 2) H_3O^+

(f) OH → H_2SO_4, heat → (g)

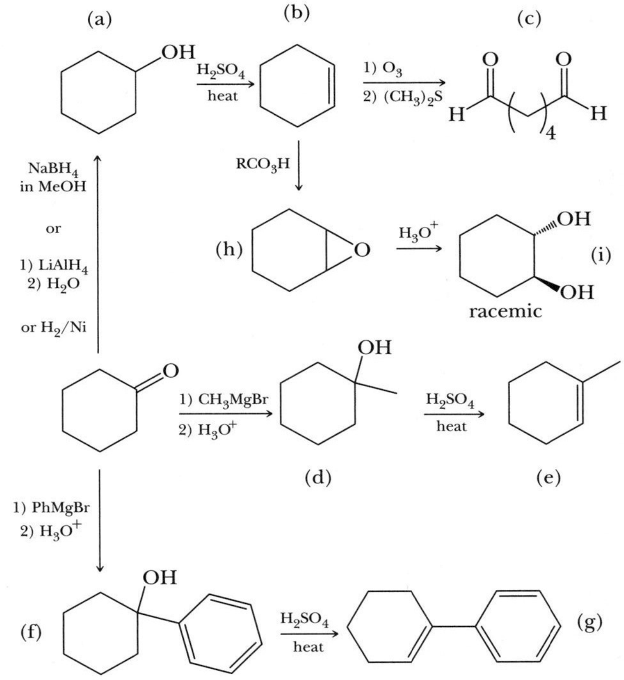

13.41 (a) 2 (b) Z

(c)

O, O, N, CH_3, CH_3 → 1) PhMgBr 2) H_3O^+ →

HO, O, N, CH_3, CH_3 → H_2SO_4, heat →

O, N, CH_3, CH_3

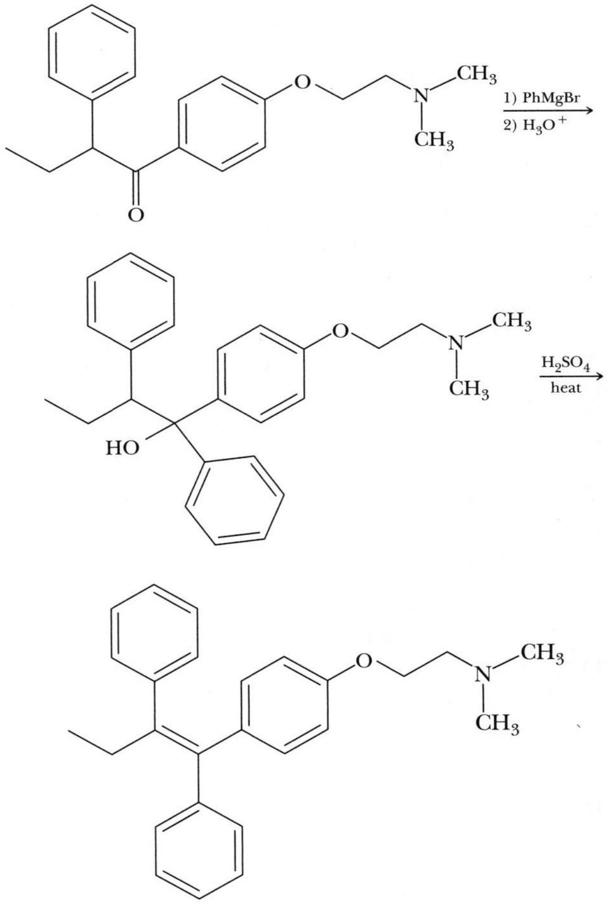

13.43 (a) Cyclopropylmagnesium bromide followed by acid treatment.

(b)

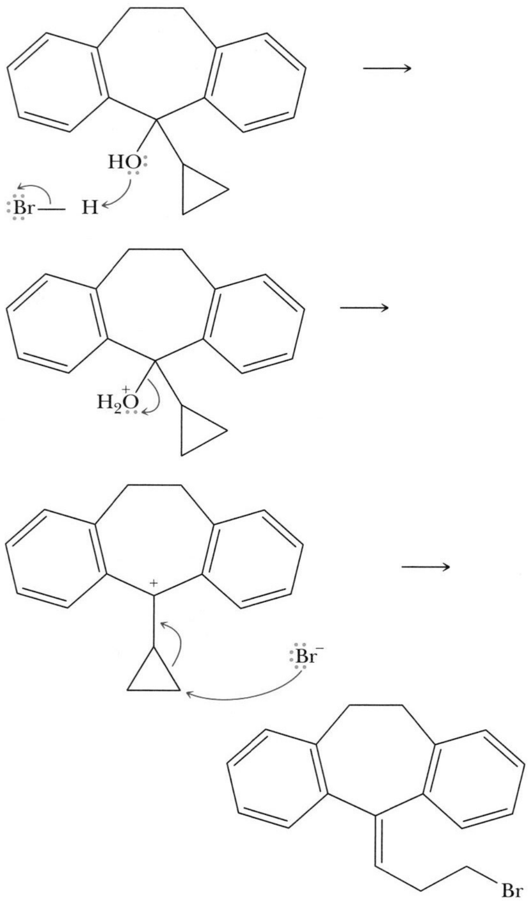

(c) $HN(CH_3)_2$

13.45 (a) Step 1 is a Friedel-Crafts acylation and involves the use of acetyl chloride (CH_3COCl) and a Lewis acid catalyst, such as $AlCl_3$.

(b) Step 2 involves the chlorination of the α-carbon, and Cl_2 in acetic acid can be used.

Step 3 involves the substitution of the Cl with dimethylamine. Two equivalents of dimethylamine are necessary.

(c) Step 4 involves the reduction of the ketone. Various reducing agents ($NaBH_4$ in an alcohol, $LiAlH_4$ followed by water, or catalytic hydrogenation) can be used.

Step 5 involves the substitution of the alcohol with a halogen. $SOCl_2$ in pyridine can be used.

(d) Step 6 is a Grignard reaction. The starting material is first treated with magnesium in ether to prepare the Grignard reagent, which is subsequently treated with cyclohexanone followed by acid.

13.47

3-methylbutanal

13.49 2,6-dimethyl-4-heptanone

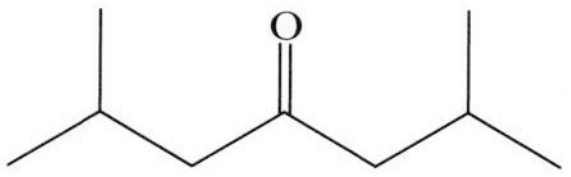

13.51

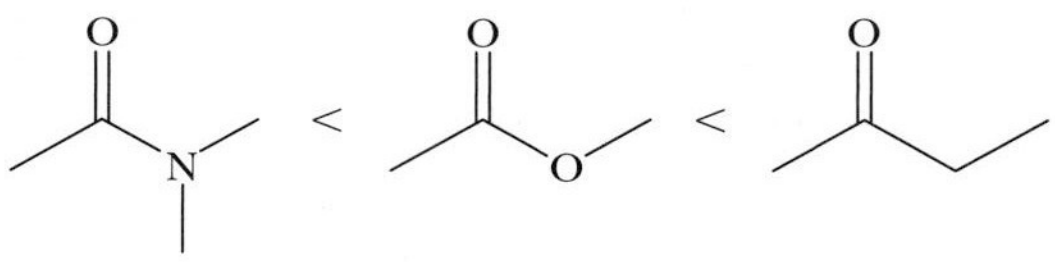

13.53 (a)

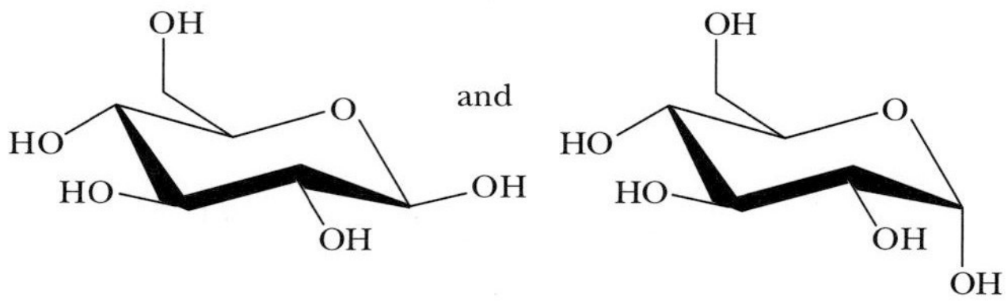

and

(b)

and

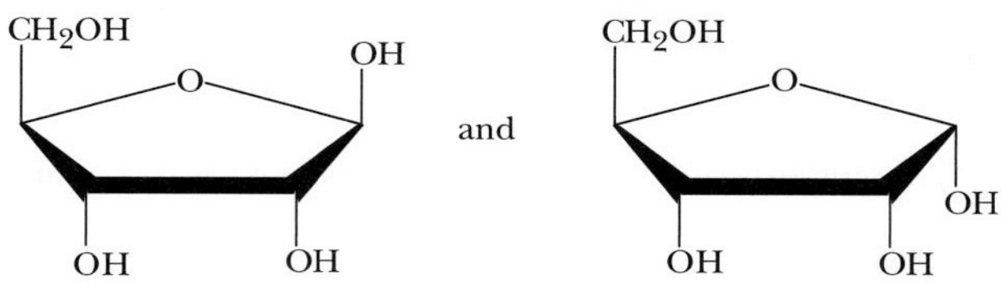

Chapter 14 Carboxylic Acids

14.1 (a) (*R*)-2,3-dihydroxypropanoic acid

(b) (*Z*)-butenedioic acid

(c) (*R*)-3,5-dihydroxy-3-methylpentanoic acid

14.2 0.22: trifluoroacetic acid

3.85: lactic acid

5.03: 2,2-dimethylpropanoic acid

14.3 (a) $COO^-NH_4^+$ ammonium butanoate

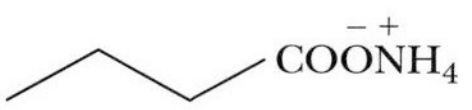

(b) $COO^-NH_4^+$ ammonium 2-hydroxypropanoate (ammonium lactate)

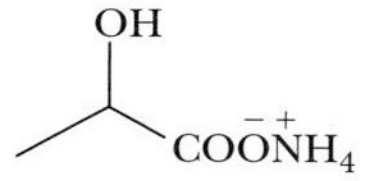

14.4 (a)

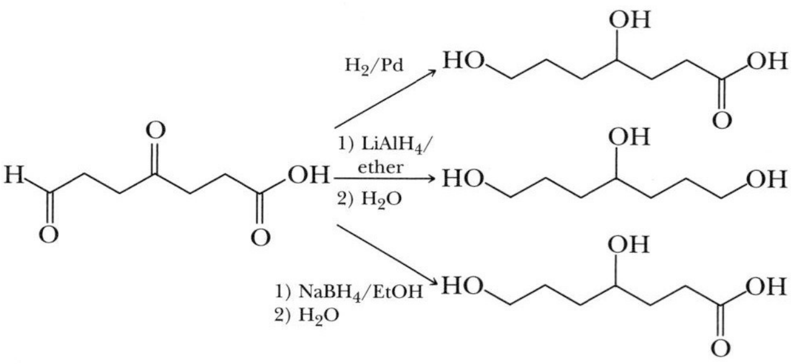

H_2/Pd

1) $LiAlH_4$/ether
2) H_2O

1) $NaBH_4$/EtOH
2) H_2O

(b)

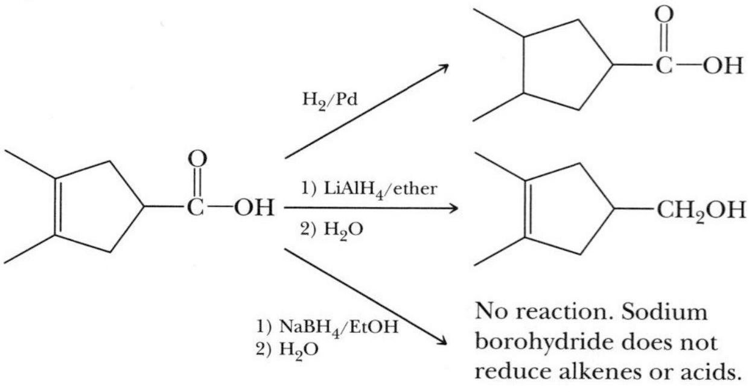

H_2/Pd

1) $LiAlH_4$/ether
2) H_2O

1) $NaBH_4$/EtOH
2) H_2O

No reaction. Sodium borohydride does not reduce alkenes or acids.

14.5 (a) 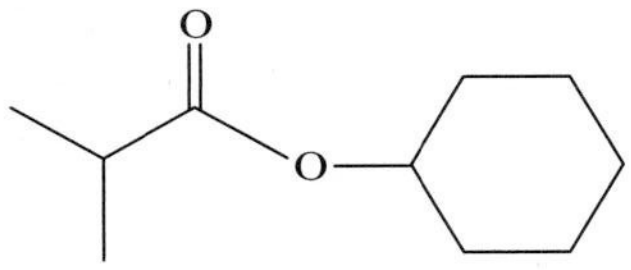 + H_2O

(b) 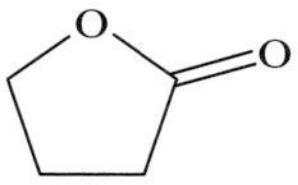 + H_2O

14.6 (a) 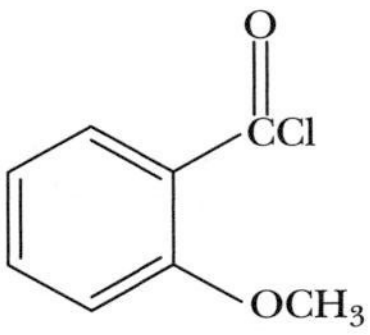 + SO_2 + HCl

(b) 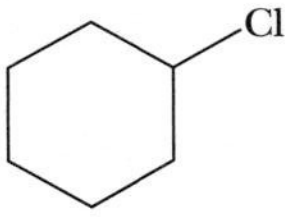 + SO_2 + HCl

14.7

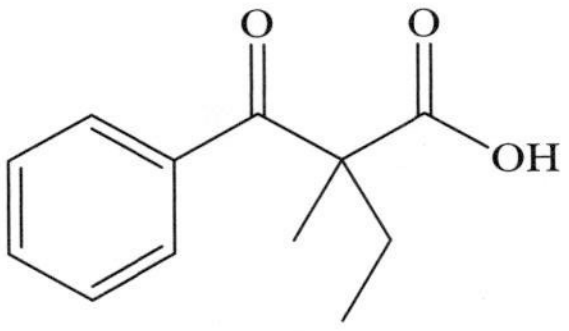

14.9 (a) 1-cyclohexenecarboxylic acid
(b) 4-hydroxypentanoic acid
(c) (*E*)-3,7-dimethyl-2,6-octadienoic acid
(d) 1-methylcyclopentanecarboxylic acid
(e) ammonium hexanoate
(f) 2-hydroxybutanedioic acid

14.11 (a) 3,5-Tetradecadienoic acid
(b) 4

14.13
(a) 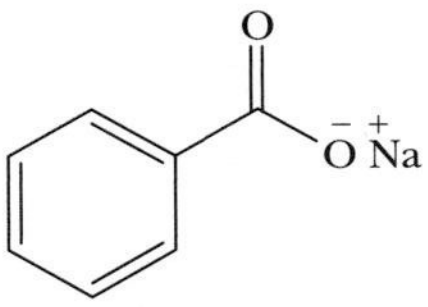 (b) 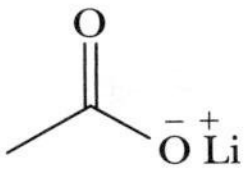 (c) 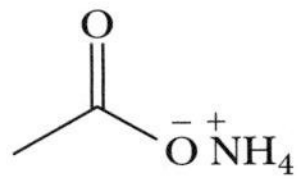 (d) 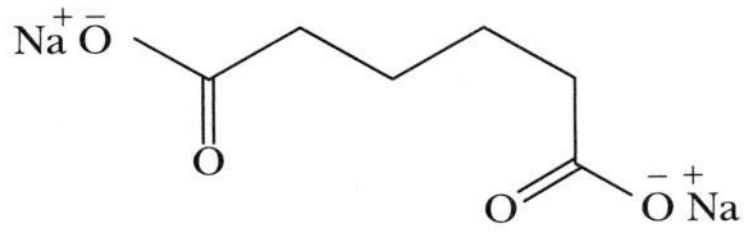 (e) 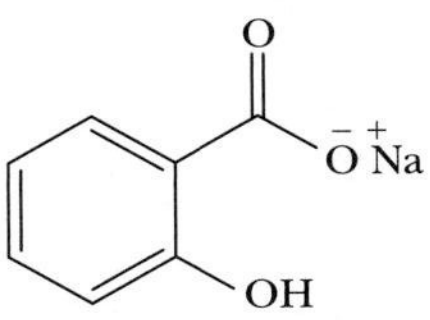 (f) 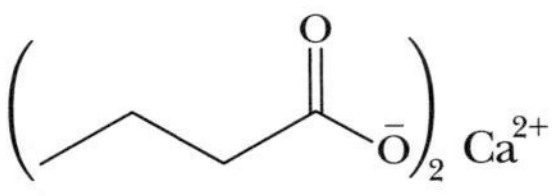 Ca^{2+}

14.15

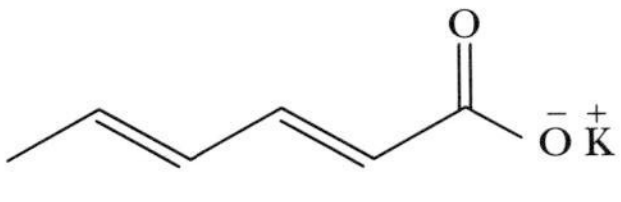

14.17
(a) $CH_3(CH_2)_6CHO < CH_3(CH_2)_6CH_2OH < CH_3(CH_2)_5COOH$
(b) $CH_3CH_2OCH_2CH_3 < CH_3CH_2CH_2CH_2OH < CH_3CH_2COOH$

14.19 (a) 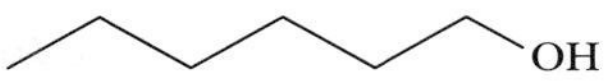 (b) 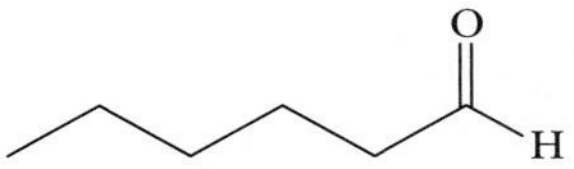 (c)

14.21 benzyl alcohol < phenol < benzoic acid

14.23 (a) 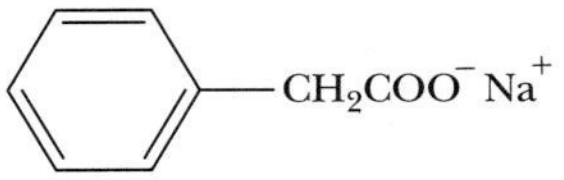 $-CH_2COO^-Na^+$ + H_2O
(b) $CH_3CH{=}CHCH_2COOO^-Na^+ + H_2O + CO_2$
(c) 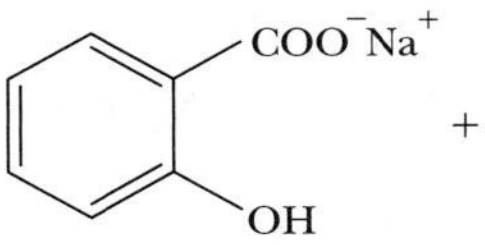 + H_2O + CO_2
(d) $CH_3CH(OH)COO^-\ H_3\overset{+}{N}CH_2CH_2OH$
(e) $CH_3CH{=}CHCH_2COOH + NaCl$

14.25 When the pH is greater than the pK_a of the acid, the acid will be deprotonated. Ascorbic acid in blood plasma exists primarily as the ascorbate anion.

14.27 Lactic acid itself

14.29 (a) $CH_3CH(\overset{+}{N}H_3)COOH$ (b) $CH_3CH(\overset{+}{N}H_3)COO^-$
(c) $CH_3CH(NH_2)COO^-$

14.31
1) $LiAlH_4$ 2) H_2O (a)
H_2/Pd H_2/Pd
1) $LiAlH_4$ 2) H_2O (b) (c)

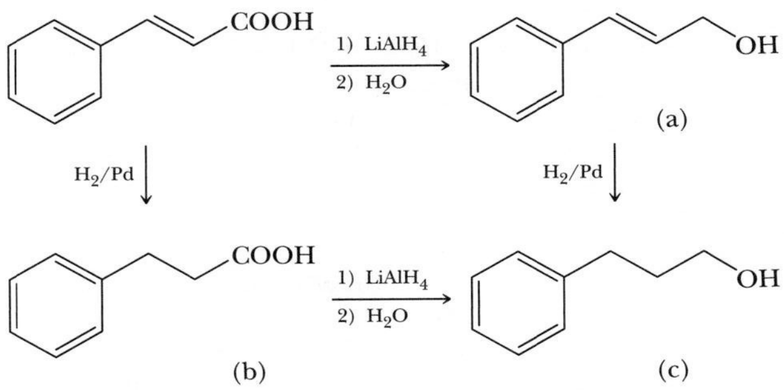

14.33 (a) 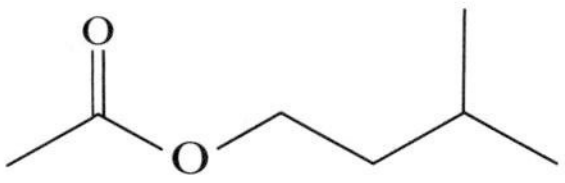 (b)

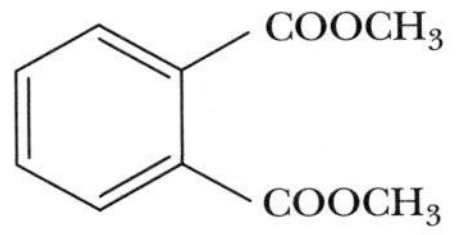

(c)

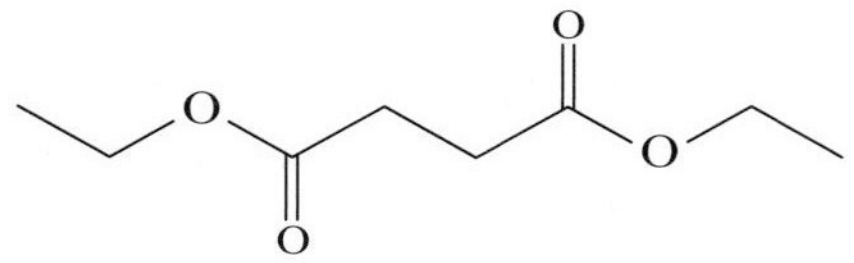

14.35

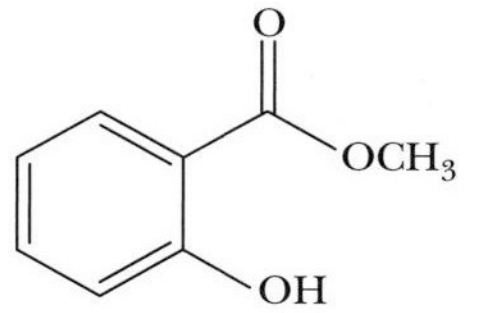

14.37 (a)

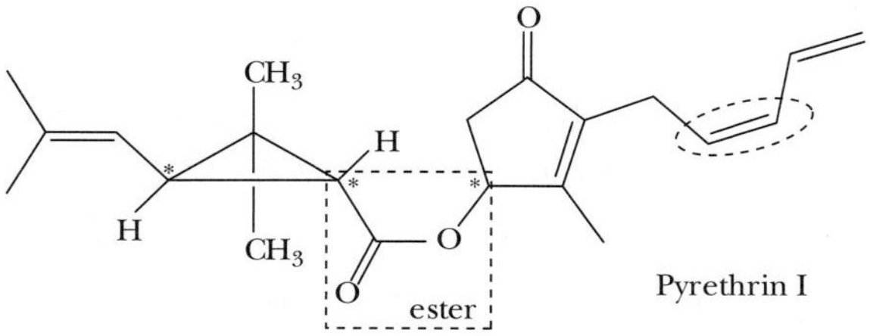

Pyrethrin I

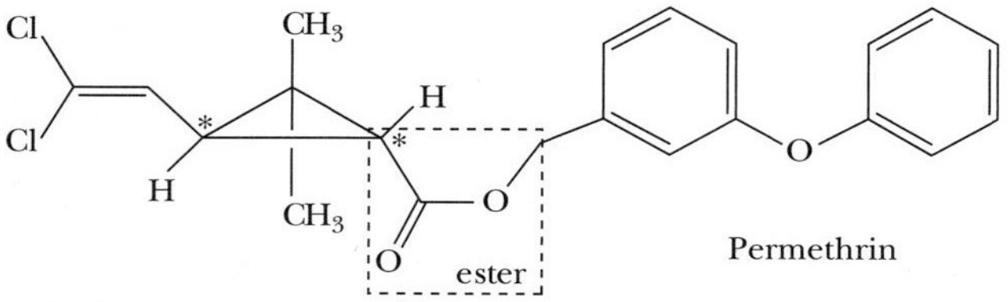

ester

Permethrin

ester

(b) Pyrethrin is chiral and the three stereocenters are labeled in the structure. In addition, *cis-trans* isomerism is possible at the circled alkene. Thus, a total of $2^4 = 16$ stereoisomers are possible.

(c) Permethrin is chiral and the two stereocenters are labeled in the structure. A total of 4 stereoisomers are possible.

14.39 (a)

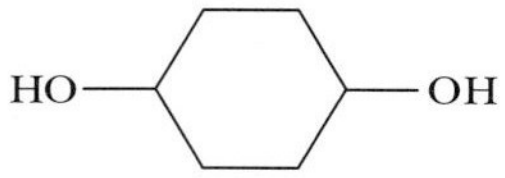

alcohol | carboxylic acid (CH_3COOH)

(b) CH_3OH alcohol; $HOCCH_2CH_2COH$ (with C=O) carboxylic acid

(c) CH_3OH alcohol; cyclohexane–COH (with C=O) carboxylic acid

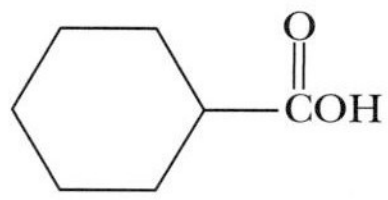

(d) $(CH_3)_2CHOH$ alcohol; $CH_3CH_2CH{=}CHCOH$ (with C=O) carboxylic acid

14.41 (a) (b)

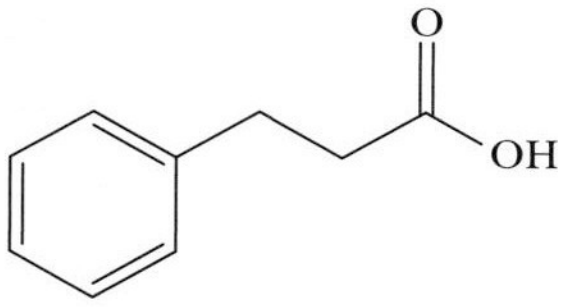

(c)

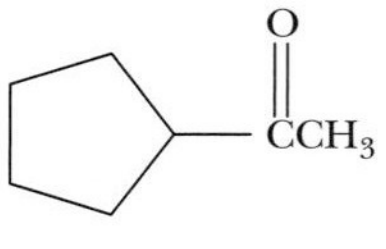

14.43

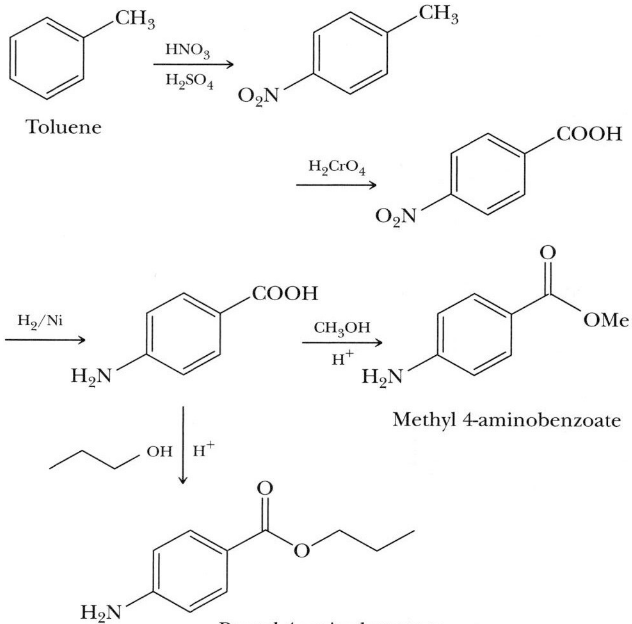

Toluene

HNO_3 / H_2SO_4

H_2CrO_4

H_2/Ni

CH_3OH / H^+

Methyl 4-aminobenzoate

Propyl 4-aminobenzoate

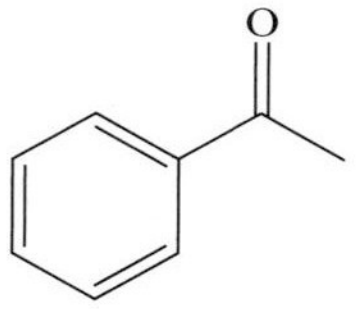

14.45 (a) $SOCl_2$ /pyridine

(b) Friedel-Crafts acylation

(c) 1. NH_3 2. H_2/Pt

(d)

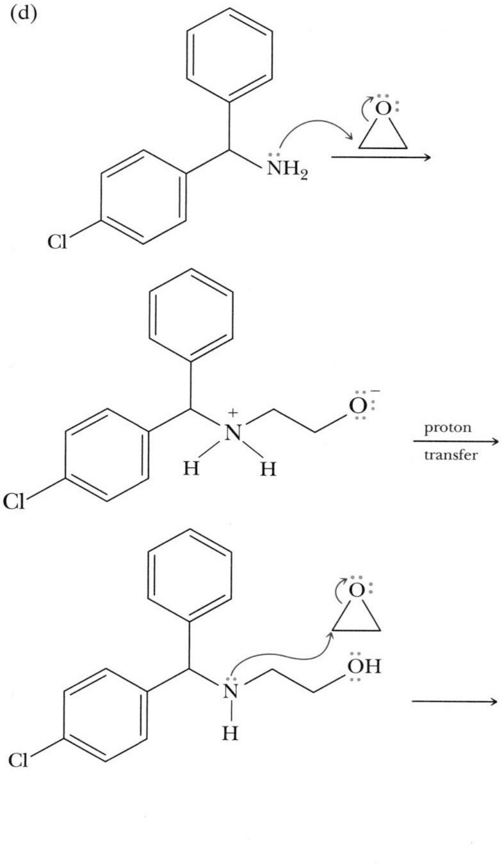

proton transfer

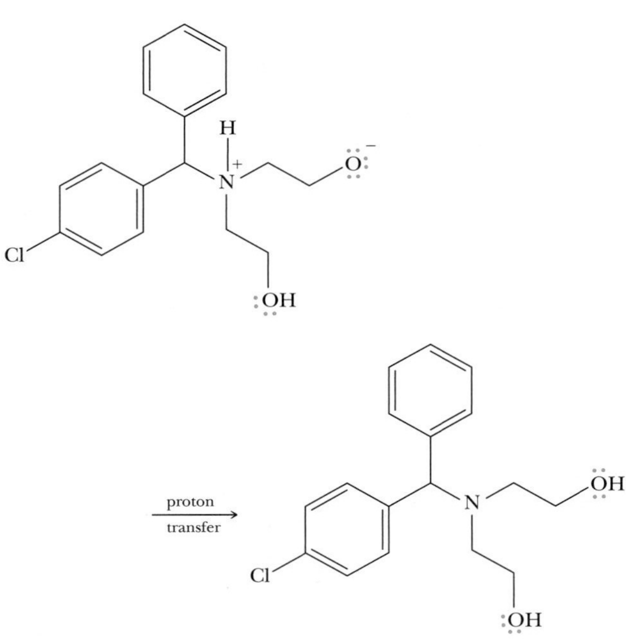

proton transfer

(e) $SOCl_2$/pyridine

(f)

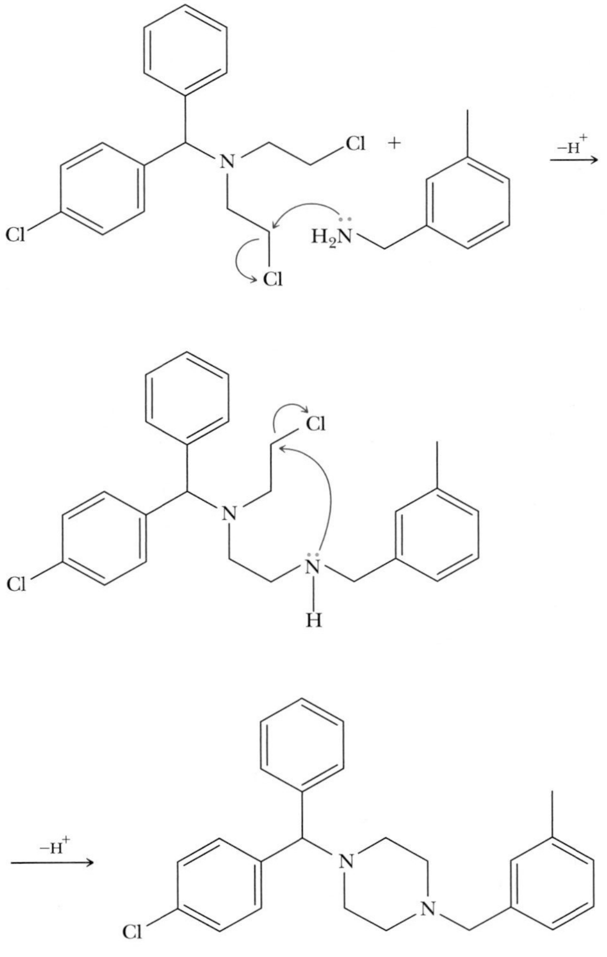

$-H^+$

$-H^+$

14.47 (a)

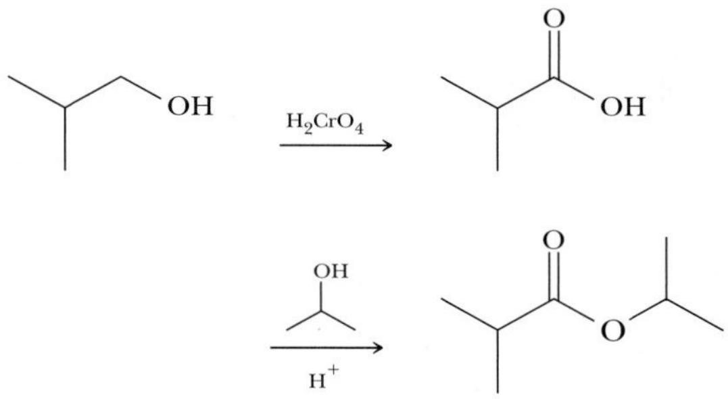

OH $\xrightarrow{H_2CrO_4}$ OH

$\xrightarrow[H^+]{\text{OH}}$ O

(b)

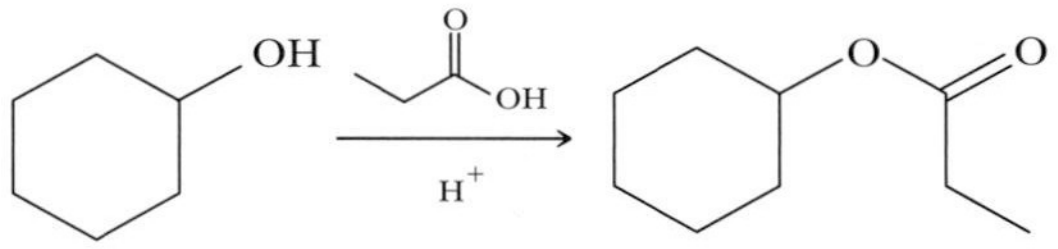

OH $\xrightarrow[H^+]{\text{OH}}$ O

(c) OH $\xrightarrow[Pt]{H_2}$ OH $\xrightarrow{H_2CrO_4}$

OH $\xrightarrow[H^+]{CH_3OH}$ O

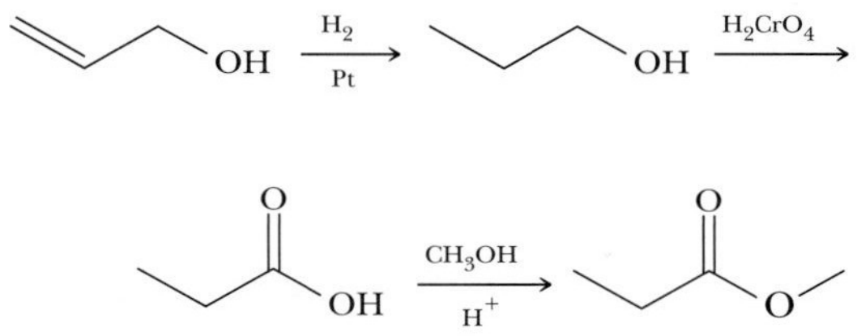

(d)

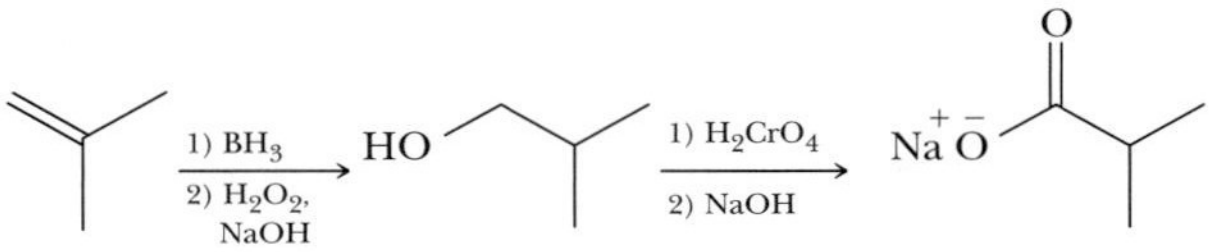

$\xrightarrow[\text{2) } H_2O_2, \text{ NaOH}]{\text{1) } BH_3}$ HO $\xrightarrow[\text{2) NaOH}]{\text{1) } H_2CrO_4}$ $Na^+ \, ^-O$

(e)

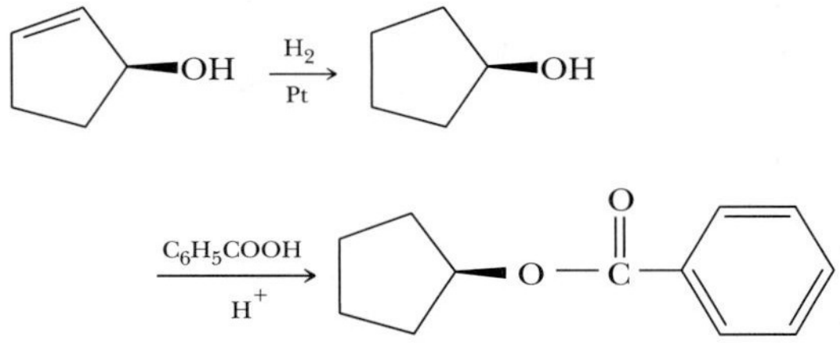

OH $\xrightarrow[Pt]{H_2}$ OH

$\xrightarrow[H^+]{C_6H_5COOH}$ O—C

(f)

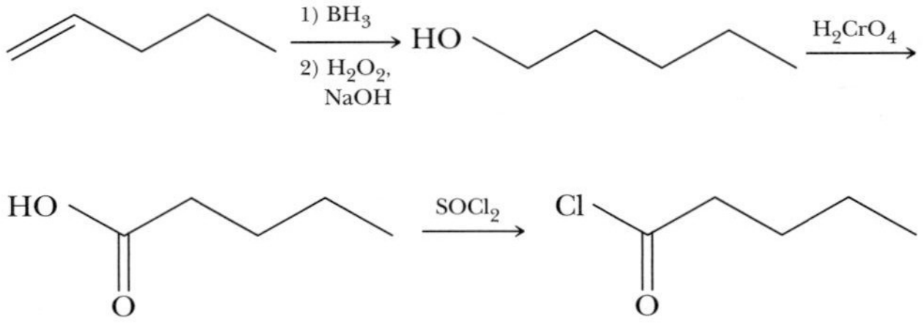

$\xrightarrow[\text{2) } H_2O_2, \text{ NaOH}]{\text{1) } BH_3}$ HO $\xrightarrow{H_2CrO_4}$

HO $\xrightarrow{SOCl_2}$ Cl

(g)

$\xrightarrow[AlCl_3]{RCl}$ R $\xrightarrow[AlCl_3]{Cl_2}$ R Cl $\xrightarrow{H_2CrO_4}$

Cl C OH $\xrightarrow[H^+]{CH_3OH}$ Cl C OCH_3

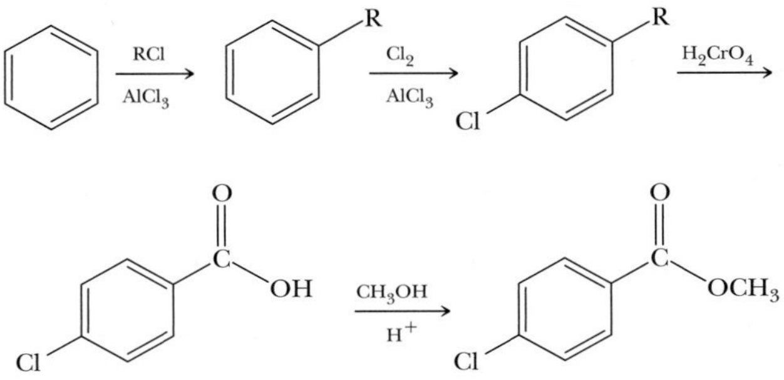

(h)

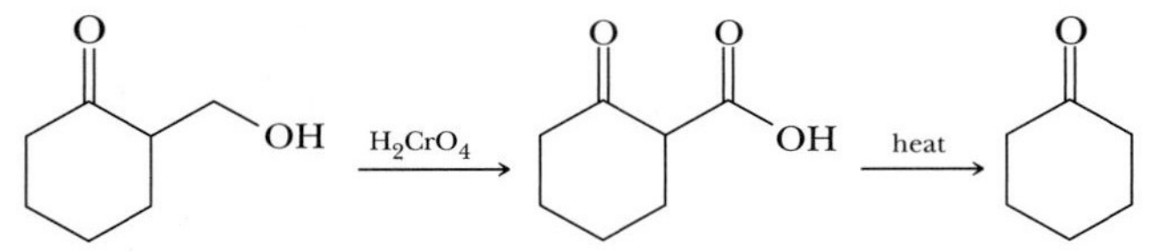

OH $\xrightarrow{H_2CrO_4}$ OH $\xrightarrow{heat}$

(i)

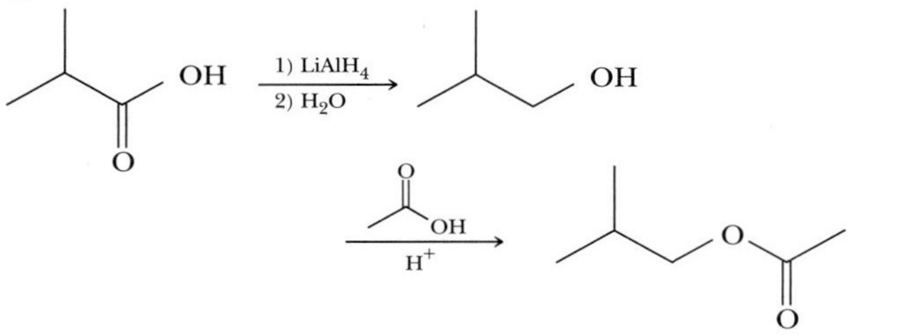

OH $\xrightarrow[2)\ H_2O]{1)\ LiAlH_4}$ OH

$\xrightarrow[H^+]{CH_3COOH}$

(j)

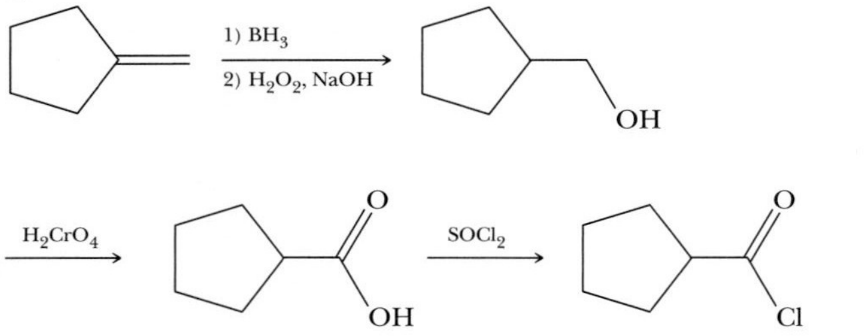

$\xrightarrow[2)\ H_2O_2,\ NaOH]{1)\ BH_3}$ OH

$\xrightarrow{H_2CrO_4}$ OH $\xrightarrow{SOCl_2}$ Cl

(k)

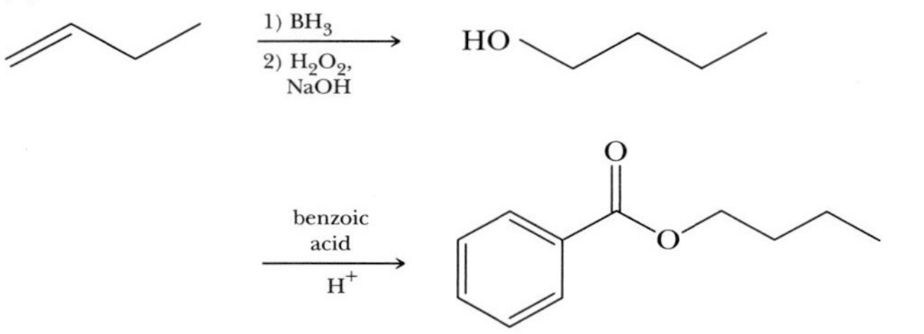

$\xrightarrow[2)\ H_2O_2,\ NaOH]{1)\ BH_3}$ HO

$\xrightarrow[H^+]{\text{benzoic acid}}$

(l)

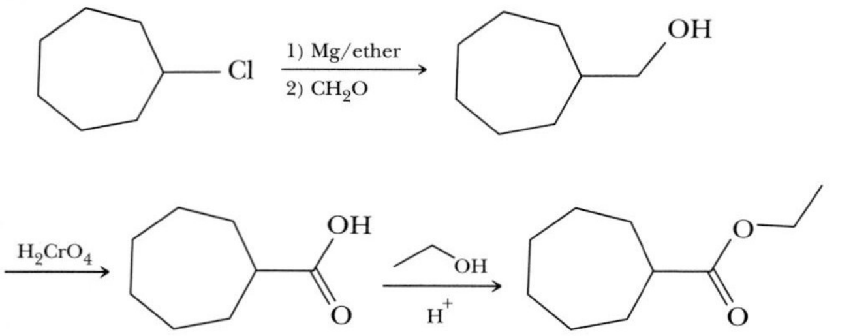

Cl $\xrightarrow[2)\ CH_2O]{1)\ Mg/ether}$ OH

$\xrightarrow{H_2CrO_4}$ OH $\xrightarrow[H^+]{CH_3CH_2OH}$

(m)

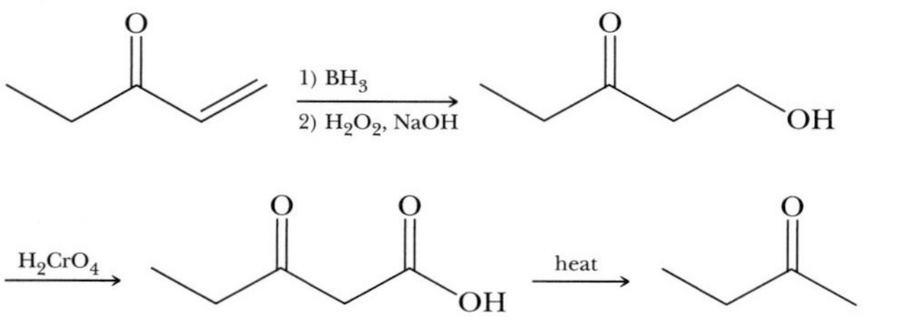

$\xrightarrow[2)\ H_2O_2,\ NaOH]{1)\ BH_3}$ OH

$\xrightarrow{H_2CrO_4}$ OH $\xrightarrow{heat}$

(n)

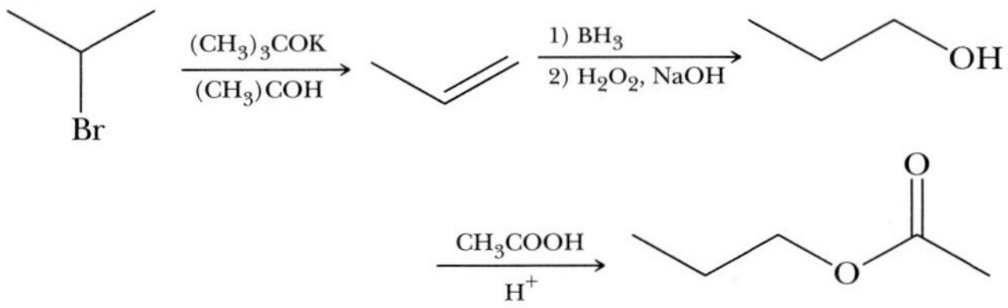

Br $\xrightarrow[(CH_3)COH]{(CH_3)_3COK}$ $\xrightarrow[2)\ H_2O_2,\ NaOH]{1)\ BH_3}$ OH

$\xrightarrow[H^+]{CH_3COOH}$

(o)

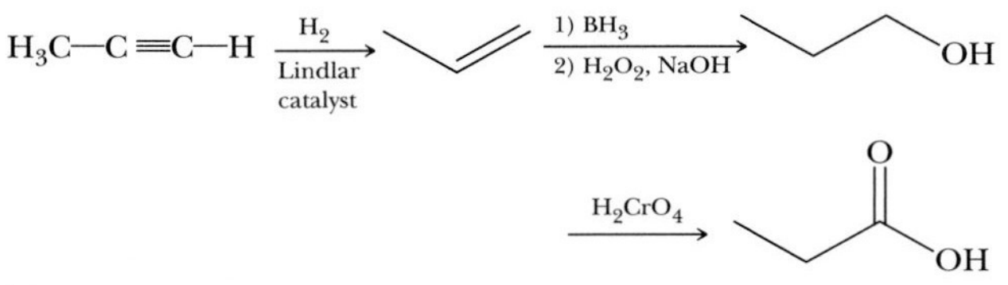

$H_3C—C{\equiv}C—H$ $\xrightarrow[\text{Lindlar catalyst}]{H_2}$ $\xrightarrow[2)\ H_2O_2,\ NaOH]{1)\ BH_3}$ OH

$\xrightarrow{H_2CrO_4}$ OH

(p)

$\xrightarrow[H_2SO_4]{HNO_3}$ O_2N $\xrightarrow[AlCl_3]{RCl}$ O_2N, R

$\xrightarrow{H_2CrO_4}$ O_2N, OH $\xrightarrow{SOCl_2}$ O_2N, Cl

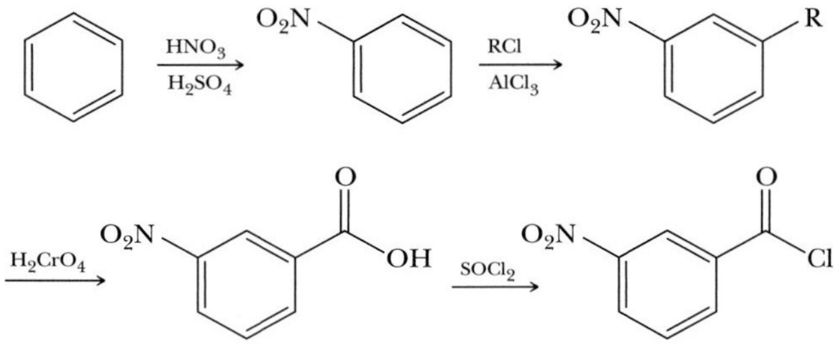

14.49 Carboxylate ion

14.51 Grignards are destroyed by acids, but will attack the carbonyl carbon of esters.

Chapter 15
Functional Derivatives of Carboxylic Acids

15.1 (a) N H (b)

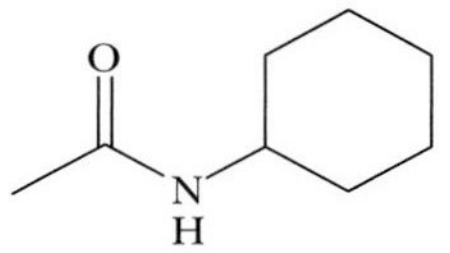

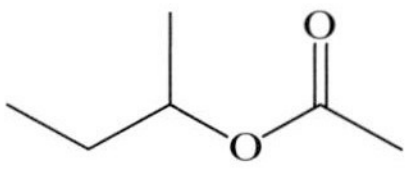

(c) 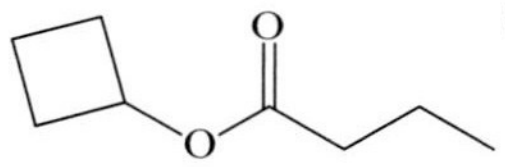 (d) H N, 5

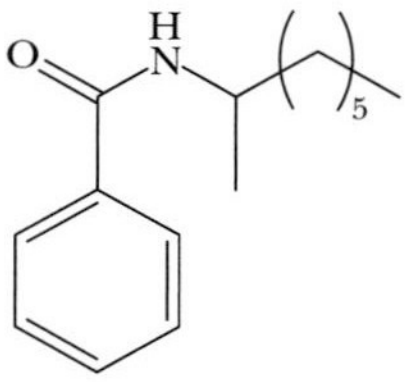

(e) 4 (f)

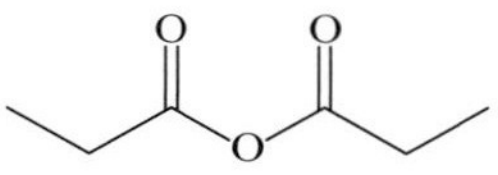

15.2 (a) COO^-Na^+, COO^-Na^+ + $2CH_3OH$

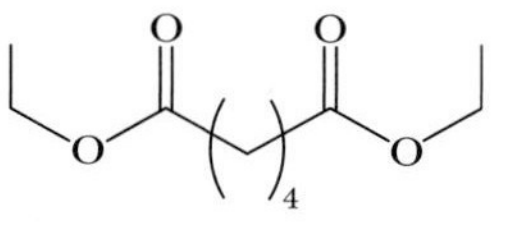

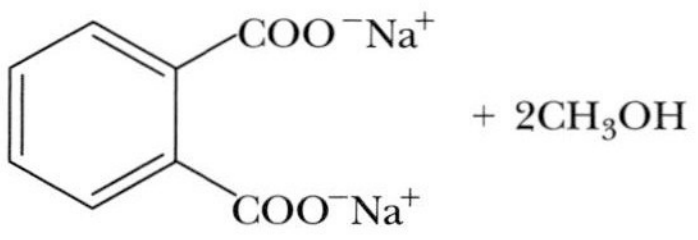

(b) OH + CH_3CH_2OH

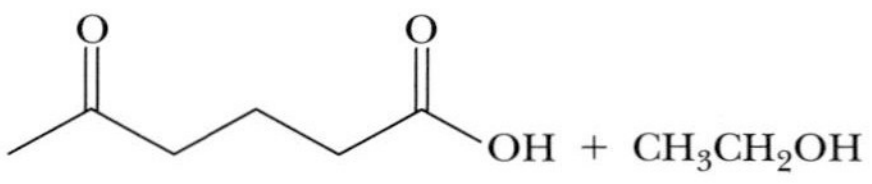

15.3 (a) $CH_3CO^-Na^+ + (CH_3)_2NH$ (with C=O)

(b) H_2N … O^-Na^+ (with C=O)

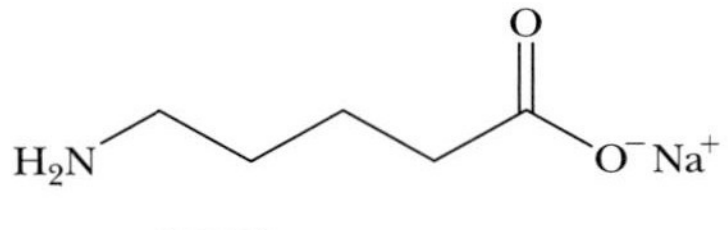

15.4 (a) + 2 HCl

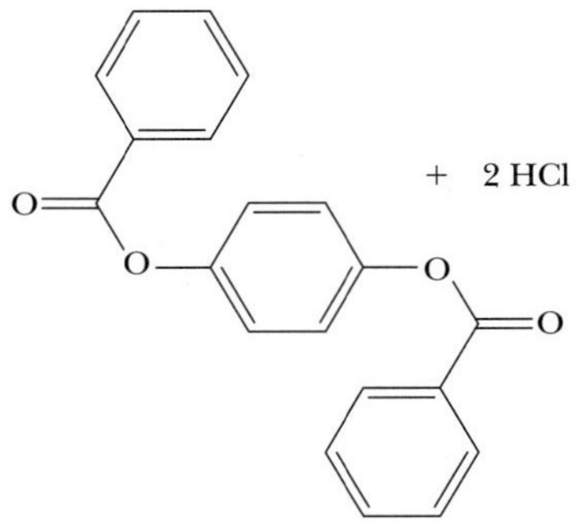

(b) HO … C=O … O

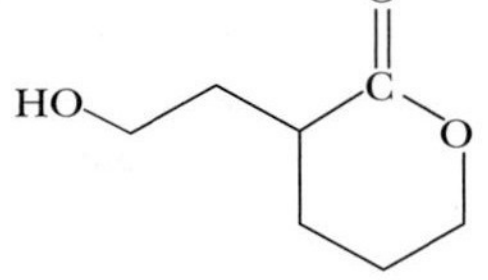

15.5 (a) HO—…—OH + $2CH_3CNH_2$ (with C=O)

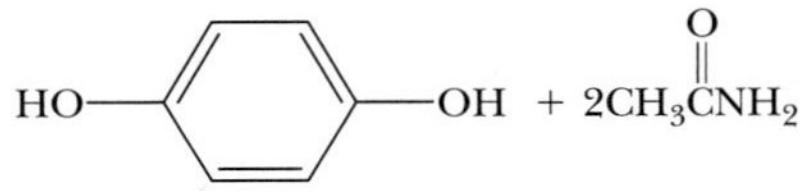

(b) HO … NH_2 (with C=O)

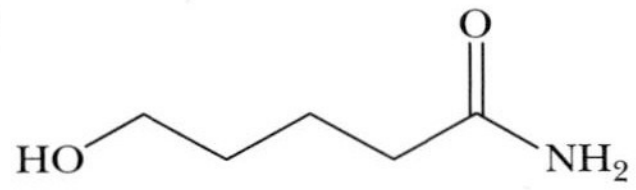

15.6 (a) H … OCH_3 (with C=O); 1) MgBr; 2) H_3O^+

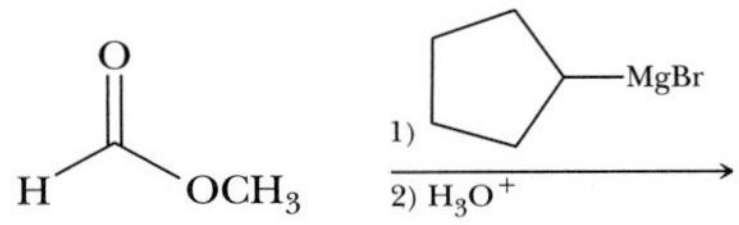

(b) Ph … OCH_3 (with C=O); 1) MgBr; 2) H_3O^+

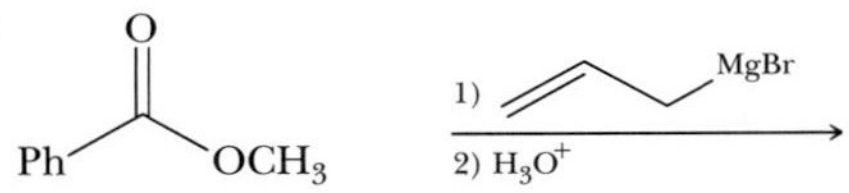

15.7 (a) OH (with C=O) → $SOCl_2$; Cl (with C=O) → $(CH_3)_2NH$; N(CH_3)(CH_3) (with C=O) → 1) $LiAlH_4$ 2) H_2O; N(CH_3)(CH_3)

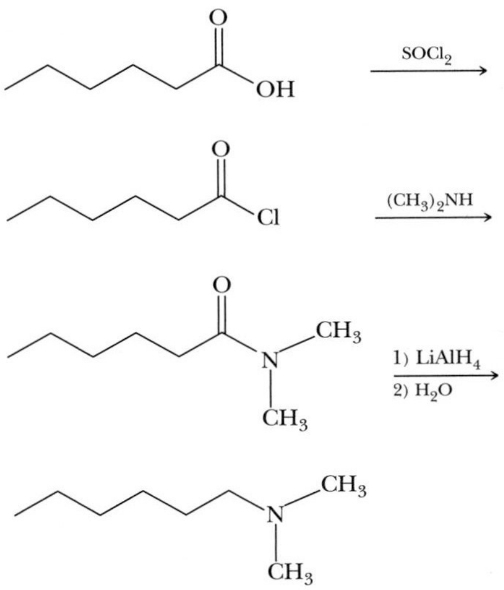

(b) OH (with C=O) → $SOCl_2$; Cl (with C=O) → NH_2; N H (with C=O) → 1) $LiAlH_4$ 2) H_2O; N H

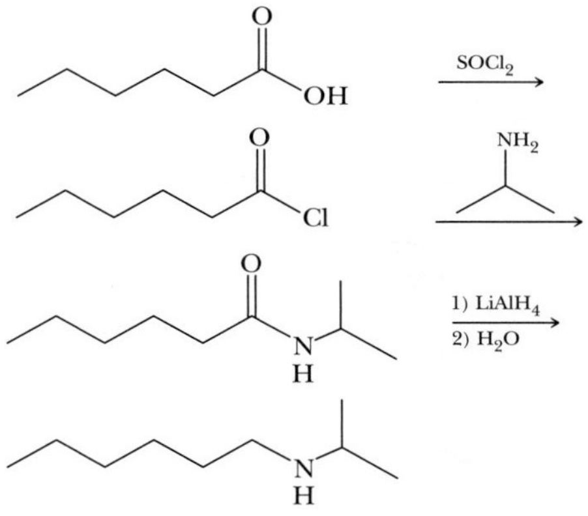

15.8 (a) H_3C, H, Ph, OH, O → 1) LiAlH4 2) H_2O → H_3C, H, Ph, OH

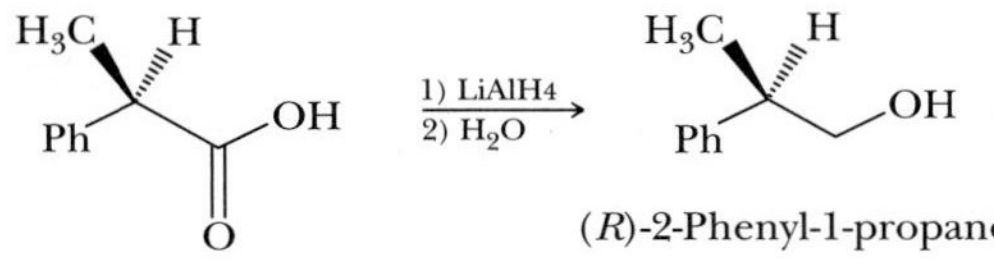

(*R*)-2-Phenyl-1-propanol

(b) H_3C, H, Ph, OH, O → $SOCl_2$ → H_3C, H, Ph, Cl, O → NH_3 → H_3C, H, Ph, NH_2, O → 1) $LiAlH_4$ 2) H_2O → H_3C, H, Ph, NH_2

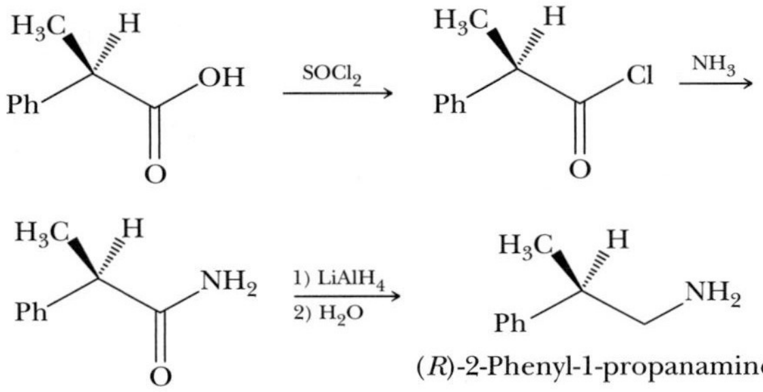

(*R*)-2-Phenyl-1-propanamine

15.9

(a) H_3CO … OCH_3 (with C=O)

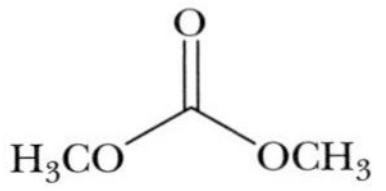

(b) O, NH_2, NO_2

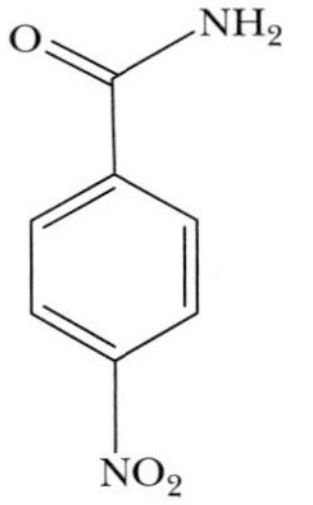

(c) O, Cl, 6

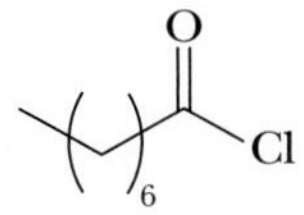

(d) O, O, O, O

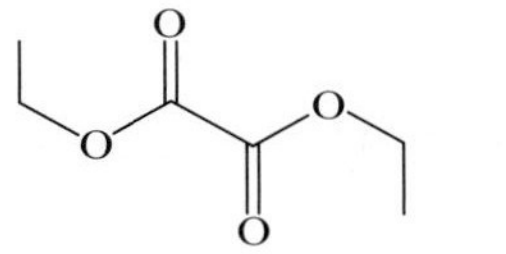

(e) O, O

(f) O, O, O

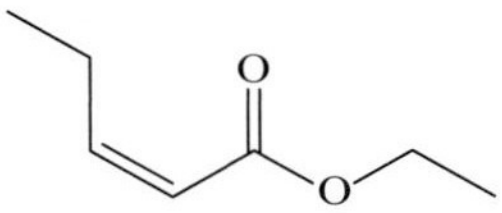

(g) O, NH 2, 10

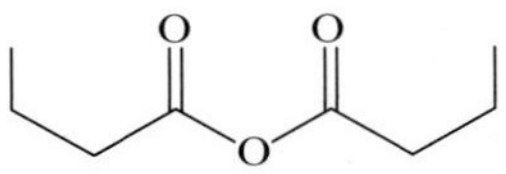

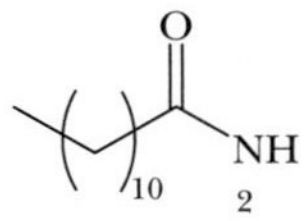

(h) OH, O, O

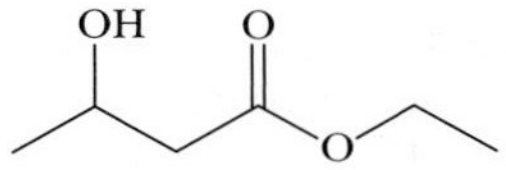

(i) O, O

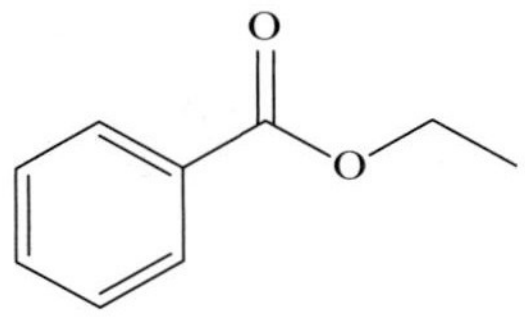

(j) O, Cl

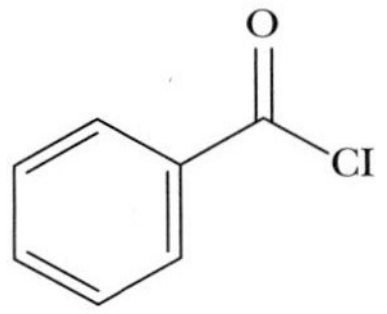

15.11

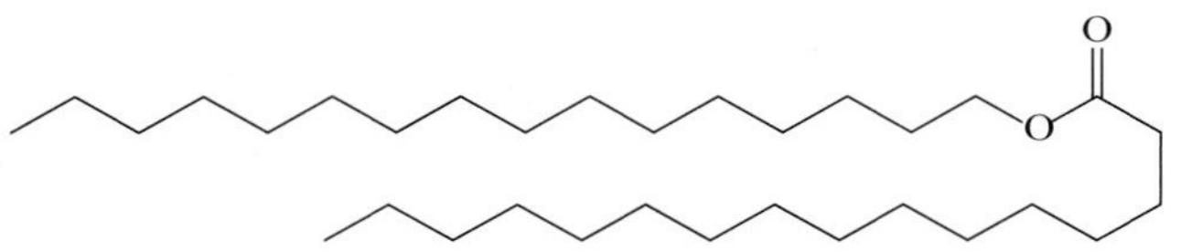

15.13 There is no intermolecular hydrogen bonding between the molecules of propyl butanoate.

15.15 $3 < 1 < 4 < 2$

15.17 (a)

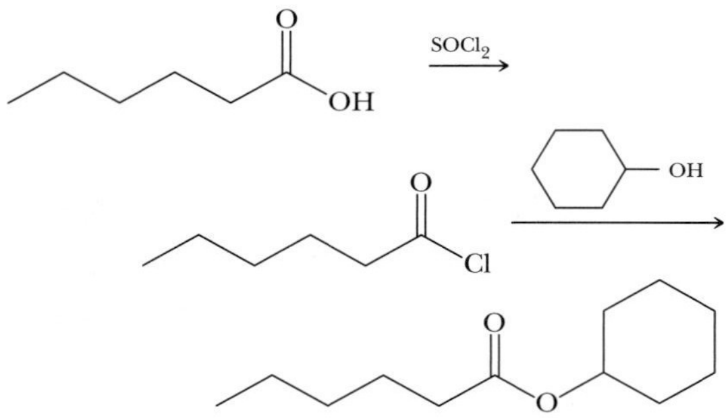

SOCl$_2$

OH

(b)

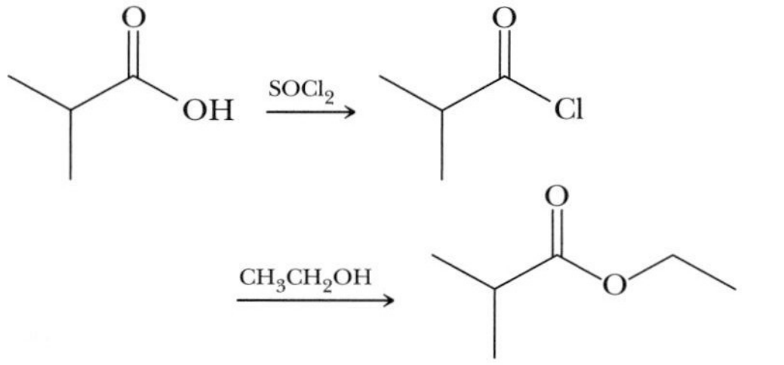

SOCl$_2$

CH_3CH_2OH

15.19 (a)

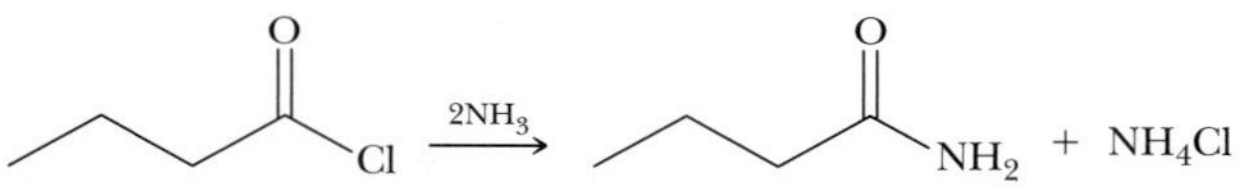

$2NH_3$

+ NH_4Cl

Step 1: Nucleophilic attack on the acyl carbon by ammonia.

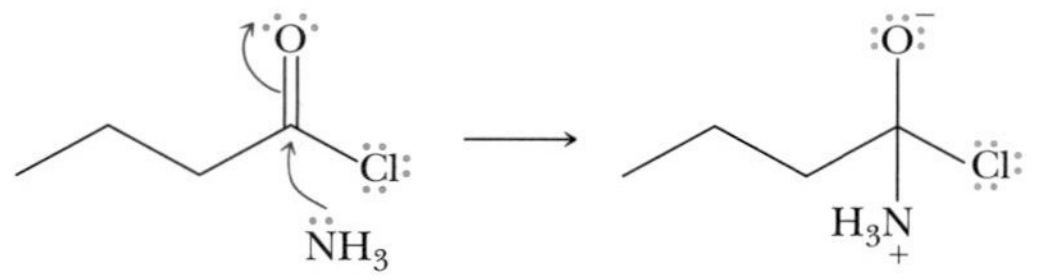

Step 2: Collapse of the tetrahedral intermediate and the loss of chloride.

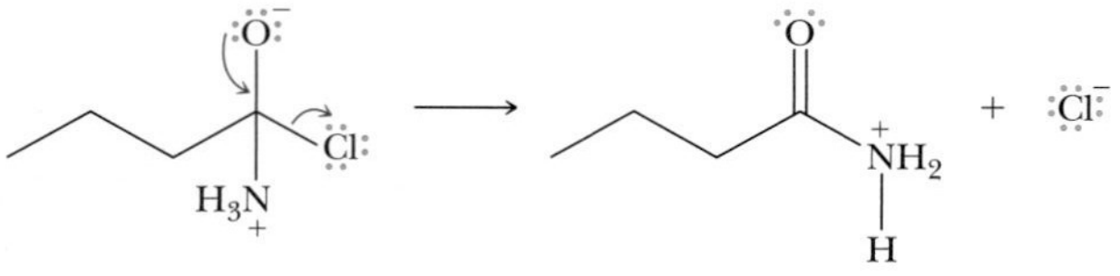

Step 3: A second molecule of ammonia deprotonates the intermediate. The ammonium ion combines with the chloride ion from the step above to form NH_4Cl.

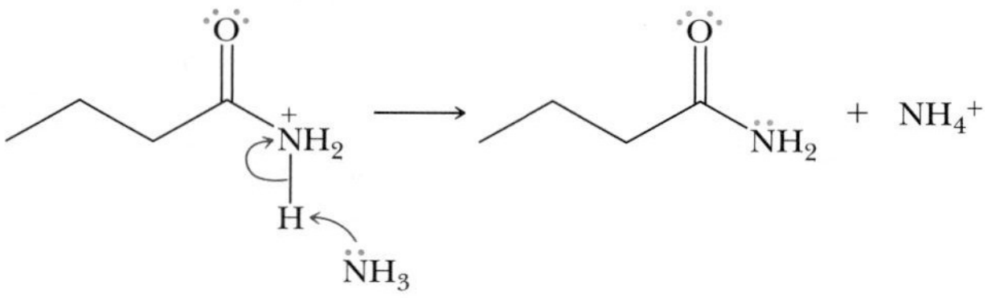

+ NH_4^+

(b)

H_3O^+

+ CH_3OH

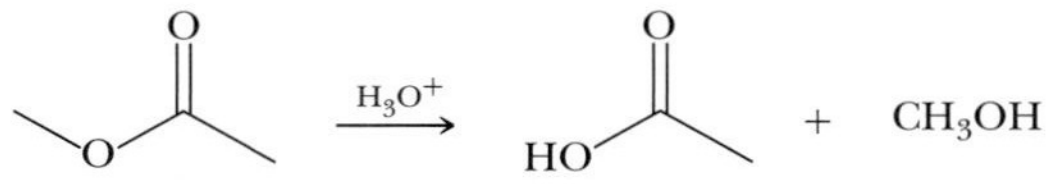

Step 1: Protonation of the acyl oxygen to form a resonance-stabilized cation.

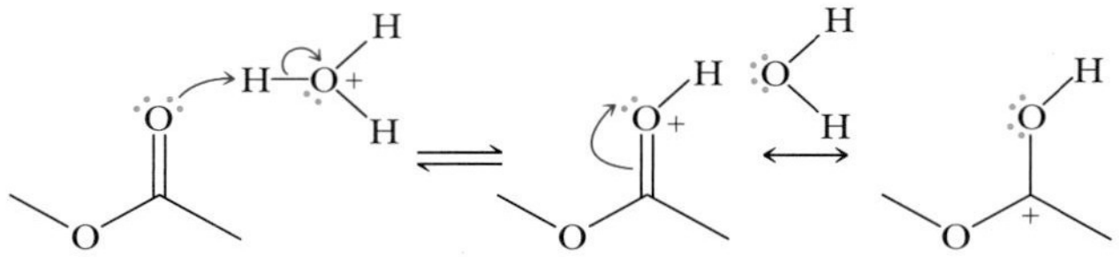

Step 2: Nucleophilic attack to form a tetrahedral intermediate.

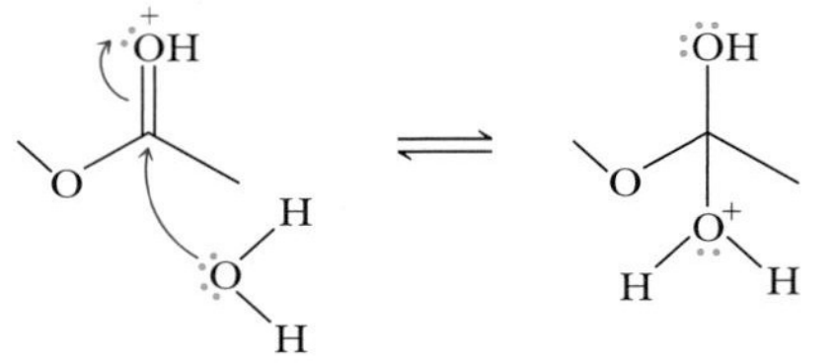

Step 3: Proton transfer from the water portion of the intermediate to the alcohol portion. This converts the alkoxy group into a good leaving group.

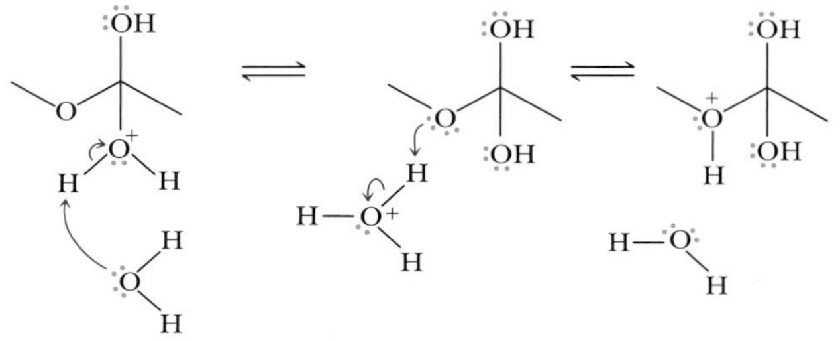

Step 4: Deprotonation and departure of the alcohol generates the carboxylic acid and the alcohol, as well as the acid catalyst. Note that either one of the two OH groups can be deprotonated.

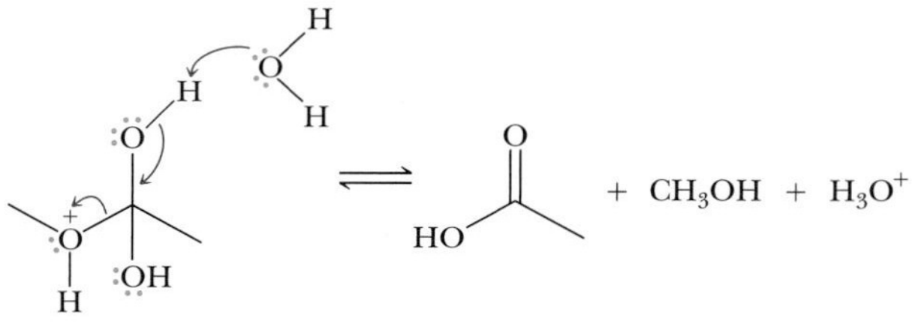

+ CH_3OH + H_3O^+

(c)

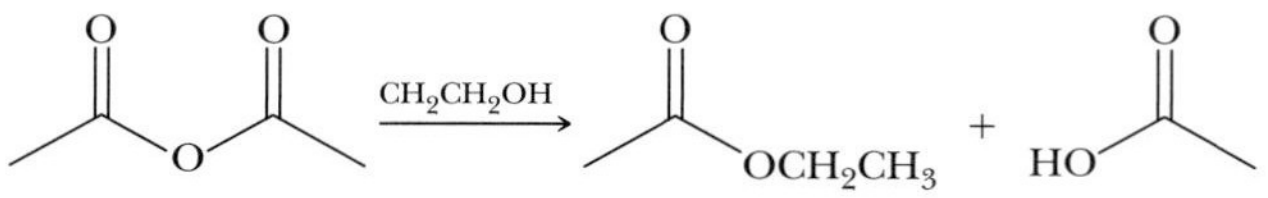

CH_3CH_2OH

OCH_2CH_3 + HO

Step 1: Nucleophilic attack on the acyl carbon by ethanol.

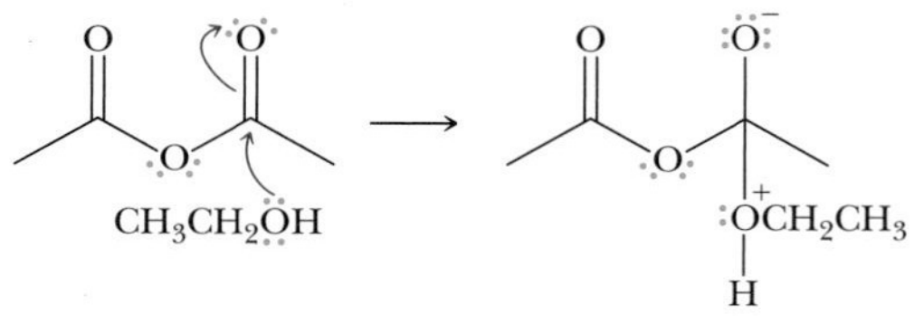

$CH_3CH_2\ddot{O}H$

$\ddot{O}CH_2CH_3$

Step 2: Collapse of the tetrahedral intermediate and the loss of acetate.

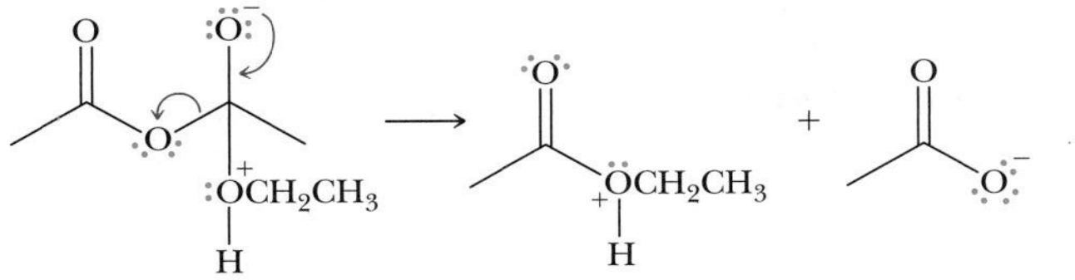

$\longrightarrow$ +

Step 3: Deprotonation of the oxonium ion generates the products.

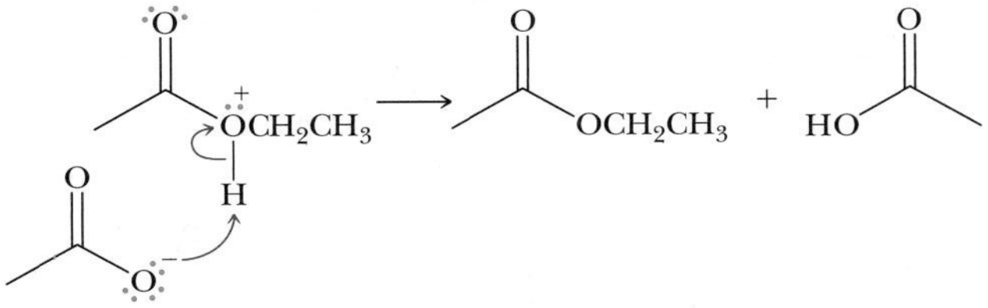

$\longrightarrow$ +

15.21 (a) +

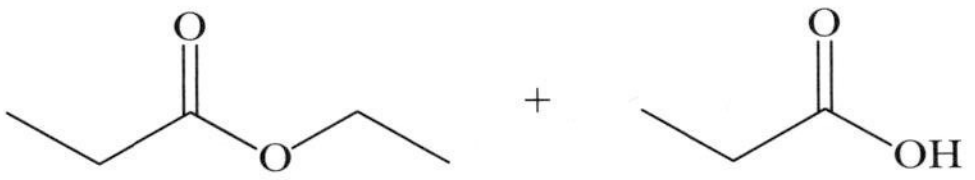

(b) +

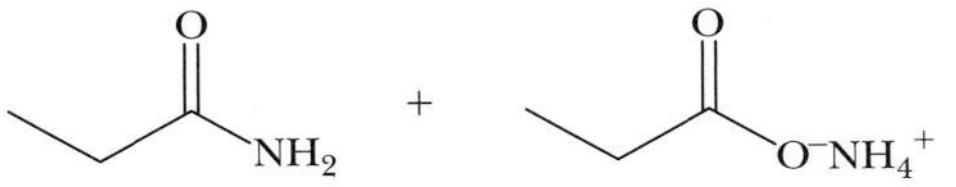

15.23

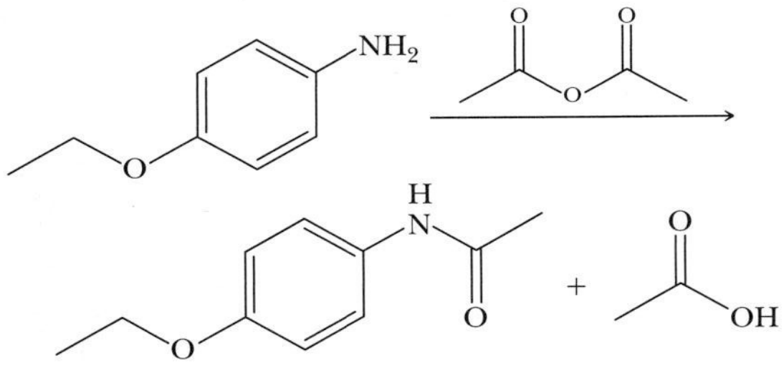

15.25 1. CH_3CH_2OH/H_2SO_4 2. Na_2CO_3/H_2O 3. NH_3

15.27 (a) + CH_3CH_2OH

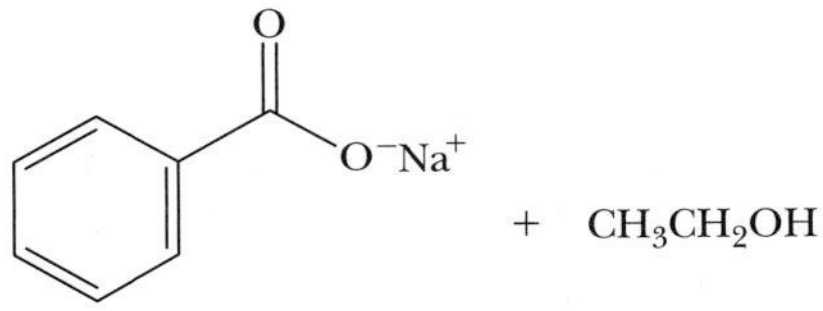

(b) + CH_3CH_2OH

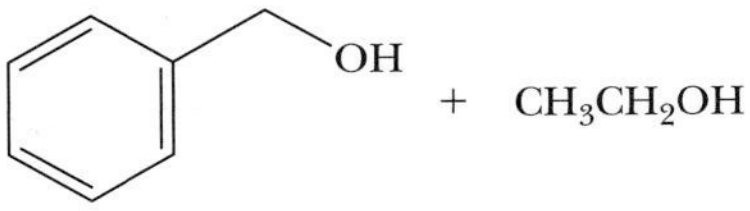

(c) + CH_3CH_2OH

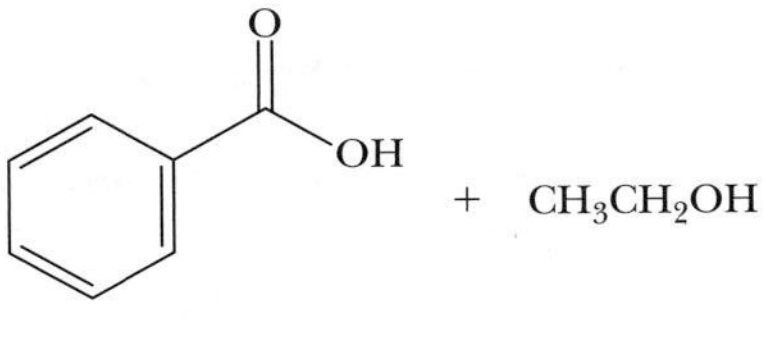

(d) + CH_3CH_2OH

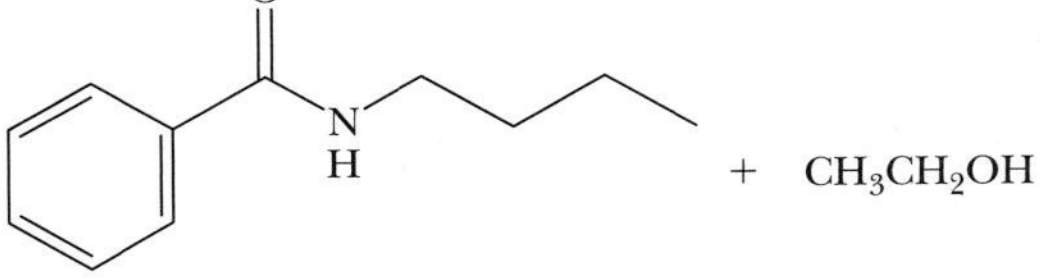

(e) $(C_6H_5)_3COH + CH_3CH_2OH$

15.29 (a) + $NH_4^+Cl^-$

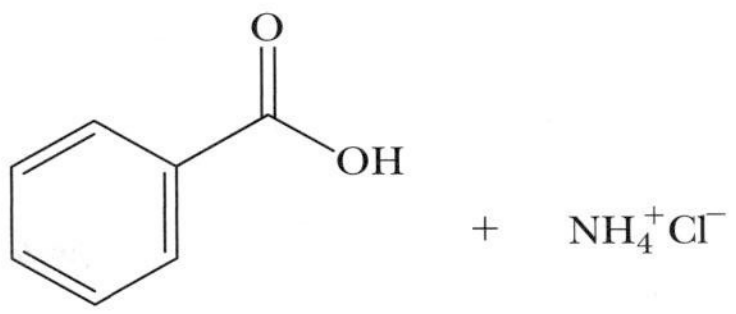

(b) + NH_3

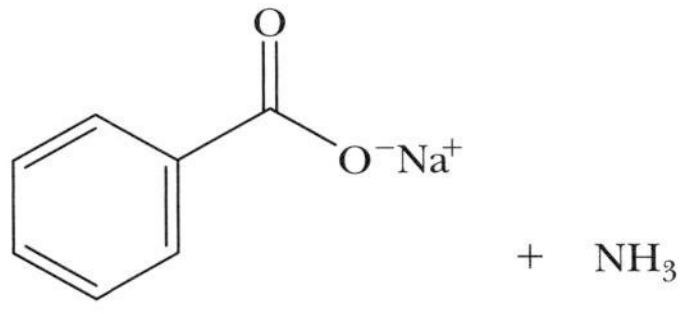

(c)

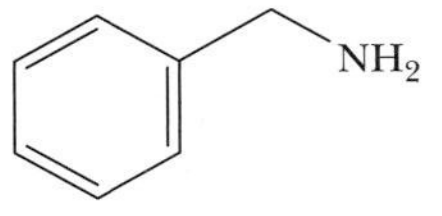

15.31 (a)

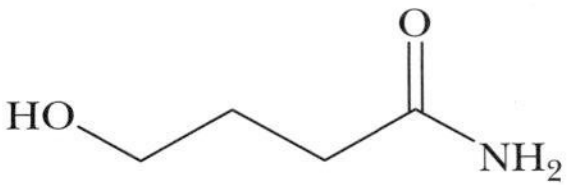

(b)

(c)

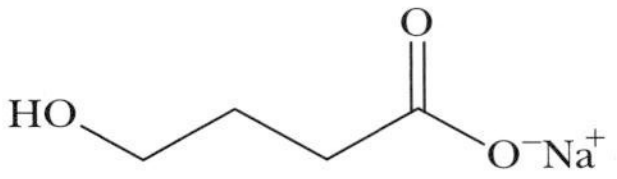

15.33 (a) + CH_3CH_2OH

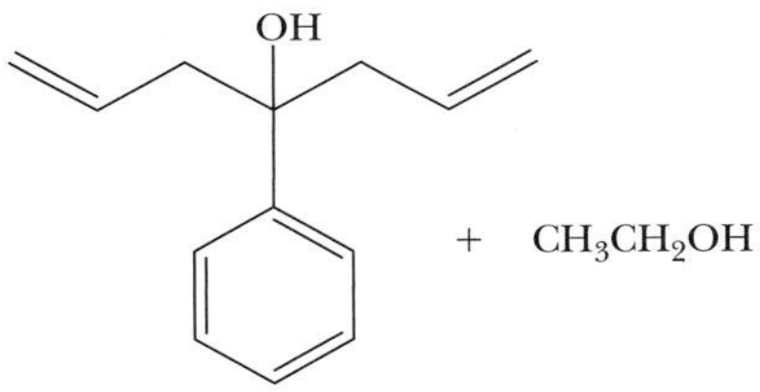

(b)

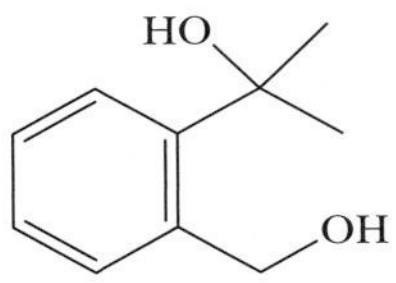

(c) + CH_3CH_2OH

15.35

Step 1: The amine attacks the carbonyl carbon to form a tetrahedral intermediate.

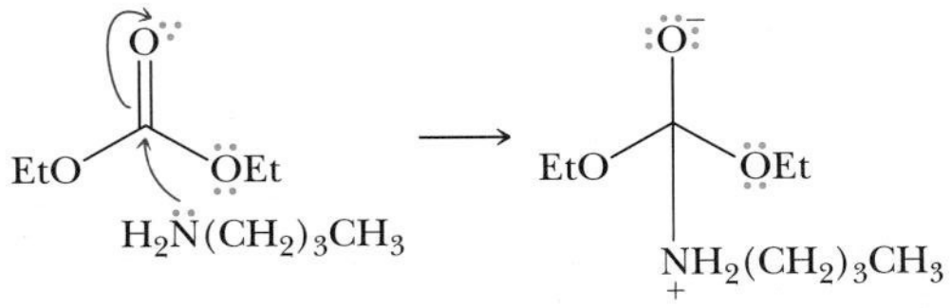

$\longrightarrow$

Step 2: Collapse of the tetrahedral intermediate forms a protonated carbamic ester.

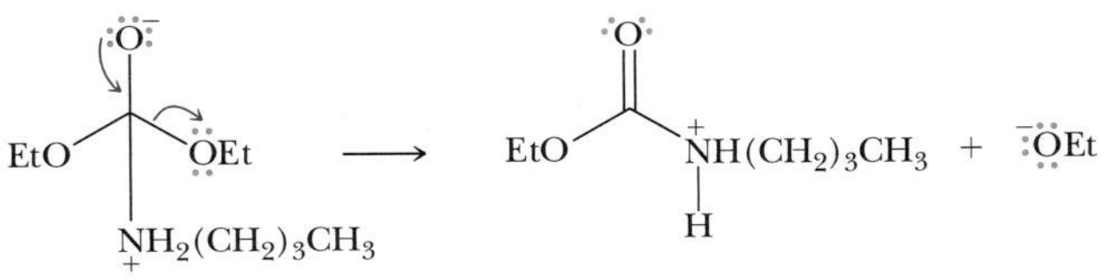

$\longrightarrow$ +

Step 3: Deprotonation results in the formation of the two products.

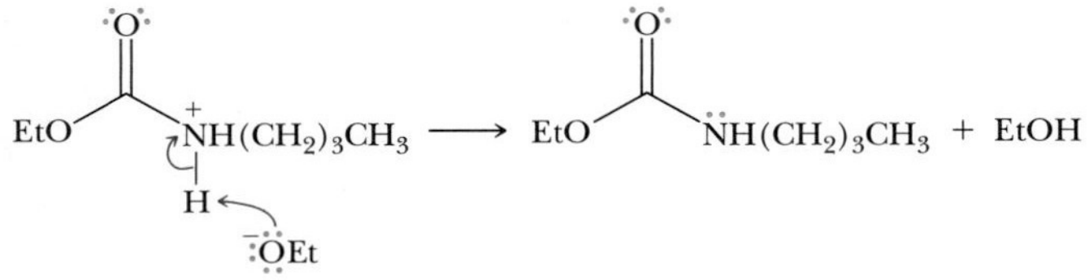

$EtO-C(=O)-\overset{+}{N}H(CH_2)_3CH_3 \longrightarrow EtO-C(=O)-NH(CH_2)_3CH_3 + EtOH$

$^{-}$:ÖEt

15.37

(a) HO, OH

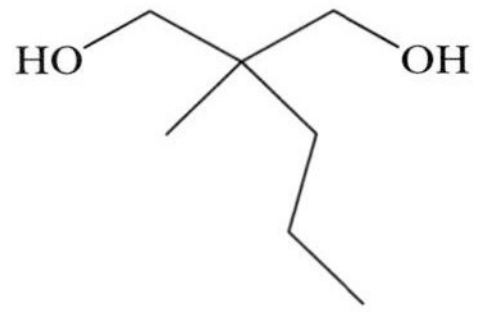

2-methyl-2-propyl-1,3-propanediol

(b) OH, O

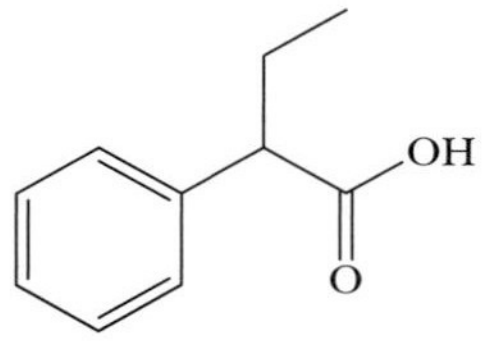

2-phenylbutanoic acid

15.39 (a) H_2/Pd (b) 1. $LiAlH_4$ 2. H_2O
(c) Br_2/CH_2Cl_2

15.41

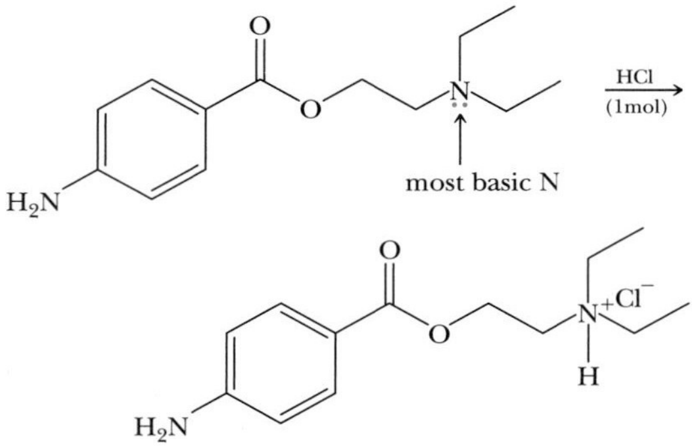

H_2N, O, O, N, HCl (1mol)

most basic N

H_2N, O, O, N^+, Cl^-, H

15.43 (a)

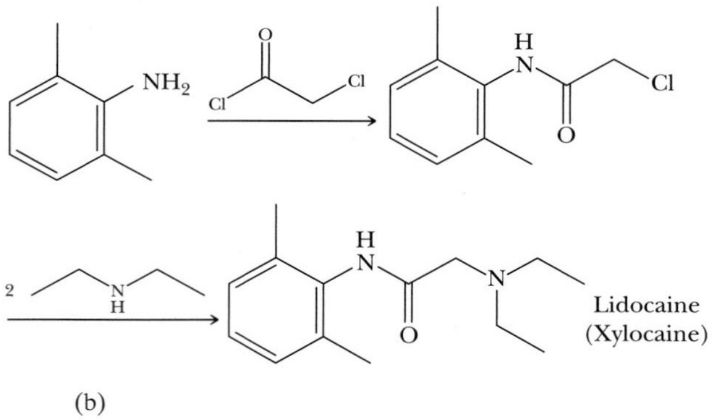

NH_2, Cl, O, Cl → H, N, O, Cl

2 (N-H) → H, N, O, N — Lidocaine (Xylocaine)

(b)

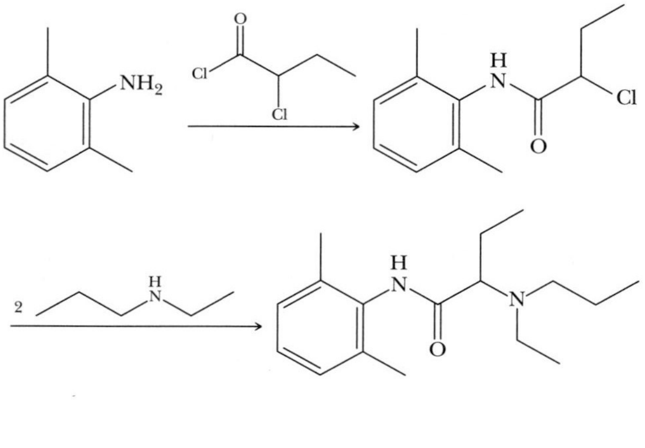

NH_2, Cl, O, Cl → H, N, O, Cl

2 (H-N) → H, N, O, N

(c)

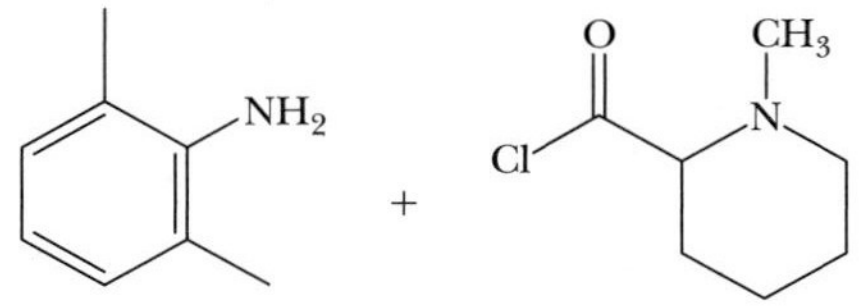

NH_2 + O, Cl, N, CH_3

15.45

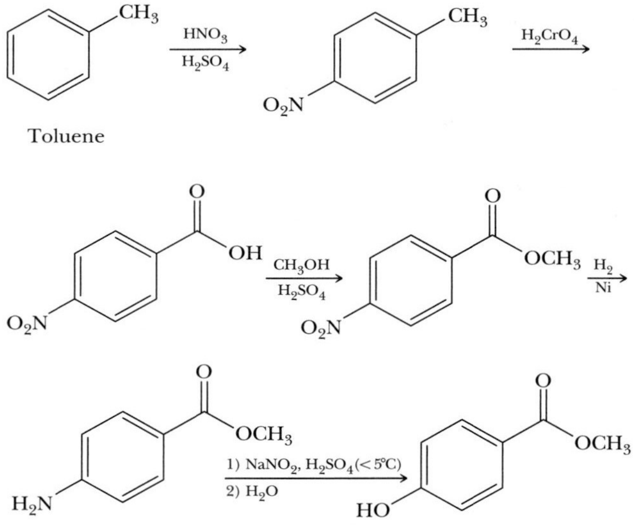

Toluene (CH_3) $\xrightarrow[H_2SO_4]{HNO_3}$ O_2N–C$_6$H$_4$–CH_3 $\xrightarrow{H_2CrO_4}$

O_2N–C$_6$H$_4$–COOH $\xrightarrow[H_2SO_4]{CH_3OH}$ O_2N–C$_6$H$_4$–$COOCH_3$ $\xrightarrow[Ni]{H_2}$

H_2N–C$_6$H$_4$–$COOCH_3$ $\xrightarrow[2)\ H_2O]{1)\ NaNO_2,\ H_2SO_4(<5°C)}$ HO–C$_6$H$_4$–$COOCH_3$

15.47

(a) O, O, H H — most acidic

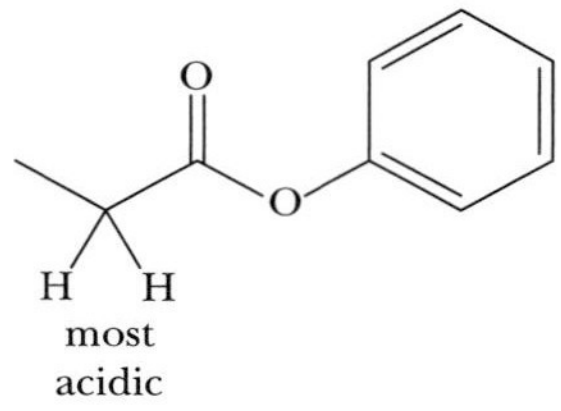

(b) O, O, O, O, H H — most acidic

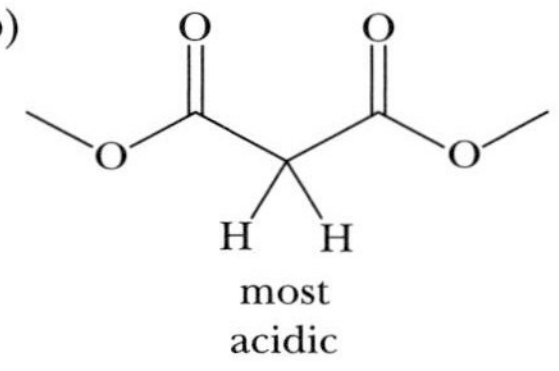

15.49 Resonance structures can be drawn such that the β-carbon is positively charged and thus electrophilic.

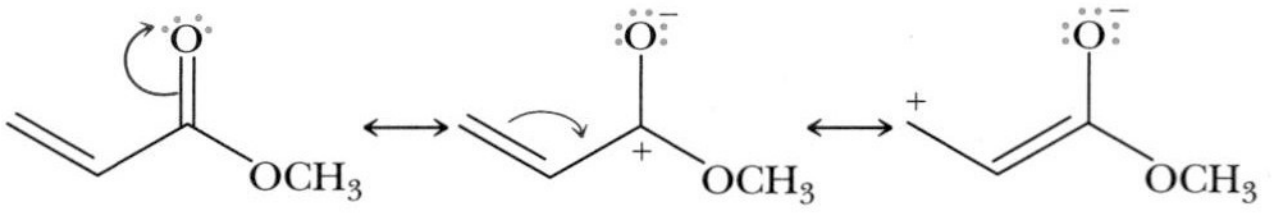

OCH_3 ⟷ OCH_3 ⟷ OCH_3

15.51 Amides have partial double-bond character between the carbon and nitrogen atoms, and at low temperature, the double bond restricts rotation, giving rise to *cis-trans* isomerism. However, at higher temperatures, there is sufficient energy to overcome the barrier of CN bond rotation.

Putting It Together

1. (d) **2.** (b)
3. (d) **4.** (b), (c), and (e)
5. (a) **6.** (e)
7. (e) **8.** (d)
9. (d) **10.** (b)

11.

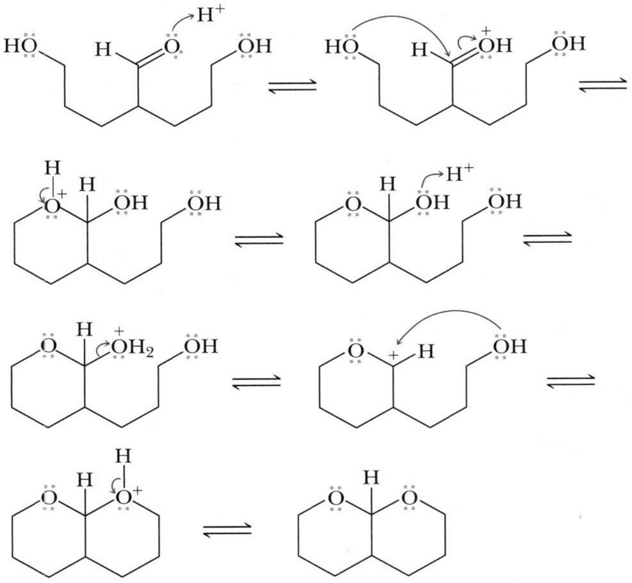

HO H O H⁺ OH ⇌ HO H OH OH ⇌

H H OH OH ⇌ O H OH H⁺ OH ⇌

O H $\overset{+}{O}H_2$ OH ⇌ O H OH ⇌

H H O⁺ ⇌ O H O

12. (c) > (a) > (d) > (b) > (e)

13. (a) (*S*)-5-ethyl-2-cyclopentenone (b) 3-phenylpropanal
(c) 3,5-dioxopentanoic acid

14.

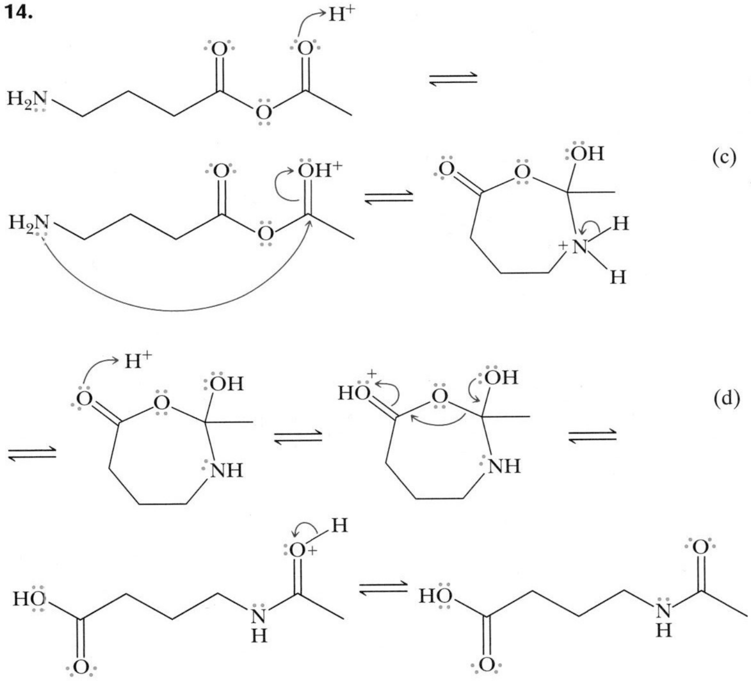

H_2N O O H⁺ ⇌

H_2N O OH^+ ⇌ O O OH N⁺ H H

⇌ O H⁺ O OH NH ⇌ HO⁺ O OH NH ⇌

HO N H O⁺ H ⇌ HO N H O

15.

(a) EtO O OEt O (b) HO OH + EtOH

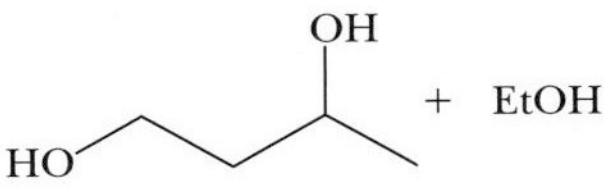

(c) O (d) HO H O OH (e) OCH_2CH_3

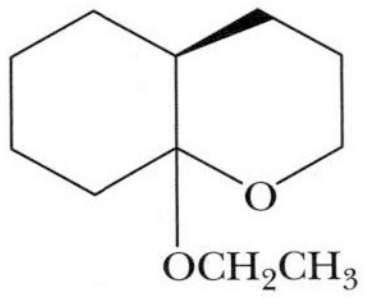

16. (a)

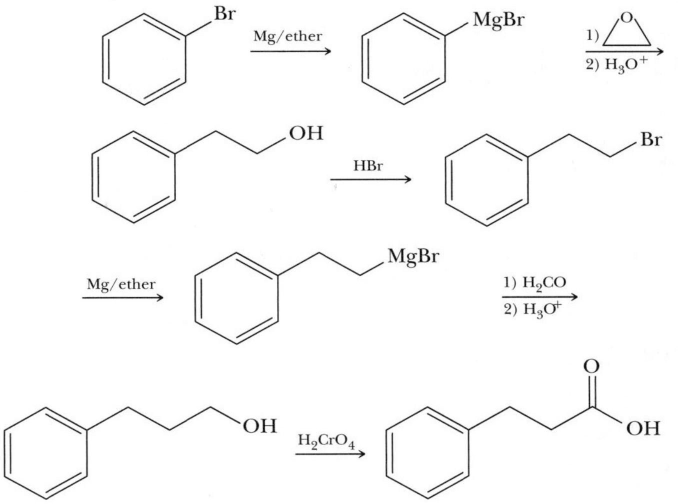

Br $\xrightarrow{\text{Mg/ether}}$ MgBr $\xrightarrow[\text{2) } H_3O^+]{\text{1) epoxide}}$

OH $\xrightarrow{\text{HBr}}$ Br

$\xrightarrow{\text{Mg/ether}}$ MgBr $\xrightarrow[\text{2) } H_3O^+]{\text{1) } H_2CO}$

OH $\xrightarrow{H_2CrO_4}$ O OH

(b)

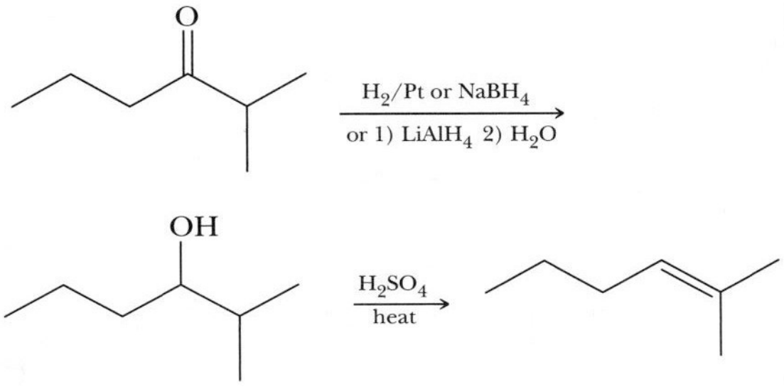

O $\xrightarrow[\text{or 1) } LiAlH_4 \text{ 2) } H_2O]{H_2/Pt \text{ or } NaBH_4}$

OH $\xrightarrow[\text{heat}]{H_2SO_4}$

(c) $\xrightarrow[H_2O]{H^+}$ OH $\xrightarrow[H_2CrO_4]{\text{PCC or}}$

O $\xrightarrow[CH_3NH_2]{H^+}$ N CH_3

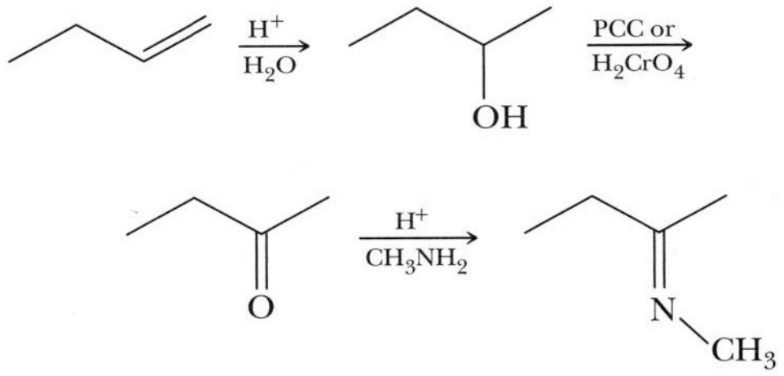

(d) OH $\xrightarrow{\text{PCC}}$ O H $\xrightarrow[\text{2) } H_3O^+]{\text{1) EtMgBr}}$

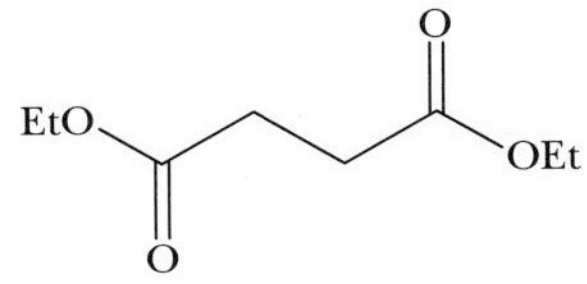

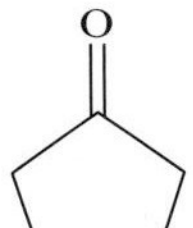

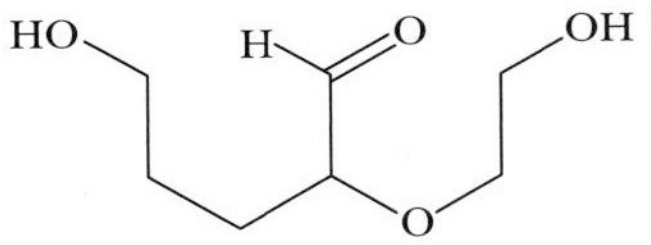

OH CH_2CH_3

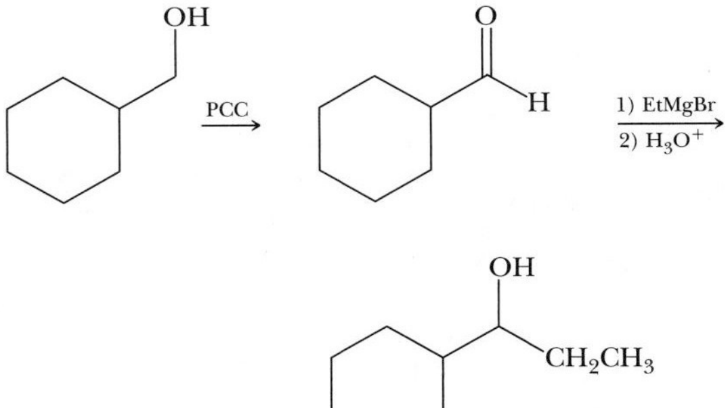

(e) OH O $\xrightarrow{SOCl_2}$ Cl O

$\xrightarrow[AlCl_3]{\text{benzene}}$ O

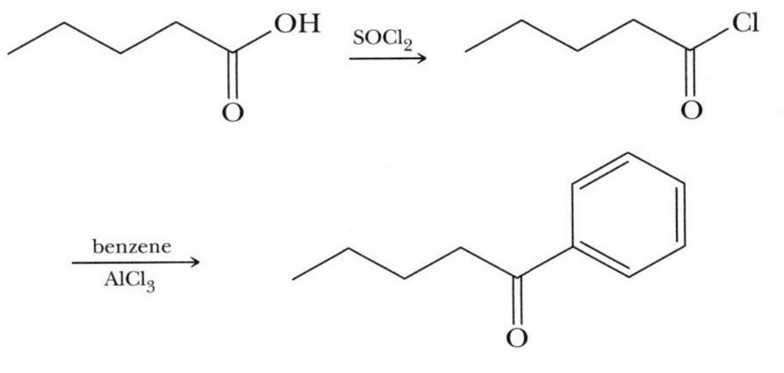

(f)

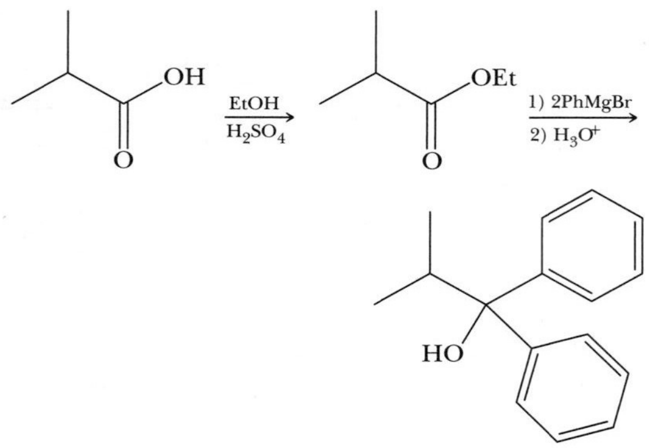

OH $\xrightarrow[H_2SO_4]{EtOH}$ OEt $\xrightarrow[2)\ H_3O^+]{1)\ 2PhMgBr}$

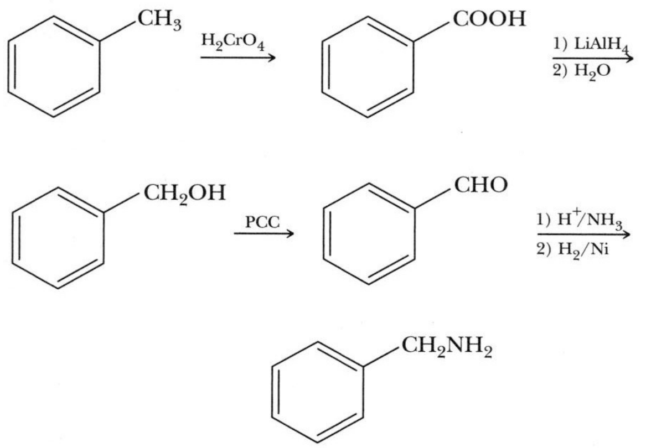

HO

(g)

CH_3 $\xrightarrow{H_2CrO_4}$ COOH $\xrightarrow[2)\ H_2O]{1)\ LiAlH_4}$

CH_2OH $\xrightarrow{PCC}$ CHO $\xrightarrow[2)\ H_2/Ni]{1)\ H^+/NH_3}$

CH_2NH_2

(h)

Br $\xrightarrow{NaOH}$ OH $\xrightarrow{H_2CrO_4}$

O OH $\xrightarrow{SOCl_2}$ O Cl

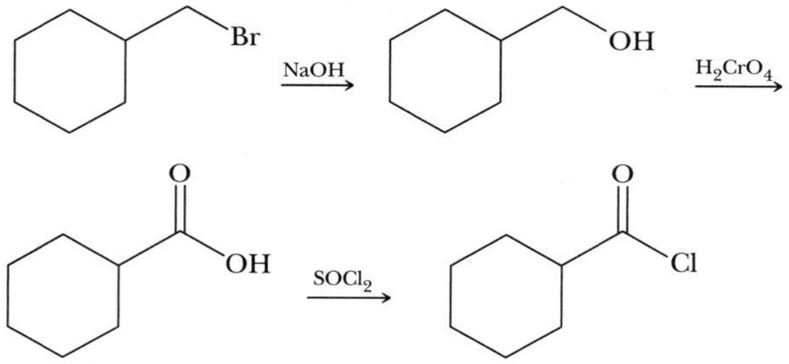

Chapter 16
Enolate Anions

16.1

(a) O H H H

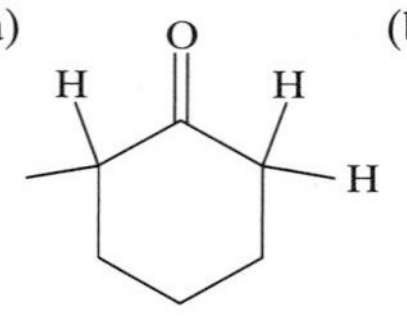

(b) O H H H

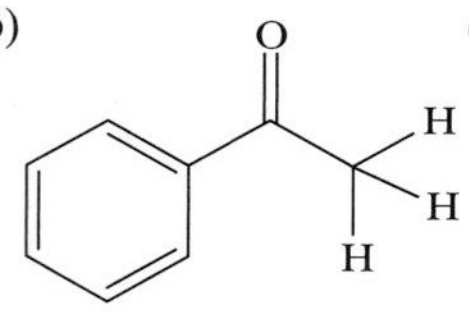

(c) O H H

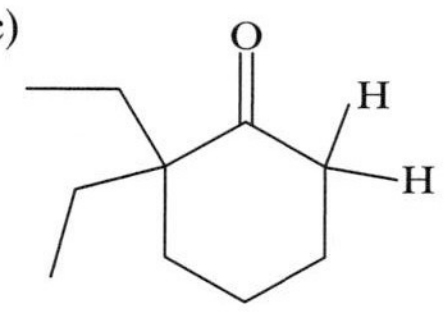

16.2

(a) O OH

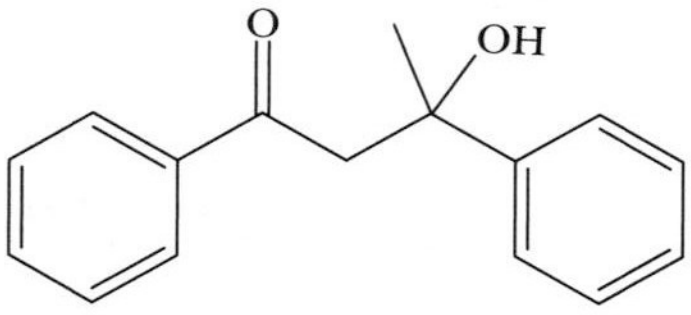

(b) O OH

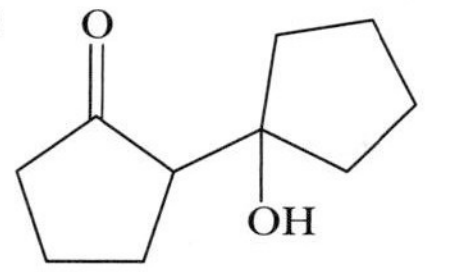

16.3

(a) O

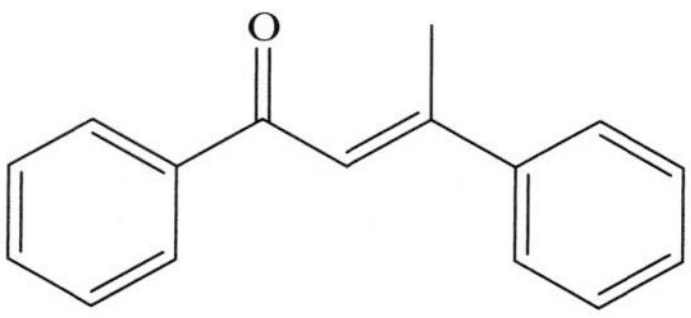

(b) O

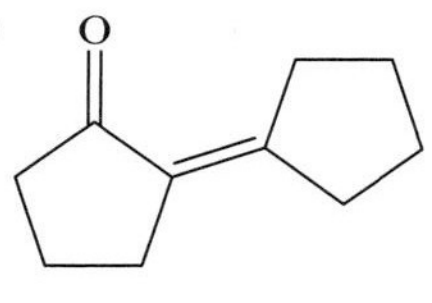

16.4

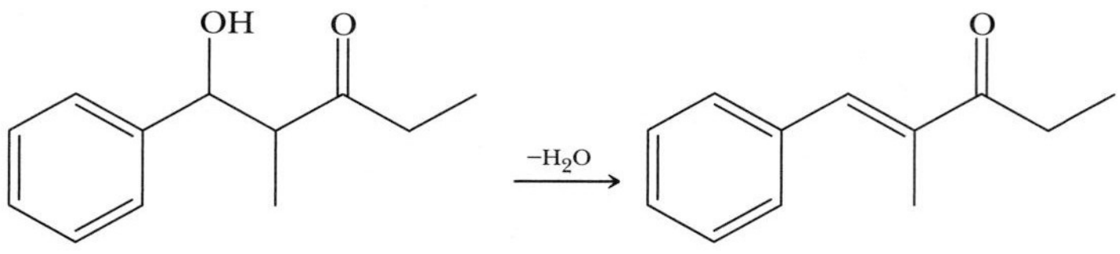

OH O $\xrightarrow{-H_2O}$ O

16.5 O

16.6 O O O

16.7 OEt O O

16.8 O O OEt

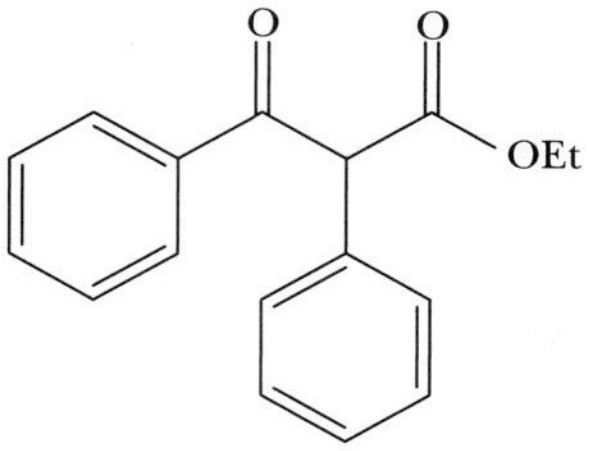

16.9

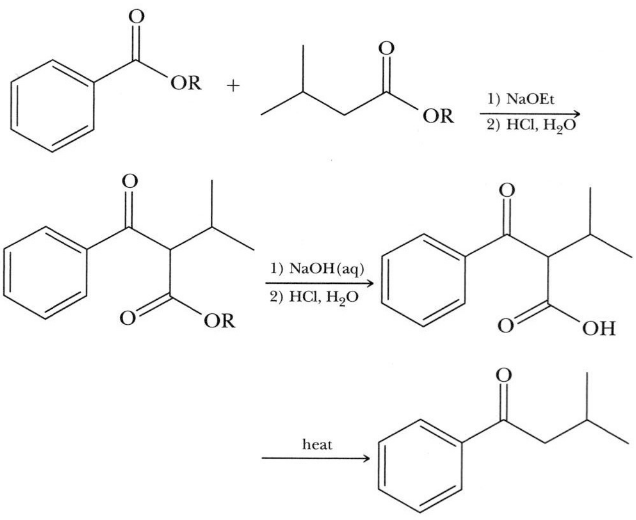

O OR + O OR $\xrightarrow[2)\ HCl,\ H_2O]{1)\ NaOEt}$

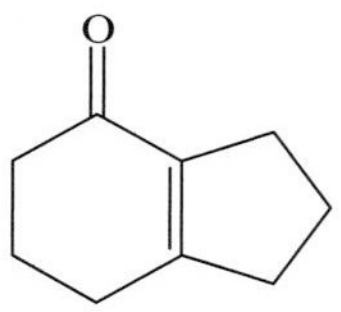

O O OR $\xrightarrow[2)\ HCl,\ H_2O]{1)\ NaOH(aq)}$ O O OH

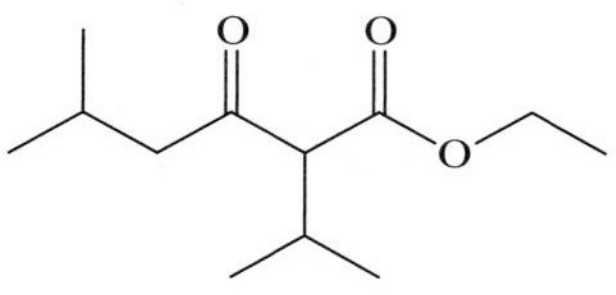

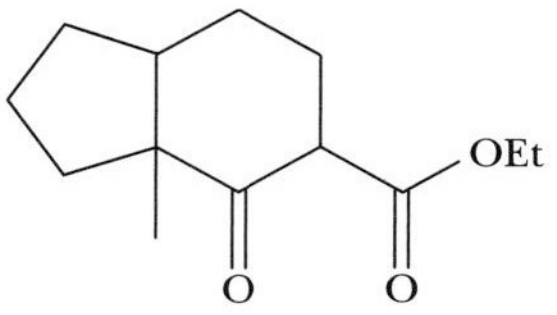

$\xrightarrow{heat}$ O

16.10 (a)

O O OEt COOEt $\xrightarrow[2)\ HCl,\ H_2O]{1)\ NaOH(aq)}$

O O OH COOH $\xrightarrow{heat}$ O O OH

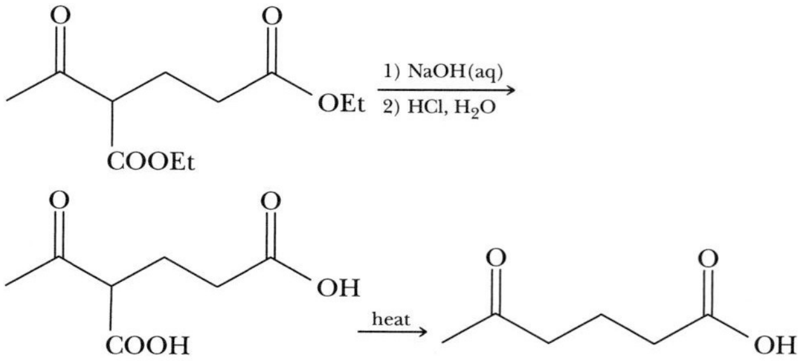

(b)

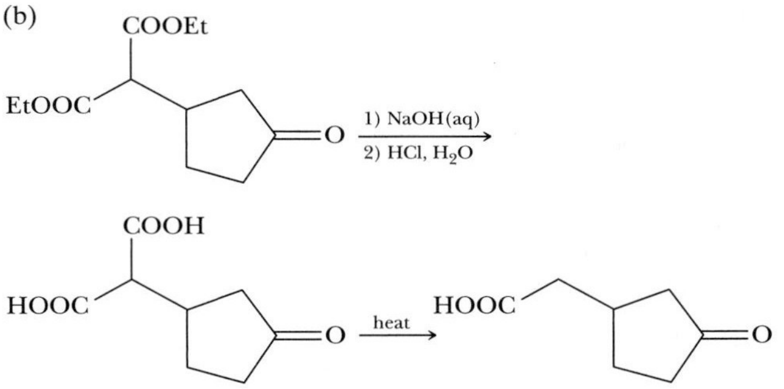

1) NaOH(aq)
2) HCl, H_2O

heat

16.11

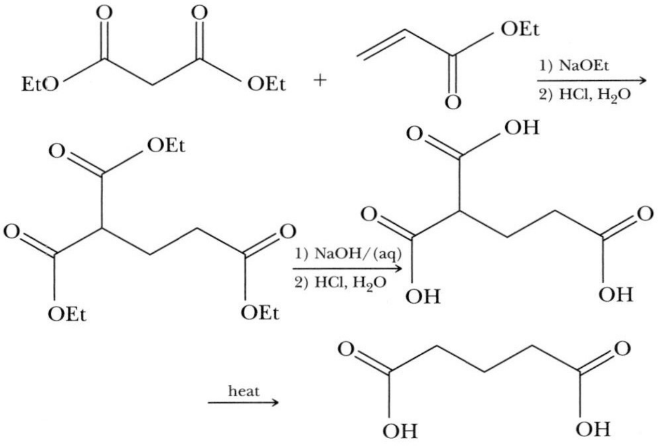

1) NaOEt
2) HCl, H_2O

1) NaOH/(aq)
2) HCl, H_2O

heat

16.12

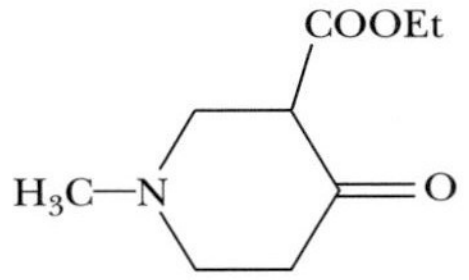

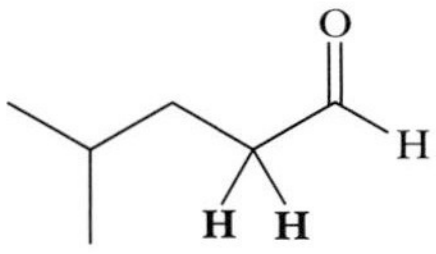

16.13

(a) (b) 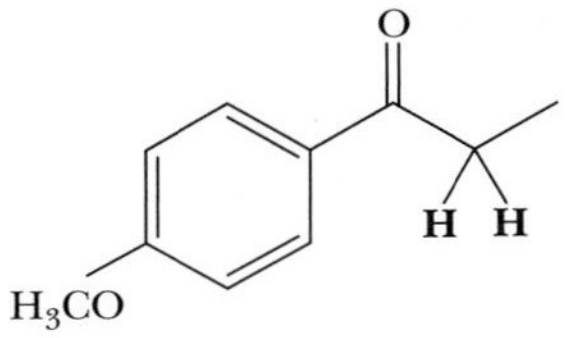 (c) 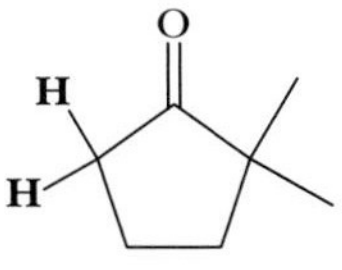 (d) 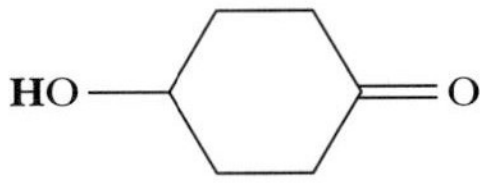 (e) 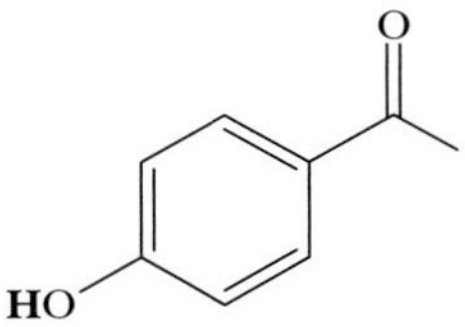 (f)

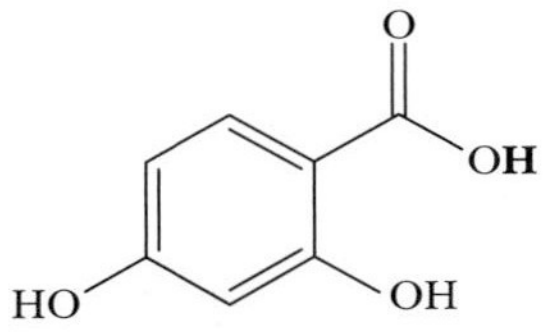

16.15 (a)

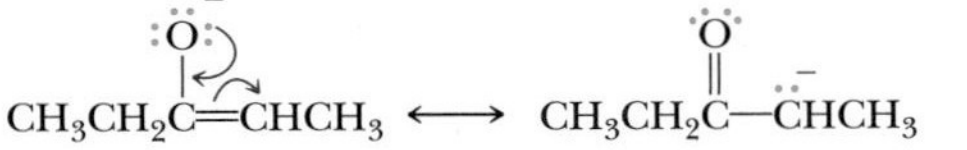

$CH_3CH_2\ddot{C}{=}CHCH_3 \longleftrightarrow CH_3CH_2C{-}\bar{C}HCH_3$

(b)

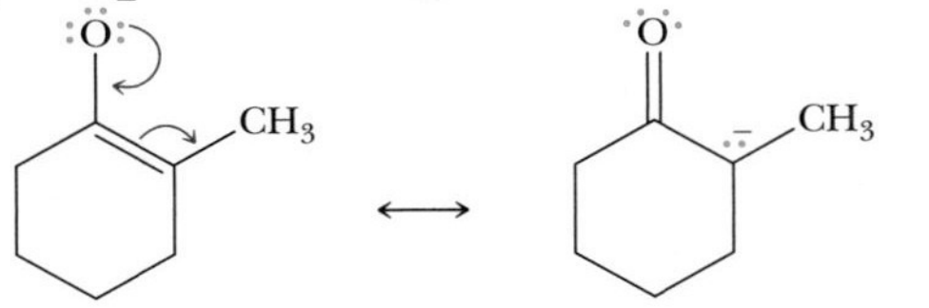

(c)

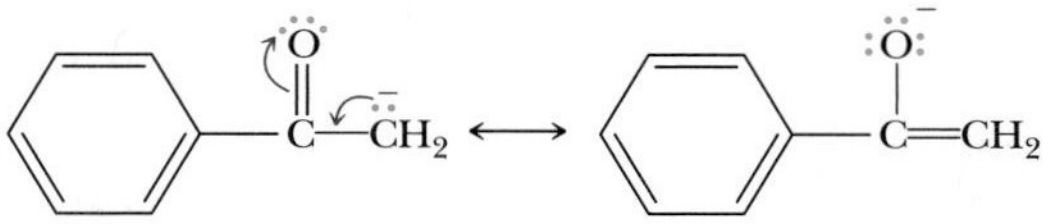

16.17 (a)

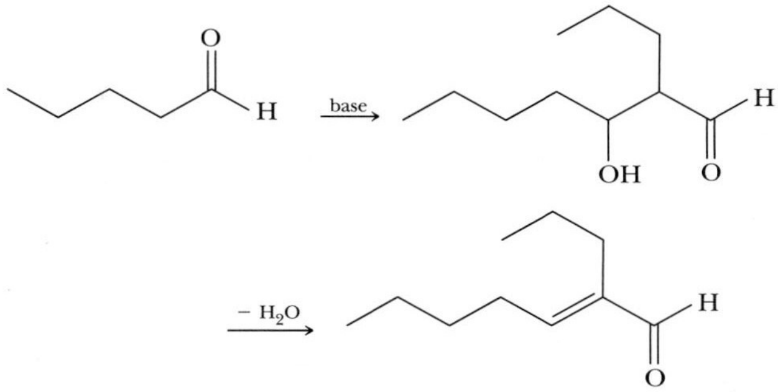

base

$-H_2O$

(b)

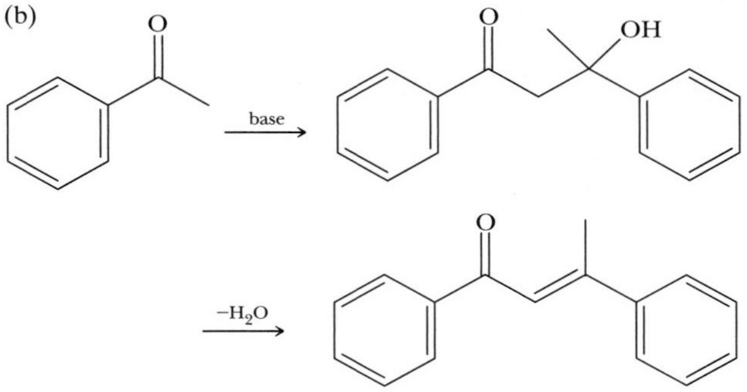

base

$-H_2O$

(c)

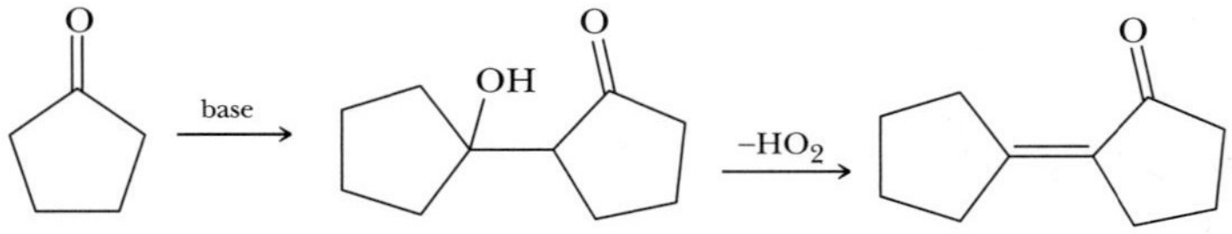

base

$-HO_2$

(d)

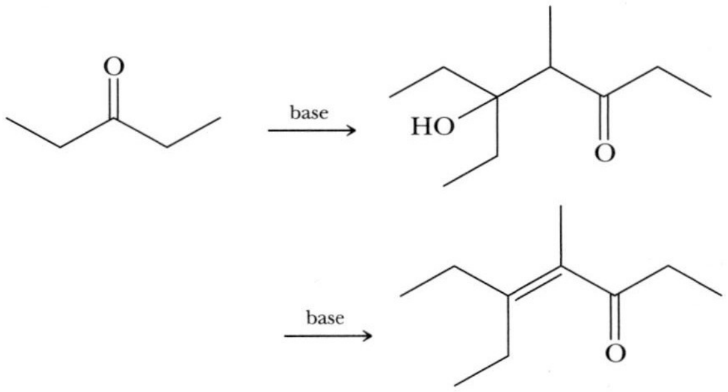

base

base

(e)

base

$-H_2O$

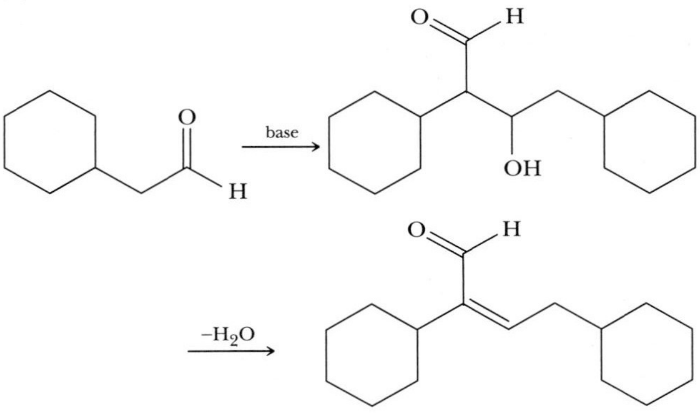

(f)

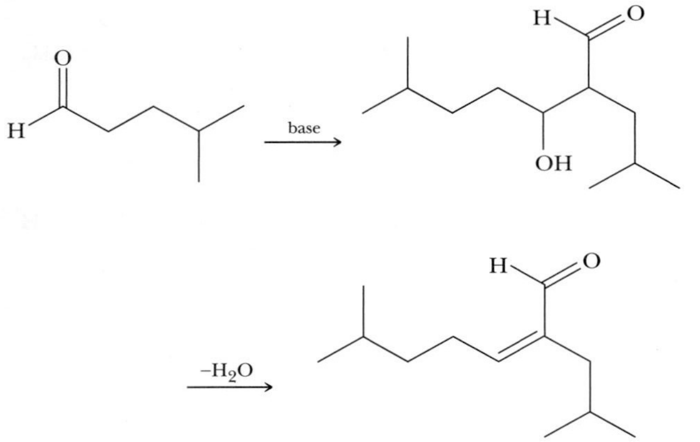

base

$-H_2O$

16.19 (a)

(b)

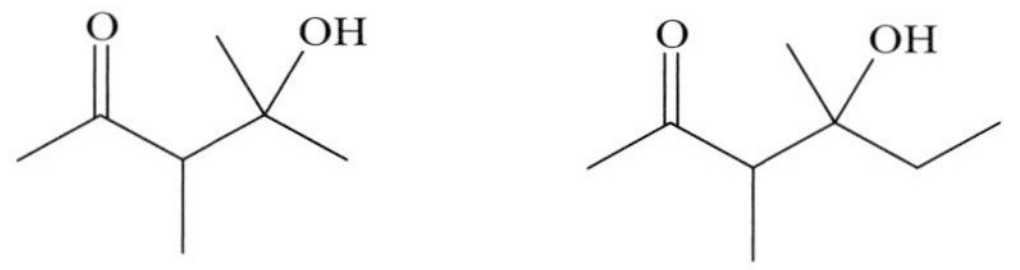

(c)

16.21 (a)

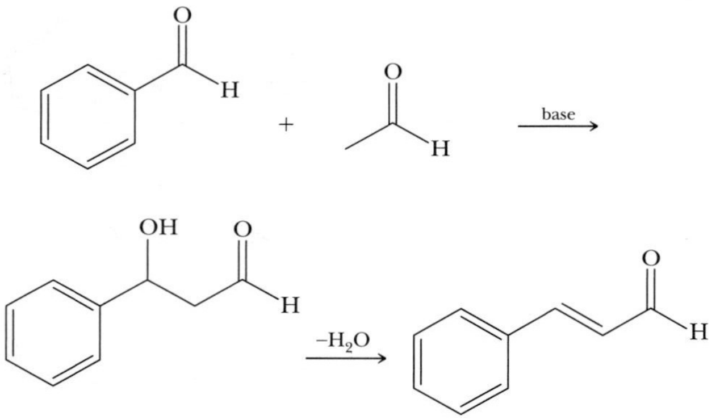

base

$-H_2O$

(b)

base

$-H_2O$

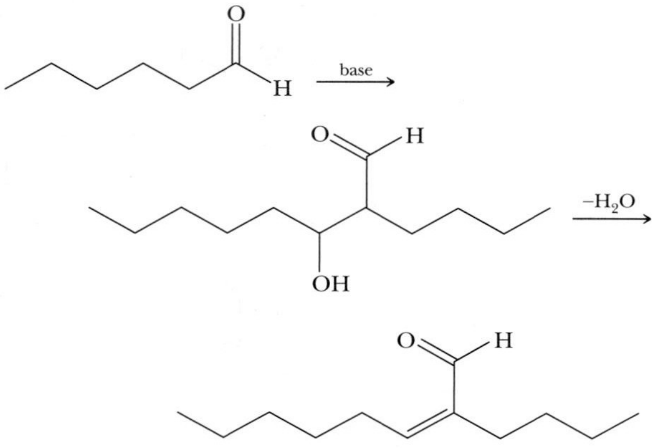

16.23

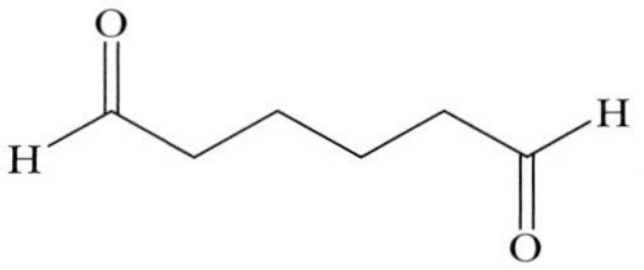

16.25 (a) 1. NaOH 2. H_2CrO_4 3. $SOCl_2$ 4. $2NH_3$ 5. RCO_3H

(b) 2 stereocenters, 4 stereoisomers

16.27 (a)

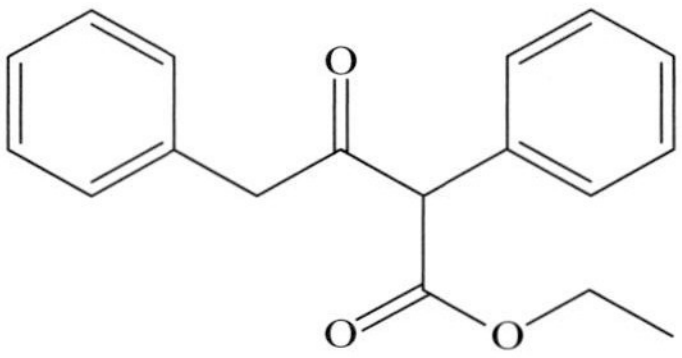

(b)

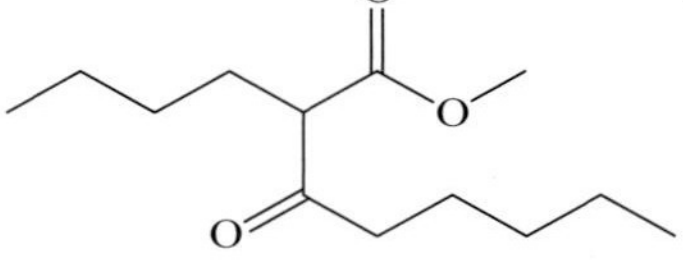

(c)

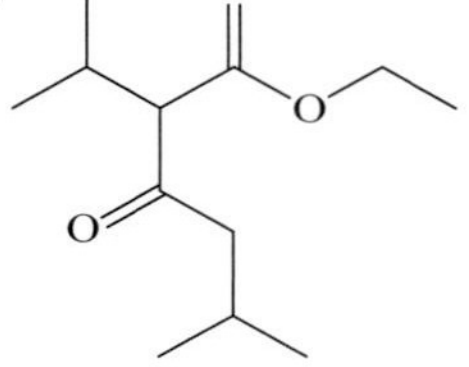

16.29 (a)

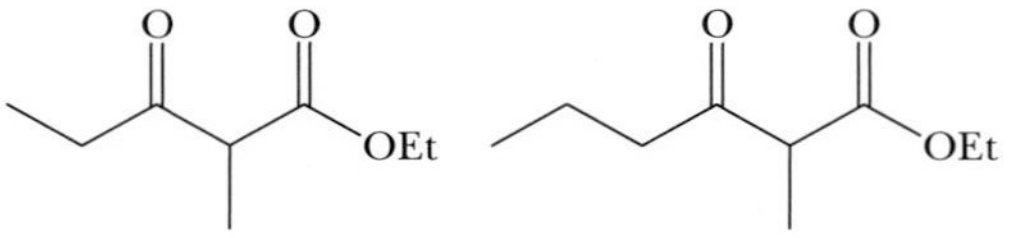

(b)

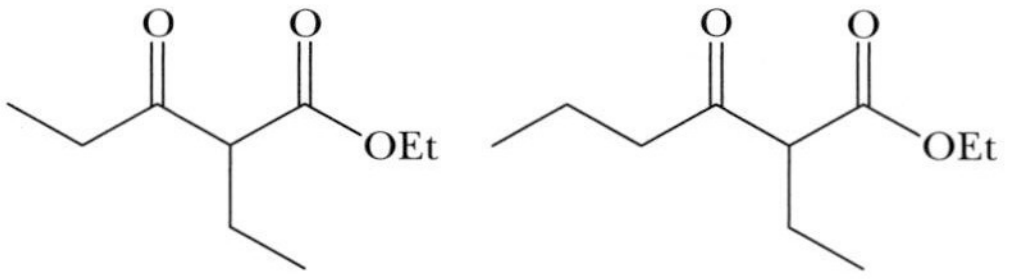

16.31

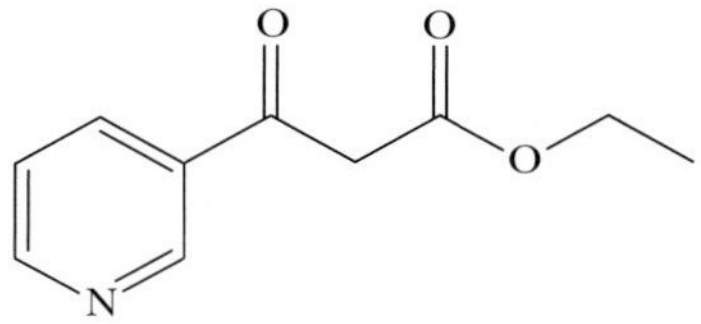

16.33 (a)

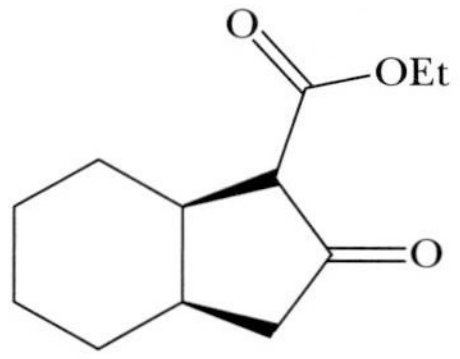

(b)

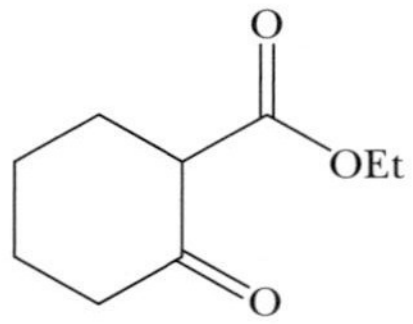

16.35

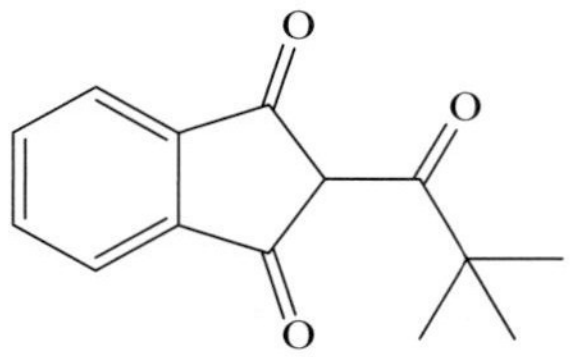

16.37

(a)

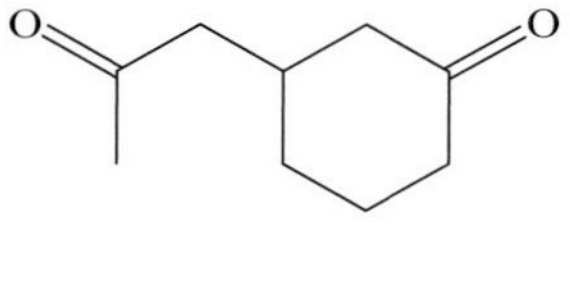

(b)

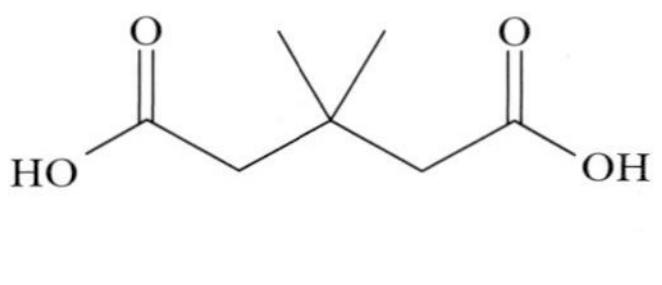

16.39 (a) Ethyl 2-propenoate, Michael addition

(b) 1. EtoNa 2. HCl, H_2O, Dieckmann condensation

(c) First, the ester can be saponified using aqueous NaOH, generating the carboxylate salt, which can then be converted to the β-ketocarboxylic acid by treatment with aqueous HCl. Heating this β-ketocarboxylic acid causes decarboxylation and forms compound C.

(d) Reagent: aniline

(E) imine

(e) H_2/Pd

(f)

(g) No. Fentanyl is achiral because it does not have any stereocenters.

16.41 (a) NaOH, aldol reaction (b) NaOH or H_2SO_4

(c) H_2/Pd

(d)

+ H_2SO_4

16.43

16.45

1) NaOH
2) H^+, heat

Chapter 17
Organic Polymer Chemistry

17.1

poly (vinyl chloride)

17.2

17.3 (a)

(b)

(c)

(d)

17.5

Poly(ethylene terephthalate) Methanol

Dimethyl terephthalate Ethylene glycol

17.7

17.9 (a) (b)

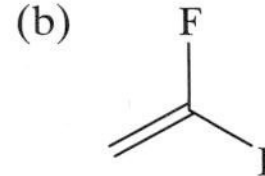

17.11 When the degree of chain branching increases, the chains are less able to pack together in the solid state. As a result, the density of the polymer decreases.

17.13

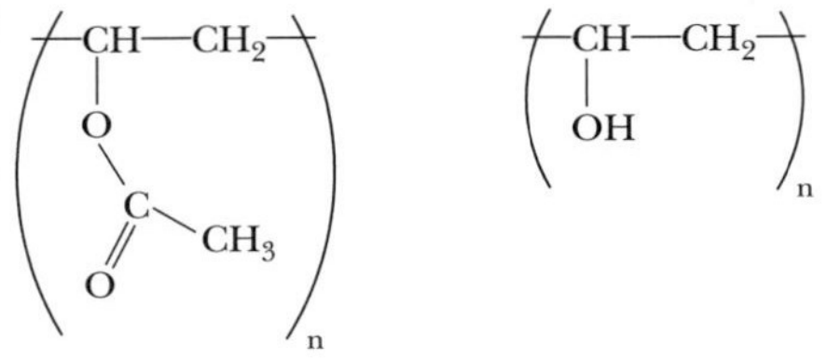

poly(vinyl acetate) poly(vinyl alcohol)

17.15 Most rigid: A and B
Most transparent: C

17.17 Yes

Chapter 18
Carbohydrates

18.1

D-ribulose L-ribulose

enantiomers

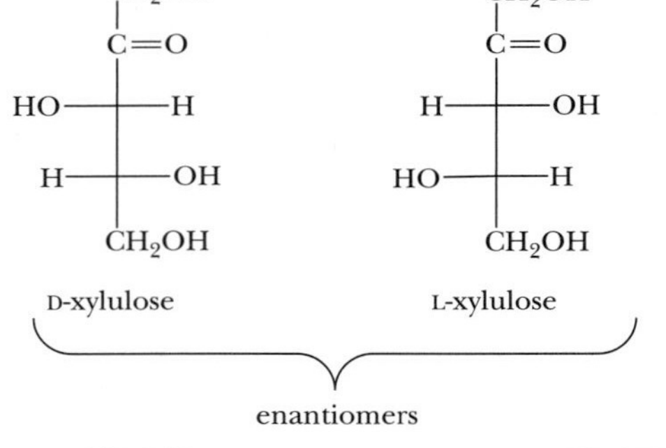

18.2

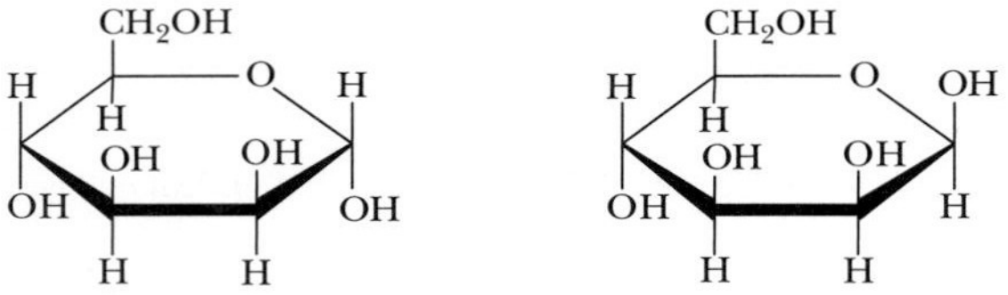

α-D-mannopyranose (α-D-mannose) β-D-mannopyranose (β-D-mannose)

18.3

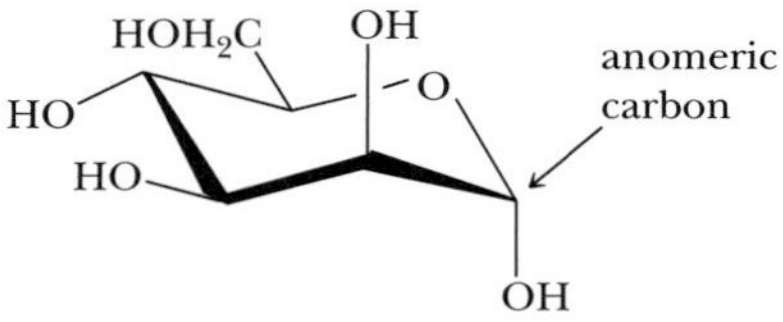

anomeric carbon

α-D-mannopyranose

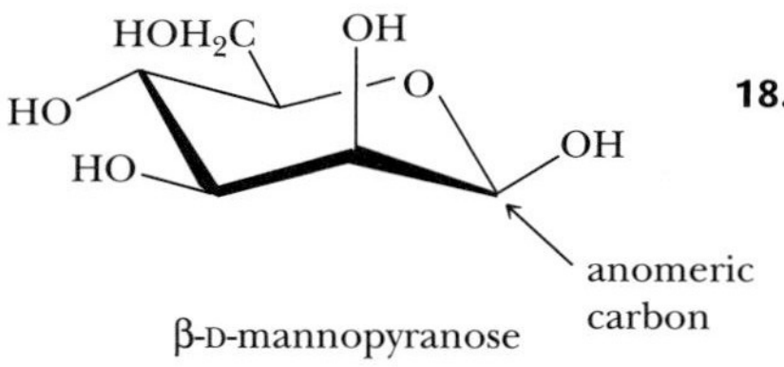

anomeric carbon

β-D-mannopyranose

18.4

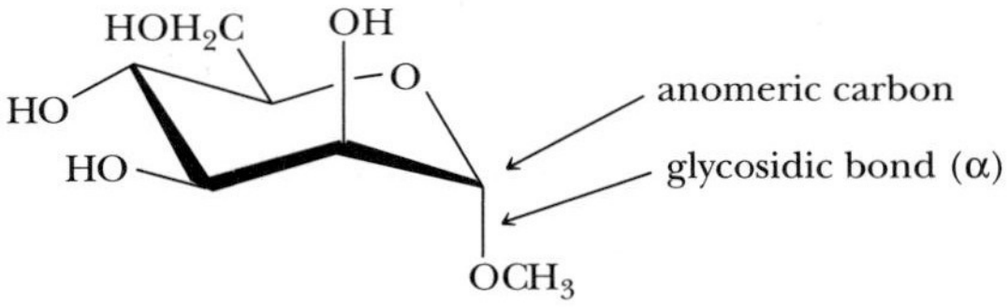

anomeric carbon

glycosidic bond (α)

18.5

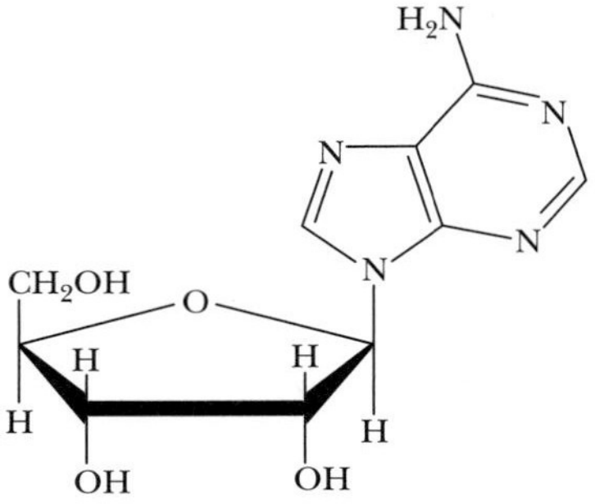

18.6 Erythritol, which is a meso compound, is optically inactive.

18.7

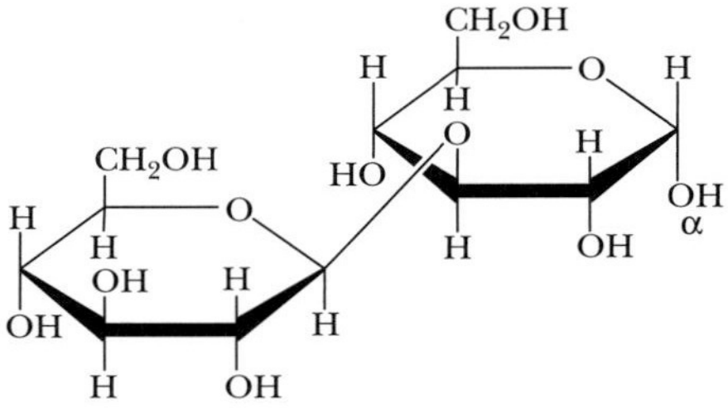

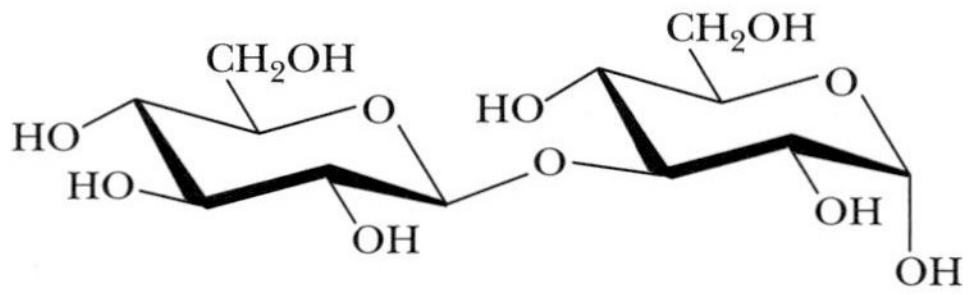

18.9 D-Glucose

18.11 The two designations refer to the configuration of the penultimate carbon, which is the stereocenter farthest away from the carbonyl group. If the monosaccharide is drawn as a Fischer projection and the OH group bonded to this carbon is on the right, the D designation is used; likewise, a monosaccharide is L if the OH group is on the left.

18.13 (a) and (c) are D-monosaccharides, while (b) is an L-monosaccharide.

18.15 Mono- and disaccharides are polar compounds and are able to participate in hydrogen bonding with water molecules.

18.17

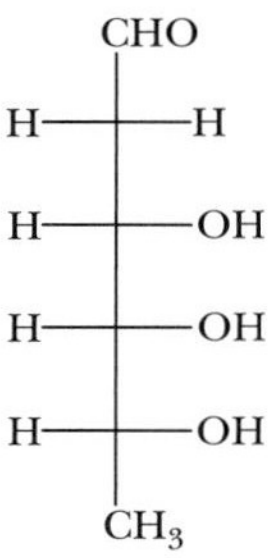

18.19 The α and β designations refer to the position of the hydroxyl ($-OH$) group on the anomeric carbon relative to the terminal $-CH_2OH$ group. If both groups of these groups are on the same side of the ring (*cis*), a β designation is assigned. If the two groups are on the opposite side of the ring (*trans*), the monosaccharide is α.

18.21 No, because anomers are compounds that differ only in the configuration of the anomeric carbon. These two compounds are enantiomers.

18.23

α-D-mannose α-D-gulose

18.25 (a)

D-galactose

(b)

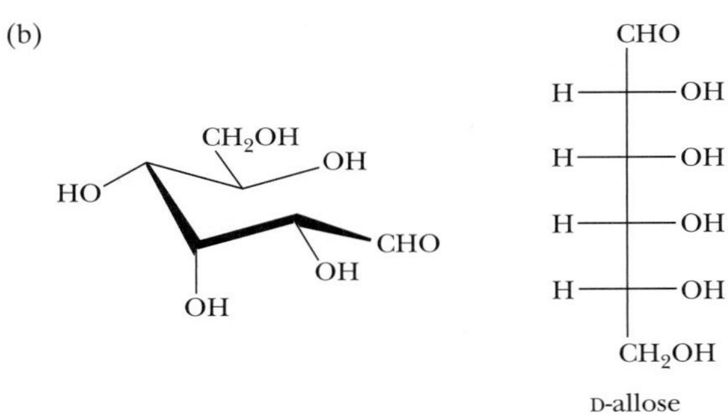

D-allose

18.27 Mutarotation involves the formation of an equilibrium mixture of the α and β anomers of a carbohydrate. This process can be detected by observing the change in the optical activity of the solution over time.

18.29 Yes, and the value will change to $-52.7°$.

18.31

glycosidic bond

anomeric carbon

18.33 (a)

N-glycosidic bond

anomeric carbon

(b)

N-glycosidic bond

anomeric carbon

18.35

D-Ribitol
optically inactive

D-Ribonic acid
optically active

18.37 D-Ribose and D-xylose yield meso compounds, which are optically inactive. D-Arabinose and D-lyxose yield optically active alditols.

18.39 D-allose and D-galactose

18.41 Ascorbic acid is a reducing agent. When ascorbic acid acts as a reducing agent, it loses two hydrogen atoms.

18.43 A glycosidic bond, also known as a glycosidic linkage, is the bond formed between the anomeric carbon of a glycoside (an acetal carbon of a carbohydrate) and an —OR group.

18.45 No.

18.47 Lactose and maltose

18.49 (a) No. (b) No.
(c) Both monosaccharides are D-glucose.

18.51 An oligosaccharide is a short polymer of about 6 to 10 monosaccharides. While there is no definite rule, polysaccharides generally contain more than 10 monosaccharides.

18.53 Both types are composed of D-glucose, and both contain α-1,4-glycosidic bonds. However, amylose is unbranched, while amylopectin contains branches that result from α-1,6-glycosidic bonds.

18.55 (a)

(b)

18.57 (a) D-galactose, D-glucose, and D-glucose
(b) α-1,4 and β-1,4
(c) With the large, nonpolar hydrocarbon chains on the right of the molecule, the molecule is expected to be relatively insoluble in water.

18.59

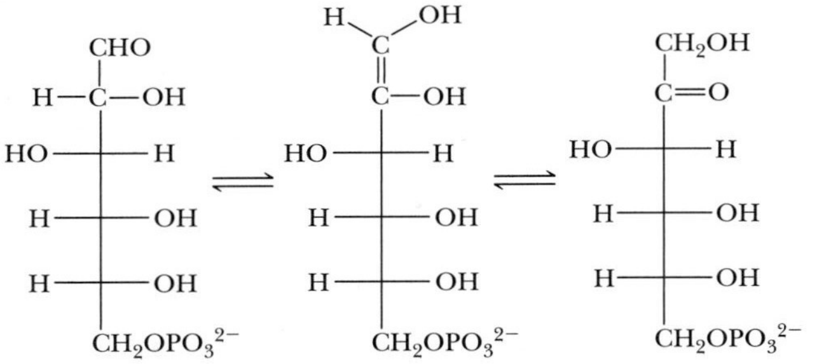

CHO, H—C—OH, HO—H, H—OH, H—OH, $CH_2OPO_3^{2-}$ ⇌ H–C(OH)=C—OH, HO—H, H—OH, H—OH, $CH_2OPO_3^{2-}$ ⇌ CH_2OH, C=O, HO—H, H—OH, H—OH, $CH_2OPO_3^{2-}$

18.61 First, it is difficult to control the stereochemistry of the glycosidic bond (α or β). In addition, it is difficult to form only the desired 1,4-linkage, as any one of the hydroxyl groups located on carbons 2, 3, 4, and 6 could be used to form the bond.

Putting It Together

1. (e) **2.** (a) **3.** (e) **4.** (d) **5.** (d) **6.** (e) **7.** (c) **8.** (b) **9.** (c) **10.** (c)

11. (a)

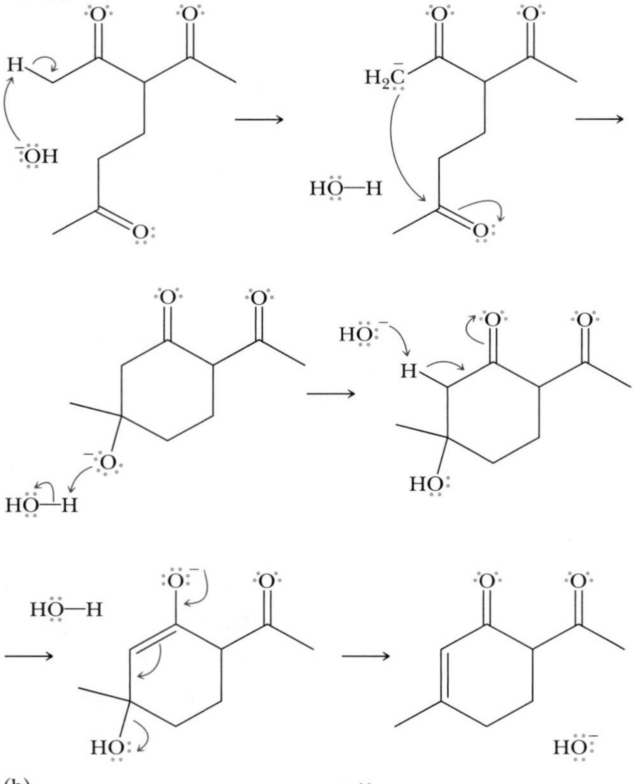

(b)

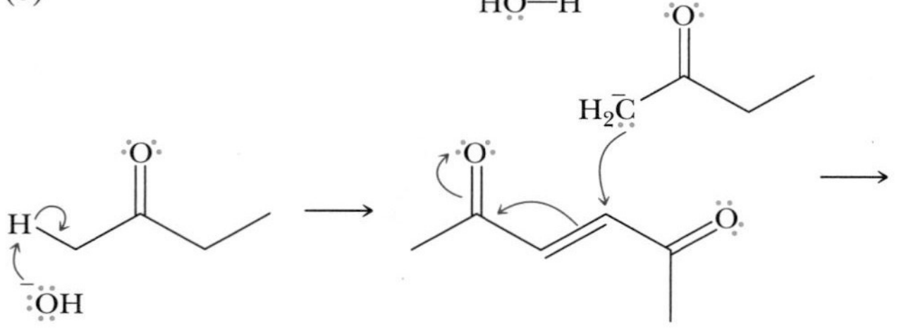

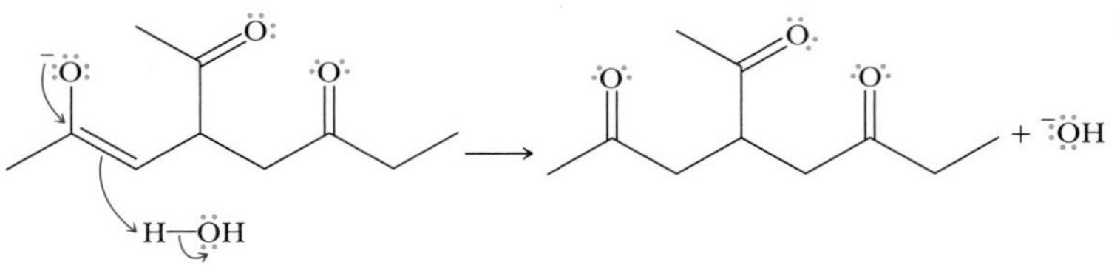

+ $^{-}$:OH

12.

glycosidic bonds

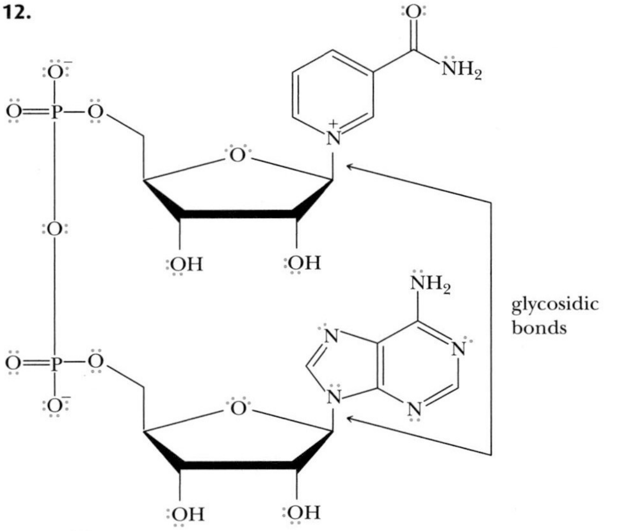

(b) No.
(d) Both carbohydrate units in NAD$^+$ are derived from D-ribose.
(e) Only β sugars are found in NAD$^+$. (f) No.

13. (a)

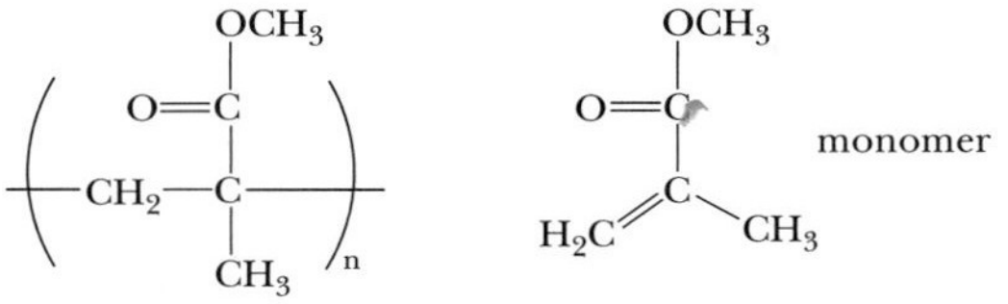

monomer

(b)

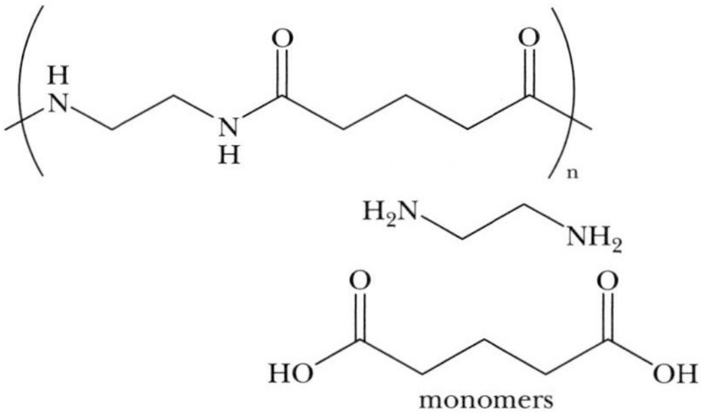

monomers

(c)

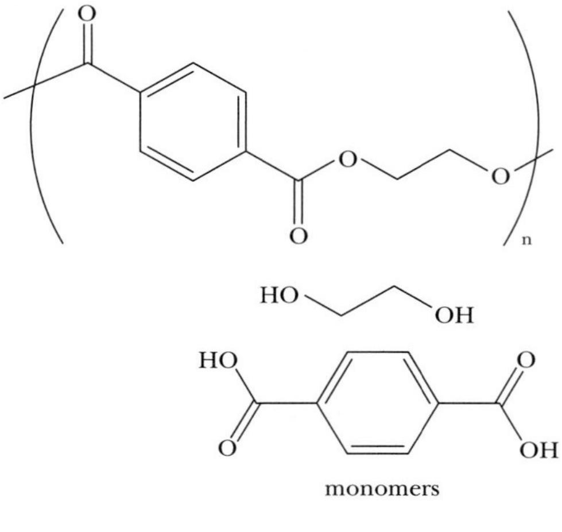

monomers

14. Step 1: Addition of ethylamine to the β-carbon, forming a resonance-stabilized enolate.

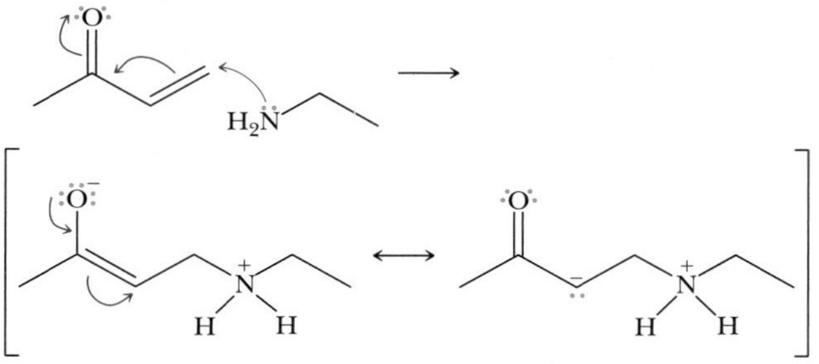

Step 2: Formation of the enol and deprotonation of the nitrogen.

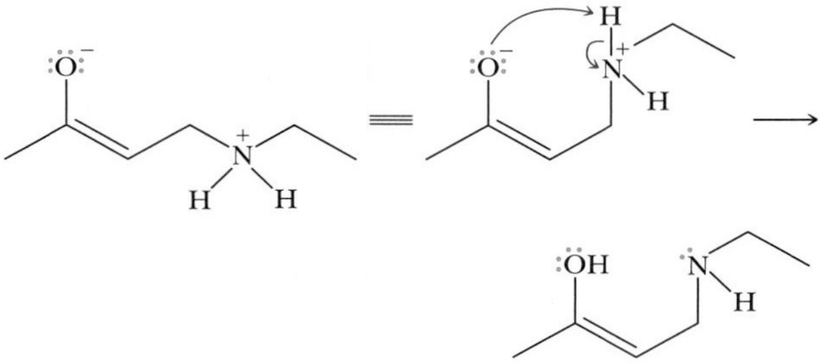

Step 3: After the addition of chloropropane to the mixture, the nitrogen deprotonates the enol, and a nucleophilic aliphatic substitution follows.

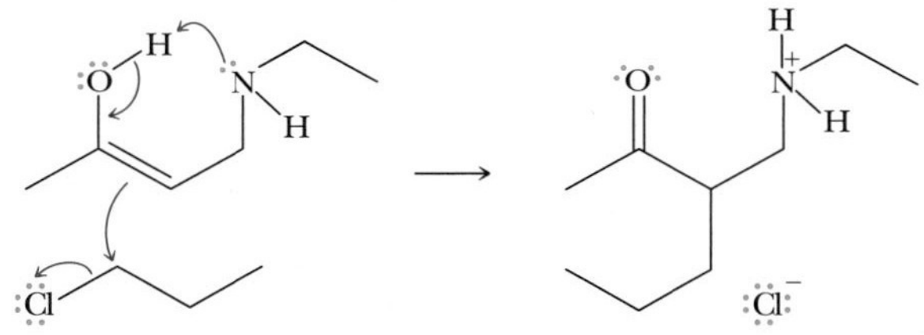

Step 4: Deprotonation of the ammonium ion to generate the product.

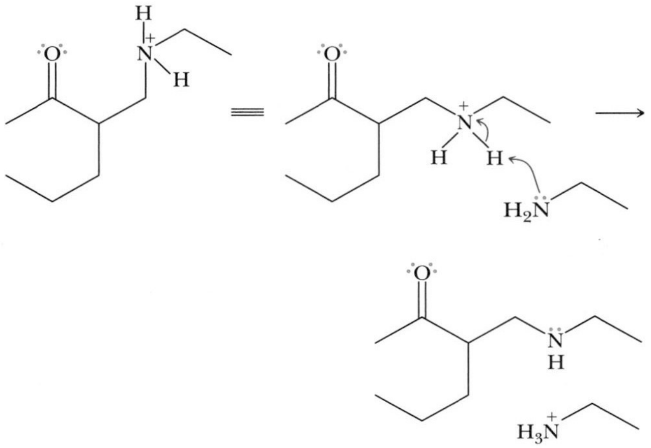

15. (a) $-(CH_2-CH(Cl))_{1000}-$ (b)

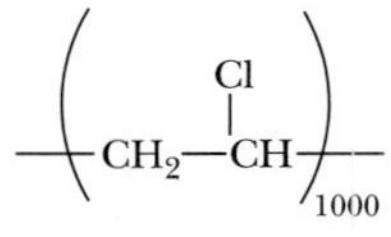

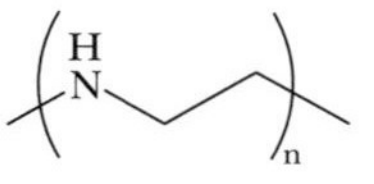

16. (a) 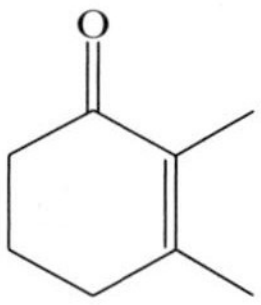 (b) 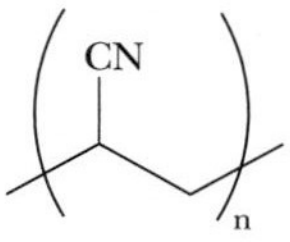 (c) 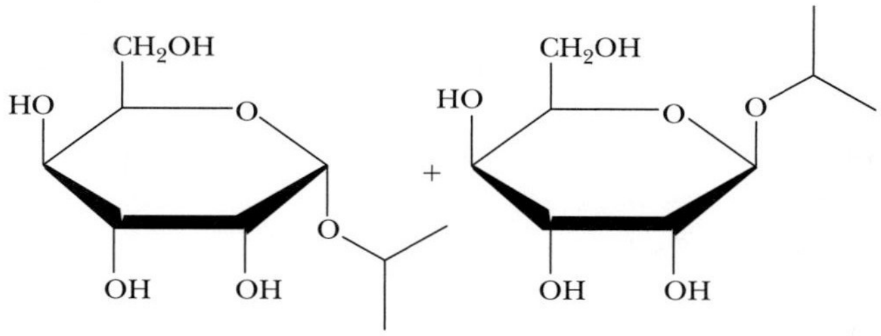 (d) $HO-C(=O)-CH_2-CH_2-C(=O)-CH_3$ (e) 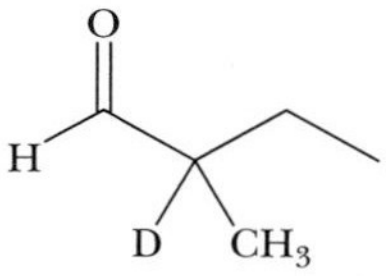 racemic mixture (f)

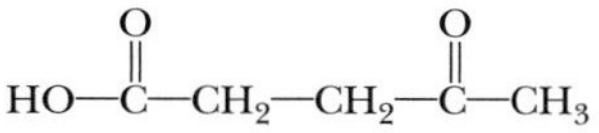

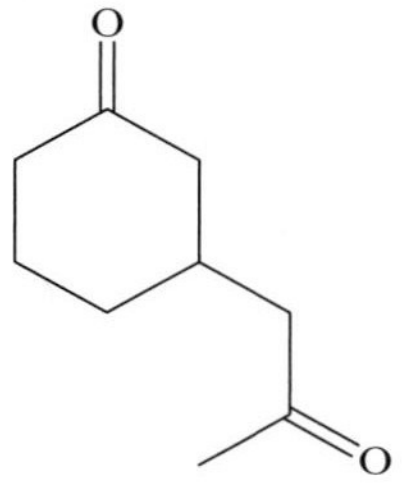

17. (a) light/benzoyl peroxide → $-(CH_2-CH(CH_2CH_2CH_3))_n-$

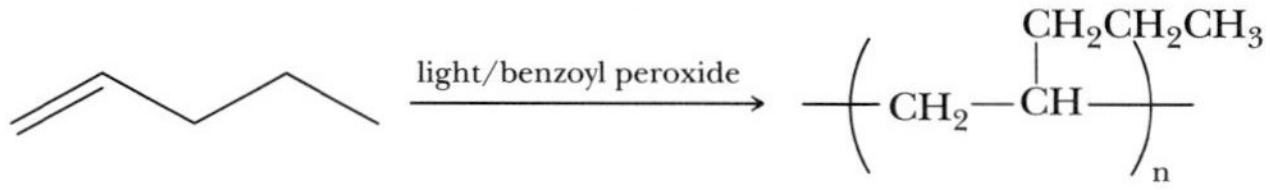

(b) H^+

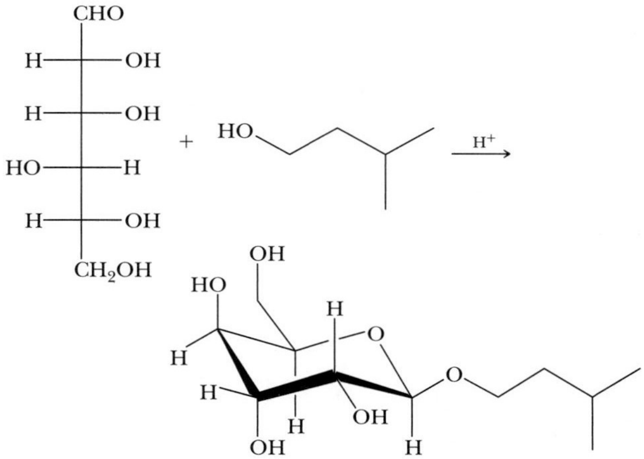

(c) EtONa / EtOH

new bond

H_3O^+

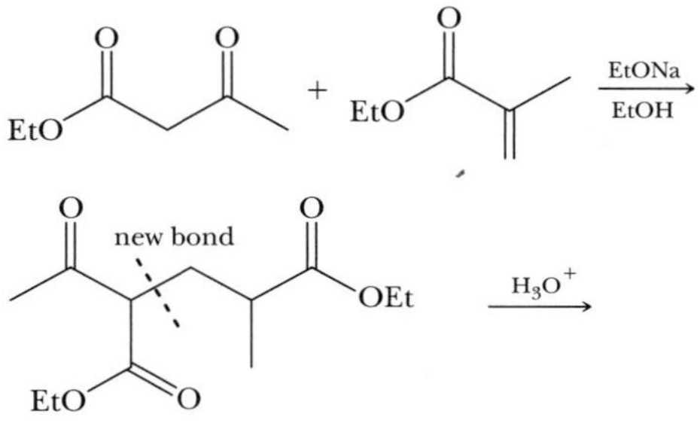

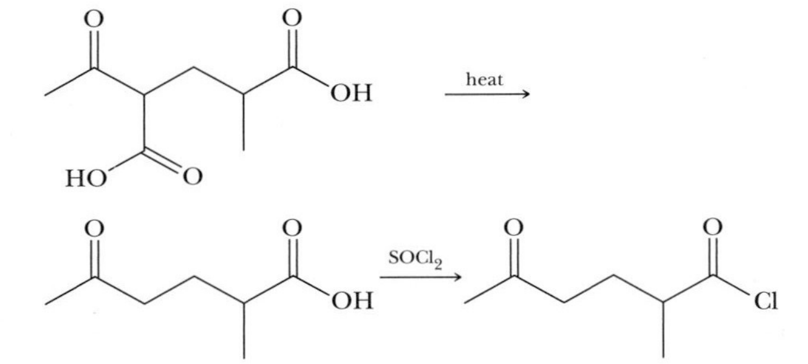

heat

$SOCl_2$

(d)

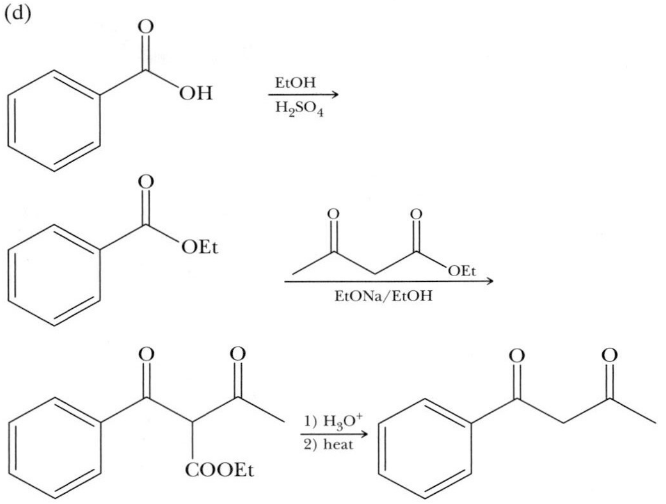

EtOH
H_2SO_4

EtONa/EtOH

1) H_3O^+
2) heat

(e)

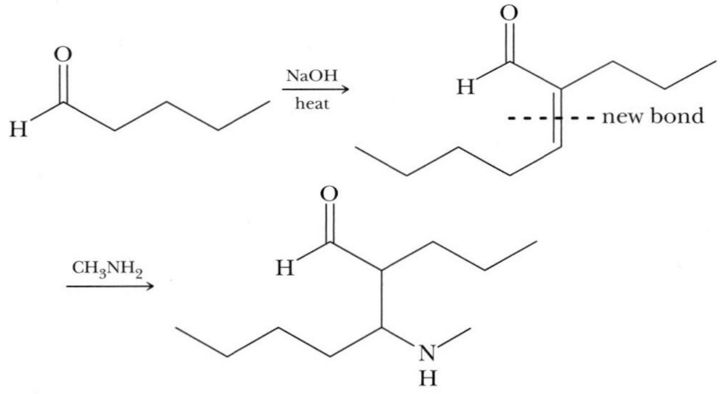

NaOH
heat

new bond

CH_3NH_2

(f)

EtONa
EtOH

H_3O^+

heat

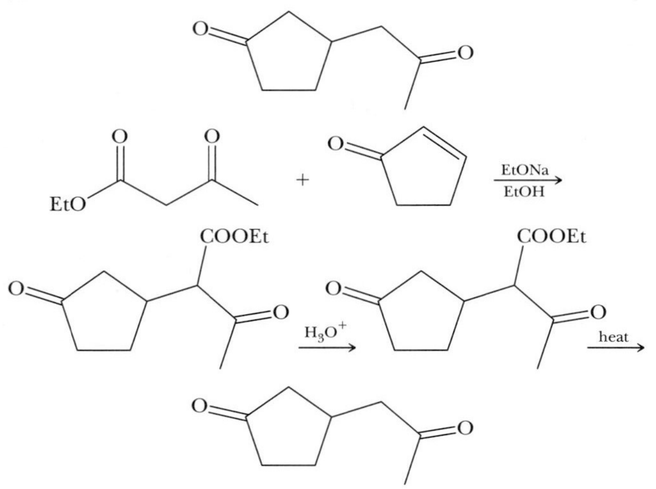

Chapter 19
Amino Acids and Proteins

19.1 Of the 20 amino acids, only glycine does not contain a stereocenter. Isoleucine and threonine each contain two stereocenters.

19.2 Negative electrode

19.3 At pH 6.0, arginine will migrate towards the negative electrode, glutamic acid will migrate towards the positive electrode, and valine will remain at the origin.

19.4 The net charge is +1.

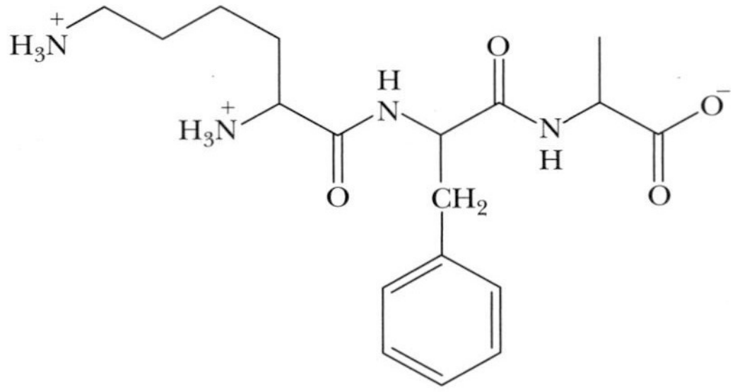

19.5 (a) Tyr-Gln-Val (b) Thr-Phe-Ser

19.6 Ala-Glu-Arg-Thr-Phe-Lys-Lys-Val-Met-Ser-Trp

19.7 Aspartic acid and glutamic acid

19.9 S

19.11

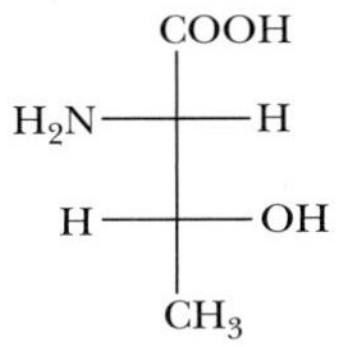

COOH
H_2N—H
H—OH
CH_3

19.13 (a) 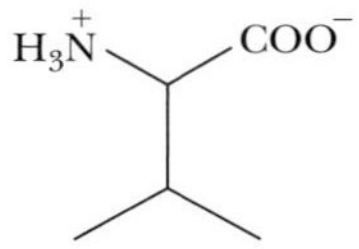 (b) 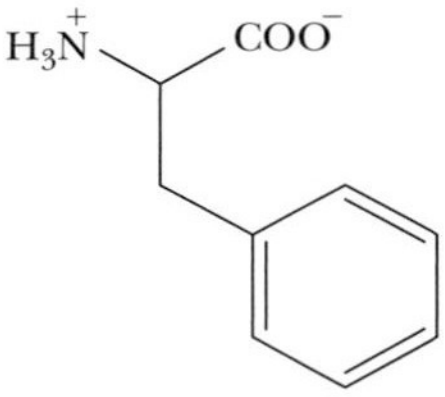 (c)

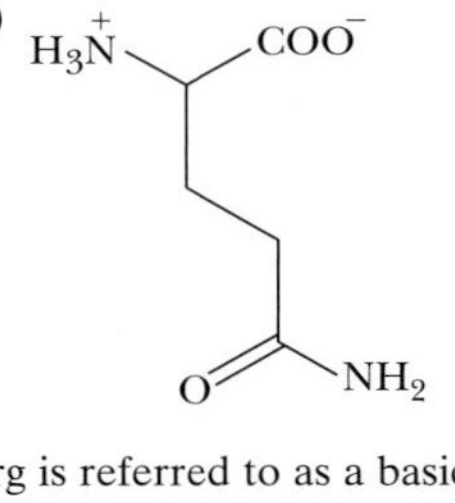

19.15 Arg is referred to as a basic amino acid because its side chain contains a basic group (the guanidinium group). Lys and His are also basic.

19.17 $H_3\overset{+}{N}CH_2CH_2COO^-$

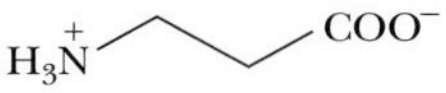

19.19 Histidine, decarboxylation

19.21 Serotonin: decarboxylation and oxidation (hydroxylation)
Melatonin: decarboxylation, oxidation (hydroxylation), methylation, and acetylation.
Both from tryptophon

19.23 (a) (b)

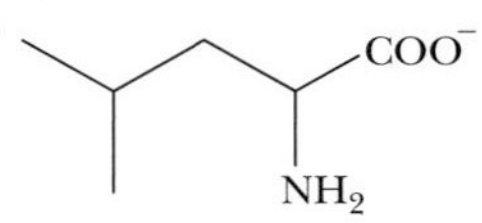

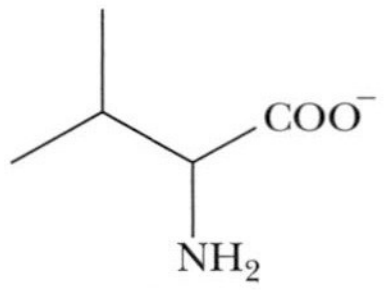

(c) (d)

19.25

1 mol NaOH

2 mol NaOH

3 mol NaOH

19.27

(a) Glutamic acid with NaOH

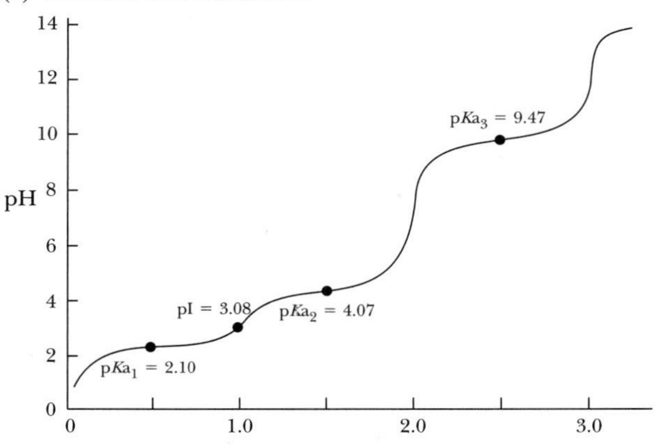

(b) Histidine with NaOH

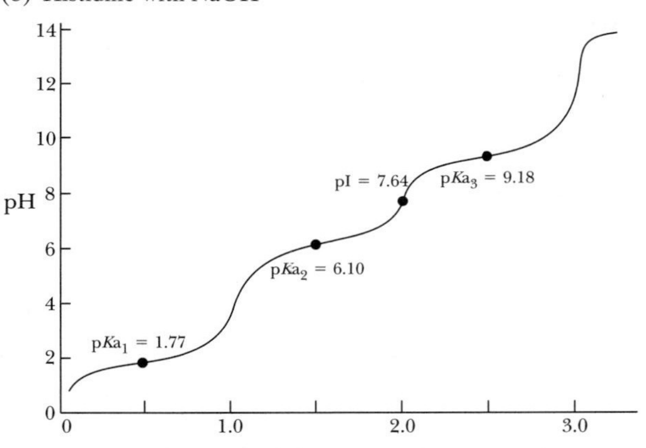

19.29 The side chain of glutamine contains an amide functional group, which is not basic or acidic. On the other hand, the side chain of glutamic acid contains a carboxyl group, which is acidic.

19.31 Guanidine and the guanidino group are very strong amine bases due to the resonance stabilization of the conjugate *acid*, the guanidinium group.

19.33 (a), (b), (d), and (f): cathode (c) and (e): anode

19.35 Neutral amino acids

19.37 (a) 24 (b) 116,280

19.39 (a) between residues 1 and 2
(b) between residues 6 and 7, 10 and 11, 13 and 14, 22 and 23, and 25 and 26
(c) between residues 12 and 13, 17 and 18, and 18 and 19.
(d) between 27 and 28

19.41 (a)

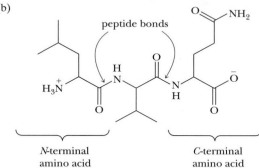

peptide bonds

N-terminal amino acid

C-terminal amino acid

(b)

peptide bonds

N-terminal amino acid

C-terminal amino acid

19.43 (a) glutamic acid, cysteine, and glycine.
(b) The carboxyl group used to form the peptide bond is not that of the α-carboxyl group but rather that of side chain.
(c) reducing agent
(d) 4 G-SH + O_2 ⟶ 2 G-S-S-G + 2 H_2O

19.45 Outside

19.47 Amino acids that are polar, acidic, and basic will prefer to be on the outside of the protein surface, in contact with the aqueous environment to maximize hydrophilic and hydrogen-bonding interactions. These amino acids include Arg, Ser, and Lys.

19.49 Denaturation is not always a reversible process.

19.51 Homoserine undergoes a cyclization to form a five-membered ring (a γ-lactone), as shown below. The same reaction with serine would form a four-membered ring (a β-lactone), which is much more strained, and higher in energy, than a five-membered ring.

19.53 Enzyme

Chapter 20 Nucleic Acids

20.1

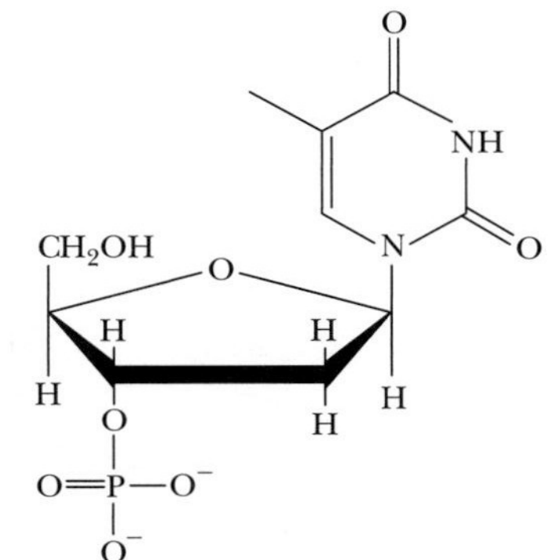

20.2

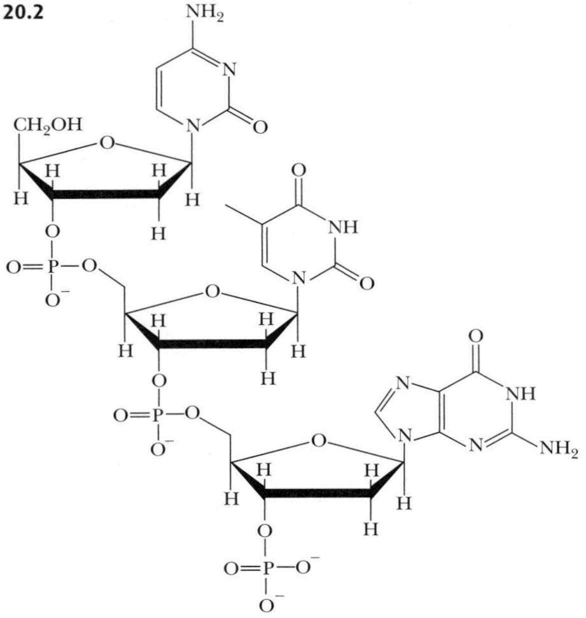

20.3 5′-TCGTACGG-3′

20.4 5′-TGGTGGACGAGTCCGGAA-3′

20.5 (a) 5′-UGC-UAU-AUU-CAA-AAU-UGC-CCU-CUU-GGU-UGA-3′

(b) Cys-Tyr-Ile-Gln-Asn-Cys-Pro-Leu-Gly

20.6 F*nu*DII and H*pa*II

20.7

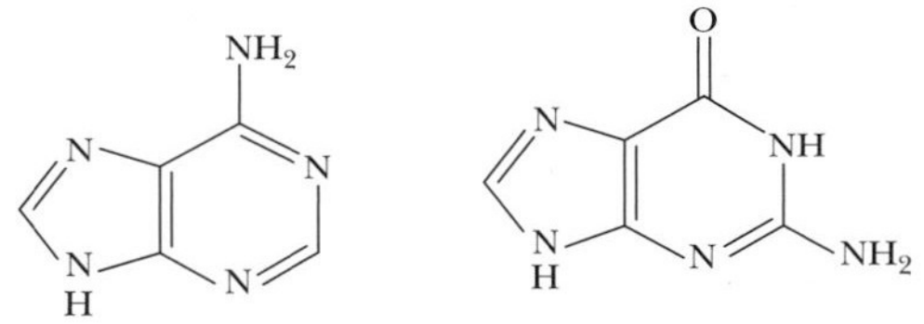

adenine

guanine

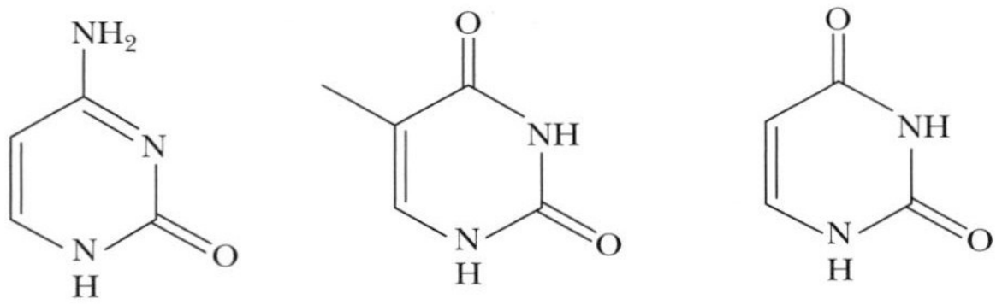

cytosine

thymine

uracil

20.9 A nucleoside consists of only a nucleobase and ribose or 2-deoxyribose. On the other hand, a nucleotide is a phosphorylated nucleoside.

20.11

(a)

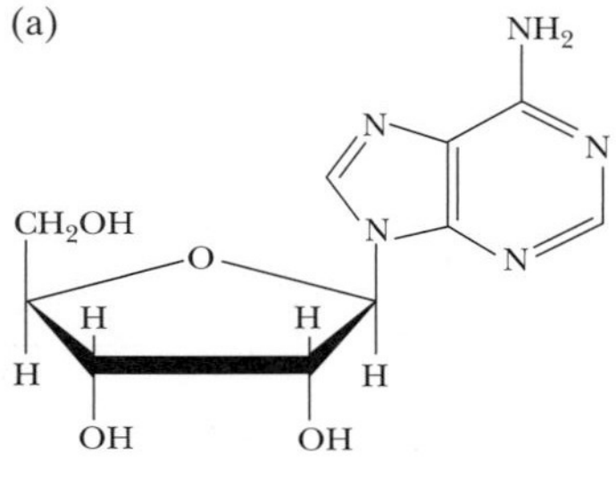

(b)

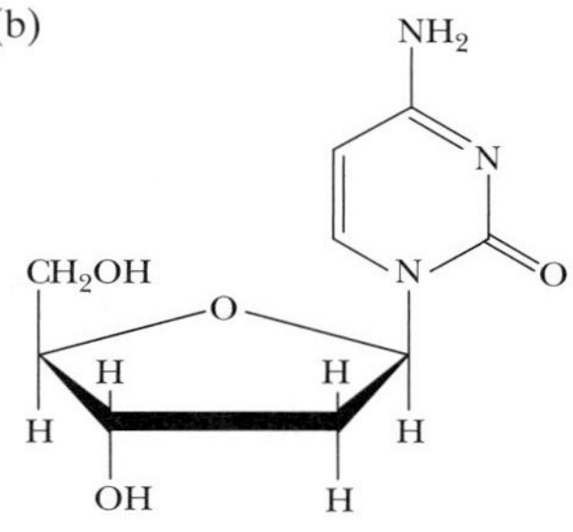

20.13 (a)

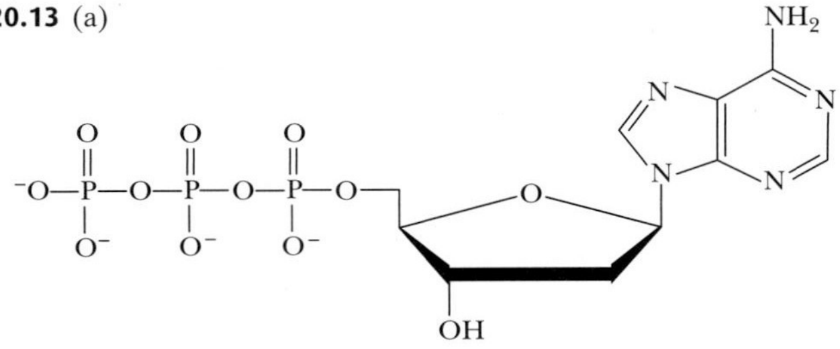

(b)

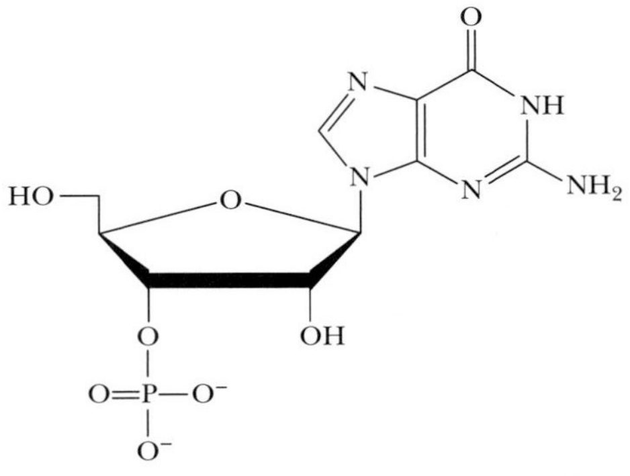

(c)

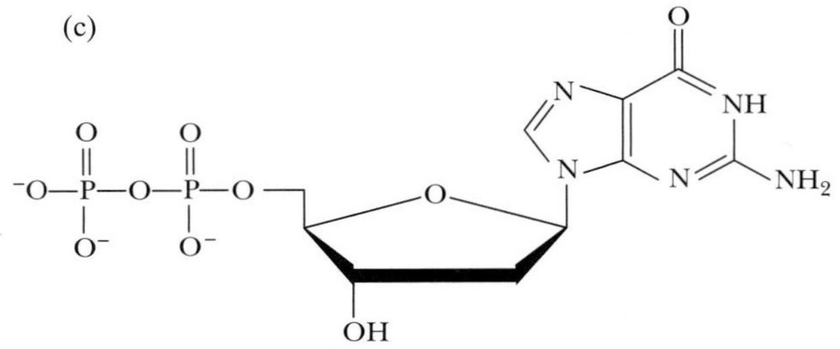

20.15 A pairs with T, so human DNA must also contain 30.4% T. Thus, A and T comprise 60.8% of human DNA. The remaining 39.2% must be equally split between 19.6% G and 19.6% C, because G pairs with C. These values agree very well with the experimental values found in Table 20.1.

20.17
- DNA consists of a two, antiparallel strands of polynucleotide that are coiled in a right-handed manner and arranged about the same axis to form a double helix.
- The nucleobases project inward towards the axis of the helix and are always paired in a very specific manner, A with T and G with C. (By projecting the bases inwards, the acid-labile *N*-glycosidic bonds are protected from the surrounding environment.)
- The base pairs are stacked with a spacing of 3.4 Å between them.
- There is one complete turn of the helix every 34 Å (ten base pairs per turn).

20.19 All nucleotides in DNA are based on 2-deoxyribose that is phosphorylated at C-5, and C-1 of the sugar is linked to the nucleobase via an *N*-glycosidic bond. The nucleotides are 2-deoxyadenosine 5′-monophosphate (dAMP), 2-deoxythymidine 5′-monophosphate (dTMP), 2-deoxyguanosine 5′-monophosphate (dGMP), and 2-deoxycytidine 5′-monophosphate (dCMP).

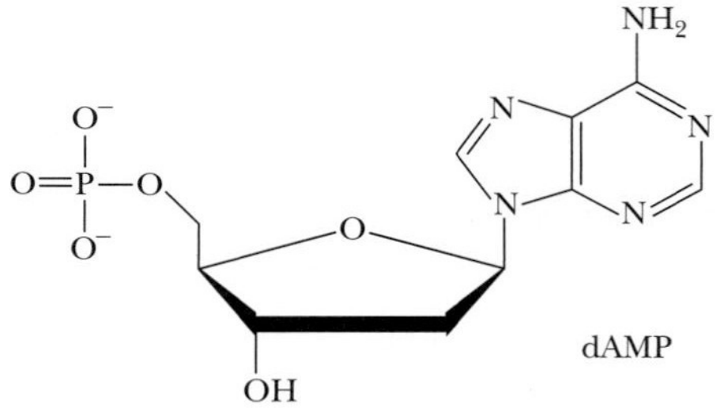

dAMP

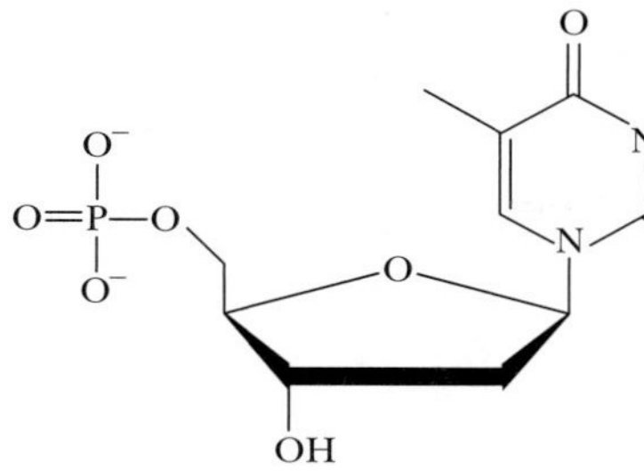

dTMP

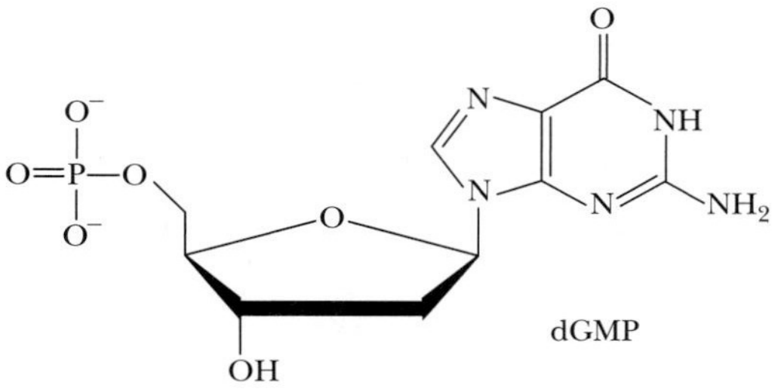

dGMP

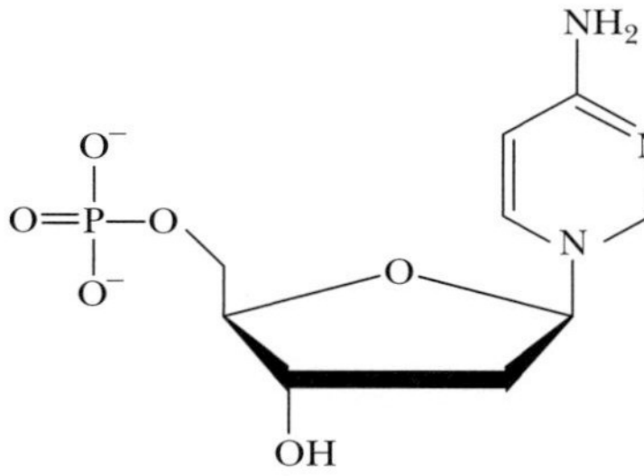

dCMP

20.21 In the structure of the double helix, the nucleobases, which are hydrophobic, are pointed inwards. This minimizes their contact with water on the outside of the helix and also allows them to stack via hydrophobic interactions.

20.23 mRNA is a carrier of protein-sequence information, tRNA carries amino acids for protein synthesis, and rRNA is a component of ribosomes.

20.25 No difference

20.27 mRNA

20.29 3′-AGUUGCUA-5′

20.31 There are 20 amino acids that are specified by the genetic code. With three nucleotides, 64 different sequence combinations are possible. If only two nucleotides were present in each unit, only 16 combinations would be possible.

20.33 Three codons, UAA, UAG, and UGA, are known as the "stop codons" and they indicate that the primary sequence of the protein is complete.

20.35 3′-TGGCAATTA-5′

20.37 Certain amino acids are coded for by more than one codon.

20.39 Both Phe and Tyr are structurally similar, except the latter contains a hydroxyl group on the aromatic ring. The codons for Phe are UUU and UUC, while those for Tyr are UAU and UAC; the codons for the two amino acids differ only in the second position.

20.41 The last base in the codons for Gly, Ala, and Val is irrelevant. Other codons in which the third base is irrelevant include those for Arg (CGX), Pro (CCX), and Thr (ACX).

20.43 With the exception of Trp and Gly, all codons with a purine in the second position code for polar, hydrophilic side chains.

20.45 426

20.47

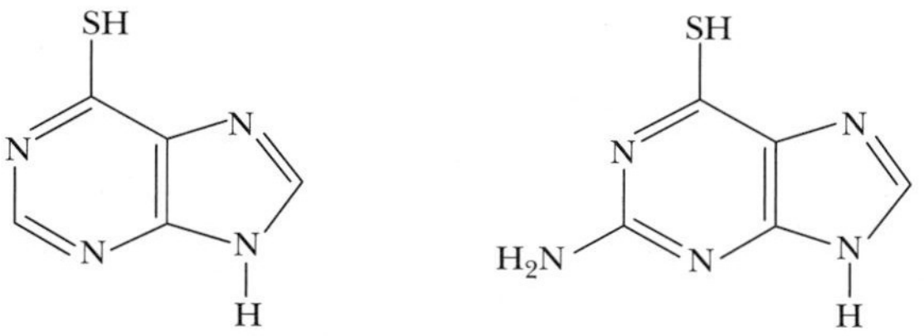

20.49 (a) In the α-helices of proteins, the repeating units are amino acids that are linked by peptide (amide) bonds. Whereas, the repeating units in DNA are 2′-deoxy-D-ribose linked via 3′,5′-phosphodiester bonds.

(b) The R groups in the α-helices of proteins point outward from the helix, whereas the nucleobases in the DNA double helix point inward and away from the aqueous environment of the cell.

20.51 (a) chain terminator

(b) interferes with transcription

(c) interferes with nucleic acid synthesis

20.53

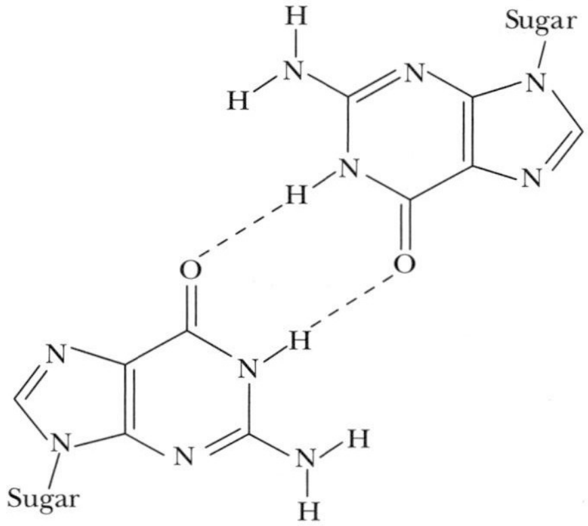

Chapter 21 Lipids

21.1 (a) 3 (b) All are chiral

21.3 Each triglyceride contains three hydrophobic regions (the hydrocarbon chains of the fatty acids) and three hydrophilic regions (the ester groups).

21.5 The melting point would increase.

21.7 Animal fat: human Plant oil: olive

21.9 A very high percentage of the saturated fatty have a low molecular weight.

21.11 Hardening refers to the catalytic hydrogenation of the C=C bonds in vegetable oil.

21.13 A good synthetic detergent should have a long, hydrophobic chain and a very hydrophilic head group (either ionic or very polar) at the end of the molecule. In addition, it should not form insoluble precipitates with the ions that are commonly found in hard water.

21.15 Anionic

21.17 (a)

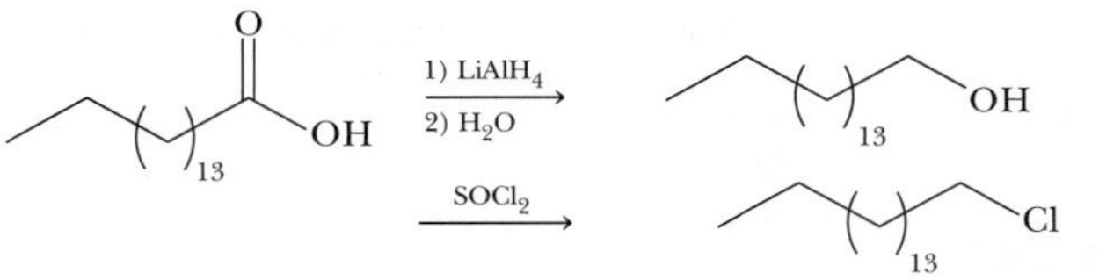

1) $LiAlH_4$ 2) H_2O

$SOCl_2$

(b)

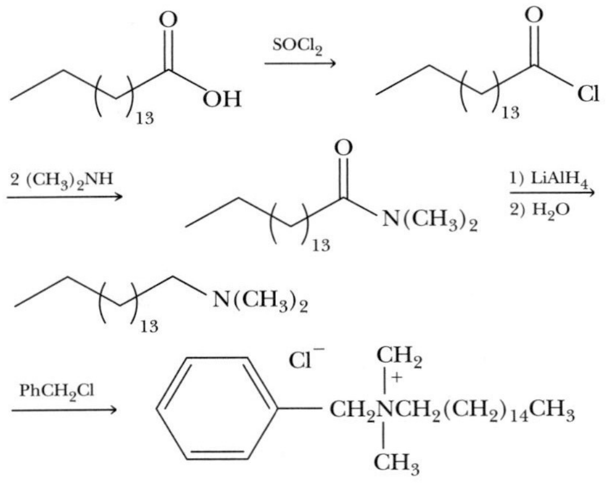

$\xrightarrow{SOCl_2}$ $\xrightarrow{2\ (CH_3)_2NH}$ $N(CH_3)_2$ $\xrightarrow{1)\ LiAlH_4,\ 2)\ H_2O}$ $N(CH_3)_2$ $\xrightarrow{PhCH_2Cl}$ $CH_2N^+(CH_3)(CH_2)_{15}CH_3\ Cl^-$

21.19 In a phospholipid, the two hydrocarbon tails are hydrophobic. The ester and the phosphate groups (as well as any other charged or polar groups that are esterified to the phosphate) are hydrophilic.

21.21 The presence of unsaturated double bonds will prevent dense packing within the hydrophobic inner layer, resulting in greater membrane fluidity.

21.23 (a)

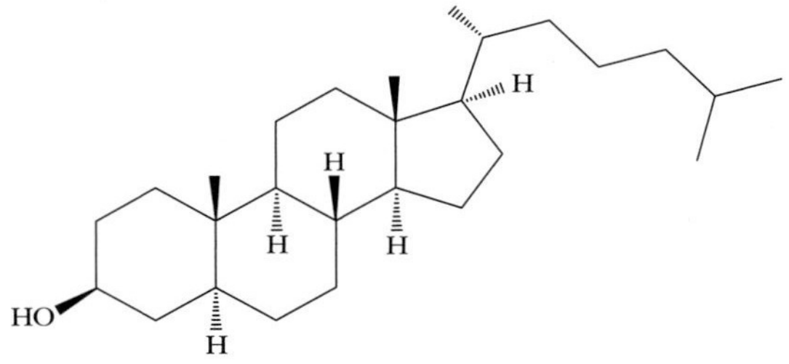

(b)

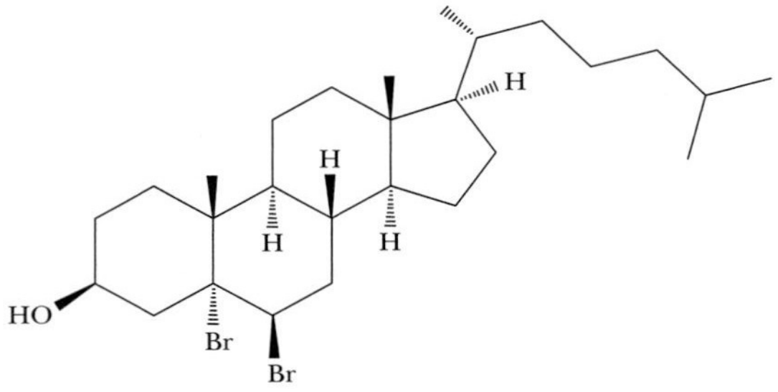

21.25

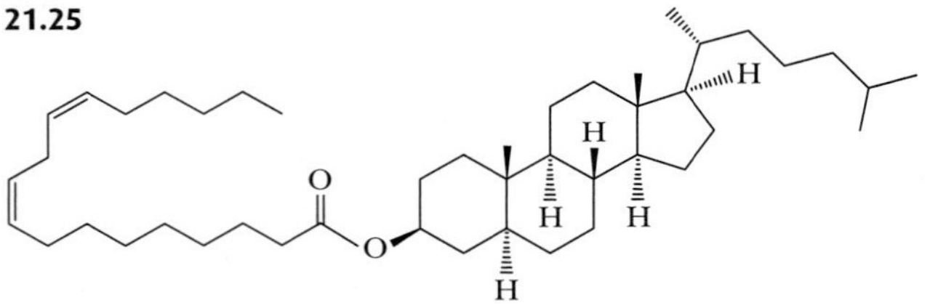

21.27 Cholic acid is able to act as an emulsifier, just like lecithin, because it contains both hydrophobic and hydrophilic groups. The hydrocarbon rings of the steroid skeleton are hydrophobic, while the carboxylate and hydroxyl groups are hydrophilic.

21.29 Methandrostenolone has an extra carbon–carbon double bond in ring A and a methyl substituent on carbon 17.

21.31 Four chemical modifications occur when progesterone is converted to estradiol: an aromatic ring is formed, a ketone is converted to a phenol, a methyl group is removed, and an acetyl group is replaced by a hydroxyl group.

21.33 The double bond between carbons 13 and 14 is reduced, carbon 15 is oxidized from an alcohol to a ketone, and there are two extra carbons (21 and 22) at the end of the molecule.

21.35 16

21.37 More soluble in dichoromethane than in water. Not soluble in blood plasma.

21.39 Fluidity increases with temperature.

21.41

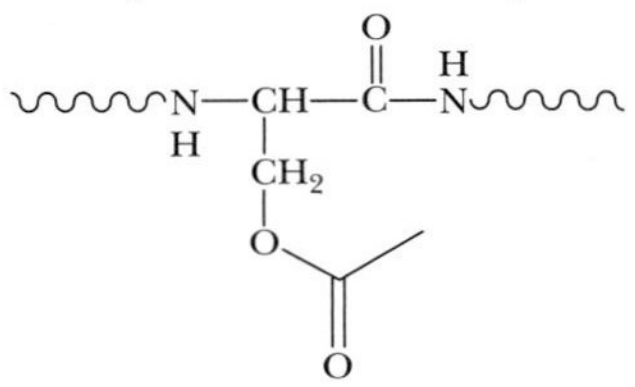

Chapter 22 The Organic Chemistry of Metabolism

22.1 neither **22.2** decrease

22.3 Asp, Glu, and protonated His, Lys, and Arg

22.5 NAD^+, niacin **22.7** 6

22.9 4 moles of each of ethanol and CO_2

22.11

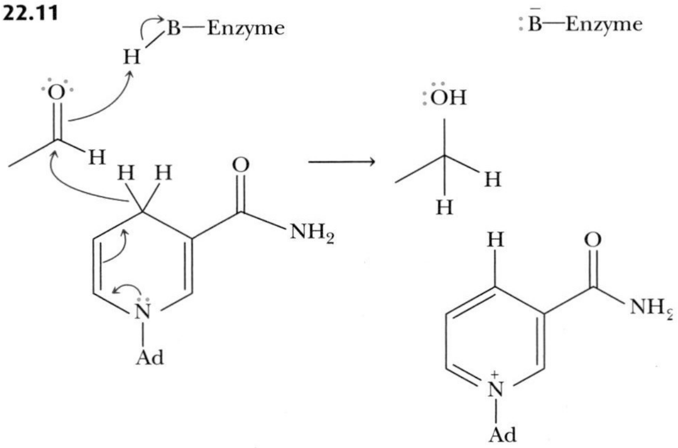

22.13 Hydrogen added to the carbonyl carbon **22.15** 3 and 4

22.17

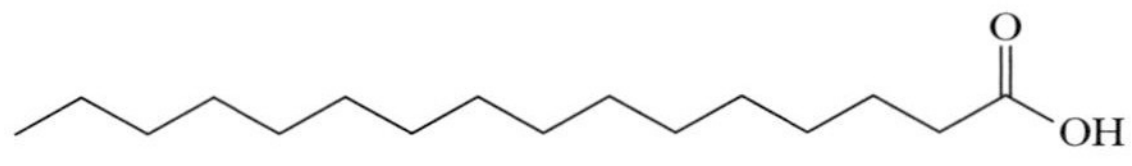

palmitic acid (C_{16})

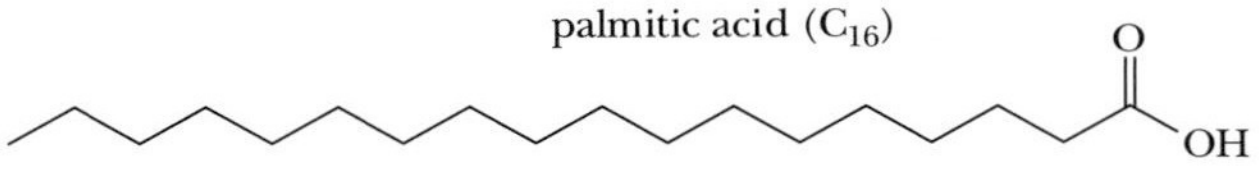

stearic acid (C_{18})

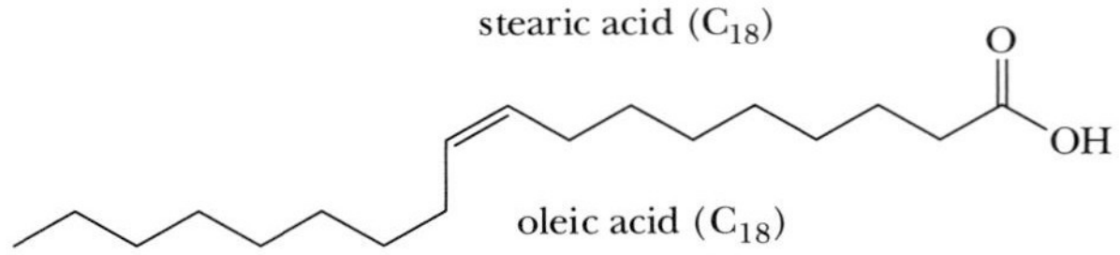

oleic acid (C_{18})

22.19 The three coenzymes and the vitamins from which they are derived are FAD (riboflavin), NAD^+ (niacin), and coenzyme A (pantothenic acid).

22.21 The main function of the citric acid cycle is to oxidize the carbon atoms of acetyl-CoA into carbon dioxide.

22.23 None of the intermediates involved in the cycle are destroyed or created in the net reaction.

22.25 (a) The oxidation of glucose into six moles of CO_2 requires six moles of O_2.

(b) RQ = 0.71 (c) decrease

22.27 Glucose: C1 and C6
Palmitic acid: all even-numbered carbons

22.29 (a) Enzyme-catalyzed reactions give 100% of the desired reactions. Laboratory reactions, which often form undesired side products, do not approach this efficiency.

(b) Enzyme-catalyzed reactions are 100% regioselective.

(c) Enzyme-catalyzed reactions are 100% stereoselective. Stereoselectivity in laboratory reactions is difficult to achieve.

22.31 Amino groups, carboxyl groups, and phosphate groups.

Index

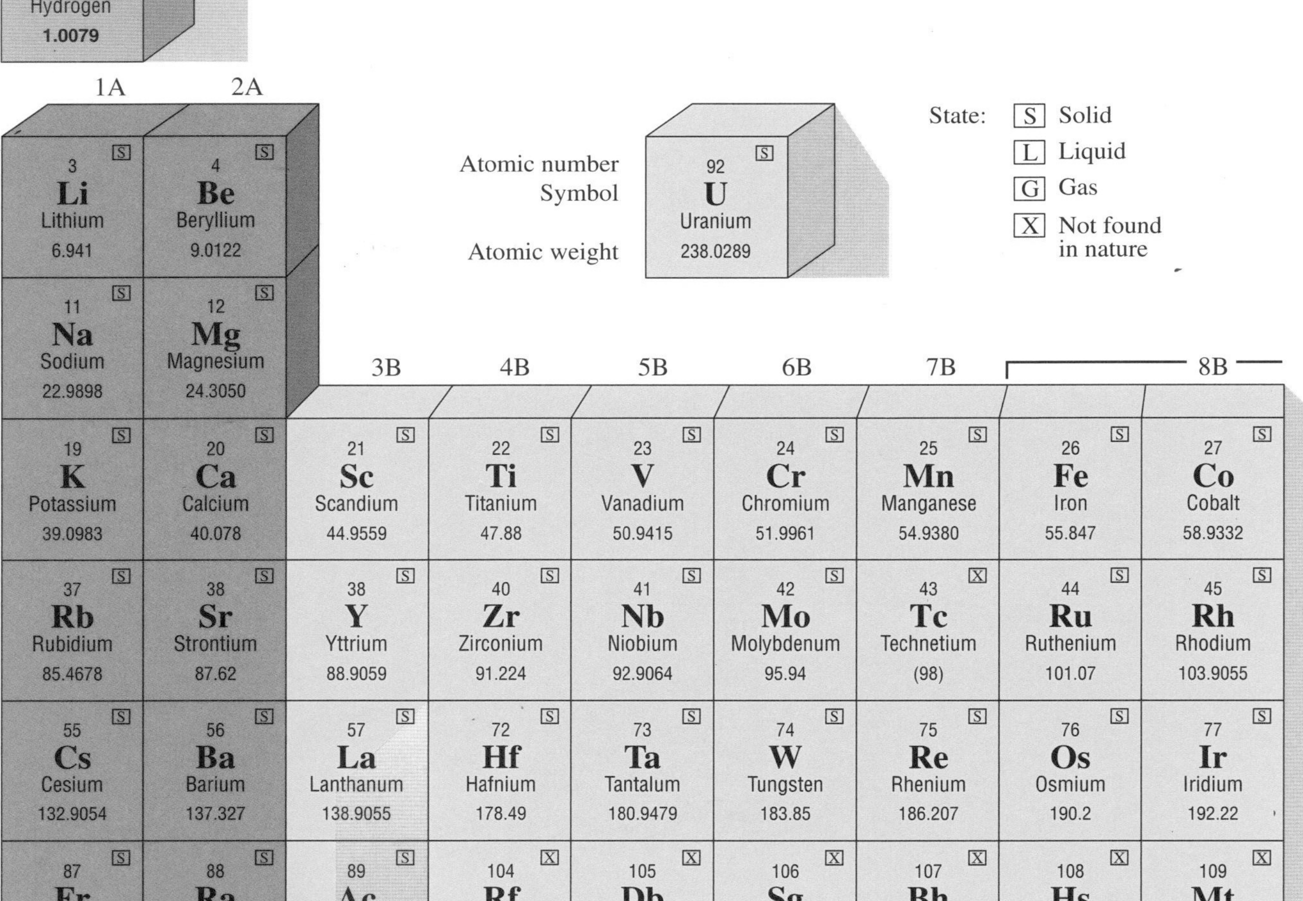

1A	2A	3B	4B	5B	6B	7B	8B	8B
1 H Hydrogen 1.0079 [G]								
3 Li Lithium 6.941 [S]	4 Be Beryllium 9.0122 [S]							
11 Na Sodium 22.9898 [S]	12 Mg Magnesium 24.3050 [S]							
19 K Potassium 39.0983 [S]	20 Ca Calcium 40.078 [S]	21 Sc Scandium 44.9559 [S]	22 Ti Titanium 47.88 [S]	23 V Vanadium 50.9415 [S]	24 Cr Chromium 51.9961 [S]	25 Mn Manganese 54.9380 [S]	26 Fe Iron 55.847 [S]	27 Co Cobalt 58.9332 [S]
37 Rb Rubidium 85.4678 [S]	38 Sr Strontium 87.62 [S]	38 Y Yttrium 88.9059 [S]	40 Zr Zirconium 91.224 [S]	41 Nb Niobium 92.9064 [S]	42 Mo Molybdenum 95.94 [S]	43 Tc Technetium (98) [X]	44 Ru Ruthenium 101.07 [S]	45 Rh Rhodium 103.9055 [S]
55 Cs Cesium 132.9054 [S]	56 Ba Barium 137.327 [S]	57 La Lanthanum 138.9055 [S]	72 Hf Hafnium 178.49 [S]	73 Ta Tantalum 180.9479 [S]	74 W Tungsten 183.85 [S]	75 Re Rhenium 186.207 [S]	76 Os Osmium 190.2 [S]	77 Ir Iridium 192.22 [S]
87 Fr Francium (223) [S]	88 Ra Radium 226.0254 [S]	89 Ac Actinium 227.0278 [S]	104 Rf Rutherfordium (261) [X]	105 Db Dubnium (262) [X]	106 Sg Seaborgium (263) [X]	107 Bh Bohrium (262) [X]	108 Hs Hassium (265) [X]	109 Mt Meitnerium (266) [X]

Lanthanides	58 Ce Cerium 140.115 [S]	59 Pr Praseodymium 140.9076 [S]	60 Nd Neodymium 144.24 [S]	61 Pm Promethium (145) [X]	62 Sm Samarium 150.36 [X]
Actinides	90 Th Thorium 232.0381 [S]	91 Pa Protactinium 231.0359 [S]	92 U Uranium 238.0289 [S]	93 Np Neptunium 237.0482 [X]	94 Pu Plutonium (244) [S]

Key: Atomic number / Symbol / Atomic weight — 92 U Uranium 238.0289 [S]

State: [S] Solid; [L] Liquid; [G] Gas; [X] Not found in nature

* Elements 110–112 have not yet been named.